Table of Atomic Masses*

Element	Symbol	Atomic Number	Atomic Mass	Element	Symbol	Atomic Number	Atomic Mass	Element	Symbol	Atomic Number	Atomic Mass
Actinium	Ac	89	(227)†	Gold	Au	79	197.0	Praseodymium	Pr	59	140.9
Aluminum	Al	13	26.98	Hafnium	Hf	72	178.5	Promethium	Pm	61	(145)
Americium	Am	95	(243)	Hassium	Hs	108	(265)	Protactinium	Pa	91	(231)
Antimony	Sb	51	121.8	Helium	He	2	4.003	Radium	Ra	88	226
Argon	Ar	18	39.95	Holmium	Ho	67	164.9	Radon	Rn	86	(222)
Arsenic	As	33	74.92	Hydrogen	H	1	1.008	Rhenium	Re	75	186.2
Astatine	At	85	(210)	Indium	In	49	114.8	Rhodium	Rh	45	102.9
Barium	Ba	56	137.3	Iodine	I	53	126.9	Rubidium	Rb	37	85.47
Berkelium	Bk	97	(247)	Iridium	Ir	77	192.2	Ruthenium	Ru	44	101.1
Beryllium	Be	4	9.012	Iron	Fe	26	55.85	Rutherfordium	Rf	104	(261)
Bismuth	Bi	83	209.0	Krypton	Kr	36	83.80	Samarium	Sm	62	150.4
Bohrium	Bh	107	(264)	Lanthanum	La	57	138.9	Scandium	Sc	21	44.96
Boron	B	5	10.81	Lawrencium	Lr	103	(260)	Seaborgium	Sg	106	(263)
Bromine	Br	35	79.90	Lead	Pb	82	207.2	Selenium	Se	34	78.96
Cadmium	Cd	48	112.4	Lithium	Li	3	6.941	Silicon	Si	14	28.09
Calcium	Ca	20	40.08	Lutetium	Lu	71	175.0	Silver	Ag	47	107.9
Californium	Cf	98	(251)	Magnesium	Mg	12	24.31	Sodium	Na	11	22.99
Carbon	C	6	12.01	Manganese	Mn	25	54.94	Strontium	Sr	38	87.62
Cerium	Ce	58	140.1	Meitnerium	Mt	109	(268)	Sulfur	S	16	32.07
Cesium	Cs	55	132.9	Mendelevium	Md	101	(258)	Tantalum	Ta	73	180.9
Chlorine	Cl	17	35.45	Mercury	Hg	80	200.6	Technetium	Tc	43	(98)
Chromium	Cr	24	52.00	Molybdenum	Mo	42	95.94	Tellurium	Te	52	127.6
Cobalt	Co	27	58.93	Neodymium	Nd	60	144.2	Terbium	Tb	65	158.9
Copper	Cu	29	63.55	Neon	Ne	10	20.18	Thallium	Tl	81	204.4
Curium	Cm	96	(247)	Neptunium	Np	93	(237)	Thorium	Th	90	232.0
Darmstadtium	Ds	110	(281)	Nickel	Ni	28	58.69	Thulium	Tm	69	168.9
Dubnium	Db	105	(262)	Niobium	Nb	41	92.91	Tin	Sn	50	118.7
Dysprosium	Dy	66	162.5	Nitrogen	N	7	14.01	Titanium	Ti	22	47.88
Einsteinium	Es	99	(252)	Nobelium	No	102	(259)	Tungsten	W	74	183.9
Erbium	Er	68	167.3	Osmium	Os	76	190.2	Uranium	U	92	238.0
Europium	Eu	63	152.0	Oxygen	O	8	16.00	Vanadium	V	23	50.94
Fermium	Fm	100	(257)	Palladium	Pd	46	106.4	Xenon	Xe	54	131.3
Fluorine	F	9	19.00	Phosphorus	P	15	30.97	Ytterbium	Yb	70	173.0
Francium	Fr	87	(223)	Platinum	Pt	78	195.1	Yttrium	Y	39	88.91
Gadolinium	Gd	64	157.3	Plutonium	Pu	94	(244)	Zinc	Zn	30	65.38
Gallium	Ga	31	69.72	Polonium	Po	84	(209)	Zirconium	Zr	40	91.22
Germanium	Ge	32	72.59	Potassium	K	19	39.10				

*The values given here are to four significant figures where possible. †A value given in parentheses denotes the mass of the longest-lived isotope.

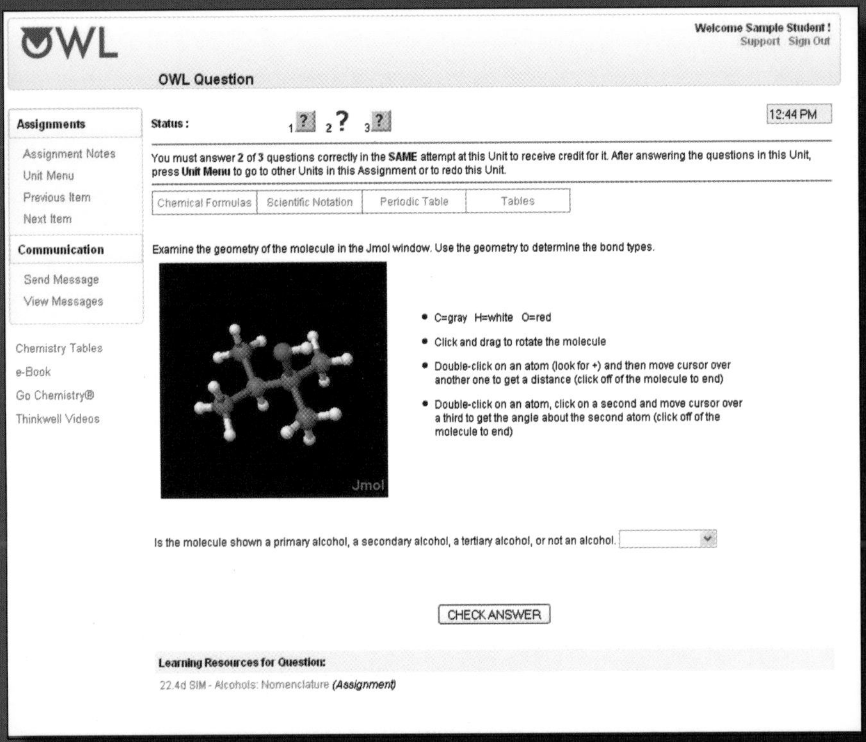

Chemical Principles

Seventh Edition

Steven S. Zumdahl

University of Illinois

Donald J. DeCoste

University of Illinois

Australia • Brazil • Japan • Korea • Mexico • Singapore • Spain • United Kingdom • United States

***Chemical Principles*, Seventh Edition**
Steven S. Zumdahl, Donald J. DeCoste

Publisher: Mary Finch
Executive Editor: Lisa Lockwood
Developmental Editor: Thomas Martin
Editorial Assistant: Krista Mastroianni
Senior Media Editor: Lisa Weber
Media Editor: Stephanie VanCamp
Marketing Manager: Nicole Hamm
Marketing Assistant: Julie Stefani
Marketing Communications Manager: Linda Yip
Content Project Manager: Teresa L. Trego
Design Director: Rob Hugel
Art Director: Maria Epes
Print Buyer: Judy Inouye
Rights Acquisitions Specialist: Tom McDonough
Production Service: Graphic World Inc.
Text Designer: Ellen Pettengill Design
Photo Researcher: Sharon Donahue
Text Researcher: Sarah D'Stair
Copy Editor: Graphic World Inc.
Illustrator: Graphic World Inc.
OWL Producers: Stephen Battisti, Cindy Stein, David Hart (Center for Educational Software Development, University of Massachusetts, Amherst)
Cover Designer: Ellen Pettengill Design
Cover Image: Sciencefoto/Oxford Scientific/PhotoLibrary
Compositor: Graphic World Inc.

For product information and technology assistance, contact us at
Cengage Learning Customer & Sales Support, 1-800-354-9706
For permission to use material from this text or product,
submit all requests online at **www.cengage.com/permissions**
Further permissions questions can be emailed to
permissionrequest@cengage.com

Library of Congress Control Number: 2011933679

ISBN-13: 978-1-111-58065-0
ISBN-10: 1-111-58065-0

Brooks/Cole
20 Davis Drive
Belmont, CA 94002-3098
USA

Cengage Learning is a leading provider of customized learning solutions with office locations around the globe, including Singapore, the United Kingdom, Australia, Mexico, Brazil, and Japan. Locate your local office at: **www.cengage.com/global**

Cengage Learning products are represented in Canada by Nelson Education, Ltd.

To learn more about Brooks/Cole, visit **www.cengage.com/brookscole**

Purchase any of our products at your local college store or at our preferred online store **www.cengagebrain.com**

Printed in the United States of America
4 5 15 14 13

Contents

3 Stoichiometry 53

4 Types of Chemical Reactions and Solution Stoichiometry 94

7 Acids and Bases 240

8 Applications of Aqueous Equilibria 292

9 Energy, Enthalpy, and Thermochemistry 360

5 Gases 148

6 Chemical Equilibrium 203

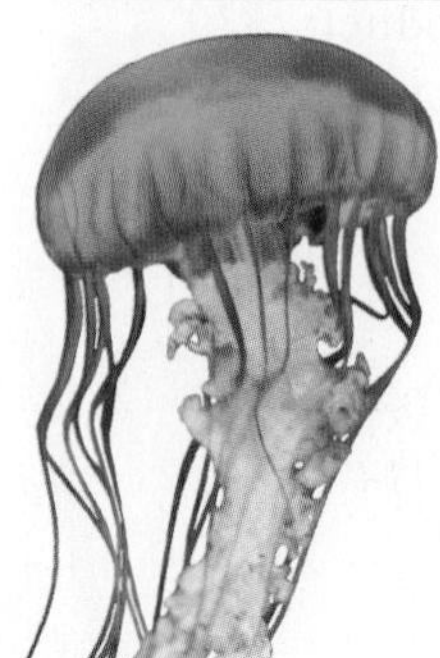

17 Properties of Solutions *851*

18 The Representative Elements *891*

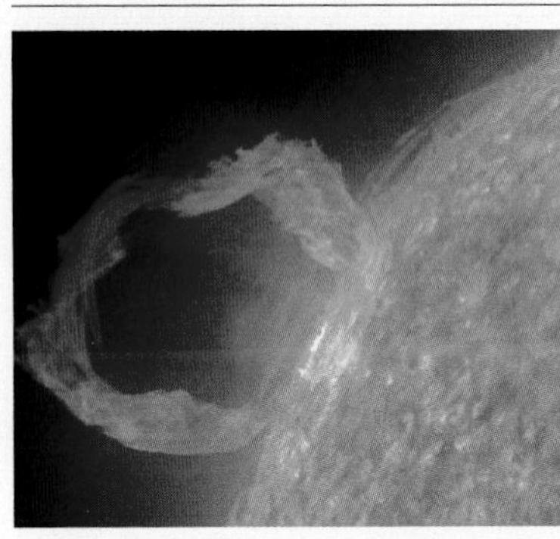

Learning to Think Like a Chemist

Chemistry is a fascinating and important subject that is challenging to teach and even more challenging to learn. Making this complex subject accessible to students without distortion is the challenge of the chemical educator, especially at the introductory level. *Chemical Principles*, Seventh Edition, provides a rigorous but understandable introduction to chemistry. It emphasizes conceptual understanding, the importance of models, and thoughtful problem solving.

Chemical Principles is based on our experiences at the University of Illinois teaching an accelerated general chemistry course for chemical sciences majors and other students who require a rigorous introductory course. These students typically have excellent credentials and a genuine aptitude for chemistry but only limited understanding of the fundamental concepts of chemistry. Although they may know how to solve stoichiometry and gas problems when they arrive in our courses, these students typically lack a thorough appreciation for the chemical principles that underlie these applications. This is not because they had inadequate preparation in high school; instead, we believe it results from the nature of chemistry itself—a subject that requires several passes before real mastery can take place.

Our mission in writing this text was to produce a book that does not assume that students already know how to think like chemists. These students will eventually do complicated and rigorous thinking, but they must be brought to that point gradually. Thus this book covers the advanced topics (in gases, atomic theory, thermodynamics, and so on) that one expects in a course for chemical sciences majors, but it starts with the fundamentals and then builds to the level required for more complete understanding. Chemistry is not the result of an inspired vision. It is the product of countless observations and many attempts, using logic and trial and error, to account for these observations. In this book we develop key chemical concepts in the same way—to show the observations first and then discuss the models that have been constructed to explain the observed behavior. We hope students will practice "thinking like a chemist" by carefully studying the observations to see if they can follow the thought process, rather than just jumping ahead to the equation or model that will follow.

In *Chemical Principles,* Seventh Edition, we take advantage of the excellent math skills that these students typically possess. As a result, there are fewer work-out examples than would be found in most mainstream books. The end-of-chapter problems cover a wide range—from drill exercises to difficult problems, some of which would challenge the average senior chemistry major. Thus instructors can tailor the problem assignments to the level appropriate for their students.

This text maintains a student-friendly approach without being patronizing. In addition, to demonstrate the importance of chemistry in real life, we have incorporated throughout the book a number of applications and recent advances in essay form.

New to This Edition

We continue to be pleased that the previous editions of the text have been well received. In response to comments from users, however, we have made some significant changes for the seventh edition.

- We have added a new section, Section 3.4, "Conceptual Problem Solving," that emphasizes the importance of conceptual problem solving in which students are shown how to think their way through a problem. We discuss how to solve problems in a flexible, creative way based on understanding the fundamental ideas of chemistry and asking and answering key questions. The students will learn that this "big picture approach" produces more long-term, meaningful learning rather than simply memorizing specific steps that are soon forgotten.
- Using the general conceptual problem-solving approach outlined in the new Section 3.4, we have introduced a series of questions into many of the in-chapter *Examples*. This more active approach helps students think their way through the solution to the problem.
- Several Chemical Insights boxes are new in the seventh edition, and many have been revised, with up-to-date topics such as geoengineering, laser cooling, graphene, and nanogenerators.
- We have replaced 10% of the end-of-chapter questions and problems and have added more than 50 new visual problems.
- We have added a list of the key terms at the end of each chapter along with an outline that reviews each chapter (termed *For Review*).
- We have revised many of the figures in the textbook to better serve visual learners.

Organization

The early chapters in this book deal with chemical reactions. Stoichiometry is covered in Chapters 3 and 4, with special emphasis on reactions in aqueous solutions. The properties of gases are treated in Chapter 5, followed by coverage of gas phase equilibria in Chapter 6. Acid–base equilibria are covered in Chapter 7, and Chapter 8 deals with additional aqueous equilibria. Thermodynamics is covered in two chapters: Chapter 9 deals with thermochemistry and the first law of thermodynamics; Chapter 10 treats the topics associated with the second law of thermodynamics. The discussion of electrochemistry follows in Chapter 11. Atomic theory and quantum mechanics are covered in Chapter 12, followed by two chapters on chemical bonding and modern spectroscopy (Chapters 13 and 14). Chemical kinetics is discussed in Chapter 15, followed by coverage of solids and liquids in Chapter 16 and the physical properties of solutions in Chapter 17. A systematic treatment of the descriptive chemistry of the representative elements is given in Chapter 18 and of the transition metals in Chapter 19. Chapter 20 covers topics in nuclear chemistry, and Chapter 21 provides an introduction to organic chemistry and to the most important biomolecules.

Flexibility of Topic Order

We recognize that the order of the chapters in this text may not fit the order of the topics in your course. Therefore, we have tried to make the order as flexible as possible. In the courses that we have taught using the text, we have successfully used it in a very different order from the one the text follows. We would encourage you to use it in whatever order that serves your purposes.

Instructors have several options for arranging the material to complement their syllabi. For example, the section on gas phase and aqueous equilibria

(Chapters 6–8) could be moved to any point later in the course. The chapters on thermodynamics can be separated: Chapter 9 can be used early in the course with Chapter 10 later. In addition, the chapters on atomic theory and bonding (Chapters 12–14) can be used near the beginning of the course. In summary, an instructor who wants to cover atomic theory early and equilibrium later might prefer the following order of chapters: 1–5, 9, 12, 13, 14, 10, 11, 6, 7, 8, 15–21. An alternative order might be: 1–5, 9, 12, 13, 14, 6, 7, 8, 10, 11, 15–21. The point is that the chapters on atomic theory and bonding (12–14), thermodynamics (9, 10), and equilibrium (6, 7, 8) can be moved around quite easily. In addition, the kinetics chapter (Chapter 15) can be covered at any time after bonding. It is also possible to use Chapter 20 (on nuclear chemistry) much earlier—after Chapter 12, for example—if desired.

Two approaches for teaching atomic theory earlier and equilibrium later in the course

APPROACH 1

Chapter 1 *Chemists and Chemistry*
Chapter 2 *Atoms, Molecules, and Ions*
Chapter 3 *Stoichiometry*
Chapter 4 *Types of Chemical Reactions and Solution Stoichiometry*
Chapter 5 *Gases*
Chapter 9 *Energy, Enthalpy, and Thermochemistry*
Chapter 12 *Quantum Mechanics and Atomic Theory*
Chapter 13 *Bonding: General Concepts*
Chapter 14 *Covalent Bonding: Orbitals*
Chapter 10 *Spontaneity, Entropy, and Free Energy*
Chapter 11 *Electrochemistry*
Chapter 6 *Chemical Equilibrium*
Chapter 7 *Acids and Bases*
Chapter 8 *Applications of Aqueous Equilibria*
Chapter 15 *Chemical Kinetics*
Chapter 16 *Liquids and Solids*
Chapter 17 *Properties of Solutions*
Chapter 18 *The Representative Elements*
Chapter 19 *Transition Metals and Coordination Chemistry*
Chapter 20 *The Nucleus: A Chemist's View*
Chapter 21 *Organic and Biochemical Molecules*

APPROACH 2

Chapter 1 *Chemists and Chemistry*
Chapter 2 *Atoms, Molecules, and Ions*
Chapter 3 *Stoichiometry*
Chapter 4 *Types of Chemical Reactions and Solution Stoichiometry*
Chapter 5 *Gases*
Chapter 9 *Energy, Enthalpy, and Thermochemistry*
Chapter 12 *Quantum Mechanics and Atomic Theory*
Chapter 13 *Bonding: General Concepts*
Chapter 14 *Covalent Bonding: Orbitals*
Chapter 6 *Chemical Equilibrium*
Chapter 7 *Acids and Bases*
Chapter 8 *Applications of Aqueous Equilibria*
Chapter 10 *Spontaneity, Entropy, and Free Energy*
Chapter 11 *Electrochemistry*
Chapter 15 *Chemical Kinetics*
Chapter 16 *Liquids and Solids*
Chapter 17 *Properties of Solutions*
Chapter 18 *The Representative Elements*
Chapter 19 *Transition Metals and Coordination Chemistry*
Chapter 20 *The Nucleus: A Chemist's View*
Chapter 21 *Organic and Biochemical Molecules*

Mathematical Level

This text assumes a solid background in algebra. All of the mathematical operations required are described in Appendix One or are illustrated in worked-out examples. A knowledge of calculus is not required for use of this text. Differential and integral notions are used only where absolutely necessary and are explained when they are used.

Alternate Versions

Hybrid Edition with Access (24 months) to OWL with Cengage YouBook
ISBN-10: 1-133-10984-5; ISBN-13: 978-1-133-10984-6
This briefer version of *Chemical Principles* does not contain the end-of-chapter problems, which can be assigned in OWL.

Supporting Materials

OWL for General Chemistry

Printed Access OWL with YouBook (24 months) ISBN-10: 1-133-04411-5; ISBN-13: 978-1-133-04411-6
Instant Access OWL with YouBook (6 months) ISBN-10: 1-133-00149-1; ISBN-13: 978-1-133-00149-2
Instant Access OWL with YouBook (24 months) ISBN-10: 1-133-00150-5; ISBN-13: 978-1-133-00150-8

By Roberta Day and Beatrice Botch of the University of Massachusetts, Amherst, and William Vining of the State University of New York at Oneonta. **OWL** Online Web Learning offers more assignable, gradable content (including end-of chapter questions specific to this textbook) and more reliability and flexibility than any other system. OWL's powerful course management tools allow instructors to control due dates, number of attempts, and whether students see answers or receive feedback on how to solve problems. OWL includes the **Cengage YouBook**, an interactive and customizable Flash-based eBook. Instructors can publish web links, modify the textbook narrative as needed with the text edit tool, quickly reorder entire sections and chapters, and hide any content they don't teach to create an eBook that perfectly matches their syllabus. The Cengage YouBook includes animated figures, video clips, highlighting, notes, and more.

Developed by chemistry instructors for teaching chemistry, OWL is the only system specifically designed to support **mastery learning**, where students work as long as they need to master each chemical concept and skill. OWL has already helped hundreds of thousands of students master chemistry through a wide range of assignment types, including tutorials, interactive simulations, and algorithmically generated homework questions that provide instant, answer-specific feedback.

OWL is continually enhanced with online learning tools to address the various learning styles of today's students such as:

- **Quick Prep** Review courses that help students learn essential skills to succeed in General and Organic Chemistry
- **Jmol** Molecular visualization program for rotating molecules and measuring bond distances and angles
- **Go Chemistry®** Mini video lectures on key concepts that students can play on their computers or download to their video iPods, smart phones, or personal video players

In addition, when you become an OWL user, you can expect service that goes far beyond the ordinary. To learn more or to see a demo, please contact your Cengage Learning representative, or visit us at www.cengage.com/owl.

For the Instructor

PowerLecture Instructor's CD/DVD Package with ExamView®

ISBN-10: 1-111-98895-1; ISBN-13: 978-1-111-98895-1

This digital library and presentation tool includes:

- **PowerPoint® lecture slides** written for this text that instructors can customize by importing their own lecture slides or other materials.
- **Image libraries** that contain digital files for figures, photographs, and numbered tables from the text, as well as multimedia animations in a variety of digital formats. Use these files to print transparencies, create your own PowerPoint slides, and supplement your lectures.
- Digital files of the complete **Instructor's Manual** and **ExamView Test Bank.**

- Sample chapters from the **Study Guide**, by Paul Kelter (Northern Illinois University)
- Sample chapters from the **Student Solutions Manual**, by Tom Hummel (University of Illinois)
- **ExamView testing software** that enables you to create, deliver, and customize tests using the more than 2000 test bank questions written specifically for this text

Instructor Companion Site

Supporting materials are available to qualified adopters. Please consult your local Cengage Learning sales representative for details. Go to login.cengage.com, find this textbook, and choose Instructor Companion Site to see samples of these materials, request a desk copy, locate your sales representative, and download the WebCT or Blackboard versions of the Test Bank.

For the Student

Visit CengageBrain.com

To access these and additional course materials, please visit www.cengagebrain.com. At the CengageBrain.com home page, search for this textbook's ISBN (from the back cover of your book). This will take you to the product page where these resources can be found. (Instructors can log in at login.cengage.com.)

OWL

OWL Quick Prep for General Chemistry

Instant Access OWL Quick Prep for General Chemistry (90 days) ISBN-10: 0-495-56030-8; ISBN-13: 978-0-495-56030-2

Quick Prep is a self-paced online short course that helps students succeed in general chemistry. Students who completed Quick Prep through an organized class or self-study averaged almost a full letter grade higher in their subsequent general chemistry course than those who did not. Intended to be taken prior to the start of the semester, Quick Prep is appropriate for both underprepared students and for students who seek a review of basic skills and concepts. Quick Prep features an assessment quiz to focus students on the concepts they need to study to be prepared for general chemistry. Quick Prep is approximately 20 hours of instruction delivered through OWL with no textbook required and can be completed at any time in the student's schedule. Professors can package a printed access card for Quick Prep with the textbook, or students can purchase instant access at www.cengagebrain.com. To view an OWL Quick Prep demonstration and for more information, visit www.cengage.com/chemistry/quickprep.

go Chemistry

Go Chemistry® for General Chemistry

ISBN-10: 0-495-38228-0; ISBN-13: 978-0-495-38228-7

Pressed for time? Missed a lecture? Need more review? Go Chemistry for General Chemistry is a set of 27 downloadable mini-video lectures. Developed by award-winning chemists, Go Chemistry helps you quickly review essential topics—whenever and wherever you want! Each video contains animations and problems and can be downloaded to your computer desktop or portable video player (like iPod or iPhone) for convenient self-study and exam review. Selected Go Chemistry videos have e-flashcards to briefly introduce a key concept and then test student understanding of the basics with a series of questions. OWL includes five Go Chemistry videos. Professors can package a printed access card for Go Chemistry with the textbook. Students can enter the ISBN above at www.cengagebrain.com to download two free videos or to purchase instant access to the 27-video set or to individual videos.

Student Companion Site

This site includes a glossary, flashcards, an interactive periodic table, and samples of the Study Guide and Student Solutions Manual, which are all accessible from www.cengagebrain.com.

Acknowledgments

The successful completion of this book is due to the efforts of many people. Mary Finch, publisher, and Lisa Lockwood, executive editor, were extremely supportive of the revision. We also wish to thank Thomas Martin, developmental editor, who has a keen eye, good ideas, and is extremely organized and helpful. We are grateful to have worked with Sharon Donahue, photo researcher, who never fails in finding just the right photo.

We greatly appreciate the efforts of Tom Hummel from the University of Illinois, who managed the revision of the end-of-chapter exercises and problems and the solutions manuals. Tom's extensive knowledge of general chemistry and high standards of accuracy ensure the quality of the problems and solutions in this text. We are deeply grateful to Gretchen Adams, who created the interactive examples and interactive end-of-chapter exercises and problems. Gretchen is extremely creative, never misses a deadline, and is a real pleasure to work with. Special thanks go to Nicole Hamm, marketing manager, who knows the market and works very hard in support of this book.

Thanks to others who provided valuable assistance on this revision: Krista Mastroianni, editorial assistant; Lisa Weber, senior media editor; Stephanie VanCamp, media editor; Julie Stefani, marketing assistant; Megan Greiner, production editor (Graphic World); Teresa L. Trego, content project manager; and Maria Epes, art director.

Our sincerest appreciation goes to all of the reviewers whose feedback and suggestions contributed to the success of this project.

Seventh Edition Reviewers

Rosemary Bartoszek-Loza, *Ohio State University*
H. Floyd Davis, *Cornell University*
Darby Feldwinn, *University of California–Santa Barbara*
Burt Goldberg, *New York University*
Kandalam V. Ramanujachary, *Rowan University*
Philip J. Reid, *University of Washington*
Christopher P. Roy, *Duke University*

Sixth Edition Reviewers

Elizabeth Day, *University of the Pacific*
Ivan J. Dmochowski, *University of Pennsylvania*
Brian Enderle, *University of California, Davis*
Regina Frey, *Washington University, St. Louis*
Brian Frost, *University of Nevada*
Derek Gragson, *California Polytechnic State University*
Keith Griffiths, *University of Western Ontario*
Carl Hoeger, *University of California, San Diego*
Robert Kerber, *State University of New York, Stony Brook*
K. C. McGill, *Georgia College and State University*
Thomas G. Minehan, *California State University, Northridge*
John H. Nelson, *University of Nevada*
Robert Price, *City College of San Francisco*
Douglas Raynie, *South Dakota State University*
Philip J. Reid, *University of Washington*
Thomas Schleich, *University of California, Santa Cruz*
Robert Sharp, *University of Michigan*

Mark Sulkes, *Tulane University*
John H. Terry, *Cornell University*
Mark Thachuk, *University of British Columbia*
Michael R. Topp, *University of Pennsylvania*
Meishan Zhao, *University of Chicago*

Fifth Edition Reviewers

Alan L. Balch, *University of California, Davis*
David Erwin, *Rose-Hulman Institute of Technology*
Michael Hecht, *Princeton University*
Rosemary Marusak, *Kenyon College*
Patricia B. O'Hara, *Amherst College*
Ruben D. Parra, *DePaul University*
Philip J. Reid, *University of Washington*
Eric Scerri, *University of California, Los Angeles*
Robert Sharp, *University of Michigan*

About the Authors

STEVEN S. ZUMDAHL received his B.S. degree in Chemistry from Wheaton College (Illinois) in 1964 and his Ph.D. in Chemistry from the University of Illinois, Urbana, in 1968.

In over 35 years of teaching he has been a faculty member at the University of Colorado, Boulder; Parkland College (Illinois); and the University of Illinois, where he served as Professor and Associate Head and Director of Undergraduate Programs in Chemistry until he became Professor Emeritus in 2003. In 1994 Dr. Zumdahl received the National Catalyst Award from the Chemical Manufacturers Association in recognition of his contribution to chemical education in the United States.

Professor Zumdahl is known at the University of Illinois for his rapport with students and for his outstanding teaching ability. During his tenure at the University, he received the University of Illinois Award for Excellence in Teaching, the Liberal Arts and Sciences College Award for Distinguished Teaching, and the School of Chemical Sciences Teaching Award (five times).

Dr. Z., as he is known to his students, greatly enjoys "mechanical things," including bicycles and cars. He collects and restores classic automobiles, having a special enthusiasm for vintage Corvettes and Packards.

DONALD J. DECOSTE is Associate Director of General Chemistry at the University of Illinois, Urbana-Champaign, and has been teaching chemistry at the high school and college levels for 24 years. He earned his B.S. degree in Chemistry and Ph.D. from the University of Illinois, Urbana-Champaign. At UIUC he has developed chemistry courses for nonscience majors, preservice secondary teachers, and preservice elementary teachers. He teaches courses in introductory chemistry and the teaching of chemistry and has received the School of Chemical Sciences Teaching Award four times. Don has led workshops for secondary teachers and graduate student teaching assistants, discussing the methods and benefits of getting students more actively involved in class. When not involved in teaching and advising, Don enjoys spending time with his wife and three children.

chapter

1

Chemists and Chemistry

A variety of chemistry glassware.

iStockphoto.com

OWL

Sign in to OWL at www.cengage.com/owl to view tutorials and simulations, develop problem-solving skills, and complete online homework assigned by your professor.

go Chemistry

Download mini lecture videos for key concept review and exam prep from OWL or purchase them from www.cengagebrain.com.

Chemistry. It is a word that evokes various, and often dramatic, responses. It is a word that is impossible to define concisely, because the field is so diverse and its practitioners perform such an incredible variety of jobs. Chemistry mainly deals with situations in which the nature of a substance is changed by altering its composition; entirely new substances are synthesized, or the properties of existing substances are enhanced.

There are many misconceptions about the practitioners of chemistry. Many people picture a chemist as a solitary figure who works in a laboratory and does not talk to anyone else for days at a time. Nothing could be further from the truth. Many chemists do indeed work in laboratories, but rarely by themselves. A typical day for a modern chemist would be spent as a member of a team solving a particular problem important to his or her company. This team might consist of chemists from various specialties, chemical engineers, development specialists, and possibly even lawyers. Figure 1.1 represents the people and organizations with which typical laboratory chemists might expect to interact in the course of their jobs.

On the other hand, many persons trained as chemists do not perform actual laboratory work but may work as patent lawyers, financial analysts, plant managers, salespeople, personnel managers, and so on. Also, it is quite common for a person trained as a chemist to have many different jobs during a career.

In Chapters 2 through 21 of this text we will concentrate on the formal discipline of chemistry—its observations, theories, and applications. The goal of Chapter 1 is to introduce some of the important aspects of chemistry not typically discussed in connection with learning chemistry. The chapter includes an introduction to the world of commercial chemistry and provides a couple of specific examples of the types of problems confronted by the practitioners of

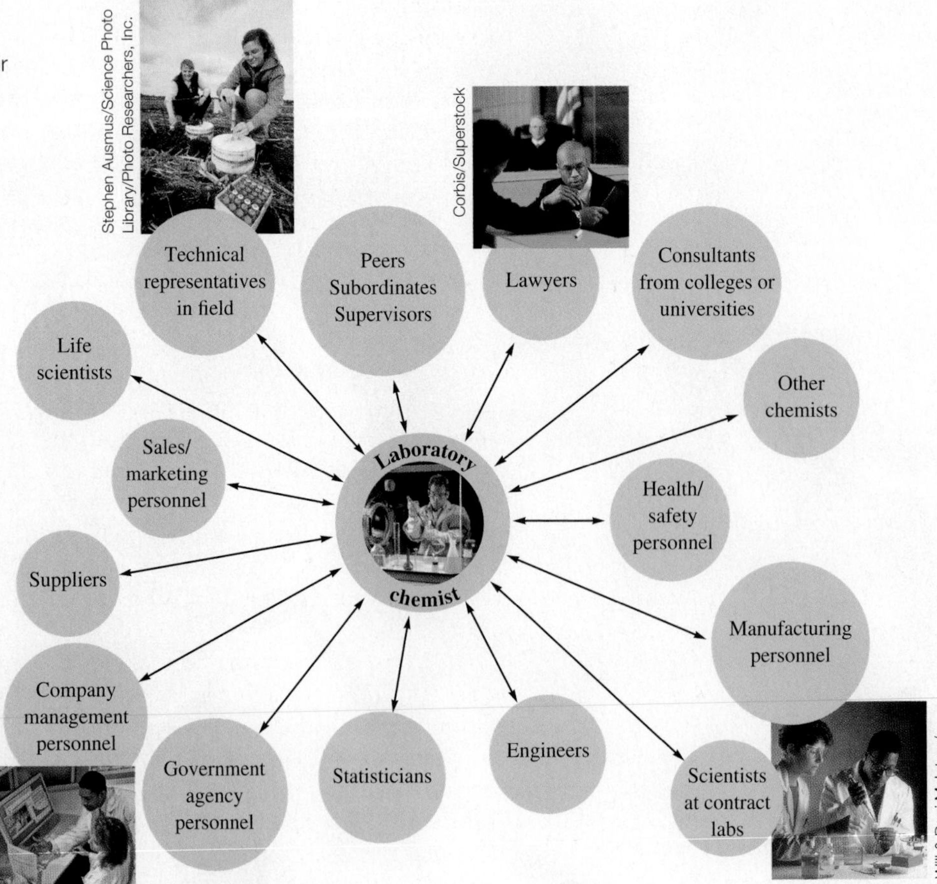

Figure 1.1
Typical chemists interact with a great variety of other people while doing their jobs. (Center photo: Lester Lefkowitz/Corbis #SC-019-0199)

the "chemical arts." We begin by considering the chemical scientist as a problem solver.

1.1 Thinking Like a Chemist

Much of your life, both personal and professional, will involve problem solving. Most likely, the more creative you are at solving problems, the more effective and successful you will be. Chemists are usually excellent problem solvers because they get a lot of practice. Chemical problems are frequently very complicated—there is usually no neat and tidy solution. Often it is difficult to know where to begin. In response to this dilemma, a chemist makes an educated guess (formulates a hypothesis) and then tests it to see if the proposed solution correctly predicts the observed behavior of the system. This process of trial and error is virtually a way of life for a chemist. Chemists rarely solve a complex problem in a straightforward, elegant manner. More commonly, they poke and prod the problem and make progress only in fits and starts.

It's very important to keep this in mind as you study chemistry. Although "plug and chug" exercises are necessary to familiarize you with the relationships that govern chemical behavior, your ultimate goal should be to advance beyond this stage to true problem solving. Unfortunately, it is impossible to give a formula for becoming a successful problem solver. Creative problem solving is a rather mysterious activity that defies simple analysis. However, it is clear that practice helps. That's why we will make every attempt in this text to challenge you to be creative with the knowledge of chemistry you will be acquiring. Although this process can be frustrating at times, it is definitely worth the struggle—both because it is one of the most valuable skills you can develop and because it helps you test your understanding of chemical concepts. If your understanding of these concepts is not sufficient to allow you to solve problems involving "twists" that you have never encountered before, your knowledge is not very useful to you. The only way to develop your creativity is to expose yourself to new situations in which you need to make new connections. A substantial part of creative problem solving involves developing the confidence necessary to think your way through unfamiliar situations. You must recognize that the entire solution to a complex problem is almost never visible in the beginning. Typically, one tries first to understand pieces of the problem and then puts those pieces together to form the solution.

1.2 A Real-World Chemistry Problem

Kristen Brochmann, Fundamental Photographs

Acid-damaged paper.

As discussed, the professional chemist is primarily a problem solver—one who daily confronts tough, but fascinating, situations that must be understood. To illustrate, we will consider an important current problem that requires chemical expertise to solve: the crumbling of the paper in many of the books published in the past century. The pages of many of these books are literally falling apart. To give some perspective on the magnitude of the problem, if the books in the New York Public Library were lined up, they would stretch for almost 100 miles. Currently, about 40 miles of these books are quietly crumbling to dust.

Because of the magnitude of this problem, the company that develops a successful preservation process will reap considerable financial rewards, in addition to performing an important service to society. Assume that you work for a company that is interested in finding a method for saving the crumbling paper in books and that you are put in charge of your company's efforts to develop such a process. What do you know about paper? Probably not much. So the first step is to go to the library to learn all you can about paper. Because paper manufacturing is a mature industry, a great deal of information is avail-

Chemistry Explorers

Alison Williams's Focus: The Structure of Nucleic Acids

Alison Williams started her scientific career as a high school student when she worked part-time at the Ohio State Agricultural Research and Development Center in Wooster, Ohio. She subsequently received her undergraduate degree from Wesleyan University, and then her master's degree and Ph.D. in biophysical chemistry. Dr. Williams has taught at Swarthmore College, Wesleyan University, Princeton University, and is now a senior lecturer at Barnard College.

Dr. Williams's primary interest is to understand the thermodynamic and kinetic behavior of nucleic acid structure. Nucleic acids, in the form of the huge polymers DNA and RNA, are central to the genetic machinery of cells. Dr. Williams and her research group are studying shorter nucleic acids, with the goal of determining how conditions such as the presence of ions in the cellular solutions affect the structures of those nucleic acids. This is significant because ion concentrations in cells vary depending on the type of cell and its growth rate. Dr. Williams's work in this area and related areas should shed light on many important biological processes.

Denise Applewhite/Office of Communications/Princeton University

Alison Williams.

able. Research at the library will show that paper is made of cellulose obtained from wood pulp and that the finished paper is "sized" to give it a smooth surface that prevents ink from "fuzzing." The agent typically used for sizing is alum [$Al_2(SO_4)_3$], which is the cause of the eventual decomposition of the paper. This happens as follows: In the presence of moisture, the Al^{3+} ions from alum become hydrated, forming $Al(H_2O)_6{}^{3+}$. The $Al(H_2O)_6{}^{3+}$ ion acts as an acid because the very strong Al^{3+}—O bond causes changes in the O—H bonds of the attached water molecules, thus allowing H^+ ions to be produced by the following reaction:

$$Al(H_2O)_6{}^{3+} \rightleftharpoons [Al(OH)(H_2O)_5]^{2+} + H^+$$

Therefore, paper sized with alum contains significant numbers of H^+ ions. This is important because the H^+ assists in the breakdown of the polymeric cellulose structure of paper. Cellulose is composed of glucose molecules ($C_6H_{12}O_6$) bonded together to form long chains. A segment of cellulose is shown in Fig. 1.2. When the long chains of glucose units in cellulose are broken into shorter pieces, the structural integrity of the paper fails and it crumbles.

Although library research helps you to understand the fundamentals of the problem, now the tough part (and the most interesting part) begins. Can you find a creative solution to the problem? Can the paper in existing books be

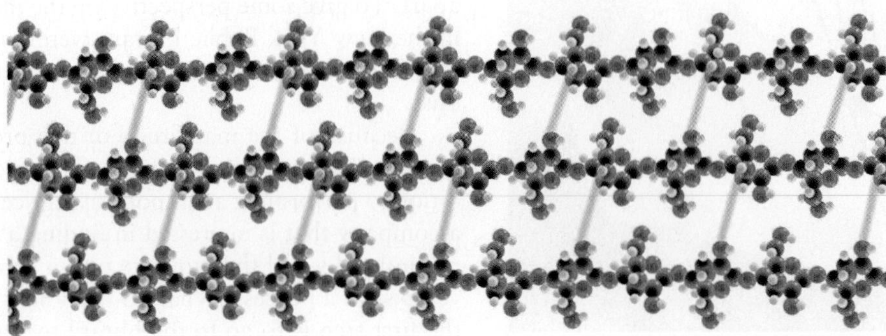

Figure 1.2
The polymer cellulose, which consists of β-D-glucose monomers. (Source: http://en.wikipedia.org/wiki/File:Cellulose_spacefilling_model.jpg)

Stephanie Burns: Chemist, Executive

Chemistry Explorers

Stephanie Burns was always interested in science, even as a little girl. This interest intensified over the years until she obtained a Ph.D. in organic chemistry from Iowa State University, where she specialized in the organic chemistry of silicon. Her career path led her to a job with Dow Corning Company, where she developed useful products containing silicon. Eventually her career path led to several positions involving product development, marketing, and business management. Her outstanding performance in these positions resulted in her appointment as an executive vice president. In early 2003, Dr. Burns, at age 48, was promoted to President and Chief Operating Officer for Dow Corning. In 2004 she became Chief Executive Officer, and in 2006 she was elected Chairman. She has repeatedly been on *Forbes*'s list of the 100 most powerful women.

Dr. Burns says "there was no magic" in reaching the position of Chairman and Chief Executive Officer of Dow Corning. "I'm driven by the science and technology of the company. It's in my blood," she says. Burns says her top priority is to encourage her company's scientists to develop innovative products and expand business built on silicon-based chemistry.

Stephanie Burns.

Courtesy Dow Corning. Photo by Jeffrey Glen.

treated to stop the deterioration in a way that is economical, permanent, and safe?

The essence of the problem seems to be the H^+ present in the paper. How can it be removed or at least rendered harmless?

Your general knowledge of chemistry tells you that some sort of base (a substance that reacts with H^+) is needed. One of the most common and least expensive bases is sodium hydroxide. Why not dip the affected books in a solution of sodium hydroxide and remove the H^+ by the reaction: $H^+ + OH^- \rightarrow H_2O$? This seems to be a reasonable first idea, but as you consider it further and discuss it with your colleagues, several problems become apparent:

1. The $NaOH(aq)$ is a strong base and is therefore quite corrosive. It will destroy the paper by breaking down the cellulose just as acid does.
2. The book bindings will be destroyed by dipping the books in water, and the pages will stick together after the books dry.
3. The process will be very labor-intensive, requiring the handling of individual books.

Some of these difficulties can be addressed. For example, a much weaker base than sodium hydroxide could be used. Also, the pages could be removed from the binding, soaked one at a time, dried, and then rebound. In fact, this process is used for some very rare and valuable books, but the labor involved makes it very expensive—much too expensive for the miles of books in the New York Public Library. Obviously, this process is not what your company is seeking.

You need to find a way to treat large numbers of books without disassembling them. How about using a gaseous base? The books could be sealed in a chamber and the gaseous base allowed to permeate them. The first candidate that occurs to you is ammonia, a readily available gaseous base that reacts with H^+ to form NH_4^+:

$$NH_3 + H^+ \longrightarrow NH_4^+$$

This seems like a very promising idea, so you decide to construct a pilot treatment chamber. To construct this chamber, you need some help from coworkers. For example, you might consult a chemical engineer for help in the design of the plumbing and pumps needed to supply ammonia to the chamber. You might also consult a mechanical engineer about the appropriate material to use for the chamber and then discuss the actual construction of the chamber with machinists and other personnel from the company's machine shop. In addition, you probably would consult a safety specialist and possibly a toxicologist about the hazards associated with ammonia.

Before the chamber is built, you also have to think carefully about how to test the effectiveness of the process. How could you evaluate, in a relatively short time, how well the process protects paper from deterioration? At this stage, you would undoubtedly do more library research and consult with other experts, such as a paper chemist your company hires as an outside consultant.

Assume now that the chamber has been constructed and that the initial tests look encouraging. At first the H^+ level is greatly reduced in the treated paper. However, after a few days the H^+ level begins to rise again. Why? The fact that ammonia is a gas at room temperature (and pressure) is an advantage because it allows you to treat many books simultaneously in a dry chamber. However, the volatility of ammonia works against you after the treatment. The process

$$NH_4^+ \longrightarrow NH_3\uparrow + H^+$$

allows the ammonia to escape after a few days. Thus this treatment is too temporary. Even though this effort failed, it was still useful because it provided an opportunity to understand what is required to solve this problem. You need a gaseous substance that *permanently* reacts with the paper and that also consumes H^+.

In discussing this problem over lunch, a colleague suggests the compound diethyl zinc [$(C_2H_5)_2Zn$], which is quite volatile (boiling point = 117°C) and which reacts with water (moisture is present in paper) as follows:

$$(C_2H_5)_2Zn + H_2O \longrightarrow ZnO + 2C_2H_6$$

The C_2H_6 (ethane) is a gas that escapes, but the white solid, ZnO, becomes an integral part of the paper. The important part of ZnO is the oxide ion, O^{2-}, which reacts with H^+ to form water:

$$O^{2-} + 2H^+ \longrightarrow H_2O$$

Thus the ZnO is a nonvolatile base that can be placed in the paper by a gaseous substance. This process seems very promising. However, the major disadvantage of this process (there are always disadvantages) is that diethyl zinc is *very* flammable and great care must be exercised in its use. This leads to another question: Is the treatment effective enough to be worth the risks involved? As it turns out, the Library of Congress used diethyl zinc until 1994, but the process was discontinued because of its risks. Since then, a process known as Bookkeeper has been used. In this process, the book is immersed into a suspension of magnesium oxide (MgO). Small particles (submicron) of MgO are deposited in the pages, and these neutralize the acid and, like ZnO formed from diethyl zinc, become an integral part of the paper. The advantages are the simplicity of the application and the safety of the method.

The type of problem solving illustrated by investigation of the acid decomposition of paper is quite typical of that which a practicing chemist confronts

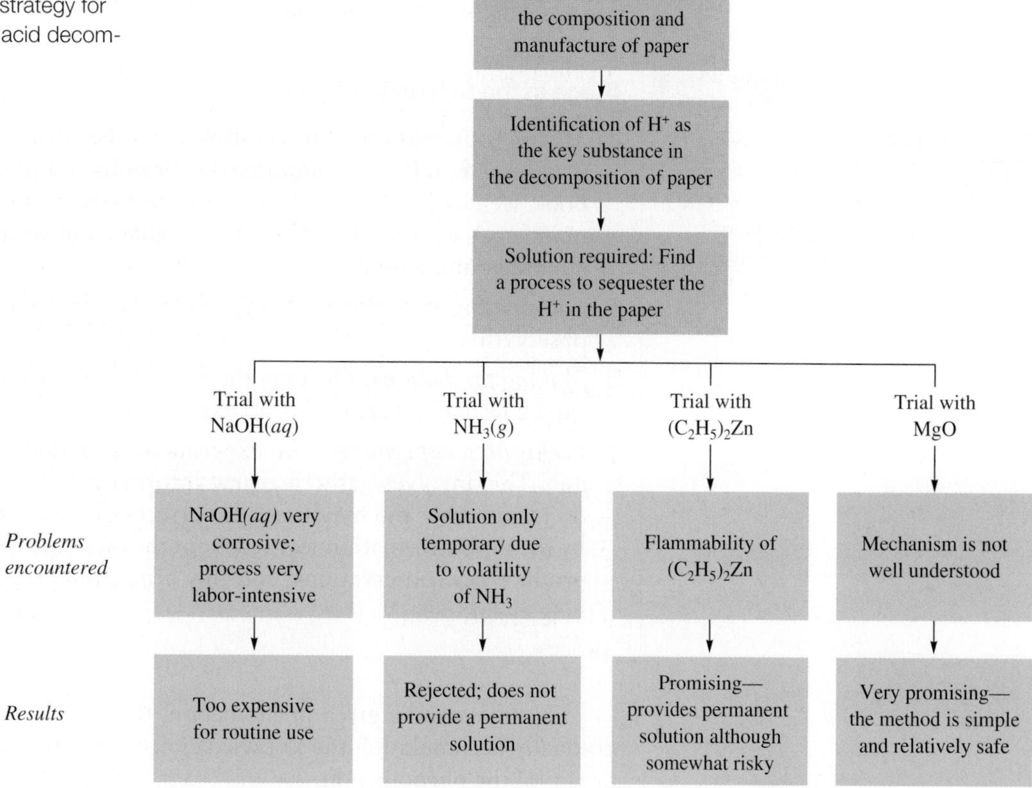

Figure 1.3
Schematic diagram of the strategy for solving the problem of the acid decomposition of paper.

daily. The first step in successful problem solving is to identify the exact nature of the problem. Although this may seem trivial, it is often the most difficult and most important part of the process. Poor problem solving often results from a fuzzy definition of the problem. You cannot efficiently solve a problem if you do not understand the essence of the problem. Once the problem is well defined, then solutions can be advanced, usually by a process of intelligent trial and error. This process typically involves starting with the simplest potential solution and iterating to a final solution as the feedback from earlier attempts is used to refine the approach. Rarely, if ever, is the solution to a complex problem obvious immediately after the problem is defined. The best solution becomes apparent only as the results from various trial solutions are evaluated. A schematic summarizing the approach for dealing with the acid decomposition of paper is shown in Fig. 1.3.

1.3 The Scientific Method

Science is a framework for gaining and organizing knowledge. Science is not simply a set of facts but is also a plan of action—a *procedure* for processing and understanding certain types of information. Scientific thinking is useful in all aspects of life, but in this text we will use it to understand how the chemical world operates. The process that lies at the center of scientific inquiry is called the **scientific method.** There are actually many scientific methods depending on the nature of the specific problem under study and on the particular investigator involved. However, it is useful to consider the following general framework for a generic scientific method:

STEPS

See Appendix A1.6 for conventions regarding the use of significant figures in connection with measurements and the calculations involving measurements. Appendix 2 discusses methods for converting among various units.

Steps in the Scientific Method

1 *Making observations.* Observations may be *qualitative* (the sky is blue; water is a liquid) or *quantitative* (water boils at 100°C; a certain chemistry book weighs 2 kilograms). A qualitative observation does not involve a number. A quantitative observation (called a **measurement**) involves both a number and a unit.

2 *Formulating hypotheses.* A hypothesis is a *possible* explanation for the observation.

3 *Making predictions.* The hypothesis then is used to make a prediction that can be tested by performing an experiment.

4 *Performing experiments.* An experiment is carried out to test the hypothesis. This involves gathering new information that enables a scientist to decide whether the hypothesis is correct—that is, whether it is supported by the new information learned from the experiment. Experiments always produce new observations, and this brings the process back to the beginning again.

To understand a given phenomenon, these steps are repeated many times, gradually accumulating the knowledge necessary to provide a possible explanation of the phenomenon.

As scientists observe nature, they often see that the same observation applies to many different systems. For example, innumerable chemical changes have shown that the total observed mass of the materials involved is the same before and after the change. Such generally observed behavior is formulated into a statement called a **natural law.** For example, the observation that the total mass of materials is not affected by a chemical change in those materials is called the law of conservation of mass. This law tells us *what* happens, but it does not tell us *why.* To try to explain why, we continue to make observations, formulate hypotheses, and test these against observations.

Once a set of hypotheses that agree with the various observations is obtained, the hypotheses are assembled into a theory. A **theory,** which is often called a *model,* is a set of tested hypotheses that gives an overall explanation of some natural phenomenon.

This portrayal of the classical scientific method probably overemphasizes the importance of observations in current scientific practice. Now that we know a great deal about the nature of matter, scientists often start with a hypothesis that they try to refute as they push forward the frontiers of science. See the writings of Karl Popper for more information on this view.

It is very important to distinguish between observations and theories. An observation is something that is witnessed and can be recorded. A theory is an *interpretation*—a possible explanation of *why* nature behaves in a particular way. For example, in Chapter 2 we will read about Dalton's atomic theory, in which John Dalton proposed that a chemical reaction is a reorganization of atoms in reacting substances to produce new substances. As we discussed, we know that mass is conserved (it is a natural law), and we can explain it by claiming that all matter is made of nonchanging atoms (the theory).

Theories inevitably change as more information becomes available. For example, we will also see in Chapter 2 that with further experimentation and observations, the atomic theory came to include subatomic particles—electrons, protons, and neutrons. The "indivisible" atom of Dalton is not indivisible after all. We see the idea of changing theories in all realms of science. For example, the motions of the sun and stars have remained virtually the same over the thousands of years during which humans have been observing them, but our explanations—our theories—for these motions have changed greatly since ancient times.

Critical Units!

Chemical Insights

How important are conversions from one unit to another? If you ask the National Aeronautics and Space Administration (NASA), very important! In 1999 NASA lost a $125 million Mars Climate Orbiter because of a failure to convert from English to metric units.

The problem arose because two teams working on the Mars mission were using different sets of units. NASA's scientists at the Jet Propulsion Laboratory in Pasadena, California, assumed that the thrust data for the rockets on the orbiter they received from Lockheed Martin Astronautics in Denver, which built the spacecraft, were in metric units. In reality, the units were English. As a result the orbiter dipped 100 kilometers lower into the Mars atmosphere than planned and the friction from the atmosphere caused the craft to burn up.

NASA's mistake refueled the controversy over whether Congress should require the United States to switch to the metric system. About 95% of the world now uses the metric system, and the United States is slowly switching from English to metric. For example, the automobile industry has adopted metric fasteners and we buy our soda in 2-liter bottles.

NASA

Artist's conception of the lost Mars Climate Orbiter.

Units can be very important. In fact, they can mean the difference between life and death on some occasions. In 1983, for example, a Canadian jetliner almost ran out of fuel when someone pumped 22,300 pounds of fuel into the aircraft instead of 22,300 kilograms. Remember to watch your units!

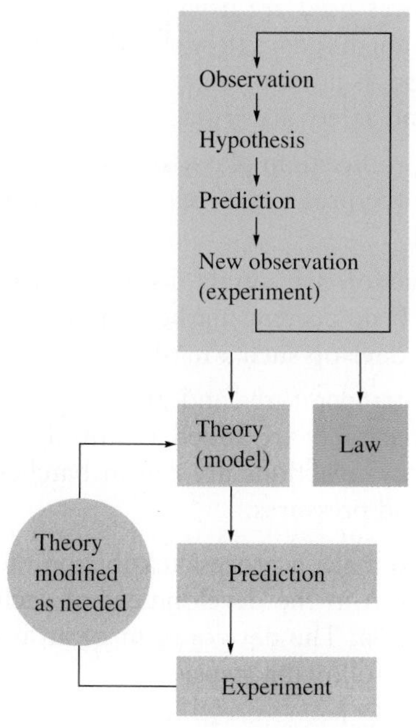

Figure 1.4
The various parts of the scientific method.

The point is that scientists do not stop asking questions just because a given theory seems to account satisfactorily for some aspect of natural behavior. They continue doing experiments to refine or replace the existing theories. This is generally done by using the currently accepted theory to make a prediction and then performing an experiment (making a new observation) to see whether the results bear out this prediction.

Always remember that theories (models) are human inventions. They represent attempts to explain observed natural behavior in terms of human experiences. A theory is actually an educated guess. We must continue to do experiments and to refine our theories (making them consistent with new knowledge) if we hope to approach a more nearly complete understanding of nature.

In this section we have described the scientific method as it might ideally be applied (Fig. 1.4). However, it is important to remember that science does not always progress smoothly and efficiently. For one thing, hypotheses and observations are not totally independent of each other, as we have assumed in the description of the idealized scientific method. The coupling of observations and hypotheses occurs because once we begin to proceed down a given theoretical path, our hypotheses are unavoidably couched in the language of those theoretical underpinnings. In other words, we tend to see what we expect to see and often fail to notice things that we do not expect. Thus the theory we are testing helps us because it focuses our questions. However, at the very same time, this focusing process may limit our ability to see other possible explanations.

It is also important to keep in mind that scientists are human. They have prejudices; they misinterpret data; they become emotionally attached to their theories and thus lose objectivity; and they play politics. Science is affected by profit motives, budgets, fads, wars, and religious beliefs. Galileo, for example, was forced to recant his astronomical observations in the face of strong religious resistance. Lavoisier, the father of modern chemistry, was beheaded because of his political affiliations. And great progress in the chemistry of nitrogen fertilizers resulted from the desire to produce explosives to fight wars. The progress of science is often affected more by the frailties of humans and their institutions than by the limitations of scientific measuring devices. The scientific methods are only as effective as the humans using them. They do not automatically lead to progress.

1.4 Industrial Chemistry

Christian Lagerek/Shutterstock.com #56046928

Industrial processes require large plants for the production of chemicals.

The impact of chemistry on our lives is due in no small measure to the many industries that process and manufacture chemicals to provide the fuels, fabrics, fertilizers, food preservatives, detergents, and many other products that affect us daily. The chemical industry can be subdivided in terms of three basic types of activities:

1. The isolation of naturally occurring substances for use as raw materials
2. The processing of raw materials by chemical reactions to manufacture commercial products
3. The use of chemicals to provide services

A given industry may participate in one, two, or all of these activities.

Producing chemicals on a large industrial scale is very different from an academic laboratory experiment. Some of the important differences are described below.

- In the academic laboratory, practicality is typically the most important consideration. Because the amounts of substances used are usually small, hazardous materials can be handled by using fume hoods, safety shields, and so on; expense, although always a consideration, is not a primary factor. However, for any industrial process, economy and safety are critical.
- In industry, containers and pipes are metal rather than glass, and corrosion is a constant problem. In addition, because the progress of reactions cannot be monitored visually, gauges must be used.
- In the laboratory, any by-products of a reaction are simply disposed of; in industry, they are usually recycled or sold. If no current market exists for a given by-product, the manufacturer tries to develop such a market.
- Industrial processes often run at very high temperatures and pressures and ideally are *continuous flow,* meaning that reactants are added and products are extracted continuously. In the laboratory, reactions are run in batches and typically at much lower temperatures and pressures.

The many criteria that must be satisfied to make a process feasible on the industrial scale require that great care be taken in the development of each process to ensure safe and economical operation. The development of an industrial chemical process typically involves the following steps:

Step 1: A need for a particular product is identified.

Step 2: The relevant chemistry is studied on a small scale in a laboratory. Various ways of producing the desired material are evaluated in terms of costs and potential hazards.

A Note-able Achievement

Chemical Insights

Post-it Notes, a product of the 3M Corporation, revolutionized casual written communications and personal reminders. Introduced in the United States in 1980, these sticky-but-not-too-sticky notes have now found countless uses in offices, cars, and homes throughout the world.

The invention of sticky notes occurred over a period of about 10 years and involved a great deal of serendipity. The adhesive for Post-it Notes was discovered by Dr. Spencer F. Silver of 3M in 1968. Silver found that when an acrylate polymer material was made in a particular way, it formed cross-linked microspheres. When suspended in a solvent and sprayed on a sheet of paper, this substance formed a "sparse monolayer" of adhesive after the solvent evaporated. Scanning electron microscope images of the adhesive show that it has an irregular surface, a little like the surface of a gravel road. In contrast, the adhesive on cellophane tape looks smooth and uniform, like a superhighway. The bumpy surface of Silver's adhesive caused it to be sticky but not so sticky as to produce permanent adhesion because the number of contact points between the binding surfaces was limited.

When he invented this adhesive, Silver had no specific ideas for its use, so he spread the word of his discovery to his fellow employees at 3M to see if anyone had an application for it. In addition, over the next several years development was carried out to improve the adhesive's properties. It was not until 1974 that the idea for Post-it Notes popped up. One Sunday, Art Fry, a chemical engineer for 3M, was singing in his church choir when he became annoyed that the bookmark in his hymnal kept falling out. He thought to himself that it would be nice if the bookmark were sticky enough to stay in place but not so sticky that it couldn't be moved. Luckily, he remembered Silver's glue—and the Post-it Note was born.

For the next three years, Fry worked to overcome the manufacturing obstacles associated with the product. By 1977 enough Post-it Notes were being produced to supply 3M's corporate headquarters, where the employees quickly became addicted to their many uses. Post-it Notes are now available in 62 colors and 25 shapes.

In the years since their introduction, 3M has heard some remarkable stories connected to the use of these notes. For example, a Post-it Note was applied to the nose of a corporate jet, where it was intended to be read by the plane's Las Vegas ground crew. Someone forgot to remove it, however. The note was still on the nose of the plane when it landed in Minneapolis, having survived a takeoff and landing and speeds of 500 miles per hour at temperatures as low as −56°F. Stories on the 3M website also describe how a Post-it Note on the front door of a home survived the 140 mile per hour winds of Hurricane Hugo and how a foreign official accepted Post-it Notes in lieu of cash when a small bribe was needed to cut through bureaucratic hassles.

Post-it Notes have definitely changed the way we communicate and remember things.

Step 3: The data are evaluated by chemists, chemical engineers, business managers, safety engineers, and others to determine which possibility is most feasible.

Step 4: A *pilot-plant test* of the process is carried out. The scale of the pilot plant is between that of the laboratory and that of a manufacturing plant. This test has several purposes: to make sure that the reaction is efficient at a larger scale, to test reactor (reaction container) designs, to determine the costs of the process, to evaluate the hazards, and to gather information on environmental impact.

1.5 Polyvinyl Chloride (PVC): Real-World Chemistry

To get a little better feel for how the world of industrial chemistry operates, we will now consider a particular product, polyvinyl chloride (PVC), to see what types of considerations have been important in making this a successful and important consumer product.

When you put on a nylon jacket, use a polyethylene wash bottle in the lab, wear contact lenses, or accidentally drop your telephone (and it doesn't break), you are benefiting from the properties of polymers. Polymers are very large molecules that are assembled from small units (called monomers). Because of their many useful properties, polymers are manufactured in huge quantities. In fact, it has been estimated that more than 50% of all industrial chemists have jobs that are directly related to polymers.

One particularly important polymer is polyvinyl chloride (PVC), which is made from the molecule commonly called vinyl chloride:

$$\begin{array}{ccc} \mathrm{H} & & \mathrm{H} \\ & \mathrm{C{=}C} & \\ \mathrm{H} & & \mathrm{Cl} \end{array}$$

When many of these units are joined together, the polymer PVC results:

$$\begin{array}{cccccc} \mathrm{H} & \mathrm{H} & \mathrm{H} & \mathrm{H} & \mathrm{H} & \mathrm{H} \\ | & | & | & | & | & | \\ \sim\mathrm{C} & -\mathrm{C} & -\mathrm{C} & -\mathrm{C} & -\mathrm{C} & -\mathrm{C}\sim \\ | & | & | & | & | & | \\ \mathrm{H} & \mathrm{Cl} & \mathrm{H} & \mathrm{Cl} & \mathrm{H} & \mathrm{Cl} \end{array}$$

This can be represented as

$$\left(\begin{array}{cc} \mathrm{H} & \mathrm{H} \\ | & | \\ -\mathrm{C} & -\mathrm{C}- \\ | & | \\ \mathrm{H} & \mathrm{Cl} \end{array} \right)_n$$

where n is usually greater than 1000.

Because the development of PVC into a useful, important material is representative of the type of problem solving encountered in industrial chemistry, we will consider it in some detail.

In pure form PVC is a hard, brittle substance that decomposes easily at the high temperatures necessary to process it. This makes it almost useless. The fact that it has become a high-volume plastic ($\approx$10 billion pounds per year produced in the United States) is a tribute to chemical innovation. Depending on the additives used, PVC can be made rigid or highly flexible, and it can be tailored for use in inexpensive plastic novelty items or for use in precision engineering applications.

A scientist inspecting a product being formed from polyvinyl plastic.

Brownie Harris/Stock Market/Corbis

The development of PVC illustrates the interplay of logic and serendipity, as well as the importance of optimizing properties both for processing and for applications. PVC production has been beset with difficulties from the beginning, but solutions have been found for each problem through a combination of chemical deduction and trial and error. For example, many additives have been found that provide temperature stability so that PVC can be processed as a melt (liquid) and so that PVC products can be used at high temperatures. However, there is still controversy among chemists about exactly how PVC decomposes thermally, and thus the reason these stabilizers work is not well understood. Also, there are approximately 100 different plasticizers (softeners) available for PVC, but the theory of its plasticization is too primitive to predict accurately which compounds might produce even better results.

PVC was discovered by a German chemical company in 1912, but its brittleness and thermal instability proved so problematical that in 1926 the company stopped paying the fees to maintain its patents. That same year Waldo Semon, a chemist at B. F. Goodrich, found that PVC could be made flexible by the addition of phosphate and phthalate esters. Semon also found that white lead [$Pb_3(OH)_2(CO_3)_2$] provided thermal stability to PVC. These advances led to the beginning of significant U.S. industrial production of PVC ($\approx$4 million pounds per year by 1936). In an attempt to further improve PVC, T. L. Gresham (also a chemist at B. F. Goodrich) tried approximately 1000 compounds, searching for a better plasticizer. The compound that he found (its identity is not important here) remains the most common plasticizer added to PVC. The types of additives commonly used in the production of PVC are listed in Table 1.1.

Although the exact mechanism of the thermal, heat-induced decomposition of PVC remains unknown, most chemists agree that the chlorine atoms present in the polymer play an important role. Lead salts are added to PVC both to provide anions less reactive than chloride and to provide lead ions to combine with the released chloride ions. As a beneficial side effect, the lead chloride formed gives PVC enhanced electrical resistance, making lead stabilizers particularly useful in producing PVC for electrical wire insulation.

One major use of PVC is for pipes in plumbing systems. Here, even though the inexpensive lead stabilizers would be preferred from an economic standpoint, the possibility that the toxic lead could be leached from the pipes into the drinking water necessitates the use of more expensive tin and antimony compounds as thermal stabilizers. Because about one-half of the annual U.S. production of PVC is formed into piping, the PVC formulation used for pipes represents a huge market for companies that manufacture additives, and the competition is very intense. A recently developed low-cost thermal stabilizer for PVC is a mixture of antimony and calcium salts. This mixture has replaced stabilizers containing tin compounds that have become increasingly costly in recent years.

Outdoor applications of PVC often require that it contain ultraviolet light absorbers to protect against damage from sunlight. For pigmented applications

Table 1.1

Types of Additives Commonly Used in the Production of PVC

Type of Additive	Effect
Plasticizer	Softens the material
Heat stabilizer	Increases resistance to thermal decomposition
Ultraviolet absorber	Prevents damage by sunlight
Flame retardant	Lowers flammability
Biocide	Prevents bacterial or fungal attack

such as vinyl siding, window frames, and building panels, titanium(IV) oxide (TiO_2) is usually used. For applications in which the PVC must be transparent, other compounds are needed.

The additives used in PVC in the largest amounts are plasticizers, but one detrimental effect of these additives is an increase in flammability. Rigid PVC, which contains little plasticizer, is quite flame resistant because of its high chloride content. However, as more plasticizer is added for flexibility, the flammability increases to the point where fire retardants must be added, the most common being antimony(III) oxide (Sb_2O_3). As the PVC is heated, this oxide forms antimony(III) chloride ($SbCl_3$), which migrates into the flame, where it inhibits the burning process. Because antimony(III) oxide is a white salt, it cannot be used for transparent or darkly colored PVC. In these cases sodium antimonate (Na_3SbO_4), a transparent salt, is used.

Once the additives have been chosen for a particular PVC application, the materials must be blended. This is often done in a dry-blending process, which produces a powder that is then used for fabrication of the final product. The powdered mixture also can be melted and formed into pellets, which are easily shipped to manufacturing plants, where they are remelted and formed into the desired products.

The production of PVC provides a good case study of an industrial process. It illustrates many of the factors that must be taken into account when any product is manufactured: effectiveness of the product, cost, ease of production, safety, and environmental impact. The last issue is becoming ever more important as our society struggles both to reduce the magnitude of the waste stream by recycling and to improve our waste disposal methods.

Key Terms

Section 1.3

scientific method
measurement
natural law
theory

For Review

Sign in at **www.cengage.com/owl** to:

- View tutorials and simulations, develop problem-solving skills, and complete online homework assigned by your professor.
- Download Go Chemistry mini lecture modules for quick review and exam prep from OWL (or purchase them at **www.cengagebrain.com**)

Thinking like a chemist

- Problem solving often requires trial and error.
- Practice helps one become a better problem solver.

Scientific method

- Make observations.
- Formulate hypotheses.
- Make predictions.
- Perform experiments.

Difference between a law and a theory

- A law summarizes what happens; it comes from generally observed behavior.
- A theory is an attempt at an explanation of why nature behaves in a particular way; it is subject to modifications over time and sometimes fails.

Three general types of activities in industrial chemistry

- Isolating naturally occurring substances for use as raw materials
- Processing raw materials into commercial products via chemical reactions
- Using chemicals to provide services

Atoms, Molecules, and Ions

2 chapter

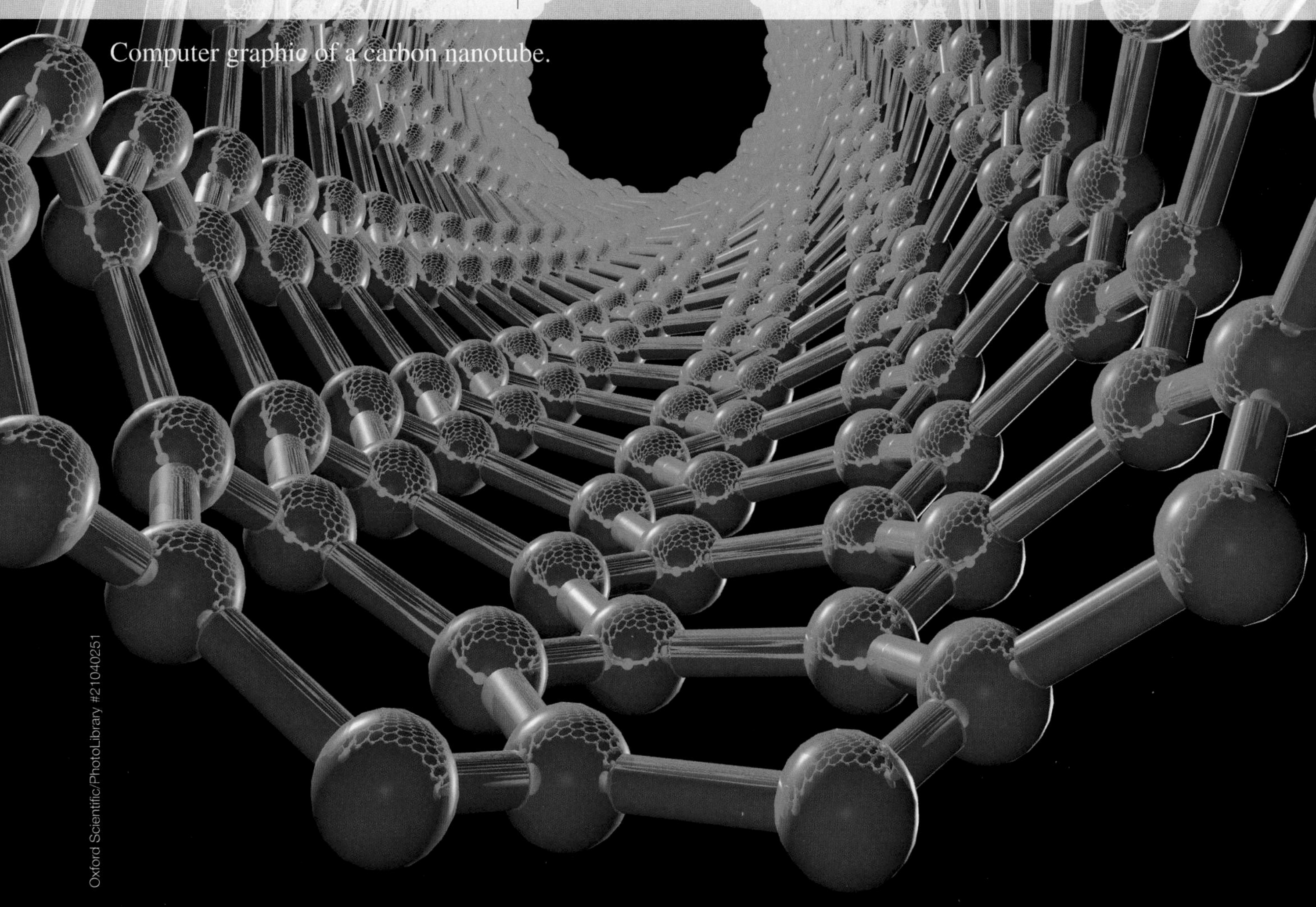

Computer graphic of a carbon nanotube.

Oxford Scientific/PhotoLibrary #21040251

In this chapter we present very briefly many of the fundamental concepts and some of the vocabulary of chemistry plus something about how the science developed. Depending on your specific background in chemistry, much of this material may be review. However, whatever your background, read this chapter carefully to be sure this material is fresh in your mind as we pursue the study of reaction chemistry in Chapters 3 and 4.

2.1 The Early History of Chemistry

OWL

Sign in to OWL at www.cengage.com/owl to view tutorials and simulations, develop problem-solving skills, and complete online homework assigned by your professor.

go Chemistry

Download mini lecture videos for key concept review and exam prep from OWL or purchase them from www.cengagebrain.com.

Chemistry has been important since ancient times. The processing of natural ores to produce metals for ornaments and weapons and the use of embalming fluids are two applications of chemical phenomena that were used before 1000 B.C.

The Greeks were the first to try to explain why chemical changes occur. By about 400 B.C. they had proposed that all matter was composed of four fundamental substances: fire, earth, water, and air. The Greeks also considered the question of whether matter is continuous, and thus infinitely divisible into smaller pieces, or composed of small indivisible particles. One supporter of the latter position was Democritus, who used the term *atomos* (which later became *atoms*) to describe these ultimate particles. However, because the Greeks had no experiments to test their ideas, no definitive conclusion about the divisibility of matter was reached.

The next 2000 years of chemical history were dominated by a pseudoscience called alchemy. Alchemists were often mystics and fakes who were obsessed with the idea of turning cheap metals into gold. However, this period also saw important discoveries: Elements such as mercury, sulfur, and antimony were discovered, and alchemists learned how to prepare the mineral acids.

Roald Hoffman/Cornell University

The Priestley Medal is the highest honor given by the American Chemical Society. It is named for **Joseph Priestley,** who was born in England on March 13, 1733. He performed many important scientific experiments, one of which led to the discovery that a gas later identified as carbon dioxide could be dissolved in water to produce *seltzer.* Also, as a result of meeting Benjamin Franklin in London in 1766, Priestley became interested in electricity and was the first to observe that graphite was an electrical conductor. However, his greatest discovery occurred in 1774, when he isolated oxygen by heating mercuric oxide.

Because of his nonconformist political views, he was forced to leave England. He died in the United States in 1804.

The foundations of modern chemistry were laid in the sixteenth century with the development of systematic metallurgy (extraction of metals from ores) by a German, Georg Bauer, and the medicinal application of minerals by the Swiss alchemist Paracelsus.

The first "chemist" to perform truly quantitative experiments was Robert Boyle (1627–1691), an Irish scientist, who carefully measured the relationship between the pressure and volume of gases. When Boyle published his book *The Sceptical Chemist* in 1661, the quantitative sciences of physics and chemistry were born. In addition to his results on the quantitative behavior of gases, Boyle's other major contribution to chemistry consisted of his ideas about the chemical elements. Boyle held no preconceived notion about the number of elements. In his view a substance was an element unless it could be broken down into two or more simpler substances. As Boyle's experimental definition of an element became generally accepted, the list of known elements began to grow, and the Greek system of four elements finally died. Although Boyle was an excellent scientist, he was not always right. For example, he clung to the alchemist's views that metals were not true elements and that a way would eventually be found to change one metal to another.

The phenomenon of combustion evoked intense interest in the seventeenth and eighteenth centuries. The German chemist Georg Stahl (1660–1734) suggested that a substance he called phlogiston flowed out of the burning material. Stahl postulated that a substance burning in a closed container eventually stopped burning because the air in the container became saturated with phlogiston. Oxygen gas, discovered by Joseph Priestley (1733–1804), an English clergyman and scientist, was found to support vigorous combustion and was thus supposed to be low in phlogiston. In fact, oxygen was originally called "dephlogisticated air." It is important to note that the observations made by

Joseph Priestley did not contradict those made by Georg Stahl. However, Priestley's theory to explain what he saw was vastly different. As we stated in Chapter 1, *what* happens doesn't change, but our ideas about *why* a phenomenon occurs can change. Such is the nature of science.

2.2 Fundamental Chemical Laws

By the late eighteenth century, combustion had been studied extensively; the gases carbon dioxide, nitrogen, hydrogen, and oxygen had been discovered; and the list of elements continued to grow. However, it was Antoine Lavoisier (1743–1794), a French chemist (Fig. 2.1), who finally explained the true nature of combustion, thus clearing the way for the tremendous progress that was made near the end of the eighteenth century. Lavoisier, like Boyle, regarded measurement as the essential operation of chemistry. His experiments, in which he carefully weighed the reactants and products of various reactions, suggested that *mass is neither created nor destroyed.* Lavoisier's discovery of this **law of conservation of mass** was the basis for the developments in chemistry in the nineteenth century.

Lavoisier's quantitative experiments showed that combustion involved oxygen (which Lavoisier named), not phlogiston. He also discovered that life was supported by a process that also involved oxygen and was similar in many ways to combustion. In 1789 Lavoisier published the first modern chemistry textbook, *Elementary Treatise on Chemistry,* in which he presented a unified picture of the chemical knowledge assembled up to that time. Unfortunately, in the same year the text was published, the French Revolution began. Lavoisier, who had been associated with collecting taxes for the government, was executed on the guillotine as an enemy of the people in 1794.

After 1800, chemistry was dominated by scientists who, following Lavoisier's lead, performed careful weighing experiments to study the course of chemical reactions and to determine the composition of various chemical compounds. One of these chemists, a Frenchman, Joseph Proust (1754–1826), showed that *a given compound always contains exactly the same proportion of elements by mass.* For example, Proust found that the substance copper car-

Figure 2.1
Antoine Lavoisier with his wife. Lavoisier was born in Paris on August 26, 1743. From the beginning of his scientific career, Lavoisier recognized the importance of accurate measurements. His careful weighings showed that mass is conserved in chemical reactions and that combustion involves reaction with oxygen. Also, he wrote the first modern chemistry textbook. He is often called the father of modern chemistry.

Because of his connection to a private tax-collecting firm, radical French revolutionaries demanded his execution, which occurred on the guillotine on May 8, 1794.

Figure 2.2
John Dalton (1766–1844), an Englishman, began teaching at a Quaker school when he was 12. His fascination with science included an intense interest in meteorology (he kept careful daily weather records for 46 years), which led to an interest in the gases of the air and their ultimate components, atoms. Dalton is best known for his atomic theory, in which he postulated that the fundamental differences among atoms are their masses. He was the first to prepare a table of relative atomic weights.

Dalton was a humble man with several apparent handicaps: He was poor; he was not articulate; he was not a skilled experimentalist; and he was color-blind, a terrible problem for a chemist. Despite these disadvantages, he helped revolutionize the science of chemistry.

bonate is always 5.3 parts copper to 4 parts oxygen to 1 part carbon (by mass). The principle of the constant composition of compounds, originally called Proust's law, is now known as the **law of definite proportion.**

Proust's discovery stimulated John Dalton (1766–1844), an English schoolteacher (Fig. 2.2), to think about atoms. Dalton reasoned that if elements were composed of tiny individual particles, a given compound should always contain the same combination of these atoms. This concept explained why the same relative masses of elements were always found in a given compound.

But Dalton discovered another principle that convinced him even more of the existence of atoms. He noted, for example, that carbon and oxygen form two different compounds that contain different relative amounts of carbon and oxygen, as shown by the following data:

	Mass of Oxygen That Combines with 1 g of Carbon
Compound I	1.33 g
Compound II	2.66 g

Dalton noted that compound II contained twice as much oxygen per gram of carbon as compound I, a fact that could be easily explained in terms of atoms. Compound I might be CO, and compound II might be CO_2. This principle, which was found to apply to compounds of other elements as well, became known as the **law of multiple proportions:** *When two elements form a series of compounds, the ratios of the masses of the second element that combine with 1 gram of the first element can always be reduced to small whole numbers.*

These ideas are also illustrated by the compounds of nitrogen and oxygen, as shown by the following data:

	Mass of Nitrogen That Combines with 1 g of Oxygen
Compound I	1.750 g
Compound II	0.8750 g
Compound III	0.4375 g

which yield the following ratios:

$$\frac{\text{I}}{\text{II}} = \frac{1.750}{0.8750} = \frac{2}{1}$$

$$\frac{\text{II}}{\text{III}} = \frac{0.8750}{0.4375} = \frac{2}{1}$$

$$\frac{\text{I}}{\text{III}} = \frac{1.750}{0.4375} = \frac{4}{1}$$

The significance of these data is that compound I contains twice as much nitrogen (N) per gram of oxygen (O) as does compound II and that compound II contains twice as much nitrogen per gram of oxygen as does compound III. In terms of the numbers of atoms combining, these data can be explained by any of the following sets of formulas:

Compound I	N_2O		NO		N_4O_2
Compound II	NO	or	NO_2	or	N_2O_2
Compound III	NO_2		NO_4		N_2O_4

In fact, an infinite number of other possibilities exists. Dalton could not deduce absolute formulas from the available data on relative masses. However, the data on the composition of compounds in terms of the relative masses of the elements supported his hypothesis that each element consisted of a certain type of atom and that compounds were formed from specific combinations of atoms.

2.3 Dalton's Atomic Theory

In 1808 Dalton published *A New System of Chemical Philosophy,* in which he presented his theory of atoms.

These statements are a modern paraphrase of Dalton's ideas.

Dalton's Model

1. Each element is made up of tiny particles called atoms.
2. The atoms of a given element are identical; the atoms of different elements are different in some fundamental way or ways.
3. Chemical compounds are formed when atoms combine with one another. A given compound always has the same relative numbers and types of atoms.
4. Chemical reactions involve reorganization of the atoms—changes in the way they are bound together. The atoms themselves are not changed in a chemical reaction.

It is instructive to consider Dalton's reasoning on the relative masses of the atoms of the various elements. In Dalton's time, water was known to be composed of the elements hydrogen and oxygen, with 8 grams of oxygen present for every 1 gram of hydrogen. If the formula for water were OH, an oxygen atom would have to have eight times the mass of a hydrogen atom. However, if the formula for water were H_2O (two atoms of hydrogen for every oxygen atom), this would mean that each atom of oxygen is 16 times as heavy as *each* atom of hydrogen (since the ratio of the mass of one oxygen to that of *two* hydrogens is 8 to 1). Because the formula for water was not then known, Dalton could not specify the relative masses of oxygen and hydrogen unambiguously. To solve the problem, Dalton made a fundamental assumption: He decided that nature would be as simple as possible. This assumption led him to conclude that the formula for water should be OH. He thus assigned hydrogen a mass of 1 and oxygen a mass of 8.

mass (m): the quantity of matter in a body

weight: $m \times g$

Using similar reasoning for other compounds, Dalton prepared the first table of **atomic masses** (formerly called atomic weights by chemists, since mass is usually determined by comparison to a standard mass—a process called *weighing**). Many of the masses were later proved to be wrong because of Dalton's incorrect assumptions about the formulas of certain compounds, but the construction of a table of masses was an important step forward.

Although not recognized as such for many years, the keys to determining absolute formulas for compounds were provided in the experimental work of the French chemist Joseph Gay-Lussac (1778–1850) and by the hypothesis of an Italian chemist named Amedeo Avogadro (1776–1856). In 1809 Gay-Lussac performed experiments in which he measured (under the same conditions of temperature and pressure) the volumes of gases that reacted with one another. For example, Gay-Lussac found that 2 volumes of hydrogen react with 1 volume of oxygen to form 2 volumes of gaseous water and that 1 vol-

*Technically, weight is the force exerted on an object by gravitational attraction to a body such as the earth (weight = mass × acceleration due to gravity). It is mass (the quantity of matter in a body), not weight, that chemists use in their measurements, although the two terms are sometimes used interchangeably.

The Granger Collection, New York

Joseph Louis Gay-Lussac (1778–1850), a French physicist and chemist, was remarkably versatile. Although he is now primarily known for his studies on the combining of volumes of gases, Gay-Lussac was instrumental in the studies of many of the other properties of gases. Some of Gay-Lussac's motivation to learn about gases arose from his passion for ballooning. In fact, he made ascents to heights of over 4 miles to collect air samples, setting altitude records that stood for approximately 50 years. Gay-Lussac also was the codiscoverer of boron and the developer of a process for manufacturing sulfuric acid. As chief assayer of the French mint, Gay-Lussac developed many techniques for chemical analysis and invented many types of glassware now used routinely in labs. Gay-Lussac spent his last 20 years as a lawmaker in the French government.

ume of hydrogen reacts with 1 volume of chlorine to form 2 volumes of hydrogen chloride.

In 1811 Avogadro interpreted these results by proposing that, *at the same temperature and pressure, equal volumes of different gases contain the same number of particles.* This assumption (called **Avogadro's hypothesis**) makes sense if the distances between the particles in a gas are very great compared with the sizes of the particles. Under these conditions the volume of a gas is determined by the number of molecules present, not by the size of the individual particles.

If Avogadro's hypothesis is correct, Gay-Lussac's result,

$$\text{2 volumes of hydrogen react with 1 volume of oxygen} \longrightarrow \text{2 volumes of water vapor}$$

can be expressed as follows:

$$\text{2 molecules of hydrogen react with 1 molecule of oxygen} \longrightarrow \text{2 molecules of water}$$

These observations can be explained best by assuming that gaseous hydrogen, oxygen, and chlorine are all composed of diatomic (two-atom) molecules: H_2, O_2, and Cl_2, respectively. Gay-Lussac's results can then be represented as shown in Fig. 2.3. (Note that this reasoning suggests that the formula for water is H_2O, not OH as Dalton believed.)

Unfortunately, Avogadro's interpretations were not accepted by most chemists. The main stumbling block seems to have been the prevailing belief that only atoms of different elements could attract each other to form molecules. Dalton and the other prominent chemists of the time assumed that identical atoms had no "affinity" for each other and thus would not form diatomic molecules.

Because no general agreement existed concerning the formulas for elements such as hydrogen, oxygen, and chlorine or for the compounds formed from these elements, chaos reigned in the first half of the nineteenth century. Although during this period chemists, such as the Swedish chemist Jöns Jakob Berzelius (1779–1848), made painstaking measurements of the masses of various elements that combined to form compounds, these results were interpreted in many different ways, depending on the assumptions about the formulas of the elements and compounds, and this led to many different tables of atomic masses. The situation was so confused that 19 different formulas for the com-

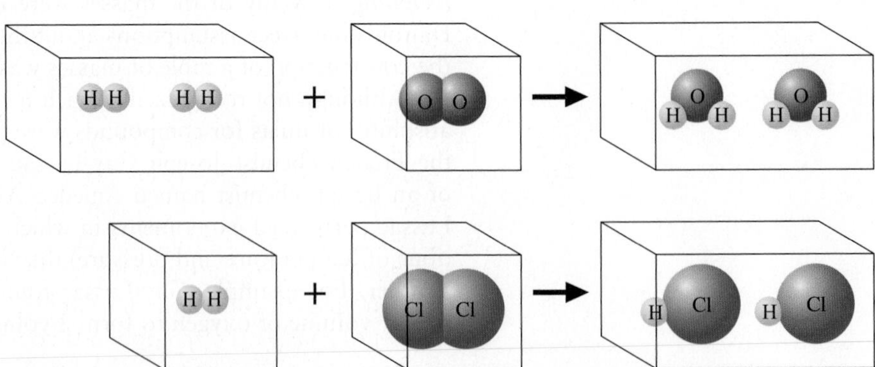

Figure 2.3

A representation of combining gases at the molecular level. The spheres represent atoms in the molecules, and the boxes represent the relative volumes of the gases.

pound acetic acid were given in a textbook written in 1861 by F. August Kekulé (1829–1896). In the next section we will see how this mess was finally cleaned up, primarily because of the leadership of the Italian chemist Stanislao Cannizzaro (1826–1910).

2.4 Cannizzaro's Interpretation

Convinced that chemists had to find a way to agree on a common set of atomic masses, the German chemist F. August Kekulé organized the First International Chemical Congress held in 1860 at Karlsruhe, Germany. At this meeting the young Italian chemist Stanislao Cannizzaro presented his ideas so clearly and forcefully, both in formal and informal talks, that a consensus about atomic masses began to develop in the chemical community. Cannizzaro was guided by two main beliefs:

1. Compounds contained whole numbers of atoms as Dalton postulated.
2. Avogadro's hypothesis was correct—equal volumes of gases under the same conditions contain the same number of molecules.

Applications of Avogadro's hypothesis to Gay-Lussac's results of combining volumes of gas convinced Cannizzaro that hydrogen gas consisted of H_2 molecules. Thus he arbitrarily assigned the relative molecular mass of hydrogen (H_2) to be 2. He then set out to measure the relative molecular masses for other gaseous substances. He did so by comparing the mass of 1 liter of a given gas with the mass of 1 liter of hydrogen gas (both gases at the same conditions of temperature and pressure). For example, the ratio of the masses of 1-liter samples of oxygen and hydrogen gas is 16:

Both gases are at the same temperature and pressure.

$$\frac{\text{Mass of 1.0 L oxygen gas}}{\text{Mass of 1.0 L hydrogen gas}} = \frac{16}{1} = \frac{32}{2}$$

Since by Avogadro's hypothesis both samples of gas contain the same number of molecules, the mass of an oxygen molecule (which he assumed to be O_2) must be 32 relative to a mass of 2 for the H_2 molecule. Since each molecule contains two atoms, the relative atomic masses for oxygen and hydrogen are then 16 and 1, respectively. Using this same method, Cannizzaro found the relative molecular mass of carbon dioxide to be 44 (relative to 2 for H_2). Chemical analysis of carbon dioxide had shown it to contain 27% carbon (by mass). This percentage corresponds to (0.27)(44 g), or 12 g, of carbon in 44 g of carbon dioxide, and 44 g − 12 g = 32 g of oxygen. Recall that the oxygen atom has a relative mass of 16. Thus if the formula of carbon dioxide is assumed to be CO_2, then the relative mass of carbon is 12 because $12 + 2(16) = 44$. However, if the formula of carbon dioxide is C_2O_2, then 12 represents the relative mass of two carbon atoms, giving carbon a relative mass of 6. Similarly, the formula C_3O_2 for carbon dioxide gives a relative mass of 4 for carbon. Thus the relative mass of the carbon atom cannot be determined from these data without knowing the formula for carbon dioxide. This is exactly the type of problem that had plagued chemists all along and was the reason for so many different mass tables.

Cannizzaro addressed this problem by obtaining the relative molecular masses of many other compounds containing carbon. For example, consider the data shown in Table 2.1. Notice from these data that the relative mass of carbon present in the compounds is always a multiple of 12. This observation strongly suggests that the relative mass of carbon is 12, which in turn would mean that the formula for carbon dioxide is CO_2.

Chemical Insights — Seeing Atoms

There are many pieces of evidence that convince us that matter is made up of atoms. Some of the most compelling evidence comes from scanning probe microscopy. This technique employs a microscopic tip, which responds to a surface to reveal its architecture. The principal methods of scanning probe microscopy are scanning tunneling microscopy (STM) and atomic force microscopy (AFM).

The scanning tunneling microscope was invented at IBM's Zurich Research Laboratory in Switzerland in the early 1980s by Gerd K. Binning and Heinrich Rohrer, who subsequently won the Nobel Prize in Physics for their work. STM uses an ultrasharp metal tip that is brought to within about 1 nm of the surface. A small voltage is applied to the tip, which produces current flow between the surface and the tip. This tunneling* current is strongly dependent on the distance of the tip from the surface. A feedback circuit, which senses the current flow, keeps the tip at a constant distance from the surface. The tip can also be used to move atoms around on the surface, as illustrated by the elliptical arrangement of cobalt atoms shown in Fig. 2.4.

AFM is similar in many ways to STM. In AFM, the attractive and repulsive forces acting on a tiny arm near the surface are measured, and a relief map is produced from the results.

Recently, IBM researcher Leo Gross and his coworkers have found that the image produced by AFM can be greatly improved by inserting a carbon monoxide (CO) molecule at the end of the gold AFM tip. This technique enables the AFM probe to produce a detailed image of an entire molecule such as pentacene ($C_{22}H_{14}$) as depicted in Figure 2.5.

*The term *tunneling* refers to the ability of electrons from the surface to escape even though they do not apparently possess enough energy to overcome the large potential energy holding them there. This quantum mechanical phenomenon is known as tunneling (the electron "tunnels through" the potential barrier).

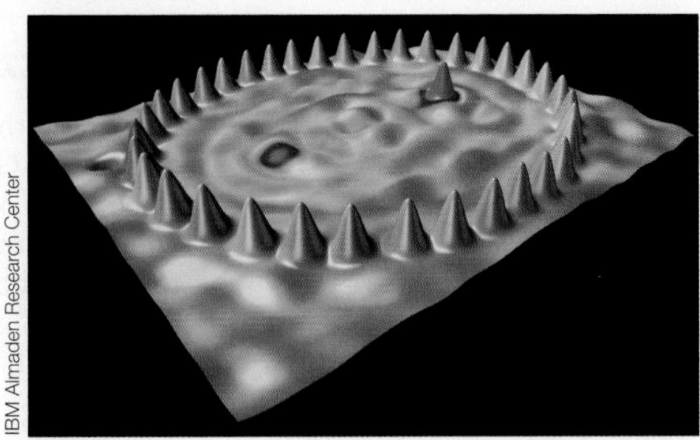
IBM Almaden Research Center

Figure 2.4
Image of a ring of cobalt atoms placed on a copper surface.

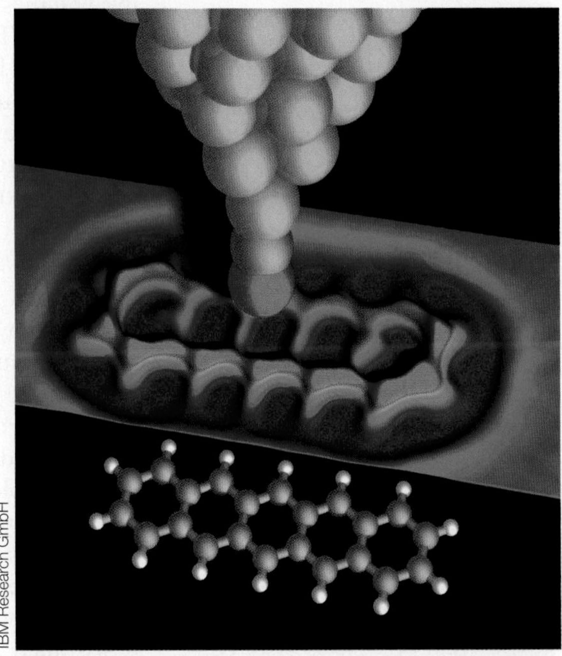
IBM Research GmbH

Figure 2.5
Top, A depiction of the EMF tip in which a CO molecule has been added to the tip (consisting of gold atoms). The image of the pentacene molecule is shown in blue, green, and red.
Bottom, A ball and stick model of the pentacene ($C_{22}H_{14}$) molecule (carbon atoms are black, hydrogen atoms are white)

Table 2.1

Relative Mass Data for Several Gases Containing Carbon

Compound	Relative Molecular Mass	Percent Carbon (by Mass)	Relative Mass of Carbon Present
Methane	16	75	12
Ethane	30	80	24
Propane	44	82	36
Butane	58	83	48
Carbon dioxide	44	27	12

EXAMPLE 2.1

The first four compounds listed in Table 2.1 contain only carbon and hydrogen. Predict the formulas for these compounds.

Solution Since the compounds contain only carbon and hydrogen, the percent hydrogen in each compound (by mass) is 100 − % carbon. We can then find the relative mass of hydrogen present as follows:

$$\text{Relative mass of hydrogen} = \frac{\text{percent hydrogen}}{100} \times \text{relative molecular mass}$$

In tabular form the results are as follows:

Compound	Relative Molecular Mass	Percent Hydrogen	Relative Mass of Hydrogen
Methane	16	25	4
Ethane	30	20	6
Propane	44	18	8
Butane	58	17	10

Combining the preceding results with those from Table 2.1, we find that methane contains relative masses of carbon and hydrogen of 12 and 4, respectively. Using the relative atomic mass values of 12 and 1 for carbon and hydrogen gives a formula of CH_4 for methane. Similarly, the relative masses of carbon and hydrogen in ethane of 24 and 6, respectively, lead to a formula of C_2H_6 for ethane. Similar reasoning gives formulas for propane and butane of C_3H_8 and C_4H_{10}, respectively.

Cannizzaro's work was so convincing because he collected data on so many compounds. Although he couldn't absolutely prove that his atomic mass values were correct (because he had no way to verify absolutely the formulas of the compounds), the consistency of the large quantity of data he had collected eventually convinced virtually everyone that his interpretation made sense and that the relative values of atomic mass that he had determined were correct. The confusion was finally over. Chemistry had the universal (relative) mass standards that it needed.

It is worthwhile to note that Cannizzaro's work led to *approximate* values of the relative atomic masses. His goal was not to determine highly precise values for atomic masses but rather to pin down the approximate values (for example, to show that oxygen's relative mass was 16 rather than 8). The most precise values for atomic masses were determined by quantitative experiments

Bettmann/Corbis

Stanislao Cannizzaro (1826–1910). Cannizzaro's work ended the confusion of atomic mass values.

in which the combining masses of elements were carefully measured, such as in the work of Berzelius.

In the next chapter we will have much more to say about atomic masses, including the origin of the very precise values used by today's chemists.

2.5 Early Experiments to Characterize the Atom

On the basis of the work of Dalton, Gay-Lussac, Avogadro, Cannizzaro, and others, chemistry was beginning to make sense. The concept of atoms was clearly a good idea. Inevitably, scientists began to wonder about the nature of the atom. What is an atom made of, and how do the atoms of the various elements differ?

The Electron

The first important experiments that led to an understanding of the composition of the atom were done by the English physicist J. J. Thomson (1856–1940), who studied electrical discharges in partially evacuated tubes called *cathode-ray tubes* (Fig. 2.6) during the period from 1898 to 1903. Thomson found that when high voltage was applied to the tube, a "ray" he called a **cathode ray** (because it emanated from the negative electrode, or cathode) was produced. Because this ray was produced at the negative electrode and was repelled by the negative pole of an applied electric field (Fig. 2.7), Thomson postulated that the ray was a stream of negatively charged particles, now called **electrons.** From experiments in which he measured the deflection of the beam of electrons in a magnetic field, Thomson determined the *charge-to-mass ratio* of an electron:

$$\frac{e}{m} = -1.76 \times 10^8 \text{ C/g}$$

where e represents the charge on the electron in coulombs and m represents the electron mass in grams.

One of Thomson's primary goals in his cathode-ray tube experiments was to gain an understanding of the structure of the atom. He reasoned that since electrons could be produced from electrodes made of various types of metals, *all* atoms must contain electrons. Since atoms were known to be electrically neutral, Thomson further assumed that atoms also must contain some positive charge. Thomson postulated that an atom consisted of a diffuse cloud of positive charge with the negative electrons embedded randomly in it. This model,

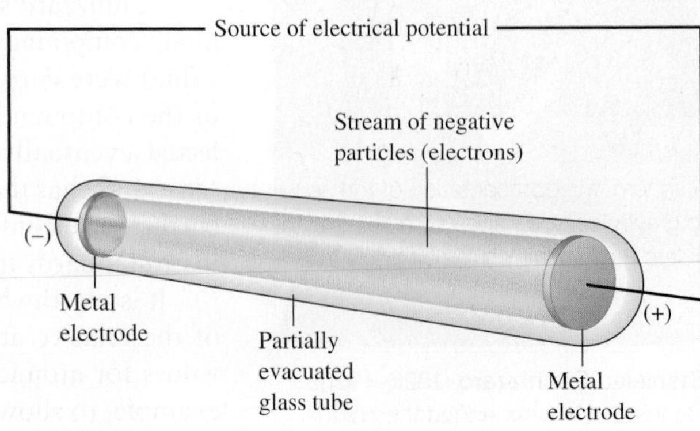

Figure 2.6
A cathode-ray tube. The fast-moving electrons excite the gas in the tube, causing a glow between the electrodes. The green color in the photo is due to the response of the screen (coated with zinc sulfide) to the electron beam.

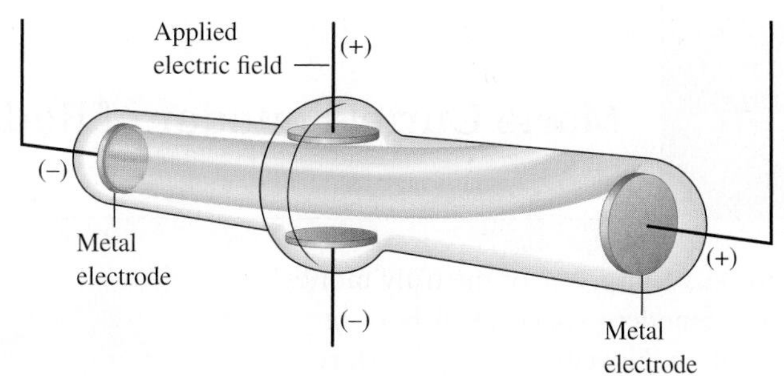

Figure 2.7
Deflection of cathode rays by an applied electric field.

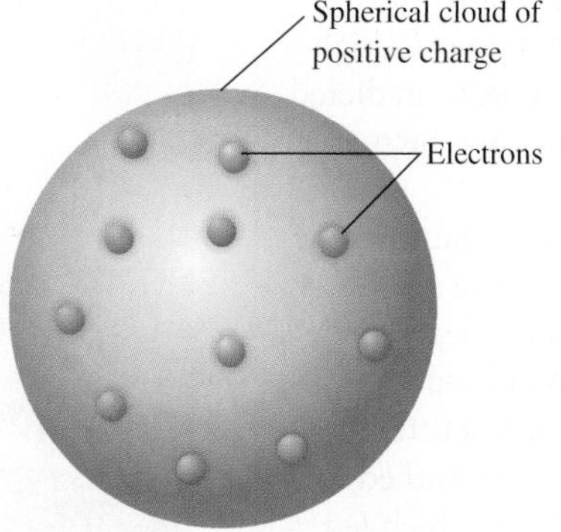

Figure 2.8
Thomson's plum pudding model.

C. Byatt-norman/
Shutterstock.com #19116133

A classic English plum pudding.

shown in Fig. 2.8, is often called the *plum pudding model* because the electrons are like raisins dispersed in a pudding (the positive-charge cloud), as in plum pudding, a favorite English dessert.*

In 1909 Robert Millikan (1868–1953), working at the University of Chicago, performed very clever experiments involving charged oil drops. These experiments allowed him to determine the magnitude of the electron charge (Fig. 2.9). With this value and the charge-to-mass ratio determined by Thomson, Millikan was able to calculate the mass of the electron as 9.11×10^{-31} kilogram.

Bettmann/Corbis #BE042343

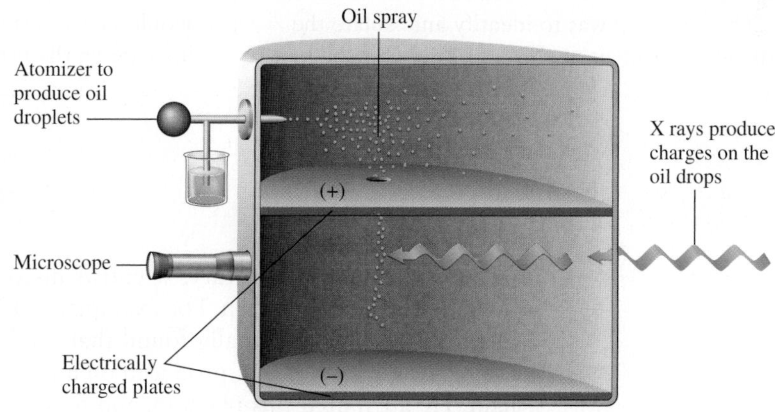

Figure 2.9
A schematic representation of the apparatus Millikan used to determine the charge on the electron. The fall of charged oil droplets due to gravity can be halted by adjusting the voltage across the two plates. The voltage and the mass of an oil drop can then be used to calculate the charge on the oil drop. Millikan's experiments showed that the charge on an oil drop is always a whole-number multiple of the electron charge.

*Although J. J. Thomson is generally given credit for this model, the idea was apparently first suggested by the English mathematician and physicist William Thomson (better known as Lord Kelvin and not related to J. J. Thomson).

Chemical Insights — Marie Curie: Founder of Radioactivity

Marie Sklodowska Curie, one of the truly monumental figures of modern science, was born in Warsaw, Poland, on November 7, 1867. Marie developed an early interest in chemistry, and it is interesting that Dmitri Mendeleev, creator of the periodic table and friend of Marie's father (a high school mathematics and physics teacher), predicted great success for the young woman when he met her in Warsaw.

To escape political persecution in Poland by the Russians, Marie emigrated in 1891 at the age of 24 to Paris, where she decided to pursue a degree in science at the Sorbonne Institute. While studying there, Marie met Pierre Curie, a well-respected physicist who, among other things, had studied the temperature dependence of magnetism, which led to the formulation of Curie's law. Marie and Pierre were married in 1895, after which Marie decided to pursue a doctorate in physics. As the subject of her doctoral thesis, she decided to study the strange radiation emitted by uranium ore, which had been accidentally discovered by Henri Becquerel. Marie was recruited for the task by Becquerel himself. As she began her studies, Madame Curie noticed that pitchblende produced more radiation than uranium, and she became convinced that an as-yet-unknown element in pitchblende was responsible for this "radioactivity"—a term that she coined.

The next step was to identify and isolate the radioactive element or elements in pitchblende.

Bettmann/Corbis

Marie Sklodowska Curie (1867–1934) in her laboratory.

Pierre interrupted his own research—he thought it would be for just a few weeks—to collaborate with his wife on the project. The Curies actually bor-

Radioactivity

In the late nineteenth century, scientists discovered that certain elements produce high-energy radiation. For example, in 1896 the French scientist Antoine Henri Becquerel accidentally found that the image of a piece of mineral containing uranium could be produced on a photographic plate in the absence of light. He attributed this phenomenon to a spontaneous emission of radiation by the uranium, which his student, Marie Curie, called **radioactivity.** Studies in the early twentieth century demonstrated three types of radioactive emission: gamma (γ) rays, beta (β) particles, and alpha (α) particles. A γ ray is high-energy "light"; a β particle is a high-speed electron; and an α particle has a 2+ charge—that is, a charge twice that of the electron and with the opposite sign. The mass of an α particle is 7300 times that of an electron. More modes of radioactivity are now known, and we will discuss them in Chapter 20. Here

rowed money to support themselves and convinced the Austrian government to send them 1 ton of pitchblende from the mines at Joachimsthal. After receiving this 5 cubic foot pile of "sand" from Austria, the Curies worked to chemically digest the ore. In this process they worked with batches as large as 40 pounds at a time in an improvised laboratory with a leaky roof. Working through the bitter winter of 1896 and all through 1897 (in which they had their first daughter, Iréne, who also became a prominent scientist), in July 1898 the Curies finally isolated a previously unknown element they named polonium after Marie's homeland. Although most people would be satisfied by discovering a new element 400 times more radioactive than uranium, the Curies kept working. By this time, the 1 ton of pitchblende had been concentrated to an amount that would fit into an ordinary flask. Marie continued to extract and crystallize increasingly smaller amounts of material. Finally, in November 1898 she obtained crystals of the salt of another new element that the Curies named radium, which turned out to be 900 times more radioactive than uranium.

For their work the Curies were awarded the Nobel Prize in Physics in 1903, sharing the award with Henri Becquerel. In 1904, Pierre was awarded a chair in physics at the Sorbonne. He was killed tragically on the streets of Paris on April 19, 1906, when he was knocked down by a cab and the wheels of a heavy van passing in the opposite direction ran over his head. After mourning for just a few weeks, Marie Curie decided to proceed with the research on radium. In an unprecedented action, she was awarded her late husband's chair and became the first woman to teach at the Sorbonne.

Marie Curie worked tirelessly to develop radioactivity as a new discipline in physics. With the help of five assistants, she studied the effects of radioactivity and developed the atomic theory of its origin. In 1911, Marie was awarded her second Nobel Prize, this time in chemistry, for the chemical processes discovered in the identification of radium and polonium and for the subsequent characterization of those elements. During World War I, she trained doctors in the new methods of radiology and, after learning to drive, personally transported medical equipment to hospitals. After the war, Madame Curie assumed leadership of the newly built Radium Institute in Paris. In 1920, a campaign was mounted in the United States to produce 1 gram of radium for Marie to support her research. She traveled to the United States to receive the precious vial of radium at the White House in 1921.

Marie Curie continued her studies of radioactivity until just before her death of leukemia in 1934. She was truly one of the greatest scientists of the twentieth century.

we will consider only α particles because they were used in some crucial early experiments.

The Nuclear Atom

In 1911 Ernest Rutherford (Fig. 2.10), who performed many of the pioneering experiments to explore radioactivity, carried out an experiment to test Thomson's plum pudding model. The experiment involved directing α particles at a thin sheet of metal foil, as illustrated in Fig. 2.11. Rutherford reasoned that if Thomson's model were accurate, the massive α particles should crash through the thin foil like cannonballs through gauze, as shown in Fig. 2.12(a). He expected the α particles to travel through the foil with, at the most, very minor deflections in their paths. The results of the experiment were very different from those Rutherford anticipated. Although most of the α particles passed

Figure 2.10
Ernest Rutherford (1871–1937) was born on a farm in New Zealand. In 1895 he placed second in a scholarship competition to attend Cambridge University but was awarded the scholarship when the winner decided to stay home and get married. As a scientist in England, Rutherford did much of the early work on characterizing radioactivity. He named the α and β particles and the γ ray and coined the term *half-life* to describe an important attribute of radioactive elements. His experiments on the behavior of α particles striking thin metal foils led him to postulate the nuclear atom. He also invented the name *proton* for the nucleus of the hydrogen atom. He received the Nobel Prize in Chemistry in 1908.

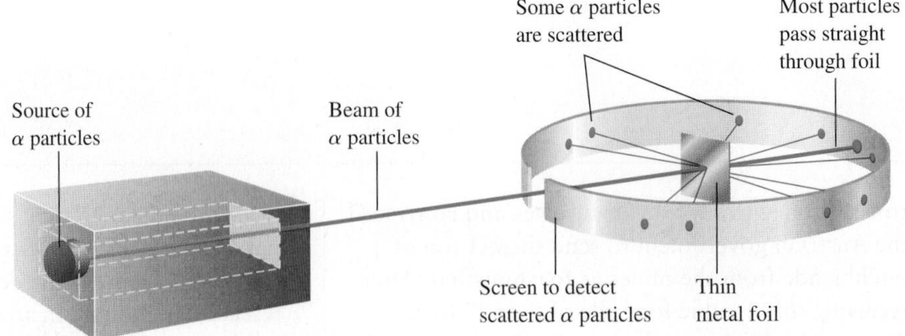

Figure 2.11
Rutherford's experiment on α-particle bombardment of metal foil. (Gold foil was used in the original experiments because it can be hammered into extremely thin sheets.)

straight through, many of the particles were deflected at large angles, as shown in Fig. 2.12(b), and some were reflected, never hitting the detector. This outcome was a great surprise to Rutherford. (He wrote that this result was comparable to shooting a howitzer at a piece of paper and having the shell reflected back.)

Rutherford knew from these results that the plum pudding model for the atom could not be correct. The large deflections of the α particles could be caused only by a center of concentrated positive charge that contains most of the atom's mass, as illustrated in Fig. 2.12(b). Most of the α particles pass directly through the foil because the atom is mostly open space. The deflected α particles are those that had a "close encounter" with the massive positive center of the atom, and the few reflected α particles are those that made a "direct hit" on the much more massive positive center.

In Rutherford's mind these results could be explained only in terms of a **nuclear atom**—an atom with a dense center of positive charge (the **nucleus**) and electrons moving around the nucleus at a distance that is large relative to the nuclear radius.

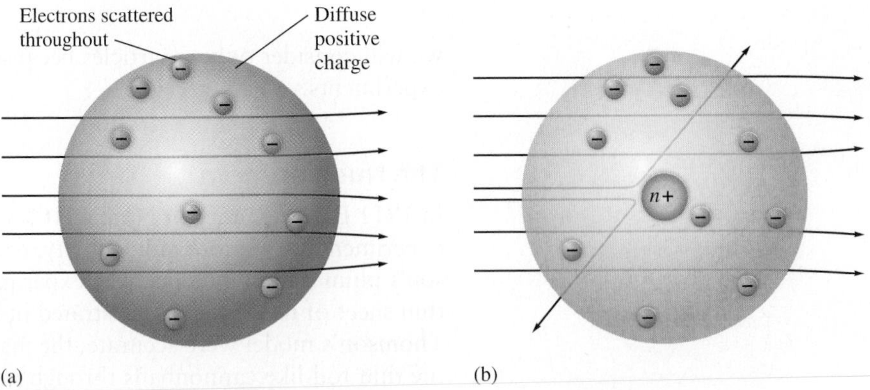

Figure 2.12
(a) The expected results of the metal foil experiment if Thomson's model were correct.
(b) Actual results.

2.6 The Modern View of Atomic Structure: An Introduction

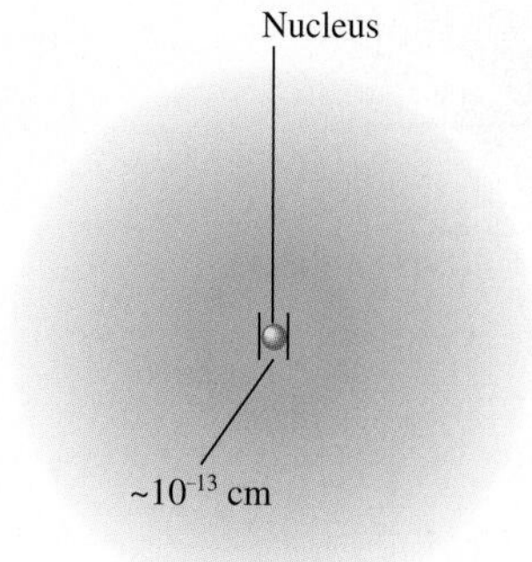

Figure 2.13
A nuclear atom viewed in cross section.

In the years since Thomson and Rutherford, a great deal has been learned about atomic structure. Because much of this material will be covered in detail in later chapters, only an introduction will be given here. The simplest view of the atom is that it consists of a tiny nucleus with a diameter of about 10^{-13} cm and electrons that move about the nucleus at an average distance of about 10^{-8} cm away from it (Fig. 2.13).

As we will see later, the chemistry of an atom mainly results from its electrons. For this reason chemists can be satisfied with a relatively crude nuclear model. The nucleus is assumed to contain **protons,** which have a positive charge equal in magnitude to the electron's negative charge, and **neutrons,** which have virtually the same mass as a proton but no charge. The masses and charges of the electron, proton, and neutron are shown in Table 2.2.

Two striking things about the nucleus are its small size, compared with the overall size of the atom, and its extremely high density. The tiny nucleus accounts for almost all of the atom's mass. Its great density is dramatically demonstrated by the fact that a piece of nuclear material about the size of a pea would have a mass of 250 million tons!

The *chemistry* of an atom arises from its electrons.

As with any theory in science, although it provides answers to questions, it also brings about more questions. An important question to consider at this point is, *"If all atoms are composed of these same components, why do different atoms have different chemical properties?"* The answer to this question lies in the number and the arrangement of the electrons. The electrons constitute most of the atomic volume and thus are the parts that "intermingle" when atoms combine to form molecules. Therefore, the number of electrons possessed by a given atom greatly affects its ability to interact with other atoms. As a result, the atoms of different elements, which have different numbers of protons and electrons, show different chemical behavior.

A sodium atom has 11 protons in its nucleus. Since atoms have no net charge, the number of electrons must equal the number of protons. Therefore, a sodium atom has 11 electrons moving around its nucleus. It is *always* true that a sodium atom has 11 protons and 11 electrons. However, each sodium atom also has neutrons in its nucleus, and different types of sodium atoms exist that have different numbers of neutrons. For example, consider the sodium atoms represented in Fig. 2.14. These two atoms are **isotopes,** or *atoms with*

Table 2.2

The Mass and Charge of the Electron, Proton, and Neutron

Particle	Mass	Charge*
Electron	9.11×10^{-31} kg	1−
Proton	1.67×10^{-27} kg	1+
Neutron	1.67×10^{-27} kg	None

*The magnitude of the charge of the electron and the proton is 1.60×10^{-19} C.

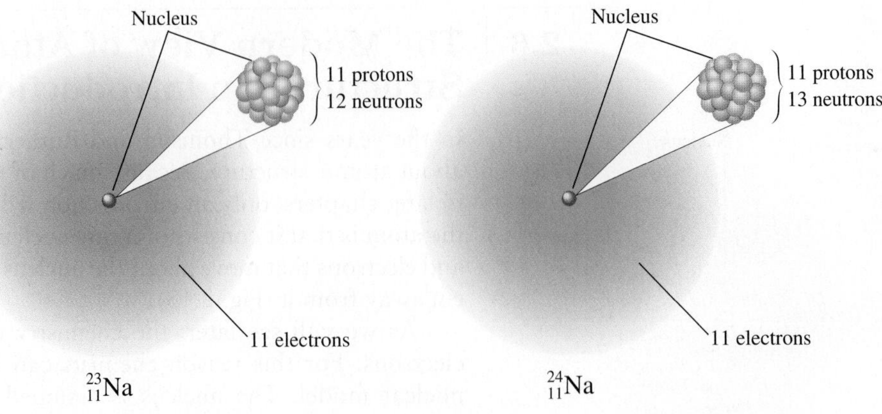

Figure 2.14 Two isotopes of sodium. Both have 11 protons and 11 electrons, but they differ in the number of neutrons in their nuclei. Sodium-23 is the only naturally occurring form of sodium. Sodium-24 does not occur naturally but can be made artificially.

the same number of protons but different numbers of neutrons. Note that the symbol for one particular type of sodium atom is written

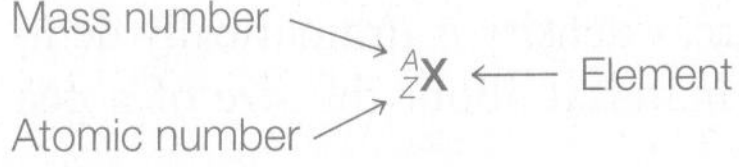

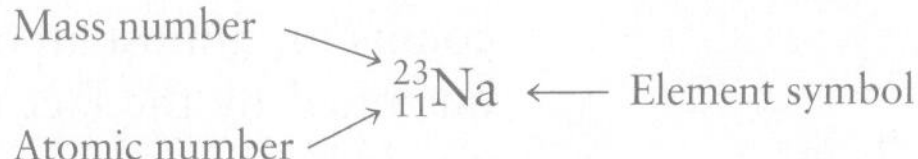

where the **atomic number,** Z (number of protons), is written as a subscript and the **mass number,** A (the total number of protons and neutrons), is written as a superscript. (The particular atom represented here is called "sodium-23." It has 11 electrons, 11 protons, and 12 neutrons.) Because the chemistry of an atom arises from its electrons, isotopes show almost identical chemical properties. In nature most elements contain a mixture of isotopes.

2.7 Molecules and Ions

From a chemist's viewpoint, the most interesting characteristic of an atom is its ability to combine with other atoms to form compounds. It was John Dalton who first recognized that chemical compounds were collections of atoms, but he could not determine the structure of atoms or their means for binding to one another. During the twentieth century, scientists have learned that atoms have electrons and that these electrons participate in the bonding of one atom to another. We will discuss bonding thoroughly in Chapters 13 and 14; here we will consider some definitions that will be useful in the next few chapters.

The forces that hold atoms together in compounds are called **chemical bonds.** One way that atoms can form bonds is by *sharing electrons.* These bonds are called **covalent bonds,** and the resulting collection of atoms is called a **molecule.** Molecules can be represented in several different ways. The simplest method is the **chemical formula,** in which the symbols for the elements are used to indicate the types of atoms present and subscripts are used to indicate the relative numbers of atoms. For example, the formula for carbon dioxide is CO_2, meaning, of course, that each molecule contains 1 atom of carbon and 2 atoms of oxygen.

Familiar examples of molecules that contain covalent bonds are hydrogen (H_2), water (H_2O), oxygen (O_2), ammonia (NH_3), and methane (CH_4). More information about a molecule is given by its **structural formula,** in which the individual bonds are shown (indicated by lines). Structural formulas may or may not indicate the actual shape of the molecule. For example, water might be represented as

```
H—O—H     or      O
                 / \
                H   H
```

The structure on the right shows the actual shape of the water molecule, based on experimental evidence. Other examples of structural formulas are

```
                                                         H
    H                N                 H                 |
    |              ⋰ ▲ ⋱               |                 C
H—N       or     H  H  H        H—C—H       or      ⋰ ▲ ⋱
    |                                  |               H  H  H
    H                                  H
        Ammonia                              Methane
```

In the actual structures on the right, the central atom and the solid lines are understood to be in the plane of the page. Atoms connected to the central atom by dashed lines are behind the plane of the page, and atoms connected to the central atom by wedges are in front of the plane of the page.

In a compound composed of molecules, the individual molecules move around as independent units. For example, a methane molecule is represented in Fig. 2.15 using a **space-filling model.** These models show the relative sizes of the atoms, as well as their relative orientation in the molecule. Figure 2.16 shows other examples. **Ball-and-stick models** are also used to represent molecules. The ball-and-stick model of methane is shown in Fig. 2.17.

A second type of chemical bonding results from attractions among ions. An **ion** is an atom or group of atoms that has a net positive or negative charge. The best-known ionic compound is common table salt, or sodium chloride, which forms when neutral chlorine and sodium react.

Figure 2.15
Space-filling model of the methane molecule. This type of model shows both the relative sizes of the atoms in the molecule and their spatial relationships.

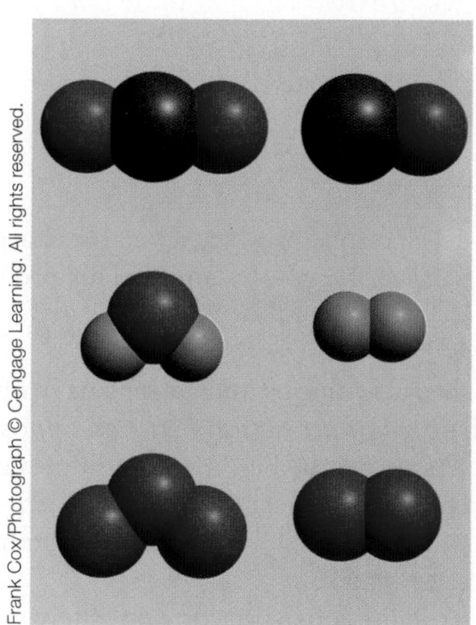

Figure 2.16
Space-filling models of various molecules.

Figure 2.17
Ball-and-stick model of methane.

To see how ions are formed, consider what happens when an electron is transferred from sodium to chlorine (the neutrons in the nuclei will be ignored):

Neutral sodium atom (Na)

Sodium ion (Na^+)

11+ — Minus 1 electron → 11+

10 electrons

11 electrons

With one electron stripped off, the sodium with its 11 protons and only 10 electrons has become a *positive ion* with a net 1+ charge. A positive ion is called a **cation.** The process can be represented in shorthand form as

$$Na \longrightarrow Na^+ + e^-$$

If an electron is added to chlorine,

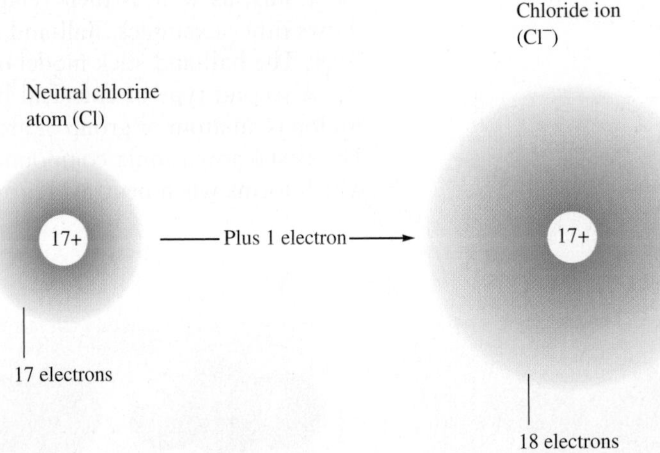

the 18 electrons produce a net 1− charge; the chlorine has become an *ion with a negative charge*—an **anion.** This process is represented as

$$Cl + e^- \longrightarrow Cl^-$$

Because anions and cations have opposite charges, they attract each other. This *force of attraction between oppositely charged ions* is called **ionic bonding.** As shown in Fig. 2.19, sodium metal and chlorine gas (a green gas composed of Cl_2 molecules) react to form solid sodium chloride, which contains many Na^+ and Cl^- ions packed together. The solid forms beautiful, colorless cubic crystals.

A solid consisting of oppositely charged ions is called an *ionic solid,* or (often) a *salt.* Ionic solids can consist of simple ions, as in sodium chloride, or of **polyatomic** (many-atom) **ions,** as in ammonium nitrate (NH_4NO_3), which contains ammonium cations (NH_4^+) and nitrate anions (NO_3^-). The ball-and-stick models of these ions are shown in Fig. 2.18.

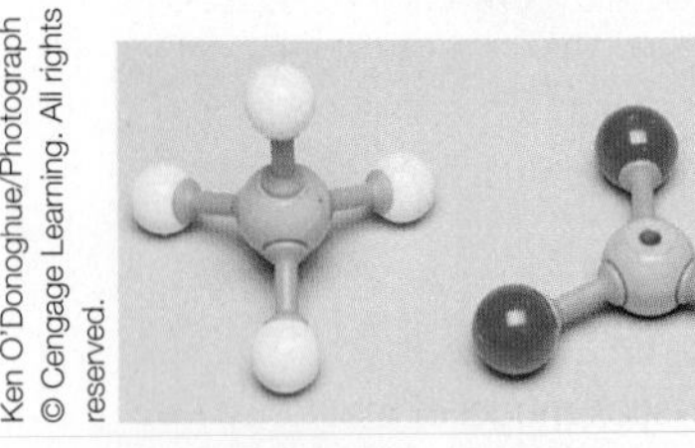

Figure 2.18
Ball-and-stick models of the ammonium ion (NH_4^+) and the nitrate ion (NO_3^-).

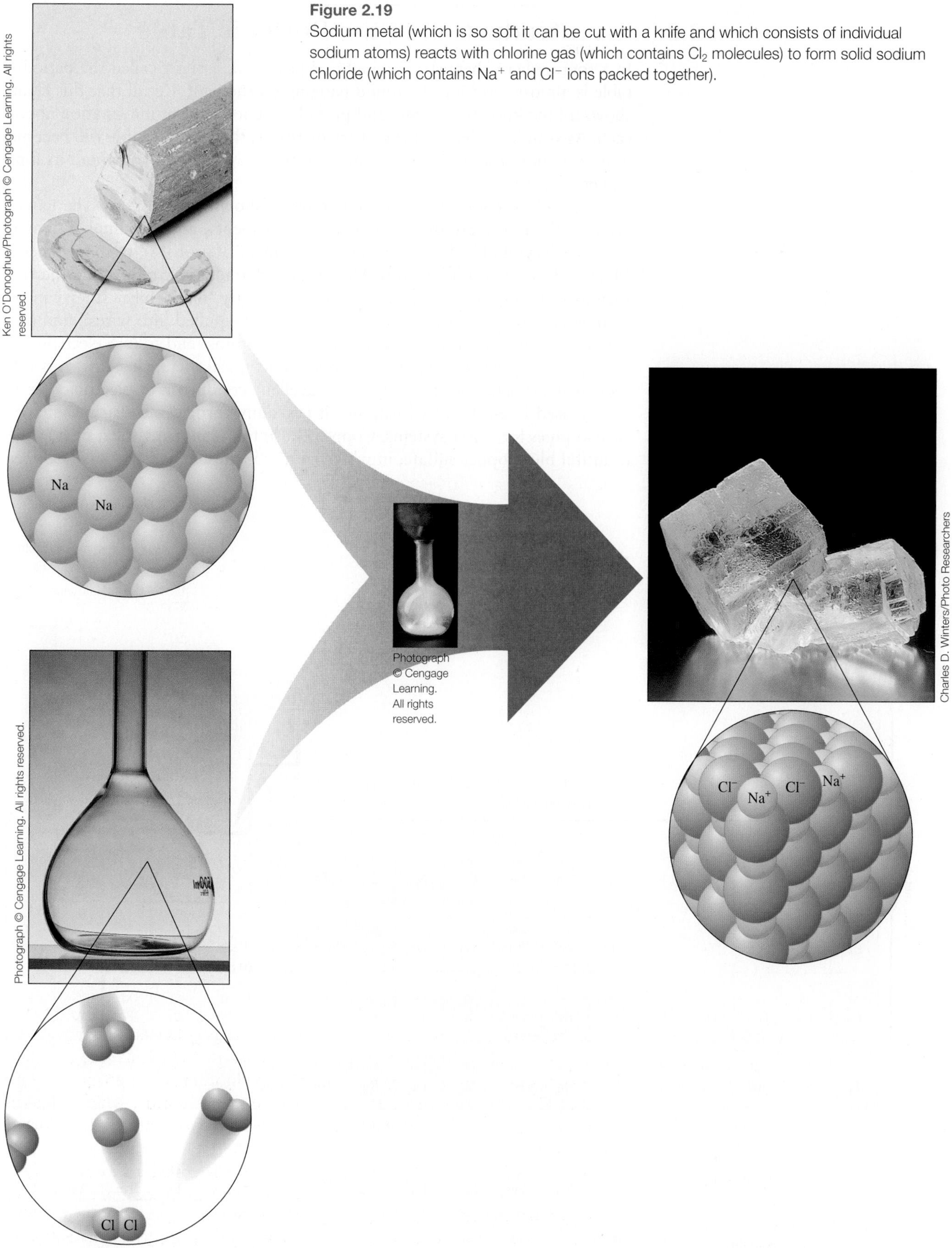

Figure 2.19
Sodium metal (which is so soft it can be cut with a knife and which consists of individual sodium atoms) reacts with chlorine gas (which contains Cl_2 molecules) to form solid sodium chloride (which contains Na^+ and Cl^- ions packed together).

2.8 An Introduction to the Periodic Table

In a room where chemistry is taught or practiced, a chart called the **periodic table** is almost certain to be found hanging on the wall. Recall that this chart shows all the known elements and provides a good deal of information about each. As your knowledge of chemistry increases, the periodic table will become more and more useful to you. In this section we will remind you about its fundamental aspects.

A simple version of the periodic table is shown in Fig. 2.20. The letters given in the boxes are the symbols for the elements, and the number shown above each symbol is the atomic number (number of protons) for that element. Most of the elements are **metals.** Metals have characteristic physical properties such as efficient conduction of heat and electricity, malleability (they can be hammered into thin sheets), ductility (they can be pulled into wires), and (often) a lustrous appearance. Chemically, metal atoms tend to *lose* electrons to form positive ions. For example, copper is a typical metal. It is lustrous (although it tarnishes readily); it is an excellent conductor of electricity (it is widely used in electrical wires); and it is readily formed into various shapes such as pipes for water systems. Copper is also found in many salts, such as the beautiful blue copper sulfate, in which copper is present as Cu^{2+} ions. Copper

Figure 2.20
The periodic table continues to expand as new elements are synthesized in particle accelerators.

Alkali metals (group 1); Alkaline earth metals (group 2); Transition metals (groups 3–12); Halogens (group 17); Noble gases (group 18)

1 1A	2 2A	3	4	5	6	7	8	9	10	11	12	13 3A	14 4A	15 5A	16 6A	17 7A	18 8A
1 H																	2 He
3 Li	4 Be											5 B	6 C	7 N	8 O	9 F	10 Ne
11 Na	12 Mg											13 Al	14 Si	15 P	16 S	17 Cl	18 Ar
19 K	20 Ca	21 Sc	22 Ti	23 V	24 Cr	25 Mn	26 Fe	27 Co	28 Ni	29 Cu	30 Zn	31 Ga	32 Ge	33 As	34 Se	35 Br	36 Kr
37 Rb	38 Sr	39 Y	40 Zr	41 Nb	42 Mo	43 Tc	44 Ru	45 Rh	46 Pd	47 Ag	48 Cd	49 In	50 Sn	51 Sb	52 Te	53 I	54 Xe
55 Cs	56 Ba	57 La*	72 Hf	73 Ta	74 W	75 Re	76 Os	77 Ir	78 Pt	79 Au	80 Hg	81 Tl	82 Pb	83 Bi	84 Po	85 At	86 Rn
87 Fr	88 Ra	89 Ac†	104 Rf	105 Db	106 Sg	107 Bh	108 Hs	109 Mt	110 Ds	111 Rg	112 Cn	113 Uut	114 Uuq	115 Uup		117 Uus	118 Uuo

*Lanthanides	58 Ce	59 Pr	60 Nd	61 Pm	62 Sm	63 Eu	64 Gd	65 Tb	66 Dy	67 Ho	68 Er	69 Tm	70 Yb	71 Lu
†Actinides	90 Th	91 Pa	92 U	93 Np	94 Pu	95 Am	96 Cm	97 Bk	98 Cf	99 Es	100 Fm	101 Md	102 No	103 Lr

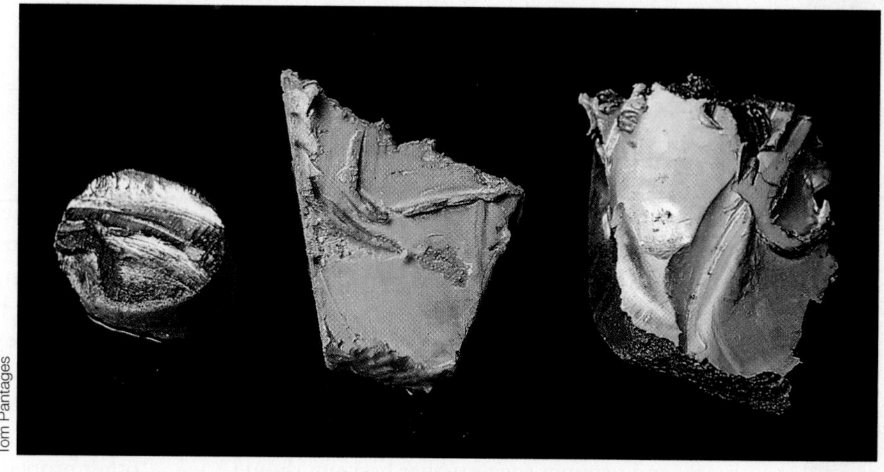

Samples of the alkali metals lithium, sodium, and potassium.

is a member of the transition metals—the metals shown in the center of the periodic table.

The relatively few **nonmetals** appear in the upper right-hand corner of the table (to the right of the heavy line in Fig. 2.20), except hydrogen, a nonmetal that is grouped with the metals. The nonmetals typically lack the physical properties that characterize the metals. Chemically, they tend to *gain* electrons to form anions in reactions with metals. Nonmetals often bond to each other by forming covalent bonds. For example, chlorine is a typical nonmetal. Under normal conditions, it exists as Cl_2 molecules; it reacts with metals to form salts containing Cl^- ions (NaCl, for example); and it forms covalent bonds with nonmetals (for example, hydrogen chloride gas, or HCl).

Metals tend to form positive ions; nonmetals tend to form negative ions.

The periodic table is arranged so that elements in the same vertical columns (called **groups** or **families**) have *similar chemical properties.* For example, all of the **alkali metals,** members of Group 1A—lithium (Li), sodium (Na), potassium (K), rubidium (Rb), cesium (Cs), and francium (Fr)—are very active elements that readily form ions with a 1+ charge when they react with nonmetals. The members of Group 2A—beryllium (Be), magnesium (Mg), calcium (Ca), strontium (Sr), barium (Ba), and radium (Ra)—are called the **alkaline earth metals.** They all form ions with a 2+ charge when they react with nonmetals. The **halogens,** the members of Group 7A—fluorine (F), chlorine (Cl), bromine (Br), iodine (I), and astatine (At)—all form diatomic molecules. Fluorine, chlorine, bromine, and iodine all react with metals to form salts containing ions with a 1− charge (F^-, Cl^-, Br^-, and I^-). The members of Group 8A—helium (He), neon (Ne), argon (Ar), krypton (Kr), xenon (Xe), and radon (Rn)—are known as the **noble gases.** They all exist under normal conditions as monatomic (single-atom) gases and have little chemical reactivity.

Three members of the halogen family: chlorine, bromine, and iodine.

The horizontal rows of elements in the periodic table are called **periods.** Horizontal row one is called the first period (it contains H and He), row two is called the second period (elements Li through Ne), and so on.

We will learn much more about the periodic table as we continue with our study of chemistry. Meanwhile, when an element is introduced in this text, you should always note its position on the periodic table.

2.9 Naming Simple Compounds

When chemistry was an infant science, there was no system for naming compounds. Names such as sugar of lead, blue vitriol, quicklime, Epsom salts, milk of magnesia, gypsum, and laughing gas were coined by early chemists. Such names are called *common names.* As chemistry grew, it became clear that using

Chemical Insights Hassium Fits Right In

Hassium, element 108, does not exist in nature but must be made in a particle accelerator. It was first created in 1984 and can be made by shooting magnesium-26 ($^{26}_{12}Mg$) atoms at curium-248 ($^{248}_{96}Cm$) atoms. The collisions between these atoms produce some hassium-265 ($^{265}_{108}Hs$) atoms. The position of hassium in the periodic table (see Fig. 2.20) in the vertical column containing iron, ruthenium, and osmium suggests that hassium should have chemical properties similar to these metals. However, it is not easy to test this prediction—only a few atoms of hassium can be made at a given time and they last for only about 9 seconds. Imagine having to get your next lab experiment done in 9 seconds!

Amazingly, a team of chemists from the Lawrence Berkeley National Laboratory in California, the Paul Scherrer Institute and the University of Bern in Switzerland, and the Institute of Nuclear Chemistry in Germany have done experiments to characterize the chemical behavior of hassium. For example, they have observed that hassium atoms react with oxygen to form a hassium oxide compound of the type expected from its position on the periodic table. The team has also measured other properties of hassium, including the energy released as it undergoes nuclear decay to another atom.

This work would have surely pleased Dmitri Mendeleev, who originally developed the periodic table and showed its power to predict chemical properties.

common names for compounds would lead to unacceptable chaos. More than 4 million chemical compounds are currently known. Memorizing common names for these compounds would be an impossible task.

The solution, of course, is to adopt a *system* for naming compounds in which the name tells something about the composition of the compound. After learning the system, a chemist given a formula should be able to name the compound, or given a name should be able to construct the compound's formula. In this section we will specify the most important rules for naming compounds other than organic compounds (those based on chains of carbon atoms).

The systematic naming of organic compounds will be discussed in Chapter 21.

We will begin with the systems for naming inorganic **binary compounds**—compounds composed of two elements—which we classify into various types for easier recognition. We will consider both ionic and covalent compounds.

Binary Compounds (Type I; Ionic)

Binary ionic compounds contain a positive ion (cation), always written first in the formula, and a negative ion (anion). In the naming of these compounds, the following rules apply:

1. The cation is always named first and the anion second.
2. A monatomic (meaning from one atom) cation takes its name from the name of the element. For example, Na^+ is called sodium in the names of compounds containing this ion.
3. A monatomic anion is named by taking the first part of the element name and adding *-ide*. Thus the Cl^- ion is called chloride.

A monatomic cation has the same name as its parent element.

Some common monatomic cations and anions and their names are given in Table 2.3.

Table 2.3

Common Monatomic Cations and Anions

Cation	Name	Anion	Name
H^+	Hydrogen	H^-	Hydride
Li^+	Lithium	F^-	Fluoride
Na^+	Sodium	Cl^-	Chloride
K^+	Potassium	Br^-	Bromide
Cs^+	Cesium	I^-	Iodide
Be^{2+}	Beryllium	O^{2-}	Oxide
Mg^{2+}	Magnesium	S^{2-}	Sulfide
Ca^{2+}	Calcium	N^{3-}	Nitride
Ba^{2+}	Barium	P^{3-}	Phosphide
Al^{3+}	Aluminum		
Ag^+	Silver		
Zn^{2+}	Zinc		

A Type I binary ionic compound contains a metal that forms only one type of cation. The rules for naming Type I compounds are illustrated by the following examples:

In formulas of ionic compounds, simple ions are represented by the element symbol: Cl means Cl^-, Na means Na^+, and so on.

Compound	*Ions Present*	*Name*
NaCl	Na^+, Cl^-	Sodium chloride
KI	K^+, I^-	Potassium iodide
CaS	Ca^{2+}, S^{2-}	Calcium sulfide
Li_3N	Li^+, N^{3-}	Lithium nitride
CsBr	Cs^+, Br^-	Cesium bromide
MgO	Mg^{2+}, O^{2-}	Magnesium oxide

Type II binary ionic compounds contain a metal that can form more than one type of cation.

Binary Compounds (Type II; Ionic)

In the ionic compounds considered previously (Type I), the metal involved forms only a single type of cation. That is, sodium forms only Na^+, calcium forms only Ca^{2+}, and so on. However, as we will see in more detail later in the text, many metals can form more than one type of positive ion and thus form more than one type of ionic compound with a given anion. For example, the compound $FeCl_2$ contains Fe^{2+} ions, and the compound $FeCl_3$ contains Fe^{3+} ions. In cases such as these, the *charge on the metal ion must be specified.* The systematic names for these two iron compounds are iron(II) chloride and iron(III) chloride, respectively, where the *Roman numeral indicates the charge of the cation.*

A compound containing a metal that forms multiple cations must have a Roman numeral in its name.

Another system for naming these ionic compounds that is seen in the older literature was used for metals that form only two ions. *The ion with the higher charge has a name ending in* -ic, *and the one with the lower charge has a name ending in* -ous. In this system, for example, Fe^{3+} is called the ferric ion, and Fe^{2+} is called the ferrous ion. The names for $FeCl_3$ and $FeCl_2$ are then ferric chloride and ferrous chloride, respectively.

Table 2.4 gives both names for many common Type II cations. The system that uses Roman numerals will be used exclusively in this text.

Note that the use of a Roman numeral in a systematic name is required only in cases in which more than one ionic compound forms between a given pair of elements. This case most commonly occurs for compounds containing transition metals, which often form more than one cation. *Elements that form only one cation do not need to be identified by a Roman numeral.* Common metals that do not require Roman numerals are the Group 1A elements, which form only 1+ ions; the Group 2A elements, which form only 2+ ions; and

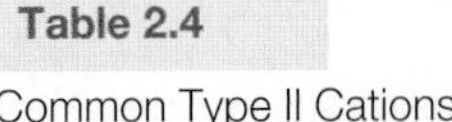

Table 2.4

Common Type II Cations

Ion	Systematic Name	Alternate Name
Fe^{3+}	Iron(III)	Ferric
Fe^{2+}	Iron(II)	Ferrous
Cu^{2+}	Copper(II)	Cupric
Cu^{+}	Copper(I)	Cuprous
Co^{3+}	Cobalt(III)	Cobaltic
Co^{2+}	Cobalt(II)	Cobaltous
Sn^{4+}	Tin(IV)	Stannic
Sn^{2+}	Tin(II)	Stannous
Pb^{4+}	Lead(IV)	Plumbic
Pb^{2+}	Lead(II)	Plumbous
Hg^{2+}	Mercury(II)	Mercuric
Hg_2^{2+}*	Mercury(I)	Mercurous

*Note that mercury(I) ions always occur bound together to form Hg_2^{2+}.

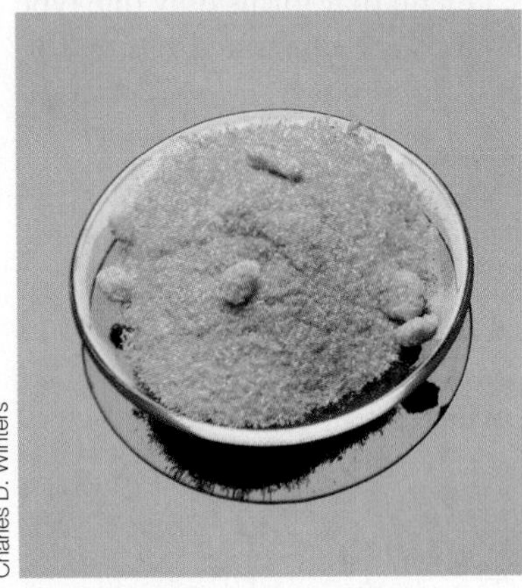

(top) Copper(I) chloride.
(bottom) Copper(II) chloride.

aluminum, which forms only Al^{3+}. Common transition metals that do not require a Roman numeral (because they form only one ion) are zinc (Zn^{2+}) and silver (Ag^{+}).

When a metal ion that forms more than one type of cation is present, the charge on the metal ion must be determined by balancing the positive and negative charges of the compound. To make this determination, you must be able to recognize the common cations and anions and know their charges (see Tables 2.3 and 2.5).

The following flowchart is useful when you are naming binary ionic compounds:

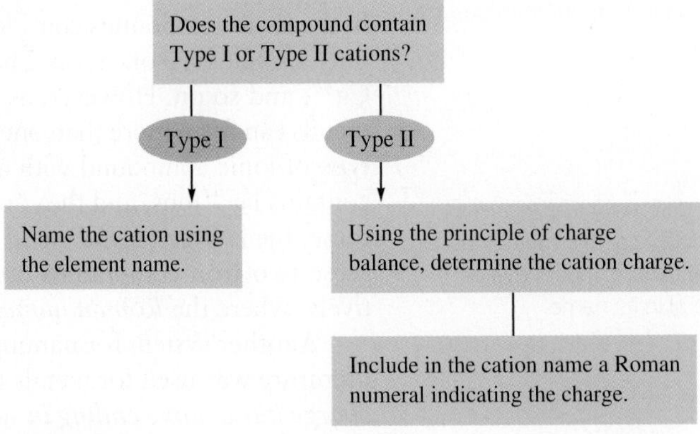

Various chromium compounds dissolved in water. From left to right: $CrCl_2$, $K_2Cr_2O_7$, $Cr(NO_3)_3$, $CrCl_3$, K_2CrO_4.

EXAMPLE 2.2

Give the systematic name of each of the following compounds.

a. $CoBr_2$ b. $CaCl_2$ c. Al_2O_3 d. $CrCl_3$

Solution

Compound	*Name*	*Comment*
a. $CoBr_2$	Cobalt(II) bromide	Cobalt is a transition metal that requires a Roman numeral. The two Br^{-} ions must be balanced by a Co^{2+} cation.

b. $CaCl_2$	Calcium chloride	Calcium, an alkaline earth metal, forms only the Ca^{2+} ion. A Roman numeral is not necessary.
c. Al_2O_3	Aluminum oxide	Aluminum forms only Al^{3+}. A Roman numeral is not necessary.
d. $CrCl_3$	Chromium(III) chloride	Chromium is a transition metal that must have a Roman numeral. $CrCl_3$ contains Cr^{3+}.

Ionic Compounds with Polyatomic Ions

Ionic compounds that contain polyatomic ions are not binary compounds.

We have not yet considered ionic compounds that contain polyatomic ions. For example, the compound ammonium nitrate (NH_4NO_3) contains the polyatomic ions NH_4^+ and NO_3^-. Polyatomic ions are assigned special names that *must be memorized* to name the compounds containing them. The most important polyatomic ions and their names are listed in Table 2.5.

Note in Table 2.5 that several series of anions contain an atom of a given element and different numbers of oxygen atoms. These anions are called **oxyanions.** When there are two members in such a series, the name of the one with the smaller number of oxygen atoms ends in *-ite,* and the name of the one with the larger number ends in *-ate*—for example, sulfite (SO_3^{2-}) and sulfate (SO_4^{2-}). When more than two oxyanions make up a series, *hypo-* (less than) and *per-* (more than) are used as prefixes to name the members of the series with the fewest and the most oxygen atoms, respectively. The best example involves the oxyanions containing chlorine, as shown in Table 2.5.

Binary Compounds (Type III; Covalent—Contain Two Nonmetals)

In binary covalent compounds, the element names follow the same rules as those for binary ionic compounds.

Binary covalent compounds are formed between *two nonmetals.* Although these compounds do not contain ions, they are named very similarly to binary ionic compounds.

Table 2.5

Common Polyatomic Ions

Ion	Name	Ion	Name
NH_4^+	Ammonium	CO_3^{2-}	Carbonate
NO_2^-	Nitrite	HCO_3^-	Hydrogen carbonate (bicarbonate is a widely used common name)
NO_3^-	Nitrate		
SO_3^{2-}	Sulfite		
SO_4^{2-}	Sulfate	$C_2H_3O_2^-$	Acetate
HSO_4^-	Hydrogen sulfate (bisulfate is a widely used common name)	MnO_4^-	Permanganate
		$Cr_2O_7^{2-}$	Dichromate
		CrO_4^{2-}	Chromate
OH^-	Hydroxide	O_2^{2-}	Peroxide
CN^-	Cyanide		
PO_4^{3-}	Phosphate	ClO^-	Hypochlorite
HPO_4^{2-}	Hydrogen phosphate	ClO_2^-	Chlorite
$H_2PO_4^-$	Dihydrogen phosphate	ClO_3^-	Chlorate
		ClO_4^-	Perchlorate

Table 2.6

Prefixes Used to Indicate Number in Chemical Names

Prefix	Number Indicated
mono-	1
di-	2
tri-	3
tetra-	4
penta-	5
hexa-	6
hepta-	7
octa-	8

In the naming of binary covalent compounds, the following rules apply:

1. The first element in the formula is named first, using the full element name.
2. The second element is named as if it were an anion.
3. Prefixes are used to denote the numbers of atoms present. These prefixes are given in Table 2.6.
4. The prefix *mono-* is never used for naming the first element. For example, CO is called carbon monoxide, *not* monocarbon monoxide.

To see how these rules apply, we will now consider the names of the several covalent compounds formed by nitrogen and oxygen:

Compound	*Systematic Name*	*Common Name*
N_2O	Dinitrogen monoxide	Nitrous oxide
NO	Nitrogen monoxide	Nitric oxide
NO_2	Nitrogen dioxide	
N_2O_3	Dinitrogen trioxide	
N_2O_4	Dinitrogen tetroxide	
N_2O_5	Dinitrogen pentoxide	

Notice from the preceding examples that to avoid awkward pronunciations, we often drop the final *o* or *a* of the prefix when the element begins with a vowel. For example, N_2O_4 is called dinitrogen tetroxide, *not* dinitrogen tetr*a*oxide; and CO is called carbon monoxide, *not* carbon monooxide.

Some compounds are always referred to by their common names. The two best examples are water and ammonia. The systematic names for H_2O and NH_3 are never used.

The rules for naming binary compounds are summarized in Fig. 2.21. Notice that prefixes to indicate the number of atoms are used only in Type III binary compounds (those containing two nonmetals). An overall strategy for naming compounds is summarized in Fig. 2.22.

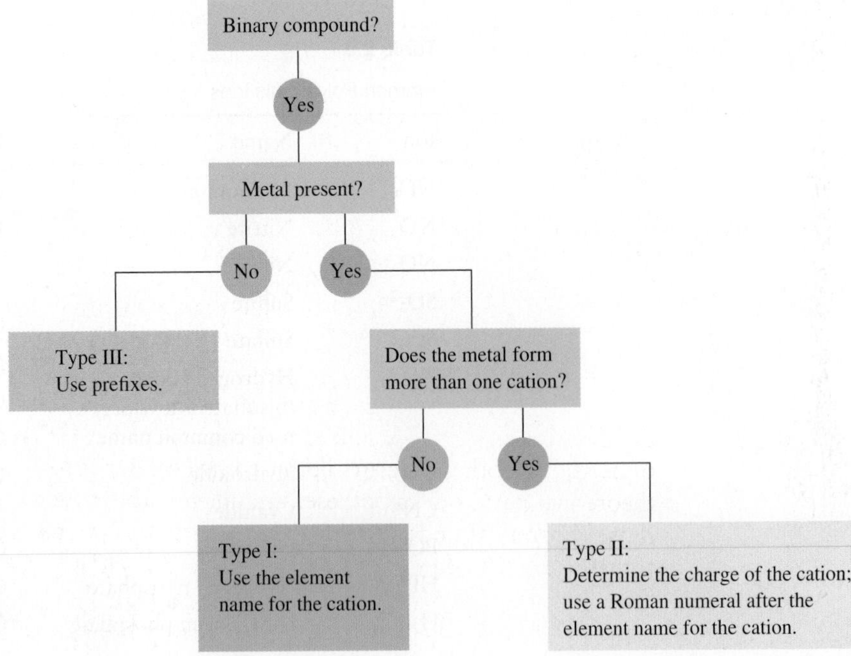

Figure 2.21
A flowchart for naming binary compounds.

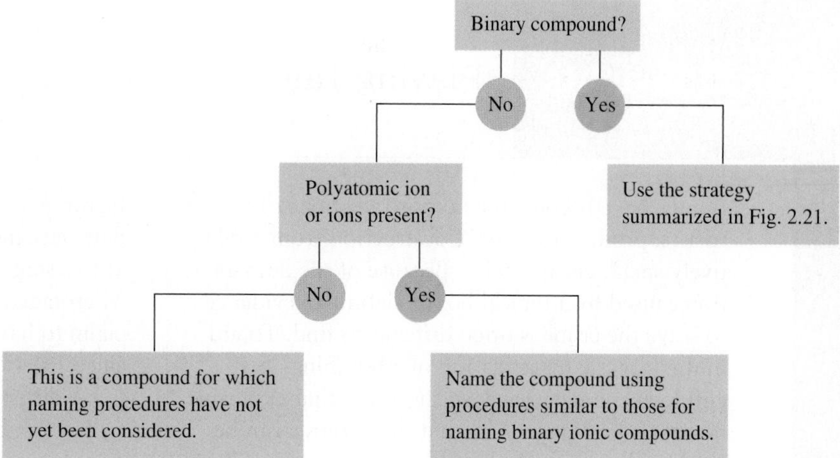

Figure 2.22
Overall strategy for naming chemical compounds.

OWL INTERACTIVE EXAMPLE 2.3

Give the systematic name of each of the following compounds.

a. Na_2SO_4 e. Na_2SO_3 i. $NaOCl$
b. KH_2PO_4 f. Na_2CO_3 j. Na_2SeO_4
c. $Fe(NO_3)_3$ g. $NaHCO_3$ k. $KBrO_3$
d. $Mn(OH)_2$ h. $CsClO_4$

Solution

Compound	*Name*	*Comment*
a. Na_2SO_4	Sodium sulfate	
b. KH_2PO_4	Potassium dihydrogen phosphate	
c. $Fe(NO_3)_3$	Iron(III) nitrate	Transition metal that requires a Roman numeral. Fe^{3+} ion balances three NO_3^- ions.
d. $Mn(OH)_2$	Manganese(II) hydroxide	Transition metal that requires a Roman numeral. Mn^{2+} ion balances two OH^- ions.
e. Na_2SO_3	Sodium sulfite	
f. Na_2CO_3	Sodium carbonate	
g. $NaHCO_3$	Sodium hydrogen carbonate	Often called sodium bicarbonate.
h. $CsClO_4$	Cesium perchlorate	
i. $NaOCl$	Sodium hypochlorite	
j. Na_2SeO_4	Sodium selenate	Atoms in the same group, such as sulfur and selenium, often form similar ions that are named similarly. Thus SeO_4^{2-} is selenate, like SO_4^{2-} (sulfate).
k. $KBrO_3$	Potassium bromate	As above BrO_3^- is bromate, like ClO_3^- (chlorate).

Chemical Insights — Playing Tag

Bombs have become the favorite weapons of terrorists, producing massive destruction from a relatively small, cheap device. Because of the devastation caused by a typical bomb, definitive evidence to solve the crime is often difficult to find. To aid in the forensic investigation of a bombing, law enforcement agencies have suggested that explosive materials be "tagged" so that their source can be readily identified. One method for tagging explosives was developed in the 1970s by Richard Livesag, a research chemist at 3M Corporation. Called Microtaggant, the tagging material consists of irregularly shaped particles approximately 0.1 millimeter in diameter. Examined under a microscope, the tiny specks reveal a laminated structure of 10 layers of melamine plastic, a chemically unreactive substance that is very difficult to destroy. The colored layers of the chips act as a kind of bar code, identifying the manufacturer, the date of production, and the name of the distributor of the explosive. Fluorescent and magnetic materials added to the chips aid in their detection at the scene of an explosion.

Although it seems that the tagging of explosives should be noncontroversial, this has not proved to be the case. Manufacturers of explosive materials and gun owners' associations have opposed the use of taggants, arguing that the added substances produce instabilities in the explosives. Also, manufacturers fear that they will be held liable for the damage caused by misuse of their products. Thus, at this stage, only Switzerland routinely uses Microtaggant to label explosives, and the Swiss claim to have solved nearly 600 bombing cases since 1984 using the taggant evidence.

A different tagging method has been pioneered by Isotag of Houston, Texas. Isotag labels explosives by inserting unusual atomic isotopes of the atoms present in the ions or molecules of the explosive substance. The isotopic labeling is done in a way that makes the tagged substances distinctly different from naturally occurring substances. For example, if labeled ammonium nitrate were used in a bomb, residue at the scene could be collected and analyzed with a mass spectrometer to show the unique isotopic patterns present, which could be used to trace the origin of the compound. Although Isotag claims its system to be safe and effective, manufacturers of fertilizers containing ammonium nitrate are reluctant to allow tagging because of fears of liability.

The U.S. Antiterrorism and Effective Death Penalty Act of 1996 authorized a thorough study of taggants. The results of this study and the outcome of the political struggle over taggants will determine whether the United States will follow Switzerland and require taggants for some or all explosive substances.

Formulas from Names

So far we have started with the chemical formula of a compound and decided on its systematic name. The reverse process is also important. For example, given the name calcium hydroxide, we can write the formula as $Ca(OH)_2$, since we know that calcium forms only Ca^{2+} ions and that, since hydroxide is OH^-, two of these anions will be required to give a neutral compound. Similarly, the name iron(II) oxide implies the formula FeO, since the Roman numeral II indicates the presence of Fe^{2+} and the oxide ion is O^{2-}.

OWL INTERACTIVE EXAMPLE 2.4

Given the following systematic names, write the formula for each compound.

a. Ammonium sulfate
b. Vanadium(V) fluoride
c. Dioxygen difluoride
d. Rubidium peroxide
e. Gallium oxide

Microtrace, LLC

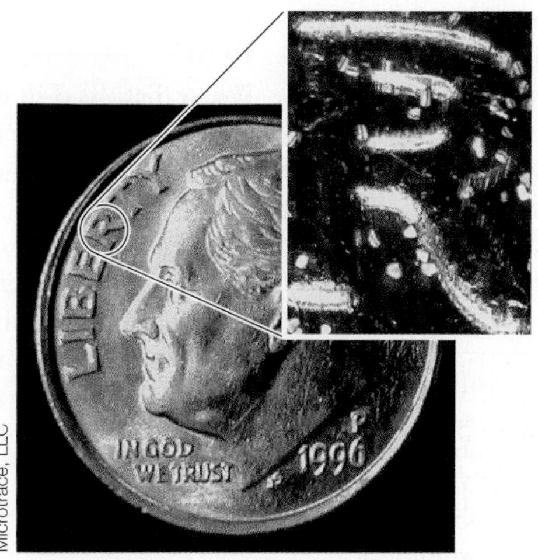

Microtrace, LLC

The multilayered plastic particles of Microtaggants (left) can produce over 37 million unique codes that can be used to identify substances. The tiny size of the taggants is shown (right) compared with the "R" on a dime.

Although taggants for explosives have proved controversial, approximately 500 consumer items already contain Microtaggants to authenticate brand-name products and discourage counterfeiting. For example, taggants are found in shampoo, paint, and carpet adhesives. The latter application enables business owners to check that a requested brand-name glue has been used to secure carpet rather than cheaper off-brand substitutes. Isotag markets taggants for substances such as gasoline and perfume, in which the plastic Microtaggant chips are impractical. For instance, isotopic tagging of perfumes enables a manufacturer to protect its product against dilution by an unethical distributor who might seek to increase profits by selling a diluted perfume.

So chemical taggants represent an increasing market. However, the question of whether explosives will be tagged remains a technical and political question mark.

Solution

Name	*Chemical Formula*	*Comment*
a. Ammonium sulfate	$(NH_4)_2SO_4$	Two ammonium ions (NH_4^+) are required for each sulfate ion (SO_4^{2-}) to achieve charge balance.
b. Vanadium(V) fluoride	VF_5	The compound contains V^{5+} ions and requires five F^- ions for charge balance.
c. Dioxygen difluoride	O_2F_2	The prefix *di-* indicates two of each atom.
d. Rubidium peroxide	Rb_2O_2	Since rubidium is in Group 1A, it forms only 1+ ions. Thus two Rb^+ ions are needed to balance the 2− charge on the peroxide ion (O_2^{2-}).

Name	*Chemical Formula*	*Comment*
e. Gallium oxide	Ga_2O_3	Gallium in Group 3A, like aluminum, forms only 3+ ions. Two Ga^{3+} ions are required to balance the charge on three O^{2-} ions.

Acids

When dissolved in water, certain molecules produce a solution containing free H^+ ions (protons). These substances, acids, will be discussed in detail in Chapters 4, 7, and 8. Here we will simply present the rules for naming acids.

An acid can be viewed as a molecule with one or more H^+ ions attached to an anion. The rules for naming acids depend on whether the anion contains oxygen. If the *anion does not contain oxygen,* the acid is named with the prefix *hydro-* and the suffix *-ic.* For example, when gaseous HCl is dissolved in water, it forms hydrochloric acid. Similarly, HCN and H_2S dissolved in water are called hydrocyanic and hydrosulfuric acid, respectively.

Table 2.7

Names of Acids That Do Not Contain Oxygen

Acid	Name
HF	Hydrofluoric acid
HCl	Hydrochloric acid
HBr	Hydrobromic acid
HI	Hydroiodic acid
HCN	Hydrocyanic acid
H_2S	Hydrosulfuric acid

When the *anion contains oxygen,* the acid name is formed from the root name of the anion with a suffix of *-ic* or *-ous.* If the anion name ends in *-ate,* the acid name ends with *-ic* (or sometimes *-ric*). For example, H_2SO_4 contains the sulfate anion (SO_4^{2-}) and is called sulfuric acid; H_3PO_4 contains the phosphate anion (PO_4^{3-}) and is called phosphoric acid; and $HC_2H_3O_2$ contains the acetate ion ($C_2H_3O_2^-$) and is called acetic acid. If the anion has an *-ite* ending, the acid name ends with *-ous.* For example, H_2SO_3, which contains sulfite (SO_3^{2-}), is named sulfurous acid; HNO_2, which contains nitrite (NO_2^-), is named nitrous acid. The application of these rules can be seen in the names of the acids of the oxyanions of chlorine:

Acid	*Anion*	*Name*
$HClO_4$	Perchlor*ate*	Perchlor*ic* acid
$HClO_3$	Chlor*ate*	Chlor*ic* acid
$HClO_2$	Chlor*ite*	Chlor*ous* acid
HClO	Hypochlor*ite*	Hypochlor*ous* acid

Table 2.8

Names of Some Oxygen-Containing Acids

Acid	Name
HNO_3	Nitric acid
HNO_2	Nitrous acid
H_2SO_4	Sulfuric acid
H_2SO_3	Sulfurous acid
H_3PO_4	Phosphoric acid
$HC_2H_3O_2$	Acetic acid

The names of the most important acids are given in Tables 2.7 and 2.8. An overall strategy for naming acids is shown in Fig. 2.23.

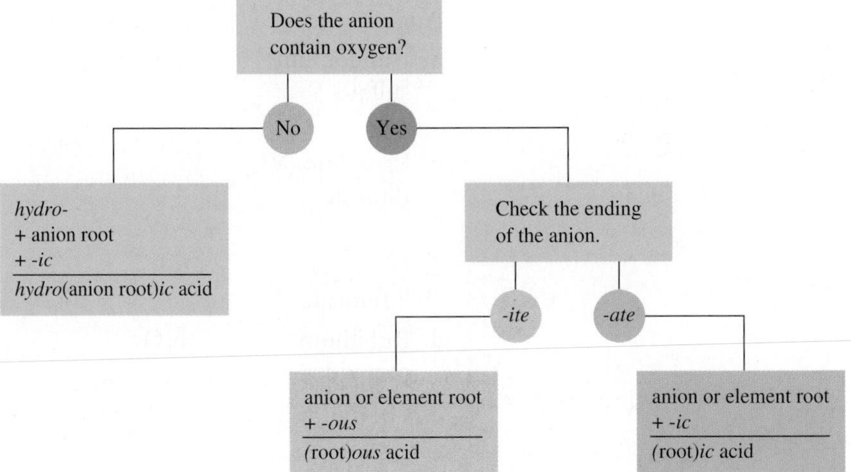

Figure 2.23
A flowchart for naming acids. The acid has one or more H^+ ions attached to an anion.

Key Terms

Section 2.2
law of conservation of mass
law of definite proportion
law of multiple proportions

Section 2.3
atomic masses
Avogadro's hypothesis

Section 2.5
cathode ray
electrons
radioactivity
nuclear atom
nucleus

Section 2.6
protons
neutrons
isotopes
atomic number
mass number

Section 2.7
chemical bonds
covalent bonds
molecule
chemical formula
structural formula
space-filling model
ball-and-stick models
ion
cation
anion
ionic bonding
polyatomic ion

Section 2.8
periodic table
metals
nonmetals
groups (families)
alkali metals
alkaline earth metals
halogens
noble gases
periods

Section 2.9
binary compounds
binary ionic compounds
oxyanions
binary covalent compounds

For Review

Sign in at **www.cengage.com/owl** to:
- View tutorials and simulations, develop problem-solving skills, and complete online homework assigned by your professor.
- Download Go Chemistry mini lecture modules for quick review and exam prep from OWL (or purchase them at **www.cengagebrain.com**)

Fundamental laws
- Conservation of mass
- Definite proportion
- Multiple proportions

Dalton's atomic theory
- All elements are composed of atoms.
- All atoms of a given element are identical.
- Chemical compounds are formed when atoms combine.

Early atomic experiments
- Thomson model: electrons in a diffuse cloud of positive charge
- Millikan experiment: determined mass of electron
- Rutherford experiment: nuclear atom

Atomic structure
- Small dense nucleus contains protons and neutrons.
  - Protons—positive charge
  - Neutrons—no charge
- Electrons reside outside the nucleus in the relatively large remaining atomic volume.
  - Electrons—negative charge, small mass (1/1840 of proton)
- Isotopes have the same atomic number but different mass numbers.

Formation of molecules
- Atoms combine to form molecules by sharing electrons to form covalent bonds.
- Molecules are described by chemical formulas.
- Chemical formulas show number and type of atoms.
  - Structural formula
  - Ball-and-stick model
  - Space-filling model

Formation of ions
- Cation—formed by a loss of an electron, positive charge
- Anion—formed by gain of an electron, negative charge
- Ionic bonds—formed by interaction of cations and anions

Periodic table
- Elements are organized in order of increasing atomic number.
- Elements with similar properties are in columns, or groups.
- Metals are in the majority and tend to form cations.
- Nonmetals tend to form anions.

Naming of compounds
- Binary compounds
  - Type I—contain a metal that always forms the same cation
  - Type II—contain a metal that can form more than one cation
  - Type III—contain two nonmetals
- Compounds containing a polyatomic ion
- Acids

Discussion Questions

These questions are designed to be considered by groups of students in class. Often these questions work well for introducing a particular topic in class.

1. You may have noticed that when water boils, you can see bubbles that rise to the surface of the water. What is inside these bubbles? Explain.
   a. air
   b. hydrogen and oxygen gas
   c. oxygen gas
   d. water vapor
   e. carbon dioxide gas
2. Which of the following is true about an individual atom?
   a. An individual atom should be considered a solid.
   b. An individual atom should be considered a liquid.
   c. An individual atom should be considered a gas.
   d. The state of the atom depends on which element it is.
   e. An individual atom cannot be considered a solid, liquid, or gas.

   Justify your choice, and for those you did not choose, explain why they are incorrect.
3. How would you go about finding the number of "chalk molecules" it takes to write your name on the board? Explain what you would need to do and provide a sample calculation.
4. These questions concern the work of J. J. Thomson:
   a. From what you know of Thomson's work, which particles do you think he would believe are most important in the formation of compounds (chemical changes) and why?
   b. Of the remaining two subatomic particles, which do you place second in importance for forming compounds and why?
   c. Propose three models that explain Thomson's findings and evaluate them. Include Thomson's findings.
5. Heat is applied to an ice cube in a closed container until only steam is present. Draw a representation of this process, assuming you can see it at an extremely high level of magnification. What happens to the size of the molecules? What happens to the total mass of the sample?
6. You have a chemical in a sealed glass container filled with air. The setup is sitting on a balance, as shown. The chemical is ignited by means of a magnifying glass focusing sunlight on the reactant. After the chemical has completely burned, which of the following is true? Explain your answer.
   a. The balance will read less than 250.0 g.
   b. The balance will read 250.0 g.
   c. The balance will read greater than 250.0 g.
   d. Cannot be determined without knowing the identity of the chemical.

7. The vitamin niacin (nicotinic acid, $C_6H_5NO_2$) can be isolated from a variety of natural sources such as liver, yeast, milk, and whole grain. It also can be synthesized from commercially available materials. From a nutritional point of view, which source of nicotinic acid is best for use in a multivitamin tablet? Why?
8. One of the best indications of a useful theory is that it raises more questions for further experimentation than it originally answered. Is this true of Dalton's atomic theory? Give examples.
9. Dalton assumed that all atoms of the same element are identical in all their properties. Explain why this assumption is not valid.
10. How does Dalton's atomic theory account for each of the following?
    a. the law of conservation of mass
    b. the law of definite proportion
    c. the law of multiple proportions
11. What refinements had to be made in Dalton's atomic theory to account for Gay-Lussac's results on the combining volumes of gases?
12. Which (if any) of the following can be determined by knowing the number of protons in a neutral element? Explain your answer.
    a. the number of neutrons in the neutral element
    b. the number of electrons in the neutral element
    c. the name of the element
13. The average mass of a carbon atom is 12.011. Assuming you were able to pick up only one carbon atom, the chance that you would randomly get one with a mass of 12.011 is
    a. 0%.
    b. 0.011%.
    c. about 12%.
    d. 12.011%.
    e. greater than 50%.
    f. None of these is true.

    Explain.
14. Which of the following explain how an ion is formed? Explain your answer.
    a. adding or subtracting protons to/from an atom
    b. adding or subtracting neutrons to/from an atom
    c. adding or subtracting electrons to/from an atom
15. The formula of water is H_2O. Which of the following is indicated by this formula? Explain your answer.
    a. the mass of hydrogen is twice that of oxygen in each molecule
    b. there are two hydrogen atoms and one oxygen atom per water molecule
    c. the mass of oxygen is twice that of hydrogen in each molecule
    d. there are two oxygen atoms and one hydrogen atom per water molecule
16. Why do we call $Ba(NO_3)_2$ barium nitrate, but we call $Fe(NO_3)_2$ iron(II) nitrate?
17. Why is calcium dichloride not the correct systematic name for $CaCl_2$?

Exercises

OWL Interactive versions of these problems may be assigned in OWL.

A blue exercise number indicates that the answer to that exercise appears at the back of this book and a solution appears in the *Solutions Guide*.

Development of the Atomic Theory

18. Explain the law of conservation of mass, the law of definite proportion, and the law of multiple proportions.

19. A reaction of 1 L of chlorine gas (Cl_2) with 5 L of fluorine gas (F_2) yields 2 L of a gaseous product. All gas volumes are at the same temperature and pressure. What is the formula of the gaseous product?

20. When mixtures of gaseous H_2 and gaseous Cl_2 react, a product forms that has the same properties regardless of the relative amounts of H_2 and Cl_2 used.
    a. How is this result interpreted in terms of the law of definite proportion?
    b. When a volume of H_2 reacts with an equal volume of Cl_2 at the same temperature and pressure, what volume of product having the formula HCl is formed?

21. Observations of the reaction between nitrogen gas and hydrogen gas show us that 1 volume of nitrogen reacts with 3 volumes of hydrogen to make 2 volumes of gaseous product, as shown below:

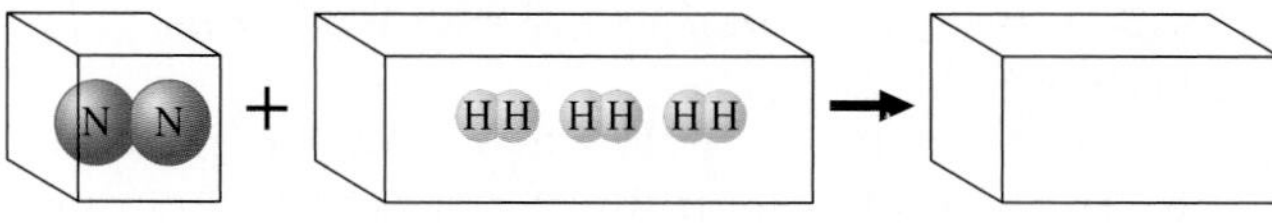

Determine the formula of the product and justify your answer.

22. The three most stable oxides of carbon are carbon monoxide (CO), carbon dioxide (CO_2), and carbon suboxide (C_3O_2). The molecules can be represented as

 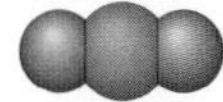 

Explain how these molecules illustrate the law of multiple proportions.

23. Hydrazine, ammonia, and hydrogen azide all contain only nitrogen and hydrogen. The mass of hydrogen that combines with 1.00 g of nitrogen for each compound is 1.44×10^{-1} g, 2.16×10^{-1} g, and 2.40×10^{-2} g, respectively. Show how these data illustrate the law of multiple proportions.

24. Consider 100.0-g samples of two different compounds consisting only of carbon and oxygen. One compound contains 27.2 g of carbon, and the other has 42.9 g of carbon. How can these data support the law of multiple proportions if 42.9 is not a multiple of 27.2? Show that these data support the law of multiple proportions.

25. Early tables of atomic weights (masses) were generated by measuring the mass of a substance that reacts with 1.00 g of oxygen. Given the following data and taking the atomic mass of hydrogen as 1.00, generate a table of relative atomic masses for oxygen, sodium, and magnesium.

Element	Mass That Combines with 1.00 g Oxygen	Assumed Formula
Hydrogen	0.126 g	HO
Sodium	2.875 g	NaO
Magnesium	1.500 g	MgO

How do your values compare with those in the periodic table? How do you account for any differences?

The Nature of the Atom

26. What evidence led to the conclusion that cathode rays had a negative charge? Is there a difference between a cathode ray and a β particle?

27. From the information in this chapter on the mass of the proton, the mass of the electron, and the sizes of the nucleus and the atom, calculate the densities of a hydrogen nucleus and a hydrogen atom.

28. A chemistry instructor makes the following claim: "Consider that if the nucleus were the size of a grape, the electrons would be about 1 *mile* away on average." Is this claim reasonably accurate? Provide mathematical support.

29. A chemist in a galaxy far, far away performed the Millikan oil drop experiment and got the following results for the charge on various drops. What is the charge of the electron in zirkombs?

2.56×10^{-12} zirkombs	7.68×10^{-12} zirkombs
3.84×10^{-12} zirkombs	6.40×10^{-13} zirkombs

30. Do the proton and the neutron have exactly the same mass? How do the masses of the proton and the neutron compare with the mass of the electron? Which particles make the greatest contribution to the mass of an atom? Which particles make the greatest contribution to the chemical properties of an atom?

31. Consider Ernest Rutherford's α-particle bombardment experiment illustrated in Fig. 2.11. How did the results of this experiment lead Rutherford away from the plum pudding model of the atom to propose the nuclear model of the atom?

Elements, Ions, and the Periodic Table

32. Distinguish between the following terms.
    a. molecule versus ion
    b. covalent bonding versus ionic bonding
    c. molecule versus compound
    d. anion versus cation
33. What is the distinction between atomic number and mass number? Between mass number and atomic mass?
34. a. Classify the following elements as metals or nonmetals.

Mg	Si	Rn
Ti	Ge	Eu
Au	B	Am
Bi	At	Br

    b. The distinction between metals and nonmetals is really not a clear one. Some elements, called *metalloids*, are intermediate in their properties. Which of these elements would you reclassify as metalloids? What other elements in the periodic table would you expect to be metalloids?
35. a. List the noble gas elements. Which of the noble gases has only radioactive isotopes? (This situation is indicated on most periodic tables by parentheses around the mass of the element. See inside front cover.)
    b. Which lanthanide element and which transition element have only radioactive isotopes?
36. Consider the elements of the carbon family: C, Si, Ge, Sn, and Pb. What is the trend in metallic character as one goes down a group in the periodic table? What is the trend in metallic character going from left to right across a period in the periodic table?
37. Identify the elements that correspond to the following atomic numbers. Label each as either a noble gas, a halogen, an alkali metal, an alkaline earth metal, a transition metal, a lanthanide metal, or an actinide metal.
    a. 17 e. 2
    b. 4 f. 92
    c. 63 g. 55
    d. 72
38. The number of protons in an atom determines the identity of the atom. What does the number and arrangement of the electrons in an atom determine? What does the number of neutrons in an atom determine?
39. For lighter, stable isotopes, the ratio of the mass number to the atomic number is close to a certain value. What is the value? What happens to the value of the mass number to atomic number ratio as stable isotopes become heavier?
40. Write the atomic symbol ($^{A}_{Z}X$) for each of the isotopes described below.
    a. number of protons = 27, number of neutrons = 31
    b. the isotope of boron with mass number 10
    c. $Z = 12$, $A = 23$
    d. atomic number 53, number of neutrons = 79
    e. $Z = 9$, number of neutrons = 10
    f. number of protons = 29, mass number 65
41. How many protons, neutrons, and electrons are in each of the following atoms or ions?
    a. $^{24}_{12}Mg$ d. $^{59}_{27}Co^{3+}$ g. $^{79}_{34}Se^{2-}$
    b. $^{24}_{12}Mg^{2+}$ e. $^{59}_{27}Co$ h. $^{63}_{28}Ni$
    c. $^{59}_{27}Co^{2+}$ f. $^{79}_{34}Se$ i. $^{59}_{28}Ni^{2+}$
42. Complete the following table.

Symbol	Number of Protons in Nucleus	Number of Neutrons in Nucleus	Number of Electrons	Net Charge
$^{238}_{92}U$	—	—	—	—
—	20	20	—	2+
—	23	28	20	—
$^{89}_{39}Y$	—	—	—	—
—	35	44	36	—
—	15	16	—	3−

43. What is the symbol for an ion with 63 protons, 60 electrons, and 88 neutrons? If an ion contains 50 protons, 68 neutrons, and 48 electrons, what is its symbol?
44. What is the symbol of an ion with 16 protons, 18 neutrons, and 18 electrons? What is the symbol for an ion that has 16 protons, 16 neutrons, and 18 electrons?
45. Would you expect each of the following atoms to gain or lose electrons when forming ions? What ion is the most likely in each case?
    a. Ra c. P e. Br
    b. In d. Te f. Rb
46. For each of the following atomic numbers, use the periodic table to write the formula (including the charge) for the simple *ion* that the element is most likely to form in ionic compounds.
    a. 13 c. 56 e. 87
    b. 34 d. 7 f. 35

Nomenclature

47. The compounds $AlCl_3$, $CrCl_3$, and ICl_3 have similar formulas, yet each follows a different set of rules to name it. Name these compounds, and then compare and contrast the nomenclature rules used in each case.
48. Each of the following compounds has three possible names listed for it. For each compound, what is the correct name and why aren't the other names used?
    a. N_2O: nitrogen oxide, nitrogen(I) oxide, dinitrogen monoxide
    b. Cu_2O: copper oxide, copper(I) oxide, dicopper monoxide
    c. Li_2O: lithium oxide, lithium(I) oxide, dilithium monoxide

49. Name each of the following compounds.

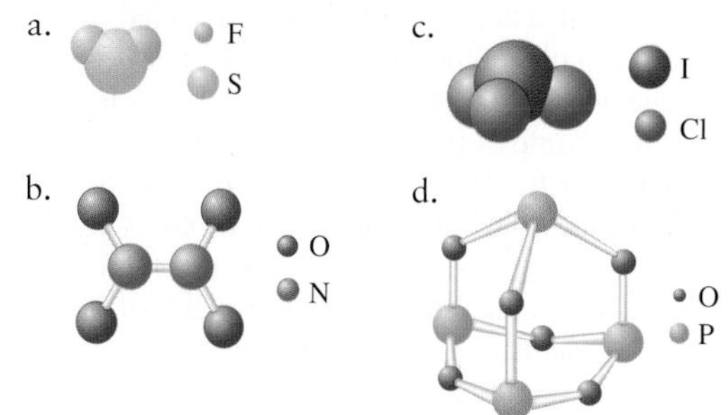

50. Name the following compounds.
   a. $NaClO_4$
   b. $Mg_3(PO_4)_2$
   c. $Al_2(SO_4)_3$
   d. SF_2
   e. SF_6
   f. Na_2HPO_4
   g. NaH_2PO_4
   h. Li_3N
   i. NaOH
   j. $Mg(OH)_2$
   k. $Al(OH)_3$
   l. Ag_2CrO_4

51. Name each of the following compounds.
   a. CuI
   b. CuI_2
   c. CoI_2
   d. Na_2CO_3
   e. $NaHCO_3$
   f. S_4N_4
   g. $SeBr_4$
   h. NaOCl
   i. $BaCrO_4$
   j. NH_4NO_3

52. Name the following compounds. Assume the potential acids are dissolved in water.
   a. $HC_2H_3O_2$
   b. NH_4NO_2
   c. Co_2S_3
   d. ICl
   e. $Pb_3(PO_4)_2$
   f. $KClO_3$
   g. H_2SO_4
   h. Sr_3N_2
   i. $Al_2(SO_3)_3$
   j. SnO_2
   k. Na_2CrO_4
   l. HClO

53. Write formulas for the following compounds.
   a. Sulfur dioxide
   b. Sulfur trioxide
   c. Sodium sulfite
   d. Potassium hydrogen sulfite
   e. Lithium nitride
   f. Chromium(III) carbonate
   g. Chromium(II) acetate
   h. Tin(IV) fluoride
   i. Ammonium hydrogen sulfate
   j. Ammonium hydrogen phosphate
   k. Potassium perchlorate
   l. Sodium hydride
   m. Hypobromous acid
   n. Hydrobromic acid

54. Write formulas for the following compounds.
   a. Sodium oxide
   b. Sodium peroxide
   c. Potassium cyanide
   d. Copper(II) nitrate
   e. Silicon tetrachloride
   f. Lead(II) oxide
   g. Lead(IV) oxide (common name lead dioxide)
   h. Copper(I) chloride
   i. Gallium arsenide
   j. Cadmium selenide
   k. Zinc sulfide
   l. Mercury(I) chloride
   m. Nitrous acid
   n. Diphosphorus pentoxide

55. The common names and formulas for several substances are given below. What are the systematic names for these substances?
   a. Sugar of lead $Pb(C_2H_3O_2)_2$
   b. Blue vitriol $CuSO_4$
   c. Quicklime CaO
   d. Epsom salts $MgSO_4$
   e. Milk of magnesia $Mg(OH)_2$
   f. Gypsum $CaSO_4$
   g. Laughing gas N_2O

56. Each of the following compounds is incorrectly named. What is wrong with each name, and what is the correct name for each compound?
   a. $FeCl_3$, iron chloride
   b. NO_2, nitrogen(IV) oxide
   c. CaO, calcium(II) monoxide
   d. Al_2S_3, dialuminum trisulfide
   e. $Mg(C_2H_3O_2)_2$, manganese diacetate
   f. $FePO_4$, iron(II) phosphide
   g. P_2S_5, phosphorus sulfide
   h. Na_2O_2, sodium oxide
   i. HNO_3, nitrate acid
   j. H_2S, sulfuric acid

57. Name the following acids.

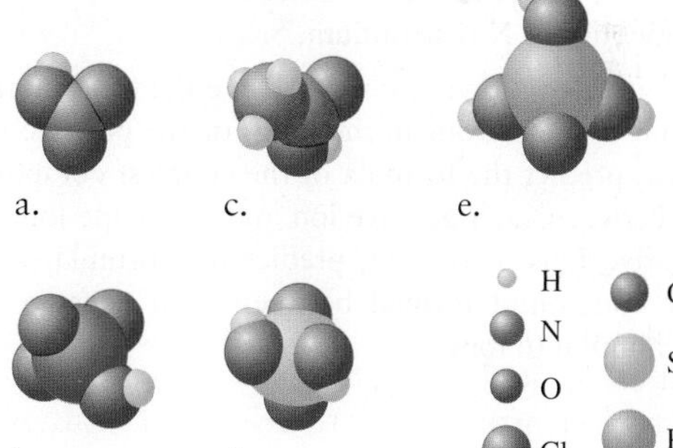

Additional Exercises

58. What discoveries were made by J. J. Thomson, Henri Becquerel, and Lord Rutherford? How did Dalton's model of the atom have to be modified to account for these discoveries?

59. Consider the chemical reaction depicted to the right. Label as much as you can using the terms *atom, molecule, element, compound, ionic, gas,* and *solid.*

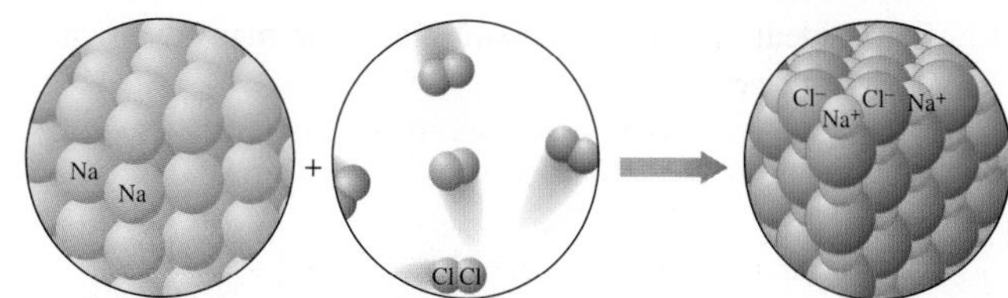

60. Section 2.3 describes the postulates of Dalton's atomic theory. With some modifications, these postulates hold up very well regarding how we view elements, compounds, and chemical reactions today. Answer the following questions concerning Dalton's atomic theory and the modifications made today.
    a. The atom can be broken down into smaller parts. What are the smaller parts?
    b. How are atoms of hydrogen identical to each other, and how can they be different from each other?
    c. How are atoms of hydrogen different from atoms of helium? How can H atoms be similar to He atoms?
    d. How is water different from hydrogen peroxide (H_2O_2) even though both compounds are composed of only hydrogen and oxygen?
    e. What happens in a chemical reaction, and why is mass conserved in a chemical reaction?

61. A sample of chloroform is found to contain 12.0 g of carbon, 106.4 g of chlorine, and 1.01 g of hydrogen. If a second sample of chloroform is found to contain 30.0 g of carbon, what is the total mass of chloroform in the second sample?

62. In a reaction, 34.0 g of chromium(III) oxide reacts with 12.1 g of aluminum to produce chromium and aluminum oxide. If 23.3 g of chromium is produced, what mass of aluminum oxide is produced?

63. The isotope of an unknown element, X, has a mass number of 79. The most stable ion of the isotope has 36 electrons and forms a binary compound with sodium having a formula of Na_2X. Which of the following statements is(are) *true?* Correct the false statements.
    a. The binary compound formed between X and fluorine will be a covalent compound.
    b. The isotope of X contains 38 protons.
    c. The isotope of X contains 41 neutrons.
    d. The identity of X is strontium, Sr.

64. For each of the following ions, indicate the total number of protons and electrons in the ion. For the positive ions in the list, predict the formula of the simplest compound formed between each positive ion and the oxide ion. For the negative ions in the list, predict the formula of the simplest compound formed between each negative ion and the aluminum ion.
    a. Fe^{2+}
    b. Fe^{3+}
    c. Ba^{2+}
    d. Cs^+
    e. S^{2-}
    f. P^{3-}
    g. Br^-
    h. N^{3-}

65. An element's most stable ion forms an ionic compound with bromine, having the formula XBr_2. If the ion of element X has a mass number of 230 and 86 electrons, what is the identity of the element, and how many neutrons does it have?

66. The early alchemists used to do an experiment in which water was boiled for several days in a sealed glass container. Eventually, some solid residue would begin to appear in the bottom of the flask. This result was interpreted to mean that some of the water in the flask had been converted into earth. When Lavoisier repeated this experiment, he found that the water weighed the same before and after heating, and the weight of the flask plus the solid residue equaled the original weight of the flask. Were the alchemists correct? Explain what really happened. (This experiment is described in the article by A. F. Scott in *Scientific American*, January 1984.)

67. Elements in the same family often form oxyanions of the same general formula. The anions are named in a similar fashion. What are the names of the oxyanions of selenium and tellurium: SeO_4^{2-}, SeO_3^{2-}, TeO_4^{2-}, TeO_3^{2-}?

68. How would you name $HBrO_4$, KIO_3, $NaBrO_2$, and HIO? Refer to Table 2.5 and the acid nomenclature discussion in the text.

69. Indium oxide contains 4.784 g of indium for every 1.000 g of oxygen. In 1869, when Mendeleev first presented his version of the periodic table, he proposed the formula In_2O_3 for indium oxide. Before that time, it was thought that the formula was InO. What values for the atomic mass of indium are obtained using these two formulas? Assume that oxygen has an atomic mass of 16.00.

70. The designations 1A through 8A used for certain families of the periodic table are helpful for predicting the charges on ions in binary ionic compounds. In these compounds, the metals generally take on a positive charge equal to the family number, and the nonmetals take on a negative charge equal to the family number minus 8. Thus the compound formed from sodium and chlorine contains Na^+ and Cl^- ions and has the formula NaCl. Predict the formula and the name of the binary compound formed from the following pairs of elements.
    a. Ca and N
    b. K and O
    c. Rb and F
    d. Mg and S
    e. Ba and I
    f. Al and Se
    g. Cs and P
    h. In and Br

71. A binary ionic compound is known to contain a cation with 51 protons and 48 electrons. The anion contains one-third the number of protons as the cation. The number of electrons in the anion is equal to the number of protons plus 1. What is the formula of this compound? What is the name of this compound?

72. Identify each of the following elements.
    a. a member of the same family as oxygen whose most stable ion contains 54 electrons
    b. a member of the alkali metal family whose most stable ion contains 36 electrons
    c. a noble gas with 18 protons in the nucleus
    d. a halogen with 85 protons and 85 electrons

73. A certain element has only two naturally occurring isotopes: one with 18 neutrons and the other with 20 neutrons. The element forms 1− charged ions when in ionic compounds. Predict the identity of the element. What number of electrons does the 1− charged ion have?

Challenge Problems

74. Reaction of 2.0 L of hydrogen gas with 1.0 L of oxygen gas yields 2.0 L of water vapor. All gases are at the same temperature and pressure. Show how these data support the idea that oxygen gas is a diatomic molecule. Must we consider hydrogen to be a diatomic molecule to explain these results?

75. Each of the statements below is true, but Dalton might have had trouble explaining some of them with his atomic theory. Give explanations for the following statements.
   a. The space-filling models for ethyl alcohol and dimethyl ether are shown below.

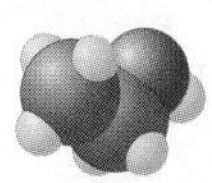

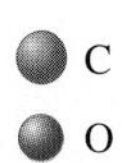

   These two compounds have the same composition by mass (52% carbon, 13% hydrogen, and 35% oxygen), yet the two have different melting points, boiling points, and solubilities in water.
   b. Burning wood leaves an ash that is only a small fraction of the mass of the original wood.
   c. Atoms can be broken down into smaller particles.
   d. One sample of lithium hydride is 87.4% lithium by mass, whereas another sample of lithium hydride is 74.9% lithium by mass. However, the two samples have the same properties.

76. You take three compounds, each consisting of two elements (X, Y, and/or Z) and decompose them to their respective elements. To determine the relative masses of X, Y, and Z, you collect and weigh the elements, obtaining the following data:

Elements in Compound	Masses of Elements
1. X and Y	X = 0.4 g, Y = 4.2 g
2. Y and Z	Y = 1.4 g, Z = 1.0 g
3. X and Y	X = 2.0 g, Y = 7.0 g

   a. What are the assumptions needed to solve this problem?
   b. What are the relative masses of X, Y, and Z?
   c. What are the chemical formulas of the three compounds?
   d. If you decompose 21 g of compound XY, how much of each element is present?

77. Two elements, R and Q, combine to form two binary compounds. In the first compound, 14.0 g of R combines with 3.00 g of Q. In the second compound, 7.00 g of R combines with 4.50 g of Q. Show that these data are in accord with the law of multiple proportions. If the formula of the second compound is RQ, what is the formula of the first compound?

78. A single molecule has a mass of 7.31×10^{-23} g. Provide an example of a real molecule that can have this mass.

79. A combustion reaction involves the reaction of a substance with oxygen gas. The complete combustion of any hydrocarbon (binary compound of carbon and hydrogen) produces carbon dioxide and water as the only products. Octane is a hydrocarbon found in gasoline. Complete combustion of octane produces 8 L of carbon dioxide for every 9 L of water vapor (both measured at the same temperature and pressure). What is the ratio of carbon atoms to hydrogen atoms in a molecule of octane?

80. You have two distinct gaseous compounds made from element X and element Y. The mass percents are as follows:

Compound I: 30.43% X, 69.57% Y

Compound II: 63.64% X, 36.36% Y

In their natural standard states, element X and element Y exist as gases. (Monatomic? Diatomic? Triatomic? That is for you to determine.) When you react "gas X" with "gas Y" to make the products, you get the following data (all at standard pressure and temperature):

$$1 \text{ volume "gas X"} + 2 \text{ volumes "gas Y"} \longrightarrow 2 \text{ volumes compound I}$$

$$2 \text{ volumes "gas X"} + 1 \text{ volume "gas Y"} \longrightarrow 2 \text{ volumes compound II}$$

Assume the simplest possible formulas for reactants and products in these chemical equations. Then determine the relative atomic masses of element X and element Y.

Marathon Problem

81. You have gone back in time and are working with Dalton on a table of relative masses. Following are his data:

0.602 g gas A reacts with 0.295 g gas B

0.172 g gas B reacts with 0.401 g gas C

0.320 g gas A reacts with 0.374 g gas C

a. Assuming simplest formulas (AB, BC, and AC), construct a table of relative masses for Dalton.

b. Knowing some history of chemistry, you tell Dalton that if he determines the volumes of the gases reacted at constant temperature and pressure, he need not assume simplest formulas. You collect the following data:

6 volumes gas A + 1 volume gas B $\longrightarrow$ 4 volumes product

1 volume gas B + 4 volumes gas C $\longrightarrow$ 4 volumes product

3 volumes gas A + 2 volumes gas C $\longrightarrow$ 6 volumes product

Write the simplest balanced equations, and find the actual relative masses of the elements. Explain your reasoning.

Stoichiometry

3

chapter

Massive eruption of the Kileuea volcano in Hawaii.

Jim Sugar/Science Faction

OWL

Sign in to OWL at www.cengage.com/owl to view tutorials and simulations, develop problem-solving skills, and complete online homework assigned by your professor.

go Chemistry

Download mini lecture videos for key concept review and exam prep from OWL or purchase them from www.cengagebrain.com.

Chemical reactions have a profound effect on our lives. There are many examples: Food is converted to energy in the human body; nitrogen and hydrogen are combined to form ammonia, which is used as a fertilizer; fuels and plastics are produced from petroleum; the starch in plants is synthesized from carbon dioxide and water using energy from sunlight; human insulin is produced in laboratories by bacteria; cancer is induced in humans by substances from our environment; and so on, in a seemingly endless list. The central activity of chemistry is to understand chemical changes such as these, and the study of reactions occupies a central place in this text. We will examine why reactions occur, how fast they occur, and the specific pathways they follow.

In this chapter we will consider the quantities of materials consumed and produced in chemical reactions. This area of study is called **chemical stoichiometry.** To understand chemical stoichiometry, you must first understand the concept of relative atomic masses.

3.1 Atomic Masses

As we saw in Chapter 2, the first quantitative information about atomic masses came from the work of Dalton, Gay-Lussac, Lavoisier, Avogadro, Cannizzaro, and Berzelius. By observing the proportions in which elements combine to form various compounds, nineteenth-century chemists calculated relative atomic masses. The modern system of atomic masses, instituted in 1961, is based on ^{12}C (carbon-12) as the standard. In this system *^{12}C is assigned a mass of exactly 12 atomic mass units* (amu), and the masses of all other atoms are given relative to this standard.

Chemist using a mass spectrometer to analyze for copper in blood plasma.

The most accurate method currently available for comparing the masses of atoms involves the use of the **mass spectrometer.** In this instrument, diagrammed in Fig. 3.1, atoms or molecules are passed into a beam of high-speed electrons. The high-speed electrons knock electrons off the atoms or molecules being analyzed and change them to positive ions. An applied electric field then accelerates these ions through a magnetic field, which deflects the paths of the ions. The amount of path deflection for each ion depends on its mass—the most massive ions are deflected the smallest amount—and this deflection causes the ions to separate, as shown in Fig. 3.1. A comparison of the positions where the ions hit the detector plate gives very accurate values of their relative masses. For example, when ^{12}C and ^{13}C are analyzed in a mass spectrometer, the ratio of their masses is found to be

$$\frac{\text{Mass } ^{13}\text{C}}{\text{Mass } ^{12}\text{C}} = 1.0836129$$

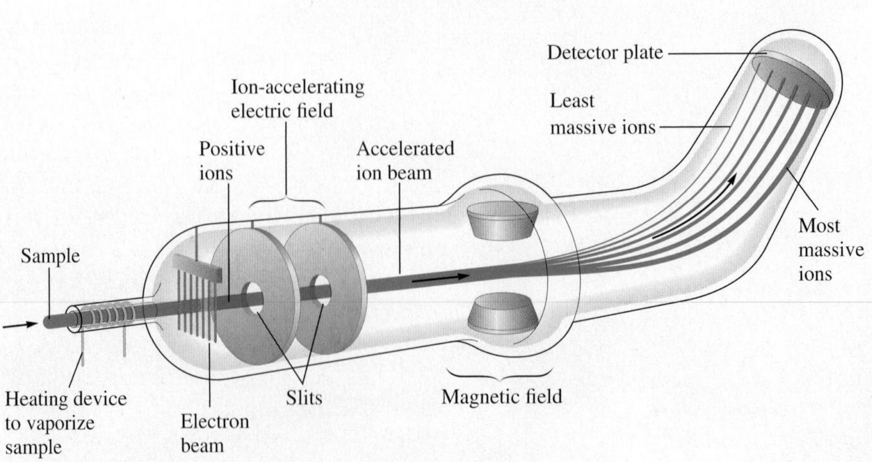

Figure 3.1
Schematic diagram of a mass spectrometer.

Elemental Analysis Catches Elephant Poachers

Chemical Insights

In an effort to combat the poaching of elephants by controlling illegal exports of ivory, scientists are now using the isotopic composition of ivory trinkets and elephant tusks to identify the region of Africa where the elephant lived. Using a mass spectrometer, scientists analyze the ivory for the relative amounts of ^{12}C, ^{13}C, ^{14}N, ^{15}N, ^{86}Sr, and ^{87}Sr to determine the diet of the elephant and thus its place of origin. For example, because grasses use a different photosynthetic pathway to produce glucose than do trees, grasses have a slightly different $^{13}C/^{12}C$ ratio than trees. They have different ratios because each time a carbon is added in going from simpler to more complex compounds, the more massive ^{13}C is disfavored relative to ^{12}C since it reacts more slowly. Because trees use more steps to build up glucose, they end up with a smaller $^{13}C/^{12}C$ ratio in their leaves relative to grasses, and this difference is then reflected in the tissues of elephants. Thus scientists can tell whether a particular tusk came from a savanna-dwelling (grass-eating) elephant or from a tree-browsing elephant.

Similarly, because the ratios of $^{15}N/^{14}N$ and $^{87}Sr/^{86}Sr$ in elephant tusks also vary depending on the region of Africa the elephant inhabited, they can be used to trace the elephant's origin. In fact, using these techniques, scientists have reported being able to discriminate between elephants living only about 100 miles apart.

A herd of savanna-dwelling elephants.

Kennan Ward/Stock Market/Corbis

There is now international concern about the dwindling elephant populations in parts of Africa, where their numbers have decreased dramatically over the past several decades. This concern has led to bans on the export of ivory from many countries in Africa. However, a few nations still allow ivory to be exported. Thus, to enforce the trade restrictions, the origin of a given piece of ivory must be established. It is hoped that the "isotope signature" of the ivory can be used for this purpose.

Since the atomic mass unit is defined such that the mass of ^{12}C is *exactly* 12 atomic mass units, then on this same scale,

$$\text{Mass of } ^{13}\text{C} = (1.0836129)(\underset{\substack{\uparrow \\ \text{Exact number,} \\ \text{by definition}}}{12\ \text{amu}}) = 13.003355\ \text{amu}$$

The masses of other atoms can be determined in a similar manner.

The mass for each element is given in the table inside the front cover of this book. This value, even though it is actually a mass, is sometimes called (for historical reasons) the atomic weight for each element.

Most elements occur in nature as mixtures of isotopes; thus atomic masses are generally reported as average values. It turns out the average mass for a given atom varies at different geographical locations, and the new values are given as intervals to reflect this. Carbon, for example, now has an official mass of [12.0096; 12.0116]. This is not due to uncertainty but takes into account variations in atomic masses based on location. This new system has recently been adopted for 10 elements (hydrogen, lithium, boron, carbon, nitrogen, oxygen, silicon, sulfur, chlorine, and thallium). In the text we will continue with the standard method of average atomic masses for these elements.

Look at the value of the atomic mass of carbon given in the table. You might expect to see 12 since we said the system of atomic masses is based on ^{12}C. However, the number given for carbon is 12.01 because the carbon found on earth (natural carbon) is a mixture of the isotopes ^{12}C, ^{13}C, and ^{14}C. All three isotopes have six protons, but they have six, seven, and eight neutrons, respectively. Because natural carbon is a mixture of isotopes, the atomic mass we use for carbon is an *average value* based on its isotopic composition.

The average atomic mass for carbon is computed as follows. Chemists know that natural carbon is composed of 98.89% ^{12}C atoms and 1.11% ^{13}C atoms. The amount of ^{14}C is negligibly small at this level of precision. Using

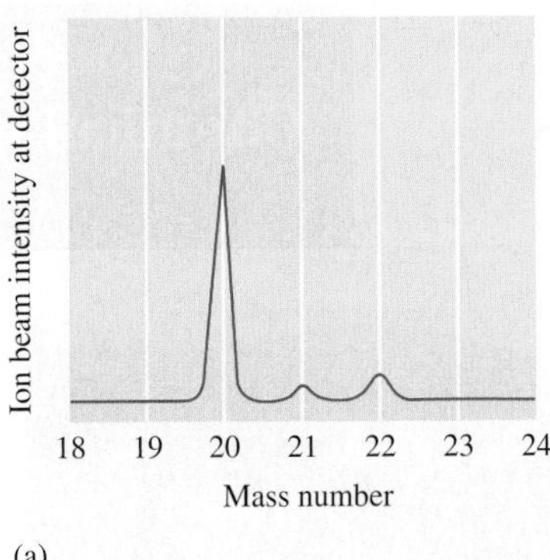

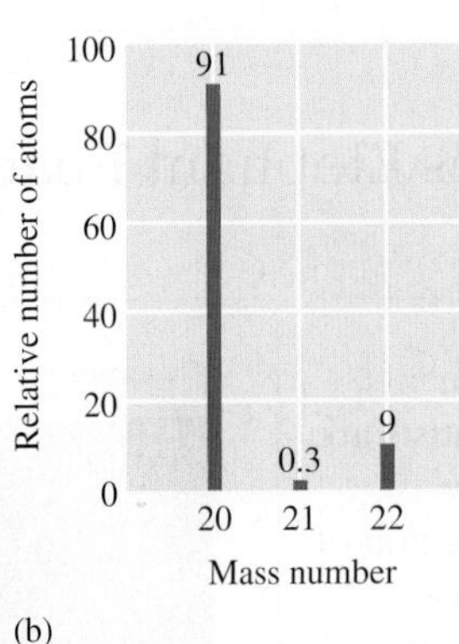

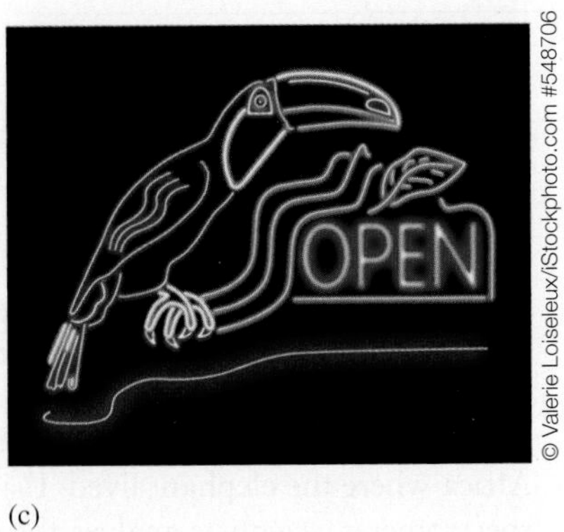

Figure 3.2
The relative intensities of the signals recorded when natural neon is injected into a mass spectrometer, represented in terms of (a) "peaks" and (b) a bar graph. The relative areas of the peaks are 0.9092 (^{20}Ne), 0.00257 (^{21}Ne), and 0.0882 (^{22}Ne); natural neon is therefore 90.92% ^{20}Ne, 0.257% ^{21}Ne, and 8.82% ^{22}Ne. (c) A neon sign.

the masses of ^{12}C (exactly 12 amu) and ^{13}C (13.003355 amu), the average atomic mass for natural carbon can be calculated.

$$98.89\% \text{ of } 12 \text{ amu} + 1.11\% \text{ of } 13.0034 \text{ amu} = (0.9889)(12 \text{ amu}) + (0.0111)(13.0034 \text{ amu}) = 12.01 \text{ amu}$$

See Appendix A1.5 for a discussion of significant figures.

This average mass is often called the atomic weight of carbon.

Even though natural carbon does not contain a single atom with mass 12.01, for stoichiometric purposes we consider carbon to be composed of only one type of atom with a mass of 12.01. We do this so that we can count atoms of natural carbon by weighing a sample of carbon. Let's consider a nonchemical example. It is much easier to weigh out 3000 grams of jelly beans (with an average mass of 3 grams per jelly bean) than to count out 1000 of them. Note that none of the jelly beans has to have a mass of 3 grams for this method to work; only the *average* mass must be 3 grams. We extend this same principle to counting atoms. For natural carbon with an average mass of 12.01 atomic mass units, to obtain 1000 atoms would require weighing out 12,010 atomic mass units of natural carbon (a mixture of ^{12}C and ^{13}C).

As in the case of carbon, the mass for each element given in the table inside the front cover of this book is an average value based on the isotopic composition of the naturally occurring element. For instance, the mass listed for hydrogen (1.008) is the average mass for natural hydrogen, which is a mixture of ^{1}H and ^{2}H (deuterium). *No* atom of hydrogen actually has the mass 1.008.

In addition to being used for determining accurate mass values for individual atoms, the mass spectrometer is used to determine the isotopic composition of a natural element. For example, when a sample of natural neon is injected into a mass spectrometer, the results shown in Fig. 3.2 are obtained. The areas of the "peaks" or the heights of the bars indicate the relative numbers of $^{22}_{10}$Ne, $^{22}_{10}$Ne, and $^{22}_{10}$Ne atoms.

Relative number of atoms: 100, 80, 60, 40, 20, 0
69.09
30.91
63 65
Mass number

Figure 3.3
Mass spectrum of natural copper.

EXAMPLE 3.1

Copper is a very important metal used for water pipes, electrical wiring, roof coverings, and other materials. When a sample of natural copper is vaporized and injected into a mass spectrometer, the results shown in Fig. 3.3 are obtained. Use these data to compute the average mass of natural copper. (The mass values for ^{63}Cu and ^{65}Cu are 62.93 amu and 64.93 amu, respectively.)

Solution As shown by the graph, of every 100 atoms of natural copper, on average 69.09 are ^{63}Cu and 30.91 are ^{65}Cu. Thus the average mass of 100 atoms of natural copper is

$$(69.09 \text{ atoms})\left(62.93 \frac{\text{amu}}{\text{atom}}\right) + (30.9 \text{ atoms})\left(64.93 \frac{\text{amu}}{\text{atom}}\right) = 6355 \text{ amu}$$

The average mass per atom is

$$\frac{6355 \text{ amu}}{100 \text{ atoms}} = 63.55 \text{ amu/atom}$$

This mass value is used in calculations involving the reactions of copper and is the value given in the table inside the front cover of this book.

3.2 The Mole

The SI definition of the mole is the amount of a substance that contains as many entities as there are in exactly 0.012 kg (12 g) of carbon-12.

Because samples of matter typically contain so many atoms, a unit of measure called the mole has been established for use in counting atoms. For our purposes it is most convenient to define the **mole** (abbreviated mol) as *the number equal to the number of carbon atoms in exactly 12 grams of pure* ^{12}C. Modern techniques that allow us to count atoms very precisely have been used to determine this number as 6.022137×10^{23} (6.022×10^{23} will be sufficient for our purposes). This number is called **Avogadro's number** to honor his contributions to chemistry. *One mole of something consists of* 6.022×10^{23} *units of that substance.* Just as a dozen eggs is 12 eggs, a mole of eggs is 6.022×10^{23} eggs.

The magnitude of the number 6.022×10^{23} is very difficult to imagine. To give you some idea, 1 mole of seconds represents a span of time 4 million times as long as the earth has already existed, and 1 mole of marbles is enough to cover the entire earth to a depth of 50 miles! However, since atoms are so tiny, a mole of atoms or molecules is a perfectly manageable quantity to use in a reaction (see Fig. 3.4).

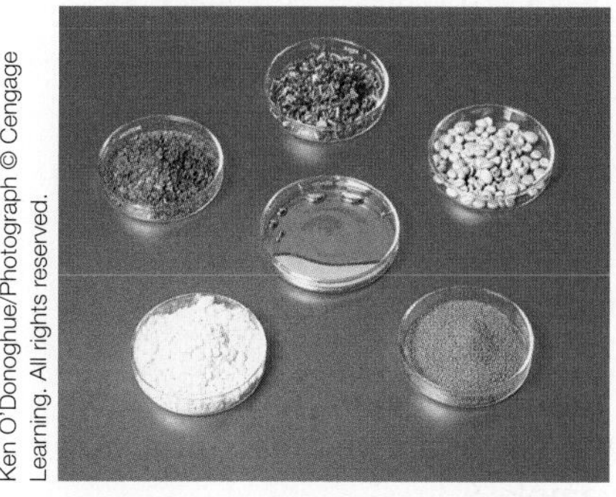

Figure 3.4
Proceeding clockwise from the top, samples containing 1 mole each of copper, aluminum, iron, sulfur, iodine, and (in the center) mercury.

How do we use the mole in chemical calculations? Recall that Avogadro's number is defined as the number of atoms in exactly 12 grams of ^{12}C. Thus 12 grams of ^{12}C contains 6.022×10^{23} atoms. Also, a 12.01-gram sample of natural carbon contains 6.022×10^{23} atoms (a mixture of ^{12}C, ^{13}C, and ^{14}C atoms, with an average mass of 12.01). Since the ratio of the masses of the samples (12 g/12.01 g) is the same as the ratio of the masses of the individual components (12 amu/12.01 amu), the two samples contain the *same number* of components.

To be sure this point is clear, think of oranges with an average mass of 0.5 pound each and grapefruit with an average mass of 1.0 pound each. Any two sacks for which the sack of grapefruit weighs twice as much as the sack of oranges will contain the same number of pieces of fruit. The same idea extends to atoms. Compare natural carbon (average mass of 12.01) and natural helium (average mass of 4.003). A sample of 12.01 grams of natural carbon contains the same number of atoms as 4.003 grams of natural helium. Both samples contain 1 mole of atoms (6.022×10^{23}). Table 3.1 gives more examples that illustrate this basic idea.

The mass of 1 mole of an element is equal to its atomic mass in grams.

Thus the mole is defined such that a sample of a natural element with a mass equal to the element's atomic mass expressed in grams contains 1 mole of atoms. This definition also fixes the relationship between the atomic mass unit

Table 3.1

Comparison of 1-Mole Samples of Various Elements

Element	Number of Atoms	Mass of Sample (g)
Aluminum	6.022×10^{23}	26.98
Gold	6.022×10^{23}	196.97
Iron	6.022×10^{23}	55.85
Sulfur	6.022×10^{23}	32.07
Boron	6.022×10^{23}	10.81
Xenon	6.022×10^{23}	131.30

and the gram. Since 6.022×10^{23} atoms of carbon (each with a mass of 12 amu) have a mass of 12 grams, then

$$(6.022 \times 10^{23} \text{ atoms})\left(\frac{12 \text{ amu}}{\text{atom}}\right) = 12 \text{ g}$$

and

$$6.022 \times 10^{23} \text{ amu} = \underset{\substack{\uparrow \\ \text{Exact number}}}{1 \text{ g}}$$

OWL INTERACTIVE EXAMPLE 3.2

Americium is an element that does not occur naturally. It can be made in very small amounts in a device called a particle accelerator. Compute the mass in grams of a sample of americium containing six atoms.

Solution From the table inside the front cover of this book, note that one americium atom has a mass of 243 amu. Thus the mass of six atoms is

$$6 \text{ atoms} \times 243 \frac{\text{amu}}{\text{atom}} = 1.46 \times 10^3 \text{ amu}$$

From the relationship 6.022×10^{23} amu = 1 g, the mass of six americium atoms in grams is

$$1.46 \times 10^3 \text{ amu} \times \frac{1 \text{ g}}{6.022 \times 10^{23} \text{ amu}} = 2.42 \times 10^{-21} \text{ g}$$

This relationship can be used to derive the factor needed to convert between atomic mass units and grams.

Refer to Appendix 2 for a discussion of units and the conversion from one unit to another.

To perform chemical calculations, you must understand what the mole means and how to determine the number of moles in a given mass of a substance. These procedures are illustrated in Example 3.3.

OWL INTERACTIVE EXAMPLE 3.3

A silicon chip used in an integrated circuit of a microcomputer has a mass of 5.68 mg. How many silicon (Si) atoms are present in this chip?

Solution The strategy for doing this problem is to convert from milligrams of silicon to grams of silicon, then to moles of silicon, and finally to atoms of silicon:

$$5.68 \text{ mg Si} \times \frac{1 \text{ g Si}}{1000 \text{ mg Si}} = 5.68 \times 10^{-3} \text{ g Si}$$

$$5.68 \times 10^{-3} \text{ g Si} \times \frac{1 \text{ mol Si}}{28.09 \text{ g Si}} = 2.02 \times 10^{-4} \text{ mol Si}$$

$$2.02 \times 10^{-4} \text{ mol Si} \times \frac{6.022 \times 10^{23} \text{ atoms}}{1 \text{ mol Si}} = 1.22 \times 10^{20} \text{ atoms}$$

It always makes sense to think about orders of magnitude as you do a calculation. In Example 3.3, the 5.68-milligram sample of silicon is clearly much less than 1 mole of silicon (which has a mass of 28.09 grams), so the final answer of 1.22×10^{20} atoms (compared with 6.022×10^{23} atoms) is at least in the right direction. Paying careful attention to units and making sure the answer is sensible can help you detect an inverted conversion factor or a number that was incorrectly entered in your calculator.

Always check to see if your answer is sensible.

3.3 Molar Mass

A chemical compound is, ultimately, a collection of atoms. For example, methane (the major component of natural gas) consists of molecules that each contain one carbon atom and four hydrogen atoms (CH_4). How can we calculate the mass of 1 mole of methane; that is, what is the mass of 6.022×10^{23} CH_4 molecules? Since each CH_4 molecule contains one carbon atom and four hydrogen atoms, 1 mole of CH_4 molecules consists of 1 mole of carbon atoms and 4 moles of hydrogen atoms. The mass of 1 mole of methane can be found by summing the masses of carbon and hydrogen present:

The average atomic mass for carbon to five significant digits is 12.011.

$$\begin{aligned} \text{Mass of 1 mol of C} &= 12.011 \text{ g} \\ \text{Mass of 4 mol of H} &= 4 \times 1.008 \text{ g} \\ \hline \text{Mass of 1 mol of } CH_4 &= 16.043 \text{ g} \end{aligned}$$

OWL INTERACTIVE EXAMPLE 3.4

Isopentyl acetate ($C_7H_{14}O_2$), the compound responsible for the scent of bananas, can be produced commercially. Interestingly, bees release about 1 μg (1×10^{-6} g) of this compound when they sting to attract other bees to join the attack. How many molecules of isopentyl acetate are released in a typical bee sting? How many atoms of carbon are present?

Solution Since we are given a mass of isopentyl acetate and want the number of molecules, we must first compute the molar mass.

The average atomic mass for hydrogen to five significant digits is 1.0079 and that for oxygen is 15.999.

$$7 \text{ mol C} \times 12.011 \frac{\text{g}}{\text{mol}} = 84.077 \text{ g C}$$

$$14 \text{ mol H} \times 1.0079 \frac{\text{g}}{\text{mol}} = 14.111 \text{ g H}$$

$$2 \text{ mol O} \times 15.999 \frac{\text{g}}{\text{mol}} = 31.998 \text{ g O}$$

$$\text{Mass of 1 mol of } C_7H_{14}O_2 = 130.186 \text{ g}$$

Thus 1 mole of isopentyl acetate (6.022×10^{23} molecules) has a mass of 130.186 g.

To find the number of molecules released in a sting, we must first determine the number of moles of isopentyl acetate in 1×10^{6} g:

$$1 \times 10^{-6} \text{ g } C_7H_{14}O_2 \times \frac{1 \text{ mol } C_7H_{14}O_2}{130.186 \text{ g } C_7H_{14}O_2} = 8 \times 10^{-9} \text{ mol } C_7H_{14}O_2$$

Since 1 mol is 6.022×10^{23} units, we can determine the number of molecules:

$$8 \times 10^{-9} \text{ mol } C_7H_{14}O_2 \times \frac{6.022 \times 10^{23} \text{ molecules}}{1 \text{ mol } C_7H_{14}O_2} = 5 \times 10^{15} \text{ molecules}$$

Chemical Insights

Measuring the Masses of Large Molecules or Making Elephants Fly

When a chemist produces a new molecule, one crucial property for making a positive identification is the molecule's mass. There are many ways to determine the molar mass of a compound, but one of the fastest and most accurate methods involves mass spectrometry. This method requires that the substance be put into the gas phase and ionized. The deflection that the resulting ion exhibits as it is accelerated through a magnetic field can be used to obtain a very precise value of its mass. One drawback of this method is that it is difficult to use with large molecules because they are difficult to vaporize. That is, substances that contain large molecules typically have very high boiling points, and these molecules are often damaged when they are vaporized at such high temperatures. A case in point involves proteins, an extremely important class of large biologic molecules that are quite fragile at high temperatures. Typical methods used to obtain the masses of protein molecules are slow and tedious.

Mass spectrometry has not previously been used to obtain protein masses because proteins decompose at the temperatures necessary to vaporize them. However, a relatively new technique called matrix-assisted laser desorption has been developed that allows mass spectrometric determination of protein molar masses. In this technique, the large "target" molecule is embedded in a matrix of smaller molecules. The matrix is then placed in a mass spectrometer and blasted with a laser beam, which causes its disintegration. Disintegration of the matrix frees the large target molecule, which is then swept into the mass spectrometer. One researcher involved in this project likened this method to an elephant on top of a tall building: "The elephant must fly if the building suddenly turns into fine grains of sand."

This technique allows scientists to determine the masses of huge molecules. So far, researchers have measured proteins with masses up to 350,000 daltons (1 dalton is equal to 1 atomic mass unit). This method probably will be extended to even larger molecules such as DNA and could be a revolutionary development in the characterization of biomolecules.

To determine the number of carbon atoms present, we must multiply the number of molecules by 7 (each molecule of isopentyl acetate contains seven carbon atoms):

To show the correct number of significant figures in each calculation, we round off after each step. In your calculations, always carry extra significant figures through to the end; then round off.

$$5 \times 10^{15} \text{ molecules} \times \frac{7 \text{ carbon atoms}}{\text{molecule}} = 4 \times 10^{16} \text{ carbon atoms}$$

Note: In keeping with our practice of always showing the correct number of significant figures, we have rounded off after each step. However, if extra digits are carried throughout this problem, the final answer rounds to 3×10^{16}.

Since the number 16.043 represents the mass of 1 mole of methane molecules, it makes sense to call it the *molar mass* for methane. However, traditionally, the term *molecular weight* has been used to describe the mass of 1 mole of a substance. Thus the terms **molar mass** and **molecular weight** mean exactly the same thing: *the mass in grams of 1 mole of a compound*. The molar mass of a known substance is obtained by summing the masses of the component atoms, as we did for methane.

A substance's molar mass (molecular weight) is the mass in grams of 1 mole of the substance.

Some substances exist as a collection of ions rather than as separate molecules. An example is ordinary table salt, sodium chloride (NaCl), which is composed of an array of Na^+ and Cl^- ions. There are no NaCl molecules present. However, in this text, for convenience, we will apply the term *molar mass* to both ionic and molecular substances. Thus we will refer to 58.44

(22.99 + 35.45) as the molar mass for NaCl. In some texts the term *formula weight* is used for ionic compounds instead of the terms *molar mass* and *molecular weight.*

3.4 Conceptual Problem Solving

One of the great rewards of studying chemistry is to become a good problem solver. Being able to solve complex problems is a talent that will serve you well in all walks of life. It is our purpose in this text to help you learn to solve problems in a flexible, creative way based on understanding the fundamental ideas of chemistry. We call this approach **conceptual problem solving.**

The ultimate goal is to be able to solve new problems (that is, problems you have not seen before) on your own. In this text we will provide problems and offer solutions by explaining how to think about the problems. Although the answers to these problems are important, it is perhaps even more important to understand the process—the thinking necessary to get the answer. While studying the solution, it is crucial that you interactively think through the problem with us. Do not skip the discussion and jump to the answer. Make sure that you understand each step in the process.

A main goal in conceptual problem solving is to get the "big picture"—a real understanding of the situation. This approach to problem solving looks within the problem for a solution. In essence, we ask a series of questions as we proceed and use our knowledge of fundamental principles to answer these questions; we then let the problem (and our questions) guide us as we solve it.

The following organizing principles will be useful to us as we proceed to solve a problem:

- First, we need to read the problem and decide on the final goal. Then we sort through the facts given, focusing on the key words and, when appropriate, drawing a diagram of the problem. In this part of the analysis we need to state the problem as simply as possible. The questions we ask at first are general and similar regardless of the problem. They are basically of the type "What are we trying to solve?" and "What does this mean?"
- We need to work backward from the final goal to decide where to start. Our questions become more specific depending on the given problem, such as, "What are the reactants and products?" "What is the balanced equation?" "What do we mean by molar mass?" and so on. Our understanding of the fundamental principles of chemistry will enable us to answer each of these questions and will eventually lead us to the final solution.
- After getting an answer we should check to see whether it is reasonable. The extent to which we can do this varies with the type of problem. If we are computing mass percents of elements in a compound, do they add up to 100%? Do we get a negative answer for a mass? Sometimes it is more difficult to judge if an answer is reasonable. However, once we get the solution of the problem, we should ask, "Does it make sense?"

In summary, instead of looking outside the problem using a memorized scheme, we will look inside the problem and let the problem help us as we proceed to a solution. Learning this approach requires some patience, but the reward for learning to solve problems this way is that you become an effective solver of any new problem that confronts you in daily life or in your work in any field. You will no longer panic when you see a problem that is different in some ways from those you have solved in the past. Although you might be frustrated at times as you learn this method, we guarantee that it will pay dividends later and should make your experience with chemistry a positive one that will prepare you for any career you choose.

We will model this approach with the examples in this text, beginning with Example 3.5. We will continue this approach throughout the examples in Chapters 3 and 4. We will not, however, discuss all example problems throughout the text to this extent because we expect you to take an increasingly active role in this process. As the problems become more complicated, the method of conceptual problem solving becomes more important, and it is crucial that you work through these on your own before reading the solution.

3.5 Percent Composition of Compounds

So far we have discussed the composition of a compound in terms of the numbers of its constituent atoms. It is often useful to know a compound's composition in terms of the masses of its elements. We can obtain this information from the formula of the compound by comparing the mass of each element present in 1 mole of the compound with the total mass of 1 mole of the compound.

For example, consider ethanol, which has the formula C_2H_5OH. The mass of each element present and the molar mass are obtained through the following procedure:

$$\text{Mass of C} = 2 \text{ mol} \times 12.011 \frac{\text{g}}{\text{mol}} = 24.022 \text{ g}$$

$$\text{Mass of H} = 6 \text{ mol} \times 1.008 \frac{\text{g}}{\text{mol}} = 6.048 \text{ g}$$

$$\text{Mass of O} = 1 \text{ mol} \times 15.999 \frac{\text{g}}{\text{mol}} = 15.999 \text{ g}$$

$$\text{Mass of 1 mol of } C_2H_5OH = 46.069 \text{ g}$$

The **mass percent** (often called the weight percent) of carbon in ethanol can be computed by comparing the mass of carbon in 1 mole of ethanol with the total mass of 1 mole of ethanol and multiplying the result by 100%:

$$\begin{aligned}\text{Mass percent of C} &= \frac{\text{mass of C in 1 mol } C_2H_5OH}{\text{mass of 1 mol } C_2H_5OH} \times 100\% \\ &= \frac{24.022 \text{ g}}{46.069 \text{ g}} \times 100\% \\ &= 52.144\%\end{aligned}$$

The mass percents of hydrogen and oxygen in ethanol are obtained in a similar manner:

$$\begin{aligned}\text{Mass percent of H} &= \frac{\text{mass of H in 1 mol } C_2H_5OH}{\text{mass of 1 mol } C_2H_5OH} \times 100\% \\ &= \frac{6.048 \text{ g}}{46.069 \text{ g}} \times 100\% \\ &= 13.13\%\end{aligned}$$

$$\begin{aligned}\text{Mass percent of O} &= \frac{\text{mass of O in 1 mol } C_2H_5OH}{\text{mass of 1 mol } C_2H_5OH} \times 100\% \\ &= \frac{15.999 \text{ g}}{46.069 \text{ g}} \times 100\% \\ &= 34.728\%\end{aligned}$$

Notice that the percentages add to 100% if rounded to two decimal places; this is the check of the calculations.

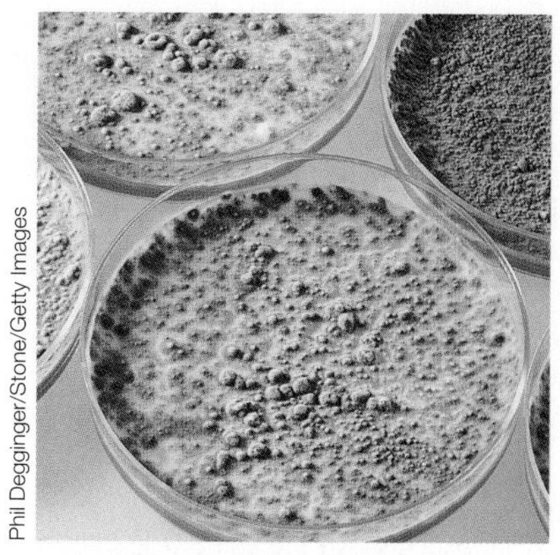
Phil Degginger/Stone/Getty Images

Penicillin is isolated from a mold.

The Nobel Foundation

Dorothy Hodgkin (1910–1994) was born in Cairo, Egypt. She became interested in chemistry and in crystals at about the age of 10, and on her 16th birthday she received a book by the Nobel Prize–winning physicist William Bragg. The subject of the book was how to use X rays to analyze crystals, and from that point on, her career path was set.

Dr. Hodgkin used X-ray analysis for three important discoveries. In 1945 she determined the structure of penicillin, which helped manufacturers create penicillin. In 1954 she determined the structure of vitamin B_{12}, which led to her winning the Nobel Prize in Chemistry in 1964. Although both of these are important and useful discoveries, Dr. Hodgkin considered her greatest scientific achievement to be the discovery of insulin (1969), now used in the treatment of diabetes.

OWL INTERACTIVE EXAMPLE 3.5

Penicillin, the first of a now large number of antibiotics (antibacterial agents), was discovered accidentally by the Scottish bacteriologist Alexander Fleming in 1928, but he was never able to isolate it as a pure compound. This and similar antibiotics have saved millions of lives that might have been lost to infections. Penicillin F has the formula $C_{14}H_{20}N_2SO_4$. Compute the mass percent of each element.

Solution As we discussed in Section 3.4, your solution to a problem should begin with questions, the first of which is:

- What is the problem asking us to solve?

In this case you are asked to determine the mass percent of each element in penicillin F. This leads directly to the next question:

- What does this mean?

The mass percent for a given element in a compound is the mass of the element in the compound as a percentage of the mass of the compound. We know the formula for penicillin F is $C_{14}H_{20}N_2SO_4$. The mass percent of carbon, for example, is

$$\text{Mass percent of carbon} = \frac{\text{Mass of carbon in 1 mole of } C_{14}H_{20}N_2SO_4}{\text{Mass of 1 mole of } C_{14}H_{20}N_2SO_4} \times 100\%$$

Notice how we are letting the problem guide us. The original problem was to find the mass percent of each element in a compound. By using the definition of mass percent, we see that we will solve for each element individually. We also have changed the problem to two new and more specific questions:

1. What is the mass of carbon in 1 mole of penicillin F? (the numerator above)
2. What is the mass of 1 mole of penicillin F? (the denominator above)

From the discussion in Section 3.3, we see that the second question is asking for the molar mass of penicillin F. This shows us that although allowing the problem to guide us is important, we also need knowledge of fundamental principles. The molar mass of penicillin F is computed as follows:

$$\text{C:} \quad 14 \text{ mol} \times 12.011 \frac{\text{g}}{\text{mol}} = 168.15 \text{ g}$$

$$\text{H:} \quad 20 \text{ mol} \times 1.008 \frac{\text{g}}{\text{mol}} = 20.16 \text{ g}$$

$$\text{N:} \quad 2 \text{ mol} \times 14.007 \frac{\text{g}}{\text{mol}} = 28.014 \text{ g}$$

$$\text{S:} \quad 1 \text{ mol} \times 32.07 \frac{\text{g}}{\text{mol}} = 32.07 \text{ g}$$

$$\text{O:} \quad 4 \text{ mol} \times 15.999 \frac{\text{g}}{\text{mol}} = \underline{63.996 \text{ g}}$$

$$\text{Mass of 1 mol of } C_{14}H_{20}N_2SO_4 = 312.39 \text{ g}$$

Notice that in solving for the molar mass of penicillin F, we also solved for the mass of each element in 1 mole of the compound. With these masses we can determine the mass percent of each element:

$$\text{Mass percent of C} = \frac{168.15 \text{ g C}}{312.39 \text{ g } C_{14}H_{20}N_2SO_4} \times 100\% = 53.827\%$$

$$\text{Mass percent of H} = \frac{20.16 \text{ g H}}{312.39 \text{ g } C_{14}H_{20}N_2SO_4} \times 100\% = 6.453\%$$

$$\text{Mass percent of N} = \frac{28.014 \text{ g N}}{312.39 \text{ g } C_{14}H_{20}N_2SO_4} \times 100\% = 8.968\%$$

$$\text{Mass percent of S} = \frac{32.07 \text{ g S}}{312.39 \text{ g } C_{14}H_{20}N_2SO_4} \times 100\% = 10.27\%$$

$$\text{Mass percent of O} = \frac{63.996 \text{ g O}}{312.39 \text{ g } C_{14}H_{20}N_2SO_4} \times 100\% = 20.486\%$$

Finally, we can check that the answer is reasonable because the percentages add up to 100%.

Let's summarize the approach:

1. We started with the general question, "What are we trying to solve?"
2. Once we decided on the answer to question 1, we explained what this meant, in this case with a formula.
3. We let the answers guide us to new questions, which were more specific to the problem (such as, "What is the mass of 1 mole of $C_{14}H_{20}N_2SO_4$?").
4. We answered these specific questions using knowledge of fundamental principles.
5. We came to the solution.

This approach will guide you in solving problems throughout the textbook. No matter how complicated the problems become, you should always know specifically what you are trying to solve and then ask (and answer) questions to get to the solution. Your knowledge is important as well, and in chemistry this knowledge builds. Solving this particular problem, for example, would not have been possible without knowing what a chemical formula means and without an understanding of molar mass.

3.6 Determining the Formula of a Compound

When a new compound is prepared, one of the first items of interest is its formula. The formula is often determined by taking a weighed sample of the compound and either decomposing it into its component elements or reacting it with oxygen to produce substances such as CO_2, H_2O, and N_2, which are then collected and weighed. A device for doing this type of analysis is shown in Fig. 3.5. The results of such analyses provide the mass of each type of element in the compound, which can be used to determine the mass percent of each element present.

We will see how information of this type can be used to compute the formula of a compound. Suppose a substance has been prepared that is composed of carbon, hydrogen, and nitrogen. When 0.1156 gram of this compound is reacted with oxygen, 0.1638 gram of carbon dioxide (CO_2) and 0.1676 gram of water (H_2O) are collected. Assuming that all of the carbon in the compound

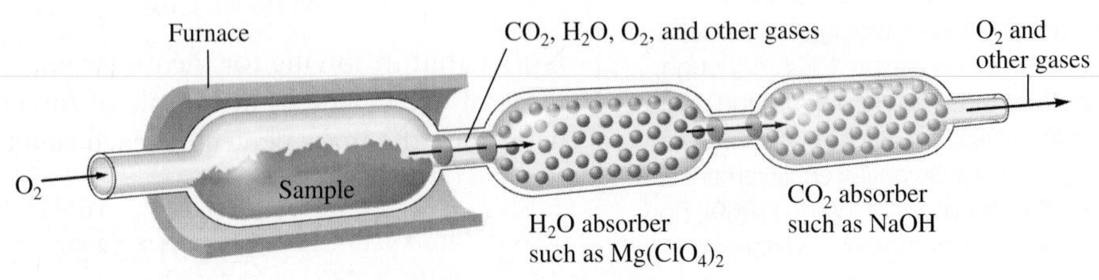

Figure 3.5
A schematic diagram of the combustion device used to analyze substances for carbon and hydrogen. The sample is burned in the presence of excess oxygen, which converts all of its carbon to carbon dioxide and all of its hydrogen to water. These compounds are collected by absorption using appropriate materials, and their amounts are determined by measuring the increase in weights of the absorbents.

is converted to CO_2, we can determine the mass of carbon originally present in the 0.1156-gram sample. To do so, we must use the fraction (by mass) of carbon in CO_2. The molar mass of CO_2 is 12.011 g/mol plus 2(15.999) g/mol, or 44.009 g/mol. The fraction of carbon present by mass (12.011 grams C/44.009 grams CO_2) can now be used to determine the mass of carbon in 0.1638 gram of CO_2:

$$0.1638 \text{ g CO}_2 \times \frac{12.011 \text{ g C}}{44.009 \text{ g CO}_2} = 0.04470 \text{ g C}$$

Remember that this carbon originally came from the 0.1156-gram sample of the unknown compound. Thus the mass percent of carbon in this compound is

$$\frac{0.04470 \text{ g C}}{0.1156 \text{ g compound}} \times 100\% = 38.67\% \text{ C}$$

The same procedure can be used to find the mass percent of hydrogen in the unknown compound. We assume that all of the hydrogen present in the original 0.1156 gram of compound was converted to H_2O. The molar mass of H_2O is 18.015 grams, and the fraction of hydrogen by mass in H_2O is 2.016 grams H/18.015 grams H_2O. Therefore, the mass of hydrogen in 0.1676 gram of H_2O is

$$0.1676 \text{ g H}_2\text{O} \times \frac{2.016 \text{ g H}}{18.015 \text{ g H}_2\text{O}} = 0.01876 \text{ g H}$$

And the mass percent of hydrogen in the compound is

$$\frac{0.01876 \text{ g H}}{0.1156 \text{ g compound}} \times 100\% = 16.23\% \text{ H}$$

The unknown compound contains only carbon, hydrogen, and nitrogen. So far, we have determined that it is 38.67% carbon and 16.23% hydrogen. The remainder must be nitrogen:

$$100.00\% - (\underset{\substack{\uparrow \\ \%\text{ C}}}{38.67\%} + \underset{\substack{\uparrow \\ \%\text{ H}}}{16.23\%}) = 45.10\% \text{ N}$$

We have determined that the compound contains 38.67% carbon, 16.23% hydrogen, and 45.10% nitrogen. Next, we use these data to obtain the formula.

Since the formula of a compound indicates the *numbers* of atoms in the compound, we must convert the masses of the elements to numbers of atoms. The easiest way to do this is to work with 100.00 grams of the compound. In the present case 38.67% carbon by mass means 38.67 grams of carbon per 100.00 grams of compound; 16.23% hydrogen means 16.23 grams of hydrogen per 100.00 grams of compound; and so on. To determine the formula, we must calculate the number of carbon atoms in 38.67 grams of carbon, the number of hydrogen atoms in 16.23 grams of hydrogen, and the number of nitrogen atoms in 45.10 grams of nitrogen. We can do this as follows:

$$38.67 \text{ g C} \times \frac{1 \text{ mol C}}{12.011 \text{ g C}} = 3.220 \text{ mol C}$$

$$16.23 \text{ g H} \times \frac{1 \text{ mol H}}{1.008 \text{ g H}} = 16.10 \text{ mol H}$$

$$45.10 \text{ g N} \times \frac{1 \text{ mol N}}{14.007 \text{ g N}} = 3.220 \text{ mol N}$$

Thus 100.00 grams of this compound contains 3.220 moles of carbon atoms, 16.10 moles of hydrogen atoms, and 3.220 moles of nitrogen atoms.

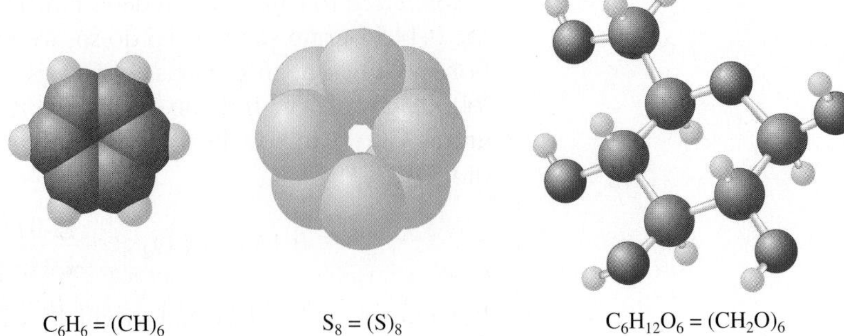

Figure 3.6
Examples of substances whose empirical and molecular formulas differ. Notice that molecular formula = (empirical formula)$_x$, where x is an integer.

We can find the smallest *whole-number ratio* of atoms in this compound by dividing each of the mole values above by the smallest of the three:

$$\text{C:} \quad \frac{3.220}{3.220} = 1$$

$$\text{H:} \quad \frac{16.10}{3.220} = 5$$

$$\text{N:} \quad \frac{3.220}{3.220} = 1$$

Thus the formula of this compound can be written CH_5N. This formula is called the **empirical formula.** It represents the *simplest whole-number ratio of the various types of atoms in a compound.*

If this compound is molecular, then the formula might well be CH_5N. It might also be $C_2H_{10}N_2$, or $C_3H_{15}N_3$, and so on—that is, some multiple of the simplest whole-number ratio. Each of these alternatives also has the correct relative numbers of atoms. Any molecule that can be represented as $(CH_5N)_x$, where x is an integer, has the empirical formula CH_5N. To be able to specify the exact formula of the molecule involved, the **molecular formula,** we must know the molar mass.

Molecular formula = (empirical formula)$_x$, where x is an integer.

Suppose we know that this compound with empirical formula CH_5N has a molar mass of 31.06. How do we determine which of the possible choices represents the molecular formula? Since the molecular formula is always a whole-number multiple of the empirical formula, we must first find the empirical formula mass for CH_5N:

$$\begin{array}{lll} \text{1 C:} & 1 \times 12.011 \text{ g} = & 12.011 \text{ g} \\ \text{5 H:} & 5 \times 1.008 \text{ g} = & 5.040 \text{ g} \\ \text{1 N:} & 1 \times 14.007 \text{ g} = & \underline{14.007 \text{ g}} \\ & \text{Formula mass of } CH_5N = & 31.058 \text{ g} \end{array}$$

This value is the same as the known molar mass of the compound. Thus, in this case, the empirical formula and the molecular formula are the same; this substance consists of molecules with the formula CH_5N. It is quite common for the empirical and molecular formulas to be different; some examples in which this is the case are shown in Fig. 3.6.

OWL INTERACTIVE EXAMPLE 3.6

A white powder is analyzed and found to contain 43.64% phosphorus and 56.36% oxygen by mass. The compound has a molar mass of 283.88 g. What are the compound's empirical and molecular formulas?

Solution

■ What are we trying to solve?

In this problem we are asked to solve for the empirical and molecular formulas for a compound.

■ What does this mean?

The empirical formula is the simplest whole-number ratio of the atoms in the compound, and the molecular formula is the actual formula for the compound. In both cases the formula will look like P_xO_y, and we are trying to solve for x and y.

We know x and y represent numbers of atoms (or relative moles of atoms), but we are given mass percents of the elements in the compound. So now we have a new question:

■ How can we convert a mass percent of an element to the moles of atoms?

We know we can convert mass to moles of atoms using atomic masses. So, we must convert a mass percent to a mass, and a mass to a number.

One way—although not the only way—is to realize that mass percents have the same value as masses if we assume 100.00 g of the sample. Thus, in 100.00 g of this compound, there are 43.64 g of phosphorus and 56.36 g of oxygen. If we convert these masses to moles, the ratio of the numbers of moles of each element represents the ratio of x to y in the formula. In terms of moles, in 100.00 g of compound we have

$$43.64 \text{ g P} \times \frac{1 \text{ mol P}}{30.97 \text{ g P}} = 1.409 \text{ mol P}$$

$$56.36 \text{ g O} \times \frac{1 \text{ mol O}}{15.999 \text{ g O}} = 3.523 \text{ mol O}$$

These numbers give us a ratio of moles of atoms of P/O, but the empirical formula is the *simplest whole-number ratio* of these atoms. Dividing both mole values by the smaller one gives

$$\frac{1.409}{1.409} = 1 \text{ P} \qquad \text{and} \qquad \frac{3.523}{1.409} = 2.5 \text{ O}$$

This yields the formula $PO_{2.5}$. Since compounds must contain whole numbers of atoms, the empirical formula should contain only whole numbers. To obtain the simplest set of whole numbers, we multiply both numbers by 2 to give the empirical formula P_2O_5.

To obtain the molecular formula, we must compare the empirical formula mass with the molar mass. The empirical formula mass for P_2O_5 is 141.94.

$$\frac{\text{Molar mass}}{\text{Empirical formula mass}} = \frac{283.88}{141.94} = 2$$

The molecular formula is $(P_2O_5)_2$, or P_4O_{10}.

■ Does the answer make sense?

The molar mass of the compound is a whole-number multiple of the empirical molar mass, which should make us more confident about the answer. The structural formula of this interesting compound is given in Fig. 3.7.

Figure 3.7
The structural formula of P_4O_{10}. Note that some of the oxygen atoms act as "bridges" between the phosphorus atoms. This compound has a great affinity for water and is often used as a desiccant, or drying agent.

In Example 3.6 we found the molecular formula by comparing the empirical formula mass with the molar mass. There is an alternative way to obtain the molecular formula. The molar mass and the percentages (by mass) of each element present can be used to compute the moles of each element present in one mole of compound. These numbers of moles then represent directly the subscripts in the molecular formula. This procedure is illustrated in Example 3.7.

OWL INTERACTIVE EXAMPLE 3.7

Caffeine, a stimulant found in coffee, tea, chocolate, and some medications, contains 49.48% carbon, 5.15% hydrogen, 28.87% nitrogen, and 16.49% oxygen by mass and has a molar mass of 194.2. Determine the molecular formula of caffeine.

Solution

- What are we trying to solve?

In this problem we are asked to solve for the molecular formula for caffeine.

- What does this mean?

The molecular formula will look like $C_aH_bN_cO_d$, where *a*, *b*, *c*, and *d* are whole numbers. We need to solve for *a*, *b*, *c*, and *d*.

This problem is similar to Example 3.6, but in this case, we are going to determine the molecular formula without the empirical formula.

- How can we use percent by mass data of an element and the molar mass?

By multiplying the molar mass of the compound by the individual percents by mass of each element, we can determine the mass of each element in 1 mole (194.2 g) of caffeine:

$$\frac{49.48 \text{ g C}}{100.0 \text{ g caffeine}} \times \frac{194.2 \text{ g}}{\text{mol}} = \frac{96.09 \text{ g C}}{\text{mol caffeine}}$$

$$\frac{5.15 \text{ g H}}{100.0 \text{ g caffeine}} \times \frac{194.2 \text{ g}}{\text{mol}} = \frac{10.0 \text{ g H}}{\text{mol caffeine}}$$

$$\frac{28.87 \text{ g N}}{100.0 \text{ g caffeine}} \times \frac{194.2 \text{ g}}{\text{mol}} = \frac{56.07 \text{ g N}}{\text{mol caffeine}}$$

$$\frac{16.49 \text{ g O}}{100.0 \text{ g caffeine}} \times \frac{194.2 \text{ g}}{\text{mol}} = \frac{32.02 \text{ g O}}{\text{mol caffeine}}$$

We have masses and we need moles, which we can compute using atomic masses.

$$\text{C:} \quad \frac{96.09 \text{ g C}}{\text{mol caffeine}} \times \frac{1 \text{ mol C}}{12.011 \text{ g C}} = \frac{8.000 \text{ mol C}}{\text{mol caffeine}}$$

$$\text{H:} \quad \frac{10.0 \text{ g H}}{\text{mol caffeine}} \times \frac{1 \text{ mol H}}{1.008 \text{ g H}} = \frac{9.92 \text{ mol H}}{\text{mol caffeine}}$$

$$\text{N:} \quad \frac{56.07 \text{ g N}}{\text{mol caffeine}} \times \frac{1 \text{ mol N}}{14.01 \text{ g N}} = \frac{4.002 \text{ mol N}}{\text{mol caffeine}}$$

$$\text{O:} \quad \frac{32.02 \text{ g O}}{\text{mol caffeine}} \times \frac{1 \text{ mol O}}{16.00 \text{ g O}} = \frac{2.001 \text{ mol O}}{\text{mol caffeine}}$$

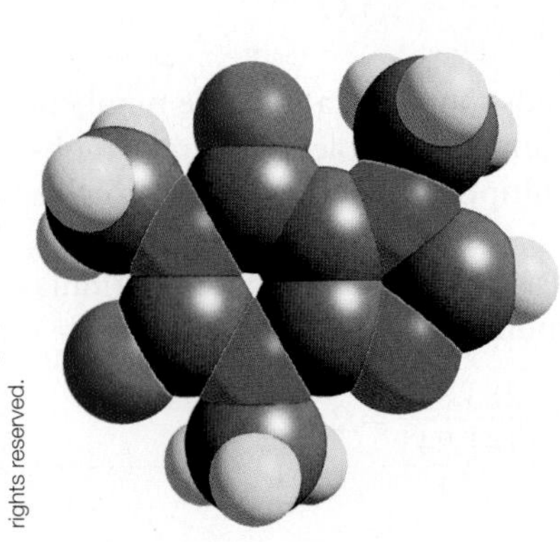

Computer-generated molecule of caffeine.

Rounding the numbers to integers gives the molecular formula for caffeine: $C_8H_{10}N_4O_2$.

- Does the answer make sense?

The answers to the number of moles should be whole numbers, because, unlike the first part of the solution to Example 3.6, we are solving for the actual number of moles of each element in 1 mole of the compound. There is not much rounding needed to get whole numbers for the subscripts. This is a good sign.

The methods for obtaining empirical and molecular formulas are summarized below.

STEPS

Determination of the Empirical Formula

Numbers very close to whole numbers, such as 9.92 and 1.08, should be rounded to whole numbers. Numbers such as 2.25, 4.33, and 2.72 should not be rounded to whole numbers.

1 Since mass percent gives the number of grams of a particular element per 100 grams of compound, base the calculation on 100 grams of compound. Each percent will then represent the mass in grams of that element present in the compound.
2 Determine the number of moles of each element present in 100 grams of compound using the atomic weights (masses) of the elements present.
3 Divide each value of the number of moles by the smallest of the values. If each resulting number is a whole number (after appropriate rounding), these numbers represent the subscripts of the elements in the empirical formula.
4 If the numbers obtained in the previous step are not whole numbers, multiply each number by an integer so that the results are all whole numbers.

STEPS

Determination of the Molecular Formula

Method 1

1 Obtain the empirical formula.
2 Compute the empirical formula mass.
3 Calculate the ratio:

$$\frac{\text{Molar mass}}{\text{Empirical formula mass}}$$

4 The integer from the previous step represents the number of empirical formula units in one molecule. When the empirical formula subscripts are multiplied by this integer, we obtain the molecular formula.

Method 2

1 Using the mass percents and the molar mass, determine the mass of each element present in 1 mole of compound.
2 Determine the number of moles of each element present in 1 mole of compound.
3 The integers from the previous step represent the subscripts in the molecular formula.

3.7 Chemical Equations

Chemical Reactions

A chemical change involves reorganization of the atoms in one or more substances. For example, when the methane (CH_4) in natural gas combines with oxygen (O_2) in the air and burns, carbon dioxide (CO_2) and water (H_2O) are formed. This process is represented by a **chemical equation** with the **reactants** (here methane and oxygen) on the left side of an arrow and the **products** (carbon dioxide and water) on the right side:

$$\underset{\text{Reactants}}{CH_4 + O_2} \longrightarrow \underset{\text{Products}}{CO_2 + H_2O}$$

Courtesy, SaskEnergy

A flare in a natural gas field is an example of a chemical reaction.

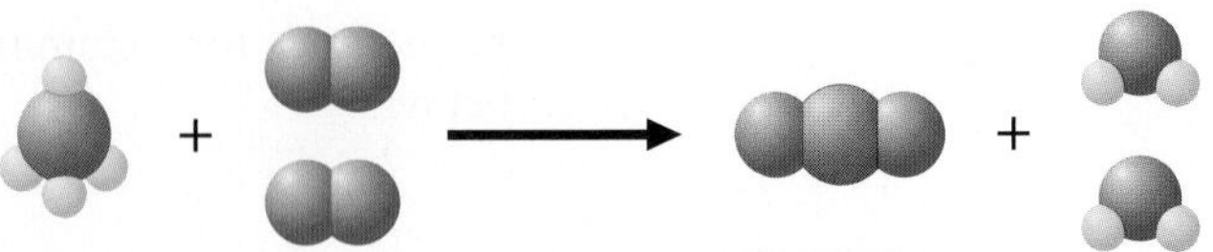

Figure 3.8
The reaction between methane and oxygen to give water and carbon dioxide. Note that no atoms have been gained or lost in the reaction. The reaction simply reorganizes the atoms.

Notice that the atoms have been reorganized. *Bonds have been broken and new ones formed.* Remember that *in a chemical reaction, atoms are neither created nor destroyed. All atoms present in the reactants must be accounted for among the products.* In other words, there must be the same number of each type of atom on the product side as there is on the reactant side of the arrow. Making sure that this rule is followed is called **balancing a chemical equation** for a reaction. The equation just shown for the reaction between CH_4 and O_2 is not balanced. As we will see in the next section, the equation can be balanced to produce

$$CH_4 + 2O_2 \longrightarrow CO_2 + 2H_2O$$

This reaction is shown graphically in Fig. 3.8. We can check whether the equation is balanced by comparing the number of each type of atom on both sides:

$$CH_4 + 2O_2 \longrightarrow CO_2 + 2H_2O$$

CH_4	$2O_2$	CO_2	$2H_2O$
1 C, 4 H	4 O	1 C, 2 O	4 H, 2 O

The Meaning of a Chemical Equation

The chemical equation for a reaction provides two important types of information: the nature of the reactants and products and the relative numbers of each. The reactants and products in a specific reaction must be identified by experiment. Besides specifying the compounds involved in the reaction, the equation often includes the *physical states* of the reactants and products:

State	Symbol
Solid	(*s*)
Liquid	(*l*)
Gas	(*g*)
Dissolved in water (in aqueous solution)	(*aq*)

For example, when hydrochloric acid in aqueous solution is added to solid sodium hydrogen carbonate, the products carbon dioxide gas, liquid water, and sodium chloride (which dissolves in the water) are formed:

$$HCl(aq) + NaHCO_3(s) \longrightarrow CO_2(g) + H_2O(l) + NaCl(aq)$$

The relative numbers of reactants and products in a reaction are indicated by the *coefficients* in the balanced equation. (The coefficients can be determined since we know that the same number of each type of atom must occur on both sides of the equation.) For example, the balanced equation

$$CH_4(g) + 2O_2(g) \longrightarrow CO_2(g) + 2H_2O(g)$$

can be interpreted in several equivalent ways, as shown in Table 3.2. Note that the total mass is 80 grams for both reactants and products. We should expect this result, since chemical reactions involve only a rearrangement of atoms. Atoms, and therefore mass, are conserved in a chemical reaction.

Table 3.2

Information Conveyed by the Balanced Equation for the Combustion of Methane

Reactants		Products
$CH_4(g) + 2O_2(g)$	$\longrightarrow$	$CO_2(g) + 2H_2O(g)$
1 molecule CH_4 + 2 molecules O_2	$\longrightarrow$	1 molecule CO_2 + 2 molecules H_2O
1 mole of CH_4 molecules + 2 moles of O_2 molecules	$\longrightarrow$	1 mole of CO_2 molecules + 2 moles of H_2O molecules
6.022×10^{23} CH_4 molecules + $2(6.022 \times 10^{23})$ O_2 molecules	$\longrightarrow$	6.022×10^{23} CO_2 molecules + $2(6.022 \times 10^{23})$ H_2O molecules
16 g CH_4 + 2(32 g) O_2	$\longrightarrow$	44 g CO_2 + 2(18 g) H_2O
80 g reactants	$\longrightarrow$	80 g products

From this discussion you can see that a balanced chemical equation gives you a great deal of information.

3.8 Balancing Chemical Equations

An unbalanced chemical equation is of limited use. Whenever you see an equation, you should ask yourself whether it is balanced. The principle that lies at the heart of the balancing process is that atoms are conserved in a chemical reaction. The same number of each type of atom must be found among the reactants and products. Also, remember that the identities of the reactants and products of a reaction are determined by experimental observation. For example, when liquid ethanol is burned in the presence of sufficient oxygen gas, the products will always be carbon dioxide and water. When the equation for this reaction is balanced, the *identities* of the reactants and products must not be changed. *The formulas of the compounds must never be changed when balancing a chemical equation.* That is, the subscripts in a formula cannot be changed, nor can atoms be added or subtracted from a formula.

In balancing equations, start with the most complicated molecule.

Most chemical equations can be balanced by inspection—that is, by trial and error. It is always best to start with the most complicated molecules (those containing the greatest number of atoms). For example, consider the reaction of ethanol with oxygen, given by the unbalanced equation

$$C_2H_5OH(l) + O_2(g) \longrightarrow CO_2(g) + H_2O(g)$$

The most complicated molecule here is C_2H_5OH. We will begin by balancing the products that contain the atoms in C_2H_5OH. Since C_2H_5OH contains two carbon atoms, we place a 2 before the CO_2 to balance the carbon atoms:

$$\underset{\text{2 C atoms}}{C_2H_5OH(l)} + O_2(g) \longrightarrow \underset{\text{2 C atoms}}{2CO_2(g)} + H_2O(g)$$

Since C_2H_5OH contains six hydrogen atoms, the hydrogen atoms can be balanced by placing a 3 before the H_2O:

$$\underset{(5+1)\text{ H}}{C_2H_5OH(l)} + O_2(g) \longrightarrow 2CO_2(g) + \underset{(3\times 2)\text{ H}}{3H_2O(g)}$$

Last, we balance the oxygen atoms. Note that the right side of the preceding equation contains seven oxygen atoms, whereas the left side has only three.

We can correct this by putting a 3 before the O_2 to produce the balanced equation:

$$\underset{1\ \mathrm{O}}{C_2H_5OH(l)} + \underset{6\ \mathrm{O}}{3O_2(g)} \longrightarrow \underset{(2\times 2)\ \mathrm{O}}{2CO_2(g)} + \underset{3\ \mathrm{O}}{3H_2O(g)}$$

Now we check:

$$C_2H_5OH(l) + 3O_2(g) \longrightarrow 2CO_2(g) + 3H_2O(g)$$

Reactants	Products
2 C atoms	2 C atoms
6 H atoms	6 H atoms
7 O atoms	7 O atoms

The balanced equation can be represented by space-filling models as follows:

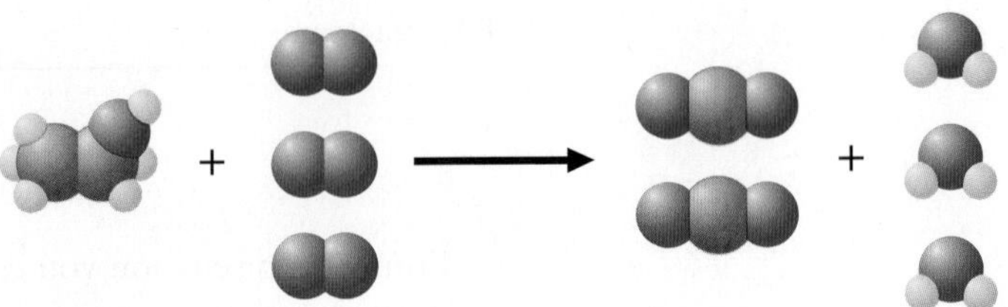

STEPS

Writing and Balancing the Equation for a Chemical Reaction

1 Determine what reaction is occurring. What are the reactants, the products, and the states involved?

2 Write the *unbalanced* equation that summarizes the preceding information.

3 Balance the equation by inspection, starting with the most complicated molecule(s). Determine what coefficients are necessary to ensure that the same number of each type of atom appears on both reactant and product sides. Do not change the identities (formulas) of any of the reactants or products.

OWL INTERACTIVE EXAMPLE 3.8

Chromate and dichromate compounds are suspected carcinogens (cancer-inducing agents) and should be handled carefully.

Chromium compounds exhibit a variety of bright colors. When solid ammonium dichromate, $(NH_4)_2Cr_2O_7$, a vivid orange compound, is ignited, a spectacular reaction occurs, as shown in the two photographs on the next page. Although the reaction is somewhat more complex, let's assume here that the products are solid chromium(III) oxide, nitrogen gas (consisting of N_2 molecules), and water vapor. Balance the equation for this reaction.

Solution From the description given, the reactant is solid ammonium dichromate, $(NH_4)_2Cr_2O_7(s)$, and the products are nitrogen gas, $N_2(g)$; water vapor, $H_2O(g)$; and solid chromium(III) oxide, $Cr_2O_3(s)$. The formula for chromium (III) oxide can be determined by recognizing that the Roman numeral III means that Cr^{3+} ions are present. For a neutral compound the formula must then be Cr_2O_3, since each oxide ion is O^{2-}.

The unbalanced equation is

$$(NH_4)_2Cr_2O_7(s) \longrightarrow Cr_2O_3(s) + N_2(g) + H_2O(g)$$

Note that nitrogen and chromium are balanced (two nitrogen atoms and two chromium atoms on each side), but hydrogen and oxygen are not. A coefficient of 4 for H_2O balances the hydrogen atoms:

$$\underset{(4 \times 2)\ \text{H}}{(NH_4)_2Cr_2O_7(s)} \longrightarrow Cr_2O_3(s) + N_2(g) + \underset{(4 \times 2)\ \text{H}}{4H_2O(g)}$$

Note that in balancing the hydrogen, we have also balanced the oxygen since there are seven oxygen atoms in the reactants and in the products.

Check: $\underset{\text{Reactant atoms}}{2\ \text{N}, 8\ \text{H}, 2\ \text{Cr}, 7\ \text{O}} \longrightarrow \underset{\text{Product atoms}}{2\ \text{N}, 8\ \text{H}, 2\ \text{Cr}, 7\ \text{O}}$

The equation is balanced.

3.9 Stoichiometric Calculations: Amounts of Reactants and Products

Recall that the coefficients in chemical equations represent *numbers* of molecules, not masses of molecules. However, in the laboratory or chemical plant, when a reaction is to be run, the amounts of substances needed cannot be determined by counting molecules directly. Counting is always done by weighing. In this section we will see how chemical equations can be used to deal with *masses* of reacting chemicals.

To develop the principles involved in dealing with the stoichiometry of reactions, we will consider the combustion of propane (C_3H_8), a hydrocarbon used for gas barbecue grills and often used as a fuel in rural areas where natural gas pipelines are not available. Propane reacts with oxygen to produce carbon dioxide and water. We will consider the question, "What mass of oxygen will react with 96.1 grams of propane?" The first thing that must always be done when performing calculations involving chemical reactions is to *write the balanced chemical equation* for the reaction. In this case the balanced equation is

Before doing any calculations involving a chemical reaction, be sure the equation for the reaction is balanced.

$$C_3H_8(g) + 5O_2(g) \longrightarrow 3CO_2(g) + 4H_2O(g)$$

Recall that this equation means that 1 mole of C_3H_8 will react with 5 moles of O_2 to produce 3 moles of CO_2 and 4 moles of H_2O. To use this equation to find the masses of reactants and products, we must be able to convert between masses and moles of substances. Thus we must first ask, "How many moles of propane are present in 96.1 grams of propane?" The molar mass of propane to three significant figures is 44.1 g/mol. The moles of propane can be calculated as follows:

$$96.1\ \text{g}\ C_3H_8 \times \frac{1\ \text{mol}\ C_3H_8}{44.1\ \text{g}\ C_3H_8} = 2.18\ \text{mol}\ C_3H_8$$

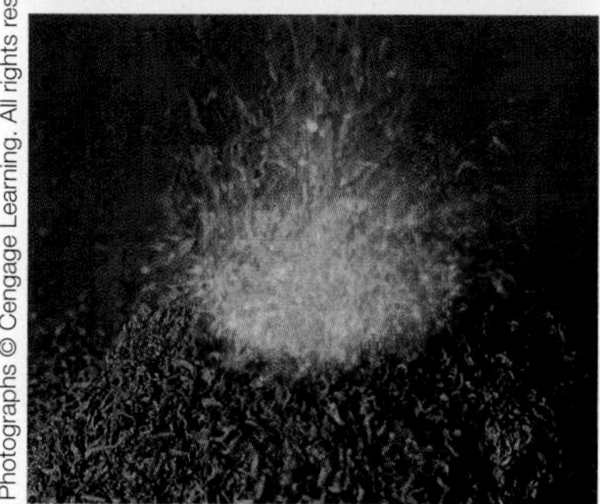

Decomposition of ammonium dichromate.

Next, we must take into account that each mole of propane reacts with 5 moles of oxygen. The best way to do this is to use the balanced equation to construct a **mole ratio.** In this case we want to convert from moles of propane to moles of oxygen. From the balanced equation we see that 5 moles of O_2 is required for each mole of C_3H_8, so the appropriate ratio is

$$\frac{5\ \text{mol}\ O_2}{1\ \text{mol}\ C_3H_8}$$

which can be used to calculate the number of moles of O_2 required:

$$2.18\ \text{mol}\ C_3H_8 \times \frac{5\ \text{mol}\ O_2}{1\ \text{mol}\ C_3H_8} = 10.9\ \text{mol}\ O_2$$

Chemical Insights High Mountains—Low Octane

The next time you visit a gas station, take a moment to note the octane rating that accompanies the grade of gasoline that you are purchasing. The gasoline is priced according to its octane rating—a measure of the fuel's antiknock properties. In a conventional internal combustion engine, gasoline vapors and air are drawn into the combustion cylinder on the downward stroke of the piston. This air–fuel mixture is compressed on the upward piston stroke (compression stroke), and a spark from the sparkplug ignites the mix. The rhythmic combustion of the air–fuel mix occurring sequentially in several cylinders furnishes the power to propel the vehicle down the road. Excessive heat and pressure (or poor-quality fuel) within the cylinder may cause the premature combustion of the mixture—commonly known as engine "knock" or "ping." Over time, this engine knock can damage the engine, resulting in inefficient performance and costly repairs.

A consumer typically is faced with three choices of gasoline, with octane ratings of 87 (regular), 89 (midgrade), and 93 (premium). But if you happen to travel or live in the higher elevations of the Rocky Mountain states, you might be surprised to find different octane ratings at the gasoline pumps. The reason for this provides a lesson in stoichiometry. At higher elevations the air is less dense—the volume of oxygen per unit volume of air is smaller. Most engines are designed to achieve a 14:1 oxygen-to-fuel ratio in the cylinder prior to combustion. If less oxygen is available, then less fuel is required to achieve this optimal ratio. In turn, the lower volumes of oxygen and fuel result in a lower pressure in the cylinder. Because high pressure tends to promote knocking, the lower pressure within engine cylinders at higher elevations promotes a more controlled combustion of the air–fuel mixture, and therefore, octane requirements are lower. Although consumers in the Rocky Mountain states can purchase three grades of gasoline, the octane ratings of these fuel blends are different from those in the rest of the United States. In Denver, Colorado, regular gasoline is 85 octane, midgrade is 87 octane, and premium is 91 octane—2 points lower than gasoline sold in most of the rest of the country.

Since the original question asked for the mass of oxygen needed to react with 96.1 grams of propane, the 10.9 moles of O_2 must be converted to *grams*, using the molar mass of O_2:

$$10.9 \text{ mol } O_2 \times \frac{32.0 \text{ g } O_2}{1 \text{ mol } O_2} = 349 \text{ g } O_2$$

Therefore, 349 grams of oxygen is required to burn 96.1 grams of propane.

This example can be extended by asking, "What mass of carbon dioxide is produced when 96.1 grams of propane is combusted with oxygen?" In this case we must convert between moles of propane and moles of carbon dioxide. This conversion can be done by inspecting the balanced equation, which shows that 3 moles of CO_2 is produced for each mole of C_3H_8 reacted:

$$2.18 \text{ mol } C_3H_8 \times \frac{3 \text{ mol } CO_2}{1 \text{ mol } C_3H_8} = 6.54 \text{ mol } CO_2$$

Then we use the molar mass of CO_2 (44.0 g/mol) to calculate the mass of CO_2 produced:

$$6.54 \text{ mol } CO_2 \times \frac{44.0 \text{ g } CO_2}{1 \text{ mol } CO_2} = 288 \text{ g } CO_2$$

The process for finding the mass of carbon dioxide produced from 96.1 grams of propane is summarized here:

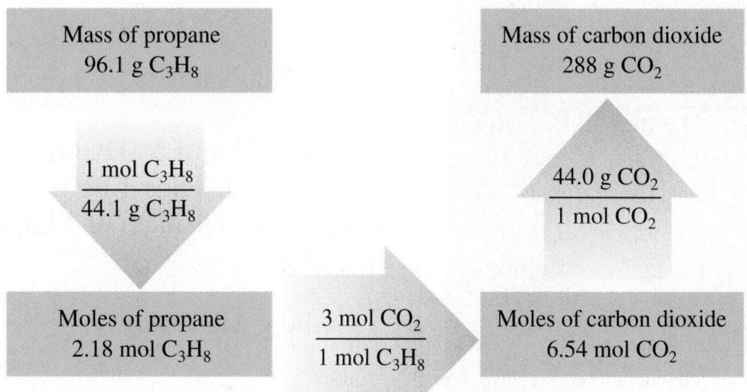

STEPS

Calculation of Masses of Reactants and Products in Chemical Reactions

1 Balance the equation for the reaction.
2 Convert the known masses of the substances to moles.
3 Use the balanced equation to set up the appropriate mole ratios.
4 Use the appropriate mole ratios to calculate the number of moles of the desired reactant or product.
5 Convert from moles back to grams if required by the problem.

OWL INTERACTIVE EXAMPLE 3.9

Baking soda ($NaHCO_3$) is often used as an antacid. It neutralizes excess hydrochloric acid secreted by the stomach:

$$NaHCO_3(s) + HCl(aq) \longrightarrow NaCl(aq) + H_2O(l) + CO_2(aq)$$

Milk of magnesia, which is an aqueous suspension of magnesium hydroxide, is also used as an antacid:

$$Mg(OH)_2(s) + 2HCl(aq) \longrightarrow 2H_2O(l) + MgCl_2(aq)$$

Which is the more effective antacid per gram, $NaHCO_3$ or $Mg(OH)_2$?

Solution

■ What are we trying to solve?

In this problem we are asked to determine which antacid is more effective.

■ What does this mean?

The more effective antacid will react with (neutralize) more acid. So to answer this question, we must answer the following:

1. How much HCl is neutralized per gram of $NaHCO_3$?
2. How much HCl is neutralized per gram of $Mg(OH)_2$?

We will then compare the answers and choose the more effective antacid.

■ How can we determine the amount of base (antacid) that reacts with an amount of acid?

We have balanced chemical equations, which give us the mole ratios of $NaHCO_3$/HCl and $Mg(OH)_2$/HCl. Thus we must convert 1.00 g of each

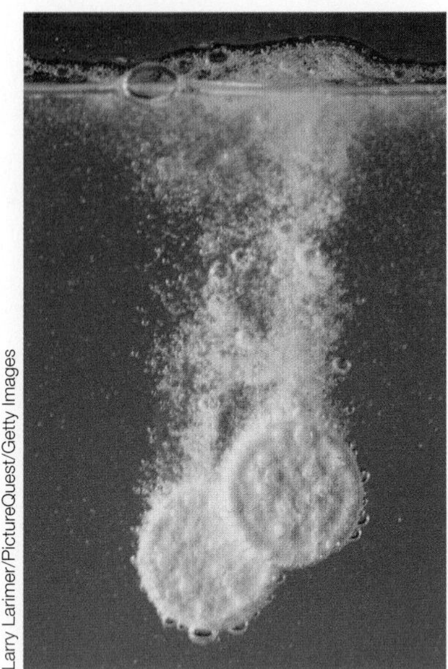
Larry Larimer/PictureQuest/Getty Images

Two antacid tablets containing HCO_3^- dissolve to produce CO_2 gas.

antacid to moles of antacid and then to moles of HCl. The antacid that reacts with the greater number of moles of HCl is the more effective one.

Using the molar mass of $NaHCO_3$, we determine the moles of $NaHCO_3$ in 1.00 g of $NaHCO_3$:

$$1.00 \text{ g NaHCO}_3 \times \frac{1 \text{ mol NaHCO}_3}{84.01 \text{ g NaHCO}_3} = 1.19 \times 10^{-2} \text{ mol NaHCO}_3$$

Because HCl and $NaHCO_3$ react 1:1, this answer also represents the moles of HCl required. Thus 1.00 g of $NaHCO_3$ will neutralize 1.19×10^{-2} mol HCl. Using the molar mass of $Mg(OH)_2$, we next determine the moles of $Mg(OH)_2$ in 1.00 g:

$$1.00 \text{ g Mg(OH)}_2 \times \frac{1 \text{ mol Mg(OH)}_2}{58.32 \text{ g Mg(OH)}_2} = 1.71 \times 10^{-2} \text{ mol Mg(OH)}_2$$

Using the balanced equation, we determine the moles of HCl that will react with this amount of $Mg(OH)_2$:

$$1.71 \times 10^{-2} \text{ mol Mg(OH)}_2 \times \frac{2 \text{ mol HCl}}{1 \text{ mol Mg(OH)}_2} = 3.42 \times 10^{-2} \text{ mol HCl}$$

Thus 1.00 g of $Mg(OH)_2$ will neutralize 3.42×10^{-2} mol HCl. It is a better antacid per gram than $NaHCO_3$.

3.10 Calculations Involving a Limiting Reactant

When chemicals are mixed together to undergo a reaction, they are often mixed in **stoichiometric quantities**—that is, in exactly the correct amounts so that all reactants "run out" (are used up) at the same time. To clarify this concept, let's consider the production of hydrogen for use in the manufacture of ammonia by the Haber process. Ammonia, a very important fertilizer itself and a starting material for other fertilizers, is made by combining nitrogen from the air with hydrogen according to the equation

$$N_2(g) + 3H_2(g) \longrightarrow 2NH_3(g)$$

The hydrogen for this process is produced from the reaction of methane with water:

$$CH_4(g) + H_2O(g) \longrightarrow 3H_2(g) + CO(g)$$

Now consider the following question: What mass of water is required to react with *exactly* 2.50×10^3 kilograms of methane? That is, how much water will just use up all of the 2.50×10^3 kilograms of methane, leaving no methane or water remaining? Using the principles developed in the preceding section, we can calculate that if 2.50×10^3 kilograms of methane is mixed with 2.81×10^3 kilograms of water, both reactants will run out at the same time. The reactants have been mixed in stoichiometric quantities.

If, however, 2.50×10^3 kilograms of methane is mixed with 3.00×10^3 kilograms of water, the methane will be consumed before the water runs out. The water will be in *excess*. In this case the quantity of products formed will be determined by the quantity of methane present. Once the methane is consumed, no more products can be formed, even though some water still remains. In this situation, because the amount of methane *limits* the amount of products that can be formed, it is called the **limiting reactant,** or **limiting reagent.** In any stoichiometry problem it is essential to determine which reactant is the limiting one to calculate correctly the amounts of products that will be formed.

Figure 3.9
Hydrogen and nitrogen react to form ammonia according to the equation $N_2 + 3H_2 \longrightarrow 2NH_3$.

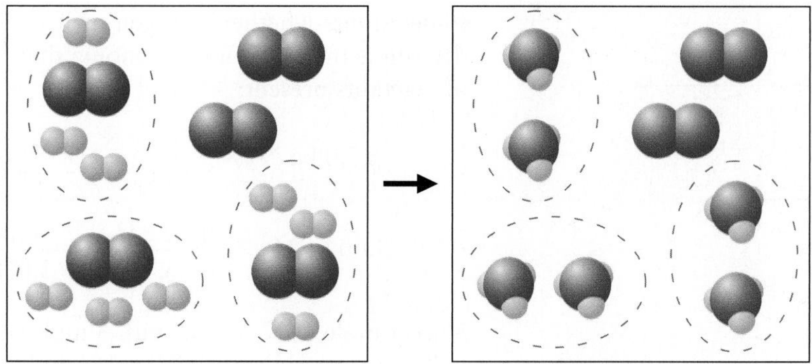

To further explore the idea of a limiting reactant, consider the ammonia synthesis reaction:

$$N_2(g) + 3H_2(g) \longrightarrow 2NH_3(g)$$

Assume that 5 N_2 molecules and 9 H_2 molecules are placed in a flask. Is this a stoichiometric mixture of reactants, or will one of them be consumed before the other runs out? From the balanced equation we know that each N_2 molecule requires 3 H_2 molecules for the reaction to occur:

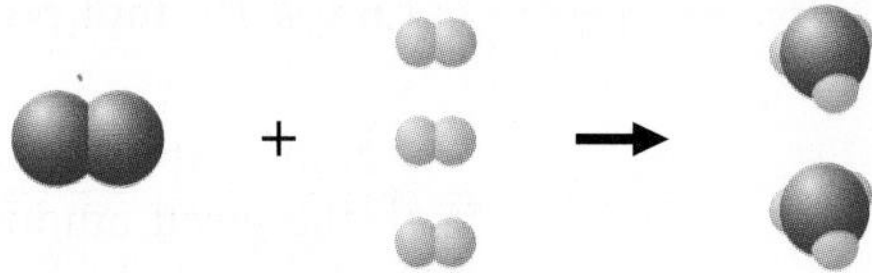

Thus the required H_2/N_2 ratio is $3H_2/1N_2$. In our experiment, we have 9 H_2 and 5 N_2, or a ratio of $9H_2/5N_2 = 1.8H_2/1N_2$.

Since the actual ratio (1.8:1) of H_2/N_2 is less than the ratio required by the balanced equation (3:1), there is not enough hydrogen to react with all the nitrogen. That is, the hydrogen will run out first, leaving some unreacted N_2 molecules. We can visualize this as shown in Fig. 3.9, which shows that 3 of the N_2 molecules react with the 9 H_2 molecules to produce 6 NH_3 molecules:

$$3N_2 + 9H_2 \longrightarrow 6NH_3$$

This leaves 2 N_2 molecules unreacted—the nitrogen is in excess.

What we have shown here is that in this experiment the hydrogen is the limiting reactant. The amount of H_2 initially present determines the amount of NH_3 that can form. The reaction was not able to use up all the N_2 molecules because the H_2 molecules were all consumed by the first 3 N_2 molecules to react.

In the laboratory or chemical plant we work with much larger quantities than the few molecules of the preceding example. Therefore we must use moles to deal with limiting reactants. The ideas are exactly the same, except that we are using moles of molecules instead of individual molecules. For example, suppose 25.0 kilograms of nitrogen and 5.00 kilograms of hydrogen are mixed and reacted to form ammonia. How do we calculate the mass of ammonia produced when this reaction is run to completion (until one of the reactants is completely consumed)?

As in the preceding example, we must use the balanced equation

$$N_2(g) + 3H_2(g) \longrightarrow 2NH_3(g)$$

to determine whether nitrogen or hydrogen is a limiting reactant and then to determine the amount of ammonia that is formed. We first calculate the moles of reactants present:

$$25.0 \text{ kg } N_2 \times \frac{1000 \text{ g } N_2}{1 \text{ kg } N_2} \times \frac{1 \text{ mol } N_2}{28.0 \text{ g } N_2} = 8.93 \times 10^2 \text{ mol } N_2$$

$$5.00 \text{ kg } H_2 \times \frac{1000 \text{ g } H_2}{1 \text{ kg } H_2} \times \frac{1 \text{ mol } H_2}{2.016 \text{ g } H_2} = 2.48 \times 10^3 \text{ mol } H_2$$

Since 1 mole of N_2 reacts with 3 moles of H_2, the number of moles of H_2 that will react exactly with 8.93×10^2 moles of N_2 is

$$8.93 \times 10^2 \text{ mol } N_2 \times \frac{3 \text{ mol } H_2}{1 \text{ mol } N_2} = 2.68 \times 10^3 \text{ mol } H_2$$

Thus 8.93×10^2 moles of N_2 requires 2.68×10^3 moles of H_2 to react completely. However, in this case only 2.48×10^3 moles of H_2 is present. Thus the hydrogen will be consumed before the nitrogen. Therefore, hydrogen is the *limiting reactant* in this particular situation, and we must use the amount of hydrogen to compute the quantity of ammonia formed:

Always check to see which reactant is limiting.

$$2.48 \times 10^3 \text{ mol } H_2 \times \frac{2 \text{ mol } NH_3}{3 \text{ mol } H_2} = 1.65 \times 10^3 \text{ mol } NH_3$$

$$1.65 \times 10^3 \text{ mol } NH_3 \times \frac{17.0 \text{ g } NH_3}{1 \text{ mol } NH_3} = 2.80 \times 10^4 \text{ g } NH_3 = 28.0 \text{ kg } NH_3$$

Note that to determine the limiting reactant, we could have started instead with the given amount of hydrogen and calculated the moles of nitrogen required:

$$2.48 \times 10^3 \text{ mol } H_2 \times \frac{1 \text{ mol } N_2}{3 \text{ mol } H_2} = 8.27 \times 10^2 \text{ mol } N_2$$

Thus 2.48×10^3 moles of H_2 requires 8.27×10^2 moles of N_2. Since 8.93×10^2 moles of N_2 is actually present, the nitrogen is in excess. The hydrogen will run out first, and thus again we find that hydrogen limits the amount of ammonia formed.

A related but simpler way to determine which reactant is limiting is to compare the mole ratio of the substances required by the balanced equation with the mole ratio of reactants actually present. For example, in this case the mole ratio of H_2 to N_2 required by the balanced equation is

$$\frac{3 \text{ mol } H_2}{1 \text{ mol } N_2}$$

That is,

$$\frac{\text{mol } H_2}{\text{mol } N_2} \text{ (required)} = \frac{3}{1} = 3$$

In this experiment we have 2.48×10^3 moles of H_2 and 8.93×10^2 moles of N_2. Thus the ratio

$$\frac{\text{mol } H_2}{\text{mol } N_2} \text{ (actual)} = \frac{2.48 \times 10^3}{8.93 \times 10^2} = 2.78$$

Since 2.78 is less than 3, the actual mole ratio of H_2 to N_2 is too small, and H_2 must be limiting. If the actual H_2/N_2 mole ratio had been greater than 3, then the H_2 would have been in excess and the N_2 would have been limiting.

OWL INTERACTIVE EXAMPLE 3.10

Nitrogen gas can be prepared by passing gaseous ammonia over solid copper(II) oxide at high temperatures. The other products of the reaction are solid copper and water vapor. If 18.1 g of NH_3 is reacted with 90.4 g of CuO, which is the limiting reactant? How many grams of N_2 will be formed?

Solution

- What are we trying to solve?

In this example we are asked to solve two problems: to determine the limiting reactant and to determine the mass of N_2 that will be formed from a given amount of reactants.

We first need to answer the following two questions: What is meant by the term limiting reactant? and What do we need to do to solve for an amount of N_2?

- What is meant by the term limiting reactant?

It is the reactant that is completely used up in the reaction. Since a chemical equation gives us the mole ratio of the reactants, we can determine how many moles of a given reactant we need to have compared with what we actually have. To do this we need to have a balanced equation, which we can get from the description of the problem.

$$2NH_3(g) + 3CuO(s) \longrightarrow N_2(g) + 3Cu(s) + 3H_2O(g)$$

To use this equation we have to compare the amounts of reactants in moles to the balanced equation, so we must compute the moles of NH_3 and CuO:

$$18.1 \text{ g NH}_3 \times \frac{1 \text{ mol NH}_3}{17.0 \text{ g NH}_3} = 1.06 \text{ mol NH}_3$$

$$90.4 \text{ g CuO} \times \frac{1 \text{ mol CuO}}{79.5 \text{ g CuO}} = 1.14 \text{ mol CuO}$$

We can now use either one of these to determine the number of moles of the other reactant that would be required to react completely. For example, if all of the NH_3 reacts, what number of moles of CuO is required?

$$1.06 \text{ mol NH}_3 \times \frac{3 \text{ mol CuO}}{2 \text{ mol NH}_3} = 1.59 \text{ mol CuO}$$

Thus 1.59 moles of CuO is required to react with 1.06 moles of NH_3. Since only 1.14 moles of CuO is actually present, the amount of CuO is limiting.

We can verify this conclusion by comparing the mole ratio of CuO to NH_3 required by the balanced equation,

$$\frac{\text{mol CuO}}{\text{mol NH}_3} \text{(required)} = \frac{3}{2} = 1.5$$

with the mole ratio actually present,

$$\frac{\text{mol CuO}}{\text{mol NH}_3} \text{(actual)} = \frac{1.14}{1.06} = 1.08$$

Since the actual ratio is too small (smaller than 1.5), CuO is the limiting reactant.

- What do we need to do to solve for an amount of N_2?

To determine this we need to use the limiting reactant to calculate the moles, then mass, of N_2.

Because CuO is the limiting reactant, we must use the amount of CuO to calculate the amount of N_2 formed:

$$1.14 \text{ mol CuO} \times \frac{1 \text{ mol N}_2}{3 \text{ mol CuO}} = 0.380 \text{ mol N}_2$$

Using the molar mass of N_2, we can calculate the mass of N_2 produced:

$$0.380 \text{ mol N}_2 \times \frac{28.0 \text{ g N}_2}{1 \text{ mol N}_2} = 10.6 \text{ g N}_2$$

Note: As with most problems, there is more than one way to solve this. We could have answered, "What is meant by the term limiting reactant?" with: the reactant that limits the amount of product. So, another way to solve this is to solve for the number of moles of N_2 produced assuming all of the NH_3 reacts and solve again for moles of N_2 produced assuming all of the CuO reacts. The reactant that results in the smaller amount of product limits the reaction. Once we know the limiting reactant, we can solve for the mass of N_2 produced. We leave this up to you to try.

The amount of a given product formed when the limiting reactant is completely consumed is called the **theoretical yield** of that product. In Example 3.10, 10.6 grams of nitrogen is the theoretical yield. This is the *maximum amount* of nitrogen that can be produced from the quantities of reactants used. Actually, the amount of product predicted by the theoretical yield is seldom obtained because of side reactions (other reactions that involve one or more of the reactants or products) and other complications. The *actual yield* of product is often given as a percentage of the theoretical yield. This value is called the **percent yield:**

Percent yield is important as an indicator of the efficiency of a particular laboratory or industrial reaction.

$$\frac{\text{Actual yield}}{\text{Theoretical yield}} \times 100\% = \text{percent yield}$$

For example, if the reaction considered in Example 3.10 actually produced 6.63 grams of nitrogen instead of the predicted 10.6 grams, the percent yield of nitrogen would be

$$\frac{6.63 \text{ g N}_2}{10.6 \text{ g N}_2} \times 100\% = 62.5\%$$

STEPS

Solving a Stoichiometry Problem Involving Masses of Reactants and Products

1 Write and balance the equation for the reaction.
2 Convert the known masses of substances to moles.
3 By comparing the mole ratio of reactants required by the balanced equation with the mole ratio of reactants actually present, determine which reactant is limiting.
4 Using the amount of the limiting reactant and the appropriate mole ratios, compute the number of moles of the desired product.
5 Convert from moles to grams using the molar mass.

OWL INTERACTIVE EXAMPLE 3.11

Potassium chromate, a bright yellow solid, is produced by the reaction of solid chromite ore ($FeCr_2O_4$) with solid potassium carbonate and gaseous oxygen at high temperatures. The other products of the reaction are solid iron(III) oxide and gaseous carbon dioxide. In a particular experiment, 169 kg of chromite ore, 298 kg of potassium carbonate, and 75.0 kg of oxygen were sealed in a reaction vessel and reacted at a high temperature. The amount of potassium chromate obtained was 194 kg. Calculate the percent yield of potassium chromate.

Solution

■ What are we trying to solve?

This problem asks us to determine percent yield.

■ What does this mean?

$$\text{Percent yield} = \frac{\text{actual yield}}{\text{theoretical yield}} \times 100\%$$

We are given the actual yield of potassium chromate as 194 kg; thus we have

$$\text{Percent yield} = \frac{194 \text{ kg}}{\text{theoretical yield}} \times 100\%$$

So we have now changed the problem to, What is the theoretical yield? Once we have this, we can solve for percent yield. This leads us to:

■ What do we mean by theoretical yield?

The theoretical yield is the amount of product we can expect if the reaction is allowed to run to completion. It is the amount of product that we have been calculating using the ideas in Sections 3.8 and 3.9.

We are given the masses (and names) of the three reactants and are told the names of the products. First, we will need a balanced chemical equation.

The unbalanced equation, which can be written from the preceding description of the reaction, is

$$FeCr_2O_4(s) + K_2CO_3(s) + O_2(g) \longrightarrow K_2CrO_4(s) + Fe_2O_3(s) + CO_2(g)$$

The balanced equation is

$$4FeCr_2O_4(s) + 8K_2CO_3(s) + 7O_2(g) \longrightarrow 8K_2CrO_4(s) + 2Fe_2O_3(s) + 8CO_2(g)$$

To determine the theoretical yield, we have to decide on which reactant is limiting and then use this reactant to compute the mass of the product formed. To determine the limiting reactant, we will compare the mole ratios of the reactants required by the balanced equation with the actual mole ratios. The numbers of moles of the various reactants are obtained as follows:

$$169 \text{ kg } FeCr_2O_4 \times \frac{1000 \text{ g } FeCr_2O_4}{1 \text{ kg } FeCr_2O_4} \times \frac{1 \text{ mol } FeCr_2O_4}{223.84 \text{ g } FeCr_2O_4} = 7.55 \times 10^2 \text{ mol } FeCr_2O_4$$

$$298 \text{ kg } K_2CO_3 \times \frac{1000 \text{ g } K_2CO_3}{1 \text{ kg } K_2CO_3} \times \frac{1 \text{ mol } K_2CO_3}{138.21 \text{ g } K_2CO_3} = 2.16 \times 10^3 \text{ mol } K_2CO_3$$

$$75.0 \text{ kg } O_2 \times \frac{1000 \text{ g } O_2}{1 \text{ kg } O_2} \times \frac{1 \text{ mol } O_2}{32.00 \text{ g } O_2} = 2.34 \times 10^3 \text{ mol } O_2$$

Now we must determine which of the three reactants is limiting. For the reactants K_2CO_3 and $FeCr_2O_4$ the required mole ratio is

$$\frac{\text{mol } K_2CO_3}{\text{mol } FeCr_2O_4} \text{ (required)} = \frac{8}{4} = 2$$

The actual mole ratio is

$$\frac{\text{mol } K_2CO_3}{\text{mol } FeCr_2O_4}\text{ (actual)} = \frac{2.16 \times 10^3}{7.55 \times 10^2} = 2.86$$

Since the actual mole ratio is greater than that required, the K_2CO_3 is in excess compared with $FeCr_2O_4$. Thus either $FeCr_2O_4$ or O_2 must be limiting. To determine which of these will limit the amounts of products, we compare the required mole ratio,

$$\frac{\text{mol } O_2}{\text{mol } FeCr_2O_4}\text{ (required)} = \frac{7}{4} = 1.75$$

with the actual mole ratio,

$$\frac{\text{mol } O_2}{\text{mol } FeCr_2O_4}\text{ (actual)} = \frac{2.34 \times 10^3}{7.55 \times 10^2} = 3.10$$

Thus more K_2CO_3 and O_2 are present than required. These reactants are in excess, so $FeCr_2O_4$ is the limiting reactant.

We must use the amount of $FeCr_2O_4$ to calculate the maximum amount of K_2CrO_4 that can be formed:

$$7.55 \times 10^2 \text{ mol } FeCr_2O_4 \times \frac{8 \text{ mol } K_2CrO_4}{4 \text{ mol } FeCr_2O_4} = 1.51 \times 10^3 \text{ mol } K_2CrO_4$$

Using the molar mass of K_2CrO_4, we can determine the mass:

$$1.51 \times 10^3 \text{ mol } K_2CrO_4 \times \frac{194.19 \text{ g } K_2CrO_4}{1 \text{ mol } K_2CrO_4} = 2.93 \times 10^5 \text{ g } K_2CrO_4$$

This value represents the theoretical yield of K_2CrO_4. The actual yield was 194 kg, or 1.94×10^5 g. Thus the percent yield is

$$\frac{1.94 \times 10^5 \text{ g } K_2CrO_4}{2.93 \times 10^5 \text{ g } K_2CrO_4} \times 100\% = 66.2\%$$

Key Terms

chemical stoichiometry

Section 3.1
mass spectrometer

Section 3.2
mole
Avogadro's number

Section 3.3
molar mass

Section 3.4
conceptual problem solving

Section 3.5
molecular weight
mass percent

Section 3.6
empirical formula
molecular formula

Section 3.7
chemical equation
reactants
products
balancing a chemical equation

For Review

Stoichiometry

- Deals with the amounts of substances consumed and/or produced in a chemical reaction.
- We count atoms by measuring the mass of the sample.
- To relate mass and the number of atoms, the average atomic mass is required.

Mole

- The number of carbon atoms in exactly 12 g of pure ^{12}C
- 6.022×10^{23} units of a substance
- The mass of 1 mole of an element = the atomic mass in grams

Molar mass

- Mass (g) of 1 mole of a compound or element
- Obtained for a compound by finding the sum of the average masses of its constituents

Section 3.9
mole ratio
Section 3.10
stoichiometric quantities
Haber process
limiting reactant (reagent)
theoretical yield
percent yield
	Percent composition
	■ The mass percent on each element in a compound
	■ Mass percent = $\frac{\text{mass of element in 1 mole of substance}}{\text{mass of 1 mole of substance}} \times 100\%$
	Empirical formula
	■ The simplest whole-number ratio of the various types of atoms in a compound
	■ Can be obtained from the mass percent of elements in a compound
	Molecular formula
	■ For molecular substance:
	  ■ The formula of the constituent molecules
	  ■ Always an integer multiple of an empirical formula
	■ For ionic substances:
	  ■ The same as the empirical formula
	Chemical reactions
	■ Reactants are turned into products.
	■ Atoms are neither created nor destroyed.
	■ All of the atoms present in the reactants must also be present in the products.
	Characteristics of a chemical equation
	■ Represents a chemical reaction
	■ Reactants on the left side of the arrow; products on the right side
	■ When balanced, gives the relative numbers of reactant and product molecules or ions
	Stoichiometry calculations
	■ Amounts of reactants consumed and products formed can be determined from the balanced chemical equation.
	■ The limiting reactant is the one consumed first, thus limiting the amount of product that can form.
	Yield
	■ The theoretical yield is the maximum amount that can be produced from a given amount of the limiting reactant.
	■ The actual yield, the amount of product actually obtained, is always less than the theoretical yield.
	■ Percent yield = $\frac{\text{actual yield (g)}}{\text{theoretical yield (g)}} \times 100\%$

Discussion Questions

These questions are designed to be considered by groups of students in class. Often these questions work well for introducing a particular topic in class.

1. The following are actual student responses to the question: Why is it necessary to balance chemical equations?
   a. The chemicals will not react until you have added the correct mole ratios.
   b. The correct products will not be formed unless the right amount of reactants have been added.
   c. A certain number of products cannot be formed without a certain number of reactants.
   d. The balanced equation tells you how much reactant you need and allows you to predict how much product you will make.
   e. A mole-to-mole ratio must be established for the reaction to occur as written.

   What is the best choice? For those you did not choose, explain why they are incorrect. State the fundamental reason why an equation for a reaction must be balanced.

2. Consider the equation $A + 2B \rightarrow AB_2$. Imagine that 10 moles of A is reacted with 26 moles of B. Use a scale from 0 to 10 to express your level of agreement with each of the following statements. Justify and discuss your responses.
   a. There will be some As left over.
   b. There will be some Bs left over.
   c. Because of leftover As, some A_2 molecules will be formed.
   d. Because of leftover Bs, some B_2 molecules will be formed.

e. Even if A is not limiting, A_2 molecules will be formed.
f. Even if B is not limiting, B_2 molecules will be formed.
g. Along with the molecule AB_2, molecules with the formula A_xB_y (other than AB_2) will be formed.

3. What information do we get from a formula? From an equation?

4. A sample of liquid heptane (C_7H_{16}) weighing 11.50 g is reacted with 1.300 moles of oxygen gas. The heptane is burned completely (heptane reacts with oxygen to form both carbon monoxide and water and carbon dioxide and water). After the reaction is complete, the amount of gas present is 1.050 moles (assume that all of the water formed is liquid).
   a. How many moles of CO are produced?
   b. How many moles of CO_2 are produced?
   c. How many moles of O_2 are left over?

5. Nitrogen (N_2) and hydrogen (H_2) react to form ammonia (NH_3). Consider the mixture of N_2( ) and H_2( ) in a closed container as illustrated:

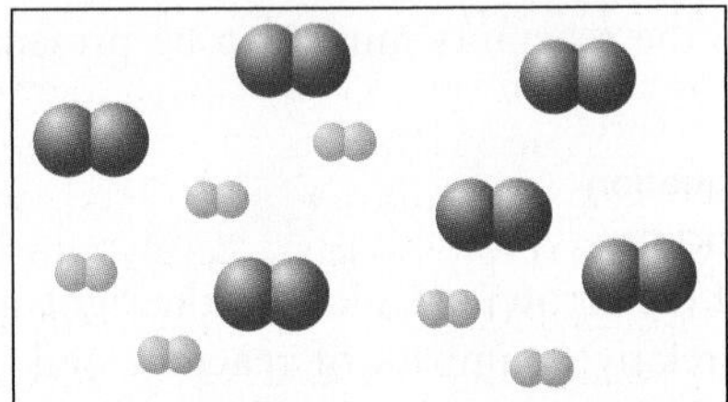

Assuming that the reaction goes to completion, draw a representation of the product mixture. Explain how you arrived at this representation.

6. For the preceding question, which of the following equations best represents the reaction?
   a. $6N_2 + 6H_2 \rightarrow 4NH_3 + 4N_2$
   b. $N_2 + H_2 \rightarrow NH_3$
   c. $N + 3H \rightarrow NH_3$
   d. $N_2 + 3H_2 \rightarrow 2NH_3$
   e. $2N_2 + 6H_2 \rightarrow 4NH_3$

   Justify your choice. For those you did not choose, explain why they are incorrect.

7. You know that chemical A reacts with chemical B. You react 10.0 g A with 10.0 g B. What information do you need to have to determine the amount of product that will be produced? Explain.

8. A kerosene lamp has a mass of 1.5 kg. You put 0.5 kg of kerosene in the lamp. You burn all the kerosene until the lamp has a mass of 1.5 kg. What is the mass of the gases that are given off? Explain.

9. Consider an iron bar on a balance as shown.

As the iron bar rusts, which of the following is true? Explain your answer.
   a. The balance will read less than 75.0 g.
   b. The balance will read 75.0 g.
   c. The balance will read greater than 75.0 g.
   d. The balance will read greater than 75.0 g, but if the bar is removed, the rust scraped off, and the bar replaced, the balance will read 75.0 g.

10. You may have noticed that water sometimes drips from an exhaust pipe of a car as it is running. Is this evidence that there is at least a small amount of water originally present in the gasoline? Explain.

Questions 11 and 12 deal with the following situation: You react chemical A with chemical B to make one product. It takes 100 g of A to react completely with 20 g B.

11. What is the mass of the product?
   a. less than 20 g
   b. between 20 and 100 g
   c. between 100 and 120 g
   d. exactly 120 g
   e. more than 120 g

12. What is true about the chemical properties of the product?
   a. The properties are more like those of chemical A.
   b. The properties are more like those of chemical B.
   c. The properties are an average of those of chemical A and chemical B.
   d. The properties are not necessarily like those of either chemical A or chemical B.

   Justify your choice. For those you did not choose, explain why they are incorrect.

13. What is the difference between the empirical and molecular formulas of a compound? Can they ever be the same? Explain.

14. Atoms of three different elements are represented by O, □, and Δ. Which compound is left over when three molecules of OΔ and three molecules of □□Δ react to form O□Δ and OΔΔ?

15. One way of determining the empirical formula is to burn a compound in air and weigh the amounts of carbon dioxide and water given off. For what types of compounds does this work? Explain the assumptions that are made. Why is the formula an empirical formula and not necessarily a molecular formula?

16. In chemistry, what is meant by the term *mole*? What is the importance of the mole concept?

17. Which (if any) of the following is true regarding the limiting reactant in a chemical reaction?
   a. The limiting reactant has the lowest coefficient in a balanced equation.
   b. The limiting reactant is the reactant for which you have the fewest number of moles.
   c. The limiting reactant has the lowest ratio of moles available/coefficient in the balanced equation.
   d. The limiting reactant has the lowest ratio of coefficient in the balanced equation/moles available.

   Justify your choice. For those you did not choose, explain why they are incorrect.

18. Consider the equation $3A + B \rightarrow C + D$. You react 4 moles of A with 2 moles of B. Which of the following is true?
   a. The limiting reactant is the one with the higher molar mass.
   b. A is the limiting reactant because you need 6 moles of A and have 4 moles.

c. B is the limiting reactant because you have fewer moles of B than A.
d. B is the limiting reactant because three A molecules react with each B molecule.
e. Neither reactant is limiting.

Justify your choice. For those you did not choose, explain why they are incorrect.

19. Chlorine exists mainly as two isotopes, ^{37}Cl and ^{35}Cl. Which is more abundant? How do you know?

20. According to the law of conservation of mass, mass cannot be gained or destroyed in a chemical reaction. Why can't you simply add the masses of two reactants to determine the total mass of product?

21. The atomic mass of boron (B) is given in the periodic table as 10.81, yet no single atom of boron has a mass of 10.81 amu. Explain.

22. Why is the actual yield of a reaction often less than the theoretical yield?

Exercises

OWL Interactive versions of these problems may be assigned in OWL.

A blue exercise number indicates that the answer to that exercise appears at the back of this book and a solution appears in the *Solutions Guide.*

Atomic Masses and the Mass Spectrometer

23. An element X has five major isotopes, which are listed below along with their abundances. Calculate the average atomic mass, and identify the element.

Isotope	Percent Natural Abundance	Mass (amu)
^{46}X	8.00	45.95269
^{47}X	7.30	46.951764
^{48}X	73.80	47.947947
^{49}X	5.50	48.947841
^{50}X	5.40	49.944792

24. The stable isotopes of iron are ^{54}Fe, ^{56}Fe, ^{57}Fe, and ^{58}Fe. The mass spectrum of iron looks like the following:

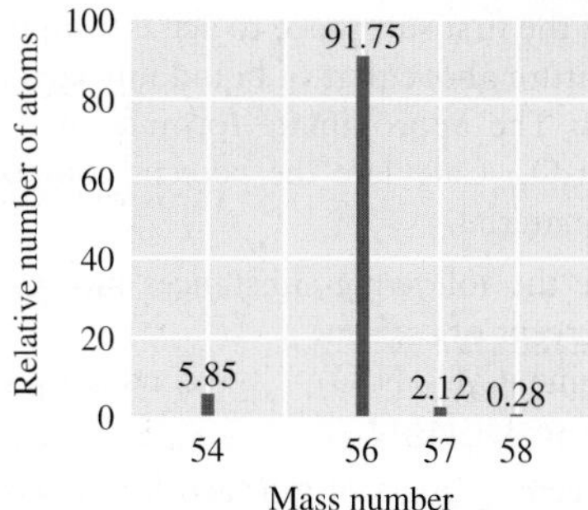

Use the data on the mass spectrum to estimate the average atomic mass of iron and compare it with the value given in the table inside the front cover of this book.

25. The element silver (Ag) has two naturally occurring isotopes: ^{109}Ag and ^{107}Ag with a mass of 106.905 amu. Silver consists of 51.82% ^{107}Ag and has an average atomic mass of 107.868 amu. Calculate the mass of ^{109}Ag.

26. The element europium exists in nature as two isotopes: ^{151}Eu has a mass of 150.9196 amu, and ^{153}Eu has a mass of 152.9209 amu. The average atomic mass of europium is 151.96 amu. Calculate the relative abundance of the two europium isotopes.

27. The element rhenium (Re) has two naturally occurring isotopes, ^{185}Re and ^{187}Re, with an average atomic mass of 186.207 amu. Rhenium is 62.60% ^{187}Re, and the atomic mass of ^{187}Re is 186.956 amu. Calculate the mass of ^{185}Re.

28. An element consists of 1.40% of an isotope with mass 203.973 amu, 24.10% of an isotope with mass 205.9745 amu, 22.10% of an isotope with mass 206.9759 amu, and 52.40% of an isotope with mass 207.9766 amu. Calculate the average atomic mass and identify the element.

29. The mass spectrum of bromine (Br_2) consists of three peaks with the following relative sizes:

Mass (amu)	Relative Size
157.84	0.2534
159.84	0.5000
161.84	0.2466

How do you interpret these data?

30. Naturally occurring tellurium (Te) has the following isotopic abundances:

Isotope	Abundance	Mass (amu)
^{120}Te	0.09%	119.90
^{122}Te	2.46%	121.90
^{123}Te	0.87%	122.90
^{124}Te	4.61%	123.90
^{125}Te	6.99%	124.90
^{126}Te	18.71%	125.90
^{128}Te	31.79%	127.90
^{130}Te	34.48%	129.91

Draw the mass spectrum of H_2Te, assuming that the only hydrogen isotope present is ^1H (mass 1.008).

31. Gallium arsenide (GaAs) has gained widespread use in semiconductor devices that interconvert light and electrical signals in fiber-optic communications systems. Gallium

consists of 60.% ^{69}Ga and 40.% ^{71}Ga. Arsenic has only one naturally occurring isotope, ^{75}As. Gallium arsenide is a polymeric material, but its mass spectrum shows fragments with formulas GaAs and Ga_2As_2. What would the distribution of peaks look like for these two fragments?

Moles and Molar Masses

32. Ascorbic acid, or vitamin C ($C_6H_8O_6$), is an essential vitamin. It cannot be stored by the body and must be present in the diet. What is the molar mass of ascorbic acid? Vitamin C tablets are taken as a dietary supplement. If a typical tablet contains 500.0 mg vitamin C, what amount (moles) and what number of molecules of vitamin C does it contain?

33. The molecular formula of acetylsalicylic acid (aspirin), one of the most commonly used pain relievers, is $C_9H_8O_4$.
    a. Calculate the molar mass of aspirin.
    b. A typical aspirin tablet contains 500. mg $C_9H_8O_4$. What amount (moles) of $C_9H_8O_4$ molecules and what number of molecules of acetylsalicylic acid are in a 500.-mg tablet?

34. Complete the following table.

Mass of Sample	Moles of Sample	Molecules in Sample	Total Atoms in Sample
4.24 g C_6H_6	_____	_____	_____
_____	0.224 mol H_2O	_____	_____
_____	_____	2.71×10^{22} molecules CO_2	_____
_____	_____	_____	3.35×10^{22} total atoms in CH_3OH sample

35. What amount (moles) is represented by each of these samples?
    a. 20.0 mg caffeine, $C_8H_{10}N_4O_2$
    b. 2.72×10^{21} molecules of ethanol, C_2H_5OH
    c. 1.50 g of dry ice, CO_2

36. How many atoms of nitrogen are present in 5.00 g of each of the following?
    a. glycine, $C_2H_5O_2N$
    b. magnesium nitride
    c. calcium nitrate
    d. dinitrogen tetroxide

37. Consider the following gas samples: 4.0 g of hydrogen gas, 4.0 g of helium gas, 1.0 mole of fluorine gas, 44.0 g of carbon dioxide gas, and 146 g of sulfur hexafluoride gas. Arrange the gas samples in order of increasing number of total atoms present.

38. Aspartame is an artificial sweetener that is 160 times sweeter than sucrose (table sugar) when dissolved in water. It is marketed as NutraSweet. The molecular formula of aspartame is $C_{14}H_{18}N_2O_5$.
    a. Calculate the molar mass of aspartame.
    b. How many moles of molecules are in 10.0 g of aspartame?
    c. What is the mass in grams of 1.56 moles of aspartame?
    d. How many molecules are in 5.0 mg of aspartame?
    e. How many atoms of nitrogen are in 1.2 g of aspartame?
    f. What is the mass in grams of 1.0×10^9 molecules of aspartame?
    g. What is the mass in grams of one molecule of aspartame?

39. Chloral hydrate ($C_2H_3Cl_3O_2$) is a drug formerly used as a sedative and hypnotic. It is the compound used to make "Mickey Finns" in detective stories.
    a. Calculate the molar mass of chloral hydrate.
    b. How many moles of $C_2H_3Cl_3O_2$ molecules are in 500.0 g of chloral hydrate?
    c. What is the mass in grams of 2.0×10^{-2} mol chloral hydrate?
    d. How many chlorine atoms are in 5.0 g chloral hydrate?
    e. What mass of chloral hydrate would contain 1.0 g Cl?
    f. What is the mass of exactly 500 molecules of chloral hydrate?

40. In the spring of 1984, concern arose over the presence of ethylene dibromide, or EDB, in grains and cereals. EDB has the molecular formula $C_2H_4Br_2$ and until 1984 was commonly used as a plant fumigant. The federal limit for EDB in finished cereal products is 30.0 parts per billion (ppb), where 1.0 ppb = 1.0×10^{-9} g of EDB for every 1.0 g of sample. How many molecules of EDB are in 1.0 lb of flour if 30.0 ppb of EDB is present?

Percent Composition

41. Anabolic steroids are performance enhancement drugs whose use has been banned from most major sporting activities. One anabolic steroid is fluoxymesterone ($C_{20}H_{29}FO_3$). Calculate the percent composition by mass of fluoxymesterone.

42. Calculate the percent composition by mass of the following compounds that are important starting materials for synthetic polymers:
    a. $C_3H_4O_2$ (acrylic acid, from which acrylic plastics are made)
    b. $C_4H_6O_2$ (methyl acrylate, from which Plexiglas is made)
    c. C_3H_3N (acrylonitrile, from which Orlon is made)

43. In 1987 the first substance to act as a superconductor at a temperature above that of liquid nitrogen (77 K) was discovered. The approximate formula of this substance is $YBa_2Cu_3O_7$. Calculate the percent composition by mass of this material.

44. Arrange the following substances in order of increasing mass percent of carbon.
    a. caffeine, $C_8H_{10}N_4O_2$
    b. sucrose, $C_{12}H_{22}O_{11}$
    c. ethanol, C_2H_5OH

45. The percent by mass of nitrogen for a compound is found to be 46.7%. Which of the following could be this species?

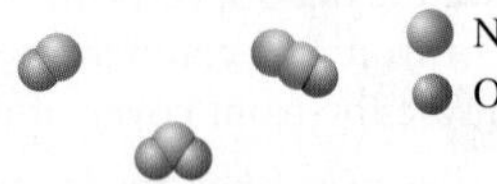

46. Vitamin B_{12}, cyanocobalamin, is essential for human nutrition. It is concentrated in animal tissue but not in higher plants. Although nutritional requirements for the vitamin are quite low, people who abstain completely from animal

products may develop a deficiency anemia. Cyanocobalamin is the form used in vitamin supplements. It contains 4.34% cobalt by mass. Calculate the molar mass of cyanocobalamin, assuming that there is one atom of cobalt in every molecule of cyanocobalamin.

47. Fungal laccase, a blue protein found in wood-rotting fungi, is 0.390% Cu by mass. If a fungal laccase molecule contains four copper atoms, what is the molar mass of fungal laccase?

48. Portland cement acts as the binding agent in concrete. A typical Portland cement has the following composition:

Formula	Name	Mass Percent
Ca_3SiO_5	Tricalcium silicate	50.
Ca_2SiO_4	Dicalcium silicate	25
$Ca_3Al_2O_6$	Tricalcium aluminate	12
Ca_2AlFeO_5	Calcium aluminoferrite	8.0
$CaSO_4 \cdot 2H_2O$	Calcium sulfate dihydrate	3.5
Other substances, mostly MgO		1.5

Assuming that the impurities contain no Ca, Al, or Fe, calculate the mass percent of these elements in this Portland cement.

Empirical and Molecular Formulas

49. Express the composition of each of the following compounds as the mass percent of its elements.
   a. formaldehyde, CH_2O
   b. glucose, $C_6H_{12}O_6$
   c. acetic acid, $HC_2H_3O_2$

   Considering your answers, which type of formula—empirical or molecular—can be obtained from elemental analysis that gives mass percent composition? Explain.

50. Give the empirical formula of each of these compounds.

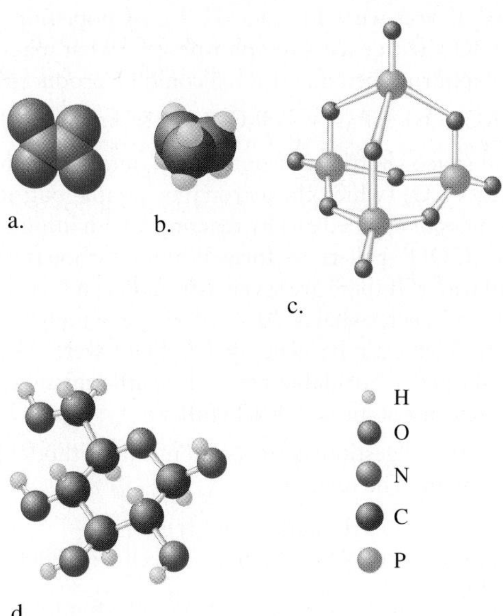

51. Determine the molecular formulas to which the following empirical formulas and molar masses pertain.
   a. SNH (188.35 g/mol)
   b. $NPCl_2$ (347.64 g/mol)
   c. CoC_4O_4 (341.94 g/mol)
   d. SN (184.32 g/mol)

52. A sample of urea contains 1.121 g N, 0.161 g H, 0.480 g C, and 0.640 g O. What is the empirical formula of urea?

53. There are two binary compounds of mercury and oxygen. Heating either of them results in the decomposition of the compound, with oxygen gas escaping into the atmosphere while leaving a residue of pure mercury. Heating 0.6498 g of one of the compounds leaves a residue of 0.6018 g. Heating 0.4172 g of the other compound results in a mass loss of 0.016 g. Determine the empirical formula of each compound.

54. The compound adrenaline contains 56.79% C, 6.56% H, 28.37% O, and 8.28% N by mass. What is the empirical formula of adrenaline?

55. A compound contains only carbon, hydrogen, nitrogen, and oxygen. Combustion of 0.157 g of the compound produced 0.213 g of CO_2 and 0.0310 g of H_2O. In another experiment, 0.103 g of the compound produced 0.0230 g of NH_3. What is the empirical formula of the compound? *Hint:* Combustion involves reacting with excess O_2. Assume that all the carbon ends up in CO_2 and all the hydrogen ends up in H_2O. Also assume that all the nitrogen ends up in the NH_3 in the second experiment.

56. Maleic acid is an organic compound composed of 41.39% C, 3.47% H, and the rest oxygen. If 0.129 mole of maleic acid has a mass of 15.0 g, what are the empirical and molecular formulas of maleic acid?

57. Determine the molecular formula of a compound that contains 26.7% P, 12.1% N, and 61.2% Cl, and has a molar mass of 580 g/mol.

58. Terephthalic acid is an important chemical used in the manufacture of polyesters and plasticizers. It contains only C, H, and O. Combustion of 19.81 mg terephthalic acid produces 41.98 mg CO_2 and 6.45 mg H_2O. If 0.250 mole of terephthalic acid has a mass of 41.5 g, determine the molecular formula of terephthalic acid.

59. A compound contains only carbon, hydrogen, and oxygen. Combustion of 10.68 mg of the compound yields 16.01 mg CO_2 and 4.37 mg H_2O. The molar mass of the compound is 176.1 g/mol. What are the empirical and molecular formulas of the compound?

60. ABS plastic is a tough, hard plastic used in applications requiring shock resistance. (See Chapter 21.) The polymer consists of three monomer units: acrylonitrile (C_3H_3N), butadiene (C_4H_6), and styrene (C_8H_8).
   a. A sample of ABS plastic contains 8.80% N by mass. It took 0.605 g of Br_2 to react completely with a 1.20-g sample of ABS plastic. Bromine reacts 1:1 (by moles) with the butadiene molecules in the polymer and

nothing else. What is the percent by mass of acrylonitrile and butadiene in this polymer?

b. What are the relative numbers of each of the monomer units in this polymer?

Balancing Chemical Equations

61. The reaction of an element X with element Y is represented in the following diagram. Which of the equations best describes this reaction?

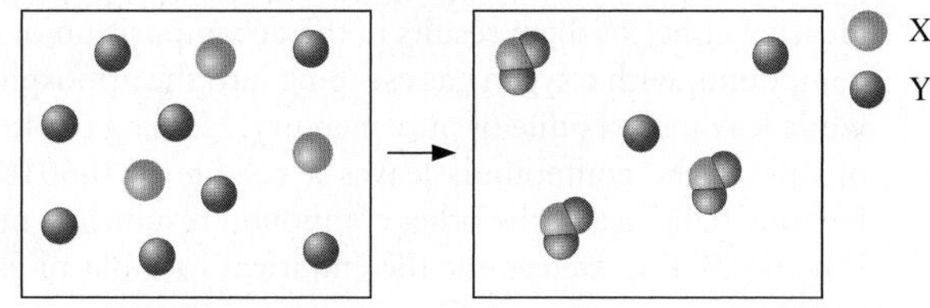

a. $3X + 8Y \rightarrow X_3Y_8$
b. $3X + 6Y \rightarrow X_3Y_6$
c. $X + 2Y \rightarrow XY_2$
d. $3X + 8Y \rightarrow 3XY_2 + 2Y$

62. Silicon is produced for the chemical and electronics industries by the following reactions. Give the balanced equation for each reaction.
   a. $SiO_2(s) + C(s) \xrightarrow[\text{arc furnace}]{\text{Electric}} Si_2(s) + CO(g)$
   b. Silicon tetrachloride is reacted with very pure magnesium, producing silicon and magnesium chloride.
   c. $Na_2SiF_6(s) + Na(s) \longrightarrow Si(s) + NaF(s)$

63. Give the balanced equation for each of the following chemical reactions.
   a. Glucose ($C_6H_{12}O_6$) reacts with oxygen gas to produce gaseous carbon dioxide and water vapor.
   b. Solid iron(III) sulfide reacts with gaseous hydrogen chloride to form solid iron(III) chloride and hydrogen sulfide gas.
   c. Carbon disulfide liquid reacts with ammonia gas to produce hydrogen sulfide gas and solid ammonium thiocyanate (NH_4SCN).

64. Give the balanced equation for each of the following.
   a. The combustion of ethanol (C_2H_5OH) forms carbon dioxide and water vapor. A combustion reaction refers to a reaction of a substance with oxygen gas.
   b. Aqueous solutions of lead(II) nitrate and sodium phosphate are mixed, resulting in the precipitate formation of lead(II) phosphate with aqueous sodium nitrate as the other product.

65. Balance the following equations.
   a. $Cr(s) + S_8(s) \longrightarrow Cr_2S_3(s)$
   b. $NaHCO_3(s) \xrightarrow{\text{Heat}} Na_2CO_3(s) + CO_2(g) + H_2O(g)$
   c. $KClO_3(s) \xrightarrow{\text{Heat}} KCl(s) + O_2(g)$
   d. $Eu(s) + HF(g) \longrightarrow EuF_3(s) + H_2(g)$
   e. $C_6H_6(l) + O_2(g) \longrightarrow CO_2(g) + H_2O(g)$

66. Balance each of the following chemical equations.
   a. $KO_2(s) + H_2O(l) \rightarrow KOH(aq) + O_2(g) + H_2O_2(aq)$
   b. $Fe_2O_3(s) + HNO_3(aq) \rightarrow Fe(NO_3)_3(aq) + H_2O(l)$
   c. $NH_3(g) + O_2(g) \rightarrow NO(g) + H_2O(g)$
   d. $PCl_5(l) + H_2O(l) \rightarrow H_3PO_4(aq) + HCl(g)$
   e. $CaO(s) + C(s) \rightarrow CaC_2(s) + CO_2(g)$
   f. $MoS_2(s) + O_2(g) \rightarrow MoO_3(s) + SO_2(g)$
   g. $FeCO_3(s) + H_2CO_3(aq) \rightarrow Fe(HCO_3)_2(aq)$

Reaction Stoichiometry

67. The reusable booster rockets of the U.S. space shuttle use a mixture of aluminum and ammonium perchlorate for fuel. A possible equation for this reaction is

$$3Al(s) + 3NH_4ClO_4(s) \longrightarrow Al_2O_3(s) + AlCl_3(s) + 3NO(g) + 6H_2O(g)$$

What mass of NH_4ClO_4 should be used in the fuel mixture for every kilogram of Al?

68. Nitric acid is produced commercially by the Ostwald process. The three steps of the Ostwald process are shown in the following equations:

$$4NH_3(g) + 5O_2(g) \longrightarrow 4NO(g) + 6H_2O(g)$$
$$2NO(g) + O_2(g) \longrightarrow 2NO_2(g)$$
$$3NO_2(g) + H_2O(l) \longrightarrow 2HNO_3(aq) + NO(g)$$

What mass of NH_3 must be used to produce 1.0×10^6 kg of HNO_3 by the Ostwald process, assuming 100% yield in each reaction and assuming the NO produced in the third stage is not recycled?

69. Over the years, the thermite reaction has been used for welding railroad rails, in incendiary bombs, and to ignite solid-fuel rocket motors. The reaction is

$$Fe_2O_3(s) + 2Al(s) \longrightarrow 2Fe(l) + Al_2O_3(s)$$

What masses of iron(III) oxide and aluminum must be used to produce 15.0 g of iron? What is the maximum mass of aluminum oxide that could be produced?

70. The reaction between potassium chlorate and red phosphorus takes place when you strike a match on a matchbox. If you were to react 52.9 g of potassium chlorate ($KClO_3$) with excess red phosphorus, what mass of tetraphosphorus decoxide (P_4O_{10}) could be produced?

$$KClO_3(s) + P_4(s) \longrightarrow P_4O_{10}(s) + KCl(s) \quad \text{(unbalanced)}$$

71. The space shuttle environmental control system handles excess CO_2 (which the astronauts breathe out; it is 4.0% by mass of exhaled air) by reacting it with lithium hydroxide (LiOH) pellets to form lithium carbonate (Li_2CO_3) and water. If there are seven astronauts on board the shuttle, and each exhales 20. L of air per minute, how long could clean air be generated if there were 25,000 g of LiOH pellets available for each shuttle mission? Assume the density of air is 0.0010 g/mL.

72. Bacterial digestion is an economical method of sewage treatment. The reaction

$$5CO_2(g) + 55NH_4^+(aq) + 76O_2(g) \xrightarrow{\text{Bacteria}} \underset{\text{Bacterial tissue}}{C_5H_7O_2N(s)} + 54NO_2^-(aq) + 52H_2O(l) + 109H^+(aq)$$

is an intermediate step in the conversion of the nitrogen in organic compounds into nitrate ions. How much bacterial tissue is produced in a treatment plant for every 1.0×10^4 kg of wastewater containing 3.0% NH_4^+ ions by mass? Assume that 95% of the ammonium ions are consumed by the bacteria.

73. Phosphorus can be prepared from calcium phosphate by the following reaction:

$$2Ca_3(PO_4)_2(s) + 6SiO_2(s) + 10C(s) \longrightarrow 6CaSiO_3(s) + P_4(s) + 10CO(g)$$

Phosphorite is a mineral that contains $Ca_3(PO_4)_2$ plus other non-phosphorus-containing compounds. What is the maximum amount of P_4 that can be produced from 1.0 kg of phosphorite if the phosphorite sample is 75% $Ca_3(PO_4)_2$ by mass? Assume an excess of the other reactants.

74. In the production of printed circuit boards for the electronics industry, a 0.60-mm layer of copper is laminated onto an insulating plastic board. Next, a circuit pattern made of a chemically resistant polymer is printed on the board. The unwanted copper is removed by chemical etching and the protective polymer is finally removed by solvents. One etching reaction is

$$Cu(NH_3)_4Cl_2(aq) + 4NH_3(aq) + Cu(s) \longrightarrow 2Cu(NH_3)_4Cl(aq)$$

A plant needs to manufacture 10,000 printed circuit boards, each 8.0×16.0 cm in area. An average of 80.% of the copper is removed from each board (density of copper = 8.96 g/cm^3). What masses of $Cu(NH_3)_4Cl_2$ and NH_3 are needed to do this? Assume 100% yield.

Limiting Reactants and Percent Yield

75. Consider the reaction between NO(g) and O_2(g) represented below.

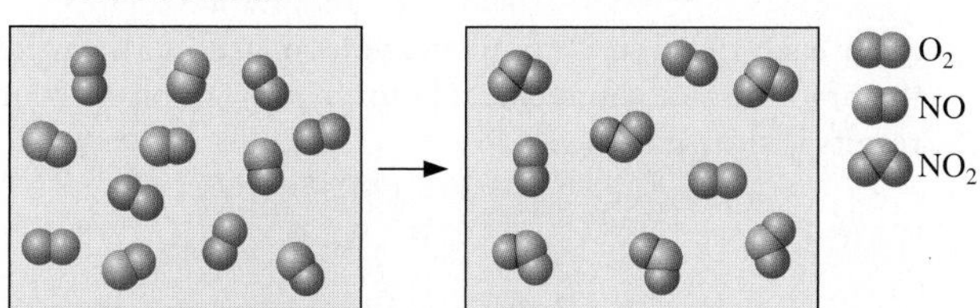

What is the balanced equation for this reaction and what is the limiting reactant?

76. Consider the following reaction:

$$4NH_3(g) + 5O_2(g) \longrightarrow 4NO(g) + 6H_2O(g)$$

If a container were to have 10 molecules of O_2 and 10 molecules of NH_3 initially, how many total molecules (reactants plus products) would be present in the container after this reaction goes to completion?

77. Hydrogen peroxide is used as a cleaning agent in the treatment of cuts and abrasions for several reasons. It is an oxidizing agent that can directly kill many microorganisms; it decomposes upon contact with blood, releasing elemental oxygen gas (which inhibits the growth of anaerobic microorganisms); and it foams upon contact with blood, which provides a cleansing action. In the laboratory, small quantities of hydrogen peroxide can be prepared by the action of an acid on an alkaline earth metal peroxide, such as barium peroxide:

$$BaO_2(s) + 2HCl(aq) \longrightarrow H_2O_2(aq) + BaCl_2(aq)$$

What amount of hydrogen peroxide should result when 1.50 g of barium peroxide is treated with 25.0 mL of hydrochloric acid solution containing 0.0272 g of HCl per mL? What mass of which reagent is left unreacted?

78. Silver sulfadiazine burn-treating cream creates a barrier against bacterial invasion and releases antimicrobial agents directly into the wound. If 25.0 g of Ag_2O is reacted with 50.0 g of $C_{10}H_{10}N_4SO_2$, what mass of silver sulfadiazine ($AgC_{10}H_9N_4SO_2$) can be produced, assuming 100% yield?

$$Ag_2O(s) + 2C_{10}H_{10}N_4SO_2(s) \longrightarrow 2AgC_{10}H_9N_4SO_2(s) + H_2O(l)$$

79. Bornite (Cu_3FeS_3) is a copper ore used in the production of copper. When heated, the following reaction occurs:

$$2Cu_3FeS_3(s) + 7O_2(g) \longrightarrow 6Cu(s) + 2FeO(s) + 6SO_2(g)$$

If 2.50 metric tons of bornite is reacted with excess O_2 and the process has an 86.3% yield of copper, how much copper is produced?

80. DDT, an insecticide harmful to fish, birds, and humans, is produced by the following reaction:

$$\underset{\text{Chlorobenzene}}{2C_6H_5Cl} + \underset{\text{Chloral}}{C_2HOCl_3} \longrightarrow \underset{\text{DDT}}{C_{14}H_9Cl_5} + H_2O$$

In a government lab, 1142 g of chlorobenzene is reacted with 485 g of chloral.

a. What mass of DDT is formed, assuming 100% yield?
b. Which reactant is limiting? Which is in excess?
c. What mass of the excess reactant is left over?
d. If the actual yield of DDT is 200.0 g, what is the percent yield?

81. Hydrogen cyanide is produced industrially from the reaction of gaseous ammonia, oxygen, and methane:

$$2NH_3(g) + 3O_2(g) + 2CH_4(g) \longrightarrow 2HCN(g) + 6H_2O(g)$$

If 5.00×10^3 kg each of NH_3, O_2, and CH_4 are reacted, what mass of HCN and of H_2O will be produced, assuming 100% yield?

82. The production capacity for acrylonitrile (C_3H_3N) in the United States exceeds 2 million pounds per year. Acrylonitrile, the building block for polyacrylonitrile fibers and a variety of plastics, is produced from gaseous propylene, ammonia, and oxygen:

$$2C_3H_6(g) + 2NH_3(g) + 3O_2(g) \longrightarrow 2C_3H_3N(g) + 6H_2O(g)$$

a. What mass of acrylonitrile can be produced from a mixture of 1.00 kg of propylene (C_3H_6), 1.50 kg of ammonia, and 2.00 kg of oxygen, assuming 100% yield?
b. What mass of water is produced, and what masses of which starting materials are left in excess?

83. Consider the following unbalanced reaction:

$$P_4(s) + F_2(g) \longrightarrow PF_3(g)$$

How many grams of F_2 are needed to produce 120. g of PF_3 if the reaction has a 78.1% yield?

84. The aspirin substitute acetaminophen ($C_8H_9O_2N$) is produced by the following three-step synthesis:

I. $C_6H_5O_3N(s) + 3H_2(g) + HCl(aq) \longrightarrow C_6H_8ONCl(s) + 2H_2O(l)$

II. $C_6H_8ONCl(s) + NaOH(aq) \longrightarrow C_6H_7ON(s) + H_2O(l) + NaCl(aq)$

III. $C_6H_7ON(s) + C_4H_6O_3(l) \longrightarrow C_8H_9O_2N(s) + HC_2H_3O_2(l)$

The first two reactions have percent yields of 87% and 98% by mass, respectively. The overall reaction yields 3 moles of acetaminophen product for every 4 moles of $C_6H_5O_3N$ reacted.

a. What is the percent yield by mass for the overall process?
b. What is the percent yield by mass of step III?

Additional Exercises

85. A sample of a hydrocarbon (a compound consisting of only carbon and hydrogen) contains 2.59×10^{23} atoms of hydrogen and is 17.3% hydrogen by mass. If the molar mass of the hydrocarbon is between 55 and 65 g/mol, how many moles of compound are present, and what is the mass of the sample?

86. A binary compound created by the reaction of an unknown element E and hydrogen contains 91.27% E and 8.73% H by mass. If the formula of the compound is E_3H_8, calculate the atomic mass of E.

87. An ionic compound MX_3 is prepared according to the following unbalanced chemical equation:

$$M + X_2 \longrightarrow MX_3$$

A 0.105-g sample of X_2 contains 8.92×10^{20} molecules. The compound MX_3 consists of 54.47% X by mass. What are the identities of M and X, and what is the correct name for MX_3? Starting with 1.00 g each of M and X_2, what mass of MX_3 can be prepared?

88. The empirical formula of styrene is CH; the molar mass of styrene is 104.14 g/mol. How many H atoms are present in a 2.00-g sample of styrene?

89. A 0.755-g sample of hydrated copper(II) sulfate ($CuSO_4 \cdot xH_2O$) was heated carefully until it had changed completely to anhydrous copper(II) sulfate ($CuSO_4$) with a mass of 0.483 g. Determine the value of x. [This number is called the "number of waters of hydration" of copper(II) sulfate. It specifies the number of water molecules per formula unit of $CuSO_4$ in the hydrated crystal.]

90. Many cereals are made with high moisture content so that the cereal can be formed into various shapes before it is dried. A cereal product containing 58% H_2O by mass is produced at the rate of 1000. kg/h. How much water must be evaporated per hour if the final product contains only 20.% water?

91. When aluminum metal is heated with an element from Group 6A of the periodic table, an ionic compound forms. When the experiment is performed with an unknown Group 6A element, the product is 18.56% Al by mass. What is the formula of the compound?

92. A salt contains only barium and one of the halide ions. A 0.158-g sample of the salt was dissolved in water, and an excess of sulfuric acid was added to form barium sulfate ($BaSO_4$), which was filtered, dried, and weighed. Its mass was found to be 0.124 g. What is the formula of the barium halide?

93. A sample of LSD (D-lysergic acid diethylamide, $C_{24}H_{30}N_3O$) is added to some table salt (sodium chloride) to form a mixture. Given that a 1.00-g sample of the mixture undergoes combustion to produce 1.20 g of CO_2, what is the mass percentage of LSD in the mixture?

94. Consider the following unbalanced equation:

$$Ca_3(PO_4)_2(s) + H_2SO_4(aq) \longrightarrow CaSO_4(s) + H_3PO_4(aq)$$

What masses of calcium sulfate and phosphoric acid can be produced from the reaction of 1.0 kg of calcium phosphate with 1.0 kg of concentrated sulfuric acid (98% H_2SO_4 by mass)?

95. A 0.4230-g sample of impure sodium nitrate was heated, converting all the sodium nitrate to 0.2864 g of sodium nitrite and oxygen gas. Determine the percent of sodium nitrate in the original sample.

96. You have seven closed containers, each with equal masses of chlorine gas (Cl_2). You add 10.0 g of sodium to the first sample, 20.0 g of sodium to the second sample, and so on (adding 70.0 g of sodium to the seventh sample). Sodium and chlorine react to form sodium chloride according to the equation

$$2Na(s) + Cl_2(g) \longrightarrow 2NaCl(s)$$

After each reaction is complete, you collect and measure the amount of sodium chloride formed. A graph of your results is shown below.

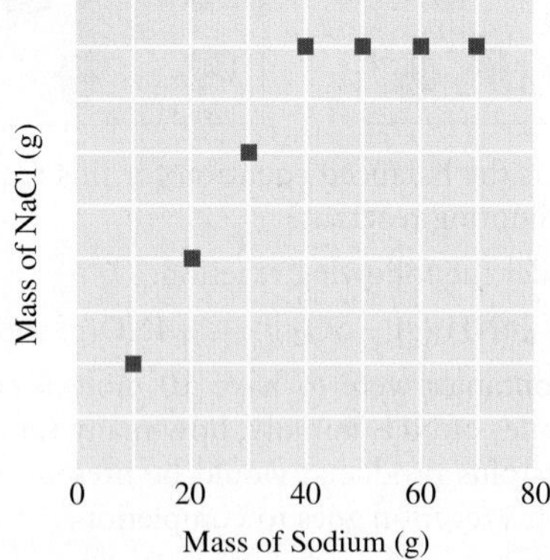

Anwer the following questions:

a. Explain the shape of the graph.
b. Calculate the mass of NaCl formed when 20.0 g of sodium is used.
c. Calculate the mass of Cl_2 in each container.
d. Calculate the mass of NaCl formed when 50.0 g of sodium is used.
e. Identify the leftover reactant and determine its mass for parts b and d.

97. An iron ore sample contains Fe_2O_3 plus other impurities. A 752-g sample of impure iron ore is heated with excess carbon, producing 453 g of pure iron by the following reaction:

$$Fe_2O_3(s) + 3C(s) \longrightarrow 2Fe(s) + 3CO(g)$$

What is the mass percent of Fe_2O_3 in the impure iron ore sample? Assume that Fe_2O_3 is the only source of iron and that the reaction is 100% efficient.

98. In using a mass spectrometer, a chemist sees a peak at a mass of 30.0106. Of the choices $^{12}C_2{}^1H_6$, $^{12}C^1H_2{}^{16}O$, and $^{14}N^{16}O$, which is responsible for this peak? Pertinent masses are ^1H, 1.007825; ^{16}O, 15.994915; and ^{14}N, 14.003074.

99. Natural rubidium has the average mass 85.4678 amu and is composed of isotopes ^{85}Rb (mass = 84.9117 amu) and ^{87}Rb. The ratio of atoms $^{85}Rb/^{87}Rb$ in natural rubidium is 2.591. Calculate the mass of ^{87}Rb.

100. Tetrodotoxin is a toxic chemical found in fugu pufferfish, a popular but rare delicacy in Japan. This compound has an LD_{50} (the amount of substance that is lethal to 50.% of a population sample) of 10. μg per kg of body mass. Tetrodotoxin is 41.38% carbon by mass, 13.16% nitrogen by mass, and 5.37% hydrogen by mass, with the remaining amount consisting of oxygen. What is the empirical formula of tetrodotoxin? If three molecules of tetrodotoxin have a mass of 1.59×10^{-21} g, what is the molecular formula of tetrodotoxin? What number of molecules of tetrodotoxin would be the LD_{50} dosage for a person weighing 165 lb?

101. Consider the following data for three binary compounds of hydrogen and nitrogen:

	% H (by Mass)	% N (by Mass)
I	17.75	82.25
II	12.58	87.42
III	2.34	97.66

When 1.00 L of each gaseous compound is decomposed to its elements, the following volumes of $H_2(g)$ and $N_2(g)$ are obtained:

	H_2 (L)	N_2 (L)
I	1.50	0.50
II	2.00	1.00
III	0.50	1.50

Use these data to determine the molecular formulas of compounds I, II, and III and to determine the relative values for the atomic masses of hydrogen and nitrogen.

102. A 0.200-g sample of protactinium(IV) oxide is converted to another oxide of protactinium by heating in the presence of oxygen to give 0.2081 g of the new oxide, Pa_xO_y. Determine the values of x and y.

103. A 1.000-g sample of XI_2 is dissolved in water, and excess silver nitrate is added to precipitate all of the iodide as AgI. The mass of the dry AgI is found to be 1.375 g. Calculate the atomic weight (mass) of X.

104. A substance X_2Z has the composition (by mass) of 40.0% X and 60.0% Z. What is the composition (by mass) of the compound XZ_2?

105. Vitamin A has a molar mass of 286.4 g and has a general molecular formula of C_xH_yE, where E is an unknown element. If vitamin A is 83.86% C and 10.56% H by mass, what is the molecular formula of vitamin A?

106. Boron consists of two isotopes, ^{10}B and ^{11}B. Chlorine also has two isotopes, ^{35}Cl and ^{37}Cl. Consider the mass spectrum of BCl_3. How many peaks would be present, and what approximate mass would each peak correspond to in the BCl_3 mass spectrum?

Challenge Problems

107. In a mass spectrometer, positive ions are produced when a gaseous mixture is ionized by electron bombardment produced by an electric discharge. When the electric-discharge voltage is low, singly positive ions are produced and the following peaks are observed in the mass spectrum:

Mass (amu)	Relative Intensity
32	0.3743
34	0.0015
40	1.0000

When the electric discharge is increased, still only singly charged ions are produced, but now the peaks observed in the mass spectrum are

Mass (amu)	Relative Intensity
16	0.7500
18	0.0015
40	1.0000

What does the gas mixture consist of, and what is the percent composition by isotope of the mixture?

108. When the supply of oxygen is limited, iron metal reacts with oxygen to produce a mixture of FeO and Fe_2O_3. In a certain experiment, 20.00 g of iron metal was reacted with 11.20 g of oxygen gas. After the experiment, the iron was totally consumed and 3.24 g of oxygen gas remained. Calculate the amounts of FeO and Fe_2O_3 formed in this experiment.

109. Element X forms both a dichloride (XCl_2) and a tetrachloride (XCl_4). Treatment of 10.00 g of XCl_2 with excess chlorine forms 12.55 g of XCl_4. Calculate the atomic weight (mass) of X and identify X.

110. Zinc and magnesium metal each react with hydrochloric acid to make chloride salts of the respective metals and hydrogen gas. A 10.00-g mixture of zinc and magnesium produces 0.5171 g of hydrogen gas upon being mixed with an excess of hydrochloric acid. Determine the percent magnesium by mass in the original mixture.

111. An unknown binary compound containing hydrogen (XH_n) has a density as a gas that is 2.393 times that of oxygen gas under the same conditions. When 2.23×10^{-2} mole of this compound reacts with excess oxygen gas, 0.803 g of water is produced. Identify the element X in this compound.

112. A 2.25-g sample of scandium metal is reacted with excess hydrochloric acid to produce 0.1502 g hydrogen gas. What is the formula of the scandium chloride produced in the reaction?

113. When $M_2S_3(s)$ is heated in air, it is converted to $MO_2(s)$. A 4.000-g sample of $M_2S_3(s)$ shows a decrease in mass of 0.277 g when it is heated in air. What is the average atomic mass of M?

114. Consider a gaseous binary compound with a molar mass of 62.09 g/mol. When 1.39 g of this compound is completely burned in excess oxygen, 1.21 g of water is formed. Determine the formula of the compound.

115. Pure carbon was burned in an excess of oxygen. The gaseous products were

CO_2	72.0 mol%
CO	16.0 mol%
O_2	12.0 mol%

How many moles of O_2 were present in the initial reaction mixture for every mole of carbon?

116. You take 1.00 g of an aspirin tablet (a compound consisting solely of carbon, hydrogen, and oxygen), burn it in air, and collect 2.20 g CO_2 and 0.400 g H_2O. You know that the molar mass of aspirin is between 170 and 190 g/mol. Reacting 1 mole of salicylic acid with 1 mole of acetic anhydride ($C_4H_6O_3$) gives you 1 mole of aspirin and 1 mole of acetic acid ($C_2H_4O_2$). Use this information to determine the molecular formula of salicylic acid.

117. Lanthanum was reacted with hydrogen in a given experiment to produce the nonstoichiometric compound $LaH_{2.90}$. Assuming that the compound contains H^-, La^{2+}, and La^{3+}, calculate the fraction of La^{2+} and La^{3+} present.

118. A 9.780-g gaseous mixture contains ethane (C_2H_6) and propane (C_3H_8). Complete combustion to form carbon dioxide and water requires 1.120 moles of oxygen gas. Calculate the mass percent of ethane in the original mixture.

119. Consider a mixture of potassium chloride and potassium nitrate that is 43.2% potassium by mass. What is the percent KCl by mass of the original mixture?

120. A 2.077-g sample of an element, which has an atomic mass between 40 and 55, reacts with oxygen to form 3.708 g of an oxide. Determine the formula of the oxide and identify the element.

121. Ammonia reacts with O_2 to form either $NO(g)$ or $NO_2(g)$ according to these unbalanced equations:

$$NH_3(g) + O_2(g) \longrightarrow NO(g) + H_2O(g)$$
$$NH_3(g) + O_2(g) \longrightarrow NO_2(g) + H_2O(g)$$

In a certain experiment, 2.00 moles of $NH_3(g)$ and 10.00 moles of $O_2(g)$ are contained in a closed flask. After the reaction is complete, 6.75 moles of $O_2(g)$ remains. Calculate the number of moles of $NO(g)$ in the product mixture. (*Hint:* You cannot do this problem by adding the balanced equations, because you cannot assume that the two reactions will occur with equal probability.)

122. A gas contains a mixture of $NH_3(g)$ and $N_2H_4(g)$, both of which react with $O_2(g)$ to form $NO_2(g)$ and $H_2O(g)$. The gaseous mixture (with an initial mass of 61.00 g) is reacted with 10.00 moles O_2, and after the reaction is complete, 4.062 moles of O_2 remains. Calculate the mass percent of $N_2H_4(g)$ in the original gaseous mixture.

Marathon Problems*

123. From the information that follows, determine the mass of substance C that will be formed if 45.0 g of substance A reacts with 23.0 g of substance B. (Assume that the reaction between A and B goes to completion.)
    a. Substance A is a gray solid that consists of an alkaline earth metal and carbon (37.5% by mass). It reacts with substance B to produce substances C and D. Forty million trillion formula units of A have a mass of 4.26 mg.
    b. 47.9 g of substance B contains 5.36 g of hydrogen and 42.5 g of oxygen.
    c. When 10.0 g of substance C is burned in excess oxygen, 33.8 g of carbon dioxide and 6.92 g of water are produced. A mass spectrum of substance C shows a parent molecular ion with a mass-to-charge ratio of 26.
    d. Substance D is the hydroxide of the metal in substance A.

124. Consider the following balanced chemical equation:

$$A + 5B \longrightarrow 3C + 4D$$

*From James H. Burness, "The Use of "Marathon" Problems as Effective Vehicles for the Presentation of General Chemistry Lectures," Journal of Chemical Education, 68(11).

a. Equal masses of A and B are reacted. Complete each of the following with either "A is the limiting reactant because ____"; "B is the limiting reactant because ____"; or "We cannot determine the limiting reactant because ____."
   i. If the molar mass of A is greater than the molar mass of B, then
   ii. If the molar mass of B is greater than the molar mass of A, then

b. The products of the reaction are carbon dioxide (C) and water (D). Compound A has a similar molar mass to carbon dioxide. Compound B is a diatomic molecule. Identify compound B and support your answer.

c. Compound A is a hydrocarbon that is 81.71% carbon by mass. Determine its empirical and molecular formulas.

4 chapter
Types of Chemical Reactions and Solution Stoichiometry

When zinc reacts with iodine, the heat produces a cloud of excess iodine.

OWL

Sign in to OWL at www.cengage.com/owl to view tutorials and simulations, develop problem-solving skills, and complete online homework assigned by your professor.

go Chemistry

Download mini lecture videos for key concept review and exam prep from OWL or purchase them from www.cengagebrain.com.

Much of the chemistry that affects each of us occurs among substances dissolved in water. For example, virtually all of the chemistry that makes life possible occurs in an aqueous environment. Also, various tests for illnesses involve aqueous reactions. Modern medical practice depends heavily on analyses of blood and other body fluids. In addition to the common tests for sugar, cholesterol, and iron, analyses for specific chemical markers allow detection of many diseases before more obvious symptoms occur.

Aqueous chemistry is also important in our environment. In recent years contamination of the groundwater by substances such as chloroform and nitrates has been widely publicized. Water is essential for life, and the maintenance of an ample supply of clean water is crucial to all civilization.

To understand the chemistry that occurs in such diverse places as the human body, the groundwater, the oceans, the local water treatment plant, your hair as you shampoo it, and so on, we must understand how substances dissolved in water react with one another.

However, before we can understand solution reactions, we need to discuss the nature of solutions in which water is the dissolving medium, or *solvent.* These solutions are called **aqueous solutions.** In this chapter we will study the nature of materials after they are dissolved in water and various types of reactions that occur among these substances. You will see that the procedures developed in Chapter 3 to deal with chemical reactions work very well for reactions that take place in aqueous solutions. To understand the types of reactions that occur in aqueous solutions, we must first explore the types of species present. This requires an understanding of the nature of water.

4.1 Water, the Common Solvent

Water is one of the most important substances on earth. It is, of course, crucial for sustaining the reactions that keep us alive, but it also affects our lives in many indirect ways. Water helps moderate the earth's temperature; it cools automobile engines, nuclear power plants, and many industrial processes; it provides a means of transportation on the earth's surface and a medium for the growth of a myriad of creatures we use as food; and much more.

One of the most valuable functions of water involves its ability to dissolve many different substances. For example, salt "disappears" when you sprinkle it into the water used to cook vegetables, as does sugar when you add it to your iced tea. In each case the disappearing substance is obviously still present—you can taste it. What happens when a solid dissolves? To understand this process, we need to consider the nature of water. Liquid water consists of a collection of H_2O molecules. An individual H_2O molecule is "bent" or V-shaped, with an H—O—H angle of about 105°:

H 105° H
O

The O—H bonds in the water molecule are covalent bonds formed by electron sharing between the oxygen and hydrogen atoms. However, the electrons of the bond are not shared equally between these atoms. For reasons we will discuss in later chapters, oxygen has a greater attraction for electrons than does hydrogen. If the electrons were shared equally between the two atoms, both would be electrically neutral because, on average, the number of electrons around each would equal the number of protons in that nucleus. However, because the oxygen atom has a greater attraction for electrons, the shared electrons tend to spend more time close to the oxygen than to either of the hydro-

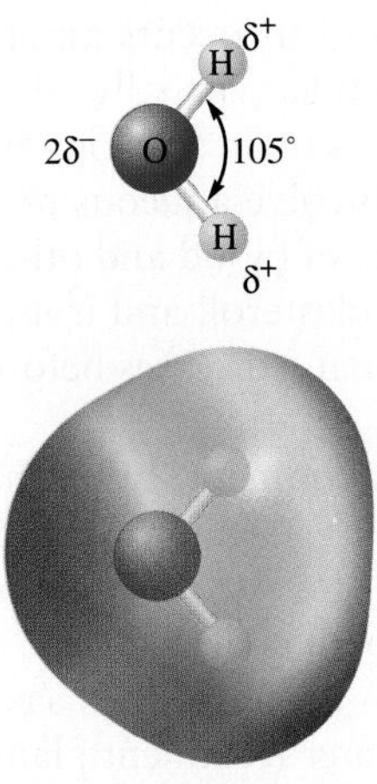

Figure 4.1
(top) The water molecule is polar. (bottom) The electrons in the water molecule are not shared equally between hydrogen and oxygen. This can be represented with a colored map of electrostatic potential. Red areas indicate high electron density, and blue areas represent low electron density. The colors in between indicate varying degrees of electron density.

gens. Thus the oxygen atom gains a slight excess of negative charge, and the hydrogen atoms become slightly positive. This is shown in Fig. 4.1, where δ (delta) indicates a *partial* charge *(less than one unit of charge)*. Because of this unequal charge distribution, water is said to be a **polar molecule.** It is this polarity that gives water its great ability to dissolve compounds.

A schematic of an ionic solid dissolving in water is shown in Fig. 4.2. Note that the "positive ends" of the water molecules are attracted to the negatively charged anions and that the "negative ends" are attracted to the positively charged cations. This process is called **hydration.** The hydration of its ions tends to cause a salt to "fall apart" in the water, or to dissolve. The strong forces present among the positive and negative ions of the solid are replaced by strong water–ion interactions.

It is very important to recognize that when ionic substances (salts) dissolve in water, they break up into the *individual* cations and anions. For instance, when ammonium nitrate (NH_4NO_3) dissolves in water, the resulting solution contains NH_4^+ and NO_3^- ions floating around independently. This process can be represented as

$$NH_4NO_3(s) \xrightarrow{H_2O(l)} NH_4^+(aq) + NO_3^-(aq)$$

where (*aq*) indicates that the ions are hydrated by unspecified numbers of water molecules.

The solubility of ionic substances in water varies greatly. For example, sodium chloride is quite soluble in water, whereas silver chloride (contains Ag^+ and Cl^- ions) is only very slightly soluble. The differences in the solubilities of ionic compounds in water typically depend on the relative affinities of the ions for each other (these forces hold the solid together) and the affinities of the ions for water molecules [which cause the solid to disperse (dissolve) in water]. Solubility is a complex issue that we will explore in much more detail in Chapter 17. However, the most important thing to remember at this point is that when an ionic solid does dissolve in water, the ions are dispersed and are assumed to move around independently.

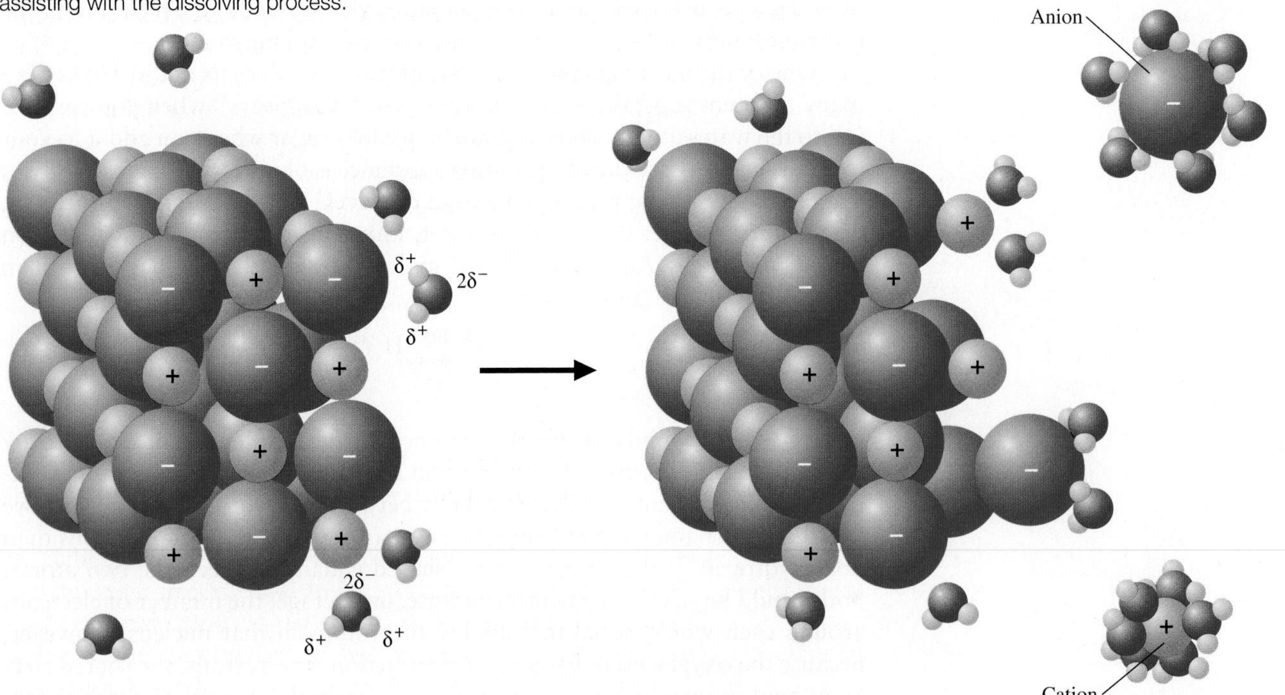

Figure 4.2
Polar water molecules interact with the positive and negative ions of a salt, assisting with the dissolving process.

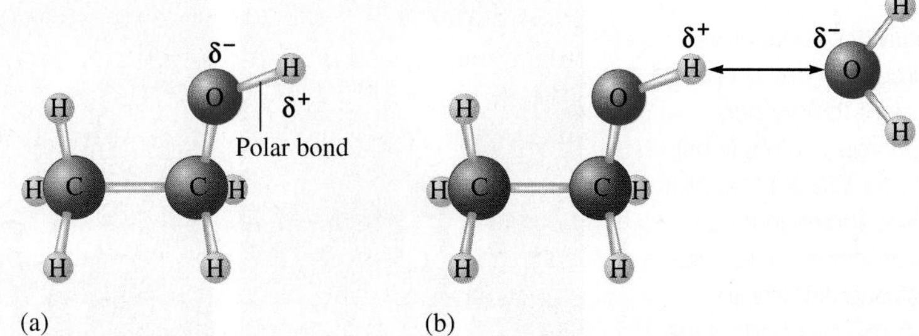

Figure 4.3
(a) The ethanol molecule contains a polar O—H bond similar to those in the water molecule. (b) The polar water molecule interacts strongly with the polar O—H bond in ethanol. This is a case of "like dissolving like."

Water also dissolves many nonionic substances. Ethanol (C_2H_5OH), for example, is very soluble in water. Wine, beer, and mixed drinks are aqueous solutions of alcohol and other substances. Why is ethanol so soluble in water? The answer lies in the structure of the alcohol molecule, which is shown in Fig. 4.3(a). The molecule contains a polar O—H bond like those in water, which makes it very compatible with water. The interaction of water with ethanol is represented in Fig. 4.3(b).

Many substances do not dissolve in water. Pure water will not, for example, dissolve animal fat because fat molecules are nonpolar and do not interact effectively with polar water molecules. In general, polar and ionic substances are expected to be more soluble in water than nonpolar substances. "Like dissolves like" is a useful rule for predicting solubility. We will explore the basis for this generalization when we discuss the details of solution formation in Chapter 17.

4.2 The Nature of Aqueous Solutions: Strong and Weak Electrolytes

Recall that a solution is a homogeneous mixture. It is the same throughout (the first sip of a cup of coffee is the same as the last), but its composition can be varied by changing the amount of dissolved substances (one can make weak or strong coffee). In this section we will consider what happens when a substance, the **solute,** is dissolved in liquid water, the **solvent.**

One useful property for characterizing a solution is its **electrical conductivity,** its ability to conduct an electric current. This characteristic can be checked conveniently by using an apparatus like the one shown in Fig. 4.4. If the solution in the container conducts electricity, the bulb lights. Some solutions conduct current very efficiently, and the bulb shines very brightly; these solutions contain **strong electrolytes.** Other solutions conduct only a small current, and the bulb glows dimly; these solutions contain **weak electrolytes.** Some solutions permit no current to flow, and the bulb remains unlit; these solutions contain **nonelectrolytes.**

The basis for the conductivity properties of solutions was first correctly identified by Svante Arrhenius, then a Swedish graduate student in physics, who carried out research on the nature of solutions at the University of Uppsala in the early 1880s. Arrhenius came to believe that the conductivity of solutions arose from the presence of ions, an idea that was at first scorned by the majority of the scientific establishment. However, in the late 1890s when atoms were found to contain charged particles, the ionic theory suddenly made sense and became widely accepted.

As Arrhenius postulated, the extent to which a solution can conduct an electric current depends directly on the number of ions present. Some materials, such as sodium chloride, readily produce ions in aqueous solution and are thus strong electrolytes. Other substances, such as acetic acid, produce relatively few ions when dissolved in water and are weak electrolytes. A third class

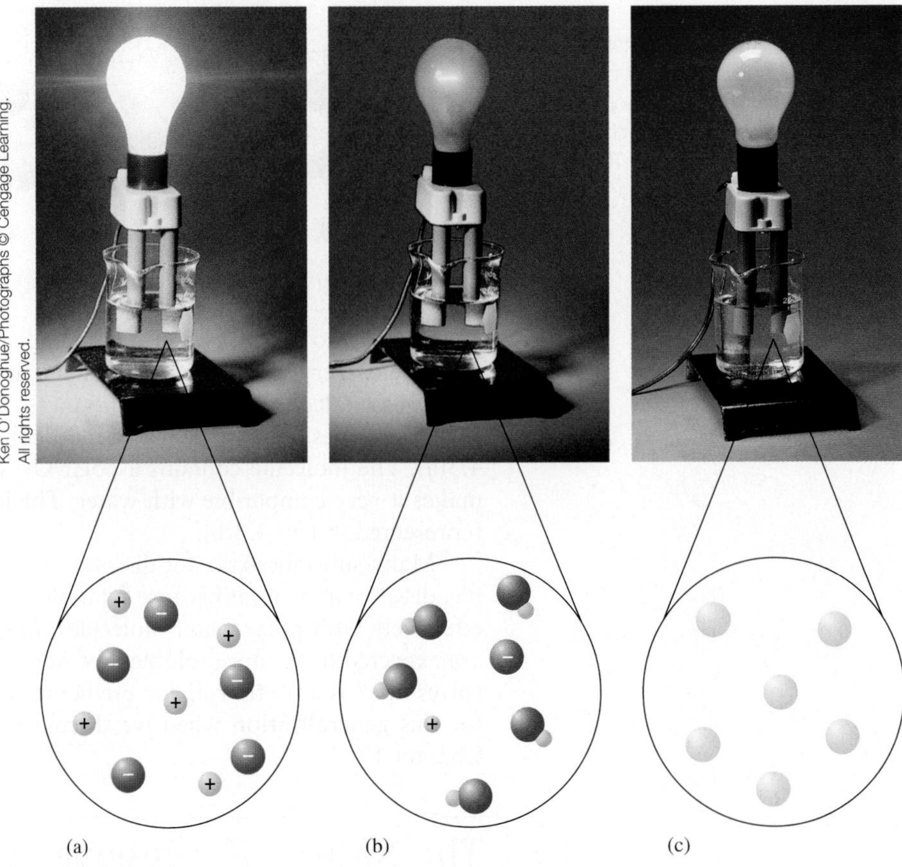

Figure 4.4
Electrical conductivity of aqueous solutions. The circuit will be completed and will allow current to flow only when there are charge carriers (ions) in the solution. *Note:* Water molecules are present but not shown in these pictures. (a) A hydrochloric acid solution, which is a strong electrolyte, contains ions that readily conduct the current and give a brightly lit bulb. (b) An acetic acid solution, which is a weak electrolyte, contains only a few ions and does not conduct as much current as a strong electrolyte. The bulb is only dimly lit. (c) A sucrose solution, which is a nonelectrolyte, contains no ions and does not conduct a current. The bulb remains unlit.

of materials, such as sugar, form virtually no ions when dissolved in water and are nonelectrolytes.

Strong Electrolytes

We will consider several classes of strong electrolytes: (1) soluble salts, (2) strong acids, and (3) strong bases.

As shown in Fig. 4.2, a salt consists of an array of cations and anions that separate and become hydrated when the salt dissolves. **Solubility** is usually measured in terms of the mass (grams) of solute that dissolves per given volume of solvent *or* in terms of the number of moles of solute that dissolve in a given volume of solution. Some salts, such as NaCl, KCl, and NH_4Cl, are very soluble in water. For example, approximately 357 grams of NaCl will dissolve in a liter of water at 25°C. On the other hand, many salts are only very slightly soluble in water; for example, silver chloride (AgCl) dissolves in water only to a slight extent (approximately 2×10^{-3} g/L at 25°C). We will consider only soluble salts at this point.

One of Arrhenius's most important discoveries concerned the nature of **acids.** Acidic behavior was first associated with the sour taste of citrus fruits. In fact, the word *acid* comes directly from the Latin word *acidus,* meaning "sour." The *mineral acids* sulfuric acid (H_2SO_4) and nitric acid (HNO_3), so named because they were originally obtained by the treatment of minerals, were discovered around 1300.

Acids were known to exist for hundreds of years before the time of Arrhenius, but no one had recognized their essential nature. In his studies of solutions, Arrhenius found that when the substances HCl, HNO_3, and H_2SO_4 were dissolved in water, they behaved as strong electrolytes. He postulated that this was the result of ionization reactions in water, for example:

$$HCl \xrightarrow{H_2O} H^+(aq) + Cl^-(aq)$$

$$HNO_3 \xrightarrow{H_2O} H^+(aq) + NO_3^-(aq)$$

$$H_2SO_4 \xrightarrow{H_2O} H^+(aq) + HSO_4^-(aq)$$

The Arrhenius definition of an acid: a substance that produces H^+ ions in solution.

Perchloric acid, $HClO_4(aq)$, is another strong acid.

Thus Arrhenius proposed that an *acid is a substance that produces H^+ ions (protons) when it is dissolved in water.*

Studies of conductivity show that when HCl, HNO_3, and H_2SO_4 are placed in water, *virtually every molecule* dissociates to give ions. These substances are strong electrolytes and are thus called **strong acids.** All three are very important chemicals, and much more will be said about them as we proceed. However, at this point the following facts are important:

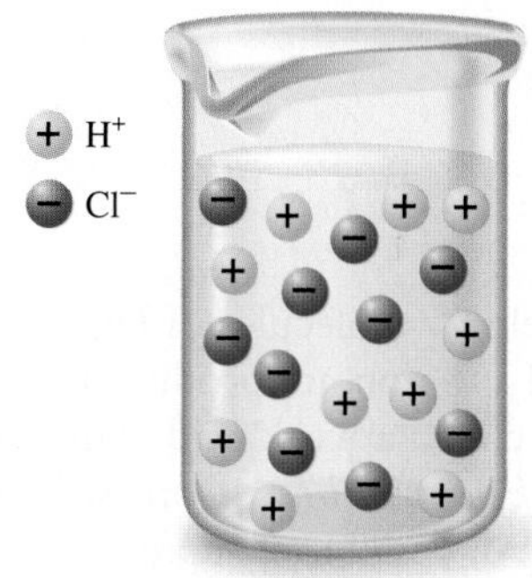

Figure 4.5
HCl(*aq*) is completely ionized.

Hydrochloric acid, nitric acid, and sulfuric acid are aqueous solutions and should be written in chemical equations as HCl(*aq*), $HNO_3(aq)$, and $H_2SO_4(aq)$, respectively, although they often appear without the (*aq*) symbol.

A strong acid is one that completely dissociates into its ions. Thus, if 100 molecules of HCl are dissolved in water, 100 H^+ ions and 100 Cl^- ions are produced. Virtually no HCl molecules exist in aqueous solution (see Fig. 4.5).

Sulfuric acid is a special case. The formula H_2SO_4 indicates that this acid can produce two H^+ ions per molecule when dissolved in water. However, only the first H^+ ion is completely dissociated. The second H^+ ion can be pulled off under certain conditions, which we will discuss later. Thus a solution of H_2SO_4 dissolved in water contains mostly H^+ ions and HSO_4^- ions.

Strong electrolytes dissociate completely in aqueous solution.

Another important class of strong electrolytes is the **strong bases,** soluble compounds containing the *hydroxide ion* (OH^-) that completely dissociate when dissolved in water. Solutions containing bases have a bitter taste and a slippery feel. The most common basic solutions are those produced when solid sodium hydroxide (NaOH) or potassium hydroxide (KOH) is dissolved in water to produce ions, as follows (Fig. 4.6):

$$NaOH(s) \xrightarrow{H_2O} Na^+(aq) + OH^-(aq)$$

$$KOH(s) \xrightarrow{H_2O} K^+(aq) + OH^-(aq)$$

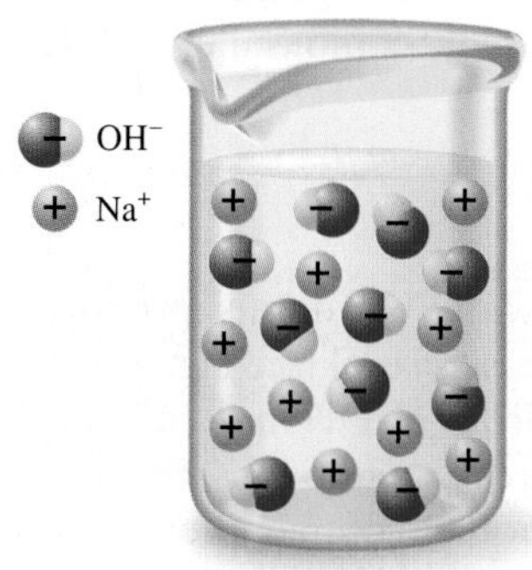

Figure 4.6
An aqueous solution of sodium hydroxide.

Weak Electrolytes

Weak electrolytes are substances that produce relatively few ions when dissolved in water, as shown in Fig. 4.4(b). The most common weak electrolytes are weak acids and weak bases.

The main acidic component of vinegar is acetic acid ($HC_2H_3O_2$). The formula is written to indicate that acetic acid has two chemically distinct types of hydrogen atoms. Formulas for acids are often written with the acidic hydrogen atom or atoms (any that will produce H^+ ions in solution) listed first. If any nonacidic hydrogens are present, they are written later in the formula. Thus the formula $HC_2H_3O_2$ indicates one acidic and three nonacidic hydrogen atoms. The dissociation reaction for acetic acid in water can be written as follows:

$$HC_2H_3O_2(aq) \xrightarrow{H_2O} H^+(aq) + C_2H_3O_2^-(aq)$$

Acetic acid is very different from the strong acids in that only about 1% of its molecules dissociate in aqueous solution (Fig. 4.7). Thus, when 100 molecules

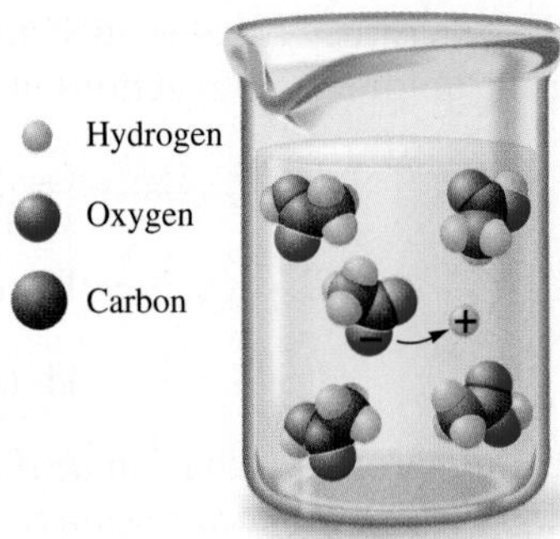

Figure 4.7
Acetic acid ($HC_2H_3O_2$) exists in water mostly as undissociated molecules. Only a small percentage of the molecules are ionized.

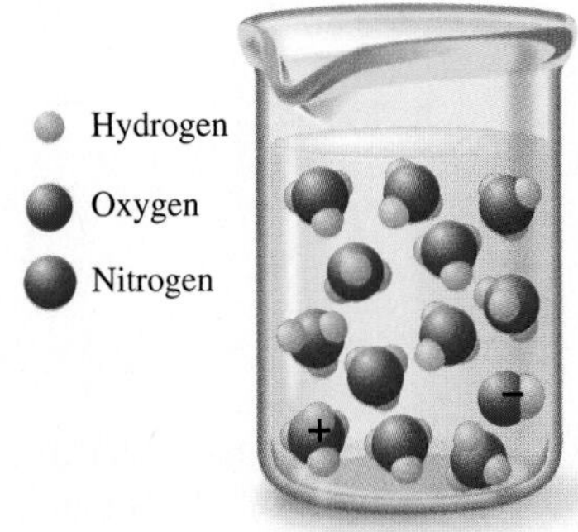

Figure 4.8
The reaction of NH_3 in water. The NH_4^+ and OH^- ions are formed by reaction of an NH_3 molecule with an H_2O molecule.

of $HC_2H_3O_2$ are dissolved in water, approximately 99 molecules of $HC_2H_3O_2$ remain intact, and only one H^+ ion and one $C_2H_3O_2^-$ ion are produced.

Weak electrolytes dissociate only to a small extent in aqueous solution.

Because acetic acid is a weak electrolyte, it is called a **weak acid.** Any acid, such as acetic acid, that *dissociates only to a slight extent in aqueous solution is called a weak acid.* We will explore weak acids in detail in Chapter 7.

The most common **weak base** is ammonia (NH_3). When ammonia is dissolved in water, it reacts as follows:

$$NH_3(aq) + H_2O(l) \longrightarrow NH_4^+(aq) + OH^-(aq)$$

The solution is *basic* since OH^- ions are produced. Ammonia is called a **weak base** because *the resulting solution is a weak electrolyte*—very few ions are present (Fig. 4.8). In fact, for every 100 molecules of NH_3 that are dissolved, only one NH_4^+ ion and one OH^- ion are produced; 99 molecules of NH_3 remain unreacted.

Nonelectrolytes

Nonelectrolytes are substances that dissolve in water but do not produce any ions [Fig. 4.4(c)]. An example of a nonelectrolyte is ethanol (see Fig. 4.3 for the structural formula). When ethanol dissolves, entire C_2H_5OH molecules are dispersed in the water. Since the molecules do not break up into ions, the resulting solution does not conduct an electric current. Another common nonelectrolyte is table sugar (sucrose, $C_{12}H_{22}O_{11}$), which is very soluble in water but produces no ions when it dissolves.

4.3 The Composition of Solutions

Chemical reactions often take place when two solutions are mixed. To perform stoichiometric calculations in such cases, we must know two things: (1) the *nature of the reaction,* which depends on the exact forms the chemicals take when dissolved, and (2) the *amounts of chemicals* present in the solutions, that is, the composition of each solution.

The composition of a solution can be described in many different ways, as we will see in Chapter 17. At this point we will consider only the most commonly used expression of concentration, **molarity** (M), which is defined as *moles of solute per volume of solution (expressed in liters):*

$$M = \text{molarity} = \frac{\text{moles of solute}}{\text{liters of solution}}$$

A solution that is 1.0 molar (written as 1.0 M) contains 1.0 mole of solute per liter of solution.

OWL INTERACTIVE EXAMPLE 4.1

Calculate the molarity of a solution prepared by bubbling 1.56 g of gaseous HCl into enough water to make 26.8 mL of solution.

Solution

■ What are we trying to solve?

We are asked to solve for the concentration of a solution in units of molarity.

■ What does this mean?

$$\text{Molarity} = \frac{\text{moles of solute}}{\text{liters of solution}}$$

We are given the volume of the solution in milliliters, and we can convert this to liters:

$$26.8 \text{ mL} \times \frac{1 \text{ L}}{1000 \text{ mL}} = 2.68 \times 10^{-2} \text{ L}$$

Thus we have

$$\text{Molarity} = \frac{\text{moles of solute}}{2.68 \times 10^{-2} \text{ L}}$$

To calculate the molarity, we need to determine the moles of solute, so we change the original question to *"How many moles of HCl are in 1.56 g?"* Recall from Chapter 3 that to do this we use the molar mass as follows:

$$1.56 \text{ g HCl} \times \frac{1 \text{ mol HCl}}{36.5 \text{ g HCl}} = 4.27 \times 10^{-2} \text{ mol HCl}$$

We can now determine the concentration as follows:

$$\text{Molarity} = \frac{4.27 \times 10^{-2} \text{ mol HCl}}{2.68 \times 10^{-2} \text{ L solution}} = 1.59\ M \text{ HCl}$$

Note that the description of a solution's composition may not accurately reflect the true chemical nature of the solution. Solution concentration is always given in terms of the form of the solute *before* it dissolves. For example, consider 1.0 liter of a solution labeled as 1.0 M NaCl. This solution was prepared by dissolving 1.0 mole of solid NaCl in enough water to make 1.0 liter of solution. The label 1.0 M does not mean that the solution contains 1.0 mole of NaCl units. Actually, the solution contains 1.0 mole of Na^+ ions and 1.0 mole of Cl^- ions.

Often we need to determine the number of moles of solute present in a given volume of a solution of known molarity. The procedure for doing so is easily derived from the definition of molarity:

$$M = \frac{\text{moles of solute}}{\text{liters of solution}}$$

$$\text{Liters of solution} \times \text{molarity} = \text{liters of solution} \times \frac{\text{moles of solute}}{\text{liters of solution}}$$

$$= \text{moles of solute}$$

OWL INTERACTIVE EXAMPLE 4.2

Calculate the number of moles of Cl^- ions in 1.75 L of 1.0×10^{-3} M $AlCl_3$.

Solution

■ What are we trying to solve?

We are looking for the number of moles of Cl^- ions, and we are given the volume of the solution and the concentration in molarity, along with the formula for the ionic compound.

We know that molarity is a ratio of moles of solute to liters of solution, or

$$M = \frac{\text{moles of solute}}{\text{liters of solution}}$$

Thus

$$\text{Moles} = M \times \text{volume}$$

This would give us the moles of $AlCl_3$ that would dissolve, but we are looking for the moles of Cl^- ions in solution.

When solid $AlCl_3$ dissolves, it produces ions as follows:

$$AlCl_3(s) \xrightarrow{H_2O} Al^{3+}(aq) + 3Cl^-(aq)$$

Thus a 1.0×10^{-3} M $AlCl_3$ solution contains 1.0×10^{-3} M Al^{3+} ions and 3.0×10^{-3} M Cl^- ions.

To calculate the moles of Cl^- ions in 1.75 L of the 1.0×10^{-3} M $AlCl_3$ solution, we must multiply the volume by the molarity:

$$1.75 \text{ L solution} \times 3.0 \times 10^{-3}\ M\ Cl^-$$

$$= 1.75 \text{ L solution} \times \frac{3.0 \times 10^{-3} \text{ mol } Cl^-}{\text{L solution}}$$

$$= 5.3 \times 10^{-3} \text{ mol } Cl^-$$

OWL INTERACTIVE EXAMPLE 4.3

Typical blood serum is about 0.14 M NaCl. What volume of blood contains 1.0 mg of NaCl?

Solution

■ What are we trying to solve?

We want to determine the volume of blood containing 1.0 mg of NaCl, and we know the concentration of NaCl in molarity. Thus we are looking for a conversion between volume and mass.

We know that molarity is a ratio of moles of solute to liters of solution, so we have a conversion between volume and moles:

$$M = \frac{\text{moles of solute}}{\text{liters of solution}}$$

Thus

$$\text{Volume} = \frac{\text{moles of solute}}{M}$$

We also have a conversion between mass and moles (molar mass). Thus we can convert mass to moles (using molar mass) and moles to volume (using molarity):

We must first determine the number of moles represented by 1.0 mg of NaCl:

$$1.0 \text{ mg NaCl} \times \frac{1 \text{ g NaCl}}{1000 \text{ mg NaCl}} \times \frac{1 \text{ mol NaCl}}{58.45 \text{ g NaCl}} = 1.7 \times 10^{-5} \text{ mol NaCl}$$

Next, we must determine what volume of 0.14 *M* NaCl solution contains 1.7×10^{-5} mole of NaCl. There is some volume *(V)* that when multiplied by the molarity of this solution yields 1.7×10^{-5} mole of NaCl. That is,

$$V \times \frac{0.14 \text{ mol NaCl}}{\text{L solution}} = 1.7 \times 10^{-5} \text{ mol NaCl}$$

Solving for the volume gives

$$V = \frac{1.7 \times 10^{-5} \text{ mol NaCl}}{\dfrac{0.14 \text{ mol NaCl}}{\text{L solution}}} = 1.2 \times 10^{-4} \text{ L solution}$$

Thus 0.12 mL of blood contains 1.7×10^{-5} mole of NaCl, or 1.0 mg of NaCl.

A **standard solution** is a solution *whose concentration is accurately known.* Standard solutions, often used in chemical analysis, can be prepared as shown in Fig. 4.9 and in Example 4.4.

OWL INTERACTIVE EXAMPLE 4.4

To analyze the alcohol content of a certain wine, a chemist needs 1.00 L of an aqueous 0.200 *M* $K_2Cr_2O_7$ (potassium dichromate) solution. How much solid $K_2Cr_2O_7$ must be weighed out to make this solution?

Solution

■ What are we trying to solve?

We are asked to determine the mass of K_2CrO_7 solute required to make 1.00 L of a 0.200 *M* solution. Knowing that molarity is a ratio of moles of solute to

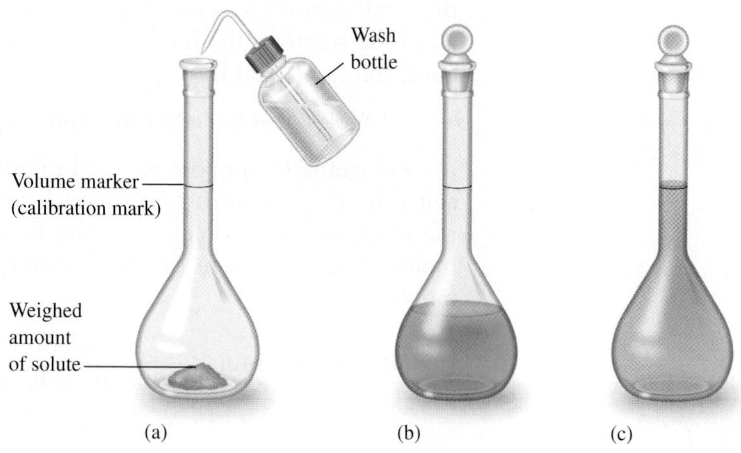

Figure 4.9
Steps involved in the preparation of a standard aqueous solution. (a) Put a weighed amount of a substance (the solute) into the volumetric flask, and add a small quantity of water. (b) Dissolve the solid in the water by gently swirling the flask *(with the stopper in place).* (c) Add more water (with gentle swirling) until the level of the solution just reaches the mark etched on the neck of the flask. Then mix the solution thoroughly by inverting the flask several times.

liters of solution, we can use molarity and volume to determine the moles of solute.

$$M = \frac{\text{moles of solute}}{\text{liters of solution}}$$

Thus

$$\text{Moles} = M \times \text{volume}$$

However, we are asked for mass, so we will need to convert the moles to a mass, and we can do this using the molar mass of the solute.

First, determine the moles of $K_2Cr_2O_7$ required:

$$1.00 \text{ L solution} \times \frac{0.200 \text{ mol } K_2Cr_2O_7}{\text{L solution}} = 0.200 \text{ mol } K_2Cr_2O_7$$

This amount can be converted to grams by using the molar mass of $K_2Cr_2O_7$:

$$0.200 \text{ mol } K_2Cr_2O_7 \times \frac{294.2 \text{ g } K_2Cr_2O_7}{\text{mol } K_2Cr_2O_7} = 58.8 \text{ g } K_2Cr_2O_7$$

Thus, to make 1.00 L of 0.200 M $K_2Cr_2O_7$, the chemist must weigh out 58.8 g of $K_2Cr_2O_7$, put it in a 1.00-L volumetric flask, and add water up to the mark on the flask.

Note: In looking back at the solutions to Examples 4.1 through 4.4, you should notice that the problems are solved similarly. For example, in all cases we need to know what molarity means and how to manipulate the formula for concentration. The problems vary in the details. In Examples 4.1, 4.3, and 4.4, we need to use molar mass = mass/mol. In Example 4.2, we need to understand how an ionic solid dissolves. As discussed in Section 3.4, we need to ask questions that allow the problem to guide us, but we also need a bank of knowledge to draw from.

Dilution

To save time and space in the laboratory, routinely used solutions are often purchased or prepared in concentrated form (these are called *stock solutions*). In a process called **dilution,** water is then added to achieve the molarity desired for a particular solution. For example, the common acids are purchased as concentrated solutions and diluted as needed. A typical dilution calculation involves determining how much water must be added to an amount of stock solution to achieve a solution of the desired concentration. The key to doing these calculations is to remember that since only water is added in the dilution, all of the solute in the final dilute solution must come from the concentrated stock solution. That is,

Dilution with water doesn't alter the number of moles of solute present.

$$\text{Moles of solute after dilution} = \text{moles of solute before dilution}$$

For example, suppose we need to prepare 500. milliliters of 1.00 M acetic acid ($HC_2H_3O_2$) from a 17.5 M stock solution of acetic acid. What volume of the stock solution is required? The first step is to determine the number of moles of acetic acid in the final solution by multiplying the volume by the molarity:

$$500. \text{ mL solution} \times \frac{1 \text{ L solution}}{1000 \text{ mL solution}} \times \frac{1.00 \text{ mol } HC_2H_3O_2}{\text{L solution}}$$
$$= 0.500 \text{ mol } HC_2H_3O_2$$

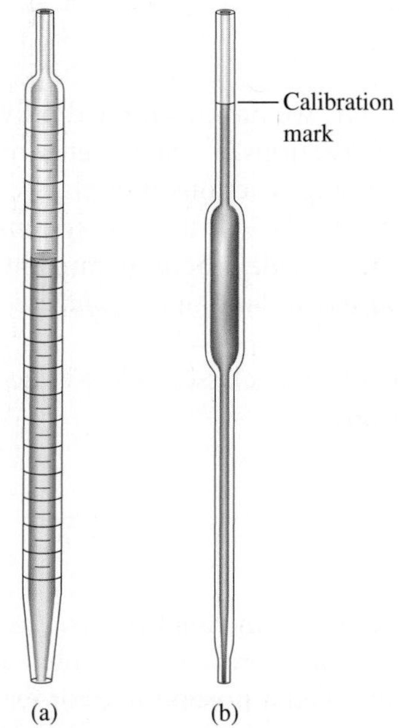

Figure 4.10
(a) A measuring pipet is graduated and can be used to measure various volumes of liquid accurately. (b) A volumetric pipet is designed to measure *one* volume accurately. When filled to the calibration mark, it delivers the volume indicated on the pipet.

Thus we need to use a volume of 17.5 *M* acetic acid that contains 0.500 mole of $HC_2H_3O_2$. That is,

$$V \times \frac{17.5 \text{ mol } HC_2H_3O_2}{\text{L solution}} = 0.500 \text{ mol } HC_2H_3O_2$$

Solving for *V* gives

$$V = \frac{0.500 \text{ mol } HC_2H_3O_2}{\dfrac{17.5 \text{ mol } HC_2H_3O_2}{\text{L solution}}} = 0.0286 \text{ L, or } 28.6 \text{ mL solution}$$

Thus, to make 500. milliliters of a 1.00 *M* acetic acid solution, we can take 28.6 milliliters of 17.5 *M* acetic acid and dilute it to a total volume of 500. milliliters.

A dilution procedure typically involves two types of glassware: a pipet and a volumetric flask. A pipet is a device for accurately measuring and transferring a given volume of solution. There are two common types of pipets: *measuring pipets* and *volumetric pipets,* as shown in Fig. 4.10. Measuring pipets are used to measure out volumes for which a volumetric pipet is not available. For example, we would use a measuring pipet as shown in Fig. 4.11 to deliver 28.6 milliliters of 17.5 *M* acetic acid into a 500-milliliter volumetric flask and then add water to the mark to perform the dilution described above.

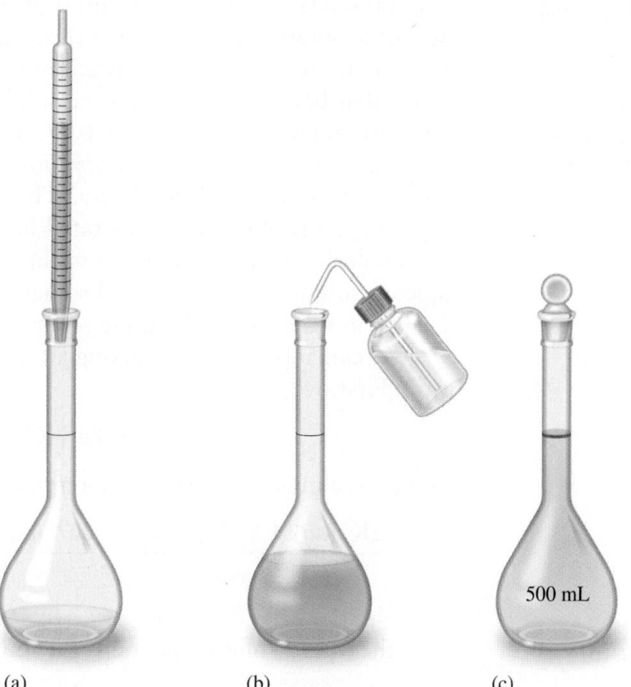

Figure 4.11
(a) A measuring pipet is used to transfer 28.7 mL of 17.4 *M* acetic acid solution to a volumetric flask. (b) Water is added to the flask to the calibration mark. (c) The resulting solution is 1.00 *M* acetic acid.

4.4 Types of Chemical Reactions

Types of Solution Reactions

Precipitation reactions
Acid–base reactions
Oxidation–reduction reactions

Although we have considered many reactions so far, we have examined only a tiny fraction of the millions of possible chemical reactions. To make sense of all these reactions, we need some system for grouping reactions into classes. Although there are many different ways to do this, we will use the system most commonly used by practicing chemists. They divide reactions into the following groups: *precipitation reactions, acid–base reactions,* and *oxidation–reduction reactions.*

Virtually all reactions can be placed into one of these classes. We will define and illustrate each type in the following sections.

4.5 Precipitation Reactions

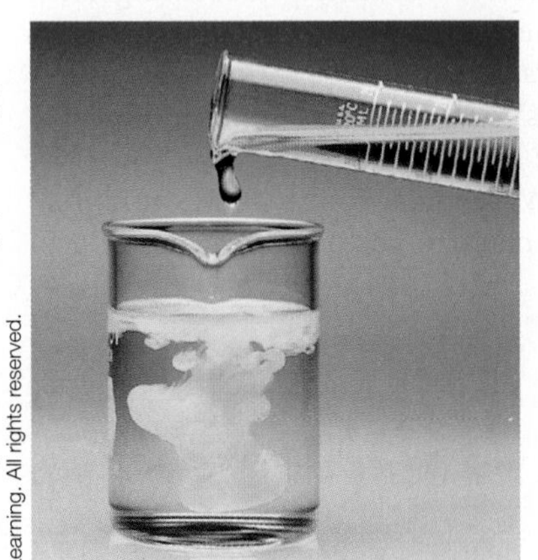

Figure 4.12
When yellow aqueous potassium chromate is added to a colorless barium nitrate solution, yellow barium chromate precipitates.

When two solutions are mixed, an insoluble substance sometimes forms; that is, a solid forms and separates from the solution. Such a reaction is called a **precipitation reaction,** and the solid that forms is called a **precipitate.** For example, a precipitation reaction occurs when an aqueous solution of potassium chromate [$K_2CrO_4(aq)$], which is yellow, is mixed with a colorless aqueous solution containing barium nitrate [$Ba(NO_3)_2(aq)$]. As shown in Fig. 4.12, when these solutions are mixed, a yellow solid forms. What is the equation that describes this chemical change? To write the equation, we must know the identities of the reactants and products. The reactants have already been described: $K_2CrO_4(aq)$ and $Ba(NO_3)_2(aq)$. Is there some way we can predict the identities of the products? In particular, what is the yellow solid?

The best way to predict the identity of this solid is to think carefully about what products are possible. To do so, we need to know what species are present in the solution formed when the reactant solutions are mixed. First, let's think about the nature of each reactant solution. The designation $Ba(NO_3)_2(aq)$ means that barium nitrate (a white solid) has been dissolved in water. Notice that barium nitrate contains the Ba^{2+} and NO_3^- ions. *Remember: In virtually every case, when a solid containing ions dissolves in water, the ions separate and move around independently.* That is, $Ba(NO_3)_2(aq)$ does not contain $Ba(NO_3)_2$ units; it contains separated Ba^{2+} and NO_3^- ions (Fig. 4.13).

Similarly, since solid potassium chromate contains K^+ and CrO_4^{2-} ions, an aqueous solution of potassium chromate (which is prepared by dissolving solid K_2CrO_4 in water) contains these separated ions (Fig. 4.13).

When ionic compounds dissolve in water, the *resulting solution contains the separated ions.*

We can represent the mixing of $K_2CrO_4(aq)$ and $Ba(NO_3)_2(aq)$ in two ways. First, we can write

$$K_2CrO_4(aq) + Ba(NO_3)_2(aq) \longrightarrow \text{products}$$

However, a much more accurate representation is

$$\underbrace{2K^+(aq) + CrO_4^{2-}(aq)}_{\substack{\text{The ions in} \\ K_2CrO_4(aq)}} + \underbrace{Ba^{2+}(aq) + 2NO_3^-(aq)}_{\substack{\text{The ions in} \\ Ba(NO_3)_2(aq)}} \longrightarrow \text{products}$$

Thus the mixed solution contains the ions

$$K^+,\ CrO_4^{2-},\ Ba^{2+},\ \text{and}\ NO_3^-$$

How can some or all of these ions combine to form the yellow solid observed when the original solutions are mixed? This is not an easy question to answer. In fact, predicting the products of a chemical reaction is one of the hardest things a beginning chemistry student is asked to do. Even an experi-

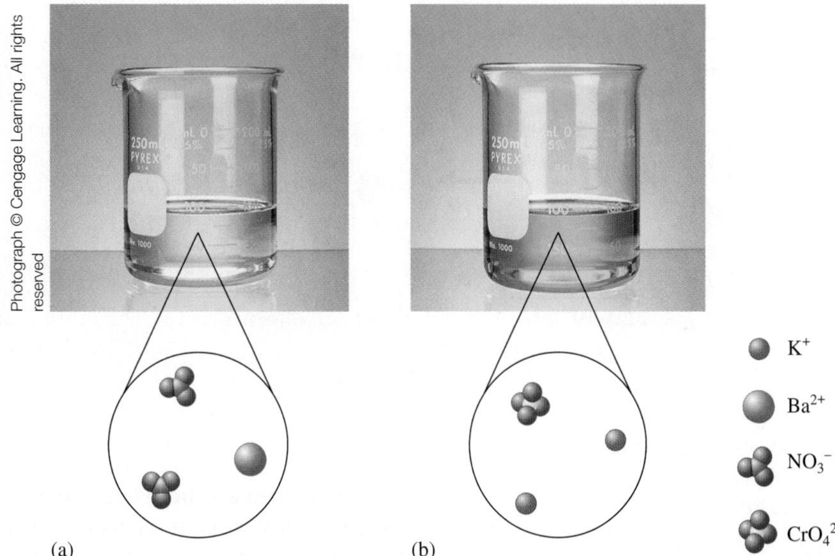

Figure 4.13
Reactant solutions: (a) $Ba(NO_3)_2(aq)$ and (b) $K_2CrO_4(aq)$.

enced chemist, when confronted with a new reaction, is often not sure what will happen. The chemist tries to think of the various possibilities, considers the likelihood of each possibility, and then makes a prediction (an educated guess). Only after identifying each product *experimentally* is the chemist sure what reaction has taken place. However, an educated guess is useful because it provides a place to start. It tells us what kinds of products we are most likely to find.

We already know some things that will help us predict the products:

1. When ions form a solid compound, the compound must have a zero net charge. Thus the products of this reaction must contain *both anions and cations*. For example, K^+ and Ba^{2+} could not combine to form the solid, nor could CrO_4^{2-} and NO_3^-.
2. Most ionic materials contain only two types of ions: one type of cation and one type of anion (for example, NaCl, KOH, Na_2SO_4, K_2CrO_4, $Co(NO_3)_2$, NH_4Cl, and Na_2CO_3).

The possible combinations of a given cation and a given anion from the list of ions K^+, CrO_4^{2-}, Ba^{2+}, and NO_3^- are

$$K_2CrO_4,\ KNO_3,\ BaCrO_4,\ \text{and}\ Ba(NO_3)_2$$

Which of these possibilities is most likely to represent the yellow solid? We know it's not K_2CrO_4 or $Ba(NO_3)_2$. They are the reactants. They were present (dissolved) in the separate solutions that were mixed. The only real possibilities for the solid that formed are

$$KNO_3 \quad \text{and} \quad BaCrO_4$$

To decide which of these possibilities most likely represents the yellow solid, we need more facts. An experienced chemist knows that the K^+ ion and the NO_3^- ion are both colorless. Thus, if the solid is KNO_3, it should be white, not yellow. On the other hand, the CrO_4^{2-} ion is yellow [note in Fig. 4.12 that $K_2CrO_4(aq)$ is yellow]. Thus the yellow solid is almost certainly $BaCrO_4$. Further tests show that this is the case.

So far we have determined that one product of the reaction between $K_2CrO_4(aq)$ and $Ba(NO_3)_2(aq)$ is $BaCrO_4(s)$, but what happened to the K^+ and NO_3^- ions? The answer is that these ions are left dissolved in the solution. That

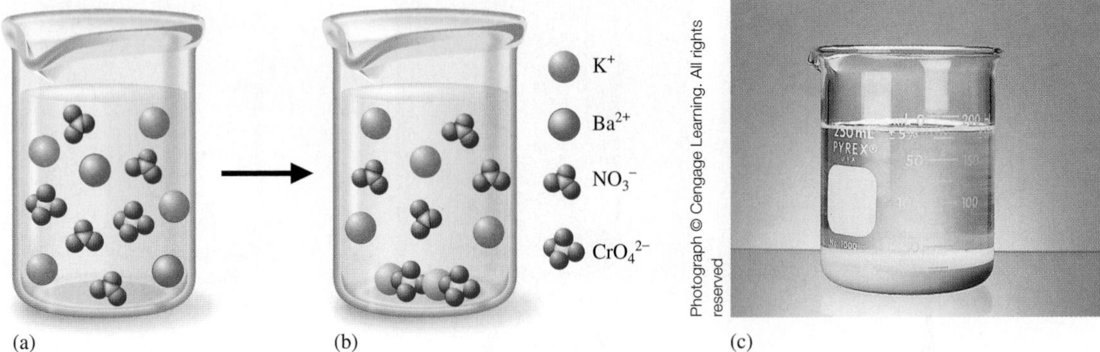

Figure 4.14
The reaction of $K_2CrO_4(aq)$ and $Ba(NO_3)_2(aq)$. (a) The molecular-level "picture" of the mixed solution before any reaction has occurred. (b) The molecular-level "picture" of the solution after the reaction has occurred to form $BaCrO_4(s)$. (c) A photo of the solution after the reaction has occurred, showing the solid $BaCrO_4$ on the bottom.

is, KNO_3 does not form a solid when the K^+ and NO_3^- ions are present in this much water. In other words, if we took the white solid, $KNO_3(s)$, and put it in the same quantity of water as is present in the mixed solution, it would dissolve. Thus, when we mix $K_2CrO_4(aq)$ and $Ba(NO_3)_2(aq)$, $BaCrO_4(s)$ forms, but KNO_3 is left behind in solution [we write it as $KNO_3(aq)$]. This reaction is illustrated in Fig. 4.14. Therefore, the equation for this precipitation reaction is

$$K_2CrO_4(aq) + Ba(NO_3)_2(aq) \longrightarrow BaCrO_4(s) + 2KNO_3(aq)$$

If we removed the solid $BaCrO_4$ by filtration and then evaporated the water, the white solid, KNO_3, would be obtained.

Now let's consider another example. When an aqueous solution of silver nitrate is added to an aqueous solution of potassium chloride, a white precipitate forms. We can represent what we know so far as

$$AgNO_3(aq) + KCl(aq) \longrightarrow \text{unknown white solid}$$

Doing chemistry requires both understanding ideas and remembering facts.

Remembering that when ionic substances dissolve in water, the ions separate, we can write

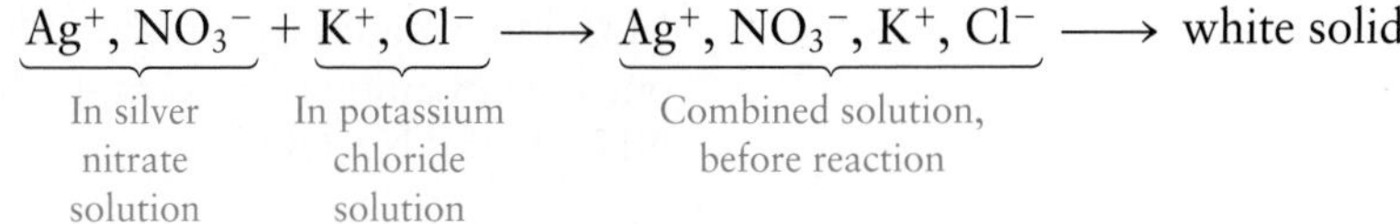

$$\underbrace{Ag^+, NO_3^-}_{\text{In silver nitrate solution}} + \underbrace{K^+, Cl^-}_{\text{In potassium chloride solution}} \longrightarrow \underbrace{Ag^+, NO_3^-, K^+, Cl^-}_{\text{Combined solution, before reaction}} \longrightarrow \text{white solid}$$

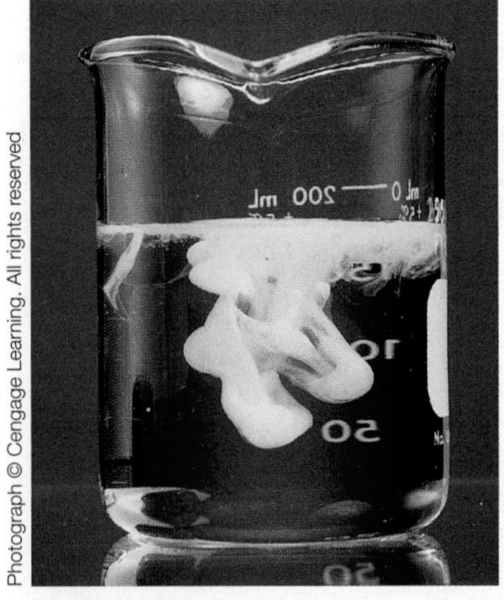

Figure 4.15
Precipitation of silver chloride by mixing solutions of silver nitrate and potassium chloride. The K^+ and NO_3^- ions remain in solution.

Since we know that the white solid must contain both positive and negative ions, the possible compounds that can be assembled from this collection of ions are

$$AgNO_3, KCl, AgCl, \text{and } KNO_3$$

Since $AgNO_3$ and KCl are the substances dissolved in the reactant solutions, we know that they do not represent the white solid product. The only real possibilities are

$$AgCl \quad \text{and} \quad KNO_3$$

From the example considered above, we know that KNO_3 is quite soluble in water. Thus solid KNO_3 will not form when the reactant solutions are mixed. The product must be $AgCl(s)$ (which can be proved by experiment). The equation for the reaction now can be written

$$AgNO_3(aq) + KCl(aq) \longrightarrow AgCl(s) + KNO_3(aq)$$

Figure 4.15 shows the result of mixing aqueous solutions of $AgNO_3$ and KCl. Figure 4.16 provides a visualization of the reaction.

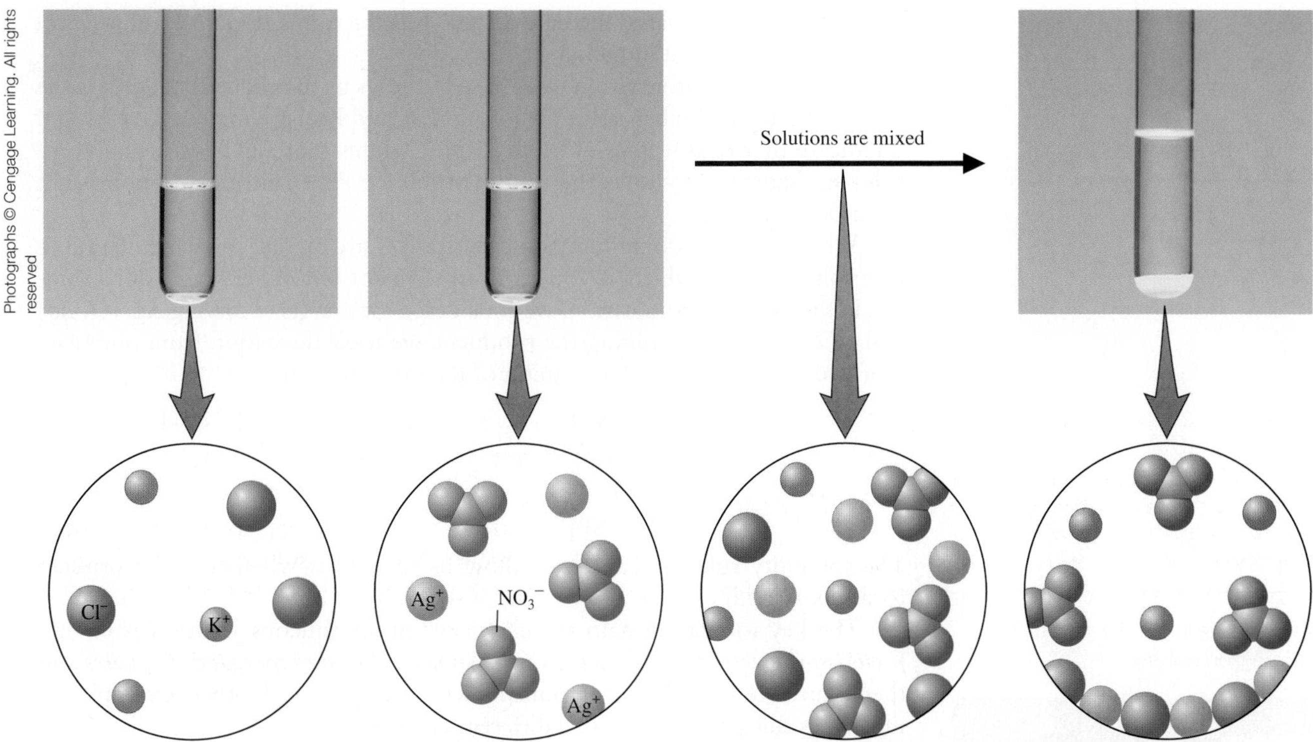

Figure 4.16
Photos and accompanying molecular-level representations illustrating the reaction of KCl(*aq*) with $AgNO_3$(*aq*) to form AgCl(*s*). Note that it is not possible to have a photo of the mixed solution before the reaction occurs because it is an imaginary step that we use to help visualize the reaction. Actually, the reaction occurs immediately when the two solutions are mixed.

Notice that to do these two examples, we had to know both concepts (solids always have a zero net charge) and facts (KNO_3 is very soluble in water, the CrO_4^{2-} is yellow, and so on).

Predicting the identity of the solid product in a precipitation reaction requires knowledge of the solubilities of common ionic substances. As an aid in predicting the products of precipitation reactions, some simple solubility rules are given in Table 4.1. You should memorize these rules.

The phrase "slightly soluble" used in the solubility rules in Table 4.1 means that the tiny amount of solid that dissolves is not noticeable. The solid

Table 4.1

Simple Rules for Solubility of Salts in Water

1. Most nitrate (NO_3^-) salts are soluble.
2. Most salts of Na^+, K^+, and NH_4^+ are soluble.
3. Most chloride salts are soluble. Notable exceptions are AgCl, $PbCl_2$, and Hg_2Cl_2.
4. Most sulfate salts are soluble. Notable exceptions are $BaSO_4$, $PbSO_4$, and $CaSO_4$.
5. Most hydroxide salts are only slightly soluble. The important soluble hydroxides are NaOH, KOH, and $Ca(OH)_2$ (marginally soluble).
6. Most sulfide (S^{2-}), carbonate (CO_3^{2-}), and phosphate (PO_4^{3-}) salts are only slightly soluble.

appears to be insoluble to the naked eye. Thus the terms *insoluble* and *slightly soluble* are often used interchangeably.

Note that the information in Table 4.1 allows us to predict that AgCl is the white solid formed when solutions of $AgNO_3$ and KCl are mixed; Rules 1 and 2 indicate that KNO_3 is soluble, and Rule 3 states that AgCl is (virtually) insoluble. Figure 4.15 shows the results of mixing silver nitrate and potassium chloride solutions.

When solutions containing ionic substances are mixed, it will be helpful in determining the products if you think in terms of *ion interchange.* For example, in the preceding discussion, we considered the results of mixing $AgNO_3(aq)$ and $KCl(aq)$. In determining the products, we took the cation from one reactant and combined it with the anion of the other reactant:

$$Ag^+ + NO_3^- + K^+ + Cl^- \longrightarrow$$

Possible solid products

The solubility rules in Table 4.1 allow us to predict whether either product forms as a solid.

Focus on the ions in solution before any reaction occurs.

The key to dealing with the chemistry of an aqueous solution is to first *focus on the actual components of the solution before any reaction occurs* and then figure out how those components will react with each other. Example 4.5 illustrates this process for three different reactions.

The precipitation of lead(II) sulfate by mixing solutions of lead(II) nitrate and sodium sulfate.

OWL INTERACTIVE EXAMPLE 4.5

Using the solubility rules in Table 4.1, predict what will happen when the following pairs of solutions are mixed.

a. $KNO_3(aq)$ and $BaCl_2(aq)$
b. $Na_2SO_4(aq)$ and $Pb(NO_3)_2(aq)$
c. $KOH(aq)$ and $Fe(NO_3)_3(aq)$

Solution

a. $KNO_3(aq)$ stands for an aqueous solution obtained by dissolving solid KNO_3 in water to form a solution containing the hydrated ions $K^+(aq)$ and $NO_3^-(aq)$. Likewise, $BaCl_2(aq)$ is a solution formed by dissolving solid $BaCl_2$ in water to produce $Ba^{2+}(aq)$ and $Cl^-(aq)$. When these two solutions are mixed, the resulting solution contains the ions K^+, NO_3^-, Ba^{2+}, and Cl^-. All will be hydrated, but (aq) is omitted for simplicity. To look for possible solid products, combine the cation from one reactant with the anion from the other:

$$K^+ + NO_3^- + Ba^{2+} + Cl^- \longrightarrow$$

Possible solid products

Note from Table 4.1 that the rules predict that both KCl and $Ba(NO_3)_2$ are soluble in water. Thus no precipitate will form when $KNO_3(aq)$ and $BaCl_2(aq)$ are mixed. All the ions will remain dissolved in the solution. No reaction occurs.

b. Using the same procedures as in part a, we find that the ions present in the combined solution before any reaction occurs are Na^+, SO_4^{2-}, Pb^{2+}, and NO_3^-. The possible salts that could form precipitates are

$$Na^+ + SO_4^{2-} + Pb^{2+} + NO_3^- \longrightarrow$$

Solid $Fe(OH)_3$ forms when aqueous KOH and $Fe(NO_3)_3$ are mixed.

The compound $NaNO_3$ is soluble, but $PbSO_4$ is insoluble (see Rule 4 in Table 4.1). When these solutions are mixed, $PbSO_4$ will precipitate from the solution. The balanced equation is

$$Na_2SO_4(aq) + Pb(NO_3)_2(aq) \longrightarrow PbSO_4(s) + 2NaNO_3(aq)$$

c. The combined solution (before any reaction occurs) contains the ions K^+, OH^-, Fe^{3+}, and NO_3^-. The salts that might precipitate are KNO_3 and $Fe(OH)_3$. The solubility rules in Table 4.1 indicate that both K^+ and NO_3^- salts are soluble. However, $Fe(OH)_3$ is only slightly soluble (Rule 5) and hence will precipitate. The balanced equation is

$$3KOH(aq) + Fe(NO_3)_3(aq) \longrightarrow Fe(OH)_3(s) + 3KNO_3(aq)$$

4.6 Describing Reactions in Solution

In this section we will consider the types of equations used to represent reactions in solution. For example, when we mix aqueous potassium chromate with aqueous barium nitrate, a reaction occurs to form a precipitate ($BaCrO_4$) and dissolved potassium nitrate. So far we have written the **molecular equation** for this reaction:

$$K_2CrO_4(aq) + Ba(NO_3)_2(aq) \longrightarrow BaCrO_4(s) + 2KNO_3(aq)$$

Although this equation shows the reactants and products of the reaction, it does not give a very clear picture of what actually occurs in solution. As we have seen, aqueous solutions of potassium chromate, barium nitrate, and potassium nitrate contain the individual ions, not molecules, as is implied by the molecular equation. Thus the **complete ionic equation**

$$2K^+(aq) + CrO_4^{2-}(aq) + Ba^{2+}(aq) + 2NO_3^-(aq) \longrightarrow BaCrO_4(s) + 2K^+(aq) + 2NO_3^-(aq)$$

A strong electrolyte is a substance that completely breaks apart into ions when dissolved in water.

better represents the actual forms of the reactants and products in solution. *In a complete ionic equation, all substances that are strong electrolytes are represented as ions.*

The complete ionic equation reveals that only some of the ions participate in the reaction. The K^+ and NO_3^- ions are present in solution both before and after the reaction. Ions such as these that do not participate directly in a reaction in solution are called **spectator ions.** The ions that participate in this reaction are the Ba^{2+} and CrO_4^{2-} ions, which combine to form solid $BaCrO_4$:

$$Ba^{2+}(aq) + CrO_4^{2-}(aq) \longrightarrow BaCrO_4(s)$$

Net ionic equations include only those components that undergo changes in the reaction.

This equation, called the **net ionic equation,** includes only those solution components directly involved in the reaction. Chemists usually write the net ionic equation for a reaction in solution because it gives the actual forms of the reactants and products and includes only the species that undergo a change.

EXAMPLE 4.6

For each of the following reactions, write the molecular equation, the complete ionic equation, and the net ionic equation.

a. Aqueous potassium chloride is added to aqueous silver nitrate to form a silver chloride precipitate plus aqueous potassium nitrate.

b. Aqueous potassium hydroxide is mixed with aqueous iron(III) nitrate to form a precipitate of iron(III) hydroxide and aqueous potassium nitrate.

Solution

a. *Molecular:*

$$KCl(aq) + AgNO_3(aq) \longrightarrow AgCl(s) + KNO_3(aq)$$

Complete ionic (remember that any ionic compound dissolved in water will be present as the separated ions):

$$K^+(aq) + Cl^-(aq) + Ag^+(aq) + NO_3^-(aq) \longrightarrow$$

↑ Spectator ion (under K^+) ↑ Spectator ion (under NO_3^-)

$$AgCl(s) + K^+(aq) + NO_3^-(aq)$$

↑ Solid, not written as separate ions (under AgCl) ↑ Spectator ion (under K^+) ↑ Spectator ion (under NO_3^-)

Net ionic: Canceling the spectator ions,

$$\cancel{K^+}(aq) + Cl^-(aq) + Ag^+(aq) + \cancel{NO_3^-}(aq) \longrightarrow AgCl(s) + \cancel{K^+}(aq) + \cancel{NO_3^-}(aq)$$

gives the following net ionic equation:

$$Cl^-(aq) + Ag^+(aq) \longrightarrow AgCl(s)$$

b. *Molecular:*

$$3KOH(aq) + Fe(NO_3)_3(aq) \longrightarrow Fe(OH)_3(s) + 3KNO_3(aq)$$

Complete ionic:

$$3K^+(aq) + 3OH^-(aq) + Fe^{3+}(aq) + 3NO_3^-(aq) \longrightarrow Fe(OH)_3(s) + 3K^+(aq) + 3NO_3^-(aq)$$

Net ionic:

$$3OH^-(aq) + Fe^{3+}(aq) \longrightarrow Fe(OH)_3(s)$$

Three Types of Equations Used to Describe Reactions in Solution

1. The *molecular equation* gives the overall reaction stoichiometry but not necessarily the actual forms of the reactants and products in solution.
2. The *complete ionic equation* represents as ions all reactants and products that are strong electrolytes.
3. The *net ionic equation* includes only those solution components undergoing a change. Spectator ions are not included.

4.7 Selective Precipitation

We can use the fact that salts have different solubilities to separate mixtures of ions. For example, suppose we have an aqueous solution containing the cations Ag^+, Ba^{2+}, and Fe^{3+}, and the anion NO_3^-. We want to separate the cations by precipitating them one at a time, a process called **selective precipitation.**

How can the separation of these cations be accomplished? We can perform some preliminary tests and observe the reactivity of each cation toward the anions Cl^-, SO_4^{2-}, and OH^-. For example, to test the reactivity of Ag^+ toward Cl^-, we can mix the $AgNO_3$ solution with aqueous KCl or NaCl. As we have

Chemical Analysis of Cockroaches

Chemical Insights

Cockroaches can be a big problem. Not only are these hardy pests unpleasant to live with, but they also consume significant quantities of the world's precious food and grain supplies. Because the many different species of cockroaches require different control measures, determining which species is causing a particular problem is important. Careful examination of a cockroach can reveal its species, but this process is very time-consuming. However, a new method of cockroach identification, based on gas chromatography, has been developed at the U.S. Department of Agriculture by D. A. Carlson and R. J. Brenner. In gas chromatography, the compounds to be separated are dispersed in a carrier gas that passes through a porous solid. Because different substances have differing tendencies to adhere to the solid, the components of the mixture travel at different rates through the system, causing them to spread out so that they can be separated and identified.

In the cockroach identification study, Carlson and Brenner found that the composition of the outer, waxy layer of a roach is distinct to the particular species. Thus, by dissolving this waxy coating and injecting it into the gas stream of a gas chromatograph, scientists can identify the cockroach unambiguously in less than half an hour. This technique is particularly useful for identifying hybrid Asian–German cockroaches, which have become a major problem for the food industry.

Although biologists might argue that the gas chromatographic method takes the fun and the challenge out of identifying cockroaches, this technique should lead to significant advances in the control of these insects.

seen, this produces a precipitate. When we carry out tests of this type using all the possible combinations, we obtain the results in Table 4.2.

After studying these results, we might proceed to separate the cations as follows:

Step 1

Add an aqueous solution of NaCl to the solution containing the Ag^+, Ba^{2+}, and Fe^{3+} ions. Solid AgCl will form and can be removed, leaving Ba^{2+} and Fe^{3+} ions in solution.

Table 4.2

Testing the Reactivity of the Cations Ag^+, Ba^{2+}, and Fe^{3+} with the Anions Cl^-, SO_4^{2-}, and OH^-

	Test Solution (anion)		
Cation	$NaCl(aq)$ (Cl^-)	$Na_2SO_4(aq)$ (SO_4^{2-})	$NaOH(aq)$ (OH^-)
Ag^+	White precipitate (AgCl)	No reaction	White precipitate that turns brown ($AgOH \longrightarrow Ag_2O$; White → Brown)
Ba^{2+}	No reaction	White precipitate ($BaSO_4$)	No reaction
Fe^{3+}	Yellow color but no solid	No reaction	Reddish brown precipitate [$Fe(OH)_3$]

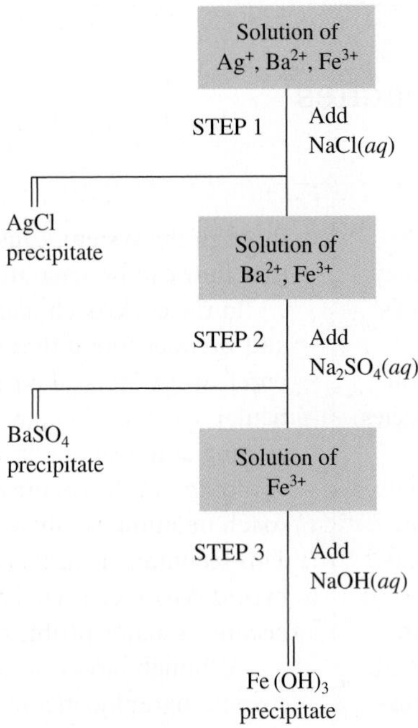

Figure 4.17 Selective precipitation of Ag^+, Ba^{2+}, and Fe^{3+} ions. In this schematic representation, a double line means that a solid forms, and a single line designates a solution.

Step 2

Add an aqueous solution of Na_2SO_4 to the solution containing the Ba^{2+} and Fe^{3+} ions. Solid $BaSO_4$ will form and can be removed, leaving only Fe^{3+} ions in solution.

Step 3

Add an aqueous solution of NaOH to the solution containing the Fe^{3+} ions. Solid $Fe(OH)_3$ will form and can be removed.

Steps 1 through 3 are represented schematically in Fig. 4.17.

Note that adding the anions in this order precipitates the cations one at a time and thus separates them. The process whereby mixtures of ions are separated and identified is called **qualitative analysis.** In this example the qualitative analysis was carried out by selective precipitation, but it can also be accomplished by using other separation techniques that will not be discussed here.

4.8 Stoichiometry of Precipitation Reactions

In Chapter 3 we covered the principles of chemical stoichiometry: the procedures for calculating quantities of reactants and products involved in a chemical reaction. Recall that in performing these calculations, we first convert all quantities to moles and then use the coefficients of the balanced equation to assemble the appropriate molar ratios. In cases in which reactants are mixed, we must determine which reactant is limiting, since the reactant that is consumed first will limit the amounts of products formed. *These same principles apply to reactions that take place in solutions.* However, two points about solution reactions need special emphasis. The first is that it is sometimes difficult to tell immediately which reaction will occur when two solutions are mixed. Usually we must think about the various possibilities and then decide what will happen. The first step in this process *always* should be to write down the species that are actually present in the solution, as we did in Section 4.5.

The second special point about solution reactions is that to obtain the moles of reactants, we must use the volume of a particular solution and its molarity. This procedure was covered in Section 4.3.

We will introduce stoichiometric calculations for reactions in solution in Example 4.7.

OWL INTERACTIVE EXAMPLE 4.7

Calculate the mass of solid NaCl that must be added to 1.50 L of a 0.100 *M* $AgNO_3$ solution to precipitate all the Ag^+ ions in the form of AgCl.

Solution

- What are we trying to solve?

We want to determine the mass of NaCl to add to a given amount of $AgNO_3(aq)$ to precipitate all of the Ag^+ ions and form AgCl.

- What does this mean?

We need to recognize that a reaction occurs to form AgCl; thus we will need a balanced equation. First, we must consider what happens chemically.

When added to the $AgNO_3$ solution (which contains Ag^+ and NO_3^- ions), the solid NaCl dissolves to yield Na^+ and Cl^- ions. Thus the mixed solution contains the ions

$$Ag^+, NO_3^-, Na^+, \text{and } Cl^-$$

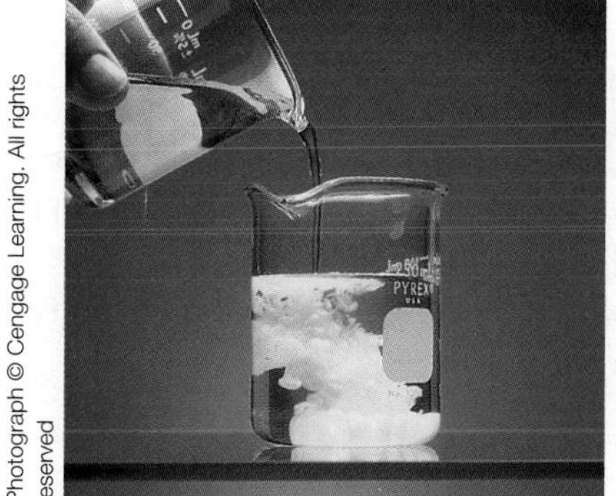

When a solution of NaCl(*aq*) is added to a solution of $AgNO_3$, the white solid AgCl forms.

$NaNO_3$ is soluble and AgCl is insoluble (Table 4.1), so solid AgCl forms according to the following net ionic reaction:

$$Ag^+(aq) + Cl^-(aq) \longrightarrow AgCl(s)$$

In this case enough Cl^- ions must be added to react with all the Ag^+ ions present. Thus we must calculate the moles of Ag^+ ions present in 1.50 L of a 0.100 *M* $AgNO_3$ solution (remember that a 0.100 *M* $AgNO_3$ solution contains 0.100 *M* Ag^+ ions and 0.100 *M* NO_3^- ions):

$$1.50 \text{ L} \times \frac{0.100 \text{ mol } Ag^+}{\text{L}} = 0.150 \text{ mol } Ag^+$$

Since Ag^+ and Cl^- react in a 1:1 ratio, 0.150 mole of Cl^- ions and thus 0.150 mole of NaCl are required. We calculate the mass of NaCl required as follows:

$$0.150 \text{ mol NaCl} \times \frac{58.4 \text{ g NaCl}}{\text{mol NaCl}} = 8.76 \text{ g NaCl}$$

Notice from Example 4.7 that the procedures for doing stoichiometric calculations for solution reactions are very similar to those for other types of reactions. It is useful to think in terms of the following steps for reactions in solution.

Species present
↓ Write the reaction
Balanced net ionic equation
↓ Determine moles of reactants
Identify limiting reactant
↓ Determine moles of products
Check units of products

STEPS

Solving a Stoichiometry Problem Involving Reactions in Solution

1 Identify the species present in the combined solution and determine which reaction occurs.
2 Write the balanced equation for the reaction.
3 Calculate the moles of reactants.
4 Determine which reactant is limiting.
5 Calculate the moles of product or products, as required.
6 Convert to grams or other units, as required.

OWL INTERACTIVE EXAMPLE 4.8

When aqueous solutions of Na_2SO_4 and $Pb(NO_3)_2$ are mixed, $PbSO_4$ precipitates. Calculate the mass of $PbSO_4$ formed when 1.25 L of 0.0500 *M* $Pb(NO_3)_2$ and 2.00 L of 0.0250 *M* Na_2SO_4 are mixed.

Solution

- What are we trying to solve?

We are asked to determine the mass of the precipitant given the amounts of the reactants.

This problem is complex in the sense that it incorporates much of what we have learned in Chapters 3 and 4. The list of what we need to know/do is extensive:

- We need to know what the solutions "look like" at a molecular level.
- We need to be able to write a net ionic equation.
- We must use volume and molarity to determine the number of moles of each reactant.
- Since we are given amounts of both reactants, we will have to determine which reactant is limiting.
- We need to convert moles to mass using molar mass.

However, by now each step of this solution should be familiar to you. Thus, although there are many parts to the solution to this problem, taking it one step at a time makes it less daunting.

When the aqueous solutions of Na_2SO_4 (containing Na^+ and SO_4^{2-} ions) and $Pb(NO_3)_2$ (containing Pb^{2+} and NO_3^- ions) are mixed, the mixed solution contains the ions Na^+, SO_4^{2-}, Pb^{2+}, and NO_3^-. Since $NaNO_3$ is soluble and $PbSO_4$ is insoluble (Table 4.1), solid $PbSO_4$ will form.

The net ionic equation is

$$Pb^{2+}(aq) + SO_4^{2-}(aq) \longrightarrow PbSO_4(s)$$

Since 0.0500 *M* $Pb(NO_3)_2$ contains 0.0500 *M* Pb^{2+} ions, we can calculate the moles of Pb^{2+} ions in 1.25 L of this solution as follows:

$$1.25 \text{ L} \times \frac{0.0500 \text{ mol } Pb^{2+}}{\text{L}} = 0.0625 \text{ mol } Pb^{2+}$$

The 0.0250 *M* Na_2SO_4 solution contains 0.0250 *M* SO_4^{2-} ions, and the number of moles of SO_4^{2-} ions in 2.00 L of this solution is

$$2.00 \text{ L} \times \frac{0.0250 \text{ mol } SO_4^{2-}}{\text{L}} = 0.0500 \text{ mol } SO_4^{2-}$$

Because Pb^{2+} and SO_4^{2-} react in a 1:1 ratio, the amount of SO_4^{2-} will be limiting.

Since the Pb^{2+} ions are present in excess, only 0.0500 mole of solid $PbSO_4$ will be formed. The mass of $PbSO_4$ formed can be calculated by using the molar mass of $PbSO_4$ (303.3):

$$0.0500 \text{ mol } PbSO_4 \times \frac{303.3 \text{ g } PbSO_4}{1 \text{ mol } PbSO_4} = 15.2 \text{ g } PbSO_4$$

One method for determining the amount of a given substance present in a solution is to form a precipitate that includes the substance. The precipitate is then filtered, dried, and weighed. This process, called gravimetric analysis, is illustrated in Example 4.9.

Jeff Daly/Visuals Unlimited

Phosphate rock is used in the manufacture of fertilizer.

EXAMPLE 4.9

Phosphorite, also called *phosphate rock,* is a mineral containing PO_4^{3-} and OH^- anions and Ca^{2+} cations. It is treated with sulfuric acid in the manufacture of phosphate fertilizers (see Chapter 3). A chemist finds the calcium content in an impure sample of phosphate rock by weighing out a 0.4367-g sample, dissolving it in water, and precipitating the Ca^{2+} ions as the insoluble hydrated salt* $CaC_2O_4 \cdot H_2O$ ($C_2O_4^{2-}$ is called the oxalate ion). After being filtered and dried (at a temperature of about 100°C so that the extraneous water is driven off but not the water of hydration), the $CaC_2O_4 \cdot H_2O$ precipitate weighed 0.2920 g. Calculate the mass percent of calcium in the sample of phosphate rock.

Solution

■ What are we trying to solve?

We are asked to determine the mass percent of calcium in a sample of phosphate rock.

■ What does this mean?

$$\text{Mass percent of calcium in the rock} = \frac{\text{mass of calcium}}{\text{mass of rock sample}} \times 100\%$$

We know that the sample of rock has a mass of 0.4367 g, so to determine the mass percent of calcium we change the question to "*What is the mass of calcium?*"

The gravimetric procedure can be summarized as follows:

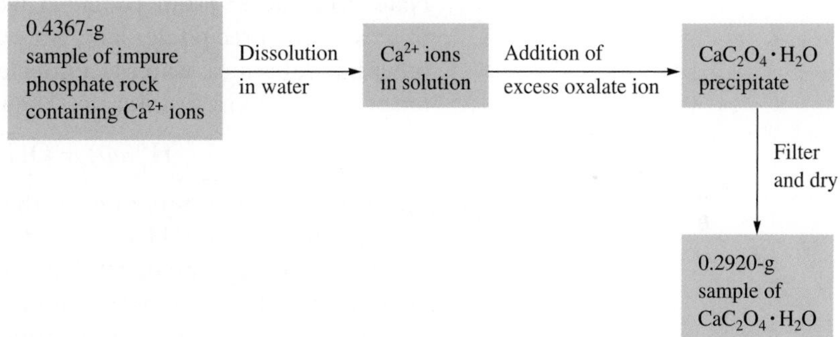

In this analysis, excess $C_2O_4^{2-}$ ions are added to ensure that all Ca^{2+} ions are precipitated. Thus the number of moles of Ca^{2+} ions in the original sample determines the number of moles of $CaC_2O_4 \cdot H_2O$ formed. Using the molar mass of $CaC_2O_4 \cdot H_2O$, we can calculate the moles of $CaC_2O_4 \cdot H_2O$:

$$0.2920 \text{ g } CaC_2O_4 \cdot H_2O \times \frac{1 \text{ mol } CaC_2O_4 \cdot H_2O}{146.12 \text{ g } CaC_2O_4 \cdot H_2O}$$

$$= 1.998 \times 10^{-3} \text{ mol } CaC_2O_4 \cdot H_2O$$

Thus the original sample of impure phosphate rock contained 1.998×10^{-3} mole of Ca^{2+} ions, which we convert to grams:

$$1.998 \times 10^{-3} \text{ mol } Ca^{2+} \times \frac{40.08 \text{ g } Ca^{2+}}{1 \text{ mol } Ca^{2+}} = 8.009 \times 10^{-2} \text{ g } Ca^{2+}$$

*Hydrated salts contain one or more H_2O molecules per formula unit in addition to the cations and anions. A dot is used in the formula of these salts.

The mass percent of calcium in the original sample is then

$$\frac{8.009 \times 10^{-2}\ \text{g}}{0.4367\ \text{g}} \times 100\% = 18.34\%$$

4.9 Acid–Base Reactions

Earlier in this chapter we considered Arrhenius's concept of acids and bases: An acid is a substance that produces H^+ ions when dissolved in water, and a base is a substance that produces OH^- ions. Although these ideas are fundamentally correct, it is convenient to have a more general definition of a base, which covers substances that do not produce OH^- ions. Such a definition was provided by Brønsted and Lowry, who defined acids and bases as follows:

The Brønsted–Lowry concept of acids and bases will be discussed in detail in Chapter 7.

An acid is a proton donor.
A base is a proton acceptor.

How do we recognize acid–base reactions? One of the most difficult tasks for someone inexperienced in chemistry is to predict which reaction might occur when two solutions are mixed. With precipitation reactions, we found that the best way to deal with this problem is to focus on the species actually present in the mixed solution. This also applies to acid–base reactions. For example, when an aqueous solution of hydrogen chloride (HCl) is mixed with an aqueous solution of sodium hydroxide (NaOH), the combined solution contains the ions H^+, Cl^-, Na^+, and OH^-, since HCl is a strong acid and NaOH is a strong base. How can we predict what reaction occurs, if any? First, will NaCl precipitate? From Table 4.1 we know that NaCl is soluble in water and thus will not precipitate. The Na^+ and Cl^- ions are spectator ions. On the other hand, because water is a nonelectrolyte, large quantities of H^+ and OH^- ions cannot coexist in solution. They will react to form H_2O molecules:

$$H^+(aq) + OH^-(aq) \longrightarrow H_2O(l)$$

This is the net ionic equation for the reaction that occurs when aqueous solutions of HCl and NaOH are mixed.

Next, consider mixing an aqueous solution of acetic acid ($HC_2H_3O_2$) with an aqueous solution of potassium hydroxide (KOH). In our earlier discussion of conductivity, we said that an aqueous solution of acetic acid is a weak electrolyte. Thus acetic acid does not dissociate into ions to any great extent. In fact, in aqueous solution 99% of the $HC_2H_3O_2$ molecules remain undissociated. However, when solid KOH is dissolved in water, it dissociates completely to produce K^+ and OH^- ions. So, in the solution formed by mixing aqueous solutions of $HC_2H_3O_2$ and KOH, *before any reaction occurs* the principal species are H_2O, $HC_2H_3O_2$, K^+, and OH^-. Which reaction will occur? A possible precipitation reaction involves K^+ and OH^-, but we know that KOH is soluble. Another possibility is a reaction involving the hydroxide ion and a proton donor. Is there a source of protons in the solution? The answer is yes—the $HC_2H_3O_2$ molecules. The OH^- ion has such a strong affinity for protons that it can strip them from the $HC_2H_3O_2$ molecules. Thus the net ionic equation for the reaction is

$$OH^-(aq) + HC_2H_3O_2(aq) \longrightarrow H_2O(l) + C_2H_3O_2^-(aq)$$

This reaction illustrates a very important general principle: *The hydroxide ion is such a strong base that for purposes of stoichiometry it is assumed to react completely with any weak acid dissolved in water.* Of course, OH^- ions also react completely with the H^+ ions in the solutions of strong acids.

We will now deal with the stoichiometry of acid–base reactions in aqueous solutions. The procedure is fundamentally the same as that used previously.

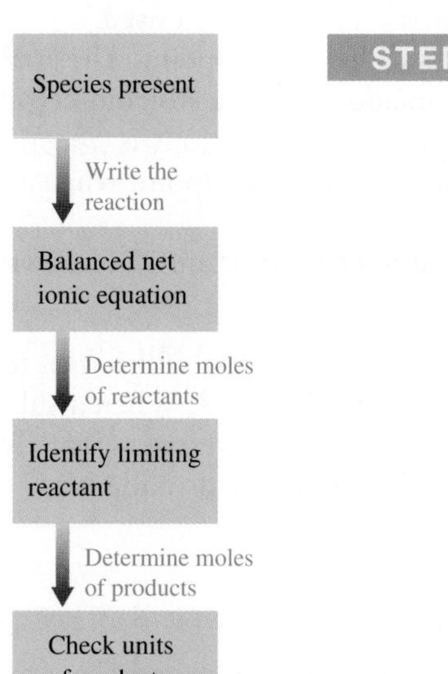

STEPS

Calculations for Acid–Base Reactions

1. List the species present in the combined solution *before reaction,* and decide which reaction will occur.
2. Write the balanced net ionic equation for this reaction.
3. Change the given quantities of reactants to moles. For reactions in solution, use the volumes of the original solutions and their molarities.
4. Determine the limiting reactant where appropriate.
5. Calculate the moles of the required reactant or product.
6. Convert to grams or a volume of solution, as required by the problem.

An acid–base reaction is often called a **neutralization reaction.** When just enough base is added to react exactly with all of the acid in a solution, we say the acid has been *neutralized.*

Acid–Base Titrations

Acid–base **titrations** are an example of **volumetric analysis,** a technique in which one solution is used to analyze another. The solution used to carry out the analysis is called the **titrant** and is delivered from a device called a buret, which measures the volume accurately. The point in the titration at which enough titrant has been added to react exactly with the substance being determined is called the **equivalence point,** or the **stoichiometric point.** This point is often marked by the change in color of a chemical called an **indicator.** The titration procedure is illustrated in Fig. 4.18.

The following requirements must be met for a titration to be successful:

The concentration of the titrant must be known. Such a titrant is called a *standard solution.*

(a)

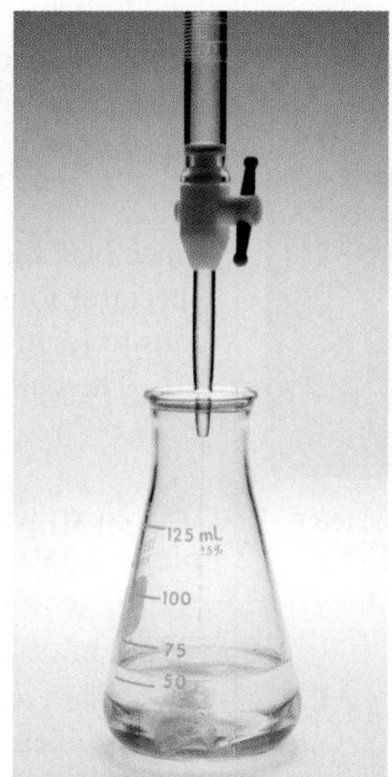
(b)

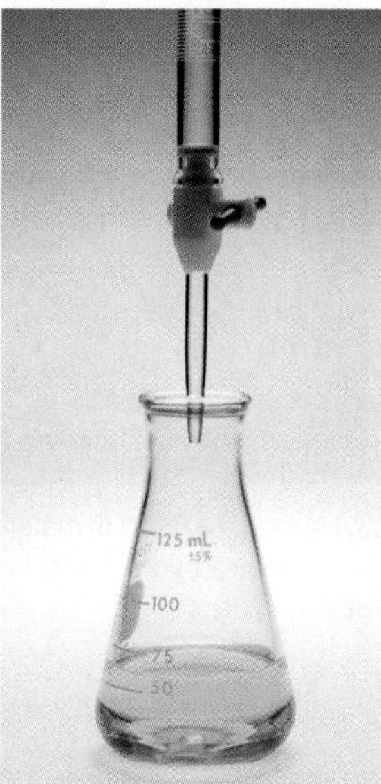
(c)

Figure 4.18
The titration of an acid with a base. (a) The titrant (the base) is in the buret, and the beaker contains the acid solution along with a small amount of indicator. (b) As base is added drop by drop to the acidic solution in the beaker during the titration, the indicator changes color, but the color disappears on mixing. (c) The stoichiometric (equivalence) point is marked by a permanent indicator color change. The volume of the base added is the difference between the final and initial buret readings.

The exact reaction between titrant and substance being analyzed must be known.

The stoichiometric (equivalence) point must be known. An indicator that changes color at, or very near, the stoichiometric point is often used.

The point at which the indicator changes color is called the **endpoint.** The goal is to choose an indicator whose endpoint coincides with the stoichiometric point. An indicator very commonly used by acid–base titrations is *phenolphthalein,* which is colorless in acid and turns pink at the endpoint when an acid is titrated with a base.

Ideally, the endpoint and stoichiometric point should coincide.

The volume of titrant required to reach the stoichiometric point must be known as accurately as possible.

We will deal with acid–base titrations only briefly here but will return to the topic of titrations and indicators in more detail in Chapter 8. When a substance being analyzed contains an acid, the amount of acid present is usually determined by titration with a standard solution containing hydroxide ions.

OWL INTERACTIVE EXAMPLE 4.10

What volume of a 0.100 *M* HCl solution is needed to neutralize 25.0 mL of a 0.350 *M* NaOH solution?

Solution

■ What are we trying to solve?

We are asked to determine the volume of a solution of HCl required to neutralize a given volume of a solution of NaOH.

■ What does this mean?

The HCl and NaOH react with each other, so we need to use a balanced equation to determine the mole ratio between the two reactants. Since we have volumes and molarities of both, we can determine the moles of each reactant. To do this, we must think about the components in the reactant solutions.

The species present in the mixed solutions before any reaction occurs are

$$\underbrace{H^+, Cl^-,}_{\text{From HCl}(aq)} \quad \underbrace{Na^+, OH^-,}_{\text{From NaOH}(aq)} \quad \text{and} \quad H_2O$$

Which reaction will occur? The two possibilities are

$$Na^+(aq) + Cl^-(aq) \longrightarrow NaCl(s)$$

$$H^+(aq) + OH^-(aq) \longrightarrow H_2O(l)$$

AP Photo/The Herald Bulletin, John P. Cleary

A Hazmat (hazardous materials) team neutralizes a hydrochloric acid (HCl) spill.

Since NaCl is soluble, the first reaction does not take place (Na^+ and Cl^- are spectator ions). However, as we have seen before, the reaction of H^+ and OH^- ions to form H_2O does occur.

The balanced net ionic equation for the reaction is

$$H^+(aq) + OH^-(aq) \longrightarrow H_2O(l)$$

Next, we calculate the number of moles of OH^- ions in the 25.0-mL sample of 0.350 *M* NaOH:

$$25.0 \text{ mL NaOH} \times \frac{1 \text{ L}}{1000 \text{ mL}} \times \frac{0.350 \text{ mol OH}^-}{\text{L NaOH}} = 8.75 \times 10^{-3} \text{ mol OH}^-$$

This problem requires the addition of just enough H^+ ions to react exactly with the OH^- ions present. Thus we need not be concerned with determining a limiting reactant.

Since H^+ and OH^- ions react in a 1:1 ratio, 8.75×10^{-3} mole of H^+ ions is required to neutralize the OH^- ions present.

The volume (V) of 0.100 M HCl required to furnish this amount of H^+ ions can be calculated as follows:

$$V \times \frac{0.100 \text{ mol H}^+}{\text{L}} = 8.75 \times 10^{-3} \text{ mol H}^+$$

Solving for V gives

$$V = 8.75 \times 10^{-2} \text{ L}$$

Thus 8.75×10^{-2} L (87.5 mL) of 0.100 M HCl is required to neutralize 25.0 mL of 0.350 M NaOH.

■ Is the answer reasonable?

The mole ratio between the reactants is 1:1. The NaOH is more concentrated than the HCl, so it makes sense that the volume of HCl is larger than the volume of NaOH.

OWL INTERACTIVE EXAMPLE 4.11

In a certain experiment, 28.0 mL of 0.250 M HNO_3 and 53.0 mL of 0.320 M KOH are mixed. Calculate the amount of water formed in the resulting reaction. What is the concentration of H^+ or OH^- ions in excess after the reaction goes to completion?

Solution

■ What are we trying to solve?

We are asked to determine the amount of water formed and the amount of excess reactant when an acid and base react.

■ What does this mean?

The acid and base react with each other, so we need to use a balanced equation to determine the mole ratio between the two reactants. Since we have volumes and molarities of both, we can determine the moles of each reactant. To do this, we must think about the components in the reactant solutions.

The species available for reaction are

$$\underbrace{H^+, NO_3^-,}_{\text{From HNO}_3\text{ solution}} \quad \underbrace{K^+, OH^-,}_{\text{From KOH solution}} \quad \text{and} \quad H_2O$$

Since KNO_3 is soluble, K^+ and NO_3^- are spectator ions, so the net ionic equation is

$$H^+(aq) + OH^-(aq) \longrightarrow H_2O(l)$$

We next compute the amounts of H^+ and OH^- ions present.

$$28.0 \text{ mL HNO}_3 \times \frac{1 \text{ L}}{1000 \text{ mL}} \times \frac{0.250 \text{ mol H}^+}{\text{L}} = 7.00 \times 10^{-3} \text{ mol H}^+$$

$$53.0 \text{ mL KOH} \times \frac{1 \text{ L}}{1000 \text{ mL}} \times \frac{0.320 \text{ mol OH}^-}{\text{L}} = 1.70 \times 10^{-2} \text{ mol OH}^-$$

Since H^+ and OH^- react in a 1:1 ratio, the limiting reactant is H^+. Thus 7.00×10^{-3} mole of H^+ ions will react with 7.00×10^{-3} mole of OH^- ions to form 7.00×10^{-3} mole of H_2O.

The amount of OH^- ions in excess is obtained from the following difference:

$$\text{Original amount} - \text{amount consumed} = \text{amount in excess}$$

$$1.70 \times 10^{-2} \text{ mol OH}^- - 7.00 \times 10^{-3} \text{ mol OH}^- = 1.00 \times 10^{-2} \text{ mol OH}^-$$

The volume of the combined solution is the sum of the individual volumes:

$$\text{Original volume of HNO}_3 + \text{original volume of KOH} = \text{total volume}$$

$$28.0 \text{ mL} + 53.0 \text{ mL} = 81.0 \text{ mL} = 8.10 \times 10^{-2} \text{ L}$$

Thus the molarity of OH^- ions in excess is

$$M = \frac{\text{mol OH}^-}{\text{L solution}} = \frac{1.00 \times 10^{-2} \text{ mol OH}^-}{8.10 \times 10^{-2} \text{ L}} = 0.123\ M \text{ OH}^-$$

■ Is the answer reasonable?

The mole ratio between the reactants is 1:1. The KOH is more concentrated than the HNO_3, and a greater volume of KOH is used. It makes sense that the OH^- is in excess.

EXAMPLE 4.12

An environmental chemist analyzed the effluent (the released waste material) from an industrial process known to produce the compounds carbon tetrachloride (CCl_4) and benzoic acid ($HC_7H_5O_2$), a weak acid that has one acidic hydrogen atom per molecule. A sample of this effluent weighing 0.3518 g was placed in water and shaken vigorously to dissolve the benzoic acid. The resulting aqueous solution required 10.59 mL of 0.1546 *M* NaOH for neutralization. Calculate the mass percent of $HC_7H_5O_2$ in the original sample.

Solution

■ What are we trying to solve?

We are asked to determine the mass percent of an acid in a mixture (effluent).

■ What does this mean?

$$\text{Mass percent of benzoic acid} = \frac{\text{mass of benzoic acid}}{\text{mass of mixture}} \times 100\%$$

We know the mass of the mixture is 0.3518 g, so we change the question to "*What is the mass of benzoic acid in the mixture?*"

In this case the sample was a mixture containing CCl_4 and $HC_7H_5O_2$, and it was titrated with OH^- ions. Clearly, CCl_4 is not an acid (it contains no hydrogen atoms), so we can assume it does not react with OH^- ions. However, $HC_7H_5O_2$ is an acid. It donates one H^+ ion per molecule to react with an OH^- ion as follows:

$$HC_7H_5O_2(aq) + OH^-(aq) \longrightarrow H_2O(l) + C_7H_5O_2^-(aq)$$

Although $HC_7H_5O_2$ is a weak acid, the OH^- ion is such a strong base that we can assume that each OH^- ion added will react with a $HC_7H_5O_2$ molecule until all the benzoic acid is consumed.

We must first determine the number of moles of OH^- ions required to react with all of the $HC_7H_5O_2$:

$$10.59 \text{ mL NaOH} \times \frac{1 \text{ L}}{1000 \text{ mL}} \times \frac{0.1546 \text{ mol OH}^-}{\text{L NaOH}}$$

$$= 1.637 \times 10^{-3} \text{ mol OH}^-$$

This number is also the number of moles of $HC_7H_5O_2$ present. The number of grams of the acid is calculated by using its molar mass:

$$1.637 \times 10^{-3} \text{ mol HC}_7\text{H}_5\text{O}_2 \times \frac{122.125 \text{ g HC}_7\text{H}_5\text{O}_2}{1 \text{ mol HC}_7\text{H}_5\text{O}_2} = 0.1999 \text{ g HC}_7\text{H}_5\text{O}_2$$

The mass percent of $HC_7H_5O_2$ in the original sample is

$$\frac{0.1999 \text{ g}}{0.3518 \text{ g}} \times 100\% = 56.82\%$$

The first step in the analysis of a complex solution is to write down the components present and to focus on the chemistry of each one.

Chemical systems often seem difficult to deal with simply because there are many components. Solving a problem involving a solution in which several components are present is simplified if you *think* about the *chemistry* involved. *The key to success is to write down all the components in the solution and to focus on the chemistry of each one.* We have been emphasizing this approach in dealing with the reactions between ions in solution. Make it a habit to write down the components of solutions before trying to decide which reaction(s) might take place.

4.10 Oxidation–Reduction Reactions

As we have seen, many important substances are ionic. Sodium chloride, for example, can be formed by the reaction of elemental sodium and chlorine:

$$2\text{Na}(s) + \text{Cl}_2(g) \longrightarrow 2\text{NaCl}(s)$$

In this reaction, solid sodium, which contains neutral sodium atoms, reacts with chlorine gas, which contains diatomic Cl_2 molecules, to form the ionic solid NaCl, which contains Na^+ and Cl^- ions. This process is represented in Fig. 4.19. *Reactions like this one, in which one or more electrons are transferred, are called* ***oxidation–reduction reactions*** *or* ***redox reactions.***

Many important chemical reactions involve oxidation and reduction. In fact, most reactions used for energy production are redox reactions. In humans the oxidation of sugars, fats, and proteins provides the energy necessary for life. Combustion reactions, which provide most of the energy to power our civilization, also involve oxidation and reduction. An example is the reaction of methane with oxygen:

$$\text{CH}_4(g) + 2\text{O}_2(g) \longrightarrow \text{CO}_2(g) + 2\text{H}_2\text{O}(g) + \text{energy}$$

Even though none of the reactants or products in this reaction is ionic, the reaction is still assumed to involve a transfer of electrons from carbon to oxygen. To explain this, we must introduce the concept of oxidation states.

Oxidation States

The concept of **oxidation states** (also called **oxidation numbers**) provides a way to keep track of electrons in oxidation–reduction reactions. Oxidation states are defined by a set of rules, most of which describe how to divide up the

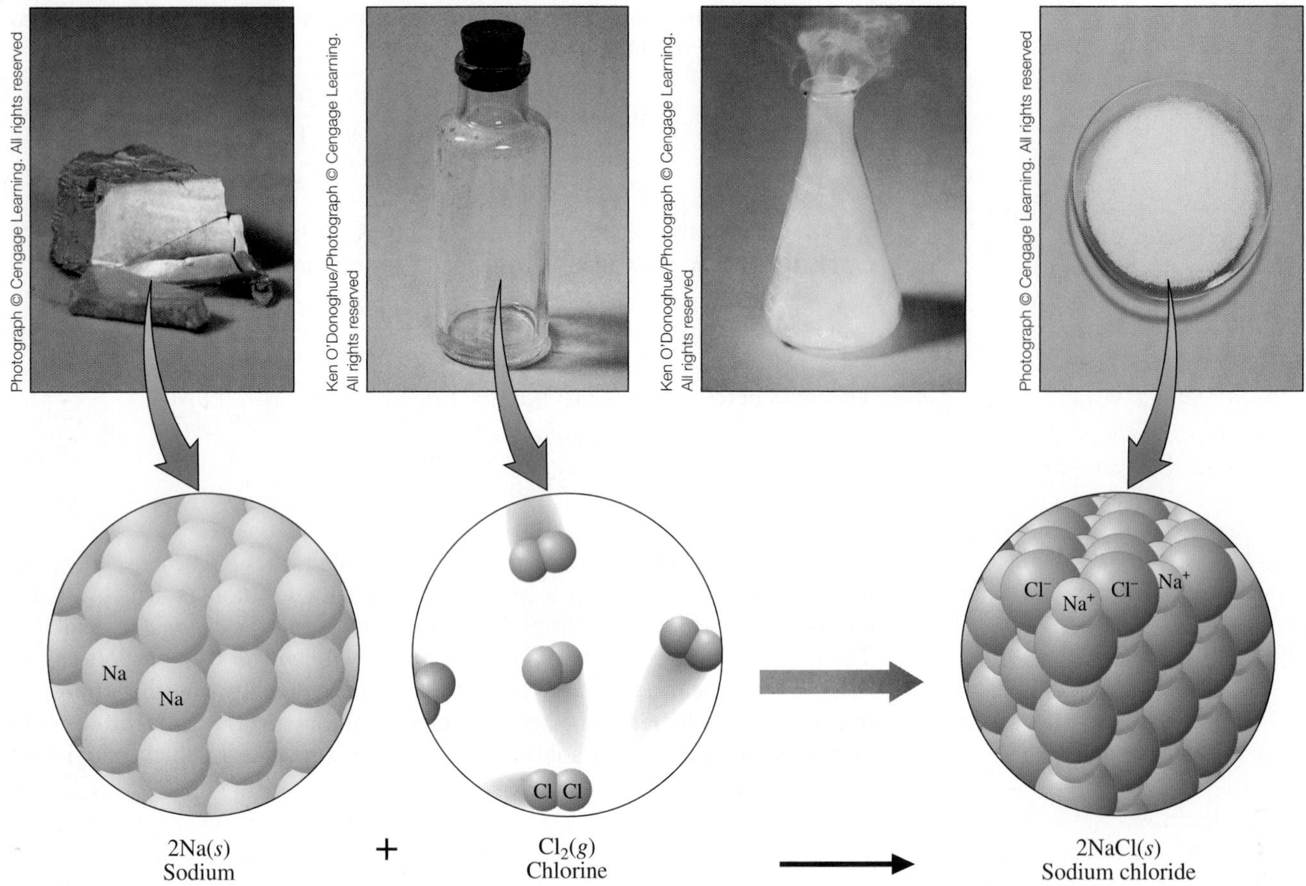

Figure 4.19
The reaction of solid sodium and gaseous chlorine to form solid sodium chloride.

Oxidation of copper metal by nitric acid. The copper atoms lose two electrons to form Cu^{2+} ions, which give a deep green color that becomes turquoise when diluted with water. The brown gas NO_2 is also evolved.

shared electrons in compounds containing covalent bonds. However, before we discuss these rules, we need to discuss the distribution of electrons in a bond.

Recall from the discussion of the water molecule in Section 4.1 that oxygen has a greater attraction for electrons than does hydrogen, causing the O—H bonds in the water molecule to be polar. This phenomenon occurs in other bonds as well, and we will discuss the topic of polarity in detail in Chapter 13. For now we will be satisfied with some general guidelines to help us keep track of electrons in oxidation–reduction reactions. The nonmetals with the highest attraction for shared electrons are in the upper right-hand corner of the periodic table. They are fluorine, oxygen, nitrogen, and chlorine. The relative ability of these atoms to attract shared electrons is

$$\mathrm{F} > \mathrm{O} > \mathrm{N} \approx \mathrm{Cl}$$

↗ Greatest attraction for electrons ↖ Least attraction for electrons

That is, fluorine attracts shared electrons to the greatest extent, followed by oxygen, then nitrogen and chlorine.

The rules for assigning oxidation states are given in Table 4.3. Application of these rules allows the assignment of oxidation states in most compounds. The principles are illustrated in Example 4.13.

State-of-the-Art Analysis

Chemical Insights

The real world of chemical analysis is often quite different from what students do in the typical university laboratory. In the real world, chemical analysis must be done quickly, accurately, economically, and often outside the laboratory setting. Analytical accuracy is crucial. A career can hinge on accuracy when a drug test is involved, and sometimes accuracy is truly a life-or-death matter, as in the screening of air travelers' luggage for explosives.

Chemical analysis can turn up in unexpected places. Modern engines in automobiles have been made much more fuel-efficient and less polluting by the inclusion of a sensor to analyze the oxygen (O_2) concentration in the exhaust gases. The signal from this sensor is sent to the computer that controls engine function so that instantaneous adjustments can be made in spark timing and air–fuel mixtures.

The automated screening of luggage for explosives is a very difficult and important analysis problem. One method being developed for luggage screening is called thermal neutron analysis (TNA), in which the substance to be analyzed is bombarded with neutrons. When nuclei in the sample absorb neutrons, they release gamma rays that are characteristic of a specific nucleus. For example, after the nucleus of a nitrogen atom absorbs a neutron, it emits a gamma ray that is unique to nitrogen, whereas an oxygen atom would produce a different gamma ray unique to oxygen, and so on. Thus, when a sample is bombarded by neutrons and the resulting gamma rays are analyzed by a detector connected to a computer, the atoms present in the sample can be specified. In an airport the luggage would pass through the TNA instrument on a conveyor belt and be bombarded by neutrons from californium-252. The detector is set up to look for unusually large quantities of nitrogen because most chemical explosives are based on nitrogen compounds. Although this system is still under development, the Federal Aviation Administration is optimistic that it will work.

Analytical chemists have always admired the supersensitive natural detection devices built into organisms as part of elaborate control systems used to regulate the levels of various crucial chemicals, such as enzymes, hormones, and neurotransmitters. Because these "biosensors" are so sensitive, chemists are now attaching them to their instruments. For example, the sensory hairs from Hawaiian red swimming crabs can be connected to electrical analyzers and used to detect hormones at concentrations lower than 10^{-12} *M*. Also, slices from the tissues of pineapple cores can be used to detect hydrogen peroxide at levels of $\approx 10^{-6}$ *M*.

Tammy Peluso/Island Effects

A Hawaiian red swimming crab.

Another state-of-the-art detection system contains a surface acoustic wave (SAW) device, which is based on a piezoelectric crystal whose resonant frequency is sensitive to tiny changes in its mass—it can sense a change of 10^{-10} g/cm . In one use of this device as a detector, it was coated with a thin film of zeolite, a silicate mineral. Zeolite has intricate passages of a very uniform size. Thus, it can act as a "molecular sieve," allowing only molecules of a certain size to pass through onto the detector, where their accumulation changes the mass and therefore alters the detector frequency. This sensor has been used to detect amounts of methyl alcohol (CH_3OH) as low as 10^{-9} g.

The face of chemical analysis is changing rapidly. In fact, although wet chemical analyses (titrations, for example) are still quite important in the chemical industry, increasingly these routine analyses are done by robots, which not only perform the analyses automatically but also send the results to a computer for interpretation.

Table 4.3

Rules for Assigning Oxidation States

1. The oxidation state of an atom in an element is 0. For example, the oxidation state of each atom in the substances Na(*s*), $O_2(g)$, $O_3(g)$, and Hg(*l*) is 0.
2. The oxidation state of a monatomic ion is the same as its charge. For example, the oxidation state of the Na^+ ion is +1.
3. In its covalent compounds with nonmetals, hydrogen is assigned an oxidation state of +1. For example, in the compounds HCl, NH_3, H_2O, and CH_4, hydrogen is assigned an oxidation state of +1.
4. Oxygen is assigned an oxidation state of −2 in its covalent compounds, such as CO, CO_2, SO_2, and SO_3. The exception to this rule occurs in peroxides (compounds containing the O_2^{2-} group), where each oxygen is assigned an oxidation state of −1. The best-known example of a peroxide is hydrogen peroxide (H_2O_2).
5. In binary compounds, the element with the greater attraction for the electrons in the bond is assigned a negative oxidation state equal to its charge in its ionic compounds. For example, fluorine is always assigned an oxidation state of −1. That is, for purposes of counting electrons, fluorine is assumed to be F^-. Nitrogen is usually assigned −3. For example, in NH_3, nitrogen is assigned an oxidation state of −3; in H_2S, sulfur is assigned an oxidation state of −2; in HI, iodine is assigned an oxidation state of −1; and so on.
6. The sum of the oxidation states must be zero for an electrically neutral compound and must be equal to the overall charge for an ionic species. For example, the sum of the oxidation states for the hydrogen and oxygen atoms in water is 0; the sum of the oxidation states for the carbon and oxygen atoms in CO_3^{2-} is −2; and the sum of oxidation states for the nitrogen and hydrogen atoms in NH_4^+ is +1.

OWL INTERACTIVE EXAMPLE 4.13

Assign oxidation states to all of the following atoms.

a. CO_2 b. SF_6 c. NO_3^-

Solution

a. The rule that takes precedence here is that oxygen is assigned an oxidation state of −2. The oxidation state for carbon can be determined by recognizing that since CO_2 has no charge, the sum of the oxidation states for oxygen and carbon must be 0. Since each oxygen is −2 and there are two oxygen atoms, the carbon atom must be assigned an oxidation state of +4:

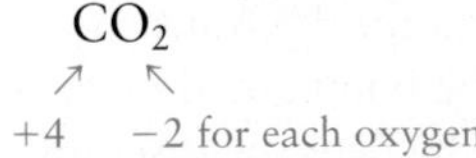

b. Since fluorine has the greater attraction for the shared electrons, we assign its oxidation state first. Because its charge in ionic compounds is 1−, we assign −1 as the oxidation state of each fluorine atom. The sulfur must then be assigned an oxidation state of +6 to balance the total of −6 from the fluorine atoms:

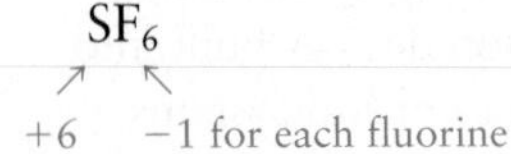

c. Since oxygen has a greater attraction than does nitrogen for the shared electrons, we assign its oxidation state of −2 first. Because the sum of the

Actual charges are given as *n*− or *n*+. Oxidation states *(not actual charges)* are given as −*n* or +*n*. For example, for NO_3^-, the overall charge is 1−; the oxidation state for N is +5 and for O is −2.

oxidation states of the three oxygens is −6 and the net charge on the NO_3^- ion is 1−, the nitrogen must have an oxidation state of +5:

$$\underset{+5}{N}\underset{-2 \text{ for each oxygen}}{O_3^-}$$

Next, let's consider the oxidation states of the atoms in Fe_3O_4, which is the main component in magnetite, an iron ore that accounts for the reddish color of many types of rocks and soils. We assign each oxygen atom its usual oxidation state of −2. The three iron atoms must yield a total of +8 to balance the total of −8 from the four oxygens. Thus each iron atom has an oxidation state of $+\frac{8}{3}$. A noninteger value for an oxidation state may seem strange since charge is expressed in whole numbers. However, although they are rare, noninteger oxidation states do occur because of the rather arbitrary way that the electrons are divided up by the rules in Table 4.3. In the compound Fe_3O_4, for example, the rules assume that all of the iron atoms are equal when in fact this compound can best be viewed as containing four O^{2-} ions, two Fe^{3+} ions, and one Fe^{2+} per formula unit. (Note that the "average" charge on iron works out to be $\frac{8}{3}+$, which is equal to the oxidation state we determined above.) Noninteger oxidation states should not intimidate you. They serve the same purpose as integer oxidation states—for keeping track of electrons.

The Characteristics of Oxidation–Reduction Reactions

Oxidation–reduction reactions are characterized by a transfer of electrons. In some cases the transfer occurs in a literal sense to form ions, such as in this reaction:

$$2Na(s) + Cl_2(g) \longrightarrow 2NaCl(s)$$

However, sometimes the transfer occurs in a more formal sense, such as in the combustion of methane (the oxidation state for each atom is given):

$$CH_4(g) + 2O_2(g) \longrightarrow CO_2(g) + 2H_2O(g)$$

Oxidation state: C −4, H +1 (each H); O in O_2 0; C in CO_2 +4, O −2 (each O); H in H_2O +1 (each H), O −2

Note that the oxidation state of oxygen in O_2 is 0 because it is in elemental form. In this reaction there are no ionic compounds, but we can still describe the process in terms of a transfer of electrons. Note that carbon undergoes a change in oxidation state from −4 in CH_4 to +4 in CO_2. Such a change can be accounted for by a loss of eight electrons (the symbol e^- stands for an electron):

$$\underset{-4}{CH_4} \longrightarrow \underset{+4}{CO_2} + 8e^-$$

On the other hand, each oxygen changes from an oxidation state of 0 in O_2 to −2 in H_2O and CO_2, signifying a gain of two electrons per atom. Since four oxygen atoms are involved, this is a gain of eight electrons:

$$\underset{0}{2O_2} + 8e^- \longrightarrow \underset{4(-2) = -8}{CO_2 + 2H_2O}$$

No change occurs in the oxidation state of hydrogen, so it is not formally involved in the electron transfer process.

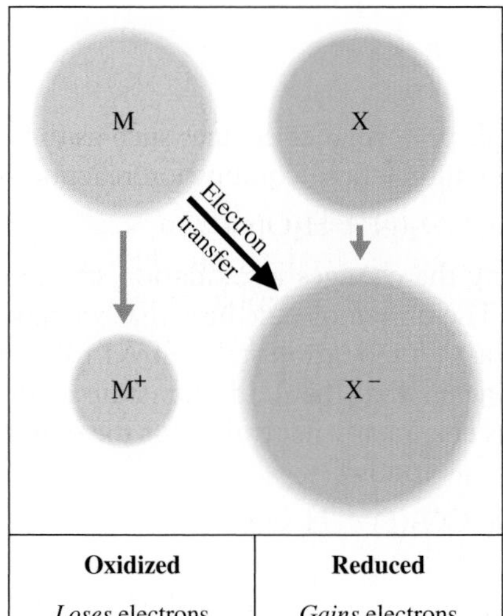

Oxidized	Reduced
Loses electrons	*Gains* electrons
Oxidation state *increases*	Oxidation state *decreases*
Reducing agent	*Oxidizing* agent

Figure 4.20
A summary of an oxidation–reduction process, in which M is oxidized and X is reduced.

With this background, we can now define some important terms. **Oxidation** is an *increase* in oxidation state (a loss of electrons). **Reduction** is a *decrease* in oxidation state (a gain of electrons). Thus in the reaction

$$\underset{0}{2\text{Na}(s)} + \underset{0}{\text{Cl}_2(g)} \longrightarrow 2\underset{+1}{\text{Na}}\underset{-1}{\text{Cl}}(s)$$

sodium is oxidized and chlorine is reduced. In addition, Cl_2 is called the **oxidizing agent (electron acceptor)**, and Na is called the **reducing agent (electron donor)**. These concepts are summarized in Fig. 4.20.

Concerning the reaction

$$\underset{-4}{\text{C}}\underset{+1}{\text{H}_4}(g) + 2\underset{0}{\text{O}_2}(g) \longrightarrow \underset{+4}{\text{C}}\underset{-2}{\text{O}_2}(g) + 2\underset{+1}{\text{H}_2}\underset{-2}{\text{O}}(g)$$

Oxidation is an increase in oxidation state. Reduction is a decrease in oxidation state.

A helpful mnemonic device is OIL RIG (*O*xidation *I*nvolves *L*oss; *R*eduction *I*nvolves *G*ain).

An oxidizing agent is reduced and a reducing agent is oxidized in a redox reaction.

we can say the following:

Carbon is oxidized because there is an increase in its oxidation state (carbon has formally lost electrons).

Oxygen is reduced as shown by the decrease in its oxidation state (oxygen has formally gained electrons).

CH_4 is the reducing agent.

O_2 is the oxidizing agent.

Note that when the oxidizing or reducing agent is named, the *whole compound* is specified, not just the element that undergoes the change in oxidation state.

4.11 Balancing Oxidation–Reduction Equations

Oxidation–reduction reactions are often complicated, which means that it can be difficult to balance their equations by simple inspection. Two methods for balancing redox reactions will be considered here: (1) the oxidation states method and (2) the half-reaction method.

The Oxidation States Method

Methanol (CH_3OH) is used as a fuel in high-performance engines such as those in the race cars in the Indianapolis 500. The unbalanced combustion reaction is

$$\text{CH}_3\text{OH}(l) + \text{O}_2(g) \longrightarrow \text{CO}_2(g) + \text{H}_2\text{O}(g)$$

We want to balance this equation by using the changes in oxidation state, so we must first specify all oxidation states. The only molecule here that we have not previously considered is CH_3OH. We assign oxidation states of $+1$ to each hydrogen and -2 to the oxygen, which means that the oxidation state of the carbon must be -2, since the compound is electrically neutral. Thus the oxidation states for the reaction participants are as follows:

$$\underset{-2}{\text{C}}\underset{\substack{+1\\(\text{each H})}}{\text{H}_3}\underset{-2}{\text{O}}\underset{+1}{\text{H}}(l) + \underset{0}{\text{O}_2}(g) \longrightarrow \underset{+4}{\text{C}}\underset{\substack{-2\\(\text{each O})}}{\text{O}_2}(g) + \underset{\substack{+1\\(\text{each H})}}{\text{H}_2}\underset{-2}{\text{O}}(g)$$

Note that the oxidation state of carbon changes from -2 to $+4$, an increase of 6. On the other hand, the oxidation state of oxygen changes from 0 to -2, a decrease of 2. This means that three oxygen atoms are needed to balance the increase in the oxidation state of the single carbon atom. We can write this relationship as follows:

$$\underset{\text{1 carbon atom}}{\text{CH}_3\text{OH}(l)} + \underset{\text{3 oxygen atoms}}{\tfrac{3}{2}\text{O}_2(g)} \longrightarrow \text{products}$$

The rest of the equation can be balanced by inspection:

$$CH_3OH(l) + \tfrac{3}{2}O_2(g) \longrightarrow CO_2(g) + 2H_2O(g)$$

We then write it in conventional format (multiply through by 2):

$$2CH_3OH(l) + 3O_2(g) \longrightarrow 2CO_2(g) + 4H_2O(g)$$

In using the oxidation states method to balance an oxidation–reduction equation, we find the coefficients for the reactants that will make the total increase in oxidation state balance the total decrease. The remainder of the equation is then balanced by inspection.

EXAMPLE 4.14

Because metals are so reactive, very few are found in pure form in nature. Metallurgy involves reducing the metal ions in ores to the elemental form. The production of manganese from the ore pyrolusite, which contains MnO_2, uses aluminum as the reducing agent. Using oxidation states, balance the equation for this process.

$$MnO_2(s) + Al(s) \longrightarrow Mn(s) + Al_2O_3(s)$$

Solution First, we assign oxidation states:

$$\underset{+4\quad -2\,(\text{each O})}{MnO_2(s)} + \underset{0}{Al(s)} \longrightarrow \underset{0}{Mn(s)} + \underset{+3\,(\text{each Al})\quad -2\,(\text{each O})}{Al_2O_3(s)}$$

Each Mn atom undergoes a decrease in oxidation state of 4 (from +4 to 0), whereas each Al atom undergoes an increase of 3 (from 0 to +3).

Thus we need three Mn atoms for every four Al atoms to balance the increase and decrease in oxidation states:

$$\text{Increase} = 4(3) = \text{decrease} = 3(4)$$

$$3MnO_2(s) + 4Al(s) \longrightarrow \text{products}$$

We balance the rest of the equation by inspection:

$$3MnO_2(s) + 4Al(s) \longrightarrow 3Mn(s) + 2Al_2O_3(s)$$

The procedures for balancing an oxidation–reduction equation by the oxidation states method are summarized below.

STEPS

Balancing an Oxidation–Reduction Equation by the Oxidation States Method

1. Assign the oxidation states of all atoms.
2. Decide which element is oxidized, and determine the increase in oxidation state.
3. Decide which element is reduced, and determine the decrease in oxidation state.
4. Choose coefficients for the species containing the atom oxidized and the atom reduced such that the total increase in oxidation state equals the total decrease in oxidation state.
5. Balance the remainder of the equation by inspection.

The Half-Reaction Method

For oxidation–reduction reactions that occur in aqueous solution, it is often useful to separate the reaction into two **half-reactions:** one involving oxidation and the other involving reduction. For example, consider the unbalanced equation for the oxidation–reduction reaction between cerium(IV) ion and tin(II) ion:

$$Ce^{4+}(aq) + Sn^{2+}(aq) \longrightarrow Ce^{3+}(aq) + Sn^{4+}(aq)$$

This reaction can be separated into a half-reaction involving the substance being *reduced,*

$$Ce^{4+}(aq) \longrightarrow Ce^{3+}(aq)$$

and one involving the substance being *oxidized,*

$$Sn^{2+}(aq) \longrightarrow Sn^{4+}(aq)$$

The general procedure is to balance the equations for the half-reactions separately and then to add them to obtain the overall balanced equation. The half-reaction method for balancing oxidation–reduction equations differs slightly depending on whether the reaction takes place in acidic or basic solution.

STEPS

Balancing Oxidation–Reduction Equations Occurring in Acidic Solution by the Half-Reaction Method

1 Write the equations for the oxidation and reduction half-reactions.

2 For each half-reaction:
- Balance all the elements except hydrogen and oxygen.
- Balance oxygen using H_2O.
- Balance hydrogen using H^+.
- Balance the charge using electrons.

3 If necessary, multiply one or both balanced half-reactions by integers to equalize the number of electrons transferred in the two half-reactions.

4 Add the half-reactions, and cancel identical species.

5 Check to be sure that the elements and charges balance.

These steps are summarized by the following flowchart:

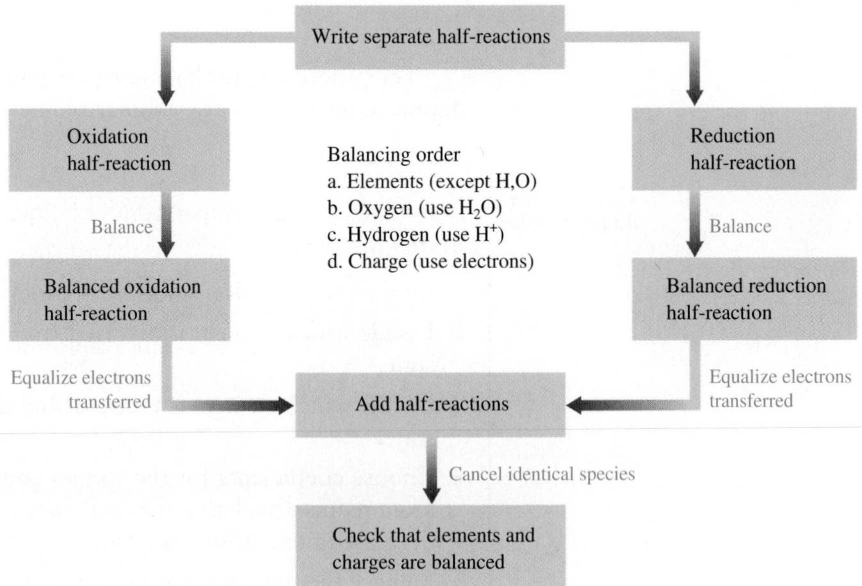

We will illustrate this method by balancing the equation for the reaction between permanganate and iron(II) ions in acidic solution:

$$MnO_4^-(aq) + Fe^{2+}(aq) \xrightarrow{\text{Acidic}} Fe^{3+}(aq) + Mn^{2+}(aq)$$

This reaction is used to analyze iron ore for its iron content.

1. **Identify and write equations for the half-reactions.**

   The oxidation states for the half-reaction involving the permanganate ion show that manganese is reduced:

$$\underset{+7\quad -2\text{ (each O)}}{MnO_4^-} \longrightarrow \underset{+2}{Mn^{2+}}$$

   This is the *reduction half-reaction.* The other half-reaction involves the oxidation of iron(II) to iron(III) ion and is the *oxidation half-reaction:*

$$\underset{+2}{Fe^{2+}} \longrightarrow \underset{+3}{Fe^{3+}}$$

2. **Balance each half-reaction.**

   For the reduction reaction, we have

$$MnO_4^-(aq) \longrightarrow Mn^{2+}(aq)$$

   a. The manganese is balanced.

   b. We balance oxygen by adding $4H_2O$ to the right side of the equation:

$$MnO_4^-(aq) \longrightarrow Mn^{2+}(aq) + 4H_2O(l)$$

   c. Next, we balance hydrogen by adding $8H^+$ to the left side:

$$8H^+(aq) + MnO_4^-(aq) \longrightarrow Mn^{2+}(aq) + 4H_2O(l)$$

   d. All the elements have been balanced, but we need to balance the charge by using electrons. At this point we have the following charges for reactants and products in the reduction half-reaction:

$$\underbrace{8H^+(aq) + MnO_4^-(aq)}_{\substack{8+\quad+\quad 1-\\ 7+}} \longrightarrow \underbrace{Mn^{2+}(aq) + 4H_2O(l)}_{\substack{2+\quad+\quad 0\\ 2+}}$$

   We can equalize the charges by adding five electrons to the left side:

$$\underbrace{5e^- + 8H^+(aq) + MnO_4^-(aq)}_{2+} \longrightarrow \underbrace{Mn^{2+}(aq) + 4H_2O(l)}_{2+}$$

   Both the *elements* and the *charges* are now balanced, so this represents the balanced reduction half-reaction. The fact that five electrons appear on the reactant side of the equation makes sense, since five electrons are required to reduce MnO_4^- (in which Mn has an oxidation state of +7) to Mn^{2+} (in which Mn has an oxidation state of +2).

   For the oxidation reaction,

$$Fe^{2+}(aq) \longrightarrow Fe^{3+}(aq)$$

   the elements are balanced, and we must simply balance the charge:

$$\underbrace{Fe^{2+}(aq)}_{2+} \longrightarrow \underbrace{Fe^{3+}(aq)}_{3+}$$

One electron is needed on the right side to give a net 2+ charge on both sides:

$$\underbrace{Fe^{2+}(aq)}_{2+} \longrightarrow \underbrace{Fe^{3+}(aq) + e^-}_{2+}$$

The number of electrons gained in the reduction half-reaction must equal the number of electrons lost in the oxidation half-reaction.

3. **Equalize the electron transfer in the two half-reactions.**
Since the reduction half-reaction involves a transfer of five electrons and the oxidation half-reaction involves a transfer of only one electron, the oxidation half-reaction must be multiplied by 5:

$$5Fe^{2+}(aq) \longrightarrow 5Fe^{3+}(aq) + 5e^-$$

4. **Add the half-reactions.**
The half-reactions are added to give

$$5e^- + 5Fe^{2+}(aq) + MnO_4^-(aq) + 8H^+(aq) \longrightarrow 5Fe^{3+}(aq) + Mn^{2+}(aq) + 4H_2O(l) + 5e^-$$

Note that the electrons cancel (as they must) to give the final balanced equation:

$$5Fe^{2+}(aq) + MnO_4^-(aq) + 8H^+(aq) \longrightarrow 5Fe^{3+}(aq) + Mn^{2+}(aq) + 4H_2O(l)$$

5. **Check that the elements and charges balance.**

Elements balance: 5 Fe, 1 Mn, 4 O, 8 H $\longrightarrow$ 5 Fe, 1 Mn, 4 O, 8 H
Charges balance: $5(2+) + (1-) + 8(1+) = 17+ \longrightarrow 5(3+) + (2+) + 0 = 17+$

The equation is balanced.

Oxidation–reduction reactions can occur in basic as well as in acidic solutions. The half-reaction method for balancing equations is slightly different in such cases.

STEPS

Balancing Oxidation–Reduction Equations Occurring in Basic Solution by the Half-Reaction Method

1 Use the half-reaction method as specified for acidic solutions to obtain the final balanced equation *as if H^+ ions were present.*

2 To both sides of the equation obtained by the procedure in Step 1, add the number of OH^- ions that is equal to the number of H^+ ions. (We want to eliminate H^+ by forming H_2O.)

3 Form H_2O on the side containing both H^+ and OH^- ions, and eliminate the number of H_2O molecules that appear on both sides of the equation.

4 Check that the elements and charges balance.

This method is summarized by the following flowchart:

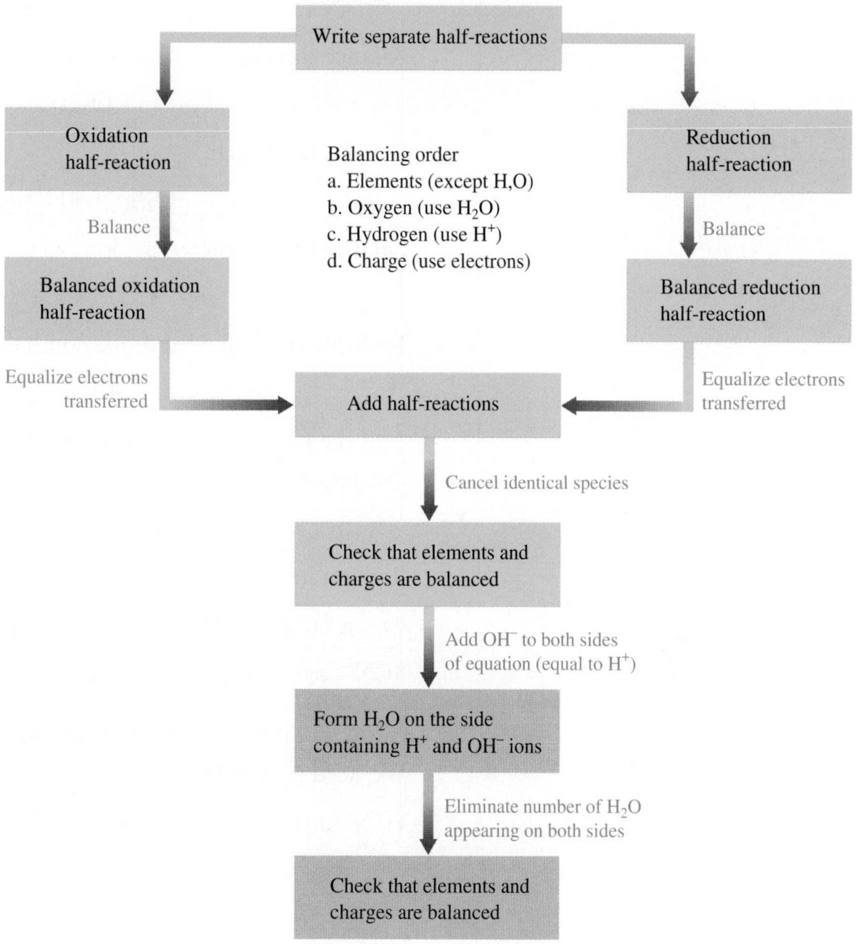

Write separate half-reactions

Oxidation half-reaction

Balance

Balanced oxidation half-reaction

Equalize electrons transferred

Reduction half-reaction

Balance

Balanced reduction half-reaction

Equalize electrons transferred

Balancing order
a. Elements (except H,O)
b. Oxygen (use H_2O)
c. Hydrogen (use H^+)
d. Charge (use electrons)

Add half-reactions

Cancel identical species

Check that elements and charges are balanced

Add OH^- to both sides of equation (equal to H^+)

Form H_2O on the side containing H^+ and OH^- ions

Eliminate number of H_2O appearing on both sides

Check that elements and charges are balanced

OWL INTERACTIVE EXAMPLE 4.15

Silver is sometimes found in nature as large nuggets; more often it is found mixed with other metals and their ores. Cyanide ion is often used to extract the silver by the following reaction that occurs in basic solution:

$$\mathrm{Ag}(s) + \mathrm{CN}^-(aq) + \mathrm{O}_2(g) \xrightarrow{\text{Basic}} \mathrm{Ag(CN)}_2{}^-(aq)$$

Balance this equation by using the half-reaction method.

Solution

1. **Balance the equation as if H^+ ions were present.**

   Balance the oxidation half-reaction:

$$\mathrm{CN}^-(aq) + \mathrm{Ag}(s) \longrightarrow \mathrm{Ag(CN)}_2{}^-(aq)$$

   Balance carbon and nitrogen:

$$2\mathrm{CN}^-(aq) + \mathrm{Ag}(s) \longrightarrow \mathrm{Ag(CN)}_2{}^-(aq)$$

   Balance the charge:

$$2\mathrm{CN}^-(aq) + \mathrm{Ag}(s) \longrightarrow \mathrm{Ag(CN)}_2{}^-(aq) + e^-$$

Balance the reduction half-reaction:

$$O_2(g) \longrightarrow$$

Balance oxygen:

$$O_2(g) \longrightarrow 2H_2O(l)$$

Balance hydrogen:

$$O_2(g) + 4H^+(aq) \longrightarrow 2H_2O(l)$$

Balance the charge:

$$4e^- + O_2(g) + 4H^+(aq) \longrightarrow 2H_2O(l)$$

Multiply the balanced oxidation half-reaction by 4:

$$8CN^-(aq) + 4Ag(s) \longrightarrow 4Ag(CN)_2^-(aq) + 4e^-$$

Add the half-reactions, and cancel identical species:

Oxidation half-reaction:

$$8CN^-(aq) + 4Ag(s) \longrightarrow 4Ag(CN)_2^-(aq) + 4e^-$$

Reduction half-reaction:

$$4e^- + O_2(g) + 4H^+(aq) \longrightarrow 2H_2O(l)$$

$$8CN^-(aq) + 4Ag(s) + O_2(g) + 4H^+(aq) \longrightarrow 4Ag(CN)_2^-(aq) + 2H_2O(l)$$

2. **Add OH^- ions to both sides of the balanced equation.**

   We need to add $4OH^-$ to each side:

$$8CN^-(aq) + 4Ag(s) + O_2(g) + \underbrace{4H^+(aq) + 4OH^-(aq)}_{4H_2O(l)}$$

$$\longrightarrow 4Ag(CN)_2^-(aq) + 2H_2O(l) + 4OH^-(aq)$$

3. **Eliminate as many H_2O molecules as possible.**

$$8CN^-(aq) + 4Ag(s) + O_2(g) + 2H_2O(l) \longrightarrow 4Ag(CN)_2^-(aq) + 4OH^-(aq)$$

4. **Check that elements and charges balance.**

OWL INTERACTIVE EXAMPLE 4.16

Cerium(IV) ion is a strong oxidizing agent that accepts one electron to produce cerium(III) ion:

$$Ce^{4+}(aq) + e^- \longrightarrow Ce^{3+}(aq)$$

A solution containing an unknown concentration of Sn^{2+} ions was titrated with a solution containing Ce^{4+} ions, which oxidize the Sn^{2+} ions to Sn^{4+} ions. In one titration, 1.00 L of the unknown solution required 46.45 mL of a 0.1050 *M* Ce^{4+} solution to reach the stoichiometric point. Calculate the concentration of Sn^{2+} ions in the unknown solution.

Solution

■ What are we trying to solve?

We are asked to determine the concentration of 1.00 L of a solution of Sn^{2+} required to react with a given volume and molarity of a Ce^{4+} solution.

■ What does this mean?

The Sn^{2+} and Ce^{4+} react with each other, so we need to use a balanced equation to determine the mole ratio between the two reactants. From the volume and molarity of the Ce^{4+} solution, we can determine the number of moles that, using the balanced equation, can be converted to moles of the Sn^{2+} solution. With moles and volume of Sn^{2+}, we can compute the concentration in terms of molarity.

The unbalanced equation for the titration reaction is

$$Ce^{4+}(aq) + Sn^{2+}(aq) \longrightarrow Ce^{3+}(aq) + Sn^{4+}(aq)$$

The balanced equation is

$$2Ce^{4+}(aq) + Sn^{2+}(aq) \longrightarrow 2Ce^{3+}(aq) + Sn^{4+}(aq)$$

We can obtain the number of moles of Ce^{4+} ions from the volume and molarity of the Ce^{4+} solution used as the titrant:

$$46.45 \text{ mL} \times \frac{1 \text{ L}}{1000 \text{ mL}} \times \frac{0.1050 \text{ mol } Ce^{4+}}{\text{L}} = 4.877 \times 10^{-3} \text{ mol } Ce^{4+}$$

The number of moles of Sn^{2+} ions can be obtained by applying the appropriate mole ratio from the balanced equation:

$$4.877 \times 10^{-3} \text{ mol } Ce^{4+} \times \frac{1 \text{ mol } Sn^{2+}}{2 \text{ mol } Ce^{4+}} = 2.439 \times 10^{-3} \text{ mol } Sn^{2+}$$

This value represents the quantity of Sn^{2+} ions in 1.00 L of solution. Thus the concentration of Sn^{2+} in the unknown solution is

$$\text{Molarity} = \frac{\text{mol } Sn^{2+}}{\text{L solution}} = \frac{2.439 \times 10^{-3} \text{ mol } Sn^{2+}}{1.00 \text{ L}} = 2.44 \times 10^{-3} \text{ M}$$

■ Is the answer reasonable?

The volume of the Sn^{2+} solution is much greater (factor of ~20) than the volume of the Ce^{4+} solution. It makes sense that the concentration of the Sn^{2+} solution is much lower than the concentration of the Ce^{4+} solution.

4.12 Simple Oxidation–Reduction Titrations

Oxidation–reduction reactions are commonly used as a basis for volumetric analytical procedures. For example, a reducing substance can be titrated with a solution of a strong oxidizing agent, or vice versa. Three of the most frequently used oxidizing agents are aqueous solutions of *potassium permanganate* ($KMnO_4$), *potassium dichromate* ($K_2Cr_2O_7$), and *cerium hydrogen sulfate* [$Ce(HSO_4)_4$].

The strong oxidizing agent, the permanganate ion (MnO_4^-), can undergo several different reactions. The reaction that occurs in acidic solution is the one most commonly used:

$$MnO_4^-(aq) + 8H^+(aq) + 5e^- \longrightarrow Mn^{2+}(aq) + 4H_2O(l)$$

Permanganate has the advantage of being its own indicator—the MnO_4^- ion is intensely purple, and the Mn^{2+} ion is almost colorless. As long as some reducing agent remains in the solution being titrated, the solution remains colorless (assuming all other species present are colorless), since the purple MnO_4^- ion being added is converted to the essentially colorless Mn^{2+} ion. However, when all the reducing agent has been consumed, the next drop of permanganate titrant will turn the solution being titrated light purple (pink). Thus the endpoint (where the color change indicates the titration should stop) occurs approximately one drop beyond the stoichiometric point (the actual point at which all the reducing agent has been consumed).

Example 4.17 describes a typical volumetric analysis using permanganate.

EXAMPLE 4.17

Iron ores often involve a mixture of oxides and contain both Fe^{2+} and Fe^{3+} ions. Such an ore can be analyzed for its iron content by dissolving it in acidic solution, reducing all the iron to Fe^{2+} ions, and then titrating with a standard solution of potassium permanganate. In the resulting solution, MnO_4^- is reduced to Mn^{2+}, and Fe^{2+} is oxidized to Fe^{3+}. A sample of iron ore weighing 0.3500 g was dissolved in acidic solution, and all the iron was reduced to Fe^{2+}. Then the solution was titrated with a 1.621×10^{-2} *M* $KMnO_4$ solution. The titration required 41.56 mL of the permanganate solution to reach the light purple (pink) endpoint. Determine the mass percent of iron in the iron ore.

Solution

- What are we trying to solve?

We are asked to determine the mass percent of iron in an iron ore.

- What does this mean?

$$\text{Mass percent of iron} = \frac{\text{mass of iron}}{\text{mass of iron ore}} \times 100\%$$

We know that the mass of the mixture is 0.3500 g, so we change the question to "*What is the mass of the iron?*"

All of the iron metal is converted to Fe^{2+}, which is reacted with a known volume and molarity of MnO_4^-. From volume and molarity we can get moles, and by using the mole ratio in a balanced equation, we can determine the moles of iron. We convert from moles to mass using the atomic mass of iron.

From the problem it is obvious that this is a redox reaction, so we will need to balance the equation accordingly.

First, we write the unbalanced equation for the reaction:

$$H^+(aq) + MnO_4^-(aq) + Fe^{2+}(aq) \longrightarrow Fe^{3+}(aq) + Mn^{2+}(aq) + H_2O(l)$$

Using the half-reaction method, we balance the equation:

$$8H^+(aq) + MnO_4^-(aq) + 5Fe^{2+}(aq) \longrightarrow 5Fe^{3+}(aq) + Mn^{2+}(aq) + 4H_2O(l)$$

The number of moles of MnO_4^- ion required in the titration is found from the volume and concentration of permanganate solution used:

$$41.56 \text{ mL} \times \frac{1 \text{ L}}{1000 \text{ mL}} \times \frac{1.621 \times 10^{-2} \text{ mol } MnO_4^-}{\text{L}} = 6.737 \times 10^{-4} \text{ mol } MnO_4^-$$

The balanced equation shows that five times as much Fe^{2+} as MnO_4^- is required:

$$6.737 \times 10^{-4} \text{ mol } MnO_4^- \times \frac{5 \text{ mol } Fe^{2+}}{1 \text{ mol } MnO_4^-} = 3.368 \times 10^{-3} \text{ mol } Fe^{2+}$$

Thus the 0.3500-g sample of iron ore contained 3.368×10^{-3} mol of iron. The mass of iron present is

$$3.368 \times 10^{-3} \text{ mol Fe} \times \frac{55.85 \text{ g Fe}}{1 \text{ mol Fe}} = 0.1881 \text{ g Fe}$$

The mass percent of iron in the iron ore is

$$\frac{0.1881 \text{ g}}{0.3500 \text{ g}} \times 100\% = 53.74\%$$

Key Terms

aqueous solutions

Section 4.1
polar molecule
hydration

Section 4.2
solute
solvent
electrical conductivity
strong electrolyte
weak electrolyte
nonelectrolyte
solubility
acid
strong acid
strong base
weak acid
weak base

Section 4.3
molarity
standard solution
dilution

Section 4.5
precipitation reaction
precipitate

Section 4.6
molecular equation
complete ionic equation
spectator ion
net ionic equation

Section 4.7
selective precipitation
qualitative analysis

Section 4.9
neutralization reaction
volumetric analysis
titrant
equivalence (stoichiometric) point
indicator, endpoint

Section 4.10
oxidation-reduction (redox) reaction
oxidation state (oxidation number)
oxidation
reduction
oxidizing agent (electron acceptor)
reducing agent (electron donor)

Section 4.11
half-reaction

For Review

OWL and go Chemistry

Sign in at **www.cengage.com/owl** to:

- View tutorials and simulations, develop problem-solving skills, and complete online homework assigned by your professor.
- Download Go Chemistry mini lecture modules for quick review and exam prep from OWL (or purchase them at **www.cengagebrain.com**)

Chemical reactions in solution are very important in everyday life.

Water is a polar solvent that dissolves many ionic and polar substances.

- Many ionic substances dissolve in water, although solubility varies greatly.
- Many polar substances dissolve in water.

Electrolytes

- Strong electrolyte: 100% dissociated to produce separate ions; strongly conducts an electric current
- Weak electrolyte: Only a small percentage of dissolved molecules produce ions; weakly conducts an electric current
- Nonelectrolyte: Dissolved substance produces no ions; does not conduct an electric current

Acids and bases

- Arrhenius model
  - Acid: produces H^+
  - Base: produces OH^-
- Brønsted–Lowry model
  - Acid: proton donor
  - Base: proton acceptor
- Strong acid: completely dissociates into separated H^+ and anions
- Strong base: completely dissociates into separated OH^- and cation
- Weak acid: dissociates to a slight extent
- Weak base: results in a solution that is a weak electrolyte and contains OH^- ions

Molarity

- One way to describe solution composition
- Molarity $(M) = \dfrac{\text{moles of solute}}{\text{volume of solution (L)}}$
- Standard solution: Molarity is accurately known.

Dilution

- Solvent is added to reduce the molarity.
- Moles of solute after dilution = moles of solute before dilution
- $M_1V_1 = M_2V_2$

Types of equations that describe solution reactions

- Molecular equation: All reactants and products are written as complete formulas.
- Complete ionic equation: All reactants and products that are strong electrolytes are written as separate ions.
- Net ionic equation: Only those compounds that undergo a change are written; spectator ions are not included.

Solubility rules

- Based on experiment observation
- Help predict the outcomes of precipitation reactions

Qualitative analysis

- A mixture of ions can be separated by selective precipitation

Important types of solution reactions
- Acid–base reactions: involve a transfer of H^+ ions
- Precipitation reactions: formation of a solid occurs
- Oxidation–reduction reactions: involve electron transfer

Titrations
- Measures the volume of a standard solution (titrant) needed to react with a substance in solution
- Stoichiometric (equivalence) point: the point at which the required amount of titrant has been added to exactly react with the substance being analyzed
- Endpoint: the point at which a chemical indicator changes color

Oxidation–reduction reactions
- Oxidation states are assigned using a set of rules to keep track of electron flow
- Oxidation: increase in oxidation state (a loss of electrons)
- Reduction: decrease in oxidation state (a gain of electrons)
- Oxidizing agent: gains electrons (is reduced)
- Reducing agent: loses electrons (is oxidized)
- Balancing oxidation–reduction reactions:
  - Oxidation state method
  - Half-reaction method
    - In acidic solutions
    - In basic solutions
- Can be used for volumetric analytical procedures
  - Titrations

Discussion Questions

These questions are designed to be considered by groups of students in class. Often these questions work well for introducing a particular topic in class.

1. Assume you have a highly magnified view of a solution of HCl that allows you to "see" the HCl. Draw this magnified view. If you dropped in a piece of magnesium, the magnesium would disappear, and hydrogen gas would be released. Represent this change using symbols for the elements, and write out the balanced equation.
2. You have a solution of table salt in water. What happens to the salt concentration (increases, decreases, or stays the same) as the solution boils? Draw pictures to explain your answer.
3. You have a sugar solution (solution A) with concentration x. You pour one-third of this solution into a beaker and add an equivalent volume of water (solution B).
   a. What is the ratio of sugar in solutions A and B?
   b. Compare the volumes of solutions A and B.
   c. What is the ratio of the concentrations of sugar in solutions A and B?
4. You add an aqueous solution of lead nitrate to an aqueous solution of potassium iodide. Draw highly magnified views of each solution individually and the mixed solution, including any product that forms. Write the balanced equation for the reaction.
5. You need to make 150.0 mL of a 0.10 M NaCl solution. You have solid NaCl and your lab partner has a 2.5 M NaCl solution. Explain how you each make the 0.10 M NaCl solution.
6. The exposed electrodes of a light bulb are placed in a solution of H_2SO_4 in an electrical circuit such that the light bulb is glowing. You add a dilute salt solution, and the bulb dims. Which of the following could be the salt in the solution?
   a. $Ba(NO_3)_2$
   b. $NaNO_3$
   c. K_2SO_4
   d. $Ca(NO_3)_2$

   Justify your choices. For those you did not choose, explain why they are incorrect.
7. You have two solutions of chemical A. To determine which has the highest concentration of A (molarity), which of the following must you know (there may be more than one answer)?
   a. the mass in grams of A in each solution
   b. the molar mass of A
   c. the volume of water added to each solution
   d. the total volume of the solution

   Explain.

8. Which of the following must be known to calculate the molarity of a salt solution (there may be more than one answer)?
   a. the mass of salt added
   b. the molar mass of the salt
   c. the volume of water added
   d. the total volume of the solution

   Explain.

9. Consider separate aqueous solutions of HCl and H_2SO_4, each with the same molar concentration. An aqueous solution of NaOH is added to each solution to neutralize the acid. Which acid solution requires the largest volume of NaOH solution to react completely with the acid present? Explain.

Exercises

OWL Interactive versions of these problems may be assigned in OWL.

A blue exercise number indicates that the answer to that exercise appears at the back of this book and a solution appears in the *Solutions Guide*.

Aqueous Solutions: Strong and Weak Electrolytes

10. Characterize strong electrolytes versus weak electrolytes versus nonelectrolytes. Give examples of each. How do you experimentally determine whether a soluble substance is a strong electrolyte, weak electrolyte, or nonelectrolyte.

11. The figures below are molecular-level representations of four aqueous solutions of the same solute. Arrange the solutions from most to least concentrated.

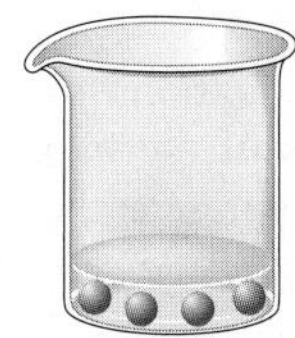
Solution A (1.0 L)

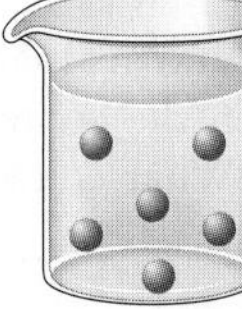
Solution B (4.0 L)

Solution C (2.0 L)

Solution D (2.0 L)

12. Which of the following statements is (are) true? Correct the false statements.
   a. A concentrated solution in water will always contain a strong or weak electrolyte.
   b. A strong electrolyte will break up into ions when dissolved in water.
   c. An acid is a strong electrolyte.
   d. All ionic compounds are strong electrolytes in water.

13. Differentiate between what happens when the following are dissolved in water.
   a. polar solute versus nonpolar solute
   b. KF versus $C_6H_{12}O_6$
   c. RbCl versus AgCl
   d. HNO_3 versus CO

14. Commercial cold packs and hot packs are available for treating athletic injuries. Both types contain a pouch of water and a dry chemical. When the pack is struck, the pouch of water breaks, dissolving the chemical, and the solution becomes either hot or cold. Many hot packs use magnesium sulfate, and many cold packs use ammonium nitrate. Write reaction equations to show how these strong electrolytes break apart in water.

15. Match each name below with the following microscopic pictures of that compound in aqueous solution.

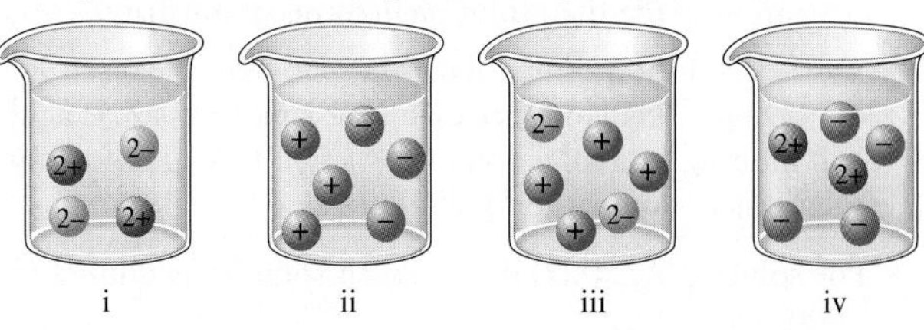

   a. barium nitrate
   b. sodium chloride
   c. potassium carbonate
   d. magnesium sulfate

Solution Concentration: Molarity

16. A solution of ethanol (C_2H_5OH) in water is prepared by dissolving 75.0 mL of ethanol (density = 0.79 g/cm^3) in enough water to make 250.0 mL of solution. What is the molarity of the ethanol in this solution?

17. Describe how you would prepare 2.00 L of each of the following solutions.
   a. 0.250 *M* NaOH from solid NaOH
   b. 0.250 *M* NaOH from 1.00 *M* NaOH stock solution
   c. 0.100 *M* K_2CrO_4 from solid K_2CrO_4
   d. 0.100 *M* K_2CrO_4 from 1.75 *M* K_2CrO_4 stock solution

18. How would you prepare 1.00 L of a 0.50 *M* solution of each of the following?
   a. H_2SO_4 from "concentrated" (18 *M*) sulfuric acid
   b. HCl from "concentrated" (12 *M*) reagent
   c. $NiCl_2$ from the salt $NiCl_2 \cdot 6H_2O$
   d. HNO_3 from "concentrated" (16 *M*) reagent
   e. Sodium carbonate from the pure solid

19. What mass of NaOH is contained in 250.0 mL of a 0.400 *M* sodium hydroxide solution?

20. If 10. g of $AgNO_3$ is available, what volume of 0.25 *M* $AgNO_3$ solution can be prepared?

21. Which of the following solutions of strong electrolytes contains the largest number of moles of chloride ions: 100.0 mL of 0.30 *M* $AlCl_3$, 50.0 mL of 0.60 *M* $MgCl_2$, or 200.0 mL of 0.40 *M* NaCl?

22. Calculate the concentration of all ions present in each of the following solutions of strong electrolytes.
   a. 0.100 mole of $Ca(NO_3)_2$ in 100.0 mL of solution
   b. 2.5 moles of Na_2SO_4 in 1.25 L of solution
   c. 5.00 g of NH_4Cl in 500.0 mL of solution
   d. 1.00 g of K_3PO_4 in 250.0 mL of solution

23. Calculate the sodium ion concentration when 70.0 mL of 3.0 *M* sodium carbonate is added to 30.0 mL of 1.0 *M* sodium bicarbonate.

24. A solution is prepared by dissolving 25.0 g of ammonium sulfate in enough water to make 100.0 mL of stock solution. A 10.00-mL sample of this stock solution is added to 50.00 mL of water. Calculate the concentration of ammonium ions and sulfate ions in the final solution.

25. A standard solution is prepared for the analysis of fluoxymesterone ($C_{20}H_{29}FO_3$), an anabolic steroid. A stock solution is first prepared by dissolving 10.0 mg of fluoxymesterone in enough water to give a total volume of 500.0 mL. A 100.0-μL aliquot (portion) of this solution is diluted to a final volume of 100.0 mL. Calculate the concentration of the final solution in terms of molarity.

26. A stock solution containing Mn^{2+} ions is prepared by dissolving 1.584 g of pure manganese metal in nitric acid and diluting to a final volume of 1.000 L. The following solutions are prepared by dilution.

    For solution A, 50.00 mL of stock solution is diluted to 1000.0 mL.

    For solution B, 10.00 mL of A is diluted to 250.0 mL.

    For solution C, 10.00 mL of B is diluted to 500.0 mL.

    Calculate the molar concentrations of the stock solution and solutions A, B, and C.

27. The units of parts per million (ppm) and parts per billion (ppb) are commonly used by environmental chemists. In general, 1 ppm means 1 part of solute for every 10^6 parts of solution. (Both solute and solution are measured using the same units.) Mathematically, by mass:

$$\text{ppm} = \frac{\mu\text{g solute}}{\text{g solution}} = \frac{\text{mg solute}}{\text{kg solution}}$$

    In the case of very dilute aqueous solutions, a concentration of 1.0 ppm is equal to 1.0 μg of solute per 1.0 mL of solution, which equals 1.0 g of solution. Parts per billion is defined in a similar fashion. Calculate the molarity of each of the following aqueous solutions.
    a. 5.0 ppb Hg in H_2O
    b. 1.0 ppb $CHCl_3$ in H_2O
    c. 10.0 ppm As in H_2O
    d. 0.10 ppm DDT ($C_{14}H_9Cl_5$) in H_2O

28. In the spectroscopic analysis of many substances, a series of standard solutions of known concentration are measured to generate a calibration curve. How would you prepare standard solutions containing 10.0, 25.0, 50.0, 75.0, and 100. ppm of copper from a commercially produced 1000.0-ppm solution? Assume each solution has a final volume of 100.0 mL. (See Exercise 27 for definitions.)

Precipitation Reactions

29. List the formulas of three soluble bromide salts and three insoluble bromide salts. Do the same exercise for sulfate salts, hydroxide salts, and phosphate salts (list three soluble salts and three insoluble salts). List the formulas for six insoluble Pb^{2+} salts and one soluble Pb^{2+} salt.

30. When 1.0 mole of solid lead nitrate is added to 2.0 moles of aqueous potassium iodide, a yellow precipitate forms. After the precipitate settles to the bottom, does the solution above the precipitate conduct electricity? Explain. Write the complete ionic equation to help you answer this question.

31. When the following solutions are mixed together, what precipitate (if any) will form?
    a. $Hg_2(NO_3)_2(aq) + CuSO_4(aq)$
    b. $Ni(NO_3)_2(aq) + CaCl_2(aq)$
    c. $K_2CO_3(aq) + MgI_2(aq)$
    d. $Na_2CrO_4(aq) + AlBr_3(aq)$

32. For the reactions in Exercise 31, write the balanced molecular equation, complete ionic equation, and net ionic equation. If no precipitate forms, write "No reaction."

33. Write the balanced molecular, complete, and net ionic equations for the reaction, if any, that occurs when aqueous solutions of the following are mixed.
    a. ammonium sulfate and barium nitrate
    b. lead(II) nitrate and sodium chloride
    c. sodium phosphate and potassium nitrate
    d. sodium bromide and rubidium chloride
    e. copper(II) chloride and sodium hydroxide

34. How would you separate the following ions in aqueous solution by selective precipitation?
    a. Ag^+, Ba^{2+}, and Cr^{3+}
    b. Ag^+, Pb^{2+}, and Cu^{2+}
    c. Hg_2^{2+} and Ni^{2+}

35. Write the balanced molecular and net ionic equations for the reaction that occurs when the contents of the two beakers are added together. What colors represent the spectator ions in each reaction?

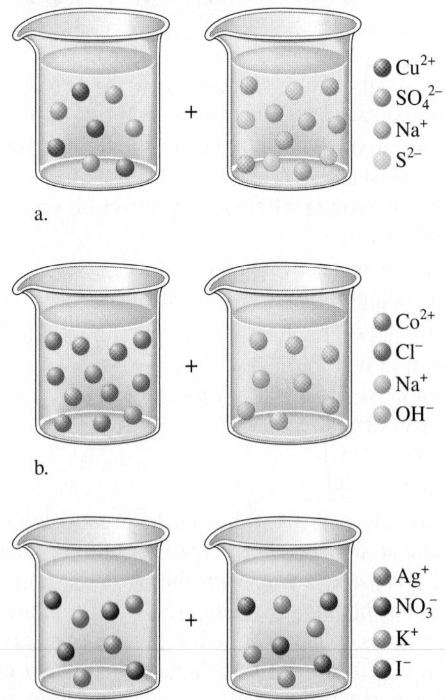

36. Give an example of how each of the following insoluble ionic compounds could be produced using a precipitation reaction. Write the balanced molecular equation for each reaction.
   a. $Fe(OH)_3(s)$
   b. $Hg_2Cl_2(s)$
   c. $PbSO_4(s)$
   d. $BaCrO_4(s)$

37. Separate samples of a solution of an unknown soluble ionic compound are treated with KCl, Na_2SO_4, and NaOH. A precipitate forms only when Na_2SO_4 is added. Which cations could be present in the unknown soluble ionic compound?

38. What volume of 0.100 *M* Na_3PO_4 is required to precipitate all of the lead(II) ions from 150.0 mL of 0.250 *M* $Pb(NO_3)_2$?

39. How many grams of silver chloride can be prepared by the reaction of 100.0 mL of 0.20 *M* silver nitrate with 100.0 mL of 0.15 *M* calcium chloride? Calculate the concentrations of each ion remaining in solution after precipitation is complete.

40. The following drawings represent aqueous solutions. Solution A is 2.00 L of a 2.00 *M* aqueous solution of copper(II) nitrate. Solution B is 2.00 L of a 3.00 *M* aqueous solution of potassium hydroxide.

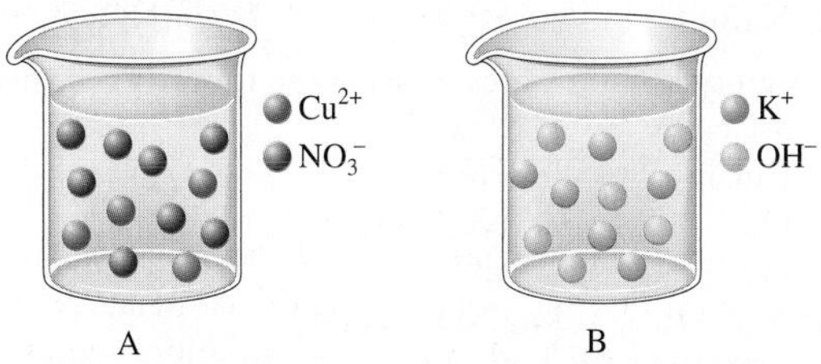

   a. Draw a picture of the solution made by mixing solutions A and B together after the precipitation reaction takes place. Make sure this picture shows the correct relative volume compared with solutions A and B and the correct relative number of ions, along with the correct relative amount of solid formed.
   b. Determine the concentrations (in *M*) of all ions left in solution (from part a) and the mass of solid formed.

41. What mass of Na_2CrO_4 is required to precipitate all of the silver ions from 75.0 mL of a 0.100 *M* solution of $AgNO_3$?

42. A 1.00-g sample of an alkaline earth metal chloride is treated with excess silver nitrate. All of the chloride is recovered as 1.38 g of silver chloride. Identify the metal.

43. A mixture contains only NaCl and $Al_2(SO_4)_3$. A 1.45-g sample of the mixture is dissolved in water, and an excess of NaOH is added, producing a precipitate of $Al(OH)_3$. The precipitate is filtered, dried, and weighed. The mass of the precipitate is 0.107 g. What is the mass percent of $Al_2(SO_4)_3$ in the sample?

44. The thallium (present as Tl_2SO_4) in a 9.486-g pesticide sample was precipitated as thallium(I) iodide. Calculate the mass percent of Tl_2SO_4 in the sample if 0.1824 g of TlI was recovered.

45. Saccharin ($C_7H_5NO_3S$) is sometimes dispensed in tablet form. Ten tablets with a total mass of 0.5894 g were dissolved in water. This solution was then oxidized to convert all the sulfur to sulfate ion, which was precipitated by adding an excess of barium chloride solution. The mass of $BaSO_4$ obtained was 0.5032 g. What is the average mass of saccharin per tablet? What is the average mass percent of saccharin in the tablets?

46. Douglasite is a mineral with the formula $2KCl \cdot FeCl_2 \cdot 2H_2O$. Calculate the mass percent of douglasite in a 455.0-mg sample if it took 37.20 mL of a 0.1000 *M* $AgNO_3$ solution to precipitate all the Cl^- as AgCl. Assume the douglasite is the only source of chloride ion.

47. A 1.42-g sample of a pure compound with formula M_2SO_4 was dissolved in water and treated with an excess of aqueous calcium chloride, resulting in the precipitation of all the sulfate ions as calcium sulfate. The precipitate was collected, dried, and found to weigh 1.36 g. Determine the atomic mass of M and identify M.

Acid–Base Reactions

48. Write balanced equations (all three types) for the reactions that occur when the following aqueous solutions are mixed.
   a. ammonia (aqueous) and nitric acid
   b. barium hydroxide (aqueous) and hydrochloric acid
   c. perchloric acid [$HClO_4(aq)$] and solid iron(III) hydroxide
   d. solid silver hydroxide and hydrobromic acid

49. What acid and what base would react in aqueous solution so that the following salts appear as products in the molecular equation? Write the balanced molecular equation for each reaction.
   a. potassium perchlorate
   b. cesium nitrate
   c. calcium iodide

50. Carminic acid, a naturally occurring red pigment extracted from the cochineal insect, contains only carbon, hydrogen, and oxygen. It was commonly used as a dye in the first half of the nineteenth century. It is 53.66% C and 4.09% H by mass. A titration required 18.02 mL of 0.0406 *M* NaOH to neutralize 0.3602 g of carminic acid. Assuming that there is only one acidic hydrogen per molecule, what is the molecular formula of carminic acid?

51. What volume of each of the following acids will react completely with 50.00 mL of 0.100 *M* NaOH?
   a. 0.100 *M* HCl
   b. 0.100 *M* H_2SO_3 (two acidic hydrogens)
   c. 0.200 *M* H_3PO_4 (three acidic hydrogens)
   d. 0.150 *M* HNO_3
   e. 0.200 *M* $HC_2H_3O_2$ (one acidic hydrogen)
   f. 0.300 *M* H_2SO_4 (two acidic hydrogens)

52. A 30.0-mL sample of an unknown strong base is neutralized after the addition of 12.0 mL of a 0.150 *M* HNO_3 solution. If the unknown base concentration is 0.0300 *M*, give some possible identities for the unknown base.

53. A student had 1.00 L of a 1.00 *M* acid solution. Much to the surprise of the student, it took 2.00 L of 1.00 *M* NaOH solution to react completely with the acid. Explain why it took twice as much NaOH to react with all of the acid.

    In a different experiment, a student had 10.0 mL of 0.020 *M* HCl. Again, much to the surprise of the student, it took only 5.00 mL of 0.020 *M* strong base to react completely with the HCl. Explain why it took only half as much strong base to react with all of the HCl.

54. Sodium hydroxide solution is usually standardized by titrating a pure sample of potassium hydrogen phthalate ($KHC_8H_4O_4$, often abbreviated KHP), an acid with one acidic hydrogen and a molar mass of 204.22 g/mol. It takes 34.67 mL of a sodium hydroxide solution to titrate a 0.1082-g sample of KHP. What is the molarity of the sodium hydroxide?

55. A 0.500-L sample of H_2SO_4 solution was analyzed by taking a 100.0-mL aliquot and adding 50.0 mL of 0.213 *M* NaOH. After the reaction occurred, an excess of OH^- ions remained in the solution. The excess base required 13.21 mL of 0.103 *M* HCl for neutralization. Calculate the molarity of the original sample of H_2SO_4. (Sulfuric acid has two acidic hydrogens.)

56. What volume of 0.0521 *M* $Ba(OH)_2$ is required to neutralize exactly 14.20 mL of 0.141 *M* H_3PO_4? Phosphoric acid contains three acidic hydrogens.

57. A 10.00-mL sample of vinegar, an aqueous solution of acetic acid ($HC_2H_3O_2$), is titrated with 0.5062 *M* NaOH, and 16.58 mL is required to reach the endpoint.
    a. What is the molarity of the acetic acid?
    b. If the density of the vinegar is 1.006 g/cm^3, what is the mass percent of acetic acid in the vinegar?

58. A student titrates an unknown amount of potassium hydrogen phthalate (KHP) with 20.46 mL of a 0.1000 *M* NaOH solution. KHP (molar mass = 204.22 g/mol) has one acidic hydrogen. How many grams of KHP were titrated (reacted completely) by the sodium hydroxide solution?

59. A student mixes four reagents together, thinking that the solutions will neutralize each other. The solutions mixed together are 50.0 mL of 0.100 *M* hydrochloric acid, 100.0 mL of 0.200 *M* of nitric acid, 500.0 mL of 0.0100 *M* calcium hydroxide, and 200.0 mL of 0.100 *M* rubidium hydroxide. Is the resulting solution neutral? If not, calculate the concentration of excess H^+ or OH^- ions left in solution.

60. A 50.00-mL sample of an ammonia solution is analyzed by titration with HCl. The reaction is

$$NH_3(aq) + H^+(aq) \longrightarrow NH_4^+(aq)$$

    It took 39.47 mL of 0.0984 *M* HCl to titrate (react completely with) the ammonia. What is the concentration of the original ammonia solution?

61. Hydrochloric acid (75.0 mL of 0.250 *M*) is added to 225.0 mL of 0.0550 *M* $Ba(OH)_2$ solution. What is the concentration of the excess H^+ or OH^- left in this solution?

62. A 2.20-g sample of an unknown acid (empirical formula = $C_3H_4O_3$) is dissolved in 1.0 L of water. A titration required 25.0 mL of 0.500 *M* NaOH to react completely with all the acid present. Assuming that the unknown acid has one acidic proton per molecule, what is the molecular formula of the unknown acid?

Oxidation–Reduction Reactions

63. Differentiate between the following terms.
    a. species reduced versus the reducing agent
    b. species oxidized versus the oxidizing agent
    c. oxidation state versus actual charge

64. How do you balance redox reactions by the oxidation states method?

65. Assign oxidation states to all atoms in each compound.
    a. $KMnO_4$
    b. NiO_2
    c. $K_4Fe(CN)_6$ (Fe only)
    d. $(NH_4)_2HPO_4$
    e. P_4O_6
    f. Fe_3O_4
    g. $XeOF_4$
    h. SF_4
    i. CO
    j. $C_6H_{12}O_6$

66. Assign oxidation states to all of the following atoms.
    a. UO_2^{2+}
    b. As_2O_3
    c. $NaBiO_3$
    d. As_4
    e. $HAsO_2$
    f. $Mg_2P_2O_7$
    g. $Na_2S_2O_3$
    h. Hg_2Cl_2
    i. $Ca(NO_3)_2$

67. Assign oxidation states to all of the following atoms.
    a. $SrCr_2O_7$
    b. $CuCl_2$
    c. O_2
    d. H_2O_2
    e. $MgCO_3$
    f. Ag
    g. $PbSO_3$
    h. PbO_2
    i. $Na_2C_2O_4$
    j. CO_2
    k. $(NH_4)_2Ce(SO_4)_3$
    l. Cr_2O_3

68. Tell which of the following are oxidation–reduction reactions. For those that are, identify the oxidizing agent, the reducing agent, the substance being oxidized, and the substance being reduced.
    a. $CH_4(g) + 2O_2(g) \longrightarrow CO_2(g) + 2H_2O(g)$
    b. $Zn(s) + 2HCl(aq) \longrightarrow ZnCl_2(aq) + H_2(g)$
    c. $Cr_2O_7^{2-}(aq) + 2OH^-(aq) \longrightarrow 2CrO_4^{2-}(aq) + H_2O(l)$
    d. $O_3(g) + NO(g) \longrightarrow O_2(g) + NO_2(g)$
    e. $2H_2O_2(l) \longrightarrow 2H_2O(l) + O_2(g)$
    f. $2CuCl(aq) \longrightarrow CuCl_2(aq) + Cu(s)$
    g. $HCl(g) + NH_3(g) \longrightarrow NH_4Cl(s)$
    h. $SiCl_4(l) + 2H_2O(l) \longrightarrow 4HCl(aq) + SiO_2(s)$
    i. $SiCl_4(l) + 2Mg(s) \longrightarrow 2MgCl_2(s) + Si(s)$

69. Many oxidation–reduction reactions can be balanced by inspection. Try to balance the following reactions by inspection. In each reaction, identify the substance reduced and the substance oxidized.
    a. $Al(s) + HCl(aq) \longrightarrow AlCl_3(aq) + H_2(g)$
    b. $CH_4(g) + S(s) \longrightarrow CS_2(l) + H_2S(g)$
    c. $C_3H_8(g) + O_2(g) \longrightarrow CO_2(g) + H_2O(l)$
    d. $Cu(s) + Ag^+(aq) \longrightarrow Ag(s) + Cu^{2+}(aq)$

70. Balance each of the following oxidation–reduction reactions by using the oxidation states method.
    a. $C_2H_6(g) + O_2(g) \longrightarrow CO_2(g) + H_2O(g)$
    b. $Mg(s) + HCl(aq) \longrightarrow Mg^{2+}(aq) + Cl^-(aq) + H_2(g)$
    c. $Cu(s) + Ag^+(aq) \longrightarrow Cu^{2+}(aq) + Ag(s)$
    d. $Zn(s) + H_2SO_4(aq) \longrightarrow ZnSO_4(aq) + H_2(g)$

71. Balance the following oxidation–reduction reactions, which occur in acidic solution, using the half-reaction method.
   a. $Cu(s) + NO_3^-(aq) \longrightarrow Cu^{2+}(aq) + NO(g)$
   b. $Cr_2O_7^{2-}(aq) + Cl^-(aq) \longrightarrow Cr^{3+}(aq) + Cl_2(g)$
   c. $Pb(s) + PbO_2(s) + H_2SO_4(aq) \longrightarrow PbSO_4(s)$
   d. $Mn^{2+}(aq) + NaBiO_3(s) \longrightarrow Bi^{3+}(aq) + MnO_4^-(aq)$
   e. $H_3AsO_4(aq) + Zn(s) \longrightarrow AsH_3(g) + Zn^{2+}(aq)$
   f. $As_2O_3(s) + NO_3^-(aq) \longrightarrow H_3AsO_4(aq) + NO(g)$
   g. $Br^-(aq) + MnO_4^-(aq) \longrightarrow Br_2(l) + Mn^{2+}(aq)$
   h. $CH_3OH(aq) + Cr_2O_7^{2-}(aq) \longrightarrow CH_2O(aq) + Cr^{3+}(aq)$

72. Balance the following oxidation–reduction reactions, which occur in basic solution, using the half-reaction method.
   a. $Al(s) + MnO_4^-(aq) \longrightarrow MnO_2(s) + Al(OH)_4^-(aq)$
   b. $Cl_2(g) \longrightarrow Cl^-(aq) + ClO^-(aq)$
   c. $NO_2^-(aq) + Al(s) \longrightarrow NH_3(g) + AlO_2^-(aq)$
   d. $MnO_4^-(aq) + S^{2-}(aq) \longrightarrow MnS(s) + S(s)$
   e. $CN^-(aq) + MnO_4^-(aq) \longrightarrow CNO^-(aq) + MnO_2(s)$

73. Balance the following equations by the half-reaction method.
   a. $Fe(s) + HCl(aq) \longrightarrow HFeCl_4(aq) + H_2(g)$
   b. $IO_3^-(aq) + I^-(aq) \xrightarrow{\text{Acidic}} I_3^-(aq)$
   c. $Cr(NCS)_6^{4-}(aq) + Ce^{4+}(aq) \xrightarrow{\text{Acidic}} Cr^{3+}(aq) + Ce^{3+}(aq) + NO_3^-(aq) + CO_2(g) + SO_4^{2-}(aq)$
   d. $CrI_3(s) + Cl_2(g) \xrightarrow{\text{Basic}} CrO_4^{2-}(aq) + IO_4^-(aq) + Cl^-(aq)$
   e. $Fe(CN)_6^{4-}(aq) + Ce^{4+}(aq) \xrightarrow{\text{Basic}} Ce(OH)_3(s) + Fe(OH)_3(s) + CO_3^{2-}(aq) + NO_3^-(aq)$

74. One of the classic methods for the determination of the manganese content in steel involves converting all the manganese to the deeply colored permanganate ion and then measuring the absorption of light. The steel is first dissolved in nitric acid, producing the manganese(II) ion and nitrogen dioxide gas. This solution is then reacted with an acidic solution containing periodate ion; the products are the permanganate and iodate ions. Write balanced chemical equations for both of these steps.

75. A solution of permanganate is standardized by titration with oxalic acid ($H_2C_2O_4$). It required 28.97 mL of the permanganate solution to react completely with 0.1058 g of oxalic acid. The unbalanced equation for the reaction is

$$MnO_4^-(aq) + H_2C_2O_4(aq) \xrightarrow{\text{Acidic}} Mn^{2+}(aq) + CO_2(g)$$

What is the molarity of the permanganate solution?

76. A 50.00-mL sample of solution containing Fe^{2+} ions is titrated with a 0.0216 *M* $KMnO_4$ solution. It required 20.62 mL of $KMnO_4$ solution to oxidize all the Fe^{2+} ions to Fe^{3+} ions by the reaction

$$MnO_4^-(aq) + Fe^{2+}(aq) \xrightarrow{\text{Acidic}} Mn^{2+}(aq) + Fe^{3+}(aq) \quad \text{(Unbalanced)}$$

   a. What was the concentration of Fe^{2+} ions in the sample solution?
   b. What volume of 0.0150 *M* $K_2Cr_2O_7$ solution would it take to do the same titration? The reaction is

$$Cr_2O_7^{2-}(aq) + Fe^{2+}(aq) \xrightarrow{\text{Acidic}} Cr^{3+}(aq) + Fe^{3+}(aq) \quad \text{(Unbalanced)}$$

77. The iron content of iron ore can be determined by titration with a standard $KMnO_4$ solution. The iron ore is dissolved in HCl, and all the iron is reduced to Fe^{2+} ions. This solution is then titrated with $KMnO_4$ solution, producing Fe^{3+} and Mn^{2+} ions in acidic solution. If it required 38.37 mL of 0.0198 *M* $KMnO_4$ to titrate a solution made from 0.6128 g of iron ore, what is the mass percent of iron in the iron ore?

78. The vanadium in a sample of ore is converted to VO^{2+}. The VO^{2+} ion is subsequently titrated with MnO_4^- in acidic solution to form $V(OH)_4^+$ and manganese(II) ion. To titrate the solution, 26.45 mL of 0.02250 *M* MnO_4^- was required. If the mass percent of vanadium in the ore was 58.1%, what was the mass of the ore sample?

79. When hydrochloric acid reacts with magnesium metal, hydrogen gas and aqueous magnesium chloride are produced. What volume of 5.0 *M* HCl is required to react completely with 3.00 g of magnesium?

80. Triiodide ions are generated in solution by the following (unbalanced) reaction in acidic solution:

$$IO_3^-(aq) + I^-(aq) \longrightarrow I_3^-(aq)$$

Triiodide ion is determined by titration with a sodium thiosulfate ($Na_2S_2O_3$) solution. The products are iodide ion and tetrathionate ion ($S_4O_6^{2-}$).
   a. Balance the equation for the reaction of IO_3^- with I^- ions.
   b. A sample of 0.6013 g of potassium iodate was dissolved in water. Hydrochloric acid and solid potassium iodide were then added in excess. What is the minimum mass of solid KI and the minimum volume of 3.00 *M* HCl required to convert all of the IO_3^- ions to I_3^- ions?
   c. Write and balance the equation for the reaction of $S_2O_3^{2-}$ with I_3^- in acidic solution.
   d. A 25.00-mL sample of a 0.0100 *M* solution of KIO_3 is reacted with an excess of KI. It requires 32.04 mL of $Na_2S_2O_3$ solution to titrate the I_3^- ions present. What is the molarity of the $Na_2S_2O_3$ solution?
   e. How would you prepare 500.0 mL of the KIO_3 solution in part d, using pure, dry KIO_3?

Additional Exercises

81. A 230.-mL sample of a 0.275 *M* $CaCl_2$ solution is left on a hot plate overnight; the following morning, the solution is 1.10 *M*. What volume of water evaporated from the 0.275 *M* $CaCl_2$ solution?

82. Using the general solubility rules given in Table 4.1, name three reagents that would form precipitates with each of the following ions in aqueous solution. Write the net ionic equation for each of your suggestions.
    a. chloride ion
    b. calcium ion
    c. iron(III) ion
    d. sulfate ion
    e. mercury(I) ion, Hg_2^{2+}
    f. silver ion

83. Consider a 1.50-g mixture of magnesium nitrate and magnesium chloride. After dissolving this mixture in water, 0.500 *M* silver nitrate is added dropwise until precipitate formation is complete. This mass of the white precipitate formed is 0.641 g.
    a. Calculate the mass percent of magnesium chloride in the mixture.
    b. Determine the minimum volume of silver nitrate that must have been added to ensure complete formation of the precipitate.

84. What mass of solid aluminum hydroxide can be produced when 50.0 mL of 0.200 *M* $Al(NO_3)_3$ is added to 200.0 mL of 0.100 *M* KOH?

85. In most of its ionic compounds, cobalt is either Co(II) or Co(III). One such compound, containing chloride ion and waters of hydration, was analyzed, and the following results were obtained. A 0.256-g sample of the compound was dissolved in water, and excess silver nitrate was added. The silver chloride was filtered, dried, and weighed, and it had a mass of 0.308 g. A second sample of 0.416 g of the compound was dissolved in water, and an excess of sodium hydroxide was added. The hydroxide salt was filtered and heated in a flame, forming cobalt(III) oxide. The mass of the cobalt(III) oxide formed was 0.145 g.
    a. What is the percent composition, by mass, of the compound?
    b. Assuming the compound contains one cobalt ion per formula unit, what is the formula?
    c. Write balanced equations for the three reactions described.

86. A mixture contains only NaCl and $Fe(NO_3)_3$. A 0.456-g sample of the mixture is dissolved in water, and an excess of NaOH is added, producing a precipitate of $Fe(OH)_3$. The precipitate is filtered, dried, and weighed. Its mass is 0.107 g. Calculate the following.
    a. the mass of iron in the sample
    b. the mass of $Fe(NO_3)_3$ in the sample
    c. the mass percent of $Fe(NO_3)_3$ in the sample

87. A mixture contains only sodium chloride and potassium chloride. A 0.1586-g sample of the mixture was dissolved in water. It took 22.90 mL of 0.1000 *M* $AgNO_3$ to completely precipitate all the chloride present. What is the composition (by mass percent) of the mixture?

88. Tris(pentafluorophenyl)borane, commonly known by its acronym BARF, is frequently used to initiate polymerization of ethylene or propylene in the presence of a catalytic transition metal compound. It is composed solely of C, F, and B; it is 42.23% C and 55.66% F by mass.
    a. What is the empirical formula of BARF?
    b. A 2.251-g sample of BARF dissolved in 347.0 mL of solution produces a 0.01267 *M* solution. What is the molecular formula of BARF?

89. A student added 50.0 mL of an NaOH solution to 100.0 mL of 0.400 *M* HCl. The solution was then treated with an excess of aqueous chromium(III) nitrate, resulting in formation of 2.06 g of precipitate. Determine the concentration of the NaOH solution.

90. In a 1-L beaker, 203 mL of 0.307 *M* ammonium chromate was mixed with 137 mL of 0.269 *M* chromium(III) nitrite to produce ammonium nitrite and chromium(III) chromate. Write the balanced chemical equation for the reaction occurring here. If the percent yield of the reaction was 88.0%, how much chromium(III) chromate was isolated?

91. It took 25.06 ± 0.05 mL of a sodium hydroxide solution to titrate a 0.4016-g sample of KHP (see Exercise 54). Calculate the concentration and uncertainty in the concentration of the sodium hydroxide solution. (See Appendix Section A1.5.) Neglect any uncertainty in the mass.

92. You wish to prepare 1 L of a 0.02 *M* potassium iodate solution. You require that the final concentration be within 1% of 0.02 *M* and that the concentration must be known accurately to the fourth decimal place. How would you prepare this solution? Specify the glassware you would use, the accuracy needed for the balance, and the ranges of acceptable masses of KIO_3 that can be used.

93. Citric acid, which can be obtained from lemon juice, has the molecular formula $C_6H_8O_7$. A 0.250-g sample of citric acid dissolved in 25.0 mL of water requires 37.2 mL of 0.105 *M* NaOH for complete neutralization. How many acidic hydrogens per molecule does citric acid have?

94. Acetylsalicylic acid is the active ingredient in aspirin. It took 35.17 mL of 0.5065 *M* sodium hydroxide to react completely with 3.210 g of acetylsalicylic acid. Acetylsalicylic acid has one acidic hydrogen. What is the molar mass of acetylsalicylic acid?

95. When organic compounds containing sulfur are burned, sulfur dioxide is produced. The amount of SO_2 formed can be determined by reaction with hydrogen peroxide:

$$H_2O_2(aq) + SO_2(g) \longrightarrow H_2SO_4(aq)$$

The resulting sulfuric acid is then titrated with a standard NaOH solution. A 1.325-g sample of coal is burned and the SO_2 collected in a solution of hydrogen peroxide. It took 28.44 mL of 0.1000 *M* NaOH to neutralize the resulting sulfuric acid. Calculate the mass percent of sulfur in the coal sample. Sulfuric acid has two acidic hydrogens.

Challenge Problems

96. The blood alcohol (C_2H_5OH) level can be determined by titrating a sample of blood plasma with an acidic potassium dichromate solution, resulting in the production of $Cr^{3+}(aq)$ and carbon dioxide. The reaction can be monitored because the dichromate ion ($Cr_2O_7^{2-}$) is orange in solution, and the Cr^{3+} ion is green. The unbalanced redox equation is

$$Cr_2O_7^{2-}(aq) + C_2H_5OH(aq) \longrightarrow Cr^{3+}(aq) + CO_2(g)$$

If 31.05 mL of 0.0600 M potassium dichromate solution is required to titrate 30.0 g of blood plasma, determine the mass percent of alcohol in the blood.

97. Zinc and magnesium metal each react with hydrochloric acid according to the following equations:

$$Zn(s) + 2HCl(aq) \longrightarrow ZnCl_2(aq) + H_2(g)$$
$$Mg(s) + 2HCl(aq) \longrightarrow MgCl_2(aq) + H_2(g)$$

A 10.00-g mixture of zinc and magnesium is reacted with the stoichiometric amount of hydrochloric acid. The reaction mixture is then reacted with 156 mL of 3.00 M silver nitrate to produce the maximum possible amount of silver chloride.
   a. Determine the percent magnesium by mass in the original mixture.
   b. If 78.0 mL of HCl was added, what was the concentration of the HCl?

98. A 10.00-g sample consisting of a mixture of sodium chloride and potassium sulfate is dissolved in water. This aqueous mixture then reacts with excess aqueous lead(II) nitrate to form 21.75 g of solid. Determine the mass percent of sodium chloride in the original mixture.

99. Consider the reaction of 19.0 g of zinc with excess silver nitrite to produce silver metal and zinc nitrite. The reaction is stopped before all the zinc metal has reacted and 29.0 g of solid metal is present. Calculate the mass of each metal in the 29.0-g mixture.

100. Consider an experiment in which two burets, Y and Z, are simultaneously draining into a beaker that initially contained 275.0 mL of 0.300 M HCl. Buret Y contains 0.150 M NaOH and buret Z contains 0.250 M KOH. The stoichiometric point in the titration is reached 60.65 minutes after Y and Z were started simultaneously. The total volume in the beaker at the stoichiometric point is 655 mL. Calculate the flow rates of burets Y and Z. Assume the flow rates remain constant during the experiment.

101. A sample is a mixture of KCl and KBr. When 0.1024 g of the sample is dissolved in water and reacted with excess silver nitrate, 0.1889 g of solid is obtained. What is the composition by mass percent of the mixture?

102. You made 100.0 mL of a lead(II) nitrate solution for lab but forgot to cap it. The next lab session you noticed that there was only 80.0 mL left (the rest had evaporated). In addition, you forgot the initial concentration of the solution. You decide to take 2.00 mL of the solution and add an excess of a concentrated sodium chloride solution. You obtain a solid with a mass of 3.407 g. What was the concentration of the original lead(II) nitrate solution?

103. Polychlorinated biphenyls (PCBs) have been used extensively as dielectric materials in electrical transformers. Because PCBs have been shown to be potentially harmful, analysis for their presence in the environment has become very important. PCBs are manufactured according to the following generic reaction:

$$C_{12}H_{10} + nCl_2 \xrightarrow[\text{catalyst}]{\text{Fe}} C_{12}H_{10-n}Cl_n + nHCl$$

This reaction results in a mixture of PCB products. The mixture is analyzed by decomposing the PCBs and then precipitating the resulting Cl^- as AgCl.
   a. Develop a general equation that relates the average value of n to the mass of a given mixture of PCBs and the mass of AgCl produced.
   b. A 0.1947-g sample of a commercial PCB yielded 0.4971 g of AgCl. What is the average value of n for this sample?

104. Consider reacting copper(II) sulfate with iron. Two possible reactions can occur, as represented by the following equations.

copper(II) sulfate(aq) + iron(s) $\longrightarrow$
copper(s) + iron(II) sulfate(aq)

copper(II) sulfate(aq) + iron(s) $\longrightarrow$
copper(s) + iron(III) sulfate(aq)

You place 87.7 mL of a 0.500 M solution of copper(II) sulfate in a beaker. You then add 2.00 g of iron filings to the copper(II) sulfate solution. After the reaction occurs, you isolate 2.27 g of copper. Which equation above describes the reaction that occurred? Support your answer.

105. A stream flows at a rate of 5.00×10^4 liters per second (L/s) upstream of a manufacturing plant. The plant discharges 3.50×10^3 L/s of water that contains 65.0 ppm HCl into the stream. (See Exercise 27 for definitions.)
   a. Calculate the stream's total flow rate downstream from this plant.
   b. Calculate the concentration of HCl in ppm downstream from this plant.
   c. Further downstream, another manufacturing plant diverts 1.80×10^4 L/s of water from the stream for its own use. This plant must first neutralize the acid and does so by adding lime:

$$CaO(s) + 2H^+(aq) \longrightarrow Ca^{2+}(aq) + H_2O(l)$$

   What mass of CaO is consumed in an 8.00-h work day by this plant?
   d. The original stream water contained 10.2 ppm Ca^{2+}. Although no calcium was in the waste water from the first plant, the waste water of the second plant contains Ca^{2+} from the neutralization process. If 90.0% of the water used by the second plant is returned to the stream, calculate the concentration of Ca^{2+} in ppm downstream of the second plant.

106. Chromium has been investigated as a coating for steel cans. The thickness of the chromium film is determined by dissolving a sample of a can in acid and oxidizing the resulting Cr^{3+} to $Cr_2O_7^{2-}$ with the peroxydisulfate ion:

$$S_2O_8^{2-}(aq) + Cr^{3+}(aq) + H_2O(l) \longrightarrow Cr_2O_7^{2-}(aq) + SO_4^{2-}(aq) + H^+(aq) \text{ (Unbalanced)}$$

After removal of unreacted $S_2O_8^{2-}$, an excess of ferrous ammonium sulfate $[Fe(NH_4)_2(SO_4)_2 \cdot 6H_2O]$ is added, reacting with $Cr_2O_7^{2-}$ produced from the first reaction. The unreacted Fe^{2+} from the excess ferrous ammonium sulfate is titrated with a separate $K_2Cr_2O_7$ solution. The reaction is:

$$H^+(aq) + Fe^{2+}(aq) + Cr_2O_7^{2-}(aq) \longrightarrow Fe^{3+}(aq) + Cr^{3+}(aq) + H_2O(l) \text{ (Unbalanced)}$$

a. Write balanced chemical equations for the two reactions.
b. In one analysis, a 40.0-cm^2 sample of a chromium-plated can was treated according to this procedure. After dissolution and removal of excess $S_2O_8^{2-}$, 3.000 g of $Fe(NH_4)_2(SO_4)_2 \cdot 6H_2O$ was added. It took 8.58 mL of 0.0520 *M* $K_2Cr_2O_7$ solution to completely react with the excess Fe^{2+}. Calculate the thickness of the chromium film on the can. (The density of chromium is 7.19 g/cm^3.)

107. One high-temperature superconductor has the general formula $YBa_2Cu_3O_x$. The copper is a mixture of Cu(II) and Cu(III) oxidation states. This mixture of oxidation states appears vital for high-temperature superconductivity to occur. A simple method for determining the average copper oxidation state has been reported [D. C. Harris, M. E. Hillis, and T. A. Hewston, *J. Chem. Educ.* 64, 847(1987)]. The described analysis takes place in two steps:

i. One superconductor sample is treated directly with I^-:

$$Cu^{2+}(aq) + I^-(aq) \longrightarrow CuI(s) + I_3^-(aq) \text{ (Unbalanced)}$$

$$Cu^{3+}(aq) + I^-(aq) \longrightarrow CuI(s) + I_3^-(aq) \text{ (Unbalanced)}$$

ii. A second superconductor sample is dissolved in acid, converting all copper to Cu(II). This solution is then treated with I^-:

$$Cu^{2+}(aq) + I^-(aq) \longrightarrow CuI(s) + I_3^-(aq) \text{ (Unbalanced)}$$

In both steps the I_3^- is determined by titrating with a standard sodium thiosulfate ($Na_2S_2O_3$) solution:

$$I_3^-(aq) + S_2O_3^{2-}(aq) \longrightarrow S_4O_6^{2-}(aq) + I^-(aq) \text{ (Unbalanced)}$$

a. Calculate the average copper oxidation states for materials with the formulas $YBa_2Cu_3O_{6.5}$, $YBa_2Cu_3O_7$, and $YBa_2Cu_3O_8$. Interpret your results in terms of a mixture of Cu(II) and Cu(III) ions, assuming that only Y^{3+}, Ba^{2+}, and O^{2-} are present in addition to the copper ions.
b. Balance the equations involved in the copper analysis.
c. A superconductor sample was analyzed by the above procedure. In step i, it took 37.77 mL of 0.1000 *M* $Na_2S_2O_3$ to react completely with the I_3^- generated from a 562.5-mg sample. In step ii, it took 22.57 mL of 0.1000 *M* $Na_2S_2O_3$ to react with the I_3^- generated by a 504.2-mg sample. Determine the formula of this superconductor sample (that is, find the value of x in $YBa_2Cu_3O_x$). Calculate the average oxidation state of copper in this material.

108. You are given a solid that is a mixture of Na_2SO_4 and K_2SO_4. A 0.205-g sample of the mixture is dissolved in water. An excess of an aqueous solution of $BaCl_2$ is added. The $BaSO_4$ that is formed is filtered, dried, and weighed. Its mass is 0.298 g. What mass of SO_4^{2-} ion is in the sample? What is the mass percent of SO_4^{2-} ion in the sample? What are the percent compositions by mass of Na_2SO_4 and K_2SO_4 in the sample?

109. A sample is a mixture of $AgNO_3$, $CuCl_2$, and $FeCl_3$. When a 1.0000-g sample of the mixture is dissolved in water and reacted with excess silver nitrate, 1.7809 g of precipitate forms. When a separate 1.0000-g sample of the mixture is treated with a reducing agent, all the metal ions in the mixture are reduced to pure metals. The total mass of pure metals produced is 0.4684 g. Calculate the mass percent of $AgNO_3$, $CuCl_2$, and $FeCl_3$ in the original mixture.

110. Three students were asked to find the identity of the metal in a particular sulfate salt. They dissolved a 0.1472-g sample of the salt in water and treated it with excess barium chloride, resulting in the precipitation of barium sulfate. After the precipitate had been filtered and dried, it weighed 0.2327 g.

Each student analyzed the data independently and came to different conclusions. Pat decided that the metal was titanium. Chris thought it was sodium. Randy reported that it was gallium. What formula did each student assign to the sulfate salt?

Look for information on the sulfates of gallium, sodium, and titanium in this text and reference books such as the *CRC Handbook of Chemistry and Physics*. What further tests would you suggest to determine which student is most likely correct?

Marathon Problems*

111. The formate ion, CHO_2^-, forms ionic compounds with many metal ions. Assume that 9.7416 g $M(CHO_2)_2$ (where M represents the atomic symbol for a particular metal) is dissolved in water. When a solution of 0.200 *M* sodium sulfate is added, a white precipitate forms. The sodium sulfate solution is added until no more precipitate forms; then a few milliliters are added in excess. The precipitate is filtered, dried, and weighed. It has a mass of 9.9392 g. The filtrate is saved for further use.

    A potassium permanganate solution is standardized by dissolving 0.9234 g sodium oxalate in dilute sulfuric acid and then titrating with the potassium permanganate solution. The principal products of the reaction are manganese(II) ion and carbon dioxide gas. The titration requires 18.55 mL of the potassium permanganate solution to reach the endpoint, which is indicated by the first permanent, but barely perceptible, pink color of the permanganate ion.

    The filtrate from the original reaction is diluted by pouring it into a 250-mL volumetric flask, diluting to the mark with water, and then mixing thoroughly. An aliquot consisting of 10.00 mL of this diluted solution is pipetted into a 125-mL Erlenmeyer flask, approximately 25 mL of water is added, and the solution is made basic. What volume of the standard permanganate solution will be needed to titrate this solution to the equivalence point? The principal products of the reaction are carbonate ion and manganese(IV) oxide. Identify M.

112. You have two 500.0-mL aqueous solutions. Solution A is a solution of a metal nitrate that is 8.246% nitrogen by mass. The ionic compound in solution B consists of potassium, chromium, and oxygen; chromium has an oxidation state of +6, and there are 2 potassiums and 1 chromium in the formula. The masses of the solutes in each of the solutions are the same. When the solutions are added together, a blood-red precipitate forms. After the reaction has gone to completion, you dry the solid and find that it has a mass of 331.8 g.
    a. Identify the ionic compounds in solution A and solution B.
    b. Identify the blood-red precipitate.
    c. Calculate the concentration (molarity) of all ions in the original solutions.
    d. Calculate the concentration (molarity) of all ions in the final solution.

*From James H. Burness, "The Use of "Marathon" Problems as Effective Vehicles for the Presentation of General Chemistry Lectures," Journal of Chemical Education, 68(11).

5 Gases

chapter

Hot air balloon taking off from the ski resort of Chateau d'Oex in the Swiss Alps.

Matter exists in three distinct physical states: gas, liquid, and solid. Of these, the gaseous state is the easiest to describe both experimentally and theoretically. In particular, the study of gases provides an excellent example of the scientific method in action. It illustrates how observations lead to natural laws, which in turn can be accounted for by models. Then, as more accurate measurements become available, the models are modified.

In addition to providing a good illustration of the scientific method, gases are important in their own right. For example, gases are often produced in chemical reactions and thus must be dealt with in stoichiometric calculations. Also, the earth's atmosphere is a mixture of gases, primarily elemental nitrogen and oxygen; it both supports life and acts as a waste receptacle for the exhaust gases that accompany many industrial processes.

For these reasons, it is important to understand the behavior of gases. We will pursue this goal by considering the properties of gases, the laws and models that describe the behavior of gases, and finally the reactions that occur among the gases in the atmosphere.

5.1 Early Experiments

Even though the Greeks considered "air" to be one of the four fundamental elements and various alchemists obtained "airs," or "vapors," in their experiments, careful study of these elusive substances proved difficult. The first person to attempt a scientific study of the "vapors" produced in chemical reactions was the Flemish physician Jan Baptista Van Helmont (1577–1644). Thinking that air and similar substances must be akin to the "chaos" from which, according to Greek myth, the universe was created, Van Helmont described these substances using the Flemish word for *chaos,* which was *gas.*

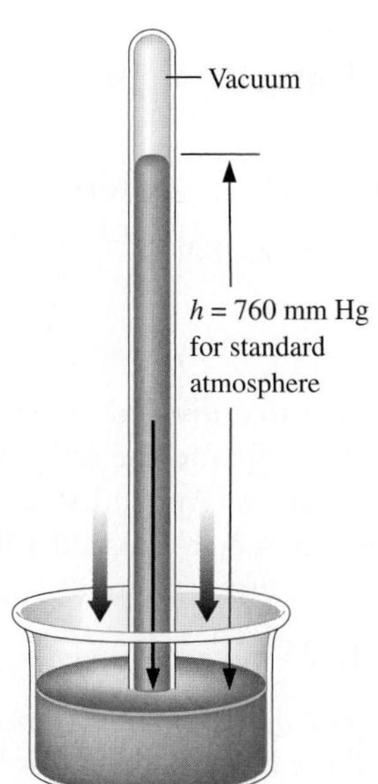

Figure 5.1
A torricellian barometer. The tube, completely filled with mercury, is inverted in a dish of mercury. Mercury flows out of the tube until the pressure of the column of mercury (shown by black arrow) "standing on the surface" of the mercury in the dish is equal to the pressure of the air (shown by purple arrows) on the rest of the surface of the mercury in the dish.

Van Helmont extensively studied a gas he obtained from burning wood, which he called "gas sylvestre" and which we now know as carbon dioxide, and noted that this substance was similar in many ways but not identical to air. By the end of his life, the importance of gases, especially air, was becoming more apparent. In 1643 the Italian physicist Evangelista Torricelli (1608–1647), who had been a student of Galileo, performed experiments that showed that *the air in the atmosphere exerts pressure.* (In fact, as we will see, all gases exert pressure.) Torricelli designed the first **barometer** by filling a tube that was closed at one end with mercury and then inverting it in a dish of mercury (Fig. 5.1). He observed that a column of mercury approximately 760 millimeters long always remained in the tube as a result of the pressure of the atmosphere.

A few years later Otto von Guericke, a German physicist, invented an air pump, often called a vacuum pump, that he used in a famous demonstration for the king of Prussia in 1654. Guericke placed two hemispheres together and pumped the air out of the resulting sphere through a valve, which was subsequently closed. He then dramatically showed that teams of horses could not pull the hemispheres apart. However, after secretly opening the valve to let air in, Guericke was able to separate the hemispheres easily by hand. The king of Prussia was so impressed by Guericke's cleverness that he awarded him a lifetime pension.

Units of Pressure

Because instruments used for measuring pressure, such as the **manometer** (Fig. 5.2), often use columns of mercury because of its high density, the most commonly used units for pressure are based on the height of the mercury column (in millimeters) the gas pressure can support. The unit **millimeters of**

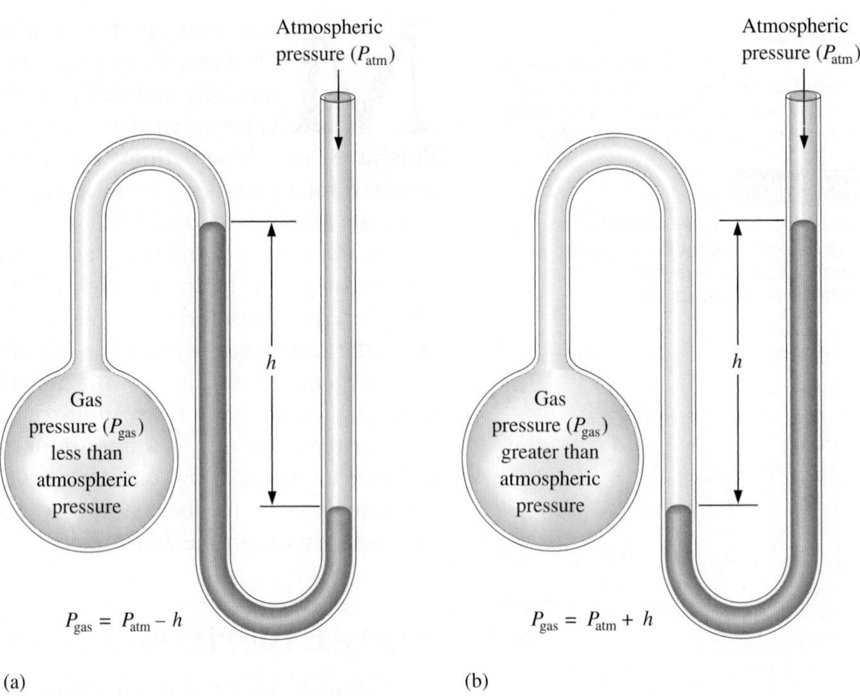

Figure 5.2
A simple manometer, a device for measuring the pressure of a gas in a container. The pressure of the gas is given by h (the difference in mercury levels) in units of torr (equivalent to mm Hg). (a) Gas pressure = atmospheric pressure − h. (b) Gas pressure = atmospheric pressure + h.

mercury (mm Hg) is called the **torr** in honor of Torricelli. A related unit for pressure is the **standard atmosphere:**

1 atm: 760 mm Hg, 760 torr, 101,325 Pa, 29.92 in Hg, 14.7 lb/in^2

$$1 \text{ standard atmosphere} = 1 \text{ atm} = 760 \text{ mm Hg} = 760 \text{ torr}$$

The International Union of Pure and Applied Chemists (IUPAC) has adopted 1 bar (100,000. Pa) as the standard pressure instead of 1 atm (101,325 Pa). Both standards are now widely used.

However, since pressure is defined as force per unit area, that is,

$$\text{Pressure} = \frac{\text{force}}{\text{area}}$$

the fundamental units of pressure involve units of force divided by units of area. In the SI system, the unit of force is the newton (N) and the unit of area is meters squared (m^2). (For a review of the SI system, see Appendix 2.) Thus the unit of pressure in the SI system is newtons per meter squared (N/m^2), called the **pascal (Pa)**. In terms of pascals the standard atmosphere is

$$1 \text{ standard atmosphere} = 101{,}325 \text{ Pa}$$

Thus 1 atm is approximately 10^5 pascals. Since the pascal is so small, and because it is not commonly used in the United States, we will use it sparingly in this book. However, converting from torrs or atmospheres to pascals is straightforward.

5.2 The Gas Laws of Boyle, Charles, and Avogadro

Boyle's Law

The first quantitative experiments on gases were performed by the Irish chemist Robert Boyle (1627–1691). Using a J-shaped tube closed at one end (Fig. 5.3), which he reportedly set up in the multistory entryway of his house, Boyle studied the relationship between the pressure of the trapped gas and its volume. Representative values from Boyle's experiments are given in Table 5.1. These data show that the product of the pressure and volume for the trapped air sample is constant within the accuracies of Boyle's measurements (note the third column in Table 5.1). This behavior can be represented by the equation

$$PV = k$$

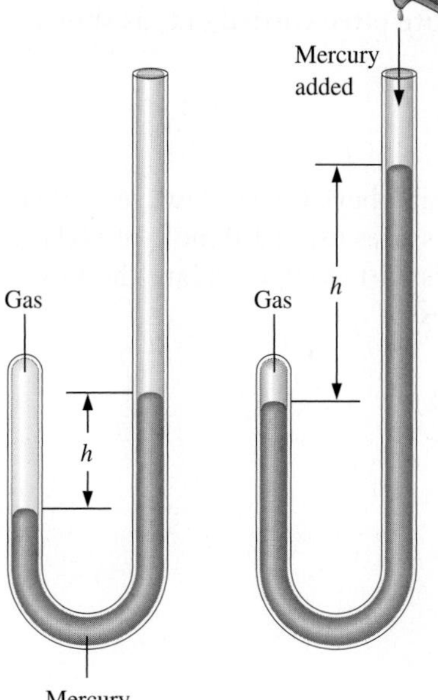

Figure 5.3
A J-tube similar to the one used by Boyle.

Table 5.1

Actual Data from Boyle's Experiments

Volume (in^3)	Pressure (in Hg)	Pressure × Volume (in Hg × in^3)
48.0	29.1	14.0×10^2
40.0	35.3	14.1×10^2
32.0	44.2	14.1×10^2
24.0	58.8	14.1×10^2
20.0	70.7	14.1×10^2
16.0	87.2	14.0×10^2
12.0	117.5	14.1×10^2

which is called **Boyle's law,** where k is a constant at a specific temperature for a given sample of air.

It is convenient to represent the data in Table 5.1 by using two different plots. Figure 5.4(a) shows a plot of P versus V, which produces a hyperbola. Notice that as the pressure drops by half, the volume doubles. Thus there is an *inverse relationship* between pressure and volume. The second type of plot can be obtained by rearranging Boyle's law to give

$$V = \frac{k}{P}$$

which is the equation for a straight line of the type

$$y = mx + b$$

Graphing is reviewed in Appendix A1.3.

where m represents the slope and b is the intercept of the straight line. In this case, $y = V$, $x = 1/P$, $m = k$, and $b = 0$. Thus a plot of V versus $1/P$ using Boyle's data gives a straight line with an intercept of zero, as shown in Fig. 5.4(b).

Boyle's law only approximately describes the relationship between pressure and volume for a gas. Highly accurate measurements on various gases at a constant temperature have shown that the product PV is not quite constant but changes with pressure. Results for several gases are shown in Fig. 5.5. Note the small changes that occur in the product PV as the pressure is varied. Such changes become very significant at pressures much higher than normal atmospheric pressure. We will discuss these deviations and the reasons for them in detail in Section 5.10. *A gas that obeys Boyle's law is called an* **ideal gas.** We will describe the characteristics of an ideal gas more completely in Section 5.3.

Boyle's law: $V \propto 1/P$ at constant temperature.

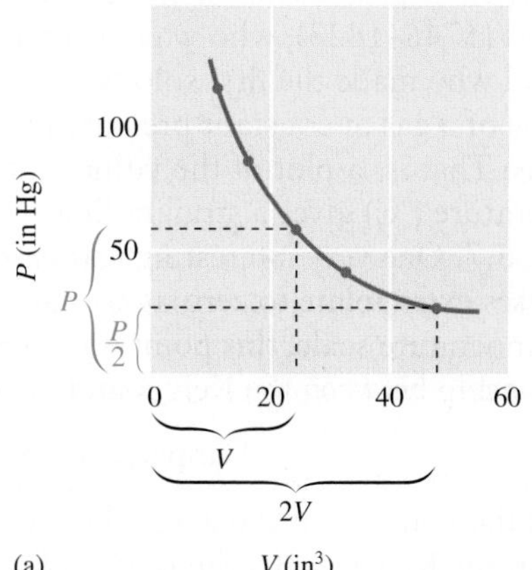

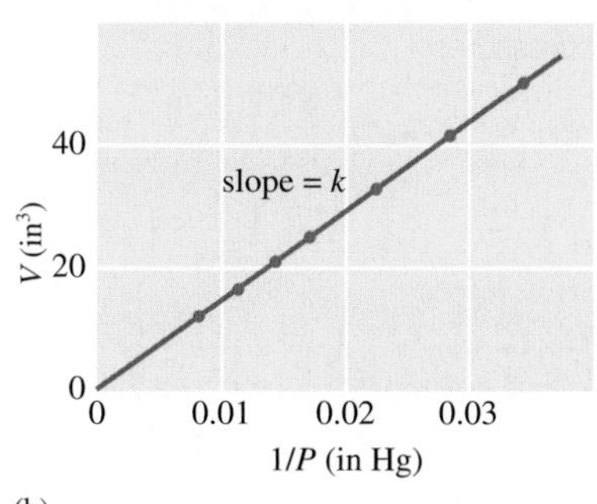

Figure 5.4
Plotting Boyle's data from Table 5.1. (a) A plot of P versus V shows that the volume doubles as the pressure is halved. (b) A plot of V versus $1/P$ gives a straight line. The slope of this line equals the value of the constant k.

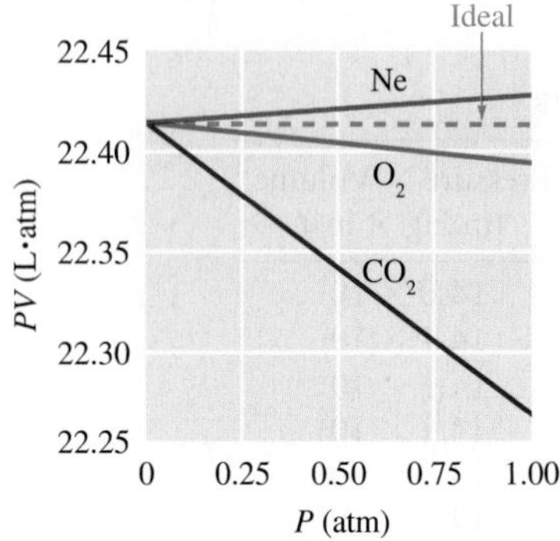

Figure 5.5
A plot of *PV* versus *P* for several gases. An ideal gas is expected to have a constant value of *PV*, as shown by the dashed line. Carbon dioxide shows the largest change in *PV*, and this change is actually quite small: *PV* changes from approximately 22.39 L atm at 0.25 atm to 22.26 L atm at 1.00 atm. Thus Boyle's law is a good approximation at these relatively low pressures.

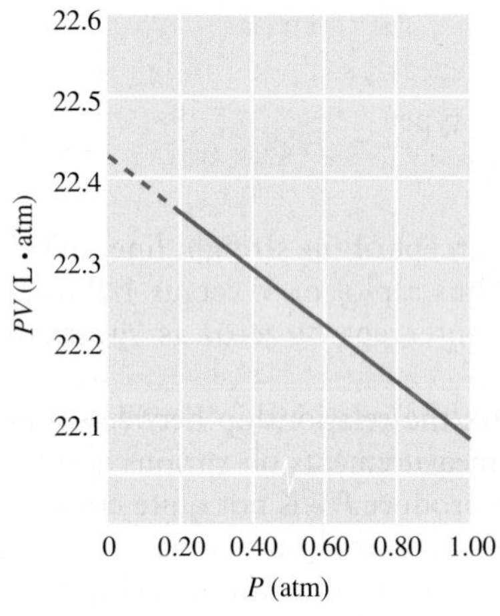

Figure 5.6
A plot of *PV* versus *P* for 1 mole of ammonia. The dashed line shows the extrapolation of the data to zero pressure to give the "ideal" value of *PV* of 22.41 L atm.

As with Boyle's law, Charles's law is obeyed exactly only at relatively low pressures.

One common use of Boyle's law is to predict the new volume of a gas when the pressure is changed (at constant temperature), or vice versa.

We mentioned earlier that Boyle's law is only approximately followed for real gases. To determine the significance of the deviations, studies of the effect of changing pressure on the volume of a gas are often carried out, as shown in Example 5.1.

EXAMPLE 5.1

In a study to see how closely gaseous ammonia obeys Boyle's law, several volume measurements were made at various pressures using 1.0 mole of NH_3 gas at a temperature of 0°C. Using the results listed below, calculate the Boyle's law constant for NH_3 at the various pressures.

Experiment	Pressure (atm)	Volume (L)
1	0.1300	172.1
2	0.2500	89.28
3	0.3000	74.35
4	0.5000	44.49
5	0.7500	29.55
6	1.000	22.08

Solution To determine how closely NH_3 gas follows Boyle's law under these conditions, we calculate the value of k (in L atm) for each set of values:

Experiment	1	2	3	4	5	6
$k = PV$	22.37	22.32	22.31	22.25	22.16	22.08

Although the deviations from true Boyle's law behavior are quite small at these low pressures, the value of k changes regularly in one direction as the pressure is increased. Thus, to calculate the "ideal" value of k for NH_3, plot PV versus P, as shown in Fig. 5.6, and extrapolate (extend the line beyond the experimental points) back to zero pressure, where, for reasons we will discuss later, a gas behaves most ideally. The value of k obtained by this extrapolation is 22.41 L atm. This is the same value obtained from similar plots for the gases CO_2, O_2, and Ne at 0°C, as shown in Fig. 5.5.

Charles's Law

In the century following Boyle's findings, scientists continued to study the properties of gases. One of these scientists was the French physicist Jacques Charles (1746–1823), who was the first person to fill a balloon with hydrogen gas and who made the first solo balloon flight. In 1787 Charles found that the volume of a gas at constant pressure increases *linearly* with the temperature of the gas. That is, a plot of the volume of a gas (at constant pressure) versus its temperature (°C) gives a straight line. This behavior is shown for several gases in Fig. 5.7. One very interesting feature of these plots is that the volumes of all the gases extrapolate to zero at the same temperature, −273.2°C. On the Kelvin temperature scale, this point is defined as 0 K, which leads to the following relationship between the Kelvin and Celsius scales:

$$\text{Temperature (K)} = 0°\text{C} + 273$$

When the volumes of the gases shown in Fig. 5.7 are plotted versus temperature on the Kelvin scale, the plots in Fig. 5.8 result. In this case the volume of

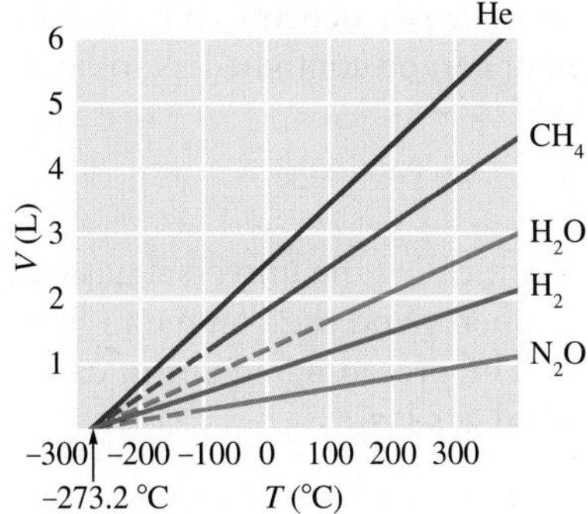

Figure 5.7
Plots of V versus T (°C) for several gases. The solid lines represent experimental measurements on gases. The dashed lines represent extrapolation of the data into regions where these gases would become liquids or solids. Note that the samples of the various gases contain different numbers of moles.

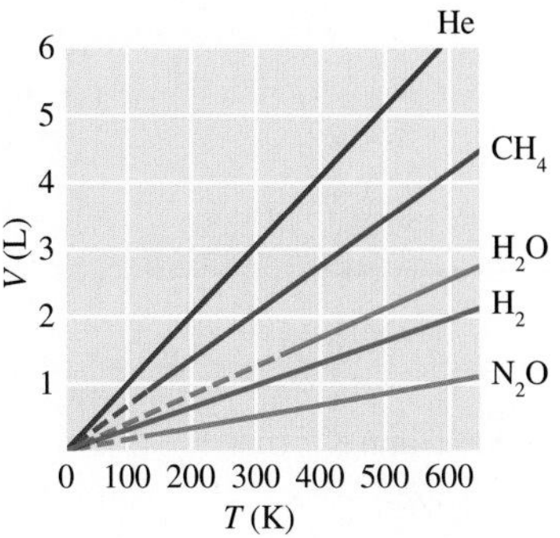

Figure 5.8
Plots of V versus T as in Fig. 5.7, except that here the Kelvin scale is used for temperature.

each gas is *directly proportional to temperature* and extrapolates to zero when the temperature is 0 K. This behavior is represented by the equation known as **Charles's law,**

$$V = bT$$

where T is the temperature (in kelvins) and b is a proportionality constant.

Before we illustrate the uses of Charles's law, let's consider the importance of 0 K. At temperatures below this point, the extrapolated volumes would become negative. The fact that a gas cannot have a negative volume suggests that 0 K has a special significance. In fact, 0 K is called **absolute zero,** and much evidence suggests that this temperature cannot be attained. Temperatures of approximately 10^{-6} K have been produced in laboratories, but 0 K has never been reached.

Avogadro's Law

In Chapter 2 we noted that in 1811 the Italian chemist Avogadro postulated that equal volumes of gases at the same temperature and pressure contain the same number of "particles." This observation is called **Avogadro's law** (or **hypothesis**), which can be stated mathematically as

$$V = an$$

where V is the volume of the gas, n is the number of moles, and a is a proportionality constant. This equation states that *for a gas at constant temperature and pressure, the volume is directly proportional to the number of moles of gas.* This relationship is obeyed closely by gases at low pressures.

5.3 The Ideal Gas Law

We have considered three laws that describe the behavior of gases as revealed by experimental observations:

Boyle's law:	$V = \frac{k}{P}$	(at constant T and n)
Charles's law:	$V = bT$	(at constant P and n)
Avogadro's law:	$V = an$	(at constant T and P)

These relationships showing how the volume of a gas depends on pressure, temperature, and number of moles of gas present can be combined as follows:

$$V = R\left(\frac{Tn}{P}\right)$$

where R is the combined proportionality constant called the **universal gas constant.** When the pressure is expressed in atmospheres and the volume in liters, R has the value of 0.08206 L atm K^{-1} mol^{-1}. The preceding equation can be rearranged to the more familiar form of the **ideal gas law:**

$R = 0.08206$ L atm K^{-1} mol^{-1}

$$PV = nRT$$

The ideal gas law is an *equation of state* for a gas, where the state of the gas is its condition at a given time. A particular *state* of a gas is described by its pressure, volume, temperature, and number of moles. Knowledge of any three of these properties is enough to completely define the state of a gas since the fourth property can then be determined from the equation for the ideal gas law.

It is important to recognize that the ideal gas law is an empirical equation—it is based on experimental measurements of the properties of gases. A gas that obeys this equation is said to behave *ideally*. That is, this equation defines the behavior of an ideal gas, which is a hypothetical substance. The ideal gas equation is best regarded as a limiting law—it expresses behavior that real gases *approach* at low pressures and high temperatures. Most gases obey this equation closely enough at pressures below 1 atm that only minimal errors result from assuming ideal behavior. Unless you are given information to the contrary, you should assume ideal gas behavior when solving problems involving gases in this text.

The ideal gas law applies best at pressures below 1 atm.

OWL INTERACTIVE EXAMPLE 5.2

A sample of hydrogen gas (H_2) has a volume of 8.56 L at a temperature of 0°C and a pressure of 1.5 atm. Calculate the moles of H_2 present in this gas sample.

Solution Solving the ideal gas law for n gives

$$n = \frac{PV}{RT}$$

In this case $P = 1.5$ atm, $V = 8.56$ L, $T = 0°C + 273 = 273$ K, and $R = 0.08206$ L atm K^{-1} mol^{-1}. Thus

$$n = \frac{(1.5\text{ atm})(8.56\text{ L})}{\left(0.08206\,\frac{\text{L atm}}{\text{K mol}}\right)(273\text{ K})} = 0.57\text{ mol}$$

The ideal gas law is often used to calculate the changes that will occur when the conditions of a gas are changed, as described below.

OWL INTERACTIVE EXAMPLE 5.3

Suppose we have a sample of ammonia gas with a volume of 3.5 L at a pressure of 1.68 atm. The gas is compressed to a volume of 1.35 L at a constant temperature. Use the ideal gas law to calculate the final pressure.

Solution The basic assumption we make when using the ideal gas law to describe a change in state for a gas is that the equation applies equally well to both the initial and the final states. In dealing with a change in state, we always *place the variables on one side of the equals sign and the constants on the other.* In this case the pressure and volume change, whereas the temperature and the number of moles remain constant (as does R, by definition). Thus we write the ideal gas law as

$$\underset{\text{Change}}{PV} = \underset{\text{Remain constant}}{nRT}$$

Since n and T remain the same in this case, we can write $P_1V_1 = nRT$ and $P_2V_2 = nRT$. Combining these equations gives

$$P_1V_1 = nRT = P_2V_2 \qquad \text{or} \qquad P_1V_1 = P_2V_2$$

We are given $P_1 = 1.68$ atm, $V_1 = 3.5$ L, $V_2 = 1.35$ L. Solving for P_2 gives

$$P_2 = \left(\frac{V_1}{V_2}\right)P_1 = \left(\frac{3.5\text{ L}}{1.35\text{ L}}\right)1.68\text{ atm} = 4.4\text{ atm}$$

Check: Does this answer make sense? The volume decreased (at constant temperature), so the pressure should increase, as the result of the calculation indicates. Note that the calculated final pressure is 4.4 atm. Because most gases do not behave ideally above 1 atm, we might find that if we *measured* the pressure of this gas sample, the observed pressure would differ slightly from 4.4 atm.

OWL INTERACTIVE EXAMPLE 5.4

A sample of methane gas that has a volume of 3.8 L at 5°C is heated to 86°C at constant pressure. Calculate its new volume.

Solution To solve this problem, we take the ideal gas law and segregate the changing variables and the constants by placing them on opposite sides of the equation. In this case volume and temperature change, and number of moles and pressure (and of course R) remain constant. Thus $PV = nRT$ becomes

$$\frac{V}{T} = \frac{nR}{P}$$

which leads to

$$\frac{V_1}{T_1} = \frac{nR}{P} \qquad \text{and} \qquad \frac{V_2}{T_2} = \frac{nR}{P}$$

Combining these equations gives

$$\frac{V_1}{T_1} = \frac{nR}{P} = \frac{V_2}{T_2} \qquad \text{or} \qquad \frac{V_1}{T_1} = \frac{V_2}{T_2}$$

We are given

$$T_1 = 5°\text{C} + 273 = 278\text{ K} \qquad T_2 = 86°\text{C} + 273 = 359\text{ K}$$

$$V_1 = 3.8\text{ L} \qquad V_2 = ?$$

Thus

$$V_2 = \frac{T_2V_1}{T_1} = \frac{(359\text{ K})(3.8\text{ L})}{278\text{ K}} = 4.9\text{ L}$$

Check: Is the answer sensible? In this case the temperature was increased (at constant pressure), so the volume should increase. Thus the answer makes sense.

The problem in Example 5.4 can be described as a Charles's law problem, whereas the problem in Example 5.3 can be said to be a Boyle's law problem. In both cases, however, we started with the ideal gas law. The real advantage of using the ideal gas law is that it applies to virtually any problem dealing with gases and is easy to remember.

OWL INTERACTIVE EXAMPLE 5.5

A sample of diborane gas (B_2H_6), a substance that bursts into flames when exposed to air, has a pressure of 345 torr at a temperature of $-15°C$ and a volume of 3.48 L. If conditions are changed so that the temperature is 36°C and the pressure is 468 torr, what will be the volume of the sample?

Solution Since, for this sample, pressure, temperature, and volume all change while the number of moles remains constant, we use the ideal gas law in the form

$$\frac{PV}{T} = \text{nR}$$

which leads to

$$\frac{P_1V_1}{T_1} = nR = \frac{P_2V_2}{T_2} \quad \text{or} \quad \frac{P_1V_1}{T_1} = \frac{P_2V_2}{T_2}$$

Then

$$V_2 = \frac{T_2P_1V_1}{T_1P_2}$$

We have

$$P_1 = 345 \text{ torr} \qquad P_2 = 468 \text{ torr}$$

$$T_1 = -15°\text{C} + 273 = 258 \text{ K} \qquad T_2 = 36°\text{C} + 273 = 309 \text{ K}$$

$$V_1 = 3.48 \text{ L} \qquad V_2 = ?$$

Thus

$$V_2 = \frac{(309 \text{ K})(345 \text{ torr})(3.48 \text{ L})}{(258 \text{ K})(468 \text{ torr})} = 3.07 \text{ L}$$

Since the equation used in Example 5.5 involved a *ratio* of pressures, it was unnecessary to convert pressures to units of atmospheres. The units of torr cancel. (You will obtain the same answer by inserting $P_1 = \frac{345}{760}$ and $P_2 = \frac{468}{760}$ into the equation.) However, temperature *must always* be converted to the Kelvin scale; since this conversion involves *addition* of 273, the conversion factor does not cancel. Be careful.

Always convert the temperature to the Kelvin scale when applying the ideal gas law.

5.4 Gas Stoichiometry

STP: 0°C and 1 atm

Suppose we have 1 mole of an ideal gas at 0°C (273.2 K) and 1 atm. From the ideal gas law, the volume of the gas is given by

$$V = \frac{nRT}{P} = \frac{(1.000 \text{ mol})(0.08206 \text{ L atm K}^{-1} \text{ mol}^{-1})(273.2 \text{ K})}{1.000 \text{ atm}} = 22.42 \text{ L}$$

This volume of 22.42 liters is called the **molar volume** of an ideal gas. The measured molar volumes of several gases are listed in Table 5.2. Note that the molar volumes of some of the gases are very close to the ideal value, but others deviate significantly. Later in this chapter we will discuss some of the reasons for the deviations.

The conditions 0°C and 1 atm, called **standard temperature and pressure** (abbreviated **STP**), are common reference conditions for the properties of gases. For example, the molar volume of an ideal gas is 22.42 L at STP.

Table 5.2

Molar Volumes for Various Gases at 0°C and 1 atm

Gas	Molar Volume (L)
Oxygen (O_2)	22.397
Nitrogen (N_2)	22.402
Hydrogen (H_2)	22.433
Helium (He)	22.434
Argon (Ar)	22.397
Carbon dioxide (CO_2)	22.260
Ammonia (NH_3)	22.079

Many chemical reactions involve gases. By assuming ideal behavior for these gases, we can carry out stoichiometric calculations if the pressure, volume, and temperature of the gases are known.

OWL INTERACTIVE EXAMPLE 5.6

Quicklime (CaO) is produced by the thermal decomposition of calcium carbonate ($CaCO_3$). Calculate the volume of CO_2 produced at STP from the decomposition of 152 g of $CaCO_3$ according to the reaction

$$CaCO_3(s) \longrightarrow CaO(s) + CO_2(g)$$

Solution

■ What are we trying to solve?

We are asked to solve for the volume of $CO_2(g)$ produced at STP. We are given the pressure and temperature, so to determine the volume we will need to determine the number of moles of gas produced.

We use the same strategy we used in the stoichiometry problems earlier in the book. That is, we compute the number of moles of $CaCO_3$ consumed and the number of moles of CO_2 produced. The moles of CO_2 can then be converted to volume by using the molar volume of an ideal gas.

Using the molar mass of $CaCO_3$, we can calculate the number of moles of $CaCO_3$:

$$152 \text{ g } CaCO_3 \times \frac{1 \text{ mol } CaCO_3}{100.1 \text{ g } CaCO_3} = 1.52 \text{ mol } CaCO_3$$

Since each mole of $CaCO_3$ produces 1 mole of CO_2, 1.52 moles of CO_2 will be formed. We can compute the volume of CO_2 at STP by using the molar volume:

$$1.52 \text{ mol } CO_2 \times \frac{22.42 \text{ L } CO_2}{1 \text{ mol } CO_2} = 34.1 \text{ L } CO_2$$

Thus the decomposition of 152 g of $CaCO_3$ will produce 34.1 L of CO_2 at STP.

Remember that the molar volume of an ideal gas is 22.42 L at STP.

Note that in Example 5.6 the final step involved calculation of the volume of gas from the number of moles. Since the conditions were specified as STP, we were able to use the molar volume of a gas at STP. If the conditions of a problem are different from STP, the ideal gas law must be used to calculate the volume.

Molar Mass

One very important use of the ideal gas law is in the calculation of the molar mass (molecular weight) of a gas from its measured density. To understand the relationship between gas density and molar mass, note that the number of moles of gas n can be expressed as

$$n = \frac{\text{grams of gas}}{\text{molar mass}} = \frac{\text{mass}}{\text{molar mass}} = \frac{m}{\text{molar mass}}$$

Substitution into the ideal gas equation gives

$$P = \frac{nRT}{V} = \frac{(m/\text{molar mass})RT}{V} = \frac{m(RT)}{V(\text{molar mass})}$$

But m/V is the gas density d in units of grams per liter. Thus

$$P = \frac{dRT}{\text{molar mass}}$$

or

$$\text{Molar mass} = \frac{dRT}{P} \tag{5.1}$$

Thus, if the density of a gas at a given temperature and pressure is known, its molar mass can be calculated.

You can memorize the equation involving gas density and molar mass, but it is better simply to remember the ideal gas equation, the definition of density, and the relationship between number of moles and molar mass. You can then derive this equation when you need it. This approach proves that you understand the concepts and means one less equation to memorize.

5.5 Dalton's Law of Partial Pressures

Among the experiments that led John Dalton to propose the atomic theory were his studies of mixtures of gases. In 1803 Dalton summarized his observations as follows: *For a mixture of gases in a container, the total pressure exerted is the sum of the pressures that each gas would exert if it were alone.* This statement, known as **Dalton's law of partial pressures,** can be expressed as follows:

$$P_{\text{Total}} = P_1 + P_2 + P_3 + \cdots$$

where the subscripts refer to the individual gases (gas 1, gas 2, and so on). The pressures P_1, P_2, P_3, and so on are called **partial pressures;** that is, each one is the pressure that gas would exert if it were alone in the container.

Assuming that each gas behaves ideally, the partial pressure of each gas can be calculated from the ideal gas law:

$$P_1 = \frac{n_1RT}{V}, \qquad P_2 = \frac{n_2RT}{V}, \qquad P_3 = \frac{n_3RT}{V}, \qquad \ldots$$

The total pressure of the mixture P_{Total} can be represented as

$$\begin{aligned} P_{\text{Total}} = P_1 + P_2 + P_3 + \cdots &= \frac{n_1RT}{V} + \frac{n_2RT}{V} + \frac{n_3RT}{V} + \cdots \\ &= (n_1 + n_2 + n_3 + \cdots)\left(\frac{RT}{V}\right) = n_{\text{Total}}\left(\frac{RT}{V}\right) \end{aligned}$$

where n_{Total} is the sum of the numbers of moles of the various gases. Thus, for a mixture of ideal gases, it is the *total number of moles of particles* that is im-

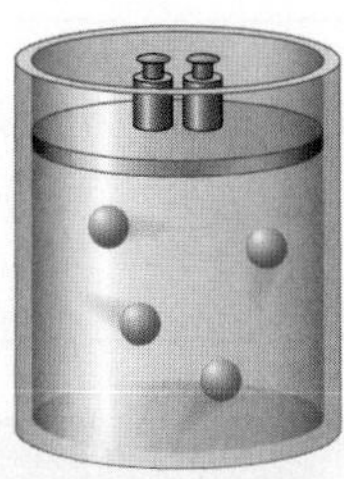

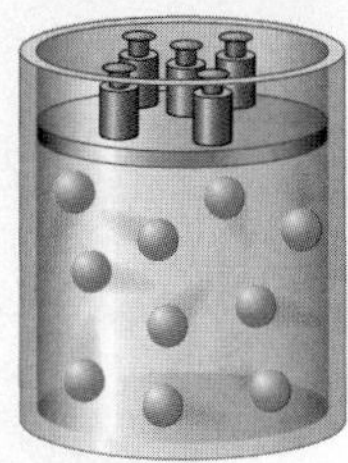

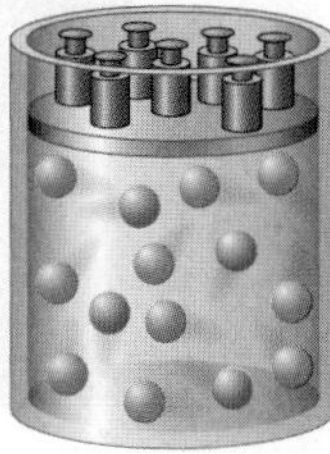

Figure 5.9
The partial pressure of each gas in a mixture of gases depends on the number of moles of that gas. The total pressure is the sum of the partial pressures and depends on the total moles of gas particles present, no matter what their identities.

portant, not the identity or composition of the individual gas particles. This idea is illustrated in Fig. 5.9.

This important result indicates some fundamental characteristics of an ideal gas. The fact that the pressure exerted by an ideal gas is not affected by the identity (structure) of the gas particles reveals two things about ideal gases: (1) the volume of the individual gas particle must not be important, and (2) the forces among the particles must not be important. If these factors were important, the pressure exerted by the gas would depend on the nature of the individual particles. These observations will strongly influence the model that we will eventually construct to explain ideal gas behavior.

At this point we need to define the **mole fraction:** *the ratio of the number of moles of a given component in a mixture to the total number of moles in the mixture.* The Greek letter chi (χ) is used to symbolize the mole fraction. For a given component in a mixture, the mole fraction χ_1 is

$$\chi_1 = \frac{n_1}{n_{\text{Total}}} = \frac{n_1}{n_1 + n_2 + n_3 + \cdots}$$

From the ideal gas equation, we know that the number of moles of a gas is directly proportional to the pressure of the gas, since

$$n = P\left(\frac{V}{RT}\right)$$

That is, for each component in the mixture,

$$n_1 = P_1\left(\frac{V}{RT}\right), \qquad n_2 = P_2\left(\frac{V}{RT}\right), \qquad \cdots$$

Therefore, we can represent the mole fraction in terms of pressures:

$$\chi_1 = \frac{n_1}{n_{\text{Total}}} = \frac{\overbrace{P_1(V/RT)}^{n_1}}{\underbrace{P_1(V/RT)}_{n_1} + \underbrace{P_2(V/RT)}_{n_2} + \underbrace{P_3(V/RT)}_{n_3} + \cdots}$$

$$= \frac{(V/RT)P_1}{(V/RT)(P_1 + P_2 + P_3 + \cdots)}$$

$$= \frac{P_1}{P_1 + P_2 + P_3 + \cdots} = \frac{P_1}{P_{\text{Total}}}$$

Similarly,
$$\chi_2 = \frac{n_2}{n_{\text{Total}}} = \frac{P_2}{P_{\text{Total}}}$$

and so on. Thus the mole fraction of a particular component in a mixture of ideal gases is directly related to its partial pressure.

The expression for the mole fraction,

$$\chi_1 = \frac{P_1}{P_{\text{Total}}}$$

can be rearranged:
$$P_1 = \chi_1 \times P_{\text{Total}}$$

That is, *the partial pressure of a particular component of a gaseous mixture is equal to the mole fraction of that component times the total pressure.*

A mixture of gases occurs whenever a gas is collected by displacement of water. For example, Fig. 5.10 shows the collection of oxygen gas produced by

Chemical Insights The Chemistry of Air Bags

Most experts agree that air bags represent a very important advance in automobile safety. First patented by American inventor John W. Hetrick in 1953, air bags are now required in all cars and trucks in the United States. These bags, which are stored in the auto's steering wheel or dash, are designed to inflate rapidly (within about 40 ms) in the event of a crash, cushioning the front-seat occupants against impact. The bags then deflate immediately to allow vision and movement after the crash. Air bags are activated when a severe deceleration (an impact) causes a steel ball to compress a spring and electrically ignite a detonator cap, which, in turn, causes sodium azide (NaN_3) to decompose explosively, forming sodium and nitrogen gas:

$$2NaN_3(s) \longrightarrow 2Na(s) + 3N_2(g)$$

This system works very well and requires only a relatively small amount of sodium azide [100 g yields 56 L of $N_2(g)$ at 25°C and 1 atm].

In addition to being located in the steering wheel and dash, air bags are now found in many other sites in motor vehicles. Air bags to protect against side impacts are located above the windows (called *curtain air bags* to protect against head injuries) and in the doors (called *torso air bags*). Some vehicles also have bags to prevent knee injuries. Because the explosive deployment of air bags can cause serious injuries, especially to children, variable force air bags have been developed that depend on the weight of the person occupying the front seat.

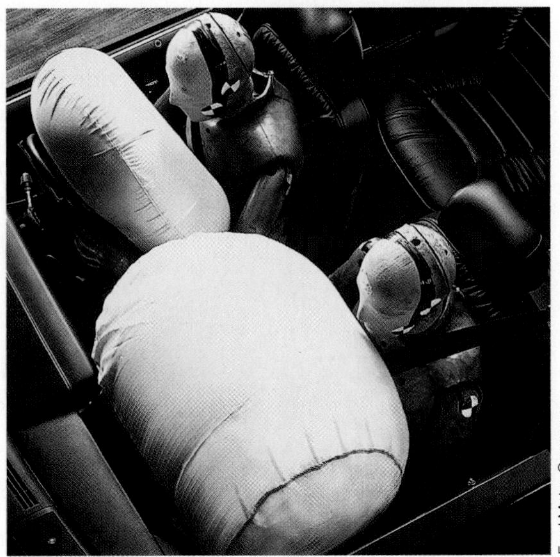

Inflated dual air bags.

When a vehicle containing air bags reaches the end of its useful life, the sodium azide present in the activators must be given proper disposal. Besides being explosive, sodium azide has a toxicity roughly equal to that of sodium cyanide. It also forms hydrazoic acid (HN_3), a toxic and explosive liquid, when treated with acid.

The air bag represents an application of chemistry that has already saved thousands of lives.

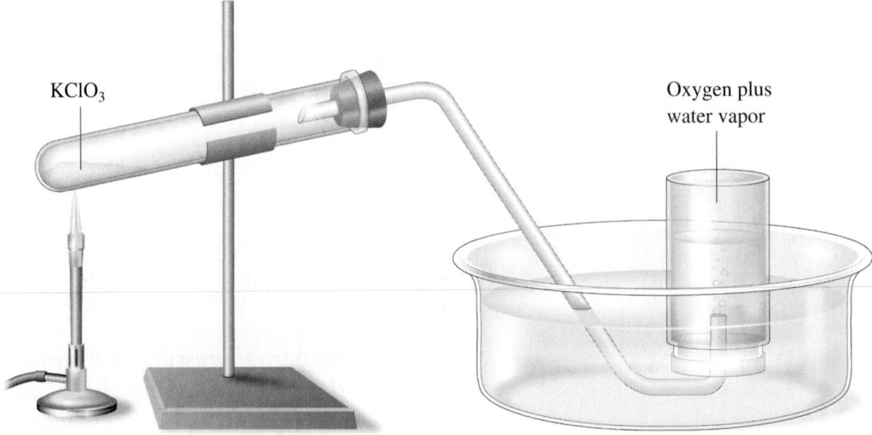

Figure 5.10
The production of oxygen by thermal decomposition of $KClO_3$. The MnO_2 catalyst is mixed with the $KClO_3$ to make the reaction faster.

Vapor pressure will be discussed in detail in Chapter 16.

the decomposition of solid potassium chlorate. In this situation the gas in the bottle is a mixture of water vapor and the gas being collected. Water vapor is present because molecules of water escape from the surface of the liquid and collect in the space above the liquid. Molecules of water also return to the liquid. When the rate of escape equals the rate of return, the number of water molecules in the vapor state remains constant, and thus the pressure of water vapor remains constant. This pressure, which depends on temperature, is called the *vapor pressure of water.*

OWL INTERACTIVE EXAMPLE 5.7

The mole fraction of nitrogen in air is 0.7808. Calculate the partial pressure of N_2 in air when the atmospheric pressure is 760. torr.

Solution The partial pressure of N_2 can be calculated as follows:

$$P_{N_2} = \chi_{N_2} \times P_{\text{Total}} = 0.7808 \times 760.\ \text{torr} = 593\ \text{torr}$$

OWL INTERACTIVE EXAMPLE 5.8

A sample of solid potassium chlorate ($KClO_3$) was heated in a test tube (Fig. 5.10) and decomposed according to the following reaction:

$$2KClO_3(s) \longrightarrow 2KCl(s) + 3O_2(g)$$

The oxygen produced was collected by displacement of water at 22°C at a total pressure of 754 torr. The volume of the gas collected was 0.650 L, and the vapor pressure of water at 22°C is 21 torr. Calculate the partial pressure of O_2 in the gas collected and the mass of $KClO_3$ in the sample that was decomposed.

Solution

- What are we trying to solve?

We are asked to find the partial pressure of O_2 in the gas collected and the mass of reactant ($KClO_3$) in the original sample.

We have the balanced equation and are given the volume, temperature, total pressure of the gas mixture, and vapor pressure of the water at that temperature. Since oxygen and water vapor are the only gases in the mixture, we can determine the partial pressure of the oxygen gas from Dalton's law of partial pressures:

$$P_{\text{Total}} = P_{O_2} + P_{H_2O} = P_{O_2} + 21\ \text{torr} = 754\ \text{torr}$$

Thus

$$P_{O_2} = 754\ \text{torr} - 21\ \text{torr} = 733\ \text{torr}$$

We now have to determine the mass of $KClO_3$ from the partial pressure of oxygen gas. Recall our strategy used in stoichiometry problems. That is, we can determine the mass of $KClO_3$ from the number of moles of $KClO_3$, which we can determine from the number of moles of O_2. We know the partial pressure of O_2. Along with volume and temperature, we can use the ideal gas law to find the number of moles of O_2:

$$n_{O_2} = \frac{P_{O_2}V}{RT}$$

In this case

$$P_{O_2} = 733 \text{ torr} = \frac{733 \text{ torr}}{760 \text{ torr/atm}} = 0.964 \text{ atm}$$

$$V = 0.650 \text{ L}$$

$$T = 22°\text{C} + 273 = 295 \text{ K}$$

$$R = 0.08206 \text{ L atm K}^{-1} \text{ mol}^{-1}$$

Thus

$$n_{O_2} = \frac{(0.964 \text{ atm})(0.650 \text{ L})}{(0.08206 \text{ L atm K}^{-1} \text{ mol}^{-1})(295 \text{ K})} = 2.59 \times 10^{-2} \text{ mol}$$

Next, we calculate the moles of $KClO_3$ needed to produce this quantity of O_2 using the mole ratio from the balanced equation for the decomposition of $KClO_3$:

$$2.59 \times 10^{-2} \text{ mol } O_2 \times \frac{2 \text{ mol } KClO_3}{3 \text{ mol } O_2} = 1.73 \times 10^{-2} \text{ mol } KClO_3$$

Using the molar mass of $KClO_3$, we calculate the grams of $KClO_3$:

$$1.73 \times 10^{-2} \text{ mol } KClO_3 \times \frac{122.6 \text{ g } KClO_3}{1 \text{ mol } KClO_3} = 2.12 \text{ g } KClO_3$$

Thus the original sample contained 2.12 g of $KClO_3$.

5.6 The Kinetic Molecular Theory of Gases

So far we have considered the behavior of gases from an experimental point of view. On the basis of observations from different types of experiments, we know that at pressure less than 1 atm most gases closely approach the behavior described by the ideal gas law. Now we want to construct a model to explain this behavior.

Before we construct the model, we will briefly review the scientific method. Recall that a law is a way of generalizing behavior that has been observed in many experiments. Laws are very useful since they allow us to predict the behavior of similar systems. For example, if a chemist prepares a new gaseous compound, a measurement of the gas density at known pressure and temperature can provide a reliable value for the compound's molar mass.

However, although laws summarize observed behavior, they do not tell us *why* nature behaves in the observed fashion. This is the central question for scientists. To try to answer this question, we construct theories (build models). The models in chemistry consist of speculations about what the individual atoms or molecules (microscopic particles) might be doing to cause the observed behavior of the macroscopic systems (collections of very large numbers of atoms and molecules).

A model is considered successful if it explains the observed behavior in question and predicts correctly the results of future experiments. Note that a model can never be proved to be absolutely true. In fact, *any model is an approximation* by its very nature and is bound to fail at some point. Models range from the simple to the extraordinarily complex. We use simple models to predict approximate behavior and more complicated models to account very precisely for observed quantitative behavior. In this text we will stress simple models that provide an approximate picture of what might be happening and that fit the most important experimental results.

Separating Gases

Chemical Insights

Assume you work for an oil company that owns a huge natural gas reservoir containing a mixture of methane and nitrogen gases. In fact, the gas mixture contains so much nitrogen that it is unusable as a fuel. Your job is to separate the nitrogen (N_2) from the methane (CH_4). How might you accomplish this task? You clearly need some sort of "molecular filter" that will stop the slightly larger methane molecules (size ≈430 pm) and allow the nitrogen molecules (size ≈410 pm) to pass through. To accomplish the separation of molecules so similar in size will require a very precise "filter."

The good news is that such a filter exists. Recent work by Steven Kuznicki and Valerie Bell at Engelhard Corporation in New Jersey and Michael Tsapatsis at the University of Massachusetts has produced a "molecular sieve" in which the pore (passage) sizes can be adjusted precisely enough to separate N_2 molecules from CH_4 molecules. The material involved is a special hydrated titanosilicate (contains H_2O, Ti, Si, O, and Sr) compound patented by Engelhard and known as ETS-4 (Engelhard TitanoSilicate-4). When sodium ions are substituted for the strontium ions in ETS-4 and the new material is carefully dehydrated, a uniform and controllable pore-size reduction occurs (see figure). The researchers have shown that the material can be used to separate N_2 (≈410 pm) from O_2 (≈390 pm). They have also shown that it is possible to reduce the nitrogen content of natural gas from 18% to less than 5% with a 90% recovery of methane.

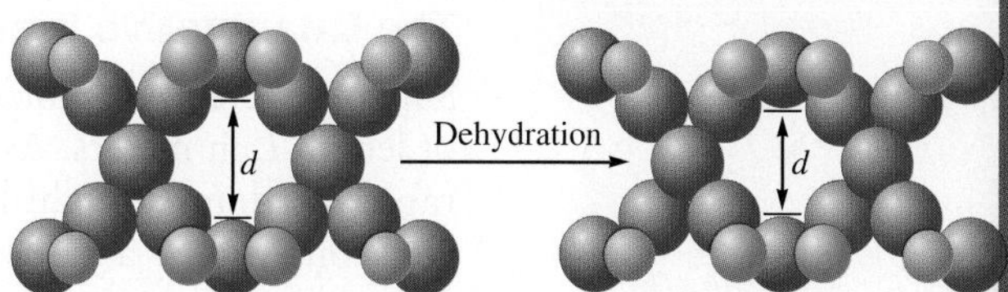

Molecular sieve framework of titanium (blue), silicon (green), and oxygen (red) atoms contracts on heating—at room temperature (left), $d = 4.27$ Å; at 250°C (right), $d = 3.94$ Å.

An example of this type of model is the **kinetic molecular theory,** a simple model that attempts to explain the properties of an ideal gas. This model is based on speculations about the behavior of the individual gas particles (atoms or molecules). The postulates of the kinetic molecular theory can be stated as follows:

Kinetic Molecular Theory

- The particles are so small compared with the distances between them that *the volume of the individual particles can be assumed to be negligible* (zero).
- *The particles are in constant motion. The collisions of the particles with the walls of the container are the cause of the pressure exerted by the gas.*
- *The particles are assumed to exert no forces on each other;* they are assumed to neither attract nor repel each other.
- *The average kinetic energy of a collection of gas particles is assumed to be directly proportional to the Kelvin temperature of the gas.*

Of course, real gas particles do have a finite volume and do exert forces on each other. Thus they do not conform exactly to these assumptions. But we will see that these postulates do indeed explain *ideal* gas behavior.

The true test of a model is how well its predictions fit the experimental observations. The postulates of the kinetic molecular model picture an ideal gas as consisting of particles having no volume and no attraction for each other, and the model assumes that the gas produces pressure on its container by collisions with the walls. To test the validity of this model, we need to consider the question: "When we apply the principles of physics to a collection of these gas particles, can we derive an expression for pressure that agrees with the ideal gas law?" The answer is, "Yes, we can." We will now consider this derivation in detail.

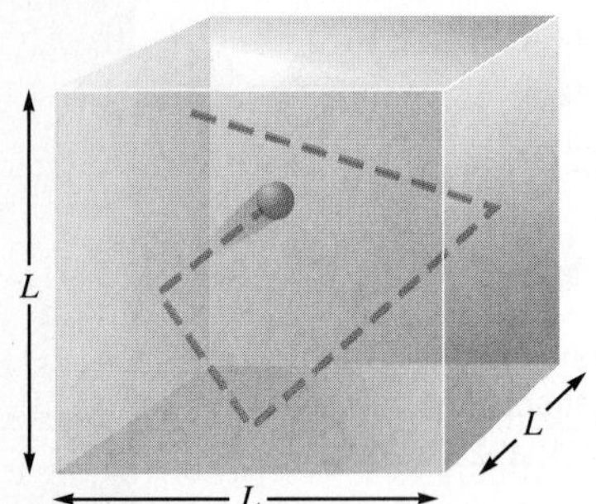

Figure 5.11
An ideal gas particle in a cube whose sides are of length L (in meters). The particle collides elastically with the walls in a random, straight-line motion.

The Quantitative Kinetic Molecular Model

Suppose there are n moles of an ideal gas in a cubical container with sides each of length L in meters. Assume each gas particle has a mass m and that it is in rapid, random, straight-line motion colliding with the walls (Fig. 5.11). The collisions will be assumed to be *elastic*—no loss of kinetic energy occurs. We want to compute the force on the walls from the colliding gas particles and then, since pressure is force per unit area, to obtain an expression for the pressure of the gas.

Before we can derive the expression for the pressure of a gas, we must first discuss some characteristics of velocity. Each particle in the gas has a particular velocity u, which can be broken into components u_x, u_y, and u_z (Fig. 5.12). First, using u_x and u_y and the Pythagorean theorem, we can obtain u_{xy} (Fig. 5.12):

$$u_{xy}^2 = u_x^2 + u_y^2$$

Hypotenuse of right triangle (u_{xy}^2); Sides of right triangle (u_x^2, u_y^2)

Then, constructing another triangle as shown in Fig. 5.12, we find

$$u^2 = u_{xy}^2 + u_z^2$$

or

$$u^2 = \underbrace{u_x^2 + u_y^2}_{u_{xy}^2} + u_z^2$$

Now let's consider how an individual gas particle moves. For example, how often does this particle strike the two walls of the box that are perpendicular to the x axis? Note that only the x component of the velocity affects the particle's impacts on these two walls (Fig. 5.13). The larger the x component of the velocity, the faster the particle travels between these two walls, thus producing more impacts per unit of time on these walls. Remember that the pressure of the gas is caused by these collisions with the walls.

The collision frequency (collisions per unit of time) with the two walls that are perpendicular to the x axis is given by

$$(\text{Collision frequency})_x = \frac{\text{velocity in the } x \text{ direction}}{\text{distance between the walls}} = \frac{u_x}{L}$$

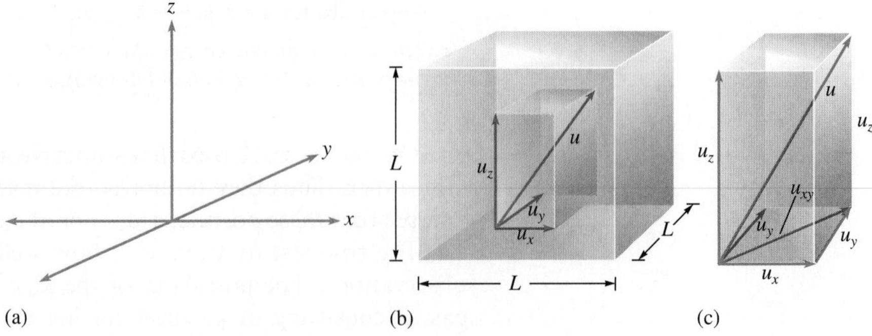

Figure 5.12
(a) The Cartesian coordinate axes.
(b) The velocity u of any gas particle can be broken down into three mutually perpendicular components, u_x, u_y, and u_z. This can be represented as a rectangular solid with sides u_x, u_y, and u_z and body diagonal u.
(c) In the xy plane,

$$u_x^2 + u_y^2 = u_{xy}^2$$

by the Pythagorean theorem. Since u_{xy} and u_z are also perpendicular,

$$u^2 = u_{xy}^2 + u_z^2 = u_x^2 + u_y^2 + u_z^2$$

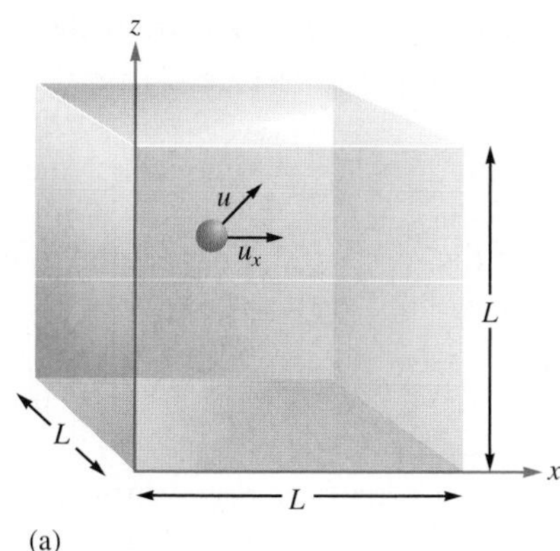

(a)

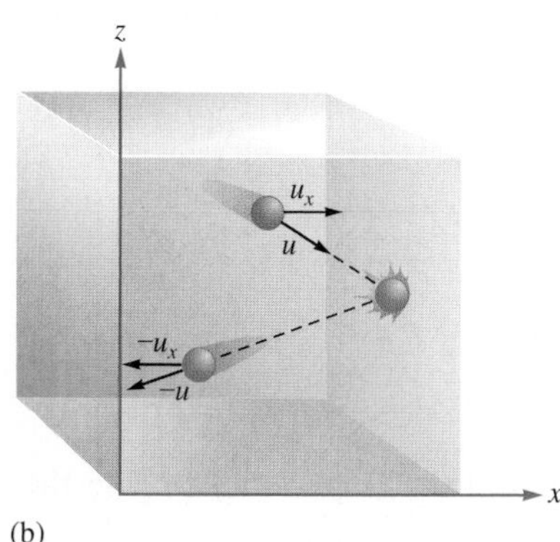

(b)

Figure 5.13
(a) Only the x component of the gas particle's velocity affects the frequency of impacts on the shaded walls, the walls that are perpendicular to the x axis. (b) For an elastic collision, there is an exact reversal of the x component of the velocity and of the total velocity. The change in momentum (final − initial) is then

$$-mu_x - mu_x = -2mu_x$$

Next, what is the force of a collision? Force is defined as mass times acceleration (change in velocity per unit of time):

$$F = ma = m\left(\frac{\Delta u}{\Delta t}\right)$$

where F represents force, a represents the acceleration, Δu represents a change in velocity, and Δt represents a given length of time.

Since we assume that the particle has constant mass, we can write

$$F = \frac{m\Delta u}{\Delta t} = \frac{\Delta(mu)}{\Delta t}$$

The quantity mu is the momentum of the particle (momentum is the product of mass and velocity), and the expression $F = \Delta(mu)/\Delta t$ means that force is the change in momentum per unit of time. When a particle hits a wall perpendicular to the x axis, as shown in Fig. 5.13, an elastic collision occurs, resulting in an *exact reversal* of the x component of velocity. That is, the *sign,* or direction, of u_x reverses when the particle collides with one of the walls perpendicular to the x axis. Thus the final momentum is the *negative,* or opposite, of the initial momentum. Remember that an elastic collision means that there is no change in the *magnitude* of the velocity. The change in momentum in the x direction is

$$\begin{aligned}\text{Change in momentum} &= \Delta(mu_x)\\ &= \text{final momentum} - \text{initial momentum}\\ &= -mu_x - mu_x\end{aligned}$$

↗ Final momentum in x direction ↖ Initial momentum in x direction

$$= -2mu_x$$

We are interested in the magnitude of the force the gas particle exerts on the walls of the box. Since we know that every action produces an equal but opposite reaction, the change in momentum with respect to the wall on impact is $-(-2mu_x)$, or $2mu_x$.

Recall that since force is the change in momentum per unit of time, then

$$\text{Force}_x = \frac{\Delta(mu_x)}{\Delta t}$$

for the walls perpendicular to the x axis.

This expression can be obtained by multiplying the change in momentum per impact by the number of impacts per unit of time:

$$\text{Force}_x = (2mu_x)\left(\frac{u_x}{L}\right) = \text{change in momentum per unit of time}$$

↗ Change in momentum per impact ↖ Impacts per unit of time

That is,

$$\text{Force}_x = \frac{2mu_x^2}{L}$$

So far we have considered only the two walls of the box perpendicular to the x axis. We can assume that the magnitude of the force on the two walls perpendicular to the y axis is given by

$$\text{Force}_y = \frac{2mu_y^2}{L}$$

and that the magnitude of the force on the two walls perpendicular to the z axis is given by

$$\text{Force}_z = \frac{2mu_z^2}{L}$$

Pressure is force per unit area, or the sum of the forces of all of the molecules striking the walls divided by the area of the wall. The pressure is

$$\text{Pressure} = \frac{\dfrac{2mu_x^2}{L} + \dfrac{2mu_y^2}{L} + \dfrac{2mu_z^2}{L}}{6L^2}$$

The 6 sides of the cube ↗ ↖ Area of each side

Since we have shown that

$$u^2 = u_x^2 + u_y^2 + u_z^2$$

the pressure is equal to

$$\text{Pressure} = \frac{\dfrac{2m}{L}(u_x^2 + u_y^2 + u_z^2)}{6L^2} = \frac{\dfrac{2m}{L}(u^2)}{6L^2}$$

Now, since we want the pressure caused by an average particle, we use the average of the square of the velocity $\overline{u^2}$ to obtain

$$\text{Pressure caused by "average" particle} = \frac{2m\overline{u^2}/L}{6L^2} = \frac{m\overline{u^2}}{3L^2}$$

Since the volume V of the cube is equal to L^3, we can write

$$\text{Pressure} = P = \frac{m\overline{u^2}}{3V}$$

So far we have considered the pressure on the walls caused by a single, "average" particle. Of course, we want the pressure caused by the entire gas

Photos Ken O'Donoghue

A balloon filled with air at room temperature.

The balloon is dipped into liquid nitrogen at 77 K.

The balloon collapses as the molecules inside slow down because of the decreased temperature. Slower molecules produce a lower pressure.

sample. The number of particles in a given gas sample can be expressed as follows:

$$\text{Number of gas particles} = nN_A$$

where n is the number of moles and N_A is Avogadro's number.

The total pressure on the box caused by n moles of a gas is therefore

$$P = nN_A \frac{m\overline{u^2}}{3V}$$

Next, we want to express the pressure in terms of the kinetic energy of the gas molecules. Kinetic energy (the energy caused by motion) is given by $\frac{1}{2}mu^2$, where m is the mass and u is the velocity. Since we are using the average of the velocity squared ($\overline{u^2}$), and since $m\overline{u^2} = 2(\frac{1}{2}m\overline{u^2})$, we have

$$P = \frac{2}{3}\left[\frac{nN_A(\frac{1}{2}m\overline{u^2})}{V}\right]$$

Recall that P is the pressure of the gas, n is the number of moles of gas, N_A is Avogadro's number, m is the mass of each particle, $\overline{u^2}$ is the average of the squares of the velocities of the particles, and V is the volume of the container.

KE = $\frac{1}{2}mu^2$, the energy caused by the motion of a particle.

The quantity $\frac{1}{2}m\overline{u^2}$ represents the average kinetic energy of a gas particle. If the average kinetic energy of an individual particle is multiplied by N_A, the number of particles in a mole, we get the average kinetic energy for a mole of gas particles:

$$(\text{KE})_{\text{avg}} = N_A(\tfrac{1}{2}m\overline{u^2})$$

Using this definition, we can rewrite the expression for pressure as

$$P = \frac{2}{3}\left[\frac{n(\text{KE})_{\text{avg}}}{V}\right] \quad \text{or} \quad \frac{PV}{n} = \frac{2}{3}(\text{KE})_{\text{avg}}$$

The fourth postulate of the kinetic molecular theory is that the average kinetic energy of the particles in the gas sample is directly proportional to the temperature in Kelvins. Thus, since $(\text{KE})_{\text{avg}} \propto T$, we can write

$$\frac{PV}{n} = \frac{2}{3}(\text{KE})_{\text{avg}} \propto T \quad \text{or} \quad \frac{PV}{n} \propto T$$

Note that this expression, has been *derived* from the assumptions of the kinetic molecular theory. How does it compare with the ideal gas law—the equation obtained from experiment? Compare the ideal gas law,

$$\frac{PV}{n} = RT \qquad \text{From experiment}$$

with the result from the kinetic molecular theory,

$$\frac{PV}{n} \propto T \qquad \text{From theory}$$

These expressions have exactly the same form if R, the universal gas constant, is considered the proportionality constant in the second case.

The agreement between the ideal gas, law and the kinetic molecular theory gives us confidence in the validity of the model. The characteristics we have assumed for ideal gas particles must agree, at least under certain conditions, with their actual behavior.

The Meaning of Temperature

We have seen from the kinetic molecular theory that the Kelvin temperature is a measure of the average kinetic energy of the gas particles. The exact relationship between temperature and average kinetic energy can be obtained by combining the equations

$$\frac{PV}{n} = RT = \frac{2}{3}(\text{KE})_{\text{avg}}$$

which yields the expression

$$(\text{KE})_{\text{avg}} = \tfrac{3}{2}RT$$

This is a very important relationship. It summarizes the meaning of the Kelvin temperature of a gas: The Kelvin temperature is an index of the random motions of the particles of a gas, with higher temperature meaning greater motion.

Root Mean Square Velocity

In the equation from the kinetic molecular theory, the average velocity of the gas particles is a special kind of average. The symbol $\overline{u^2}$ means the average of the *squares* of the particle velocities. The square root of $\overline{u^2}$ is called the **root mean square velocity** and is symbolized by u_{rms}:

$$u_{\text{rms}} = \sqrt{\overline{u^2}}$$

We can obtain an expression for u_{rms} from the equations

$$(\text{KE})_{\text{avg}} = N_A(\tfrac{1}{2}m\overline{u^2}) \qquad \text{and} \qquad (\text{KE})_{\text{avg}} = \tfrac{3}{2}RT$$

Combination of these equations gives

$$N_A(\tfrac{1}{2}m\overline{u^2}) = \tfrac{3}{2}RT \qquad \text{or} \qquad \overline{u^2} = \frac{3RT}{N_A m}$$

Taking the square root of both sides of the last equation produces

$$\sqrt{\overline{u^2}} = u_{\text{rms}} = \sqrt{\frac{3RT}{N_A m}}$$

In this expression, m represents the mass in kilograms of a single gas particle. When N_A, the number of particles in a mole, is multiplied by m, the product is the mass of a *mole* of gas particles in *kilograms*. We will call this quantity M. Substituting M for $N_A m$ in the equation for u_{rms}, we obtain

$$u_{\text{rms}} = \sqrt{\frac{3RT}{M}}$$

Before we can use this equation, we need to consider the units for R. So far we have used 0.08206 L atm K^{-1} mol^{-1} as the value of R. But to obtain the desired units (meters per second) for u_{rms}, R must be expressed in different units. As we will see in more detail in Chapter 9, the energy unit most often used in the SI system is the joule (J). A **joule** is defined as a kilogram meter squared per second squared (kg m^2/s^2). When R is converted from liter atmospheres to joules, it has the value 8.3145 J K^{-1} mol^{-1}. When R with these units is used in the expression $\sqrt{3RT/M}$, u_{rms} has units of meters per second, as desired.

So far we have said nothing about the range of velocities actually found in a gas sample. In a real gas, there are large numbers of collisions between particles. For example, when an odorous gas such as ammonia is released in a room, it takes some time for the odor to permeate the air, as we will see in Section 5.7. This delay results from collisions between the NH_3 molecules and O_2 and N_2 molecules in the air, which greatly slow the mixing process.

$R = 0.08206$ L atm K^{-1} mol^{-1}
$R = 8.3145$ J K^{-1} mol^{-1}

If the path of a particular gas particle could be monitored, it would probably look very erratic, something like that shown in Fig. 5.14. The average distance a particle travels between collisions in a particular gas sample is called the **mean free path.** It is typically a very small distance (1×10^{-7} m for O_2 at STP). One effect of the many collisions among gas particles is to produce a large range of velocities as the particles collide and exchange kinetic energy. Although u_{rms} for oxygen gas at STP is approximately 500 m/s, the majority of O_2 molecules do not have this velocity. The actual distribution of molecular velocities for oxygen gas at STP is shown in Fig. 5.15. This figure shows the relative number of gas molecules having each particular velocity.

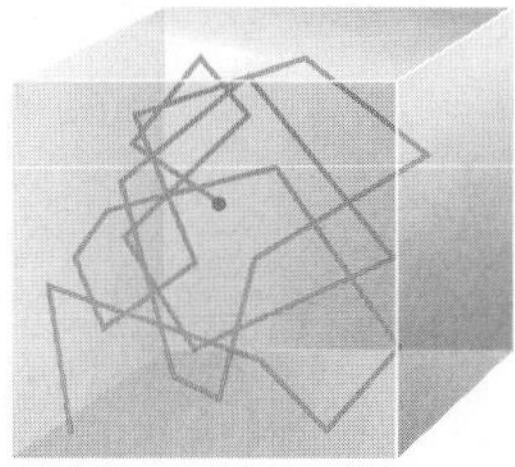

Figure 5.14
Path of one particle in a gas. Any given particle will continuously change its course as a result of collisions with other particles as well as with the walls of the container.

We are also interested in the effect of *temperature* on the velocity distribution in a gas. Figure 5.16 shows the velocity distribution for nitrogen gas at three temperatures. Note that as the temperature is increased, the curve maximum, which reflects the average velocity, moves toward higher values and the range of velocities becomes much larger.

The distribution of velocities of the particles in an ideal gas is described by the Maxwell–Boltzmann distribution law:

$$f(u) = 4\pi\left(\frac{m}{2k_BT}\right)^{3/2} u^2 e^{-mu^2/2k_BT}$$

where

u = velocity in m/s

m = mass of a gas particle in kg

k_B = Boltzmann's constant = 1.38066×10^{-23} J/K

T = temperature in K

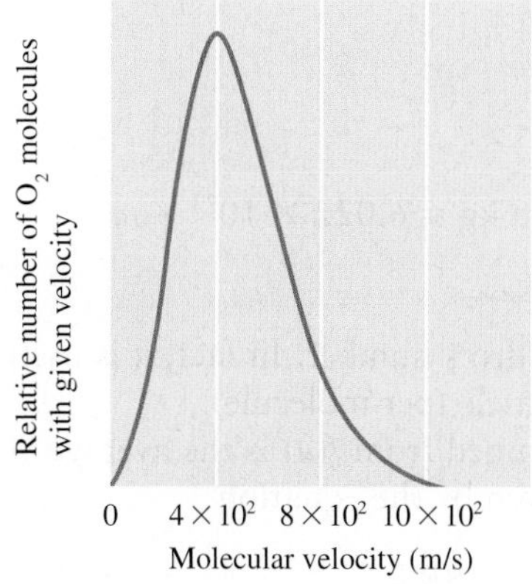

Figure 5.15
A plot of the relative number of O_2 molecules that have a given velocity at STP.

This equation was derived independently by James C. Maxwell, a Scottish physicist, and Ludwig E. Boltzmann, an Austrian physicist who did much of the fundamental theoretical work on the kinetic molecular description of an ideal gas. The product of $f(u)du$ represents the fraction of gas molecules with velocities between u and $u + du$, where du represents an infinitesimal velocity increment. This function is the one plotted in Figs. 5.15 and 5.16.

Analysis of the expression for $f(u)$ yields the following equation for the most probable velocity u_{mp} (the velocity possessed by the greatest number of gas particles):

$$u_{mp} = \sqrt{\frac{2k_BT}{m}} = \sqrt{\frac{2RT}{M}}$$

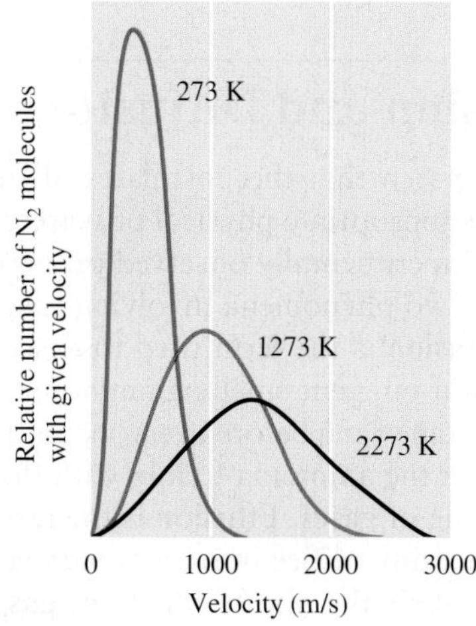

Figure 5.16
A plot of the relative number of N_2 molecules that have a given velocity at three temperatures. Note that as the temperature increases, both the average velocity (reflected by the curve's peak) and the spread of velocities increase.

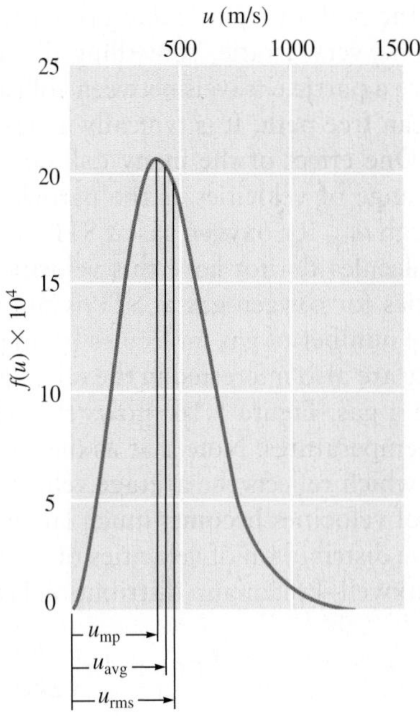

Figure 5.17
The velocity distribution for nitrogen gas at 273 K, with the values of most probable velocity u_{mp} (the velocity at the curve maximum), the average velocity u_{avg}, and the root mean square velocity u_{rms} indicated.

where

$$M = \text{molar mass of the gas particles in kg} = 6.022 \times 10^{23} \times m$$

$$R = \text{gas constant} = 6.022 \times 10^{23} \times k_B$$

Note that R and k_B are related by Avogadro's number. In fact, it is useful to think of k_B as the gas law constant per particle (per molecule).

Another type of velocity that can be obtained from $f(u)$ is the average velocity u_{avg} (sometimes written $\bar{u}$), which is given by the equation

$$u_{avg} = \bar{u} = \sqrt{\frac{8k_BT}{\pi m}} = \sqrt{\frac{8RT}{\pi M}}$$

Thus we have three ways to describe a "typical" velocity for the particles in an ideal gas: the root mean square velocity, the most probable velocity, and the average velocity. As can be seen from the equations for u_{rms}, u_{mp}, and u_{avg}, these velocities are not the same. In fact, they stand in the ratios

$$u_{mp} : u_{avg} : u_{rms} = 1.000 : 1.128 : 1.225$$

This relationship is shown for nitrogen gas at 0°C in Fig. 5.17.

5.7 Effusion and Diffusion

We have seen that the postulates of the kinetic molecular theory, combined with the appropriate physical principles, produce an equation that successfully fits the experimentally observed properties of gases as they approach ideal behavior. Two phenomena involving gases provide further tests of this model.

Diffusion is the term used to describe the mixing of gases. When a small amount of pungent-smelling ammonia is released at the front of a classroom, it takes some time before everyone in the room can smell it because time is required for the ammonia to mix with the air. The rate of diffusion is the rate of the mixing of gases. **Effusion** is the term used to describe the passage of a gas through a tiny orifice into an evacuated chamber (Fig. 5.18). The rate of effusion measures the rate at which the gas is transferred into the chamber.

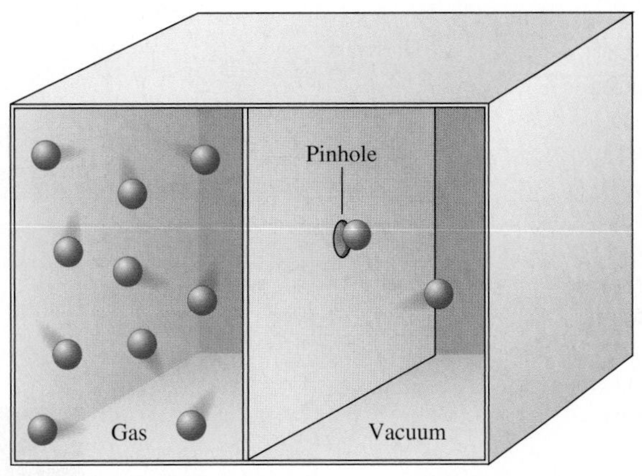

Figure 5.18
The effusion of a gas into an evacuated chamber. The rate of effusion (the rate at which the gas is transferred across the barrier through the pin hole) is inversely proportional to the square root of the mass of the gas molecules.

Effusion

Thomas Graham (1805–1869), a Scottish chemist, found experimentally that the rate of effusion of a gas is inversely proportional to the square root of the mass of its particles. Stated in another way, the relative rates of effusion of two gases at the same temperature and pressure are given by the inverse ratio of the square roots of the masses of the gas particles:

$$\frac{\text{Rate of effusion for gas 1}}{\text{Rate of effusion for gas 2}} = \frac{\sqrt{M_2}}{\sqrt{M_1}}$$

where M_1 and M_2 represent the molar masses of the gases. This equation is called **Graham's law of effusion.**

In Graham's law, the units for molar mass can be g/mol or kg/mol, since the units cancel in the ratio $\sqrt{M_2}/\sqrt{M_1}$.

Does the kinetic molecular model for gases correctly predict the relative effusion rates of gases summarized by Graham's law? To answer this question, we must recognize that the effusion rate for a gas depends directly on the average velocity of its particles. The faster the gas particles are moving, the more likely they are to pass through the effusion orifice. This reasoning leads to the following *prediction* for two gases at the same temperature T:

$$\frac{\text{Effusion rate for gas 1}}{\text{Effusion rate for gas 2}} = \frac{u_{\text{avg}} \text{ for gas 1}}{u_{\text{avg}} \text{ for gas 2}} = \frac{\sqrt{8RT/\pi M_1}}{\sqrt{8RT/\pi M_2}} = \frac{\sqrt{M_2}}{\sqrt{M_1}}$$

This equation is Graham's law, and thus the kinetic molecular model fits the experimental results for the effusion of gases.

Diffusion

Diffusion is frequently illustrated by the lecture demonstration represented in Fig. 5.19, in which two cotton plugs, one soaked in ammonia and the other in hydrochloric acid, are simultaneously placed at the ends of a long tube. A white ring of ammonium chloride (NH_4Cl) forms where the NH_3 and HCl molecules meet several minutes later:

$$NH_3(g) + HCl(g) \longrightarrow \underset{\text{White solid}}{NH_4Cl(s)}$$

The progress of the gases through the tube is surprisingly slow in light of the fact that the velocities of the HCl and NH_3 molecules at 25°C are approximately 450 and 660 m/s, respectively. Why does it take several minutes for the NH_3 and HCl molecules to meet? The answer is that the tube contains air and thus the NH_3 and HCl molecules undergo many collisions with O_2 and N_2 molecules as they travel through the tube. Although these collisions greatly slow their progress through the tube, it still seems reasonable to expect the

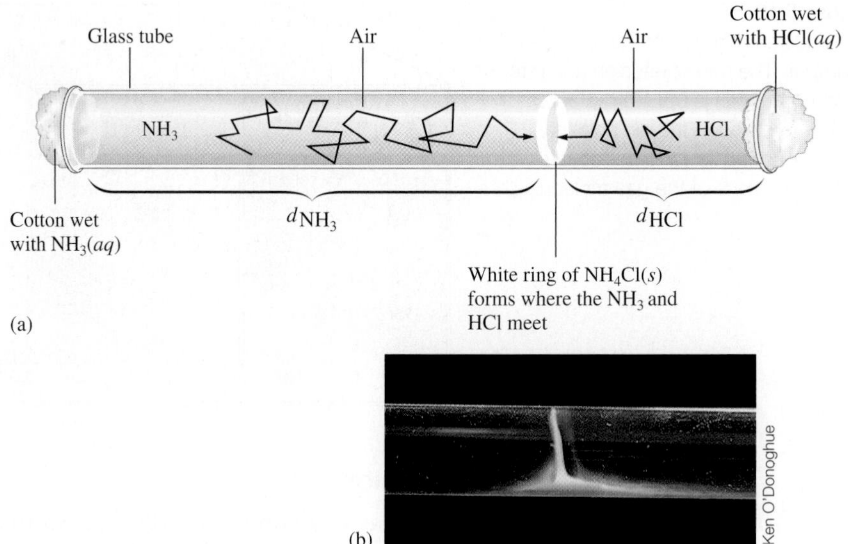

Figure 5.19
(a) A demonstration of the relative diffusion rates of NH_3 and HCl molecules through air. Two cotton plugs, one dipped in HCl(*aq*) and one dipped in NH_3(*aq*), are simultaneously inserted into the ends of the tube. Gaseous NH_3 and HCl vaporizing from the cotton plugs diffuse toward each other and, where they meet, react to form NH_4Cl(*s*).
(b) When HCl(*g*) and NH_3(*g*) meet in the tube, a white ring of NH_4Cl(*s*) forms.

relative distances traveled by the NH_3 and HCl molecules to be related to their velocities:

$$\frac{d_{NH_3}}{d_{HCl}} = \frac{\text{distance traveled by NH}_3}{\text{distance traveled by HCl}} = \frac{u_{avg(NH_3)}}{u_{avg(HCl)}} = \sqrt{\frac{M_{HCl}}{M_{NH_3}}} = \sqrt{\frac{36.5}{17}} = 1.5$$

However, careful experiments show that this prediction is not borne out—the observed ratio of distances is 1.3, not 1.5 as predicted by Graham's law. This discrepancy is not due to a failure of the kinetic molecular theory or of Graham's law; it exists because this "diffusion" experiment does not involve a simple diffusion process. Rather, it involves a *flow* of ammonia and hydrogen chloride gases through the air in the tube. Because the NH_3 and HCl molecules suffer many collisions with the N_2 and O_2 molecules in the tube, the flow rates of NH_3 and HCl are not directly proportional to their molecular velocities. Higher velocities lead to a higher number of intermolecular collisions, which in turn impedes the flow of the gas. Because of its smaller mass (and thus higher average velocity), the flow of the ammonia gas is impeded more than the flow of the hydrogen chloride gas. Therefore, the NH_3(*g*) travels a smaller distance to meet the HCl(*g*) than is expected from Graham's law (the distance ratio is smaller than 1.5).

If no air were present in the tube, the ratio of distances would be 1.5 as predicted from Graham's law.

Although we have given only a qualitative treatment here, the phenomena accompanying the mixing of gases are well understood, and the results of this experiment can be described very accurately by quantitative theories.

Although other technologies are now coming into use for this purpose, gaseous diffusion has played an important role in the enrichment of uranium for use in nuclear reactors. Natural uranium is mostly $^{238}_{92}U$, which cannot be fissioned to produce energy. It contains only about 0.7% of the fissionable nuclide $^{235}_{92}U$. For uranium to be useful as a nuclear fuel, the relative amount of $^{235}_{92}U$ must be increased to about 3%. In the gas diffusion enrichment process, the natural uranium (containing $^{238}_{92}U$ and a small amount of $^{235}_{92}U$) reacts with fluorine to form a mixture of $^{238}UF_6$ and $^{235}UF_6$. Because these molecules have slightly different masses, they will have slightly different velocities at a given temperature, which allows them to be separated by a multistage diffusion process. To understand how this process works, imagine a series of chambers separated by semiporous walls that allow passage of some UF_6 molecules but prevent bulk flow of gas. In effect, each porous wall acts much like a tiny hole in an effusion cell. Assume that the UF_6 from natural uranium is placed in chamber 1. Thus chamber 1 contains 99.3% $^{238}UF_6$ and 0.7% $^{235}UF_6$ (that is, 993 molecules of $^{238}UF_6$ for every 7 molecules of $^{235}UF_6$). Some molecules of this UF_6 diffuse through the semiporous barrier into chamber 2, which was initially empty. Because of its smaller mass, $^{235}UF_6$, which has a slightly greater velocity than $^{238}UF_6$, diffuses at a slightly greater rate. Thus chamber 2 will contain a ratio of $^{235}UF_6$ to $^{238}UF_6$ that is slightly greater than 7 to 993.

Although the process is called *gaseous diffusion* because the chambers are separated by barriers that effectively allow only individual UF_6 molecules to pass through, it behaves like an effusion process. Thus we can find the actual ratio of the two types of UF_6 in chamber 2 from Graham's law:

$$\frac{\text{Diffusion rate for } ^{235}\text{UF}_6}{\text{Diffusion rate for } ^{238}\text{UF}_6} = \sqrt{\frac{\text{mass}(^{238}\text{UF}_6)}{\text{mass}(^{235}\text{UF}_6)}} = \sqrt{\frac{352.05 \text{ g/mol}}{349.03 \text{ g/mol}}} = 1.0043$$

We can use this factor to calculate the ratio of $^{235}UF_6/^{238}UF_6$ in chamber 2:

$$\underset{\text{Chamber 2}}{\underset{\uparrow}{\frac{^{235}\text{UF}_6}{^{238}\text{UF}_6}}} = 1.0043 \times \underset{\text{Chamber 1}}{\underset{\uparrow}{\frac{^{235}\text{UF}_6}{^{238}\text{UF}_6}}} = 1.0043\left(\frac{7}{993}\right)$$

$$= 1.0043(7.0493 \times 10^{-3})$$

$$= 7.0797 \times 10^{-3}$$

This very slight increase represents a change from the ratio of 70,493 $^{235}UF_6$ molecules per 10,000,000 $^{238}UF_6$ molecules in chamber 1 to the ratio of 70,797 $^{235}UF_6$ molecules per 10,000,000 $^{238}UF_6$ molecules in chamber 2.

This enrichment process (in $^{235}UF_6$) continues as the slightly enriched gas in chamber 2 diffuses into chamber 3 and is again enriched by a factor of 1.0043. The same process is repeated until sufficient enrichment occurs. Obviously, this process will take many stages. For example, to calculate the number of steps required to enrich from 0.700% ^{235}U to 3.00% ^{235}U, we have the following equation:

$$\underset{\text{Original ratio}}{\underset{\uparrow}{\frac{0.700\ ^{235}\text{UF}_6}{99.3\ ^{238}\text{UF}_6}}} \times (1.0043)^N = \underset{\text{Desired ratio}}{\underset{\uparrow}{\frac{3.00\ ^{235}\text{UF}_6}{97.0\ ^{238}\text{UF}_6}}}$$

where N represents the number of stages. This equation follows from the fact that each stage produces an enrichment by the factor 1.0043. Thus

$$\text{Original ratio} \times \underset{\substack{\text{First}\\\text{stage}}}{\underset{\uparrow}{1.0043}} \times \underset{\substack{\text{Second}\\\text{stage}}}{\underset{\uparrow}{1.0043}} \times \underset{\substack{\text{Third}\\\text{stage}}}{\underset{\uparrow}{1.0043}} \times \cdots = \text{final ratio}$$

Solving this equation for N yields 345. Thus we predict that 345 stages are required to obtain the desired enrichment.

Although we have greatly oversimplified* the actual enrichment process here, this discussion gives you an idea of how it is accomplished. A photo of actual diffusion cells is shown in Fig. 5.20.

U.S. Department of Energy

Figure 5.20
Uranium-enrichment converters from the Paducah gaseous diffusion plant in Kentucky.

*For a more detailed description, see W. Spindel and T. Ishida, "Isotope Separation," *J. Chem. Ed.* 68 (1991): 312.

Enrichment of uranium by gaseous diffusion has become obsolete. Increasingly, a process using gas centrifuges has become the method of choice. Such a process uses a large number of rotating cylinders (centrifuges) that cause the more massive $^{238}UF_6$ to move to the outside of each cylinder relative to the $^{235}UF_6$, which remains closer to the center. Gas centrifuge technology requires much less energy than gaseous diffusion, and each step has a 1.3 separation factor compared to 1.0043 for gaseous diffusion. Other technologies are also being explored for uranium enrichment, but we will not discuss them here.

5.8 Collisions of Gas Particles with the Container Walls

In the analysis of the kinetic molecular model that led to the ideal gas equation, we assumed that the pressure a gas exerts is caused by the collisions of its particles with the walls of its container. In this section we will consider the details of that phenomenon.

Our goal is to obtain an equation that describes the number of particles that collide per second with a given area of the wall. Although a rigorous derivation of such an equation can be carried out from the details of the kinetic molecular theory, we will not do that. Instead, we will pursue a qualitative strategy, trying to obtain the fundamental relationships from our conceptual understanding of how an ideal gas is expected to behave. We will define the quantity we are looking for as Z_A, the collision rate (per second) of the gas particles with a section of wall that has an area A (in m^2). We expect Z_A to depend on the following factors:

1. The average velocity of the gas particles
2. The size of the area being considered
3. The number of particles in the container

How is Z_A expected to depend on the average velocity of the gas particles? For example, if we double the average velocity, we double the number of wall impacts, so Z_A should double. Thus Z_A depends directly on u_{avg}:

$$Z_A \propto u_{avg}$$

Similarly, Z_A depends directly on A, the area of the wall under consideration. That is, if we double the area being considered, we will double the number of impacts per second that occur within that section of the wall. Thus $Z_A \propto A$.

Likewise, if the number of particles in the container is doubled, the impacts with the wall will double. For a general case, we need to consider not the absolute number of particles but the number of particles per unit volume (the number density of particles), which can be represented by N/V, the number of particles N divided by the volume V (in m^3). Thus Z_A is expected to depend directly on N/V. That is, $Z_A \propto N/V$.

In summary, Z_A should be directly proportional to u_{avg}, A, and N/V:

$$Z_A \propto u_{avg} \times A \times \frac{N}{V}$$

Note that the units for Z_A expected from this relationship are

$$\frac{\text{m}}{\text{s}} \times \text{m}^2 \times \frac{\text{(particles)}}{\text{m}^3} \longrightarrow \frac{\text{(particles)}}{\text{s}} \quad \text{or} \quad \frac{\text{(collisions)}}{\text{s}}$$

The parentheses are used here because particles and collisions are understood and are not actual units. The correct units for Z_A are 1/s, or s^{-1}. The fact that the product $u_{avg} \times A \times N/V$ gives the units expected for Z_A indicates that we are considering all the gas properties that influence Z_A. Substituting the expression for u_{avg} gives

$$Z_A \propto \frac{N}{V}A\sqrt{\frac{8RT}{\pi M}}$$

A more detailed analysis of the situation shows that the proportionality constant is $\frac{1}{4}$. Thus the exact equation for Z_A is

$$Z_A = \frac{1}{4}\frac{N}{V}A\sqrt{\frac{8RT}{\pi M}} = A\frac{N}{V}\sqrt{\frac{RT}{2\pi M}}$$

OWL INTERACTIVE EXAMPLE 5.9

Calculate the impact rate on a 1.00-cm² section of a vessel containing oxygen gas at a pressure of 1.00 atm and 27°C.

Solution To calculate Z_A, we must identify the values of the variables in the equation

$$Z_A = A\frac{N}{V}\sqrt{\frac{RT}{2\pi M}}$$

In this case, A is given as 1.00 cm². However, to be inserted into the expression for Z_A, A must have the units m². The appropriate conversion gives $A = 1.00 \times 10^{-4}\ m^2$.

The quantity N/V can be obtained from the ideal gas law by solving for n/V and then converting to the appropriate units:

$$\frac{n}{V} = \frac{P}{RT} = \frac{1.00\ \text{atm}}{\left(0.08206\ \frac{\text{L atm}}{\text{K mol}}\right)(300.\ \text{K})} = 4.06 \times 10^{-2}\ \text{mol/L}$$

To obtain N/V, which has the units (molecules)/m³, from n/V, we make the following conversion:

$$\frac{N}{V} = 4.06 \times 10^{-2}\ \frac{\text{mol}}{\text{L}} \times 6.022 \times 10^{23}\ \frac{\text{(molecules)}}{\text{mol}} \times \frac{1000\ \text{L}}{\text{m}^3}$$

$$= 2.44 \times 10^{25}\ \text{(molecules)/m}^3$$

The quantity M represents the molar mass of O_2 in kg. Thus

$$M = 32.0\ \frac{\text{g}}{\text{mol}} \times \frac{1\ \text{kg}}{1000\ \text{g}} = 3.20 \times 10^{-2}\ \text{kg/mol}$$

Next, we insert these quantities into the expression for Z_A:

$$Z_A = A\frac{N}{V}\sqrt{\frac{RT}{2\pi M}} = (1.00 \times 10^{-4}\ \text{m}^2)(2.44 \times 10^{25}\ \text{m}^{-3})$$

$$\times \sqrt{\frac{\left(8.3145\frac{\text{J}}{\text{K mol}}\right)(300.\ \text{K})}{(2)(3.14)\left(3.20 \times 10^{-2}\ \frac{\text{kg}}{\text{mol}}\right)}} = 2.72 \times 10^{23}\ \text{s}^{-1}$$

That is, in this gas 2.72×10^{23} collisions per second occur on each 1.00-cm² area of the container.

5.9 Intermolecular Collisions

Recall that the postulates of the kinetic molecular model do not take into account collisions between gas particles. Since this model correctly fits ideal gas behavior (that is, the behavior approached by real gases at high T and low P), our conclusion is that intermolecular collisions apparently do not have an important influence on the pressure, volume, or temperature of a gas behaving ideally. That is, the effects of the collisions must somehow "cancel out" relative to the properties P, V, and T of an ideal gas. However, there is much evidence to suggest that collisions do occur among the gas particles in a real gas. For example, a gas that is somehow disturbed from a Maxwell–Boltzmann distribution of velocities will rapidly change until it again reaches a Maxwell–Boltzmann distribution. This behavior must be caused by energy exchanges through collisions.

In this section we will consider the collision frequency of the particles in a gas. We will start by considering a single spherical gas particle with diameter d (in meters) that is moving with velocity u_{avg}. As this particle moves through the gas in a straight line, it will collide with another particle only if the other particle has its center in a cylinder with radius d, as shown in Fig. 5.21.

Any particle with its center outside this cylinder will not be hit by our particle. Thus our particle "sweeps out" a cylinder of radius d and length $u_{avg} \times 1$ second during every second of its flight. Therefore, the volume of the cylinder swept out per second is

$$V = \text{volume} = \underbrace{(\pi d^2)}_{\substack{\text{Area of}\\ \text{cylinder}\\ \text{slice}}} \underbrace{(u_{avg})(1\text{ s})}_{\substack{\text{Length of}\\ \text{cylinder}}}$$

As the particle travels through this cylinder, the number of collisions depends on the number of gas particles in that volume. To specify the number of gas particles, we use the number density of the gas N/V, which indicates the number of gas particles per unit volume. Thus we can write

$$\begin{aligned}\text{Number of collisions per second} &= \begin{pmatrix}\text{volume}\\ \text{swept out}\end{pmatrix} \times \frac{N}{V} = \pi d^2 (u_{avg}) \left(\frac{N}{V}\right) \\ &= \pi d^2 \left(\sqrt{\frac{8RT}{\pi M}}\right)\left(\frac{N}{V}\right) = \frac{N}{V} d^2 \sqrt{\frac{8\pi RT}{M}}\end{aligned}$$

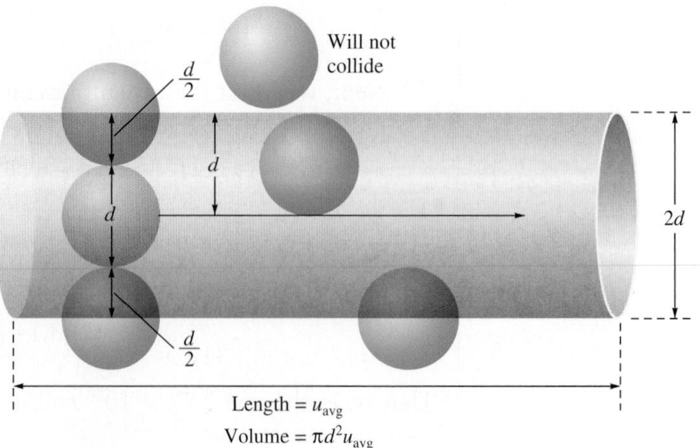

Figure 5.21
The cylinder swept out by a gas particle of diameter *d*.

This equation is not quite correct. If you are thinking carefully about this situation, you may be asking yourself the question, "What about the motions of the other particles?" That is, we have said that the primary particle has velocity u_{avg}, but we have assumed that the other particles are stationary. Of course, they are not really stationary. They are moving in various directions with various velocities. When the motions of the other particles are accounted for (a derivation we will not show here), the *relative velocity* of the primary particle becomes $\sqrt{2}\, u_{avg}$ rather than the value u_{avg} that we have been using. Thus the expression for the collision rate becomes

$$\text{Collision rate (per second)} = Z = \sqrt{2}\,\frac{N}{V}\,d^2\sqrt{\frac{8\pi RT}{M}} = 4\frac{N}{V}\,d^2\sqrt{\frac{\pi RT}{M}}$$

OWL INTERACTIVE EXAMPLE 5.10

Calculate the collision frequency for an oxygen molecule in a sample of pure oxygen gas at 27°C and 1.0 atm. Assume that the diameter of an O_2 molecule is 300 pm.

Solution To obtain the collision frequency, we must identify the quantities in the expression

$$Z = 4\frac{N}{V}\,d^2\sqrt{\frac{\pi RT}{M}}$$

that are appropriate to this case. We can obtain the value of *N*/*V* for this sample of oxygen by assuming ideal behavior. From the ideal gas law,

$$\frac{n}{V} = \frac{P}{RT} = \frac{1.0 \text{ atm}}{\left(0.08206\,\frac{\text{L atm}}{\text{K mol}}\right)(300.\text{ K})} = 4.1 \times 10^{-2}\text{ mol/L}$$

Thus

$$\frac{N}{V} = \left(4.1 \times 10^{-2}\,\frac{\text{mol}}{\text{L}}\right)\left(6.022 \times 10^{23}\,\frac{\text{molecules}}{\text{mol}}\right)\left(\frac{1000\text{ L}}{\text{m}^3}\right)$$

$$= 2.5 \times 10^{25}\text{ (molecules)/m}^3$$

From the given information, we know that

$$d = 300\text{ pm} = 300 \times 10^{-12}\text{ m or } 3 \times 10^{-10}\text{ m}$$

Also, for O_2, $M = 3.20 \times 10^{-2}$ kg/mol. Thus

$$Z = 4(2.5 \times 10^{25}\text{ m}^{-3})(3 \times 10^{-10}\text{ m})^2 \times \sqrt{\frac{\pi(8.3145\text{ J K}^{-1}\text{ mol}^{-1})(300\text{ K})}{3.20 \times 10^{-2}\text{ kg/mol}}}$$

$$= 4 \times 10^9\text{ (collisions)/s} = 4 \times 10^9\text{ s}^{-1}$$

Notice how large this number is. Each O_2 molecule undergoes approximately 4 billion collisions per second in this gas sample.

Mean Free Path

As we saw earlier, the collision frequency *Z* represents the number of collisions per second that occur in a given gas sample. On the other hand, the reciprocal of *Z* gives the time (in seconds) between collisions. Thus, if $Z = 4 \times 10^9$ (collisions) per second, then $1/Z = 2.5 \times 10^{-10}$ seconds between collisions. Now if we multiply $1/Z$ by the average velocity, we obtain the **mean free path λ:**

$$\lambda = \underset{\substack{\nearrow \\ \text{Time between} \\ \text{collisions (s)}}}{\frac{1}{Z}} \times \underset{\substack{\uparrow \\ \text{Distance traveled} \\ \text{per second}}}{u_{avg}} = \text{distance between collisions}$$

Substituting the expressions for $1/Z$ and u_{avg} gives

$$\lambda = \left[\frac{1}{4(N/V)(d^2)\sqrt{RT/M}}\right]\left(\sqrt{\frac{8RT}{\pi M}}\right) = \frac{1}{\sqrt{2}(N/V)(\pi d^2)}$$

EXAMPLE 5.11

Calculate the mean free path in a sample of oxygen gas at 27°C and 1.0 atm.

Solution Using data from the preceding example, we have

$$\lambda = \frac{1}{\sqrt{2}(2.5 \times 10^{25}\ \text{m}^{-3})(\pi)(3 \times 10^{-10}\ \text{m})^2} = 1 \times 10^{-7}\ \text{m}$$

Note that an O_2 molecule travels only a very short distance before it collides with another O_2 molecule. This produces a path for a given O_2 molecule like the one represented in Fig. 5.14, where the length of each straight line is $\sim 10^{-7}$ m.

5.10 Real Gases

An ideal gas is a hypothetical concept. No gas *exactly* follows the ideal gas law, although many gases come very close at low pressures and/or high temperatures. Thus ideal gas behavior can best be thought of as the behavior *approached by real gases* under certain conditions.

We have seen that a very simple model, the kinetic molecular theory, by making some rather drastic assumptions (no interparticle interactions and zero volume for the gas particles), successfully explains ideal behavior. However, it is important that we examine real gas behavior to see how it differs from that predicted by the ideal gas law and to determine what modifications of the kinetic molecular theory are needed to explain the observed behavior. Since a model is an approximation and will inevitably fail, we must be ready to learn from such failures. In fact, we often learn more about nature from the failures of our models than from their successes.

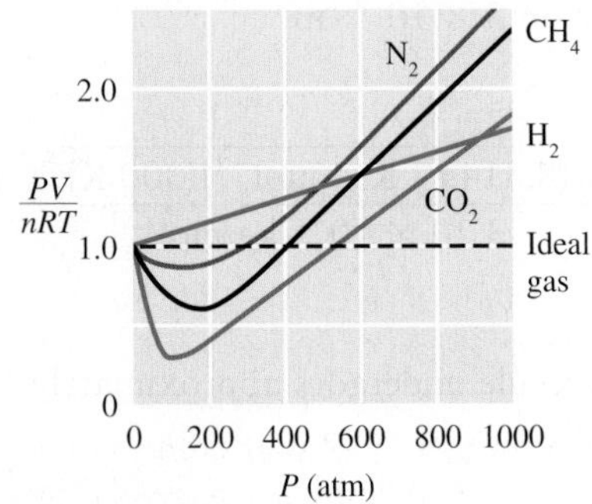

Figure 5.22
Plots of *PV/nRT* versus *P* for several gases (200 K). Note the significant deviations from ideal behavior (*PV/nRT* = 1). The behavior is close to ideal only at low pressures (less than 1 atm).

We will examine the experimentally observed behavior of real gases by measuring the pressure, volume, temperature, and number of moles for a gas and noting how the quantity PV/nRT depends on pressure. Plots of PV/nRT versus P are shown for several gases in Fig. 5.22. For an ideal gas, PV/nRT equals 1 under all conditions, but notice that for real gases, PV/nRT approaches 1 only at low pressures (typically 1 atm). To illustrate the effect of temperature, we have plotted PV/nRT versus P for nitrogen gas at several temperatures in Fig. 5.23. Notice that the behavior of the gas appears to become more nearly ideal as the temperature is increased. The most important conclusion to be drawn from these plots is that a real gas typically exhibits behavior that is closest to ideal behavior at *low pressures* and *high temperatures*.

How can we modify the assumptions of the kinetic molecular theory to fit the behavior of real gases? An equation for real gases was developed in 1873 by Johannes van der Waals, a physics professor at the University of Amster-

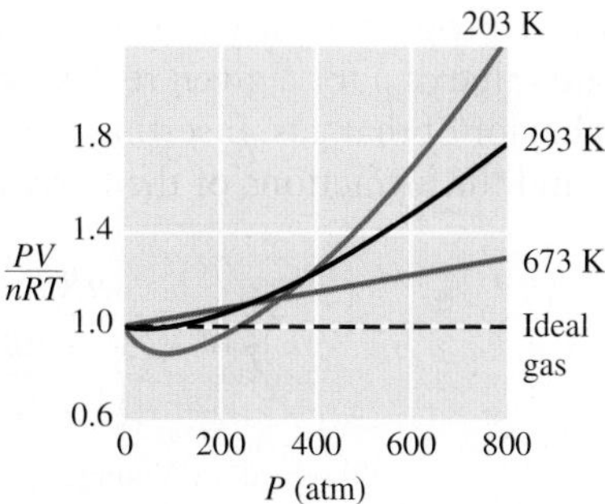

Figure 5.23
Plots of PV/nRT versus P for nitrogen gas at three temperatures. Note that although nonideal behavior is evident in each case, the deviations are smaller at the higher temperatures.

PV/nRT is also 1 at high pressures for many gases because of a canceling of nonideal effects.

dam who in 1910 received a Nobel Prize for his work. To follow his analyses, we start with the ideal gas law,

$$P = \frac{nRT}{V}$$

P' is corrected for the finite volume of the particles. The attractive forces have not yet been taken into account.

P_{obs} is usually called just P.

We have now corrected for both the finite volume and the attractive forces of the particles.

Remember that this equation describes the behavior of a hypothetical gas consisting of volumeless entities that do not interact with each other. In contrast, a real gas consists of atoms or molecules that have finite volumes. Thus the volume available to a given particle in a real gas is less than the volume of the container because the gas particles themselves take up some of the space. To account for this discrepancy, van der Waals represented the actual volume as the volume of the container V minus a correction factor for the volume of the molecules nb, where n is the number of moles of gas and b is an empirical constant (one determined by fitting the equation to the experimental results). Thus the volume *actually available* to a given gas molecule is given by the difference $V - nb$.

This modification of the ideal gas equation leads to the expression

$$P' = \frac{nRT}{(V - nb)}$$

The volume of the gas particles has now been taken into account.

The next step is to account for the attractions that occur among the particles in a real gas. The effect of these attractions is to make the observed pressure P_{obs} smaller than it would be if the gas particles did not interact:

$$P_{obs} = (P' - \text{correction factor}) = \left(\frac{nRT}{V - nb} - \text{correction factor}\right)$$

This effect can be understood by using the following model. When gas particles come close together, attractive forces occur, which cause the particles to hit the wall slightly less often than they would in the absence of these interactions (Fig. 5.24).

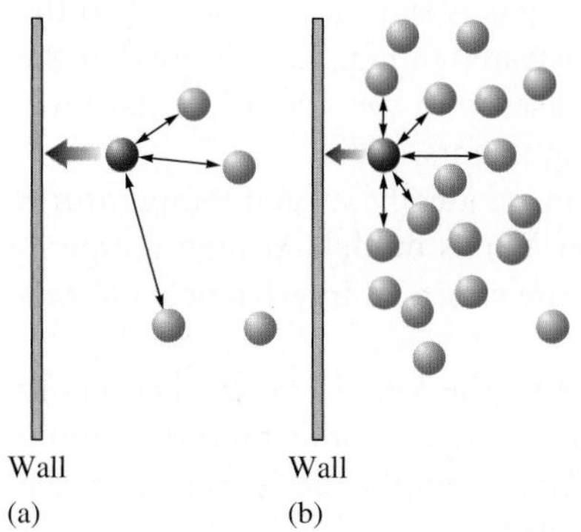

Figure 5.24
(a) Gas at low concentration—relatively few interactions between particles. The indicated gas particle exerts a pressure on the wall close to that predicted for an ideal gas.
(b) Gas at high concentration—many more interactions between particles. Because of these interactions the collision frequency with the walls is lowered, thus causing the observed pressure to be smaller than if the gas were behaving ideally.

The size of the correction factor depends on the concentration of gas molecules defined in terms of moles of gas particles per liter (n/V). The higher the concentration, the more likely a pair of gas particles will be close enough to attract each other. For large numbers of particles, the number of interacting *pairs* of particles depends on the square of the number of particles and thus on the square of the concentration, or $(n/V)^2$. This reasoning can be justified as follows: In a gas sample containing N particles, there are $N - 1$ partners available for each particle, as shown in Fig. 5.25. Since the $1 \cdots 2$ pair is the same as the $2 \cdots 1$ pair, this analysis counts each pair twice. Thus for N particles there are $N(N - 1)/2$ pairs. If N is a very large number, $N - 1$ approximately equals N, giving $N^2/2$ possible pairs. Thus the correction to the ideal pressure for the attractions of the particles has the form

$$P_{obs} = P' - a\left(\frac{n}{V}\right)^2$$

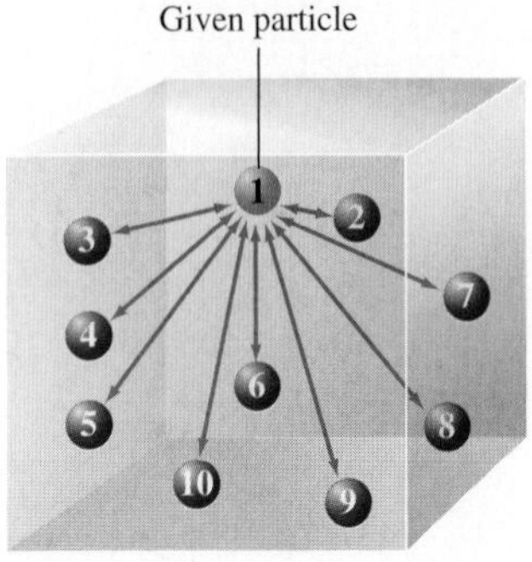

Figure 5.25
Illustration of pairwise interactions among gas particles. In a sample with 10 particles, each particle has 9 possible partners, to give 10(9)/2 = 45 distinct pairs. The factor of $\frac{1}{2}$ arises because when particle 1 is the particle of interest, we count the 1 ··· 2 pair, and when particle 2 is the particle of interest, we count the 2 ··· 1 pair. However, 1 ··· 2 and 2 ··· 1 are the same pair, which we thus have counted twice. Therefore, we must divide by 2 to get the correct number of pairs.

where a is a proportionality constant (which includes the factor of $1\frac{1}{2}$ from $N^2/2$). The value of a for a given real gas can be determined from observing the actual behavior of that gas. Inserting the corrections for both the volume of the particles and the attractions of the particles gives the equation

$$P_{\text{obs}} = \frac{nRT}{V - nb} - a\left(\frac{n}{V}\right)^2$$

Observed pressure (P_{obs}); Volume of the container (V); Volume correction (nb); Pressure correction ($a(n/V)^2$)

This equation can be rearranged to give the **van der Waals equation:**

$$\underbrace{\left[P_{\text{obs}} + a\left(\frac{n}{V}\right)^2\right]}_{\substack{\text{Corrected pressure} \\ P_{\text{ideal}}}} \underbrace{(V - nb)}_{\substack{\text{Corrected volume} \\ V_{\text{ideal}}}} = nRT$$

The values of the weighting factors, a and b, are determined for a given gas by fitting experimental behavior. That is, a and b are varied until the best fit of the observed pressure is obtained under all conditions. The values of a and b for various gases are given in Table 5.3.

Experimental studies indicate that the changes van der Waals made in the basic assumptions of the kinetic molecular theory corrected the major flaws in the model. First, consider the effects of volume. For a gas at low pressure (large volume), the volume of the container is very large compared with the volumes of the gas particles. That is, the volume available to the gas is essentially equal to the volume of the container, so the gas behaves ideally. On the other hand, for a gas at high pressure (small volume), the volume of the particles becomes significant so that the volume available to the gas is significantly less than the container volume. These observations are illustrated in Fig. 5.26. Note that the volume-correction constant b generally increases with the size of the gas molecule, which gives further support to these arguments.

The fact that a real gas tends to behave more ideally at high temperatures can also be explained in terms of the van der Waals model. At high temperatures the particles are moving so rapidly that the effects of interparticle interactions are not very important.

The corrections made by van der Waals to the kinetic molecular theory make physical sense, which makes us confident that we understand the fundamentals of gas behavior at the particle level. This is significant because so much important chemistry takes place in the gas phase.

Table 5.3

Values of van der Waals Constants for Some Common Gases

Gas	$a\left(\frac{\text{atm L}^2}{\text{mol}^2}\right)$	$b\left(\frac{\text{L}}{\text{mol}}\right)$
He	0.034	0.0237
Ne	0.211	0.0171
Ar	1.35	0.0322
Kr	2.32	0.0398
Xe	4.19	0.0511
H_2	0.244	0.0266
N_2	1.39	0.0391
O_2	1.36	0.0318
Cl_2	6.49	0.0562
CO_2	3.59	0.0427
CH_4	2.25	0.0428
NH_3	4.17	0.0371
H_2O	5.46	0.0305

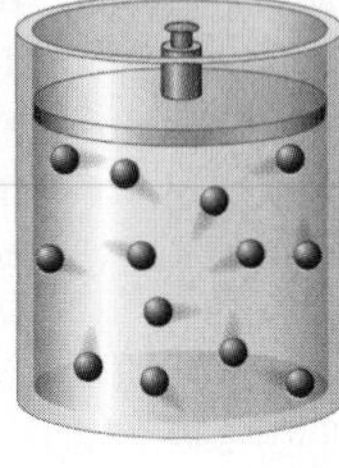

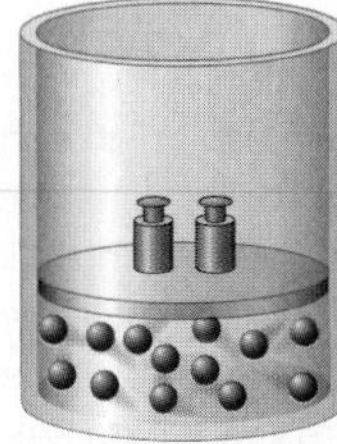

Figure 5.26
The volume occupied by the gas particles themselves is less important at (a) large container volumes (low pressure) than at (b) small container volumes (high pressure).

Kenneth Suslick Practices Sound Chemistry

Chemistry Explorers

Professor Kenneth S. Suslick, who received his B.S. degree from the California Institute of Technology and his Ph.D. from Stanford University, has spent his entire academic career at the University of Illinois at Urbana-Champaign. Although his research interests have spanned the traditional areas of chemistry, Professor Suslick and his group have been especially interested in sonochemistry, the chemical effects of ultrasound in which sound waves (pitched above human hearing) are applied to solutions.

The effects of high-intensity sound waves on solutions come from cavitation: the formation, growth, and implosive collapse of bubbles in a liquid. Cavitational collapse produces very high pressures that lead to local heating so intense that they cause reactions requiring high energies to occur, often accompanied by the emission of light.

In fact, Suslick's research has just shown that as the bubble implodes, the interior can reach temperatures higher than 15,000 K—about three times the temperature of the surface of the sun. At these temperatures atoms come apart to form a plasma, which contains ions and electrons. Thus cavitation can create extraordinarily hot spots in an otherwise cold liquid.

Besides its use for studying high-energy reactions, ultrasound has also proved valuable in the synthesis of nanostructured materials. One example is the formation of liquid-containing protein microspheres that could be injected into the body's circulatory system to deliver drugs. Another application involves surface chemistry. When cavitation occurs near a solid surface, the high-speed jets of liquid and the associated sound waves can alter the surface properties of the solid. This has proved useful in activating the surfaces of metals used as catalysts.

Sonochemistry is indeed sound chemistry.

Jim Gray/Courtesy of Kenneth S. Suslick

A humorous photograph of Professor Kenneth S. Suslick. The fog apparently emanating from his head is produced by a very cold vapor from liquid nitrogen, which freezes moisture from the air.

5.11 Characteristics of Several Real Gases

We can understand gas behavior more completely if we examine the characteristics of several common gases. Note from Fig. 5.22 that the gases H_2, N_2, CH_4, and CO_2 show different behavior when the compressibility PV/nRT is plotted versus P. For example, notice that the plot for $H_2(g)$ never drops below the ideal value (1.0) in contrast to all the other gases. What is special about H_2 compared to these other gases? Recall from Section 5.8 that the reason that the compressibility of a real gas falls below 1.0 is that the actual (observed) pressure is lower than the pressure expected for an ideal gas due to the intermolecular attractions that occur in real gases. This must mean that H_2 molecules have very low attractive forces for each other. This idea is borne out by looking at the van der Waals a value for H_2 in Table 5.3. Note that H_2 has the lowest value among the gases H_2, N_2, CH_4, and CO_2. Remember that the value of a reflects how much of a correction must be made to adjust the observed pressure up to the expected ideal pressure:

$$P_{\text{ideal}} = P_{\text{obs}} + a\left(\frac{n}{V}\right)^2$$

A low value for a reflects weak intermolecular forces among the gas molecules.

Also notice that although the compressibility for N_2 dips below 1.0, it does not show as much deviation as that for CH_4, which in turn does not show as much deviation as the compressibility for CO_2. Based on this behavior, we can surmise that the importance of intermolecular interactions increases in this order:

$$H_2 < N_2 < CH_4 < CO_2$$

This order is reflected by the relative *a* values for these gases in Table 5.3. In Section 16.1, we will see how these variations in intermolecular interactions can be explained. The main point to be made here is that real gas behavior can tell us about the relative importance of intermolecular attractions among gas molecules.

5.12 Chemistry in the Atmosphere

Table 5.4

Atmospheric Composition Near Sea Level (dry air)*

Component	Mole Fraction
N_2	0.78084
O_2	0.20946
Ar	0.00934
CO_2	0.000345
Ne	0.00001818
He	0.00000524
CH_4	0.00000168
Kr	0.00000114
H_2	0.0000005
NO	0.0000005
Xe	0.000000087

*The atmosphere contains various amounts of water vapor, depending on conditions.

The gases that are most important to us are located in the **atmosphere** that surrounds the earth's surface. The principal components are N_2 and O_2, but many other important gases, such as H_2O and CO_2, are also present. The average composition of the earth's atmosphere near sea level, with the water vapor removed, is shown in Table 5.4. Because of gravitational effects, the composition of the earth's atmosphere is not constant: Heavier molecules tend to be near the earth's surface, and light molecules tend to migrate to higher altitudes and eventually to escape into space. The atmosphere is a highly complex and dynamic system, but for convenience, we divide it into several layers based on the way the temperature changes with altitude. (The lowest layer, called the *troposphere*, is shown in Fig. 5.27.) Note that in contrast to the complex temperature profile of the atmosphere in general, the pressure decreases in a regular way with increasing altitude in the troposphere.

The chemistry occurring in the higher levels of the atmosphere is mostly determined by the effects of high-energy radiation and particles from the sun and other sources in space. In fact, the upper atmosphere serves as an important shield to prevent this high-energy radiation from reaching the earth, where it would damage the relatively fragile molecules sustaining life. In particular, the ozone in the upper atmosphere helps prevent high-energy ultraviolet radiation from penetrating to the earth. Intensive research is in progress to determine the natural factors that control the ozone concentration and to understand how it is affected by chemicals released into the atmosphere.

The chemistry occurring in the troposphere is strongly influenced by human activities. Millions of tons of gases and particulates are released into the tropo-

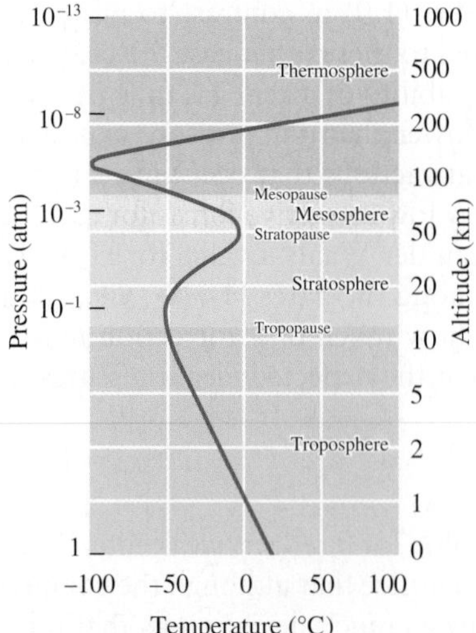

Figure 5.27
The variation of temperature and pressure with altitude. Note that the pressure steadily decreases with increasing altitude but that the temperature does not change monotonically.

sphere by our highly industrial civilization. Actually, it is amazing that the atmosphere can absorb so much material with relatively small permanent changes.

Significant changes, however, are occurring. Severe **air pollution** is found around many large cities, and it is probable that long-range changes in the planet's weather are taking place. We will deal only with the short-term, localized effects of pollution.

The two main sources of pollution are transportation and the production of electricity. The combustion of petroleum in vehicles produces CO, CO_2, NO, and NO_2, along with unburned molecules from petroleum. When this mixture is trapped close to the ground in stagnant air, reactions occur, producing chemicals that are potentially irritating and harmful to living systems.

Although represented here as O_2, the actual oxidant is an organic peroxide such as CH_3COO, formed by reaction of O_2 with organic pollutants.

The complex chemistry of polluted air appears to center on ozone and the nitrogen oxides (NO_x). At the high temperatures in the gasoline and diesel engines of cars and trucks, N_2 and O_2 react to form a small quantity of NO, which is emitted into the air with the exhaust gases (Fig. 5.28). This NO is oxidized in air to NO_2, which in turn absorbs energy from sunlight and breaks up into nitric oxide and free oxygen atoms:

$$NO_2(g) \xrightarrow{\text{Radiant energy}} NO(g) + O(g)$$

Oxygen atoms are very reactive and can combine with O_2 to form *ozone:*

$$O(g) + O_2(g) \longrightarrow O_3(g)$$

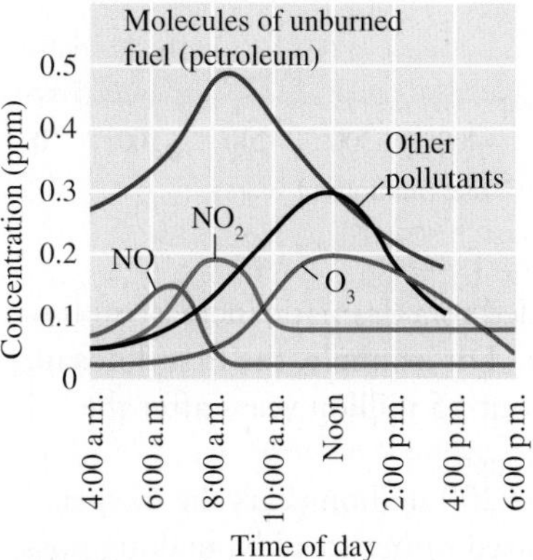

Figure 5.28
Concentration (in molecules per million molecules of "air") of some smog components versus time of day. After P. A. Leighton's classic experiment, "Photochemistry of Air Pollution," in *Physical Chemistry: A Series of Monographs,* ed. Eric Hutchinson and P. Van Rysselberghe, Vol. IX. New York: Academic Press, 1961.

Ozone is also very reactive. It can react with the unburned hydrocarbons in the polluted air to produce chemicals that cause the eyes to water and burn and are harmful to the respiratory system.

The end product of this whole process is often referred to as **photochemical smog,** so called because light is required to initiate some of the reactions. The production of photochemical smog can be more clearly understood by examining as a group the preceding reactions:

$$\begin{aligned} NO_2(g) &\longrightarrow NO(g) + O(g) \\ O(g) + O_2(g) &\longrightarrow O_3(g) \\ NO(g) + \tfrac{1}{2}O_2(g) &\longrightarrow NO_2(g) \\ \hline \text{Net reaction:} \quad \tfrac{3}{2}O_2(g) &\longrightarrow O_3(g) \end{aligned}$$

Note that the NO_2 molecules assist in the formation of ozone without being consumed themselves. The ozone then produces other pollutants.

We can observe this process by analyzing polluted air at various times during a day (Fig. 5.28). As people drive to work between 6 and 8 a.m., the amounts of NO, NO_2, and unburned molecules from petroleum increase. Later, as the decomposition of NO_2 occurs, the concentration of ozone and other pollutants builds up. Current efforts to combat the formation of photochemical smog are focused on cutting down the amounts of molecules from unburned fuel in automobile exhaust and designing engines that produce less nitric oxide (Fig. 5.29).

John Lawlor

Figure 5.29
Our various modes of transportation produce large amounts of nitrogen oxides, which facilitate the formation of photochemical smog.

The other major source of pollution results from burning coal to produce electricity. Much of the coal found in the Midwest contains significant quantities of sulfur, which, when burned, produces sulfur dioxide:

$$S\ (\text{In coal}) + O_2(g) \longrightarrow SO_2(g)$$

A further oxidation reaction occurs when sulfur dioxide is changed to sulfur trioxide in the air:

$$2SO_2(g) + O_2(g) \longrightarrow 2SO_3(g)$$

Chemical Insights

The Importance of Oxygen

Oxygen has been present only for the last half of the earth's history—appearing about 2.5 billion years ago in what is called *The Great Oxidation Event*. Before this time the earth's atmosphere contained large quantities of methane, which reacted with oxygen and prevented its buildup. Now there is geological evidence that about 2.7 billion years ago the amount of dissolved nickel in the oceans began to decrease. This was important because the organisms that produced methane required nickel ions to exist. As the methane concentrations in the atmosphere decreased, oxygen produced by chlorophyll-containing organisms began to build up. In contrast, oxygen was removed from the atmosphere by mountain building and erosion as freshly exposed rocks combined with oxygen to form oxygen-containing minerals.

The concentration of oxygen in the atmosphere has varied greatly over the last 600 million years, as shown in the accompanying graph. Note that 300 million years ago (at the end of the Carboniferous period) the air consisted of about 35% oxygen. The fossil record indicates that during this time insects and other arthropods that absorb oxygen through holes in their exoskeletons were extraordinarily large. It is thought that mayflies as big as today's robins and dragonflies as big as modern hawks were commonplace in this period.

About 255 million years ago, the oxygen concentration in the atmosphere was about 30%, but for some reason the oxygen content plunged to about 13% (about the concentration at an elevation of 15,000 feet in today's world) in the relatively short geological time span of 10 million years. Die-offs during this period claimed as many as 95% of the species living in the ocean and about 70% of those living on land. The oxygen content then began to rebound (to about 16%) 200 million years ago, which led to a dramatic increase in biological innovation. For example, the first dinosaurs appeared only about 15 million years after the mass die-offs.

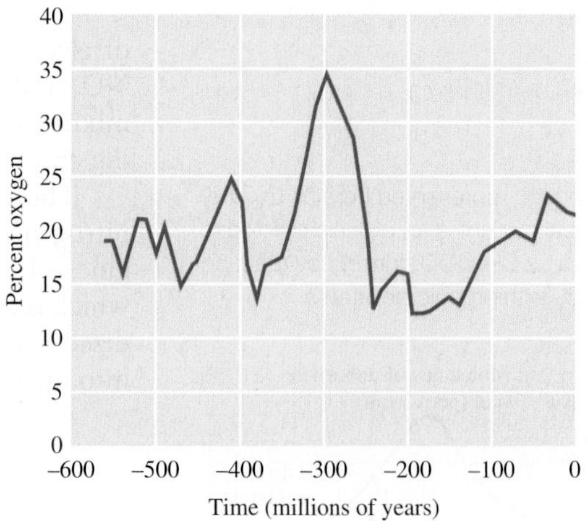

During the last 200 million years the oxygen content has increased rather steadily, making possible the existence of fuel-intense species such as mammals. In fact about 25 million years ago when oxygen concentration maximized at 23%, many mammals had become gigantic. For example, the relatives of today's rhino stood almost 5 m tall and weighed 15 metric tons—the largest ever land mammals. After peaking at 23%, the oxygen levels dropped to today's level of 21% and the "megamammals" disappeared. If history is any indication, the oxygen levels in the atmosphere will continue to change significantly over time, but this is obviously not a high-priority problem for us in the twenty-first century.

This equation describes only the overall stoichiometry of the process; many different oxidants actually participate in the oxidation of sulfur dioxide (see Chapter 15 for a further discussion). The production of sulfur trioxide is significant because it can combine with droplets of water in the air to form sulfuric acid:

$$SO_3(g) + H_2O(l) \longrightarrow H_2SO_4(aq)$$

Kristie A. Boering and Ronald C. Cohen Study the Earth's Atmosphere

Chemistry Explorers

One of the most important areas of research is the study of the chemistry of the earth's atmosphere and how human activities are changing it. Two researchers who are exploring these issues are Kristie A. Boering and Ronald C. Cohen, both faculty members at the University of California at Berkeley. Professor Boering and her research group are studying the earth's atmospheric chemistry through observations from aircraft, balloons, and ground-based platforms; computer simulations; and laboratory experiments. Dr. Boering is particularly interested in the exchange of gases between the biosphere and the atmosphere in modern times and over the past millennium.

Professor Cohen's overall goal is to develop a model for the ways human activity can cause global changes in the atmosphere. His group is particularly interested in what chemical reactions control ozone formation and depletion.

Photo by Michael Barnes/University of California at Berkeley

Kristie A. Boering.

Photo by Michael Barnes/University of California at Berkeley

Ronald C. Cohen.

Professors Boering and Cohen hope their efforts to understand the changes in atmospheric chemistry will lead to solutions to problems such as global warming.

Sulfuric acid is very corrosive to both living things and building materials. Another result of this type of pollution is **acid rain** (Fig. 5.30). In many parts of the northeastern United States and southeastern Canada, acid rain has caused some freshwater lakes to become too acidic for fish to live.

M. Edwards/Peter Arnold, Inc.

Figure 5.30
A helicopter dropping lime in a lake in Sweden to neutralize excess acid from acid rain.

Chemical Insights

Acid Rain: An Expensive Problem

Rainwater, even in pristine wilderness areas, is slightly acidic because some of the carbon dioxide present in the atmosphere dissolves in the raindrops to produce H^+ ions by the following reaction:

$$H_2O(l) + CO_2(g) \longrightarrow H^+(aq) + HCO_3^-(aq)$$

This process produces only very small concentrations of H^+ ions in the rainwater. However, gases such as NO_2 and SO_2, which are by-products of energy use, can produce significantly higher H^+ concentrations. Nitrogen dioxide reacts with water to give a mixture of nitrous acid and nitric acid:

$$2NO_2(g) + H_2O(l) \longrightarrow HNO_2(aq) + HNO_3(aq)$$

Sulfur dioxide is oxidized to sulfur trioxide, which then reacts with water to form sulfuric acid:

$$2SO_2(g) + O_2(g) \longrightarrow 2SO_3(g)$$

$$SO_3(g) + H_2O(l) \longrightarrow H_2SO_4(aq)$$

The damage caused by the acid formed in polluted air is a growing worldwide problem. Lakes are dying in Norway, the forests are sick in Germany, and buildings and statues are deteriorating all over the world.

For example, the Field Museum in Chicago contains more white Georgia marble than any other structure in the world. But more than 70 years of exposure to the elements has taken such a toll on it that the building underwent a multimillion-dollar renovation to replace the damaged marble with freshly quarried material.

What is the chemistry of the deterioration of marble by sulfuric acid? Marble is produced by geological processes at high temperatures and pressures from limestone, a sedimentary rock formed by slow deposition of calcium carbonate from the shells of marine organisms. Limestone and marble are chemically identical ($CaCO_3$) but differ in physical properties because limestone is composed of smaller particles of calcium carbonate and is thus more porous and more workable. Although both limestone and marble are used for buildings, marble can be polished to a higher sheen and is often preferred for decorative purposes.

Both marble and limestone react with sulfuric acid to form calcium sulfate. The process can be represented most simply as

$$CaCO_3(s) + H_2SO_4(aq) \longrightarrow Ca^{2+}(aq) + SO_4^{2-}(aq) + H_2O(l) + CO_2(g)$$

The problem of sulfur dioxide pollution is further complicated by the energy crisis. As petroleum supplies dwindle and the price increases, our dependence on coal will grow. As supplies of low-sulfur coal are used up, high-sulfur coal will be used. One way to use high-sulfur coal without further harming the air quality is to remove the sulfur dioxide from the exhaust gas by means of a system called a *scrubber* before it is emitted from the power plant stack. A common method of *scrubbing* involves blowing powdered limestone ($CaCO_3$) into the combustion chamber, where it is decomposed to lime and carbon dioxide:

$$CaCO_3(s) \longrightarrow CaO(s) + CO_2(g)$$

The lime then combines with the sulfur dioxide to form calcium sulfite:

$$CaO(s) + SO_2(g) \longrightarrow CaSO_3(s)$$

The calcium sulfite and any remaining unreacted sulfur dioxide are removed by injecting an aqueous suspension of lime into the combustion chamber and the stack, producing a *slurry* (a thick suspension), as shown in Fig. 5.31.

Unfortunately, there are many problems associated with scrubbing. The systems are complicated and expensive and consume a great deal of energy. The large quantities of calcium sulfite produced in the process present a disposal problem. For a typical scrubber approximately 1 ton of calcium sulfite

In this equation the calcium sulfate is represented by separate hydrated ions because calcium sulfate dissolves in rainwater. Thus in areas bathed in rainwater, the marble slowly dissolves away.

In areas of the building protected from the rain, the calcium sulfate can form the mineral gypsum, $CaSO_4 \cdot 2H_2O$. The $\cdot\ 2H_2O$ in the formula of gypsum indicates the presence of two water molecules (called waters of hydration) for each $CaSO_4$ formula unit in the solid. The smooth surface of the marble is thus replaced by a thin layer of gypsum, a more porous material that binds soot and dust.

What can be done to protect limestone and marble structures from this kind of damage? Of course, one approach is to lower sulfur dioxide and nitrogen oxide emissions from power plants (Fig. 5.31). In addition, scientists are experimenting with coatings to protect marble from the acidic atmosphere. However, a coating can do more harm than good unless it "breathes." If moisture trapped beneath the coating freezes, the expanding ice can fracture the marble. Needless to say, it is difficult to find a coating that will allow water to pass but not allow acid to pass, so the search continues.

Photos Field Museum, Chicago #CSGN40263 and #GN83213_6C.

The damaging effects of acid rain can be seen by comparing these photos of a decorative statue at the Field Museum in Chicago. The photo on the left was taken c. 1920; the photo on the right was taken in 1990. Recent renovation has since replaced the deteriorating marble.

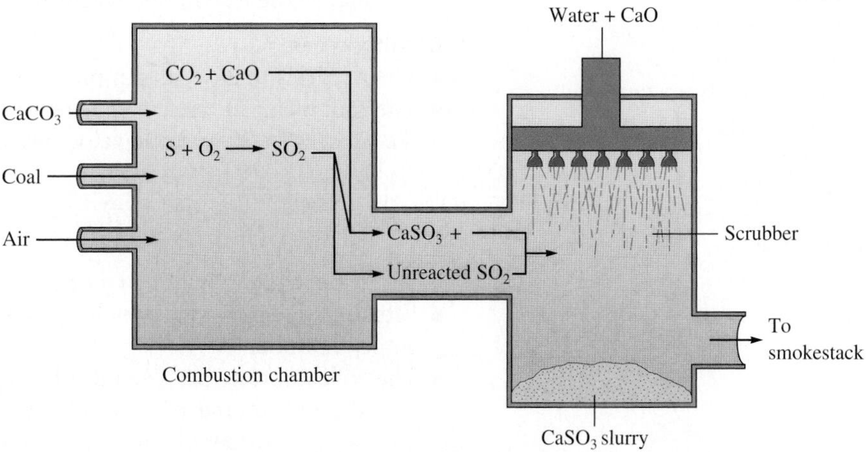

Figure 5.31
Schematic diagram of the process for scrubbing sulfur dioxide from stack gases in power plants.

per year is produced per person served by the power plant. Since no use has yet been found for this calcium sulfite, it is usually buried in a landfill. As a result of these difficulties, air pollution by sulfur dioxide continues to be a major problem, one that is expensive in terms of damage to the environment and to human health, as well as in monetary terms.

Key Terms

Section 5.1
barometer
manometer
mm Hg
torr
standard atmosphere
pascal

Section 5.2
Boyle's law
ideal gas
Charles's law
absolute zero
Avogadro's law

Section 5.3
universal gas constant
ideal gas law

Section 5.4
molar volume
standard temperature and pressure (STP)

Section 5.5
Dalton's law of partial pressures
partial pressure
mole fraction

Section 5.6
kinetic molecular theory
root mean square velocity
joule
mean free path

Section 5.7
diffusion
effusion
Graham's law of effusion

Section 5.9
mean free path (λ)

Section 5.10
van der Waals equation

Section 5.12
atmosphere
air pollution
photochemical smog
acid rain

For Review

State of a gas

- The state of a gas can be described completely by specifying its pressure (P), volume (V), temperature (T), and the amount (moles) of gas present (n).
- Pressure
  - Common units
    - 1 torr = 1 mm Hg
    - 1 atm = 760 torr
  - SI unit: pascal
    - 1 atm = 101,325 Pa

Gas laws

- Discovered by observing the properties of gases
- Boyle's law: $PV = k$
- Charles's law: $V = bT$
- Avogadro's law: $V = an$
- Ideal gas law: $PV = nRT$
- Dalton's law of partial pressures: $P_{\text{Total}} = P_1 + P_2 + P_3 + \cdots$, where P_n represents the partial pressure of component n in a mixture of gases

Kinetic molecular theory (KMT)

- Model that accounts for ideal gas behavior
- Postulates of the KMT:
  - Volume of gas particles is negligible (zero)
  - No particle interactions
  - Particles are in constant motion, colliding with the container walls to produce pressure
  - The average kinetic energy of the gas particles is directly proportional to the Kelvin temperature of the gas

Gas properties

- The particles in any gas sample have a range of velocities
- The root mean square (rms) velocity for a gas represents the average of the squares of the particle velocities

$$\mu_{\text{rms}} = \sqrt{\frac{3RT}{M}}$$

- Diffusion: the mixing of two or more gases
- Effusion: the process in which a gas passes through a small hole into an empty chamber
- The collision rate (per second) of gas particles with the container walls (Z_A) depends on the average velocity of the gas particles, the size of the area being considered, and the number of particles in the container

$$Z_A = A\frac{N}{V}\sqrt{\frac{RT}{1\pi M}}$$

- The intermolecular collision rate (per second) of gas particles (Z) depends on the average velocity of the gas particles, the size of the gas particles, and the number of particles in the container

$$Z = 4\frac{N}{V}d^2\sqrt{\frac{\pi RT}{M}}$$

Real gas behavior

- Real gases approach ideal behavior at high temperatures and low pressures
- Understanding how the ideal gas law must be modified to account for real gas behavior helps us understand how gases behave on a molecular level
- Van der Waals found that to describe real gas behavior we must consider particle interactions and particle volumes

Discussion Questions

These questions are designed to be considered by groups of students in class. Often these questions work well for introducing a particular topic in class.

1. Consider the following apparatus: a test tube covered with a nonpermeable elastic membrane inside a container that is closed with a cork. A syringe goes through the cork.

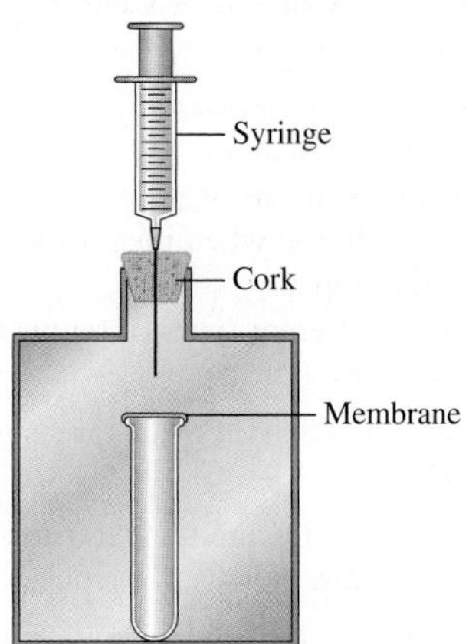

   a. As you push down on the syringe, how does the membrane covering the test tube change?
   b. You stop pushing the syringe but continue to hold it down. In a few seconds, what happens to the membrane?

2. Figure 5.1 shows a picture of a barometer. Which of the following statements is the best explanation of how this barometer works?
   a. Air pressure outside the tube causes the mercury to move in the tube until the air pressure inside and outside the tube is equal.
   b. Air pressure inside the tube causes the mercury to move in the tube until the air pressure inside and outside the tube is equal.
   c. Air pressure outside the tube counterbalances the weight of the mercury in the tube.
   d. Capillary action of the mercury causes the mercury to go up the tube.
   e. The vacuum that is formed at the top of the tube holds up the mercury.

   Justify your choice. For choices that you did not pick, explain why they are incorrect. Pictures help!

3. The barometer on the left in the following diagram shows the level of mercury at a given atmospheric pressure. Fill all the other barometers with mercury for that same atmospheric pressure. Explain your answer.

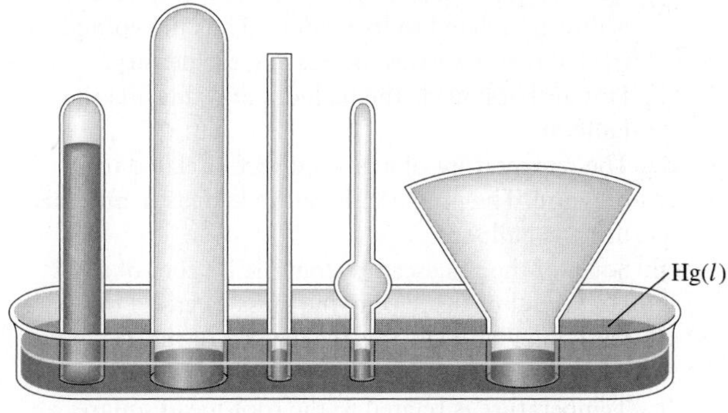

4. As you increase the temperature of a gas in a sealed, rigid container, what happens to the density of the gas? Would the results be the same if you did the same experiment in a container with a piston at constant pressure? Explain.

5. A diagram in a chemistry book shows a magnified view of a flask of air as follows:

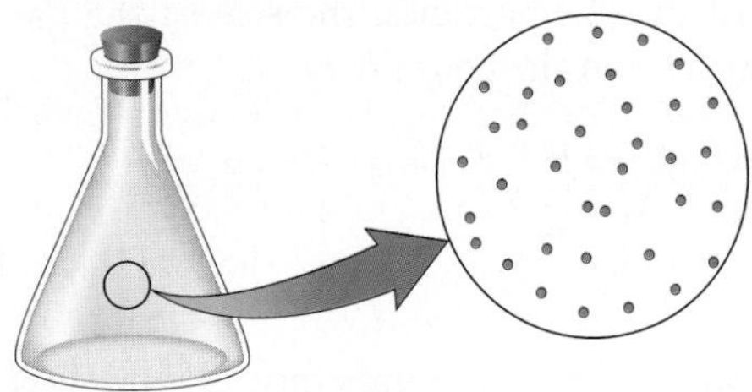

What do you suppose is between the dots (the dots represent air molecules)?

a. air
b. dust
c. pollutants
d. oxygen
e. nothing

6. If you put a drinking straw in water, place your finger over the opening, and lift the straw out of the water, some water stays in the straw. Explain.

7. A chemistry student relates the following story: "I noticed my tires were a bit low and went to the gas station. As I was filling the tires, I thought about the kinetic molecular theory (KMT), and I realized that I was increasing both the pressure and volume of the tires as I filled the tires with air. 'Hmmm,' I thought, 'that goes against what I learned in chemistry, where I was told pressure and volume are inversely proportional.'" What is the fault of the logic of the chemistry student in this situation? Explain *why* we think pressure and volume to be inversely related (draw pictures and use the KMT).

8. Chemicals X and Y (both gases) react to form the gas XY, but it takes a bit of time for the reaction to occur. Both X and Y are placed in a container with a piston (free to move), and you note the volume. As the reaction occurs, what happens to the volume of the container? Explain.

9. Which statement best explains why a hot-air balloon rises when the air in the balloon is heated?
   a. According to Charles's law, the temperature of a gas is directly related to its volume. Thus the volume of the balloon increases, decreasing the density.
   b. Hot air rises inside the balloon, and this lifts the balloon.
   c. The temperature of a gas is directly related to its pressure. The pressure therefore increases, and this lifts the balloon.
   d. Some of the gas escapes from the bottom of the balloon, thus decreasing the mass of gas in the balloon. This decreases the density of the gas in the balloon, and this lifts the balloon.
   e. Temperature is related to the root mean square velocity of the gas molecules. Thus the molecules are moving faster, hitting the balloon more often, and thus lifting the balloon.

   Justify your choice. For those you did not choose, explain why they are incorrect.

10. Draw a highly magnified view of a sealed, rigid container filled with a gas. Then draw what it would look like if you cooled the gas significantly, but kept the temperature above the boiling point of the substance in the container. Also draw what it would look like if you heated the gas significantly. Finally, draw what each situation would look like if you evacuated enough of the gas to decrease the pressure by a factor of 2.

11. If you release a helium balloon, it soars upward and eventually pops. Explain.

12. If you have any two gases in different containers that are the same size at the same pressure and same temperature, what is true about the moles of each gas? Why is this true?

13. Explain the following seeming contradiction: You have two gases, A and B, in two separate containers of equal volume and at equal pressure and temperature. Therefore, you must have the same number of moles of each gas. Because the two temperatures are equal, the average kinetic energies of the two samples are equal. Therefore, since the energy of such a system corresponds to translational motion, the root mean square velocities of the two are equal, and thus the particles in each sample move, on average, with the same relative speed. Since A and B are different gases, each must have a different molar mass. If A has a higher molar mass than B, the particles of A must be hitting the sides of the container with more force. Thus the pressure in the container of gas A must be higher than that in the container with gas B. However, one of our initial assumptions was that the pressures were equal. Explain.

14. Using postulates of the kinetic molecular theory, give a molecular interpretation of Boyle's law, Charles's law, and Dalton's law of partial pressures.

15. Rationalize the following observations.
    a. Aerosol cans will explode if heated.
    b. You can drink through a soda straw.
    c. A thin-walled can will collapse when the air inside is removed by a vacuum pump.
    d. Manufacturers produce different types of tennis balls for high and low elevations.

16. Show how Boyle's law and Charles's law are special cases of the ideal gas law.

17. At the same conditions of pressure and temperature, ammonia gas is less dense than air. Why is this true?

18. For each of the quantities (a–f) listed below, explain which of the following properties (mass of the molecule, density of the gas sample, temperature of the gas sample, size of the molecule, and number of moles of gas) must be known to calculate the quantity.
    a. average kinetic energy
    b. average number of collisions per second with other gas molecules
    c. average force of each impact with the wall of the container
    d. root mean square velocity
    e. average number of collisions with a given area of the container
    f. distance between collisions

19. You have two containers each with 1 mole of xenon gas at 15°C. Container A has a volume of 3.0 L, and container B has a volume of 1.0 L. Explain how the following quantities compare between the two containers.
    a. the average kinetic energy of the Xe atoms
    b. the force with which the Xe atoms collide with the container walls
    c. the root mean square velocity of the Xe atoms
    d. the collision frequency of the Xe atoms (with other atoms)
    e. the pressure of the Xe sample

20. You have a balloon covering the mouth of a flask filled with air at 1 atm. You apply heat to the bottom of the flask until the volume of the balloon is equal to that of the flask.
    a. Which has more air in it—the balloon or the flask? Or do both contain the same amount of air? Explain.
    b. In which is the pressure greater—the balloon or the flask? Or is the pressure the same? Explain.

Exercises

OWL Interactive versions of these problems may be assigned in OWL.

A blue exercise number indicates that the answer to that exercise appears at the back of this book and a solution appears in the *Solutions Guide*.

Pressure

21. A sealed-tube manometer as shown below can be used to measure pressures below atmospheric pressure. The tube above the mercury is evacuated. When there is a vacuum in the flask, the mercury levels in both arms of the U-tube are equal. If a gaseous sample is introduced into the flask, the mercury levels are different. The difference h is a measure of the pressure of the gas inside the flask. If h is equal to 4.75 cm, calculate the pressure in the flask in torr, pascals, and atmospheres.

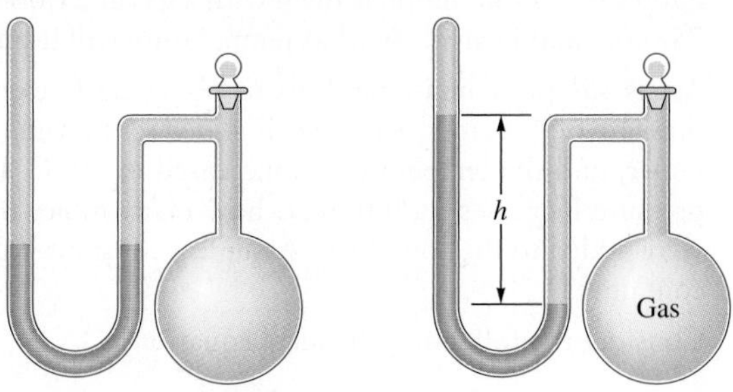

22. A diagram for an open-tube manometer is shown below.

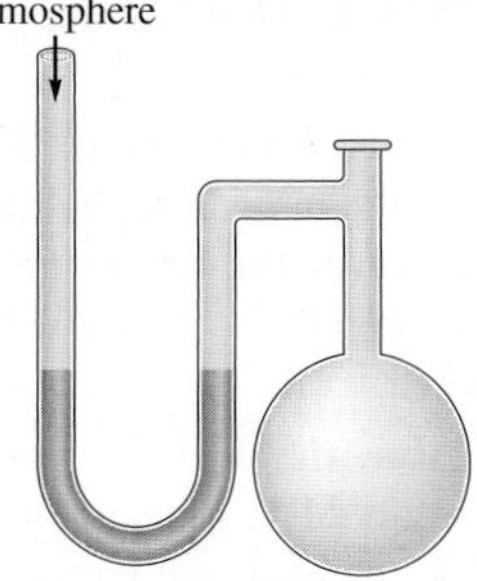

If the flask is open to the atmosphere, the mercury levels are equal. For each of the following situations in which a gas is contained in the flask, calculate the pressure in the flask in torr, atmospheres, and pascals.

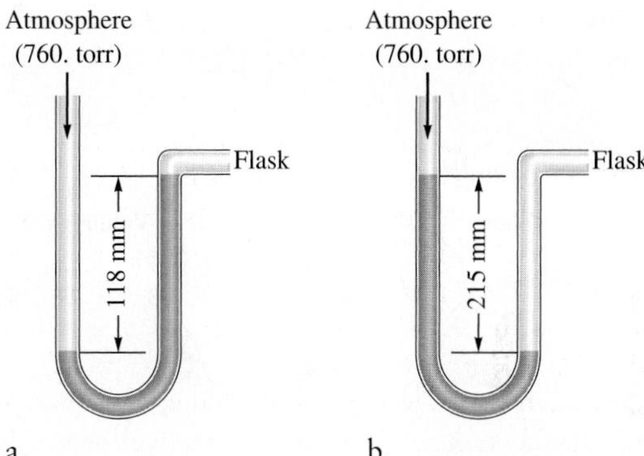

    c. Calculate the pressures in the flask in parts a and b (in torr) if the atmospheric pressure is 635 torr.

23. The gravitational force exerted by an object is given by

$$F = mg$$

where F is the force in newtons, m is the mass in kilograms, and g is the acceleration due to gravity, 9.81 m/s^2. Calculate the force exerted per unit of area by a column of mercury (density = 13.59 g/cm^3) that is 76.0 cm high. How high would a column of water (density = 1.00 g/cm^3) have to be to exert the same force?

24. a. If the open-tube manometer in Exercise 22 contains a nonvolatile silicone oil (density = 1.30 g/cm^3) instead of mercury (density = 13.6 g/cm^3), what are the pressures in the flask as shown in parts a and b in torr, atmospheres, and pascals?
    b. What advantage would there be in using a less dense fluid than mercury in a manometer used to measure relatively small differences in pressure?

25. Freon-12 (CF_2Cl_2) is commonly used as the refrigerant in central home air conditioners. The system is initially charged to a pressure of 4.8 atm. Express this pressure in each of the following units (1 atm = 14.7 psi).
    a. mm Hg b. torr c. Pa d. psi

Gas Laws

26. Draw a qualitative graph to show how the first property varies with the second in each of the following (assume 1 mole of an ideal gas and T in kelvins).
    a. PV versus V with constant T
    b. P versus T with constant V

c. T versus V with constant P
d. P versus V with constant T
e. P versus $1/V$ with constant T
f. PV/T versus P

27. As weather balloons rise from the earth's surface, the pressure of the atmosphere becomes less, tending to cause the volume of the balloons to expand. However, the temperature is much lower in the upper atmosphere than at sea level. Would this temperature effect tend to make such a balloon expand or contract? Weather balloons do, in fact, expand as they rise. What does this tell you?

28. Consider the flasks in the following diagrams.

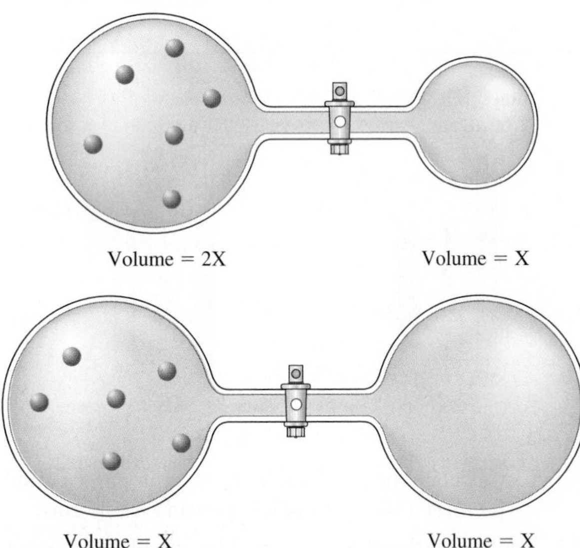

Assuming the connecting tube has negligible volume, draw what each diagram will look like after the stopcock between the two flasks is opened. Also, solve for the final pressure in each case, in terms of the original pressure. Assume temperature is constant.

29. Consider the flask diagramed below. What are the final partial pressures of H_2 and N_2 after the stopcock between the two flasks is opened? (Assume the final volume is 3.00 L.) What is the total pressure (in torr)?

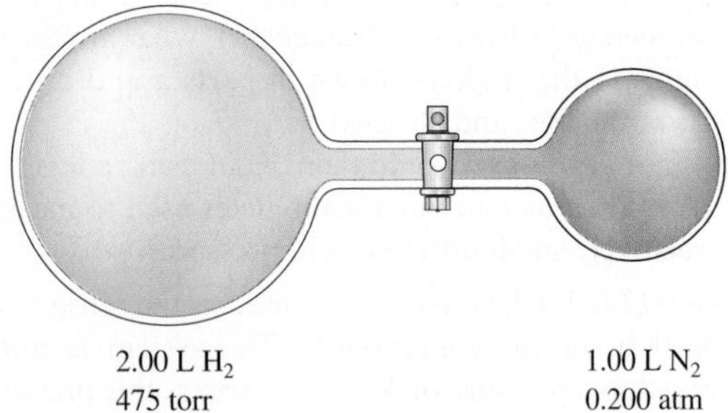

30. Consider the flask apparatus in Exercise 29, which contains 2.00 L of H_2 at a pressure of 360. torr and 1.00 L of N_2 at an unknown pressure. If the total pressure in the flasks is 320. torr after the stopcock is opened, determine the initial pressure of N_2 in the 1.00-L flask.

31. A compressed-gas cylinder contains 1.00×10^3 g of argon gas. The pressure inside the cylinder is 2050. psi (pounds per square inch) at a temperature of 18°C. How much gas remains in the cylinder if the pressure is decreased to 650. psi at a temperature of 26°C?

32. A sealed balloon is filled with 1.00 L of helium at 23°C and 1.00 atm. The balloon rises to a point in the atmosphere where the pressure is 220. torr and the temperature is −31°C. What is the change in the volume of the balloon as it ascends from 1.00 atm to a pressure of 220. torr?

33. A piece of solid carbon dioxide, with a mass of 22.0 g, is placed in an otherwise empty 4.00-L container at 27°C. What is the pressure in the container after all the carbon dioxide vaporizes? If 22.0 g of solid carbon dioxide was placed in a similar container already containing air at 740. torr, what would be the partial pressure of carbon dioxide and the total pressure in the container after the carbon dioxide had vaporized?

34. An ideal gas is in a cylinder with a volume of 5.0×10^2 mL at a temperature of 30.°C and a pressure of 710 torr. The gas is compressed to a volume of 25 mL, and the temperature is raised to 820°C. What is the new pressure?

35. Suppose two 200.0-L tanks are to be filled separately with the gases helium and hydrogen. What mass of each gas is needed to produce a pressure of 135 atm in its respective tank at 24°C?

36. An ideal gas at 7°C is in a spherical flexible container having a radius of 1.00 cm. The gas is heated at constant pressure to 88°C. Determine the radius of the spherical container after the gas is heated. (Volume of a sphere = $4/3\pi r^3$.)

37. A flask that can withstand an internal pressure of 2500 torr, but no more, is filled with a gas at 21.0°C and 758 torr and heated. At what temperature will it burst?

38. A gas sample containing 1.50 moles at 25°C exerts a pressure of 400. torr. Some gas is *added* to the same container, and the temperature is increased to 50.°C. If the pressure increases to 800. torr, how many moles of gas were added to the container? Assume a constant-volume container.

39. Consider the following chemical equation:

$$2NO_2(g) \longrightarrow N_2O_4(g)$$

If 25.0 mL of NO_2 gas is completely converted to N_2O_4 gas under the same conditions, what volume will the N_2O_4 occupy?

40. A bicycle tire is filled with air to a pressure of 75 psi at a temperature of 19°C. Riding the bike on asphalt on a hot day increases the temperature of the tire to 58°C. The volume of the tire increases by 4.0%. What is the new pressure in the bicycle tire?

41. A hot-air balloon is filled with air to a volume of 4.00×10^3 m^3 at 745 torr and 21°C. The air in the balloon is then heated to 62°C, causing the balloon to expand to a volume of 4.20×10^3 m^3. What is the ratio of the number of moles of air in the heated balloon to the original number of moles of air in the balloon? (*Hint:* Openings in the balloon allow air to flow in and out. Thus the pressure in the balloon is always the same as that of the atmosphere.)

42. Determine the partial pressure of each gas as shown in the figure below. *Note:* The relative numbers of each type of gas are depicted in the figure.

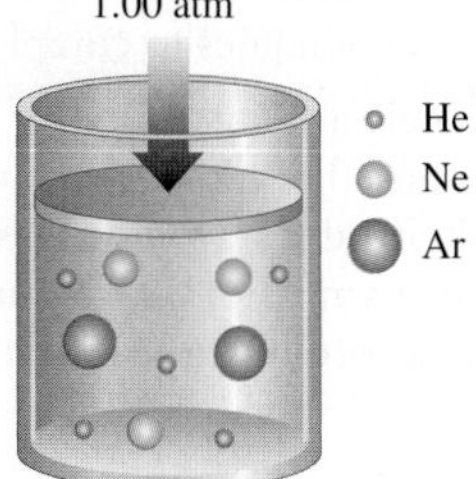

43. Consider the flasks in the following diagrams.

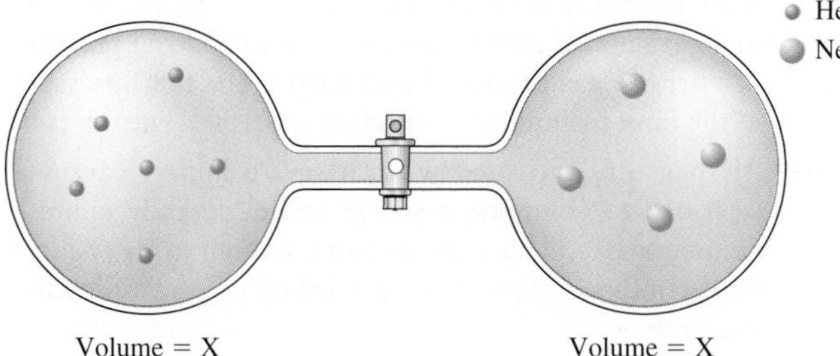

a. Which is greater, the initial pressure of helium or initial pressure of neon? How much greater?
b. Assuming the connecting tube has negligible volume, draw what each diagram will look like after the stopcock between the two flasks is opened.
c. Solve for the final pressure in terms of the original pressures of helium and neon. Assume temperature is constant.
d. Solve for the final partial pressures of helium and neon in terms of their original pressures. Assume the temperature is constant.

44. A sample of nitrogen gas was collected over water at 20.°C and a total pressure of 1.00 atm. A total volume of 2.50×10^2 mL was collected. What mass of nitrogen was collected? (At 20.°C the vapor pressure of water is 17.5 torr.)

45. Helium is collected over water at 25°C and 1.00 atm total pressure. What total volume of gas must be collected to obtain 0.586 g of helium? (At 25°C the vapor pressure of water is 23.8 torr.)

46. A 2.00-L sample of $O_2(g)$ was collected over water at a total pressure of 785 torr and 25°C. When the $O_2(g)$ was dried (water vapor removed), the gas had a volume of 1.94 L at 25°C and 785 torr. Calculate the vapor pressure of water at 25°C.

47. In a mixture of the two gases, the partial pressures of $CH_4(g)$ and $O_2(g)$ are 0.175 atm and 0.250 atm, respectively.
a. What is the mole fraction of each gas in the mixture?
b. If the mixture occupies a volume of 10.5 L at 65°C, calculate the total number of moles of gas in the mixture.
c. Calculate the number of grams of each gas in the mixture.

48. A 1.00-L gas sample at 100.°C and 600. torr contains 50.0% helium and 50.0% xenon by mass. What are the partial pressures of the individual gases?

49. At 0°C a 1.0 L flask contains 5.0×10^{-2} mole of N_2, 1.5×10^2 mg O_2, and 5.0×10^{21} molecules of NH_3. What is the partial pressure of each gas, and what is the total pressure in the flask?

Gas Density, Molar Mass, and Reaction Stoichiometry

50. Given that a sample of air is made up of nitrogen, oxygen, and argon in the mole fractions 78% N_2, 21% O_2, and 1.0% Ar, what is the density of air at standard temperature and pressure?

51. Consider two different containers, each filled with 2 moles of Ne(g). One of the containers is rigid and has constant volume. The other container is flexible (like a balloon) and is capable of changing its volume to keep the external pressure and internal pressure equal to each other. If you raise the temperature in both containers, what happens to the pressure and density of the gas inside each container? Assume a constant external pressure.

52. An unknown diatomic gas has a density of 3.164 g/L at STP. What is the identity of the gas?

53. A compound contains only nitrogen and hydrogen and is 87.4% nitrogen by mass. A gaseous sample of the compound has a density of 0.977 g/L at 710. torr and 100.°C. What is the molecular formula of the compound?

54. A compound has the empirical formula CHCl. A 256-mL flask, at 373 K and 750. torr, contains 0.800 g of the gaseous compound. Give the molecular formula.

55. One of the chemical controversies of the nineteenth century concerned the element beryllium (Be). Berzelius originally claimed that beryllium was a trivalent element (forming Be^{3+} ions) and that it formed an oxide with the formula Be_2O_3. This assumption resulted in a calculated atomic mass of 13.5 for beryllium. In formulating his periodic table, Mendeleev proposed that beryllium was divalent (forming Be^{2+} ions) and that it gave an oxide with the formula BeO. This assumption gives an atomic mass of 9.0. In 1894 A. Combes (*Comptes Rendes*, 1894, p. 1221) reacted beryllium with the anion $C_5H_7O_2^-$ and measured the density of the gaseous product. Combes's data for two different experiments are as follows:

	I	II
Mass	0.2022 g	0.2224 g
Volume	22.6 cm^3	26.0 cm^3
Temperature	13°C	17°C
Pressure	765.2 torr	764.6 torr

If beryllium is a divalent metal, the molecular formula of the product will be $Be(C_5H_7O_2)_2$; if it is trivalent, the formula will be $Be(C_5H_7O_2)_3$. Show how Combes's data help to confirm that beryllium is a divalent metal.

56. Discrepancies in the experimental values of the molar mass of nitrogen provided some of the first evidence for

the existence of the noble gases. If pure nitrogen is collected from the decomposition of ammonium nitrite,

$$NH_4NO_2(s) \xrightarrow{\text{Heat}} N_2(g) + 2H_2O(g)$$

its measured molar mass is 28.01. If O_2, CO_2, and H_2O are removed from air, the remaining gas has an average molar mass of 28.15. Assuming this discrepancy is solely a result of contamination with argon (atomic mass = 39.95), calculate the ratio of moles of Ar to moles of N_2 in air.

57. A sample of methane (CH_4) gas contains a small amount of helium. Calculate the volume percentage of helium if the density of the sample is 0.70902 g/L at 0.0°C and 1.000 atm.

58. Metallic molybdenum can be produced from the mineral molybdenite (MoS_2). The mineral is first oxidized in air to molybdenum trioxide and sulfur dioxide. Molybdenum trioxide is then reduced to metallic molybdenum using hydrogen gas. The balanced equations are

$$MoS_2(s) + 7/2O_2(g) \longrightarrow MoO_3(s) + 2SO_2(g)$$

$$MoO_3(s) + 3H_2(g) \longrightarrow Mo(s) + 3H_2O(l)$$

Calculate the volumes of air and hydrogen gas at 17°C and 1.00 atm that are necessary to produce 1.00×10^3 kg of pure molybdenum from MoS_2. Assume air contains 21% oxygen by volume and assume 100% yield for each reaction.

59. In 1897 the Swedish explorer Andreé tried to reach the North Pole in a balloon. The balloon was filled with hydrogen gas. The hydrogen gas was prepared from iron splints and diluted sulfuric acid. The reaction is

$$Fe(s) + H_2SO_4(aq) \longrightarrow FeSO_4(aq) + H_2(g)$$

The volume of the balloon was 4800 m^3, and the loss of hydrogen gas during filling was estimated at 20.%. What mass of iron splints and 98% (by mass) H_2SO_4 were needed to ensure the complete filling of the balloon? Assume a temperature at 0°C, a pressure of 1.0 atm during filling, and 100% yield.

60. Urea (H_2NCONH_2) is used extensively as a nitrogen source in fertilizers. It is produced commercially from the reaction of ammonia and carbon dioxide:

$$2NH_3(g) + CO_2(g) \xrightarrow[\text{Pressure}]{\text{Heat}} H_2NCONH_2(s) + H_2O(g)$$

Ammonia gas at 223°C and 90. atm flows into a reactor at a rate of 500. L/min. Carbon dioxide at 223°C and 45 atm flows into the reactor at a rate of 600. L/min. What mass of urea is produced per minute by this reaction assuming 100% yield?

61. Methanol (CH_3OH) can be produced by the following reaction:

$$CO(g) + 2H_2(g) \longrightarrow CH_3OH(g)$$

Hydrogen at STP flows into a reactor at a rate of 16.0 L/min. Carbon monoxide at STP flows into the reactor at a rate of 25.0 L/min. If 5.30 g of methanol is produced per minute, what is the percent yield of the reaction?

62. Consider the reaction between 50.0 mL of liquid methanol (CH_3OH; density = 0.850 g/mL) and 22.8 L of O_2 at 27°C and a pressure of 2.00 atm. The products of the reaction are $CO_2(g)$ and $H_2O(g)$. Calculate the number of moles of H_2O formed if the reaction goes to completion.

63. Some very effective rocket fuels are composed of lightweight liquids. The fuel composed of dimethylhydrazine [$(CH_3)_2N_2H_2$] mixed with dinitrogen tetroxide was used to power the lunar lander in its missions to the moon. The two components react according to the following equation:

$$(CH_3)_2N_2H_2(l) + 2N_2O_4(l) \longrightarrow 3N_2(g) + 4H_2O(g) + 2CO_2(g)$$

If 150 g of dimethylhydrazine reacts with excess dinitrogen tetroxide and the product gases are collected at 27°C in an evacuated 250-L tank, what is the partial pressure of nitrogen gas produced and what is the total pressure in the tank assuming the reaction has 100% yield?

64. Air bags are activated when a severe impact causes a steel ball to compress a spring and electrically ignite a detonator cap. This action causes sodium azide (NaN_3) to decompose explosively according to the following reaction:

$$2NaN_3(s) \longrightarrow 2Na(s) + 3N_2(g)$$

What mass of $NaN_3(s)$ must be reacted to inflate an air bag to 70.0 L at STP?

65. At elevated temperatures, sodium chlorate decomposes to produce sodium chloride and oxygen gas. A 0.8765-g sample of impure sodium chlorate was heated until the production of oxygen gas ceased. The oxygen gas collected over water occupied 57.2 mL at a temperature of 22°C and a pressure of 734 torr. Calculate the mass percent of $NaClO_3$ in the original sample. (At 22°C the vapor pressure of water is 19.8 torr.)

66. Xenon and fluorine will react to form binary compounds when a mixture of these two gases is heated to 400°C in a nickel reaction vessel. A 100.0-mL nickel container is filled with xenon and fluorine giving partial pressures of 1.24 atm and 10.10 atm, respectively, at a temperature of 25°C. The reaction vessel is heated to 400°C to cause a reaction to occur and then cooled to a temperature at which F_2 is a gas and the xenon fluoride is a nonvolatile solid. The remaining F_2 gas is transferred to another 100.0-mL nickel container where the pressure of F_2 at 25°C is 7.62 atm. Assuming all of the xenon has reacted, what is the formula of the product?

67. The nitrogen content of organic compounds can be determined by the Dumas method. The compound in question is first reacted by passage over hot $CuO(s)$:

$$\text{Compound} \xrightarrow[\text{CuO}(s)]{\text{Hot}} N_2(g) + CO_2(g) + H_2O(g)$$

The gaseous products are then passed through a concentrated solution of KOH to remove the CO_2. After passage through the KOH solution, the gas contains N_2 and is saturated with water vapor. In a given experiment, a 0.253-g sample of a compound produced 31.8 mL of N_2 saturated with water vapor at 25°C and 726 torr. What is the mass percent of nitrogen in the compound? (The vapor pressure of water at 25°C is 23.8 torr.)

68. An organic compound contains C, H, N, and O. Combustion of 0.1023 g of the compound in excess oxygen yielded 0.2766 g of CO_2 and 0.0991 g of H_2O. A sample of 0.4831 g of the compound was analyzed for nitrogen by the Dumas method (see Exercise 67). At STP, 27.6 mL of dry N_2 was obtained. In a third experiment, the density of the compound as a gas was found to be 4.02 g/L at 127°C and 256 torr. What are the empirical formula and the molecular formula of the compound?

69. Nitric acid is produced commercially by the Ostwald process. In the first step, ammonia is oxidized to nitric oxide:

$$4NH_3(g) + 5O_2(g) \longrightarrow 4NO(g) + 6H_2O(g)$$

Assume this reaction is carried out in the apparatus diagramed below.

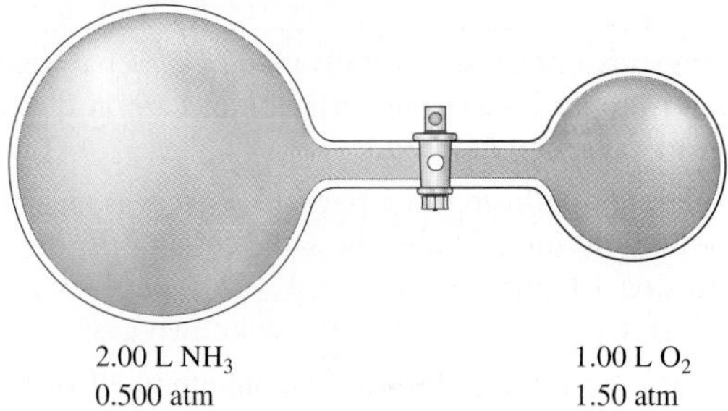

The stopcock between the two reaction containers is opened, and the reaction proceeds using proper catalysts. Calculate the partial pressure of NO after the reaction is complete. Assume 100% yield for the reaction, assume the final container volume is 3.00 L, and assume the temperature is constant.

70. Consider the following balanced equation in which gas X forms gas X_2:

$$2X(g) \longrightarrow X_2(g)$$

Equal moles of X are placed in two separate containers. One container is rigid, so the volume cannot change; the other container is flexible, so the volume changes to keep the internal pressure equal to the external pressure. The above reaction is run in each container. What happens to the pressure and density of the gas inside each container as reactants are converted to products?

71. As $NH_3(g)$ is decomposed into nitrogen gas and hydrogen gas at constant pressure and temperature, the volume of the product gases collected is twice the volume of NH_3 reacted. Explain. As $NH_3(g)$ is decomposed into nitrogen gas and hydrogen gas at constant volume and temperature, the total pressure increases by some factor. Why does the increase in pressure occur, and by what factor does the total pressure increase when reactants are completely converted into products? How do the partial pressures of the product gases compare to each other and to the initial pressure of NH_3?

Kinetic Molecular Theory and Real Gases

72. Use the postulates of the kinetic molecular theory (KMT) to explain why Boyle's law, Charles's law, Avogadro's law, and Dalton's law of partial pressures hold true for ideal gases. Use the KMT to explain the P versus n (at constant V and T) relationship and the P versus T (at constant V and n) relationship.

73. You have a gas in a container fitted with a piston and you change one of the conditions of the gas such that a change takes place, as shown below:

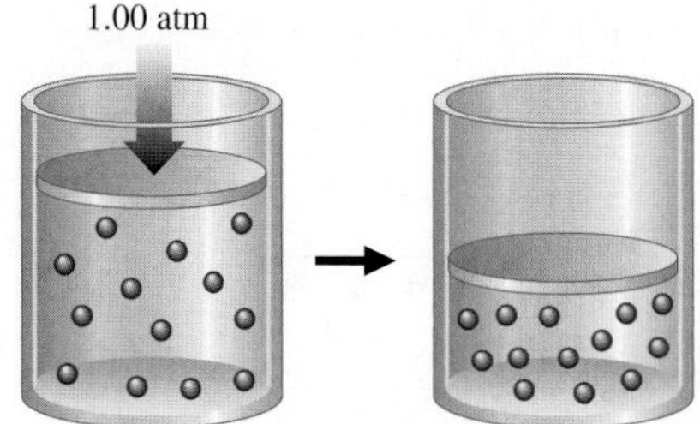

State two distinct changes you can make to accomplish this, and explain why each would work.

74. You have a gas in a container fitted with a piston and you change one of the conditions of the gas such that a change takes place, as shown below:

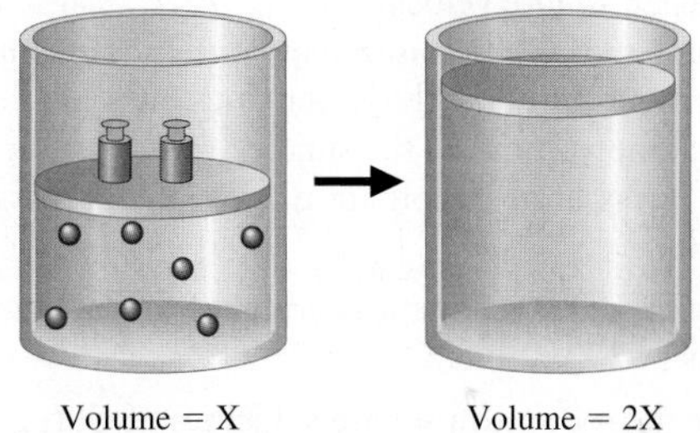

State three distinct changes you can make to accomplish this, and explain why each would work.

75. Consider two gases, A and B, each in a 1.0-L container with both gases at the same temperature and pressure. The mass of gas A in the container is 0.34 g, and the mass of gas B in the container is 0.48 g.

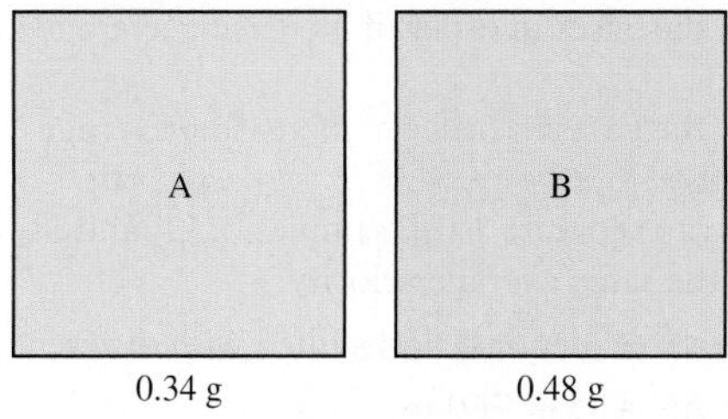

a. Which gas sample has the most molecules present? Explain.
b. Which gas sample has the largest average kinetic energy? Explain.
c. Which gas sample has the fastest average velocity? Explain.

d. How can the pressure in the two containers be equal to each other since the larger gas B molecules collide with the container walls more forcefully?

76. Consider the following samples of gases at the same temperature.

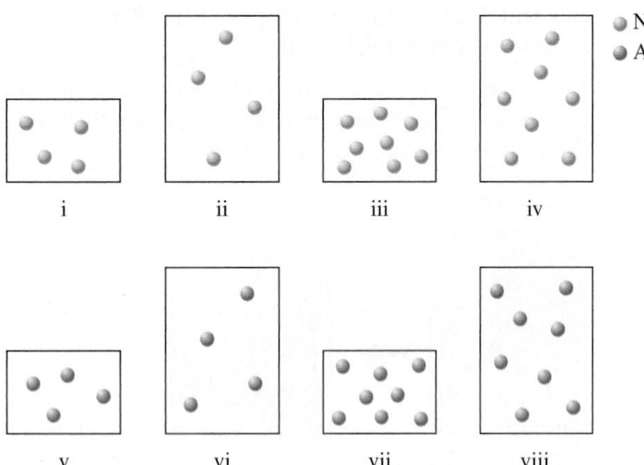

Arrange each of these samples in order from lowest to highest.
a. pressure
b. average kinetic energy
c. density
d. root mean square velocity
Note: Some samples of gases may have equal values for these attributes. Assume the larger containers have a volume twice the volume of the smaller containers and assume the mass of an argon atom is twice the mass of a neon atom.

77. Calculate the average kinetic energies of the CH_4 and N_2 molecules at 273 K and 546 K.

78. Calculate the root mean square velocities of CH_4 and N_2 molecules at 273 K and 546 K.

79. Do all the molecules in a 1-mole sample of $CH_4(g)$ have the same kinetic energy at 273 K? Do all the molecules in a 1-mole sample of $N_2(g)$ have the same velocity at 546 K? Explain.

80. Consider separate 1.0-L gaseous samples of H_2, Xe, Cl_2, and O_2, all at STP.
a. Rank the gases in order of increasing average kinetic energy.
b. Rank the gases in order of increasing average velocity.
c. How can separate 1.0-L samples of O_2 and H_2 both have the same average velocity?

81. Consider three identical flasks filled with different gases.

Flask A: CO at 760 torr and 0°C

Flask B: N_2 at 250 torr and 0°C

Flask C: H_2 at 100 torr and 0°C

a. In which flask will the molecules have the greatest average kinetic energy?
b. In which flask will the molecules have the greatest root mean square velocity?
c. Which flask will have the greatest number of collisions per second with the walls of the container?

82. Consider a 1.0-L container of neon gas at STP. Will the average kinetic energy, root mean square velocity, frequency of collisions of gas molecules with each other, frequency of collisions of gas molecules with the walls of the container, and energy of impact of gas molecules with the container increase, decrease, or remain the same under each of the following conditions?
a. The temperature is increased to 100°C.
b. The temperature is decreased to −50°C.
c. The volume is decreased to 0.5 L.
d. The number of moles of neon is doubled.

83. Freon-12 is used as a refrigerant in central home air conditioners. The rate of effusion of Freon-12 to Freon-11 (molar mass = 137.4 g/mol) is 1.07:1. The formula of Freon-12 is one of the following: CF_4, CF_3Cl, CF_2Cl_2, $CFCl_3$, or CCl_4. Which formula is correct for Freon-12?

84. One way of separating oxygen isotopes is by gaseous diffusion of carbon monoxide. The gaseous diffusion process behaves like an effusion process. Calculate the relative rates of effusion of $^{12}C^{16}O$, $^{12}C^{17}O$, and $^{12}C^{18}O$. List some advantages and disadvantages of separating oxygen isotopes by gaseous diffusion of carbon dioxide instead of carbon monoxide.

85. The rate of effusion of a particular gas was measured to be 24.0 mL/min. Under the same conditions, the rate of effusion of pure methane gas (CH_4) is 47.8 mL/min. What is the molar mass of the unknown gas?

86. It took 4.5 minutes for 1.0 L of helium to effuse through a porous barrier. How long will it take for 1.0 L of Cl_2 gas to effuse under identical conditions?

87. Calculate the pressure exerted by 0.5000 mole of N_2 in a 1.0000-L container at 25.0°C. (See Table 5.3.)
a. Use the ideal gas law.
b. Use the van der Waals equation.
c. Compare the results from parts a and b.

88. Calculate the pressure exerted by 0.5000 mole of N_2 in a 10.000-L container at 25.0°C. (See Table 5.3.)
a. Use the ideal gas law.
b. Use the van der Waals equation.
c. Compare the results from parts a and b.
d. Compare the results with those in Exercise 87.

89. Why do real gases not always behave ideally? Under what conditions does a real gas behave most ideally? Why?

90. Consider the following velocity distribution curves *A* and *B*.

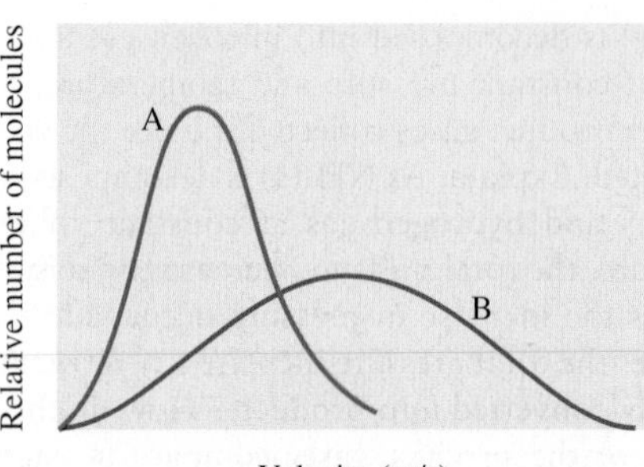

a. If the plots represent the velocity distribution of 1.0 L of He(*g*) at STP versus 1.0 L of $Cl_2(g)$ at STP, which plot corresponds to each gas? Explain your reasoning.

b. If the plots represent the velocity distribution of 1.0 L of $O_2(g)$ at temperatures of 273 K versus 1273 K, which plot corresponds to each temperature? Explain your reasoning. Under which temperature condition would the $O_2(g)$ sample behave most ideally? Explain.

91. In the van der Waals equation, why is a term added to the observed pressure and why is a term subtracted from the container volume to correct for nonideal gas behavior?

92. Without looking at tables of values, which of the following gases would you expect to have the largest value of the van der Waals constant b: H_2, N_2, CH_4, C_2H_6, or C_3H_8?

93. From the values in Table 5.3 for the van der Waals constant a for the gases H_2, CO_2, N_2, and CH_4, predict which molecule shows the strongest intermolecular attractions.

94. The Maxwell–Boltzmann distribution function $f(u)$ increases at small values of u and decreases at large values of u. Identify the parts of the function responsible for this behavior.

95. Calculate the root mean square, the most probable, and the average velocities for $N_2(g)$ at 227°C.

96. Calculate the kinetic energy possessed by 1.00×10^{20} molecules of methane gas (CH_4) at $T = 27°C$, assuming ideal behavior.

97. A flask contains $\frac{1}{3}$ mole of H_2 and $\frac{2}{3}$ mole of He. Compare the force on the wall per impact of H_2 relative to that for He.

98. A certain sample of uranium is reacted with fluorine to form a mixture of $^{235}UF_6(g)$ and $^{238}UF_6(g)$. After 100 diffusion steps, the gas contains 1526 $^{235}UF_6$ molecules per 1.000×10^5 total number of molecules in the gas ($^{235}UF_6 + {}^{238}UF_6$). What is the ratio of ^{235}U to ^{238}U atoms in the original sample of uranium?

99. Consider separate 1.0-L samples of $O_2(g)$ and He(g), both at 25°C and the same pressure. Compare the change in momentum per impact and the number of impacts per second in the two samples.

100. Consider separate 1.00-L samples of Ar(g), both containing the same number of moles, one at 27°C and the other at 77°C. Compare the change in momentum per impact and the number of impacts per second in the two samples.

101. Calculate the intermolecular collision frequency and the mean free path in a sample of helium gas with a volume of 5.0 L at 27°C and 3.0 atm. Assume that the diameter of a helium atom is 50. pm.

Atmosphere Chemistry

102. Use the data in Table 5.4 to calculate the partial pressure of He in dry air assuming that the total pressure is 1.0 atm. Assuming a temperature of 25°C, calculate the number of He atoms per cubic centimeter.

103. Atmospheric scientists often use mixing ratios to express the concentrations of trace compounds in air. Mixing ratios are often expressed as ppmv (parts per million volume):

$$\text{ppmv of } X = \frac{\text{vol of X at STP}}{\text{total vol of air at STP}} \times 10^6$$

On a certain November day, the concentration of carbon monoxide in the air in downtown Denver, Colorado, reached 3.0×10^2 ppmv. The atmospheric pressure at that time was 628 torr and the temperature was 0°C.

a. What was the partial pressure of CO?

b. What was the concentration of CO in molecules per cubic meter?

c. What was the concentration of CO in molecules per cubic centimeter?

104. Write reactions to show how nitric and sulfuric acids are produced in the atmosphere. Write reactions to show how the nitric and sulfuric acids in acid rain react with marble and limestone. (Both marble and limestone are primarily calcium carbonate.)

105. Trace organic compounds in the atmosphere are first concentrated and then measured by gas chromatography. In the concentration step, several liters of air are pumped through a tube containing a porous substance that traps organic compounds. The tube is then connected to a gas chromatograph and heated to release the trapped compounds. The organic compounds are separated in the column and the amounts are measured. In an analysis for benzene and toluene in air, a 3.00-L sample of air at 748 torr and 23°C was passed through the trap. The gas chromatography analysis showed that this air sample contained 89.6 ng of benzene (C_6H_6) and 153 ng of toluene (C_7H_8). Calculate the mixing ratio (see Exercise 103) and number of molecules per cubic centimeter for both benzene and toluene.

Additional Exercises

106. A form of Boyle's law is $PV = k$ (at constant T and n). Table 5.1 contains actual data from pressure–volume experiments conducted by Robert Boyle. The value of k in most experiments is 14.1×10^2 in Hg · in^3. Express k in units of atm · L. In Example 5.1, k was determined for NH_3 at various pressures and volumes. Give some reasons why the k values differ so dramatically between Example 5.1 and Table 5.1.

107. A glass vessel contains 28 g nitrogen gas. Assuming ideal behavior, which of the processes listed below would double the pressure exerted on the walls of the vessel?

a. Adding enough mercury to fill one-half the container.

b. Raising the temperature of the container from 30.°C to 60.°C.

c. Raising the temperature of the container from −73°C to 127°C.

d. Adding 28 g nitrogen gas.

108. Which of the following statements is(are) true? For the false statements, correct them.
   a. At constant temperature, the lighter the gas molecules, the faster the average velocity of the gas molecules.
   b. At constant temperature, the heavier the gas molecules, the larger the average kinetic energy of the gas molecules.
   c. A real gas behaves most ideally when the container volume is relatively large and the gas molecules are moving relatively quickly.
   d. As temperature increases, the effect of interparticle interactions on gas behavior is increased.
   e. At constant V and T, as gas molecules are added into a container, the number of collisions per unit area increases, resulting in a higher pressure.
   f. The kinetic molecular theory predicts that pressure is inversely proportional to temperature at constant volume and moles of gas.

109. A person accidentally swallows a drop of liquid oxygen, $O_2(l)$, which has a density of 1.149 g/mL. Assuming the drop has a volume of 0.050 mL, what volume of gas will be produced in the person's stomach at body temperature (37°C) and a pressure of 1.0 atm?

110. Hydrogen azide, HN_3, decomposes on heating by the following *unbalanced* reaction:

$$HN_3(g) \longrightarrow N_2(g) + H_2(g)$$

If 3.0 atm of pure $HN_3(g)$ is decomposed initially, what is the final total pressure in the reaction container? What are the partial pressures of nitrogen and hydrogen gas? Assume the volume and temperature of the reaction container are constant.

111. A 20.0 L stainless steel container at 25°C was charged with 2.00 atm of hydrogen gas and 3.00 atm of oxygen gas. A spark ignited the mixture, producing water. What is the pressure in the tank at 25°C? If the same experiment were performed, but the temperature was 125°C instead of 25°C, what would be the pressure in the tank?

112. In the "Méthode Champenoise," grape juice is fermented in a wine bottle to produce sparkling wine. The reaction is

$$C_6H_{12}O_6(aq) \longrightarrow 2C_2H_5OH(aq) + 2CO_2(g)$$

Fermentation of 750. mL of grape juice (density = 1.0 g/cm^3) is allowed to take place in a bottle with a total volume of 825 mL until 12% by volume is ethanol (C_2H_5OH). Assuming that the CO_2 is insoluble in H_2O (actually a wrong assumption), what would be the pressure of CO_2 inside the wine bottle at 25°C? (The density of ethanol is 0.79 g/cm^3.)

113. A 2.747-g sample of manganese metal is reacted with excess HCl gas to produce 3.22 L of $H_2(g)$ at 373 K and 0.951 atm and a manganese chloride compound ($MnCl_x$). What is the formula of the manganese chloride compound produced in the reaction?

114. The total mass that can be lifted by a balloon is given by the difference between the mass of air displaced by the balloon and the mass of the gas inside the balloon. Consider a hot-air balloon that approximates a sphere 5.00 m in diameter and contains air heated to 65°C. The surrounding air temperature is 21°C. The pressure in the balloon is equal to the atmospheric pressure, which is 745 torr.
   a. What total mass can the balloon lift? Assume the average molar mass of air is 29.0 g/mol. (*Hint:* Heated air is less dense than cool air.)
   b. If the balloon is filled with enough helium at 21°C and 745 torr to achieve the same volume as in part a, what total mass can the balloon lift?
   c. What mass could the hot-air balloon (from part a) lift if it were on the ground in Denver, Colorado, where a typical atmospheric pressure is 630. torr?
   d. What mass could the hot-air balloon (from part a) lift if it were a cold day with a temperature of −8°C?

115. At STP, 1.0 L Br_2 reacts completely with 3.0 L F_2, producing 2.0 L of a product. What is the formula of the product? (All substances are gases.)

116. Natural gas is a mixture of hydrocarbons, primarily methane (CH_4) and ethane (C_2H_6). A typical mixture might have $\chi_{methane} = 0.915$ and $\chi_{ethane} = 0.085$. What are the partial pressures of the two gases in a 15.00-L container of natural gas at 20.°C and 1.44 atm? Assuming complete combustion of both gases in the natural gas sample, what is the total mass of water formed?

117. An important process for the production of acrylonitrile (C_3H_3N) (annual U.S. production is greater than 10^9 lb) is given by the following equation:

$$2C_3H_6(g) + 2NH_3(g) + 3O_2(g) \longrightarrow 2C_3H_3N(g) + 6H_2O(g)$$

A 150.-L reactor is charged to the following partial pressures at 25°C:

$$P_{C_3H_6} = 0.500 \text{ MPa}$$
$$P_{NH_3} = 0.800 \text{ MPa}$$
$$P_{O_2} = 1.500 \text{ MPa}$$

What mass of acrylonitrile can be produced from this mixture (MPa = 10^6 Pa)?

118. The oxides of Group 2A metals (symbolized by M here) react with carbon dioxide according to the following reaction:

$$MO(s) + CO_2(g) \longrightarrow MCO_3(s)$$

A 2.85-g sample containing only MgO and CuO is placed in a 3.00-L container. The container is filled with CO_2 to a pressure of 740. torr at 20.°C. After the reaction has gone to completion, the pressure inside the flask is 390. torr at 20.°C. What is the mass percent of MgO in the mixture? Assume that only the MgO reacts with CO_2.

119. Small quantities of hydrogen gas can be prepared in the laboratory by the addition of aqueous hydrochloric acid to metallic zinc.

$$Zn(s) + 2HCl(aq) \longrightarrow ZnCl_2(aq) + H_2(g)$$

Typically, the hydrogen gas is bubbled through water for collection and becomes saturated with water vapor. Suppose 240. mL of hydrogen gas is collected at 30.°C and has a total pressure of 1.032 atm by this process. What is the partial pressure of hydrogen gas in the sample? How many grams of zinc must have reacted to produce this quantity of hydrogen? (The vapor pressure of water is 32 torr at 30°C.)

120. Nitrogen gas (N_2) reacts with hydrogen gas (H_2) to form ammonia gas (NH_3). You have nitrogen and hydrogen gases in a 15.0-L container fitted with a movable piston (the piston allows the container volume to change so as to keep the pressure constant inside the container). Initially, the partial pressure of each reactant gas is 1.00 atm. Assume the temperature is constant and the reaction goes to completion.
   a. Calculate the partial pressure of ammonia in the container after the reaction has reached completion.
   b. Calculate the volume of the container after the reaction has reached completion.

121. Consider the three flasks in the diagram below. Assuming the connecting tubes have negligible volume, what is the partial pressure of each gas and the total pressure after all the stopcocks are opened?

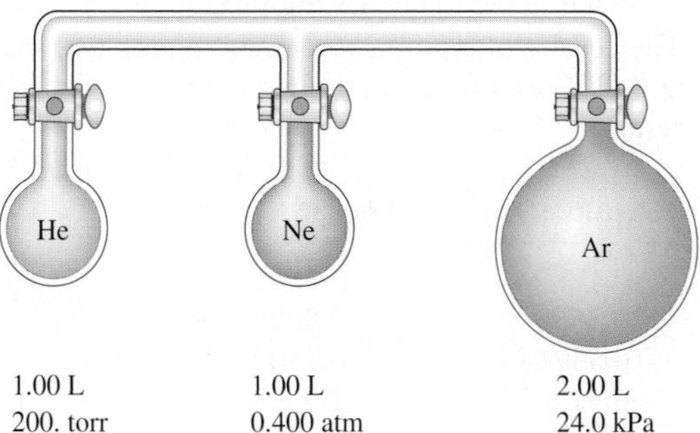

122. Equal moles of sulfur dioxide gas and oxygen gas are mixed in a flexible reaction vessel and then sparked to initiate the formation of gaseous sulfur trioxide. Assuming that the reaction goes to completion, what is the ratio of the final volume of the gas mixture to the initial volume of the gas mixture if both volumes are measured at the same temperature and pressure?

123. Silane (SiH_4) is the silicon analogue of methane (CH_4). It is prepared industrially according to the following equations:

$$Si(s) + 3HCl(g) \longrightarrow HSiCl_3(l) + H_2(g)$$

$$4HSiCl_3(l) \longrightarrow SiH_4(g) + 3SiCl_4(l)$$

   a. If 156 mL of $HSiCl_3$ (d = 1.34 g/mL) is isolated when 15.0 L of HCl at 10.0 atm and 35°C is used, what is the percent yield of $HSiCl_3$?
   b. When 156 mL of $HSiCl_3$ is heated, what volume of SiH_4 at 10.0 atm and 35°C will be obtained if the percent yield of the reaction is 93.1%?

124. A compound containing only C, H, and N yields the following data.
   i. Complete combustion of 35.0 mg of the compound produced 33.5 mg of CO_2 and 41.1 mg of H_2O.
   ii. A 65.2-mg sample of the compound was analyzed for nitrogen by the Dumas method (see Exercise 67), giving 35.6 mL of dry N_2 at 740. torr and 25°C.
   iii. The effusion rate of the compound as a gas was measured and found to be 24.6 mL/min. The effusion rate of argon gas, under identical conditions, is 26.4 mL/min.

   What is the formula of the compound?

125. A 15.0-L tank is filled with H_2 to a pressure of 2.00×10^2 atm. How many balloons (each 2.00 L) can be inflated to a pressure of 1.00 atm from the tank? Assume that there is no temperature change and that the tank cannot be emptied below 1.00 atm pressure.

126. Consider the following diagram.

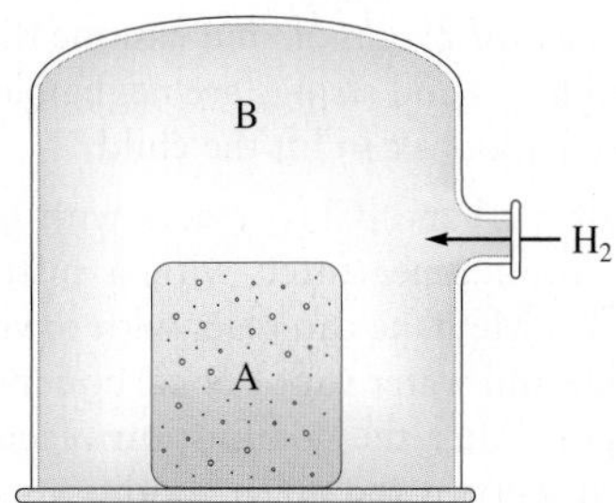

A porous container (A), filled with air at STP, is contained in a large enclosed container (B), which is flushed with $H_2(g)$. What will happen to the pressure inside container A? Explain your answer.

127. A 100.-L flask contains a mixture of methane (CH_4) and argon at 25°C. The mass of argon present is 228 g and the mole fraction of methane in the mixture is 0.650. Calculate the total kinetic energy of the gaseous mixture.

128. Represent the following plots.
   a. PV/n (y axis) versus P (x axis) for a real gas that obeys the equation $PV/n = \alpha + \beta P$
   b. change in momentum per impact versus mass of an individual gas particle for a series of ideal gases all at the same temperature
   c. P versus T (°C) for an ideal gas where n and V are constant

129. A spherical glass container of unknown volume contains helium gas at 25°C and 1.960 atm. When a portion of the helium is withdrawn and adjusted to 1.00 atm at 25°C, it is found to have a volume of 1.75 cm^3. The gas remaining in the first container shows a pressure of 1.710 atm. Calculate the volume of the spherical container.

130. A compound Z is known to have a composition of 34.38% Ni, 28.13% C, and 37.48% O. In an experiment 1.00 L of gaseous Z is mixed with 1.00 L of argon, where each gas is at P = 2.00 atm and T = 25°C. When this mixture of gases is put in an effusion chamber, the ratio of Z molecules to Ar molecules in the effused mixture is 0.4837. Using these data, calculate the following.
   a. the empirical formula for Z
   b. the molar mass for Z
   c. the molecular formula for Z
   d. the mole ratio of Z to argon in a sample of gas obtained by five effusion steps (starting with the original mixture)

131. Hydrogen cyanide gas is commercially prepared by the reaction of methane [$CH_4(g)$], ammonia [$NH_3(g)$], and oxygen [$O_2(g)$] at a high temperature. The other product is gaseous water.
  a. Write a balanced chemical equation for the reaction.
  b. Methane and ammonia gases flow into a reactor at a rate of 20.0 L/s. Oxygen gas is introduced at a flow rate of 40.0 L/s. All the reactant gases are at 1.00 atm and 150.°C. What mass of HCN is produced per second by this reaction assuming 100% yield?

Challenge Problems

132. Consider a children's cartoon illustrating a child holding the strings of several helium balloons and being lifted into the sky.
  a. Estimate the minimum number of 10.-L balloons it would take to lift a 50.-lb child. Assume air has an average molar mass of 29 g/mol, and assume the masses of the balloons and strings are negligible.
  b. Explain why the balloons can lift the child.

133. A 16.0-g sample of methane (CH_4) reacts with 64.0 g of oxygen gas in a container fitted with a piston (at 1.00 atm and 425 K). Methane can react with oxygen to form carbon dioxide and water vapor or carbon monoxide and water vapor. After the combustion reaction is complete, the gas density at the given conditions is observed to be 0.7282 g/L. Calculate the mole fraction of methane that reacts to form carbon monoxide rather than carbon dioxide.

134. You have two samples of helium gas at the same pressure in separate steel containers of the same volume. You want the number of collisions of helium atoms with the walls of container 1 to be twice the number of collisions of helium atoms with the walls of container 2. Assume ideal behavior.
  a. How does the temperature in container 1 compare to the temperature in container 2? That is, which temperature is larger and by what factor? Explain your answer and support it mathematically.
  b. If the number of collisions is different in each container, how can the pressure be the same? Provide a written explanation with mathematical support.

135. A mixture of chromium and zinc weighing 0.362 g was reacted with an excess of hydrochloric acid. After all the metals in the mixture reacted, 225 mL of dry hydrogen gas was collected at 27°C and 750. torr. Determine the mass percent of Zn in the metal sample. [Zinc reacts with hydrochloric acid to produce zinc chloride and hydrogen gas; chromium reacts with hydrochloric acid to produce chromium(III) chloride and hydrogen gas.]

136. You have a sealed, flexible balloon filled with argon gas. The atmospheric pressure is 1.00 atm and the temperature is 25°C. The air has a mole fraction of nitrogen of 0.79, the rest being oxygen.
  a. Explain why the balloon would float when heated. Make sure to discuss which factors change and which remain constant, and why this matters. Be complete.
  b. Above what temperature would you heat the balloon so that it would float?

137. Derive a linear relationship between gas density and temperature, and use it to estimate the value of absolute zero temperature (in °C to the nearest 0.1°C) from an air sample whose density is 1.2930 g/L at 0.0°C and 0.9460 g/L at 100.0°C. Assume air obeys the ideal gas law and that the pressure is held constant.

138. A chemist weighed out 5.14 g of a mixture containing unknown amounts of $BaO(s)$ and $CaO(s)$ and placed the sample in a 1.50-L flask containing $CO_2(g)$ at 30.0°C and 750. torr. After the reaction to form $BaCO_3(s)$ and $CaCO_3(s)$ was completed, the pressure of $CO_2(g)$ remaining was 230. torr. Calculate the mass percents of $CaO(s)$ and $BaO(s)$ in the mixture.

139. The density of a pure gaseous compound was measured at 0.00°C as a function of pressure to give the following results:

Density (g/L)	Pressure (atm)
0.17893	0.2500
0.35808	0.5000
0.53745	0.7500
0.71707	1.000

Calculate the molar mass of this compound, corrected for any nonideal behavior of the gas. Assume the nonideal gas obeys the equation $PV/nRT = 1 + \beta P$. (*Hint:* Derive an equation for P/d and plot P/d versus P.)

140. Consider separate 1.0-L samples of $He(g)$ and $UF_6(g)$, both at 1.00 atm and containing the same number of moles. What ratio of temperatures for the two samples would produce the same collision frequency with the vessel walls?

141. The most probable velocity u_{mp} is the velocity possessed by the greatest number of gas particles. At a certain temperature, the probability that a gas particle has the most probable velocity is equal to one-half the probability that the same gas particle has the most probable velocity at 300. K. Is the temperature higher or lower than 300. K? Calculate the temperature.

142. Derive Dalton's law of partial pressures from the kinetic molecular theory of gases. What assumptions are necessary?

143. One of the assumptions of the kinetic molecular theory is that the volume of a gas particle is negligible. If this were the case, the ratio of the number of collisions of gas particles with the walls of the container compared to the

number of collisions a given gas particle experiences with other gas particles should be quite high. Determine the volume of a cube (in L) filled with helium such that the ratio of the number of collisions of helium atoms with the container walls to the number of intermolecular collisions for a given helium atom is 1 quintillion (1 quintillion = 1.00×10^{18}). The atomic radius of helium is 3.2×10^{-11} m.

144. Consider a sample of a hydrocarbon (a compound consisting of only carbon and hydrogen) at 0.959 atm and 298 K. Upon combusting the entire sample in oxygen, you collect a mixture of gaseous carbon dioxide and water vapor at 1.51 atm and 375 K. This mixture has a density of 1.391 g/L and occupies a volume four times as large as that of the pure hydrocarbon. Determine the molecular formula of the hydrocarbon.

145. A steel cylinder contains 5.00 moles of graphite (pure carbon) and 5.00 moles of O_2. The mixture is ignited and all the graphite reacts. Combustion produces a mixture of CO gas and CO_2 gas. After the cylinder has cooled to its original temperature, it is found that the pressure of the cylinder has increased by 17.0%. Calculate the mole fractions of CO, CO_2, and O_2 in the final gaseous mixture.

146. You have an equimolar mixture of the gases SO_2 and O_2, along with some He, in a container fitted with a piston. The density of this mixture at STP is 1.924 g/L. Assume ideal behavior and constant temperature.
   a. What is the mole fraction of He in the original mixture?
   b. The SO_2 and O_2 react to completion to form SO_3. What is the density of the gas mixture after the reaction is complete?

147. Methane (CH_4) gas flows into a combustion chamber at a rate of 200. L/min at 1.50 atm and ambient temperature. Air is added to the chamber at 1.00 atm and the same temperature, and the gases are ignited.
   a. To ensure complete combustion of CH_4 to $CO_2(g)$ and $H_2O(g)$, three times as much oxygen as is necessary is reacted. Assuming air is 21 mole percent O_2 and 79 mole percent N_2, calculate the flow rate of air necessary to deliver the required amount of oxygen.
   b. Under the conditions in part a, combustion of methane was not complete as a mixture of $CO_2(g)$ and $CO(g)$ was produced. It was determined that 95.0% of the carbon in the exhaust gas was present in the CO_2. The remainder was present as carbon in the CO. Calculate the composition of the exhaust gas in terms of mole fractions of CO, CO_2, O_2, N_2, and H_2O. Assume CH_4 is completely reacted and N_2 is unreacted.
   c. Assuming a total pressure of the exhaust gas of 1.00 atm, calculate the partial pressures of the gases in part b.

148. A spherical vessel with a volume of 1.00 L was evacuated and sealed. Twenty-four hours later the pressure of air in the vessel was found to be 1.20×10^{-6} atm. During this 24-h period, the vessel had been surrounded by air at 27°C and 1.00 atm. Assuming that air is 78 mole percent nitrogen and that the remainder is oxygen, calculate the diameter of the tiny circular hole in the vessel that allowed the air to leak in.

149. Calculate the number of stages needed to change a mixture of $^{13}CO_2$ and $^{12}CO_2$ that is originally 0.10% (by moles) $^{13}CO_2$ to a mixture that is 0.010% $^{13}CO_2$ by a gaseous diffusion process. (The mass of ^{13}C is 13.003355 amu.)

150. Two samples of gas are separated in two rectangular 1.00-L chambers by a thin metal wall. One sample is pure helium and the other is pure radon. Both samples are at 27°C and show a pressure of 2.00×10^{-6} atm. Assuming that the metal wall separating the gases suddenly develops a circular hole of radius 1.00×10^{-6} m, calculate the pressure in each chamber after 10.0 h have passed.

151. You have a helium balloon at 1.00 atm and 25°C. You want to make a hot-air balloon with the same volume and same lift as the helium balloon. Assume air is 79.0% nitrogen and 21.0% oxygen by volume. The "lift" of a balloon is given by the difference between the mass of air displaced by the balloon and the mass of gas inside the balloon.
   a. Will the temperature in the hot-air balloon have to be higher or lower than 25°C? Explain.
   b. Calculate the temperature of the air required for the hot-air balloon to provide the same lift as the helium balloon at 1.00 atm and 25°C. Assume atmospheric conditions are 1.00 atm and 25°C.

152. Consider an equimolar mixture (equal number of moles) of two diatomic gases (A_2 and B_2) in a container fitted with a piston. The gases react to form one product (which is also a gas) with the formula A_xB_y. The density of the sample after the reaction is complete (and the temperature returns to its original state) is 1.50 times greater than the density of the reactant mixture.
   a. Specify the formula of the product, and explain if more than one answer is possible based on the given data.
   b. Can you determine the molecular formula of the product with the information given or only the empirical formula?

153. You are given an unknown gaseous binary compound (that is, a compound consisting of two different elements). When 10.0 g of the compound is burned in excess oxygen, 16.3 g of water is produced. The compound has a density 1.38 times that of oxygen gas at the same conditions of temperature and pressure. Give a possible identity for the unknown compound.

Marathon Problem

154.* Use the following information to identify element A and compound B, then answer questions a and b.

An empty glass container has a mass of 658.572 g. It has a mass of 659.452 g after it has been filled with nitrogen gas at a pressure of 790. torr and a temperature of 15°C. When the container is evacuated and refilled with a certain element (A) at a pressure of 745 torr and a temperature of 26°C, it has a mass of 660.59 g.

Compound B, a gaseous organic compound that consists of 85.6% carbon and 14.4% hydrogen by mass, is placed in a stainless steel vessel (10.68 L) with excess oxygen gas. The vessel is placed in a constant-temperature bath at 22°C. The pressure in the vessel is 11.98 atm. In the bottom of the vessel is a container that is packed with Ascarite and a desiccant. Ascarite is asbestos impregnated with sodium hydroxide; it quantitatively absorbs carbon dioxide:

$$2NaOH(s) + CO_2(g) \longrightarrow Na_2CO_3(s) + H_2O(l)$$

The desiccant is anhydrous magnesium perchlorate, which quantitatively absorbs the water produced by the combustion reaction as well as the water produced by the preceding reaction. Neither the Ascarite nor the desiccant reacts with compound B or oxygen. The total mass of the container with the Ascarite and desiccant is 765.3 g.

The combustion reaction of compound B is initiated by a spark. The pressure immediately rises, then begins to decrease, and finally reaches a steady value of 6.02 atm. The stainless steel vessel is carefully opened, and the mass of the container inside the vessel is found to be 846.7 g.

A and B react quantitatively in a 1:1 mole ratio to form 1 mole of gas C.

a. How many grams of C will be produced if 10.0 L of A and 8.60 L of B (each at STP) are reacted by opening a stopcock connecting the two samples?
b. What will be the total pressure in the system?

*From James H. Burness, "The Use of "Marathon" Problems as Effective Vehicles for the Presentation of General Chemistry Lectures," Journal of Chemical Education, 68(11).

Chemical Equilibrium

6

chapter

Black sea nettle in the Monterey Bay Aquarium in California. Jellyfish use statocysts to maintain their physical equilibrium. A statocyst consists of a fluid-filled sac containing statoliths that stimulate sensory cells and help indicate position when the animal moves.

Lee Foster/Taxi/Getty Images

In doing stoichiometry calculations, we assume that reactions proceed to completion—that is, until one of the reactants is consumed. Many reactions *do* proceed essentially to completion. For such reactions it can be assumed that the reactants are quantitatively converted to products and that the amount of limiting reactant that remains is negligible. On the other hand, there are many chemical reactions that stop far short of completion. An example is the dimerization of nitrogen dioxide:

$$2NO_2(g) \longrightarrow N_2O_4(g)$$

The reactant NO_2 is a reddish brown gas, and the product N_2O_4 is a colorless gas. When NO_2 is placed in an evacuated, sealed glass vessel at 25°C, the initial dark brown color decreases in intensity as the NO_2 is converted to colorless N_2O_4. However, even over a long period of time, the contents of the reaction vessel do not become colorless. Instead, the intensity of the brown color eventually becomes constant, which means that the concentration of NO_2 is no longer changing. This is illustrated on the molecular level in Fig. 6.1. This observation is a clear indication that the reaction has stopped short of completion. In fact, the system has reached **chemical equilibrium**, *the state in which the concentrations of all reactants and products remain constant with time.*

Any chemical reaction carried out in a closed vessel will reach equilibrium. For some reactions the equilibrium position so favors the products that the reaction appears to have gone to completion. We say that the equilibrium position for such a reaction lies *far to the right,* in the direction of the products. For example, when gaseous hydrogen and oxygen are mixed in stoichiometric quantities and react to form water vapor, the reaction proceeds essentially to completion. The amounts of the reactants that remain when the system reaches equilibrium are so tiny that they are negligible. In contrast, some reactions occur only to a slight extent. For example, when solid CaO is placed in a closed vessel at 25°C, the decomposition to solid Ca and gaseous O_2 is virtually undetectable. In cases like this, the equilibrium position is said to lie *far to the left,* in the direction of the reactants.

In this chapter we will discuss how and why a chemical system comes to equilibrium and the characteristics of a system at equilibrium. In particular, we will discuss how to calculate the concentrations of the reactants and products present for a given system at equilibrium.

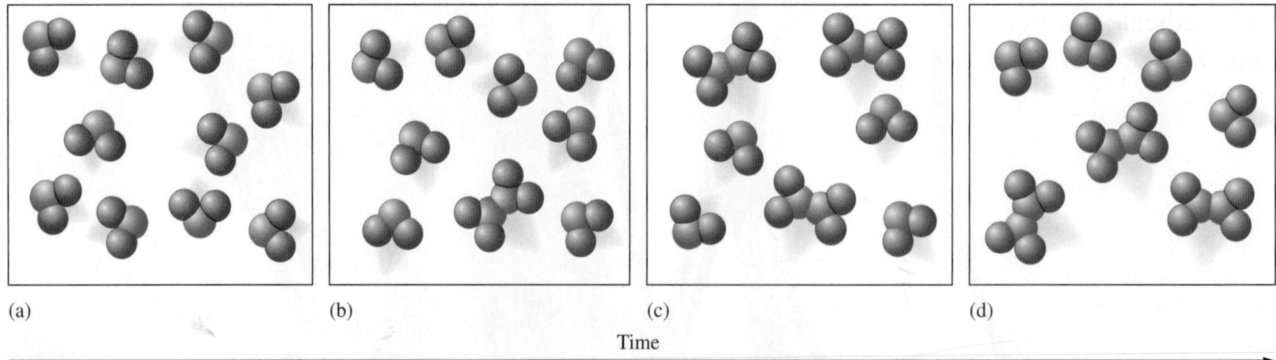

Figure 6.1
A molecular representation of the reaction $2NO_2(g) \longrightarrow N_2O_4(g)$ over time in a closed vessel. Note that the numbers of NO_2 and N_2O_4 in the container become constant (c) and (d) after sufficient time has passed.

6.1 The Equilibrium Condition

Since no changes occur in the concentrations of reactants or products in a reaction system at equilibrium, it may appear that everything has stopped. However, this is not the case. On the molecular level, there is frenetic activity. Equilibrium is not static; it is a highly *dynamic* situation. The concept of chemical equilibrium is analogous to two island cities connected by a bridge. Suppose the traffic flow on the bridge is the same in both directions. It is obvious that there is motion, since one can see the cars traveling across the bridge, but the number of cars in each city is not changing because equal numbers are entering and leaving. The result is no *net* change in the car population.

Reddish brown nitrogen dioxide gas streaming from a flask where copper is reacting with concentrated nitric acid.

To see how this concept applies to chemical reactions, consider the reaction between steam and carbon monoxide in a closed vessel at a high temperature, where the reaction takes place rapidly:

$$H_2O(g) + CO(g) \rightleftharpoons H_2(g) + CO_2(g)$$

Assume that the same number of moles of gaseous CO and gaseous H_2O are placed in a closed vessel and allowed to react. The plots of the concentrations of reactants and products versus time are shown in Fig. 6.2. Note that since CO and H_2O were originally present in equal molar quantities, and since they react in a 1:1 ratio, the concentrations of the two gases are always equal. Also, since H_2 and CO_2 are formed in equal amounts, they are always present at the same concentrations.

Figure 6.2 is a profile of the progress of the reaction. When CO and H_2O are mixed, they immediately begin reacting to form H_2 and CO_2. This leads to a decrease in the concentrations of the reactants, but the concentrations of the products, which were initially at zero, are increasing. Beyond a certain time, indicated by the dashed line in Fig. 6.2, the concentrations of reactants and products no longer change—equilibrium has been reached. Unless the system is somehow disturbed, no further changes in concentrations will occur. Note that although the equilibrium position lies far to the right, the concentrations of reactants never reach zero; the reactants will always be present in small but constant concentrations. This is shown on the microscopic level in Fig. 6.3.

Equilibrium is a dynamic situation.

Why does equilibrium occur? As we will see in much more detail in Chapter 15, chemical reactions occur via collisions of the reacting molecules. The energy associated with a collision can break bonds in the reactant molecules, allowing them to rearrange to form the products. Since the collision rate of the molecules in a gas depends on the concentration of molecules present, the rate of a reaction depends on the concentrations of the reactants. Thus, as reactants collide and react to form products

$$H_2O(g) + CO(g) \longrightarrow H_2(g) + CO_2(g)$$

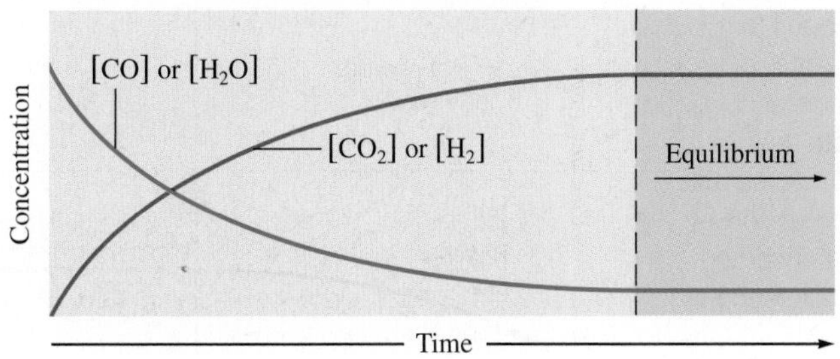

Figure 6.2
The changes in concentrations with time for the reaction $H_2O(g) + CO(g) \rightleftharpoons H_2(g) + CO_2(g)$ when equal molar quantities of $H_2O(g)$ and $CO(g)$ are mixed.

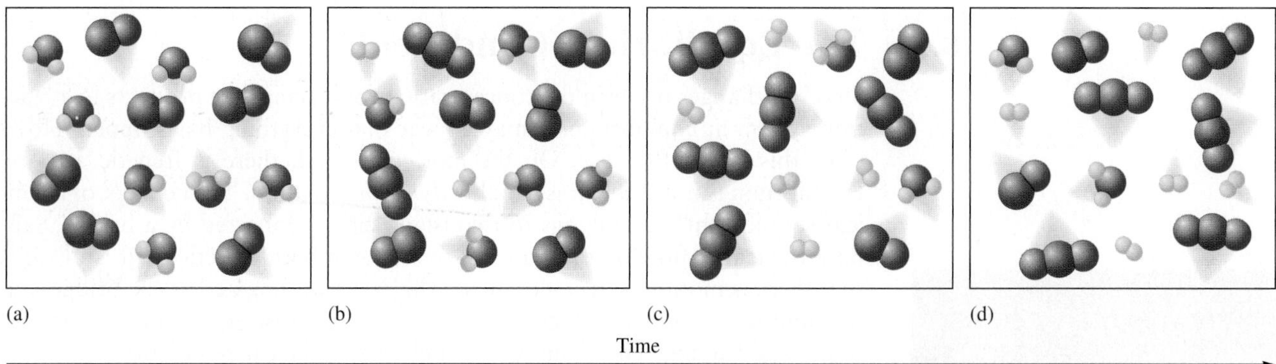

Figure 6.3
(a) H_2O and CO are mixed in equal numbers and begin to react (b) to form CO_2 and H_2. After time has passed, equilibrium is reached (c), and the numbers of reactant and product molecules then remain constant over time (d).

the concentrations of the reactants decrease, causing the rate of this reaction (the *forward* reaction) to decrease—that is, the reaction slows down (Fig. 6.4).

As in the bridge-traffic analogy, there is also a reverse direction:

$$H_2(g) + CO_2(g) \longrightarrow H_2O(g) + CO(g)$$

Initially in this experiment no H_2 and CO_2 were present, and this reverse reaction could not occur. However, as the forward reaction proceeds, the concentrations of H_2 and CO_2 build up, and the rate of the reverse reaction increases (Fig. 6.4) as the forward reaction slows down. Eventually, the concentrations reach levels where the rate of the forward reaction equals the rate of the reverse reaction. That is, the concentrations of the reactants and products achieve values such that H_2O and CO are being consumed by the forward reaction at exactly the same rate as they are being produced by the reverse reaction. The system has reached equilibrium.

What would happen to the gaseous equilibrium mixture of reactants and products represented in Fig. 6.3, parts (c) and (d), if we injected some $H_2O(g)$ into the box? To answer this question, we need to be sure we understand the equilibrium condition: The concentrations of reactants and products remain constant at equilibrium because the forward and reverse reaction rates are equal. If we inject some H_2O molecules, what will happen to the forward reaction: $H_2O + CO \rightarrow H_2 + CO_2$? It will speed up because when there are more H_2O molecules there will be more collisions between H_2O and CO molecules. This in turn will form more products and will cause the reverse reaction $H_2O + CO \leftarrow H_2 + CO_2$ to speed up. Thus the system will change until the forward and reverse reaction rates again become equal. Will this new equilibrium position contain more or fewer product molecules than are shown in Fig. 6.3(c) and (d)? Think about this carefully. If you are not sure of the answer

Figure 6.4
The changes with time in the rates of forward and reverse reactions for $H_2O(g) + CO(g) \rightleftharpoons H_2(g) + CO_2(g)$ when equal molar quantities of $H_2O(g)$ and $CO(g)$ are mixed. Note that the rates for the forward and reverse reactions do not change in the same way with time. We will not be concerned with the reasons for this difference at this point.

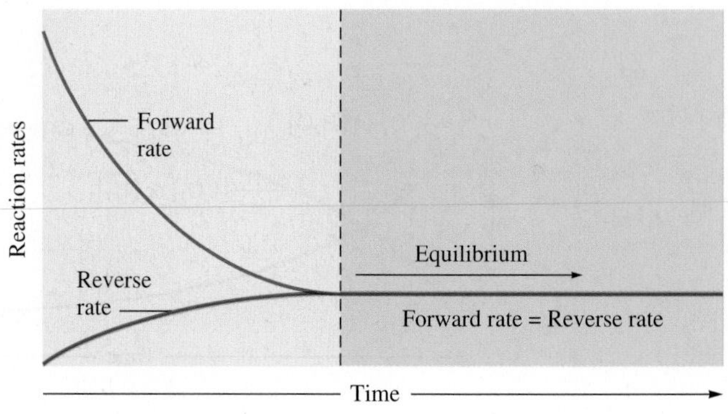

now, keep reading. We will consider this type of situation in more detail later in this chapter.

The equilibrium position of a reaction—left, right, or somewhere in between—is determined by many factors: the initial concentrations, the relative energies of the reactants and products, and the relative "degree of organization" of the reactants and products. Energy and organization come into play because nature tries to achieve minimum energy and maximum disorder. For now, we will simply view the equilibrium phenomenon as an experimentally verified fact. The theoretical origins of equilibrium will be explored in detail in Chapter 10.

The relationship between equilibrium and thermodynamics is explored in Section 10.11.

The Characteristics of Chemical Equilibrium

To explore the important characteristics of chemical equilibrium, we will consider the synthesis of ammonia from elemental nitrogen and hydrogen:

$$N_2(g) + 3H_2(g) \rightleftharpoons 2NH_3(g)$$

The United States produces almost 20 million tons of ammonia annually.

This process (called the Haber process) is of great commercial value because ammonia is an important fertilizer for the growth of corn and other crops. Ironically, this beneficial process was discovered in Germany just before World War I in a search for ways to produce nitrogen-based explosives. In the course of this work, German chemist Fritz Haber (1868–1934) pioneered the large-scale production of ammonia.

When gaseous nitrogen, hydrogen, and ammonia are mixed in a closed vessel at 25°C, no apparent change in the concentrations occurs over time, regardless of the original amounts of the gases. It would seem that equilibrium has been attained. However, this is not necessarily true.

There are two possible reasons why the concentrations of the reactants and products of a given chemical reaction remain unchanged when mixed:

1. The system is at chemical equilibrium.
2. The forward and reverse reactions are so slow that the system moves toward equilibrium at an undetectable rate.

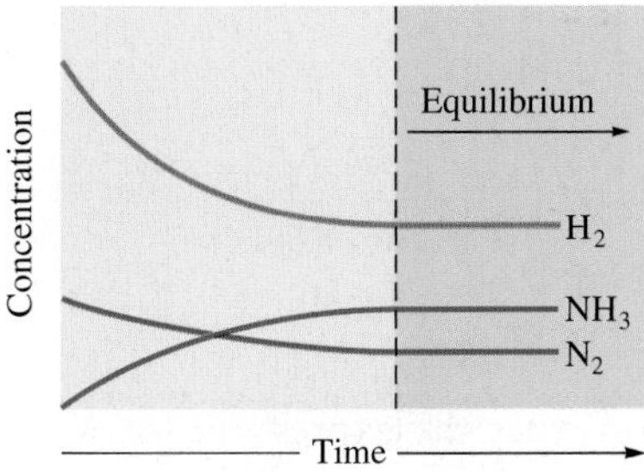

Figure 6.5
A concentration profile for the reaction $N_2(g) + 3H_2(g) \rightleftharpoons 2NH_3(g)$ when only $N_2(g)$ and $H_2(g)$ are mixed initially.

The second reason applies to the nitrogen, hydrogen, and ammonia mixture at 25°C. Because the molecules involved have strong chemical bonds, mixtures of N_2, H_2, and NH_3 at 25°C can exist with no apparent change over long periods of time. However, under appropriate conditions the system does reach equilibrium, as shown in Fig. 6.5. Note that because of the reaction stoichiometry, H_2 disappears three times as fast as N_2 does, and NH_3 forms twice as fast as N_2 disappears.

6.2 The Equilibrium Constant

Science is fundamentally empirical—it is based on experiment. The development of the equilibrium concept is typical. From observations of many chemical reactions, two Norwegian chemists, Cato Maximilian Guldberg (1836–1902) and Peter Waage (1833–1900), proposed the **law of mass action** in 1864 as a general description of the equilibrium condition. For a reaction of the type

The law of mass action is based on experimental observations.

$$jA + kB \rightleftharpoons lC + mD$$

where A, B, C, and D represent chemical species and j, k, l, and m are their coefficients in the balanced equation, the law of mass action is represented by the following **equilibrium expression:**

$$K = \frac{[C]^l[D]^m}{[A]^j[B]^k}$$

The square brackets indicate the concentrations of the chemical species *at equilibrium,* and K is a constant called the **equilibrium constant.**

Guldberg and Waage found that the equilibrium concentrations for every reaction system that they studied obeyed this relationship. That is, when the observed equilibrium concentrations are inserted into the equilibrium expression constructed from the law of mass action for a given reaction, the result is a constant (at a given temperature and assuming ideal behavior). Thus the value of the equilibrium constant for a given reaction system can be calculated from the measured concentrations of reactants and products present at equilibrium, a procedure illustrated in Example 6.1.

The K's used in this section might best be called $K^{observed}$ since they are calculated from "observed" concentrations that are not corrected for the effects of nonideality. Only K^{obs} values have units.

OWL INTERACTIVE EXAMPLE 6.1

The following equilibrium concentrations were observed for the Haber process at 127°C:

$$[NH_3] = 3.1 \times 10^{-2} \text{ mol/L}$$
$$[N_2] = 8.5 \times 10^{-1} \text{ mol/L}$$
$$[H_2] = 3.1 \times 10^{-3} \text{ mol/L}$$

a. Calculate the value of K at 127°C for this reaction.

b. Calculate the value of the equilibrium constant at 127°C for the reaction

$$2NH_3(g) \rightleftharpoons N_2(g) + 3H_2(g)$$

c. Calculate the value of the equilibrium constant at 127°C for the reaction given by the equation

$$\tfrac{1}{2}N_2(g) + \tfrac{3}{2}H_2(g) \rightleftharpoons NH_3(g)$$

Solution

a. The balanced equation for the Haber process is

$$N_2(g) + 3H_2(g) \rightleftharpoons 2NH_3(g)$$

Thus, using the law of mass action to construct the expression for K, we have

$$K = \frac{[NH_3]^2}{[N_2][H_2]^3} = \frac{(3.1 \times 10^{-2} \text{ mol/L})^2}{(8.5 \times 10^{-1} \text{ mol/L})(3.1 \times 10^{-3} \text{ mol/L})^3}$$
$$= 3.8 \times 10^4 \text{ L}^2/\text{mol}^2$$

b. This reaction is written in the reverse order of the equation given in part a. This leads to the equilibrium expression

$$K' = \frac{[N_2][H_2]^3}{[NH_3]^2}$$

which is the reciprocal of the expression used in part a. So

$$K' = \frac{[N_2][H_2]^3}{[NH_3]^2} = \frac{1}{K} = \frac{1}{3.8 \times 10^4 \text{ L}^2/\text{mol}^2} = 2.6 \times 10^{-5} \text{ mol}^2/\text{L}^2$$

c. We use the law of mass action: $K'' = \dfrac{[NH_3]}{[N_2]^{1/2}[H_2]^{3/2}}$

If we compare this expression with the one obtained in part a, we see that since

$$\frac{[NH_3]}{[N_2]^{1/2}[H_2]^{3/2}} = \left(\frac{[NH_3]^2}{[N_2][H_2]^3}\right)^{1/2}$$

then

$$K'' = K^{1/2}$$

Thus

$$K'' = K^{1/2} = (3.8 \times 10^4 \text{ L}^2/\text{mol}^2)^{1/2} = 1.9 \times 10^2 \text{ L/mol}$$

We can draw some important conclusions from the results of Example 6.1. For a reaction of the form

$$j\text{A} + k\text{B} \rightleftharpoons l\text{C} + m\text{D}$$

the equilibrium expression is

$$K = \frac{[\text{C}]^l[\text{D}]^m}{[\text{A}]^j[\text{B}]^k}$$

If this reaction is reversed, the new equilibrium expression is

$$K' = \frac{[\text{A}]^j[\text{B}]^k}{[\text{C}]^l[\text{D}]^m} = \frac{1}{K}$$

If the original reaction is multiplied by some factor n to give

$$nj\text{A} + nk\text{B} \rightleftharpoons nl\text{C} + nm\text{D}$$

the equilibrium expression becomes

$$K'' = \frac{[\text{C}]^{nl}[\text{D}]^{nm}}{[\text{A}]^{nj}[\text{B}]^{nk}} = K^n$$

Some Characteristics of the Equilibrium Expression

1. The equilibrium expression for a reaction written in reverse is the reciprocal of that for the original reaction.
2. When the balanced equation for a reaction is multiplied by a factor n, the equilibrium expression for the new reaction is the original expression raised to the nth power. Thus $K_{\text{new}} = (K_{\text{original}})^n$.
3. The apparent units for K are determined by the powers of the various concentration terms. The (apparent) units for K therefore depend on the reaction being considered. We will have more to say about the units for K in Section 6.4.

The K's referred to here are K^{obs} values.

The law of mass action applies to solution and gaseous equilibria.

The law of mass action is widely applicable. It correctly describes the equilibrium behavior of all chemical reaction systems whether they occur in solution or in the gas phase. Although, as we will see later, corrections for nonideal behavior must be applied in certain cases, such as for concentrated aqueous solutions and for gases at high pressures, the law of mass action provides a remarkably accurate description of all types of chemical equilibria. For example, consider again the ammonia synthesis reaction. At 500°C the value of K for this reaction is 6.0×10^{-2} L^2/mol^2. Whenever N_2, H_2, and NH_3 are mixed together at this temperature, the system will always come to an equilibrium position such that

$$\frac{[\text{NH}_3]^2}{[\text{N}_2][\text{H}_2]^3} = 6.0 \times 10^{-2} \text{ L}^2/\text{mol}^2$$

This expression has the same value at 500°C, *regardless of the amounts of the gases that are mixed together initially.*

Danny Eilers/Alamy

Applying anhydrous ammonia to soybean stubble prior to planting corn.

Although the special ratio of products to reactants defined by the equilibrium expression is constant for a given reaction system at a given temperature, the *equilibrium concentrations will not always be the same.* Table 6.1 gives

Table 6.1

Results of Three Experiments for the Reaction $N_2(g) + 3H_2(g) \rightleftharpoons 2NH_3(g)$

Experiment	Initial Concentrations	Equilibrium Concentrations	$K = \frac{[NH_3]^2}{[N_2][H_2]^3}$
I	$[N_2]_0 = 1.000\ M$ $[H_2]_0 = 1.000\ M$ $[NH_3]_0 = 0$	$[N_2] = 0.921\ M$ $[H_2] = 0.763\ M$ $[NH_3] = 0.157\ M$	$K = 6.02 \times 10^{-2}\ L^2/mol^2$
II	$[N_2]_0 = 0$ $[H_2]_0 = 0$ $[NH_3]_0 = 1.000\ M$	$[N_2] = 0.399\ M$ $[H_2] = 1.197\ M$ $[NH_3] = 0.203\ M$	$K = 6.02 \times 10^{-2}\ L^2/mol^2$
III	$[N_2]_0 = 2.00\ M$ $[H_2]_0 = 1.00\ M$ $[NH_3]_0 = 3.00\ M$	$[N_2] = 2.59\ M$ $[H_2] = 2.77\ M$ $[NH_3] = 1.82\ M$	$K = 6.02 \times 10^{-2}\ L^2/mol^2$

three sets of data for the synthesis of ammonia, showing that even though the individual sets of equilibrium concentrations are quite different for the different situations, the *equilibrium constant, which depends on the ratio of the concentrations, remains the same* (within experimental error). Note that subscripts of zero indicate initial concentrations.

Each *set of equilibrium concentrations* is called an **equilibrium position.** It is essential to distinguish between the equilibrium constant and the equilibrium positions for a given reaction system. There is only *one* equilibrium constant for a particular system at a particular temperature, but there are an *infinite* number of equilibrium positions. The specific equilibrium position adopted by a system depends on the initial concentrations, but the equilibrium constant does not.

For a reaction at a given temperature, there are many equilibrium positions but only one value for *K*.

6.3 Equilibrium Expressions Involving Pressures

PV = nRT

P = n/V RT

So far we have been describing equilibria involving gases in terms of concentrations. Equilibria involving gases can also be described in terms of pressures. The relationship between the pressure and the concentration of a gas can be seen from the ideal gas equation:

$$PV = nRT \quad \text{or} \quad P = \left(\frac{n}{V}\right)RT = CRT$$

where C equals n/V, or the number of moles of gas n per unit volume V. Thus C represents the *molar concentration of the gas.*

For the ammonia synthesis reaction, the equilibrium expression can be written in terms of concentrations,

$$K = \frac{[NH_3]^2}{[N_2][H_2]^3} = \frac{C_{NH_3}{}^2}{(C_{N_2})(C_{H_2}{}^3)}$$

or in terms of the *equilibrium partial pressures of the gases,*

$$K_p = \frac{P_{NH_3}{}^2}{(P_{N_2})(P_{H_2}{}^3)}$$

K involves concentrations; K_p involves pressures.

In this book K denotes an equilibrium constant in terms of concentrations, and K_p represents an equilibrium constant in terms of partial pressures.

The relationship between K and K_p for a particular reaction follows from the fact that for an ideal gas, $C = P/RT$. For example, for the ammonia synthesis reaction,

$$K = \frac{[NH_3]^2}{[N_2][H_2]^3} = \frac{C_{NH_3}^{\ 2}}{(C_{N_2})(C_{H_2}^{\ 3})}$$

$$= \frac{\left(\frac{P_{NH_3}}{RT}\right)^2}{\left(\frac{P_{N_2}}{RT}\right)\left(\frac{P_{H_2}}{RT}\right)^3} = \frac{P_{NH_3}^{\ 2}}{(P_{N_2})(P_{H_2}^{\ 3})} \times \frac{\left(\frac{1}{RT}\right)^2}{\left(\frac{1}{RT}\right)^4}$$

$$= \frac{P_{NH_3}^{\ 2}}{(P_{N_2})(P_{H_2}^{\ 3})}(RT)^2 = K_p(RT)^2$$

However, for the synthesis of hydrogen fluoride from its elements,

$$H_2(g) + F_2(g) \rightleftharpoons 2HF(g)$$

the relationship between K and K_p is

$$K = \frac{[HF]^2}{[H_2][F_2]} = \frac{C_{HF}^{\ 2}}{(C_{H_2})(C_{F_2})} = \frac{\left(\frac{P_{HF}}{RT}\right)^2}{\left(\frac{P_{H_2}}{RT}\right)\left(\frac{P_{F_2}}{RT}\right)} = \frac{P_{HF}^{\ 2}}{(P_{H_2})(P_{F_2})} = K_p$$

Thus for this reaction K is equal to K_p. This equality occurs because the sum of the coefficients on either side of the balanced equation is identical, so the terms in RT cancel. In the equilibrium expression for the ammonia synthesis reaction, the sum of the powers in the numerator is different from that in the denominator, so K does not equal K_p.

For the general reaction

$$jA + kB \rightleftharpoons lC + mD$$

the relationship between K and K_p is

$$K_p = K(RT)^{\Delta n}$$

where Δn is the sum of the coefficients of the *gaseous* products minus the sum of the coefficients of the *gaseous* reactants. This equation is easy to derive from the definitions of K and K_p and the relationship between pressure and concentration. For the preceding general reaction,

$$K_p = \frac{(P_C^{\ l})(P_D^{\ m})}{(P_A^{\ j})(P_B^{\ k})} = \frac{(C_C \times RT)^l (C_D \times RT)^m}{(C_A \times RT)^j (C_B \times RT)^k}$$

$$= \frac{(C_C^{\ l})(C_D^{\ m})}{(C_A^{\ j})(C_B^{\ k})} \times \frac{(RT)^{l+m}}{(RT)^{j+k}} = K(RT)^{(l+m)-(j+k)} = K(RT)^{\Delta n}$$

Δn always involves products minus reactants.

where $\Delta n = (l + m) - (j + k)$, the difference in the sums of the coefficients for the gaseous products and reactants.

We have seen that the (apparent) units of the equilibrium constant depend on the specific reaction. For example, for the reaction

$$H_2(g) + F_2(g) \rightleftharpoons 2HF(g)$$

the units for K and K_p can be found as follows:

$$K = \frac{C_{HF}^{\ 2}}{(C_{H_2})(C_{F_2})} = \frac{(\text{mol/L})^2}{(\text{mol/L})(\text{mol/L})} \Rightarrow \text{no units}$$

$$K_p = \frac{P_{HF}^{\ 2}}{(P_{H_2})(P_{F_2})} = \frac{(\text{atm})^2}{(\text{atm})(\text{atm})} \Rightarrow \text{no units}$$

For this reaction, neither K nor K_p has units, and K is equal to K_p. Because there are identical powers in the numerator and denominator, the units cancel. For equilibrium expressions in which the powers in the numerator and denominator are not the same, the equilibrium constants will have (apparent) units and K will not equal K_p.

Note that in the preceding discussion we used the term "apparent units" when referring to equilibrium constants. This term was used because the theoretical foundation for the concept of equilibrium based on thermodynamics includes a *reference state* for each substance, which always causes the units of concentration or pressure to cancel. We will explore this situation thoroughly in Chapter 10, but we will introduce this concept in Section 6.4.

6.4 The Concept of Activity

As we will see in Chapter 10, the "true" equilibrium constant expression does not simply involve the observed equilibrium pressure or the concentration for a substance but involves the *ratio* of the equilibrium pressure (or concentration) for a given substance to a *reference* pressure (or concentration) for that substance. This ratio is defined as the **activity** of the substance, which in terms of pressures is defined as

In 1982, the International Union of Pure and Applied Chemists (IUPAC) set the standard pressure as 1 bar (exactly). Both conventions are now used, although most thermodynamic data are now referenced to 1 bar instead of 1 atm.

$$\text{Activity } (i\text{th component}) = a_i = \frac{P_i}{P_{\text{reference}}}$$

where P_i = partial pressure of the ith gaseous component

$P_{\text{reference}}$ = 1 atm (exactly)

and where ideal behavior is assumed.

Using the concept of activities, the equilibrium expression for the reaction

$$j\text{A}(g) + k\text{B}(g) \rightleftharpoons l\text{C}(g) + m\text{D}(g)$$

is written as

$$K = \frac{(a_\text{C})^l(a_\text{D})^m}{(a_\text{A})^j(a_\text{B})^k} = \frac{\left(\frac{P_\text{C}}{P_\text{ref}}\right)^l\left(\frac{P_\text{D}}{P_\text{ref}}\right)^m}{\left(\frac{P_\text{A}}{P_\text{ref}}\right)^j\left(\frac{P_\text{B}}{P_\text{ref}}\right)^k}$$

With all the pressures expressed in atmospheres, we have

$$K_\text{p} = \frac{\left(\frac{P_\text{C}\,\cancel{(\text{atm})}}{1\,\cancel{\text{atm}}}\right)^l\left(\frac{P_\text{D}\,\cancel{(\text{atm})}}{1\,\cancel{\text{atm}}}\right)^m}{\left(\frac{P_\text{A}\,\cancel{(\text{atm})}}{1\,\cancel{\text{atm}}}\right)^j\left(\frac{P_\text{B}\,\cancel{(\text{atm})}}{1\,\cancel{\text{atm}}}\right)^k} = \frac{P_\text{C}{}^l P_\text{D}{}^m}{P_\text{A}{}^j P_\text{B}{}^k}$$

where K_p is unitless as shown.

When the equilibrium composition of a system is expressed in units of moles per liter, the reference state is (exactly) 1 mol/L.

Because of the difference in reference states, in general,

$$K \neq K_\text{p}$$

K: Equilibrium composition expressed in concentration units

K_p: Equilibrium composition expressed as pressures

The only exception to this principle occurs when the sum of the powers in the numerator and the denominator are the same (as discussed previously for $H_2 + F_2 \rightleftharpoons 2HF$). In such a case $K = K_p$.

Because of the ideas discussed in this section, it is customary to give the values of equilibrium constants without units. This is the practice we will follow in the remainder of the text.

6.5 Heterogeneous Equilibria

So far we have discussed equilibria only for systems in the gas phase, where all reactants and products are gases. These situations represent **homogeneous equilibria.** However, many equilibria involve more than one phase and are called **heterogeneous equilibria.** For example, the thermal decomposition of calcium carbonate in the commercial preparation of lime occurs by a reaction involving both solid and gas phases:

Lime is among the top six chemicals manufactured in the United States in terms of amount produced.

$$CaCO_3(s) \rightleftharpoons \underset{\text{Lime}}{\underset{\uparrow}{CaO(s)}} + CO_2(g)$$

Larry Andre Maslennikov/Peter Arnold/PhotoLibrary

Helicopter liming acidic wetland in Sweden.

Straightforward application of the law of mass action leads to the equilibrium expression

$$K' = \frac{[CO_2][CaO]}{[CaCO_3]}$$

However, experimental results show that the *position of a heterogeneous equilibrium does not depend on the amounts of pure solids or liquids present* (Fig. 6.6). This result makes sense when the meaning of an activity for a pure liquid or solid is understood. For a pure liquid or solid, the reference state is the pure liquid or solid. Thus, for the composition of $CaCO_3$ considered above, we do not insert $[CaCO_3]$ or $[CaO]$ into the equilibrium expression but rather into the activity of each:

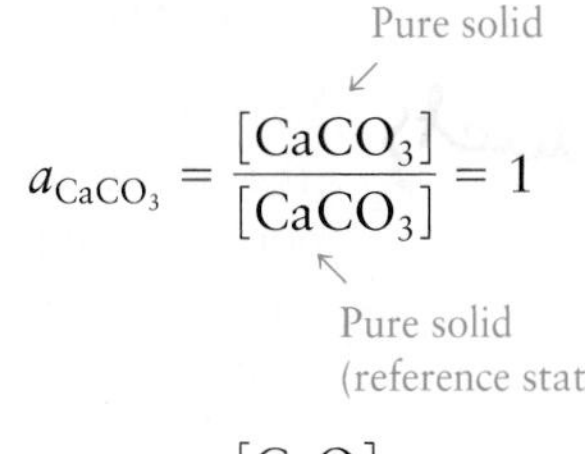

$$a_{CaCO_3} = \frac{[CaCO_3]}{[CaCO_3]} = 1$$

and

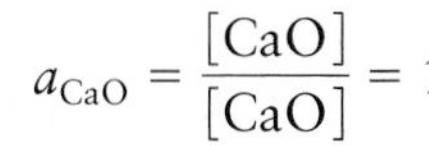

$$a_{CaO} = \frac{[CaO]}{[CaO]} = 1$$

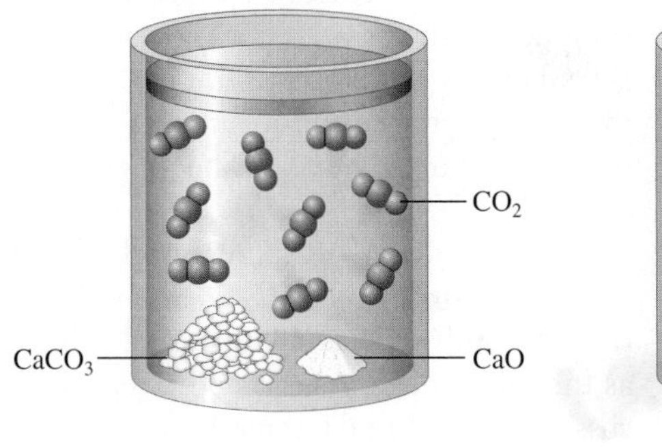

Figure 6.6
The position of the equilibrium $CaCO_3(s) \rightleftharpoons CaO(s) + CO_2(g)$ does not depend on the amounts of $CaCO_3(s)$ and $CaO(s)$ present.

Thus the equilibrium expressions for the decomposition of solid $CaCO_3$ are

$$K = \frac{[CO_2](1)}{1} = [CO_2] \qquad \text{and} \qquad K_p = \frac{P_{CO_2}(1)}{1} = P_{CO_2}$$

In summary, we can make the following general statement: *The activity of a pure solid or liquid is always 1.*

Note that the net effect of inserting an activity of 1 into the equilibrium expression for each pure solid or liquid in the reaction has the same effect as simply disregarding them. If pure solids or pure liquids are involved in a chemical reaction, their concentrations *are not included in the equilibrium expression* for the reaction. This simplification occurs *only* with pure solids or liquids, not with solutions or gases, because in these last two cases the activity cannot be assumed to be 1.

For example, in the decomposition of liquid water to gaseous hydrogen and oxygen,

$$2H_2O(l) \rightleftharpoons 2H_2(g) + O_2(g)$$

where $\qquad K = [H_2]^2[O_2] \qquad \text{and} \qquad K_p = (P_{H_2}{}^2)(P_{O_2})$

water is not included in either equilibrium expression because it is present as a pure liquid ($a_{H_2O(l)} = 1$). However, if the reaction were carried out under conditions in which the water is a gas rather than a liquid,

$$2H_2O(g) \rightleftharpoons 2H_2(g) + O_2(g)$$

then $\qquad K = \dfrac{[H_2]^2[O_2]}{[H_2O]^2} \qquad \text{and} \qquad K_p = \dfrac{(P_{H_2}{}^2)(P_{O_2})}{P_{H_2O}{}^2}$

because the concentration or pressure of water vapor can assume different values, depending on the conditions. That is, we cannot assume an activity of 1 in such an instance.

6.6 Applications of the Equilibrium Constant

Knowing the equilibrium constant for a reaction allows us to predict several important features of the reaction: the tendency of the reaction to occur (but not the speed of the reaction), whether a given set of concentrations represents an equilibrium condition, and the equilibrium position that will be achieved from a given set of initial concentrations.

The Extent of a Reaction

The inherent tendency for a reaction to occur is indicated by the magnitude of the equilibrium constant. A value of K that is much larger than 1 means that at equilibrium the reaction system will consist of mostly products—the equilibrium lies to the right. That is, reactions with very large equilibrium constants go essentially to completion. On the other hand, a very small value of K means that the system at equilibrium will consist of mostly reactants—the equilibrium position is far to the left. The given reaction does not occur to any significant extent.

It is important to understand that *the size of K and the time required to reach equilibrium are not directly related.* The time required to achieve equilibrium depends on the reaction rate. The size of K is determined by factors such as the difference in energy between products and reactants, which will be discussed in detail in Chapter 10.

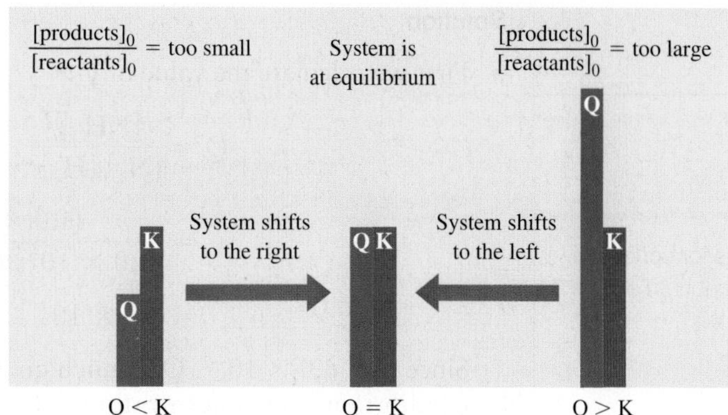

Figure 6.7
Comparing the values of Q and K allows us to determine the direction the system will shift to reach equilibrium.

Reaction Quotient

When the reactants and products of a given chemical reaction are mixed, it is useful to know whether the mixture is at equilibrium and, if it is not, in which direction the system will shift to reach equilibrium. If the concentration of one of the reactants or products is zero, the system will shift in the direction that produces the missing component. However, if all the initial concentrations are not zero, it is more difficult to determine the direction of the move toward equilibrium. To determine the shift in such cases, we use the **reaction quotient,** Q. The reaction quotient is obtained by applying the law of mass action, but using *initial concentrations* instead of equilibrium concentrations. For example, for the synthesis of ammonia,

$$N_2(g) + 3H_2(g) \rightleftharpoons 2NH_3(g)$$

the expression for the reaction quotient is

$$Q = \frac{[NH_3]_0^2}{[N_2]_0[H_2]_0^3}$$

where the zero subscripts indicate initial concentrations.

To determine in which direction a system will shift to reach equilibrium, we compare the values of Q and K (Fig. 6.7). There are three possible situations:

1. *Q is equal to K.* The system is at equilibrium; no shift will occur.
2. *Q is greater than K.* In this case the ratio of initial concentrations of products to initial concentrations of reactants is too large. For the system to reach equilibrium, a net change of products to reactants must occur. The system *shifts to the left,* consuming products and forming reactants, until equilibrium is achieved.
3. *Q is less than K.* In this case the ratio of initial concentrations of products to initial concentrations of reactants is too small. The system *must shift to the right,* consuming reactants and forming products, to attain equilibrium.

OWL INTERACTIVE EXAMPLE 6.2

For the synthesis of ammonia at 500°C, the equilibrium constant is 6.0×10^{-2}. Predict the direction in which the system will shift to reach equilibrium in each of the following cases.

a. $[NH_3]_0 = 1.0 \times 10^{-3}\ M$; $[N_2]_0 = 1.0 \times 10^{-5}\ M$; $[H_2]_0 = 2.0 \times 10^{-3}\ M$

b. $[NH_3]_0 = 2.00 \times 10^{-4}\ M$; $[N_2]_0 = 1.50 \times 10^{-5}\ M$; $[H_2]_0 = 3.54 \times 10^{-1}\ M$

c. $[NH_3]_0 = 1.0 \times 10^{-4}\ M$; $[N_2]_0 = 5.0\ M$; $[H_2]_0 = 1.0 \times 10^{-2}\ M$

We will not include units for concentrations that are in the equilibrium expression or reaction quotient.

Solution

a. First we calculate the value of Q:

$$Q = \frac{[NH_3]_0^2}{[N_2]_0[H_2]_0^3}$$

$$= \frac{(1.0 \times 10^{-3})^2}{(1.0 \times 10^{-5})(2.0 \times 10^{-3})^3}$$

$$= 1.3 \times 10^7$$

Since $K = 6.0 \times 10^{-2}$, Q is much greater than K. For the system to attain equilibrium, the concentrations of the products must be decreased and the concentrations of the reactants increased. The system will shift to the left:

$$N_2 + 3H_2 \longleftarrow 2NH_3$$

b. We calculate the value of Q:

$$Q = \frac{[NH_3]_0^2}{[N_2]_0[H_2]_0^3} = \frac{(2.00 \times 10^{-4})^2}{(1.50 \times 10^{-5})(3.54 \times 10^{-1})^3}$$

$$= 6.01 \times 10^{-2}$$

In this case $Q = K$, so the system is at equilibrium. No shift will occur.

c. The value of Q is

$$Q = \frac{[NH_3]_0^2}{[N_2]_0[H_2]_0^3} = \frac{(1.0 \times 10^{-4})^2}{(5.0)(1.0 \times 10^{-2})^3}$$

$$= 2.0 \times 10^{-3}$$

Here Q is less than K, so the system will shift to the right, attaining equilibrium by increasing the concentration of the product and decreasing the concentrations of the reactants:

$$N_2 + 3H_2 \longrightarrow 2NH_3$$

Calculating Equilibrium Pressures and Concentrations

A typical equilibrium problem involves finding the equilibrium concentrations (or pressures) of reactants and products given the value of the equilibrium constant and the initial concentrations (or pressures).

OWL INTERACTIVE EXAMPLE 6.3

Assume that the reaction for the formation of gaseous hydrogen fluoride from hydrogen and fluorine has an equilibrium constant of 1.15×10^2 at a certain temperature. In a particular experiment at this temperature 3.000 moles of each component was added to a 1.500-liter flask. Calculate the equilibrium concentrations of all species.

Solution The balanced equation for the reaction is

$$H_2(g) + F_2(g) \rightleftharpoons 2HF(g)$$

The equilibrium expression is

$$K = 1.15 \times 10^2 = \frac{[HF]^2}{[H_2][F_2]}$$

We first calculate the initial concentrations:

$$[HF]_0 = [H_2]_0 = [F_2]_0 = \frac{3.000 \text{ mol}}{1.500 \text{ L}} = 2.000\ M$$

From the value of Q,

$$Q = \frac{[HF]_0^2}{[H_2]_0[F_2]_0} = \frac{(2.000)^2}{(2.000)(2.000)} = 1.000$$

which is much less than K, we know that the system must shift to the right to reach equilibrium.

What change in the concentrations is necessary? Since the answer to this question is presently unknown, we will define the change needed in terms of x. Let x equal the number of moles per liter of H_2 consumed to reach equilibrium. The stoichiometry of the reaction shows that x mol/L of F_2 will also be consumed and that $2x$ mol/L of HF will be formed:

$$H_2(g) + F_2(g) \longrightarrow 2HF(g)$$

$$x \text{ mol/L} + x \text{ mol/L} \longrightarrow 2x \text{ mol/L}$$

Now the equilibrium concentrations can be expressed in terms of x:

Initial Concentration (mol/L)	Change (mol/L)	Equilibrium Concentration (mol/L)
$[H_2]_0 = 2.000$	$-x$	$[H_2] = 2.000 - x$
$[F_2]_0 = 2.000$	$-x$	$[F_2] = 2.000 - x$
$[HF]_0 = 2.000$	$+2x$	$[HF] = 2.000 + 2x$

These concentrations can be represented in a shorthand table as follows:

This shorthand representation is sometimes called an *ICE table* from the first letters of the labels.

	$H_2(g)$	+	$F_2(g)$	$\rightleftharpoons$	$2HF(g)$
Initial:	2.000		2.000		2.000
Change:	$-x$		$-x$		$+2x$
Equilibrium:	$2.000 - x$		$2.000 - x$		$2.000 + 2x$

To solve for x, we substitute the equilibrium concentrations into the equilibrium expression:

$$K = 1.15 \times 10^2 = \frac{[HF]^2}{[H_2][F_2]} = \frac{(2.000 + 2x)^2}{(2.000 - x)^2}$$

The right side of this equation is a perfect square, so taking the square root of both sides gives

$$\sqrt{1.15 \times 10^2} = \frac{2.000 + 2x}{2.000 - x}$$

which yields $x = 1.528$. The equilibrium concentrations can now be calculated:

$$[H_2] = [F_2] = 2.000\ M - x = 0.472\ M$$

$$[HF] = 2.000\ M + 2x = 5.056\ M$$

Check: Checking these values by substituting them into the equilibrium expression gives

$$\frac{[HF]^2}{[H_2][F_2]} = \frac{(5.056)^2}{(0.472)^2} = 1.15 \times 10^2$$

which agrees with the given value of K.

6.7 Solving Equilibrium Problems

We have already considered most of the strategies needed to solve equilibrium problems. The typical procedure for analyzing a chemical equilibrium problem can be summarized as shown below.

STEPS

Solving Equilibrium Problems

1. Write the balanced equation for the reaction.
2. Write the equilibrium expression using the law of mass action.
3. List the initial concentrations.
4. Calculate Q, and determine the direction of the shift to equilibrium.
5. Define the change needed to reach equilibrium, and define the equilibrium concentrations by applying the change to the initial concentrations.
6. Substitute the equilibrium concentrations into the equilibrium expression, and solve for the unknown.
7. Check your calculated equilibrium concentrations by making sure that they give the correct value of K.

In Example 6.3 we were able to solve for the unknown by taking the square root of both sides of the equation. However, this situation is not very common, so we must now consider a more typical problem. Suppose that for a synthesis of hydrogen fluoride from hydrogen and fluorine, 3.000 moles of H_2 and 6.000 moles of F_2 are mixed in a 3.000-L flask. The equilibrium constant for the synthesis reaction at this temperature is 1.15×10^2. We calculate the equilibrium concentration of each component as follows:

- We begin as usual by writing the balanced equation for the reaction:

$$H_2(g) + F_2(g) \rightleftharpoons 2HF(g)$$

- The equilibrium expression is

$$K = 1.15 \times 10^2 = \frac{[HF]^2}{[H_2][F_2]}$$

- The initial concentrations are

$$[H_2]_0 = \frac{3.000 \text{ mol}}{3.000 \text{ L}} = 1.000\ M$$

$$[F_2]_0 = \frac{6.000 \text{ mol}}{3.000 \text{ L}} = 2.000\ M$$

$$[HF]_0 = 0$$

- There is no need to calculate Q; since no HF is initially present, we know that the system must shift to the right to reach equilibrium.
- If we let x represent the number of moles per liter of H_2 consumed to reach equilibrium, we can represent the concentrations as follows:

Initial Concentration (mol/L)	Change (mol/L)	Equilibrium Concentration (mol/L)
$[H_2]_0 = 1.000$	$-x$	$[H_2] = 1.000 - x$
$[F_2]_0 = 2.000$	$-x$	$[F_2] = 2.000 - x$
$[HF]_0 = 0$	$+2x$	$[HF] = 0 + 2x$

Or in shorthand form:

	$H_2(g)$	+	$F_2(g)$	$\rightleftharpoons$	$2HF(g)$
Initial:	1.000		2.000		0
Change:	$-x$		$-x$		$+2x$
Equilibrium:	$1.000 - x$		$2.000 - x$		$2x$

- Substituting the equilibrium concentrations into the equilibrium expression gives

$$K = 1.15 \times 10^2 = \frac{[HF]^2}{[H_2][F_2]} = \frac{(2x)^2}{(1.000 - x)(2.000 - x)}$$

To solve for x, we perform the indicated multiplication,

$$(1.000 - x)(2.000 - x)(1.15 \times 10^2) = (2x)^2$$

to give

$$(1.15 \times 10^2)x^2 - 3.000(1.15 \times 10^2)x + 2.000(1.15 \times 10^2) = 4x^2$$

and collect terms,

$$(1.11 \times 10^2)x^2 - (3.45 \times 10^2)x + 2.30 \times 10^2 = 0$$

This expression is a quadratic equation of the general form

$$ax^2 + bx + c = 0$$

where the roots can be obtained from the quadratic formula:

$$x = \frac{-b \pm \sqrt{b^2 - 4ac}}{2a}$$

In this example $a = 1.11 \times 10^2$, $b = -3.45 \times 10^2$, and $c = 2.30 \times 10^2$. Substituting these values into the quadratic formula gives two values for x:

$$x = 2.14 \text{ mol/L} \quad \text{and} \quad x = 0.968 \text{ mol/L}$$

Both of these results cannot be valid (because a *given* set of initial concentrations leads to only *one* equilibrium position). How can we choose between them? Since the expression for the equilibrium concentration of H_2 is

$$[H_2] = 1.000\ M - x$$

the value of x cannot be 2.14 mol/L (because subtracting 2.14 M from 1.000 M gives a negative concentration for H_2, which is physically impossible). Thus the correct value for x is 0.968 mol/L, and the equilibrium concentrations are as follows:

$$[H_2] = 1.000\ M - 0.968\ M = 3.2 \times 10^{-2}\ M$$

$$[F_2] = 2.000\ M - 0.968\ M = 1.032\ M$$

$$[HF] = 2(0.968\ M) = 1.936\ M$$

- We can check these concentrations by substituting them into the equilibrium expression:

$$\frac{[HF]^2}{[H_2][F_2]} = \frac{(1.936)^2}{(3.2 \times 10^{-2})(1.032)} = 1.13 \times 10^2$$

This value is in close agreement with the given value for K (1.15×10^2), so the calculated equilibrium concentrations are correct.

Note that although we used the quadratic formula to solve for x in this problem, other methods are also available. For example, trial and error is always a possibility. However, use of successive approximations (see Appendix A1.4) is often preferable. For example, in this case successive approximations can be carried out conveniently by starting with the quadratic equation

$$(1.11 \times 10^2)x^2 - (3.45 \times 10^2)x + 2.30 \times 10^2 = 0$$

and dividing it by 1.11×10^2 to give

$$x^2 - 3.11x + 2.07 = 0$$

which can then be rearranged to

$$x^2 = 3.11x - 2.07$$

or

$$x = \sqrt{3.11x - 2.07}$$

Now we can proceed by guessing a value of x, which is then inserted into the square root expression. Next, we calculate a "new" value of x from the expression

$$\underset{\text{"New" value calculated}}{x} = \sqrt{3.11\underset{\text{Guessed value of } x \text{ inserted}}{x} - 2.07}$$

When the calculated value (the new value) of x agrees with the guessed value, the equation has been solved.

To solve the algebraic equations you encounter when doing chemistry problems, use whatever method is convenient and comfortable for you.

Treating Systems That Have Small Equilibrium Constants

We have seen that fairly complicated calculations are often necessary to solve equilibrium problems. However, under certain conditions we can make simplifications that greatly reduce the mathematical difficulties. For example, consider gaseous NOCl, which decomposes to form the gases NO and Cl_2. At 35°C the equilibrium constant is 1.6×10^{-5} mol/L. In an experiment in which 1.0 mole of NOCl is placed in a 2.0-L flask, what are the equilibrium concentrations?

The balanced equation is

$$2NOCl(g) \rightleftharpoons 2NO(g) + Cl_2(g)$$

and

$$K = \frac{[NO]^2[Cl_2]}{[NOCl]^2} = 1.6 \times 10^{-5}$$

The initial concentrations are

$$[NOCl]_0 = \frac{1.0 \text{ mol}}{2.0 \text{ L}} = 0.50\ M, \qquad [NO]_0 = 0, \qquad \text{and} \qquad [Cl_2]_0 = 0$$

Since there are no products initially, the system will move to the right to reach equilibrium. We will define x as the change in concentration of Cl_2 needed to reach equilibrium. The changes in the concentrations of NOCl and NO can then be obtained from the balanced equation:

$$2NOCl(g) \longrightarrow 2NO(g) + Cl_2(g)$$

$$2x \longrightarrow 2x + x$$

The concentrations can be summarized as follows:

Initial Concentration (mol/L)	Change (mol/L)	Equilibrium Concentration (mol/L)
$[NOCl]_0 = 0.50$	$-2x$	$[NOCl] = 0.50 - 2x$
$[NO]_0 = 0$	$+2x$	$[NO] = 0 + 2x = 2x$
$[Cl_2]_0 = 0$	$+x$	$[Cl_2] = 0 + x = x$

Or in shorthand form:

	$2NOCl(g)$	$\rightleftharpoons$	$2NO(g)$	+	$Cl_2(g)$
Initial:	0.50		0		0
Change:	$-2x$		$+2x$		$+x$
Equilibrium:	$0.50 - 2x$		$2x$		x

The equilibrium concentrations must satisfy the equilibrium expression:

$$K = 1.6 \times 10^{-5} = \frac{[NO]^2[Cl_2]}{[NOCl]^2} = \frac{(2x)^2(x)}{(0.50 - 2x)^2}$$

Multiplying and collecting terms results in an equation that requires complicated methods to solve directly. However, we can avoid this situation by recognizing that since K is so small (1.6×10^{-5} mol/L), the system will not proceed far to the right to reach equilibrium. That is, *x represents a relatively small number.* Consequently, the term $(0.50 - 2x)$ can be approximated by 0.50. That is, when x is small,

$$0.50 - 2x \approx 0.50$$

Approximations can simplify complicated math, but their validity should be carefully checked.

Making this approximation allows us to simplify the equilibrium expression:

$$1.6 \times 10^{-5} = \frac{(2x)^2(x)}{(0.50 - 2x)^2} \approx \frac{(2x)^2(x)}{(0.50)^2} = \frac{4x^3}{(0.50)^2}$$

Solving for x^3 gives

$$x^3 = \frac{(1.6 \times 10^{-5})(0.50)^2}{4} = 1.0 \times 10^{-6}$$

and $x = 1.0 \times 10^{-2}$ mol/L.

Next, we must check the validity of the approximation. If $x = 1.0 \times 10^{-2}$, then

$$0.50 - 2x = 0.50 - 2(1.0 \times 10^{-2}) = 0.48$$

A good way to assess whether a 4% error is acceptable here is to examine the precision of the data given. For example, note that the value of K is 1.6×10^{-5}, which can be interpreted as $(1.6 \pm 0.1) \times 10^{-5}$. Thus the uncertainty in K is at least 1 part in 16, or about 6%. Therefore, a 4% error in [NOCl] is acceptable.

The difference between 0.50 and 0.48 is 0.02, or 4% of the initial concentration of NOCl, a relatively small discrepancy that will have little effect on the outcome. That is, since $2x$ is very small compared with 0.50, the value of x obtained in the approximate solution should be very close to the exact value. We use this approximate value of x to calculate the equilibrium concentrations:

$$[NOCl] = 0.50 - 2x = 0.48\ M \approx 0.50\ M$$

$$[NO] = 2x = 2(1.0 \times 10^{-2}\ M) = 2.0 \times 10^{-2}\ M$$

$$[Cl_2] = x = 1.0 \times 10^{-2}\ M$$

Check: $$\frac{[NO]^2[Cl_2]}{[NOCl]^2} = \frac{(2.0 \times 10^{-2})^2(1.0 \times 10^{-2})}{(0.50)^2} = 1.6 \times 10^{-5}$$

Since the given value of K is 1.6×10^{-5}, these calculated concentrations are correct.

This problem turned out to be relatively easy to solve because the *small value of K and the resulting small shift to the right to reach equilibrium allowed simplification.*

6.8 Le Châtelier's Principle

It is important to understand the factors that control the *position* of a chemical equilibrium. For example, when a chemical is manufactured, the chemists and chemical engineers in charge of production want to choose conditions that favor the desired product as much as possible. In other words, they want the equilibrium to lie far to the right. When Fritz Haber was developing the process for the synthesis of ammonia, he did extensive studies on how the temperature and pressure affect the equilibrium concentration of ammonia. Some of his results are given in Table 6.2. Note that the amount of NH_3 at equilibrium increases with an increase in pressure but decreases with an increase in temperature. Thus the amount of NH_3 present at equilibrium is favored by conditions of low temperature and high pressure.

However, this is not the whole story. Carrying out the process at low temperatures is not feasible because then the reaction is too slow. Even though the equilibrium tends to shift to the right as the temperature is lowered, the attainment of equilibrium is much too slow at low temperatures to be practical. This observation emphasizes once again that we must study both the thermodynamics (Chapter 10) and the kinetics (Chapter 15) of a reaction before we really understand the factors that control it.

Science Photo Library/Photo Researchers

Henri Louis Le Châtelier (1850–1936), the French physical chemist and metallurgist, seen here while a student at the École Polytechnique.

We can qualitatively predict the effects of changes in concentration, pressure, and temperature on a system at equilibrium by using **Le Châtelier's principle,** which states that *if a change in conditions (a "stress") is imposed on a system at equilibrium, the equilibrium position will shift in a direction that tends to reduce that change in conditions.* Although this rule, put forth by Henri Le Châtelier in 1884, sometimes oversimplifies the situation, it works remarkably well.

The Effect of a Change in Concentration

To see how we can predict the effects of a change in concentration on a system at equilibrium, we will consider the ammonia synthesis reaction. Suppose there is an equilibrium position described by these concentrations:

$$[N_2] = 0.399\ M, \qquad [H_2] = 1.197\ M, \qquad \text{and} \qquad [NH_3] = 0.202\ M$$

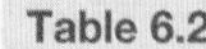

Table 6.2

Percent by Mass of NH_3 at Equilibrium in a Mixture of N_2, H_2, and NH_3 as a Function of Temperature and Total Pressure*

	Total Pressure		
Temperature (°C)	300 atm	400 atm	500 atm
400	48% NH_3	55% NH_3	61% NH_3
500	26% NH_3	32% NH_3	38% NH_3
600	13% NH_3	17% NH_3	21% NH_3

*Each experiment was begun with a 3:1 mixture of H_2 and N_2.

What will happen if 1.000 mole per liter of N_2 is suddenly injected into the system at constant volume? We can answer this question by calculating the value of Q. The concentrations before the system adjusts are

$$[N_2]_0 = 0.399\ M + 1.000\ M = 1.399\ M$$

$$\uparrow \text{ Added } N_2$$

$$[H_2]_0 = 1.197\ M$$

$$[NH_3]_0 = 0.202\ M$$

Note that we label these as "initial concentrations" because the system is no longer at equilibrium. Then

$$Q = \frac{[NH_3]_0{}^2}{[N_2]_0[H_2]_0{}^3} = \frac{(0.202)^2}{(1.399)(1.197)^3} = 1.70 \times 10^{-2}$$

Since we are not given the value of K, we must calculate it from the first set of equilibrium concentrations:

$$K = \frac{[NH_3]^2}{[N_2][H_2]^3} = \frac{(0.202)^2}{(0.399)(1.197)^3} = 5.96 \times 10^{-2}$$

As we might have expected, because the concentration of N_2 was increased, Q is less than K. The system will shift to the right to arrive at the new equilibrium position. Rather than do the calculations, we simply summarize the results:

Blue anhydrous cobalt(II) chloride and pink hydrated cobalt(II) chloride. Since the reaction $CoCl_2(s) + 6H_2O(g) \longrightarrow CoCl_2 \cdot 6H_2O(s)$ is shifted to the right by water vapor, $CoCl_2$ is often used in novelty devices to detect humidity.

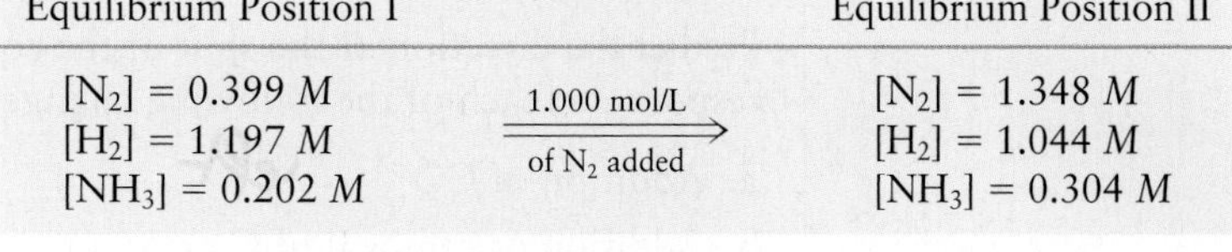
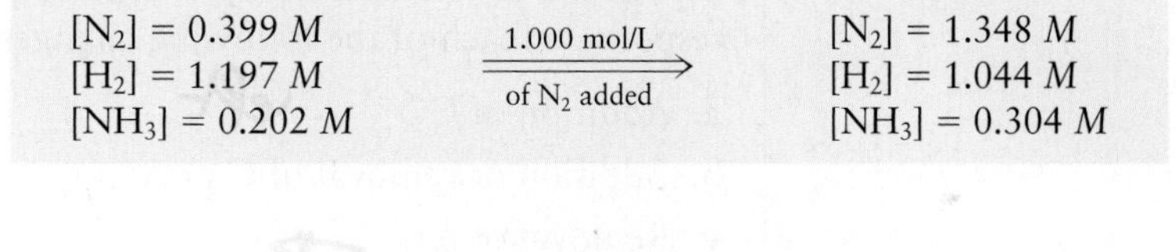

Equilibrium Position I		Equilibrium Position II
$[N_2] = 0.399\ M$ $[H_2] = 1.197\ M$ $[NH_3] = 0.202\ M$	1.000 mol/L of N_2 added $\Longrightarrow$	$[N_2] = 1.348\ M$ $[H_2] = 1.044\ M$ $[NH_3] = 0.304\ M$

These data reveal that the equilibrium position does in fact shift to the right: The concentration of H_2 decreases; the concentration of NH_3 increases; and, of course, since nitrogen is added, the concentration of N_2 shows an increase relative to the amount present at the original equilibrium position. (However, note that the nitrogen decreased from the amount present immediately after addition of the 1.000 mole of N_2 because the reaction shifted to the right.)

We can predict this shift qualitatively by using Le Châtelier's principle. Since the stress imposed is the addition of nitrogen, Le Châtelier's principle predicts that the system will shift in a direction that consumes nitrogen. This reduces the effect of the addition. Thus Le Châtelier's principle correctly predicts that adding nitrogen causes the equilibrium to shift to the right (Fig. 6.8).

The system shifts in the direction that compensates for the imposed change in conditions.

If ammonia had been added instead of nitrogen, the system would have shifted to the left to consume ammonia. So we can paraphrase Le Châtelier's principle for this case as follows: *If a gaseous reactant or product is added to a system at equilibrium, the system will shift away from the added component. If a gaseous reactant or product is removed, the system will shift toward the removed component.*

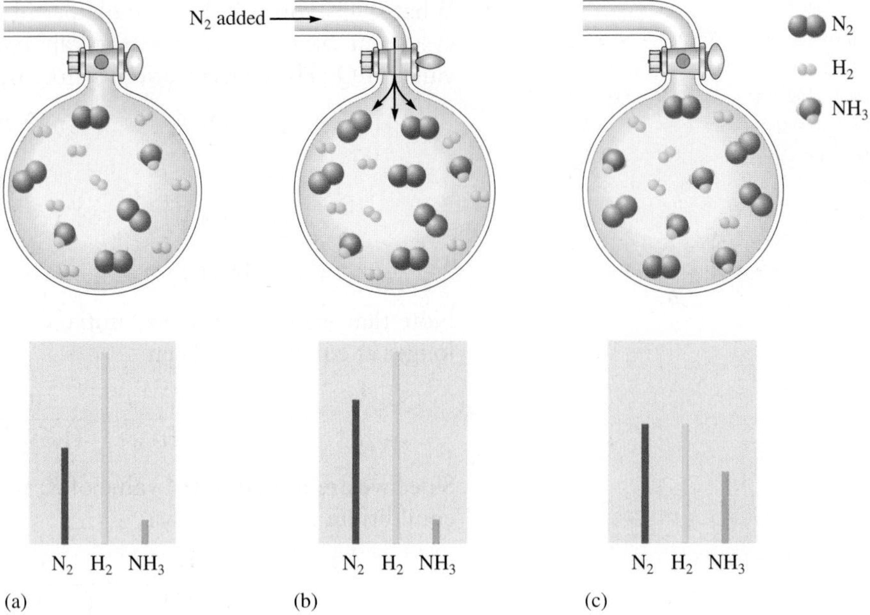

Figure 6.8
(a) The initial equilibrium mixture of N_2, H_2, and NH_3. (b) Addition of N_2. (c) The new equilibrium position for the system containing more N_2 (due to addition of N_2), less H_2, and more NH_3 than the mixture in (a).

OWL INTERACTIVE EXAMPLE 6.4

Arsenic can be extracted from its ores by first reacting the ore with oxygen (a process called *roasting*) to form solid As_4O_6, which is then reduced with carbon:

$$As_4O_6(s) + 6C(s) \rightleftharpoons As_4(g) + 6CO(g)$$

Predict the direction of the shift of the equilibrium position for this reaction in response to each of the following changes in conditions.

a. Addition of CO

b. Addition or removal of C or As_4O_6

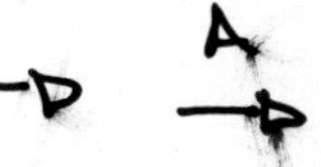

c. Removal of As_4

Solution

a. Le Châtelier's principle predicts that the shift will be away from the substance whose concentration is increased. The equilibrium position will shift to the left when CO is added.

b. Since the amount of a pure solid has no effect on the equilibrium position, changing the amount of C or As_4O_6 will have no effect.

c. If gaseous As_4 is removed, the equilibrium position will shift to the right to form more products. In industrial processes the desired product is often continuously removed from the reaction system to increase the yield.

The Effect of a Change in Pressure

Basically, there are three ways to change the pressure of a reaction system involving gaseous components at a given temperature:

1. Add or remove a gaseous reactant or product at constant volume.
2. Add an inert gas (one not involved in the reaction) at constant volume.
3. Change the volume of the container.

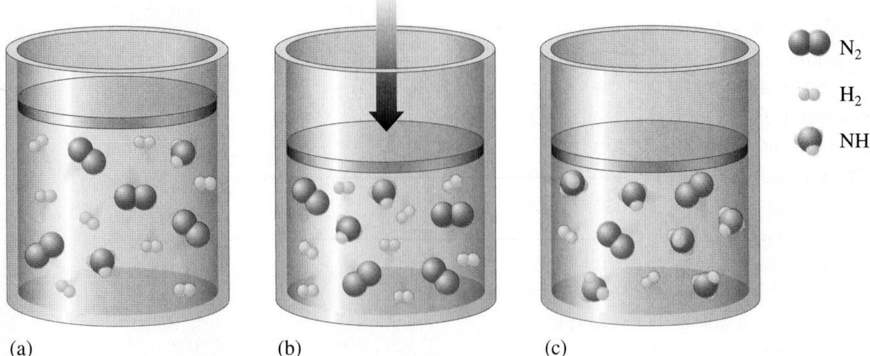

Figure 6.9
(a) A mixture of $NH_3(g)$, $N_2(g)$, and $H_2(g)$ at equilibrium. (b) The volume is suddenly decreased. (c) The new equilibrium position for the system containing more NH_3, less N_2, and less H_2. The reaction $N_2(g) + 3H_2(g) \rightleftharpoons 2NH_3(g)$ shifts to the right (toward the side with fewer molecules) when the container volume is decreased.

We have already considered the addition or removal of a reactant or product. When an inert gas is added at constant volume, there is no effect on the equilibrium position. *The addition of an inert gas increases the total pressure but has no effect on the concentrations or partial pressures of the reactants or products (assuming ideal gas behavior).* Thus the system remains at the original equilibrium position.

When the volume of the container is changed, the concentrations (and thus the partial pressures) of both reactants and products are changed. We could calculate Q and predict the direction of the shift. However, for systems involving gaseous components, there is an easier way: We focus on the volume. The central idea is that *when the volume of the container holding a gaseous system is reduced, the system responds by reducing its own volume. This is done by decreasing the total number of gaseous molecules in the system.*

To see how this works, we can rearrange the ideal gas law to give

$$V = \left(\frac{RT}{P}\right)n$$

or at constant T and P

$$V \propto n$$

That is, at constant temperature and pressure, the volume of a gas is directly proportional to the number of moles of gas present.

Suppose we have a mixture of gaseous nitrogen, hydrogen, and ammonia at equilibrium (Fig. 6.9). If we suddenly reduce the volume, what will happen to the equilibrium position? The reaction system can reduce its volume by reducing the number of molecules present. Consequently, the reaction

$$N_2(g) + 3H_2(g) \rightleftharpoons 2NH_3(g)$$

will shift to the right, since in this direction four molecules (one of nitrogen and three of hydrogen) react to produce two molecules (of ammonia), thus *reducing the total number of gaseous molecules present.* The new equilibrium position will be further to the right than the original one. That is, the equilibrium position will shift toward the side of the reaction involving the smaller number of gaseous molecules in the balanced equation. This phenomenon is illustrated in Fig. 6.10.

The opposite is also true. When the container volume is increased, the system will shift in the direction that increases its volume. An increase in volume in the ammonia synthesis system will produce a shift to the left to increase the total number of gaseous molecules present.

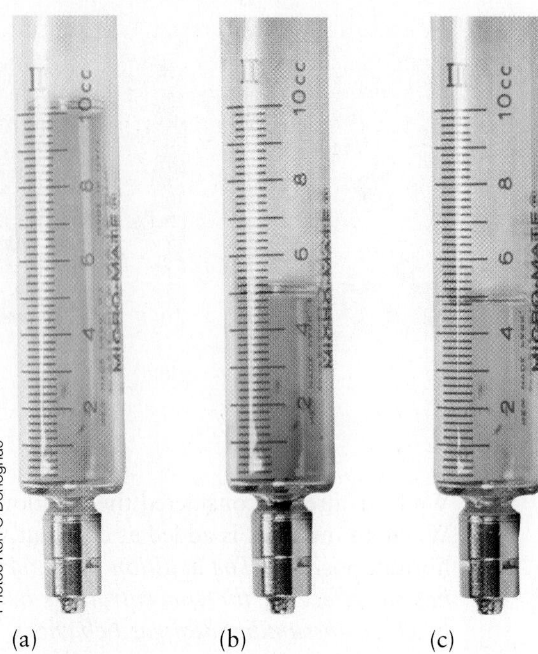

Photos Ken O'Donoghue

Figure 6.10
(a) Brown $NO_2(g)$ and colorless $N_2O_4(g)$ at equilibrium in a syringe. (b) The volume is suddenly decreased, giving a greater concentration of both N_2O_4 and NO_2 (indicated by the darker brown color). (c) A few seconds after the sudden volume decrease, the color becomes a much lighter brown as the equilibrium shifts from brown $NO_2(g)$ to colorless $N_2O_4(g)$. This is predicted by Le Châtelier's principle, since in the equilibrium

$$2NO_2(g) \rightleftharpoons N_2O_4(g)$$

the product side has the smaller number of molecules.

OWL INTERACTIVE EXAMPLE 6.5

Predict the shift in equilibrium position that will occur for each of the following processes when the volume is reduced.

a. The preparation of liquid phosphorus trichloride:

$$P_4(s) + 6Cl_2(g) \rightleftharpoons 4PCl_3(l)$$

b. The preparation of gaseous phosphorus pentachloride:

$$PCl_3(g) + Cl_2(g) \rightleftharpoons PCl_5(g)$$

c. The reaction of phosphorus trichloride with ammonia:

$$PCl_3(g) + 3NH_3(g) \rightleftharpoons P(NH_2)_3(g) + 3HCl(g)$$

Solution

a. Since P_4 and PCl_3 are a pure solid and a pure liquid, respectively, we need to consider only the effect of the decrease in volume on Cl_2. The position of the equilibrium will shift to the right, since the reactant side contains six gaseous molecules and the product side has none.

b. Decreasing the volume will shift the given reaction to the right, since the product side contains only one gaseous molecule and the reactant side has two.

c. Both sides of the balanced reaction equation have four gaseous molecules. A change in volume will have no effect on the equilibrium position. There is no shift in this case.

The Effect of a Change in Temperature

It is important to recognize that although the changes we have just discussed may alter the equilibrium *position,* they do not alter the equilibrium *constant* (assuming ideal behavior). For example, the addition of a reactant shifts the equilibrium position to the right but has no effect on the value of the equilib-

Table 6.3

Observed Value of K for the Ammonia Synthesis Reaction as a Function of Temperature*

Temperature (K)	K
500	90
600	3
700	0.3
800	0.04

*For this exothermic reaction, the value of K decreases as the temperature increases, as predicted by Le Châtelier's principle.

rium constant; the new equilibrium concentrations satisfy the original equilibrium constant.

The effect of temperature on equilibrium is different, however, because *the value of K changes with temperature.* We can use Le Châtelier's principle to predict the direction of the change.

The synthesis of ammonia from nitrogen and hydrogen releases energy (is *exothermic*). We can represent this situation by treating energy as a product:

$$N_2(g) + 3H_2(g) \rightleftharpoons 2NH_3(g) + \text{energy}$$

If energy in the form of heat is added to this system at equilibrium, Le Châtelier's principle predicts that the shift will be in the direction that consumes energy, in this case to the left. Note that this shift decreases the concentration of NH_3 and increases the concentrations of N_2 and H_2, thus *decreasing the value of K*. The experimentally observed change in K with temperature for this reaction is indicated in Table 6.3. The value of K decreases with increased temperature, as predicted.

On the other hand, for a reaction that consumes energy (an *endothermic* reaction), such as the decomposition of calcium carbonate,

$$\text{Energy} + CaCO_3(s) \rightleftharpoons CaO(s) + CO_2(g)$$

an increase in temperature will cause the equilibrium to shift to the right and the value of K to increase.

In summary, to use Le Châtelier's principle to describe the effect of a temperature change on a system at equilibrium, treat energy as a reactant (in an endothermic process) or as a product (in an exothermic process), and predict the direction of the shift as if an actual reactant or product is added or removed. Although Le Châtelier's principle cannot predict the size of the change in K, it can correctly predict the direction of the change.

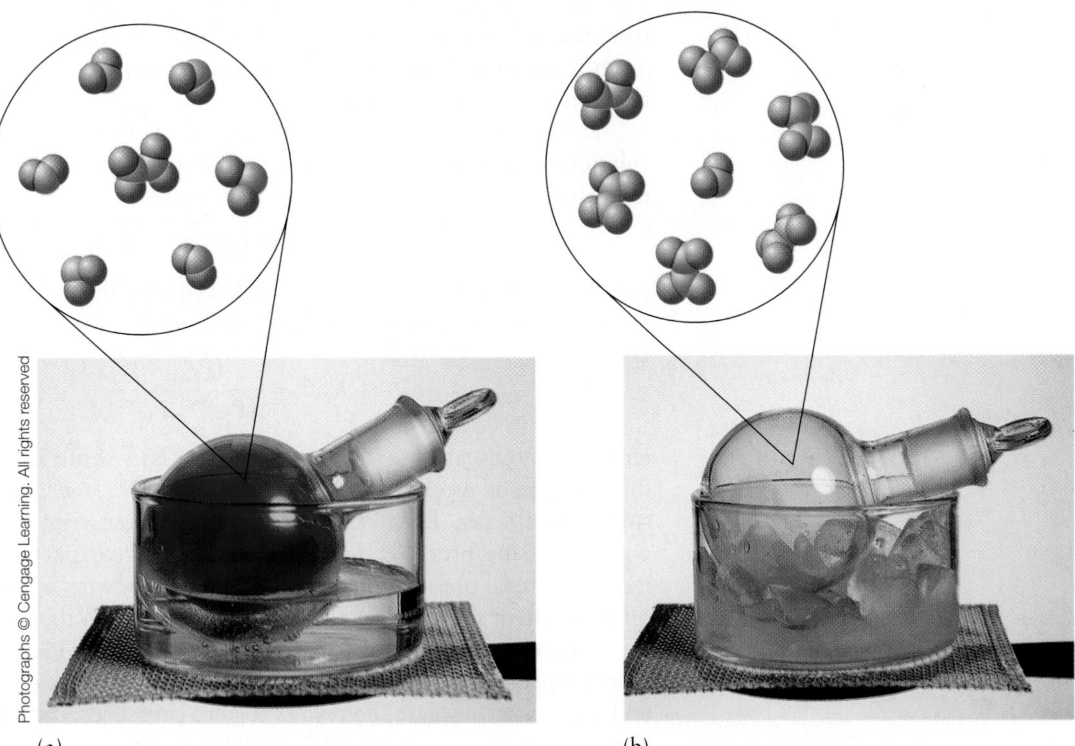

Shifting the $N_2O_4(g) \longrightarrow 2NO_2(g)$ equilibrium by changing the temperature. (a) At 100°C the flask is definitely reddish brown due to a large amount of NO_2 present. (b) At 0°C the equilibrium is shifted toward colorless $N_2O_4(g)$.

Table 6.4

Shifts in the Equilibrium Position for the Reaction $N_2O_4(g) \rightleftharpoons 2NO_2(g)$

Change	Shift
Addition of $N_2O_4(g)$	Right
Addition of $NO_2(g)$	Left
Removal of $N_2O_4(g)$	Left
Removal of $NO_2(g)$	Right
Addition of He(g)	None
Decrease in container volume	Left
Increase in container volume	Right
Increase in temperature	Right
Decrease in temperature	Left

We have seen how Le Châtelier's principle can be used to predict the effect of several types of changes on a system at equilibrium. As a summary of these ideas, Table 6.4 shows how various changes affect the equilibrium position of the endothermic reaction

$$N_2O_4(g) \rightleftharpoons 2NO_2(g)$$

6.9 Equilibria Involving Real Gases

Up to this point in our discussion of the equilibrium phenomenon, we have assumed ideal behavior for all substances. In fact, the value of K calculated from the law of mass action is the true value of the equilibrium constant for a given reaction system only if the observed pressures (concentrations) are corrected for any nonideal behavior.

To gain some appreciation for the effect of nonideal behavior on the calculation of equilibrium constants, consider the data in Table 6.5, which show the values of K_p (at 723 K) for the reaction

$$N_2(g) + 3H_2(g) \rightleftharpoons 2NH_3(g)$$

calculated from the (uncorrected) observed equilibrium pressures (P^{obs}) at various total pressures. Note that K_p^{obs}, defined as

$$K_p^{obs} = \frac{(P_{NH_3}^{obs})^2}{(P_{N_2}^{obs})(P_{H_2}^{obs})^3}$$

increases significantly with total pressure. This result makes sense in view of the fact that, as we discussed in Section 5.10, $P^{obs} < P^{ideal}$ for a real gas at pressures above 1 atm. Recall that the discrepancy between P^{obs} and P^{ideal} increases with increasing pressure. Thus, for this case, we expect K_p^{obs} to increase with increasing total pressure because the excess of powers in the denominator magnifies the error in pressures there as compared with the numerator.

One common method for finding the limiting value (the "true" value) of K_p is to measure K_p at various values of total pressure (constant temperature) and then to extrapolate the results to zero pressure. Another way to obtain the

Table 6.5

Values of K_p^{obs} at 723 K for the Reaction $N_2(g) + 3H_2(g) \rightleftharpoons 2NH_3(g)$ as a Function of Total Pressure (at equilibrium)

Total Pressure (atm)	K_p^{obs}
10	4.4×10^{-5}
50	4.6×10^{-5}
100	5.2×10^{-5}
300	7.7×10^{-5}
600	1.7×10^{-4}
1000	5.3×10^{-4}

true value of K_p is to correct the observed equilibrium pressures for any nonideal behavior. For example, we might represent the activity of the *i*th gaseous component as

$$a_i = \frac{\gamma_i P_i^{obs}}{P_{ref}}$$

where γ_i is called the activity coefficient for correcting P_i^{obs} to the ideal value. Obtaining the values of the activity coefficients is a complex process, which will not be treated here.

In general, for equilibrium pressures of 1 atm or less, the value of K_p calculated from the observed equilibrium pressures is expected to be within about 1% of the true value. However, at high pressures the deviations can be quite severe, as illustrated by the data in Table 6.5.

Key Terms

chemical equilibrium

Section 6.1
law of mass action
equilibrium expression
equilibrium constant
equilibrium position

Section 6.4
activity

Section 6.5
homogeneous equilibria
heterogeneous equilibria

Section 6.6
reaction quotient, *Q*

Section 6.8
LeChâtelier's principle

For Review

Chemical equilibrium

- When a reaction takes place in a closed system, it reaches a condition where the concentrations of the reactants and products remain constant over time
- Dynamic state: reactants and products are interconverted continually
  - Forward rate = reverse rate
- The law of mass action: for the reaction

$$jA + kB \rightleftharpoons mC + nD$$

$$K = \frac{[C]^m[D]^n}{[A]^j[B]^k} = \text{equilibrium constant}$$

- A pure liquid or solid is never included in the equilibrium expression
- For a gas-phase reaction the reactants and products can be described in terms of their partial pressures and the equilibrium constant called K_p:

$$K_p = K(RT)^{\Delta n}$$

where Δn is the sum of the coefficients of the gaseous products minus the sum of the coefficients of the gaseous reactants.

Activity

- An equilibrium constant expression involves the ratio of the equilibrium concentration (pressure) to a reference concentration (pressure) for that substance
  - Equilibrium constants are therefore unitless

Equilibrium position

- A set of reactant and product concentrations that satisfies the equilibrium constant expression
  - There is one value of K for a given system at a given temperature
  - There are an infinite number of equilibrium positions at a given temperature depending on the initial concentrations
- A small value of K means the equilibrium lies to the left; a large value of K means the equilibrium lies to the right.
  - The magnitude of K has no relationship to the speed at which equilibrium is achieved
- Q, the reaction quotient, applies the law of mass action to initial concentrations rather than equilibrium concentrations
  - If $Q > K$, the system will shift to the left to achieve equilibrium.
  - If $Q < K$, the system will shift to the right to achieve equilibrium.
- Finding the concentrations that characterize a given equilibrium position:
  - Start with the given initial concentrations (pressures)
  - Define the change needed to reach equilibrium
  - Apply the change to the initial concentrations (pressures) and solve for the equilibrium concentrations (pressures)

LeChâtelier's principle

- Enables qualitative prediction of the effects of changes in concentration, pressure, and temperature on a system at equilibrium
- If a change in conditions is imposed on a system at equilibrium, the system will shift in a direction that compensates for the imposed change

Discussion Questions

These questions are designed to be considered by groups of students in class. Often these questions work well for introducing a particular topic in class.

1. Consider an equilibrium mixture of four chemicals (A, B, C, and D, all gases) reacting in a closed flask according to the equation

$$A + B \rightleftharpoons C + D$$

   a. You add more A to the flask. How does the concentration of each chemical compare with its original concentration after equilibrium is reestablished? Justify your answer.

   b. You have the original setup at equilibrium, and add more D to the flask. How does the concentration of each chemical compare with its original concentration after equilibrium is reestablished? Justify your answer.

2. The boxes shown below represent a set of initial conditions for the reaction:

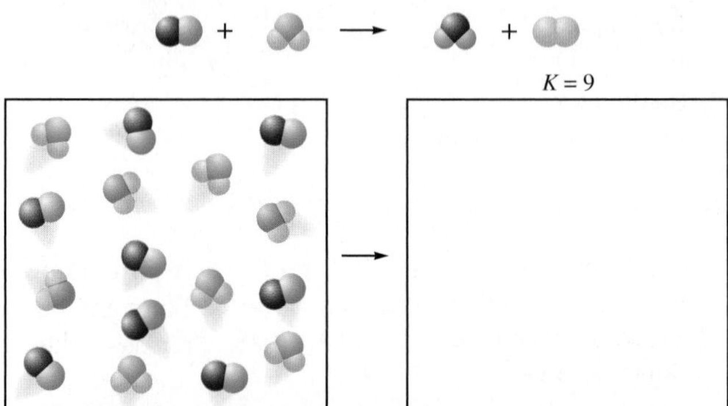

Draw a quantitative molecular picture that shows what this system looks like after the reactants are mixed in one of the boxes and the system reaches equilibrium. Support your answer with calculations.

3. For the reaction $H_2(g) + I_2(g) \rightleftharpoons 2HI(g)$, consider two possibilities using the same-sized, rigid container: (a) you add 0.5 mole of each reactant, allow the system to come

to equilibrium, then add another mole of H_2, and allow the system to reach equilibrium again, or (b) you add 1.5 moles of H_2 and 0.5 mole of I_2 and allow the system to come to equilibrium. Will the final equilibrium mixture be different for the two procedures? Explain.

4. Given the reaction $A(g) + B(g) \rightleftharpoons C(g) + D(g)$, consider the following situations:
   i. You have 1.3 *M* A and 0.8 *M* B initially.
   ii. You have 1.3 *M* A, 0.8 *M* B, and 0.2 *M* C initially.
   iii. You have 2.0 *M* A and 0.8 *M* B initially.

   After equilibrium has been reached, order i–iii in terms of increasing equilibrium concentrations of D. Explain your sequence. Then give the order in terms of increasing equilibrium concentration of B and explain.

5. Consider the reaction $A(g) + 2B(g) \rightleftharpoons C(g) + D(g)$ in a 1.0-L rigid flask. Answer the following questions for each situation (a–d):
   i. Estimate a range (as small as possible) for the requested substance. For example, [A] could be between 95 *M* and 100 *M*.
   ii. Explain how you decided on the limits for the estimated range.
   iii. Indicate what other information would enable you to narrow your estimated range.
   iv. Compare the estimated concentrations for a through d, and explain any differences.

   a. If at equilibrium [A] = 1 *M*, and then 1 mole of C is added, estimate the value for [A] once equilibrium is reestablished.
   b. If at equilibrium [B] = 1 *M*, and then 1 mole of C is added, estimate the value for [B] once equilibrium is reestablished.
   c. If at equilibrium [C] = 1 *M*, and then 1 mole of C is added, estimate the value for [C] once equilibrium is reestablished.
   d. If at equilibrium [D] = 1 *M*, and then 1 mole of C is added, estimate the value for [D] once equilibrium is reestablished.

6. Consider the reaction $A(g) + B(g) \rightleftharpoons C(g) + D(g)$. A friend asks the following: "I know we have been told that if a mixture of A, B, C, and D is at equilibrium and more of A is added, more C and D will form. But how can more C and D form if we do not add more B?" What do you tell your friend?

7. Consider the following statements: "Consider the reaction $A(g) + B(g) \rightleftharpoons C(g)$, at equilibrium in a 1-L container, with [A] = 2 *M*, [B] = 1 *M*, and [C] = 4 *M*. To this 1-L container you add 3 moles of B. A possible new equilibrium condition is [A] = 1 *M*, [B] = 3 *M*, and [C] = 6 *M* because in both cases *K* = 2." Indicate everything you think is correct in these statements and everything that is incorrect. Correct the incorrect statements, and explain.

8. Le Châtelier's principle is stated (Section 6.8) as "If a gaseous reactant or product is added to a system at equilibrium, the system will shift away from the added component." The system $N_2(g) + 3H_2(g) \rightleftharpoons 2NH_3(g)$ is used as an example in which the addition of nitrogen gas at equilibrium results in a decrease in H_2 concentration and an increase in NH_3 concentration as equilibrium is reestablished. In this experiment the volume is assumed to be constant. On the other hand, if N_2 is added to the reaction system in a container with a piston so that the pressure can be held constant, the concentration of NH_3 could actually *decrease* and the concentration of H_2 would *increase* as equilibrium is reestablished. Explain how this is possible. Also, if you consider this same system at equilibrium, the addition of an inert gas, at constant pressure, *does* affect the equilibrium position. How is the equilibrium position affected? Explain.

9. The value of the equilibrium constant *K* depends on which of the following (there may be more than one answer)?
   a. the initial concentrations of the reactants
   b. the initial concentrations of the products
   c. the temperature of the system
   d. the nature of the reactants and products

   Explain.

Exercises

OWL Interactive versions of these problems may be assigned in OWL.

A blue exercise number indicates that the answer to that exercise appears at the back of this book and a solution appears in the *Solutions Guide*.

Characteristics of Chemical Equilibrium

10. Consider an initial mixture of N_2 and H_2 gases that can be represented as follows.

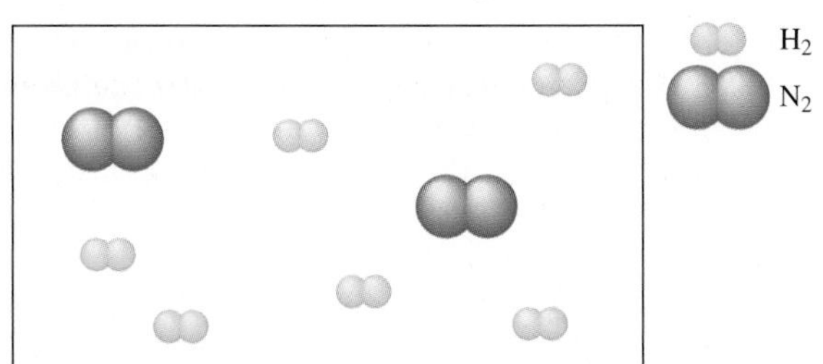

The gases react to form ammonia gas (NH_3) as represented by the following concentration profile.

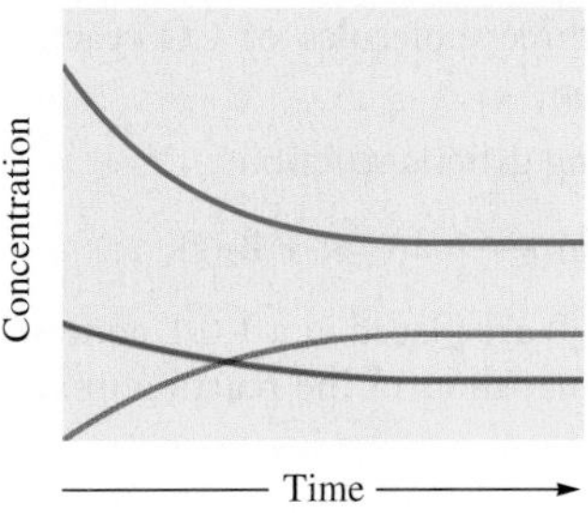

a. Label each plot on the graph as N_2, H_2, or NH_3 and explain your answers.
b. Explain the relative shapes of the plots.
c. When is equilibrium reached? How do you know?

11. Consider the following reactions at some temperature:

$$2NOCl(g) \rightleftharpoons 2NO(g) + Cl_2(g) \qquad K = 1.6 \times 10^{-5}$$

$$2NO(g) \rightleftharpoons N_2(g) + O_2(g) \qquad K = 1 \times 10^{31}$$

For each reaction some quantities of the reactants were placed in separate containers and allowed to come to equilibrium. Describe the relative amounts of reactants and products that are present at equilibrium. At equilibrium, which is faster, the forward or reverse reaction in each case?

12. Consider the following reaction:

$$H_2O(g) + CO(g) \rightleftharpoons H_2(g) + CO_2(g)$$

Amounts of H_2O, CO, H_2, and CO_2 are put into a flask so that the composition corresponds to an equilibrium position. If the CO placed in the flask is labeled with radioactive ^{14}C, will ^{14}C be found only in CO molecules for an indefinite period of time? Why or why not?

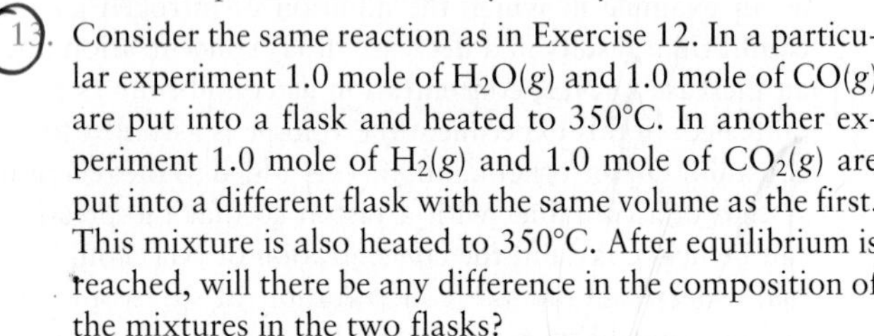

13. Consider the same reaction as in Exercise 12. In a particular experiment 1.0 mole of $H_2O(g)$ and 1.0 mole of $CO(g)$ are put into a flask and heated to 350°C. In another experiment 1.0 mole of $H_2(g)$ and 1.0 mole of $CO_2(g)$ are put into a different flask with the same volume as the first. This mixture is also heated to 350°C. After equilibrium is reached, will there be any difference in the composition of the mixtures in the two flasks?

14. Consider the following reaction at some temperature:

$$H_2O(g) + CO(g) \rightleftharpoons H_2(g) + CO_2(g) \qquad K = 2.0$$

Some molecules of H_2O and CO are placed in a 1.0-L container as shown below.

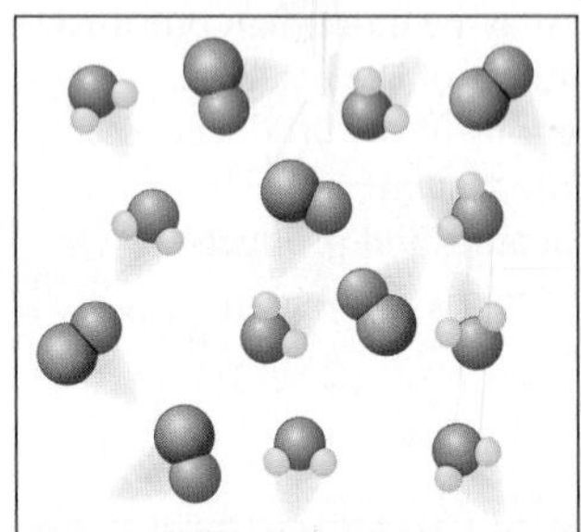

When equilibrium is reached, how many molecules of H_2O, CO, H_2, and CO_2 are present? Do this problem by trial and error—that is, if two molecules of CO react, is this equilibrium; if three molecules of CO react, is this equilibrium; and so on.

15. Consider the following generic reaction:

$$2A_2B(g) \rightleftharpoons 2A_2(g) + B_2(g)$$

Some molecules of A_2B are placed in a 1.0-L container. As time passes, several snapshots of the reaction mixture are taken as illustrated below.

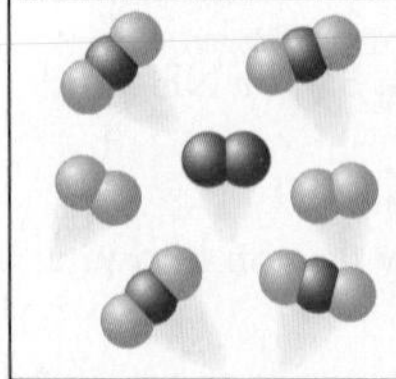

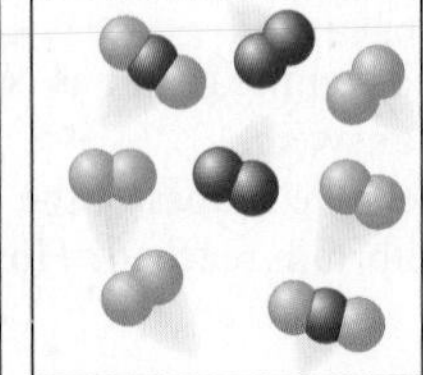

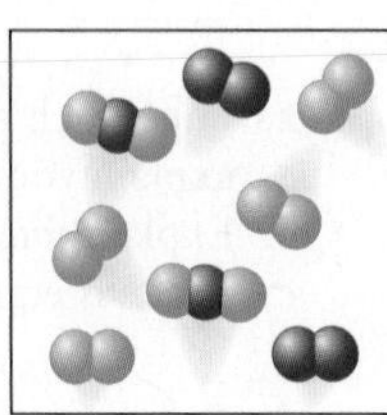

Which illustration is the first to represent an equilibrium mixture? Explain. How many molecules of A_2B were initially placed in the container?

The Equilibrium Constant

16. There is only one value of the equilibrium constant for a particular system at a particular temperature, but there are an infinite number of equilibrium positions. Explain.

17. Explain the difference between K, K_p, and Q.

18. What are homogeneous equilibria? Heterogeneous equilibria? What is the difference in writing K expressions for homogeneous versus heterogeneous reactions? Summarize which species are included in K expressions and which species are not included.

19. Write expressions for K and K_p for the following reactions.
   a. $2NH_3(g) + CO_2(g) \rightleftharpoons N_2CH_4O(s) + H_2O(g)$
   b. $2NBr_3(s) \rightleftharpoons N_2(g) + 3Br_2(g)$
   c. $2KClO_3(s) \rightleftharpoons 2KCl(s) + 3O_2(g)$
   d. $CuO(s) + H_2(g) \rightleftharpoons Cu(l) + H_2O(g)$

20. For which reactions in Exercise 19 is K_p equal to K?

21. At a particular temperature, a 3.0-L flask contains 2.4 moles of Cl_2, 1.0 mole of NOCl, and 4.5×10^{-3} mole of NO. Calculate K at this temperature for the following reaction.

$$2NOCl(g) \rightleftharpoons 2NO(g) + Cl_2(g)$$

22. At 1100 K, $K_p = 0.25$ for the reaction

$$2SO_2(g) + O_2(g) \rightleftharpoons 2SO_3(g)$$

What is the value of K at this temperature?

23. At 327°C, the equilibrium concentrations are $[CH_3OH] = 0.15\ M$, $[CO] = 0.24\ M$, and $[H_2] = 1.1\ M$ for the reaction

$$CH_3OH(g) \rightleftharpoons CO(g) + 2H_2(g)$$

Calculate K_p at this temperature.

24. For the reaction

$$H_2(g) + Br_2(g) \rightleftharpoons 2HBr(g)$$

$K_p = 3.5 \times 10^4$ at 1495 K. What is the value of K_p for the following reactions at 1495 K?
   a. $HBr(g) \rightleftharpoons \frac{1}{2}H_2(g) + \frac{1}{2}Br_2(g)$
   b. $2HBr(g) \rightleftharpoons H_2(g) + Br_2(g)$
   c. $\frac{1}{2}H_2(g) + \frac{1}{2}Br_2(g) \rightleftharpoons HBr(g)$

25. At a particular temperature, a 2.00-L flask at equilibrium contains 2.80×10^{-4} mole of N_2, 2.50×10^{-5} mole of O_2, and 2.00×10^{-2} mole of N_2O. Calculate K at this temperature for the reaction

$$2N_2(g) + O_2(g) \rightleftharpoons 2N_2O(g)$$

If $[N_2] = 2.00 \times 10^{-4}\ M$, $[N_2O] = 0.200\ M$, and $[O_2] = 0.00245\ M$, does this represent a system at equilibrium?

26. The following equilibrium pressures were observed at a certain temperature for the reaction

$$N_2(g) + 3H_2(g) \rightleftharpoons 2NH_3(g)$$

$$P_{NH_3} = 3.1 \times 10^{-2} \text{ atm}$$

$$P_{N_2} = 8.5 \times 10^{-1} \text{ atm}$$

$$P_{H_2} = 3.1 \times 10^{-3} \text{ atm}$$

Calculate the value for the equilibrium constant K_p at this temperature. If $P_{N_2} = 0.525$ atm, $P_{NH_3} = 0.0167$ atm, and $P_{H_2} = 0.00761$ atm, does this represent a system at equilibrium?

27. For the reaction

$$2H_2O(g) \rightleftharpoons 2H_2(g) + O_2(g)$$

$K = 2.4 \times 10^{-3}$ at a given temperature. At equilibrium in a 2.0-L container, it is found that $[H_2O(g)] = 1.1 \times 10^{-1}$ *M* and $[H_2(g)] = 1.9 \times 10^{-2}$ *M*. Calculate the moles of $O_2(g)$ present under these conditions.

28. The reaction

$$2NO(g) + Br_2(g) \rightleftharpoons 2NOBr(g)$$

as $K_p = 109$ at 25°C. If the equilibrium partial pressure of Br_2 is 0.0159 atm and the equilibrium partial pressure of NOBr is 0.0768 atm, calculate the partial pressure of NO at equilibrium.

29. In a study of the reaction

$$3Fe(s) + 4H_2O(g) \rightleftharpoons Fe_3O_4(s) + 4H_2(g)$$

at 1200 K, it was observed that when the equilibrium partial pressure of water vapor is 15.0 torr, the total pressure at equilibrium is 36.3 torr. Calculate K_p for this reaction at 1200 K.

30. Nitrogen gas (N_2) reacts with hydrogen gas (H_2) to form ammonia (NH_3). At 200°C in a closed container, 1.00 atm of nitrogen gas is mixed with 2.00 atm of hydrogen gas. At equilibrium, the total pressure is 2.00 atm. Calculate the partial pressure of hydrogen gas at equilibrium, and calculate the value of K_p for this reaction.

31. A sample of gaseous PCl_5 was introduced into an evacuated flask so that the pressure of pure PCl_5 would be 0.50 atm at 523 K. However, PCl_5 decomposes to gaseous PCl_3 and Cl_2, and the actual pressure in the flask was found to be 0.84 atm. Calculate K_p for the decomposition reaction

$$PCl_5(g) \rightleftharpoons PCl_3(g) + Cl_2(g)$$

at 523 K. Also calculate *K* at this temperature.

32. A sample of $S_8(g)$ is placed in an otherwise empty, rigid container at 1325 K at an initial pressure of 1.00 atm, where it decomposes to $S_2(g)$ by the reaction

$$S_8(g) \rightleftharpoons 4S_2(g)$$

At equilibrium, the partial pressure of S_8 is 0.25 atm. Calculate K_p for this reaction at 1325 K.

33. At a particular temperature, 12.0 moles of SO_3 is placed into a 3.0-L rigid container, and the SO_3 dissociates by the reaction

$$2SO_3(g) \rightleftharpoons 2SO_2(g) + O_2(g)$$

At equilibrium, 3.0 moles of SO_2 is present. Calculate *K* for this reaction.

34. At a particular temperature, 8.0 moles of NO_2 is placed into a 1.0-L container and the NO_2 dissociates by the reaction

$$2NO_2(g) \rightleftharpoons 2NO(g) + O_2(g)$$

At equilibrium the concentration of NO(*g*) is 2.0 *M*. Calculate *K* for this reaction.

Equilibrium Calculations

35. The equilibrium constant is 0.0900 at 25°C for the reaction

$$H_2O(g) + Cl_2O(g) \rightleftharpoons 2HOCl(g)$$

For which of the following sets of conditions is the system at equilibrium? For those which are not at equilibrium, in which direction will the system shift?

a. A 1.0-L flask contains 1.0 mole of HOCl, 0.10 mole of Cl_2O, and 0.10 mole of H_2O.
b. A 2.0-L flask contains 0.084 mole of HOCl, 0.080 mole of Cl_2O, and 0.98 mole of H_2O.
c. A 3.0-L flask contains 0.25 mole of HOCl, 0.0010 mole of Cl_2O, and 0.56 mole of H_2O.

36. The equilibrium constant is 0.0900 at 25°C for the reaction

$$H_2O(g) + Cl_2O(g) \rightleftharpoons 2HOCl(g)$$

For which of the following sets of conditions is the system at equilibrium? For those which are not at equilibrium, in which direction will the system shift?

a. $P_{H_2O} = 1.00$ atm, $P_{Cl_2O} = 1.00$ atm, $P_{HOCl} = 1.00$ atm
b. $P_{H_2O} = 200.$ torr, $P_{Cl_2O} = 49.8$ torr, $P_{HOCl} = 21.0$ torr
c. $P_{H_2O} = 296$ torr, $P_{Cl_2O} = 15.0$ torr, $P_{HOCl} = 20.0$ torr

37. At 900.°C, $K_p = 1.04$ for the reaction

$$CaCO_3(s) \rightleftharpoons CaO(s) + CO_2(g)$$

At a low temperature dry ice (solid CO_2), calcium oxide, and calcium carbonate are introduced into a 50.0-L reaction chamber. The temperature is raised to 900.°C. For the following mixtures, will the initial amount of calcium oxide increase, decrease, or remain the same as the system moves toward equilibrium?

a. 655 g of $CaCO_3$, 95.0 g of CaO, 58.4 g of CO_2
b. 780 g of $CaCO_3$, 1.00 g of CaO, 23.76 g of CO_2
c. 0.14 g of $CaCO_3$, 5000 g of CaO, 23.76 g of CO_2
d. 715 g of $CaCO_3$, 813 g of CaO, 4.82 g of CO_2

38. At a particular temperature, $K = 3.75$ for the reaction

$$SO_2(g) + NO_2(g) \rightleftharpoons SO_3(g) + NO(g)$$

If all four gases had initial concentrations of 0.800 *M*, calculate the equilibrium concentrations of the gases.

39. At 25°C, $K = 0.090$ for the reaction

$$H_2O(g) + Cl_2O(g) \rightleftharpoons 2HOCl(g)$$

Calculate the concentrations of all species at equilibrium for each of the following cases.

a. 1.0 g of H_2O and 2.0 g of Cl_2O are mixed in a 1.0-L flask.
b. 1.0 mole of pure HOCl is placed in a 2.0-L flask.

40. For the reaction below at a certain temperature, it is found that the equilibrium concentrations in a 5.00-L rigid container are $[H_2] = 0.0500\ M$, $[F_2] = 0.0100\ M$, and $[HF] = 0.400\ M$.

$$H_2(g) + F_2(g) \rightleftharpoons 2HF(g)$$

If 0.200 mole of F_2 is added to this equilibrium mixture, calculate the concentrations of all gases once equilibrium is reestablished.

41. At 1100 K, $K_p = 0.25$ for the following reaction:

$$2SO_2(g) + O_2(g) \rightleftharpoons 2SO_3(g)$$

Calculate the equilibrium partial pressures of SO_2, O_2, and SO_3 produced from an initial mixture in which $P_{SO_2} = P_{O_2} = 0.50$ atm and $P_{SO_3} = 0$.

42. At 2200°C, $K = 0.050$ for the reaction

$$N_2(g) + O_2(g) \rightleftharpoons 2NO(g)$$

What is the partial pressure of NO at equilibrium assuming the N_2 and O_2 had initial pressures of 0.80 atm and 0.20 atm, respectively?

43. A type of reaction we will study is that having a very small K value ($K << 1$). Solving for equilibrium concentrations in an equilibrium problem usually requires many mathematical operations to be performed. However, the math involved in solving equilibrium problems for reactions having small K values ($K << 1$) is simplified. What assumption is made when solving equilibrium concentrations for reactions having small K values? Whenever assumptions are made, they must be checked for validity. In general, the "5% rule" is used to check the validity of assuming that x (or $2x$, $3x$, and so on) is very small compared to some number. When x (or $2x$, $3x$, and so on) is less than 5% of the number the assumption was made against, then the assumption is said to be valid. If the 5% rule fails, what do you do to solve for the equilibrium concentrations?

44. At 35°C, $K = 1.6 \times 10^{-5}$ for the reaction

$$2NOCl(g) \rightleftharpoons 2NO(g) + Cl_2(g)$$

Calculate the concentrations of all species at equilibrium for each of the following original mixtures.

a. 2.0 moles of pure NOCl in a 2.0-L flask
b. 2.0 moles of NO and 1.0 mole of Cl_2 in a 1.0-L flask
c. 1.0 mole of NOCl and 1.0 mole of NO in a 1.0-L flask
d. 3.0 moles of NO and 1.0 mole of Cl_2 in a 1.0-L flask
e. 2.0 moles of NOCl, 2.0 moles of NO, and 1.0 mole of Cl_2 in a 1.0-L flask
f. 1.00 mol/L concentration of all three gases

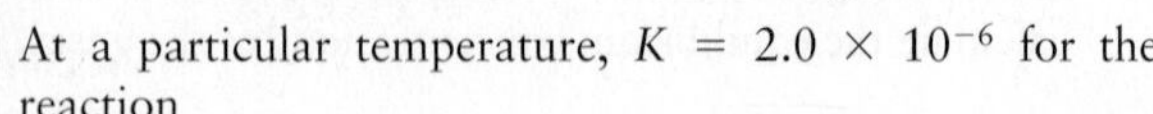

45. At a particular temperature, $K = 2.0 \times 10^{-6}$ for the reaction

$$2CO_2(g) \rightleftharpoons 2CO(g) + O_2(g)$$

If 2.0 moles of CO_2 is initially placed into a 5.0-L vessel, calculate the equilibrium concentrations of all species.

46. At a particular temperature, $K_p = 0.25$ for the reaction

$$N_2O_4(g) \rightleftharpoons 2NO_2(g)$$

a. A flask containing only N_2O_4 at an initial pressure of 4.5 atm is allowed to reach equilibrium. Calculate the equilibrium partial pressures of the gases.
b. A flask containing only NO_2 at an initial pressure of 9.0 atm is allowed to reach equilibrium. Calculate the equilibrium partial pressures of the gases.
c. From your answers to parts a and b, does it matter from which direction an equilibrium position is reached?
d. The volume of the container in part a is decreased to one-half the original volume. Calculate the new equilibrium partial pressures.

47. For the reaction below, $K_p = 1.16$ at 800.°C.

$$CaCO_3(s) \rightleftharpoons CaO(s) + CO_2(g)$$

If a 20.0-g sample of $CaCO_3$ is put into a 10.0-L container and heated to 800.°C, what percentage by mass of the $CaCO_3$ will react to reach equilibrium?

48. At 25°C, $K_p = 2.9 \times 10^{-3}$ for the reaction

$$NH_4OCONH_2(s) \rightleftharpoons 2NH_3(g) + CO_2(g)$$

In an experiment carried out at 25°C, a certain amount of NH_4OCONH_2 is placed in an evacuated rigid container and allowed to come to equilibrium. Calculate the total pressure in the container at equilibrium.

49. Lexan is a plastic used to make compact discs, eyeglass lenses, and bullet-proof glass. One of the compounds used to make Lexan is phosgene ($COCl_2$), a poisonous gas. Phosgene is produced by the reaction

$$CO(g) + Cl_2(g) \rightleftharpoons COCl_2(g)$$

for which $K = 4.5 \times 10^9$ at 100.°C.

a. Calculate K_p at 100.°C.
b. Equal moles of CO and Cl_2 are reacted at 100.°C. If the total pressure at equilibrium is 5.0 atm, calculate the equilibrium partial pressures of all the gases.

50. At a certain temperature, $K = 1.1 \times 10^3$ for the reaction

$$Fe^{3+}(aq) + SCN^-(aq) \rightleftharpoons FeSCN^{2+}(aq)$$

Calculate the concentrations of Fe^{3+}, SCN^-, and $FeSCN^{2+}$ at equilibrium if 0.020 mole of $Fe(NO_3)_3$ is added to 1.0 L of 0.10 M KSCN. (Neglect any volume change.)

Le Châtelier's Principle

51. Which of the following statements is(are) true? Correct the false statement(s).

a. When a reactant is added to a system at equilibrium at a given temperature, the reaction will shift right to reestablish equilibrium.
b. When a product is added to a system at equilibrium at a given temperature, the value of K for the reaction will increase when equilibrium is reestablished.

c. When temperature is increased for a reaction at equilibrium, the value of K for the reaction will increase.
d. Addition of a catalyst (a substance that increases the speed of the reaction) has no effect on the equilibrium position.

52. How will the equilibrium position of a gas-phase reaction be affected if the volume of the reaction vessel changes? Are there reactions that will not have their equilibria shifted by a change in volume? Explain. Why does changing the pressure in a rigid container by adding an inert gas not shift the equilibrium position for a gas-phase reaction?

53. Suppose the reaction system

$$UO_2(s) + 4HF(g) \rightleftharpoons UF_4(g) + 2H_2O(g)$$

has already reached equilibrium. Predict the effect that each of the following changes will have on the equilibrium position. Tell whether the equilibrium will shift to the right, will shift to the left, or will not be affected.
a. More $UO_2(s)$ is added to the system.
b. The reaction is performed in a glass reaction vessel; $HF(g)$ attacks and reacts with glass.
c. Water vapor is removed.

54. Consider the reaction

$$Fe^{3+}(aq) + SCN^-(aq) \rightleftharpoons FeSCN^{2+}(aq)$$

How will the equilibrium position shift if
a. water is added, doubling the volume?
b. $AgNO_3(aq)$ is added? (AgSCN is insoluble.)
c. $NaOH(aq)$ is added? [$Fe(OH)_3$ is insoluble.]
d. $Fe(NO_3)_3(aq)$ is added?

55. Chromium(VI) forms two different oxyanions, the orange dichromate ion ($Cr_2O_7^{2-}$), and the yellow chromate ion (CrO_4^{2-}). (See the photos below.) The equilibrium reaction between the two ions is

$$Cr_2O_7^{2-}(aq) + H_2O(l) \rightleftharpoons 2CrO_4^{2-}(aq) + 2H^+(aq)$$

Explain why orange dichromate solutions turn yellow when sodium hydroxide is added.

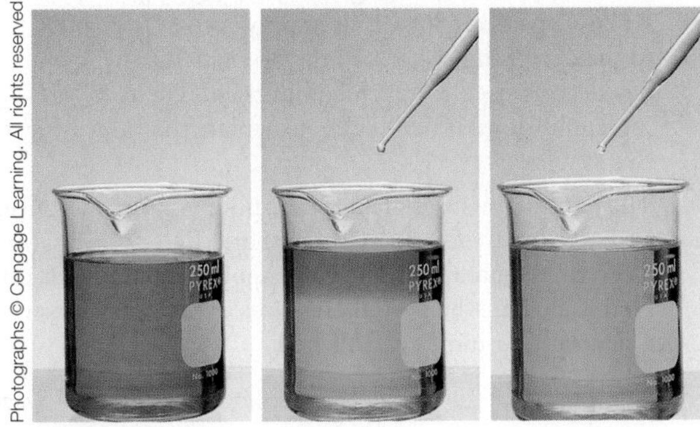

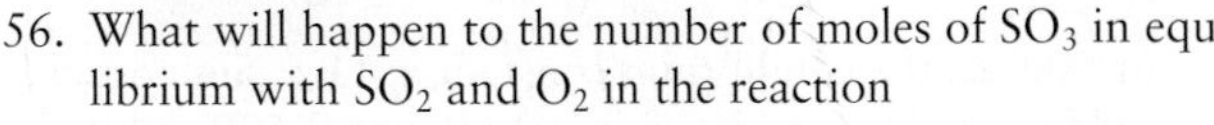

56. What will happen to the number of moles of SO_3 in equilibrium with SO_2 and O_2 in the reaction

$$2SO_3(g) \rightleftharpoons 2SO_2(g) + O_2(g)$$

in each of the following cases?
a. Oxygen gas is added.
b. The pressure is increased by decreasing the volume of the reaction container.
c. In a rigid reaction container, the pressure is increased by adding argon gas.
d. The temperature is decreased (the reaction is endothermic).
e. Gaseous sulfur dioxide is removed.

57. An important reaction in the commercial production of hydrogen is

$$CO(g) + H_2O(g) \rightleftharpoons H_2(g) + CO_2(g)$$

How will this system at equilibrium shift in each of the five following cases?
a. Gaseous carbon dioxide is removed.
b. Water vapor is added.
c. In a rigid reaction container, the pressure is increased by adding helium gas.
d. The temperature is increased (the reaction is exothermic).
e. The pressure is increased by decreasing the volume of the reaction container.

58. Hydrogen for use in ammonia production is produced by the reaction

$$CH_4(g) + H_2O(g) \xrightleftharpoons[750°C]{\text{Ni catalyst}} CO(g) + 3H_2(g)$$

What will happen to a reaction mixture at equilibrium if
a. $H_2O(g)$ is removed?
b. the temperature is increased (the reaction is endothermic)?
c. an inert gas is added to a rigid reaction container?
d. $CO(g)$ is removed?
e. the volume of the container is tripled?

59. In which direction will the position of the equilibrium

$$2HI(g) \rightleftharpoons H_2(g) + I_2(g)$$

be shifted for each of the following changes?
a. $H_2(g)$ is added.
b. $I_2(g)$ is removed.
c. $HI(g)$ is removed.
d. In a rigid reaction container, some $Ar(g)$ is added.
e. The volume of the container is doubled.
f. The temperature is decreased (the reaction is exothermic).

60. Predict the shift in the equilibrium position that will occur for each of the following reactions when the volume of the reaction container is increased.
a. $N_2(g) + 3H_2(g) \rightleftharpoons 2NH_3(g)$
b. $PCl_5(g) \rightleftharpoons PCl_3(g) + Cl_2(g)$
c. $H_2(g) + F_2(g) \rightleftharpoons 2HF(g)$
d. $COCl_2(g) \rightleftharpoons CO(g) + Cl_2(g)$
e. $CaCO_3(s) \rightleftharpoons CaO(s) + CO_2(g)$

61. Old-fashioned "smelling salts" consist of ammonium carbonate [$(NH_4)_2CO_3$]. The reaction for the decomposition of ammonium carbonate

$$(NH_4)_2CO_3(s) \rightleftharpoons 2NH_3(g) + CO_2(g) + H_2O(g)$$

is endothermic. Would the smell of ammonia increase or decrease as the temperature is increased?

62. Ammonia is produced by the Haber process, in which nitrogen and hydrogen are reacted directly using an iron mesh impregnated with oxides as a catalyst. For the reaction

$$N_2(g) + 3H_2(g) \rightleftharpoons 2NH_3(g)$$

equilibrium constants as a function of temperature are

300°C, 4.34×10^{-3}

500°C, 1.45×10^{-5}

600°C, 2.25×10^{-6}

Is the reaction exothermic or endothermic? Explain.

Additional Exercises

63. At 25°C, $K_p \approx 1 \times 10^{-31}$ for the reaction

$$N_2(g) + O_2(g) \rightleftharpoons 2NO(g)$$

a. Calculate the concentration of NO (in molecules/cm^3) that can exist in equilibrium in air at 25°C. In air $P_{N_2} = 0.8$ atm and $P_{O_2} = 0.2$ atm.
b. Typical concentrations of NO in relatively pristine environments range from 10^8 molecules/cm^3 to 10^{10} molecules/cm^3. Why is there a discrepancy between these values and your answer to part a?

64. Given the following equilibrium constants at 427°C,

$Na_2O(s) \rightleftharpoons 2Na(l) + \frac{1}{2}O_2(g)$ $\quad K_1 = 2 \times 10^{-25}$

$NaO(g) \rightleftharpoons Na(l) + \frac{1}{2}O_2(g)$ $\quad K_2 = 2 \times 10^{-5}$

$Na_2O_2(s) \rightleftharpoons 2Na(l) + O_2(g)$ $\quad K_3 = 5 \times 10^{-29}$

$NaO_2(s) \rightleftharpoons Na(l) + O_2(g)$ $\quad K_4 = 3 \times 10^{-14}$

determine the values for the equilibrium constants for the following reactions.
a. $Na_2O(s) + \frac{1}{2}O_2(g) \rightleftharpoons Na_2O_2(s)$
b. $NaO(g) + Na_2O(s) \rightleftharpoons Na_2O_2(s) + Na(l)$
c. $2NaO(g) \rightleftharpoons Na_2O_2(s)$
(*Hint:* When reaction equations are added, the equilibrium expressions are multiplied.)

65. Calculate a value for the equilibrium constant for the reaction

$$O_2(g) + O(g) \rightleftharpoons O_3(g)$$

given that

$NO_2(g) \overset{h\nu}{\rightleftharpoons} NO(g) + O(g)$ $\quad K = 6.8 \times 10^{-49}$

$O_3(g) + NO(g) \rightleftharpoons NO_2(g) + O_2(g)$ $\quad K = 5.8 \times 10^{-34}$

(See the hint in Exercise 64.)

66. Given $K = 3.50$ at 45°C for the reaction

$$A(g) + B(g) \rightleftharpoons C(g)$$

and $K = 7.10$ at 45°C for the reaction

$$2A(g) + D(g) \rightleftharpoons C(g)$$

what is the value of K at the same temperature for the reaction

$$C(g) + D(g) \rightleftharpoons 2B(g)$$

What is the value of K_p at 45°C for the reaction? Starting with 1.50 atm partial pressures of both C and D, what is the mole fraction of B once equilibrium is reached?

67. An initial mixture of nitrogen gas and hydrogen gas is reacted in a rigid container at a certain temperature as follows:

$$3H_2(g) + N_2(g) \rightleftharpoons 2NH_3(g)$$

At equilibrium, the concentrations are $[H_2] = 5.0$ M, $[N_2] = 8.0$ M, and $[NH_3] = 4.0$ M. What were the concentrations of nitrogen gas and hydrogen gas that were reacted initially?

68. At 25°C, $K_p = 5.3 \times 10^5$ for the reaction

$$N_2(g) + 3H_2(g) \rightleftharpoons 2NH_3(g)$$

When a certain partial pressure of $NH_3(g)$ is put into an otherwise empty rigid vessel at 25°C, equilibrium is reached when 50.0% of the original ammonia has decomposed. What was the original partial pressure of ammonia before any decomposition occurred?

69. A 2.4156-g sample of PCl_5 was placed in an empty 2.000-L flask and allowed to decompose to PCl_3 and Cl_2 at 250.0°C:

$$PCl_5(g) \rightleftharpoons PCl_3(g) + Cl_2(g)$$

At equilibrium the total pressure inside the flask was observed to be 358.7 torr.
a. Calculate the partial pressure of each gas at equilibrium and the value of K_p at 250.0°C.
b. What are the new equilibrium pressures if 0.250 mole of Cl_2 gas is added to the flask?

70. At 25°C, gaseous SO_2Cl_2 decomposes to $SO_2(g)$ and $Cl_2(g)$ to the extent that 12.5% of the original SO_2Cl_2 (by moles) has decomposed to reach equilibrium. The total pressure (at equilibrium) is 0.900 atm. Calculate the value of K_p for this system.

71. The partial pressures of an equilibrium mixture of $N_2O_4(g)$ and $NO_2(g)$ are $P_{N_2O_4} = 0.34$ atm and $P_{NO_2} = 1.20$ atm at a certain temperature. The volume of the container is doubled. Calculate the partial pressures of the two gases when a new equilibrium is established.

72. For the reaction

$$PCl_5(g) \rightleftharpoons PCl_3(g) + Cl_2(g)$$

at 600. K, the equilibrium constant is 11.5. Suppose that 2.450 g of PCl_5 is placed in an evacuated 500.-mL bulb, which is then heated to 600. K.
a. What would the pressure of PCl_5 be if it did not dissociate?
b. What is the partial pressure of PCl_5 at equilibrium?

c. What is the total pressure in the bulb at equilibrium?
d. What is the degree of dissociation of PCl_5 at equilibrium?

73. At 125°C, $K_p = 0.25$ for the reaction

$$2NaHCO_3(s) \rightleftharpoons Na_2CO_3(s) + CO_2(g) + H_2O(g)$$

A 1.00-L flask containing 10.0 g of $NaHCO_3$ is evacuated and heated to 125°C.
a. Calculate the partial pressures of CO_2 and H_2O after equilibrium is established.
b. Calculate the masses of $NaHCO_3$ and Na_2CO_3 present at equilibrium.
c. Calculate the minimum container volume necessary for all the $NaHCO_3$ to decompose.

74. The gas arsine (AsH_3) decomposes as follows:

$$2AsH_3(g) \rightleftharpoons 2As(s) + 3H_2(g)$$

In an experiment pure $AsH_3(g)$ was placed in an empty, rigid, sealed flask at a pressure of 392.0 torr. After 48 h the pressure in the flask was observed to be constant at 488.0 torr.
a. Calculate the equilibrium pressure of $H_2(g)$.
b. Calculate K_p for this reaction.

75. For the reaction

$$NH_3(g) + H_2S(g) \rightleftharpoons NH_4HS(s)$$

$K = 400.$ at 35.0°C. If 2.00 moles each of NH_3, H_2S, and NH_4HS are placed in a 5.00-L vessel, what mass of NH_4HS will be present at equilibrium? What is the pressure of H_2S at equilibrium?

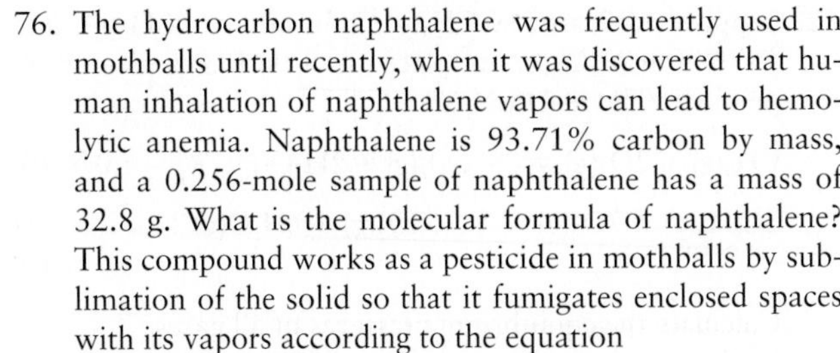

76. The hydrocarbon naphthalene was frequently used in mothballs until recently, when it was discovered that human inhalation of naphthalene vapors can lead to hemolytic anemia. Naphthalene is 93.71% carbon by mass, and a 0.256-mole sample of naphthalene has a mass of 32.8 g. What is the molecular formula of naphthalene? This compound works as a pesticide in mothballs by sublimation of the solid so that it fumigates enclosed spaces with its vapors according to the equation

$$\text{Naphthalene}(s) \rightleftharpoons \text{naphthalene}(g) \qquad K = 4.29 \times 10^{-6} \text{ (at 298 K)}$$

If 3.00 g of solid naphthalene is placed in an enclosed space with a volume of 5.00 L at 25°C, what percentage of the naphthalene will have sublimed once equilibrium has been established?

77. Consider the decomposition of the compound $C_5H_6O_3$ as follows:

$$C_5H_6O_3(g) \longrightarrow C_2H_6(g) + 3CO(g)$$

When a 5.63-g sample of pure $C_5H_6O_3(g)$ was sealed in an otherwise empty 2.50-L flask and heated to 200.°C, the pressure in the flask gradually rose to 1.63 atm and remained at that value. Calculate K for this reaction.

Challenge Problems

78. A sample of $N_2O_4(g)$ is placed in an empty cylinder at 25°C. After equilibrium is reached, the total pressure is 1.5 atm, and 16% (by moles) of the original $N_2O_4(g)$ has dissociated to $NO_2(g)$.
a. Calculate the value of K_p for this dissociation reaction at 25°C.
b. If the volume of the cylinder is increased until the total pressure is 1.0 atm (the temperature of the system remains constant), calculate the equilibrium pressure of $N_2O_4(g)$ and $NO_2(g)$.
c. What percentage (by moles) of the original $N_2O_4(g)$ is dissociated at the new equilibrium position (total pressure = 1.00 atm)?

79. Nitric oxide and bromine at initial partial pressures of 98.4 torr and 41.3 torr, respectively, were allowed to react at 300. K. At equilibrium the total pressure was 110.5 torr. The reaction is

$$2NO(g) + Br_2(g) \rightleftharpoons 2NOBr(g)$$

a. Calculate the value of K_p.
b. What would be the partial pressures of all species if NO and Br_2, both at an initial partial pressure of 0.30 atm, were allowed to come to equilibrium at this temperature?

80. Consider the decomposition equilibrium for dinitrogen pentoxide:

$$2N_2O_5(g) \rightleftharpoons 4NO_2(g) + O_2(g)$$

At a certain temperature and a total pressure of 1.00 atm, the N_2O_5 is 0.50% decomposed (by moles) at equilibrium.
a. If the volume is increased by a factor of 10.0, will the mole percent of N_2O_5 decomposed at equilibrium be greater than, less than, or equal to 0.50%? Explain your answer.
b. Calculate the mole percent of N_2O_5 that will be decomposed at equilibrium if the volume is increased by a factor of 10.0.

81. Consider the reaction

$$P_4(g) \longrightarrow 2P_2(g)$$

where $K_p = 1.00 \times 10^{-1}$ at 1325 K. In an experiment where $P_4(g)$ was placed in a container at 1325 K, the equilibrium mixture of $P_4(g)$ and $P_2(g)$ has a total pressure of 1.00 atm. Calculate the equilibrium pressures of $P_4(g)$ and $P_2(g)$. Calculate the fraction (by moles) of $P_4(g)$ that has dissociated to reach equilibrium.

82. Suppose 1.50 atm of $CH_4(g)$, 2.50 atm of $C_2H_6(g)$, and 15.00 atm of $O_2(g)$ are placed in a flask at a given temperature. The reactions are

$$CH_4(g) + 2O_2(g) \rightleftharpoons CO_2(g) + 2H_2O(g) \quad K_p = 1.0 \times 10^4$$

$$2C_2H_6(g) + 7O_2(g) \rightleftharpoons 4CO_2(g) + 6H_2O(g) \quad K_p = 1.0 \times 10^8$$

Calculate the equilibrium pressures of all gases.

83. Consider the reaction

$$3O_2(g) \rightleftharpoons 2O_3(g)$$

At 175°C and a pressure of 128 torr, an equilibrium mixture of O_2 and O_3 has a density of 0.168 g/L. Calculate K_p for the above reaction at 175°C.

84. A mixture of N_2, H_2 and NH_3 is at equilibrium according to the equation $N_2(g) + 3H_2(g) \rightleftharpoons 2NH_3(g)$ as depicted below.

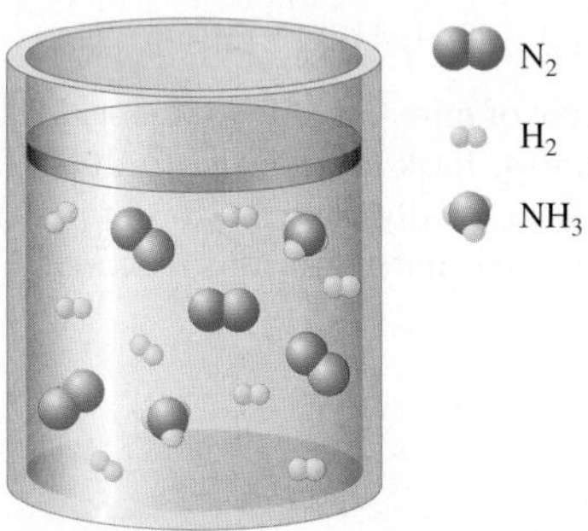

The volume is suddenly decreased (by increasing the external pressure), and a new equilibrium is established as depicted below.

a. If the volume of the final equilibrium mixture is 1.00 L, determine the value of the equilibrium constant K for the reaction. Assume temperature is constant.
b. Determine the volume of the initial equilibrium mixture assuming a final equilibrium volume of 1.00 L and assuming a constant temperature.

85. A 4.72-g sample of methanol (CH_3OH) was placed in an otherwise empty 1.00-L flask and heated to 250.°C to vaporize the methanol. Over time the methanol vapor decomposed by the following reaction:

$$CH_3OH(g) \rightleftharpoons CO(g) + 2H_2(g)$$

After the system has reached equilibrium, a tiny hole is drilled in the side of the flask allowing gaseous compounds to effuse out of the flask. Measurements of the effusing gas show that it contains 33.0 times as much $H_2(g)$ as $CH_3OH(g)$. Calculate K for this reaction at 250.°C.

86. The compound $SbCl_5(g)$ decomposes at high temperatures to gaseous antimony trichloride and chlorine gas. When 89.7 g of $SbCl_5(g)$ is placed in a 15.0-L container at 180°C, the $SbCl_5(g)$ is 29.2% decomposed (by moles) after the system has reached equilibrium.
a. Calculate the value of K for this reaction at 180°C.
b. Determine the number of moles of chlorine gas that must be injected into the flask to make the new equilibrium pressure of antimony trichloride half that of the original equilibrium pressure of antimony trichloride in the original experiment.

87. At 207°C, $K_p = 0.267$ for the reaction

$$PCl_5(g) \rightleftharpoons PCl_3(g) + Cl_2(g)$$

a. If 0.100 mole of $PCl_5(g)$ is placed in an otherwise empty 12.0-L vessel at 207°C, calculate the partial pressures of $PCl_5(g)$, $PCl_3(g)$, and $Cl_2(g)$ at equilibrium.
b. In another experiment the total pressure of an equilibrium mixture is 2.00 atm at 207°C. What mass of PCl_5 was introduced into a 5.00-L vessel to reach this equilibrium position?

88. A 1.604-g sample of methane (CH_4) gas and 6.400 g of oxygen gas are sealed in a 2.50-L vessel at 411°C and are allowed to reach equilibrium. Methane can react with oxygen to form gaseous carbon dioxide and water vapor, or methane can react with oxygen to form gaseous carbon monoxide and water vapor. At equilibrium the pressure of oxygen is 0.326 atm, and the pressure of water vapor is 4.45 atm. Calculate the pressures of carbon monoxide and carbon dioxide present at equilibrium.

89. At 1000 K the $N_2(g)$ and $O_2(g)$ in air (78% N_2, 21% O_2, by moles) react to form a mixture of $NO(g)$ and $NO_2(g)$. The values of the equilibrium constants are 1.5×10^{-4} and 1.0×10^{-5} for the formation of $NO(g)$ and $NO_2(g)$, respectively. At what total pressure will the partial pressures of $NO(g)$ and $NO_2(g)$ be equal in an equilibrium mixture of $N_2(g)$, $O_2(g)$, $NO(g)$, and $NO_2(g)$?

90. The equilibrium constant K_p for the reaction

$$CCl_4(g) \rightleftharpoons C(s) + 2Cl_2(g)$$

at 700°C is 0.76. Determine the initial pressure of carbon tetrachloride that will produce a total equilibrium pressure of 1.20 atm at 700°C.

91. An 8.00-g sample of SO_3 was placed in an evacuated container, where it decomposed at 600.°C according to the following reaction:

$$SO_3(g) \rightleftharpoons SO_2(g) + \tfrac{1}{2}O_2(g)$$

At equilibrium the total pressure and the density of the gaseous mixture were 1.80 atm and 1.60 g/L, respectively. Calculate K_p for this reaction.

92. A sample of iron(II) sulfate was heated in an evacuated container to 920 K, where the following reactions occurred:

$$2FeSO_4(s) \rightleftharpoons Fe_2O_3(s) + SO_3(g) + SO_2(g)$$

$$SO_3(g) \rightleftharpoons SO_2(g) + \tfrac{1}{2}O_2(g)$$

After equilibrium was reached, the total pressure was 0.836 atm, and the partial pressure of oxygen was 0.0275 atm. Calculate K_p for each of the above reactions.

93. At 450°C, $K_p = 6.5 \times 10^{-3}$ for the ammonia synthesis reaction. Assume that a reaction vessel with a movable piston initially contains 3.0 moles of $H_2(g)$ and 1.0 mole of $N_2(g)$. Make a plot to show how the partial pressure of $NH_3(g)$ present at equilibrium varies for the total pressures of 1.0 atm, 10.0 atm, 100. atm, and 1000. atm (assuming that K_p remains constant). [*Note:* Assume these total pressures represent the initial total pressure of $H_2(g)$ plus $N_2(g)$, where $P_{NH_3} = 0$.]

94. A sample of gaseous nitrosyl bromide (NOBr) was placed in a container fitted with a frictionless, massless piston, where it decomposed at 25°C according to the following equation:

$$2NOBr(g) \rightleftharpoons 2NO(g) + Br_2(g)$$

The initial density of the system was recorded as 4.495 g/L. After equilibrium was reached, the density was noted to be 4.086 g/L.

a. Determine the value of the equilibrium constant *K* for the reaction.

b. If Ar(*g*) is added to the system at equilibrium at constant temperature, what will happen to the equilibrium position? What happens to the value of *K*? Explain each answer.

95. A gaseous material XY(*g*) dissociates to some extent to produce X(*g*) and Y(*g*):

$$XY(g) \rightleftharpoons X(g) + Y(g)$$

A 2.00-g sample of XY (molar mass = 165 g/mol) is placed in a container with a movable piston at 25°C. The pressure is held constant at 0.967 atm. As XY begins to dissociate, the piston moves until 35.0 mole percent of the original XY has dissociated and then remains at a constant position. Assuming ideal behavior, calculate the density of the gas in the container after the piston has stopped moving, and determine the value of *K* for this reaction at 25°C.

Marathon Problem

96.* Consider the reaction

$$A(g) + B(g) \rightleftharpoons C(g)$$

for which $K = 1.30 \times 10^2$. Assume that 0.406 mole of C(*g*) is placed in the cylinder represented here. The temperature is 300.0 K, and the barometric pressure on the piston (which is assumed to be massless and frictionless) is constant at 1.00 atm. The original volume [before the 0.406 mole of C(*g*) begins to decompose] is 10.00 L. What is the volume in the cylinder at equilibrium?

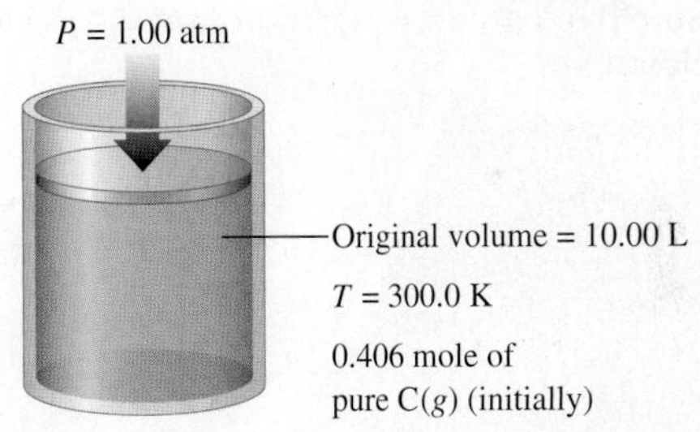

*From James H. Burness, "The Use of "Marathon" Problems as Effective Vehicles for the Presentation of General Chemistry Lectures," Journal of Chemical Education, 68(11). Copyright © 1991 American Chemical Society. Reprinted by permission.

7 Acids and Bases

chapter

A polarized light micrograph of tartaric acid crystals. Tartaric acid is found in grapes as well as other fruits.

Sign in to OWL at www.cengage.com/owl to view tutorials and simulations, develop problem-solving skills, and complete online homework assigned by your professor.

go Chemistry

Download mini lecture videos for key concept review and exam prep from OWL or purchase them from www.cengagebrain.com.

In this chapter we reencounter two very important classes of compounds, acids and bases. We will explore their interactions and apply the fundamentals of chemical equilibria discussed in Chapter 6 to systems involving proton transfer reactions.

Acid–base chemistry is important in a wide variety of everyday applications. There are complex systems in our bodies that carefully control the acidity of our blood, and even small deviations may lead to serious illness and death. The same sensitivity exists in other life forms. If you have ever owned tropical fish or goldfish, you know how important it is to monitor and control the acidity of the water in the aquarium.

Acids and bases are also important industrially. For example, the vast quantity of sulfuric acid manufactured in the United States each year is needed to produce fertilizers, polymers, steel, and many other materials.

The influence of acids on living things has assumed special importance in the United States, Canada, and Europe in recent years as a result of the phenomenon of acid rain. This problem is complex, and its diplomatic and economic overtones make it all the more difficult to solve.

7.1 The Nature of Acids and Bases

Acids were first recognized as a class of substances that taste sour. Vinegar tastes sour because it is a dilute solution of acetic acid; citric acid is responsible for the sour taste of a lemon. Bases, sometimes called *alkalis,* are characterized by their bitter taste and slippery feel. Commercial preparations for unclogging drains are highly basic.

The first person to recognize the essential nature of acids and bases was Svante Arrhenius. Based on his experiments with electrolytes, Arrhenius postulated that *acids produce hydrogen ions in aqueous solution, and bases produce hydroxide ions.* At the time of its discovery, the **Arrhenius concept** of acids and bases was a major step forward in quantifying acid–base chemistry, but this concept is limited because it applies only to aqueous solutions and allows for only one kind of base—the hydroxide ion. A more general definition of acids and bases was suggested independently by Danish chemist Johannes N. Brønsted (1879–1947) and English chemist Thomas M. Lowry (1874–1936) in 1923. In terms of the **Brønsted–Lowry definition,** *an acid is a proton (H^+) donor, and a base is a proton acceptor.* For example, when gaseous HCl dissolves in water, each HCl molecule donates a proton to a water molecule, and so HCl qualifies as a Brønsted–Lowry acid. The molecule that accepts the proton—water in this case—is a Brønsted–Lowry base.

To understand how water can act as a base, we need to recognize that the oxygen of the water molecule has two unshared electron pairs, either of which can form a covalent bond with an H^+ ion. When gaseous HCl dissolves, the following reaction occurs:

$$\text{H—}\ddot{\text{O}}\text{:}(\text{—H}) + \text{H—Cl} \longrightarrow \left[\text{H—}\ddot{\text{O}}\text{—H}(\text{—H})\right]^+ + \text{Cl}^-$$

Note that the proton is transferred from the HCl molecule to the water molecule to form H_3O^+, which is called the **hydronium ion.**

The general reaction that occurs when an acid is dissolved in water can best be represented as

Recall that (*aq*) means the substance is hydrated.

$$\underset{\text{Acid}}{HA(aq)} + \underset{\text{Base}}{H_2O(l)} \rightleftharpoons \underset{\text{Conjugate acid}}{H_3O^+(aq)} + \underset{\text{Conjugate base}}{A^-(aq)} \tag{7.1}$$

This representation emphasizes the significant role of the polar water molecule in pulling the proton from the acid. Note that the **conjugate base** is everything that remains of the acid molecule after a proton is lost. The **conjugate acid** is formed when the proton is transferred to the base. A **conjugate acid–base pair** consists of two substances related to each other by the donating and accepting of a single proton. In Equation (7.1) there are two conjugate acid–base pairs: HA and A^-, and H_2O and H_3O^+.

It is important to note that Equation (7.1) really represents *a competition for the proton between the two bases H_2O and A^-*. If H_2O is a much stronger base than A^-—that is, if H_2O has a much greater affinity for H^+ than A^- does—the equilibrium position will be far to the right; most of the acid dissolved will be in the ionized form. Conversely, if A^- is a much stronger base than H_2O, the equilibrium position will lie far to the left. In this instance most of the acid dissolved will be present at equilibrium as HA.

The equilibrium expression for the reaction given in Equation (7.1) is

$$K_a = \frac{[H_3O^+][A^-]}{[HA]} = \frac{[H^+][A^-]}{[HA]} \tag{7.2}$$

In this chapter we will always represent an acid as simply dissociating. This does not mean we are using the Arrhenius model for acids. Since water does not affect the equilibrium position, we leave it out of the acid dissociation reaction for simplicity.

where K_a is called the **acid dissociation constant**. Both $H_3O^+(aq)$ and $H^+(aq)$ are commonly used to represent the hydrated proton. In this book we will often use simply H^+, but you should remember that it is hydrated in aqueous solutions.

In Chapter 6 we saw that pure solids and liquids are always omitted from the equilibrium expression because they have unit activities. In a dilute solution containing an acid, we can assume that the activity of water is 1. Thus the term $[H_2O]$ is not included in Equation (7.2), and the equilibrium expression for K_a has the same form as that for the simple dissociation

$$HA(aq) \rightleftharpoons H^+(aq) + A^-(aq)$$

You should not forget, however, that water plays an important role in causing the acid to dissociate, as represented below.

Note that K_a is the equilibrium constant for the reaction in which a proton is removed from HA to form the conjugate base A^-. We use K_a to represent *only* this type of reaction. With this information you can write the K_a expression for any acid, even one that is totally unfamiliar to you.

The Brønsted–Lowry definition is not limited to aqueous solutions; it can be extended to reactions in the gas phase. For example, consider the reaction between gaseous hydrogen chloride and ammonia that we discussed when we studied diffusion (Chapter 5):

Lewis structures, which represent the electron arrangements in molecules, will be discussed fully in Chapter 13.

$$NH_3(g) + HCl(g) \rightleftharpoons NH_4Cl(s)$$

In this reaction a proton is donated by the hydrogen chloride to the ammonia, as shown by the following Lewis structures:

$$\text{H}-\underset{\text{H}}{\overset{\text{H}}{\text{N}}}\text{:} \quad \text{H}-\ddot{\underset{..}{\text{Cl}}}\text{:} \rightleftharpoons \left[\text{H}-\underset{\text{H}}{\overset{\text{H}}{\text{N}}}-\text{H}\right]^+ + \left[:\ddot{\underset{..}{\text{Cl}}}:\right]^-$$

Note that this reaction is not considered an acid–base reaction according to the Arrhenius concept.

7.2 Acid Strength

The strength of an acid is defined by the equilibrium position of its dissociation reaction:

$$HA(aq) + H_2O(l) \rightleftharpoons H_3O^+(aq) + A^-(aq)$$

A **strong acid** is one for which *this equilibrium lies far to the right.* This means that almost all the original HA is dissociated at equilibrium [Fig. 7.1(a)]. There is an important connection between the strength of an acid and that of its conjugate base. *A strong acid yields a weak conjugate base*—one that has a low affinity for a proton. A strong acid can also be described as an acid whose conjugate base is a much weaker base than water (Fig. 7.2). In this case the water molecules win the competition for the H^+ ions.

Conversely, a **weak acid** is one for which *the equilibrium lies far to the left.* Most of the acid originally placed in the solution is still present as HA at equilibrium. That is, a weak acid dissociates only to a very small extent in aqueous solution [Fig. 7.1(b)]. In contrast to a strong acid, a weak acid has a conjugate base that is a much stronger base than water. In this case a water molecule is not very successful in pulling an H^+ ion from the conjugate base. *A weak acid yields a relatively strong conjugate base.*

The various ways of describing the strength of an acid are summarized in Table 7.1.

The common strong acids are sulfuric acid [$H_2SO_4(aq)$], hydrochloric acid [$HCl(aq)$], nitric acid [$HNO_3(aq)$], and perchloric acid [$HClO_4(aq)$]. Sulfuric

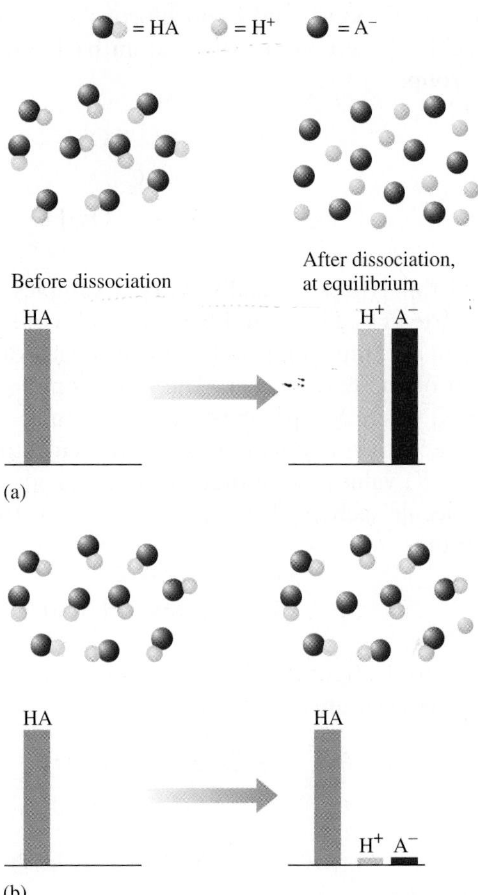

Figure 7.1
Graphical representation of the behavior of acids of different strengths in aqueous solution. (a) A strong acid. (b) A weak acid.

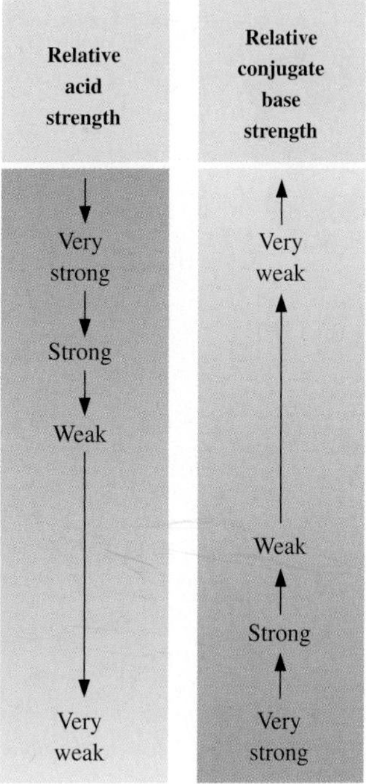

Figure 7.2
The relationship of acid strength and conjugate base strength for the reaction

$$\underset{\text{Acid}}{HA(aq)} + H_2O(l) \rightleftharpoons H_3O^+(aq) + \underset{\text{Conjugate base}}{A^-(aq)}$$

Table 7.1

Various Ways to Describe Acid Strength

Property	Strong Acid	Weak Acid
K_a value	K_a is large	K_a is small
Position of the dissociation equilibrium	Far to the right	Far to the left
Equilibrium concentration of H^+ compared with original concentration of HA	$[H^+] \approx [HA]_0$	$[H^+] \ll [HA]_0$
Strength of conjugate base compared with that of water	A^- much weaker base than H_2O	A^- much stronger base than H_2O

Perchloric acid can explode if handled improperly.

acid is actually a **diprotic acid,** an acid having two acidic protons. The acid H_2SO_4 is a strong acid, virtually 100% dissociated in water:

$$H_2SO_4(aq) \longrightarrow H^+(aq) + HSO_4^-(aq)$$

The HSO_4^- ion is a weak acid:

$$HSO_4^-(aq) \rightleftharpoons H^+(aq) + SO_4^{2-}(aq)$$

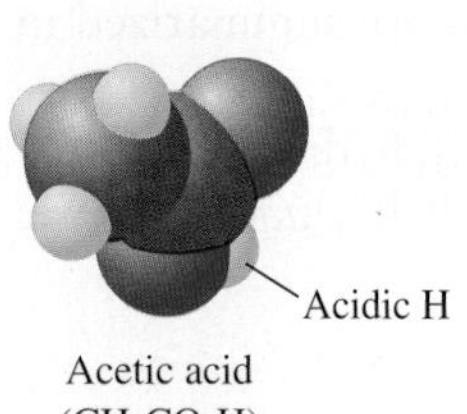

Acetic acid (CH_3CO_2H)

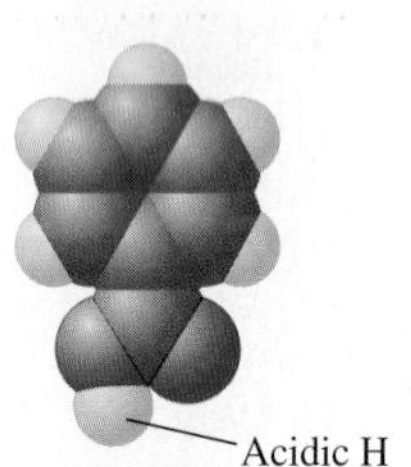

Benzoic acid ($C_6H_5CO_2H$)

Most acids are **oxyacids,** in which the acidic proton is attached to an oxygen atom. The strong acids mentioned above, except hydrochloric acid, are typical examples. Many common weak acids, such as phosphoric acid (H_3PO_4), nitrous acid (HNO_2), and hypochlorous acid (HOCl), are also oxyacids. **Organic acids,** those with a carbon atom backbone, commonly contain the **carboxyl group:**

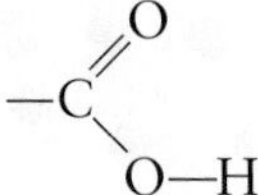

Acids of this type are usually weak. Examples are acetic acid (CH_3COOH), often written $HC_2H_3O_2$, and benzoic acid (C_6H_5COOH).

There are some important acids in which the acidic proton is attached to an atom other than oxygen. The most common of these are the hydrohalic acids HX, where X represents a halogen atom.

Table 7.2 lists common **monoprotic acids** (those having *one* acidic proton) and their K_a values. Note that the strong acids are not listed. When a strong acid molecule such as HCl is placed in water, the position of the dissociation equilibrium

$$HCl(aq) \rightleftharpoons H^+(aq) + Cl^-(aq)$$

lies so far to the right that [HCl] cannot be measured accurately. This situation prevents an accurate calculation of K_a:

$$K_a = \frac{[H^+][Cl^-]}{[HCl]}$$

↑ Very small and highly uncertain

Appendix Table A5.1 contains K_a values.

Table 7.2

Values of K_a for Some Common Monoprotic Acids

Formula	Name	Value of K_a
HSO_4^-	Hydrogen sulfate ion	1.2×10^{-2}
$HClO_2$	Chlorous acid	1.2×10^{-2}
$HC_2H_2ClO_2$	Monochloracetic acid	1.35×10^{-3}
HF	Hydrofluoric acid	7.2×10^{-4}
HNO_2	Nitrous acid	4.0×10^{-4}
$HC_2H_3O_2$	Acetic acid	1.8×10^{-5}
$[Al(H_2O)_6]^{3+}$	Hydrated aluminum(III) ion	1.4×10^{-5}
HOCl	Hypochlorous acid	3.5×10^{-8}
HCN	Hydrocyanic acid	6.2×10^{-10}
NH_4^+	Ammonium ion	5.6×10^{-10}
HOC_6H_5	Phenol	1.6×10^{-10}

Increasing acid strength ↑

OWL INTERACTIVE EXAMPLE 7.1

Using Table 7.2, arrange the following species according to their strength as bases: H_2O, F^-, Cl^-, NO_2^-, and CN^-.

Solution Remember that water is a stronger base than the conjugate base of a strong acid but a weaker base than the conjugate base of a weak acid. This rule leads to the following order:

$$Cl^- < H_2O < \text{conjugate bases of weak acids}$$

Weakest bases ⟶ Strongest bases

We can order the remaining conjugate bases by recognizing that the strength of an acid is *inversely related* to the strength of its conjugate base. From Table 7.2 we have

$$K_a \text{ for HF} > K_a \text{ for } HNO_2 > K_a \text{ for HCN}$$

Thus the base strengths increase as follows:

$$F^- < NO_2^- < CN^-$$

The combined order of increasing base strength is

$$Cl^- < H_2O < F^- < NO_2^- < CN^-$$

Water as an Acid and a Base

A substance is said to be *amphoteric* if it can behave either as an acid or as a base. Water is the most common **amphoteric substance.** We see this behavior in the **autoionization** of water, which involves the transfer of a proton from one water molecule to another to produce a hydroxide ion and a hydronium ion:

$$H{-}\ddot{O}{-}H + H{-}\ddot{O}{-}H \rightleftharpoons \left[H_3O\right]^+ + \left[:\ddot{O}{-}H\right]^-$$

In this reaction one water molecule acts as an acid by furnishing a proton, and the other acts as a base by accepting the proton. This reaction also can be represented as follows:

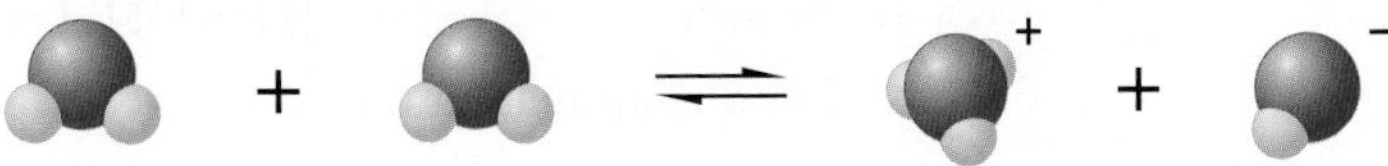

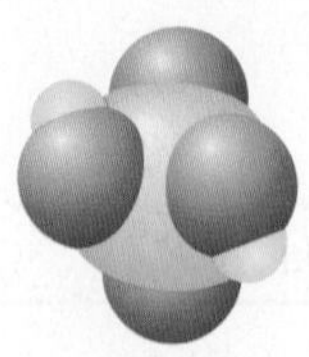
Sulfuric acid (H_2SO_4)

Autoionization can occur in other liquids as well. For example, in liquid ammonia the autoionization reaction is

$$NH_3 + NH_3 \rightleftharpoons [NH_4]^+ + [NH_2]^-$$

The autoionization reaction for water

$$2H_2O(l) \rightleftharpoons H_3O^+(aq) + OH^-(aq)$$

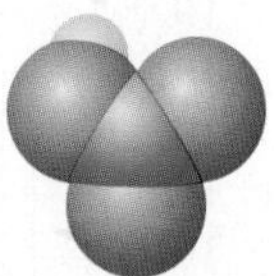
Nitric acid (HNO_3)

leads to the equilibrium expression

$$K_w = [H_3O^+][OH^-] = [H^+][OH^-]$$

where K_w, called the **ion-product constant** (or the *dissociation constant*), always refers to the autoionization of water.

Experiment shows that at 25°C

$$[H^+] = [OH^-] = 1.0 \times 10^{-7} M$$

Perchloric acid ($HClO_4$)

which means that at 25°C

$$K_w = [H^+][OH^-] = (1.0 \times 10^{-7} \text{ mol/L})(1.0 \times 10^{-7} \text{ mol/L})$$
$$= 1.0 \times 10^{-14} \text{ mol}^2/\text{L}^2$$

The units are customarily omitted, for reasons discussed in Section 6.3.

It is important to recognize the meaning of K_w. In any aqueous solution at 25°C, *no matter what it contains,* the product of $[H^+]$ and $[OH^-]$ must always equal 1.0×10^{-14}. There are three possible situations:

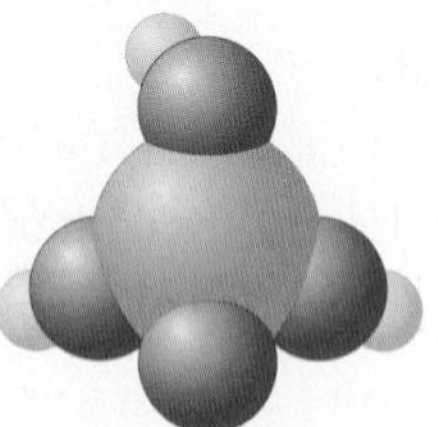
Phosphoric acid (H_3PO_4)

1. A neutral solution, where $[H^+] = [OH^-]$
2. An acidic solution, where $[H^+] > [OH^-]$
3. A basic solution, where $[OH^-] > [H^+]$

In each case, however, at 25°C

$$K_w = [H^+][OH^-] = 1.0 \times 10^{-14}$$

Nitrous acid (HNO_2)

Hypochlorous acid ($HOCl$)

EXAMPLE 7.2

At 60°C the value of K_w is 1×10^{-13}.

a. Using Le Châtelier's principle, predict whether the reaction

$$2H_2O(l) \rightleftharpoons H_3O^+(aq) + OH^-(aq)$$

is exothermic (releases energy) or endothermic (absorbs energy).

b. Calculate $[H^+]$ and $[OH^-]$ in a neutral solution at 60°C.

Solution

a. K_w *increases* from 1×10^{-14} at 25°C to 1×10^{-13} at 60°C. Le Châtelier's principle states that if a system at equilibrium is heated, it will adjust to consume energy. Since the value of K_w increases with temperature, we think of energy as a reactant, and so the process must be endothermic.

b. At 60°C $\quad [H^+][OH^-] = 1 \times 10^{-13}$

For a neutral solution

$$[H^+] = [OH^-] = \sqrt{1 \times 10^{-13}} = 3 \times 10^{-7} M$$

7.3 The pH Scale

Because $[H^+]$ in an aqueous solution is typically quite small, the pH scale provides a convenient way to represent solution acidity. The pH is a log scale based on 10, where

$$pH = -\log[H^+]$$

Thus, for a solution in which

$$[H^+] = 1.0 \times 10^{-7}\ M$$

then

$$pH = -(-7.00) = 7.00$$

The pH scale is a compact way to represent solution acidity. It involves base-10 logs, not natural logs (ln).

The definition of pH in terms of $[H^+]$ neglects any correction for nonideality of the solutions.

At this point we need to discuss significant figures for logarithms. The rule is that *the number of decimal places in the log is equal to the number of significant figures in the original number.* Thus

$$[H^+] = 1.0 \times 10^{-9}\ M \quad \text{(2 significant figures)}$$
$$pH = 9.00 \quad \text{(2 decimal places)}$$

Similar log scales are used for representing other quantities. For example,

$$pOH = -\log[OH^-]$$
$$pK = -\log K$$

Since pH is a log scale based on 10, *the pH changes by 1 for every power of 10 change in $[H^+]$.* For example, a solution of pH 3 has an H^+ concentration 10 times that of a solution of pH 4 and 100 times that of a solution of pH 5. Also note that because pH is defined as $-\log[H^+]$, *the pH decreases as $[H^+]$ increases.* The pH scale and the pH values for several common substances are shown in Fig. 7.3.

The pH of a solution is usually measured by using a pH meter, an electronic device with a probe that is inserted into a solution of unknown pH. The probe contains an acidic aqueous solution enclosed by a special glass membrane that allows migration of H^+ ions. If the unknown solution has a different pH from that of the solution in the probe, an electrical potential results, which is registered on the meter (Fig. 7.4).

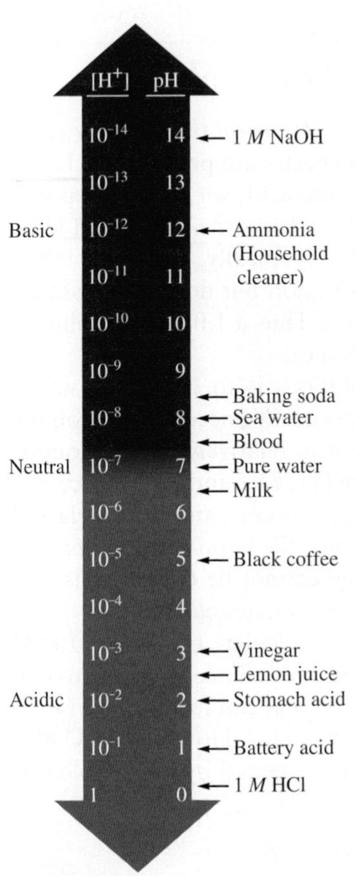

Figure 7.3
The pH scale and pH values of some common substances.

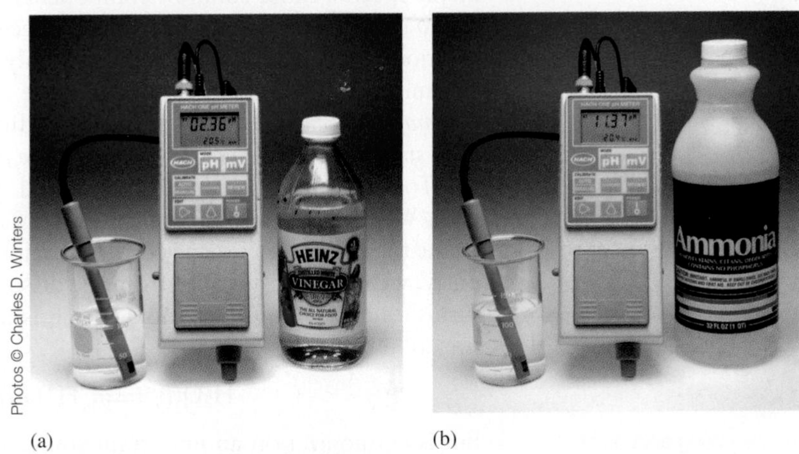

Figure 7.4
(a) Measuring the pH of vinegar. (b) Measuring the pH of aqueous ammonia.

Now that we have considered all the fundamental definitions relevant to acid–base solutions, we can proceed to a quantitative description of the equilibria present in these solutions. The main reason that acid–base problems sometimes seem difficult is that because a typical aqueous solution contains many components, the problems tend to be complicated. However, you can deal with these problems successfully if you use the following general strategies.

General Strategies for Solving Acid–Base Problems

1. *Think chemistry.* Focus on the solution components and their reactions. It will almost always be possible to choose one reaction that is the most important.
2. *Be systematic.* Acid–base problems require a step-by-step approach.
3. *Be flexible.* Although all acid–base problems are similar in many ways, important differences do occur. Treat each problem as a separate entity. Do not try to force a given problem to match any you have solved before. Look for both the similarities and the differences.
4. *Be patient.* The complete solution to a complicated problem cannot be seen immediately in all its detail. Pick the problem apart into workable steps.
5. *Be confident.* Look within the problem for the solution, and let the problem guide you. Assume that you can think it out. Do not rely on memorizing solutions to problems. In fact, memorizing solutions is usually detrimental because you tend to try to force a new problem to be the same as one you have seen before. *Understand and think; don't just memorize.*

7.4 Calculating the pH of Strong Acid Solutions

When we deal with acid–base equilibria, *we must focus on the solution components and their chemistry.* For example, what species are present in a 1.0 *M* solution of HCl? Since hydrochloric acid is a strong acid, we assume that it is completely dissociated. Thus, although the label on the bottle says 1.0 *M* HCl, the solution contains virtually no HCl molecules. Typically, container labels indicate the substance(s) used to make up the solution but do not necessarily describe the solution components after dissolution. Thus a 1.0 *M* HCl solution contains H^+ and Cl^- ions rather than HCl molecules.

The next step in dealing with aqueous solutions is to determine which components are significant and which can be ignored. We need to focus on the **major species,** those solution components present in relatively large amounts. In 1.0 *M* HCl, for example, the major species are H^+, Cl^-, and H_2O. Since this solution is very acidic, OH^- is present only in tiny amounts and thus is classed as a minor species. In attacking acid–base problems, the importance of *writing the major species in the solution* as the first step cannot be overemphasized. *This single step is the key to solving these problems successfully.*

To illustrate the main ideas involved, we will calculate the pH of 1.0 *M* HCl. We first list the major species: H^+, Cl^-, and H_2O. Since we want to calculate the pH, we will focus on those major species that can furnish H^+. Obviously, we must consider H^+ from the dissociation of HCl. However, H_2O also furnishes H^+ by autoionization, which is often represented by the simple dissociation reaction

Always write the major species present in the solution.

$$H_2O(l) \rightleftharpoons H^+(aq) + OH^-(aq)$$

The H^+ from the strong acid drives the equilibrium $H_2O \rightleftharpoons H^+ + OH^-$ to the left.

But is autoionization an important source of H^+ ions? In pure water at 25°C, $[H^+]$ is 10^{-7} *M*. In 1.0 *M* HCl the water will produce even less than 10^{-7} *M* H^+, since by Le Châtelier's principle the H^+ from the dissociated HCl will drive the position of the water equilibrium to the left. Thus the amount of H^+ contributed by water is negligible compared with the 1.0 *M* H^+ from the dissociation of HCl. Therefore we can say that $[H^+]$ in the solution is 1.0 *M* and that

$$pH = -\log[H^+] = -\log(1.0) = 0.00$$

7.5 Calculating the pH of Weak Acid Solutions

$HF + H_2O \rightarrow H_3O^+ + F^-$

Since a weak acid dissolved in water can be viewed as a prototype of almost any equilibrium occurring in aqueous solution, we will proceed carefully and systematically. Although some of the procedures we develop here may seem superfluous, they will become essential as the problems become more complicated. We will develop the necessary strategies by calculating the pH of a 1.00 *M* solution of HF ($K_a = 7.2 \times 10^{-4}$).

First, as always, write the major species present in the solution.

The first step, as always, is to *write the major species in the solution.* From its small K_a value, we know that hydrofluoric acid is a weak acid and will be dissociated only to a slight extent. Thus when we write the major species, the hydrofluoric acid will be represented in its dominant form, as HF. The major species in solution are HF and H_2O.

Major Species

 HF

 H_2O

The next step is to decide which of the major species can furnish H^+ ions. Actually, both major species can do so:

$$HF(aq) \rightleftharpoons H^+(aq) + F^-(aq) \qquad K_a = 7.2 \times 10^{-4}$$

$$H_2O(l) \rightleftharpoons H^+(aq) + OH^-(aq) \qquad K_w = 1.0 \times 10^{-14}$$

In aqueous solutions typically one source of H^+ can be singled out as dominant. By comparing K_a for HF with K_w for H_2O, we see that hydrofluoric acid, although weak, is still a much stronger acid than water. Thus we will assume that hydrofluoric acid will be the dominant source of H^+. We will ignore the tiny contribution expected from water.

Just as with gas-phase equilibria, the equilibrium constant expressions for reactions in solution actually require activities. In all our calculations, we will assume that $a_x = [x]$, where *x* represents a species in solution.

Therefore, it is the dissociation of HF that will determine the equilibrium concentration of H^+ and hence the pH:

$$HF(aq) \rightleftharpoons H^+(aq) + F^-(aq)$$

The equilibrium expression is

$$K_a = 7.2 \times 10^{-4} = \frac{[H^+][F^-]}{[HF]}$$

To solve the equilibrium problem, we follow the procedures developed in Chapter 6 for gas-phase equilibria. First, we list the initial concentrations, the *concentrations before the reaction of interest has proceeded to equilibrium.* Before any HF dissociates, the concentrations of the species in the equilibrium are

$$[HF]_0 = 1.00\ M, \qquad [F^-]_0 = 0, \qquad \text{and} \qquad [H^+]_0 = 10^{-7}\ M \approx 0$$

(Note that the zero value for $[H^+]_0$ is an approximation, since we are neglecting the H^+ ions from the autoionization of water.)

The next step is to determine the change required to reach equilibrium. Since some HF will dissociate to come to equilibrium (but that amount is presently unknown), we let *x* be the change in the concentration of HF that is required to achieve equilibrium. That is, we assume that *x* moles per liter of HF will dissociate to produce *x* moles per liter of H^+ and *x* moles per liter of F^- as the system adjusts to its equilibrium position. Now the equilibrium concentrations can be defined in terms of *x*:

$$[HF] = [HF]_0 - x = 1.00 - x$$

$$[F^-] = [F^-]_0 + x = 0 + x = x$$

$$[H^+] = [H^+]_0 + x \approx 0 + x = x$$

Substituting these equilibrium concentrations into the equilibrium expression gives

$$K_a = 7.2 \times 10^{-4} = \frac{[H^+][F^-]}{[HF]} = \frac{(x)(x)}{1.00 - x}$$

This expression produces a quadratic equation that can be solved by using the quadratic formula, as for the gas-phase systems in Chapter 6. However, since K_a for HF is so small, HF will dissociate only slightly; thus x is expected to be small. This will allow us to simplify the calculation. If x is very small compared with 1.00, the term in the denominator can be approximated as follows:

$$1.00 - x \approx 1.00$$

The equilibrium expression then becomes

$$7.2 \times 10^{-4} = \frac{(x)(x)}{1.00 - x} \approx \frac{(x)(x)}{1.00}$$

which yields

$$x^2 \approx (7.2 \times 10^{-4})(1.00) = 7.2 \times 10^{-4}$$

$$x \approx \sqrt{7.2 \times 10^{-4}} = 2.7 \times 10^{-2}$$

The validity of an approximation should always be checked.

How valid is the approximation that $[HF] = 1.00\ M$? Because this question will arise often in connection with acid–base equilibrium calculations, we will consider it carefully. *The validity of the approximation depends on how much accuracy we demand for the calculated value of* $[H^+]$. Typically, the K_a values for acids are known to an accuracy of only about ±5%. Therefore, it is reasonable to apply this figure when determining the validity of the approximation

Note that although the 5% rule is an arbitrary choice, it makes sense because of the typical uncertainty in K_a values. Be aware that the precision of the data for a particular situation should be used to evaluate which approximations are reasonable.

$$[HA]_0 - x \approx [HA]_0$$

We will use the following test.

First calculate the value of x by making the approximation

$$K_a = \frac{x^2}{[HA]_0 - x} \approx \frac{x^2}{[HA]_0}$$

where

$$x^2 \approx K_a[HA]_0 \quad \text{and} \quad x \approx \sqrt{K_a[HA]_0}$$

Then compare the sizes of x and $[HA]_0$. If the expression

$$\frac{x}{[HA]_0} \times 100\%$$

is less than or equal to 5%, the value of x is small enough for the approximation

$$[HA]_0 - x \approx [HA]_0$$

to be considered valid.

In our example

$$x = 2.7 \times 10^{-2}\ \text{mol/L}$$

$$[HA]_0 = [HF]_0 = 1.00\ \text{mol/L}$$

and

$$\frac{x}{[HA]_0} \times 100\% = \frac{2.7 \times 10^{-2}}{1.00} \times 100\% = 2.7\%$$

The approximation we made is considered valid, so the value of x calculated by using that approximation is acceptable. Thus

$$x = [H^+] = 2.7 \times 10^{-2}\ M \qquad \text{and} \qquad pH = -\log(2.7 \times 10^{-2}) = 1.57$$

This problem illustrates all the important steps required for solving a typical equilibrium problem involving a weak acid. These steps are summarized below.

STEPS

Solving Weak Acid Equilibrium Problems

1 List the major species in the solution.

2 Choose the species that can produce H^+, and write balanced equations for the reactions producing H^+.

3 Comparing the values of the equilibrium constants for the reactions you have written, decide which reaction will dominate in the production of H^+.

4 Write the equilibrium expression for the dominant reaction.

5 List the initial concentrations of the species participating in the dominant reaction.

6 Define the change needed to achieve equilibrium; that is, define x.

7 Write the equilibrium concentrations in terms of x.

8 Substitute the equilibrium concentrations into the equilibrium expression.

9 Solve for x the "easy" way—that is, by assuming that $[HA]_0 - x \approx [HA]_0$.

10 Verify whether the approximation is valid (the 5% rule is the test in this case).

11 Calculate $[H^+]$ and pH.

The pH of a Mixture of Weak Acids

A table of K_a values for various weak acids is given in Appendix Table A5.1.

Sometimes a solution contains two weak acids of very different strengths. This situation is considered in Example 7.3. Note that the usual steps are followed (although not labeled).

EXAMPLE 7.3

Calculate the pH of a solution that contains 1.00 M HCN ($K_a = 6.2 \times 10^{-10}$) and 5.00 M HNO_2 ($K_a = 4.0 \times 10^{-4}$). Also calculate the concentration of cyanide ion (CN^-) in this solution at equilibrium.

Major Species

HCN

HNO_2

H_2O

Solution Since HCN and HNO_2 are both weak acids and are thus primarily undissociated, the major species in the solution are

$$HCN, \qquad HNO_2, \qquad \text{and} \qquad H_2O$$

All three components produce H^+:

$$HCN(aq) \rightleftharpoons H^+(aq) + CN^-(aq) \qquad K_a = 6.2 \times 10^{-10}$$

$$HNO_2(aq) \rightleftharpoons H^+(aq) + NO_2^-(aq) \qquad K_a = 4.0 \times 10^{-4}$$

$$H_2O(l) \rightleftharpoons H^+(aq) + OH^-(aq) \qquad K_w = 1.0 \times 10^{-14}$$

A mixture of three acids might lead to a very complicated problem. However, the situation is greatly simplified by the fact that even though HNO_2 is a weak acid, it is much stronger than the other two acids present (as revealed by the K values). Thus HNO_2 can be assumed to be the dominant producer of H^+, so we will focus on the equilibrium expression:

$$K_a = 4.0 \times 10^{-4} = \frac{[H^+][NO_2^-]}{[HNO_2]}$$

The initial concentrations, the definition of x, and the equilibrium concentrations are as follows:

Initial Concentration (mol/L)		Equilibrium Concentration (mol/L)
$[HNO_2]_0 = 5.00$ $[NO_2^-]_0 = 0$ $[H^+]_0 \approx 0$	$\xrightarrow[\text{dissociates}]{x \text{ mol/L } HNO_2}$	$[HNO_2] = 5.00 - x$ $[NO_2^-] = x$ $[H^+] = x$

It is often convenient to represent these concentrations in the following shorthand form:

	$HNO_2(aq)$	$\rightleftharpoons$	$H^+(aq)$	+	$NO_2^-(aq)$
Initial:	5.00		0		0
Change:	$-x$		$+x$		$+x$
Equilibrium:	$5.00 - x$		x		x

Substituting the equilibrium concentrations into the equilibrium expression and making the approximation that $5.00 - x = 5.00$ gives

$$K_a = 4.0 \times 10^{-4} = \frac{(x)(x)}{5.00 - x} \approx \frac{x^2}{5.00}$$

We solve for x: $$x = 4.5 \times 10^{-2}$$

Using the 5% rule we show that the approximation is valid:

$$\frac{x}{[HNO_2]_0} \times 100\% = \frac{4.5 \times 10^{-2}}{5.00} \times 100\% = 0.90\%$$

Therefore,

$$[H^+] = x = 4.5 \times 10^{-2}\ M \quad \text{and} \quad \text{pH} = 1.35$$

We also want to calculate the equilibrium concentration of cyanide ion in this solution. The CN^- ions in this solution come from the dissociation of HCN:

$$HCN(aq) \rightleftharpoons H^+(aq) + CN^-(aq)$$

Although the position of this equilibrium lies far to the left and does not contribute *significantly* to $[H^+]$, HCN is the *only source* of CN^-. Thus we must consider the extent of the dissociation of HCN to calculate $[CN^-]$. The equilibrium expression for the preceding reaction is

$$K_a = 6.2 \times 10^{-10} = \frac{[H^+][CN^-]}{[HCN]}$$

We know $[H^+]$ for this solution from the results for the first part of this problem. Note that *there is only one kind of* H^+ *in this solution.* It does not matter from which acid the H^+ ions originate. The equilibrium value of $[H^+]$ for the HCN dissociation is $4.5 \times 10^{-2}\ M$, even though the H^+ was contributed almost entirely by the dissociation of HNO_2. What is [HCN] at equilibrium? We know $[HCN]_0 = 1.00\ M$, and since K_a for HCN is so small, a negligible amount of HCN will dissociate.

Thus
$$[HCN] = [HCN]_0 - \text{amount of HCN dissociated}$$
$$\approx [HCN]_0 = 1.00\ M$$

Since $[H^+]$ and [HCN] are known, we can find $[CN^-]$ from the equilibrium expression:

$$K_a = 6.2 \times 10^{-10} = \frac{[H^+][CN^-]}{[HCN]} = \frac{(4.5 \times 10^{-2})[CN^-]}{1.00}$$

$$[CN^-] = \frac{(6.2 \times 10^{-10})(1.00)}{4.5 \times 10^{-2}} = 1.4 \times 10^{-8}\ M$$

Note the significance of this result. Since $[CN^-] = 1.4 \times 10^{-8}\ M$, and since HCN is the only source of CN^-, only 1.4×10^{-8} mol/L of HCN has dissociated. This is a very small amount compared with the initial concentration of HCN, which is exactly what we would expect from its very small K_a value. Thus $[HCN] = 1.00\ M$ as assumed. Also, this result confirms that HNO_2 is the only significant source of H^+.

Percent Dissociation

It is often useful to specify the amount of weak acid that has dissociated in achieving equilibrium in an aqueous solution. The **percent dissociation** is defined as follows:

$$\text{Percent dissociation} = \frac{\text{amount dissociated (mol/L)}}{\text{initial concentration (mol/L)}} \times 100\% \qquad (7.3)$$

For example, we found earlier that in a 1.00 *M* solution of HF, $[H^+] = 2.7 \times 10^{-2}\ M$. For the system to reach equilibrium, 2.7×10^{-2} moles per liter of the original 1.00 *M* HF dissociates, so

$$\text{Percent dissociation} = \frac{2.7 \times 10^{-2}\ \text{mol/L}}{1.00\ \text{mol/L}} \times 100\% = 2.7\%$$

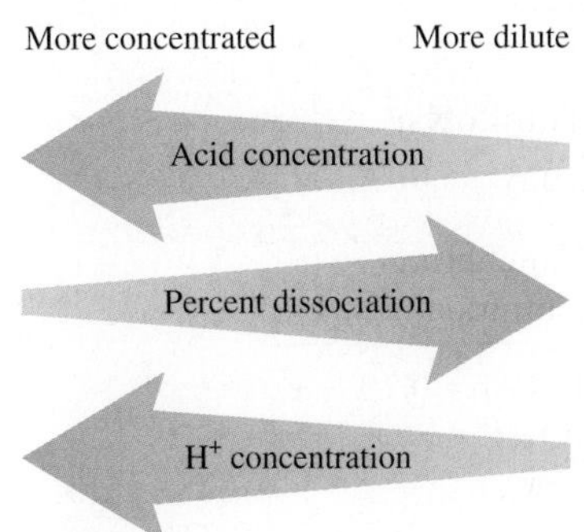

Figure 7.5
The effect of dilution on the percent dissociation and $[H^+]$ of a weak acid solution.

For a given weak acid, the percent dissociation increases as the acid becomes more dilute. For example, the percent dissociation of acetic acid ($HC_2H_3O_2$, $K_a = 1.8 \times 10^{-5}$) is significantly greater in a 0.10 *M* solution than in a 1.0 *M* solution.

Demonstrate for yourself (by doing the calculations) that even though the concentration of H^+ ion at equilibrium is smaller in the 0.10 *M* acetic acid solution than in the 1.0 *M* acetic acid solution, the percent dissociation is significantly greater in the 0.10 *M* solution (1.3%) than in the 1.0 *M* solution (0.42%). This is a general result. *For solutions of any weak acid HA, $[H^+]$ decreases as $[HA]_0$ decreases, but the percent dissociation increases as $[HA]_0$ decreases.*

This phenomenon can be explained in the following way. Consider the weak acid HA with the initial concentration $[HA]_0$. At equilibrium

$$[HA] = [HA]_0 - x \approx [HA]_0$$
$$[H^+] = [A^-] = x$$

Thus
$$K_a = \frac{[H^+][A^-]}{[HA]} \approx \frac{(x)(x)}{[HA_0]}$$

The more dilute the weak acid solution, the greater the percent dissociation.

Now suppose enough water is added to dilute the solution by a factor of 10. The new concentrations before any adjustment occurs are

$$[A^-]_{new} = [H^+]_{new} = \frac{x}{10}$$

$$[HA]_{new} = \frac{[HA]_0}{10}$$

and Q, the reaction quotient, is

$$Q = \frac{(x/10)(x/10)}{[HA]_0/10} = \frac{1(x)(x)}{10[HA]_0} = \frac{1}{10}K_a$$

Since Q is less than K_a, the system must adjust to the right to reach the new equilibrium position. Thus the percent dissociation increases as the acid becomes more dilute. This behavior is summarized in Fig. 7.5. In Example 7.4 we see how the percent dissociation can be used to calculate the K_a value for a weak acid.

Patrick Ward/Corbis

A runner struggles to the top of a hill during a cross-country race in the hills near Wasdale Head, England.

OWL INTERACTIVE EXAMPLE 7.4

Lactic acid ($HC_3H_5O_3$) is a waste product that accumulates in muscle tissue during exertion, leading to pain and a feeling of fatigue. In a 0.100 *M* aqueous solution, lactic acid is 3.7% dissociated. Calculate the value of K_a for this acid.

Solution The small value for the percent dissociation clearly indicates that $HC_3H_5O_3$ is a weak acid. Thus the major species in the solution are the undissociated acid and water:

Major Species

$HC_3H_5O_3$

H_2O

$$HC_3H_5O_3 \quad \text{and} \quad H_2O$$

Although $HC_3H_5O_3$ is a weak acid, it is much stronger than water and thus will be the dominant source of H^+ in the solution. The dissociation reaction is

$$HC_3H_5O_3(aq) \rightleftharpoons H^+(aq) + C_3H_5O_3^-(aq)$$

and the equilibrium expression is

$$K_a = \frac{[H^+][C_3H_5O_3^-]}{[HC_3H_5O_3]}$$

The initial and equilibrium concentrations are as follows:

Initial Concentration (mol/L)		Equilibrium Concentration (mol/L)
$[HC_3H_5O_3]_0 = 0.10$ $[C_3H_5O_3^-]_0 = 0$ $[H^+]_0 \approx 0$	$\xrightarrow[\text{HC}_3\text{H}_5\text{O}_3\text{ dissociates}]{x\text{ mol/L}}$	$[HC_3H_5O_3] = 0.10 - x$ $[C_3H_5O_3^-] = x$ $[H^+] = x$

The change needed to reach equilibrium can be obtained from the percent dissociation and Equation (7.3). For this acid

$$\text{Percent dissociation} = 3.7\% = \frac{x}{[HC_3H_5O_3]_0} \times 100\% = \frac{x}{0.10} \times 100\%$$

and

$$x = \frac{3.7}{100}(0.10) = 3.7 \times 10^{-3} \text{ mol/L}$$

Now we can calculate the equilibrium concentrations:

$$[HC_3H_5O_3] = 0.10 - x = 0.10\ M \quad \text{(to the correct number of significant figures)}$$

$$[C_3H_5O_3^-] = [H^+] = x = 3.7 \times 10^{-3}\ M$$

These concentrations can now be used to calculate the value of K_a for lactic acid:

$$K_a = \frac{[H^+][C_3H_5O_3^-]}{[HC_3H_5O_3]} = \frac{(3.7 \times 10^{-3})(3.7 \times 10^{-3})}{0.10} = 1.4 \times 10^{-4}$$

7.6 Bases

In a basic solution pH > 7.

According to the Arrhenius concept, a base is a substance that produces OH^- ions in aqueous solution. According to the Brønsted–Lowry definition, a base is a proton acceptor. The bases sodium hydroxide (NaOH) and potassium hydroxide (KOH) fulfill both criteria. They contain OH^- ions in the solid lattice and behave as strong electrolytes, dissociating completely when dissolving in water:

$$NaOH(s) \xrightarrow{H_2O} Na^+(aq) + OH^-(aq)$$

Thus a 1.0 M NaOH solution actually contains 1.0 M Na^+ and 1.0 M OH^-. Because of their complete dissociation, NaOH and KOH are called **strong bases** in the same sense as we defined strong acids.

All the hydroxides of the Group 1A elements (LiOH, NaOH, KOH, RbOH, and CsOH) are strong bases, but only NaOH and KOH are common laboratory reagents because the lithium, rubidium, and cesium compounds are expensive. The alkaline earth (Group 2A) hydroxides—$Ca(OH)_2$, $Ba(OH)_2$, and $Sr(OH)_2$—are also strong bases. For these compounds, 2 moles of hydroxide ion are produced for every 1 mole of metal hydroxide dissolved in aqueous solution.

The alkaline earth hydroxides are not very soluble and are used only when the solubility factor is not important. In fact, the low solubility of these bases

Rick Poley/Visuals Unlimited

Tanks in Miami, Florida, used to soften, filter, and disinfect the public water supply.

can be an advantage. For example, many antacids are suspensions of metal hydroxides such as aluminum hydroxide and magnesium hydroxide. The low solubility of these compounds prevents the formation of a large hydroxide ion concentration that would harm the tissues of the mouth, esophagus, and stomach. Yet these suspensions furnish plenty of hydroxide ion to react with stomach acid, since the salts dissolve as this reaction proceeds.

Calcium hydroxide [$Ca(OH)_2$], often called **slaked lime,** is widely used in industry because it is inexpensive and plentiful. For example, slaked lime is used in scrubbing stack gases to remove sulfur dioxide from the exhaust of power plants and factories. In the scrubbing process, a suspension of slaked lime is sprayed into the stack gases to react with sulfur dioxide gas according to the following equations:

$$SO_2(g) + H_2O(l) \rightleftharpoons H_2SO_3(aq)$$

$$Ca(OH)_2(aq) + H_2SO_3(aq) \rightleftharpoons CaSO_3(s) + 2H_2O(l)$$

Hard water contains Ca^{2+} and Mg^{2+} ions, among others, which are detrimental to detergent action.

Slaked lime is also widely used in water treatment plants for softening hard water, which involves the removal of ions such as Ca^{2+} and Mg^{2+}, ions that hamper the action of detergents. The softening method most often used in water treatment plants is the **lime–soda process,** in which *lime* (CaO) and *soda ash* (Na_2CO_3) are added to the water. As we will see in more detail later in this chapter, the CO_3^{2-} ion from soda ash reacts with water to produce the HCO_3^- ion. When lime is added to hard water, it forms slaked lime,

$$CaO(s) + H_2O(l) \longrightarrow Ca(OH)_2(aq)$$

which then reacts with the HCO_3^- ion and a Ca^{2+} ion to produce calcium carbonate:

$$Ca(OH)_2(aq) + \underset{\text{From hard water}}{\underbrace{Ca^{2+}(aq)}} + 2HCO_3^-(aq) \longrightarrow 2CaCO_3(s) + 2H_2O(l)$$

Thus, for every mole of $Ca(OH)_2$ consumed, 1 mole of Ca^{2+} is removed from the hard water, thereby softening it. Some hard water naturally contains bicarbonate ions. In this case no soda ash is needed—simply adding lime accomplishes the softening.

Calculating the pH of a strong base solution is relatively simple, as illustrated in Example 7.5.

OWL INTERACTIVE EXAMPLE 7.5

Calculate the pH of a 5.0×10^{-2} *M* NaOH solution.

Major Species

Na^+

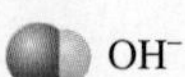
OH^-

H_2O

Solution The major species in this solution are

$$\underbrace{Na^+, \quad OH^-,}_{\text{From NaOH}} \quad \text{and} \quad H_2O$$

Although the autoionization of water also produces OH^- ions, the pH will be determined by the OH^- ions from the dissolved NaOH. Thus in the solution

$$[OH^-] = 5.0 \times 10^{-2}\ M$$

The concentration of H^+ can be calculated from K_w:

$$[H^+] = \frac{K_w}{[OH^-]} = \frac{1.0 \times 10^{-14}}{5.0 \times 10^{-2}} = 2.0 \times 10^{-13}\ M$$

$$pH = 12.70$$

Note that this solution is basic:

$$[OH^-] > [H^+] \qquad \text{and} \qquad pH > 7$$

The added OH^- from the salt has shifted the water autoionization equilibrium

$$H_2O(l) \rightleftharpoons H^+(aq) + OH^-(aq)$$

to the left, significantly lowering the $[H^+]$ compared with that in pure water.

A base does not have to contain the hydroxide ion.

Many types of proton acceptors (bases) do not contain the hydroxide ion. However, when dissolved in water, these substances increase the concentration of hydroxide ion by reacting with water. For example, ammonia reacts with water as follows:

$$NH_3(aq) + H_2O(l) \rightleftharpoons NH_4^+(aq) + OH^-(aq)$$

The ammonia molecule accepts a proton and thus functions as a base. Water is the acid in this reaction. Note that even though the base ammonia contains no hydroxide ion, it still increases the concentration of hydroxide ion to yield a basic solution.

Bases like ammonia typically have at least one unshared pair of electrons that is capable of forming a bond with a proton. The reaction of an ammonia molecule with a water molecule can be represented as follows:

$$H_3N{:} \; + \; H{-}\ddot{O}{:}{-}H \rightleftharpoons H{-}N^+(H)_2{-}H \; + \; {:}\ddot{O}{-}H^-$$

There are many bases like ammonia that produce hydroxide ions by reaction with water. In most of these bases, the lone pair is located on a nitrogen atom. Some examples are

$H_3C-\ddot{N}H_2$	$(H_3C)_2\ddot{N}H$	$(H_3C)_3\ddot{N}$	$C_2H_5\ddot{N}H_2$	$C_5H_5\ddot{N}$
Methylamine	Dimethylamine	Trimethylamine	Ethylamine	Pyridine

Note that the first four bases can be thought of as substituted ammonia molecules where hydrogen atoms are replaced by methyl (CH_3) or ethyl (C_2H_5) groups. The pyridine molecule is like benzene

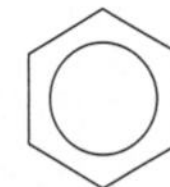

except that a nitrogen atom has replaced one of the carbon atoms in the ring.

Benzene (C_6H_6) is often represented by the symbol

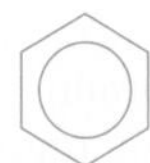

where each vertex represents a carbon atom. The hydrogen atoms are not shown.

The general reaction between a base (B) and water is given by

$$\underset{\text{Base}}{B(aq)} + \underset{\text{Acid}}{H_2O(l)} \rightleftharpoons \underset{\text{Conjugate acid}}{BH^+(aq)} + \underset{\text{Conjugate base}}{OH^-(aq)} \tag{7.4}$$

Appendix Table A5.3 contains K_b values.

The equilibrium reaction for this general solution is

$$K_b = \frac{[BH^+][OH^-]}{[B]}$$

where K_b *always refers to the reaction of a base with water to form the conjugate acid and the hydroxide ion.*

Table 7.3

Values of K_b for Some Common Weak Bases

Name	Formula	Conjugate Acid	K_b
Ammonia	NH_3	NH_4^+	1.8×10^{-5}
Methylamine	CH_3NH_2	$CH_3NH_3^+$	4.38×10^{-4}
Ethylamine	$C_2H_5NH_2$	$C_2H_5NH_3^+$	5.6×10^{-4}
Aniline	$C_6H_5NH_2$	$C_6H_5NH_3^+$	3.8×10^{-10}
Pyridine	C_5H_5N	$C_5H_5NH^+$	1.7×10^{-9}

Bases of the type represented by B in Equation (7.4) compete with OH^-, a very strong base, for the H^+ ion. Thus their K_b values tend to be small (for example, for ammonia $K_b = 1.8 \times 10^{-5}$), and they are called **weak bases.** The K_b values for some common weak bases are listed in Table 7.3.

Typically, pH calculations for solutions of weak bases are very similar to those for weak acids.

OWL INTERACTIVE EXAMPLE 7.6

Calculate the pH of a 1.0 M solution of methylamine ($K_b = 4.38 \times 10^{-4}$).

Major Species

CH_3NH_2

H_2O

Solution Since methylamine (CH_3NH_2) is a weak base, the major species in solution are

$$CH_3NH_2 \quad \text{and} \quad H_2O$$

Both are bases; however, since water can be neglected as a source of OH^-, the dominant equilibrium is

$$CH_3NH_2(aq) + H_2O(l) \rightleftharpoons CH_3NH_3^+(aq) + OH^-(aq)$$

and

$$K_b = 4.38 \times 10^{-4} = \frac{[CH_3NH_3^+][OH^-]}{[CH_3NH_2]}$$

The concentrations are as follows:

Initial Concentration (mol/L)		Equilibrium Concentration (mol/L)
$[CH_3NH_2]_0 = 1.0$	x mol/L CH_3NH_2 reacts with H_2O	$[CH_3NH_2] = 1.0 - x$
$[CH_3NH_3^+]_0 = 0$	$\longrightarrow$	$[CH_3NH_3^+] = x$
$[OH^-]_0 \approx 0$	to reach equilibrium	$[OH^-] = x$

Or in shorthand form:

	$CH_3NH_2(aq)$	+	$H_2O(l)$	$\rightleftharpoons$	$CH_3NH_3^+(aq)$	+	$OH^-(aq)$
Initial:	1.0		—		0		0
Change:	$-x$		—		$+x$		$+x$
Equilibrium:	$1.0 - x$		—		x		x

Amines

Chemical Insights

We have seen that many bases have nitrogen atoms with one lone pair of electrons. These bases can be viewed as substituted ammonia molecules, with the general formula $R_xNH_{(3-x)}$. Compounds of this type are called **amines.**

Amines are widely distributed in animals and plants, often serving as messengers or regulators. For example, in the human nervous system there are two amine stimulants, *norepinephrine* and *adrenaline:*

HO, HO on benzene ring; $CHCH_2NH_2$ with OH

Norepinephrine

HO, HO on benzene ring; $CHCH_2NHCH_3$ with OH

Adrenaline

Ephedrine, widely used as a decongestant, was a known drug in China over 2000 years ago. People from cultures in Mexico and the Southwest have used the hallucinogen *mescaline,* extracted from peyote cactus, for centuries.

A peyote cactus in bloom.

R. Konig/Jacana/Photo Researchers

Ephedrine: phenyl–C(H)(O–H)–C(H)(CH$_3$)–N(H)–CH$_3$

Ephedrine

$CH_2CH_2NH_2$ on benzene ring with three $O{-}CH_3$ groups

Mescaline

Many other drugs, such as codeine and quinine, are amines, but they are rarely used in their pure amine forms. Instead, they are treated with an acid to become acid salts. An example of an acid salt is ammonium chloride, obtained by the reaction

$$NH_3 + HCl \longrightarrow NH_4Cl$$

Amines can also be protonated in this way. The resulting acid salt, written as AHCl (where A represents the amine), contains AH^+ and Cl^-. In general, the acid salts are more stable and more soluble in water than the parent amines. For instance, the parent amine of the well-known local anaesthetic *novocaine* is water-insoluble, but the acid salt is much more soluble.

H_2N–C$_6$H$_4$–C(=O)–O–CH$_2$–CH$_2$–N$^+$H(CH_2CH_3)$_2$ Cl^-

Novocaine hydrochloride

Substituting the equilibrium concentrations into the equilibrium expression and making the usual approximation gives

$$K_b = 4.38 \times 10^{-4} = \frac{[CH_3NH_3^+][OH^-]}{[CH_3NH_2]} = \frac{(x)(x)}{1.0 - x} \approx \frac{x^2}{1.0}$$

$$x \approx 2.1 \times 10^{-2}$$

The approximation is valid by the 5% rule, so

$$[OH^-] = x = 2.1 \times 10^{-2}\ M \text{ and pOH} = 1.68$$

Note that since $[H^+][OH^-] = 1.0 \times 10^{-14}$, pH + pOH = 14. Thus pH = 14.00 − 1.68 = 12.32.

7.7 Polyprotic Acids

Some important acids, such as sulfuric acid (H_2SO_4) and phosphoric acid (H_3PO_4), can furnish more than one proton per molecule and are called **polyprotic acids.** A polyprotic acid always dissociates in a *stepwise* manner, one proton at a time.

Carbonic Acid

The diprotic (two-proton) *carbonic acid* (H_2CO_3), which is vital for maintaining a constant pH in human blood, dissociates in the following steps:

$$H_2CO_3(aq) \rightleftharpoons H^+(aq) + HCO_3^-(aq) \qquad K_{a_1} = \frac{[H^+][HCO_3^-]}{[H_2CO_3]} = 4.3 \times 10^{-7}$$

$$HCO_3^-(aq) \rightleftharpoons H^+(aq) + CO_3^{2-}(aq) \qquad K_{a_2} = \frac{[H^+][CO_3^{2-}]}{[HCO_3^-]} = 4.8 \times 10^{-11}$$

The successive K_a values for the dissociation equilibria are designated K_{a_1} and K_{a_2}. Note that the conjugate base HCO_3^- of the first dissociation equilibrium becomes the acid in the second step.

Carbonic acid is formed when carbon dioxide gas is dissolved in water. In fact, the first dissociation step for carbonic acid is best represented by the reaction

$$CO_2(aq) + H_2O(l) \rightleftharpoons H^+(aq) + HCO_3^-(aq)$$

since relatively little H_2CO_3 actually exists in solution. However, it is convenient to consider CO_2 in water as H_2CO_3 so that we can treat such solutions by using the familiar dissociation reactions for weak acids.

Any solution of carbonic acid contains various amounts of all three species—$H_2CO_3(CO_2, H_2O)$, HCO_3^-, and CO_3^{2-}—depending on the pH of the solution. At a given pH, the fractions of the various species can be calculated from the dissociation equilibrium constants as shown in Example 7.7.

OWL INTERACTIVE EXAMPLE 7.7

Calculate the fractions of H_2CO_3, HCO_3^-, and CO_3^{2-} at pH 9.00.

Solution The fraction of each species present is the concentration of that species divided by the total concentrations of all three species. For example, for HCO_3^- we have

$$\text{Fraction } HCO_3^- = \frac{[HCO_3^-]}{[H_2CO_3] + [HCO_3^-] + [CO_3^{2-}]} = f_{HCO_3^-}$$

Dividing the numerator and denominator by $[HCO_3^-]$ gives

$$f_{HCO_3^-} = \frac{1}{\dfrac{[H_2CO_3]}{[HCO_3^-]} + 1 + \dfrac{[CO_3^{2-}]}{[HCO_3^-]}}$$

We can calculate the $[H_2CO_3]/[HCO_3^-]$ ratio from K_{a_1}:

$$K_{a_1} = \frac{[H^+][HCO_3^-]}{[H_2CO_3]}$$

or

$$\frac{[H_2CO_3]}{[HCO_3^-]} = \frac{[H^+]}{K_{a_1}}$$

Since $K_{a_1} = 4.3 \times 10^{-7}$ and pH = 9.00 ($[H^+] = 1.00 \times 10^{-9}$),

$$\frac{[H_2CO_3]}{[HCO_3^-]} = \frac{1.00 \times 10^{-9}}{4.3 \times 10^{-7}} = 2.3 \times 10^{-3}$$

We can use K_{a_2} to calculate the $[CO_3^{2-}]/[HCO_3^-]$ ratio:

$$K_{a_2} = \frac{[H^+][CO_3^{2-}]}{[HCO_3^-]} = 4.8 \times 10^{-11}$$

so

$$\frac{[CO_3^{2-}]}{[HCO_3^-]} = \frac{K_{a_2}}{[H^+]} = \frac{4.8 \times 10^{-11}}{1.00 \times 10^{-9}} = 4.8 \times 10^{-2}$$

Now we can calculate the fraction of $[HCO_3^-]$ present:

$$f_{HCO_3^-} = \frac{1}{\dfrac{[H_2CO_3]}{[HCO_3^-]} + 1 + \dfrac{[CO_3^{2-}]}{[HCO_3^-]}} = \frac{1}{2.3 \times 10^{-3} + 1 + 4.8 \times 10^{-2}}$$

$$= \frac{1}{1.05} = 0.95$$

Note that at pH 9, HCO_3^- represents 95% of the carbonate-containing species in the solution.

Similar calculations can be used to find the fraction of each species over the entire pH range. A graph of the results of these calculations is shown in Fig. 7.6.

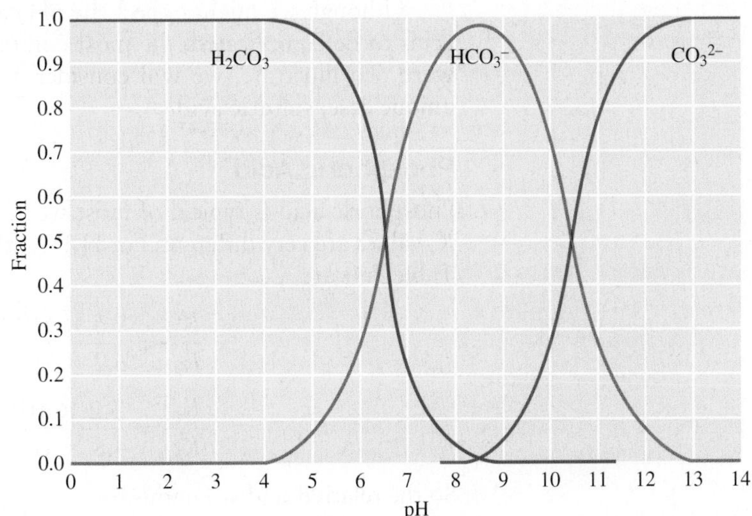

Figure 7.6
A plot of the fractions of H_2CO_3, HCO_3^-, and CO_3^{2-} in aqueous solution as a function of pH.

Table 7.4

Stepwise Dissociation Constants for Several Common Polyprotic Acids

Name	Formula	K_{a_1}	K_{a_2}	K_{a_3}
Phosphoric acid	H_3PO_4	7.5×10^{-3}	6.2×10^{-8}	4.8×10^{-13}
Arsenic acid	H_3AsO_4	5×10^{-3}	8×10^{-8}	6×10^{-10}
Carbonic acid*	H_2CO_3	4.3×10^{-7}	4.8×10^{-11}	
Sulfuric acid	H_2SO_4	Large	1.2×10^{-2}	
Sulfurous acid	H_2SO_3	1.5×10^{-2}	1.0×10^{-7}	
Hydrosulfuric acid†	H_2S	1.0×10^{-7}	$\approx 10^{-19}$	
Oxalic acid	$H_2C_2O_4$	6.5×10^{-2}	6.1×10^{-5}	
Ascorbic acid (vitamin C)	$H_2C_6H_6O_6$	7.9×10^{-5}	1.6×10^{-12}	

*This is really $CO_2(aq)$.

†The K_{a_2} value for H_2S is quite uncertain. Its small size makes it very difficult to measure.

Phosphoric acid is a **triprotic acid** (three protons) that dissociates in the following steps:

$$H_3PO_4(aq) \rightleftharpoons H^+(aq) + H_2PO_4^{2-}(aq) \qquad K_{a_1} = \frac{[H^+][H_2PO_4^-]}{[H_3PO_4]} = 7.5 \times 10^{-3}$$

$$H_2PO_4^{2-}(aq) \rightleftharpoons H^+(aq) + HPO_4^{2-}(aq) \qquad K_{a_2} = \frac{[H^+][HPO_4^{2-}]}{[H_2PO_4^-]} = 6.2 \times 10^{-8}$$

$$HPO_4^{2-}(aq) \rightleftharpoons H^+(aq) + PO_4^{3-}(aq) \qquad K_{a_3} = \frac{[H^+][PO_4^{3-}]}{[HPO_4^{2-}]} = 4.8 \times 10^{-13}$$

For a typical, weak polyprotic acid,

$$K_{a_1} > K_{a_2} > K_{a_3}$$

That is, the acid involved in each successive step of the dissociation is weaker. This is shown by the stepwise dissociation constants given in Table 7.4. These values indicate that the loss of a second or third proton occurs less readily than loss of the first proton. This result is not surprising; the greater the negative charge on the acid, the more difficult it becomes to remove the positively charged proton.

Although we might expect the pH calculations for solutions of polyprotic acids to be complicated, the most common cases are surprisingly straightforward. To illustrate, we will consider a typical case, phosphoric acid, and a unique case, sulfuric acid.

Phosphoric Acid

Phosphoric acid is typical of most weak polyprotic acids in that its successive K_a values are very different. For H_3PO_4, the ratios of successive K_a values (from Table 7.4) are

$$\frac{K_{a_1}}{K_{a_2}} = \frac{7.5 \times 10^{-3}}{6.2 \times 10^{-8}} = 1.2 \times 10^5$$

$$\frac{K_{a_2}}{K_{a_3}} = \frac{6.2 \times 10^{-8}}{4.8 \times 10^{-13}} = 1.3 \times 10^5$$

So the relative acid strengths are

$$H_3PO_4 \gg H_2PO_4^- \gg HPO_4^{2-}$$

For a typical polyprotic acid in water, only the first dissociation step is important in determining the pH.

This means that in a solution prepared by dissolving H_3PO_4 in water, *only the first dissociation step makes an important contribution to [H^+].* This greatly simplifies the pH calculations for phosphoric acid solutions, as is illustrated in Example 7.8.

EXAMPLE 7.8

Calculate the pH of a 5.0 *M* H_3PO_4 solution and determine equilibrium concentrations of the species H_3PO_4, $H_2PO_4^-$, HPO_4^{2-}, and PO_4^{3-}.

Major Species

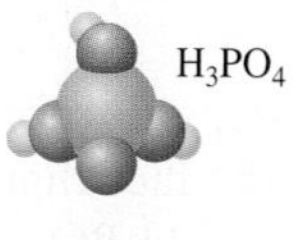

Solution The major species in solution are

$$H_3PO_4 \quad \text{and} \quad H_2O$$

None of the dissociation products of H_3PO_4 is written, since the K_a values are all so small that they will be minor species. The dominant equilibrium will be the dissociation of H_3PO_4:

$$H_3PO_4(aq) \rightleftharpoons H^+(aq) + H_2PO_4^-(aq)$$

where $$K_{a_1} = 7.5 \times 10^{-3} = \frac{[H^+][H_2PO_4^-]}{[H_3PO_4]}$$

The concentrations are as follows:

Initial Concentration (mol/L)		Equilibrium Concentration (mol/L)
$[H_3PO_4]_0 = 5.0$	x mol/L H_3PO_4	$[H_3PO_4] = 5.0 - x$
$[H_2PO_4^-]_0 = 0$	$\longrightarrow$	$[H_2PO_4^-] = x$
$[H^+]_0 \approx 0$	dissociates	$[H^+] = x$

Or in shorthand form:

	$H_3PO_4(aq)$	$\rightleftharpoons$	$H^+(aq)$	+	$H_2PO_4^-(aq)$
Initial:	5.0		0		0
Change:	$-x$		$+x$		$+x$
Equilibrium:	$5.0 - x$		x		x

Substituting the equilibrium concentrations into the expression for K_{a_1} and making the usual approximation gives

$$K_{a_1} = 7.5 \times 10^{-3} = \frac{[H^+][H_2PO_4^-]}{[H_3PO_4]} = \frac{(x)(x)}{5.0 - x} \approx \frac{x^2}{5.0}$$

Thus $$x \approx 1.9 \times 10^{-1}$$

Since 1.9×10^{-1} is less than 5% of 5.0, the approximation is acceptable, and

$$[H^+] = x = 0.19\ M \quad \text{and} \quad \text{pH} = 0.72$$

So far we have determined that

$$[H^+] = [H_2PO_4^-] = 0.19\ M$$

and $$[H_3PO_4] = 5.0 - x = 4.8\ M$$

The concentration of HPO_4^{2-} can be obtained by using the expression for K_{a_2}:

$$K_{a_2} = 6.2 \times 10^{-8} = \frac{[H^+][HPO_4^{2-}]}{[H_2PO_4^-]}$$

where $$[H^+] = [H_2PO_4^-] = 0.19\ M$$

Thus $$[HPO_4^{2-}] = K_{a_2} = 6.2 \times 10^{-8}\ M$$

To calculate $[PO_4^{3-}]$, we use the expression for K_{a_3} and the values of $[H^+]$ and $[HPO_4^{2-}]$ calculated previously:

$$K_{a_3} = \frac{[H^+][PO_4^{3-}]}{[HPO_4^{2-}]} = 4.8 \times 10^{-13} = \frac{0.19[PO_4^{3-}]}{6.2 \times 10^{-8}}$$

$$[PO_4^{3-}] = \frac{(4.8 \times 10^{-13})(6.2 \times 10^{-8})}{0.19} = 1.6 \times 10^{-19}\ M$$

These results show that the second and third dissociation steps do not make an important contribution to $[H^+]$. This is apparent from the fact that $[HPO_4^{2-}]$ is $6.2 \times 10^{-8}\ M$, indicating that only 6.2×10^{-8} mol/L of $H_2PO_4^-$ has dissociated. The value of $[PO_4^{3-}]$ shows that the dissociation of HPO_4^{2-} is even smaller. But we must use the second and third dissociation steps to calculate $[HPO_4^{2-}]$ and $[PO_4^{3-}]$, since these steps are the only sources of these ions.

EXAMPLE 7.9

A 0.200-mole sample of sodium phosphate is dissolved in water, and hydrochloric acid is added to give a total volume of 1.00 L and a final pH of 4.630. Calculate the concentrations of all phosphate-containing species in this solution.

Solution Before we do the complete calculation, we can determine the ratios of the various phosphate-containing species from the various K_a expressions:

$$\frac{[H_3PO_4]}{[H_2PO_4^-]} = \frac{[H^+]}{K_{a_1}} = \frac{2.34 \times 10^{-5}}{7.5 \times 10^{-3}} = 3.1 \times 10^{-3}$$

The inverse ratio is

$$\frac{[H_2PO_4^-]}{[H_3PO_4]} = \frac{1}{3.1 \times 10^{-3}} = 3.2 \times 10^2$$

$$\frac{[H_2PO_4^-]}{[HPO_4^{2-}]} = \frac{[H^+]}{K_{a_2}} = \frac{2.34 \times 10^{-5}}{6.2 \times 10^{-8}} = 3.8 \times 10^2$$

$$\frac{[HPO_4^{2-}]}{[PO_4^{3-}]} = \frac{[H^+]}{K_{a_3}} = \frac{2.34 \times 10^{-5}}{4.8 \times 10^{-13}} = 4.9 \times 10^7$$

These ratios show that $H_2PO_4^-$ is the dominant species at a pH of 4.630. Notice that this is exactly what is predicted by the plot shown in Fig. 7.7. This

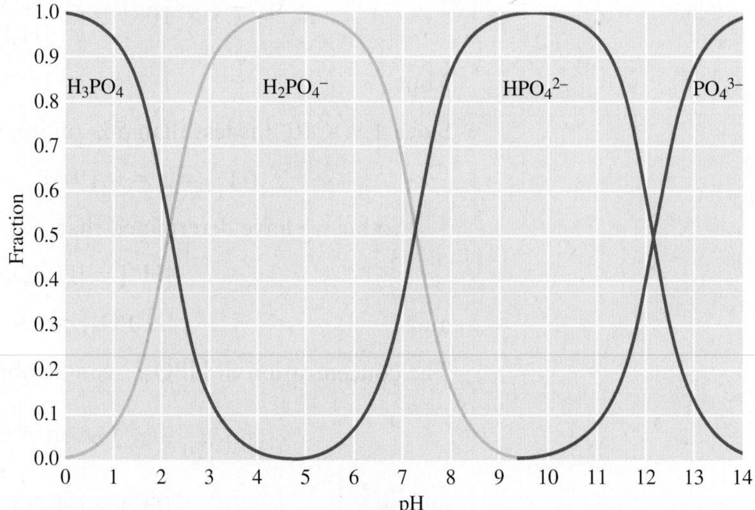

Figure 7.7
A plot of the fractions of H_3PO_4, $H_2PO_4^-$, HPO_4^{2-}, and PO_4^{3-} in aqueous solution as a function of pH.

plot shows the fractions of phosphate-containing species as a function of pH. Note that the fraction of $H_2PO_4^-$ is maximum at about pH 5.

Now we will proceed to determine the concentrations of H_3PO_4, $H_2PO_4^-$, HPO_4^{2-}, and PO_4^{3-}. We know that the total concentrations of these species must be 0.200 *M*, since this represents the concentration of PO_4^{3-} originally added to the solution.

Thus

$$[PO_4^{3-}] + [HPO_4^{2-}] + [H_2PO_4^-] + [H_3PO_4] = 0.200\ M$$

Also note that a pH of 4.630 means that $[H^+] = 2.34 \times 10^{-5}$.

To solve this problem we need to reduce the equation to one variable. For example, we can use the K_{a_1}, K_{a_2}, and K_{a_3} expressions to represent each species in terms of $[H_3PO_4]$ as follows:

$$[H_2PO_4^-] = \frac{[H_3PO_4]K_{a_1}}{[H^+]} = \frac{[H_3PO_4]7.5 \times 10^{-3}}{2.34 \times 10^{-5}} = 3.21 \times 10^2[H_3PO_4]$$

$$[HPO_4^{2-}] = \frac{[H_2PO_4^-]K_{a_2}}{[H^+]} = \frac{3.21 \times 10^2[H_3PO_4]K_{a_2}}{[H^+]} = 8.49 \times 10^{-1}[H_3PO_4]$$

Since $K_{a_1} \times K_{a_2} \times K_{a_3} = \dfrac{[H^+]^3[PO_4^{3-}]}{[H_3PO_4]}$,

$$[PO_4^{3-}] = \frac{(K_{a_1} \times K_{a_2} \times K_{a_3})[H_3PO_4]}{[H^+]^3} = 1.74 \times 10^{-8}[H_3PO_4]$$

Substituting these expressions into the equation

$$[PO_4^{3-}] + [HPO_4^{2-}] + [H_2PO_4^-] + [H_3PO_4] = 0.200\ M$$

gives

$$1.74 \times 10^{-8}[H_3PO_4] + 8.49 \times 10^{-1}[H_3PO_4] + 3.21 \times 10^2[H_3PO_4] + [H_3PO_4] = 0.200\ M$$

$$[H_3PO_4] = \frac{0.200\ M}{322.8} = 6.20 \times 10^{-4}\ M$$

$$[H_2PO_4^-] = (3.21 \times 10^2)(6.20 \times 10^{-4}\ M) = 0.199\ M$$

$$[HPO_4^{2-}] = (8.49 \times 10^{-1})(6.20 \times 10^{-4}\ M) = 5.26 \times 10^{-4}\ M$$

$$[PO_4^{3-}] = (1.74 \times 10^{-8})(6.20 \times 10^{-4}\ M) = 1.08 \times 10^{-11}\ M$$

Notice that at this pH the dominant species by far is $H_2PO_4^-$, as we predicted earlier.

Sulfuric Acid

Sulfuric acid is unique among the common acids because it is *a strong acid in its first dissociation step and a weak acid in its second step:*

$$H_2SO_4(aq) \longrightarrow H^+(aq) + HSO_4^-(aq) \qquad K_{a_1} \text{ is very large}$$

$$HSO_4^-(aq) \rightleftharpoons H^+(aq) + SO_4^{2-}(aq) \qquad K_{a_2} = 1.2 \times 10^{-2}$$

Example 7.10 illustrates how to calculate the pH for sulfuric acid solutions.

EXAMPLE 7.10

Major Species

H^+

HSO_4^-

H_2O

Calculate the pH of a 1.0 M H_2SO_4 solution.

Solution The major species in the solution are

$$H^+, \quad HSO_4^-, \quad \text{and} \quad H_2O$$

where the first two ions are produced by the complete first dissociation step of H_2SO_4. The concentration of H^+ in this solution will be at least 1.0 M, since this amount is produced by the first dissociation step of H_2SO_4. We must now answer this question: "Does the HSO_4^- ion dissociate enough to make a significant contribution to the concentration of H^+?" This question can be answered by calculating the equilibrium concentrations for the dissociation reaction of HSO_4^-:

$$HSO_4^-(aq) \rightleftharpoons H^+(aq) + SO_4^{2-}(aq)$$

where

$$K_{a_2} = 1.2 \times 10^{-2} = \frac{[H^+][SO_4^{2-}]}{[HSO_4^-]}$$

The concentrations are as follows:

Initial Concentration (mol/L)		Equilibrium Concentration (mol/L)
$[HSO_4^-]_0 = 1.0$ $[SO_4^{2-}]_0 = 0$ $[H^+]_0 = 1.0$	x mol/L HSO_4^- dissociates $\longrightarrow$ to reach equilibrium	$[HSO_4^-] = 1.0 - x$ $[SO_4^{2-}] = x$ $[H^+] = 1.0 + x$

Or in shorthand form:

	$HSO_4^-(aq)$	$\rightleftharpoons$	$H^+(aq)$	+	$SO_4^{2-}(aq)$
Initial:	1.0		1.0		0
Change:	$-x$		$+x$		$+x$
Equilibrium:	$1.0 - x$		$1.0 + x$		x

Note that $[H^+]_0$ is not equal to zero, as is usually the case for a weak acid, because the first dissociation step has already produced some H^+.

Substituting the equilibrium concentrations into the expression for K_{a_2} and making the usual approximation gives

$$K_{a_2} = 1.2 \times 10^{-2} = \frac{[H^+][SO_4^{2-}]}{[HSO_4^-]} = \frac{(1.0 + x)(x)}{1.0 - x} \approx \frac{(1.0)(x)}{(1.0)}$$

Thus

$$x \approx 1.2 \times 10^{-2}$$

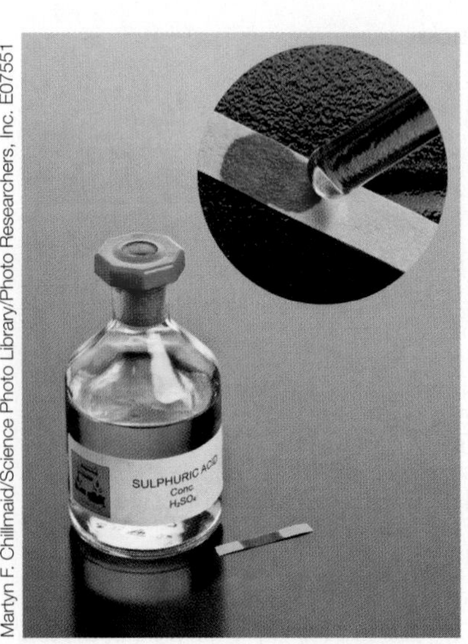

Martyn F. Chillmaid/Science Photo Library/Photo Researchers, Inc. E07551

A bottle of sulfuric acid. A drop of sulfuric acid turns Universal indicator paper bright red, showing it is very acidic.

Since 1.2×10^{-2} is 1.2% of 1.0, the approximation is valid according to the 5% rule. Note that x is not equal to $[H^+]$ in this case. Instead,

$$[H^+] = 1.0\ M + x = 1.0\ M + (1.2 \times 10^{-2})\ M$$

$$= 1.0\ M \quad \text{(to the correct number of significant figures)}$$

Thus, since the dissociation of HSO_4^- does not make a significant contribution to the concentration of H^+,

$$[H^+] = 1.0\ M \quad \text{and} \quad \text{pH} = 0.00$$

Only in dilute H_2SO_4 solutions does the second dissociation step contribute significantly to $[H^+]$.

Example 7.10 illustrates the most common case for sulfuric acid in which only the first dissociation makes an important contribution to the concentration of H^+. In solutions more dilute than 1.0 M (for example, 0.10 M H_2SO_4), the dissociation of HSO_4^- is important. Solving this type of problem requires use of the quadratic formula, as shown in Example 7.11.

EXAMPLE 7.11

Calculate the pH of a 1.00×10^{-2} M H_2SO_4 solution.

Solution The major species in solution are

$$H^+, \quad HSO_4^-, \quad \text{and} \quad H_2O$$

Proceeding as in Example 7.10, we consider the dissociation of HSO_4^-, which leads to the following concentrations:

Initial Concentration (mol/L)		Equilibrium Concentration (mol/L)
$[HSO_4^-]_0 = 0.0100$	x mol/L HSO_4^- dissociates → to reach equilibrium	$[HSO_4^-] = 0.0100 - x$
$[SO_4^{2-}]_0 = 0$		$[SO_4^{2-}] = x$
$[H^+]_0 = 0.0100$ ↗ From dissociation of H_2SO_4		$[H^+] = 0.0100 + x$

Substituting the equilibrium concentrations into the expression for K_{a_2} gives

$$1.2 \times 10^{-2} = K_{a_2} = \frac{[H^+][SO_4^{2-}]}{[HSO_4^-]} = \frac{(0.0100 + x)(x)}{(0.0100 - x)}$$

If we make the usual approximation, then $0.010 + x \approx 0.010$ and $0.010 - x \approx 0.010$, and we have

$$1.2 \times 10^{-2} = \frac{(0.0100 + x)(x)}{(0.0100 - x)} \approx \frac{(0.0100)x}{(0.0100)}$$

The calculated value of x is

$$x = 1.2 \times 10^{-2} = 0.012$$

This value is larger than 0.010, clearly a ridiculous result. Thus we cannot make the usual approximation and must instead solve the quadratic equation. The expression

$$1.2 \times 10^{-2} = \frac{(0.0100 + x)(x)}{(0.0100 - x)}$$

leads to

$$(1.2 \times 10^{-2})(0.0100 - x) = (0.0100 + x)(x)$$

$$(1.2 \times 10^{-4}) - (1.2 \times 10^{-2})x = (1.0 \times 10^{-2})x + x^2$$

$$x^2 + (2.2 \times 10^{-2})x - (1.2 \times 10^{-4}) = 0$$

This equation can be solved by using the quadratic formula,

$$x = \frac{-b \pm \sqrt{b^2 - 4ac}}{2a}$$

where $a = 1$, $b = 2.2 \times 10^{-2}$, and $c = -1.2 \times 10^{-4}$. Use of the quadratic formula gives one negative root (which cannot be correct) and one positive root,

$$x = 4.5 \times 10^{-3}$$

Thus $$[H^+] = 0.0100 + x = 0.0100 + 0.0045 = 0.0145$$

and $$pH = 1.84$$

Note that in this case the second dissociation step produces about half as many H^+ ions as the initial step does.

This problem can also be solved by successive approximations, a method illustrated in Appendix A1.4.

Characteristics of Weak Polyprotic Acids

1. Typically, successive K_a values are so much smaller than the first value that only the first dissociation step makes a significant contribution to the equilibrium concentration of H^+. This means that the calculation of the pH for a solution of a weak polyprotic acid is identical to that for a solution of a weak monoprotic acid.
2. Sulfuric acid is unique in being a strong acid in its first dissociation step and a weak acid in its second step. For relatively concentrated solutions of sulfuric acid (1.0 *M* or higher), the large concentration of H^+ from the first dissociation step represses the second step, which can be neglected as a contributor of H^+ ions. For dilute solutions of sulfuric acid, the second step does make a significant contribution and must be considered in obtaining the total H^+ concentration.

7.8 Acid–Base Properties of Salts

The term **salt** is often used by chemists as simply another name for *ionic compound.* When a salt dissolves in water, we assume that it breaks up into its ions, which move about independently, at least in dilute solutions. Under certain conditions these ions can behave as acids or bases. In this section we explore such reactions.

Salts That Produce Neutral Solutions

Recall that the conjugate base of a strong acid has virtually no affinity for protons as compared with that of the water molecule. For this reason strong acids completely dissociate in aqueous solution. Thus, when anions such as Cl^- and NO_3^- are placed in water, they do not combine with H^+ and therefore have no effect on the pH. Cations such as K^+ and Na^+ from strong bases have no affinity for H^+ and no ability to produce H^+, so they too have no effect on the pH of an aqueous solution. *Salts that consist of the cations of strong bases and the anions of strong acids have no effect on [H⁺] when dissolved in water.* This means that aqueous solutions of salts such as KCl, NaCl, $NaNO_3$, and KNO_3 are neutral (have a pH of 7).

The salt of a strong acid and a strong base gives a neutral solution.

Salts That Produce Basic Solutions

In an aqueous solution of sodium acetate ($NaC_2H_3O_2$), the major species are

$$Na^+, \quad C_2H_3O_2^-, \quad \text{and} \quad H_2O$$

Major Species

Na^+

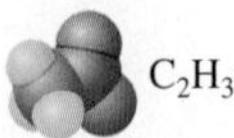
$C_2H_3O_2^-$

H_2O

What are the acid–base properties of each component? The Na^+ ion has neither acid nor base properties. The $C_2H_3O_2^-$ ion is the conjugate base of acetic acid, a weak acid. This means that $C_2H_3O_2^-$ has a significant affinity for a proton and acts as a base. Finally, water is neither a strong enough acid nor a strong enough base to affect $[H^+]$.

Thus the pH of this solution will be controlled by the $C_2H_3O_2^-$ ion. Since $C_2H_3O_2^-$ is a base, it will react with the best proton donor available. In this case water is the *only* source of protons, and the reaction is

$$C_2H_3O_2^-(aq) + H_2O(l) \rightleftharpoons HC_2H_3O_2(aq) + OH^-(aq) \quad (7.5)$$

Note that this reaction, which yields a basic solution, involves a *base reacting with water to produce the hydroxide ion and a conjugate acid.* We have defined K_b as the equilibrium constant for such a reaction. In this instance

$$K_b = \frac{[HC_2H_3O_2][OH^-]}{[C_2H_3O_2^-]}$$

The value of K_a for acetic acid is well known (1.8×10^{-5}). But how can we obtain the K_b value for the acetate ion? The answer lies in the relationships among K_a, K_b, and K_w. Note that when the K_a expression for acetic acid is multiplied by the K_b expression for the acetate ion, the result is K_w:

$$K_a \times K_b = \frac{[H^+][C_2H_3O_2^-]}{[HC_2H_3O_2]} \times \frac{[HC_2H_3O_2][OH^-]}{[C_2H_3O_2^-]} = [H^+][OH^-] = K_w$$

This is a very important result. For any weak acid and its conjugate base,

$$K_a \times K_b = K_w$$

Thus, when either K_a or K_b is known, the other constant can be calculated. For the acetate ion,

$$K_b = \frac{K_w}{K_a \text{ (for } HC_2H_3O_2)} = \frac{1.0 \times 10^{-14}}{1.8 \times 10^{-5}} = 5.6 \times 10^{-10}$$

This is the K_b value for the reaction described by Equation (7.5). Note that it is obtained from the K_a value of the parent weak acid, in this case acetic acid.

A basic solution is formed if the anion of the salt is the conjugate base of a weak acid.

The sodium acetate solution is an example of an important general case. *For any salt whose cation has neutral properties (such as Na^+ or K^+) and whose anion is the conjugate base of a weak acid, the aqueous solution will be basic.* The K_b value for the anion can be obtained from the relationship $K_b = K_w/K_a$. Equilibrium calculations of this type are illustrated in Example 7.12.

OWL INTERACTIVE EXAMPLE 7.12

Calculate the pH of a 0.30 *M* NaF solution. The K_a value for HF is 7.2×10^{-4}.

Major Species

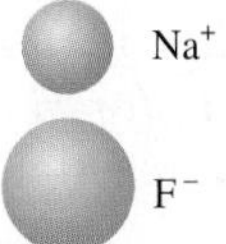

Solution The major species in solution are

$$Na^+, \quad F^-, \quad \text{and} \quad H_2O$$

Since HF is a weak acid, the F^- ion must have a significant affinity for protons. Therefore the dominant reaction will be

$$F^-(aq) + H_2O(l) \rightleftharpoons HF(aq) + OH^-(aq)$$

which yields the K_b expression

$$K_b = \frac{[HF][OH^-]}{[F^-]}$$

The value of K_b can be calculated from K_w and the K_a value for HF:

$$K_b = \frac{K_w}{K_a \text{ (for HF)}} = \frac{1.0 \times 10^{-14}}{7.2 \times 10^{-4}} = 1.4 \times 10^{-11}$$

The concentrations are as follows:

Initial Concentration (mol/L)		Equilibrium Concentration (mol/L)
$[F^-]_0 = 0.30$ $[HF]_0 = 0$ $[OH^-]_0 \approx 0$	$\xrightarrow[\text{H}_2\text{O to reach equilibrium}]{x \text{ mol/L } F^- \text{ reacts with}}$	$[F^-] = 0.30 - x$ $[HF] = x$ $[OH^-] = x$

Thus

$$K_b = 1.4 \times 10^{-11} = \frac{[HF][OH^-]}{[F^-]} = \frac{(x)(x)}{0.30 - x} \approx \frac{x^2}{0.30}$$

$$x \approx 2.0 \times 10^{-6}$$

The approximation is valid by the 5% rule, so

$$[OH^-] = x = 2.0 \times 10^{-6}\ M$$

$$pOH = 5.69$$

$$pH = 14.00 - 5.69 = 8.31$$

As expected, the solution is basic.

Base Strength in Aqueous Solution

To emphasize the concept of base strength, consider the basic properties of the cyanide ion. One relevant reaction is the dissociation of hydrocyanic acid in water:

$$HCN(aq) + H_2O(l) \rightleftharpoons H_3O^+(aq) + CN^-(aq) \qquad K_a = 6.2 \times 10^{-10}$$

Since HCN is such a weak acid, CN^- appears to be a *strong* base, showing a very high affinity for H^+ *compared with* H_2O, with which it is competing. However, we also need to look at the reaction in which cyanide ion reacts with water:

$$CN^-(aq) + H_2O(l) \rightleftharpoons HCN(aq) + OH^-(aq)$$

where

$$K_b = \frac{K_w}{K_a} = \frac{1.0 \times 10^{-14}}{6.2 \times 10^{-10}} = 1.6 \times 10^{-5}$$

In this reaction CN^- appears to be a weak base; the K_b value is only 1.6×10^{-5}. What accounts for this apparent difference in base strength? The key idea is that in the reaction of CN^- with H_2O, *CN^- is competing with OH^- for H^+, instead of competing with H_2O*, as it does in the HCN dissociation reaction. These equilibria show the following relative base strengths:

$$OH^- > CN^- > H_2O$$

Similar arguments can be made for other "weak" bases, such as ammonia, the acetate ion, and the fluoride ion.

Salts That Produce Acidic Solutions

Some salts produce acidic solutions when dissolved in water. For example, when solid NH_4Cl is dissolved in water, NH_4^+ and Cl^- ions are released, with NH_4^+ behaving as a weak acid:

$$NH_4^+(aq) \rightleftharpoons NH_3(aq) + H^+(aq)$$

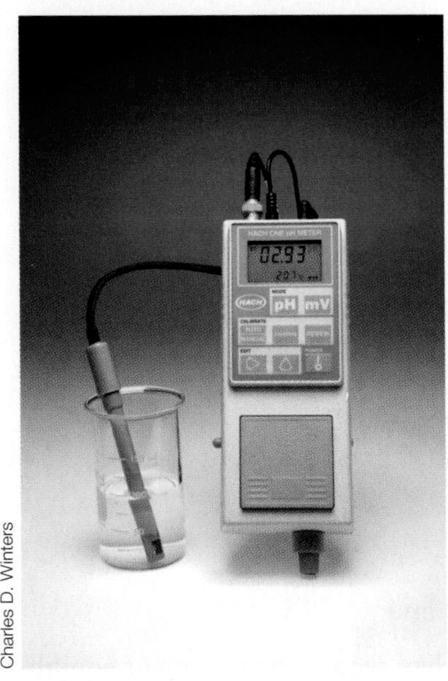

Charles D. Winters

A pH meter showing that the pH of 0.1 *M* $AlCl_3$ is 2.93.

The Cl^- ion, having virtually no affinity for H^+ in water, does not affect the pH of the solution.

In general, *a salt whose cation is the conjugate acid of a weak base produces an acidic solution.*

A second type of salt that produces an acidic solution is one that contains a *highly charged metal ion.* For example, when solid aluminum chloride ($AlCl_3$) is dissolved in water, the resulting solution is significantly acidic. Although the Al^{3+} ion is not itself a Brønsted–Lowry acid, the hydrated ion $Al(H_2O)_6^{3+}$ formed in water is a weak acid:

$$Al(H_2O)_6^{3+}(aq) \rightleftharpoons Al(OH)(H_2O)_5^{2+}(aq) + H^+(aq)$$

The high charge on the metal ion polarizes the O—H bonds in the attached water molecules, making the hydrogens in these water molecules more acidic than those in free water molecules. Typically, the higher the charge on the metal ion, the stronger is the acidity of the hydrated ion.

OWL INTERACTIVE EXAMPLE 7.13

Calculate the pH of a 0.10 *M* NH_4Cl solution. The K_b value for NH_3 is 1.8×10^{-5}.

Major Species

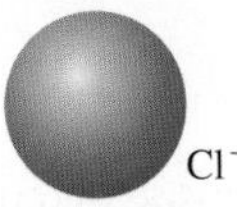

Solution The major species in solution are

$$NH_4^+, \quad Cl^-, \quad \text{and} \quad H_2O$$

Note that both NH_4^+ and H_2O can produce H^+. The dissociation reaction for the NH_4^+ ion is

$$NH_4^+(aq) \rightleftharpoons NH_3(aq) + H^+(aq)$$

for which

$$K_a = \frac{[NH_3][H^+]}{[NH_4^+]}$$

Note that although the K_b value for NH_3 is given, the reaction corresponding to K_b is not appropriate here, since NH_3 is not a major species in the solution. Instead, the given value of K_b is used to calculate K_a for NH_4^+ from the relationship

$$K_a \times K_b = K_w$$

Thus

$$K_a \text{ (for } NH_4^+) = \frac{K_w}{K_b \text{ (for } NH_3)} = \frac{1.0 \times 10^{-14}}{1.8 \times 10^{-5}} = 5.6 \times 10^{-10}$$

Although NH_4^+ is a very weak acid, as indicated by its K_a value, it is stronger than H_2O and thus will dominate in the production of H^+. Therefore, we will focus on the dissociation reaction of NH_4^+ to calculate the pH of this solution.

We solve the weak acid problem in the usual way:

Initial Concentration (mol/L)		Equilibrium Concentration (mol/L)
$[NH_4^+]_0 = 0.10$	x mol/L NH_4^+ dissociates	$[NH_4^+] = 0.10 - x$
$[NH_3]_0 = 0$	$\longrightarrow$	$[NH_3] = x$
$[H^+]_0 \approx 0$	to reach equilibrium	$[H^+] = x$

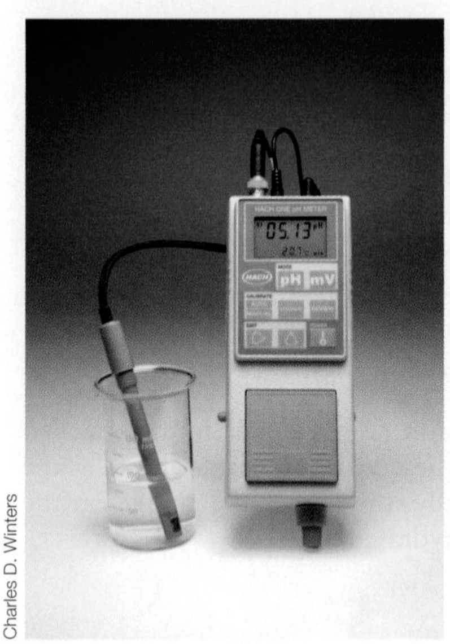

Charles D. Winters

A pH meter showing that the pH of 0.1 M NH_4Cl is 5.13.

Thus

$$5.6 \times 10^{-10} = K_a = \frac{[H^+][NH_3]}{[NH_4^+]} = \frac{(x)(x)}{0.10 - x} = \frac{x^2}{0.10}$$

$$x \approx 7.5 \times 10^{-6}$$

The approximation is valid by the 5% rule, so

$$[H^+] = x = 7.5 \times 10^{-6}\ M \qquad \text{and} \qquad \text{pH} = 5.13$$

EXAMPLE 7.14

Calculate the pH of a 0.010 M $AlCl_3$ solution. The K_a value for $Al(H_2O)_6^{3+}$ is 1.4×10^{-5}.

Solution The major species in solution are

$$Al(H_2O)_6^{3+}, \qquad Cl^-, \qquad \text{and} \qquad H_2O$$

Major Species

Cl^-

$Al(H_2O)_6^{3+}$

H_2O

Since the $Al(H_2O)_6^{3+}$ ion is a stronger acid than water, the dominant equilibrium is

$$Al(H_2O)_6^{3+}(aq) \rightleftharpoons Al(OH)(H_2O)_5^{2+}(aq) + H^+(aq)$$

and

$$1.4 \times 10^{-5} = K_a = \frac{[Al(OH)(H_2O)_5^{2+}][H^+]}{[Al(H_2O)_6^{3+}]}$$

This is a typical weak acid problem, which we can solve with the usual procedures.

Initial Concentration (mol/L)		Equilibrium Concentration (mol/L)
$[Al(H_2O)_6^{3+}]_0 = 0.010$ $[Al(OH)(H_2O)_5^{2+}]_0 = 0$ $[H^+]_0 \approx 0$	x mol/L $Al(H_2O)_6^{3+}$ dissociates to reach equilibrium $\longrightarrow$	$[Al(H_2O)_6^{3+}] = 0.010 - x$ $[Al(OH)(H_2O)_5^{2+}] = x$ $[H^+] = x$

Thus

$$1.4 \times 10^{-5} = K_a = \frac{[Al(OH)(H_2O)_5^{2+}][H^+]}{[Al(H_2O)_6^{3+}]} = \frac{(x)(x)}{0.010 - x} \approx \frac{x^2}{0.010}$$

$$x \approx 3.7 \times 10^{-4}$$

Since the approximation is valid by the 5% rule,

$$[H^-] = x = 3.7 \times 10^{-4}\ M \qquad \text{and} \qquad \text{pH} = 3.43$$

So far we have considered salts containing only one ion that has acidic or basic properties. For many salts, such as ammonium acetate ($NH_4C_2H_3O_2$), both ions affect the pH of the aqueous solution. First, we will consider the qualitative aspects of such problems. We can predict whether the solution will be basic, acidic, or neutral by comparing the K_a value for the acidic ion with the K_b value for the basic ion. If the K_a value for the acidic ion is larger than the K_b value for the basic ion, the solution will be acidic. If the K_b value is larger than the K_a value, the solution is basic. When the K_a and K_b values are equal, the solution is neutral. These facts are summarized in Table 7.5. Table 7.6 summarizes the acid–base properties of aqueous solutions of various salts.

Table 7.5

Qualitative Prediction of pH for Solutions of Salts for Which Both Cation and Anion Have Acidic or Basic Properties

$K_a > K_b$	pH < 7 (acidic)
$K_b > K_a$	pH > 7 (basic)
$K_a = K_b$	pH = 7 (neutral)

Table 7.6

Acid–Base Properties of Aqueous Solutions of Various Types of Salts

Type of Salt	Examples	Comments	pH of Solution
Cation is from strong base; anion is from strong acid	KCl, KNO_3, $NaCl$, $NaNO_3$	Acts as neither an acid nor a base	Neutral
Cation is from strong base; anion is from weak acid	$NaC_2H_3O_2$, KCN, NaF	Anion acts as a base; cation has no effect on pH	Basic
Cation is conjugate acid of weak base; anion is from strong acid	NH_4Cl, NH_4NO_3	Cation acts as an acid; anion has no effect on pH	Acidic
Cation is conjugate acid of weak base; anion is conjugate base of weak acid	$NH_4C_2H_3O_2$, NH_4CN	Cation acts as an acid; anion acts as a base	Acidic if $K_a > K_b$, basic if $K_b > K_a$, neutral if $K_a = K_b$
Cation is highly charged metal ion; anion is from strong acid	$Al(NO_3)_3$, $FeCl_3$	Hydrated cation acts as an acid; anion has no effect on pH	Acidic

OWL INTERACTIVE EXAMPLE 7.15

Predict whether an aqueous solution of each of the following salts will be acidic, basic, or neutral.

a. $NH_4C_2H_3O_2$ **b.** NH_4CN **c.** $Al_2(SO_4)_3$

Solution

a. The ions in solution are NH_4^+ and $C_2H_3O_2^-$. As we mentioned previously, K_a for NH_4^+ is 5.6×10^{-10}, and K_b for $C_2H_3O_2^-$ is 5.6×10^{-10}. Thus, since K_a for NH_4^+ is equal to K_b for $C_2H_3O_2^-$, the solution will be neutral (pH = 7).

b. The solution will contain NH_4^+ and CN^- ions. The K_a value for NH_4^+ is 5.6×10^{-10}, and

$$K_b\ (\text{for CN}^-) = \frac{K_w}{K_a\ (\text{for HCN})} = 1.6 \times 10^{-5}$$

Since K_b for CN^- is much larger than K_a for NH_4^+, this solution will be basic.

c. The solution will contain $Al(H_2O)_6^{3+}$ and SO_4^{2-} ions. The K_a value for $Al(H_2O)_6^{3+}$ is 1.4×10^{-5}, as given in Example 7.14. We must calculate K_b for SO_4^{2-}. The HSO_4^- ion is the conjugate acid of SO_4^{2-}, and its K_a value is K_{a_2} for sulfuric acid, or 1.2×10^{-2}. Therefore,

$$K_b\ (\text{for SO}_4^{2-}) = \frac{K_w}{K_{a_2}\ (\text{for sulfuric acid})}$$

$$= \frac{1.0 \times 10^{-14}}{1.2 \times 10^{-2}} = 8.3 \times 10^{-13}$$

This solution will be acidic, since K_a for $Al(H_2O)_6^{3+}$ is much greater than K_b for SO_4^{2-}.

We have seen that it is possible to make a qualitative prediction of the acidity or basicity of an aqueous solution containing a dissolved salt. We also can give a quantitative description of these solutions by using the procedures we have developed for treating acid–base equilibria. Example 7.16 illustrates this technique.

EXAMPLE 7.16

Calculate the pH of a 0.100 *M* solution of NH_4CN.

Solution The major species in solution are

$$NH_4^+, \quad CN^-, \quad \text{and} \quad H_2O$$

The familiar reactions involving these species are

$$NH_4^+(aq) \rightleftharpoons NH_3(aq) + H^+(aq) \qquad K_a = 5.6 \times 10^{-10}$$

$$CN^-(aq) + H_2O(l) \rightleftharpoons HCN(aq) + OH^-(aq) \qquad K_b = 1.6 \times 10^{-5}$$

$$H_2O(l) + H_2O(l) \rightleftharpoons H_3O^+(aq) + OH^-(aq) \qquad K_w = 1.0 \times 10^{-14}$$

However, noting that NH_4^+ is an acid and CN^- is a base, it is also sensible to consider the reaction

$$NH_4^+(aq) + CN^-(aq) \rightleftharpoons HCN(aq) + NH_3(aq)$$

To evaluate the importance of this reaction, we need the value of its equilibrium constant. Note that

$$K = \frac{[NH_3][HCN]}{[NH_4^+][CN^-]} = \frac{[H^+][NH_3]}{[NH_4^+]} \times \frac{[HCN]}{[H^+][CN^-]}$$

$$= K_a(NH_4^+) \times \frac{1}{K_a\,(HCN)}$$

Thus the value of the equilibrium constant we need is

$$K = \frac{K_a\,(NH_4^+)}{K_a\,(HCN)} = \frac{5.6 \times 10^{-10}}{6.2 \times 10^{-10}} = 0.90$$

Notice that this equilibrium constant is much larger than those for the other possible reactions. Thus we expect this reaction to be dominant in this solution.

Following the usual procedures, we have the concentrations listed below.

Initial Concentration (mol/L)	Equilibrium Concentration (mol/L)
$[NH_4^+]_0 = 0.100$	$[NH_4^+] = 0.100 - x$
$[CN^-]_0 = 0.100$	$[CN^-] = 0.100 - x$
$[NH_3]_0 = 0$	$[NH_3] = x$
$[HCN]_0 = 0$	$[HCN] = x$

Then

$$K = 0.90 = \frac{x^2}{(0.100 - x)^2}$$

Taking the square root of both sides yields

$$0.95 = \frac{x}{0.100 - x}$$

and

$$x = 4.9 \times 10^{-2}\ M = [NH_3] = [HCN]$$

Notice that the reaction under consideration does not involve H^+ or OH^- directly. Thus, to obtain the pH, we must consider the position of the HCN or NH_4^+ dissociation equilibrium. For example, for HCN

$$K_a = 6.2 \times 10^{-10} = \frac{[H^+][CN^-]}{[HCN]}$$

From the preceding calculations,

$$[CN^-] = 0.100 - x = 0.100 - 0.049 = 0.051\ M$$

$$[HCN] = x = 4.9 \times 10^{-2}\ M$$

Substituting these values into the K_a expression for HCN gives

$$[H^+] = 6.0 \times 10^{-10}\ M$$

and

$$pH = 9.22$$

Note that this solution is basic, just as we predicted in Example 7.15.

7.9 Acid Solutions in Which Water Contributes to the H^+ Concentration

In the typical case involving a weak acid HA in water, the $[H^+]$ produced by the acid is much greater than that produced by water, so

$$[H^+] = [H^+]_{HA} + [H^+]_{H_2O} = [H^+]_{HA}$$

The HA and H_2O are simultaneously in equilibrium, but in this case we can ignore water as a source of H^+.

In most solutions containing an acid, we can assume that the acid dominates in the production of H^+ ions. That is, we typically can assume that the acid produces so much H^+ in comparison with the amount of H^+ produced by water that water can be ignored as a source of H^+. However, in certain cases water must be taken into account when the pH of an aqueous solution is calculated. For example, consider a $1.0 \times 10^{-4}\ M$ solution of a very weak acid HA ($K_a = 1.0 \times 10^{-10}$). A quick calculation shows that if water is ignored, the $[H^+]$ produced by this acid is $1.0 \times 10^{-7}\ M$. This value cannot be the correct $[H^+]$ in this solution at equilibrium because in pure water $[H^+] = 1.0 \times 10^{-7}\ M$. In this instance perhaps the thing to do to get the total $[H^+]$ is to add the two H^+ concentrations:

$$\underset{\substack{\uparrow \\ \text{From} \\ 10^{-4}\ M\ \text{HA}}}{1.0 \times 10^{-7}\ M} + \underset{\substack{\uparrow \\ \text{From} \\ H_2O}}{1.0 \times 10^{-7}\ M} = 2.0 \times 10^{-7}\ M$$

However, this procedure is not correct because the two sources of H^+ will affect each other. That is, because both of the reactions

$$H_2O(l) \rightleftharpoons H^+(aq) + OH^-(aq)$$

$$HA(aq) \rightleftharpoons H^+(aq) + A^-(aq)$$

involve H^+, the equilibrium position of each will be affected by the other. Thus we must solve these equilibria simultaneously. This procedure will lead to a concentration of H^+ such that

$$1.0 \times 10^{-7}\ M < [H^+] < 2.0 \times 10^{-7}\ M$$

Note that in these two equilibria there are four unknown concentrations:

$$[H^+], \quad [OH^-], \quad [HA], \quad \text{and} \quad [A^-]$$

To solve for these concentrations, we need four independent equations that relate them. Two of these equations are provided by the two equilibrium expressions:

$$K_w = [H^+][OH^-] \quad \text{and} \quad K_a = \frac{[H^+][A^-]}{[HA]}$$

A third equation can be derived from the principle of **charge balance:** The positive and negative charges carried by the ions in an aqueous solution must balance. That is, the "concentration of positive charge" must equal the "concentration of negative charge." In this case H^+ is the only positive ion, and the negative ions are A^- and OH^-. Thus the *charge balance* expression is

$$[H^+] = [A^-] + [OH^-]$$

Another relationship can be obtained by recognizing that all the HA originally dissolved must be present at equilibrium as either A^- or HA. This observation leads to the equation

This expression conserves A.

$$[HA]_0 = [HA] + [A^-]$$

↑ Original concentration of HA dissolved

which is called the **material balance equation.**

These four equations can be used to derive an equation involving only $[H^+]$. We will start with the K_a expression,

$$K_a = \frac{[H^+][A^-]}{[HA]}$$

and use the other relationships to express $[A^-]$ and [HA] in terms of $[H^+]$. Recall that the charge balance equation is

$$[H^+] = [A^-] + [OH^-]$$

Using the K_w expression, we have

$$[OH^-] = \frac{K_w}{[H^+]}$$

and the charge balance equation becomes

$$[H^+] = [A^-] + \frac{K_w}{[H^+]} \quad \text{or} \quad [A^-] = [H^+] - \frac{K_w}{[H^+]}$$

This equation gives $[A^-]$ in terms of $[H^+]$.

The material balance equation is

$$[HA]_0 = [HA] + [A^-] \quad \text{or} \quad [HA] = [HA]_0 - [A^-]$$

Since
$$[A^-] = [H^+] - \frac{K_w}{[H^+]}$$

we have
$$[HA] = [HA]_0 - \left([H^+] - \frac{K_w}{[H^+]}\right)$$

Now we substitute the expressions for $[A^-]$ and [HA] into the K_a expression:

$$K_a = \frac{[H^+][A^-]}{[HA]} = \frac{[H^+]\left([H^+] - \frac{K_w}{[H^+]}\right)}{[HA]_0 - \left([H^+] - \frac{K_w}{[H^+]}\right)} = \frac{[H^+]^2 - K_w}{[HA]_0 - \frac{[H^+]^2 - K_w}{[H^+]}}$$

This expression permits the calculation of the $[H^+]$ in a solution containing a weak acid. That is, it gives the correct $[H^+]$ for any solution made by dissolving a weak acid in pure water.

See Appendix A1.4 for information on using successive approximations.

The equation can be solved by simple trial and error or by the more systematic method of successive approximations. Recall that the usual way of doing successive approximations is to substitute a guessed value of the variable of interest ($[H^+]$ in this case) into the equation everywhere it appears except in one place. The equation is then solved to obtain a new value of the variable, which becomes the "guessed value" in the next round. The process is continued until the calculated value equals the guessed value.

Even though the full equation can be solved in this manner, the process is tedious and time-consuming. We would certainly like to use the simpler method (ignoring the contribution of water to the $[H^+]$) whenever possible. Thus a key question arises: "Under what conditions can problems involving a weak acid be done in the simple way?"

Notice that the term $[H^+]^2 - K_w$ appears twice in the full equation:

$$K_a = \frac{[H^+]^2 - K_w}{[HA]_0 - \dfrac{[H^+]^2 - K_w}{[H^+]}}$$

Now, assume that the condition

$$[H^+]^2 \gg K_w$$

applies, which means that

$$[H^+]^2 - K_w \approx [H^+]^2$$

Under this condition the full equation can be simplified as follows:

$$K_a = \frac{[H^+]^2 - K_w}{[HA]_0 - \dfrac{[H^+]^2 - K_w}{[H^+]}} \approx \frac{[H^+]^2}{[HA]_0 - \dfrac{[H^+]^2}{[H^+]}} = \frac{[H^+]^2}{[HA]_0 - [H^+]}$$

$$= \frac{x^2}{[HA]_0 - x}$$

where $x = [H^+]$ at equilibrium.

This is an important result: If $[H^+]^2 \gg K_w$, the full equation reduces to the typical expression for a weak acid, which originates from ignoring water as a source of H^+. Because uncertainties in K_a values are typically greater than 1%, we can safely assume that "much greater than" means at least 100 times greater. Then since $K_w = 1.0 \times 10^{-14}$, the $[H^+]^2$ must be at least 100×10^{-14}, or 10^{-12}, which corresponds to $[H^+] = 10^{-6}$. Thus, if $[H^+]$ is greater than or equal to 10^{-6} *M*, the complicated question reduces to the simple equation—that is, you get the *same answer* by using either equation.

How do we decide when we must use the complicated equation? The best way to proceed is as follows: Calculate the $[H^+]$ in the normal way, ignoring any contribution from H_2O. If $[H^+]$ from this calculation is greater than or equal to 10^{-6} *M*, the answer is correct—that is, the complicated equation will give the same answer. If the $[H^+]$ calculated from the simple equation is less than 10^{-6} *M*, you must use the full equation—that is, water must be considered as a source of H^+.

EXAMPLE 7.17

Calculate the $[H^+]$ in

a. 1.0 *M* HCN ($K_a = 6.2 \times 10^{-10}$).

b. 1.0×10^{-4} *M* HCN ($K_a = 6.2 \times 10^{-10}$).

Solution

a. First, do the weak acid problem the "normal" way. This technique leads to the expression

$$\frac{x^2}{1.0 - x} = 6.2 \times 10^{-10} \approx \frac{x^2}{1.0}$$

$$x = 2.5 \times 10^{-5}\, M = [H^+]$$

Note that the $[H^+]$ from the dissociation of HCN is greater than $10^{-6}\, M$, so we are finished. Water makes no important contribution to the $[H^+]$ in this solution.

b. First, do the weak acid problem the "normal" way. This procedure leads to the expression

$$K_a = 6.2 \times 10^{-10} = \frac{x^2}{1.0 \times 10^{-4} - x} \approx \frac{x^2}{1.0 \times 10^{-4}}$$

$$x = 2.5 \times 10^{-7}\, M$$

In this very dilute solution of HCN, the $[H^+]$ from HCN alone is less than $10^{-6}\, M$, so the full equation must be used to obtain the correct $[H^+]$ in the solution:

$$6.2 \times 10^{-10} = K_a = \frac{[H^+]^2 - 10^{-14}}{1.0 \times 10^{-4} - \dfrac{[H^+]^2 - 10^{-14}}{[H^+]}}$$

We will now solve for $[H^+]$ by use of successive approximations. First we must determine a reasonable guess for $[H^+]$. Note from the preceding simple calculation that $[H^+]$, ignoring the contribution from water, is $2.5 \times 10^{-7}\, M$. Will the actual $[H^+]$ be larger or smaller than this value? It will be a little larger because of the contribution from H_2O. So a reasonable guess for $[H^+]$ is $3.0 \times 10^{-7}\, M$.

We now substitute this value for $[H^+]$ into the denominator of the equation, to give

$$K_a = 6.2 \times 10^{-10} = \frac{[H^+]^2 - 1.0 \times 10^{-14}}{1.0 \times 10^{-4} - \dfrac{(3.0 \times 10^{-7})^2 - 1.0 \times 10^{-14}}{3.0 \times 10^{-7}}}$$

$$6.2 \times 10^{-10} = \frac{[H^+]^2 - 1.0 \times 10^{-14}}{1.0 \times 10^{-4} - 2.67 \times 10^{-7}}$$

Now, rearrange this equation so that a value for $[H^+]$ can be calculated:

$$[H^+]^2 = 6.2 \times 10^{-14} - 1.66 \times 10^{-16} + 1.0 + 10^{-14}$$

$$= 7.2 \times 10^{-14}$$

$$[H^+] = \sqrt{7.2 \times 10^{-14}} = 2.68 \times 10^{-7}$$

Recall that the original guessed value of $[H^+]$ was 3.0×10^{-7}. Since the calculated value and the guessed value do not agree, use 2.68×10^{-7} as the new guessed value:

$$K_a = \frac{[H^+]^2 - 1.0 \times 10^{-14}}{1.0 \times 10^{-4} - \dfrac{(2.68 \times 10^{-7})^2 - 1.0 \times 10^{-14}}{2.68 \times 10^{-7}}}$$

Solving for $[H^+]$, we have

$$[H^+] = 2.68 \times 10^{-7} = 2.7 \times 10^{-7}\ M$$

Since the guessed value and the newly calculated value agree, this answer is correct, and it takes into account both contributors to $[H^+]$ (water and HCN). Thus for this solution,

$$pH = -\log(2.7 \times 10^{-7}) = 6.57$$

As you followed the preceding procedure, you may have noticed that there was an opportunity to simplify the math, but we did not take advantage of it. Note that the term

$$[HA]_0 - \frac{[H^+]^2 - K_w}{[H^+]}$$

occurs in the denominator of the overall equation. Because $[H^+]$ will be between 10^{-6} and 10^{-7} (otherwise, we would be using the simple equation), we can see that the value of the term

$$\frac{[H^+]^2 - K_w}{[H^+]}$$

will be between 0 (if $[H^+] = 10^{-7}\ M$) and 10^{-6} (if $[H^+] = 10^{-6}\ M$). Thus, if $[HA]_0 > 2 \times 10^{-5}\ M$ in a given acid solution, then

$$[HA]_0 - \frac{[H^+]^2 - K_w}{[H^+]} \approx [HA]_0$$

within the limits of the 5% rule. Under these conditions the equation

$$K_a = \frac{[H^+]^2 - K_w}{[HA]_0 - \dfrac{[H^+]^2 - K_w}{[H^+]}}$$

becomes

$$K_a \approx \frac{[H^+]^2 - K_w}{[HA]_0}$$

which can be readily solved for $[H^+]$:

$$[H^+] \approx \sqrt{K_a[HA]_0 + K_w}$$

Note that this equation gives the same $[H^+]$ for $1.0 \times 10^{-4}\ M$ HCN as the full equation.

This simplified equation applies to all cases except for very dilute weak acid solutions.

We will now summarize the conclusions of this section.

The pH Calculations for an Aqueous Solution of a Weak Acid HA (Major Species HA and H_2O)

1. The full equation for this case is

$$K_a = \frac{[H^+]^2 - K_w}{[HA]_0 - \dfrac{[H^+]^2 - K_w}{H^+}}$$

2. When the weak acid by itself produces $[H^+] \geq 10^{-6}\ M$, the full equation becomes

$$K_a = \frac{[H^+]^2}{[HA]_0 - [H^+]}$$

This corresponds to the typical weak acid case.

Continued

3. When

$$[HA]_0 \gg \frac{[H^+]^2 - K_w}{[H^+]}$$

the full equation becomes

$$K_a = \frac{[H^+]^2 - K_w}{[HA]_0}$$

which gives

$$[H^+] = \sqrt{K_a[HA]_0 + K_w}$$

7.10 Strong Acid Solutions in Which Water Contributes to the H^+ Concentration

Although in a typical strong acid solution (for example, 0.1 *M* HCl) the $[H^+]$ is determined by the amount of strong acid present, there are circumstances in which the contribution of water must be taken into account. For example, consider a 1.0×10^{-7} *M* HNO_3 solution. For this very dilute solution, the strong acid and the water make comparable contributions to $[H^+]$ at equilibrium. Because the H^+ from HNO_3 will affect the position of the water equilibrium, the total $[H^+]$ in this solution will not be simply $1.0 \times 10^{-7}\ M + 1.0 \times 10^{-7}\ M = 2.0 \times 10^{-7}\ M$. Rather, $[H^+]$ will be between $1.0 \times 10^{-7}\ M$ and $2.0 \times 10^{-7}\ M$. We can calculate the exact $[H^+]$ by using the principle of charge balance:

$$[\text{Positive charge}] = [\text{negative charge}]$$

In this case we have

$$[H^+] = [NO_3^-] + [OH^-]$$

which, from $K_w = [H^+][OH^-]$, can be written as

$$[H^+] = [NO_3^-] + \frac{K_w}{[H^+]} \qquad \text{or} \qquad \frac{[H^+]^2 - K_w}{[H^+]} = [NO_3^-]$$

The fact that the solution contains 1.0×10^{-7} *M* HNO_3 means that $[NO_3^-] = 1.0 \times 10^{-7}\ M$. Inserting this value, we can solve the preceding equation to give $[H^+] = 1.6 \times 10^{-7}\ M$.

This approach applies for any strong acid (although it is unnecessary for more typical concentrations). It can also be adapted to calculate the pH of a very dilute strong base solution. (Try your hand at this problem by calculating the pH of a 5.0×10^{-8} *M* KOH solution.)

7.11 Strategy for Solving Acid–Base Problems: A Summary

In this chapter we have encountered many different situations involving aqueous solutions of acids and bases, and in the next chapter we will encounter still more. In solving for the equilibrium concentrations in these aqueous solutions, you may be tempted to create a pigeonhole for each possible situation and to memorize the procedures necessary to deal with each particular situation. This approach is just not practical and usually leads to frustration: Too many pigeonholes are required, because there seems to be an infinite number of cases. But you can handle any case successfully by taking a systematic, patient, and thoughtful approach. When analyzing an acid–base equilibrium problem, do *not* ask yourself how a memorized solution can be used to solve the problem. Instead, ask yourself this question: "*What are the major species in the solution, and how does each behave chemically?*"

The most important part of doing a complicated acid–base equilibrium problem is the analysis you do at the beginning of a problem:

Which major species are present?
Does a reaction occur that can be assumed to go to completion?
Which equilibrium dominates the solution?

Let the problem guide you. Be patient.

The following steps outline a general strategy for solving problems involving acid–base equilibria.

STEPS

Solving Acid–Base Equilibria Problems

1 List the major species in solution.

2 Look for reactions that can be assumed to go to completion, such as a strong acid dissociating or H^+ reacting with OH^-.

3 For a reaction that can be assumed to go to completion:
   a. Determine the concentrations of the products.
   b. Write down the major species in solution after the reaction.

4 Look at each major component of the solution, and decide whether it is an acid or a base.

5 Pick the equilibrium that will control the pH. Use known values of the dissociation constants for the various species to determine the dominant equilibrium.
   a. Write the equation for the reaction and the equilibrium expression.
   b. Compute the initial concentrations (assuming that the dominant equilibrium has not yet occurred—for example, there has been no acid dissociation).
   c. Define x.
   d. Compute the equilibrium concentrations in terms of x.
   e. Substitute the concentrations into the equilibrium expression, and solve for x.
   f. Check the validity of the approximation.
   g. Calculate the pH and other concentrations as required.

Although these procedures may seem somewhat cumbersome, especially for simpler problems, they will become increasingly helpful as the aqueous solutions become more complicated. If you develop the habit of approaching acid–base problems systematically, the more complex cases will be much easier to manage.

Key Terms

Section 7.1
Arrhenius concept
Brønsted–Lowry model
hydronium ion
conjugate base
conjugate acid
conjugate acid–base pair
acid dissociation constant

Section 7.2
strong acid
weak acid
diprotic acid
oxyacid
organic acid
carboxyl group
monoprotic acid
amphoteric substance
autoionization
ion-product (dissociating) constant

Section 7.4
major species

Section 7.5
percent dissociation

Section 7.6
strong bases
slaked lime
lime-soda process
weak bases
amine

Section 7.7
polyprotic acid
triprotic acid

Section 7.8
salt

Section 7.9
charge balance
material balance equation

For Review

OWL and go Chemistry

Sign in at **www.cengage.com/owl** to:

- View tutorials and simulations, develop problem-solving skills, and complete online homework assigned by your professor.
- Download Go Chemistry mini lecture modules for quick review and exam prep from OWL (or purchase them at **www.cengagebrain.com**)

Models for acids and bases

- Arrhenius model
  - Acids produce H^+ in solution
  - Bases produce OH^- in solution
- Brønsted–Lowry model
  - An acid is a proton donor
  - A base is a proton acceptor
  - In this model an acid molecule reacts with a water molecule, which behaves as a base:

$$\underset{\text{Acid}}{HA(aq)} + \underset{\text{Base}}{H_2O(l)} \rightleftharpoons \underset{\text{Conjugate acid}}{H_3O^+(aq)} + \underset{\text{Conjugate base}}{A^-(aq)}$$

    to form a new acid (conjugate acid) and a new base (conjugate base).
- Lewis model
  - A Lewis acid is an electron-pair acceptor
  - A Lewis base is an electron-pair donor

Acid–base equilibrium

- The equilibrium constant for an acid dissociating (ionizing) in water is called K_a
- The K_a expression is

$$K_a = \frac{[H_3O^+][A^-]}{[HA]}$$

  which is often simplified as

$$K_a = \frac{[H^+][A^-]}{[HA]}$$

  - $[H_2O]$ is never included because it is assumed to be constant

Acid strength

- A strong acid has a very large K_a value
  - The acid completely dissociates (ionizes) in water
  - The dissociation (ionization) equilibrium position lies all the way to the right
  - Strong acids have very weak conjugate bases
  - The common strong acids are nitric acid [$HNO_3(aq)$], hydrochloric acid [$HCl(aq)$], sulfuric acid [$H_2SO(aq)$], and perchloric acid [$HClO_4(aq)$]
- A weak acid has a small K_a value
  - The acid dissociates (ionizes) to only a slight extent
  - The dissociation (ionization) equilibrium position lies far to the left
  - Weak acids have relatively strong conjugate bases
  - Percent dissociation of a weak acid

$$\%\text{ dissociation} = \frac{\text{amount dissociated (mol/L)}}{\text{initial concentration (mol/L)}} \times 100\%$$

  - The smaller the percent dissociation, the weaker the acid
  - Dilution of a weak acid increases its percent dissociation

Autoionization of water

- Water is an amphoteric substance: it behaves as both an acid and a base
- Water reacts with itself in an acid–base reaction

$$H_2O(l) + H_2O(l) \rightleftharpoons H_3O^+(aq) + OH^-(aq)$$

which leads to the equilibrium expression

$$K_w = [H_3O^+][OH^-] \quad \text{or} \quad [H^+][OH^-] = K_w$$

  - K_w is the ion-product constant for water
  - At 25°C in pure water $[H^+] = [OH^-] = 1.0 \times 10^{-7}$, so $K_w = 1.0 \times 10^{-14}$
- Acidic solution: $[H^+] > [OH^-]$
- Basic solution: $[OH^-] > [H^+]$
- Neutral solution: $[H^+] = [OH^-]$

The pH scale

- $pH = -\log [H^+]$
- Since pH is a log scale, the pH changes by 1 for every 10-fold change in $[H^+]$
- The log scale is also used for $[OH^-]$ and for K_a values

$$pOH = -\log[OH^-]$$

$$pK_a = -\log K_a$$

Bases

- Strong bases are hydroxide salts, such as NaOH and KOH
- Weak bases react with water to produce OH^-

$$B(aq) + H_2O(l) \rightleftharpoons BH^+(aq) + OH^-(aq)$$

  - The equilibrium constant for this reaction is called K_b where

$$K_b = \frac{[BH^+][OH^-]}{[B]}$$

  - In water a base B is always competing with OH^- for a proton (H^+), so K_b values tend to be very small, thus making B a weak base (compared to OH^-)

Polyprotic acids

- A polyprotic acid has more than one acidic proton
- Polyprotic acids dissociate one proton at a time
  - Each step has a characteristic K_a value
  - Typically for a weak polyprotic acid, $K_{a_1} > K_{a_2} > K_{a_3}$
- Sulfuric acid is unique
  - It is a strong acid in the first dissociation step (K_{a_1} is very large)
  - It is a weak acid in the second step

Acid–base properties of salts

- Can produce acidic, basic, or neutral solutions
- Salts that contain:
  - Cations of strong bases and anions of strong acids produce neutral solutions
  - Cations of strong bases and anions of weak acids produce basic solutions
  - Cations of weak bases and anions of strong acids produce acidic solutions
- Acidic solutions are produced by salts containing a highly charged metal cation—for example, Al^{3+} and Fe^{3+}

Solutions for which water contributes significantly to the H^+ concentration

- Weak acid solutions
  - $$K_a = \frac{[H^+]^2 - K_w}{[HA]_0 - \dfrac{[H^+]^2 - K_w}{[H^+]}}$$
- Strong acid solutions
  - Charge balance
  - $[H^+] = [A^-] + [OH^-]$

Discussion Questions

These questions are designed to be considered by groups of students in class. Often these questions work well for introducing a particular topic in class.

1. Consider two beakers of pure water at different temperatures. How do their pH values compare? Which is more acidic? More basic? Explain.
2. Differentiate between the terms *strength* and *concentration* as they apply to acids and bases. When is HCl strong? Weak? Concentrated? Dilute? Answer the same questions for ammonia.
3. Sketch two graphs: (a) percent dissociation of weak acid HA versus initial concentration of HA ($[HA]_0$), and (b) H^+ concentration versus $[HA]_0$. Explain both.
4. Consider a solution prepared by mixing a weak acid HA and HCl. What are the major species? Explain what is occurring in solution. How would you calculate the pH? What if you added NaA to this solution? Then added NaOH?
5. Explain why salts can be acidic, basic, or neutral, and show examples. Do so without specific numbers.
6. Consider two separate aqueous solutions: one of a weak acid HA and one of HCl. Assuming you started with 10 molecules of each:
   a. Draw a picture of what each looks like at equilibrium.
   b. What are the major species in each beaker?
   c. From your pictures, calculate the K_a values of each acid.
   d. Order the following from strongest to weakest base: H_2O, A^-, and Cl^-. Explain your sequence.
7. You are asked for the H^+ concentration in a solution of NaOH(*aq*). Because sodium hydroxide is a strong base, can we say there is no H^+, since having H^+ would imply that the solution is acidic?
8. Consider a solution prepared by mixing equal moles of a weak acid HA, HCl, and NaA. Which of the following best describes what happens?
   a. The H^+ from the HCl reacts completely with the A^- from the NaA. Then the HA dissociates to some extent.
   b. The H^+ from the HCl reacts with the A^- from the NaA to make HA, whereas the HA is dissociating. Eventually you have equal amounts of everything.
   c. The H^+ from the HCl reacts with the A^- from the NaA to make HA, whereas the HA is dissociating. Eventually all the reactions have equal rates.
   d. The H^+ from the HCl reacts completely with the A^- from the NaA. Then the HA dissociates until "too much" H^+ and A^- are formed, so the H^+ and A^- react to form HA, and so on. Eventually equilibrium is reached.

   Justify the best choice. For those you did not choose, explain why they are incorrect.
9. Consider a solution formed by mixing 100.0 mL of 0.10 *M* HA ($K_a = 1.0 \times 10^{-6}$), 100.0 mL of 0.10 *M* NaA, and 100.0 mL of 0.05 *M* HCl. In calculating the pH for the final solution, you would make some assumptions about the order in which various reactions occur to simplify the calculations. State these assumptions. Does it matter whether the reactions actually occur in the assumed order? Relate this to Question 8. Explain.
10. A certain sodium compound is dissolved in water to liberate Na^+ ions and a particular negative ion. What evidence would you look for to determine whether the anion is behaving as an acid or a base (without measuring the pH of the solution)? Explain how the anion could behave simultaneously as an acid and a base.
11. Acids and bases can be thought of as chemical opposites (acids are proton donors, and bases are proton acceptors). Therefore, one might think that $K_a = 1/K_b$. Why isn't this the case? What is the relationship between K_a and K_b? Prove it with a derivation.
12. You have two solutions of the salts NaX(*aq*) and NaY(*aq*) at equal concentrations. What would you need to know to determine which solution has the higher pH? Explain how you would decide (perhaps even provide a sample calculation).
13. Is the conjugate base of a weak acid a strong base? Explain. Explain why Cl^- does not affect the pH of an aqueous solution.

14. Match the following pH values: 1, 2, 5, 6, 6.5, 8, 11, 11, and 13 with the following chemicals (of equal concentration): HBr, NaOH, NaF, NaCN, NH_4F, CH_3NH_3F, HF, HCN, and NH_3. Answer this question without performing calculations.

15. The salt BX, when dissolved in water, produces an acidic solution. Which of the following could be true? (There may be more than one correct answer.)
    a. The acid HX is a weak acid.
    b. The acid HX is a strong acid.
    c. The cation B^+ is a weak acid.
    Explain.

Exercises

OWL Interactive versions of these problems may be assigned in OWL.

A blue exercise number indicates that the answer to that exercise appears at the back of this book and a solution appears in the *Solutions Guide.*

Nature of Acids and Bases

16. Consider the autoionization of liquid ammonia:

Label each of the species in the equation as an acid or a base and explain your answer.

17. The following are representations of acid–base reactions:

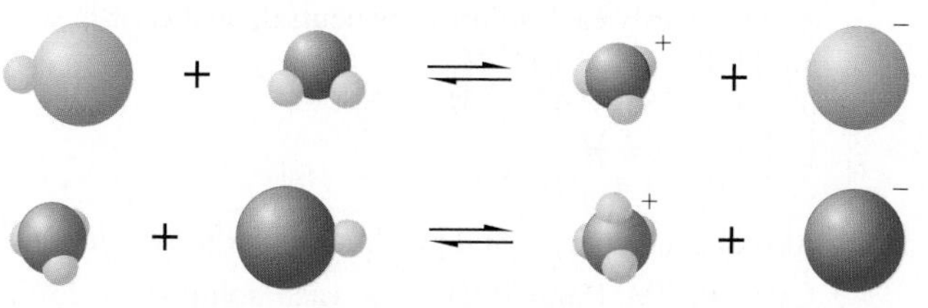

    a. Label each of the species in both equations as an acid or a base and explain.
    b. For those species that are acids, which labels apply: Arrhenius acid and/or Brønsted-Lowery acid? What about the bases?

18. Consider the following statements. Write out an example reaction and K expression that are associated with each statement.
    a. The autoionization of water.
    b. An acid reacts with water to produce the conjugate base of the acid and the hydronium ion.
    c. A base reacts with water to produce the conjugate acid of the base and the hydroxide ion.

19. For each of the following aqueous reactions, identify the acid, the base, the conjugate base, and the conjugate acid.
    a. $H_2O + H_2CO_3 \rightleftharpoons H_3O^+ + HCO_3^-$
    b. $C_5H_5NH^+ + H_2O \rightleftharpoons C_5H_5N + H_3O^+$
    c. $HCO_3^- + C_5H_5NH^+ \rightleftharpoons H_2CO_3 + C_5H_5N$

20. Write balanced equations that describe the following reactions.
    a. The dissociation of perchloric acid in water.
    b. The dissociation of propanoic acid ($CH_3CH_2CO_2H$) in water.
    c. The dissociation of ammonium ion in water.

21. Write the dissociation reaction and the corresponding K_a equilibrium expression for each of the following acids in water.
    a. $HC_2H_3O_2$ b. $Co(H_2O)_6^{3+}$ c. $CH_3NH_3^+$

22. Classify each of the following as a strong acid or a weak acid.

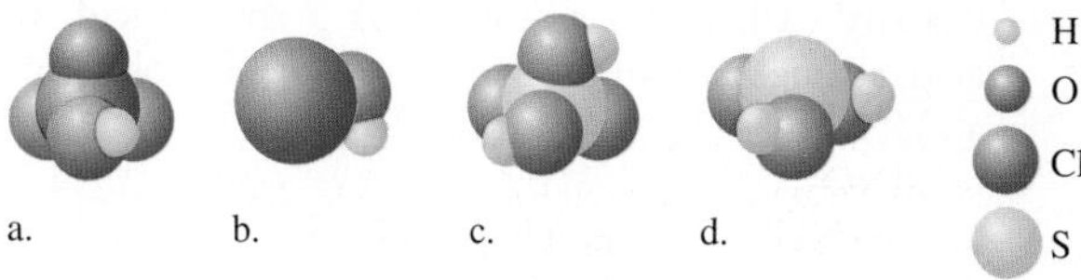

23. Consider the following illustrations:

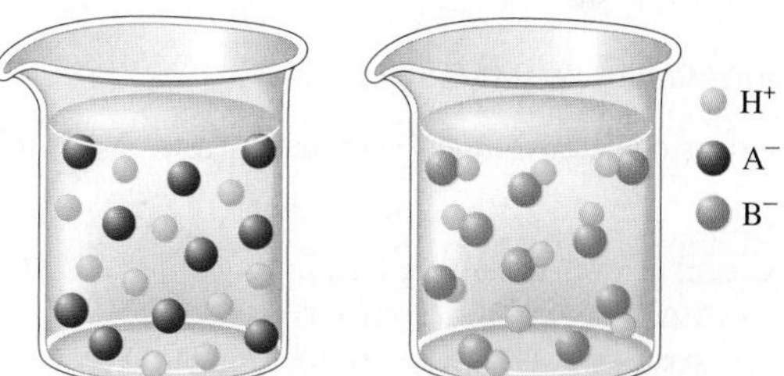

Which beaker best illustrates what happens when the following acids are dissolved in water?
    a. HNO_2 d. HF
    b. HNO_3 e. $HC_2H_3O_2$
    c. HCl

24. Write the reaction and the corresponding K_b equilibrium expression for each of the following substances (acting as bases in water).
    a. NH_3 c. pyridine, C_5H_5N
    b. CN^- d. aniline, $C_6H_5NH_2$

25. Use Table 7.2 to order the following from the strongest to the weakest acid.

$$HClO_2,\ H_2O,\ NH_4^+,\ HClO_4$$

26. Use Table 7.2 to order the following from the strongest to the weakest base.

$$ClO_2^-,\ H_2O,\ NH_3,\ ClO_4^-$$

27. You may need Table 7.2 to answer the following questions.
    a. Which is the stronger acid, HCl or H_2O?
    b. Which is the stronger acid, H_2O or HNO_2?
    c. Which is the stronger acid, HCN or HOC_6H_5?

28. You may need Table 7.2 to answer the following questions.
    a. Which is the stronger base, Cl^- or H_2O?
    b. Which is the stronger base, H_2O or NO_2^-?
    c. Which is the stronger base, CN^- or $OC_6H_5^-$?

29. Consider the reaction of acetic acid in water

$$CH_3CO_2H(aq) + H_2O(l) \rightleftharpoons CH_3CO_2^-(aq) + H_3O^+(aq)$$

where $K_a = 1.8 \times 10^{-5}$.
    a. Which two bases are competing for the proton?
    b. Which is the stronger base?
    c. In light of your answer to part b, why do we classify the acetate ion ($CH_3CO_2^-$) as a weak base? Use an appropriate reaction to justify your answer.

30. In general, as base strength increases, conjugate acid strength decreases. Explain why the conjugate acid of the weak base NH_3 is a weak acid.

31. Classify each of the following as a strong acid, weak acid, strong base, or weak base in aqueous solution.
    a. HNO_2
    b. HNO_3
    c. CH_3NH_2
    d. NaOH
    e. NH_3
    f. HF
    g. $HC(=O)—OH$
    h. $Ca(OH)_2$
    i. H_2SO_4

Autoionization of Water and pH Scale

32. Values of K_w as a function of temperature are as follows:

Temp (°C)	K_w
0	1.14×10^{-15}
25	1.00×10^{-14}
35	2.09×10^{-14}
40.	2.92×10^{-14}
50.	5.47×10^{-14}

    a. Is the autoionization of water exothermic or endothermic?
    b. What is the pH of pure water at 50.°C?
    c. From a plot of $\ln(K_w)$ versus $1/T$ (using the Kelvin scale), estimate K_w at 37°C, normal physiological temperature.
    d. What is the pH of a neutral solution at 37°C?

33. At 40.°C the value of K_w is 2.92×10^{-14}.
    a. Calculate the $[H^+]$ and $[OH^-]$ in pure water at 40.°C.
    b. What is the pH of pure water at 40.°C?
    c. If the hydroxide ion concentration in a solution is 0.10 *M*, what is the pH at 40.°C?

34. Give the conditions for a neutral aqueous solution at 25°C, in terms of $[H^+]$, pH, and the relationship between $[H^+]$ and $[OH^-]$. Do the same for an acidic solution and for a basic solution. As a solution becomes more acidic, what happens to pH, pOH, $[H^+]$, and $[OH^-]$? As a solution becomes more basic, what happens to pH, pOH, $[H^+]$, and $[OH^-]$?

35. Calculate the $[H^+]$ of each of the following solutions at 25°C. Identify each solution as neutral, acidic, or basic.
    a. $[OH^-] = 1.5$ *M*
    b. $[OH^-] = 3.6 \times 10^{-15}$ *M*
    c. $[OH^-] = 1.0 \times 10^{-7}$ *M*
    d. $[OH^-] = 7.3 \times 10^{-4}$ *M*

    Also calculate the pH and pOH of each of these solutions.

36. Calculate the $[OH^-]$ of each of the following solutions at 25°C. Identify each solution as neutral, acidic, or basic.
    a. $[H^+] = 1.0 \times 10^{-7}$ *M*
    b. $[H^+] = 8.3 \times 10^{-16}$ *M*
    c. $[H^+] = 12$ *M*
    d. $[H^+] = 5.4 \times 10^{-5}$ *M*

    Also calculate the pH and pOH of each of these solutions.

37. Calculate $[H^+]$ and $[OH^-]$ for each solution at 25°C. Identify each solution as neutral, acidic, or basic.
    a. pH = 7.40 (the normal pH of blood)
    b. pH = 15.3
    c. pH = −1.0
    d. pH = 3.20
    e. pOH = 5.0
    f. pOH = 9.60

38. Fill in the missing information in the following table.

	pH	pOH	$[H^+]$	$[OH^-]$	Acidic, Basic, or Neutral?
Solution a	9.63	____	____	____	____
Solution b	____	____	____	3.9×10^{-6} *M*	____
Solution c	____	____	0.027 *M*	____	____
Solution d	____	12.2	____	____	____

Solutions of Acids

39. Calculate the pH of each of the following solutions of a strong acid in water.
    a. 0.10 *M* HCl
    b. 5.0 *M* $HClO_4$
    c. 1.0×10^{-11} *M* HI

40. A solution is prepared by adding 50.0 mL of 0.050 *M* HBr to 150.0 mL of 0.10 *M* HI. Calculate $[H^+]$ and the pH of this solution. HBr and HI are both considered strong acids.

41. How would you prepare 1600 mL of a pH = 1.50 solution using concentrated (12 *M*) HCl?

42. What mass of HNO_3 is present in 250.0 mL of a nitric acid solution having a pH = 5.10?

43. What are the major species present in 0.250 *M* solutions of each of the following acids? Calculate the pH of each of these solutions.
    a. HNO_2 b. CH_3CO_2H ($HC_2H_3O_2$)

44. What are the major species present in 0.250 *M* solutions of each of the following acids? Calculate the pH of each of these solutions.
    a. HOC_6H_5 b. HCN

45. Calculate the concentration of all species present and the pH of a 0.020 *M* HF solution.

46. Calculate the percent dissociation for a 0.22 *M* solution of chlorous acid ($HClO_2$, $K_a = 1.2 \times 10^{-2}$).

47. The following illustration displays the relative number of species when an acid, HA, is added to water.

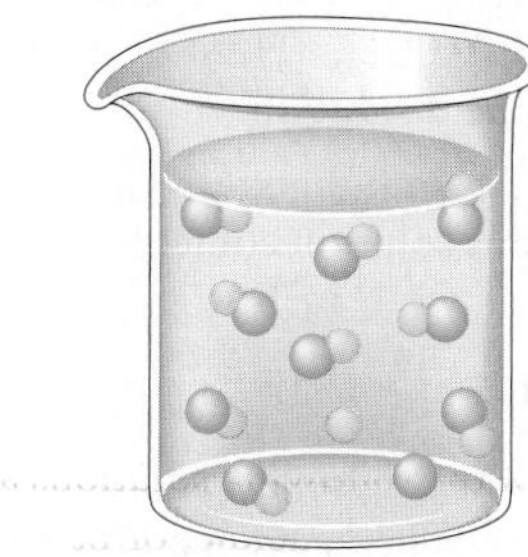

    a. Is HA a weak or strong acid? How can you tell?
    b. Using the relative numbers given in the illustration, determine the value for K_a and the percent dissociation of the acid. Assume the initial acid concentration is 0.20 *M*.

48. Monochloroacetic acid ($HC_2H_2ClO_2$) is a skin irritant that is used in "chemical peels" intended to remove the top layer of dead skin from the face and ultimately improve the complexion. The value of K_a for monochloroacetic acid is 1.35×10^{-3}. Calculate the pH of a 0.10 *M* solution of monochloroacetic acid.

49. Calculate the pH of a 0.010 *M* solution of iodic acid (HIO_3, $K_a = 0.17$).

50. For propanoic acid ($HC_3H_5O_2$, $K_a = 1.3 \times 10^{-5}$), determine the concentration of all species present, the pH, and the percent dissociation of a 0.100 *M* solution.

51. A solution is prepared by dissolving 0.56 g of benzoic acid ($C_6H_5CO_2H$, $K_a = 6.4 \times 10^{-5}$) in enough water to make 1.0 L of solution. Calculate $[C_6H_5CO_2H]$, $[C_6H_5CO_2^-]$, $[H^+]$, $[OH^-]$, and the pH of this solution.

52. At 25°C a saturated solution of benzoic acid (see Exercise 51) has a pH of 2.80. Calculate the water solubility of benzoic acid in moles per liter and grams per 100. mL.

53. A typical aspirin tablet contains 325 mg of acetylsalicylic acid ($HC_9H_7O_4$). Calculate the pH of a solution that is prepared by dissolving two aspirin tablets in one cup (237 mL) of solution. Assume the aspirin tablets are pure acetylsalicylic acid, $K_a = 3.3 \times 10^{-4}$.

54. Calculate the pH of a solution that contains 1.0 *M* HF and 1.0 *M* HOC_6H_5. Also calculate the concentration of $OC_6H_5^-$ in this solution at equilibrium.

55. Calculate the pH of each of the following.
    a. a solution containing 0.10 *M* HCl and 0.10 *M* HOCl
    b. a solution containing 0.050 *M* HNO_3 and 0.50 *M* $HC_2H_3O_2$

56. A 0.15 *M* solution of a weak acid is 3.0% dissociated. Calculate K_a.

57. An acid HX is 25% dissociated in water. If the equilibrium concentration of HX is 0.30 *M*, calculate the K_a value for HX.

58. Calculate the percent dissociation of the acid in each of the following solutions.
    a. 0.50 *M* acetic acid
    b. 0.050 *M* acetic acid
    c. 0.0050 *M* acetic acid
    d. Use Le Châtelier's principle to explain why percent dissociation increases as the concentration of a weak acid decreases.
    e. Even though the percent dissociation increases from solutions a to c, the $[H^+]$ decreases. Explain.

59. The pH of a 1.00×10^{-2} *M* solution of cyanic acid (HOCN) is 2.77 at 25°C. Calculate K_a for HOCN from this result.

60. Trichloroacetic acid (CCl_3CO_2H) is a corrosive acid that is used to precipitate proteins. The pH of a 0.050 *M* solution of trichloroacetic acid is the same as the pH of a 0.040 *M* $HClO_4$ solution. Calculate K_a for trichloroacetic acid.

61. A typical sample of vinegar has a pH of 3.0. Assuming that vinegar is only an aqueous solution of acetic acid ($K_a = 1.8 \times 10^{-5}$), calculate the concentration of acetic acid in vinegar.

62. You have 100.0 g of saccharin, a sugar substitute, and you want to prepare a pH = 5.75 solution. What volume of solution can be prepared? For saccharin ($HC_7H_4NSO_3$), $pK_a = 11.70$ ($pK_a = -\log K_a$).

Solutions of Bases

63. Using Table 7.3, order the following bases from strongest to weakest.

    NO_3^-, H_2O, NH_3, and C_5H_5N

64. Using Table 7.3, order the following acids from strongest to weakest.

    HNO_3, H_2O, NH_4^+, and $C_5H_5NH^+$

65. Use Table 7.3 to help answer the following questions.
    a. Which is the stronger base, ClO_4^- or $C_6H_5NH_2$?
    b. Which is the stronger base, H_2O or $C_6H_5NH_2$?
    c. Which is the stronger base, OH^- or $C_6H_5NH_2$?
    d. Which is the stronger base, $C_6H_5NH_2$ or CH_3NH_2?

66. Use Table 7.3 to help answer the following questions.
    a. Which is the stronger acid, $HClO_4$ or $C_6H_5NH_3^+$?
    b. Which is the stronger acid, H_2O or $C_6H_5NH_3^+$?
    c. Which is the stronger acid, $C_6H_5NH_3^+$ or $CH_3NH_3^+$?

67. Calculate the pH of the following solutions.
a. 0.10 M NaOH
c. 2.0 M NaOH
b. 1.0×10^{-10} M NaOH

68. Calculate $[OH^-]$, pOH, and pH for each of the following.
a. 0.00040 M $Ca(OH)_2$
b. a solution containing 25 g of KOH per liter
c. a solution containing 150.0 g of NaOH per liter

69. Calculate the concentration of an aqueous $Ba(OH)_2$ solution that has pH = 10.50.

70. What mass of KOH is necessary to prepare 800.0 mL of a solution having a pH = 11.56?

71. For the reaction of hydrazine (N_2H_4) in water.

$$H_2NNH_2(aq) + H_2O(l) \rightleftharpoons H_2NNH_3^+(aq) + OH^-(aq)$$

K_b is 3.0×10^{-6}. Calculate the concentrations of all species and the pH of a 2.0 M solution of hydrazine in water.

72. Calculate the percentage of pyridine (C_5H_5N) that forms pyridinium ion, $C_5H_5NH^+$, in a 0.10 M aqueous solution of pyridine ($K_b = 1.7 \times 10^{-9}$).

73. The presence of what element most commonly results in basic properties for an organic compound? What is present on this element in compounds that allows it to accept a proton?

74. Calculate $[OH^-]$, $[H^+]$, and the pH of 0.40 M solutions of each of the following amines (the K_b values are found in Table 7.3).
a. aniline
b. methylamine

75. Calculate the pH of a 0.20 M $C_2H_5NH_2$ solution ($K_b = 5.6 \times 10^{-4}$).

76. Calculate the pH of a 0.050 M $(C_2H_5)_2NH$ solution ($K_b = 1.3 \times 10^{-3}$).

77. Codeine is a derivative of morphine that is used as an analgesic, narcotic, or antitussive. It was once commonly used in cough syrups but is now available only by prescription because of its addictive properties. The formula of codeine is $C_{18}H_{21}NO_3$, and the pK_b is 6.05. Calculate the pH of a 10.0-mL solution containing 5.0 mg of codeine ($pK_b = -\log K_b$).

78. A codeine-containing cough syrup lists codeine sulfate as a major ingredient instead of codeine. *The Merck Index* gives $C_{36}H_{44}N_2O_{10}S$ as the formula for codeine sulfate. Describe the composition of codeine sulfate (see Exercise 77). Why is codeine sulfate used instead of codeine?

79. What is the percent ionization in each of the following solutions?
a. 0.10 M NH_3
c. 0.10 M CH_3NH_2
b. 0.010 M NH_3

80. Quinine ($C_{20}H_{24}N_2O_2$) is the most important alkaloid derived from cinchona bark. It is used as an antimalarial drug. For quinine $pK_{b_1} = 5.1$ and $pK_{b_2} = 9.7$ ($pK_b = -\log K_b$). Only 1.0 g of quinine will dissolve in 1900.0 mL of solution. Calculate the pH of a saturated aqueous solution of quinine. Consider only the reaction $Q + H_2O \rightleftharpoons QH^+ + OH^-$ described by pK_{b_1}, where Q = quinine.

81. The pH of a 0.016 M aqueous solution of p-toluidine ($CH_3C_6H_4NH_2$) is 8.60. Calculate K_b.

82. Calculate the mass of $HONH_2$ required to dissolve in enough water to make 250.0 mL of solution having a pH of 10.00 ($K_b = 1.1 \times 10^{-8}$).

Polyprotic Acids

83. Write out the stepwise K_a reactions for citric acid ($H_3C_6H_5O_7$), a triprotic acid.

84. Consider a 0.10 M H_2CO_3 solution and a 0.10 M H_2SO_4 solution. Without doing any detailed calculations, choose one of the following statements that best describes the $[H^+]$ of each solution and explain your answer.
a. The $[H^+]$ is less than 0.10 M.
b. The $[H^+]$ is 0.10 M.
c. The $[H^+]$ is between 0.10 M and 0.20 M.
d. The $[H^+]$ is 0.20 M.

85. Arsenic acid (H_3AsO_4) is a triprotic acid with $K_{a_1} = 5 \times 10^{-3}$, $K_{a_2} = 8 \times 10^{-8}$, and $K_{a_3} = 6 \times 10^{-10}$. Calculate $[H^+]$, $[OH^-]$, $[H_3AsO_4]$, $[H_2AsO_4^-]$, $[HAsO_4^{2-}]$, and $[AsO_4^{3-}]$ in a 0.20 M arsenic acid solution.

86. Calculate $[CO_3^{2-}]$ in a 0.010 M solution of CO_2 in water (H_2CO_3). If all the CO_3^{2-} in this solution comes from the reaction

$$HCO_3^-(aq) \rightleftharpoons H^+(aq) + CO_3^{2-}(aq)$$

what percentage of the H^+ ions in the solution is a result of the dissociation of HCO_3^-? When acid is added to a solution of sodium hydrogen carbonate ($NaHCO_3$), vigorous bubbling occurs. How is this reaction related to the existence of carbonic acid (H_2CO_3) molecules in aqueous solution?

87. A typical vitamin C tablet (containing pure ascorbic acid, $H_2C_6H_6O_6$) weighs 500. mg. One vitamin C tablet is dissolved in enough water to make 200.0 mL of solution. Calculate the pH of this solution. Ascorbic acid is a diprotic acid.

88. Calculate the pH and $[S^{2-}]$ in a 0.10 M H_2S solution. Assume $K_{a_1} = 1.0 \times 10^{-7}$; $K_{a_2} = 1.0 \times 10^{-19}$.

89. Calculate the pH of a 2.0 M solution of H_2SO_4.

90. Calculate the pH of a 5.0×10^{-3} M solution of H_2SO_4.

Acid–Base Properties of Salts

91. Give three example solutions that fit each of the following descriptions.
a. a strong electrolyte solution that is very acidic
b. a strong electrolyte solution that is slightly acidic
c. a strong electrolyte solution that is very basic
d. a strong electrolyte solution that is slightly basic
e. a strong electrolyte solution that is neutral

92. Derive an expression for the relationship between pK_a and pK_b for a conjugate acid–base pair.

93. Rank the following 0.10 M solutions in order of increasing pH.
a. HI, HF, NaF, NaI
b. NH_4Br, HBr, KBr, NH_3
c. $C_6H_5NH_3NO_3$, $NaNO_3$, NaOH, HOC_6H_5, KOC_6H_5, $C_6H_5NH_2$, HNO_3

94. Arrange the following 0.10 *M* solutions in order from most acidic to most basic.

$CaBr_2$, KNO_2, $HClO_4$, HNO_2, $HONH_3ClO_4$, and NH_4NO_2

95. Are solutions of the following salts acidic, basic, or neutral? For those that are not neutral, explain why the solution is acidic or basic. The relevant K_a and K_b values are found in Tables 7.2, 7.3, and 7.4.
a. $Sr(NO_3)_2$
b. $C_2H_5NH_3CN$
c. C_5H_5NHF
d. $NH_4C_2H_3O_2$
e. $NaHCO_3$

96. Is an aqueous solution of $NaHSO_4$ acidic, basic, or neutral? What reaction occurs with water? Calculate the pH of a 0.10 *M* solution of $NaHSO_4$.

97. Determine $[OH^-]$, $[H^+]$, and the pH of each of the following solutions.
a. 1.0 *M* KCl
b. 1.0 *M* KF

98. Calculate the concentrations of all species present in a 0.25 *M* solution of ethylammonium chloride ($C_2H_5NH_3Cl$).

99. A 0.050 *M* solution of the salt NaB has a pH of 9.00. Calculate the pH of a 0.010 *M* solution of HB.

100. Calculate the pH of each of the following solutions.
a. 0.10 *M* CH_3NH_3Cl
b. 0.050 *M* NaCN

101. Calculate the pH of each of the following solutions.
a. 0.12 *M* KNO_2
b. 0.45 *M* NaOCl
c. 0.40 *M* NH_4ClO_4

102. Sodium azide (NaN_3) is sometimes added to water to kill bacteria. Calculate the concentration of all species in a 0.010 *M* solution of NaN_3. The K_a value for hydrazoic acid (HN_3) is 1.9×10^{-5}.

103. An unknown salt is either NaCN, $NaC_2H_3O_2$, NaF, NaCl, or NaOCl. When 0.100 mole of the salt is dissolved in 1.00 L of solution, the pH of the solution is 8.07. What is the identity of the salt?

104. Consider a solution of an unknown salt having the general formula BHCl, where B is one of the weak bases in Table 7.3. A 0.10 *M* solution of the unknown salt has a pH of 5.82. What is the actual formula of the salt?

105. Calculate the pH of a 0.10 *M* solution of $CoCl_3$. The K_a value for $Co(H_2O)_6^{3+}$ is 1.0×10^{-5}.

106. Calculate the pH of a 0.200 *M* solution of C_5H_5NHF.

107. Determine the pH of a 0.50 *M* solution of NH_4OCl.

108. Calculate the pH of a 0.10 *M* solution of sodium phosphate.

Solutions of Dilute Acids and Bases

109. Using the assumptions we ordinarily make in calculating the pH of an aqueous solution of a weak acid, calculate the pH of a 1.0×10^{-6} *M* solution of hypobromous acid (HBrO, $K_a = 2 \times 10^{-9}$). What is wrong with your answer? Why is it wrong? Without trying to solve the problem, tell what has to be included to solve the problem correctly.

110. Calculate the pH of 4.0×10^{-5} *M* phenol ($K_a = 1.6 \times 10^{-10}$).

111. Calculate the pH of 5.0×10^{-4} *M* HCN.

112. Calculate the pH of 5.0×10^{-8} *M* HNO_3.

113. Calculate the pH of a 7.0×10^{-7} *M* HCl solution.

114. Calculate the pH of a 1.0×10^{-7} *M* solution of NaOH in water.

Additional Exercises

115. Calculate the value for the equilibrium constant for each of the following aqueous reactions.
a. $NH_3 + H_3O^+ \rightleftharpoons NH_4^+ + H_2O$
b. $NO_2^- + H_3O^+ \rightleftharpoons HNO_2 + H_2O$
c. $NH_4^+ + CH_3CO_2^- \rightleftharpoons NH_3 + CH_3CO_2H$
d. $H_3O^+ + OH^- \rightleftharpoons 2H_2O$
e. $NH_4^+ + OH^- \rightleftharpoons NH_3 + H_2O$
f. $HNO_2 + OH^- \rightleftharpoons H_2O + NO_2^-$

116. A solution is prepared by adding 50.0 mL concentrated hydrochloric acid and 20.0 mL concentrated nitric acid to 300 mL water. More water is added until the final volume is 1.00 L. Calcuate $[H^+]$, $[OH^-]$, and the pH for this solution. [*Hint:* Concentrated HCl is 38% HCl (by mass) and has a density of 1.19 g/mL; concentrated HNO_3 is 70.% HNO_3 (by mass) and has a density of 1.42 g/mL.]

117. A 10.0-mL sample of an HCl solution has a pH of 2.000. What volume of water must be added to change the pH to 4.000?

118. Phosphoric acid is a common ingredient in traditional cola drinks. It is added to provide the drinks with a pleasantly tart taste. Although phosphoric acid is a triprotic acid, its protons are lost one at a time. Assuming that in cola drinks the concentration of phosphoric acid is 0.007 *M*, calculate the pH in this solution.

119. Hemoglobin (abbreviated Hb) is a protein that is responsible for the transport of oxygen in the blood of mammals. Each hemoglobin molecule contains four iron atoms that are the binding sites for O_2 molecules. The oxygen binding is pH-dependent. The relevant equilibrium reaction is

$$HbH_4^{4+}(aq) + 4O_2(g) \rightleftharpoons Hb(O_2)_4(aq) + 4H^+(aq)$$

Use Le Châtelier's principle to answer the following.
a. What form of hemoglobin, HbH_4^{4+} or $Hb(O_2)_4$, is favored in the lungs? What form is favored in the cells?
b. When a person hyperventilates, the concentration of CO_2 in the blood is decreased. How does this affect the oxygen-binding equilibrium? How does breathing into a paper bag help to counteract this effect? (See Exercise 120.)

c. When a person has suffered a cardiac arrest, injection of a sodium bicarbonate solution is given. Why is this necessary? (*Hint:* CO_2 blood levels increase during cardiac arrest.)

120. The pH of human blood is steady at a value of approximately 7.4 owing to the following equilibrium reactions:

$$CO_2(aq) + H_2O(l) \rightleftharpoons H_2CO_3(aq) \rightleftharpoons HCO_3^-(aq) + H^+(aq)$$

Acids formed during normal celluar respiration react with the HCO_3^- to form carbonic acid, which is in equilibrium with $CO_2(aq)$ and $H_2O(l)$. During vigorous exercise, a person's H_2CO_3 blood levels were 26.3 m*M*, whereas his CO_2 levels were 1.63 m*M*. On resting, the H_2CO_3 levels declined to 24.9. What was the CO_2 blood level at rest?

121. A solution is tested for pH and conductivity as pictured below:

The solution contains one of the following substances: HCl, NaOH, NH_4Cl, HCN, NH_3, HF, or NaCN. If the solute concentration is about 1.0 *M*, what is the identity of the solute?

122. A 0.20 *M* sodium chlorobenzoate ($NaC_7H_4ClO_2$) solution has a pH of 8.65. Calculate the pH of a 0.20 *M* chlorobenzoic acid ($HC_7H_4ClO_2$) solution.

123. A 0.25-g sample of lime (CaO) is dissolved in enough water to make 1500 mL of solution. Calculate the pH of the solution.

124. Isocyanic acid (HNCO) can be prepared by heating sodium cyanate in the presence of solid oxalic acid according to the equation

$$2NaOCN(s) + H_2C_2O_4(s) \longrightarrow 2HNCO(l) + Na_2C_2O_4(s)$$

Upon isolating pure HNCO(*l*), an aqueous solution of HNCO can be prepared by dissolving the liquid HNCO in water. What is the pH of a 100.-mL solution of HNCO prepared from the reaction of 10.0 g each of NaOCN and $H_2C_2O_4$, assuming all of the HNCO produced is dissolved in solution? (K_a of HNCO $= 1.2 \times 10^{-4}$.)

125. Papaverine hydrochloride (abbreviated papH$^+$Cl$^-$; molar mass = 378.85 g/mol) is a drug that belongs to a group of medicines called vasodilators, which cause blood vessels to expand, thereby increasing blood flow. This drug is the conjugate acid of the weak base papaverine (abbreviated pap; $K_b = 8.33 \times 10^{-9}$ at 35.0°C). Calculate the pH of a 30.0 mg/mL aqueous dose of papH$^+$Cl$^-$ prepared at 35.0°C. (K_w at 35.0°C is 2.1×10^{-14}.)

126. Acrylic acid ($CH_2{=}CHCO_2H$) is a precursor for many important plastics. (K_a for acrylic acid is 5.6×10^{-5}.)
   a. Calculate the pH of a 0.10 *M* solution of acrylic acid.
   b. Calculate the percent dissociation of a 0.10 *M* solution of acrylic acid.
   c. Calculate the [H^+] necessary to ensure that the percent dissociation of a 0.10 *M* solution of acrylic acid is less than 0.010%.
   d. Calculate the pH of a 0.050 *M* solution of sodium acrylate ($NaC_3H_3O_2$).

127. The equilibrium constant K_a for the reaction

$$Fe(H_2O)_6^{3+}(aq) + H_2O(l) \rightleftharpoons Fe(H_2O)_5(OH)^{2+}(aq) + H_3O^+(aq)$$

is 6.0×10^{-3}.
   a. Calculate the pH of a 0.10 *M* solution of $Fe(H_2O)_6^{3+}$.
   b. Calculate the pH necessary for 99.90% of the iron(III) to be in the form $Fe(H_2O)_6^{3+}$.
   c. Will a 1.0 *M* solution of iron(II) nitrate have a higher or lower pH than a 1.0 *M* solution of iron(III) nitrate? Explain.

128. How many moles of HCl(*g*) must be added to 1.0 L of 2.0 *M* NaOH to achieve a pH of 0.00? (Neglect any volume changes.)

129. A solution contains a mixture of acids: 0.50 *M* HA ($K_a = 1.0 \times 10^{-3}$), 0.20 *M* HB ($K_a = 1.0 \times 10^{-10}$), and 0.10 *M* HC ($K_a = 1.0 \times 10^{-12}$). Calculate the [$H^+$] in this solution.

130. One mole of a weak acid HA was dissolved in 2.0 L of solution. After the system had come to equilibrium, the concentration of HA was found to be 0.45 *M*. Calculate K_a for HA.

131. Calculate [OH^-] in a solution obtained by adding 0.0100 mole of solid NaOH to 1.00 L of 15.0 *M* NH_3.

132. Calculate the pH of an aqueous solution containing 1.0×10^{-2} *M* HCl, 1.0×10^{-2} *M* H_2SO_4, and 1.0×10^{-2} *M* HCN.

133. A solution is made by adding 50.0 mL of 0.200 *M* acetic acid ($K_a = 1.8 \times 10^{-5}$) to 50.0 mL of 1.00×10^{-3} *M* HCl.
   a. Calculate the pH of the solution.
   b. Calculate the acetate ion concentration.

134. Will 0.10 *M* solutions of the following salts be acidic, basic, or neutral? See Appendix 5 for K_a values.
   a. ammonium bicarbonate
   b. sodium dihydrogen phosphate
   c. sodium hydrogen phosphate
   d. ammonium dihydrogen phosphate
   e. ammonium formate

135. Citric acid ($H_3C_6H_5O_7$) is a triprotic acid with $K_{a_1} = 8.4 \times 10^{-4}$, $K_{a_2} = 1.8 \times 10^{-5}$, and $K_{a_3} = 4.0 \times 10^{-6}$. Calculate the pH of 0.15 *M* citric acid.

Challenge Problems

136. Consider 1000. mL of a 1.00×10^{-4} M solution of a certain acid HA that has a K_a value equal to 1.00×10^{-4}. How much water must be added or removed (by evaporation) so that a solution remains in which 25.0% of HA is dissociated at equilibrium? Assume that HA is nonvolatile.

137. a. The principal equilibrium in a solution of $NaHCO_3$ is

$$HCO_3^-(aq) + HCO_3^-(aq) \rightleftharpoons H_2CO_3(aq) + CO_3^{2-}(aq)$$

Calculate the value of the equilibrium constant for this reaction.

b. At equilibrium, what is the relationship between $[H_2CO_3]$ and $[CO_3^{2-}]$?

c. Using the equilibrium

$$H_2CO_3(aq) \rightleftharpoons 2H^+(aq) + CO_3^{2-}(aq)$$

and the result from part b, derive an expression for the pH of the solution in terms of K_{a_1} and K_{a_2}.

d. What is the pH of the solution of $NaHCO_3$?

138. A typical solution of baking soda (sodium bicarbonate, $NaHCO_3$) has a pH that is independent of concentration. The pH of a solution of sodium bisulfate ($NaHSO_4$) does depend on the concentration.

a. Explain *why* the pH of a typical solution of sodium bicarbonate is concentration-independent, whereas the pH of a solution of sodium bisulfate is concentration-dependent.

b. What is the pH of a solution of $NaHCO_3$?

c. Determine the pH of a solution of 0.010 M $NaHSO_4$.

139. Calculate the mass of sodium hydroxide that must be added to 1.00 L of 1.00 M $HC_2H_3O_2$ to double the pH of the solution (assume that the added NaOH does not change the volume of the solution).

140. A certain acid, HA, has a vapor density of 5.11 g/L when in the gas phase at a temperature of 25°C and a pressure of 1.00 atm. When 1.50 g of this acid is dissolved in enough water to make 100.0 mL of solution, the pH is found to be 1.80. Calculate K_a for HA.

141. A 0.100 M solution of the salt BHX has a pH of 8.00, where B is a weak base and X^- is the anion of the weak acid HX. Calculate the K_a value for HX if the K_b value for B is 1.0×10^{-3}.

142. Determine the pH of a 0.100 M solution of $(NH_4)_2C_2O_4$.

143. An aqueous solution contains a mixture of 0.0500 M HCOOH ($K_a = 1.77 \times 10^{-4}$) and 0.150 M CH_3CH_2COOH ($K_a = 1.34 \times 10^{-5}$). Calculate the pH of this solution.

144. Consider 50.0 mL of a solution of weak acid HA ($K_a = 1.00 \times 10^{-6}$), which has a pH of 4.000. What volume of water must be added to make the pH = 5.000?

145. A 50.00-mL solution of a weak acid HA ($K_a = 5.00 \times 10^{-10}$) in water has a pH = 5.650. Calculate the amount of water that must be added to reach a pH value of 6.650.

146. Calculate the pH of a solution initially with 0.10 M $NaHSO_4$ and 0.10 M NH_3.

147. A chemist dissolves 0.135 mole of $CO_2(g)$ in 2.50 L of 0.105 M Na_2CO_3. Calculate the pH of the resulting solution.

148. Derive an equation to solve for the pH for a dilute weak base in which water is a significant contributor to the pH. Use this formula to calculate the pH of a 100.0-mL sample of 2.0×10^{-5} M B (where B is a weak base with a K_b value of 6.1×10^{-11}).

149. Calculate the pH of 6.0×10^{-4} M $NaNO_2$.

150. Calculate the resulting pH when 1.00 L of 1.00 M H_2SO_4 is added to a tank containing 1.00×10^7 L of pure water.

151. Calculate the pH of a solution prepared by mixing equal volumes of 1.0×10^{-4} M NH_3 and 1.0×10^{-4} M HCl.

152. Calculate $[OH^-]$ in a 3.0×10^{-7} M solution of $Ca(OH)_2$.

Marathon Problems

153.* Captain Kirk, of the Starship *Enterprise*, has been told by his superiors that only a chemist can be trusted with the combination to the safe containing the dilithium crystals that power the ship. The combination is the pH of solution A described below, followed by the pH of solution C. (Example: If the pH of solution A is 3.47 and that of solution C is 8.15, then the combination to the safe is 3-47-8-15). The chemist must determine the combination using only the information below (all solutions are at 25°C):

Solution A is 50.0 mL of a 0.100 M solution of the weak monoprotic acid HX.

Solution B is a 0.0500 M solution of the salt NaX. It has a pH of 10.02.

Solution C is made by adding 15.0 mL of 0.250 M KOH to solution A.

What is the combination to the safe?

154. Mix equal volumes of one solution from Group I with one solution from Group II to achieve the results below. Calculate the pH of each solution.

Group I: 0.20 M NH_4Cl, 0.20 M HCl, 0.20 M $C_6H_5NH_3Cl$, 0.20 M $(C_2H_5)_3NHCl$

Group II: 0.20 M KOI, 0.20 M NaCN, 0.20 M KOCl, 0.20 M $NaNO_2$

a. the solution with the lowest pH
b. the solution with the highest pH
c. the solution with the pH closest to 7.00

*From James H. Burness, "The Use of "Marathon" Problems as Effective Vehicles for the Presentation of General Chemistry Lectures," Journal of Chemical Education, 68(11).

8 chapter Applications of Aqueous Equilibria

Stalactites and stalagmites in Carlsbad Caverns, New Mexico.

OWL

Sign in to OWL at www.cengage.com/owl to view tutorials and simulations, develop problem-solving skills, and complete online homework assigned by your professor.

go Chemistry

Download mini lecture videos for key concept review and exam prep from OWL or purchase them from www.cengagebrain.com.

Much important chemistry, including most of the chemistry of the natural world, occurs in aqueous solution. We have already introduced one very significant class of aqueous equilibria, acid–base reactions. In this chapter we consider more applications of acid–base chemistry and introduce two additional types of aqueous equilibria, those involving the solubility of salts and the formation of complex ions.

The interplay of acid–base, solubility, and complex ion equilibria is often important in natural processes, such as the weathering of minerals, the uptake of nutrients by plants, and tooth decay. For example, limestone ($CaCO_3$) will dissolve in water made acidic by dissolved carbon dioxide:

$$CO_2(aq) + H_2O(l) \rightleftharpoons H^+(aq) + HCO_3^-(aq)$$

$$H^+(aq) + CaCO_3(s) \rightleftharpoons Ca^{2+}(aq) + HCO_3^-(aq)$$

This process and its reverse account for the formation of limestone caves and the stalactites and stalagmites found there. The acidic water (containing carbon dioxide) dissolves the underground limestone deposits, thereby forming a cavern. As the water drips from the ceiling of the cave, the carbon dioxide is lost to the air, and solid calcium carbonate forms by the reverse of the preceding process to produce stalactites on the ceiling and stalagmites where the drops hit the cave floor.

Before we consider the other types of aqueous equilibria, we will deal with acid–base equilibria in more detail.

8.1 Solutions of Acids or Bases Containing a Common Ion

In Chapter 7 we were concerned with calculating the equilibrium concentrations of species (particularly H^+ ions) in solutions containing an acid or a base. In this section we discuss solutions that contain not only the weak acid HA but also its salt NaA. Although this case appears to be a new type of problem, it can be handled rather easily by using the procedures developed in Chapter 7.

Suppose we have a solution containing weak-acid hydrofluoric acid (HF, $K_a = 7.2 \times 10^{-4}$) and its salt, sodium fluoride (NaF). Recall that when a salt dissolves in water, it breaks up completely into its ions—it is a strong electrolyte:

$$NaF(s) \xrightarrow{H_2O(l)} Na^+(aq) + F^-(aq)$$

Since hydrofluoric acid is a weak acid and only slightly dissociated, the major species in the solution are HF, Na^+, F^-, and H_2O. The **common ion** in this solution is F^-, since it is produced by both hydrofluoric acid and sodium fluoride. What effect does the presence of the dissolved sodium fluoride have on the dissociation equilibrium of hydrofluoric acid?

To answer this question, we compare the extent of dissociation of hydrofluoric acid in two different solutions, the first containing 1.0 *M* HF and the second containing 1.0 *M* HF and 1.0 *M* NaF. According to Le Châtelier's principle, the dissociation equilibrium for HF

$$HF(aq) \rightleftharpoons H^+(aq) + F^-(aq)$$

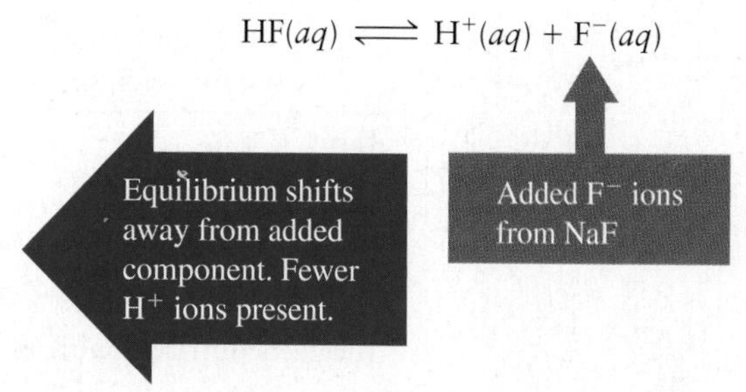

in the second solution will be *driven to the left by the presence of the F^- ions from the NaF*. Thus the extent of dissociation of HF will be less in the presence of dissolved NaF: The shift in equilibrium position that occurs because of the addition of an ion already involved in the equilibrium reaction is called the **common ion effect.** This effect makes a solution of NaF and HF less acidic than a solution of HF alone.

The common ion effect is quite general. For example, when solid NH_4Cl is dissolved in a 1.0 *M* NH_3 solution

$$NH_4Cl(s) \longrightarrow NH_4^+(aq) + Cl^-(aq)$$

the added ammonium ions cause the position of the ammonia–water equilibrium

$$NH_3(aq) + H_2O(l) \rightleftharpoons NH_4^+(aq) + OH^-(aq)$$

to shift to the left, reducing the concentration of OH^- ions.

The common ion effect is also important in solutions of polyprotic acids. The production of protons by the first dissociation step greatly inhibits the succeeding dissociation steps, which also produce protons, the common ion in this case. We will see later in this chapter that the common ion effect is also important in dealing with the solubility of salts.

OWL INTERACTIVE EXAMPLE 8.1

In Section 7.5 we found that the equilibrium concentration of H^+ in a 1.0 *M* HF solution is 2.7×10^{-2} *M* and the percent dissociation of HF is 2.7%. Calculate $[H^+]$ and the percent dissociation of HF in a solution containing both 1.0 *M* HF ($K_a = 7.2 \times 10^{-4}$) and 1.0 *M* NaF.

Major Species

Na^+

Solution As the aqueous solutions we consider become more complex, it becomes increasingly important to be systematic and to *focus on the chemistry* occurring in the solution before thinking about mathematical procedures. *Always* write the major species first and consider the chemical properties of each component.

In a solution containing 1.0 *M* HF and 1.0 *M* NaF, the major species are

$$HF, \quad F^-, \quad Na^+, \quad \text{and} \quad H_2O$$

Since Na^+ ions have neither acidic nor basic properties, and since water is such a weak acid or base, the important species are HF and F^-; they participate in the acid dissociation equilibrium that controls $[H^+]$ in this solution. That is, the position of the equilibrium

$$HF(aq) \rightleftharpoons H^+(aq) + F^-(aq)$$

will determine $[H^+]$ in the solution. The equilibrium expression is

$$K_a = \frac{[H^+][F^-]}{[HF]} = 7.2 \times 10^{-4}$$

The important concentrations are listed in the following table.

Initial Concentration (mol/L)		Equilibrium Concentration (mol/L)
$[HF]_0 = 1.0$ (from dissolved HF)		$[HF] = 1.0 - x$
$[F^-]_0 = 1.0$ (from dissolved NaF)	x mol/L HF dissociates $\longrightarrow$ to reach equilibrium	$[F^-] = 1.0 + x$
$[H^+]_0 = 0$ (neglect contribution from H_2O)		$[H^+] = x$

Note that $[F^-]_0 = 1.0\ M$ from the dissolved sodium fluoride and that the equilibrium $[F^-] > 1.0\ M$ because when the acid dissociates, it produces F^- as well as H^+. Then

$$K_a = 7.2 \times 10^{-4} = \frac{[H^+][F^-]}{[HF]} = \frac{(x)(1.0 + x)}{1.0 - x} \approx \frac{(x)(1.0)}{1.0}$$

since x is expected to be small.

Solving for x gives

$$x = \frac{1.0}{1.0}(7.2 \times 10^{-4}) = 7.2 \times 10^{-4}$$

Noting that x is small compared with 1.0, we conclude that this result is acceptable. Thus

$$[H^+] = x = 7.2 \times 10^{-4}\ M \qquad \text{(The pH is 3.14.)}$$

The percent dissociation of HF in this solution is

$$\frac{[H^+]}{[HF]_0} \times 100\% = \frac{7.2 \times 10^{-4}\ M}{1.0\ M} \times 100\% = 0.072\%$$

Compare these values for $[H^+]$ and percent dissociation of HF with those for a 1.0 *M* HF solution, where $[H^+] = 2.7 \times 10^{-2}\ M$ and the percent dissociation is 2.7%. The large difference clearly shows that the presence of the F^- ions from the dissolved NaF greatly inhibits the dissociation of HF. The position of the acid dissociation equilibrium has been shifted to the left by the presence of F^- ions from NaF.

8.2 Buffered Solutions

The most important buffering system in the blood involves HCO_3^- and H_2CO_3.

The most important application of acid–base solutions containing a common ion is buffering. A **buffered solution** is one that *resists a change in pH* when either hydroxide ions or protons are added. The most important practical example of a buffered solution is human blood, which can absorb the acids and bases produced by biological reactions without changing its pH. A constant pH for blood is vital because cells can survive only in a very narrow pH range around 7.4.

A buffered solution may contain a weak acid and its salt (for example, HF and NaF) or a weak base and its salt (for example, NH_3 and NH_4Cl). By choosing the appropriate components, a solution can be buffered at virtually any pH.

In treating buffered solutions in this chapter, we will start by considering the equilibrium calculations. We will then use these results to show how buffering works. That is, we will answer the question: "How does a buffered solution resist changes in pH when an acid or base is added?"

As you do the calculations associated with buffered solutions, keep in mind that they are merely solutions containing weak acids or bases and that the procedures required are the same ones we have already developed. Be sure to use the systematic approach introduced in Chapter 7.

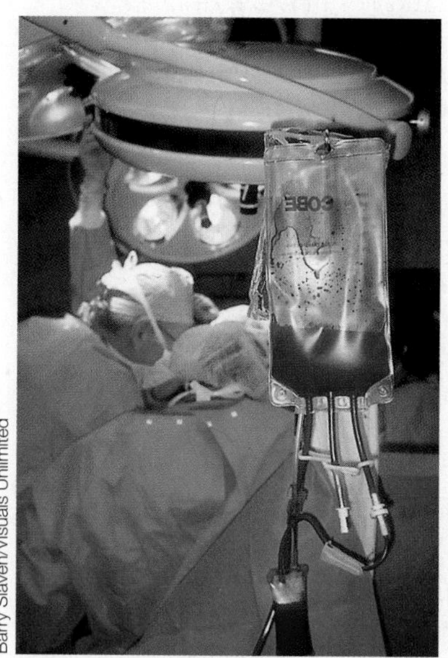

Barry Slaven/Visuals Unlimited

Human blood is a buffered solution.

OWL INTERACTIVE EXAMPLE 8.2

A buffered solution contains 0.50 *M* acetic acid ($HC_2H_3O_2$, $K_a = 1.8 \times 10^{-5}$) and 0.50 *M* sodium acetate ($NaC_2H_3O_2$).

a. Calculate the pH of this solution.

Major Species

$HC_2H_3O_2$

$C_2H_3O_2^-$

Na^+

H_2O

Solution The major species in the solution are

$HC_2H_3O_2$,	Na^+,	$C_2H_3O_2^-$,	and	H_2O
↑	↑	↑		↑
Weak acid	Neither acid nor base	Base (conjugate base of $HC_2H_3O_2$)		Very weak acid or base

Examination of the solution components leads to the conclusion that the acetic acid dissociation equilibrium, which involves both $HC_2H_3O_2$ and $C_2H_3O_2^-$, will control the pH of the solution:

$$HC_2H_3O_2(aq) \rightleftharpoons H^+(aq) + C_2H_3O_2^-(aq)$$

$$K_a = 1.8 \times 10^{-5} = \frac{[H^+][C_2H_3O_2^-]}{[HC_2H_3O_2]}$$

The concentrations are as follows:

Initial Concentration (mol/L)		Equilibrium Concentration (mol/L)
$[HC_2H_3O_2]_0 = 0.50$ $[C_2H_3O_2^-]_0 = 0.50$ $[H^+]_0 \approx 0$	x mol/L of $HC_2H_3O_2$ dissociates to reach equilibrium ⟶	$[HC_2H_3O_2] = 0.50 - x$ $[C_2H_3O_2^-] = 0.50 + x$ $[H^+] = x$

Then

$$K_a = 1.8 \times 10^{-5} = \frac{[H^+][C_2H_3O_2^-]}{[HC_2H_3O_2]} = \frac{(x)(0.50 + x)}{0.50 - x} \approx \frac{(x)(0.50)}{0.50}$$

and

$$x = 1.8 \times 10^{-5}$$

The approximation is valid (by the 5% rule), so

$$[H^+] = x = 1.8 \times 10^{-5}\ M \quad \text{and} \quad \text{pH} = 4.74$$

A digital pH meter shows the pH of the buffered solution to be 4.74.

b. Calculate the change in pH that occurs when 0.010 mole of solid NaOH is added to 1.0 L of the buffered solution. Compare this pH change with the change that occurs when 0.010 mole of solid NaOH is added to 1.0 L of water.

Solution Since the added solid NaOH will completely dissociate, the major species in solution *before any reaction occurs* are $HC_2H_3O_2$, Na^+, $C_2H_3O_2^-$, OH^-, and H_2O. Note that the solution contains a relatively large amount of the very strong base, the hydroxide ion, which has a great affinity for protons. The best source of protons is the acetic acid, so the reaction that will occur is

$$OH^- + HC_2H_3O_2 \longrightarrow H_2O + C_2H_3O_2^-$$

Although acetic acid is a weak acid, the hydroxide ion is such a strong base that the preceding reaction will *proceed essentially to completion* (until the OH^- ions are consumed).

The best approach to this problem involves two distinct steps: (1) assume the reaction goes to completion and carry out the stoichiometric calculations, and then (2) carry out the equilibrium calculations.

■ *The stoichiometry problem.* The reaction occurs as shown below.

	$HC_2H_3O_2$	+ OH^-	$\longrightarrow$ $C_2H_3O_2^-$	+ H_2O
Before reaction:	$1.0\text{ L} \times 0.50\ M = 0.50$ mole	0.010 mole	$1.0\text{ L} \times 0.50\ M = 0.50$ mole	
After reaction:	$0.50 - 0.01 = 0.49$ mole	$0.010 - 0.010 = 0$ mole	$0.50 + 0.01 = 0.51$ mole	

Note that 0.01 mole of $HC_2H_3O_2$ has been converted to 0.01 mole of $C_2H_3O_2^-$ by the added OH^-.

■ *The equilibrium problem.* After the reaction between OH^- and $HC_2H_3O_2$ has run to completion, the major species in solution are

$$HC_2H_3O_2, \quad Na^+, \quad C_2H_3O_2^-, \quad \text{and} \quad H_2O$$

The dominant equilibrium involves the dissociation of acetic acid.

This problem is very similar to that in part a. The only difference is that the addition of 0.01 mole of OH^- has consumed some $HC_2H_3O_2$ and produced some $C_2H_3O_2^-$, yielding the following concentrations:

Initial Concentration (mol/L)		Equilibrium Concentration (mol/L)
$[HC_2H_3O_2]_0 = 0.49$ $[C_2H_3O_2^-]_0 = 0.51$ $[H^+]_0 \approx 0$	x mol/L of $HC_2H_3O_2$ dissociates to reach equilibrium $\longrightarrow$	$[HC_2H_3O_2] = 0.49 - x$ $[C_2H_3O_2^-] = 0.51 + x$ $[H^+] = x$

Note that the initial concentrations are defined after the reaction with OH^- is complete but before the system adjusts to equilibrium.

Following the usual procedures gives

$$K_a = 1.8 \times 10^{-5} = \frac{[H^+][C_2H_3O_2^-]}{[HC_2H_3O_2]} = \frac{(x)(0.51 + x)}{0.49 - x} \approx \frac{(x)(0.51)}{0.49}$$

$$x \approx 1.7 \times 10^{-5}$$

The approximations are valid (by the 5% rule), so

$$[H^+] = x = 1.7 \times 10^{-5} \quad \text{and} \quad \text{pH} = 4.76$$

The change in pH produced by the addition of 0.01 mole of OH^- to this buffered solution is then

$$\underset{\text{New solution}}{\underset{\uparrow}{4.76}} - \underset{\text{Original solution}}{\underset{\uparrow}{4.74}} = +0.02$$

The pH has increased by 0.02 pH units.

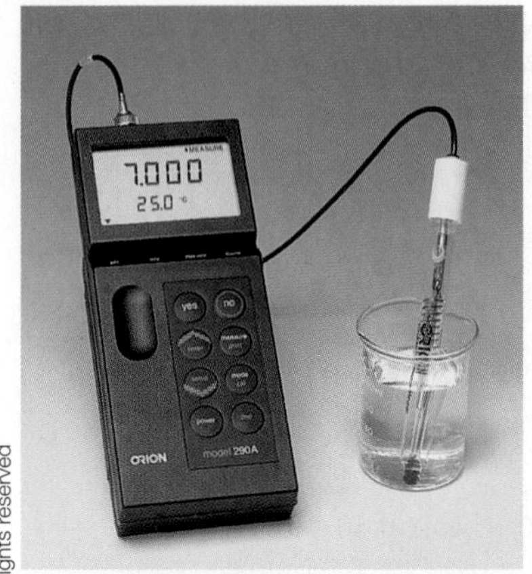

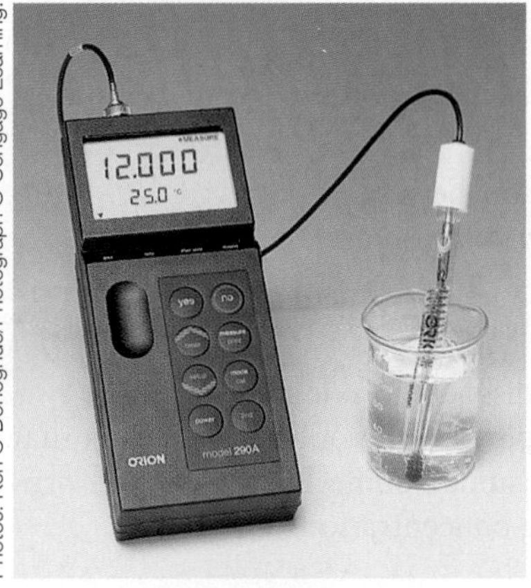

(top) Pure water at pH 7.00.
(bottom) When 0.01 mole of NaOH is added to 1.0 L of pure water, the pH jumps to 12.00.

Now compare this result with what happens when 0.01 mole of solid NaOH is added to 1.0 L of water to give 0.01 *M* NaOH. In this case, $[OH^-] = 0.01\ M$ and

$$[H^+] = \frac{K_w}{[OH^-]} = \frac{1.0 \times 10^{-14}}{1.0 \times 10^{-2}} = 1.0 \times 10^{-12}$$

$$pH = 12.00$$

Thus the change in pH is

$$\underset{\text{New solution}}{\underset{\uparrow}{12.00}} - \underset{\text{Pure water}}{\underset{\uparrow}{7.00}} = +5.00$$

The increase is 5.00 pH units. Note how well the buffered solution resists a change in pH compared with pure water.

Example 8.2 is a typical buffer problem. It contains all the concepts necessary for handling the calculations for buffered solutions containing weak acids. Pay special attention to the following points:

1. Buffered solutions are simply solutions of weak acids or bases containing a common ion. The pH calculations for buffered solutions require exactly the same procedures previously introduced in Chapter 7. *This is not a new type of problem.*
2. When a strong acid or base is added to a buffered solution, it is best to deal with the stoichiometry of the resulting reaction first. After the stoichiometric calculations are completed, then consider the equilibrium calculations. This procedure can be represented as follows:

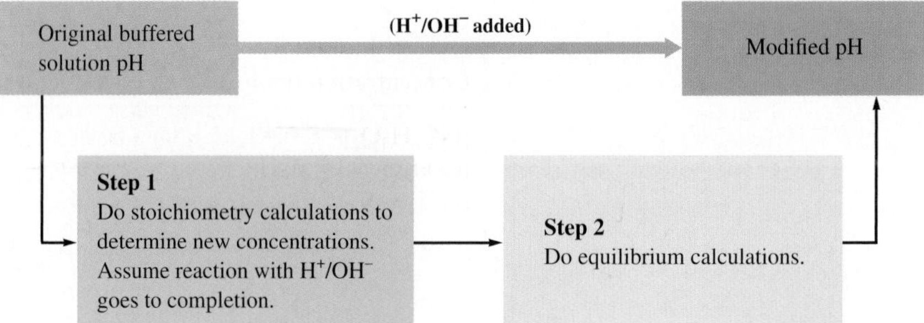

Buffering: How Does It Work?

Example 8.2 demonstrates the ability of a buffered solution to absorb hydroxide ions without a significant change in pH. *But how does a buffer work?* Suppose a buffered solution contains relatively large quantities of a weak acid HA and its conjugate base A^-. Since the weak acid represents the best source of protons, the following reaction occurs when hydroxide ions are added to the solution:

$$OH^- + HA \longrightarrow A^- + H_2O$$

The net result is that OH^- ions are not allowed to accumulate but are replaced by A^- ions.

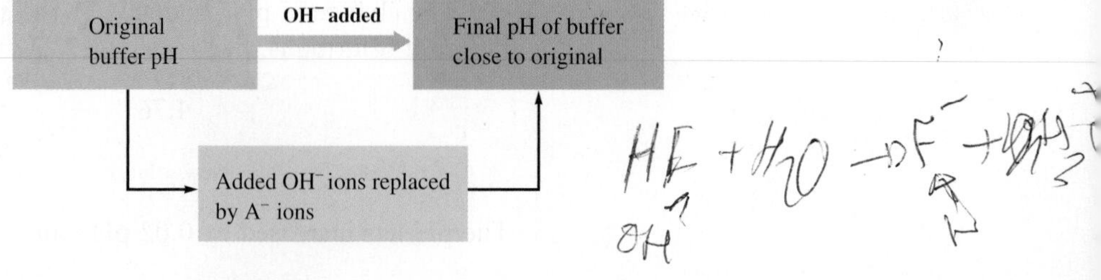

The stability of the pH under these conditions can be understood by examining the equilibrium expression for the dissociation of HA:

$$K_a = \frac{[H^+][A^-]}{[HA]} \quad \text{or rearranging,} \quad [H^+] = K_a\frac{[HA]}{[A^-]}$$

In a buffered solution the pH is governed by the ratio [HA]/[A⁻].

In other words, the *equilibrium concentration of* H^+ *and thus the pH are determined by the ratio [HA]/[*A^-*]*. When OH^- ions are added, HA is converted to A^-, causing the ratio [HA]/[A^-] to decrease. However, *if the amounts of HA and* A^- *originally present are very large compared with the amount of* OH^- added, the change in the [HA]/[A^-] ratio is small.

In Example 8.2

$$\frac{[HA]}{[A^-]} = \frac{0.50}{0.50} = 1.0 \qquad \text{Initially}$$

$$\frac{[HA]}{[A^-]} = \frac{0.49}{0.51} = 0.96 \qquad \text{After adding 0.01 mol/L of } OH^-$$

The change in the ratio [HA]/[A^-] is very small. Thus the [H^+] and the pH remain essentially constant.

The essence of buffering, then, is that [HA] and [A^-] are large compared with the amount of OH^- added. Thus, when the OH^- is added, the concentrations of HA and A^- change, but only by small amounts. Under these conditions, the [HA]/[A^-] ratio and thus the [H^+] stay virtually constant.

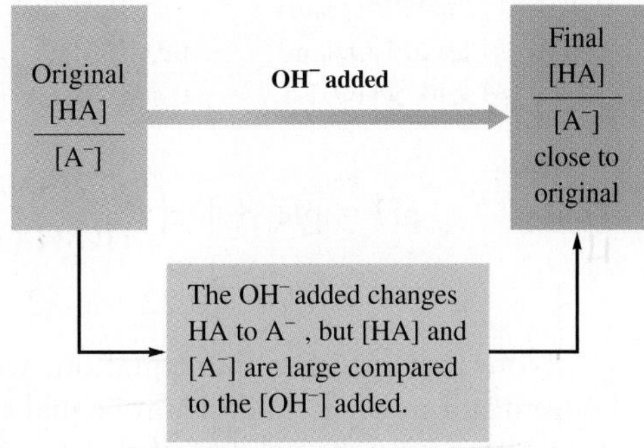

Similar reasoning applies when protons are added to a buffered solution containing a weak acid and a salt of its conjugate base. Because the A^- ion has a high affinity of H^+, the added H^+ ions react with A^- to form the weak acid:

$$H^+ + A^- \longrightarrow HA$$

Thus free H^+ ions do not accumulate. In this case there will be a net change of A^- to HA. However, if [A^-] and [HA] are large compared with the [H^+] added, only a slight change in the pH occurs.

An alternative form of the acid dissociation equilibrium expression,

$$[H^+] = K_a\frac{[HA]}{[A^-]} \tag{8.1}$$

is often useful for calculating [H^+] in buffered solutions, since [HA] and [A^-] are usually known. For example, to calculate [H^+] in a buffered solution containing 0.10 *M* HF ($K_a = 7.2 \times 10^{-4}$) and 0.30 *M* NaF, we simply substitute the respective concentrations into Equation (8.1):

$$[H^+] = \underset{K_a}{(7.2 \times 10^{-4})}\frac{\overset{[HF]}{0.10}}{\underset{[F^-]}{0.30}} = 2.4 \times 10^{-4}\ M$$

Another useful form of Equation (8.1) can be obtained by taking the negative log of both sides:

$$-\log[H^+] = -\log(K_a) - \log\left(\frac{[HA]}{[A^-]}\right)$$

That is,

$$pH = pK_a - \log\left(\frac{[HA]}{[A^-]}\right)$$

or inverting the log term and reversing the sign,

$$pH = pK_a + \log\left(\frac{[A^-]}{[HA]}\right) = pK_a + \log\left(\frac{[\text{base}]}{[\text{acid}]}\right) \tag{8.2}$$

This log form of the expression for K_a is called the **Henderson–Hasselbalch equation** and is useful for calculating the pH of solutions when the ratio $[HA]/[A^-]$ is known.

For a particular buffering system (acid–conjugate base pair), all solutions that have the same ratio of $[A^-]/[HA]$ have the same pH. For example, a buffered solution containing 5.0 M $HC_2H_3O_2$ and 3.0 M $NaC_2H_3O_2$ has the same pH as one containing 0.050 M $HC_2H_3O_2$ and 0.030 M $NaC_2H_3O_2$. This result can be shown as follows:

System	$[A^-]/[HA]$
5.0 *M* $HC_2H_3O_2$ and 3.0 *M* $NaC_2H_3O_2$	$\frac{3.0\ M}{5.0\ M} = 0.60$
0.050 *M* $HC_2H_3O_2$ and 0.030 *M* $NaC_2H_3O_2$	$\frac{0.030\ M}{0.050\ M} = 0.60$

Thus

$$pH = pK_a + \log\left(\frac{[C_2H_3O_2^-]}{[HC_2H_3O_2]}\right) = 4.74 + \log(0.60)$$

$$= 4.74 - 0.22 = 4.52$$

Note that in using this equation, we have assumed that the equilibrium concentrations of A^- and HA are equal to their initial concentrations. That is, we are assuming the validity of the approximations

$$[A^-] = [A^-]_0 + x \approx [A^-]_0 \quad \text{and} \quad [HA] = [HA]_0 - x \approx [HA]_0$$

where x represents the amount of acid that dissociates. Since the initial concentrations of HA and A^- are expected to be relatively large in a buffered solution, this assumption is generally acceptable.

OWL INTERACTIVE EXAMPLE 8.3

A buffered solution contains 0.25 *M* NH_3 ($K_b = 1.8 \times 10^{-5}$) and 0.40 *M* NH_4Cl.

a. Calculate the pH of this solution.

Solution The major species in the solution are

$$NH_3, \quad \underbrace{NH_4^+, \quad Cl^-,}_{\text{From the dissolved } NH_4Cl} \quad \text{and} \quad H_2O$$

Major Species

H₂O

Since Cl^- is such a weak base and water is a weak acid or base, the important equilibrium is

$$NH_3(aq) + H_2O(l) \rightleftharpoons NH_4^+(aq) + OH^-(aq)$$

and

$$K_b = 1.8 \times 10^{-5} = \frac{[NH_4^+][OH^-]}{[NH_3]}$$

The appropriate concentrations are as follows:

Initial Concentration (mol/L)		Equilibrium Concentration (mol/L)
$[NH_3]_0 = 0.25$ $[NH_4^+]_0 = 0.40$ $[OH^-] \approx 0$	$\xrightarrow[\text{reacts with } H_2O]{x \text{ mol/L } NH_3}$	$[NH_3] = 0.25 - x$ $[NH_4^+] = 0.40 + x$ $[OH^-] = x$

Thus

$$K_b = 1.8 \times 10^{-5} = \frac{[NH_4^+][OH^-]}{[NH_3]} = \frac{(0.40 + x)(x)}{0.25 - x} \approx \frac{(0.40)(x)}{0.25}$$

$$x = 1.1 \times 10^{-5}$$

The approximations are valid (by the 5% rule), so

$$[OH^-] = x = 1.1 \times 10^{-5}$$

$$pOH = 4.95$$

$$pH = 14.00 - 4.95 = 9.05$$

This case is typical of a buffered solution in that the initial and equilibrium concentrations of buffering materials are essentially the same.

Alternative Solution. There is another way to solve this problem. Since the solution contains relatively large quantities of *both* NH_4^+ and NH_3, we can use the equilibrium

$$NH_3(aq) + H_2O(l) \rightleftharpoons NH_4^+(aq) + OH^-(aq)$$

to calculate $[OH^-]$ and then calculate $[H^+]$ from K_w, as we have just done. Or we can use the dissociation equilibrium for NH_4^+,

$$NH_4^+(aq) \rightleftharpoons NH_3(aq) + H^+(aq)$$

to calculate $[H^+]$ directly. *Either choice will give the same answer,* since the same equilibrium concentrations of NH_3 and NH_4^+ must satisfy both equilibria.

We can obtain the K_a value for NH_4^+ from the given K_b value for NH_3, since $K_a \times K_b = K_w$:

$$K_a = \frac{K_w}{K_b} = \frac{1.0 \times 10^{-14}}{1.8 \times 10^{-5}} = 5.6 \times 10^{-10}$$

Then using the Henderson–Hasselbalch equation, we have

$$pH = pK_a + \log\left(\frac{[\text{base}]}{[\text{acid}]}\right)$$

$$= 9.25 + \log\left(\frac{0.25\ M}{0.40\ M}\right) = 9.25 - 0.20 = 9.05$$

b. Calculate the pH of the solution that results when 0.10 mole of gaseous HCl is added to 1.0 L of the buffered solution from part a.

Solution *Before any reaction occurs,* the solution contains the following major species:

$$NH_3, \quad NH_4^+, \quad Cl^-, \quad H^+, \quad \text{and} \quad H_2O$$

Remember: Think about the chemistry first. Ask yourself if a reaction will occur among the major species.

What reaction can occur? We know that H^+ will not react with Cl^- to form HCl. In contrast to Cl^-, the NH_3 molecule has a great affinity for protons [this is demonstrated by the fact that NH_4^+ is such a weak acid ($K_a = 5.6 \times 10^{-10}$)]. Thus NH_3 will react with H^+ to form NH_4^+:

$$NH_3(aq) + H^+(aq) \longrightarrow NH_4^+(aq)$$

Since we can assume that this reaction goes essentially to completion, forming the very weak acid NH_4^+, we will do the stoichiometric calculations before we consider the equilibrium calculations. That is, we will let the reaction run to completion and then consider the equilibrium.

The stoichiometric calculations for this process are shown below.

	NH_3	+	H^+	$\longrightarrow$	NH_4^+
Before reaction:	(1.0 L)(0.25 *M*) = 0.25 mole		0.10 mole ↑ Limiting reactant		(1.0 L)(0.40 *M*) = 0.40 mole
After reaction:	0.25 − 0.10 = 0.15 mole		0		0.40 + 0.10 = 0.50 mole

After the reaction goes to completion, the solution contains the major species

$$NH_3, \quad NH_4^+, \quad Cl^-, \quad \text{and} \quad H_2O$$

and

$$[NH_3]_0 = \frac{0.15 \text{ mol}}{1.0 \text{ L}} = 0.15\ M$$

$$[NH_4^+]_0 = \frac{0.50 \text{ mol}}{1.0 \text{ L}} = 0.50\ M$$

We can use the Henderson–Hasselbalch equation, where

$$[\text{Base}] = [NH_3] \approx [NH_3]_0 = 0.15\ M$$

$$[\text{Acid}] = [NH_4^+] \approx [NH_4^+]_0 = 0.50\ M$$

Then

$$\text{pH} = \text{p}K_a + \log\left(\frac{[NH_3]}{[NH_4^+]}\right) = 9.25 + \log\left(\frac{0.15\ M}{0.50\ M}\right)$$

$$= 9.25 - 0.52 = 8.73$$

Note that the addition of HCl only slightly decreases the pH (from 9.05 to 8.73), as we would expect in a buffered solution.

We can now summarize the most important characteristics of buffered solutions.

Characteristics of Buffered Solutions

1. Buffered solutions contain relatively large concentrations of a weak acid and its corresponding weak base. They can involve a weak acid HA and the conjugate base A^- or a weak base B and the conjugate acid BH^+.
2. When H^+ is added to a buffered solution, it reacts essentially to completion with the weak base present:

$$H^+ + A^- \longrightarrow HA \quad \text{or} \quad H^+ + B \longrightarrow BH^+$$

3. When OH^- is added to a buffered solution, it reacts essentially to completion with the weak acid present:

$$OH^- + HA \longrightarrow A^- + H_2O \quad \text{or} \quad OH^- + BH^+ \longrightarrow B + H_2O$$

4. The pH of the buffered solution is determined by the ratio of the concentrations of the weak base and weak acid. As long as this ratio remains virtually constant, the pH will remain virtually constant. This will be the case as long as the concentrations of the buffering materials (HA and A^- or B and BH^+) are large compared with the amounts of H^+ or OH^- added.

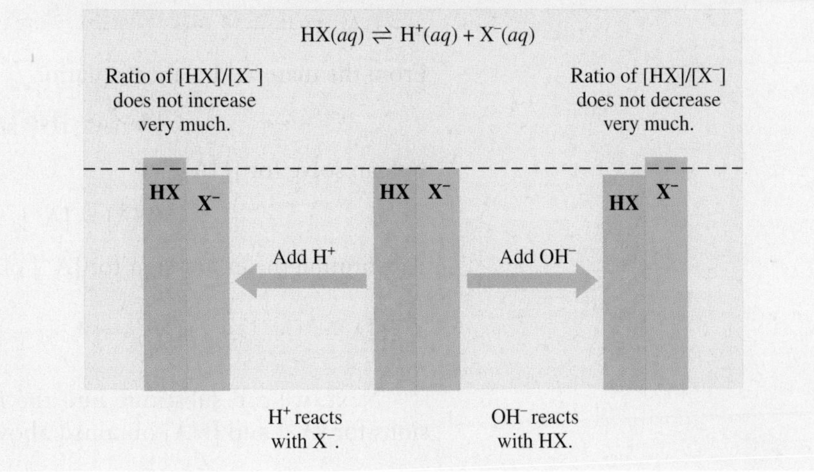

8.3 Exact Treatment of Buffered Solutions

Because buffered solutions can have a pH value of 7, the question naturally arises about the possible importance of water as a contributor to the $[H^+]$ under these conditions. To derive a general equation for buffered solutions that includes any possible contribution from water, we will use the approach developed in Section 7.9.

In an aqueous buffered solution containing the weak acid HA and its salt NaA, the equilibria of interest are

$$HA(aq) \rightleftharpoons H^+(aq) + A^-(aq)$$

and

$$H_2O(l) \rightleftharpoons H^+(aq) + OH^-(aq)$$

The charge balance relationship for this solution is

$$[Na^+] + [H^+] = [A^-] + [OH^-]$$

in which we must include the Na^+ added to the solution.

The material balance equation based on the conservation of A is

$$\underset{\text{From added NaA}}{[A^-]_0} + \underset{\text{Original HA concentration}}{[HA]_0} = [HA] + [A^-]$$

Since $[A^-]_0 = [Na^+]$, because all of the Na^+ comes from the added salt (NaA), and

$$[OH^-] = \frac{K_w}{[H^+]},$$

we can rewrite the charge balance equation

$$[Na^+] + [H^+] = [A^-] + [OH^-]$$

as follows:

$$[A^-]_0 + [H^+] = [A^-] + \frac{K_w}{[H^+]}$$

We can now solve for $[A^-]$:

$$[A^-] = [A^-]_0 + [H^+] - \frac{K_w}{[H^+]}$$
$$= [A^-]_0 + \frac{[H^+]^2 - K_w}{[H^+]}$$

From the material balance equation

$$[A^-]_0 + [HA]_0 = [HA] + [A^-]$$

we can solve for [HA]:

$$[HA] = [A^-]_0 + [HA]_0 - [A^-]$$

Substituting the expression for $[A^-]$ obtained above, we have

$$[HA] = [A^-]_0 + [HA]_0 - [A^-]_0 - \frac{[H^+]^2 - K_w}{[H^+]} = [HA]_0 - \frac{[H^+]^2 - K_w}{[H^+]}$$

Next, we can substitute into the K_a expression for HA using the expressions for $[A^-]$ and [HA] obtained above:

$$K_a = \frac{[H^+][A^-]}{[HA]} = \frac{[H^+]\left([A^-]_0 + \dfrac{[H^+]^2 - K_w}{[H^+]}\right)}{[HA]_0 - \dfrac{[H^+]^2 - K_w}{[H^+]}}$$

Notice that when $[A^-]_0$ is zero—that is, when no salt is added—this expression turns into the equation obtained in Section 7.9 for a weak acid HA dissolved in water. Thus the strategies useful for solving this equation are the same as those presented in Section 7.9.

OWL INTERACTIVE EXAMPLE 8.4

Calculate the $[H^+]$ of a buffered solution containing 3.0×10^{-4} *M* HOCl ($K_a = 3.5 \times 10^{-8}$) and 1.0×10^{-4} *M* NaOCl.

Solution We recognize that this problem involves a buffered solution containing the weak acid HOCl and its salt NaOCl.

First let's solve this problem using the regular procedures:

$$K_a = \frac{[H^+][OCl^-]}{[HOCl]} = 3.5 \times 10^{-8}$$

If we let $x = [H^+]$, then

$$[OCl^-] = 1.0 \times 10^{-4} + x$$
$$[HOCl] = 3.0 \times 10^{-4} - x$$

and we have

$$3.5 \times 10^{-8} = \frac{[H^+][OCl^-]}{[HOCl]} = \frac{(x)(1.0 \times 10^{-4} + x)}{(3.0 \times 10^{-4} - x)}$$

Assuming x is small compared to 1.0×10^{-4}, we have

$$3.5 \times 10^{-8} \approx \frac{(x)(1.0 \times 10^{-4})}{(3.0 \times 10^{-4})}$$

$$[H^+] = x = \frac{1.05 \times 10^{-11}}{1.0 \times 10^{-4}} = 1.05 \times 10^{-7}\,M = 1.1 \times 10^{-7}\,M$$

Note that the $[H^+]$ value obtained in this way is very close to 1.0×10^{-7} *M*. Since the calculated $[H^+]$ is so close to 10^{-7} *M*, must we consider the contribution of water to the pH?

To answer that question, we will use the complete equation derived earlier, which does include water's contribution to the pH. This equation is

$$K_a = 3.5 \times 10^{-8} = \frac{[H^+]\left([OCl^-]_0 + \dfrac{[H^+]^2 - 1.0 \times 10^{-14}}{[H^+]}\right)}{[HOCl]_0 - \dfrac{[H^+]^2 - 1.0 \times 10^{-14}}{[H^+]}}$$

where

$$[OCl^-]_0 = 1.0 \times 10^{-4}\ M$$

and

$$[HOCl]_0 = 3.0 \times 10^{-4}\ M$$

We expect the final $[H^+]$ to be very close to 1.0×10^{-7}, so the term

$$\frac{[H^+]^2 - 1.0 \times 10^{-14}}{[H^+]}$$

will be very small since $[H^+]^2$ will be very close to 1.0×10^{-14}. Thus, as a first approximation, we will assume that

$$[OCl^-]_0 = 1.0 \times 10^{-4}\ M >>> \frac{[H^+]^2 - 1.0 \times 10^{-14}}{[H^+]}$$

and

$$[HOCl]_0 = 3.0 \times 10^{-4}\ M >>> \frac{[H^+]^2 - 1.0 \times 10^{-14}}{[H^+]}$$

The expression then becomes

$$3.5 \times 10^{-8} = \frac{[H^+][OCl^-]_0}{[HOCl]_0} = \frac{[H^+](1.0 \times 10^{-4})}{(3.0 \times 10^{-4})}$$

and

$$[H^+] = 1.05 \times 10^{-7}\ M = 1.1 \times 10^{-7}\ M$$

as we found before.

Using this result we can check the magnitude of the neglected term

$$\frac{[H^+]^2 - 1.0 \times 10^{-14}}{[H^+]} = \frac{(1.05 \times 10^{-7})^2 - 1.0 \times 10^{-14}}{1.05 \times 10^{-7}} = 9.8 \times 10^{-9}$$

This term is clearly negligible as compared with 1.0×10^{-4} and 3.0×10^{-4}.

In this case water does not have to be explicitly taken into account. Notice that the rigorous equation produces a different result from the normal procedure only when the buffering materials are very dilute (on the order of 10^{-6} *M* or smaller). For buffered solutions with "normal" concentrations of HA and A^-, the simple approach used in this text to treat buffers will suffice. However, the rigorous equation always gives the correct answer and has the advantage of emphasizing explicitly the approximations being made in the simpler approach.

8.4 Buffer Capacity

A buffer with a large capacity contains large concentrations of the buffering components.

The **buffer capacity** of a buffered solution is defined in terms of the amount of protons or hydroxide ions it can absorb without a significant change in pH. A buffer with a large capacity contains large concentrations of buffering components and so can absorb a relatively large amount of protons or hydroxide ions and show little pH change. *The pH of a buffered solution is determined by the ratio [A⁻]/[HA]. The capacity of a buffered solution is determined by the magnitudes of [HA] and [A⁻].*

EXAMPLE 8.5

Calculate the change in pH that occurs when 0.010 mole of gaseous HCl is added to 1.0 L of each of the following solutions.

Solution A: 5.00 *M* $HC_2H_3O_2$ and 5.00 *M* $NaC_2H_3O_2$

Solution B: 0.050 *M* $HC_2H_3O_2$ and 0.050 *M* $NaC_2H_3O_2$

For acetic acid, $K_a = 1.8 \times 10^{-5}$.

Solution For both solutions the initial pH can be determined from the Henderson–Hasselbalch equation:

$$pH = pK_a + \log\left(\frac{[C_2H_3O_2^-]}{[HC_2H_3O_2]}\right)$$

In each case $[C_2H_3O_2^-] = [HC_2H_3O_2]$. Thus, for both A and B,

$$pH = pK_a + \log(1) = pK_a = -\log(1.8 \times 10^{-5}) = 4.74$$

After the addition of HCl to each of these solutions, the major species *before any reaction occurs* are

$$HC_2H_3O_2, \quad Na^+, \quad C_2H_3O_2^-, \quad \underbrace{H^+, \quad Cl^-,}_{\text{From the added HCl}} \quad \text{and} \quad H_2O$$

Will any reactions occur among these species? Note that we have a relatively large quantity of H^+ that will readily react with any effective base. We know that Cl^- will not react with H^+ to form HCl in water. However, $C_2H_3O_2^-$ will react with H^+ to form the weak acid $HC_2H_3O_2$:

$$H^+(aq) + C_2H_3O_2^-(aq) \longrightarrow HC_2H_3O_2(aq)$$

Because $HC_2H_3O_2$ is a weak acid, we assume that this reaction runs to completion; the 0.010 mole of added H^+ will convert 0.010 mole of $C_2H_3O_2^-$ to 0.010 mole of $HC_2H_3O_2$.

For solution A (since the solution volume is 1.0 L, the number of moles equals the molarity), we have the following concentrations:

	H^+	+ $C_2H_3O_2^-$	$\longrightarrow$ $HC_2H_3O_2$
Before reaction:	0.010 *M*	5.00 *M*	5.00 *M*
After reaction:	0	4.99 *M*	5.01 *M*

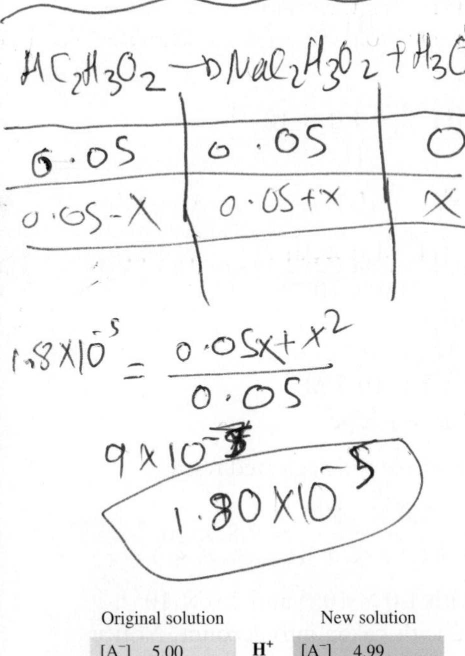

The new pH can be obtained by substituting the new concentrations into the Henderson–Hasselbalch equation:

$$pH = pK_a + \log\left(\frac{[C_2H_3O_2^-]}{[HC_2H_3O_2]}\right) = 4.74 + \log\left(\frac{4.99}{5.01}\right)$$

$$= 4.74 - 0.0017 = 4.74$$

There is virtually no change in pH for solution A when 0.010 mole of gaseous HCl is added.

For solution B we have the following calculations:

	H^+	+ $C_2H_3O_2^-$	$\longrightarrow$ $HC_2H_3O_2$
Before reaction:	0.010 *M*	0.050 *M*	0.050 *M*
After reaction:	0	0.040 *M*	0.060 *M*

Original solution

$\frac{[A^-]}{[HA]} = \frac{.050}{.050} = 1.0$ $\xrightarrow{H^+ \text{ added}}$ New solution $\frac{[A^-]}{[HA]} = \frac{0.040}{0.060} = 0.67$

The new pH is

$$pH = 4.74 + \log\left(\frac{0.040}{0.060}\right) = 4.74 - 0.18 = 4.56$$

Although the pH change for solution B is small, a change did occur, in contrast to the case for solution A.

These results show that solution A, which contains much larger quantities of buffering components, has a much higher buffering capacity than solution B.

We have seen that the pH of a buffered solution depends on the ratio of the concentrations of buffering components. When this ratio is least affected by added protons or hydroxide ions, the solution is the most resistant to a change in pH. To find the ratio that gives optimum buffering, suppose that we have a buffered solution containing a large concentration of acetate ion and only a small concentration of acetic acid. Addition of protons to form acetic acid will produce a relatively large *percentage* change in the concentration of acetic acid and so will produce a relatively large change in the ratio $[C_2H_3O_2^-]/[HC_2H_3O_2]$ (Table 8.1). Similarly, if hydroxide ions are added to remove some acetic acid, the percentage change in the concentration of acetic acid is again large. The same effects are seen if the initial concentration of acetic acid is large and that of acetate ion is small.

Because large changes in the ratio $[A^-]/[HA]$ will produce large changes in pH, we want to avoid this situation for the most effective buffering. This type of reasoning leads us to the general conclusion that optimum buffering will occur when [HA] is equal to $[A^-]$. It is under this condition that the ratio $[A^-]/[HA]$ is most resistant to change when H^+ or OH^- is added. Thus, when choosing the buffering components for a specific application, we want $[A^-]/[HA]$ to equal 1. It follows that since

$$pH = pK_a + \log\left(\frac{[A^-]}{[HA]}\right) = pK_a + \log(1) = pK_a$$

the *pK_a of the weak acid selected for the buffer should be as close as possible to the desired pH.* For example, suppose we need a buffered solution with a pH of 4.00. The most effective buffering will occur when [HA] is equal to $[A^-]$. From the Henderson–Hasselbalch equation,

$$pH = pK_a + \log\left(\frac{[A^-]}{[HA]}\right)$$

pH ↑ 4.00; log ratio ↑ Ratio = 1 for most effective buffer

That is,

$$4.00 = pK_a + \log(1) = pK_a + 0 \quad \text{and} \quad pK_a = 4.00$$

Thus in this case the best choice is a weak acid that has $pK_a = 4.00$, or $K_a = 1.0 \times 10^{-4}$.

Solution A

Original A: $\frac{[Ac^-]}{[HAc]} = 1.00$ $\xrightarrow{H^+}$ Final A: $\frac{[Ac^-]}{[HAc]} = 0.980$

2% change

Solution B

Original B: $\frac{[Ac^-]}{[HAc]} = 100.$ $\xrightarrow{H^+}$ Final B: $\frac{[Ac^-]}{[HAc]} = 49.5$

50.5% change

Table 8.1

Change in $[C_2H_3O_2^-]/[HC_2H_3O_2]$ for Two Solutions When 0.01 mole of H^+ Is Added to 1.0 L of Each Solution

Solution	$\left(\frac{[C_2H_3O_2^-]}{[HC_2H_3O_2]}\right)_{orig}$	$\left(\frac{[C_2H_3O_2^-]}{[HC_2H_3O_2]}\right)_{new}$	Change	Percent Change
A	$\frac{1.00\ M}{1.00\ M} = 1.00$	$\frac{0.99\ M}{1.01\ M} = 0.98$	$1.00 \longrightarrow 0.98$	2.00%
B	$\frac{1.00\ M}{0.01\ M} = 100$	$\frac{0.99\ M}{0.02\ M} = 49.5$	$100 \longrightarrow 49.5$	50.5%

OWL INTERACTIVE EXAMPLE 8.6

A chemist needs to prepare a solution buffered at pH 4.30 using one of the following acids (and its sodium salt):

a. chloroacetic acid ($K_a = 1.35 \times 10^{-3}$)
b. propanoic acid ($K_a = 1.3 \times 10^{-5}$)
c. benzoic acid ($K_a = 6.4 \times 10^{-5}$)
d. hypochlorous acid ($K_a = 3.5 \times 10^{-8}$)

Calculate the ratio of $[HA]/[A^-]$ required for each system to yield a pH of 4.30. Which system will work best?

Solution A pH of 4.30 corresponds to

$$[H^+] = 10^{-4.30} = \text{antilog}(-4.30) = 5.0 \times 10^{-5}\ M$$

Since K_a values rather than pK_a values are given for the various acids, we use Equation (8.1),

$$[H^+] = K_a \frac{[HA]}{[A^-]}$$

instead of the Henderson–Hasselbalch equation. We substitute the required $[H^+]$ and K_a for each acid into Equation (8.1) to calculate each ratio of $[HA]/[A^-]$ needed. The results are as follows:

Acid	$[H^+] = K_a \frac{[HA]}{[A^-]}$	$\frac{[HA]}{[A^-]}$
a. Chloroacetic	$5.0 \times 10^{-5} = 1.35 \times 10^{-3}\left(\frac{[HA]}{[A^-]}\right)$	3.7×10^{-2}
b. Propanoic	$5.0 \times 10^{-5} = 1.3 \times 10^{-5}\left(\frac{[HA]}{[A^-]}\right)$	3.8
c. Benzoic	$5.0 \times 10^{-5} = 6.4 \times 10^{-5}\left(\frac{[HA]}{[A^-]}\right)$	0.78
d. Hypochlorous	$5.0 \times 10^{-5} = 3.5 \times 10^{-8}\left(\frac{[HA]}{[A^-]}\right)$	1.4×10^{-3}

Since $[HA]/[A^-]$ for benzoic acid is closest to 1, the system of benzoic acid and its sodium salt is the best choice among those given for buffering a solution at pH 4.30. This example demonstrates the principle that the optimum buffering system has a pK_a value close to the desired pH. The pK_a for benzoic acid is 4.19.

8.5 Titrations and pH Curves

As we saw in Chapter 4, a titration is commonly used to analyze the amount of acid or base in a solution. This process involves delivering a solution of known concentration (the titrant) from a buret into the unknown solution until the substance being analyzed is just consumed. The stoichiometric or equivalence point is usually signaled by the color change of an indicator. In this section we will discuss the pH changes that occur during an acid–base titration. We will use this information later in this chapter to show how an appropriate indicator can be chosen for a particular titration.

The progress of an acid–base titration is often monitored by plotting the pH of the solution being analyzed as a function of the amount of titrant added. Such a plot is called a **pH curve, or titration curve.**

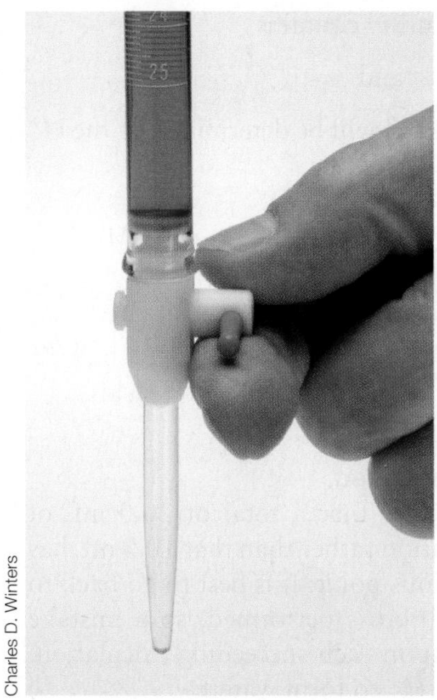
Charles D. Winters

A buret valve.

$1 \text{ millimole} = 1 \times 10^{-3} \text{ mol}$

$1 \text{ mL} = 1 \times 10^{-3} \text{ L}$

$\frac{\text{mmol}}{\text{mL}} = \frac{\text{mol}}{\text{L}} = M$

Strong Acid–Strong Base Titrations

The reaction for a strong acid–strong base titration is

$$H^+(aq) + OH^-(aq) \longrightarrow H_2O(l)$$

To compute $[H^+]$ at a given point in the titration, we must determine the moles of H^+ remaining at that point and divide by the total volume of the solution. Before we proceed, we need to consider a new unit, which is especially convenient for titrations. Since titrations usually involve small quantities (burets are typically graduated in milliliters), the mole is inconveniently large. Therefore, we will use the **millimole** (abbreviated **mmol**), where

$$1 \text{ mmol} = \frac{1 \text{ mol}}{1000} = 10^{-3} \text{ mol}$$

and

$$\text{Molarity} = \frac{\text{mol of solute}}{\text{L of solution}} = \frac{\dfrac{\text{mol of solute}}{1000}}{\dfrac{\text{L of solution}}{1000}} = \frac{\text{mmol of solute}}{\text{mL of solution}}$$

A 1.0 M solution thus contains 1.0 mole of solute per liter of solution or, *equivalently*, 1.0 millimole of solute per milliliter of solution. Just as we obtain the number of moles of solute from the product of the volume in liters and the molarity, we obtain the number of millimoles of solute from the product of the volume in milliliters and the molarity.

We will illustrate the calculations involved in a strong acid–strong base titration by considering the titration of 50.0 mL of 0.200 M HNO_3 with 0.100 M NaOH. We will calculate the pH of the solution at selected points during the course of the titration where specific volumes of 0.100 M NaOH have been added.

A. No NaOH has been added.

Since HNO_3 is a strong acid (is completely dissociated), the solution contains the major species

$$H^+, \quad NO_3^-, \quad \text{and} \quad H_2O$$

The pH is determined by the H^+ from the nitric acid. Since 0.200 M HNO_3 contains 0.200 M H^+,

$$[H^+] = 0.200\ M \quad \text{and} \quad \text{pH} = 0.699$$

B. 10.0 mL of 0.100 *M* NaOH has been added.

In the mixed solution *before any reaction occurs,* the major species are

$$H^+, \quad NO_3^-, \quad Na^+, \quad OH^-, \quad \text{and} \quad H_2O$$

Note that large quantities of both H^+ and OH^- are present. The 1.0 mmol (10.0 mL × 0.10 M) of added OH^- will react with 1.0 mmol of H^+ to form water:

	H^+	+	OH^-	$\longrightarrow$	H_2O
Before reaction:	50.0 mL × 0.200 M = 10.0 mmol		10.0 mL × 0.100 M = 1.00 mmol		
After reaction:	10.0 − 1.00 = 9.0 mmol		1.00 − 1.00 = 0 mmol		

After the reaction is complete, the solution contains

$$H^+, \quad NO_3^-, \quad Na^+, \quad \text{and} \quad H_2O$$

(The OH^- ions have been consumed.) The pH will be determined by the H^+ remaining:

The final solution volume is the sum of the original volume of HNO_3 and the volume of added NaOH.

$$[H^+] = \frac{\text{mmol } H^+ \text{ left}}{\text{volume of solution (mL)}} = \frac{9.0 \text{ mmol}}{(50.0 + 10.0) \text{ mL}} = 0.15\ M$$

(50.0: Original volume of HNO_3 solution; 10.0: Volume of NaOH added)

$$pH = -\log(0.15) = 0.82$$

C. 20.0 mL (total) of 0.100 *M* NaOH has been added.
We consider this point from the perspective that a total of 20.0 mL of NaOH has been added to the *original* solution rather than that 10.0 mL has been added to the solution from the previous point. It is best to go back to the original solution each time a calculation is performed, so a mistake made at an earlier point does not show up in each succeeding calculation. As before, the added OH^- will react with H^+ to form water:

Go back to the original solution each time you perform a calculation to avoid mistakes.

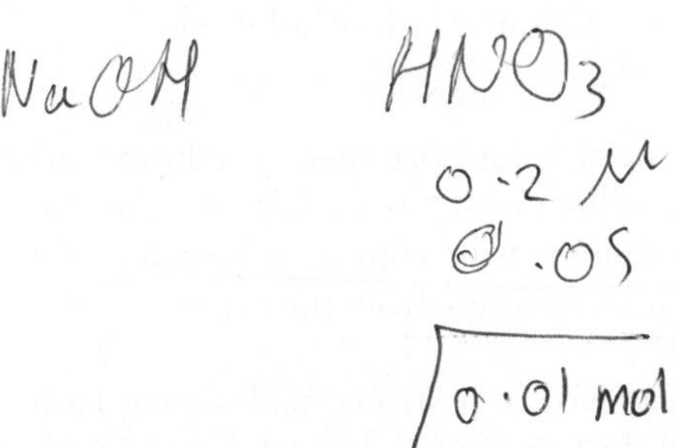

	H^+	+	OH^-	$\longrightarrow$	H_2O
Before reaction:	50.0 mL × 0.200 *M* = 10.0 mmol		20.0 mL × 0.100 *M* = 2.00 mmol		
After reaction:	10.0 − 2.00 = 8.0 mmol		2.00 − 2.00 = 0 mmol		

After the reaction,

$$[H^+] = \frac{8.0 \text{ mmol}}{(50.0 + 20.0) \text{ mL}} = 0.11\ M$$

(8.0 mmol: H^+ remaining)

$$pH = 0.96$$

D. 50.0 mL (total) of 0.100 *M* NaOH has been added.
Proceeding exactly as for the previous two points, we find the pH to be 1.30.

E. 100.0 mL (total) of 0.100 *M* NaOH has been added.
At this point the amount of NaOH that has been added is

$$100.0 \text{ mL} \times 0.100\ M = 10.0 \text{ mmol}$$

The original amount of nitric acid was

$$50.0 \text{ mL} \times 0.200\ M = 10.0 \text{ mmol}$$

Enough OH^- has been added to react exactly with all of the H^+ from the nitric acid. This is the **stoichiometric point,** or **equivalence point,** of the titration. At this point the major species in the solution are

$$Na^+, \quad NO_3^-, \quad \text{and} \quad H_2O$$

Since Na^+ has no acid or base properties and NO_3^- is the anion of the strong acid HNO_3 and is therefore a very weak base, neither NO_3^- nor Na^+ affects the pH. Thus the solution is neutral (the pH is 7.00).

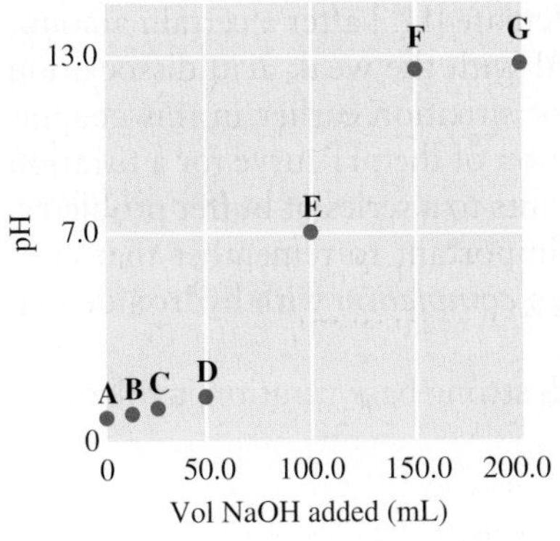

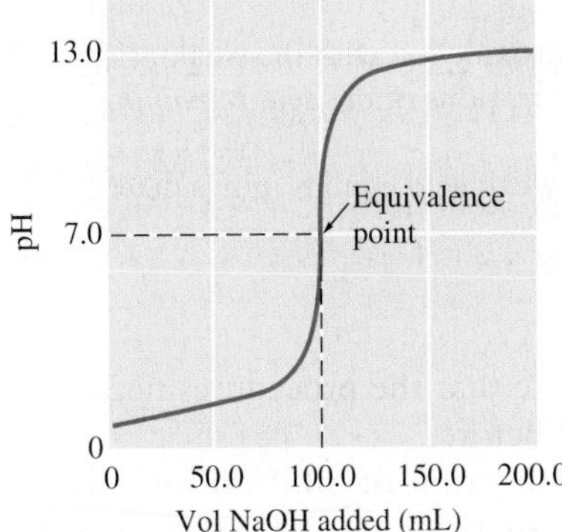

Figure 8.1
The pH curve for the titration of 50.0 mL of 0.200 *M* HNO_3 with 0.100 *M* NaOH. Note that the equivalence point occurs when 100.0 mL of NaOH has been added, the point where exactly enough OH^- has been added to react with all the H^+ originally present. The pH of 7 at the equivalence point is characteristic of a strong acid–strong base titration.

F. 150.0 mL (total) of 0.100 *M* NaOH has been added.
The titration reaction is as follows:

	H^+	+	OH^-	$\longrightarrow$	H_2O
Before reaction:	50.0 mL × 0.200 *M* = 10.0 mmol		150.00 mL × 0.100 *M* = 15.0 mmol		
After reaction:	10.0 − 10.0 = 0 mmol		15.0 − 10.0 = 5.0 mmol ↑ Excess OH^- added		

Now the OH^- is *in excess* and thus will determine the pH:

$$[OH^-] = \frac{\text{mmol OH}^- \text{ in excess}}{\text{volume (mL)}} = \frac{5.0 \text{ mmol}}{(50.0 + 150.0) \text{ mL}} = \frac{5.0 \text{ mmol}}{200.0 \text{ mL}}$$

$$= 0.025\ M$$

Since $[H^+][OH^-] = 1.0 \times 10^{-14}$,

$$[H^+] = \frac{1.0 \times 10^{-14}}{2.5 \times 10^{-2}} = 4.0 \times 10^{-13}\ M \quad \text{and} \quad \text{pH} = 12.40$$

G. 200.0 mL (total) of 0.100 *M* NaOH has been added.
Proceeding as for the previous point, we find the pH to be 12.60.

The results of these calculations are summarized by the pH curve shown in Fig. 8.1. Note that the pH changes very gradually until the titration is close to the equivalence point, where a dramatic change occurs. This behavior occurs because early in the titration there is a relatively large amount of H^+ in the solution, and the addition of a given amount of OH^- thus produces only a small change in pH. However, near the equivalence point $[H^+]$ is relatively small, and the addition of a small amount of OH^- produces a large change.

The pH curve in Fig. 8.1, typical of the titration of a strong acid with a strong base, has the following characteristics.

Before the equivalence point, $[H^+]$ (and hence the pH) can be calculated by dividing the number of millimoles of H^+ remaining at that point by the total volume of the solution in milliliters.

At the equivalence point, the pH is 7.00.

After the equivalence point, $[OH^-]$ can be calculated by dividing the number of millimoles of excess OH^- by the total volume of the solution. Then $[H^+]$ is obtained from K_w.

The titration of a strong base with a strong acid requires reasoning very similar to that used above, except, of course, that OH^- is in excess before the equivalence point, and H^+ is in excess after the equivalence point. The pH curve for the titration of 100.0 mL of 0.50 *M* NaOH with 1.0 *M* HCl is shown in Fig. 8.2.

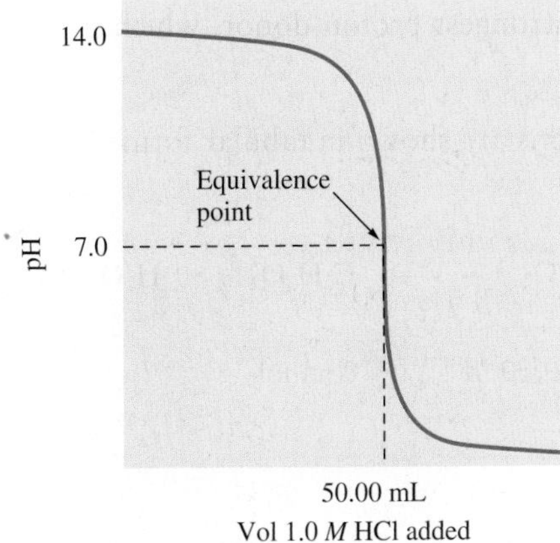

Figure 8.2
The pH curve for the titration of 100.0 mL of 0.50 *M* NaOH with 1.0 *M* HCl. The equivalence point occurs when 50.00 mL of HCl has been added, since at this point 5.0 mmol of H^+ ions has been added to react with the original 5.0 mmol of OH^- ions.

Titrations of Weak Acids with Strong Bases

We have seen that since strong acids and strong bases are completely dissociated, the calculations required to obtain the pH curves for titrations involving the two are quite straightforward. However, when the acid being titrated is a

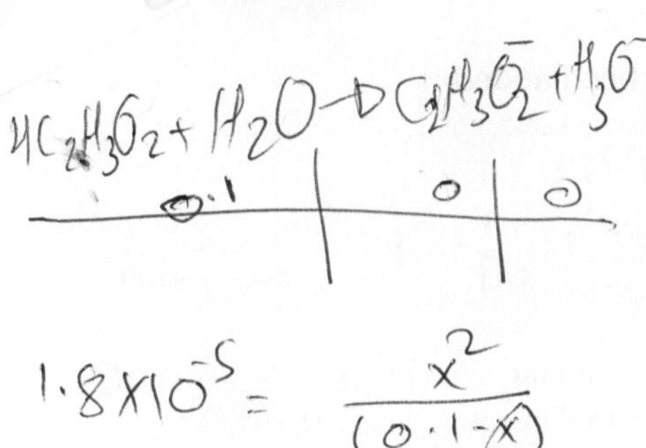

weak acid, there is a major difference: To calculate $[H^+]$ after a certain amount of strong base has been added, we must deal with the weak acid dissociation equilibrium. We dealt with this same type of situation earlier in this chapter when we treated buffered solutions. Calculation of the pH curve for a titration of a weak acid with a strong base really amounts to a series of buffer problems. In performing these calculations, it is very important to remember that even though the acid is weak, it *reacts essentially to completion* with hydroxide ion, a very strong base.

Calculating the pH curve for a weak acid–strong base titration involves the following two-step procedure.

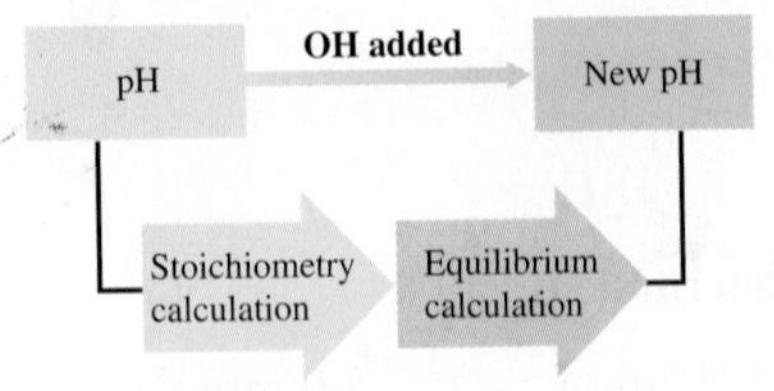

Titration Curve Calculations

1. *A stoichiometry problem.* The reaction of hydroxide ion with the weak acid is assumed to run to completion, and the concentrations of the acid *remaining* and the conjugate base *formed* are determined.
2. *An equilibrium problem.* The position of the weak acid equilibrium is determined, and the pH is calculated.

Treat the stoichiometry and equilibrium problems separately.

It is *essential* to do these steps *separately*. Note that the procedures necessary for solving these problems have all been used before.

As an illustration, we will consider the titration of 50.0 mL of 0.10 *M* acetic acid ($HC_2H_3O_2$, $K_a = 1.8 \times 10^{-5}$) with 0.10 *M* NaOH. As before, we will calculate the pH at various points representing volumes of added NaOH.

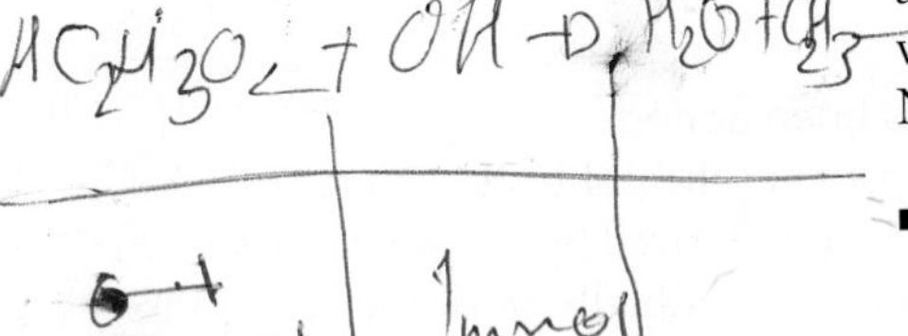

- **No NaOH has been added.**

  This is a typical weak acid calculation of the type introduced in Chapter 7. The pH is 2.87. (Check this value yourself.)

- **10.0 mL of 0.10 *M* NaOH has been added.**

  The major species in the mixed solution *before any reaction takes place* are

$$HC_2H_3O_2, \quad OH^-, \quad Na^+, \quad \text{and} \quad H_2O$$

The strong base OH^- will react with the strongest proton donor, which in this case is $HC_2H_3O_2$.

The Stoichiometry Problem. The calculations are shown in tabular form.

	OH^-	+	$HC_2H_3O_2$	$\longrightarrow$	$C_2H_3O_2^-$ + H_2O
Before reaction:	10.0 mL × 0.10 *M* = 1.0 mmol		50.0 mL × 0.10 *M* = 5.0 mmol		0 mmol
After reaction:	1.0 − 1.0 = 0 mmol ↑ Limiting reactant		5.0 − 1.0 = 4.0 mmol		1.0 mmol

The Equilibrium Problem. We examine the major components left in the solution *after the reaction takes place* to select the dominant equilibrium. The major species are

$$HC_2H_3O_2, \quad C_2H_3O_2^-, \quad Na^+, \quad \text{and} \quad H_2O$$

Since $HC_2H_3O_2$ is a much stronger acid than H_2O, and since $C_2H_3O_2^-$ is the conjugate base of $HC_2H_3O_2$, the pH will be determined by the position of the acetic acid dissociation equilibrium:

$$HC_2H_3O_2(aq) \rightleftharpoons H^+(aq) + C_2H_3O_2^-(aq)$$

where
$$K_a = \frac{[H^+][C_2H_3O_2^-]}{[HC_2H_3O_2]}$$

We follow the usual steps to complete the equilibrium calculations.

The initial concentrations are defined after the reaction with OH^- has gone to completion but before any dissociation of $HC_2H_3O_2$ has occurred.

Initial Concentration		Equilibrium Concentration
$[HC_2H_3O_2]_0 = \frac{4.0 \text{ mmol}}{(50.0 + 10.0) \text{ mL}} = \frac{4.0}{60.0}$	x mmol/mL $HC_2H_3O_2$ dissociates $\longrightarrow$	$[HC_2H_3O_2] = \frac{4.0}{60.0} - x$
$[C_2H_3O_2^-]_0 = \frac{1.0 \text{ mmol}}{(50.0 + 10.0) \text{ mL}} = \frac{1.0}{60.0}$		$[C_2H_3O_2^-] = \frac{1.0}{60.0} + x$
$[H^+]_0 \approx 0$		$[H^+] = x$

Then

$$1.8 \times 10^{-5} = K_a = \frac{[H^+][C_2H_3O_2^-]}{[HC_2H_3O_2]} = \frac{x\left(\frac{1.0}{60.0} + x\right)}{\frac{4.0}{60.0} - x} \approx \frac{x\left(\frac{1.0}{60.0}\right)}{\frac{4.0}{60.0}}$$

$$= \left(\frac{1.0}{4.0}\right)x$$

$$x = \left(\frac{4.0}{1.0}\right)(1.8 \times 10^{-5}) = 7.2 \times 10^{-5} = [H^+]$$

Note that the approximations made are well within 5% uncertainty limits.

and
$$\text{pH} = 4.14$$

- **25.0 mL (total) of 0.10 M NaOH has been added.**

The procedure here is very similar to that used for the previous point and will be summarized briefly. The stoichiometry problem is as follows:

	OH^-	+	$HC_2H_3O_2$	$\longrightarrow$	$C_2H_3O_2^- + H_2O$
Before reaction:	25.0 mL × 0.10 *M* = 2.5 mmol		50.0 mL × 0.10 *M* = 5.0 mmol		0 mmol
After reaction:	2.5 − 2.5 = 0		5.0 − 2.5 = 2.5 mmol		2.5 mmol

After the reaction the major species in solution are

$$HC_2H_3O_2, \quad C_2H_3O_2^-, \quad Na^+, \quad \text{and} \quad H_2O$$

The equilibrium that will control the pH is

$$HC_2H_3O_2(aq) \rightleftharpoons H^+(aq) + C_2H_3O_2^-(aq)$$

Initial Concentration		Equilibrium Concentration
$[HC_2H_3O_2]_0 = \frac{2.5 \text{ mmol}}{(50.0 + 25.0) \text{ mL}}$	x mmol/mL $HC_2H_3O_2$ dissociates $\longrightarrow$	$[HC_2H_3O_2] = \frac{2.5}{75.0} - x$
$[C_2H_3O_2^-]_0 = \frac{2.5 \text{ mmol}}{(50.0 + 25.0) \text{ mL}}$		$[C_2H_3O_2^-] = \frac{2.5}{75.0} + x$
$[H^+]_0 \approx 0$		$[H^+] = x$

Therefore,

$$1.8 \times 10^{-5} = K_a = \frac{[H^+][C_2H_3O_2^-]}{[HC_2H_3O_2]} = \frac{x\left(\frac{2.5}{75.0} + x\right)}{\frac{2.5}{75.0} - x} \approx \frac{x\left(\frac{2.5}{75.0}\right)}{\frac{2.5}{75.0}}$$

$$x = 1.8 \times 10^{-5} = [H^+] \quad \text{and} \quad pH = 4.74$$

This is a special point in the titration because it is *halfway to the equivalence point*. The original solution, 50.0 mL of 0.10 M $HC_2H_3O_2$, contained 5.0 mmol of $HC_2H_3O_2$. Thus 5.0 mmol of OH^- is required to reach the equivalence point. This corresponds to 50.0 mL of NaOH, since

$$(50.0 \text{ mL})(0.10\ M) = 5.0 \text{ mmol}$$

At this point half of the acid has been used up, so $[HC_2H_3O_2] = [C_2H_3O_2^-]$.

After 25.0 mL of NaOH has been added, half of the original $HC_2H_3O_2$ has been converted to $C_2H_3O_2^-$. At this point in the titration, $[HC_2H_3O_2]_0$ is equal to $[C_2H_3O_2^-]_0$. We can neglect the effect of dissociation; that is,

$$[HC_2H_3O_2] = [HC_2H_3O_2]_0 - x \approx [HC_2H_3O_2]_0$$

$$[C_2H_3O_2^-] = [C_2H_3O_2^-]_0 + x \approx [C_2H_3O_2^-]_0$$

The expression for K_a at the halfway point is

$$K_a = \frac{[H^+][C_2H_3O_2^-]}{[HC_2H_3O_2]} = \frac{[H^+][C_2H_3O_2^-]_0}{[HC_2H_3O_2]_0} = [H^+]$$

Equal at the halfway point

Thus, *at the halfway point* in the titration,

$$[H^+] = K_a \quad \text{and} \quad pH = pK_a$$

- **40.0 mL (total) of 0.10 M NaOH has been added.**

The procedures required here are the same as those used for the previous two points. The pH is 5.35. (Check this value yourself.)

- **50.0 mL (total) of 0.10 M NaOH has been added.**

This point is the equivalence point of the titration; 5.0 mmol of OH^- has been added, which is just enough to react with the 5.0 mmol of $HC_2H_3O_2$ originally present. At this point the solution contains these major species:

$$Na^+, \quad C_2H_3O_2^-, \quad \text{and} \quad H_2O$$

Note that the solution contains $C_2H_3O_2^-$, which is a base. Remember that a base will combine with a proton, and the only source of protons in this solution is water. Thus the reaction will be

$$C_2H_3O_2^-(aq) + H_2O(l) \rightleftharpoons HC_2H_3O_2(aq) + OH^-(aq)$$

This is a *weak base* reaction characterized by K_b:

$$K_b = \frac{[HC_2H_3O_2][OH^-]}{[C_2H_3O_2^-]} = \frac{K_w}{K_a} = \frac{1.0 \times 10^{-14}}{1.8 \times 10^{-5}} = 5.6 \times 10^{-10}$$

The relevant concentrations are

Initial Concentration (before any $C_2H_3O_2^-$ reacts with H_2O)		Equilibrium Concentration
$[C_2H_3O_2^-]_0 = \frac{5.0 \text{ mmol}}{(50.0 + 50.0) \text{ mL}} = 0.050\ M$	x mmol/mL $C_2H_3O_2^-$ reacts with H_2O →	$[C_2H_3O_2^-] = 0.050 - x$
$[OH^-]_0 \approx 0$		$[OH^-] = x$
$[HC_2H_3O_2]_0 = 0$		$[HC_2H_3O_2] = x$

Then

$$5.6 \times 10^{-10} = K_b = \frac{[HC_2H_3O_2][OH^-]}{[C_2H_3O_2^-]} = \frac{(x)(x)}{0.050 - x} \approx \frac{x^2}{0.050}$$

$$x = 5.3 \times 10^{-6}$$

The approximation is valid (by the 5% rule), so

$$[OH^-] = 5.3 \times 10^{-6}\ M$$

and

$$[H^+][OH^-] = K_w = 1.0 \times 10^{-14}$$

$$[H^+] = 1.9 \times 10^{-9}\ M \quad \text{and} \quad pH = 8.72$$

The pH at the equivalence point of a titration of a weak acid with a strong base is always greater than 7.

This is another important result: *The pH at the equivalence point of a titration of a weak acid with a strong base is always greater than 7.* This result occurs because the anion of the acid, which remains in solution at the equivalence point, is a base. In contrast, for the titration of a strong acid with a strong base, the pH at the equivalence point is 7 because the anion remaining in this case is *not* an effective base.

- **60.0 mL (total) of 0.10 *M* NaOH has been added.**

At this point excess OH^- has been added:

	OH^-	+	$HC_2H_3O_2$	$\longrightarrow$	$C_2H_3O_2^- + H_2O$
Before reaction:	60.0 mL × 0.10 *M* = 6.0 mmol		50.0 mL × 0.10 *M* = 5.0 mmol		0 mmol
After reaction:	6.0 − 5.0 = 1.0 mmol in excess		5.0 − 5.0 = 0 mmol		5.0 mmol

After the reaction is complete, the solution contains these major species:

$$Na^+, \quad C_2H_3O_2^-, \quad OH^-, \quad \text{and} \quad H_2O$$

There are two bases in this solution, OH^- and $C_2H_3O_2^-$. However, $C_2H_3O_2^-$ is a weak base compared with OH^-. Therefore, the amount of OH^- produced by the reaction of $C_2H_3O_2^-$ with H_2O will be small compared with the excess OH^- already in solution. You can verify this conclusion by looking at the previous point, where only 5.3×10^{-6} *M* OH^- was produced by $C_2H_3O_2^-$. The amount in this case will be even smaller, since the excess OH^- will push the K_b equilibrium to the left.

Thus the pH is determined by the excess OH^-:

$$[OH^-] = \frac{\text{mmol of } OH^- \text{ in excess}}{\text{volume (in mL)}} = \frac{1.0 \text{ mmol}}{(50.0 + 60.0) \text{ mL}}$$

$$= 9.1 \times 10^{-3}\ M$$

and

$$[H^+] = \frac{1.0 \times 10^{-14}}{9.1 \times 10^{-3}} = 1.1 \times 10^{-12}\ M \quad \text{and} \quad pH = 11.96$$

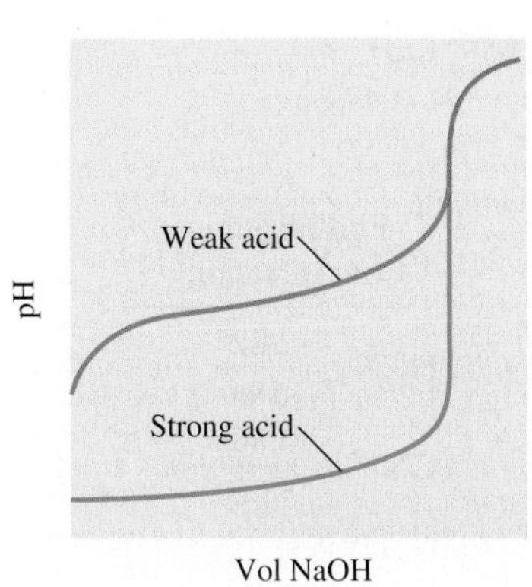

- **75.0 mL (total) of 0.10 *M* NaOH has been added.**

The procedure needed here is very similar to that for the previous point. The pH is 12.30. (Check this value.)

The pH curve for this titration is shown in Fig. 8.3. Note the differences between this curve and the curve in Fig. 8.1. For example, the shapes of the plots are quite different before the equivalence point, but they are very similar after that point. (The shapes of the strong and weak acid curves are the same after the equivalence points because excess OH^- controls the pH in this region in both cases.) Near the beginning of the titration of the weak acid, the pH increases more rapidly than it does in the strong acid case. It levels off near the halfway point and then increases rapidly again. The leveling off near the halfway point is caused by buffering effects. Earlier in this chapter we saw that optimum buffering occurs when [HA] is equal to [A^-]. This is exactly the case at the halfway point of the titration. As we can see from the curve, the pH changes least around this point in the titration.

The other notable difference between the curves for strong and weak acids is the value of the pH at the equivalence point. For the titration of a strong acid, the equivalence point occurs at a pH of 7. For the titration of a weak acid, the pH at the equivalence point is greater than 7 because of the basicity of the conjugate base of the weak acid.

The equivalence point is defined by the stoichiometry, not by the pH.

It is important to understand that the equivalence point in an acid–base titration is *defined by the stoichiometry, not by the pH*. The equivalence point occurs when enough titrant has been added to react exactly with all of the acid or base being titrated.

OWL INTERACTIVE EXAMPLE 8.7

Hydrogen cyanide gas (HCN) is a powerful respiratory inhibitor that is highly toxic. It is a very weak acid ($K_a = 6.2 \times 10^{-10}$) when dissolved in water. If a 50.0-mL sample of 0.100 *M* HCN is titrated with 0.100 *M* NaOH, calculate the pH of the solution at the following points.

a. after 8.00 mL of 0.100 *M* NaOH has been added

Solution

The Stoichiometry Problem. After 8.00 mL of 0.100 *M* NaOH has been added, we have the following:

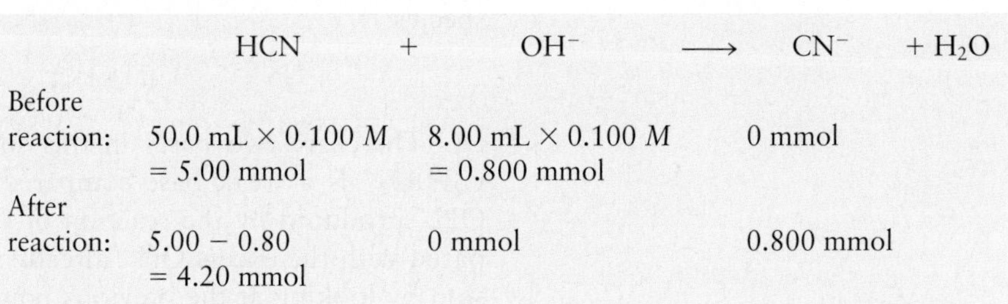

	HCN	+	OH^-	$\longrightarrow$	CN^-	$+ H_2O$
Before reaction:	$50.0 \text{ mL} \times 0.100\ M = 5.00$ mmol		$8.00 \text{ mL} \times 0.100\ M = 0.800$ mmol		0 mmol	
After reaction:	$5.00 - 0.80 = 4.20$ mmol		0 mmol		0.800 mmol	

The Equilibrium Problem. Since the solution contains the major species

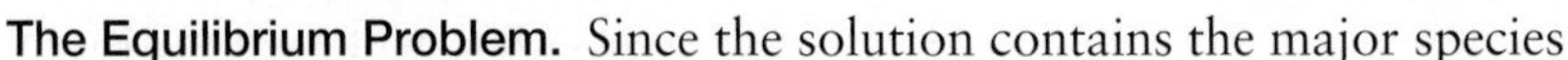

$$HCN, \quad CN^-, \quad Na^+, \quad \text{and} \quad H_2O$$

the position of the acid dissociation equilibrium

$$HCN(aq) \rightleftharpoons H^+(aq) + CN^-(aq)$$

will determine the pH.

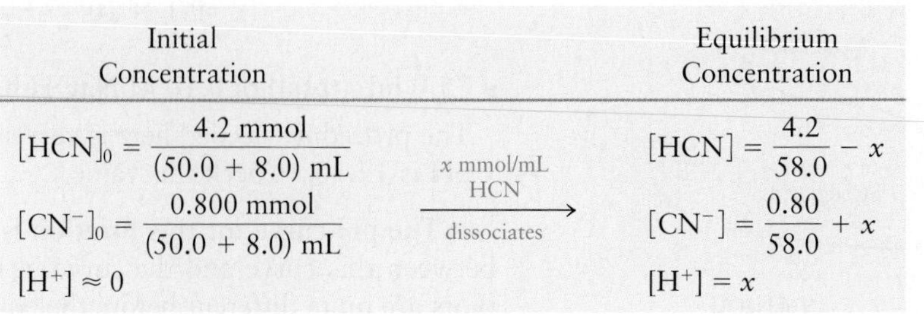

Initial Concentration		Equilibrium Concentration
$[HCN]_0 = \dfrac{4.2 \text{ mmol}}{(50.0 + 8.0) \text{ mL}}$	$\xrightarrow[\text{dissociates}]{x \text{ mmol/mL HCN}}$	$[HCN] = \dfrac{4.2}{58.0} - x$
$[CN^-]_0 = \dfrac{0.800 \text{ mmol}}{(50.0 + 8.0) \text{ mL}}$		$[CN^-] = \dfrac{0.80}{58.0} + x$
$[H^+] \approx 0$		$[H^+] = x$

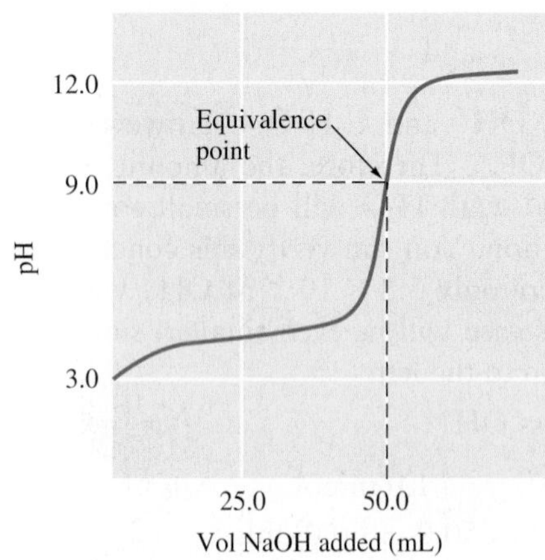

Figure 8.3
The pH curve for the titration of 50.0 mL of 0.10 *M* $HC_2H_3O_2$ with 0.10 *M* NaOH. Note that the equivalence point occurs when 50.0 mL of NaOH has been added, where the amount of added OH^- ions exactly equals the original amount of acid. The pH at the equivalence point is greater than 7 because the $C_2H_3O_2^-$ ion present at this point is a base and reacts with water to produce OH^-.

Substituting into the expression for K_a gives

$$6.2 \times 10^{-10} = K_a = \frac{[H^+][CN^-]}{[HCN]} = \frac{x\left(\frac{0.80}{58.0} + x\right)}{\frac{4.2}{58.0} - x} \approx \frac{x\left(\frac{0.80}{58.0}\right)}{\left(\frac{4.2}{58.0}\right)}$$

$$= x\left(\frac{0.80}{4.2}\right)$$

The approximations made here are well within 5% uncertainty limits.

$$x = 3.3 \times 10^{-9}\ M = [H^+] \quad \text{and} \quad pH = 8.49$$

b. at the halfway point in the titration

Solution The amount of HCN originally present can be obtained from the original volume and molarity:

$$50.0\ mL \times 0.100\ M = 5.00\ mmol$$

Thus the halfway point will occur when 2.50 mmol of OH^- has been added:

$$\text{Volume of NaOH (in mL)} \times 0.100\ M = 2.50\ mmol\ OH^-$$

or

$$\text{Volume of NaOH} = 25.0\ mL$$

As was pointed out previously, at the halfway point [HCN] is equal to $[CN^-]$ and pH is equal to pK_a. Thus, after 25.0 mL of 0.100 M NaOH has been added,

$$pH = pK_a = -\log(6.2 \times 10^{-10}) = 9.21$$

c. at the equivalence point

Solution The equivalence point will occur when a total of 5.00 mmol of OH^- has been added. Since the NaOH solution is 0.100 M, the equivalence point occurs when 50.0 mL of NaOH has been added. This results in the formation of 5.00 mmol of CN^-. The major species in solution at the equivalence point are

$$CN^-, \quad Na^+, \quad \text{and} \quad H_2O$$

Thus the reaction that will control the pH involves the basic cyanide ion extracting a proton from water:

$$CN^-(aq) + H_2O(l) \rightleftharpoons HCN(aq) + OH^-(aq)$$

and

$$K_b = \frac{K_w}{K_a} = \frac{1.0 \times 10^{-14}}{6.2 \times 10^{-10}} = 1.6 \times 10^{-5} = \frac{[HCN][OH^-]}{[CN^-]}$$

Initial Concentration		Equilibrium Concentration
$[CN^-]_0 = \frac{5.00\ mmol}{(50.0 + 50.0)\ mL} = 5.00 \times 10^{-2}\ M$	x mmol/mL of CN^- reacts with H_2O →	$[CN^-] = (5.00 \times 10^{-2}) - x$
$[HCN]_0 = 0$		$[HCN] = x$
$[OH^-]_0 \approx 0$		$[OH^-] = x$

Substituting the equilibrium concentrations into the expression for K_b and solving in the usual way gives

$$[OH^-] = x = 8.9 \times 10^{-4}$$

Then from K_w we have

$$[H^+] = 1.1 \times 10^{-11} \quad \text{and} \quad pH = 10.96$$

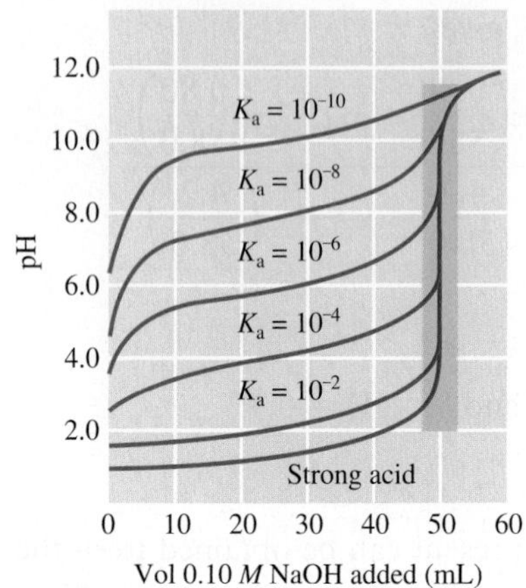

Figure 8.4
The pH curves for the titrations of 50.0-mL samples of 0.10 *M* solutions of various acids with 0.10 *M* NaOH.

Two important conclusions can be drawn from a comparison of the titration of 50.0 mL of 0.1 *M* acetic acid covered earlier in this section with that of 50.0 mL of 0.1 *M* hydrocyanic acid (Example 8.7). First, the same amount of 0.1 *M* NaOH is required to reach the equivalence point in both cases. The fact that HCN is a much weaker acid than $HC_2H_3O_2$ has no bearing on the amount of base required. It is the *amount* of acid, not its strength, that determines the equivalence point. Second, the *pH value* at the equivalence point *is* affected by the acid strength. For the titration of acetic acid, the pH at the equivalence point is 8.72; for the titration of hydrocyanic acid, the pH at the equivalence point is 10.96. This difference occurs because the CN^- ion is a much stronger base than the $C_2H_3O_2^-$ ion. Also, the pH at the halfway point of the titration is much higher for HCN than for $HC_2H_3O_2$, again because of the greater base strength of the CN^- ion (or equivalently, the smaller acid strength of HCN).

The strength of a weak acid has a significant effect on the shape of its pH curve. Figure 8.4 shows pH curves for 50-mL samples of 0.10 *M* solutions of various acids titrated with 0.10 *M* NaOH. Note that the equivalence point occurs at the same volume of 0.10 *M* NaOH for each case but that the shapes of the curves are dramatically different. The weaker the acid, the greater is the pH value at the equivalence point. In particular, note that the vertical region surrounding the equivalence point becomes shorter as the acid being titrated becomes weaker. We will see in the next section how this affects the choice of an indicator for such a titration.

The amount of acid present, not its strength, determines the equivalence point.

In addition to being used to analyze the amount of acid or base in a solution, titrations can be used to determine the values of equilibrium constants, as shown in Example 8.8.

OWL INTERACTIVE EXAMPLE 8.8

A chemist has synthesized a monoprotic weak acid and wants to determine its K_a value. To do so, the chemist dissolves 2.00 mmol of the solid acid in 100.0 mL of water and titrates the resulting solution with 0.0500 *M* NaOH. After 20.0 mL of NaOH has been added, the pH is 6.00. What is the K_a value for the acid?

Solution

The Stoichiometry Problem. We represent the monoprotic acid as HA. The stoichiometry for the titration reaction is shown below.

HA	+ OH^-	$\longrightarrow$ A^-	+ H_2O
Before reaction: 2.00 mmol	20.0 mL × 0.0500 *M* = 1.00 mmol	0 mmol	
After reaction: 2.00 − 1.00 = 1.00 mmol	1.00 − 1.00 = 0 mmol	1.00 mmol	

2.00 mmol HA
↓ add OH^-
1.00 mmol HA
1.00 mmol A^-

The Equilibrium Problem. After the reaction, the solution contains the major species

$$HA, \quad A^-, \quad Na^+, \quad \text{and} \quad H_2O$$

The pH will be determined by the equilibrium

$$HA(aq) \rightleftharpoons H^+(aq) + A^-(aq)$$

$$K_a = \frac{[H^+][A^-]}{[HA]}$$

Initial Concentration		Equilibrium Concentration
$[HA]_0 = \frac{1.00 \text{ mmol}}{(100.0 + 20.0) \text{ mL}} = 8.33 \times 10^{-3}\ M$		$[HA] = 8.33 \times 10^{-3} = x$
$[A^-]_0 = \frac{1.00 \text{ mmol}}{(100.0 + 20.0) \text{ mL}} = 8.33 \times 10^{-3}\ M$	$\xrightarrow[\text{dissociates}]{x \text{ mmol/mL HA}}$	$[A^-] = 8.33 \times 10^{-3} + x$
$[H^+]_0 \approx 0$		$[H^+] = x$

Note that x is actually known here because the pH at this point is known to be 6.00. Thus

$$x = [H^+] = \text{antilog}(-\text{pH}) = 1.0 \times 10^{-6}\ M$$

Substituting the equilibrium concentrations into the expression for K_a allows calculation of the K_a value:

$$K_a = \frac{[H^+][A^-]}{[HA]} = \frac{x(8.33 \times 10^{-3} + x)}{(8.33 \times 10^{-3}) - x}$$

$$= \frac{(1.0 \times 10^{-6})(8.33 \times 10^{-3} + 1.0 \times 10^{-6})}{(8.33 \times 10^{-3}) - (1.0 \times 10^{-6})}$$

$$= \frac{(1.0 \times 10^{-6})(8.33 \times 10^{-3})}{8.33 \times 10^{-3}} = 1.0 \times 10^{-6}$$

There is an easier way to think about this problem. The original solution contained 2.00 mmol of HA, and since 20.0 mL of added 0.0500 M NaOH contains 1.00 mmol of OH^-, this is the halfway point in the titration (where [HA] is equal to $[A^-]$). Thus

$$[H^+] = K_a = 1.0 \times 10^{-6}$$

Titrations of Weak Bases with Strong Acids

Titrations of weak bases with strong acids can be treated using the procedures we have introduced previously. As always, you should *first think about the major species in solution* and decide whether a reaction occurs that runs essentially to completion. If such a reaction does occur, let it run to completion and then do the stoichiometric calculations. Finally, choose the dominant equilibrium and calculate the pH.

The calculations for the titration of a weak base with a strong acid are illustrated by the following titration of 100.0 mL of 0.050 M NH_3 with 0.10 M HCl. The strategies needed at several key areas in the titration will be described qualitatively. The actual calculations are summarized in Table 8.2.

- **Before the addition of any HCl.**

1. The major species are

$$NH_3 \quad \text{and} \quad H_2O$$

   NH_3 is a base and will seek a source of protons. In this case H_2O is the only available source of protons.

2. No reactions occur that go to completion, since NH_3 cannot readily take a proton from H_2O. This is evidenced by the small K_b value for NH_3.

3. The equilibrium that controls the pH involves the reaction of ammonia with water:

$$NH_3(aq) + H_2O(l) \rightleftharpoons NH_4^+(aq) + OH^-(aq)$$

Table 8.2

Summary of Results for the Titration of 100.0 mL 0.050 *M* NH_3 with 0.10 *M* HCl

Volume of 0.10 *M* HCl Added (mL)	$[NH_3]_0$	$[NH_4^+]_0$	$[H^+]$	pH
0	0.050 *M*	0	1.1×10^{-11} *M*	10.96
10.0	$\frac{4.0 \text{ mmol}}{(100 + 10) \text{ mL}}$	$\frac{1.0 \text{ mmol}}{(100 + 10) \text{ mL}}$	1.4×10^{-10} *M*	9.85
25.0*	$\frac{2.5 \text{ mmol}}{(100 + 25) \text{ mL}}$	$\frac{2.5 \text{ mmol}}{(100 + 25) \text{ mL}}$	5.6×10^{-10} *M*	9.25
50.0†	0	$\frac{5.0 \text{ mmol}}{(100 + 50) \text{ mL}}$	4.3×10^{-6} *M*	5.36
60.0‡	0	$\frac{5.0 \text{ mmol}}{(100 + 60) \text{ mL}}$	$\frac{1.0 \text{ mmol}}{160 \text{ mL}} = 6.2 \times 10^{-3}$ *M*	2.21

*Halfway point.
†Equivalence point.
‡$[H^+]$ determined by the 1.0 mmol of excess H^+.

Use K_b to calculate $[OH^-]$. Although NH_3 is a weak base (compared with OH^-), it produces much more OH^- in this reaction than is produced from the autoionization of H_2O.

- **Before the equivalence point.**

1. The major species (before any reaction occurs) are

$$NH_3, \quad \underbrace{H^+, \quad Cl^-,}_{\text{From added HCl}} \quad \text{and} \quad H_2O$$

2. The NH_3 will react with H^+ from the added HCl:

$$NH_3(aq) + H^+(aq) \rightleftharpoons NH_4^+(aq)$$

This reaction proceeds essentially to completion because the NH_3 readily reacts with a free proton. This case is much different from the previous case, where H_2O was the only source of protons. The stoichiometric calculations are then carried out using the known volume of 0.10 *M* HCl added.

3. After the reaction of NH_3 with H^+ is run to completion, the solution contains the following major species:

$$NH_3, \quad \underset{\substack{\uparrow \\ \text{Formed in} \\ \text{titration reaction}}}{NH_4^+,} \quad Cl^-, \quad \text{and} \quad H_2O$$

The solution contains NH_3 and NH_4^+; thus the equilibria involving these species will determine $[H^+]$. The $[H^+]$ can be calculated using either the dissociation reaction of NH_4^+,

$$NH_4^+(aq) \rightleftharpoons NH_3(aq) + H^+(aq)$$

or the reaction of NH_3 with H_2O,

$$NH_3(aq) + H_2O(l) \rightleftharpoons NH_4^+(aq) + OH^-(aq)$$

to calculate the pH.

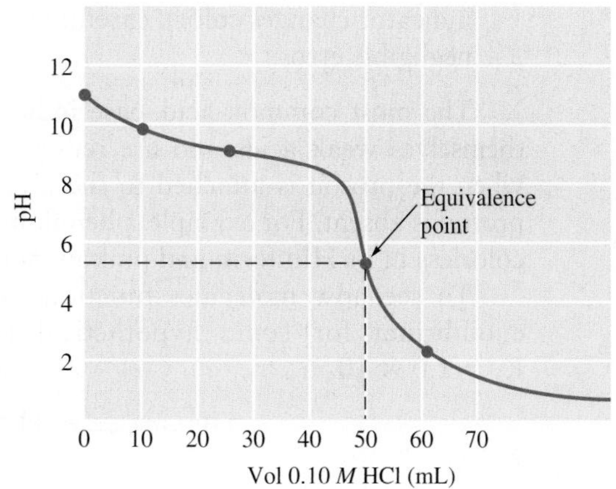

Figure 8.5
The pH curve for the titration of 100.0 mL of 0.050 M NH_3 with 0.10 M HCl. Note that the pH at the equivalence point is less than 7, since the solution contains the weak acid NH_4^+.

- **At the equivalence point.**
  1. By definition, the equivalence point occurs when all the original NH_3 is converted to NH_4^+. Thus the major species in solution are
  $$NH_4^+, \quad Cl^-, \quad \text{and} \quad H_2O$$
  2. No reaction goes to completion.
  3. The equilibrium that controls $[H^+]$ is the dissociation of the weak acid NH_4^+, for which
  $$K_a = \frac{K_w}{K_b(\text{for } NH_3)}$$

- **Beyond the equivalence point.**
  1. Excess HCl has been added and the major species are
  $$H^+, \quad NH_4^+, \quad Cl^-, \quad \text{and} \quad H_2O$$
  2. No reaction goes to completion.
  3. Although NH_4^+ will dissociate, it is such a weak acid that $[H^+]$ will be determined simply by the excess H^+:
  $$[H^+] = \frac{\text{mmol of } H^+ \text{ in excess}}{\text{mL of solution}}$$

The results of these calculations are shown in Table 8.2. The pH curve is shown in Fig. 8.5.

8.6 Acid–Base Indicators

There are two common methods for determining the equivalence point of an acid–base titration:

1. Use a pH meter to monitor the pH and then plot a titration curve. The center of the vertical region of the pH curve indicates the equivalence point (for example, see Figs. 8.1 through 8.5).
2. Use an **acid–base indicator,** which marks the endpoint of a titration by changing color. Although the *equivalence point of a titration, defined by the stoichiometry, is not necessarily the same as the endpoint* (where the

The *endpoint* is defined by the change in color of the indicator. The *equivalence point* is defined by the reaction stoichiometry.

Photos: Charles D. Winters

Figure 8.6
The indicator phenolphthalein is pink in basic solution (top) and colorless in acidic solution (bottom).

indicator changes color), careful selection of the indicator will ensure only negligible error.

The most common acid–base indicators are complex molecules that are themselves weak acids and are represented by HIn. They exhibit one color when the proton is attached to the molecule and a different color when the proton is absent. For example, **phenolphthalein,** a commonly used indicator, is colorless in its HIn form and pink in its In^-, or basic, form (Fig. 8.6).

To see how molecules function as indicators, consider the following equilibrium for some hypothetical indicator HIn, a weak acid with $K_a = 1.0 \times 10^{-8}$:

$$\underset{\text{Red}}{\mathrm{HIn}(aq)} \rightleftharpoons \mathrm{H^+}(aq) + \underset{\text{Blue}}{\mathrm{In^-}(aq)}$$

$$K_a = \frac{[\mathrm{H^+}][\mathrm{In^-}]}{[\mathrm{HIn}]}$$

By rearranging, we get

$$\frac{K_a}{[\mathrm{H^+}]} = \frac{[\mathrm{In^-}]}{[\mathrm{HIn}]}$$

Suppose we add a few drops of this indicator to an acidic solution whose pH is 1.0 ($[H^+] = 1.0 \times 10^{-1}$). Then

$$\frac{K_a}{[\mathrm{H^+}]} = \frac{1. \times 10^{-8}}{1.0 \times 10^{-1}} = 10^{-7} = \frac{[\mathrm{In^-}]}{[\mathrm{HIn}]} = \frac{1}{10{,}000{,}000}$$

This ratio shows that the predominant form of the indicator is HIn, resulting in a red solution. As OH^- is added to this solution in a titration, $[H^+]$ decreases, and the equilibrium shifts to the right, changing HIn to In^-. At some point in the titration, enough of the In^- form will be present in the solution so that a purple tint will be noticeable. That is, a color change from red to reddish purple will occur.

How much In^- must be present in the solution for the human eye to detect that the color is different from the original one? For most indicators, about one-tenth of the initial form must be converted to the other form before a new color is apparent. We will assume, then, that in the titration of an acid with a base, the color change will be apparent at a pH where

$$\frac{[\mathrm{In^-}]}{[\mathrm{HIn}]} = \frac{1}{10}$$

EXAMPLE 8.9

Bromthymol blue, an indicator with a K_a value of 1.0×10^{-7}, is yellow in its HIn form and blue in its In^- form. Suppose we put a few drops of this indicator in a strongly acidic solution. If the solution is then titrated with NaOH, at what pH will the indicator color change first be visible?

Solution For bromthymol blue,

$$K_a = 1.0 \times 10^{-7} = \frac{[\mathrm{H^+}][\mathrm{In^-}]}{[\mathrm{HIn}]}$$

We assume the color change is visible when

$$\frac{[\mathrm{In^-}]}{[\mathrm{HIn}]} = \frac{1}{10}$$

(a) (b) (c)

Photos: Charles D. Winters

Figure 8.7
(a) Yellow acid form of bromthymol blue; (b) a greenish tint is seen when the solution contains 1 part blue and 10 parts yellow; (c) blue basic form.

That is, we assume we can see the first hint of a greenish tint (yellow plus a little blue) when the solution contains 1 part blue and 10 parts yellow (Fig. 8.7). Thus

$$K_a = 1.0 \times 10^{-7} = \frac{[H^+](1)}{10}$$

$$[H^+] = 1.0 \times 10^{-6} \quad \text{or} \quad pH = 6.0$$

The color change is first visible at a pH of 6.0.

The Henderson–Hasselbalch equation is very useful in determining the pH at which an indicator changes color. For example, application of Equation (8.2) to the K_a expression for the general indicator HIn yields

$$pH = pK_a + \log\left(\frac{[In^-]}{[HIn]}\right)$$

where K_a is the dissociation constant for the acid form [HIn]. Since we assume that the color change is visible when

$$\frac{[In^-]}{[HIn]} = \frac{1}{10}$$

we have the following equation for determining the pH at which the color change occurs:

$$pH = pK_a + \log(\tfrac{1}{10}) = pK_a - 1$$

For bromthymol blue ($K_a = 1 \times 10^{-7}$, or $pK_a = 7$), the pH at the color change is

$$pH = 7 - 1 = 6$$

as we calculated in Example 8.9.

When a basic solution is titrated, the indicator HIn will initially exist as In^- in solution, but as acid is added, more HIn will form. In this case the color change will be visible when there is a mixture of 10 parts In^- to 1 part HIn. That is, a color change from blue to blue-green will occur (Fig. 8.7) because of the presence of some of the yellow HIn molecules. This color change will be first visible when

$$\frac{[In^-]}{[HIn]} = \frac{10}{1}$$

Note that this expression is the reciprocal of the ratio for the titration of an acid. Substituting this ratio into the Henderson–Hasselbalch equation gives

$$pH = pK_a + \log(\tfrac{10}{1}) = pK_a + 1$$

For bromthymol blue ($pK_a = 7$), we have a color change at

$$pH = 7 + 1 = 8$$

In summary, when bromthymol blue is used for the titration of an acid, the starting form will be HIn (yellow), and the color change occurs at a pH of about 6. When bromthymol blue is used for the titration of a base, the starting form is In^- (blue), and the color change occurs at a pH of about 8. Thus the useful pH range for bromthymol blue is

$$pK_a \text{ (bromthymol blue)} \pm 1 = 7 \pm 1$$

or from 6 to 8. The useful pH ranges for several common indicators are shown in Fig. 8.8 on page 275.

When we choose an indicator for a titration, we want the indicator endpoint (where the color changes) and the titration equivalence point to be as close as possible. Choosing an indicator is easier if there is a large change in pH near the equivalence point of the titration. The dramatic change in pH near the equivalence point in a strong acid–strong base titration (Figs. 8.1 and 8.2) produces a sharp endpoint; that is, the complete color change (from the acid-to-base or the base-to-acid colors) usually occurs over one drop of added titrant.

What indicator should we use for the titration of 100.00 mL of 0.100 *M* HCl with 0.100 *M* NaOH? We know that the equivalence point occurs at pH 7.00. In the initially acidic solution, the indicator will exist predominantly in the HIn form. As OH^- ions are added, the pH will increase rather slowly at first (Fig. 8.1) and then will rise rapidly at the equivalence point. This sharp change causes the indicator dissociation equilibrium

$$HIn \rightleftharpoons H^+ + In^-$$

to shift suddenly to the right, producing enough In^- ions to cause a color change. Since we are titrating an acid, the indicator is predominantly in the acid form initially. Therefore the first observable color change will occur at a pH where

$$\frac{[In^-]}{[HIn]} = \frac{1}{10}$$

Thus $$pH = pK_a + \log(\tfrac{1}{10}) = pK_a - 1$$

If we want an indicator that will change color at pH 7, we can use this relationship to find the pK_a value of a suitable indicator:

$$pH = 7 = pK_a - 1 \qquad \text{or} \qquad pK_a = 7 + 1 = 8$$

Thus an indicator with a pK_a value of 8 ($K_a = 1 \times 10^{-8}$) will change color at about a pH of 7 and will be ideal to mark the endpoint for a strong acid–strong base titration.

How crucial is it in a strong acid–strong base titration that the indicator change color exactly at a pH of 7? We can answer this question by examining the pH change near the equivalence point of the titration of 100.00 mL of 0.100 *M* HCl with 0.100 *M* NaOH. The data for a few points at or near the equivalence point are shown in Table 8.3. Note that in going from 99.99 mL to 100.01 mL of added NaOH solution (about half of a drop), the pH changes from 5.3 to 8.7—a very dramatic change. This behavior leads to the following general conclusions about indicators for a strong acid–strong base titration:

Table 8.3

Selected pH Values Near the Equivalence Point in the Titration of 100.0 mL of 0.100 *M* HCl with 0.100 *M* NaOH

NaOH Added (mL)	pH
99.99	5.3
100.00	7.0
100.01	8.7

Indicator color change will be sharp, occurring with the addition of a single drop of titrant.

There is a wide choice of suitable indicators. The results will agree within one drop of titrant, using indicators with endpoints as far apart as pH = 5 and pH = 9 (Fig. 8.9).

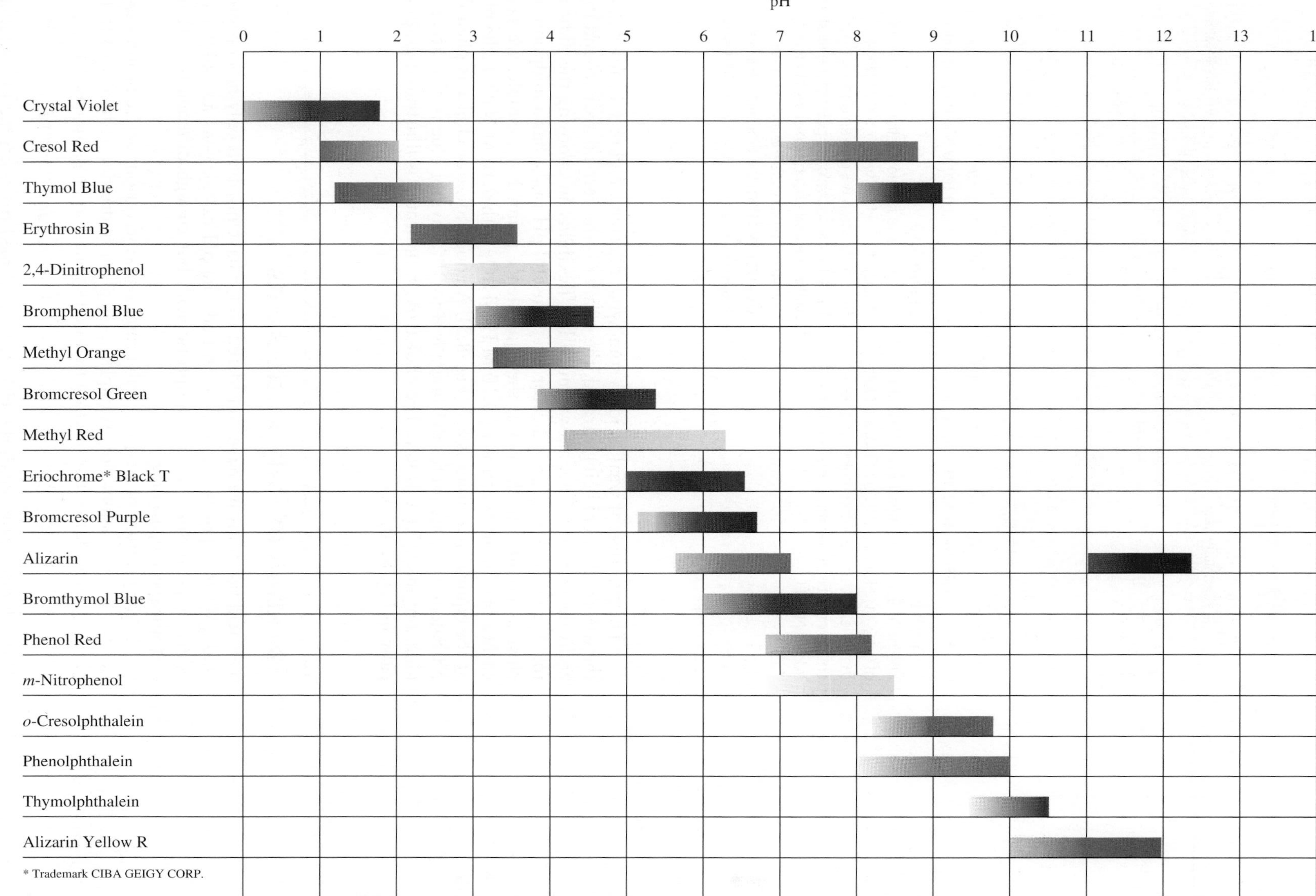

Figure 8.8
The useful pH ranges for several common indicators. Note that most indicators have a useful range of about two pH units, as predicted by the expression $pK_a \pm 1$.

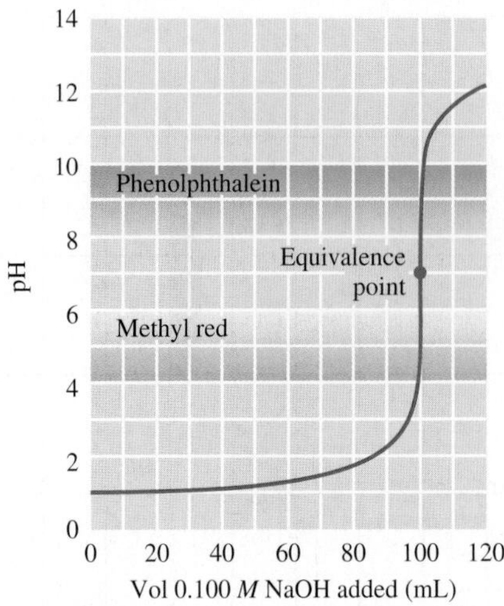

Figure 8.9
The pH curve for the titration of 100.0 mL of 0.100 *M* HCl with 0.100 *M* NaOH. Note that phenolphthalein and methyl red have endpoints at virtually the same amounts of added NaOH.

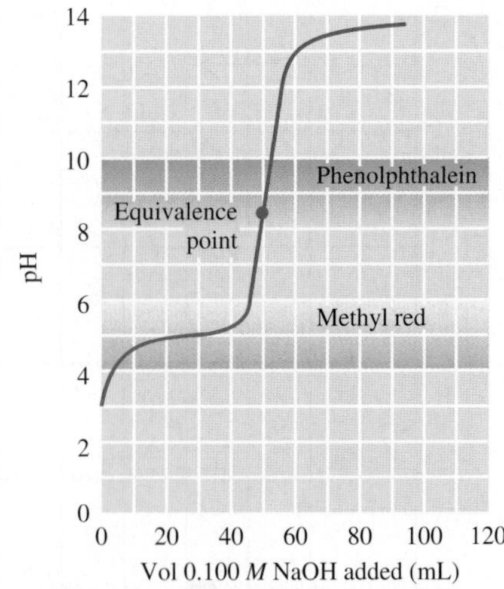

Figure 8.10
The pH curve for the titration of 50 mL of 0.100 *M* $HC_2H_3O_2$ with 0.100 *M* NaOH. Phenolphthalein will give an endpoint very close to the equivalence point of the titration. Methyl red will change color well before the equivalence point (so the endpoint will be very different from the equivalence point) and is not a suitable indicator for this titration.

The titration of weak acids is somewhat different. Figure 8.4 shows that the weaker the acid being titrated, the smaller is the vertical area around the equivalence point. This allows much less flexibility in choosing the indicator. We must choose an indicator whose useful pH range has a midpoint as close as possible to the pH at the equivalence point. For example, we saw earlier that in the titration of 0.10 *M* $HC_2H_3O_2$ with 0.10 *M* NaOH, the pH at the equivalence point is 8.7 (Fig. 8.3). A good indicator choice is phenolphthalein, since its useful pH range is 8 to 10. Thymol blue is also acceptable, but methyl red is not. The choice of an indicator is illustrated graphically in Fig. 8.10.

8.7 Titration of Polyprotic Acids

The acid titrations we have considered so far have involved only monoprotic acids. When a polyprotic acid is titrated, the pH calculations are similar in many ways to those for a monoprotic acid, but enough differences exist to warrant special coverage.

In the titration of a typical polyprotic acid, the various acidic protons are titrated in succession. For example, as sodium hydroxide is used to titrate phosphoric acid, the first reaction that takes place can be represented as

$$H_3PO_4(aq) + OH^-(aq) \longrightarrow H_2PO_4^-(aq) + H_2O(l)$$

This reaction occurs until the H_3PO_4 is consumed (to reach the first equivalence point). Therefore, at the first equivalence point the solution contains the major species Na^+, $H_2PO_4^-$, and H_2O. Then, as more sodium hydroxide is added, the reaction

$$H_2PO_4^-(aq) + OH^-(aq) \longrightarrow HPO_4^{2-}(aq) + H_2O(l)$$

Table 8.4

A Summary of Various Points in the Titration of a Triprotic Acid

Point in the Titration	Major Species Present	Equilibrium Expression Used to Obtain the pH
No base added	H_3A, H_2O	$K_{a_1} = \frac{[H^+][H_2A^-]}{[H_3A]}$
Base added		
Before the first equivalence point	H_3A, H_2A^-, H_2O	$K_{a_1} = \frac{[H^+][H_2A^-]}{[H_3A]}$
At the first equivalence point	H_2A^-, H_2O	See discussion in text
Between the first and second equivalence points	H_2A^-, HA^{2-}, H_2O	$K_{a_2} = \frac{[H^+][HA^{2-}]}{[H_2A^-]}$
At the second equivalence point	HA^{2-}, H_2O	See discussion in text
Between the second and third equivalence points	HA^{2-}, A^{3-}, H_2O	$K_{a_3} = \frac{[H^+][A^{3-}]}{[HA^{2-}]}$
At the third equivalence point	A^{3-}, H_2O	$K_b = \frac{K_w}{K_{a_3}} = \frac{[HA^{2-}][OH^-]}{[A^{3-}]}$
Beyond the third equivalence point	A^{3-}, OH^-, H_2O	pH determined by excess OH^-

occurs to give a solution that contains Na^+, HPO_4^{2-}, and H_2O as the major species at the second equivalence point. As sodium hydroxide is added beyond the second equivalence point, the reaction that occurs can be represented as

$$HPO_4^{2-}(aq) + OH^-(aq) \longrightarrow PO_4^{3-}(aq) + H_2O(l)$$

As mentioned earlier, the calculations involved in obtaining the pH curve for the titration of a polyprotic acid are closely related to those for a monoprotic acid. The same principles apply, but we must be very careful in identifying which of the various equilibria is appropriate to use in a given case. The secret to success here is, as always, identifying the major species in solution at any given point in the titration. We summarize the various cases in Table 8.4 for a triprotic acid H_3A with dissociation constants K_{a_1}, K_{a_2}, and K_{a_3}.

A point that cannot be overemphasized is that the appropriate equilibrium expression is chosen by knowing what major species are present. Thus, if the solution contains HA^{2-} and A^{3-}, we must use K_{a_3} to determine the pH, and so on. Note that in two instances in Table 8.4 we did not specify the equilibrium expression to be used. These two cases, which need to be considered in more detail, are discussed in the section that follows.

Solutions Containing Amphoteric Anions as the Only Acid–Base Major Species

At the first equivalence point in the titration of the acid H_3A with sodium hydroxide, the major species are H_2A^- and H_2O. What equilibrium will control the $[H^+]$ in such a solution?

The key to answering this question is to recognize that H_2A^- is an amphoteric species: It is both an acid,

$$H_2A^-(aq) \rightleftharpoons H^+(aq) + HA^{2-}(aq)$$

and a base,

$$H_2A^-(aq) + H^+(aq) \rightleftharpoons H_3A(aq)$$

What is the best source of protons for H_2A^- behaving as a base? There are two possible sources: H_2O and other H_2A^- ions in the solution. By now we realize that H_2O is a very weak acid. In a typical case H_2A^- will be a much stronger acid than H_2O. Thus, in a solution containing H_2A^- and H_2O as the major species, we expect the reaction

$$H_2A^-(aq) + H_2A^-(aq) \rightleftharpoons H_3A(aq) + HA^{2-}(aq)$$

to be the most important acid–base reaction. Notice that this reaction expresses both the acidic and basic properties of H_2A^- (in the preceding equation, one H_2A^- is behaving as the acid and the other H_2A^- as the base). This reaction leads to the equilibrium expression

$$K = \frac{[H_3A][HA^{2-}]}{[H_2A^-]^2}$$

We can obtain the value for K by recognizing that

$$\frac{[H_3A][HA^{2-}]}{[H_2A^-][H_2A^-]} = \frac{[H_3A]}{[H^+][H_2A^-]} \times \frac{[H^+][HA^{2-}]}{[H_2A^-]} = \frac{1}{K_{a_1}} \times K_{a_2}$$

Thus
$$K = \frac{K_{a_2}}{K_{a_1}}$$

The position of this equilibrium will determine the concentrations of H_3A, H_2A^-, and HA^{2-} in the solution and thus will determine the pH. We can obtain an expression for $[H^+]$ from the equation

$$\frac{K_{a_2}}{K_{a_1}} = \frac{[H_3A][HA^{2-}]}{[H_2A^-]^2}$$

by recognizing that if the reaction

$$H_2A^-(aq) + H_2A^-(aq) \rightleftharpoons H_3A(aq) + HA^{2-}(aq)$$

is the only important reaction involving these species, then

$$[H_3A] = [HA^{2-}]$$

This condition allows us to write

$$\frac{K_{a_2}}{K_{a_1}} = \frac{[H_3A][HA^{2-}]}{[H_2A^-]^2} = \frac{[H_3A]^2}{[H_2A^-]^2}$$

From the expression for K_{a_1} we have

$$\frac{[H^+]}{K_{a_1}} = \frac{[H_3A]}{[H_2A^-]}$$

Thus
$$\frac{K_{a_2}}{K_{a_1}} = \frac{[H_3A]^2}{[H_2A^-]^2} = \frac{[H^+]^2}{K_{a_1}{}^2}$$

or
$$[H^+]^2 = K_{a_1}{}^2 \times \frac{K_{a_2}}{K_{a_1}} = K_{a_1}K_{a_2}$$

and
$$[H^+] = K_{a_1}K_{a_2}$$

In terms of pH, this equation becomes

$$\text{pH} = \frac{\text{p}K_{a_1} + \text{p}K_{a_2}}{2}$$

Note that this equation applies only to a solution containing the major species H_2A^- and H_2O. It does not apply, for example, to a solution that contains the major species H_2A^-, HA^{2-}, and H_2O. In the former case H_2A^- is simultaneously the best acid and the best base, and thus the pH is determined by the reaction of H_2A^- with itself. In the latter case the solution contains the acid H_2A^- and its conjugate base HA^{2-}. In this case H_2A^- is the best acid and HA^{2-} is the best base, and the equilibrium that controls the pH is that described by K_{a_2}:

$$H_2A^-(aq) \rightleftharpoons H^+(aq) + HA^{2-}(aq)$$

Example 8.10 shows how the pH is calculated at the first equivalence point of the titration of phosphoric acid with sodium hydroxide, where the solution contains $H_2PO_4^-$. Note that the pH of the solution does not depend on the concentration of $H_2PO_4^-$ in the solution. Thus the pH is 4.67 at the first equivalence point in any typical titration of H_3PO_4. Likewise, at the second equivalence point, where the major species are HPO_4^{2-} and H_2O, the pH is calculated from the expression

$$\text{pH} = \frac{\text{p}K_{a_2} + \text{p}K_{a_3}}{2}$$

OWL INTERACTIVE EXAMPLE 8.10

Calculate the pH of a 1.0 *M* solution of NaH_2PO_4. (For H_3PO_4, $K_{a_1} = 7.5 \times 10^{-3}$, $K_{a_2} = 6.2 \times 10^{-8}$, and $K_{a_3} = 4.8 \times 10^{-13}$.)

Solution The major species in solution are

$$Na^+, \quad H_2PO_4^-, \quad \text{and} \quad H_2O$$

This is an example of a solution containing the amphoteric anion $H_2PO_4^-$, which is simultaneously the best acid and the best base. Both properties must be considered to calculate the pH correctly. Use the formula involving the average of the p*K*'s:

$$\text{pH} = \frac{\text{p}K_{a_1} + \text{p}K_{a_2}}{2} = \frac{2.12 + 7.21}{2} = 4.67$$

Titration of a Polyprotic Acid with Sodium Hydroxide—A Summary

At this point it is useful to summarize the pH calculations associated with the titration of a triprotic acid H_3A. Figure 8.11 shows which expression should be used for the major species in the solution at a given point in the titration.

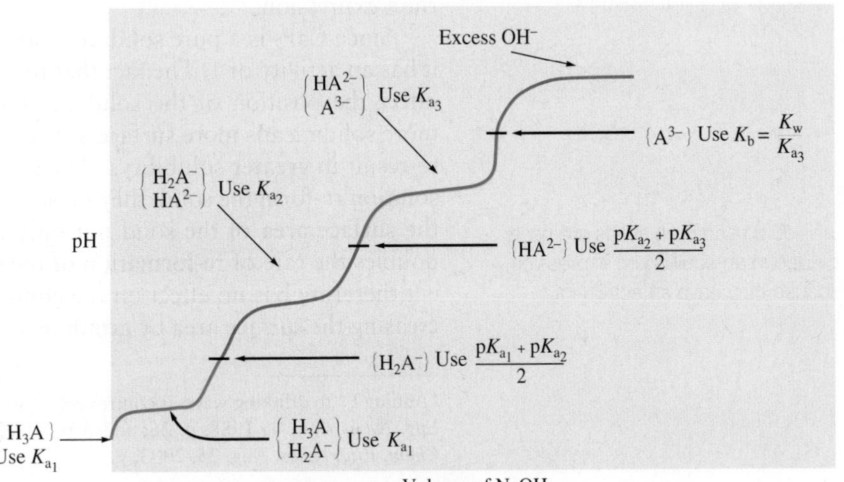

Figure 8.11 A summary of the important equilibria at various points in the titration of a triprotic acid.

8.8 Solubility Equilibria and the Solubility Product

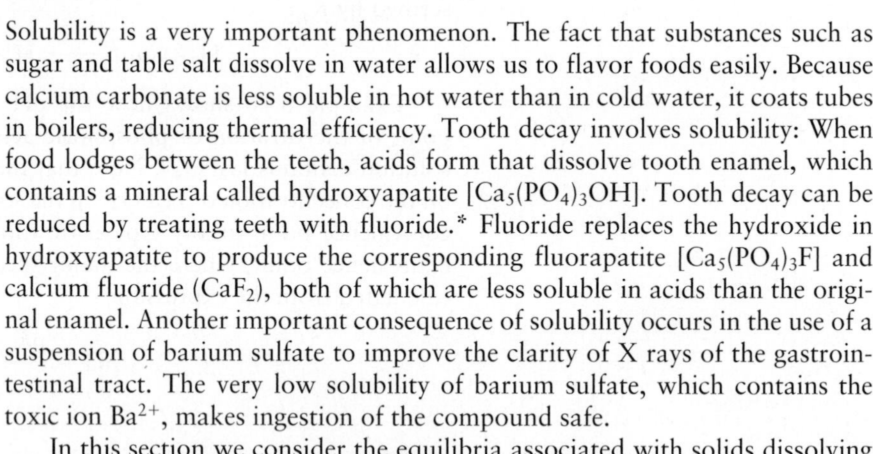

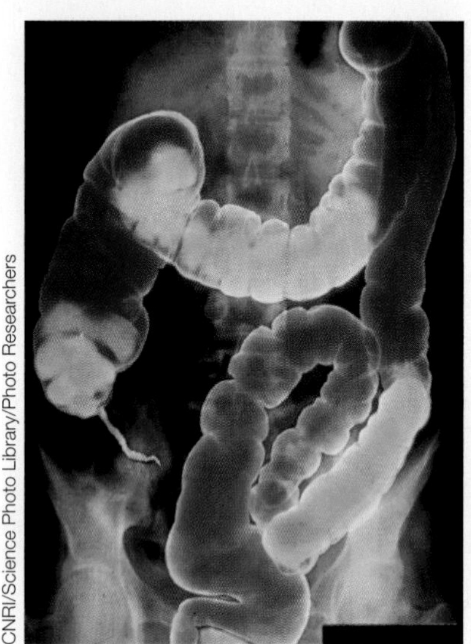

CNRI/Science Photo Library/Photo Researchers

An X ray of the upper gastrointestinal tract clarified by barium sulfate.

Solubility is a very important phenomenon. The fact that substances such as sugar and table salt dissolve in water allows us to flavor foods easily. Because calcium carbonate is less soluble in hot water than in cold water, it coats tubes in boilers, reducing thermal efficiency. Tooth decay involves solubility: When food lodges between the teeth, acids form that dissolve tooth enamel, which contains a mineral called hydroxyapatite [$Ca_5(PO_4)_3OH$]. Tooth decay can be reduced by treating teeth with fluoride.* Fluoride replaces the hydroxide in hydroxyapatite to produce the corresponding fluorapatite [$Ca_5(PO_4)_3F$] and calcium fluoride (CaF_2), both of which are less soluble in acids than the original enamel. Another important consequence of solubility occurs in the use of a suspension of barium sulfate to improve the clarity of X rays of the gastrointestinal tract. The very low solubility of barium sulfate, which contains the toxic ion Ba^{2+}, makes ingestion of the compound safe.

In this section we consider the equilibria associated with solids dissolving in water to form aqueous solutions. When an ionic solid dissolves in water, we typically assume that it dissociates into separate hydrated cations and anions. For example, when calcium fluoride dissolves in water, we typically represent the situation as follows:

$$CaF_2(s) \xrightarrow{H_2O} Ca^{2+}(aq) + 2F^-(aq)$$

When the solid salt is first added to the water, no Ca^{2+} or F^- ions are present. However, as the dissolution proceeds, the concentrations of Ca^{2+} and F^- increase, making it more and more likely that these ions will collide and re-form the solid phase. Thus two competing processes are occurring, the preceding reaction and the reverse reaction:

$$Ca^{2+}(aq) + 2F^-(aq) \longrightarrow CaF_2(s)$$

Ultimately, dynamic equilibrium is reached:

$$CaF_2(s) \rightleftharpoons Ca^{2+}(aq) + 2F^-(aq)$$

At this point, no more solid can dissolve (the solution is said to be *saturated*).

We can write an equilibrium expression for this process according to the law of mass action:

$$K_{sp} = [Ca^{2+}][F^-]^2$$

This analysis ignores the presence of ion pairs such as CaF^-, which may be present in significant amounts.

where $[Ca^{2+}]$ and $[F^-]$ are expressed in mol/L. The constant K_{sp} is called the **solubility product constant,** or simply the **solubility product** for the equilibrium expression.

Pure liquids and pure solids are never included in an equilibrium expression because they have an activity of 1.

Since CaF_2 is a pure solid, it is not included in the equilibrium expression; it has an activity of 1. The fact that the amount of excess solid present does not affect the position of the solubility equilibrium might seem strange at first; more solid means more surface area exposed to the solvent, which would seem to result in greater solubility. This is not the case, however. When the ions in solution re-form the solid, they do so on the surface of the solid. Thus doubling the surface area of the solid not only doubles the rate of dissolving but also doubles the rate of re-formation of the solid. The amount of excess solid present therefore has no effect on the equilibrium position. Similarly, although increasing the surface area by grinding up the solid or stirring the solution speeds

*Adding F^- to drinking water is controversial. See Bette Hileman, "Fluoridation of Water," *Chem. Eng. News,* Aug. 1, 1988, p. 26; and Bette Hileman, "Fluoride Concerns Surface Once Again," *Chem. Eng. News,* Aug. 25, 2003, p. 22.

Table 8.5

K_{sp} Values at 25°C for Common Ionic Solids

Ionic Solid	K_{sp} (at 25°C)
Fluorides	
BaF_2	2.4×10^{-5}
MgF_2	6.4×10^{-9}
PbF_2	4×10^{-8}
SrF_2	7.9×10^{-10}
CaF_2	4.0×10^{-11}
Chlorides	
$PbCl_2$	1.6×10^{-5}
$AgCl$	1.6×10^{-10}
Hg_2Cl_2*	1.1×10^{-18}
Bromides	
$PbBr_2$	4.6×10^{-6}
$AgBr$	5.0×10^{-13}
Hg_2Br_2*	1.3×10^{-22}
Iodides	
PbI_2	1.4×10^{-8}
AgI	1.5×10^{-16}
Hg_2I_2*	4.5×10^{-29}
Sulfates	
$CaSO_4$	6.1×10^{-5}
Ag_2SO_4	1.2×10^{-5}
$SrSO_4$	3.2×10^{-7}
$PbSO_4$	1.3×10^{-8}
$BaSO_4$	1.5×10^{-9}
Chromates	
$SrCrO_4$	3.6×10^{-5}

Ionic Solid	K_{sp} (at 25°C)
Hg_2CrO_4*	2×10^{-9}
$BaCrO_4$	8.5×10^{-11}
Ag_2CrO_4	9.0×10^{-12}
$PbCrO_4$	2×10^{-16}
Carbonates	
$NiCO_3$	1.4×10^{-7}
$CaCO_3$	8.7×10^{-9}
$BaCO_3$	1.6×10^{-9}
$SrCO_3$	7×10^{-10}
$CuCO_3$	2.5×10^{-10}
$ZnCO_3$	2×10^{-10}
$MnCO_3$	8.8×10^{-11}
$FeCO_3$	2.1×10^{-11}
Ag_2CO_3	8.1×10^{-12}
$CdCO_3$	5.2×10^{-12}
$PbCO_3$	1.5×10^{-15}
$MgCO_3$	1×10^{-15}
Hg_2CO_3*	9.0×10^{-15}
Hydroxides	
$Ba(OH)_2$	5.0×10^{-3}
$Sr(OH)_2$	3.2×10^{-4}
$Ca(OH)_2$	1.3×10^{-6}
$AgOH$	2.0×10^{-8}
$Mg(OH)_2$	8.9×10^{-12}
$Mn(OH)_2$	2×10^{-13}
$Cd(OH)_2$	5.9×10^{-15}
$Pb(OH)_2$	1.2×10^{-15}
$Fe(OH)_2$	1.8×10^{-15}

Ionic Solid	K_{sp} (at 25°C)
$Co(OH)_2$	2.5×10^{-16}
$Ni(OH)_2$	1.6×10^{-16}
$Zn(OH)_2$	4.5×10^{-17}
$Cu(OH)_2$	1.6×10^{-19}
$Hg(OH)_2$	3×10^{-26}
$Sn(OH)_2$	3×10^{-27}
$Cr(OH)_3$	6.7×10^{-31}
$Al(OH)_3$	2×10^{-32}
$Fe(OH)_3$	4×10^{-38}
$Co(OH)_3$	2.5×10^{-43}
Sulfides	
MnS	2.3×10^{-13}
FeS	3.7×10^{-19}
NiS	3×10^{-21}
CoS	5×10^{-22}
ZnS	2.5×10^{-22}
SnS	1×10^{-26}
CdS	1.0×10^{-28}
PbS	7×10^{-29}
CuS	8.5×10^{-45}
Ag_2S	1.6×10^{-49}
HgS	1.6×10^{-54}
Phosphates	
Ag_3PO_4	1.8×10^{-18}
$Sr_3(PO_4)_2$	1×10^{-31}
$Ca_3(PO_4)_2$	1.3×10^{-32}
$Ba_3(PO_4)_2$	6×10^{-39}
$Pb_3(PO_4)_2$	1×10^{-54}

*Contains Hg_2^{2+} ions. $K_{sp} = [Hg_2^{2+}][X^-]^2$ for Hg_2X_2 salts.

up the attainment of equilibrium, neither procedure changes the amount of solid dissolved at equilibrium. Neither the amount of excess solid nor the size of the particles will shift the *position* of the solubility equilibrium.

K_{sp} is an equilibrium constant; solubility is an equilibrium position.

It is very important to distinguish between the *solubility* of a given solid and its *solubility product*. The solubility product is an *equilibrium constant* and thus has only *one* value for a given solid at a given temperature. Solubility, on the other hand, is an *equilibrium position* and has an *infinite number* of possible values at a given temperature, depending on the other conditions (such as the presence of a common ion). The K_{sp} values at 25°C for many common ionic solids are listed in Table 8.5. The units are customarily omitted.

OWL INTERACTIVE EXAMPLE 8.11

Calculate the K_{sp} value for bismuth sulfide (Bi_2S_3), which has a solubility of 1.0×10^{-15} mol/L at 25°C.

Solution The system initially contains H_2O and solid Bi_2S_3. In the simplest treatment, we assume the solid dissolves as follows:

$$Bi_2S_3(s) \rightleftharpoons 2Bi^{3+}(aq) + 3S^{2-}(aq)$$

Precipitation of bismuth sulfide.

This treatment ignores ion pairs between the Bi^{3+} and S^{2-} ions.

Therefore $$K_{sp} = [Bi^{3+}]^2[S^{2-}]^3$$

Since no Bi^{3+} or S^{2-} ions are present in solution before the Bi_2S_3 dissolves,

$$[Bi^{3+}]_0 = [S^{2-}]_0 = 0$$

Thus the equilibrium concentrations of these ions will be determined by the amount of salt that dissolves to reach equilibrium, which in this case is 1.0×10^{-15} mol/L. Since each Bi_2S_3 unit contains $2Bi^{3+}$ and $3S^{2-}$ ions,

$$1.0 \times 10^{-15} \text{ mol/L } Bi_2S_3(s) \longrightarrow 2(1.0 \times 10^{-15} \text{ mol/L}) \; Bi^{3+}(aq) + 3(1.0 \times 10^{-15} \text{ mol/L}) \; S^{2-}(aq)$$

The equilibrium concentrations are

$$[Bi^{3+}] = [Bi^{3+}]_0 + \text{change} = 0 + 2.0 \times 10^{-15} \text{ mol/L}$$

$$[S^{2-}] = [S^{2-}]_0 + \text{change} = 0 + 3.0 \times 10^{-15} \text{ mol/L}$$

Then

$$K_{sp} = [Bi^{3+}]^2[S^{2-}]^3 = (2.0 \times 10^{-15})^2(3.0 \times 10^{-15})^3 = 1.1 \times 10^{-73}$$

In Example 8.11 we used the solubility of an ionic solid to calculate its K_{sp} value. The reverse is also possible: The solubility of an ionic solid can be calculated if its K_{sp} value is known.

OWL INTERACTIVE EXAMPLE 8.12

The K_{sp} value for copper(II) iodate $[Cu(IO_3)_2]$ is 1.4×10^{-7} at 25°C. Calculate its solubility at 25°C.

Solution The system initially contains H_2O and solid $Cu(IO_3)_2$. The solid dissolves according to the following equilibrium:

$$Cu(IO_3)_2(s) \rightleftharpoons Cu^{2+}(aq) + 2IO_3^-(aq)$$

Therefore $$K_{sp} = [Cu^{2+}][IO_3^-]^2$$

A more detailed analysis of this situation would include ion pairs such as $[CuIO_3]^+$, but this is beyond the level of this text.

To find the solubility of $Cu(IO_3)_2$, we must find the equilibrium concentrations of the Cu^{2+} and IO_3^- ions. We do this in the usual way by specifying the initial concentrations (before any solid has dissolved) and then defining the change required to reach equilibrium. Since in this case we do not know the solubility, we will assume that x mol/L of the solid dissolves to reach equilibrium. The 1:2 stoichiometry of the salt means that

$$x \text{ mol/L } Cu(IO_3)_2(s) \longrightarrow x \text{ mol/L } Cu^{2+}(aq) + 2x \text{ mol/L } IO_3^-(aq)$$

The concentrations are as follows:

Initial Concentration (mol/L) [before any $Cu(IO_3)_2$ dissolves]		Equilibrium Concentration (mol/L)
$[Cu^{2+}]_0 = 0$ $[IO_3^-]_0 = 0$	x mol/L dissolves ⟶ to reach equilibrium	$[Cu^{2+}] = x$ $[IO_3^-] = 2x$

Substituting the equilibrium concentrations into the expression for K_{sp} gives

$$1.4 \times 10^{-7} = K_{sp} = [Cu^{2+}][IO_3^-]^2 = (x)(2x)^2 = 4x^3$$

Then $$x = \sqrt[3]{3.5 \times 10^{-8}} = 3.3 \times 10^{-3} \text{ mol/L}$$

Thus the solubility of solid $Cu(IO_3)_2$ is 3.3×10^{-3} mol/L.

Relative Solubilities

A salt's K_{sp} value provides information about its solubility. However, we must be careful in using K_{sp} values to predict the *relative* solubilities of a group of salts. There are two possible situations.

1. The salts being compared produce the same number of ions. For example, consider

$$\text{AgI}(s) \quad K_{sp} = 1.5 \times 10^{-16}$$
$$\text{CuI}(s) \quad K_{sp} = 5.0 \times 10^{-12}$$
$$\text{CaSO}_4(s) \quad K_{sp} = 6.1 \times 10^{-5}$$

Each of these solids dissolves to produce two ions:

$$\text{Salt} \rightleftharpoons \text{cation} + \text{anion}$$
$$K_{sp} = [\text{cation}][\text{anion}]$$

If x is the solubility in mol/L, then at equilibrium

$$[\text{Cation}] = x$$
$$[\text{Anion}] = x$$
$$K_{sp} = [\text{cation}][\text{anion}] = x^2$$
$$x = \sqrt{K_{sp}} = \text{solubility}$$

Thus in this case we can compare the solubilities of these solids by comparing their K_{sp} values:

$$\underset{\substack{\text{Most soluble;}\\ \text{largest } K_{sp}}}{\text{CaSO}_4(s)} > \text{CuI}(s) > \underset{\substack{\text{Least soluble;}\\ \text{smallest } K_{sp}}}{\text{AgI}(s)}$$

2. The salts being compared produce different numbers of ions. For example, consider

$$\text{CuS}(s) \quad K_{sp} = 8.5 \times 10^{-45}$$
$$\text{Ag}_2\text{S}(s) \quad K_{sp} = 1.6 \times 10^{-49}$$
$$\text{Bi}_2\text{S}_3(s) \quad K_{sp} = 1.1 \times 10^{-73}$$

Since these salts produce different numbers of ions when they dissolve, the K_{sp} values cannot be compared *directly* to determine the relative solubilities. In fact, if we calculate the solubilities (using the procedure in Example 8.12), we obtain the results summarized in Table 8.6. The order of solubilities is

$$\underset{\text{Most soluble}}{\text{Bi}_2\text{S}_3(s)} > \text{Ag}_2\text{S}(s) > \underset{\text{Least soluble}}{\text{CuS}(s)}$$

which is opposite to the order of the K_{sp} values.

Table 8.6

Calculated Solubilities for CuS, Ag_2S, and Bi_2S_3 at 25°C

Salt	K_{sp}	Calculated Solubility (mol/L)
CuS	8.5×10^{-45}	9.2×10^{-23}
Ag_2S	1.6×10^{-49}	3.4×10^{-17}
Bi_2S_3	1.1×10^{-73}	1.0×10^{-15}

Remember that relative solubilities can be predicted by comparing K_{sp} values *only* for salts that produce the same total number of ions.

Common Ion Effect

So far we have considered ionic solids dissolved in pure water. We will now see what happens when the water contains an ion in common with the dissolving salt. For example, consider the solubility of solid silver chromate (Ag_2CrO_4, $K_{sp} = 9.0 \times 10^{-12}$) in a 0.100 M solution of $AgNO_3$. Before any Ag_2CrO_4 dissolves, the solution contains the major species Ag^+, NO_3^-, and H_2O. Since

A potassium chromate solution being added to aqueous silver nitrate, forming silver chromate.

NO_3^- is not found in Ag_2CrO_4, we can ignore it. The relevant initial concentrations (before any Ag_2CrO_4 dissolves) are

$$[Ag^+]_0 = 0.100\ M \qquad \text{(from the dissolved } AgNO_3\text{)}$$
$$[CrO_4^{2-}]_0 = 0$$

The system comes to equilibrium as Ag_2CrO_4 dissolves according to the reaction

$$Ag_2CrO_4(s) \rightleftharpoons 2Ag^+(aq) + CrO_4^{2-}(aq)$$

for which

$$K_{sp} = [Ag^+]^2[CrO_4^{2-}] = 9.0 \times 10^{-12}$$

We assume that x mol/L of Ag_2CrO_4 dissolves to reach equilibrium, which means that

$$x \text{ mol/L } Ag_2CrO_4(s) \longrightarrow 2x \text{ mol/L } Ag^+(aq) + x \text{ mol/L } CrO_4^{2-}(aq)$$

Now we can specify the equilibrium concentrations in terms of x:

$$[Ag^+] = [Ag^+]_0 + \text{change} = 0.100 + 2x$$
$$[CrO_4^{2-}] = [CrO_4^{2-}]_0 + \text{change} = 0 + x = x$$

Substituting these concentrations into the expression for K_{sp} gives

$$9.0 \times 10^{-12} = [Ag^+]^2[CrO_4^{2-}] = (0.100 + 2x)^2(x)$$

The mathematics required here appears to be complicated, since the right-hand side of the equation produces an expression that contains an x^3 term. However, as is usually the case, we can make simplifying assumptions. Since the K_{sp} value for Ag_2CrO_4 is small (the position of the equilibrium lies far to the left), x is expected to be small compared with 0.100 M. Therefore, $0.100 + 2x \approx 0.100$, which allows simplification of the expression:

$$9.0 \times 10^{-12} = (0.100 + 2x)^2(x) \approx (0.100)^2(x)$$

Then

$$x \approx \frac{9.0 \times 10^{-12}}{(0.100)^2} = 9.0 \times 10^{-10} \text{ mol/L}$$

Since x is much less than 0.100 M, the approximation is valid. Thus

$$\text{Solubility of } Ag_2CrO_4 \text{ in } 0.100\ M\ AgNO_3 = x = 9.0 \times 10^{-10} \text{ mol/L}$$

and the equilibrium concentrations are

$$[Ag^+] = 0.100 + 2x = 0.100 + 2(9.0 \times 10^{-10}) = 0.100\ M$$
$$[CrO_4^{2-}] = x = 9.0 \times 10^{-10}\ M$$

Now we compare the solubilities of Ag_2CrO_4 in pure water and in 0.100 M $AgNO_3$:

$$\text{Solubility of } Ag_2CrO_4 \text{ in pure water} = 1.3 \times 10^{-4} \text{ mol/L}$$
$$\text{Solubility of } Ag_2CrO_4 \text{ in } 0.100\ M\ AgNO_3 = 9.0 \times 10^{-10} \text{ mol/L}$$

Note that the solubility of Ag_2CrO_4 is much less in the presence of the Ag^+ ions from $AgNO_3$. This is another example of the common ion effect. The solubility of a solid is lowered if the solution already contains ions common to the solid.

Complications Inherent in Solubility Calculations

So far in this section we have assumed a direct relationship between the observed solubility of a salt and the concentrations of the component ions in the solution. However, this procedure is fraught with difficulties. For example, when we calculated the K_{sp} for Bi_2S_3 earlier, no allowance was made for the fact that S^{2-} is an excellent base, causing a significant amount of the reaction

$$S^{2-}(aq) + H_2O(l) \rightleftharpoons HS^-(aq) + OH^-(aq)$$

to occur in aqueous solution. Thus, although we assumed in the earlier calculation that all of the sulfide from dissolving Bi_2S_3 exists as S^{2-} in solution, this is clearly not the case. The solubility of a salt containing a basic anion can be calculated accurately by simultaneously considering the K_{sp} and K_b equilibria. However, we will not show that calculation here.

Another complication that clouds the relationship between the measured solubility of a salt and its K_{sp} value is ion pairing. For example, when $CaSO_4$ dissolves in water, the solution contains

$$Ca^{2+}(s), \quad SO_4^{2-}(aq), \quad \text{and} \quad CaSO_4(aq)$$

the last representing an ion pair surrounded by water molecules:

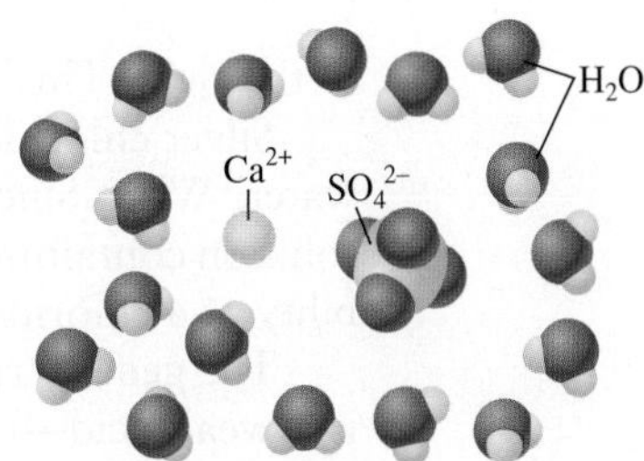

Therefore, from a measured solubility of $CaSO_4$ of $\approx 10^{-3}\ M$, we cannot safely assume that

$$[Ca^{2+}] = [SO_4^{2-}] = 10^{-3}\ M$$

since significant numbers of Ca^{2+} and SO_4^{2-} might be present as ion pairs. In very accurate solubility calculations, the activities of the ions are used instead of the stoichiometric concentrations that we have used in this chapter. In obtaining the ion activities, corrections are made for effects such as ion pairing. However, these calculations are beyond the scope of our treatment of solubility.

Yet another complication in K_{sp} calculations involves the formation of complex ions. For example, when AgCl is dissolved in water, some of the ions combine to form $AgCl_2^-$, a complex ion. Thus a saturated solution of AgCl will contain at least the ions Ag^+, Cl^-, and $AgCl_2^-$. In addition, other complex ions such as $AgCl_3^{2-}$ may exist as well as AgCl ion pairs. Again, the assumption that the concentrations of Ag^+ and Cl^- ions can be obtained directly from the measured solubility of AgCl is suspect. These effects can be corrected for by treating the solubility and complex ion equilibria simultaneously.

The point is this: The assumption that the ion concentrations for a particular solid can be obtained directly from its observed solubility causes our results to be, at best, approximations. We will not deal here with the procedures for correcting solubility calculations for these various effects.*

pH and Solubility

The pH of a solution can affect a salt's solubility quite significantly. For example, magnesium hydroxide dissolves according to the equilibrium

$$Mg(OH)_2(s) \rightleftharpoons Mg^{2+}(aq) + 2OH^-(aq)$$

*For additional information, see J. M. Bonicamp, et al., *J. Chem. Ed.* 75 (1998): 1182.

Addition of OH^- ions (an increase in pH) will, by the common ion effect, force the equilibrium to the left, decreasing the solubility of $Mg(OH)_2$. On the other hand, an addition of H^+ ions (a decrease in pH) increases the solubility because OH^- ions are removed from solution by reacting with the added H^+ ions. In response to the lower concentration of OH^-, the equilibrium position moves to the right. This explains how a suspension of solid $Mg(OH)_2$, known as milk of magnesia, dissolves in the stomach to combat excess acidity.

This concept also applies to salts with other types of anions. For example, the solubility of silver phosphate (Ag_3PO_4) is greater in acid than in pure water because the PO_4^{3-} ion is a strong base that reacts with H^+ to form the HPO_4^{2-} ion. The reaction

$$H^+ + PO_4^{3-} \longrightarrow HPO_4^{2-}$$

occurs in acidic solution, thus lowering the concentration of PO_4^{3-} and shifting the solubility equilibrium

$$Ag_3PO_4(s) \rightleftharpoons 3Ag^+(aq) + PO_4^{3-}(aq)$$

to the right. This in turn increases the solubility of silver phosphate.

Silver chloride (AgCl), however, has the same solubility in acid as in pure water. Why? Since the Cl^- ion is a very weak base, the addition of H^+ to a solution containing Cl^- does not affect $[Cl^-]$ and thus has no effect on the solubility of a chloride salt.

The general rule is that if the anion X^- is an effective base—that is, if HX is a weak acid—the salt MX will show increased solubility in an acidic solution. Examples of common anions that are effective bases are OH^-, S^{2-}, CO_3^{2-}, $C_2O_4^{2-}$, and CrO_4^{2-}. Salts involving these anions are much more soluble in an acidic solution than in pure water.

As mentioned at the beginning of this chapter, one practical result of the increased solubility of carbonates in acid is the formation of huge limestone caves such as Mammoth Cave in Kentucky and Carlsbad Caverns in New Mexico. Carbon dioxide dissolved in groundwater makes it acidic, increasing the solubility of calcium carbonate and eventually producing huge caverns. As the carbon dioxide escapes to the air, the pH of the dripping water goes up and the calcium carbonate precipitates, forming stalactites and stalagmites.

8.9 Precipitation and Qualitative Analysis

So far we have considered solids dissolving in aqueous solutions. Now we will consider the reverse process—the formation of precipitates. When solutions are mixed, various reactions can occur. We have already considered acid–base reactions in some detail. In this section we show how to predict whether a precipitate will form when two solutions are mixed. We will use the **ion product**, which is defined just like the K_{sp} expression for a given solid except that *initial concentrations are used* instead of equilibrium concentrations. For solid CaF_2 the expression for the ion product Q is written

$$Q = [Ca^{2+}]_0[F^-]_0^2$$

If we add a solution containing Ca^{2+} ions to a solution containing F^- ions, a precipitate may or may not form, depending on the concentrations of these ions in the mixed solution. To predict whether precipitation will occur, we consider the relationship between Q and K_{sp}:

If Q *is greater than* K_{sp}, precipitation occurs and will continue until the concentrations of ions remaining in solution satisfy K_{sp}.

If Q *is less than* K_{sp}, no precipitation occurs.

Sometimes we will want to do more than simply predict whether precipitation occurs; we will want to calculate the equilibrium concentrations in the solution after precipitation is complete. For example, we will next calculate the equilibrium concentrations of Pb^{2+} and I^- ions in a solution formed by mixing 100.0 mL of 0.0500 M $Pb(NO_3)_2$ and 200.0 mL of 0.100 M NaI. First, we must determine whether solid PbI_2 ($K_{sp} = 1.4 \times 10^{-8}$) forms when the solutions are mixed. To do so, we first calculate $[Pb^{2+}]_0$ and $[I^-]_0$ before any reaction occurs:

$$[Pb^{2+}]_0 = \frac{\text{mmol of } Pb^{2+}}{\text{mL of solution}} = \frac{(100.0 \text{ mL})(0.0500 \text{ mmol/mL})}{300.0 \text{ mL}} = 1.67 \times 10^{-2}\ M$$

$$[I^-]_0 = \frac{\text{mmol of } I^-}{\text{mL of solution}} = \frac{(200.0 \text{ mL})(0.100 \text{ mmol/mL})}{300.0 \text{ mL}} = 6.67 \times 10^{-2}\ M$$

The ion product for PbI_2 is

$$Q = [Pb^{2+}]_0[I^-]_0^2 = (1.67 \times 10^{-2})(6.67 \times 10^{-2})^2 = 7.43 \times 10^{-5}$$

Since Q is greater than K_{sp}, a precipitate of PbI_2 will form.

Note that since the K_{sp} for PbI_2 is quite small (1.4×10^{-8}), only very small quantities of Pb^{2+} and I^- can coexist in aqueous solution. In other words, when Pb^{2+} and I^- are mixed, most of these ions will precipitate out as PbI_2. That is, the reaction

The equilibrium constant for formation of solid PbI_2 is $1/K_{sp}$, or 7×10^7, so this equilibrium lies far to the right.

$$Pb^{2+}(aq) + 2I^-(aq) \longrightarrow PbI_2(s)$$

which is the reverse of the dissolution reaction, goes essentially to completion.

If, when two solutions are mixed, a reaction that goes virtually to completion occurs, it is essential to do the stoichiometric calculations before considering the equilibrium calculations. In this case we let the system go completely in the direction toward which it tends. Then we let it adjust back to equilibrium. If we let Pb^{2+} and I^- react to completion, we have the following calculations:

	Pb^{2+}	+	$2I^-$	$\longrightarrow$	PbI_2
Before reaction:	(100.0 mL)(0.0500 M) = 5.00 mmol		(200.0 mL)(0.100 M) = 20.0 mmol		The amount of PbI_2 formed does not influence the equilibrium
After reaction:	0 mmol		20.0 − 2(5.00) = 10.0 mmol		

Next we must allow the system to reach equilibrium. At equilibrium $[Pb^{2+}]$ is not zero, because the reaction really does not quite go to completion. The best way to think about this is that once the PbI_2 is formed, a very small amount redissolves to reach equilibrium. Since I^- is in excess, this PbI_2 is dissolving into a solution that contains 10.0 mmol of I^- per 300.0 mL of solution, or $3.33 \times 10^{-2}\ M\ I^-$.

We can state the resulting problem as follows: What is the solubility of solid PbI_2 in a $3.33 \times 10^{-2}\ M$ NaI solution? The lead iodide dissolves according to the equation

$$PbI_2(s) \rightleftharpoons Pb^{2+}(aq) + 2I^-(aq)$$

The concentrations are as follows:

Initial Concentration (mol/L)		Equilibrium Concentration (mol/L)
$[Pb^{2+}]_0 = 0$ $[I^-]_0 = 3.33 \times 10^{-2}$	$\xrightarrow[\text{dissolves}]{x\text{ mol/L } PbI_2(s)}$	$[Pb^{2+}] = x$ $[I^-] = 3.33 \times 10^{-2} + 2x$

Substituting into the expression for K_{sp} gives

$$K_{sp} = 1.4 \times 10^{-8} = [Pb^{2+}][I^-]^2 = (x)(3.33 \times 10^{-2} + 2x)^2$$
$$\approx (x)(3.33 \times 10^{-2})^2$$

Then

$$[Pb^{2+}] = x = 1.3 \times 10^{-5}\ M$$
$$[I^-] = 3.33 \times 10^{-2}\ M$$

Note that $3.33 \times 10^{-2} \gg 2x$, so the approximation is valid. These Pb^{2+} and I^- concentrations thus represent the equilibrium concentrations in the solution formed by mixing 100.0 mL of 0.0500 *M* $Pb(NO_3)_2$ and 200.0 mL of 0.100 *M* NaI.

Selective Precipitation

Mixtures of metal ions in aqueous solution are often separated by **selective precipitation** — that is, by using a reagent whose anion forms a precipitate with only one of the metal ions in the mixture. For example, suppose we have a solution containing both Ba^{2+} and Ag^+ ions. If NaCl is added to the solution, AgCl precipitates as a white solid, but since $BaCl_2$ is soluble, the Ba^{2+} ions remain in solution.

EXAMPLE 8.13

A solution contains 1.0×10^{-4} *M* Cu^+ and 2.0×10^{-3} *M* Pb^{2+}. If a source of I^- is added to this solution gradually, will PbI_2 ($K_{sp} = 1.4 \times 10^{-8}$) or CuI ($K_{sp} = 5.3 \times 10^{-12}$) precipitate first? Specify the concentration of I^- necessary to begin precipitation of each salt.

Solution For PbI_2 the K_{sp} expression is

$$1.4 \times 10^{-8} = K_{sp} = [Pb^{2+}][I^-]^2$$

Since $[Pb^{2+}]$ in this solution is known to be 2.0×10^{-3} *M*, the greatest concentration of I^- that can be present without causing precipitation of PbI_2 can be calculated from the K_{sp} expression:

$$1.4 \times 10^{-8} = [Pb^{2+}][I^-]^2 = (2.0 \times 10^{-3})[I^-]^2$$
$$[I^-] = 2.6 \times 10^{-3}\ M$$

Any I^- in excess of this concentration will cause solid PbI_2 to form.

Similarly, for CuI the K_{sp} expression is

$$5.3 \times 10^{-12} = K_{sp} = [Cu^+][I^-] = (1.0 \times 10^{-4})[I^-]$$

and

$$[I^-] = 5.3 \times 10^{-8}\ M$$

A concentration of I^- in excess of 5.3×10^{-8} *M* will cause formation of solid CuI.

When $KI(aq)$ is added to a solution containing 1.0×10^{-4} *M* Cu^+ and 2.0×10^{-3} *M* Pb^{2+}, white CuI(*s*) precipitates first.

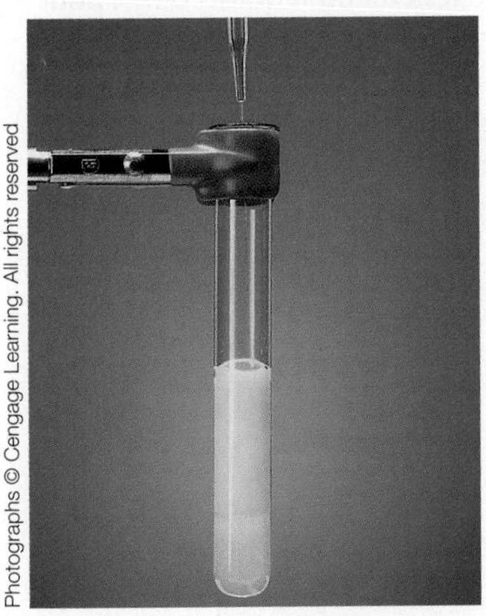

Yellow $PbI_2(s)$ precipitates next.

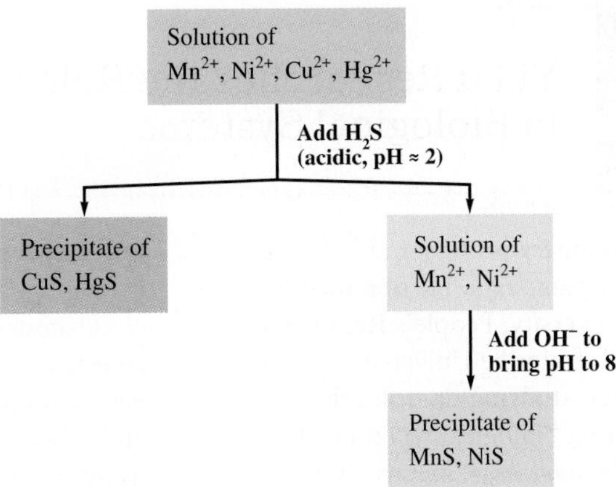

Figure 8.12
The separation of Cu^{2+} and Hg^{2+} from Ni^{2+} and Mn^{2+} using H_2S. At a low pH, $[S^{2-}]$ is relatively low and only the very insoluble HgS and CuS precipitate. When OH^- is added to lower $[H^+]$, the value of $[S^{2-}]$ increases, and MnS and NiS precipitate.

As I^- is added to the mixed solution, CuI will precipitate first, since the $[I^-]$ required is less. Therefore, Cu^+ can be separated from Pb^{2+} by using this reagent.

Since metal sulfide salts differ dramatically in their solubilities, the sulfide ion is often used to separate metal ions by selective precipitation. For example, consider a solution containing a mixture of 10^{-3} *M* Fe^{2+} and 10^{-3} *M* Mn^{2+}. Since FeS ($K_{sp} = 3.7 \times 10^{-19}$) is much less soluble than MnS ($K_{sp} = 2.3 \times 10^{-13}$), careful addition of S^{2-} to the mixture will precipitate Fe^{2+} as FeS, leaving Mn^{2+} in solution.

We can directly compare K_{sp} values to find relative solubilities because FeS and MnS produce the same number of ions in solution.

One real advantage of using the sulfide ion as a precipitating reagent is that because it is basic, its concentration can be controlled by regulating the pH of the solution. H_2S is a diprotic acid that dissociates in two steps, as shown in the following reactions:

$$H_2S(aq) \rightleftharpoons H^+(aq) + HS^-(aq) \qquad K_{a_1} = 1.0 \times 10^{-7}$$

$$HS^-(aq) \rightleftharpoons H^+(aq) + S^{2-}(aq) \qquad K_{a_2} \approx 10^{-19}$$

Note from the small K_{a_2} value that the S^{2-} ion has a high affinity for protons. In an acidic solution (large $[H^+]$), $[S^{2-}]$ will be relatively small, since under these conditions the dissociation equilibria will lie far to the left. On the other hand, in a basic solution, $[S^{2-}]$ will be relatively large, since the very small value of $[H^+]$ will pull both equilibria to the right, producing relatively large amounts of S^{2-}.

Thus the most insoluble sulfide salts, such as CuS ($K_{sp} = 8.5 \times 10^{-45}$) and HgS ($K_{sp} = 1.6 \times 10^{-54}$), can be precipitated from an acidic solution, leaving the more soluble ones, such as MnS ($K_{sp} = 2.3 \times 10^{-13}$) and NiS ($K_{sp} = 3 \times 10^{-21}$), still dissolved. The more soluble sulfides can then be precipitated by making the solution slightly basic. This procedure is diagramed in Fig. 8.12.

Qualitative Analysis

The classic scheme for **qualitative analysis** of a mixture containing all the common cations (listed in Fig. 8.13) involves first separating the cations into five major groups based on solubilities. (These groups are not directly related to the groups of the periodic table.) Each group is then treated further to separate and

Chemistry Explorers

Yi Lu Researches the Role of Metals in Biological Systems

Professor Lu is a chemistry professor at University of Illinois, Urbana–Champaign. He obtained his B.S. from Beijing University, People's Republic of China, and his Ph.D. from the University of California at Los Angeles, studying the roles that metals play in assisting biological reactions. His main goal is to understand how metals containing catalysts work. This knowledge could lead to the design of more efficient catalysts for processes involved in renewable energy, destroying environmental pollutants, fighting diseases such as AIDS, among others.

Based on their knowledge of metal ion chemistry, Professor Lu and his coworkers have designed a quick and accurate method to test for lead in paint and water. Detection of lead is important because lead is very toxic, especially to children. Studies have shown that children exposed to lead can suffer behavioral problems, learning disabilities, and, in severe cases, even death. Lead exposure is a particular problem in old houses where lead-based paint was used.

Yi Lu

Courtesy, Yi Lu/University of Illinois at Urbana

Professor Lu and his colleagues have found new ways to test for lead and other toxic metals using single-strand DNA molecules that change color when they interact with metals such as lead. The colorimetric test for lead involves dipping a treated "indicator" paper into the water to be tested and requires only about 2 minutes to show the result. The test for paint requires extraction of lead from the paint; the whole process takes about 20 minutes. Professor Lu has founded a company (DzymeTech), which is planning to develop a low-cost, easy-to-use kit that can be used to test for possible lead contamination. This should significantly lower the incidence of lead poisoning.

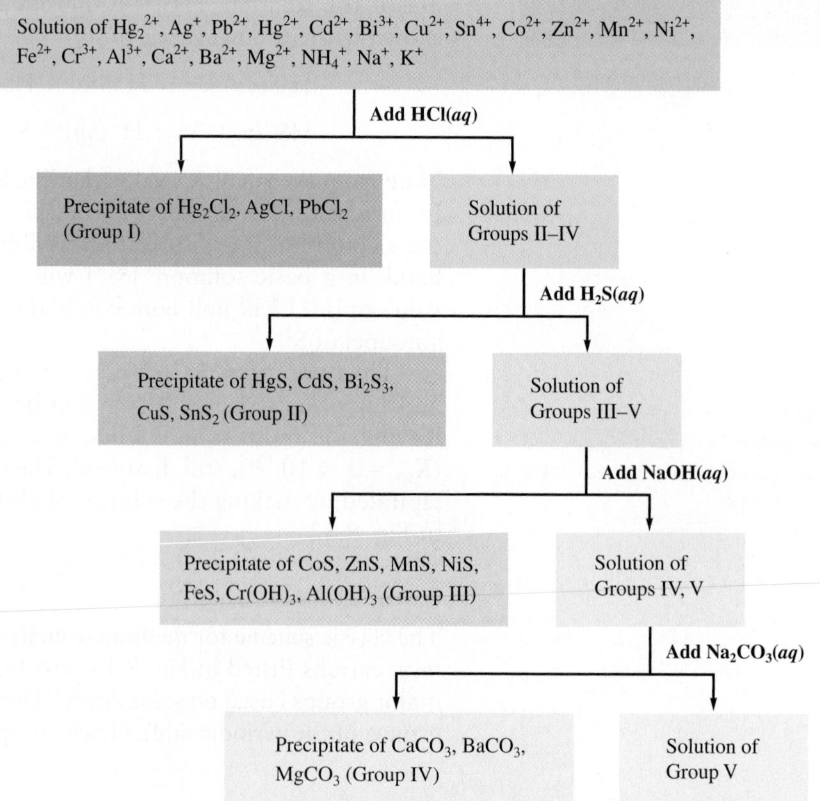

Figure 8.13
A schematic diagram of the classic method for separating the common cations by selective precipitation.

identify the individual ions. We will be concerned here only with separation of the major groups.

- **Group I—insoluble chlorides.**

  When dilute aqueous HCl is added to a solution containing a mixture of the common cations, only Ag^+, Pb^{2+}, and Hg_2^{2+} will precipitate as insoluble chlorides. All other chlorides are soluble and remain in solution. The Group I precipitate is removed, leaving the other ions in solution for treatment with sulfide ion.

- **Group II—sulfides insoluble in acid solution.**

  After the insoluble chlorides are removed, the solution is still acidic, since HCl was added. If H_2S is added to this solution, only the most insoluble sulfides (those of Hg^{2+}, Cd^{2+}, Bi^{3+}, Cu^{2+}, and Sn^{4+}) will precipitate, since $[S^{2-}]$ is relatively low because of the high concentration of H^+. The more soluble sulfides will remain dissolved under these conditions. The precipitate of the insoluble salts is removed.

- **Group III—sulfides insoluble in basic solution.**

  The solution is made basic at this stage and more H_2S is added. As we saw earlier, a basic solution produces a higher $[S^{2-}]$, which leads to precipitation of the more soluble sulfides. The cations precipitated as sulfides at this stage are Co^{2+}, Zn^{2+}, Mn^{2+}, Ni^{2+}, and Fe^{2+}. If any Cr^{3+} and Al^{3+} ions are present, they will also precipitate, but as insoluble hydroxides (remember that the solution is now basic). The precipitate is separated from the solution containing the rest of the ions.

- **Group IV—insoluble carbonates.**

  At this point all the cations have been precipitated except those from Groups 1A and 2A of the periodic table. The Group 2A cations form insoluble carbonates and can be precipitated by the addition of CO_3^{2-}. For example, Ba^{2+}, Ca^{2+}, and Mg^{2+} form solid carbonates and can be removed from the solution.

- **Group V—alkali metal and ammonium ions.**

  The only ions remaining in solution at this point are the Group 1A cations and the NH_4^+ ion, all of which form soluble salts with the common anions.

(top) Flame test for potassium.
(bottom) Flame test for sodium.

From left to right, cadmium sulfide, chromium(III) hydroxide, aluminum hydroxide, and nickel(II) hydroxide.

The Group 1A cations are usually identified by the characteristic colors they produce when heated in a flame. These colors are caused by the emission spectra of these ions.

The qualitative analysis scheme for cations based on the selective precipitation procedure described above is summarized in Fig. 8.13.

8.10 Complex Ion Equilibria

A **complex ion** is a charged species consisting of a metal ion surrounded by *ligands.* A ligand is a molecule or an ion having a lone pair of electrons that can be donated to the metal ion to form a covalent bond. Some common ligands are H_2O, NH_3, Cl^-, and CN^-. The number of ligands attached to a metal ion is called the *coordination number.* The most common coordination numbers are 6, for example, in $Co(H_2O)_6^{2+}$ and $Ni(NH_3)_6^{2+}$; 4, for example, in $CoCl_4^{2-}$ and $Cu(NH_3)_4^{2+}$; and 2, for example, in $Ag(NH_3)_2^+$; but others are known.

The properties of complex ions will be discussed in more detail in Chapter 19. For now we will just look at the equilibria involving these species. Metal ions add ligands one at a time in steps characterized by equilibrium constants called **formation constants,** or **stability constants.** For example, when solutions containing Ag^+ ions and NH_3 molecules are mixed, the following reactions take place:

$$Ag^+(aq) + NH_3(aq) \rightleftharpoons Ag(NH_3)^+(aq) \qquad K_1 = 2.1 \times 10^3$$

$$Ag(NH_3)^+(aq) + NH_3(aq) \rightleftharpoons Ag(NH_3)_2^+(aq) \qquad K_2 = 8.2 \times 10^3$$

where K_1 and K_2 are the formation constants for the two steps. In a solution containing Ag^+ and NH_3, all the species—NH_3, Ag^+, $Ag(NH_3)^+$, and $Ag(NH_3)_2^+$—exist at equilibrium. Calculating the concentrations of all these components can be complicated. However, usually the total concentration of the ligand is much larger than the total concentration of the metal ion, and approximations can greatly simplify the problems.

For example, consider a solution prepared by mixing 100.0 mL of 2.0 *M* NH_3 with 100.0 mL of 1.0×10^{-3} *M* $AgNO_3$. *Before any reaction occurs,* the mixed solution contains the major species Ag^+, NO_3^-, NH_3, and H_2O. What reaction or reactions will occur in this solution? From our discussions of acid–base chemistry, we know that one reaction is

$$NH_3(aq) + H_2O(l) \rightleftharpoons NH_4^+(aq) + OH^-(aq)$$

However, we are interested in the reaction between NH_3 and Ag^+ to form complex ions, and since the position of the preceding equilibrium lies far to the left (K_b for NH_3 is 1.8×10^{-5}), we can neglect the amount of NH_3 consumed in the reaction with water. Thus, before any complex ion formation occurs, the concentrations in the mixed solution are

$$[Ag^+]_0 = \frac{(100.0\text{ mL})(1.0 \times 10^{-3}\ M)}{200.0\text{ mL}} = 5.0 \times 10^{-4}\ M$$

↖ Total volume

$$[NH_3]_0 = \frac{(100.0\text{ mL})(2.0\ M)}{200.0\text{ mL}} = 1.0\ M$$

As mentioned already, the Ag^+ ion reacts with NH_3 in a stepwise manner to form $AgNH_3^+$ and then $Ag(NH_3)_2^+$:

$$Ag^+(aq) + NH_3(aq) \rightleftharpoons Ag(NH_3)^+(aq) \qquad K_1 = 2.1 \times 10^3$$

$$AgNH_3^+(aq) + NH_3(aq) \rightleftharpoons Ag(NH_3)_2^+(aq) \qquad K_2 = 8.2 \times 10^3$$

Since both K_1 and K_2 are large, and because there is a large excess of NH_3, *both reactions can be assumed to go essentially to completion.* This is equivalent to writing the net reaction in the solution as follows:

$$Ag^+ + 2NH_3 \longrightarrow Ag(NH_3)_2^+$$

The stoichiometric calculations are summarized below.

	Ag^+	+	$2NH_3$	$\longrightarrow$	$Ag(NH_3)_2^+$
Before reaction:	$5.0 \times 10^{-4}\ M$		$1.0\ M$		0
After reaction:	0		$1.0 - 2(5.0 \times 10^{-4}) \approx 1.0\ M$		$5.0 \times 10^{-4}\ M$
			↗ Twice as much NH_3 as Ag^+ is required		

Note that in this case we have used molarities when performing the calculations and we have assumed this reaction to be complete, using all the original Ag^+ to form $Ag(NH_3)_2^+$. In reality, a *very small* amount of the $Ag(NH_3)_2^+$ formed will dissociate to produce small amounts of $Ag(NH_3)^+$ and Ag^+. However, since the amount of $Ag(NH_3)_2^+$ dissociating will be so small, we can safely assume that $[Ag(NH_3)_2^+]$ is $5.0 \times 10^{-4}\ M$ at equilibrium. Also, we know that since so little NH_3 has been consumed, $[NH_3]$ is essentially $1.0\ M$ at equilibrium. We can use these concentrations to calculate $[Ag^+]$ and $[Ag(NH_3)^+]$ using the K_1 and K_2 expressions.

To calculate the equilibrium concentration of $Ag(NH_3)^+$, we use

$$K_2 = 8.2 \times 10^3 = \frac{[Ag(NH_3)_2^+]}{[Ag(NH_3)^+][NH_3]}$$

since $[Ag(NH_3)_2^+]$ and $[NH_3]$ are known. Rearranging and solving for $[Ag(NH_3)^+]$ gives

$$[Ag(NH_3)^+] = \frac{[Ag(NH_3)_2^+]}{K_2[NH_3]} = \frac{5.0 \times 10^{-4}}{(8.2 \times 10^{-3})(1.0)} = 6.1 \times 10^{-8}\ M$$

Now the equilibrium concentration of Ag^+ can be calculated by using K_1:

$$K_1 = 2.1 \times 10^3 = \frac{[Ag(NH_3)^+]}{[Ag^+][NH_3]} = \frac{6.1 \times 10^{-8}}{[Ag^+](1.0)}$$

$$[Ag^+] = \frac{6.1 \times 10^{-8}}{(2.1 \times 10^3)(1.0)} = 2.9 \times 10^{-11}\ M$$

So far we have assumed that $Ag(NH_3)_2^+$ is the dominant silver-containing species in solution. Is this a valid assumption? The calculated concentrations are

$$[Ag(NH_3)_2^+] = 5.0 \times 10^{-4}\ M$$

$$[AgNH_3^+] = 6.1 \times 10^{-8}\ M$$

$$[Ag^+] = 2.9 \times 10^{-11}\ M$$

These values clearly support the conclusion that

$$[Ag(NH_3)_2^+] \gg [AgNH_3^+] \gg [Ag^+]$$

Essentially all the Ag^+ ions originally present end up in $Ag(NH_3)_2^+$ at equilibrium.

Thus the assumption that $[Ag(NH_3)_2^+]$ is dominant is valid, and the calculated concentrations are correct.

This analysis shows that although complex ion equilibria have many species present and look complicated, the calculations are actually quite straightforward, especially if the ligand is present in large excess.

EXAMPLE 8.14

Calculate the concentrations of Ag^+, $Ag(S_2O_3)^-$, and $Ag(S_2O_3)_2^{3-}$ in a solution prepared by mixing 150.0 mL of 1.00×10^{-3} *M* $AgNO_3$ with 200.0 mL of 5.00 *M* $Na_2S_2O_3$. The stepwise formation equilibria are

$$Ag^+(aq) + S_2O_3^{2-}(aq) \rightleftharpoons Ag(S_2O_3)^-(aq) \qquad K_1 = 7.4 \times 10^8$$

$$Ag(S_2O_3)^-(aq) + S_2O_3^{2-}(aq) \rightleftharpoons Ag(S_2O_3)_2^{3-}(aq) \qquad K_2 = 3.9 \times 10^4$$

Solution The concentrations of the ligand and metal ion in the mixed solution *before any reaction occurs* are

$$[Ag^+]_0 = \frac{(150.0 \text{ mL})(1.00 \times 10^{-3}\ M)}{150.0 \text{ mL} + 200.0 \text{ mL}} = 4.29 \times 10^{-4}\ M$$

$$[S_2O_3^{2-}]_0 = \frac{(200.0 \text{ mL})(5.00\ M)}{150.0 \text{ mL} + 200.0 \text{ mL}} = 2.86\ M$$

Since $[S_2O_3^{2-}]_0 \gg [Ag^+]_0$, and because K_1 and K_2 are large, both formation reactions can be assumed to go to completion. The net reaction in the solution is as follows:

	Ag^+	+	$2S_2O_3^{2-}$	$\longrightarrow$	$Ag(S_2O_3)_2^{3-}$
Before reaction:	$4.29 \times 10^{-4}\ M$		$2.86\ M$		0
After reaction:	≈ 0		$2.86 - 2(4.29 \times 10^{-4})$ $\approx 2.86\ M$		$4.29 \times 10^{-4}\ M$

Note that Ag^+ is limiting and that the amount of $S_2O_3^{2-}$ consumed is negligible. Also note that since all these species are in the same solution, the molarities can be used to do the stoichiometry problem.

Of course, the concentrations calculated earlier do not represent the equilibrium concentrations. For example, the concentration of Ag^+ is not zero at equilibrium, and there is some $Ag(S_2O_3)^-$ in the solution. To calculate the equilibrium concentrations of these species, we must use the K_1 and K_2 expressions. We can calculate the concentration of $Ag(S_2O_3)^-$ from K_2:

$$3.9 \times 10^4 = K_2 = \frac{[Ag(S_2O_3)_2^{3-}]}{[Ag(S_2O_3)^-][S_2O_3^{2-}]} = \frac{4.29 \times 10^{-4}}{[Ag(S_2O_3)^-](2.86)}$$

$$[Ag(S_2O_3)^-] = 3.8 \times 10^{-9}\ M$$

We can calculate $[Ag^+]$ from K_1:

$$7.4 \times 10^8 = K_1 = \frac{[Ag(S_2O_3)^-]}{[Ag^+][S_2O_3^{2-}]} = \frac{3.8 \times 10^{-9}}{[Ag^+](2.86)}$$

$$[Ag^+] = 1.8 \times 10^{-18}\ M$$

These results show that

$$[Ag(S_2O_3)_2^{3-}] \gg [Ag(S_2O_3)^-] \gg [Ag^+]$$

Thus the assumption that essentially all the original Ag^+ is converted to $Ag(S_2O_3)_2^{3-}$ at equilibrium is valid.

Complex Ions and Solubility

Often ionic solids that are only slightly soluble in water must be dissolved in aqueous solutions. For example, when the various qualitative analysis groups are precipitated, the precipitates must be redissolved to separate the ions within

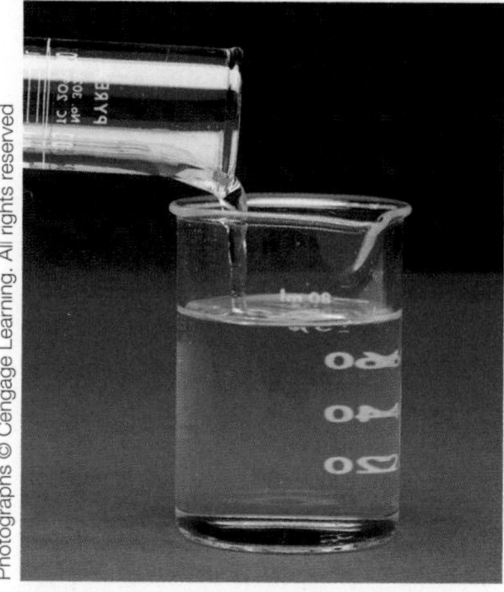

(top) Aqueous ammonia is added to silver chloride (white). (bottom) The silver chloride, insoluble in water, dissolves to form $Ag(NH_3)_2^+(aq)$ and $Cl^-(aq)$.

each group. Consider a solution of cations that contains Ag^+, Pb^{2+}, and Hg_2^{2+}, among others. When dilute aqueous HCl is added to this solution, the Group I ions will form the insoluble chlorides AgCl, $PbCl_2$, and Hg_2Cl_2. Once this mixed precipitate is separated from the solution, it must be redissolved to identify the cations individually. How can this be done? We know that some solids are more soluble in acidic than in neutral solutions. What about chloride salts? For example, can AgCl be dissolved by using a strong acid? The answer is no, because Cl^- ions have virtually no affinity for H^+ ions in aqueous solution. The position of the dissolution equilibrium

$$AgCl(s) \rightleftharpoons Ag^+(aq) + Cl^-(aq)$$

is not affected by the presence of H^+.

How can we pull the dissolution equilibrium to the right, even though Cl^- is an extremely weak base? The key is to lower the concentration of Ag^+ in solution by forming complex ions. For example, Ag^+ reacts with excess NH_3 to form the stable complex ion $Ag(NH_3)_2^+$. As a result, AgCl is quite soluble in concentrated ammonia solutions. The relevant reactions are

$$AgCl(s) \rightleftharpoons Ag^+(aq) + Cl^-(aq) \qquad K_{sp} = 1.6 \times 10^{-10}$$

$$Ag^+(aq) + NH_3(aq) \rightleftharpoons Ag(NH_3)^+(aq) \qquad K_1 = 2.1 \times 10^3$$

$$Ag(NH_3)^+(aq) + NH_3(aq) \rightleftharpoons Ag(NH_3)_2^+(aq) \qquad K_2 = 8.2 \times 10^3$$

The Ag^+ ion produced by dissolving solid AgCl combines with NH_3 to form $Ag(NH_3)_2^+$. This causes more AgCl to dissolve until the point at which

$$[Ag^+][Cl^-] = K_{sp} = 1.6 \times 10^{-10}$$

Here $[Ag^+]$ refers only to the Ag^+ ion that is present as a separate species in solution. It does *not* represent the total silver content of the solution, which is

$$[Ag]_{\text{total dissolved}} = [Ag^+] + [AgNH_3^+] + [Ag(NH_3)_2^+]$$

As we saw in the preceding section, virtually all the Ag^+ from the dissolved AgCl ends up in the complex ion $Ag(NH_3)_2^+$, so we can represent the dissolving of solid AgCl in excess NH_3 by the equation

$$AgCl(s) + 2NH_3(aq) \rightleftharpoons Ag(NH_3)_2^+(aq) + Cl^-(aq)$$

Since this equation is the *sum of the three stepwise reactions* given above, the equilibrium constant for this reaction is the product of the constants for the three reactions. (Demonstrate this result to yourself by multiplying the three expressions for K_{sp}, K_1, and K_2.) The equilibrium expression is

When reactions are added, the equilibrium constant for the overall process is the product of the constants for the individual reactions.

$$\begin{aligned} K &= \frac{[Ag(NH_3)_2^+][Cl^-]}{[NH_3]^2} \\ &= K_{sp} \times K_1 \times K_2 \\ &= (1.6 \times 10^{-10})(2.1 \times 10^3)(8.2 \times 10^3) \\ &= 2.8 \times 10^{-3} \end{aligned}$$

Using this expression, we will now calculate the solubility of solid AgCl in a 10.0 *M* NH_3 solution. If we let x be the solubility (in mol/L) of AgCl in this solution, we can then write the following expressions for the equilibrium concentrations of the pertinent species:

$[Cl^-] = x$

$[Ag(NH_3)_2^+] = x$ ← x mol/L of AgCl dissolves to produce x mol/L of Cl^- and x mol/L of $Ag(NH_3)_2^+$

$[NH_3] = 10.0 - 2x$ ← Formation of x mol/L of $Ag(NH_3)_2^+$ requires $2x$ mol/L of NH_3, since each complex ion contains two NH_3 ligands

Substituting these concentrations into the equilibrium expression gives

$$K = 2.8 \times 10^{-3} = \frac{[Ag(NH_3)_2^+][Cl^-]}{[NH_3]^2} = \frac{(x)(x)}{(10.0 - 2x)^2} = \frac{x^2}{(10.0 - 2x)^2}$$

No approximations are necessary here. Taking the square root of both sides of the equation gives

$$\sqrt{2.8 \times 10^{-3}} = \frac{x}{10.0 - 2x}$$

$$x = 0.48 \text{ mol/L} = \text{solubility of AgCl}(s) \text{ in } 10.0\ M\ NH_3$$

Thus the solubility of AgCl in 10.0 M NH_3 is much greater than in pure water, which is

$$\sqrt{K_{sp}} = 1.3 \times 10^{-5} \text{ mol/L}$$

In this chapter we have considered two strategies for dissolving a water-insoluble ionic solid. If the *anion* of the solid is a good base, the solubility is greatly increased by acidifying the solution. In cases where the anion is not sufficiently basic, the ionic solid can often be dissolved in a solution containing a ligand that forms stable complex ions with its *cation.*

Sometimes solids are so insoluble that combinations of reactions are needed to dissolve them. For example, to dissolve the extremely insoluble HgS ($K_{sp} = 1.6 \times 10^{-54}$), we must use a mixture of concentrated HCl and concentrated HNO_3, called *aqua regia.* The H^+ ions in the aqua regia react with the S^{2-} ions to form H_2S, and Cl^- reacts with Hg^{2+} to form various complex ions such as $HgCl_4^{2-}$. In addition, NO_3^- oxidizes S^{2-} to elemental sulfur. These processes lower the concentrations of Hg^{2+} and S^{2-} and thus promote the solubility of HgS.

Since the solubility of many salts increases with temperature, simple heating is sometimes enough to make a salt sufficiently soluble. For example, earlier in this section we considered the mixed chloride precipitates of the Group I ions—$PbCl_2$, AgCl, and Hg_2Cl_2. The effect of temperature on the solubility of $PbCl_2$ is such that we can precipitate $PbCl_2$ with cold aqueous HCl and then redissolve it by heating the solution to near boiling. In this way, $PbCl_2$ can be separated from the silver and mercury(I) chlorides, since they are not significantly soluble in hot water. Subsequently, solid AgCl can be dissolved and separated from Hg_2Cl_2 by using aqueous ammonia. The solid Hg_2Cl_2 reacts with NH_3 to form a mixture of elemental mercury and $HgNH_2Cl$:

$$Hg_2Cl_2(s) + 2NH_3(aq) \longrightarrow \underset{\text{White}}{HgNH_2Cl(s)} + \underset{\text{Black}}{Hg(l)} + NH_4^+(aq) + Cl^-(aq)$$

The mixed precipitate appears gray. This is an oxidation–reduction reaction in which one mercury(I) ion in Hg_2Cl_2 is oxidized to Hg^{2+} in $HgNH_2Cl$ and the other mercury(I) ion is reduced to Hg, or elemental mercury.

The treatment of the Group I ions is summarized in Fig. 8.14. Note that the presence of Pb^{2+} is confirmed by adding CrO_4^{2-}, which forms bright yellow lead(II) chromate ($PbCrO_4$). Also note that H^+ added to a solution containing $Ag(NH_3)_2^+$ and Cl^- reacts with the NH_3 to form NH_4^+, thus destroying the $Ag(NH_3)_2^+$ complex. Silver chloride then re-forms:

$$2H^+(aq) + Ag(NH_3)_2^+(aq) + Cl^-(aq) \longrightarrow 2NH_4^+(aq) + AgCl(s)$$

Note that the qualitative analysis of cations by selective precipitation involves all the types of reactions we have discussed and represents an excellent application of the principles of chemical equilibrium.

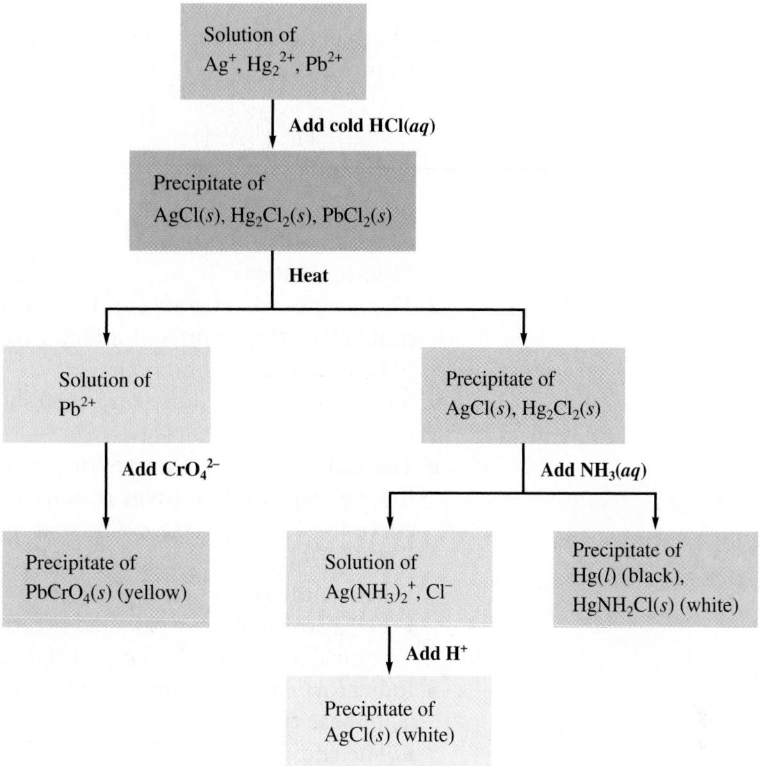

Figure 8.14
The separation of the Group I ions in the classic scheme of qualitative analysis.

Key Terms

Section 8.1
common ion
common ion effect

Section 8.2
buffered solution
Henderson–Hasselbalch equation

Section 8.4
buffer capacity

Section 8.5
pH curve (titration curve)
millimole (mmol)
equivalence point (stoichiometric point)

Section 8.6
acid–base indicator
phenolphthalein

Section 8.8
solubility product constant (solubility product)

Section 8.9
ion product
selective precipitation
qualitative analysis

Section 8.10
complex ion
formation (stability) constant

For Review

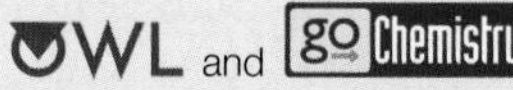

Buffered solutions

- Contains a weak acid (HA) and its salt (NaA) or a weak base (B) and its salt (BHCl)
- Resists a change in its pH when H^+ or OH^- is added
- For a buffered solution containing HA and A^-
  - The Henderson–Hasselbalch equation is useful:

$$\text{pH} = \text{p}K_\text{a} + \log\left(\frac{[\text{A}^-]}{[\text{HA}]}\right)$$

  - The capacity of the buffered solution depends on the amounts of HA and A^- present
- The most efficient buffering occurs when the $\frac{[\text{A}^-]}{[\text{HA}]}$ ratio is close to 1
- Buffering works because the amounts of HA (which reacts with added OH^-) and A^- (which reacts with added H^+) are large enough that the $\frac{[\text{A}^-]}{[\text{HA}]}$ ratio does not change significantly when strong acids or bases are added

- The general equation for buffered solutions that includes contributions from water is
  - $$K_a = \frac{[H^+]\left([A^-]_0 + \dfrac{[H^+]^2 - K_w}{[H^+]}\right)}{[HA]_0 - \dfrac{[H^+]^2 - K_w}{[H^+]}}$$

Acid–base titrations

- The progress of a titration is represented by plotting the pH of the solution versus the volume of added titrant; the resulting graph is called a pH curve or titration curve
- Strong acid–strong base titrations show a sharp change in pH near the equivalence point
- The shape of the pH curve for a strong base–strong acid titration before the equivalence point is quite different from the shape of the pH curve for a strong base–weak acid titration
  - The strong base–weak acid pH curve shows the effects of buffering before the equivalence point
  - For a strong base–weak acid titration, the pH is greater than 7 at the equivalence point because of the basic properties of A^-
- Indicators are sometimes used to mark the equivalence point of an acid–base titration
  - The end point is where the indicator changes color
  - The goal is to have the end point and the equivalence point be as close as possible

Solids dissolving in water

- For a slightly soluble salt, an equilibrium is set up between the excess solid (MX) and the ions in solution:

$$MX(s) \rightleftharpoons M^+(aq) + X^-(aq)$$

- The corresponding constant is called K_{sp}:

$$K_{sp} \rightleftharpoons [M^+][X^-]$$

  - The solubility of $MX(s)$ is decreased by the presence of another source of either M^+ or X^-; this is called the common ion effect
- Predicting whether precipitation will occur when two solutions are mixed involves calculating Q for the initial concentrations
  - If $Q > K_{sp}$, precipitation occurs
  - If $Q \leq K_{sp}$, no precipitation occurs

Qualitative analysis

- A mixture of ions can be separated by selective precipitation
  - The ions are first separated into groups by adding $HCl(aq)$, then $H_2S(aq)$, then $NaOH(aq)$, and finally $Na_2CO_3(aq)$
  - The ions in the groups are separated and identified by further selective dissolution and precipitation

Complex ions

- Complex ions consist of a metal ion surrounded by attached ligands
  - A ligand is a Lewis base
  - The number of ligands is called the *coordination number,* which is commonly 2, 4, or 6
- Complex ion equilibria in solution are described by formation (stability) constants
- The formation of complex ions can be used to selectively dissolve solids in the qualitative analysis scheme

Discussion Questions

These questions are designed to be considered by groups of students in class. Often these questions work well for introducing a particular topic in class.

1. What are the major species in solution after $NaHSO_4$ is dissolved in water? What happens to the pH of the solution as more $NaHSO_4$ is added? Why? Would the results vary if baking soda ($NaHCO_3$) were used instead? Explain.
2. A friend asks the following: "Consider a buffered solution made up of the weak acid HA and its salt NaA. If a strong base such as NaOH is added, the HA reacts with the OH^- to make A^-. Thus the amount of acid (HA) is decreased, and the amount of base (A^-) is increased. Analogously, adding HCl to the buffered solution forms more of the acid (HA) by reacting with the base (A^-). Thus how can we claim that a buffered solution resists changes in the pH of the solution?" How would you explain buffering to your friend?
3. Mixing together solutions of acetic acid and sodium hydroxide can make a buffered solution. Explain. How does the amount of each solution added change the effectiveness of the buffer? Would a buffered solution made by mixing HCl and NaOH be effective? Explain.
4. Sketch two pH curves, one for the titration of a weak acid with a strong base, and one for the titration of a strong acid with a strong base. How are they similar? How are they different? Account for the similarities and the differences.
5. Sketch a pH curve for the titration of a weak acid (HA) with a strong base (NaOH). List the major species, and explain how you would calculate the pH of the solution at various points, including the halfway point and the equivalence point.
6. You have a solution of the weak acid HA and add some HCl to it. What are the major species in the solution? What do you need to know to calculate the pH of the solution, and how would you use this information? How does the pH of the solution of just the HA compare with that of the final mixture? Explain.
7. You have a solution of the weak acid HA and add some of the salt NaA to it. What are the major species in the solution? What do you need to know to calculate the pH of the solution, and how would you use this information? How does the pH of the solution of just the HA compare with that of the final mixture? Explain.
8. Devise as many ways as you can to experimentally determine the K_{sp} value of a solid. Explain why each of these would work.
9. You are browsing through the *Handbook of Hypothetical Chemistry* when you come across a solid that is reported to have a K_{sp} value of zero in water at 25°C. What does this mean?
10. A friend tells you: "The constant K_{sp} of a salt is called the solubility product constant and is calculated from the concentrations of ions in the solution. Thus, if salt A dissolves to a greater extent than salt B, salt A must have a higher K_{sp} than salt B." Do you agree with your friend? Explain.
11. What happens to the K_{sp} value of a solid as the temperature of the solution changes? Consider both increasing and decreasing temperatures, and explain your answer.
12. Which is more likely to dissolve in an acidic solution, silver sulfide or silver chloride? Why?
13. You have two salts AgX and AgY with very similar K_{sp} values. You know that the K_a value for HX is much greater than the K_a value for HY. Which salt is more soluble in an acidic solution? Explain.
14. Under what circumstances can the relative solubilities of two salts be compared by directly comparing values of their solubility products?

Exercises

OWL Interactive versions of these problems may be assigned in OWL.

A blue exercise number indicates that the answer to that exercise appears at the back of this book and a solution appears in the *Solutions Guide*.

Buffers

15. Define a buffered solution. What makes up a buffered solution? Explain how buffers absorb added H^+ or OH^- with little pH change. A certain buffer is made by dissolving $NaHCO_3$ and Na_2CO_3 in some water. Write equations to show how this buffer neutralizes added H^+ and OH^-.
16. A good buffer generally contains relatively equal concentrations of a weak acid and its conjugate base. If you wanted to buffer a solution at pH = 4.00 or pH = 10.00, how would you decide which weak acid–conjugate base or weak base–conjugate acid pair to use? The second characteristic of a good buffer is good buffering capacity. What is the *capacity* of a buffer? How do the following buffers differ in capacity? How do they differ in pH?

    0.01 *M* acetic acid/0.01 *M* sodium acetate

    0.1 *M* acetic acid/0.1 *M* sodium acetate

    1.0 *M* acetic acid/1.0 *M* sodium acetate
17. How many of the following are buffered solutions? Explain your answer. *Note:* Counter-ions and water molecules have been omitted from the illustrations for clarity.

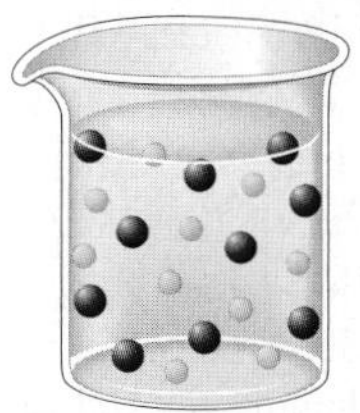

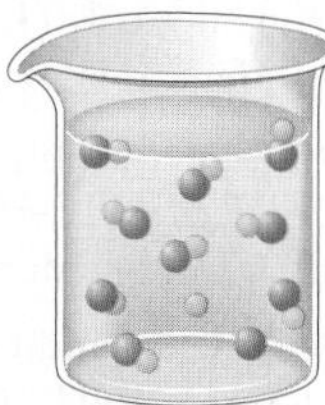

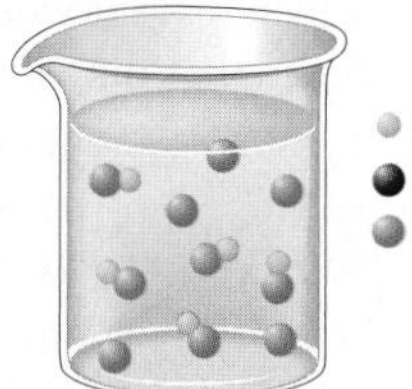

18. Which of the following can be classified as buffer solutions?
    a. 0.25 *M* HBr + 0.25 *M* HOBr
    b. 0.15 *M* $HClO_4$ + 0.20 *M* RbOH
    c. 0.50 *M* HOCl + 0.35 *M* KOCl
    d. 0.70 *M* KOH + 0.70 *M* $HONH_2$
    e. 0.85 *M* H_2NNH_2 + 0.60 *M* $H_2NNH_3NO_3$

19. Consider a buffered solution where [weak acid] > [conjugate base]. How is the pH of the solution related to the pK_a value of the weak acid? If [conjugate base] > [weak acid], how is pH related to pK_a?

20. Derive an equation analogous to the Henderson–Hasselbalch equation that relates pOH and pK_b of a buffered solution composed of a weak base and its conjugate acid, such as NH_3 and NH_4^+.

21. Calculate the pH of each of the following solutions.
    a. 0.100 *M* propanoic acid ($HC_3H_5O_2$, $K_a = 1.3 \times 10^{-5}$)
    b. 0.100 *M* sodium propanoate ($NaC_3H_5O_2$)
    c. pure H_2O
    d. 0.100 *M* $HC_3H_5O_2$ and 0.100 *M* $NaC_3H_5O_2$

22. Calculate the pH after 0.020 mole of HCl is added to 1.00 L of each of the four solutions in Exercise 21.

23. Calculate the pH after 0.020 mole of NaOH is added to 1.00 L of each of the four solutions in Exercise 21.

24. The results of Exercises 21–23 illustrate an important property of buffered solutions. Which solution in Exercise 21 is the buffered solution and what important property is illustrated by the results?

25. Calculate the pH of a solution that is 0.60 *M* HF and 1.00 *M* KF.

26. Calculate the pH of a solution that is 0.100 *M* $HONH_2$ and 0.100 *M* $HONH_3Cl$.

27. Calculate the pH after 0.10 mole of NaOH is added to 1.00 L of the solution in Exercise 25, and calculate the pH after 0.20 mole of HCl is added to 1.00 L of the solution in Exercise 25.

28. Calculate the pH after 0.020 mole of NaOH is added to 1.00 L of the solution in Exercise 26, and calculate the pH after 0.020 mole of HCl is added to 1.00 L of the solution in Exercise 26.

29. Calculate the pH of a solution that is 0.40 *M* H_2NNH_2 and 0.80 *M* $H_2NNH_3NO_3$. In order for this buffer to have pH = pK_a, would you add HCl or NaOH? What quantity (moles) of which reageant would you add to 1.0 L of the original buffer so that the resulting solution has pH = pK_a?

30. Calculate the pH of a solution that is 0.20 *M* HOCl and 0.90 *M* KOCl. In order for this buffer to have pH = pK_a, would you add HCl or NaOH? What quantity (moles) of which reagent would you add to 1.0 L of the original buffer so that the resulting solution has pH = pK_a?

31. Calculate the pH of a buffered solution prepared by dissolving 21.5 g of benzoic acid ($HC_7H_5O_2$) and 37.7 g of sodium benzoate in 200.0 mL of solution.

32. A buffered solution is made by adding 50.0 g NH_4Cl to 1.00 L of a 0.75 *M* solution of NH_3. Calculate the pH of the final solution. (Assume no volume change.)

33. Consider a solution that contains both C_5H_5N and $C_5H_5NHNO_3$. Calculate the ratio $[C_5H_5N]/[C_5H_5NH^+]$ if the solution has the following pH values.
    a. pH = 4.50 c. pH = 5.23
    b. pH = 5.00 d. pH = 5.50

34. How many moles of NaOH must be added to 1.0 L of 2.0 *M* $HC_2H_3O_2$ to produce a solution buffered at each pH?
    a. pH = pK_a b. pH = 4.00 c. pH = 5.00

35. Calculate the number of moles of HCl(*g*) that must be added to 1.0 L of 1.0 *M* $NaC_2H_3O_2$ to produce a solution buffered at each pH.
    a. pH = pK_a b. pH = 4.20 c. pH = 5.00

36. You make 1.00 L of a buffered solution (pH = 4.00) by mixing acetic acid and sodium acetate. You have 1.00 *M* solutions of each component of the buffered solution. What volume of each solution do you mix to make such a buffered solution?

37. Calculate the mass of sodium acetate that must be added to 500.0 mL of 0.200 *M* acetic acid to form a pH = 5.00 buffered solution.

38. Calculate the pH after 0.010 mole of gaseous HCl is added to 250.0 mL of each of the following buffered solutions.
    a. 0.050 *M* NH_3 and 0.15 *M* NH_4Cl
    b. 0.50 *M* NH_3 and 1.50 *M* NH_4Cl

39. An aqueous solution contains dissolved $C_6H_5NH_3Cl$ and $C_6H_5NH_2$. The concentration of $C_6H_5NH_2$ is 0.50 *M* and pH is 4.20.
    a. Calculate the concentration of $C_6H_5NH_3^+$ in this buffered solution.
    b. Calculate the pH after 4.0 g of NaOH(*s*) is added to 1.0 L of this solution. (Neglect any volume change.)

40. What volumes of 0.50 *M* HNO_2 and 0.50 *M* $NaNO_2$ must be mixed to prepare 1.00 L of a solution buffered at pH = 3.55?

41. Phosphate buffers are important in regulating the pH of intracellular fluids at pH values generally between 7.1 and 7.2.
    a. What is the concentration ratio of $H_2PO_4^-$ to HPO_4^{2-} in intracellular fluid at pH = 7.15?

    $$H_2PO_4^-(aq) \rightleftharpoons HPO_4^{2-}(aq) + H^+(aq) \quad K_a = 6.2 \times 10^{-8}$$

    b. Why is a buffer composed of H_3PO_4 and $H_2PO_4^-$ ineffective in buffering the pH of intracellular fluid?

    $$H_3PO_4(aq) \rightleftharpoons H_2PO_4^-(aq) + H^+(aq) \quad K_a = 7.5 \times 10^{-3}$$

42. Carbonate buffers are important in regulating the pH of blood at 7.40. If the carbonic acid concentration in a sample of blood is 0.0012 *M*, determine the bicarbonate ion concentration required to buffer the pH of blood at pH = 7.40.

    $$H_2CO_3(aq) \rightleftharpoons HCO_3^-(aq) + H^+(aq) \quad K_a = 4.3 \times 10^{-7}$$

43. When a person exercises, muscle contractions produce lactic acid. Moderate increases in lactic acid can be handled by the blood buffers without decreasing the pH of blood. However, excessive amounts of lactic acid can overload the blood buffer system, resulting in a lowering of the blood pH. A condition called *acidosis* is diagnosed if the blood pH falls to 7.35 or lower. Assume the primary

blood buffer system is the carbonate buffer system described in Exercise 42. Calculate what happens to the $[H_2CO_3]/[HCO_3^-]$ ratio in blood when the pH decreases from 7.40 to 7.35.

44. Which of the following mixtures would result in a buffered solution when 1.0 L of each of the two solutions are mixed?
    a. 0.1 *M* KOH and 0.1 *M* CH_3NH_3Cl
    b. 0.1 *M* KOH and 0.2 *M* CH_3NH_2
    c. 0.2 *M* KOH and 0.1 *M* CH_3NH_3Cl
    d. 0.1 *M* KOH and 0.2 *M* CH_3NH_3Cl

45. Which of the following mixtures would result in a buffered solution when 1.0 L of each of the two solutions are mixed?
    a. 0.2 *M* HNO_3 and 0.4 *M* $NaNO_3$
    b. 0.2 *M* HNO_3 and 0.4 *M* HF
    c. 0.2 *M* HNO_3 and 0.4 *M* NaF
    d. 0.2 *M* HNO_3 and 0.4 *M* NaOH

46. Calculate the pH of a solution formed by mixing 100.0 mL of 0.100 *M* NaF and 100.0 mL of 0.025 *M* HCl.

47. Consider the acids in Table 7.2. Which acid would be the best choice for preparing a pH = 7.00 buffer? Explain how to make 1.0 L of this buffer.

48. Consider the bases in Table 7.3. Which base would be the best choice for preparing a pH = 5.00 buffer? Explain how to make 1.0 L of this buffer.

49. A solution contains 1.0×10^{-6} *M* HOCl and an unknown concentration of KOCl. If the pH of the solution is 7.20, calculate the KOCl concentration. (*Hint:* The contribution of water to the $[H^+]$ cannot be ignored.)

50. In Section 8.3 an equation was derived for the exact treatment of HA/NaA-type buffers. What would be the expression for B/BHCl-type buffers stated in terms of K_b, $[OH^-]$, [B], and $[BH^+]$? Would it be necessary to use this exact expression to solve for the pH of a solution containing 1.0×10^{-4} *M* $HONH_2$ and 1.0×10^{-5} *M* $HONH_3Cl$? Explain.

51. Consider a weak acid HA with a K_a value of 1.6×10^{-7}. Calculate the pH of a solution that is 5.0×10^{-7} *M* HA and 5.0×10^{-7} *M* NaA.

Acid–Base Titrations

52. Consider the following pH curves for 100.0 mL of two different acids with the same intital concentration each titrated by 0.10 *M* NaOH:

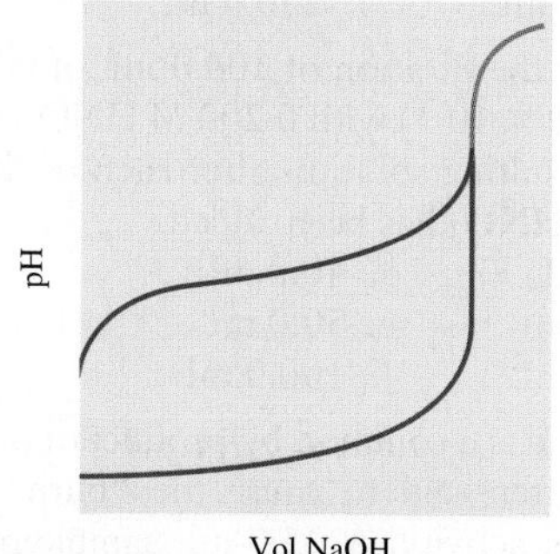

    a. Which plot represents a pH curve of a weak acid, and which plot is for a strong acid? How can you tell? Cite three differences between the plots that help you decide.
    b. In both cases the pH is relatively constant before the pH changes greatly. Does this mean that at some point in each titration each solution was a buffered solution?
    c. True or false? The equivalence point volume for each titration is the same. Explain your answer.
    d. True or false? The pH at the equivalence point for each titration is the same. Explain your answer.

53. An acid is titrated with NaOH. The following beakers are illustrations of the contents of the beaker at various times during the titration. These are presented out of order. *Note:* Counter-ions and water molecules have been omitted from the illustrations for clarity.

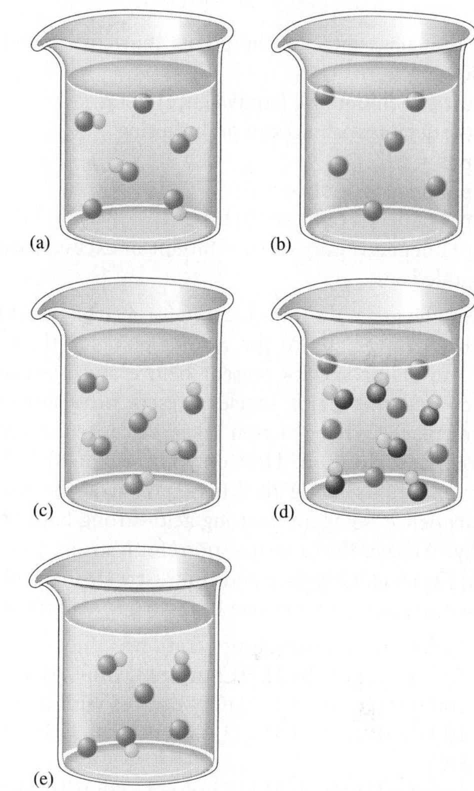

    a. Is the acid a weak or strong acid? How can you tell?
    b. Arrange the beakers in order of what the contents would look like as the titration progresses.
    c. For which beaker would pH = pK_a? Explain your answer.
    d. Which beaker represents the equivalence point of the titration? Explain your answer.
    e. For which beaker would the K_a value for the acid not be necessary to determine the pH? Explain your answer.

54. Consider the titration of a generic weak acid HA with a strong base that gives the following titration curve:

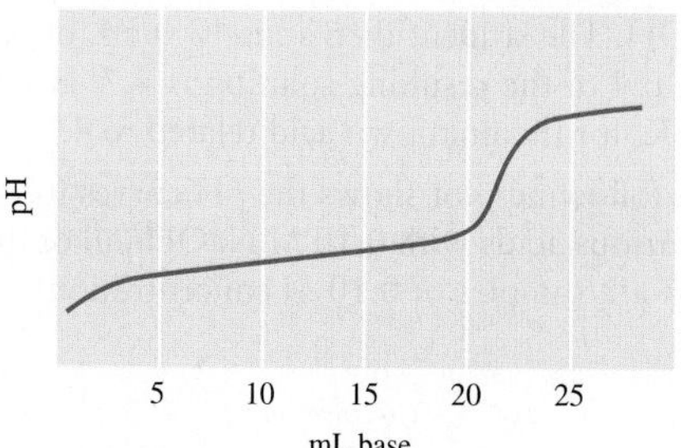

On the curve indicate the points that correspond to the following.

a. the equivalence point
b. the maximum buffering region
c. pH = pK_a
d. pH depends only on [HA]
e. pH depends only on [A^-]
f. pH depends only on the amount of excess strong base added

55. Sketch the titration curve for the titration of a generic weak base B with a strong acid. The titration reaction is

$$B + H^+ \rightleftharpoons BH^+$$

On the curve indicate the points that correspond to the following.

a. the stoichiometric (equivalence) point
b. the region with maximum buffering
c. pH = pK_a
d. pH depends only on [B]
e. pH depends only on [BH^+]
f. pH depends only on the amount of excess strong acid added

56. Draw the general titration curve for a strong acid titrated with a strong base. At the various points in the titration, list the major species present before any reaction takes place and the major species present after any reaction takes place. What reaction takes place in a strong acid–strong base titration? How do you calculate the pH at the various points along the curve? What is the pH at the equivalence point for a strong acid–strong base titration? Why? Answer the same questions for a strong base–strong acid titration. Compare and contrast a strong acid–strong base titration with a strong base–strong acid titration.

57. Consider the following four titrations:

i. 100.0 mL of 0.10 *M* HCl titrated with 0.10 *M* NaOH
ii. 100.0 mL of 0.10 *M* NaOH titrated with 0.10 *M* HCl
iii. 100.0 mL of 0.10 *M* CH_3NH_2 titrated with 0.10 *M* HCl
iv. 100.0 mL of 0.10 *M* HF titrated with 0.10 *M* NaOH

Rank the titrations in order of

a. increasing volume of titrant added to reach the equivalence point.
b. increasing pH initially before any titrant has been added.
c. increasing pH at the halfway point in equivalence.
d. increasing pH at the equivalence point.

How would the rankings change if C_5H_5N replaced CH_3NH_2 and if HOC_6H_5 replaced HF?

58. A student titrates an unknown weak acid HA to a pale-pink phenolphthalein endpoint with 25.0 mL of 0.100 *M* NaOH. The student then adds 13.0 mL of 0.100 *M* HCl. The pH of the resulting solution is 4.7. How is the value of pK_a for the unknown acid related to 4.7?

59. The following plot shows the pH curves for the titrations of various acids with 0.10 *M* NaOH (all of the acids were 50.0-mL samples of 0.10 *M* concentration).

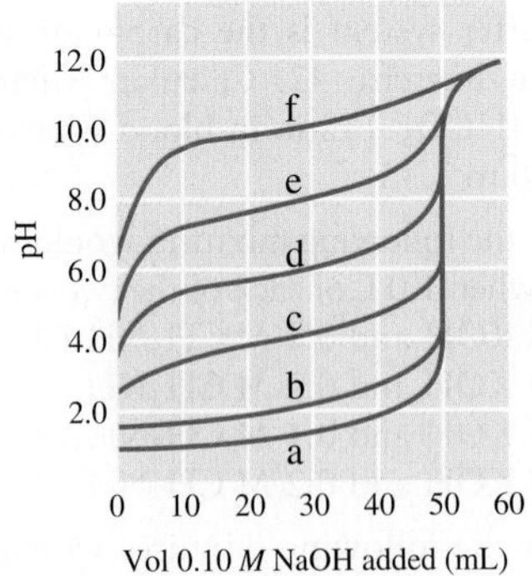

a. Which pH curve corresponds to the weakest acid?
b. Which pH curve corresponds to the strongest acid? Which point on the pH curve would you examine to see if this acid is a strong acid or a weak acid (assuming you did not know the initial concentration of the acid)?
c. Which pH curve corresponds to an acid with $K_a \approx 1 \times 10^{-6}$?

60. The figure in the preceding exercise shows the pH curves for the titrations of six different acids with NaOH. Make a similar plot for the titration of three different bases with 0.10 *M* HCl. Assume 50.0 mL of 0.20 *M* of the bases, and assume the three bases are a strong base (KOH), a weak base with $K_b = 1 \times 10^{-5}$, and another weak base with $K_b = 1 \times 10^{-10}$.

61. Consider the titration of 40.0 mL of 0.200 *M* $HClO_4$ with 0.100 *M* KOH. Calculate the pH of the resulting solution after the following volumes of KOH have been added.

a. 0.0 mL
b. 10.0 mL
c. 40.0 mL
d. 80.0 mL
e. 100.0 mL

62. Consider the titration of 80.0 mL of 0.100 *M* $Ba(OH)_2$ with 0.400 *M* HCl. Calculate the pH of the resulting solution after the following volumes of HCl have been added.

a. 0.0 mL
b. 20.0 mL
c. 30.0 mL
d. 40.0 mL
e. 80.0 mL

63. Consider the titration of 100.0 mL of 0.200 *M* acetic acid ($K_a = 1.8 \times 10^{-5}$) with 0.100 *M* KOH. Calculate the pH of the resulting solution after each of the following volumes of KOH has been added.

a. 0.0 mL
b. 50.0 mL
c. 100.0 mL
d. 150.0 mL
e. 200.0 mL
f. 250.0 mL

64. Consider the titration of 100.0 mL of 0.100 *M* H_2NNH_2 ($K_b = 3.0 \times 10^{-6}$) with 0.200 *M* HNO_3. Calculate the pH of the resulting solution after each of the following volumes of HNO_3 has been added.

a. 0.0 mL
b. 20.0 mL
c. 25.0 mL
d. 40.0 mL
e. 50.0 mL
f. 100.0 mL

65. Lactic acid is a common by-product of cellular respiration and is often said to cause the "burn" associated with strenuous activity. A 25.0-mL sample of 0.100 *M* lactic acid ($HC_3H_5O_3$, pK_a = 3.86) is titrated with 0.100 *M* NaOH solution. Calculate the pH after the addition of 0.0 mL, 4.0 mL, 8.0 mL, 12.5 mL, 20.0 mL, 24.0 mL, 24.5 mL, 24.9 mL, 25.0 mL, 25.1 mL, 26.0 mL, 28.0 mL,

and 30.0 mL of the NaOH. Plot the results of your calculations as pH versus milliliters of NaOH added.

66. Repeat the procedure in Exercise 65 for the titration of 25.0 mL of 0.100 *M* propanoic acid ($HC_3H_5O_2$, $K_a = 1.3 \times 10^{-5}$) with 0.100 *M* KOH.

67. Repeat the procedure in Exercise 65 for the titration of 25.0 mL of 0.100 *M* NH_3 ($K_b = 1.8 \times 10^{-5}$) with 0.100 *M* HCl.

68. Repeat the procedure in Exercise 65 for the titration of 25.0 mL of 0.100 *M* pyridine ($K_b = 1.7 \times 10^{-9}$) with 0.100 *M* hydrochloric acid. Do not do the points at 24.9 mL and 25.1 mL.

69. Calculate the pH at the halfway point and at the equivalence point for each of the following titrations.
    a. 100.0 mL of 0.10 *M* $HC_7H_5O_2$ ($K_a = 6.4 \times 10^{-5}$) titrated with 0.10 *M* NaOH
    b. 100.0 mL of 0.10 *M* $C_2H_5NH_2$ ($K_b = 5.6 \times 10^{-4}$) titrated with 0.20 *M* HNO_3
    c. 100.0 mL of 0.50 *M* HCl titrated with 0.25 *M* NaOH

70. You have 75.0 mL of 0.10 *M* HA. After adding 30.0 mL of 0.10 *M* NaOH, the pH is 5.50. What is the K_a value of HA?

71. A student dissolves 0.0100 mole of an unknown weak base in 100.0 mL water and titrates the solution with 0.100 *M* HNO_3. After 40.0 mL of 0.100 *M* HNO_3 was added, the pH of the resulting solution was 8.00. Calculate the K_b value for the weak base.

Indicators

72. What is an acid–base indicator? Define the equivalence (stoichiometric) point and the endpoint of a titration. Why should you choose an indicator so that the two points coincide? Do the pH values of the two points have to be within ±0.01 pH unit of each other? Explain. Why does an indicator change from its acid color to its base color over a range of pH values? In general, when do color changes start to occur for indicators? Can the indicator thymol blue contain only a single —CO_2H group and no other acidic or basic functional group? Explain.

73. Two drops of indicator HIn ($K_a = 1.0 \times 10^{-9}$), where HIn is yellow and In^- is blue, are placed in 100.0 mL of 0.10 *M* HCl.
    a. What color is the solution initially?
    b. The solution is titrated with 0.10 *M* NaOH. At what pH will the color change (yellow to greenish yellow) occur?
    c. What color will the solution be after 200.0 mL of NaOH has been added?

74. A certain indicator HIn has a pK_a of 3.00 and a color change becomes visible when 7.00% of the indicator has been converted to In^-. At what pH is this color change visible?

75. Estimate the pH of a solution in which bromcresol green is blue and thymol blue is yellow (see Fig. 8.8).

76. A solution has a pH of 7.0. What would be the color of the solution if each of the following indicators were added? (See Fig. 8.8.)
    a. thymol blue
    b. bromthymol blue
    c. methyl red
    d. crystal violet

77. Which of the indicators in Fig. 8.8 could be used for doing the titrations in Exercises 61 and 63?

78. Which of the indicators in Fig. 8.8 could be used for doing the titrations in Exercises 62 and 64?

79. Which of the indicators in Fig. 8.8 could be used for doing the titrations in Exercises 65 and 67?

80. Which of the indicators in Fig. 8.8 could be used for doing the titrations in Exercises 66 and 68?

81. Methyl red ($K_a = 5.0 \times 10^{-6}$) undergoes a color change from red to yellow as the solution gets more basic. Calculate an approximate pH range for which methyl red is useful. What color change occurs and what is the pH at the color change when a weak acid is titrated with a strong base using methyl red as an indicator? What color change occurs and what is the pH at the color change when a weak base is titrated with a strong acid using methyl red as an indicator? For which of these two types of titrations is methyl red a possible indicator?

82. Indicators can be used to estimate the pH values of solutions. To determine the pH of a 0.01 *M* weak acid (HX) solution, a few drops of three different indicators are added to separate portions of 0.01 *M* HX. The resulting colors of the HX solution are summarized in the last column of the accompanying table. What is the approximate pH of the 0.01 *M* HX solution? What is the approximate K_a value for HX?

Indicator (HIn)	Color of HIn	Color of In^-	pK_a of HIn	Color of 0.01 *M* HX
Bromphenol blue	Yellow	Blue	4.0	Blue
Bromcresol purple	Yellow	Purple	6.0	Yellow
Bromcresol green	Yellow	Blue	4.8	Green

Polyprotic Acid Titrations

83. When a diprotic acid, H_2A, is titrated with NaOH, the protons on the diprotic acid are generally removed one at a time, resulting in a pH curve that has the following generic shape:

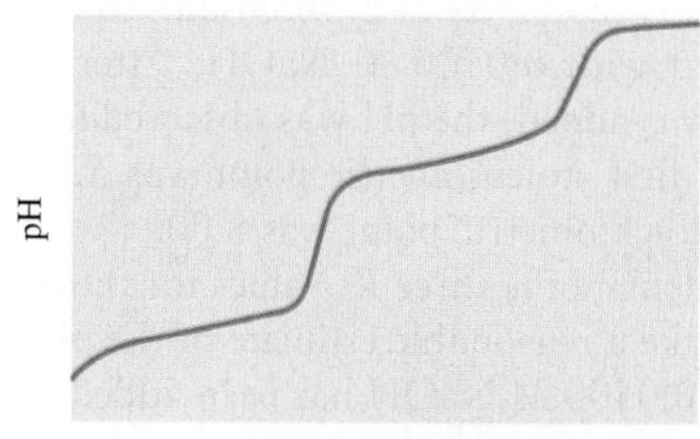

a. Notice that the plot has essentially two titration curves. If the first equivalence point occurs at 100.0 mL NaOH added, what volume of NaOH added corresponds to the second equivalence point?

b. For the following volumes of NaOH added, list the major species present after the OH^- reacts completely.
   i. 0 mL NaOH added
   ii. beween 0 and 100.0 mL NaOH added
   iii. 100.0 mL NaOH added
   iv. between 100.0 and 200.0 mL NaOH added
   v. 200.0 mL NaOH added
   vi. after 200.0 mL NaOH added

c. If the pH at 50.0 mL NaOH added is 4.0 and the pH at 150.0 mL NaOH added is 8.0, determine the values K_{a_1} and K_{a_2} for the diprotic acid.

84. A student was given a 0.10 *M* solution of an unknown diprotic acid H_2A and asked to determine the K_{a_1} and K_{a_2} values for the diprotic acid. The student titrated 50.0 mL of the 0.10 *M* H_2A with 0.10 *M* NaOH. After 25.0 mL of NaOH was added, the pH of the resulting solution was 6.70. After 50.0 mL of NaOH was added, the pH of the resulting solution was 8.00. What are the values of K_{a_1} and K_{a_2} for the diprotic acid?

85. Consider the titration of 100.0 mL of a 0.0500 *M* solution of the hypothetical weak acid H_3X ($K_{a_1} = 1.0 \times 10^{-3}$, $K_{a_2} = 1.0 \times 10^{-7}$, $K_{a_3} = 1.0 \times 10^{-12}$) with 0.100 *M* KOH. Calculate the pH of the solution under the following conditions.
   a. before any KOH has been added
   b. after 10.0 mL of 0.100 *M* KOH has been added
   c. after 25.0 mL of 0.100 *M* KOH has been added
   d. after 50.0 mL of 0.100 *M* KOH has been added
   e. after 60.0 mL of 0.100 *M* KOH has been added
   f. after 75.0 mL of 0.100 *M* KOH has been added
   g. after 100.0 mL of 0.100 *M* KOH has been added
   h. after 125.0 mL of 0.100 *M* KOH has been added
   i. after 150.0 mL of 0.100 *M* KOH has been added
   j. after 200.0 mL of 0.100 *M* KOH has been added

86. Consider 100.0 mL of a 0.100 *M* solution of H_3A ($K_{a_1} = 1.5 \times 10^{-4}$, $K_{a_2} = 3.0 \times 10^{-8}$, $K_{a_3} = 5.0 \times 10^{-12}$).
   a. Calculate the pH of this solution.
   b. Calculate the pH of the solution after 10.0 mL of 1.00 *M* NaOH has been added to the original solution.
   c. Calculate the pH of the solution after 25.0 mL of 1.00 *M* NaOH has been added to the original solution.

87. A 0.200-g sample of a triprotic acid (molar mass = 165.0 g/mol) is dissolved in a 50.00-mL aqueous solution and titrated with 0.0500 *M* NaOH. After 10.50 mL of the base was added, the pH was observed to be 3.73. The pH at the first stoichiometric point was 5.19 and at the second stoichiometric point was 8.00.
   a. Calculate the three K_a values for the acid.
   b. Make a reasonable estimate of the pH after 59.0 mL of 0.0500 *M* NaOH has been added. Explain your answer.
   c. Calculate the pH after 59.0 mL of 0.0500 *M* NaOH has been added.

88. Consider the titration of 100.0 mL of 0.100 *M* H_3A ($K_{a_1} = 5.0 \times 10^{-4}$, $K_{a_2} = 1.0 \times 10^{-8}$, $K_{a_3} = 1.0 \times 10^{-11}$) with 0.0500 *M* NaOH.
   a. Calculate the pH after 100.0 mL of 0.0500 *M* NaOH has been added.
   b. What total volume of 0.0500 *M* NaOH is required to reach a pH of 8.67?

89. The titration of Na_2CO_3 with HCl has the following qualitative profile:

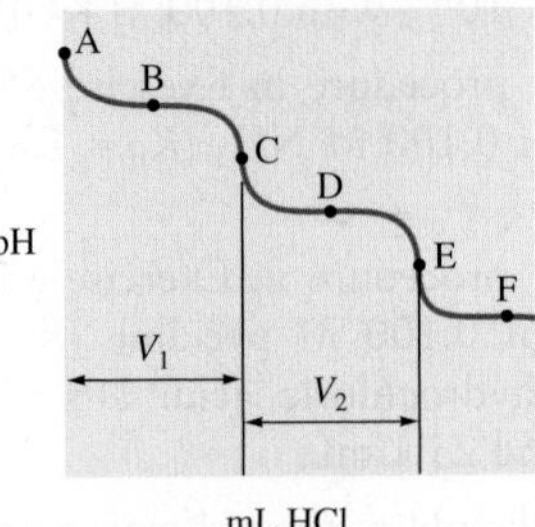

   a. Identify the major species in solution as points A–F.
   b. For the titration of 25.00 mL of 0.100 *M* Na_2CO_3 with 0.100 *M* HCl, calculate the pH at points A–E. (B and D are halfway points to equivalence.)

90. Consider 100.0 mL of a solution of 0.200 *M* Na_2A, where A^{2-} is a base with corresponding acids H_2A ($K_a = 1.0 \times 10^{-3}$) and HA^- ($K_a = 1.0 \times 10^{-8}$).
   a. What volume of 1.00 *M* HCl must be added to this solution to reach pH = 8.00?
   b. Calculate the pH at the second stoichiometric point of the titration of 0.200 *M* Na_2A, with 1.00 *M* HCl.

Solubility Equilibria

91. For which of the following is the K_{sp} value of the ionic compound the largest? The smallest? Explain your answer.

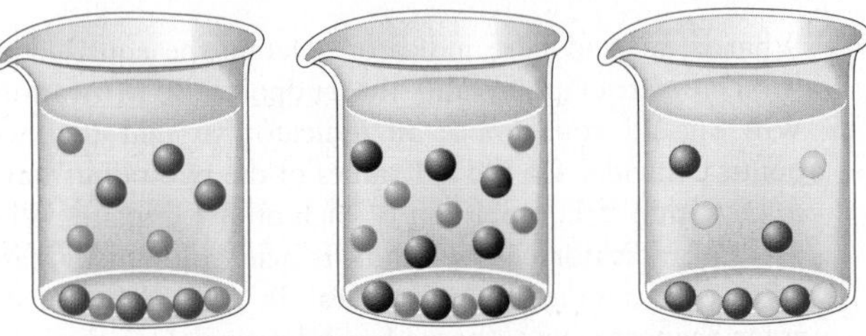

92. $Ag_2S(s)$ has a larger molar solubility than CuS even though Ag_2S has the smaller K_{sp} value. Explain how this is possible.

93. Calculate the solubility of each of the following compounds in moles per liter and grams per liter. (Ignore any acid–base properties.)
   a. Ag_3PO_4, $K_{sp} = 1.8 \times 10^{-18}$
   b. $CaCO_3$, $K_{sp} = 8.7 \times 10^{-9}$
   c. Hg_2Cl_2, $K_{sp} = 1.1 \times 10^{-18}$ (Hg_2^{2+} is the cation in solution.)

94. Calculate the solubility of each of the following compounds in moles per liter. Ignore any acid–base properties.
   a. PbI_2, $K_{sp} = 1.4 \times 10^{-8}$
   b. $CdCO_3$, $K_{sp} = 5.2 \times 10^{-12}$
   c. $Sr_3(PO_4)_2$, $K_{sp} = 1 \times 10^{-31}$

95. Use the following data to calculate the K_{sp} value for each solid.
   a. The solubility of CaC_2O_4 is 6.1×10^{-3} g/L.
   b. The solubility of BiI_3 is 1.32×10^{-5} mol/L.

96. The concentration of Pb^{2+} in a solution saturated with $PbBr_2(s)$ is 2.14×10^{-2} *M*. Calculate K_{sp} for $PbBr_2$.

97. The concentration of Ag^+ in a solution saturated with $Ag_2C_2O_4(s)$ is 2.2×10^{-4} *M*. Calculate K_{sp} for $Ag_2C_2O_4$.

98. The solubility of the ionic compound M_2X_3, having a molar mass of 288 g/mol, is 3.60×10^{-7} g/L. Calculate the K_{sp} of the compound.

99. For each of the following pairs of solids, determine which solid has the smallest molar solubility.
   a. $CaF_2(s)$, $K_{sp} = 4.0 \times 10^{-11}$ or $BaF_2(s)$, $K_{sp} = 2.4 \times 10^{-5}$
   b. $Ca_3(PO_4)_2(s)$, $K_{sp} = 1.3 \times 10^{-32}$ or $FePO_4(s)$, $K_{sp} = 1.0 \times 10^{-22}$

100. The K_{sp} for silver sulfate (Ag_2SO_4) is 1.2×10^{-5}. Calculate the solubility of silver sulfate in each of the following.
   a. water b. 0.10 *M* $AgNO_3$ c. 0.20 *M* K_2SO_4

101. Calculate the solubility (in mol/L) of $Fe(OH)_3$ ($K_{sp} = 4 \times 10^{-38}$) in each of the following.
   a. water (assume pH is 7.0 and constant)
   b. a solution buffered at pH = 5.0
   c. a solution buffered at pH = 11.0

102. The solubility of $Ce(IO_3)_3$ in a 0.20 *M* KIO_3 solution is 4.4×10^{-8} mol/L. Calculate K_{sp} for $Ce(IO_3)_3$.

103. What mass of ZnS ($K_{sp} = 2.5 \times 10^{-22}$) will dissolve in 300.0 mL of 0.050 *M* $Zn(NO_3)_2$? Ignore the basic properties of S^{2-}.

104. The concentration of Mg^{2+} in seawater is 0.052 *M*. At what pH will 99% of the Mg^{2+} be precipitated as the hydroxide salt? [K_{sp} for $Mg(OH)_2 = 8.9 \times 10^{-12}$.]

105. For the substances in Exercises 93 and 94, which will show increased solubility as the pH of the solution becomes more acidic? Write equations for the reactions that occur to increase the solubility.

106. Explain the following phenomenon: You have a test tube with an aqueous solution of silver nitrate as shown in test tube 1 below. A few drops of aqueous sodium chromate solution was added with the end result shown in test tube 2. A few drops of aqueous sodium chloride solution was then added with the end result shown in test tube 3.

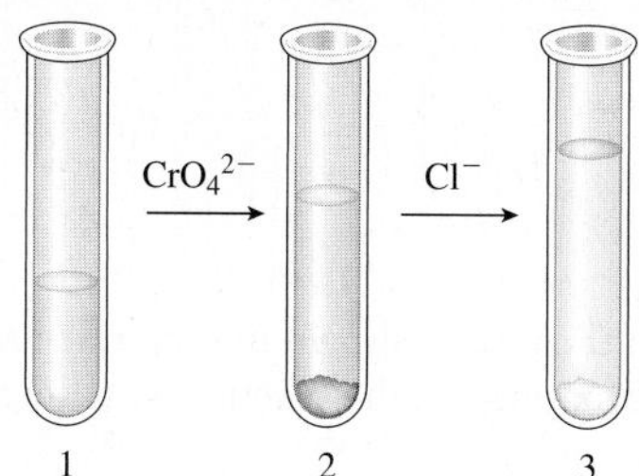

Use the K_{sp} values in the book to support your explanation, and include the balanced equations. Also, list the ions that are present in solution in each test tube.

107. For which salt in each of the following groups will the solubility depend on pH?
   a. AgF, AgCl, AgBr
   b. $Pb(OH)_2$, $PbCl_2$
   c. $Sr(NO_3)_2$, $Sr(NO_2)_2$
   d. $Ni(NO_3)_2$, $Ni(CN)_2$

108. A solution is prepared by mixing 75.0 mL of 0.020 *M* $BaCl_2$ and 125 mL of 0.040 *M* K_2SO_4. What are the concentrations of barium and sulfate ions in this solution? Assume only SO_4^{2-} ions (no HSO_4^-) are present.

109. Calculate the final concentrations of $K^+(aq)$, $C_2O_4^{2-}(aq)$, $Ba^{2+}(aq)$, and $Br^-(aq)$ in a solution prepared by adding 0.100 L of 6.0×10^{-4} *M* $K_2C_2O_4$ to 0.150 L of 1.0×10^{-4} *M* $BaBr_2$. (For BaC_2O_4, $K_{sp} = 2.3 \times 10^{-8}$.)

110. A solution is prepared by mixing 50.0 mL of 0.10 *M* $Pb(NO_3)_2$ with 50.0 mL of 1.0 *M* KCl. Calculate the concentrations of Pb^{2+} and Cl^- at equilibrium. [K_{sp} for $PbCl_2(s) = 1.6 \times 10^{-5}$.]

111. The K_{sp} of $Al(OH)_3$ is 2×10^{-32}. At what pH will a 0.2 *M* Al^{3+} solution begin to show precipitation of $Al(OH)_3$?

112. A solution is 1×10^{-4} *M* in NaF, Na_2S, and Na_3PO_4. What would be the order of precipitation as a source of Pb^{2+} is added gradually to the solution? The relevant K_{sp} values are $K_{sp}(PbF_2) = 4 \times 10^{-8}$, $K_{sp}(PbS) = 7 \times 10^{-29}$, and $K_{sp}[Pb_3(PO_4)_2] = 1 \times 10^{-54}$.

113. A solution contains 1.0×10^{-5} *M* Na_3PO_4. What is the minimum concentration of $AgNO_3$ that would cause precipitation of solid Ag_3PO_4 ($K_{sp} = 1.8 \times 10^{-18}$)?

114. A solution contains 0.25 *M* $Ni(NO_3)_2$ and 0.25 *M* $Cu(NO_3)_2$. Can the metal ions be separated by slowly adding Na_2CO_3? Assume that for successful separation, 99% of the metal ion must be precipitated before the other metal ion begins to precipitate, and assume that no volume change occurs on addition of Na_2CO_3.

115. Describe how you could separate the ions in each of the following groups by selective precipitation.
   a. Ag^+, Mg^{2+}, Cu^{2+}
   b. Pb^{2+}, Ca^{2+}, Fe^{2+}
   c. Cl^-, Br^-, I^-
   d. Pb^{2+}, Bi^{3+}

116. If a solution contains either $Pb^{2+}(aq)$ or $Ag^+(aq)$, how can temperature be manipulated to help identify the ion in solution?

117. Sulfide precipitates are generally grouped as sulfides insoluble in acidic solution and sulfides insoluble in basic solution. Explain why there is a difference between the two groups of sulfide precipitates.

Complex Ion Equilibria

118. Nanotechnology has become an important field, with applications ranging from high-density data storage to the design of "nano machines." One common building block of nanostructured architectures is manganese oxide nanoparticles. The particles can be formed from manganese oxalate nanorods, the formation of which can be described as follows:

$$Mn^{2+}(aq) + C_2O_4^{2-}(aq) \rightleftharpoons MnC_2O_4(aq) \quad K_1 = 7.9 \times 10^3$$

$$MnC_2O_4(aq) + C_2O_4^{2-}(aq) \rightleftharpoons Mn(C_2O_4)_2^{2-}(aq) \quad K_2 = 7.9 \times 10^1$$

Calculate the value for the overall formation constant for $Mn(C_2O_4)_2^{2-}$:

$$K = \frac{[Mn(C_2O_4)_2^{2-}]}{[Mn^{2+}][C_2O_4^{2-}]^2}$$

119. When aqueous KI is added gradually to mercury(II) nitrate, an orange precipitate forms. Continued addition of KI causes the precipitate to dissolve. Write balanced

equations to explain these observations. (*Hint:* Hg^{2+} reacts with I^- to form HgI_4^{2-}.)

120. As a sodium chloride solution is added to a solution of silver nitrate, a white precipitate forms. Ammonia is added to the mixture and the precipitate dissolves. When potassium bromide solution is then added, a pale yellow precipitate appears. When a solution of sodium thiosulfate is added, the yellow precipitate dissolves. Finally, potassium iodide is added to the solution and a yellow precipitate forms. Write reactions for all the changes mentioned above. What conclusions can you draw concerning the sizes of the K_{sp} values for AgCl, AgBr, and AgI? What can you say about the relative values of the formation constants of $Ag(NH_3)_2^+$ and $Ag(S_2O_3)_2^{3-}$?

121. The overall formation constant for HgI_4^{2-} is 1.0×10^{30}. That is,

$$1.0 \times 10^{30} = \frac{[HgI_4^{2-}]}{[Hg^{2+}][I^-]^4}$$

What is the concentration of Hg^{2+} in 500.0 mL of a solution that was originally 0.010 *M* Hg^{2+} and had 65 g of KI added to it? The reaction is

$$Hg^{2+}(aq) + 4I^-(aq) \rightleftharpoons HgI_4^{2-}(aq)$$

122. A solution is prepared by adding 0.090 mole of $K_3[Fe(CN)_6]$ to 0.60 L of 2.0 *M* NaCN. Assuming no volume change, calculate the concentrations of $Fe(CN)_6^{3-}$ and Fe^{3+} in this solution. The *K* (overall) for the formation of $Fe(CN)_6^{3-}$ is 1×10^{42}.

123. A solution is prepared by mixing 100.0 mL of 1.0×10^{-4} *M* $Be(NO_3)_2$ and 100.0 mL of 8.0 *M* NaF.

$$Be^{2+}(aq) + F^-(aq) \rightleftharpoons BeF^+(aq) \qquad K_1 = 7.9 \times 10^4$$

$$BeF^+(aq) + F^-(aq) \rightleftharpoons BeF_2(aq) \qquad K_2 = 5.8 \times 10^3$$

$$BeF_2(aq) + F^-(aq) \rightleftharpoons BeF_3^-(aq) \qquad K_3 = 6.1 \times 10^2$$

$$BeF_3^-(aq) + F^-(aq) \rightleftharpoons BeF_4^{2-}(aq) \qquad K_4 = 2.7 \times 10^1$$

Calculate the equilibrum concentrations of F^-, Be^{2+}, BeF^+, BeF_2, BeF_3^-, and BeF_4^{2-} in this solution.

124. K_f for the complex ion $Ag(NH_3)_2^+$ is 1.7×10^7. K_{sp} for AgCl is 1.6×10^{-10}. Calculate the molar solubility of AgCl in 1.0 *M* NH_3.

125. a. Using the K_{sp} for $Cu(OH)_2$ (1.6×10^{-19}) and the overall formation constant for $Cu(NH_3)_4^{2+}$ (1.0×10^{13}), calculate a value for the equilibrium constant for the reaction

$$Cu(OH)_2(s) + 4NH_3(aq) \rightleftharpoons Cu(NH_3)_4^{2+}(aq) + 2OH^-(aq)$$

b. Use the value of the equilibrium constant you calculated in part a to calculate the solubility (in mol/L) of $Cu(OH)_2$ in 5.0 *M* NH_3. In 5.0 *M* NH_3, the concentration of OH^- is 0.0095 *M*.

126. The copper(I) ion forms a chloride salt that has $K_{sp} = 1.2 \times 10^{-6}$. Copper(I) also forms a complex ion with Cl^-:

$$Cu^+(aq) + 2Cl^-(aq) \rightleftharpoons CuCl_2^-(aq) \qquad K = 8.7 \times 10^4$$

a. Calculate the solubility of copper(I) chloride in pure water. (Ignore $CuCl_2^-$ formation for part a.)
b. Calculate the solubility of copper(I) chloride in 0.10 *M* NaCl.

127. Solutions of sodium thiosulfate are used to dissolve unexposed AgBr in the developing process for black-and-white film. What mass of AgBr can dissolve in 1.00 L of 0.500 *M* $Na_2S_2O_3$? Assume the overall formation constant for $Ag(S_2O_3)_2^{3-}$ is 2.9×10^{13} and K_{sp} for AgBr is 5.0×10^{-13}.

128. a. Calculate the molar solubility of AgI in pure water. K_{sp} for AgI is 1.5×10^{-16}.
b. Calculate the molar solubility of AgI in 3.0 *M* NH_3. The overall formation constant for $Ag(NH_3)_2^+$ is 1.7×10^7.
c. Compare the calculated solubilities from parts a and b. Explain any differences.

129. A series of chemicals was added to some $AgNO_3(aq)$. $NaCl(aq)$ was added first to the silver nitrate solution, with the end result shown below in test tube 1; $NH_3(aq)$ was then added, with the end result shown in test tube 2; and $HNO_3(aq)$ was added last, with the end result shown in test tube 3.

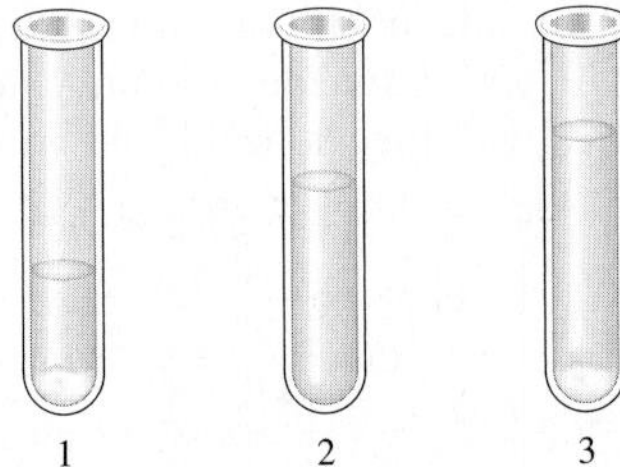

Explain the results shown in each test tube. Include a balanced equation for the reaction(s) taking place.

Additional Exercises

130. Will a precipitate of $Cd(OH)_2$ form if 1.0 mL of 1.0 *M* $Cd(NO_3)_2$ is added to 1.0 L of 5.0 *M* NH_3?

$$Cd^{2+}(aq) + 4NH_3(aq) \rightleftharpoons Cd(NH_3)_4^{2+}(aq) \qquad K = 1.0 \times 10^7$$

$$Cd(OH)_2(s) \rightleftharpoons Cd^{2+}(aq) + 2OH^-(aq) \qquad K_{sp} = 5.9 \times 10^{-15}$$

131. Tris(hydroxymethyl)aminomethane, commonly called TRIS or Trizma, is often used as a buffer in biochemical studies. Its buffering range is from pH 7 to 9, and K_b is 1.19×10^{-6} for the reaction

$$\underset{\text{TRIS}}{(HOCH_2)_3CNH_2(aq)} + H_2O(l) \rightleftharpoons \underset{\text{TRISH}^+}{(HOCH_2)_3CNH_3^+(aq)} + OH^-(aq)$$

a. What is the optimum pH for TRIS buffers?
b. Calculate the ratio [TRIS]/[TRISH$^+$] at pH = 7.00 and at pH = 9.00.
c. A buffer is prepared by diluting 50.0 g of TRIS base and 65.0 g of TRIS hydrochloride (written as TRISHCl) to a total volume of 2.0 L. What is the pH of this buffer? What is the pH after 0.50 mL of 12 *M* HCl is added to a 200.0-mL portion of the buffer?

132. The salts in Table 8.5, with the possible exception of the hydroxide salts, have one of the following mathematical relationships between the K_{sp} value and the molar solubility *s*.
i. $K_{sp} = s^2$
ii. $K_{sp} = 4s^3$
iii. $K_{sp} = 27s^4$
iv. $K_{sp} = 108s^5$

For each mathematical relationship, give an example of a salt in Table 8.5 that exhibits that relationship.

133. You have the following reagents on hand:

Solids (pK_a of Acid Form Is Given)	Solutions
Benzoic acid (4.19)	5.0 *M* HCl
Sodium acetate (4.74)	1.0 *M* acetic acid (4.74)
Potassium fluoride (3.14)	2.6 *M* NaOH
Ammonium chloride (9.26)	1.0 *M* HOCl (7.46)

What combinations of reagents would you use to prepare buffers at the following pH values?
a. 3.0 b. 4.0 c. 5.0 d. 7.0 e. 9.0

134. a. Calculate the pH of a buffered solution that is 0.100 *M* in $C_6H_5CO_2H$ (benzoic acid, $K_a = 6.4 \times 10^{-5}$) and 0.100 *M* in $C_6H_5CO_2Na$.
b. Calculate the pH after 20.0% (by moles) of the benzoic acid is converted to benzoate anion by addition of base. Use the dissociation equilibrium:

$$C_6H_5CO_2H(aq) \rightleftharpoons C_6H_5CO_2^-(aq) + H^+(aq)$$

c. Do the same calculation as in part b, but use the following equilibrium to calculate the pH:

$$C_6H_5CO_2^-(aq) + H_2O(l) \rightleftharpoons C_6H_5CO_2H(aq) + OH^-(aq)$$

d. Do your answers in parts b and c agree? Why or why not?

135. One method for determining the purity of aspirin (empirical formula $C_9H_8O_4$) is to hydrolyze it with NaOH solution and then to titrate the remaining NaOH. The reaction of aspirin with NaOH is as follows:

$$\underset{\text{Aspirin}}{C_9H_8O_4(s)} + 2OH^-(aq) \xrightarrow[\text{10 min}]{\text{Boil}} \underset{\text{Salicylate ion}}{C_7H_5O_3^-(aq)} + \underset{\text{Acetate ion}}{C_2H_3O_2^-(aq)} + H_2O(l)$$

A sample of aspirin with a mass of 1.427 g was boiled in 50.00 mL of 0.500 *M* NaOH. After the solution was cooled, it took 31.92 mL of 0.289 *M* HCl to titrate the excess NaOH. Calculate the purity of the aspirin. What indicator should be used for this titration? Why?

136. Another way to treat data from a pH titration is to graph the absolute value of the change in pH per change in milliliters added versus milliliters added (ΔpH/ΔmL versus mL added). Make this graph using your results from Exercise 65. What advantage might this method have over the traditional method for treating titration data?

137. Potassium hydrogen phthalate, known as KHP (molar mass = 204.22 g/mol), can be obtained in high purity and is used to determine the concentration of solutions of strong bases by the reaction

$$HP^-(aq) + OH^-(aq) \longrightarrow H_2O(l) + P^{2-}(aq)$$

If a typical titration experiment begins with approximately 0.5 g of KHP and has a final volume of about 100 mL, what is an appropriate indicator to use? The pK_a for HP^- is 5.51.

138. A 10.00-g sample of the ionic compound NaA, where A^- is the anion of a weak acid, was dissolved in enough water to make 100.0 mL of solution and was then titrated with 0.100 *M* HCl. After 500.0 mL HCl was added, the pH was 5.00. The experimenter found that 1.00 L of 0.100 *M* HCl was required to reach the stoichiometric point of the titration.
a. What is the molar mass of NaA?
b. Calculate the pH of the solution at the stoichiometric point of the titration.

139. What mass of $Ca(NO_3)_2$ must be added to 1.0 L of a 1.0 *M* HF solution to begin precipitation of $CaF_2(s)$? For CaF_2, $K_{sp} = 4.0 \times 10^{-11}$ and K_a for HF = 7.2×10^{-4}. Assume no volume change on addition of $Ca(NO_3)_2(s)$.

140. The equilibrium constant for the following reaction is 1.0×10^{23}:

$$Cr^{3+}(aq) + H_2EDTA^{2-}(aq) \rightleftharpoons CrEDTA^-(aq) + 2H^+(aq)$$

$$EDTA^{4-} = \begin{matrix} ^-O_2C\text{—}CH_2 & & CH_2\text{—}CO_2^- \\ & N\text{—}CH_2\text{—}CH_2\text{—}N & \\ ^-O_2C\text{—}CH_2 & & CH_2\text{—}CO_2^- \end{matrix}$$

Ethylenediaminetetraacetate

EDTA is used as a complexing agent in chemical analysis. Solutions of EDTA, usually containing the disodium salt Na_2H_2EDTA, are used to treat heavy metal poisoning. Calculate [Cr^{3+}] at equilibrium in a solution originally 0.0010 *M* in Cr^{3+} and 0.050 *M* in H_2EDTA^{2-} and buffered at pH = 6.00.

141. Calculate the concentration of Pb^{2+} in each of the following.
a. a saturated solution of $Pb(OH)_2$; $K_{sp} = 1.2 \times 10^{-15}$
b. a saturated solution of $Pb(OH)_2$ buffered at pH = 13.00
c. 0.010 mole of $Pb(NO_3)_2$ added to 1.0 L of aqueous solution, buffered at pH = 13.00 and containing 0.050 *M* Na_4EDTA. Does $Pb(OH)_2$ precipitate from this solution? For the reaction,

$$Pb^{2+}(aq) + EDTA^{4-}(aq) \rightleftharpoons PbEDTA^{2-}(aq) \qquad K = 1.1 \times 10^{18}$$

142. The solubility rules outlined in Chapter 4 say that $Ba(OH)_2$, $Sr(OH)_2$, and $Ca(OH)_2$ are marginally soluble hydroxides. Calculate the pH of a saturated solution of each of these marginally soluble hydroxides.

143. A certain acetic acid solution has pH = 2.68. Calculate the volume of 0.0975 *M* KOH required to "neutralize" 25.0 mL of this solution.

144. Calculate the volume of 1.50×10^{-2} *M* NaOH that must be added to 500.0 mL of 0.200 *M* HCl to give a solution that has pH = 2.15.

145. A 0.400 *M* solution of ammonia was titrated with hydrochloric acid to the equivalence point, where the total volume was 1.50 times the original volume. At what pH does the equivalence point occur?

146. A student intends to titrate a solution of a weak monoprotic acid with a sodium hydroxide solution but reverses the two solutions and places the weak acid solution in the buret. After 23.75 mL of the weak acid solution has been added to 50.0 mL of the 0.100 *M* NaOH solution, the pH of the resulting solution is 10.50. Calculate the original concentration of the solution of weak acid.

147. The active ingredient in aspirin is acetylsalicylic acid. A 2.51-g sample of acetylsalicylic acid required 27.36 mL of 0.5106 *M* NaOH for complete reaction. Addition of 15.44 mL of 0.4524 *M* HCl to the flask containing the aspirin and the sodium hydroxide produced a mixture with pH = 3.48. Find the molar mass of acetylsalicylic acid and its *K*a value. Acetylsalicylic acid is a monoprotic acid.

148. A solution is formed by mixing 50.0 mL of 10.0 *M* NaX with 50.0 mL of 2.0×10^{-3} *M* $CuNO_3$. Assume that Cu(I) forms complex ions with X^- as follows:

$$Cu^+(aq) + X^-(aq) \rightleftharpoons CuX(aq) \qquad K_1 = 1.0 \times 10^2$$
$$CuX(aq) + X^-(aq) \rightleftharpoons CuX_2^-(aq) \qquad K_2 = 1.0 \times 10^4$$
$$CuX_2^-(aq) + X^-(aq) \rightleftharpoons CuX_3^{2-}(aq) \qquad K_3 = 1.0 \times 10^3$$

Calculate the following concentrations at equilibrium.
a. CuX_3^{2-} b. CuX_2^- c. Cu^+

149. When phosphoric acid is titrated with a NaOH solution, only two stoichiometric points are seen. Why?

150. Consider the following two acids:

[Structure: benzene ring with OH and CO_2H substituents] $pK_{a_1} = 2.98$; $pK_{a_2} = 13.40$

Salicylic acid

$HO_2CCH_2CH_2CH_2CH_2CO_2H$
Adipic acid $pK_{a_1} = 4.41$; $pK_{a_2} = 5.28$

In two separate experiments, the pH was measured during the titration of 5.00 mmol of each acid with 0.200 *M* NaOH. Each experiment showed only one stoichiometric point when the data were plotted. In one experiment the stoichiometric point was at 25.00 mL added NaOH, and in the other experiment the stoichiometric point was at 50.00 mL NaOH. Explain these results. (See Exercise 83.)

Challenge Problems

151. A buffer is made using 45.0 mL of 0.750 *M* $HC_3H_5O_2$ ($K_a = 1.3 \times 10^{-5}$) and 55.0 mL of 0.700 *M* $NaC_3H_5O_2$. What volume of 0.10 *M* NaOH must be added to change the pH of the original buffer solution by 2.5%?

152. Consider a solution prepared by mixing the following:
50.0 mL of 0.100 *M* Na_3PO_4
100.0 mL of 0.0500 *M* KOH
200.0 mL of 0.0750 *M* HCl
50.0 mL of 0.150 *M* NaCN

Determine the volume of 0.100 *M* HNO_3 that must be added to this mixture to achieve a final pH value of 7.21.

153. For solutions containing salts of the form NH_4X, the pH is determined by using the equation

$$pH = \frac{pK_a(NH_4^+) + pK_a(HX)}{2}$$

a. Derive this equation. (*Hint:* Review Section 8.7 on the pH of solutions containing amphoteric species.)
b. Use this equation to calculate the pH of the following solutions: ammonium formate, ammonium acetate, and ammonium bicarbonate. See Appendix 5 for K_a values.
c. Solutions of ammonium acetate are commonly used as pH = 7.0 buffers. Write equations to show how an ammonium acetate solution neutralizes added H^+ and OH^-.

154. Consider the titration of 100.0 mL of a solution that contains a mixture of 0.050 *M* H_2SO_4 and 0.20 *M* $H_2C_6H_6O_6$. Calculate the pH
a. before any 0.10 *M* NaOH has been added.
b. after a total of 100.0 mL of 0.10 *M* NaOH has been added.
c. after a total of 300.0 mL of 0.10 *M* NaOH has been added.
d. after a total of 500.0 mL of 0.10 *M* NaOH has been added.

155. The copper(I) ion forms a complex ion with CN^- according to the following equation:

$$Cu^+(aq) + 3CN(aq)^- \rightleftharpoons Cu(CN)_3^{2-}(aq) \qquad K_f = 1.0 \times 10^{11}$$

a. Calculate the solubility of CuBr(*s*) ($K_{sp} = 1.0 \times 10^{-5}$) in 1.0 L of 1.0 *M* NaCN.
b. Calculate the concentration of Br^- at equilibrium.
c. Calculate the concentration of CN^- at equilibrium.

156. Aluminum ions react with the hydroxide ion to form the precipitate $Al(OH)_3(s)$, but can also react to form the soluble complex ion $Al(OH)_4^-$. In terms of solubility, $Al(OH)_3(s)$ will be more soluble in very acidic solutions as well as more soluble in very basic solutions.
a. Write equations for the reactions that occur to increase the solubility of $Al(OH)_3(s)$ in very acidic solutions and in very basic solutions.

b. Show that the solubility of $Al(OH)_3$, as a function of $[H^+]$, obeys the equation

$$S = [H^+]^3K_{sp}/K_w^3 + KK_w/[H^+]$$

where S = solubility = $[Al^{3+}] + [Al(OH)_4^-]$ and K is the equilibrium constant for

$$Al(OH)_3(s) + OH^-(aq) \rightleftharpoons Al(OH)_4^-(aq)$$

c. The value of K is 40.0, and K_{sp} for $Al(OH)_3$ is 2×10^{-32}. Plot the solubility of $Al(OH)_3$ in the pH range 4–12.

157. a. Calculate the molar solubility of SrF_2 in water, ignoring the basic properties of F^-. (For SrF_2, $K_{sp} = 7.9 \times 10^{-10}$.)
b. Would the measured molar solubility of SrF_2 be greater than or less than the value calculated in part a? Explain.
c. Calculate the molar solubility of SrF_2 in a solution buffered at pH = 2.00. (K_a for HF is 7.2×10^{-4}.)

158. What is the maximum possible concentration of Ni^{2+} ion in water at 25°C that is saturated with 0.10 M H_2S and maintained at pH 3.0 with HCl?

159. A mixture contains 1.0×10^{-3} M Cu^{2+} and 1.0×10^{-3} M Mn^{2+} and is saturated with 0.10 M H_2S. Determine a pH where CuS precipitates but MnS does not precipitate. K_{sp} for CuS = 8.5×10^{-45} and K_{sp} for MnS = 2.3×10^{-13}.

160. Consider 1.0 L of an aqueous solution that contains 0.10 M sulfuric acid to which 0.30 mole barium nitrate is added. Assuming no change in volume of the solution, determine the pH, the concentration of barium ions in the final solution, and the mass of solid formed.

161. Calculate the solubility of AgCN(s) ($K_{sp} = 2.2 \times 10^{-12}$) in a solution containing 1.0 M H^+. (K_a for HCN is 6.2×10^{-10}.)

162. Calcium oxalate (CaC_2O_4) is relatively insoluble in water ($K_{sp} = 2 \times 10^{-9}$). However, calcium oxalate is more soluble in acidic solution. How much more soluble is calcium oxalate in 0.10 M H^+ than in pure water? In pure water, ignore the basic properties of $C_2O_4^{2-}$.

163. What volume of 0.0100 M NaOH must be added to 1.00 L of 0.0500 M HOCl to achieve a pH of 8.00?

164. Consider the titration of 100.0 mL of a 1.00×10^{-4} M solution of an acid HA ($K_a = 5.0 \times 10^{-10}$) with 1.00×10^{-3} M NaOH. Calculate the pH for the following conditions.
a. before any NaOH has been added
b. after 5.00 mL of NaOH has been added
c. at the stoichiometric point

165. Consider a solution formed by mixing 200.0 mL of 0.250 M Na_3PO_4, 135.0 mL of 1.000 M HCl, and 100.0 mL of 0.100 M NaCN.
a. Calculate the pH of this solution.
b. Calculate the concentration of HCN in this solution.

166. Consider a solution formed by mixing 50.0 mL of 0.100 M H_2SO_4, 30.0 mL of 0.100 M HOCl, 25.0 mL of 0.200 M NaOH, 25.0 mL of 0.100 M $Ba(OH)_2$, and 10.0 mL of 0.150 M KOH. Calculate the pH of this solution.

167. Calculate the pH of a solution prepared by mixing 500.0 mL of 0.50 M Na_3PO_4 and 500.0 mL of 0.10 M H_2SO_4.

168. Consider the titration of 100.0 mL of 0.10 M phosphoric acid with 0.10 M NaOH.
a. Determine the pH at the third half-equivalence point by assuming it is a "special point" (see Fig. 8.11).
b. Calculate the pH at the third equivalence point.
c. Why must the answer to part a be incorrect? Why can't we use the "special point" on the graph? (Explain the assumption made in using the special point and why it is not valid in this case.)
d. Calculate the pH at the third half-equivalence point.

169. In the titration of 100.0 mL of a 0.0500 M solution of acid H_3A ($K_{a_1} = 1.0 \times 10^{-3}$, $K_{a_2} = 5.0 \times 10^{-8}$, $K_{a_3} = 2.0 \times 10^{-12}$), calculate the volume of 1.00 M NaOH required to reach pH values of 9.50 and 4.00.

170. Consider the titration curve in Exercise 89 for the titration of Na_2CO_3 with HCl.
a. If a mixture of $NaHCO_3$ and Na_2CO_3 was titrated, what would be the relative sizes of V_1 and V_2?
b. If a mixture of NaOH and Na_2CO_3 was titrated, what would be the relative sizes of V_1 and V_2?
c. A sample contains a mixture of $NaHCO_3$ and Na_2CO_3. When this sample was titrated with 0.100 M HCl, it took 18.9 mL to reach the first stoichiometric point and an additional 36.7 mL to reach the second stoichiometric point. What is the composition in mass percent of the sample?

Marathon Problem

171.* A 225-mg sample of a diprotic acid is dissolved in enough water to make 250. mL of solution. The pH of this solution is 2.06. A saturated solution of calcium hydroxide ($K_{sp} = 1.3 \times 10^{-6}$) is prepared by adding excess calcium hydroxide to pure water and then removing the undissolved solid by filtration. Enough of the calcium hydroxide solution is added to the solution of the acid to reach the second equivalence point. The pH at the second equivalence point (as determined by a pH meter) is 7.96. The first dissociation constant for the acid (K_{a_1}) is 5.90×10^{-2}. Assume that the volumes of the solutions are additive, that all solutions are at 25°C, and that K_{a_1} is at least 1000 times greater than K_{a_2}.
a. Calculate the molar mass of the acid.
b. Calculate the second dissociation constant for the acid (K_{a_2}).

*From James H. Burness, "The Use of "Marathon" Problems as Effective Vehicles for the Presentation of General Chemistry Lectures," Journal of Chemical Education, 68(11). Copyright © 1991 American Chemical Society. Reprinted by permission.

9 Energy, Enthalpy, and Thermochemistry

chapter

Rafflesia arnoldii in Sumatra.

Photo by Jeremy Holden

OWL

Sign in to OWL at www.cengage.com/owl to view tutorials and simulations, develop problem-solving skills, and complete online homework assigned by your professor.

go Chemistry

Download mini lecture videos for key concept review and exam prep from OWL or purchase them from www.cengagebrain.com.

Energy is the essence of our very existence as individuals and as a society. The food that we eat furnishes the energy to live, work, and play, just as the coal and oil consumed by manufacturing and transportation systems power our modern, industrialized civilization.

Huge quantities of carbon-based fossil fuels have been available for the taking. This abundance of fuels has led to a world society with a voracious appetite for energy, consuming millions of barrels of petroleum every day. We are now dangerously dependent on the dwindling supplies of oil, and this dependence is an important source of tension among nations in today's world. In an incredibly short time, we have moved from a period of ample and cheap supplies of petroleum to one of high prices and uncertain supplies. If our present standard of living is to be maintained, we must find alternatives to petroleum. To do so, we need to know the relationship between chemistry and energy, which we explore in this chapter.

There are additional problems with fossil fuels. The waste products from burning fossil fuels significantly affect our environment. For example, when a carbon-based fuel is burned, the carbon reacts with oxygen to form carbon dioxide, which is released into the atmosphere. Although much of this carbon dioxide is consumed in various natural processes such as photosynthesis and the formation of carbonate minerals, the amount of carbon dioxide in the atmosphere is steadily increasing. This increase is significant because atmospheric carbon dioxide absorbs heat radiated from the earth's surface and radiates some of it back toward the earth. Since this is an important mechanism for controlling the earth's temperature, many scientists fear that an increase in the concentration of carbon dioxide will warm the earth, causing significant changes in climate. In addition, impurities in the fuels react with components of air to produce air pollution.

In this chapter we will cover the fundamental concepts of energy and take a brief look at the practical aspects of the energy supply and pollution. Additional theoretical aspects of energy will be presented in Chapter 10.

9.1 The Nature of Energy

The total energy content of the universe is constant.

Although the concept of energy is quite familiar, energy is rather difficult to define precisely. We will define **energy** as the *capacity to do work or to produce heat*. In this chapter we will concentrate on the transfer of energy via heat flow that accompanies chemical processes.

One of the most important characteristics of energy is that it is conserved. The **law of conservation of energy** states that *energy can be converted from one form to another but can be neither created nor destroyed*. That is, the energy of the universe is constant. Energy can be classified as either potential energy or kinetic energy. **Potential energy** is energy due to position or composition. For example, water behind a dam has potential energy that can be converted to work when the water flows down through turbines, thereby creating electricity. Attractive and repulsive forces also lead to potential energy. The energy released when gasoline is burned results from differences in the attractive forces between nuclei and electrons in the reactants and products. The **kinetic energy** of an object is due to the motion of the object and depends on the mass of the object (m) and its velocity (v): $KE = \frac{1}{2}mv^2$.

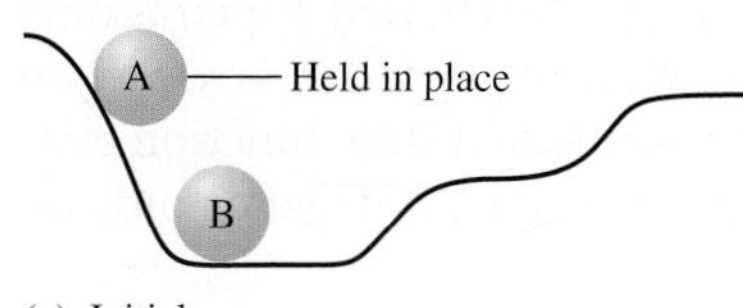

(a) Initial

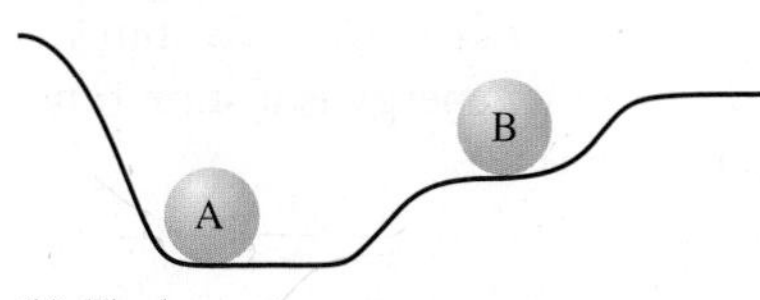

(b) Final

Figure 9.1
(a) In the initial positions, ball A has a higher potential energy than ball B. (b) After A has rolled down the hill, the potential energy lost by A has been converted to random motions of the components of the hill (frictional heating) and to the increase in the potential energy of B.

Energy can be converted from one form to another easily. For example, consider the two balls in Fig. 9.1(a). Ball A, because of its higher position, initially has more potential energy than ball B. When A is released, it moves down the hill and strikes B. Eventually the arrangement shown in Fig. 9.1(b) is achieved. What has happened in going from the initial to the final arrangement? The potential energy of A has decreased, but since energy is conserved, all the energy lost by A must be accounted for. How is this energy distributed?

Initially, the potential energy of A is changed to kinetic energy as the ball rolls down the hill. Part of this kinetic energy has been transferred to B, causing it to be raised to a higher final position. Thus the potential energy of B has increased. However, since the final position of B is lower than the original position of A, some of the energy is still unaccounted for. Both balls are at rest in their final positions, so the missing energy cannot be due to their motions. What has happened to the remaining energy?

The answer lies in the interaction between the hill's surface and the ball. As A rolls down the hill, some of its kinetic energy is transferred to the surface of the hill as heat. This transfer of energy is called *frictional heating*. The temperature of the hill increases very slightly as the ball rolls down.

Heat involves a *transfer* of energy.

At this point it is important to recognize that heat and temperature are decidedly different. Recall that temperature is a property that reflects the random motions of the particles in a particular substance. **Heat**, on the other hand, involves the *transfer* of energy between two objects due to a temperature difference. Heat is not a substance contained in an object, although we often talk of heat as if this were true.

Note that in going from the initial to the final arrangements in Fig. 9.1, ball B gains potential energy because ball A has done work on B. **Work** is defined as *a force acting over a distance.* Work is required to raise B from its original position to a higher one. Part of the original energy stored as potential energy in A has been transferred through work to B, thereby increasing B's potential energy. Thus there are two ways to transfer energy: through work and through heat.

In rolling to the bottom of the hill as shown in Fig. 9.1, ball A always loses the same amount of potential energy. However, the way that this energy transfer is divided between work and heat depends on the specific conditions—the **pathway.** For example, the surface of the hill might be so rough that the energy of A is expended completely through frictional heating; A is moving so slowly when it hits B that it cannot move B to the next level. In this case no work is done. Regardless of the condition of the hill's surface, the *total energy* transferred will be constant. However, the amounts of heat and work will differ. Energy change is independent of the pathway; however, work and heat are both dependent on the pathway.

This brings us to a very important concept: the **state function** or *state property*. A state function refers to a property of the system that depends only on its *present state*. A state function (property) does not depend in any way on the system's past (or future). In other words, the value of a state function does not depend on how the system arrived at the present state; it depends only on the characteristics of the present state.

Stated more precisely, one very important characteristic of a state function is that a change in this function (property) in going from one state to another state is independent of the particular pathway taken between the two states.

Energy is a state function; work and heat are not.

Of the functions considered in our present example, energy is a state function, but work and heat are not state functions.

Chemical Energy

The ideas we have just illustrated using mechanical examples also apply to chemical systems. The combustion of methane, for example,

$$CH_4(g) + 2O_2(g) \longrightarrow CO_2(g) + 2H_2O(g) + \text{energy (heat)}$$

is used to heat many homes in the United States. To discuss this reaction, we divide the universe into two parts: the system and the surroundings. The **system** is the part of the universe on which we wish to focus attention; the **surroundings** include everything else in the universe. In this case we define the system as the reactants and products of the reaction. The surroundings consist of the reaction container, the room, and everything else other than the reactants and products.

When a reaction results in the evolution of heat, it is said to be **exothermic** (*exo-* is a prefix meaning "out of"); that is, energy flows *out of the system.* For example, in the combustion of methane, energy flows out of the system as heat. Reactions that absorb energy from the surroundings are said to be **endothermic.** That is, when the heat flow is *into a system,* the process is endothermic. For example, the formation of nitric oxide from nitrogen and oxygen is endothermic:

$$N_2(g) + O_2(g) + \text{energy (heat)} \longrightarrow 2NO(g)$$

A familiar endothermic physical process is the vaporization of water:

$$H_2O(l) + \text{energy} \longrightarrow H_2O(g)$$

Where does the energy, released as heat, come from in an exothermic reaction? The answer lies in the difference in potential energy between the products and the reactants. In an exothermic reaction, which has lower potential energy, the reactants or the products? We know that total energy is conserved and that energy flows from the system into the surroundings in an exothermic reaction. This means that *the energy gained by the surroundings must be equal to the energy lost by the system.* For methane combustion, the energy content of the system decreases, which means that 1 mole of CO_2 and 2 moles of H_2O molecules (the products) possess less potential energy than do 1 mole of CH_4 and 2 moles of O_2 molecules (the reactants). The heat flow into the surroundings results from a lowering of the potential energy of the reaction system. This always holds true. *In any exothermic reaction, the potential energy stored in the chemical bonds is being converted to thermal energy (random kinetic energy) via heat.*

The energy diagram for the combustion of methane is shown in Fig. 9.2, where Δ(PE) represents the *change* in potential energy stored in the bonds of the products as compared with the bonds of the reactants. In other words, this quantity represents the difference between the energy required to break the bonds in the reactants and the energy released when the bonds in the products are formed. In an exothermic process, the bonds in the products are stronger (on average) than those of the reactants. That is, more energy is released in forming the new bonds in the products than is consumed in breaking the bonds in the reactants. The net result is that the quantity of energy Δ(PE) is transferred to the surroundings through heat.

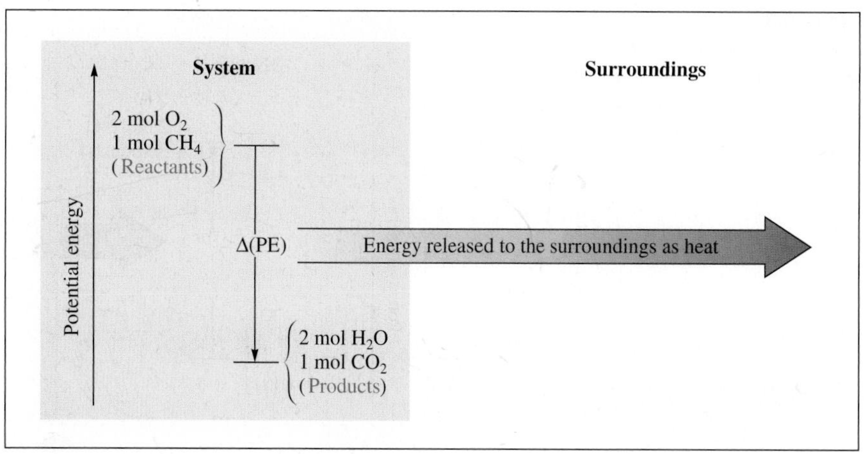

Figure 9.2
The combustion of methane releases the quantity of energy Δ(PE) to the surroundings via heat flow. This is an exothermic process.

Chemical Insights Bees Are Hot

A recent study at Yunnan Agricultural University in China has shown how certain species of Asian honeybees protect their colonies. Large wasps (*Vespa velutina*) with wingspans as large as 5 cm specialize in attacking social-insect nests, and they carry off the larvae as food for young wasps back at their home colony. The wasps can defeat colonies of honeybees with thousands of residents. The invader wasp sits at the entrance to the colony and kills the guard honeybees as they emerge to defend their nest. When the guard bees are all dead, the wasp then mines the honeybee nest for larvae. However, Asian honeybees (*Aspis carama*) have developed a special defense against invading wasps. They engulf the wasp in a living ball of defenders and cook the predator to death. Studies show that within about 5 minutes the center of the ball of defenders reaches a temperature of about 45°C. Separate studies of the temperature tolerances of the bees showed that the wasps die at 45.7°C, but the Asian honeybees can survive to 50.7°C. Thus the defenders come within about 5°C of a temperature that would be fatal to all of them.

For an endothermic reaction, the situation is reversed, as shown in Fig. 9.3. Energy that flows into the system as heat is used to increase the potential energy of the system. In this case the products have higher potential energy (weaker bonds on average) than the reactants.

The study of energy and its interconversions is called **thermodynamics.** The law of conservation of energy is often called the **first law of thermodynamics** and is stated as follows: *The energy of the universe is constant.*

The **internal energy** (E) of a system can be defined most precisely as the sum of the kinetic and potential energies of all of the "particles" in the system. The internal energy of a system can be changed by a flow of work, heat, or both. That is,

$$\Delta E = q + w$$

where ΔE represents the change in the system's internal energy, q represents heat, and w represents work.

Thermodynamic quantities always consist of two parts: a *number,* giving the magnitude of the change; and a *sign,* indicating the direction of the flow. *The sign reflects the system's point of view.* For example, if a quantity of en-

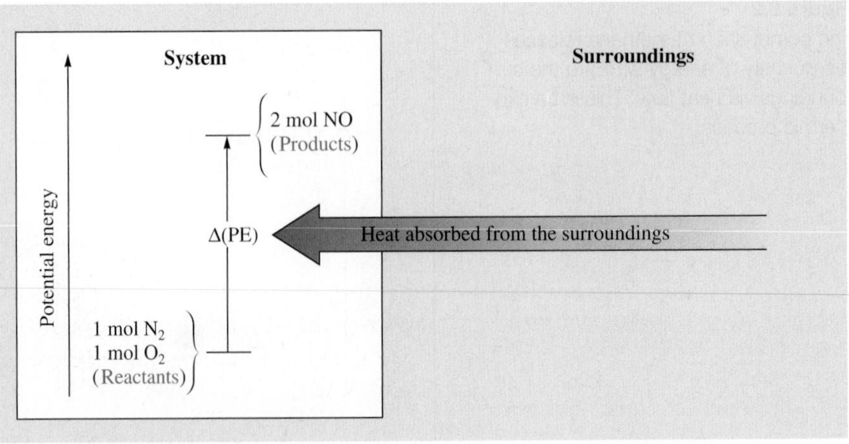

Figure 9.3
The energy diagram for the reaction of nitrogen and oxygen to form nitric oxide. This is an endothermic process.

ergy flows *into* the system via heat (an endothermic process), q is equal to $+x$, where the *positive* sign indicates that the *system's energy is increasing.* On the other hand, when energy flows *out of* the system via heat (an exothermic process), q is equal to $-x$, where the *negative* sign indicates that the *system's energy is decreasing.*

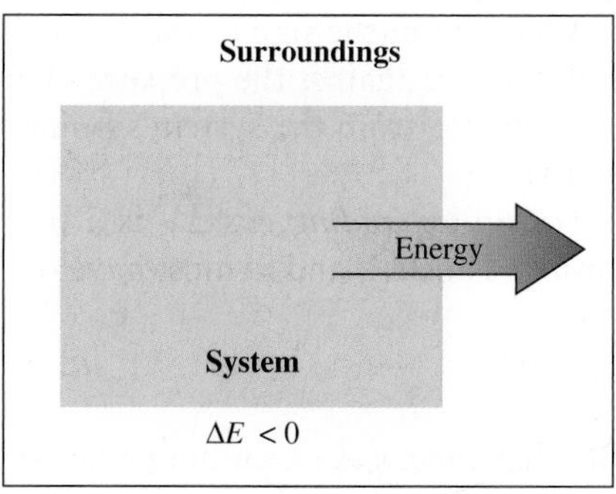

Exothermic

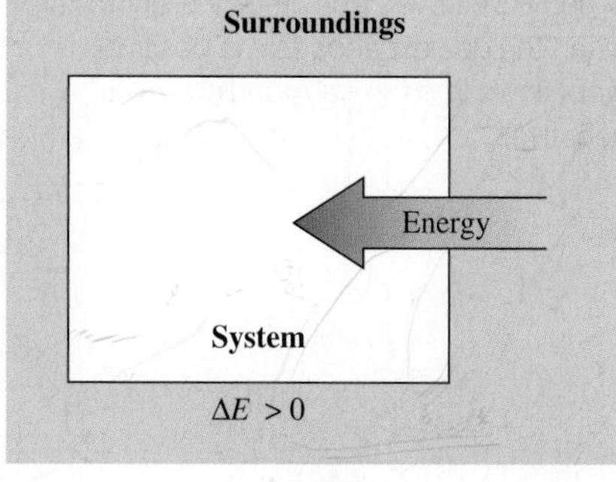

Endothermic

The convention in this text is to take the system's point of view: $q = -x$ denotes an exothermic process, and $q = +x$ denotes an endothermic one.

In this text the same conventions also apply to the flow of work. If the system does work on the surroundings (energy flows out of the system), w is negative. If the surroundings do work on the system (energy flows into the system), w is positive. We define work from the system's point of view to be consistent for all thermodynamic quantities. That is, in this convention the signs of both q and w reflect what happens to the system; thus we use $\Delta E = q + w$.

In this text we *always* take the system's point of view. This convention is not followed in every area of science. For example, engineers are in the business of designing machines to do work—that is, to make the system (the machine) transfer energy to its surroundings through work. Consequently, engineers define work from the surroundings' point of view. In their convention, work that flows out of the system is treated as positive because the energy of the surroundings has increased. The first law of thermodynamics is then written $\Delta E = q - w'$, where w' signifies work from the surroundings' point of view.

A common type of work associated with chemical processes is work done by a gas (through *expansion*) or work done to a gas (through *compression*). For example, in an automobile engine, the heat from the combustion of the gasoline expands the gases in the cylinder, pushing back the piston. This motion is then translated into the motion of the car.

Suppose we have a gas confined to a cylindrical container with a movable piston, as shown in Fig. 9.4, where F is the force acting on a piston of area A. Since pressure is defined as force per unit area, the pressure of the gas is

$$P = \frac{F}{A}$$

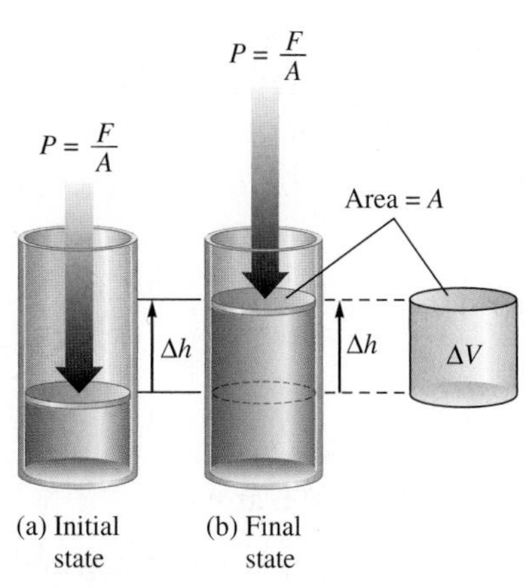

Figure 9.4
(a) The piston, moving a distance Δh against a pressure P, does work on the surroundings. (b) Since the volume of a cylinder is the area of the base times its height, the change in volume of the gas is given by $\Delta h \times A = \Delta V$.

Work is defined as a force applied over a given distance, so if the piston moves a distance Δh, as shown in Fig. 9.4, then the magnitude of the work is

$$|\text{Work}| = |\text{force} \times \text{distance}| = |F \times \Delta h|$$

Since $P = F/A$, or $F = P \times A$, then

$$|\text{Work}| = |F \times \Delta h| = |P \times A \times \Delta h|$$

Since the volume of the cylinder equals the area of the piston times the height of the cylinder (Fig. 9.4), the change in volume ΔV resulting from the piston moving a distance Δh is

$$\Delta V = \text{final volume} - \text{initial volume} = A \times \Delta h$$

Substituting $\Delta V = A \times \Delta h$ into the expression for the magnitude of the work gives

$$|\text{Work}| = |P \times A \times \Delta h| = |P\Delta V|$$

Note that this expression gives the *magnitude* (size) of the work required to expand a gas by ΔV against a pressure P.

w and $P\Delta V$ have opposite signs since when the gas expands (ΔV is positive), work flows into the surroundings (*w* is negative).

What about the sign of the work? The gas (the system) is expanding, moving the piston against the pressure. Thus the system is doing work on the surroundings, so from the system's point of view, the sign of the work should be negative.

For an *expanding gas* ΔV is a positive quantity because the volume is increasing. Thus ΔV and w must have opposite signs, which leads to the equation

$$w = -P\Delta V$$

Note that for a gas expanding against an external pressure P, w is a negative quantity as required, since work flows out of the system. When a gas is *compressed,* ΔV is a negative quantity (the volume decreases), which makes w a positive quantity (work flows into the system).

The work accompanying a change in volume of a gas is often called "*PV* work."

In dealing with "PV work," keep in mind that the P in $P\Delta V$ always refers to the external pressure—the pressure that causes a compression or that resists an expansion.

OWL INTERACTIVE EXAMPLE 9.1

A balloon is inflated to its full extent by heating the air inside it. In the final stages of this process, the volume of the balloon changes from 4.00×10^6 L to 4.50×10^6 L by addition of 1.3×10^8 J of energy as heat. Assuming the balloon expands against a constant pressure of 1.0 atm, calculate ΔE for the process.

Solution To calculate ΔE we use the equation

$$\Delta E = q + w$$

The joule (J) is the fundamental SI unit for energy:

$$\text{J} = \frac{\text{kg m}^2}{\text{s}^2}$$

Since the problem states that 1.3×10^8 J of energy is *added* as heat,

$$q = +1.3 \times 10^8 \text{ J}$$

The work done can be calculated from the expression

$$w = -P\Delta V$$

In this case $P = 1.0$ atm (the external pressure) and

$$\begin{aligned}\Delta V &= V_{\text{final}} - V_{\text{initial}} \\ &= 4.50 \times 10^6 \text{ L} - 4.00 \times 10^6 \text{ L} = 0.50 \times 10^6 \text{ L} \\ &= 5.0 \times 10^5 \text{ L}\end{aligned}$$

Thus

$$w = -1.0 \text{ atm} \times 5.0 \times 10^5 \text{ L} = -5.0 \times 10^5 \text{ L atm}$$

The conversion factor between L atm and J can be obtained from the values of *R*:

$$0.08206 \frac{\text{L atm}}{\text{K mol}}$$

and

$$8.3145 \frac{\text{J}}{\text{K mol}}$$

Note that the negative sign for w makes sense, since the gas is expanding and thus doing work on the surroundings.

To calculate ΔE, we must add q and w. However, since q is given in units of J and w is given in units of L atm, we must change the work to units of joules:

$$w = -5.0 \times 10^5 \text{ L atm} \times \frac{101.3 \text{ J}}{\text{L atm}} = -5.1 \times 10^7 \text{ J}$$

Then

$$\Delta E = q + w = (+1.3 \times 10^8\ \text{J}) + (-5.1 \times 10^7\ \text{J}) = 8 \times 10^7\ \text{J}$$

Since more energy is added through heating than the gas expends doing work, there is a net increase in the energy of the gas in the balloon. Hence ΔE is positive.

9.2 Enthalpy

So far we have discussed the internal energy of a system. A less familiar property of a system is its **enthalpy** (H), which is defined as

$$H = E + PV$$

where E is the internal energy of the system, P is the pressure of the system, and V is the volume of the system.

Enthalpy is a state function. A change in enthalpy does not depend on the pathway between two states.

Since internal energy, pressure, and volume are all state functions, *enthalpy is also a state function.* But what exactly is enthalpy? To help answer this question, consider a process carried out at constant pressure, where the only work allowed is pressure-volume work ($w = -P\Delta V$). Under these conditions the expression

$$\Delta E = q_{\text{P}} + w$$

becomes

$$\Delta E = q_{\text{P}} - P\Delta V$$

or

$$q_{\text{P}} = \Delta E + P\Delta V$$

where q_{P} is the heat at constant pressure.

We will now relate q_{P} to the change in enthalpy. The definition of enthalpy is $H = E + PV$. Therefore,

$$(\text{Change in } H) = (\text{change in } E) + (\text{change in } PV)$$

or

$$\Delta H = \Delta E + \Delta(PV)$$

Since P is constant, the change in PV is caused only by a change in volume. Thus

$$\Delta(PV) = P\Delta V$$

and

$$\Delta H = \Delta E + P\Delta V$$

This expression is identical to the one we obtained for q_{P}:

$$q_{\text{P}} = \Delta E + P\Delta V$$

$\Delta H = q$ at constant pressure, where only "PV work" is allowed.

Thus, for a process carried out at constant pressure, where the only work allowed is that from a volume change,

$$\Delta H = q_{\text{P}}$$

The change in enthalpy of a system has no easily interpreted meaning except at constant pressure, where ΔH = heat.

At constant pressure (where only PV work is allowed), the change in enthalpy (ΔH) of the system is equal to the energy flow as heat. This means that for a reaction studied at constant pressure, the flow of heat is a measure of the change in enthalpy for the system. For this reason, the terms *heat of reaction* and *change in enthalpy* are used interchangeably for reactions studied at constant pressure.

For a chemical reaction the enthalpy change is given by the equation

$$\Delta H = H_{\text{products}} - H_{\text{reactants}}$$

At constant pressure exothermic means ΔH is negative; endothermic means ΔH is positive.

In a case in which the products of a reaction have greater enthalpy than the reactants, ΔH will be positive. Thus heat is absorbed by the system, and the reaction is endothermic. On the other hand, if the enthalpy of the products is less than that of the reactants, ΔH is negative. In this case the overall decrease in enthalpy is achieved by the generation of heat, and the reaction is exothermic.

9.3 Thermodynamics of Ideal Gases

In developing the concepts of thermodynamics, we often find it useful to refer to the properties of matter in the simplest possible context. For this reason we often start with the thermodynamic characteristics of the ideal gas—the hypothetical condition approached by real gases at high temperatures and low pressures such that they obey the relationship $PV = nRT$.

To this point, we have assumed that the ideal gas particles have no internal structure—they are monatomic.

Recall from Chapter 5 that for an ideal gas

$$(\text{KE})_{\text{avg}} = \tfrac{3}{2}RT$$

where $(\text{KE})_{\text{avg}}$ represents the average, random, translational energy for 1 mole of gas at a given temperature T (in kelvins). The only way to change the kinetic energy of an ideal gas is to change its temperature. The energy ("heat") required to change the energy of 1 mole of an ideal gas by ΔT is

$$\text{Energy ("heat") required} = \tfrac{3}{2}R\Delta T$$

Note that for a temperature change of 1 K ($\Delta T = 1$), the energy required is $\frac{3}{2}R$.

The **molar heat capacity** of a substance is defined as the energy required to raise the temperature of 1 mole of that substance by 1 K. Thus we might conclude that the molar heat capacity of an ideal gas is $\frac{3}{2}R$. However, we will have to qualify this conclusion when we consider the implications of the PV work that can occur when a gas is heated.

Heating an Ideal Gas at Constant Volume

For an ideal gas, work occurs only when its volume changes. Thus, if a gas is heated at constant volume, the pressure increases but no work occurs.

When an ideal gas is heated in a rigid container in which no change in volume occurs, there can be no PV work ($\Delta V = 0$). Under these conditions all the energy that flows into the gas is used to increase the translational energies of the gas molecules. Thus C_v, the molar heat capacity of an ideal gas *at constant volume,* is $\frac{3}{2}R$, the result anticipated in the preceding discussion:

$$C_v = \tfrac{3}{2}R = \begin{array}{c}\text{"heat" required to change the temperature} \\ \text{of 1 mol of gas by 1 K at constant volume}\end{array}$$

Heating an Ideal Gas at Constant Pressure

When an ideal gas is heated at constant pressure, its volume increases and PV work occurs. Thus, when a gas is heated at constant pressure, energy must be supplied both to change the translational energy of the gas and to provide the work the gas does as it expands:

$$\text{Energy required} = \text{"heat"} = \begin{array}{c}\text{energy needed} \\ \text{to change the} \\ \text{translational energy}\end{array} + \begin{array}{c}\text{energy needed to} \\ \text{do the } PV \text{ work}\end{array}$$

The heat needed to increase the translational energy is $\frac{3}{2}R$, as we concluded above.

The *quantity* of work done as the gas expands by ΔV is $P\Delta V$. Using the ideal gas law, we see that

$$P\Delta V = nR\Delta T = R\Delta T \qquad \text{(per mole)}$$

Thus for a 1 K change in temperature ($\Delta T = 1$ K) the work is R, so

$$\begin{array}{c}\text{Heat required to increase the temperature} \\ \text{of 1 mol of gas by 1 K (constant } P\text{)}\end{array} = \tfrac{3}{2}R + R = \tfrac{5}{2}R$$

$$= C_v + R = C_p$$

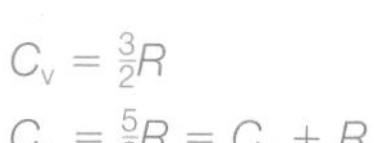

$C_v = \frac{3}{2}R$

$C_p = \frac{5}{2}R = C_v + R$

Therefore, we have shown that Cp, the molar heat capacity of an ideal gas at constant pressure, is $\frac{5}{2}R$ or $C_v + R$.

Heating a Polyatomic Gas

We have established that for an ideal gas the molar heat capacity at constant volume is $\frac{3}{2}R$. This value of C_v assumes that an ideal gas consists of "particles" that have no structure. That is, we assume that the particles are monatomic (consisting of a single atom). Monatomic real gases, such as helium, have measured values of C_v very close to $\frac{3}{2}R$. However, gases such as SO_2 and $CHCl_3$ that contain polyatomic molecules have observed values for C_v that are significantly greater than $\frac{3}{2}R$. For example, the value of C_v for SO_2 is almost $4R$ at 25°C. This larger value for C_v results because polyatomic molecules absorb energy to excite rotational and vibrational motions in addition to translational motions. That is, at 25°C the molecules in such a gas are rotating, and the atoms in the molecule are vibrating, as if the bonds were springs.

As a polyatomic gas is heated, the gas molecules absorb energy to increase their rotational and vibrational motions as well as to move through space (translate) at higher speeds. Recall from our previous discussions that the temperature of a monatomic ideal gas is an index of the average random *translational* energy of the gas. Thus, when a gas is heated, the temperature only increases to the extent that the translational energies of the molecules increase. Any energy that is absorbed to increase the vibrational and rotational energies does not contribute directly to the translational kinetic energy; so, for a gas that consists of diatomic or polyatomic molecules, much of the heat absorbed is used in processes that do not directly increase the temperature. Thus its heat capacity (the energy required to change its *temperature* by 1 K) is greater than $\frac{3}{2}R$.

C_v is greater than $\frac{3}{2}R$ for a gas in which the individual particles are molecules because some of the energy added via heat flow is "stored" in motions that do not directly raise the temperature of the gas. This effect is not related to whether or not the gas is behaving ideally.

Note that the elevated value of C_v for a gas whose particles are molecules is not caused by nonideal behavior. That is, it does not depend on whether the gas obeys the ideal gas law. Rather, it is simply that the internal structure of the molecules enables them to absorb energy for processes other than translational motions.

Recall that C_p is greater than C_v because of the work done by the heated gas as it expands at constant pressure. Thus, if we assume that a given polyatomic gas obeys the ideal gas law, the expression

$$C_p = C_v + R$$

can be used to calculate C_p if the value of C_v for the gas is known.

The observed heat capacities of several gases are shown in Table 9.1. Notice that the monatomic gases have values of C_v equal to $\frac{3}{2}R$ (12.47 J K^{-1} mol^{-1}). Note also that as the molecules become more complex (more atoms), C_v increases. This result is expected because the presence of more atoms means that more nontranslational motions are available to absorb energy. Finally, notice that in all cases $C_p - C_v = R$, as expected for gases that closely obey the ideal gas law.

Heating a Gas: Energy and Enthalpy

Recall that the average translational energy of an ideal gas E is given by the expression

$$E = \tfrac{3}{2}RT \qquad \text{(per mole)}$$

Table 9.1

Molar Heat Capacities of Various Gases at 298 K

Gas	$C_v\left(\frac{J}{K\ mol}\right)$	$C_p\left(\frac{J}{K\ mol}\right)$	$C_p - C_v$
He, Ne, Ar	12.47	20.80	8.33
H_2	20.54	28.86	8.32
N_2	20.71	29.03	8.32
N_2O	30.38	38.70	8.32
CO_2	28.95	37.27	8.32
C_2H_6	44.60	52.92	8.32

for a monatomic ideal gas. The energy of an ideal gas can be changed only by changing the temperature:

$$\Delta E = \tfrac{3}{2}R\Delta T \qquad \text{(per mole)}$$

Note that this expression corresponds to

$$\Delta E = C_v\Delta T \qquad \text{(per mole)}$$

or

$$\Delta E = nC_v\Delta T \qquad (n \text{ moles})$$

The constant-volume heat capacity appears in this expression because when a gas is heated at constant volume, all the input energy (heat) goes toward increasing E (no heat is needed to do work).

On the other hand, when a gas is heated at constant pressure, the volume changes and work occurs. In this case (for n moles of gas),

$$\begin{aligned} \text{“Heat” required} &= q_p = nC_p\Delta T \\ &= n(C_v + R)\Delta T \\ &= \underbrace{nC_v\Delta T}_{\Delta E} + \underbrace{nR\Delta T}_{P\Delta V = \text{work required}} \end{aligned}$$

Notice that although this process is carried out at constant pressure, ΔE is still given by $nC_v\Delta T$. This result seems contradictory at first glance, but actually it makes good sense. Because E for an ideal gas depends only on T (it does not depend on pressure or volume, for example), $\Delta E = nC_v\Delta T$ when an ideal gas is heated whether the process occurs at constant volume or constant pressure.

Next, consider the change in enthalpy when a gas is heated. Recall that by definition

$$H = E + PV$$

Thus, in general, a change in enthalpy is given by the expression

$$\Delta H = \Delta E + \Delta(PV)$$

which (using the ideal gas law) becomes

$$\Delta H = \Delta E + \Delta(nRT) = \Delta E + nR\Delta T$$

for a sample of ideal gas containing n moles. Substituting $\Delta E = nC_v\Delta T$, we have

$$\begin{aligned} \Delta H &= nC_v\Delta T + nR\Delta T \\ &= n(C_v + R)\Delta T = nC_p\Delta T \end{aligned}$$

Note that we have shown that

$$\Delta H = nC_p\Delta T$$

Table 9.2

Thermodynamic Properties of an Ideal Gas

Expression	Application
$C_v = \frac{3}{2}R$	Monatomic ideal gas
$C_v > \frac{3}{2}R$	Polyatomic ideal gas (value must be measured experimentally)
$C_p = C_v + R$	All ideal gases
$C_p = \frac{5}{2}R = \frac{3}{2}R + R$	Monatomic ideal gas
$C_p > \frac{5}{2}R$	Polyatomic ideal gas (specific value depends on the value of C_v)
$\Delta E = nC_v\Delta T$	All ideal gases
$\Delta H = nC_p\Delta T$	All ideal gases

The only way to change H and E for an ideal gas is to change the temperature of the gas. Thus, for any process involving *an ideal gas at constant temperature,* $\Delta H = 0$ and $\Delta E = 0$.

even though we have not assumed constant pressure (or volume). Thus we have shown that for an ideal gas we can always use the expression $nC_p\Delta T$ to calculate the change in enthalpy when n moles of an ideal gas are heated, regardless of any conditions on pressure or volume.

Again, it may seem contradictory that C_p appears in this expression for ΔH even though the pressure may or may not be constant in the process. However, note that the enthalpy ($H = E + PV$) of an ideal gas depends on E and the product PV. We have seen that E depends directly on T, and from the ideal gas law we can easily show that PV depends directly on T ($PV = nRT$) for a given sample of ideal gas (containing n moles). Thus both the energy E and the enthalpy H of an ideal gas depend only on T, not on P or V (individually):

$$E \propto T \qquad \text{and} \qquad H \propto T$$

For energy (E), the proportionality constant is C_v (per mole), and for enthalpy (H), the proportionality constant is C_p (per mole).

In considering these ideas, we must distinguish among q, ΔH, and ΔE. In the calculation of the heat flow for an ideal gas, the equation

$$q = nC\Delta T$$

applies, where C_v or C_p is used depending on the conditions. In contrast, $\Delta H = nC_p\Delta T$ and $\Delta E = nC_v\Delta T$ for a temperature change of an ideal gas regardless of whether pressure or volume (or neither) is constant. Also, note that the heat flow equals ΔE at constant volume ($\Delta E = q_v$) but the heat flow equals ΔH at constant pressure ($\Delta H = q_p$). These results are summarized in Table 9.2.

We will illustrate these concepts in Example 9.2.

EXAMPLE 9.2

Consider 2.00 moles of a monatomic ideal gas that is taken from state A ($P_A = 2.00$ atm, $V_A = 10.0$ L) to state B ($P_B = 1.00$ atm, $V_B = 30.0$ L) by two different pathways:

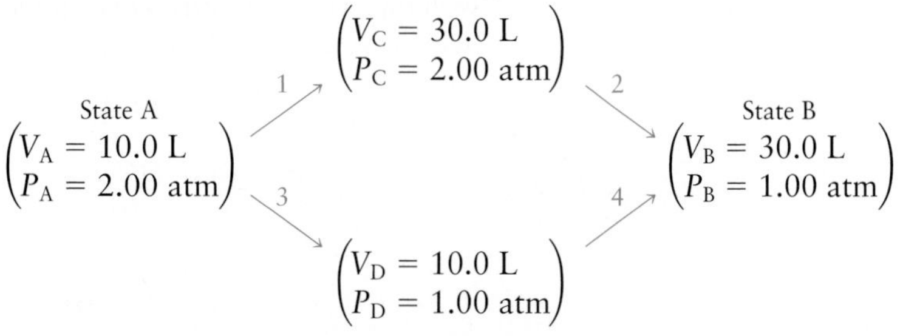

$$\begin{pmatrix} V_C = 30.0\text{ L} \\ P_C = 2.00\text{ atm} \end{pmatrix}$$

State A $\begin{pmatrix} V_A = 10.0\text{ L} \\ P_A = 2.00\text{ atm} \end{pmatrix}$ →(1) C →(2) State B $\begin{pmatrix} V_B = 30.0\text{ L} \\ P_B = 1.00\text{ atm} \end{pmatrix}$

State A →(3) D →(4) State B

$$\begin{pmatrix} V_D = 10.0\text{ L} \\ P_D = 1.00\text{ atm} \end{pmatrix}$$

Calculate q, w, ΔE, and ΔH for both pathways.

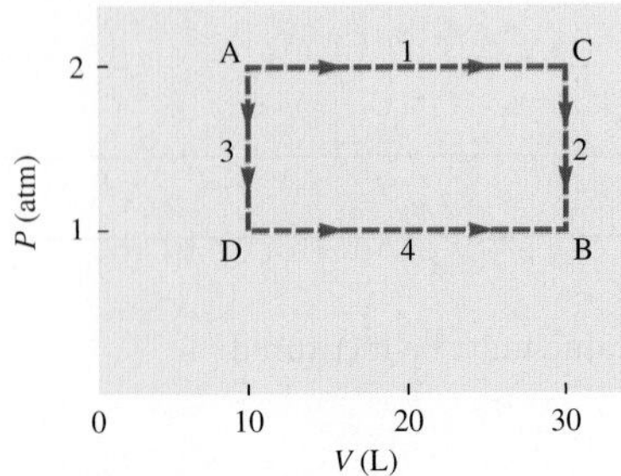

Figure 9.5
Summary of the two pathways discussed in Example 9.2.

Solution Before we do any calculations, it is useful to summarize the processes described above by using the "PV diagram" shown in Fig. 9.5.

Step 1

Notice from Fig. 9.5 that this step corresponds to an expansion from 10.0 to 30.0 L at a constant pressure of 2.00 atm. This process must occur by heating the gas to produce some temperature change ΔT (not specified in the given data). From the ideal gas law we know that

$$P\Delta V = nR\Delta T$$

In this case $\Delta V = 30.0 \text{ L} - 10.0 \text{ L}$, so

$$P\Delta V = (2.00 \text{ atm})(20.0 \text{ L}) = 4.00 \times 10^1 \text{ L atm}$$

or if we convert to joules,

$$P\Delta V = 4.00 \times 10^1 \text{ L atm} \times \frac{101.3 \text{ J}}{\text{L atm}} = 4.05 \times 10^3 \text{ J}$$

It follows that

$$\Delta T = \frac{P\Delta V}{nR} = \frac{4.05 \times 10^3 \text{ J}}{nR}$$

We know that $w = -P\Delta V$, and because in this case the gas expands against a constant external pressure of 2.00 atm, we have

$$w_1 = -(2.00 \text{ atm})(30.0 \text{ L} - 10.0 \text{ L}) = -4.00 \times 10^1 \text{ L atm}$$
$$= -4.05 \times 10^3 \text{ J}$$

Also, in this case (constant P)

$$q_1 = q_\text{P} = nC_\text{p}\Delta T$$
$$= n\left(\frac{5}{2}R\right)\left(\frac{4.05 \times 10^3 \text{ J}}{nR}\right) = 1.01 \times 10^4 \text{ J}$$

(Note that the signs of w and q are as expected: The gas expands, so work flows out of the system; the gas is heated, so heat flows into the system.)

We can calculate ΔE_1 and ΔH_1 as follows:

$$\Delta E_1 = nC_\text{v}\Delta T = n\left(\frac{3}{2}R\right)\left(\frac{4.05 \times 10^3 \text{ J}}{nR}\right) = 6.08 \times 10^3 \text{ J}$$

$$\Delta H_1 = nC_\text{p}\Delta T = n\left(\frac{5}{2}R\right)\left(\frac{4.05 \times 10^3 \text{ J}}{nR}\right) = 1.01 \times 10^4 \text{ J}$$

Note that in this case $q_1(q_\text{P})$ equals ΔH_1, as expected, because this process is carried out at constant pressure.

Step 2

In this step the gas pressure decreases from 2.00 atm to 1.00 atm at constant volume. This step must correspond to the cooling of the gas by a quantity ΔT, which we can find from the ideal gas law:

$$\Delta PV = nR\Delta T$$

$$\Delta T = \frac{\Delta PV}{nR} = \frac{(1.00 \text{ atm} - 2.00 \text{ atm})(30.0 \text{ L})}{nR}$$
$$= \frac{-30.0 \text{ L atm}}{nR} = \frac{-3.04 \times 10^3 \text{ J}}{nR}$$

Note that ΔT is negative, as expected for a cooling process.

Because in this step $\Delta V = 0$, thus $w_2 = 0$. In this case

$$q_2 = q_v = nC_v\Delta T = n\left(\frac{3}{2}R\right)\left(\frac{-3.04 \times 10^3 \text{ J}}{nR}\right)$$

$$= -4.56 \times 10^3 \text{ J}$$

Also,
$$\Delta E_2 = nC_v\Delta T = n\left(\frac{3}{2}R\right)\left(\frac{-3.04 \times 10^3 \text{ J}}{nR}\right)$$

$$= -4.56 \times 10^3 \text{ J} = q_v$$

and
$$\Delta H_2 = nC_p\Delta T = n\left(\frac{5}{2}R\right)\left(\frac{-3.04 \times 10^3 \text{ J}}{nR}\right)$$

$$= -7.60 \times 10^3 \text{ J}$$

Notice that in this case $q_2 = q_v = \Delta E$, as expected for a constant-volume process.

Using similar reasoning, we can compute the required quantities for steps 3 and 4.

Step 3

$$\Delta T = \frac{\Delta PV}{nR} = \frac{(-1.00 \text{ atm})(10.0 \text{ L})}{nR} = \frac{-10.0 \text{ L atm}}{nR}$$

$$= \frac{-1.01 \times 10^3 \text{ J}}{nR}$$

$$w_3 = 0 \qquad (\Delta V = 0)$$

$$q_3 = q_v = nC_v\Delta T = n\left(\frac{3}{2}R\right)\left(\frac{-1.01 \times 10^3 \text{ J}}{nR}\right)$$

$$= -1.52 \times 10^3 \text{ J}$$

$$\Delta E_3 = q_v = -1.52 \times 10^3 \text{ J}$$

$$\Delta H_3 = nC_p\Delta T = n\left(\frac{5}{2}R\right)\left(\frac{-1.01 \times 10^3 \text{ J}}{nR}\right) = -2.53 \times 10^3 \text{ J}$$

Step 4

$$\Delta T = \frac{P\Delta V}{nR} = \frac{(1.00 \text{ atm})(20.0 \text{ L})}{nR} = \frac{20.0 \text{ L atm}}{nR}$$

$$= \frac{2.03 \times 10^3 \text{ J}}{nR}$$

$$w_4 = -P\Delta V = -(1.00 \text{ atm})(20.0 \text{ L}) = -20.0 \text{ L atm}$$

$$= -2.03 \times 10^3 \text{ J}$$

$$q_4 = q_p = nC_p\Delta T = n\left(\frac{5}{2}R\right)\left(\frac{2.03 \times 10^3 \text{ J}}{nR}\right)$$

$$= 5.08 \times 10^3 \text{ J}$$

$$\Delta E_4 = nC_vT = n\left(\frac{3}{2}R\right)\left(\frac{2.03 \times 10^3 \text{ J}}{nR}\right)$$

$$= 3.05 \times 10^3 \text{ J}$$

$$\Delta H_4 = nC_p\Delta T = n\left(\frac{5}{2}R\right)\left(\frac{2.03 \times 10^3 \text{ J}}{nR}\right)$$

$$= 5.08 \times 10^3 \text{ J} = q_p$$

Summary

■ Pathway one (steps 1 and 2):

$$q_{one} = q_1 + q_2 = 1.01 \times 10^4\ J - 4.56 \times 10^3\ J$$
$$= 5.5 \times 10^3\ J$$
$$w_{one} = w_1 + w_2 = -4.05 \times 10^3\ J$$
$$q_{one} + w_{one} = 1.5 \times 10^3\ J = \Delta E_{one}$$
$$\Delta H_{one} = \Delta H_1 + \Delta H_2$$
$$= 1.01 \times 10^4\ J - 7.60 \times 10^3\ J$$
$$= 2.5 \times 10^3\ J$$

■ Pathway two (steps 3 and 4):

$$q_{two} = q_3 + q_4 = -1.52 \times 10^3\ J + 5.08 \times 10^3\ J$$
$$= 3.56 \times 10^3\ J$$
$$w_{two} = w_3 + w_4 = -2.03 \times 10^3\ J$$
$$q_{two} + w_{two} = 3.55 \times 10^3\ J - 2.03 \times 10^3\ J$$
$$= 1.52 \times 10^3\ J = \Delta E_{two}$$
$$\Delta H_{two} = \Delta H_3 + \Delta H_4$$
$$= -2.53 \times 10^3\ J + 5.08 \times 10^3\ J$$
$$= 2.55 \times 10^3\ J$$

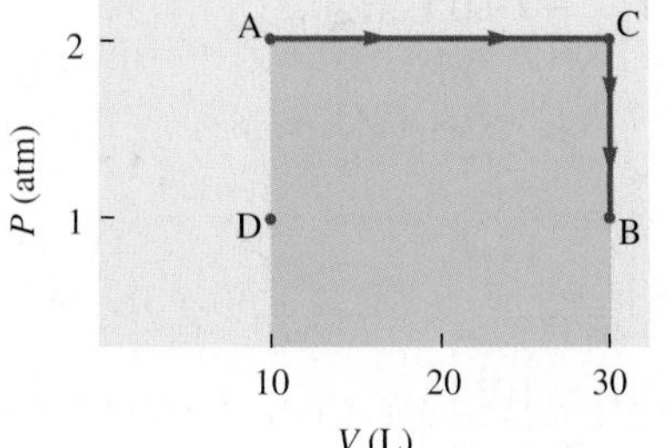

(a)

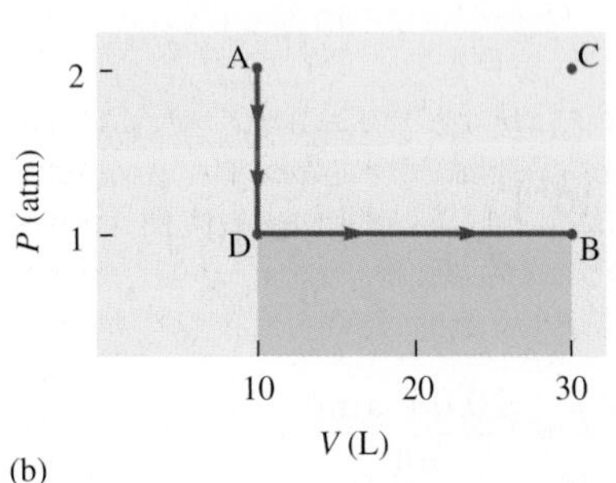

(b)

Figure 9.6
The magnitude of the work for pathway one (a) and pathway two (b) is shown by the colored areas: $|w| = |P\Delta V|$.

Notice from the results of Example 9.2 that the work and heat are different for the two pathways between states A and B. For example, in pathway one the expansion (by 20.0 L) is carried out at a pressure of 2.00 atm, and in pathway two the expansion (by 20.0 L) is carried out at 1.00 atm. Thus, since $|w| = |P\Delta V|$, twice as much work is obtained by way of pathway one as via pathway two. This result is shown graphically in Fig. 9.6, where the darker areas represent $|P\Delta V|$. These results again emphasize that heat and work are both pathway-dependent. On the other hand, note that the sum of q and w is the same for both pathways (within round-off differences). This is expected. Recall that

$$\Delta E = q + w$$

and that E is a state function. Note also that the overall ΔH value for pathway one equals that for pathway two (within rounding errors), as expected, because enthalpy is a state function as well.

9.4 Calorimetry

We can determine the heat associated with a chemical reaction experimentally by using a device called a **calorimeter. Calorimetry,** the science of measuring heat, is based on observing the temperature change when a body absorbs or discharges energy as heat. Substances respond differently to being heated. We have already discussed the response of ideal gases to heating. Now we expand that discussion to include other substances. In general terms, the **heat capacity** (C) of a substance is defined as

$$C = \frac{\text{heat absorbed}}{\text{increase in temperature}}$$

Table 9.3

The Specific Heat Capacities of Some Common Substances

Substance	Specific Heat Capacity (J °C^{-1} g^{-1})
$H_2O(l)$	4.18
$H_2O(s)$	2.03
Al(s)	0.89
Fe(s)	0.45
Hg(l)	0.14
C(s) (graphite)	0.71

Specific heat capacity: The energy required to raise the temperature of 1 g of a substance by 1°C.

Molar heat capacity: The energy required to raise the temperature of 1 mole of a substance by 1°C.

When an element or a compound is heated, the energy required to reach a certain temperature will depend on the amount of the substance present (for example, it takes twice as much energy to raise the temperature of 2 g of water by 1°C as it takes to raise the temperature of 1 g of water by 1°C). Thus, in defining the heat capacity of a substance, the amount of substance must be specified. If the heat capacity is given *per gram* of substance, it is called the **specific heat capacity** with units of J K^{-1} g^{-1} or J °C^{-1} g^{-1}. If the heat capacity is given *per mole* of the substance, it is called the *molar heat capacity,* which has the units J K^{-1} mol^{-1} or J °C^{-1} mol^{-1}. The specific heat capacities of some common substances are given in Table 9.3.

Although the calorimeters used for highly accurate work are precision instruments, a very simple calorimeter can be used to examine the fundamentals of calorimetry. All we need are two nested Styrofoam cups with a Styrofoam cover through which a stirrer and thermometer can be inserted, as shown in Fig. 9.7. This device is called a *coffee cup calorimeter*. The outer cup is used to provide extra insulation. The inner cup holds the solution in which the reaction occurs.

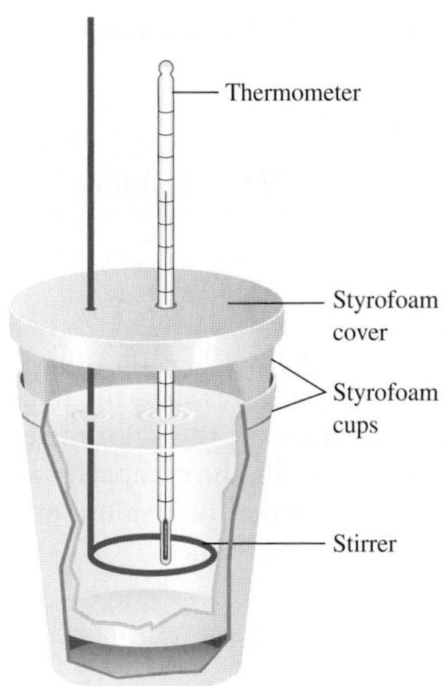

Figure 9.7
A coffee cup calorimeter made of two Styrofoam cups.

The measurement of heat using a simple calorimeter such as that shown in Fig. 9.7 is an example of **constant-pressure calorimetry,** since the pressure (atmospheric pressure) remains constant during the process. Constant-pressure calorimetry is used in determining the changes in enthalpy occurring in solution. Recall that under these conditions the change in enthalpy equals the heat.

For example, suppose we mix 50.0 mL of 1.0 *M* HCl at 25.0°C with 50.0 mL of 1.0 *M* NaOH also at 25°C in a calorimeter. After the reactants are mixed, the temperature is observed to increase to 31.9°C. As we saw in Chapter 4, the net ionic equation for this reaction is

$$H^+(aq) + OH^-(aq) \longrightarrow H_2O(l)$$

If two reactants at the same temperature are mixed and the resulting solution gets warmer, it means the reaction taking place is exothermic. An endothermic reaction cools the solution.

When these reactants (both originally at the same temperature) are mixed, the temperature of the mixed solution is observed to increase. Thus the chemical reaction must be releasing energy as heat. This increases the random motions of the solution components, which in turn increases the temperature. The quantity of energy released can be determined from the temperature increase, the mass of the solution, and the specific heat capacity of the solution. For an approximate result, we will assume that the calorimeter does not absorb or leak any heat and that the solution can be treated as if it were pure water with a density of 1.0 g/mL.

We also need to know the heat required to raise the temperature of a given amount of water by 1°C. Table 9.3 lists the specific heat capacity of water as 4.18 J °C^{-1} g^{-1}. This means that 4.18 J of energy is required to raise the temperature of 1 g of water by 1°C.

Note that the typical units for molar heat capacity ($\Delta E/\Delta T$ per mole) are

$$\frac{\text{J}}{\text{K mol}}$$

Because 1 K equals 1°C, the units

$$\frac{\text{J}}{\text{°C mol}}$$

are used as well.

From these assumptions and definitions, we can calculate the heat (change in enthalpy) for the neutralization reaction:

$$\begin{aligned}\text{Energy released by the reaction} &= \text{energy absorbed by the solution}\\ &= \text{specific heat capacity} \times \text{mass of solution}\\ &\quad \times \text{increase in temperature}\end{aligned}$$

where the increase in temperature = 31.9°C − 25.0°C = 6.9°C, and where

$$\text{Mass of solution} = 100.0 \text{ mL} \times 1.0 \text{ g/mL} = 1.0 \times 10^2 \text{ g}$$

Thus

$$\text{Energy released} = \left(4.18 \frac{\text{J}}{\text{°C g}}\right)(1.0 \times 10^2 \text{ g})(6.9\text{°C}) = 2.9 \times 10^3 \text{ J}$$

Enthalpies of reaction are often expressed in terms of moles of reacting substances. The number of moles of H^+ ions consumed in the preceding experiment is

$$50.0 \text{ mL} \times \frac{1 \text{ L}}{1000 \text{ mL}} \times \frac{1.0 \text{ mol}}{\text{L}} \text{H}^+ = 5.0 \times 10^{-2} \text{ mol H}^+$$

Thus 2.9×10^3 J of heat was released when 5.0×10^{-2} moles of H^+ ions reacted. Thus

$$\frac{2.9 \times 10^3 \text{ J}}{5.0 \times 10^{-2} \text{ mol H}^+} = 5.8 \times 10^4 \text{ J}$$

Notice that in this example we mentally keep track of the direction of the energy flow and assign the correct sign at the end of the calculation.

is the heat released per 1.0 mole of H^+ ions neutralized. The *magnitude* of the enthalpy of reaction per mole for

$$\text{H}^+(aq) + \text{OH}^-(aq) \longrightarrow \text{H}_2\text{O}(l)$$

at constant pressure is 58 kJ/mol. Since heat is *evolved*, $\Delta H = -58$ kJ/mol.

OWL INTERACTIVE EXAMPLE 9.3

When 1.00 L of 1.00 *M* $Ba(NO_3)_2$ at 25.0°C is mixed with 1.00 L of 1.00 *M* Na_2SO_4 at 25°C in a calorimeter, the white solid $BaSO_4$ forms, and the temperature of the mixture increases to 28.1°C. Assuming that the calorimeter absorbs only a negligible quantity of heat, that the specific heat capacity of the solution is 4.18 J °C^{-1} g^{-1}, and that the density of the final solution is 1.0 g/mL, calculate the enthalpy change per mole of $BaSO_4$ formed.

Solution

■ What are we trying to solve?

In this problem we are asked to solve for the enthalpy change per mole of $BaSO_4$ formed. To solve this, we need to first determine the balanced equation for the reaction that occurs.

The ions present before any reaction occurs are Ba^{2+}, NO_3^-, Na^+, and SO_4^{2-}. The Na^+ and NO_3^- ions are spectator ions, since $NaNO_3$ is very soluble in water and will not precipitate under these conditions. The net ionic equation of the reaction is therefore

$$\text{Ba}^{2+}(aq) + \text{SO}_4^{2-}(aq) \longrightarrow \text{BaSO}_4(s)$$

From the balanced chemical equation, we can see there is a 1:1:1 mole ratio for Ba^{2+}, SO_4^{2-}, and $BaSO_4$. Since 1.00 L of 1.00 *M* $Ba(NO_3)_2$ contains 1.00 mole of Ba^{2+} ions, and 1.00 L of 1.00 *M* Na_2SO_4 contains 1.00 mole of SO_4^{2-} ions, 1.00 mole of solid $BaSO_4$ is formed in this experiment. Thus the enthalpy of

Solutions of $Ba(NO_3)_2$ and Na_2SO_4 both initially at 25.0°C.

When barium nitrate and sodium sulfate are mixed in an insulated container, they form a white solid $BaSO_4$ (barium sulfate), and the temperature rises to 28.1°C.

this reaction is the enthalpy per mole of $BaSO_4$ formed. This leads to the next question:

- How do we determine the enthalpy for the reaction?

Since the temperature increases, the formation of solid $BaSO_4$ must be exothermic; ΔH will be negative.

$$\text{Heat evolved by reaction} = \text{heat absorbed by solution}$$
$$= \text{specific heat capacity} \times \text{mass of solution} \times \text{increase in temperature}$$

Since 1.00 L of each solution is used, the total solution volume is 2.00 L, and

$$\text{Mass of solution} = 2.00 \text{ L} \times \frac{1000 \text{ mL}}{1 \text{ L}} \times \frac{1.0 \text{ g}}{\text{mL}}$$
$$= 2.0 \times 10^3 \text{ g}$$
$$\text{Temperature increase} = 28.1°\text{C} - 25.0°\text{C} = 3.1°\text{C}$$
$$\text{Heat evolved} = (4.18 \text{ J °C}^{-1} \text{ g}^{-1})(2.0 \times 10^3 \text{ g})(3.1°\text{C})$$
$$= 2.6 \times 10^4 \text{ J}$$

Thus
$$q = q_\text{p} = \Delta H = -2.6 \times 10^4 \text{ J}$$

Thus the enthalpy change per mole of $BaSO_4$ formed is

$$\Delta H = -2.6 \times 10^4 \text{ J/mol} = -26 \text{ kJ/mol}$$

Calculation of ΔH and ΔE for Cases in Which *PV* Work Occurs

In the examples of constant-pressure calorimetry we have considered so far, the reactions have occurred in solution, where no appreciable volume changes occur (that is, the total volume of the reactant solution is the sum of the volumes of the solutions that are mixed and remains constant as the reaction proceeds). Under these conditions no work occurs (since $\Delta V = 0$, $P\Delta V = 0$, and $w = 0$). Thus, since $\Delta H = q_\text{p}$ (constant pressure) and $w = 0$,

$$\Delta E = q_\text{p} + w = \Delta H + 0$$

At constant pressure, where $\Delta V = 0$, no work is done and $\Delta E = \Delta H = q_\text{p}$.

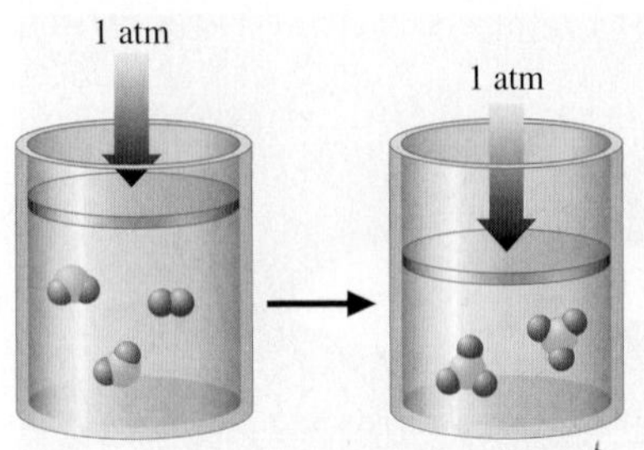

Figure 9.8
A schematic to show the change in volume for the reaction

$$2SO_2(g) + O_2(g) \longrightarrow 2SO_3(g)$$

However, when a reaction involving gases is studied at constant pressure, ΔE may not equal ΔH. The reaction

$$2SO_2(g) + O_2(g) \longrightarrow 2SO_3(g)$$

provides an example of this case. Picture this reaction being carried out at constant temperature and pressure, as shown in Fig. 9.8.

In going from reactants to products, the volume of the system decreases because the number of moles of gas decreases. Since ΔV $(= V_{\text{final}} - V_{\text{initial}})$ is negative, w is positive:

$$w = -\underbrace{P\Delta V}_{\substack{\uparrow \\ \text{Negative in this case}}}$$

Thus work flows into the system. In addition, because

$$\Delta E = q + w$$

and at constant pressure

$$\Delta H = q_{\text{P}}$$

then

$$\Delta E = q_{\text{P}} + w = \Delta H + w$$

In this case $w \neq 0$, so ΔE and ΔH are different. This case is illustrated in Example 9.4.

OWL INTERACTIVE EXAMPLE 9.4

When 2.00 moles of $SO_2(g)$ reacts completely with 1.00 mole of $O_2(g)$ to form 2.00 moles of $SO_3(g)$ at 25°C and a constant pressure of 1.00 atm, 198 kJ of energy is released as heat. Calculate ΔH and ΔE for this process.

Solution Because the pressure is constant for this process, $\Delta H = q_{\text{P}}$. The description of the experiment states that 198 kJ of heat is *released*. Thus $\Delta H = q_{\text{P}} = -198$ kJ, where the negative sign indicates that energy flows *out of* the system.

The value of ΔE can be calculated from the relationship

$$\Delta E = q + w$$

Since q is known (-198 kJ), we need only the value for w. We know that

$$w = -P\Delta V$$

Solving the ideal gas law for ΔV gives

$$\Delta V = \Delta n\left(\frac{RT}{P}\right)$$

where only n changes (T and P are constant) as the reaction occurs.

In this case

$$\Delta n = n_{\text{final}} - n_{\text{initial}}$$

$$n_{\text{final}} = \underset{\substack{\uparrow \\ \text{Moles of SO}_3}}{2\text{ mol}}$$

$$n_{\text{initial}} = \underset{\substack{\uparrow \\ \text{Moles of SO}_2}}{2\text{ mol}} + \underset{\substack{\uparrow \\ \text{Moles of O}_2}}{1\text{ mol}}$$

So

$$\Delta n = 2\text{ mol} - 3\text{ mol} = -1\text{ mol}$$

Now we can calculate w:

$$w = -P\Delta V = -P\underbrace{\left(\Delta n \times \frac{RT}{P}\right)}_{\Delta V} = -\Delta nRT$$

where $\Delta n = -1 \text{ mol}$

$R = 8.3145 \text{ J K}^{-1} \text{ mol}^{-1}$

$T = 25°\text{C} + 273 = 298 \text{ K}$

Thus $$w = -(-1 \text{ mol})\left(8.3145 \frac{\text{J}}{\text{K mol}}\right)(298 \text{ K}) = 2.48 \text{ kJ}$$

Using the values of q and w, we can calculate ΔE:

$$\Delta E = q + w = \Delta H + w = -198 \text{ kJ} + 2.48 \text{ kJ} = -196 \text{ kJ}$$

Note that ΔE and ΔH are different for this case because the volume changes (and therefore work occurs).

Calorimetry experiments can also be performed at **constant volume.** For example, when a photographic flashbulb flashes, the bulb becomes very hot, since the reaction of the zirconium or magnesium wire with the oxygen inside the bulb is exothermic. The reaction occurs inside the flashbulb, which is rigid (does not change volume). Under these conditions no work is done, since the volume must change for pressure-volume work to be performed. To study the energy changes in reactions under conditions of constant volume, a **bomb calorimeter** (Fig. 9.9) is often used. Weighed reactants are placed inside a rigid steel container (the "bomb") and ignited. The energy change is determined by measuring the increase in the temperature of the water and other calorimeter parts. For a constant-volume process, the change in volume (ΔV) is equal to zero, so the work is also equal to zero. Therefore,

$$\Delta E = q + w = q = q_v \qquad \text{(constant volume)}$$

Suppose we wish to measure the energy of combustion of octane (C_8H_{18}), a component of gasoline. A 0.5269-g sample of octane is placed in a bomb

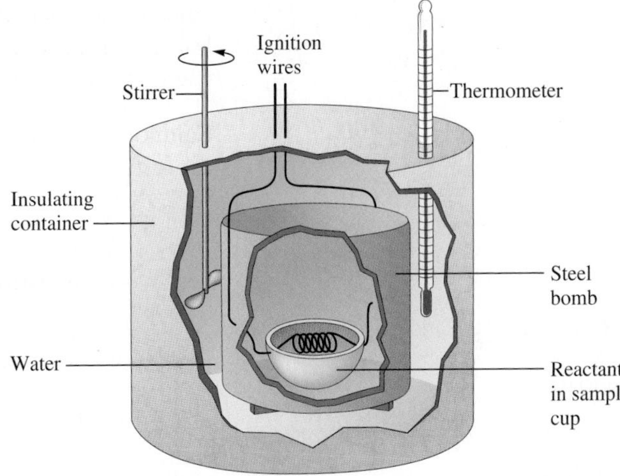

Figure 9.9
(left) A commercial "bomb." (right) Schematic of a bomb calorimeter. The reaction is carried out inside a rigid steel "bomb," and the heat evolved is absorbed by the surrounding water and the other calorimeter parts. The quantity of energy produced by the reaction can be calculated from the temperature increase.

calorimeter known to have a heat capacity of 11.3 kJ/°C. This means that 11.3 kJ of energy is required to raise the temperature of the water and other parts of the calorimeter by 1°C. The octane is ignited in the presence of excess oxygen, causing the temperature of the calorimeter to increase by 2.25°C. The amount of energy released by the reaction is calculated as follows:

$$\begin{aligned}\text{Energy released by the reaction} &= \text{temperature increase} \times \text{energy required to change the temperature by } 1^\circ\text{C} \\ &= \Delta T \times \text{heat capacity of calorimeter} \\ &= 2.25^\circ\text{C} \times 11.3 \text{ kJ/}^\circ\text{C} = 25.4 \text{ kJ}\end{aligned}$$

This means that 25.4 kJ of energy was released by the combustion of 0.5269 g of octane. The number of moles of octane is

$$0.5269 \text{ g octane} \times \frac{1 \text{ mol octane}}{114.2 \text{ g octane}} = 4.614 \times 10^{-3} \text{ mol octane}$$

Since 25.4 kJ of energy was released for 4.614×10^{-3} mole of octane, the energy released per mole is

$$\frac{25.4 \text{ kJ}}{4.614 \times 10^{-3} \text{ mol}} = 5.50 \times 10^3 \text{ kJ/mol}$$

Since the reaction is exothermic, ΔE is negative:

$$\Delta E_{\text{combustion}} = -5.50 \times 10^3 \text{ kJ/mol}$$

Note that since no work is done in this case, ΔE is equal to the heat:

$$\begin{aligned}\Delta E &= q + w = q \\ &= -5.50 \times 10^3 \text{ kJ/mol}\end{aligned}$$

OWL INTERACTIVE EXAMPLE 9.5

Hydrogen's potential as a fuel is discussed in Section 9.8.

It has been suggested that hydrogen gas obtained from the decomposition of water might be a substitute for natural gas (principally methane). To compare the energies of combustion of these fuels, the following experiment was carried out using a bomb calorimeter with a heat capacity of 11.3 kJ/°C. When a 1.50-g sample of methane gas was burned with excess oxygen in the calorimeter, the temperature increased by 7.3°C. When a 1.15-g sample of hydrogen gas was burned with excess oxygen, the temperature increase was 14.3°C. Calculate the energy of combustion (per gram) for hydrogen and methane.

Solution We calculate the energy of combustion for methane using the heat capacity of the calorimeter (11.3 kJ/°C) and the observed temperature increase of 7.3°C:

$$\text{Energy } \textit{released} \text{ in the combustion of 1.50 g of } CH_4 = (11.3 \text{ kJ/}^\circ\text{C})(7.3^\circ\text{C}) = 83 \text{ kJ}$$

$$\text{Energy released in the combustion of 1 g of } CH_4 = \frac{83 \text{ kJ}}{1.50 \text{ g}} = 55 \text{ kJ/g}$$

The direction of the energy flow is indicated by words in this example. Using signs, we have

$$\Delta E_{\text{combustion}} = -55 \text{ kJ/g}$$

for methane and

$$\Delta E_{\text{combustion}} = -141 \text{ kJ/g}$$

for hydrogen.

Similarly, for hydrogen,

$$\begin{aligned}\text{Energy } \textit{released} \text{ in the combustion of 1.15 g of } H_2 &= (11.3 \text{ kJ/}^\circ\text{C})(14.3^\circ\text{C}) \\ &= 162 \text{ kJ}\end{aligned}$$

$$\text{Energy released in the combustion of 1 g of } H_2 = \frac{162 \text{ kJ}}{1.15 \text{ g}} = 141 \text{ kJ/g}$$

The energy released by the combustion of 1 g of hydrogen is approximately 2.5 times that for 1 g of methane, indicating that hydrogen gas is a potentially useful fuel.

9.5 Hess's Law

ΔH is not dependent on the reaction pathway.

Since enthalpy is a state function, the change in enthalpy in going from some initial state to some final state is independent of the pathway. This means that *in going from a particular set of reactants to a particular set of products, the change in enthalpy is the same whether the reaction takes place in one step or in a series of steps.* This principle is known as **Hess's law** and can be illustrated by examining the oxidation of nitrogen to produce nitrogen dioxide. The overall reaction can be written in one step, where the enthalpy change is represented by ΔH_1:

$$N_2(g) + 2O_2(g) \longrightarrow 2NO_2(g) \qquad \Delta H_1 = 68 \text{ kJ}$$

This reaction can also be carried out in two distinct steps, with enthalpy changes designated by ΔH_2 and ΔH_3:

$$N_2(g) + O_2(g) \longrightarrow 2NO(g) \qquad \Delta H_2 = 180 \text{ kJ}$$

$$2NO(g) + O_2(g) \longrightarrow 2NO_2(g) \qquad \Delta H_3 = -112 \text{ kJ}$$

Net reaction: $$N_2(g) + 2O_2(g) \longrightarrow 2NO_2(g) \qquad \Delta H_2 + \Delta H_3 = 68 \text{ kJ}$$

Note that the sum of these two steps gives the net, or overall, reaction and that

$$\Delta H_1 = \Delta H_2 + \Delta H_3 = 68 \text{ kJ}$$

The principle of Hess's law is shown schematically in Fig. 9.10.

Characteristics of Enthalpy Changes

To use Hess's law to compute enthalpy changes for reactions, we must understand two characteristics of ΔH for a reaction:

Reversing a reaction changes the sign of ΔH.

1. If a reaction is reversed, the sign of ΔH is also reversed.
2. The magnitude of ΔH is directly proportional to the quantities of reactants and products in a reaction. If the coefficients in a balanced reaction are multiplied by an integer, the value of ΔH is multiplied by the same integer.

Both of these rules follow in a straightforward way from the properties of enthalpy changes. The first rule can be explained by recalling that the *sign* of

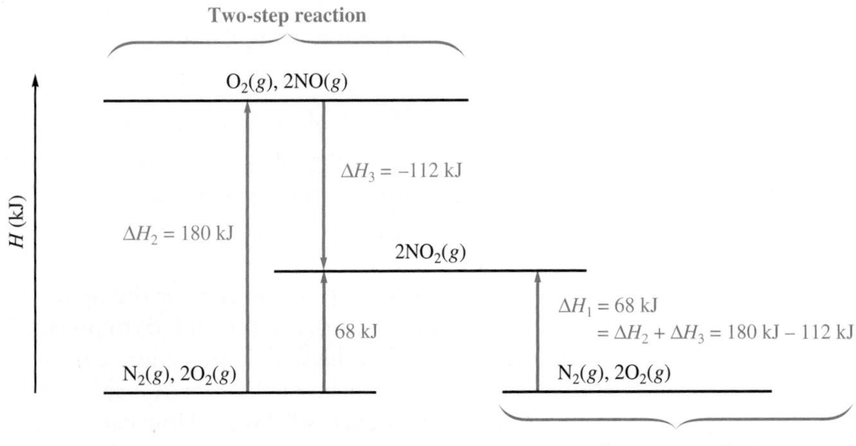

Figure 9.10
The principle of Hess's law. The same change in enthalpy occurs when nitrogen and oxygen react to form nitrogen dioxide, regardless of whether the reaction occurs in one (red) or two (blue) steps.

Crystals of xenon tetrafluoride, the first reported binary compound containing a noble gas element.

ΔH indicates the *direction* of the heat flow at constant pressure. If the direction of the reaction is reversed, the direction of the heat flow is also reversed. To see this, consider the preparation of xenon tetrafluoride, which was the first reported binary compound made from a noble gas:

$$Xe(g) + 2F_2(g) \longrightarrow XeF_4(s) \qquad \Delta H = -251 \text{ kJ}$$

This reaction is exothermic, and 251 kJ of energy flows into the surroundings as heat. On the other hand, if the colorless XeF_4 crystals are decomposed into the elements according to the equation

$$XeF_4(s) \longrightarrow Xe(g) + 2F_2(g)$$

the opposite energy flow occurs because 251 kJ of energy has to be added to the system in this case. Thus, for this reaction, $\Delta H = +251$ kJ.

The second rule comes from the fact that ΔH is an extensive property depending on the amount of substances reacting. For example, 251 kJ of energy is evolved for the reaction

$$Xe(g) + 2F_2(g) \longrightarrow XeF_4(s)$$

Thus, for a preparation involving twice the quantities of reactants and products,

$$2Xe(g) + 4F_2(g) \longrightarrow 2XeF_4(s)$$

twice as much heat would be evolved:

$$\Delta H = 2(-251 \text{ kJ}) = -502 \text{ kJ}$$

Hints for Using Hess's Law

Calculations involving Hess's law typically require that several reactions be manipulated and combined to finally give the reaction of interest. In doing this procedure, you should work *backward* from the required reaction, using the reactants and products to decide how to manipulate the other reactions at your disposal. Reverse any reactions as needed to give the required reactants and products, and then multiply the reactions to give the correct numbers of reactants and products. This process involves some trial and error but can be very systematic if you always allow the final reaction to guide you.

OWL INTERACTIVE EXAMPLE 9.6

Diborane (B_2H_6) is a highly reactive boron hydride that was once considered as a possible rocket fuel for the U.S. space program. Calculate ΔH for the synthesis of diborane from its elements, according to the equation

$$2B(s) + 3H_2(g) \longrightarrow B_2H_6(g)$$

using the following data:

Reaction	ΔH
(a) $2B(s) + \frac{3}{2}O_2(g) \rightarrow B_2O_3(s)$	−1273 kJ
(b) $B_2H_6(g) + 3O_2(g) \rightarrow B_2O_3(s) + 3H_2O(g)$	−2035 kJ
(c) $H_2(g) + \frac{1}{2}O_2(g) \rightarrow H_2O(l)$	−286 kJ
(d) $H_2O(l) \rightarrow H_2O(g)$	44 kJ

Solution To obtain ΔH for the required reaction, we must somehow combine equations (a), (b), (c), and (d) to produce that reaction and add the corresponding ΔH values. This procedure can best be done by focusing on the reactants and products of the required reaction. The reactants are $B(s)$ and $H_2(g)$, and the product is $B_2H_6(g)$. How can we obtain the correct equation? Reaction (a) has $B(s)$ as a reactant, as needed in the required equation. Thus reaction (a) will

Firewalking: Magic or Science?

Chemical Insights

For millennia people have been amazed at the ability of Eastern mystics to walk across beds of glowing coals without any apparent discomfort. Even in the United States thousands of people have performed feats of firewalking as part of motivational seminars. How is this feat possible? Do firewalkers have supernatural powers?

Actually, there are sound scientific explanations of why firewalking is possible. The first important factor concerns the heat capacity of feet. Because human tissue is mainly composed of water, it has a relatively large specific heat capacity. This means that a large amount of energy must be transferred from the coals to significantly change the temperature of the feet. During the brief contact between feet and coals, there is relatively little time for energy flow, so the feet do not reach a high enough temperature to cause damage.

Second, although the surface of the coals has a very high temperature, the red-hot layer is very thin. Therefore, the quantity of energy available to heat the feet is smaller than might be expected. This factor points out the difference between temperature and heat. Temperature reflects the *intensity* of the random kinetic energy in a given sample of matter. The amount of energy available for heat flow, on the other hand, depends on the quantity of matter at a given temperature—10 g of matter at a given temperature contains ten times as much thermal energy as 1 g of the same matter. For example, the tiny spark from a sparkler does not hurt when it hits your hand. The spark has a very high temperature but has so little mass that no significant energy transfer occurs to your hand. This same argument applies to the very thin hot layer on the coals.

Itsuo Inouye/AP/Wide World Photos

A group of firewalkers in Japan.

A third factor that aids firewalkers is the presence of moisture from the perspiration on the feet of the presumably tense firewalker. In addition, because firewalking is often done at night, with moist grass surrounding the bed of coals, the firewalker's feet are probably damp before the walk. Vaporization of this moisture consumes some of the energy from the hot coals.

Thus, although firewalking is an impressive feat, there are several sound scientific reasons why anyone should be able to do it with the proper training and a properly prepared bed of coals.

be used as it is. Reaction (b) has $B_2H_6(g)$ as a reactant, but this substance is needed as a product. Thus reaction (b) must be reversed, and the sign of ΔH changed accordingly. Up to this point we have

$$\text{(a)} \quad 2B(s) + \tfrac{3}{2}O_2(g) \longrightarrow B_2O_3(s) \qquad \Delta H = -1273\ \text{kJ}$$

$$-\text{(b)} \quad B_2O_3(s) + 3H_2O(g) \longrightarrow B_2H_6(g) + 3O_2(g) \qquad \Delta H = -(-2035\ \text{kJ})$$

$$\text{Sum:} \quad B_2O_3(s) + 2B(s) + \tfrac{3}{2}O_2(g) + 3H_2O(g) \longrightarrow B_2O_3(s) + B_2H_6(g) + 3O_2(g) \qquad \Delta H = 762\ \text{kJ}$$

Deleting the species that occur on both sides gives

$$2B(s) + 3H_2O(g) \longrightarrow B_2H_6(g) + \tfrac{3}{2}O_2(g) \qquad \Delta H = 762\ \text{kJ}$$

We are closer to the required reaction, but we still need to remove $H_2O(g)$ and $O_2(g)$ and introduce $H_2(g)$ as a reactant. We can do so by using reactions

(c) and (d). If we multiply reaction (c) and its ΔH value by 3 and add the result to the preceding equation, we have

$$\begin{array}{llll} & 2B(s) + 3H_2O(g) \longrightarrow B_2H_6(g) + \frac{3}{2}O_2(g) & & \Delta H = 762 \text{ kJ} \\ 3 \times (c) & 3[H_2(g) + \frac{1}{2}O_2(g) \longrightarrow H_2O(l)] & & \Delta H = 3(-286 \text{ kJ}) \\ \hline \text{Sum:} & 2B(s) + 3H_2(g) + \frac{3}{2}O_2(g) + 3H_2O(g) \longrightarrow B_2H_6(g) + \frac{3}{2}O_2(g) + 3H_2O(l) & & \Delta H = -96 \text{ kJ} \end{array}$$

We can cancel the $\frac{3}{2}O_2(g)$ on both sides, but we cannot cancel the H_2O because it is gaseous on one side and liquid on the other. This problem can be solved by adding reaction (d), multiplied by 3:

$$\begin{array}{lll} & 2B(s) + 3H_2(g) + 3H_2O(g) \longrightarrow B_2H_6(g) + 3H_2O(l) & \Delta H = -96 \text{ kJ} \\ 3 \times (d) & 3[H_2O(l) \longrightarrow H_2O(g)] & \Delta H = 3(44 \text{ kJ}) \\ \hline & 2B(s) + 3H_2(g) + 3H_2O(g) + 3H_2O(l) \longrightarrow B_2H_6(g) + 3H_2O(l) + 3H_2O(g) & \Delta H = +36 \text{ kJ} \end{array}$$

This step gives the reaction required by the problem:

$$2B(s) + 3H_2(g) \longrightarrow B_2H_6(g) \qquad \Delta H = +36 \text{ kJ}$$

Thus ΔH for the synthesis of 1 mole of diborane is +36 kJ.

9.6 Standard Enthalpies of Formation

For a reaction studied under conditions of constant pressure, we can obtain the enthalpy change by using a calorimeter. However, this process can be very difficult. In fact, in some cases it is impossible, since certain reactions do not lend themselves to such study. An example is the conversion of solid carbon from its graphite form to its diamond form:

$$C_{\text{graphite}}(s) \longrightarrow C_{\text{diamond}}(s)$$

The value of ΔH for this process cannot be obtained readily by measurement in a calorimeter. We will show next how to *calculate* ΔH for chemical reactions and physical changes by using standard enthalpies of formation.

The **standard enthalpy of formation** (ΔH_f°) of a compound is defined as the *change in enthalpy that accompanies the formation of 1 mole of a compound from its elements with all substances in their standard states.*

The *superscript zero* on a thermodynamic function (for example, ΔH°) indicates that the corresponding process has been carried out under standard conditions. The **standard state** for a substance is a precisely defined reference state. Because thermodynamic functions often depend on the concentrations (or pressures) of the substances involved, we must use a common reference state to properly compare the thermodynamic properties of two substances. This is especially important because for most thermodynamic properties we can measure only *changes* in that property. For example, we have no method for determining absolute values of enthalpy. We can measure only enthalpy changes (ΔH values) by performing heat flow experiments.

Standard states are defined as follows:

Definitions of Standard States

1. For a gas the standard state is a pressure of exactly 1 atm.
2. For a substance present in a solution, the standard state is a concentration of exactly 1 *M* at an applied pressure of 1 atm.
3. For a pure substance in a condensed state (liquid or solid), the standard state is the pure liquid or solid.
4. For an element the standard state is the form in which the element exists (is most stable) under conditions of 1 atm and the temperature of interest (usually 25°C).

Standard state is *not* the same as standard temperature and pressure (STP) for a gas.

The standard state for oxygen is $O_2(g)$ at a pressure of 1 atm; the standard state for sodium is Na(*s*); the standard state for mercury is Hg(*l*); and so on.

The International Union of Pure and Applied Chemistry (IUPAC) has adopted 1 bar (100,000. Pa) as the standard pressure instead of 1 atm (101,325 Pa). Both standards are now widely used.

Several important characteristics of the definition of enthalpy of formation will become clearer if we again consider the formation of nitrogen dioxide from the elements in their standard states:

$$\tfrac{1}{2}N_2(g) + O_2(g) \longrightarrow NO_2(g) \qquad \Delta H_f^\circ = 34 \text{ kJ/mol}$$

Note that the reaction is written so that both elements are in their standard states, and 1 mole of product is formed. Enthalpies of formation are *always* given per mole of product with the product in its standard state.

The formation reaction for methanol is written as

$$C(s) + 2H_2(g) + \tfrac{1}{2}O_2(g) \longrightarrow CH_3OH(l) \qquad \Delta H_f^\circ = -239 \text{ kJ/mol}$$

The standard state of carbon is graphite, the standard states for oxygen and hydrogen are the diatomic gases, and the standard state for methanol is the liquid.

The ΔH_f° values for some common substances are shown in Table 9.4. More values are found in Appendix 4. The importance of tabulated ΔH_f° values is that enthalpies for many reactions can be calculated using these numbers. To see how this is done, we will calculate the standard enthalpy change for the combustion of methane:

Table 9.4

Standard Enthalpies of Formation for Several Compounds at 25°C

Compound	ΔH_f° (kJ/mol)
$NH_3(g)$	−46
$NO_2(g)$	34
$H_2O(l)$	−286
$Al_2O_3(s)$	−1676
$Fe_2O_3(s)$	−826
$CO_2(g)$	−394
$CH_3OH(l)$	−239
$C_8H_{18}(l)$	−269

$$CH_4(g) + 2O_2(g) \longrightarrow CO_2(g) + 2H_2O(l)$$

Enthalpy is a state function, so we can invoke Hess's law and choose *any* convenient pathway from reactants to products and then sum the enthalpy changes. A convenient pathway, shown in Fig. 9.11, involves taking the reactants apart into the respective elements in their standard states in reactions (a) and (b) and then forming the products from these elements in reactions (c) and (d). This general pathway will work for any reaction, since atoms are conserved in a chemical reaction.

Note from Fig. 9.11 that reaction (a), where methane is taken apart into its elements,

$$CH_4(g) \longrightarrow C(s) + 2H_2(g)$$

is just the reverse of the formation reaction for methane:

$$C(s) + 2H_2(g) \longrightarrow CH_4(g) \qquad \Delta H_f^\circ = -75 \text{ kJ/mol}$$

Note that although the tabulated values of ΔH_f° usually correspond to a temperature of 25°C, values of ΔH_f° can be obtained at any temperature.

Since reversing a reaction means changing the sign of ΔH but keeping the same magnitude, ΔH for reaction (a) is $-\Delta H_f^\circ$, or 75 kJ. Thus $\Delta H^\circ_{(a)} = 75$ kJ.

Next, we consider reaction (b). Here oxygen is already an element in its standard state, so no change is necessary. Thus $\Delta H^\circ_{(b)} = 0$.

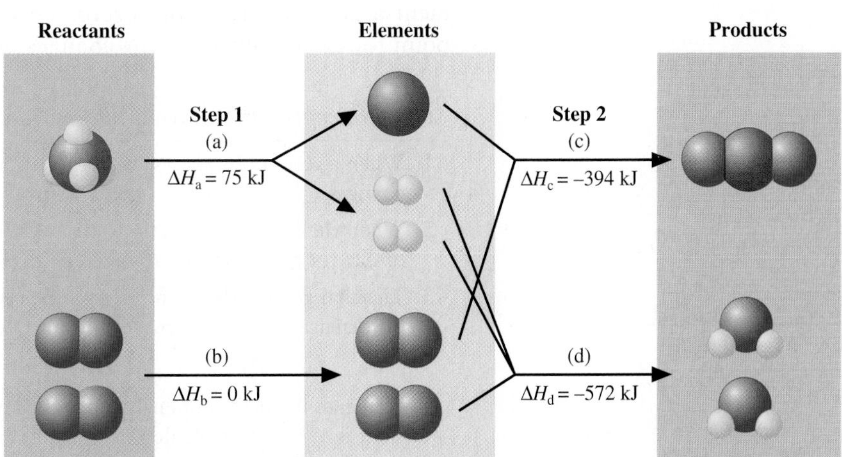

Figure 9.11
A schematic diagram of the energy changes for the reaction $CH_4(g) + 2O_2(g) \rightarrow CO_2(g) + 2H_2O(l)$.

The next steps, reactions (c) and (d), use the elements formed in reactions (a) and (b) to form the products. Note that reaction (c) is simply the formation reaction for carbon dioxide:

$$C(s) + O_2(g) \longrightarrow CO_2(g) \qquad \Delta H_f^\circ = -394 \text{ kJ/mol}$$

and

$$\Delta H^\circ_{(c)} = \Delta H_f^\circ \text{ [for } CO_2(g)] = -394 \text{ kJ}$$

Reaction (d) is the formation reaction for water:

$$H_2(g) + \tfrac{1}{2}O_2(g) \longrightarrow H_2O(l) \qquad \Delta H_f^\circ = -286 \text{ kJ/mol}$$

However, since 2 moles of water are required in the balanced equation, we must form 2 moles of water from the elements:

$$2H_2(g) + O_2(g) \longrightarrow 2H_2O(l)$$

Thus

$$\Delta H^\circ_{(d)} = 2 \times \Delta H_f^\circ \text{ [for } H_2O(l)] = 2(-286 \text{ kJ}) = -572 \text{ kJ}$$

We have now completed the pathway from the reactants to the products. The change in enthalpy for the overall reaction is the sum of the ΔH values (including their signs) for the steps:

$$\begin{aligned}\Delta H^\circ_{\text{reaction}} &= \Delta H^\circ_{(a)} + \Delta H^\circ_{(b)} + \Delta H^\circ_{(c)} + \Delta H^\circ_{(d)} \\ &= -\Delta H_f^\circ \text{ [for } CH_4(g)] + 0 + \Delta H_f^\circ \text{ [for } CO_2(g)] \\ &\quad + 2 \times \Delta H_f^\circ \text{ [for } H_2O(l)] \\ &= -(-75 \text{ kJ}) + 0 + (-394 \text{ kJ}) + (-572 \text{ kJ}) \\ &= -891 \text{ kJ}\end{aligned}$$

Let's examine carefully the pathway we used in this example. First, the reactants were broken down into the elements in their standard states. This step involved reversing the formation reactions and thus switching the signs of the respective enthalpies of formation. The products were then constructed from these elements. This step involved formation reactions and thus enthalpies of formation. We can summarize the entire process as follows: *The enthalpy change for a given reaction can be calculated by subtracting the enthalpies of formation of the reactants from the enthalpies of formation of the products.* Remember to multiply the enthalpies of formation by integers as required by the balanced equation. This procedure can be represented symbolically as follows:

$$\Delta H^\circ_{\text{reaction}} = \Sigma \Delta H_f^\circ \text{ (products)} - \Sigma \Delta H_f^\circ \text{ (reactants)} \qquad (9.1)$$

where the symbol Σ (sigma) means "to take the sum of the terms."

Elements are not included in the calculation since elements require no change in form. We have in effect *defined* the enthalpy of formation of an element in its standard state as zero, since we have chosen this as our reference point for calculating enthalpy changes in reactions.

Key Concepts for Doing Enthalpy Calculations

1. When a reaction is reversed, the magnitude of ΔH remains the same, but the sign changes.
2. When the balanced equation for a reaction is multiplied by an integer, the value of ΔH for that reaction must be multiplied by the same integer.
3. The change in enthalpy for a given reaction can be calculated from the enthalpies of formation of the reactants and products:

$$\Delta H^\circ_{\text{reaction}} = \Sigma \Delta H_f^\circ \text{ (products)} - \Sigma \Delta H_f^\circ \text{ (reactants)}$$

4. Elements in their standard states are not included in the $\Delta H_{\text{reaction}}$ calculations. That is, ΔH_f° for an element in its standard state is zero.

OWL INTERACTIVE EXAMPLE 9.7

Using the standard enthalpies of formation listed in Table 9.4, calculate the standard enthalpy change for the overall reaction that occurs when ammonia is burned in air to form nitrogen dioxide and water. This is the first step in the manufacture of nitric acid.

$$4NH_3(g) + 7O_2(g) \longrightarrow 4NO_2(g) + 6H_2O(l)$$

Solution We will use the pathway in which the reactants are broken down into elements in their standard states and then are used to form the products (Fig. 9.12).

Step 1

Decomposition of $NH_3(g)$ into elements [reaction (a) in Fig. 9.12]. The first step involves decomposing 4 moles of NH_3 into N_2 and H_2:

$$4NH_3(g) \longrightarrow 2N_2(g) + 6H_2(g)$$

This reaction is 4 times the *reverse* of the formation reaction for NH_3:

$$\tfrac{1}{2}N_2(g) + \tfrac{3}{2}H_2(g) \longrightarrow NH_3(g) \qquad \Delta H_f^\circ = -46 \text{ kJ/mol}$$

Thus

$$\Delta H^\circ_{(a)} = 4 \text{ mol}[-(-46 \text{ kJ/mol})] = 184 \text{ kJ}$$

Step 2

Elemental oxygen [reaction (b) in Fig. 9.12]. Since $O_2(g)$ is an element in its standard state, $\Delta H^\circ_{(b)} = 0$.

We now have the elements $N_2(g)$, $H_2(g)$, and $O_2(g)$, which can be combined to form the products of the overall reaction.

Step 3

Synthesis of $NO_2(g)$ from the elements [reaction (c) in Fig. 9.12]. The overall reaction equation has 4 moles of NO_2. Thus the required reaction is 4 times the formation reaction for NO_2:

$$4 \times [\tfrac{1}{2}N_2(g) + O_2(g) \longrightarrow NO_2(g)]$$

and

$$\Delta H^\circ_{(c)} = 4 \times \Delta H_f^\circ \text{ [for } NO_2(g)]$$

From Table 9.4, ΔH_f° [for $NO_2(g)$] = 34 kJ/mol, so

$$\Delta H^\circ_{(c)} = 4 \text{ mol} \times 34 \text{ kJ/mol} = 136 \text{ kJ}$$

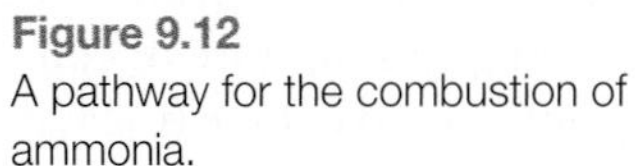

Figure 9.12
A pathway for the combustion of ammonia.

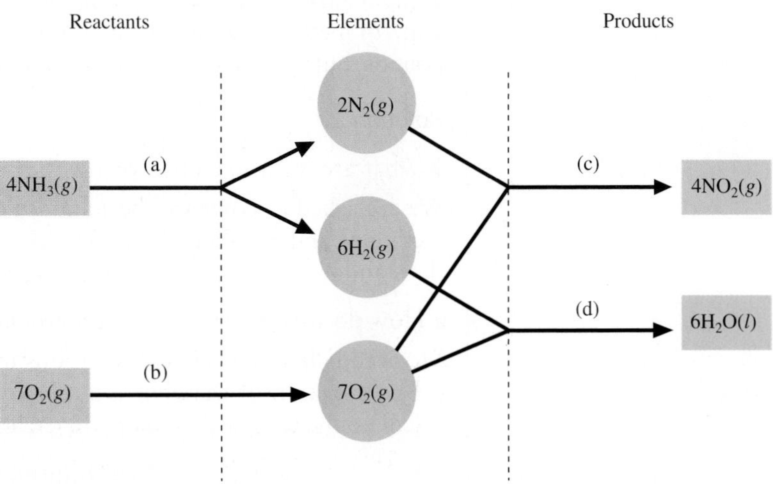

Step 4

Synthesis of $H_2O(l)$ from the elements [reaction (d) in Fig. 9.12]. Since the overall reaction equation has 6 moles of $H_2O(l)$, the required reaction is 6 times the formation reaction of $H_2O(l)$:

$$6 \times [H_2(g) + \tfrac{1}{2}O_2(g) \longrightarrow H_2O(l)]$$

and

$$\Delta H^\circ_{(d)} = 6 \times \Delta H_f^\circ \text{ [for } H_2O(l)]$$

From Table 9.4, ΔH_f° [for $H_2O(l)$] $= -286$ kJ/mol, so

$$\Delta H^\circ_{(d)} = 6 \text{ mol}(-286 \text{ kJ/mol}) = -1716 \text{ kJ}$$

To summarize, we have done the following:

$$\begin{array}{lcl} 4NH_3(g) \xrightarrow{\Delta H^\circ_{(a)}} & \left\{\begin{array}{l} 2N_2(g) + 6H_2(g) \\ 7O_2(g) \end{array}\right\} & \xrightarrow{\Delta H^\circ_{(c)}} 4NO_2(g) \\ 7O_2(g) \xrightarrow{\Delta H^\circ_{(b)} = 0} & & \xrightarrow{\Delta H^\circ_{(d)}} 6H_2O(l) \end{array}$$

Elements in their standard states

We now add the ΔH° values for the steps to obtain ΔH° for the overall reaction:

$$\begin{aligned} \Delta H^\circ_{\text{reaction}} &= \Delta H^\circ_{(a)} + \Delta H^\circ_{(b)} + \Delta H^\circ_{(c)} + \Delta H^\circ_{(d)} \\ &= 4 \times -\Delta H_f^\circ \text{ [for } NH_3(g)] + 0 + 4 \times \Delta H_f^\circ \text{ [for } NO_2(g)] \\ &\quad + 6 \times \Delta H_f^\circ \text{ [for } H_2O(l)] \\ &= 4 \times \Delta H_f^\circ \text{ [for } NO_2(g)] + 6 \times \Delta H_f^\circ \text{ [for } H_2O(l)] \\ &\quad - 4 \times \Delta H_f^\circ \text{ [for } NH_3(g)] \\ &= \Delta H_f^\circ \text{ (products)} - \Delta H_f^\circ \text{ (reactants)} \end{aligned}$$

Remember that elemental reactants and products do not need to be included, since ΔH_f° for an element in its standard state is zero. Note that we have again obtained Equation (9.1). The final solution is

$$\begin{aligned} \Delta H^\circ_{\text{reaction}} &= 6 \times (-286 \text{ kJ}) + 4 \times (34 \text{ kJ}) - 4 \times (-46 \text{ kJ}) \\ &= -1396 \text{ kJ} \end{aligned}$$

Now that we have shown the basis for Equation (9.1), we will make direct use of it to calculate ΔH for reactions in succeeding examples.

EXAMPLE 9.8

Methanol (CH_3OH) is sometimes used as a fuel in high-performance engines. Using the data in Table 9.4, compare the standard enthalpy of combustion per gram of methanol with that of gasoline. Gasoline is actually a mixture of compounds, but assume for this problem that gasoline is pure liquid octane (C_8H_{18}).

Solution

- What are we trying to solve?

We are asked to compare the standard enthalpies of combustion per gram for methanol and gasoline. To make this comparison we will have to determine the standard enthalpy for each reaction.

- How do we determine the standard enthalpy of combustion for each?

We need balanced chemical equations for the combustion of methanol and for the combustion of gasoline (octane). We then use the data in Table 9.4 and calculate the standard enthalpy of each reaction using

$$\Delta H^\circ_{\text{reactants}} = \Sigma\Delta H_f^\circ \text{ (products)} - \Sigma\Delta H_f^\circ \text{ (reactants)}$$

Once we determine each of these, we will have to determine the standard enthalpy per gram of methanol and of octane.

The combustion reaction for methanol is

$$2CH_3OH(l) + 3O_2(g) \longrightarrow 2CO_2(g) + 4H_2O(l)$$

Using the standard enthalpies of formation from Table 9.4 and Equation (9.1), we have

$$\begin{aligned}\Delta H^\circ_{\text{reaction}} &= 2 \times \Delta H^\circ_f\,[\text{for } CO_2(g)] + 4 \times \Delta H^\circ_f\,[\text{for } H_2O(l)] \\ &\quad - 2 \times \Delta H^\circ_f\,[\text{for } CH_3OH(l)] \\ &= 2 \times (-394\text{ kJ}) + 4 \times (-286\text{ kJ}) - 2 \times (-239\text{ kJ}) \\ &= -1454\text{ kJ}\end{aligned}$$

Thus 1454 kJ of heat is evolved when 2 moles of methanol burn. The molar mass of methanol is 32.0 g/mol. This means that 1454 kJ of energy is produced when 64.0 g of methanol burn. The enthalpy of combustion per gram of methanol is

$$\frac{-1454\text{ kJ}}{64.0\text{ g}} = -22.7\text{ kJ/g}$$

The combustion reaction for octane is

$$2C_8H_{18}(l) + 25O_2(g) \longrightarrow 16CO_2(g) + 18H_2O(l)$$

Using the standard enthalpies of formation from Table 9.4 and Equation (9.1), we have

$$\begin{aligned}\Delta H^\circ_{\text{reaction}} &= 16 \times \Delta H^\circ_f\,[\text{for } CO_2(g)] + 18 \times \Delta H^\circ_f\,[\text{for } H_2O(l)] \\ &\quad - 2 \times \Delta H^\circ_f\,[\text{for } C_8H_{18}(l)] \\ &= 16 \times (-394\text{ kJ}) + 18 \times (-286\text{ kJ}) - 2 \times (-269\text{ kJ}) \\ &= -1.09 \times 10^4\text{ kJ}\end{aligned}$$

This value is the amount of heat evolved when 2 moles of octane burn. Since the molar mass of octane is 114.2 g/mol, the enthalpy of combustion per gram of octane is

$$\frac{-1.09 \times 10^4\text{ kJ}}{2(114.2\text{ g})} = -47.7\text{ kJ/g}$$

The enthalpy of combustion per gram of octane is about twice that per gram of methanol. On this basis, gasoline appears to be superior to methanol for use in a racing car, where weight considerations are usually very important. Why, then, is methanol used in racing cars? The answer is that methanol burns much more smoothly than gasoline in high-performance engines, and this advantage more than compensates for its weight disadvantage.

9.7 Present Sources of Energy

Woody plants, coal, petroleum, and natural gas provide a vast resource of energy that originally came from the sun. By the process of photosynthesis, plants store energy that can be claimed by burning the plants themselves or the decay products that have been converted to **fossil fuels.** Although the United States currently depends heavily on petroleum for energy, this dependency is a relatively recent phenomenon, as shown in Fig. 9.13. In this section we discuss some sources of energy and their effects on the environment.

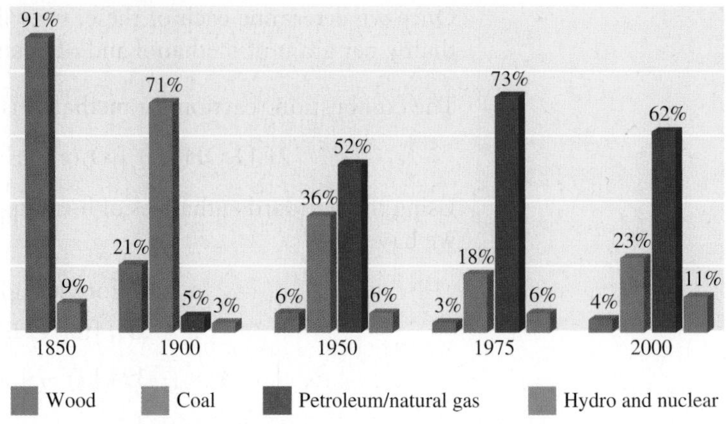

Figure 9.13 Energy sources used in the United States.

Petroleum and Natural Gas

Although how they were produced is not completely understood, petroleum and natural gas were most likely formed from the remains of marine organisms that lived about 500 million years ago. **Petroleum** is a thick, dark liquid composed mostly of compounds called *hydrocarbons* that contain carbon and hydrogen. Table 9.5 lists the formulas and names of several common hydrocarbons. **Natural gas,** usually associated with petroleum deposits, consists mostly of methane but also contains significant amounts of ethane, propane, and butane.

Table 9.5 Formulas and Names for Some Common Hydrocarbons

Formula	Name
CH_4	Methane
C_2H_6	Ethane
C_3H_8	Propane
C_4H_{10}	Butane
C_5H_{12}	Pentane
C_6H_{14}	Hexane
C_7H_{16}	Heptane
C_8H_{18}	Octane

The composition of petroleum varies somewhat, but it consists mostly of hydrocarbons having chains that contain from 5 to more than 25 carbons. To be used efficiently, the petroleum must be separated into fractions by boiling. The lighter molecules (having the lowest boiling points) can be boiled off, leaving the heavier ones behind. The uses of various petroleum fractions are shown in Table 9.6.

Table 9.6 Uses of the Various Petroleum Fractions

Petroleum Fraction in Terms of Numbers of Carbon Atoms	Major Uses
C_5–C_{10}	Gasoline
C_{10}–C_{18}	Kerosene Jet fuel
C_{15}–C_{25}	Diesel fuel Heating oil Lubricating oil
$>C_{25}$	Asphalt

The petroleum era began when the demand for lamp oil during the Industrial Revolution outstripped the traditional sources, animal fats and whale oil. In response to this increased demand, Edwin Drake drilled the first oil well in 1859 at Titusville, Pennsylvania. The petroleum from this well was refined to produce *kerosene* (fraction C_{10}–C_{18}), which served as an excellent lamp oil. *Gasoline* (fraction C_5–C_{10}) had limited use and was often discarded. However, the development of the electric light decreased the need for kerosene, and the advent of the "horseless carriage" with its gasoline-powered engine signaled the birth of the gasoline age.

As gasoline became more important, new ways were sought to increase the yield of gasoline obtained from each barrel of petroleum. William Burton invented a process at Standard Oil of Indiana called *pyrolytic (high-temperature) cracking*. In this process the heavier molecules of the kerosene fraction are heated to about 700°C, causing them to break (crack) into the smaller molecules of hydrocarbons in the gasoline fraction. As cars became larger, more efficient internal combustion engines were designed. Because of the uneven burning of the gasoline then available, these engines "knocked," producing unwanted noise and even engine damage. Intensive research to find additives that would promote smoother burning produced tetraethyl lead [$(C_2H_5)_4Pb$], a very effective "antiknock" agent.

The addition of tetraethyl lead to gasoline became a common practice, and by 1960 gasoline contained as much as 3 g of lead per gallon. As we have discovered so often in recent years, technological advances can produce environmental problems. To prevent air pollution from automobile exhaust, manufacturers have added catalytic converters to car exhaust systems. The effectiveness of these converters, however, is destroyed by lead. The use of leaded gasoline has also greatly increased the amount of lead in the environment, where it can be ingested by animals and humans. For these reasons, the use of lead in gasoline has been phased out, which has required extensive (and expensive) modifications of engines and of the gasoline-refining process.

The estimates of natural gas reserves in the United States have risen sharply in recent years, mainly due to new technologies that enable access to deep pockets of natural gas in shale rock formations. A new technique called *hydraulic fracturing,* in which water is injected into wells at high pressures to shatter rocks deep underground, frees the trapped natural gas. This shale gas represents about one-third of U.S. reserves of natural gas. Although this shale gas currently provides only a small portion of U.S. total gas production, experts believe that natural gas will be substituted for other fossil fuels over the coming years. An added benefit of natural gas is that it produces less carbon dioxide per unit of energy generated than either oil or coal. Present estimates predict that the United States has about a 100-year supply of natural gas.

Coal

Coal has variable composition depending on both its age and where it was formed.

Coal was formed from the remains of plants that were buried and subjected to pressure and heat over long periods of time. Plant materials have a high content of cellulose, a complex molecule whose empirical formula is CH_2O but whose molar mass is around 500,000. After the plants and trees that flourished on the earth at various times and places died and were buried, chemical changes gradually lowered the oxygen and hydrogen content of the cellulose molecules. Coal "matures" through four stages: lignite, subbituminous, bituminous, and anthracite. Each stage has higher carbon-to-oxygen and carbon-to-hydrogen ratios; that is, the relative carbon content gradually increases. Typical elemental compositions of the various coals are given in Table 9.7. The energy available from the combustion of a given mass of coal increases as the carbon content increases. Anthracite is the most valuable coal and lignite the least valuable.

Klaus Andrews/Argus Fotoarchiv/Peter Arnold/PhotoLibrary

Use of coal can pose pollution problems.

Coal is an important and plentiful fuel in the United States, currently furnishing about 20% of our energy. As the supply of petroleum dwindles, the share of the energy supply from coal is expected to increase to about 30%. However, coal is expensive and dangerous to mine underground, and the strip mining of fertile farmland in the Midwest or of scenic land in the West causes obvious problems. In addition, the burning of coal, especially high-sulfur coal, yields air pollutants such as sulfur dioxide, which in turn can lead to acid rain, as we learned in Chapter 5. However, even if coal were pure carbon, the carbon dioxide produced when it was burned would still have significant effects on the earth's climate.

Effects of Carbon Dioxide on Climate

The electromagnetic spectrum including visible and infrared radiation is discussed in Chapter 12.

The earth receives a tremendous quantity of radiant energy from the sun, about 30% of which is reflected into space by the earth's atmosphere. The remaining energy passes through the atmosphere to the earth's surface. Some of this en-

Table 9.7

Elemental Composition of Various Types of Coal

	Mass Percent of Each Element				
Type of Coal	C	H	O	N	S
Lignite	71	4	23	1	1
Subbituminous	77	5	16	1	1
Bituminous	80	6	8	1	5
Anthracite	92	3	3	1	1

Chemical Insights Hiding Carbon Dioxide

Global warming now seems to be a reality. At the heart of this issue is the carbon dioxide produced by society's widespread use of fossil fuels. For example, in the United States CO_2 makes up 81% of greenhouse gas emissions. Thirty percent of this CO_2 comes from coal-fired power plants used to produce electricity. One way to solve this problem would be to phase out coal-fired power plants. However, this outcome is not likely because the United States possesses so much coal (at least a 250-year supply) and coal is so cheap (about $0.01 per pound). Recognizing this fact, the U.S. government has instituted a research program to see if the CO_2 produced at power plants can be captured and sequestered (stored) underground in deep geological formations. The factors that need to be explored to determine whether sequestration is feasible are the capacities of underground storage sites and the chances that the sites will leak.

The injection of CO_2 into the earth's crust is already being undertaken by various oil companies. Since 1996, the Norwegian oil company Statoil has separated more than 1 million tons of CO_2 annually from natural gas and pumped it into a saltwater aquifer beneath the floor of the North Sea. In western Canada a group of oil companies has injected CO_2 from a North Dakota synthetic fuels plant into oil fields in an effort to increase oil recovery. The oil companies expect to store 22 million tons of CO_2 there and to produce 130 million barrels of oil over the next 20 years.

Sequestration of CO_2 has great potential as one method for decreasing the rate of global warming. Only time will tell whether it will work.

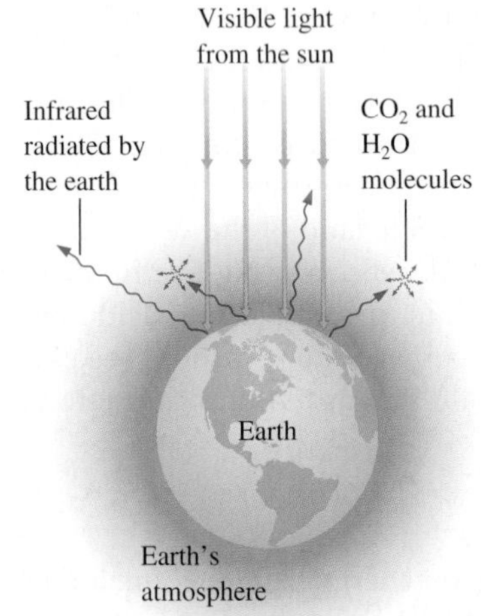

Figure 9.14
The earth's atmosphere is transparent to visible light from the sun. This visible light strikes the earth, and part of it is changed to infrared radiation. The infrared radiation from the earth's surface is strongly absorbed by CO_2, H_2O, and other molecules present in smaller amounts (for example, CH_4 and N_2O) in the atmosphere. In effect, the atmosphere traps some of the energy, acting like the glass in a greenhouse and keeping the earth warmer than it would otherwise be.

ergy is absorbed by plants to drive photosynthesis and some by the oceans to evaporate water, but most of it is absorbed by soil, rock, and water, resulting in an increase in the temperature of the earth's surface. This energy is in turn radiated from the heated surface mainly as *infrared radiation,* often called heat radiation.

The atmosphere, like window glass, is transparent to visible light but does not allow all the infrared radiation to pass through. Molecules in the atmosphere, principally H_2O and CO_2, strongly absorb infrared radiation, trapping it in the earth's atmosphere, as shown in Fig. 9.14. A net amount of thermal energy is thus retained by the earth's atmosphere, which causes the earth to be much warmer than it would be without its atmosphere. In a way the atmosphere acts like the glass of a greenhouse, which is transparent to visible light but absorbs infrared radiation, thus raising the temperature inside the building. This **greenhouse effect** is seen even more spectacularly on Venus, where the dense atmosphere is mainly responsible for the high surface temperature of that planet.

Thus the temperature of the earth's surface is controlled to a significant extent by the carbon dioxide and water content of the atmosphere. The effect of atmospheric moisture (humidity) is apparent in the Midwest. In summer when the humidity is high, the heat of the sun is retained well into the night, giving very high nighttime temperatures. On the other hand, in winter the coldest temperatures always occur on clear nights, when the low humidity allows efficient radiation of energy back into space.

The atmosphere's water content is controlled by the water cycle (evaporation and precipitation), and the average content remains constant over the years. However, as fossil fuels have come into more extensive use, the carbon dioxide concentration has increased significantly. This increase, which was 16% from 1880 to 1980, has continued in the past two decades. Some projections indicate that the carbon dioxide content of the atmosphere may be dou-

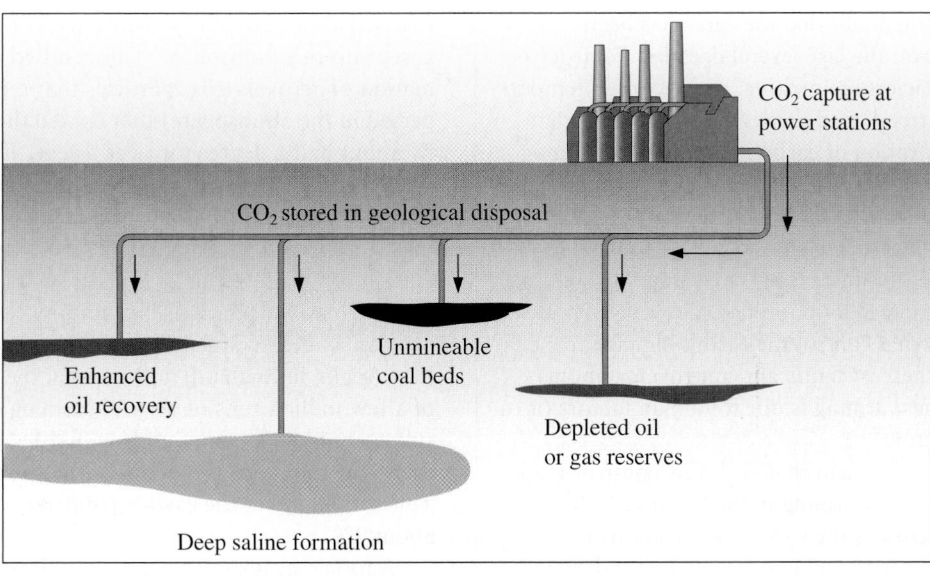

The average temperature of the earth's surface is 288 K. It would be ≈255 K without the "greenhouse gases."

ble in the twenty-first century what it was in 1880. As a result, the earth's average temperature could increase by as much as 3°C, causing dramatic changes in climate and greatly affecting the growth of food crops.

Most atmospheric scientists believe the warming trend has already started. For example, in 2005 the global average temperature of 14.6°C (58.3°F) was the warmest since records have been kept starting in the late 1800s. Global average temperatures have risen 0.6°C in the past 30 years and 0.8°C in the past century.

How well can we predict long-term effects? Because weather has been studied for a period of time that is miniscule compared with the age of the earth, the factors that control the earth's climate in the long range are not clearly understood. For example, we do not understand what causes the earth's periodic ice ages. So, indeed, it is difficult to estimate the impact of the increasing carbon dioxide levels.

In fact, the variation in the earth's average temperature over the past century is somewhat confusing. In the northern latitudes during the past century, the average temperature rose by 0.8°C over a period of 60 years, then cooled by 0.5°C during the next 25 years, and finally warmed by 0.2°C in the past 20 years. Such fluctuations do not match the steady increase in carbon dioxide. However, in southern latitudes and in areas near the equator, the average temperature showed a steady increase totaling 0.4°C over the past century. This figure is in reasonable agreement with the predicted effect of the increasing carbon dioxide concentration over that period.

Another significant fact is that the last 10 years of the twentieth century appear to be the warmest decade on record. Although the exact relationship between the carbon dioxide concentration in the atmosphere and the earth's temperature is not known at present, one thing is clear: The increase in the atmospheric concentration of carbon dioxide is quite dramatic. We must consider the implications of this increase as we consider our future energy needs.

Chemical Insights Geoengineering

There is little doubt that the earth has been warming over the last several decades. Polar ice is melting, glaciers are melting, and recorded temperatures are trending. There is also agreement that the concentration of carbon dioxide in the atmosphere has increased significantly since the Industrial Revolution in the 1700s, and most of the increase has occurred since the 1950s. Atmospheric CO_2 has increased from about 280 ppm in 1750 to nearly 390 ppm today. However, because the factors that control the earth's climate are so complex, there is significant controversy about whether the warming is due to human factors or to natural factors.

One development that fuels this controversy is that the rate of warming in the last decade has decreased even as the CO_2 concentrations have continued to increase. Recent research indicates that the decrease in the rate of warming is due to a sudden unexplained drop in the amount of water vapor in the lower stratosphere in late 2000 and early 2001. The water vapor concentration decreased by 10% and has been stable at that level ever since. Because water is a potent greenhouse gas, this decline has slowed the increase in global temperature.

There are many other natural factors that influence the earth's temperature, such as changes in solar activity and cloud formation. So the earth is warming, but we are not 100% sure why. How do we respond to global warming? One strategy we could pursue is to find ways to cool the earth by geoengineering. For example, it is well known that tiny particles or liquid droplets in the atmosphere block enough of the sunlight to produce significant cooling. We know this from previous major volcanic eruptions, such as the eruption of Mount Pinatubo in the Philippines in 1991, which released about 20 million metric tons of sulfurous gases into the atmosphere. This resulted in the formation of aerosols (tiny particles that remain dispersed in the atmosphere) that cooled the earth by about half a degree for over a year. These types of effects cause scientists to wonder if we can produce aerosols artificially to instigate the cooling of the earth. In fact, Nobel Prize–winning chemist Paul J. Crutzen has suggested that such a plan might work. Calculations by Timothy Lenton, a professor of earth systems at the University of East Anglia in England, indicate that the addition of a few million tons of aerosol-forming sulfur or sulfur-containing compounds would be required each year to reflect about 2% of the sun's rays. This should lower the earth's temperature by about 2°C.

Another strategy for decreasing the earth's warming would be to remove CO_2 from the atmosphere. Klaus Lackner from Columbia University and David Keith of the University of Calgary have each designed air capture systems for CO_2. Professor Keith plans to build a prototype that could remove 100,000 tons of CO_2 per year at a cost of \$100 per ton. At that rate, absorbing all of the CO_2 produced in the United States annually would require 58,000 air capture units and would cost more than \$580 billion. Professor Lackner is developing a device based on different technology that he projects will capture a ton of CO_2 per day at a cost of less than \$100 per ton.

Although various schemes put forth to control global warming may seem impractical at present, there is growing interest in the United States and the United Kingdom. In fact, the U.S. House of Representatives has already held several hearings to learn more about geoengineering. Geoengineering seems to be cool these days.

9.8 New Energy Sources

As we search for the energy sources of the future, we need to consider economic, climatic, and supply factors. There are several potential energy sources: the sun (solar), nuclear processes (fission and fusion), biomass (plants), and synthetic fuels. Direct use of the sun's radiant energy to heat our homes and run our factories and transportation systems seems a sensible long-term goal. But what do we do now? Conservation of fossil fuels is one obvious step, but substitutes for fossil fuels must be found eventually. We will discuss some alternative sources of energy here. Nuclear power will be considered in Chapter 20.

Coal Conversion

One alternative energy source involves using a traditional fuel—coal—in new ways. Since transportation costs for solid coal are high, more energy-efficient fuels are being developed from coal. One possibility is to produce a gaseous fuel. Substances like coal that contain large molecules have high boiling points and tend to be solids or thick liquids. To convert coal from a solid to a gas therefore requires reducing the size of the molecules; the coal structure must be broken down in a process called *coal gasification.* This process is carried out by treating the coal with oxygen and steam at high temperatures to break many of the carbon–carbon bonds. These bonds are replaced by carbon–hydrogen and carbon–oxygen bonds as the coal fragments react with the water and oxygen. This process is represented in Fig. 9.15. The desired fuel consists of a mixture of carbon monoxide and hydrogen called *synthetic gas,* or **syngas,** and methane (CH_4) gas. Since all the components of this product can react with oxygen to release heat in a combustion reaction, this gas is a useful fuel.

An industrial process must be energy-efficient.

One of the most important considerations in designing an industrial process is the efficient use of energy. In coal gasification some of the reactants are exothermic:

$$C(s) + 2H_2(g) \longrightarrow CH_4(g) \qquad \Delta H° = -75 \text{ kJ}$$

$$C(s) + \tfrac{1}{2}O_2(g) \longrightarrow CO(g) \qquad \Delta H° = -111 \text{ kJ}$$

$$C(s) + O_2(g) \longrightarrow CO_2(g) \qquad \Delta H° = -394 \text{ kJ}$$

Other gasification reactions are endothermic. For example,

$$C(s) + H_2O(g) \longrightarrow H_2(g) + CO(g) \qquad \Delta H° = 131 \text{ kJ}$$

If the rate at which the coal, air, and steam are combined is carefully controlled, the correct temperature can be maintained in the process without using any external energy source. That is, an energy balance is attained.

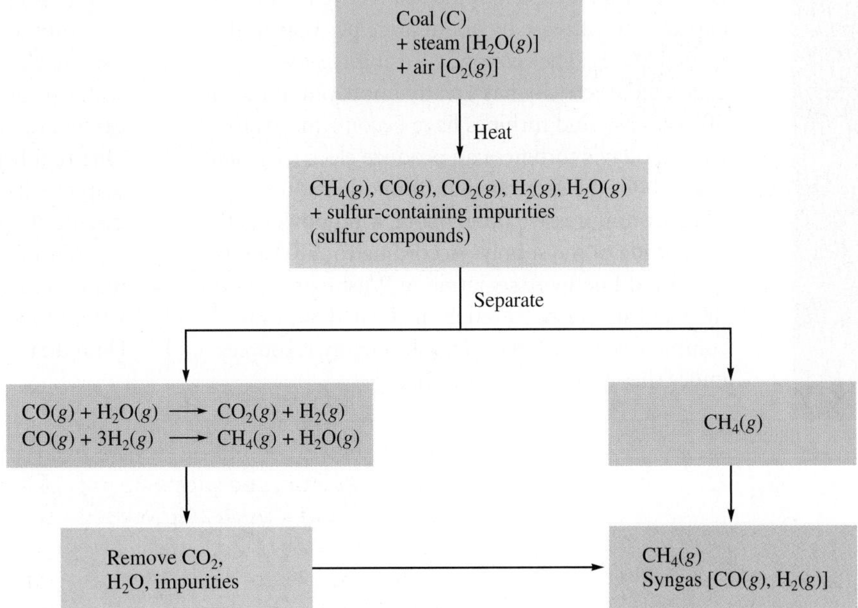

Figure 9.15
Coal gasification. Reaction of coal with a mixture of steam and air breaks down the large hydrocarbon molecules in the coal to smaller gaseous molecules, which can be used as fuels.

Chemical Insights Farming the Wind

In the Midwest the wind blows across fields of corn, soybeans, wheat, and wind turbines—wind turbines? It turns out that the wind that seems to blow almost continuously across the plains is now becoming the latest cash crop. One of these new-breed wind farmers is Daniel Juhl, who recently erected 17 wind turbines on six acres of land near Woodstock, Minnesota. These turbines can generate as much as 10 megawatts (MW) of electricity, which Juhl sells to the local electrical utility.

The largest wind farm in the world, located on the Oregon–Washington border, generates 300 MW. A controversial wind farm called Cape Wind, in the ocean five miles off the coast of Cape Cod, is planned to produce more than 400 MW of power.

There is plenty of untapped wind power in the United States. Wind mappers rate regions on a scale from 1 to 6 (with 6 being the best) to indicate the quality of the wind resource. Wind farms are now being developed in areas rated from 4 to 6. The farmers who own the land welcome the increased income derived from the wind blowing across their land. Economists estimate that each acre devoted to wind turbines can pay royalties to the farmers of as much as $8000 per year, or many times the revenue from growing corn on that same land. Daniel Juhl claims that farmers who construct the turbines themselves can realize as much as $20,000 per year per turbine. Globally, wind generation of electricity has nearly quadrupled in the last five years and is expected to increase by about 60% per year in the United States. The economic feasibility of wind-generated electricity has greatly improved in the last 30 years as wind turbines have become more efficient. Today's turbines can produce electricity that costs about the same as that from other sources. The most impressive thing about wind power is the magnitude of the supply. According to the American Wind Energy Association in Washington, D.C., the wind-power potential in the United States is comparable to or larger than the energy resources under the sands of Saudi Arabia.

Stateline Wind Generating Project in Walla Walla, Washington.

The biggest hurdle that must be overcome before wind power can become a significant electricity producer in the United States is construction of the transmission infrastructure—the power lines needed to move the electricity from the rural areas to the cities where most of the power is used. For example, the hundreds of turbines planned in southwest Minnesota in a development called Buffalo Ridge could supply enough electricity to power 1 million homes if transmission problems can be solved.

Another possible scenario for wind farms is to use the electrical power generated to decompose water to produce hydrogen gas that could be carried to cities by pipelines and used as a fuel. One real benefit of hydrogen is that it produces water as its only combustion product. Thus it is essentially pollution-free.

Within a few years wind power could be a major source of electricity. There could be a fresh wind blowing across the energy landscape of the United States in the near future.

As we stated earlier, syngas can be used directly as a fuel, but it is also important as a raw material for producing other fuels. For example, syngas can be directly converted to methanol:

$$CO(g) + 2H_2(g) \longrightarrow CH_3OH(l)$$

Methanol is used in the production of synthetic fibers and plastics and can also be used as a fuel. In addition, it can be converted directly to gasoline. About half of South Africa's gasoline supply comes from methanol produced from syngas.

In addition to coal gasification, the formation of *coal slurries* is another new use of coal. A slurry is a suspension of fine particles in a liquid. Coal must be pulverized and mixed with water to form a slurry. The resulting slurry can be handled, stored, and burned in ways similar to those used for *residual oil,* a heavy fuel oil from petroleum accounting for 13% of U.S. petroleum imports. One hope is that coal slurries might replace solid coal and residual oil as fuels for electricity-generating power plants. However, the water needed for slurries might place an unacceptable burden on water resources, especially in the western states.

Hydrogen as a Fuel

If you have ever seen a lecture demonstration where hydrogen–oxygen mixtures were ignited, you have witnessed a demonstration of hydrogen's potential as a fuel. The combustion reaction is

$$H_2(g) + \tfrac{1}{2}O_2(g) \longrightarrow H_2O(l) \qquad \Delta H^\circ = -286 \text{ kJ}$$

As we saw in Example 9.5, the heat of combustion of $H_2(g)$ per gram is about 2.5 times that of natural gas. In addition, hydrogen has a real advantage over fossil fuels in that the only product of hydrogen combustion is water; fossil fuels also produce carbon dioxide. But even though it appears that hydrogen is a very logical choice for a major future fuel, three main problems are associated with its use: the costs of production, storage, and transport.

First, let's look at the production problem. Although hydrogen is very abundant on earth, virtually none of it exists as the free gas. Currently, the main source of hydrogen gas is from the treatment of natural gas with steam:

$$CH_4(g) + H_2O(g) \longrightarrow 3H_2(g) + CO(g)$$

We can calculate ΔH for this reaction by using Equation (9.1):

$$\begin{aligned}\Delta H^\circ &= \Sigma\Delta H_f^\circ\,(\text{products}) - \Sigma\Delta H_f^\circ\,(\text{reactants})\\ &= \Delta H_f^\circ\,[\text{for } CO(g)] - \Delta H_f^\circ\,[\text{for } CH_4(g)] - \Delta H_f^\circ\,[\text{for } H_2O(g)]\\ &= -111 \text{ kJ} - (-75 \text{ kJ}) - (-242 \text{ kJ}) = 206 \text{ kJ}\end{aligned}$$

Note that this reaction is highly endothermic; treating methane with steam is not an efficient way to obtain hydrogen for fuel. It would be much more economical to burn the methane directly.

A virtually inexhaustible supply of hydrogen exists in the waters of the world's oceans. However, the reaction

$$H_2O(l) \longrightarrow H_2(g) + \tfrac{1}{2}O_2(g)$$

requires 286 kJ of energy per mole of liquid water, and under current circumstances large-scale production of hydrogen from water is not economically feasible. However, several methods for such production are currently being studied: electrolysis of water, thermal decomposition of water, and biological decomposition of water.

Electrolysis will be discussed in Chapter 11.

Electrolysis of water involves passing an electric current through it. The present cost of electricity makes the hydrogen produced by electrolysis too expensive to be competitive as a fuel. However, if in the future we develop more efficient sources of electricity, this situation could change.

Thermal decomposition is another method for producing hydrogen from water. This method involves heating the water to several thousand degrees, where it spontaneously decomposes into hydrogen and oxygen. However, attaining temperatures in this range would be very expensive even if a practical heat source and a suitable reaction container were available.

In the thermochemical decomposition of water, chemical reactions, as well as heat, are used to "split" water into its components. One such system involves the following reactions (the temperature required for each is given in parentheses):

$$2HI \longrightarrow I_2 + H_2 \qquad (425°C)$$

$$2H_2O + SO_2 + I_2 \longrightarrow H_2SO_4 + 2HI \qquad (90°C)$$

$$H_2SO_4 \longrightarrow SO_2 + H_2O + \tfrac{1}{2}O_2 \qquad (825°C)$$

$$\text{Net reaction:} \quad H_2O \longrightarrow H_2 + \tfrac{1}{2}O_2$$

Note that the HI is not consumed in this reaction. Note also that the maximum temperature required is 825°C, a temperature that is feasible if a nuclear reactor is used as a heat source. A current research goal is to find a system for which the required temperatures are low enough that sunlight can be used as the energy source.

But what about the systems on earth that biologically decompose water without the aid of electricity or high temperatures? In the process of photosynthesis, green plants absorb carbon dioxide and water and use them along with energy from the sun to produce the substances needed for growth. Scientists have studied photosynthesis for years, hoping to get answers to humanity's food and energy shortages. At present much of this research involves attempts to modify the photosynthetic process so that plants will release hydrogen gas from water instead of using the hydrogen to produce complex compounds. Small-scale experiments have shown that under certain conditions plants do produce hydrogen gas, but the yields are far from being commercially useful. Thus the economical production of hydrogen gas remains unrealized.

The storage and transportation of hydrogen present two problems. First, on metal surfaces the H_2 molecule decomposes to atoms. Since the atoms are so small, they can migrate into the metal, causing structural changes that make it brittle. This might lead to a pipeline failure if hydrogen were pumped under high pressure.

A second problem is the relatively small amount of energy that is available *per unit volume* of hydrogen. Although the energy available per gram of hydrogen is significantly greater than that per gram of methane, the energy available per given volume of hydrogen is about one-third that available from the same volume of methane. Could hydrogen be considered as a potential fuel for automobiles? This is an intriguing question. The internal combustion engines in automobiles can be easily adapted to burn hydrogen. However, the primary difficulty is the storage of enough hydrogen to give an automobile a reasonable range. This is illustrated in Example 9.9.

EXAMPLE 9.9

Assuming that the combustion of hydrogen gas provides three times as much energy per gram as gasoline, calculate the volume of liquid H_2 (density = 0.0710 g/mL) required to furnish the energy contained in 80.0 L (about 20 gal) of gasoline (density = 0.740 g/mL). Calculate also the volume that this hydrogen would occupy as a gas at 1.00 atm and 25°C.

Solution

■ What are we trying to solve?

We are asked to solve for the volume of liquid H_2 required to release the same amount of energy as about 20 gal of gasoline.

We are told to assume the combustion of hydrogen gas provides three times as much energy per gram as gasoline, but we are working with volumes. Thus, we will have to convert volume of gasoline to mass of gasoline.

The mass of 80.0 L of gasoline is

$$80.0\ \text{L} \times \frac{1000\ \text{mL}}{1\ \text{L}} \times \frac{0.740\ \text{g}}{\text{mL}} = 59{,}200\ \text{g}$$

Since H_2 furnishes three times as much energy per gram as gasoline, only one-third as much liquid hydrogen is needed to furnish the same energy:

$$\text{Mass of } H_2(l) \text{ needed} = \frac{59{,}200\ \text{g}}{3} = 19{,}700\ \text{g}$$

Since density = mass/volume, volume = mass/density, and the volume of $H_2(l)$ needed is

$$V = \frac{19{,}700\ \text{g}}{0.0710\ \text{g/mL}} = 2.77 \times 10^5\ \text{mL} = 277\ \text{L}$$

Thus 277 L of liquid H_2 is needed to furnish the same energy of combustion as 80.0 L of gasoline.

To calculate the volume that this hydrogen would occupy as a gas at 1.00 atm and 25°C, we use the ideal gas law:

$$PV = nRT$$

In this case $P = 1.00$ atm, $T = 273 + 25°\text{C} = 298$ K, and $R = 0.08206$ L atm K^{-1} mol^{-1}. Also,

$$n = 19{,}700\ \text{g}\ H_2 \times \frac{1\ \text{mol}\ H_2}{2.02\ \text{g}\ H_2} = 9.75 \times 10^3\ \text{mol}\ H_2$$

Thus $$V = \frac{nRT}{P} = \frac{(9.75 \times 10^3\ \text{mol})(0.08206\ \text{L atm K}^{-1}\ \text{mol}^{-1})(298\ \text{K})}{1.00\ \text{atm}}$$

$$= 2.38 \times 10^5\ \text{L} = 238{,}000\ \text{L}$$

At 1 atm and 25°C, the hydrogen gas needed to replace 20 gal of gasoline occupies a volume of 238,000 L.

You can see from Example 9.9 that an automobile would need a huge tank to hold enough hydrogen gas to have a typical mileage range. Clearly, hydrogen must be stored in some other way, possibly as a liquid. Is this feasible? Because of hydrogen's very low boiling point (20 K), storage of liquid hydrogen requires a superinsulated container that can withstand high pressures. Storage in this manner would be both expensive and hazardous because of the potential for explosion. Thus storage of hydrogen in the individual automobile as a liquid does not seem practical.

Metal hydrides are discussed in Chapter 18.

A much better alternative seems to be the use of metals that absorb hydrogen to form solid metal hydrides:

$$H_2(g) + M(s) \longrightarrow MH_2(s)$$

In this method of storage, hydrogen gas would be pumped into a tank containing the solid metal, where it would be absorbed to form a hydride, whose volume would be little more than that of the metal. This hydrogen would then

Chemical Insights Heat Packs

A skier is trapped by a sudden snowstorm. After building a snow cave for protection, she realizes her hands and feet are freezing; she is in danger of frostbite. Then she remembers the four small packs in her pocket. She removes the plastic cover from each one to reveal a small paper packet. She places one packet in each boot and one in each mitten. Soon her hands and feet are toasty warm.

These "magic" packets of energy contain a mixture of powdered iron, activated carbon, sodium chloride, cellulose (sawdust), and zeolite, all moistened by a little water. The paper cover is permeable to air.

The exothermic reaction that produces the heat is a very common one—the rusting of iron. The overall reaction can be represented as

$$4Fe(s) + 3O_2(g) \longrightarrow 2Fe_2O_3(s)$$

$$\Delta H° = -1652 \text{ kJ}$$

although in reality it is somewhat more complicated. When the plastic envelope is removed, O_2 molecules penetrate the paper, causing the reaction to begin.

The oxidation of iron by oxygen occurs naturally. Any steel surface exposed to the atmosphere inevitably rusts. But this oxidation process is quite slow—much too slow to be useful in hot packs.

Courtesy, Grabber Inc.

However, if the iron is ground into a fine powder, the resulting increase in surface area causes the reaction with oxygen to be fast enough to warm hands and feet. The packet can produce heat for up to six hours.

be available for combustion in the engine by release of $H_2(g)$ from the hydride as needed:

$$MH_2(s) \longrightarrow M(s) + H_2(g)$$

Several types of solids that absorb hydrogen to form hydrides are being studied for use in hydrogen-powered vehicles.

Other Energy Alternatives

Many other energy sources are being considered for future use. The western states, especially Colorado, contain huge deposits of *oil shale,* which consists of a complex carbon-based material called *kerogen* contained in porous rock formations. These deposits have the potential of being a larger energy source than the vast petroleum deposits of the Middle East. However, the main problem with oil shale is that the trapped fuel is not fluid and cannot be pumped. For recovery of the fuel, the rock must be heated to a temperature of 250°C or higher to decompose the kerogen to smaller molecules that produce gaseous and liquid products. This process is expensive and yields large quantities of waste rock, which have a negative environmental impact.

Lukiyanova Natalia/frenta/Shutterstock.com #5558134O

Sunflower oil can be used as motor fuel.

Ethanol (C_2H_5OH) is another fuel with the potential to supplement, if not replace, gasoline. The most common method of producing ethanol is fermentation, a process in which sugar is changed to alcohol by the action of yeast. The sugar can come from virtually any source, including fruits and grains, although fuel-grade ethanol would probably come mostly from corn. Car engines can burn pure alcohol or *gasohol,* an alcohol–gasoline mixture (10% ethanol in gasoline), with little modification. Gasohol is now widely available in the United States. The use of pure alcohol as a motor fuel is not feasible in most of the United States because it does not vaporize easily when temperatures are low. As a result, a fuel called E85, which is 85% ethanol and 15% gasoline, is now being used in so-called flex-fuel cars, which can run on either E85 or gasoline. Pure ethanol could be a very practical fuel in warm climates. For example, in Brazil large quantities of ethanol fuel are produced for cars.

Another potential source of liquid fuel is oil squeezed from seeds (*seed oil*). For example, some farmers in North Dakota, South Africa, and Australia are now using sunflower oil to replace diesel fuel. Oil seeds, found in a wide variety of plants, can be processed to produce a "biodiesel" oil composed mainly of carbon and hydrogen, which of course reacts with oxygen to produce carbon dioxide, water, and heat. The main advantage of seed oil as a fuel is that it is renewable. It is hoped that oil seed plants can be developed that will thrive under soil and climatic conditions unsuitable for corn and wheat. Ideally, fuel would be grown just like food crops. One major advantage of biodiesel is that its production is quite energy-efficient. When the energy required to produce biodiesel is compared to the energy gained from its use as a fuel, there is a 93% net energy gain. The net energy gain for ethanol is only about 25% because its production, which includes distillation, is relatively energy-intensive.

Key Terms

Section 9.1
energy
law of conservation of energy
potential energy
kinetic energy
heat
work
pathway
state function
system
surroundings
exothermic
endothermic
thermodynamics
first law of thermodynamics
internal energy

Section 9.2
enthalpy

Section 9.3
molar heat capacity

Section 9.4
calorimeter
calorimetry
specific heat capacity

For Review

OWL and 

Sign in at **www.cengage.com/owl** to:

- View tutorials and simulations, develop problem-solving skills, and complete online homework assigned by your professor.
- Download Go Chemistry mini lecture modules for quick review and exam prep from OWL (or purchase them at **www.cengagebrain.com**)

Energy

- The capacity to do work or produce heat
- Is conserved (first law of thermodynamics)
- Can be converted from one form to another
- Is a state function
- Potential energy: stored energy
- Kinetic energy: energy due to motion
- The internal energy for a system is the sum of its potential and kinetic energies
- The internal energy of a system can be changed by work and heat:

$$\Delta E = q + w$$

Work

- Force applied over a distance
- For an expanding/contracting gas
- Not a state function

$$w = -P\Delta V$$

constant-pressure calorimetry
constant-volume calorimetry
bomb calorimeter
Section 9.5
Hess's law
Section 9.6
standard enthalpy of formation
standard state
Section 9.7
fossil fuels
petroleum
natural gas
coal
greenhouse effect
Section 9.8
syngas
	Heat
	■ Energy flow due to a temperature difference
	■ Exothermic: energy as heat flows out of a system
	■ Endothermic: energy as heat flows into a system
	■ Not a state function
	■ Measured for chemical reactions by calorimetry
	Enthalpy
	■ $H = E + PV$
	■ Is a state function
	■ Hess's law: the change in enthalpy in going from a given set of reactants to a given set of products is the same whether the process takes place in one step or a series of steps
	■ Standard enthalpies of formation (ΔH_f°) can be used to calculate ΔH for a chemical reaction
	$\Delta H^\circ_{\text{reaction}} = \Sigma\, n_p \Delta H_f^\circ(\text{products}) - \Sigma\, n_r \Delta H^\circ(\text{reactants})$
	Thermodynamics of ideal gases
	■ Heating an ideal gas at constant volume
	  ■ No PV work is done; all energy is used to increase the translational energies of the gas molecules.
	  ■ $C_v = \frac{3}{2}R$
	■ Heating an ideal gas at constant pressure
	  ■ PV work is done; energy is used to increase the translational energies of the gas molecules and do work.
	  ■ $C_p = C_v + R$
	■ $\Delta H = nC_p\Delta T$
	Energy use
	■ Energy sources from fossil fuels are associated with difficult supply and environmental impact issues
	■ The greenhouse effect results from release into the atmosphere of gases, including carbon dioxide, that strongly absorb infrared radiation, thus warming the earth
	■ Alternative fuels are being sought to replace fossil fuels:
	  ■ Hydrogen
	  ■ Syngas from coal
	  ■ Biofuels from plants such as corn and certain seed-producing plants

Discussion Questions

These questions are designed to be considered by groups of students in class. Often these questions work well for introducing a particular topic in class.

1. Objects placed together eventually reach the same temperature. When you go into a room and touch a piece of metal in that room, it feels colder than a piece of plastic. Explain.

2. What is meant by the term *lower in energy?* Which is lower in energy, a mixture of hydrogen and oxygen gases or liquid water? How do you know? Which of the two is more stable? How do you know?

3. A fire is started in a fireplace by striking a match and lighting crumpled paper under some logs. Explain the energy transfers in this scenario using the terms *exothermic, endothermic, system, surroundings, potential energy,* and *kinetic energy.*

4. Liquid water turns to ice. Is this process endothermic or exothermic? Explain what is occurring using the terms *system, surroundings, heat, potential energy,* and *kinetic energy.*

5. Consider the following statements: "Heat is a form of energy, and energy is conserved. The heat lost by the

system must be equal to the amount of heat gained by the surroundings. Therefore, heat is conserved." Indicate everything you think is correct in these statements and everything you think is incorrect in these statements. Correct the incorrect statements and explain.

6. Photosynthetic plants use the following reaction to produce glucose, cellulose, and more:

$$6CO_2(g) + 6H_2O(l) \longrightarrow C_6H_{12}O_6(s) + 6O_2(g)$$

How might extensive destruction of forests exacerbate the greenhouse effect?

7. Explain why oceanfront areas generally have smaller temperature fluctuations than inland areas.
8. Predict the signs of q and w for the process of boiling water.
9. Hess's law is really just another statement of the first law of thermodynamics. Explain.
10. In the equation $w = -P\Delta V$, why is there a negative sign?
11. Why is C_p larger than C_v? Provide a conceptual rationale.
12. You have an ideal gas with an initial volume of 1.0 L and initial pressure of 1.0 atm. You decide to change the conditions such that $P_{final} = 2.0$ atm and $V_{final} = 2.0$ L. To make things more interesting, you and a friend each carry out this change in two steps. You first change the volume so that at one point $P = 1.0$ atm and $V = 2.0$ atm. Your friend first changes the pressure so that at one point $P = 2.0$ atm and $V = 1.0$ atm. Both of you, of course, end up with the same final conditions after the second step.
    a. How should your ΔE's, ΔH's, q's, and w's compare with those of your friend? Why?
    b. Calculate ΔE, ΔH, q, and w for each case. Do they make sense? Explain.
13. When is $\Delta H = \frac{5}{2}RT$? When is $\Delta E = \frac{5}{2}RT$? When is $\Delta H = \frac{3}{2}RT$? When is $\Delta E = \frac{3}{2}RT$? When is $\Delta H = \Delta E$? What does this say, if anything, about ΔE and ΔH as state functions?
14. For a liquid, which would you expect to be larger: $\Delta H_{vaporization}$ or ΔH_{fusion}? Explain.

Exercises

OWL Interactive versions of these problems may be assigned in OWL.

A blue exercise number indicates that the answer to that exercise appears at the back of this book and a solution appears in the *Solutions Guide.*

The Nature of Energy

15. Consider the accompanying diagram. Ball A is allowed to fall and strike ball B. Assume that all of ball A's energy is transferred to ball B at point I and that there is no loss of energy to other sources. Calculate the kinetic energy and the potential energy of ball B at point II. For a falling object, the potential energy is given by PE = mgz, where m is the mass in kilograms, g is the gravitational constant (9.81 m/s^2), and z is the distance in meters.

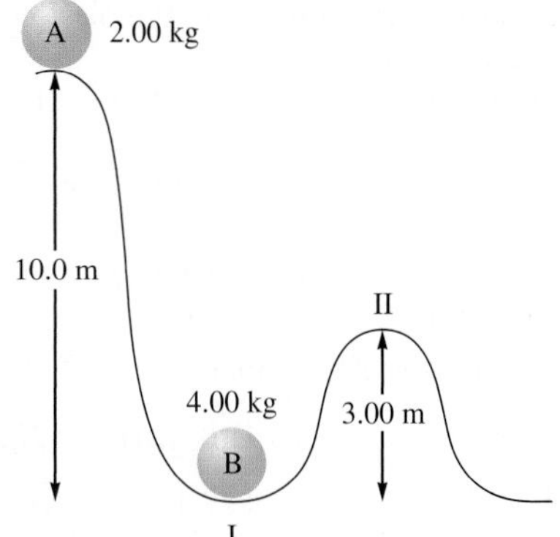

16. Consider the following potential energy diagrams for two different reactions.

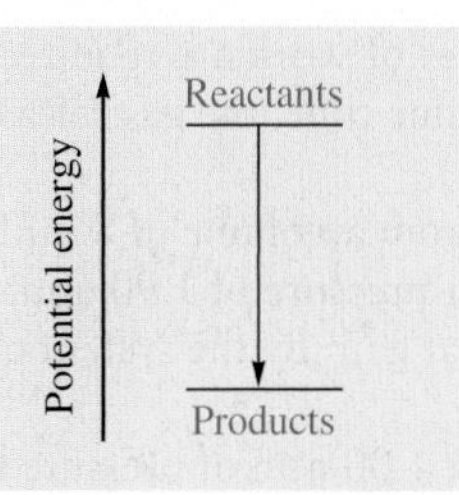

a.

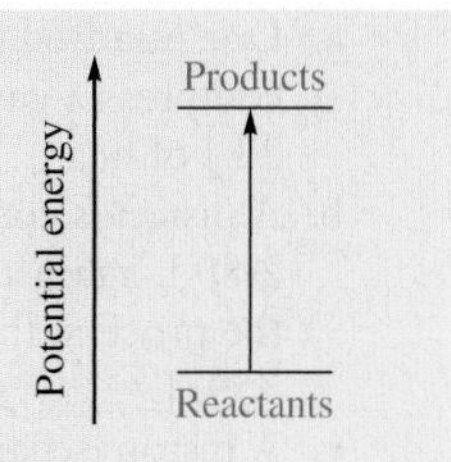

b.

Which plot represents an exothermic reaction? In plot a, do the reactants on average have stronger or weaker bonds than the products? In plot b, reactants must gain potential energy to convert to products. How does this occur?

17. Consider an airplane trip from Chicago, Illinois, to Denver, Colorado. List some path-dependent functions and some state functions for the plane trip.
18. Consider 2.00 moles of an ideal gas that is taken from state A ($P_A = 2.00$ atm, $V_A = 10.0$ L) to state B ($P_B = 1.00$ atm, $V_B = 30.0$ L) by two different pathways.

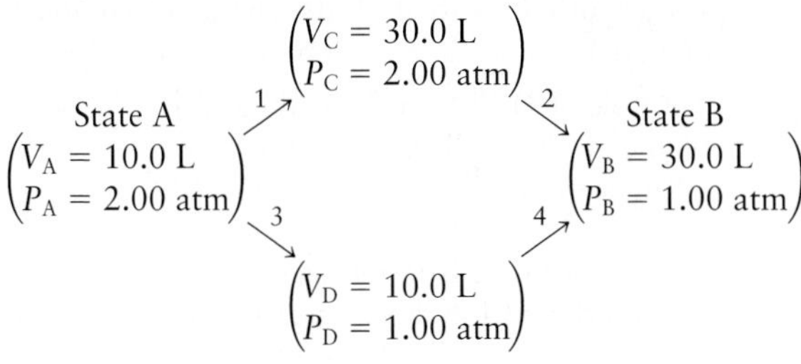

These pathways are summarized in the following graph of P versus V:

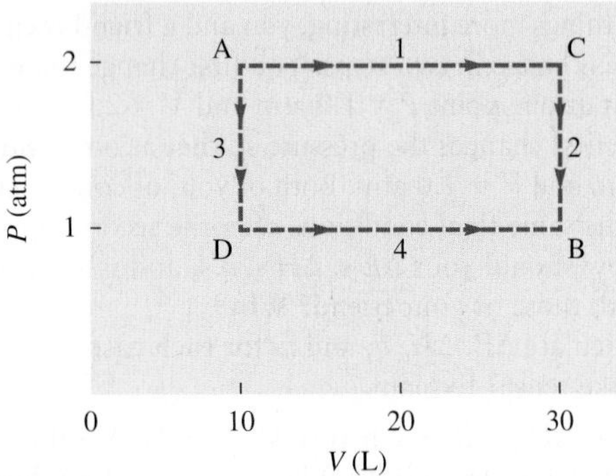

Calculate the work (in units of J) associated with the two pathways. Is work a state function? Explain.

19. A system undergoes a process consisting of the following two steps:

Step 1: The system absorbs 72 J of heat while 35 J of work is done on it.

Step 2: The system absorbs 35 J of heat while performing 72 J of work.

Calculate ΔE for the overall process.

20. Calculate the internal energy change for each of the following.
    a. One hundred (100.) joules of work are required to compress a gas. At the same time, the gas releases 23 J of heat.
    b. A piston is compressed from a volume of 8.30 L to 2.80 L against a constant pressure of 1.90 atm. In the process, there is a heat gain by the system of 350. J.
    c. A piston expands against 1.00 atm of pressure from 11.2 L to 29.1 L. In the process, 1037 J of heat is absorbed.

21. A balloon filled with 39.1 moles of helium has a volume of 876 L at 0.0°C and 1.00 atm pressure. At constant pressure, the temperature of the balloon is increased to 38.0°C, causing the balloon to expand to a volume of 998 L. Calculate q, w, and ΔE for the helium in the balloon. (The molar heat capacity for helium gas is 20.8 J $°C^{-1}$ mol^{-1}.)

22. Consider a mixture of air and gasoline vapor in a cylinder with a piston. The original volume is 40. cm^3. If the combustion of this mixture releases 950. J of energy, to what volume will the gases expand against a constant pressure of 650. torr if all the energy of combustion is converted into work to push back the piston?

23. One mole of $H_2O(g)$ at 1.00 atm and 100.°C occupies a volume of 30.6 L. When 1 mole of $H_2O(g)$ is condensed to 1 mole of $H_2O(l)$ at 1.00 atm and 100.°C, 40.66 kJ of heat is released. If the density of $H_2O(l)$ at this temperature and pressure is 0.996 g/cm^3, calculate ΔE for the condensation of 1 mole of water at 1.00 atm and 100.°C.

24. As a system increases in volume, it absorbs 52.5 J of energy in the form of heat from the surroundings. The piston is working against a pressure of 0.500 atm. The final volume of the system is 58.0 L. What was the initial volume of the system if the internal energy of the system decreased by 102.5 J?

Properties of Enthalpy

25. What is the difference between ΔH and ΔE at constant P?

26. Are the following processes exothermic or endothermic?
    a. the combustion of gasoline in a car engine
    b. water condensing on a cold pipe
    c. $CO_2(s) \longrightarrow CO_2(g)$
    d. $F_2(g) \longrightarrow 2F(g)$

27. The reaction

$$SO_3(g) + H_2O(l) \longrightarrow H_2SO_4(aq)$$

is the last step in the commercial production of sulfuric acid. The enthalpy change for this reaction is −227 kJ. In the design of a sulfuric acid plant, is it necessary to provide for heating or cooling of the reaction mixture? Explain your answer.

28. Consider the following reaction:

$$2H_2(g) + O_2(g) \longrightarrow 2H_2O(l) \qquad \Delta H = -572 \text{ kJ}$$

    a. How much heat is evolved for the production of 1.00 mole $H_2O(l)$?
    b. How much heat is evolved when 4.03 g hydrogen is reacted with excess oxygen?
    c. How much heat is evolved when 186 g oxygen is reacted with excess hydrogen?
    d. The total volume of hydrogen gas needed to fill the *Hindenburg* was 2.0×10^8 L at 1.0 atm and 25°C. How much heat was evolved when the *Hindenburg* exploded, assuming all of the hydrogen reacted?

29. The overall reaction in a commercial heat pack can be represented as

$$4Fe(s) + 3O_2(g) \longrightarrow 2Fe_2O_3(s) \qquad \Delta H = -1652 \text{ kJ}$$

    a. How much heat is released when 4.00 moles of iron is reacted with excess O_2?
    b. How much heat is released when 1.00 mole of Fe_2O_3 is produced?
    c. How much heat is released when 1.00 g iron is reacted with excess O_2?
    d. How much heat is released when 10.0 g Fe and 2.00 g O_2 are reacted?

30. Consider the combustion of propane:

$$C_3H_8(g) + 5O_2(g) \longrightarrow 3CO_2(g) + 4H_2O(l) \quad \Delta H = -2221 \text{ kJ}$$

Assume that all the heat in Example 9.1 comes from the combustion of propane. What mass of propane must be burned to furnish this amount of energy, assuming the heat transfer process is 60.% efficient?

31. For the process $H_2O(l) \longrightarrow H_2O(g)$ at 298 K and 1.0 atm, ΔH is more positive than ΔE by 2.5 kJ/mol. What does the 2.5 kJ/mol quantity represent?

32. For the following reactions at constant pressure, predict if $\Delta H > \Delta E$, $\Delta H < \Delta E$, or $\Delta H = \Delta E$.
   a. $2HF(g) \longrightarrow H_2(g) + F_2(g)$
   b. $N_2(g) + 3H_2(g) \longrightarrow 2NH_3(g)$
   c. $4NH_3(g) + 5O_2(g) \longrightarrow 4NO(g) + 6H_2O(g)$

The Thermodynamics of Ideal Gases

33. Calculate the energy required to heat 1.00 kg of ethane gas (C_2H_6) from 25.0°C to 75.0°C first under conditions of constant volume and then at a constant pressure of 2.00 atm. Calculate ΔE, ΔH, and w for these processes also. (See Table 9.1 for relevant data.)

34. Calculate q, w, ΔE, and ΔH for the process in which 88.0 g of nitrous oxide gas (N_2O) is cooled from 165°C to 55°C at a constant pressure of 5.00 atm. (See Table 9.1.)

35. Consider a sample containing 5.00 moles of a monatomic ideal gas that is taken from state A to state B by the following two pathways:

Pathway one: $P_A = 3.00$ atm, $V_A = 15.0$ L $\xrightarrow{1}$ $P_C = 3.00$ atm, $V_C = 55.0$ L $\xrightarrow{2}$ $P_B = 6.00$ atm, $V_B = 20.0$ L

Pathway two: $P_A = 3.00$ atm, $V_A = 15.0$ L $\xrightarrow{3}$ $P_D = 6.00$ atm, $V_D = 15.0$ L $\xrightarrow{4}$ $P_B = 6.00$ atm, $V_B = 20.0$ L

For each step, assume that the external pressure is constant and equals the final pressure of the gas for that step. Calculate q, w, ΔE, and ΔH for each step, and calculate overall values for each pathway. Explain how the overall values for the two pathways illustrate that ΔE and ΔH are state functions, whereas q and w are path functions.

36. Consider a sample containing 2.00 moles of a monatomic ideal gas that undergoes the following changes:

$P_A = 10.0$ atm, $V_A = 10.0$ L $\xrightarrow{1}$ $P_B = 10.0$ atm, $V_B = 5.0$ L $\xrightarrow{2}$ $P_C = 20.0$ atm, $V_C = 5.0$ L $\xrightarrow{3}$ $P_D = 20.0$ atm, $V_D = 25.0$ L

For each step, assume that the external pressure is constant and equals the final pressure of the gas for that step. Calculate q, w, ΔE, and ΔH for each step and for the overall change from state A to state D.

Calorimetry and Heat Capacity

37. Explain how calorimetry works to calculate ΔH or ΔE for a reaction. Does the temperature of the calorimeter increase or decrease for an endothermic reaction? How about for an exothermic reaction? Explain why ΔH is obtained directly from a coffee cup calorimeter, whereas ΔE is obtained directly from a bomb calorimeter.

38. The specific heat capacity of silver is 0.24 J °C^{-1} g^{-1}.
   a. Calculate the energy required to raise the temperature of 150.0 g Ag from 273 K to 298 K.
   b. Calculate the energy required to raise the temperature of 1.0 mole of Ag by 1.0°C (called the *molar heat capacity* of silver).
   c. It takes 1.25 kJ of energy to heat a sample of pure silver from 12.0°C to 15.2°C. Calculate the mass of the sample of silver.

39. Consider the substances in Table 9.3. Which substance requires the largest amount of energy to raise the temperature of 25.0 g of the substance from 15.0°C to 37.0°C? Calculate the energy. Which substance in Table 9.3 has the largest temperature change when 550. g of the substance absorbs 10.7 kJ of energy? Calculate the temperature change.

40. A 150.0-g sample of a metal at 75.0°C is added to 150.0 g of H_2O at 15.0°C. The temperature of the water rises to 18.3°C. Calculate the specific heat capacity of the metal, assuming that all the heat lost by the metal is gained by the water.

41. A biology experiment requires the preparation of a water bath at 37.0°C (body temperature). The temperature of the cold tap water is 22.0°C, and the temperature of the hot tap water is 55.0°C. If a student starts with 90.0 g of cold water, what mass of hot water must be added to reach 37.0°C?

42. A 5.00-g sample of aluminum pellets (specific heat capacity = 0.89 J °C^{-1} g^{-1}) and a 10.00-g sample of iron pellets (specific heat capacity = 0.45 J °C^{-1} g^{-1}) are heated to 100.0°C. The mixture of hot iron and aluminum is then dropped into 97.3 g of water at 22.0°C. Calculate the final temperature of the metal and water mixture, assuming no heat loss to the surroundings.

43. A sample of nickel is heated to 99.8°C and placed in a coffee cup calorimeter containing 150.0 g water at 23.5°C. After the metal cools, the final temperature of metal and water mixture is 25.0°C. If the specific heat capacity of nickel is 0.444 J/°C · g, what mass of nickel was originally heated? Assume no heat loss to the surroundings.

44. Hydrogen gives off 120. J/g of energy when burned in oxygen, and methane gives off 50. J/g under the same circumstances. If a mixture of 5.0 g hydrogen and 10. g methane is burned, and the heat released is transferred to 50.0 g water at 25.0°C, what final temperature will be reached by the water?

45. In a coffee cup calorimeter, 50.0 mL of 0.100 M $AgNO_3$ and 50.0 mL of 0.100 M HCl are mixed. The following reaction occurs:

$$Ag^+(aq) + Cl^-(aq) \longrightarrow AgCl(s)$$

If the two solutions are initially at 22.60°C, and if the final temperature is 23.40°C, calculate ΔH for the reaction in kJ/mol of AgCl formed. Assume a mass of 100.0 g for the combined solution and a specific heat capacity of 4.18 J °C^{-1} g^{-1}.

46. In a coffee cup calorimeter, 100.0 mL of 1.0 *M* NaOH and 100.0 mL of 1.0 *M* HCl are mixed. Both solutions were orignially at 24.6°C. After the reaction, the final temperature is 31.3°C. Assuming that all the solutions have a density of 1.0 g/cm^3 and a specific heat capacity of 4.18 J °C^{-1} g^{-1}, calculate the enthalpy change for the neutralization of HCl by NaOH. Assume that no heat is lost to the surroundings or to the calorimeter.

47. A coffee cup calorimeter initially contains 125 g water at 24.2°C. Potassium bromide (10.5 g), also at 24.2°C, is added to the water, and after the KBr dissolves, the final temperature is 21.1°C. Calculate the enthalpy change for dissolving the salt in J/g and kJ/mol. Assume that the specific heat capacity of the solution is 4.18 J °C^{-1} g^{-1} and that no heat is transferred to the surroundings or to the calorimeter.

48. In a coffee cup calorimeter, 1.60 g of NH_4NO_3 is mixed with 75.0 g of water at an initial temperature of 25.00°C. After dissolution of the salt, the final temperature of the calorimeter contents is 23.34°C. Assuming the solution has a heat capacity of 4.18 J °C^{-1} g^{-1} and assuming no heat loss to the calorimeter, calculate the enthalpy change for the dissolution of NH_4NO_3 in units of kJ/mol.

49. Consider the dissolution of $CaCl_2$:

$$CaCl_2(s) \longrightarrow Ca^{2+}(aq) + 2Cl^-(aq) \qquad \Delta H = -81.5 \text{ kJ}$$

An 11.0-g sample of $CaCl_2$ is dissolved in 125 g of water, with both substances at 25.0°C. Calculate the final temperature of the solution assuming no heat loss to the surroundings and assuming the solution has a specific heat capacity of 4.18 J °C^{-1} g^{-1}.

50. Consider the reaction

$$2HCl(aq) + Ba(OH)_2(aq) \longrightarrow BaCl_2(aq) + 2H_2O(l) \qquad \Delta H = -118 \text{ kJ}$$

Calculate the heat when 100.0 mL of 0.500 *M* HCl is mixed with 300.0 mL of 0.100 *M* $Ba(OH)_2$. Assuming that the temperature of both solutions was initially 25.0°C and that the final mixture has a mass of 400.0 g and a specific heat capacity of 4.18 J °C^{-1} g^{-1}, calculate the final temperature of the mixture.

51. The heat capacity of a bomb calorimeter was determined by burning 6.79 g of methane (energy of combustion = −802 kJ/mol CH_4) in the bomb. The temperature changed by 10.8°C.
    a. What is the heat capacity of the bomb?
    b. A 12.6-g sample of acetylene (C_2H_2) produced a temperature increase of 16.9°C in the same calorimeter. What is the energy of combustion of acetylene (in kJ/mol)?

52. The combustion of 0.1584 g benzoic acid increases the temperature of a bomb calorimeter by 2.54°C. Calculate the heat capacity of this calorimeter. (The energy released by combustion of benzoic acid is 26.42 kJ/g.) A 0.2130-g sample of vanillin ($C_8H_8O_3$) is then burned in the same calorimeter, and the tempearture increases by 3.25°C. What is the energy of combustion per gram of vanillin? Per mole of vanillin?

53. Combustion of table sugar produces $CO_2(g)$ and $H_2O(l)$. When 1.46 g of table sugar is combusted in a constant-volume (bomb) calorimeter, 24.00 kJ of heat is liberated.
    a. Assuming that table sugar is pure sucrose [$C_{12}H_{22}O_{11}(s)$], write the balanced equation for the combustion reaction.
    b. Calculate ΔE in kJ/mol $C_{12}H_{22}O_{11}$ for the combustion reaction of sucrose.
    c. Calculate ΔH in kJ/mol $C_{12}H_{22}O_{11}$ for the combustion reaction of sucrose at 25°C.

54. Calculate w and ΔE when 1 mole of a liquid is vaporized at its boiling point (80.°C) and 1.00 atm pressure. ΔH_{vap} for the liquid is 30.7 kJ mol^{-1} at 80.°C.

Hess's Law

55. Given the following data:

(g) ⟶ (g) + (g) $\qquad \Delta H = 92$ kJ

(g) + (g) ⟶ (g) $\qquad \Delta H = -484$ kJ

N H O

calculate ΔH for the reaction

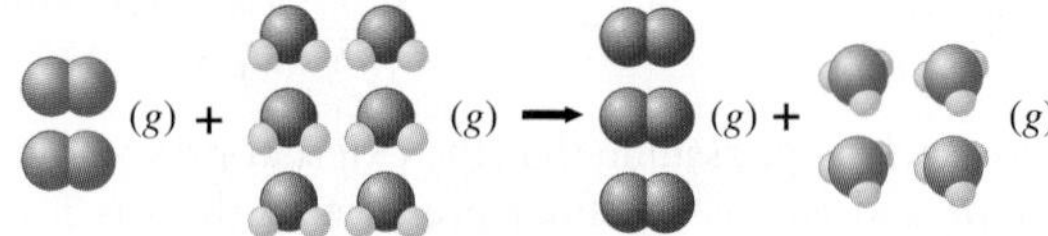

On the basis of enthalpy change, is this a useful reaction for the synthesis of ammonia?

56. Calculate ΔH for the reaction

$$N_2H_4(l) + O_2(g) \longrightarrow N_2(g) + 2H_2O(l)$$

given the following data:

$$2NH_3(g) + 3N_2O(g) \longrightarrow 4N_2(g) + 3H_2O(l) \quad \Delta H = -1010. \text{ kJ}$$
$$N_2O(g) + 3H_2(g) \longrightarrow N_2H_4(l) + H_2O(l) \quad \Delta H = -317 \text{ kJ}$$
$$2NH_3(g) + \tfrac{1}{2}O_2(g) \longrightarrow N_2H_4(l) + H_2O(l) \quad \Delta H = -143 \text{ kJ}$$
$$H_2(g) + \tfrac{1}{2}O_2(g) \longrightarrow H_2O(l) \quad \Delta H = -286 \text{ kJ}$$

57. Given the following data:

$$C_2H_2(g) + \tfrac{5}{2}O_2(g) \longrightarrow 2CO_2(g) + H_2O(l) \quad \Delta H = -1300. \text{ kJ}$$
$$C(s) + O_2(g) \longrightarrow CO_2(g) \quad \Delta H = -394 \text{ kJ}$$
$$H_2(g) + \tfrac{1}{2}O_2(g) \longrightarrow H_2O(l) \quad \Delta H = -286 \text{ kJ}$$

calculate ΔH for the reaction

$$2C(s) + H_2(g) \longrightarrow C_2H_2(g)$$

58. Given the following data:

$$2ClF(g) + O_2(g) \longrightarrow Cl_2O(g) + F_2O(g) \quad \Delta H = 167.4 \text{ kJ}$$
$$2ClF_3(g) + 2O_2(g) \longrightarrow Cl_2O(g) + 3F_2O(g) \quad \Delta H = 341.4 \text{ kJ}$$
$$2F_2(g) + O_2(g) \longrightarrow 2F_2O(g) \quad \Delta H = -43.4 \text{ kJ}$$

calculate ΔH for the reaction

$$ClF(g) + F_2(g) \longrightarrow ClF_3(g)$$

59. Given the following data:

$$\begin{aligned} Ca(s) + 2C(\textit{graphite}) &\longrightarrow CaC_2(s) & \Delta H &= -62.8 \text{ kJ} \\ Ca(s) + \tfrac{1}{2}O_2(g) &\longrightarrow CaO(s) & \Delta H &= -635.5 \text{ kJ} \\ CaO(s) + H_2O(l) &\longrightarrow Ca(OH)_2(aq) & \Delta H &= -653.1 \text{ kJ} \\ C_2H_2(g) + \tfrac{5}{2}O_2(g) &\longrightarrow 2CO_2(g) + H_2O(l) & \Delta H &= -1300. \text{ kJ} \\ C(\textit{graphite}) + O_2(g) &\longrightarrow CO_2(g) & \Delta H &= -393.5 \text{ kJ} \end{aligned}$$

calculate ΔH for the reaction

$$CaC_2(s) + 2H_2O(l) \longrightarrow Ca(OH)_2(aq) + C_2H_2(g)$$

60. Given the following data:

$$\begin{aligned} Fe_2O_3(s) + 3CO(g) &\longrightarrow 2Fe(s) + 3CO_2(g) & \Delta H &= -23 \text{ kJ} \\ 3Fe_2O_3(s) + CO(g) &\longrightarrow 2Fe_3O_4(s) + CO_2(g) & \Delta H &= -39 \text{ kJ} \\ Fe_3O_4(s) + CO(g) &\longrightarrow 3FeO(s) + CO_2(g) & \Delta H &= 18 \text{ kJ} \end{aligned}$$

calculate ΔH for the reaction

$$FeO(s) + CO(g) \longrightarrow Fe(s) + CO_2(g)$$

61. Combustion reactions involve reacting a substance with oxygen. When compounds containing carbon and hydrogen are combusted, carbon dioxide and water are the products. Using the enthalpies of combustion for C_4H_4 (−2341 kJ/mol), C_4H_8 (−2755 kJ/mol), and H_2 (−286 kJ/mol), calculate ΔH for the reaction

$$C_4H_4(g) + 2H_2(g) \longrightarrow C_4H_8(g)$$

62. Given the following data:

$$\begin{aligned} 2O_3(g) &\longrightarrow 3O_2(g) & \Delta H &= -427 \text{ kJ} \\ O_2(g) &\longrightarrow 2O(g) & \Delta H &= 495 \text{ kJ} \\ NO(g) + O_3(g) &\longrightarrow NO_2(g) + O_2(g) & \Delta H &= -199 \text{ kJ} \end{aligned}$$

calculate ΔH for the reaction

$$NO(g) + O(g) \longrightarrow NO_2(g)$$

63. The bombardier beetle uses an explosive discharge as a defensive measure. The chemical reaction involved is the oxidation of hydroquinone by hydrogen peroxide to produce quinone and water:

$$C_6H_4(OH)_2(aq) + H_2O_2(aq) \longrightarrow C_6H_4O_2(aq) + 2H_2O(l)$$

Calculate ΔH for this reaction from the following data:

$$\begin{aligned} C_6H_4(OH)_2(aq) &\longrightarrow C_6H_4O_2(aq) + H_2(g) & \Delta H &= 177.4 \text{ kJ} \\ H_2(g) + O_2(g) &\longrightarrow H_2O_2(aq) & \Delta H &= -191.2 \text{ kJ} \\ H_2(g) + \tfrac{1}{2}O_2(g) &\longrightarrow H_2O(g) & \Delta H &= -241.8 \text{ kJ} \\ H_2O(g) &\longrightarrow H_2O(l) & \Delta H &= -43.8 \text{ kJ} \end{aligned}$$

64. Given the following data:

$$\begin{aligned} P_4(s) + 6Cl_2(g) &\longrightarrow 4PCl_3(g) & \Delta H &= -1225.6 \text{ kJ} \\ P_4(s) + 5O_2(g) &\longrightarrow P_4O_{10}(s) & \Delta H &= -2967.3 \text{ kJ} \\ PCl_3(g) + Cl_2(g) &\longrightarrow PCl_5(g) & \Delta H &= -84.2 \text{ kJ} \\ PCl_3(g) + \tfrac{1}{2}O_2(g) &\longrightarrow Cl_3PO(g) & \Delta H &= -285.7 \text{ kJ} \end{aligned}$$

calculate ΔH for the reaction

$$P_4O_{10}(s) + 6PCl_5(g) \longrightarrow 10Cl_3PO(g)$$

Standard Enthalpies of Formation

65. Given the definition of the standard enthalpy of formation for a substance, write separate reactions for the formation of NaCl, H_2O, $C_6H_{12}O_6$, and $PbSO_4$ that have $\Delta H°$ values equal to $\Delta H_f°$ for each compound.

66. The combustion of methane can be represented as follows:

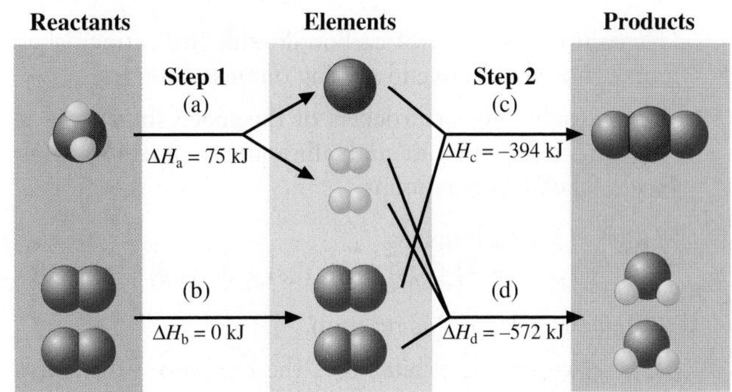

a. Use the information given above to determine the value of ΔH for the combustion of methane to form $CO_2(g)$ and $2H_2O(l)$.
b. What is $\Delta H_f°$ for an element in its standard state? Why is this? Use the figure above to support your answer.
c. How does ΔH for the reaction $CO_2(g) + 2H_2O(l) \longrightarrow CH_4(g) + O_2(g)$ compare to that of the combustion of methane? Why is this?

67. Use the values of $\Delta H_f°$ in Appendix 4 to calculate $\Delta H°$ for the following reactions.

a.

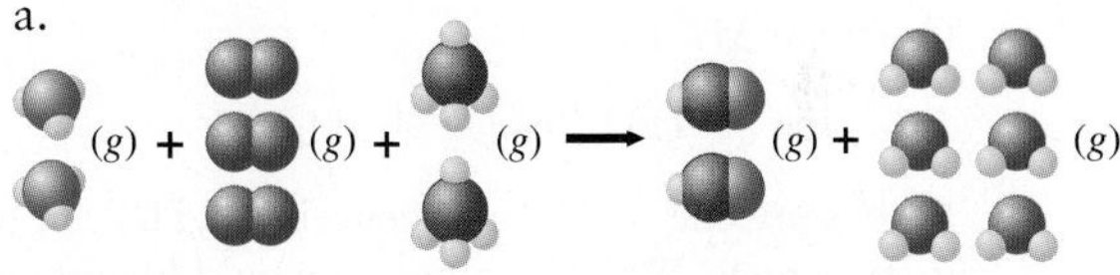

N H O C

b. $Ca_3(PO_4)_2(s) + 3H_2SO_4(l) \longrightarrow 3CaSO_4(s) + 2H_3PO_4(l)$
c. $NH_3(g) + HCl(g) \longrightarrow NH_4Cl(s)$
d. (l) + (g) → (g) + (g)
e. $SiCl_4(l) + 2H_2O(l) \longrightarrow SiO_2(s) + 4HCl(aq)$
f. $MgO(s) + H_2O(l) \longrightarrow Mg(OH)_2(s)$

68. The Ostwald process for the commercial production of nitric acid from ammonia and oxygen involves the following steps:

$$\begin{aligned} 4NH_3(g) + 5O_2(g) &\longrightarrow 4NO(g) + 6H_2O(g) \\ 2NO(g) + O_2(g) &\longrightarrow 2NO_2(g) \\ 3NO_2(g) + H_2O(l) &\longrightarrow 2HNO_3(aq) + NO(g) \end{aligned}$$

a. Use the values of ΔH_f° in Appendix 4 to calculate the value of ΔH° for each of the preceding reactions.
b. Write the overall equation for the production of nitric acid by the Ostwald process by combining the preceding equations. (Water is also a product.) Is the overall reaction exothermic or endothermic?

69. Calculate ΔH° for each of the following reactions using the data in Appendix 4:

$$4Na(s) + O_2(g) \longrightarrow 2Na_2O(s)$$
$$2Na(s) + 2H_2O(l) \longrightarrow 2NaOH(aq) + H_2(g)$$
$$2Na(s) + CO_2(g) \longrightarrow Na_2O(s) + CO(g)$$

Explain why a water or carbon dioxide fire extinguisher might not be effective in putting out a sodium fire.

70. The reusable booster rockets of the space shuttle use a mixture of aluminum and ammonium perchlorate as fuel. A possible reaction is

$$3Al(s) + 3NH_4ClO_4(s) \longrightarrow Al_2O_3(s) + AlCl_3(s) + 3NO(g) + 6H_2O(g)$$

Calculate ΔH° for this reaction.

71. The space shuttle orbiter uses the oxidation of methyl hydrazine by dinitrogentetroxide for propulsion. The balanced reaction is

$$5N_2O_4(l) + 4N_2H_3CH_3(l) \longrightarrow 12H_2O(g) + 9N_2(g) + 4CO_2(g)$$

Calculate ΔH° for this reaction.

72. Does the reaction in Exercise 70 or that in Exercise 71 produce more energy per kilogram of reactant mixture (stoichiometric amounts)?

73. At 298 K, the standard enthalpies of formation for $C_2H_2(g)$ and $C_6H_6(l)$ are 227 kJ/mol and 49 kJ/mol, respectively.
a. Calculate ΔH° for

$$C_6H_6(l) \longrightarrow 3C_2H_2(g)$$

b. Both acetylene (C_2H_2) and benzene (C_6H_6) can be used as fuels. Which compound would liberate more energy per gram when combusted in air?

74. Calculate ΔH° for each of the following reactions, which occur in the atmosphere.
a. $C_2H_4(g) + O_3(g) \rightarrow CH_3CHO(g) + O_2(g)$
b. $O_3(g) + NO(g) \rightarrow NO_2(g) + O_2(g)$
c. $SO_3(g) + H_2O(l) \rightarrow H_2SO_4(aq)$
d. $2NO(g) + O_2(g) \rightarrow 2NO_2(g)$

75. Use the reaction

$$2ClF_3(g) + 2NH_3(g) \longrightarrow N_2(g) + 6HF(g) + Cl_2(g) \qquad \Delta H^\circ = -1196 \text{ kJ}$$

to calculate ΔH_f° for $ClF_3(g)$.

76. The standard enthalpy of combustion of ethene gas [$C_2H_4(g)$] is −1411.1 kJ/mol at 298 K. Given the following enthalpies of formation, calculate ΔH_f° for $C_2H_4(g)$.

$CO_2(g)$	−393.5 kJ/mol
$H_2O(l)$	−285.8 kJ/mol

Energy Consumption and Sources

77. The complete combustion of acetylene [$C_2H_2(g)$] produces 1300. kJ of energy per mole of acetylene consumed. How many grams of acetylene must be burned to produce enough heat to raise the temperature of 1.00 gal of water by 10.0°C if the process is 80.0% efficient? Assume the density of water is 1.00 g/cm^3.

78. Assume that 4.19×10^6 kJ of energy is needed to heat a home. If this energy is derived from the combustion of methane (CH_4), what volume of methane, measured at STP, must be burned? ($\Delta H^\circ_{\text{combustion}}$ for $CH_4 = -891$ kJ/mol)

79. Syngas can be burned directly or converted to methanol. Calculate ΔH° for the reaction

$$CO(g) + 2H_2(g) \longrightarrow CH_3OH(l)$$

80. Ethanol (C_2H_5OH) has been proposed as an alternative fuel. Calculate the standard enthalpy of combustion per gram of liquid ethanol.

81. Methanol (CH_3OH) has also been proposed as an alternative fuel. Calculate the standard enthalpy of combustion per gram of liquid methanol, and compare this answer to that for ethanol in Exercise 80.

82. Some automobiles and buses have been equipped to burn propane (C_3H_8) as a fuel. Compare the amount of energy that can be obtained per gram of $C_3H_8(g)$ with that per gram of gasoline, assuming that gasoline is octane [$C_8H_{18}(l)$]. (See Example 9.8.) Look up the physical properties of propane. What disadvantages are there to using propane instead of gasoline as a fuel?

Additional Exercises

83. Consider the following cyclic process carried out in two steps on a gas:

Step 1: 45 J of heat is added to the gas, and 10. J of expansion work is performed.

Step 2: 60. J of heat is removed from the gas as the gas is compressed back to the initial state.

Calculate the work for the gas compression in step 2.

84. Determine ΔE for the process $H_2O(l) \longrightarrow H_2O(g)$ at 25°C and 1 atm.

85. The standard enthalpy of formation of $H_2O(l)$ at 298 K is −285.8 kJ/mol. Calculate the change in internal energy for the following process at 298 K and 1 atm:

$$H_2O(l) \longrightarrow H_2(g) + \tfrac{1}{2}O_2(g) \qquad \Delta E^\circ = ?$$

86. A piece of chocolate cake contains about 400 Calories. A nutritional Calorie is equal to 1000 calories (thermochemical calories). How many 8-in-high steps must a 180-lb man climb to expend the 400 Cal from the piece of cake? See Exercise 15 for the formula for potential energy.

87. In a bomb calorimeter, the bomb is surrounded by water that must be added for each experiment. Since the amount of water is not constant from experiment to experiment, mass must be measured in each case. The heat capacity of the calorimeter is broken down into two parts: the water and the calorimeter components. If a calorimeter contains 1.00 kg of water and has a total heat capacity of 10.84 kJ/°C, what is the heat capacity of the calorimeter components?

88. The bomb calorimeter in Exercise 87 is filled with 987 g of water. The initial temperature of the calorimeter contents is 23.32°C. A 1.056-g sample of benzoic acid ($\Delta E_{comb} = -26.42$ kJ/g) is combusted in the calorimeter. What is the final temperature of the calorimeter contents?

89. When 1.00 L of 2.00 *M* Na_2SO_4 solution at 30.0°C is added to 2.00 L of 0.750 *M* $Ba(NO_3)_2$ solution at 30.0°C in a calorimeter, a white solid ($BaSO_4$) forms. The temperature of the mixture increases to 42.0°C. Assuming that the specific heat capacity of the solution is 6.37 J $°C^{-1}$ g^{-1} and that the density of the final solution is 2.00 g/mL, calculate the enthalpy change per mole of $BaSO_4$ formed.

90. If a student performs an endothermic reaction in a calorimeter, how does the calculated value of ΔH differ from the actual value if the heat exchanged with the calorimeter is not taken into account?

91. The enthalpy of neutralization for the reaction of a strong acid with a strong base is −56 kJ/mol of water produced. How much energy will be released when 200.0 mL of 0.400 *M* HNO_3 is mixed with 150.0 mL of 0.500 *M* KOH?

92. Three gas-phase reactions were run in a constant-pressure piston apparatus as illustrated. For each reaction, give the balanced reaction and predict the sign of *w* (the work done) for the reaction.

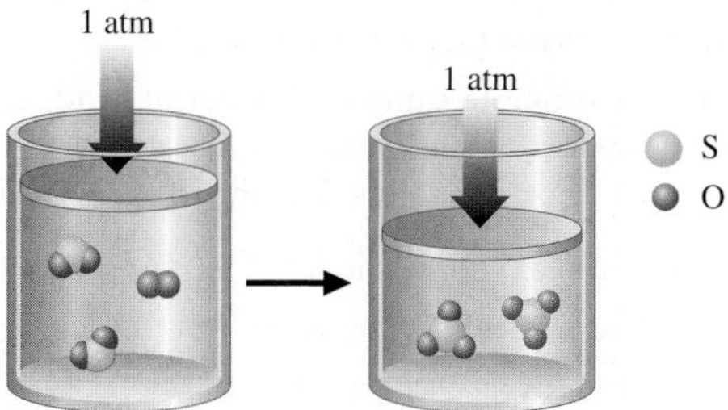

a.

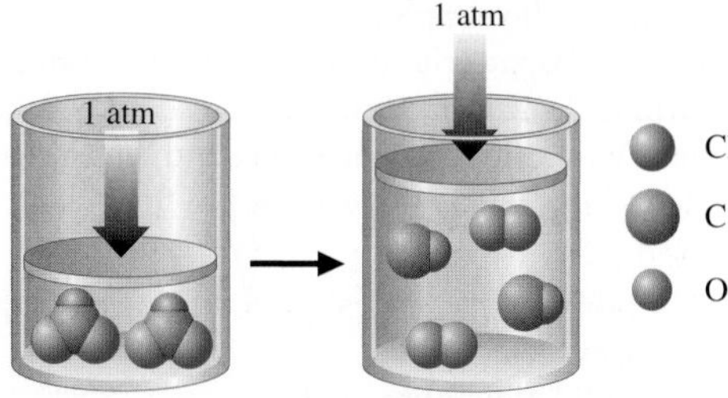

b.

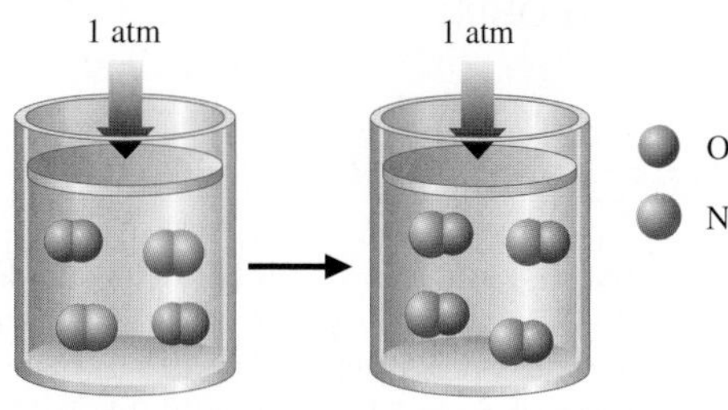

c.

If just the balanced reactions were given, how could you predict the sign of *w* for a reaction?

93. Consider the following changes:
a. $N_2(g) \rightarrow N_2(l)$
b. $CO(g) + H_2O(g) \rightarrow H_2(g) + CO_2(g)$
c. $Ca_3P_2(s) + 6H_2O(l) \rightarrow 3Ca(OH)_2(s) + 2PH_3(g)$
d. $2CH_3OH(l) + 3O_2(g) \rightarrow 2CO_2(g) + 4H_2O(l)$
e. $I_2(s) \rightarrow I_2(g)$

At constant temperature and pressure, in which of these changes is work done by the system on the surroundings? By the surroundings on the system? In which of them is no work done?

94. Nitrogen gas reacts with hydrogen gas to form ammonia gas. Consider the reaction between nitrogen and hydrogen as depicted below:

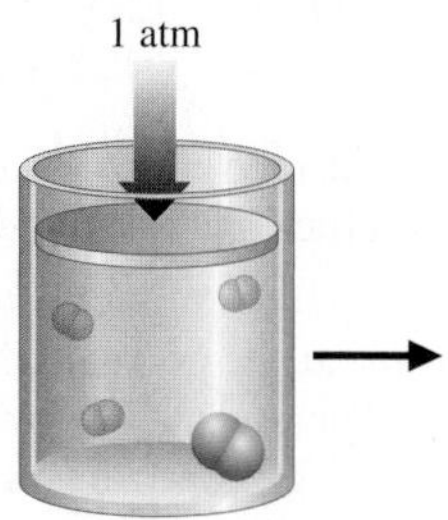

a. Draw what the container will look like after the reaction has gone to completion. Assume a constant pressure of 1 atm.
b. Is the sign of work positive or negative, or is the value of work equal to zero for the reaction? Explain your answer.

95. Using the following data, calculate the standard heat of formation of ICl(g) in kJ/mol:

$$Cl_2(g) \longrightarrow 2Cl(g) \qquad \Delta H° = 242.3 \text{ kJ}$$
$$I_2(g) \longrightarrow 2I(g) \qquad \Delta H° = 151.0 \text{ kJ}$$
$$ICl(g) \longrightarrow I(g) + Cl(g) \qquad \Delta H° = 211.3 \text{ kJ}$$
$$I_2(s) \longrightarrow I_2(g) \qquad \Delta H° = 62.8 \text{ kJ}$$

96. High-quality audio amplifiers generate large amounts of heat. To dissipate the heat and prevent damage to the electronic devices, manufacturers use heat-radiating metal fins. Would it be better to make these fins out of iron or aluminum? Why? (See Table 9.3 for specific heat capacities.)

97. Write reactions that correspond to the following enthalpy changes:
 a. ΔH_f° for solid aluminum oxide
 b. the standard enthalpy of combustion of liquid ethanol [$C_2H_5OH(l)$]
 c. the standard enthalpy of neutralization of barium hydroxide solution by hydrochloric acid
 d. ΔH_f° for gaseous vinyl chloride [$C_2H_3Cl(g)$]
 e. the enthalpy of combustion of liquid benzene [$C_6H_6(l)$]
 f. the enthalpy of solution of solid ammonium bromide

Challenge Problems

98. The heat required to raise the temperature from 300.0 K to 400.0 K for 1 mole of a gas at constant volume is 2079 J. The internal energy required to heat the same gas at constant pressure from 550.0 K to 600.0 K is 1305 J. The gas does 150. J of work during this expansion at constant pressure. Is this gas behaving ideally? Is the gas a monatomic gas? Explain.

99. When water is supercooled, it freezes at a temperature below 0.0°C. If 10.9 kJ of heat is released when 2.00 moles of supercooled water at −15.0°C freezes, calculate the molar enthalpy of fusion for ice at 0.0°C and 1 atm. Assume the molar heat capacities for $H_2O(s)$ and $H_2O(l)$ are 37.5 J K^{-1} mol^{-1} and 75.3 J K^{-1} mol^{-1}, respectively, and are temperature independent.

100. The sun supplies energy at a rate of about 1.0 kilowatt per square meter of surface area (1 watt = 1 J/s). The plants in an agricultural field produce the equivalent of 20. kg of sucrose ($C_{12}H_{22}O_{11}$) per hour per hectare (1 ha = 10,000 m^2). Assuming that sucrose is produced by the reaction

$$12CO_2(g) + 11H_2O(l) \longrightarrow C_{12}H_{22}O_{11}(s) + 12O_2(g) \qquad \Delta H = 5640 \text{ kJ}$$

calculate the percentage of sunlight used to produce the sucrose—that is, determine the efficiency of photosynthesis.

101. The heat of vaporization of water at the normal boiling point, 373.2 K, is 40.66 kJ/mol. The specific heat capacity of liquid water is 4.184 J K^{-1} g^{-1} and of gaseous water is 2.02 J K^{-1} g^{-1}. Assume that these values are independent of temperature. What is the heat of vaporization of water at 298.2 K? Does this result agree with Appendix 4 data?

102. Consider the following reaction at 248°C and 1.00 atm:

$$CH_3Cl(g) + H_2(g) \longrightarrow CH_4(g) + HCl(g)$$

For this reaction, the enthalpy change at 248°C is −83.3 kJ/mol. At constant pressure the molar heat capacities (C_p) for the compounds are as follows: CH_3Cl (48.5 J K^{-1} mol^{-1}), H_2 (28.9 J K^{-1} mol^{-1}), CH_4 (41.3 J K^{-1} mol^{-1}), and HCl (29.1 J K^{-1} mol^{-1}).
 a. Assuming that the C_p values are independent of temperature, calculate ΔH° for this reaction at 25°C.
 b. Calculate ΔH_f° for CH_3Cl using data from Appendix 4 and the result from part a.

103. The best solar panels currently available are about 15% efficient in converting sunlight to electricity. A typical home will use about 40. kWh of electricity per day (1 kWh = 1 kilowatt hour; 1 kW = 1000 J/s). Assuming 8.0 hours of useful sunlight per day, calculate the minimum solar panel surface area necessary to provide all of a typical home's electricity. (See Exercise 100 for the energy rate supplied by the sun.)

104. You have 2.4 moles of a gas contained in a 4.0-L bulb at a temperature of 32°C. This bulb is connected to a 20.0-L sealed, initially evacuated bulb via a valve. Assume the temperature remains constant.
 a. What should happen to the gas when you open the valve? Calculate any changes of conditions.
 b. Calculate ΔH, ΔE, q, and w for the process you described in part a.
 c. Given your answer to part b, what is the driving force for the process?

105. An isothermal process is one in which the temperatures of the system and surroundings remain constant at all times. With this in mind, what is wrong with the following statement: "For an isothermal expansion of an ideal gas against a constant pressure, $\Delta T = 0$, so $q = 0$"? What is q equal to in an isothermal expansion of an ideal gas against a constant external pressure?

106. You have a 1.00-mole sample of water at −30.°C, and you heat it until you have gaseous water at 140.°C. Calculate q for the entire process. Use the following data:

Specific heat capacity of ice = 2.03 J °C^{-1} g^{-1}
Specific heat capacity of water = 4.18 J °C^{-1} g^{-1}
Specific heat capacity of steam = 2.02 J °C^{-1} g^{-1}

$H_2O(s) \longrightarrow H_2O(l)$ $\qquad \Delta H_{fusion} = 6.01$ kJ/mol (at 0°C)
$H_2O(l) \longrightarrow H_2O(g)$ $\qquad \Delta H_{vaporization} = 40.7$ kJ/mol (at 100.°C)

Marathon Problems

107.* A sample consisting of 22.7 g of a nongaseous, unstable compound X is placed inside a metal cylinder with a radius of 8.00 cm, and a piston is carefully placed on the surface of the compound so that, for all practical purposes, the distance between the bottom of the cylinder and the piston is zero. (A hole in the piston allows trapped air to escape as the piston is placed on the compound; then this hole is plugged so that nothing inside the cylinder can escape.) The piston-and-cylinder apparatus is carefully placed in 10.00 kg of water at 25.00°C. The barometric pressure is 778 torr.

When the compound spontaneously decomposes, the piston moves up, the temperature of the water reaches a maximum of 29.52°C, and then it gradually decreases as the water loses heat to the surrounding air. The distance between the piston and the bottom of the cylinder, at the maximum temperature, is 59.8 cm. Chemical analysis shows that the cylinder contains 0.300 mole of carbon dioxide, 0.250 mole of liquid water, 0.025 mole of oxygen gas, and an undetermined amount of a gaseous element A.

It is known that the enthalpy change for the decomposition of X, according to the reaction described above, is −1893 kJ/mol X. The heat capacity for water is 4.184 J $°C^{-1}$ g^{-1}. The conversion factor between L atm and J can be determined from the two values for the gas constant *R*—namely, 0.08206 L atm K^{-1} mol^{-1} and 8.3145 J K^{-1} mol^{-1}. The vapor pressure of water at 29.5°C is 31 torr. Assume that the heat capacity of the piston-and-cylinder apparatus is negligible and that the piston has negligible mass.

Given the preceding information, determine

a. the formula for X.
b. the pressure-volume work (in kJ) for the decomposition of the 22.7-g sample of X.
c. the *molar* change in internal energy for the decomposition of X.

*From James H. Burness, "The Use of "Marathon" Problems as Effective Vehicles for the Presentation of General Chemistry Lectures," Journal of Chemical Education, 68(11). Copyright © 1991 American Chemical Society. Reprinted by permission.

108. A gaseous hydrocarbon reacts completely with oxygen gas to form carbon dioxide and water vapor. Given the following data, determine ΔH_f° for the hydrocarbon:

$\Delta H_{rxn} = -2044.5$ kJ/mol

$\Delta H_f^\circ\ (CO_2) = -393.5$ kJ/mol

$\Delta H_f^\circ\ (H_2O) = -242$ kJ/mol

Density of CO_2 and H_2O mixture at 1 atm, 200.°C = 0.751 g/L

The density of the hydrocarbon is less than the density of Kr at the same conditions.

10 Spontaneity, Entropy, and Free Energy

chapter

Methane burns spontaneously in air to release energy as heat and light.

Sign in to OWL at www.cengage.com/owl to view tutorials and simulations, develop problem-solving skills, and complete online homework assigned by your professor.

go Chemistry

Download mini lecture videos for key concept review and exam prep from OWL or purchase them from www.cengagebrain.com.

The *first law of thermodynamics* is a statement of the law of conservation of energy: Energy can be neither created nor destroyed. In other words, *the energy of the universe is constant.* Although the total energy remains constant, the various forms of energy can be interchanged through physical and chemical processes. For example, if you drop a book, some of the initial potential energy of the book is changed to kinetic energy, which is transferred to the atoms in the air and the floor as random motion. The net effect of this process is to change a given quantity of potential energy to exactly the same quantity of thermal energy. Energy has been converted from one form to another, but the same quantity of energy exists before and after the process.

Now we will consider a chemical example. When methane is burned in excess oxygen, the major reaction is

$$CH_4(g) + 2O_2(g) \longrightarrow CO_2(g) + 2H_2O(g) + \text{energy}$$

This reaction produces a quantity of energy that is released as heat. This energy flow results from a lowering of the potential energy stored in the bonds of CH_4 and O_2 as they react to form CO_2 and H_2O (Fig. 10.1). Potential energy is converted to thermal energy, but the energy content of the universe remains constant.

The first law of thermodynamics is used mainly for energy bookkeeping—that is, to answer questions such as the following:

How much energy is involved in the change?

Does energy flow into or out of the system?

What form does the energy finally assume?

The first law of thermodynamics: The energy of the universe is constant.

Although the first law of thermodynamics provides the means to account for energy changes, it gives no hint as to *why* a particular process occurs in a given direction. What are the driving forces that cause a process to occur? This is the main question to be considered in this chapter.

Also you might wonder why there is an energy supply crisis in the world given that the first law of thermodynamics states that the energy of the universe is constant. It turns out that the problem is not the *quantity* of energy in the universe but the *quality* of that energy. The key question is, What happens to the usefulness of energy when we convert it from one form to another? We will see in this chapter that when we "use" energy, the amount of energy is unchanged but the new form of the energy is less useful than the original form.

10.1 Spontaneous Processes

Spontaneous does not mean fast.

A process is said to be *spontaneous* if it *occurs without outside intervention.* **Spontaneous processes** may be fast or slow. As we will see in this chapter, thermodynamics can tell us the *direction* in which a process will occur but can say nothing about the *speed* (rate) of the process. As we will explore in detail in Chapter 15, the rate of a reaction depends on many factors, including temperature and concentration. In describing a chemical reaction, the discipline of

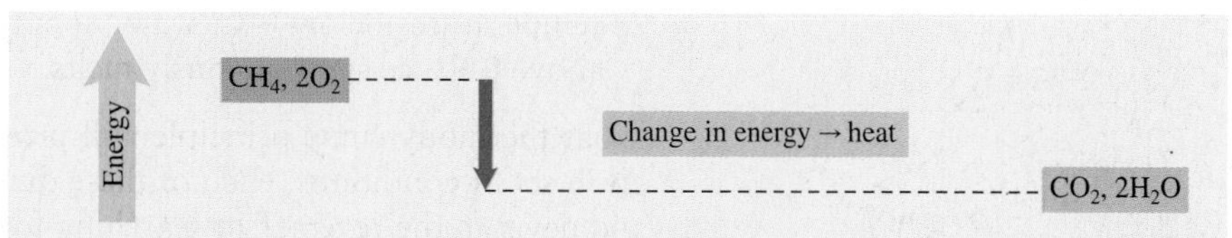

Figure 10.1
When methane and oxygen react to form carbon dioxide and water, the products have lower potential energy than the reactants. This change in potential energy results in energy flow (heat) to the surroundings.

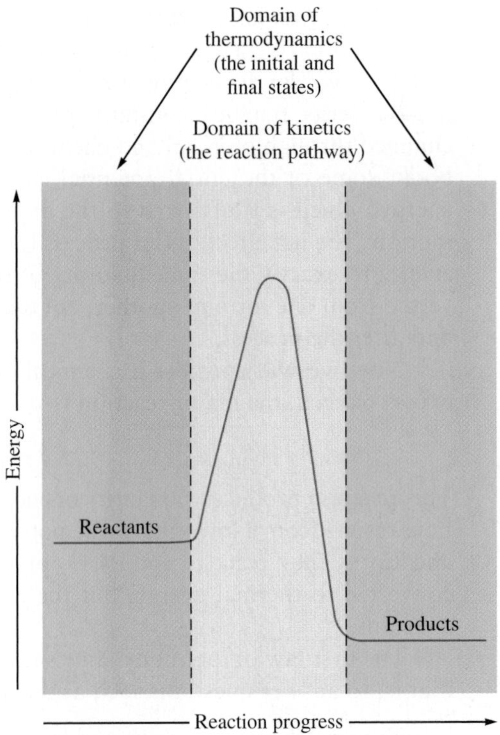

Figure 10.2
The rate of a reaction depends on the pathway from reactants to products; this is the domain of kinetics. Thermodynamics tells us whether a reaction is spontaneous based only on the properties of the reactants and products. The predictions of thermodynamics do not require knowledge of the pathway between reactants and products.

chemical kinetics (the study of reaction rates) focuses on the pathway between reactants and products; in contrast, thermodynamics considers only the initial and final states and does not require knowledge of the pathway between the reactants and products (Fig. 10.2).

In summary, thermodynamics lets us predict whether a process will occur but gives no information about the amount of time required. For example, according to the principles of thermodynamics, a diamond should change spontaneously to graphite at 25°C and 1 atm pressure. The fact that we do not observe this process does not mean the prediction is wrong; it simply means the process is too slow to observe. Thus we need both thermodynamics and kinetics to describe reactions fully.

To explore the idea of spontaneity, consider the following physical and chemical processes:

A ball rolls down a hill but never spontaneously rolls back up the hill.

If exposed to air and moisture, steel rusts spontaneously. However, the iron oxide in rust does not spontaneously change back to iron metal and oxygen gas.

A gas fills its container uniformly. It never spontaneously collects at one end of the container.

Heat flow always occurs from a hot object to a cooler one. The reverse process never occurs spontaneously.

Wood burns spontaneously in an exothermic reaction to form carbon dioxide and water, but wood is not formed when carbon dioxide and water are heated together.

At temperatures below 0°C, water spontaneously freezes; and at temperatures above 0°C, ice spontaneously melts.

What thermodynamic principle will provide an explanation for why, under a given set of conditions, each of these diverse processes occurs in one direction and never in the reverse? In searching for an answer, we could explain the be-

Chaos, Keep It Coming!

Can you imagine how life would be
If there were no entropy?
Or, making matters even worse,
The laws of entropy were reversed?
Books would get straighter on their shelves,
And children's rooms would clean themselves!
And every rock or stick or tree
Would form a crystal, perfectly.
There'd be no anarchy or war
For everyone would know the score.
Every thing and every face
Would have its certain time and place.
Replacing every beach would pass
An endless stretch of flawless glass.
The sea would be the brightest blue,
And every day the sky would too.
How beautiful would be our world
If order did command it.
If all were straight and never curled:
Perhaps we should demand it.

You'd think a world sans entropy
Would be a lovely place to be.
I said this recently myself,
As all my books fell off their shelf.
Yet pondering this ordered bliss,
I noticed things that I would miss,
Like rolling waves upon the sea,
Or sugar for my morning tea:
The sugar won't dissolve, it's true,
That anti-entropy holds like glue.
And after that, I saw with grief,
There'd be no fractaled maple leaf:
No beauty in the summer wood,
Should chaos disappear for good.
What a bore, to know each day
Would turn out in the same old way.
If entropy would disappear
There'd be no fortune, fate or luck
And even after many years,
Vegas wouldn't make a buck.

Heather Ryphemi Stregay

havior of a ball on a hill in terms of gravity. But what does gravity have to do with the rusting of a nail or the freezing of water? Early developers of thermodynamics thought that exothermicity might be the key, that a process would be spontaneous if it were exothermic. Although this factor does appear to be important, since many spontaneous processes are exothermic, it is not the only factor. For example, the melting of ice, which occurs spontaneously at temperatures greater than 0°C, is an endothermic process.

What common characteristic causes the processes listed earlier to be spontaneous in one direction only? After many years of observation, scientists have concluded that the characteristic common to all spontaneous processes is an increase in a property called **entropy** (S). *The change in the entropy of the universe for a given process is a measure of the driving force behind that process.*

What is entropy? Ultimately, entropy is about how energy is distributed among the energy levels in the "particles" that constitute a given system. How-

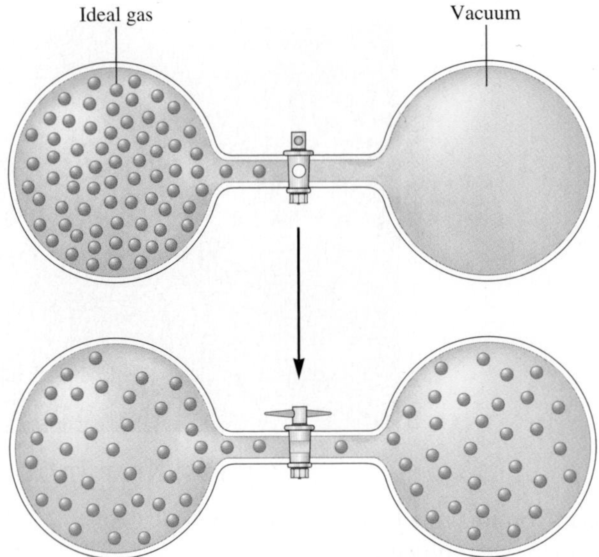

Figure 10.3
The expansion of an ideal gas into an evacuated bulb.

ever, in beginning to understand how entropy operates, it is useful to think about how probability is a driving force in the macroscopic world around us. The macroscopic world illustrates that the natural progression of things is from order to disorder—that is, from lower probability to higher probability. You merely have to think about the condition of your room to be convinced of this. Your room naturally tends to get messy (disordered) because an ordered room requires everything to be in its place. There are simply many more ways for things to be out of place than to be in place, as the poem by Heather Ryphemi Stregay, written while she was a student in general chemistry, observes.*

As another example, suppose you have a deck of playing cards ordered in some particular way. You throw these cards into the air and pick them all up at random. Looking at the new sequence of the cards, you would be very surprised to find that it matched the original order. Such an event would be possible but *very improbable*. There are billions of ways for the deck to be disordered but only one way to be ordered according to your definition. Thus the chances of picking the cards up out of order are much greater than of picking them up in order. It is natural for disorder to increase.

Entropy is closely associated with probability. The key concept is that the more ways a particular state can be achieved, the greater is the likelihood (probability) that that state will occur. In other words, *nature spontaneously proceeds toward the states that have the highest probabilities of existing*. This conclusion is not surprising at all. The difficulty comes in connecting this concept to real-life processes. For example, what does the spontaneous rusting of steel have to do with probability?

Understanding the connection between entropy and spontaneity will allow us to answer such questions. We will begin to explore this connection by considering a very simple process, the expansion of an ideal gas into a vacuum, as represented in Fig. 10.3. Why is this process spontaneous? What causes the gas to expand to a uniform state? The driving force can be explained in terms of simple probability. Because there are more ways of having the gas evenly spread throughout the container than there are ways for it to be in any other possible state, the gas spontaneously attains the uniform distribution.

To understand this conclusion, we will greatly simplify the system and consider some of the possible arrangements of only four gas molecules in a

---

*In her poem Ms. Stregay takes poetic license in her use of the term *entropy*.

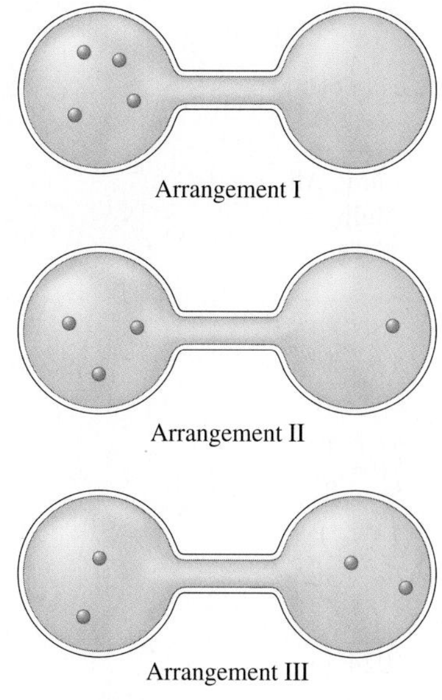

Figure 10.4
Three possible arrangements (states) of four molecules in a two-bulbed flask.

Table 10.1

The Microstates That Give a Particular Arrangement (State)

Arrangement	Microstates	
I		A B C D | —
II	B D C | A	A B D | C
	A C D | B	A B C | D
III	A B | C D	B C | A D
	A C | B D	B D | A C
	A D | B C	C D | A B

two-bulbed container* (Fig. 10.4). How many ways can each arrangement (state) be achieved? Arrangement I can be achieved in only one way—all the molecules must be in one end. Arrangement II can be achieved in four ways, as shown in Table 10.1. Each configuration that gives a particular arrangement is called a *microstate.* Arrangement I has one microstate, and arrangement II has four microstates. Arrangement III can be achieved in six ways (six microstates), as shown in Table 10.1. *Which arrangement is most likely to occur?* The one that can be achieved in the greatest number of ways is the most probable. Thus arrangement III is most probable, and the relative probabilities of arrangements III, II, and I are 6:4:1. We have discovered an important principle: *The probability of occurrence of a particular arrangement (state) depends on the number of ways (microstates) in which that arrangement can be achieved.*

The consequences of this principle are dramatic for large numbers of molecules. One gas molecule in the flask in Fig. 10.4 has one chance in two of being in the left bulb. We say that the probability of finding the molecule in the left bulb is $\frac{1}{2}$. For two molecules in the flask, there is one chance in two of finding each molecule in the left bulb, so there is one chance in four ($\frac{1}{2} \times \frac{1}{2} = \frac{1}{4}$) that *both* molecules will be in the left bulb. As the number of molecules increases, the relative probability of finding all of them in the left bulb decreases, as shown in Table 10.2. For 1 mole of gas, the probability of finding all the molecules in the left bulb is so small that this arrangement would "never" occur.

Thus a gas placed in one end of a container will spontaneously expand to fill the entire vessel evenly because for a large number of gas molecules there is a huge number of microstates corresponding to equal numbers of molecules in both ends. On the other hand, the opposite process,

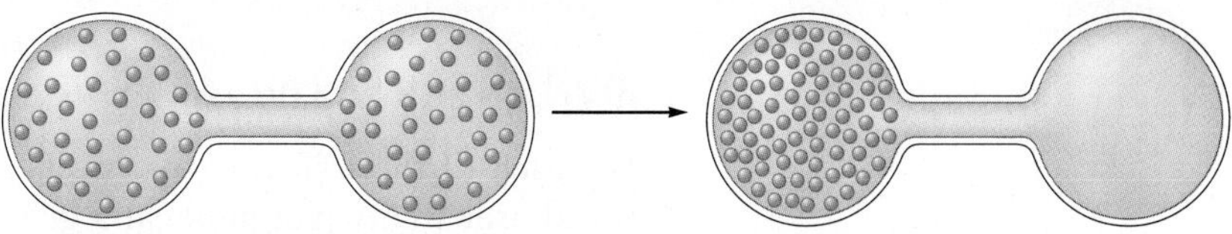

*Note that this treatment is oversimplified. In reality, the molecules of an ideal gas are indistinguishable—we can't really label them as in this example. The general idea, however, is correct.

Table 10.2

Probability of Finding All the Molecules in the Left Bulb as a Function of the Total Number of Molecules

Number of Molecules	Relative Probability of Finding All Molecules in the Left Bulb
1	$\frac{1}{2}$
2	$\frac{1}{2} \times \frac{1}{2} = \frac{1}{2^2} = \frac{1}{4}$
3	$\frac{1}{2} \times \frac{1}{2} \times \frac{1}{2} = \frac{1}{2^3} = \frac{1}{8}$
5	$\frac{1}{2} \times \frac{1}{2} \times \frac{1}{2} \times \frac{1}{2} \times \frac{1}{2} = \frac{1}{2^5} = \frac{1}{32}$
10	$\frac{1}{2^{10}} = \frac{1}{1024}$
n	$\frac{1}{2^n} = \left(\frac{1}{2}\right)^n$
6×10^{23} (1 mole)	$\left(\frac{1}{2}\right)^{6 \times 10^{23}} = 10^{-(2 \times 10^{23})}$

For two molecules in the flask, there are four possible microstates:

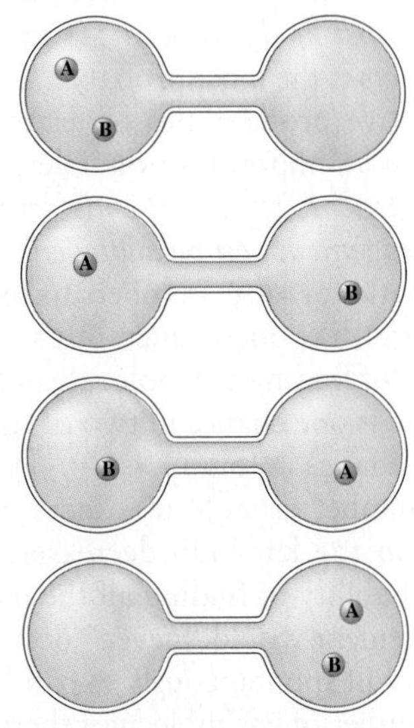

Thus there is one chance in four of finding

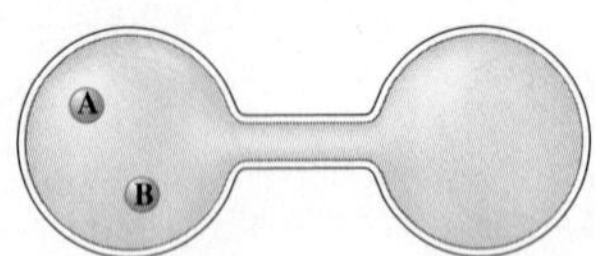

although not impossible, is *highly* improbable since only one microstate leads to this arrangement. Therefore, this process does not occur spontaneously.

The type of probability we have been considering in this example is called **positional probability** because it depends on the number of configurations in space (positional microstates) that yield a particular state. A gas expands into a vacuum to give a uniform distribution because the expanded state has the highest positional probability of the states available to the system.

Positional probability is also illustrated by changes of state. In general, positional probability increases in going from solid to liquid to gas. A mole of a substance has a much smaller volume in the solid state than in the gaseous state. In the solid state the molecules are close together, with relatively few positions available to them; in the gaseous state the molecules are far apart, with many more positions available to them.

Positional probability can also be invoked to explain the formation of solutions. The change in positional probability associated with the mixing of two pure substances is expected to be positive. There are many more microstates for the mixed condition than for the separated condition because of the increased volume available to the particles of each component of the mixture. For example, when two liquids are mixed, the molecules of each liquid have more available space and thus more available positions. This will be discussed in detail in Chapter 17.

OWL INTERACTIVE EXAMPLE 10.1

For each of the following pairs, choose the substance with the higher positional probability (per mole) at a given temperature.

a. solid CO_2 and gaseous CO_2

b. N_2 gas at 1 atm and N_2 gas at 1.0×10^{-2} atm

Solid iodine being heated slightly, causing it to sublime (turn to a vapor). The vapor recrystallizes on a cool evaporating dish.

Solution

a. Since a mole of gaseous CO_2 has the greater volume, the molecules have many more available positions than in a mole of solid CO_2. Thus gaseous CO_2 has the higher positional probability.

b. A mole of N_2 gas at 1×10^{-2} atm has a volume 100 times that (at a given temperature) of a mole of N_2 gas at 1 atm. Thus N_2 gas at 1×10^{-2} atm has the higher positional probability.

OWL INTERACTIVE EXAMPLE 10.2

Predict the sign of the change in positional probability for each of the following processes.

a. Solid sugar is added to water to form a solution.

b. Iodine vapor condenses on a cold surface to form crystals.

Solution

a. The sugar molecules become randomly dispersed in the water when the solution forms. The sugar molecules have access to a larger volume and therefore have more positions available to them. Thus the positional disorder increases.

b. Gaseous iodine is forming a solid. This process involves a change from a relatively large volume to a much smaller volume, which results in lower positional probability.

10.2 The Isothermal Expansion and Compression of an Ideal Gas

In this section we will lay the groundwork for several fundamental concepts of thermodynamics by considering the isothermal expansion and compression of an ideal gas. An **isothermal process** is one in which the temperatures of the system and the surroundings remain constant at all times. Recall that the energy of an ideal gas can be changed only by changing its temperature. Therefore, for any isothermal process *involving an ideal gas,*

$$\Delta E = 0$$

and since

$$\Delta E = q + w = 0$$

then

$$q = -w$$

To illustrate the work and heat effects that accompany the expansion or compression of an ideal gas, consider the apparatus shown in Fig. 10.5. Assume that the pulley is frictionless and that the cable and pan have zero mass.

Initially, assume that the gas occupies a volume V_1 at pressure P_1, where P_1 is just balanced by a mass M_1 on the pan. Thus

$$P_1 = \frac{\text{force}}{\text{area}} = \frac{M_1 g}{A}$$

where A is the area of the piston and g is the gravitational constant. Thus state 1 of the gas is defined by P_1, V_1, n, and T, where n and T remain constant as the expansion occurs.

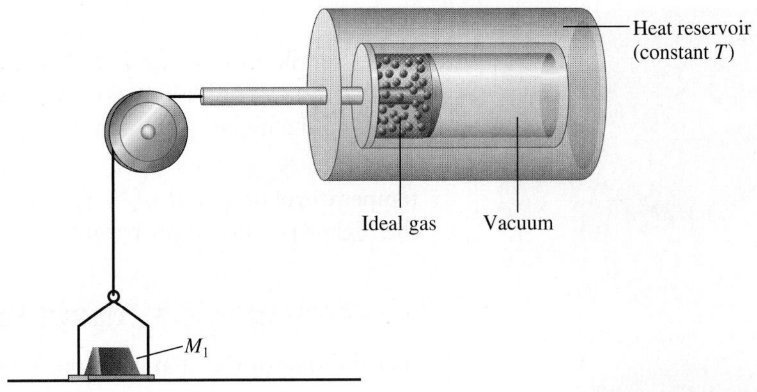

Figure 10.5
A device for the isothermal expansion/compression of an ideal gas.

One-Step Expansion—No Work. If mass M_1 is removed from the pan, the gas will expand, moving the piston to the right end of the cylinder. After expansion, the gas occupies a volume $V_2 = 4V_1$ and pressure $P_2 = P_1/4$.

When the process goes from state 1 (P_1, V_1) to state 2 ($P_1/4$, $4V_1$) with no mass on the pan, no heat flows into or out of the gas because T is constant, and no work is done (no mass is lifted). Thus work $= w_0 = 0$. This is called a *free expansion.*

P_{external} (the pressure against which the gas expands) is zero in a free expansion.

One-Step Expansion. Now consider an experiment with the gas initially at state 1 where the mass M_1 is replaced by a mass $M_1/4$. The gas will now expand against the pressure (P_{external}):

$$P_{\text{ex}} = \frac{\left(\frac{M_1}{4}\right)g}{A} = \frac{P_1}{4}$$

The mass is lifted and the gas expands until the pressure is $P_1/4$. The new volume is $4V_1$. In this case work is performed:

$$|\text{Work}| = |w_1| = \left(\frac{M_1}{4}\right)gh$$

where h is the *change* in height of the mass.

This work can also be expressed in terms of the external pressure (P_{ex}) on the gas as it expands and the change in volume (ΔV). Recall from Chapter 9 that

$$w = -P_{\text{ex}}\Delta V$$

In this expansion the magnitude (absolute value) of the work is

$$|w_1| = P_{\text{ex}}\Delta V = \frac{P_1}{4}(V_2 - V_1) = \frac{P_1}{4}(4V_1 - V_1) = \frac{3}{4}P_1V_1$$

Since in this case work flows out of the system into the surroundings, the correct sign is

$$w_1 = -P_{\text{ex}}\Delta V = -\frac{3}{4}P_1V_1$$

Two-Step Expansion. Next, we will expand the gas in two steps by using two different weights on the pan. First, we put a weight with mass $M_1/2$ on the pan. In this case

$$P_{\text{ex}} = \frac{\left(\frac{M_1}{2}\right)g}{A} = \frac{P_1}{2}$$

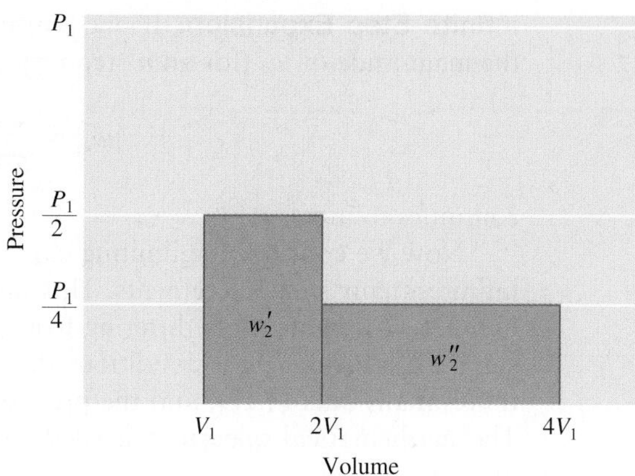

Figure 10.6
The *PV* diagram for a two-step expansion.

and the gas expands until $P_2 = P_1/2$ and $V_2 = 2V_1$. The magnitude of the work is

$$|w'_2| = \frac{P_1}{2}(V_2 - V_1) = \frac{P_1}{2}(2V_1 - V_1) = \frac{P_1V_1}{2}$$

Next, replace the mass $M_1/2$ with a mass $M_1/4$. The gas expands again until $P_3 = P_1/4$ and $V_3 = 4V_1$. The quantity of work in this step is

$$|w''_2| = \frac{P_1}{4}(4V_1 - 2V_1) = \frac{P_1V_1}{2}$$

The total quantity of work in this two-step expansion is

$$|w_2| = \frac{P_1V_1}{2} + \frac{P_1V_1}{2} = P_1V_1$$

With its correct sign $w_2 = -P_1V_1$. This process is diagramed in Fig. 10.6. Note that $|w_2| > |w_1| > |w_0|$, even though in each case the gas is taken from state 1 (P_1, V_1) to state 2 $(P_1/4, 4V_1)$. This result illustrates a property of work we have discussed before: Work is pathway-dependent—it is not a state function.

Six-Step Expansion. Next, we will consider the expansion of the gas from state 1 to state 2 in six steps using several masses between M_1 and $M_1/4$. This process is summarized in Fig. 10.7. In this case $|w_6|$, which is the sum of these six steps, is clearly greater than $|w_2|$.

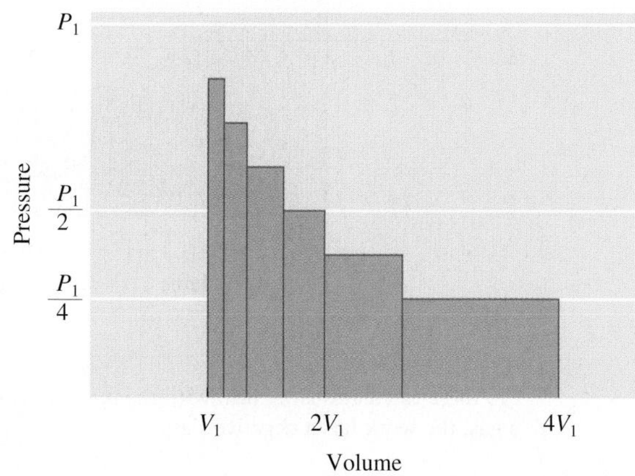

Figure 10.7
The *PV* diagram for a six-step expansion.

Infinite-Step Expansion. If one continues to increase the number of steps, the magnitude of w_n (for an n-step process),

$$|w_n| = \sum_{i=1}^{n} P_i \Delta V_i$$

continues to increase.

Now we consider the limiting case—a process in which P_{ex} is changed by infinitesimally small increments. This case corresponds to the use of an infinite number of weights, each differing from the previous one by an infinitesimally small mass. Under these conditions the successive volume changes become infinitesimally small (dV), and the process requires an infinite number of steps. The mathematical operation needed to sum the steps in this instance is the integral:

$$|\text{Work}| = \int_{V_1}^{V_2} P_{ex} dV$$

The diagram corresponding to this process is given in Fig. 10.8.

It is important to recognize that when the expansion of the gas is carried out in an infinite number of steps, the *external pressure is always almost exactly equal to the pressure produced by the gas.* That is, at any given time P_{ex} is only less than $P_{gas}(= P)$ by an infinitesimally small amount dP. Thus one can assume that at any point in the process $P \approx P_{ex}$.

A process like this one, carried out so that the system is always at equilibrium, is called a *reversible process.* The expansions described earlier that were carried out in a finite number of steps were not done reversibly because in these processes P_{ex} was smaller than P by a finite amount during each step. We will have more to say presently about what the term *reversible* means in this context.

Since $P_{ex} \approx P_{gas} = P$ in the reversible expansion, by use of the ideal gas law,

$$P_{ex} \approx P = \frac{nRT}{V}$$

and

$$|\text{Total work}|^* = |w_\infty| = |w_{rev}| = \int_{V_1}^{V_2} \frac{nRT\,dV}{V}$$

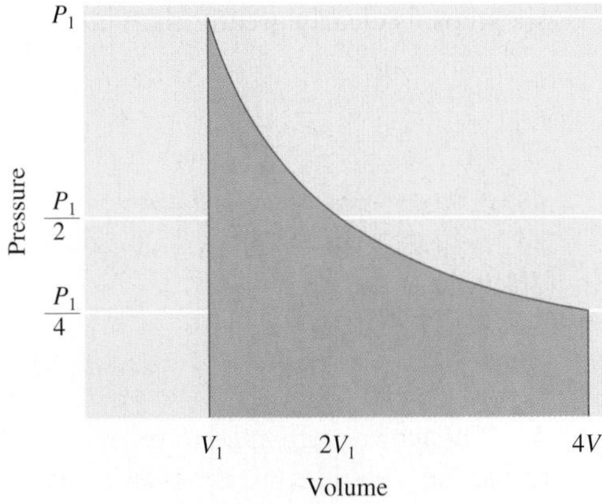

Figure 10.8
The *PV* diagram for the reversible expansion.

*In these calculations we deal with the magnitude for convenience. Actually, for the expansion of a gas, the work has a negative sign.

Since n and T are held constant in this experiment,

$$|w_{\text{rev}}| = nRT\int_{V_1}^{V_2} \frac{dV}{V} = nRT(\ln V_2 - \ln V_1) = nRT \ln\left(\frac{V_2}{V_1}\right)$$

In this specific experiment $V_2 = 4V_1$. Therefore,

$$|w_{\text{rev}}| = nRT \ln 4 = 1.4nRT$$

And since $P_1V_1 = nRT$,

$$|w_{\text{rev}}| = 1.4P_1V_1$$

for this particular expansion.

Note that as the number of steps in the expansion increases, the amount of work the gas performs also increases. The maximum work that a given amount of gas can perform in going from V_1 to V_2 at constant temperature occurs in the reversible expansion. Thus

$$|w_{\text{max}}| = |w_{\text{rev}}| = nRT \ln\left(\frac{V_2}{V_1}\right)$$

for the isothermal expansion of n moles of an ideal gas.

In doing this hypothetical experiment, we have assumed that the gas behaves ideally. When a given amount of an ideal gas expands at constant temperature, the internal energy of the gas remains constant; that is, $\Delta E = 0$. As mentioned earlier, this condition means that $q = -w$. Thus a quantity of energy q (equal in magnitude to w) flows into the gas (as heat) as the expansion occurs, and the work (w) is performed. Therefore, the surroundings furnished the energy (through heat flow) necessary to perform the work.

Since work flows out of the system in the reversible expansion, it has a negative sign:

$$w_{\text{rev}} = -nRT \ln\left(\frac{V_2}{V_1}\right)$$

and

$$q_{\text{rev}} = -w_{\text{rev}} = nRT \ln\left(\frac{V_2}{V_1}\right)$$

The Isothermal Compression of an Ideal Gas

Now let's consider the opposite experiment: compressing a gas at pressure $P_1/4$ and volume $4V_1$ to pressure P_1 and volume V_1. This experiment can be done in one step or a number of steps.

One-Step Compression. Initially, the gas is at pressure $P_1/4$ and volume $4V_1$. When mass M_1 is placed on the pan, the gas will be rapidly compressed to the state described by P_1 and V_1. To see how this process works, consider the diagram for the gas at $P_1/4$ and $4V_1$ (Fig. 10.9). We raise the mass M_1 to the pan, which then causes the gas to be compressed to P_1 and V_1 as the pan returns to the original level.

The work performed to recompress the gas is equal to the work performed to lift the weight up to the pan. It also can be expressed in terms of $P_{\text{ex}}\Delta V$:

$$|w_1'| = M_1gh = P_1\Delta V = P_1(4V_1 - V_1) = 3P_1V_1$$

Two-Step Compression. We can compress the gas in two steps. For example, a mass $M_1/2$ is lifted onto the pan, and after the gas has been partially

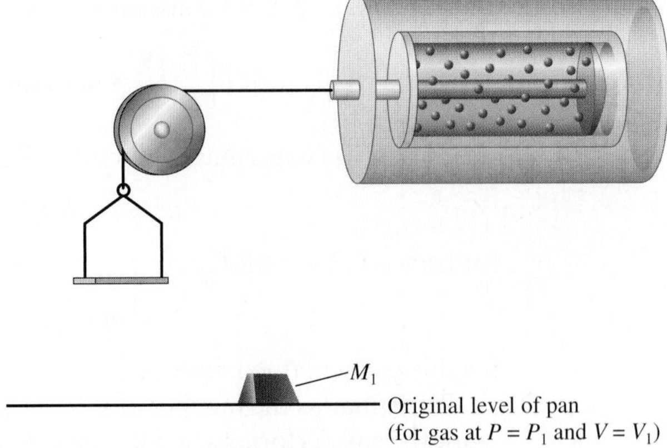

Figure 10.9
The situation before the one-step compression.

compressed, M_1 is put onto the pan in place of $M_1/2$ to finish the compression. In this case the total work required for the compression is

$$|w'_2| = \frac{P_1}{2}(4V_1 - 2V_1) + P_1(2V_1 - V_1) = 2P_1V_1$$

Infinite-Step Compression. Notice that in compressing the gas isothermally, as the number of steps increases, the work required to compress the gas decreases.

If we compress the gas in an infinite number of steps (in which, at all times, $P_{ex} \approx P$), the work required is

$$|w'_\infty| = \int_{V_1}^{V_2} P dV = nRT \ln\left(\frac{V_2}{V_1}\right)$$
$$= nRT \ln\left(\frac{V_1}{4V_1}\right) = 1.4P_1V_1$$

Because $P_{ex} \approx P$ throughout the process, this is a *reversible* compression. In this case w has a positive sign because we are performing work on the system. Thus, in the reversible, isothermal compression of the gas, $w'_\infty = 1.4P_1V_1$, and since $\Delta E = 0$,

$$q_\infty = -w'_\infty = -1.4P_1V_1$$

As the gas is compressed reversibly and isothermally, it releases $1.4P_1V_1$ (L atm) of energy as heat to the surroundings. In other words, the same quantity of energy flows into the gas as work and flows out of the gas as heat to produce no net change in E as the compression occurs.

In general terms, for a reversible compression from V_1 to V_2,

$$w_{rev} = -q_{rev} = -nRT \ln\left(\frac{V_2}{V_1}\right)$$

or in terms of pressures,

$$w_{rev} = -q_{rev} = -nRT \ln\left(\frac{P_1}{P_2}\right)$$

Note that since this process is a compression ($V_2 < V_1$), $\ln(V_2/V_1)$ has a negative sign, making w_{rev} positive and q_{rev} negative, as expected.

Summary

We summarize the results of these expansion and compression experiments in Table 10.3. The most important conclusion that can be drawn from these re-

Table 10.3

Summary of the Isothermal Expansion and Compression Experiments

	Number of Steps	w	q
Expansion (constant T)	0 (no mass)	0	0
	1	$-0.75P_1V_1$	$0.75P_1V_1$
	2	$-1P_1V_1$	$1P_1V_1$
	4	$-1.16P_1V_1$	$1.16P_1V_1$
	∞	$-1.4P_1V_1$	$1.4P_1V_1$
Compression (constant T)	1	$3P_1V_1$	$-3P_1V_1$
	2	$2P_1V_1$	$-2P_1V_1$
	4	$1.67P_1V_1$	$-1.67P_1V_1$
	∞	$1.4P_1V_1$	$-1.4P_1V_1$

sults can be stated as follows: Only when the expansion and compression *are both done reversibly* (by an infinite number of steps) is the universe the same after the cyclic process (the expansion and the subsequent compression of the gas back to its original state). That is, only for the reversible processes is the heat absorbed during expansion exactly equal to the heat released during compression. In all the processes carried out using a finite number of steps, more heat is released into the surroundings than is absorbed in the comparable expansion (same number of steps).

For example, in the one-step expansion and compression, the net work done is

$$w_{\text{net}} = \underset{\substack{\uparrow \\ \text{Expansion}}}{-0.75P_1V_1} + \underset{\substack{\uparrow \\ \text{Compression}}}{3P_1V_1} = 2.25P_1V_1$$

and the net heat flow is $-2.25P_1V_1$.

In terms of the *system,* this one-step expansion/compression is cyclic. That is, the system starts at state 1 before the expansion process and ends up back at state 1 after the compression process. However, although the system is returned to the same state in this cyclic expansion–compression process, the *surroundings* are not the same. In fact, we might say that in the surroundings "work has been changed to heat." In this one-step cyclic process, $2.25P_1V_1$ (L atm) of work is changed to $2.25P_1V_1$ (L atm) of heat (thermal energy) in the surroundings. This amount represents the net extra work required to lift the weights onto the pan to compress the gas compared with the work that was obtained as the gas expanded.

This is a general result. In any finite-step, cyclic expansion–compression process, "work is always converted to heat":

$$\underset{\substack{\text{Ordered} \\ \text{energy}}}{\text{Work}} \longrightarrow \underset{\substack{\text{Disordered} \\ \text{energy}}}{\text{Heat}}$$

This result applies whenever the process is carried out in a nonreversible (irreversible) manner. In other words, in an *irreversible* cyclic process, more work must be input to the system than the system produces. In all the finite gas compressions, the work required is greater than $1.4P_1V_1$, which is the maximum work available from the expansion.

All real processes are irreversible.

Of course, all real processes are irreversible, because they cannot be carried out in an infinite number of steps without taking an infinite amount of time. In other words, *all real processes are irreversible* (in a thermodynamic sense).

The maximum work corresponds to the reversible expansion.

Another important conclusion to be drawn from the preceding example is that the *maximum work obtainable from the gas occurs when the expansion is carried out reversibly* ($w_{max} = w_{rev}$). This result is always true for PV work, as well as for any other type of work, such as electrical work performed by an electrochemical cell. We will examine this latter example in the next chapter.

The final point that this experiment reemphasizes is that work and heat are pathway-dependent and thus are not state functions. Energy, on the other hand, is a state function. In each of these isothermal expansions and compressions between (P_1, V_1) and ($P_1/4$, $4V_1$), ΔE is always zero, regardless of the number of steps, since T is constant.

At this point we can precisely define the terms *reversible* and *irreversible* in a thermodynamic sense. In a reversible cyclic process, *both the system and the surroundings are returned exactly to their original conditions.* As it turns out, this process is hypothetical. On the other hand, an irreversible process is one in which, even when the system is cycled (state 1 $\rightarrow$ state 2 $\rightarrow$ state 1) and thus returned to its original state, the surroundings are changed in a permanent way. All real processes are irreversible.

10.3 The Definition of Entropy

Ludwig Boltzmann's tomb in Vienna. Notice Boltzmann's equation on the monument.

So far we have discussed entropy in a very qualitative way. Here we want to give a precise, quantitative definition of entropy.

We have said that entropy is related to probability. If a system has several states available to it, the one that can be achieved in the greatest number of ways (has the largest number of microstates) is the one most likely to occur. That is, the state with the greatest probability has the highest entropy.

How does entropy depend on the number of microstates? To explain this idea, we will consider the following simple system.

Suppose we have two vessels each containing a single gas particle at the same temperature. There is entropy associated with each vessel because the velocity of even one particle is an *average* velocity at a given temperature. That is, there is a range of actual particle velocities. To make matters simple, let's suppose there are three possible velocities (microstates): lower than average (L), average (A), and higher than average (H). The two vessels, therefore, have the same entropy, and the total entropy of both vessels is simply the sum of the entropies of each vessel—entropy is additive. The total entropy of the two vessels is twice that of each individual vessel.

But what about the number of microstates? There were three microstates in each of the individual vessels (L, A, H). But when we consider the two vessels together, there are nine microstates (LL, LA, LH, AL, AA, AH, HL, HA, HH). So while entropy is additive, the number of microstates is multiplicative ($3 \times 3 = 9$). How can we state this relationship mathematically?

Because $\log(x)^2 = 2\log x$, we can quantify the relationship between entropy and the number of microstates as follows:

$$S \; \alpha \; \log(\text{microstates})$$

We can see that this relationship makes sense because if we square the number of microstates, the entropy is doubled. Now let's define the entropy of each individual vessel as S_1 and S_2.

$$S_1 \; \alpha \; \log(3) = k_1\log(3)$$

$$S_2 \; \alpha \; \log(3) = k_2\log(3)$$

$$S_{total} = S_1 + S_2 = k_1\log(3) + k_2\log(3)$$

If $k_1 = k_2 = k$, we have

$$S_{total} = S_1 + S_2 = k\log(3) + k\log(3) = 2k\log(3) = k\log(3)^2$$

More formally, we can connect entropy and probability quantitatively by defining the entropy function S as follows:

$$S = k_B \ln \Omega$$

where

k_B = Boltzmann's constant, the gas constant per molecule (R/N_A)

Ω = the number of microstates corresponding to a given state (including both position and energy)

This definition of entropy shows its exact relationship to probability. However, it is not useful in a practical sense for the typical types of samples used by chemists because those samples contain so many components. For example, a mole of gas contains 6.022×10^{23} individual particles. In addition, according to one estimate, describing the positions and velocities of this mole of particles would require a stack of paper 10 *light-years* tall—and this description would apply for only an instant. Clearly, we cannot deal directly with this definition of entropy for typical-sized samples. We must find a way to connect entropy to the macroscopic properties of matter. To do so, we will consider an ideal gas that expands isothermally from volume V_1 to volume $2V_1$ (Fig. 10.10).

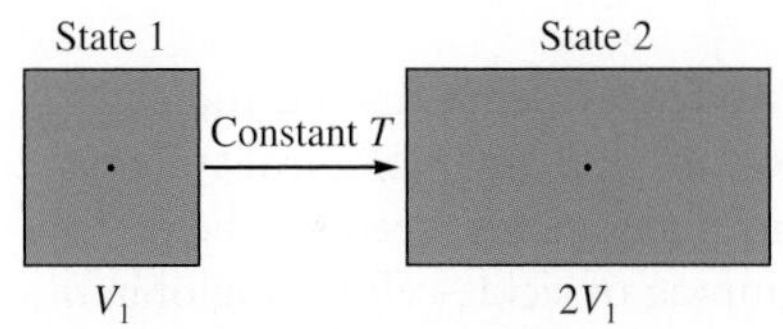

Figure 10.10
A particle in a gas that expands from volume V_1 to volume $2V_1$.

Focus on an individual particle in this gas. When the gas goes from volume V_1 to $2V_1$, each particle has double the number of positions* available to it. That is, $\Omega_2 = 2\Omega_1$.

Now we will use the definition of entropy to calculate ΔS for this expansion. In this case $S_1 = k_B \ln \Omega_1$ and $S_2 = k_B \ln \Omega_2$, so

$$\Delta S = S_2 - S_1 = k_B \ln \Omega_2 - k_B \ln \Omega_1 = k_B \ln\left(\frac{\Omega_2}{\Omega_1}\right)$$

$$= k_B \ln\left(\frac{2\,\Omega_1}{\Omega_1}\right) = k_B \ln 2$$

This quantity represents ΔS for each particle in the gas. If the gas contains five particles, then the ratio of Ω_2 to Ω_1 is

$$\frac{\Omega_2}{\Omega_1} = 2 \times 2 \times 2 \times 2 \times 2 = 2^5$$

where each 2 represents the fact that each particle has twice as many positions available to it in $2V_1$ as in V_1. This argument can be readily extended to a sample with 1 mole of particles:

$$\frac{\Omega_2}{\Omega_1} = 2^{6 \times 10^{23}}$$

and

$$\Delta S = k_B \ln(2^{6 \times 1023}) = (6 \times 10^{23})(k_B \ln 2)$$

$$= N_A k_B \ln 2 = R \ln 2$$

where R is the gas constant.

---

*Technically, when we double the volume, we are doubling the "density" (the closeness) of energy microstates. However, this requires an understanding of concepts that are beyond the scope of this text.

Chemical Insights

Entropy: An Organizing Force?

In this text we have emphasized the meaning of the second law of thermodynamics—that the entropy of the universe is always increasing. Although the results of all our experiments support this conclusion, this does not mean that order cannot appear spontaneously in a given part of the universe. The best example of this phenomenon involves the assembly of cells in living organisms. Of course, when a process that creates an ordered system is examined in detail, it is found that other parts of the process involve an increase in disorder such that the sum of all the entropy changes is positive. In fact, scientists are now finding that the search for maximum entropy in one part of a system can be a powerful force for organization in another part of the system.

To understand how entropy can be an organizing force, look at the accompanying figure. In a system containing large and small "balls" as shown in the figure, the small balls can "herd" the large balls into clumps in the corners and near the walls. This clears out the maximum space for the small balls so that they can move more freely, thus maximizing the entropy of the system, as demanded by the second law of thermodynamics.

In essence, the ability to maximize entropy by sorting different-sized objects creates a kind of attractive force, called a *depletion,* or *excluded-volume, force*. These "entropic forces" operate for objects in the size range of approximately 10^{-8} m to approximately 10^{-6} m. For entropy-induced ordering to occur, the particles must be constantly jostling each other and must be constantly agitated by solvent molecules, thus making gravity unimportant.

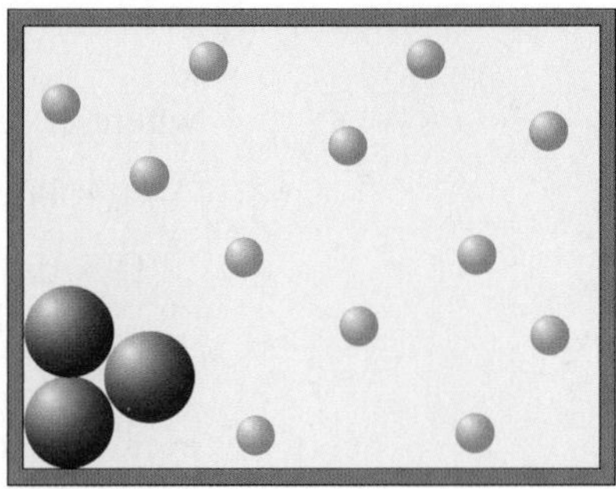

There is increasing evidence that entropic ordering is important in many biological systems. For example, this phenomenon seems to be responsible for the clumping of sickle-cell hemoglobin in the presence of much smaller proteins that act as the "smaller balls." Entropic forces also have been linked to the clustering of DNA in cells without nuclei. Allen Minton of the National Institutes of Health in Bethesda, Maryland, is studying the role of entropic forces in the binding of proteins to cell membranes.

Entropic ordering also appears in nonbiological settings, especially in the ways polymer molecules clump together. For example, polymers added to paint to improve the flow characteristics of the paint actually caused it to coagulate because of depletion forces.

Thus, as you probably have concluded already, entropy is a complex issue. As entropy drives the universe to its ultimate death of maximum chaos, it provides some order along the way.

Now for a general case in which a gas is expanded for V_1 to V_2, the ratio of Ω_2 to Ω_1 is

$$\frac{\Omega_2}{\Omega_1} = \frac{V_2}{V_1}$$

and for 1 mole of gas,

$$\Delta S_{V_1 \to V_2} = R \ln\left(\frac{V_2}{V_1}\right)$$

For n moles of gas we have

$$\Delta S_{V_1 \to V_2} = nR \ln\left(\frac{V_2}{V_1}\right)$$

What we have accomplished here is to use the definition of entropy in terms of probability to derive an expression for ΔS that depends on volume, a macroscopic property of the gas. We can now relate the change in entropy to heat flow by noting the striking similarity between the preceding equation for ΔS and the one derived in Section 10.2 describing q_{rev} for the isothermal expansion–compression of an ideal gas. Compare

$$\Delta S = nR \ln\left(\frac{V_2}{V_1}\right) \qquad \text{with} \qquad q_{rev} = nRT \ln\left(\frac{V_2}{V_1}\right)$$

Combining these equations gives

$$\Delta S = \frac{q_{rev}}{T}$$

This very important relationship is the macroscopic (thermodynamic) definition of ΔS. In our treatment we started with the definition of entropy based on probability because that definition better emphasizes the fundamental character of entropy. However, it is also very important to know how entropy changes relate to changes in macroscopic properties, such as volume and heat, because these changes are relatively easy to measure.

10.4 Entropy and Physical Changes

Although chemists deal primarily with the chemical changes of matter, physical changes are also very important. In this section we will consider how the entropy of a substance depends on its temperature and on its physical state.

Temperature Dependence of Entropy

For an isothermal process, we have seen that the change in entropy is defined by the relationship

$$\Delta S = \frac{q_{rev}}{T}$$

We can calculate ΔS for a change in temperature from T_1 to T_2 by summing infinitesimal increments in entropy at each temperature T:

$$dS = \frac{dq_{rev}}{T}$$

Using integration, we have

$$\Delta S_{T_1 \to T_2} = \int_{T_1}^{T_2} dS = \int_{T_1}^{T_2} \frac{dq_{rev}}{T}$$

If the process is carried out at constant pressure, then

$$dq_{rev} = nC_p \, dT$$

for n moles of substance. Thus

$$\Delta S_{T_1 \to T_2} = \int_{T_1}^{T_2} \frac{dq_{rev}}{T} = nC_p \int_{T_1}^{T_2} \frac{dT}{T}$$

assuming that C_p is constant between T_1 and T_2. Performing the integration gives

$$\Delta S_{T_1 \to T_2} = nC_p \ln\left(\frac{T_2}{T_1}\right)$$

Similarly, for a process carried out at constant volume,

$$\Delta S_{T_1 \to T_2} = nC_v \ln\left(\frac{T_2}{T_1}\right)$$

Entropy Changes Associated with Changes of State

At the normal melting point or boiling point of a substance, the two states of matter present at that temperature and at 1 atm pressure are in equilibrium. That is, the two states can coexist indefinitely if the system is isolated (left totally undisturbed). Recall that a reversible process can occur only at equilibrium. Thus, since a change of state from solid to liquid at the substance's melting point is a reversible process, we can calculate the change in entropy for this process by using the equation

$$\Delta S = \frac{q_{rev}}{T}$$

where

$$q_{rev} = \Delta H_{fusion} = \begin{array}{c}\text{energy required to melt 1 mole}\\ \text{of solid at the melting point}\end{array}$$

$$T = \text{melting point in } K$$

The same reasoning applies to a change from liquid to gas at the boiling point, except in this case $q_{rev} = \Delta H_{vaporization}$ and T = boiling point.

OWL INTERACTIVE EXAMPLE 10.3

Calculate the change in entropy that occurs when a sample containing 2.00 moles of water is heated from 50.°C to 150.°C at 1 atm pressure. The molar heat capacities for $H_2O(l)$ and $H_2O(g)$ are 75.3 J K^{-1} mol^{-1} and 36.4 J K^{-1} mol^{-1}, respectively, and the enthalpy of vaporization for water is 40.7 kJ/mol at 100°C.

Solution

■ What are we trying to solve?

In this problem we are asked to solve for the change in entropy when water is heated from 50°C to 150°C. We predict the entropy will increase with an increase in temperature, but we also need to realize that water changes from a liquid to a gas at 100°C.

So we must ask the question: *What happens as we heat water from 50°C to 150°C?*

The temperature of the liquid water will increase from 50°C to 100°C. Then the liquid will change to a gas at a constant temperature of 100°C. Finally, the temperature of the water vapor will increase from 100°C to 150°C

We will do this calculation in three (reversible) steps:

1. $\Delta S(l)_{50°C \to 100°C}$ (heat the liquid)
2. $\Delta S^{100°C}_{(l) \to (g)}$ (change the liquid to a gas)
3. $\Delta S(g)_{100°C \to 150°C}$ (heat the gas)

and total the results.

1. The entropy change for heating 2.00 moles of liquid water from 50°C to 100°C is

$$\Delta S_{(1)} = (2.00 \text{ mol})\left(75.3 \frac{\text{J}}{\text{K mol}}\right) \ln\left(\frac{373}{323}\right) = 21.7 \text{ J/K}$$

2. The entropy change for the vaporization of 2.00 moles of water at 100°C can be calculated from the equation

$$\Delta S = \frac{q_{\text{rev}}}{T} = \frac{\Delta H_{\text{vap}}}{T_{\text{bp}}}$$

$$= \frac{4.07 \times 10^4 \text{ J/mol}}{373 \text{ K}} = 1.09 \times 10^2 \text{ J K}^{-1} \text{ mol}^{-1}$$

In this case 2.00 moles of water are vaporized, so

$$\Delta S_{(2)} = \left(1.09 \times 10^2 \frac{\text{J}}{\text{K mol}}\right)(2.00 \text{ mol}) = 2.18 \times 10^2 \text{ J/K}$$

3. The entropy change for heating 2.00 moles of gaseous water from 100°C to 150°C is

$$\Delta S_{(3)} = (2.00 \text{ mol})\left(36.4 \frac{\text{J}}{\text{K mol}}\right) \ln\left(\frac{423}{373}\right) = 9.16 \text{ J/K}$$

The total entropy change is

$$\Delta S_{50°\text{C} \rightarrow 150°\text{C}} = \Delta S_{(1)} + \Delta S_{(2)} + \Delta S_{(3)}$$
$$= 21.7 \text{ J/K} + 218 \text{ J/K} + 9.16 \text{ J/K} = 249 \text{ J/K}$$

■ Does our answer make sense?

The total entropy change is positive as we predicted. Notice as well that the vaporization of the water contributes most to the total entropy change.

10.5 Entropy and the Second Law of Thermodynamics

We have seen that processes are spontaneous when they result in an increase in disorder. Nature always moves toward the most probable state available to it. We can state this principle in terms of entropy: *In any spontaneous process there is always an increase in the entropy of the universe.* This is the **second law of thermodynamics.** Contrast this law with the first law of thermodynamics, which tells us that the energy of the universe is constant. Energy is conserved in the universe, but entropy is not. In fact, the second law can be paraphrased as follows: *The entropy of the universe is increasing.*

The total energy of the universe is constant, but the entropy is increasing.

As in Chapter 9, we find it convenient to divide the universe into a system and the surroundings. Thus we can represent the change in the entropy of the universe as

$$\Delta S_{\text{univ}} = \Delta S_{\text{sys}} + \Delta S_{\text{surr}}$$

where ΔS_{sys} and ΔS_{surr} represent the changes in entropy that occur in the system and in the surroundings, respectively.

To predict whether a given process will be spontaneous, we must know the sign of ΔS_{univ}. If ΔS_{univ} is positive, the entropy of the universe increases, and the process is spontaneous in the direction written. If ΔS_{univ} is negative, the process is spontaneous in the *opposite* direction. If ΔS_{univ} is zero, the process has no tendency to occur, indicating that the system is at equilibrium. To predict whether a process is spontaneous, we must consider the entropy changes that occur both in the system and in the surroundings.

10.6 The Effect of Temperature on Spontaneity

To explore the interplay of ΔS_{sys} and ΔS_{surr} in determining the sign of ΔS_{univ}, we will first discuss the change of state for 1 mole of water from liquid to gas,

$$H_2O(l) \longrightarrow H_2O(g)$$

considering the water to be the system and everything else the surroundings.

What happens to the entropy of water in this process? A mole of liquid water (18 g) has a volume of approximately 18 mL. A mole of gaseous water at 1 atm and 100°C occupies a volume of approximately 31 L. Clearly, there are many more positions available to the water molecules in a volume of 31 L than in 18 mL; thus the vaporization of water is favored by this increase in positional probability. That is, for this process, the entropy of the system increases; ΔS_{sys} has a positive sign.

What about the entropy change in the surroundings? Although we will not prove it here, entropy changes in the surroundings are determined primarily by the flow of energy into or out of the system as heat. To understand this observation, suppose an exothermic process transfers 50 joules of energy as heat to the surroundings, where it becomes thermal energy—that is, kinetic energy associated with the random motions of atoms. This flow of energy into the surroundings increases the random motions of atoms there and so increases the entropy of the surroundings. The sign of ΔS_{surr} is positive. When an endothermic process occurs in the system, it produces the opposite effect. Heat flows from the surroundings to the system, causing the random motions of the atoms in the surroundings to decrease, decreasing the entropy of the surroundings. The vaporization of water is an endothermic process. Thus, for this change of state, ΔS_{surr} is negative.

In an endothermic process heat flows from the surroundings into the system. In an exothermic process heat flows into the surroundings.

Remember it is the sign of ΔS_{univ} that tells us whether the vaporization of water is spontaneous. We have seen that ΔS_{sys} is positive and favors the process and that ΔS_{surr} is negative and unfavorable. Thus the components of ΔS_{univ} are in opposition. Which one controls the situation? The answer *depends on the temperature*. We know that at a pressure of 1 atm water changes spontaneously from liquid to gas at all temperatures above 100°C. Below 100°C the opposite process (condensation) is spontaneous.

Inga Spence/Visuals Unlimited

Steam rising from a thermal pool in Yellowstone.

Since ΔS_{sys} and ΔS_{surr} are in opposition for the vaporization of water, the temperature must have an effect on the relative importance of these two terms. To understand why, we must discuss in more detail the factors that control the entropy changes in the surroundings. The central idea is that *the entropy changes in the surroundings are primarily determined by heat flow.* An exothermic process in the system increases the entropy of the surroundings because the energy flow increases the random motions in the surroundings. This means that exothermicity is an important driving force for spontaneity. In earlier chapters we have seen that a system tends to undergo changes that lower its energy. We now understand the reason for that tendency. When a system at constant temperature moves to a lower energy, the energy it gives up is transferred to the surroundings. Some of this energy is transferred as heat, thus leading to an increase in entropy in the surroundings.

The significance of exothermicity as a driving force *depends on the temperature at which the process occurs.* That is, the magnitude of ΔS_{surr} depends on the temperature at which the heat is transferred. We will not attempt to prove this fact here. Instead, we offer an analogy. Suppose that you have \$50 to give away. Giving it to a millionaire will not create much of an impression—a millionaire has money to spare. However, to a poor college student, \$50 represents a significant sum and will be received with considerable joy. The same principle can be applied to energy transfer via the flow of heat. If 50 joules of energy is transferred to the surroundings, the impact of that event depends greatly on the temperature. If the temperature of the surroundings is very high, the atoms there are in rapid motion. The 50 joules of energy will not make a large percentage change in these motions. On the other hand, if 50 joules of energy is transferred to the surroundings at a very low temperature, where the motions are slow, the energy will cause a large percentage change in the motions. Thus the impact of the transfer of a given quantity of energy as heat to or from the surroundings is greatest at low temperatures.

For our purposes the entropy changes that occur in the surroundings have two important characteristics.

In a process occurring at constant temperature, the tendency for the system to lower its energy is due to the resulting positive ΔS_{surr}.

1. *The sign of ΔS_{surr} depends on the direction of the heat flow.* At constant temperature an exothermic process in the system causes heat to flow into the surroundings, increasing the random motions and thereby increasing the entropy of the surroundings. For this case ΔS_{surr} is positive. This principle is often stated in terms of energy. An important driving force in nature results from the tendency of a system to achieve the lowest possible energy by transferring energy as heat to the surroundings.
2. *The magnitude of ΔS_{surr} depends on the temperature.* The transfer of a given quantity of energy as heat produces a much greater percentage change in the randomness of the surroundings at a low temperature than it does at a high temperature. Thus ΔS_{surr} depends directly on the quantity of heat transferred and inversely on temperature. In other words, the tendency for the system to lower its energy becomes a more important driving force at lower temperatures:

$$\begin{array}{c}\text{Driving force provided by}\\\text{the energy flow (heat)}\end{array} = \begin{array}{c}\text{magnitude of the}\\\text{entropy change of}\\\text{the surroundings}\end{array} = \frac{\text{quantity of heat (J)}}{\text{temperature (K)}}$$

These ideas are summarized as follows:

Exothermic process:

ΔS_{surr} = positive

Endothermic process:

ΔS_{surr} = negative

$$\text{Exothermic process:} \quad \Delta S_{surr} = +\frac{\text{quantity of heat (J)}}{\text{temperature (K)}}$$

$$\text{Endothermic process:} \quad \Delta S_{surr} = -\frac{\text{quantity of heat (J)}}{\text{temperature (K)}}$$

We can express ΔS_{surr} in terms of the change in enthalpy (ΔH) for a process occurring at constant pressure (where only PV work is allowed), since under these conditions

$$\text{Heat flow (constant } P) = \text{change in enthalpy} = \Delta H$$

When no subscript is present, the quantity (for example, ΔH) refers to the system.

Recall that ΔH consists of two parts: a sign and a number. The *sign* indicates the direction of flow; a plus sign means "into the system" (endothermic) and a minus sign means "out of the system" (exothermic). The *number* indicates the quantity of energy.

Combining all these concepts yields the following definition of ΔS_{surr} for a reaction that takes place under conditions of constant temperature (K) and pressure:

$$\Delta S_{surr} = -\frac{\Delta H}{T}$$

The minus sign is necessary because the sign of ΔH is determined with respect to the reaction system, and this equation expresses a property of the surroundings. This means that if the reaction is exothermic, ΔH has a negative sign, but since heat flows into the surroundings, ΔS_{surr} is positive.

Stibnite contains Sb_2S_3.

OWL INTERACTIVE EXAMPLE 10.4

In the metallurgy of antimony, the pure metal is recovered by different reactions, depending on the composition of the ore. For example, iron is used to reduce antimony in sulfide ores:

$$Sb_2S_3(s) + 3Fe(s) \longrightarrow 2Sb(s) + 3FeS(s) \qquad \Delta H = -125 \text{ kJ}$$

Carbon is used as the reducing agent in oxide ores:

$$Sb_4O_6(s) + 6C(s) \longrightarrow 4Sb(s) + 6CO(g) \qquad \Delta H = 778 \text{ kJ}$$

Calculate ΔS_{surr} for each of these reactions at 25°C and 1 atm.

Solution We use

$$\Delta S_{surr} = -\frac{\Delta H}{T}$$

where $T = 25 + 273 = 298$ K

For the sulfide ore reaction,

$$\Delta S_{surr} = -\frac{-125 \text{ kJ}}{298 \text{ K}} = 0.419 \text{ kJ/K} = 419 \text{ J/K}$$

Note that ΔS_{surr} is positive, as expected, since this reaction is exothermic; energy flows to the surroundings as heat, increasing the randomness of the surroundings.

For the oxide ore reaction,

$$\Delta S_{surr} = -\frac{778 \text{ kJ}}{298} = -2.61 \text{ kJ/K} = -2.61 \times 10^3 \text{ J/K}$$

In this case ΔS_{surr} is negative because heat flows from the surroundings to the system.

We have seen that the spontaneity of a process is determined by the entropy change it produces in the universe. We have also seen that ΔS_{univ} has two components, ΔS_{sys} and ΔS_{surr}. If for some process both ΔS_{sys} and ΔS_{surr} are positive, then ΔS_{univ} is positive, and the process is spontaneous. If, on the other hand, both ΔS_{sys} and ΔS_{surr} are negative, the process does not occur in the direction indicated but is spontaneous in the opposite direction. Finally, if ΔS_{sys} and ΔS_{surr} have opposite signs, the spontaneity of the process depends on the sizes of the opposing terms. These cases are summarized in Table 10.4.

We can now understand why spontaneity often depends on temperature and thus why water spontaneously freezes below 0°C and melts above 0°C. The term ΔS_{surr} is temperature-dependent. Since

$$\Delta S_{surr} = -\frac{\Delta H}{T}$$

Table 10.4

Interplay of ΔS_{sys} and ΔS_{surr} in Determining the Sign of ΔS_{univ}

Signs of Entropy Changes			
ΔS_{sys}	ΔS_{surr}	ΔS_{univ}	Process Spontaneous?
+	+	+	Yes
−	−	−	No (process will occur in opposite direction)
+	−	?	Yes, if ΔS_{sys} has a larger magnitude than ΔS_{surr}
−	+	?	Yes, if ΔS_{surr} has a larger magnitude than ΔS_{sys}

at constant pressure, the value of ΔS_{surr} changes markedly with temperature. The magnitude of ΔS_{surr} is very small at high temperatures and increases as the temperature decreases. That is, exothermicity is most important as a driving force at low temperatures.

10.7 Free Energy

So far we have used ΔS_{univ} to predict the spontaneity of a process. Now we will define another thermodynamic function that is also related to spontaneity and is especially useful in dealing with the temperature dependence of spontaneity. This function is called **free energy** (G) and is defined as

The symbol *G* for free energy honors Josiah Willard Gibbs (1839–1903), who was Professor of Mathematical Physics at Yale University from 1871 to 1903. He laid the foundations of many areas of thermodynamics, particularly as they apply to chemistry.

$$G = H - TS$$

where H is the enthalpy, T is the kelvin temperature, and S is the entropy.

For a process that occurs at constant temperature, the change in free energy (ΔG) is given by the equation

$$\Delta G = \Delta H - T\Delta S$$

Note that all quantities here refer to the system. From this point on we will follow the usual convention that when no subscript is included, the quantity refers to the system.

To see how this equation relates to spontaneity, we divide both sides of the equation by $-T$ to produce

$$-\frac{\Delta G}{T} = -\frac{\Delta H}{T} + \Delta S$$

Remember that at constant temperature and pressure

$$\Delta S_{surr} = -\frac{\Delta H}{T}$$

So we can write $-\frac{\Delta G}{T} = -\frac{\Delta H}{T} + \Delta S = \Delta S_{surr} + \Delta S = \Delta S_{univ}$

We have shown that $\Delta S_{univ} = -\frac{\Delta G}{T}$ at constant T and P

This result is very important. It means that a process carried out at constant temperature and pressure will be spontaneous only if ΔG is negative. That is, *a process (at constant T and P) is spontaneous in the direction in which the free energy decreases* ($-\Delta G$ means $+\Delta S_{univ}$).

Now we have two functions that can be used to predict spontaneity: the entropy of the universe, which applies to all processes, and free energy, which can be used for processes carried out at constant temperature and pressure. Since so many chemical reactions occur under the latter conditions, the free energy function is more useful to chemists.

Let's use the free energy equation to predict the spontaneity of the melting of ice:

The superscript degree symbol (°) indicates that all substances are in their standard states.

$$H_2O(s) \longrightarrow H_2O(l)$$

for which $\Delta H^\circ = 6.03 \times 10^3$ J/mol and $\Delta S^\circ = 22.1$ J K^{-1} mol^{-1}

Results for the calculations of ΔS_{univ} and ΔG° at -10°C, 0°C, and 10°C are shown in Table 10.5. These data predict that the process is spontaneous at 10°C; that is, ice melts at this temperature since ΔS_{univ} is positive and ΔG° is negative. The opposite is true at -10°C, at which water freezes spontaneously.

Table 10.5

Results of the Calculations of ΔS_{univ} and $\Delta G°$ for the Process $H_2O(s) \rightarrow H_2O(l)$ at −10°C, 0°C, and 10°C*

T (°C)	T (K)	$\Delta H°$ (J/mol)	$\Delta S°$ (J K^{-1} mol^{-1})	$\Delta S_{surr} = -\frac{\Delta H°}{T}$ (J K^{-1} mol^{-1})	$\Delta S_{univ} = \Delta S° + \Delta S_{surr}$ (J K^{-1} mol^{-1})	$T\Delta S°$ (J/mol)	$\Delta G° = \Delta H° - T\Delta S°$ (J/mol)
−10	263	6.03×10^3	22.1	−22.9	−0.8	5.81×10^3	$+2.2 \times 10^2$
0	273	6.03×10^3	22.1	−22.1	0	6.03×10^3	0
10	283	6.03×10^3	22.1	−21.3	+0.8	6.25×10^3	-2.2×10^2

*Note that at 10°C, $\Delta S°$ (ΔS_{sys}) controls, and the process occurs even though it is endothermic. At −10°C, the magnitude of ΔS_{surr} is larger than that of $\Delta S°$, so the process is spontaneous in the opposite (exothermic) direction. In these calculations we are assuming that $\Delta H°$ and $\Delta S°$ are temperature-independent.

To be exact, the calculations shown in Table 10.5 should account for the temperature dependences of ΔH and ΔS. For example,

$$\begin{array}{ccc} H_2O(l) & \xrightarrow[\Delta H_2]{0°C} & H_2O(s) \\ \uparrow \Delta H_1° & & \uparrow \Delta H_3° \\ H_2O(l) & \xrightarrow[\Delta H_4]{-10°C} & H_2O(s) \end{array}$$

$$\Delta H_4° = \Delta H_1° + \Delta H_2° + \Delta H_3°$$

However, since $\Delta H_1°$ and $\Delta H_3°$ are small compared to $\Delta H_2°$,

$$\Delta H_4° \approx \Delta H_2°$$

The same ideas hold for $\Delta S°$.

Why is this so? The answer lies in the fact that ΔS_{sys} ($\Delta S°$) and ΔS_{surr} oppose each other. The term $\Delta S°$ favors the melting of ice because of the increase in positional entropy, and ΔS_{surr} favors the freezing of water because it is an exothermic process. At temperatures below 0°C the change of state occurs in the exothermic direction because ΔS_{surr} is larger in magnitude than ΔS_{sys}. But above 0°C the change occurs in the direction in which ΔS_{sys} is favorable, since in this case ΔS_{sys} is larger in magnitude than ΔS_{surr}. At 0°C the *opposing tendencies just balance,* and the two states coexist; there is no driving force in either direction. An equilibrium exists between the two states of water. Note that ΔS_{univ} is equal to 0 at 0°C.

We can reach the same conclusions by examining $\Delta G°$. At −10°C, $\Delta G°$ is positive because the $\Delta H°$ term is larger than the $T\Delta S°$ term. The opposite is true at 10°C. At 0°C, $\Delta H°$ is equal to $T\Delta S°$ and $\Delta G°$ is equal to 0. This means that solid H_2O and liquid H_2O have the same free energy at 0°C ($\Delta G° = G_{(l)} - G_{(s)}$), and the system is at equilibrium.

We can understand the temperature dependence of spontaneity by examining the behavior of ΔG. For a process occurring at constant temperature and pressure,

$$\Delta G = \Delta H - T\Delta S$$

If ΔH and ΔS favor opposite processes, spontaneity will depend on temperature in such a way that the exothermic direction will be favored at low temperatures. For example, for the process

$$H_2O(s) \longrightarrow H_2O(l)$$

ΔH and ΔS are both positive. The natural tendency for this system to lower its energy is in opposition to its natural tendency to increase its positional randomness. At low temperatures ΔH dominates, and at high temperatures ΔS dominates. The various cases are summarized in Table 10.6.

Table 10.6

Various Possible Combinations of ΔH and ΔS for a Process and the Resulting Dependence of Spontaneity on Temperature

Case	Result
ΔS positive, ΔH negative	Spontaneous at all temperatures
ΔS positive, ΔH positive	Spontaneous at high temperatures (where exothermicity is relatively unimportant)
ΔS negative, ΔH negative	Spontaneous at low temperatures (where exothermicity is dominant)
ΔS negative, ΔH positive	Process not spontaneous at *any* temperature (reverse process is spontaneous at *all* temperatures)

EXAMPLE 10.5

At what temperatures is the following process spontaneous at 1 atm?

$$Br_2(l) \longrightarrow Br_2(g)$$

where $\Delta H^\circ = 31.0$ kJ/mol and $\Delta S^\circ = 93.0 \text{ J K}^{-1} \text{ mol}^{-1}$

What is the normal boiling point of liquid Br_2?

Solution The vaporization process will be spontaneous at all temperatures at which ΔG° is negative. Note that ΔS° favors the vaporization process because of the increase in positional entropy, and ΔH° favors the *opposite* process, which is exothermic. These opposite tendencies will exactly balance at the boiling point of liquid Br_2, since at this temperature liquid and gaseous Br_2 are in equilibrium ($\Delta G^\circ = 0$). We can find this temperature by setting $\Delta G^\circ = 0$ in the equation

$$\Delta G^\circ = \Delta H^\circ - T\Delta S^\circ$$

Thus we have

$$0 = \Delta H^\circ - T\Delta S^\circ$$

$$\Delta H^\circ = T\Delta S^\circ$$

and

$$T = \frac{\Delta H^\circ}{\Delta S^\circ} = \frac{3.10 \times 10^4 \text{ J/mol}}{93.0 \text{ J K}^{-1} \text{ mol}^{-1}} = 333 \text{ K}$$

At temperatures above 333 K, $T\Delta S^\circ$ has a larger magnitude than ΔH°, and ΔG° (or $\Delta H^\circ - T\Delta S^\circ$) is negative. Above 333 K the vaporization process is spontaneous; the opposite process occurs spontaneously below this temperature. At 333 K liquid and gaseous Br_2 will coexist in equilibrium. These observations can be summarized as follows (the pressure is 1 atm in each case):

1. $T > 333$ K. The term ΔS° controls. The increase in entropy occurring when liquid Br_2 is vaporized is dominant.
2. $T < 333$ K. The process is spontaneous in the direction in which it is exothermic. The term ΔH° controls.
3. $T = 333$ K. The opposing driving forces are just balanced ($\Delta G^\circ = 0$), and the liquid and gaseous phases of Br_2 coexist. This is the normal boiling point.

10.8 Entropy Changes in Chemical Reactions

The second law of thermodynamics tells us that a process will be spontaneous if the entropy of the universe increases when the process occurs. We saw in Section 10.7 that for a process at constant temperature and pressure, we can use the change in free energy of the system to predict the sign of ΔS_{univ} and thus the direction in which it is spontaneous. So far we have applied these ideas only to physical processes, such as changes of state and the formation of solutions. However, the main business of chemistry is studying chemical reactions, and therefore, we want to apply the second law to reactions.

First, we will consider the entropy changes accompanying chemical reactions that occur under conditions of constant temperature and pressure. As for the other types of processes we have considered, the entropy changes in the *surroundings* are determined by the heat flow that occurs as the reaction takes place. However, the entropy changes in the *system* (the reactants and products of the reaction) can be predicted by considering the change in positional probability.

For example, in the ammonia synthesis reaction

$$N_2(g) + 3H_2(g) \longrightarrow 2NH_3(g)$$

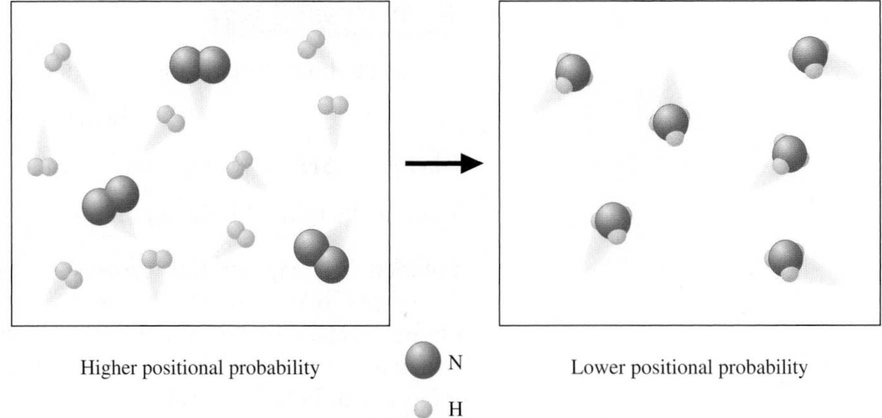

four reactant molecules are changed to two product molecules, lowering the number of independent units in the system and thus leading to lower positional disorder.

Fewer gaseous molecules means fewer possible configurations. To help clarify this result, consider a special container with a million compartments, each large enough to hold a hydrogen molecule. There are a million ways one H_2 molecule can be placed in this container. But suppose we break the H — H bond and place the two independent H atoms in the same container. A little thought will convince you that there are *many* more than a million ways to place the two separate atoms in the container. The number of arrangements possible for the two independent atoms is much greater than the number for the molecule. Thus for the process

$$H_2 \longrightarrow 2H$$

positional probability increases.

Does positional probability increase or decrease when the following reaction takes place?

$$4NH_3(g) + 5O_2(g) \longrightarrow 4NO(g) + 6H_2O(g)$$

In this case 9 gaseous molecules are changed to 10 gaseous molecules, and the positional probability increases. There are more independent units as products than as reactants. In general, when a reaction involves gaseous molecules, *the change in positional probability is dominated by the relative numbers of molecules of gaseous reactants and products.* If the number of molecules of the gaseous products is greater than the number of molecules of the gaseous reactants, positional probability typically increases, and ΔS is positive for the reaction.

EXAMPLE 10.6

Predict the sign of $\Delta S°$ for each of the following reactions.

a. the thermal decomposition of solid calcium carbonate:

$$CaCO_3(s) \longrightarrow CaO(s) + CO_2(g)$$

b. the oxidation of SO_2 in air:

$$2SO_2(g) + O_2(g) \longrightarrow 2SO_3(g)$$

Solution

a. Since in this reaction a gas is produced from a solid reactant, the positional probability increases, and $\Delta S°$ is positive.

b. Here three molecules of gaseous reactants become two molecules of gaseous products. Since the number of gas molecules decreases, positional probability decreases, and $\Delta S°$ is negative.

Absolute Entropies and the Third Law of Thermodynamics

In thermodynamics it is the *change* in a certain function that is usually important. The change in enthalpy determines whether a reaction is exothermic or endothermic at constant pressure. The change in free energy determines whether a process is spontaneous at constant temperature and pressure. It is fortunate that changes in thermodynamic functions are sufficient for most purposes because absolute values for many thermodynamic characteristics of a system (such as enthalpy or free energy) cannot be determined.

However, we can assign *absolute* entropy values. Consider a solid at 0 K, at which molecular motion virtually ceases. If it is a perfect crystal, its internal arrangement is absolutely regular [Fig. 10.11(a)]. There is only *one way* to achieve this perfect order; every particle must be in its place. For example, with N coins there is only one way to achieve the state of all heads. Thus a perfect crystal represents the lowest possible entropy; that is, *the entropy of a perfect crystal at 0 K is zero.* This is a statement of the **third law of thermodynamics.**

As the temperature of a perfect crystal is increased, the random vibrational motions increase, and disorder increases within the crystal [Fig. 10.11(b)]. Thus the entropy of a substance increases with temperature. Since S is zero for a perfect crystal at 0 K, the entropy value for a substance at a particular temperature can be calculated if we know the temperature dependence of entropy.

We have shown that the change in entropy that accompanies a change in temperature of a substance from T_1 to T_2 can be calculated from the expression

$$\Delta S_{T_1 \to T_2} = nC \ln\left(\frac{T_2}{T_1}\right)$$

where C is C_p or C_v depending on the conditions. In using this equation, we are assuming that the heat capacity is independent of temperature. Unfortunately, this assumption is often not true, especially over a large temperature range. Thus the dependence of C on temperature must be taken into account for accurate calculations. However, we will not be concerned with these procedures here.

When a substance is heated from temperature T_1 to T_2, a change of state may occur. If it does, as we saw in Section 10.4, we use the expression

$$\Delta S = \frac{\Delta H}{T}$$

at the melting point or boiling point to account for the entropy change that accompanies the change of state.

The standard entropy values represent the increase in entropy that occurs when a substance is heated from 0 K to 298 K at 1 atm.

The *standard entropy values* ($S°$) of many common substances at 298 K and 1 atm are listed in Appendix 4. From these values you will see that the entropy of a substance does indeed increase in going from solid to liquid to gas.

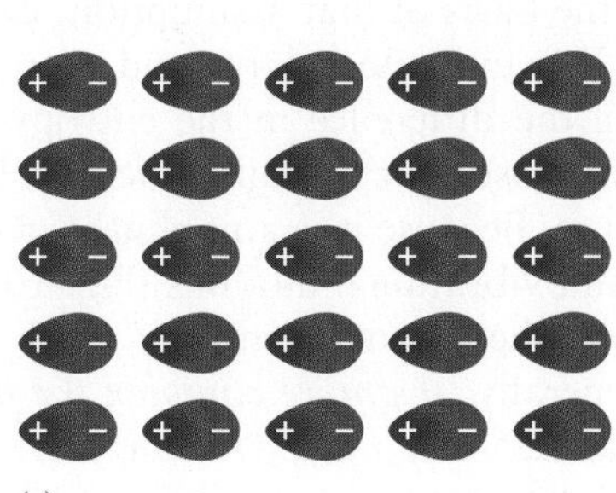
(a)

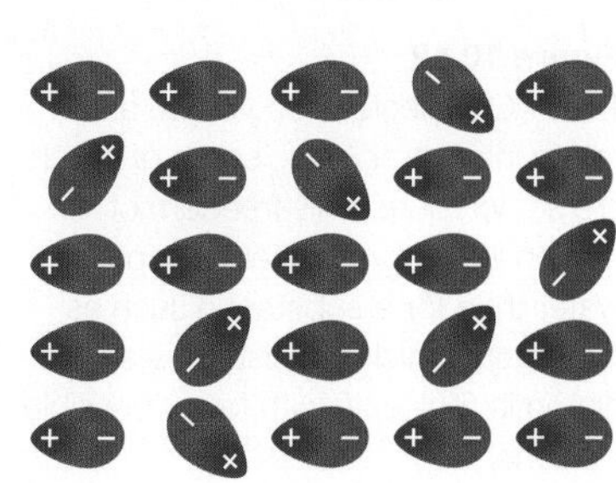
(b)

Figure 10.11
(a) An idealized perfect crystal of hydrogen chloride at 0 K; the dipolar HCl molecules are represented by (+ −). The entropy is zero ($S = 0$) for this crystal at 0 K. (b) As the temperature rises above 0 K, lattice vibrations allow some dipoles to change their orientations, producing some disorder and an increase in entropy ($S > 0$).

Because *entropy is a state function of the system* (it is not pathway-dependent), the entropy change for a given chemical reaction can be calculated by taking the difference between the standard entropy values of the products and those of the reactants:

$$\Delta S^\circ_{\text{reaction}} = \Sigma S^\circ_{\text{products}} - \Sigma S^\circ_{\text{reactants}}$$

where, as usual, Σ represents the sum of the terms. It is important to note that entropy is an extensive property (it depends on the amount of substance present). This means that *the number of moles of a given reactant or product must be taken into account.*

EXAMPLE 10.7

Calculate ΔS° for the reduction of aluminum oxide by hydrogen gas

$$Al_2O_3(s) + 3H_2(g) \longrightarrow 2Al(s) + 3H_2O(g)$$

using the following standard entropy values.

Substance	S° (J K^{-1} mol^{-1})
$Al_2O_3(s)$	51
$H_2(g)$	131
$Al(s)$	28
$H_2O(g)$	189

Solution

$$\begin{aligned}\Delta S^\circ &= \Sigma S^\circ_{\text{products}} - \Sigma S^\circ_{\text{reactants}} \\ &= 2S^\circ_{Al(s)} + 3S^\circ_{H_2O(g)} - 3S^\circ_{H_2(g)} - S^\circ_{Al_2O_3(s)} \\ &= 2\text{ mol}\left(28\frac{\text{J}}{\text{K mol}}\right) + 3\text{ mol}\left(189\frac{\text{J}}{\text{K mol}}\right) \\ &\quad -3\text{ mol}\left(131\frac{\text{J}}{\text{K mol}}\right) - 1\text{ mol}\left(51\frac{\text{J}}{\text{K mol}}\right) \\ &= 56\text{ J/K} + 567\text{ J/K} - 393\text{ J/K} - 51\text{ J/K} \\ &= 179\text{ J/K}\end{aligned}$$

The reaction considered in Example 10.7 involves 3 moles of hydrogen gas on the reactant side and 3 moles of water vapor on the product side. Would you expect ΔS to be large or small for such a case? We have assumed that ΔS depends on the relative numbers of molecules of gaseous reactants and products. On the basis of that assumption, ΔS should be near zero for the present reaction. However, ΔS is large and positive. Why? The large value for ΔS results from the difference in the entropy values for hydrogen gas and water vapor. The reason for this difference can be traced to the difference in molecular structure. Because it is a nonlinear, triatomic molecule, H_2O has more rotational and vibrational motions (Fig. 10.12) than does the diatomic H_2 molecule. Thus the standard entropy value for $H_2O(g)$ is greater than that for $H_2(g)$. Generally, *the more complex the molecule, the higher the standard entropy value.*

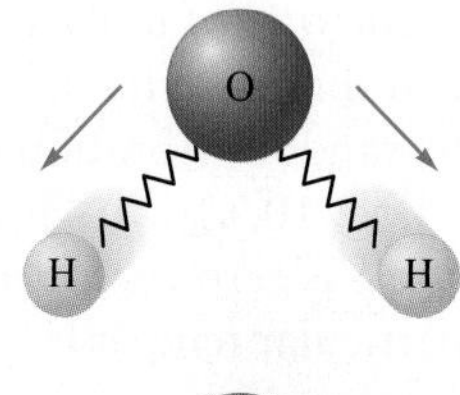

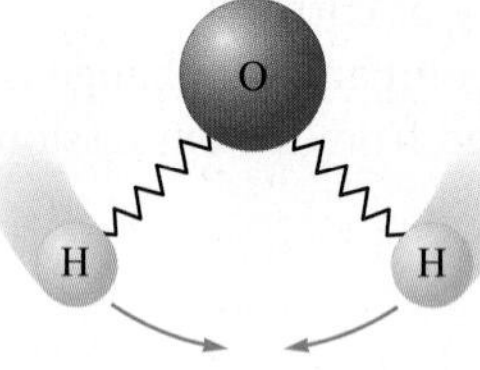

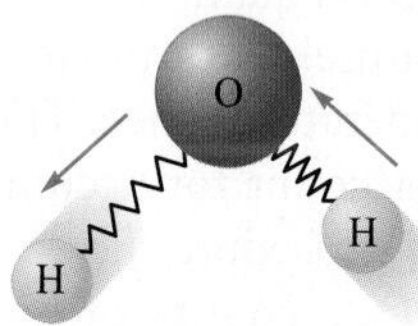

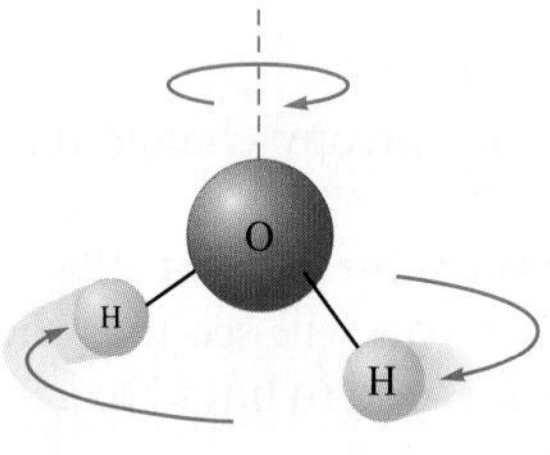

Figure 10.12
The H_2O molecule can vibrate and rotate in several ways, some of which are shown here. This freedom of motion leads to a higher entropy for water than for a substance such as hydrogen, which consists of a simple diatomic molecule with fewer possible motions.

10.9 Free Energy and Chemical Reactions

For chemical reactions we are often interested in the **standard free energy change** ($\Delta G°$), *the change in free energy that occurs if the reactants in their standard states are converted to the products in their standard states.* For example, for the ammonia synthesis reaction at 25°C,

$$N_2(g) + 3H_2(g) \longrightarrow 2NH_3(g) \qquad \Delta G° = -33.3 \text{ kJ} \qquad (10.1)$$

This $\Delta G°$ value represents the change in free energy that occurs when 1 mole of nitrogen gas at 1 atm reacts with 3 moles of hydrogen gas at 1 atm to produce 2 moles of gaseous ammonia at 1 atm.

It is important to recognize that the standard free energy change for a reaction is not measured directly. For example, we can measure heat flow in a calorimeter to determine $\Delta H°$, but we cannot measure $\Delta G°$ this way. The value of $\Delta G°$ for the ammonia synthesis in Equation (10.1) was *not* obtained by mixing 1 mole of N_2 with 3 moles of H_2 in a flask and measuring the change in free energy as 2 moles of NH_3 formed. For one thing, if we mixed 1 mole of N_2 and 3 moles of H_2 in a flask, the system would go to equilibrium rather than to completion. Also, we have no instrument that directly measures free energy. Although we cannot directly measure $\Delta G°$ for a reaction, we can calculate it from other measured quantities, as we will see later in this section.

Why is it useful to know $\Delta G°$ for a reaction? As we will see in more detail later, knowing the $\Delta G°$ values for several reactions allows us to compare the relative tendency of these reactions to occur. The more negative the value of $\Delta G°$, the further a reaction will go to the right to reach equilibrium. We must use standard-state free energies to make this comparison because free energy depends on pressure or concentration. Thus, to get an accurate comparison of reaction tendencies, we must compare all reactions under the same pressure or concentration conditions. We will have more to say about the significance of $\Delta G°$ later.

The value of $\Delta G°$ tells us nothing about the rate of a reaction, only its eventual equilibrium position.

There are several ways to calculate $\Delta G°$. One common method uses the equation

$$\Delta G° = \Delta H° - T\Delta S°$$

which applies to a reaction carried out at constant temperature. For example, consider the reaction

$$C(s) + O_2(g) \longrightarrow CO_2(g)$$

The values of $\Delta H°$ and $\Delta S°$ are known to be -393.5 kJ and 3.05 J/K, respectively, and $\Delta G°$ can be calculated at 298 K as follows:

$$\Delta G° = \Delta H° - T\Delta S° = -3.935 \times 10^5 \text{ J} - (298 \text{ K})(3.05 \text{ J/K})$$
$$= -3.944 \times 10^5 \text{ J} = -394.4 \text{ kJ} \qquad (\text{per mole of } CO_2)$$

OWL INTERACTIVE EXAMPLE 10.8

Consider the reaction

$$2SO_2(g) + O_2(g) \longrightarrow 2SO_3(g)$$

carried out at 25°C and 1 atm. Calculate $\Delta H°$, $\Delta S°$, and $\Delta G°$ using the following data:

Substance	$\Delta H_f°$ (kJ/mol)	$S°$ (J K^{-1} mol^{-1})
$SO_2(g)$	−297	248
$SO_3(g)$	−396	257
$O_2(g)$	0	205

Solution The value of ΔH° can be calculated from the enthalpies of formation using the equation we discussed in Section 9.6:

$$\Delta H^\circ = \Sigma \Delta H_f^\circ \text{ (products)} - \Sigma \Delta H_f^\circ \text{ (reactants)}$$

Then

$$\begin{aligned}\Delta H^\circ &= 2\Delta H^\circ_{f\,[SO_3(g)]} - 2\Delta H^\circ_{f\,[SO_2(g)]} - \Delta H^\circ_{f\,[O_2(g)]} \\ &= 2 \text{ mol}(-396 \text{ kJ/mol}) - 2 \text{ mol}(-297 \text{ kJ/mol}) - 0 \\ &= -792 \text{ kJ} + 594 \text{ kJ} = -198 \text{ kJ}\end{aligned}$$

The value of ΔS° can be calculated by using the standard entropy values and the equation discussed in Section 10.8:

$$\Delta S^\circ = \Sigma S^\circ_{products} - \Sigma S^\circ_{reactants}$$

Thus

$$\begin{aligned}\Delta S^\circ &= 2S^\circ_{SO_3(g)} - 2S^\circ_{SO_2(g)} - S^\circ_{O_2(g)} \\ &= 2 \text{ mol}(257 \text{ J K}^{-1} \text{ mol}^{-1}) - 2 \text{ mol}(248 \text{ J K}^{-1} \text{ mol}^{-1}) \\ &\quad -1 \text{ mol}(205 \text{ J K}^{-1} \text{ mol}^{-1}) \\ &= 514 \text{ J/K} - 496 \text{ J/K} - 205 \text{ J/K} = -187 \text{ J/K}\end{aligned}$$

We expect ΔS° to be negative since three molecules of gaseous reactants give two molecules of gaseous products.

The value of ΔG° can now be calculated from the equation

$$\Delta G^\circ = \Delta H^\circ - T\Delta S^\circ$$

So

$$\begin{aligned}\Delta G^\circ &= -198 \text{ kJ} - (298 \text{ K})\left(-187 \frac{\text{J}}{\text{K}}\right)\left(\frac{1 \text{ kJ}}{1000 \text{ J}}\right) \\ &= -198 \text{ kJ} + 55.7 \text{ kJ} = -142 \text{ kJ}\end{aligned}$$

Like enthalpy and entropy, *free energy is a state function.* Thus we can use procedures for finding ΔG that are similar to those for finding ΔH using Hess's law.

To illustrate this second method for calculating the free energy change, we will obtain ΔG° for the reaction

$$2CO(g) + O_2(g) \longrightarrow 2CO_2(g) \qquad (10.2)$$

from the following data:

$$2CH_4(g) + 3O_2(g) \longrightarrow 2CO(g) + 4H_2O(g) \qquad \Delta G^\circ = -1088 \text{ kJ} \qquad (10.3)$$

$$CH_4(g) + 2O_2(g) \longrightarrow CO_2(g) + 2H_2O(g) \qquad \Delta G^\circ = -801 \text{ kJ} \qquad (10.4)$$

Note that $CO(g)$ is a reactant in Equation (10.2), so Equation (10.3) must be reversed, since $CO(g)$ is a product in the reaction as written. When a reaction is reversed, the sign of ΔG° is also reversed. In Equation (10.4), $CO_2(g)$ is a product, as it is in Equation (10.2), but only one molecule of CO_2 is formed. Thus Equation (10.4) must be multiplied by 2, which means the ΔG° value for Equation (10.4) must also be multiplied by 2. Free energy is an extensive property since it is defined by two extensive properties, H and S.

Reversed Equation (10.3)

$$2CO(g) + 4H_2O(g) \longrightarrow 2CH_4(g) + 3O_2(g) \qquad \Delta G^\circ = -(-1088 \text{ kJ})$$

2 × Equation (10.4)

$$2CH_4(g) + 4O_2(g) \longrightarrow 2CO_2(g) + 4H_2O(g) \qquad \Delta G^\circ = 2(-801 \text{ kJ})$$

$$2CO(g) + O_2(g) \longrightarrow 2CO_2(g) \qquad \Delta G^\circ = -(-1088 \text{ kJ}) + 2(-801 \text{ kJ}) = -514 \text{ kJ}$$

This example shows that ΔG values for reactions are manipulated in exactly the same way as ΔH values.

EXAMPLE 10.9

Using the following data (at 25°C),

$$C_{(s)}^{\text{diamond}} + O_2(g) \longrightarrow CO_2(g) \qquad \Delta G^\circ = -397\text{ kJ} \qquad (10.5)$$

$$C_{(s)}^{\text{graphite}} + O_2(g) \longrightarrow CO_2(g) \qquad \Delta G^\circ = -394\text{ kJ} \qquad (10.6)$$

calculate ΔG° for the reaction

$$C_{(s)}^{\text{diamond}} \longrightarrow C_{(s)}^{\text{graphite}}$$

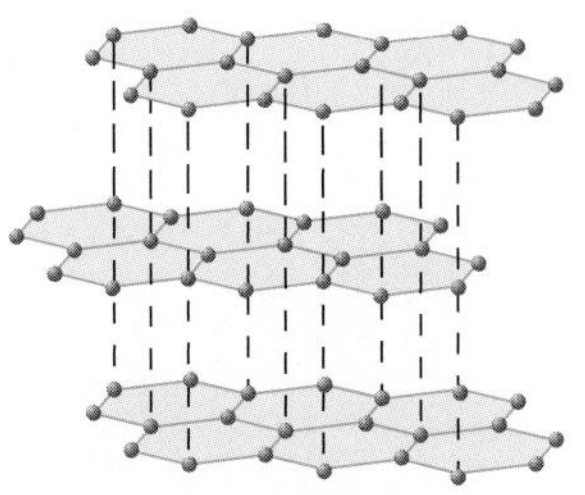

Graphite

Solution We reverse Equation (10.6) to make graphite a product, as required, and then add the new equation to Equation (10.5):

$$C_{(s)}^{\text{diamond}} + O_2(g) \longrightarrow CO_2(g) \qquad \Delta G^\circ = -397\text{ kJ}$$

Reversed Equation (10.6)

$$CO_2(g) \longrightarrow C_{(s)}^{\text{graphite}} + O_2(g) \qquad \Delta G^\circ = -(-394\text{ kJ})$$

$$C_{(s)}^{\text{diamond}} \longrightarrow C_{(s)}^{\text{graphite}} \qquad \Delta G^\circ = -397\text{ kJ} + 394\text{ kJ} = -3\text{ kJ}$$

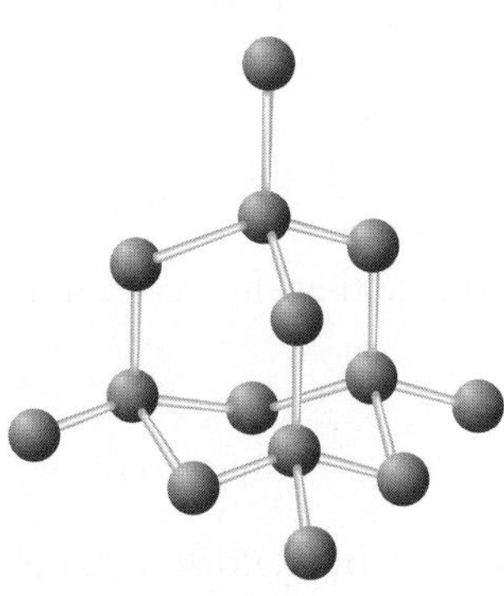

Diamond

Since ΔG° is negative for this process, diamond should spontaneously change to graphite at 25°C and 1 atm. However, the reaction is so slow under these conditions that we do not observe the process. This example shows *kinetic* rather than *thermodynamic* control of a reaction. That is, thermodynamically, diamond should change to graphite, but this spontaneous change is not observed because its rate is so slow. We say that diamond is kinetically stable with respect to graphite, even though it is thermodynamically unstable.

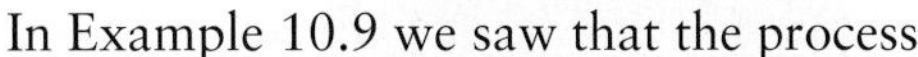

In Example 10.9 we saw that the process

$$C_{(s)}^{\text{diamond}} \longrightarrow C_{(s)}^{\text{graphite}}$$

is spontaneous but very slow at 25°C and 1 atm. The reverse process can be forced to occur at high temperatures and pressures. Diamond has a more compact structure and thus a higher density than graphite, so the exertion of very high pressure causes it to become thermodynamically favored. If high temperatures are also used to make the process fast enough to be feasible, diamonds can be made from graphite. The conditions usually involve temperatures greater than 1000°C and pressures of about 10^5 atm. About half of all industrial diamonds are made this way. We will discuss this process in more detail in Chapter 16.

Phil A. Harrington/Peter Arnold/PhotoLibrary

Synthetic diamonds.

A third method for calculating the free energy change for a reaction uses standard free energies of formation. The **standard free energy of formation** (ΔG_f°) of a substance is defined as the *change in free energy that accompanies the formation of 1 mole of that substance from its constituent elements with all reactants and products in their standard states.* For example, for the formation of glucose ($C_6H_{12}O_6$), the appropriate reaction is

$$6C(s) + 6H_2(g) + 3O_2(g) \longrightarrow C_6H_{12}O_6(s)$$

The standard free energy associated with this process is called the free energy of formation of glucose. Values of the standard free energy of formation are useful in calculating ΔG° for specific chemical reactions using the equation

$$\Delta G^\circ = \Sigma\Delta G_f^\circ\text{ (products)} - \Sigma\Delta G_f^\circ\text{ (reactants)}$$

Values of ΔG_f° for many common substances are listed in Appendix 4. Note that, analogous to the enthalpy of formation, *the standard free energy of formation of an element in its standard state is zero.*

The standard state of an element is its most stable state at the temperature of interest (usually 25°C) and 1 atm.

EXAMPLE 10.10

Calculating ΔG° from free energies of formation is very similar to calculating ΔH°, as shown in Section 9.6.

Methanol is a high-octane fuel used in high-performance racing engines. Calculate ΔG° for the reaction

$$2CH_3OH(g) + 3O_2(g) \longrightarrow 2CO_2(g) + 4H_2O(g)$$

given the following free energies of formation:

Substance	ΔG_f° (kJ/mol)
$CH_3OH(g)$	−163
$O_2(g)$	0
$CO_2(g)$	−394
$H_2O(g)$	−229

Solution We use

$$\begin{aligned}\Delta G_f^\circ &= \Sigma\Delta G_f^\circ\text{ (products)} - \Sigma\Delta G_f^\circ\text{ (reactants)}\\ &= 2\Delta G^\circ_{f\,[CO_2(g)]} + 4\Delta G^\circ_{f\,[H_2O(g)]} - 3\Delta G^\circ_{f\,[O_2(g)]} - 2\Delta G^\circ_{f\,[CH_3OH(g)]}\\ &= 2\text{ mol}(-394\text{ kJ/mol}) + 4\text{ mol}(-229\text{ kJ/mol}) - 3(0)\\ &\quad - 2\text{ mol}(-163\text{ kJ/mol})\\ &= -1378\text{ kJ}\end{aligned}$$

The large magnitude and the negative sign of ΔG° indicate that this reaction is very favorable thermodynamically.

EXAMPLE 10.11

A chemical engineer wants to determine the feasibility of making ethanol (C_2H_5OH) by reacting water with ethylene (C_2H_4) according to the equation

$$C_2H_4(g) + H_2O(l) \longrightarrow C_2H_5OH(l)$$

Is this reaction spontaneous under standard conditions?

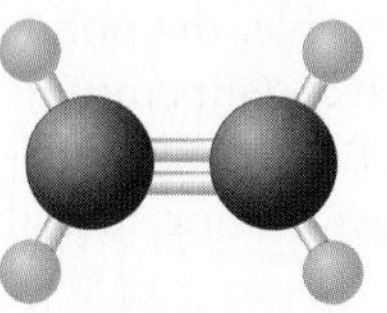
Ethylene

Solution To determine the spontaneity of this reaction under standard conditions, we must determine ΔG° for the reaction by using the appropriate standard free energies of formation at 25°C from Appendix 4:

$$\Delta G^\circ_{f\,[C_2H_5OH(l)]} = -175\text{ kJ/mol}$$

$$\Delta G^\circ_{f\,[H_2O(l)]} = -237\text{ kJ/mol}$$

$$\Delta G^\circ_{f\,[C_2H_4(g)]} = 68\text{ kJ/mol}$$

Thus

$$\begin{aligned}\Delta G^\circ &= \Delta G^\circ_{f\,[C_2H_5OH(l)]} - \Delta G^\circ_{f\,[H_2O(l)]} - \Delta G^\circ_{f\,[C_2H_4(g)]}\\ &= -175\text{ kJ} - (-237\text{ kJ}) - 68\text{ kJ} = -6\text{ kJ}\end{aligned}$$

and the process is spontaneous under standard conditions at 25°C.

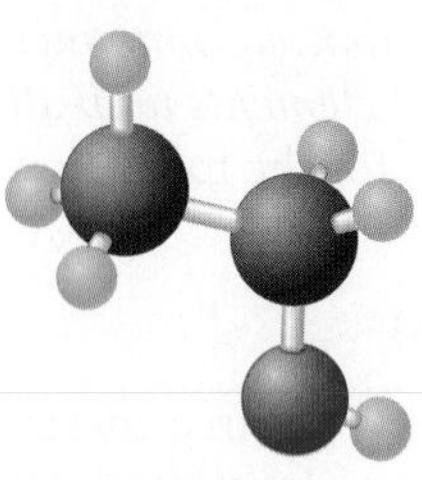
Ethanol

Although the reaction considered in Example 10.11 is spontaneous, other features of the reaction must be studied to see whether the process is feasible. For example, the chemical engineer will need to study the kinetics of the reaction to determine whether it is fast enough to be useful and, if it is not, whether a catalyst can be found to enhance the rate. In doing these studies, the engineer must remember that ΔG° depends on temperature:

$$\Delta G^\circ = \Delta H^\circ - T\Delta S^\circ$$

Thus, if the process must be carried out at high temperatures to be fast enough to be feasible, ΔG° must be recalculated at that temperature using the ΔH° and ΔS° values for the reaction.

10.10 The Dependence of Free Energy on Pressure

In this chapter we have seen that a reaction system at constant temperature and pressure will proceed spontaneously in the direction that lowers its free energy. For this reason reactions proceed until they reach equilibrium. As we will see later in this section, the equilibrium position represents the lowest free energy value available to a particular reaction system. The free energy of a reaction system changes as the reaction proceeds because free energy depends on the pressure of a gas (or on the concentration of species in solution). We will deal only with the pressure dependence of the free energy of an ideal gas. The dependence of free energy on concentration can be developed using similar reasoning.

To understand the pressure dependence of free energy, we need to know how pressure affects the thermodynamic functions that constitute free energy—that is, enthalpy and entropy (recall that $G = H - TS$). For an ideal gas, enthalpy is not pressure-dependent. However, entropy *does* depend on pressure because of its dependence on volume. Consider 1 mole of an ideal gas at a given temperature. At a volume of 10.0 L, the gas has many more positions available for the molecules than if its volume is 1.0 L. The positional probability is greater for the larger volume. In summary, at a given temperature for 1 mole of ideal gas

$$S_{\text{large volume}} > S_{\text{small volume}}$$

or since pressure and volume are inversely related,

$$S_{\text{low pressure}} > S_{\text{high pressure}}$$

We have shown qualitatively that the entropy and therefore the free energy of an ideal gas depend on its pressure. From a more detailed argument, which we will not consider here, one can show that

$$G = G^\circ + RT\ln(P)$$

where G° is the free energy of the gas at a pressure of 1 atm, G is the free energy of the gas at a pressure of P atm, R is the universal gas constant, and T is the kelvin temperature.

To see how the change in free energy for a reaction depends on pressure, we will consider the ammonia synthesis reaction

$$N_2(g) + 3H_2(g) \longrightarrow 2NH_3(g)$$

The absolute value of the free energy of a substance cannot be obtained. We use it symbolically here to show that it is the change in free energy that is really significant.

In general, $$\Delta G = \Sigma G_{\text{products}} - \Sigma G_{\text{reactants}} \qquad (10.7)$$

For this reaction $$\Delta G = 2G_{NH_3} - G_{N_2} - 3G_{H_2}$$

where $$G_{NH_3} = G^\circ_{NH_3} + RT\ln(P_{NH_3})$$

$$G_{N_2} = G^\circ_{N_2} + RT\ln(P_{N_2})$$

$$G_{H_2} = G^\circ_{H_2} + RT\ln(P_{H_2})$$

Substituting these values into Equation (10.7) gives

$$\Delta G = 2[G^\circ_{NH_3} + RT\ln(P_{NH_3})] - [G^\circ_{N_2} + RT\ln(P_{N_2})] - 3[G^\circ_{H_2} + RT\ln(P_{H_2})]$$
$$= 2G^\circ_{NH_3} - G^\circ_{N_2} - 3G^\circ_{H_2} + 2RT\ln(P_{NH_3}) - RT\ln(P_{N_2}) - 3RT\ln(P_{H_2})$$
$$= \underbrace{(2G^\circ_{NH_3} - G^\circ_{N_2} - 3G^\circ_{H_2})}_{\Delta G^\circ_{\text{reaction}}} + RT[2\ln(P_{NH_3}) - \ln(P_{N_2}) - 3\ln(P_{H_2})]$$

The first term in parentheses is ΔG° for the reaction. Thus we have

$$\Delta G = \Delta G^\circ_{\text{reaction}} + RT[2\ln(P_{NH_3}) - \ln(P_{N_2}) - 3\ln(P_{H_2})]$$

Since

$$2\ln(P_{NH_3}) = \ln(P_{NH_3}{}^2)$$

$$-\ln(P_{N_2}) = \ln\left(\frac{1}{P_{N_2}}\right)$$

and

$$-3\ln(P_{H_2}) = \ln\left(\frac{1}{P_{H_2}{}^3}\right)$$

the equation becomes

$$\Delta G = \Delta G^\circ + RT\ln\left[\frac{P_{NH_3}{}^2}{(P_{N_2})(P_{H_2}{}^3)}\right]$$

But the term

$$\frac{P_{NH_3}{}^2}{(P_{N_2})(P_{H_2}{}^3)}$$

is the reaction quotient Q discussed in Section 6.6. Therefore, we have

$$\Delta G = \Delta G^\circ + RT\ln(Q)$$

where Q is the reaction quotient (from the law of mass action), T is the temperature (K), R is the gas law constant, ΔG° is the free energy change for the reaction with all reactants and products at a pressure of 1 atm, and ΔG is the free energy change for the reaction at the specified pressures of reactants and products.

OWL INTERACTIVE EXAMPLE 10.12

One method for synthesizing methanol (CH_3OH) involves reacting gaseous carbon monoxide and hydrogen:

$$CO(g) + 2H_2(g) \longrightarrow CH_3OH(l)$$

Calculate ΔG at 25°C for this reaction, in which carbon monoxide gas at 5.0 atm and hydrogen gas at 3.0 atm are converted to liquid methanol.

Solution

■ What are we trying to solve?

We are asked to solve for ΔG of the reaction between $CO(g)$ and $H_2(g)$. We are asked to determine ΔG and not ΔG° because we have these gases at pressures other than 1.00 atm; in other words, not at standard conditions. To calculate ΔG for this process, we use the equation

$$\Delta G = \Delta G^\circ + RT\ln(Q)$$

We must first compute $\Delta G°$ from standard free energies of formation (see Appendix 4). Since

$$\Delta G°_{f\,[CH_3OH(l)]} = -166 \text{ kJ}$$
$$\Delta G°_{f\,[H_2(g)]} = 0$$
$$\Delta G°_{f\,[CO(g)]} = -137 \text{ kJ}$$

then

$$\Delta G° = -166 \text{ kJ} - (-137 \text{ kJ}) - 0 = -29 \text{ kJ} = -2.9 \times 10^4 \text{ J}$$

Note that this is the value of $\Delta G°$ for the reaction of 1 mole of CO with 2 moles of H_2 to produce 1 mole of CH_3OH. We might call this the value of $\Delta G°$ for one "unit" of the reaction or for "1 mole of the reaction." Thus the $\Delta G°$ value might be written more accurately as -2.9×10^4 J/mol of reaction, or -2.9×10^4 J/mol rxn.

Note that in this case ΔG is defined for 1 mole of the reaction—that is, for 1 mole of $CO(g)$ reacting with 2 moles of $H_2(g)$ to form 1 mole of $CH_3OH(l)$. Thus ΔG, $\Delta G°$, and $RT \ln(Q)$ all have units of J/mol of reaction. In this case the units of R are actually J K^{-1} (mol of reaction)$^{-1}$, although they are usually not written this way.

We can now calculate ΔG, where

$$\Delta G° = -2.9 \times 10^4 \text{ J/mol rxn}$$
$$R = 8.3145 \text{ J K}^{-1} \text{ mol}^{-1}$$
$$T = 273 + 25 = 298 \text{ K}$$
$$Q = \frac{1}{(P_{CO})(P_{H_2}{}^2)} = \frac{1}{(5.0)(3.0)^2} = 2.2 \times 10^{-2}$$

Note that the pure liquid methanol is not included in the calculation of Q because a pure liquid has an activity of 1, as discussed in Chapter 6. Thus

$$\begin{aligned}\Delta G &= \Delta G° + RT \ln(Q) \\ &= (-2.9 \times 10^4 \text{ J/mol rxn}) \\ &\quad + [8.3145 \text{ J K}^{-1} (\text{mol rxn})^{-1}](298 \text{ K})[\ln(2.2 \times 10^{-2})] \\ &= (-2.9 \times 10^4 \text{ J/mol rxn}) - (9.4 \times 10^3 \text{ J/mol rxn}) \\ &= -3.8 \times 10^4 \text{ J/mol rxn} \\ &= -38 \text{ kJ/mol rxn}\end{aligned}$$

Note that ΔG is significantly more negative than $\Delta G°$, implying that the reaction is more spontaneous at reactant pressures greater than 1 atm. We might expect this result from Le Châtelier's principle.

The Meaning of ΔG for a Chemical Reaction

In this section we have learned to calculate ΔG for chemical reactions under various conditions. For example, in Example 10.12 the calculations show that the formation of $CH_3OH(l)$ from $CO(g)$ at 5.0 atm reacting with $H_2(g)$ at 3.0 atm is spontaneous. What does this result mean? Does it mean that if we mixed 1.0 mole of $CO(g)$ and 2.0 moles of $H_2(g)$ together at pressures of 3.0 atm and 5.0 atm, respectively, that 1.0 mole of $CH_3OH(l)$ would form in the reaction flask? The answer is no. This answer may surprise you in view of what has been said in this section. It is true that 1.0 mole of $CH_3OH(l)$ has a lower free energy than 1.0 mole of $CO(g)$ at 5.0 atm plus 2.0 moles of $H_2(g)$ at 3.0 atm. However, when $CO(g)$ and $H_2(g)$ are mixed under these conditions, there is *an even lower free energy available to this system than 1.0 mole of pure $CH_3OH(l)$.* For reasons we will discuss shortly, *the system can achieve the lowest possible free energy by going to equilibrium, not by going to completion.* At the equilibrium position, some of the $CO(g)$ and $H_2(g)$ will remain in the reaction flask. So even though 1.0 mole of pure $CH_3OH(l)$ is at a lower free energy than

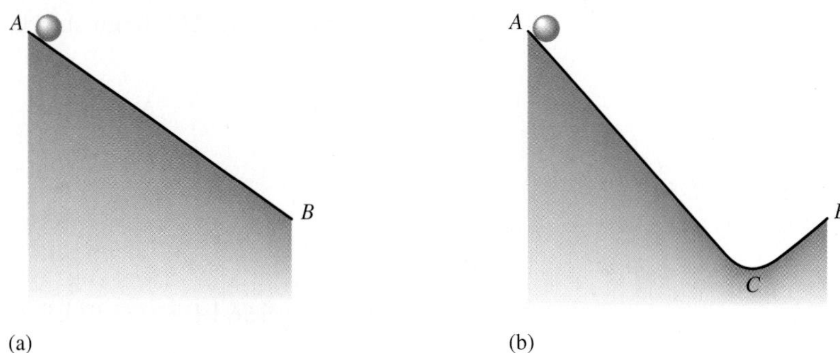

Figure 10.13
Schematic representations of balls rolling down two types of hills.

1.0 mole of $CO(g)$ and 2.0 moles of $H_2(g)$ at 5.0 atm and 3.0 atm, respectively, the reaction system will stop short of forming 1.0 mole of $CH_3OH(l)$. The reaction stops short of completion because the equilibrium mixture of $CH_3OH(l)$, $CO(g)$, and $H_2(g)$ exists at the lowest possible free energy available to the system.

To illustrate this point, we will explore a mechanical example. Consider balls rolling down the two hills shown in Fig. 10.13. Note that in both cases point *B* has a lower potential energy than point *A*.

In Fig. 10.13(a) the ball will roll to point *B*. This diagram is analogous to a phase change. For example, at 25°C ice will spontaneously change completely to liquid water because the latter has the lowest free energy. In this case liquid water is the only choice. There is no intermediate substance with lower free energy.

The situation is different for a chemical reaction system, as illustrated in Fig. 10.13(b). In Fig. 10.13(b) the ball will not get to point *B* because there is a lower potential energy at point *C*. Like the ball, a chemical system will seek the *lowest possible* free energy, which, for reasons we will discuss in the next section, is the equilibrium position.

Therefore, although the value of ΔG for a given reaction system tells us whether the products or reactants are favored under a given set of conditions, it does not mean that the system will proceed to pure products (if ΔG is negative) or remain at pure reactants (if ΔG is positive). Instead, the system will spontaneously go to the equilibrium position, the lowest possible free energy available to it. In the next section, we will see that the value of $\Delta G°$ for a particular reaction tells us exactly where this position will be.

10.11 Free Energy and Equilibrium

When the components of a given chemical reaction are mixed, they will proceed, rapidly or slowly depending on the kinetics of the process, to the equilibrium position. In Chapter 6 we defined the equilibrium position as the point at which the forward and reverse reaction rates are equal. In this chapter we look at equilibrium from a thermodynamic point of view, and we find that *the equilibrium point occurs at the lowest value of free energy available to the reaction system.* As it turns out, the two definitions give the same equilibrium state, which must be the case for both the kinetic and thermodynamic models to be valid.

To understand the relationship between free energy and equilibrium, let's consider the following simple hypothetical reaction:

$$A(g) \rightleftharpoons B(g)$$

where 1.0 mole of gaseous A is initially placed in a reaction vessel at a pressure of 2.0 atm. The free energies for A and B are diagramed as shown in Fig. 10.14(a).

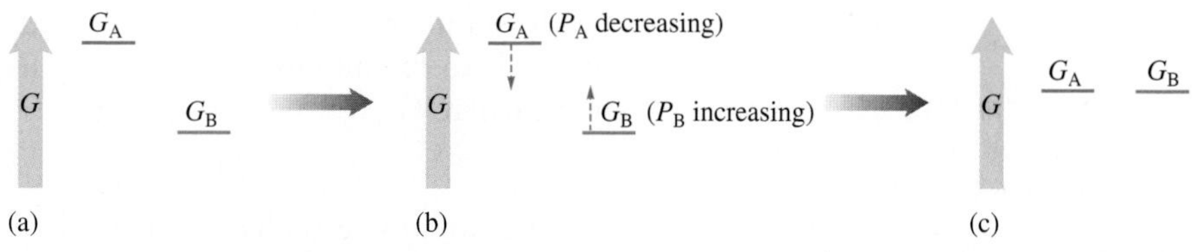

Figure 10.14
(a) The initial free energies of A and B. (b) As A(*g*) changes to B(*g*), the free energy of A decreases and that of B increases. (c) Eventually, pressures of A and B are achieved such that $G_A = G_B$, the equilibrium position.

As A reacts to form B, the total free energy of the system changes, yielding the following results:

$$\text{Total free energy of A} = G_A = n_A[G_A^\circ + RT\ln(P_A)]$$

$$\text{Total free energy of B} = G_B = n_B[G_B^\circ + RT\ln(P_B)]$$

where G_A and G_B are the total free energies of A and B, and n_A and n_B are the moles of A and B, respectively.

$$\text{Total free energy of system} = G = G_A + G_B$$

As A changes to B, G_A decreases because P_A is decreasing [Fig. 10.14(b)]. In contrast, G_B increases because P_B is increasing. The reaction proceeds to the right as long as the total free energy of the system decreases (as long as G_B is less than G_A). At some point the pressures of A and B reach the values P_A^e and P_B^e that make G_A equal to G_B. *The system has reached equilibrium* [Fig. 10.14(c)]. Since A at pressure P_A^e and B at pressure P_B^e have the same free energy (G_A equals G_B), ΔG is zero for A at pressure P_A^e changing to B at pressure P_B^e. *The system has reached minimum free energy.* There is no longer any driving force to change A to B or B to A, so the system remains at this position (the pressures of A and B remain constant).

Note that G_A and G_B are defined here as the total free energies of A and B and are dependent on the moles of A and B as well as the pressures of A and B.

Suppose that for the experiment just described, the plot of free energy versus the mole fraction of A reacted is defined as shown in Fig. 10.15(a). In this experiment minimum free energy is reached when 75% of A has been changed to B. At this point the pressure of A is 0.25 times the original pressure, or

$$(0.25)(2.0 \text{ atm}) = 0.50 \text{ atm}$$

The pressure of B is

$$(0.75)(2.0 \text{ atm}) = 1.5 \text{ atm}$$

For the reaction $A(g) \rightleftharpoons B(g)$, the pressure is constant during the reaction, since the same number of gas molecules is always present.

Since this is the equilibrium position, we can use the equilibrium pressures to calculate a value for K for the reaction in which A is converted to B:

$$K = \frac{P_B^e}{P_A^e} = \frac{1.5 \text{ atm}}{0.50 \text{ atm}} = 3.0$$

Exactly the same equilibrium point will be achieved if we place 1.0 mole of pure B(*g*) in the flask at a pressure of 2.0 atm. In this case B will change to A until equilibrium ($G_B = G_A$) is reached [Fig. 10.15(b)].

The overall free energy curve for this system is shown in Fig. 10.15(c). Note that any mixture of A(*g*) and B(*g*) containing 1.0 mole of A plus B at a total pressure of 2.0 atm will react until it reaches the minimum on the curve.

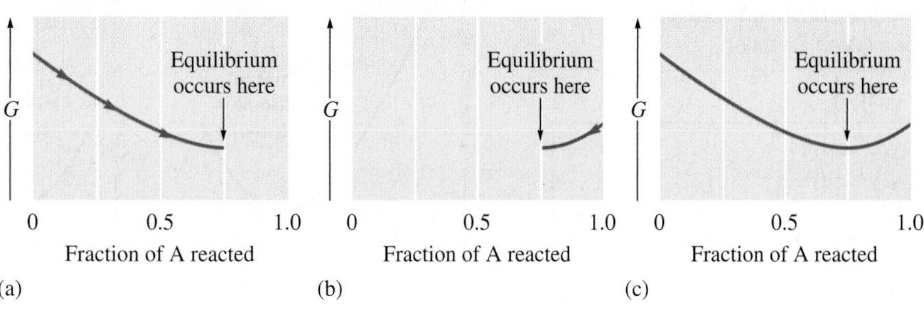

Figure 10.15
(a) The change in free energy to reach equilibrium, beginning with 1.0 mole of A(*g*) at $P_A = 2.0$ atm. (b) The change in free energy to reach equilibrium, beginning with 1.0 mole of B(*g*) at $P_B = 2.0$ atm. (c) The free energy profile for $A(g) \rightleftharpoons B(g)$ in a system containing 1.0 mole (A plus B) at $P_{Total} = 2.0$ atm. Each point on the curve corresponds to the total free energy for a given combination of A and B.

Here $G_{products}$ and $G_{reactants}$ represent the sums for all products and all reactants, respectively.

In summary, when substances undergo a chemical reaction, the reaction proceeds to the minimum free energy (equilibrium), which corresponds to the point where $G_{products} = G_{reactants}$, or

$$\Delta G = G_{products} - G_{reactants} = 0$$

We can now establish a quantitative relationship between free energy and the value of the equilibrium constant. We have seen that

$$\Delta G = \Delta G^\circ + RT \ln(Q)$$

and at equilibrium ΔG equals 0 and Q equals K. So

$$\Delta G = 0 = \Delta G^\circ + RT \ln(K)$$

or

$$\Delta G^\circ = -RT \ln(K)$$

We must note the following characteristics of this very important equation:

Case 1: $\Delta G^\circ = 0$. When ΔG° equals zero for a particular reaction, the free energies of the reactants and products are equal when all components are in the standard states (1 atm for gases). The system is at equilibrium when the pressures of all reactants and products are 1 atm, which means that *K equals 1.*

Case 2: $\Delta G^\circ < 0$. In this case ΔG° ($G^\circ_{products} - G^\circ_{reactants}$) is negative, which means that

$$G^\circ_{products} < G^\circ_{reactants}$$

If a flask contains the reactants and products, all at 1 atm, the system is *not* at equilibrium. Since $G^\circ_{products}$ is less than $G^\circ_{reactants}$, the system adjusts to the right to reach equilibrium. In this case K is *greater than 1,* since the pressures of the products at equilibrium are greater than 1 atm and the pressures of the reactants at equilibrium are less than 1 atm.

Case 3: $\Delta G^\circ > 0$. Since ΔG° ($G^\circ_{products} - G^\circ_{reactants}$) is positive,

$$G^\circ_{reactants} < G^\circ_{products}$$

If a flask contains the reactants and products, all at 1 atm, the system is *not* at equilibrium. In this case the system adjusts to the left (toward the reactants, which have a lower free energy) to reach equilibrium. The value of K is *less than 1,* since at equilibrium the pressures of the reactants are greater than 1 atm and the pressures of the products are less than 1 atm.

These results are summarized in Table 10.7. The value of K for a specific reaction can be calculated from the equation

$$\Delta G^\circ = -RT \ln(K)$$

as is shown in Examples 10.13 and 10.14.

Table 10.7

Qualitative Relationship Between the Change in Standard Free Energy and the Equilibrium Constant for a Given Reaction

ΔG°	K
$\Delta G^\circ = 0$	$K = 1$
$\Delta G^\circ < 0$	$K > 1$
$\Delta G^\circ > 0$	$K < 1$

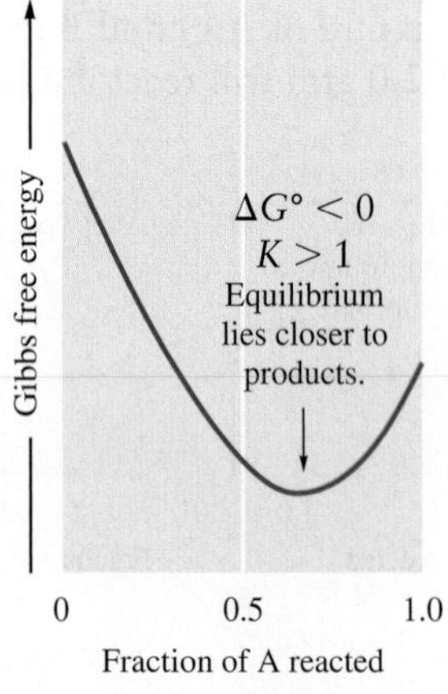

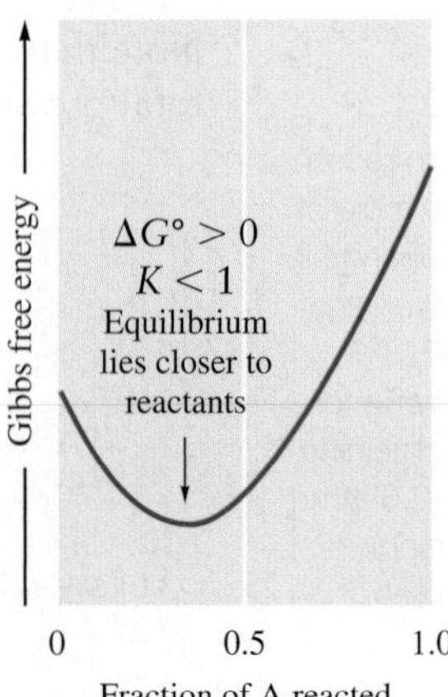

EXAMPLE 10.13

Consider the ammonia synthesis reaction

$$N_2(g) + 3H_2(g) \rightleftharpoons 2NH_3(g)$$

where $\Delta G° = -33.3$ kJ per mole of N_2 consumed at 25°C. For each of the following mixtures of reactants and products at 25°C, predict the direction in which the system will shift to reach equilibrium.

a. $P_{NH_3} = 1.00$ atm, $P_{N_2} = 1.47$ atm, and $P_{H_2} = 1.00 \times 10^{-2}$ atm

b. $P_{NH_3} = 1.00$ atm, $P_{N_2} = 1.00$ atm, and $P_{H_2} = 1.00$ atm

Solution

a. We can predict the direction of the shift to equilibrium by calculating the value of ΔG using the equation

$$\Delta G = \Delta G° + RT\ln(Q)$$

The units of ΔG, $\Delta G°$, and $RT\ln(Q)$ are all per "mole of reaction," although the "per mole" is indicated only for R (as is customary).

$$\text{where } Q = \frac{P_{NH_3}{}^2}{(P_{N_2})(P_{H_2}{}^3)} = \frac{(1.00)^2}{(1.47)[(1.00 \times 10^{-2})^3]} = 6.80 \times 10^5$$

$$T = 25 + 273 = 298 \text{ K}$$

$$R = 8.3145 \text{ J K}^{-1}\text{ mol}^{-1}$$

$$\text{and} \quad \Delta G° = -33.3 \text{ kJ/mol} = -3.33 \times 10^4 \text{ J/mol}$$

Thus

$$\Delta G = (-3.33 \times 10^4 \text{ J/mol}) + (8.3145 \text{ J K}^{-1}\text{ mol}^{-1})(298 \text{ K})\ln(6.80 \times 10^5)$$

$$= (-3.33 \times 10^4 \text{ J/mol}) + (3.33 \times 10^4 \text{ J/mol}) = 0$$

Since $\Delta G = 0$, the reactants and products have the same free energies at the given partial pressures. The system is already at equilibrium, and no shift occurs.

b. The partial pressures given here are all 1.00 atm, indicating that the system is in the standard state. That is,

$$\Delta G = \Delta G° + RT\ln(Q) = \Delta G° + RT\ln\frac{(1.00)^2}{(1.00)(1.00)^3}$$

$$= \Delta G° + RT\ln(1.00) = \Delta G° + 0 = \Delta G°$$

For this reaction at 25°C,

$$\Delta G° = -33.3 \text{ kJ/mol}$$

The negative value for $\Delta G°$ means that in their standard states the products have a lower free energy than the reactants. Thus the system moves to the right to reach equilibrium. That is, K is greater than 1.

OWL INTERACTIVE EXAMPLE 10.14

The overall reaction for the corrosion (rusting) of iron by oxygen is

$$4Fe(s) + 3O_2(g) \rightleftharpoons 2Fe_2O_3(s)$$

Using the following data, calculate the equilibrium constant for this reaction at 25°C.

Substance	$\Delta H_f°$ (kJ/mol)	$S°$ (J K^{-1} mol^{-1})
$Fe_2O_3(s)$	−826	90
$Fe(s)$	0	27
$O_2(g)$	0	205

Michael S. Yamashita/Corbis

Rusted warships in Micronesia.

Solution To calculate K for this reaction, we will use the equation

$$\Delta G^\circ = -RT\ln(K)$$

We must first calculate ΔG° from

$$\Delta G^\circ = \Delta H^\circ - T\Delta S^\circ$$

where

$$\begin{aligned}\Delta H^\circ &= 2\Delta H^\circ_{\text{f [Fe}_2\text{O}_3(s)]} - 3\Delta H^\circ_{\text{f [O}_2(g)]} - 4\Delta H^\circ_{\text{f [Fe}(s)]}\\ &= 2\text{ mol}(-826\text{ kJ/mol}) - 0 - 0\\ &= -1652\text{ kJ} = -1.652 \times 10^6\text{ J}\end{aligned}$$

$$\begin{aligned}\Delta S^\circ &= 2S^\circ_{\text{Fe}_2\text{O}_3} - 3S^\circ_{\text{O}_2} - 4S^\circ_{\text{Fe}}\\ &= 2\text{ mol}(90\text{ J K}^{-1}\text{ mol}^{-1}) - 3\text{ mol}(205\text{ J K}^{-1}\text{ mol}^{-1})\\ &\quad - 4\text{ mol}(27\text{ J K}^{-1}\text{ mol}^{-1})\\ &= -543\text{ J/K}\end{aligned}$$

and

$$T = 273 + 25 = 298\text{ K}$$

The units of ΔG, ΔG°, and $RT\ln(Q)$ are all per "mole of reaction," although the "per mole" is indicated only for R (as is customary).

Then

$$\begin{aligned}\Delta G^\circ &= \Delta H^\circ - T\Delta S^\circ = (-1.652 \times 10^6\text{ J}) - (298\text{ K})(-543\text{ J/K})\\ &= -1.490 \times 10^6\text{ J}\end{aligned}$$

and

$$\begin{aligned}\Delta G^\circ &= -RT\ln(K) = -1.490 \times 10^6\text{ J}\\ &= -(8.3145\text{ J K}^{-1}\text{ mol}^{-1})(298\text{ K})\ln(K)\end{aligned}$$

Thus

$$\ln(K) = \frac{1.490 \times 10^6}{2.48 \times 10^3} = 601 \quad \text{and} \quad K = e^{601}$$

In terms of base = 10, $K = 10^{261}$

This is a very large equilibrium constant. The rusting of iron is clearly very favorable from a thermodynamic point of view.

The Temperature Dependence of K

In Chapter 6 we used Le Châtelier's principle to predict qualitatively how the value of K for a given reaction would change with a change in temperature. Now we can specify the quantitative dependence of the equilibrium constant on temperature from the relationship

$$\Delta G^\circ = -RT\ln(K) = \Delta H^\circ - T\Delta S^\circ$$

We can rearrange this equation to give

$$\ln(K) = -\frac{\Delta H^\circ}{RT} + \frac{\Delta S^\circ}{R} = -\frac{\Delta H^\circ}{R}\left(\frac{1}{T}\right) + \frac{\Delta S^\circ}{R}$$

Note that this is a linear equation of the form $y = mx + b$, where $y = \ln(K)$, $m = -\Delta H^\circ/R$ = slope, $x = 1/T$, and $b = \Delta S^\circ/R$ = intercept. This means that if values of K for a given reaction are determined at various temperatures, a plot of $\ln(K)$ versus $1/T$ will be linear, with slope $-\Delta H^\circ/R$ and intercept $\Delta S^\circ/R$. This result assumes that both ΔH° and ΔS° are independent of temperature over the temperature range considered. This assumption is valid only over a relatively small temperature range.

An important conclusion that can be drawn from this equation is that the sign of the slope of the plot of $\ln(K)$ versus $1/T$ depends on the sign of ΔH° for the reaction. Note that an exothermic reaction ($\Delta H^\circ < 0$) will show a positive slope (ΔH° is negative, so $-\Delta H^\circ/R$ is positive) for the $\ln(K)$ versus $1/T$ plot. In this case $\ln(K)$ will increase as $1/T$ increases (T decreases). Thus K increases as

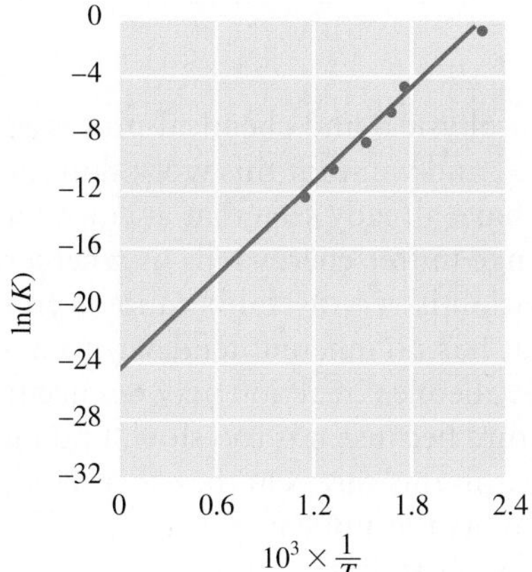

Figure 10.16
Experimental data showing the dependence of K on T (in kelvins) for the ammonia synthesis reaction. In this case, as T decreases, K increases.

T is decreased or, conversely, K decreases as T is increased. This is exactly the temperature dependence of K predicted for an exothermic reaction by Le Châtelier's principle (see Section 6.8). This effect can be shown quantitatively by examining how the value of K for the ammonia synthesis reaction

$$N_2(g) + 3H_2(g) \rightleftharpoons 2NH_3(g)$$

depends on temperature. Figure 10.16 shows a plot of $\ln(K)$ versus $1/T$ for this exothermic reaction ($\Delta H° = -92$ kJ for the reaction as written). Note that $\ln(K)$ increases as $1/T$ increases, meaning that K increases as T is decreased, as expected. Of course, the reverse applies for the equilibrium constant for an endothermic reaction: The value of K increases as the temperature is increased.

Once the temperature dependence of K for a given reaction is known, this relationship can be used to predict the value of K at any temperature (assuming $\Delta H°$ is constant with T). This can be seen most easily as follows. Assuming that K_1 and K_2 are the equilibrium constants for a given reaction at temperatures T_1 and T_2, we can write

$$\ln(K_2) = \frac{-\Delta H°}{RT_2} + \frac{\Delta S°}{R}$$

and

$$\ln(K_1) = \frac{-\Delta H°}{RT_1} + \frac{\Delta S°}{R}$$

Subtracting the second equation from the first gives the combined equation:

$$\ln\left(\frac{K_2}{K_1}\right) = \frac{-\Delta H°}{R}\left[\frac{1}{T_2} - \frac{1}{T_1}\right]$$

This is called the *van't Hoff equation* after Dutch chemist Jacobus van't Hoff. This equation can be used to calculate K at any temperature once $\Delta H°$ and K are known at a given temperature. The accuracy of this calculation depends on whether $\Delta H°$ and $\Delta S°$ are constant over the temperature range considered.

EXAMPLE 10.15

The value of K_p is 3.7×10^{-6} at 900. K for the ammonia synthesis reaction. Assuming the value of $\Delta H°$ for this reaction is -92 kJ, calculate the value of K_p at 550. K.

Solution We use the van't Hoff equation

$$\ln\left(\frac{K_2}{K_1}\right) = \frac{-\Delta H°}{R}\left(\frac{1}{T_2} - \frac{1}{T_1}\right)$$

where $K_1 = 3.7 \times 10^{-6}$, $T_1 = 900.$ K, and $T_2 = 550.$ K.

$$\ln\left(\frac{K_2}{3.7 \times 10^{-6}}\right) = -\left(\frac{-92{,}000 \text{ J/mol}}{8.3145 \dfrac{\text{J}}{\text{K mol}}}\right)\left(\frac{1}{550.} - \frac{1}{900.}\right)$$

$$\ln(K_2) - \ln(3.7 \times 10^{-6}) = 1.1 \times 10^4 \text{ K } (1.8 \times 10^{-3} - 1.1 \times 10^{-3})$$

Solving this equation gives

$$\ln(K_2) = -4.8$$

$$K_2 = 8.2 \times 10^{-3} = K_p \text{ at } 550. \text{ K}$$

Notice that the value of K_p increased as the temperature decreased, as expected for an exothermic reaction.

10.12 Free Energy and Work

One of the main reasons we are interested in physical and chemical processes is that we want to use them to do work for us, and we want this work done as efficiently and economically as possible. We have already seen that at constant temperature and pressure the sign of the change in free energy tells us whether a given process is spontaneous. This information is very useful because it prevents us from wasting effort on a process that has no inherent tendency to occur. Although a thermodynamically favorable chemical reaction may not occur to any appreciable extent at a given temperature because it is too slow, finding a catalyst to speed up the reaction makes sense in this case. On the other hand, if the reaction is prevented from occurring by its thermodynamic characteristics, we would be wasting our time looking for a catalyst.

In addition to being important qualitatively (telling us whether a process is spontaneous), the change in free energy is important quantitatively because it can tell us how much work can be done through a given process. In fact, as we will show, the *maximum possible useful work obtainable from a process at constant temperature and pressure is equal to the change in free energy:*

$$w_{\text{useful}}^{\text{max}} = \Delta G$$

This relationship explains why this function is called the *free* energy. Under certain conditions ΔG for a spontaneous process represents the energy that is *free to do useful work.* On the other hand, for a process that is not spontaneous, the value of ΔG tells us the minimum amount of work that must be *expended* to make the process occur.

Recall that the maximum work would occur only along the hypothetical reversible pathway and is thus unattainable (although it can be approached closely in some situations). In any case, knowing the maximum work for a process is still important because then we can evaluate the efficiency of any machine that might be based on the process.

We will now prove the preceding relationship between ΔG and $w_{\text{useful}}^{\text{max}}$. First, we define the total work w:

$$w = w_{\text{useful}} + \underset{\substack{\uparrow \\ PV\text{ work}}}{w_{\text{useless}}} = w_{\text{useful}} + w_{\text{pv}}$$

Note that PV work is related to the expansion or contraction of the system and is not counted as useful work. From the definition of ΔE, and assuming constant P and T, we have

$$\begin{aligned}\Delta E &= q_{\text{p}} + w = q_{\text{p}} + w_{\text{useful}} + w_{\text{pv}} \\ &= q_{\text{p}} + w_{\text{useful}} - P\Delta V\end{aligned}$$

From the definition of enthalpy,

$$H = E + PV$$

we have

$$\begin{aligned}\Delta H &= \Delta E + P\Delta V \\ &= \underbrace{q_{\text{p}} + w_{\text{useful}} - P\Delta V}_{\Delta E} + P\Delta V \\ &= q_{\text{p}} + w_{\text{useful}}\end{aligned}$$

Next, from the definition of free energy,

$$G = H - TS$$

we have

$$\Delta G = \Delta H - T\Delta S = \underbrace{q_p + w_{useful}}_{\Delta H} - T\Delta S$$

For the reversible pathway

$$w_{useful} = w_{useful}^{max} \qquad \text{and} \qquad q_p = q_p^{rev}$$

Thus for the reversible pathway

$$\Delta G = q_p^{rev} + w_{useful}^{max} - T\Delta S$$

and since
$$\Delta S = \frac{q_p^{rev}}{T}$$

then
$$q_p^{rev} = T\Delta S$$

So we have

$$\Delta G = T\Delta S + w_{useful}^{max} - T\Delta S \qquad \text{or} \qquad \Delta G = w_{useful}^{max}$$

Thus we have shown that at constant temperature and pressure the change in free energy for a process gives the maximum useful work available from that process.

Let's consider a few more points in connection with these relationships. If a process is carried out so that $w_{useful} = 0$, then the expression

$$\Delta G = q_p + w_{useful} - T\Delta S$$

becomes
$$\Delta G = q_p - T\Delta S$$

And since $\Delta G = \Delta H - T\Delta S$, we have

$$\Delta H - T\Delta S = q_p - T\Delta S$$

$$\Delta H = q_p$$

This relationship between ΔH and q_p is used frequently in thermochemical studies. We bring it up again to emphasize that $\Delta H = q_p$ only at constant pressure *and when no useful work is done* (only PV work is allowed). This last condition is often neglected.

If a process is carried out so that w_{useful} is at a maximum (the hypothetical reversible pathway where $\Delta G = w_{useful}$), then from the expression

$$\Delta G = q_p + w_{useful} - T\Delta S$$

we have
$$q_p = T\Delta S$$

Thus q_p, which is pathway-dependent, varies between ΔH (when $w_{useful} = 0$) and $T\Delta S$ (when $w_{useful} = w_{useful}^{max}$). The quantity $T\Delta S$ represents the minimum heat flow that must accompany the process under consideration. That is, $T\Delta S$ represents the minimum energy that must be "wasted" through heat flow as the process occurs.

In summary, at constant T and P,

$$q_p = \Delta H \quad \text{if} \quad w_{useful} = 0$$

$$q_p = T\Delta S \quad \text{if} \quad w_{useful} = w_{useful}^{max}$$

10.13 Reversible and Irreversible Processes: A Summary

As we demonstrated in the analysis of the isothermal expansion–compression of an ideal gas in Section 10.2, the amount of work we actually obtain from a spontaneous process is *always* less than the maximum possible amount.

To explore this idea more fully in a more realistic context than that of an ideal gas, let's consider an electric current flowing through the starter motor of a car. The current is generated from a chemical change in a battery. Since we can calculate ΔG for the battery reaction, we can determine the energy available to do work. Can we use all of this energy to do work? No, because a current flowing through a wire causes frictional heating, and the greater the current, the greater the heat. This heat represents wasted energy—it is not useful for running the starter motor. We can minimize this energy waste by running very low currents through the motor circuit. However, zero current flow would be necessary to eliminate frictional heating entirely, and we cannot derive any work from the motor if no current flows. This example shows the difficulty nature places us in. Using a process to do work requires that some of the energy be wasted, and usually, the faster we run the process, the more energy we waste.

Achieving the maximum work available from a spontaneous process can occur only via a hypothetical pathway. Any real pathway wastes energy in the sense that the maximum work is not obtained. If we could discharge the battery infinitely slowly by an infinitesimally small current flow, we could achieve the maximum useful work. Also, if we could then recharge the battery by using an infinitesimally small current, exactly the same amount of energy would be used to return the battery to its original state as was obtained in the infinitesimally slow discharge. After we cycle the battery in this way, the universe (the system and surroundings) is exactly the same as it was before the cyclic process. Therefore this is a reversible process (Fig. 10.17).

However, if the battery is discharged to run the starter motor and then recharged by using a *finite* current flow, as is actually the case, *more* work will always be required to recharge the battery than the battery produces as it discharges. Thus, even though the battery (the system) has returned to its original state, the surroundings have not because the surroundings had to furnish a net amount of work as the battery was cycled. The *universe is different* after this cyclic process is performed, and this process is irreversible. *All real processes are irreversible.*

Recall that after any real cyclic process is carried out in a system, the surroundings have less ability to do work and contain more thermal energy. In other words, in any real cyclic process, work is changed to heat in the surroundings, and the entropy of the universe increases. This is another way of stating the second law of thermodynamics.

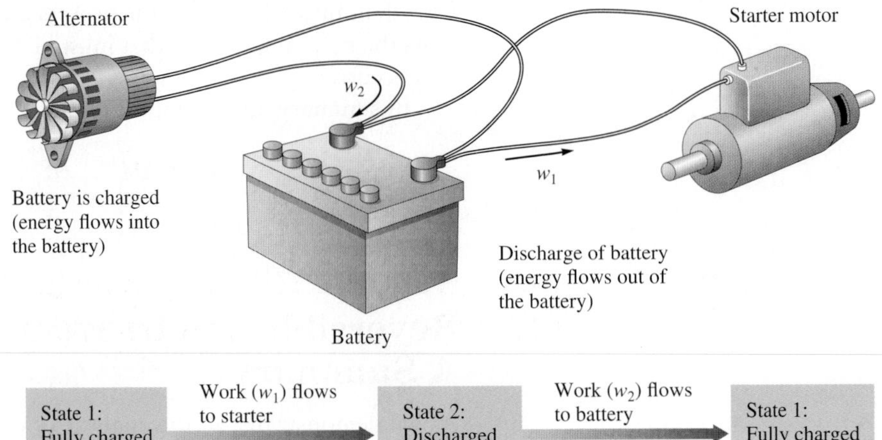

Figure 10.17
A battery can do work by sending current to a starter motor. The battery can then be recharged by forcing current through it. If the current flow in both processes is infinitesimally small, $|w_1| = |w_2|$. This is a *reversible process.* But if the current flow is finite, as it would be in any real case, $|w_2| > |w_1|$. This is an *irreversible process* (the *universe is different* after the cyclic process occurs). All real processes are irreversible.

Thus thermodynamics tells us the work potential of a process and then tells us that we can never achieve this potential. In this spirit, thermodynamicist Henry Bent paraphrased the first two laws of thermodynamics as follows:

First law: You can't win; you can only break even.

Second law: You can't break even.

When energy is used to do work, it becomes less organized and less concentrated and thus less useful.

The ideas we have discussed in this section are applicable to the energy crisis that will probably increase in severity over the next 25 years. The crisis is obviously not one of supply; the first law tells us that the universe contains a constant supply of energy. The problem is the availability of *useful* energy. *As we use energy, we degrade its usefulness.* For example, when gasoline reacts with oxygen in the combustion reaction to run our cars, the change in potential energy results in heat flow. Thus the energy concentrated in the bonds of the gasoline and oxygen molecules ends up *spread* over the surroundings as thermal energy, where it is much more difficult to harness for useful work. In this way the entropy of the universe increases: Concentrated energy becomes spread out—more disordered and less useful. Therefore, the crux of the energy problem is that we are rapidly consuming the concentrated energy found in fossil fuels. It took millions of years to concentrate the sun's energy in these fuels, which we will consume in a few hundred years. Thus we must use these energy sources as wisely as possible.

10.14 Adiabatic Processes

So far in this chapter we have focused on isothermal (constant-temperature) processes for ideal gases. In this section we introduce the **adiabatic process**—*a process in which no energy as heat flows into or out of the system.* That is, an adiabatic process occurs when the system is thermally isolated (insulated) from the surroundings. For an adiabatic process

$$q = 0$$

and

$$\Delta E = q + w = w$$

Figure 10.18
An ideal gas confined in an insulated container with a movable piston. No heat flow with the surroundings can occur.

Consider an ideal gas confined to a cylinder with a movable piston, as shown in Fig. 10.18. Initially, the pressure of the gas (P_{gas}) equals the external pressure (P_{ext}) and the piston is stationary. If P_{ext} is decreased, the gas will expand by doing PV work. What will happen to the temperature of the gas? (Remember that $q = 0$.) Because the expanding gas does work on the surroundings,

$$w = -P_{ext}\Delta V$$

and energy flows out of the system:

$$\Delta E = w = -P_{ext}\Delta V$$

What is the source of this energy? In an isothermal expansion, energy as heat enters the system to just balance the outflow of energy as work (see Section 10.2). Since $q = 0$ for an adiabatic process, the energy to do the work must come from the thermal energy of the gas. That is, in an adiabatic expansion, the temperature of the gas decreases (the average kinetic energy of sample decreases) to furnish the energy to do the work.

Recall that the energy of an ideal gas depends only on its temperature:

$$E = nC_vT$$

So for an adiabatic process,

$$\Delta E = w = -P_{ext}\Delta V = nC_v\Delta T$$

For an infinitesimal adiabatic change,

$$dE = -P_{ext}dV = nC_v dT$$

Assume that the adiabatic expansion or compression is carried out reversibly. That is, P_{ext} is only infinitesimally different from P_{gas} ($P_{ext} \approx P_{gas}$). Then

$$P_{ext} = P_{gas} = \frac{nRT}{V}$$

Thus, for a reversible, adiabatic expansion–compression, we have

$$dE = nC_v dT = -P_{ext}dV = -P_{gas}dV = -\frac{nRT}{V}dV$$

and

$$-\frac{nRT}{V}dV = nC_v dT$$

which can be rearranged to

$$\frac{C_v}{T}dT = -\frac{R}{V}dV$$

We can derive an expression for a reversible, adiabatic change from V_1 to V_2 and from T_1 to T_2 by summing (integrating) the infinitesimal changes required:

$$C_v\int_{T_1}^{T_2}\frac{1}{T}dT = -R\int_{V_1}^{V_2}\frac{1}{V}dV$$

where C_v is assumed to be independent of temperature over the interval T_1 to T_2. Evaluating the integrals gives

$$C_v \ln\left(\frac{T_2}{T_1}\right) = -R\ln\left(\frac{V_2}{V_1}\right) = R\ln\left(\frac{V_1}{V_2}\right)$$

Taking the antilog of each side, we have

$$\left(\frac{T_2}{T_1}\right)^{C_v} = \left(\frac{V_1}{V_2}\right)^{R}$$

Since $C_p = C_v + R$, we can write

$$\left(\frac{T_2}{T_1}\right)^{C_v} = \left(\frac{V_1}{V_2}\right)^{(C_p - C_v)}$$

or

$$\left(\frac{T_2}{T_1}\right) = \left(\frac{V_1}{V_2}\right)^{\left(\frac{C_p}{C_v}-1\right)} = \left(\frac{V_1}{V_2}\right)^{(\gamma-1)}$$

where

$$\gamma = \frac{C_p}{C_v}$$

Thus

$$\frac{T_2}{T_1} = \frac{V_1^{\gamma-1}}{V_2^{\gamma-1}}$$

or

$$T_1V_1^{\gamma-1} = T_2V_2^{\gamma-1}$$

Using the ideal gas law, we can also express this result in terms of pressure. Since in this case,

$$\frac{P_1V_1}{T_1} = \frac{P_2V_2}{T_2}$$

then $$\frac{T_2}{T_1} = \frac{P_2V_2}{P_1V_1} = \frac{V_1^{\gamma-1}}{V_2^{\gamma-1}}$$

and $$P_1V_1^{\gamma} = P_2V_2^{\gamma}$$

We can use these equations to calculate the changes in various properties of an ideal gas undergoing a reversible, adiabatic expansion or compression. This is illustrated in Example 10.16.

EXAMPLE 10.16

Consider a sample containing 5.00 moles of a monatomic ideal gas at 25.0°C and an initial pressure of 10.0 atm. Suppose the external pressure is lowered to 1.00 atm in a reversible manner. Calculate the final pressure and volume of the gas sample and compute the work for the process.

Solution In this case we know the initial and final pressures, so we will use the equation

$$P_1V_1^{\gamma} = P_2V_2^{\gamma}$$

where $$\gamma = \frac{C_p}{C_v} = \frac{\frac{5}{2}R}{\frac{3}{2}R} = \frac{5}{3}$$

for a monatomic gas. Thus we have

$$P_1V_1^{5/3} = P_2V_2^{5/3}$$

We can calculate V_1 from the ideal gas law:

$$V_1 = \frac{nRT_1}{P_1} = \frac{(5.00\ \text{mol})\left(0.08206\ \frac{\text{L atm}}{\text{K mol}}\right)(298\ \text{K})}{10.0\ \text{atm}} = 12.2\ \text{L}$$

Now we can solve for V_2:

$$V_2^{5/3} = \frac{P_1V_1^{5/3}}{P_2} = \frac{(10.0\ \text{atm})(12.2\ \text{L})^{5/3}}{1.00\ \text{atm}}$$

$$V_2 = 48.6\ \text{L}$$

Thus the final volume is 48.6 L. We can calculate the work from the expression

$$\Delta E = w = nC_v\Delta T = (5.00)\left(\frac{3}{2}R\right)(T_2 - T_1)$$

but first we must compute T_2 from the ideal gas law:

$$T_2 = \frac{P_2V_2}{nR} = \frac{(1.00\ \text{atm})(48.6\ \text{L})}{(5.00\ \text{mol})\left(0.08206\frac{\text{L atm}}{\text{K mol}}\right)} = 118\ \text{K}$$

So

$$w = \Delta E = (5.00\ \text{mol})\left(\frac{3}{2}\right)\left(8.3145\frac{\text{J}}{\text{K mol}}\right)(118\ \text{K} - 298\ \text{K}) = -11{,}200\ \text{J}$$

Note that ΔE and w have negative signs because energy as work flows out of the system in the expansion.

Notice for the reversible adiabatic expansion considered in Example 10.16 that the temperature of the sample changed from 298 K to 118 K—a very dramatic temperature decrease. Because of this significant temperature decrease, the final volume of the gas is much smaller than if the expansion were carried out isothermally, where the temperature would remain at 298 K. For a reversible isothermal expansion at 298 K from $P_1 = 10.0$ atm and $V_1 = 12.2$ L to $P_2 = 1.00$ atm, the final volume is

$$V_2 = \frac{P_1 V_1}{P_2} = \frac{(10.0 \text{ mol})(12.2 \text{ L})}{1.00 \text{ atm}} = 122 \text{ L}$$

For this expansion the work is

$$w = -nRT \ln\left(\frac{V_2}{V_1}\right) = -(5.00 \text{ mol})\left(8.3145 \frac{\text{J}}{\text{K mol}}\right)(298 \text{ K}) \ln\left(\frac{122 \text{ L}}{12.2 \text{ L}}\right)$$
$$= -28{,}500 \text{ J}$$

As expected from the much greater volume change, the work delivered to the surroundings in the reversible isothermal expansion is much greater than for the reversible adiabatic expansion. The two types of expansions starting at $P_1 = 10.0$ atm and $V_1 = 12.2$ L are compared in Fig. 10.19. Note that for reversible isothermal expansion

$$P_1 V_1 = P_2 V_2$$

or

$$PV = \text{constant}$$

On the other hand, for the reversible adiabatic expansion

$$P_1 V_1^{\gamma} = P_2 V_2^{\gamma}$$

or

$$PV^{\gamma} = \text{constant}$$

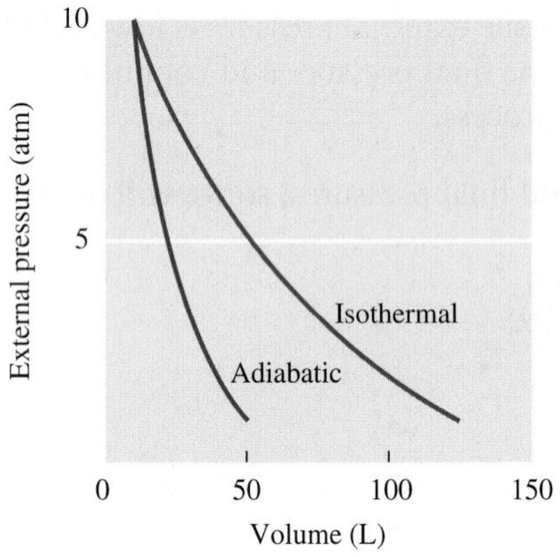

Figure 10.19
Comparison of the adiabatic and isothermal expansions for ideal gas samples in which $n = 5$, $P_1 = 10.0$ atm, and $V_1 = 12.2$ L.

Key Terms

Section 10.1
spontaneous process
entropy
positional probability

Section 10.2
isothermal process

Section 10.5
second law of thermodynamics

Section 10.7
free energy

Section 10.8
third law of thermodynamics

Section 10.9
standard free energy change
standard free energy of formation

Section 10.14
adiabatic process

For Review

Sign in at **www.cengage.com/owl** to:
- View tutorials and simulations, develop problem-solving skills, and complete online homework assigned by your professor.
- Download Go Chemistry mini lecture modules for quick review and exam prep from OWL (or purchase them at **www.cengagebrain.com**)

First law of thermodynamics
- States that the energy of the universe is constant
- Provides a way to keep track of energy as it changes form
- Gives no information about why a particular process occurs in a given direction

Second law of thermodynamics
- States that for any spontaneous process there is always an increase in the entropy of the universe
- Entropy(S) is a thermodynamic function that describes the number of arrangements (positions and/or energy levels) available to a system existing in a given state
  - $\Delta S = \frac{q_{rev}}{T}$
  - Nature spontaneously proceeds toward states that have the highest probability of occurring

- Using entropy, thermodynamics can predict the direction in which a process will occur spontaneously:

$$\Delta S_{univ} = \Delta S_{sys} + \Delta S_{surr}$$

- For a spontaneous process, ΔS_{univ} must be positive
- For a process at constant temperature and pressure:
  - ΔS_{sys} is dominated by "positional" entropy
  - For a chemical reaction, ΔS_{sys} is dominated by changes in the number of gaseous molecules
  - ΔS_{surr} is determined by heat:

$$\Delta S_{surr} = -\frac{\Delta H}{T}$$

  - ΔS_{surr} is positive for an exothermic process (ΔH is negative)
  - Because ΔS_{surr} depends inversely on T, exothermicity becomes a more important driving force at low temperatures
- Thermodynamics cannot predict the rate at which a system will spontaneously change; the principles of kinetics are necessary to do this

Third law of thermodynamics

- States that the entropy of a perfect crystal at 0 K is zero

Free energy (*G*)

- Free energy is a state function:

$$G = H - TS$$

- A process occurring at constant temperature and pressure is spontaneous in the direction in which its free energy decreases ($\Delta G < 0$)
- For a reaction the standard free energy change (ΔG°) is the change in free energy that occurs when reactants in their standard states are converted to products in their standard states
- The standard free energy change for a reaction can be determined from the standard free energies of formation (ΔG_f°) of the reactants and products:

$$\Delta G^\circ = \Sigma n_p \Delta G_f^\circ(\text{products}) - \Sigma n_r \Delta G_f^\circ(\text{reactants})$$

- Free energy depends on temperature and pressure:

$$G = G^\circ + RT \ln P$$

- This relationship can be used to derive the relationship between ΔG° for a reaction and the value of its equilibrium constant K:

$$\Delta G^\circ = -RT \ln K$$

  - For $\Delta G^\circ = 0, K = 1$
  - For $\Delta G^\circ < 0, K > 1$
  - For $\Delta G^\circ > 0, K < 1$
- The maximum possible useful work obtainable from a process at constant temperature and pressure is equal to the change in free energy:

$$w_{max} = \Delta G$$

  - In any real process, $w < w_{max}$
  - When energy is used to do work in a real process, the energy of the universe remains constant but the usefulness of the energy decreases
    - Concentrated energy is spread out in the surroundings as thermal energy

Adiabatic process

- A process in which no energy as heat flows into or out of the system ($q = 0$)
- $\Delta E = w = -P_{ext}\Delta V$

Discussion Questions

These questions are designed to be considered by groups of students in class. Often these questions work well for introducing a particular topic in class.

1. For the process $A(l) \rightarrow A(g)$, which direction is favored by changes in energy probability? Positional probability? Explain your answer. If you wanted to favor the process as written, would you raise or lower the temperature of the system? Explain.
2. For a liquid, which would you expect to be larger: ΔS_{fusion} or $\Delta S_{evaporation}$? Why?
3. Gas A_2 reacts with gas B_2 to form gas AB at constant temperature. The bond energy of AB is much greater than that of either reactant. What can be said about the sign of ΔH? ΔS_{surr}? ΔS? Explain how potential energy changes for this process. Explain how random kinetic energy changes during the process.
4. What types of experiments can be carried out to determine if a reaction is spontaneous? Does spontaneity have any relationship to the final equilibrium position of a reaction? Explain.
5. A friend tells you "Free energy G and pressure P are directly related by the equation $G = G° + RT\ln(P)$. Also, G is related to the equilibrium constant K in that when $G_{products} = G_{reactants}$, the system is at equilibrium. Therefore it must be true that a system is at equilibrium when all pressures are equal." Do you agree with this friend? Explain.
6. You remember that $\Delta G°$ is related to $RT\ln(K)$ but can't remember if it is $RT\ln(K)$ or $-RT\ln(K)$. Realizing what $\Delta G°$ and K mean, how can you figure out the correct sign?
7. Predict the sign of ΔS for each of the following and explain.
   a. the evaporation of alcohol
   b. the freezing of water
   c. compressing an ideal gas at constant temperature
   d. heating an ideal gas at constant pressure
   e. dissolving NaCl in water
8. Which is larger: ΔS at constant pressure or ΔS at constant volume? Provide a conceptual rationale.
9. Is ΔS_{surr} favorable or unfavorable for exothermic reactions? endothermic reactions? Explain.
10. At 1 atm, liquid water is heated above 100°C. For this process which of the following choices (i–iv) is correct for ΔS_{surr}? ΔS? ΔS_{univ}? Explain each answer.
    i. greater than zero
    ii. less than zero
    iii. equal to zero
    iv. cannot be determined
11. High temperatures are favorable to a reaction kinetically but may be unfavorable to a reaction thermodynamically. Explain.

Exercises

OWL Interactive versions of these problems may be assigned in OWL.

A blue exercise number indicates that the answer to that exercise appears at the back of this book and a solution appears in the *Solutions Guide*.

Spontaneity and Entropy

12. Define the following.
    a. spontaneous process
    b. entropy
    c. positional probability
    d. system
    e. surroundings
    f. universe
13. Table 10.1 shows the possible arrangements of four molecules in a two-bulbed flask. What are the possible arrangements if there is one molecule in this two-bulbed flask or two molecules or three molecules? For each, what arrangement is most likely?
14. Consider the following illustration of six molecules of gas in a two-bulbed flask

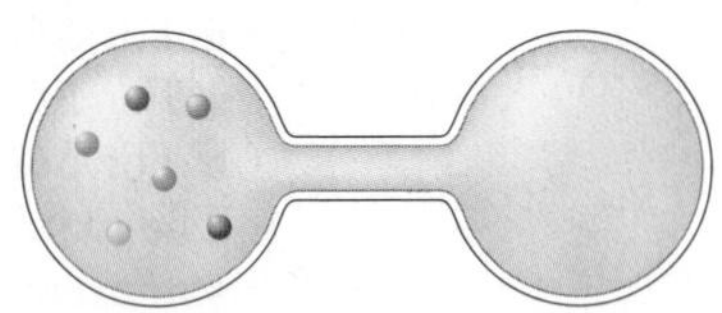

    a. What is the most likely arrangement of molecules? How many microstates are there for this arrangement?
    b. Determine the probability of finding the gas in its most likely arrangement.

15. Consider the following energy levels, each capable of holding two objects:

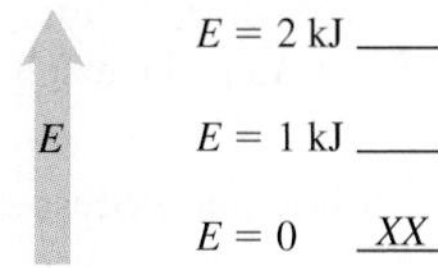

Draw all the possible arrangements of the two identical particles (represented by X) in the three energy levels. What total energy is most likely, that is, occurs the greatest number of times? Assume that the particles are indistinguishable from each other.

16. Do Exercise 15 with two particles A and B that can be distinguished from each other.

17. Which of the following processes require energy as they occur?
    a. Salt dissolves in H_2O.
    b. A clear solution becomes a uniform color after a few drops of dye are added.
    c. A cell produces proteins from amino acids.
    d. Iron rusts.
    e. A house is built.
    f. A satellite is launched into orbit.
    g. A satellite falls back to earth.

18. Which of the following involve an increase in the entropy of the system under consideration?
    a. melting of a solid
    b. evaporation of a liquid
    c. sublimation
    d. freezing
    e. mixing
    f. separation
    g. diffusion

19. Describe how the following changes affect the positional probability of a substance.
    a. increase in volume of a gas at constant T
    b. increase in temperature of a gas at constant V
    c. increase in pressure of a gas at constant T

20. Choose the substance with the larger positional probability in each case.
    a. 1 mole of H_2 (at STP) or 1 mole of H_2 (at 100°C, 0.5 atm)
    b. 1 mole of N_2 (at STP) or 1 mole of N_2 (at 100 K, 2.0 atm)
    c. 1 mole of $H_2O(s)$ (at 0°C) or 1 mole of $H_2O(l)$ (at 20°C)

21. In the roll of two dice, what *total* number is the most likely to occur? Is there an energy reason why this number is favored? Would energy have to be spent to increase the probability of getting a particular number (that is, to cheat)?

22. Entropy can be calculated by a relationship proposed by Ludwig Boltzmann:

$$S = k_B \ln \Omega$$

where $k_B = 1.38 \times 10^{-23}$ J/K and Ω is the number of ways a particular state can be obtained. (This equation is engraved on Boltzmann's tombstone.) Calculate S for the three arrangements of particles in Table 10.1.

Energy, Enthalpy, and Entropy Changes Involving Ideal Gases and Physical Changes

23. Calculate the energy required to change the temperature of 1.00 kg of ethane (C_2H_6) from 25.0°C to 73.4°C in a rigid vessel. (C_v for C_2H_6 is 44.60 J K^{-1} mol^{-1}.) Calculate the energy required for this same temperature change at constant pressure. Calculate the change in internal energy of the gas in each of these processes.

24. For nitrogen gas the values of C_v and C_p at 25°C are 20.8 J K^{-1} mol^{-1} and 29.1 J K^{-1} mol^{-1}, respectively. When a sample of nitrogen is heated at constant pressure, what fraction of the energy is used to increase the internal energy of the gas? How is the remainder of the energy used? How much energy is required to raise the temperature of 100.0 g N_2 from 25.0°C to 85.0°C in a vessel having a constant volume?

25. Consider a rigid, insulated box containing 0.400 mole of He(g) at 20.0°C and 1.00 atm in one compartment and 0.600 mole of $N_2(g)$ at 100.0°C and 2.00 atm in the other compartment. These compartments are connected by a partition that transmits heat. What is the final temperature in the box at thermal equilibrium? [For He(g), C_v = 12.5 J K^{-1} mol^{-1}; for $N_2(g)$, C_v = 20.7 J K^{-1} mol^{-1}.]

26. One mole of an ideal gas is contained in a cylinder with a movable piston. The temperature is constant at 77°C. Weights are removed suddenly from the piston to give the following sequence of three pressures:
    a. P_1 = 5.00 atm (initial state)
    b. P_2 = 2.24 atm
    c. P_3 = 1.00 atm (final state)

    What is the total work (in joules) in going from the initial to the final state by way of the preceding two steps? What would be the total work if the process were carried out reversibly?

27. One mole of an ideal gas with a volume of 1.0 L and a pressure of 5.0 atm is allowed to expand isothermally into an evacuated bulb to give a total volume of 2.0 L. Calculate w and q. Also calculate q_{rev} for this change of state.

28. A cylinder with an initial volume of 10.0 L is fitted with a frictionless piston and is filled with 1.00 mole of an ideal gas at 25°C. Assume that the surroundings are large enough so that if heat is withdrawn from or added to it, the temperature does not change.
    a. The gas expands isothermally and reversibly from 10.0 L to 20.0 L. Calculate the work and the heat.
    b. The gas expands isothermally and irreversibly from 10.0 L to 20.0 L as the external pressure changes instantaneously from 2.46 atm to 1.23 atm. Calculate the work and the heat.

29. The molar heat capacities for carbon dioxide at 298.0 K are

$$C_v = 28.95 \text{ J K}^{-1} \text{ mol}^{-1}$$

$$C_p = 37.27 \text{ J K}^{-1} \text{ mol}^{-1}$$

The molar entropy of carbon dioxide gas at 298.0 K and 1.000 atm is 213.64 J K^{-1} mol^{-1}.
    a. Calculate the energy required to change the temperature of 1.000 mole of carbon dioxide gas from

298.0 K to 350.0 K, both at constant volume and at constant pressure.

b. Calculate the molar entropy of $CO_2(g)$ at 350.0 K and 1.000 atm.

c. Calculate the molar entropy of $CO_2(g)$ at 350.0 K and 1.174 atm.

30. The molar entropy of helium gas at 25°C and 1.00 atm is 126.1 J K^{-1} mol^{-1}. Assuming ideal behavior, calculate the entropy of the following.
    a. 0.100 mole of He(*g*) at 25°C and a volume of 5.00 L
    b. 3.00 moles of He(*g*) at 25°C and a volume of 3000.0 L

31. Consider the process

$$\underset{75°C}{A(l)} \longrightarrow \underset{155°C}{A(g)}$$

which is carried out at constant pressure. The total ΔS for this process is known to be 75.0 J K^{-1} mol^{-1}. For A(*l*) and A(*g*), the C_p values are 75.0 J K^{-1} mol^{-1} and 29.0 J K^{-1} mol^{-1}, respectively, and are not dependent on temperature. Calculate $\Delta H_{vaporization}$ for A(*l*) at 125°C (its boiling point).

32. A sample of ice weighing 18.02 g, initially at −30.0°C, is heated to 140.0°C at a constant pressure of 1.00 atm. Calculate q, w, ΔE, ΔH, and ΔS for this process. The molar heat capacities (C_p) for solid, liquid, and gaseous water—37.5 J K^{-1} mol^{-1}, 75.3 J K^{-1} mol^{-1}, and 36.4 J K^{-1} mol^{-1}, respectively—are assumed to be temperature-independent. The enthalpies of fusion and vaporization are 6.01 kJ/mol and 40.7 kJ/mol, respectively. Assume ideal gas behavior.

33. Calculate the entropy change for a process in which 3.00 moles of liquid water at 0°C is mixed with 1.00 mole of water at 100.°C in a perfectly insulated container. (Assume that the molar heat capacity of water is constant at 75.3 J K^{-1} mol^{-1}.)

34. Calculate the change in entropy that occurs when 18.02 g of ice at −10.0°C is placed in 54.05 g of water at 100.0°C in a perfectly insulated vessel. Assume that the molar heat capacities for $H_2O(s)$ and $H_2O(l)$ are 37.5 J K^{-1} mol^{-1} and 75.3 J K^{-1} mol^{-1}, respectively, and the molar enthalpy of fusion for ice is 6.01 kJ/mol.

Entropy and the Second Law of Thermodynamics: Free Energy

35. The synthesis of glucose directly from CO_2 and H_2O and the synthesis of proteins directly from amino acids are both nonspontaneous processes under standard conditions. Yet these processes must occur for life to exist. In light of the second law of thermodynamics, how can life exist?

36. A green plant synthesizes glucose by photosynthesis as shown in the reaction

$$6CO_2(g) + 6H_2O(l) \longrightarrow C_6H_{12}O_6(s) + 6O_2(g)$$

Animals use glucose as a source of energy:

$$C_6H_{12}O_6(s) + 6O_2(g) \longrightarrow 6CO_2(g) + 6H_2O(l)$$

If we were to assume that both of these processes occur to the same extent in a cyclic process, what thermodynamic property must have a nonzero value?

37. What determines ΔS_{surr} for a process? To calculate ΔS_{surr} at constant pressure and temperature, we use the following equation: $\Delta S_{surr} = -\Delta H/T$. Why does a minus sign appear in the equation, and why is ΔS_{surr} inversely proportional to temperature?

38. Predict the sign of ΔS_{surr} for the following processes.
    a. $H_2O(l) \longrightarrow H_2O(g)$
    b. $I_2(g) \longrightarrow I_2(s)$

39. Calculate ΔS_{surr} for the following reactions at 25°C and 1 atm.
    a. $C_3H_8(g) + 5O_2(g) \longrightarrow 3CO_2(g) + 4H_2O(l)$ $\quad \Delta H° = -2221$ kJ
    b. $2NO_2(g) \longrightarrow 2NO(g) + O_2(g)$ $\quad \Delta H° = 112$ kJ

40. For each of the following pairs of substances, which substance has the greater value of $S°$ at 25°C and 1 atm?
    a. $C_{graphite}(s)$ or $C_{diamond}(s)$
    b. $C_2H_5OH(l)$ or $C_2H_5OH(g)$
    c. $CO_2(s)$ or $CO_2(g)$
    d. $N_2O(g)$ or $He(g)$
    e. $HF(g)$ or $HCl(g)$

41. Predict the sign of $\Delta S°$ for each of the following changes.
    a.

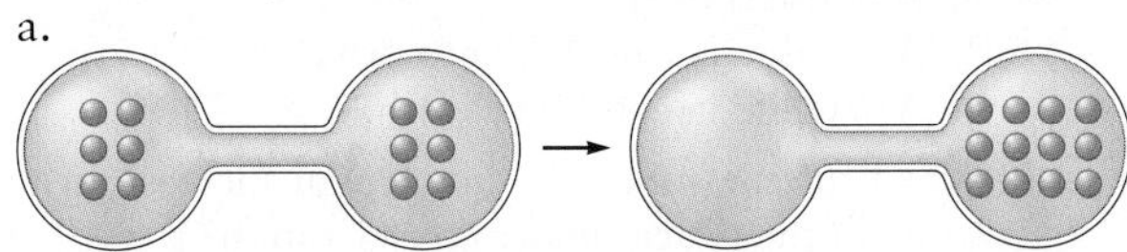

    b. $AgCl(s) \longrightarrow Ag^+(aq) + Cl^-(aq)$
    c. $2H_2(g) + O_2(g) \longrightarrow 2H_2O(l)$
    d. $Na(s) + \frac{1}{2}Cl_2(g) \longrightarrow NaCl(s)$
    e. $HCl(g) \longrightarrow H^+(aq) + Cl^-(aq)$
    f. $KBr(s) \longrightarrow K^+(aq) + Br^-(aq)$

42. Predict the sign of $\Delta S°$ and then calculate $\Delta S°$ for each of the following reactions.
    a. $2H_2S(g) + SO_2(g) \longrightarrow 3S_{rhombic}(s) + 2H_2O(g)$
    b. $2SO_3(g) \longrightarrow 2SO_2(g) + O_2(g)$
    c. $Fe_2O_3(s) + 3H_2(g) \longrightarrow 2Fe(s) + 3H_2O(g)$

43. For the reaction

$$C_2H_2(g) + 4F_2(g) \longrightarrow 2CF_4(g) + H_2(g)$$

$\Delta S°$ is equal to −358 J/K. Use this value and data from Appendix 4 to calculate the value of $S°$ for $CF_4(g)$.

44. Using Appendix 4 and the following data, determine $S°$ for $Fe(CO)_5(g)$.

$$Fe(s) + 5CO(g) \longrightarrow Fe(CO)_5(g) \qquad \Delta S° = ?$$

$$Fe(CO)_5(l) \longrightarrow Fe(CO)_5(g) \qquad \Delta S° = 107 \text{ J/K}$$

$$Fe(s) + 5CO(g) \longrightarrow Fe(CO)_5(l) \qquad \Delta S° = -677 \text{ J/K}$$

45. For the reaction

$$2Al(s) + 3Br_2(l) \longrightarrow 2AlBr_3(s)$$

$\Delta S°$ is equal to −144 J/K. Use this value and data from Appendix 4 to calculate the value of $S°$ for solid aluminum bromide.

46. Ethanethiol (C_2H_5SH; also called ethyl mercaptan) is commonly added to natural gas to provide the "rotten egg" smell of a gas leak. The boiling point of ethanethiol is 35°C and its heat of vaporization is 27.5 kJ/mol. What is the entropy of vaporization for this substance?

47. For mercury at 1 atm, the enthalpy of vaporization is 58.51 kJ/mol and the entropy of vaporization is 92.92 J K^{-1} mol^{-1}. What is the boiling point of mercury?

48. The enthalpy of vaporization of ethanol is 38.7 kJ/mol at its boiling point (78°C). Determine ΔS_{sys}, ΔS_{surr}, and ΔS_{univ} when 1.00 mole of ethanol is vaporized at 78°C and 1.00 atm.

49. For ammonia (NH_3) the enthalpy of fusion is 5.65 kJ/mol, and the entropy of fusion is 28.9 J K^{-1} mol^{-1}.
    a. Will $NH_3(s)$ spontaneously melt at 200. K?
    b. What is the approximate melting point of ammonia?

50. It is quite common for a solid to change from one structure to another at a temperature below its melting point. For example, sulfur undergoes a phase change from the rhombic crystal structure to the monoclinic crystal form at temperatures above 95°C.
    a. Predict the signs of ΔH and ΔS for the process $S_{rhombic} \longrightarrow S_{monoclinic}$.
    b. Which form of sulfur has the more ordered crystalline structure (has the smaller positional probability)?

51. As $O_2(l)$ is cooled at 1 atm, it freezes at 54.5 K to form solid I. At a lower temperature, solid I rearranges to solid II, which has a different crystal structure. Thermal measurements show that ΔH for the I $\longrightarrow$ II phase transition is -743.1 J/mol, and ΔS for the same transition is -17.0 J K^{-1} mol^{-1}. At what temperature are solids I and II in equilibrium?

Free Energy and Chemical Reactions

52. From data in Appendix 4, calculate $\Delta H°$, $\Delta S°$, and $\Delta G°$ for each of the following reactions at 25°C.
    a. $CH_4(g) + 2O_2(g) \longrightarrow CO_2(g) + 2H_2O(g)$
    b. $6CO_2(g) + 6H_2O(l) \longrightarrow \underset{\text{Glucose}}{C_6H_{12}O_6(s)} + 6O_2(g)$
    c. $P_4O_{10}(s) + 6H_2O(l) \longrightarrow 4H_3PO_4(s)$
    d. $HCl(g) + NH_3(g) \longrightarrow NH_4Cl(s)$

53. The value of $\Delta G°$ for the reaction

$$2C_4H_{10}(g) + 13O_2(g) \longrightarrow 8CO_2(g) + 10H_2O(l)$$

is $-5490.$ kJ. Use this value and data from Appendix 4 to calculate the standard free energy of formation for $C_4H_{10}(g)$.

54. Of the functions $\Delta H°$, $\Delta S°$, and $\Delta G°$, which depends most strongly on temperature? When $\Delta G°$ is calculated at temperatures other than 25°C, what assumptions are generally made concerning $\Delta H°$ and $\Delta S°$?

55. For the reaction at 298 K,

$$2NO_2(g) \rightleftharpoons N_2O_4(g)$$

the values of $\Delta H°$ and $\Delta S°$ are -58.03 kJ and -176.6 J/K, respectively. What is the value of $\Delta G°$ at 298 K? Assuming that $\Delta H°$ and $\Delta S°$ do not depend on temperature, at what temperature is $\Delta G° = 0$? Is $\Delta G°$ negative above or below this temperature?

56. Acrylonitrile is the starting material used in the manufacture of acrylic fibers (U.S. annual production capacity is more than 2 million pounds). Three industrial processes for the production of acrylonitrile are given below. Using data from Appendix 4, calculate $\Delta S°$, $\Delta H°$, and $\Delta G°$ for each process. For part a, assume that $T = 25°C$; for part b, $T = 70.°C$; and for part c, $T = 700.°C$. Assume that $\Delta H°$ and $\Delta S°$ do not depend on temperature.
    a. $CH_2\!-\!CH_2(g)$ (ring with O) $+ HCN(g)$
       Ethylene oxide
       $\longrightarrow \underset{\text{Acrylonitrile}}{CH_2\!=\!CHCN(g)} + H_2O(l)$
    b. $HC\!\equiv\!CH(g) + HCN(g) \xrightarrow[70°C\text{–}90°C]{CaCl_2 \cdot HCl} CH_2\!=\!CHCN(g)$
    c. $4CH_2\!=\!CHCH_3(g) + 6NO(g)$
       $\xrightarrow[Ag]{700°C} 4CH_2\!=\!CHCN(g) + 6H_2O(g) + N_2(g)$

57. Consider the reaction

$$2POCl_3(g) \longrightarrow 2PCl_3(g) + O_2(g)$$

    a. Calculate $\Delta G°$ for this reaction. The $\Delta G_f°$ values for $POCl_3(g)$ and $PCl_3(g)$ are -502 kJ/mol and $-270.$ kJ/mol, respectively.
    b. Is this reaction spontaneous under standard conditions at 298 K?
    c. The value of $\Delta S°$ for this reaction is 179 J/K. At what temperatures is this reaction spontaneous at standard conditions? Assume that $\Delta H°$ and $\Delta S°$ do not depend on temperature.

58. Consider two reactions for the production of ethanol:

$$C_2H_4(g) + H_2O(g) \longrightarrow CH_3CH_2OH(l)$$

$$C_2H_6(g) + H_2O(g) \longrightarrow CH_3CH_2OH(l) + H_2(g)$$

Which would be more thermodynamically feasible? Why? Assume standard conditions and assume that $\Delta H°$ and $\Delta S°$ are temperature-independent.

59. Using data from Appendix 4, calculate $\Delta H°$, $\Delta S°$, and $\Delta G°$ for the following reactions that produce acetic acid:

$$CH_4(g) + CO_2(g) \longrightarrow CH_3\overset{\overset{\displaystyle O}{\|}}{C}\!-\!OH(l)$$

$$CH_3OH(g) + CO(g) \longrightarrow CH_3\overset{\overset{\displaystyle O}{\|}}{C}\!-\!OH(l)$$

Which reaction would you choose as a commercial method for producing acetic acid (CH_3CO_2H)? What temperature conditions would you choose for the reaction? Assume standard conditions and assume that $\Delta H°$ and $\Delta S°$ are temperature-independent.

60. Given the following data:

$$2C_6H_6(l) + 15O_2(g) \longrightarrow 12CO_2(g) + 6H_2O(l) \qquad \Delta G° = -6399 \text{ kJ}$$

$$C(s) + O_2(g) \longrightarrow CO_2(g) \qquad \Delta G° = -394 \text{ kJ}$$

$$H_2(g) + \tfrac{1}{2}O_2(g) \longrightarrow H_2O(l) \qquad \Delta G° = -237 \text{ kJ}$$

calculate $\Delta G°$ for the reaction

$$6C(s) + 3H_2(g) \longrightarrow C_6H_6(l)$$

61. When most biological enzymes are heated, they lose their catalytic activity. The change

$$\text{Original enzyme} \longrightarrow \text{new form}$$

that occurs upon heating is endothermic and spontaneous. Is the structure of the original enzyme or its new form more ordered (has the smaller positional probability)? Explain your answer.

62. For the reaction

$$2O(g) \longrightarrow O_2(g)$$

a. predict the signs of ΔH and ΔS.
b. would the reaction be more spontaneous at high or low temperatures?

63. Hydrogen cyanide is produced industrially by the following exothermic reaction:

$$2NH_3(g) + 3O_2(g) + 2CH_4(g) \xrightarrow[\text{Pt-Rh}]{1000°C} 2HCN(g) + 6H_2O(g)$$

Is the high temperature needed for thermodynamic or for kinetic reasons?

Free Energy: Pressure Dependence and Equilibrium

64. A reaction at constant T and P is spontaneous as long as ΔG is negative; that is, reactions always proceed as long as the products have a lower free energy than the reactants. What is so special about equilibrium? Why don't reactions move away from equilibrium?

65. ΔG predicts spontaneity for a reaction at constant T and P, whereas $\Delta G°$ predicts the equilibrium position. Explain what this statement means. Under what conditions can you use $\Delta G°$ to determine the spontaneity of a reaction?

66. Using thermodynamic data from Appendix 4, calculate $\Delta G°$ at 25°C for the process

$$2SO_2(g) + O_2(g) \longrightarrow 2SO_3(g)$$

where all gases are at 1.00 atm pressure. Also calculate $\Delta G°$ at 25°C for this same reaction but with all gases at 10.0 atm pressure.

67. Consider the reaction

$$2NO_2(g) \rightleftharpoons N_2O_4(g)$$

For each of the following mixtures of reactants and products at 25°C, predict the direction in which the reaction will shift to reach equilibrium. Use thermodynamic data in Appendix 4.

a. $P_{NO_2} = P_{N_2O_4} = 1.0$ atm
b. $P_{NO_2} = 0.21$ atm, $P_{N_2O_4} = 0.50$ atm
c. $P_{NO_2} = 0.29$ atm, $P_{N_2O_4} = 1.6$ atm

68. Using data from Appendix 4, calculate ΔG for the reaction

$$2H_2S(g) + SO_2(g) \rightleftharpoons 3S(s) + 2H_2O(g)$$

for the following conditions at 25°C:

$$P_{H_2S} = 1.0 \times 10^{-4} \text{ atm}$$
$$P_{SO_2} = 1.0 \times 10^{-2} \text{ atm}$$
$$P_{H_2O} = 3.0 \times 10^{-2} \text{ atm}$$

69. Using data from Appendix 4, calculate $\Delta H°$, $\Delta S°$, and K (at 298 K) for the synthesis of ammonia by the Haber process:

$$N_2(g) + 3H_2(g) \rightleftharpoons 2NH_3(g)$$

Calculate ΔG for this reaction under the following conditions (assume an uncertainty of ±1 in all quantities):

a. $T = 298$ K, $P_{N_2} = P_{H_2} = 200$ atm, $P_{NH_3} = 50$ atm
b. $T = 298$ K, $P_{N_2} = 200$ atm, $P_{H_2} = 600$ atm, $P_{NH_3} = 200$ atm
c. $T = 100$ K, $P_{N_2} = 50$ atm, $P_{H_2} = 200$ atm, $P_{NH_3} = 10$ atm
d. $T = 700$ K, $P_{N_2} = 50$ atm, $P_{H_2} = 200$ atm, $P_{NH_3} = 10$ atm

Assume that $\Delta H°$ and $\Delta S°$ do not depend on temperature.

70. One of the reactions that destroys ozone in the upper atmosphere is

$$NO(g) + O_3(g) \rightleftharpoons NO_2(g) + O_2(g)$$

Using data from Appendix 4, calculate $\Delta G°$ and K (at 298 K) for this reaction.

71. Hydrogen sulfide can be removed from natural gas by the reaction

$$2H_2S(g) + SO_2(g) \rightleftharpoons 3S(s) + 2H_2O(g)$$

Calculate $\Delta G°$ and K (at 298 K) for this reaction. Would this reaction be favored at a high or low temperature?

72. Consider the autoionization of water at 25°C:

$$H_2O(l) \rightleftharpoons H^+(aq) + OH^-(aq) \qquad K_w = 1.00 \times 10^{-14}$$

a. Calculate $\Delta G°$ for this process at 25°C.
b. At 40.°C, $K_w = 2.92 \times 10^{-14}$. Calculate $\Delta G°$ at 40.°C.

73. How can one estimate the value of K at temperatures other than 25°C for a reaction? How can one estimate the temperature where $K = 1$ for a reaction? Do all reactions have a specific temperature where $K = 1$?

74. The standard free energies of formation and the standard enthalpies of formation at 298 K for difluoroacetylene (C_2F_2) and hexafluorobenzene (C_6F_6) are

	$\Delta G_f°$ (kJ/mol)	$\Delta H_f°$ (kJ/mol)
$C_2F_2(g)$	191.2	241.3
$C_6F_6(g)$	78.2	132.8

For the following reaction:

$$C_6F_6(g) \rightleftharpoons 3C_2F_2(g)$$

a. calculate $\Delta S°$ at 298 K.
b. calculate K at 298 K.
c. estimate K at 3000. K, assuming $\Delta H°$ and $\Delta S°$ do not depend on temperature.

75. Consider the reaction

$$Fe_2O_3(s) + 3H_2(g) \longrightarrow 2Fe(s) + 3H_2O(g)$$

Assuming $\Delta H°$ and $\Delta S°$ do not depend on temperature, calculate the temperature where $K = 1.00$ for this reaction.

76. The Ostwald process for the commercial production of nitric acid involves three steps:

$$4NH_3(g) + 5O_2(g) \xrightarrow[825°C]{Pt} 4NO(g) + 6H_2O(g)$$

$$2NO(g) + O_2(g) \longrightarrow 2NO_2(g)$$

$$3NO_2(g) + H_2O(l) \longrightarrow 2HNO_3(l) + NO(g)$$

a. Calculate $\Delta H°$, $\Delta S°$, $\Delta G°$, and K (at 298 K) for each of the three steps in the Ostwald process (see Appendix 4).
b. Calculate the equilibrium constant for the first step at 825°C. Assume that $\Delta H°$ and $\Delta S°$ are temperature-independent.
c. Is there a thermodynamic reason for the high temperature in the first step assuming standard conditions?

77. Consider the following reaction at 800. K:

$$N_2(g) + 3F_2(g) \longrightarrow 2NF_3(g)$$

An equilibrium mixture contains the following partial pressures: $P_{N_2} = 0.021$ atm, $P_{F_2} = 0.063$ atm, and $P_{NF_3} =$ 0.48 atm. Calculate $\Delta G°$ for the reaction at 800. K.

78. Consider the following reaction at 298 K:

$$2SO_2(g) + O_2(g) \longrightarrow 2SO_3(g)$$

An equilibrium mixture contains $O_2(g)$ and $SO_3(g)$ at partial pressures of 0.50 atm and 2.0 atm, respectively. Using data from Appendix 4, determine the equilibrium partial pressure of SO_2 in the mixture. Will this reaction be most favored at a high or a low temperature, assuming standard conditions?

79. For the reaction

$$A(g) + 2B(g) \rightleftharpoons C(g)$$

the initial partial pressures of gases A, B, and C are all 0.100 atm. Once equilibrium has been established, it is found that [C] = 0.040 atm. What is $\Delta G°$ for this reaction at 25°C?

80. Consider the following diagram of free energy (G) versus fraction of A reacted in terms of moles for the reaction $2A(g) \longrightarrow B(g)$.

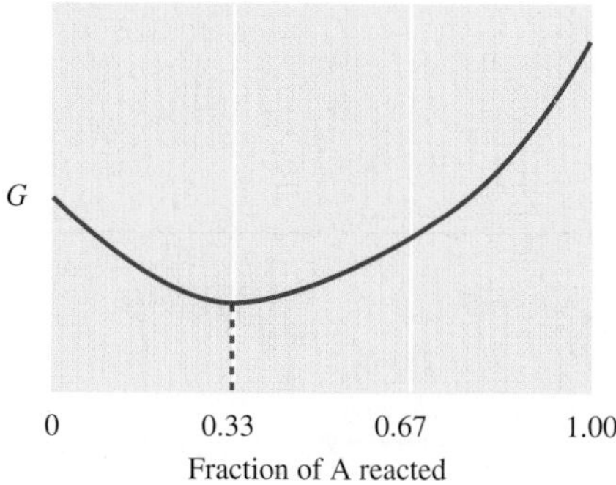

Before any A has reacted, $P_A = 3.0$ atm and $P_B = 0$. Determine the sign of $\Delta G°$ and the value of K for this reaction.

81. Calculate $\Delta G°$ for $H_2O(g) + \frac{1}{2}O_2(g) \rightleftharpoons H_2O_2(g)$ at 600. K, using the following data:

$$H_2(g) + O_2(g) \rightleftharpoons H_2O_2(g) \quad K = 2.3 \times 10^6 \text{ at 600. K}$$

$$2H_2(g) + O_2(g) \rightleftharpoons 2H_2O(g) \quad K = 1.8 \times 10^{37} \text{ at 600. K}$$

82. Cells use the hydrolysis of adenosine triphosphate, abbreviated ATP, as a source of energy. Symbolically, this reaction can be represented as

$$ATP(aq) + H_2O(l) \longrightarrow ADP(aq) + H_2PO_4^-(aq)$$

where ADP represents adenosine diphosphate. For this reaction $\Delta G° = -30.5$ kJ/mol.
a. Calculate K at 25°C.
b. If all the free energy from the metabolism of glucose

$$C_6H_{12}O_6(s) + 6O_2(g) \longrightarrow 6CO_2(g) + 6H_2O(l)$$

goes into the production of ATP, how many ATP molecules can be produced for every molecule of glucose?

83. Carbon monoxide is toxic because it bonds much more strongly to the iron in hemoglobin (Hgb) than does O_2. Consider the following reactions and approximate standard free energy changes:

$$Hgb + O_2 \longrightarrow HgbO_2 \quad \Delta G° = -70 \text{ kJ}$$

$$Hgb + CO \longrightarrow HgbCO \quad \Delta G° = -80 \text{ kJ}$$

Using these data, estimate the equilibrium constant value at 25°C for the following reaction:

$$HgbO_2 + CO \rightleftharpoons HgbCO + O_2$$

84. One reaction that occurs in human metabolism is

$$HO_2CCH_2CH_2\underset{\displaystyle NH_2}{\underset{|}{C}}HCO_2H(aq) + NH_3(aq) \rightleftharpoons$$

Glutamic acid

$$H_2N\overset{\displaystyle O}{\overset{\|}{C}}CH_2CH_2\underset{\displaystyle NH_2}{\underset{|}{C}}HCO_2H(aq) + H_2O(l)$$

Glutamine

For this reaction $\Delta G° = 14$ kJ at 25°C.
a. Calculate K for this reaction at 25°C.
b. In a living cell this reaction is coupled with the hydrolysis of ATP. (See Exercise 82.) Calculate $\Delta G°$ and K at 25°C for the following reaction:

$$\text{Glutamic acid}(aq) + ATP(aq) + NH_3(aq) \rightleftharpoons \text{Glutamine}(aq) + ADP(aq) + H_2PO_4^-(aq)$$

85. At 25.0°C, for the reaction

$$2NO_2(g) \rightleftharpoons N_2O_4(g)$$

the values of $\Delta H°$ and $\Delta S°$ are -58.03 kJ/mol and -176.6 J K^{-1} mol^{-1}, respectively. Calculate the value of K at 25.0°C. Assuming $\Delta H°$ and $\Delta S°$ are temperature-independent, estimate the value of K at 100.0°C.

86. Consider the relationship

$$\ln(K) = \frac{-\Delta H°}{RT} + \frac{\Delta S°}{R}$$

The equilibrium constant for some hypothetical process was determined as a function of temperature (in kelvins) with the results plotted below.

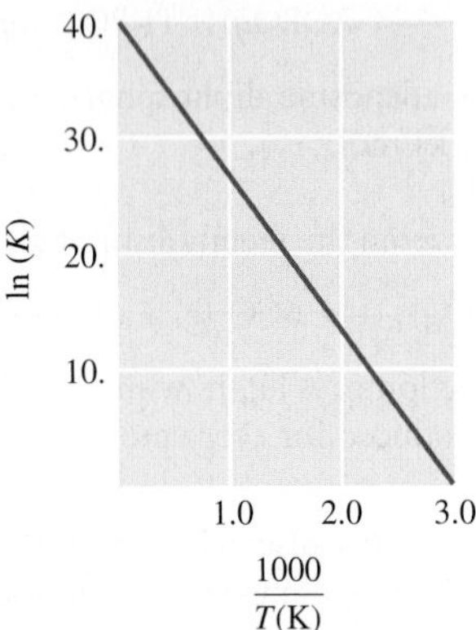

From the plot, determine the values of $\Delta H°$ and $\Delta S°$ for this process. What would be the major difference in the ln(*K*) versus 1/*T* plot for an endothermic process as compared to an exothermic process?

87. a. Use the equation in Exercise 86 to determine $\Delta H°$ and $\Delta S°$ for the autoionization of water:

$$H_2O(l) \rightleftharpoons H^+(aq) + OH^-(aq)$$

T (°C)	K
0	1.14×10^{-15}
25	1.00×10^{-14}
35	2.09×10^{-14}
40.	2.92×10^{-14}
50.	5.47×10^{-14}

b. Estimate the value of $\Delta G°$ for the autoionization of water at its critical temperature, 374°C.

88. The equilibrium constant *K* for the reaction

$$2Cl(g) \rightleftharpoons Cl_2(g)$$

was measured as a function of temperature (in kelvins). A graph of ln(*K*) versus 1/*T* for this reaction gives a straight line with a slope of 1.352×10^4 K and a *y* intercept of −14.51. Determine the values of $\Delta H°$ and $\Delta S°$ for this reaction. (See Exercise 86.)

89. The equilibrium constant for a certain reaction decreases from 8.84 to 3.25×10^{-2} when the temperature increases from 25°C to 75°C. Estimate the temperature where $K = 1.00$ for this reaction. Estimate the value of $\Delta S°$ for this reaction. (*Hint:* Manipulate the equation given in Exercise 86.)

Adiabatic Processes

90. A sample of a monatomic ideal gas at 1.00 atm and 25°C expands adiabatically and reversibly from 5.00 L to 12.5 L. Calculate the final temperature and pressure of the gas, the work associated with this process, and the change in internal energy.

91. A sample of 1.75 moles of H_2 ($C_v = 20.5$ J K^{-1} mol^{-1}) at 21°C and 1.50 atm undergoes a reversible adiabatic compression until the final pressure is 4.50 atm. Calculate the final volume of the gas sample and the work associated with this process. Assume that the gas behaves ideally.

92. A 1.50-mole sample of an ideal gas is allowed to expand adiabatically and reversibly to twice its original volume. In the expansion the temperature dropped from 296 K to 239 K. Calculate ΔE and ΔH for the gas expansion.

93. Consider 1.00 mole of $CO_2(g)$ at 300. K and 5.00 atm. The gas expands until the final pressure is 1.00 atm. For each of the following conditions describing the expansion, calculate q, w, and ΔE. C_p for CO_2 is 37.1 J K^{-1} mol^{-1}, and assume that the gas behaves ideally.
   a. The expansion occurs isothermally and reversibly.
   b. The expansion occurs isothermally against a constant external pressure of 1.00 atm.
   c. The expansion occurs adiabatically and reversibly.

94. Consider 1.00 mole of $CO_2(g)$ at 300. K and 5.00 atm. The gas expands until the final pressure is 1.00 atm. For each of the following conditions describing the expansion, calculate ΔS, ΔS_{surr}, and ΔS_{univ}. C_p for CO_2 is 37.1 J K^{-1} mol^{-1}, and assume that the gas behaves ideally.
   a. The expansion occurs isothermally and reversibly.
   b. The expansion occurs isothermally against a constant external pressure of 1.00 atm.
   c. The expansion occurs adiabatically and reversibly.

Additional Exercises

95. Monochloroethane (C_2H_5Cl) can be produced by the direct reaction of ethane gas (C_2H_6) with chlorine gas or by the reaction of ethylene gas (C_2H_4) with hydrogen chloride gas. The second reaction gives almost a 100% yield of pure C_2H_5Cl at a rapid rate without catalysis. The first method requires light as an energy source or the reaction would not occur. Yet $\Delta G°$ for the first reaction is considerably more negative than $\Delta G°$ for the second reaction. Explain how this can be so.

96. Given the following illustration, what can be said about the sign of ΔS for the process of solid NaCl dissolving in water? What can be said for ΔH about this process?

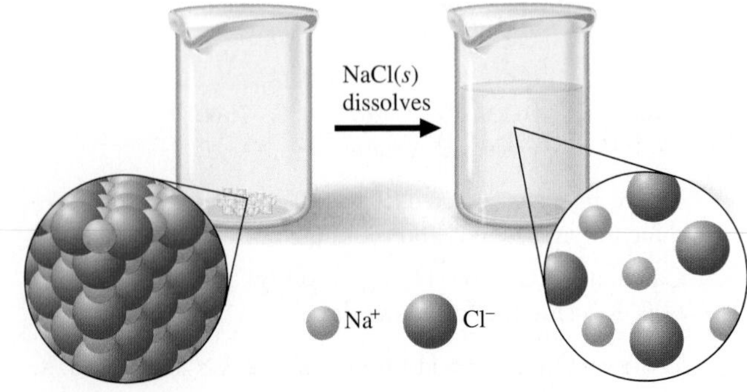

97. Some water is placed in a coffee cup calorimeter. When 1.0 g of an ionic solid is added, the temperature of the solution increases from 21.5°C to 24.2°C as the solid dissolves. For the dissolving process, what are the signs for ΔS_{sys}, ΔS_{surr}, and ΔS_{univ}?

98. Entropy has been described as "time's arrow." Interpret this view of entropy.

99. Discuss the relationship between w_{max} and the magnitude and sign of the free energy change for a reaction. Also discuss w_{max} for real processes.

100. Human DNA contains almost twice as much information as is needed to code for all the substances produced in the body. Likewise, the digital data sent from *Voyager 2* contain one redundant bit out of every two bits of information. The Hubble space telescope transmits three redundant bits for every bit of information. How is entropy related to the transmission of information? What do you think is accomplished by having so many redundant bits of information in both DNA and the space probes?

101. The enthalpy of vaporization of chloroform ($CHCl_3$) is 31.4 kJ/mol at its boiling point (61.7°C). Determine ΔS_{sys}, ΔS_{surr}, and ΔS_{univ} when 1.00 mole of chloroform is vaporized at 61.7°C and 1.00 atm.

102. Two crystalline forms of white phosphorus are known. Both forms contain P_4 molecules, but the molecules are packed together in different ways. The α form is always obtained when the liquid freezes. However, below −76.9°C, the α form spontaneously converts to the β form:

$$P_4(s, \alpha) \longrightarrow P_4(s, \beta)$$

a. Predict the signs of ΔH and ΔS for this process.
b. Predict which form of phosphorus has the more ordered crystalline structure (has the smaller positional probability).

103. In the text the equation

$$\Delta G = \Delta G° + RT\ln(Q)$$

was derived for gaseous reactions where the quantities in Q were expressed in units of pressure. We also can use units of mol/L for the quantities in Q—specifically for aqueous reactions. With this in mind, consider the reaction

$$HF(aq) \rightleftharpoons H^+(aq) + F^-(aq)$$

for which $K_a = 7.2 \times 10^{-4}$ at 25°C. Calculate ΔG for the reaction under the following conditions at 25°C:

a. $[HF] = [H^+] = [F^-] = 1.0\ M$
b. $[HF] = 0.98\ M$, $[H^+] = [F^-] = 2.7 \times 10^{-2}\ M$
c. $[HF] = [H^+] = [F^-] = 1.0 \times 10^{-5}\ M$
d. $[HF] = [F^-] = 0.27\ M$, $[H^+] = 7.2 \times 10^{-4}\ M$
e. $[HF] = 0.52\ M$, $[F^-] = 0.67\ M$, $[H^+] = 1.0 \times 10^{-3}\ M$

Based on the calculated ΔG values, in which direction will the reaction shift to reach equilibrium for each of the five sets of conditions?

104. Many biochemical reactions that occur in cells require relatively high concentrations of potassium ion (K^+). The concentration of K^+ in muscle cells is about 0.15 M. The concentration of K^+ in blood plasma is about 0.0050 M. The high internal concentration in cells is maintained by pumping K^+ from the plasma. How much work must be done to transport 1.0 mole of K^+ from the blood to the inside of a muscle cell at 37°C (normal body temperature)? When 1.0 mole of K^+ is transferred from blood to the cells, do any other ions have to be transported? Why or why not? Much of the ATP (see Exercise 82) formed from metabolic processes is used to provide energy for transport of cellular components. How much ATP must be hydrolyzed to provide the energy for the transport of 1.0 mole of K^+?

105. Consider the following system at equilibrium at 25°C:

$$PCl_3(g) + Cl_2(g) \rightleftharpoons PCl_5(g) \quad \Delta G° = -92.50\ \text{kJ}$$

What will happen to the ratio of partial pressure of PCl_5 to partial pressure of PCl_3 if the temperature is raised? Explain completely.

106. Consider the reaction

$$H_2(g) + Br_2(g) \rightleftharpoons 2HBr(g)$$

where $\Delta H° = -103.8$ kJ. In a particular experiment, 1.00 atm of $H_2(g)$ and 1.00 atm of $Br_2(g)$ were mixed in a 1.00-L flask at 25°C and allowed to reach equilibrium. Then the molecules of H_2 were counted by using a very sensitive technique, and 1.10×10^{13} molecules were found. For this reaction, calculate the values of K, $\Delta G°$, and $\Delta S°$.

107. At 1500 K the process

$$\underset{10\ \text{atm}}{I_2(g)} \longrightarrow \underset{10\ \text{atm}}{2I(g)}$$

is not spontaneous. However, the process

$$\underset{0.10\ \text{atm}}{I_2(g)} \longrightarrow \underset{0.10\ \text{atm}}{2I(g)}$$

is spontaneous at 1500 K. Explain.

108. Using the following data, calculate the value of K_{sp} for $Ba(NO_3)_2$, one of the least soluble of the common nitrate salts.

Species	$\Delta G_f°$
$Ba^{2+}(aq)$	−561 kJ/mol
$NO_3^-(aq)$	−109 kJ/mol
$Ba(NO_3)_2(s)$	−797 kJ/mol

109. Sodium chloride is added to water (at 25°C) until it is saturated. Calculate the Cl^- concentration in such a solution.

Species	$\Delta G°$(kJ/mol)
$NaCl(s)$	−384
$Na^+(aq)$	−262
$Cl^-(aq)$	−131

110. What is the pH of a 0.125 M solution of the weak base B if $\Delta H° = -28.0$ kJ and $\Delta S° = -175$ J/K for the following equilibrium reaction at 25°C?

$$B(aq) + H_2O(l) \rightleftharpoons BH^+(aq) + OH^-(aq)$$

111. Consider the reactions

$$Ni^{2+}(aq) + 6NH_3(aq) \longrightarrow Ni(NH_3)_6^{2+}(aq) \quad (1)$$

$$Ni^{2+}(aq) + 3en(aq) \longrightarrow Ni(en)_3^{2+}(aq) \quad (2)$$

where

$$en = H_2N{-}CH_2{-}CH_2{-}NH_2$$

The ΔH values for the two reactions are quite similar, yet $K_{reaction\ 2} > K_{reaction\ 1}$. Explain.

112. The deciding factor on why HF is a weak acid and not a strong acid like the other hydrogen halides is entropy. What occurs when HF dissociates in water as compared to the other hydrogen halides?

113. The third law of thermodynamics states that the entropy of a perfect crystal at 0 K is zero. In Appendix 4, $F^-(aq)$, $OH^-(aq)$, and $S^{2-}(aq)$ all have negative standard entropy values. How can $S°$ values be less than zero?

114. Calculate the entropy change for the vaporization of liquid methane and hexane using the following data:

	Boiling Point (1 atm)	ΔH_{vap}
Methane	112 K	8.20 kJ/mol
Hexane	342 K	28.9 kJ/mol

Compare the molar volume of gaseous methane at 112 K with that of gaseous hexane at 342 K. How do the differences in molar volume affect the values of ΔS_{vap} for these liquids?

115. The standard entropy values ($S°$) for $H_2O(l)$ and $H_2O(g)$ are 70. J K^{-1} mol^{-1} and 189 J K^{-1} mol^{-1}, respectively. Calculate the ratio of Ω_g to Ω_l for water using Boltzmann's equation. (See Exercise 22.)

116. Calculate the values of ΔS and ΔG for each of the following processes at 298 K:

$$H_2O(l, 298\ K) \longrightarrow H_2O(g, V = 1000.\ L/mol)$$

$$H_2O(l, 298\ K) \longrightarrow H_2O(g, V = 100.\ L/mol)$$

The standard enthalpy of vaporization for water at 298 K is 44.02 kJ/mol. Does either of these processes occur spontaneously?

117. Calculate the changes in free energy, enthalpy, and entropy when 1.00 mole of Ar(*g*) at 27°C is compressed isothermally from 100.0 L to 1.00 L.

118. Consider the isothermal expansion of 1.00 mole of ideal gas at 27°C. The volume increases from 30.0 L to 40.0 L. Calculate q, w, ΔE, ΔH, ΔS, and ΔG for two situations:
   a. a free expansion
   b. a reversible expansion

119. A 1.00-mole sample of an ideal gas in a vessel with a movable piston initially occupies a volume of 5.00 L at an external pressure of 5.00 atm.
   a. If P_{ex} is suddenly lowered to 2.00 atm and the gas is allowed to expand isothermally, calculate the following quantities for the system: ΔE, ΔH, ΔS, ΔG, w, and q.
   b. Show by the second law that this process will occur spontaneously.

120. One mole of an ideal gas with a volume of 6.67 L and a pressure of 1.50 atm is contained in a vessel with a movable piston. The external pressure is suddenly increased to 5.00 atm and the gas is compressed isothermally (T = 122 K). Calculate ΔE, ΔH, ΔS, w, q, ΔS_{surr}, ΔS_{univ}, and ΔG.

Challenge Problems

121. Consider a 2.00-mole sample of Ar at 2.00 atm and 298 K.
   a. If the gas sample expands adiabatically and reversibly to a pressure of 1.00 atm, calculate the final temperature of the gas sample assuming ideal gas behavior.
   b. If the gas sample expands adiabatically and irreversibly against a constant 1.00 atm pressure, calculate the final temperature of the gas sample assuming ideal gas behavior.

122. Consider 1.0 mole of a monatomic ideal gas in a container fitted with a piston. The initial conditions are 5.0 L and P = 5.0 atm at some constant T.
   a. If the external pressure is suddenly changed to 2.0 atm, show that expansion of the gas is spontaneous.
   b. If the external pressure suddenly changes back to 5.0 atm, show that compression of the gas is spontaneous.
   c. Calculate and compare signs of ΔG for each case (parts a and b) and discuss why this sign cannot be used to predict spontaneity.

123. One mole of an ideal gas undergoes an isothermal reversible expansion at 25°C. During this process, the system absorbs 855 J of heat from the surroundings. When this gas is compressed to the original state in one step (isothermally), *twice* as much work is done on the system as was performed on the surroundings in the expansion.
   a. What is ΔS for the one-step isothermal compression?
   b. What is ΔS_{univ} for the overall process (expansion and compression)?

124. At least some of what is in the following quoted statement is false. Change the incorrect statements so that they are correct and defend your answer. What is correct in the statements? What is wrong? Discuss a real-world situation that supports your position.

"The magnitude of ΔS is always larger than the magnitude of ΔS_{surr}. This is so because ΔS is related to q_{rev}, whereas ΔS_{surr} is related to q_{actual}, and the magnitude of q_{rev} is always larger than the magnitude of q_{actual}."

125. You have a 1.00-L sample of hot water (90.°C) sitting open in a 25°C room. Eventually the water cools to 25°C, whereas the temperature of the room remains unchanged.

Calculate ΔS_{univ} for this process. Assume the density of water is 1.00 g/mL over this temperature range and that the heat capacity of water is constant over this temperature range and equal to 75.3 J K^{-1} mol^{-1}.

126. Consider two perfectly insulated vessels. Vessel 1 initially contains an ice cube at 0°C and water at 0°C. Vessel 2 initially contains an ice cube at 0°C and a saltwater solution at 0°C. Consider the process $H_2O(s) \rightarrow H_2O(l)$.
   a. Determine the sign of ΔS, ΔS_{surr}, and ΔS_{univ} for the process in vessel 1.
   b. Determine the sign of ΔS, ΔS_{surr}, and ΔS_{univ} for the process in vessel 2.

   (*Hint:* Think about the effect that a salt has on the freezing point of a solvent.)

127. If wet silver carbonate is dried in a stream of hot air, the air must have a certain concentration level of carbon dioxide to prevent silver carbonate from decomposing by the reaction

$$Ag_2CO_3(s) \rightleftharpoons Ag_2O(s) + CO_2(g)$$

$\Delta H°$ for this reaction is 79.14 kJ/mol in the temperature range of 25°C–125°C. Given that the partial pressure of carbon dioxide in equilibrium with pure solid silver carbonate is 6.23×10^{-3} torr at 25°C, calculate the partial pressure of CO_2 necessary to prevent decomposition of Ag_2CO_3 at 110.°C.

128. Consider a weak acid HX. If a 0.10 *M* solution of HX has a pH of 5.83 at 25°C, what is $\Delta G°$ for the acid's dissociation reaction at 25°C?

129. Using data from Appendix 4, calculate $\Delta H°$, $\Delta G°$, and *K* (at 298 K) for the production of ozone from oxygen:

$$3O_2(g) \rightleftharpoons 2O_3(g)$$

At 30 km above the surface of the earth, the temperature is about 230. K and the partial pressure of oxygen is about 1.0×10^{-3} atm. Estimate the partial pressure of ozone in equilibrium with oxygen at 30 km above the earth's surface. Is it reasonable to assume that the equilibrium between oxygen and ozone is maintained under these conditions? Explain.

130. One mole of a monatomic ideal gas (for which $S° =$ 8.00 J K^{-1} mol^{-1} at −73.0°C) was heated at a constant pressure of 2.00 atm from −73.0°C to 27.0°C. Calculate ΔH, ΔE, w, q, ΔS due to the change in volume, ΔS due to the change in temperature, and ΔG.

131. Consider the system

$$A(g) \longrightarrow B(g)$$

at 25°C.
   a. Assuming that $G_A^\circ = 8996$ J/mol and $G_B^\circ =$ 11,718 J/mol, calculate the value of the equilibrium constant for this reaction.
   b. Calculate the equilibrium pressures that result if 1.00 mole of A(*g*) at 1.00 atm and 1.00 mole of B(*g*) at 1.00 atm are mixed at 25°C.
   c. Show by calculations that $\Delta G = 0$ at equilibrium.

132. Liquid water at 25°C is introduced into an evacuated, insulated vessel. Identify the signs of the following thermodynamic functions for the process that occurs: ΔH, ΔS, ΔG, ΔT_{water}, ΔS_{surr}, ΔS_{univ}.

133. Consider 1.00 mole of an ideal gas that is expanded isothermally at 25°C from 2.45×10^{-2} atm to 2.45×10^{-3} atm in the following three irreversible steps:

   Step 1: from 2.45×10^{-2} atm to 9.87×10^{-3} atm
   Step 2: from 9.87×10^{-3} atm to 4.93×10^{-3} atm
   Step 3: from 4.93×10^{-3} atm to 2.45×10^{-3} atm

   Calculate q, w, ΔE, ΔS, ΔH, and ΔG for each step and for the overall process.

134. Consider 1.00 mole of an ideal gas at 25°C.
   a. Calculate q, w, ΔE, ΔS, ΔH, and ΔG for the expansion of this gas isothermally and irreversibly from 2.45×10^{-2} atm to 2.45×10^{-3} atm in one step.
   b. Calculate q, w, ΔE, ΔS, ΔH, and ΔG for the same change of pressure as in part a but performed isothermally and reversibly.
   c. Calculate q, w, ΔE, ΔS, ΔH, and ΔG for the one-step isothermal, irreversible compression of 1.00 mole of an ideal gas at 25°C from 2.45×10^{-3} atm to 2.45×10^{-2} atm.
   d. Construct the *PV* diagrams for the processes described in parts a, b, and c.
   e. Calculate the entropy change in the surroundings for the processes described in parts a, b, and c.

135. Consider the reaction

$$2CO(g) + O_2(g) \longrightarrow 2CO_2(g)$$

   a. Using data from Appendix 4, calculate *K* at 298 K.
   b. What is ΔS for this reaction at $T = 298$ K if the reactants, each at 10.0 atm, are changed to products at 10.0 atm? (*Hint:* Construct a thermodynamic cycle and consider how entropy depends on pressure.)

136. Calculate $\Delta H°$ and $\Delta S°$ at 25°C for the reaction

$$2SO_2(g) + O_2(g) \longrightarrow 2SO_3(g)$$

at a constant pressure of 1.00 atm using thermodynamic data in Appendix 4. Also calculate $\Delta H°$ and $\Delta S°$ at 227°C and 1.00 atm, assuming that the constant-pressure molar heat capacities for $SO_2(g)$, $O_2(g)$, and $SO_3(g)$ are 39.9 J K^{-1} mol^{-1}, 29.4 J K^{-1} mol^{-1}, and 50.7 J K^{-1} mol^{-1}, respectively. (*Hint:* Construct a thermodynamic cycle, and consider how enthalpy and entropy depend on temperature.)

137. Although we often assume that the heat capacity of a substance is not temperature-dependent, this is not strictly true, as shown by the following data for ice:

Temperature (°C)	C_p (J K^{-1} mol^{-1})
−200.	12
−180.	15
−160.	17
−140.	19
−100.	24
−60.	29
−30.	33
−10.	36
0	37

Use these data to calculate graphically the change in entropy for heating ice from −200.°C to 0°C. (*Hint:* Recall that

$$\Delta S_{T_1 \rightarrow T_2} = \int_{T_1}^{T_2} \frac{C_p\, dT}{T}$$

and that integration from T_1 to T_2 sums the area under the curve of a plot of C_p/T versus T from T_1 to T_2.)

138. Consider the following C_p values for $N_2(g)$:

C_p (J K^{-1} mol^{-1})	T (K)
28.7262	300.0
29.2937	400.0
29.8545	500.0

Assume that C_p can be expressed in the form

$$C_p = a + bT + cT^2$$

Estimate the value of C_p for $N_2(g)$ at 900. K. Assuming that C_p shows this temperature dependence over the range 100 K to 900 K, calculate ΔS for heating 1.00 mole of $N_2(g)$ from 100. K to 900. K.

139. Benzene (C_6H_6) has a melting point of 5.5°C and an enthalpy of fusion of 10.04 kJ/mol at 25.0°C. The molar heat capacities at constant pressure for solid and liquid benzene are 100.4 J K^{-1} mol^{-1} and 133.0 J K^{-1} mol^{-1}, respectively. For the reaction

$$C_6H_6(l) \rightleftharpoons C_6H_6(s)$$

calculate ΔS_{sys} and ΔS_{surr} at 10.0°C.

Marathon Problems

140. Impure nickel, refined by smelting sulfide ores in a blast furnace, can be converted into metal from 99.90% to 99.99% purity by the Mond process. The primary reaction involved in the Mond process is

$$Ni(s) + 4CO(g) \rightleftharpoons Ni(CO)_4(g)$$

a. Without referring to Appendix 4, predict the sign of $\Delta S°$ for the preceding reaction. Explain.

b. The spontaneity of the preceding reaction is temperature-dependent. Predict the sign of ΔS_{surr} for this reaction. Explain.

c. For $Ni(CO)_4(g)$, $\Delta H_f^\circ = -607$ kJ/mol and $S° =$ 417 J K^{-1} mol^{-1} at 298 K. Using these values and data in Appendix 4, calculate $\Delta H°$ and $\Delta S°$ for the preceding reaction.

d. Calculate the temperature at which $\Delta G° = 0$ ($K = 1$) for the preceding reaction, assuming that $\Delta H°$ and $\Delta S°$ do not depend on temperature.

e. The first step of the Mond process involves equilibrating impure nickel with $CO(g)$ and $Ni(CO)_4(g)$ at about 50°C. The purpose of this step is to convert as much nickel as possible into the gas phase. Calculate the equilibrium constant for the preceding reaction at 50.°C.

f. In the second step of the Mond process, the gaseous $Ni(CO)_4$ is isolated and heated at 227°C. The purpose of this step is to deposit as much nickel as possible as pure solid (the reverse of the preceding reaction). Calculate the equilibrium constant for the preceding reaction at 227°C.

g. Why is temperature increased for the second step of the Mond process?

h. The Mond process relies on the volatility of $Ni(CO)_4$ for its success. Only pressures and temperatures at which $Ni(CO)_4$ is a gas are useful. A recently developed variation of the Mond process carries out the first step at higher pressures and a temperature of 152°C. Estimate the maximum pressure of $Ni(CO)_4(g)$ that can be attained before the gas will liquefy at 152°C. The boiling point for $Ni(CO)_4$ is 42°C, and the enthalpy of vaporization is 29.0 kJ/mol. [*Hint:* The phase-change reaction and the corresponding equilibrium expression are

$$Ni(CO)_4(l) \rightleftharpoons Ni(CO)_4(g) \qquad K = P_{Ni(CO)_4}$$

$Ni(CO)_4(g)$ will liquefy when the pressure of $Ni(CO)_4$ is greater than the K value.]

141. The initial state of an ideal gas is 2.00 atm, 2.00 L. The final state is 1.00 atm, 4.00 L. The expansion is accomplished isothermally.

a. If the expansion is a free expansion, calculate w, q, ΔE, and ΔH.

b. If the expansion is done in one step, calculate w, q, ΔE, and ΔH.

c. If the expansion is done in two steps (with $V =$ 3.00 L as the intermediate step), calculate w, q, ΔE, and ΔH.

d. If the expansion is reversible, calculate w, q, ΔE, and ΔH.

You have the new state of an ideal gas at 1.00 atm, 4.00 L. You take the gas back to conditions of 2.00 atm, 2.00 L. The compression is accomplished isothermally.

e. If the compression is done in one step, calculate w, q, ΔE, and ΔH.

f. If the compression is done in two steps (with $V =$ 3.00 L as the intermediate step), calculate w, q, ΔE, and ΔH.

g. If the compression is reversible, calculate w, q, ΔE, and ΔH. Explain.

Compare your answers for the expansion and compression. Discuss the implications, especially considering the changes to the system and the changes to the surroundings that have occurred even though the system was brought back to its initial state.

Electrodynamics

11

chapter

A rusted car from the 1930s.

Sign in to OWL at www.cengage.com/owl to view tutorials and simulations, develop problem-solving skills, and complete online homework assigned by your professor.

go Chemistry

Download mini lecture videos for key concept review and exam prep from OWL or purchase them from www.cengagebrain.com.

Electrochemistry constitutes one of the most important interfaces between chemistry and everyday life. Every time you start your car, turn on your calculator, look at your digital watch, or listen to a radio at the beach, you are depending on electrochemical reactions. Our society sometimes seems to run almost entirely on batteries. Certainly, the advent of small, dependable batteries along with silicon chip technology has made possible the tiny calculators, tape recorders, and clocks that we take for granted.

Electrochemistry is important in other less obvious ways. For example, the corrosion of iron, which has tremendous economic implications, is an electrochemical process. In addition, many important industrial materials such as aluminum, chlorine, and sodium hydroxide are prepared by electrolytic processes. In analytical chemistry, electrochemical techniques use electrodes that are specific for a given molecule or ion, including H^+ (pH meters), F^-, Cl^-, and many others. These increasingly important methods are used to analyze for trace pollutants in natural waters or for the tiny quantities of chemicals in human blood that may signal the development of a specific disease.

Electrochemistry is best defined as *the study of the interchange of chemical and electrical energy*. It is primarily concerned with two processes that involve oxidation–reduction reactions: the generation of an electric current from a chemical reaction and the opposite process, the use of a current to produce chemical change.

11.1 Galvanic Cells

Recall from Chapter 4 that an **oxidation–reduction (redox) reaction** involves a transfer of electrons from the **reducing agent** to the **oxidizing agent** and that **oxidation** involves a *loss of electrons* (an increase in oxidation number) and **reduction** involves a *gain of electrons* (a decrease in oxidation number).

To understand how a redox reaction can be used to generate a current, we will consider the reaction between MnO_4^- and Fe^{2+}:

$$8H^+(aq) + MnO_4^-(aq) + 5Fe^{2+}(aq) \longrightarrow Mn^{2+}(aq) + 5Fe^{3+}(aq) + 4H_2O(l)$$

In this reaction Fe^{2+} is oxidized and MnO_4^- is reduced; electrons are transferred from Fe^{2+} (the reducing agent) to MnO_4^- (the oxidizing agent).

Balancing half-reactions is discussed in Section 4.11.

It is useful to break a redox reaction into two **half-reactions,** one involving oxidation and the other involving reduction. For the preceding reaction, the half-reactions are

$$8H^+ + MnO_4^- + 5e^- \longrightarrow Mn^{2+} + 4H_2O$$

$$5(Fe^{2+} \longrightarrow Fe^{3+} + e^-)$$

Note that the second half-reaction must occur five times for each time the first reaction occurs. The balanced overall reaction is the sum of the half-reactions.

When MnO_4^- and Fe^{2+} are present in the same solution, the electrons are transferred directly as the reactants collide. Under these conditions, no useful work is obtained from the chemical energy associated with the reaction, which instead is released as heat. How can we harness this energy? The key is to separate the oxidizing agent from the reducing agent, thus requiring the electron transfer to occur through a wire. The current produced in the wire by the electron flow can then be directed through a device, such as an electric motor, to provide useful work.

For example, consider the system illustrated in Fig. 11.1. If our reasoning has been correct, electrons should flow through the wire from Fe^{2+} to MnO_4^-. However, when we construct the apparatus as shown, no electron flow is ap-

Dental Resistance

Chemical Insights

A trip to the dentist is a necessary but often disquieting event. Because dealing with dental cavities can be expensive and unpleasant, dentists are searching for ways to find budding cavities at the very earliest stages of their development, well before they can be detected by X rays. One such method is being developed by Chris Longbottom, a dentist at the University of Dundee Dental School in Scotland. Longbottom is experimenting with measuring the electrical resistance of teeth using tiny, 10-millivolt currents. This approach takes advantage of the changes in a tooth that occur as a cavity begins to form. Bacteria in the mouth produce acids that dissolve tooth mineral, a process that starts in the pores of the teeth. As a given pore is enlarged by the action of acid, the electrical resistance of the tooth changes because the electrolyte solution that collects in the enlarged pore is a better electrical conductor than the tooth mineral. Using special electrodes that fit over the teeth at points where the teeth touch (and are thus most vulnerable to trapped food that stimulates bacterial growth), Longbottom and his colleagues have noted changes in resistance as cavities start to form. This method of detection is so sensitive that the tiny cavities can be treated with fluoride or antibiotics, thus checking the growth of cavities long before drilling is necessary. Although the technique needs further development, Longbottom hopes it will join dentists' arsenal of tools soon.

parent. Why? Careful observation would show that when we connect the wires from the two compartments, current flows for an instant and then ceases. The current stops flowing because of charge buildups in the two compartments. If electrons flowed from the right to the left compartment in the apparatus as shown, the left compartment (receiving electrons) would become negatively charged, and the right compartment (losing electrons) would become positively charged. Creating a charge separation of this type requires a large amount of energy. Thus sustained electron flow cannot occur under these conditions.

We can, however, solve this problem very simply. The solutions must be connected so that ions can flow to maintain the net charge of zero in each compartment. This connection might involve a **salt bridge** (a U-tube filled with an electrolyte) or a **porous disk** in a tube connecting the two solutions (Fig. 11.2). Either of these devices allows ion flow without extensive mixing of the solutions. When we make the provision for ion flow, the circuit is complete. Electrons flow through the wire from reducing agent to oxidizing agent, and ions flow between the compartments to keep the net charge zero in each.

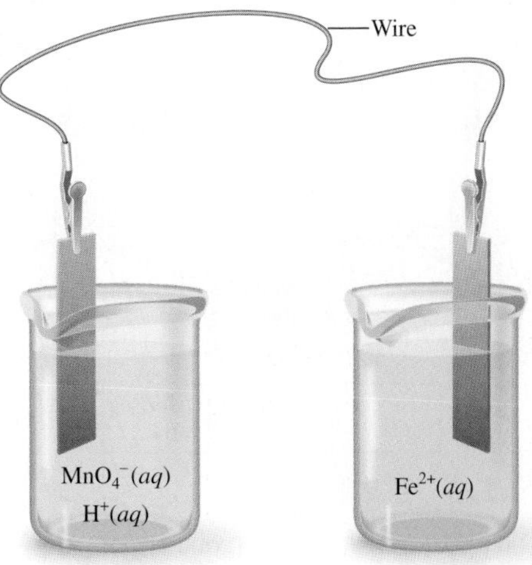

Figure 11.1
Schematic of a method to separate the oxidizing and reducing agents of a redox reaction. (The solutions also contain counter ions to balance the charge.)

Figure 11.2
Galvanic cells can contain a salt bridge as in (a) or a porous-disk connection as in (b). A salt bridge contains a strong electrolyte held in a Jello-like matrix. A porous disk contains tiny passages that allow hindered flow of ions.

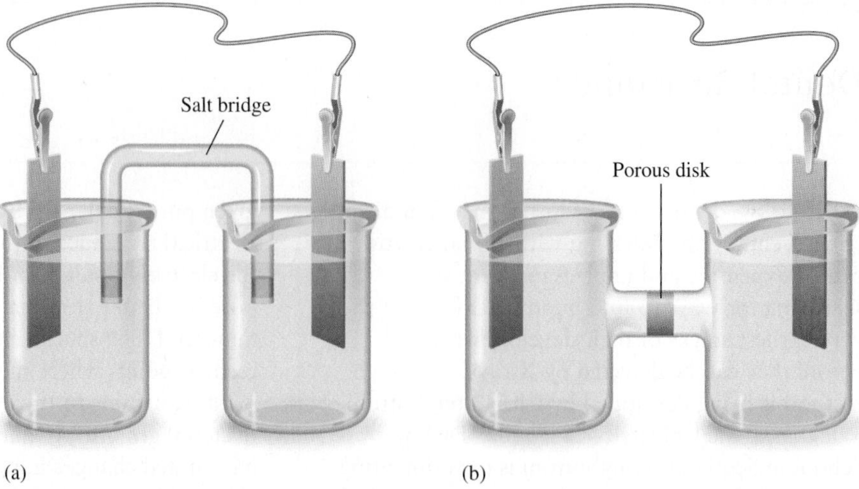

A galvanic cell uses a spontaneous redox reaction to produce a current that can be used to do work.

We now have covered all the essential characteristics of a **galvanic cell,** *a device in which chemical energy is changed to electrical energy.* (The opposite process, *electrolysis,* will be considered in Section 11.7.)

Oxidation occurs at the anode. Reduction occurs at the cathode.

The reaction in an electrochemical cell occurs at the interface between an electrode and the solution where the electron transfer occurs. The electrode at which *oxidation* occurs is called the **anode;** the electrode at which *reduction* occurs is called the **cathode** (Fig. 11.3).

Cell Potential

A galvanic cell consists of an oxidizing agent in one compartment that pulls electrons through a wire from a reducing agent in the other compartment. The "pull," or driving force, on the electrons is called the **cell potential** ($\mathscr{E}_{cell}$), or the **electromotive force** (emf), of the cell. The unit of electrical potential is the **volt** (abbreviated V), which is defined as 1 joule of work per coulomb of charge transferred.

A volt is 1 joule of work per coulomb of charge transferred: 1 V = 1 J/C.

How can we measure the cell potential? One possible instrument is a crude **voltmeter,** which works by drawing current through a known resistance. However, when current flows through a wire, the frictional heating that occurs wastes some of the useful energy of the cell. A traditional voltmeter will therefore measure a potential that is lower than the maximum cell potential. The

Figure 11.3
An electrochemical process involves electron transfer at the interface between the electrode and the solution. (a) The species in the solution acting as the reducing agent supplies electrons to the anode. (b) The species in the solution acting as the oxidizing agent receives electrons from the cathode.

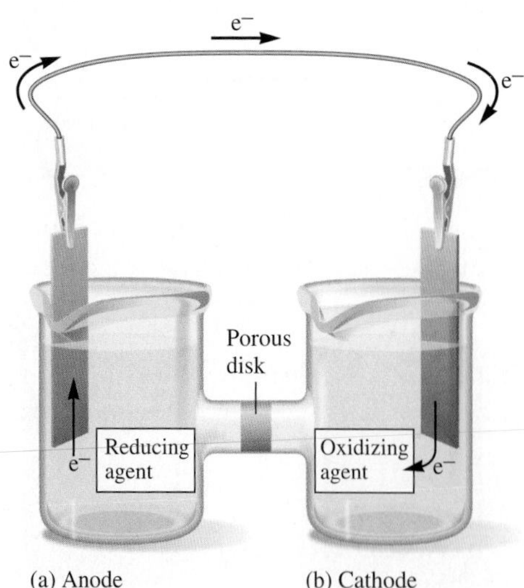

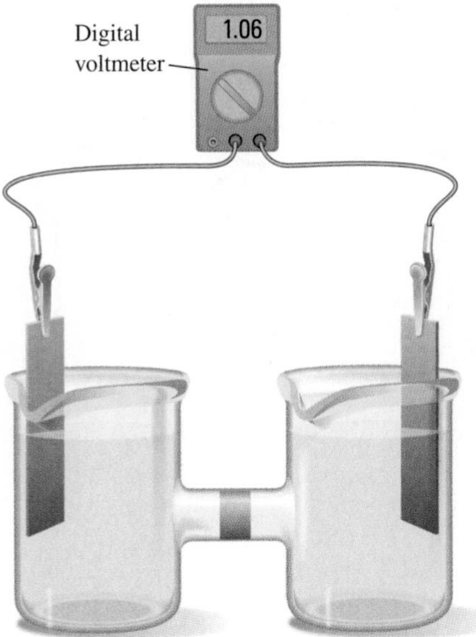

Figure 11.4
Digital voltmeters draw only a negligible current and are convenient to use.

key to determining the maximum potential is to perform the measurement under conditions of zero current so that no energy is wasted. Traditionally, this measurement has been accomplished by inserting a variable-voltage device (powered from an external source) in *opposition* to the cell potential. The voltage on this instrument, called a **potentiometer,** is adjusted until no current flows in the cell circuit. Under such conditions the cell potential is equal in magnitude and opposite in sign to the voltage setting of the potentiometer. This value represents the *maximum* cell potential, since no energy is wasted heating the wire. Advances in electronic technology have allowed the design of *digital voltmeters* that draw only a negligible amount of current (Fig. 11.4). Since these instruments are more convenient to use, they have replaced potentiometers in the modern laboratory.

11.2 Standard Reduction Potentials

The name *galvanic cell* honors Luigi Galvani (1737–1798), an Italian scientist generally credited with the discovery of electricity. Galvanic cells are sometimes called *voltaic cells* after Alessandro Volta (1745–1827), another Italian, who first constructed cells of this type around 1800.

The reaction in a galvanic cell is always an oxidation–reduction reaction that can be broken down into half-reactions. It would be convenient to assign a potential to *each* half-reaction so that when we construct a cell from a given pair of half-reactions, we can obtain the cell potential by summing the half-cell potentials. For example, the observed potential for the cell shown in Fig. 11.5(a) is 0.76 volt, and the cell reaction* is

$$2H^+(aq) + Zn(s) \longrightarrow Zn^{2+}(aq) + H_2(g)$$

For this cell the anode compartment contains a zinc metal electrode with Zn^{2+} and SO_4^{2-} ions in an aqueous solution that bathes the electrode. The anode reaction is the oxidation half-reaction:

$$Zn \longrightarrow Zn^{2+} + 2e^-$$

Each zinc atom loses two electrons to produce a Zn^{2+} ion that enters the solution. The electrons flow through the wire. For now we will assume that all cell

*In this text we will follow the convention of indicating the physical states of the reactants and products only in the overall redox reaction. For simplicity, half-reactions will *not* include the physical states.

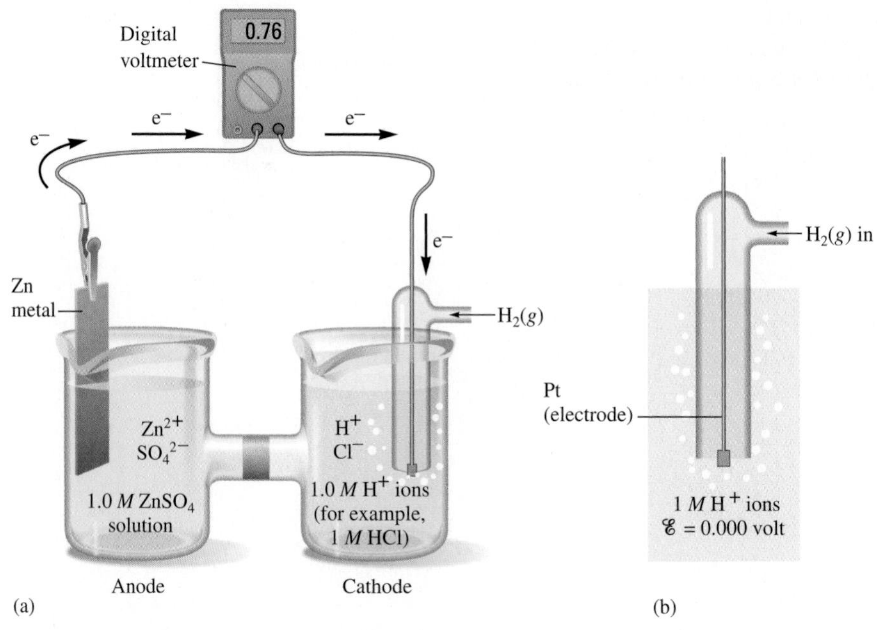

Figure 11.5
(a) A galvanic cell involving the reactions $Zn \longrightarrow Zn^{2+} + 2e^-$ (at the anode) and $2H^+ + 2e^- \longrightarrow H_2$ (at the cathode) has a potential of 0.76 V.
(b) The standard hydrogen electrode where $H_2(g)$ at 1 atm is passed over a platinum electrode in contact with 1 *M* H^+ ions. This electrode process (assuming ideal behavior) is arbitrarily assigned a value of exactly zero volts.

components are in their standard states, so in this case the solution in the anode compartment will contain 1 *M* Zn^{2+}. The cathode reaction of this cell is

$$2H^+ + 2e^- \longrightarrow H_2$$

The standard hydrogen potential is the reference potential against which all half-reaction potentials are assigned.

The cathode consists of a platinum electrode (used because it is a chemically inert conductor) in contact with 1 *M* H^+ ions and bathed by hydrogen gas at 1 atm. Such an electrode, called the **standard hydrogen electrode,** is shown in Fig. 11.5(b).

Although we can measure the *total* potential of this cell (0.76 volt), there is no way to measure the potentials of the individual electrodes. Thus, if we desire potentials for half-reactions (half-cells), we must arbitrarily divide up the total cell potential. For example, if we assign the reaction

$$2H^+ + 2e^- \longrightarrow H_2$$

where

$$[H^+] = 1\ M \qquad \text{and} \qquad P_{H_2} = 1 \text{ atm}$$

as having a potential of exactly 0 volts, then the reaction

$$Zn \longrightarrow Zn^{2+} + 2e^-$$

will have a potential of 0.76 volt, since

$$\underset{0.76\text{ V}}{\underset{\uparrow}{\mathscr{E}^\circ_{\text{cell}}}} = \underset{0\text{ V}}{\underset{\uparrow}{\mathscr{E}^\circ_{H^+\to H_2}}} + \underset{0.76\text{ V}}{\underset{\uparrow}{\mathscr{E}^\circ_{Zn\to Zn^{2+}}}}$$

Standard states were discussed in Section 9.6.

Recall that the superscript ° indicates that *standard states* are used.

By setting the standard potential for the half-reaction $2H^+ + 2e^- \rightarrow H_2$ equal to zero, we can assign values to all other half-reactions. For example, the measured potential for the cell shown in Fig. 11.6 is 1.10 volt. The cell reaction is

$$Zn(s) + Cu^{2+}(aq) \longrightarrow Zn^{2+}(aq) + Cu(s)$$

which can be divided into the half-reactions

$$\text{Anode:} \quad Zn \longrightarrow Zn^{2+} + 2e^-$$

$$\text{Cathode:} \quad Cu^{2+} + 2e^- \longrightarrow Cu$$

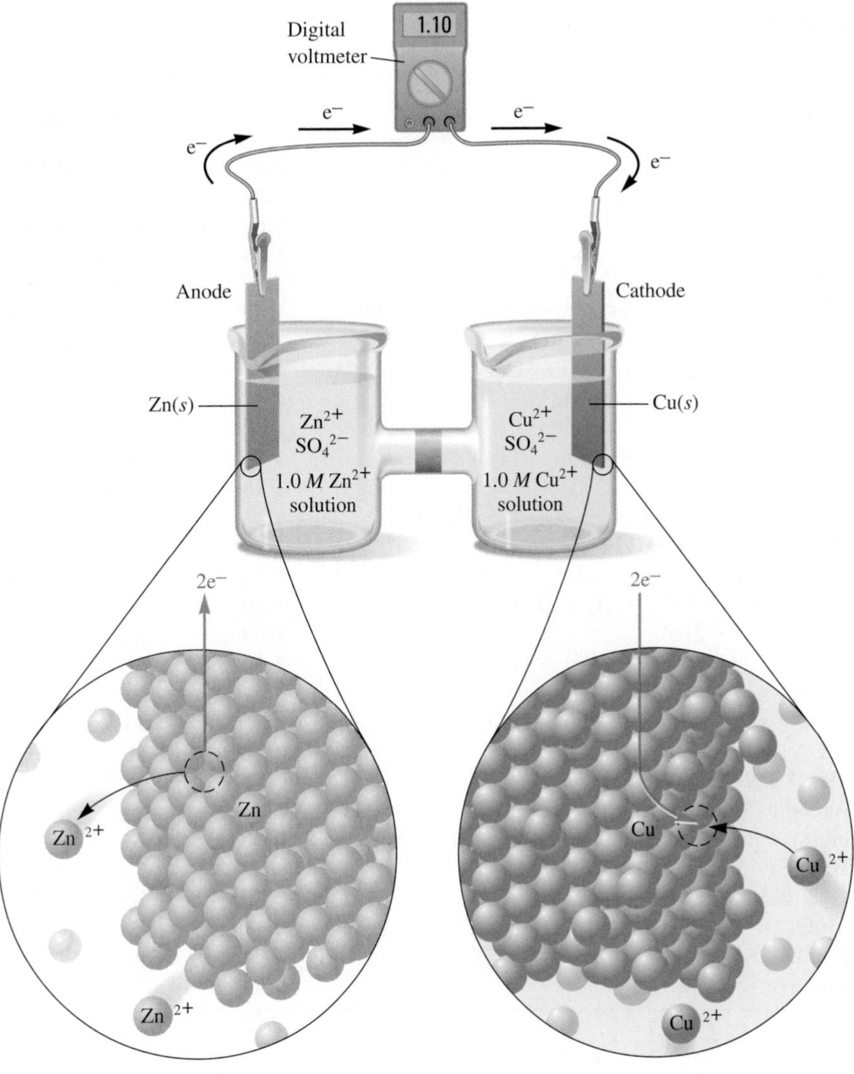

Figure 11.6
A galvanic cell involving the half-reactions $Zn \longrightarrow Zn^{2+} + 2e^-$ (anode) and $Cu^{2+} + 2e^- \longrightarrow Cu$ (cathode), with $\mathscr{E}^\circ_{cell} = 1.10$ V.

Thus
$$\mathscr{E}^\circ_{cell} = \mathscr{E}^\circ_{Zn\rightarrow Zn^{2+}} + \mathscr{E}^\circ_{Cu^{2+}\rightarrow Cu}$$

Since $\mathscr{E}^\circ_{Zn\rightarrow Zn^{2+}}$ is assigned a value of 0.76 volt, the value of $\mathscr{E}^\circ_{Cu^{2+}\rightarrow Cu}$ must be 0.34 volt:

$$1.10 \text{ V} = 0.76 \text{ V} + 0.34 \text{ V}$$

Copper being plated onto the copper metal cathode on the right, and zinc dissolving from the zinc metal anode on the left. Note the porous disk in the tube connecting the two solutions, which allows the ion flow to balance the electron flow through the wire.

All half-reactions are given as reduction processes in standard tables.

The scientific community has universally accepted the values for half-reaction potentials based on the assignment of 0 volts to the process $2H^+ + 2e^- \rightarrow H_2$ (under standard conditions where ideal behavior is assumed). However, before we can use these values, we need to understand several essential characteristics of half-cell potentials.

The accepted convention is to give the potentials of half-reactions as *reduction* processes. For example,

$$2H^+ + 2e^- \longrightarrow H_2$$
$$Cu^{2+} + 2e^- \longrightarrow Cu$$
$$Zn^{2+} + 2e^- \longrightarrow Zn$$

The $\mathscr{E}^\circ$ values corresponding to these half-reactions are called **standard reduction potentials.** Standard reduction potentials for the most common half-reactions are given in Table 11.1 and Appendix A5.5.

Table 11.1

Standard Reduction Potentials at 25°C (298 K) for Many Common Half-reactions

Half-reaction	$\mathscr{E}°$ (V)	Half-reaction	$\mathscr{E}°$ (V)
$F_2 + 2e^- \rightarrow 2F^-$	2.87	$O_2 + 2H_2O + 4e^- \rightarrow 4OH^-$	0.40
$Ag^{2+} + e^- \rightarrow Ag^+$	1.99	$Cu^{2+} + 2e^- \rightarrow Cu$	0.34
$Co^{3+} + e^- \rightarrow Co^{2+}$	1.82	$Hg_2Cl_2 + 2e^- \rightarrow 2Hg + 2Cl^-$	0.27
$H_2O_2 + 2H^+ + 2e^- \rightarrow 2H_2O$	1.78	$AgCl + e^- \rightarrow Ag + Cl^-$	0.22
$Ce^{4+} + e^- \rightarrow Ce^{3+}$	1.70	$SO_4^{2-} + 4H^+ + 2e^- \rightarrow H_2SO_3 + H_2O$	0.20
$PbO_2 + 4H^+ + SO_4^{2-} + 2e^- \rightarrow PbSO_4 + 2H_2O$	1.69	$Cu^{2+} + e^- \rightarrow Cu^+$	0.16
$MnO_4^- + 4H^+ + 3e^- \rightarrow MnO_2 + 2H_2O$	1.68	$2H^+ + 2e^- \rightarrow H_2$	0.00
$IO_4^- + 2H^+ + 2e^- \rightarrow IO_3^- + H_2O$	1.60	$Fe^{3+} + 3e^- \rightarrow Fe$	−0.036
$MnO_4^- + 8H^+ + 5e^- \rightarrow Mn^{2+} + 4H_2O$	1.51	$Pb^{2+} + 2e^- \rightarrow Pb$	−0.13
$Au^{3+} + 3e^- \rightarrow Au$	1.50	$Sn^{2+} + 2e^- \rightarrow Sn$	−0.14
$PbO_2 + 4H^+ + 2e^- \rightarrow Pb^{2+} + 2H_2O$	1.46	$Ni^{2+} + 2e^- \rightarrow Ni$	−0.23
$Cl_2 + 2e^- \rightarrow 2Cl^-$	1.36	$PbSO_4 + 2e^- \rightarrow Pb + SO_4^{2-}$	−0.35
$Cr_2O_7^{2-} + 14H^+ + 6e^- \rightarrow 2Cr^{3+} + 7H_2O$	1.33	$Cd^{2+} + 2e^- \rightarrow Cd$	−0.40
$O_2 + 4H^+ + 4e^- \rightarrow 2H_2O$	1.23	$Fe^{2+} + 2e^- \rightarrow Fe$	−0.44
$MnO_2 + 4H^+ + 2e^- \rightarrow Mn^{2+} + 2H_2O$	1.21	$Cr^{3+} + e^- \rightarrow Cr^{2+}$	−0.50
$IO_3^- + 6H^+ + 5e^- \rightarrow \frac{1}{2}I_2 + 3H_2O$	1.20	$Cr^{3+} + 3e^- \rightarrow Cr$	−0.73
$Br_2 + 2e^- \rightarrow 2Br^-$	1.09	$Zn^{2+} + 2e^- \rightarrow Zn$	−0.76
$VO_2^+ + 2H^+ + e^- \rightarrow VO^{2+} + H_2O$	1.00	$2H_2O + 2e^- \rightarrow H_2 + 2OH^-$	−0.83
$AuCl_4^- + 3e^- \rightarrow Au + 4Cl^-$	0.99	$Mn^{2+} + 2e^- \rightarrow Mn$	−1.18
$NO_3^- + 4H^+ + 3e^- \rightarrow NO + 2H_2O$	0.96	$Al^{3+} + 3e^- \rightarrow Al$	−1.66
$ClO_2 + e^- \rightarrow ClO_2^-$	0.954	$H_2 + 2e^- \rightarrow 2H^-$	−2.23
$2Hg^{2+} + 2e^- \rightarrow Hg_2^{2+}$	0.91	$Mg^{2+} + 2e^- \rightarrow Mg$	−2.37
$Ag^+ + e^- \rightarrow Ag$	0.80	$La^{3+} + 3e^- \rightarrow La$	−2.37
$Hg_2^{2+} + 2e^- \rightarrow 2Hg$	0.80	$Na^+ + e^- \rightarrow Na$	−2.71
$Fe^{3+} + e^- \rightarrow Fe^{2+}$	0.77	$Ca^{2+} + 2e^- \rightarrow Ca$	−2.76
$O_2 + 2H^+ + 2e^- \rightarrow H_2O_2$	0.68	$Ba^{2+} + 2e^- \rightarrow Ba$	−2.90
$MnO_4^- + e^- \rightarrow MnO_4^{2-}$	0.56	$K^+ + e^- \rightarrow K$	−2.92
$I_2 + 2e^- \rightarrow 2I^-$	0.54	$Li^+ + e^- \rightarrow Li$	−3.05
$Cu^+ + e^- \rightarrow Cu$	0.52		

Combining two half-reactions to obtain a balanced oxidation–reduction reaction often requires two manipulations:

When a half-reaction is reversed, the sign of $\mathscr{E}°$ is reversed.

1. One of the reduction half-reactions must be reversed (since redox reactions must involve a substance being oxidized and a substance being reduced). The half-reaction with the largest positive potential will run as written (as a reduction), and the other half-reaction will be forced to run in reverse (will be the oxidation reaction). The net potential of the cell will be the *difference* between the two. Since the reduction process occurs at the cathode and the oxidation process occurs at the anode, we can write

$$\mathscr{E}°_{cell} = \mathscr{E}° \text{ (cathode)} - \mathscr{E}° \text{ (anode)}$$

Because subtraction means "change the sign and add," in the examples done here we will change the sign of the oxidation (anode) reaction when we reverse it and add it to the reduction (cathode) reaction.

When a half-reaction is multiplied by an integer, $\mathscr{E}°$ remains the same.

2. Since the number of electrons lost must equal the number gained, the half-reactions must be multiplied by integers as necessary to achieve electron balance. However, the *value of* $\mathscr{E}°$ *is not changed* when a half-reaction is multiplied by an integer. Since a standard reduction potential is an *intensive property* (it does not depend on how many times the reaction occurs), the potential is *not* multiplied by the integer required to balance the cell reaction.

Consider a galvanic cell based on the redox reaction

$$Fe^{3+}(aq) + Cu(s) \longrightarrow Cu^{2+}(aq) + Fe^{2+}(aq)$$

The pertinent half-reactions are

$$Fe^{3+} + e^- \longrightarrow Fe^{2+} \qquad \mathscr{E}° = 0.77\ V \qquad (1)$$

$$Cu^{2+} + 2e^- \longrightarrow Cu \qquad \mathscr{E}° = 0.34\ V \qquad (2)$$

Since the Cu^{2+}/Cu half-reaction has the lower positive $\mathscr{E}°$ value, it will be forced to run in reverse:

$$Cu \longrightarrow Cu^{2+} + 2e^-$$

It is the anode reaction.

Then, since each Cu atom produces two electrons but each Fe^{3+} ion accepts only one electron, reaction (1) must be multiplied by 2:

$$2Fe^{3+} + 2e^- \longrightarrow 2Fe^{2+}$$

Now we can obtain the balanced cell reaction by summing the appropriately modified half-reactions:

$$2Fe^{3+} + 2e^- \longrightarrow 2Fe^{2+} \qquad \mathscr{E}\ (\text{cathode}) = 0.77\ V$$

$$Cu \longrightarrow Cu^{2+} + 2e^- \qquad -\mathscr{E}\ (\text{anode}) = -0.34\ V$$

Cell reaction: $Cu(s) + 2Fe^{3+}(aq) \longrightarrow Cu^{2+}(aq) + 2Fe^{2+}(aq)$ $\qquad \mathscr{E}°_{cell} = \mathscr{E}°\ (\text{cathode}) - \mathscr{E}°\ (\text{anode}) = 0.77\ V - 0.34\ V = 0.43\ V$

A galvanic cell runs spontaneously in the direction that gives a positive value for $\mathscr{E}°_{cell}$.

Next, we want to consider how to describe a galvanic cell fully, given just its half-reactions. This description will include the cell reaction, the cell potential, and the physical setup of the cell. Let's consider a galvanic cell based on the following half-reactions:

$$Fe^{2+} + 2e^- \longrightarrow Fe \qquad \mathscr{E}° = -0.44\ V$$

$$MnO_4^- + 5e^- + 8H^+ \longrightarrow Mn^{2+} + 4H_2O \qquad \mathscr{E}° = 1.51\ V$$

In a working galvanic cell, one of these reactions must run in reverse—the one with the least positive potential.

This will lead to a positive overall cell potential: *A cell will always run spontaneously in the direction that produces a positive cell potential.* Thus in the present case the half-reaction involving iron must be reversed.

$$Fe \longrightarrow Fe^{2+} + 2e^- \qquad \text{Anode reaction}$$

$$MnO_4^- + 5e^- + 8H^+ \longrightarrow Mn^{2+} + 4H_2O \qquad \text{Cathode reaction}$$

where

$$\mathscr{E}°_{cell} = \mathscr{E}°\ (\text{cathode}) - \mathscr{E}°\ (\text{anode}) = 1.51\ V - 0.44\ V$$

The balanced cell reaction is obtained as follows:

$$5(Fe \longrightarrow Fe^{2+} + 2e^-)$$

$$2(MnO_4^- + 5e^- + 8H^+ \longrightarrow Mn^{2+} + 4H_2O)$$

$$2MnO_4^-(aq) + 5Fe(s) + 16H^+(aq) \longrightarrow 5Fe^{2+}(aq) + 2Mn^{2+}(aq) + 8H_2O(l)$$

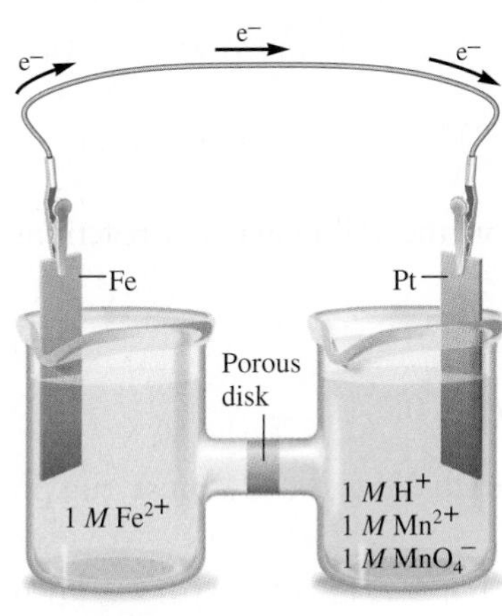

Figure 11.7
The schematic of a galvanic cell based on the half-reactions

$$Fe \longrightarrow Fe^{2+} + 2e^-$$

$$MnO_4^- + 5e^- + 8H^+ \longrightarrow Mn^{2+} + 4H_2O$$

Now consider the physical setup of the cell, shown schematically in Fig. 11.7. In the left compartment, the active components in their standard states are pure metallic iron (Fe) and 1.0 M Fe^{2+}. The anion present depends on the iron salt used. In this compartment the anion does not participate in the reaction but simply balances the charge. The half-reaction that takes place at this electrode is

$$Fe \longrightarrow Fe^{2+} + 2e^-$$

which is an oxidation reaction, so this compartment is the anode. The electrode consists of pure iron metal.

In the right compartment, the active components in their standard states are 1.0 M MnO_4^-, 1.0 M H^+, and 1.0 M Mn^{2+}, with appropriate unreacting ions (often called *counter ions*) to balance the charge. The half-reaction in this compartment is

$$MnO_4^- + 5e^- + 8H^+ \longrightarrow Mn^{2+} + 4H_2O$$

which is a reduction reaction, so this compartment is the cathode. Since neither MnO_4^- nor Mn^{2+} ion can serve as the electrode, a nonreacting conductor must be used. The usual choice is platinum.

The next step is to determine the direction of electron flow. In the left compartment, the half-reaction is the oxidation of iron:

$$Fe \longrightarrow Fe^{2+} + 2e^-$$

In the right compartment, the half-reaction is the reduction of MnO_4^-:

$$MnO_4^- + 5e^- + 8H^+ \longrightarrow Mn^{2+} + 4H_2O$$

Thus the electrons flow from Fe to MnO_4^- in this cell, or from the anode to the cathode, as is always the case.

Items Needed for a Description of a Galvanic Cell

1. The cell potential (always positive for a galvanic cell) and the balanced cell reaction.
2. The direction of electron flow, obtained by inspecting the half-reactions and using the directions that give a positive $\mathscr{E}^\circ_{cell}$.
3. Designation of the anode and the cathode.
4. The nature of each electrode and the ions present in each compartment. A chemically inert conductor is required if none of the substances participating in the half-reaction is a conducting solid.

EXAMPLE 11.1

Describe completely the galvanic cell based on the following half-reactions under standard conditions:

$$Ag^+ + e^- \longrightarrow Ag \qquad \mathscr{E}^\circ = 0.80 \text{ V} \qquad (1)$$

$$Fe^{3+} + e^- \longrightarrow Fe^{2+} \qquad \mathscr{E}^\circ = 0.77 \text{ V} \qquad (2)$$

Solution Since a positive $\mathscr{E}^\circ_{cell}$ value is required, reaction (2) must run in reverse:

$$\begin{array}{lll} Ag^+ + e^- \longrightarrow Ag & \mathscr{E}^\circ \text{ (cathode)} = & 0.80 \text{ V} \\ Fe^{2+} \longrightarrow Fe^{3+} + e^- & -\mathscr{E}^\circ \text{ (anode)} = & -0.77 \text{ V} \\ \hline \text{Cell reaction: } Ag^+(aq) + Fe^{2+}(aq) \longrightarrow Fe^{3+}(aq) + Ag(s) & \mathscr{E}^\circ_{cell} = & 0.03 \text{ V} \end{array}$$

Since Ag^+ receives electrons and Fe^{2+} loses electrons in the cell reaction, the electrons flow from the compartment containing Fe^{2+} to the compartment containing Ag^+.

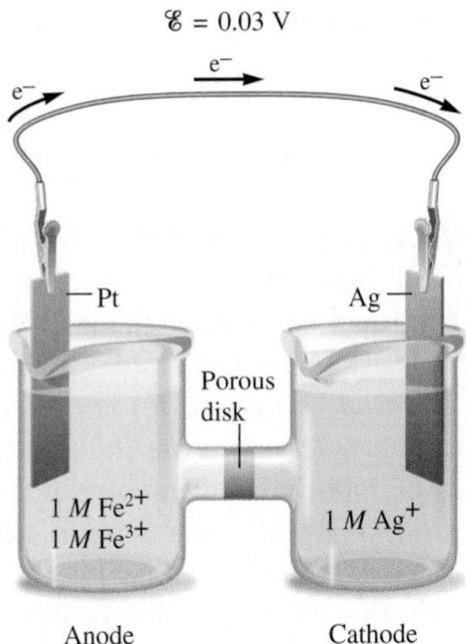

Figure 11.8
Schematic diagram for the galvanic cell based on the half-reactions

$$Ag^+ + e^- \longrightarrow Ag$$
$$Fe^{2+} \longrightarrow Fe^{3+} + e^-$$

Oxidation occurs in the compartment containing Fe^{2+}. Hence this compartment functions as the anode. Reduction occurs in the compartment containing Ag^+, so this compartment is the cathode.

The electrode in the Ag/Ag^+ compartment is silver metal, and an inert conductor, such as platinum, must be used in the Fe^{2+}/Fe^{3+} compartment. Appropriate counter ions are assumed to be present. The diagram for this cell is shown in Fig. 11.8.

11.3 Cell Potential, Electrical Work, and Free Energy

So far we have considered electrochemical cells in a very practical manner without much theoretical background. The next step will be to explore the relationship between thermodynamics and electrochemistry.

The work that can be accomplished when electrons are transferred through a wire depends on the "push" (the thermodynamic driving force) behind the electrons. This driving force (the emf) is defined in terms of a *potential* difference (in volts) between two points in the circuit. Recall that a volt represents a joule of work per coulomb of charge transferred:

$$\text{emf} = \text{potential difference (V)} = \frac{\text{work (J)}}{\text{charge (C)}}$$

Thus 1 joule of work is produced or required (depending on the direction) when 1 coulomb of charge is transferred between two points in the circuit that differ by a potential of 1 volt.

In this book *work is viewed from the point of view of the system.* Thus work flowing out of the system is indicated by a minus sign. When a cell produces a current, the cell potential is positive, and the current can be used to do work—to run a motor, for instance. Thus the cell potential ($\mathscr{E}$) and the work (w) have opposite signs:

$$\mathscr{E} = \frac{-w}{q}$$

(Work: $-w$; Charge: q)

Thus $$-w = q\mathscr{E}$$

From this equation we can see that the maximum work in a cell is obtained at the maximum cell potential:

$$-w_{max} = q\mathscr{E}_{max} \qquad \text{or} \qquad w_{max} = -q\mathscr{E}_{max}$$

The work is never the maximum possible if any current is flowing.

However, there is a problem. For the system to perform electrical work, current must flow. When current flows, some energy is inevitably wasted through frictional heating, so the maximum work is not obtained. This observation reflects the important general principle introduced in Section 10.12: *In any real, spontaneous process, some energy is always wasted—the actual work realized is always less than the calculated maximum.* This principle is a consequence of the fact that the entropy of the universe must increase in any spontaneous process. Recall from Section 10.12 that the only process from which maximum work could be realized is the hypothetical reversible process. For a galvanic cell, this would involve an infinitesimally small current flow and thus an infinite amount of time to do the work. Even though we can never achieve the maximum work through the actual discharge of a galvanic cell, we can measure the maximum potential. There is negligible current flow when a cell potential is measured with a potentiometer or an efficient digital voltmeter. No current flow implies no waste of energy, so the measured potential is the maximum.

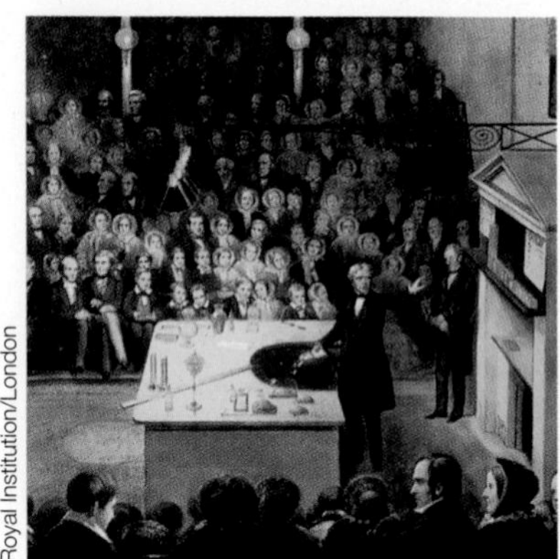
Royal Institution/London

Michael Faraday lecturing at the Royal Institution before Prince Albert and others (1855).

Although we can never actually realize the maximum work from a cell reaction, its value is still useful for evaluating the efficiency of a real process based on the cell reaction. For example, suppose a certain galvanic cell has a maximum potential of 2.50 V. In a particular experiment, 1.33 moles of electrons pass through this cell at an average actual potential of 2.10 V. The actual work done is

$$w = -q\mathscr{E}$$

where $\mathscr{E}$ represents the actual potential difference at which the current flowed (2.10 V or 2.10 J/C) and q is the quantity of charge transferred (in coulombs). The charge on 1 mole of electrons is called the **faraday** (abbreviated F): *96,485 coulombs of charge per mole of electrons.* Thus q equals the number of moles of electrons times the charge per mole of electrons:

$$q = nF = 1.33 \text{ mol e}^- \times 96{,}485 \text{ C/mol e}^-$$

For the experiment above, the actual work is

$$w = -q\mathscr{E} = -\left(1.33 \text{ mol e}^- \times 96{,}485 \frac{\text{C}}{\text{mol e}^-}\right) \times \left(2.10 \frac{\text{J}}{\text{C}}\right)$$
$$= -2.69 \times 10^5 \text{ J}$$

For the maximum possible work, the calculation is similar, except that the maximum potential is used:

$$w_{max} = -q\mathscr{E}_{max} = -\left(1.33 \text{ mol e}^- \times 96{,}485 \frac{\text{C}}{\text{mol e}^-}\right)\left(2.50 \frac{\text{J}}{\text{C}}\right)$$
$$= -3.21 \times 10^5 \text{ J}$$

Thus, in its actual operation, the efficiency of this cell is

$$\frac{w}{w_{max}} \times 100\% = \frac{-2.69 \times 10^5 \text{ J}}{-3.21 \times 10^5 \text{ J}} \times 100\% = 83.8\%$$

Next we want to relate the potential of a galvanic cell to free energy. In Section 10.12 we saw that for a process carried out at constant temperature and pressure, the change in free energy equals the maximum useful work obtainable from that process:

$$w_{max} = \Delta G$$

For a galvanic cell,

$$w_{max} = -q\mathscr{E}_{max} = \Delta G$$

Since $$q = nF$$

we have $$\Delta G = -q\mathscr{E}_{max} = -nF\mathscr{E}_{max}$$

From now on the subscript on $\mathscr{E}_{max}$ will be deleted, with the understanding that any potential given in this book is the maximum potential. Thus

$$\Delta G = -nF\mathscr{E}$$

For standard conditions

$$\Delta G^\circ = -nF\mathscr{E}^\circ$$

This equation states that *the maximum cell potential is directly related to the free energy difference between the reactants and the products in the cell.* This relationship is important because it provides an experimental means of obtaining ΔG for a reaction. It also confirms that a galvanic cell runs in the direction that gives a positive value for $\mathscr{E}_{cell}$; a positive $\mathscr{E}_{cell}$ value corresponds to a negative ΔG value, which is the condition for spontaneity.

OWL INTERACTIVE EXAMPLE 11.2

Using the data in Table 11.1, calculate ΔG° for the reaction

$$Cu^{2+}(aq) + Fe(s) \longrightarrow Cu(s) + Fe^{2+}(aq)$$

Is this reaction spontaneous?

Solution

■ What are we trying to solve?

We want to know if the reaction is spontaneous. Which leads to the question: *What do we mean by spontaneous?*

The reaction is spontaneous if it proceeds without outside intervention, and we can tell this by looking at the signs of both $\mathscr{E}^\circ_{cell}$ (positive for a spontaneous process) and ΔG° (negative for a spontaneous process).

The half-reactions are

$$Cu^{2+} + 2e^- \longrightarrow Cu \qquad \mathscr{E}^\circ \text{ (cathode)} = 0.34 \text{ V}$$

$$Fe \longrightarrow Fe^{2+} + 2e^- \qquad -\mathscr{E}^\circ \text{ (anode)} = 0.44 \text{ V}$$

$$Cu^{2+} + Fe \longrightarrow Fe^{2+} + Cu \qquad \mathscr{E}^\circ_{cell} = 0.78 \text{ V}$$

We can calculate ΔG° from the equation

$$\Delta G^\circ = -nF\mathscr{E}^\circ$$

Since two electrons are transferred in the reaction, 2 moles of electrons are required per mole of reactants and products. Thus $n = 2$ moles of e^-, $F = 96{,}485$ C/mol e^-, and $\mathscr{E}^\circ = 0.78$ V $= 0.78$ J/C. Therefore,

$$\Delta G^\circ = -(2 \text{ mol e}^-)\left(96{,}485\frac{\text{C}}{\text{mol e}^-}\right)\left(0.78\frac{\text{J}}{\text{C}}\right)$$

$$= -1.5 \times 10^5 \text{ J}$$

The process is spontaneous, as indicated both by the negative sign of $\Delta G°$ and by the positive sign of $\mathscr{E}°_{cell}$.

This reaction is used industrially to deposit copper metal from solutions containing dissolved copper ores.

EXAMPLE 11.3

Using the data from Table 11.1, predict whether 1 *M* HNO_3 will dissolve gold metal to form a 1 *M* Au^{3+} solution.

A gold ring does not dissolve in nitric acid.

Solution

■ What are we trying to solve?

We want to determine if gold will dissolve in 1 *M* nitric acid (HNO_3) to form a 1 *M* Au^{3+} solution. To make this determination, we need to see if the reaction between Au and HNO_3 to form Au^{3+} is spontaneous or has a standard potential greater than zero. We must use the appropriate half-reactions. For this process to occur, gold would be oxidized, so the HNO_3 would act as an oxidizing agent.

The half-reaction for HNO_3 acting as an oxidizing agent is

$$NO_3^- + 4H^+ + 3e^- \longrightarrow NO + 2H_2O \qquad \mathscr{E}° \text{ (cathode)} = 0.96 \text{ V}$$

The reaction for the oxidation of solid gold to Au^{3+} ions is

$$Au \longrightarrow Au^{3+} + 3e^- \qquad -\mathscr{E}° \text{ (anode)} = -1.50 \text{ V}$$

The sum of these half-reactions gives the required reaction:

$$Au(s) + NO_3^-(aq) + 4H^+(aq) \longrightarrow Au^{3+}(aq) + NO(g) + 2H_2O(l)$$

and

$$\mathscr{E}°_{cell} = 0.96 \text{ V} - 1.50 \text{ V} = -0.54 \text{ V}$$

Since the $\mathscr{E}°$ value is negative, the process will *not* occur under standard conditions. That is, gold will not dissolve in 1 *M* HNO_3 to give 1 *M* Au^{3+}. In fact, a mixture (1 : 3 by volume) of concentrated nitric and hydrochloric acid, called *aqua regia,* is required to dissolve gold.

11.4 Dependence of the Cell Potential on Concentration

So far we have described cells under standard conditions. In this section we consider the dependence of the cell potential on concentration. For example, under standard conditions (all concentrations 1 *M*), the cell with the reaction

$$Cu(s) + 2Ce^{4+}(aq) \longrightarrow Cu^{2+}(aq) + 2Ce^{3+}(aq)$$

has a potential of 1.36 V. What will the cell potential be if $[Ce^{4+}]$ is greater than 1.0 *M*? This question can be answered qualitatively in terms of Le Châtelier's principle. An increase in the concentration of Ce^{4+} will favor the forward reaction and thus increase the driving force on the electrons. The cell potential will increase. On the other hand, an increase in the concentration of a product (Cu^{2+} or Ce^{3+}) will oppose the forward reaction, thus decreasing the cell potential.

These ideas are illustrated in Example 11.4.

EXAMPLE 11.4

For the cell reaction

$$2Al(s) + 3Mn^{2+}(aq) \longrightarrow 2Al^{3+}(aq) + 3Mn(s) \qquad \mathscr{E}°_{cell} = 0.48 \text{ V}$$

predict whether $\mathscr{E}_{cell}$ will be larger or smaller than $\mathscr{E}°_{cell}$ for the following cases.

a. $[Al^{3+}] = 2.0\ M$, $[Mn^{2+}] = 1.0\ M$
b. $[Al^{3+}] = 1.0\ M$, $[Mn^{2+}] = 3.0\ M$

Solution

a. A product concentration has been raised above 1.0 *M*. This will oppose the cell reaction, causing $\mathscr{E}_{cell}$ to be less than $\mathscr{E}^\circ_{cell}$ ($\mathscr{E}_{cell} < 0.48$ V).

b. A reactant concentration has been increased above 1.0 *M*, and $\mathscr{E}_{cell}$ will be greater than $\mathscr{E}^\circ_{cell}$ ($\mathscr{E}_{cell} > 0.48$ V).

The Nernst Equation

Photo by Trans-Ocean, courtesy AIP Emilio Segre Visual Archives, Sawyer Collection

Professor Walter Hermann von Nernst (1864–1941) was one of the pioneers in the development of electrochemical theory and is generally given credit for first stating the third law of thermodynamics. He won the Nobel Prize in chemistry in 1920 for his contributions to our understanding of thermodynamics.

The dependence of the cell potential on concentration results directly from the dependence of free energy on concentration. Recall from Chapter 10 that the equation

$$\Delta G = \Delta G^\circ + RT\ln(Q)$$

where *Q* is the reaction quotient, was used to calculate the effect of concentration on ΔG. Since $\Delta G = -nF\mathscr{E}$ and $\Delta G^\circ = -nF\mathscr{E}^\circ$, the equation becomes

$$-nF\mathscr{E} = -nF\mathscr{E}^\circ + RT\ln(Q)$$

Dividing each side of the equation by $-nF$ gives

$$\mathscr{E} = \mathscr{E}^\circ - \frac{RT}{nF}\ln(Q) \qquad (11.1)$$

Equation (11.1), which gives the relationship between the cell potential and the concentrations of the cell components, is commonly called the **Nernst equation** after German chemist Walter Hermann von Nernst.

The Nernst equation is often given in terms of a log (base-10) form that is valid at 25°C:

$$\mathscr{E} = \mathscr{E}^\circ - \frac{0.0591}{n}\log(Q)$$

Using this relationship we can calculate the potential of a cell in which some or all of the components are not in their standard states.

For example, $\mathscr{E}^\circ_{cell}$ is 0.48 V for the galvanic cell based on the reaction

$$2Al(s) + 3Mn^{2+}(aq) \longrightarrow 2Al^{3+}(aq) + 3Mn(s)$$

Consider a cell in which

$$[Mn^{2+}] = 0.50\ M \qquad \text{and} \qquad [Al^{3+}] = 1.50\ M$$

The cell potential at 25°C for these concentrations must be calculated using the Nernst equation:

$$\mathscr{E}_{cell} = \mathscr{E}^\circ_{cell} - \frac{0.0591}{n}\log(Q)$$

We know that

$$\mathscr{E}^\circ_{cell} = 0.48\text{ V} \qquad \text{and} \qquad Q = \frac{[Al^{3+}]^2}{[Mn^{2+}]^3} = \frac{(1.50)^2}{(0.50)^3} = 18$$

Since the half-reactions are

$$2Al \longrightarrow 2Al^{3+} + 6e^- \qquad \text{and} \qquad 3Mn^{2+} + 6e^- \longrightarrow 3Mn$$

we know that $\qquad n = 6$

Thus

$$\mathscr{E}_{\text{cell}} = 0.48 - \frac{0.0591}{6}\log(18) = 0.47 \text{ V}$$

Note that the cell voltage decreases slightly because of the nonstandard concentrations. This change is consistent with the predictions of Le Châtelier's principle (see Example 11.4). In this case, since the reactant concentration is lower than 1.0 M and the product concentration is higher than 1.0 M, $\mathscr{E}_{\text{cell}}$ is less than $\mathscr{E}^\circ_{\text{cell}}$.

The potential calculated from the Nernst equation is the maximum potential before any current flow occurs. As the cell discharges and current flows from anode to cathode, the concentrations change; as a result, $\mathscr{E}_{\text{cell}}$ changes. In fact, *the cell will spontaneously discharge until it reaches equilibrium,* at which point

$$Q = K \text{ (the equilibrium constant)} \qquad \text{and} \qquad \mathscr{E}_{\text{cell}} = 0$$

A "dead" battery is one in which the cell reaction has reached equilibrium; there is no longer any chemical driving force to push electrons through the wire. In other words, *at equilibrium the components in the two cell compartments have the same free energy;* that is, $\Delta G = 0$ for the cell reaction at the equilibrium concentrations. The cell no longer has the ability to do work.

EXAMPLE 11.5

Describe the cell based on the following half-reactions:

$$VO_2^+ + 2H^+ + e^- \longrightarrow VO^{2+} + H_2O \qquad \mathscr{E}^\circ = 1.00 \text{ V} \qquad (1)$$

$$Zn^{2+} + 2e^- \longrightarrow Zn \qquad \mathscr{E}^\circ = -0.76 \text{ V} \qquad (2)$$

where

$$T = 25°\text{C} \qquad [VO_2^+] = 2.0\ M \qquad [VO^{2+}] = 1.0 \times 10^{-2}\ M$$

$$[H^+] = 0.50\ M \qquad [Zn^{2+}] = 1.0 \times 10^{-1}\ M$$

Solution The balanced cell reaction is obtained by reversing reaction (2) and multiplying reaction (1) by 2:

2 × reaction (1)

$$2VO_2^+ + 4H^+ + 2e^- \longrightarrow 2VO^{2+} + 2H_2O \qquad \mathscr{E}^\circ \text{ (cathode)} = 1.00 \text{ V}$$

Reaction (2) reversed

$$Zn \longrightarrow Zn^{2+} + 2e^- \qquad -\mathscr{E}^\circ \text{ (anode)} = 0.76 \text{ V}$$

Cell reaction: $2VO_2^+(aq) + 4H^+(aq) + Zn(s)$

$$\longrightarrow 2VO^{2+}(aq) + 2H_2O(l) + Zn^{2+}(aq) \qquad \mathscr{E}^\circ_{\text{cell}} = 1.76 \text{ V}$$

Since the cell contains components at concentrations other than 1 M, we must use the Nernst equation, where $n = 2$ (since two electrons are transferred), to calculate the cell potential. At 25°C we can use the equation in the form

$$\mathscr{E} = \mathscr{E}^\circ - \frac{0.0591}{n}\log(Q)$$

Thus

$$\mathscr{E} = 1.76 - \frac{0.0591}{2}\log\left(\frac{[Zn^{2+}][VO^{2+}]^2}{[VO_2^+]^2[H^+]^2[H^+]^4}\right)$$

$$= 1.76 - \frac{0.0591}{2}\log\left[\frac{(1.0 \times 10^{-1})(1.0 \times 10^{-2})^2}{(2.0)^2(0.50)^4}\right]$$

$$= 1.76 - \frac{0.0591}{2}\log(4 \times 10^{-5}) = 1.76 + 0.13 = 1.89 \text{ V}$$

The cell diagram is given in Fig. 11.9.

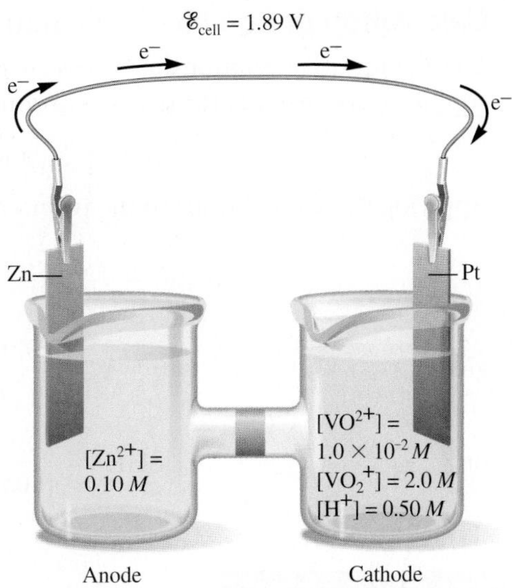

Figure 11.9
Schematic diagram of the cell described in Example 11.5.

Ion-Selective Electrodes

Because the cell potential is sensitive to the concentrations of the reactants and products involved in the cell reaction, measured potentials can be used to determine the concentration of an ion. A pH meter is a familiar example of an instrument that measures concentration from an observed potential. The pH meter has three main components: a standard electrode of known potential, a special **glass electrode** that changes potential depending on the concentration of H^+ ion in the solution into which it is dipped, and a potentiometer that measures the potential between the two electrodes. The potentiometer reading is automatically converted electronically to a direct reading of the pH of the solution being tested.

The glass electrode (Fig. 11.10) contains a reference solution of dilute hydrochloric acid in contact with a thin glass membrane. The electrical potential of the glass electrode depends on the difference in $[H^+]$ between the reference solution and the solution into which the electrode is dipped. Thus the electrical potential varies with the pH of the solution being tested.

Electrodes that are sensitive to the concentration of a particular ion are called **ion-selective electrodes,** of which the glass electrode for pH measurement is just one example. Glass electrodes can be made sensitive to ions such as Na^+, K^+, or NH_4^+ by changing the composition of the glass. Other ions can be detected if an appropriate crystalline solid replaces the glass membrane. For example, a crystal of lanthanum(III) fluoride (LaF_3) can be used in an electrode to measure $[F^-]$. Solid silver sulfide (Ag_2S) can be used to measure $[Ag^+]$ and $[S^{2-}]$. Some of the ions that can be detected by ion-selective electrodes are listed in Table 11.2.

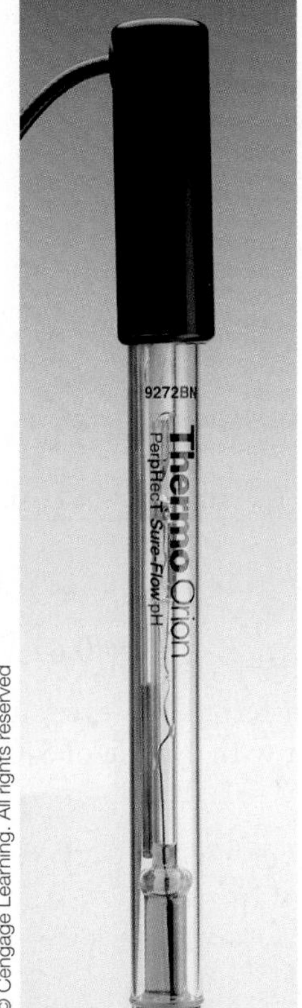

Figure 11.10
A glass electrode contains a reference solution of dilute hydrochloric acid in contact with a thin glass membrane, in which a silver wire coated with silver chloride has been embedded. When the electrode is dipped into a solution containing H^+ ions, the electrode potential is determined by the difference in $[H^+]$ between the two solutions.

Table 11.2
Some Ions Whose Concentrations Can Be Detected by Ion-Selective Electrodes

Cations	Anions
H^+	Br^-
Cd^{2+}	Cl^-
Ca^{2+}	CN^-
Cu^{2+}	F^-
K^+	NO_3^-
Ag^+	S^{2-}
Na^+	

Calculation of Equilibrium Constants for Redox Reactions

The quantitative relationship between $\mathscr{E}°$ and $\Delta G°$ allows the calculation of equilibrium constants for redox reactions. For a cell at equilibrium,

$$\mathscr{E}_{cell} = 0 \qquad \text{and} \qquad Q = K$$

Applying these conditions to the form of the Nernst equation valid at 25°C,

$$\mathscr{E} = \mathscr{E}° - \frac{0.0591}{n}\log(Q)$$

gives

$$0 = \mathscr{E}° - \frac{0.0591}{n}\log(K)$$

or

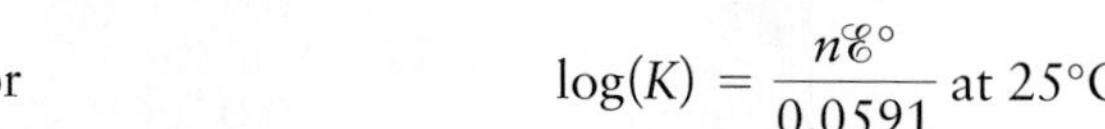

$$\log(K) = \frac{n\mathscr{E}°}{0.0591} \quad \text{at } 25°\text{C}$$

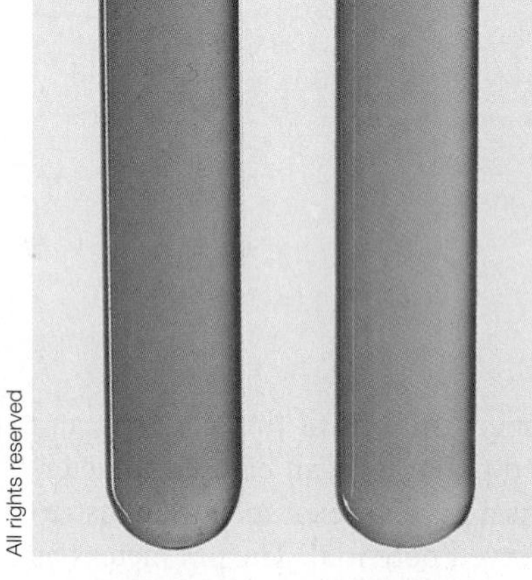

The blue solution contains Cr^{2+} ions, and the green solution contains Cr^{3+} ions.

EXAMPLE 11.6

For the oxidation–reduction reaction

$$S_4O_6^{2-}(aq) + Cr^{2+}(aq) \longrightarrow Cr^{3+}(aq) + S_2O_3^{2-}(aq)$$

the appropriate half-reactions are

$$S_4O_6^{2-} + 2e^- \longrightarrow 2S_2O_3^{2-} \qquad \mathscr{E}° = 0.17\text{ V} \qquad (1)$$

$$Cr^{3+} + e^- \longrightarrow Cr^{2+} \qquad \mathscr{E}° = -0.50\text{ V} \qquad (2)$$

Balance the redox reaction and calculate $\mathscr{E}°$ and K (at 25°C).

Solution To obtain the balanced reaction, we must reverse reaction (2), multiply it by 2, and add it to reaction (1):

$$S_4O_6^{2-} + 2e^- \longrightarrow 2S_2O_3^{2-} \qquad \mathscr{E}°\text{ (cathode)} = 0.17\text{ V}$$

$$2(Cr^{2+} \longrightarrow Cr^{3+} + e^-) \qquad -\mathscr{E}°\text{ (anode)} = -(-0.50)\text{ V}$$

Cell reaction: $2Cr^{2+}(aq) + S_4O_6^{2-}(aq) \longrightarrow 2Cr^{3+}(aq) + 2S_2O_3^{2-}(aq) \qquad \mathscr{E}° = 0.67\text{ V}$

In this reaction 2 moles of electrons are transferred for every unit of reaction—that is, for every 2 moles of Cr^{2+} reacting with 1 mole of $S_4O_6^{2-}$ to form 2 moles of Cr^{3+} and 2 moles of $S_2O_3^{2-}$. Thus $n = 2$. Then

$$\log(K) = \frac{n\mathscr{E}°}{0.0591} = \frac{2(0.67)}{0.0591} = 22.6$$

$$K = 10^{22.6} = 4 \times 10^{22}$$

This very large equilibrium constant is not unusual for a redox reaction.

Figure 11.11
A concentration cell that contains a silver electrode and aqueous silver nitrate in both compartments. Because the right compartment contains 1 M Ag^+ and the left compartment contains 0.1 M Ag^+, there will be a driving force to transfer electrons from left to right. Silver will be deposited on the right electrode, thus lowering the concentration of Ag^+ in the right compartment. In the left compartment, the silver electrode dissolves (producing Ag^+ ions) to raise the concentration of Ag^+ in solution.

Concentration Cells

Because cell potentials depend on concentration, we can construct galvanic cells in which both compartments contain the same components but at different concentrations. For example, in the cell in Fig. 11.11, both compartments contain aqueous $AgNO_3$, but with different molarities. Let's consider the potential of this cell and the direction of electron flow. The half-reaction relevant to both compartments of this cell is

$$Ag^+ + e^- \longrightarrow Ag \qquad \mathscr{E}° = 0.80\text{ V}$$

If the cell had 1 M Ag^+ in both compartments,

$$\mathscr{E}^\circ_{cell} = 0.80 \text{ V} - 0.80 \text{ V} = 0 \text{ V}$$

However, in the cell shown in Fig. 11.11, the concentrations of Ag^+ in the two compartments are 1 M and 0.1 M. Because the concentrations of Ag^+ are unequal, the actual half-cell potentials will not be identical. Thus the cell will exhibit a positive voltage. In which direction will the electrons flow in this cell? The best way to think about this question is to recognize that nature will try to equalize the concentrations of Ag^+ in the two compartments. This can be done by transferring electrons from the compartment containing 0.1 M Ag^+ to the one containing 1 M Ag^+ (left to right in Fig. 11.11). This electron transfer will produce Ag^+ in the left compartment and consume Ag^+ (to form Ag) in the right compartment.

To calculate the potential at 25°C for the cell shown in Fig. 11.11, we use the Nernst equation in the form

$$\mathscr{E} = \mathscr{E}^\circ - \frac{0.0591}{n}\log(Q)$$

In this case $n = 1$ because the reaction in both compartments is $Ag^+ + e^- \rightarrow Ag$ but running in opposite directions. What is $\mathscr{E}^\circ$ for this cell? Because $\mathscr{E}^\circ$ refers to standard conditions, this corresponds to a cell like the one in Fig. 11.11, *except* that $[Ag^+] = 1$ M in both compartments. Such a cell has a potential of zero. That is, $\mathscr{E}^\circ = 0$ for this cell, as noted previously. Next, we need to consider the form of Q. Recall that Q is a ratio of product to reactant concentrations. We can represent the process taking place in this cell as

$$Ag^+(1\ M) \longrightarrow Ag^+(0.1\ M)$$

Thus we can write the Nernst equation for this case:

$$\mathscr{E} = \mathscr{E}^\circ - \frac{0.0591}{n}\log(Q) = 0 - \frac{0.0591}{1}\log\left(\frac{0.10}{1.0}\right)$$

$$= -\frac{0.0591}{1}(-1.00) = 0.0591 \text{ V}$$

Thus the cell potential is 0.0591 V.

A cell in which both compartments have the same components but at different concentrations is called a **concentration cell.** The difference in concentration is the only factor that produces a cell potential in this case, and the voltages are typically small.

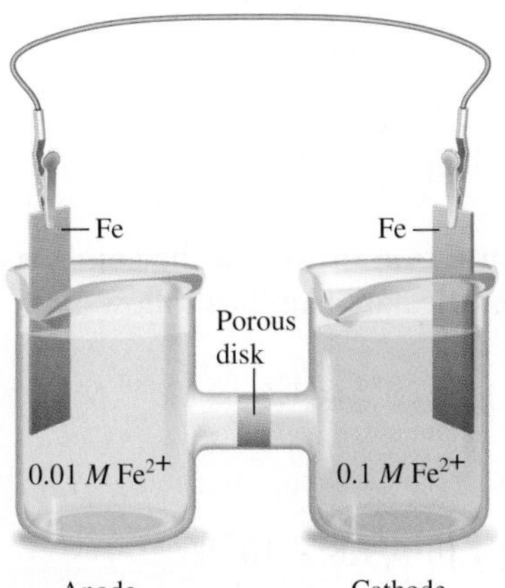

Figure 11.12
A concentration cell containing iron electrodes and different concentrations of Fe^{2+} ion in the two compartments.

EXAMPLE 11.7

Determine the direction of electron flow, designate the anode and cathode, and calculate the potential at 25°C for the cell represented in Fig. 11.12.

Solution The concentrations of Fe^{2+} ion in the compartments can eventually be equalized by transferring electrons from the left compartment to the right. This will cause Fe^{2+} to be formed in the left compartment, and iron metal will be deposited on the right electrode, thus consuming Fe^{2+} ions. Therefore, electron flow is from left to right, oxidation occurs in the left compartment (the anode), and reduction occurs in the right compartment (the cathode).

To calculate the cell potential, we use the Nernst equation in the form

$$\mathscr{E} = \mathscr{E}^\circ - \frac{0.0591}{n}\log(Q)$$

where $n = 2$ because the cell half-reaction is $Fe^{2+} + 2e^- \rightarrow Fe$ or its opposite. Also, $\mathscr{E}° = 0$ (as always) for a concentration cell, and $Q = 0.01/0.10$ because Fe^{2+} is being formed in the compartment with the lower concentration. Thus

$$\mathscr{E} = 0 - \frac{0.0591}{2}\log\left(\frac{0.01}{0.10}\right) = -0.0296(-1.00) = 0.0296 \text{ V}$$

Because the potential of an electrochemical cell depends on the concentrations of the participating ions, the observed potential can be used as a sensitive method for measuring ion concentrations in solution. We have already mentioned the ion-selective electrodes that work by this principle. Another application of the relationship between cell potential and concentration is the determination of equilibrium constants for reactions that are not redox reactions. For example, consider a modified version of the silver concentration cell shown in Fig. 11.11. If the 0.10 *M* $AgNO_3$ solution in the left-hand compartment is replaced by 1.0 *M* NaCl and an excess of solid AgCl is added to the cell, the observed cell potential can be used to determine the concentration of Ag^+ in equilibrium with the AgCl(*s*). In other words, at 25°C we can write the Nernst equation as

Ag^+ in equilibrium with AgCl(*s*)
↓

$$\mathscr{E} = 0 - \frac{0.0591}{1}\log\left(\frac{[Ag^+]}{1.0}\right)$$

Note that the measurement of $\mathscr{E}$ for the cell allows the $[Ag^+]$ in equilibrium with AgCl(*s*) to be calculated, which, in turn, allows the K_{sp} for AgCl to be calculated:

$$K_{sp} = [Ag^+][Cl^-]$$

In this case $[Cl^-] = 1.0$ *M* from the 1.0 *M* NaCl, and $[Ag^+]$ can be obtained from the measured value for $\mathscr{E}$. We will illustrate this process in Example 11.8.

EXAMPLE 11.8

A silver concentration cell similar to the one shown in Fig. 11.11 is set up at 25°C with 1.0 *M* $AgNO_3$ in the left compartment and 1.0 *M* NaCl along with excess AgCl(*s*) in the right compartment. The measured cell potential is 0.58 V. Calculate the K_{sp} value for AgCl at 25°C.

Solution In this case at 25°C

$$\mathscr{E} = 0.58 \text{ V} = 0 - \frac{0.0591}{1}\log\left(\frac{[Ag^+]}{1.0}\right)$$

where $[Ag^+]$ represents the equilibrium concentration of Ag^+ in the compartment containing 1.0 *M* NaCl and AgCl(*s*). We calculate $[Ag^+]$ as follows:

$$\log[Ag^+] = -\frac{0.58}{0.0591} = -9.80 \quad \text{and} \quad [Ag^+] = 1.6 \times 10^{-10}\ M$$

Thus

$$K_{sp} = [Ag^+][Cl^-] = (1.6 \times 10^{-10})(1.0) = 1.6 \times 10^{-10}$$

This calculation neglects any complications from complex ions and ion pairs.

Because the measured potential of an electrochemical cell provides a very sensitive method for the experimental determination of equilibrium concentrations, the values of equilibrium constants are often determined from electrochemical measurements.

11.5 Batteries

A **battery** is a galvanic cell or, more commonly, *a group of galvanic cells connected in series,* where the potentials of the individual cells add to give the total battery potential. Batteries are a source of direct current and have become an essential source of portable power in our society. In this section we examine the most common types of batteries. Some new batteries currently being developed are described at the end of the chapter.

Lead Storage Battery

Since about 1912 when self-starters were first used in automobiles, the **lead storage battery** has been a major factor in making the automobile a practical means of transportation. This type of battery can function for several years under temperature extremes from −30°F to 100°F and under incessant punishment from rough roads.

In this battery, lead serves as the anode, and lead coated with lead dioxide serves as the cathode. The electrodes dip into an electrolyte solution of sulfuric acid. The electrode reactions are as follows:

Anode reaction: $Pb + HSO_4^- \longrightarrow PbSO_4 + H^+ + 2e^-$

Cathode reaction: $PbO_2 + HSO_4^- + 3H^+ + 2e^- \longrightarrow PbSO_4 + 2H_2O$

Cell reaction: $Pb(s) + PbO_2(s) + 2H^+(aq) + 2HSO_4^-(aq) \longrightarrow 2PbSO_4(s) + 2H_2O(l)$

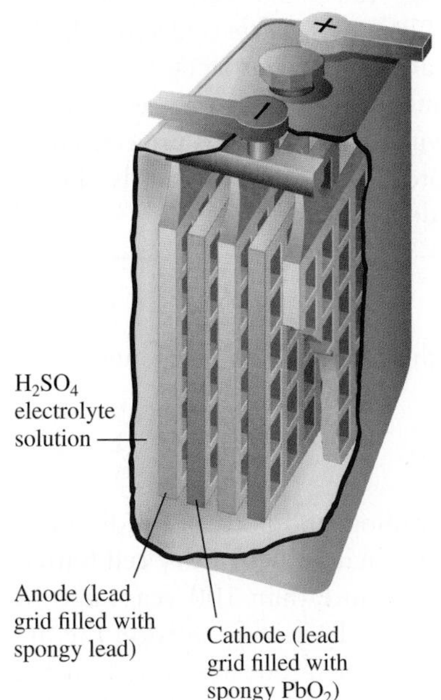

Figure 11.13
One of the six cells in a 12-V lead storage battery. The anode consists of a lead grid filled with spongy lead, and the cathode is a lead grid filled with lead dioxide. The cell also contains 38% (by mass) sulfuric acid.

The typical automobile lead storage battery has six cells connected in series. Each cell contains multiple electrodes in the form of grids (Fig. 11.13) and produces about 2 volts, to give a total battery potential of about 12 volts. Note from the cell reaction that sulfuric acid is consumed as the battery discharges, which lowers the density of the electrolyte solution from its initial value of about 1.28 g/cm^3 in the fully charged battery. As a result, the condition of the battery can be monitored by measuring the density of the sulfuric acid solution. The solid lead sulfate formed in the cell reaction during discharge adheres to the grid surfaces of the electrodes. The battery is recharged by forcing current through the battery in the opposite direction to reverse the cell reaction. A car's battery is continuously charged by an alternator driven by the automobile's engine.

An automobile with a dead battery can be jump-started by connecting its battery to the battery in a running automobile. This process can be dangerous, however, because the resulting flow of current causes electrolysis of water in the dead battery, producing hydrogen and oxygen gases (see Section 11.7 for details). Disconnecting the jumper cables after the disabled car starts causes an arc that can ignite the gaseous mixture. If this happens, the battery may explode, ejecting corrosive sulfuric acid. This problem can be averted by connecting the ground jumper cable to a part of the engine remote from the battery. Any arc produced when this cable is disconnected will then be harmless.

Traditional types of storage batteries require periodic "topping off" because the water in the electrolyte solution is depleted by the electrolysis that accompanies the charging process. Recent types of batteries have electrodes made of an alloy of calcium and lead that inhibits the electrolysis of water. These batteries can be sealed since they require no addition of water.

It is rather amazing that in the 85 years that lead storage batteries have been used, no better system has been found. Although a lead storage battery does provide excellent service, it has a useful lifetime of only 3–5 years in an automobile. Although it might seem that the battery could undergo an indefinite number of discharge–charge cycles, physical damage from road shock,

Chemical Insights — Electrochemical Window Shades

Everybody likes large windows in their homes and workplaces. However, windows present major problems in making buildings energy efficient—they admit sunlight in the summer, increasing air-conditioning demands, and they allow heat to flow out of the building in the winter. To combat these problems, researchers from Electro-Optics Technology Center at Tufts University are developing a "smart" window that uses a short pulse of electricity to vary the transparency of the window in response to changing climate conditions.

The Tufts smart window consists of seven thin layers as shown in the accompanying illustration. Production of the incredibly thin layers (the seven layers have a total thickness of about 10^{-3} mm) has a lot in common with the manufacture of computer chips and has proved to be very difficult. The center layer of the window contains an electrolyte that allows ion flow to both of the surrounding layers. One of the sandwiching layers contains tungsten(VI) oxide (WO_3) and the other contains lithium cobalt(III) oxide ($LiCoO_2$). Because the transparency and reflectivity of these compounds change in response to electron and ion flow, they can be used to regulate the passage of light through the window. Depending on the size of the current pulse delivered across the window layers, a continuum of optical properties can be produced ranging from completely transparent to opaque.

A somewhat simpler electrochemical system is being developed by Claes-Goran Granquist and coworkers at the University of Uppsala in Sweden. This system involves two transparent polyester foils that are first coated with a conductive indium–tin oxide layer. One of the foils is then coated with tungsten oxide and the other with nickel oxide. An ion-conducting polymer is then sandwiched between the foils. The system works similar to a battery, with the foils acting as the anode and the cathode with the sandwiched polymer acting as the electrolyte. When 1.4 V is applied to the system, the direction that charges the tungsten oxide coating (turns dark) and discharges the nickel oxide coating (turns dark as it discharges) causes the system to block the passage of light. When the voltage is reversed, both oxides become transparent. The voltage can be controlled with a common switch, photo detector, or voice-sensitive switch. The product is projected to be useful in motorcycle helmets, face shields, ski goggles, and windows in buildings.

which tends to shake the solid $PbSO_4$ from the electrodes, and chemical side reactions eventually cause it to fail.

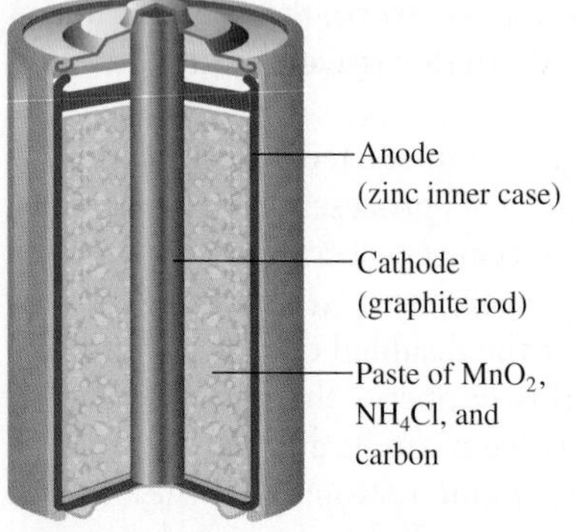

Figure 11.14
A common dry cell battery.

Dry Cell Batteries

The calculators, electronic watches, portable radios, and portable audio players that are so familiar to us are all powered by small, efficient **dry cell batteries.** The common dry cell battery was invented more than 100 years ago by George Leclanché (1839–1882), a French chemist. In its *acid version* the dry cell battery contains a zinc inner case that acts as the anode and a carbon rod in contact with a moist paste of solid MnO_2, solid NH_4Cl, and carbon that acts as the cathode (Fig. 11.14). The half-cell reactions are complex but can be approximated as follows:

Anode reaction: $$Zn \longrightarrow Zn^{2+} + 2e^-$$

Cathode reaction: $$2NH_4^+ + 2MnO_2 + 2e^- \longrightarrow Mn_2O_3 + 2NH_3 + H_2O$$

This cell produces a potential of about 1.5 volts.

In the *alkaline version* of the dry cell battery, the solid NH_4Cl is replaced with KOH or NaOH. In this case the half-reactions can be approximated as follows:

Anode reaction: $$Zn + 2OH^- \longrightarrow ZnO + H_2O + 2e^-$$

Cathode reaction: $$2MnO_2 + H_2O + 2e^- \longrightarrow Mn_2O_3 + 2OH^-$$

Although many practical problems remain to be solved, there is real hope that smart windows will be available for buildings and cars in the near future. This could lead to a significant change in the energy demands of our society—it is estimated that smart windows can reduce a building's energy usage by as much as 50%.

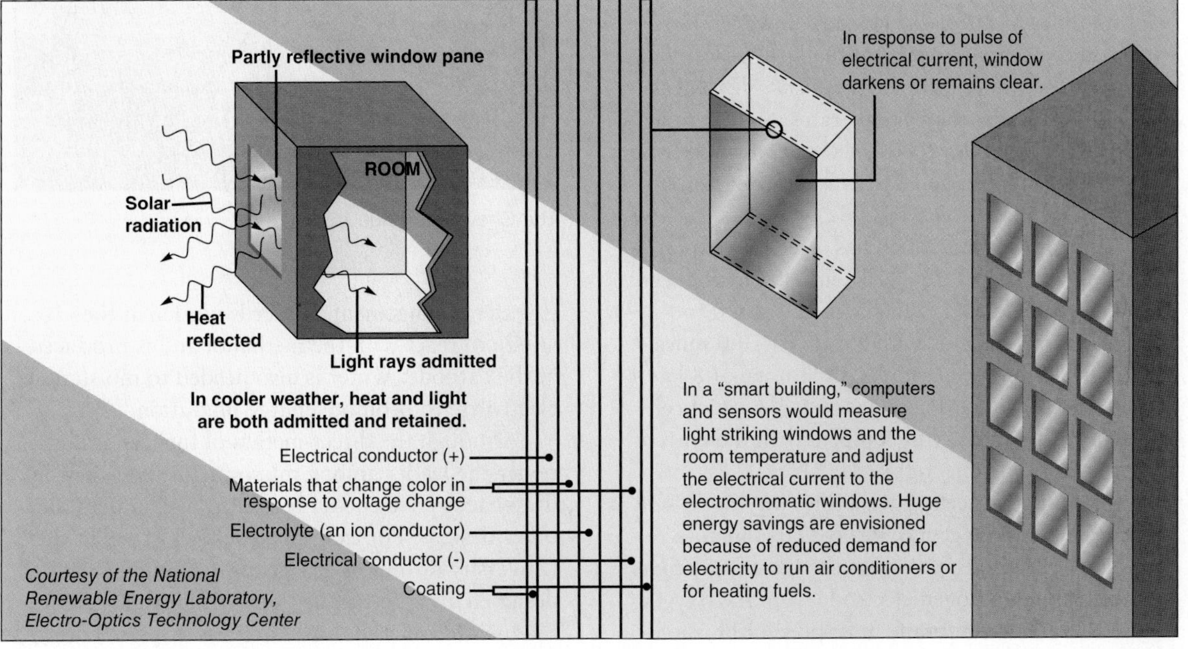

Courtesy of the National Renewable Energy Laboratory, Electro-Optics Technology Center

The alkaline dry cell lasts longer than the acidic cell mainly because the zinc anode corrodes less rapidly under basic conditions than under acidic conditions.

Other types of dry cell batteries include the *silver cell,* which has a Zn anode. Its cathode uses Ag_2O as the oxidizing agent in a basic environment. *Mercury cells,* often used in calculators, also have a Zn anode. The cathode uses HgO as the oxidizing agent in a basic medium (Fig. 11.15).

An especially important type of dry cell is the *nickel–cadmium battery,* in which the electrode reactions are as follows:

Anode reaction: $$Cd + 2OH^- \longrightarrow Cd(OH)_2 + 2e^-$$

Cathode reaction: $$NiO_2 + 2H_2O + 2e^- \longrightarrow Ni(OH)_2 + 2OH^-$$

As in the lead storage battery, the products adhere to the electrodes. Therefore, a nickel–cadmium battery can be recharged an indefinite number of times.

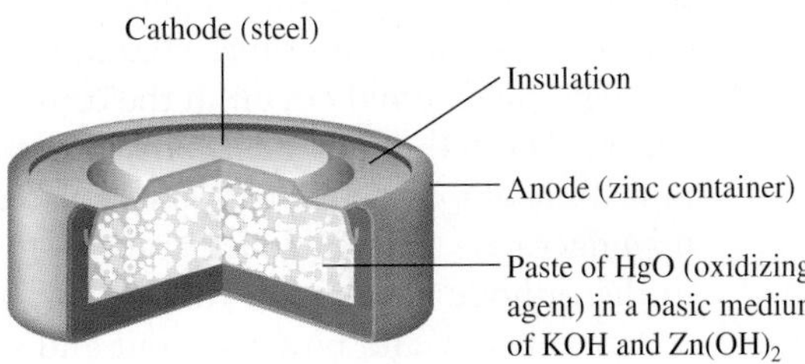

Figure 11.15
A mercury battery of the type used in calculators.

Chemical Insights Fuel Cells—Portable Energy

The promise of an energy-efficient, environmentally sound source of electrical power has spurred an intense interest in fuel cells in recent years. Although fuel cells have long been used in the U.S. space program, no practical fuel cell for powering automobiles has been developed. However, we are now on the verge of practical fuel-cell–powered cars. For example, DaimlerChrysler's NECAR 5 was driven across the United States, a 3000-mile trip that took 16 days. NECAR 5 is powered by a H_2/O_2 fuel cell that generates its H_2 from decomposition of methanol (CH_3OH).

General Motors, which has also been experimenting with H_2/O_2 fuel cells, in 2005 introduced the Chevrolet Sequel, which can go from 0 to 60 mph in 10 seconds with a range of 300 miles. The car has composite tanks for storage of 8 kg of liquid hydrogen. GM's goal is to build a fuel-cell system by 2010 that can compete with the current internal combustion engine.

In reality, fuel cells have a long way to go before they can be economically viable in automobiles. The main problem is the membrane that separates the hydrogen electrode from the oxygen electrode. This membrane must prevent H_2 molecules from passing through and still allow ions to pass between the electrodes. Current membranes cost over $3000 for an automobile-size fuel cell. The hope is to reduce this by a factor of 10 in the next few years.

Besides providing power for automobiles, fuel cells are being considered for powering small electronic devices such as cameras, cell phones, and laptop computers. Many of these micro fuel cells currently use methanol as the fuel (reducing agent) rather than H_2. However, these direct-methanol fuel cells are rife with problems. A major difficulty is water management. Water is needed at the anode to react with the methanol and is produced at the cathode. Water is also needed to moisten the electrolyte to promote charge migration.

The Chevrolet Sequel car.

Although the direct-methanol fuel cell is currently the leader among micro fuel-cell designs, its drawbacks have encouraged the development of other designs. For example, Richard Masel at the University of Illinois at Urbana–Champaign has designed a micro fuel cell that uses formic acid as the fuel. Masel and others are also experimenting with mini hot chambers external to the fuel cell that break down hydrogen-rich fuels into hydrogen gas, which is then fed into the tiny fuel cells.

To replace batteries, fuel cells must be demonstrated to be economically feasible, safe, and dependable. Today, rapid progress is being made to overcome the current problems. A recent estimate indicates that by late in this decade annual sales of the little power plants may reach 200 million units per year. It appears that after years of hype about the virtues of fuel cells, we are finally going to realize their potential.

Lithium-ion batteries involve the migration of Li^+ ions from the cathode to the anode, where they intercolate (enter the interior) as the battery is charged. At the same time, charge-balancing electrons travel to the anode through the external circuit in the charger. On discharge, the opposite process occurs. The cathode of the first successful lithium-ion batteries originally contained $LiCoO_2$ and a lithium-intercolated carbon (LiC_6) anode. More recently manufacturers have included transition metals such as nickel and manganese in the cathode in addition to cobalt. The mixed-metal cathodes have greater charge capacity and power output and shorter recharge times.

Lithium-ion batteries are used in a wide variety of applications, including cell phones, laptop computers, power tools, and even electric drive systems in automobiles and motorcycles.

Fuel Cells

A **fuel cell** is *a galvanic cell in which the reactants are continuously supplied.* To illustrate the principles of fuel cells, we will consider the exothermic redox reaction of methane with oxygen:

$$CH_4(g) + 2O_2(g) \longrightarrow CO_2(g) + 2H_2O(g) + \text{energy}$$

Usually, the energy from this reaction is released as heat to warm homes and to run machines. However, in a fuel cell designed to use this reaction, the energy is used to produce an electric current; the electrons flow from the reducing agent (CH_4) to the oxidizing agent (O_2) through a conductor.

The U.S. space program has supported extensive research to develop fuel cells. The space shuttle missions use a fuel cell based on the reaction of hydrogen and oxygen to form water:

$$2H_2(g) + O_2(g) \longrightarrow 2H_2O(l)$$

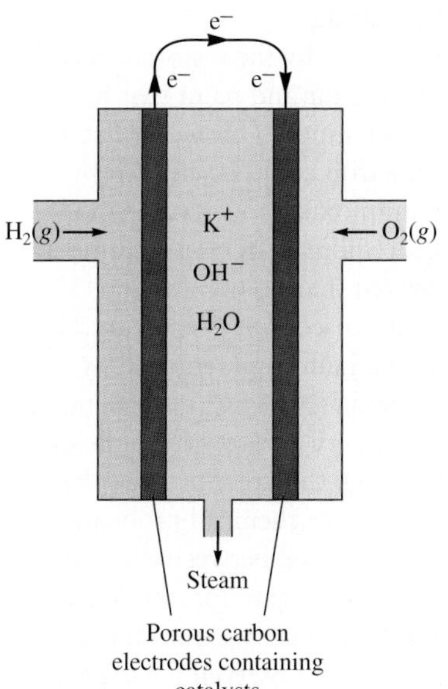

Figure 11.16
Schematic of the hydrogen–oxygen fuel cell.

A schematic of a fuel cell that uses this reaction is shown in Fig. 11.16. The half-cell reactions are as follows:

$$\text{Anode reaction:} \quad 2H_2 + 4OH^- \longrightarrow 4H_2O + 4e^-$$

$$\text{Cathode reaction:} \quad 4e^- + O_2 + 2H_2O \longrightarrow 4OH^-$$

Fuel cells are finding use as permanent power sources. A power plant built in New York City contains stacks of hydrogen–oxygen fuel cells, which can be rapidly put online in response to fluctuating power demands. The hydrogen gas is obtained by decomposing the methane in natural gas. A plant of this type has also been constructed in Tokyo.

Although the H_2/O_2 fuel cells developed for the U.S. space program are too expensive for commercial use, progress is being made toward achieving more widespread use of fuel cells for transportation. For example, the cities of Chicago and Vancouver (Canada) each have three city buses that are powered by H_2/O_2 fuel cells. These buses are more powerful and much quieter than their diesel-powered counterparts and emit only water in their exhausts. The major problem with these buses is their prohibitive cost—$1.4 million each. The large automobile manufacturers are also working on a fuel-cell–powered vehicle. DaimlerChrysler is now experimenting with a fourth-generation fuel-cell car and expects to have a practical version of this car available to the public very soon.

Because of the problems associated with the storage of hydrogen gas, scientists are also working to develop fuel cells that use gasoline or methanol to generate the H_2 needed for the fuel cell.

11.6 Corrosion

Corrosion can be viewed as the process of returning metals to their natural state—the ores from which they were originally obtained. Corrosion involves oxidation of the metal. Since corroded metal often loses its structural integrity and attractiveness, this spontaneous process has great economic impact. About one-fifth of the iron and steel produced annually is used to replace rusted metal.

Some metals, such as copper, gold, silver, and platinum, are relatively difficult to oxidize. They are often called noble metals.

Metals corrode because they oxidize easily. Table 11.1 shows that, with the exception of gold, metals commonly used for structural and decorative purposes all have standard reduction potentials less positive than that of oxy-

Chemical Insights Refurbishing the Lady

The restoration of the Statue of Liberty in New York harbor more than 30 years ago represents a fascinating blend of science, technology, and art. The statue consists of copper sheets attached to a framework of iron, which had become so weakened by corrosion during its 100 years of exposure to the elements that it was in danger of collapsing.

Gustave Eiffel, the French engineer who designed the ingenious support structure, knew from experience that if the copper touched the iron framework, the more active iron would corrode very rapidly. Why does this happen? It is apparent from the reduction potentials

$$Cu^{2+} + 2e^- \longrightarrow Cu \qquad \mathscr{E}° = 0.34\ V$$

$$Fe^{2+} + 2e^- \longrightarrow Fe \qquad \mathscr{E}° = -0.44\ V$$

that Cu^{2+} will spontaneously oxidize iron. However, as with many electrochemical processes, the situation is more complex than it first appears. Since we are talking about two metal strips touching each other, the question is: Where do the Cu^{2+} ions (the oxidizing agents) come from? In fact, research on this process suggests that the copper simply acts as a conductor for electrons and that the oxidizing agent is not Cu^{2+} at all but probably O_2 or oxides of N or S.

Whatever the mechanism, Eiffel attempted to combat the corrosion problem by inserting asbestos pads between the copper sheets and the frame. However, this idea did not work, probably because copper is such a good conductor that *any* contact between the two metals anywhere on the statue totally thwarted the effect of the insulation. In fact, workers found that the iron framework was so corroded it had to be completely replaced with stainless steel, which is much more corrosion-resistant.

Stainless steel has its own problems, however. In being bent to achieve the intricate shapes needed, the steel bars often became brittle. Flexibility was restored by heating each bar to a very high temperature, using a current of 30,000 amperes, and then cooling the bar suddenly. Unfortunately, this process also destroyed the corrosion resistance of the stainless steel. That resistance had to be restored by soaking the bars in nitric acid, an oxidizing acid that re-forms the protective oxide coating removed by heating.

Another problem faced by the restorers was the removal of layers of coal tar and paint that had been applied in vain attempts to protect the statue's interior. Although the iron could be cleaned by blasting with aluminum oxide powder, the more fragile copper sheets required a gentler treatment. The restorers discovered that liquid nitrogen (77 K) cracked the paint and caused it to peel away. The coal-tar layer under the paint was removed by blasting with baking soda ($NaHCO_3$), a substance sometimes used by museum curators for polishing dinosaur bones. Although this treatment worked very well, it created another chemical problem: Where the baking soda seeped between the seams in the copper plates onto the exterior, the statue turned from green to light blue when it rained. After this phenomenon was observed for the first time, workers stationed on the exterior scaffolding immediately cleaned off any leaking $NaHCO_3$.

gen gas. When any one of these half-reactions is reversed (to show oxidation of the metal) and combined with the reduction half-reaction for oxygen, the result is a positive $\mathscr{E}°$ value. Thus the oxidation of most metals by oxygen is spontaneous (although we cannot tell from the potential how fast it will occur).

In view of the large differences in the reduction potentials between oxygen and most metals, it is surprising that the problem of corrosion does not prevent the use of metals in air. However, most metals develop a thin oxide coating that tends to protect their internal atoms against further oxidation. The metal that best demonstrates this phenomenon is aluminum. With a reduction potential of -1.7 volts, aluminum should be easily oxidized by O_2. According to the apparent thermodynamics of the reaction, an aluminum airplane could dissolve in a rainstorm. This very active metal can be used as a structural material because the formation of a thin, adherent layer of aluminum oxide [Al_2O_3,

One of the most interesting aspects of the chemistry of the Statue of Liberty is the green patina on its surface. Copper metal exposed to the atmosphere changes from its bright reddish brown metallic luster first to an almost black color and then to the familiar green patina. The initial blackening is mainly caused by formation of copper oxide and copper sulfide. The green patina that forms consists of thin layers of two types of basic copper sulfates: brochantite, $CuSO_4 \cdot 3Cu(OH)_2$, and antlerite, $CuSO_4 \cdot 2Cu(OH)_2$. Crystals of these compounds seem to be cemented onto the copper surface by organic molecules from the air.

In recent years large areas of the statue's left side have been observed to darken as if the patina were being removed. Scientists speculated that this darkening may be the result of acid rain, which converts brochantite to the more soluble antlerite, which is then washed off by rainwater.

To make any new copper sheets look like the old ones, the restorers transplanted the patina from a weathered piece of copper to the new surface. Applying acetone, an organic solvent, to the weathered surface and scraping with an abrasive cloth caused tiny flakes of patina to fall off. These flakes, applied to the new copper, attached themselves permanently in one to three weeks of exposure to the atmosphere.

The restoration of the Statue of Liberty made use of the latest advances in chemistry as well as facts known to most general chemistry students. It's an example of the fascinating and varied problems faced by chemists as they pursue their profession.

The restored Statue of Liberty in New York harbor.

Suggested Reading

Ivars Peterson, "Lessons Learned from a Lady," *Science News* 130 (1986): 392.

more properly represented as $Al_2(OH)_6$] greatly inhibits further corrosion. The potential of the "passive," oxide-coated aluminum is -0.6 volt, a value that causes it to behave much like a noble metal.

Iron can also form a protective oxide coating. This coating is not an infallible shield against corrosion, however; when steel is exposed to oxygen in moist air, the oxide that forms tends to scale off, exposing new metal surfaces to corrosion.

The corrosion products of noble metals such as copper and silver are complex and affect the use of these metals as decorative materials. Under normal atmospheric conditions, copper forms an external layer of greenish copper carbonate called *patina*. *Silver tarnish* is silver sulfide (Ag_2S), which in thin layers gives the silver surface a richer appearance. Gold, with a positive standard reduction potential (1.50 volts) significantly larger than that for oxygen (1.23 volts), shows no appreciable corrosion in air.

Corrosion of Iron

Since steel is the main structural material for bridges, buildings, and automobiles, controlling its corrosion is extremely important. To do so, we must understand the corrosion mechanism. Instead of being a direct oxidation process, as we might expect, the corrosion of iron is an electrochemical reaction, as illustrated in Fig. 11.17.

Steel has a nonuniform surface because its chemical composition is not completely homogeneous. In addition, physical strains leave stress points in the metal. These nonuniformities produce areas where the iron is more easily oxidized (*anodic regions*) than it is at others (*cathodic regions*). In the anodic regions, each iron atom gives up two electrons to form the Fe^{2+} ion:

$$Fe \longrightarrow Fe^{2+} + 2e^-$$

The electrons that are released flow through the steel, as they do through the wire of a galvanic cell, to a cathodic region where they react with oxygen:

$$O_2 + 2H_2O + 4e^- \longrightarrow 4OH^-$$

The Fe^{2+} ions formed in the anodic regions travel to the cathodic regions through the moisture on the surface of the steel, just as ions travel through a salt bridge in a galvanic cell. In the cathodic regions, Fe^{2+} ions react with oxygen to form rust, which is hydrated iron(III) oxide of variable composition:

$$4Fe^{2+}(aq) + O_2(g) + (4 + 2n)H_2O(l) \longrightarrow \underset{\text{Rust}}{2Fe_2O_3 \cdot nH_2O(s)} + 8H^+(aq)$$

Because of the migration of ions and electrons, rust often forms at sites that are remote from those where the iron dissolved to form pits in the steel. Also the degree of hydration of the iron oxide affects the color of the rust, which may vary from black to yellow to the familiar reddish brown.

The electrochemical nature of the rusting of iron explains the importance of moisture in the corrosion process. Moisture must be present to act as a "salt bridge" between anodic and cathodic regions. Steel does not rust in dry air, a fact that explains why cars last much longer in the arid Southwest than in the relatively humid Midwest. Salt also accelerates rusting, a fact all too easily recognized by car owners in the colder parts of the United States, where salt is used on roads to melt snow and ice. The severity of rusting is greatly increased because the dissolved salt on the moist steel surface increases the conductivity of the aqueous solution formed there and thus accelerates the electrochemical corrosion process. Chloride ions also form very stable complex ions with Fe^{3+}, and this factor tends to encourage the dissolving of the iron, further accelerating the corrosion.

Prevention of Corrosion

Prevention of corrosion is an important way of conserving our natural resources of energy and metals. The primary means of protection is the application of a coating, most commonly paint or metal plating, to protect the metal

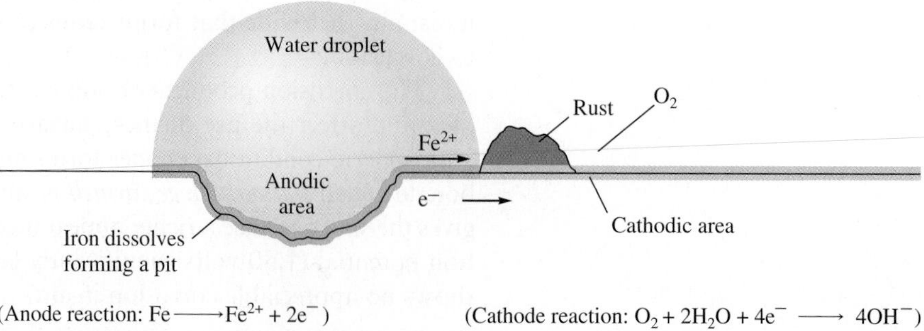

Figure 11.17
The electrochemical corrosion of iron.

Paint That Stops Rust—Completely

Chemical Insights

Traditionally, paint has provided the most economical method for protecting steel against corrosion. However, as people who live in the Midwest know well, paint cannot prevent a car from rusting indefinitely. Eventually, flaws develop in the paint that allow the ravages of rusting to take place.

This situation may soon change. Chemists at Glidden Research Center in Ohio have developed a paint called Rustmaster Pro that worked so well to prevent rusting in its initial tests that the scientists did not believe their results. Steel coated with the new paint showed no signs of rusting after an astonishing 10,000 hours of exposure in a salt spray chamber at 38°C.

Rustmaster is a water-based polymer formulation that prevents corrosion in two different ways. First, the polymer layer that cures in air forms a barrier impenetrable to both oxygen and water vapor. Second, the chemicals in the coating react with the steel surface to produce an interlayer between the metal and the polymer coating. This interlayer is a complex mineral called pyroaurite that contains cations of the form $[M_{1-x}Z_x(OH)_2]^{x+}$, where M is a 2+ ion (Mg^{2+}, Fe^{2+}, Zn^{2+}, Co^{2+}, or Ni^{2+}), Z is a 3+ ion (Al^{3+}, Fe^{3+}, Mn^{3+}, Co^{3+}, or Ni^{3+}), and x is a number between 0 and 1. The anions in pyroaurite are typically CO_3^{2-}, Cl^-, and/or SO_4^{2-}.

This pyroaurite interlayer is the real secret of the paint's effectiveness. Because the corrosion of steel has an electrochemical mechanism, motion of ions must be possible between the cathodic and anodic areas on the surface of the steel for rusting to occur. However, the pyroaurite interlayer grows into the neighboring polymer layer, thus preventing this crucial movement of ions. In effect, this layer prevents corrosion in the same way that removing the salt bridge prevents current from flowing in a galvanic cell.

In addition to having an extraordinary corrosion-fighting ability, Rustmaster yields an unusually small quantity of volatile solvents as it dries. A typical paint can produce from 1 to 5 kg of volatiles per gallon; Rustmaster produces only 0.05 kg. This paint may signal a new era in corrosion prevention.

---

from oxygen and moisture. Chromium and tin are often used to plate steel (see Section 11.8) because they react with oxygen to form a durable, effective oxide coating. Zinc, also used to coat steel in a process called **galvanizing,** does not form an oxide coating. However, since it is a more active metal than iron, as the potentials for the oxidation half-reactions show,

$$Fe \longrightarrow Fe^{2+} + 2e^- \qquad -\mathscr{E}° = 0.44 \text{ V}$$

$$Zn \longrightarrow Zn^{2+} + 2e^- \qquad -\mathscr{E}° = 0.76 \text{ V}$$

any oxidation that occurs dissolves zinc rather than iron. Recall that the reaction with the most positive standard potential has the greatest thermodynamic tendency to occur. Thus zinc acts as a "sacrificial" coating on steel.

Alloying is also used to prevent corrosion. *Stainless steel* contains chromium and nickel, both of which form oxide coatings that change steel's reduction potential to one characteristic of the noble metals. A new technology is now being developed to create surface alloys. Instead of forming a metal alloy such as stainless steel, which has the same composition throughout, a cheaper carbon steel is treated by ion bombardment to produce a thin layer of stainless steel or other desirable alloy on the surface. In this process a "plasma" or "ion gas" of the alloying ions is formed at high temperatures and is then directed onto the surface of the metal.

Cathodic protection is a method most often used to protect steel in buried fuel tanks and pipelines. An active metal, such as magnesium, is connected by a wire to the pipeline or tank to be protected (Fig. 11.18). Because magnesium is a better reducing agent than iron, electrons are furnished by the magnesium,

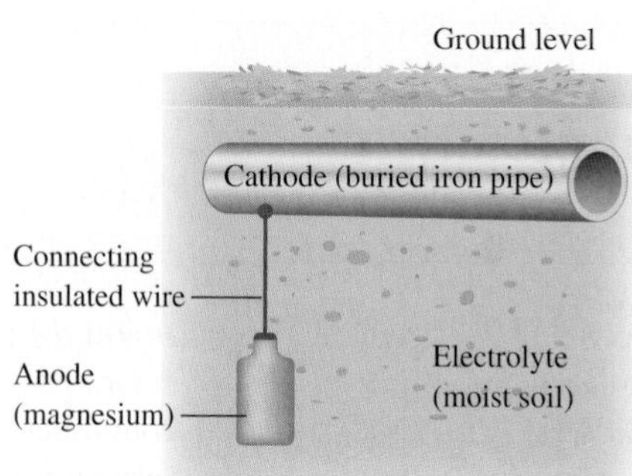

Figure 11.18
Cathodic protection of an underground pipe.

keeping the iron from being oxidized. As oxidation occurs, the magnesium anode dissolves, and so it must be replaced periodically.

11.7 Electrolysis

A galvanic cell produces current when an oxidation–reduction reaction proceeds spontaneously. A similar apparatus, an **electrolytic cell,** uses electrical energy to produce chemical change. The process of **electrolysis** involves *forcing a current through a cell to produce a chemical change for which the cell potential is negative;* that is, electrical work causes an otherwise nonspontaneous chemical reaction to occur. Electrolysis has great practical importance; for example, charging a battery, producing aluminum metal, and chrome plating an object are all done electrolytically.

To illustrate the difference between a galvanic cell and an electrolytic cell, consider the cell shown in Fig. 11.19(a) as it runs spontaneously to produce 1.10 volts. In this *galvanic cell* the reaction at the anode is

$$Zn \longrightarrow Zn^{2+} + 2e^-$$

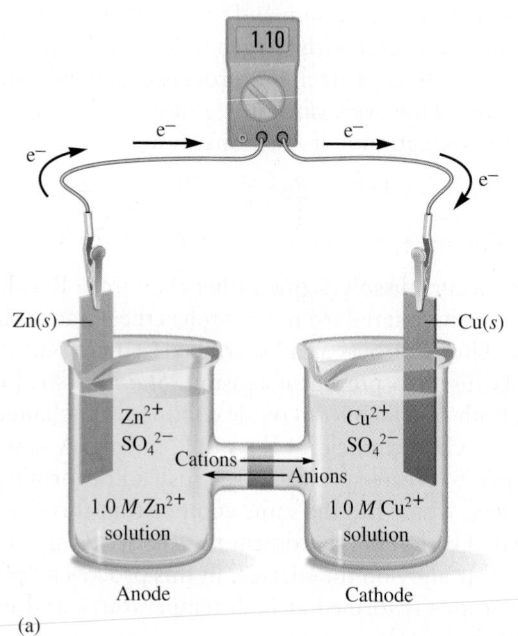

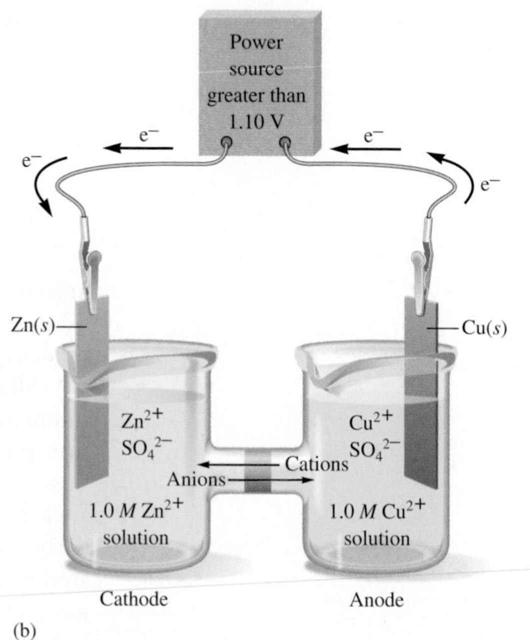

Figure 11.19
(a) A standard galvanic cell based on the spontaneous reaction

$$Zn + Cu^{2+} \longrightarrow Zn^{2+} + Cu$$

(b) A standard electrolytic cell. A power source forces the opposite reaction

$$Cu + Zn^{2+} \longrightarrow Cu^{2+} + Zn$$

and the cathode reaction is

$$Cu^{2+} + 2e^- \longrightarrow Cu$$

Figure 11.19(b) shows an external power source forcing electrons through the cell in the *opposite* direction to that in (a). This requires an external potential greater than 1.10 V, which must be applied in opposition to the natural cell potential. This device is an *electrolytic cell.* Notice that since electron flow is opposite in the two cases, the anode and cathode are reversed in (a) and (b). Also, ion flow through the salt bridge is opposite in the two cells.

An electrolytic cell uses electrical energy to produce a chemical change that would otherwise not occur spontaneously.

Now we will consider the stoichiometry of electrolytic processes—that is, *how much chemical change occurs with the flow of a given current for a specified time.* Suppose we wish to determine the mass of copper that is plated out when a current of 10.0 amperes [an **ampere** (amp) is *1 coulomb of charge per second*] is passed for 30.0 minutes through a solution containing Cu^{2+}. *Plating* means depositing the neutral metal on the electrode surface by reducing the metal ions in solution. In this case each Cu^{2+} ion requires two electrons to become an atom of copper metal:

1 A = 1 C/s

$$Cu^{2+}(aq) + 2e^- \longrightarrow Cu(s)$$

This reduction process occurs at the cathode of the electrolytic cell.

To solve this stoichiometry problem, we use the following steps:

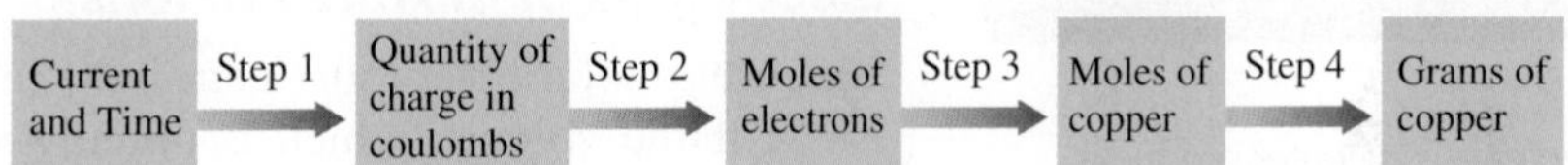

1. Since an amp is a coulomb of charge per second, we multiply the current by the time in seconds to obtain the total coulombs of charge passed into the solution at the cathode:

$$\text{Coulombs of charge} = \text{amps} \times \text{seconds} = \frac{C}{s} \times s$$

$$= 10.0\,\frac{C}{s} \times 30.0\text{ min} \times 60.0\,\frac{s}{\text{min}}$$

$$= 1.80 \times 10^4\text{ C}$$

2. Since 1 mole of electrons carries a charge of 1 faraday, or 96,485 coulombs, we can calculate the number of moles of electrons required to carry 1.80×10^4 coulombs of charge:

$$1.80 \times 10^4\text{ C} \times \frac{1\text{ mol e}^-}{96{,}485\text{ C}} = 1.87 \times 10^{-1}\text{ mol e}^-$$

This means that 0.187 mole of electrons flowed into the solution containing Cu^{2+}.

3. Each Cu^{2+} ion requires two electrons to become a copper atom. Thus each mole of electrons produces $\frac{1}{2}$ mole of copper metal:

$$1.87 \times 10^{-1}\text{ mol e}^- \times \frac{1\text{ mol Cu}}{2\text{ mol e}^-} = 9.35 \times 10^{-2}\text{ mol Cu}$$

4. Since we now know the number of moles of copper metal plated onto the cathode, we can calculate the mass of copper formed:

$$9.35 \times 10^{-2}\text{ mol Cu} \times \frac{63.546\text{ g}}{\text{mol Cu}} = 5.94\text{ g Cu}$$

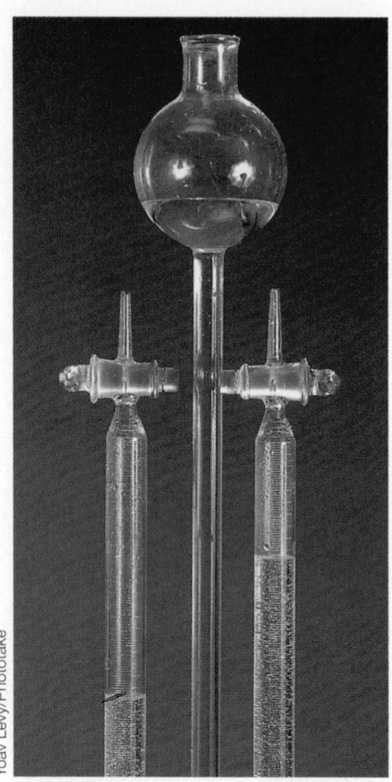

Yoav Levy/Phototake

Figure 11.20
The electrolysis of water produces hydrogen gas at the cathode (on the left) and oxygen gas at the anode (on the right). Note that twice as much hydrogen gas is produced as oxygen gas.

Electrolysis of Water

We have seen that hydrogen and oxygen combine spontaneously to form water and that the accompanying decrease in free energy can be used to run a fuel cell to produce electricity. The reverse process, which is, of course, nonspontaneous, can be forced by electrolysis:

Anode reaction:	$2H_2O \longrightarrow O_2 + 4H^+ + 4e^-$	$-\mathscr{E}° = -1.23\ V$
Cathode reaction:	$4H_2O + 4e^- \longrightarrow 2H_2 + 4OH^-$	$\mathscr{E}° = -0.83\ V$
Net reaction:	$6H_2O \longrightarrow 2H_2 + O_2 + \underbrace{4(H^+ + OH^-)}_{4H_2O}$	
or	$2H_2O \longrightarrow 2H_2 + O_2$	$\mathscr{E}° = -2.06\ V$

Note that these potentials assume an anode chamber with 1 *M* H^+ and a cathode chamber with 1 *M* OH^-. In pure water, where $[H^+] = [OH^-] = 10^{-7}\ M$, the potential for the overall process is -1.23 V.

In practice, however, if platinum electrodes connected to a 6-volt battery are dipped into pure water, no reaction is observed; pure water contains so few ions that only a negligible current can flow. However, addition of even a small amount of a soluble salt causes an immediate evolution of bubbles of hydrogen and oxygen, as illustrated in Fig. 11.20.

Electrolysis of Mixtures of Ions

Suppose a solution in an electrolytic cell contains the ions Cu^{2+}, Ag^+, and Zn^{2+}. If the voltage, which is initially very low, is gradually turned up, in which order will the metals be plated out onto the cathode? This question can be answered by looking at the standard reduction potentials of these ions:

$$Ag^+ + e^- \longrightarrow Ag \qquad \mathscr{E}° = 0.80\ V$$

$$Cu^{2+} + 2e^- \longrightarrow Cu \qquad \mathscr{E}° = 0.34\ V$$

$$Zn^{2+} + 2e^- \longrightarrow Zn \qquad \mathscr{E}° = -0.76\ V$$

Remember that the more *positive* the $\mathscr{E}°$ value, the more the reaction has a tendency to proceed in the direction indicated. Of the three reactions listed, the order of oxidizing ability is

$$Ag^+ > Cu^{2+} > Zn^{2+}$$

This means that silver will plate out first as the potential is increased, followed by copper, and finally zinc.

The principle described in this section is very useful, but it must be applied with some caution. For example, in the electrolysis of an aqueous solution of sodium chloride, we should be able to use $\mathscr{E}°$ values to predict which products are expected. Of the major species in the solution (Na^+, Cl^-, and H_2O), only Cl^- and H_2O can be readily oxidized. The half-reactions (written as oxidation processes) are

$$2Cl^- \longrightarrow Cl_2 + 2e^- \qquad -\mathscr{E}° = -1.36\ V$$

$$2H_2O \longrightarrow O_2 + 4H^+ + 4e^- \qquad -\mathscr{E}° = -1.23\ V$$

Since water has the more positive potential, we would expect to see O_2 produced at the anode. However, this does not happen. As the voltage is increased in the cell, the Cl^- ion is the first to be oxidized. A much higher potential than expected is required to oxidize water. The voltage required in excess of the expected value (called the *overvoltage*) is much greater for the production of O_2 than for Cl_2, which explains why chlorine is produced at the lower voltage.

The Chemistry of Sunken Treasure

Chemical Insights

When the galleon *Atocha* was destroyed on a reef by a hurricane in 1622, it was bound for Spain carrying about 47 tons of copper, gold, and silver from the New World. The bulk of the treasure was silver bars and coins packed in wooden chests. When treasure hunter Mel Fisher salvaged the silver in 1985, corrosion and marine growth had transformed the shiny metal into something that looked like coral. Restoring the silver to its original condition required an understanding of the chemical changes that had occurred in 350 years of being submerged in the ocean. Much of this chemistry we have already considered at various places in this text.

As the wooden chests containing the silver decayed over the years, the oxygen supply was depleted. This favored the growth of certain bacteria that use the sulfate ion rather than oxygen as an oxidizing agent to generate energy. As these bacteria consume sulfate ions, they release hydrogen sulfide gas that reacts with silver to form black silver sulfide:

$$2Ag(s) + H_2S(aq) \longrightarrow Ag_2S(s) + H_2(g)$$

Thus over the years the surface of the silver became covered with a tightly adhering layer of corrosion, which fortunately protected the silver underneath, thus preventing total conversion of the silver to silver sulfide.

Another change that took place as the wood decomposed was the formation of carbon dioxide. This shifted the equilibrium that is present in the ocean,

$$CO_2(aq) + H_2O(l) \rightleftharpoons HCO_3^-(aq) + H^+(aq)$$

to the right, producing higher concentrations of HCO_3^-. In turn, the HCO_3^- reacted with Ca^{2+} ions present in the seawater to form calcium carbonate:

$$Ca^{2+}(aq) + HCO_3^-(aq) \rightleftharpoons CaCO_3(s) + H^+(aq)$$

Calcium carbonate is the main component of limestone. Thus over time the corroded silver coins and bars became encased in limestone.

Both the limestone formation and the corrosion had to be dealt with. Since $CaCO_3$ contains the basic anion CO_3^{2-}, acid dissolves limestone:

$$2H^+(aq) + CaCO_3(s) \longrightarrow Ca^{2+}(aq) + CO_2(g) + H_2O(l)$$

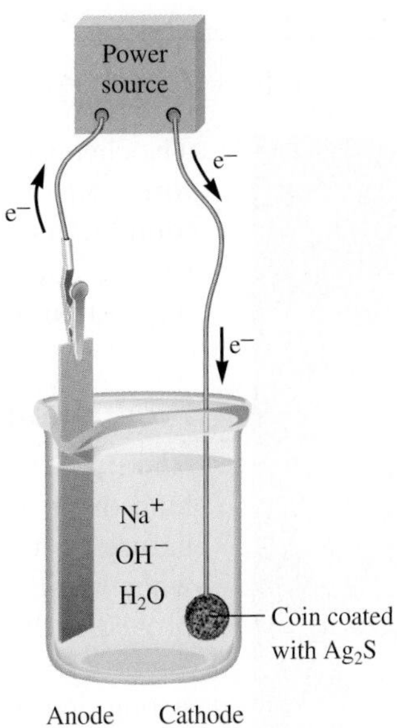

Soaking the mass of coins in a buffered acidic bath for several hours allowed the individual pieces to be separated, and the black Ag_2S on the surfaces was revealed. An abrasive could not have been used to remove this corrosion; it would have destroyed the details of the engraving—a very valuable feature of the coins to a historian or a collector—and it would have washed away some of the silver. Instead, the corrosion reaction was reversed through electrolytic reduction. The coins were connected to the cathode of an electrolytic cell in a dilute sodium hydroxide solution, as represented in the figure.

As electrons flow, the Ag^+ ions in the silver sulfide are reduced to silver metal,

$$Ag_2S + 2e^- \longrightarrow Ag + S^{2-}$$

As a by-product, bubbles of hydrogen gas from the reduction of water,

$$2H_2O + 2e^- \longrightarrow H_2 + 2OH^-$$

form on the surface of the coins.

The agitation caused by the bubbles loosens the flakes of metal sulfide and helps clean the coins.

Using these procedures, technicians have been able to restore the treasure to very nearly the same condition it was in when the *Atocha* sailed many years ago.

The causes of overvoltage are very complex. Basically, the phenomenon is caused by difficulties in transferring electrons from the species in the solution to the atoms on the electrode across the electrode–solution interface. Because of this situation, $\mathscr{E}°$ values must be used cautiously in predicting the actual order of oxidation or reduction of species in an electrolytic cell.

11.8 Commercial Electrolytic Processes

Oberlin College Archives/Oberlin College

Figure 11.21
Charles Martin Hall (1863–1914) was a student at Oberlin College in Ohio when he first became interested in aluminum. One of his professors commented that anyone who could manufacture aluminum cheaply would make a fortune, so Hall decided to give it a try. The 21-year-old Hall worked in a wooden shed near his house with an iron frying pan as a container, a blacksmith's forge as a heat source, and galvanic cells constructed from fruit jars. Using these crude galvanic cells, Hall found that he could produce aluminum by passing a current through a molten Al_2O_3/Na_3AlF_6 mixture. By a strange coincidence, Paul Heroult, a Frenchman who was born and died in the same years as Hall, made the same discovery at about the same time.

The chemistry of metals is characterized by their ability to donate electrons to form ions. Because metals are typically very good reducing agents, most are found in nature in *ores*, mixtures of ionic compounds often containing oxide, sulfide, and silicate anions. The noble metals, such as gold, silver, and platinum, are more difficult to oxidize and are often found as pure metals.

Production of Aluminum

Aluminum is one of the most abundant elements on earth, ranking third behind oxygen and silicon. Since aluminum is a very active metal, it is found in nature as its oxide in an ore called *bauxite* (named after Les Baux, France, where it was discovered in 1821). Production of aluminum metal from its ore proved to be more difficult than production of most other metals. In 1782 Lavoisier recognized aluminum as a metal "whose affinity for oxygen is so strong that it cannot be overcome by any known reducing agent." As a result, pure aluminum metal remained unknown. Finally, in 1854 a process was found for producing metallic aluminum using sodium, but aluminum remained a very expensive rarity. In fact, it is said that Napoleon III served his most honored guests with aluminum forks and spoons, while the others had to settle for gold and silver utensils.

The breakthrough came in 1886 when two men, Charles M. Hall (Fig. 11.21) in the United States and Paul Heroult in France, almost simultaneously discovered a practical electrolytic process for producing aluminum. The key factor in the *Hall–Heroult process* is the use of molten cryolite (Na_3AlF_6) as the solvent for the aluminum oxide.

Electrolysis is possible only if ions can move to the electrodes. A common method for producing ion mobility is dissolving the substance to be electrolyzed in water. This method cannot be used for aluminum because water is more easily reduced than Al^{3+}, as the following standard reduction potentials show:

$$Al^{3+} + 3e^- \longrightarrow Al \qquad \mathscr{E}° = -1.66 \text{ V}$$

$$2H_2O + 2e^- \longrightarrow H_2 + 2OH^- \qquad \mathscr{E}° = -0.83 \text{ V}$$

Thus aluminum metal cannot be plated out of an aqueous solution of Al^{3+}.

Ion mobility can also be produced by melting the salt. But the melting point of solid Al_2O_3 is much too high (2050°C) to allow practical electrolysis of the molten oxide. A mixture of Al_2O_3 and Na_3AlF_6, however, has a melting point of 1000°C, and the resulting molten mixture can be used to obtain aluminum metal electrolytically. Because of this discovery by Hall and Heroult, the price of aluminum plunged (Table 11.3), and its use became economically feasible.

Bauxite is not pure aluminum oxide (called *alumina*) but also contains the oxides of iron, silicon, and titanium, and various silicate materials. The pure hydrated alumina ($Al_2O_3 \cdot nH_2O$) is obtained by treating the crude bauxite with aqueous sodium hydroxide. Being amphoteric, alumina dissolves in the basic solution:

$$Al_2O_3(s) + 2OH^-(aq) \longrightarrow 2AlO_2^-(aq) + H_2O(l)$$

Table 11.3

The Price of Aluminum Over the Past Century

Date	Price of Aluminum ($/lb)*
1855	100,000
1885	100
1890	2
1895	0.50
1970	0.30
1980	0.80
1990	0.74

*Note the precipitous drop in price after the discovery of the Hall–Heroult process.

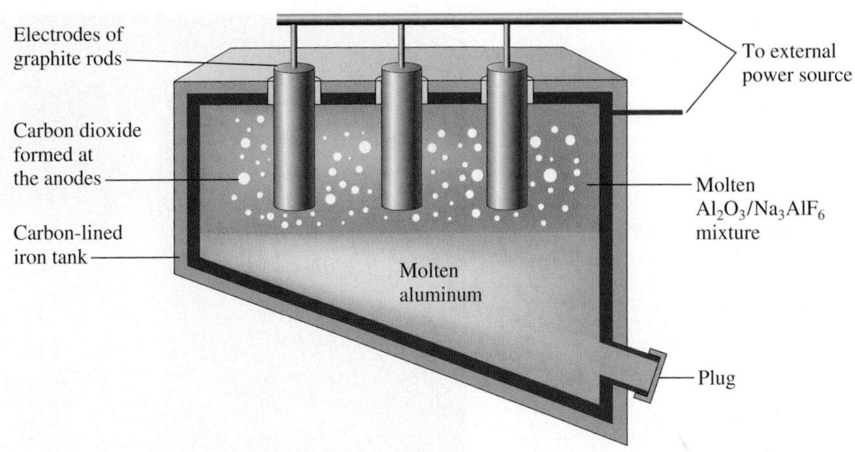

Figure 11.22
A schematic diagram of an electrolytic cell for producing aluminum by the Hall–Heroult process. Because molten aluminum is more dense than the mixture of molten cryolite and alumina, it settles to the bottom of the cell and is drawn off periodically. The graphite electrodes are gradually eaten away and must be replaced from time to time. The cell operates at a current flow of up to 250,000 A.

The other metal oxides, which are basic, remain as solids. The solution containing the aluminate ion (AlO_2^-) is separated from the sludge of other oxides and is acidified with carbon dioxide gas, causing the hydrated alumina to reprecipitate:

$$2CO_2(g) + 2AlO_2^-(aq) + (n + 1)H_2O(l) \longrightarrow 2HCO_3^-(aq) + Al_2O_3 \cdot nH_2O(s)$$

The purified alumina is then mixed with cryolite and melted, and the aluminum ion is reduced to aluminum metal in an electrolytic cell of the type shown in Fig. 11.22. Because the electrolyte solution contains a large number of aluminum-containing ions, the chemistry is not completely understood. However, the alumina probably reacts with the cryolite anion as follows:

$$Al_2O_3 + 4AlF_6^{3-} \longrightarrow 3Al_2OF_6^{2-} + 6F^-$$

The electrode reactions are thought to be the following:

Cathode reaction: $$AlF_6^{3-} + 3e^- \longrightarrow Al + 6F^-$$

Anode reaction: $$2Al_2OF_6^{2-} + 12F^- + C \longrightarrow 4AlF_6^{3-} + CO_2 + 4e^-$$

The overall cell reaction can be written as

$$2Al_2O_3 + 3C \longrightarrow 4Al + 3CO_2$$

The aluminum produced in this electrolytic process is 99.5% pure. To be useful as a structural material, aluminum is alloyed with metals such as zinc (used for trailer and aircraft construction) and manganese (used for cooking utensils, storage tanks, and highway signs). The production of aluminum consumes almost 5% of all electricity used in the United States.

Electrorefining of Metals

Purification of metals is another important application of electrolysis. For example, impure copper from the chemical reduction of copper ore is cast into large slabs that serve as the anodes for electrolytic cells. Aqueous copper sulfate is the electrolyte, and thin sheets of ultrapure copper function as the cathodes (Fig. 11.23).

Figure 11.23
Ultrapure copper sheets (serving as cathodes) are lowered between slabs of impure copper (serving as anodes) into a tank containing an aqueous solution of copper sulfate ($CuSO_4$). It takes about four weeks for the anodes to dissolve and for the pure copper to be deposited on the cathodes.

The main reaction at the anode is

$$Cu \longrightarrow Cu^{2+} + 2e^-$$

Other metals such as iron and zinc are also oxidized from the impure anode:

$$Zn \longrightarrow Zn^{2+} + 2e^-$$

$$Fe \longrightarrow Fe^{2+} + 2e^-$$

Noble metal impurities in the anode are not oxidized at the voltage used; they fall to the bottom of the cell to form a sludge, which is processed to remove the valuable silver, gold, and platinum.

The Cu^{2+} ions from the solution are deposited onto the cathode,

$$Cu^{2+} + 2e^- \longrightarrow Cu$$

producing copper that is 99.95% pure.

Metal Plating

Metals that readily corrode can often be protected by the application of a thin coating of a metal that resists corrosion. Examples are "tin" cans, which are actually steel cans with a thin coating of tin, and chrome-plated steel bumpers for automobiles.

An object can be plated by making it the cathode in a tank containing ions of the plating metal. The silver plating of a spoon is shown schematically in Fig. 11.24. In an actual plating process, the solution also contains ligands that form complexes with the silver ion. When the concentration of Ag^+ is lowered in this way, a smooth, even coating of silver is obtained.

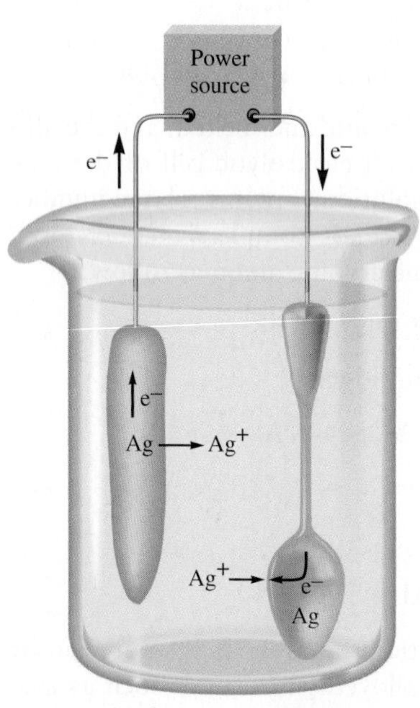

Figure 11.24
Schematic of the electroplating of a spoon. The item to be plated is the cathode, and the anode is a silver bar. Silver is plated out at the cathode:

$$Ag^+ + e^- \longrightarrow Ag$$

Note that a salt bridge is not needed here since Ag^+ ions are involved at both electrodes.

Electrolysis of Sodium Chloride

Sodium metal is produced mainly by the electrolysis of molten sodium chloride. Because solid NaCl has a rather high melting point (800°C), it is usually mixed with solid $CaCl_2$ to lower the melting point to about 600°C. The mixture is then electrolyzed in a **Downs cell,** as illustrated in Fig. 11.25, where the reactions are as follows:

Anode reaction: $2Cl^- \longrightarrow Cl_2 + 2e^-$

Cathode reaction: $Na^+ + e^- \longrightarrow Na$

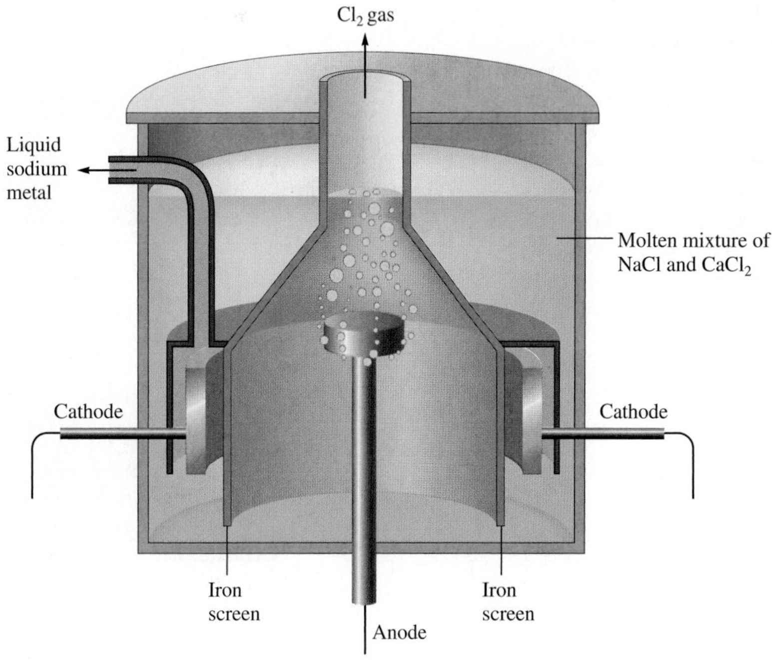

Figure 11.25
The Downs cell for the electrolysis of molten sodium chloride. The cell is designed so that the sodium and chlorine produced cannot come into contact with each other to re-form NaCl.

At the temperatures in the Downs cell, the sodium is liquid and can be drained off, cooled, and cast into blocks. Because it is so reactive, sodium must be stored in an inert solvent, such as mineral oil, to prevent its oxidation.

Electrolysis of aqueous sodium chloride (brine) is an important industrial process for the production of chlorine and sodium hydroxide. In fact, this process is second only to the production of aluminum as a consumer of electricity in the United States. Sodium is not produced in this process under normal circumstances because H_2O is more easily reduced than Na^+, as the standard reduction potentials show:

$$Na^+ + e^- \longrightarrow Na \qquad \mathscr{E}° = -2.71 \text{ V}$$

$$2H_2O + 2e^- \longrightarrow H_2 + 2OH^- \qquad \mathscr{E}° = -0.83 \text{ V}$$

Hydrogen, not sodium, is produced at the cathode.

For the reasons we discussed in Section 11.7, chlorine gas is produced at the anode. Thus the electrolysis of brine produces hydrogen and chlorine:

$$\text{Anode reaction:} \qquad 2Cl^- \longrightarrow Cl_2 + 2e^-$$

$$\text{Cathode reaction:} \qquad 2H_2O + 2e^- \longrightarrow H_2 + 2OH^-$$

It leaves a solution containing dissolved NaOH and NaCl.

The contamination of the sodium hydroxide by NaCl can be virtually eliminated by using a special **mercury cell** for electrolyzing brine (Fig. 11.26). In this cell mercury is the conductor at the cathode, and because hydrogen gas has an extremely high overvoltage with a mercury electrode, Na^+ is reduced instead of H_2O. The resulting sodium metal dissolves in the mercury, forming a liquid alloy, which is then pumped to a chamber where the dissolved sodium is reacted with water to produce hydrogen:

$$2Na(s) + 2H_2O(l) \longrightarrow 2Na^+(aq) + 2OH^-(aq) + H_2(g)$$

Relatively pure solid NaOH is recovered from the aqueous solution, and the regenerated mercury is then pumped back to the electrolysis cell. This process, called the **chlor-alkali process,** has often resulted in significant mercury contamination of the environment; the waste solutions from this process are now carefully treated to remove mercury.

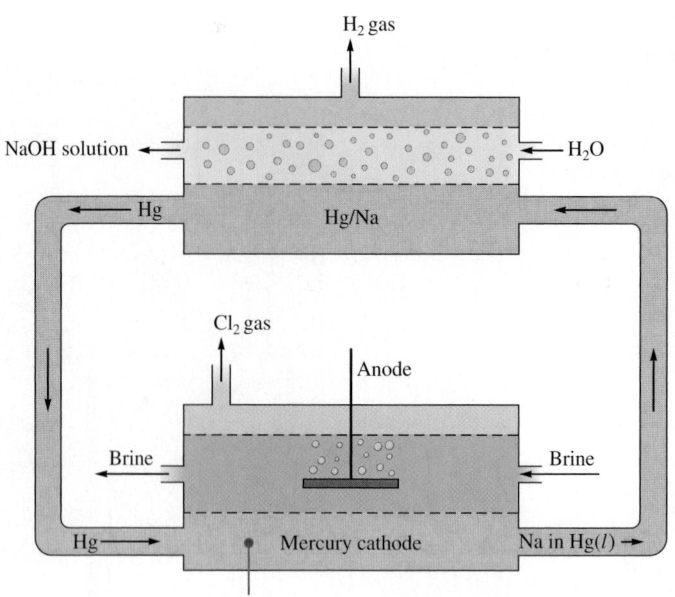

Figure 11.26
The mercury cell for production of chlorine and sodium hydroxide. The large overvoltage required to produce hydrogen at a mercury electrode means that Na^+ ions are reduced rather than water. The sodium formed dissolves in the liquid mercury and is then pumped to a chamber, where it reacts with water.

Because of the environmental problems associated with the mercury cell, it has been primarily displaced in the chlor-alkali industry by other technologies. In the United States nearly 75% of chlor-alkali production is now carried out in diaphragm cells. In a diaphragm cell the cathode and the anode are separated by a diaphragm that allows passage of H_2O molecules, Na^+ ions, and, to a limited extent, Cl^- ions. The diaphragm does not allow OH^- ions to pass through it. Thus the H_2 and OH^- formed at the cathode are kept separate from the Cl_2 formed at the anode. The major disadvantage of this process is that the aqueous effluent pumped from the cathode compartment contains a mixture of sodium hydroxide and unreacted sodium chloride, which must be separated if pure sodium hydroxide is a desired product.

In the past 25 years, a new process has been developed in the chlor-alkali industry that uses a membrane to separate the anode and cathode compartments in brine electrolysis cells. The membrane is superior to the diaphragm used in diaphragm cells because the membrane is impermeable to anions. Only cations can flow through the membrane. Because neither Cl^- nor OH^- ions can pass through the membrane separating the anode and cathode compartments, NaCl contamination of the NaOH formed at the cathode is not a problem. Although membrane technology is only now becoming prominent in the United States, it is already the dominant method for chlor-alkali production in Japan.

Key Terms

electrochemistry

Section 11.1

oxidation-reduction (redox) reaction
reducing agent
oxidizing agent
oxidation
reduction
half-reactions

For Review

OWL and go Chemistry

Sign in at **www.cengage.com/owl** to:

- View tutorials and simulations, develop problem-solving skills, and complete online homework assigned by your professor.
- Download Go Chemistry mini lecture modules for quick review and exam prep from OWL (or purchase them at **www.cengagebrain.com**)

Electrochemistry

- The study of the interchange of chemical and electrical energy
- Employs oxidation–reduction reactions

salt bridge
porous disk
galvanic cell
anode
cathode
cell potential (electromotive force)
voltmeter
potentiometer

Section 11.2
standard hydrogen electrode
standard reduction potential

Section 11.3
faraday

Section 11.4
Nernst equation
glass electrode
ion-selective electrode
concentration cell

Section 11.5
battery
lead storage battery
dry cell battery
fuel cell

Section 11.6
corrosion
galvanizing
cathodic protection

Section 11.7
electrolytic cell
electrolysis
ampere

Section 11.8
Downs cell
mercury cell
chlor-alkali process

- Galvanic cell: chemical energy is transformed into electrical energy by separating the oxidizing and reducing agents and forcing the electrons to travel through a wire
- Electrolytic cell: electrical energy is used to produce a chemical change

Galvanic cell

- Anode: the electrode where oxidation occurs
- Cathode: the electrode where reduction occurs
- The driving force behind the electron transfer is called the cell potential ($\mathscr{E}_{cell}$)
  - The potential is measured in units of volts (V), defined as a joule of work per coulomb of charge:

$$\mathscr{E}(\mathrm{V}) = \frac{-\text{work (J)}}{\text{charge (C)}} = -\frac{w}{q}$$

  - A system of half-reactions, called standard reduction potentials, can be used to calculate the potentials of various cells
    - The half-reaction $2H^+ + 2e^- \longrightarrow H_2$ is arbitrarily assigned a potential of 0 V

Free energy and work

- The maximum work that a cell can perform is

$$-w_{max} = q\mathscr{E}_{max}$$

  where $\mathscr{E}_{max}$ represents the cell potential when no current is flowing
- The actual work obtained from a cell is always less than the maximum because energy is lost through frictional heating of the wire when current flows
- For a process carried out at constant temperature and pressure, the change in free energy equals the maximum useful work obtainable from that process:

$$\Delta G = w_{max} = -q\mathscr{E}_{max} = -nF\mathscr{E}$$

  where F (faraday) equals 96,485 C and n is the number of moles of electrons transferred in the process

Concentration cell

- A galvanic cell in which both compartments have the same components but at different concentrations
- The electrons flow in the direction that tends to equalize the concentrations

Nernst equation

- Shows how the cell potential depends on the concentrations of the cell components:

$$\mathscr{E} = \mathscr{E}_0 - \frac{0.0591}{n}\log Q \quad \text{at } 25°\text{C}$$

- When a galvanic cell is at equilibrium, $\mathscr{E} = 0$ and $Q = K$

Batteries

- A battery consists of a galvanic cell or group of cells connected in series that serve as a source of direct current
- Lead storage battery
  - Anode: lead
  - Cathode: lead coated with PbO_2
  - Electrolyte: $H_2SO_4(aq)$

- Dry cell battery
  - Contains a moist paste instead of a liquid electrolyte
  - Anode: usually Zn
  - Cathode: carbon rod in contact with an oxidizing agent (which varies depending on the application)

Fuel cells

- Galvanic cells in which the reactants are continuously supplied
- The H_2/O_2 fuel cell is based on the reaction between H_2 and O_2 to form water

Corrosion

- Involves the oxidation of metals to form mainly oxides and sulfides
- Some metals, such as aluminum and chromium, form a thin, protective oxide coating that prevents further corrosion
- The corrosion of iron to form rust is an electrochemical process
  - The Fe^{2+} ions formed at anodic areas of the surface migrate through the moisture layer to cathodic regions, where they react with oxygen from the air
  - Iron can be protected from corrosion by coating it with paint or with a thin layer of metal such as chromium, tin, or zinc; by alloying; and by cathodic protection

Electrolysis

- Used to place a thin coating of metal onto steel
- Used to produce pure metals such as aluminum and copper

Discussion Questions

These questions are designed to be considered by groups of students in class. Often these questions work well for introducing a particular topic in class.

1. Sketch a galvanic cell, and explain how it works. Look at Figs. 11.1 and 11.2. Explain what is occurring in each container and why the cell in Fig. 11.2 "works" but the one in Fig. 11.1 does not.
2. In making a specific galvanic cell, explain how one determines which electrodes and solutions to use in the cell.
3. You want to "plate out" nickel metal from a nickel nitrate solution onto a piece of metal inserted into the solution. Should you use copper or zinc (or can you use either)? Explain.
4. A copper penny can be dissolved in nitric acid but not in hydrochloric acid. Using reduction potentials given in the book, show why this is so. What are the products of the reaction? Newer pennies contain a mixture of zinc and copper. What happens to the zinc in the penny when placed in nitric acid? Hydrochloric acid? Support your explanations with the data from the book, and include balanced equations for all reactions.
5. Sketch a cell that forms iron metal from iron(II) while changing chromium metal to chromium(III). Calculate the voltage, show the electron flow, label the anode and cathode, and balance the overall cell equation.
6. Which of the following is the best reducing agent: F_2, H^+, Na, Na^+, or F^-? Explain. Order as many of these species as possible from the best to the worst oxidizing agent. Why can't you order all of them? From Table 11.1 choose the species that is the best oxidizing agent. Choose the best reducing agent. Explain.
7. You are told that metal A is a better reducing agent than metal B. What, if anything, can be said about A^+ compared with B^+? Explain.
8. Explain the following relationships: ΔG and w, cell potential and w, cell potential and ΔG, cell potential and Q. Using these relationships, explain how you could make a cell in which both electrodes are the same metal and both solutions contain the same compound, but at different concentrations. How could this cell run spontaneously?
9. Explain why cell potentials are not multiplied by the coefficients in the balanced equation. (*Hint:* Use the relationship between ΔG and cell potential.)
10. What is the difference between $\mathscr{E}$ and $\mathscr{E}°$? When is $\mathscr{E}$ equal to zero? When is $\mathscr{E}°$ equal to zero? (Consider "regular" galvanic cells as well as concentration cells.)
11. Consider the following galvanic cell:

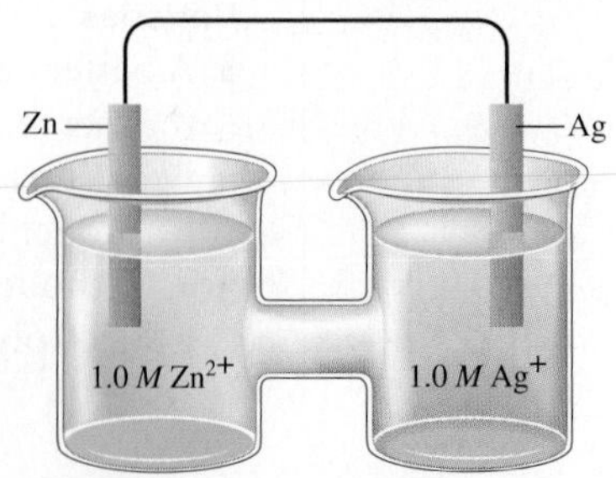

What happens to $\mathscr{E}$ as the concentration of Zn^{2+} is increased? as the concentration of Ag^+ is increased? What happens to $\mathscr{E}°$ in these cases?

12. Look up the reduction potential for Fe^{3+} to Fe^{2+}. Look up the reduction potential for Fe^{2+} to Fe. Finally, look up the reduction potential for Fe^{3+} to Fe. You should notice that adding the reduction potentials for the first two does not give you the potential for the third. Why not? Show how you can use the first two potentials to calculate the third potential.

13. If the cell potential is proportional to work and the standard reduction potential for the hydrogen ion is zero, does this mean that the reduction of the hydrogen ion requires no work?

14. Is the following statement true or false? Concentration cells work because standard reduction potentials are dependent on concentration. Explain.

Exercises

OWL Interactive versions of these problems may be assigned in OWL.

A blue exercise number indicates that the answer to that exercise appears at the back of this book and a solution appears in the *Solutions Guide*.

Galvanic Cells, Cell Potentials, and Standard Reduction Potentials

15. What is electrochemistry? What are redox reactions? Explain the difference between a galvanic cell and an electrolytic cell.

16. When magnesium metal is added to a beaker of HCl(*aq*), a gas is produced. Knowing that magnesium is oxidized and that hydrogen is reduced, write the balanced equation for the reaction. How many electrons are transferred in the balanced equation? What quantity of useful work can be obtained when Mg is added directly to the beaker of HCl? How can you harness this reaction to do useful work?

17. Consider the following galvanic cell:

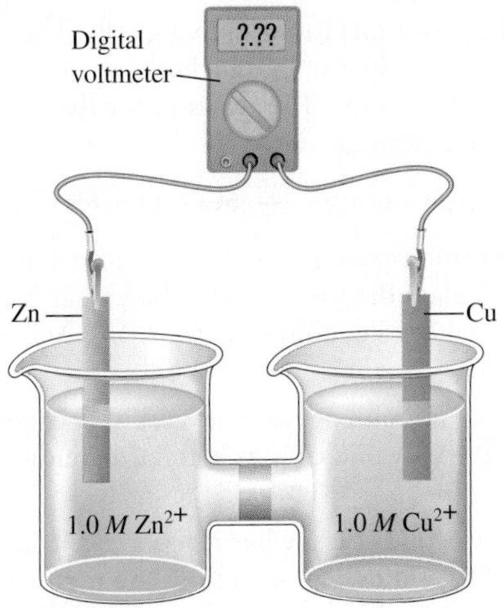

a. Label the reducing agent and the oxidizing agent, and describe the direction of the electron flow.
b. Determine the standard cell potential.
c. Which electrode increases in mass as the reaction proceeds and which electrode decreases in mass?

18. How can you construct a galvanic cell from two substances, each having a negative standard reduction potential?

19. Sketch the galvanic cells based on the following overall reactions. Calculate $\mathscr{E}°$, show the direction of electron flow and the direction of ion migration through the salt bridge, identify the cathode and anode, and give the overall balanced equation. Assume that all concentrations are 1.0 *M* and that all partial pressures are 1.0 atm. Standard reduction potentials are found in Table 11.1.
a. $Cr^{3+}(aq) + Cl_2(g) \rightleftharpoons Cr_2O_7^{2-}(aq) + Cl^-(aq)$
b. $Cu^{2+}(aq) + Mg(s) \rightleftharpoons Mg^{2+}(aq) + Cu(s)$
c. $IO_3^-(aq) + Fe^{2+}(aq) \rightleftharpoons Fe^{3+}(aq) + I_2(s)$
d. $Zn(s) + Ag^+(aq) \rightleftharpoons Zn^{2+}(aq) + Ag(s)$

20. Calculate $\mathscr{E}°$ values for the following cells. Which reactions are spontaneous as written (under standard conditions)? Balance the equations. Standard reduction potentials are found in Table 11.1.
a. $MnO_4^-(aq) + I^-(aq) \rightleftharpoons I_2(aq) + Mn^{2+}(aq)$
b. $MnO_4^-(aq) + F^-(aq) \rightleftharpoons F_2(g) + Mn^{2+}(aq)$
c. $H_2(g) \rightleftharpoons H^+(aq) + H^-(aq)$
d. $Au^{3+}(aq) + Ag(s) \rightleftharpoons Ag^+(aq) + Au(s)$

21. Sketch the galvanic cells based on the following half-reactions. Calculate $\mathscr{E}°$, show the direction of electron flow and the direction of ion migration through the salt bridge, identify the cathode and anode, and give the overall balanced equation. Assume that all concentrations are 1.0 *M* and that all partial pressures are 1.0 atm.
a. $Cl_2 + 2e^- \longrightarrow 2Cl^-$ $\quad \mathscr{E}° = 1.36$ V
$Br_2 + 2e^- \longrightarrow 2Br^-$ $\quad \mathscr{E}° = 1.09$ V
b. $MnO_4^- + 8H^+ + 5e^- \longrightarrow Mn^{2+} + 4H_2O$ $\quad \mathscr{E}° = 1.51$ V
$IO_4^- + 2H^+ + 2e^- \longrightarrow IO_3^- + H_2O$ $\quad \mathscr{E}° = 1.60$ V
c. $H_2O_2 + 2H^+ + 2e^- \longrightarrow 2H_2O$ $\quad \mathscr{E}° = 1.78$ V
$O_2 + 2H^+ + 2e^- \longrightarrow H_2O_2$ $\quad \mathscr{E}° = 0.68$ V
d. $Mn^{2+} + 2e^- \longrightarrow Mn$ $\quad \mathscr{E}° = -1.18$ V
$Fe^{3+} + 3e^- \longrightarrow Fe$ $\quad \mathscr{E}° = -0.036$ V

22. Consider the following galvanic cells:

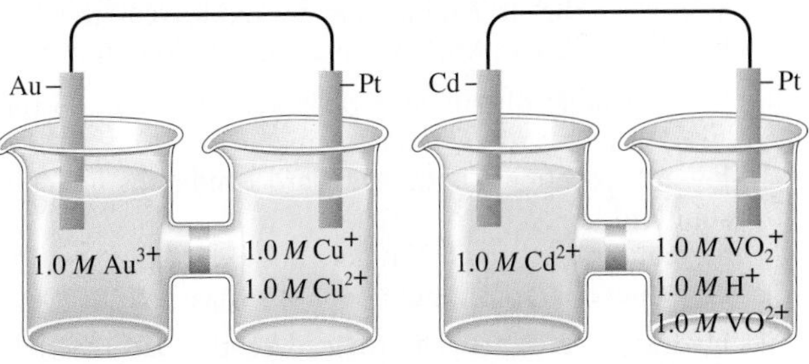

For each galvanic cell, give the balanced cell equation and determine $\mathscr{E}°$. Standard reduction potentials are found in Table 11.1.

23. Give the standard line notation for each cell in Exercise 21.

24. The saturated calomel electrode, abbreviated SCE, is often used as a reference electrode in making electrochemical measurements. The SCE is composed of mercury in contact with a saturated solution of calomel (Hg_2Cl_2). The electrolyte solution is saturated KCl. $\mathscr{E}_{SCE}$ is +0.242 V relative to the standard hydrogen electrode. Calculate the potential for each of the following galvanic cells containing a saturated calomel electrode and the given half-cell components as standard conditions. In each case indicate whether the SCE is the cathode or the anode. Standard reduction potentials are found in Table 11.1.
    a. $Cu^{2+} + 2e^- \longrightarrow Cu$
    b. $Fe^{3+} + e^- \longrightarrow Fe^{2+}$
    c. $AgCl + e^- \longrightarrow Ag + Cl^-$
    d. $Al^{3+} + 3e^- \longrightarrow Al$
    e. $Ni^{2+} + 2e^- \longrightarrow Ni$

25. Answer the following questions using data from Table 11.1 (all under standard conditions).
    a. Is $H^+(aq)$ capable of oxidizing $Cu(s)$ to $Cu^{2+}(aq)$?
    b. Is $Fe^{3+}(aq)$ capable of oxidizing $I^-(aq)$?
    c. Is $H_2(g)$ capable of reducing $Ag^+(aq)$?
    d. Is $Fe^{2+}(aq)$ capable of reducing $Cr^{3+}(aq)$ to $Cr^{2+}(aq)$?

26. Using data from Table 11.1, place the following in order of increasing strength as oxidizing agents (all under standard conditions).

$$Cd^{2+}, IO_3^-, K^+, H_2O, AuCl_4^-, \text{and } I_2$$

27. Using data from Table 11.1, place the following in order of increasing strength as reducing agents (all under standard conditions).

$$Cu^+, F^-, H^-, H_2O, I_2, \text{and } K$$

28. Consider only the species (at standard conditions)

$$Br^-, Br_2, H^+, H_2, La^{3+}, Ca, Cd$$

in answering the following questions. Give reasons for your answers.
    a. Which is the strongest oxidizing agent?
    b. Which is the strongest reducing agent?
    c. Which species can be oxidized by MnO_4^- in acid?
    d. Which species can be reduced by $Zn(s)$?

29. Use the table of standard reduction potentials (Table 11.1) to pick a reagent that is capable of each of the following oxidations (under standard conditions in acidic solution).
    a. oxidizes Br^- to Br_2 but does not oxidize Cl^- to Cl_2
    b. oxidizes Mn to Mn^{2+} but does not oxidize Ni to Ni^{2+}

30. Use the table of standard reduction potentials (Table 11.1) to pick a reagent that is capable of each of the following reductions (under standard conditions in acidic solution).
    a. reduces Cu^{2+} to Cu but does not reduce Cu^{2+} to Cu^+
    b. reduces Br_2 to Br^- but does not reduce I_2 to I^-

31. Hydrazine is somewhat toxic. Use the half-reactions shown below to explain why household bleach (a highly alkaline solution of sodium hypochlorite) should not be mixed with household ammonia or glass cleansers that contain ammonia.

$$ClO^- + H_2O + 2e^- \longrightarrow 2OH^- + Cl^- \qquad \mathscr{E}° = 0.90 \text{ V}$$

$$N_2H_4 + 2H_2O + 2e^- \longrightarrow 2NH_3 + 2OH^- \qquad \mathscr{E}° = -0.10 \text{ V}$$

32. A patent attorney has asked for your advice concerning the merits of a patent application claiming the invention of an aqueous single galvanic cell capable of producing a 12-V potential. Comment.

Cell Potential, Free Energy, and Equilibrium

33. The free energy change for a reaction ΔG is an extensive property. What is an extensive property? Surprisingly, one can calculate ΔG from the cell potential $\mathscr{E}$ for the reaction. This is surprising because $\mathscr{E}$ is an intensive property. How can the extensive property ΔG be calculated from the intensive property $\mathscr{E}$?

34. The equation $\Delta G° = -nF\mathscr{E}°$ also can be applied to half-reactions. Use standard reduction potentials to estimate $\Delta G_f°$ for $Fe^{2+}(aq)$ and $Fe^{3+}(aq)$. ($\Delta G_f°$ for $e^- = 0$.)

35. Estimate $\mathscr{E}°$ for the half-reaction.

$$2H_2O + 2e^- \longrightarrow H_2 + 2OH^-$$

given the following values of $\Delta G_f°$:

$$H_2O(l) = -237 \text{ kJ/mol}$$
$$H_2(g) = 0.0$$
$$OH^-(aq) = -157 \text{ kJ/mol}$$
$$e^- = 0.0$$

Compare this value of $\mathscr{E}°$ with the value of $\mathscr{E}°$ given in Table 11.1.

36. Glucose is the major fuel for most living cells. The oxidative breakdown of glucose by our body to produce energy is called *respiration*. The reaction for the complete combustion of glucose is

$$C_6H_{12}O_6(s) + 6O_2(g) \longrightarrow 6CO_2(g) + 6H_2O(l)$$

If this combustion reaction could be harnessed as a galvanic cell, calculate the theoretical voltage that could be produced at standard conditions. (*Hint:* Use $\Delta G_f°$ values from Appendix 4.)

37. Calculate the maximum amount of work that can be obtained from the galvanic cells at standard conditions in Exercise 22.

38. Under standard conditions, what reaction occurs, if any, when each of the following operations are performed?
    a. Crystals of I_2 are added to a solution of NaCl.
    b. Cl_2 gas is bubbled into a solution of NaI.
    c. A silver wire is placed in a solution of $CuCl_2$.
    d. An acidic solution of $FeSO_4$ is exposed to air.

For the reactions that occur, write a balanced equation and calculate $\mathscr{E}°$, $\Delta G°$, and K at 25°C.

39. Calculate $\Delta G°$ and K at 25°C for the galvanic cell reactions in Exercise 21.

40. Chlorine dioxide (ClO_2), which is produced by the reaction

$$2NaClO_2(aq) + Cl_2(g) \longrightarrow 2ClO_2(g) + 2NaCl(aq)$$

has been tested as a disinfectant for municipal water treatment.

a. Using data from Table 11.1, calculate $\mathscr{E}°$, $\Delta G°$, and K at 25°C for the production of ClO_2.

b. One of the concerns in using ClO_2 as a disinfectant is that the carcinogenic chlorate ion (ClO_3^-) might be a by-product. It can be formed from the reaction

$$ClO_2(g) \rightleftharpoons ClO_3^-(aq) + Cl^-(aq)$$

Balance the equation for the decomposition of ClO_2.

41. The amount of manganese in steel is determined by changing it to permanganate ion. The steel is first dissolved in nitric acid, producing Mn^{2+} ions. These ions are then oxidized to the deeply colored MnO_4^- ions by periodate ion (IO_4^-) in acid solution.

a. Complete and balance an equation describing each of the above reactions.

b. Calculate $\mathscr{E}°$, $\Delta G°$, and K at 25°C for each reaction.

42. The overall reaction and equilibrium constant value for a hydrogen–oxygen fuel cell at 298 K is

$$2H_2(g) + O_2(g) \longrightarrow 2H_2O(l) \qquad K = 1.28 \times 10^{83}$$

a. Calculate $\mathscr{E}°$ and $\Delta G°$ at 298 K for the fuel-cell reaction.

b. Predict the signs of $\Delta H°$ and $\Delta S°$ for the fuel-cell reaction.

c. As temperature increases, does the maximum amount of work obtained from the fuel-cell reaction increase, decrease, or remain the same? Explain.

43. Combine the equations

$$\Delta G° = -nF\mathscr{E}° \quad \text{and} \quad \Delta G° = \Delta H° - T\Delta S°$$

to derive an expression for $\mathscr{E}°$ as a function of temperature. Describe how one can graphically determine $\Delta H°$ and $\Delta S°$ from measurements of $\mathscr{E}°$ at different temperatures, assuming that $\Delta H°$ and $\Delta S°$ do not depend on temperature. What property would you look for in designing a reference half-cell that would produce a potential relatively stable with respect to temperature?

44. Calculate $\mathscr{E}°$ for the reaction

$$CH_3OH(l) + \tfrac{3}{2}O_2(g) \longrightarrow CO_2(g) + 2H_2O(l)$$

using values of $\Delta G_f°$ in Appendix 4. Will $\mathscr{E}°$ increase or decrease with an increase in temperature? (See Exercise 43 for the dependence of $\mathscr{E}°$ on temperature.)

45. A disproportionation reaction involves a substance that acts as both an oxidizing agent and a reducing agent, producing higher and lower oxidation states of the same element in the products. Which of the following disproportionation reactions are spontaneous under standard conditions? Calculate $\Delta G°$ and K at 25°C for those reactions that are spontaneous under standard conditions.

a. $2Cu^+(aq) \longrightarrow Cu^{2+}(aq) + Cu(s)$

b. $3Fe^{2+}(aq) \longrightarrow 2Fe^{3+}(aq) + Fe(s)$

c. $HClO_2(aq) \longrightarrow ClO_3^-(aq) + HClO(aq)$ (unbalanced)

Use the half-reactions:

$$ClO_3^- + 3H^+ + 2e^- \longrightarrow HClO_2 + H_2O \qquad \mathscr{E}° = 1.21\text{ V}$$

$$HClO_2 + 2H^+ + 2e^- \longrightarrow HClO + H_2O \qquad \mathscr{E}° = 1.65\text{ V}$$

46. Calculate $\mathscr{E}°$ for the following half-reaction:

$$AgI(s) + e^- \longrightarrow Ag(s) + I^-$$

(*Hint:* Reference the K_{sp} value for AgI and the standard reduction potential for Ag^+.)

47. For the following half-reaction, $\mathscr{E}° = -2.07$ V:

$$AlF_6^{3-} + 3e^- \longrightarrow Al + 6F^-$$

Using data from Table 11.1, calculate the equilibrium constant at 25°C for the reaction

$$Al^{3+}(aq) + 6F^-(aq) \rightleftharpoons AlF_6^{3-}(aq)$$

48. Calculate K_{sp} for iron(II) sulfide given the following data:

$$FeS(s) + 2e^- \longrightarrow Fe(s) + S^{2-}(aq) \qquad \mathscr{E}° = -1.01\text{ V}$$

$$Fe^{2+}(aq) + 2e^- \longrightarrow Fe(s) \qquad \mathscr{E}° = -0.44\text{ V}$$

49. The solubility product for CuI(s) is 1.1×10^{-12}. Calculate the value of $\mathscr{E}°$ for the half-reaction

$$CuI + e^- \longrightarrow Cu + I^-$$

Galvanic Cells: Concentration Dependence

50. Explain the following statement: $\mathscr{E}$ determines spontaneity, while $\mathscr{E}°$ determines the equilibrium position. Under what conditions can you use $\mathscr{E}°$ to predict spontaneity?

51. Calculate the pH of the cathode compartment for the following reaction given $\mathscr{E}_{cell} = 3.01$ V when $[Cr^{3+}] = 0.15\ M$, $[Al^{3+}] = 0.30\ M$, and $[Cr_2O_7^{2-}] = 0.55\ M$.

$$2Al(s) + Cr_2O_7^{2-}(aq) + 14H^+(aq) \longrightarrow 2Al^{3+}(aq) + 2Cr^{3+}(aq) + 7H_2O(l)$$

52. Consider the galvanic cell based on the following half-reactions:

$$Zn^{2+} + 2e^- \longrightarrow Zn \qquad \mathscr{E}° = -0.76\text{ V}$$

$$Fe^{2+} + 2e^- \longrightarrow Fe \qquad \mathscr{E}° = -0.44\text{ V}$$

a. Determine the overall cell reaction and calculate $\mathscr{E}°_{cell}$.

b. Calculate $\Delta G°$ and K for the cell reaction at 25°C.

c. Calculate $\mathscr{E}_{cell}$ at 25°C when $[Zn^{2+}] = 0.10\ M$ and $[Fe^{2+}] = 1.0 \times 10^{-5}\ M$.

53. Consider the galvanic cell based on the following half-reactions:

$$Au^{3+} + 3e^- \longrightarrow Au \qquad \mathscr{E}° = 1.50\text{ V}$$

$$Tl^+ + e^- \longrightarrow Tl \qquad \mathscr{E}° = -0.34\text{ V}$$

a. Determine the overall cell reaction and calculate $\mathscr{E}°_{cell}$.

b. Calculate $\Delta G°$ and K for the cell reaction at 25°C.

c. Calculate $\mathscr{E}_{cell}$ at 25°C when $[Au^{3+}] = 1.0 \times 10^{-2}\ M$ and $[Tl^+] = 1.0 \times 10^{-4}\ M$.

54. Consider the following galvanic cell at 25°C:

$$Pt \mid Cr^{2+}(0.30\ M), Cr^{3+}(2.0\ M) \parallel Co^{2+}(0.20\ M) \mid Co$$

The overall reaction and equilibrium constant value are

$$2Cr^{2+}(aq) + Co^{2+}(aq) \longrightarrow 2Cr^{3+}(aq) + Co(s) \quad K = 2.79 \times 10^7$$

Calculate the cell potential $\mathscr{E}$ for this galvanic cell and ΔG for the cell reaction at these conditions.

55. Consider the cell described below:

$$Al\,|\,Al^{3+}(1.00\ M)\,||\,Pb^{2+}\ (1.00\ M)\,|\,Pb$$

Calculate the cell potential after the reaction has operated long enough for the $[Al^{3+}]$ to have changed by 0.60 mol/L. (Assume $T = 25°C$.)

56. The Nernst equation can be applied to half-reactions. Calculate the reduction potential at 25°C of each of the following half-cells.
    a. Cu/Cu^{2+} (0.10 M)
       (The half-reaction is $Cu^{2+} + 2e^- \rightarrow Cu$.)
    b. Cu/Cu^{2+} (2.0 M)
    c. $Cu/Cu^{2+}(1.0 \times 10^{-4}\ M)$
    d. MnO_4^- (0.10 M)/Mn^{2+} (0.010 M) at pH = 3.00
       (The half-reaction is $MnO_4^- + 8H^+ + 5e^- \rightarrow Mn^{2+} + 4H_2O$.)
    e. MnO_4^- (0.10 M)/Mn^{2+} (0.010 M) at pH = 1.00

57. The overall reaction in the lead storage battery is

$$Pb(s) + PbO_2(s) + 2H^+(aq) + 2HSO_4^-(aq) \longrightarrow 2PbSO_4(s) + 2H_2O(l)$$

   a. Calculate $\mathscr{E}$ at 25°C for this battery when $[H_2SO_4] = 4.5\ M$; that is, $[H^+] = [HSO_4^-] = 4.5\ M$. At 25°C, $\mathscr{E}° = 2.04$ V for the lead storage battery.
   b. For the cell reaction $\Delta H° = -315.9$ kJ and $\Delta S° = 263.5$ J/K. Calculate $\mathscr{E}°$ at −20.°C. (See Exercise 43.)
   c. Calculate $\mathscr{E}$ at −20.°C when $[H_2SO_4] = 4.5\ M$.
   d. Based on your previous answers, why does it seem that batteries fail more often on cold days than on warm days?

58. A chemist wishes to determine the concentration of CrO_4^{2-} electrochemically. A cell is constructed consisting of a saturated calomel electrode (SCE; see Exercise 24) and a silver wire coated with Ag_2CrO_4. The $\mathscr{E}°$ value for the following half-reaction is +0.446 V relative to the standard hydrogen electrode:

$$Ag_2CrO_4 + 2e^- \longrightarrow 2Ag + CrO_4^{2-}$$

   a. Calculate $\mathscr{E}_{cell}$ and ΔG at 25°C for the cell reaction when $[CrO_4^{2-}] = 1.00$ mol/L.
   b. Write the Nernst equation for the cell. Assume that the SCE concentrations are constant.
   c. If the coated silver wire is placed in a solution (at 25°C) in which $[CrO_4^{2-}] = 1.00 \times 10^{-5}\ M$, what is the expected cell potential?
   d. The measured cell potential at 25°C is 0.504 V when the coated wire is dipped into a solution of unknown $[CrO_4^{2-}]$. What is the $[CrO_4^{2-}]$ for this solution?
   e. Using data from this problem and from Table 11.1, calculate the solubility product (K_{sp}) for Ag_2CrO_4.

59. What are concentration cells? What is $\mathscr{E}°$ in a concentration cell? What is the driving force for a concentration cell to produce a voltage? Is the higher or the lower ion concentration solution present at the anode? When the anode ion concentration is decreased and/or the cathode ion concentration is increased, both give rise to larger cell potentials. Why?

60. A silver concentration cell is set up at 25°C as shown below:

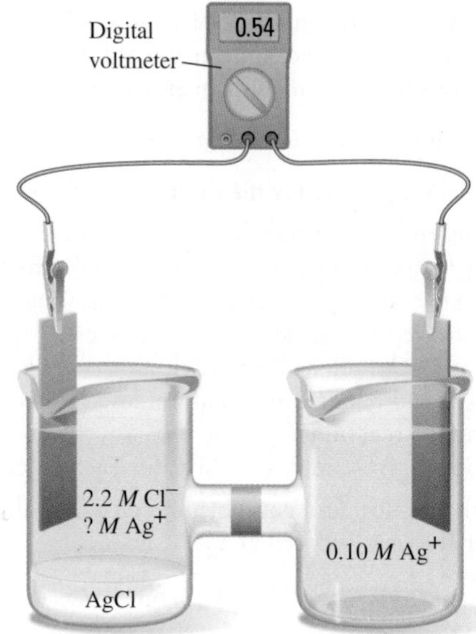

The AgCl(s) is in excess in the left compartment.
   a. Label the anode and cathode, and describe the direction of the electron flow.
   b. Determine the value of K_{sp} for AgCl at 25°C.

61. Consider the concentration cell shown below. Calculate the cell potential at 25°C when the concentration of Ni^{2+} in the compartment on the right has each of the following values.

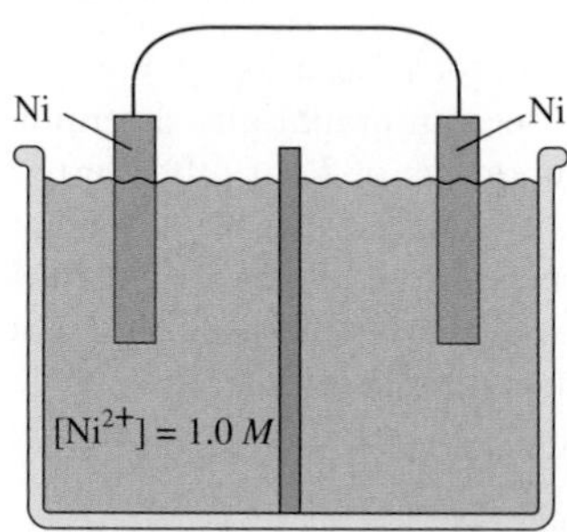

   a. 1.0 M
   b. 2.0 M
   c. 0.10 M
   d. $4.0 \times 10^{-5}\ M$
   e. Calculate the potential when both solutions are 2.5 M in Ni^{2+}.

For each case, identify the cathode, the anode, and the direction in which electrons flow.

62. Consider a concentration cell that has both electrodes made of some metal M. Solution A in one compartment of the cell contains 1.0 M M^{2+}. Solution B in the other cell compartment has a volume of 1.00 L. At the beginning of the experiment, 0.0100 mole of $M(NO_3)_2$ and 0.0100 mole of Na_2SO_4 are dissolved in solution B (ignore volume changes), where the reaction

$$M^{2+}(aq) + SO_4^{2-}(aq) \rightleftharpoons MSO_4(s)$$

occurs. For this reaction equilibrium is rapidly established, whereupon the cell potential is found to be +0.44 V at 25°C. Assume that the process

$$M^{2+} + 2e^- \longrightarrow M$$

has a standard reduction potential of +0.80 V and that no other redox process occurs in the cell. Calculate the value of K_{sp} for $MSO_4(s)$ at 25°C.

63. An electrochemical cell consists of a standard hydrogen electrode and a copper metal electrode.
    a. What is the potential of the cell at 25°C if the copper electrode is placed in a solution in which $[Cu^{2+}] = 2.5 \times 10^{-4}$ *M*?
    b. If the copper electrode is placed in a solution of 0.10 *M* NaOH that is saturated with $Cu(OH)_2$, what is the cell potential at 25°C? For $Cu(OH)_2$, $K_{sp} = 1.6 \times 10^{-19}$.
    c. The copper electrode is placed in a solution of unknown $[Cu^{2+}]$. The measured potential at 25°C is 0.195 V. What is $[Cu^{2+}]$? (Assume that Cu^{2+} is reduced.)
    d. If you wish to construct a calibration curve to show how the cell potential varies with $[Cu^{2+}]$, what should you plot to obtain a straight line? What will the slope of this line be?

64. An electrochemical cell consists of a nickel metal electrode immersed in a solution with $[Ni^{2+}] = 1.0$ *M* separated by a porous disk from an aluminum metal electrode immersed in a solution with $[Al^{3+}] = 1.0$ *M*. Sodium hydroxide is added to the aluminum compartment, causing $Al(OH)_3(s)$ to precipitate. After precipitation of $Al(OH)_3$ has ceased, the concentration of OH^- is 1.0×10^{-4} *M*, and the measured cell potential is 1.82 V. Calculate the K_{sp} value for $Al(OH)_3$.

$$Al(OH)_3(s) \rightleftharpoons Al^{3+}(aq) + 3OH^-(aq) \qquad K_{sp} = ?$$

65. You have a concentration cell in which the cathode has a silver electrode with 0.10 *M* Ag^+. The anode also has a silver electrode with $Ag^+(aq)$, 0.050 *M* $S_2O_3^{2-}$, and 1.0×10^{-3} *M* $Ag(S_2O_3)_2^{3-}$. You read the voltage to be 0.76 V.
    a. Calculate the concentration of Ag^+ at the anode.
    b. Determine the value of the equilibrium constant for the formation of $Ag(S_2O_3)_2^{3-}$.

$$Ag^+(aq) + 2S_2O_3^{2-}(aq) \rightleftharpoons Ag(S_2O_3)_2^{3-}(aq) \qquad K = ?$$

66. Consider a galvanic cell at standard conditions based on the following half-reactions:

$$Au^{3+} + 3e^- \longrightarrow Au \qquad \mathscr{E}° = 1.50 \text{ V}$$

$$Fe^{3+} + e^- \longrightarrow Fe^{2+} \qquad \mathscr{E}° = 0.77 \text{ V}$$

When enough NaCl(*s*) is added to the compartment containing gold to make the $[Cl^-] = 0.10$ *M*, the cell potential is observed to be 0.31 V. Assume that Au^{3+} is reduced, and assume that the reaction in the compartment containing gold is

$$Au^{3+}(aq) + 4Cl^-(aq) \rightleftharpoons AuCl_4^-(aq)$$

Calculate the value of *K* for this reaction at 25°C.

67. An electrochemical cell consists of a silver metal electrode immersed in a solution with $[Ag^+] = 1.0$ *M* separated by a porous disk from a copper metal electrode. If the copper electrode is placed in a solution of 5.0 *M* NH_3 that is also 0.010 *M* in $Cu(NH_3)_4^{2+}$, what is the cell potential at 25°C.

$$Cu^{2+}(aq) + 4NH_3(aq) \rightleftharpoons Cu(NH_3)_4^{2+}(aq) \qquad K = 1.0 \times 10^{13}$$

Electrolysis

68. How long will it take to plate out each of the following with a current of 100.0 A?
    a. 1.0 kg of Al from aqueous Al^{3+}
    b. 1.0 g of Ni from aqueous Ni^{2+}
    c. 5.0 moles of Ag from aqueous Ag^+

69. What mass of each of the following substances can be produced in 1.0 h with a current of 15 A?
    a. Co from aqueous Co^{2+}
    b. Hf from aqueous Hf^{4+}
    c. I_2 from aqueous KI
    d. Cr from molten CrO_3

70. It took 2.30 min with a current of 2.00 A to plate out all the silver from 0.250 L of a solution containing Ag^+. What was the original concentration of Ag^+ in the solution?

71. The electrolysis of BiO^+ produces pure bismuth. How long would it take to produce 10.0 g of Bi by the electrolysis of a BiO^+ solution using a current of 25.0 A?

72. Why is the electrolysis of molten salts much easier to predict in terms of what occurs at the anode and cathode than the electrolysis of aqueous dissolved salts?

73. What reactions take place at the cathode and the anode when each of the following is electrolyzed? Assume standard conditions.
    a. molten $NiBr_2$
    b. molten AlF_3
    c. molten MnI_2
    d. 1.0 *M* $NiBr_2$ solution
    e. 1.0 *M* AlF_3 solution
    f. 1.0 *M* MnI_2 solution

74. a. In the electrolysis of an aqueous solution of Na_2SO_4, what reactions occur at the anode and the cathode (assuming standard conditions)?

	$\mathscr{E}°$
$S_2O_8^{2-} + 2e^- \longrightarrow 2SO_4^{2-}$	2.01 V
$O_2 + 4H^+ + 4e^- \longrightarrow 2H_2O$	1.23 V
$2H_2O + 2e^- \longrightarrow H_2 + 2OH^-$	−0.83 V
$Na^+ + e^- \longrightarrow Na$	−2.71 V

    b. When water containing a small amount (~0.01 *M*) of sodium sulfate is electrolyzed, measurement of the volume of gases generated consistently gives a result that the volume ratio of hydrogen to oxygen is not quite 2:1. To what do you attribute this discrepancy? Predict whether the measured ratio is greater than or less than 2:1.

75. A solution at 25°C contains 1.0 *M* Cd^{2+}, 1.0 *M* Ag^+, 1.0 *M* Au^{3+}, and 1.0 *M* Ni^{2+} in the cathode compartment of an electrolytic cell. Predict the order in which the metals will plate out as the voltage is gradually increased.

76. An aqueous solution of an unknown salt of ruthenium is electrolyzed by a current of 2.50 A passing for 50.0 min. If 2.618 g Ru is produced at the cathode, what is the charge on the ruthenium ions in solution?

77. Consider the following half-reactions:

$$IrCl_6^{3-} + 3e^- \longrightarrow Ir + 6Cl^- \quad \mathscr{E}° = 0.77\text{ V}$$
$$PtCl_4^{2-} + 2e^- \longrightarrow Pt + 4Cl^- \quad \mathscr{E}° = 0.73\text{ V}$$
$$PdCl_4^{2-} + 2e^- \longrightarrow Pd + 4Cl^- \quad \mathscr{E}° = 0.62\text{ V}$$

A hydrochloric acid solution contains platinum, palladium, and iridium as chloro-complex ions. The solution is a constant 1.0 *M* in chloride ion and 0.020 *M* in each complex ion. Is it feasible to separate the three metals from this solution by electrolysis? (Assume that 99% of a metal must be plated out before another metal begins to plate out.)

78. An unknown metal M is electrolyzed. It took 74.1 s for a current of 2.00 A to plate out 0.107 g of the metal from a solution containing $M(NO_3)_3$. Identify the metal.

79. Electrolysis of an alkaline earth metal chloride using a current of 5.00 A for 748 seconds deposits 0.471 g of metal at the cathode. What is the identity of the alkaline earth metal chloride?

80. One of the few industrial-scale processes that produces organic compounds electrochemically is used by the Monsanto Company to produce 1,4-dicyanobutane. The reduction reaction is

$$2CH_2{=}CHCN + 2H^+ + 2e^- \longrightarrow NC{-}(CH_2)_4{-}CN$$

The $NC{-}(CH_2)_4{-}CN$ is then chemically reduced by hydrogen to $H_2N{-}(CH_2)_6{-}NH_2$, which is used in the production of nylon. What current must be used to produce 150. kg of $NC{-}(CH_2)_4{-}CN$ per hour?

81. What volume of F_2 gas, at 25°C and 1.00 atm, is produced when molten KF is electrolyzed by a current of 10.0 A for 2.00 h? What mass of potassium metal is produced? At which electrode does each reaction occur?

82. It takes 15 kWh (kilowatt hours) of electrical energy to produce 1.0 kg of aluminum metal from aluminum oxide by the Hall–Heroult process. Compare this value with the amount of energy necessary to melt 1.0 kg of aluminum metal. Why is it economically feasible to recycle aluminum cans? (The enthalpy of fusion for aluminum metal is 10.7 kJ/mol and 1 watt = 1 J/s.)

83. In the electrolysis of a sodium chloride solution, what volume of $Cl_2(g)$ is produced in the same time it takes to produce 6.00 L of $H_2(g)$, both volumes measured at 0°C and 1.00 atm?

84. What volumes of $H_2(g)$ and $O_2(g)$ at STP are produced from the electrolysis of water by a current of 2.50 A in 15.0 min?

85. Copper can be plated onto a spoon by placing the spoon in an acidic solution of $CuSO_4(aq)$ and connecting it to a copper strip via a power source as illustrated below:

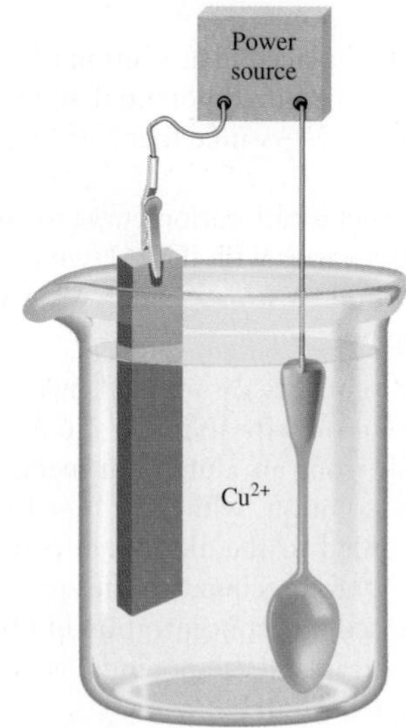

a. Label the anode and cathode, and describe the direction of the electron flow.
b. Write out the chemical equations for the reactions that occur at each electrode.

Additional Exercises

86. Consider the following electrochemical cell:

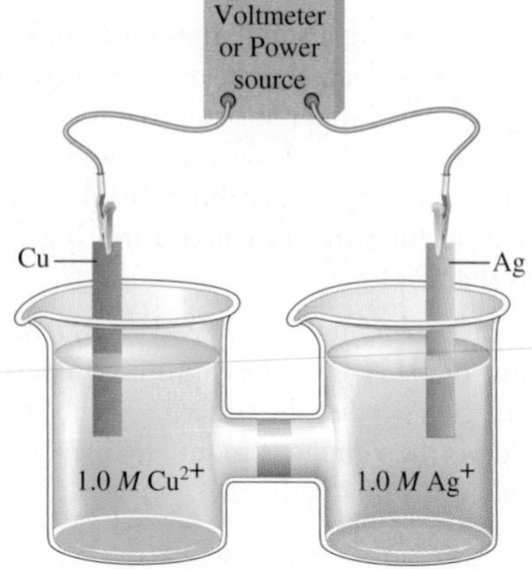

a. If silver metal is a product of the reaction, is the cell a galvanic cell or electrolytic cell? Label the cathode and anode, and describe the direction of the electron flow.
b. If copper metal is a product of the reaction, is the cell a galvanic cell or electrolytic cell? Label the cathode and anode, and describe the direction of the electron flow.
c. If the above cell is a galvanic cell, determine the standard cell potential.
d. If the above cell is an electrolytic cell, determine the minimum external potential needed to cause the reaction to occur.

87. An experimental fuel cell has been designed that uses carbon monoxide as fuel. The overall reaction is

$$2CO(g) + O_2(g) \longrightarrow 2CO_2(g)$$

The two half-cell reactions are

$$CO + O^{2-} \longrightarrow CO_2 + 2e^-$$

$$O_2 + 4e^- \longrightarrow 2O^{2-}$$

The two half-reactions are carried out in separate compartments connected with a solid mixture of CeO_2 and Gd_2O_3. Oxide ions can move through this solid at high temperatures (about 800°C). ΔG for the overall reaction at 800°C under certain concentration conditions is −380 kJ. Calculate the cell potential for this fuel cell at the same temperature and concentration conditions.

88. Batteries are galvanic cells. What happens to $\mathscr{E}_{cell}$ as a battery discharges? Does a battery represent a system at equilibrium? Explain. What is $\mathscr{E}_{cell}$ when a battery reaches equilibrium? How are batteries and fuel cells alike? How are they different? The U.S. space program uses hydrogen–oxygen fuel cells to produce power for its spacecraft. What is a hydrogen–oxygen fuel cell?

89. A fuel cell designed to react grain alcohol with oxygen has the following net reaction:

$$C_2H_5OH(l) + 3O_2(g) \longrightarrow 2CO_2(g) + 3H_2O(l)$$

The maximum work 1 mole of alcohol can yield by this process is 1320 kJ. What is the theoretical maximum voltage this cell can achieve?

90. What is the maximum work that can be obtained from a hydrogen–oxygen fuel cell at standard conditions that produces 1.00 kg of water at 25°C? Why do we say that this is the maximum work that can be obtained? What are the advantages and disadvantages in using fuel cells rather than the corresponding combustion reactions to produce electricity?

91. The overall reaction and standard cell potential at 25°C for the rechargeable nickel–cadmium alkaline battery is

$$Cd(s) + NiO_2(s) + 2H_2O(l) \longrightarrow Ni(OH)_2(s) + Cd(OH)_2(s) \qquad \mathscr{E}° = 1.10 \text{ V}$$

For every mole of Cd consumed in the cell, what is the maximum useful work that can be obtained at standard conditions?

92. Not all spontaneous redox reactions produce wonderful results. Corrosion is an example of a spontaneous redox process that has negative effects. What happens in the corrosion of a metal such as iron? What must be present for the corrosion of iron to take place? How can moisture and salt increase the severity of corrosion?

93. Explain how the following protect metals from corrosion.
a. paint
b. durable oxide coatings
c. galvanizing
d. sacrificial metal
e. alloying
f. cathodic protection

94. In theory, most metals should easily corrode in air. Why? A group of metals called the noble metals are relatively difficult to corrode in air. Some noble metals include gold, platinum, and silver. Reference Table 11.1 to come up with a possible reason why the noble metals are relatively difficult to corrode.

95. In 1973 the wreckage of the Civil War ironclad USS *Monitor* was discovered near Cape Hatteras, North Carolina. [The *Monitor* and the CSS *Virginia* (formerly the USS *Merrimack*) fought the first battle between iron-armored ships.] In 1987 investigations were begun to see whether the ship could be salvaged. *Time* reported (June 22, 1987) that scientists were considering adding sacrificial anodes of zinc to the rapidly corroding metal hull of the *Monitor*. Describe how attaching zinc to the hull would protect the *Monitor* from further corrosion.

96. A standard galvanic cell is constructed so that the overall cell reaction is

$$2Al^{3+}(aq) + 3M(s) \longrightarrow 3M^{2+}(aq) + 2Al(s)$$

where M is an unknown metal. If $\Delta G° = -411$ kJ for the overall cell reaction, identify the metal used to construct the standard cell.

97. Consider the following half-reactions:

$$Pt^{2+} + 2e^- \longrightarrow Pt \qquad \mathscr{E}° = 1.188 \text{ V}$$

$$PtCl_4^{2-} + 2e^- \longrightarrow Pt + 4Cl^- \qquad \mathscr{E}° = 0.755 \text{ V}$$

$$NO_3^- + 4H^+ + 3e^- \longrightarrow NO + 2H_2O \qquad \mathscr{E}° = 0.96 \text{ V}$$

Explain why platinum metal will dissolve in aqua regia (a mixture of hydrochloric and nitric acids) but not in either concentrated nitric or concentrated hydrochloric acid individually.

98. Consider the following reduction potentials:

$$Co^{3+} + 3e^- \longrightarrow Co \qquad \mathscr{E}° = 1.26 \text{ V}$$

$$Co^{2+} + 2e^- \longrightarrow Co \qquad \mathscr{E}° = -0.28 \text{ V}$$

a. When cobalt metal dissolves in 1.0 *M* nitric acid, will Co^{3+} or Co^{2+} be the primary product (assuming standard conditions)?
b. Is it possible to change the concentration of HNO_3 to get a different result in part a?

99. The measurement of pH using a glass electrode obeys the Nernst equation. The typical response of a pH meter at 25.00°C is given by the equation

$$\mathscr{E}_{meas} = \mathscr{E}_{ref} + 0.05916 \text{ pH}$$

where $\mathscr{E}_{ref}$ contains the potential of the reference electrode and all other potentials that arise in the cell that are not related to the hydrogen ion concentration. Assume that $\mathscr{E}_{ref} = 0.250$ V and that $\mathscr{E}_{meas} = 0.480$ V.
a. What is the uncertainty in the values of pH and $[H^+]$ if the uncertainty in the measured potential is ±1 mV (±0.001 V)?
b. To what accuracy must the potential be measured for the uncertainty in pH to be ±0.02 pH unit?

100. The black silver sulfide discoloration of silverware can easily be removed by heating the silver article in a sodium carbonate solution in an aluminum pan. The reaction is

$$3Ag_2S(s) + 2Al(s) \rightleftharpoons 6Ag(s) + 3S^{2-}(aq) + 2Al^{3+}(aq)$$

a. Using data in Appendix 4, calculate $\Delta G°$, K, and $\mathscr{E}°$ for the above reaction. [For $Al^{3+}(aq)$, $\Delta G_f° =$ $-480.$ kJ/mol.]
b. Calculate the value of the standard reduction potential for the following half-reaction:

$$2e^- + Ag_2S(s) \longrightarrow 2Ag(s) + S^{2-}(aq)$$

101. Consider the standard galvanic cell based on the following half-reactions

$$Cu^{2+} + 2e^- \longrightarrow Cu$$
$$Ag^+ + e^- \longrightarrow Ag$$

The electrodes in this cell are Ag(*s*) and Cu(*s*). Does the cell potential increase, decrease, or remain the same when the following changes occur to the standard cell?
a. $CuSO_4(s)$ is added to the copper half-cell compartment (assume no volume change).
b. $NH_3(aq)$ is added to the copper half-cell compartment. [*Hint:* Cu^{2+} reacts with NH_3 to form $Cu(NH_3)_4^{2+}(aq)$.]
c. NaCl(*s*) is added to the silver half-cell compartment. [*Hint:* Ag^+ reacts with Cl^- to form AgCl(*s*).]
d. Water is added to both half-cell compartments until the volume of solution is doubled.
e. The silver electrode is replaced with a platinum electrode.

$$Pt^{2+} + 2e^- \longrightarrow Pt \qquad \mathscr{E}° = 1.19 \text{ V}$$

102. It took 150. s for a current of 1.25 A to plate out 0.109 g of a metal from a solution containing its cations. Show that it is not possible for the cations to have a charge of 1+.

Challenge Problems

103. Three electrochemical cells were connected in series so that the same quantity of electrical current passes through all three cells. In the first cell, 1.15 g of chromium metal was deposited from a chromium(III) nitrate solution. In the second cell, 3.15 g of osmium was deposited from a solution made of Os^{n+} and nitrate ions. What is the name of the salt? In the third cell, the electrical charge passed through a solution containing X^{2+} ions caused deposition of 2.11 g of metallic X. Identify X.

104. An electrochemical cell is set up using the following unbalanced reaction:

$$M^{a+}(aq) + N(s) \longrightarrow N^{2+}(aq) + M(s)$$

The standard reduction potentials are

$$M^{a+} + ae^- \longrightarrow M \qquad \mathscr{E}° = +0.400 \text{ V}$$
$$N^{2+} + 2e^- \longrightarrow N \qquad \mathscr{E}° = +0.240 \text{ V}$$

The cell contains 0.10 *M* N^{2+} and produces a voltage of 0.180 V. If the concentration of M^{a+} is such that the value of the reaction quotient *Q* is 9.32×10^{-3}, calculate $[M^{a+}]$. Calculate w_{max} for this electrochemical cell.

105. A zinc–copper battery is constructed as follows:

$$Zn\,|\,Zn^{2+}(0.10\ M)\,||\,Cu^{2+}(2.50\ M)\,|\,Cu$$

The mass of each electrode is 200. g.
a. Calculate the cell potential when this battery is first connected.
b. Calculate the cell potential after 10.0 A of current has flowed for 10.0 h. (Assume each half-cell contains 1.00 L of solution.)
c. Calculate the mass of each electrode after 10.0 h.
d. How long can this battery deliver a current of 10.0 A before it goes dead?

106. The measurement of F^- ion concentration by ion-selective electrodes at 25.00°C obeys the equation

$$\mathscr{E}_{meas} = \mathscr{E}_{ref} - 0.05916 \log[F^-]$$

a. For a given solution, $\mathscr{E}_{meas}$ is 0.4462 V. If $\mathscr{E}_{ref}$ is 0.2420 V, what is the concentration of F^- in the solution?
b. Hydroxide ion interferes with the measurement of F^-. Therefore, the response of a fluoride electrode is

$$\mathscr{E}_{meas} = \mathscr{E}_{ref} - 0.05916 \log([F^-] + k[OH^-])$$

where $k = 1.00 \times 10^1$ and is called the selectivity factor for the electrode response. Calculate $[F^-]$ for the data in part a if the pH is 9.00. What is the percent error introduced in the $[F^-]$ if the hydroxide interference is ignored?
c. For the $[F^-]$ in part b, what is the maximum pH such that $[F^-]/k[OH^-] = 50.$?
d. At low pH, F^- is mostly converted to HF. The fluoride electrode does not respond to HF. What is the minimum pH at which 99% of the fluoride is present as F^- and only 1% is present as HF?
e. Buffering agents are added to solutions containing fluoride before making measurements with a fluoride-selective electrode. Why?

107. When copper reacts with nitric acid, a mixture of NO(*g*) and $NO_2(g)$ is evolved. The volume ratio of the two product gases depends on the concentration of the nitric acid according to the equilibrium

$$2H^+(aq) + 2NO_3^-(aq) + NO(g) \rightleftharpoons 3NO_2(g) + H_2O(l)$$

Consider the following standard reduction potentials at 25°C:

$$3e^- + 4H^+(aq) + NO_3^-(aq) \longrightarrow NO(g) + 2H_2O(l) \qquad \mathscr{E}° = 0.957 \text{ V}$$

$$e^- + 2H^+(aq) + NO_3^-(aq) \longrightarrow NO_2(g) + H_2O(l) \qquad \mathscr{E}° = 0.775 \text{ V}$$

a. Calculate the equilibrium constant for this reaction.
b. What concentration of nitric acid will produce an NO and NO_2 mixture with only 0.20% NO_2 (by

moles) at 25°C and 1.00 atm? Assume that no other gases are present and that the change in acid concentration can be neglected.

108. Consider the following galvanic cell:

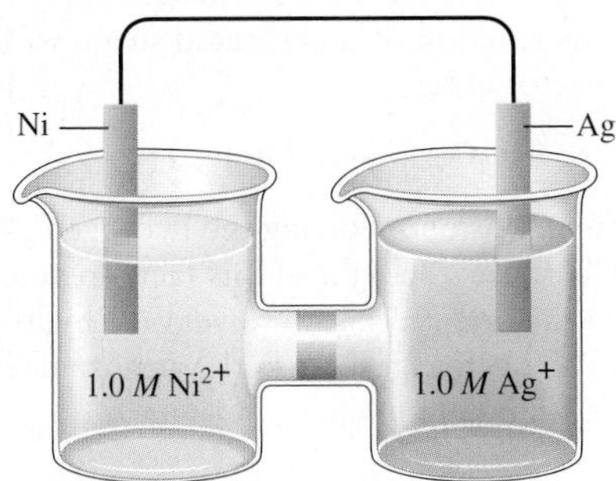

Calculate the concentrations of $Ag^+(aq)$ and $Ni^{2+}(aq)$ once the cell is "dead."

109. A galvanic cell is based on the following half-reactions:

$$Fe^{2+} + 2e^- \longrightarrow Fe(s) \qquad \mathscr{E}° = -0.440\text{ V}$$

$$2H^+ + 2e^- \longrightarrow H_2(g) \qquad \mathscr{E}° = 0.000\text{ V}$$

In this cell the iron compartment contains an iron electrode and $[Fe^{2+}] = 1.00 \times 10^{-3}$ *M*, and the hydrogen compartment contains a platinum electrode, $P_{H_2} = 1.00$ atm and a weak acid HA at an initial concentration of 1.00 *M*. If the observed cell potential is 0.333 V at 25°C, calculate the K_a value for the weak acid HA at 25°C.

110. You have a concentration cell with Cu electrodes and $[Cu^{2+}] = 1.00$ *M* (right side) and 1.0×10^{-4} *M* (left side).
a. Calculate the potential for this cell at 25°C.
b. The Cu^{2+} ion reacts with NH_3 to form $Cu(NH_3)_4^{2+}$, where the stepwise formation constants are $K_1 = 1.0 \times 10^3$, $K_2 = 1.0 \times 10^4$, $K_3 = 1.0 \times 10^3$, and $K_4 = 1.0 \times 10^3$. Calculate the new cell potential after enough NH_3 is added to the left cell compartment such that at equilibrium $[NH_3] = 2.0$ *M*.

111. A galvanic cell is based on the following half-reactions:

$$Ag^+ + e^- \longrightarrow Ag(s) \qquad \mathscr{E}° = 0.80\text{ V}$$

$$Cu^{2+} + 2e^- \longrightarrow Cu(s) \qquad \mathscr{E}° = 0.34\text{ V}$$

In this cell the silver compartment contains a silver electrode and excess $AgCl(s)$ ($K_{sp} = 1.6 \times 10^{-10}$), and the copper compartment contains a copper electrode and $[Cu^{2+}] = 2.0$ *M*.
a. Calculate the potential for this cell at 25°C.
b. Assuming 1.0 L of 2.0 *M* Cu^{2+} in the copper compartment, calculate the moles of NH_3 that would have to be added to give a cell potential of 0.52 V at 25°C (assume no volume change on addition of NH_3).

$$Cu^{2+}(aq) + 4NH_3(aq) \rightleftharpoons Cu(NH_3)_4^{2+}(aq)$$

$$K = 1.0 \times 10^{13}$$

112. Consider the following galvanic cell:

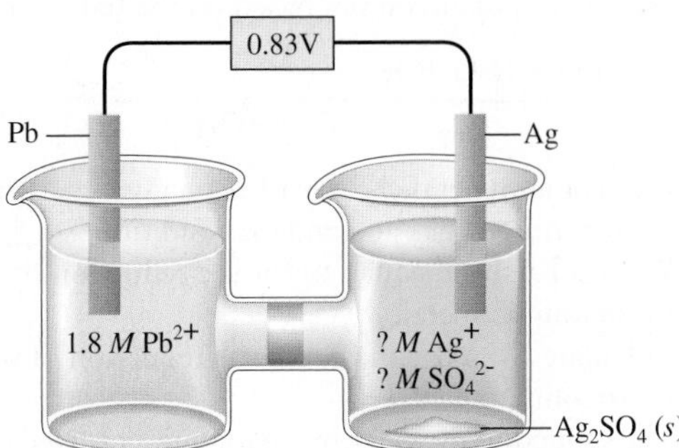

Calculate the K_{sp} value for $Ag_2SO_4(s)$. Note that to obtain silver ions in the right compartment (the cathode compartment), excess solid Ag_2SO_4 was added and some of the salt dissolved.

113. Consider the following galvanic cell:

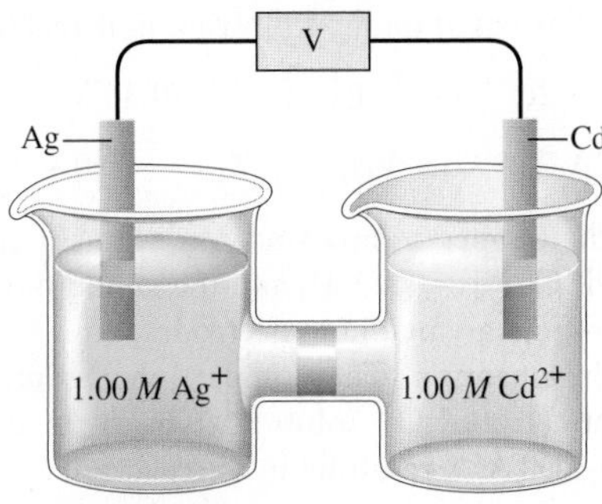

A 15.0-mole sample of NH_3 is added to the Ag compartment (assume 1.00 L of total solution after the addition). The silver ion reacts with ammonia to form complex ions as shown:

$$Ag^+(aq) + NH_3(aq) \rightleftharpoons AgNH_3^+(aq) \qquad K_1 = 2.1 \times 10^3$$

$$AgNH_3^+(aq) + NH_3(aq) \rightleftharpoons Ag(NH_3)_2^+(aq) \qquad K_2 = 8.2 \times 10^3$$

Calculate the cell potential after the addition of 15.0 moles of NH_3.

114. Given the following two standard reduction potentials,

$$M^{3+} + 3e^- \longrightarrow M \qquad \mathscr{E}° = -0.10\text{ V}$$

$$M^{2+} + 2e^- \longrightarrow M \qquad \mathscr{E}° = -0.50\text{ V}$$

determine for the standard reduction potential of the half-reaction

$$M^{3+} + e^- \longrightarrow M^{2+}$$

(*Hint:* You must use the extensive property $\Delta G°$ to determine the standard reduction potential.)

115. Zirconium is one of the few metals that retains its structural integrity upon exposure to radiation. For this reason, the fuel rods in most nuclear reactors are made of

zirconium. Answer the following questions about the redox properties of zirconium based on the half-reaction

$$ZrO_2 \cdot H_2O + H_2O + 4e^- \longrightarrow Zr + 4OH^- \qquad \mathscr{E}° = -2.36 \text{ V}$$

a. Is zirconium metal capable of reducing water to form hydrogen gas at standard conditions?
b. Write a balanced equation for the reduction of water by zirconium metal.
c. Calculate $\mathscr{E}°$, $\Delta G°$, and K for the reduction of water by zirconium metal.
d. The reduction of water by zirconium occurred during the accident at Three Mile Island, Pennsylvania, in 1979. The hydrogen produced was successfully vented and no chemical explosion occurred. If 1.00×10^3 kg of Zr reacts, what mass of H_2 is produced? What volume of H_2 at 1.0 atm and 1000.°C is produced?
e. At Chernobyl, USSR, in 1986, hydrogen was produced by the reaction of superheated steam with the graphite reactor core:

$$C(s) + H_2O(g) \longrightarrow CO(g) + H_2(g)$$

A chemical explosion involving the hydrogen gas did occur at Chernobyl. In light of this fact, do you think it was a correct decision to vent the hydrogen and other radioactive gases into the atmosphere at Three Mile Island? Explain.

Marathon Problems

116. A galvanic cell is based on the following half-reactions:

$$Cu^{2+} + 2e^- \longrightarrow Cu(s) \qquad \mathscr{E}° = 0.34 \text{ V}$$

$$V^{2+} + 2e^- \longrightarrow V(s) \qquad \mathscr{E}° = -1.20 \text{ V}$$

In this cell the copper compartment contains a copper electrode and $[Cu^{2+}] = 1.00$ *M*, and the vanadium compartment contains a vanadium electrode and V^{2+} at an unknown concentration. The compartment containing the vanadium (1.00 L of solution) was titrated with 0.0800 *M* H_2EDTA^{2-}, resulting in the reaction

$$H_2EDTA^{2-}(aq) + V^{2+}(aq) \rightleftharpoons VEDTA^{2-}(aq) + 2H^+(aq) \qquad K = ?$$

The potential of the cell was monitored to determine the stoichiometric point for the process, which occurred at a volume of 500.0 mL of H_2EDTA^{2-} solution added. At the stoichiometric point, $\mathscr{E}_{cell}$ was observed to be 1.98 V. The solution was buffered at a pH of 10.00.

a. Calculate $\mathscr{E}_{cell}$ before the titration was carried out.
b. Calculate the value of the equilibrium constant K for the titration reaction.
c. Calculate $\mathscr{E}_{cell}$ at the halfway point in the titration.

117. The table below lists the cell potentials for the 10 possible galvanic cells assembled from the metals A, B, C, D, and E and their respective 1.00 *M* 2+ ions in solution. Using the data in the table, establish a standard reduction potential table similar to Table 11.1 in the text. Assign a reduction potential of 0.00 V to the half-reaction that falls in the middle of the series. You should get two different tables. Explain why, and discuss what you could do to determine which table is correct.

	$A(s)$ in $A^{2+}(aq)$	$B(s)$ in $B^{2+}(aq)$	$C(s)$ in $C^{2+}(aq)$	$D(s)$ in $D^{2+}(aq)$
$E(s)$ in $E^{2+}(aq)$	0.28 V	0.81 V	0.13 V	1.00 V
$D(s)$ in $D^{2+}(aq)$	0.72 V	0.19 V	1.13 V	—
$C(s)$ in $C^{2+}(aq)$	0.41 V	0.94 V	—	—
$B(s)$ in $B^{2+}(aq)$	0.53 V	—	—	—

Quantum Mechanics and Atomic Theory

12

chapter

White light passing through two prisms.

Tetra Images/Alamy

Sign in to OWL at www.cengage.com/owl to view tutorials and simulations, develop problem-solving skills, and complete online homework assigned by your professor.

go Chemistry

Download mini lecture videos for key concept review and exam prep from OWL or purchase them from www.cengagebrain.com.

In the past 200 years, a great deal of experimental evidence has accumulated to support the atomic model. This theory has proved to be both extremely useful and physically reasonable. When atoms were first suggested by the Greek philosophers Democritus and Leucippus about 400 B.C., the concept was based mostly on intuition. In fact, for the following 20 centuries, no convincing experimental evidence was available to support the existence of atoms. The first real scientific data were gathered by Lavoisier and others from quantitative measurements of chemical reactions. The results of these stoichiometric experiments led John Dalton to propose the first systematic atomic theory. Dalton's theory, although crude, has stood the test of time extremely well.

Once we came to "believe in" atoms, it was logical to ask: What is the nature of an atom? Does an atom have parts, and if so, what are they? In Chapter 2 we considered some of the experiments most important for shedding light on the nature of the atom. Now we will see how the atomic theory has evolved to its present state.

One of the most striking things about the chemistry of the elements is the periodic repetition of properties. There are several groups of elements that show great similarities in chemical behavior. As we saw in Chapter 2, these similarities led to the development of the periodic table of the elements. In this chapter we will see that the modern theory of atomic structure accounts for periodicity in terms of the electron arrangements in atoms.

However, before we examine atomic structure, we must consider the revolution that took place in physics in the first 30 years of the twentieth century. During that time, experiments were carried out, the results of which could not be explained by the theories of classical physics developed by Isaac Newton and many others who followed him. A radical new theory called quantum mechanics was developed to account for the behavior of light and atoms. This "new physics" provides many surprises for people who are used to the macroscopic world, but it seems to account flawlessly (within the bounds of necessary approximations) for the behavior of matter.

As the first step in our exploration of this revolution in science, we will consider the properties of light, more properly called electromagnetic radiation.

12.1 Electromagnetic Radiation

One of the ways that energy travels through space is by **electromagnetic radiation.** The light from the sun, the energy used to cook food in a microwave oven, the X rays used by dentists, and the radiowaves used by physicians to make MRI maps of body tissues are all examples of electromagnetic radiation. Although these forms of radiant energy seem quite different, they all exhibit the same type of wavelike behavior and travel at the speed of light in a vacuum. Electromagnetic radiation is so named because it has electrical and magnetic fields that simultaneously oscillate in planes mutually perpendicular to each other and to the direction of propagation through space (Fig. 12.1).

Waves are characterized by wavelength, frequency, and speed. As shown in Fig. 12.2, **wavelength** (symbolized by the Greek letter lambda, λ) is the *distance between two consecutive peaks or troughs in a wave*. The **frequency** (symbolized by the Greek letter nu, ν) is defined as the *number of waves (cycles) per second that pass a given point in space*. Since all types of electromagnetic radiation travel at the speed of light, short-wavelength radiation must have a high frequency. You can see this in Fig. 12.2, where three waves are shown traveling between two points at constant speed. Note that the wave with the shortest wavelength (λ_3) has the highest frequency, and the wave with

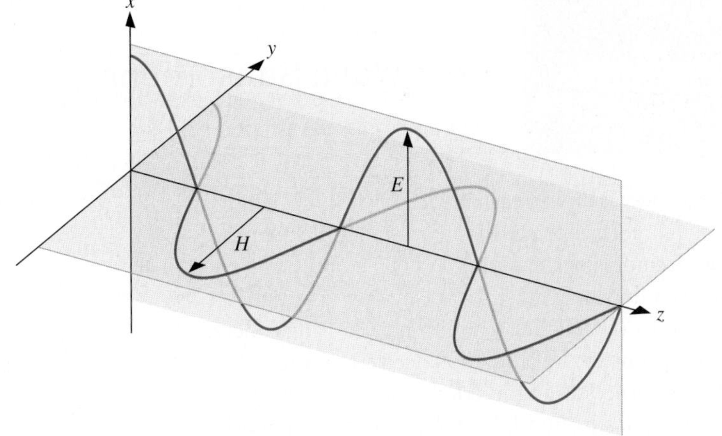

Figure 12.1
Electromagnetic radiation has oscillating electric (E) and magnetic (H) fields in planes perpendicular to each other and to the direction of propagation.

the longest wavelength (λ_1) has the lowest frequency. This implies an inverse relationship between wavelength and frequency; that is, $\lambda \propto 1/\nu$, or

$$\lambda\nu = c$$

Wavelength (λ) and frequency (ν) are inversely related.

In this equation λ is the wavelength in meters, ν is the frequency in cycles per second, and c is the speed of light, a defined quantity with the exact value of 2.99792458×10^8 m/s. In the SI system, *cycles* is understood, and the unit cycles per second becomes 1/s, or s^{-1}, which is called the *hertz* (abbreviated Hz).

Electromagnetic radiation is classified as shown in Fig. 12.3. Radiation provides an important means of energy transfer. For example, the energy from the sun reaches the earth mainly in the forms of visible and ultraviolet radiation, and the glowing coals of a fireplace transmit heat energy by infrared radiation. In a microwave oven, the water molecules in food absorb microwave radiation, which

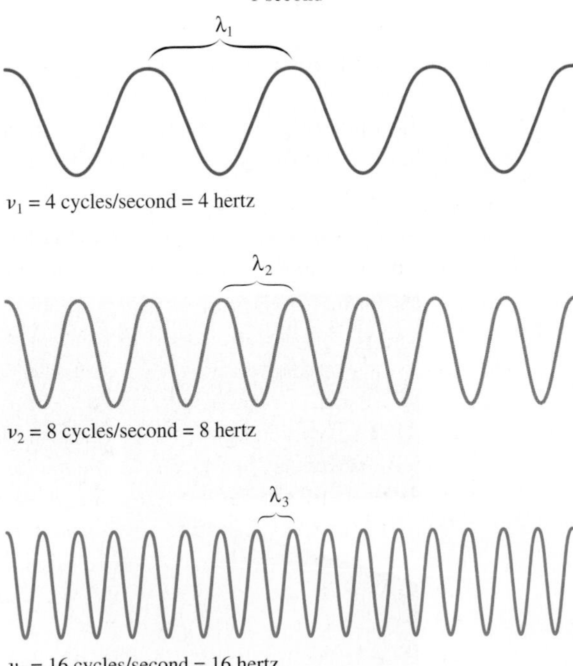

Figure 12.2
The nature of waves. Note that the radiation with the shortest wavelength has the highest frequency. This diagram can represent the oscillating electric or magnetic field of the wave.

Chemical Insights New-Wave Sunscreens

Skin cancer is an increasingly important problem. Recent studies indicate that basal cell and squamous cell carcinomas have more than tripled in the last 30 years due in greatest part to overexposure to ultraviolet (UV) light. In addition to causing cancer, overexposure to UV radiation accelerates the "aging" processes in the skin because it damages proteins, elastin, and DNA in the upper and middle layers of the skin. Ultraviolet radiation (UVR) is classified as UVB (wavelengths from 280–320 nm) and UVA (wavelengths from 320–400 nm). Studies show that UVA accounts for over 80% of the detrimental effects of exposure to UVR, including DNA damage and ultimately skin cancer. UVA rays, which penetrate deeper into the skin, can pass through clouds, windows, and even clothing.

Although it's impossible to eliminate UVR exposure to the skin, it can be minimized by wearing protective clothing and by applying sunblocks and sunscreens. Sunblocks, which are physical agents that act as barriers to UVR, often contain zinc oxide (ZnO), titanium(IV) oxide (TiO_2), or both, which are highly reflective white powders. Traditional ZnO sunblocks worn by lifeguards are opaque white, but by reducing the particle sizes of the white oxide to about 100 nm, they can be made invisible on the skin while retaining their UV-blocking ability. Traditional chemical sunscreens, most of which are based on *p*-aminobenzoic acid, act primarily by binding to skin proteins and absorbing UVB photons. Recently, the Food and Drug Administration (FDA) has approved a sunscreen formula for use in the United States that has long been used abroad. The facial moisturizer Anthelios SX, marketed by L'Oreal, contains ecamsule (trade name Mexoryl SX), a sunscreen ingredient that is particularly effective at blocking short UVA radiation (320–340 nm). The product also contains other ingredients that block longer UVA rays and UVB rays to make it a broad-spectrum sunscreen. Although there are already some sunscreens, such as avobenzene, that offer UVA protection, the advantage of ecamsule is that it does not decompose on exposure to sunlight. A major disadvantage is that ecamsule is water soluble and thus is not useful at the beach.

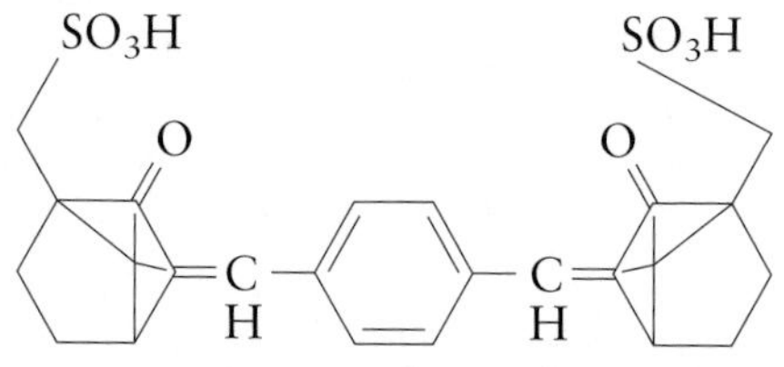

Structure of ecamsule

increases their motions. This energy is then transferred to other types of molecules via collisions, causing an increase in the food's temperature. As we proceed in the study of chemistry, we will consider many of the classes of electromagnetic radiation and the ways in which they affect matter.

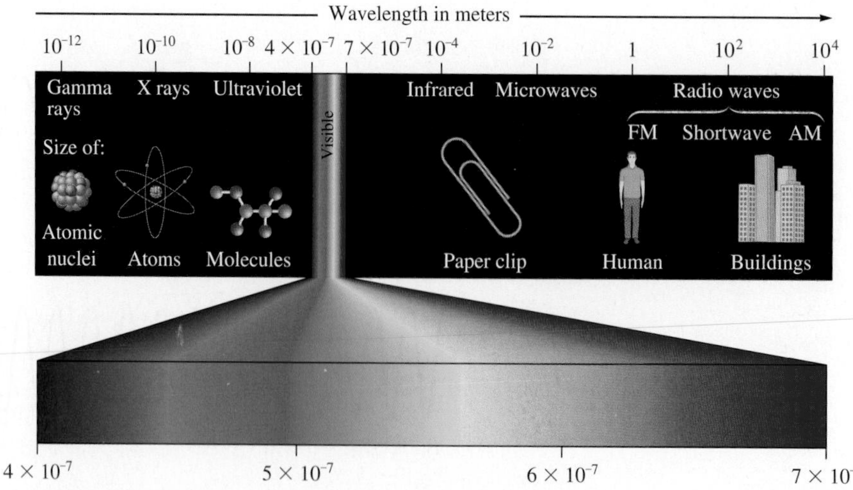

Figure 12.3 Classification of electromagnetic radiation.

12.2 The Nature of Matter

It is probably fair to say that at the end of the nineteenth century physicists were feeling rather smug. Available theories could explain phenomena ranging from the motions of the planets to the dispersion of visible light by a prism. Rumor has it that students were being discouraged from pursuing physics as a career because it was believed that all the major problems had been solved or at least described in terms of the current physical theories.

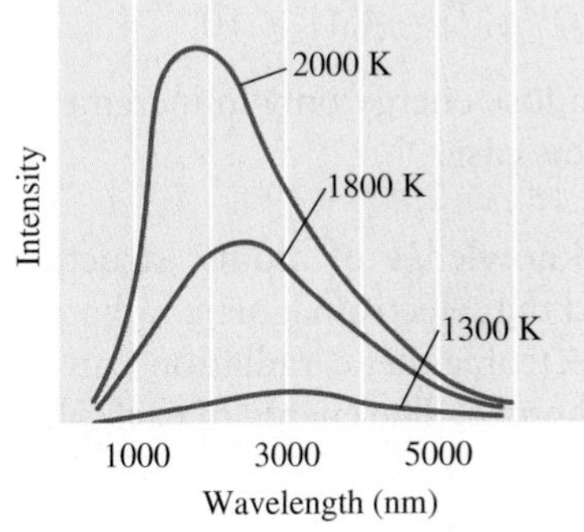

Figure 12.4
The profile of radiation emitted from a blackbody. Note that the maximum shifts to shorter wavelengths as the temperature is increased, in agreement with the observed change from a reddish to a white glow as iron is heated to higher temperatures.

At the end of the nineteenth century, the idea prevailed that matter and energy were distinct. Matter was thought to consist of particles, whereas energy in the form of light (electromagnetic radiation) was described as a wave. Particles were things that had mass and whose positions in space could be specified. Waves were described as massless and delocalized; that is, their positions in space could not be specified. It was also assumed that there was no intermingling of matter and light. Everything known before 1900 seemed to fit neatly into this view.

At the beginning of the twentieth century, however, certain experimental results suggested that this picture was incorrect. The first important advance came in 1901 from German physicist Max Planck, who studied the profile (intensity versus wavelength) of the electromagnetic radiation emitted from a solid body heated to incandescence. (An example is a piece of iron that glows red and then white as it is heated to higher and higher temperatures.) The profiles shown in Fig. 12.4 are for so-called blackbody radiation. This may seem like a contradiction in terms, since a blackbody is an idealized object that absorbs all the radiation incident on it. The term is used in this context to mean radiation that originates from the thermal energy of the body only. It does not include radiation reflected from the object and does not depend on the material composing the object. Blackbody radiation is closely approximated by the radiation emitted through a tiny hole from a cavity inside an object. The main point to be made here is that the radiation profiles shown in Fig. 12.4 are not the ones expected from classical physics. The classical theory of matter, which assumes that matter can absorb or emit any quantity of energy, predicts a radiation profile that has no maximum and goes to infinite intensity at very short wavelengths (an effect often called the **ultraviolet catastrophe**).

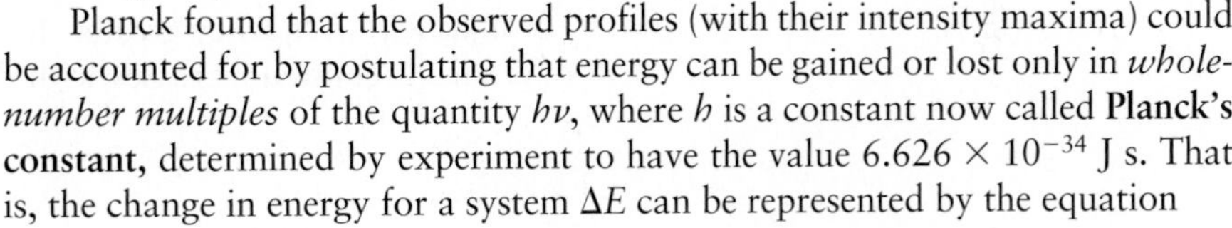

Energy can be gained or lost only in integer multiples of $h\nu$.

Planck's constant = 6.626×10^{-34} J s.

Planck found that the observed profiles (with their intensity maxima) could be accounted for by postulating that energy can be gained or lost only in *whole-number multiples* of the quantity $h\nu$, where h is a constant now called **Planck's constant,** determined by experiment to have the value 6.626×10^{-34} J s. That is, the change in energy for a system ΔE can be represented by the equation

$$\Delta E = nh\nu$$

where n is an integer (1, 2, 3, . . .), h is Planck's constant, and ν is the frequency of the electromagnetic radiation absorbed or emitted.

Planck's result was a real surprise. Physicists had always assumed that the energy of matter was continuous, which meant that the transfer of any quantity of energy was possible. Now it seemed clear that energy is in fact **quantized** and can be transferred only in discrete units of size $h\nu$. Each of these small "packets" of energy is called a *quantum*. A system can transfer energy only in whole *quanta*. Thus energy seems to have particulate properties.

OWL INTERACTIVE EXAMPLE 12.1

The blue color in fireworks is often achieved by heating copper(I) chloride (CuCl) to about 1200°C. The hot compound emits blue light having a wavelength of 450 nm. What is the increment of energy (the quantum) that is emitted at 4.50×10^2 nm by CuCl?

CuCl produces a blue flame when heated in a burner.

G. N. Lewis (see Section 13.10) actually coined the term "photon" in 1926.

Solution The quantum of energy can be calculated from the equation

$$\Delta E = h\nu$$

The frequency ν for this case can be calculated as follows:

$$\nu = \frac{c}{\lambda} = \frac{2.9979 \times 10^8 \text{ m/s}}{4.50 \times 10^{-7}\text{m}} = 6.66 \times 10^{14} \text{ s}^{-1}$$

So

$$\Delta E = h\nu = (6.626 \times 10^{-34} \text{ J s})(6.66 \times 10^{14} \text{ s}^{-1}) = 4.41 \times 10^{-19} \text{ J}$$

A sample of CuCl emitting light at 450 nm can lose energy only in increments of 4.41×10^{-19} J, the size of the quantum in this case.

The next important development in the knowledge of atomic structure came when Albert Einstein (Fig. 12.5) proposed that electromagnetic radiation is itself quantized. Einstein suggested that electromagnetic radiation can be viewed as a stream of "particles" now called **photons.** The energy of each photon is given by the expression

$$E_{\text{photon}} = h\nu = \frac{hc}{\lambda}$$

where h is Planck's constant, ν is the frequency of the radiation, and λ is the wavelength of the radiation.

Einstein arrived at this conclusion through his analysis of the **photoelectric effect** (for which he later was awarded the Nobel Prize). The photoelectric effect refers to the phenomenon in which electrons are emitted from the surface of a metal when light strikes it. The following observations characterize the photoelectric effect.

1. Studies in which the frequency of the light is varied show that no electrons are emitted by a given metal below a specific threshold frequency ν_0.
2. For light with frequency lower than the threshold frequency, no electrons are emitted regardless of the intensity of the light.
3. For light with frequency greater than the threshold frequency, the number of electrons emitted increases with the intensity of the light.

Figure 12.5
Albert Einstein (1879–1955) was born in Germany. Nothing in his early development suggested genius; even at the age of 9 he did not speak clearly, and his parents feared that he might have a handicap. When asked what profession Einstein should follow, his school principal replied, "It doesn't matter; he'll never make a success of anything." When he was 10, Einstein entered the Luitpold Gymnasium (high school), which was typical of German schools of that time in being harshly disciplinarian. There he developed a deep suspicion of authority and a skepticism that encouraged him to question and doubt—valuable qualities in a scientist. In 1905, while a patent clerk in Switzerland, Einstein published a paper explaining the photoelectric effect via the quantum theory. For this revolutionary thinking, he received a Nobel Prize in 1921. Highly regarded by this time, he worked in Germany until 1933, when Hitler's persecution of Jews forced him to come to the United States. He worked at the Institute for Advanced Study at Princeton University until his death in 1955.

Einstein was undoubtedly the greatest physicist of our age. Even if someone else had derived the theory of relativity, his other work would have ensured his ranking as the second greatest physicist of his time. Our concepts of space and time were radically changed by ideas he first proposed when he was 26 years old. From then until the end of his life, he attempted unsuccessfully to find a single unifying theory that would explain all physical events.

4. For light with frequency greater than the threshold frequency, the kinetic energy of the emitted electrons increases linearly with the frequency of the light.

These observations can be explained by assuming that electromagnetic radiation is quantized (consists of photons) and that the threshold frequency represents the minimum energy required to remove the electron from the metal's surface.

$$\text{Minimum energy required to remove an electron} = E_0 = h\nu_0$$

Because a photon with energy less than E_0 ($\nu < \nu_0$) cannot remove an electron, light with a frequency less than the threshold frequency produces no electrons. On the other hand, for light where $\nu > \nu_0$, the energy in excess of that required to remove the electron is given to the electron as kinetic energy (KE):

$$\text{KE}_{\text{electron}} = \frac{1}{2}mv^2 = h\nu - h\nu_0$$

m: Mass of electron; v: Velocity of electron; $h\nu$: Energy of incident photon; $h\nu_0$: Energy required to remove electron from metal's surface

Because in this picture the intensity of light is a measure of the number of photons present in a given part of the beam, a greater intensity means that more photons are available to release electrons (as long as $\nu > \nu_0$ for the radiation).

At about the same time that Einstein was performing his analysis of the photoelectric effect, he was also constructing the theory of special relativity. In connection with this work, Einstein derived the famous equation

$$E = mc^2$$

which he published in 1905. This equation points out the close relationship between energy and mass. In fact, we have learned that mass is a form of energy. When a system loses energy, it also loses mass. This result is more apparent if we rearrange the equation to the following form:

$$m = \frac{E}{c^2}$$

m: Mass; E: Energy; c: Speed of light

Using this form of the equation, we can calculate the mass associated with a given quantity of energy.

If a beam of light can be considered to be a stream of "particles" (photons), do the photons have mass? The answer to this question is "no." Photons do not exhibit mass in the same way as classical particles do. Einstein's equations, however, predict that a photon has momentum, which is best thought of as an intrinsic property of the photon that does not depend separately on mass and velocity, unlike the case for a classical particle. In 1922 American physicist Arthur Compton performed experiments involving collisions of X rays with electrons. These experiments showed that photons do exhibit the momentum calculated from Einstein's equation. Also, photons do seem to be affected by gravity, as Einstein postulated in his general theory of relativity. However, it is important to recognize that the photon is in no sense a typical particle. A photon has mass only in a relativistic sense—it has no rest mass.

We can summarize the important conclusions from the work of Planck and Einstein as follows:

Energy is quantized. It can be transferred only in discrete units called quanta.

Electromagnetic radiation, which was previously thought to exhibit only wave properties, seems to show certain characteristics of particulate matter as well. This phenomenon, illustrated in Fig. 12.6, is sometimes referred to as the **dual nature of light.**

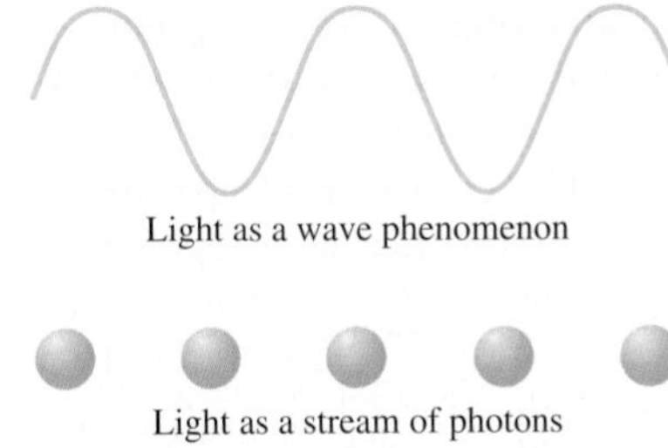

Figure 12.6
Electromagnetic radiation exhibits wave properties and particulate properties. The energy of each photon of the radiation is related to the wavelength and frequency by the equation $E_{\text{photon}} = h\nu = hc/\lambda$.

Do not confuse ν (frequency) with v (velocity).

Thus light, which was previously thought to be purely wavelike, was found to have certain characteristics of particulate matter. But is the opposite also true? That is, does matter that is normally assumed to be particulate exhibit wave properties? This question was raised in 1923 by a young French physicist named Louis de Broglie (1892–1987), who derived the following relationship for the wavelength of a particle with momentum mv:

$$\lambda = \frac{h}{mv}$$

This equation, called de Broglie's equation, allows us to calculate the wavelength for a particle, as shown in Example 12.2.

OWL INTERACTIVE EXAMPLE 12.2

Compare the wavelength for an electron (mass = 9.11×10^{-31} kg) traveling at a speed of 1.0×10^7 m/s with that for a ball (mass = 0.10 kg) traveling at 35 m/s.

Solution We use the equation $\lambda = h/mv$, where

$$h = 6.626 \times 10^{-34} \text{ J s} \quad \text{or} \quad 6.626 \times 10^{-34} \text{ kg m}^2/\text{s}$$

since

$$1 \text{ J} = 1 \text{ kg m}^2/\text{s}^2$$

For the electron,

$$\lambda_e = \frac{6.626 \times 10^{-34} \frac{\text{kg m}^2}{\text{s}}}{(9.11 \times 10^{-31} \text{ kg})(1.0 \times 10^7 \text{ m/s})} = 7.3 \times 10^{-11} \text{ m}$$

For the ball,

$$\lambda_b = \frac{6.626 \times 10^{-34} \frac{\text{kg m}^2}{\text{s}}}{(0.10 \text{ kg})(35 \text{ m/s})} = 1.9 \times 10^{-34} \text{ m}$$

Lester V. Bergman/Corbis

Spectrum obtained by shining a beam of light at an angle to a diffraction grating.

Notice from Example 12.2 that the wavelength associated with the ball is incredibly short. On the other hand, the wavelength of the electron, although quite small, happens to be of the same order as the spacing between the atoms in a typical crystal. This is important because, as we will see presently, it provides a means for testing de Broglie's equation.

Diffraction results when light is scattered from a regular array of points or lines. The diffraction of light from a diffraction grating is shown in the accompanying photograph. The colors result because the various wavelengths of visible light are not all scattered in the same way. The colors are "separated," giving the same effect as light passing through a prism. Just as a regular arrangement of ridges and grooves produces diffraction, so does a regular array of atoms or ions in a crystal. For example, when X rays are directed onto a crystal of sodium chloride with its regular array of Na^+ and Cl^- ions, the scattered radiation produces a **diffraction pattern** of bright areas and dark spots on a photographic plate, as shown in Fig. 12.7(a). This pattern occurs because the scattered light can interfere constructively (the peaks and troughs of the beams are in phase) to produce a bright area [Fig. 12.7(b)] or destructively (the peaks and troughs are out of phase) to produce a dark spot [Fig. 12.7(c)].

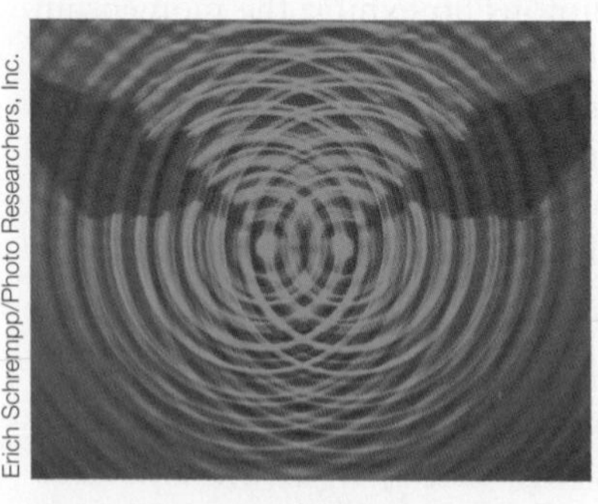
Erich Schrempp/Photo Researchers, Inc.

Constructive and destructive interference with water waves.

A diffraction pattern can be explained only in terms of waves. Thus this phenomenon provides a test for the postulate that particles such as electrons

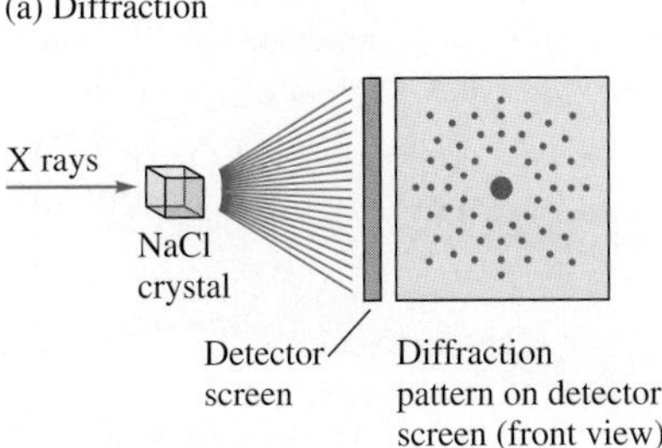

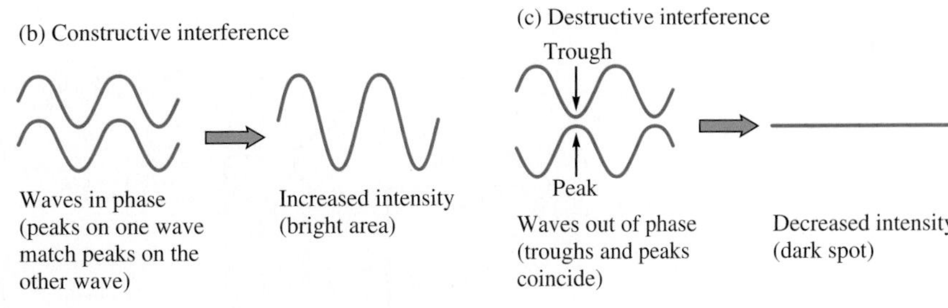

Figure 12.7
(a) Diffraction occurs when electromagnetic radiation is scattered from a regular array of objects, such as the ions in a crystal of sodium chloride. The large spot in the center is from the main incident beam of X rays. (b) Bright areas in the diffraction pattern result from *constructive interference* of waves. The waves are in phase; that is, their peaks match. (c) Dark spots result from *destructive interference* of waves. The waves are out of phase; the peaks of one wave coincide with the troughs of another wave.

have wave properties. As we saw in Example 12.2, an electron with a velocity of 10^7 m/s (easily achieved by acceleration of the electron in an electric field) has a wavelength of about 10^{-10} m, roughly the distance between the components in a typical crystal. This is important because diffraction occurs most efficiently when the spacing between the scattering points is about the same as the wavelength. Thus, if electrons actually do have an associated wavelength, a crystal should diffract electrons. An experiment to test this idea was carried out in 1927 by Davisson and Germer at Bell Laboratories. When they directed a beam of electrons at a nickel crystal, they observed a diffraction pattern similar to that seen from the diffraction of X rays. This result verified de Broglie's relationship, at least for electrons. Larger chunks of matter, such as balls, have wavelengths (Example 12.2) too small to verify experimentally. However, we believe that all matter obeys de Broglie's equation.

Now we have come full circle. Electromagnetic radiation, which at the turn of the twentieth century was thought to be a pure waveform, was found to exhibit particulate properties. Conversely, electrons, which were thought to be particles, were found to have a wavelength associated with them. The significance of these results is that matter and energy are not distinct. Energy is really a form of matter, and all matter shows the same types of properties. That is, *all matter exhibits both particulate and wave properties.* Large "pieces" of matter, such as baseballs, exhibit predominantly particulate properties. The associated wavelength is so small that it is not observed. Very small "pieces" of matter, such as photons, while showing some particulate properties through relativistic effects, exhibit predominantly wave properties. Pieces of matter with intermediate mass, such as electrons, show both the particulate and wave properties of matter.

12.3 The Atomic Spectrum of Hydrogen

Recall from Chapter 2 that key information about the structure of the atom came from several experiments carried out in the early twentieth century. Particularly important were Thomson's discovery of the electron and Rutherford's discovery of the nucleus. Another important experiment concerned the study of the emission of light by excited hydrogen atoms. When a high-energy discharge is passed through a sample of hydrogen gas, the H_2 molecules absorb energy, causing some of the H—H bonds to break. The resulting hydrogen atoms are *excited;* that is, they contain excess energy, which they release by emitting light of various wavelengths to produce what is called the *emission spectrum* of the hydrogen atom.

To understand the significance of the hydrogen emission spectrum, we must first describe the **continuous spectrum** that results when white light is passed through a prism, as shown in Fig. 12.8(a). This spectrum, like the rainbow produced when sunlight is dispersed by raindrops, contains *all* the wave-

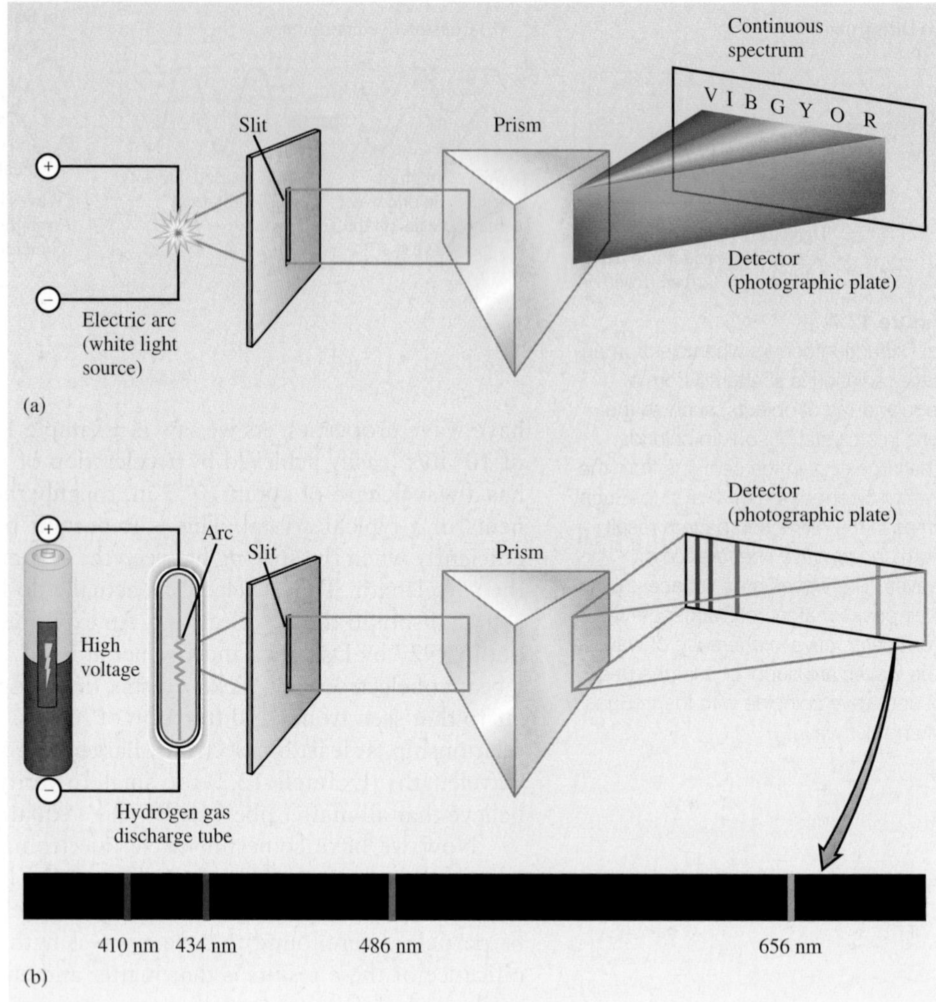

Figure 12.8
(a) A continuous spectrum containing all wavelengths of visible light (indicated by the first letters of the colors of the rainbow). (b) The hydrogen line spectrum contains only a few discrete wavelengths.

lengths of visible light. In contrast, when the hydrogen emission spectrum in the visible region is passed through a prism, as shown in Fig. 12.8(b), we see only a few lines, each corresponding to a discrete wavelength. The hydrogen emission spectrum is called a **line spectrum.**

The energy of the electron in the hydrogen atom is quantized.

What is the significance of the line spectrum of hydrogen? It indicates that *only certain energies are allowed for the electron in the hydrogen atom.* In other words, the energy of the electron in the hydrogen atom is *quantized.* Changes in energy between discrete energy levels in hydrogen produce only certain wavelengths of emitted light, as shown in Fig. 12.9. For example, a given change in energy from a high to a lower level gives a wavelength of light that can be calculated from Einstein's equation:

$$\Delta E = h\nu = \frac{hc}{\lambda}$$

(↑ Change in energy; ↖ Frequency of light emitted; ← Wavelength of light emitted)

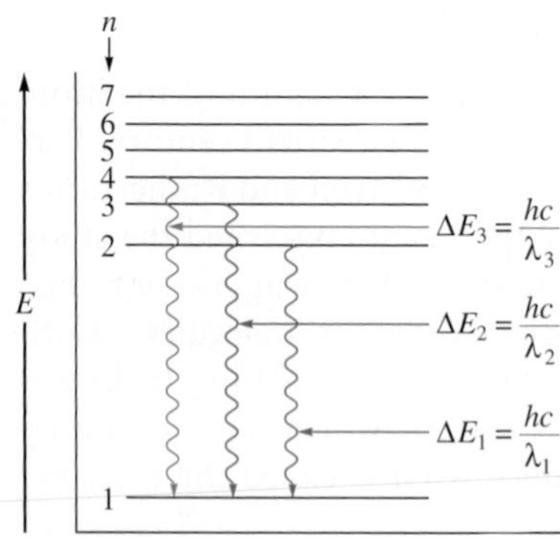

Figure 12.9
A change between two discrete energy levels emits a photon of light.

The discrete line spectrum of hydrogen shows that only certain energies are possible; that is, the electron energy levels are quantized. In contrast, if any energy level were allowed, the emission spectrum would be continuous.

12.4 The Bohr Model

The Granger Collection, New York

Niels Bohr at 37 years of age, photographed in 1922 when he received the Nobel Prize for physics.

In 1913 a Danish physicist named Niels Bohr, aware of most of the experimental results we have just discussed, developed a **quantum model** for the hydrogen atom. Bohr proposed a model that included the idea that the *electron in a hydrogen atom moves around the nucleus only in certain allowed circular orbits.* He calculated the radii for these allowed orbits by using the theories of classical physics and by making some new assumptions.

From classical physics Bohr knew that a particle in motion tends to move in a straight line and can be made to travel in a circle only by application of a force toward the center of the circle. Thus Bohr reasoned that the tendency of the revolving electron to fly off the atom must be exactly balanced by its attraction for the positively charged nucleus. But classical physics also decreed that a charged particle under acceleration should radiate energy. Since an electron revolving around the nucleus constantly changes its direction, it is constantly accelerating. Therefore, the electron should emit light and lose energy and thus be drawn into the nucleus. This conclusion, of course, does not correlate with the existence of stable atoms.

Clearly, an atomic model based solely on the theories of classical physics was untenable. Bohr dealt with the problem of the collapse of the classical atom by simply assuming that the hydrogen electron could exist only in stationary, nonradiating orbits. Ultimately, the correct model had to account for the experimental spectrum of hydrogen, which showed that only certain electron energies were allowed. The experimental data were absolutely clear on this point. Bohr's model fits the experimental results by assuming that the angular momentum of the electron could occur only in certain increments. It wasn't clear why this should be true, but with this assumption Bohr's model gave hydrogen atom energy levels consistent with the hydrogen emission spectrum. The model is represented pictorially in Fig. 12.10.

Angular momentum equals the product of mass, velocity, and orbital radius.

Although we will not discuss its origins here, the expression for the *energy levels available to the electron in the hydrogen atom is*

$$E = -2.178 \times 10^{-18}\,\text{J}\left(\frac{Z^2}{n^2}\right) \tag{12.1}$$

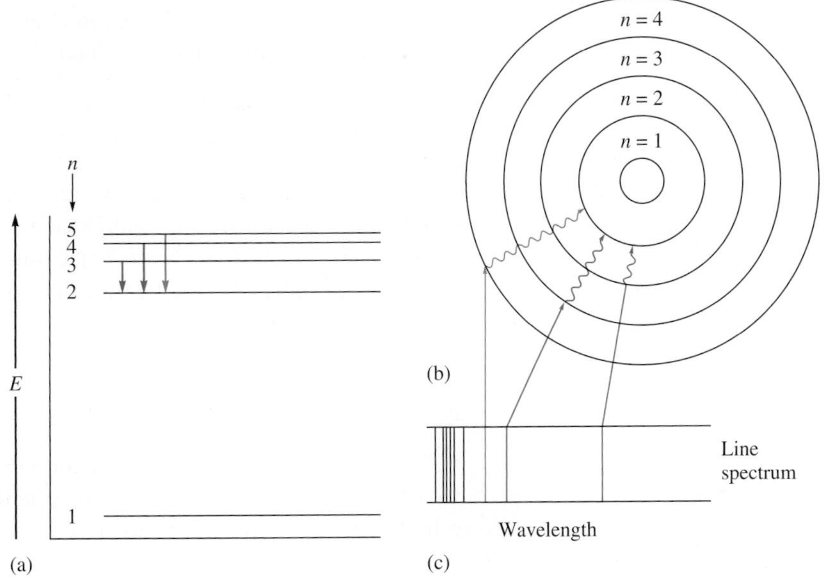

Figure 12.10
Electronic transitions in the Bohr model for the hydrogen atom. (a) An energy-level diagram for electronic transitions. (b) An orbit-transition diagram, which accounts for the experimental spectrum. (Note that the orbits shown are schematic. They are not drawn to scale.) (c) The resulting line spectrum on a photographic plate. Note that the lines in the visible region of the spectrum correspond to transitions from higher levels to the $n = 2$ level.

where n is an integer (the larger the value of n, the larger is the orbit radius) and Z is the atomic number ($Z = 1$ for hydrogen). The negative sign in Equation (12.1) simply means that the energy of the electron bound to the nucleus is lower than it would be if the electron were at an infinite distance ($n = \infty$) from the nucleus, where there is no interaction and the energy is zero:

The "J" in Equation (12.1) stands for joules.

$$E = -2.178 \times 10^{-18}\,\text{J}\left(\frac{Z^2}{\infty}\right) = 0$$

The energy of the electron in any orbit is negative relative to this reference state.

Equation (12.1) applies to all one-electron species. In addition to being used to calculate the energy levels in the hydrogen atom, it can also be used for He^{+} ($Z = 2$), Li^{2+} ($Z = 3$), Be^{3+} ($Z = 4$), and so on.

Equation (12.1) can be used to calculate the change in energy when the electron changes orbits. For example, suppose the electron in level $n = 6$ of an excited hydrogen atom falls back to level $n = 1$ as the hydrogen atom returns to its lowest possible energy state, its **ground state.** We use Equation (12.1) with $Z = 1$ since the hydrogen nucleus contains a single proton. The energies corresponding to the two states are

For $n = 6$: $$E_6 = -2.178 \times 10^{-18}\,\text{J}\left(\frac{1^2}{6^2}\right) = -6.05 \times 10^{-20}\,\text{J}$$

For $n = 1$: $$E_1 = -2.178 \times 10^{-18}\,\text{J}\left(\frac{1^2}{1^2}\right) = -2.178 \times 10^{-18}\,\text{J}$$

Note that for $n = 1$ the electron has a more negative energy than it does for $n = 6$, which means that the electron is more tightly bound in the smallest allowed orbit.

The change in energy ΔE when the electron falls from $n = 6$ to $n = 1$ is

$$\begin{aligned}\Delta E &= \text{energy of final state} - \text{energy of initial state}\\ &= E_1 - E_6 = (-2.178 \times 10^{-18}\,\text{J}) - (-6.05 \times 10^{-20}\,\text{J})\\ &= -2.118 \times 10^{-18}\,\text{J}\end{aligned}$$

The negative sign for the *change* in energy indicates that the atom has *lost* energy and is now in a more stable state. The energy is carried away from the atom by the production (emission) of a photon.

The wavelength of the emitted photon can be calculated from the equation

$$\Delta E = h\left(\frac{c}{\lambda}\right) \quad \text{or} \quad \lambda = \frac{hc}{\Delta E}$$

where ΔE represents the change in energy of the atom and thus equals the energy of the emitted photon. We have

$$\lambda = \frac{hc}{\Delta E} = \frac{(6.626 \times 10^{-34}\,\text{J s})(2.9979 \times 10^{8}\,\text{m/s})}{2.18 \times 10^{-18}\,\text{J}} = 9.379 \times 10^{-8}\,\text{m}$$

Note that for this calculation the absolute value of ΔE is used. In this case the direction of energy flow is indicated by saying that a photon of wavelength 9.379×10^{-8} m has been *emitted* from the hydrogen atom. Simply plugging the negative value of ΔE into the equation would produce a negative value for λ, which is physically meaningless.

OWL INTERACTIVE EXAMPLE 12.3

Calculate the energy required to excite the hydrogen electron from level $n = 1$ to level $n = 2$. Also calculate the wavelength of light that must be absorbed by a hydrogen atom in its ground state to reach this excited state.

Solution Using Equation (12.1) with $Z = 1$, we have

$$E_1 = -2.178 \times 10^{-18}\,\text{J}\left(\frac{1^2}{1^2}\right) = -2.178 \times 10^{-18}\,\text{J}$$

$$E_2 = -2.178 \times 10^{-18}\,\text{J}\left(\frac{1^2}{2^2}\right) = -5.445 \times 10^{-19}\,\text{J}$$

$$\Delta E = E_2 - E_1 = (-5.445 \times 10^{-19}\,\text{J}) - (-2.178 \times 10^{-18}\,\text{J})$$
$$= 1.634 \times 10^{-18}\,\text{J}$$

The positive value for ΔE indicates that the system has gained energy. The wavelength of light that must be *absorbed* to produce this change is

Note from Fig. 12.3 that the light required to produce the transition from the $n = 1$ to $n = 2$ level in hydrogen lies in the ultraviolet region.

$$\lambda = \frac{hc}{\Delta E} = \frac{(6.626 \times 10^{-34}\,\text{J s})(2.9979 \times 10^{8}\,\text{m/s})}{1.634 \times 10^{-18}\,\text{J}}$$
$$= 1.216 \times 10^{-7}\,\text{m}$$

At this time we must emphasize two important points about the Bohr model:

1. The model correctly fits the quantized energy levels of the hydrogen atom as inferred from its emission spectrum. These energy levels correspond to certain allowed circular orbits for the electrons.
2. As the electron becomes more tightly bound, its energy becomes more negative relative to the zero-energy reference state (corresponding to the electron being an infinite distance from the nucleus). That is, as the electron is brought closer to the nucleus, energy is released from the system.

Using Equation (12.1), we can derive a general equation for the electron moving from one level (n_{initial}) to another level (n_{final}):

$$\Delta E = \text{energy of level } n_{\text{final}} - \text{energy of level } n_{\text{initial}}$$
$$= E_{\text{final}} - E_{\text{initial}}$$
$$= (-2.178 \times 10^{-18}\,\text{J})\left(\frac{1^2}{n_{\text{final}}^2}\right) - (-2.178 \times 10^{-18}\,\text{J})\left(\frac{1^2}{n_{\text{initial}}^2}\right)$$
$$= -2.178 \times 10^{-18}\,\text{J}\left(\frac{1}{n_{\text{final}}^2} - \frac{1}{n_{\text{initial}}^2}\right) \quad (12.2)$$

EXAMPLE 12.4

Calculate the minimum energy required to remove the electron from a hydrogen atom in its ground state.

Solution

■ What are we trying to solve?

We want to determine the energy needed to remove a ground state electron from a hydrogen atom. We know how to determine the energy required for the electron moving from one level to another. The equation is

$$\Delta E = -2.178 \times 10^{-18}\,\text{J}\left(\frac{1}{n_{\text{final}}^2} - \frac{1}{n_{\text{initial}}^2}\right)$$

Since the electron starts in the ground state, $n_{\text{initial}} = 1$. To determine ΔE we only need n_{final}. So what is n_{final} if we are removing an electron? Recall that

Chemical Insights — The New, Improved Atomic Clock

To celebrate the turn of the century, the National Institute of Standards and Technology (NIST) in Boulder, Colorado, gave the world a new precision timepiece—a clock so accurate that it will neither gain nor lose a second in 20 million years! Called NIST F-1, the new cesium atomic clock is classified as a fountain clock because it tosses spheres of cesium atoms upward inside the device.

The timepiece works as follows: Gaseous cesium atoms are introduced into a vacuum chamber where six infrared laser beams are positioned to push the cesium atoms into a spherical arrangement. In this process the cesium atoms are slowed almost to the point of absolute zero. Two vertical lasers then "toss" the sphere upward for about a meter in a microwave-filled cavity. At this point all of the lasers are turned off. The sphere of atoms then falls through the cavity under the influence of gravity, and some of the atoms are excited by the microwaves. At the bottom the cesium atoms, which were excited by the microwaves, are induced (by a laser beam) to emit light (a process called fluorescence). The entire process is repeated many times during which the frequency of the microwave energy in the cavity is adjusted until the maximum fluorescence is achieved. The microwave frequency that produces the greatest fluorescence is the natural resonance frequency for the cesium atom. This frequency (9.192631770×10^9 cycles per second) is used to define the second. The NIST F-1 clock is much more accurate than its predecessor, NIST-7, which fired heated cesium atoms at high speeds through a horizontal microwave chamber. In NIST F-1 the slower movement of the atoms gives more time for the microwaves to interact with the atoms so that the characteristic frequency can be more precisely determined.

NIST F-1, which is now the official clock in the United States, provides the extremely accurate timekeeping necessary for modern technology-based operations. NIST F-1 is also in the pool of atomic clocks used to define Coordinated Universal Time (known as UTC), the official world time.

However, the reign of the cesium-based superclocks may be over very soon. A team led by physicist John C. Berquist at NIST has developed an atomic clock, called the NIST optical clock, that is based on the interactions between ultraviolet light and a single mercury atom. Studies by the Berquist group show that the new optical clock is 10 times more precise than the cesium clock. This means that the optical clock would have an error of only 0.1 second in 70 million years of continuous operation. The search is now underway for an atom that might provide even more precise time measurements than mercury.

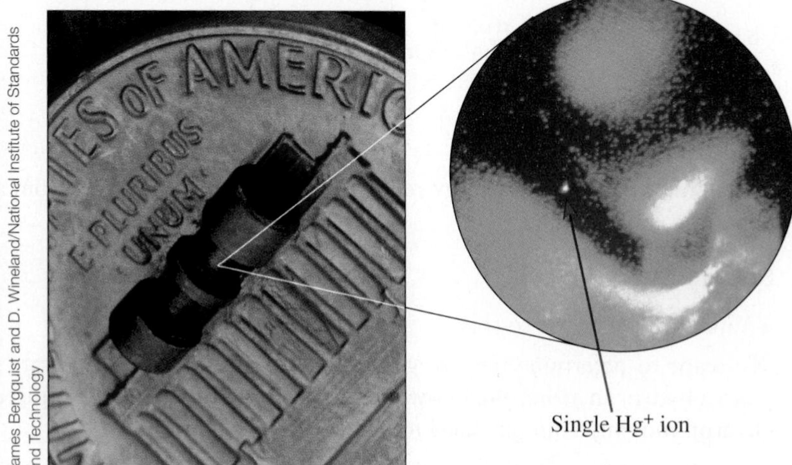

James Bergquist and D. Wineland/National Institute of Standards and Technology

When installed in an atomic clock, this molybdenum structure (left) traps a mercury ion (right) at its center. The clock uses the ion to keep time with unprecedented precision.

$E = 0$ when there is no interaction between the nucleus and the electron. This is true if the electron were at an infinite distance from the nucleus or $n_{\text{final}} = \infty$.

So we have

$$\Delta E = -2.178 \times 10^{-18}\ \text{J}\left(\frac{1}{\infty} - \frac{1}{1^2}\right)$$

$$\Delta E = -2.178 \times 10^{-18}\ \text{J}(0 - 1) = 2.178 \times 10^{-18}\ \text{J}$$

Thus the energy required to remove the electron from a hydrogen atom in its ground state is 2.178×10^{-18} J.

Equation (12.2) can be used to calculate the energy change between *any* two energy levels in a hydrogen atom, as shown in Example 12.4.

Although Bohr's model fits the energy levels for hydrogen, it is a fundamentally incorrect model for the hydrogen atom.

At first, Bohr's model appeared to be very promising. The energy levels calculated by Bohr closely agreed with the values obtained from the hydrogen emission spectrum. However, when Bohr's model was applied to atoms other than hydrogen, it did not work at all. Although some attempts were made to adapt the model using elliptical orbits, it was concluded that Bohr's model is fundamentally incorrect. The model is, however, very important historically because it shows that the observed quantization of energy in atoms can be explained by making rather simple assumptions. Bohr's model paved the way for later theories. It is important to realize, however, that the current theory of atomic structure is in no way derived from the Bohr model. Electrons do *not* move around the nucleus in circular orbits, as we will see later in this chapter.

12.5 The Quantum Mechanical Description of the Atom

By the mid-1920s it had become apparent that the Bohr model was not a valid one. A totally new approach was needed. Three physicists were at the forefront of this effort: Werner Heisenberg, Louis de Broglie, and Erwin Schrödinger. The approach developed by de Broglie and Schrödinger became known as **wave mechanics** or, more commonly, **quantum mechanics.** As we have already seen, de Broglie originated the idea that the electron, previously considered to be a particle, also shows wave properties. Pursuing this line of reasoning, Schrödinger, an Austrian physicist, decided to attack the problem of atomic structure by giving emphasis to the wave properties of the electron. To Schrödinger and de Broglie, the electron bound to the nucleus seemed similar to a standing wave, and they began research on a wave mechanical description of the atom.

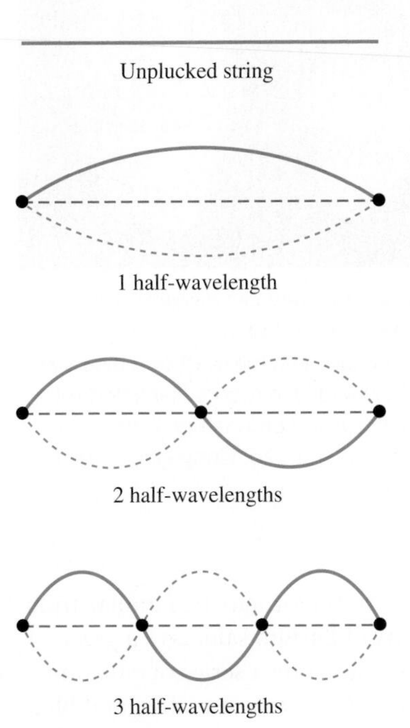

Figure 12.11
The standing wave produced by the vibration of a guitar string fastened at both ends. Each dot represents a node (a point of zero displacement).

The most familiar example of standing waves occurs in association with musical instruments such as guitars or violins, where a string attached at both ends vibrates to produce the musical tone. The waves are described as standing since they are stationary; the waves do not travel along the length of the string. The motions of the string can be explained as a combination of simple waves of the type shown in Fig. 12.11. The dots in this figure indicate the nodes, or points of zero lateral (sideways) displacement for a given wave. Note that there are limitations on the allowed wavelengths of the standing wave. Since each end of the string is fixed, there is always a node at each end. This means that there must be a whole number of *half*-wavelengths in any of the allowed motions of the string (Fig. 12.11).

The wave model was applied by de Broglie to the Bohr atom by imagining the electron in the hydrogen atom to be a standing wave. As shown in

Chemical Insights Fireworks

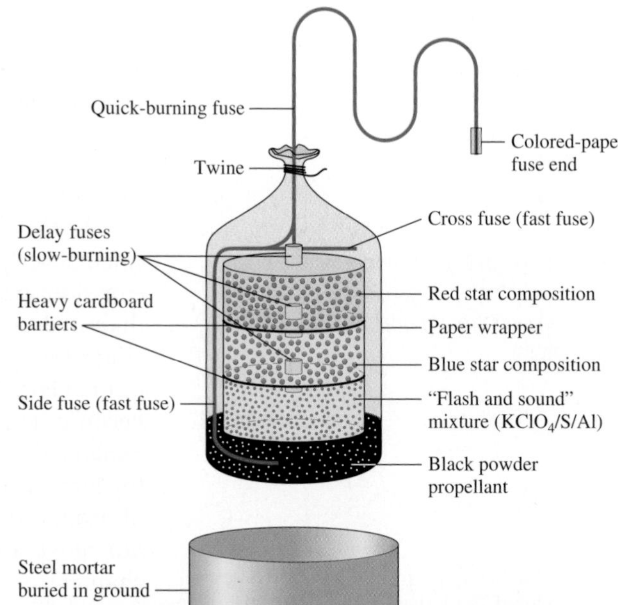

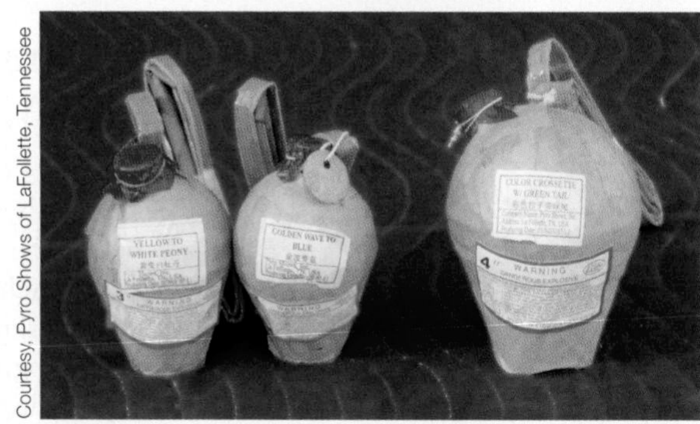
Courtesy, Pyro Shows of LaFollette, Tennessee

A typical aerial shell used in fireworks displays. Time-delayed fuses cause a shell to explode in stages. In this case a red starburst occurs first, followed by a blue starburst, and finally a flash and loud report. Reprinted with permission from *Chemical and Engineering News,* Vol. 59, Issue 26, June 29, 1981, p. 25. Copyright © 1981 American Chemical Society.

The art of using mixtures of chemicals to produce explosives is an ancient one. Black powder—a mixture of potassium nitrate, charcoal, and sulfur—was being used in China well before 1000 A.D. and has been used through the centuries in military explosives, in construction blasting, and in fireworks. The DuPont Company, now a major chemical manufacturer, started out as a manufacturer of black powder. In fact, the founder, Eleuthère du Pont, learned the manufacturing technique from none other than Lavoisier.

Before the nineteenth century, fireworks were mainly confined to rockets and loud bangs. Orange and yellow colors came from the presence of charcoal and iron filings. However, with the great advances in chemistry in the nineteenth century, new compounds found their way into fireworks. Salts of copper, strontium, and barium added brilliant colors. Magnesium and aluminum metals gave a dazzling white light. Fireworks, in fact, have changed very little since then.

How do fireworks produce their brilliant colors and loud bangs? Actually, only a handful of different chemicals are responsible for most of the spectacular effects. The noise and flashes are produced by an oxidizer (an oxidizing agent) and a fuel (a reducing agent). A common mixture involves potassium perchlorate ($KClO_4$) as the oxidizer and aluminum and sulfur as the fuel. The perchlorate compound oxidizes the fuel in a very exothermic reaction, which produces a brilliant flash, caused by the aluminum, and a loud report from the rapidly expanding gases produced. For a color effect, an element with a colored emission spectrum is included. Recall that the electrons in atoms can be raised to higher energy levels when the atoms absorb energy. The excited atoms can then release this excess energy by emitting light of specific wavelengths, often in the visible region. In fireworks the energy to excite the electrons comes from the reaction between the oxidizer and fuel.

Yellow colors in fireworks are caused by the 589-nm emission of sodium atoms. Red colors come from strontium salts emitting at 606 nm and 636–688 nm. (This red color may be familiar from highway safety flares.) Barium salts give a green color in fireworks, caused by a series of emission lines between 505 and 535 nm. A really good blue

Chemicals Commonly Used in the Manufacture of Fireworks

Oxidizers	Fuels	Special Effects
Potassium nitrate	Aluminum	Red flame: strontium nitrate, strontium carbonate
Potassium chlorate	Magnesium	Green flame: barium nitrate, barium chlorate
Potassium perchlorate	Titanium	Blue flame: copper carbonate, copper sulfate, copper oxide
Ammonium perchlorate	Charcoal	Yellow flame: sodium oxalate, cryolite (Na_3AlF_6)
Barium nitrate	Sulfur	White flame: magnesium, aluminum
Barium chlorate	Antimony sulfide	Gold sparks: iron filings, charcoal
Strontium nitrate	Dextrin	White sparks: aluminum, magnesium, aluminum–magnesium alloy, titanium
	Red gum	Whistle effect: potassium benzoate or sodium salicylate
	Polyvinyl chloride	White smoke: mixture of potassium nitrate and sulfur
		Colored smoke: mixture of potassium chlorate, sulfur, and an organic dye

color, however, is hard to obtain. Copper salts give a blue color, emitting in the 420–460-nm region. But difficulties occur because another commonly used oxidizing agent, potassium chlorate ($KClO_3$), reacts with copper salts to form copper chlorate, a highly explosive compound that is dangerous to store. Paris green, a copper salt containing arsenic, was once extensively used but is now considered to be too toxic.

A typical aerial shell is shown in the figure. The shell is launched from a mortar (a steel cylinder), using black powder as the propellant. Time-delayed fuses are used to fire the shell in stages. A list of chemicals commonly used in fireworks is given in the table.

Although you might think that the chemistry of fireworks is simple, the achievement of the vivid white flashes and the brilliant colors requires complex combinations of chemicals. For example, because the white flashes produce high flame temperatures, the colors tend to wash out. Thus oxidizers, such as $KClO_4$, are sometimes used with fuels that produce relatively low flame temperatures. An added difficulty, however, is that perchlorates are very prone to accidental ignition and are therefore quite hazardous. Another problem arises from the use of sodium salts. Because sodium produces an extremely bright yellow emission, sodium salts cannot be used when other colors are desired. Carbon-based fuels also give a yellow flame that masks other colors, therefore limiting the use of organic compounds as fuels. You can see that the manufacture of fireworks that produce the desired effects and are also safe to handle requires careful selection of chemicals. And, of course, there is still the dream of that special deep blue flame.

Carlos E. Santa Maria/Shutterstock.com

Fireworks are a beautiful illustration of chemistry in action.

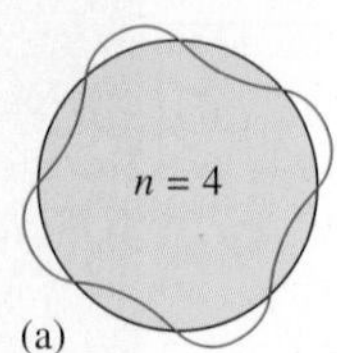

(a)

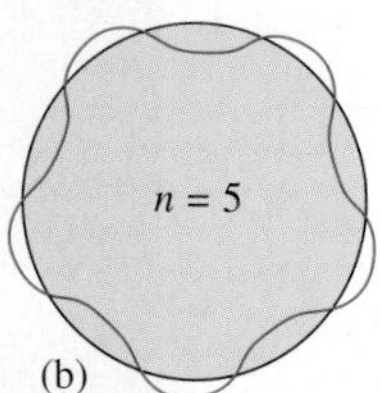

(b)

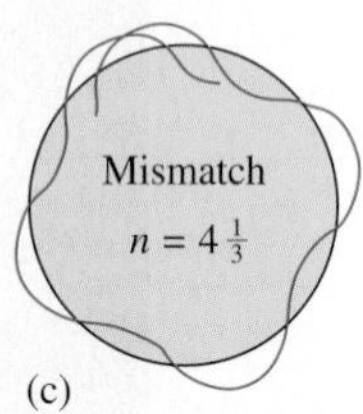

(c)

Figure 12.12
The hydrogen electron visualized as a standing wave around the nucleus. The circumference of a particular circular orbit has to correspond to a whole number of wavelengths, as shown in (a) and (b), or else destructive interference occurs, as shown in (c). This model is consistent with the fact that only certain electron energies are allowed; the atom is quantized. (Although this idea has encouraged scientists to use a wave theory, it does not mean that the electron really travels in circular orbits.)

Fig. 12.12, only certain circular orbits have a circumference into which a whole number of wavelengths of the standing electron wave will "fit." All other orbits would produce destructive interference of the standing electron wave and are not allowed. This seemed like a possible explanation for the observed quantization of the hydrogen atom. The mathematical formalism that Schrödinger developed in 1925 to describe the hydrogen electron as a wave was heavily based on the classical descriptions of wave phenomena. We will first give an overview of this approach before considering it in more detail.

The form of Schrödinger's equation is

$$\hat{H}\psi = E\psi$$

where ψ, called the **wave function,** is a function of the coordinates (x, y, and z) of the electron's position in three-dimensional space, and where $\hat{H}$ represents a set of mathematical instructions called an *operator.* An operator is a mathematical tool that acts on a function to produce another function. In some special cases, the operator gives back the original function simply multiplied by a constant. Note that the Schrödinger equation corresponds to such a special case. In the Schrödinger equation, the operator $\hat{H}$, called the Hamiltonian, acts to give back the wave function multiplied by the constant E, which represents the total energy of the atom (the sum of the potential energy due to the attraction between the proton and the electron and the kinetic energy of the moving electron). When this equation is analyzed, many solutions are found. Each solution consists of a wave function ψ that is characterized by a particular value of E. A specific wave function for a given electron is often called an **orbital.** It is important to recognize that Schrödinger could not be sure that treating an electron as a wave makes any sense: The test would be whether the model could correctly fit the experimental data for hydrogen and other atoms.

To illustrate the most important ideas of the wave mechanical (quantum mechanical) model of the atom, we will first concentrate on the wave function corresponding to the lowest energy for the hydrogen atom. This wave function is called the 1*s* orbital. The first point of interest is the meaning of the word *orbital.* One thing is clear: An orbital is *not* a Bohr orbit. The electron in the hydrogen 1*s* orbital is not moving around the nucleus in a circular orbit. How, then, is the electron moving? The answer is somewhat surprising: *We do not know.* The wave function gives us no information about the movements of the electron. This observation is somewhat disturbing. When we solve problems involving the motions of particles in the macroscopic world, we are able to predict their trajectories. For example, when two billiard balls with known velocities collide, we can predict their motions after the collision. However, we cannot predict the electron's motion using the 1*s* orbital function. Does this mean that the theory is useless? Not necessarily: We have already learned that an electron does not behave much like a billiard ball, so we must examine the situation closely before we discard the theory.

Werner Heisenberg, who was also involved in the development of the quantum mechanical model for the atom, discovered a very important principle in 1927 that helps us to understand the meaning of orbitals—the **Heisenberg uncertainty principle.** Heisenberg's mathematical analysis led him to a surprising conclusion: *There is a fundamental limitation to just how precisely we can know both the position and the momentum of a particle at a given time.* Stated mathematically, the uncertainty principle is

$$\Delta x \cdot \Delta p \geq \frac{\hbar}{2}$$

where Δx is the uncertainty in a particle's position, Δp is the uncertainty in a particle's momentum, and $\hbar$ is Planck's constant divided by 2π ($\hbar = h/2\pi$).

Thus the minimum uncertainty in the product $\Delta x \cdot \Delta p$ is $h/4\pi$. This relationship means that the more precisely we know a particle's position, the less precisely we can know its momentum, and vice versa. This limitation is so small for large particles such as baseballs or billiard balls that it is unnoticed. However, for a small particle such as the electron, the limitation becomes quite important. Applied to the electron, the uncertainty principle implies that we cannot know the exact path of the electron as it moves around the nucleus. It is therefore not appropriate to assume that the electron is moving around the nucleus in a well-defined orbit as in the Bohr model.

EXAMPLE 12.5

The uncertainty principle is sometimes called the indeterminacy principle.

The hydrogen atom has a radius on the order of 0.05 nm. Assuming that we know the position of an electron to an accuracy of 1% of the hydrogen radius, calculate the uncertainty in the velocity of the electron using the Heisenberg uncertainty principle. Then compare this value with the uncertainty in the velocity of a ball of mass 0.2 kg and radius 0.05 m whose position is known to an accuracy of 1% of its radius.

Solution From Heisenberg's uncertainty principle, the smallest possible uncertainty in the product $\Delta x \cdot \Delta p$ is $\hbar/2$; that is,

$$\Delta x \cdot \Delta p = \frac{\hbar}{2} = \frac{h}{4\pi}$$

For the electron the uncertainty in position (Δx) is 1% of 0.05 nm, or

$$\Delta x = (0.01)(0.05 \text{ nm}) = 5. \times 10^{-4} \text{ nm}$$

Converting to meters gives

$$5 \times 10^{-4} \text{ nm} \times \frac{10^{-9} \text{ m}}{1 \text{ nm}} = 5 \times 10^{-13} \text{ m}$$

The values of the constants are

$$m = \text{mass of the electron} = 9.11 \times 10^{-31} \text{ kg}$$

$$h = 6.626 \times 10^{-34} \text{ J s} = 6.626 \times 10^{-34} \frac{\text{kg m}^2}{\text{s}}$$

$$\pi = 3.14$$

We can now solve for the uncertainty in momentum:

$$\Delta p = \frac{\hbar}{2 \cdot \Delta x} = \frac{h}{4\pi \cdot \Delta x} = \frac{6.626 \times 10^{-34} \frac{\text{kg m}^2}{\text{s}}}{4(3.14)(5 \times 10^{-13} \text{ m})}$$

$$= 1.05 \times 10^{-22} \text{ kg m/s} \quad \text{(keeping extra significant figures)}$$

Recalling that $p = mv$ and assuming that the electron mass is constant (ignoring any relativistic corrections), we have

$$\Delta p = \Delta(mv) = m\Delta v$$

and the uncertainty in velocity is

$$\Delta v = \frac{\Delta p}{m} = \frac{1.05 \times 10^{-22} \text{ kg m/s}}{9.11 \times 10^{-31} \text{ kg}} = 1.15 \times 10^{8} \text{ m/s} = 1 \times 10^{8} \text{ m/s}$$

Thus, if we know the electron's position with a minimum uncertainty of 5×10^{-13} m, the uncertainty in the electron's velocity is at least 1×10^{8} m/s. This is a very large number; in fact, it is the same magnitude as the speed of

Chemical Insights

Electrons as Waves

Although scientists talk about the dual wave and particle properties of electrons, many nonscientists still believe that electrons are only tiny particles. Rooted as we are in the macroscopic world, it can be difficult for some to picture a particle as also being a wave. One look at the accompanying picture, however, should help change that. What looks like ripples surrounding two barely submerged pebbles in a pool of water is really the surface of a copper crystal.

Although they are true believers in the wave nature of electrons, the physicists at the IBM Almaden Research Center in San Jose, California, were genuinely surprised when their scanning tunneling microscope (STM) produced this image of the copper surface. "We looked at the surface with all these waves and thought, 'Is our machine broken?'" says Michael Crommie, one of the IBM physicists. But the researchers soon realized that the waves were produced by electrons confined to the metal's surface that bounced off impurities (the two pits). Because the electrons are waves, they form interference patterns after reflecting off the impurities, producing standing waves.

To further explore this behavior, the IBM scientists constructed a "quantum corral" by using their STM to place 48 iron atoms on a copper surface in a circle approximately 14 nm in diameter. Then using the STM to study electron behavior on the copper surface inside the corral, they observed the standing electron waves shown in the photo on the right. This image provides a unique visual confirmation of what the Schrödinger equation predicts. Electrons are wavelike. Seeing is believing!

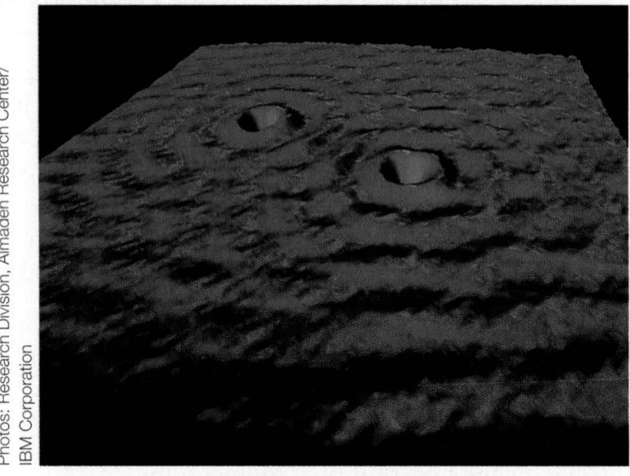

Photos: Research Division, Almaden Research Center/ IBM Corporation

(left) The electrons form interference patterns, the ripples shown here, and produce standing waves.
(right) Iron atoms in a circular "corral" cause electron standing waves.

light (3×10^8 m/s). At this level of uncertainty, we have virtually no idea of the velocity of the electron.

For the ball, the uncertainty in position (Δx) is 1% of 0.05 m, or 5×10^{-4} m. Thus the minimum uncertainty in velocity is

$$\Delta v = \frac{\Delta p}{m} = \frac{h}{\Delta x \cdot m \cdot 4\pi} = \frac{6.626 \times 10^{-34} \dfrac{\text{kg m}^2}{\text{s}}}{(5 \times 10^{-4}\ \text{m})(0.2\ \text{kg})(4)(3.14)}$$
$$= 5 \times 10^{-31}\ \text{m/s}$$

This means there is a very small (undetectable) uncertainty in our measurements of the speed of a ball. Note that this uncertainty is not caused by the limitations of measuring instruments; Δv is an *inherent* uncertainty.

Thus the uncertainty principle is negligible in the world of macroscopic objects but is very important for objects with small masses, such as the electron.

12.6 The Particle in a Box

Overview

The Schrödinger wave equation, $\hat{H}\psi = E\psi$, lies at the heart of the quantum mechanical description of atoms. Recall from the preceding discussion that $\hat{H}$ represents an operator (the Hamiltonian) that "extracts" the total energy E (the sum of the potential and kinetic energies) from the wave function. The wave function ψ depends on the x, y, and z coordinates of the electron's position in space.

Note that the Schrödinger equation requires that when ψ is operated on by $\hat{H}$, the result is ψ multiplied by a constant, E, that represents the total energy of the particular state described by ψ. As we will see, there are many possible solutions to the Schrödinger equation for a given system. For example, for the hydrogen atom there are many functions that satisfy the Schrödinger equation, each one corresponding to a particular energy for hydrogen's electron. Each of these specific wave functions for the hydrogen atom is called an orbital.

Although the detailed solution of the Schrödinger equation for the hydrogen atom is not appropriate in this text, we will illustrate some of the properties of wave mechanics and wave functions by using the wave equation to describe a very simple, hypothetical system commonly called "the particle in a box," a situation in which a particle is trapped in a one-dimensional box that has infinitely high "sides." It is important to recognize that this situation is not an accurate physical model for the hydrogen atom. That is, the hydrogen atom is really not much like this particle in a box. The reasons for treating the particle in a box are that (1) it illustrates the mathematics of wave mechanics, (2) it gives an indication of the characteristics of wave functions, and (3) it shows how energy quantization arises. Thus this treatment of a particle in a box illustrates the "flavor" of the wave mechanical description of the hydrogen atom, but it should not be taken to be an accurate representation of the hydrogen atom itself.

The Particle in a Box as a Model

Consider a particle with mass m that is free to move back and forth along one dimension (we arbitrarily choose x) between the values $x = 0$ and $x = L$ (that is, we are considering a one-dimensional "box" of size L meters). We will assume that the potential energy $V(x)$ of the particle is zero at all points along its path, except at the endpoints $x = 0$ and $x = L$, where $V(x)$ is infinitely large. In effect, we have a repulsive barrier of infinite strength at each end of the box. Thus the particle is trapped in a one-dimensional box with impenetrable walls (Fig. 12.13).

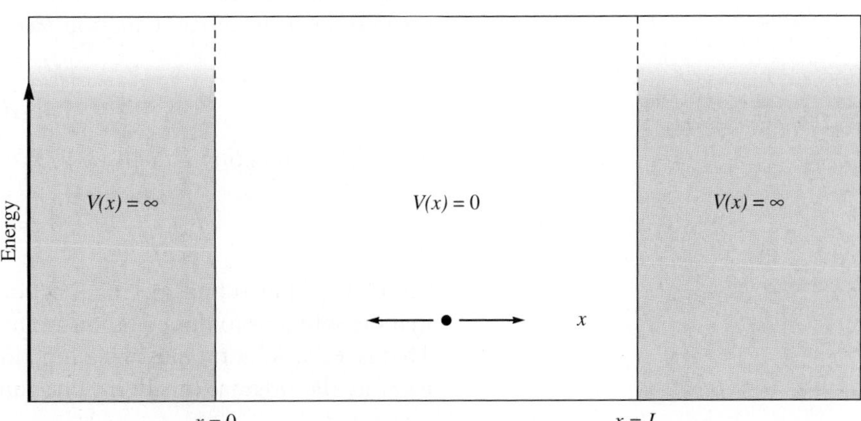

Figure 12.13
A schematic representation of a particle in a one-dimensional box with infinitely high potential walls.

Chemistry Explorers Charles Sykes Researches Surface Architecture

Charles Sykes, Assistant Professor of Chemistry at Tufts University in Medford, Massachusetts, did his undergraduate work at Oxford University (United Kingdom) and obtained his Ph.D. from Cambridge University (United Kingdom). His interests focus on how atoms and molecules interact on surfaces and on building novel, nanoscale surface structures using scanning tunneling microscopy (STM), which enables control of single atoms and molecules. The figure shows a "quantum corral," which is made of approximately 650 gold atoms on a metallic surface. The peaks inside the triangle represent electron density maxima that are caused by constructive interference of the electron waves as they are reflected by the edges of the box. Simple particle-in-a-box calculations can predict these maxima.

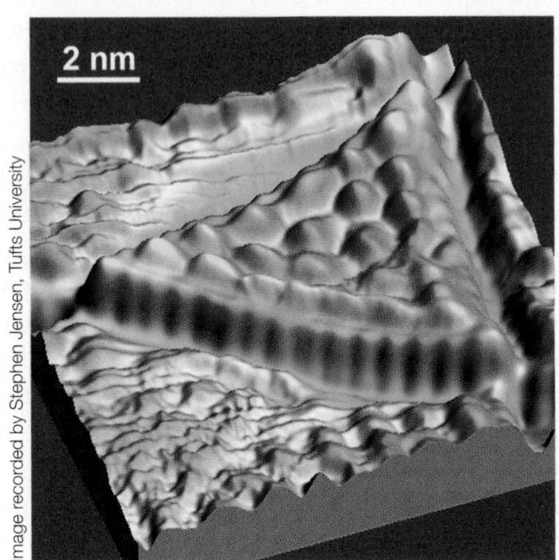

Surface architecture showing electrons forming an interference pattern.

Charles Sykes.

A goal of Professor Sykes's research group is to understand how molecules approaching such a surface will interact with the electron density on the surface and to explore how to tailor surfaces and better assist molecules to self-assemble in new surface architectures. Professor Sykes is also interested in studying the effect of surface structure on the ability of the surface to catalyze various chemical reactions.

As we mentioned before, the Schrödinger equation contains the energy operator $\hat{H}$. In this case, since the potential energy is zero inside the box, the only energy possible is the kinetic energy of the particle as it moves back and forth along the x axis. The operator for this kinetic energy is

$$-\frac{\hbar^2}{2m}\frac{d^2}{dx^2}$$

where $\hbar$ is Planck's constant divided by 2π, m is the mass of the particle, and d^2/dx^2 is the second derivative with respect to x. The form of this operator comes from the description of waves in classical physics. Inserting this operator into the Schrödinger equation $\hat{H}\psi = E\psi$ gives

$$-\frac{\hbar^2}{2m}\frac{d^2\psi}{dx^2} = E\psi$$

where ψ is a function of x $\{[\psi(x)]\}$. We can rearrange this equation to give

$$\frac{d^2\psi}{dx^2} = -\frac{2mE}{\hbar^2}\psi$$

Our goal is to find specific functions $\psi(x)$ that satisfy this equation. Notice that the solutions to this equation are functions such that $d^2\psi/dx^2 = (\text{constant})\psi$. That is, each solution must be a function whose second derivative has the same form as the original function. One function that behaves this way is the sine

Substituting $\hbar = h/2\pi$ gives

$$E = \frac{n^2h^2}{8mL^2} \quad \text{where} \quad n = 1, 2, 3, 4, \ldots$$

Note that this analysis leads to a series of solutions to the Schrödinger equation, where each function corresponds to a given energy state:

n	Function	Energy
1	$\psi_1 = \sqrt{\frac{2}{L}}\sin\left(\frac{\pi}{L}x\right)$	$E_1 = \frac{h^2}{8mL^2}$
2	$\psi_2 = \sqrt{\frac{2}{L}}\sin\left(\frac{2\pi}{L}x\right)$	$E_2 = \frac{4h^2}{8mL^2} = \frac{h^2}{2mL^2}$
3	$\psi_3 = \sqrt{\frac{2}{L}}\sin\left(\frac{3\pi}{L}x\right)$	$E_3 = \frac{9h^2}{8mL^2}$
4	$\psi_4 = \sqrt{\frac{2}{L}}\sin\left(\frac{4\pi}{L}x\right)$	$E_4 = \frac{16h^2}{8mL^2} = \frac{2h^2}{mL^2}$
⋮	⋮	⋮

Notice something very important about these results. The application of the boundary conditions has led to a series of *quantized* energy levels. That is, only certain energies are allowed for the particle bound in the box. This result fits very nicely with the experimental evidence, such as the hydrogen emission spectrum, that nature does not allow continuous energy levels for *bound* systems, as classical physics had led us to expect. Note that the energies are quantized, because the boundary conditions require that n assume only integer values. Consequently, we call n the quantum number for this system.

We can diagram the solutions to the particle-in-a-box problem conveniently by showing a plot of the wave function that corresponds to each energy level. The energy level, wave function, and probability distribution are shown in Fig. 12.14 for the first three levels.

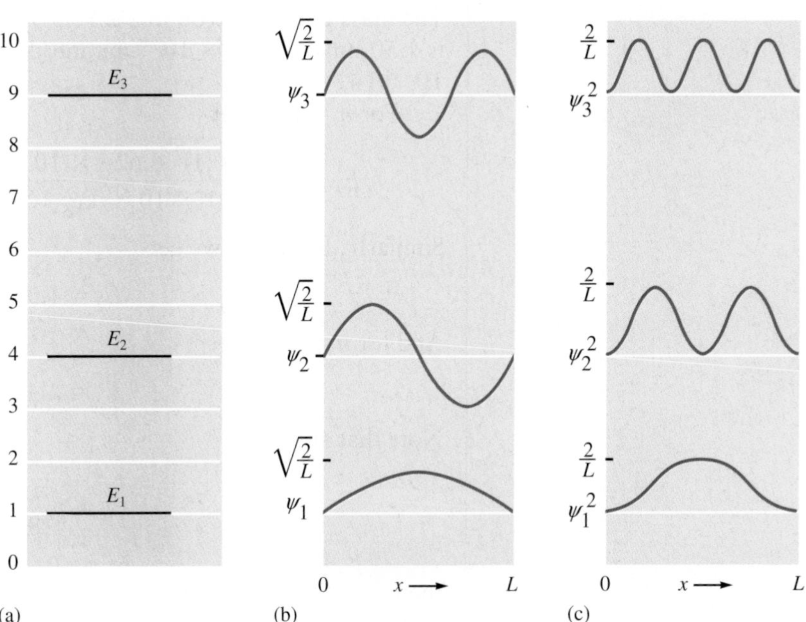

Figure 12.14
(a) The first three energy levels for a particle in a one-dimensional box in increments of $h^2/(8mL^2)$. (b) The wave functions for the first three levels plotted as a function of x. Note that the maximum value is $\sqrt{2/L}$ in each case. (c) The square of the wave functions for the first three levels plotted as a function of x. Note that the maximum value is $2/L$ in each case.

Note that each wave function goes to zero at the edges of the box, as required by the boundary conditions. Another way to say this is that the standing waves that represent the particle must have wavelengths such that an *integral number of half-wavelengths exactly equals the size of the box*. Waves with any other wavelengths could not exist because they would destructively interfere. Also note from Fig. 12.14 that the probability distribution is significantly different for the three levels. For $n = 1$ (the lowest energy or ground state), the particle is most likely to be found near the center of the box. In contrast, for $n = 2$ the particle has zero probability of being found in the center of the box. This zero point is called a node. Notice that the number of nodes increases with n.

Another interesting characteristic of the particle in a box is that the particle cannot have zero energy (that is, n cannot equal zero). For example, if n were equal to zero, ψ_0 would be zero everywhere in the box ($\sin 0 = 0$). This would mean that ψ_0^2 would also be zero. In this case there could be no particle in the box, which contradicts the boundary conditions. This fact that the particle must have a nonzero energy in its ground state is a characteristic of all particles with quantized energies. In addition, for the particle in a box, a value of zero for the energy would mean that the particle was sitting still (zero kinetic energy). This condition would violate the uncertainty principle because we would simultaneously know the exact values of the momentum (zero) and the position of the particle. For similar reasons all quantized particles must possess a minimum energy, often called the *zero-point energy*.

EXAMPLE 12.6

Assume that an electron is confined to a one-dimensional box 1.50 nm in length. Calculate the lowest three energy levels for this electron, and calculate the wavelength of light necessary to promote the electron from the ground state to the first excited state.

Solution To solve this problem, we need to substitute appropriate values into the general expression for energy:

$$E = \frac{n^2h^2}{8mL^2}$$

The mass of an electron (m) is 9.11×10^{-31} kg; the dimension of the box (L) is 1.50 nm, or 1.50×10^{-9} m; and the value of Planck's constant is 6.626×10^{-34} J s.

For $n = 1$ we get

$$E_1 = \frac{(1)^2(6.626 \times 10^{-34}\text{ J s})^2}{(8)(9.11 \times 10^{-31}\text{ kg})(1.50 \times 10^{-9}\text{ m})^2} = 2.68 \times 10^{-20}\text{ J}$$

Similarly, for $n = 2$ we get

$$E_2 = 1.07 \times 10^{-19}\text{ J}$$

And for $n = 3$ we get

$$E_3 = 2.41 \times 10^{-19}\text{ J}$$

Note that since

$$E_n = n^2\frac{h^2}{8mL^2} = n^2E_1$$

then $$E_2 = (2)^2\frac{h^2}{8mL^2} = 4E_1 \quad \text{and} \quad E_3 = 9E_1$$

To calculate the wavelength of light necessary to excite the electron from level 1 to level 2 (the first *excited* state), we first need to obtain the energy difference between the two levels:

$$\Delta E = E_2 - E_1 = (n_2^2 - n_1^2)\frac{h^2}{8mL^2}$$

$$= (3)(2.68 \times 10^{-20}\text{ J}) = 8.04 \times 10^{-20}\text{ J}$$

Then we find the wavelength required from the equation

$$\Delta E = \frac{hc}{\lambda}$$

Inserting the appropriate values gives

$$\lambda = \frac{hc}{\Delta E} = \frac{(6.626 \times 10^{-34}\text{ J s})(2.9979 \times 10^{8}\text{ m/s})}{8.04 \times 10^{-20}\text{ J}}$$

$$= 2.47 \times 10^{-6}\text{ m} = 2470\text{ nm}$$

12.7 The Wave Equation for the Hydrogen Atom

Unlike the particle in a one-dimensional box, the electron of the hydrogen atom moves in three dimensions and has potential energy because of its attraction to the positive nucleus at the atom's center. These differences can be easily accounted for by including the second derivatives with respect to all three of the Cartesian coordinates and by inserting a term that specifies the dependence of the electron's potential energy on its position in space.

Because it is more convenient mathematically, the coordinate system is changed from Cartesian to spherical polar coordinates (Fig. 12.15) before the Schrödinger equation is solved. In the system of spherical polar coordinates, a given point in space, specified by values of the Cartesian coordinates x, y, and z, is described by specific values of r, θ, and ϕ.

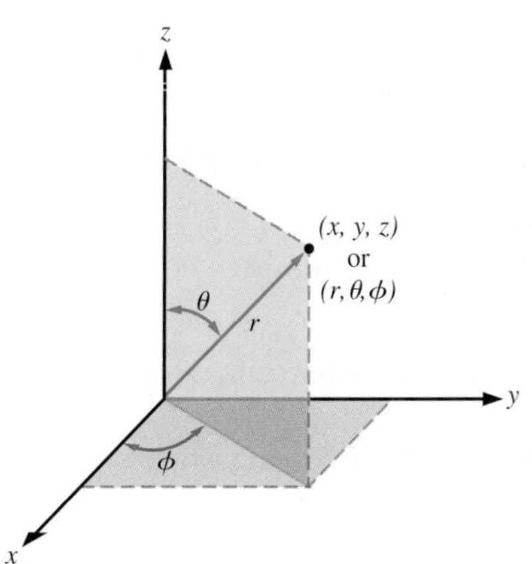

Figure 12.15
The spherical polar coordinate system.

In the spherical polar coordinate system, the wave function $\psi(r, \theta, \phi)$ can be written as a product of one function depending on r, one depending on θ, and one depending on ϕ:

$$\psi(r, \theta, \phi) = R(r)\Theta(\theta)\Phi(\phi)$$

This separation of variables allows an exact solution to the Schrödinger equation

$$\hat{H}\psi = E\psi$$

for the hydrogen atom.

In spherical polar coordinates, the potential energy (in cgs units) of the electron is

$$V(r) = -\frac{(Ze)(e)}{r}$$

To convert the potential energy from cgs units to SI units (joules), the expression shown must be multiplied by $1/4\pi\epsilon_0$, where ϵ_0 (the permittivity of the vacuum) is 8.854×10^{-12} C^2/J m.

where Ze represents the nuclear charge ($Z = 1$ for the hydrogen atom). As with the particle in a box, when the Schrödinger equation for the hydrogen atom is solved and the boundary conditions are applied, a series of wave functions is obtained, each function corresponding to a particular energy. In contrast to the particle in a one-dimensional box, where one quantum number emerges from the mathematics, the three-dimensional hydrogen atom gives rise to three quantum numbers.

Chemical Insights 0.035 Femtometer Is a Big Deal

The long accepted value for the radius of the proton is 0.8768 fm (fm = 10^{-15} m). However, new experiments suggest that this value might be about 4% too large. This doesn't sound like much, but if proved correct it will raise havoc in the scientific community.

The possible problem with the currently accepted value of the proton radius results from work done by Randolf Pohl and his coworkers at the Max Planck Institute of Quantum Optics in Garching, Germany. These scientists created an exotic form of hydrogen in which they replaced the hydrogen electron by a muon. A muon is a particle that has the same charge as an electron but is 200 times more massive than the electron. Because of its greater mass, the muon has orbitals with smaller average radii than those of an electron. The German team attempted to excite the muon to higher energy levels using laser pulses. They set their laser to detect muon transitions assuming the proton radius was in the range from 0.87 to 0.91 femtometers (fm), the value expected from current theories. After years of failing to see the expected energy transitions, the German scientists were within weeks of shutting down the experiment when they decided to assume a smaller value for the proton radius. They saw the expected transition when they tuned their laser at a value corresponding to a proton radius of 0.84184 fm. Although this value doesn't seem too different from 0.8768 fm, it has tremendous implications. A change in the radius of the proton will affect the value of the charge density of the proton, a value that affects the values of all of the fundamental physical constants since all of these constants are interrelated. Thus, more experiments are needed to determine the correct value of the proton radius. If the new, smaller value proves to be correct, it has important implications for the fundamental theory (quantum electrodynamics) of matter. Perhaps this will lead to a better understanding of nature.

The boundary conditions, which differ in some important aspects from those of the particle in the box because of the very different nature of the physical system, will not be discussed here.

The conventional symbols for these quantum numbers are as follows:

n	the principal quantum number
ℓ	the angular momentum quantum number
m_ℓ	the magnetic quantum number

We will have more to say in succeeding sections about what values these quantum numbers can assume and their physical meanings.

The mathematics of wave mechanics leads to the following expression for the allowed energies of hydrogen's electron:

$$E_n = -\frac{Z^2}{n^2}\left(\frac{me^4}{8\epsilon_0{}^2h^2}\right) = -2.178 \times 10^{-18}\,\mathrm{J}\left(\frac{Z^2}{n^2}\right)$$

where $Z = 1$ for hydrogen and where n can assume only integer values (1, 2, 3, . . .). Several characteristics of this equation are worth emphasizing. First, note that the energy of the electron depends only on the principal quantum number (this is true only for one-electron species). Also note that because n is restricted to integer values, hydrogen's electron can assume only discrete energy values—the energy levels are quantized. Finally, note that this is exactly the same equation for energy as obtained in the Bohr model.

So that you have an idea of what they look like, the first few wave functions for hydrogen are shown in Table 12.1, along with the three quantum numbers n, ℓ, and m_ℓ.

When we solve the Schrödinger equation for the hydrogen atom, some of the solutions contain complex numbers (that is, they contain $i = \sqrt{-1}$). Because it is more convenient physically to deal with orbitals that contain only real numbers, the complex orbitals are usually combined (added and sub-

Table 12.1

Solutions of the Schrödinger Wave Equation for a One-Electron Atom

n	ℓ	m_ℓ	Orbital	Solution
1	0	0	$1s$	$\psi_{1s} = \frac{1}{\sqrt{\pi}}\left(\frac{Z}{a_0}\right)^{3/2} e^{-\sigma}$
2	0	0	$2s$	$\psi_{2s} = \frac{1}{4\sqrt{2\pi}}\left(\frac{Z}{a_0}\right)^{3/2} (2 - \sigma)e^{-\sigma/2}$
2	1	0	$2p_z$	$\psi_{2p_z} = \frac{1}{4\sqrt{2\pi}}\left(\frac{Z}{a_0}\right)^{3/2} \sigma e^{-\sigma/2} \cos\theta$
2	1	±1	$2p_x$	$\psi_{2p_x} = \frac{1}{4\sqrt{2\pi}}\left(\frac{Z}{a_0}\right)^{3/2} \sigma e^{-\sigma/2} \sin\theta\cos\phi$
			$2p_y$	$\psi_{2p_y} = \frac{1}{4\sqrt{2\pi}}\left(\frac{Z}{a_0}\right)^{3/2} \sigma e^{-\sigma/2} \sin\theta\sin\phi$
3	0	0	$3s$	$\psi_{3s} = \frac{1}{81\sqrt{3\pi}}\left(\frac{Z}{a_0}\right)^{3/2} (27 - 18\sigma + 2\sigma^2)e^{-\sigma/3}$
3	1	0	$3p_z$	$\psi_{3p_z} = \frac{\sqrt{2}}{81\sqrt{\pi}}\left(\frac{Z}{a_0}\right)^{3/2} (6\sigma - \sigma^2)e^{-\sigma/3} \cos\theta$
3	1	±1	$3p_x$	$\psi_{3p_x} = \frac{\sqrt{2}}{81\sqrt{\pi}}\left(\frac{Z}{a_0}\right)^{3/2} (6\sigma - \sigma^2)e^{-\sigma/3} \sin\theta\cos\phi$
			$3p_y$	$\psi_{3p_y} = \frac{\sqrt{2}}{81\sqrt{\pi}}\left(\frac{Z}{a_0}\right)^{3/2} (6\sigma - \sigma^2)e^{-\sigma/3} \sin\theta\sin\phi$
3	2	0	$3d_{z^2}$	$\psi_{3d_{z^2}} = \frac{1}{81\sqrt{6\pi}}\left(\frac{Z}{a_0}\right)^{3/2} \sigma^2 e^{-\sigma/3} (3\cos^2\theta - 1)$
3	2	±1	$3d_{xz}$	$\psi_{3d_{xz}} = \frac{\sqrt{2}}{81\sqrt{\pi}}\left(\frac{Z}{a_0}\right)^{3/2} \sigma^2 e^{-\sigma/3} \sin\theta\cos\theta\cos\phi$
			$3d_{yz}$	$\psi_{3d_{yz}} = \frac{\sqrt{2}}{81\sqrt{\pi}}\left(\frac{Z}{a_0}\right)^{3/2} \sigma^2 e^{-\sigma/3} \sin\theta\cos\theta\sin\phi$
3	2	±2	$3d_{xy}$	$\psi_{3d_{xy}} = \frac{1}{81\sqrt{2\pi}}\left(\frac{Z}{a_0}\right)^{3/2} \sigma^2 e^{-\sigma/3} \sin^2\theta\sin 2\phi$
			$3d_{x^2-y^2}$	$\psi_{3d_{x^2-y^2}} = \frac{1}{81\sqrt{2\pi}}\left(\frac{Z}{a_0}\right)^{3/2} \sigma^2 e^{-\sigma/3} \sin^2\theta\cos 2\phi$

Note: $\sigma = Zr/a_0$, where $Z = 1$ for hydrogen; $a_0 = \epsilon_0 h^2/\pi m e^2 = 5.29 \times 10^{-11}$ m.

tracted) to remove the complex portions. For example, the p_x and p_y orbitals shown in Table 12.1 are combinations of the complex orbitals that correspond to values of m_ℓ of $+1$ and -1. These orbitals are indicated with a brace in Table 12.1. The last four *d* orbitals listed are also obtained by combination of complex orbitals, as indicated by braces in Table 12.1.

12.8 The Physical Meaning of a Wave Function

Now that we have examined some of the mathematical details of the quantum mechanical treatment of the hydrogen atom, we need to consider what it all means. What is a wave function, and what does it tell us about the electron to which it applies? First, a warning: There is always danger in taking a mathematical description of nature and using our human experiences to interpret it.

Although our attempts to attach physical significance to mathematical descriptions are quite useful to us as we try to understand how nature operates, they must be viewed with caution. Simple pictorial models of a particular natural phenomenon always oversimplify the phenomenon and should not be taken too literally. With this caveat we will proceed to try to picture what the "quantum mechanical atom" is like.

Recall that the uncertainty principle indicates that there is no way of knowing the detailed movements of the electron in a hydrogen atom. Given this severe limitation, what then is the physical meaning of a wave function for an electron? Although the function itself has no easily visualized meaning, as we mentioned in the treatment of the particle in a box, the square of the wave function does have a physical significance. *The square of the function evaluated at a particular point in space indicates the probability of finding an electron near that point.* For example, suppose we have two positions in space: one defined by the coordinates r_1, θ_1, and ϕ_1 and the other defined by the coordinates r_2, θ_2, and ϕ_2. The relative probability of finding the electron near positions 1 and 2 is determined by substituting the values of r, θ, and ϕ for the two positions into the wave function, squaring the function value, and computing the following ratio:

The square of the function here means the square of the magnitude, $|\psi|^2$. This distinction is important when orbitals with complex numbers are being considered: $|\psi|^2$ = (real part)^2 + (imaginary part)^2.

$$\frac{[\psi(r_1, \theta_1, \phi_1)]^2\,dv}{[\psi(r_2, \theta_2, \phi_2)]^2\,dv} = \frac{N_1}{N_2}$$

The quotient N_1/N_2 is the ratio of the probabilities of finding the electron in the infinitesimally small volume elements dv around points 1 and 2. For example, if the value of the ratio N_1/N_2 is 100, the electron is 100 times more likely to be found at position 1 than at position 2. The model gives no information concerning when the electron will be at either position or how it moves between the positions. This vagueness is consistent with the concept of the Heisenberg uncertainty principle.

The square of the wave function is most conveniently represented as a **probability distribution,** in which the intensity of color is used to indicate the probability value at a given point in space. The probability distribution for the hydrogen 1*s* orbital is shown in Fig. 12.16(a). The best way to think about this diagram is as a three-dimensional time exposure, with the electron as a tiny moving light. The more times the electron visits a particular point, the darker the negative becomes. Thus the darkness (intensity) of a point indicates the probability of finding an electron at that position. This diagram is sometimes known as an *electron density map;* electron density and electron probability mean the same thing.

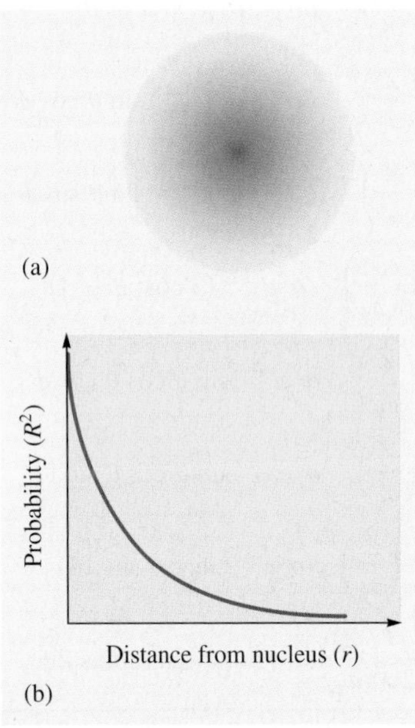

Figure 12.16
(a) The probability distribution for the hydrogen 1*s* orbital in three-dimensional space. (b) The probability density of the electron at points along a line drawn outward from the nucleus in any direction for the hydrogen 1*s* orbital.

Another way of representing the electron probability distribution for the 1*s* orbital is to calculate the probability at points along a line drawn outward in any direction from the nucleus. The result is shown in Fig. 12.16(b), where R^2 (the square of the radial part—the part that depends on r—of the 1*s* orbital) is plotted versus r. Note that the probability of finding the electron at a particular position is greatest close to the nucleus and that it drops off rapidly as the distance from the nucleus increases.

We are also interested in knowing the *total* probability of finding the electron in the hydrogen atom at a particular *distance* from the nucleus. Imagine that the space around the hydrogen nucleus is made up of a series of thin spherical shells (rather like layers in an onion), as shown in Fig. 12.17(a). When the total probability of finding the electron in each spherical shell is plotted versus the distance from the nucleus, the plot in Fig. 12.17(b) is obtained. This graph is called the **radial probability distribution,** which is a plot of $4\pi r^2R^2$ versus r, where R represents the radial part of the wave function.

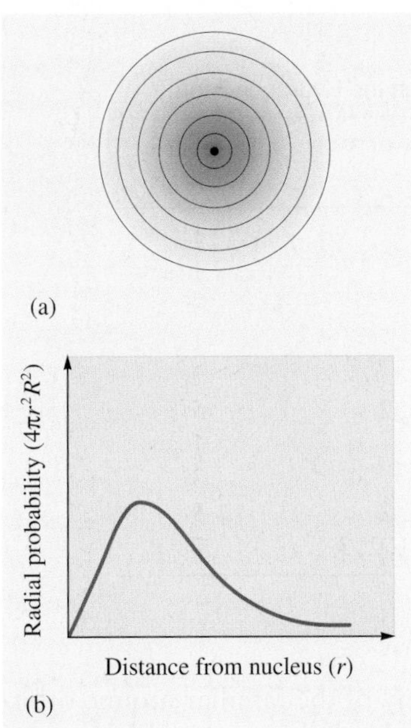

Figure 12.17
(a) Cross section of the hydrogen 1*s* orbital probability distribution divided into successive thin spherical shells. (b) The radial probability distribution. A plot of the total probability of finding the electron in each thin spherical shell as a function of distance from the nucleus.

1 Å = 10^{-10} m; the angstrom is often used as the unit for atomic radius because of its convenient size. Another convenient unit is the picometer (1 pm = 10^{-12} m).

The maximum in the curve occurs because of two opposing effects. The probability of finding an electron at a particular position is greatest near the nucleus, but the volume of the spherical shell increases with the distance from the nucleus. Therefore, as we move away from the nucleus, the probability of finding the electron at a given position decreases. However, we are summing more positions. Thus the total probability increases to a certain radius and then decreases as the electron probability at each position becomes very small. Mathematically, the maximum occurs because in the function $4\pi r^2R^2$, r^2 increases with r while R^2 decreases with r [Fig. 12.16(b)]. For the hydrogen 1*s* orbital, the maximum radial probability (the distance at which the electron is most likely to be found) occurs at a distance of 5.29×10^{-2} nm, or 0.529 Å (angstrom), from the nucleus. Interestingly, this distance is exactly the radius of the innermost orbit in the Bohr model and thus is called the Bohr radius, denoted by a_0. Note that in Bohr's model the electron is assumed to have a circular path and so is *always* found at this distance. In the wave mechanical model the specific electron motions are unknown; therefore, this is the *most probable* distance at which the electron is found.

One more characteristic of the hydrogen 1*s* orbital that we must consider is its size. As we can see from Fig. 12.16, the size of this orbital cannot be precisely defined, since the probability never becomes zero (although it drops to an extremely small value at large values of r). Therefore, the hydrogen 1*s* orbital has no distinct size. However, it is useful to have a definition of relative orbital size. *The normally accepted arbitrary definition of the size of the hydrogen 1s orbital is the radius of the sphere that encloses 90% of the total electron probability.* That is, 90% of the time the electron is found inside this sphere. Application of this rule to the hydrogen atom 1*s* orbital gives a sphere with radius 2.6 a_0, or 1.4×10^{-10} m (140 pm).

So far we have described only the lowest-energy wave function in the hydrogen atom, the 1*s* orbital. Hydrogen has many other orbitals, which are described in the next section.

12.9 The Characteristics of Hydrogen Orbitals

Quantum Numbers

As we have seen, when we solve the Schrödinger equation for the hydrogen atom, we find many wave functions (orbitals) that satisfy it. Each of these orbitals is characterized by a set of **quantum numbers** that arise when the boundary conditions are applied. Now we will systematically describe these quantum numbers in terms of the values they can assume and their physical meanings.

Table 12.2

The Angular Momentum Quantum Numbers and Corresponding Letter Symbols

Value	Letter Used
0	*s*
1	*p*
2	*d*
3	*f*
4	*g*

The **principal quantum number** (n), which can have integral values (1, 2, 3, . . .), is related to the size and energy of the orbital. As n increases, the orbital becomes larger and the electron spends more time farther from the nucleus. An increase in n also means higher energy because the electron is less tightly bound to the nucleus, and the energy is less negative. The **angular momentum quantum number** (ℓ) can have integral values from 0 to $n - 1$ for each value of n. This quantum number relates to the angular momentum of an electron in a given orbital. The dependence of the wave functions on ℓ determines the shapes of the atomic orbitals. The value of ℓ for a particular orbital is commonly assigned a letter: $\ell = 0$ is called *s*, $\ell = 1$ is called *p*, $\ell = 2$ is called *d*, and $\ell = 3$ is called *f* (Table 12.2). The **magnetic quantum number** (m_ℓ) can have integral values between ℓ and $-\ell$, including zero. The value of m_ℓ relates to the orientation in space of the angular momentum associated

Table 12.3

Quantum Numbers for the First Four Levels of Orbitals in the Hydrogen Atom

n	ℓ	Orbital Designation	m_ℓ	Number of Orbitals
1	0	1*s*	0	1
2	0	2*s*	0	1
	1	2*p*	−1, 0, +1	3
3	0	3*s*	0	1
	1	3*p*	−1, 0, 1	3
	2	3*d*	−2, −1, 0, 1, 2	5
4	0	4*s*	0	1
	1	4*p*	−1, 0, 1	3
	2	4*d*	−2, −1, 0, 1, 2	5
	3	4*f*	−3, −2, −1, 0, 1, 2, 3	7

$n = 1, 2, 3, \ldots$
$\ell = 0, 1, \ldots, (n - 1)$
$m_\ell = -\ell, \ldots, 0, \ldots, +\ell$

with the orbital. As we mentioned earlier, many of the familiar atomic orbitals are actually a combination of a complex orbital characterized by m_ℓ and one characterized by $-m_\ell$.

The labels *s*, *p*, *d*, and *f* are used for historical reasons. They originally referred to characteristics of lines observed in the atomic spectra: *s* (sharp), *p* (principal), *d* (diffuse), and *f* (fundamental). Beyond *f* the letters become alphabetic: *g*, *h*, . . . , skipping *j*, which is reserved as a symbol for angular momentum.

OWL INTERACTIVE EXAMPLE 12.7

For principal quantum level $n = 5$, determine the number of subshells (different values of ℓ) and give the designation of each.

Solution For $n = 5$ the allowed values of ℓ run from 0 to 4 ($n - 1 = 5 - 1$). Thus the subshells and their designations are

$\ell = 0$	$\ell = 1$	$\ell = 2$	$\ell = 3$	$\ell = 4$
5*s*	5*p*	5*d*	5*f*	5*g*

Number of Orbitals per Subshell
s = 1
p = 3
d = 5
f = 7
g = 9

The first four levels of orbitals in the hydrogen atom are listed with their quantum numbers in Table 12.3. Note that each set of orbitals with a given value of ℓ (sometimes called a **subshell**) is designated by giving the value of n and the letter for ℓ. Thus an orbital where $n = 2$ and $\ell = 1$ is symbolized as 2*p*. There are three 2*p* orbitals, which have different orientations in space. We will describe these orbitals in the next section.

Orbital Shapes and Energies

We have seen that the meaning of an orbital is illustrated most clearly by a probability distribution. Each orbital in the hydrogen atom has a unique probability distribution. We also have seen that another means of representing an orbital is by the surface that surrounds 90% of the total electron probability. These three types of representations for the hydrogen 1*s*, 2*s*, and 3*s* orbitals are shown in Fig. 12.18. Note the characteristic spherical shape of each of the *s* orbitals. Note also that the 2*s* and 3*s* orbitals contain areas of high probability separated by areas of zero probability. These latter areas are called **nodal surfaces**, or simply **nodes**. The number of nodes increases as n increases. For *s* orbitals, the number of nodes is given by $n - 1$. For our purposes, however, we will think of *s* orbitals only in terms of their overall spherical shape, which becomes larger as the value of n increases.

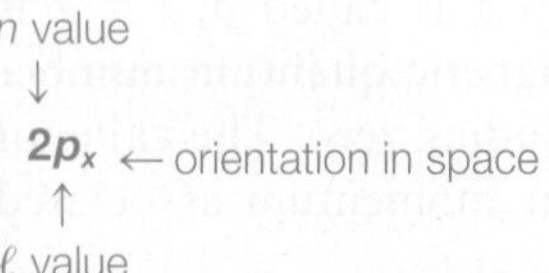

Two types of representations for the 2*p* orbitals (there are no 1*p* orbitals) are shown in Fig. 12.19. Note that the *p* orbitals are not spherical, like *s* orbitals, but have two **lobes** separated by a node at the nucleus. The *p* orbitals are

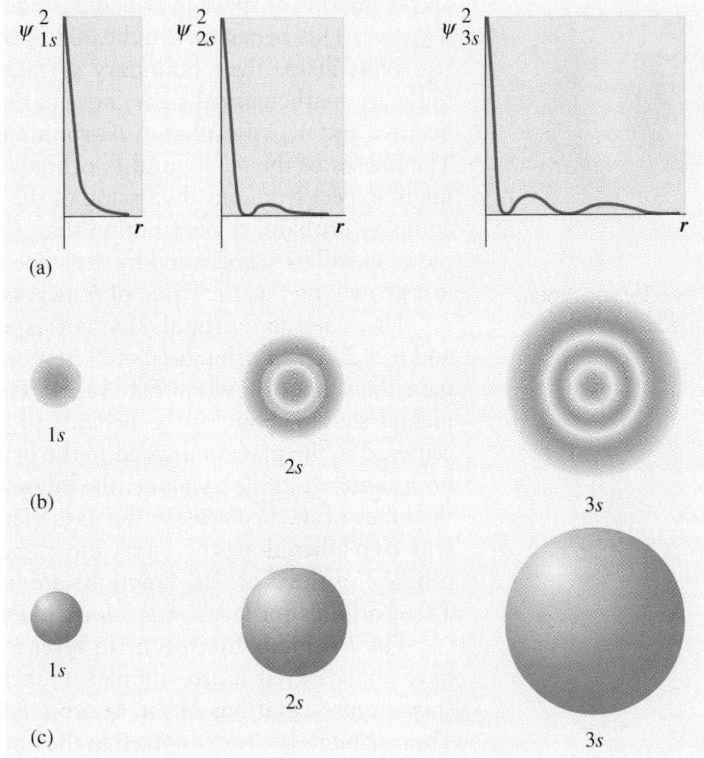

Figure 12.18
Three representations of the hydrogen 1*s*, 2*s*, and 3*s* orbitals. (a) The square of the wave function. (b) "Slices" of the three-dimensional electron density. (c) The surfaces that contain 90% of the total electron probability (the "sizes" of the orbitals).

labeled according to the axis of the Cartesian coordinate system along which the lobes lie. For example, the 2*p* orbital with lobes along the *x* axis is called the $2p_x$ orbital.

At this point it is useful to remember that mathematical functions have signs. For example, a simple sine wave (Fig. 12.1) oscillates from positive to negative and repeats this pattern. Atomic orbital functions also have signs. The functions for *s* orbitals are positive everywhere in three-dimensional space. That is, when the *s* orbital function is evaluated at any point in space, it results in a positive number. In contrast, the *p* orbital functions have different signs in different regions of space. For example, the p_z orbital has a positive sign in

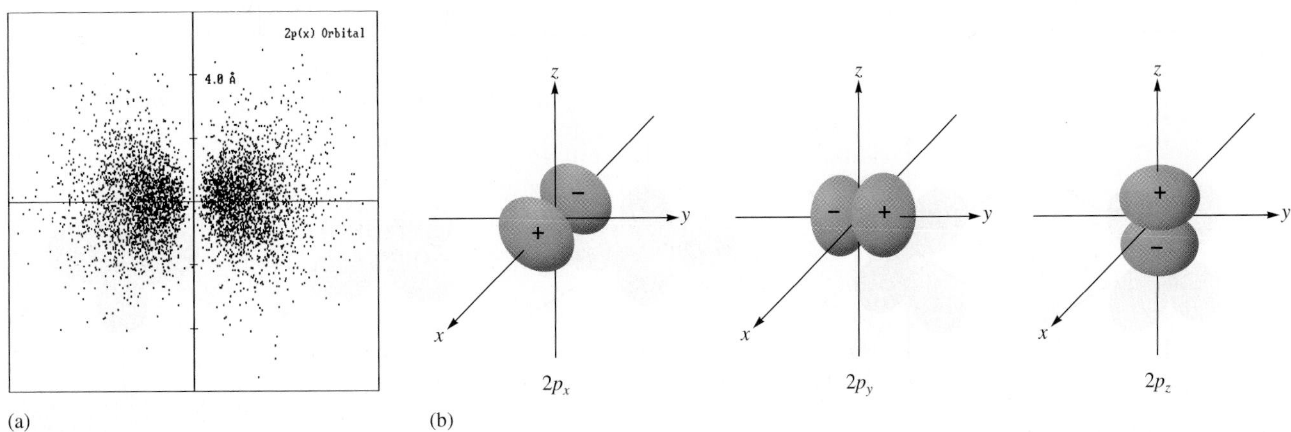

Figure 12.19
Representation of the 2*p* orbitals. (a) The electron probability distribution for a 2*p* orbital. Generated from a program by Robert Allendoerfer on Project SERAPHIM disk PC 2402; reprinted with permission. (b) The boundary surface representations of all three 2*p* orbitals. Note that the signs inside the surface indicate the phases (signs) of the orbital in that region of space.

Figure 12.20
A cross section of the electron probability distribution for a 3*p* orbital.

all the regions of space in which z is positive and has a negative sign when z is negative. This behavior is indicated in Fig. 12.19(b) by the positive and negative signs inside their boundary surfaces. It is important to understand that these are mathematical signs, not charges. Just as a sine wave has alternating positive and negative phases, p orbitals also have positive and negative phases. The phases of the p_x, p_y, and p_z orbitals are indicated in Fig. 12.19(b). As you might expect from our discussion of the s orbitals, the $3p$ orbitals have a more complex probability distribution than that of the $2p$ orbitals (Fig. 12.20), but they can still be represented by the same boundary surface shapes. The surfaces just grow larger as the value of n increases.

There are no d orbitals that correspond to principal quantum levels $n = 1$ and $n = 2$. The d orbitals ($\ell = 2$) first occur in level $n = 3$. The five $3d$ orbitals have the shapes shown in Fig. 12.21. The d orbitals have two different fundamental shapes. Four of the orbitals (d_{xz}, d_{yz}, d_{xy}, and $d_{x^2-y^2}$) have four lobes centered in the plane indicated in the orbital label. Note that d_{xy} and $d_{x^2-y^2}$ are both centered in the xy plane; the lobes of $d_{x^2-y^2}$ lie *along* the x and y axes, but the lobes of d_{xy} lie *between* the axes. The fifth orbital, d_{z^2}, has a unique shape with two lobes along the z axis and a "belt" centered in the xy plane. The signs (phases) of the d orbital functions are indicated inside the boundary surfaces. The d orbitals for levels $n > 3$ look like the $3d$ orbitals but have larger lobes.

The f orbitals first occur in level $n = 4$, and as might be expected, they have shapes even more complex than those of the d orbitals. Figure 12.22 shows representations of the $4f$ orbitals ($\ell = 3$) along with their designations. These orbitals are not involved in the bonding in any of the compounds we will consider in this text. Their shapes and labels are included here for complete-

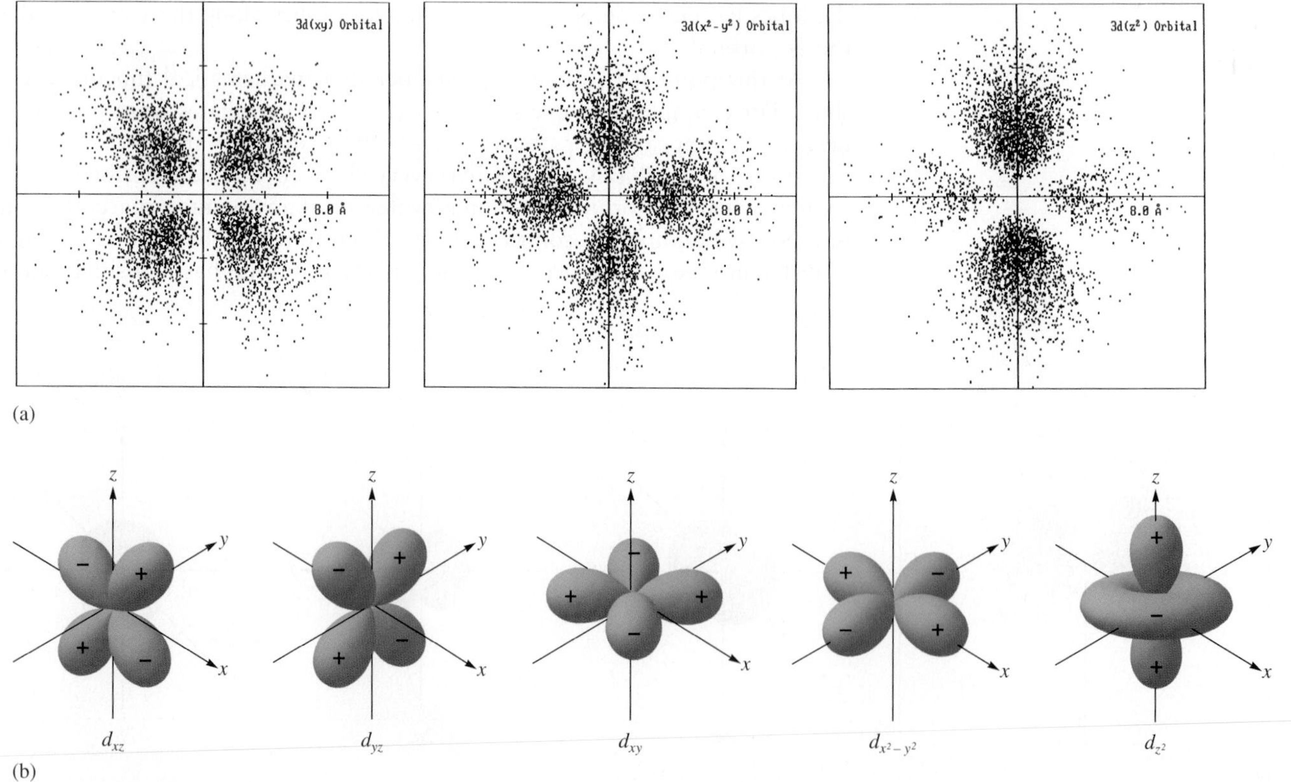

Figure 12.21
Representation of the 3*d* orbitals. (a) Electron density plots of selected 3*d* orbitals. Generated from a program by Robert Allendoerfer on Project SERAPHIM disk PC 2402; reprinted with permission. (b) The boundary surfaces of all five 3*d* orbitals, with the signs (phases) indicated.

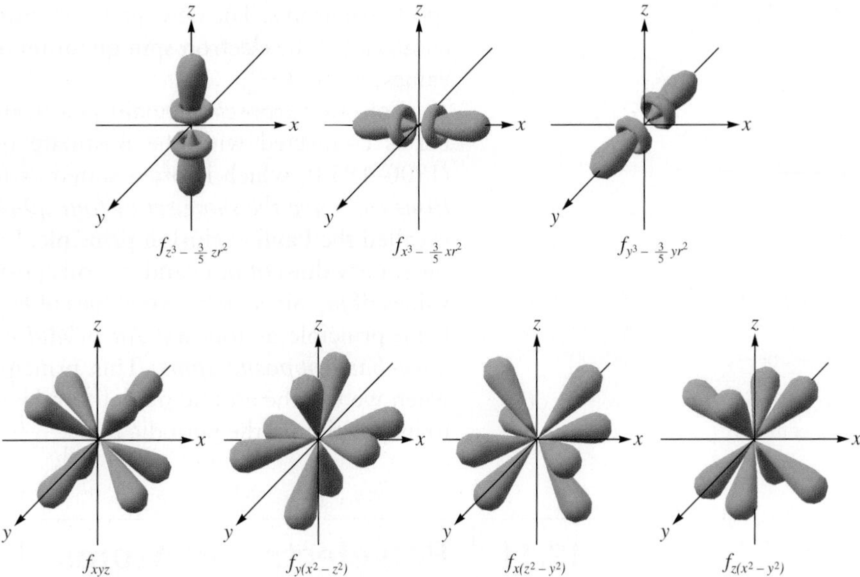

Figure 12.22
Representation of the 4*f* orbitals in terms of their boundary surfaces.

ness. Because of their complexity, the phases of the *f* orbital functions are not represented in this diagram.

So far we have talked about the shapes of the hydrogen atomic orbitals but not about their energies. For the hydrogen atom, the energy of a particular orbital is determined by its value of n. Thus *all* orbitals with the same value of n have the *same energy*—they are said to be **degenerate.** This feature is illustrated in Fig. 12.23, where the energies for the orbitals in the first three quantum levels for hydrogen are shown.

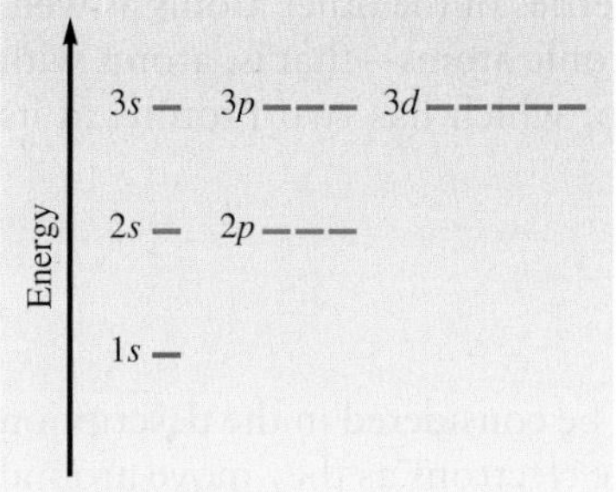

Figure 12.23
Orbital energy levels for the hydrogen atom.

Hydrogen's single electron can occupy any of its atomic orbitals. However, in the lowest energy state, the ground state, the electron resides in the 1*s* orbital. If energy is put into the atom, the electron can be transferred to a higher-energy orbital, producing an excited state.

The Hydrogen Atom

1. In the quantum mechanical model, the electron is described as a wave. This representation leads to a series of wave functions (orbitals) that describe the possible energies and spatial distributions available to the electron.
2. In agreement with the Heisenberg uncertainty principle, the model cannot specify the detailed electron motions. Instead, the square of the wave function represents the probability distribution of the electron in that orbital. This approach allows us to picture orbitals in terms of probability distributions, or electron density maps.
3. The size of an orbital is arbitrarily defined as the surface that contains 90% of the total electron probability.
4. The hydrogen atom has many types of orbitals. In the ground state, the single electron resides in the 1*s* orbital. The electron can be excited to higher-energy orbitals if the atom absorbs energy.

12.10 Electron Spin and the Pauli Principle

The electron does not literally spin. The term "spin" is just a name for the intrinsic angular momentum of the electron.

The concept of **electron spin** was developed by Samuel Goudsmit and George Uhlenbeck in 1925 while they were graduate students at the University of Leyden in the Netherlands. They found that a fourth quantum number (in addition to n, ℓ, and m_ℓ) was necessary to account for the details of the emission spectra of atoms. The new quantum number adopted to describe this phenomenon, called the **electron spin quantum number** (m_s), can have only one of two values, $+\frac{1}{2}$ and $-\frac{1}{2}$.

For our purposes the main significance of the electron spin quantum number is connected with the postulate of Austrian physicist Wolfgang Pauli (1900–1958), which is often stated as follows: *In a given atom no two electrons can have the same set of four quantum numbers* (n, ℓ, m_ℓ, and m_s). This is called the **Pauli exclusion principle.** Since electrons in the same orbital have the same values of n, ℓ, and m_ℓ, this postulate requires that they have different values of m_s. Since only two values of m_s are allowed, we might paraphrase the Pauli principle as follows: *An orbital can hold only two electrons, and they must have opposite spins.* This principle will have important consequences when we use the atomic model to relate the electron arrangement of an atom to its position in the periodic table.

$m_s = +\frac{1}{2}$ or $-\frac{1}{2}$

Each orbital can hold a maximum of two electrons.

12.11 Polyelectronic Atoms

The quantum mechanical model provides a description of the hydrogen atom that agrees very well with experimental data. However, the model would not be very useful if it did not account for the properties of the other atoms as well.

To see how the model applies to **polyelectronic atoms**—that is, atoms with more than one electron—let's consider helium, which has two protons in its nucleus and two electrons:

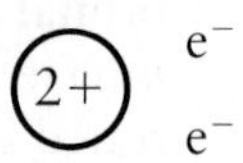

There are three energy contributions that must be considered in the description of the helium atom: (1) the kinetic energy of the electrons as they move around the nucleus, (2) the potential energy of attraction between the nucleus and the electrons, and (3) the potential energy of repulsion between the two electrons.

Although this atom can be readily described in terms of the quantum mechanical model, the Schrödinger equation that results cannot be solved exactly. The difficulty arises in dealing with the repulsion between the electrons. This so-called *electron correlation problem* refers to the fact that we cannot rigorously account for the effect a given electron has on the motions of the other electrons in an atom.

The electron correlation problem occurs with all polyelectronic atoms. To treat these systems using the quantum mechanical model, we must make approximations. The simplest approximation involves treating each electron as if it were moving in a *field of charge that is the net result of the nuclear attraction and the average repulsions of all the other electrons.* To see how this is done, let's compare the neutral helium atom and the He^+ ion:

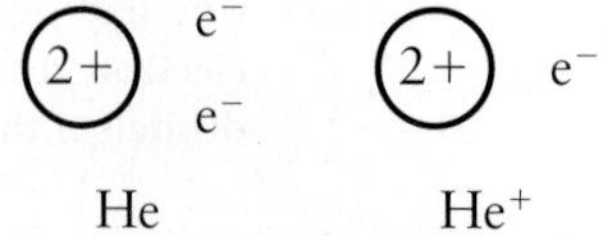

What energy is required to remove an electron from each of these species? Experiments show that 2372 kJ of energy is required to remove one electron from all the atoms in a mole of helium. Removing the one electron from each ion in a mole of He^+ ions requires 5248 kJ of energy. Thus it takes more than twice as much energy to remove an electron from a He^+ ion than from a He atom.

Why is there such a large difference? In both cases the nucleus has a 2+ charge. However, in the helium atom, there are two electrons that repel each other, but in the He^+ ion there is only one electron and thus no electron–electron repulsion. That is, the large difference in the energies required to remove one electron must arise from the electron–electron repulsions in the neutral atom. Each electron in the He atom is much less tightly bound to the nucleus than the electron in the He^+ ion. In other words, the effectiveness of the positively charged nucleus in binding the electrons has been decreased by the repulsions between the electrons. Thus the *effect of the electron repulsions can be thought of as reducing the nuclear charge* to an apparent value of less than 2+ toward a particular electron, as shown below:

(2+) e^- e^-	(Z_{eff}^+) e^-
Actual He atom	Hypothetical He atom

The *apparent* nuclear charge, or the **effective nuclear charge,** is designated Z_{eff}. For a helium atom Z_{eff}, the charge "experienced" by each electron, is less than 2. In general,

$$\text{Effective nuclear charge} = Z_{eff} = Z_{actual} - (\text{effect of electron repulsions})$$

$Z_{eff} = Z -$ effect of electron repulsions

where $Z_{actual} = Z$, the atomic number (number of protons).

This simplification allows us to treat each electron individually, where each electron is viewed as moving under the influence of a positive nuclear charge Z_{eff}. This simplified atom has one electron like hydrogen, but with a positive nuclear charge of Z_{eff} instead of 1. We therefore can find the energy and wave function for each helium electron by substituting Z_{eff} in place of $Z = 1$ in the hydrogen wave mechanical equations. When we do this, we find that both helium electrons reside in a modified 1*s* orbital that is spherical, like that for the hydrogen atom, but smaller because Z_{eff} is greater than 1. The larger nuclear charge draws each of the electrons closer to the nucleus, therefore binding each more tightly than the electron in hydrogen is bound. The increased nuclear charge of the helium atom is more important than the repulsions between the two electrons, so that each of the electrons in helium is bound more tightly than the electron in the hydrogen atom.

The model we have just described so greatly oversimplifies the structure of polyelectronic atoms that, although it produces some qualitatively useful ideas about polyelectronic atoms, it is not satisfactory for the description of quantitative atomic properties. To get an accurate description of polyelectronic atoms, we must take into account the electron–electron interactions in a much more detailed manner than simply assuming that they reduce the nuclear charge.

Nothing we do will allow us to solve the problem exactly, because the electron motions are correlated. That is, because electrons repel each other, the movement of a given electron will affect the movements of all the others. This correlation problem is reflected in the Schrödinger equation for polyelectronic atoms in the following way. Because the equation contains energy terms that simultaneously involve two different electrons, it cannot be separated rigorously into equations that involve only one electron. Thus the Schrödinger equation for polyelectronic atoms cannot be solved exactly.

One approach for dealing with this problem is to solve the equation numerically. That is, a computer is used to find the numerical values of the wave functions at each point in space that produce the lowest overall energy for the atom. Although this approach allows accurate calculation of atomic properties, it suffers from two major disadvantages: It is prohibitively time-consuming for any but the simplest of atoms, and the results are very difficult to interpret physically.

A more practical approach, the **self-consistent field (SCF) method,** is now used almost universally to treat polyelectronic atoms. In this method a given electron is assumed to be moving in a potential energy field that is a result of both the nucleus and the average "electron density" of all the other electrons in the atom (residing in their various orbitals). This approximation allows the many-electron Schrödinger equation to be separated into a set of one-electron equations that can be solved by computers. The orbitals (one-electron functions) that result from this approach have angular properties exactly the same as those of the hydrogen orbitals but have radial characteristics somewhat different from those of the hydrogen orbitals. Although the quantum numbers obtained in the description of the hydrogen atom do not apply exactly to the orbitals obtained from the self-consistent field approach, we still use them as convenient labels for the atomic orbitals in polyelectronic atoms.

We will have more to say later about the self-consistent field approach, but first we will see how the atomic orbitals for polyelectronic atoms can be used to account for the form of the periodic table of the elements.

12.12 The History of the Periodic Table

The modern periodic table contains a tremendous amount of useful information. In this section we will discuss the origin of this valuable tool; later, we will see how the quantum mechanical model for the atom explains the periodicity of chemical properties. Certainly one of the greatest successes of the quantum mechanical model is its ability to account for the arrangement of the elements on the periodic table.

The periodic table was originally constructed to represent the patterns observed in the chemical properties of the elements. As chemistry progressed during the eighteenth and nineteenth centuries, it became evident that the earth is composed of a great many elements with very different properties. Things are much more complicated than the simple model of earth, air, fire, and water suggested by the ancients. At first, the array of elements and properties was bewildering. Gradually, however, patterns were noticed.

The first chemist to recognize patterns was Johann Dobereiner, who found several groups of three elements with similar properties—for example, chlorine, bromine, and iodine. However, as Dobereiner attempted to expand this model of **triads** (as he called them) to the rest of the known elements, it became clear that this concept was severely limited.

The next notable attempt was made by English chemist John Newlands, who in 1864 suggested that elements should be arranged in **octaves.** He noticed that certain properties seemed to repeat for every eighth element in a way similar to the musical scale, which repeats for every eighth tone. Although this model managed to group several elements with similar properties, it was not generally successful.

The present form of the periodic table was conceived independently by two chemists in 1869: German Julius Lothar Meyer and Russian Dmitri Ivanovich Mendeleev (Fig. 12.24). Usually, Mendeleev is given most of the credit because it was he who showed how useful the table could be in predicting the existence and properties of yet unknown elements. For example, in 1872 when Mendeleev first published his table (Fig. 12.25), the elements gallium, scandium, and germanium were unknown. Mendeleev correctly predicted the existence and properties of these elements from gaps in his periodic table. The data for germanium (which Mendeleev called *ekasilicon*) are shown in Table 12.4. Note the excellent agreement between the actual values and Mendeleev's predictions, which were based on the properties of other members in the group of elements similar to germanium.

Image Select/Art Resource, NY

Figure 12.24

Dmitri Ivanovich Mendeleev (1834–1907), born in Siberia as the youngest of 17 children, taught chemistry at the University of St. Petersburg. In 1860 Mendeleev heard Italian chemist Cannizzaro lecture on a reliable method for determining the correct atomic masses of the elements. This important development paved the way for Mendeleev's own brilliant contribution to chemistry—the periodic table. In 1861 Mendeleev returned to St. Petersburg, where he wrote a book on organic chemistry. Later Mendeleev also wrote a book on inorganic chemistry, and he was struck by the fact that the systematic approach characterizing organic chemistry was lacking in inorganic chemistry. In attempting to systematize inorganic chemistry, he eventually arranged the elements in the form of the periodic table.

Mendeleev was a versatile genius who was interested in many fields of science. He worked on many problems associated with Russia's natural resources, such as coal, salt, and various metals. Being particularly interested in the petroleum industry, he visited the United States in 1876 to study the Pennsylvania oil fields. His interests also included meteorology and hot-air balloons. In 1887 he made an ascent in a balloon to study a total eclipse of the sun.

Tabelle II.

Reihen	Gruppe I. — R^2O	Gruppe II. — RO	Gruppe III. — R^2O^3	Gruppe IV. RH^4 RO^2	Gruppe V. RH^3 R^2O^5	Gruppe VI. RH^2 RO^3	Gruppe VII. RH R^2O^7	Gruppe VIII. — RO^4
1	H=1							
2	Li=7	Be=9,4	B=11	C=12	N=14	O=16	F=19	
3	Na=23	Mg=24	Al=27,3	Si=28	P=31	S=32	Cl=35,5	
4	K=39	Ca=40	—=44	Ti=48	V=51	Cr=52	Mn=55	Fe=56, Co=59, Ni=59, Cu=63.
5	(Cu=63)	Zn=65	—=68	—=72	As=75	Se=78	Br=80	
6	Rb=85	Sr=87	?Yt=88	Zr=90	Nb=94	Mo=96	—=100	Ru=104, Rh=104, Pd=106, Ag=108.
7	(Ag=108)	Cd=112	In=113	Sn=118	Sb=122	Te=125	J=127	
8	Cs=133	Ba=137	?Di=138	?Ce=140	—	—	—	— — — —
9	(—)	—	—	—	—	—	—	
10	—	—	?Er=178	?La=180	Ta=182	W=184	—	Os=195, Ir=197, Pt=198, Au=199.
11	(Au=199)	Hg=200	Tl=204	Pb=207	Bi=208	—	—	
12	—	—	—	Th=231	—	U=240	—	— — — —

Annalen der Chemie und Pharmacia, VIII, Supplementary Volume for 1872

Figure 12.25

Mendeleev's early periodic table, published in 1872. Note the spaces left for missing elements with atomic masses 44, 68, 72, and 100. From *Annalen der Chemie und Pharmacia,* VIII, Supplementary Volume for 1872, page 511.

Table 12.4

Comparison of the Properties of Germanium as Predicted by Mendeleev and as Actually Observed

Properties of Germanium	Predicted in 1871	Observed in 1886
Atomic mass	72	72.3
Density	5.5 g/cm^3	5.47 g/cm^3
Specific heat	0.31 J °C^{-1} g^{-1}	0.32 J °C^{-1} g^{-1}
Melting point	Very high	960°C
Oxide formula	RO_2	GeO_2
Oxide density	4.7 g/cm^3	4.70 g/cm^3
Chloride formula	RCl_4	$GeCl_4$
Boiling point of chloride	100°C	86°C

Table 12.5

Predicted Properties of Elements 113 and 114

Property	Element 113	Element 114
Chemically like	Thallium	Lead
Atomic mass	297	298
Density	16 g/mL	14 g/mL
Melting point	430°C	70°C
Boiling point	1100°C	150°C

Using his table, Mendeleev was also able to correct several values of atomic masses. For example, the original atomic mass of 76 for indium was based on the assumption that indium oxide had the formula InO. This atomic mass placed indium, which has metallic properties, among the nonmetals. Mendeleev assumed that the atomic mass was probably incorrect and proposed that the formula of indium oxide was really In_2O_3. On the basis of this (correct) formula, indium has an atomic mass of about 113, placing the element among the metals. Mendeleev also corrected the atomic masses of beryllium and uranium.

Because of its obvious usefulness, Mendeleev's periodic table was almost universally adopted, and it remains one of the most valuable tools at the chemist's disposal. For example, it is still used to predict the properties of elements yet to be discovered, as shown in Table 12.5.

A current version of the periodic table is shown inside the front cover of this book. The only fundamental difference between this table and that of Mendeleev is that the current table lists the elements in order of atomic number rather than atomic mass. The reason for this will become clear later in this chapter as we explore the electron arrangements of the atom.

12.13 The Aufbau Principle and the Periodic Table

We can use the quantum mechanical model of the atom to show how the electron arrangements in the atomic orbitals of the various atoms account for the organization of the periodic table. Our main assumption here is that all atoms have orbitals similar to those that have been described for the hydrogen atom. *As protons are added one by one to the nucleus to build up the elements, electrons are similarly added to these atomic orbitals.* This is called the **aufbau principle.**

Aufbau is German for "building up."

Hydrogen has one electron, which occupies the 1*s* orbital in its ground state. The configuration for hydrogen is written as $1s^1$, which can be represented by the following *orbital diagram:*

		1*s*	2*s*	2*p*
H:	$1s^1$	[↑]	[ ]	[ ][ ][ ]

The arrow represents a particular electron spin state.

The next element, *helium*, has two electrons. Since two electrons with opposite spins can occupy an orbital, according to the Pauli exclusion principle, the electrons for helium are in the 1*s* orbital with opposite spins. This yields a $1s^2$ configuration:

		1*s*	2*s*	2*p*
He:	$1s^2$	[↑↓]	[ ]	[ ][ ][ ]

We will see in Section 12.14 why the 2*s* orbital is lower in energy than the 2*p* orbital.

Lithium has three electrons, two of which can go into the 1*s* orbital before the orbital is filled. Since the 1*s* orbital is the only orbital with $n = 1$, the third

The Chemistry of Copernicium

Chemical Insights

Although element 112 is not exactly a newborn (it was discovered in 1996), it has just been given a name. The International Union of Pure and Applied Chemistry announced on February 19, 2010, that the official name of 112 is Copernicium (pronounced koh-pur-NEE-see-um) and has the symbol Cn. The name honors Nicolaus Copernicus, who is best known for first recognizing that the planetary system is sun-centered rather than Earth-centered.

Now scientists from the Paul Sherrer Institute in Switzerland and the Joint Institute for Nuclear Research in Russia have studied the properties of Cn. By firing a beam of ^{48}Ca projectiles at a target of ^{242}Pu (doped with Nd_2O_3), these scientists have produced just two atoms of Cn. Given the relatively long lifetime of Cn atoms (a few seconds), the researchers have been able to compare the properties of Cn, Hg, and Ar atoms. When these atoms were injected into a series of gold-covered detectors, it was found that, similar to Hg (and in contrast to Ar), Cn atoms were "mildly" volatile and readily formed bonds with gold. Thus, as predicted by the periodic table, the element Cn, which has an expected electron configuration of $[Rn]7s^2 6d^{10}$, fits with the Zn, Cd, and Hg group of elements, all of which have the $ns^2(n-1)d^{10}$ configuration.

electron will occupy the lowest-energy orbital with $n = 2$, or the 2*s* orbital, giving a $1s^2 2s^1$ configuration:

		1*s*	2*s*	2*p*
Li:	$1s^2 2s^1$	[↑↓]	[↑]	[ ][ ][ ]

The next element, *beryllium,* has four electrons, which occupy the 1*s* and 2*s* orbitals:

		1*s*	2*s*	2*p*
Be:	$1s^2 2s^2$	[↑↓]	[↑↓]	[ ][ ][ ]

Boron has five electrons, four of which occupy the 1*s* and 2*s* orbitals. The fifth electron goes into the second type of orbital with $n = 2$, the 2*p* orbitals:

		1*s*	2*s*	2*p*
B:	$1s^2 2s^2 2p^1$	[↑↓]	[↑↓]	[↑][ ][ ]

Since all the 2*p* orbitals are equivalent, it does not matter which 2*p* orbital the electron occupies.

Carbon has six electrons: Two electrons occupy the 1*s* orbital, two occupy the 2*s* orbital, and two occupy 2*p* orbitals. Since there are three equivalent 2*p* orbitals, the electrons will occupy *separate* 2*p* orbitals.

For an atom with unfilled subshells, the lowest energy is achieved by electrons occupying separate orbitals, as allowed by the Pauli exclusion principle.

This behavior is summarized by **Hund's rule** (named for German physicist F. H. Hund), which states that *the lowest-energy configuration for an atom is the one having the maximum number of unpaired electrons allowed by the Pauli principle in a particular set of degenerate orbitals.*

The configuration for carbon could be written $1s^2 2s^2 2p^1 2p^1$ to indicate that the electrons occupy separate 2*p* orbitals. However, the configuration is usually given as $1s^2 2s^2 2p^2$, and it is understood that the electrons are in different 2*p* orbitals. The orbital diagram for carbon is

		1*s*	2*s*	2*p*
C:	$1s^2 2s^2 2p^2$	[↑↓]	[↑↓]	[↑][↑][ ]

Note the unpaired electrons in the 2*p* orbitals, as required by Hund's rule.

The configuration for *nitrogen,* which has seven electrons, is $1s^22s^22p^3$. The three electrons in $2p$ orbitals occupy separate orbitals:

		$1s$	$2s$	$2p$
N:	$1s^22s^22p^3$	⇅	⇅	↑ ↑ ↑

The configuration for *oxygen,* which has eight electrons, is $1s^22s^22p^4$. One of the $2p$ orbitals is now occupied by a pair of electrons with opposite spins, as required by the Pauli exclusion principle:

		$1s$	$2s$	$2p$
O:	$1s^22s^22p^4$	⇅	⇅	⇅ ↑ ↑

The orbital diagrams and electron configurations for *fluorine* (nine electrons) and *neon* (ten electrons) are given below:

		$1s$	$2s$	$2p$
F:	$1s^22s^22p^5$	⇅	⇅	⇅ ⇅ ↑
Ne:	$1s^22s^22p^6$	⇅	⇅	⇅ ⇅ ⇅

With neon, the orbitals with $n = 1$ and $n = 2$ are now completely filled.

For *sodium* the first ten electrons occupy the $1s$, $2s$, and $2p$ orbitals, and the eleventh electron must occupy the first orbital with $n = 3$, the $3s$ orbital. The electron configuration for sodium is $1s^22s^22p^63s^1$. To avoid writing the inner-level electrons, we often abbreviate this configuration as [Ne]$3s^1$, where [Ne] represents the electron configuration of neon, $1s^22s^22p^6$.

[Ne] is shorthand for $1s^22s^22p^6$.

The next element, *magnesium,* has the configuration $1s^22s^22p^63s^2$, or [Ne]$3s^2$. Then the next six elements, *aluminum* through *argon,* have configurations obtained by filling the $3p$ orbitals one electron at a time. Figure 12.26 summarizes the electron configurations of the first 18 elements by giving the number of electrons in the type of orbital occupied last.

At this point it is useful to introduce the concept of **valence electrons,** *the electrons in the outermost principal quantum level of an atom.* The valence electrons of the nitrogen atom, for example, are the $2s$ and $2p$ electrons. For the sodium atom, the valence electron is the electron in the $3s$ orbital, and so on. Valence electrons are the most important electrons to chemists because they are involved in bonding, as we will see in the next two chapters. The inner electrons are known as **core electrons.**

Note in Fig. 12.26 that a very important pattern is developing: *The elements in the same group (vertical column of the periodic table) have the same valence electron configuration.* Remember that Mendeleev originally placed the elements in groups based on similarities in chemical properties. Now we understand the reason behind these groupings. Elements with the same valence electron configuration often show similar chemical behavior.

The element after argon is *potassium.* Since the $3p$ orbitals are fully occupied in argon, we might expect the next electron to go into a $3d$ orbital (recall

Figure 12.26
The electron configurations in the type of orbital occupied last for the first 18 elements.

H $1s^1$							He $1s^2$
Li $2s^1$	Be $2s^2$	B $2p^1$	C $2p^2$	N $2p^3$	O $2p^4$	F $2p^5$	Ne $2p^6$
Na $3s^1$	Mg $3s^2$	Al $3p^1$	Si $3p^2$	P $3p^3$	S $3p^4$	Cl $3p^5$	Ar $3p^6$

that for $n = 3$ the orbitals are 3*s*, 3*p*, and 3*d*). However, the chemistry of potassium is clearly very similar to that of lithium and sodium, indicating that the last electron in potassium occupies the 4*s* orbital instead of one of the 3*d* orbitals, a conclusion confirmed by many types of experiments. The electron configuration of potassium is

[Ar] is shorthand for $1s^22s^22p^63s^23p^6$.

$$\text{K:} \quad 1s^22s^22p^63s^23p^64s^1 \quad \text{or} \quad [\text{Ar}]4s^1$$

The next element is *calcium:*

$$\text{Ca:} \quad [\text{Ar}]4s^2$$

The next element, *scandium,* begins a series of ten elements (scandium through zinc) called the **transition metals,** whose configurations are obtained by adding electrons to the five 3*d* orbitals. The configuration of scandium is

$$\text{Sc:} \quad [\text{Ar}]4s^23d^1$$

When an electron configuration is given in this text, the orbitals are listed in the order in which they fill.

That of *titanium* is Ti: $[\text{Ar}]4s^23d^2$

And that of *vanadium* is V: $[\text{Ar}]4s^23d^3$

Chromium is the next element. The expected configuration is $[\text{Ar}]4s^23d^4$. However, the observed configuration is

$$\text{Cr:} \quad [\text{Ar}]4s^13d^5$$

The explanation for this configuration of chromium is beyond the scope of this book. In fact, chemists are still disagreeing over the exact cause of this anomaly.* Note, however, that the observed configuration has both the 4*s* and 3*d* orbitals half-filled. This is a good way to remember the correct configuration.

The next four elements, *manganese* through *nickel,* have the expected configurations:

Mn: $[\text{Ar}]4s^23d^5$ Co: $[\text{Ar}]4s^23d^7$

Fe: $[\text{Ar}]4s^23d^6$ Ni: $[\text{Ar}]4s^23d^8$

The configuration for *copper* is expected to be $[\text{Ar}]4s^23d^9$. However, the observed configuration is

$$\text{Cu:} \quad [\text{Ar}]4s^13d^{10}$$

In this case a half-filled 4*s* orbital and a filled set of 3*d* orbitals characterize the actual configuration.

Zinc has the expected configuration:

$$\text{Zn:} \quad [\text{Ar}]4s^23d^{10}$$

The configurations of the transition metals are shown in Fig. 12.27. The next six elements, *gallium* through *krypton,* have configurations that correspond to filling the 4*p* orbitals (Fig. 12.27).

The entire periodic table is represented in Fig. 12.28 in terms of which orbitals are being filled. The valence electron configurations are given in Fig. 12.29 on page 567.

From these two figures, note the following additional points:

The $(n + 1)s$ orbitals fill before the *nd* orbitals.

1. The $(n + 1)s$ orbitals always fill before the *nd* orbitals. For example, the 5*s* orbitals fill in rubidium and strontium before the 4*d* orbitals fill in the second row of transition metals (yttrium through cadmium).

Lanthanides are elements in which the 4*f* orbitals are being filled.

2. After lanthanum, which has the configuration $[\text{Xe}]6s^25d^1$, a group of 14 elements called the **lanthanide series,** or the lanthanides, occurs. This series

---

*See M. P. Melrose and E. R. Scerri, *J. Chem. Educ.* 73 (1996): 498, for more information.

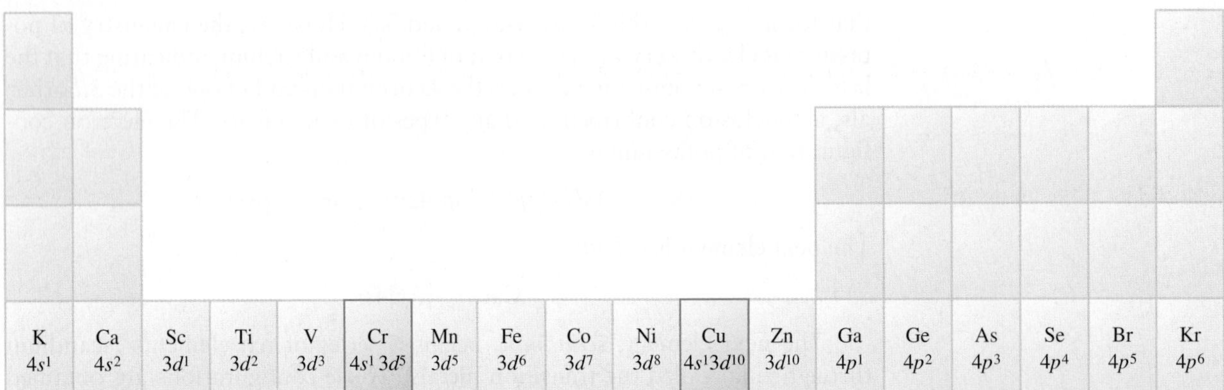

Figure 12.27
Electron configurations for potassium through krypton. The transition metals (scandium through zinc) have the general configuration $[Ar]4s^2 3d^n$, except for chromium and copper.

of elements corresponds to the filling of the seven 4*f* orbitals. Note that sometimes one electron occupies a 5*d* instead of a 4*f* orbital. This occurs because the energies of the 4*f* and 5*d* orbitals are very similar.

Actinides are elements in which the 5*f* orbitals are being filled.

3. After actinium, which has the configuration $[Rn]7s^2 6d^1$, a group of 14 elements called the **actinide series,** or the actinides, occurs. This series corresponds to the filling of the seven 5*f* orbitals. Note that sometimes one or two electrons occupy the 6*d* orbitals instead of the 5*f* orbitals because these orbitals have very similar energies.

The group label tells the total number of valence electrons for that group.

4. The group labels for the Groups 1A, 2A, 3A, 4A, 5A, 6A, 7A, and 8A indicate the *total number* of valence electrons for the atoms in these groups. For example, all the elements in Group 5A have the configuration ns^2np^3. (The *d* electrons fill one period late and are usually not counted as valence electrons.) The meaning of the group labels for the transition metals is not as clear as for the A group elements, so these will not be used in this text.
5. The groups labeled 1A, 2A, 3A, 4A, 5A, 6A, 7A, and 8A are often called the **main-group,** or **representative, elements.** Remember that every member of these groups has the same valence electron configuration.

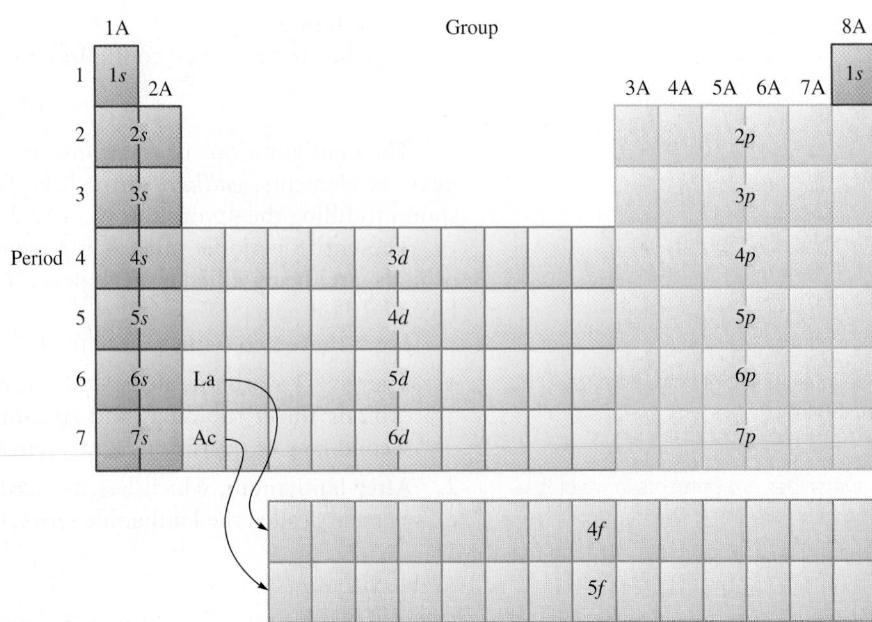

Figure 12.28
The orbitals being filled for elements in various parts of the periodic table. Note that when we move along a horizontal row (a period), the $(n + 1)s$ orbital fills before the *nd* orbital. The group labels indicate the number of valence electrons (*ns* plus *np* electrons) for the elements in each group.

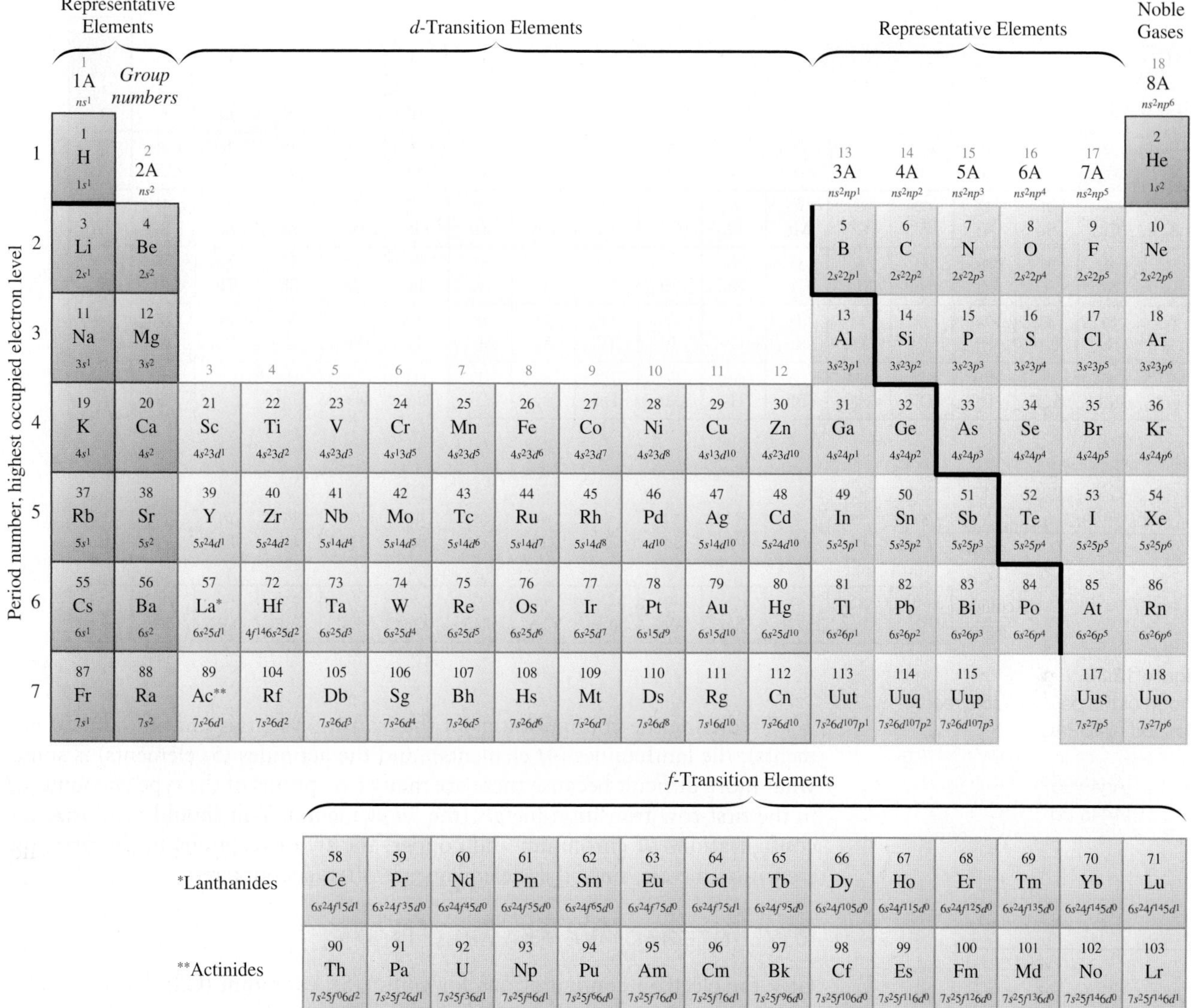

Figure 12.29
The periodic table with atomic symbols, atomic numbers, and partial electron configurations.

In 1985 the International Union of Pure and Applied Chemistry (IUPAC), a body of scientists organized to standardize scientific conventions, recommended a new form for the periodic table, which the American Chemical Society has adopted (Fig. 12.30 on page 568). In this new version, the group number indicates the number of *s*, *p*, and *d* electrons added since the last noble gas. We will not use the new format in this book, but you should be aware that the familiar periodic table may soon be replaced by this or a similar format.

The results considered in this section are very important. We have seen that the wave mechanical model can be used to explain the arrangement of elements in the periodic table. This model allows us to understand that the similar chemistry exhibited by the members of a given group arises from the fact that they all have the same valence electron configuration. Only the principal quantum number of the occupied orbitals changes in going down a particular group.

It is important to be able to give the electron configuration for each of the main-group elements. This is most easily done by using the periodic table. If you understand how the table is organized, it is not necessary to memorize the order in which the orbitals fill. Review Fig. 12.28 and Fig. 12.29 to make sure that you understand the correspondence among the orbitals and the periods and groups.

1	2	3	4	5	6	7	8	9	10	11	12	13	14	15	16	17	18
1 H																	2 He
3 Li	4 Be											5 B	6 C	7 N	8 O	9 F	10 Ne
11 Na	12 Mg											13 Al	14 Si	15 P	16 S	17 Cl	18 Ar
19 K	20 Ca	21 Sc	22 Ti	23 V	24 Cr	25 Mn	26 Fe	27 Co	28 Ni	29 Cu	30 Zn	31 Ga	32 Ge	33 As	34 Se	35 Br	36 Kr
37 Rb	38 Sr	39 Y	40 Zr	41 Nb	42 Mo	43 Tc	44 Ru	45 Rh	46 Pd	47 Ag	48 Cd	49 In	50 Sn	51 Sb	52 Te	53 I	54 Xe
55 Cs	56 Ba	57 La*	72 Hf	73 Ta	74 W	75 Re	76 Os	77 Ir	78 Pt	79 Au	80 Hg	81 Tl	82 Pb	83 Bi	84 Po	85 At	86 Rn
87 Fr	88 Ra	89 Ac**	104 Rf	105 Db	106 Sg	107 Bh	108 Hs	109 Mt	110 Ds	111 Rg	112 Cn	113 Uut	114 Uuq	115 Uup		117 Uus	118 Uuo

*Lanthanide series	58 Ce	59 Pr	60 Nd	61 Pm	62 Sm	63 Eu	64 Gd	65 Tb	66 Dy	67 Ho	68 Er	69 Tm	70 Yb	71 Lu
**Actinide series	90 Th	91 Pa	92 U	93 Np	94 Pu	95 Am	96 Cm	97 Bk	98 Cf	99 Es	100 Fm	101 Md	102 No	103 Lr

Figure 12.30
A form of the periodic table recommended by IUPAC.

Cr: [Ar]$4s^1 3d^5$
Cu: [Ar]$4s^1 3d^{10}$

Predicting the configurations of the transition metals (3*d*, 4*d*, and 5*d* elements), the lanthanides (4*f* elements), and the actinides (5*f* elements) is somewhat more difficult because there are many exceptions of the type encountered in the first-row transition metals (the 3*d* elements). You should memorize the configurations of chromium and copper, the two exceptions in the first-row transition metals, since these elements are often encountered.

OWL INTERACTIVE EXAMPLE 12.8

Give the electron configurations for sulfur (S), cadmium (Cd), hafnium (Hf), and radium (Ra), using the periodic table inside the front cover of this book.

Solution *Sulfur*, element 16, resides in Period 3, where the 3*p* orbitals are being filled (Fig. 12.31). Since sulfur is the fourth among the 3*p* elements, it must have four 3*p* electrons. Its configuration is

$$\text{S:} \quad 1s^2 2s^2 2p^6 3s^2 3p^4 \quad \text{or} \quad [\text{Ne}]3s^2 3p^4$$

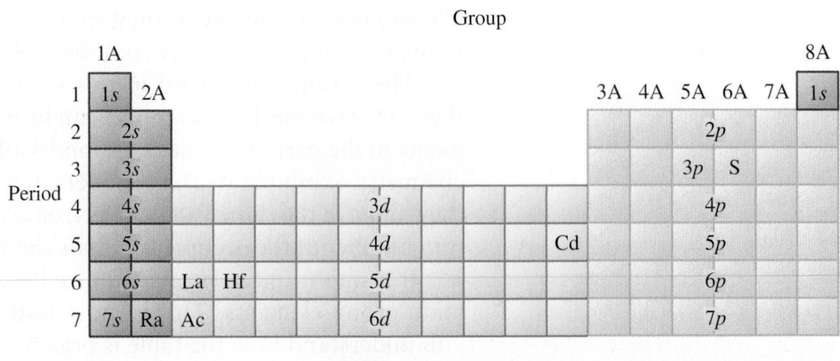

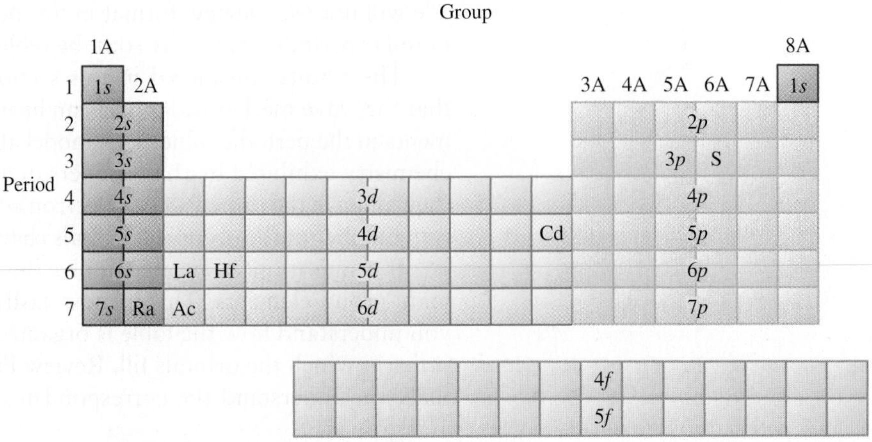

Figure 12.31
The positions of the elements considered in Example 12.8.

Cadmium, element 48, is located in Period 5 at the end of the $4d$ transition metals, as shown in Fig. 12.31. It is the tenth element in the series and thus has ten electrons in the $4d$ orbitals (in addition to the two electrons in the $5s$ orbital). The configuration is

$$\text{Cd:} \quad 1s^2 2s^2 2p^6 3s^2 3p^6 4s^2 3d^{10} 4p^6 5s^2 4d^{10} \quad \text{or} \quad [\text{Kr}]5s^2 4d^{10}$$

Hafnium, element 72, is found in Period 6, as shown in Fig. 12.31. Note that it occurs just after the lanthanide series. Thus the $4f$ orbitals are already filled. Hafnium is the second member of the $5d$ transition series and has two $5d$ electrons. The configuration is

$$\text{Hf:} \quad 1s^2 2s^2 2p^6 3s^2 3p^6 4s^2 3d^{10} 4p^6 5s^2 4d^{10} 5p^6 6s^2 4f^{14} 5d^2$$

$$\text{or} \quad [\text{Xe}]6s^2 4f^{14} 5d^2$$

Radium, element 88, is in Period 7 (and Group 2A), as shown in Fig. 12.31. Thus radium has two electrons in the $7s$ orbital, and the configuration is

$$\text{Ra:} \quad 1s^2 2s^2 2p^6 3s^2 3p^6 4s^2 3d^{10} 4p^6 5s^2 4d^{10} 5p^6 6s^2 4f^{14} 5d^{10} 6p^6 7s^2$$

$$\text{or} \quad [\text{Rn}]7s^2$$

12.14 Further Development of the Polyelectronic Model

Before we proceed with further discussion of polyelectronic atoms, we should summarize some of the most important things that have been said about the quantum mechanical description of atoms to this point. Most important, there is a fundamental difference between the solution of the Schrödinger equation for the hydrogen atom and the solutions for all polyelectronic atoms. The Schrödinger equation for the hydrogen atom can be solved exactly to yield the now-familiar hydrogen orbitals. These orbitals are characterized by the quantum numbers n, ℓ, m_ℓ, and m_s, and the energy levels corresponding to these orbitals depend only on n (all orbitals with the same value of n are degenerate). Recall that some of the orbitals directly obtained from the solution to the Schrödinger equation are complex (contain $\sqrt{-1}$). For example, of the three orbitals corresponding to $n = 2$ and $\ell = 1$ ($2p$ orbitals), the orbital corresponding to $m_\ell = 0$ is real (the $2p_z$ orbital), but the orbitals corresponding to the values of m_ℓ of $+1$ and -1 are complex. For ease of physical interpretation, these latter two orbitals are combined to produce two real orbitals ($2p_x$ and $2p_y$). These same procedures apply to all the p orbitals (corresponding to higher values of n). Similarly, for the $3d$ orbitals ($n = 3$, $\ell = 2$), the orbital corresponding to $m_\ell = 0$ is real (d_{z^2}), whereas the orbitals corresponding to m_ℓ values of ± 1 and ± 2 are complex and are used to construct the familiar real orbitals.

In contrast to the Schrödinger equation for the hydrogen atom, the Schrödinger equation for a polyelectronic atom cannot be solved exactly. For example, although the hydrogen and helium atoms are similar in many respects, the mathematical descriptions of these atoms are fundamentally different. Because electrons repel each other, the motions of the two helium electrons are correlated (coupled), and this fact prevents the exact separation of the Schrödinger equation for helium into independent, solvable equations for each electron. Thus solving the Schrödinger equation for helium (or any other polyelectronic atom) requires approximations. The approach most commonly used, the self-consistent field (SCF) method, was developed by Hartree and is applied as follows: For an atom containing N electrons, a wave function (orbital) is guessed for each electron except electron 1. For example, assume that orbitals are guessed for electrons 2, 3, 4, . . . , N: $\psi_2, \psi_3, \psi_4, \ldots, \psi_N$. The next step involves solving the Schrödinger equation for electron 1, which is moving in a potential field created by the nucleus and the electrons in orbitals $\psi_2, \psi_3, \ldots, \psi_N$. The repulsions between electron 1 and the other electrons are computed at each point in space from the sum of the average electron densities (probabilities) corresponding to $|\psi_2|^2, |\psi_3|^2, \ldots, |\psi_N|^2$ in volume element dv around that point. With the aid of a computer, the problem is solved to yield the wave function for electron 1, which we will label ψ_1'.

The next step is to do the same type of calculation to obtain a new wave function for electron 2 moving in a field of electrons described by the wave functions $\psi_1', \psi_3, \psi_4, \ldots, \psi_N$. This step leads to a new function ψ_2' for electron 2. Now the process is carried out for electron 3 interacting with electrons described by the wave functions $\psi_1', \psi_2', \psi_4, \ldots, \psi_N$ to produce a new function ψ_3'. This procedure continues until all electrons have been covered to yield the wave functions $\psi_1', \psi_2', \psi_3', \ldots, \psi_N'$. Then the entire process starts again with electron 1 and continues through electron N to give the new functions $\psi_1'', \psi_2'', \psi_3'', \ldots, \psi_N''$. The procedure is diagramed in Fig. 12.32. When a given cycle produces a set of wave functions that are virtually identical to the previous set, a self-consistent field is achieved and the procedure is terminated.

The orbitals that arise from the SCF method are quite similar to hydrogen orbitals. They have the same angular characteristics (same type of boundary surfaces) as do the orbitals of hydrogen. However, the radial parts of the orbitals are different from those of the hydrogen orbitals. Although the n quantum number from the treatment of hydrogen does not apply exactly to the SCF orbitals, it is still convenient to retain it as a label. It is important to note that the energies of the SCF orbitals for polyelectronic atoms depend on both n and ℓ, not just n, as for hydrogen.

Finally, although it is not precisely correct to assume that the N electrons in an atom occupy N independent one-electron orbitals, this remains a very useful idea for understanding many atomic properties, including the organization of the periodic table. Recall that for us to account for the arrangement of the atoms on the periodic table, the orbitals that correspond to a given value of n must fill in the order ns, then np, then nd, and finally, nf. From this observation we would expect the energies of the one-electron SCF orbitals to vary in the order

$$E_{ns} < E_{np} < E_{nd} < E_{nf}$$

and this ordering is borne out by the calculations.

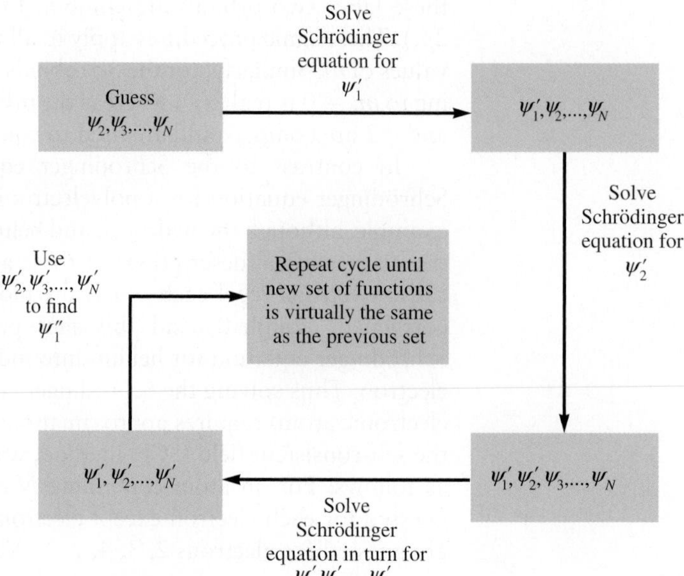

Figure 12.32
A schematic of the SCF method for obtaining the orbitals of a polyelectronic atom.

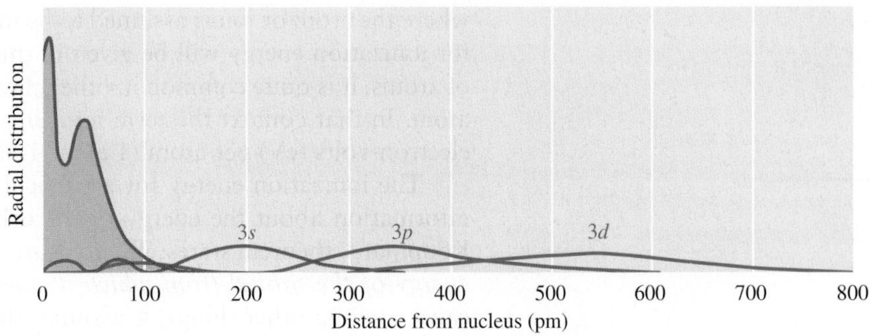

Figure 12.33
The radial distribution of electron probability density for the sodium atom. The shaded area represents the 10 core electrons. The radial distributions of the 3*s*, 3*p*, and 3*d* orbitals are also shown. Note the difference in the penetration effects of an electron in these three orbitals.

We can understand the observed order of orbital energies in a qualitative sense by considering the so-called **penetration effect.** To get an appreciation for this concept, consider the radial probability plots for the 3*s*, 3*p*, and 3*d* orbitals in a sodium atom, as shown in Fig. 12.33. Note that although an electron in the 3*s* orbital spends most of its time far from the nucleus and outside the core electrons (the electrons in the 1*s*, 2*s*, and 2*p* orbitals), which **shields** it from the nuclear charge, it has a small but significant probability of being quite close to the nucleus. We say it significantly *penetrates* the shielding core electron "cloud" and thus "feels" more of the nuclear charge. On the other hand, an electron in a 3*p* orbital does not have a probability maximum close to the nucleus. Thus we can say that an electron in a 3*p* orbital penetrates the core electrons to a lesser extent than an electron in the 3*s* orbital. Similarly, an electron in a 3*d* orbital (Fig. 12.33) shows much less penetration than a 3*p* electron does. These ideas help us to understand why an electron "prefers" the 3*s* orbital to the 3*p* or 3*d* and why, after the 3*s* orbital is filled, the next electron occupies the 3*p* rather than the 3*d* orbital. That is, the penetration effect helps us to understand qualitatively the order

$$E_{3s} < E_{3p} < E_{3d}$$

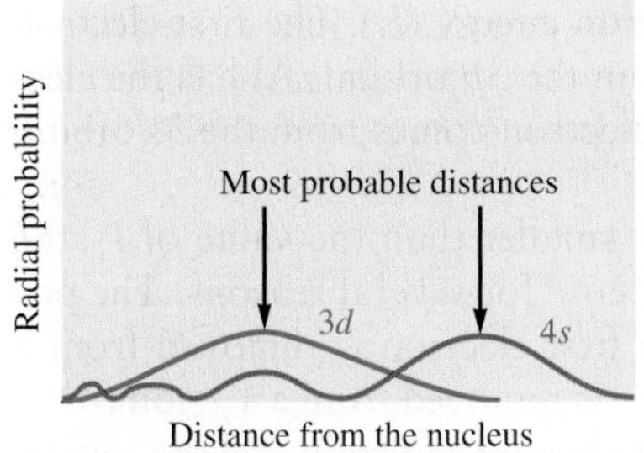

Figure 12.34
Radial probability distributions for the 3*d* and 4*s* orbitals. Note that the most probable distance of the electron from the nucleus for the 3*d* orbital is less than that for the 4*s* orbital. However, the 4*s* orbital allows more electron penetration close to the nucleus and thus is preferred over the 3*d* orbital.

The penetration effect also helps to explain why the 4*s* orbital fills before the 3*d* orbital. Recall that potassium has the electron configuration $1s^22s^22p^63s^23p^64s^1$ rather than the expected $1s^22s^22p^63s^23p^63d^1$. We can explain this result by observing that an electron in a 4*s* orbital penetrates much more than an electron in a 3*d* orbital, as shown graphically in Fig. 12.34. Note that although the most probable distance from the nucleus for a 3*d* electron is less than that for a 4*s* electron, the 4*s* electron has a significant probability of penetrating close to the nucleus. This explains why the potassium atom in its lowest-energy state has its last electron in the 4*s* orbital rather than in the 3*d* orbital.

Although the rigorous description of polyelectronic atoms is quite complicated, our simple qualitative ideas about electrons in independent orbitals are often very useful when we try to understand why atoms behave the way they do. We will consider some specific atomic properties in the next section.

12.15 Periodic Trends in Atomic Properties

We have developed a fairly complete picture of polyelectronic atoms that is quite successful in accounting for the periodic table of elements. We will next use the model to account for the observed trends in several important atomic properties: ionization energy, electron affinity, and atomic size.

Ionization Energy

Ionization energy is the energy required to remove an electron from a gaseous atom or ion,

$$X(g) \longrightarrow X^+(g) + e^-$$

where the atom or ion is assumed to be in its ground state. Although the values for ionization energy will be given in this text in terms of kilojoules per mole of atoms, it is quite common in other chemical literature to see values given per atom. In that context the term *ionization potential* is used, and the units are electron-volts (eV) per atom (1 eV = 1.602×10^{-19} J).

The ionization energy for a particular electron in an atom is a source of information about the energy of the orbital it occupies in the atom. In fact, Koopmans' theorem states: *The ionization energy of an electron is equal to the energy of the orbital from which it came.* This rule is an approximation because, among other things, it assumes that the electrons left behind in the resulting ion will not reorganize in response to the removal of an electron. However, ionization energies do provide information that is quite useful in testing the orbital model of the atom.

To introduce some of the characteristics of ionization energy, we will consider the energy required to remove several electrons in succession from aluminum atoms in the gaseous state. The ionization energies are

$$Al(g) \longrightarrow Al^{+}(g) + e^{-} \qquad I_1 = 580 \text{ kJ/mol}$$

$$Al^{+}(g) \longrightarrow Al^{2+}(g) + e^{-} \qquad I_2 = 1815 \text{ kJ/mol}$$

$$Al^{2+}(g) \longrightarrow Al^{3+}(g) + e^{-} \qquad I_3 = 2740 \text{ kJ/mol}$$

$$Al^{3+}(g) \longrightarrow Al^{4+}(g) + e^{-} \qquad I_4 = 11{,}600 \text{ kJ/mol}$$

Several important points can be illustrated from these results. In a stepwise ionization process, it is always the highest-energy electron (the one bound least tightly) that is removed first. The energy required to remove the highest-energy electron of an atom is called the **first ionization energy** (I_1). The first electron removed from the aluminum atom comes from the $3p$ orbital (Al has the electron configuration [Ne]$3s^2 3p^1$). The second electron comes from the $3s$ orbital (since Al^+ has the configuration [Ne]$3s^2$).

Note that the value of I_1 is considerably smaller than the value of I_2, the **second ionization energy.** This result makes sense for several reasons. The primary factor is simply charge. Note that the first electron is removed from a neutral atom (Al), whereas the second electron is removed from a 1+ ion (Al^+). The increase in positive charge binds the electrons more firmly and the ionization energy increases. The same trend shows up in the third (I_3) and fourth (I_4) ionization energies, where the electron is removed from the Al^{2+} and Al^{3+} ions, respectively.

The increase in successive ionization energies for an atom also makes sense in terms of relative orbital energies. The increase from I_1 to I_2 is expected because the first electron is removed from a $3p$ orbital that is higher in energy than the $3s$ orbital from which the second electron is removed. The largest jump in ionization energy by far occurs in going from the third ionization energy (I_3) to the fourth (I_4). This large jump occurs because I_4 corresponds to removing a core electron (Al^{3+} has the configuration $1s^2 2s^2 2p^6$), and core electrons are bound much more tightly than valence electrons.

Table 12.6 gives the values of ionization energies for all the Period 3 elements. Note the large jump in energy in each case in going from the removal of valence electrons to the removal of core electrons.

The values of the first ionization energy for the elements in the first five periods of the periodic table are graphed in Fig. 12.35. Note that, in general, as we go *across a period from left to right, the first ionization energy increases.* To account for this trend qualitatively, we need to consider more fully the concept of electrons shielding each other from the nuclear charge. Shielding occurs because electrons repel each other. Our simple pictures of the atom lead us to expect that core electrons should be quite effective in shielding outer

Table 12.6

Successive Ionization Energies in Kilojoules per Mole for the Elements in Period 3

Element	I_1	I_2	I_3	I_4	I_5	I_6	I_7
Na	495	4560					
Mg	735	1445	7730	Core electrons*			
Al	580	1815	2740	11,600			
Si	780	1575	3220	4350	16,100		
P	1060	1890	2905	4950	6270	21,200	
S	1005	2260	3375	4565	6950	8490	27,000
Cl	1255	2295	3850	5160	6560	9360	11,000
Ar	1527	2665	3945	5770	7230	8780	12,000

General increase → (across); General decrease → (down)

*Note the large jump in ionization energy in going from removal of valence electrons to removal of core electrons.

Electrons in the same principal quantum level do not shield each other as well as core electrons shield outer electrons.

electrons from the nuclear charge, since the core electrons are between the nucleus and the outer electrons. On the other hand, electrons in the same principal quantum level, which on the average are all at about the same distance from the nucleus, are not expected to shield each other very well. Thus, as we move from left to right in a given period of the periodic table, we do not expect the electrons to completely shield each other from the increasing nuclear charge as the number of protons in the nucleus increases. The electrons are expected to be bound more firmly in going from left to right across a given period, which means that the ionization energy should increase.

First ionization energy increases across a period and decreases down a group.

On the other hand, *the first ionization energy values decrease in going down a group*. This can be seen most clearly by focusing on the Group 1A elements (the alkali metals) and the Group 8A elements (the noble gases), as shown in Table 12.7. The main reason for the decrease in going down a group

Table 12.7

First Ionization Energies for the Alkali Metals and Noble Gases

Atom	I_1 (kJ/mol)
Group 1A	
Li	520.
Na	495
K	419
Rb	409
Cs	382
Group 8A	
He	2377
Ne	2088
Ar	1527
Kr	1356
Xe	1176
Rn	1042

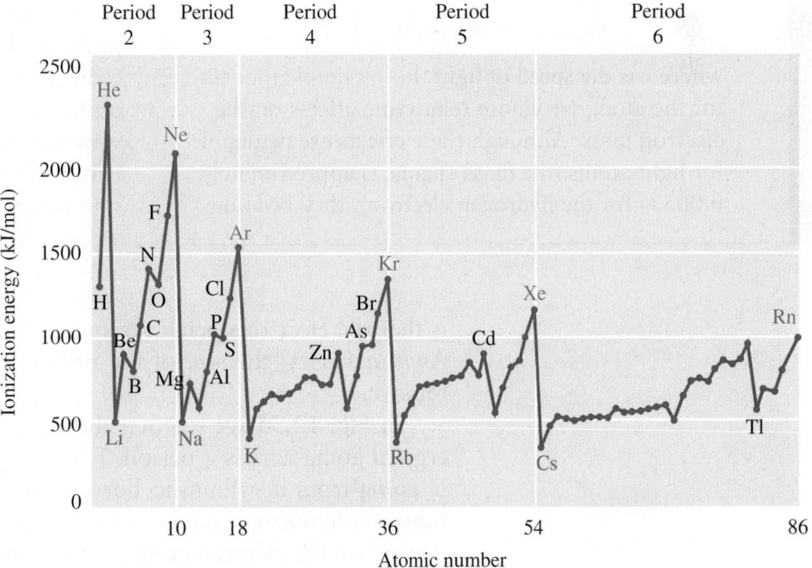

Figure 12.35
The values of first ionization energy for the elements in the first five periods. In general, ionization energy decreases in going down a group. For example, note the decrease in values for Group 1A and Group 8A. In general, ionization energy increases in going from left to right across a period. For example, note the sharp increase going across Period 2 from lithium through neon.

Chemical Insights — Why Is Mercury a Liquid?

The silver liquid called mercury has been known since ancient times. In fact, the symbol for mercury (Hg) comes from its Greek name *Hydrargyrum,* which means "watery silver." Although elements in the liquid state at ambient temperature and pressure are quite rare (Br_2 is another example), the liquid nature of mercury is especially confounding. For example, compare the properties of mercury and gold:

	Mercury	Gold
Melting point	−39°C	1064°C
Density	13.6 g cm^{-3}	19.3 g cm^{-3}
Enthalpy of fusion	2.30 kJ mol^{-1}	12.8 kJ mol^{-1}
Conductivity	10.4 kS m^{-1}	426 kS m^{-1}

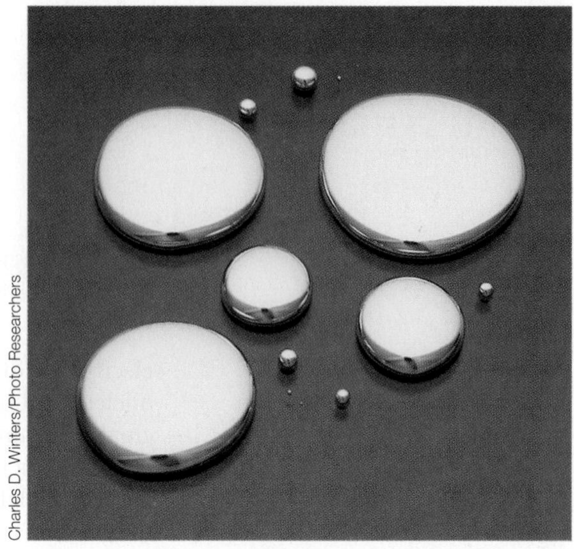

Charles D. Winters/Photo Researchers

Liquid mercury forms flat drops because of its lack of surface tension.

It is quite apparent that these metals, which are neighbors on the periodic table, have strikingly different properties. Why? The answer is not at all straightforward—but very interesting. It seems to hinge on relativity.

Recall that Einstein postulated in his theory of special relativity in 1905 that the mass (m) of a moving object increases with its velocity (v):

$$m_{\text{relativistic}} = m_{\text{rest}}/\sqrt{1 - (v/c)^2}$$

where c is the speed of light. In the simple models for the atom, we ignore relativistic effects on the electron mass. Although these effects are negligible for light atoms (the mass change is approximately 0.003% for the hydrogen electron), they become important for heavy elements such as gold and mercury. For example, the relativistic mass for a 1s electron in mercury is approximately 1.23 times its rest mass, and this effect leads to a very significant contraction in the radius of the 1s orbital.

It turns out that relativity has an even more profound impact on atomic theory than the preceding calculations suggest. A relativistic treatment of the atom fundamentally changes the way we view the electrons in atoms. In fact, as shown by British physicist Paul Dirac, the concept of electron spin occurs naturally in a relativistic treatment of

is that the electrons being removed are, on average, farther from the nucleus. As n increases, the size of the orbital increases, and the electron is easier to remove.

In Fig. 12.35 we see that there are some discontinuities in ionization energy in going across a period. For example, discontinuities occur in Period 2, in going from beryllium to boron and from nitrogen to oxygen. These exceptions to the normal trend can be explained in terms of how electron repulsions depend on the electron configuration. We will discuss the elements in Period 2 individually to further develop the concept of shielding.

The increase in the ionization energy in going from lithium ($1s^22s^1$) to beryllium ($1s^22s^2$) is expected, since the 2s electrons do not shield each other completely. The decrease in the ionization energy in going from beryllium ($1s^22s^2$) to boron ($1s^22s^22p^1$) suggests that the electrons in the 2s orbital effectively shield the 2p electron. This is sensible in view of the greater penetration

the atom and does not have to be "tacked on" as in the wave mechanical treatment. The point here is not to explain these very complex ideas but to alert you to concepts you will be learning more about in higher-level courses.

How does relativity explain why mercury has a melting point of $-39°C$, whereas that of neighboring gold is $1064°C$? The first step in answering this question involves considering the electron configurations of these atoms:

$$\text{Au: [Xe]}4f^{14}5d^{10}6s^1$$

$$\text{Hg: [Xe]}4f^{14}5d^{10}6s^2$$

Notice that gold has an unfilled 6*s* subshell, but the 6*s* level is filled in mercury. Because of its configuration, a gold atom can use its half-filled 6*s* orbital to form a bond to another gold atom. In fact, the metal–metal bond in the Au_2 molecule is an astonishingly strong 221 kJ mol^{-1}, a value very close to the bond energy of the Cl_2 molecule (239 kJ mol^{-1}) and greater than the bond energy of I_2 (149 kJ mol^{-1}). In addition, gold has an electron affinity (-220 kJ mol^{-1}) that is higher than that of oxygen and sulfur. Further, gold forms a compound with cesium (CsAu) that exhibits the CsCl crystal structure (see Fig. 16.41 and Section 16.8) in which gold atoms take the place of Cl^- ions. Thus a gold atom seems to behave a lot like a halogen atom.

What causes gold to emulate many properties of the nonmetallic halogens? The apparent answer lies in the dramatic contraction of the gold 6*s* orbital because of relativistic effects. The unexpectedly small radius of the 6*s* orbital of a gold atom results in a much lower energy than is predicted in the absence of relativistic effects. It is this very-low-energy, unfilled 6*s* orbital that causes gold atoms to form very stable Au_2 molecules in the gas phase and to bind strongly to each other in the solid state, producing its high melting point. This same low-energy 6*s* orbital also leads to gold's unexpectedly high electron affinity and to its unusual color. Gold is not the silvery color exhibited by most metals because of the absorption of blue light to transfer an electron between the 5*d* and 6*s* orbitals in gold atoms.

So why are gold and mercury so different? The answer lies in the different electron configurations of the two atoms. Unlike gold, the low-energy 6*s* orbital in mercury is filled, and these two electrons are very tightly bound to the mercury atom. In fact, one can think of Hg as being analogous to He. That is, the low-energy pair of 6*s* electrons on mercury causes it to behave like a noble gas atom—it cannot bond to another mercury atom. This explains why mercury is unique among metals in that it is almost entirely monomeric in its gas phase. In contrast, the species Hg_2^{2+} is extremely stable even in aqueous solution. This fact is not surprising once it is realized that Hg^+ is isoelectronic with Au.

Thus it is the unusually low energy of the 6*s* orbitals apparently caused by relativistic effects that causes gold to behave like a halogen atom and mercury to behave like a noble gas.

by the 2*s* electrons as compared with the 2*p* electron. The 2*s* electrons spend more time closer to the nucleus than the 2*p* electrons, where they provide effective shielding. The steady increase in the ionization energy in going from boron ($1s^22s^22p^1$) to carbon ($1s^22s^22p^2$) to nitrogen ($1s^22s^22p^3$) is expected because the 2*p* electrons are not very effective in shielding each other from the increasing nuclear charge. The drop in the ionization energy in going from nitrogen ($1s^22s^22p^3$) to oxygen ($1s^22s^22p^4$) is usually explained as follows: In nitrogen each 2*p* electron is in a separate orbital. When the extra electron for oxygen is added, one 2*p* orbital becomes doubly occupied. The electron repulsions between the electrons in the doubly occupied orbital make either of these electrons easier to remove.* As we move from oxygen ($1s^22s^22p^4$) to fluorine

*This explanation is greatly oversimplified. A more complete explanation involves a detailed analysis of the electron interactions, which is beyond the scope of this text.

($1s^22s^22p^5$) to neon ($1s^22s^22p^6$), the ionization energy increases because in each case the electron being considered is in a doubly occupied orbital. Note that the ionization energy values for oxygen, fluorine, and neon are different from those of boron, carbon, and nitrogen by an amount that represents the extra electron repulsions in doubly occupied orbitals, as shown in Fig. 12.35.

EXAMPLE 12.9

The first ionization energy for phosphorus is 1060 kJ/mol, and that for sulfur is 1005 kJ/mol. Why?

Solution Phosphorus and sulfur are neighboring elements in Period 3 of the periodic table and have the following valence electron configurations: Phosphorus is $3s^23p^3$, and sulfur is $3s^23p^4$.

Ordinarily, the first ionization energy increases as we go across a period, so we might expect sulfur to have a greater ionization energy than phosphorus. However, in this case the fourth *p* electron in sulfur must be placed in an already occupied orbital. The electron–electron repulsions that result cause this electron to be more easily removed than might be expected.*

OWL INTERACTIVE EXAMPLE 12.10

Consider atoms with the following electron configurations:

$$1s^22s^22p^6$$
$$1s^22s^22p^63s^1$$
$$1s^22s^22p^63s^2$$

Which atom has the largest first ionization energy, and which has the smallest second ionization energy? Explain your choices.

Solution The atom with the largest value of I_1 is the one with the configuration $1s^22s^22p^6$ (this is the neon atom) because this element is found at the right end of Period 2. Since the $2p$ electrons do not shield each other very effectively, I_1 will be large. The other configurations given include $3s$ electrons. These electrons are effectively shielded by the core electrons and are farther from the nucleus than the $2p$ electrons in neon. Thus I_1 for these atoms is smaller than I_1 for neon.

The atom with the smallest value of I_2 is the one with the configuration $1s^22s^22p^63s^2$ (the magnesium atom). For magnesium both I_1 and I_2 involve valence electrons. For the atom with the configuration $1s^22s^22p^63s^1$ (sodium), the second electron lost (corresponding to I_2) is a core electron from a $2p$ orbital).

Electron Affinity

Electron affinity *is the energy change associated with the addition of an electron to a gaseous atom:*

$$X(g) + e^- \longrightarrow X^-(g)$$

The sign convention for electron affinity values follows the convention for energy changes used in Chapters 9 and 10.

Because two different conventions have been used, there is a good deal of confusion in the chemical literature about the signs for electron affinity values. Electron affinity has been defined in many textbooks as the energy *released* when an electron is added to a gaseous atom. This convention requires that a positive sign be attached to an exothermic addition of an electron to an atom,

*See page 573.

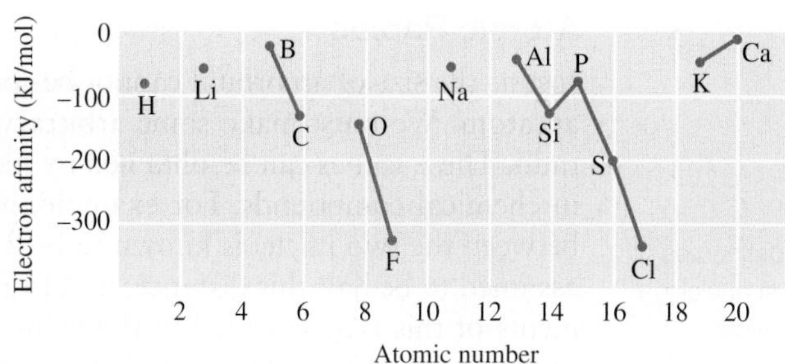

Figure 12.36
The electron affinity values for atoms among the first 20 elements that form stable, isolated X^- ions. The lines shown connect adjacent elements. The absence of a line indicates missing elements (He, Be, N, Ne, Mg, and Ar) whose atoms do not add an electron exothermically and thus do not form stable, isolated X^- ions.

which opposes normal thermodynamic conventions. Therefore, in this book we define electron affinity as a *change* in energy. This means that if the addition of the electron is exothermic, the corresponding value for electron affinity will carry a negative sign.

Figure 12.36 shows the electron affinity values for the atoms among the first 20 elements that form stable, isolated, negative ions—that is, the atoms that undergo the addition of an electron as shown above. As expected, all these elements have negative (exothermic) electron affinities. Note that the *more negative* the energy, the greater is the quantity of energy released. Although electron affinities generally become more negative from left to right across a period, there are several exceptions to this rule in each period. The dependence of electron affinity on atomic number can be explained by considering the changes in electron repulsions as a function of electron configurations. For example, the fact that the nitrogen atom does not form a stable, isolated $N^-(g)$ ion, whereas carbon forms $C^-(g)$, reflects the difference in the electron configurations of these atoms. An electron added to nitrogen ($1s^22s^22p^3$) to form the $N^-(g)$ ion ($1s^22s^22p^4$) would have to occupy a $2p$ orbital that already contains one electron. The extra repulsion between the electrons in this doubly occupied orbital causes $N^-(g)$ to be unstable. When an electron is added to carbon ($1s^22s^22p^2$) to form the $C^-(g)$ ion ($1s^22s^22p^3$), no such extra repulsions occur.

In contrast to the nitrogen atom, the oxygen atom can add one electron to form the stable $O^-(g)$ ion. Presumably, oxygen's greater nuclear charge, compared with that of nitrogen, is sufficient to overcome the repulsion associated with putting a second electron into an already occupied $2p$ orbital. However, it should be noted that a second electron *cannot* be added to an oxygen atom [$O^-(g) + e^- \not\rightarrow O^{2-}(g)$] to form an isolated oxide ion. This outcome seems strange in view of the many stable oxide compounds (MgO, Fe_2O_3, and so on) that are known. As we will discuss in detail in Chapter 13, the O^{2-} ion is stabilized in ionic compounds by the large attractions that occur among the positive ions and the oxide ions.

When we go down a group, electron affinity should become more positive (less energy released), since the electron is added at increasing distances from the nucleus. Although this is generally the case, the changes in electron affinity in going down most groups are relatively small, and numerous exceptions occur. This behavior is demonstrated by the electron affinities of the Group 7A elements (the halogens) shown in Table 12.8. Note that the range of values is quite small compared with the changes that typically occur across a period. Also note that although chlorine, bromine, and iodine show the expected trend, the energy released when an electron is added to fluorine is smaller than might be expected. This smaller energy release has been attributed to the small size of the $2p$ orbitals. Because the electrons must be very close together in these orbitals, there are unusually large electron–electron repulsions. In the other halogens with their larger orbitals, the repulsions are not as severe.

Table 12.8
Electron Affinities of the Halogens

Atom	Electron Affinity (kJ/mol)
F	−327.8
Cl	−348.7
Br	−324.5
I	−295.2

Figure 12.37
The radius of an atom (r) is defined as half the distance between the nuclei in a molecule consisting of identical atoms.

Atomic Radius

Just as the size of an orbital cannot be specified exactly, neither can the size of an atom. We must make some arbitrary choices to obtain values for **atomic radii.** These values can be obtained by measuring the distances between atoms in chemical compounds. For example, in the bromine molecule, the distance between the two nuclei is known to be 228 pm. The bromine atomic radius is assumed to be half this distance, or 114 pm, as shown in Fig. 12.37. Measurements of this type have led to the values of the atomic radii for the elements shown in Fig. 12.38. These radii are often called *covalent atomic radii* because of the way they are determined. These values are significantly smaller than might be expected from the 90% electron density volumes of isolated atoms because when atoms form bonds, their electron "clouds" interpenetrate. However, these values form a self-consistent data set that can be used to discuss the trends in atomic radii. The radii for metal atoms are obtained by halving the distance between metal atoms in solid metal crystals. Note from Fig. 12.38 that the atomic radius decreases in going from left to right across a period. This decrease can be explained in terms of the increasing effective nuclear charge (decreasing shielding) in going from left to right. This means that the valence electrons are drawn closer to the nucleus, decreasing the size of the atom.

Atomic radius decreases across a period and increases down a group.

Atomic radius increases down a group because of the increases in the orbital sizes in successive principal quantum levels.

Figure 12.38
Atomic radii (in picometers) for selected atoms. Note that atomic radius decreases going across a period and increases going down a group. The values for the noble gases are estimated because data from bonded atoms are lacking.

Atomic radius decreases →

Atomic radius increases ↓

1A	2A	3A	4A	5A	6A	7A	8A
H 37							He 32
Li 152	Be 113	B 88	C 77	N 70	O 66	F 64	Ne 69
Na 186	Mg 160	Al 143	Si 117	P 110	S 104	Cl 99	Ar 97
K 227	Ca 197	Ga 122	Ge 122	As 121	Se 117	Br 114	Kr 110
Rb 247	Sr 215	In 163	Sn 140	Sb 141	Te 143	I 133	Xe 130
Cs 265	Ba 217	Tl 170	Pb 175	Bi 155	Po 167	At 140	Rn 145

OWL INTERACTIVE EXAMPLE 12.11

Predict the trend in radius of the following ions: Be^{2+}, Mg^{2+}, Ca^{2+}, and Sr^{2+}.

Solution All these ions are formed by removing two electrons from an atom of a Group 2A element. In going from beryllium to strontium, we are going down the group, so the sizes increase.

$$Be^{2+} < Mg^{2+} < Ca^{2+} < Sr^{2+}$$

↑ Smallest radius (Be²⁺) ↑ Largest radius (Sr²⁺)

12.16 The Properties of a Group: The Alkali Metals

We have seen that the periodic table originated as a way to portray the systematic properties of the elements. Mendeleev was primarily responsible for first showing its usefulness in correlating and predicting the elemental properties. In this section we will summarize much of the information available from the table. We will also illustrate the usefulness of the table by discussing the properties of a representative group, the alkali metals.

Information Contained in the Periodic Table

1. The essence of the periodic table is that members of each group of representative elements exhibit similar chemical properties that change in a regular way. The quantum mechanical model has allowed us to understand that the similarity of properties of the atoms in a group arises from the identical valence electron configurations shared by group members. *It is the number and type of valence electrons that primarily determine an atom's chemistry.*
2. One of the most valuable types of information available from the periodic table is the electron configuration of any representative element. If you understand the organization of the table, you do not need to memorize electron configurations for the elements. Although the predicted electron configurations for transition metals are sometimes incorrect, this is not a serious problem. You should, however, memorize the configuration of two exceptions, chromium and copper, since these 3*d* transition elements are found in many important compounds.
3. As we mentioned in Chapter 2, certain groups in the periodic table have special names. These names are summarized in Fig. 12.39.
4. The most fundamental classification of the elements is into metals and nonmetals. The essential chemical property of a metal is the tendency to give up electrons to form a positive ion; metals tend to have low ionization energies. The metallic elements are found on the left side of the table, as shown in Fig. 12.39. The most reactive metals are found on the lower left-hand portion of the table where the ionization energies are smallest. The distinctive chemical property of a nonmetal is the ability to gain electrons to form an anion when reacting with a metal. The nonmetals have large ionization energies and most have negative electron affinities. The nonmetals are found on the right side of the table. The most reactive ones are located in the upper right-hand corner, excluding the noble gas elements, which are quite unreactive. The division between metals and nonmetals shown in Fig. 12.39 is only approximate. Many elements along the division line exhibit both metallic and nonmetallic properties under certain circumstances. These elements are called **metalloids,** or **semimetals.**

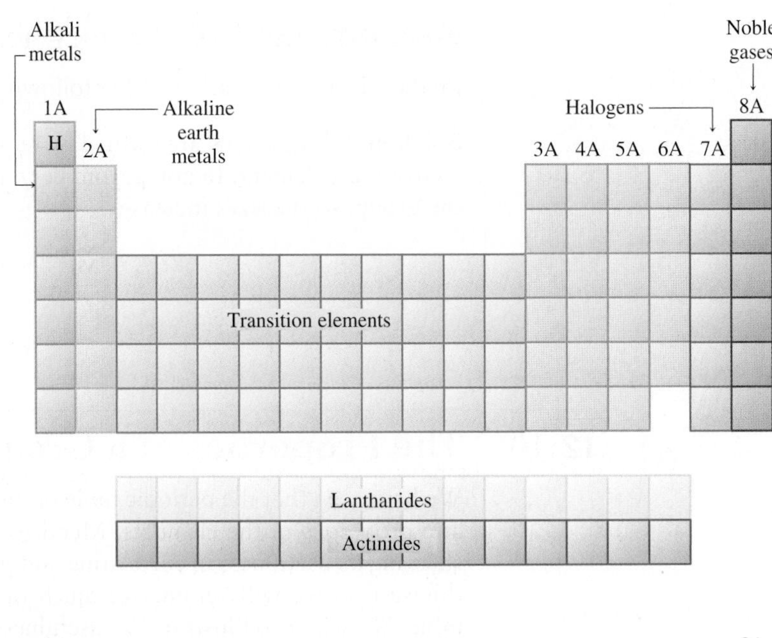

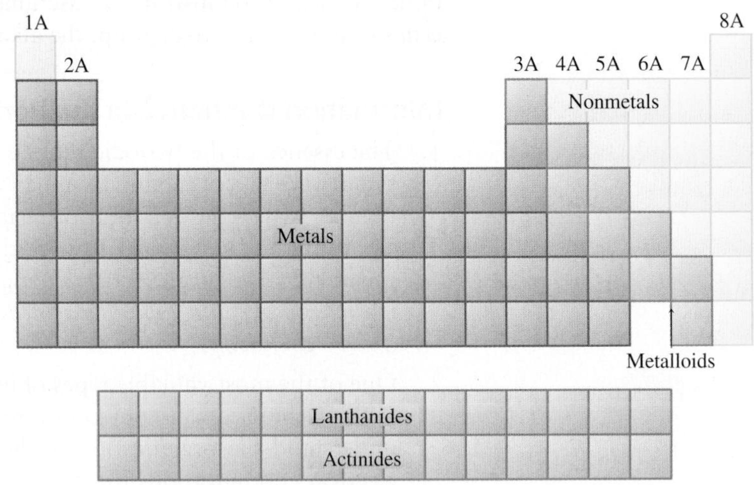

Figure 12.39
Special names for groups in the periodic table.

The Alkali Metals

The metals of Group 1A, the alkali metals, illustrate well the relationships among the properties of the elements in a group. Lithium, sodium, potassium, rubidium, cesium, and francium are the most reactive of the metals. We will not discuss francium here since it occurs in nature in only very small quantities. Although hydrogen is found in Group 1A, it behaves as a nonmetal, in contrast to the other members of that group. The fundamental reason for hydrogen's nonmetallic character is its very small size (see Fig. 12.38). The electron in the small 1*s* orbital is bound very tightly to the nucleus.

Some important properties of the first five alkali metals are shown in Table 12.9. The data in Table 12.9 show that when we move down the group, the first ionization energy decreases and the atomic radius increases. This agrees with the general trends discussed in Section 12.15.

The overall increase in density in going down Group 1A is typical of all groups. It occurs because atomic mass generally increases more rapidly than atomic size.

The smooth decrease in melting point (mp) and boiling point (bp) in going down Group 1A is not typical; in most other groups, more complicated behavior occurs. Note that the melting point of cesium is only 29°C. Cesium can be

Table 12.9

Properties of Five Alkali Metals

Element	Valence Electron Configuration	Density at 25°C (g/cm^3)	mp (°C)	bp (°C)	First Ionization Energy (kJ/mol)	Atomic (covalent) Radius (pm)	Ionic (M^+) Radius (pm)
Li	$2s^1$	0.53	180	1330	520.	152	60
Na	$3s^1$	0.97	98	892	495	186	95
K	$4s^1$	0.86	64	760	419	227	133
Rb	$5s^1$	1.53	39	688	409	247	148
Cs	$6s^1$	1.87	29	690	382	265	169

melted readily from the heat of your hand. Cesium's low melting point is very unusual—metals typically have high melting points. For example, tungsten melts at 3410°C. The only other metals with low melting points are mercury (mp = −39°C) and gallium (mp = 30°C).

Recall that the chemical property most characteristic of a metal is the ability to lose its valence electrons. The Group 1A elements are very reactive. They have low ionization energies and react readily with nonmetals to form ionic solids. A typical example involves the reaction of sodium with chlorine to form sodium chloride:

$$2Na(s) + Cl_2(g) \longrightarrow 2NaCl(s)$$

This is an oxidation–reduction reaction in which chlorine oxidizes sodium to form Na^+ and Cl^- ions. In the reactions between metals and nonmetals, it is typical for the nonmetal to behave as the oxidizing agent and the metal to behave as the reducing agent, as shown by the following reactions:

$$2Na(s) + S(s) \longrightarrow Na_2S(s)$$

Contains Na^+ and S^{2-} ions

$$6Li(s) + N_2(g) \longrightarrow 2Li_3N(s)$$

Contains Li^+ and N^{3-} ions

$$4Na(s) + O_2(g) \longrightarrow 2Na_2O(s)$$

Contains Na^+ and O^{2-} ions

For reactions of this type, the relative reducing powers of the alkali metals can be predicted from the first ionization energies listed in Table 12.9. Since it is much easier to remove an electron from a cesium atom than from a lithium atom, cesium should be the better reducing agent. The expected trend in reducing ability is

$$Cs > Rb > K > Na > Li$$

This order is observed experimentally for direct reactions between the solid alkali metals and nonmetals. However, this order of reducing ability is not observed when the alkali metals react in aqueous solution. For example, the reduction of water by an alkali metal is very vigorous and exothermic:

$$2M(s) + 2H_2O(l) \longrightarrow H_2(g) + 2M^+(aq) + 2OH^-(aq) + \text{energy}$$

The order of reducing abilities observed for this reaction (for the first three group members) is

$$Li > K > Na$$

which is not the order expected from the relative ionization energies of these metals.

Chemical Insights — Lithium: Behavior Medicine

More and more people in our society seem to be suffering from the debilitating effects of mania and depression, but the alkali metal lithium can provide help for many. In fact, more than three million prescriptions for lithium carbonate are filled annually by retail pharmacies.

Although the details are not well understood, the lithium ion seems to alleviate mood disorders by affecting the way that brain cells respond to neurotransmitters, a class of molecules that facilitate the transmission of nerve impulses.

Specifically, physiologists think that the lithium ion may interfere with a complex cycle of reactions that relays and amplifies messages carried to the cells by neurotransmitters and hormones. They theorize that exaggerated forms of behavior, such as mania or depression, arise from the overactivity of this cycle. Thus the fact that lithium inhibits this cycle may be responsible for its moderating effect on behavior.

There is a growing collection of evidence that violent behavior may result at least partially from the improper regulation of neurotransmitters and hormones. For example, a study in Finland showed that violent criminals, especially arsonists, often had low levels of serotonin, a common neurotransmitter. Studies are now under way to determine whether lithium might also be effective for treating these and other aberrant forms of behavior.

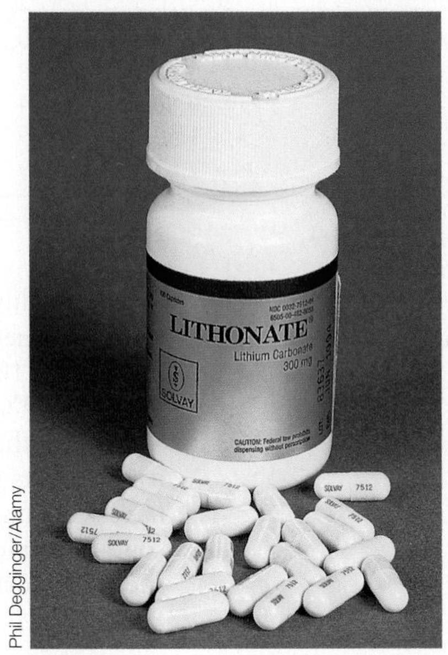

Phil Degginger/Alamy

Capsules of lithium carbonate.

This unexpected order occurs because the formation of the M^+ ions in aqueous solution is strongly influenced by the hydration of these ions by the polar water molecules. The hydration energy of an ion represents the change in energy that occurs when water molecules attach to the M^+ ion. The hydration energies for the Li^+, Na^+, and K^+ ions shown in Table 12.10 indicate that the process is exothermic in each case. However, nearly twice as much energy is released by the hydration of the Li^+ ion as compared with that of the K^+ ion. This difference is caused by size effects; the Li^+ ion is much smaller than the K^+ ion, and thus its *charge density* (charge per unit volume) is much greater. This means that the polar water molecules are more strongly attracted to the small Li^+ ion. Because the Li^+ ion is so strongly hydrated, its formation from the lithium atom occurs more readily than the formation of the K^+ ion from the potassium atom. Although a potassium atom in the gas phase loses its valence electron more readily than a lithium atom in the gas phase, the opposite is true in aqueous solution. This anomaly is an example of the importance of the polarity of the water molecule in aqueous reactions.

Table 12.10

Hydration Energies for Li^+, Na^+, and K^+ Ions

Ion	Hydration Energy (kJ/mol)
Li^+	−500
Na^+	−400
K^+	−300

There is one more surprise involving the highly exothermic reactions of the alkali metals with water. Experiments show that lithium is the best reducing agent in water, so we might expect lithium to react most violently with water. However, it does not. Sodium and potassium react much more vigorously. Why? The answer lies in the relatively high melting point for lithium. When sodium and potassium react with water, the heat evolved causes them to melt, giving a larger area of contact with water. Lithium, on the other hand, does not melt under these conditions and thus reacts more slowly. This example illustrates the important principle (which we will discuss in detail in Chapter 15) that the energy of a reaction and the rate at which it occurs are not necessarily related.

In this section we have seen that the trends in atomic properties summarized by the periodic table can be a great help in understanding the chemical behavior of the elements. This fact will be emphasized over and over as we proceed in our study of chemistry.

Key Terms

Section 12.1
electromagnetic radiation
wavelength
frequency

Section 12.2
ultraviolet catastrophe
Planck's constant
quantized
photon
photoelectric effect
dual nature of light
diffraction
diffraction pattern

Section 12.3
continuous spectrum
line spectrum

Section 12.4
quantum model
ground state

Section 12.5
wave mechanics (quantum mechanics)
wave function
orbital
Heisenberg uncertainty principle

Section 12.8
probability distribution
radial probability distribution

Section 12.9
quantum numbers
principal quantum number (n)
angular momentum quantum number (l)
magnetic quantum number (m_l)
subshell
nodal surfaces (nodes)
lobe
degenerate orbital

Section 12.10
electron spin
electron spin quantum number
Pauli exclusion principle

Section 12.11
polyelectronic atom
effective nuclear charge
self-consistent field (SCF) method

For Review

OWL and go Chemistry

Sign in at **www.cengage.com/owl** to:

- View tutorials and simulations, develop problem-solving skills, and complete online homework assigned by your professor.
- Download Go Chemistry mini lecture modules for quick review and exam prep from OWL (or purchase them at **www.cengagebrain.com**)

Electromagnetic radiation

- Characterized by its wavelength (λ), frequency (ν), and speed ($c = 2.9979 \times 10^8$ m/s)

$$\lambda\nu = c$$

- Can be viewed as a stream of "particles" called photons, each with energy $h\nu$, where h is Planck's constant (6.626×10^{-34} J · s)

Photoelectric effect

- When light strikes a metal surface, electrons are emitted
- Analysis of the kinetic energy and numbers of the emitted electrons led Einstein to suggest that electromagnetic radiation can be viewed as a stream of photons

Hydrogen spectrum

- The emission spectrum of hydrogen shows discrete wavelengths
- Indicates that hydrogen has discrete energy levels

Bohr model of the hydrogen atom

- Using the data from the hydrogen spectrum and assuming angular momentum to be quantized, Bohr devised a model in which the electron traveled in circular orbits
- Although an important pioneering effort, this model proved to be entirely incorrect

Wave (quantum) mechanical model

- An electron is described as a standing wave
- The square of the wave function (often called an orbital) gives a probability distribution for the electron position
- The exact position of the electron is never known, which is consistent with the Heisenberg uncertainty principle: it is impossible to know accurately both the position and the momentum of a particle simultaneously
- Probability maps are used to define orbital shapes
- Orbitals are characterized by the quantum numbers n, ℓ, and m_ℓ

Particle in a box

- Not an accurate physical model of the hydrogen atom
- Illustrates the mathematics of wave mechanics
- $E = \dfrac{n^2h^2}{8mL^2}$

Section 12.12
triads
octaves

Section 12.13
aufbau principle
Hund's rule
valence electron
core electron
transition metals
lanthanide series
actinide series
main-group (representative) elements

Section 12.14
penetration effect
shielding

Section 12.15
ionization energy
first ionization energy
second ionization energy
electron affinity
atomic radii

Section 12.16
metalloids (semimetals)

Electron spin
- Described by the spin quantum number m_s, which can have values of $\pm\frac{1}{2}$
- Pauli exclusion principle: no two electrons in a given atom can have the same set of quantum numbers n, ℓ, m_ℓ, and m_s
- Only two electrons with opposite spins can occupy a given orbital

Periodic table
- By populating the orbitals from the wave mechanical model (the aufbau principle), the form of the periodic table can be explained
- According to the wave mechanical model, atoms in a given group have the same valence (outer) electron configuration
- The trends in properties such as ionization energies and atomic radii can be explained in terms of the concepts of nuclear attraction, electron repulsions, shielding, and penetration

Discussion Questions

These questions are designed to be considered by groups of students in class. Often these questions work well for introducing a particular topic in class.

1. Explain what it means for something to have wavelike properties; for something to have particulate properties. Electromagnetic radiation can be discussed in terms of both particles and waves. Explain the experimental verification for each of these views.
2. Defend and criticize Bohr's model. Why was it reasonable that such a model was proposed, and what evidence was there that it "works"? Why do we no longer "believe" in it?
3. The first ionization energy for magnesium is 735 kJ/mol. Which electron is this for? Estimate Z_{eff} for this electron, and explain your reasoning. Calculate Z_{eff} for this electron, and compare it to your estimate.
4. The first four ionization energies for elements X and Y are shown below. The units are not kJ/mol.

	X	Y
First	170	200
Second	350	400
Third	1800	3500
Fourth	2500	5000

Identify the elements X and Y. There may be more than one answer, so explain completely.

5. Compare the first ionization energy of helium with its second ionization energy, remembering that both electrons come from the 1*s* orbital. Explain the difference without using actual numbers from the text.
6. Which has a larger second ionization energy, lithium or beryllium? Why?
7. Explain why a graph of ionization energy versus atomic number (across a row) is not linear. Where are the exceptions? Explain why they occur.
8. Without referring to your text, predict the trend of second ionization energies for the elements sodium through argon. Compare your answer with the data in Table 12.6. Explain any differences.
9. Account for the fact that the line that separates the metals from the nonmetals on the periodic table is diagonal downward to the right instead of horizontal or vertical.
10. Explain the term *electron* from a quantum mechanical perspective, including a discussion of atomic radii, probabilities, and orbitals.
11. Choose the best response for the following. The ionization energy for the chlorine atom is equal in magnitude to the electron affinity for
    a. the Cl atom
    b. the Cl^- ion
    c. the Cl^+ ion
    d. the F atom
    e. none of these

    Explain.

12. Consider the following statement: "The ionization energy for the potassium atom is negative because when K loses an electron to become K^+ it achieves a noble gas electron configuration." Indicate what is incorrect. Explain.

13. What is the difference between Z_{eff} and Z? When are they the same? Explain.

14. In going across a row of the periodic table, electrons are added and ionization energy generally increases. In going down a column of the periodic table, electrons are also being added but ionization energy generally decreases. Explain.

15. Explain the difference between the probability density distribution for an orbital and its radial probability.

16. How does the energy of a hydrogen 1*s* orbital compare with that of a lithium 1*s* orbital? Why? What is meant by the term *energy of the orbital?* What is its sign? Why? What is meant by the term *lower in energy?*

17. Which is larger, the hydrogen 1*s* orbital or the lithium 1*s* orbital? Why? Which has the larger radius, the hydrogen atom or the lithium atom? Why?

18. Is the following statement true or false: The hydrogen atom has a 3*s* orbital. Explain.

19. Which is higher in energy: the 2*s* or 2*p* orbital in hydrogen? Is this also true for helium? Explain.

20. Prove mathematically that it is more energetically favorable for a fluorine atom to take an electron from a sodium atom than for a fluorine atom to take an electron from another fluorine atom.

Exercises

OWL Interactive versions of these problems may be assigned in OWL.

A blue exercise number indicates that the answer to that exercise appears at the end of this book and a solution appears in the *Solutions Guide*.

Light and Matter

21. Microwave radiation has a wavelength on the order of 1.0 cm. Calculate the frequency and the energy of a single photon of this radiation. Calculate the energy of an Avogadro's number of photons (called an einstein) of this electromagnetic radiation.

22. Consider the following waves representing electromagnetic radiation:

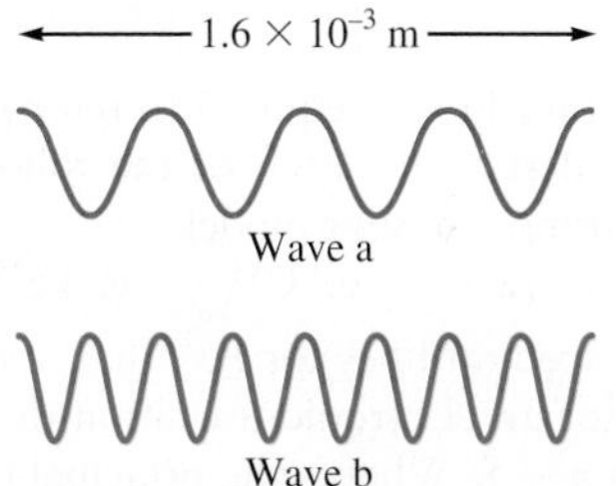

Which wave has the longer wavelength? Calculate the wavelength. Which wave has the higher frequency and larger photon energy? Calculate these values. Which wave has the greater velocity? What type of electromagnetic radiation does each wave represent?

23. One type of electromagnetic radiation has a frequency of 107.1 MHz, another type has a wavelength of 2.12×10^{-10} m, and another type of electromagnetic radiation has photons with energy equal to 3.97×10^{-19} J/photon. Identify each type of electromagnetic radiation, and place them in order of increasing photon energy and increasing frequency.

24. Carbon absorbs energy at a wavelength of 150. nm. The total amount of energy emitted by a carbon sample is 1.98×10^5 J. Calculate the number of carbon atoms present in the sample, assuming that each atom emits one photon.

25. A carbon–oxygen double bond in a certain organic molecule absorbs radiation that has a frequency of 6.0×10^{13} s^{-1}.
    a. What is the wavelength of this radiation?
    b. To what region of the spectrum does this radiation belong?
    c. What is the energy of this radiation per photon? Per mole of photons?
    d. A carbon–oxygen bond in a different molecule absorbs radiation with frequency equal to 5.4×10^{13} s^{-1}. Is this radiation more or less energetic?

26. X rays have wavelengths on the order of 1×10^{-10} m. Calculate the energy of 1.0×10^{-10} m X rays in units of kilojoules per mole of X rays. AM radio waves have wavelengths on the order of 1×10^4 m. Calculate the energy of 1.0×10^4 m radio waves in units of kilojoules per mole of radio waves. Consider that the bond energy of a carbon–carbon single bond found in organic compounds is 347 kJ/mol. Would X rays and/or radio waves be able to disrupt organic compounds by breaking carbon–carbon single bonds?

27. The work function of an element is the energy required to remove an electron from the surface of the solid. The work function for lithium is 279.7 kJ/mol (that is, it takes 279.7 kJ of energy to remove 1 mole of electrons from 1 mole of Li atoms on the surface of Li metal). What is the maximum wavelength of light that can remove an electron from an atom in lithium metal?

28. Ionization energy is the energy required to remove an electron from an atom in the gas phase. The ionization energy of gold is 890.1 kJ/mol. Is light with a wavelength of 225 nm capable of ionizing a gold atom (removing an electron) in the gas phase?

29. It takes 208.4 kJ of energy to remove 1 mole of electrons from the atoms on the surface of rubidium metal. If rubidium metal is irradiated with 254-nm light, what is the maximum kinetic energy the released electrons can have?

30. What experimental evidence supports the quantum theory of light? Explain the wave–particle duality of all matter. For what size particle must one consider both the wave and the particle properties?

31. Explain the photoelectric effect.

32. Calculate the de Broglie wavelength for each of the following.
    a. an electron with a velocity 10.% of the speed of light
    b. a tennis ball (55 g) served at 35 m/s (~80 mi/h)

33. Neutron diffraction is used in determining the structures of molecules.
    a. Calculate the de Broglie wavelength of a neutron moving at 1.00% of the speed of light.
    b. Calculate the velocity of a neutron with a wavelength of 75 pm (1 pm = 10^{-12} m).

34. Calculate the velocities of electrons with de Broglie wavelengths of 1.0×10^2 nm and 1.0 nm, respectively.

35. An atom of a particular element is traveling at 1% of the speed of light. The de Broglie wavelength is found to be 3.31×10^{-3} pm. Which element is this?

Hydrogen Atom: The Bohr Model

36. Characterize the Bohr model of the atom. In the Bohr model, what do we mean when we say something is quantized? How does the Bohr model of the hydrogen atom explain the hydrogen emission spectrum? Why is the Bohr model fundamentally incorrect?

37. The following is an energy-level diagram illustrating three different electronic transitions in the Bohr hydrogen atom.

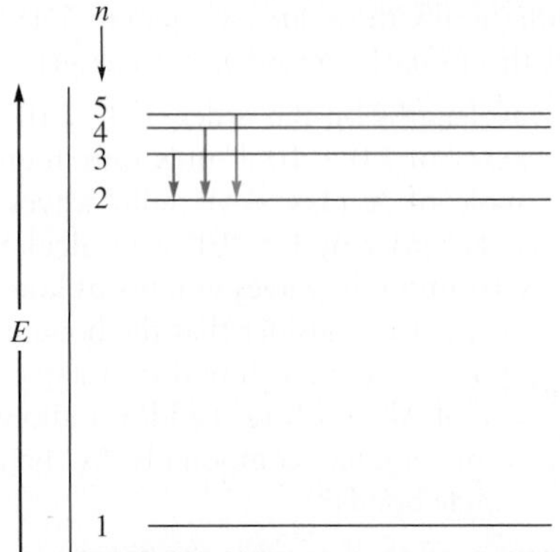

    a. Explain why the energy levels get closer together as they increase. Provide mathematical support for this.
    b. Verify that the colors given in the diagram are correct. Provide mathematical support.

38. Calculate the wavelength of light emitted when each of the following transitions occur in the hydrogen atom. What type of electromagnetic radiation is emitted in each transition?
    a. $n = 4 \rightarrow n = 3$
    b. $n = 5 \rightarrow n = 4$
    c. $n = 5 \rightarrow n = 3$

39. Assume that a hydrogen atom's electron has been excited to the $n = 5$ level. How many different wavelengths of light can be emitted as this excited atom loses energy?

40. What is the maximum wavelength of light capable of removing an electron from a hydrogen atom in the energy states characterized by $n = 1$ and $n = 3$?

41. An electron is excited from the ground state to the $n = 3$ state in a hydrogen atom. Which of the following statements are true? Correct any false statements.
    a. It takes more energy to ionize (remove) the electron from $n = 3$ than from the ground state.
    b. The electron is farther from the nucleus on average in the $n = 3$ state than in the ground state.
    c. The wavelength of light emitted if the electron drops from $n = 3$ to $n = 2$ is shorter than the wavelength of light emitted if the electron falls from $n = 3$ to $n = 1$.
    d. The wavelength of light emitted when the electron returns to the ground state from $n = 3$ is the same as the wavelength of light absorbed to go from $n = 1$ to $n = 3$.
    e. The first excited state corresponds to $n = 3$.

42. Does a photon of visible light (λ = 400–700 nm) have sufficient energy to excite an electron in a hydrogen atom from the $n = 1$ to the $n = 5$ energy state? From the $n = 2$ to the $n = 6$ energy state?

43. An excited hydrogen atom emits light with a wavelength of 397.2 nm to reach the energy level for which $n = 2$. In which principal quantum level did the electron begin?

44. An excited hydrogen atom with an electron in the $n = 5$ state emits light having a frequency of 6.90×10^{14} s^{-1}. Determine the principal quantum level for the final state in this electronic transition.

45. Consider an electron for a hydrogen atom in an excited state. The maximum wavelength of electromagnetic radiation that can completely remove (ionize) the electron from the H atom is 1460 nm. Determine the initial excited state for the electron (n = ?).

46. Calculate the energy (in kJ/mol) required to remove the electron in the ground state for each of the following one-electron species using the Bohr model.
    a. H b. He^+ c. Li^{2+} d. C^{5+} e. Fe^{25+}

47. One of the emission spectral lines for Be^{3+} has a wavelength of 253.4 nm for an electronic transition that begins in the state with $n = 5$. What is the principal quantum number of the lower-energy state corresponding to this emission?

Wave Mechanics and Particle in a Box

48. The Heisenberg uncertainty principle can be expressed in the form

$$\Delta E \cdot \Delta t \geq \frac{\hbar}{2}$$

where E represents energy and t represents time. Show that the units for this form are the same as the units for the form used in this chapter:

$$\Delta x \cdot \Delta p \geq \frac{\hbar}{2}$$

49. Using the Heisenberg uncertainty principle, calculate Δx for each of the following.
a. an electron with $\Delta v = 0.100$ m/s
b. a baseball (mass = 145 g) with $\Delta v = 0.100$ m/s
How does the answer in part a compare with the size of a hydrogen atom? How does the answer in part b correspond to the size of a baseball?

50. We can represent both probability and radial probability versus distance from the nucleus for a hydrogen 1*s* orbital as depicted below.

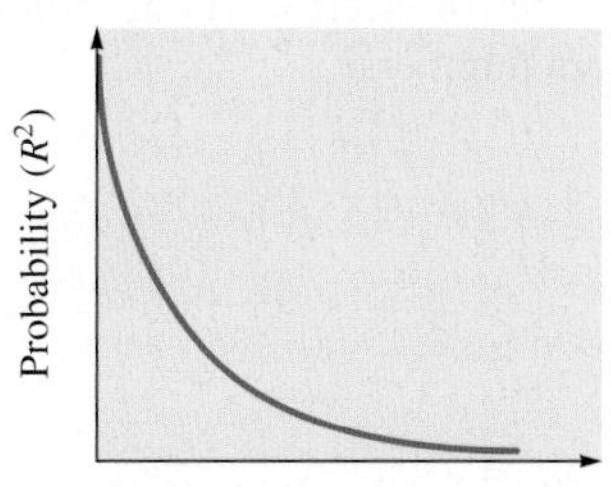

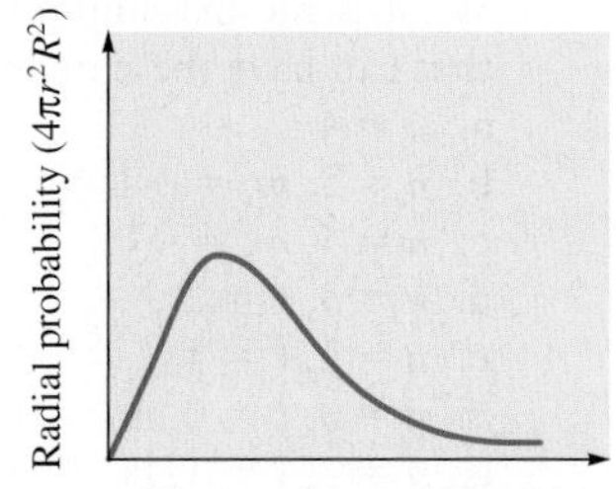

What does each graph tell us about the electron in a hydrogen 1*s* orbital? Describe the significance of the radial probability distribution.

51. Discuss why a function of the type $A \cos(Lx)$ is not an appropriate solution for the particle in a one-dimensional box.

52. Calculate the wavelength of the electromagnetic radiation required to excite an electron from the ground state to the level with $n = 5$ in a one-dimensional box 40.0 pm in length.

53. An electron in a one-dimensional box requires a wavelength of 8080 nm to excite an electron from the $n = 2$ to the $n = 3$ energy level. Calculate the length of this box.

54. An electron in a 10.0-nm one-dimensional box is excited from the ground state into a higher-energy state by absorbing a photon of electromagnetic radiation with a wavelength of 1.374×10^{-5} m. Determine the final energy state for this transition.

55. Discuss what happens to the energy levels for an electron trapped in a one-dimensional box as the length of the box increases.

56. What is the total probability of finding a particle in a one-dimensional box in level $n = 3$ between $x = 0$ and $x = L/6$?

57. Which has the lowest (ground-state) energy, an electron trapped in a one-dimensional box of length 10^{-6} m or one with length 10^{-10} m?

Orbitals and Quantum Numbers

58. What are quantum numbers? What information do we get from the quantum numbers n, ℓ, and m_ℓ? We define a spin quantum number (m_s), but do we know that an electron literally spins?

59. How do 2*p* orbitals differ from each other? How do 2*p* and 3*p* orbitals differ from each other? What is a nodal surface in an atomic orbital?

60. Identify each of the following orbitals, and determine the n and l quantum numbers. Explain your answers.

a.

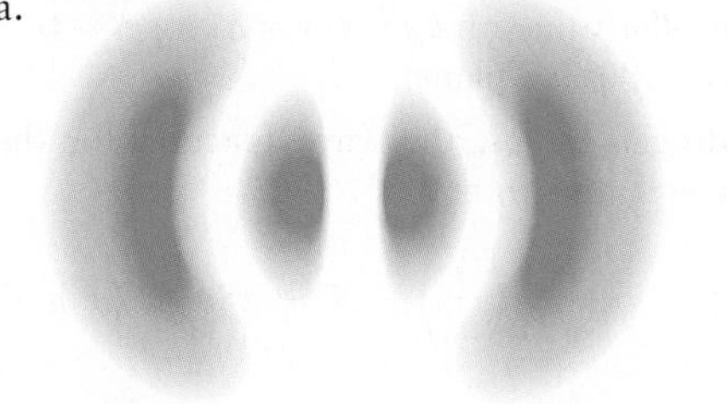

b.

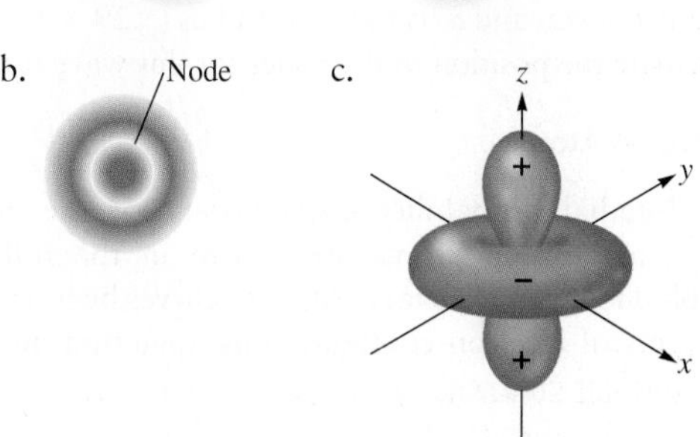

61. Which of the following orbital designations are incorrect: 1*s*, 1*p*, 7*d*, 9*s*, 3*f*, 4*f*, 2*d*?

62. Which of the following sets of quantum numbers are not allowed in the hydrogen atom? For the sets of quantum numbers that are incorrect, state what is wrong in each set.
a. $n = 3, \ell = 2, m_\ell = 2$
b. $n = 4, \ell = 3, m_\ell = 4$
c. $n = 0, \ell = 0, m_\ell = 0$
d. $n = 2, \ell = -1, m_\ell = 1$

63. Which of the following sets of quantum numbers are not allowed? For each incorrect set, state why it is incorrect.
a. $n = 3, \ell = 3, m_\ell = 0, m_s = -\frac{1}{2}$
b. $n = 4, \ell = 3, m_\ell = 2, m_s = -\frac{1}{2}$
c. $n = 4, \ell = 1, m_\ell = 1, m_s = +\frac{1}{2}$
d. $n = 2, \ell = 1, m_\ell = -1, m_s = -1$
e. $n = 5, \ell = -4, m_\ell = 2, m_s = +\frac{1}{2}$
f. $n = 3, \ell = 1, m_\ell = 2, m_s = -\frac{1}{2}$

64. How many orbitals can have the designation 5*p*, $3d_{z^2}$, 4*d*, $n = 5$, and $n = 4$?

65. How many electrons in an atom can have the designation 1*p*, $6d_{x^2-y^2}$, 4*f*, $7p_y$, 2*s*, and $n = 3$?

66. What is the physical significance of the value of ψ^2 at a particular point in an atomic orbital?

67. In defining the sizes of orbitals, why must we use an arbitrary value, such as 90% of the probability of finding an electron in that region?

68. From the diagrams of 2*p* and 3*p* orbitals in Fig. 12.19 and Fig. 12.20, respectively, draw a rough graph of the square of the wave function for these orbitals in the direction of one of the lobes.

69. The wave function for the $2p_z$ orbital in the hydrogen atom is

$$\psi_{2p_z} = \frac{1}{4\sqrt{2\pi}}\left(\frac{Z}{a_0}\right)^{3/2} \sigma e^{-\sigma/2} \cos\theta$$

where a_0 is the value for the radius of the first Bohr orbit in meters (5.29×10^{-11}), σ is Zr/a_0, r is the value for the distance from the nucleus in meters, and θ is an angle. Calculate the value of $\psi_{2p_z}{}^2$ at $r = a_0$ for $\theta = 0$ (z axis) and for $\theta = 90°$ (xy plane).

70. For hydrogen atoms, the wave function for the state $n = 3$, $\ell = 0$, and $m_\ell = 0$ is

$$\psi_{300} = \frac{1}{81\sqrt{3\pi}}\left(\frac{1}{a_0}\right)^{3/2}(27 - 18\sigma + 2\sigma^2)e^{-\sigma/3}$$

where $\sigma = r/a_0$ and a_0 is the Bohr radius (5.29×10^{-11} m). Calculate the position of the nodes for this wave function.

Polyelectronic Atoms

71. Total radial probability distributions for the helium, neon, and argon atoms are shown in the following graph. How can the shapes of these curves be interpreted in terms of electron configurations, quantum numbers, and nuclear charges?

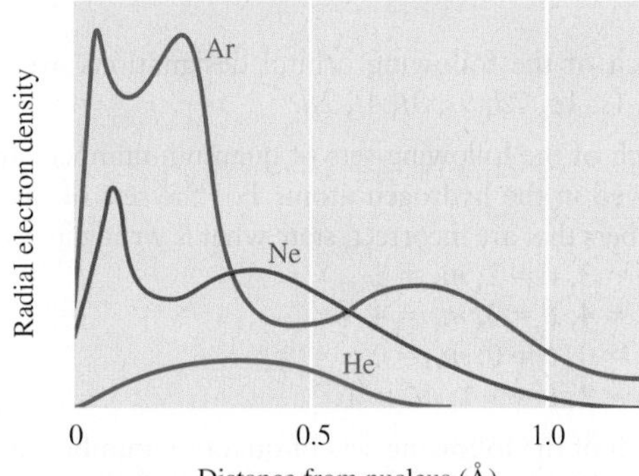

72. The relative orbital levels for the hydrogen atom can be represented as

E

3s — 3p — — — 3d — — — — —

2s — 2p — — —

1s —

Draw the relative orbital energy levels for atoms with more than one electron, and explain your answer. Also explain how the following radial probability distributions support your answer.

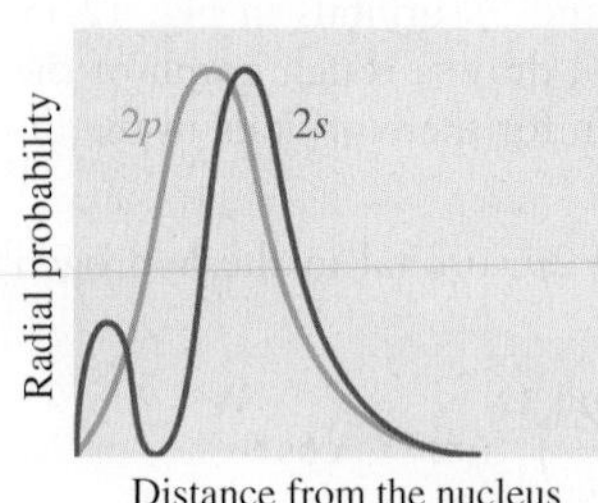

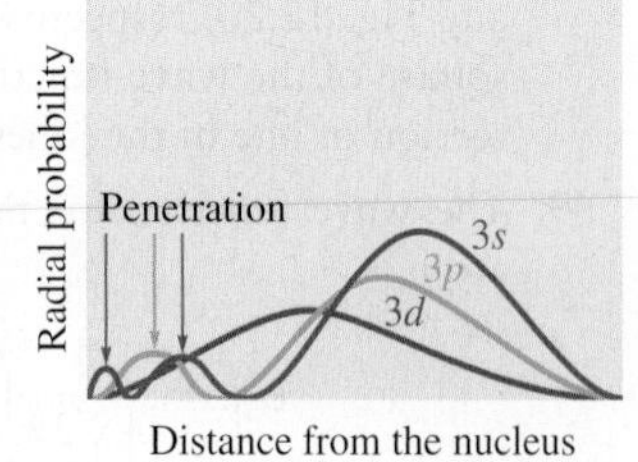

73. What is the difference between core electrons and valence electrons? Why do we emphasize the valence electrons in an atom when discussing atomic properties? What is the relationship between valence electrons and elements in the same group of the periodic table?

74. The periodic table consists of four blocks of elements that correspond to s, p, d, and f orbitals being filled. After f orbitals come g and h orbitals. In theory, if a g block and an h block of elements existed, how long would the rows of g and h elements be in this theoretical periodic table?

75. What is the maximum number of electrons in an atom that can have these quantum numbers?
    a. $n = 4$
    b. $n = 5$, $m_\ell = +1$
    c. $n = 5$, $m_s = +\frac{1}{2}$
    d. $n = 3$, $\ell = 2$
    e. $n = 2$, $\ell = 1$
    f. $n = 0$, $\ell = 0$, $m_\ell = 0$
    g. $n = 2$, $\ell = 1$, $m_\ell = -1$, $m_s = -\frac{1}{2}$
    h. $n = 3$, $m_s = +\frac{1}{2}$
    i. $n = 2$, $\ell = 2$
    j. $n = 1$, $\ell = 0$, $m_\ell = 0$

76. The elements of Si, Ga, As, Ge, Al, Cd, S, and Se are all used in the manufacture of various semiconductor devices. Write the expected electron configurations for these atoms.

77. Write the expected electron configurations for the following atoms: Sc, Fe, P, Cs, Eu, Pt, Xe, and Br.

78. Write the expected electron configurations for each of the following atoms: Cl, As, Sr, W, Pb, and Cf.

79. Using Fig. 12.29, list elements (ignore the lanthanides and actinides) that have ground-state electron configurations that differ from those we would expect from their positions in the periodic table.

80. Write the expected ground-state electron configuration for the following.
    a. the element with one unpaired $5p$ electron that forms a covalent compound with fluorine
    b. the (as yet undiscovered) alkaline earth metal after radium
    c. the noble gas with electrons occupying $4f$ orbitals
    d. the first-row transition metal with the most unpaired electrons

81. For elements 1–36, there are two exceptions to the filling order as predicted from the periodic table. Draw the atomic orbital diagrams for the two exceptions, and indicate how many unpaired electrons are present.

82. Given the valence electron orbital level diagram and the description, identify the element or ion.
    a. A ground state atom

    3s: [↑↓] 3p: [↑↓][↑][↑]

    b. An atom in an excited state (assume two electrons occupy the 1s orbital)

    2s: [↑] 2p: [↑↓][↑][↑]

c. A ground state ion with a charge of -1

4s: [↑↓] 4p: [↑↓|↑↓|↑]

83. How many valence electrons do each of the following elements have, and what are the specific valence electrons for each element?

a. Ca
b. O
c. element 117
d. In
e. Ar
f. Bi

84. In the ground state of mercury (Hg),

a. how many electrons occupy atomic orbitals with $n = 3$?
b. how many electrons occupy d atomic orbitals?
c. how many electrons occupy p_z atomic orbitals?
d. how many electrons have spin "up" ($m_s = +\frac{1}{2}$)?

85. In the ground state of element 115, Uup,

a. how many electrons have $n = 5$ as one of their quantum numbers?
b. how many electrons have $\ell = 3$ as one of their quantum numbers?
c. how many electrons have $m_\ell = 1$ as one of their quantum numbers?
d. how many electrons have $m_s = -\frac{1}{2}$ as one of their quantum numbers?

86. Give possible values for the quantum numbers of the valence electrons in an atom of titanium (Ti).

87. One bit of evidence that the quantum mechanical model is "correct" lies in the magnetic properties of matter. Atoms with unpaired electrons are attracted by magnetic fields and thus are said to exhibit *paramagnetism*. The degree to which this effect is observed is directly related to the number of unpaired electrons present in the atom. Consider the ground-state electron configurations for Li, N, Ni, Te, Ba, and Hg. Which of these atoms would be expected to be paramagnetic, and how many unpaired electrons are present in each paramagnetic atom?

88. Which of elements 1–36 have two unpaired electrons in the ground state?

89. Which of elements 1–36 have one unpaired electron in the ground state?

90. A certain oxygen atom has the electron configuration $1s^22s^22p_x^22p_y^2$. How many unpaired electrons are present? Is this an excited state for oxygen? In going from this state to the ground state, would energy be released or absorbed?

91. How many unpaired electrons are present in each of the following in the ground state: O, O^+, O^-, Os, Zr, S, F, Ar?

92. Which of the following electron configurations correspond to an excited state? Identify the atoms, and write the ground-state electron configuration where appropriate.

a. $1s^22s^23p^1$
b. $1s^22s^22p^6$
c. $1s^22s^22p^43s^1$
d. $[Ar]4s^23d^54p^1$

The Periodic Table and Periodic Properties

93. Using the element phosphorus as an example, write equations for the processes in which the energy change will correspond to the ionization energy and to the electron affinity.

94. Explain why the first ionization energy tends to increase as one proceeds from left to right across a period. Why is the first ionization energy of aluminum lower than that of magnesium and the first ionization energy of sulfur lower than that of phosphorus?

95. Why do the successive ionization energies of an atom always increase? Note the successive ionization energies for silicon given in Table 12.6. Would you expect to see any large jumps between successive ionization energies of silicon as you removed all the electrons, one by one, beyond those shown in the table?

96. The radius trend and the ionization energy trend are exact opposites. Does this make sense? Define electron affinity. Electron affinity values are both exothermic (negative) and endothermic (positive). However, ionization energy values are always endothermic (positive). Explain.

97. Arrange the following groups of atoms in order of increasing size.

a. Te, S, Se
b. K, Br, Ni
c. Ba, Si, F
d. Rb, Na, Be
e. Sr, Se, Ne
f. Fe, P, O

98. Arrange the atoms in Exercise 97 in order of increasing first ionization energy.

99. In each of the following sets, which atom or ion has the smallest ionization energy?

a. Ca, Sr, Ba
b. K, Mn, Ga
c. N, O, F
d. S^{2-}, S, S^{2+}
e. Cs, Ge, Ar

100. In each of the following sets, which atom or ion has the smallest radius?

a. H, He
b. Cl, In, Se
c. element 120, element 119, element 117
d. Nb, Zn, Si
e. Na^-, Na, Na^+

101. The first ionization energies of As and Se are 0.947 MJ/mol and 0.941 MJ/mol, respectively. Rationalize these values in terms of electron configurations.

102. Rank the elements Be, B, C, N, and O in order of increasing first ionization energy. Explain your reasoning.

103. We expect the atomic radius to increase down a group in the periodic table. Can you suggest why the atomic radius of hafnium breaks this rule? (See the following data.)

Element	Atomic Radius (Å)	Element	Atomic Radius (Å)
Sc	1.57	Ti	1.477
Y	1.693	Zr	1.593
La	1.915	Hf	1.476

104. Three elements have the electron configurations $1s^22s^22p^63s^23p^6$, $1s^22s^22p^63s^2$, and $1s^22s^22p^63s^23p^64s^1$. The first ionization energies of these elements (not in the same order) are 0.419, 0.735, and 1.527 MJ/mol. The atomic radii are 1.60, 0.98, and 2.35 Å. Identify the

three elements, and match the appropriate values of ionization energy and atomic radius to each configuration.

105. Predict some of the properties of element 117 (symbol Uus following conventions proposed by the International Union of Pure and Applied Chemistry [IUPAC]).
   a. What will be its electron configuration?
   b. What element will it most resemble chemically?
   c. What will be the formulas of the neutral binary compounds it forms with sodium, magnesium, carbon, and oxygen?
   d. What oxyanions would you expect Uus to form?

106. Order each of the following sets from the least exothermic electron affinity to the most.
   a. F, Cl, Br, I   b. N, O, F

107. In the second row of the periodic table, Be, N, and Ne all have endothermic (unfavorable) electron affinities, whereas the other second-row elements have exothermic (favorable) electron affinities. Rationalize why Be, N, and Ne have unfavorable electron affinities.

108. Which has the more negative electron affinity, the oxygen atom or the O^- ion? Explain your answer.

109. The electron affinity for sulfur is more exothermic than that for oxygen. How do you account for this?

110. The electron affinities of the elements from aluminum to chlorine are −44 kJ/mol, −120 kJ/mol, −74 kJ/mol, −200.4 kJ/mol, and −348.7 kJ/mol, respectively. Rationalize the trend in these values.

111. Use data in this chapter to determine the following.
   a. the electron affinity of Mg^{2+}
   b. the electron affinity of Al^+
   c. the ionization energy of Cl^-
   d. the ionization energy of Cl
   e. the electron affinity of Cl^+

112. For each of the following pairs of elements,

   (C and N)   (Ar and Br)   (Mg and K)   (F and Cl)

   pick the one with
   a. the more favorable (exothermic) electron affinity
   b. the higher ionization energy
   c. the larger size

The Alkali Metals

113. Does the information on alkali metals in Table 12.9 of the text confirm the general periodic trends in ionization energy and atomic radius? Explain.

114. An ionic compound of potassium and oxygen has the empirical formula KO. Would you expect this compound to be potassium(II) oxide or potassium peroxide? Explain.

115. Complete and balance the equations for the following reactions.
   a. $Li(s) + N_2(g) \longrightarrow$
   b. $Rb(s) + S(s) \longrightarrow$
   c. $Cs(s) + H_2O(l) \longrightarrow$
   d. $Na(s) + Cl_2(g) \longrightarrow$

116. Cesium was discovered in natural mineral waters in 1860 by R. W. Bunsen and G. R. Kirchhoff, using the spectroscope they invented in 1859. The name comes from the Latin word *caesius,* meaning "sky blue," which describes the prominent blue line observed for this element at 455.5 nm. Calculate the frequency and energy of a photon of this light.

117. The bright yellow light emitted by a sodium vapor lamp consists of two emission lines at 589.0 nm and 589.6 nm. What are the frequency and the energy of a photon of light at each of these wavelengths? What are the energies in kJ/mol?

118. Give the name and formula of the binary compound formed by each of the following pairs of elements.
   a. Li and N   d. Li and P
   b. Na and Br   e. Rb and H
   c. K and S   f. Na and H

119. Predict the atomic number of the next alkali metal after francium, and give its ground-state electron configuration.

Additional Exercises

120. Spectroscopists use emission spectra to confirm the presence of an element in materials of unknown composition. How is this possible?

121. On which quantum number(s) does the energy of an electron depend in each of the following?
   a. a one-electron atom or ion
   b. an atom or ion with more than one electron

122. Elements with very large ionization energies also tend to have highly exothermic electron affinities. Explain. Which group of elements would you expect to be an exception to this statement?

123. Diagonal relationships in the periodic table exist as well as vertical relationships. For example, Be and Al are similar in some of their properties, as are B and Si. Rationalize why these diagonal relationships hold for properties such as size, ionization energy, and electron affinity.

124. A certain microwave oven delivers 750. watts (J/s) of power to a coffee cup containing 50.0 g of water at 25.0°C. If the wavelength of microwaves in the oven is 9.75 cm, how long does it take, and how many photons must be absorbed, to make the water boil? The specific heat capacity of water is 4.18 J $°C^{-1}$ g^{-1}. Assume that only the water absorbs the energy of the microwaves.

125. Mars is roughly 60 million km from earth. How long does it take for a radio signal originating from earth to reach Mars?

126. Photogray lenses incorporate small amounts of silver chloride in the glass of the lens. When light hits the AgCl particles, the following reaction occurs:

$$AgCl \xrightarrow{h\nu} Ag + Cl$$

The silver metal formed causes the lenses to darken. The enthalpy change for this reaction is 3.10×10^2 kJ/mol. Assuming that all this energy must be supplied by light, what is the maximum wavelength of light that can cause this reaction?

127. Consider the following approximate visible light spectrum:

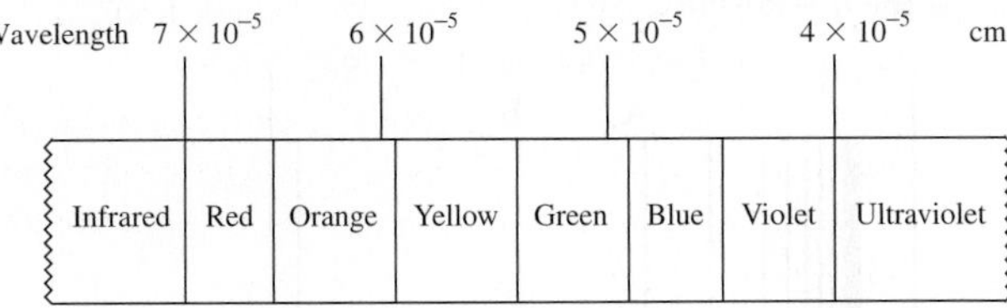

Barium emits light in the visible region of the spectrum. If each photon of light emitted from barium has an energy of 3.59×10^{-19} J, what color of visible light is emitted?

128. One of the visible lines in the hydrogen emission spectrum corresponds to the $n = 6$ to $n = 2$ electronic transition. What color light is this transition? See Exercise 127.

129. Consider the representations of the p and d atomic orbitals in Figs. 12.19 and 12.21. What do the + and − signs indicate?

130. The following graph plots the first, second, and third ionization energies for Mg, Al, and Si. Without referencing the text, which plot corresponds to which element? In one of the plots, there is a huge jump in energy between I_2 and I_3, unlike in the other two plots. Explain this phenomenon.

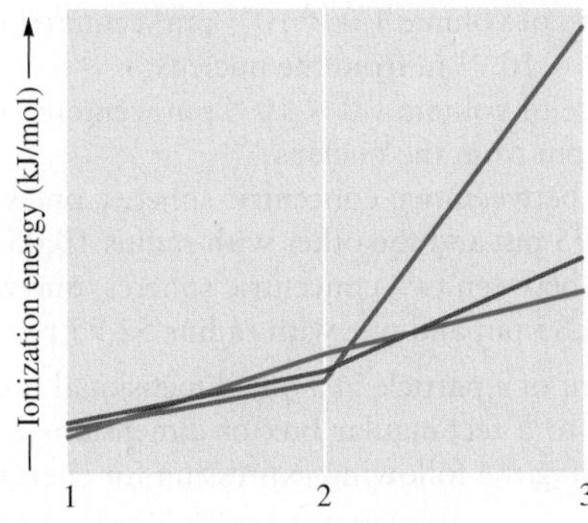

131. Using data from this chapter, calculate the change in energy expected for each of the following processes.
   a. $Na(g) + Cl(g) \longrightarrow Na^+(g) + Cl^-(g)$
   b. $Mg(g) + F(g) \longrightarrow Mg^+(g) + F^-(g)$
   c. $Mg^+(g) + F(g) \longrightarrow Mg^{2+}(g) + F^-(g)$
   d. $Mg(g) + 2F(g) \longrightarrow Mg^{2+}(g) + 2F^-(g)$

132. Write equations corresponding to the following energy terms.
   a. the fourth ionization energy of Se
   b. the electron affinity of S^-
   c. the electron affinity of Fe^{3+}
   d. the ionization energy of Mg
   e. the work function of Mg (see Exercise 27)

133. The successive ionization energies for an unknown element are

$$I_1 = 896 \text{ kJ/mol}$$
$$I_2 = 1752 \text{ kJ/mol}$$
$$I_3 = 14{,}807 \text{ kJ/mol}$$
$$I_4 = 17{,}948 \text{ kJ/mol}$$

To which family in the periodic table does the unknown element most likely belong?

134. An unknown element is a nonmetal and has a valence electron configuration of ns^2np^4.
   a. How many valence electrons does this element have?
   b. What are some possible identities for this element?
   c. What is the formula of the compound this element would form with potassium?
   d. Would this element have a larger or smaller radius than barium?
   e. Would this element have a greater or smaller ionization energy than fluorine?

135. An ion having a 4+ charge and a mass of 49.9 amu has two electrons with $n = 1$, eight electrons with $n = 2$, and ten electrons with $n = 3$. Supply the following properties for the ion. (*Hint:* In forming ions, the 4s electrons are lost before the 3d electrons.)
   a. the atomic number
   b. total number of s electrons
   c. total number of p electrons
   d. total number of d electrons
   e. the number of neutrons in the nucleus
   f. the mass of 3.01×10^{23} atoms
   g. the ground-state electron configuration of the neutral atom

136. Consider the following ionization energies for aluminum.

$$Al(g) \longrightarrow Al^+(g) + e^- \qquad I_1 = 580 \text{ kJ/mol}$$
$$Al^+(g) \longrightarrow Al^{2+}(g) + e^- \qquad I_2 = 1815 \text{ kJ/mol}$$
$$Al^{2+}(g) \longrightarrow Al^{3+}(g) + e^- \qquad I_3 = 2740 \text{ kJ/mol}$$
$$Al^{3+}(g) \longrightarrow Al^{4+}(g) + e^- \qquad I_4 = 11{,}600 \text{ kJ/mol}$$

   a. Account for the increasing trend in the values of the ionization energies.
   b. Explain the large increase between I_3 and I_4.
   c. Which one of the four ions has the greatest electron affinity? Explain.
   d. List the four aluminum ions given in the preceding reactions in order of increasing size, and explain your ordering. (*Hint:* Remember that most of the size of an atom or ion is due to its electrons.)

137. Answer the following questions, assuming that m_s has four values rather than two and that the normal rules apply for n, ℓ, and m_ℓ.
   a. How many electrons could an orbital hold?
   b. How many elements would be contained in the first and second periods of the periodic table?
   c. How many elements would be contained in the first transition metal series?
   d. How many electrons would the set of 4f orbitals be able to hold?

138. Although Mendeleev predicted the existence of several undiscovered elements, he did not predict the existence of the noble gases, the lanthanides, or the actinides. Propose reasons why Mendeleev was not able to predict the existence of the noble gases.

139. Human color vision is "produced" by the nervous system based on how three different cone receptors interact with photons of light in the eye. These three different types of cones interact with photons of different frequency light, as indicated in the following chart:

Cone Type	Range of Light Frequency Detected
S	$6.00–7.49 \times 10^{14}$ s^{-1}
M	$4.76–6.62 \times 10^{14}$ s^{-1}
L	$4.28–6.00 \times 10^{14}$ s^{-1}

What wavelength ranges (and corresponding colors) do the three types of cones detect?

140. Assume that four electrons are confined to a one-dimensional box 5.64×10^{-10} m in length. If two electrons can occupy each allowed energy level, calculate the wavelength of electromagnetic radiation necessary to promote the highest-energy electron into the first excited state.

141. The figure below represents part of the emission spectrum for a one-electron ion in the gas phase. All the lines result from electronic transitions from excited states to the $n = 3$ state.

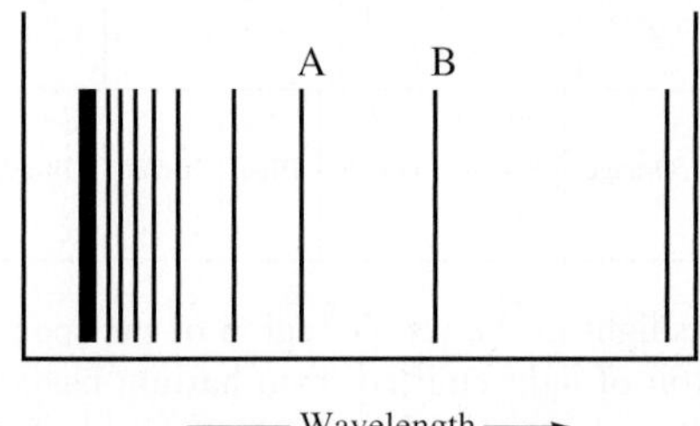

a. What electronic transitions correspond to lines A and B?
b. If the wavelength of line B is 142.5 nm, calculate the wavelength of line A.

Challenge Problems

142. An atom moving at its root mean square velocity at 100.°C has a wavelength of 2.31×10^{-11} m. Which atom is it?

143. The ground state ionization energy for the one electron ion X^{m+} is 4.72×10^4 kJ/mol. Identify X and m.

144. When the excited electron in a hydrogen atom falls from $n = 5$ to $n = 2$, a photon of blue light is emitted. If an excited electron in He^+ falls from $n = 4$, which energy level must it fall to so that a similar blue light (as with hydrogen) is emitted? Prove it.

145. The treatment of a particle in a one-dimensional box can be extended to a two-dimensional box of dimensions L_x and L_y yielding the following expression for energy:

$$E = \frac{h^2}{8m}\left(\frac{n_x^2}{L_x^2} + \frac{n_y^2}{L_y^2}\right)$$

The two quantum numbers independently can assume only integer values. Consider an electron confined to a two-dimensional box that is 8.00 nm in the x direction and 5.00 nm in the y direction.
a. What are the quantum numbers for the first three allowed energy levels?
b. Calculate the wavelength of light necessary to promote an electron from the first excited state to the second excited state.

146. The following numbers are the ratios of second ionization energy to first ionization energy:

Na:	9.2	P:	1.8
Mg:	2.0	S:	2.3
Al:	3.1	Cl:	1.8
Si:	2.0	Ar:	1.8

Explain these relative numbers.

147. For a hydrogen atom in its ground state, calculate the relative probability of finding the electron in the area described.
a. in a sphere of volume 1.0×10^{-3} pm^3 centered at the nucleus
b. in a sphere of volume 1.0×10^{-3} pm^3 centered on a point 1.0×10^{-11} m from the nucleus
c. in a sphere of volume 1.0×10^{-3} pm^3 centered on a point 53 pm from the nucleus
d. in a shell between two concentric spheres, one with radius 9.95 pm and the other with radius 10.05 pm
e. in a shell between two concentric spheres, one with radius 52.85 pm and one with radius 52.95 pm

148. The treatment of a particle in a one-dimensional box can be extended to a rectangular box of dimensions L_x, L_y, and L_z, yielding the following expression for energy:

$$E = \frac{h^2}{8m}\left(\frac{n_x^2}{L_x^2} + \frac{n_y^2}{L_y^2} + \frac{n_z^2}{L_z^2}\right)$$

The three quantum numbers n_x, n_y, and n_z independently can assume only integer values.
a. Determine the energies of the three lowest levels, assuming that the box is cubic.
b. Describe the degeneracies of all the levels that correspond to quantum numbers having values of 1 or 2. How will these degeneracies change in a box where $L_x \neq L_y \neq L_z$?

149. Assume that eight electrons are placed into the allowed energy levels of a cubic box where two electrons can occupy each allowed energy level. (See Exercise 148 for the appropriate energy equation.) Calculate the wavelength of light necessary to promote the highest-energy ground-state electron into the lowest-energy excited state assum-

ing a cubic box with dimensions 1.50 nm × 1.50 nm × 1.50 nm.

150. Assume that we are in another universe with different physical laws. Electrons in this universe are described by four quantum numbers with meanings similar to those we use. We will call these quantum numbers p, q, r, and s. The rules for these quantum numbers are as follows:

$p = 1, 2, 3, 4, 5, \ldots$

q takes on positive odd integer values and $q \leq p$.

r takes on all even integer values from $-q$ to $+q$.

(Zero is considered an even number.)

$s = +\frac{1}{2}$ or $-\frac{1}{2}$

a. Sketch what the first four periods of the periodic table will look like in this universe.
b. What are the atomic numbers of the first four elements you would expect to be least reactive?
c. Give an example, using elements in the first four rows, of ionic compounds with the formulas XY, XY_2, X_2Y, XY_3, and X_2Y_3.
d. How many electrons can have $p = 3$?
e. How many electrons can have $p = 4$, $q = 3$, $r = 2$?
f. How many electrons can have $p = 4$, $q = 3$?
g. How many electrons can have $p = 3$, $q = 0$, $r = 0$?
h. What are the possible values of q and r for $p = 5$?
i. How many electrons can have $p = 6$?

151. The ionization energy for a $1s$ electron in a silver atom is 2.462×10^6 kJ/mol.

a. Determine an approximate value for Z_{eff} for the Ag $1s$ electron. You will first have to derive an equation that relates Z_{eff} to ionization energy.
b. How does Z_{eff} from part a compare to Z for Ag? Rationalize the relative numbers.

152. Without looking at data in the text, sketch a qualitative graph of the third ionization energy versus atomic number for the elements Na through Ar, and explain your graph.

Marathon Problem*

153. From the information below, identify element X.

a. The wavelength of the radiowaves sent by an FM station broadcasting at 97.1 MHz is 30 million (3.00×10^7) times greater than the wavelength corresponding to the energy difference between a particular excited state of the hydrogen atom and the ground state.
b. Let V represent the principal quantum number for the valence shell of element X. If an electron in the hydrogen atom falls from shell V to the inner shell corresponding to the excited state mentioned in part a, the wavelength of light emitted is the same as the wavelength of an electron moving at a speed of 570. m/s.
c. The number of unpaired electrons for element X in the ground state is the same as the maximum number of electrons in an atom that can have the quantum number designations $n = 2$, $m_\ell = -1$, and $m_s = -\frac{1}{2}$
d. Let A represent the principal quantum number for the electron in an excited He^+ ion in which the single electron has the same energy as the electron in the ground state of a hydrogen atom. This value of A also represents the angular momentum quantum number for the subshell containing the unpaired electron(s) for element X.

---

*From James H. Burness, "The Use of "Marathon" Problems as Effective Vehicles for the Presentation of General Chemistry Lectures," Journal of Chemical Education, 68(11).

Bonding: General Concepts

13

chapter

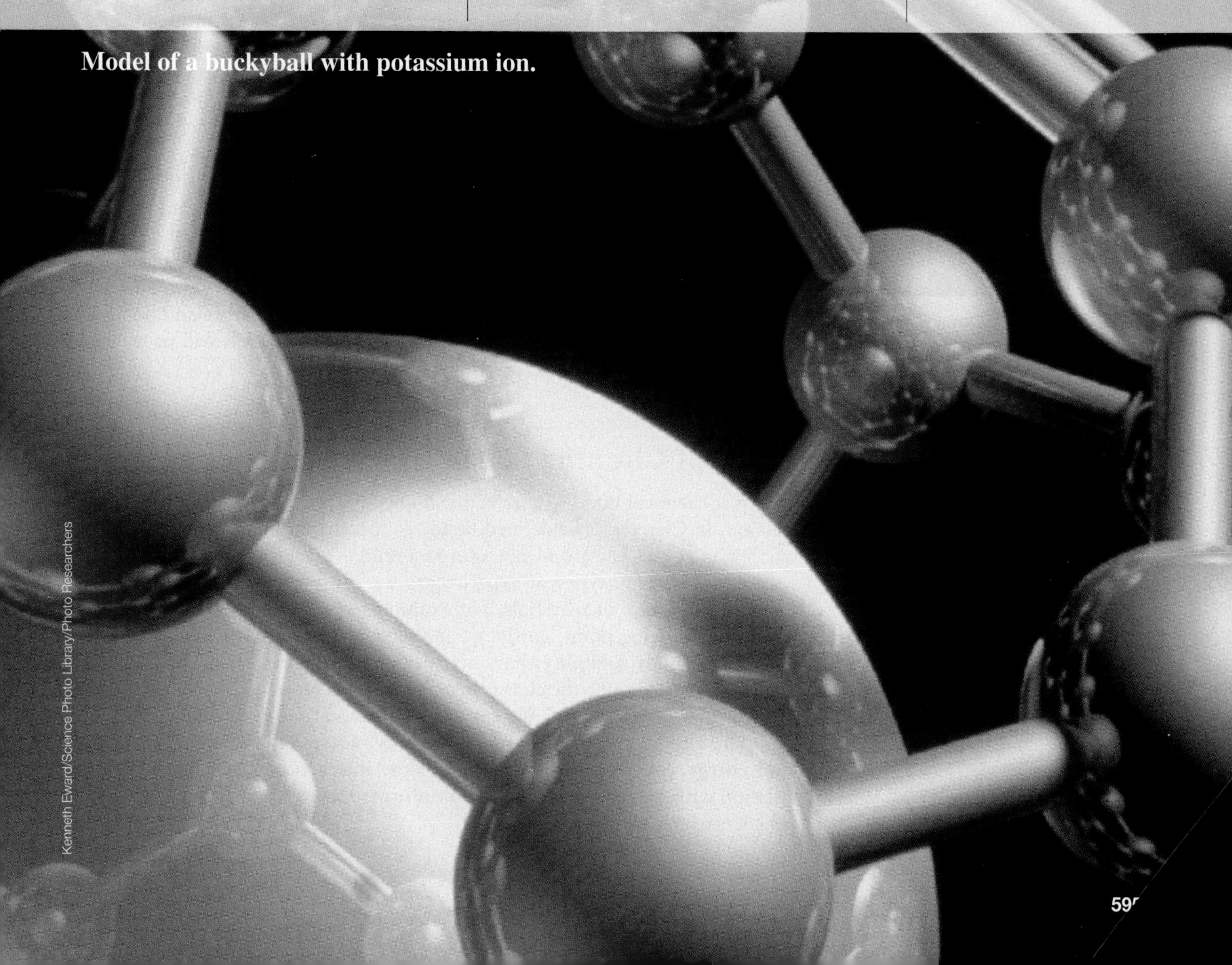

Model of a buckyball with potassium ion.

Kenneth Eward/Science Photo Library/Photo Researchers

OWL

Sign in to OWL at www.cengage.com/owl to view tutorials and simulations, develop problem-solving skills, and complete online homework assigned by your professor.

Download mini lecture videos for key concept review and exam prep from OWL or purchase them from www.cengagebrain.com.

(top) Quartz grows in beautiful, regular crystals. (bottom) Two forms of carbon: graphite and diamond.

The world around us is composed almost entirely of compounds and mixtures of compounds: Rocks, coal, soil, petroleum, trees, and human bodies are all complex mixtures of chemical compounds in which different kinds of atoms are bound together. Substances composed of unbound atoms do exist in nature, but they are very rare. Examples are the argon in the atmosphere and the helium mixed with natural gas reserves.

The manner in which atoms are bound together in a given substance has a profound effect on its chemical and physical properties. For example, graphite is a soft, slippery material used as a lubricant in locks, and diamond is one of the hardest materials known, valuable both as a gemstone and in industrial cutting tools. Why do these materials, both composed solely of carbon atoms, have such different properties? The answer, as we will see, lies in the bonding within these substances.

Silicon and carbon are next to each other in Group 4A on the periodic table. From our knowledge of periodic trends, we might expect SiO_2 and CO_2 to be very similar. But SiO_2 is the empirical formula of silica, which is found in sand and quartz, whereas carbon dioxide is a gas, a product of respiration. Why are they so different? We will be able to answer this question after we have developed models for bonding.

Bonding and structure play a central role in determining the course of all chemical reactions, many of which are vital to our survival. Later in this book we will demonstrate the importance of bonding and structure by showing how enzymes facilitate complex chemical reactions, how genetic characteristics are transferred, and how hemoglobin in the blood carries oxygen throughout the body. All these fundamental biological reactions hinge on the geometric structures of molecules, sometimes depending on very subtle differences in molecular shape to channel the chemical reaction one way rather than another.

Many of the world's current problems require fundamentally chemical answers: disease and pollution control, the search for new energy sources, the development of new fertilizers to increase crop yields, the improvement of the protein content in various staple grains, and many more. To understand the behavior of natural materials, we must understand the nature of chemical bonding and the factors that control the structures of compounds. In this chapter we will present various classes of compounds that illustrate the different types of bonds and then develop models to describe the structure and bonding that characterize materials found in nature. Later these models will prove useful in understanding chemical reactions.

13.1 Types of Chemical Bonds

What is a chemical bond? There is no simple and yet complete answer to this question. In Chapter 2 we defined bonds as forces that hold groups of atoms together and make the atoms function as a unit.

There are many types of experiments we can perform to determine the fundamental nature of materials. For example, we can study physical properties such as melting point, hardness, and electrical and thermal conductivity. We can also study solubility characteristics and the properties of the resulting solutions. To determine the charge distribution in a molecule, we can study its behavior in an electric field. We can obtain information about the strength of a bonding interaction by measuring the energy required to break the bond, the **bond energy**. Spectroscopy, the study of the interactions of electromagnetic radiation with matter, gives a wealth of information about molecular structure and energy level spacings.

There are several ways atoms can interact with one another to form aggregates. We will consider several specific examples to illustrate the various types of chemical bonds.

When solid sodium chloride is melted, it conducts electricity, a fact that convinces us that sodium chloride contains Na^+ and Cl^- ions. Thus, when sodium and chlorine react to form sodium chloride, electrons must be transferred from the sodium atoms to the chlorine atoms to form Na^+ and Cl^- ions, which then aggregate to form solid sodium chloride. Why does this happen? The best simple answer is that *the system can achieve the lowest possible energy by behaving in this way.* Part of the favorable energy change results from the attraction of a chlorine atom for an extra electron. Even more important are the very strong attractions between the oppositely charged ions. The resulting solid sodium chloride is a very sturdy material; it has a melting point of about 800°C. The bonding forces that produce this great thermal stability result from the electrostatic attractions of the closely packed, oppositely charged ions. This is an example of *ionic bonding.* Ionic substances are formed when an atom that loses electrons relatively easily reacts with an atom that has a high affinity for electrons. That is, an **ionic compound** results when a metal reacts with a nonmetal.

The energy of interaction between a pair of ions can be calculated by using **Coulomb's law:**

$$V = \frac{Q_1 Q_2}{4\pi\epsilon_0 r} = 2.31 \times 10^{-19}\ \text{J nm}\left(\frac{Q_1 Q_2}{r}\right)$$

where V has units of joules, r is the distance between the ion centers in nanometers, Q_1 and Q_2 are the numerical ion charges, and ϵ_0 is the permittivity of the vacuum. For example, in solid sodium chloride, where the distance between the centers of the Na^+ and Cl^- ions is 276 picometers (0.276 nm), the ionic energy per pair of ions is

$$V = 2.31 \times 10^{-19}\ \text{J nm}\left[\frac{(+1)(-1)}{0.276\ \text{nm}}\right] = -8.37 \times 10^{-19}\ \text{J}$$

The negative sign indicates an attractive force. That is, *the ion pair has lower energy than the separated ions.* For a mole of pairs of Na^+ and Cl^- ions, the energy of interaction is

$$V = \left(-8.37 \times 10^{-19}\ \frac{\text{J}}{\text{ion pair}}\right)\left(6.022 \times 10^{23}\ \frac{\text{ion pair}}{\text{mol}}\right)$$
$$= -504\ \text{kJ/mol}$$

Note that this energy refers to a mole of $Na^+ \cdot\cdot Cl^-$ ion pairs in the gas phase where a given pair is far from any other pair. In solid sodium chloride, which contains a large array of closely packed Na^+ and Cl^- ions, where a given ion is close to many oppositely charged ions, the energy associated with ionic bonding is much greater than 504 kJ/mol because of the larger numbers of interacting ions.

Coulomb's law can also be used to calculate the repulsive energy when two like-charged ions are brought together. In this case the calculated energy value will have a positive sign.

We have seen that a bonding force develops when two very different atoms react to form oppositely charged ions. But how does a bonding force develop between two identical atoms? Let's explore this situation from a very simple point of view by considering the energy terms that result when two hydrogen atoms are brought close together, as shown in Fig. 13.1(a). For two closely spaced hydrogen atoms, there are two unfavorable energy terms, proton–proton

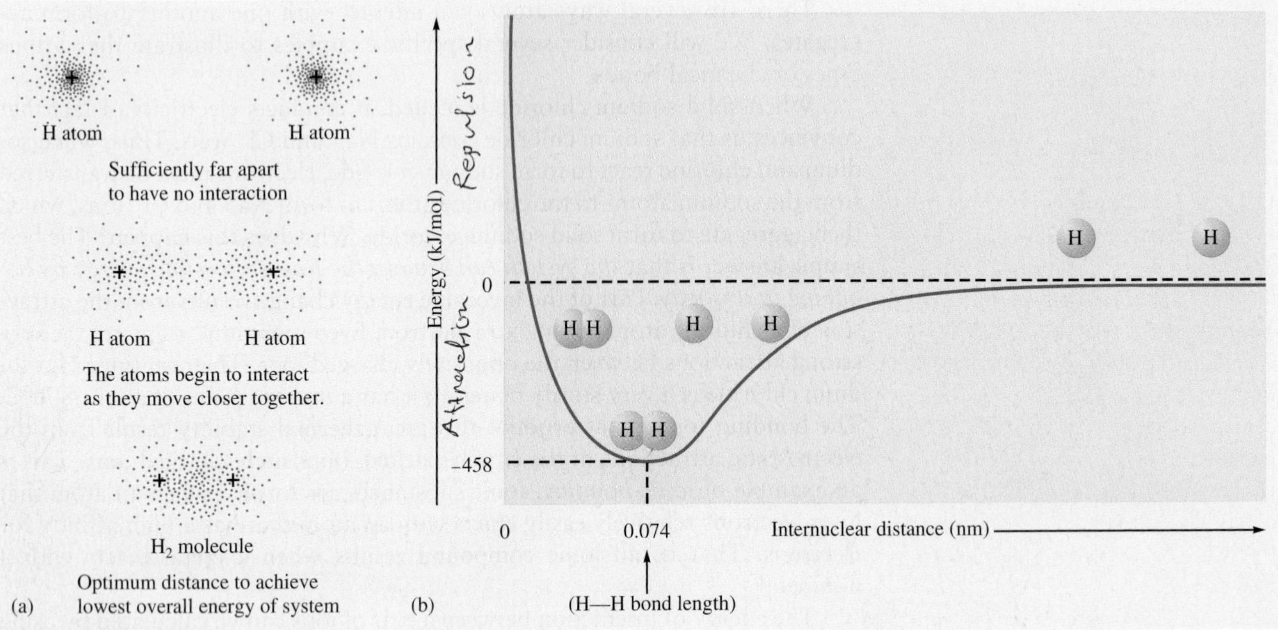

Figure 13.1
(a) The interaction of two hydrogen atoms. (b) Energy profile as a function of the distance between the nuclei of the hydrogen atoms. As the atoms approach each other, the energy decreases until the distance reaches 0.074 nm (0.74 Å) and then begins to increase again because of repulsions.

repulsion and electron–electron repulsion, and one favorable term, proton–electron attraction. Under what conditions will the H_2 molecule be favored over the separated hydrogen atoms? That is, what conditions will favor bond formation? The answer lies in nature's strong tendency to achieve the lowest possible energy. A bond will form—that is, the two hydrogen atoms will exist as a molecular unit—if the system can lower its total energy in the process.

Bonding occurs if the energy of the aggregate is lower than that of the separated atoms.

Therefore, the hydrogen atoms will assume the positions that give the lowest possible energy; the system will act to minimize the sum of the positive (repulsive) energy terms and the negative (attractive) energy terms. The distance at which the energy is minimum is called the equilibrium internuclear distance or, more commonly, the **bond length.** The total energy of this system as a function of distance between the hydrogen nuclei is shown in Fig. 13.1(b). Note four important features of this diagram:

1. The energy terms involved are the potential energy that results from the attractions and repulsions among the charged particles and the kinetic energy caused by the motions of the electrons.
2. The zero reference point for energy is defined for the atoms at infinite separation.
3. At very short distances the energy rises steeply because of the great importance of the internuclear repulsive forces at these distances.
4. The bond length is the distance at which the system has minimum energy, and the bond energy corresponds to the depth of the "well" at this distance.

The atoms in H_2 (and all other molecules) actually vibrate back and forth around the equilibrium internuclear distance.

In the H_2 molecule the electrons reside primarily in the space between the two nuclei, where they are attracted simultaneously by both protons. This positioning is precisely what leads to the stability of the H_2 molecule relative to two separated hydrogen atoms. The potential energy of each electron is lowered because of the increased attractive forces in this area. Although it is not usually discussed in connection with simple models of bonding, we should note that the kinetic energy of the electrons also changes when the individual atoms form the molecule. Thus the energy plotted in Fig. 13.1(b) is the total energy of the system, not just the potential energy. When we say that a bond is formed between the hydrogen atoms, we mean that the H_2 molecule is more stable

No Lead Pencils

Chemical Insights

Did you ever wonder why the part of a pencil that makes the mark is called the "lead"? Pencils have no lead in them now—and they never have. Apparently the association between writing and the element lead arose during the Roman Empire, when lead rods were used as writing utensils because they leave a gray mark on paper. Many centuries later, in 1564, a deposit of a black substance found to be very useful for writing was discovered in Borrowdale, England. This substance, originally called "black lead," was shown in 1879 by Swedish chemist Carl Sheele to be a form of carbon and was subsequently named graphite (after the Greek *graphein,* meaning "to write").

Originally, chunks of graphite from Borrowdale, called "marking stones," were used as writing instruments. Later, sticks of graphite were used. Because graphite is brittle, the sticks needed reinforcement. At first they were wrapped in string, which was unwound as the core wore down. Eventually, graphite rods were tied between two wooden slats or inserted into hollowed-out wooden sticks to form the first crude pencils.

Although Borrowdale graphite was pure enough to use directly, most graphite must be mixed with other materials to be useful for writing instruments. In 1795, French chemist Nicolas-Jaques Conté invented a process in which graphite is mixed with clay and water to produce pencil "lead," a recipe that is still used today. In modern pencil manufacture, graphite and clay are mixed and crushed into a fine powder to which water is added. After the gray sludge is blended for several days, it is dried, ground up again, and mixed with more water to give a gray paste. The paste is extruded through a metal tube to form thin rods, which are then cut into pencil-length pieces called "leads." These leads are heated in an oven to 1000°C until they are smooth and hard. The ratio of clay to graphite is adjusted to vary the hardness of the lead—the more clay in the mix, the harder the lead and the lighter the line it makes.

Pencils are made from a slat of wood with several grooves cut in it to hold the leads. A similar grooved slat is then placed on top and glued to form a "sandwich" from which individual pencils are cut, sanded smooth, and painted. Although many types of wood have been used over the years to make pencils, the current favorite is incense cedar from the Sierra Nevada Mountains of California.

Modern pencils are simple but amazing instruments. The average pencil can write approximately 45,000 words, which is equivalent to a line 35 miles long. The graphite in a pencil is easily transferred to paper because graphite contains layers of carbon atoms bound together in a "chicken-wire" structure. Although the bonding *within* each layer is very strong, the bonding *between* layers is weak, giving graphite its slippery, soft nature. In this way, graphite is much different from diamond, the other common elemental form of carbon. In diamond the carbon atoms are bound tightly in all three dimensions, making it extremely hard—the hardest natural substance.

Pencils are very useful—especially for doing chemistry problems—because we can erase our mistakes. Most pencils used in the United States have erasers (first attached to pencils in 1858), although most European pencils do not. Laid end to end, the number of pencils made in the United States each year would circle the earth about 15 times. Pencils illustrate how useful a simple substance like graphite can be.

than two separated hydrogen atoms by a certain quantity of energy (the bond energy).

We can also think of a bond in terms of forces. The simultaneous attraction for each electron by the two protons generates a force that pulls the protons toward each other. This attractive force just balances the proton–proton and electron–electron repulsive forces at the distance corresponding to the bond length.

The type of bonding we encounter in the hydrogen molecule and in many other molecules in which *electrons are shared by nuclei* is called **covalent bonding.**

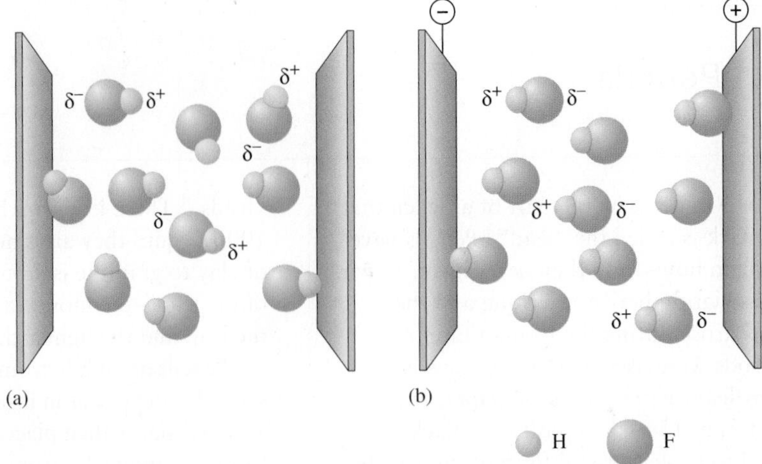

Figure 13.2
The effect of an electric field on hydrogen fluoride molecules. (a) When no electric field is present, the molecules are randomly oriented. (b) When the field is turned on, the molecules tend to line up with their negative ends toward the positive pole and their positive ends toward the negative pole. (This illustration exaggerates the effect. Actually, only a small fraction of the molecules are lined up with the field at a given instant.)

So far we have considered two extreme types of bonding. In ionic bonding the participating atoms are so different that one or more electrons are transferred to form oppositely charged ions. The bonding results from electrostatic interactions among the resulting ions. In covalent bonding two identical atoms share electrons equally. The bonding results from the mutual attraction of the two nuclei for the shared electrons. Between these extremes lie intermediate cases in which the atoms are not so different that electrons are completely transferred but are different enough so that unequal sharing results. These are called **polar covalent bonds.** An example of this type of bond occurs in the hydrogen fluoride (HF) molecule. When a sample of hydrogen fluoride gas is placed in an electric field, the molecules tend to orient themselves as shown in Fig. 13.2, with the fluoride end closest to the positive pole and the hydrogen end closest to the negative pole. This result implies that the HF molecule has the following charge distribution:

$$\underset{\delta+}{\mathrm{H}}\text{—}\underset{\delta-}{\mathrm{F}}$$

where δ (delta) is used to indicate a fractional charge. This same effect was noted in Chapter 4, where many of water's unusual properties were attributed to the polar O—H bonds in the H_2O molecule.

The most logical explanation for the development of the partial positive and negative charges on the atoms (bond polarity) in such molecules as HF and H_2O is that the electrons in the bonds are not shared equally. For example, we can account for the polarity of the HF molecule by assuming that the fluorine atom has a stronger attraction for the shared electrons than the hydrogen atom. Similarly, in the H_2O molecule the oxygen atom appears to attract the shared electrons more strongly than the hydrogen atoms. Because bond polarity has important chemical implications, we find it useful to quantify the ability of an atom to attract shared electrons. In the next section we show how this is done.

13.2 Electronegativity

The different affinities of atoms for the electrons in a bond are described by a property called **electronegativity:** *the ability of an atom in a molecule to attract shared electrons to itself.*

The most widely accepted method for determining electronegativity values is that of Linus Pauling (1901–1995), an American scientist who won Nobel Prizes for both chemistry and peace. To understand Pauling's model, consider a hypothetical molecule HX. The relative electronegativities of the H and X atoms are determined by comparing the measured H—X bond energy with the

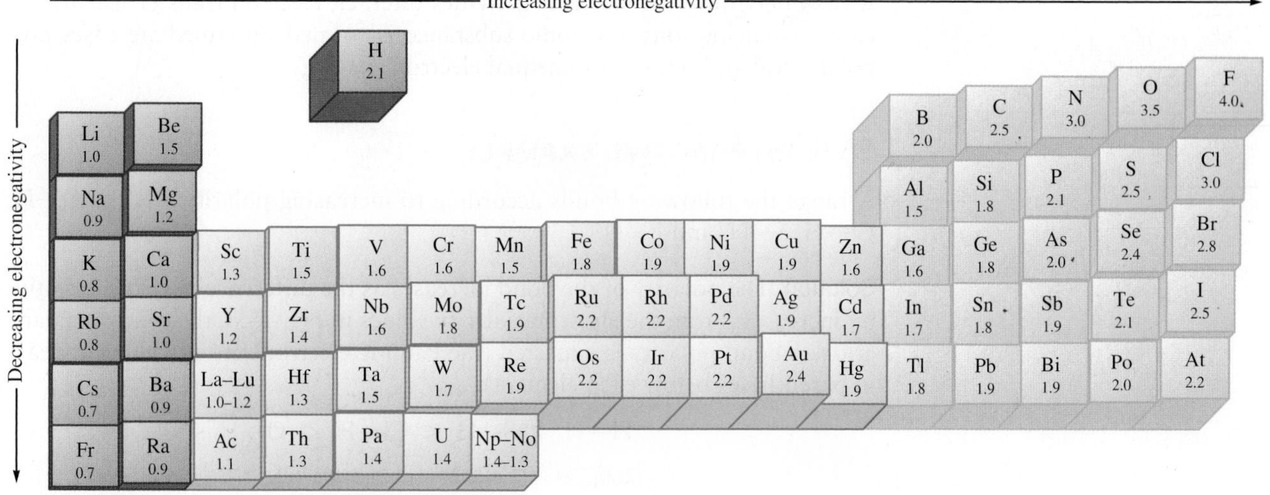

Figure 13.3
The Pauling electronegativity values. Electronegativity generally increases across a period and decreases down a group.

"expected" H—X bond energy. The expected bond energy is an "average" (actually the geometric mean) of the H—H and X—X bond energies:

$$\text{Expected H—X bond energy} = [(\text{H—H bond energy})(\text{X—X bond energy})]^{1/2}$$

The difference (Δ) between the actual (measured) and expected bond energies is

$$\Delta = (\text{H—X})_{\text{act}} - (\text{H—X})_{\text{exp}}$$

If H and X have identical electronegativities, $(\text{H—X})_{\text{act}}$ and $(\text{H—X})_{\text{exp}}$ are the same and Δ is 0. On the other hand, if X has a greater electronegativity than H, the shared electron(s) will tend to be closer to the X atom. The molecule will be polar, with the following charge distribution:

$$\underset{\delta+\ \ \delta-}{\text{H—X}}$$

Note that this bond can be viewed as having an ionic as well as a covalent component. The electrostatic attraction between the partially charged H and X atoms will lead to a greater bond strength. Thus $(\text{H—X})_{\text{act}}$ will be larger than $(\text{H—X})_{\text{exp}}$. The greater the difference in the electronegativities of the atoms, the greater is the ionic component of the bond and the greater is the value of Δ. Thus the relative electronegativities of H and X can be assigned from the Δ values.

The factor of 0.102 is a conversion factor between kJ and eV (the units originally used by Pauling).

The actual formula Pauling used to calculate electronegativity (EN) differences is

$$\text{EN(X)} - \text{EN(H)} = 0.102\sqrt{\Delta}$$

where all bond energies are in units of kJ/mol. Pauling then obtained absolute electronegativity values for the elements by assigning a value of 4.0 to fluorine (the element with the highest electronegativity).

Electronegativity values have been determined by this process for virtually all the elements; the results are given in Fig. 13.3. Note that for the representative elements, electronegativity generally increases from left to right across a period and decreases down a group. The range of electronegativity values is from 4.0 for fluorine to 0.7 for francium.

Table 13.1

The Relationship Between Electronegativity and Bond Type

Electronegativity Difference in the Bonding Atoms	Bond Type
Zero	Covalent
|	|
Intermediate	Polar covalent
↓	↓
Large	Ionic

The relationship between electronegativity and bond type is shown in Table 13.1. For identical atoms (electronegativity difference of zero), the electrons in the bond are shared equally and no polarity occurs. When two atoms with widely differing electronegativities interact, electron transfer usually occurs, producing ions—an ionic substance is formed. Intermediate cases give polar covalent bonds with unequal electron sharing.

OWL INTERACTIVE EXAMPLE 13.1

Arrange the following bonds according to increasing polarity: H—H, O—H, Cl—H, S—H, and F—H.

Solution The polarity of the bond increases as the difference in electronegativity increases. From the electronegativity values in Fig. 13.3, the following variation in bond polarity is expected (the Pauling electronegativity value appears in parentheses below each element):

	H—H	< S—H	< Cl—H	< O—H	< F—H
	(2.2)(2.2)	(2.6)(2.2)	(3.2)(2.2)	(3.4)(2.2)	(4.0)(2.2)
Electronegativity difference	0	0.4	1.1	1.2	1.8

Covalent bond ⟶ (Polarity increases) polar covalent bond

Table 13.2

A Comparison of Pauling's and Allen's Electronegativity Values for Selected Representative Elements

Atom	Pauling	Allen
H	2.20	2.300
Li	0.98	0.912
Be	1.57	1.576
B	2.04	2.051
C	2.55	2.544
N	3.04	3.066
O	3.44	3.610
F	3.98	4.193
Ne	—	4.787
Cl	3.16	2.869
Br	2.96	2.685
I	2.66	2.359

Although Pauling's electronegativity values are the ones most commonly found in textbooks, several other systems for obtaining electronegativity values have been proposed. Recently, Leland C. Allen of Princeton University has pioneered an electronegativity scale based on the average ionization energies of the valence electrons for a given atom. Allen's system allows calculation of an electronegativity value for an atom that is independent of its bonding environment. In this system electronegativity becomes a property of the isolated atom. This approach is fundamentally different from Pauling's system, which is derived from the bond energies of atoms attached to one another.

Although we will not consider the details of Allen's quantum mechanical analysis, it is useful to compare Allen's and Pauling's values for some of the more important representative elements (Table 13.2). While many values are similar, some significant differences appear.

Allen has made a strong case that his system for obtaining electronegativities is more meaningful than that created by Pauling. Allen's values are now becoming accepted by the chemical community.

13.3 Bond Polarity and Dipole Moments

We have seen that when hydrogen fluoride is placed in an electric field, the molecules have a preferential orientation (Fig. 13.2). This follows from the charge distribution in the HF molecule, which has a positive end and a negative end. A molecule such as HF that has a center of positive charge and a center of negative charge is said to be *dipolar,* or to have a **dipole moment.** A molecule that has a positive center of charge of magnitude Q and a negative center of charge of magnitude Q separated by a distance R has a dipole moment given by the expression

$$\text{Dipole moment} = \mu = QR$$

The debye is named after Peter Debye, who pioneered in the measurement of dipole moments.

which has SI units of coulomb meter (C m) but is most often given in units of debye [1 debye (D) = 3.336×10^{-30} C m]. The dipolar character of a molecule is often represented by an arrow pointing to the negative charge center, with the tail of the arrow indicating the positive center of charge:

$$\underset{\delta+\quad\;\delta-}{\longmapsto}$$

The dipole moment of a molecule gives useful information about its bonding and electron distribution. For example, the observed dipole moment for HF is 1.83 D. If HF were totally ionic (H^+F^-), the expected dipole moment (symbolized by μ) would be

$$\mu = \underset{\text{Electron charge}}{(1.60 \times 10^{-19}\ \text{C})}\underset{\text{H—F bond distance}}{(9.17 \times 10^{-11}\ \text{m})}$$

$$= 1.47 \times 10^{-29}\ \text{C m} = 4.40\ \text{D}$$

This calculation shows that HF is not fully ionic, since the measured dipole moment is much less than 4.40 D. We can estimate the ionic character of HF by assuming that the hydrogen has a charge $\delta+$ and the fluorine has a charge $\delta-$. Using the measured dipole moment, we have

$$1.83\ \text{D} = (\delta)(9.17 \times 10^{-11}\ \text{m}) \times \frac{1\ \text{D}}{3.336 \times 10^{-30}\ \text{C m}}$$

Solving for δ gives 6.66×10^{-20} C. Since the charge on an electron is 1.60×10^{-19} C, each atom in HF has a fractional charge of

$$\frac{6.66 \times 10^{-20}\ \text{C}}{1.60 \times 10^{-19}\ \text{C}} = 0.416$$

From this argument we might say that HF has 42% ionic bonding. Although this analysis is somewhat oversimplified (it assumes that charge distributions can be represented as point charges, for example), it does provide useful information about bonding. Recall that the dipolar character of a molecule is often represented by an arrow pointing to the negative charge center with the tail of the arrow indicating the positive center of charge:

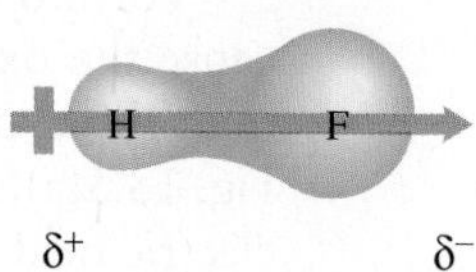

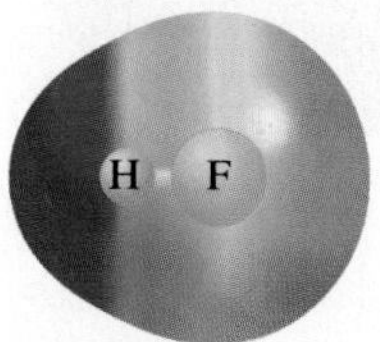

Figure 13.4
An electrostatic potential diagram of HF. Red indicates the most electron-rich area (the fluorine atom) and blue indicates the most electron-poor region (the hydrogen atom).

Another way to represent the charge distribution in HF is by an electrostatic potential diagram (Fig. 13.4). For this representation the colors of visible light are used to show the variation in charge distribution. Red indicates the most electron-rich region of the molecule, and blue indicates the most electron-poor region.

From what has been said so far, we would expect any diatomic molecule with a polar bond (between atoms with different electronegativities) to exhibit a dipole moment. Although this is generally true, the observed dipole moments of diatomic molecules are sometimes smaller than expected. For example, the carbon monoxide molecule CO has a dipole moment of only 0.11 D, much smaller than expected from the polarity of the CO bond. This discrepancy is most likely caused by the lone pairs of electrons on the atoms, which make large contributions to the dipole moment in opposition to that from the bond polarity. We will not explore the details of this situation here. The dipole moments of some representative diatomic molecules are listed in Table 13.3.

Table 13.3

The Dipole Moments of Some Diatomic Molecules (gas phase)

Molecule	Dipole Moment (D)
CO	0.112
HF	1.83
HCl	1.11
HBr	0.78
HI	0.38
NaCl	9.00
LiF	6.33
KF	8.60
KBr	10.41

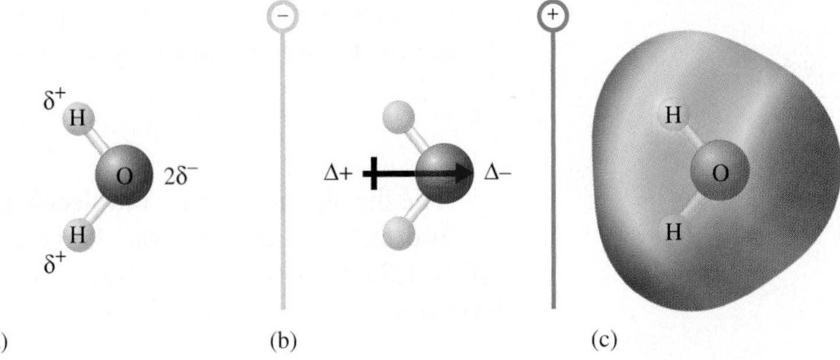

Figure 13.5
(a) The charge distribution in the water molecule. (b) The water molecule in an electric field. (c) The electrostatic potential diagram of the water molecule.

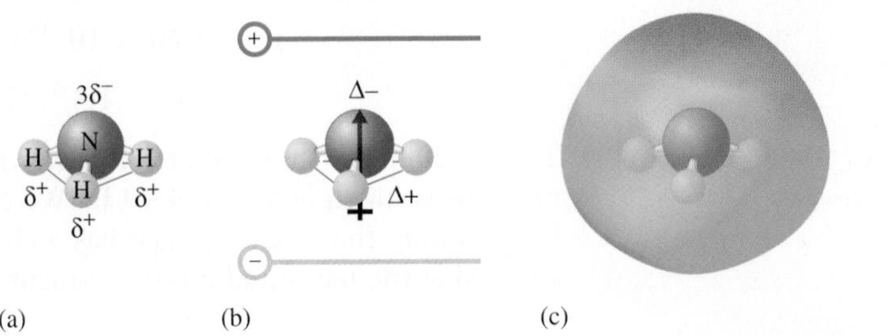

Figure 13.6
(a) The structure and charge distribution of the ammonia molecule. The polarity of the N—H bonds occurs because nitrogen has a greater electronegativity than hydrogen. (b) The dipole moment of the ammonia molecule oriented in an electric field. (c) The electrostatic potential diagram for ammonia.

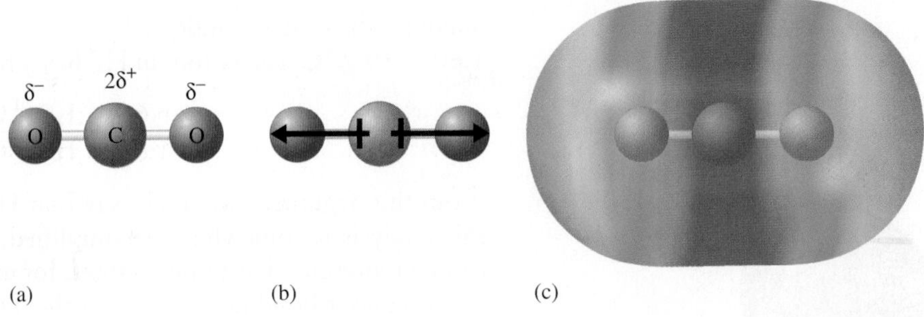

Figure 13.7
(a) The carbon dioxide molecule. (b) The opposed bond polarities cancel out, and the carbon dioxide molecule has no dipole moment. (c) The electrostatic potential diagram for carbon dioxide.

Polyatomic molecules can also exhibit dipolar behavior. For example, because the oxygen atom in the water molecule has a greater electronegativity than the hydrogen atoms, the molecular charge distribution is that shown in Fig. 13.5(a). This charge distribution causes the water molecule to behave in an electric field as if it had two centers of charge—one positive and one negative—as shown in Fig. 13.5(b). Thus the water molecule has a dipole moment. Similar behavior is observed for the NH_3 molecule (Fig. 13.6). Some molecules have polar bonds but do not have a dipole moment. This occurs when the individual bond polarities are arranged in such a way that they cancel. An example is the CO_2 molecule, a linear molecule that has the charge distribution shown in Fig. 13.7. In this case, since the opposing bond polarities cancel, the carbon dioxide molecule does not have a dipole moment. There is no preferential way for this molecule to line up in an electric field. (Try to find a preferred orientation.)

There are many cases where the bond polarities in molecules oppose and exactly cancel each other. Some common types of molecules with polar bonds but without dipole moments are shown in Table 13.4.

EXAMPLE 13.2

For each of the following molecules, show the direction of the bond polarities. Also indicate which ones have dipole moments: HCl, Cl_2, SO_3 (planar), CH_4 (tetrahedral), and H_2S (V-shaped).

Table 13.4

Types of Molecules with Polar Bonds but No Resulting Dipole Moment

Type		Cancellation of Polar Bonds	Example	Ball-and-Stick Model
Linear molecules with two identical bonds	B—A—B		CO_2	
Planar molecules with three identical bonds 120 degrees apart	B, A, B, B, 120°		SO_3	
Tetrahedral molecules with four identical bonds 109.5 degrees apart	B, A, B, B, B		CCl_4	

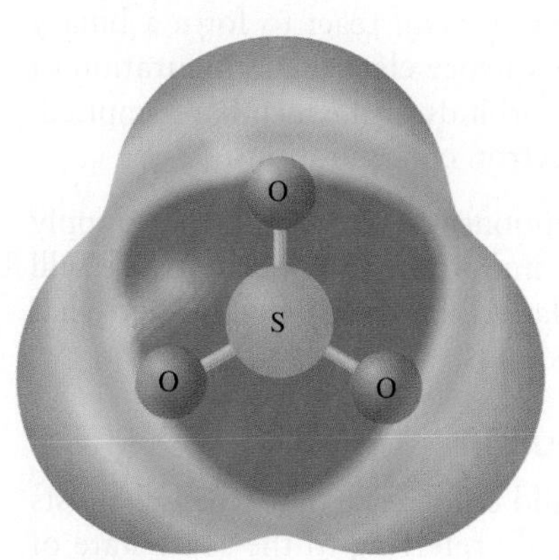

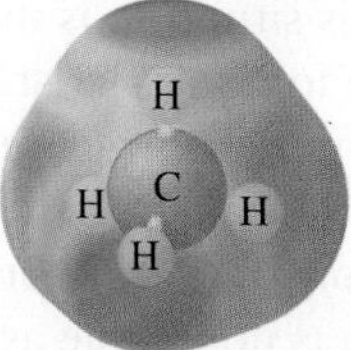

Solution The HCl molecule: Because the electronegativity of chlorine (3.2) is greater than that of hydrogen (2.2), the chlorine is partially negative, and the hydrogen is partially positive. The HCl molecule has a dipole moment oriented as follows:

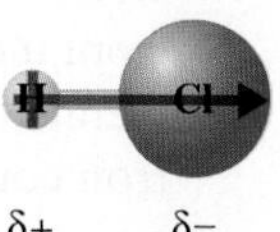

The Cl_2 molecule: Because the two chlorine atoms share the electrons equally, no bond polarity occurs. The Cl_2 molecule has no dipole moment.

The SO_3 molecule: Because the electronegativity of oxygen (3.4) is greater than that of sulfur (2.6), each oxygen has a partial negative charge, and the sulfur has a partial positive charge:

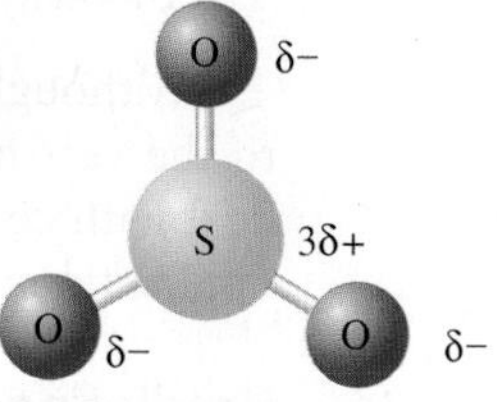

However, the bond polarities cancel, and the molecule has no dipole moment.

The CH_4 molecule: Carbon has a slightly higher electronegativity (2.6) than hydrogen (2.2). This leads to small partial positive charges on the hydrogen atoms and a small partial negative charge on the carbon:

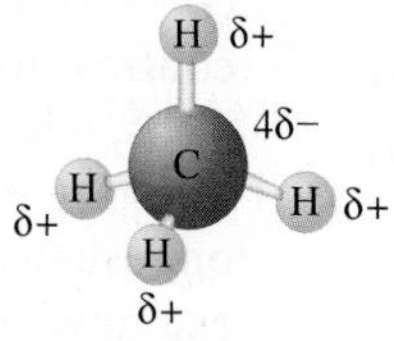

This case is similar to the third type in Table 13.4. Since the bond polarities cancel, the molecule has no dipole moment.

The H_2S molecule: Since the electronegativity of sulfur (2.6) is greater than that of hydrogen (2.2), the sulfur has a partial negative charge, and the hydrogen atoms have a partial positive charge, which can be represented as follows:

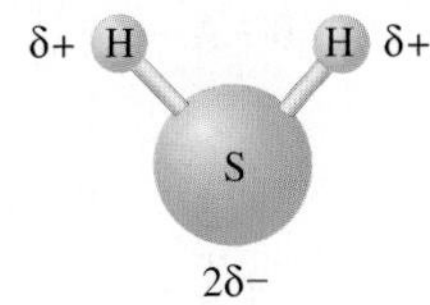

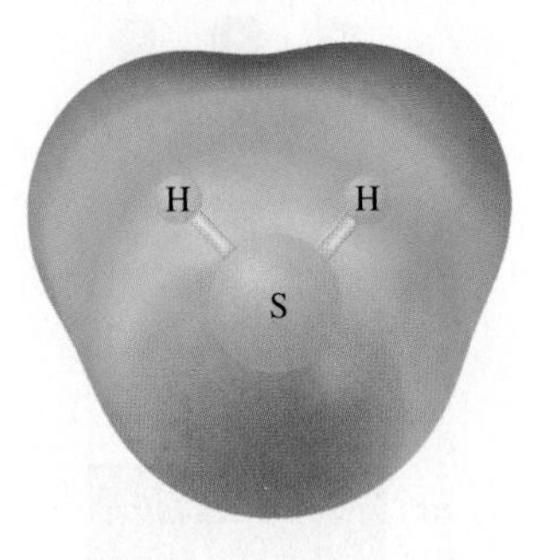

This case is analogous to the water molecule. The polar bonds result in a dipole moment oriented as shown.

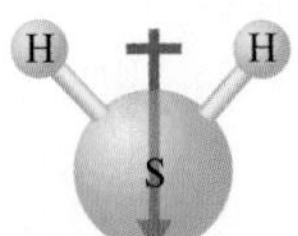

13.4 Ions: Electron Configurations and Sizes

The description of the electron arrangements in atoms that emerged from the quantum mechanical model has helped a great deal in our understanding of what constitutes a stable compound. For example, in a very large number of stable compounds, the atoms have noble gas arrangements of electrons. Nonmetallic elements achieve a noble gas electron configuration either by sharing electrons with other nonmetals to form covalent bonds or by taking electrons from metals to form ions. In the latter case the nonmetals form anions and the metals form cations. The following generalizations can be applied to the electron configurations in most stable compounds:

Atoms in stable compounds usually have a noble gas electron configuration.

As we will see later, there are exceptions to these rules, but they remain a useful place to start.

- When *two nonmetals* react to form a covalent bond, they share electrons in a way that completes the valence electron configurations of both atoms. That is, both nonmetals attain noble gas electron configurations.
- When *a nonmetal and a representative group metal* react to form a binary ionic compound, the ions form so that the valence electron configuration of the nonmetal is completed and the valence orbitals of the metal are emptied. In this way both ions achieve noble gas electron configurations.

Although there are some important exceptions, these generalizations apply to the vast majority of compounds and are important to remember. We will deal with covalent bonds more thoroughly later. Next, we will consider what implications these rules hold for ionic compounds.

Predicting Formulas of Ionic Compounds

In the solid state of an ionic compound, the ions are relatively close together, and many ions are simultaneously interacting:

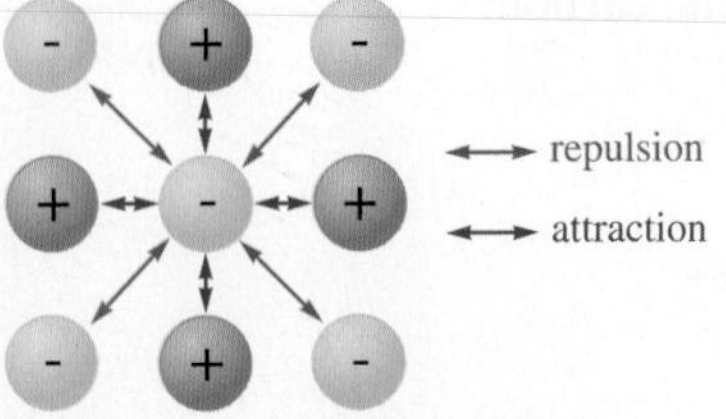

At the beginning of this discussion, we should emphasize that when chemists use the term *ionic compound,* they are usually referring to the solid state of that compound. Solid ionic compounds contain a large collection of positive and negative ions packed together in a way that minimizes the ⊖ · · ⊖ and ⊕ · · ⊕ repulsions and maximizes the ⊕ · · ⊖ attractions. This situation is in contrast to the gas phase of an ionic substance, where discrete ion pairs exist. Thus, when we speak in this text of the stability of an ionic compound, we are referring to the solid state, where the large attractive forces present among the oppositely charged ions tend to stabilize (favor the formation of) the ions. For example, as we mentioned in the previous chapter, the O^{2-} ion is not stable as

In the gas phase of an ionic substance, the ions would be relatively far apart and would not contain large groups of ions:

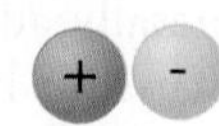

an isolated gas-phase species but, of course, is very stable in many solid ionic compounds. That is, MgO(*s*), which contains Mg^{2+} and O^{2-} ions, is very stable, but the isolated gas-phase ion pair $Mg^{2+} \cdot\cdot O^{2-}$ is not energetically favorable in comparison with the separate neutral gaseous atoms. Thus you should keep in mind that in this section, and in most other cases in which we are describing the nature of ionic compounds, the discussion usually refers to the solid state, in which many ions are simultaneously interacting.

To illustrate the principles of electron configurations in stable, solid ionic compounds, we will consider the formation of an ionic compound from calcium and oxygen. We can predict what compound will form by considering the valence electron configurations of the two atoms:

$$\text{Ca:} \quad [\text{Ar}]4s^2$$

$$\text{O:} \quad [\text{He}]2s^22p^4$$

From Fig. 13.3 we see that the electronegativity of oxygen (3.4) is much greater than that of calcium (1.0). Because of this large difference, electrons will be transferred from calcium to oxygen to form oxygen anions and calcium cations in the compound. How many electrons are transferred? We can base our prediction on the observation that noble gas configurations are generally the most stable. Note that oxygen needs two electrons to fill its 2*s* and 2*p* valence orbitals and to achieve the configuration of neon ($1s^22s^22p^6$). And by losing two electrons, calcium can achieve the configuration of argon. Two electrons are therefore transferred:

$$\text{Ca} + \text{O} \longrightarrow \text{Ca}^{2+} + \text{O}^{2-} \quad (2e^- \text{ transferred from Ca to O})$$

To predict the formula of the ionic compound, we simply recognize that chemical compounds are always electrically neutral—they have the same quantities of positive and negative charges. In this case we must have equal numbers of Ca^{2+} and O^{2-} ions, and the empirical formula of the compound is CaO.

The same principles can be applied to many other cases. For example, consider the compound formed between aluminum and oxygen. Because aluminum has the configuration $[\text{Ne}]3s^23p^1$, it must lose three electrons to form the Al^{3+} ion and thus achieve the neon configuration. Therefore, the Al^{3+} and O^{2-} ions form in this case. Since the compound must be electrically neutral, there must be three O^{2-} ions for every two Al^{3+} ions, and the compound has the empirical formula Al_2O_3.

Table 13.5 shows common elements that form ions with noble gas electron configurations in ionic compounds. In losing electrons to form cations, metals in Group 1A lose one electron, those in Group 2A lose two electrons, and those in Group 3A lose three electrons. In gaining electrons to form anions, nonmetals in Group 7A (the halogens) gain one electron, and those in Group 6A gain

Table 13.5

Common Ions with Noble Gas Electron Configurations in Ionic Compounds

Group 1A	Group 2A	Group 3A	Group 6A	Group 7A	Electron Configuration
H^-, Li^+	Be^{2+}				[He]
Na^+	Mg^{2+}	Al^{3+}	O^{2-}	F^-	[Ne]
K^+	Ca^{2+}		S^{2-}	Cl^-	[Ar]
Rb^+	Sr^{2+}		Se^{2-}	Br^-	[Kr]
Cs^+	Ba^{2+}		Te^{2-}	I^-	[Xe]

two electrons. Hydrogen typically behaves as a nonmetal and can gain one electron to form the hydride ion (H^-), which has the electron configuration of helium.

There are some important exceptions to the rules discussed here. For example, tin forms both Sn^{2+} and Sn^{4+} ions, and lead forms both Pb^{2+} and Pb^{4+} ions. Also, bismuth forms Bi^{3+} and Bi^{5+} ions, and thallium forms Tl^+ and Tl^{3+} ions. There are no simple explanations for the behavior of these ions. For now, just note them as exceptions to the very useful rule that ions generally adopt noble gas electron configurations in ionic compounds. Our discussion here refers to representative metals. The transition metals exhibit more complicated behavior, forming a variety of ions that will be considered in Chapter 19.

Sizes of Ions

Ion size plays an important role in determining the structure and stability of ionic solids, the properties of ions in aqueous solution, and the biological effects of ions. As with atoms, it is impossible to define precisely the sizes of ions. Most often, ionic radii are determined from the measured distances between ion centers in ionic compounds. This method, of course, involves an assumption about how the distance should be divided up between the two ions. Thus you will note considerable disagreement among ionic sizes given in various sources. Here we are mainly interested in trends and will be less concerned with absolute ion sizes.

Various factors influence ionic size. We will first consider the relative sizes of an ion and its parent atom. Since a positive ion is formed by removing electrons from a neutral atom, the resulting cation is smaller than its parent atom. The opposite is true for negative ions; the addition of electrons to a neutral atom produces an anion significantly larger than its parent atom.

It is also important to know how the sizes of ions vary depending on the positions of the parent elements in the periodic table. Figure 13.8 shows the sizes of the most important ions (each with a noble gas configuration) and their position in the periodic table. Note that ion size increases down a group. The changes that occur horizontally are complicated because of the change from predominantly metals on the left-hand side of the periodic table to nonmetals on the right-hand side. A given period thus contains both elements that give up electrons to form cations and ones that accept electrons to form anions.

One trend worth noting involves the relative sizes of a set of **isoelectronic ions**—*ions containing the same number of electrons.* Consider the ions O^{2-}, F^-, Na^+, Mg^{2+}, and Al^{3+}. Each of these ions has the neon electron configuration. How do the sizes of these ions vary? In general, there are two important facts to consider in predicting the relative sizes of ions: the number of electrons and the number of protons. Since these ions are isoelectronic, the number of electrons is 10 in each case. Electron repulsions should therefore be about the same in all cases. However, the number of protons increases from 8 to 13 as we go from the O^{2-} ion to the Al^{3+} ion. Thus, in going from O^{2-} to Al^{3+}, the 10 electrons experience a greater attraction as the positive charge on the nucleus increases. This causes the ions to become smaller. You can confirm this by looking at Fig. 13.8. In general, for a series of isoelectronic ions, the size decreases as the nuclear charge (Z) increases.

For isoelectronic ions, size generally decreases as Z increases.

OWL INTERACTIVE EXAMPLE 13.3

Arrange the ions Se^{2-}, Br^-, Rb^+, and Sr^{2+} in order of decreasing size.

Solution This is an isoelectronic series of ions with the krypton electron configuration. Since these ions all have the same number of electrons, their sizes

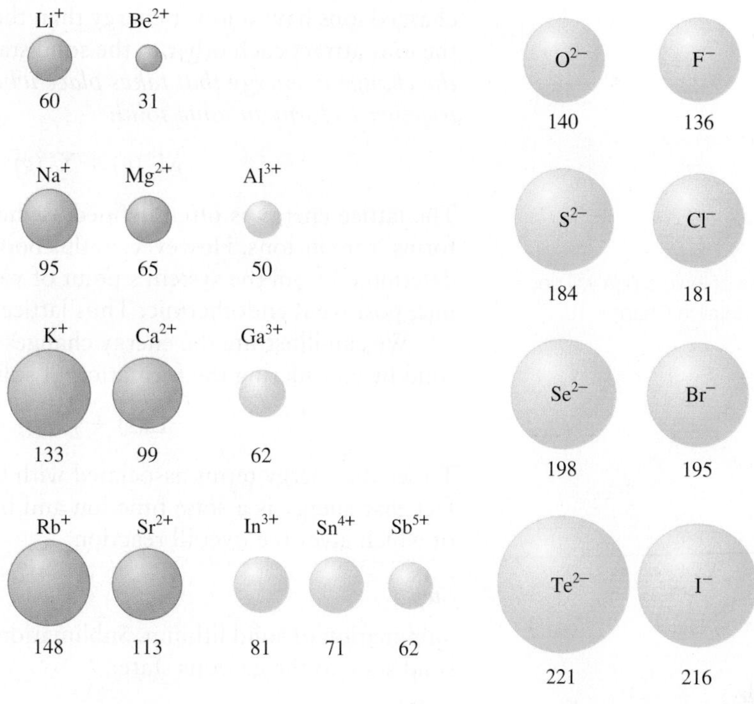

Figure 13.8 Sizes of ions related to positions of elements in the periodic table. Note that size generally increases down a group. Also note that in a series of isoelectronic ions, size decreases with increasing atomic number. The ionic radii are given in units of picometers.

will depend on nuclear charge. The Z values are 34 for Se^{2-}, 35 for Br^-, 37 for Rb^+, and 38 for Sr^{2+}. Since the nuclear charge is greatest for Sr^{2+}, it is the smallest of these ions. The Se^{2-} ion is largest.

$$\underset{\text{Largest}}{\overset{}{Se^{2-}}} > Br^- > Rb^+ > \underset{\text{Smallest}}{Sr^{2+}}$$

OWL INTERACTIVE EXAMPLE 13.4

Choose the largest ion in each of the following groups.

a. $Li^+, Na^+, K^+, Rb^+, Cs^+$ **b.** $Ba^{2+}, Cs^+, I^-, Te^{2-}$

Solution

Ion size generally increases down a group.

a. The ions are all from Group 1A elements. Since size increases down a group (the ion with the greatest number of electrons is the largest), Cs^+ is the largest ion.

b. This is an isoelectronic series of ions, all of which have the xenon electron configuration. The ion with the smallest nuclear charge is the largest ion.

$$\underset{Z=52}{Te^{2-}} > \underset{Z=53}{I^-} > \underset{Z=55}{Cs^+} > \underset{Z=56}{Ba^{2+}}$$

13.5 Formation of Binary Ionic Compounds

In this section we will introduce the factors that influence the stability and the structures of solid binary ionic compounds. We know that metals and nonmetals react by transferring electrons to form cations and anions that are mutually attractive. The resulting ionic solid forms because the aggregated oppositely charged ions have a lower energy than the original elements. Just how strongly the ions attract each other in the solid state is indicated by the **lattice energy**—*the change in energy that takes place when separated gaseous ions are packed together to form an ionic solid:*

$$M^+(g) + X^-(g) \longrightarrow MX(s)$$

The structures of ionic solids will be discussed in detail in Chapter 16.

The lattice energy is often defined as the energy *released* when an ionic solid forms from its ions. However, in this book the sign of an energy term is always determined from the system's point of view: negative if the process is exothermic; positive if endothermic. Thus lattice energy has a negative sign.

We can illustrate the energy changes involved in the formation of an ionic solid by considering the formation of solid lithium fluoride from its elements:

$$Li(s) + \tfrac{1}{2}F_2(g) \longrightarrow LiF(s)$$

To see the energy terms associated with this process, we take advantage of the fact that energy is a state function and break this reaction into steps, the sum of which gives the overall reaction.

Step 1

Sublimation of solid lithium. Sublimation involves taking a substance from the solid state to the gaseous state:

$$Li(s) \longrightarrow Li(g)$$

The enthalpy of sublimation for Li(s) is 161 kJ/mol.

Step 2

Ionization of lithium atoms to form Li^+ ions in the gas phase:

$$Li(g) \longrightarrow Li^+(g) + e^-$$

This process corresponds to the first ionization energy for lithium, which is 520 kJ/mol.

Step 3

Dissociation of fluorine molecules. We need to form 1 mole of fluorine atoms by breaking the F—F bonds in 0.5 mole of F_2 molecules:

$$\tfrac{1}{2}F_2(g) \longrightarrow F(g)$$

The energy required to break this bond is 154 kJ/mol. In this case we are breaking the bonds in a half mole of fluorine, so the energy required for this step is 154 kJ/2, or 77 kJ.

Step 4

Formation of F^- ions from fluorine atoms in the gas phase:

$$F(g) + e^- \longrightarrow F^-(g)$$

The energy change for this process corresponds to the electron affinity of fluorine, which is −328 kJ/mol.

Step 5

Formation of solid lithium fluoride from the gaseous Li^+ and F^- ions:

$$Li^+(g) + F^-(g) \longrightarrow LiF(s)$$

This corresponds to the lattice energy for LiF, which is −1047 kJ/mol.

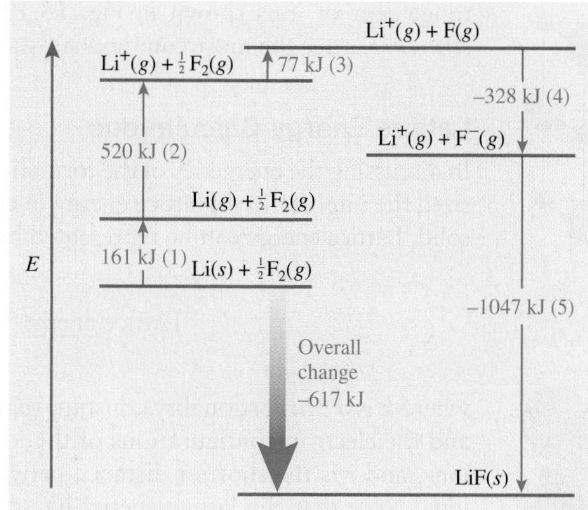

Figure 13.9
The energy changes involved in the formation of solid lithium fluoride from its elements. The numbers in parentheses refer to the reaction steps discussed in the text.

Since the sum of these five processes yields the desired overall reaction, the sum of the individual energy changes gives the overall energy change:

	Process	Energy Change (kJ)
	$Li(s) \rightarrow Li(g)$	161
	$Li(g) \rightarrow Li^+(g) + e^-$	520
	$\frac{1}{2}F_2(g) \rightarrow F(g)$	77
	$F(g) + e^- \rightarrow F^-(g)$	−328
	$Li^+(g) + F^-(g) \rightarrow LiF(s)$	−1047
Overall:	$Li(s) + \frac{1}{2}F_2(g) \rightarrow LiF(s)$	−617 kJ (per mole of LiF)

In doing this calculation, we have ignored the small difference between ΔH_{sub} and ΔE_{sub}.

This process is summarized by the energy diagram in Fig. 13.9. Note that the formation of solid lithium fluoride from its elements is highly exothermic mainly because of the very large negative lattice energy. A great deal of energy is released when the ions combine to form the solid. In fact, note that the energy released when an electron is added to a fluorine atom to form the F^- ion (328 kJ/mol) is not enough to remove an electron from lithium (520 kJ/mol). That is, when a metallic lithium atom reacts with a nonmetallic fluorine atom to form *separated* ions,

$$Li(g) + F(g) \longrightarrow Li^+(g) + F^-(g)$$

the process is endothermic and thus unfavorable. Clearly, then, the main impetus for the formation of the ionic compound rather than a covalent compound results from the strong mutual attractions among the Li^+ and F^- ions in the solid. The lattice energy is the dominant energy term.

The structure of the solid lithium fluoride is represented in Fig. 13.10. Note the alternating arrangement of the Li^+ and F^- ions. Also note that each Li^+ is surrounded by six F^- ions and each F^- ion is surrounded by six Li^+ ions. This structure can be rationalized by assuming that the ions behave as hard spheres that pack together in a way that both maximizes the attractions among the oppositely charged ions and minimizes the repulsions among the identically charged ions.

All the binary ionic compounds formed by an alkali metal and a halogen have the structure shown in Fig. 13.10, except for the cesium salts. The ar-

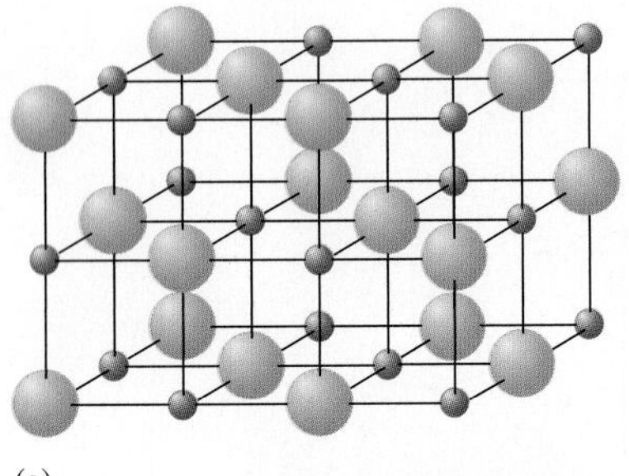

(a)

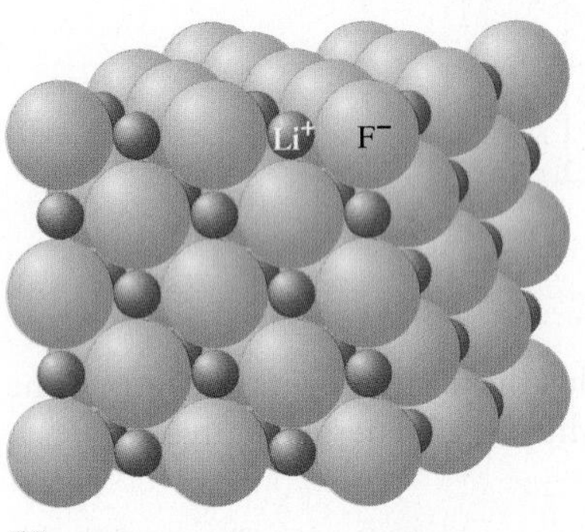

(b)

Figure 13.10
The structure of lithium fluoride.
(a) Represented by a ball-and-stick model. Note that each Li^+ ion is surrounded by six F^- ions, and each F^- ion is surrounded by six Li^+ ions.
(b) Represented with the ions shown as spheres. The structure is determined by packing the spherical ions in a way that both maximizes the ionic attractions and minimizes the ionic repulsions.

rangement of ions shown in Fig. 13.10 is often called the *sodium chloride structure*, after the most common substance that possesses it.

Lattice Energy Calculations

In discussing the energetics of the formation of solid lithium fluoride, we emphasized the importance of lattice energy in contributing to the stability of the ionic solid. Lattice energy can be represented by a modified form of Coulomb's law,

$$\text{Lattice energy} = k\left(\frac{Q_1Q_2}{r}\right)$$

where k is a proportionality constant that depends on the structure of the solid and the electron configurations of the ions, Q_1 and Q_2 are the charges on the ions, and r is the shortest distance between the centers of the cations and anions. Note that the lattice energy has a negative sign when Q_1 and Q_2 have opposite signs. This result is expected, since bringing cations and anions together is an exothermic process. Also note that the process becomes more exothermic as the ionic charges increase and as the distances between the ions in the solid decrease.

The importance of the charges in ionic solids can be illustrated by comparing the energies involved in the formation of NaF(*s*) and MgO(*s*). These solids contain the isoelectric ions Na^+, F^-, Mg^{2+}, and O^{2-}. The energy diagram for the formation of the two solids is given in Fig. 13.11. Note several important features:

The energy released when the gaseous Mg^{2+} and O^{2-} ions combine to form solid MgO is much greater (more than four times greater) than that released when the gaseous Na^+ and F^- ions combine to form solid NaF.

The energy required to remove two electrons from the magnesium atom (735 kJ/mol for the first and 1445 kJ/mol for the second, yielding a total of 2180 kJ/mol) is much greater than the energy required to remove an electron from a sodium atom (495 kJ/mol).

Energy (737 kJ/mol) is required to add two electrons to the oxygen atom in the gas phase. Addition of the first electron is exothermic (−141 kJ/mol), but addition of the second electron is quite endothermic (878 kJ/mol). This latter energy must be obtained indirectly, since the $O^{2-}(g)$ is not stable.

Since the equation for lattice energy contains the product Q_1Q_2, the lattice energy for a solid with 2+ and 2− ions should be four times that for a solid with 1+ and 1− ions. That is,

$$\frac{(+2)(-2)}{(+1)(-1)} = 4$$

For MgO and NaF, the observed ratio of lattice energies (Fig. 13.11) is

$$\frac{-3916 \text{ kJ}}{-923 \text{ kJ}} = 4.24$$

Because twice as much energy is required to remove the second electron from magnesium as to remove the first, and because addition of an electron to the gaseous O^- ion is quite endothermic, it seems puzzling that magnesium oxide contains Mg^{2+} and O^{2-} ions rather than Mg^+ and O^- ions. The answer lies in the lattice energy. Note that the lattice energy for combining gaseous Mg^{2+} and O^{2-} ions to form MgO(*s*) is 3000 kJ/mol more negative than that for combining gaseous Na^+ and F^- ions to form NaF(*s*). Thus the energy released in forming a solid containing Mg^{2+} and O^{2-} ions rather than Mg^+ and O^- ions more than compensates for the energies required for the processes that produce the Mg^{2+} and O^{2-} ions.

If there is so much lattice energy to be gained in going from singly charged to doubly charged ions in the case of magnesium oxide, why then does solid sodium fluoride contain Na^+ and F^- ions rather than Na^{2+} and F^{2-} ions? We can answer this question by recognizing that both Na^+ and F^- ions have the neon electron configuration. Removal of an electron from Na^+ requires an extremely large quantity of energy (4560 kJ/mol) because a 2*p* electron must be removed. Conversely, the addition of an electron to F^- would require use of the relatively high-energy 3*s* orbital, which is also an unfavorable process. Thus we

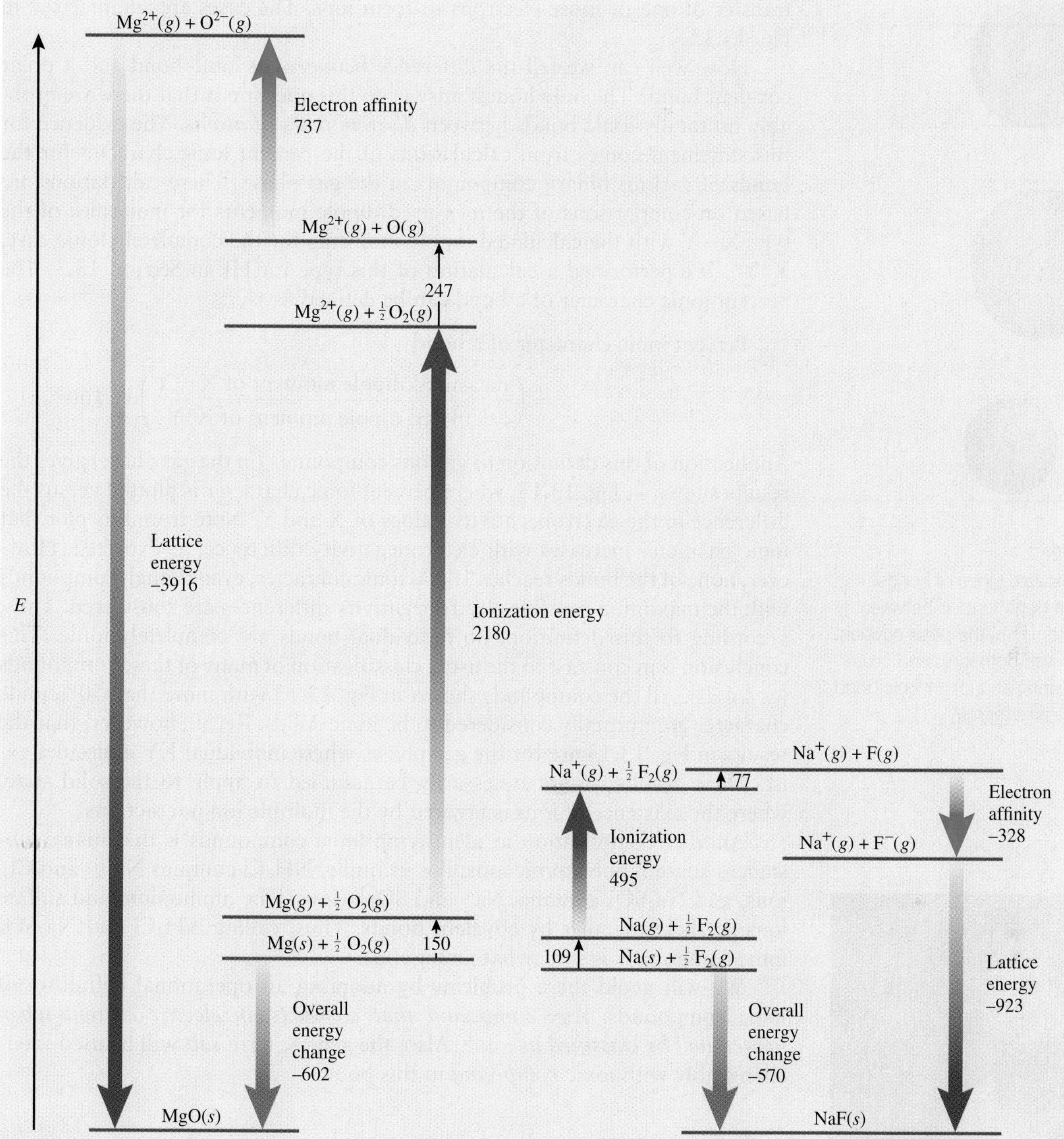

Figure 13.11
Comparison of the energy changes involved in the formation of solid sodium fluoride and solid magnesium oxide. Note the large lattice energy for magnesium oxide (where doubly charged ions are combining) compared with that for sodium fluoride (where singly charged ions are combining).

can say that for sodium fluoride the extra energy required to form the doubly charged ions is greater than the gain in lattice energy that would result.

This discussion of the energies involved in the formation of solid ionic compounds illustrates that a variety of factors operate to determine the composition and structure of these compounds. The most important of these factors involve the balancing of the energies required to form highly charged ions and the energy released when highly charged ions combine to form the solid.

13.6 Partial Ionic Character of Covalent Bonds

Recall that when atoms with different electronegativities react to form molecules, the electrons are not shared equally. The possible result is a polar covalent bond or, in the case of a large electronegativity difference, a complete

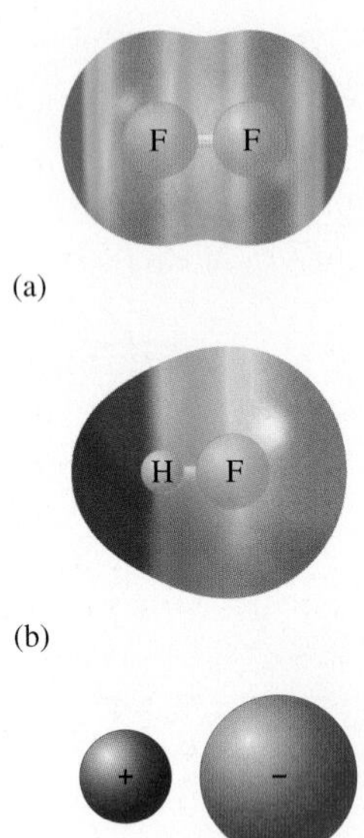

Figure 13.12
The three possible types of bonds: (a) a covalent bond formed between identical F atoms; (b) the polar covalent bond of HF, with both ionic and covalent components; and (c) an ionic bond with no electron sharing.

transfer of one or more electrons to form ions. The cases are summarized in Fig. 13.12.

How well can we tell the difference between an ionic bond and a polar covalent bond? The only honest answer to this question is that there are probably no totally ionic bonds between *discrete pairs of atoms*. The evidence for this statement comes from calculations of the percent ionic character for the bonds of various binary compounds in the gas phase. These calculations are based on comparisons of the measured dipole moments for molecules of the type X—Y with the calculated dipole moments for the completely ionic case, X^+Y^-. We performed a calculation of this type for HF in Section 13.3. The percent ionic character of a bond can be defined as

$$\text{Percent ionic character of a bond} = \left(\frac{\text{measured dipole moment of X—Y}}{\text{calculated dipole moment of } X^+Y^-}\right) \times 100\%$$

Application of this definition to various compounds (in the gas phase) gives the results shown in Fig. 13.13, where percent ionic character is plotted versus the difference in the electronegativity values of X and Y. Note from this plot that ionic character increases with electronegativity difference, as expected. However, none of the bonds reaches 100% ionic character, even though compounds with the maximum possible electronegativity differences are considered. Thus, according to this definition, no individual bonds are completely ionic. This conclusion is in contrast to the usual classification of many of these compounds (as solids). All the compounds shown in Fig. 13.13 with more than 50% ionic character are normally considered to be ionic solids. Recall, however, that the results in Fig. 13.13 are for the gas phase, where individual XY molecules exist. These results cannot necessarily be assumed to apply to the solid state, where the existence of ions is favored by the multiple ion interactions.

Another complication in identifying ionic compounds is that many substances contain polyatomic ions. For example, NH_4Cl contains NH_4^+ and Cl^- ions, and Na_2SO_4 contains Na^+ and SO_4^{2-} ions. The ammonium and sulfate ions are held together by covalent bonds. Thus, calling NH_4Cl and Na_2SO_4 ionic compounds is somewhat ambiguous.

We will avoid these problems by adopting an operational definition of ionic compounds: *Any compound that conducts an electric current when melted will be classified as ionic.* Also, the generic term *salt* will be used interchangeably with *ionic compound* in this book.

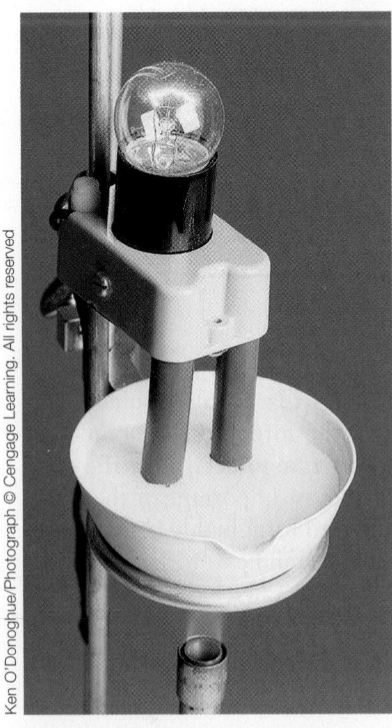

Molten NaCl conducts an electric current, indicating the presence of mobile Na^+ and Cl^- ions.

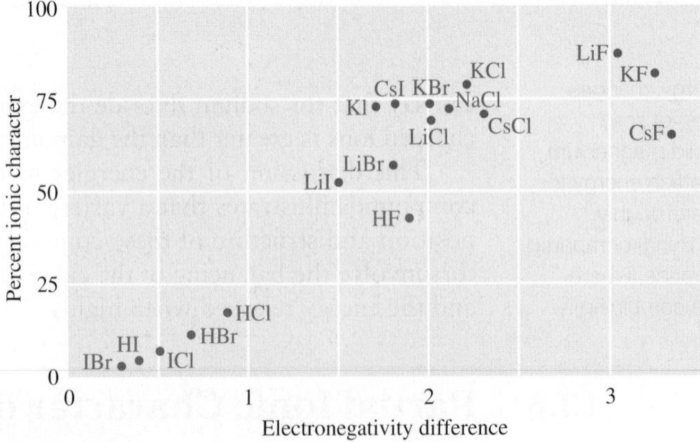

Figure 13.13
The relationship between the ionic character of a covalent bond and the electronegativity difference of the bonded atoms. The compounds normally considered to be ionic in the solid phase are shown in red.

13.7 The Covalent Chemical Bond: A Model

Before we develop specific models for covalent chemical bonding, it will be helpful to summarize some of the concepts introduced in this chapter.

What is a chemical bond? Chemical bonds can be viewed as forces that cause a group of atoms to behave as a unit.

Why do chemical bonds occur? There is no principle of nature that states that bonds are favored or disfavored. Bonds are neither inherently "good" nor inherently "bad" as far as nature is concerned; they result from the tendency of a system to seek its lowest possible energy. From a simplistic point of view, bonds occur when collections of atoms are more stable (lower in energy) than the separate atoms. For example, about 1652 kJ of energy is required to break a mole of methane (CH_4) molecules into separate C and H atoms. Or, taking the opposite view, 1652 kJ of energy is released when 1 mole of methane is formed from 1 mole of gaseous C atoms and 4 moles of gaseous H atoms. Thus we can say that 1 mole of CH_4 molecules in the gas phase is 1652 kJ lower in energy than 1 mole of carbon atoms plus 4 moles of hydrogen atoms. Methane is therefore a stable molecule relative to its separated atoms.

We find it useful to interpret molecular stability in terms of a model called a chemical bond. To help understand why this model was invented, let's continue with methane, which consists of four hydrogen atoms arranged at the corners of a tetrahedron around a carbon atom:

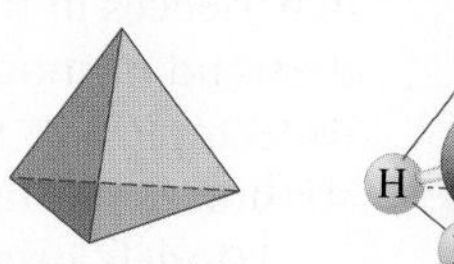

A tetrahedron has four equal triangular faces.

Given this structure, it is natural to envision four individual C—H interactions (we call them bonds). The energy of stabilization of CH_4 is divided equally among the four bonds to give an average C—H bond energy per mole of C—H bonds:

$$\frac{1652 \text{ kJ}}{4} = 413 \text{ kJ}$$

Next, consider methyl chloride, which consists of CH_3Cl molecules having the structure

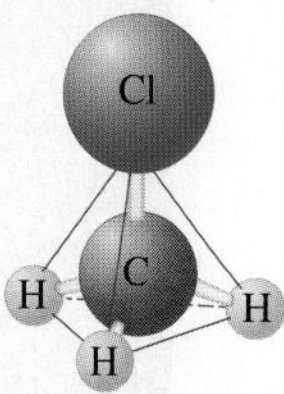

Experiments have shown that about 1578 kJ of energy is required to break down 1 mole of gaseous CH_3Cl molecules into gaseous carbon, chlorine, and hydrogen atoms. The reverse process can be represented as

$$C(g) + Cl(g) + 3H(g) \longrightarrow CH_3Cl(g) + 1578 \text{ kJ/mol}$$

A mole of gaseous methyl chloride is lower in energy by 1578 kJ than its separate gaseous atoms. Thus a mole of methyl chloride is held together by 1578 kJ of energy. Again, it is very useful to divide this energy into individual bonds. Methyl chloride can be visualized as containing one C—Cl bond and three C—H bonds. If we assume arbitrarily that a C—H interaction represents the same quantity of energy in any situation (that is, the strength of a

C—H bond is independent of its molecular environment), we can do the following bookkeeping:

$$1 \text{ mol of C—Cl bonds plus 3 mol of C—H bonds} = 1578 \text{ kJ}$$

$$\text{C—Cl bond energy} + 3(\text{average C—H bond energy}) = 1578 \text{ kJ}$$

$$\text{C—Cl bond energy} + 3(413 \text{ kJ/mol}) = 1578 \text{ kJ}$$

$$\text{C—Cl bond energy} = 1578 - 1239 = 339 \text{ kJ/mol}$$

These assumptions allow us to associate given quantities of energy with C—H and C—Cl bonds.

It is important to note that the bond concept is a human invention. Bonds provide a method for dividing up the energy evolved when a stable molecule is formed from its component atoms. Thus in this context *a bond represents a quantity of energy* obtained from the molecular energy of stabilization in a rather arbitrary way. This is not to say that the concept of individual bonds is a bad idea. In fact, the modern concept of the chemical bond, conceived by American chemists G. N. Lewis and Linus Pauling, is one of the most useful ideas chemists have ever developed.

Models: An Overview

The framework of chemistry, like that of any science, consists of models—attempts to explain how nature operates on the microscopic level based on experiences in the macroscopic world. To understand chemistry, one must understand its models and how they are used. We will use the concept of bonding to reemphasize the important characteristics of models, including their origin, structure, and uses.

Bonding is a model proposed to explain molecular stability.

Models originate from our observations of the properties of nature. For example, the concept of bonds arose from the observations that most chemical processes involve collections of atoms and that chemical reactions involve rearrangements of the ways in which the atoms are grouped. So to understand reactions, we must understand the forces that bind atoms together.

In natural processes there is a tendency toward lower energy. Collections of atoms therefore occur because the aggregated state has lower energy than

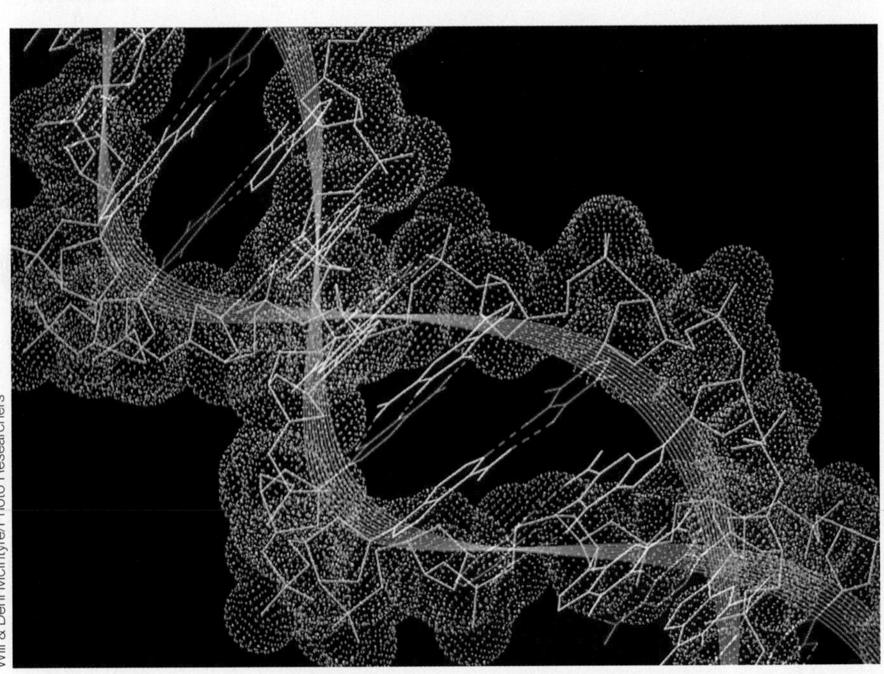

Will & Deni McIntyre/Photo Researchers

The concept of individual bonds makes it much easier to deal with complex molecules such as DNA. A small segment of a DNA molecule is shown here.

Ken Eward/Science Source/ Photo Researchers

A ball-and-stick model of a protein segment illustrating the alpha helix.

the separated atoms. Why? As we have seen earlier in this chapter, the best explanations for the energy change involve atoms sharing electrons or atoms transferring electrons to become ions. In the case of electron sharing, we find it convenient to assume that individual bonds occur between pairs of atoms. Let's explore the validity of this assumption and see how it is useful.

In a diatomic molecule such as H_2, it is natural to assume that a bond exists between the atoms, holding them together. It is also useful to assume that individual bonds are present in polyatomic molecules such as CH_4. So instead of thinking of CH_4 as a unit with a stabilization energy of 1652 kJ per mole, we choose to think of CH_4 as containing four C—H bonds, each worth 413 kJ of energy per mole of bonds. Without this concept of individual bonds in molecules, chemistry would be hopelessly complicated. There are millions of different chemical compounds, and if each of these compounds had to be considered as an entirely new entity, the task of understanding chemical behavior would be overwhelming.

The bonding model provides a framework to systematize chemical behavior by enabling us to think of molecules as collections of common fundamental components. For example, a typical biomolecule, such as a protein, contains hundreds of atoms and might seem discouragingly complex. However, if we think of a protein as constructed of individual bonds, C—C, C—H, C—N, C—O, N—H, and so on, it helps tremendously in predicting and understanding the protein's behavior. The essential idea is that we expect a given bond to behave about the same in any molecular environment. Used in this way, the model of the chemical bond has helped chemists to systematize the reactions of the millions of existing compounds.

Fundamental Properties of Models

1. Models are human inventions, always based on an incomplete understanding of how nature works. *A model does not equal reality.*
2. Models are often wrong. This property derives from the first property. Models are based on speculation and are always oversimplifications.
3. Models tend to become more complicated as they age. As flaws are discovered in our models, we "patch" them and thus add more detail.
4. It is very important to understand the assumptions that are inherent in a particular model before you use it to interpret observations or to make predictions. Simple methods usually involve very restrictive assumptions and can be expected to yield only qualitative information. Asking for a sophisticated explanation from a simple model is like expecting to get an accurate mass for a diamond by using a bathroom scale.

   For a model to be used effectively, we must understand its strengths and weaknesses and ask only appropriate questions. An illustration of this point is the simple aufbau principle used to explain the electron configurations of the elements. Although this model correctly predicts the configuration for most atoms, chromium and copper do not agree with the predictions. Detailed studies show that the configurations of chromium and copper result from complex electron interactions that are not taken into account in the simple model. However, this does not mean that we should discard the simple model that is so useful for most atoms. Instead, we must apply it with caution and not expect it to be correct in every case.
5. When a model is wrong, we often learn much more than when it is right. If a model makes a wrong prediction, it usually means we do not understand some fundamental characteristics of nature. We often learn by making mistakes. (Try to remember that when you get back your next chemistry test.)

In addition to being very useful, the bonding model is physically sensible. It makes sense that atoms can form stable groups by sharing electrons; shared electrons give a lower energy state because they are simultaneously attracted by two nuclei.

Also, as we will see in the next section, bond energy data support the existence of discrete bonds that are relatively independent of the molecular environment. It is very important to remember, however, that the chemical bond is only a model. Although our concept of discrete bonds in molecules agrees with many of our observations, some molecular properties require that we think of a molecule as a whole, with the electrons free to move through the entire molecule. This is called *delocalization* of the electrons, a concept that will be discussed more completely in the next chapter.

13.8 Covalent Bond Energies and Chemical Reactions

In this section we will consider the energies associated with various types of bonds and see how the bonding concept is useful in dealing with the energies of chemical reactions. One important consideration is to establish the sensitivity of a particular type of bond to its molecular environment. For example, consider the stepwise decomposition of methane:

Process	*Energy Required (kJ/mol)*
$CH_4(g) \rightarrow CH_3(g) + H(g)$	435
$CH_3(g) \rightarrow CH_2(g) + H(g)$	453
$CH_2(g) \rightarrow CH(g) + H(g)$	425
$CH(g) \rightarrow C(g) + H(g)$	339
	Total = 1652

$$\text{Average} = \frac{1652}{4} = 413$$

Although a C—H bond is broken in each case, the energy required varies in a nonsystematic way. This example shows that the C—H bond is somewhat sensitive to its environment. We use the *average* of these individual bond dissociation energies even though this quantity only approximates the energy associated with a C—H bond in a particular molecule. The degree of sensitivity of a bond to its environment can also be seen from experimental measurements of the energy required to break the C—H bond in the following molecules:

Molecule	Measured C—H Bond Energy (kJ/mol)
$HCBr_3$	380
$HCCl_3$	380
HCF_3	430
C_2H_6	410

These data show that the C—H bond strength varies significantly with its environment, but the concept of an average C—H bond strength remains useful to chemists. The average values of bond energies for various types of bonds are listed in Table 13.6.

So far we have discussed bonds in which one pair of electrons is shared. This type of bond is called a **single bond.** As we will see in more detail later, atoms sometimes share two pairs of electrons, forming a **double bond,** or share

Table 13.6

Average Bond Energies (kJ/mol)

Single Bonds						Multiple Bonds	
H—H 432		N—H 391		I—I 149		C=C	614
H—F	565	N—N	160	I—Cl	208	C≡C	839
H—Cl	427	N—F	272	I—Br	175	O=O	495
H—Br	363	N—Cl	200			C=O*	745
H—I	295	N—Br	243	S—H	347	C≡O	1072
		N—O	201	S—F	327	N=O	607
C—H	413	O—H	467	S—Cl	253	N=N	418
C—C	347	O—O	146	S—Br	218	N≡N	941
C—N	305	O—F	190	S—S	266	C=N	615
C—O	358	O—Cl	203			C≡N	891
C—F	485	O—I	234	Si—Si	340		
C—Cl	339			Si—H	393		
C—Br	276	F—F	154	Si—C	360		
C—I	240	F—Cl	253	Si—O	452		
C—S	259	F—Br	237				
		Cl—Cl	239				
		Cl—Br	218				
		Br—Br	193				

*C=O (CO_2) = 799

three pairs of electrons, forming a **triple bond.** The bond energies for these *multiple bonds* are also given in Table 13.6.

A relationship also exists between the number of shared electron pairs and the bond length. As the number of shared electrons increases, the bond length shortens. This relationship is shown for selected bonds in Table 13.7.

Bond Energy and Enthalpy

Bond energy values can be used to calculate approximate energies for reactions. To illustrate how this is done, we will calculate the change in energy that accompanies the following reaction:

$$H_2(g) + F_2(g) \longrightarrow 2HF(g)$$

This reaction involves breaking one H—H and one F—F bond and forming two H—F bonds. For bonds to be broken, energy must be *added* to the system—an endothermic process. Consequently, the energy terms associated with bond breaking have *positive* signs. The formation of a bond *releases* energy, an exo-

Table 13.7

Bond Lengths for Selected Bonds

Bond	Bond Type	Bond Length (pm)	Bond Energy (kJ/mol)
C—C	Single	154	347
C=C	Double	134	614
C≡C	Triple	120	839
C—O	Single	143	358
C=O	Double	123	745
C—N	Single	143	305
C=N	Double	138	615
C≡N	Triple	116	891

thermic process, and the energy terms associated with bond making carry a *negative* sign. We can write the enthalpy change for a reaction as follows:

ΔH = sum of the energies required to break old bonds (positive signs) plus the sum of the energies released in the formation of new bonds (negative signs)

This leads to the expression

$$\Delta H = \underbrace{\Sigma D \text{ (bonds broken)}}_{\text{Energy required}} - \underbrace{\Sigma D \text{ (bonds formed)}}_{\text{Energy released}}$$

where Σ represents the sum of terms and D represents the bond energy per mole of bonds. (D *always* has a positive sign.)

In the case of the formation of HF,

$$\begin{aligned}\Delta H &= D_{H-H} + D_{F-F} - 2D_{H-F} \\ &= 1 \text{ mol} \times \frac{432 \text{ kJ}}{\text{mol}} + 1 \text{ mol} \times \frac{154 \text{ kJ}}{\text{mol}} - 2 \text{ mol} \times \frac{565 \text{ kJ}}{\text{mol}} \\ &= -544 \text{ kJ}\end{aligned}$$

Thus, when 1 mole of $H_2(g)$ and 1 mole of $F_2(g)$ react to form 2 moles of $HF(g)$, 544 kJ of energy should be released.

This result can be compared with the calculation of ΔH for this reaction from the standard enthalpy of formation for HF (−271 kJ/mol):

$$\Delta H = 2 \text{ mol} \times (-271 \text{ kJ/mol}) = -542 \text{ kJ}$$

Thus the use of bond energies to calculate ΔH works quite well in this case.

Since bond energies are typically averages taken from several compounds, the ΔH calculated from bond energies is not expected to agree exactly with that calculated from enthalpies of formation.

OWL INTERACTIVE EXAMPLE 13.5

Using the bond energies listed in Table 13.6, calculate ΔH for the reaction of methane with chlorine and fluorine to give Freon-12 (CF_2Cl_2).

$$CH_4(g) + 2Cl_2(g) + 2F_2(g) \longrightarrow CF_2Cl_2(g) + 2HF(g) + 2HCl(g)$$

Solution The idea here is to break the bonds in the reactants to give individual atoms and then assemble these atoms into the products by forming new bonds:

$$\text{Reactants} \xrightarrow[\text{required}]{\text{Energy}} \text{atoms} \xrightarrow[\text{released}]{\text{Energy}} \text{products}$$

We then combine the energy changes to calculate ΔH:

$$\Delta H = \begin{matrix}\text{energy required} \\ \text{to break bonds}\end{matrix} - \begin{matrix}\text{energy released} \\ \text{when bonds form}\end{matrix}$$

where the minus sign gives the correct sign to the energy terms for the exothermic processes.

Reactant bonds broken:

CH_4	4 mol C—H	$4 \text{ mol} \times \frac{413 \text{ kJ}}{\text{mol}}$	= 1652 kJ
$2Cl_2$	2 mol Cl—Cl	$2 \text{ mol} \times \frac{239 \text{ kJ}}{\text{mol}}$	= 478 kJ
$2F_2$	2 mol F—F	$2 \text{ mol} \times \frac{154 \text{ kJ}}{\text{mol}}$	= 308 kJ
		Total energy required	= 2438 kJ

Product bonds formed:

CF_2Cl_2	2 mol C—F	$2 \text{ mol} \times \frac{485 \text{ kJ}}{\text{mol}}$	= 970 kJ
	and		
	2 mol C—Cl	$2 \text{ mol} \times \frac{339 \text{ kJ}}{\text{mol}}$	= 678 kJ
HF	2 mol H—F	$2 \text{ mol} \times \frac{565 \text{ kJ}}{\text{mol}}$	= 1130 kJ
HCl	2 mol H—Cl	$2 \text{ mol} \times \frac{427 \text{ kJ}}{\text{mol}}$	= 854 kJ
		Total energy released	= 3632 kJ

We now can calculate ΔH:

$$\Delta H = \begin{matrix}\text{energy required} \\ \text{to break bonds}\end{matrix} - \begin{matrix}\text{energy released} \\ \text{when bonds form}\end{matrix}$$

$$= 2438 \text{ kJ} - 3632 \text{ kJ} = -1194 \text{ kJ}$$

Since the sign of the value for the enthalpy change is negative, this means that 1194 kJ of energy is released per mole of CF_2Cl_2 formed. The value of ΔH calculated for this reaction using enthalpies of formation is -1126 kJ.

In performing the calculations in this section, we have made several approximations. Of course, we made the usual assumption that average bond energies apply regardless of the specific molecular environment. We also ignored the difference between enthalpy and internal energy in these calculations. Recall from Chapter 9 that at constant pressure $\Delta E = \Delta H - P\Delta V$. Thus, if a reaction involves a change in volume, a correction should be applied to the calculations of reaction enthalpies from bond energies. However, this correction is usually small compared with the uncertainties inherent in this method.

It should also be noted that the use of bond energies to estimate enthalpies of reaction should be limited to cases in which all the reactants and products are in the gas phase, where intermolecular interactions are expected to be minimal. Bond energies do not take these interactions into account.

13.9 The Localized Electron Bonding Model

So far we have discussed the general characteristics of the chemical bonding model and have seen that properties such as bond strength and polarity can be assigned to individual bonds. In this section we introduce a specific model used to describe covalent bonds. We need a simple model that can be easily applied even to very complicated molecules and that can be used routinely by chemists to interpret and organize the wide variety of chemical phenomena. The model that serves this purpose is often called the **localized electron (LE) model.** This model assumes that *a molecule is composed of atoms that are bound together by using atomic orbitals to share electron pairs.* The electron pairs in the molecule are assumed to be localized on a particular atom or in the space between two atoms. Those pairs of electrons localized on an atom are called **lone pairs,** and those found in the space between the atoms are called **bonding pairs.**

As we will apply it, the LE model has three parts:

1. Description of the valence electron arrangement in the molecule using Lewis structures (will be discussed in the next section).
2. Prediction of the geometry of the molecule, using the valence shell electron-pair repulsion (VSEPR) model (will be discussed in Section 13.13).
3. Description of the types of atomic orbitals used by the atoms to share electrons or hold lone pairs (will be discussed in Chapter 14).

13.10 Lewis Structures

The **Lewis structure** of a molecule represents the arrangement of valence electrons among the atoms in the molecule. These representations are named after G. N. Lewis (Fig. 13.14). The rules for writing Lewis structures are based on the observations of thousands of molecules, which show that *in most stable compounds the atoms achieve noble gas electron configurations.* Although this is not always the case, it is so common that it provides a very useful place to start.

Lewis structures show only valence electrons.

We have already seen that when metals and nonmetals react to form solid binary ionic compounds, electrons are transferred, and the resulting ions typically have noble gas electron configurations. An example is the formation of KBr, where the K^+ ion has the [Ar] electron configuration and the Br^- ion has the [Kr] electron configuration. In writing Lewis structures, the rule is that *only the valence electrons are included.* Using dots to represent electrons, the Lewis structure for KBr is

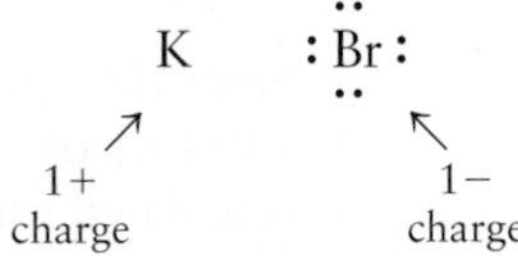

No dots are shown in the K^+ ion since it has no valence electrons. The Br^- ion is shown with eight electrons since it has a filled valence shell.

The Bancroft Library

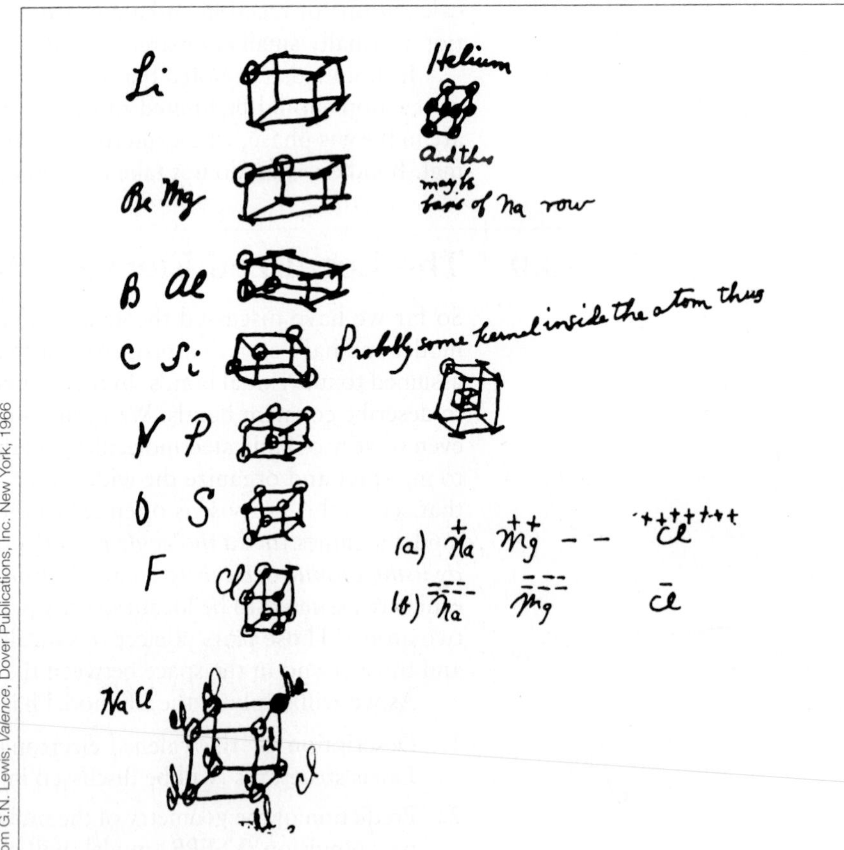

From G.N. Lewis, *Valence*, Dover Publications, Inc. New York, 1966

Figure 13.14
G. N. Lewis (above) conceived the octet rule while lecturing to a class of general chemistry students in 1902. He was also one of the two authors of a now classic work on thermodynamics, Lewis and Randall, *Thermodynamics and the Free Energy of Chemical Substances* (1923). (right) This is his original sketch. From G. N. Lewis, *Valence*, Dover Publications, Inc., New York, 1966.

Next, we will consider Lewis structures for molecule
involving elements in the first and second periods. The pri
noble gas electron configuration applies to these elements a

Hydrogen forms stable molecules where it shares two electro lows a **duet rule.** For example, when two hydrogen atoms, electron, combine to form the H_2 molecule, we have

$$H\cdot \quad \cdot H \quad \searrow \quad \swarrow \quad H:H$$

By sharing electrons, each hydrogen in H_2 has two electrons. This gives hydrogen a filled valence shell.

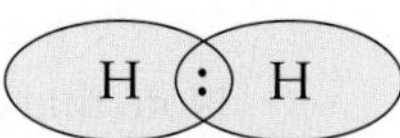

Helium does not form bonds because its valence orbitals are already filled; it is a noble gas. Helium has the electron configuration $1s^2$ and can be represented by the Lewis structure

$$He:$$

Carbon, nitrogen, oxygen, and fluorine almost always obey the octet rule in stable molecules.

The second-row nonmetals (carbon through fluorine) form stable molecules when they are surrounded by enough electrons to fill the valence orbitals—the $2s$ and the three $2p$ orbitals. Since eight electrons are required to fill the $2s$ and $2p$ orbitals, these elements typically obey the **octet rule;** they are surrounded by eight electrons. An example is the F_2 molecule, which has the following Lewis structure:

$$:\ddot{\underset{\cdot\cdot}{F}}\cdot \longrightarrow :\ddot{\underset{\cdot\cdot}{F}}:\ddot{\underset{\cdot\cdot}{F}}: \longleftarrow \cdot\ddot{\underset{\cdot\cdot}{F}}:$$

F atom with seven valence electrons — F_2 molecule — F atom with seven valence electrons

Note that each fluorine atom in F_2 is, in effect, surrounded by eight electrons, two of which are shared with the other atom. Recall that the shared pair of electrons is called a *bonding pair.* Each fluorine atom also has three pairs of electrons not involved in bonding. These are the *lone pairs.*

Neon does not form bonds since it already has an octet of valence electrons (it is a noble gas). The Lewis structure is

$$:\ddot{\underset{\cdot\cdot}{Ne}}:$$

Note that only the valence electrons of the neon atom ($2s^22p^6$) are represented by the Lewis structure. The $1s^2$ electrons are core electrons and take no part in chemical reactions.

From the discussion above, we can formulate the following rules for writing Lewis structures of molecules containing atoms from the first two periods.

STEPS

Writing Lewis Structures

1. Sum the valence electrons from all the atoms. Do not worry about keeping track of which electrons come from which atoms. It is the *total* number of electrons that is important.
2. Use a pair of electrons to form a bond between each pair of bound atoms.
3. Arrange the remaining electrons to satisfy the duet rule for hydrogen and the octet rule for the second-row elements.

To see how these rules are applied, we will construct the Lewis structures for a few molecules. We will first consider the water molecule and follow the rules above.

Step 1

We sum the *valence* electrons for H_2O as shown:

$$\underset{\text{H}}{1} + \underset{\text{H}}{1} + \underset{\text{O}}{6} = 8 \text{ valence electrons}$$

Step 2

Using one pair of electrons per bond, we draw in the two O—H single bonds:

H—O—H

Note that *a line instead of a pair of dots is used to indicate each pair of bonding electrons.* This is the standard notation.

Step 3

We distribute the remaining electrons around the atoms to achieve a noble gas electron configuration for each atom. Since four electrons have been used in forming the two bonds, four electrons (8 − 4) remain to be distributed. Hydrogen is satisfied with two electrons (duet rule), but oxygen needs eight electrons to achieve a noble gas configuration. Thus the remaining four electrons are added to oxygen as two lone pairs. Dots are used to represent the lone pairs:

H—Ö—H (with a second lone pair below O) Lone pairs

This is the correct Lewis structure for the water molecule. Each hydrogen has two electrons and the oxygen has eight:

H—O—H represents H:O:H

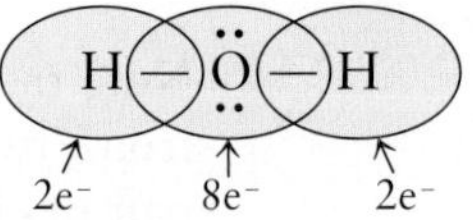

As a second example, we will write the Lewis structure for carbon dioxide. Summing the valence electrons gives

$$\underset{\text{C}}{4} + \underset{\text{O}}{6} + \underset{\text{O}}{6} = 16$$

After forming a bond between the carbon and each oxygen,

O—C—O

the remaining electrons are distributed to achieve noble gas configurations on each atom. In this case we have 12 electrons (16 − 4) remaining after the bonds are drawn. The distribution of these electrons is determined by a trial-and-error process. We have six pairs of electrons to distribute. Suppose we try three pairs on each oxygen to give

:Ö—C—Ö: (with a lone pair below each O)

To see if this structure is correct, we need to check two things:

1. The total number of electrons. There are 16 valence electrons in this structure, which is the correct number.
2. The octet rule for each atom. Each oxygen has eight electrons, but the carbon has only four. This cannot be the correct Lewis structure.

How can we arrange the 16 available electrons to achieve an octet for each atom? Suppose that there are two shared pairs between the carbon and each oxygen:

O=C=O represents O::C::O

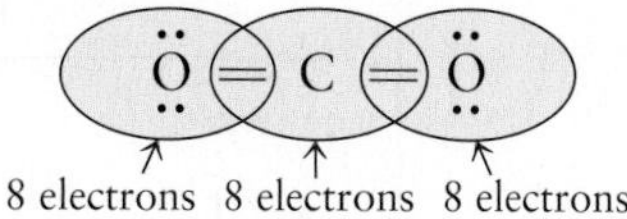

Now each atom is surrounded by 8 electrons, and the total number of electrons is 16, as required. Thus the correct Lewis structure for carbon dioxide has two double bonds.

Finally, let's consider the Lewis structure of the CN^- (cyanide) ion. Summing the valence electrons, we have

$$\underset{4}{C}\ \underset{5}{N}\ \underset{1}{^-} \quad 4 + 5 + 1 = 10$$

Note that the negative charge requires that an extra electron be added. After drawing a single bond (C—N), we distribute the remaining electrons to achieve a noble gas configuration for each atom. Eight electrons remain to be distributed. We can try various possibilities, for example:

:C̈—N̈:

This structure is incorrect because C and N have only six electrons each instead of eight. The correct arrangement is

[:C≡N:]⁻

(Satisfy yourself that both carbon and nitrogen have eight electrons.)

OWL INTERACTIVE EXAMPLE 13.6

Give the Lewis structure for each of the following.

a. HF b. N_2 c. NH_3 d. CH_4 e. CF_4 f. NO^+

Solution In each case we apply the three rules for writing Lewis structures. Recall that lines are used to indicate shared electron pairs and that dots are used to indicate nonbonding pairs (lone pairs). We have the following tabulated results:

	Total Valence Electrons	Draw Single Bonds	Calculate Number of Electrons Remaining	Use Remaining Electrons to Achieve Noble Gas Configurations
a. HF	1 + 7 = 8	H—F	6	H—F̤̈:
b. N_2	5 + 5 = 10	N—N	8	:N≡N:
c. NH_3	5 + 3(1) = 8	H—N(—H)—H	2	H—N̈(—H)—H
d. CH_4	4 + 4(1) = 8	H—C(H)(H)—H	0	H—C(H)(H)—H

(continued)

	Total Valence Electrons	Draw Single Bonds	Calculate Number of Electrons Remaining	Use Remaining Electrons to Achieve Noble Gas Configurations
e. CF_4	$4 + 4(7) = 32$	F—C—F (with F above and below C)	24	:F—C—F: (each F with three lone pairs; F above and below C)
f. NO^+	$5 + 6 - 1 = 10$	N—O	8	$[:N{\equiv}O:]^+$

When writing Lewis structures, don't worry about which electrons came from which atoms. The best way to look at a molecule is to regard it as a new entity that uses all of the available valence electrons of the atoms to achieve the lowest possible energy.* The valence electrons belong to the molecule rather than to the individual atoms. Simply distribute all valence electrons so that the various rules are satisfied, without regard to the origin of each particular electron.

13.11 Resonance

A valid Lewis structure is one that obeys the rules we have outlined.

Sometimes more than one valid Lewis structure is possible for a given molecule. For example, consider the Lewis structure for the nitrate ion (NO_3^-), which has 24 valence electrons. So that an octet of electrons surrounds each atom, a structure like the following is required:

$$\left[\begin{array}{c} \ddot{O} \\ \| \\ N \\ \ddot{O} \quad \ddot{O} \end{array}\right]^-$$

If this structure accurately represents the bonding in NO_3^-, there should be two types of N—O bonds observed in the molecule: one shorter bond (the double bond) and two identical longer ones (the two single bonds). However, experiments clearly show that NO_3^- exhibits only *one* type of N—O bond with a length and strength between those expected for a single bond and a double bond. Thus, although the structure we have shown above is a valid Lewis structure, it does *not* correctly represent the bonding in NO_3^-. This is a serious problem, and it means that the model must be modified.

Look again at the proposed Lewis structure for NO_3^-. Because there is no reason for choosing a particular oxygen atom to have the double bond, there are really three valid Lewis structures:

$$\left[\begin{array}{c} O \\ \| \\ N \\ O \quad O \end{array}\right]^- \quad \left[\begin{array}{c} O \\ | \\ N \\ O \;\!=\; \quad O \end{array}\right]^- \quad \left[\begin{array}{c} O \\ | \\ N \\ O \quad =O \end{array}\right]^-$$

Is any of these structures a correct description of the bonding in NO_3^-? No, because NO_3^- does not have one double and two single bonds—it has three

*In a sense this approach corrects for the fact that the LE model overemphasizes the point that a molecule is simply a sum of its parts—that is, that the atoms retain their individual identities in the molecule.

equivalent bonds. We can solve this problem by making the following assumption: The correct description of NO_3^- is *not given by any one* of the three Lewis structures individually but is given only by the *superposition of all three.*

The nitrate ion does not exist as any of the three extreme forms indicated by the individual Lewis structures but instead exists as an average of all three. **Resonance** *is invoked when more than one valid Lewis structure can be written for a particular molecule.* The resulting electron structure of the molecule is given by the average of these **resonance structures.** This situation is usually represented by double-headed arrows as follows:

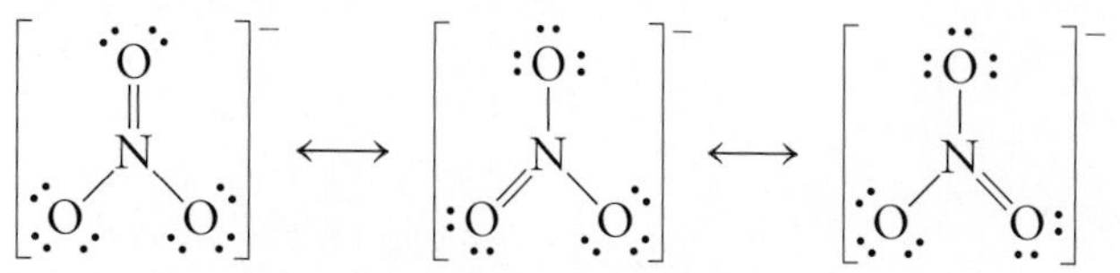

Note that in all these resonance structures the arrangement of the nuclei is the same. Only the placement of the electrons differs. The arrows do not indicate that the molecule "flips" from one resonance structure to another. They simply show that the *actual structure is an average of the three resonance structures.*

The concept of resonance is necessary because the LE model postulates that electrons are localized between a given pair of atoms. However, nature doesn't really operate this way. Electrons are actually delocalized—they can move around the entire molecule. The valence electrons in the NO_3^- molecule distribute themselves to provide equivalent N—O bonds. Resonance is necessary to compensate for this defective assumption of the LE model. However, because this model is so useful, we retain the concept of localized electrons and add resonance to accommodate species like NO_3^-.

EXAMPLE 13.7

Describe the electron arrangement in the nitrite anion (NO_2^-) using the LE model.

Equivalent Lewis structures contain the same numbers of single and multiple bonds. For example, the resonance structures for O_3,

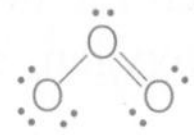

and

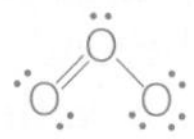

are equivalent Lewis structures. They are equally important in describing the bonding in O_3. Nonequivalent Lewis structures contain different numbers of single and multiple bonds.

Solution We will follow the usual procedure for obtaining the Lewis structure for the NO_2^- ion.

In NO_2^- there are $5 + 2(6) + 1 = 18$ valence electrons.
Indicating the single bonds gives the structure

O—N—O

The remaining 14 electrons $(18 - 4)$ can be distributed to produce these structures:

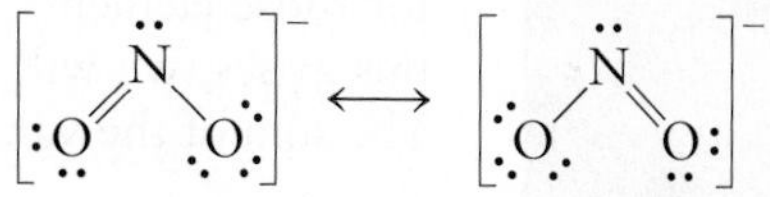

This is a resonance situation. Two equivalent Lewis structures can be drawn. *The electronic structure of the molecule is not correctly represented by either resonance structure but by the average of the two.* There are two equivalent N—O bonds, each one intermediate between a single and a double bond.

13.12 Exceptions to the Octet Rule

The LE model is a simple but very successful model, and the rules we have used for Lewis structures apply to most molecules. To implement this model, we have relied heavily on the octet rule. So far we have treated molecules for which this rule is easily applied. However, inevitably, cases arise where the importance of an octet of electrons is called into question. Boron, for example, tends to form compounds in which the boron atom has fewer than eight electrons around it—it does not have a complete octet. Boron trifluoride (BF_3), a gas at normal temperatures and pressures, reacts very energetically with molecules such as water and ammonia that have available lone pairs. The violent reactivity of BF_3 with electron-rich molecules occurs because the boron atom is electron-deficient. Boron trifluoride has 24 valence electrons. The Lewis structure that seems most consistent with the properties of BF_3 is

:F:
|
B
F   F

Note that in this structure boron has only six electrons around it. The octet rule for boron can be satisfied by drawing a structure with a double bond, such as

F
‖
B
F   F

and there are some theoretical studies that support such a structure for BF_3. However, since fluorine is so much more electronegative than boron, this structure seems questionable. In fact, some experiments indicate that each B—F bond is probably best described by the first Lewis structure, which is also consistent with the reactivity of BF_3 toward electron-rich molecules such as NH_3 with which it reacts to form H_3NBF_3:

H—N: (with two H) + B—F: (with two F) ⟶ H—N—B—F: (H₃N–BF₃)

In this stable compound, boron has an octet of electrons.

It is characteristic of boron to form molecules in which the boron atom is electron-deficient. On the other hand, carbon, nitrogen, oxygen, and fluorine do obey the octet rule.

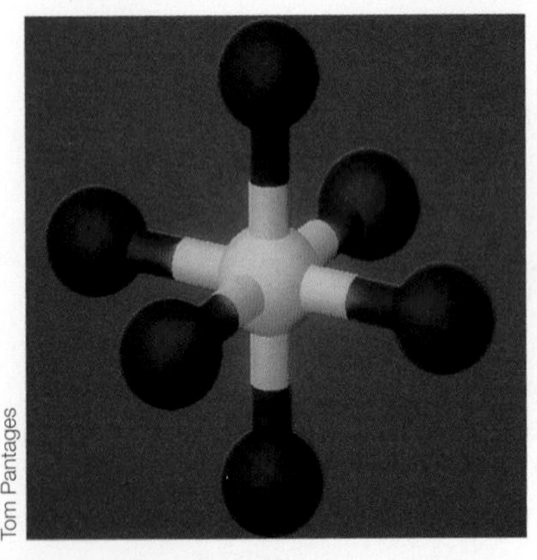

Tom Pantages

A computer-generated representation of sulfur hexafluoride.

Some atoms appear to exceed the octet rule. This behavior is observed only for those elements in Period 3 of the periodic table and beyond. To see how this arises, we will consider the Lewis structure for sulfur hexafluoride (SF_6). The sum of the valence electrons for SF_6 is

$$6 + 6(7) = 48 \text{ electrons}$$

Indicating the single bonds gives the structure on the left below:

F F F / S / F F F   (S bonded to six F; right: each F with three lone pairs)

We have used 12 electrons to form the S—F bonds, which leaves 36 electrons. Since fluorine always follows the octet rule, we complete the six fluorine octets to give the structure on the right above. This structure uses all 48 valence electrons for SF_6, but sulfur has 12 electrons around it; that is, sulfur *exceeds* the octet rule. How can this happen? There are several ways to approach this situation. The classical explanation for molecules like SF_6 involves using the empty $3d$ orbitals on the third-period elements. Recall that the second-row elements have only $2s$ and $2p$ valence orbitals, whereas the third-row elements have $3s$, $3p$, and $3d$ orbitals. The $3s$ and $3p$ orbitals fill with electrons in going from sodium to argon, but the $3d$ orbitals remain empty. For example, the valence-orbital diagram for a sulfur atom is

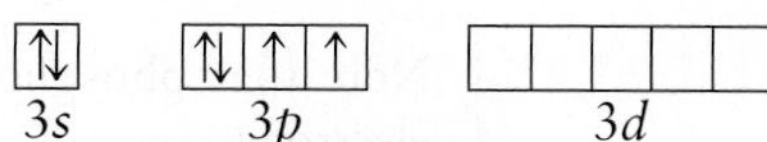

Thus one way to account for the electronic structure of SF_6 is to assume that the empty $3d$ orbitals on sulfur can be used to accommodate extra electrons: The sulfur atom can have 12 electrons around it by using the $3s$ and $3p$ orbitals to hold 8 electrons, with the extra 4 electrons in the formerly empty $3d$ orbitals.

Lewis Structures and the Octet Rule

1. The second-row elements C, N, O, and F should always be assumed to obey the octet rule.
2. The second-row elements B and Be often have fewer than eight electrons around them in their compounds. These electron-deficient compounds are very reactive.
3. The second-row elements never exceed the octet rule, since their valence orbitals ($2s$ and $2p$) can accommodate only eight electrons.
4. Third-row and heavier elements often satisfy the octet rule but are assumed in the simplest model to exceed the octet rule by using their empty valence d orbitals.
5. When writing the Lewis structure for a molecule, first draw single bonds between all bonded atoms, and then satisfy the octet rule for all the atoms. If electrons remain after the octet rule has been satisfied, place them on the elements having available d orbitals (elements in the third period or beyond).

OWL INTERACTIVE EXAMPLE 13.8

Write the Lewis structure for PCl_5.

Solution We can follow the same stepwise procedure we used previously for sulfur hexafluoride.

Step 1

Sum the valence electrons.

$$\underset{\text{P}}{\underset{\uparrow}{5}} + 5(\underset{\text{Cl}}{\underset{\uparrow}{7}}) = 40 \text{ electrons}$$

Step 2

Indicate single bonds between bound atoms.

```
      Cl
      |  Cl
      | /
Cl—P
      | \
      |  Cl
      Cl
```

Step 3

Distribute the remaining electrons. In this case, 30 electrons (40 − 10) remain. These are used to satisfy the octet rule for each chlorine atom. The final Lewis structure is

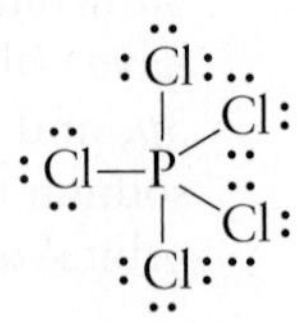

Note that phosphorus, a third-row element, exceeds the octet rule by two electrons.

In the PCl_5 and SF_6 molecules, the third-row central atoms (P and S, respectively) are assigned the extra electrons. However, in molecules having more than one atom that can exceed the octet rule, it is not always clear which atom should have the extra electrons. Consider the Lewis structure for the triiodide ion (I_3^-), which has

$$\underset{\text{I}}{3(7)} + \underset{-1 \text{ charge}}{1} = 22 \text{ valence electrons}$$

Indicating the single bonds gives I—I—I. At this point 18 electrons (22 − 4) remain. Trial and error will convince you that one of the iodine atoms must exceed the octet rule, but *which* one?

The rule we will follow is that *when it is necessary to exceed the octet rule for one of several third-row (or higher) elements, assume that the extra electrons are placed on the central atom.*

Thus for I_3^- the Lewis structure is

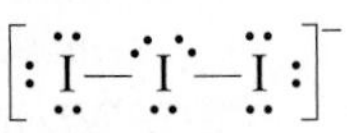

$$[:\ddot{\text{I}}—\ddot{\text{I}}—\ddot{\text{I}}:]^-$$

where the central iodine exceeds the octet rule. This structure agrees with known properties of I_3^-.

EXAMPLE 13.9

Write the Lewis structure for each molecule or ion.

a. ClF_3 **b.** XeO_3 **c.** $RnCl_2$ **d.** $BeCl_2$ **e.** ICl_4^-

Solution

a. The chlorine atom (third row) accepts the extra electrons.

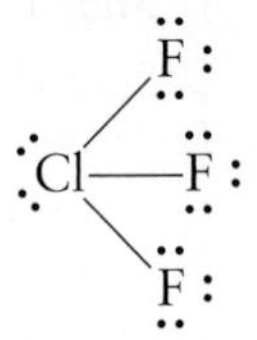

b. All atoms obey the octet rule.

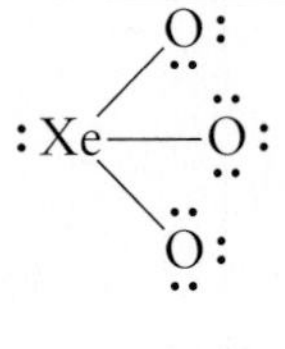

c. Radon, a noble gas in Period 6, accepts the extra electrons.

$$:\ddot{\underset{..}{Cl}}—\ddot{\underset{..}{Rn}}—\ddot{\underset{..}{Cl}}:$$

d. Beryllium is electron-deficient.

$$:\ddot{\underset{..}{Cl}}—Be—\ddot{\underset{..}{Cl}}:$$

e. Iodine exceeds the octet rule.

$$\left[\begin{array}{ccc} :\ddot{\underset{..}{Cl}} & & \ddot{\underset{..}{Cl}}: \\ & \ddot{\underset{..}{I}} & \\ :\ddot{\underset{..}{Cl}} & & \ddot{\underset{..}{Cl}}: \end{array}\right]^-$$

So far we have assumed that third-row and heavier atoms can exceed the octet rule by using their valence *d* orbitals to accommodate the extra electrons. However, recent calculations* indicate that because the 3*d* orbitals are so much higher in energy than the 3*s* and 3*p* orbitals for a given atom, it is not feasible to use them. Researchers argue that satisfying the octet rule is a high priority for molecules and that there are ways to describe the bonding in molecules like SF_6 and PCl_5 without formally exceeding the octet rule. One such explanation depends on a concept called *hyperconjugation,* which is explained in the special feature "Hyperconjugation—The Octet Rules" on p. 632.

The main point that should be made here is that chemists disagree on the best way to explain molecules such as SF_6 and PCl_5. Researchers continue to try to find the best answers.

Odd-Electron Molecules

In addition to the question about the use of *d* orbitals in third-row and heavier atoms, another problem for the simple LE model involves molecules with odd numbers of electrons. Although relatively few molecules formed from nonmetals contain odd numbers of electrons, there are some notable examples. One such molecule is nitric oxide (NO), which is formed when nitrogen and oxygen gases react at the high temperatures present in automobile engines. Nitric oxide is emitted into the air, where it reacts with oxygen to form gaseous nitrogen dioxide (NO_2), another odd-electron molecule.

Since the LE model is based on pairs of electrons, it does not handle odd-electron cases in a natural way, although Lewis structures are sometimes written for these species. To treat odd-electron molecules accurately, we need a more sophisticated model.

Formal Charge

Molecules or polyatomic ions often have many nonequivalent Lewis structures, all which obey the rules for writing Lewis structures. For example, as we will see in detail, the sulfate ion has a Lewis structure with all single bonds and several Lewis structures that contain double bonds. How do we decide which of the many possible Lewis structures best describes the actual bonding in sulfate? One method involves estimating the charge on each atom in the various possible Lewis structures and using these charges to select the most appropriate structure(s). We will see below how this is done, but first we must decide on a method to evaluate atomic charges in molecules.

*L. Suidan, J. K. Bodenhopp, E. D. Glendening, and F. Weinhold, *J. Chem. Ed.* **72** (1995): 583.

Chemical Insights — Hyperconjugation—The Octet Rules

All our observations tell us that molecules such as SF_6 and PCl_5 are very stable, but what is the bonding like in these molecules? There are two extreme points of view: (1) The S and P atoms in these molecules exceed the octet rule, placing the extra electrons in 3*d* orbitals, and (2) the S and P atoms in these molecules obey the octet rule, resorting to hyperconjugation. *Hyperconjugation* involves binding *n* atoms to a central atom using fewer than *n* electron pairs around the central atom. To illustrate this model, consider SF_6. We know that there are six fluorine atoms around the central sulfur atom in this molecule. Can we explain the bonding without exceeding the octet rule? The answer is yes, by using Lewis structures such as the following:

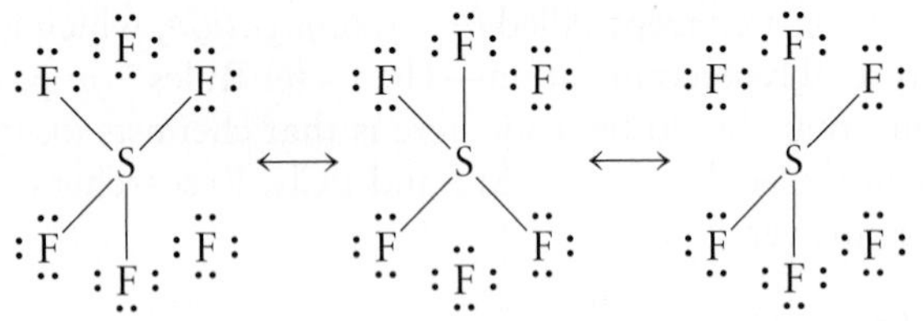

Note that in each of these resonance structures (and the many other similar ones that can be drawn) the sulfur atom always has an octet of electrons around it. The "true" structure is a composite of these equivalent resonance structures. The overall bonding is described as a combination of covalent and ionic contributions. The covalent contribution to the bonding involves four electron pairs "spread out" over the six sulfur–fluorine bonds. The ionic contribution to the bonding arises as follows. Note that for each Lewis structure two of the fluorines have −1 formal charges and the sulfur has a +2 formal charge,

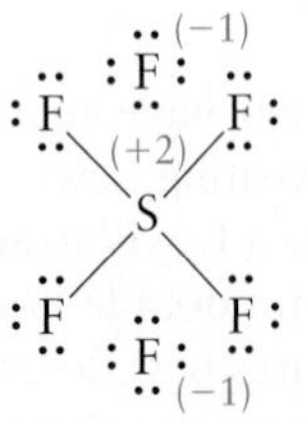

leading to ionic attractive forces. It is important to recall at this point that the description of the bonding in SF_6 by this model involves *all* the resonance structures similar to the one shown. Thus the F atoms are bound to the S atom by a combination of covalent bonding,

$$\frac{\text{4 electron-pair bonds}}{\text{6 S—F interactions}} = \tfrac{2}{3} \text{ covalent bond per S—F interaction}$$

plus the ionic bonding described above.

A similar treatment can be given for PCl_5, using the following resonance structures,

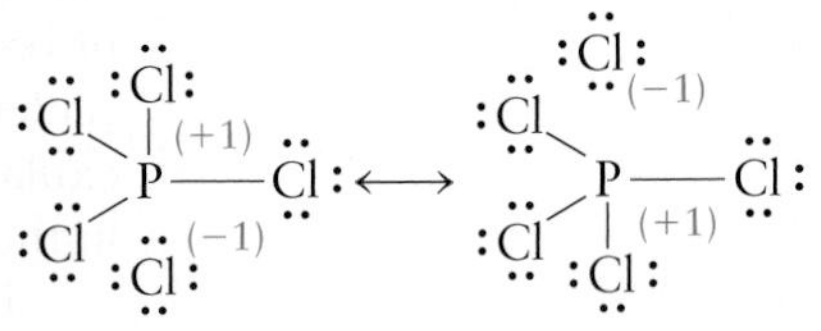

each of which satisfies the octet rule.

Thus the use of hyperconjugation preserves the octet rule and does not require the central atom to use *d* orbitals, which this model's proponents argue are too high in energy to participate in the bonding of these types of molecules.

So honest disagreements exist among chemists as to the best Lewis structures for molecules that, at least at first glance, appear to exceed the octet rule. This uncertainty shows the limitations of the Lewis model with its localized electron pairs. Note, however, that even with its limitations, it is still very useful because of its simplicity. The ability to obtain a reasonable bonding picture with a "back-of-the-envelope" model has led to the enduring influence of the Lewis model.

In Chapter 4 we discussed one system for obtaining charges for atoms in molecules—oxidation states. However, in assigning oxidation states, we always count *both* of the shared electrons as belonging to the more electronegative atom in a bond. This practice leads to highly exaggerated estimates of charge. In other words, although oxidation states are useful for bookkeeping electrons in redox reactions, they are not realistic estimates of the actual charges on individual atoms in a molecule, and so they are not suitable for judging the appropriateness of Lewis structures. A second definition of atomic charges in a molecule, the **formal charge,** is more suitable for evaluating Lewis structures.

The concept of formal charge requires that we compare

1. the number of valence electrons on the free neutral atom (which has a charge of zero because the number of electrons equals the number of protons) and
2. the number of valence electrons "belonging" to a given atom in a molecule.

If an atom in a molecule has the same number of valence electrons as it does in the free state, the positive and negative charges just balance, and the atom has a formal charge of 0. If an atom has one more valence electron in a molecule than it has on a free atom, it has a formal charge of -1, and so on. Thus the formal charge on an atom in a molecule is defined as

$$\text{Formal charge} = \begin{pmatrix}\text{number of valence} \\ \text{electrons on a free atom}\end{pmatrix} - \begin{pmatrix}\text{number of valence electrons} \\ \text{assigned to the atom in the} \\ \text{molecule}\end{pmatrix}$$

To compute the formal charge of an atom in a molecule, we assign the valence electrons to the various atoms by making the following assumptions:

1. Lone pair electrons belong entirely to the atom in question.
2. Shared electrons are *divided equally* between the two sharing atoms.

Thus the number of valence electrons assigned to a given atom is calculated as follows:

$$(\text{Valence electrons})_{\text{assigned}} = \begin{pmatrix}\text{number of lone} \\ \text{pair electrons}\end{pmatrix} + \tfrac{1}{2}\begin{pmatrix}\text{number of} \\ \text{shared electrons}\end{pmatrix}$$

We will illustrate the procedures for calculating formal charges by considering two of the possible Lewis structures for the sulfate ion, which has 32 valence electrons. For the Lewis structure

```
  [      ··      ]2−
  [     :O:      ]
  [  ··  |   ··  ]
  [ :O — S — O:  ]
  [  ··  |   ··  ]
  [     :O:      ]
  [      ··      ]
```

each oxygen atom has six lone pair electrons and shares two electrons with the sulfur atom. Thus, according to the preceding assumptions, each oxygen is assigned seven valence electrons:

$$\text{Valence electrons assigned to each oxygen} = \underset{\substack{\uparrow \\ \text{Lone} \\ \text{pair} \\ \text{electrons}}}{6} \text{ plus } \underset{\substack{\uparrow \\ \text{Shared} \\ \text{electrons}}}{\tfrac{1}{2}(2)} = 7$$

$$\text{Formal charge on oxygen} = 6 - 7 = -1$$

(6: Valence electrons on a free O atom; 7: Valence electrons assigned to each O in SO_4^{2-})

The formal charge on each oxygen is -1.

For the sulfur atom, there are no lone pair electrons, and eight electrons are shared with the oxygen atoms. Thus for sulfur,

$$\text{Valence electrons assigned to sulfur} = 0 + \tfrac{1}{2}(8) = 4$$

(0: Lone pair electrons; 8: Shared electrons)

$$\text{Formal charge on sulfur} = 6 - 4 = 2$$

(6: Valence electrons on free S atom; 4: Valence electrons assigned to S in SO_4^{2-})

A second possible Lewis structure is

```
   [     ..      ]2-
   [     O       ]
   [ ..  ‖   ..  ]
   [ :O—S—O:     ]
   [ ..  ‖   ..  ]
   [     O       ]
   [     ..      ]
```

In this case the formal charges are as calculated below.

For oxygen atoms with single bonds:

$$\text{Valence electrons assigned} = 6 + \tfrac{1}{2}(2) = 7$$

$$\text{Formal charge} = 6 - 7 = -1$$

For oxygen atoms with double bonds:

$$\text{Valence electrons assigned} = 4 + \tfrac{1}{2}(4) = 6$$

(Each double bond has 4 electrons)

$$\text{Formal charge} = 6 - 6 = 0$$

For the sulfur atom:

$$\text{Valence electrons assigned} = 0 + \tfrac{1}{2}(12) = 6$$

$$\text{Formal charge} = 6 - 6 = 0$$

Now, having determined the formal charges for the various Lewis structures of the sulfate ion, can we use them to identify which of the resonance structures is closest to the actual electronic structure of the ion? There are two schools of thought on this issue. One position assumes that the atoms in a molecule or ion will try to achieve minimum formal charges. In other words, the assumption is that electrons will naturally flow from negatively charged parts of the molecule to positively charged parts, thus minimizing the charges on the atoms. From this point of view, the resonance structures of SO_4^{2-} with two double bonds (and minimum formal charges)

$$\left[\text{O(0)}=\text{S(0)}=\text{O(0)};\ \text{S}-\text{O}^{(-1)},\ \text{S}-\text{O}^{(-1)}\right]^{2-} \longleftrightarrow \left[\text{O}^{(0)}=\text{S}^{(0)}=\text{O}^{(0)};\ \text{S}-\text{O}^{(-1)},\ \text{S}-\text{O}^{(-1)}\right]^{2-}$$

would be favored. Note that these structures exceed the octet rule, thus requiring the sulfur to use its $3d$ orbitals to hold electrons.

The other school of thought argues for the primacy of the octet rule and against the use of $3d$ orbitals. This position favors the single-bonded resonance structure of SO_4^{2-}

$$\left[\text{S}^{(+2)} \text{ singly bonded to four } \text{O}^{(-1)}\right]^{2-}$$

with its octet of electrons around the sulfur and its high formal charges.

Which point of view is correct? What do we know about the sulfate ion that might help us decide? One pertinent fact is that the sulfur–oxygen bonds in SO_4^{2-} are known by experiment to be shorter than expected for normal single bonds. This would seem to favor the resonance structures with double bonds. However, one can also argue that the high formal charges on the atoms in the single-bonded structure cause ionic attractions that pull the atoms closer together. Thus both schools of thought can adequately explain the short sulfur–oxygen bond lengths.

At this point the dispute continues about which position is correct. For second-row elements where the octet rule is never exceeded, it seems clear that the "best" Lewis structures conform to the following rules:

1. Atoms in molecules try to achieve formal charges as close to zero as possible.
2. Any negative formal charges are expected to reside on the most electronegative atoms.

However, for molecules or ions containing third-row or heavier atoms, the situation is still unclear.

Rules Governing Formal Charge

1. To calculate the formal charge on an atom:
   - Take the sum of the lone pair electrons and one-half of the shared electrons. This is the number of valence electrons assigned to a given atom in the molecule.
   - Subtract the number of assigned electrons from the number of valence electrons on the free, neutral atom to obtain the formal charge.
2. The sum of the formal charges of all atoms in a given molecule or ion must equal the overall charge on that species.
3. If nonequivalent Lewis structures exist for a species containing second-row atoms, those with formal charges closest to zero and with any negative formal charges on the most electronegative atoms are considered to best describe the bonding in the molecule or ion.

EXAMPLE 13.10

Give possible Lewis structures for XeO_3, an explosive compound of xenon. Determine the formal charges of each atom in the various Lewis structures.

Solution For XeO_3 (26 valence electrons) we can draw the following possible Lewis structures (formal charges are indicated in parentheses):

Xe (+3), O (−1), O (−1), O (−1) — Xe (+2), O (0), O (−1), O (−1) — Xe (+2), O (−1), O (0), O (−1) — Xe (+2), O (−1), O (−1), O (0)

Xe (+1), O (−1), O (0), O (0) — Xe (+1), O (0), O (−1), O (0) — Xe (+1), O (0), O (0), O (−1) — Xe (0), O (0), O (0), O (0)

The concept of formal charge is most often used to evaluate the importance of various Lewis structures for molecules that exhibit resonance. However, formal charge arguments also can be helpful in predicting which, among a given group of atoms, is the central atom in a simple molecule. For example, why is carbon dioxide O—C—O rather than C—O—O? Although this question can be pursued at many different levels of sophistication, the simplest approach involves considering the formal charges in the two possible structures. Note that in the Lewis structure given previously for carbon dioxide,

$$\ddot{O}=C=\ddot{O}$$

all atoms have formal charges of 0. However, if the atoms are arranged as follows,

C—O—O

all the Lewis structures give unreasonable formal charges. Consider the following possibilities, where the formal charges are listed below each atom:

:C≡O—Ö:	C=O=O	:C—O≡O:
−1 +2 −1	−2 +2 0	−3 +2 +1

None of these Lewis structures (with their resulting formal charges) agrees with our observation that oxygen has a significantly greater electronegativity than carbon. That is, it doesn't make sense that a compound would contain a negatively charged carbon atom next to a positively charged oxygen atom.

As a final note, there are several cautions about formal charge to keep in mind. First, although formal charges are closer to actual atomic charges in molecules than are oxidation states, formal charges still are only estimates of charge—they should not be taken as actual atomic charges. Second, the evaluation of Lewis structures using formal charge ideas can lead to erroneous predictions.

In this same vein, note the difference between a "correct," or valid, Lewis structure and an electronic structure that accurately accounts for a molecule's observed properties. A valid Lewis structure is one that obeys the rules we have established for Lewis structures. However, this Lewis structure may or may not give an accurate picture of the molecule and its properties. Experiments must be carried out to make the final decisions on the correct description of the bonding in a molecule or polyatomic ion.*

*For a discussion of this issue, see Gordon H. Purser, *J. Chem. Ed.* **76** (1999): 1013.

13.13 Molecular Structure: The VSEPR Model

The structures of molecules play a very important role in determining their chemical properties. As we will see later, structure is particularly important for biological molecules; a slight change in the structure of a large biomolecule can completely destroy its usefulness to a cell or may even change the cell from normal to cancerous.

Many accurate methods now exist for determining **molecular structure,** the three-dimensional arrangement of the atoms in a molecule. These methods must be used if precise information about structure is required. However, it is often useful to predict the approximate molecular structure of a molecule. In this section we consider a simple model that allows us to do this. This model, called the **valence shell electron-pair repulsion (VSEPR) model,** is useful in predicting the geometries of molecules formed from nonmetals. The main postulate of this model is that *the structure around a given atom is determined principally by minimizing electron-pair repulsions.* The idea is that the bonding and nonbonding pairs around a given atom should be positioned as far apart as possible. To see how this model works, we will first consider the molecule $BeCl_2$, which has the Lewis structure

The origin of the "repulsions" among electron pairs probably results more from the operation of the Pauli exclusion principle than from electrostatic effects, but we will not be concerned with that in this text.

:C̤̈l—Be—C̤̈l:

Note that there are two pairs of electrons around the beryllium atom. What arrangement of these electron pairs allows them to be as far apart as possible to minimize the repulsions? Clearly, the best arrangement places the pairs on opposite sides of the beryllium atom at 180 degrees:

This is the maximum possible separation for two electron pairs. Once we have determined the optimal arrangement of the electron pairs around the central atom, we can specify the molecular structure of $BeCl_2$ — that is, the positions of the atoms. Since each electron pair on beryllium is shared with a chlorine atom, the molecule has a **linear structure** with a bond angle of 180 degrees:

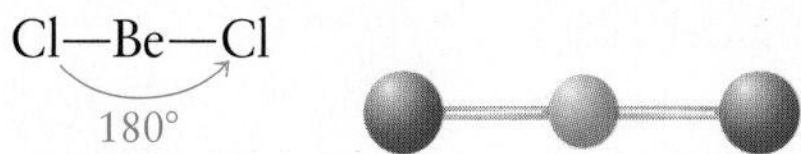

Next, let's consider BF_3, which has the Lewis structure

:F̤̈:
|
:F̤̈—B—F̤̈:

Here the boron atom is surrounded by three pairs of electrons. What arrangement will minimize the repulsions? The electron pairs are farthest apart at angles of 120 degrees:

Since each electron pair is shared with a fluorine atom, the molecular structure is

F
B 120°
F F

This is a planar (flat) and triangular molecule, which is commonly described as **trigonal planar.**

Next, let's consider the methane molecule, which has the Lewis structure

H
H—C—H
H

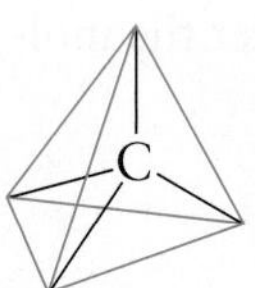

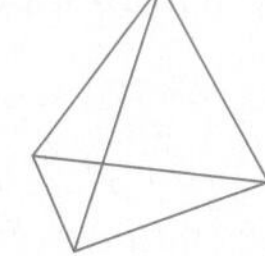

There are four pairs of electrons around the central carbon atom. What arrangement of these electron pairs best minimizes the repulsions? First, let's try a square planar arrangement:

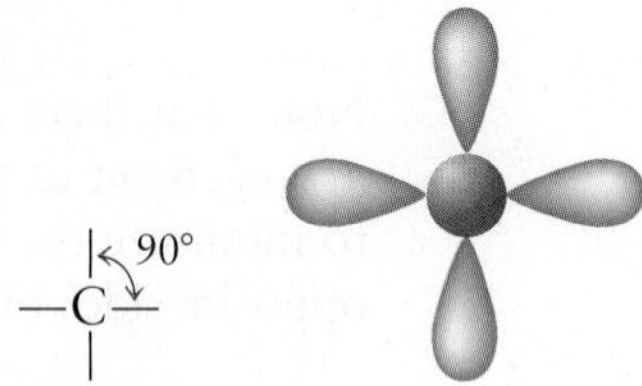

When four uniform balloons are tied together, they naturally form a tetrahedral shape.

The carbon atom and the electron pairs are centered in the plane of the paper, and the angles between the pairs are all 90 degrees.

Is there another arrangement with angles greater than 90 degrees that would put the electron pairs even farther away from each other? The answer is yes. The **tetrahedral arrangement** has angles of 109.5 degrees:

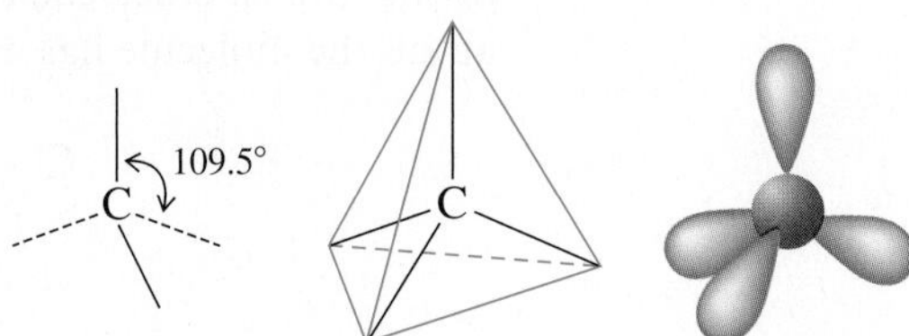

It can be shown that this is the maximum possible separation of four pairs around a given atom. This means that *whenever four pairs of electrons are present around an atom, they should always be arranged tetrahedrally.*

Now that we have obtained the electron-pair arrangement that gives the least repulsions, we can determine the positions of the atoms and thus the molecular structure of CH_4. In methane each of the four electron pairs is shared between the carbon atom and a hydrogen atom. Thus the hydrogen atoms are placed as shown in Fig. 13.15, giving the molecule a tetrahedral structure with the carbon atom at the center.

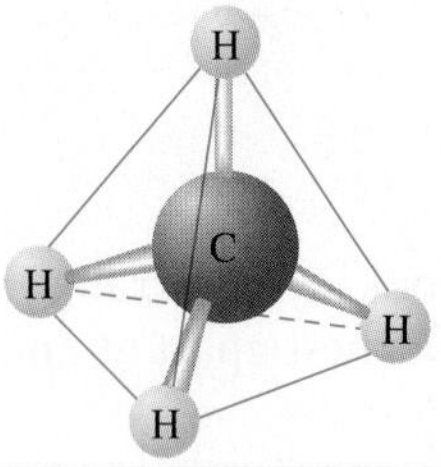

Figure 13.15
The molecular structure of methane. The tetrahedral arrangement of electron pairs produces a tetrahedral arrangement of hydrogen atoms.

Recall that the fundamental idea of the VSEPR model is to find the arrangement of electron pairs around the central atom that minimizes the electron repulsions. Then we can determine the molecular structure from knowing how the electron pairs are shared with the peripheral atoms.

STEPS

Steps for Using the VSEPR Model

1. Draw the Lewis structure for the molecule.
2. Count the electron pairs around the central atom, and arrange them in the way that minimizes repulsions (that is, put the pairs as far apart as possible).
3. Determine the positions of the atoms from the ways the electron pairs are shared.
4. Name the molecular structure from the positions of the *atoms*.

We will now predict the structure of ammonia (NH_3) using this approach.

- **Draw the Lewis structure.**

$$\mathrm{H{-}\ddot{N}{-}H}$$
$$\mathrm{|}$$
$$\mathrm{H}$$

- **Count the pairs of electrons, and arrange them to minimize repulsions.**

  The NH_3 molecule has four pairs of electrons: three bonding pairs and one nonbonding pair. From the discussion of the methane molecule, we know that the best arrangement of four electron pairs is a tetrahedral array, as shown in Fig. 13.16(a).

- **Determine the positions of the atoms.**

  The three H atoms share electron pairs, as shown in Fig. 13.16(b).

- **Name the molecular structure.**

  It is very important to recognize that the *name* of the molecular structure is always based on the *positions of the atoms*. The placement of the electron pairs determines the structure, but the name is based on the positions of the atoms. Thus it is incorrect to say that the NH_3 molecule is tetrahedral. It has a tetrahedral arrangement of electron pairs but not a tetrahedral arrangement of atoms. The molecular structure of ammonia is a **trigonal pyramid** (one triangular side is different from the other three), rather than a tetrahedron, as shown in Fig. 13.16(c).

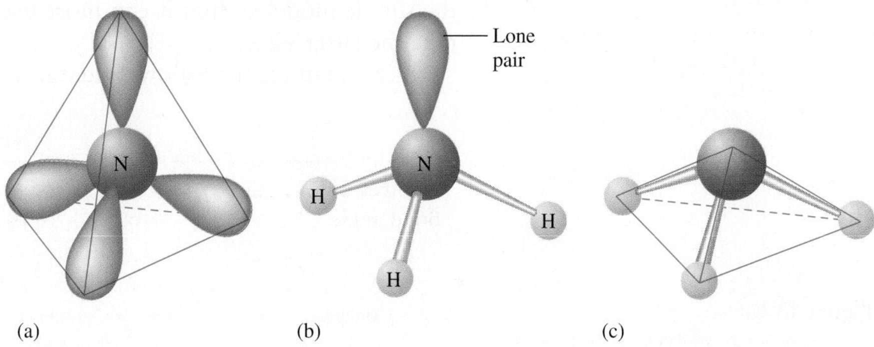

Figure 13.16

(a) The tetrahedral arrangement of electron pairs around the nitrogen atom in the ammonia molecule. (b) Three of the electron pairs around nitrogen are shared with hydrogen atoms, as shown, and the fourth is a lone pair. Although the arrangement of *electron pairs* is tetrahedral, as in the methane molecule, the hydrogen atoms in the ammonia molecule occupy only three corners of the tetrahedron. A lone pair occupies the fourth corner. (c) Note that molecular geometry is trigonal pyramidal, not tetrahedral.

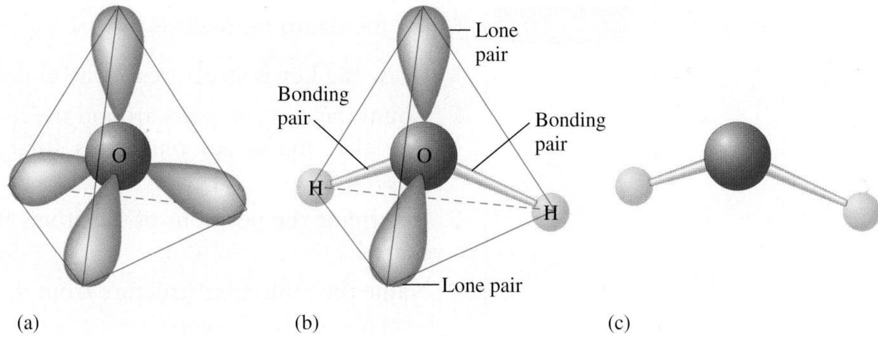

Figure 13.17
(a) The tetrahedral arrangement of the four electron pairs around oxygen in the water molecule. (b) Two of the electron pairs are shared between oxygen and the hydrogen atoms, and two are lone pairs. (c) The V-shaped molecular structure of the water molecule.

EXAMPLE 13.11

Describe the molecular structure of the water molecule.

Solution The Lewis structure for water is

$$\mathrm{H{-}\ddot{\underset{\cdot\cdot}{O}}{-}H}$$

There are four pairs of electrons: two bonding pairs and two nonbonding pairs. To minimize repulsions, these are best arranged in a tetrahedral array, as shown in Fig. 13.17(a). Although H_2O has a tetrahedral arrangement of electron pairs, it is not a tetrahedral molecule. The atoms in the H_2O molecule form a V-shape, as shown in Fig. 13.17(b) and (c).

From Example 13.11 we see that the H_2O molecule is V-shaped, or bent, because of the presence of the lone pairs. If no lone pairs were present, the molecule would be linear, the polar bonds would cancel, and the molecule would have no dipole moment. This would make water very different from the polar substance so familiar to us.

From the previous discussion, we would predict that the H—X—H bond angle (where X is the central atom) in CH_4, NH_3, and H_2O should be the tetrahedral angle (109.5 degrees). Experiments, however, show that the actual bond angles are those given in Fig. 13.18. What significance do these results have for the VSEPR model? One possible point of view is that the observed angles are close enough to the tetrahedral angle to be satisfactory. The opposite view is that the deviations are significant enough to require modification of the simple model so that it can more accurately handle similar cases. We will take the latter view.

Let's examine the following data:

	CH_4	NH_3	H_2O
Number of lone pairs	0	1	2
Bond angle	109.5°	107°	104.5°

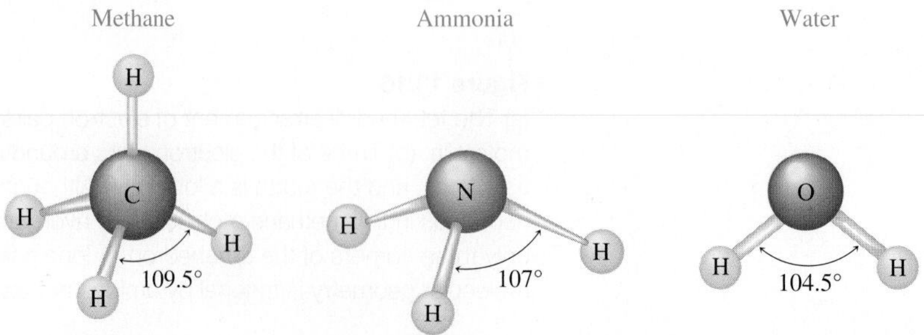

Figure 13.18
The bond angles in the CH_4, NH_3, and H_2O molecules. Note that the bond angle between bonding pairs decreases as the number of lone pairs increases.

(a)

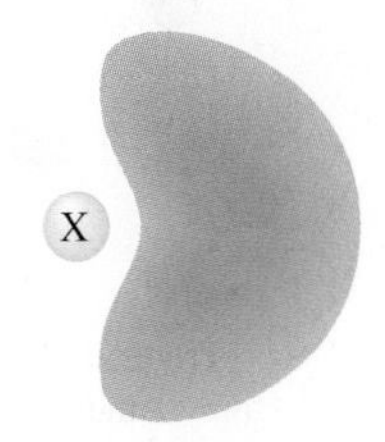

(b)

Figure 13.19
(a) In a bonding pair of electrons, the electrons are shared by two nuclei.
(b) In a lone pair, since both electrons must be close to a single nucleus, they tend to take up more of the space around that atom.

One interpretation of the trend observed here is that lone pairs require more space than bonding pairs; in other words, as the number of lone pairs increases, the bonding pairs are increasingly squeezed together.

This interpretation seems to make physical sense if we think in the following terms: A bonding pair is shared between two nuclei, and the electrons can be close to either nucleus. Therefore they are relatively confined between the two nuclei. A lone pair is localized on only one nucleus, so both electrons are close to that nucleus only, as shown schematically in Fig. 13.19. These pictures help us to understand why a lone pair may require more space near an atom than a bonding pair.

As a result of these observations, we make the following addition to the original postulate of the VSEPR model: *Lone pairs require more room than bonding pairs and tend to compress the angles between the bonding pairs.*

So far we have considered cases with two, three, and four electron pairs around the central atom. These are summarized in Table 13.8. Table 13.9 summarizes the structures possible for molecules in which there are four electron pairs around the central atom with various numbers of atoms bonded to it. Note that molecules with four pairs of electrons around the central atom can be tetrahedral (AB_4), trigonal pyramidal (AB_3), and V-shaped (AB_2). For five pairs

Table 13.8

Arrangements of Electron Pairs Around an Atom Yielding Minimum Repulsion

Number of Electron Pairs	Arrangement of Electron Pairs		Example
2	Linear	180°	
3	Trigonal planar	120°	
4	Tetrahedral	109.5°	
5	Trigonal bipyramidal	90°, 120°	
6	Octahedral	90°, 90°	

Table 13.9

Structures of Molecules That Have Four Electron Pairs Around the Central Atom

Electron-Pair Arrangement	Molecular Structure
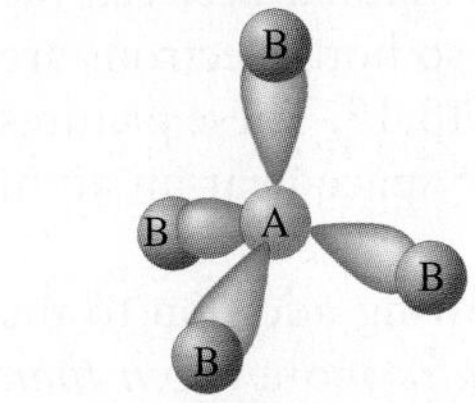	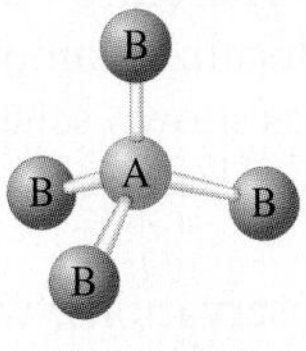 Tetrahedral
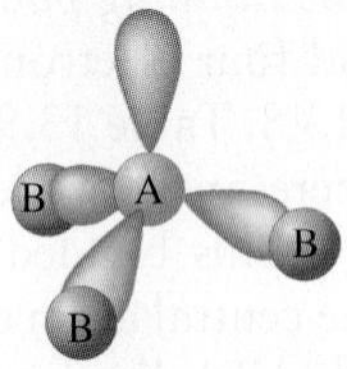	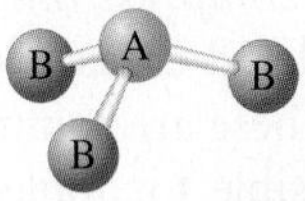 Trigonal pyramid
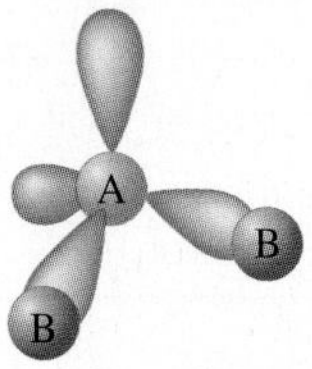	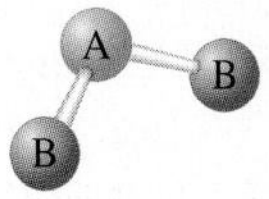 V-shaped (bent)

Table 13.10

Structures of Molecules with Five Electron Pairs Around the Central Atom

Electron-Pair Arrangement	Molecular Structure
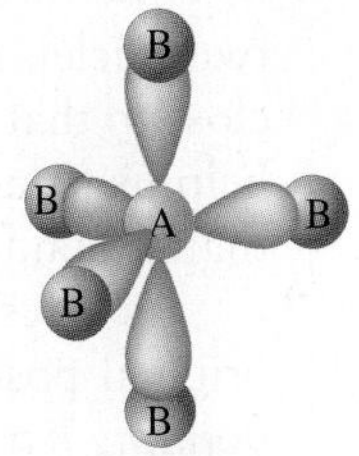	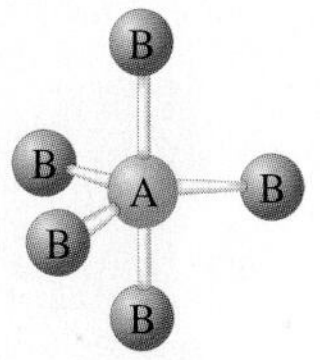 Trigonal bipyramidal
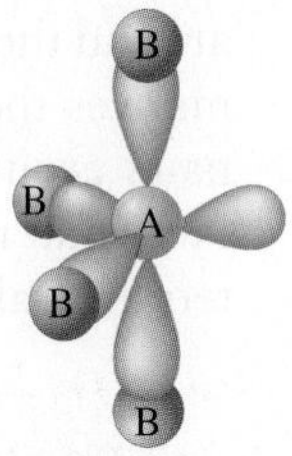	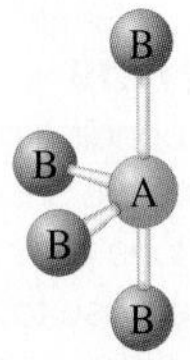 See-saw
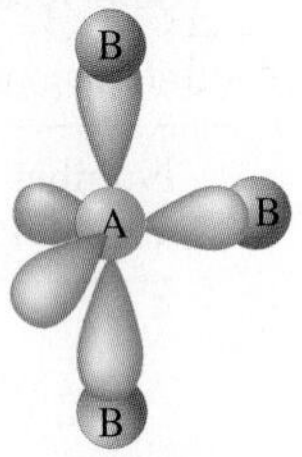	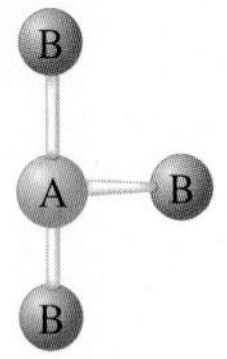 T-shaped
	 Linear

of electrons, there are several possible electron-pair arrangements. The one that produces minimum repulsion is a **trigonal bipyramid.** Note from Table 13.8 that this arrangement has two different angles, 90 degrees and 120 degrees. As the name suggests, the structure formed by this arrangement of pairs consists of two trigonal-based pyramids that share a common base. Table 13.10 summarizes the structures possible for molecules in which there are five electron pairs around the central atom with various numbers of atoms bonded to it. Note that molecules with five pairs of electrons around the central atom can be trigonal bipyramidal (AB_5), see-saw (AB_4), T-shaped (AB_3), and linear (AB_2). Six pairs of electrons can best be arranged around a given atom to form an **octahedral structure** with 90-degree angles, as shown in Table 13.8.

To use the VSEPR model to determine the geometric structures of molecules, you should memorize the relationships between the number of electron pairs and their best arrangement.

OWL INTERACTIVE EXAMPLE 13.12

When phosphorus reacts with excess chlorine gas, the compound phosphorus pentachloride (PCl_5) is formed. In the gaseous and liquid states, this substance consists of PCl_5 molecules, but in the solid state, it consists of a 1:1 mixture of PCl_4^+ and PCl_6^- ions. Predict the geometric structures of PCl_5, PCl_4^+, and PCl_6^-.

Lewis structure for PCl_5

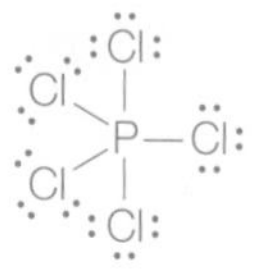

Solution The traditional Lewis structure for PCl_5 is shown in the margin. Five pairs of electrons around the phosphorus atom require a trigonal bipyramidal arrangement (see Table 13.8). When the chlorine atoms are included, a trigonal bipyramidal molecule results.

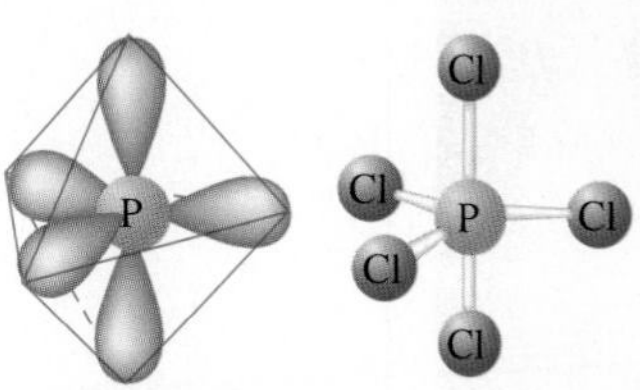

Lewis structure for PCl_4^+

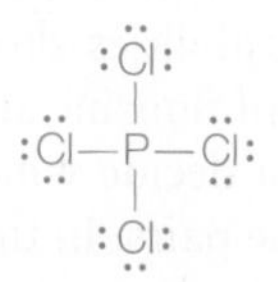

The Lewis structure for the PCl_4^+ ion [$5 + 4(7) - 1 = 32$ valence electrons] is shown in the margin. There are four pairs of electrons surrounding the phosphorus atom in the PCl_4^+ ion. This requires a tetrahedral arrangement of the pairs, as shown in the figure in the margin. Since each pair is shared with a chlorine atom, a tetrahedral PCl_4^+ cation results.

The traditional Lewis structure for PCl_6^- [$5 + 6(7) + 1 = 48$ valence electrons] is

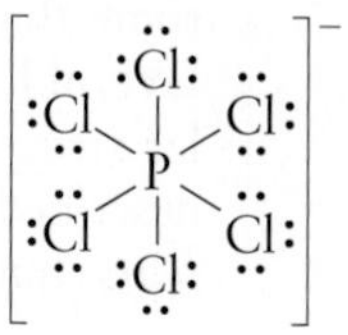

Since phosphorus is surrounded by six pairs of electrons, an octahedral arrangement is required to minimize repulsions, as shown below on the left. Since each electron pair is shared with a chlorine atom, an octahedral PCl_6^- anion is predicted.

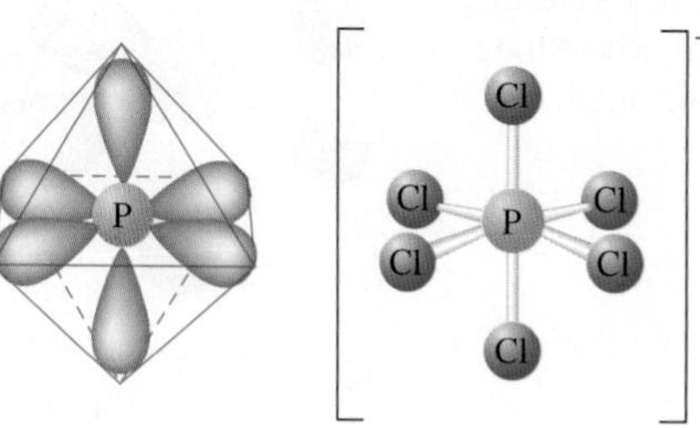

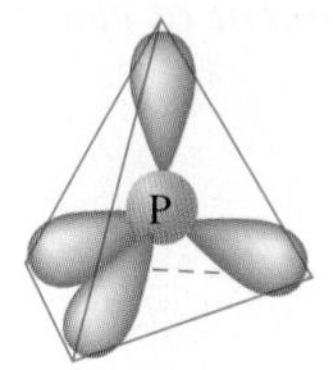

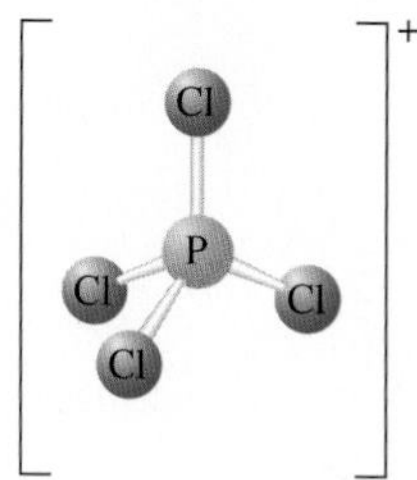

Tetrahedral PCl_4^+ cation

OWL INTERACTIVE EXAMPLE 13.13

Because the noble gases have filled *s* and *p* valence orbitals, they are not expected to be chemically reactive. In fact, for many years these elements were called *inert gases* because of this supposed inability to form any compounds. However, in the early 1960s, several compounds of krypton, xenon, and radon were synthesized. For example, a team at the Argonne National Laboratory produced the stable colorless compound xenon tetrafluoride (XeF_4). Predict its structure and determine whether it has a dipole moment.

Xenon tetrafluoride crystals.

Solution The traditional Lewis structure for XeF_4 is

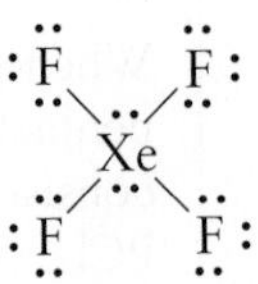

The xenon atom in this molecule is surrounded by six pairs of electrons, requiring an octahedral arrangement:

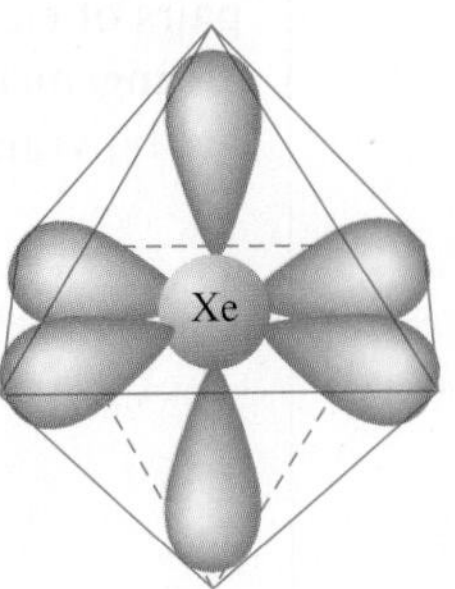

The structure predicted for this molecule depends on how the lone pairs and bonding pairs are arranged. Consider the two possibilities shown in Fig. 13.20. The bonding pairs are indicated by the presence of fluorine atoms. Since the structure predicted differs in the two cases, we must decide which of these arrangements is preferable. The key is to look at the lone pairs. In the structure in part (a), the lone pair–lone pair angle is 90 degrees; in the structure in part (b), the lone pairs are separated by 180 degrees. Since lone pairs require more room than bonding pairs, a structure with two lone pairs at 90 degrees is unfavorable. Thus the arrangement in Fig. 13.20(b) is preferred, and the molecular structure is predicted to be square planar. Note that this molecule is *not* described as being octahedral. There is an *octahedral arrangement of electron pairs*, but the *atoms* form a **square planar** structure.

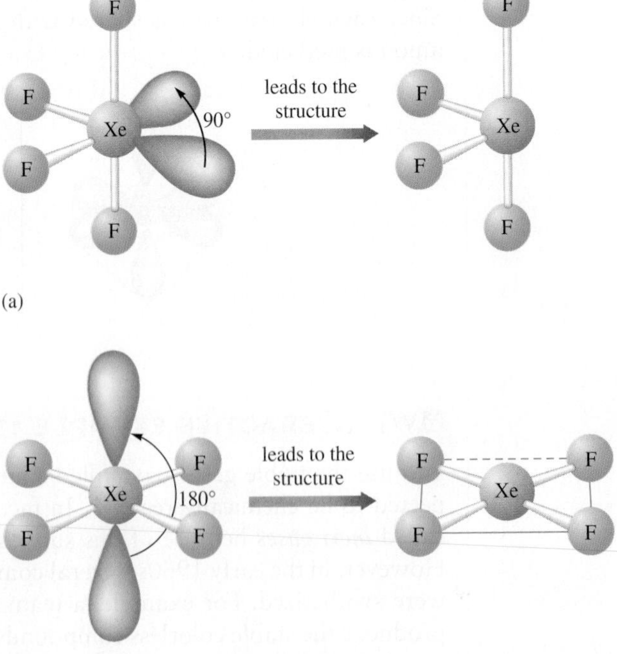

Figure 13.20
Possible electron-pair arrangement for XeF_4. Since arrangement (a) has lone pairs 90 degrees apart, it is less favorable than arrangement (b), where the lone pairs are 180 degrees apart.

Although each Xe—F bond is polar (fluorine has a greater electronegativity than xenon), the square planar arrangement of these bonds causes the polarities to cancel.

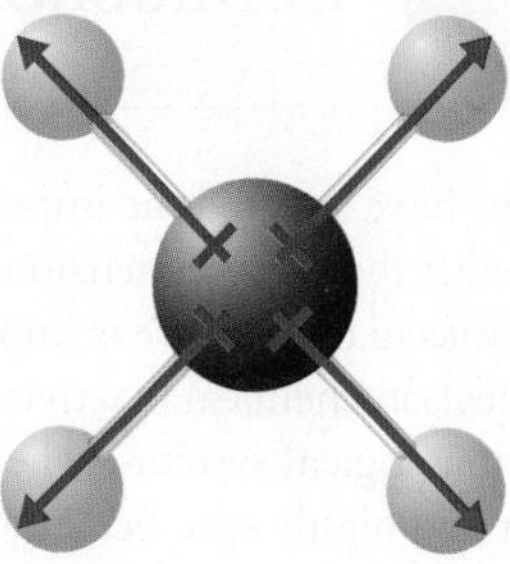

Thus XeF_4 has no dipole moment.

We can further illustrate the use of the VSEPR model for molecules or ions with lone pairs by considering the triiodide ion (I_3^-). The Lewis structure for I_3^- is

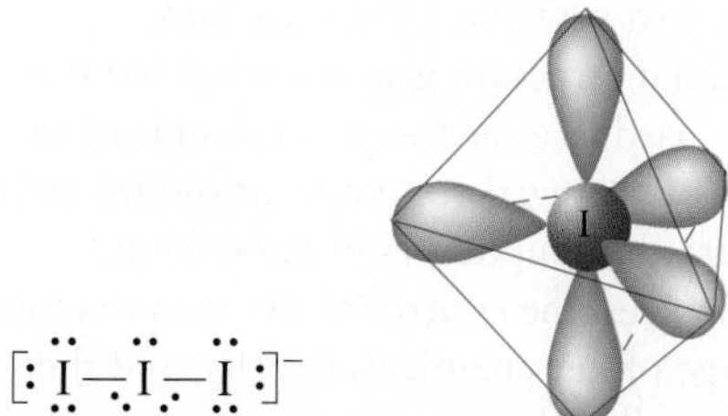

The central iodine atom has five pairs around it, requiring a trigonal bipyramidal arrangement. Several possible arrangements of the lone pairs are shown in Fig. 13.21. Note that structures (a) and (b) have lone pairs at 90 degrees, whereas in (c) all lone pairs are at 120 degrees. Thus structure (c) is preferred. The resulting molecular structure for I_3^- is linear.

$$[\text{I—I—I}]^-$$

The VSEPR Model and Multiple Bonds

So far in our treatment of the VSEPR model, we have not considered any molecules with multiple bonds. To see how these molecules are handled by this model, let's consider the NO_3^- ion, which requires three resonance structures to describe its electronic structure:

$$\left[\begin{array}{c}\text{O}\\ \| \\ \text{N}\\ \text{O}\quad\text{O}\end{array}\right]^- \longleftrightarrow \left[\begin{array}{c}\text{O}\\ | \\ \text{N}\\ \text{O}\;/\!\!/\quad\text{O}\end{array}\right]^- \longleftrightarrow \left[\begin{array}{c}\text{O}\\ | \\ \text{N}\\ \text{O}\quad\backslash\!\!\backslash\;\text{O}\end{array}\right]^-$$

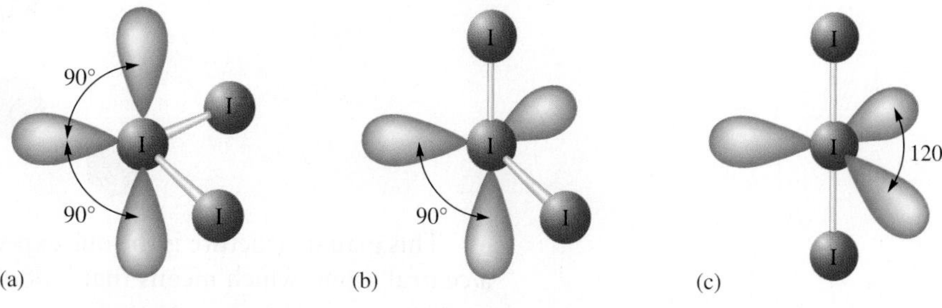

Figure 13.21
Three possible arrangements of the electron pairs in the I_3^- ion. Arrangement (c) is preferred because there are no 90-degree lone pair–lone pair interactions.

Chemical Insights

Chemical Structure and Communication: Semiochemicals

In this chapter we have stressed the importance of being able to predict the three-dimensional structure of a molecule. Molecular structure is important because of its effect on chemical reactivity. This is especially true in biological systems, where reactions must be efficient and highly specific. Among the hundreds of types of molecules in the fluids of a typical biological system, the appropriate reactants must find and react only with each other—they must be very discriminating. This specificity depends primarily on structure. The molecules are constructed so that only the appropriate partners can approach each other in a way that allows reaction.

Molecular structure is also central for those molecules used as a means of communication. Examples of chemical communication occurring in humans are the conduction of nerve impulses across synapses, the control of the manufacture and storage of key chemicals in cells, and the senses of smell and taste. Plants and animals also use chemical communication. For example, ants lay down a chemical trail so that other ants can find a certain food supply. Ants also warn their fellow workers of approaching danger by emitting certain chemicals.

Molecules convey messages by fitting into appropriate receptor sites in a very specific way, which is determined by their structure. When a molecule occupies a receptor site, chemical processes that produce the appropriate response are stimulated. Sometimes receptors can be fooled, as in the use of artificial sweeteners—molecules fit the sites on the taste buds that stimulate a "sweet" response in the brain, but they are not metabolized in the same way as natural sugars. Similar deception is useful in insect control. If an area is sprayed with synthetic female sex attractant molecules, the males of that species become so confused that mating does not occur.

A *semiochemical* is a molecule that delivers a message between members of the same or different species of plant or animal. There are three groups of these chemical messengers: allomones, kairomones, and pheromones. Each is of great ecological importance.

An *allomone* is defined as a chemical that gives adaptive advantage to the producer. For example, leaves of the black walnut tree contain an herbicide, juglone, that appears after the leaves fall to the ground. Juglone is not toxic to grass or certain grains, but it is effective against plants such as apple trees that would compete for the available water and food supplies.

Antibiotics are also allomones, since the microorganisms produce them to inhibit other species from growing near them.

Many plants produce bad-tasting chemicals to protect themselves from plant-eating insects and animals. The familiar compound nicotine deters animals from eating the tobacco plant. The millipede sends an unmistakable "back off" message by squirting a predator with benzaldehyde and hydrogen cyanide.

Defense is not the only use of allomones, however. Flowers use scent to attract pollinating insects. Honeybees, for instance, are guided to alfalfa flowers by a series of sweet-scented compounds.

Kairomones are chemical messengers that bring advantageous news to the receiver. For example,

The NO_3^- ion is known to be planar with 120-degree bond angles:

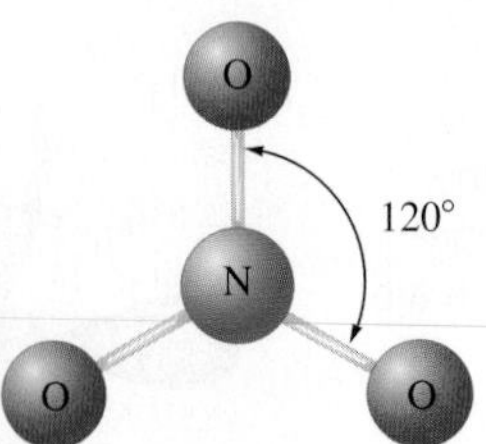

This planar structure is the one expected for three pairs of electrons around a central atom, which means that *a double bond should be counted as one ef-*

the floral scents are kairomones from the honeybees' viewpoint. Many predators are guided by kairomones emitted by their food. For example, apple skins exude a chemical that attracts the codling moth larva. In some cases kairomones help the underdog. Certain marine mollusks can pick up the "scent" of their predators, the sea stars, and make their escape.

Pheromones are chemicals that affect receptors of the same species as the donor. That is, they are specific within a species. *Releaser pheromones* cause an immediate reaction in the receptor, whereas *primer pheromones* cause long-term effects. Examples of releaser pheromones are the sex attractants of insects, generated in some species by the males and in others by the females. Sex pheromones have also been found in plants and mammals.

Alarm pheromones are highly volatile compounds (ones easily changed to a gas) released to warn of danger. Honeybees produce isoamyl acetate ($C_7H_{14}O_2$) in their sting glands. Because of its high volatility, this compound does not linger after the state of alert is over. Social behavior in insects is characterized by the use of *trail pheromones,* which are used to indicate a food source. Social insects such as bees, ants, wasps, and termites use these substances. Since trail pheromones are less volatile compounds, the indicators persist for some time.

Primer pheromones, which cause long-term behavioral changes, are harder to isolate and identify. One example, however, is the "queen substance" produced by queen honeybees. All the eggs in a colony are laid by one queen bee. If she is removed from the hive or dies, the worker bees are activated by the absence of the queen substance and begin to feed royal jelly to bee larvae to raise a new queen. The queen substance also prevents the development of the workers' ovaries so that only the queen herself can produce eggs.

Scott Camazine/Photo Researchers

Honeybees are attracted to these alfalfa flowers by sweet-scented compounds the flowers emit.

Many studies of insect pheromones are now under way in the hope that they will provide a method of controlling insects that is more efficient and safer than the current chemical pesticides.

fective pair in using the VSEPR model. This makes sense because the two pairs of electrons involved in the double bond are *not* independent pairs. Both of the electron pairs must be in the space between the nuclei of the two atoms to form the double bond. In other words, the double bond acts as one center of electron density to repel the other pairs of electrons. The same holds true for triple bonds. This leads us to another general rule: *For the VSEPR model, multiple bonds count as one effective electron pair.*

The molecular structure of nitrate also illustrates one more important point: *When a molecule exhibits resonance, any one of the resonance structures can be used to predict the molecular structure using the VSEPR model.* These rules are illustrated in Example 13.14.

Chemical Insights Smelling and Tasting Electronically

The human nose and tongue are excellent quality-control sensors. For example, we can tell whether food is spoiled by its disagreeable odor and taste. Because it's impractical to use humans as sensors in industrial settings, several companies are now developing electronic noses.

Cyrano Sciences has developed a hand-held electronic nose (called a Cyranose) that can identify specific odors by the "smell print" they create in the 32 sensors of the instrument. Each sensor is composed of conductive carbon black combined with a nonconducting polymer. In the presence of the chemicals associated with an odor, the polymer swells, thereby disrupting the conductive pathways and increasing the resistance. The device is "trained" to detect particular odors by subjecting it to known sources of those odors and storing the electrical signals in the machine's database. When the "nose" is then exposed to an odor in an industrial setting, it identifies the odor by comparing the input signals with its database. The Cyranose is used in the food and beverage industries to monitor the condition of raw materials and the quality of the finished product. In the chemical industry, it is used to pinpoint leaks and to identify manufacturing odors.

Other artificial noses are now being developed that can furnish early detection of diseases. Various researchers are now developing "noses" that can detect different types of bacteria and can provide early diagnosis of lung cancer.

A company called Alpha M.O.S. has recently introduced what it calls the "world's first commercial electronic tongue," a device used to identify various tastes associated with liquids. Like the Cyranose, the electronic tongue must first be "trained" by exposing its probe to known liquid characteristics to record their "fingerprints." Once programmed, the "tongue" can recognize tastes such as sweetness, sourness, bitterness, and saltiness and can detect rancidity and various types of contamination. The electronic tongue is of particular interest to the pharmaceutical industry, where it is hoped it can reduce the need for human tasting panels.

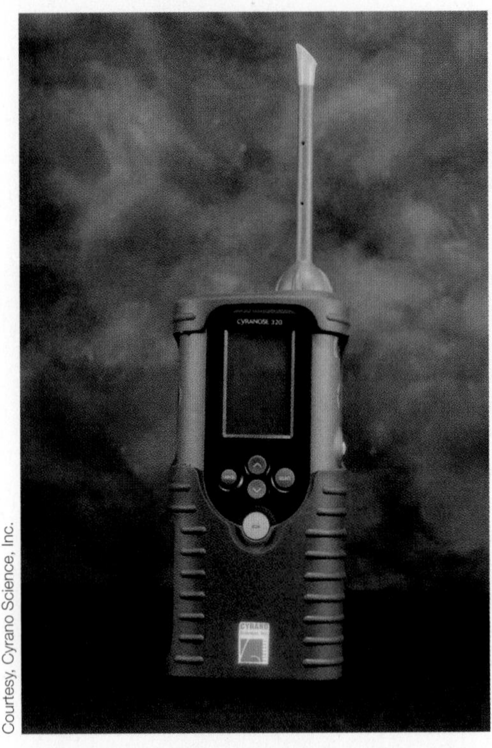

Courtesy, Cyrano Science, Inc.

A Cyranose sensor (an "electronic nose").

Artificial tasters are also being developed for sensing diseases. For example, Eric V. Anslyn and his coworkers at the University of Texas at Austin have reported a system that can recognize the "taste of heart disease" by analyzing the "taste" of blood samples.

Humans and animals are still the best "smellers" and tasters, but computers are rapidly catching up.

OWL INTERACTIVE EXAMPLE 13.14

Predict the molecular structure of the sulfur dioxide molecule. Is this molecule expected to have a dipole moment?

Solution First, we must determine the Lewis structure for the SO_2 molecule, which has 18 valence electrons. The expected resonance structures are

$$:\ddot{O}=\ddot{S}-\ddot{O}: \longleftrightarrow :\ddot{O}-\ddot{S}=\ddot{O}:$$

To determine the molecular structure, we must count the electron pairs around the sulfur atom. In each resonance structure, the sulfur has one lone pair, one pair in a single bond, and one double bond. Counting the double bond as one pair yields three effective pairs around the sulfur. According to Table 13.8, a trigonal planar arrangement is required, yielding a V-shaped molecule:

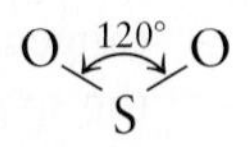

O 120° O
S

Thus the structure of the SO_2 molecule is expected to be V-shaped with a 120-degree bond angle. The molecule has a dipole moment directed as shown:

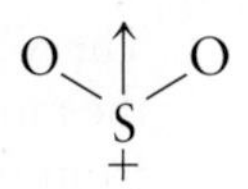

O ↑ O
S
+

Since the molecule is V-shaped, the polar bonds do not cancel.

It should be noted at this point that lone pairs oriented at least 120 degrees from other pairs do not produce significant distortions of bond angles. For example, the angle in the SO_2 molecule is actually quite close to 120 degrees. We will follow the general principle that *an angle of 120 degrees provides lone pairs with enough space so that distortions do not occur. Angles less than 120 degrees are distorted when lone pairs are present.*

Molecules Containing No Single Central Atom

So far we have considered molecules consisting of one central atom surrounded by other atoms. The VSEPR model can be readily extended to more complicated molecules, such as methanol (CH_3OH). This molecule is represented by the following Lewis structure:

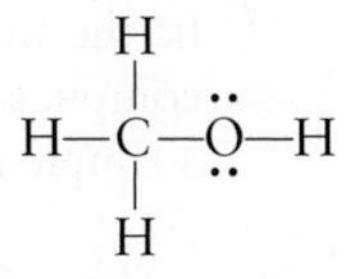

```
    H
    |   ..
H—C—O—H
    |   ..
    H
```

The VSEPR Model

The following rules are helpful in using the VSEPR model to predict molecular structure.

1. Determine the Lewis structure(s) for the molecule.
2. For molecules with resonance structures, use any of the structures to predict the molecular structure.
3. Sum the electron pairs around the central atom.

(continued)

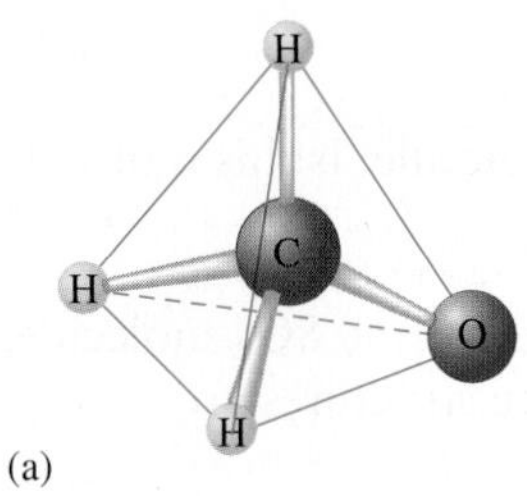

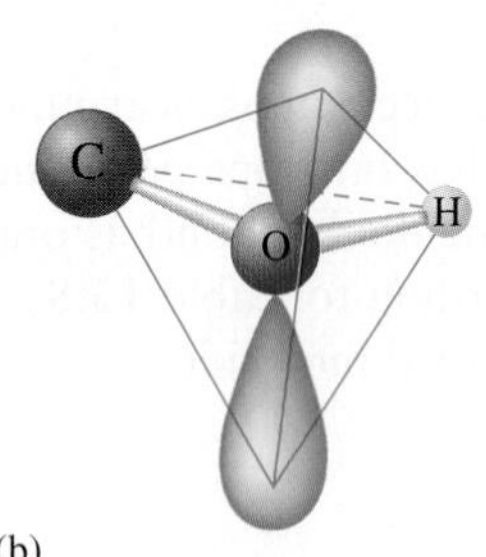

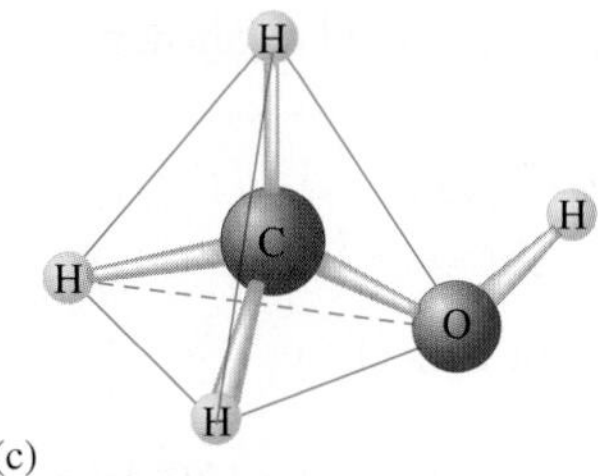

Figure 13.22
The molecular structure of methanol. (a) The arrangement of electron pairs and atoms around the carbon atom. (b) The arrangement of bonding and lone pairs around the oxygen atom. (c) The molecular structure.

The VSEPR Model (Continued)

4. When counting pairs, count each multiple bond as a single effective pair.
5. Determine the arrangement of the pairs that minimizes electron-pair repulsions. These arrangements are shown in Table 13.8.
6. Lone pairs require more space than bonding pairs. Choose an arrangement that gives the lone pairs as much room as possible, although it appears that an angle of at least 120 degrees between lone pairs provides enough space. Recognize that lone pairs at angles less than 120 degrees may produce distortions from the idealized structure.

The molecular structure can be predicted from the arrangement of pairs around the carbon and oxygen atoms. Note that there are four pairs of electrons around the carbon, which calls for a tetrahedral arrangement, as shown in Fig. 13.22(a). The oxygen also has four pairs, requiring a tetrahedral arrangement. However, in this case the tetrahedron will be slightly distorted by the space requirements of the lone pairs [Fig. 13.22(b)]. The overall geometric arrangement for the molecule is shown in Fig. 13.22(c).

The VSEPR Model—How Well Does It Work?

The VSEPR model is very simple. There are only a few rules to remember, yet the model correctly predicts the molecular structures of most molecules formed from nonmetallic elements. Molecules of any size can be treated by applying the VSEPR model to each appropriate atom (those bonded to at least two other atoms) in the molecule. Thus we can use this model to predict the structures of molecules with hundreds of atoms. It does, however, fail in a few instances. For example, phosphine (PH_3), which has a Lewis structure analogous to that of ammonia,

```
  ..          ..
H—P—H       H—N—H
  |           |
  H           H
```

would be predicted to have a molecular structure similar to that for NH_3 with bond angles of about 107 degrees. However, the bond angles of phosphine are actually 94 degrees. There are ways of explaining this structure, but more rules have to be added to the model.

This example again illustrates the point that simple models will certainly have exceptions. In introductory chemistry we want to use simple models that fit the majority of cases; we are willing to accept a few failures rather than complicate the model. The amazing thing about the VSEPR model is that such a simple model correctly predicts the structures of so many molecules.

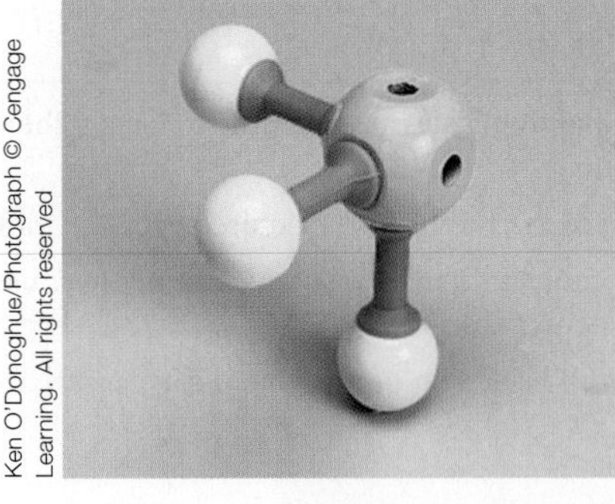

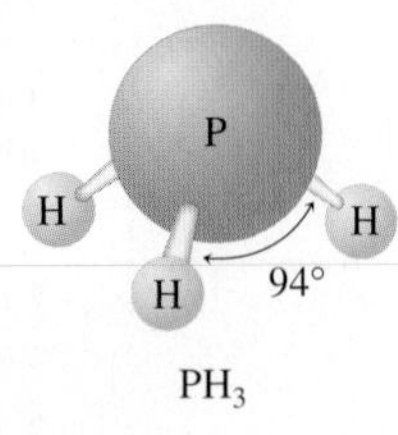

Key Terms

Section 13.1
bond energy
ionic compound
Coulomb's law
bond length
covalent bonding
polar covalent bond

Section 13.2
electronegativity

Section 13.3
dipole moment

Section 13.4
isoelectronic ions

Section 13.5
lattice energy

Section 13.8
single bond
double bond
triple bond

Section 13.9
localized electron (LE) model
lone pair
bonding pair

Section 13.10
Lewis structure
duet rule
octet rule

Section 13.11
resonance
resonance structure

Section 13.12
formal charge

Section 13.13
molecular structure
valence shell electron-pair repulsion (VSEPR) model
linear structure
trigonal planar
tetrahedral arrangement
trigonal pyramid
trigonal bipyramid
octahedral structure
square planar

For Review

Sign in at **www.cengage.com/owl** to:

- View tutorials and simulations, develop problem-solving skills, and complete online homework assigned by your professor.
- Download Go Chemistry mini lecture modules for quick review and exam prep from OWL (or purchase them at **www.cengagebrain.com**)

Chemical bonds

- Hold groups of atoms together
- Occur when a group of atoms can lower its total energy by aggregating
- Types of chemical bonds
  - Ionic: electrons are transferred to form ions
  - Covalent: equal sharing of electrons
  - Polar covalent: unequal electron sharing
- Percent ionic character of a bond X—Y

$$\frac{\text{Measured dipole moment of X—Y}}{\text{Calculated dipole moment for } X^+ Y^-} \times 100\%$$

- Electronegativity: the relative ability of an atom to attract shared electrons
  - The polarity of a bond depends on the electronegativity difference of the bonded atoms
- The spatial arrangement of polar bonds in a molecule determines whether the molecule has a dipole moment

Ionic bonding

- An ion has a different size than its parent atom
  - An anion is larger than its parent ion
  - A cation is smaller than its parent atom
- Lattice energy: the change in energy when ions are packed together to form an ionic solid

Bond energy

- The energy necessary to break a covalent bond
- Increases as the number of shared pairs increases
- Can be used to estimate the enthalpy change for a chemical reaction

Lewis structures

- Show how the valence electron pairs are arranged among the atoms in a molecule or polyatomic ion
- Stable molecules usually contain atoms that have their valence orbitals filled
  - Leads to a duet rule for hydrogen
  - Leads to an octet rule for second-row elements
  - The atoms of elements in the third row and beyond can exceed the octet rule
- Several equivalent Lewis structures can be drawn for some molecules, a concept called resonance
- When several nonequivalent Lewis structures can be drawn for a molecule, formal charge is often used to choose the most appropriate structure(s)

VSEPR model

- Based on the idea that electron pairs will be arranged around a central atom in a way that minimizes the electron repulsions
- Can be used to predict the geometric structure of most molecules

Discussion Questions

These questions are designed to be considered by groups of students in class. Often these questions work well for introducing a particular topic in class.

1. Explain the electronegativity trends across a row and down a column of the periodic table. Compare these trends with those of ionization energy and atomic radii. How are they all related?
2. The ionic compound AB is formed. The charges on the ions may be $+1, -1$; $+2, -2$; $+3, -3$; or even larger. What are the factors that determine the charge for an ion in an ionic compound?
3. Using only the periodic table, predict the most stable ion for Na, Mg, Al, S, Cl, K, Ca, and Ga. Arrange these from largest to smallest radius, and explain why the radius varies as it does. Compare your predictions with Fig. 13.8.
4. The bond energy for the C—H bond is about 413 kJ/mol in CH_4 but 380 kJ/mol in $CHBr_3$. Although these values are relatively close in magnitude, they are different. Explain why they are different. Does the fact that the C—H bond energy in $CHBr_3$ is lower make any sense? Why?
5. Consider the following statement: "Because oxygen seems to prefer a negative two charge, the second electron affinity is more negative than the first." Indicate everything that is correct in this statement. Indicate everything that is incorrect. Correct the incorrect information, and explain.
6. Which has the greater bond lengths: NO_2^- or NO_3^-? Explain.
7. The following ions are best described with resonance structures. Draw the resonance structures, and using formal charge arguments, predict the "best" Lewis structure for each ion.
   a. NCO^-
   b. CNO^-
8. Would you expect the electronegativity of titanium to be the same in the species Ti, Ti^{2+}, Ti^{3+}, and Ti^{4+}? Explain.
9. The second electron affinity values for both oxygen and sulfur are unfavorable (endothermic). Explain.
10. Arrange the following molecules from most to least polar, and explain your order: CH_4, CF_2Cl_2, CF_2H_2, CCl_4, and CCl_2H_2.

Exercises

OWL Interactive versions of these problems may be assigned in OWL.

A blue exercise number indicates that the answer to that exercise appears at the back of this book and a solution appears in the *Solutions Guide*.

Chemical Bonds and Electronegativity

11. Distinguish between the terms *electronegativity* and *electron affinity*, *covalent bond* and *ionic bond*, and *pure covalent bond* and *polar covalent bond*. Characterize the types of bonds in terms of electronegativity difference. Energetically, why do ionic and covalent bonds form?
12. The following electrostatic potential diagrams represent H_2, HCl, or NaCl. Label each, and explain your choices.

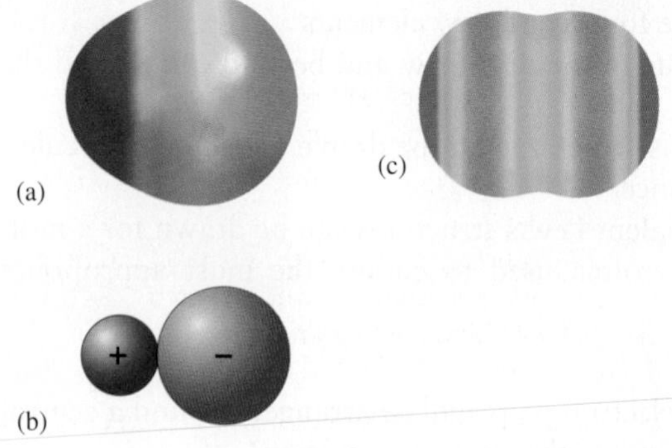

13. An alternative definition of electronegativity is

$$\text{Electronegativity} = \text{constant (I.E.} - \text{E.A.)}$$

where I.E. is the ionization energy and E.A. is the electron affinity using the sign conventions of this book. Use data in Chapter 12 to calculate the (I.E. − E.A.) term for F, Cl, Br, and I. Do these values show the same trend as the electronegativity values given in this chapter? The first ionization energies of the halogens are 1678, 1255, 1138, and 1007 kJ/mol, respectively. (*Hint:* Choose a constant so that the electronegativity of fluorine equals 4.0. Using this constant, calculate relative electronegativities for the other halogens and compare to values given in the text.)

14. Use Coulomb's law,

$$V = \frac{Q_1Q_2}{4\pi\epsilon_0 r} = 2.31 \times 10^{-19}\ \text{J nm}\left(\frac{Q_1Q_2}{r}\right)$$

to calculate the energy of interaction for the following two arrangements of charges, each having a magnitude equal to the electron charge.

a. $(+1) \xleftrightarrow{1 \times 10^{-10}\text{ m}} (-1) \longleftarrow \infty \longrightarrow (+1) \xleftrightarrow{1 \times 10^{-10}\text{ m}} (-1)$

b. Four charges arranged in a square: (-1) at top, $(+1)$ at left and right, (-1) at bottom; each side 1×10^{-10} m
1×10^{-10} m 1×10^{-10} m
1×10^{-10} m 1×10^{-10} m

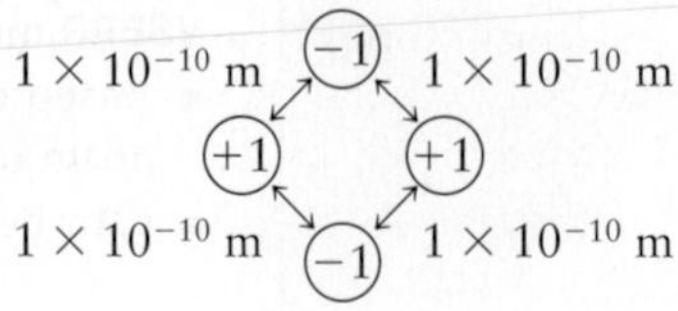

15. Without using Fig. 13.3, predict the order of increasing electronegativity in each of the following groups of elements.

a. C, N, O
b. S, Se, Cl
c. Si, Ge, Sn
d. Tl, S, Ge
e. Na, K, Rb
f. B, O, Ga

16. Without using Fig. 13.3, predict which bond in each of the following groups is the most polar.

a. C—F, Si—F, Ge—F
b. P—Cl, S—Cl
c. S—F, S—Cl, S—Br
d. Ti—Cl, Si—Cl, Ge—Cl
e. C—H, Si—H, Sn—H
f. Al—Br, Ga—Br, In—Br, Tl—Br

17. Repeat Exercises 15 and 16. This time use the values of the electronegativities of the elements given in Fig. 13.3. Are there any differences among your answers?

18. Hydrogen has an electronegativity value between boron and carbon and identical to phosphorus. With this in mind, rank the following bonds in order of decreasing polarity: P—H, O—H, N—H, F—H, C—H.

19. Rank the following bonds in order of increasing ionic character: N—O, Ca—O, C—F, Br—Br, K—F.

20. List all the possible bonds that can occur between the elements P, Cs, O, and H. Predict the type of bond (ionic, covalent, or polar covalent) one would expect to form for each bond.

21. Some plant fertilizer compounds are $(NH_4)_2SO_4$, $Ca_3(PO_4)_2$, K_2O, P_2O_5, and KCl. Which of these compounds contain both ionic and covalent bonds?

22. The following electrostatic potential diagrams represent CH_4, NH_3, or H_2O. Label each, and explain your choices.

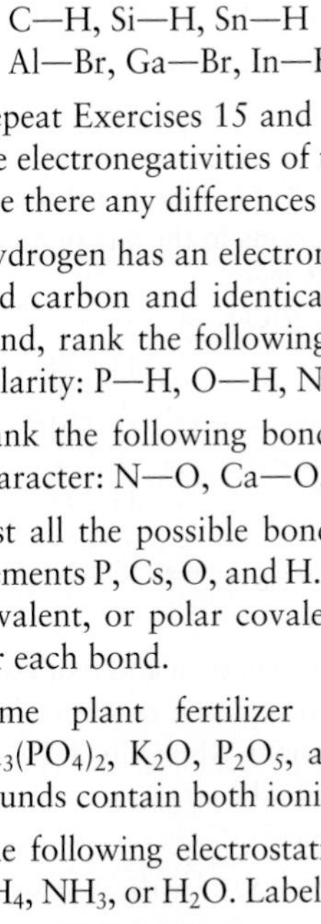

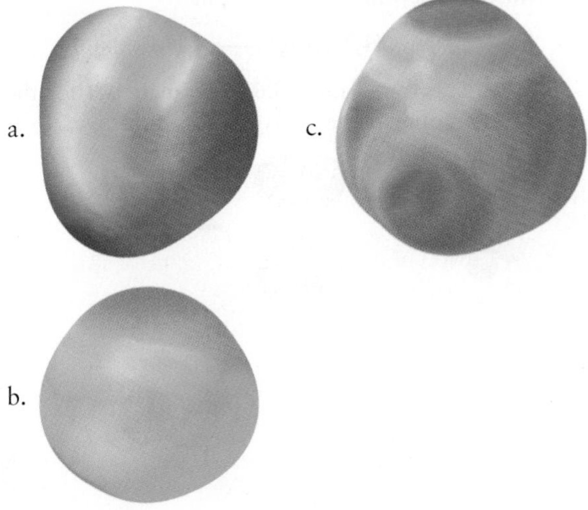

Ions and Ionic Compounds

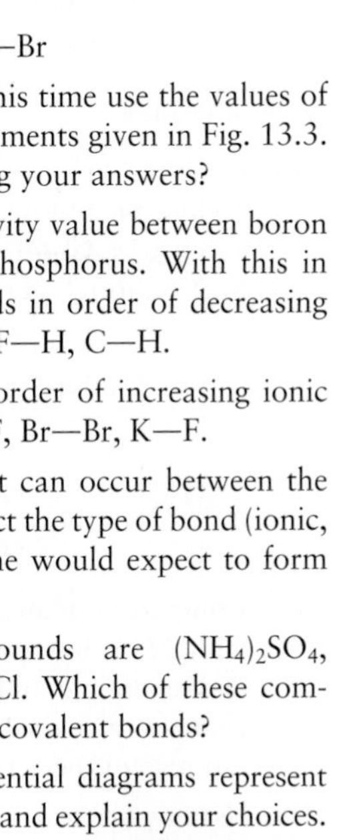

23. When an element forms an anion, what happens to the radius? When an element forms a cation, what happens to the radius? Why? Define the term *isoelectronic*. When comparing sizes of ions, which ion has the largest radius, and which ion has the smallest radius in an isoelectronic series? Why?

24. Consider the ions Sc^{3+}, Cl^-, K^+, Ca^{2+}, and S^{2-}. Match these ions to the following pictures that represent the relative sizes of the ions.

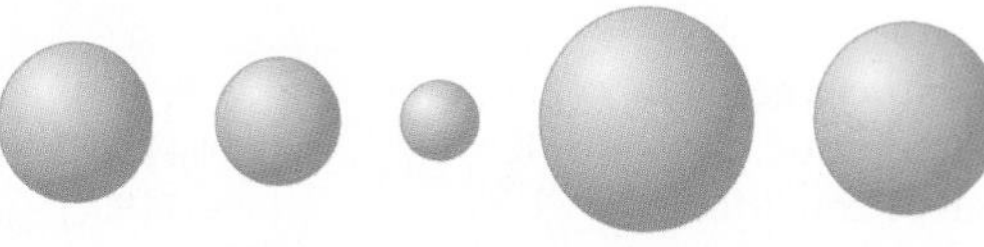

25. For each of the following groups, place the atoms and ions in order of decreasing size.

a. Cu, Cu^+, Cu^{2+}
b. Ni^{2+}, Pd^{2+}, Pt^{2+}
c. O, O^-, O^{2-}
d. La^{3+}, Eu^{3+}, Gd^{3+}, Yb^{3+}
e. Te^{2-}, I^-, Cs^+, Ba^{2+}, La^{3+}

26. Write electron configurations for each of the following.

a. the cations: Mg^{2+}, Sn^{2+}, K^+, Al^{3+}, Tl^+, As^{3+}
b. the anions: N^{3-}, O^{2-}, F^-, Te^{2-}

27. Write electron configurations for the most stable ion formed by each of the elements Rb, Ba, Se, and I (when in stable ionic compounds).

28. Give an example of an ionic compound where both the anion and the cation are isoelectronic with each of the following noble gases.

a. Ne
b. Ar
d. Kr
e. Xe

29. What noble gas has the same electron configuration as each of the ions in the following compounds?

a. cesium sulfide
b. strontium fluoride
c. calcium nitride
d. aluminum bromide

30. Which of the following ions have noble gas electron configurations?

a. Fe^{2+}, Fe^{3+}, Sc^{3+}, Co^{3+}
b. Tl^+, Te^{2-}, Cr^{3+}
c. Pu^{4+}, Ce^{4+}, Ti^{4+}
d. Ba^{2+}, Pt^{2+}, Mn^{2+}

31. Give three ions that are isoelectronic with krypton. Place these ions in order of increasing size.

32. Which compound in each of the following pairs of ionic substances has the most exothermic lattice energy? Justify your answers.

a. LiF, CsF
b. NaBr, NaI
c. $BaCl_2$, BaO
d. Na_2SO_4, $CaSO_4$
e. KF, K_2O
f. Li_2O, Na_2S

33. Predict the empirical formulas of the ionic compounds formed from the following pairs of elements. Name each compound.

a. Al and S
b. K and N
c. Mg and Cl
d. Cs and Br

34. Following are some important properties of ionic compounds:

i. low electrical conductivity as solids, and high conductivity in solution or when molten
ii. relatively high melting and boiling points
iii. brittleness

How does the concept of ionic bonding discussed in this chapter account for these properties?

35. Use the following data to estimate ΔH_f° for potassium chloride.

$$K(s) + \frac{1}{2}Cl_2(g) \longrightarrow KCl(s)$$

Lattice energy	−690. kJ/mol
Ionization energy for K	419 kJ/mol
Electron affinity of Cl	−349 kJ/mol
Bond energy of Cl_2	239 kJ/mol
Enthalpy of sublimation for K	90. kJ/mol

36. Use the following data to estimate ΔH_f° for magnesium fluoride.

$$Mg(s) + F_2(g) \longrightarrow MgF_2(s)$$

Lattice energy	−2913 kJ/mol
First ionization energy of Mg	735 kJ/mol
Second ionization energy of Mg	1445 kJ/mol
Electron affinity of F	−328 kJ/mol
Bond energy of F_2	154 kJ/mol
Enthalpy of sublimation of Mg	150. kJ/mol

37. Consider the following: $Li(s) + \frac{1}{2}I_2(g) \rightarrow LiI(s)$ $\Delta H = -292$ kJ. LiI(*s*) has a lattice energy of −753 kJ/mol. The ionization energy of Li(*g*) is 520. kJ/mol, the bond energy of $I_2(g)$ is 151 kJ/mol, and the electron affinity of I(*g*) is −295 kJ/mol. Use these data to determine the heat of sublimation of Li(*s*).

38. In general, the higher the charge on the ions in an ionic compound, the more favorable is the lattice energy. Why do some stable ionic compounds have +1 charged ions even though +4, +5, and +6 charged ions would have a more favorable lattice energy?

39. Consider the following energy changes:

	ΔE (kJ/mol)
$Mg(g) \longrightarrow Mg^+(g) + e^-$	735
$Mg^+(g) \longrightarrow Mg^{2+}(g) + e^-$	1445
$O(g) + e^- \longrightarrow O^-(g)$	−141
$O^-(g) + e^- \longrightarrow O^{2-}(g)$	878

a. Magnesium oxide exists as $Mg^{2+}O^{2-}$, not as Mg^+O^-. Explain.
b. What experiment could be done to confirm that magnesium oxide does not exist as Mg^+O^-?

40. Use the following data (in kJ/mol) to estimate ΔH for the reaction $S^-(g) + e^- \rightarrow S^{2-}(g)$. Include an estimate of uncertainty.

	ΔH_f°	Lattice Energy	IE of M	ΔH_{sub} of M
Na_2S	−365	−2203	495	109
K_2S	−381	−2052	419	90.
Rb_2S	−361	−1949	409	82
Cs_2S	−360.	−1850.	382	78

$$S(s) \longrightarrow S(g) \qquad \Delta H = 277 \text{ kJ/mol}$$

$$S(g) + e^- \longrightarrow S^-(g) \qquad \Delta H = -200. \text{ kJ/mol}$$

41. Rationalize the following lattice energy values:

Compound	Lattice Energy (kJ/mol)
CaSe	−2862
Na_2Se	−2130
CaTe	−2721
Na_2Te	−2095

42. The lattice energies of $FeCl_3$, $FeCl_2$, and Fe_2O_3 are (in no particular order) −2631 kJ/mol, −5339 kJ/mol, and −14,774 kJ/mol. Match the appropriate formula to each lattice energy.

Bond Energies

43. Use bond energy values in Table 13.6 to estimate ΔH for each of the following reactions in the gas phase.
a. $H_2(g) + Cl_2(g) \rightarrow 2HCl(g)$
b. $N{\equiv}N(g) + 3H_2(g) \rightarrow 2NH_3(g)$
c. $H{-}C{\equiv}N(g) + 2H_2(g) \longrightarrow H_2C{-}NH_2(g)$ (H–C(H)(H)–N(H)(H))
d. $H_2N{-}NH_2(g) + 2F_2(g) \longrightarrow N{\equiv}N(g) + 4HF(g)$

44. Compare your answers from parts a and b of Exercise 43 with ΔH values calculated for each reaction using standard enthalpies of formation in Appendix 4. Do enthalpy changes calculated from bond energies give a reasonable estimate of the actual values?

45. Use bond energies to predict ΔH for the isomerization of methyl isocyanide to acetonitrile.

$$CH_3N{\equiv}C(g) \longrightarrow CH_3C{\equiv}N(g)$$

46. Use data from Table 13.6 to estimate ΔH for the combustion of methane (CH_4), as shown below:

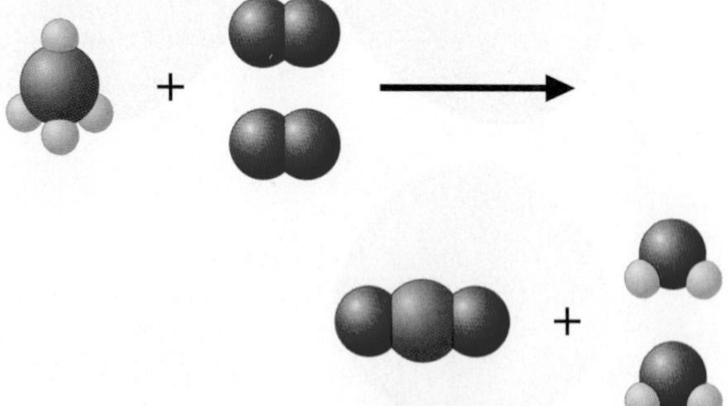

47. Use bond energies to estimate ΔH for the combustion of 1 mole of acetylene:

$$C_2H_2(g) + \frac{5}{2}O_2(g) \longrightarrow 2CO_2(g) + H_2O(g)$$

48. Consider the following reaction:

$$A_2 + B_2 \longrightarrow 2AB \qquad \Delta H = -285 \text{ kJ}$$

The bond energy for A_2 is one-half the amount of the AB bond energy. The bond energy of $B_2 = 432$ kJ/mol. What is the bond energy of A_2?

49. The space shuttle orbiter uses the oxidation of methyl hydrazine by dinitrogen tetroxide for propulsion:

$$5N_2O_4(g) + 4N_2H_3CH_3(g) \longrightarrow 12H_2O(g) + 9N_2(g) + 4CO_2(g)$$

Use bond energies to estimate ΔH for this reaction. The structures for the reactants are

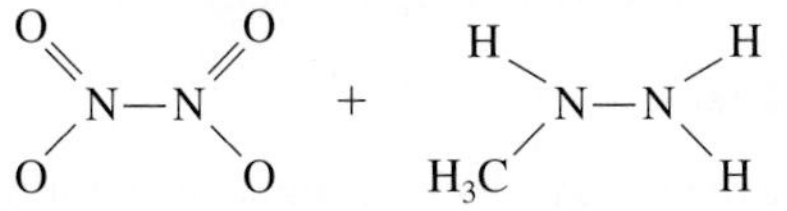

$$O_2N{-}NO_2 \quad + \quad (H_3C)(H)N{-}N(H)(H)$$

50. Following are three processes that have been used for the industrial manufacture of acrylonitrile—an important chemical used in the manufacture of plastics, synthetic rubber, and fibers. Use bond energy values (Tables 13.6 and 13.7) to estimate ΔH for each of the reactions.

a. 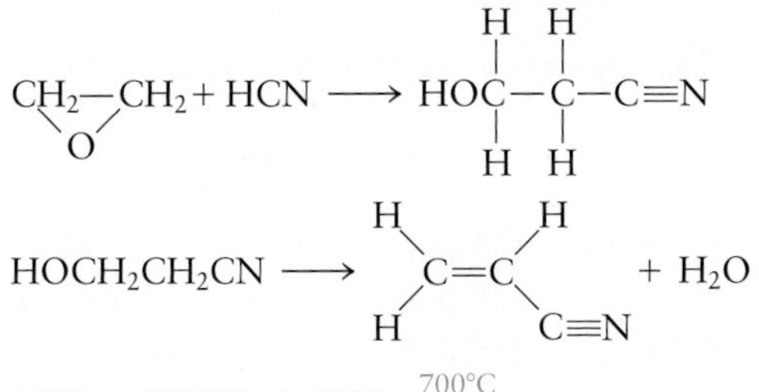 $CH_2{-}CH_2$ (epoxide, bridged by O) $+ HCN \longrightarrow HOCH_2{-}CH_2{-}C{\equiv}N$ (HOC(H)(H)—C(H)(H)—C≡N)

$$HOCH_2CH_2CN \longrightarrow H_2C{=}CH{-}C{\equiv}N + H_2O$$

b. $$4CH_2{=}CHCH_3 + 6NO \xrightarrow[\text{Ag}]{700°C} 4CH_2{=}CHCN + 6H_2O + N_2$$

The nitrogen–oxygen bond energy in nitric oxide (NO) is 630. kJ/mol.

c. $$2CH_2{=}CHCH_3 + 2NH_3 + 3O_2 \xrightarrow[425\text{–}510°C]{\text{Catalyst}} 2CH_2{=}CHCN + 6H_2O$$

51. Is the elevated temperature noted in parts b and c of Exercise 50 needed to provide energy to endothermic reactions?

52. Acetic acid is responsible for the sour taste of vinegar. It can be manufactured using the following reaction:

$$CH_3OH(g) + C{\equiv}O(g) \longrightarrow CH_3C({=}O){-}OH(l)$$

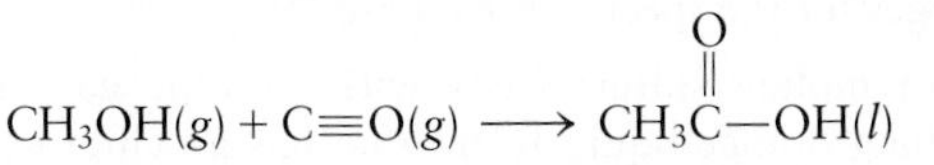

Use tabulated values of bond energies (Table 13.6) to estimate ΔH for this reaction. Compare this result to the ΔH value calculated using standard enthalpies of formation in Appendix 4. Explain any discrepancies.

53. Use bond energies (Table 13.6), values of electron affinities (Table 12.8), and the ionization energy of hydrogen (1312 kJ/mol) to estimate ΔH for each of the following reactions.
a. $HF(g) \longrightarrow H^+(g) + F^-(g)$
b. $HCl(g) \longrightarrow H^+(g) + Cl^-(g)$
c. $HI(g) \longrightarrow H^+(g) + I^-(g)$
d. $H_2O(g) \longrightarrow H^+(g) + OH^-(g)$
(Electron affinity of $OH(g) = -180.$ kJ/mol.)

54. The standard enthalpies of formation of $S(g)$, $F(g)$, $SF_4(g)$, and $SF_6(g)$ are +278.8 kJ/mol, +79.0 kJ/mol, −775 kJ/mol, and −1209 kJ/mol, respectively.
a. Use these data to estimate the energy of an S—F bond.
b. Compare the value that you calculated in part a with the value given in Table 13.6. What conclusions can you draw?
c. Why are the ΔH_f° values for $S(g)$ and $F(g)$ not equal to zero, even though sulfur and fluorine are elements?

55. Use the following standard enthalpies of formation to estimate the N—H bond energy in ammonia. Compare this with the value in Table 13.6.

$N(g)$	472.7 kJ/mol
$H(g)$	216.0 kJ/mol
$NH_3(g)$	−46.1 kJ/mol

56. The standard enthalpy of formation for $NO(g)$ is 90. kJ/mol. Use this and the values for the O=O and N≡N bond energies to estimate the bond strength in NO.

Lewis Structures and Resonance

57. Write Lewis structures that obey the octet rule for each of the following. Except for HCN and H_2CO, the first atom listed is the central atom. For HCN and H_2CO, carbon is the central atom.

a. HCN	d. NH_4^+	g. CO_2
b. PH_3	e. H_2CO	h. O_2
c. $CHCl_3$	f. SeF_2	i. HBr

58. Draw a Lewis structure that obeys the octet rule for each of the following molecules and ions. In each case the first atom listed is the central atom.
a. $POCl_3$, SO_4^{2-}, XeO_4, PO_4^{3-}, ClO_4^-
b. NF_3, SO_3^{2-}, PO_3^{3-}, ClO_3^-
c. ClO_2^-, SCl_2, PCl_2^-

59. Considering your answers to Exercise 58, what conclusions can you draw concerning the structures of species containing the same number of atoms and the same number of valence electrons?

60. Draw Lewis structures for the following. Show all resonance structures, where applicable. Carbon is the central atom in OCN^- and SCN^-.
a. NO_2^-, NO_3^-, N_2O_4(N_2O_4 exists as $O_2N{-}NO_2$.)
b. OCN^-, SCN^-, N_3^-

61. Some of the pollutants in the atmosphere are ozone, sulfur dioxide, and sulfur trioxide. Draw Lewis structures for these three molecules. Show all resonance structures.

62. Peroxyacetyl nitrate, or PAN, is present in photochemical smog. Draw Lewis structures (including resonance forms) for PAN. The skeletal arrangement is

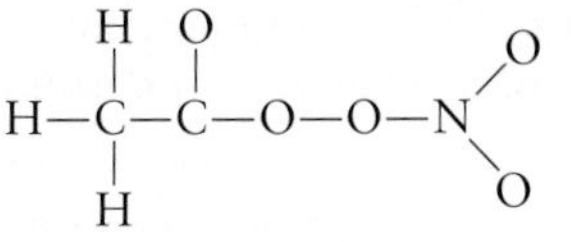

$$H{-}C(H)(H){-}C(O){-}O{-}O{-}N(O)(O)$$

63. A toxic cloud covered Bhopal, India, in December 1984 when water leaked into a tank of methyl isocyanate, and the product escaped into the atmosphere. Methyl isocya-

nate is used in the production of many pesticides. Draw the Lewis structures for methyl isocyanate, CH_3NCO, including resonance forms.

64. Explain the terms *resonance* and *delocalized electrons.* When a substance exhibits resonance, we say that none of the individual Lewis structures accurately portrays the bonding in the substance. Why do we draw resonance structures?

65. Benzene (C_6H_6) consists of a six-membered ring of carbon atoms with one hydrogen bonded to each carbon. Draw Lewis structures for benzene, including resonance structures.

66. An important observation supporting the need for resonance in the LE model is that there are only three different structures of dichlorobenzene ($C_6H_4Cl_2$). How does this fact support the need for the concept of resonance?

67. Borazine ($B_3N_3H_6$) has often been called "inorganic" benzene. Draw Lewis structures for borazine. Borazine is a six-membered ring of alternating boron and nitrogen atoms.

68. Draw all the possible Lewis structures for dimethylborazine [$(CH_3)_2B_3N_3H_4$]. (See Exercise 67.) Would there be a different number of structures if there was no resonance?

69. Which of the following statements is(are) true? Correct the false statements.
    a. It is impossible to satisfy the octet rule for all atoms in XeF_2.
    b. Because SF_4 exists, OF_4 should also exist because oxygen is in the same family as sulfur.
    c. The bond in NO^+ should be stronger than the bond in NO^-.
    d. As predicted from the two Lewis structures for ozone, one oxygen–oxygen bond is stronger than the other oxygen–oxygen bond.

70. Lewis structures can be used to understand why some molecules react in certain ways. Write the Lewis structures for the reactants and products in the reactions described below.
    a. Nitrogen dioxide dimerizes to produce dinitrogen tetroxide.
    b. Boron trihydride accepts a pair of electrons from ammonia, forming BH_3NH_3.

    Give a possible explanation for why these two reactions occur.

71. The most common type of exception to the octet rule are compounds or ions with central atoms having more than eight electrons around them. PF_5, SF_4, ClF_3, and Br_3^- are examples of this type of exception. Draw the Lewis structures for these compounds or ions. Which elements, when they have to, can have more than eight electrons around them? How is this rationalized?

72. SF_6, ClF_5, and XeF_4 are three compounds whose central atoms do not follow the octet rule. Draw Lewis structures for these compounds.

73. Consider the following bond lengths:

    C—O 1.43 Å   C=O 1.23 Å   C≡O 1.09 Å

    In the CO_3^{2-} ion, all three C—O bonds have identical bond lengths of 1.36 Å. Why?

74. Order the following species with respect to the carbon–oxygen bond length (longest to shortest):

    CO, CO_2, CO_3^{2-}, CH_3OH

    What is the order from the weakest to the strongest carbon–oxygen bond?

75. Place the species below in order of the shortest to the longest nitrogen–oxygen bond.

    H_2NOH, N_2O, NO^+, NO_2^-, NO_3^-

    (H_2NOH exists as H_2N—OH.)

Formal Charge

76. Use the formal charge arguments to rationalize why BF_3 would not follow the octet rule.

77. Use formal charge arguments to explain why CO has a much smaller dipole moment than would be expected on the basis of electronegativity.

78. Nitrous oxide (N_2O) has three possible Lewis structures:

    $$:\ddot{N}=N=\ddot{O}: \longleftrightarrow :N\equiv N-\ddot{\underset{\cdot\cdot}{O}}: \longleftrightarrow :\ddot{\underset{\cdot\cdot}{N}}-N\equiv O:$$

    Given the following bond lengths,

N—N	167 pm	N=O	115 pm
N=N	120 pm	N—O	147 pm
N≡N	110 pm		

    rationalize the observations that the N—N bond length in N_2O is 112 pm and that the N—O bond length is 119 pm. Assign formal charges to the resonance structures for N_2O. Can you eliminate any of the resonance structures on the basis of formal charges? Is this consistent with observation?

79. Draw Lewis structures that obey the octet rule for the following species. Assign the formal charge to each central atom.
    a. $POCl_3$ c. ClO_4^- e. SO_2Cl_2 g. ClO_3^-
    b. SO_4^{2-} d. PO_4^{3-} f. XeO_4 h. NO_4^{3-}

80. Draw the Lewis structures that involve minimum formal charges for the species in Exercise 79.

81. When molten sulfur reacts with chlorine gas, a vile-smelling orange liquid forms that has an empirical formula of SCl. The structure of this compound has a formal charge of zero on all elements in the compound. Draw the Lewis structure for the vile-smelling orange liquid.

82. Oxidation of the cyanide ion produces the stable cyanate ion (OCN^-). The fulminate ion (CNO^-), on the other hand, is very unstable. Fulminate salts explode when struck; $Hg(CNO)_2$ is used in blasting caps. Write the Lewis structures and assign formal charges for the cyanate and fulminate ions. Why is the fulminate ion so unstable? (C is the central atom in OCN^-, and N is the central atom in CNO^-.)

83. Write the Lewis Structure for O_2F_2 (O_2F_2 exists as F—O—O—F). Assign oxidation states and formal charges to the atoms in O_2F_2. This compound is a vigorous and

potent oxidizing and fluorinating agent. Are oxidation states or formal charges more useful in accounting for these properties of O_2F_2?

84. Benzoic acid is a food preservative. The space-filling model for benzoic acid is shown below.

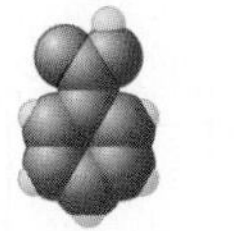

Benzoic acid
($C_6H_5CO_2H$)

Draw the Lewis structure for benzoic acid, including all resonance structures in which all atoms have a formal charge of zero.

Molecular Structure and Polarity

85. Write the name of each of the following molecular structures.

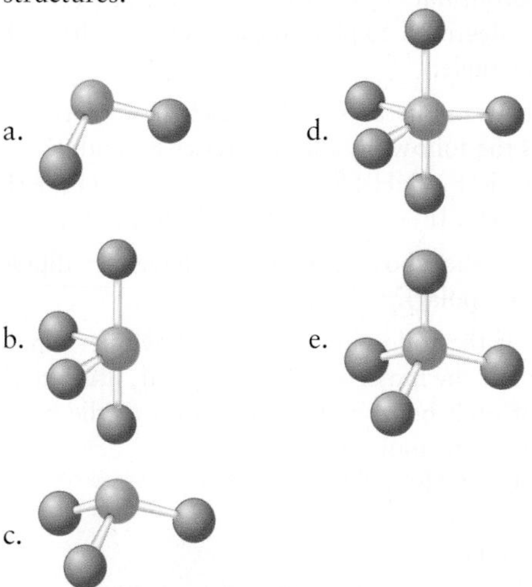

86. State whether or not each of the following has a permanent dipole moment.

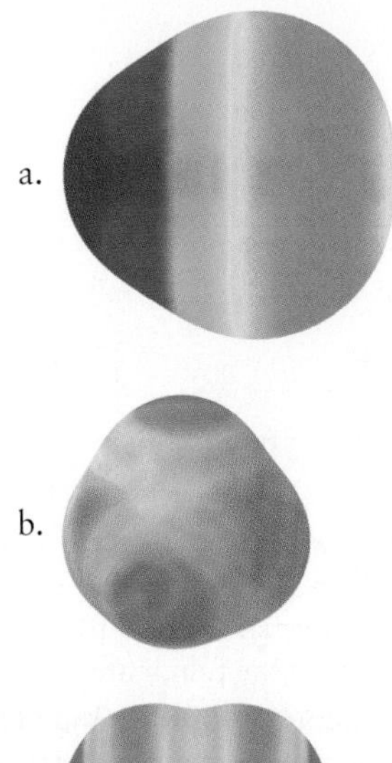

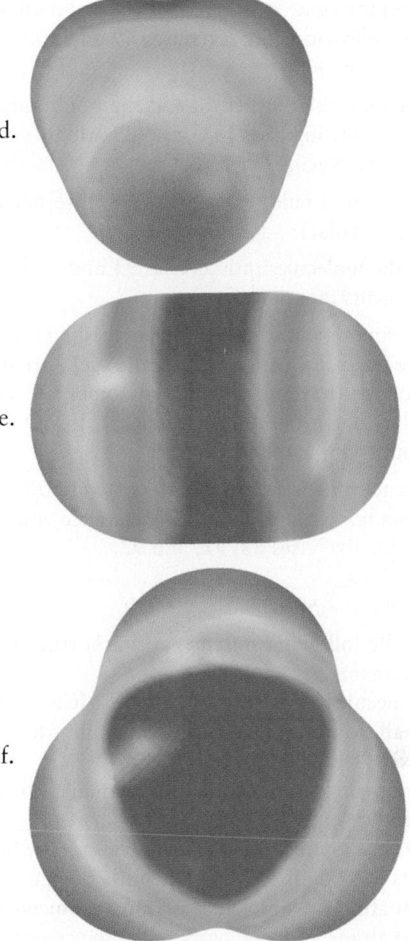

87. Predict the molecular structure and the bond angles for each molecule or ion in Exercises 57, 58, and 60.

88. Predict the molecular structure and the bond angles for each of the following.
a. SeO_3 b. SeO_2 c. PCl_3 d. SCl_2 e. SiF_4

89. There are several molecular structures based on the trigonal bipyramid geometry. Three such structures are

A, 180°, B, A — Linear
A, 90°, B, A, 90°, A — T-shaped
A, 90°, A, 120°, B, A, 90°, A — See-saw

Which of the compounds or ions in Exercises 71 and 72 have these molecular structures?

90. Two variations of the octahedral geometry are illustrated below.

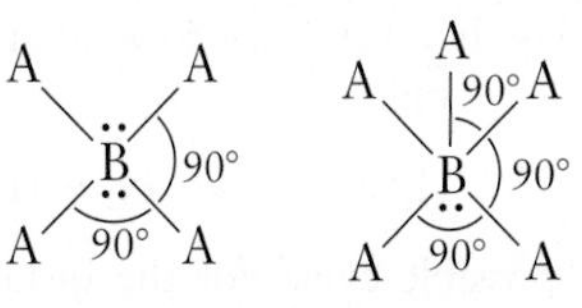

Which of the compounds or ions in Exercises 71 and 72 have these molecular structures?

91. Predict the molecular structure and the bond angles for each of the following. (See Exercises 89 and 90.)
   a. $XeCl_2$ b. ICl_3 c. TeF_4 d. PCl_5

92. Predict the molecular structure and the bond angles for each of the following. (See Exercises 89 and 90.)
   a. ICl_5 b. $XeCl_4$ c. $SeCl_6$

93. Which of the molecules in Exercise 88 have net dipole moments (are polar)?

94. Which of the molecules in Exercises 91 and 92 have net dipole moments (are polar)?

95. Give two requirements that should be satisfied for a molecule to be polar. Explain why CF_4 and XeF_4 are nonpolar compounds (have no net dipole moments), whereas SF_4 is polar (has a net dipole moment). Is CO_2 polar? What about COS? Explain.

96. What do each of the following sets of compounds/ions have in common with each other? Reference your Lewis structures for Exercises 88, 91, and 92.
   a. $XeCl_4$, $XeCl_2$
   b. ICl_5, TeF_4, ICl_3, PCl_3, SCl_2, SeO_2

97. Which of the following statements is(are) true? Correct the false statements.
   a. The molecules SeS_3, SeS_2, PCl_5, $TeCl_4$, ICl_3, and $XeCl_2$ all exhibit at least one bond angle which is approximately 120 degrees.
   b. The bond angle in SO_2 should be similar to the bond angle in CS_2 or SCl_2.
   c. Of the compounds CF_4, KrF_4, and SeF_4, only SeF_4 exhibits an overall dipole moment (is polar).
   d. Central atoms in a molecule adopt a geometry of the bonded atoms and lone pairs about the central atom in order to maximize electron repulsions.

98. Consider the following Lewis structure, where E is an unknown element:

   [:Ö—Ë—Ö:]⁻ with :Ö: bonded below E

   What are some possible identities for element E? Predict the molecular structure (including bond angles) for this ion.

99. Consider the following Lewis structure, where E is an unknown element:

   [:F̤—Ë(—Ö:)(—F̤:)]²⁻

   What are some possible identities for element E? Predict the molecular structure (including bond angles) for this ion. (See Exercises 89 and 90.)

100. Although the VSEPR model is correct in predicting that CH_4 is tetrahedral, NH_3 is pyramidal, and H_2O is bent, the model in its simplest form does not account for the fact that these molecules do not have exactly the same bond angles ($<$ HCH is 109.5 degrees, as expected for a tetrahedron, but $<$ HNH is 107.3 degrees and $<$ HOH is 104.5 degrees). Explain these deviations from the tetrahedral angle.

101. Draw Lewis structures and predict the molecular structures of the following. (See Exercises 89 and 90.)
   a. OCl_2, KrF_2, BeH_2, SO_2
   b. SO_3, NF_3, IF_3
   c. CF_4, SeF_4, KrF_4
   d. IF_5, AsF_5

   Which of the above compounds have net dipole moments (are polar)?

102. Which of the following molecules have net dipole moments? For the molecules that are polar, indicate the polarity of each bond and the direction of the net dipole moment of the molecule.
   a. CH_2Cl_2, $CHCl_3$, CCl_4
   b. CO_2, N_2O
   c. PH_3, NH_3

103. The molecules BF_3, CF_4, CO_2, PF_5, and SF_6 are all nonpolar, even though they contain polar bonds. Why?

Additional Exercises

104. Although both the Br_3^- and I_3^- ions are known, the F_3^- ion does not exist. Explain.

105. Write Lewis structures for CO_3^{2-}, HCO_3^-, and H_2CO_3. When acid is added to an aqueous solution containing carbonate or bicarbonate ions, carbon dioxide gas is formed. We generally say that carbonic acid (H_2CO_3) is unstable. Use bond energies to estimate ΔH for the reaction (in the gas phase):

$$H_2CO_3 \longrightarrow CO_2 + H_2O$$

   Specify a possible cause for the instability of carbonic acid.

106. The structure of TeF_5^- is

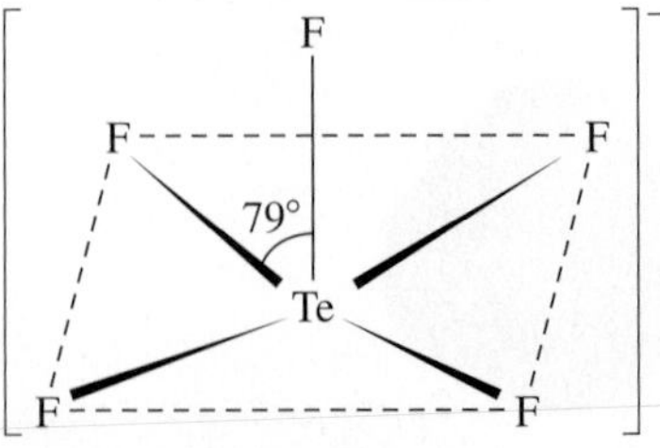

   Draw a complete Lewis structure for TeF_5^-, and explain the distortion from the ideal square pyramidal structure. (See Exercise 90.)

107. The compound NF_3 is quite stable, but NCl_3 is very unstable (NCl_3 was first synthesized in 1811 by P. L. Dulong, who lost three fingers and an eye studying its properties). The compounds NBr_3 and NI_3 are unknown, although the explosive compound $NI_3 \cdot NH_3$ is known. Account for the instability of these halides of nitrogen.

108. There are two possible structures of XeF_2Cl_2, where Xe is the central atom. Draw them, and describe how measurements of dipole moments might be used to distinguish among them.

109. Which member of the following pairs would you expect to be more energetically stable? Justify each choice.
a. $NaBr$ or $NaBr_2$ c. SO_4 or XeO_4
b. ClO_4 or ClO_4^- d. OF_4 or SeF_4

110. Many times, extra stability is characteristic of a molecule or ion in which resonance is possible. How could this feature be used to explain the acidities of the following compounds? (The acidic hydrogen is marked by an asterisk.) Part c shows resonance in the phenyl ring (C_6H_5).

```
       O
       ‖
a. H—C—OH*

          O           OH*
          ‖            |
b. CH3—C—CH=C—CH3

      OH*
       |
c.  (phenyl ring)
```

111. Arrange the following in order of increasing radius and increasing ionization energy.
a. N^+, N, N^-
b. Se, Se^-, Cl, Cl^+
c. Br^-, Rb^+, Sr^{2+}

112. Draw a Lewis structure for the *N,N*-dimethylformamide molecule. The skeletal structure is

```
      O
      |
  H—C—N—CH3
         |
        CH3
```

Various types of evidence lead to the conclusion that there is some double-bond character to one of the C—N bonds. Draw one or more resonance structures that support this observation.

113. A compound, XF_5, is 42.81% fluorine by mass. Identify the element X. What is the molecular structure of XF_5?

114. The study of carbon-containing compounds and their properties is called organic chemistry. Besides carbon atoms, *organic compounds* also can contain hydrogen, oxygen, and nitrogen atoms (as well as other types of atoms). A common trait of simple organic compounds is to have Lewis structures in which all atoms have a formal charge of zero. Consider the following incomplete Lewis structure for an organic compound called *histidine* (one of the amino acids, which are the building blocks of proteins found in human bodies):

```
        H—C—N
          |    \
          |     C—H
          |    /
          C—N—H
          |
  2    H—C—H     1
   ↘      |    ↙
   H—N—C—C—O—H
     |   |   |
     H   H   O
```

Draw a complete Lewis structure for histidine in which all atoms have a formal charge of zero. What are the approximate bond angles about the carbon atom labeled 1 and the nitrogen atom labeled 2?

115. Do the Lewis structures obtained in Exercises 79 and 80 predict the same molecular structure for each case?

116. Predict the molecular structure for each of the following. (See Exercises 89 and 90.)
a. $BrFI_2$ b. XeO_2F_2 c. $TeF_2Cl_3^-$

For each formula, there are at least two different structures that can be drawn using the same central atom. Draw the possible structures for each formula.

Challenge Problems

117. Predict the molecular structure of KrF_2. Using hyperconjugation, draw the Lewis structures for KrF_2 that obey the octet rule. Show all resonance forms.

118. Consider the following computer-generated model of caffeine.

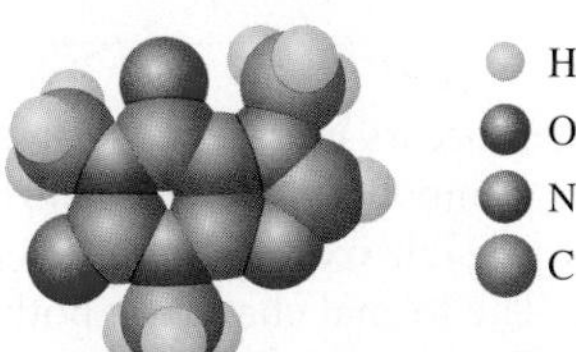

Draw a Lewis structure for caffeine in which all atoms have a formal charge of zero.

119. Given the following information:

Heat of sublimation of Li(*s*) = 166 kJ/mol
Bond energy of HCl = 427 kJ/mol
Ionization energy of Li(*g*) = 520. kJ/mol
Electron affinity of Cl(*g*) = −349 kJ/mol
Lattice energy of LiCl(*s*) = −829 kJ/mol
Bond energy of H_2 = 432 kJ/mol

Calculate the net change in energy for the following reaction:

$$2Li(s) + 2HCl(g) \longrightarrow 2LiCl(s) + H_2(g)$$

120. Use data in this chapter and Chapter 12 to discuss why MgO is an ionic compound but CO is not an ionic compound.

121. A promising new material with great potential as a fuel in solid rocket motors is ammonium dinitramide [$NH_4N(NO_2)_2$].
   a. Draw Lewis structures (including resonance forms) for the dinitramide ion [$N(NO_2)_2^-$].
   b. Predict the bond angles around each nitrogen in the dinitramide ion.
   c. Ammonium dinitramide can decompose explosively to nitrogen, water, and oxygen. Write a balanced equation for this reaction, and use bond energies to estimate ΔH for the explosive decomposition of this compound.
   d. To estimate ΔH from bond energies, you made several assumptions. What are some of your assumptions?

122. Think of forming an ionic compound as three steps (this is a simplification, as with all models): (1) removing an electron from the metal, (2) adding an electron to the nonmetal, and (3) allowing the metal cation and nonmetal anion to come together.
   a. What is the sign of the energy change for each of these three processes?
   b. In general, what is the sign of the sum of the first two processes? Use examples to support your answer.
   c. What must be the sign of the sum of the three processes?
   d. Given your answer to part c, why do ionic bonds occur?
   e. Given your explanations to part d, why is NaCl stable but not Na_2Cl_2 and $NaCl_2$? What about MgO compared to MgO_2 and Mg_2O?

123. The compound hexaazaisowurtzitane is one of the highest-energy explosives known (*C & E News*, p. 26, Jan. 17, 1994). The compound, also known as CL-20, was first synthesized in 1987. The method of synthesis and detailed performance data are still classified information because of CL-20's potential military application in rocket boosters and in warheads of "smart" weapons. The structure of CL-20 is

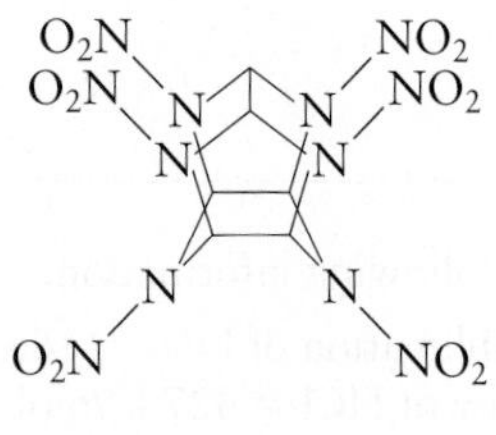

O_2N NO_2
O_2N N N NO_2
N N
N N
O_2N NO_2

In such shorthand structures, each point where lines meet represents a carbon atom. In addition, the hydrogens attached to the carbon atoms are omitted. Each of the six carbon atoms has one hydrogen atom attached. Three possible reactions for the explosive decomposition of CL-20 are

   i. $C_6H_6N_{12}O_{12}(s) \longrightarrow 6CO(g) + 6N_2(g) + 3H_2O(g) + \frac{3}{2}O_2(g)$
   ii. $C_6H_6N_{12}O_{12}(s) \longrightarrow 3CO(g) + 3CO_2(g) + 6N_2(g) + 3H_2O(g)$
   iii. $C_6H_6N_{12}O_{12}(s) \longrightarrow 6CO_2(g) + 6N_2(g) + 3H_2(g)$

   a. Use bond energies to estimate ΔH for these three reactions.
   b. Which of the above reactions releases the largest amount of energy per kilogram of CL-20?

124. In molecules of the type X—O—H, as the electronegativity of X increases, the acid strength increases. In addition, if the electronegativity of X is a very small value, the molecule acts like a base. Explain these observations and provide examples.

125. Calculate the standard heat of formation of the compound ICl(*g*) at 25°C. (*Hint:* Use Table 13.6 and Appendix 4 data.)

126. An ionic compound made from the metal M and the diatomic gas X_2 has the formula M_aX_b, in which $a = 1$ or 2 and $b = 1$ or 2. Use the data provided to determine the most likely values for a and b, along with the most likely charges for each of the ions in the ionic compound.

Data (in units of kJ/mol)
Successive ionization energies of M: 480., 4750.
Successive electron affinity values for X: −175, 920.
Enthalpy of sublimation for $M(s) \rightarrow M(g)$: 110.
Bond energy of X_2: 250.
Lattice energy for MX (M^+ and X^-): −1200. kJ/mol
Lattice energy for MX_2 (M^{2+} and X^-): −3500. kJ/mol
Lattice energy for M_2X (M^+ and X^{2-}): −3600. kJ/mol
Lattice energy for MX (M^{2+} and X^{2-}): −4800. kJ/mol

Marathon Problem

127. Identify the following five compounds of H, N, and O. For each compound, write a Lewis structure that is consistent with the information given.
   a. All the compounds are electrolytes, although not all are strong electrolytes. Compounds C and D are ionic and compound B is covalent.
   b. Nitrogen occurs in its highest possible oxidation state in compounds A and C; nitrogen occurs in its lowest possible oxidation state in compounds C, D, and E. The formal charge on both nitrogens in compound C is +1; the formal charge on the only nitrogen in compound B is 0.

c. Compounds A and E exist in solution. Both solutions give off gases. Commercially available concentrated solutions of compound A are normally 16 *M*. The commercial, concentrated solution of compound E is 15 *M*.
d. Commercial solutions of compound E are labeled with a misnomer that implies that a binary, gaseous compound of nitrogen and hydrogen has reacted with water to produce ammonium ions and hydroxide ions. Actually, this reaction occurs to only a slight extent.
e. Compound D is 43.7% N and 50.0% O by mass. If compound D were a gas at STP, it would have a density of 2.86 g/L.
f. A formula unit of compound C has one more oxygen than a formula unit of compound D. Compounds C and A have one ion in common when compound A is acting as a strong electrolyte.
g. Solutions of compound C are weakly acidic; solutions of compound A are strongly acidic; solutions of compounds B and E are basic. The titration of 0.726 g of compound B requires 21.98 mL of 1.000 *M* HCl for complete neutralization.

Covalent Bonding: Orbitals

14

chapter

A stylized model of a nitrogen molecule.

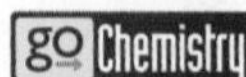

Sign in to OWL at www.cengage.com/owl to view tutorials and simulations, develop problem-solving skills, and complete online homework assigned by your professor.

go Chemistry

Download mini lecture videos for key concept review and exam prep from OWL or purchase them from www.cengagebrain.com.

In Chapter 13 we discussed the fundamental concepts of bonding and introduced the most widely used simple model for covalent bonding: the localized electron (LE) model. We saw the usefulness of a bonding model as a means for systematizing chemistry by allowing us to look at molecules in terms of individual bonds. We also saw that the approximate molecular structure can be predicted using a model that focuses on minimizing electron-pair repulsions. In this chapter we will examine bonding models in more detail, particularly focusing on the role of orbitals.

14.1 Hybridization and the Localized Electron Model

Recall that the LE model views a molecule as a collection of atoms bound together by sharing electrons between atomic orbitals. The arrangement of valence electrons is represented by the Lewis structure (or structures, where resonance occurs), and the approximate molecular geometry can be predicted using the VSEPR model. In this section we will describe what types of atomic orbitals are used by this model to share the electrons and hence to form the bonds.

*sp*³ Hybridization

Let's reconsider the bonding in methane, which has the Lewis structure and molecular geometry shown in Fig. 14.1. In general, we assume that bonding involves only the valence orbitals. This means that the hydrogen atoms in methane use 1*s* orbitals. The valence orbitals of a carbon atom are the 2*s* and 2*p* orbitals shown in Fig. 14.2. Thinking about how carbon can use these orbitals to bond to the hydrogen atoms reveals two related problems:

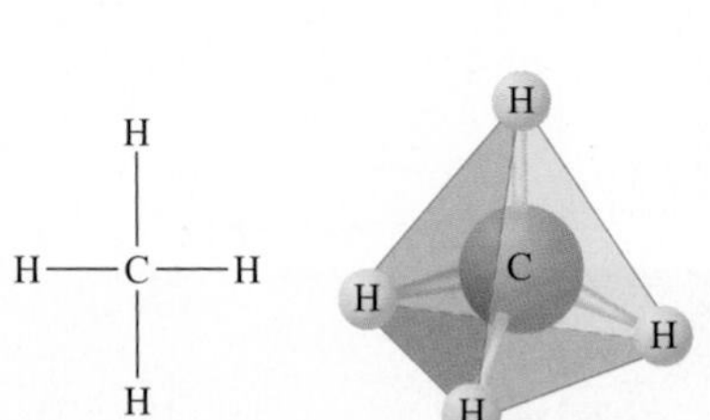

Figure 14.1
(a) The Lewis structure of the methane molecule. (b) The tetrahedral molecular geometry of the methane molecule.

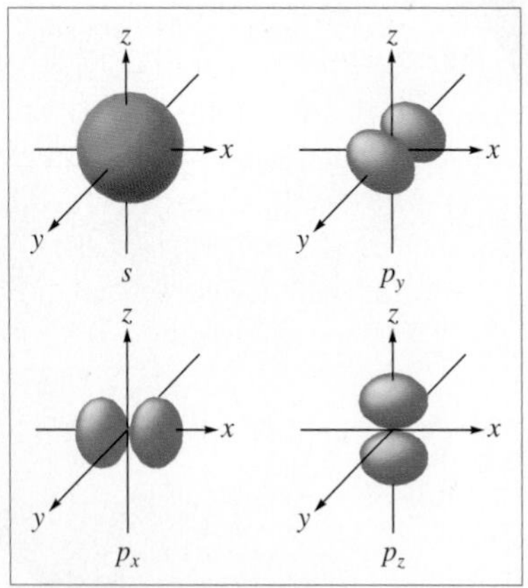

Figure 14.2
The valence orbitals on a free carbon atom: 2*s*, $2p_x$, $2p_y$, and $2p_z$.

Hybridization is a modification of the LE model to account for the observation that atoms often seem to use special atomic orbitals in forming molecules.

1. Using the 2*p* and 2*s* atomic orbitals will lead to two different types of C—H bonds: (a) those from the overlap of a 2*p* orbital of carbon and a 1*s* orbital of hydrogen (there will be three of these) and (b) those from the overlap of a 2*s* orbital of carbon and a 1*s* orbital of hydrogen (there will be one of these). This presents a problem because the experimental evidence indicates that methane has four identical C—H bonds.
2. Since the carbon 2*p* orbitals are mutually perpendicular, we might expect the three C—H bonds formed with these orbitals to be oriented at 90-degree angles:

However, the methane molecule is known by experiment to be tetrahedral with bond angles of 109.5 degrees.

This analysis suggests that carbon adopts a set of atomic orbitals other than its "native" 2*s* and 2*p* orbitals to bond to the hydrogen atoms in forming the methane molecule. In fact, it is not surprising that the 2*s* and 2*p* orbitals present on an *isolated* carbon atom might not be the best set of orbitals for bonding. That is, a different set of atomic orbitals might better serve the carbon atom to form the most stable CH_4 molecule.

To account for the known structure of methane in terms of this model, we need a set of four equivalent atomic orbitals, arranged tetrahedrally. In fact, such a set of orbitals can be obtained quite readily by combining the carbon 2*s* and 2*p* orbitals, as shown schematically in Fig. 14.3. This mixing of the na-

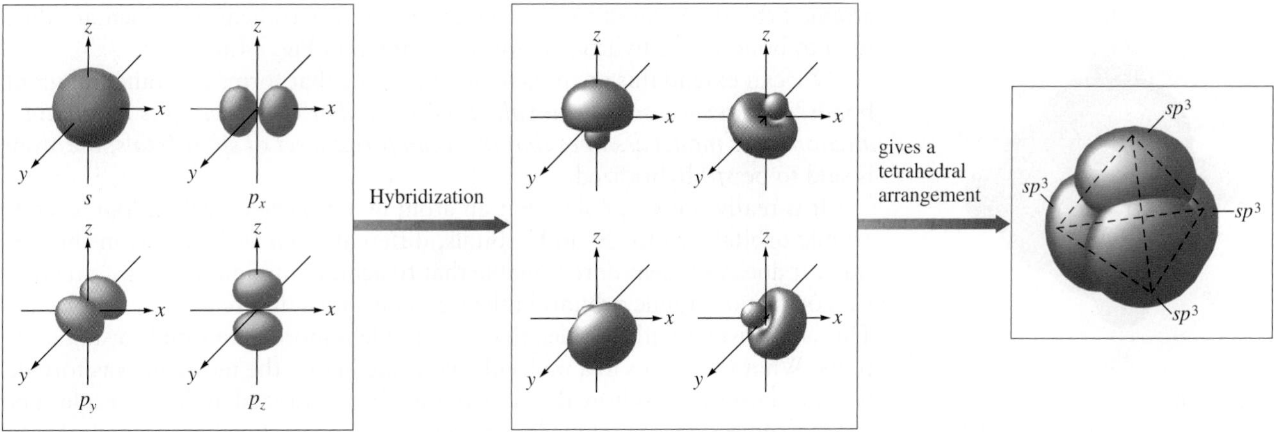

Figure 14.3
The "native" 2*s* and three 2*p* atomic orbitals characteristic of a free carbon atom are combined to form a new set of four sp^3 orbitals. The small lobes of the orbitals are usually omitted from diagrams for clarity.

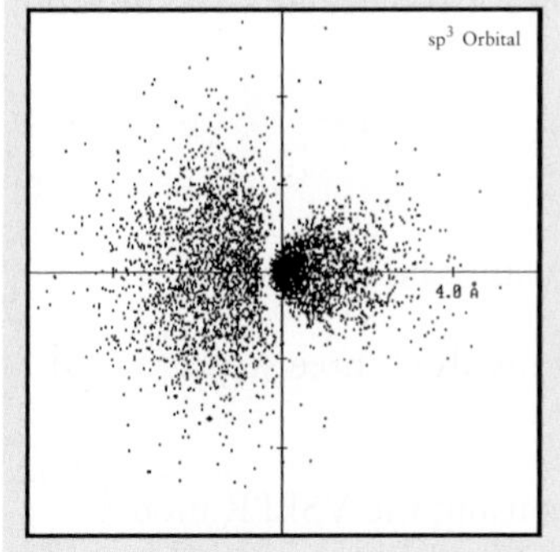

Figure 14.4
Cross section of an sp^3 orbital. Generated from a program by Robert Allendoerfer on Project SERAPHIM disk PC 2402; reprinted with permission.

sp^3 hybridization gives a tetrahedral set of orbitals.

tive atomic orbitals to form special orbitals for bonding is called **hybridization.** The four new orbitals are called sp^3 orbitals since they are formed from one 2*s* and three 2*p* orbitals (s^1p^3). Similarly, we say that the carbon atom undergoes $\boldsymbol{sp^3}$ **hybridization** or is described to be sp^3 hybridized. The four sp^3 orbitals are identical in shape, each one having a large lobe and a small lobe (Fig. 14.4). The four orbitals are oriented in space so that the large lobes form a tetrahedral arrangement, as shown in Fig. 14.3.

The linear combinations of the 2*s* and 2*p* orbitals that give the four sp^3 hybrid orbitals are listed below:

$$\phi_1 = \tfrac{1}{2}[(s) + (p_x) + (p_y) + (p_z)]$$
$$\phi_2 = \tfrac{1}{2}[(s) + (p_x) - (p_y) - (p_z)]$$
$$\phi_3 = \tfrac{1}{2}[(s) - (p_x) + (p_y) - (p_z)]$$
$$\phi_4 = \tfrac{1}{2}[(s) - (p_x) - (p_y) + (p_z)]$$

where (*s*) and (*p*) represent 2*s* and 2*p* atomic orbital functions, and where the factor of $\frac{1}{2}$ is present to satisfy the boundary condition that the total probability is 1 for each orbital. Each of the functions ϕ_1, ϕ_2, ϕ_3, and ϕ_4 represents a separate sp^3 hybrid orbital.

The hybridization of the carbon 2*s* and 2*p* orbitals can also be represented by an orbital energy-level diagram, as shown in Fig. 14.5. Note that electrons have been omitted because we are not concerned with the electron arrangements on the individual atoms—it is the total number of valence electrons and the arrangement of those electrons in the *molecule* that are important.

In summary, the experimentally known structure of methane can be explained by the LE model if we assume that carbon adopts a special set of sp^3

Figure 14.5
An energy-level diagram showing the formation of four sp^3 orbitals.

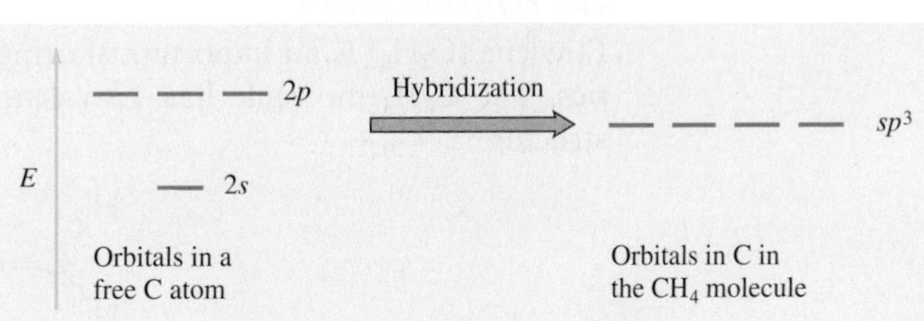

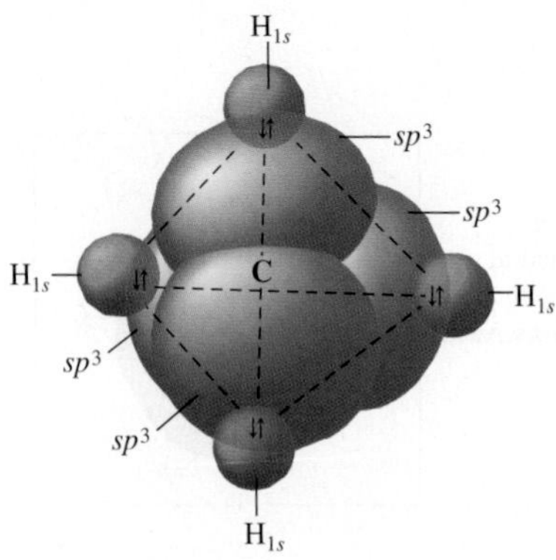

Figure 14.6
The tetrahedral set of four sp^3 orbitals on the carbon atom is used to share electron pairs with the four 1*s* orbitals of the hydrogen atoms to form the four equivalent C—H bonds. This accounts for the known tetrahedral structure of the CH_4 molecule.

atomic orbitals oriented toward the corners of a tetrahedron, which are then used to bond to the hydrogen atoms, as shown in Fig. 14.6.

We can extend this analysis to any molecule that forms a tetrahedral set of bonds: *Whenever a set of equivalent tetrahedral atomic orbitals is required by an atom, this model assumes that the atom forms a set of sp^3 orbitals;* the atom is said to be sp^3 hybridized.

It is really not surprising that an atom in a molecule might adopt a set of atomic orbitals, called **hybrid orbitals,** different from those it has in the free state. It does not seem unreasonable that to achieve minimum energy, an atom uses one set of atomic orbitals in the free state and a different set in a molecule. This is consistent with the idea that a molecule is more than simply a sum of its parts. What the atoms in a molecule were like before the molecule was formed is not as important as how the electrons are best arranged in the molecule. The individual atoms are assumed to respond as needed to achieve the minimum energy for the molecule. However, although hybridization enables us to account for the known structures of molecules, its predictions about the energies of the electrons in molecules are often incorrect. For example, the sp^3 model for methane predicts that all four pairs of electrons have identical energies. Experiments show, however, that in methane two of the electrons are present at a lower energy than the other six. This situation reminds us again that we must be cautious when we use models. Models are *always* oversimplifications and will give misleading predictions if used incorrectly. The LE model using hybridization does a very good job of accounting for molecular structures but does not accurately describe electron energies.

EXAMPLE 14.1

Describe the bonding in the ammonia molecule using the LE model.

Solution A complete description of the bonding involves three steps:

1. Write the Lewis structure.
2. Determine the arrangement of electron pairs using the VSEPR model.
3. Determine the hybrid atomic orbitals used for bonding in the molecule.

The Lewis structure for NH_3 is

```
   ..
H—N—H
   |
   H
```

The four electron pairs around the nitrogen atom require a tetrahedral arrangement. We have seen that a tetrahedral set of sp^3 hybrid orbitals is obtained by combining the 2*s* and three 2*p* orbitals. In the NH_3 molecule, three of the sp^3 orbitals are used to form bonds to the three hydrogen atoms, and the fourth sp^3 orbital is used to hold the lone pair, as shown in Fig. 14.7.

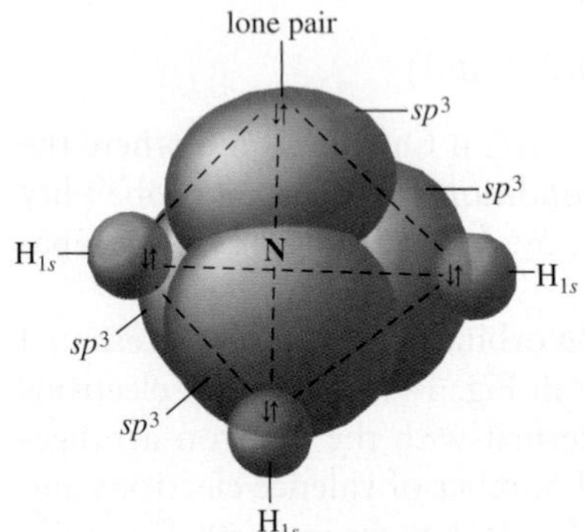

Figure 14.7
The nitrogen atom in ammonia is sp^3 hybridized.

sp^2 Hybridization

Ethylene (C_2H_4) is an important starting material in the manufacture of plastics. The C_2H_4 molecule has 12 valence electrons and the following Lewis structure:

```
H       H
 \     /
  C=C
 /     \
H       H
```

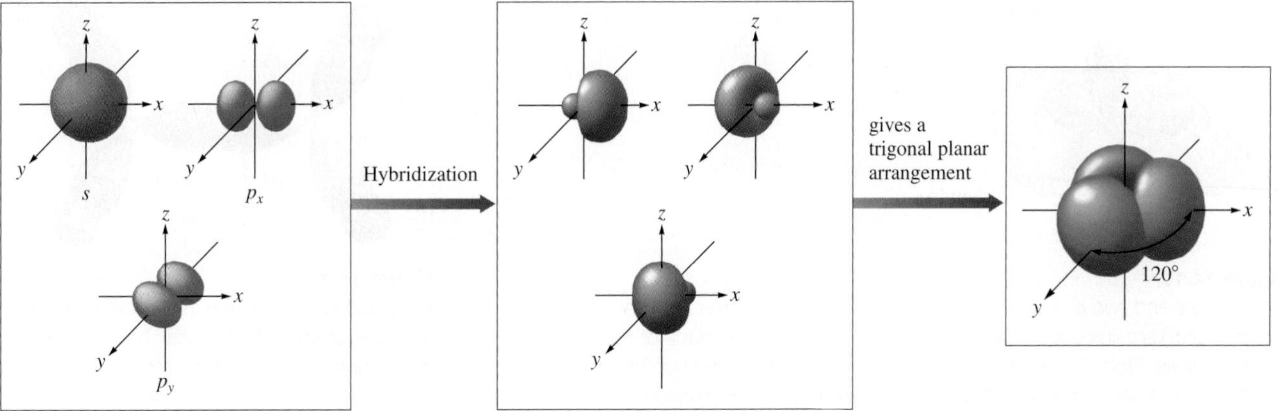

Figure 14.8
The hybridization of the s, p_x, and p_y atomic orbitals results in the formation of three sp^2 orbitals centered in the xy plane. The large lobes of the orbitals lie in the plane at angles of 120 degrees and point toward the corners of a triangle.

The assumption that a double bond acts as one effective electron pair, equivalent to a single bonding pair, works well to give the *approximate* molecular structure.

sp^2 hybridization gives a trigonal planar arrangement of atomic orbitals.

Charles D. Winters

The plastics shown here were manufactured with a polymer based on ethylene.

Recall that in the VSEPR model a double bond acts as one effective pair. Thus in the ethylene molecule each carbon is surrounded by three effective pairs. This model requires a trigonal planar arrangement with bond angles of 120 degrees. What orbitals do the carbon atoms use in this molecule? The molecular geometry requires a planar set of orbitals at angles of 120 degrees. Since the $2s$ and $2p$ valence orbitals of carbon do not have the required arrangement, we need a set of hybrid orbitals.

The sp^3 orbitals we have just considered will not work because they exhibit angles of 109.5 degrees rather than the required 120 degrees. Therefore, in ethylene the carbon atom must hybridize in a different manner. A set of three orbitals arranged at 120-degree angles in the same plane can be obtained by combining one s orbital and two p orbitals, as shown schematically in Fig. 14.8. The orbital energy-level diagram for this arrangement is shown in Fig. 14.9. Since one $2s$ and two $2p$ orbitals are used to form these hybrid orbitals, this is called **sp^2 hybridization.** Note from Fig. 14.8 that the plane of the sp^2 hybridized orbitals is determined by which p orbitals are used. Since in this case we have arbitrarily decided to use the p_x and p_y orbitals, the hybrid orbitals are centered in the xy plane. We will not show the detailed forms of these functions.

In the formation of the sp^2 orbitals, one $2p$ orbital on carbon has not been used. This remaining p orbital (p_z) is oriented perpendicular to the plane of the sp^2 orbitals, as shown in Fig. 14.10.

Now we will see how these orbitals can be used to account for the Lewis structure of ethylene. The three sp^2 orbitals on each carbon are used to share electrons, as shown in Fig. 14.11. In each of these bonds, the electron pair is shared in an area centered on a line running between the atoms. This type of

Figure 14.9
An orbital energy-level diagram for sp^2 hybridization. Note that one p orbital remains unchanged.

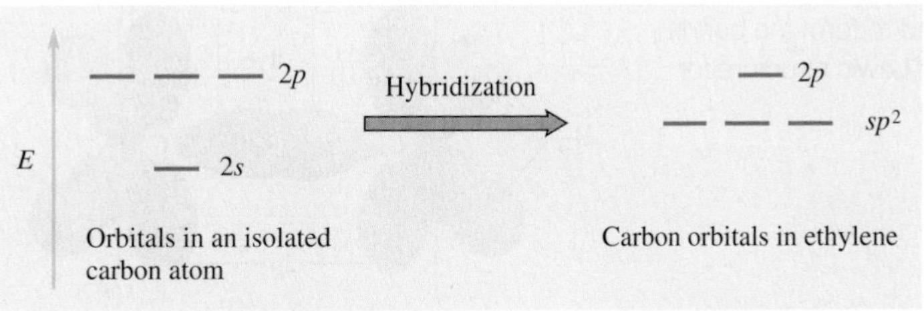

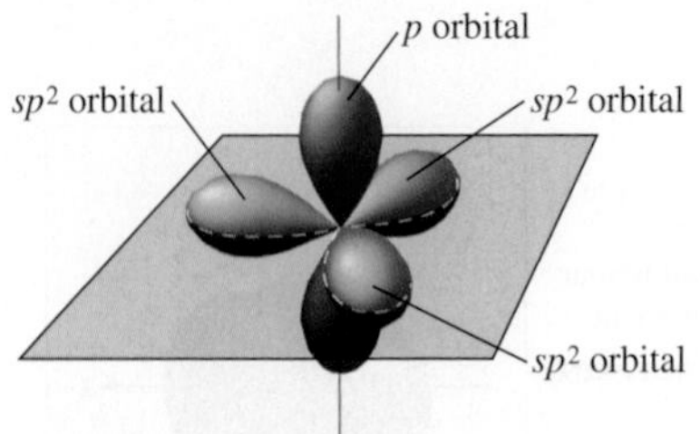

Figure 14.10
When one *s* and two *p* orbitals are mixed to form a set of three sp^2 orbitals, one *p* orbital remains unchanged and is perpendicular to the plane of the hybrid orbitals. Note that in this figure and those that follow, the hybrid orbitals are drawn with narrowed lobes to show their orientations more clearly.

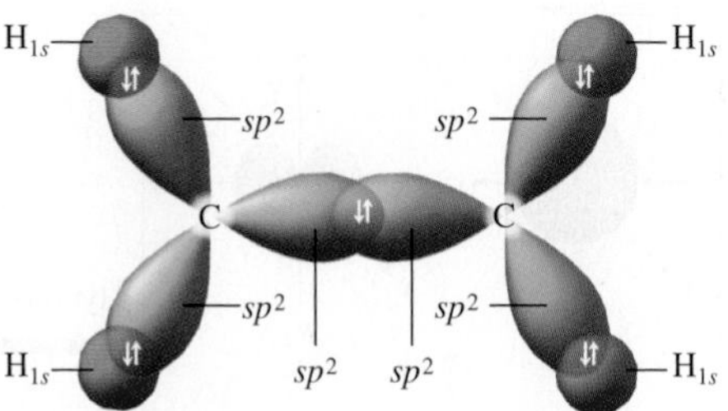

Figure 14.11
The σ bonds in ethylene. Note that for each bond the shared electron pair occupies the region directly between the atoms.

covalent bond is called a **sigma (σ) bond.** In the ethylene molecule, the σ bonds are formed using the sp^2 orbitals on each carbon atom and the 1*s* orbital on each hydrogen atom.

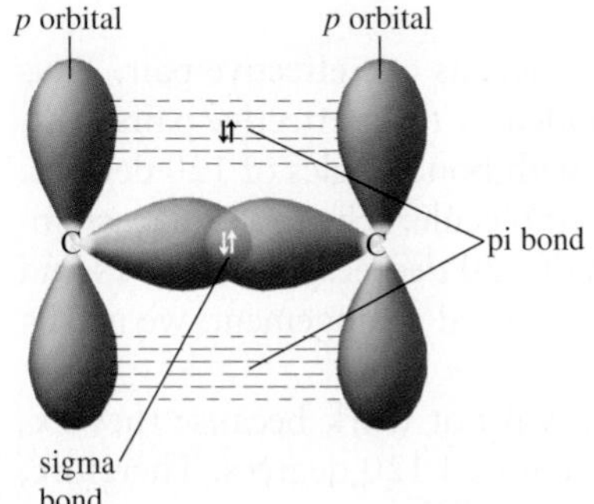

Figure 14.12
A carbon–carbon double bond consists of a σ bond and a π bond. In the σ bond, the shared electrons occupy the space directly between the atoms. The π bond, is formed from the unhybridized *p* orbitals on the two carbon atoms. In a π bond, the shared electron pair occupies the space above and below a line joining the atoms.

How can we explain the double bond between the carbon atoms? In the σ bond, the electron pair occupies the space between the carbon atoms. The second bond must therefore result from sharing an electron pair in the space *above and below* the σ bond. This type of bond can be formed using the 2*p* orbital perpendicular to the sp^2 hybrid orbitals on each carbon atom (refer to Fig. 14.10). These parallel *p* orbitals can share an electron pair, which occupies the space above and below a line joining the atoms, to form a **pi (π) bond,** as shown in Fig. 14.12.

Note that σ bonds are formed from orbitals whose lobes point toward each other, but π bonds result from parallel orbitals. A *double bond consists of one σ bond,* where the electron pair is located directly between the atoms, *and one π bond,* where the shared pair occupies the space above and below the σ bond.

We can now completely specify the orbitals used to form the bonds in the ethylene molecule. As shown in Fig. 14.13, the carbon atoms are described as using sp^2 hybrid orbitals to form the σ bonds to the hydrogen atoms and to each other and using *p* orbitals to form the π bond with each other. Note that we have accounted fully for the Lewis structure of ethylene with its carbon–carbon double bond and carbon–hydrogen single bonds.

This example illustrates an important general principle of this model: *Whenever an atom is surrounded by three effective pairs, a set of sp^2 hybrid orbitals is required.*

sp Hybridization

Another type of hybridization occurs in carbon dioxide, which has the following Lewis structure:

$$:\!\ddot{O}\!=\!C\!=\!\ddot{O}\!:$$

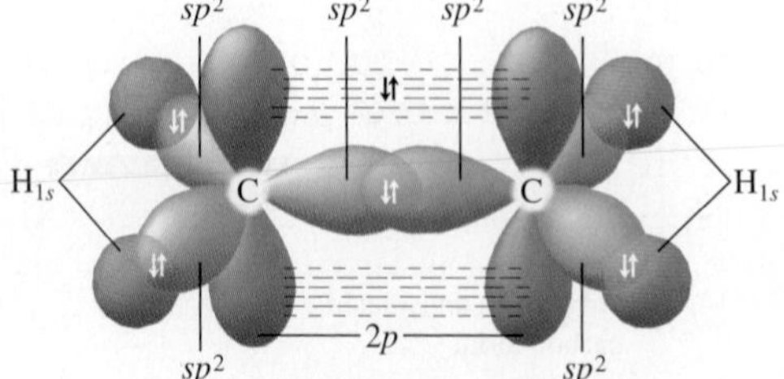

Figure 14.13
(a) The orbitals used to form the bonds in ethylene. (b) The Lewis structure for ethylene.

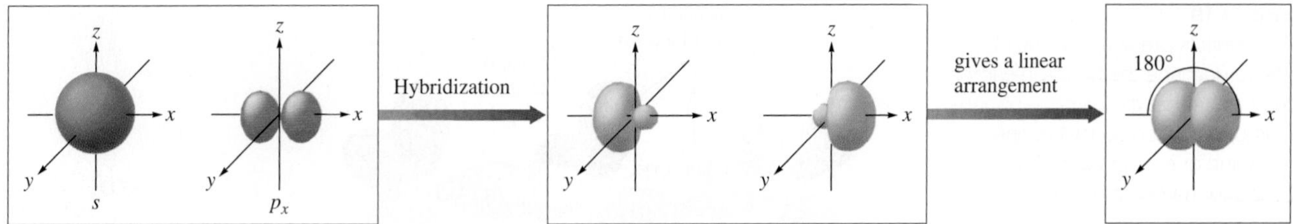

Figure 14.14
When one *s* orbital and one *p* orbital are hybridized, a set of two *sp* orbitals oriented at 180 degrees results.

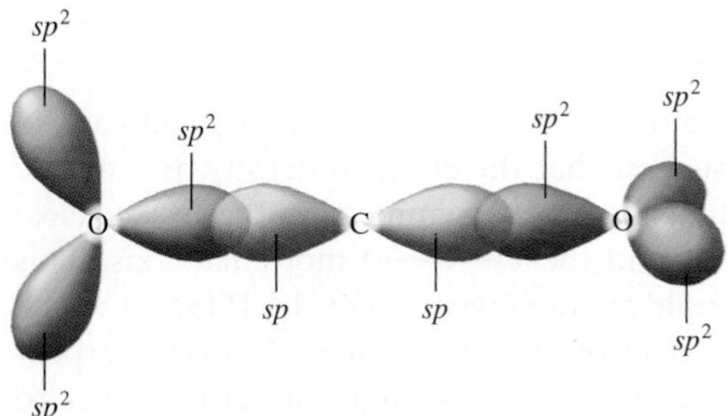

Figure 14.15
The hybrid orbitals in the CO_2 molecule.

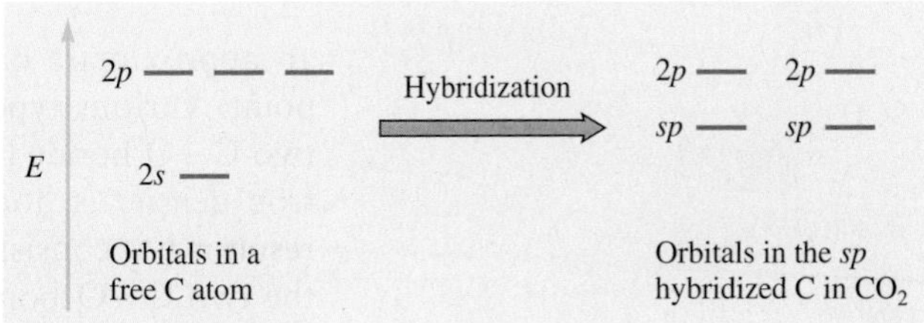

Figure 14.16
The orbital energy-level diagram for the formation of *sp* hybrid orbitals of carbon.

In the CO_2 molecule, the carbon atom has two effective pairs that are arranged at an angle of 180 degrees. We therefore need a pair of atomic orbitals oriented in opposite directions. This requires a new type of hybridization, since neither sp^3 nor sp^2 hybrid orbitals fit this case. Obtaining two hybrid orbitals arranged at 180 degrees requires ***sp* hybridization,** involving one *s* orbital and one *p* orbital, as shown schematically in Fig. 14.14.

In terms of this model, *two effective pairs around an atom will always require* sp *hybridization of that atom.* The *sp* orbitals of carbon in carbon dioxide are shown in Fig. 14.15, and the corresponding orbital energy-level diagram for their formation is given in Fig. 14.16. The *sp* hybrid orbitals are used to form the σ bonds between carbon and the oxygen atoms. Note that two $2p$ orbitals remain unchanged on the *sp* hybridized carbon. They are used to form the π bonds to the oxygen atoms.

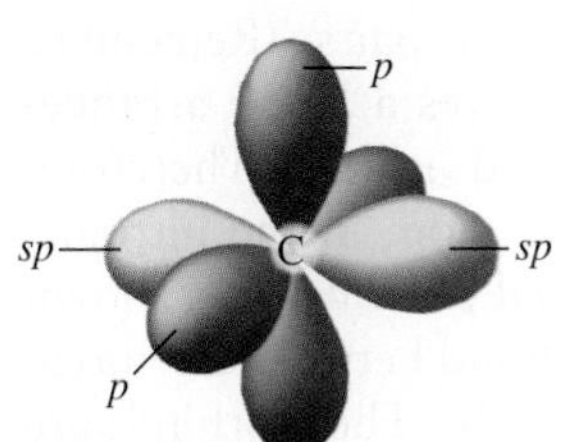

Figure 14.17
The orbitals of an *sp* hybridized carbon atom.

In the CO_2 molecule, each oxygen atom has three effective pairs around it, requiring a trigonal planar arrangement of the pairs. Since a trigonal set of hybrid orbitals corresponds to sp^2 hybridization, each oxygen atom can be assumed to be sp^2 hybridized. The orbital on each oxygen left unchanged by the hybridization process is used for the π bond to the carbon atom.

Now we are ready to account for the Lewis structure of carbon dioxide. The *sp* orbitals on carbon form σ bonds with the sp^2 orbitals on the two oxygen atoms (Fig. 14.15). The remaining sp^2 orbitals on the oxygen atoms hold lone pairs. The π bonds between the carbon atom and each oxygen atom are formed by the overlap of parallel $2p$ orbitals. The *sp* hybridized carbon atom has two unhybridized *p* orbitals, pictured in Fig. 14.17. Each of these *p* orbitals is used to form a π bond with an oxygen atom (Fig. 14.18). The total bonding picture predicted by the LE model for the CO_2 molecule is shown in Fig. 14.19. Note that this picture of the bonding neatly explains the arrangement of electrons predicted by the Lewis structure.

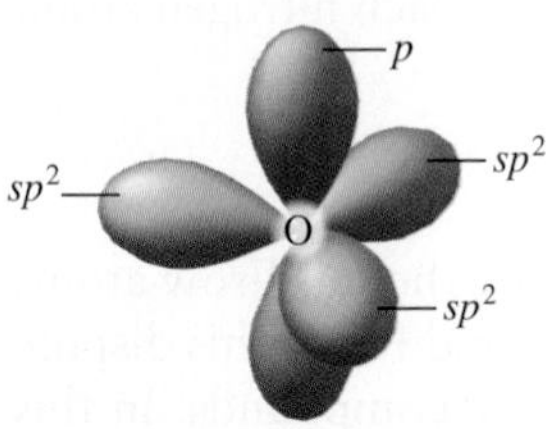

Figure 14.18
The orbital arrangement for an sp^2 hybridized oxygen atom.

It is useful to remind ourselves at this point that because we are using a very simple bonding model, the picture of a molecule we get from this model is

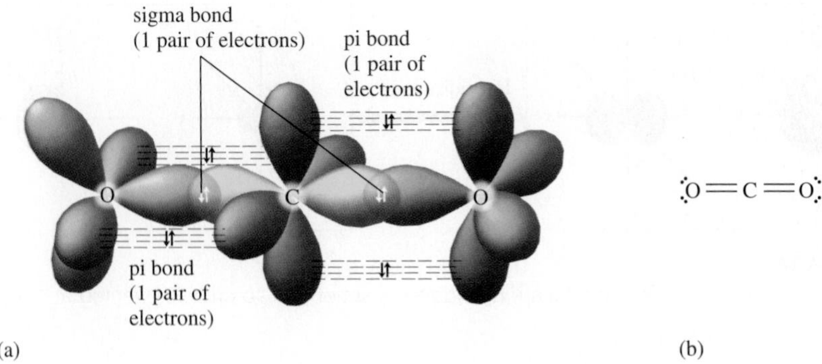

Figure 14.19
(a) The orbitals predicted by the LE model to describe the bonds in carbon dioxide. Note that the carbon–oxygen double bonds each consist of one σ bond and one π bond. (b) The Lewis structure for carbon dioxide.

an approximate one. The carbon dioxide molecule provides a good case in point. Various types of evidence suggest that the electron density around the two C—O bonds in CO_2 is actually cylindrically symmetric—that is, the electron density is homogeneous all around the O—C—O molecular axis. This result is not consistent with the simple picture given in Fig. 14.19(a), in which the two C—O bonds have their π electron densities centered in two perpendicular planes. The model can be corrected in a straightforward manner to remove this difficulty, but we will not pursue this correction here. The point is that a very simple model is quite useful to us because it is so easy to apply. Simple models help us to organize our observations and often provide a place to start as we pursue the answer to a problem. However, we must be wary of simple models—they often provide a significantly distorted picture.

EXAMPLE 14.2

Describe the bonding in the N_2 molecule.

Solution The Lewis structure for the nitrogen molecule is

$$:N \equiv N:$$

where each nitrogen atom is surrounded by two effective pairs. (Remember that multiple bonds count as one effective pair.) This gives a linear arrangement (180 degrees) requiring a pair of oppositely directed orbitals. Therefore, this situation requires *sp* hybridization. Each nitrogen atom in the nitrogen molecule has two *sp* hybrid orbitals and two unchanged *p* orbitals, as shown in Fig. 14.20(a). The *sp* orbitals are used to form the σ bond between the nitrogen atoms and to hold lone pairs, as shown in Fig. 14.20(b). The *p* orbitals are used to form the two π bonds [Fig. 14.20(c)]; each pair of overlapping, parallel *p* orbitals holds one electron pair. Such bonding accounts for the electron arrangement given by the Lewis structure. The triple bond consists of a σ bond (overlap of two *sp* orbitals) and two π bonds (each from an overlap of two *p* orbitals). In addition, a lone pair occupies an *sp* orbital on each nitrogen atom.

dsp^3 Hybridization

In Section 13.12 we considered the dispute about whether the third-row atoms actually exceed the octet rule in compounds such as SF_6 and PCl_5. This dispute hinges on the use of the 3*d* orbitals for bonding in these compounds. In this section we will describe the hybrid orbitals that are involved when the 3*d* orbitals are assumed to participate in the bonding.

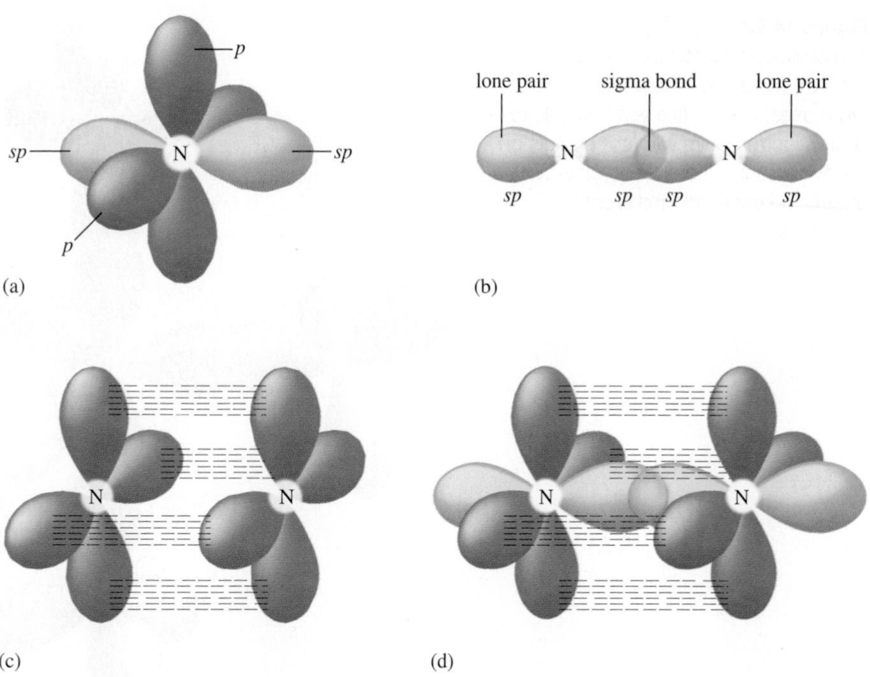

Figure 14.20
(a) An *sp* hybridized nitrogen atom. There are two *sp* hybrid orbitals and two unhybridized *p* orbitals. (b) The σ bond in the N_2 molecule. (c) The two π bonds in N_2 are formed when electron pairs are shared between two sets of parallel *p* orbitals. (d) The total bonding picture for N_2.

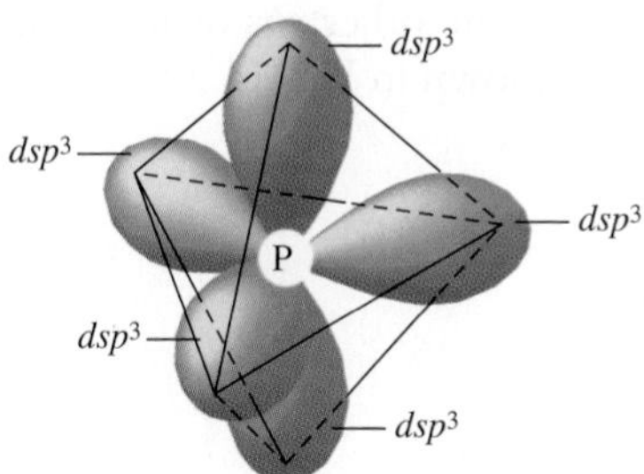

Figure 14.21
A set of dsp^3 hybrid orbitals on a phosphorus atom. Note that the set of five dsp^3 orbitals has a trigonal bipyramidal arrangement. (Each dsp^3 orbital also has a small lobe that is not shown in this diagram.)

First we will consider phosphorus pentachloride (PCl_5). The traditional Lewis structure for PCl_5 (assuming *d* orbital participation)

:Cl: :Cl:
:Cl—P
:Cl: :Cl:

shows that the phosphorus atom is surrounded by five electron pairs. Since in the VSEPR model five pairs require a trigonal bipyramidal arrangement, we need a trigonal bipyramidal set of atomic orbitals on phosphorus. Such a set of orbitals is formed by ***dsp*³ hybridization** of one *d* orbital, one *s* orbital, and three *p* orbitals, as shown in Fig. 14.21.

Although it will not be important for our purposes, the dsp^3 hybrid orbital set is different from the hybrids we have considered so far in that the hybrid orbitals pointing to the vertices of the triangle (often called the three equatorial hybrid orbitals) are slightly different in shape than the other two (the *axial* orbitals). This situation stands in contrast to the *sp*, sp^2, and sp^3 hybrid sets in which each orbital in a particular set is identical in shape to the others.

The dsp^3 hybridized phosphorus atom in the PCl_5 molecule uses its five dsp^3 orbitals to share electrons with the five chlorine atoms. Note that according to this model *a set of five effective pairs around a given atom always requires a trigonal bipyramidal arrangement, which in turn involves dsp^3 hybridization of that atom.*

The Lewis structure for PCl_5 shows that each chlorine atom is surrounded by four electron pairs. This requires a tetrahedral arrangement, which in turn requires a set of four sp^3 orbitals on each chlorine atom.

Now we can describe the bonding in the PCl_5 molecule. The five P—Cl σ bonds are formed by sharing electrons between a dsp^3 orbital on the phos-

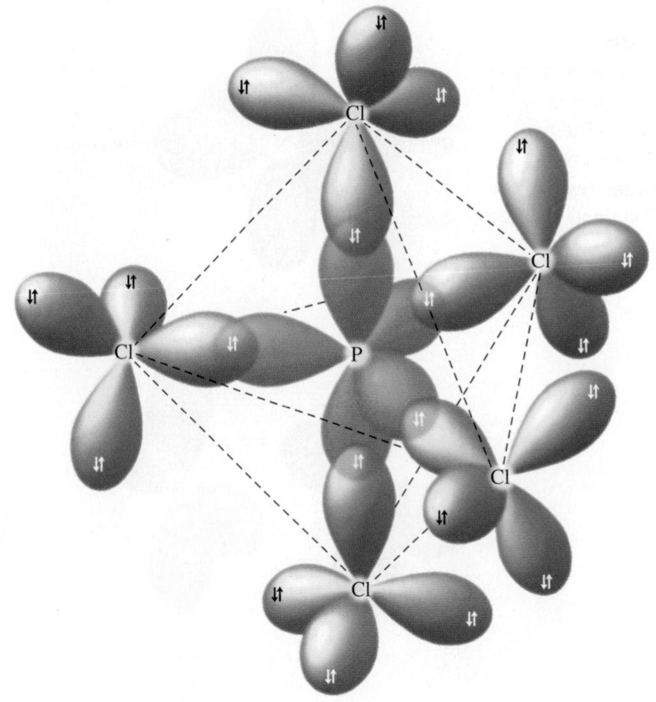

Figure 14.22
The orbitals used to form the bonds in the PCl_5 molecule. The phosphorus atom uses a set of five dsp^3 orbitals to share electron pairs with sp^3 orbitals on the five chlorine atoms. The other sp^3 orbitals on each chlorine atom hold lone pairs.

phorus atom and an sp^3 orbital on each chlorine.* The other sp^3 orbitals on each chlorine hold lone pairs. This arrangement is shown in Fig. 14.22.

Although in its simplest form the LE model of bonding invokes the use of hybrids involving *d* orbitals for trigonal bipyramidal and octahedral structures, there is considerable disagreement about the actual participation of *d* orbitals in such molecules. For a discussion, see A. E. Reed and F. Weinhold, *J. Am. Chem. Soc.* 108 (1986): 3586, and references therein.

EXAMPLE 14.3

Describe the bonding in the triiodide ion (I_3^-).

Solution The traditional Lewis structure for I_3^-

$$\left[:\ddot{\underset{\cdot\cdot}{I}}-\ddot{\underset{\cdot\cdot}{I}}-\ddot{\underset{\cdot\cdot}{I}}:\right]^-$$

shows that the central iodine atom has five pairs of electrons. A set of five pairs requires a trigonal bipyramidal arrangement, which in turn requires a set of dsp^3 orbitals. The outer iodine atoms have four pairs of electrons, which calls for a tetrahedral arrangement and sp^3 hybridization.

Thus the central iodine is dsp^3 hybridized. Three of these hybrid orbitals hold lone pairs, and two of them overlap with sp^3 orbitals from the other two iodine atoms to form σ bonds.

d^2sp^3 Hybridization

Next, we consider sulfur hexafluoride (SF_6), which has the traditional Lewis structure

```
     ..
 .. :F: ..
:F   |   F:
 ..\ | / ..
     S
 ../ | \ ..
:F   |   F:
 .. :F: ..
     ..
```

*Although we have no way of proving conclusively that each chlorine atom is sp^3 hybridized, we assume that minimizing electron-pair repulsions is as important for peripheral atoms as for the central atom. Thus we will apply the VSEPR model and hybridization to both central and peripheral atoms.

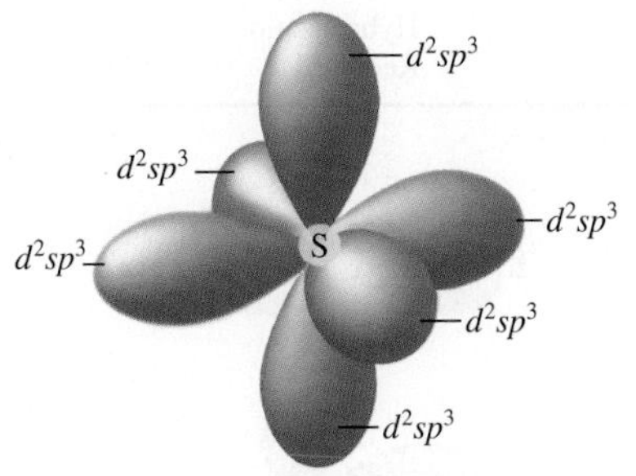

Figure 14.23
An octahedral set of d^2sp^3 orbitals on a sulfur atom. The small lobe of each hybrid orbital has been omitted for clarity.

This requires an octahedral arrangement of pairs and, in turn, an octahedral set of six hybrid orbitals. This leads to **d^2sp^3 hybridization,** in which two *d* orbitals, one *s* orbital, and three *p* orbitals are combined (Fig. 14.23). Note that *six electron pairs around an atom are always arranged octahedrally, requiring d^2sp^3 hybridization of the atom.* Each d^2sp^3 orbital on the sulfur atom is used to bond to a fluorine atom. Since there are four pairs on each fluorine atom, the fluorine atoms are assumed to be sp^3 hybridized.

EXAMPLE 14.4

How is the xenon atom in XeF_4 hybridized?

Solution The Lewis structure for XeF_4 has six pairs of electrons around xenon that are arranged octahedrally to minimize repulsions. An octahedral set of six atomic orbitals is required to hold these electrons, and the xenon atom is assumed to be d^2sp^3 hybridized.

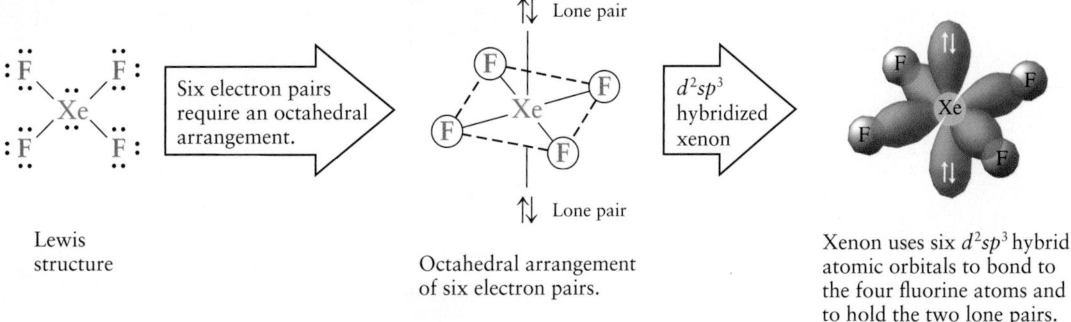

The description of a molecule using the LE model involves three distinct steps.

STEPS

Describing a Molecule with the Localized Electron Model

1. Draw the Lewis structure(s).
2. Determine the arrangement of electron pairs using the VSEPR model.
3. Specify the hybrid orbitals needed to accommodate the electron pairs.

It is important to perform the steps in this order. For a model to be successful, it must follow nature's priorities. In the case of bonding, it seems clear that the tendency for a molecule to minimize its energy is more important than maintenance of the characteristics of atoms as they exist in the free state. The atoms adjust to meet the "needs" of the molecule. When considering the bonding in a particular molecule, therefore, we always start with the molecule rather than the component atoms. In the molecule the electrons are arranged to give each atom a noble gas configuration where possible and to minimize electron-pair repulsions. We then assume that the atoms adjust their orbitals by hybridization to allow the molecule to adopt the structure that gives the minimum energy.

In applying the LE model, we must not overemphasize the characteristics of the separate atoms. The particular atom on which the valence electrons originated is not important; what is important is where they are needed in the molecule to achieve maximum stability. In the same vein, the forms of the orbitals on the isolated atoms are relatively unimportant; what really matters are

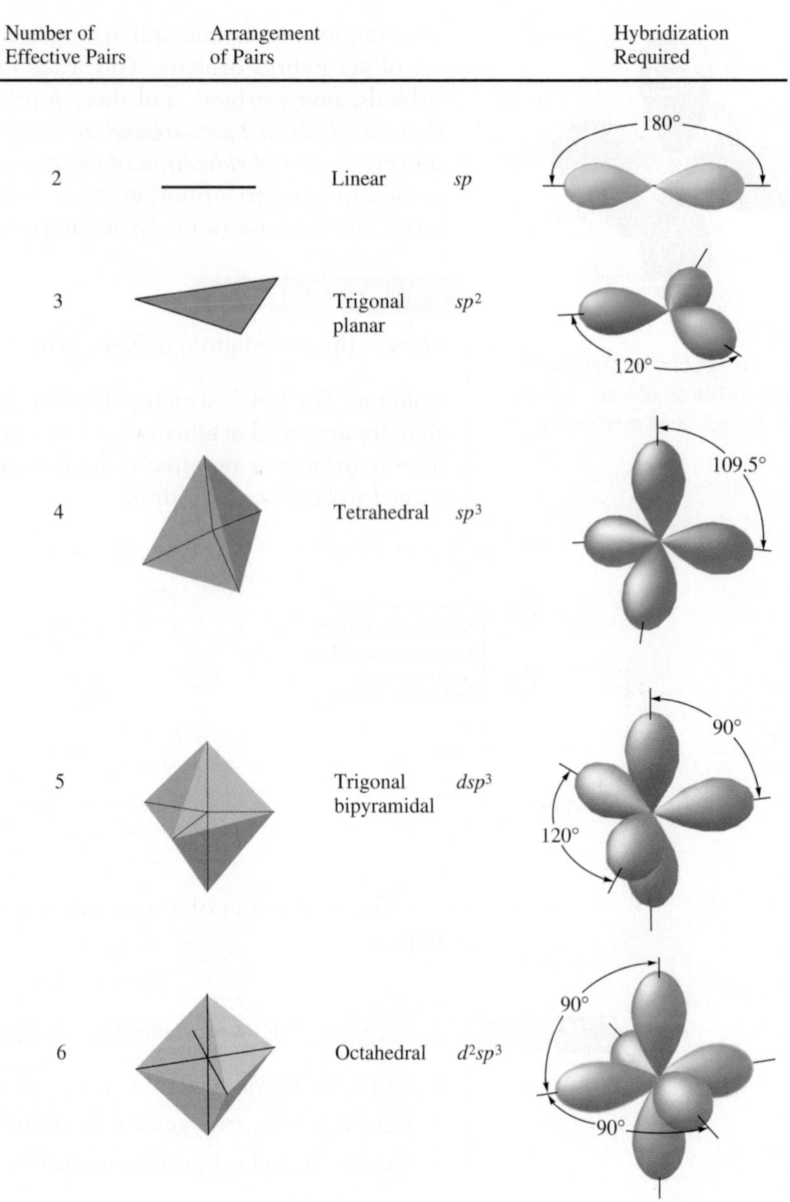

Figure 14.24 The relationship among the number of effective pairs, their spatial arrangement, and the hybrid orbital set required.

the forms needed by the molecule to achieve minimum energy. The requirements for the various types of hybridization are summarized in Fig. 14.24.

OWL INTERACTIVE EXAMPLE 14.5

For each of the following molecules or ions, describe the molecular structure, and predict the hybridization of each atom.

a. CO **b.** BF_4^- **c.** XeF_2

Solution

a. The CO molecule has 10 valence electrons, and its Lewis structure is

$$:C{\equiv}O:$$

Each atom has two effective pairs, which means that both are *sp* hybridized. The triple bond consists of a σ bond produced by the overlap of an *sp* orbital from each atom and two π bonds produced by the overlap of

$2p$ orbitals from each atom. The lone pairs are in sp orbitals. Since the CO molecule has only two atoms, it must be linear.

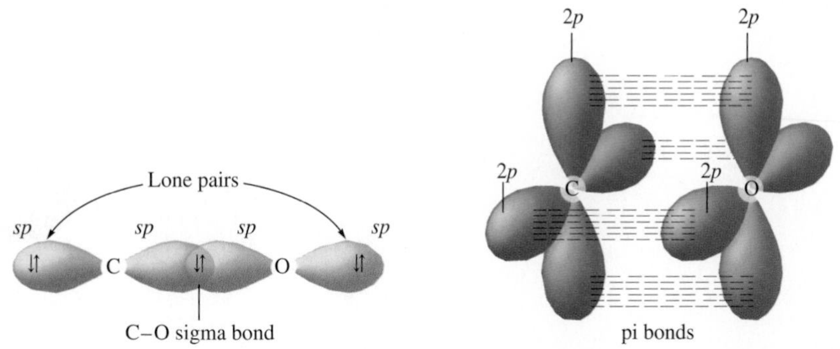

b. The BF_4^- ion has 32 valence electrons. The Lewis structure shows four pairs of electrons around the boron atom, requiring a tetrahedral arrangement:

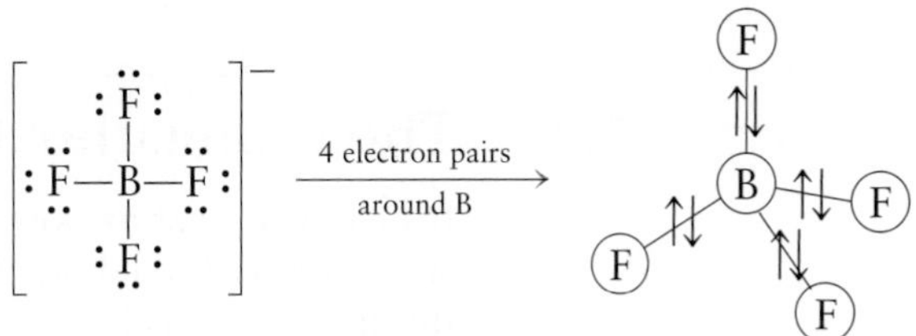

This requires sp^3 hybridization of the boron atom. Each fluorine atom also has four electron pairs and can be assumed to be sp^3 hybridized (only one sp^3 orbital is shown for each fluorine atom). The BF_4^- ion's molecular structure is tetrahedral.

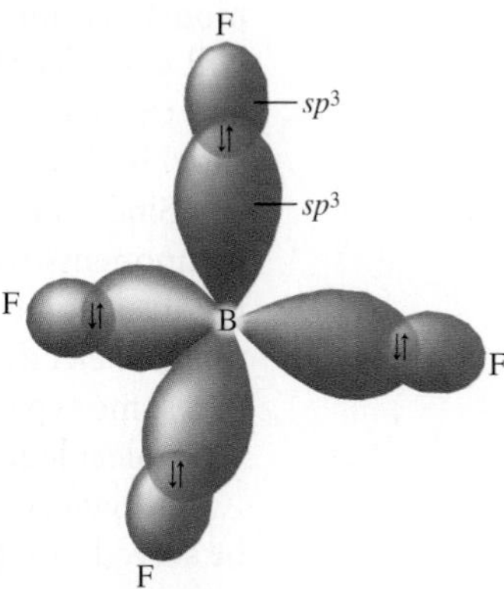

c. The XeF_2 molecule has 22 valence electrons. The Lewis structure shows five electron pairs on the xenon atom, which requires a trigonal bipyramidal arrangement:

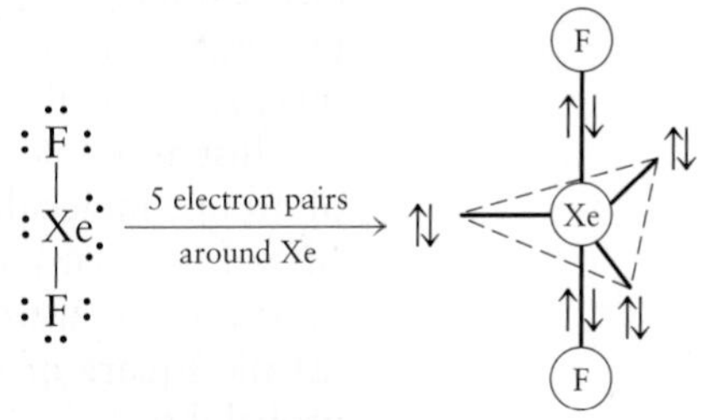

5 electron pairs
around Xe

Note that the lone pairs are placed in the plane where they are 120 degrees apart. Accommodating five pairs at the vertices of a trigonal bipyramid can be explained if the xenon atom adopts a set of five dsp^3 orbitals. Each fluorine atom has four electron pairs and can be assumed to be sp^3 hybridized. The XeF_2 molecule has a linear arrangement of atoms.

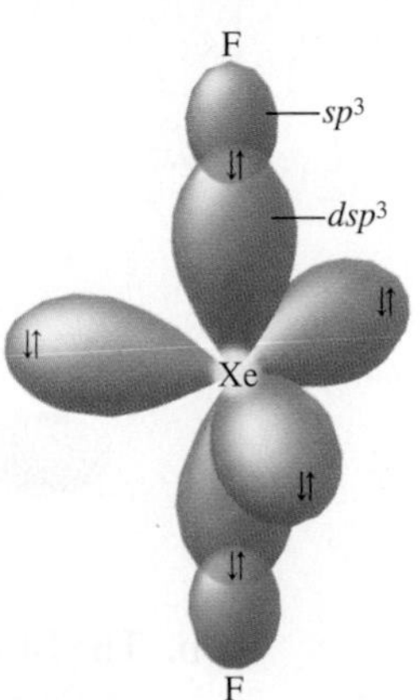

14.2 The Molecular Orbital Model

It should be clear by now that the LE model is of great value in interpreting the structures and bonding of molecules. However, there are some problems with this model at this level of approximation. For example, since it incorrectly assumes that electrons are localized, the concept of resonance must be added. Also, the model does not deal easily with molecules containing unpaired electrons. And finally, the model in this form gives no direct information about bond energies.

Another model often used to describe bonding is the **molecular orbital model.** To introduce the assumptions, methods, and results of this model, we will consider the simplest of all molecules, H_2, which consists of two protons and two electrons. A very stable molecule, H_2 is lower in energy than the separated hydrogen atoms by 432 kJ/mol.

Since the hydrogen molecule consists of protons and electrons, the same components found in separated hydrogen atoms, it seems reasonable to use a theory similar to the atomic theory discussed in Chapter 12, which assumes that the electrons in an atom exist in orbitals of a given energy. Can we apply this same type of model to the hydrogen molecule? Yes; in fact, describing the H_2 molecule in terms of quantum mechanics is quite straightforward.

However, even though it is formulated rather easily, this problem cannot be solved exactly. The difficulty is the same as that encountered in dealing with polyelectronic atoms—the electron correlation problem. Since we cannot account for the details of the electron movements, we cannot deal with the electron–electron interactions in a specific way. We need to make approximations that allow the solution of the problem but that do not destroy the model's physical integrity. The success of these approximations can be measured only by comparing predictions from the theory with experimental observations. In this case we will see that the simplified model works well.

Just as atomic orbitals are solutions to the quantum mechanical treatment of atoms, **molecular orbitals (MOs)** are solutions to the molecular problem. MOs have many of the same characteristics as atomic orbitals. Two of the most important are (1) they can hold two electrons with opposite spins and (2) the square of the molecular orbital wave function indicates the electron probability.

As in the application of quantum mechanics to isolated atoms, the MO orbital treatment can be carried out at various levels of sophistication. In our description of the model, we will assume that the MOs for H_2 are constructed

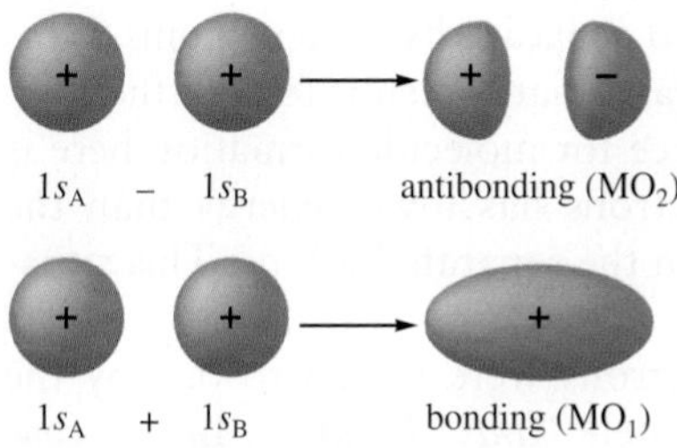

Figure 14.25
The combination of hydrogen 1*s* atomic orbitals to form MOs. The phases of the orbitals are shown by signs inside the boundary surfaces. When the orbitals are added, the matching phases produce constructive interference, which gives enhanced electron probability between the nuclei. This results in a bonding molecular orbital. When one orbital is subtracted from the other, destructive interference occurs between the opposite phases, leading to a node between the nuclei. This is an antibonding MO.

using hydrogen 1*s* orbitals. We say that the 1*s* orbitals form the "basis set" for the MOs. A more detailed treatment would use a different basis set—one in which the radial part of the atomic orbitals would be allowed to vary to achieve the lowest-energy MOs for the hydrogen molecule. However, to avoid as many complications as possible, as we discuss the fundamental ideas of the MO description of molecules, we will use the simplest version of this model.

We will now describe the bonding in the hydrogen molecule using the MO model. The first step is to obtain the hydrogen molecule's orbitals, a process that is greatly simplified if we assume that the MOs can be constructed from the hydrogen 1*s* orbitals.

In this approximate treatment of the hydrogen molecule, two MOs result:

$$MO_1 = 1s_A + 1s_B$$

$$MO_2 = 1s_A - 1s_B$$

where $1s_A$ and $1s_B$ represent the 1*s* orbitals from the two separated hydrogen atoms. This process is shown schematically in Fig. 14.25 along with the phases of the orbitals.

The orbital properties of greatest interest are size, shape (described by the electron probability distribution), and energy. These properties for the hydrogen MOs are represented in Fig. 14.26. From this figure we can note several important points:

1. The electron probability of both MOs is centered along the line passing through the two nuclei. For MO_1, the greatest electron probability is *between* the nuclei. In this case the matching phases of the orbitals produce constructive interference and enhanced electron probability between the two nuclei. For MO_2, it is centered along the molecular axis but outside the area between the two nuclei. In this case the mismatched phases produce destructive interference leading to a node in the electron probability between the two nuclei. In both MOs the electron density has cylindrical symmetry with respect to the molecular axis. That is, the electron probability is the same along any line drawn perpendicular to the bond axis at a given point on the axis. This type of cylindrically symmetric electron distribution is described as *sigma* (σ), as in the localized electron model. Accordingly, we refer to MO_1 and MO_2 as **sigma (σ) MOs.**
2. In the molecule, only the MOs are available for occupation by electrons. The 1*s* atomic orbitals of the hydrogen atoms no longer exist because the H_2 molecule—a new entity—has its own set of new orbitals.
3. MO_1 is lower in energy than the 1*s* orbitals of free hydrogen atoms, but MO_2 is higher in energy than the 1*s* orbitals. This fact has very important implications for the stability of the H_2 molecule: If the two electrons (one from each hydrogen atom) occupy the lower-energy MO, they will have

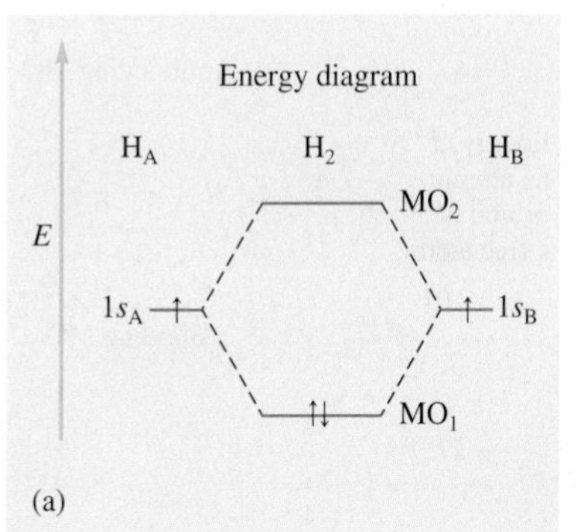

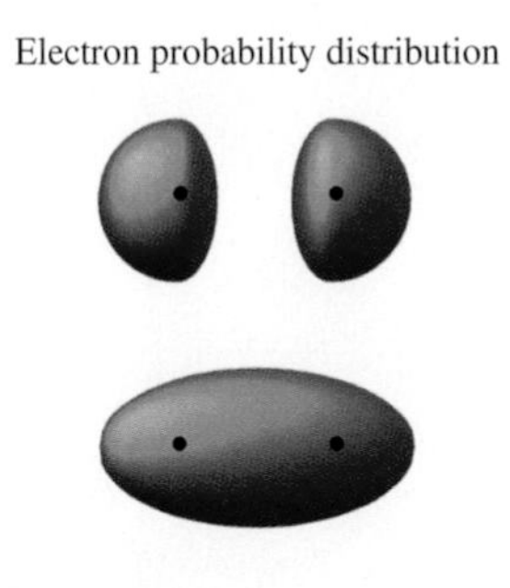

Figure 14.26
(a) The MO energy-level diagram for the H_2 molecule. (b) The shapes of the MOs are obtained by squaring the wave functions for MO_1 and MO_2. The positions of the nuclei are indicated by •.

lower energy than they have in the two separate hydrogen atoms. This situation favors molecule formation because nature tends to seek the lowest energy state. That is, the driving force for molecule formation here is that the MO available to the two electrons has lower energy than the atomic orbitals these electrons occupy in the separated atoms. This situation is *"pro" bonding.*

Bonding will result if the molecule has lower energy than the separated atoms.

On the other hand, if the two electrons were forced to occupy the higher-energy MO, they would be definitely *"anti" bonding.* In this case these electrons would have lower energy in the separated atoms than in the molecule; thus the separated state would be favored. Of course, since the lower-energy MO_1 *is* available, the two electrons occupy that MO and the resulting molecule is stable.

We have seen that the MOs of the hydrogen molecule fall into two classes: bonding and antibonding. A **bonding MO** is *lower in energy than the atomic orbitals of which it is composed.* Electrons in this type of orbital favor the molecule; that is, they will favor bonding. An **antibonding MO** is *higher in energy than the atomic orbitals of which it is composed.* Electrons in this type of orbital will favor the separated atoms (they are antibonding). Figure 14.27 illustrates these ideas.

4. Figure 14.26 shows that for the bonding MO in the H_2 molecule, the electrons have the greatest probability of being between the nuclei. This is exactly what we would expect, since the electrons can lower their energies by being simultaneously attracted by both nuclei. On the other hand, the electron distribution for the antibonding MO is such that the electrons are mainly outside the space between the nuclei. This type of distribution is not expected to provide any bonding force. In fact, it causes the electrons to be higher in energy than in the separated atoms. Thus the MO model produces electron distributions and energies that agree with our basic ideas of bonding. This fact reassures us that the model is physically reasonable.
5. The labels on MOs indicate their symmetries (shapes), their parent atomic orbitals, and whether they are bonding or antibonding. Antibonding character is indicated by an asterisk. For the H_2 molecule, both MOs have σ symmetry and both are constructed from hydrogen 1*s* atomic orbitals. The molecular orbitals for H_2 are therefore labeled as follows:

$$MO_1 = \sigma_{1s}$$
$$MO_2 = \sigma_{1s}{}^*$$

6. Molecular electron configurations can be written in much the same way as atomic configurations. Since the H_2 molecule has two electrons in the σ_{1s} molecular orbital, the electron configuration is $\sigma_{1s}{}^2$.

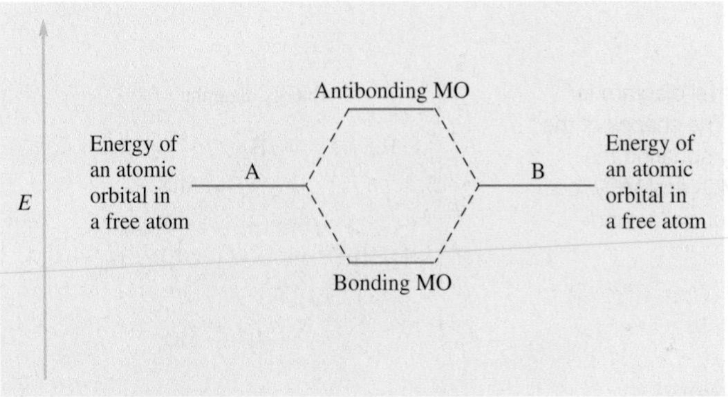

Figure 14.27
Bonding and antibonding MOs.

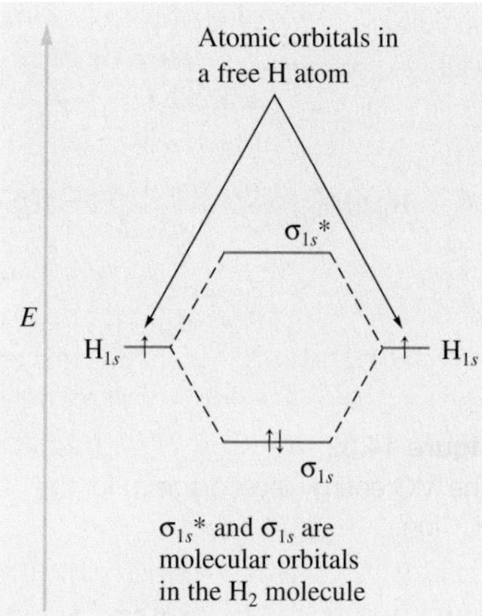

Figure 14.28
MO energy-level diagram for the H_2 molecule.

7. Each MO can hold two electrons, but the electrons' spins must be opposite.
8. Orbitals are conserved. The number of MOs is always the same as the number of atomic orbitals used to construct them.

Many of these points are summarized in Fig. 14.28.

Bond Order

Fundamentally, a molecule forms because it has lower energy than the separated atoms. In the simple MO model, this is reflected by the number of bonding electrons (those that achieve lower energy in going from the free atoms to the molecule) versus the number of antibonding electrons (those that are higher in energy in the molecule than in the free atoms). If the number of bonding electrons is greater than the number of antibonding electrons in a given molecule, the molecule is predicted to be stable.

The quantitative indicator of molecular stability (bond strength) for a diatomic molecule is the **bond order:** *the difference between the number of bonding electrons and the number of antibonding electrons, divided by 2.*

$$\text{Bond order} = \frac{\text{number of bonding electrons} - \text{number of antibonding electrons}}{2}$$

We divide by 2 because we are used to thinking of bonds in terms of *pairs* of electrons.

Bond order is an indication of bond strength because it reflects the difference between the number of bonding electrons and the number of antibonding electrons, which in turn reflects the quantity of energy released when the molecule is formed from its atoms. *Therefore, larger bond order indicates greater bond strength.*

Since the H_2 molecule has two bonding electrons and no antibonding electrons, the bond order is 1:

$$\text{Bond order} = \frac{2 - 0}{2} = 1$$

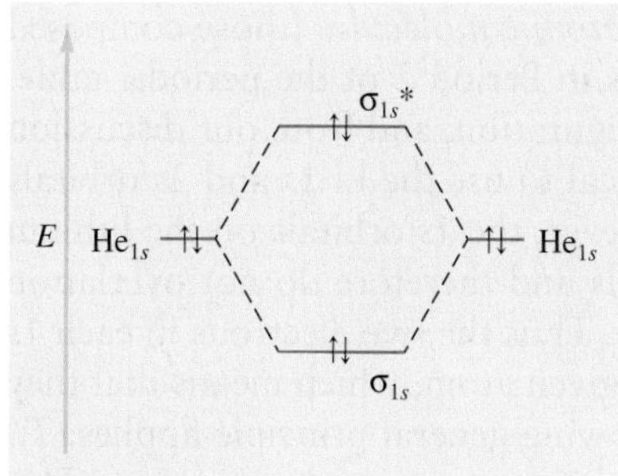

Figure 14.29
The MO energy-level diagram for the He_2 molecule.

We will now apply the MO model to the helium molecule (He_2). Does this model predict that this molecule is stable? Since the He atom has a $1s^2$ configuration, $1s$ orbitals are used to construct the MOs. Therefore, the molecules will have four electrons. From the diagram shown in Fig. 14.29, it is apparent that two electrons are raised in energy and two are lowered in energy. Thus the bond order is zero:

$$\frac{2 - 2}{2} = 0$$

This implies that the He_2 molecule is *not* stable with respect to the two free He atoms, which agrees with the observation that helium gas consists of individual He atoms.

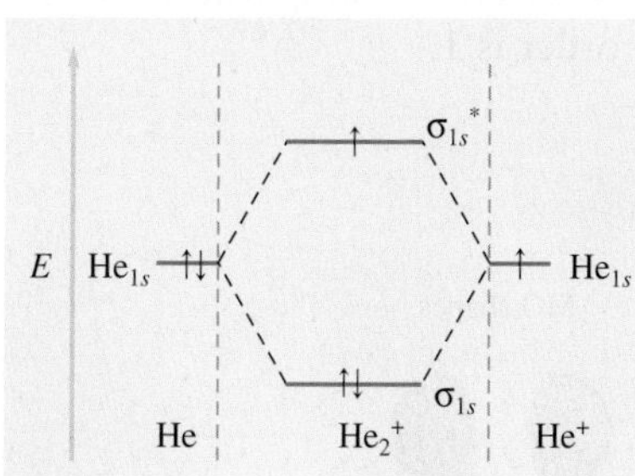

Figure 14.30
The MO energy-level diagram for the He_2^+ ion. Note that in this case the components from which He_2^+ is "constructed" are He and He^+.

Next, we will apply the simple MO model to diatomic ions, starting with He_2^+. Is this ion expected to be stable? Examination of Fig. 14.30 shows that when He is combined with He^+ to form He_2^+, two electrons are lowered in energy and only one is raised in energy. That is, the bond order is $(2 - 1)/2 = \frac{1}{2}$, and He_2^+ is predicted to be stable. Experiments have shown that He_2^+ does indeed exist with a bond energy of 250 kJ/mol.

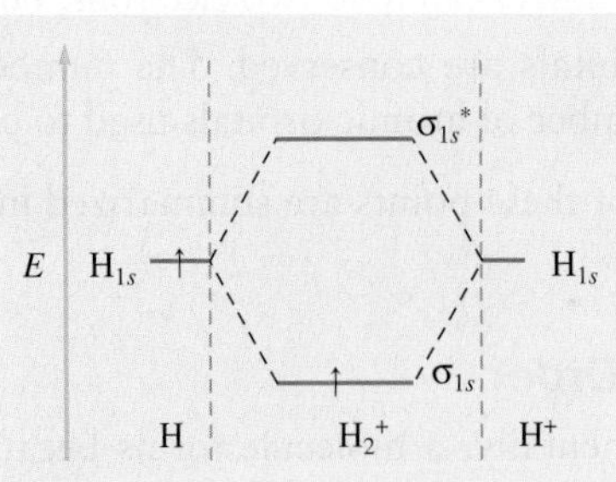

Figure 14.31
The MO energy-level diagram for the H_2^+ ion.

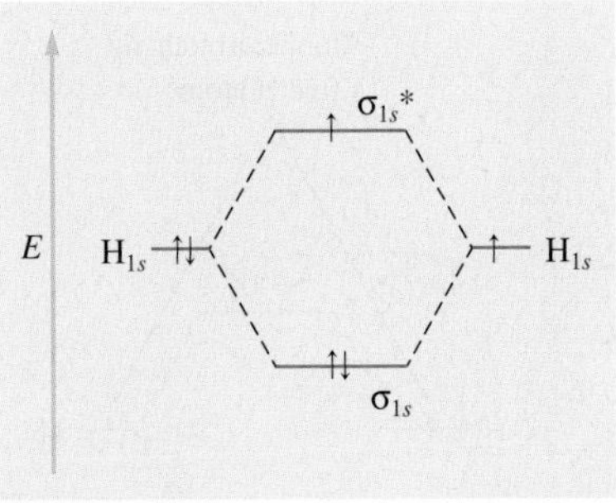

Figure 14.32
The MO energy-level diagram for the H_2^- ion.

Similarly, the H_2^+ ion is known to exist with a bond energy of 255 kJ/mol. This result is in agreement with the simple MO model (Fig. 14.31), which predicts a bond order of $(1 - 0)/2 = \frac{1}{2}$.

Figure 14.32 shows the MO diagram for the H_2^- ion, which is predicted to have a bond order of $(2 - 1)/2 = \frac{1}{2}$. However, in this case the prediction of the model does not agree with experimental results. The H_2^- ion is not known, and all indications are that it is unstable, immediately decomposing to H_2 and a free electron. Thus, again, we are reminded about the perils of simple models. The reasons for the instability of the H_2^- ion are beyond the scope of this treatment.

14.3 Bonding in Homonuclear Diatomic Molecules

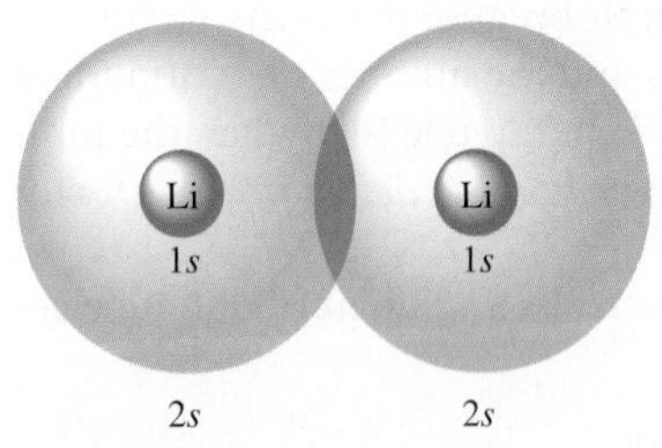

Figure 14.33
The relative sizes of the lithium 1*s* and 2*s* atomic orbitals.

In this section we consider *homonuclear diatomic molecules* (those composed of two identical atoms) formed by elements in Period 2 of the periodic table. The lithium atom has a $1s^22s^1$ electron configuration, and from our discussion in the preceding section, it would seem logical to use the Li 1*s* and 2*s* orbitals to form the MOs of the Li_2 molecule. However, the 1*s* orbitals on the lithium atoms are much smaller than the 2*s* orbitals and therefore do not overlap in space to any appreciable extent (Fig. 14.33). Thus the two electrons in each 1*s* orbital can be assumed to be localized on a given atom, which means that they do not participate in the bonding. The following general principle applies: *To participate in molecular orbitals, atomic orbitals must overlap in space.* This means that only the valence orbitals of atoms contribute significantly to the MOs of a particular molecule.

The MO diagram of the Li_2 molecule and the shapes of its bonding and antibonding MOs are shown in Fig. 14.34. The electron configuration for Li_2 (valence electrons only) is $\sigma_{2s}{}^2$, and the bond order is 1:

$$\frac{2 - 0}{2} = 1$$

Figure 14.34
The MO energy-level diagram for the Li_2 molecule.

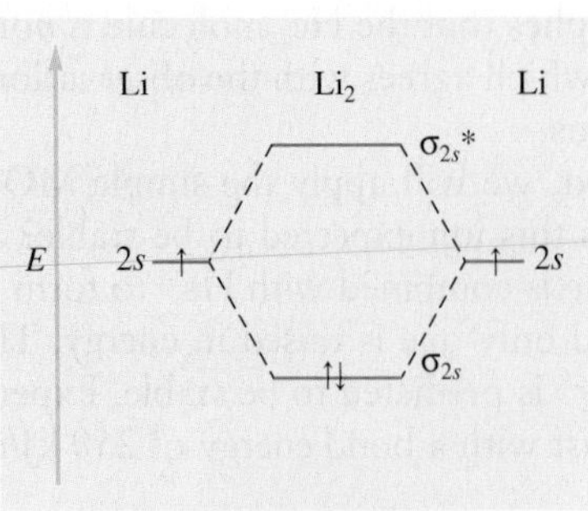

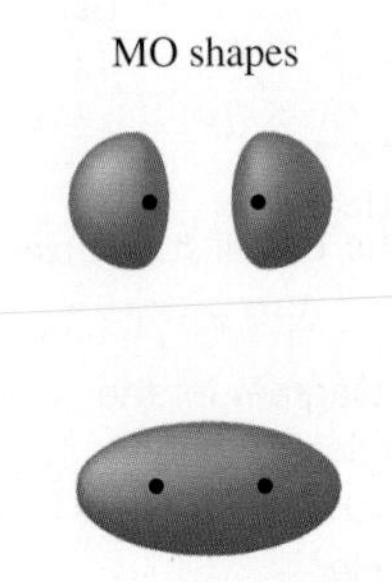

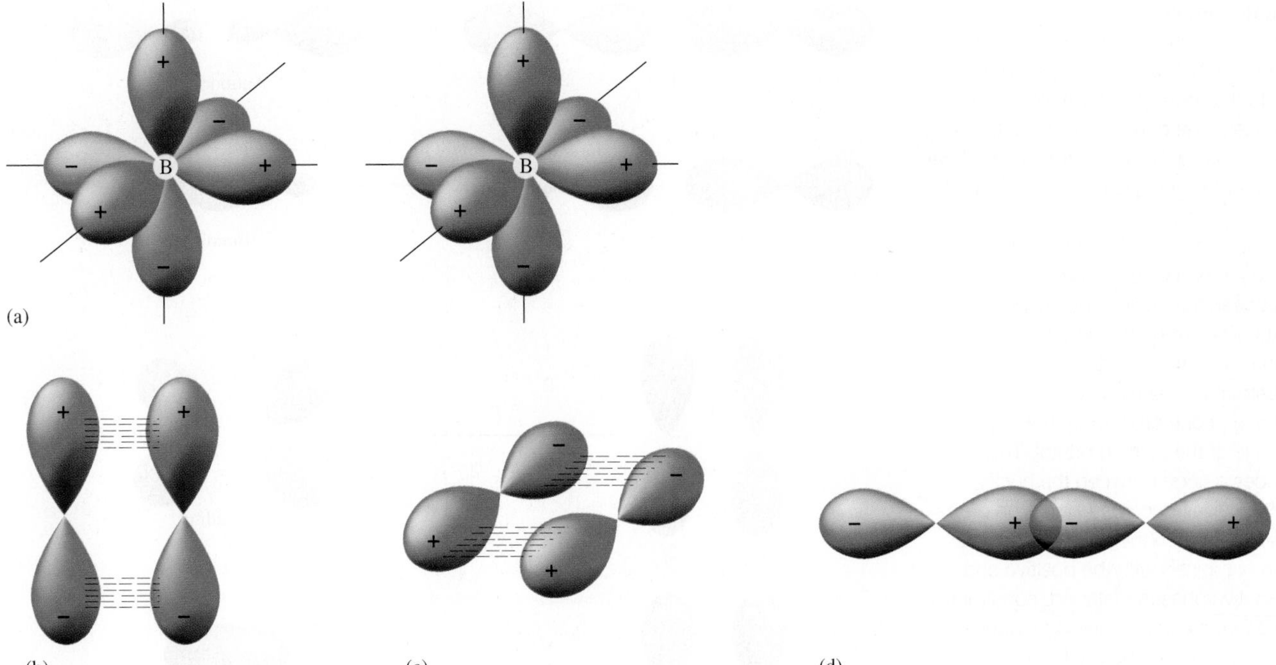

Figure 14.35
(a) The three mutually perpendicular $2p$ orbitals on two adjacent boron atoms. The signs indicate the orbital phases. Two pairs of parallel p orbitals can overlap, as shown in (b) and (c), and the third pair can overlap head-on, as shown in (d).

Only valence atomic orbitals contribute significantly to MOs.

Thus Li_2 is expected to be a stable molecule (has lower energy than two separated lithium atoms). However, this does not mean that Li_2 is the most stable form of elemental lithium. In fact, at normal temperature and pressure, lithium exists as a solid containing many lithium atoms bonded together.

For the beryllium molecule (Be_2), the bonding and antibonding orbitals both contain two electrons. In this case the bond order is $(2 - 2)/2 = 0$. Thus the model predicts that since Be_2 is not more stable than two separated Be atoms, no molecule should form. However, experiments indicate that $Be_2(g)$ does exist, although it has a *very* weak bond (bond energy $\approx$ 10 kJ/mol). In contrast, beryllium metal contains many beryllium atoms bonded to each other and is stable for reasons we will discuss in Chapter 16.

Since the boron atom has a $1s^22s^22p^1$ configuration, we describe the B_2 molecule by considering how p atomic orbitals combine to form MOs. Recall that p orbitals have two lobes and that they occur in sets of three mutually perpendicular orbitals [Fig. 14.35(a)]. When two B atoms approach each other, two pairs of p orbitals can overlap in a parallel manner [Figs. 14.35(b) and (c)] and one pair can overlap head-on [Fig. 14.35(d)].

First, let's consider the MOs formed by the head-on overlap, as shown in Fig. 14.36(a). If the two orbitals are combined directly (added together), the positive phase of the left orbital overlaps with the negative phase of the right orbital. This destructive interference of the orbitals produces a node between the nuclei that leads to decreased electron probability. Therefore, this results in an antibonding MO. On the other hand, when the phase of the right orbital is reversed (its sign is reversed by subtracting it from the left orbital), constructive interference results, giving enhanced electron probability between the nuclei. This is a bonding MO. Note that the electrons in the bonding MO are, as expected, concentrated between the nuclei, and the electrons in the antibonding MO are concentrated outside the area between the two nuclei. Both of these MOs are σ MOs.

The p orbitals that overlap in a parallel manner also produce bonding and antibonding orbitals [Fig. 14.36(b)]. When the parallel p orbitals are added

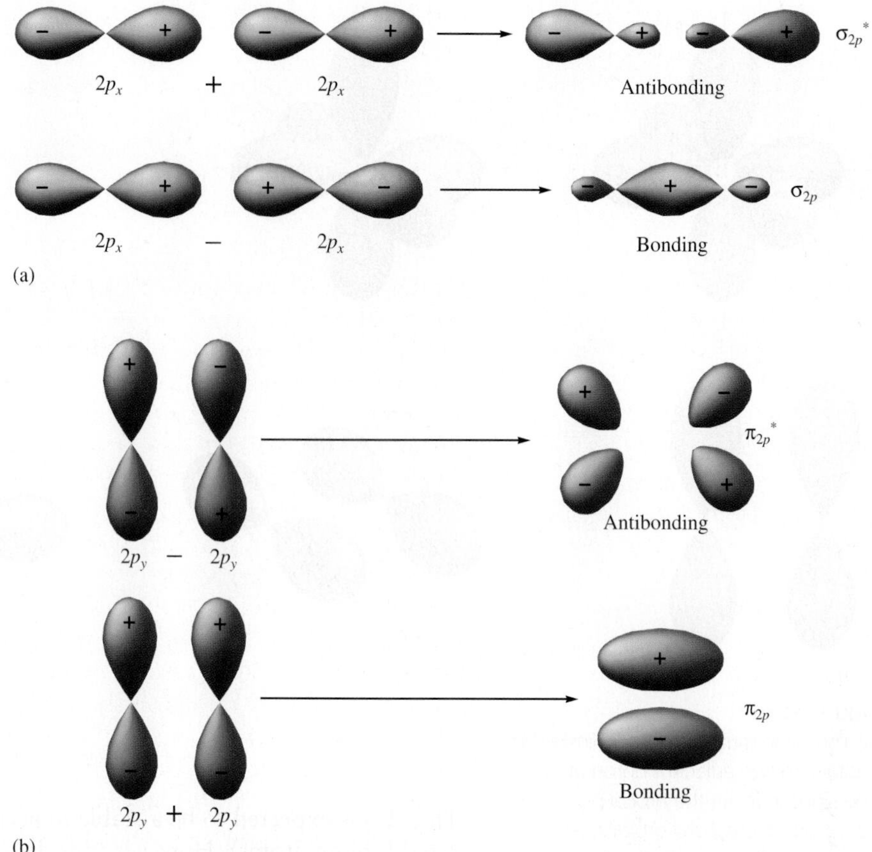

Figure 14.36
(a) The two *p* orbitals on the boron atoms that overlap head-on combine to form σ bonding and antibonding orbitals. The bonding orbital is formed by reversing the sign of the right orbital so that the positive phases of both orbitals match between the nuclei to produce constructive interference. This leads to enhanced electron probability between the nuclei. The antibonding orbital is formed by the direct combination of the orbitals, which gives destructive interference of the positive phase of one orbital with the negative phase of the second orbital. This produces a node between the nuclei, which gives decreased electron probability. (b) When the parallel *p* orbitals are combined with the positive and negative phases matched, constructive interference occurs, giving a bonding π orbital. When the orbitals have opposite phases (the signs of one orbital are reversed), destructive interference occurs, resulting in an antibonding π orbital.

together, the phases match, and constructive interference occurs, giving a bonding orbital. When the phase of the right orbital is reversed and combined with the left orbital, destructive interference occurs, giving an antibonding orbital. Since the electron probability lies above and below the line between the nuclei, both the orbitals are **pi (π) MOs.** They are designated as π_{2p} for the bonding MO and $\pi_{2p}{}^*$ for the antibonding MO.

Let's try to make an educated guess about the relative energies of the σ and π MOs formed from the 2*p* atomic orbitals. Would we expect the electrons to prefer the σ bonding orbital (where the electron probability is concentrated in the area between the nuclei) or the π bonding orbital? The σ orbital would seem to have the lower energy since the electrons are closest to the two nuclei. This agrees with the observation that σ interactions are typically stronger than π interactions.

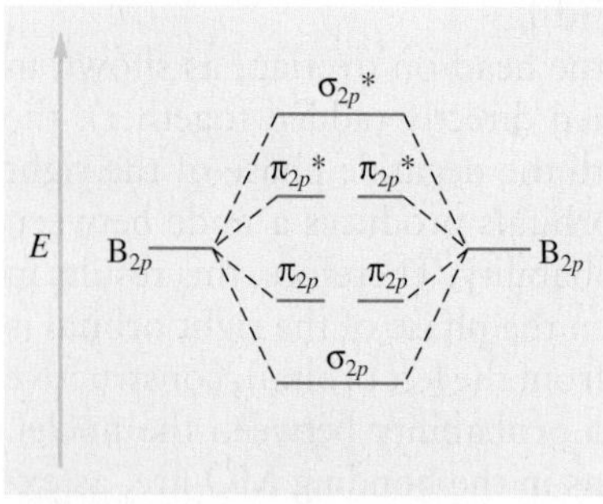

Figure 14.37
The *expected* MO energy-level diagram for the combination of the 2*p* orbitals on two boron atoms.

Figure 14.37 gives the MO energy-level diagram *expected* when the two sets of 2*p* orbitals on the two borom atoms combine to form MOs. Note that there are two π bonding orbitals at the same energy (degenerate orbitals) formed from the two pairs of parallel *p* orbitals, and there are two degenerate π antibonding orbitals. The energy of the π_{2p} orbitals is expected to be higher than that of the σ_{2p} orbital because σ interactions are generally stronger than π interactions.

To construct the total MO diagram for the B_2 molecule, we make the assumption that the 2*s* and 2*p* orbitals combine separately (in other words, there is no 2*s*–2*p* mixing). The resulting diagram is shown in Fig. 14.38. Note that B_2 has six *valence* electrons. (Remember that the 1*s* orbitals and electrons are

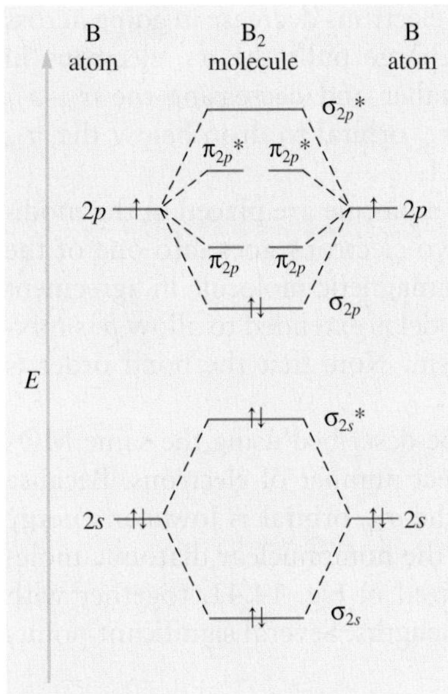

Figure 14.38
The *expected* MO energy-level diagram for the B_2 molecule.

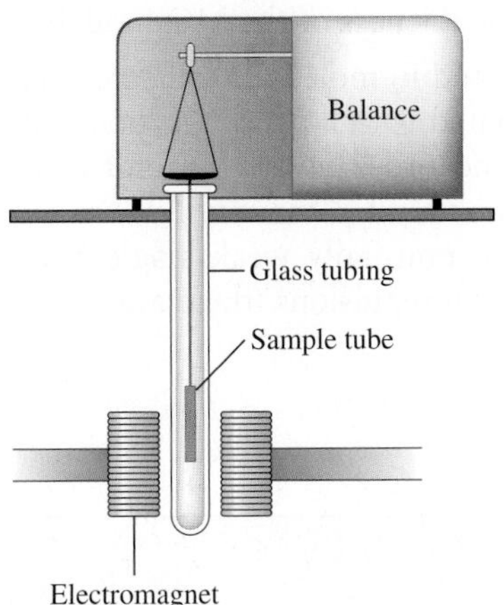

Figure 14.39
An apparatus used to measure the paramagnetism of a sample. A paramagnetic sample will appear heavier when the electromagnet is turned on because the sample is attracted into the inducing magnetic field.

assumed not to participate in the bonding.) This diagram predicts a bond order of 1:

$$\frac{4 - 2}{2} = 1$$

Therefore, B_2 should be a stable molecule.

Paramagnetism

At this point we need to discuss an additional molecular property—magnetism. Most materials have no magnetism until they are placed in a magnetic field. However, in the presence of such a field, magnetism of two types can be induced. **Paramagnetism** causes the substance to be attracted toward the inducing magnetic field. **Diamagnetism** causes the substance to be repelled from the inducing magnetic field. Figure 14.39 illustrates how paramagnetism is measured. The sample is weighed with the electromagnet turned off and then weighed again with the electromagnet turned on. An increase in weight when the field is turned on indicates that the sample is paramagnetic. Studies have shown that *paramagnetism is associated with unpaired electrons,** *and diamagnetism is associated with paired electrons.* Any substance that has both paired and unpaired electrons will exhibit net paramagnetism, since the effect of paramagnetism is much stronger than that of diamagnetism. Thus the phenomenon of paramagnetism provides a ready means for testing whether a substance contains unpaired electrons.

The MO energy-level diagram given in Fig. 14.38 predicts that the B_2 molecule will be diamagnetic, since the MOs contain only paired electrons. However, experiments show that B_2 is actually paramagnetic with two unpaired electrons. Why does the model yield the wrong prediction? This is yet another illustration of how models are developed and used. In general, we try to use the simplest possible model that accounts for all of the important observations. In this case, although the simplest model has been relatively successful in describing all the diatomic molecules up to B_2, it certainly is suspect if it cannot describe the B_2 molecule correctly.

Let's reconsider one of our previous assumptions. In our treatment of B_2, we have assumed that the *s* and *p* orbitals combine separately to form MOs. Calculations show that when the *s* and *p* orbitals are allowed to mix (participate) in the same MO, a different energy-level diagram results for B_2 (Fig. 14.40). Note that even though the *s* and *p* contributions to the MOs are no longer separate, we retain the simple orbital designations. The mixing of *p* and *s* atomic orbitals occurs only in the σ MOs (σ_{2s}, σ_{2s}*, σ_{2p}, and σ_{2p}*). Because the energy of the σ_{2p} orbital is changed by *p–s* mixing, the energies of π_{2p} and σ_{2p} orbitals are reversed. Also, the *p–s* mixing changes the energies of the σ_{2s} and σ_{2s}* such that they are no longer equally spaced relative to the energy of the free 2*s* orbital.

Perhaps an easier way to understand the effects of *p–s* mixing is to think in terms of electron repulsions. Note from Figs. 14.34 and 14.36 that both the σ_{2s} and the σ_{2p} orbitals show significant electron probability in the area between the nuclei of the bonded atoms. This means that electrons in the σ_{2s} orbital will repel any electrons placed in the σ_{2p} orbital, thus making this orbital "less attractive" (higher in energy) to electrons. Therefore, the σ_{2p} orbital is pushed to a higher energy level than the π_{2p} orbitals, which do not have electron probability between the nuclei and are not affected by electrons in the σ_{2s} orbital. This situation applies to B_2, C_2, and N_2, as shown in Fig. 14.41.

*This is an oversimplification but one that will not concern us here.

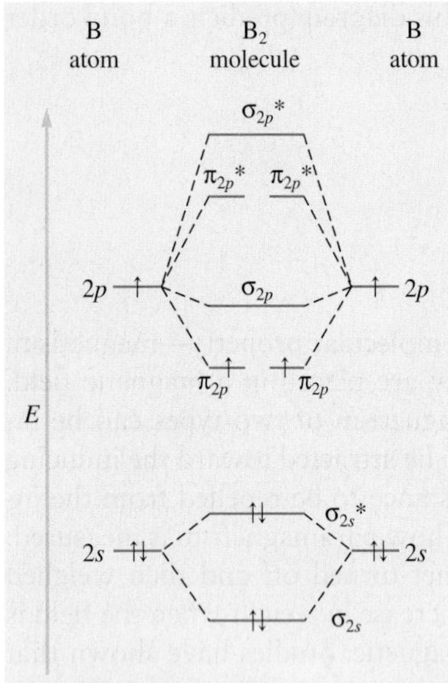

Figure 14.40
The *correct* MO energy-level diagram for the B_2 molecule. When *p–s* mixing is allowed, the energies of the σ_{2p} and π_{2p} orbitals are reversed. The two electrons from the B 2*p* orbitals now occupy separate, degenerate π_{2p} MOs and thus have parallel spins. Therefore, this diagram explains the observed paramagnetism of B_2.

Bond energy increases with bond order; bond length decreases with increasing bond order.

The repulsion effects between σ_{2s} and σ_{2p} electrons decrease in going across the period because the increasing nuclear charge pulls the σ_{2s} electrons in closer, effectively making the σ_{2s} orbital smaller and decreasing the σ_{2s}–σ_{2p} interactions. This causes the energy of the σ_{2p} orbital to drop below the π_{2p} orbital in O_2 and F_2, as indicated in Fig. 14.41.

When the six valence electrons of the B_2 molecule are placed in the modified energy-level diagram, each of the last two electrons goes into one of the degenerate π_{2p} orbitals. This produces a paramagnetic molecule in agreement with experimental results. Thus, when the model is extended to allow *p–s* mixing in MOs, it predicts the correct magnetism. Note that the bond order is $(4 - 2)/2 = 1$, as before.

The diatomic molecules C_2 and N_2 can be described using the same MOs as for B_2 (Fig. 14.40) and inserting the correct number of electrons. Because *p–s* mixing is less important in O_2 and F_2, the σ_{2p} orbital is lower in energy than the π_{2p} orbitals. The MO diagrams for the homonuclear diatomic molecules for the Period 2 elements are summarized in Fig. 14.41, together with experimentally obtained bond strengths and lengths. Several significant points arise from these results.

1. There are definite correlations between bond order, bond energy, and bond length. As the bond order predicted by the MO model increases, the bond energy increases, and the bond length decreases. This is a clear indication that the bond order predicted by the model accurately reflects bond strength, and it strongly supports the reasonableness of the MO model.
2. Comparison of the bond energies of the B_2 and F_2 molecules indicates that bond order cannot automatically be associated with a particular bond energy. Although both molecules have a bond order of 1, the bond in B_2 appears to be about twice as strong as the bond in F_2. As we will see in our later discussion of the halogens, F_2 has an unusually weak single bond because of larger-than-usual electron–electron repulsions (there are 14 valence electrons on the small F_2 molecule).

Figure 14.41
The MO energy-level diagrams, bond orders, bond energies, and bond lengths for the diatomic molecules B_2 through F_2. Note that for O_2 and F_2 the σ_{2p} orbital is lower in energy than the π_{2p} orbitals.

E (↑)	B_2	C_2	N_2		O_2	F_2
σ_{2p}*	—	—	—	σ_{2p}*	—	—
π_{2p}*	— —	— —	— —	π_{2p}*	↑ ↑	↑↓ ↑↓
σ_{2p}	—	—	↑↓	π_{2p}	↑↓ ↑↓	↑↓ ↑↓
π_{2p}	↑ ↑	↑↓ ↑↓	↑↓ ↑↓	σ_{2p}	↑↓	↑↓
σ_{2s}*	↑↓	↑↓	↑↓	σ_{2s}*	↑↓	↑↓
σ_{2s}	↑↓	↑↓	↑↓	σ_{2s}	↑↓	↑↓
Magnetism	Para–magnetic	Dia–magnetic	Dia–magnetic		Para–magnetic	Dia–magnetic
Bond order	1	2	3		2	1
Observed bond dissociation energy (kJ/mol)	290	620	942		495	154
Observed bond length (pm)	159	131	110		121	143

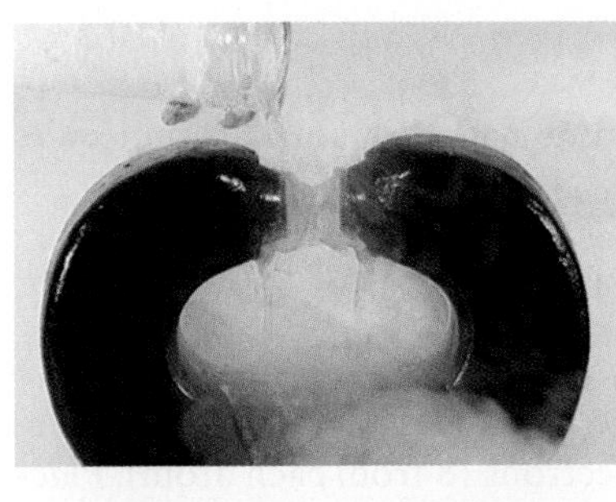

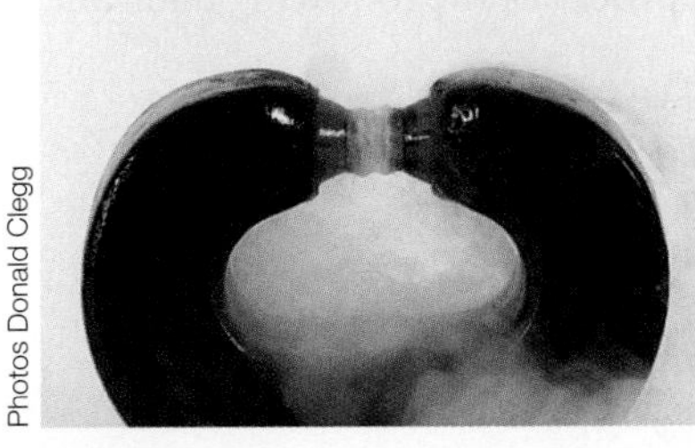

Photos Donald Clegg

Figure 14.42
When liquid oxygen is poured into the space between the poles of a strong magnet, it remains there until it boils away. This attraction of liquid oxygen for the magnetic field demonstrates the paramagnetism of the O_2 molecule.

3. Note the very large bond energy associated with the N_2 molecule, which the MO model predicts will have a bond order of 3, a triple bond. The very strong bond in N_2 is the principal reason that many nitrogen-containing compounds are used as explosives. The reactions involving these explosives give the very stable N_2 molecule as a product, thus releasing large quantities of energy.
4. The O_2 molecule is known to be paramagnetic. This can be very convincingly demonstrated by pouring liquid oxygen between the poles of a strong magnet, as shown in Fig. 14.42. Because of its paramagnetism, the oxygen is attracted to the magnet gap, where it remains until it evaporates. Significantly, the MO model correctly predicts oxygen's paramagnetism, but the simplest form of the LE model predicts a diamagnetic molecule.

OWL INTERACTIVE EXAMPLE 14.6

For the species O_2, O_2^+, and O_2^-, give the electron configuration and the bond order for each. Which has the strongest bond?

Solution The O_2 molecule has 12 valence electrons (6 + 6), O_2^+ has 11 valence electrons (6 + 6 − 1), and O_2^- has 13 valence electrons (6 + 6 + 1). We will assume that the ions can be treated using the same MO diagram used for the neutral diatomic molecule:

	O_2	O_2^+	O_2^-
σ_{2p}^*	—	—	—
π_{2p}^*	↑ ↑	↑ —	↑↓ ↑
π_{2p}	↑↓ ↑↓	↑↓ ↑↓	↑↓ ↑↓
σ_{2p}	↑↓	↑↓	↑↓
σ_{2s}^*	↑↓	↑↓	↑↓
σ_{2s}	↑↓	↑↓	↑↓

The electron configuration for each species can then be taken from the diagram:

$$O_2 \quad (\sigma_{2s})^2(\sigma_{2s}^*)^2(\sigma_{2p})^2(\pi_{2p})^4(\pi_{2p}^*)^2$$

$$O_2^+ \quad (\sigma_{2s})^2(\sigma_{2s}^*)^2(\sigma_{2p})^2(\pi_{2p})^4(\pi_{2p}^*)^1$$

$$O_2^- \quad (\sigma_{2s})^2(\sigma_{2s}^*)^2(\sigma_{2p})^2(\pi_{2p})^4(\pi_{2p}^*)^3$$

The bond orders are as follows:

$$\text{For } O_2\text{:} \quad \frac{8-4}{2} = 2$$

$$\text{For } O_2^+\text{:} \quad \frac{8-3}{2} = 2.5$$

$$\text{For } O_2^-\text{:} \quad \frac{8-5}{2} = 1.5$$

Thus O_2^+ is expected to have the strongest bond of the three species.

Experimental evidence supports these predictions. The bond energies for O_2^+, O_2, and O_2^- are 643, 495, and 395 kJ/mol, respectively.

OWL INTERACTIVE EXAMPLE 14.7

Use the MO model to predict the bond order and magnetism of each molecule.

a. Ne_2 **b.** P_2

Solution

a. The valence orbitals for Ne are $2s$ and $2p$. Thus we can use the MOs we have already constructed for the diatomic molecules of the Period 2 elements. The Ne_2 molecule has 16 valence electrons (8 from each atom). Placing these electrons in the appropriate MOs produces the following diagram:

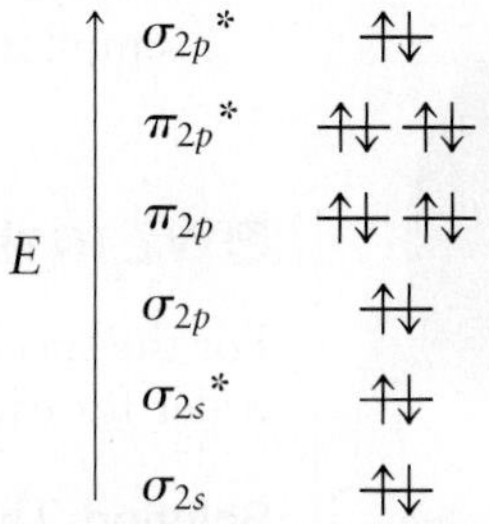

E ↑	
$\sigma_{2p}{}^*$	↑↓
$\pi_{2p}{}^*$	↑↓ ↑↓
π_{2p}	↑↓ ↑↓
σ_{2p}	↑↓
$\sigma_{2s}{}^*$	↑↓
σ_{2s}	↑↓

The bond order is $(8 - 8)/2 = 0$, so Ne_2 should not exist.

b. The P_2 molecule found in "brown phosphorus," which is stable only at very low temperatures, contains phosphorus atoms from the third row of the periodic table. We will assume that diatomic molecules of the Period 3 elements can be treated very similarly to those from Period 2. The only change will be that the MOs are formed from $3s$ and $3p$ atomic orbitals. The P_2 molecule has 10 valence electrons (5 from each phosphorus atom). The resulting molecular orbital diagram follows.

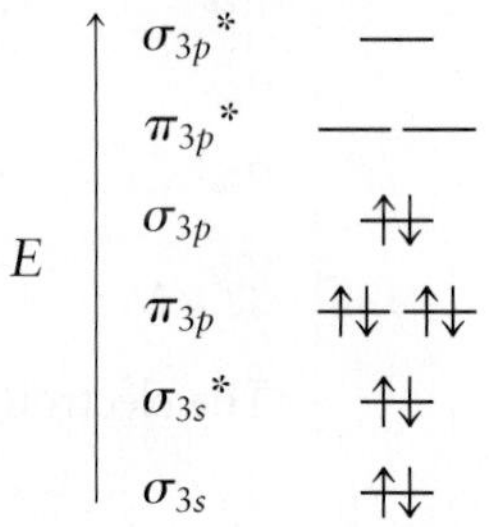

E ↑	
$\sigma_{3p}{}^*$	—
$\pi_{3p}{}^*$	— —
σ_{3p}	↑↓
π_{3p}	↑↓ ↑↓
$\sigma_{3s}{}^*$	↑↓
σ_{3s}	↑↓

The molecule has a bond order of 3 and is expected to be diamagnetic.

It should be noted that phosphorus exists in nature as P_4 molecules (and as other, more complex forms).

E ↑	
$\sigma_{2p}{}^*$	—
$\pi_{2p}{}^*$	↑ —
σ_{2p}	↑↓
π_{2p}	↑↓ ↑↓
$\sigma_{2s}{}^*$	↑↓
σ_{2s}	↑↓

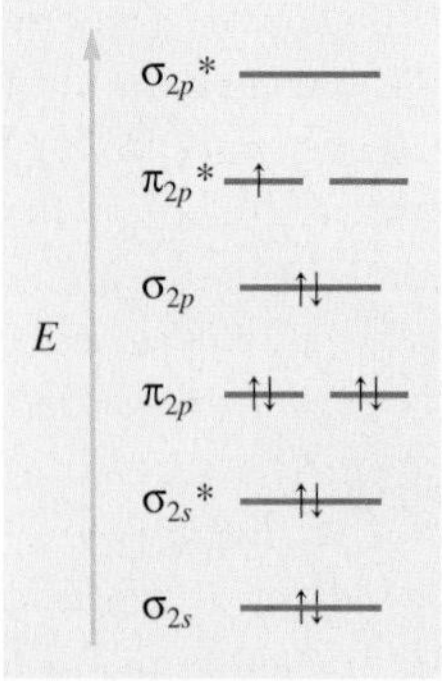

Figure 14.43
The MO energy-level diagram for the NO molecule. We assume that orbital order is the same as that for N_2. The bond order is 2.5.

14.4 Bonding in Heteronuclear Diatomic Molecules

In this section we will deal with selected examples of **heteronuclear diatomic molecules**—those containing two different atoms. A special case involves molecules containing atoms adjacent to each other in the periodic table. Since the atoms involved in such a molecule are so similar, we can use the MO diagram for homonuclear molecules. For example, we can predict the bond order and magnetism of nitric oxide (NO) by placing its 11 valence electrons (5 from nitrogen and 6 from oxygen) in the MO energy-level diagram shown in Fig. 14.43. The molecule should be paramagnetic with a bond order of

$$\frac{8 - 3}{2} = 2.5$$

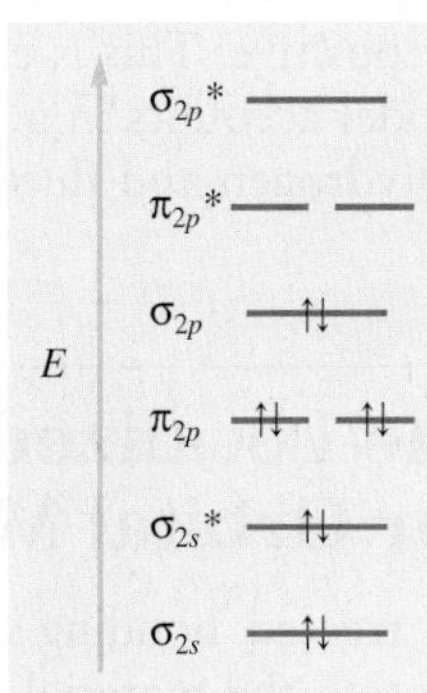

Figure 14.44
The MO energy-level diagram for both the NO^+ and CN^- ions.

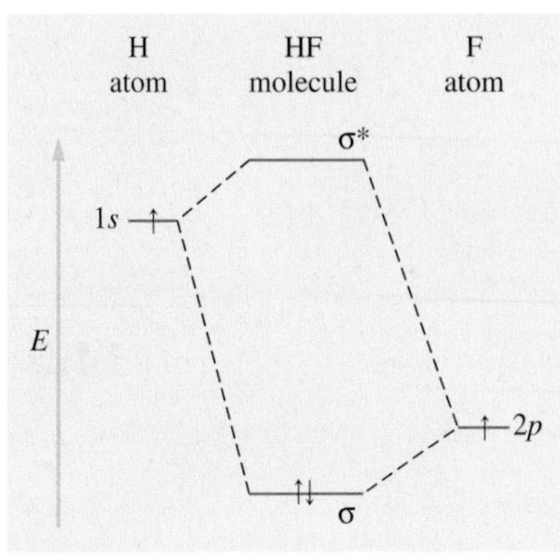

Figure 14.45
A partial MO energy-level diagram for the HF molecule.

Experimentally, nitric oxide is indeed found to be paramagnetic. Note that this odd-electron molecule is described very naturally by the MO model. In contrast, the LE model, in the simple form used in this text, does not readily describe such molecules.

OWL INTERACTIVE EXAMPLE 14.8

Use the MO model to predict the magnetism and bond order of the NO^+ and CN^- ions.

Solution The NO^+ ion has 10 valence electrons $(5 + 6 - 1)$. The CN^- ion also has 10 valence electrons $(4 + 5 + 1)$. Both ions are therefore diamagnetic and have a bond order of 3:

$$\frac{8 - 2}{2} = 3$$

The MO diagram for these two ions is the same (Fig. 14.44).

When the two atoms of a diatomic molecule are very different, the energy-level diagram for homonuclear molecules can no longer be used. A new diagram must be devised for each molecule. We will illustrate such a case by considering the hydrogen fluoride (HF) molecule. The electron configurations of the hydrogen and fluorine atoms are $1s^1$ and $1s^22s^22p^5$, respectively. To keep things as simple as possible, we will assume that fluorine uses only one of its $2p$ orbitals to bond to hydrogen. Thus the MOs for HF will be composed of fluorine $2p$ and hydrogen $1s$ orbitals. Figure 14.45 gives the partial MO energy-level diagram for HF, focusing only on the orbitals involved in the bonding. We are assuming that fluorine's other valence electrons remain localized on the fluorine atom. The $2p$ orbital of fluorine is shown at a lower energy than the $1s$ orbital of hydrogen on the diagram because fluorine binds its valence electrons more tightly. Thus the $2p$ electron on a free fluorine atom is at lower energy than the $1s$ electron on a free hydrogen atom. The diagram predicts that the HF molecule should be stable since both electrons are lowered in energy relative to their energies in the free hydrogen and fluorine atoms, and this is the driving force for bond formation.

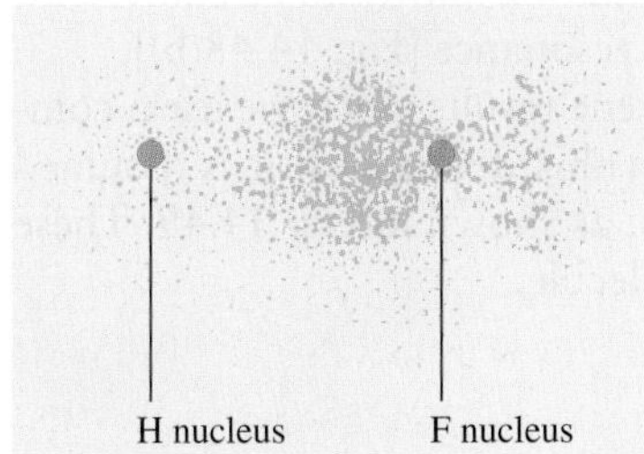

Figure 14.46
The electron probability distribution in the bonding MO of the HF molecule. Note the greater electron density close to the fluorine atom.

Because the fluorine $2p$ orbital is lower in energy than the hydrogen $1s$ orbital, the electrons prefer to be closer to the fluorine atom. That is, the σ MO containing the bonding electron pair shows greater electron probability close to the fluorine (Fig. 14.46). The electron pair is not shared equally. This causes the fluorine atom to have a slight excess of negative charge and leaves the hydrogen atom partially positive. This is exactly the bond polarity observed for HF. Thus the MO model accounts in a straightforward way for the different electronegativities of hydrogen and fluorine and the resulting unequal charge distribution.

14.5 Combining the Localized Electron and Molecular Orbital Models

In this text we have treated bonding in terms of simple models not only because we are presenting the material at an introductory level but also because a lot of first-order thinking by chemists uses these simple "back-of-the-envelope" models. Thus, as long as we are aware of the pitfalls of oversimplified models, we can use them to our benefit.

In this section we will attempt to make these simple models even more useful by addressing a particular shortcoming of the LE model—its assumption that electrons are localized (restricted to the space between a given pair of atoms). This problem is most apparent for molecules where several valid Lewis structures can be drawn. Recall that none of the resonance structures taken alone adequately describes the electronic structure of the molecule. The concept of resonance was invented to solve this problem. However, even with resonance included, the LE model does not describe molecules and ions such as O_3 and NO_3^- in a very satisfying way.

It would seem that the ideal bonding model would be one with the simplicity of the LE model but with the delocalization characteristics of the MO model. We can achieve this by combining the two models to describe molecules that require resonance. Note that for species such as O_3 and NO_3^- the double bond changes position in the resonance structures (Fig. 14.47). Since a double bond involves one σ bond and one π bond, there is a σ bond between all bound atoms in each resonance structure. It is really the π bond that has different locations in the various resonance structures.

In molecules that require resonance, it is the π bonding that is most clearly delocalized.

We conclude that the σ electrons in a molecule can be described as being localized with no apparent problems. It is the π electrons that must be treated as being delocalized. Thus, for molecules that require resonance, we will use the LE model to describe the σ bonding and the MO model to describe the π bonding. This allows us to keep the bonding model as simple as possible and yet give a more physically accurate description of such molecules.

We will illustrate this procedure by considering the bonding in benzene, an important industrial chemical that must be handled carefully because it is a known carcinogen. The benzene molecule (C_6H_6) consists of a planar hexagon of carbon atoms with one hydrogen atom bound to each carbon atom, as shown in Fig. 14.48(a). In the molecule all six C—C bonds are known to be equivalent. To explain this fact, the LE model must invoke resonance [Fig. 14.48(b)].

A better description of the bonding in benzene results when we use a combination of the models, as described above. In this description it is assumed that the σ bonds to carbon involve sp^2 orbitals, as shown in Fig. 14.49. These σ bonds are all centered in the plane of the molecule.

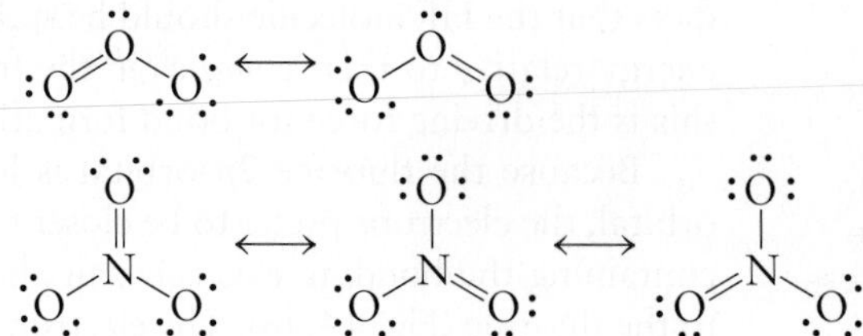

Figure 14.47
The resonance structures for O_3 and NO_3^-. Note that it is the double bond that occupies various positions in the resonance structures.

The Always Interesting NO

Chemical Insights

Nitrogen monoxide (commonly called nitric oxide) is a compound with a dual personality. Produced by the reaction of N_2 and O_2 at the high temperatures associated with combustion in automobiles and conventional power plants, NO immediately reacts with O_2 to form NO_2, a major air pollutant. Nitric oxide has also been implicated as a cause of brain damage in people with strokes. During a stroke, relatively large amounts of NO are produced by the body, leading to nerve damage in the brain. On the other hand, NO has in recent years been shown to be an essential regulator of many body processes, including controlling blood pressure, aiding in penile erection, assisting digestion, and battling tumor cells.

Recent findings indicate additional important roles for NO. For example, researchers in the United States and Tanzania have discovered that the presence of NO in the blood seems to aid in the body's battle against infectious diseases. These results indicate that patients with relatively high blood levels of NO are much more successful in fighting off malaria. However, the situation is complicated: NO's dual personality may again be evident. The studies indicate that while NO may bolster the body's resistance to malaria, the release of excess NO in the brain in people with certain forms of malaria may in fact be responsible for major brain damage. The role of NO in malaria and other infectious diseases continues to be the focus of much current research.

On a related front, studies at the Karolinska Institute in Stockholm, Sweden, seem to indicate that NO furnishes protection against sinus infections. Unlike the nose, where bacteria thrive, the sinuses—the air-filled cavities in the head surrounding the nose—are surprisingly microbe-free. In trying to discern the reason for the lack of bacteria in the sinuses—a warm, moist, seemingly bacteria-friendly environment—Swedish scientists found that the cavities contain NO concentrations close to the maximum levels allowed in polluted air. They discovered that cells in the walls of the sinuses produce NO, presumably to protect against bacterial infection—the NO binds to vital enzymes in microbes, stopping their growth and reproduction. But the importance of NO goes beyond its antibacterial role. It also behaves as a powerful vasodilator, a substance that dilates blood vessels. The researchers postulate that some of the NO in the sinuses travels through the airways into the lungs, where it dilates blood vessels in areas well supplied with O_2, thus increasing the amount of oxygen absorbed into the blood.

The powers of NO are also apparent in the operation of hemoglobin, the protein that delivers O_2 to the tissues and returns CO_2 to be exhaled in the lungs. Recent studies at Duke University indicate that hemoglobin has a dangling side chain that binds NO. As the hemoglobin releases O_2 to the tissues it also sheds some of its bound NO, which dilates blood vessels and helps the oxygen get to the tissues. The hemoglobin then picks up CO_2 and any circulating NO and returns them to the lungs to be exhaled. Thus hemoglobin not only exchanges O_2 for CO_2 at the tissues but also regulates blood pressure in the process.

Thus NO continues to show its dual personality—sometimes positive, sometimes negative, always interesting.

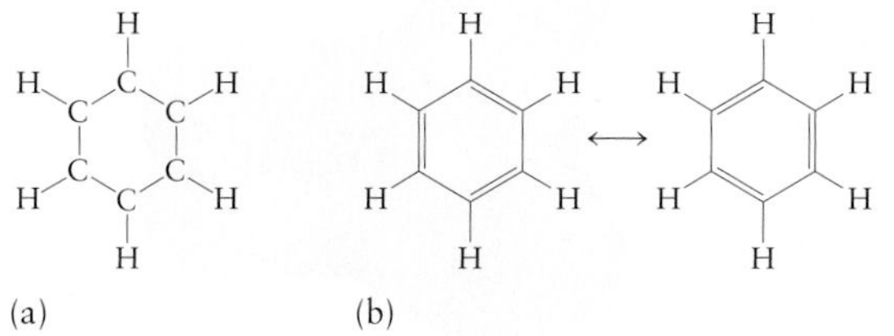

(a) (b)

Figure 14.48
(a) The benzene molecule consists of a ring of six carbon atoms with one hydrogen atom bound to each carbon; all atoms are in the same plane. All the C—C bonds are known to be equivalent. (b) Two of the resonance structures for the benzene molecule. The LE model must invoke resonance to account for the six equal C—C bonds.

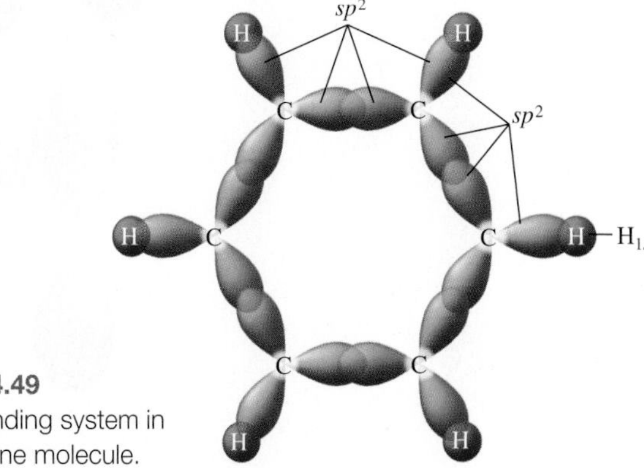

Figure 14.49
The σ bonding system in the benzene molecule.

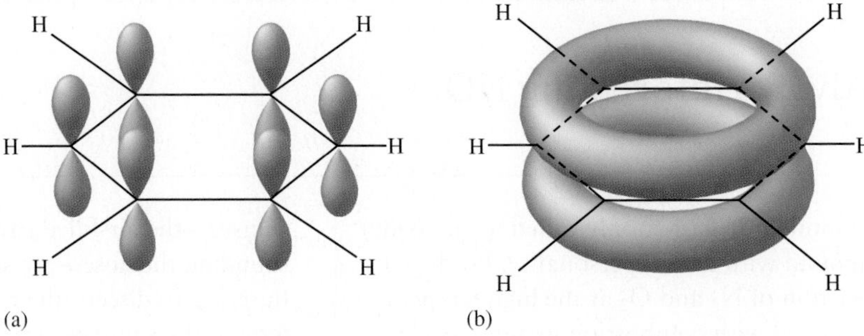

Figure 14.50
(a) The π MO system in benzene is formed by combining the six *p* orbitals from the six sp^2 hybridized carbon atoms. (b) The electrons in the resulting π MO are delocalized over the entire ring of carbon atoms, giving six equivalent bonds. A composite of these orbitals is represented here.

Since each carbon atom is sp^2 hybridized, a *p* orbital perpendicular to the plane of the ring remains on each carbon atom. These six *p* orbitals can be used to form π molecular orbitals, as shown in Fig. 14.50(a). The electrons in the resulting π molecular orbitals are delocalized above and below the plane of the ring, as shown in Fig. 14.50(b). This gives six equivalent C—C bonds, as required by the known structure of the benzene molecule. The benzene structure is often written as

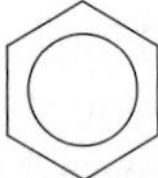

to indicate the **delocalized π bonding** in the molecule.

Very similar treatments can be applied to other planar molecules for which resonance is required by the LE model. For example, the NO_3^- ion can be described using the π MO system shown in Fig. 14.51. In this molecule each atom is assumed to be sp^2 hybridized. This leaves one *p* orbital on each atom perpendicular to the plane of the ion. These *p* orbitals can combine to form the π MO system.

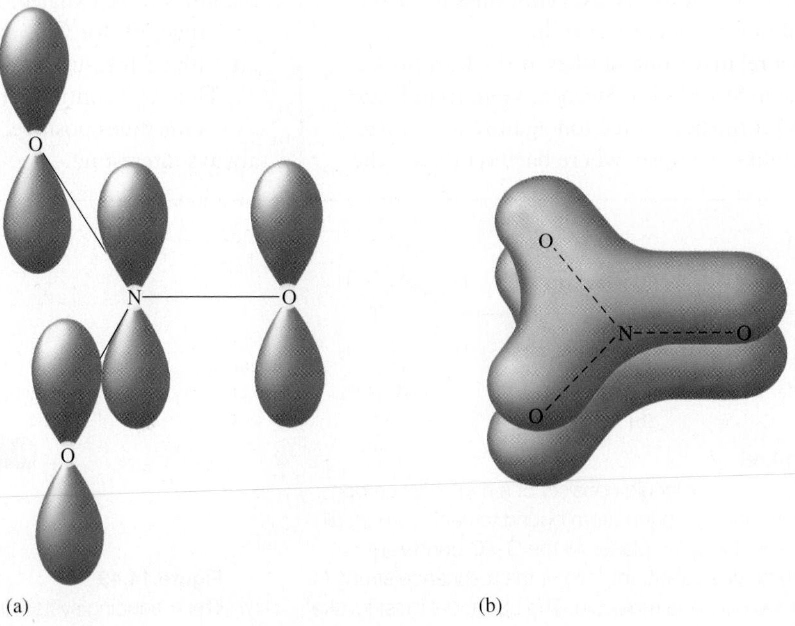

Figure 14.51
(a) The *p* orbitals used to form the π bonding system in the NO_3^- ion. (b) A representation of the delocalization of the electrons in the π MO system of the NO_3^- ion.

14.6 Orbitals: Human Inventions

In the treatment of the bonding and structure in molecules we have stressed that the descriptions we have given are models—and simple ones, at that. These models are extraordinarily valuable because they provide pictures in our minds of what molecules might look like and how they might behave. This helps us decide what questions to ask (what experiments to do) as we try to understand more completely the complicated behavior of matter. However, as valuable as these pictures are, we must remember that they greatly oversimplify a very complex situation. We must realize that these models can give us a rather superficial view that is often highly misleading.

For example, after all of the talk about orbitals in the last several chapters, you probably have the distinct impression that orbitals actually exist. The truth is, they do not—at least not in the physical sense. Orbitals are mathematical functions—solutions to a modified Schrödinger equation (the changes made to allow separation into independent electron equations).

Thus orbitals are mathematical tools we use to describe atoms and molecules. However, they have no physical reality in the sense that if you could look at a water molecule, you would probably not "see" anything like the pictures we have shown in this book. These pictures merely help us visualize the theoretical information we have accumulated for water.

We should also make clear that much more sophisticated treatments of atoms and molecules than we have considered here do exist and are carried out quite routinely by chemists who specialize in this area. These quantum mechanical calculations involve many fewer approximations than are made to obtain the models we have discussed here. However, although these treatments produce accurate mathematical descriptions of atomic and molecular properties, they are usually very difficult to interpret physically.

In contrast, the simple, highly pictorial models we have considered in this text are extremely useful because they help us do first-order chemical thinking and organize the information about matter that we gather from experiments. For example, there is a great deal of experimental evidence to suggest that methane has a tetrahedral structure. We have seen that the VSEPR model predicts this structure and that sp^3 hybrid orbitals on the carbon atom are consistent with this structure. Of course, this does not mean that methane really has four localized pairs of electrons, all with the same energy. In fact, the electrons in methane are delocalized and do not all have the same energy. However, the fact that methane is more complicated than these models indicate does not destroy the usefulness of the models. They still help us greatly as we organize the vast amount of information we collect about molecules. For example, chemists have learned to expect certain behaviors of "sp^3 carbon atoms" no matter what molecule contains them. A similar statement also applies to "sp^2 carbon atoms," "*sp* carbon atoms," and so on.

There is another point that should be made in connection with the theories of atoms and molecules. Even the experts in the field disagree (sometimes violently) about what the mathematical descriptions of atoms and molecules mean.*

*References to some interesting and provocative papers on this subject follow. These articles are probably too sophisticated to be read with great understanding now, but you can try to read them and then put copies of them in your files to be reread as your knowledge of chemistry increases. They raise some interesting questions.

L. Pauling, "The Nature of the Chemical Bond—1992," *J. Chem. Ed.* **69** (1992): 519.

J. F. Ogilvie, "The Nature of the Chemical Bond—1990," *J. Chem. Ed.* **67** (1990): 280.

M. Laing, "No Rabbit Ears on Water," *J. Chem. Ed.* **64** (1987): 124.

R. B. Martin, "Localized and Spectroscopic Orbitals: Squirrel Ears on Water," *J. Chem. Ed.* **65** (1988): 688.

E. R. Scerri, "Have Orbitals Really Been Observed?" *J. Chem. Ed.* **77** (2000): 1492.

Chemistry Explorers

Jonathan Kenny Looks at Spectroscopy in the Real World

Professor Jonathan Kenny, who has been on the faculty at Tufts University in Medford, Massachusetts, since 1981, earned his B.S. degree at the University of Notre Dame and his Ph.D. at the University of Chicago.

Professor Kenny's research focuses on the use of spectroscopy to analyze samples in the real world. One project involves a vehicle that uses a hydraulic press to insert a small-diameter rod into the ground to depths of 50 m or more. A sapphire window near the tip of the rod allows spectroscopic signals to pass through. The goal is to detect and analyze groundwater contaminants. This type of analysis presents major challenges because of the number of species present and the complex nature of the soil.

Another project of Kenny's is the development of a fingerprinting method for ballast water in ships based on multidimensional fluorescence spectroscopy. Ships involved in international trade often transport aquatic nuisance species in their ballast water, and the U.S. Coast Guard must enforce existing regulations that require all ships entering U.S. ports to exchange their ballast water in the deep ocean. Kenny and his students demonstrated that their fluorescence-based method can accurately identify water samples from different ports. Professor Kenny is also interested in developing innovative ways to help students learn chemistry in the context of related sciences such as astrophysics and biology. He teaches an environmental chemistry course for liberal arts students organized around the cosmic evolution time line.

Courtesy of Tim Kenny/Jonathan E. Kenny

Jonathan Kenny.

And even the most fundamental ideas about bonding that we have discussed in this text are subject to controversy. For example, Bacskay, Reimers (University of Sydney, Australia), and Nordholm (Göteburg University, Sweden) have suggested* that the simple "electrostatic model" presented here neglects the significance of changes in electron kinetic energies and atomic orbital contractions. The point here is not to present these more complex concepts but to make sure that you appreciate the nature of models and that you recognize the difference between the information learned about matter from experiments and the information learned from theories. Because matter is so complex, we often tailor our theories to explain a limited area of nature so as to keep the theories simple enough to understand and use. However, it is very dangerous to assume that such a limited theory can then be used in a wider sense. We must be very careful to ask proper questions of our theories or we face the danger of being greatly misled.

14.7 Molecular Spectroscopy: An Introduction

Spectroscopy can be defined as the study of the interaction of electromagnetic radiation with matter. Spectroscopy provides a nondestructive and highly sensitive method for obtaining information about the identity, structure, and properties of substances. We have already discussed the emission spectrum of the hydrogen atom (Chapter 12) and the importance of the information it provides about hydrogen's quantized energy levels. Spectroscopy is also very useful for the study of molecules. Molecules can absorb electromagnetic radiation

*Bacskay, Reimers, and Nordholm, *J. Chem. Ed.* 74 (1997), 1494.

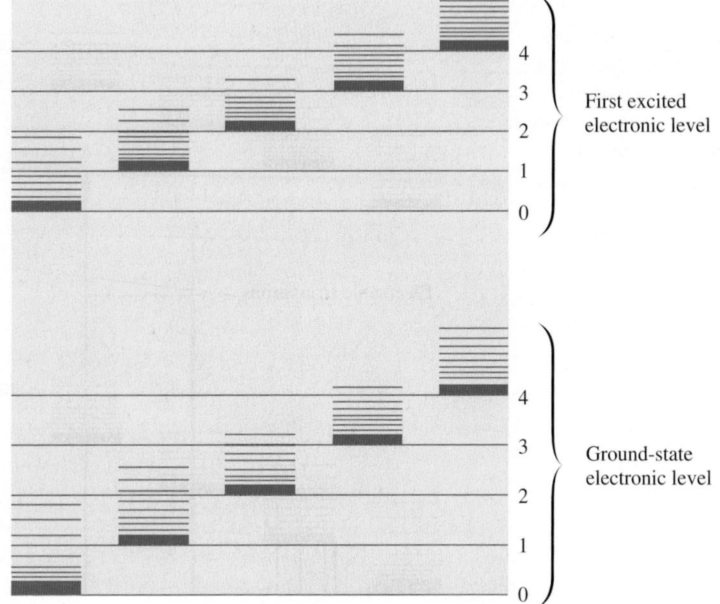

Figure 14.52
A schematic representation of two electronic energy levels in a molecule, with the vibrational (in red) and rotational (in blue) energy levels shown for each electronic state.

to furnish energy for many different processes, all which have quantized energy levels. For example, a molecule can absorb or emit a photon and go from a lower electronic energy state to a higher electronic energy state, or vice versa. This so-called *electronic transition* can be described approximately as a change from one electron arrangement to another. Typically, electronic transitions require photons in the ultraviolet or visible regions of the spectrum. In addition to undergoing electronic transitions, molecules can undergo vibrational and rotational energy transitions. For example, the atoms in a molecule vibrate around their equilibrium positions, giving rise to quantized vibrational energy levels whose spacings correspond to the energies of photons in the infrared region of the electromagnetic spectrum. Molecules also rotate and have quantized rotational energy levels whose spacings correspond to the energies of photons in the microwave region. In fact, it is the rotational excitation of the water molecules in food and the transfer of this energy to other molecules that form the basis for microwave cooking.

A schematic representation of two molecular electronic states with the accompanying vibrational and rotational energy levels is given in Fig. 14.52. Note that each electronic energy level has a set of quantized vibrational states and that each vibrational state has a set of rotational states.

Depending on the energy of the photons interacting with it, a molecule can undergo a pure rotational transition (change rotational levels but remain in the same vibrational and electronic states), a vibrational transition that may also involve a simultaneous rotational transition (Fig. 14.53), or an electronic transition that may involve a simultaneous vibrational and/or rotational change. The details of the geometric structure and electronic structure of the molecule determine exactly what types of transitions can occur. We will not deal with these issues here because they are beyond the scope of this text.

14.8 Electronic Spectroscopy

The electronic spectrum of a molecule, which typically occurs in the ultraviolet or visible region of the electromagnetic spectrum, provides information about the spacings of the electronic energy levels in the molecule. The electronic spec-

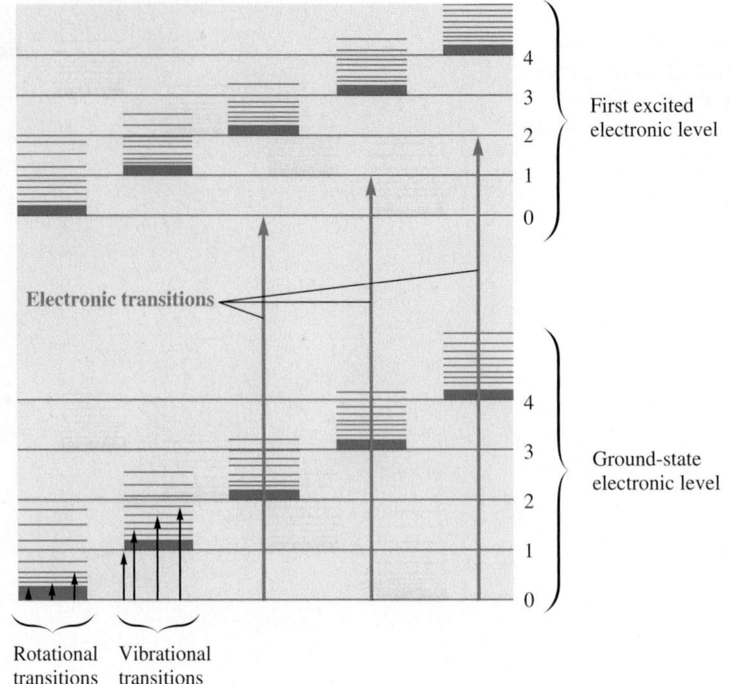

Figure 14.53
The various types of transitions are shown by vertical arrows. Note that rotational changes are lowest in energy, followed by vibrational changes and then electronic changes, which require the highest-energy photons.

trum plots the quantity of radiation absorbed versus the wavelength of the radiation, showing peaks (maxima) at wavelengths where the photons have an energy that matches an energy gap in the molecule. The electronic spectrum corresponding to the electronic transitions shown in Fig. 14.53 is represented* in Fig. 14.54.

Electronic transitions from bonding to antibonding molecular orbitals are often encountered. In this case the potential energy curve for the ground state will be quite different from that for the excited state because less bonding electron density is found in the excited state. An example is the transition from the ground state of NO^+ ion to the first excited state, as shown in Fig. 14.55.

The majority of electronic transitions in molecules occurs in the UV region of the spectrum because the energy separations of electron states typically cor-

A technician using an electron spectroscopy unit used for chemical analysis.

*In the gas phase, various rotational levels in the ground vibrational state will be populated, and transitions to various rotational levels in the excited state will occur, giving rise to a fine structure in the spectrum. This fine structure is absent in solutions because collision of the solute with a solvent molecule occurs before a rotation is completed. Rotational fine structure will not be discussed here.

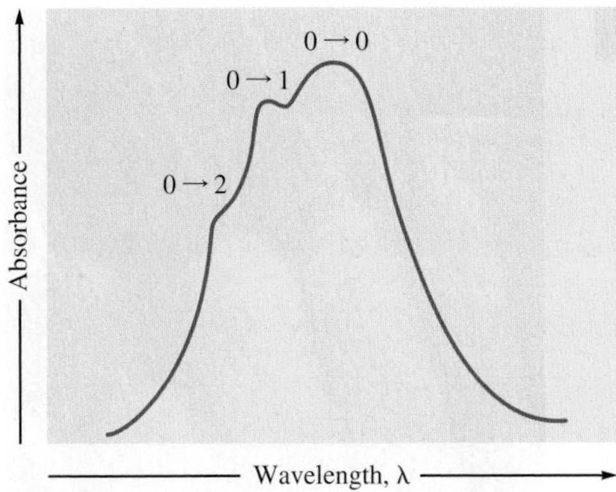

Figure 14.54
Spectrum corresponding to the changes indicated in Fig. 14.53. The transitions correspond to the vibrational changes $\nu = 0 \rightarrow \nu' = 0$, $\nu = 0 \rightarrow \nu' = 1$, and $\nu = 0 \rightarrow \nu' \rightarrow 2$.

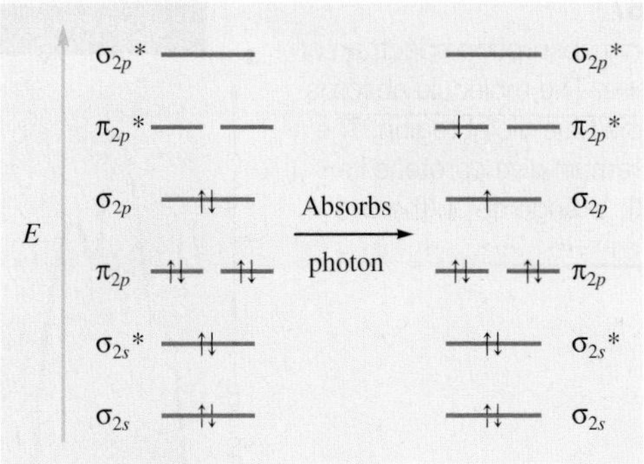

Figure 14.55
The molecular orbital diagram for the ground state of NO^+ (on the left) and the first excited state of NO^+ (on the right). This transition involves a bonding electron being transferred to an antibonding orbital.

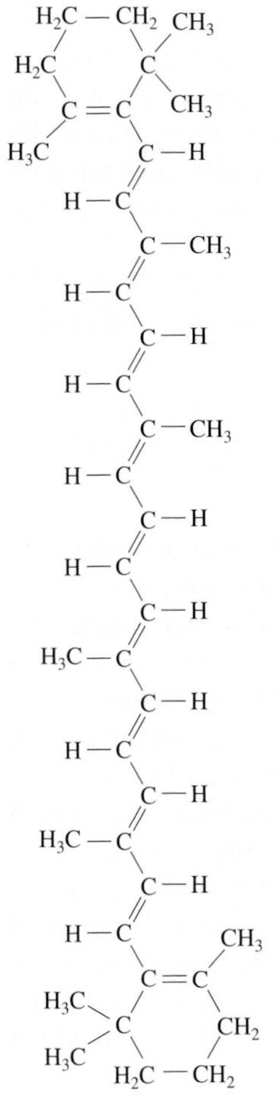

respond to the energies of photons in the UV region. However, some molecules have electronic energy separations that correspond to radiation in the visible region (Fig. 12.3). One such class of compounds includes the coordination compounds that contain transition metal ions (see Section 19.6). Another class of compounds that absorbs in the visible region of the spectrum involves molecules with long chains of carbon molecules that have alternating double bonds, such as carotene (Fig. 14.56). Substances with alternating double bonds are called conjugated molecules. As a first approximation, these molecules can be described in terms of a particle in a box (Section 12.6). Note from Section 12.6 that the energy of the particle (in this case an electron in a conjugated *pi* system) is inversely proportional to the length of the box L. As the box gets longer, the energy separations decrease. This is exactly what happens with a conjugated molecule. As the conjugated system gets longer, the electron energies get closer together, and the light absorbed corresponds to longer wavelengths.

White light consists of all of the colors of the rainbow (Fig. 12.3), so when particular colors are absorbed by a given molecule, the substance appears colored. The color that results is the one given by the "sum" of the colors that remain unabsorbed. For example, carotene absorbs visible light in the violet and blue regions (Fig. 14.57). It appears orange in color because the colors not absorbed "add up" to produce orange. In fact, carotene is the substance that give carrots their bright orange color.

Often when a molecule absorbs a photon, the resulting excited state is much more chemically reactive than is the molecule in its ground state. The study of the chemistry of these electronically excited molecules is called *photochemistry.* A particularly important area of photochemistry concerns the study of the reactions that occur in the earth's atmosphere. For example, photochemical smog depends to a large extent on the photodecomposition of nitrogen dioxide to form very reactive oxygen atoms that combine with oxygen molecules in the air to form ozone, which then goes on to react with other pollutants. Ironically, the problems of ozone *depletion* in the upper atmosphere also result from a photochemical reaction—in this case the photodecomposition of chlorofluorocarbons such as CF_2Cl_2 to produce reactive chlorine atoms, which catalyze ozone decomposition. We will discuss this situation in more detail in Chapter 15.

Figure 14.56
The molecular structure of beta-carotene. The double bonds alternate with single bonds in this molecule, called a conjugated bonding system.

Lasers, which are discussed in more detail in Chapter 16, have assumed a very important role in electronic spectroscopy in recent years for two reasons: They can deliver large quantities of energy at very precise wavelengths, and the

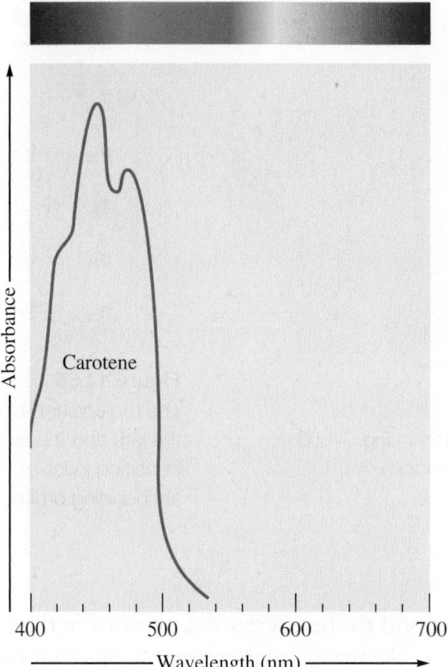

Naturelle/Alamy Images

Figure 14.57
The electronic absorption spectrum of beta-carotene. The molecule absorbs strongly in the blue-violet region. The colors that remain give carotene its characteristic orange color (the dye in carrots).

power can be delivered in very short bursts—on the order of only 10^{-15} second (one femtosecond) in duration. This allows the excitation of molecules with an ultrashort pulse of energy, after which the molecule can be observed as it decays back to the ground state by various pathways. These techniques are now being used to elucidate the mechanisms of chemical reactions, as we will discuss in Chapter 15.

Electronic spectroscopy also provides a sensitive and accurate method for determining the amount of absorbing species present in a sample (quantitative analysis). This technique is described in Appendix 3.

14.9 Vibrational Spectroscopy

As we have seen, a molecule can be approximated as a collection of atoms held together by bonds. In describing the vibrations in a molecule, we can compare the bond between a given pair of atoms to a spring attached to two masses. As the atoms move apart in a vibrational motion, the bond—like a spring—provides a restoring force that pulls the atoms back toward each other.

For example, for a diatomic molecule, the potential energy of the system as a function of the distance between the atoms can be represented as shown in Fig. 14.58. In this figure R_e represents the equilibrium bond distance, which occurs at the minimum potential energy. The potential energy increases if the atoms are moved closer together than R_e because of the repulsive forces generated by the positively charged nuclei. The increase in the potential energy as the bond is stretched represents the energy needed to overcome the bonding force generated by the shared electrons. For small displacements from R_e, the bond can be treated as a harmonic oscillator (red curve in Fig. 14.58) in which the force F is given by

$$F = -k(R - R_e)$$

where k is the force constant and R is the actual bond length. When the stretched bond is released, if the system behaves like a classical harmonic os-

Potential energy
D_e
R_e
Actual bond length (R)

Figure 14.58
The potential curve for a diatomic molecule (blue) where R_e represents the equilibrium bond distance and D_e is the bond dissociation energy. The parabolic curve (red) represents the behavior of a true harmonic oscillator.

cillator, the atoms will oscillate around R_e with a frequency v given by the equation

$$v = \frac{1}{2\pi}\sqrt{\frac{k}{\mu}}$$

in which μ is the reduced mass:

$$\mu = \frac{(m_1)(m_2)}{m_1 + m_2}$$

where m_1 and m_2 represent the masses of the atoms connected by the bond. In classical physics the oscillations could have any frequency. However, like all of the energies in a molecule, the vibrational energy levels are actually quantized. As a consequence, only certain, discrete frequencies are allowed. The energy levels that correspond to the allowed frequencies are given by the following equation:

$$E_\nu = hv_0\left(\nu + \frac{1}{2}\right)$$

where v_0 = the characteristic frequency of the vibration*

ν = the vibrational quantum number, which can assume only the values: 0, 1, 2, 3, . . .

This equation assumes harmonic behavior where the vibrational levels are equally spaced (separated by an energy equivalent to v_0). However, bonds do not behave exactly like springs (for example, the restoring force is weaker for a bond than is predicted for a spring at greatly stretched bond lengths). The actual potential energy curve (called the Morse potential) for a bond is represented by the blue curve in Fig. 14.58 and is shown in detail in Fig. 14.59. Note that the vibrational energy spacings on the Morse potential get smaller with increasing v, leading to "anharmonic" behavior.

For purposes of determining the force constant for a particular bond, the energy required for the $\nu = 0$ to $\nu = 1$ transition is used (this is assumed to be v_0). This procedure is illustrated in Example 14.9. Vibrational transitions in molecules typically require energies that correspond to the infrared region of the electromagnetic spectrum. The data are often represented in "wave numbers"; a wave number is the reciprocal of the wavelength (in cm) required to cause the vibrational transition.

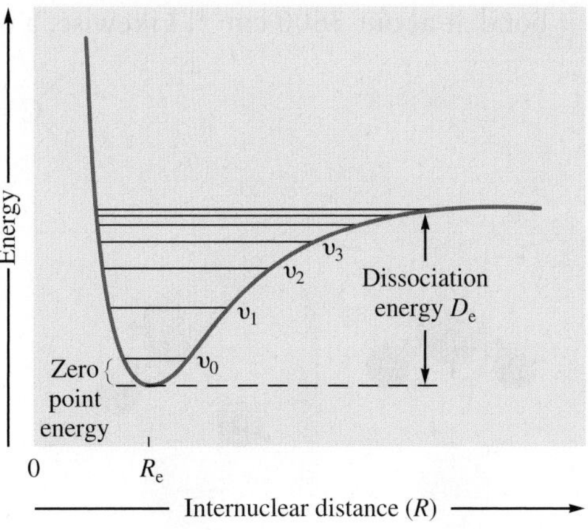

Figure 14.59
Morse energy curve for a diatomic molecule.

*Notice that the vibrational energy is not zero when $\nu = 0$. The energy $hv_0/2$ corresponding to $\nu = 0$ is the "zero-point energy" of the molecule.

EXAMPLE 14.9

The infrared spectrum of gaseous HCl ($^1H^{35}Cl$) shows the $\nu = 0$ to $\nu = 1$ transition at 2885 cm^{-1}. Calculate the vibrational force constant for the HCl molecule.

Solution The fundamental vibrational frequency for HCl is

$$\frac{1}{2\pi}\sqrt{\frac{k}{\mu}} = \upsilon_0 = \frac{c}{\lambda} = (2.998 \times 10^{10}\ \text{cm s}^{-1})(2885\ \text{cm}^{-1})$$

so

$$\frac{1}{2\pi}\sqrt{\frac{k}{\mu}} = 8.649 \times 10^{13}\ \text{s}^{-1}$$

where

$$\mu = 1.627 \times 10^{-27}\ \text{kg for } ^1\text{H}^{35}\text{Cl (See Example 14.10.)}$$

Thus

$$\begin{aligned} k &= 4\pi^2(1.627 \times 10^{-27}\ \text{kg})(8.649 \times 10^{13}\ \text{s}^{-1})^2 \\ &= 4.80 \times 10^2\ \text{kg s}^{-2} = 4.80 \times 10^2\ \text{J m}^{-2} \\ &\qquad = \text{force constant for the H—Cl bond} \end{aligned}$$

The determination of the force constant for a particular bond can give information about the bond strength. Strong springs have relatively large force constants, and the same holds true for strong bonds.

The analysis given thus far applies to diatomic molecules. For simple polyatomic molecules, several characteristic vibrational motions, called normal modes, are possible. As an example, the normal modes of vibration for the SO_2 molecule are shown in Fig. 14.60.

A particular bonded pair of atoms has a characteristic vibrational frequency (wave number) that is relatively insensitive to its molecular environment. Thus a signal that appears in the infrared (IR) spectrum at that characteristic frequency provides good evidence that this particular atom pair is present in the molecule. For example, a C—H pair in a molecule will always show a vibrational signal at about 3000 cm^{-1} (the range of wave numbers is actually 2850–3300 cm^{-1}, depending on the specific molecular environment). On the other hand, the O—H group in a molecule will show a vibrational bond at about 3600 cm^{-1}. Likewise, a carbonyl group

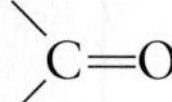

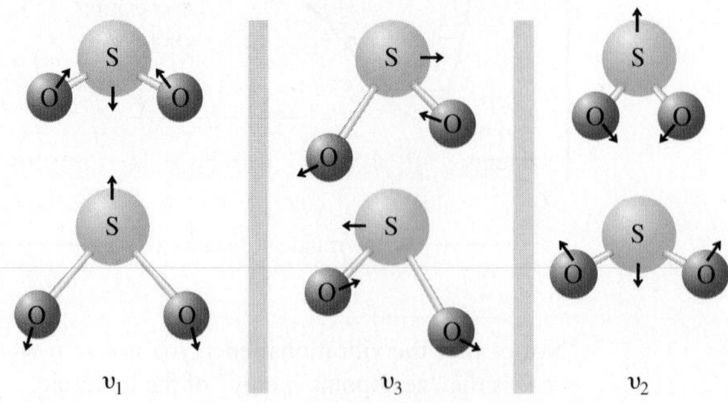

Figure 14.60
The three fundamental vibrations for sulfur dioxide. Note that the two structures for each fundamental vibration show the extremes of the motions of the atoms. (The amplitudes are exaggerated to illustrate the motion.)

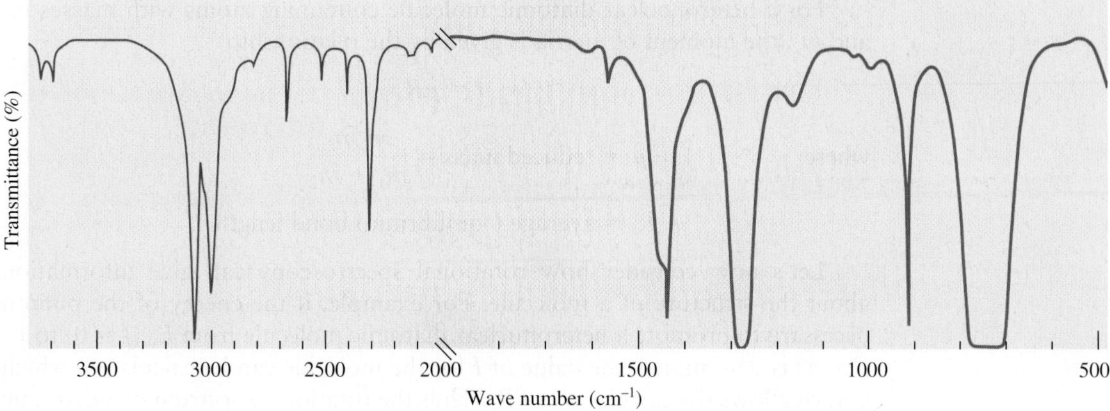

Figure 14.61
The infrared spectrum of CH_2Cl_2. (The wave number scale changes on this spectrum at 2000 cm^{-1}.)

always shows a distinctively shaped vibrational band at about 1700 cm^{-1}. A carbon–carbon double bond

$$>C=C<$$

shows a vibration at about 1650 cm^{-1}, and so on. The IR spectrum of a molecule can be a great aid in identifying which atom groupings are present in a molecule and thus can provide valuable information for identifying a specific molecule. A typical IR spectrum is shown in Fig. 14.61. Note that for this spectrum percent transmittance ($\%T$) is plotted instead of absorbance. Absorbance (A) and $\%T$ are related as follows:

$$A = -\log\left(\frac{\%T}{100}\right)$$

This causes the "peaks" in a vibrational spectrum to be pointed in the opposite direction as those for an electronic spectrum.

Typical frequency ranges for the stretching motions of several common bonds are shown in Table 14.1. The exact frequency for a given bond in a molecule depends on its molecular environment.

Table 14.1

Characteristic Stretching Frequency Ranges for Several Common Types of Bonds

Bond	Frequency Range (cm^{-1})
C—H	2850–3300
C=C	1640–1680
C≡C	2100–2260
C—O	1080–1300
C=O	1690–1760
O—H	3610–3640

14.10 Rotational Spectroscopy

Because the spacings of the quantized rotational energy levels for a molecule depend on its molecular structure, the experimental determination of a molecule's rotational energy levels provides information about its structure. That is, determination of the specific energies of microwave radiation that a given molecule absorbs yields direct information about the rotational energy spacings, which in turn gives information about the details of the molecule's structure. For example, for a linear molecule, the energies of the rotational states are given by the formula

$$E_J = \frac{h^2}{8\pi^2 I} J(J+1) = hBJ(J+1)$$

where J = the rotational quantum number, which can assume only integer values and zero ($J = 0, 1, 2, 3, \ldots$)

h = Planck's constant = 6.626×10^{-34} J s

I = the moment of inertia of the molecule

$B = h/(8\pi^2 I)$

For a heteronuclear diatomic molecule containing atoms with masses m_1 and m_2, the moment of inertia is given by the relationship

$$I = \mu R_e^2$$

where μ = reduced mass = $\dfrac{m_1 m_2}{m_1 + m_2}$

R_e = average (equilibrium) bond length

Let's now consider how rotational spectroscopy can give information about the structure of a molecule. For example, if the energy of the photon necessary to promote a heteronuclear diatomic molecule from E_0 ($J = 0$) to E_1 ($J = 1$) is determined, the value of I for the molecule can be calculated, which in turn allows the calculation of R_e. Thus the rotational spectrum of a diatomic molecule provides an accurate method for measuring its average bond length.

Rotational energy levels are very close together. As a consequence, rotational energy transitions are initiated by relatively low-energy (long-wavelength) radiation. Typically, rotational changes are produced by radiation in the microwave region. To absorb radiation and thereby change its rotational energy, a molecule must have a permanent dipole moment. Also, heteronuclear diatomic molecules can change only one rotational level at a time—for example, from a level with $J = 1$ to a level with $J = 2$, from $J = 2$ to $J = 3$, and so on. That is, for these molecules, $\Delta J = +1$.

Thus the allowed absorption frequencies can be derived (in terms of J, B, and I) as follows: Recall that for an energy-level change ΔE,

$$\Delta E = h\upsilon = E_f - E_i$$

where E_f and E_i are the final and initial energies, respectively. Using the equation above for E_J, we have

$$E_f = hBJ_f(J_f + 1)$$

and

$$E_i = hBJ_i(J_i + 1)$$

Therefore,

$$\Delta E = E_f - E_i = hB[J_f(J_f + 1) - J_i(J_i + 1)]$$

Since J_f and J_i can differ only by 1, we know that

$$J_f = J_i + 1$$

Thus

$$\Delta E = hB[(J_i + 1)(J_i + 2) - J_i(J_i + 1)]$$

and

$$\Delta E = 2hB(J_i + 1)$$

We can convert to frequency υ as follows:

$$\upsilon = \frac{\Delta E}{h} = 2B(J_i + 1)$$

Thus, for a given heteronuclear diatomic molecule, the frequencies for the various rotational transitions are separated by $2B$, as shown in Table 14.2.

Example 14.10 illustrates the use of the rotational transition energy to calculate the bond length of a diatomic molecule.

Table 14.2

Several Allowed Rotational Transitions for a Heteronuclear Diatomic Molecule

Transition	Frequency
$J = 0 \rightarrow J = 1$	$2B$
$J = 1 \rightarrow J = 2$	$4B$
$J = 2 \rightarrow J = 3$	$6B$
$J = 3 \rightarrow J = 4$	$8B$
$J = 4 \rightarrow J = 5$	$10B$

EXAMPLE 14.10

The microwave spectrum of $^1H^{35}Cl$ shows that the transition from $J = 0$ to $J = 1$ requires electromagnetic radiation with a wavelength of 4.85×10^{-4} m. Calculate the bond length of the $^1H^{35}Cl$ molecule. (The masses of ^1H and ^{35}Cl are 1.0078 amu and 34.9689 amu, respectively.)

Solution We know that

$$\Delta E = 2hB(J_i + 1) = h\nu = \frac{hc}{\lambda}$$

so

$$\frac{c}{\lambda} = 2B(J_i + 1)$$

where

$$\lambda = 4.85 \times 10^{-4}\ \text{m}$$

$$c = 2.998 \times 10^{8}\ \text{m/s}$$

$$J_i = 0$$

Then

$$\frac{c}{\lambda} = 2B(1) = \frac{2.998 \times 10^{8}\ \text{ms}^{-1}}{4.85 \times 10^{-4}\ \text{m}}$$

and

$$B = 3.09 \times 10^{11}\ \text{s}^{-1}$$

Since

$$B = \frac{h}{8\pi^2 I}$$

then

$$I = \frac{h}{8\pi^2 B} = \frac{6.626 \times 10^{-34}\ \text{J s}}{8\pi^2(3.09 \times 10^{11}\ \text{s}^{-1})} = 2.57 \times 10^{-47}\ \text{kg m}^2$$

where

$$I = \mu R_e^2$$

For $^1H^{35}Cl$,

$$\mu = \frac{m_H m_{Cl}}{m_H + m_{Cl}} = \frac{(1.0078)(34.9689)}{1.0078 + 34.9689} = 0.9796\ \text{amu}$$

Since one amu = 1.661×10^{-27} kg,

$$\mu = 0.9796\ \text{amu} \times 1.661 \times 10^{-27}\ \frac{\text{kg}}{\text{amu}} = 1.627 \times 10^{-27}\ \text{kg}$$

then

$$R_e^2 = \frac{I}{\mu} = \frac{2.57 \times 10^{-47}\ \text{kg m}^2}{1.627 \times 10^{-27}\ \text{kg}}$$

and

$$R_e = 1.26 \times 10^{-10}\ \text{m} = 126\ \text{pm}$$

This represents the measured bond length of the $^1H^{35}Cl$ molecule.

As might be expected, the rotational spectra of polyatomic molecules are more complex than those for diatomic molecules. For example, depending on their structures, polyatomic molecules can have as many as three different moments of inertia. Although more complicated, the analysis of the structures of polyatomic molecules using rotational spectra follows the same principles as we have discussed for diatomic molecules.

14.11 Nuclear Magnetic Resonance Spectroscopy

Nuclear magnetic resonance (NMR) spectroscopy has risen to the same level of importance as electronic, vibrational, and rotational spectroscopy as a tool for studying molecular properties, particularly structural properties. Certain nuclei, such as the hydrogen nucleus (proton), have a nuclear "spin." The "spinning" nucleus* generates a small magnetic field that we will call μ. When placed

*The nucleus does not literally spin, although we often picture it that way. Nuclear "spin," like electron "spin," is simply a name for a property called spin angular momentum.

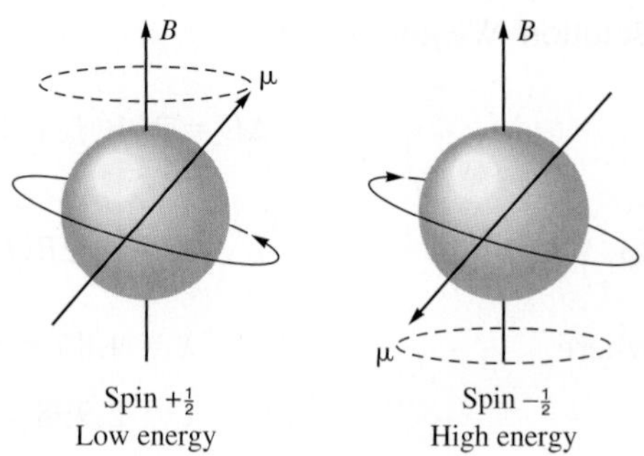

Figure 14.62
Representations of the two spin states of the proton interacting with a large, steady magnetic field *B*.

in a strong, external magnetic field (B), the nucleus can exist in two distinct states: a low-energy state in which μ is aligned with the external magnetic field B and a high-energy state in which μ is opposed to the external magnetic field (Fig. 14.62). A transition from the more stable alignment (with the field) to the less stable alignment (against the field) occurs when the nucleus absorbs electromagnetic energy that is exactly equal to the energy separation between the states (ΔE). This energy gap is very small and corresponds to the radiofrequency range of the electromagnetic spectrum. Because the net absorption energies are so small, a special technique is needed to detect the signal, a condition called resonance. We will not discuss the details of the instrumentation here.

We will focus on three features of the NMR spectrum: the number and size of the signals, the chemical shift, and spin–spin coupling.

Number and Size of Signals

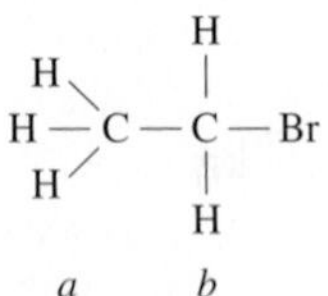

Figure 14.63
The molecular structure of bromoethane. The molecule has two distinct magnetic environments labeled *a* and *b*.

First, we will consider how the NMR spectrometer can distinguish between hydrogen nuclei in different molecular environments and produce multiple signals. Magnetically equivalent protons (hydrogen nuclei in the same local environment) produce one signal. For example, in a molecule such as bromoethane (Fig. 14.63), there are two sets of magnetically equivalent hydrogens. The hydrogens labeled *a* are the three magnetically equivalent methyl hydrogens, and the hydrogens labeled *b* are the two magnetically equivalent methylene hydrogens. Magnetically equivalent protons sense the same local magnetic environment. That is, they occupy equivalent positions in the molecule.

Because of rapid rotations about sigma bonds and molecular symmetry, the three methyl hydrogens are all equivalent to each other. For the same reasons, the two methylene protons are also equivalent to each other. In summary, the methyl hydrogens (*a*) all exist in the same magnetic environment but experience a different magnetic field than the methylene hydrogens (*b*) because of different local magnetic fields. As a result, the resonance energy ΔE corresponding to the frequency of absorption v will be different for these two groups of hydrogen nuclei, producing two distinct signals. Thus the NMR spectrometer can distinguish between groups of hydrogen nuclei that experience different local magnetic fields.

The areas of the signals are directly proportional to the number of hydrogens. In the case of bromoethane, the two peaks have an area ratio of 3:2 because there are three methyl hydrogens (*a*) and two methylene hydrogens (*b*). We would expect a spectrum that looks like the idealized version shown in Fig. 14.64. We will look at an actual spectrum presently.

Chemical Shift

When a molecule is placed in an external magnetic field B, each hydrogen nucleus experiences a total field that is the sum of B and two other local magnetic fields: one produced by the electrons (H_e) and the other produced by neighbor-

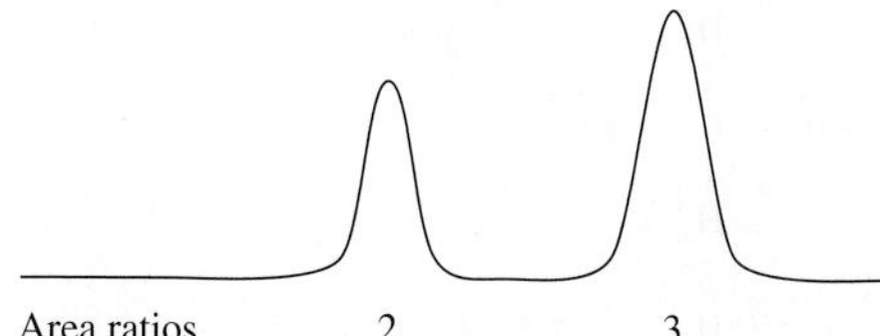

Figure 14.64
The expected NMR spectrum for bromoethane.

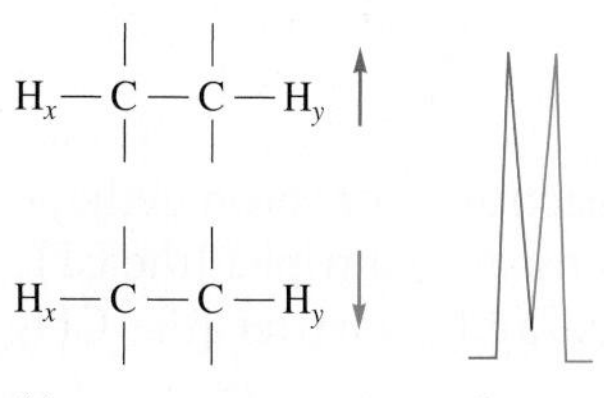

ing protons that possess a nuclear spin (H_h). Thus the applied magnetic field (B) is constant, whereas the magnetic fields H_e and H_h depend on the molecular environment. The magnetic field produced by the neighboring electrons (H_e) determines the position (relative frequency) of an NMR signal, called the chemical shift. The magnetic fields produced by neighboring nuclei (H_h) are smaller than H_e and can cause splitting of the NMR signal, called spin–spin coupling. This effect will be discussed in the next subsection.

The position of an NMR signal is typically recorded relative to the position of the signal of an internal standard. This standard is commonly tetramethyl silane (TMS) [$(CH_3)_4Si$]. Assuming the difference in frequency between TMS and the hydrogen nucleus of interest is Δ, the spectrometer will record a signal at

$$\frac{\Delta}{\upsilon_0} \times 10^6 = \delta$$

where υ_0 is the spectrometer frequency. The symbol δ (delta) represents the position of the signal, the chemical shift, expressed as parts per million (ppm) on a scale of 1–10 ppm. The signal for TMS occurs at 0.0 ppm, whereas the signals for most nuclei are "downfield" toward 10 ppm.

Figure 14.65
(a) The spin of proton H_y can be "up" or "down." (b) This leads to a "doublet" (two peaks with equal areas) as the signal for proton H_x.

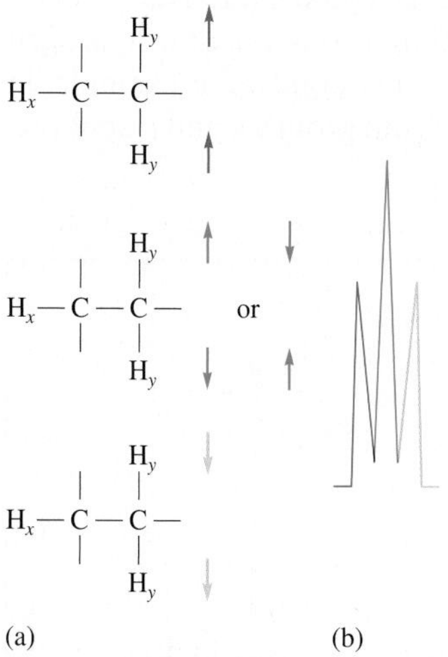

Spin–Spin Coupling

Next, we will consider how the magnetic field produced by neighboring protons (H_h) can cause the splitting of the NMR signal. First, we will consider the signal due to a hydrogen atom (H_x) with only one neighboring hydrogen atom (H_y) (Fig. 14.65). The neighboring hydrogen nucleus will have approximately equal probability of existing in either the "up" or "down" orientation. In other words, half the molecules exist with proton H_y "up" and half will exist with the H_y proton "down." As a result, two different peaks are observed, because H_h for H_x adds and subtracts from the local field "felt" by H_y. This pattern is called a doublet, and we say that the neighboring hydrogen nucleus (H_y) splits the signal of H_x into a doublet. The intensity of the two lines is equal because the probability of the neighboring hydrogen nucleus (H_y) existing in either spin state is approximately equal.

When there are two magnetically equivalent adjacent hydrogen nuclei, three possibilities exist for their combined magnetic fields (Fig. 14.66). Both can have spin "up," one can have spin "up" and the other have spin "down," (there are two ways to achieve this), or both can have spin "down" [Fig. 14.66(a)]. These three possibilities have a probability ratio of 1:2:1, which corresponds to three different signal frequencies. The appearance of the resulting NMR signal for H_x is a three-line pattern, called a triplet, with intensities 1:2:1 [Fig. 14.66(b)].

When there are three magnetically equivalent neighboring hydrogen nuclei (Fig. 14.67), the H_x signal splits into a quartet with intensity pattern 1:3:3:1 for similar reasons to those discussed for the other cases.

Figure 14.66
(a) The spins for protons H_y can both be "up," can be opposed (in two ways), or can both be "down." (b) This leads to a "triplet" signal for proton H_x, with areas 1:2:1 reflecting the ways of achieving the various states shown in (a).

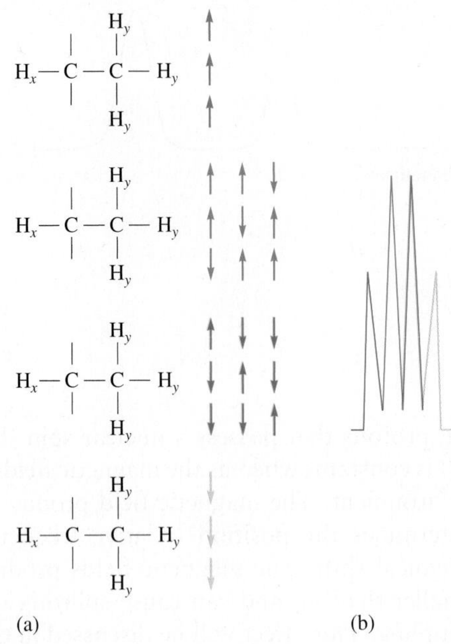

Figure 14.67
The spins for protons H_y can be arranged as shown in (a), leading to four different magnetic environments for proton H_x. This arrangement leads to a "quartet" signal (b) for proton H_x, with areas 1:3:3:1 reflecting the ways of achieving the various states shown in (a).

Next, we will examine the actual NMR spectrum for bromoethane (CH_3CH_2Br), shown in Fig. 14.68. This spectrum consists of a triplet [the CH_3 protons (*a*), split by the two adjacent CH_2 protons] and a quartet [the CH_2 protons (*b*) split by the adjacent CH_3 protons].

Spin–spin coupling is transmitted through the covalent bonds of the molecule by causing polarization of the electron spins. That is, when the proton of a hydrogen bonded to a carbon has spin "up," it causes a slight "unpairing" of the electrons in the bond between the hydrogen and the carbon. As a result, the "spin down" electron is favored in the neighborhood of the proton; the "spin up" electron spends more time near the carbon. This spin polarization, which continues through neighboring bonds, decreases as it gets farther from the original proton spin. Thus spin–spin coupling gets weaker (attenuates) with the number of bonds. In general, spin–spin coupling is observed for hydrogen nuclei that are separated by no more than three sigma bonds. For example, in the molecule 2-butanone (Fig. 14.69), hydrogens from groups *a* and *b* split one

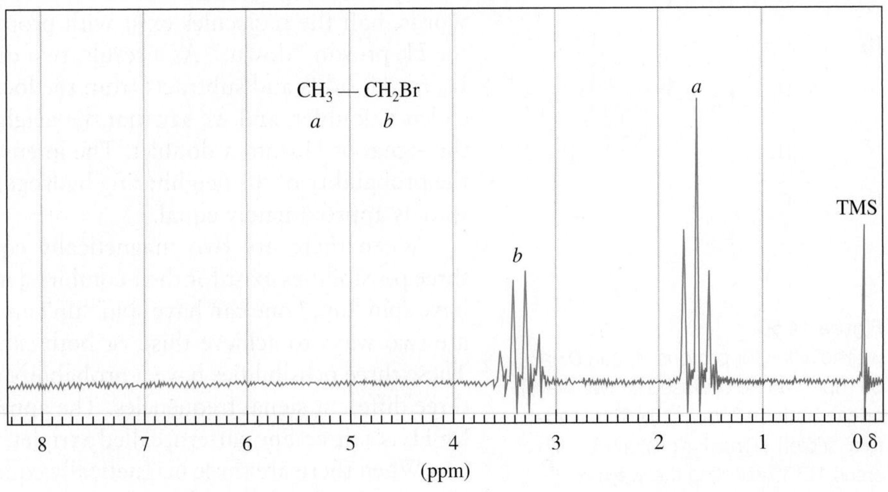

Figure 14.68
The NMR spectrum of CH_3CH_2Br (bromoethane) with TMS as reference.

NMR and Oenology

Chemical Insights

Serious wine collectors can pay many thousands of dollars for a rare vintage wine. For example, a 1946 Chateau Petrus might bring as much as $7000. Buying such a bottle of wine can be risky. It is not uncommon for a wine exposed to oxygen through a bad cork to be spoiled through decomposition of the alcohol to vinegar (acetic acid) or acetaldehyde. In the past, the only way to find out whether the wine was still good was to open the bottle and taste it. Now, however, there is a new technology based on NMR that should allow the detection of excess acetic acid without opening the bottle.

Wine collector Gene Mulvihill has funded the construction of a special NMR spectrometer by chemistry professor Matthew Augstine of the University of California at Davis. The special NMR spectrometer, which costs about $100,000, uses a superconducting magnet that operates near 0 K. The bottle of wine to be evaluated is placed in the magnetic field and the resulting NMR spectrum is analyzed for signals characteristic of the protons in the hydrogen atoms in acetic acid.

Mulvihill has formed a company, Wine Scanner, Inc., to market the services of the wine-analyzing NMR machine to wine collectors and upscale restaurants with expensive wine inventories. *In vino veritas.*

Joshua Lutz/Redux Pictures

Gene Mulvihill and his wine-analyzing NMR spectrometer.

Figure 14.69
(a) The molecule (2-butanone)

$$\mathrm{CH_3CH_2\overset{\displaystyle O}{\overset{\|}{C}}CH_3}$$

has the NMR spectrum shown in (b).

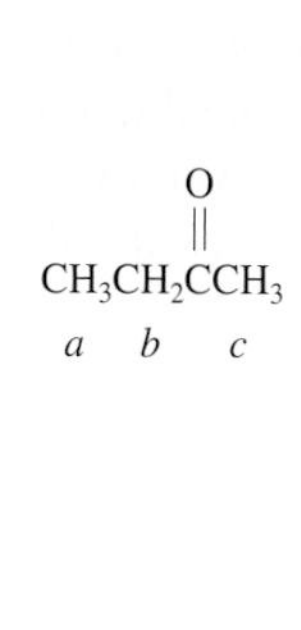

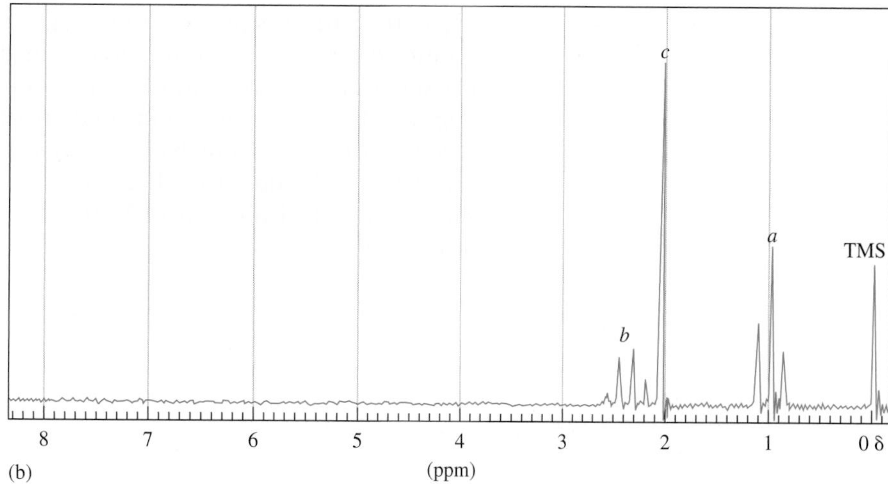

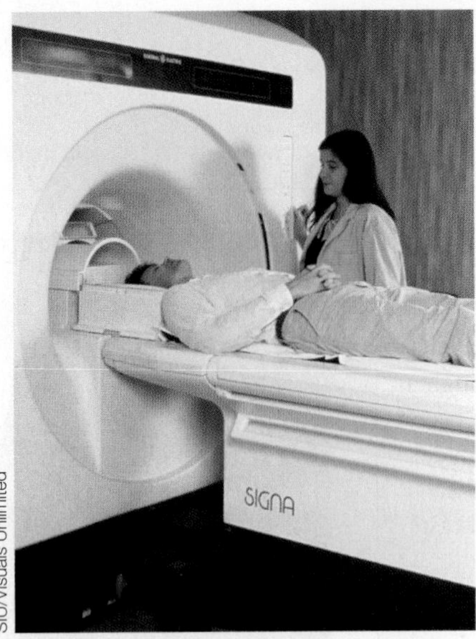

SIU/Visuals Unlimited

Figure 14.70
A technician speaks to a patient before he is moved into the cavity of a magnetic resonance imaging (MRI) machine.

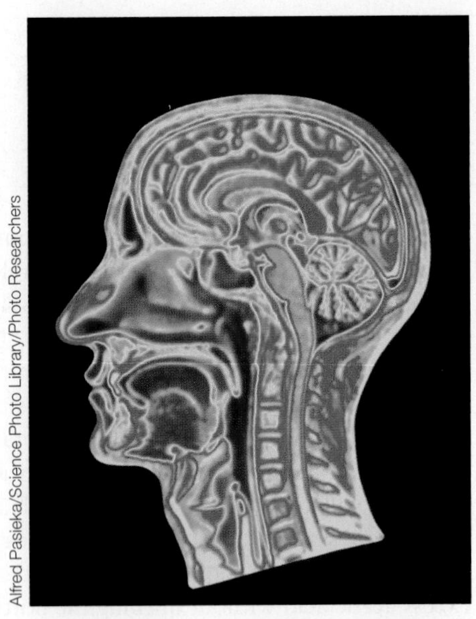

Alfred Pasieka/Science Photo Library/Photo Researchers

Figure 14.71
A colored magnetic resonance imaging (MRI) scan through a human head, showing a healthy brain in side view.

another because they are separated by only three bonds. However, hydrogens from group *c* are not split by any other hydrogens because they are separated from the other hydrogens by more than three sigma bonds. Note also that the coupling is never observed between magnetically equivalent hydrogen atoms. That is, magnetically equivalent protons do not split each other.

NMR spectroscopy has assumed a major role in our attempts to understand the structures and functions of molecules. For example, NMR has developed to the stage where it is possible to determine the three-dimensional structures of protein molecules containing thousands of atoms. One reason that this feat is possible is that other nuclei such as ^{13}C and ^{31}P have spin states that respond to an external magnetic field in much the same way as does ^{1}H. Therefore, NMR spectra of these nuclei can also be obtained, thus providing more structural information. The use of NMR to study protein structure was deemed so important that a Nobel Prize was recently awarded for this work.

Another very important application of NMR, called magnetic resonance imaging (MRI), is widely used as a medical diagnosis tool. In using this technique, the patient is placed in a strong magnetic field (Fig. 14.70) and the responses of protons in the water contained in the various types of tissues are measured and mapped into an image of the part of the body under study (Fig. 14.71). This sensitive and nondestructive method for providing images of various parts of the body has revolutionized medical diagnosis. The discoverers of this technique, Paul Lauterbur of the University of Illinois and Peter Mansfield of the University of Nottingham, were awarded the Nobel Prize for medicine in 2003.

Key Terms

Section 14.1
hybridization
sp^3 hybridization
hybrid orbitals
sp^2 hybridization
sigma (σ) bond
pi (π) bond
sp hybridization
dsp^3 hybridization
d^2sp^3 hybridization

Section 14.2
molecular orbital model
molecular orbital (MO)
sigma (σ) molecular orbital
bonding molecular orbital
antibonding molecular orbital
bond order

Section 14.3
pi (π) molecular orbital
paramagnetism
diamagnetism

Section 14.4
heteronuclear diatomic molecule

Section 14.5
delocalized π bonding

Section 14.7
spectroscopy

For Review

OWL and go Chemistry

Sign in at **www.cengage.com/owl** to:

- View tutorials and simulations, develop problem-solving skills, and complete online homework assigned by your professor.
- Download Go Chemistry mini lecture modules for quick review and exam prep from OWL (or purchase them at **www.cengagebrain.com**)

Two widely used bonding models

- Localized electron model
- Molecular orbital model

Localized electron model

- Molecule is pictured as a group of atoms sharing electron pairs between atomic orbitals
- Hybrid orbitals, which are combinations of the "native" atomic orbitals, are often required to account for the molecular structure
  - Six electron pairs (octahedral arrangement) requires d^2sp^3 orbitals
  - Five electron pairs (trigonal bipyramidal arrangement) requires dsp^3 orbitals
  - Four electron pairs (tetrahedral arrangement) require sp^3 orbitals
  - Three electron pairs (trigonal planar arrangement) require sp^2 orbitals
  - Two electron pairs (linear arrangement) requires *sp* orbitals

Two types of bonds

- Sigma: electrons are shared in the area centered on a line joining the atoms
- Pi: a shared electron pair occupies the space above and below the line joining the atoms

Molecular orbital model

- A molecule is assumed to be a new entity consisting of positively charged nuclei and electrons
- The electrons in the molecule are contained in molecular orbitals, which in the simplest form of the model are constructed from the atomic orbitals of the constituent atoms
- The model correctly predicts relative bond strength, magnetism, and bond polarity
- It correctly portrays electrons as being delocalized in polyatomic molecules
- The main disadvantage of the model is that it is difficult to apply qualitatively to polyatomic molecules

Molecular orbitals are classified in two ways: energy and shape

- Energy
  - A bonding MO is lower in energy than the atomic orbitals from which it is constructed. Electrons in this type of MO are lower in energy in the molecule than in the separated atoms and thus favor molecule formation.
  - An antibonding MO is higher in energy than the atomic orbitals from which it is constructed. Electrons in this type of MO are higher in energy in the molecule than in the separated atoms and thus do not favor molecule formation.
- Shape (symmetry)
  - Sigma (σ) MOs have their electron probability centered on a line passing through the nuclei
  - Pi (π) MOs have their electron probability concentrated above and below the line connecting the nuclei

Bond order is an index of bond strength

$$\text{Bond order} = \frac{\text{number of bonding electrons} - \text{number of antibonding electrons}}{2}$$

Molecules that require the concept of resonance in the localized electron model can be more accurately described by combining the localized electron and molecular orbital models

- The σ bonds are localized
- The π bonds are delocalized

Spectroscopy

- Provides a method for obtaining information about the identity, structure, and properties of substances.
- Electronic spectroscopy: provides information about the spacings of the electronic energy levels.
- Vibrational spectroscopy: typically in the infrared region (IR spectroscopy); gives evidence for a particular atom pair in a molecule.
- Rotational spectroscopy: provides information about the structure of a given molecule.
- Nuclear magnetic resonance (NMR) spectroscopy: allows for the understanding of the structures and functions of molecules.

Discussion Questions

These questions are designed to be considered by groups of students in class. Often these questions work well for introducing a particular topic in class.

1. What are molecular orbitals? How do they compare with atomic orbitals? Can you tell by the shape of bonding orbitals and antibonding orbitals which is lower in energy? Explain.
2. Explain the difference between the σ and π MOs for homonuclear diatomic molecules. How are bonding orbitals and antibonding orbitals different? Why are there two π MOs and one σ MO? Why are the π MOs degenerate?
3. Compare Figs. 14.38 and 14.40. Why are they different? Because B_2 is known to be paramagnetic, the π_{2p} and σ_{2p} MOs must be switched from our first prediction. What is the rationale for this? Why might one expect the σ_{2p} to be lower in energy than the π_{2p}? Why can't we use diatomic oxygen to help us decide whether the σ_{2p} or π_{2p} is lower in energy?
4. Which of the following would you expect to be more favorable energetically? Explain.
   a. An H_2 molecule in which enough energy is added to excite one electron from a bonding to an antibonding MO
   b. Two separate H atoms
5. Draw the Lewis structure for HCN. Indicate the hybrid orbitals, and draw a picture showing all the bonds between the atoms, labeling each bond as σ or π.
6. Which is the more correct statement: "The methane molecule (CH_4) is a tetrahedral molecule because it is sp^3 hybridized" or "The methane molecule (CH_4) is sp^3 hybridized because it is a tetrahedral molecule"? What, if anything, is the difference between these two statements?
7. Compare and contrast the MO model with the LE model. When is each useful?
8. What are the relationships among bond order, bond energy, and bond length? Which of these quantities can be measured?

Exercises

OWL Interactive versions of these problems may be assigned in OWL.

A blue exercise number indicates that the answer to that exercise appears at the back of this book and a solution appears in the *Solutions Guide*.

The LE Model and Hybrid Orbitals

9. Why do we hybridize atomic orbitals to explain the bonding in covalent compounds? What type of bonds form from hybrid orbitals, σ or π? Explain.

10. What hybridization is required for central atoms that have a tetrahedral arrangement of electron pairs? A trigonal planar arrangement of electron pairs? A linear arrangement of electron pairs? How many unhybridized p atomic orbitals are present when a central atom exhibits tetrahedral geometry? Trigonal planar geometry? Linear geometry? What are the unhybridized p atomic orbitals used for?

11. Why are d orbitals sometimes used to form hybrid orbitals? Which period of elements does not use d orbitals for hybridization? If necessary, which d orbitals ($3d$, $4d$, $5d$, or $6d$) would sulfur use to form hybrid orbitals requiring d atomic orbitals? Answer the same question for arsenic and for iodine.

12. The atoms in a single bond can rotate about the internuclear axis without breaking the bond. The atoms in a double bond and a triple bond cannot rotate about the internuclear axis unless the bond is broken. Why?

13. Use the LE model to describe the bonding in H_2O.

14. Use the LE model to describe the bonding in H_2CO and C_2H_2. Carbon is the central atom in H_2CO, and C_2H_2 exists as HCCH.

15. The space-filling models of ethane and ethanol are shown below.

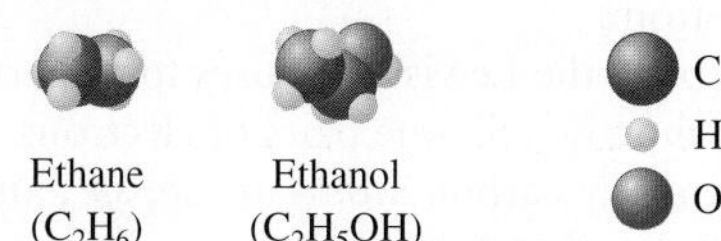

Use the LE model to describe the bonding in ethane and ethanol.

16. The space-filling models of hydrogen cyanide and phosgene are shown below.

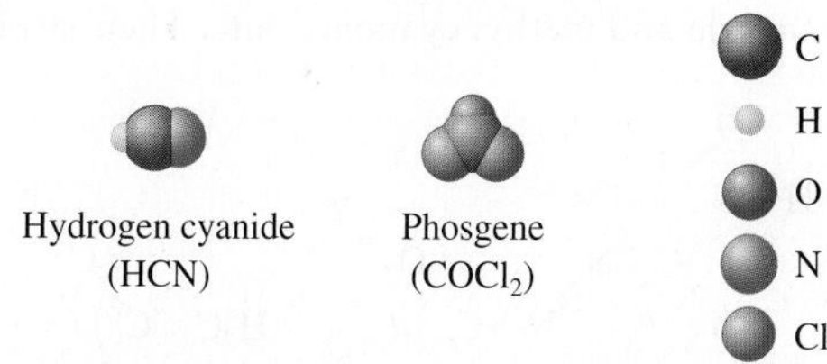

Use the LE model to describe the bonding in hydrogen cyanide and phosgene.

17. Give the expected hybridization of the central atom for the molecules or ions in Exercises 57, 58, and 60 from Chapter 13.

18. Give the expected hybridization of the central atom for the molecules in Exercises 91 and 92 from Chapter 13.

19. Urea, a compound formed in the liver, is one of the ways humans excrete nitrogen. The Lewis structure for urea is

```
   H  :O:  H
   |   ||  |
H—N—C—N—H
   ..      ..
```

Using hybrid orbitals for carbon, nitrogen, and oxygen, determine which orbitals overlap to form the various bonds in urea.

20. Give the expected hybridization for the molecular structures illustrated below.

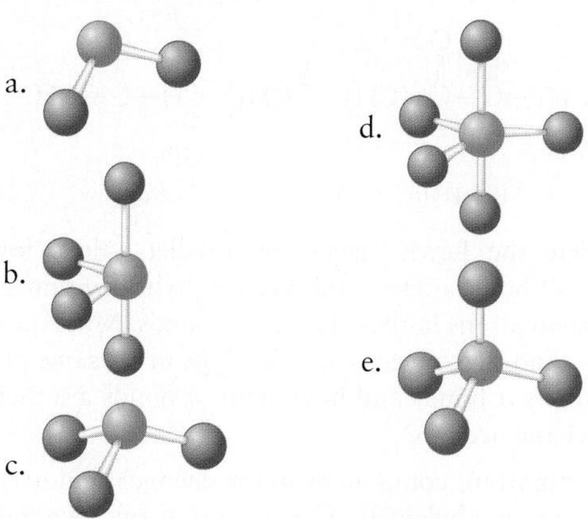

21. For each of the following molecules, write a Lewis structure, predict the molecular structure (including bond angles), give the expected hybrid orbitals on the central atom, and predict the overall polarity.

a. CF_4	g. AsF_5
b. NF_3	h. KrF_2
c. OF_2	i. KrF_4
d. BF_3	j. SeF_6
e. BeH_2	k. IF_5
f. TeF_4	l. IF_3

22. Predict the hybrid orbitals used by the sulfur atom(s) in each of the following. Also predict the molecular structure, including bond angle(s).

a. SO_2

b. SO_3

c. $S_2O_3^{2-}$

```
 [      O     ] 2-
 [      |     ]
 [  S—S—O  ]
 [      |     ]
 [      O     ]
```

d. $S_2O_8^{2-}$

```
 [      O              O      ] 2-
 [      |              |      ]
 [  O—S—O—O—S—O  ]
 [      |              |      ]
 [      O              O      ]
```

e. SO_3^{2-}	h. SF_4
f. SO_4^{2-}	i. SF_6
g. SF_2	j. $F_3S—SF$

23. Why must all six atoms in C_2H_4 be in the same plane?

24. The allene molecule has the following Lewis structure:

```
H          H
 \        /
  C=C=C
 /        \
H          H
```

Must all four hydrogen atoms lie in the same plane? If not, what is the spatial relationship among them? Why?

25. Biacetyl and acetoin are added to margarine to make it taste more like butter.

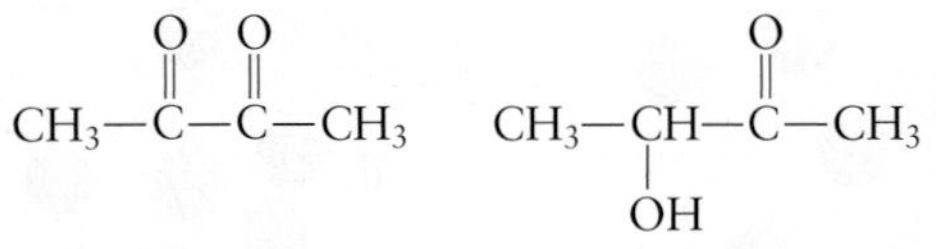

$$CH_3{-}\overset{O}{\overset{\|}{C}}{-}\overset{O}{\overset{\|}{C}}{-}CH_3 \qquad CH_3{-}\underset{OH}{\underset{|}{C}H}{-}\overset{O}{\overset{\|}{C}}{-}CH_3$$

Biacetyl Acetoin

Complete the Lewis structures, predict values for all C—C—O bond angles, and give the hybridization of all the carbon atoms in these two compounds. Must the four carbons and two oxygens in biacetyl lie in the same plane? How many σ bonds and how many π bonds are there in biacetyl and acetoin?

26. Many important compounds in the chemical industry are derivatives of ethylene (C_2H_4). Two of them are acrylonitrile and methyl methacrylate.

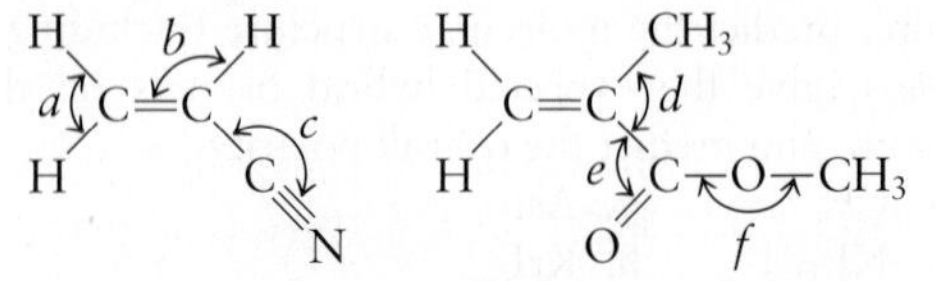

Acrylonitrile Methyl methacrylate

Complete the Lewis structures for these molecules, showing all lone pairs. Give approximate values for bond angles *a* through *f*, and give the hybridization of all carbon atoms. In acrylonitrile and methyl methacrylate indicate which atoms in each molecule must lie in the same plane. How many σ bonds and how many π bonds are there in acrylonitrile and methyl methacrylate?

27. One of the first drugs to be approved for use in treatment of acquired immune deficiency syndrome (AIDS) was azidothymidine (AZT). Complete the Lewis structure of AZT.

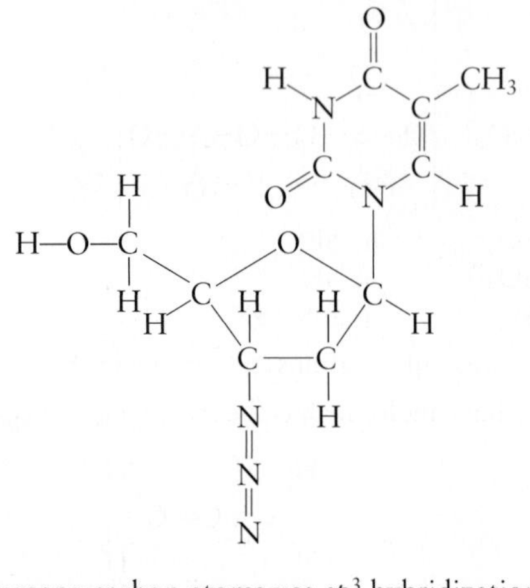

a. How many carbon atoms use sp^3 hybridization?
b. How many carbon atoms use sp^2 hybridization?
c. Which atom is *sp* hybridized?
d. How many σ bonds are in the molecule?
e. How many π bonds are in the molecule?
f. What is the N—N—N bond angle in the azide (—N_3) group?
g. What is the H—O—C bond angle in the side group attached to the five-membered ring?
h. What is the hybridization of the oxygen atom in the —CH_2OH group?

28. Hot and spicy foods contain molecules that stimulate pain-detecting nerve endings. Two such molecules are piperine and capsaicin:

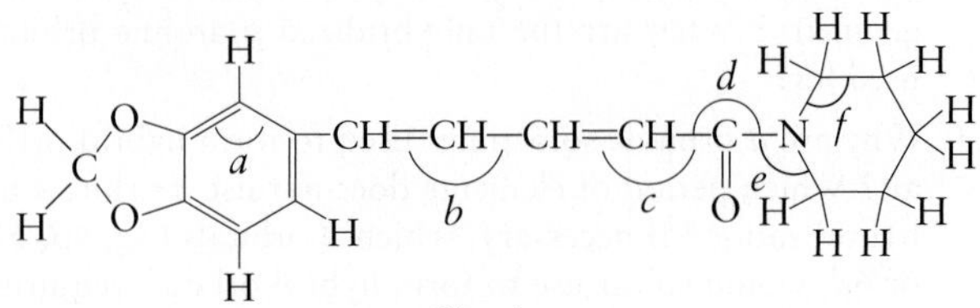

Piperine

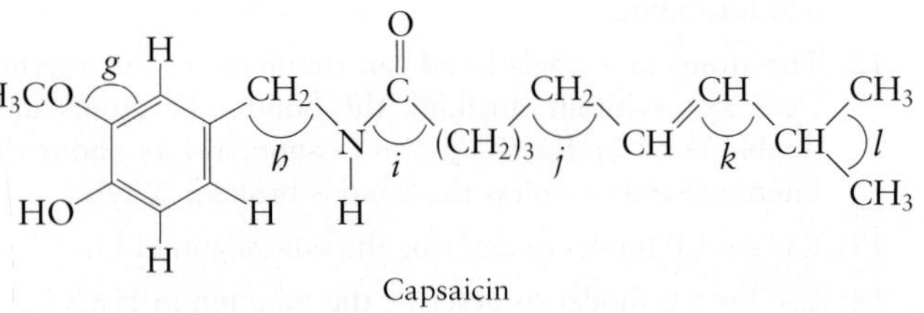

Capsaicin

Piperine is the active compound in white and black pepper, and capsaicin is the active compound in chili peppers. The ring structures in piperine and capsaicin are shorthand notation. Each point where lines meet represents a carbon atom.

a. Complete the Lewis structures for piperine and capsaicin, showing all lone pairs of electrons.
b. How many carbon atoms are *sp*, sp^2, and sp^3 hybridized in each molecule?
c. Which hybrid orbitals are used by the nitrogen atoms in each molecule?
d. Give approximate values for the bond angles marked *a* through *l* in the above structures.

29. Two molecules used in the polymer industry are azodicarbonamide and methyl cyanoacrylate. Their structures are

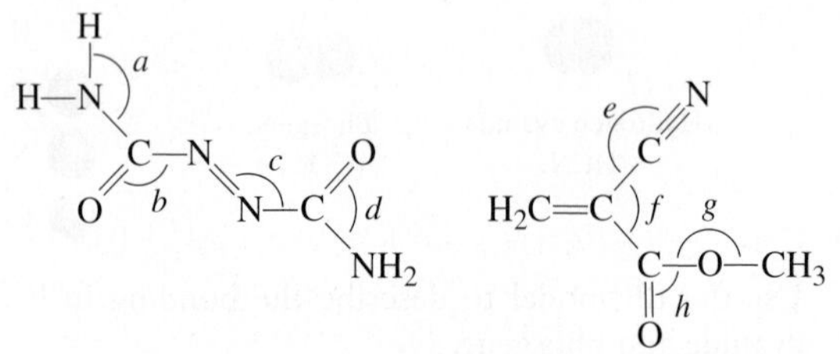

Azodicarbonamide Methyl cyanoacrylate

Azodicarbonamide is used in forming polystyrene. When added to the molten plastic, it decomposes to nitrogen, carbon monoxide, and ammonia gases, which are captured as bubbles in the molten polymer. Methyl cyanoacrylate is the main ingredient in super glue. As the glue sets, methyl cyanoacrylate polymerizes across the carbon–carbon double bond. (See Chapter 21.)

a. Complete the Lewis structures showing all lone pairs of electrons.
b. Which hybrid orbitals are used by the carbon atoms in each molecule and the nitrogen atoms in azodicarbonamide?

c. How many π bonds are present in each molecule?
d. Give approximate values for the bond angles marked *a* through *h* in the above structures.

30. The antibiotic thiarubin-A was discovered by studying the feeding habits of wild chimpanzees in Tanzania. The structure for thiarubin-A is

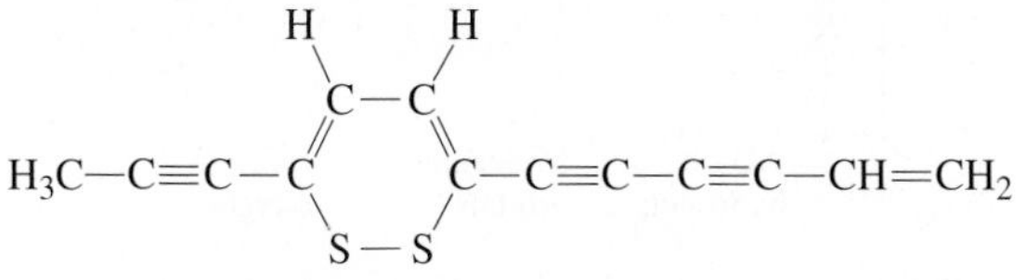

```
             H     H
              \   /
               C—C
              //   \\
H3C—C≡C—C         C—C≡C—C≡C—CH=CH2
              \   /
               S—S
```

a. Complete the Lewis structures showing all lone pairs of electrons.
b. Indicate the hybrid orbitals used by the carbon and sulfur atoms in thiarubin-A.
c. How many σ and π bonds are present in this molecule?

31. The three most stable oxides of carbon are carbon monoxide (CO), carbon dioxide (CO_2), and carbon suboxide (C_3O_2). The space-filling models for these three compounds are

For each oxide, draw the Lewis structure, predict the molecular structure, and describe the bonding (in terms of the hybrid orbitals for the carbon atoms).

The MO Model

32. Consider the following molecular orbitals formed from the combination of two hydrogen 1*s* orbitals:

  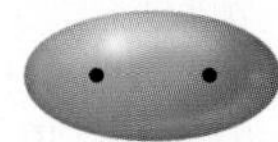

a. Which is the bonding molecular orbital, and which is the antibonding molecular orbital? Explain how you can tell by looking at their shapes.
b. Which of the two molecular orbitals is lower in energy? Why is this true?

33. Sketch the molecular orbital and label its type (σ or π; bonding or antibonding) that would be formed when the following atomic orbitals overlap. Explain your labels.

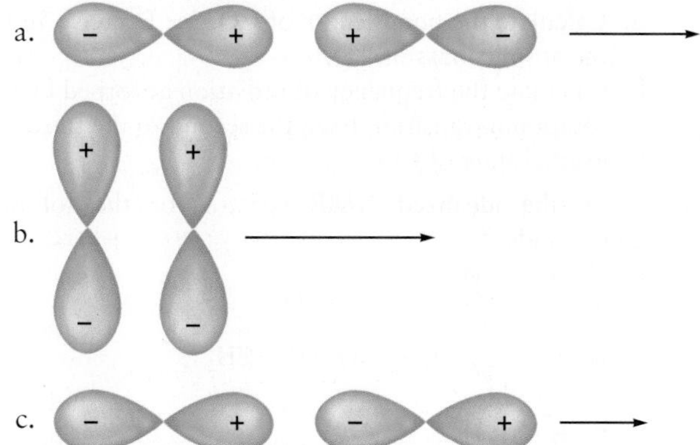

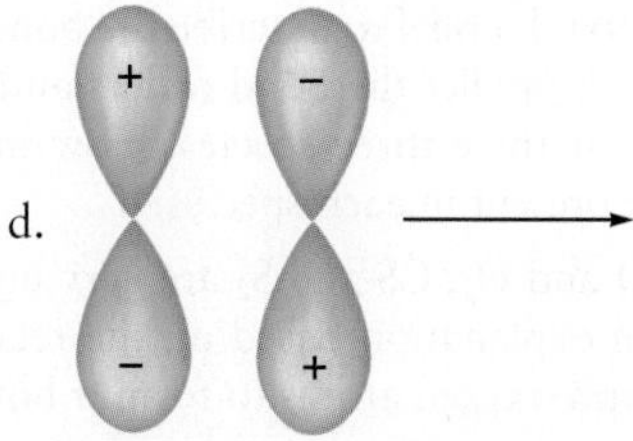

34. What are the relationships among bond order, bond energy, and bond length? Which of these can be measured? Distinguish between the terms *paramagnetic* and *diamagnetic*. What type of experiment can be done to determine if a material is paramagnetic?

35. What modification to the molecular orbital model was made from the experimental evidence that B_2 is paramagnetic?

36. A Lewis structure obeying the octet rule can be drawn for O_2 as follows:

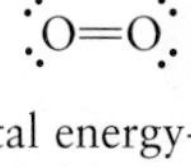

$$:\!\ddot{\text{O}}=\ddot{\text{O}}\!:$$

Use the molecular orbital energy-level diagram for O_2 to show that the above Lewis structure corresponds to an exicted state.

37. How does molecular orbital theory explain the following observations?
a. H_2 is stable, whereas He_2 is unstable.
b. B_2 and O_2 are paramagnetic, whereas C_2, N_2, and F_2 are diamagnetic.
c. N_2 has a very large bond energy associated with it.
d. NO^+ is more stable than NO^-.

38. Why does the molecular orbital model do a better job in explaining the bonding in NO^- and NO than the hybrid orbital model?

39. Show how a hydrogen 1*s* atomic orbital and a fluorine 2*p* atomic orbital overlap to form bonding and antibonding MOs in the hydrogen fluoride molecule. Are they σ or π MOs?

40. Which of the following are predicted by the molecular orbital model to be stable diatomic species?
a. H_2^+, H_2, H_2^-, H_2^{2-}
b. N_2^{2-}, O_2^{2-}, F_2^{2-}
c. Be_2, B_2, Ne_2

41. Which charge(s) for the N_2 molecule would give a bond order of 2.5?

42. In which of the following diatomic molecules would the bond strength be expected to weaken as an electron is removed?
a. H_2 b. B_2 c. C_2^{2-} d. OF

43. In terms of the molecular orbital model, which species in each of the following two pairs will most likely be the one to gain an electron? Explain.

CN or NO O_2^{2+} or N_2^{2+}

44. Using the molecular orbital model to describe the bonding in CO, CO^+, and CO^{2+}, predict the bond orders and the relative bond lengths for these three species. How many unpaired electrons are present in each species?

45. As compared with CO and O_2, CS and S_2 are very unstable molecules. Give an explanation based on the relative abilities of the sulfur and oxygen atoms to form π bonds.

46. Consider the following electron configuration:

$$(\sigma_{3s})^2(\sigma_{3s}^*)^2(\sigma_{3p})^2(\pi_{3p})^4(\pi_{3p}^*)^4$$

Give four species that, in theory, would have this electron configuration.

47. What type of molecular orbital would result from the combination of two d_{xz} atomic orbitals shown below? Assume the x axis is the internuclear axis.

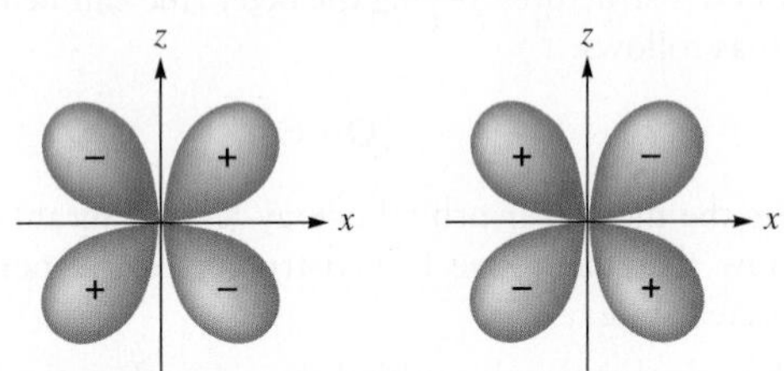

48. Using an MO energy-level diagram, would you expect O_2 to have a lower or higher first ionization energy than atomic oxygen? Why?

49. Use Figs. 14.45 and 14.46 to answer the following questions.
   a. Would the bonding MO in HF place greater electron density near the H or the F atom? Why?
   b. Would the bonding MO have greater fluorine $2p$ character, greater hydrogen $1s$ character, or an equal contribution from both? Why?
   c. Answer parts a and b for the antibonding MO in HF.

50. The diatomic molecule OH exists in the gas phase. OH plays an important part in combustion reactions and is a reactive oxidizing agent in polluted air. The bond length and bond energy have been measured to be 97.06 pm and 424.7 kJ/mol, respectively. Assume that the OH molecule is analogous to the HF molecule discussed in the chapter and that the MOs result from the overlap of a p_z orbital from oxygen and the $1s$ orbital of hydrogen (the O—H bond lies along the z axis).
   a. Draw pictures of the sigma bonding and antibonding molecular orbitals in OH.
   b. Which of the two MOs has the greater hydrogen $1s$ character?
   c. Can the $2p_x$ orbital of oxygen form MOs with the $1s$ orbital of hydrogen? Explain.
   d. Knowing that only the $2p$ orbitals of oxygen interact significantly with the $1s$ orbital of hydrogen, complete the MO energy-level diagram for OH. Place the correct number of electrons in the energy levels.
   e. Estimate the bond order for OH.
   f. Predict whether the bond order of OH^+ is greater than, less than, or the same as that of OH. Explain.

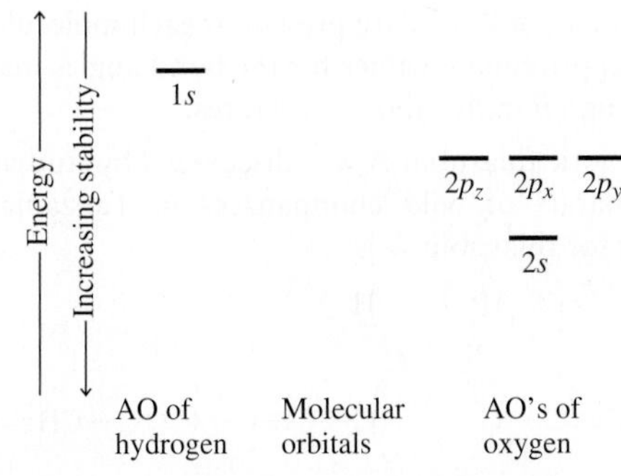

51. What is delocalized π bonding, and what does it explain? Explain the delocalized π bonding system in C_6H_6 (benzene) and SO_2.

52. Describe the bonding in the O_3 molecule and the NO_2^- ion using the LE model. How would the molecular orbital model describe the π bonding in these two species?

53. Describe the bonding in the CO_3^{2-} ion using the LE model. How would the molecular orbital model describe the π bonding in this species?

54. The space-filling model for benzoic acid is shown below.

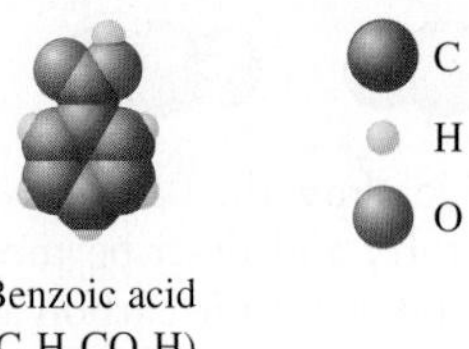

Benzoic acid ($C_6H_5CO_2H$)

Describe the bonding in benzoic acid using the LE model combined with the molecular orbital model.

Spectroscopy

55. The infrared spectrum of $^1H^{79}Br$ shows the $v = 0$ to $v = 1$ transition at 2650. cm^{-1}. Determine the vibrational force constant for the HBr molecule. The atomic mass of ^1H is 1.0078 amu, and the atomic mass of ^{79}Br is 78.918 amu.

56. If the force constant of $^{14}N^{16}O$ is 1550. N m^{-1}, determine the wave number of a line in the infrared spectrum of NO. The atomic mass of ^{14}N is 14.003 amu, and the atomic mass of ^{16}O is 15.995 amu.

57. The microwave spectrum of $^{12}C^{16}O$ shows that the transition from $J = 0$ to $J = 1$ requires electromagnetic radiation with a wavelength of 2.60×10^{-3} m.
   a. Calculate the bond length of CO. See Exercise 56 for the atomic mass of ^{16}O.
   b. Calculate the frequency of radiation absorbed in a rotational transition from the second to the third excited state of CO.

58. Draw the idealized NMR spectra for the following compounds.
   a. (benzene ring with H atoms)—CH_2—O—C(=O)—CH_3

Assume that the five hydrogen atoms in the benzene ring are equivalent with no spin–spin coupling.

b.

$$CH_3-\overset{\overset{\displaystyle O}{\|}}{C}-\underset{\underset{\displaystyle CH_3}{|}}{\overset{\overset{\displaystyle CH_3}{|}}{C}}-CH_3$$

c.

(benzene ring with substituents: CH_2CH_3, H, CH_2CH_3, H, CH_3CH_2, H)

59. Consider the following idealized NMR multiplets.

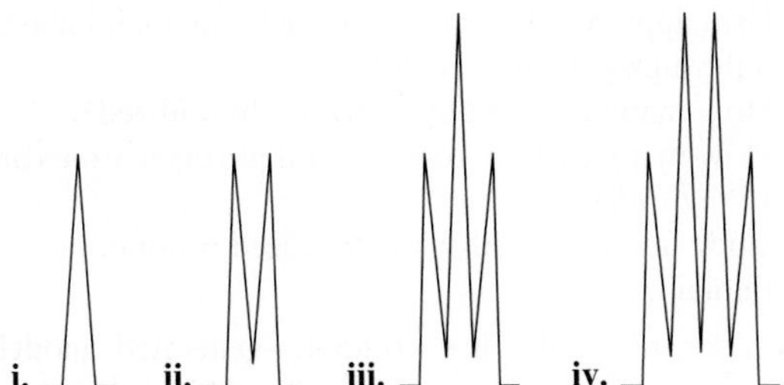

Which is the correct multiplet for the underlined group in the following molecules?

a. $CH_3-\underline{CH_2}-O-CH_2-CH_3$

b. $\underline{CH_3}-O-\overset{\overset{\displaystyle O}{\|}}{C}-CH_2-CH_3$

c. $(CH_3)_3C-O-\overset{\overset{\displaystyle O}{\|}}{C}-\underline{CH_2}-CH_2-CH_3$

d. $\underline{H_2\overset{\overset{\displaystyle F}{|}}{C}}-\overset{\overset{\displaystyle F}{|}}{C}H-CH_2-\overset{\overset{\displaystyle O}{\|}}{C}-O-CH_2-CH_3$

60. The NMR spectra below are for the organic compounds C_6H_{12} and $C_4H_{10}O$. Deduce the structures for these compounds. See Exercise 70 for a discussion of the bonding in organic compounds. The structure of C_6H_{12} has one double bond, with the rest being single bonds, and the structure of $C_4H_{10}O$ has only single bonds. Note that the TMS reference has been omitted in each spectrum.

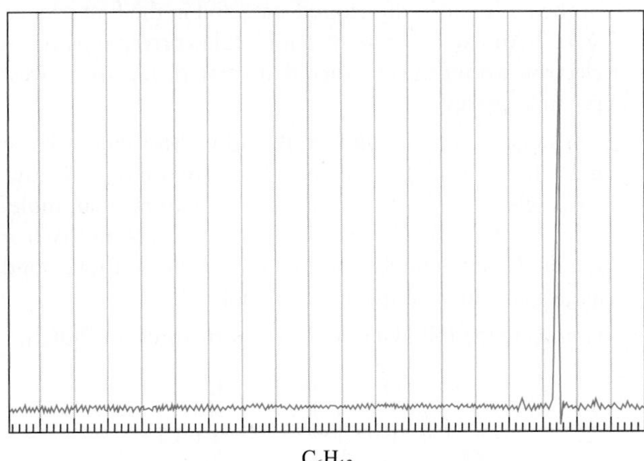

C_6H_{12}

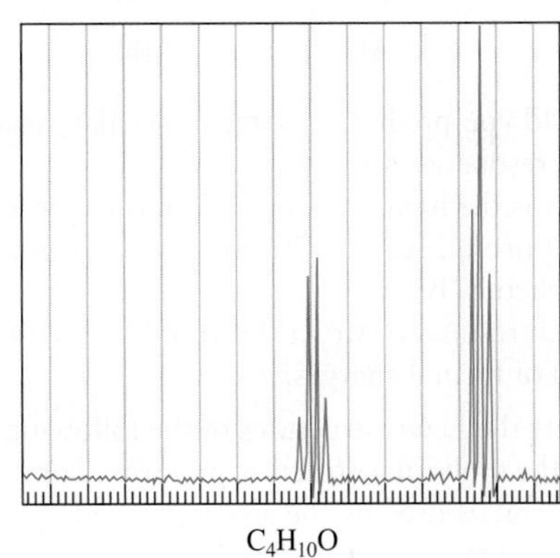

$C_4H_{10}O$

Additional Exercises

61. Draw the Lewis structures, predict the molecular structures, and describe the bonding (in terms of the hybrid orbitals for the central atom) for the following.

a. XeO_3
b. XeO_4
c. $XeOF_4$
d. $XeOF_2$
e. XeO_3F_2

62. $FClO_2$ and F_3ClO can both gain a fluoride ion to form stable anions. F_3ClO and F_3ClO_2 can also lose a fluoride ion to form stable cations. Draw Lewis structures and describe the hybrid orbitals used by chlorine in these four ions.

63. Two structures can be drawn for cyanuric acid:

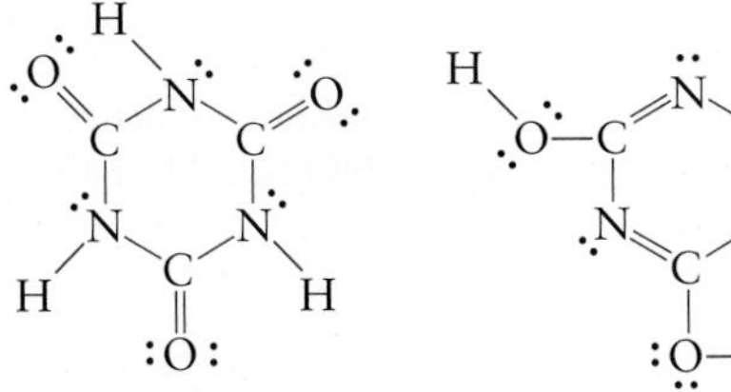

a. Are these two structures the same molecule? Why or why not?

b. Give the hybridization of the carbon and nitrogen atoms in each structure.

c. Use bond energies (Table 13.6) to predict which form is more stable; that is, which contains the strongest bonds?

64. Using bond energies from Table 13.6, estimate the barrier to rotation around a carbon–carbon double bond. To do this, consider what must happen to go from

(Cl)(H)C=C(H)(Cl) to (Cl)(H)C=C(Cl)(H)

in terms of making and breaking chemical bonds; that is, what happens to the π bond?

65. Show how a d_{xz} atomic orbital and a p_z atomic orbital combine to form a bonding molecular orbital. Assume the x axis is the internuclear axis. Is a σ or a π molecular orbital formed? Explain.

66. Describe the bonding in the first excited state of N_2 (the one closest in energy to the ground state) using the MO model. What differences do you expect in the properties of the molecule in the ground state and in the first excited state? (An excited state of a molecule corresponds to an electron arrangement other than that giving the lowest possible energy.)

67. Consider three molecules: A, B, and C. Molecule A has a hybridization of sp^3. Molecule B has two more effective pairs (electron pairs around the central atom) than molecule A. Molecule C consists of two σ bonds and two π bonds. Give the molecular structure, hybridization, bond angles, and an example for each molecule.

68. Complete the following resonance structures for $POCl_3$:

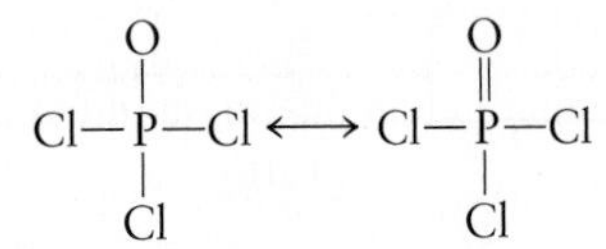

```
      O                O
      |                ‖
  Cl—P—Cl  ⟷   Cl—P—Cl
      |                |
      Cl               Cl
     (A)              (B)
```

a. Would you predict the same molecular structure from each resonance structure?
b. What is the hybridization of P in each structure?
c. What orbitals can the P atom use to form the π bond in structure B?
d. Which resonance structure would be favored on the basis of formal charges?

69. Complete the Lewis structures of the following molecules. Predict the molecular structure, polarity, bond angles, and hybrid orbitals used by the atoms marked by asterisks for each molecule.

a. BH_3 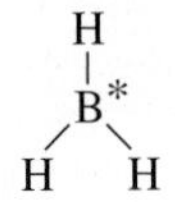
```
      H
      |
      B*
     / \
    H   H
```
b. N_2F_2 F—N*—N*—F

c. C_4H_6 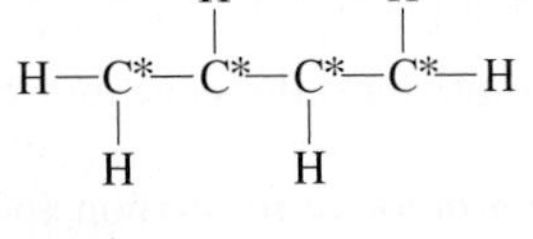
```
      H         H
      |         |
  H—C*—C*—C*—C*—H
      |         |
      H         H
```
d. ICl_3
```
  Cl—I*—Cl
      |
      Cl
```

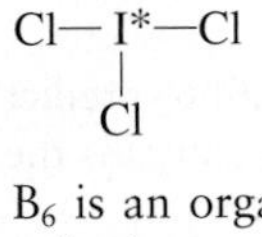

70. Vitamin B_6 is an organic compound whose deficiency in the human body can cause apathy, irritability, and an increased susceptibility to infections. Below is an incomplete Lewis structure, for vitamin B_6. Complete the Lewis structure and answer the following questions. [*Hint:* Vitamin B_6 can be classified as an organic compound (a compound based on carbon atoms). The majority of Lewis structures for simple organic compounds have all atoms with a formal charge of zero. Therefore, add lone pairs and multiple bonds to the structure below to give each atom a formal charge of zero.]

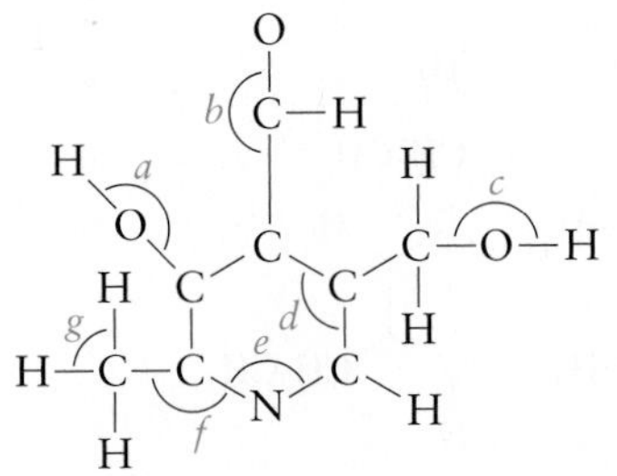

```
                 O
                 |
              b( C—H
 H  a            |         H    c
  O)             C         |
    \          /   \       C—O—H
 H   C          C         |
 g|  |        d/  e        H
 H—C—C    N    C
   |    f        \
   H              H
```

a. How many σ bonds and π bonds exist in vitamin B_6?
b. Give approximate values for the bond angles marked *a* through *g* in the structure.
c. How many carbon atoms are sp^2 hybridized?
d. How many carbon, oxygen, and nitrogen atoms are sp^3 hybridized?
e. Does vitamin B_6 exhibit delocalized π bonding? Explain.

71. Consider the following computer-generated model of caffeine:

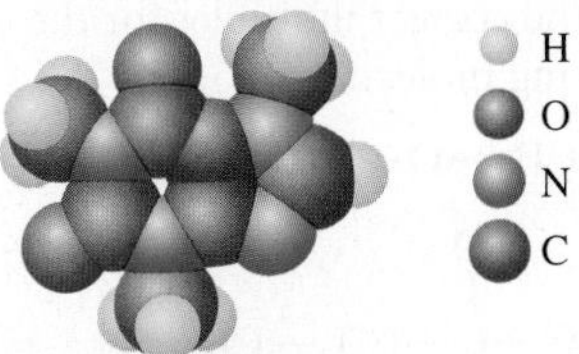

Complete a Lewis structure for caffeine in which all atoms have a formal charge of zero (as is typical with most organic compounds). How many C and N atoms are sp^2 hybridized? How many C and N atoms are sp^3 hybridized? sp hybridized? How many σ and π bonds are there?

72. Aspartame is an artificial sweetener marketed under the name Nutra-Sweet. A partial Lewis structure for aspartame is shown below:

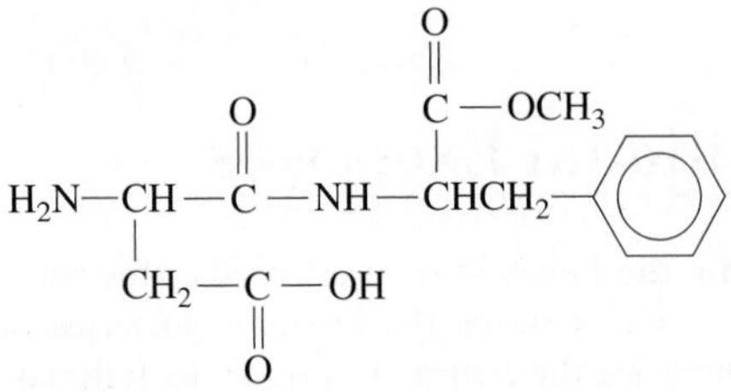

```
                          O
                          ‖
              O           C—OCH3
              ‖           |
   H2N—CH—C—NH—CHCH2—⟨benzene ring⟩
       |
       CH2—C—OH
           ‖
           O
```

Note that the six-sided ring is shorthand notation for a benzene ring (—C_6H_5). Benzene is discussed in Section 14.5. Complete the Lewis structure for aspartame. How many C and N atoms exhibit sp^2 hybridization? How many C and O atoms exhibit sp^3 hybridization? How many σ and π bonds are in aspartame? Aspartame is an organic compound, and the Lewis structure follows the guidelines outlined in Exercise 70.

73. The N_2O molecule is linear and polar.
a. On the basis of this experimental evidence, which arrangement is correct, NNO or NON? Explain your answer.

b. On the basis of your answer in part a, write the Lewis structure of N_2O (including resonance forms). Give the formal charge on each atom and the hybridization of the central atom.

c. How would the multiple bonding in $:N{\equiv}N{-}\ddot{\underset{\cdot\cdot}{O}}:$ be described in terms of orbitals?

74. Values of measured bond energies may vary greatly depending on the molecule studied. Consider the following reactions:

$$NCl_3(g) \longrightarrow NCl_2(g) + Cl(g) \qquad \Delta H = 375 \text{ kJ/mol}$$

$$ONCl(g) \longrightarrow NO(g) + Cl(g) \qquad \Delta H = 158 \text{ kJ/mol}$$

Rationalize the difference in the values of ΔH for these reactions, even though each reaction appears to involve the breaking of only one N—Cl bond. (*Hint:* Consider the bond order of the NO bond in ONCl and in NO.)

Challenge Problems

75. Given that the ionization energy of F_2^- is 290. kJ/mol, do the following.

a. Calculate the bond energy of F_2^-. You will need to look up the bond energy of F_2 and ionization energy of F^-.

b. Explain the difference in bond energy between F_2^- and F_2 using MO theory.

76. Bond energy has been defined in the text as the amount of energy required to break a chemical bond, so we have come to think of the addition of energy as breaking bonds. However, in some cases the addition of energy can cause the formation of bonds. For example, in a sample of helium gas subjected to a high-energy source, some He_2 molecules exist momentarily and then dissociate. Use MO theory (and diagrams) to explain why He_2 molecules can come to exist and why they dissociate.

77. a. A flask containing gaseous N_2 is irradiated with 25-nm light. Using the following information, indicate what species can form in this flask during irradiation.

$$N_2(g) \longrightarrow 2N(g) \qquad \Delta H = 941 \text{ kJ/mol}$$

$$N_2(g) \longrightarrow N_2^+(g) + e^- \qquad \Delta H = 1501 \text{ kJ/mol}$$

$$N(g) \longrightarrow N^+(g) + e^- \qquad \Delta H = 1402 \text{ kJ/mol}$$

b. What range of wavelengths will produce atomic nitrogen in the flask but will not produce any ions?

c. Explain why the first ionization energy of N_2 (1501 kJ/mol) is greater than the first ionization energy of atomic nitrogen (1402 kJ/mol).

78. Use the MO model to determine which of the following has the smallest ionization energy: N_2, O_2, N_2^{2-}, N_2^-, O_2^+. Explain your answer.

79. Cholesterol ($C_{27}H_{46}O$) has the following structure:

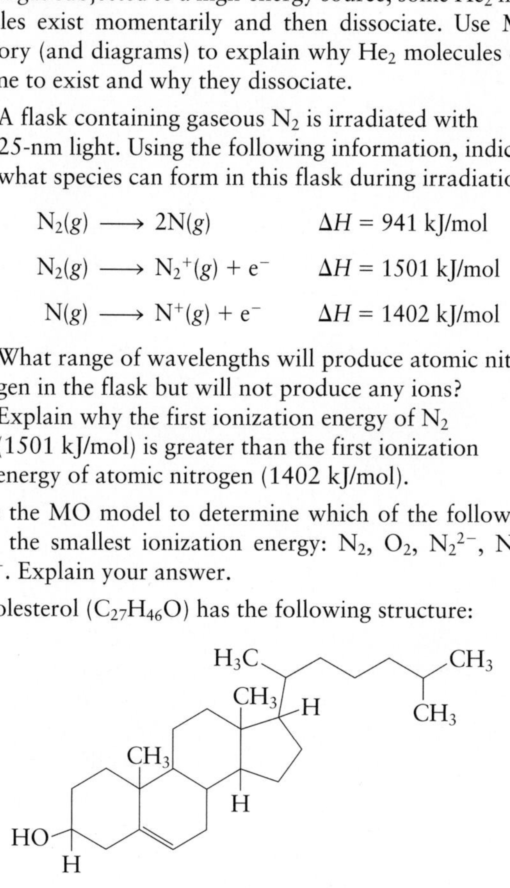

In such shorthand structures, each point where lines meet represents a carbon atom, and most H atoms are not shown. Draw the complete structure, showing all carbon and hydrogen atoms. (There will be four bonds to each carbon atom.) Indicate which carbon atoms use sp^2 or sp^3 hybrid orbitals. Are all carbon atoms in the same plane, as implied by the structure?

80. Arrange the following from lowest to highest ionization energy: O, O_2, O_2^-, O_2^+. Explain your answer.

81. Carbon monoxide (CO) forms bonds to a variety of metals and metal ions. Its ability to bond to iron in hemoglobin is the reason that CO is so toxic. The bond carbon monoxide forms to metals is through the carbon atom:

$$M{-}C{\equiv}O$$

a. On the basis of electronegativities, would you expect the carbon atom or the oxygen atom to form bonds to metals?

b. Assign formal charges to the atoms in CO. Which atom would you expect to bond to a metal on this basis?

c. In the MO model, bonding MOs place more electron density near the more electronegative atom. (See the HF molecule, Figs. 14.45 and 14.46.) Antibonding MOs place more electron density near the less electronegative atom in the diatomic molecule. Use the MO model to predict which atom of carbon monoxide should form bonds to metals.

82. Use the MO model to explain the bonding in BeH_2. When constructing the MO energy-level diagram, assume that the Be 1*s* electrons are not involved in bond formation.

83. In Exercise 65 in Chapter 13, the Lewis structures for benzene (C_6H_6) were drawn. Using one of the Lewis structures, estimate ΔH°_f for $C_6H_6(g)$ using bond energies and given the standard enthalpy of formation of C(*g*) is 717 kJ/mol. The experimental ΔH°_f value for $C_6H_6(g)$ is 83 kJ/mol. Explain the discrepancy between the experimental value and the calculated ΔH°_f value for $C_6H_6(g)$.

84. For each chemical formula below, an NMR spectrum is described, including relative overall areas (intensities) for the various signals given in parentheses. Draw the structure of a compound having the specific formula that would give the described NMR spectrum. (*Hint:* All of

these formulas represent organic compounds. Lewis structures for organic compounds typically have all atoms in the compound with a formal charge of zero. This is the case in this problem.)

a. $C_2H_3Cl_3$; NMR has one singlet signal.
b. $C_3H_6Cl_2$; NMR has a triplet (4) and a quintet (2) signal.
c. $C_3H_6O_2$; NMR has a singlet (1), a quartet (2), and a triplet (3) signal.
d. $C_5H_{10}O$; NMR has a heptet (1), a singlet (3), and a doublet (6) signal.
e. C_3H_6O; NMR has a triplet (3), a quintet (2), and a triplet (1) signal.

85. Structural isomers are compounds that have the same chemical formula but the atoms are bonded together differently giving different compounds. Consider the two structural isomers having the formula C_2H_6O. Draw the two isomers, and describe what should be seen in the NMR spectrum for each isomer (types of signals and relative overall intensities). (*Hint:* Both of the isomers are organic compounds. Lewis structures for organic compounds typically have all atoms in the compound with a formal charge of zero. This is the case in this problem.)

Marathon Problem

86. The sp^2 hybrid atomic orbitals have the following general form:

$$\Phi_1 = \sqrt{1/3}\Phi_s + 2A\Phi_{px}$$

$$\Phi_2 = \sqrt{1/3}\Phi_s - A\Phi_{px} + B\Phi_{py}$$

$$\Phi_3 = \sqrt{1/3}\Phi_s - A\Phi_{px} - B\Phi_{py}$$

where Φ_s, Φ_{px}, and Φ_{py} represent orthonormal (normalized and orthogonalized) atomic orbitals. Calculate the values of A and B.

Chemical Kinetics

15

chapter

Soccer goalie catching a ball.

Sign in to OWL at www.cengage.com/owl to view tutorials and simulations, develop problem-solving skills, and complete online homework assigned by your professor.

go Chemistry

Download mini lecture videos for key concept review and exam prep from OWL or purchase them from www.cengagebrain.com.

The applications of chemistry focus primarily on chemical reactions, and the commercial use of a reaction requires knowledge of several of its characteristics. A reaction is defined by its reactants and products, whose identities must be learned by experiment. Once the reactants and products are known, the equation for the reaction can be written and balanced, and stoichiometric calculations can be carried out. Another very important characteristic of a reaction is its spontaneity. Spontaneity refers to the *inherent tendency* for the process to occur; however, it implies nothing about speed. *Spontaneous does not mean fast.* There are many spontaneous reactions that are so slow that no apparent reaction occurs over a period of weeks or years at normal temperatures. For example, there is a strong inherent tendency for gaseous hydrogen and oxygen to combine to form water,

$$2H_2(g) + O_2(g) \longrightarrow 2H_2O(l)$$

but in fact the two gases can coexist indefinitely at 25°C. Similarly, the gaseous reactions

$$H_2(g) + Cl_2(g) \longrightarrow 2HCl(g)$$
$$N_2(g) + 3H_2(g) \longrightarrow 2NH_3(g)$$

are both highly likely to occur from a thermodynamic standpoint, but we observe no product formation under normal conditions. In addition, the process of changing diamond to graphite is spontaneous but is so slow that it is not detectable.

To be useful, reactions must occur at a reasonable rate. To produce the 20 million tons of ammonia needed each year for fertilizer, we cannot simply mix nitrogen and hydrogen gases at 25°C and wait for them to react. It is not enough to understand the stoichiometry and thermodynamics of a reaction; we must also understand the factors that govern the rate of the reaction. The area of chemistry that concerns reaction rates is called **chemical kinetics.**

One of the main goals of chemical kinetics is to understand the steps by which a reaction takes place. This series of steps is called the *reaction mechanism.* Understanding the mechanism allows us to find ways to facilitate the reaction. For example, the Haber process for the production of ammonia requires high temperatures to achieve commercially feasible reaction rates. However, even higher temperatures (and more cost) would be required without the use of iron oxide, which speeds up the reaction.

In this chapter we will consider the fundamental ideas of chemical kinetics. We will explore rate laws, reaction mechanisms, and simple models for chemical reactions.

15.1 Reaction Rates

The kinetics of air pollution is discussed in Section 15.9.

To introduce the concept of reaction rate, we will consider the decomposition of nitrogen dioxide, a gas that causes air pollution. Nitrogen dioxide decomposes to nitric oxide and oxygen as follows:

$$2NO_2(g) \longrightarrow 2NO(g) + O_2(g)$$

Suppose in a particular experiment we start with a flask of nitrogen dioxide at room temperature, where it is stable indefinitely, and quickly heat it to 300°C, where it decomposes according to the preceding equation. We then measure the concentrations of nitrogen dioxide, nitric oxide, and oxygen over time as the nitrogen dioxide decomposes. The results of this experiment are summarized in Table 15.1, and the data are plotted in Fig. 15.1.

Table 15.1

Concentrations of Reactant and Products as a Function of Time for the Reaction $2NO_2(g) \longrightarrow 2NO(g) + O_2(g)$ (at 300°C)

	Concentration (mol/L)		
Time (±1 s)	NO_2	NO	O_2
0	0.0100	0	0
50	0.0079	0.0021	0.0011
100	0.0065	0.0035	0.0018
150	0.0055	0.0045	0.0023
200	0.0048	0.0052	0.0026
250	0.0043	0.0057	0.0029
300	0.0038	0.0062	0.0031
350	0.0034	0.0066	0.0033
400	0.0031	0.0069	0.0035

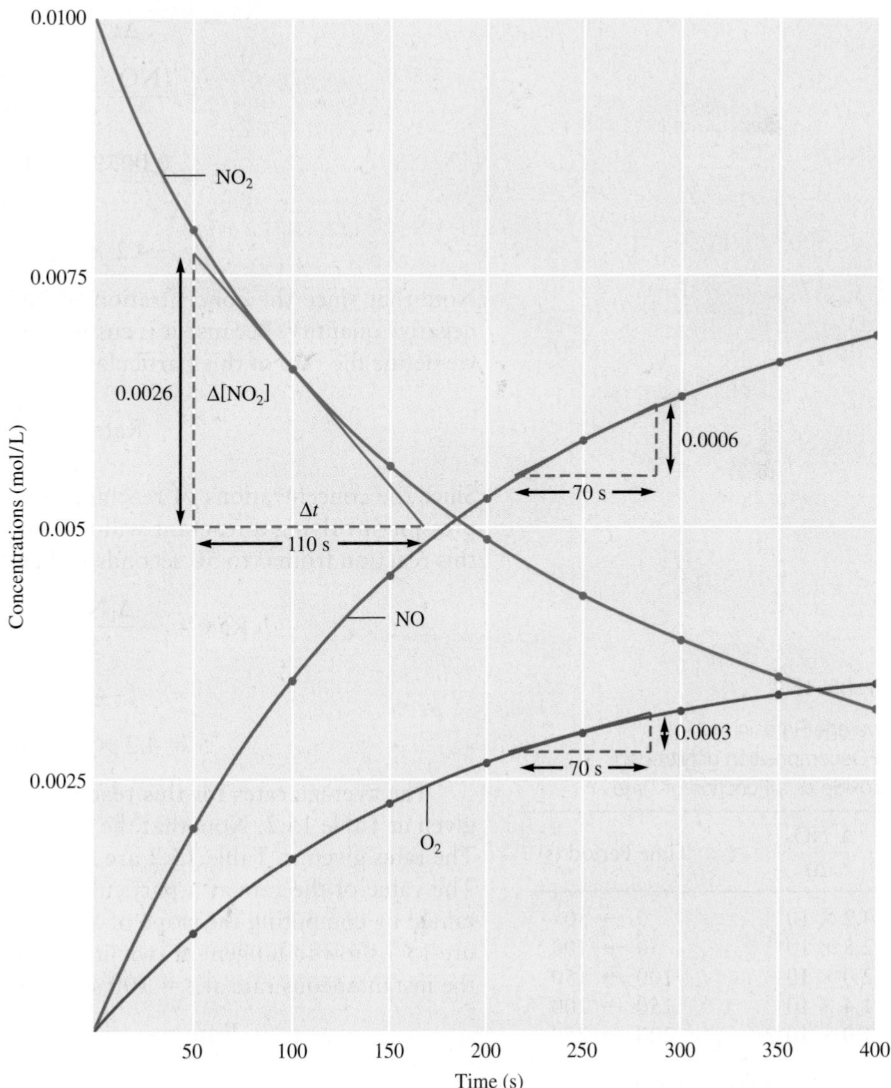

Figure 15.1
Starting with pure nitrogen dioxide at 300°C, the concentrations of nitrogen dioxide, nitric oxide, and oxygen are plotted versus time.

This definition of reaction rate assumes that the volume remains constant. A more general definition of reaction rate is the change in the number of moles per unit of volume per unit of time.

Note from these results that the concentration of the reactant (NO_2) decreases with time and that the concentrations of the products (NO and O_2) increase with time. Chemical kinetics deals with the speed at which these changes occur. The speed, or *rate,* of a process is defined as the change in a given quantity over a specific period of time. For chemical reactions the quantity that changes is the amount or concentration of a reactant or product. So the **reaction rate** of a chemical reaction is defined as the *change in concentration of a reactant or product per unit time:*

$$\text{Rate} = \frac{\text{concentration of A at time } t_2 - \text{concentration of A at time } t_1}{t_2 - t_1} = \frac{\Delta[\text{A}]}{\Delta t}$$

where A represents a specific reactant or product and the square brackets indicate concentration in mol/L. As usual, the symbol Δ indicates a *change* in a given quantity.

Now let's calculate the average rate at which the concentration of NO_2 changes over the first 50 seconds of the reaction, using the data given in Table 15.1.

$$\begin{aligned}\text{Rate} &= \frac{\Delta[\text{NO}_2]}{\Delta t} \\ &= \frac{[\text{NO}_2]_{t=50} - [\text{NO}_2]_{t=0}}{50\text{ s} - 0\text{ s}} \\ &= \frac{0.0079\text{ mol/L} - 0.0100\text{ mol/L}}{50\text{ s}} \\ &= -4.2 \times 10^{-5}\text{ mol L}^{-1}\text{ s}^{-1}\end{aligned}$$

Note that since the concentration of NO_2 decreases with time, $\Delta[NO_2]$ is a negative quantity. Because it is customary to work with *positive* reaction rates, we define the rate of this particular reaction as

$$\text{Rate} = -\frac{\Delta[\text{NO}_2]}{\Delta t}$$

Since the concentrations of reactants always decrease with time, any rate expression involving a reactant will include a negative sign. The average rate of this reaction from 0 to 50 seconds is then

$$\begin{aligned}\text{Rate} &= -\frac{\Delta[\text{NO}_2]}{\Delta t} \\ &= -(-4.2 \times 10^{-5}\text{ mol L}^{-1}\text{ s}^{-1}) \\ &= 4.2 \times 10^{-5}\text{ mol L}^{-1}\text{ s}^{-1}\end{aligned}$$

Table 15.2

Average Rate (in mol L^{-1} s^{-1}) of Decomposition of Nitrogen Dioxide as a Function of Time

$-\frac{\Delta[\text{NO}_2]}{\Delta t}$	Time Period (s)
4.2×10^{-5}	$0 \rightarrow 50$
2.8×10^{-5}	$50 \rightarrow 100$
2.0×10^{-5}	$100 \rightarrow 150$
1.4×10^{-5}	$150 \rightarrow 200$
1.0×10^{-5}	$200 \rightarrow 250$

Note: The *rate* decreases with time.

The average rates for this reaction during several other time intervals are given in Table 15.2. Note that the rate is not constant but decreases with time. The rates given in Table 15.2 are *average* rates over 50-second time intervals. The value of the rate at a particular time (the **instantaneous rate**) can be obtained by computing the slope of a line tangent to the curve at that point. Figure 15.1 shows a tangent drawn at $t = 100$ seconds. The *slope* of this line gives the instantaneous rate at $t = 100$ seconds:

$$\begin{aligned}\text{Rate} &= -(\text{slope of the tangent line}) \\ &= 2.4 \times 10^{-5}\text{ mol L}^{-1}\text{ s}^{-1}\end{aligned}$$

Chemical Insights

Femtochemistry

Some reactions occur very slowly, such as when a nail rusts. Others occur very rapidly, such as when methane is combusted in a Bunsen burner. Studying very fast reactions requires very special techniques, usually involving lasers—devices that produce high-energy bursts of light with very precise frequencies. The study of very fast reactions is one of the most important areas of chemical research, as demonstrated by the fact that the 1999 Nobel Prize in chemistry was awarded to Ahmed H. Zewail of the California Institute of Technology in Pasadena, California. Zewail's studies involve reactions that occur on the femtosecond (10^{-15} s) time scale—the time scale for molecular vibrations.

In Zewail's laboratory a strong laser flash of a few femtoseconds' duration shines on beams of molecules streaming into a vacuum chamber. The laser beam is tuned to excite all of the molecules to the same state, where they are vibrating in unison. Subsequent, weaker laser pulses monitor the concentrations of the reactants, intermediates, and products as the reaction occurs.

One reaction studied by Zewail is the decomposition of cyclobutane to two ethylene molecules:

```
 H       H          H     H     H     H
  \     /            \   /       \   /
H--C---C--H            C           C
   |   |     --->      ||    +     ||
H--C---C--H            C           C
  /     \            /   \       /   \
 H       H          H     H     H     H
```

Zewail has shown that the reaction mechanism involves the breaking of one of the carbon–carbon bonds in cyclobutane to produce a tetramethylene intermediate:

```
                        Unpaired electrons
 H       H            H    /    /H
  \     /              \  •    •/
H--C---C--H          H--C      C--H
   |   |     --->       |      |
H--C---C--H          H--C------C--H
  /     \              /        \
 H       H            H          H
```

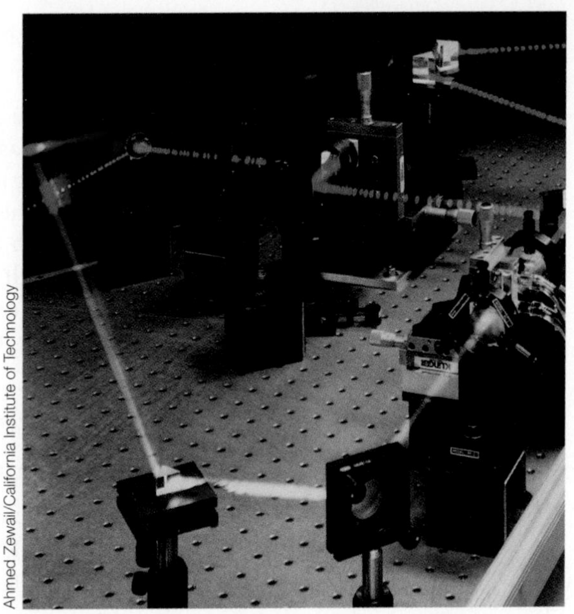
Ahmed Zewail/California Institute of Technology

A laser spectroscopy laboratory at the California Institute of Technology.

The intermediate exists for approximately 700 fs (700×10^{-15} s) and then decomposes into two ethylene molecules. Techniques similar to Zewail's approach (commonly called femtochemistry) are now widely used in chemical research and are being applied to problems such as elucidating the mechanism for energy conversion in chlorophyll (photosynthesis) and understanding the way human eyes detect light.

So far we have discussed the rate of this reaction only in terms of the reactant. The rate can also be defined in terms of the products. However, in doing so, we must take into account the coefficients in the balanced equation for the reaction because the stoichiometry determines the relative rates of the consumption of reactants and the generation of products. For example, in the reaction

$$2NO_2(g) \longrightarrow 2NO(g) + O_2(g)$$

both NO_2 and NO have a coefficient of 2, so NO is produced at the same rate as NO_2 is consumed. We can verify this from Fig. 15.1. Note that the curve for NO is the same shape as the curve for NO_2 except that it is inverted. This means that at any point in time the slope of the tangent to the curve for NO will be the negative of the slope to the curve for NO_2. (Verify this at the point $t = 100$ seconds on both curves.) In contrast, O_2 has a coefficient of 1, which means it is produced half as fast as NO, which has a coefficient of 2. That is, the rate of NO production is twice the rate of O_2 production.

We can also verify this fact from Fig. 15.1. For example, at $t = 250$ seconds,

$$\text{Slope of the tangent to the NO curve} = 8.6 \times 10^{-6} \text{ mol L}^{-1} \text{ s}^{-1}$$

$$\text{Slope of the tangent to the } O_2 \text{ curve} = 4.3 \times 10^{-6} \text{ mol L}^{-1} \text{ s}^{-1}$$

The slope at $t = 250$ seconds on the NO curve is twice the slope of that point on the O_2 curve, showing that the rate of production of NO is twice that of O_2.

The rate information can be summarized as follows:

$$\boxed{\text{Rate of consumption of } NO_2} = \boxed{\text{rate of production of NO}} = \boxed{2(\text{rate of production of } O_2)}$$

Because the reaction rate changes with time, and because the rate may be different for the various reactants and products (by factors that depend on the coefficients in the balanced equation), we must be very specific when we describe a rate for a chemical reaction.

15.2 Rate Laws: An Introduction

Note that the term *reversible* has different meanings in kinetics and thermodynamics.

Chemical reactions are *reversible.* In our discussion of the decomposition of nitrogen dioxide, we have so far considered only the forward reaction:

$$2NO_2(g) \longrightarrow 2NO(g) + O_2(g)$$

However, the reverse reaction can also occur. As NO and O_2 accumulate, they can react to re-form NO_2:

$$O_2(g) + 2NO(g) \longrightarrow 2NO_2(g)$$

When gaseous NO_2 is placed in an otherwise empty container, the dominant reaction initially is

$$2NO_2(g) \longrightarrow 2NO(g) + O_2(g)$$

and the change in the concentration of NO_2 ($\Delta[NO_2]$) depends only on the forward reaction. However, after a period of time, enough products accumulate so that the reverse reaction becomes important. Now $\Delta[NO_2]$ depends on the *difference in the rates of the forward and reverse reactions.* This complication can be avoided if we study the rate of a reaction under conditions where the reverse reaction makes only a negligible contribution. Typically, this means that we study a reaction at a point soon after the reactants are mixed, before the products have had time to build up to significant levels.

Under conditions such that the reverse reaction can be neglected, the *reaction rate depends only on the concentrations of the reactants.* For the decomposition of nitrogen dioxide, we can write

$$\text{Rate} = k[NO_2]^n \tag{15.1}$$

Such an expression, which shows how the rate depends on the concentrations of reactants, is called a **rate law.** The proportionality constant k, called the **rate constant,** and n, called the **order** of the reactant, must both be determined by experiment. The order of a reactant can be positive or negative and can be an integer or a fraction. For the relatively simple reactions we will consider in this book, the orders will generally be positive integers.

Note two important points about Equation (15.1):

1. The concentrations of the products do not appear in the rate law because the reaction rate is being studied under conditions where the reverse reaction does not contribute to the overall rate.
2. The value of the exponent n must be determined by experiment; it cannot be written from the balanced equation.

Before we go further, we must define exactly what we mean by the term *rate* in Equation (15.1). In Section 15.1 we saw that reaction rate means a change in concentration per unit time. However, which reactant or product concentration do we choose in defining the rate? For example, for the decomposition of NO_2 to produce O_2 and NO considered in Section 15.1, we could define the rate in terms of any of these three species. However, since O_2 is produced only half as fast as NO_2, we must be careful to specify which species we are talking about in a given case. For instance, we might choose to define the reaction rate in terms of the consumption of NO_2:

$$\text{Rate} = -\frac{\Delta[NO_2]}{\Delta t} = -\frac{d[NO_2]}{dt} = k[NO_2]^n$$

where d indicates an infinitesimally small change. On the other hand, we could define the rate in terms of the production of O_2:

$$\text{Rate}' = \frac{d[O_2]}{dt} = k'[NO_2]^n$$

Note that because $2NO_2$ molecules are consumed for every O_2 molecule produced,

$$\text{Rate} = 2 \times \text{rate}'$$

or

$$k[NO_2]^n = 2k'[NO_2]^n$$

and

$$k = 2 \times k'$$

Thus the value of the rate constant depends on how the rate is defined.

In this text we will always be careful to define exactly what is meant by the rate for a given reaction so that there will be no confusion about which specific rate constant is being used.

Types of Rate Laws

Notice that the rate law we have used to this point expresses rate as a function of concentration. For example, for the decomposition of NO_2 we have defined the rate as

$$\text{Rate} = -\frac{d[NO_2]}{dt} = k[NO_2]^n$$

The terms *differential rate law* and *rate law* will be used interchangeably in this text.

which tells us (once we have determined the value of n) exactly how the rate depends on the concentration of the reactant NO_2. A rate law that expresses how the *rate depends on concentration* is called the **differential rate law,** but it is often simply called the **rate law.** Thus, when we use the term *rate law* in this text, we mean the expression that gives the rate as a function of concentration.

A second kind of rate law, the **integrated rate law,** will also be important in our study of kinetics. The integrated rate law expresses how the *concentrations depend on time.* As we will see, a given differential rate law is always related to a certain type of integrated rate law, and vice versa. That is, if we determine the differential rate law for a given reaction, we automatically know the form of the integrated rate law for the reaction. This means that once we determine either type of rate law for a reaction, we also know the other one.

Which rate law we choose to determine by experiment often depends on what types of data are easiest to collect. If we can conveniently measure how the rate changes as the concentrations are changed, we can readily determine the differential (rate/concentration) rate law. On the other hand, if it is more convenient to measure the concentration as a function of time, we can determine the form of the integrated (concentration/time) rate law. We will discuss how rate laws are actually determined in the next several sections.

Why are we interested in determining the rate law for a reaction? How does it help us? It helps us because we can work backward from the rate law to find the steps by which the reaction occurs. Most chemical reactions do not take place in a single step but result from a series of sequential steps. To understand a chemical reaction, we must learn what these steps are. For example, a chemist who is designing an insecticide may study the reactions involved in the process of insect growth to see what type of molecule may interrupt this series of reactions. Or an industrial chemist may be trying to make a reaction occur faster, using less expensive conditions. To accomplish this, he or she must know which step is slowest because it is that step that must be speeded up. Thus a chemist is usually not interested in a rate law for its own sake but for what it tells about the steps by which a reaction occurs. We will develop a process for finding the reaction steps later in this chapter.

Rate Laws

1. There are two types of rate laws:
   - The differential rate law (often called simply the rate law) shows how the rate of a reaction depends on concentrations.
   - The integrated rate law shows how the concentrations of species in the reaction depend on time.
2. Because we will typically consider reactions under conditions where the reverse reaction is unimportant, our rate laws will involve concentrations of reactants.
3. Because the differential and integrated rate laws for a given reaction are related in a well-defined way, the experimental determination of *either* of the rate laws is sufficient.
4. Experimental convenience usually dictates which type of rate law is determined experimentally.
5. Knowing the rate law for a reaction is important mainly because we can usually infer the individual steps involved in the reaction from the specific form of the rate law.

15.3 Determining the Form of the Rate Law

Table 15.3

Concentration/Time Data for the Reaction $2N_2O_5(soln) \longrightarrow 4NO_2(soln) + O_2(g)$ (at 45°C)

$[N_2O_5]$ (mol/L)	Time (s)
1.00	0
0.88	200
0.78	400
0.69	600
0.61	800
0.54	1000
0.48	1200
0.43	1400
0.38	1600
0.34	1800
0.30	2000

The first step in understanding how a given chemical reaction occurs is to determine the *form* of the rate law. In this section we will explore ways to obtain the differential rate law for a reaction. First, we will consider the decomposition of dinitrogen pentoxide in carbon tetrachloride solution:

$$2N_2O_5(soln) \longrightarrow 4NO_2(soln) + O_2(g)$$

Data for this reaction at 45°C are listed in Table 15.3 and plotted in Fig. 15.2. In this reaction the oxygen gas escapes from the solution and thus does not react with the nitrogen dioxide, so we do not have to be concerned about the effects of the reverse reaction at any time over the life of the reaction. In other words, the reverse reaction is negligible at all times over the course of this reaction.

Evaluation of the reaction rates at N_2O_5 concentrations of 0.90 M and 0.45 M, by taking the slopes of the tangents to the curve at these points (Fig. 15.2), yields the following data:

$[N_2O_5]$	Rate (mol L^{-1} s^{-1})
0.90 M	5.4×10^{-4}
0.45 M	2.7×10^{-4}

Note that when $[N_2O_5]$ is halved, the rate is also halved. This means that the rate of this reaction depends on the concentration of N_2O_5 to the *first power.* In other words, the (differential) rate law for this reaction is

$$\text{Rate} = -\frac{d[N_2O_5]}{dt} = k[N_2O_5]$$

First order: Rate = k[A]. Doubling the concentration of A doubles the reaction rate.

Thus the reaction is *first order* in N_2O_5. Note that for this reaction the order is *not* the same as the coefficient of N_2O_5 in the balanced equation for the reaction. This reemphasizes the fact that the order of a particular reactant must be obtained by *observing* how the reaction rate depends on the concentration of that reactant.

We have seen that determining the instantaneous rate at two different reactant concentrations gives the following rate law for the decomposition of N_2O_5:

$$\text{Rate} = -\frac{d[A]}{dt} = k[A]$$

where A represents N_2O_5.

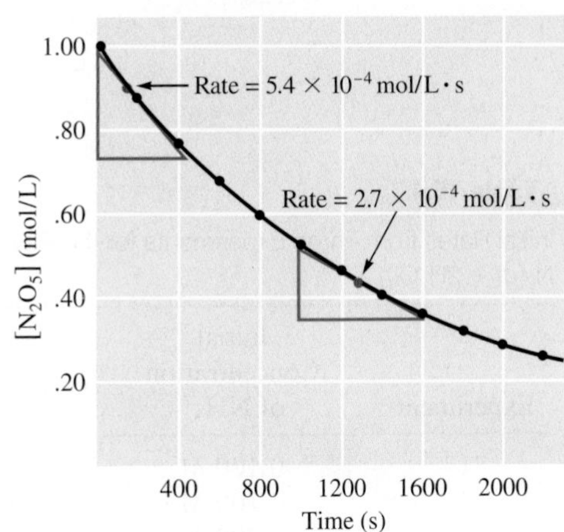

Figure 15.2
A plot of the concentration of N_2O_5 as a function of time for the reaction $2N_2O_5(soln) \longrightarrow 4NO_2(soln) + O_2(g)$ (at 45°C). Note that the reaction rate at $[N_2O_5] = 0.90\ M$ is twice that at $[N_2O_5] = 0.45\ M$.

Method of Initial Rates

The value of the initial rate is determined for each experiment at the same value of t as close to $t = 0$ as possible.

The most common method for directly determining the form of the differential rate law for a reaction is the **method of initial rates.** The **initial rate** of a reaction is the instantaneous rate determined just after the reaction begins (just after $t = 0$). The idea is to determine the instantaneous rate before the initial concentrations of reactants have changed significantly. Several experiments are carried out using different initial concentrations, and the initial rate is determined for each run. The results are then compared to see how the initial rate depends on the initial concentrations. This procedure allows the form of the rate law to be determined. We will illustrate the method of initial rates by using the following reaction:

$$NH_4^+(aq) + NO_2^-(aq) \longrightarrow N_2(g) + 2H_2O(l)$$

Table 15.4 gives initial rates obtained from three experiments involving different initial concentrations of reactants.

The general form of the rate law for this reaction is

$$\text{Rate} = -\frac{d[NH_4^+]}{dt} = k[NH_4^+]^n[NO_2^-]^m$$

We can determine the values of n and m by observing how the initial rate depends on the initial concentrations of NH_4^+ and NO_2^-. In Experiments 1 and 2, where the initial concentration of NH_4^+ remains the same but where the initial concentration of NO_2^- doubles, the observed initial rate also doubles. Since

$$\text{Rate} = k[NH_4^+]^n[NO_2^-]^m$$

we have, for Experiment 1,

$$\text{Rate} = 1.35 \times 10^{-7}\ \text{mol L}^{-1}\text{ s}^{-1} = k(0.100\ \text{mol/L})^n(0.0050\ \text{mol/L})^m$$

and for Experiment 2,

$$\text{Rate} = 2.70 \times 10^{-7}\ \text{mol L}^{-1}\text{ s}^{-1} = k(0.100\ \text{mol/L})^n(0.010\ \text{mol/L})^m$$

Rates 1, 2, and 3 were determined at the same value of t (very close to $t = 0$).

The ratio of these rates is

$$\frac{\text{Rate 2}}{\text{Rate 1}} = \underbrace{\frac{2.70 \times 10^{-7}\ \text{mol L}^{-1}\text{ s}^{-1}}{1.35 \times 10^{-7}\ \text{mol L}^{-1}\text{ s}^{-1}}}_{2.00} = \frac{\cancel{k(0.100\ \text{mol/L})^n}(0.010\ \text{mol/L})^m}{\cancel{k(0.100\ \text{mol/L})^n}(0.0050\ \text{mol/L})^m}$$

$$= \underbrace{\frac{(0.010\ \text{mol/L})^m}{(0.0050\ \text{mol/L})^m}}_{(2.0)^m}$$

Table 15.4

Initial Rates from Three Experiments for the Reaction $NH_4^+(aq) + NO_2^-(aq) \longrightarrow N_2(g) + 2H_2O(l)$

Experiment	Initial Concentration of NH_4^+	Initial Concentration of NO_2^-	Initial Rate (mol L^{-1} s^{-1})
1	0.100 M	0.0050 M	1.35×10^{-7}
2	0.100 M	0.010 M	2.70×10^{-7}
3	0.200 M	0.010 M	5.40×10^{-7}

Thus $$\frac{\text{Rate 2}}{\text{Rate 1}} = 2.00 = (2.0)^m$$

which means the value of m is 1. The rate law for this reaction is first order in the reactant NO_2^-.

A similar analysis of the results for Experiments 2 and 3 yields the following ratio:

$$\frac{\text{Rate 3}}{\text{Rate 2}} = \frac{5.40 \times 10^{-7} \text{ mol L}^{-1} \text{ s}^{-1}}{2.70 \times 10^{-7} \text{ mol L}^{-1} \text{ s}^{-1}} = \frac{(0.200 \text{ mol/L})^n}{(0.100 \text{ mol/L})^n}$$

$$= 2.00 = \left(\frac{0.200}{1.100}\right)^n = (2.00)^n$$

The value of n is also 1.

We have shown that the values of n and m are both 1. Therefore, the rate law is

$$\text{Rate} = k[NH_4^+][NO_2^-]$$

This rate law is first order in both NO_2^- and NH_4^+. Note that it is merely a coincidence that n and m have the same values as the coefficients of NH_4^+ and NO_2^- in the balanced equation for the reaction.

Overall reaction order is the sum of the orders for each reactant. For a discussion of how this term can be misleading, see John C. Reeve, "Some Provocative Opinions on the Terminology of Chemical Kinetics," *J. Chem. Ed.* 68 (1991): 278.

The **overall reaction order** is the sum of n and m. For this reaction, $n + m = 2$. The reaction is second order overall.

The value of the rate constant (k) can now be calculated by using the results of *any* of the three experiments shown in Table 15.4. From the data for Experiment 1, we know that

$$\text{Rate} = k[NH_4^+][NO_2^-]$$

$$1.35 \times 10^{-7} \text{ mol L}^{-1} \text{ s}^{-1} = k(0.100 \text{ mol/L})(0.0050 \text{ mol/L})$$

Then

$$k = \frac{1.35 \times 10^{-7} \text{ mol L}^{-1} \text{ s}^{-1}}{(0.100 \text{ mol/L})(0.0050 \text{ mol/L})} = 2.7 \times 10^{-4} \text{ L mol}^{-1} \text{ s}^{-1}$$

EXAMPLE 15.1

The reaction between bromate ions and bromide ions in acidic aqueous solution is given by the following equation:

$$BrO_3^-(aq) + 5Br^-(aq) + 6H^+(aq) \longrightarrow 3Br_2(l) + 3H_2O(l)$$

Table 15.5 gives the results of four experiments involving this reaction. Using these data, determine the orders for all three reactants, the overall reaction order, and the value of the rate constant.

Table 15.5

The Results from Four Experiments to Study the Reaction $BrO_3^-(aq) + 5Br^-(aq) + 6H^+(aq) \longrightarrow 3Br_2(l) + 3H_2O(l)$

Experiment	Initial Concentration of BrO_3^- (mol/L)	Initial Concentration of Br^- (mol/L)	Initial Concentration of H^+ (mol/L)	Measured Initial Rate (mol L^{-1} s^{-1})
1	0.10	0.10	0.10	8.0×10^{-4}
2	0.20	0.10	0.10	1.6×10^{-3}
3	0.20	0.20	0.10	3.2×10^{-3}
4	0.10	0.10	0.20	3.2×10^{-3}

A solution containing BrO_3^- ions being added to a solution containing Br^- ions to form Br_2.

Solution The general form of the rate law for this reaction is

$$\text{Rate} = k[BrO_3^-]^n[Br^-]^m[H^+]^p$$

We can determine the values of n, m, and p by comparing the rates from the various experiments. To determine the value of n, we use the results from Experiments 1 and 2, in which only $[BrO_3^-]$ changes:

$$\frac{\text{Rate 2}}{\text{Rate 1}} = \frac{1.6 \times 10^{-3} \text{ mol L}^{-1}\text{ s}^{-1}}{8.0 \times 10^{-4} \text{ mol L}^{-1}\text{ s}^{-1}}$$

$$= \frac{k(0.20 \text{ mol/L})^n(0.10 \text{ mol/L})^m(0.10 \text{ mol/L})^p}{k(0.10 \text{ mol/L})^n(0.10 \text{ mol/L})^m(0.10 \text{ mol/L})^p}$$

$$2.0 = \left(\frac{0.20 \text{ mol/L}}{0.10 \text{ mol/L}}\right)^n = (2.0)^n$$

Thus n is equal to 1.

To determine the value of m, we use the results from Experiments 2 and 3, in which only $[Br^-]$ changes:

$$\frac{\text{Rate 3}}{\text{Rate 2}} = \frac{3.2 \times 10^{-3} \text{ mol L}^{-1}\text{ s}^{-1}}{1.6 \times 10^{-3} \text{ mol L}^{-1}\text{ s}^{-1}}$$

$$= \frac{k(0.20 \text{ mol/L})^n(0.20 \text{ mol/L})^m(0.10 \text{ mol/L})^p}{k(0.20 \text{ mol/L})^n(0.10 \text{ mol/L})^m(0.10 \text{ mol/L})^p}$$

$$2.0 = \left(\frac{0.20 \text{ mol/L}}{0.10 \text{ mol/L}}\right)^m = (2.0)^m$$

Thus m is equal to 1.

To determine the value of p, we use the results from Experiments 1 and 4, in which $[BrO_3^-]$ and $[Br^-]$ are constant but $[H^+]$ changes:

$$\frac{\text{Rate 4}}{\text{Rate 1}} = \frac{3.2 \times 10^{-3} \text{ mol L}^{-1}\text{ s}^{-1}}{8.0 \times 10^{-4} \text{ mol L}^{-1}\text{ s}^{-1}}$$

$$= \frac{k(0.10 \text{ mol/L})^n(0.10 \text{ mol/L})^m(0.20 \text{ mol/L})^p}{k(0.10 \text{ mol/L})^n(0.10 \text{ mol/L})^m(0.10 \text{ mol/L})^p}$$

$$4.0 = \left(\frac{0.20 \text{ mol/L}}{0.10 \text{ mol/L}}\right)^p$$

$$4.0 = (2.0)^p = (2.0)^2$$

Thus p is equal to 2.

The rate of this reaction is first order in BrO_3^- and Br^- and second order in H^+. The overall reaction order is $n + m + p = 4$.

The rate law can now be written:

$$\text{Rate} = k[BrO_3^-][Br^-][H^+]^2$$

The value of the rate constant k can be calculated from the results of any of the four experiments. For Experiment 1 the initial rate is 8.0×10^{-4} mol L^{-1} s^{-1}, and $[BrO_3^-] = 0.100\ M$, $[Br^-] = 0.100\ M$, and $[H^+] = 0.100\ M$. Using these values in the rate law gives

$$8.00 \times 10^{-4} \text{ mol L}^{-1}\text{ s}^{-1} = k(0.10 \text{ mol/L})(0.10 \text{ mol/L})(0.10 \text{ mol/L})^2$$

$$8.00 \times 10^{-4} \text{ mol L}^{-1}\text{ s}^{-1} = k(1.0 \times 10^{-4} \text{ mol}^4/\text{L}^4)$$

$$k = \frac{8.0 \times 10^{-4} \text{ mol L}^{-1}\text{ s}^{-1}}{1.0 \times 10^{-4} \text{ mol}^4/\text{L}^4}$$

$$= 8.00 \text{ L}^3 \text{ mol}^{-3}\text{ s}^{-1}$$

Check: Verify that the same value of k can be obtained from the results of the other experiments.

15.4 The Integrated Rate Law

In the field of kinetics, the rate for this type of reaction is usually defined as

$$\text{Rate} = -\frac{1}{a}\frac{d[A]}{dt}$$

where a is the coefficient of A in the balanced equation. However, to avoid complications, we will leave out the factor of $1/a$, which simply changes the value of the rate constant by a factor of a.

The rate laws we have considered so far express the rate as a function of the reactant concentrations. It is also useful to be able to express the reactant concentrations as a function of time, given the (differential) rate law for the reaction. In this section we will show how this is done.

We will proceed by first looking at reactions involving a single reactant:

$$aA \longrightarrow \text{products}$$

all of which have a rate law of the form

$$\text{Rate} = -\frac{d[A]}{dt} = k[A]^n$$

We will develop the integrated rate laws individually for the cases $n = 1$ (first order), $n = 2$ (second order), and $n = 0$ (zero order).

First-Order Rate Laws

For the reaction

$$2N_2O_5(soln) \longrightarrow 4NO_2(soln) + O_2(g)$$

experiments show that the rate law is

$$\text{Rate} = -\frac{d[N_2O_5]}{dt} = k[N_2O_5]$$

The decomposition of N_2O_5 in solution to form NO_2 plus O_2 (bubbles). The brown gas above the solution is escaping NO_2.

Since the rate of this reaction depends on the concentration of N_2O_5 to the first power, it is a **first-order reaction.** This means that if the concentration of N_2O_5 in a flask were suddenly doubled, the rate of production of NO_2 and O_2 would also double. Using calculus, this differential rate law can be integrated, which yields the expression

$$\ln[N_2O_5] = -kt + \ln[N_2O_5]_0$$

where ln indicates the natural logarithm, t is the time, $[N_2O_5]$ is the concentration of N_2O_5 at time t, and $[N_2O_5]_0$ is the initial concentration of N_2O_5 (at $t = 0$, the start of the experiment). Note that such an equation, called the integrated rate law, expresses the *concentration of the reactant as a function of time.*

For a chemical reaction of the form

$$aA \longrightarrow \text{products}$$

where the kinetics are first order in [A], the differential rate law is of the form

$$\text{Rate} = -\frac{d[A]}{dt} = k[A]$$

An integrated rate law relates concentration to reaction time.

and the **integrated first-order rate law** is

$$\ln[A] = -kt + \ln[A]_0 \tag{15.2}$$

There are three important things to note about Equation (15.2):

1. The equation shows how the concentration of A depends on time. If the initial concentration of A and the value of the rate constant k are known, the concentration of A at any time can be calculated.
2. Equation (15.2) is of the form $y = mx + b$, where a plot of y versus x is a straight line with slope m and intercept b. In this case

$$y = \ln[A] \qquad x = t \qquad m = -k \qquad b = \ln[A]_0$$

For a first-order reaction, a plot of ln[A] versus *t* is a straight line.

Thus, for a first-order reaction, plotting the natural logarithm of concentration versus time always gives a straight line. This fact is often used to test whether a reaction is first order. For the reaction of the type

$$a\text{A} \longrightarrow \text{products}$$

the *reaction is first order in A if a plot of ln[A] versus t is a straight line.* Conversely, if the plot is not a straight line, the reaction is not first order.

3. The integrated rate law for a first-order reaction can also be expressed in terms of the *ratio* of [A] and $[A]_0$ as follows:

$$\ln\left(\frac{[A]_0}{[A]}\right) = kt$$

EXAMPLE 15.2

The decomposition of N_2O_5 in the gas phase was studied at constant temperature:

$$2N_2O_5(g) \longrightarrow 4NO_2(g) + O_2(g)$$

The following results were collected:

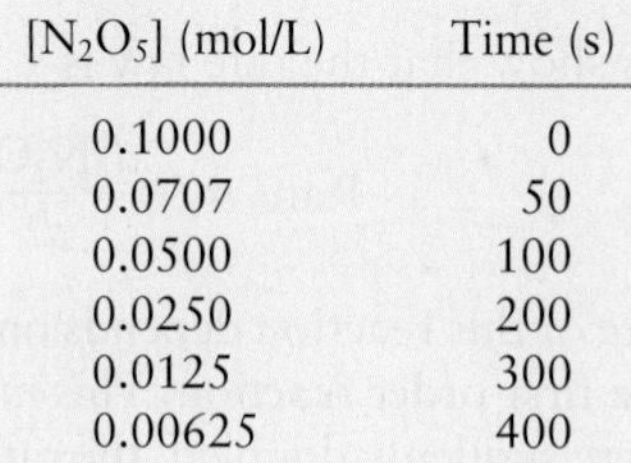

$[N_2O_5]$ (mol/L)	Time (s)
0.1000	0
0.0707	50
0.0500	100
0.0250	200
0.0125	300
0.00625	400

Using these data, verify that the rate law is first order in $[N_2O_5]$, and calculate the value of the rate constant, where the rate $= -d[N_2O_5]/dt$.

Solution We can verify that the rate law is first order in $[N_2O_5]$ by constructing a plot of $\ln[N_2O_5]$ versus time. The values of $\ln[N_2O_5]$ at various times are given below, and the plot of $\ln[N_2O_5]$ versus time is shown in Fig. 15.3.

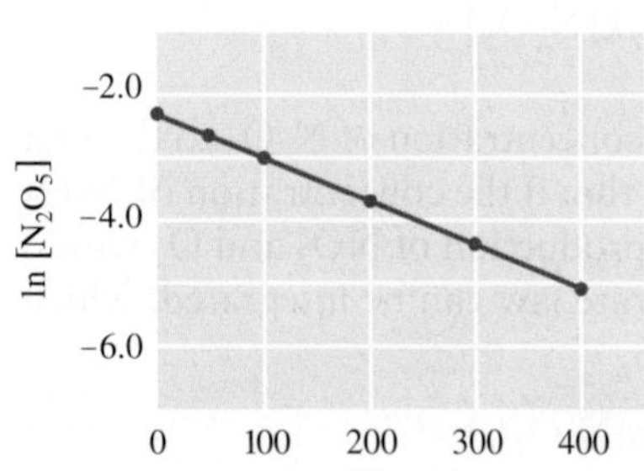

Figure 15.3
A plot of $\ln[N_2O_5]$ versus time.

$\ln[N_2O_5]$	Time (s)
−2.303	0
−2.649	50
−2.996	100
−3.689	200
−4.382	300
−5.075	400

The plot is a straight line, confirming that the reaction is first order in N_2O_5, since it follows the equation $\ln[N_2O_5] = -kt + \ln[N_2O_5]_0$.

Since the reaction is first order, the slope of the line equals $-k$. In this case

$$k = -(\text{slope}) = 6.93 \times 10^{-3}\ \text{s}^{-1}$$

EXAMPLE 15.3

Using the data given in Example 15.2, calculate $[N_2O_5]$ 150. s after the start of the reaction.

Solution We know from Example 15.2 that $[N_2O_5] = 0.0500$ mol/L at 100 s and $[N_2O_5] = 0.0250$ mol/L at 200 s. Since 150 s is halfway between 100 s and

200 s, it is tempting to assume that we can use a simple average to obtain $[N_2O_5]$ at that time. This is incorrect because it is $\ln[N_2O_5]$, not $[N_2O_5]$, that depends directly on t. To calculate $[N_2O_5]$ after 150 s, we must use Equation (15.2):

$$\ln[N_2O_5] = -kt + \ln[N_2O_5]_0$$

where $t = 150.$ s, $k = 6.93 \times 10^{-3}\ s^{-1}$ (as determined in Example 15.2), and $[N_2O_5]_0 = 0.100$ mol/L.

$$\ln([N_2O_5])_{t=150} = -(6.93 \times 10^{-3}\ s^{-1})(150.\ s) + \ln(0.100)$$
$$= -1.040 - 2.303 = -3.343$$
$$[N_2O_5]_{t=150} = \text{antilog}(-3.343) = 0.0353 \text{ mol/L}$$

Note that this value of $[N_2O_5]$ is *not* halfway between 0.0500 mol/L and 0.0250 mol/L.

Half-Life of a First-Order Reaction

The time required for a reactant to reach half of its original concentration is called the **half-life of a reaction** and is designated by $t_{1/2}$. To illustrate this idea, we can calculate the half-life of the decomposition reaction discussed in Example 15.2. The data plotted in Fig. 15.4 show that the half-life for this reaction is 100 seconds. We can see this by considering the following numbers:

$[N_2O_5]$ (mol/L)	Time (s)	
0.1000	0	
		$\Delta t = 100$ s; $\frac{[N_2O_5]_{t=100}}{[N_2O_5]_{t=0}} = \frac{0.050}{0.100} = \frac{1}{2}$
0.0500	100	
		$\Delta t = 100$ s; $\frac{[N_2O_5]_{t=200}}{[N_2O_5]_{t=100}} = \frac{0.025}{0.050} = \frac{1}{2}$
0.0250	200	
		$\Delta t = 100$ s; $\frac{[N_2O_5]_{t=300}}{[N_2O_5]_{t=200}} = \frac{0.0125}{0.0250} = \frac{1}{2}$
0.0125	300	

Note that it *always* takes 100 seconds for $[N_2O_5]$ to be halved in this reaction.

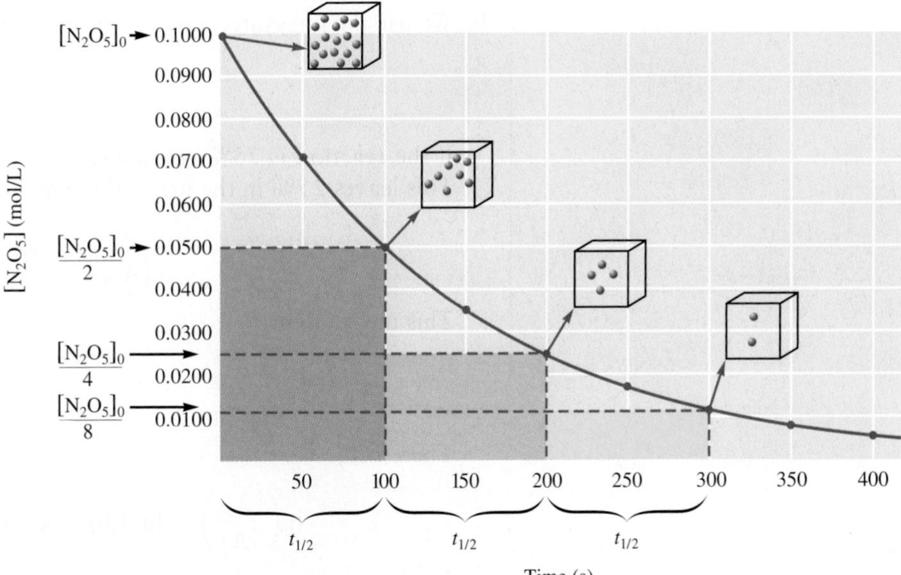

Figure 15.4
A plot of $[N_2O_5]$ versus time for the decomposition reaction of N_2O_5.

A general formula for the half-life of a first-order reaction can be derived from the integrated rate law for the general reaction,

$$a\text{A} \longrightarrow \text{products}$$

If the reaction is first order in [A],

$$\ln\left(\frac{[\text{A}]_0}{[\text{A}]}\right) = kt$$

By definition, when t = $t_{1/2}$, $$[\text{A}] = \frac{[\text{A}]_0}{2}$$

Then for $t = t_{1/2}$, the integrated rate law becomes

$$\ln\left(\frac{[\text{A}]_0}{[\text{A}]_0/2}\right) = kt_{1/2}$$

or $$\ln(2) = kt_{1/2}$$

Substituting the value of ln(2) and solving for $t_{1/2}$ gives

$$t_{1/2} = \frac{0.693}{k} \qquad (15.3)$$

For a first-order reaction, $t_{1/2}$ is independent of the initial concentration.

This is the *general equation for the half-life of a first-order reaction.* Equation (15.3) can be used to calculate $t_{1/2}$ if k is known, or k if $t_{1/2}$ is known. Note that for a first-order reaction *the half-life does not depend on concentration.*

OWL INTERACTIVE EXAMPLE 15.4

A certain first-order reaction has a half-life of 20.0 minutes.

a. Calculate the rate constant for this reaction.

b. How much time is required for this reaction to be 75% complete?

Solution

a. Solving Equation (15.3) for k gives

$$k = \frac{0.692}{t_{1/2}} = \frac{0.693}{20.0 \text{ min}} = 3.47 \times 10^{-2} \text{ min}^{-1}$$

b. We use the integrated rate law in the form

$$\ln\left(\frac{[\text{A}]_0}{[\text{A}]}\right) = kt$$

If the reaction is 75% complete, 75% of the reactant has been consumed. This leaves 25% in the original form:

$$\frac{[\text{A}]}{[\text{A}]_0} \times 100 = 25$$

This means that

$$\frac{[\text{A}]}{[\text{A}]_0} = 0.25 \quad \text{and} \quad \frac{[\text{A}]_0}{[\text{A}]} = 4.0$$

Therefore,

$$\ln\left(\frac{[\text{A}]_0}{[\text{A}]}\right) = \ln(4.0) = kt = \left(\frac{3.47 \times 10^{-2}}{\text{min}}\right)t$$

and $$t = 40.\text{ min}$$

Thus it takes 40. minutes for this particular reaction to reach 75% completion.

Let's consider another way of solving this problem by using the definition of half-life. After one half-life the reaction has gone 50% to completion. If the initial concentration were 1.0 mol/L, after one half-life the concentration would be 0.50 mol/L. One more half-life would produce a concentration of 0.25 mol/L. Comparing 0.25 mol/L with the original 1.0 mol/L shows that 25% of the reactant is left after two half-lives. This is a general result. (What percentage of reactant remains after three half-lives?) Two half-lives for this reaction is 2(20.0 min), or 40.0 minutes, which agrees with the above answer.

Second-Order Rate Laws

For a general reaction involving a single reactant,

$$a\text{A} \longrightarrow \text{products}$$

Second order: Rate = $k[\text{A}]^2$. Doubling the concentration of A quadruples the reaction rate; tripling the concentration of A increases the rate by nine times.

which is second order in A, the rate law can be defined as

$$\text{Rate} = -\frac{d[\text{A}]}{dt} = k[\text{A}]^2 \qquad (15.4)$$

Integration of this differential rate law yields the **integrated second-order rate law:**

$$\frac{1}{[\text{A}]} = kt + \frac{1}{[\text{A}]_0} \qquad (15.5)$$

Note the following characteristics of Equation (15.5):

For a second-order reaction, a plot of $1/[\text{A}]$ versus t is linear.

1. A plot of $1/[\text{A}]$ versus t will produce a straight line with a slope equal to k.
2. Equation (15.5) shows how [A] depends on time and can be used to calculate [A] at any time t, provided k and $[\text{A}]_0$ are known.

When one half-life of a second-order reaction has elapsed ($t = t_{1/2}$), by definition, $[\text{A}] = [\text{A}]_0/2$. Equation (15.5) then becomes

$$\frac{1}{\frac{[\text{A}]_0}{2}} = kt_{1/2} + \frac{1}{[\text{A}]_0}$$

and $$\frac{1}{[\text{A}]_0} = kt_{1/2}$$

Solving for $t_{1/2}$ gives *the expression for the half-life of a second-order reaction:*

$$t_{1/2} = \frac{1}{k[\text{A}]_0} \qquad (15.6)$$

EXAMPLE 15.5

When two identical molecules combine, the resulting molecule is called a *dimer*.

Butadiene reacts to form its dimer according to the equation

$$2C_4H_6(g) \longrightarrow C_8H_{12}(g)$$

The following data were collected for this reaction at a given temperature:

$[C_4H_6]$ (mol/L)	Time (±1 s)
0.01000	0
0.00625	1000
0.00476	1800
0.00370	2800
0.00313	3600
0.00270	4400
0.00241	5200
0.00208	6200

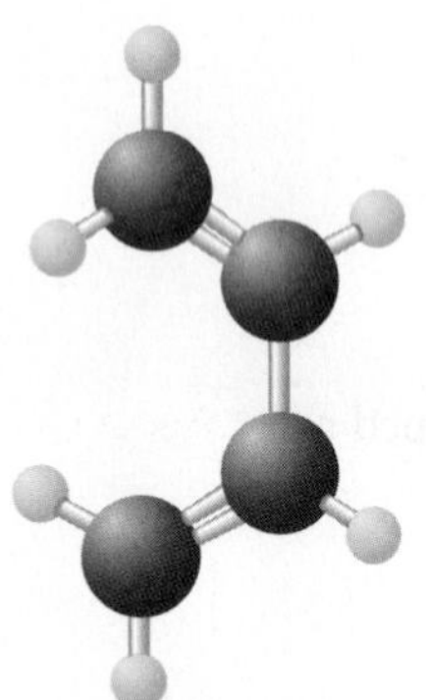

Butadiene (C_4H_6)

a. Is this reaction first order or second order?

b. What is the value of the rate constant for the reaction?

c. What is the half-life for the reaction under the conditions of this experiment?

Solution

a. To decide whether the rate law for this reaction is first order or second order, we must see whether the plot of $\ln[C_4H_6]$ versus time is a straight line (first order) or the plot of $1/[C_4H_6]$ versus time is a straight line (second order). The data necessary to make these plots are as follows:

t (s)	$\frac{1}{[C_4H_6]}$	$\ln[C_4H_6]$
0	100	−4.605
1000	160	−5.075
1800	210	−5.348
2800	270	−5.599
3600	319	−5.767
4400	370	−5.915
5200	415	−6.028
6200	481	−6.175

The resulting plots are shown in Fig. 15.5. Since the $\ln[C_4H_6]$ versus t plot is not a straight line, the reaction is *not* first order. The reaction is,

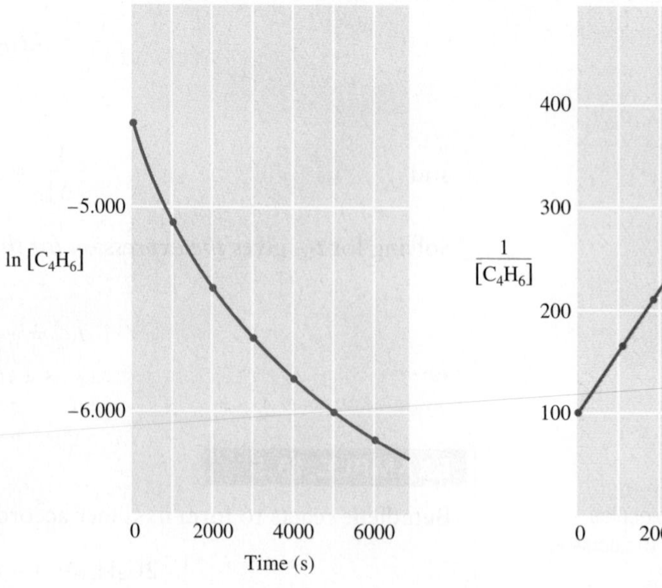

Figure 15.5
(a) A plot of $\ln[C_4H_6]$ versus t.
(b) A plot of $1/[C_4H_6]$ versus t.

however, second order, as shown by the linearity of the $1/[C_4H_6]$ versus t plot. Thus we can now write the rate law for this second-order reaction:

$$\text{Rate} = -\frac{d[C_4H_6]}{dt} = k[C_4H_6]^2$$

b. For a second-order reaction, a plot of $1/[C_4H_6]$ versus t produces a straight line with slope k. In terms of the standard equation for a straight line, $y = mx + b$, we have $y = 1/[C_4H_6]$ and $x = t$. In this case,

$$k = \text{slope} = 6.14 \times 10^{-2}\ \text{L mol}^{-1}\ \text{s}^{-1}$$

c. The expression for the half-life of a second-order reaction is

$$t_{1/2} = \frac{1}{k[A]_0}$$

In this case $k = 6.14 \times 10^{-2}\ \text{L mol}^{-1}\ \text{s}^{-1}$ (from part b) and $[A]_0 = [C_4H_6]_0 = 0.01000\ M$ (the concentration at $t = 0$). Thus

$$t_{1/2} = \frac{1}{(6.14 \times 10^{-2}\ \text{L mol}^{-1}\ \text{s}^{-1})(1.000 \times 10^{-2}\ \text{mol/L})}$$
$$= 1.63 \times 10^3\ \text{s}$$

The initial concentration of C_4H_6 is halved in 1630 s.

For a second-order reaction, $t_{1/2}$ depends on $[A]_0$. For a first-order reaction, $t_{1/2}$ is independent of $[A]_0$.

It is important to recognize the difference between the half-life for a first-order reaction and the half-life for a second-order reaction. For a second-order reaction, $t_{1/2}$ depends on both k and $[A]_0$; for a first-order reaction, $t_{1/2}$ depends only on k. For a first-order reaction, a constant time is required to reduce the concentration of the reactant by half, and then by half again, and so on, as the reaction proceeds. In Example 15.5 we saw that this is *not* true for a second-order reaction. For that second-order reaction, we found that the first half-life (the time required to go from $[C_4H_6] = 0.010\ M$ to $[C_4H_6] = 0.0050\ M$) is 1630 seconds. We can estimate the second half-life from the concentration data as a function of time. Note that to reach $0.0024\ M$ C_4H_6 (approximately 0.0050/2) requires 5200 seconds of reaction time. Thus, to get from $0.0050\ M$ C_4H_6 to $0.0024\ M$ C_4H_6 takes 3570 seconds (5200 − 1630). The second half-life is much longer than the first. This pattern is characteristic of second-order reactions. In fact, *for a second-order reaction, each successive half-life is double the preceding one* (provided the effects of the reverse reaction can be ignored, as we are assuming here). Prove this to yourself by examining the equation $t_{1/2} = 1/(k[A]_0)$.

For each successive half-life, $[A]_0$ is halved. Since $t_{1/2} = 1/k[A]_0$, $t_{1/2}$ doubles.

Zero-Order Rate Laws

Most reactions involving a single reactant show either first-order or second-order kinetics. However, sometimes such a reaction can be a **zero-order reaction.** The rate law for a zero-order reaction is

$$\text{Rate} = k[A]^0 = k(1) = k$$

A zero-order reaction has a constant rate.

For a zero-order reaction, the rate is a constant. It does not change with concentration as it does for first-order or second-order reactions.

The **integrated rate law for a zero-order reaction** is

$$[A] = -kt + [A]_0 \qquad (15.7)$$

In this case a plot of [A] versus t gives a straight line of slope $-k$, as shown in Fig. 15.6.

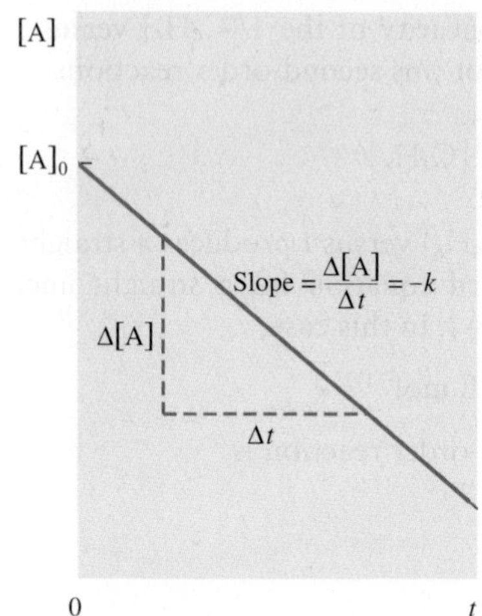

Figure 15.6
A plot of [A] versus *t* for a zero-order reaction.

The expression for the half-life of a zero-order reaction can be obtained from the integrated rate law. By definition, $[A] = [A]_0/2$ when $t = t_{1/2}$, so

$$\frac{[A]_0}{2} = -kt_{1/2} + [A]_0$$

or

$$kt_{1/2} = \frac{[A]_0}{2}$$

Solving for $t_{1/2}$ gives

$$t_{1/2} = \frac{[A]_0}{2k} \tag{15.8}$$

Zero-order reactions are most often encountered when a substance such as a metal surface or an enzyme is required for the reaction to occur. For example, the decomposition reaction

$$2N_2O(g) \longrightarrow 2N_2(g) + O_2(g)$$

occurs on a hot platinum surface. When the platinum surface is completely covered with N_2O molecules, an increase in the concentration of N_2O has no effect on the rate, since only those N_2O molecules on the surface can react. Under these conditions, *the rate is a constant* because it is controlled by what happens on the platinum surface rather than by the total concentration of N_2O, as illustrated in Fig. 15.7. This reaction can also occur at high temperatures with no platinum surface present, but under these conditions it is not zero order.

Integrated Rate Laws for Reactions with More Than One Reactant

So far we have considered the integrated rate laws for simple reactions with only one reactant. Special techniques are required to deal with more complicated reactions. For example, consider the reaction

$$BrO_3^-(aq) + 5Br^-(aq) + 6H^+(aq) \longrightarrow 3Br_2(l) + 3H_2O(l)$$

From experimental evidence we know that the rate law is

$$\text{Rate} = -\frac{d[BrO_3^-]}{dt} = k[BrO_3^-][Br^-][H^+]^2$$

Suppose we run this reaction under conditions where $[BrO_3^-]_0 = 1.0 \times 10^{-3}\ M$, $[Br^-]_0 = 1.0\ M$, and $[H^+]_0 = 1.0\ M$. As the reaction proceeds, $[BrO_3^-]$

Figure 15.7
The decomposition reaction $2N_2O(g) \longrightarrow 2N_2(g) + O_2(g)$ takes place on a platinum surface. Although $[N_2O]$ is twice as great in (b) as in (a), the rate of decomposition of N_2O is the same in both cases since the platinum surface can accommodate only a certain number of molecules. As a result, this reaction is zero order.

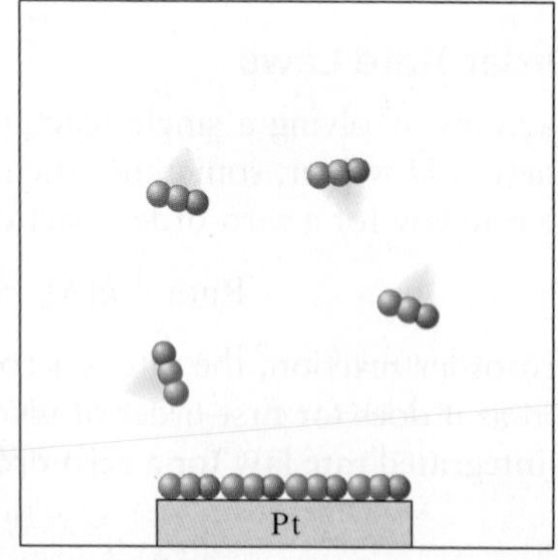

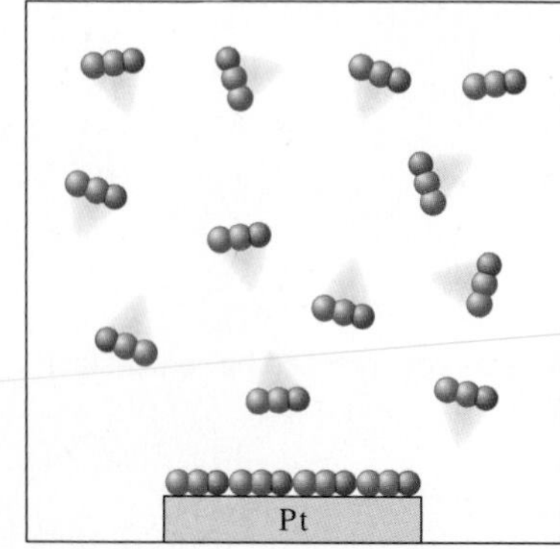

N_2O

(a) (b)

decreases significantly, but because the Br^- ion and H^+ ion concentrations are so large initially, relatively little of either of these two reactants is consumed. Thus $[Br^-]$ and $[H^+]$ remain *approximately constant.* In other words, under the conditions where the Br^- ion and H^+ ion concentrations are much larger than the BrO_3^- ion concentration, we can assume that throughout the reaction

$$[Br^-] = [Br^-]_0 \quad \text{and} \quad [H^+] = [H^+]_0$$

This means that the rate law can be written as

$$\text{Rate} = k[Br^-]_0[H^+]_0^2[BrO_3^-] = k'[BrO_3^-]$$

where, since $[Br^-]_0$ and $[H^+]_0$ are constant,

$$k' = k[Br^-]_0[H^+]_0^2$$

The rate law

$$\text{Rate} = k'[BrO_3^-]$$

is first order. However, since this rate law was obtained by simplifying a more complicated one, it is called a **pseudo-first-order rate law.** Under the conditions of this experiment, a plot of $\ln[BrO_3^-]$ versus t gives a straight line with a slope equal to $-k'$. Since $[Br^-]_0$ and $[H^+]_0$ are known, the value of k can be calculated from the equation

$$k' = k[Br^-]_0[H^+]_0^2$$

which can be rearranged to give

$$k = \frac{k'}{[Br^-]_0[H^+]_0^2}$$

Note that the kinetics of complicated reactions can be studied by observing the behavior of one reactant at a time. If the concentration of one reactant is much smaller than the concentrations of the others, then the amounts of those reactants present in large concentrations will not change significantly and can be regarded as constant. The change in concentration with time of the reactant present in a relatively small amount can then be used to determine the order of the reaction in that component. This technique allows us to determine rate laws for complex reactions.

15.5 Rate Laws: A Summary

In the last several sections, we have developed the following important points:

1. To simplify the rate laws for reactions, we have always assumed that the rate is being studied under conditions where only the forward reaction is important. This produces rate laws that contain only reactant concentrations.
2. There are two types of rate laws.
   a. The differential rate law (often called *the rate law*) shows how the rate depends on the concentrations. The forms of the rate laws for zero-order, first-order, and second-order kinetics of reactions with a single reactant are shown in Table 15.6.
   b. The integrated rate law shows how concentration depends on time. The integrated rate laws corresponding to zero-order, first-order, and second-order kinetics of reactions with a single reactant are given in Table 15.6.
3. Whether we determine the differential rate law or the integrated rate law depends on the type of data that can be collected conveniently and accurately. Once we have experimentally determined either type of rate law for a given reaction, we can write the other rate law.

Table 15.6

Summary of the Kinetics for Reactions of the Type $aA \longrightarrow$ Products That Are Zero, First, or Second Order in [A]

	Order		
	Zero	First	Second
Rate law	Rate $= k$	Rate $= k[A]$	Rate $= k[A]^2$
Integrated rate law	$[A] = -kt + [A]_0$	$\ln[A] = -kt + \ln[A]_0$	$\frac{1}{[A]} = kt + \frac{1}{[A]_0}$
Plot needed to give a straight line	[A] versus t	ln[A] versus t	$\frac{1}{[A]}$ versus t
Relationship of rate constant to the slope of the straight line	Slope $= -k$	Slope $= -k$	Slope $= k$
Half-life	$t_{1/2} = \frac{[A]_0}{2k}$	$t_{1/2} = \frac{0.693}{k}$	$t_{1/2} = \frac{1}{k[A]_0}$

4. The most common method for experimentally determining the differential rate law is the method of initial rates. In this method several experiments are run at different initial concentrations, and the instantaneous rates are determined for each at the same value of t as close to $t = 0$ as possible. The point is to evaluate the rate before the concentrations change significantly from the initial values. From a comparison of the initial rates and the initial concentrations, the dependence of the rate on the concentrations of various reactants can be obtained—that is, the order in each reactant can be determined.
5. To experimentally determine the integrated rate law for a reaction, we measure concentrations at various values of t as the reaction proceeds. Then we see which integrated rate law correctly fits the data. Typically, this is done by ascertaining which type of plot gives a straight line. This information is described for reactions with a single reactant in Table 15.6. Once the correct straight-line plot is found, the correct integrated rate law can be chosen and the value of k obtained from the slope. Also, the (differential) rate law for the reaction can then be written.
6. The integrated rate law for a reaction that involves several reactants can be treated by choosing conditions such that the concentration of only one reactant varies in a given experiment. This is done by having the concentration of one reactant be small compared with the concentrations of all the others, causing a rate law such as

$$\text{Rate} = k[A]^n[B]^m[C]^p$$

to reduce to

$$\text{Rate} = k'[A]^n$$

where $k' = k[B]_0^m[C]_0^p$, $[B]_0 \gg [A]_0$, and $[C]_0 \gg [A]_0$. The value of n is obtained by determining whether a plot of [A] versus t is linear ($n = 0$), a plot of ln[A] versus t is linear ($n = 1$), or a plot of 1/[A] versus t is linear ($n = 2$). The value of k' is determined from the slope of the appropriate plot. The values of m, p, and k are found by determining the value of k' at several different concentrations of B and C.

15.6 Reaction Mechanisms

Most chemical reactions occur by a *series of steps* called the **reaction mechanism.** To understand a reaction, we must know its mechanism, and one of the main purposes for studying kinetics is to learn as much as possible about the steps involved in a reaction. In this section we explore some of the fundamental characteristics of reaction mechanisms.

Consider the reaction between nitrogen dioxide and carbon monoxide:

$$NO_2(g) + CO(g) \longrightarrow NO(g) + CO_2(g)$$

The rate law for this reaction is known from experiment to be

$$\text{Rate} = k[NO_2]^2$$

A balanced equation does not tell us *how* the reactants become products.

As we will see below, this reaction is more complicated than it appears from the balanced equation. This is quite typical; the balanced equation for a reaction tells us the reactants, the products, and the stoichiometry but gives no direct information about the reaction mechanism.

For the reaction between nitrogen dioxide and carbon monoxide, the mechanism is thought to involve the steps shown in Fig. 15.8, where k_1 and k_2 are the rate constants of the individual reactions. In this mechanism gaseous NO_3 is an **intermediate,** a species that is neither a reactant nor a product but that is formed and consumed during the reaction sequence (Fig. 15.8).

An intermediate is formed in one step and consumed in a subsequent step and so is never seen as a product.

Each of these reactions is called an **elementary step,** *a reaction whose rate law can be written from its molecularity.* **Molecularity** is defined as the number of species that must collide to produce the reaction indicated by that step. A reaction involving one molecule is called a **unimolecular step.** Reactions involving the collision of two and three species are termed **bimolecular** and **termolecular,** respectively. Termolecular steps are quite rare because the probability of three molecules colliding simultaneously is very small. Examples of these three types of elementary steps and the corresponding rate laws are shown in Table 15.7. Note from Table 15.7 that the rate law for an elementary step follows *directly* from the molecularity of that step. For example, for a bimolecular step, the rate law is always second order, either of the form $k[A]^2$ for a step with a single reactant or of the form $k[A][B]$ for a step involving two reactants.

The prefix *uni-* means "one," *bi-* means "two," and *ter-* means "three."

A unimolecular elementary step is always first order, a bimolecular step is always second order, and so on.

We can now define a reaction mechanism more precisely. It is a *series of elementary steps that must satisfy two requirements:*

1. The sum of the elementary steps must give the overall balanced equation for the reaction.
2. The mechanism must agree with the experimentally determined rate law.

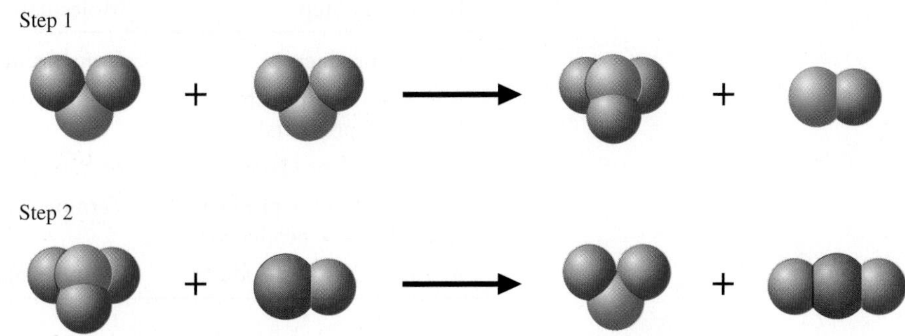

Figure 15.8
A molecular representation of the elementary steps in the reaction of NO_2 and CO.

Chemical Insights Ultracold Reactions

More than 200 years ago the work of Charles Gay-Lussac led to the suspicion that an absolute low temperature exists for matter. In recent years scientists have come close to cooling matter to 0 K. Temperatures of about 10^{-7} K have been achieved by *laser cooling* in which a laser beam is directed against a beam of atoms, dramatically slowing the movement of atoms to near zero.

It is extremely difficult to slow atoms to near absolute zero, but it is even harder to cool molecules to these temperatures. One strategy is to begin with "hot" diatomic molecules and then cool them down using lasers. However, as the velocities of these molecules decrease, they retain high levels of vibrational energy that ultimately cause the molecules to break apart. The other strategy is to cool atoms and then try to get them to form molecules. This is the approach used by physicist Jun Ye and his coworkers at the University of Colorado in Boulder. Ye and his research group were able to form molecules of potassium and rubidium at ultracold temperatures and then encourage the molecules to emit their excess vibrational energy using a laser pulse to produce stable molecules at about 2×10^{-7} K. Because the K-Rb molecule is slightly polar (the potassium atom has a slightly greater electronegativity value than the rubidium atom, which leads to a dipole $^{\delta-}$K-Rb$^{\delta+}$), the molecules can be manipulated by an electric field. Ye and his coworkers have observed the reaction

$$2\text{K-Rb} \longrightarrow K_2 + Rb_2$$

at these ultralow temperatures. Because the motions of the molecules are so slow at these temperatures, the researchers have been able to study the details of this reaction. For example, they found that the reaction occurs only when the molecules are head to tail and that the reaction is greatly speeded up when the electric field around the molecule is increased. They also observed that the molecules react with greater certainty when they are within 1 nm of each other.

Theoretical studies by Paul Julienne at the National Institute of Standards and Technology in Maryland and Zbigniew Idziaczek at the University of Warsaw predict how the molecules find each other so that they can react. At these low temperatures, the molecules behave more like diffuse waves than discrete spots, and the waves can spread beyond the individual molecules for hundreds of nanometers. These waves affect how the molecules close in on each other to react. As more of these studies are done at ultracold temperatures, we are certain to learn a lot more about how molecules react with each other.

Table 15.7

Examples of Elementary Steps and Corresponding Rate Laws

Elementary Step	Molecularity	Rate Law
A $\longrightarrow$ products	*Uni*molecular	Rate = $k[A]$
A + A $\longrightarrow$ products (2A $\longrightarrow$ products)	*Bi*molecular	Rate = $k[A]^2$
A + B $\longrightarrow$ products	*Bi*molecular	Rate = $k[A][B]$
A + A + B $\longrightarrow$ products (2A + B $\longrightarrow$ products)	*Ter*molecular	Rate = $k[A]^2[B]$
A + B + C $\longrightarrow$ products	*Ter*molecular	Rate = $k[A][B][C]$

Christopher Rose-Petruck Studies Ultrafast X-ray Spectroscopy

Chemistry Explorers

Christopher Rose-Petruck, Associate Professor of Chemistry at Brown University, was educated in Germany, receiving his Ph.D. at the Ludwig-Maximilians University in Munich.

Professor Rose-Petruck's research focuses on the detailed mechanisms of certain important reactions that occur in solution. Because molecules undergo structural changes as they react on the timescale of picoseconds (10^{-12} s) and femtoseconds (10^{-15} s), "seeing" these changes occur requires ultrafast spectroscopic techniques. Professor Rose-Petruck and his group use ultrafast X-ray spectroscopy to study reactions as they occur in solutions. As part of their research, they have designed a reliable laboratory instrument that delivers the ultrashort X-ray pulses needed for ultrafast X-ray spectroscopy. To do this Professor Rose-Petruck developed a high power laser–driven X-ray source.

One system that Professor Rose-Petruck has been investigating involves the reactions of the iron carbonyl compound $Fe(CO)_5$ in solution. He and his students have found that $Fe(CO)_5$ rearranges in solution in a way that allows a solvent molecule to directly attach to the iron atom. This knowledge is important in understanding how $Fe(CO)_5$ and similar compounds react in solution.

Christopher Rose-Petruck with his children.

Courtesy of Gabriela Petruck/Christopher Rose-Petruck

Professor Rose-Petruck plans to continue his studies of chemical reactions with the goal of "seeing" the reactants approach each other, form some sort of activated complex, and follow the products as they separate.

To see how these requirements are applied, we will consider the mechanism given above for the reaction of nitrogen dioxide with carbon monoxide. First, note that the sum of the two steps gives the overall balanced equation:

$$NO_2(g) + NO_2(g) \longrightarrow NO_3(g) + NO(g)$$
$$NO_3(g) + CO(g) \longrightarrow NO_2(g) + CO_2(g)$$
$$\cancel{NO_2}(g) + NO_2(g) + \cancel{NO_3}(g) + CO(g) \longrightarrow \cancel{NO_3}(g) + NO(g) + \cancel{NO_2}(g) + CO_2(g)$$
$$\text{Overall reaction:} \quad NO_2(g) + CO(g) \longrightarrow NO(g) + CO_2(g)$$

The first requirement for a correct mechanism is met. To see whether the mechanism meets the second requirement, we need to introduce a new concept: the **rate-determining step.*** Multistep reactions may have one step that is much slower than all the others. Reactants can become products only as fast as they can complete this slowest step. That is, the overall reaction can be no faster than the slowest or rate-determining step in the sequence. An analogy for this situation is the rapid pouring of water through a funnel into a container. The water collects in the container at a rate that is essentially determined by the size of the funnel opening and not by the rate of pouring.

A reaction is only as fast as its slowest step.

*The concept of a rate-controlling step in a reaction mechanism is complex and often misleading. For a comprehensive discussion of this issue, see "Rate-Controlling Step: A Necessary or Useful Concept," by Keith J. Laidler, *J. Chem. Ed.* 65 (1988): 250.

The rate at which this colored solution enters the flask is determined by the size of the funnel stem, not how fast the solution is poured.

Which is the rate-determining step in the reaction of nitrogen dioxide with carbon monoxide? Let's *assume* that the first step is rate determining and the second step is relatively fast:

$$NO_2(g) + NO_2(g) \longrightarrow NO_3(g) + NO(g) \quad \text{Slow (rate determining)}$$

$$NO_3(g) + CO(g) \longrightarrow NO_2(g) + CO_2(g) \quad \text{Fast}$$

What we have really assumed here is that the formation of NO_3 occurs much more slowly than its reaction with CO. The rate of CO_2 production is then controlled by the rate of formation of NO_3 in the first step. Since this is an elementary step, we can write the rate law from the molecularity. The bimolecular first step has the rate law

$$\text{Rate of formation of } NO_3 = \frac{d[NO_3]}{dt} = k_1[NO_2]^2$$

Since the overall reaction rate can be no faster than the slowest step,

$$\text{Overall rate} = k_1[NO_2]^2$$

Note that this rate law agrees with the experimentally determined rate law given earlier. Since the mechanism we assumed above satisfies the two requirements stated earlier, it *may* be the correct mechanism for the reaction.

EXAMPLE 15.6

The balanced equation for the reaction of gaseous nitrogen dioxide and fluorine is

$$2NO_2(g) + F_2(g) \longrightarrow 2NO_2F(g)$$

The experimentally determined rate law is

$$\text{Rate} = k[NO_2][F_2]$$

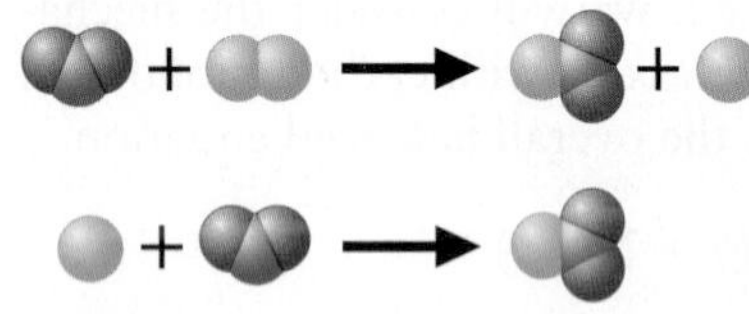

A suggested mechanism for this reaction is

$$NO_2 + F_2 \xrightarrow{k_1} NO_2F + F \quad \text{Slow}$$

$$F + NO_2 \xrightarrow{k_2} NO_2F \quad \text{Fast}$$

Is this an acceptable mechanism? That is, does it satisfy the two requirements?

Solution The first requirement for an acceptable mechanism is that the sum of the steps give the balanced equation:

$$NO_2 + F_2 \longrightarrow NO_2F + F$$

$$F + NO_2 \longrightarrow NO_2F$$

$$2NO_2 + F_2 + \cancel{F} \longrightarrow 2NO_2F + \cancel{F}$$

$$\text{Overall reaction:} \quad 2NO_2 + F_2 \longrightarrow 2NO_2F$$

The first requirement is met.

The second requirement is that the mechanism must agree with the experimentally determined rate law. Since the proposed mechanism states that the first step is rate determining, the overall reaction rate must be that of the first step. The first step is bimolecular, so the rate law is

$$\text{Rate} = k_1[NO_2][F_2]$$

This has the same form as the experimentally determined rate law. The proposed mechanism is acceptable because it satisfies both requirements. (Note that we have not proved it is *the correct* mechanism.)

How does a chemist deduce the mechanism for a given reaction? The rate law is always determined first. Then, using chemical intuition and following the two rules given above, the chemist constructs a possible mechanism. Until recently, we could not prove a mechanism absolutely. We could only say that a mechanism that satisfies the two requirements is *possibly* correct. Recent advances in spectroscopy, however, such as those pioneered by Professor Ahmed Zewail (see "Femtochemistry" on page 721) and the use of STM techniques to study reactions, have provided the means for identifying mechanisms exactly (see "Seeing Reaction Mechanisms" on page 744). Deducing the mechanism for a chemical reaction is difficult; it requires skill and experience. We will only touch on this process in this text.

Mechanisms with Fast Forward and Reverse First Steps

A common type of reaction mechanism is one involving a first step in which *both* the forward and reverse reactions are very fast compared with the reactions in the second step. An example of this type of mechanism is that for the decomposition of ozone to oxygen. The balanced reaction is

$$2O_3(g) \longrightarrow 3O_2(g)$$

The observed rate law is

$$\text{Rate} = k\frac{[O_3]^2}{[O_2]}$$

Note that this rate law is unusual in that it contains the concentration of a *product*. The mechanism proposed for this process is

$$O_3 \underset{k_{-1}}{\overset{k_1}{\rightleftharpoons}} O_2 + O$$

$$O + O_3 \xrightarrow{k_2} 2O_2$$

The double arrows in the first step indicate that both the forward and reverse reactions are important. They have the rate constants k_1 and k_{-1}, respectively.

The second step is relatively slow because of the very small concentration of O_3 molecules.

For this mechanism we will assume that *both* the forward and reverse reactions of the first step are very fast compared with the reaction in the second step. This means that the second step is rate determining. Therefore, the rate for the overall reaction is equal to the rate of the second step:

$$\text{Rate} = k_2[O][O_3]$$

This rate law does not have the same form as the experimentally determined rate law. For one thing, it contains the concentration of the intermediate, an oxygen atom. We can remove [O] and obtain a rate law that agrees with the experiment results by making an additional assumption. We assume that the rates of the forward and reverse reactions in the first step are equal. That is, we assume that the initial reversible fast step is at equilibrium. This makes sense because the rates of both the forward and reverse reactions for the first step are so much faster than the rate of the second step. For the first step,

$$\text{Rate of forward reaction} = k_1[O_3]$$

and $$\text{Rate of reverse reaction} = k_{-1}[O_2][O]$$

At equilibrium we have

$$k_1[O_3] = k_{-1}[O_2][O]$$

Chemical Insights Seeing Reaction Mechanisms

Until very recently chemists could only guess at the details of reaction mechanisms. New developments such as femtosecond spectroscopy (see "Femtochemistry" on page 721) and scanning tunneling microscopy (STM; see "Seeing Atoms" on page 22) have enabled scientists to begin to see the details of chemical reactions.

A perfect example of these new techniques is furnished by the study of the oxidation of carbon monoxide by oxygen to form carbon dioxide by Professor Wilson Ho and his coworkers at the University of California at Irvine. Taking advantage of the single-molecule dexterity of the tip of an STM instrument, Ho and his colleagues maneuvered a single CO molecule on a silver surface until it was near an adsorbed O_2 molecule (Fig. 15.9). At this point an O—CO—O intermediate formed on the surface. A little shot of energy from the STM tip caused the intermediate to form CO_2, which exited the surface while leaving a lone O atom behind. In another study Ho and his coworkers found that CO_2 could be formed by depositing a CO molecule from the STM tip onto an oxygen atom on the surface.

The new techniques of femtosecond spectroscopy and STM are revolutionizing the way chemistry is carried out. We are now at the point where we can "see" the detailed pathways of chemical reactions.

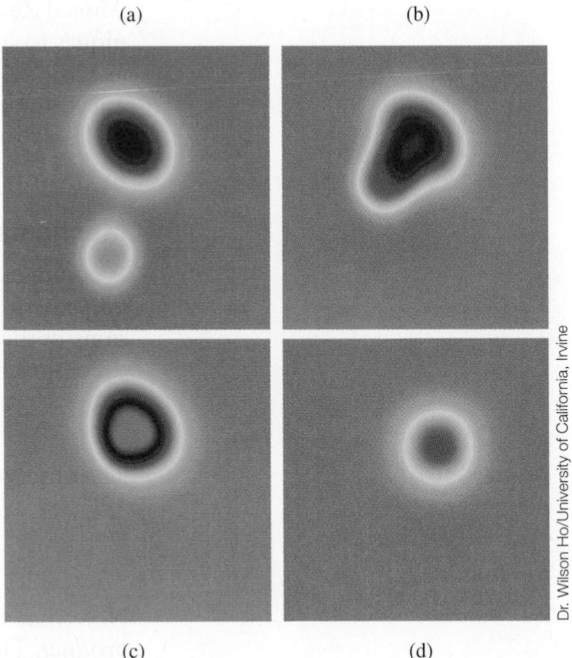

Figure 15.9
The STM images of the reaction of CO and O_2. (a) An O_2 molecule (oval) and a CO molecule (circle) on a surface. As the two molecules approach each other, a reaction occurs (b) to form an O—CO—O complex (c). After an electron pulse is applied to the complex, a newly formed CO_2 molecule exits the surface, leaving behind a single O atom (d).

We solve for [O]:

$$[O] = \frac{k_1[O_3]}{k_{-1}[O_2]}$$

Now we substitute the expression for [O] into the rate law for the second step:

$$\text{Rate} = k_2[O][O_3] = k_2\left(\frac{k_1[O_3]}{k_{-1}[O_2]}\right)[O_3] = \frac{k_2k_1[O_3]^2}{k_{-1}[O_2]}$$

$$= k\frac{[O_3]^2}{[O_2]}$$

where k is a composite constant representing k_2k_1/k_{-1}.

This rate law, *derived* by postulating the two elementary steps and making assumptions about the relative rates of these steps, agrees with the experimental rate law. Since this mechanism (the elementary steps *plus* the assumptions) also gives the correct overall stoichiometry, it is an acceptable mechanism for the decomposition of ozone to oxygen.

EXAMPLE 15.7

The gas-phase reaction of chlorine with chloroform is described by the equation

$$Cl_2(g) + CHCl_3(g) \longrightarrow HCl(g) + CCl_4(g)$$

The rate law determined from experiment has a noninteger order:

$$\text{Rate} = k[Cl_2]^{1/2}[CHCl_3]$$

A proposed mechanism for this reaction follows:

$$Cl_2(g) \underset{k_{-1}}{\overset{k_1}{\rightleftharpoons}} 2Cl(g) \qquad \text{Both fast with equal rates (fast equilibrium)}$$

$$Cl(g) + CHCl_3(g) \xrightarrow{k_2} HCl(g) + CCl_3(g) \qquad \text{Slow}$$

$$CCl_3(g) + Cl(g) \xrightarrow{k_3} CCl_4(g) \qquad \text{Fast}$$

Is this an acceptable mechanism for the reaction?

Solution Two questions must be answered. First, does the mechanism give the correct overall stoichiometry? Adding the three steps does yield the correct balanced equation:

$$Cl_2(g) \rightleftharpoons 2Cl(g)$$

$$Cl(g) + CHCl_3(g) \longrightarrow HCl(g) + CCl_3(g)$$

$$CCl_3(g) + Cl(g) \longrightarrow CCl_4(g)$$

$$Cl_2(g) + \cancel{Cl}(g) + CHCl_3(g) + \cancel{CCl_3}(g) + \cancel{Cl}(g) \longrightarrow \cancel{2Cl}(g) + HCl(g) + \cancel{CCl_3}(g) + CCl_4(g)$$

$$\text{Overall reaction:} \quad Cl_2(g) + CHCl_3(g) \longrightarrow HCl(g) + CCl_4(g)$$

Second, does the mechanism agree with the observed rate law? Since the overall reaction rate is determined by the rate of the slowest step,

$$\text{Overall rate} = \text{rate of second step} = k_2[Cl][CHCl_3]$$

Since the chlorine atom is an intermediate, we must find a way to eliminate [Cl] in the rate law. This can be done by recognizing that since the first step is at equilibrium, its forward and reverse rates are equal:

$$k_1[Cl_2] = k_{-1}[Cl]^2$$

Solving for $[Cl]^2$ gives

$$[Cl]^2 = \frac{k_1[Cl_2]}{k_{-1}}$$

Taking the square root of both sides yields

$$[Cl] = \left(\frac{k_1}{k_{-1}}\right)^{1/2} [Cl_2]^{1/2}$$

and

$$\text{Rate} = k_2[Cl][CHCl_3] = k_2\left(\frac{k_1}{k_{-1}}\right)^{1/2} [Cl_2]^{1/2}[CHCl_3] = k[Cl_2]^{1/2}[CHCl_3]$$

where

$$k = k_2\left(\frac{k_1}{k_{-1}}\right)^{1/2}$$

The rate law derived from the mechanism agrees with the experimentally observed rate law. This mechanism satisfies the two requirements and thus is an acceptable mechanism.

15.7 The Steady-State Approximation

In the simplest reaction mechanisms, one particular step is usually rate determining. However, it is not unusual in complex, multistep reaction mechanisms for different steps to be rate determining under different sets of conditions.

In cases where a specific rate-determining step cannot be chosen, an analysis called the **steady-state approximation** is often used. The central feature of this method is the assumption that the concentration of any intermediate remains constant as the reaction proceeds. An intermediate is neither a reactant nor a product but something that is formed and then consumed as the reaction proceeds.

For example, the reaction between nitric oxide and hydrogen,

$$2NO(g) + H_2(g) \longrightarrow N_2O(g) + H_2O(g)$$

may proceed via the following mechanism:

1. $2NO \underset{k_{-1}}{\overset{k_1}{\rightleftharpoons}} N_2O_2$

2. $N_2O_2 + H_2 \xrightarrow{k_2} N_2O + H_2O$

In this mechanism the intermediate is N_2O_2. To apply the steady-state approximation to this mechanism, we assume the concentration of N_2O_2 remains constant. That is,

$$\frac{d[N_2O_2]}{dt} = 0$$

Next, we will identify the steps that produce N_2O_2 and those that consume N_2O_2 and write the rate law for each. Then we will apply the condition that the concentration of N_2O_2 is constant by setting the total rate of production of N_2O_2 equal to the total rate of consumption of N_2O_2. That is, if

$$\text{Rate of production of } N_2O_2 = \text{rate of consumption of } N_2O_2$$

then

$$\frac{d[N_2O_2]}{dt} = 0$$

Rate of Production of N_2O_2

In this mechanism N_2O_2 is produced only in the forward part of the first elementary step,

$$2NO \xrightarrow{k_1} N_2O_2$$

and the rate law for this step is

$$\frac{d[N_2O_2]}{dt} = k_1[NO]^2$$

Rate of Consumption of N_2O_2

In this mechanism N_2O_2 is consumed in the reverse part of the first step,

$$2NO \xleftarrow{k_{-1}} N_2O_2$$

and in the second step,

$$N_2O_2 + H_2 \xrightarrow{k_2} N_2O + H_2O$$

The rate laws for these steps are

$$-\frac{d[N_2O_2]}{dt} = k_{-1}[N_2O_2] \quad \text{and} \quad -\frac{d[N_2O_2]}{dt} = k_2[N_2O_2][H_2]$$

The Steady-State Condition

Now we equate the rates of production and consumption of N_2O_2:

$$\underset{\text{Rate of production}}{k_1[\text{NO}]^2} = \underset{\text{Total rate of consumption}}{k_{-1}[\text{N}_2\text{O}_2] + k_2[\text{N}_2\text{O}_2][\text{H}_2]}$$

The Rate Law for the Overall Reaction

Next, we will write the rate law for the overall reaction,

$$2\text{NO} + \text{H}_2 \longrightarrow \text{N}_2\text{O} + \text{H}_2\text{O}$$

We can do this in several ways, depending on which reactant or product we use to represent the rate. In this case we will choose the decomposition of H_2 to define the rate:

$$\text{Rate of reaction} = -\frac{d[\text{H}_2]}{dt}$$

Note that H_2 is consumed only in the second step of the mechanism:

$$\text{N}_2\text{O}_2 + \text{H}_2 \xrightarrow{k_1} \text{N}_2\text{O} + \text{H}_2\text{O}$$

Thus the rate law is

$$-\frac{d[\text{H}_2]}{dt} = k_2[\text{N}_2\text{O}_2][\text{H}_2]$$

However, this is not the final form of the rate law for the overall reaction because it contains the concentration of an intermediate. We can remove this concentration from the rate law by solving the steady-state expression

$$k_1[\text{NO}]^2 = k_{-1}[\text{N}_2\text{O}_2] + k_2[\text{N}_2\text{O}_2][\text{H}_2]$$

for $[N_2O_2]$:

$$[\text{N}_2\text{O}_2] = \frac{k_1[\text{NO}]^2}{k_{-1} + k_2[\text{H}_2]}$$

We substitute this expression into the rate law,

$$\text{Rate} = -\frac{d[\text{H}_2]}{dt} = k_2[\text{N}_2\text{O}_2][\text{H}_2]$$

to give

$$\text{Rate} = -\frac{d[\text{H}_2]}{dt} = k_2[\text{H}_2]\left(\frac{k_1[\text{NO}]^2}{k_{-1} + k_2[\text{H}_2]}\right)$$

or

$$\text{Rate} = -\frac{d[\text{H}_2]}{dt} = \frac{k_2k_1[\text{H}_2][\text{NO}]^2}{k_{-1} + k_2[\text{H}_2]}$$

This is the overall rate law for the proposed mechanism based on the steady-state analysis. Note that this rate law is quite complicated, which is common for rate laws obtained by assuming steady-state conditions. The usual practice for testing the validity of a complicated rate law involves choosing concentration conditions that produce a simpler form of the rate law. For example, if the reaction between NO and H_2 is studied under conditions where the concentration of H_2 is large enough so that

$$k_2[\text{H}_2] \gg k_{-1}$$

then the full rate law

$$\frac{k_2k_1[H_2][NO]^2}{k_{-1} + k_2[H_2]}$$

reduces to

$$\frac{k_2k_1[H_2][NO]^2}{k_2[H_2]} = k_1[NO]^2$$

Thus, at sufficiently high concentrations of H_2, the reaction should show second-order dependence on the concentration of nitric oxide if the suggested mechanism is valid.

On the other hand, at low concentrations of H_2, such that

$$k_{-1} \gg k_2[H_2]$$

the rate law reduces to the form

$$\text{Rate} = \frac{k_2k_1}{k_{-1}}[H_2][NO]^2 = k[H_2][NO]^2$$

Studies of the reaction under these conditions should show first-order dependence on $[H_2]$, as well as second-order dependence on [NO], if the suggested mechanism is valid.

The ideas we have developed above for a specific mechanism can be generalized as follows.

STEPS

Analyzing a Mechanism Using the Steady-State Approximation

1 Write the proposed mechanism (the elementary steps).

2 Construct a steady-state expression for each intermediate I by applying the criterion

$$\frac{d[I]}{dt} = 0$$

which means that

Rate of production of I = rate of consumption of I

This condition is implemented by identifying each step that produces or consumes I and writing the appropriate rate law for each. The sum of the rate laws that produce I are then set equal to the sum of the rate laws that consume I.

3 From the steady-state approximation for each intermediate $I_1, I_2, \ldots,$ solve for $[I_1], [I_2], \ldots.$

4 Construct the rate law for the overall reaction in terms of one of the reactants or products. The decision about which reactant or product to use in constructing the rate law is based on convenience.

5 Use the expressions from step 3 for $[I_1], [I_2], \ldots$ to substitute for the concentrations of intermediates found in the rate law for step 4. The goal is to obtain an overall rate law that contains only reactant and/or product concentrations.

15.8 A Model for Chemical Kinetics

How do chemical reactions occur? We already have given some indications. For example, we have seen that the rates of chemical reactions depend on the concentrations of the reacting species. The initial rate for the reaction

$$aA + bB \longrightarrow \text{products}$$

can be described by the rate law

$$\text{Rate} = k[A]^n[B]^m$$

where the order of each reactant depends on the detailed reaction mechanism. This explains why reaction rates depend on concentration. But what about some of the other factors affecting reaction rates? For example, how does temperature affect the speed of a reaction?

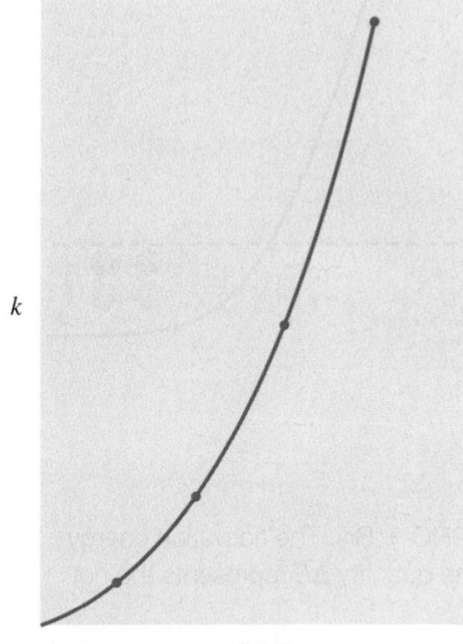

Figure 15.10
A plot showing the exponential dependence of the rate constant on absolute temperature. The exact temperature dependence of k is different for each reaction. This plot represents the behavior of a rate constant that doubles for every increase in temperature of 10 K.

We can answer this question qualitatively from our experience. We use refrigerators because food spoilage is retarded at low temperatures. The combustion of wood occurs at a measurable rate only at high temperatures. An egg cooks in boiling water much faster at sea level than in Leadville, Colorado (elevation 10,000 feet), where the boiling point of water is about 90°C. These observations and others lead us to conclude that *chemical reactions speed up when the temperature is increased.* Experiments have shown that virtually all rate constants show an exponential increase with absolute temperature, as represented in Fig. 15.10.

In this section we will introduce a model that can be used to account for the observed characteristics of reaction rates. This model, the **collision model,** is built around the central idea that *molecules must collide to react.* We have already seen that this assumption can explain the concentration dependence of reaction rates. Now we need to consider whether this model can also account for the observed temperature dependence of reaction rates.

The kinetic molecular theory of gases predicts that an increase in temperature increases molecular velocities and so increases the frequency of intermolecular collisions. This agrees with the observation that reaction rates are greater at higher temperatures. Thus there is qualitative agreement between the collision model and experimental observations. However, it is found that the rate of reaction is much smaller than the calculated collision frequency in a given collection of gas particles. This must mean that *only a small fraction of the collisions produces a reaction.* Why?

This question was first addressed in the 1880s by Svante Arrhenius. He proposed the existence of a *threshold energy,* called the **activation energy,** that must be overcome to produce a chemical reaction. We can see that this proposal makes sense by considering the decomposition of BrNO in the gas phase:

$$2BrNO(g) \longrightarrow 2NO(g) + Br_2(g)$$

In this reaction two Br—N bonds must be broken and one Br—Br bond must be formed. Breaking a Br—N bond requires considerable energy (243 kJ/mol), which must come from somewhere. The collision model postulates that the energy required to break the bonds comes from the kinetic energies possessed by the reacting molecules before the collision. This kinetic energy is changed into potential energy as the molecules are distorted during a collision, breaking bonds and rearranging the atoms into the product molecules.

The higher the activation energy, the slower is the reaction at a given temperature.

We can envision the reaction progress as shown in Fig. 15.11. The arrangement of atoms found at the top of the potential energy "hill," or barrier, is called the **activated complex,** or **transition state.** The conversion of BrNO to NO and Br_2 is exothermic, as indicated by the fact that the products have

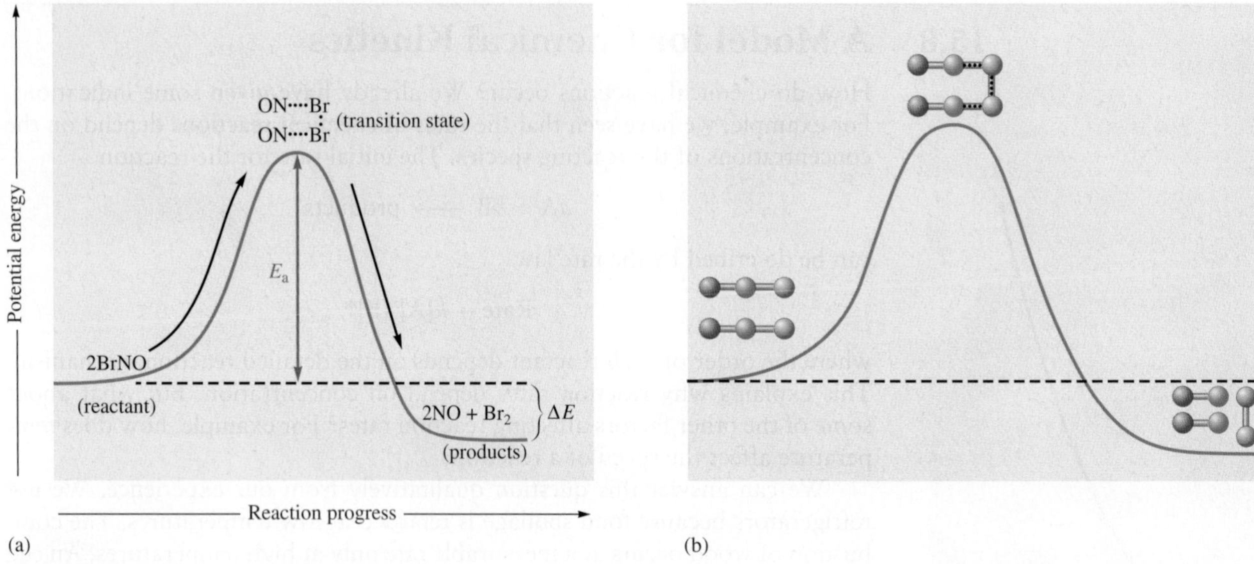

Figure 15.11
(a) The change in potential energy as a function of reaction progress for the reaction $2BrNO \longrightarrow 2NO + Br_2$. The activation energy E_a represents the energy needed to disrupt the BrNO molecules so that they can form products. The quantity ΔE represents the net change in energy in going from reactant to products. (b) A molecular representation of the reaction.

lower energy than the reactant. However, ΔE has no effect on the rate of the reaction. Rather, the rate depends on the size of the activation energy E_a.

The main point here is that a certain minimum energy is required for two BrNO molecules to "get over the hill" so that products can form. This is furnished by the collision energy. A collision between two BrNO molecules with small kinetic energies will not have enough energy to get over the barrier, and no reaction occurs. At a given temperature, only a certain fraction of the collisions possess enough energy to be effective and thus to result in product formation.

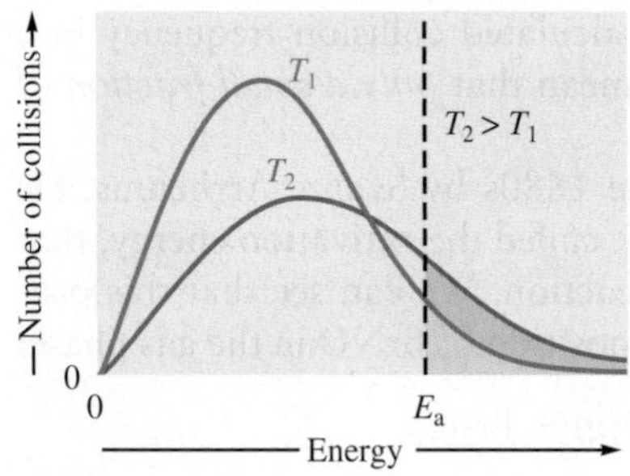

Figure 15.12
Plot showing the number of collisions with a particular energy at T_1 and T_2, where $T_2 > T_1$.

We can be more precise by recalling from Chapter 5 that a distribution of velocities occurs in a sample of gas molecules. Therefore, a distribution of collision energies also occurs, as shown in Fig. 15.12 for two different temperatures. Figure 15.12 also shows the activation energy for the reaction in question. Only collisions with energy greater than the activation energy are able to react (get over the barrier). At the lower temperature T_1, the fraction of effective collisions is quite small. However, as the temperature is increased to T_2, the fraction of collisions with the required activation energy increases dramatically. When the temperature is doubled, the fraction of effective collisions much more than doubles. In fact, the fraction of effective collisions increases *exponentially* with temperature. This agrees with the observation that rates of reactions increase exponentially with temperature.

Arrhenius postulated that the number of collisions having an energy equal to or greater than the activation energy is given by the expression

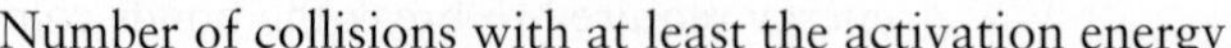

$$\text{Number of collisions with at least the activation energy} = (\text{total number of collisions})e^{-E_a/RT}$$

where E_a is the activation energy, R is the universal gas constant, and T is the Kelvin temperature. The factor $e^{-E_a/RT}$ represents the fraction of collisions with energy E_a or greater at temperature T.

We have seen that not all molecular collisions are effective in producing chemical reactions because a minimum energy is required for the reaction to occur. There is, however, another complication. Experiments show that the

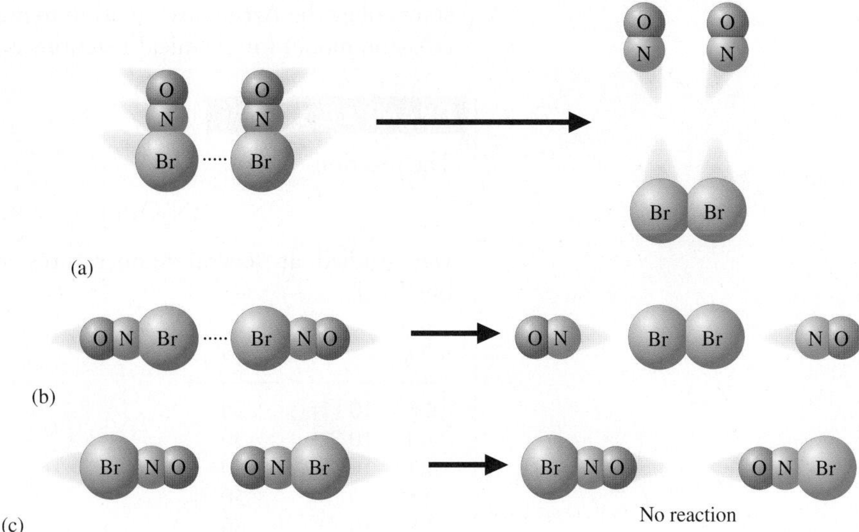

Figure 15.13
Several possible orientations for a collision between two BrNO molecules. Orientations (a) and (b) can lead to a reaction, but orientation (c) cannot.

observed reaction rate is considerably smaller than the rate of collisions with enough energy to surmount the barrier. This means that many collisions, even though they have the required energy, still do not produce a reaction. Why not?

The answer lies in the **molecular orientations** during collisions. We can illustrate this effect by using the reaction between two BrNO molecules, as shown in Fig. 15.13. Some collision orientations can lead to reaction, and others cannot. Therefore, we must include a correction factor to allow for collisions with nonproductive molecular orientations.

To summarize, two requirements must be satisfied for reactants to collide successfully (to rearrange to form products):

1. The collision must involve enough energy to produce the reaction; that is, the collision energy must equal or exceed the activation energy.
2. The relative orientations of the reactants must allow formation of any new bonds necessary to produce products.

Taking these factors into account, we can represent the rate constant as

$$k = zpe^{-E_a/RT}$$

where z is the collision frequency (the total number of collisions per second). The factor p in this expression, called the **steric factor,** reflects the fraction of collisions with effective orientations. Recall that the factor $e^{-E_a/RT}$ represents the fraction of collisions with sufficient energy to produce a reaction. This expression is most often written in the form

$$k = Ae^{-E_a/RT} \tag{15.9}$$

which is called the **Arrhenius equation.** In this equation A, which replaces zp, is called the **pre-exponential factor,** or **frequency factor,** for the reaction.

Taking the natural logarithm of each side of the Arrhenius equation yields

$$\ln(k) = -\frac{E_a}{R}\left(\frac{1}{T}\right) + \ln(A) \tag{15.10}$$

Equation (15.10) is a linear equation of the type $y = mx + b$, where $y = \ln(k)$, $m = -E_a/R$ = slope, $x = 1/T$, and $b = \ln(A)$ = intercept. Thus, for a reaction where the rate constant obeys the Arrhenius equation, a plot of $\ln(k)$ versus $1/T$ gives a straight line. The slope and intercept can be used to determine the values of E_a and A characteristic of that reaction. The fact that most rate constants obey the Arrhenius equation to a good approximation indicates that the collision model for chemical reactions is physically reasonable.

EXAMPLE 15.8

The reaction

$$2N_2O_5(g) \longrightarrow 4NO_2(g) + O_2(g)$$

was studied at several temperatures and the following values of k were obtained:

k (s^{-1})	T (°C)
2.0×10^{-5}	20
7.3×10^{-5}	30
2.7×10^{-4}	40
9.1×10^{-4}	50
2.9×10^{-3}	60

Calculate the value of E_a for this reaction.

Solution To obtain the value of E_a, we need to construct a plot of ln(k) versus 1/T. First, we must calculate values of ln(k) and 1/T:

T (°C)	T (K)	1/T (K)	k (s^{-1})	ln(k)
20	293	3.41×10^{-3}	2.0×10^{-5}	−10.82
30	303	3.30×10^{-3}	7.3×10^{-5}	−9.53
40	313	3.19×10^{-3}	2.7×10^{-4}	−8.22
50	323	3.10×10^{-3}	9.1×10^{-4}	−7.00
60	333	3.00×10^{-3}	2.9×10^{-3}	−5.84

The plot of ln(k) versus 1/T is shown in Fig. 15.14. The slope is found to be -1.2×10^4 K. Since

$$\text{Slope} = -\frac{E_a}{R}$$

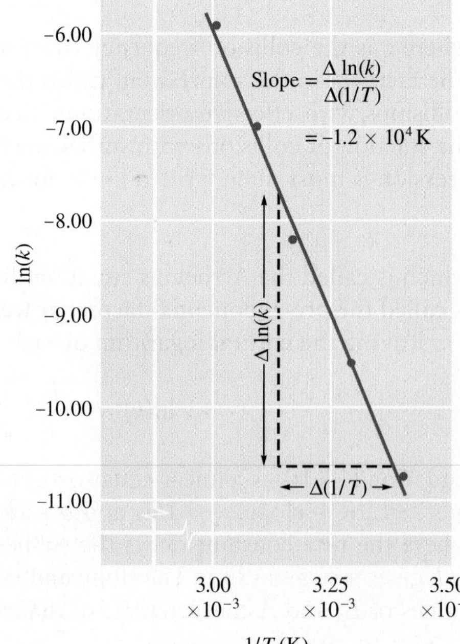

Figure 15.14
Plot of ln(k) versus 1/T for the reaction $2N_2O_5(g) \longrightarrow 4NO_2(g) + O_2(g)$. The value of the activation energy for this reaction can be obtained from the slope of the line, which equals $-E_a/R$.

then

$$E_a = -R(\text{slope}) = -(8.3145 \text{ J K}^{-1} \text{ mol}^{-1})(-1.2 \times 10^4 \text{ K})$$
$$= 1.0 \times 10^5 \text{ J/mol}$$

Thus the value of the activation energy for this reaction is 1.0×10^5 J/mol.

The most common procedure for finding E_a for a reaction involves measuring the rate constant k at several temperatures and then plotting $\ln(k)$ versus $1/T$, as shown in Example 15.8. However, E_a can also be calculated from the values of k at only two temperatures using a formula that can be derived as follows from Equation (15.10).

At temperature T_1, the rate constant is k_1; thus

$$\ln(k_1) = -\frac{E_a}{RT_1} + \ln(A)$$

At temperature T_2, the rate constant is k_2; thus

$$\ln(k_2) = -\frac{E_a}{RT_2} + \ln(A)$$

Subtracting the first equation from the second gives

$$\ln\left(\frac{k_2}{k_1}\right) = \frac{E_a}{R}\left(\frac{1}{T_1} - \frac{1}{T_2}\right) \tag{15.11}$$

Therefore, the values of k_1 and k_2 measured at temperatures T_1 and T_2 can be used to calculate E_a.

15.9 Catalysis

We have seen that the rate of a reaction increases dramatically with temperature. If a particular reaction does not occur fast enough at normal temperatures, we can speed it up by raising the temperature. However, sometimes this is not feasible. For example, since living cells can survive only in a rather narrow temperature range, the human body is designed to operate at an almost constant temperature of 98.6°F. But the many complicated biochemical reactions keeping us alive would be much too slow at this temperature without intervention. We survive only because the body contains many substances called **enzymes** that increase the rates of these reactions even though body temperature remains constant. In fact, almost every biologically important reaction is assisted by a specific enzyme. An important example involves the enzyme carbonic anhydrase, which catalyzes the reaction of carbon dioxide with water:

$$CO_2 + H_2O \rightleftharpoons HCO_3^- + H^+$$

This crucial reaction allows the carbon dioxide that forms in the cells during metabolism to be removed. If the carbon dioxide were allowed to accumulate, it would poison the cell. Carbonic anhydrase is so efficient that one molecule of enzyme can catalyze the reaction of over *600,000 carbon dioxide molecules in one second!*

Although it is possible to use higher temperatures to speed up commercially important reactions, such as the Haber process for synthesizing ammonia, this is very expensive. In a chemical plant an increase in temperature means significantly increased energy costs. The use of an appropriate catalyst allows a reaction to proceed rapidly at a relatively low temperature and therefore can hold down production costs.

Chemical Insights

TiO_2—One of Nature's Most Versatile Materials

Titanium(IV) oxide (commonly called titanium dioxide) is a compound with an amazing array of talents. It is a widely used white pigment for paint, paper, vinyl floor coverings, sunscreens, and synthetic fibers. However, it is much more than a simple pigment. TiO_2, it appears, has an almost unlimited ability to catalyze the breakdown of organic materials. In Japan, scores of companies are scrambling to take advantage of this remarkable talent. TiO_2-impregnated paint used on buildings and in automobile and rail tunnels prevents the buildup of oily dust, TiO_2-impregnated glass prevents oily buildup on chandeliers and automobile windows, and a TiO_2-impregnated fan belt can help break down grease that often clogs kitchen exhaust fans. TiO_2 catalysis can also be used to break down air pollutants such as cigarette smoke and nitrogen oxides. In addition to all these talents, TiO_2 kills bacteria.

TiO_2 is a photocatalyst—it requires light for its catalytic activity. It also requires the presence of oxygen and water. The mechanism for the catalytic behavior of TiO_2 results from its behavior as a semiconductor. When light is absorbed by TiO_2, free electrons and "holes" are created that cause adsorbed water and oxygen molecules to form hydroxyl (OH) and superoxide (O_2^-) radicals, species with unpaired electrons. These extremely reactive radicals can react with and destroy virtually all organic materials. The formation of these radicals also explains the ability of TiO_2 to catalyze the reactions of nitrogen oxides to form nitric acid. These radicals are lethal to microorganisms and explain the ability of TiO_2 to behave as a bactericide.

New uses are being found daily for the many talents of TiO_2.

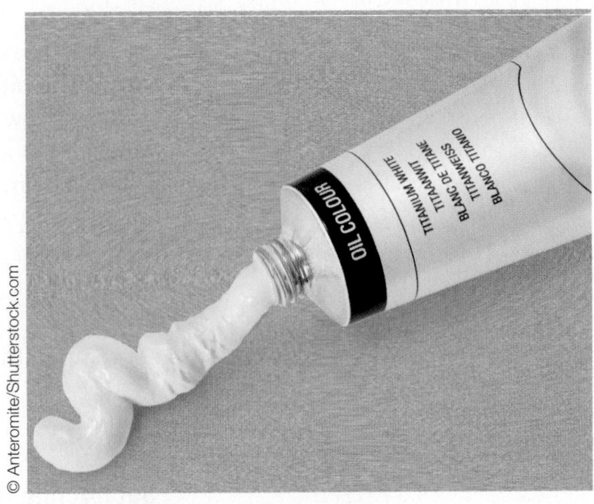

A white paint pigment that contains TiO_2.

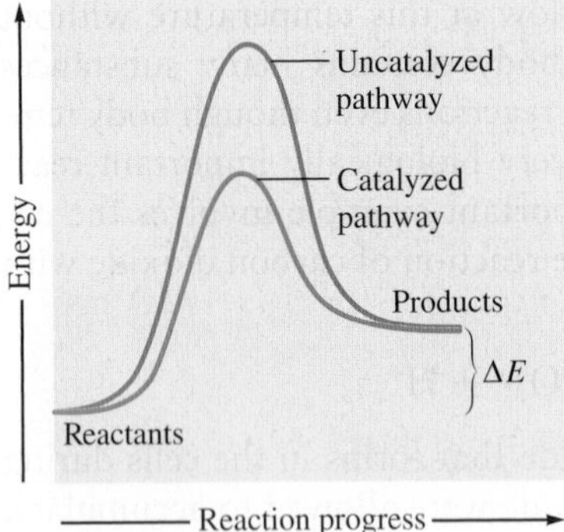

Figure 15.15
Energy plots for catalyzed and uncatalyzed pathways for a given reaction. Note that although the shapes of the catalyzed and uncatalyzed energy profiles are shown here the same for simplicity, they do not have identical shapes for actual reactions.

A **catalyst** is *a substance that speeds up a reaction without being consumed itself.* Just as virtually all vital biological reactions are assisted by enzymes (biological catalysts), almost all industrial processes also benefit from the use of catalysts. For example, the production of sulfuric acid uses vanadium(V) oxide, and the Haber process uses a mixture of iron and iron oxide.

How does a catalyst work? Remember that for each reaction a certain energy barrier must be surmounted. How can we make a reaction occur faster without raising the temperature to increase the molecular kinetic energies? The solution is to provide a new pathway for the reaction, one with a *lower activation energy*. That is what a catalyst does, as shown in Fig. 15.15. Because the catalyst allows the reaction to occur along a pathway with a lower activation energy, a much larger fraction of collisions is effective at a given temperature. Thus the reaction rate is increased. This effect is illustrated in Fig. 15.16. Note from this diagram that although a catalyst lowers the activation energy (E_a) for a reaction, it does not affect the energy difference (ΔE) between products and reactants.

Catalysts are classified as homogeneous or heterogeneous. A **homogeneous catalyst** is one that is *present in the same phase (physical state) as the reacting molecules.* A **heterogeneous catalyst** exists *in a different phase,* usually as a solid.

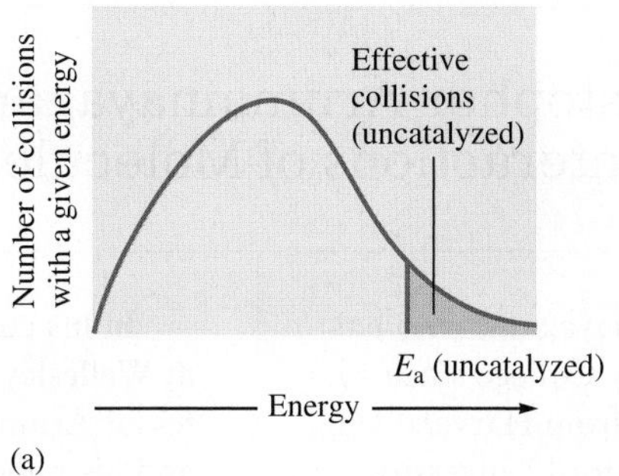

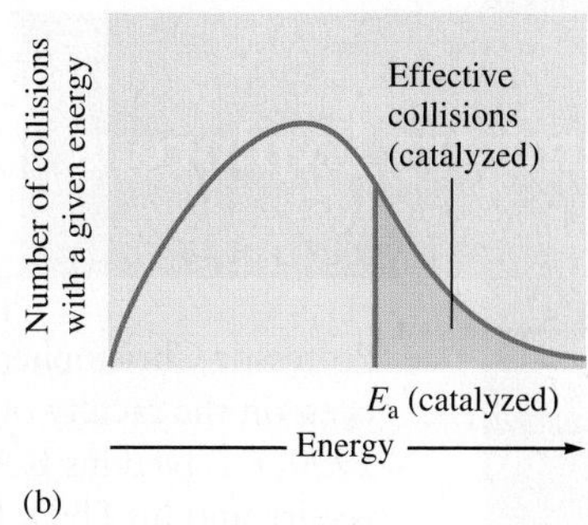

Figure 15.16
Effect of a catalyst on the number of reaction-producing collisions. Because a catalyst provides a reaction pathway with a lower activation energy, a much greater fraction of the collisions is effective for the catalyzed pathway (b) than for the uncatalyzed pathway (a) (at a given temperature). This allows reactants to become products at a much higher rate, even though there is no temperature increase.

Heterogeneous Catalysis

Heterogeneous catalysis most often involves gaseous reactants being adsorbed on the surface of a solid catalyst. **Adsorption** refers to the collection of one substance on the surface of another substance; **absorption** refers to the penetration of one substance into another. Water is *absorbed* by a sponge.

One of the earliest examples of heterogeneous catalysis involves the synthesis of ammonia from nitrogen and hydrogen. This process was developed in 1909 by German chemist Fritz Haber—who tested more than 1000 possible catalysts before settling on iron as the best choice. (Today, ammonia manufacturers use a solid catalyst consisting of a mixture of iron, potassium, and calcium that performs better than iron alone.) Although Haber had no means for determining why iron was a good catalyst, it is now understood that the key to iron's effectiveness is that the strong nitrogen–nitrogen and hydrogen–hydrogen bonds are weakened when the H_2 and N_2 molecules are bound to iron atoms on the surface of the metal. Iron turns out to be an ideal catalyst because the iron–nitrogen bond is sufficiently strong that the nitrogen atoms on the surface do not recombine to form N_2 but is weak enough to allow nitrogen and hydrogen atoms on the surface to combine to form ammonia.

An important example of heterogeneous catalysis occurs in the hydrogenation of unsaturated hydrocarbons, compounds composed mainly of carbon and hydrogen and containing some carbon–carbon double bonds. Hydrogenation is an important industrial process used to change unsaturated fats, occurring as oils, to saturated fats (solid shortenings such as Crisco). In this process the C=C bonds are converted to C—C bonds through the addition of hydrogen.

A simple example of hydrogenation involves ethylene:

$$\underset{\text{Ethylene}}{H_2C{=}CH_2}(g) + H_2(g) \longrightarrow \underset{\text{Ethane}}{H_3C{-}CH_3}(g)$$

This reaction is quite slow at normal temperatures mainly because the strong bond in the hydrogen molecule results in a large activation energy for the reaction. However, the reaction rate can be greatly increased by using a solid catalyst of platinum, palladium, or nickel. The hydrogen and ethylene adsorb on the catalyst surface, where the reaction occurs. The main function of the catalyst apparently is to allow formation of metal–hydrogen interactions that weaken the H—H bonds and facilitate the reaction. The mechanism is illustrated in Fig. 15.17.

Chemistry Explorers

Christopher Arumainayagam Researches the Interactions of Molecules with Surfaces

Professor Christopher Arumainayagam, who has been on the faculty of Wellesley College since 1990, received his B.A. degree from Harvard University and his Ph.D. from Stanford University.

One goal of Professor Arumainayagam's research is to understand the interactions of molecules with solid surfaces in order to learn more about the reaction mechanisms of heterogeneous catalysis. In experiments conducted using molecular beams (narrow streams of molecules that are undisturbed by collisions with other gas molecules), Professor Arumainayagam has studied how adsorption on a solid surface depends on the velocity and angle of approach of the incident molecule. In collaboration with Professor John Yates, Professor Arumainayagam has also studied the production and removal of surface defects (irregularities) that are thought to play a critical role in the performance of heterogeneous catalysts.

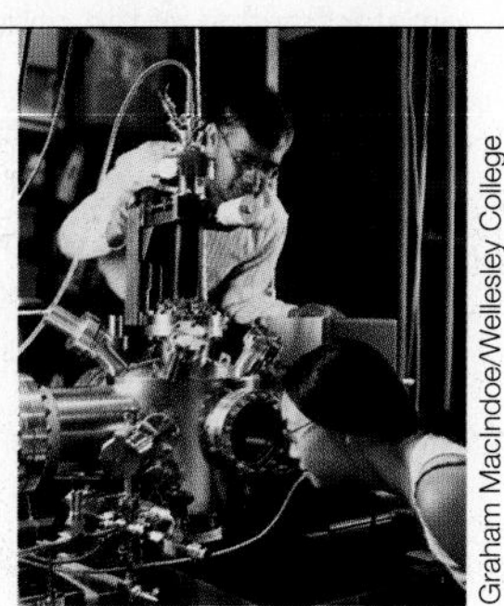

Christopher Arumainayagam and a student.

In his current research at Wellesley College, Professor Arumainayagam and his undergraduate students are using surface science techniques and low-energy electron beams to understand how high-energy particles (e.g., electrons, protons, alpha particles) and high-energy photons (e.g., X rays and gamma rays) interact with matter, causing ionization. Results of their studies may provide information that could further cost-efficient destruction of hazardous chemicals and improve our understanding of the electron-induced decomposition of feed gases used in the processing of semiconductor devices.

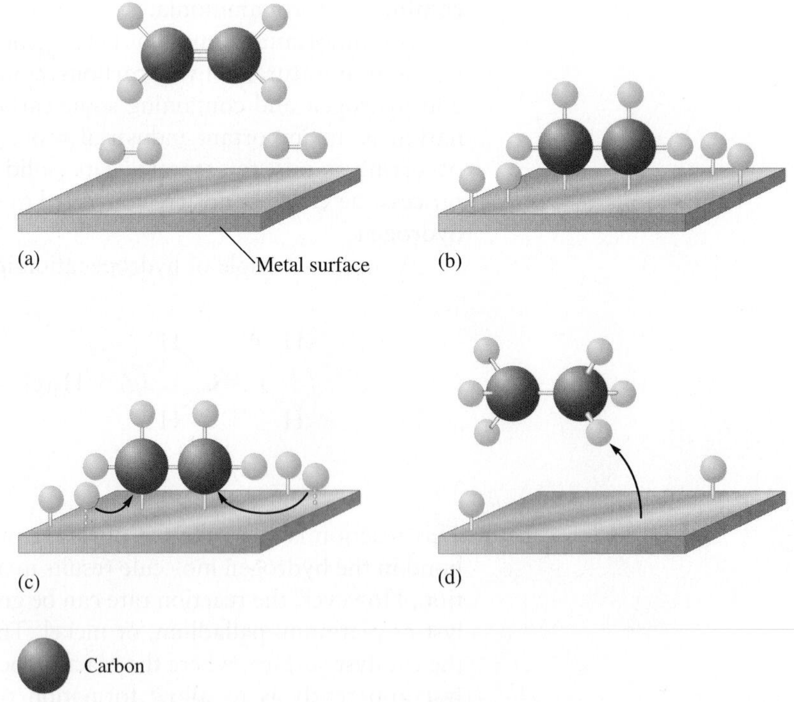

Figure 15.17
Heterogeneous catalysis of the hydrogenation of ethylene. (a) Hydrogen is adsorbed on the metal surface, forming metal–hydrogen bonds and breaking the H—H bonds. (b) During adsorption, the C—C π bond in ethylene is broken and metal–carbon bonds are formed. (c) The adsorbed molecules and atoms migrate toward each other on the metal surface, forming new C—H bonds. (d) The C atoms in ethane (C_2H_6) have completely saturated bonding capacities and so cannot bind strongly to the metal surfaces. The C_2H_6 molecule thus escapes.

Typically, heterogeneous catalysis involves four steps:

1. Adsorption and activation of the reactants
2. Migration of the adsorbed reactants on the surface
3. Reaction among the adsorbed substances
4. Escape, or *desorption,* of the products

Heterogeneous catalysis also occurs in the oxidation of gaseous sulfur dioxide to gaseous sulfur trioxide. This process is especially interesting because it illustrates both positive and negative consequences of chemical catalysis.

The negative side is the formation of damaging air pollutants. Recall that sulfur dioxide, a toxic gas with a choking odor, is formed when sulfur-containing fuels are burned. However, it is sulfur trioxide that causes most of the environmental damage, mainly through the production of acid rain. When sulfur trioxide combines with a droplet of water, sulfuric acid is formed:

$$H_2O(l) + SO_3(g) \longrightarrow H_2SO_4(aq)$$

This sulfuric acid can cause considerable damage to vegetation, buildings and statues, and fish populations.

Sulfur dioxide is *not* rapidly oxidized to sulfur trioxide in clean, dry air. Why, then, is there a problem? The answer is catalysis. Dust particles and water droplets catalyze the reaction between SO_2 and O_2 in the air.

On the positive side, the heterogeneous catalysis of the oxidation of SO_2 is used to advantage in the manufacture of sulfuric acid, where the reaction of O_2 and SO_2 to form SO_3 is catalyzed by a solid mixture of platinum and vanadium(V) oxide.

Heterogeneous catalysis is also used in the catalytic converters of automobile exhaust systems. The exhaust gases, containing compounds such as nitric oxide, carbon monoxide, and unburned hydrocarbons, are passed through a converter containing beads of solid catalyst (Fig. 15.18). The catalyst promotes the conversion of carbon monoxide to carbon dioxide, hydrocarbons to carbon dioxide and water, and nitric oxide to nitrogen gas, to lessen the environmental impact of the exhaust gases. This beneficial catalysis can, unfortunately, be accompanied by the unwanted catalysis of the oxidation of SO_2 to SO_3, the latter reacting with the moisture present to form sulfuric acid.

Because of the complex nature of the reactions that take place in the converter, a mixture of catalysts is used. The most effective catalytic materials are transition metal oxides and noble metals such as palladium and platinum. A catalytic converter typically consists of platinum and rhodium particles deposited on a ceramic honeycomb, a configuration that maximizes the contact between the metal particles and the exhaust gases. In studies performed during the last ten years, researchers at General Motors have shown that rhodium

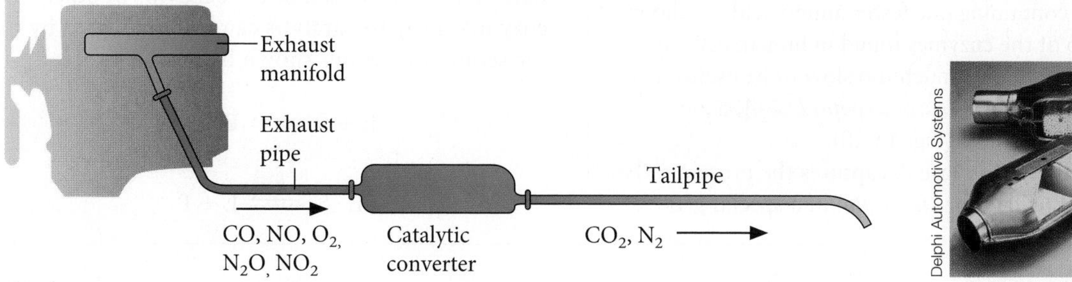

Figure 15.18
The exhaust gases from an automobile engine are passed through a catalytic converter to minimize environmental damage.

Chemical Insights — Enzymes: Nature's Catalysts

The most impressive examples of homogeneous catalysis occur in nature, where the complex reactions necessary for plant and animal life are made possible by enzymes. Enzymes are large molecules specifically tailored to facilitate a given type of reaction. Usually enzymes are proteins, an important class of biomolecules constructed from α-amino acids that have the general structure

$$\mathrm{H_2N{-}C(R)(H){-}C({=}O){-}OH}$$

where R represents any one of 20 different substituents. These amino acid molecules can be "hooked together" to form a polymer (a word meaning "many parts") called a *protein*. The general structure of a protein can be represented as follows:

$$\mathrm{H_2N\sim\sim\sim N(H){-}C(R)(H){-}C({=}O){-}N(H){-}C(R')(H){-}C({=}O){-}N(H){-}C(R'')(H){-}C({=}O){-}O{-}H}$$

Many amino acid fragments | Fragment from an amino acid with substituent R | Fragment from an amino acid with substituent R′ | Fragment from an amino acid with substituent R″

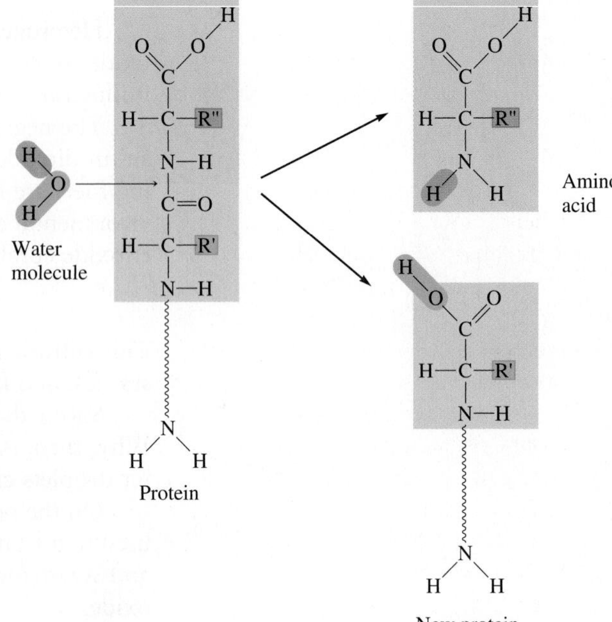

Figure 15.19
The removal of the end amino acid from a protein by reaction with a molecule of water. The products are an amino acid and a new, smaller protein.

Since specific proteins are needed by the human body, the proteins in food must be broken into their constituent amino acids, which are then used to construct new proteins in the body's cells. The reaction in which a protein is broken down one amino acid at a time is shown in Fig. 15.19. Note that in this reaction a water molecule reacts with a protein molecule to produce an amino acid and a new protein containing one fewer amino acid. Without the help of the enzymes found in human cells, this reaction would be much too slow to be useful. One of these enzymes is *carboxypeptidase-A,* a zinc-containing protein (Fig. 15.20).

Carboxypeptidase-A captures the protein to be acted on (called the *substrate*) in a special groove and positions the substrate so that the end is in the *active site,* where the catalysis occurs (Fig. 15.21). Note that the Zn^{2+} ion bonds to the oxygen of the C=O (carbonyl) group. This polarizes the electron density in the carbonyl group, allowing the neighboring C—N bond to be broken much more easily. When the reaction is completed, the remaining portion of the substrate protein and the newly formed amino acid are released by the enzyme.

The process just described for carboxypeptidase-A is characteristic of the behavior of other enzymes. Enzyme catalysis can be represented by the series of reactions shown below:

$$\mathrm{E + S} \underset{k_{-1}}{\overset{k_1}{\rightleftharpoons}} \mathrm{E\cdot S}$$

$$\mathrm{E\cdot S} \overset{k_2}{\longrightarrow} \mathrm{E + P}$$

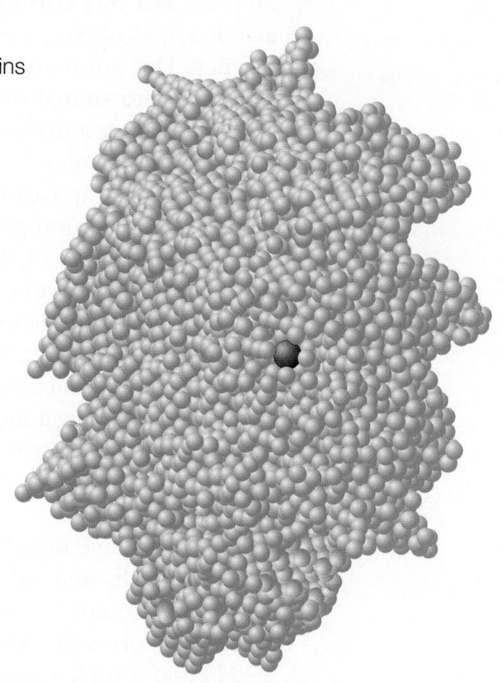

Figure 15.20
The structure of the enzyme carboxypeptidase-A, which contains 307 amino acids.

where E represents the enzyme, S represents the substrate, E · S represents the enzyme–substrate complex, and P represents the products. The enzyme and substrate form a complex where the reaction occurs. The enzyme then releases the products and is ready to repeat the process. The most amazing thing about enzymes is their efficiency. Because an enzyme plays its catalytic role over and over and very rapidly, only a tiny amount of enzyme is required. This makes the isolation of enzymes for study quite difficult.

Figure 15.21
Protein–substrate interaction. The substrate is shown in black and red, with the red representing the terminal amino acid. Blue indicates side chains from the enzyme that help bind the substrate.

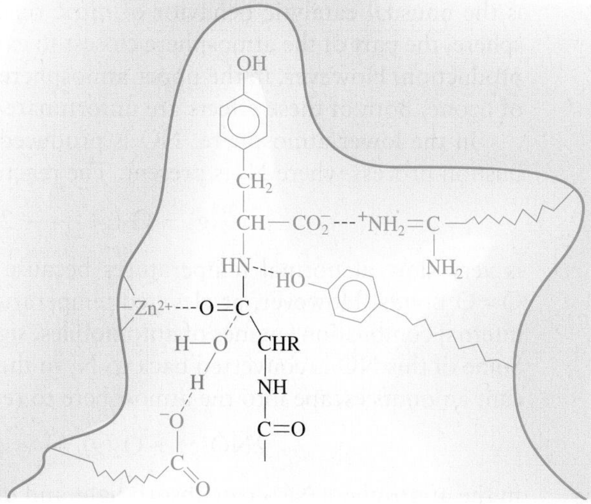

promotes the dissociation of NO molecules adsorbed on its surface, thereby enhancing the conversion of NO, a serious air pollutant, to N_2, a natural component of pure air.

One consequence of the widespread use of catalytic converters has been the need to remove lead from gasoline. Tetraethyl lead was used for more than 50 years as a very effective "octane booster" because of its antiknocking characteristics. However, lead quickly destroys much of a converter's catalytic efficiency. This poisoning effect, along with health concerns about the toxicity of lead, has necessitated the removal of lead from gasoline, which has caused a search for other antiknock additives and the redesigning of engines to run on lower-octane gasoline.

Yet another application of solid-state catalysis occurs in the desulfurization of petroleum. Natural petroleum includes various molecules that contain sulfur atoms. Combustion of this petroleum produces SO_2, which must be removed from the exhaust to prevent air pollution. One way to prevent pollution by SO_2 is to remove the sulfur from the petroleum before it is used for fuel—the desulfurization of petroleum. One type of sulfur-containing molecules found in petroleum are thiols, which can be written R—SH, where R represents a molecular fragment containing a long chain of carbon atoms. In desulfurization the goal is to remove the sulfur from this molecule to produce a hydrocarbon (R—H):

$$\text{R—SH} \longrightarrow \text{R—H} + \text{S}$$

Oil chemists have found that this process is catalyzed by a mixture of molybdenum, cobalt, and sulfur. Because this catalyst is a very complicated substance, we are not certain how it works. It is thought that the desulfurization reaction involves the thiol binding to the catalytic surface. In this process the S—H bond is broken, with the R—S and the H being bound to metal atoms on the surface of the catalyst. The H atom is then transferred to the R fragment to form R—H, which migrates away, leaving the S behind.

Homogeneous Catalysis

A homogeneous catalyst exists in the same phase as the reacting molecules. There are many examples in both the gas and liquid phases. One such example is the unusual catalytic behavior of nitric oxide toward ozone. In the troposphere, the part of the atmosphere closest to earth, nitric oxide catalyzes ozone production. However, in the upper atmosphere, it catalyzes the decomposition of ozone. Both of these effects are unfortunate environmentally.

In the lower atmosphere, NO is produced in any high-temperature combustion process where N_2 is present. The reaction

$$N_2(g) + O_2(g) \longrightarrow 2NO(g)$$

is very slow at normal temperatures because of the very strong N≡N and O=O bonds. However, at elevated temperatures, such as those found in the internal combustion engines of automobiles, significant quantities of NO form. Some of this NO is converted back to N_2 in the catalytic converter, but significant amounts escape into the atmosphere to react with oxygen:

$$2NO(g) + O_2(g) \longrightarrow 2NO_2(g)$$

Although O_2 is represented here as the oxidizing agent for NO, the actual oxidizing agent is probably some type of peroxide compound produced by the reaction of oxygen with pollutants. The direct reaction of NO with O_2 is very slow.

In the atmosphere NO_2 can absorb light and decompose as follows:

$$NO_2(g) \xrightarrow{\text{Light}} NO(g) + O(g)$$

The oxygen atom formed in this process is very reactive and can combine with oxygen molecules to form ozone:

$$O_2(g) + O(g) \longrightarrow O_3(g)$$

Hot, New Enzymes

Chemical Insights

Enzymes, nature's catalysts, are very attractive to chemical companies: Enzymes are very efficient catalysts and as natural products they tend to be easier on the environment than human-made catalysts. However, one disadvantage of most enzymes is that they can function only near room temperature and at pH values near 7. Because of these limitations, scientists are now looking at the enzymes that occur in organisms that exist in extreme conditions.

The exotic microorganisms that thrive in the scalding temperatures of hot springs, the freezing temperatures of the Arctic, or extremes in pH or salinity are called *extremophiles* and their enzymes are called *extremozymes.* Although the extremozymes are attractive to industry because they can survive in the extreme conditions that often characterize industrial processes, the parent extremophiles are difficult to grow "in captivity" because of the unusual conditions they require. Because of the difficulties in culturing these microorganisms, biotechnology companies are now "shotgun" cloning the DNA from a mix of these extremophiles. The steps in this process are represented below.

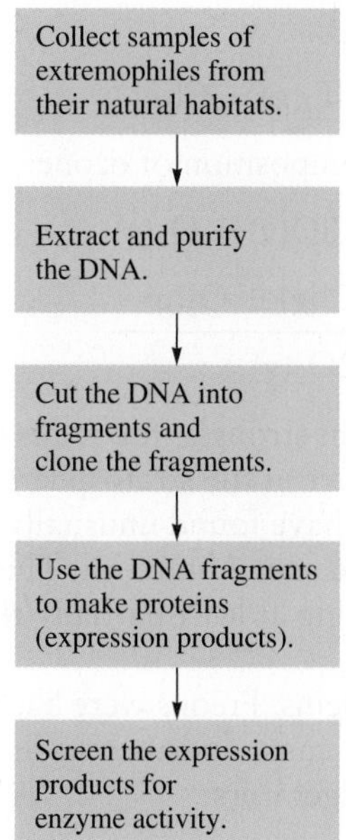

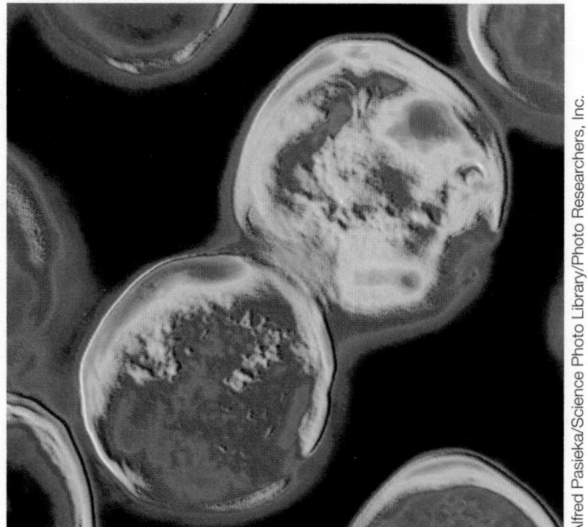

Alfred Pasieka/Science Photo Library/Photo Researchers, Inc.

A micrograph of the extremophile *Archaeoglobus fulgidis,* an organism that lives in the hot sediments near submarine hydrothermal vents.

The DNA fragments are expressed (proteins are made from the genes contained in the fragments) most commonly using *Escherichia coli.* Interestingly, even though *E. coli* must be cultured under the mild conditions necessary for them to survive, the extremozymes formed seem to have their characteristic catalytic activities, indicating that they have the same structures as when they are formed in their native extreme conditions.

The mixtures of extremozymes produced by the expression process are then tested to see if they have catalytic activity for the industrial processes of interest. If a given mixture shows catalytic activity, it is then usually subjected to random DNA mutagenesis or "molecule breeding" to see whether random evolution of the enzymes will lead to improved activities.

The potential market for the extremozymes is huge. Currently, the global market for industrial enzymes is more than $7 billion. The companies developing the extremozymes hope that their products can replace existing industrial catalysts, making processes more efficient by lowering costs and minimizing the formation of by-products.

Ozone is a powerful oxidizing agent that can react with other air pollutants to form substances irritating to the eyes and lungs.

In this series of reactions, nitric oxide is acting as a true catalyst because it assists the production of ozone without being consumed itself. This can be seen by summing the reactions:

$$
\begin{array}{rcl}
NO(g) + \frac{1}{2}O_2(g) & \longrightarrow & NO_2(g) \\
NO_2(g) & \xrightarrow{\text{Light}} & NO(g) + O(g) \\
O_2(g) + O(g) & \longrightarrow & O_3(g) \\
\hline
\frac{3}{2}O_2(g) & \longrightarrow & O_3(g)
\end{array}
$$

In the upper atmosphere, the presence of nitric oxide has the opposite effect—the depletion of ozone. The series of reactions involved is

$$
\begin{array}{rcl}
NO(g) + O_3(g) & \longrightarrow & NO_2(g) + O_2(g) \\
O(g) + NO_2(g) & \longrightarrow & NO(g) + O_2(g) \\
\hline
O(g) + O_3(g) & \longrightarrow & 2O_2(g)
\end{array}
$$

Nitric oxide is again catalytic, but here its effect is to change O_3 to O_2. This is a potential problem because O_3, which absorbs ultraviolet light, is necessary to protect us from the harmful effects of this high-energy radiation. That is, we want O_3 in the upper atmosphere to block ultraviolet radiation from the sun. However, we do not want it in the lower atmosphere where we have to breathe it and its oxidation products.

Over the last decade, research has shown that the ozone layer in the upper atmosphere is also threatened by *Freons,** a group of stable, noncorrosive compounds long used as refrigerants and propellants in aerosol cans. The most commonly used substance of this type was Freon-12 (CCl_2F_2). The chemical inertness of Freons makes them more useful but also creates a problem, since they remain for a long time in the environment. Eventually, they migrate into the upper atmosphere to be decomposed by high-energy light. Among the decomposition products are chlorine atoms:

$$
CCl_2F_2(g) \xrightarrow{\text{Light}} CClF_2(g) + Cl(g)
$$

These chlorine atoms can catalyze the decomposition of ozone:

$$
\begin{array}{rcl}
Cl(g) + O_3(g) & \longrightarrow & ClO(g) + O_2(g) \\
O(g) + ClO(g) & \longrightarrow & Cl(g) + O_2(g) \\
\hline
O(g) + O_3(g) & \longrightarrow & 2O_2(g)
\end{array}
$$

The problem of Freons has been brought strongly into focus by the discovery of a mysterious "hole" in the ozone layer in the stratosphere over Antarctica. Studies to find the reason for the hole have found unusually high levels of chlorine monoxide (ClO) in the atmosphere over Antarctica. This strongly implicates the Freons in the atmosphere as being at least partially responsible for the ozone destruction in the area.

Because they pose environmental problems, Freons were banned years ago by the U.S. government for use in aerosol cans, and they have recently been banned for use in air conditioners and refrigerators.

*For more information, see S. Elliott and F. S. Rowland, "Chlorofluorocarbons and Stratospheric Ozone," *J. Chem. Ed.* 64 (1987): 387, and P. S. Zurer, "Ozone Depletion's Recurring Surprises Challenge Atmospheric Scientists," *Chem. Eng. News,* May 24, 1993: 8.

Key Terms

chemical kinetics

Section 15.1
reaction rate
instantaneous rate

Section 15.2
rate law
rate constant
order
differential rate law (rate law)
integrated rate law

Section 15.3
method of initial rates
initial rate
overall reaction order

Section 15.4
first-order reaction
integrated first-order rate law
half-life of a reaction
integrated second-order rate law
zero-order reaction
integrated rate law for a zero-order reaction
pseudo-first-order rate law

Section 15.6
reaction mechanism
intermediate
elementary step
molecularity
unimolecular step
bimolecular step
termolecular step
rate-determining step

Section 15.7
steady-state approximation

Section 15.8
collision model
activation energy
activated complex (transition state)
molecular orientations
steric factor
Arrhenius equation
pre-exponential (frequency) factor

Section 15.9
enzyme
catalyst
homogeneous catalyst
heterogeneous catalyst
adsorption
absorption

For Review

OWL and go Chemistry

Sign in at **www.cengage.com/owl** to:

- View tutorials and simulations, develop problem-solving skills, and complete online homework assigned by your professor.
- Download Go Chemistry mini lecture modules for quick review and exam prep from OWL (or purchase them at **www.cengagebrain.com**)

Chemical kinetics

- The study of the factors that control the rate (speed) of a chemical reaction
  - Rate is defined in terms of the change in concentration of a given reaction component per unit time
  - Kinetic measurements are often made under conditions where the reverse reaction is insignificant
- The kinetic and thermodynamic properties of a reaction are not fundamentally related

Rate laws

- Differential rate law: describes the rate as a function of concentration

$$\text{Rate} = -\frac{\Delta[\text{A}]}{\Delta t} = k[\text{A}]^n$$

  - k is the rate constant
  - n is the order; not related to the coefficients in the balanced equation
- Integrated rate law: describes the concentration as a function of time
  - For a reaction of the type

$$a\text{A} \longrightarrow \text{products}$$

for which

$$\text{Rate} = k[\text{A}]^n$$

$n = 0$: $$[\text{A}] = -kt + [\text{A}]_0$$

$$t_{1/2} = \frac{[\text{A}]_0}{2k}$$

$n = 1$: $$\ln[\text{A}] = -kt + \ln[\text{A}]_0$$

$$t_{1/2} = \frac{0.693}{k}$$

$n = 2$: $$\frac{1}{[\text{A}]} = kt + \frac{1}{k[\text{A}]_0}$$

$$t_{1/2} = \frac{1}{k[\text{A}]_0}$$

  - The value of k can be determined from the plot of the appropriate function of [A] versus t

Reaction mechanism

- Series of elementary steps by which an overall reaction occurs
  - Elementary step: rate law for the step can be written from the molecularity of the reaction
- Two requirements for an acceptable mechanism:
  - The elementary steps sum to give the correct overall balanced equation
  - The mechanism agrees with the experimentally determined rate law

- Simple reactions can have an elementary step that is slower than all of the other steps; which is called the rate-determining step.

Steady state approximation
- Used when a specific rate-determining step cannot be chosen.
- Assumes the concentration of any intermediate remains constant as the reaction proceeds.

Kinetic models
- The simplest model to account for reaction kinetics is the collision model
  - Molecules must collide to react
  - The collision kinetic energy furnishes the potential energy needed to enable the reactants to rearrange to form products
  - A certain threshold energy called the activation energy (E_a) is necessary for a reaction to occur
  - The relative orientations of the colliding reactants are also a determining factor in the reaction rate
  - This model leads to the Arrhenius equation:

$$k = Ae^{-E_a/RT}$$

  - A depends on the collision frequency and relative orientation of the molecules
  - The value of E_a can be found by obtaining the values of k at several temperatures

Catalyst
- Speeds up a reaction without being consumed
- Works by providing a lower-energy pathway for the reaction
- Enzymes are biological catalysts
- Catalysts can be classified as homogeneous or heterogeneous
  - Homogeneous: exist in the same phase as the reactants
  - Heterogeneous: exist in a different phase than the reactants

Discussion Questions*

These questions are designed to be considered by groups of students in class. Often these questions work well for introducing a particular topic in class.

1. Define the term *stability* from both a kinetic and a thermodynamic perspective. Give examples to show the differences in these concepts.
2. Describe at least two experiments you could perform to determine a rate law.
3. Make a graph of [A] versus time for zero-, first-, and second-order reactions. From these graphs, compare successive half-lives.
4. How does temperature affect k, the rate constant? Explain.
5. Consider the following statements: "In general, the rate of a chemical reaction increases at first. After that the rate of the reaction decreases because its rate is dependent on the concentrations of the reactants, and these are decreasing." Indicate everything that is correct in these statements, and indicate everything that is incorrect. Correct the incorrect statements, and explain.
6. For the reaction $A + B \longrightarrow C$, explain at least two ways in which the rate law could be zero order in chemical A.
7. A friend of yours states, "A balanced equation tells us how chemicals interact. Therefore, we can determine the rate law directly from the balanced equation." What do you tell your friend?
8. The rate constant (k) depends on which of the following? (There may be more than one answer.)
   a. the concentration of the reactants
   b. the nature of the reactants
   c. the temperature
   d. the order of the reaction

   Explain.
9. Provide a conceptual rationale for the differences in the half-lives of zero-, first-, and second-order reactions.

*In the Questions and the Exercises the term *rate law* always means differential rate law.

Exercises*

OWL Interactive versions of these problems may be assigned in OWL.

A blue exercise number indicates that the answer to that exercise appears at the back of this book and a solution appears in the *Solutions Guide.*

Reaction Rates

10. Define *reaction rate.* Distinguish between the initial rate, average rate, and instantaneous rate of a chemical reaction. Which of these rates is usually fastest? The initial rate is the rate used by convention. Give a possible explanation as to why.

11. Consider the general reaction

$$aA + bB \longrightarrow cC$$

and the following average rate data over a specific time period Δt:

$$-\frac{\Delta A}{\Delta t} = 0.0080 \text{ mol L}^{-1} \text{ s}^{-1}$$

$$-\frac{\Delta B}{\Delta t} = 0.0120 \text{ mol L}^{-1} \text{ s}^{-1}$$

$$\frac{\Delta C}{\Delta t} = 0.0160 \text{ mol L}^{-1} \text{ s}^{-1}$$

Determine a set of possible coefficients to balance this general reaction.

12. Consider the reaction

$$4PH_3(g) \longrightarrow P_4(g) + 6H_2(g)$$

If, in a certain experiment, over a specific time period, 0.0048 mole of PH_3 is consumed in a 2.0-L container during each second of the reaction, what are the rates of production of P_4 and H_2 in this experiment?

13. In the Haber process for the production of ammonia,

$$N_2(g) + 3H_2(g) \longrightarrow 2NH_3(g)$$

what is the relationship between the rate of production of ammonia and the rate of consumption of hydrogen?

14. What are the units for each of the following if concentrations are expressed in moles per liter and time in seconds?
    a. rate of a chemical reaction
    b. rate constant for a zero-order rate law
    c. rate constant for a first-order rate law
    d. rate constant for a second-order rate law
    e. rate constant for a third-order rate law

15. The rate law for the reaction

$$Cl_2(g) + CHCl_3(g) \longrightarrow HCl(g) + CCl_4(g)$$

is

$$\text{Rate} = k[Cl_2]^{1/2}[CHCl_3]$$

What are the units for k assuming time in seconds?

16. The hydroxyl radical (OH) is an important oxidizing agent in the atmosphere. At 298 K the rate constant for the reaction of OH with benzene is 1.24×10^{-12} cm^3 molecule^{-1} s^{-1}. Calculate the value of the rate constant in L mol^{-1} s^{-1}.

Rate Laws from Experimental Data: Initial-Rates Method

17. The reaction

$$2NO(g) + Cl_2(g) \longrightarrow 2NOCl(g)$$

was studied at $-10°C$. The following results were obtained, where

$$\text{Rate} = -\frac{d[Cl_2]}{dt}$$

$[NO]_0$ (mol/L)	$[Cl_2]_0$ (mol/L)	Initial Rate (mol L^{-1} min^{-1})
0.10	0.10	0.18
0.10	0.20	0.36
0.20	0.20	1.45

a. What is the rate law?
b. What is the value of the rate constant?

18. The following data were obtained for the gas-phase decomposition of dinitrogen pentoxide,

$$2N_2O_5(g) \longrightarrow 4NO_2(g) + O_2(g)$$

$[N_2O_5]_0$ (mol/L)	Initial Rate (mol L^{-1} s^{-1})
0.0750	8.90×10^{-4}
0.190	2.26×10^{-3}
0.275	3.26×10^{-3}
0.410	4.85×10^{-3}

where $\text{Rate} = \frac{-d[N_2O_5]}{dt}$

Write the rate law and calculate the value of the rate constant.

19. The decomposition of nitrosyl chloride was studied:

$$2NOCl(g) \rightleftharpoons 2NO(g) + Cl_2(g)$$

The following data were obtained, where

$$\text{Rate} = \frac{-d[NOCl]}{dt}$$

$[NOCl]_0$ (molecules/cm^3)	Initial Rate (molecules cm^{-3} s^{-1})
3.0×10^{16}	5.98×10^4
2.0×10^{16}	2.66×10^4
1.0×10^{16}	6.64×10^3
4.0×10^{16}	1.06×10^5

*In the Questions and the Exercises, the term *rate law* always means differential rate law.

a. What is the rate law?
b. Calculate the rate constant.
c. Calculate the rate constant for the concentrations given in moles per liter.

20. The rate of the reaction between hemoglobin (Hb) and carbon monoxide (CO) was studied at 20°C. The following data were collected, with all concentration units in μmol/L. (A hemoglobin concentration of 2.21 μmol/L is equal to 2.21×10^{-6} mol/L.)

$[Hb]_0$ (μmol/L)	$[CO]_0$ (μmol/L)	Initial Rate (μmol L^{-1} s^{-1})
2.21	1.00	0.619
4.42	1.00	1.24
4.42	3.00	3.71

a. Determine the orders of this reaction with respect to Hb and CO.
b. Determine the rate law.
c. Calculate the value of the rate constant.
d. What would be the initial rate for an experiment with $[Hb]_0 = 3.36$ μmol/L and $[CO]_0 = 2.40$ μmol/L?

21. The following data were obtained for the reaction

$$2ClO_2(aq) + 2OH^-(aq) \longrightarrow ClO_3^-(aq) + ClO_2^-(aq) + H_2O(l)$$

where $\text{Rate} = -\frac{d[ClO_2]}{dt}$

$[ClO_2]_0$ (mol/L)	$[OH^-]_0$ (mol/L)	Initial Rate (mol L^{-1} s^{-1})
0.0500	0.100	5.75×10^{-2}
0.100	0.100	2.30×10^{-1}
0.100	0.0500	1.15×10^{-1}

a. Determine the rate law and the value of the rate constant.
b. What would be the initial rate for an experiment with $[ClO_2]_0 = 0.175$ mol/L and $[OH^-]_0 =$ 0.0844 mol/L?

22. The reaction

$$2NO(g) + O_2(g) \longrightarrow 2NO_2(g)$$

was studied, and the following data were obtained, where

$$\text{Rate} = -\frac{d[O_2]}{dt}$$

$[NO]_0$ (molecules/cm^3)	$[O_2]_0$ (molecules/cm^3)	Initial Rate (molecules cm^{-3} s^{-1})
1.00×10^{18}	1.00×10^{18}	2.00×10^{16}
3.00×10^{18}	1.00×10^{18}	1.80×10^{17}
2.50×10^{18}	2.50×10^{18}	3.13×10^{17}

What would be the initial rate for an experiment where $[NO]_0 = 6.21 \times 10^{18}$ molecules/cm^3 and $[O_2]_0 = 7.36 \times 10^{18}$ molecules/cm^3?

23. The reaction

$$H_2SeO_3(aq) + 6I^-(aq) + 4H^+(aq) \longrightarrow Se(s) + 2I_3^-(aq) + 3H_2O(l)$$

was studied at 0°C, and the following data were obtained:

$[H_2SeO_3]_0$ (mol/L)	$[H^+]_0$ (mol/L)	$[I^-]_0$ (mol/L)	Initial Rate (mol L^{-1} s^{-1})
1.0×10^{-4}	2.0×10^{-2}	2.0×10^{-2}	1.66×10^{-7}
2.0×10^{-4}	2.0×10^{-2}	2.0×10^{-2}	3.33×10^{-7}
3.0×10^{-4}	2.0×10^{-2}	2.0×10^{-2}	4.99×10^{-7}
1.0×10^{-4}	4.0×10^{-2}	2.0×10^{-2}	6.66×10^{-7}
1.0×10^{-4}	1.0×10^{-2}	2.0×10^{-2}	0.42×10^{-7}
1.0×10^{-4}	2.0×10^{-2}	4.0×10^{-2}	13.2×10^{-7}
1.0×10^{-4}	1.0×10^{-2}	4.0×10^{-2}	3.36×10^{-7}

These relationships hold only if there is an insignificant amount of I_3^- present. What is the rate law and the value of the rate constant? $\left(\text{Assume that rate} = -\frac{d[H_2SeO_3]}{dt}.\right)$

24. The initial rate of a reaction doubles as the concentration of one of the reactants is quadrupled. What is the order of this reactant? If a reactant has a −1 order, what happens to the initial rate when the concentration of that reactant increases by a factor of two?

25. A study was made of the effect of the hydroxide concentration on the rate of the reaction

$$I^-(aq) + OCl^-(aq) \longrightarrow IO^-(aq) + Cl^-(aq)$$

The following data were obtained:

$[I^-]_0$ (mol/L)	$[OCl^-]_0$ (mol/L)	$[OH^-]_0$ (mol/L)	Initial Rate (mol L^{-1} s^{-1})
0.0013	0.012	0.10	9.4×10^{-3}
0.0026	0.012	0.10	18.7×10^{-3}
0.0013	0.0060	0.10	4.7×10^{-3}
0.0013	0.018	0.10	14.0×10^{-3}
0.0013	0.012	0.050	18.7×10^{-3}
0.0013	0.012	0.20	4.7×10^{-3}
0.0013	0.018	0.20	7.0×10^{-3}

Determine the rate law and the value of the rate constant for this reaction.

Integrated Rate Laws

26. The initial rate for a reaction is equal to the slope of the tangent line at $t \approx 0$ in a plot of [A] versus time. From calculus,

$$\text{Initial rate} = \frac{-d[A]}{dt}$$

Therefore, the differential rate law for a reaction is

$$\text{Rate} = \frac{-d[A]}{dt} = k[A]^n$$

Assuming you have some calculus in your background, derive the zero-, first-, and second-order integrated rate laws using the differential rate law.

27. If the half-life for a reaction is 20. seconds, what would be the second half-life, assuming the reaction is either zero, first, or second order?

28. A certain reaction has the following general form:

$$aA \longrightarrow bB$$

At a particular temperature and $[A]_0 = 2.80 \times 10^{-3}$ *M*, concentration versus time data were collected for this reaction, and a plot of 1/[A] versus time resulted in a straight line with a slope value of $+3.60 \times 10^{-2}$ L mol^{-1} s^{-1}.
   a. Determine the rate law, the integrated rate law, and the value of the rate constant for this reaction.
   b. Calculate the half-life for this reaction.
   c. How much time is required for the concentration of A to decrease to 7.00×10^{-4} *M*?

29. A certain reaction has the following general form:

$$aA \longrightarrow bB$$

At a particular temperature and $[A]_0 = 2.00 \times 10^{-2}$ *M*, concentration versus time data were collected for this reaction, and a plot of ln[A] versus time resulted in a straight line with a slope value of -2.97×10^{-2} min^{-1}.
   a. Determine the rate law, the integrated rate law, and the value of the rate constant for this reaction.
   b. Calculate the half-life for this reaction.
   c. How much time is required for the concentration of A to decrease to 2.50×10^{-3} *M*?

30. The decomposition of ethanol (C_2H_5OH) on an alumina (Al_2O_3) surface,

$$C_2H_5OH(g) \longrightarrow C_2H_4(g) + H_2O(g)$$

was studied at 600 K. Concentration versus time data were collected for this reaction, and a plot of [A] versus time resulted in a straight line with a slope value of -4.00×10^{-5} mol L^{-1} s^{-1}.
   a. Determine the rate law, the integrated rate law, and the value of the rate constant for this reaction.
   b. If the initial concentration of C_2H_5OH was 1.25×10^{-2} *M*, calculate the half-life for this reaction.
   c. How much time is required for all of the 1.25×10^{-2} *M* C_2H_5OH to decompose?

31. The decomposition of hydrogen peroxide was studied at a particular temperature. The following data were obtained, where

$$\text{Rate} = -\frac{d[H_2O_2]}{dt}$$

Time (s)	$[H_2O_2]$ (mol/L)
0	1.00
120 ± 1	0.91
300 ± 1	0.78
600 ± 1	0.59
1200 ± 1	0.37
1800 ± 1	0.22
2400 ± 1	0.13
3000 ± 1	0.082
3600 ± 1	0.050

Determine the integrated rate law, the differential rate law, and the value of the rate constant. Calculate the $[H_2O_2]$ at 4000. s after the start of the reaction.

32. The dimerization of butadiene was studied at 500. K:

$$2C_4H_6(g) \longrightarrow C_8H_{12}(g)$$

The following data were obtained, where

$$\text{Rate} = -\frac{d[C_4H_6]}{dt}$$

Time (s)	$[C_4H_6]$ (mol/L)
195	1.6×10^{-2}
604	1.5×10^{-2}
1246	1.3×10^{-2}
2180	1.1×10^{-2}
6210	0.68×10^{-2}

Determine the forms of the integrated rate law, the differential rate law, and the value of the rate constant for this reaction.

33. The rate of the reaction

$$NO_2(g) + CO(g) \longrightarrow NO(g) + CO_2(g)$$

depends only on the concentration of nitrogen dioxide at temperatures below 225°C. At a temperature below 225°C, the following data were collected:

Time (s)	$[NO_2]$ (mol/L)
0	0.500
1.20×10^3	0.444
3.00×10^3	0.381
4.50×10^3	0.340
9.00×10^3	0.250
1.80×10^4	0.174

Determine the integrated rate law, the differential rate law, and the value of the rate constant at this temperature. Calculate $[NO_2]$ at 2.70×10^4 s after the start of the reaction.

34. The rate of the reaction

$$O(g) + NO_2(g) \longrightarrow NO(g) + O_2(g)$$

was studied at a certain temperature. This reaction is one step of the nitric oxide–catalyzed destruction of ozone in the upper atmosphere.
   a. In one experiment, NO_2 was in large excess at a concentration of 1.0×10^{13} molecules/cm^3 with the following data collected:

Time (s)	[O] (atoms/cm^3)
0	5.0×10^9
1.0×10^{-2}	1.9×10^9
2.0×10^{-2}	6.8×10^8
3.0×10^{-2}	2.5×10^8

   What is the order of the reaction with respect to oxygen atoms?

b. The reaction is known to be first order with respect to NO_2. Determine the overall rate law and the value of the rate constant.

35. At 500 K in the presence of a copper surface, ethanol decomposes according to the equation

$$C_2H_5OH(g) \longrightarrow CH_3CHO(g) + H_2(g)$$

The pressure of C_2H_5OH was measured as a function of time, and the following data were obtained:

Time (s)	$P_{C_2H_5OH}$ (torr)
0	250.
100.	237
200.	224
300.	211
400.	198
500.	185

Since the pressure of a gas is directly proportional to the concentration of the gas, we can express the rate law for a gaseous reaction in terms of partial pressures. Using the preceding data, deduce the rate law, the integrated rate law, and the value of the rate constant, all in terms of pressure units in atm and time in seconds. Predict the pressure of C_2H_5OH after 900. s from the start of the reaction. (*Hint:* To determine the order of the reaction with respect to C_2H_5OH, compare how the pressure of C_2H_5OH decreases with each time listing.)

36. Experimental data for the reaction

$$A \longrightarrow 2B + C$$

have been plotted in the following three different ways (with concentration units in mol/L):

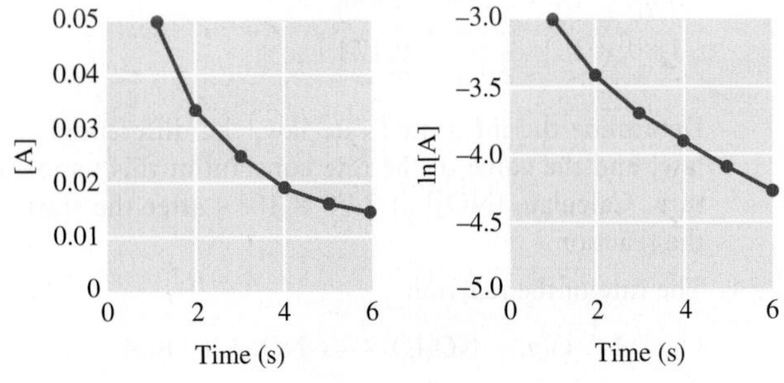

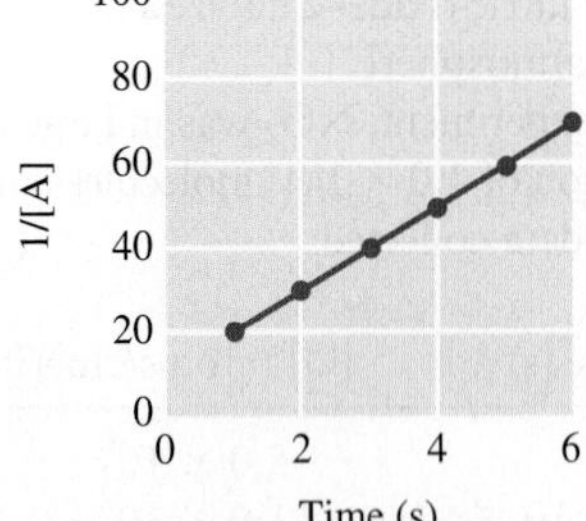

a. What is the order of the reaction with respect to A, and what is the initial concentration of A?
b. What is the concentration of A after 9 s?
c. What are the first three half-lives for this experiment?

37. The reaction

$$NO(g) + O_3(g) \longrightarrow NO_2(g) + O_2(g)$$

was studied by performing two experiments. In the first experiment (results shown in following table), the rate of disappearance of NO was followed in a large excess of O_3. (The $[O_3]$ remains effectively constant at 1.0×10^{14} molecules/cm^3.)

Time (ms)	[NO] (molecules/cm^3)
0	6.0×10^8
100 ± 1	5.0×10^8
500 ± 1	2.4×10^8
700 ± 1	1.7×10^8
1000 ± 1	9.9×10^7

In the second experiment, [NO] was held constant at 2.0×10^{14} molecules/cm^3. The data for the disappearance of O_3 were as follows:

Time (ms)	$[O_3]$ (molecules/cm^3)
0	1.0×10^{10}
50 ± 1	8.4×10^9
100 ± 1	7.0×10^9
200 ± 1	4.9×10^9
300 ± 1	3.4×10^9

a. What is the order with respect to each reactant?
b. What is the overall rate law?
c. What is the value of the rate constant obtained from each set of experiments?

$$\text{Rate} = k'[NO]^x \qquad \text{Rate} = k''[O_3]^y$$

d. What is the value of the rate constant for the overall rate law?

$$\text{Rate} = k[NO]^x[O_3]^y$$

38. Determine the forms of the integrated and the differential rate laws for the decomposition of benzene diazonium chloride,

$$C_6H_5N_2Cl(aq) \longrightarrow C_6H_5Cl(l) + N_2(g)$$

from the following data, which were collected at 50.°C and 1.00 atm:

Time (s)	N_2 Evolved (mL)
6	19.3
9	26.0
14	36.0
22	45.0
30.	50.4
∞	58.3

The total solution volume was 40.0 mL.

39. You and a coworker have developed a molecule that has shown potential as cobra antivenom (AV). This antivenom

works by binding to the venom (V), thereby rendering it nontoxic. This reaction can be desrcibed by the rate law

$$\text{Rate} = k[\text{AV}]^1[\text{V}]^1$$

You have been given the following data from your coworker:

$$[\text{V}]_0 = 0.20\ M$$

$$[\text{AV}]_0 = 1.0 \times 10^{-4}\ M$$

A plot of ln[AV] versus time gives a straight line with a slope of $-0.32\ \text{s}^{-1}$. What is the value of the rate constant (k) for this reaction?

40. Consider the following representation of the reaction $2\text{NO}_2(g) \rightarrow 2\text{NO}(g) + \text{O}_2(g)$.

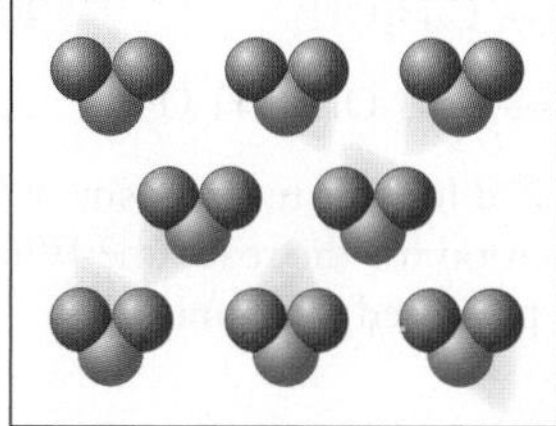

(a) time = 0 minutes

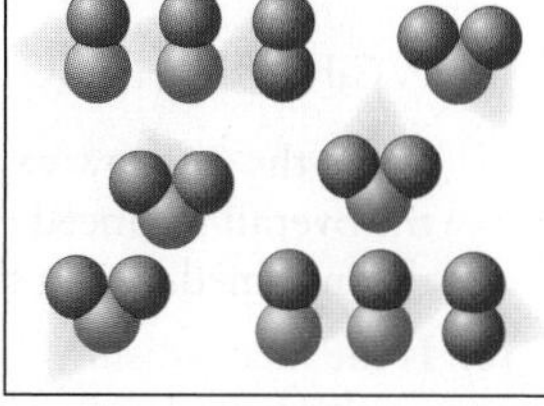

(b) time = 10 minutes

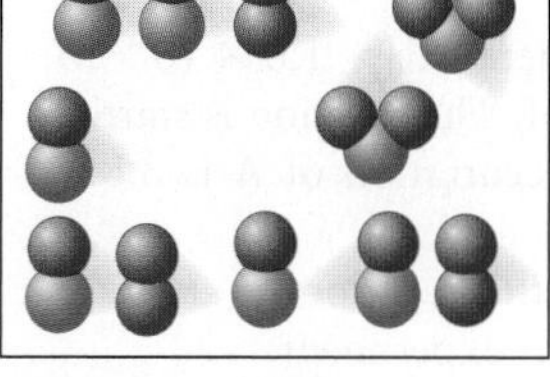

(c) time = ? minutes

Time

Determine the time for the final representation above if the reaction is
a. first order
b. second order
c. zero order

41. The rate law for the decomposition of phosphine (PH_3) is

$$\text{Rate} = -\frac{d[\text{PH}_3]}{dt} = k[\text{PH}_3]$$

It takes 120. s for the concentration of 1.00 M PH_3 to decrease to 0.250 M. How much time is required for 2.00 M PH_3 to decrease to a concentration of 0.350 M?

42. The radioactive isotope ^{32}P decays by first-order kinetics and has a half-life of 14.3 days. How long does it take for 95.0% of a given sample of ^{32}P to decay?

43. Consider the following initial rate data for the decomposition of compound AB to give A and B:

$[\text{AB}]_0$ (mol/L)	Initial Rate (mol $\text{L}^{-1}\ \text{s}^{-1}$)
0.200	3.20×10^{-3}
0.400	1.28×10^{-2}
0.600	2.88×10^{-2}

Determine the half-life for the decomposition reaction initially having 1.00 M AB present.

44. The rate law for the reaction

$$2\text{NOBr}(g) \longrightarrow 2\text{NO}(g) + \text{Br}_2(g)$$

at some temperature is

$$\text{Rate} = -\frac{d[\text{NOBr}]}{dt} = k[\text{NOBr}]^2$$

a. If the half-life for this reaction is 2.00 s when $[\text{NOBr}]_0 = 0.900\ M$, calculate the value of k for this reaction.
b. How much time is required for the concentration of NOBr to decrease to 0.100 M?

45. A first-order reaction is 75.0% complete in 320. s.
a. What are the first and second half-lives for this reaction?
b. How long does it take for 90.0% completion?

46. For the reaction A $\rightarrow$ products, successive half-lives are observed to be 10.0, 20.0, and 40.0 min for an experiment in which $[\text{A}]_0 = 0.10\ M$. Calculate the concentration of A at the following times.
a. 80.0 min b. 30.0 min

47. The decomposition of hydrogen iodide on finely divided gold at 150°C is zero order with respect to HI. The rate defined below is constant at 1.20×10^{-4} mol/L · s.

$$2\text{HI}(g) \xrightarrow{\text{Au}} \text{H}_2(g) + \text{I}_2(g)$$

$$\text{Rate} = -\frac{d[\text{HI}]}{dt} = k = 1.20 \times 10^{-4}\ \text{mol/L} \cdot \text{s}$$

a. If the initial HI concentration was 0.250 mol/L, calculate the concentration of HI at 25 minutes after the start of the reaction.
b. How long will it take for all of the 0.250 M HI to decompose?

48. Consider two reaction vessels, one containing A and the other containing B, with equal concentrations at $t = 0$. If both substances decompose by first-order kinectics, where

$$k_\text{A} = 4.50 \times 10^{-4}\ \text{s}^{-1}$$

$$k_\text{B} = 3.70 \times 10^{-3}\ \text{s}^{-1}$$

how much time must pass to reach a condition such that $[A] = 4.00[B]$?

49. Theophylline is a phamaceutical drug that is sometimes used to help with lung function. You observe a case where the initial lab results indicate that the concentration of theophylline in a patient's body decreased from 2.0×10^{-3} *M* to 1.0×10^{-3} *M* in 24 hours. In another 12 hours, the drug concentration was found to be 5.0×10^{-4} *M*. What is the value of the rate constant for the metabolism of this drug in the body?

50. Consider the hypothetical reaction

$$A + B + 2C \longrightarrow 2D + 3E$$

where the rate law is

$$\text{Rate} = -\frac{d[A]}{dt} = k[A][B]^2$$

An experiment is carried out where $[A]_0 = 1.0 \times 10^{-2}$ *M*, $[B]_0 = 3.0$ *M*, and $[C]_0 = 2.0$ *M*. The reaction is started, and after 8.0 seconds, the concentration of A is 3.8×10^{-3} *M*.

a. Calculate the value of k for this reaction.
b. Calculate the half-life for this experiment.
c. Calculate the concentration of A after 13.0 seconds.
d. Calculate the concentration of C after 13.0 seconds.

51. Consider the reaction

$$3A + B + C \longrightarrow D + E$$

where the rate law is defined as

$$-\frac{d[A]}{dt} = k[A]^2[B][C]$$

An experiment is carried out where $[B]_0 = [C]_0 = 1.00$ *M* and $[A]_0 = 1.00 \times 10^{-4}$ *M*.

a. If after 3.00 minutes $[A] = 3.26 \times 10^{-5}$ *M*, calculate the value of k.
b. Calculate the half-life for this experiment.
c. Calculate the concentration of B and the concentration of A after 10.0 minutes.

Reaction Mechanisms

52. Define each of the following.
a. elementary step
b. molecularity
c. reaction mechanism
d. intermediate
e. rate-determining step

53. Define what is meant by unimolecular and bimolecular steps. Why are termolecular steps infrequently seen in chemical reactions?

54. What two requirements must be met to call a mechanism plausible? Why say a "plausible" mechanism instead of the "correct" mechanism? Is it true that most reactions occur by a one-step mechanism? Explain.

55. Write the rate laws for the following elementary reactions.
a. $CH_3NC(g) \longrightarrow CH_3CN(g)$
b. $O_3(g) + NO(g) \longrightarrow O_2(g) + NO_2(g)$
c. $O_3(g) \longrightarrow O_2(g) + O(g)$
d. $O_3(g) + O(g) \longrightarrow 2O_2(g)$
e. ${}^{14}_{6}C \longrightarrow {}^{14}_{7}N + \beta$ particle (nuclear decay)

56. A possible mechanism for the decomposition of hydrogen peroxide is

$$H_2O_2 \longrightarrow 2OH$$
$$H_2O_2 + OH \longrightarrow H_2O + HO_2$$
$$HO_2 + OH \longrightarrow H_2O + O_2$$

Using your results fom Exercise 31, specify which step is the rate-determining step. What is the overall balanced equation for the reaction?

57. A proposed mechanism for a reaction is

$C_4H_9Br \longrightarrow C_4H_9^+ + Br^-$	Slow
$C_4H_9^+ + H_2O \longrightarrow C_4H_9OH_2^+$	Fast
$C_4H_9OH_2^+ + H_2O \longrightarrow C_4H_9OH + H_3O^+$	Fast

Write the rate law expected for this mechanism. What is the overall balanced equation for the reaction? What are the intermediates in the proposed mechanism?

58. Is the mechanism

$$NO + Cl_2 \xrightarrow{k_1} NOCl_2$$
$$NOCl_2 + NO \xrightarrow{k_2} 2NOCl$$

consistent with the results you obtained in Exercise 17? If so, which step is the rate-determining step?

59. The reaction

$$2NO(g) + O_2(g) \longrightarrow 2NO_2(g)$$

exhibits the rate law

$$\text{Rate} = k[NO]^2[O_2]$$

Which of the following mechanisms is consistent with this rate law?

a. $NO + O_2 \longrightarrow NO_2 + O$	Slow
$O + NO \longrightarrow NO_2$	Fast
b. $NO + O_2 \rightleftharpoons NO_3$	Fast equilibrium
$NO_3 + NO \longrightarrow 2NO_2$	Slow
c. $2NO \longrightarrow N_2O_2$	Slow
$N_2O_2 + O_2 \longrightarrow N_2O_4$	Fast
$N_2O_4 \longrightarrow 2NO_2$	Fast
d. $2NO \rightleftharpoons N_2O_2$	Fast equilibrium
$N_2O_2 \longrightarrow NO_2 + O$	Slow
$O + NO \longrightarrow NO_2$	Fast

60. The gas-phase reaction between Br_2 and H_2 to form HBr is assumed to proceed by the following mechanism:

$$Br_2 \underset{k_{-1}}{\overset{k_1}{\rightleftharpoons}} 2Br$$
$$Br + H_2 \underset{k_{-2}}{\overset{k_2}{\rightleftharpoons}} HBr + H$$
$$H + Br_2 \xrightarrow{k_3} HBr + Br$$
$$2Br \xrightarrow{k_4} Br_2$$

a. Under what conditions does the rate law have the form rate $= k'[Br_2]$?

b. Under what conditions does the rate law have the form rate $= k''[H_2][Br_2]^{1/2}$?

c. Give expressions for k' and k'' in terms of the rate constants used to define the mechanism.

61. The reaction

$$5Br^-(aq) + BrO_3^-(aq) + 6H^+(aq) \longrightarrow 3Br_2(l) + 3H_2O(l)$$

is expected to obey the mechanism

$$BrO_3^-(aq) + H^+(aq) \underset{k_{-1}}{\overset{k_1}{\rightleftharpoons}} HBrO_3(aq) \quad \text{Fast equilibrium}$$

$$HBrO_3(aq) + H^+(aq) \underset{k_{-2}}{\overset{k_2}{\rightleftharpoons}} H_2BrO_3^+(aq) \quad \text{Fast equilibrium}$$

$$Br^-(aq) + H_2BrO_3^+(aq) \xrightarrow{k_3} (Br\text{—}BrO_2)(aq) + H_2O(l) \quad \text{Slow}$$

$$(Br\text{—}BrO_2)(aq) + 4H^+(aq) + 4Br^-(aq) \longrightarrow \text{products} \quad \text{Fast}$$

Write the rate law for this reaction.

62. The rate law for the reaction

$$BrO_3^-(aq) + 3SO_3^{2-}(aq) \longrightarrow Br^-(aq) + 3SO_4^{2-}(aq)$$

is $\quad \text{Rate} = k[BrO_3^-][SO_3^{2-}][H^+]$

The first step in a proposed mechanism is

$$SO_3^{2-}(aq) + H^+(aq) \xrightarrow{k_1} HSO_3^-(aq) \quad \text{Fast}$$

The second step is rate determining. Write a possible second step for the mechanism.

63. The reaction

$$I^-(aq) + OCl^-(aq) \longrightarrow IO^-(aq) + Cl^-(aq)$$

is believed to occur by the following mechanism:

$$OCl^- + H_2O \underset{k_{-1}}{\overset{k_1}{\rightleftharpoons}} HOCl + OH^- \quad \text{Fast equilibrium}$$

$$I^- + HOCl \xrightarrow{k_2} HOI + Cl^- \quad \text{Slow}$$

$$HOI + OH^- \xrightarrow{k_3} H_2O + IO^- \quad \text{Fast}$$

Write the rate law for this reaction. *Note:* Since the reaction is in aqueous solution, the effective concentration of water remains constant. Thus the rate of the forward reaction in the first step can be written as

$$\text{Rate} = k[H_2O][OCl^-] = k_1[OCl^-]$$

64. In the gas phase, the production of phosgene from chlorine and carbon monoxide is assumed to proceed by the following mechanism:

$$Cl_2 \underset{k_{-1}}{\overset{k_1}{\rightleftharpoons}} 2Cl \quad \text{Fast equilibrium}$$

$$Cl + CO \underset{k_{-2}}{\overset{k_2}{\rightleftharpoons}} COCl \quad \text{Fast equilibrium}$$

$$COCl + Cl_2 \xrightarrow{k_3} COCl_2 + Cl \quad \text{Slow}$$

$$2Cl \xrightarrow{k_4} Cl_2 \quad \text{Fast}$$

Overall reaction: $CO + Cl_2 \longrightarrow COCl_2$

a. Write the rate law for this reaction.

b. Which species are intermediates?

65. The following mechanism is proposed for the reduction of NO_3^- by $MoCl_6^{2-}$:

$$MoCl_6^{2-} \underset{k_{-1}}{\overset{k_1}{\rightleftharpoons}} MoCl_5^- + Cl^-$$

$$NO_3^- + MoCl_5^- \xrightarrow{k_2} OMoCl_5^- + NO_2^-$$

a. What is the intermediate?

b. Derive an expression for the rate law (rate $= d[NO_2^-]/dt$) for the overall reaction using the steady-state approximation.

66. The following mechanism has been proposed to account for the rate law of the decomposition of ozone to $O_2(g)$:

$$O_3 + M \underset{k_{-1}}{\overset{k_1}{\rightleftharpoons}} O_2 + O + M$$

$$O + O_3 \xrightarrow{k_2} 2O_2$$

Apply the steady-state hypothesis to the concentration of atomic oxygen, and derive the rate law for the decomposition of ozone. (M stands for an atom or molecule that can exchange kinetic energy with the particles undergoing the chemical reaction.)

67. Consider the hypothetical reaction

$$B \longrightarrow E + F$$

which is assumed to occur by the mechanism

$$B + B \underset{k_{-1}}{\overset{k_1}{\rightleftharpoons}} B^* + B$$

$$B^* \xrightarrow{k_2} E + F$$

where B* represents a B molecule with enough energy to surmount the reaction energy barrier.

a. Derive the rate law for the production of E using the steady-state approximation.

b. Assume that this reaction is known to be first order. Under what conditions does your derived rate law (from part a) agree with this observation?

c. Explain how a chemical reaction can be first order, since even in a simple case ($B \longrightarrow E + F$) molecules must collide to build up enough energy to get over the energy barrier. Why aren't all reactions at least second order? In other words, explain the physical significance of the result from part b.

Temperature Dependence of Rate Constants and the Collision Model

68. How is the rate of a reaction affected by each of the following.

a. activation energy

b. temperature

c. frequency of collisions

d. orientation of collisions

69. The central idea of the collision model is that molecules must collide in order to react. Give two reasons why not all collisions of reactant molecules result in product formation.

70. Consider the following potential energy plots

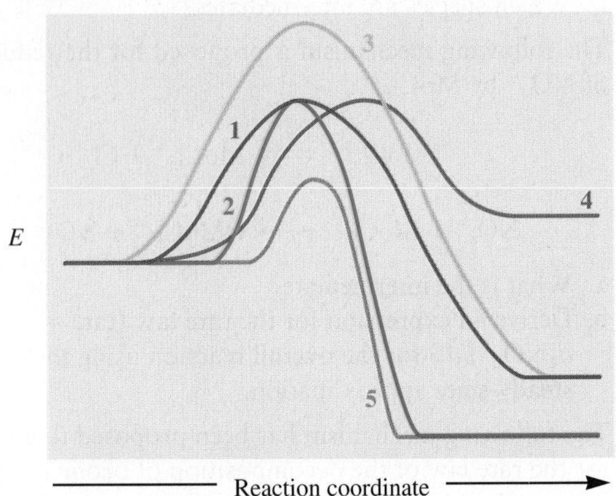

a. Rank the reactions from fastest to slowest, and explain your answer. If any reactions have equal rates, explain why.
b. Label the reactions as endothermic or exothermic, and supply your answer.
c. Rank the exothermic reactions from greatest to least change in potential energy, and support your answer.

71. The graph below shows the number of collisions with a particular energy for two different temperatures.

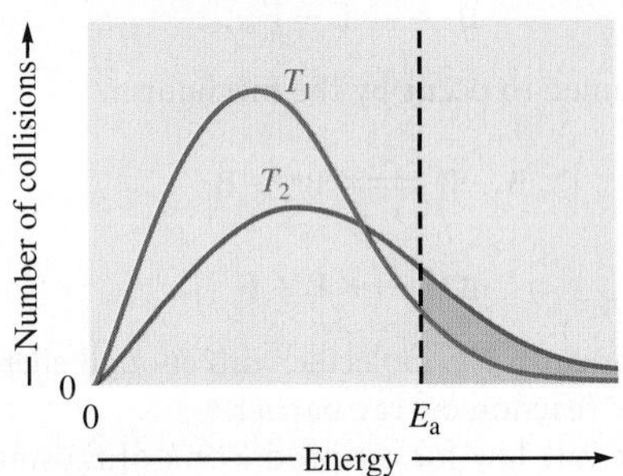

a. Which is greater, T_2 or T_1? How can you tell?
b. What does this graph tell us about the temperature dependence of the rate of a chemical reaction? Explain your answer.

72. Which of the following reactions would you expect to have the larger rate at room temperature? Why? (*Hint:* Think of which would have the lower activation energy.)

$$2Ce^{4+}(aq) + Hg_2^{2+}(aq) \longrightarrow 2Ce^{3+}(aq) + 2Hg^{2+}(aq)$$

$$H_3O^+(aq) + OH^-(aq) \longrightarrow 2H_2O(l)$$

73. The activation energy for the decomposition of $HI(g)$ to $H_2(g)$ and $I_2(g)$ is 186 kJ/mol. The rate constant at 555 K is 3.52×10^{-7} L mol^{-1} s^{-1}. What is the rate constant at 645 K?

74. The decomposition of iodoethane in the gas phase proceeds according to the following equation:

$$C_2H_5I(g) \longrightarrow C_2H_4(g) + HI(g)$$

At 660. K, $k = 7.2 \times 10^{-4}$ s^{-1}; at 720. K, $k = 1.7 \times 10^{-2}$ s^{-1}. What is the rate constant for this first-order decomposition at 325°C? If the initial pressure of iodoethane is 894 torr at 245°C, what is the pressure of iodoethane after three half-lives?

75. A certain reaction has an activation energy of 54.0 kJ/mol. As the temperature is increased from 22°C to a higher temperature, the rate constant increases by a factor of 7.00. Calculate the higher temperature.

76. Chemists commonly use a rule of thumb that an increase of 10 K in temperature doubles the rate of a reaction. What must the activation energy be for this statement to be true for a temperature increase from 25°C to 35°C?

77. The reaction

$$(CH_3)_3CBr + OH^- \longrightarrow (CH_3)_3COH + Br^-$$

in a certain solvent is first order with respect to $(CH_3)_3CBr$ and zero order with respect to OH^-. In several experiments the rate constant k was determined at different temperatures. A plot of $\ln(k)$ versus $1/T$ was constructed that resulted in a straight line with a slope of -1.10×10^4 K and a y intercept of 33.5. Assume that k has units of s^{-1}.
a. Determine the activation energy for this reaction.
b. Determine the value of the frequency factor A.
c. Calculate the value of k at 25°C.

78. The rate constant for the gas-phase decomposition of N_2O_5,

$$N_2O_5 \longrightarrow 2NO_2 + \tfrac{1}{2}O_2$$

has the following temperature dependence:

T (K)	k (s^{-1})
338	4.9×10^{-3}
318	5.0×10^{-4}
298	3.5×10^{-5}

Make the appropriate graph using these data, and determine the activation energy for this reaction.

79. Experimental values for the temperature dependence of the rate constant for the gas-phase reaction

$$NO(g) + O_3(g) \longrightarrow NO_2(g) + O_2(g)$$

are as follows:

T (K)	k (L mol^{-1} s^{-1})
195	1.08×10^9
230.	2.95×10^9
260.	5.42×10^9
298	12.0×10^9
369	35.5×10^9

Make the appropriate graph using these data, and determine the activation energy for this reaction.

80. Draw a rough sketch of the energy profile for each of the following cases.
a. $\Delta E = +10$ kJ/mol, $E_a = 25$ kJ/mol
b. $\Delta E = -10$ kJ/mol, $E_a = 50$ kJ/mol
c. $\Delta E = -50$ kJ/mol, $E_a = 50$ kJ/mol

Which reaction will have the greatest rate at 298 K? Assume the frequency factor A is the same for all three reactions.

81. For the following reaction profiles, indicate
   a. the positions of reactants and products.
   b. the activation energy.
   c. ΔE for the reaction.

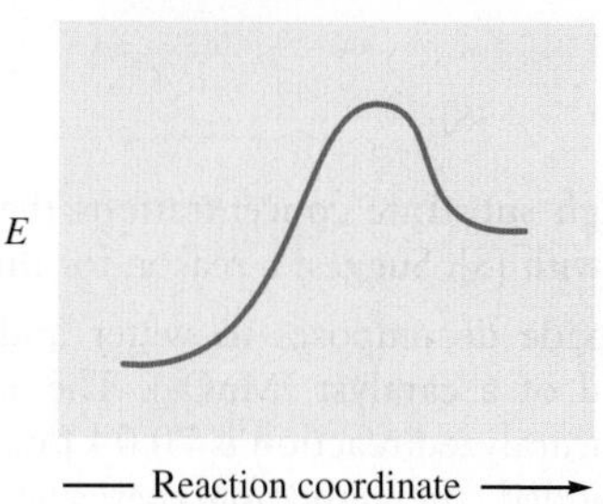

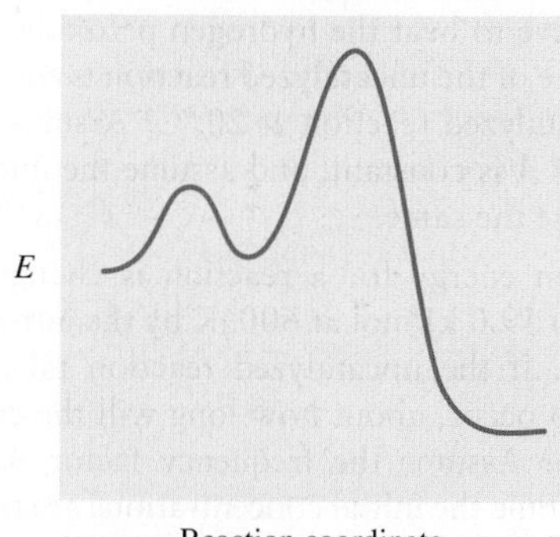

   d. The second reaction profile is representative of a reaction that occurs by a two-step mechanism. Which point on the plot represents the energy of the intermediate in the two-step reaction? Which step in the mechanism is rate determining, the first or the second step? Explain.

82. The activation energy for the reaction

$$NO_2(g) + CO(g) \longrightarrow NO(g) + CO_2(g)$$

is 125 kJ/mol, and ΔE for the reaction is −216 kJ/mol. What is the activation energy for the reverse reaction [$NO(g) + CO_2(g) \longrightarrow NO_2(g) + CO(g)$]?

83. The activation energy for the reaction

$$A_2(g) + B_2(g) \longrightarrow 2AB(g)$$

is 167 kJ/mol, and ΔE for the reaction is +28 kJ/mol. What is the activation energy for the decomposition of AB?

Catalysis

84. Why does a catalyst increase the rate of a reaction? What is the difference between a homogeneous and a heterogeneous catalyst? Would a given reaction necessarily have the same rate law for both a catalyzed and an uncatalyzed pathway? Explain.

85. Consider the following potential energy plots for a chemical reaction when answering the questions below.

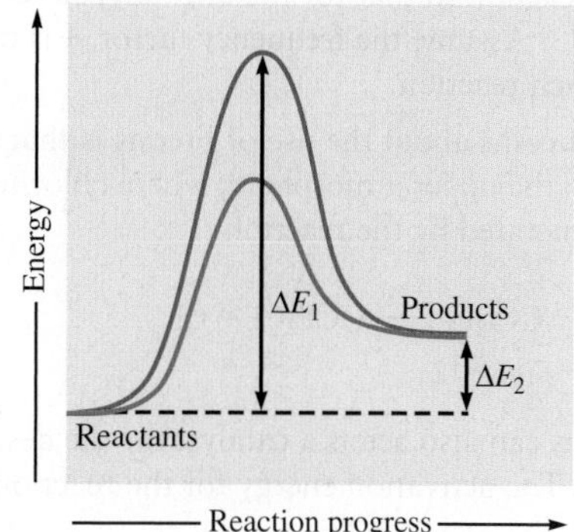

   a. Which plot (red or blue) is the catalyzed pathway? How do you know?
   b. What does ΔE_1 represent?
   c. What does ΔE_2 represent?
   d. Is the reaction endothermic or exothermic?

86. Would the slope of a ln(k) versus 1/T (K) plot for a catalyzed reaction be more or less negative than the slope of a ln(k) versus 1/T (K) plot for the uncatalyzed reaction? Assume that both rate laws are first order. Explain.

87. The decomposition of NH_3 to N_2 and H_2 was studied on two surfaces:

Surface	E_a (kJ/mol)
W	163
Os	197

Without a catalyst, the activation energy is 335 kJ/mol.
   a. Which surface is the better heterogeneous catalyst for the decomposition of NH_3? Why?
   b. How many times faster is the reaction at 298 K on the W surface compared with the reaction with no catalyst present? Assume that the frequency factor A is the same for each reaction.
   c. The decomposition reaction on the two surfaces obeys a rate law of the form

$$\text{Rate} = k\frac{[NH_3]}{[H_2]}$$

   How can you explain the inverse dependence of the rate on the H_2 concentration?

88. One pathway for the destruction of ozone in the upper atmosphere is

$O_3(g) + NO(g) \longrightarrow NO_2(g) + O_2(g)$	Slow
$NO_2(g) + O(g) \longrightarrow NO(g) + O_2(g)$	Fast

Overall reaction: $O_3(g) + O(g) \longrightarrow 2O_2(g)$

   a. Which species is a catalyst?
   b. Which species is an intermediate?
   c. E_a for the uncatalyzed reaction

$$O_3(g) + O(g) \longrightarrow 2O_2(g)$$

is 14.0 kJ. E_a for the same reaction when catalyzed is 11.9 kJ. What is the ratio of the rate constant for the catalyzed reaction to that for the uncatalyzed reaction at 25°C? Assume the frequency factor A is the same for each reaction.

89. One of the concerns about the use of Freons is that they will migrate to the upper atmosphere, where chlorine atoms can be generated by the reaction

$$\underset{\text{Freon-12}}{CCl_2F_2} \xrightarrow{h\nu} CF_2Cl + Cl$$

Chlorine atoms can also act as a catalyst for the destruction of ozone. The activation energy for the reaction

$$Cl + O_3 \longrightarrow ClO + O_2$$

is 2.1 kJ/mol. Which is the more effective catalyst for the destruction of ozone, Cl or NO? (See Exercise 88.)

90. Assuming that the mechanism for the hydrogenation of C_2H_4 given in Section 15.9 is correct, would you predict that the product of the reaction of C_2H_4 with D_2 would be $CH_2D—CH_2D$ or $CHD_2—CH_3$?

91. For enzyme-catalyzed reactions that follow the mechanism

$$E + S \rightleftharpoons E \cdot S$$
$$E \cdot S \rightleftharpoons E + P$$

a graph of the rate as a function of [S], the concentration of the substrate, has the following general appearance:

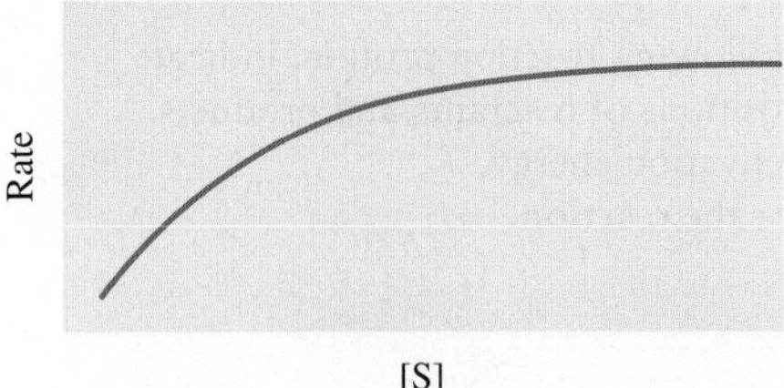

Note that at high substrate concentrations the rate no longer changes with [S]. Suggest a reason for this.

92. Hydrogen peroxide decomposes to water and oxygen gas with the aid of a catalyst (MnO_2). The activation energy of the uncatalyzed reaction is 70.0 kJ/mol. When the catalyst is added, the activation energy at 20.°C is 42.0 kJ/mol. Theoretically, to what temperature (°C) would one have to heat the hydrogen peroxide solution so that the rate of the uncatalyzed reaction is equal to the rate of the catalyzed reaction at 20.°C? Assume the frequency factor A is constant, and assume the initial concentrations are the same.

93. The activation energy for a reaction is changed from 184 kJ/mol to 59.0 kJ/mol at 600. K by the introduction of a catalyst. If the uncatalyzed reaction takes about 2400 years to occur, about how long will the catalyzed reaction take? Assume the frequency factor A is constant, and assume the initial concentrations are the same.

Additional Exercises

94. The rate law for a reaction can be determined only from experiment and not from the balanced equation. Two experimental procedures were outlined in this chapter. What are these two procedures? Explain how each method is used to determine rate laws.

95. The type of rate law for a reaction, either the differential rate law or the integrated rate law, is usually determined by which data are easiest to collect. Explain.

96. a. Using the free energy profile for a simple one-step reaction, show that at equilibrium $K = k_f/k_r$, where k_f and k_r are the rate constants for the forward and reverse reactions. *Hint:* Use the relationship $\Delta G° = -RT\ln(K)$, and represent k_f and k_r using the Arrhenius equation ($k = Ae^{-E_a/RT}$).

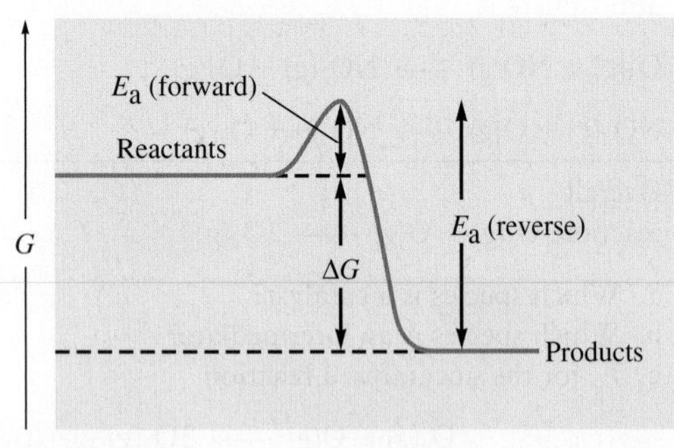

b. Why is the following statement false? "A catalyst can increase the rate of a forward reaction but not the rate of the reverse reaction."

97. Iodomethane (CH_3I) is a commonly used reagent in organic chemistry. When used properly, this reagent allows chemists to introduce methyl groups in many different, useful applications. The chemical does pose a risk as a carcinogen, possibly owing to iodomethane's ability to react with portions of the DNA strand (if they were to come in contact). Consider the following hypothetical initial rates data:

$[DNA]_0$ (μmol/L)	$[CH_3I]_0$ (μmol/L)	Initial Rate (μmol L^{-1} s^{-1})
0.100	0.100	3.20×10^{-4}
0.100	0.200	6.40×10^{-4}
0.200	0.200	1.28×10^{-3}

Which of the following could be a possible mechanism to explain the initial rate data?

Mechanism I $DNA + CH_3I \longrightarrow DNA—CH_3^+ + I^-$

Mechanism II $CH_3I \longrightarrow CH_3^+ + I^-$ Slow

$DNA + CH_3^+ \longrightarrow DNA—CH_3^+$ Fast

98. Two isomers (A and B) of a given compound dimerize as follows:

$$2A \xrightarrow{k_1} A_2$$
$$2B \xrightarrow{k_2} B_2$$

Both processes are known to be second order in the reactant, and k_1 is known to be 0.250 L mol^{-1} s^{-1} at 25°C. In a particular experiment A, and B were placed in separate containers at 25°C, where $[A]_0 = 1.00 \times 10^{-2}$ *M* and $[B]_0 = 2.50 \times 10^{-2}$ *M*. After each reaction had progressed for 3.00 min, [A] = 3.00[B]. In this case the rate laws are defined as follows:

$$\text{Rate} = -\frac{d[A]}{dt} = k_1[A]^2$$

$$\text{Rate} = -\frac{d[B]}{dt} = k_2[B]^2$$

a. Calculate the concentration of A_2 after 3.00 min.
b. Calculate the value of k_2.
c. Calculate the half-life for the experiment involving A.

99. The thermal degradation of silk was studied by Kuruppillai, Hersh, and Tucker ("Historic Textile and Paper Materials," *ACS Advances in Chemistry Series,* No. 212, 1986) by measuring the tensile strength of silk fibers at various times of exposure to elevated temperature. The loss of tensile strength follows first-order kinetics,

$$-\frac{ds}{dt} = ks$$

where *s* is the strength of the fiber retained after heating and *k* is the first-order rate constant. The effects of adding a deacidifying agent and an antioxidant to the silk were studied, and the following data were obtained:

	Strength Retained (%)		
Heating Time (days)	Untreated	Deacidifying Agent	Antioxidant
0.00	100.0	100.1	114.6
1.00	67.9	60.8	65.2
2.00	38.9	26.8	28.1
3.00	16.1	—	11.3
6.00	6.8	—	—

a. Determine the first-order rate constants for the thermal degradation of silk for each of the three experiments.
b. Does either of the two additives appear to retard the degradation of silk?
c. Calculate the half-life for the thermal degradation of silk for each of the three experiments.

100. Sulfuryl chloride (SO_2Cl_2) decomposes to sulfur dioxide (SO_2) and chlorine (Cl_2) by reaction in the gas phase. The following data were obtained when a sample containing 5.00×10^{-2} mole of sulfuryl chloride was heated to 600 K ± 1 K in a 5.00×10^{-1} L container.

Time (h)	0.00	1.00	2.00	4.00	8.00	16.00
Pressure (atm)	4.93	5.60	6.34	7.33	8.56	9.52

Define the rate as $-d[SO_2Cl_2]/dt$.
a. Determine the value of the rate constant for the decomposition of sulfuryl chloride at 600 K.
b. What is the half-life of the reaction?
c. What would be the pressure in the vessel after 0.500 h and after 12.0 h?
d. What fraction of the sulfuryl chloride remains after 20.0 h?

101. One reason suggested for the instability of long chains of silicon atoms is that the decomposition involves the following transition state:

```
   H  H
   |  |
H—Si—Si—H  ⟶
   |  |
   H  H

 ⎧  H                   H  ⎫
 ⎪  |                   |  ⎪
 ⎨H—Si - - - - - - - - -Si:⎬  ⟶  SiH4 + :SiH2
 ⎪  |   ⟍          ⟋    |  ⎪
 ⎩  H      ⟍ H ⟋        H  ⎭
```

The activation energy for such a process is 210 kJ/mol, which is less than either the Si—Si or the Si—H bond energy. Why would a similar mechanism not be expected to play a very important role in the decomposition of long chains of carbon atoms as seen in organic compounds?

102. The following results were obtained at 600 K for the decomposition of ethanol on an alumina (Al_2O_3) surface,

$$C_2H_5OH(g) \longrightarrow C_2H_4(g) + H_2O(g)$$

t (s)	P_{Total} (torr)
0	250.
10.	265
20.	280.
30.	295
40.	310.
50.	325

a. Predict P_{Total} in torr at $t = 80.$ s.
b. What is the value of the rate constant, and what are its units?
c. What is the order of the reaction?
d. Calculate P_{Total} at $t = 300.$ s.

103. At 620. K butadiene dimerizes at a moderate rate. The following data were obtained in an experiment involving this reaction:

t (s)	$[C_4H_6]$ (mol/L)
0	0.01000
1000.	0.00629
2000.	0.00459
3000.	0.00361

a. Determine the order of the reaction in butadiene.
b. In how many seconds is the dimerization 1.0% complete?
c. In how many seconds is the dimerization 2.0% complete?
d. What is the half-life for the reaction if the initial concentration of butadiene is 0.0200 *M*?
e. Use the results from this problem and Exercise 32 to calculate the activation energy for the dimerization of butadiene.

104. The decomposition of $NO_2(g)$ occurs by the following bimolecular elementary reaction:

$$2NO_2(g) \longrightarrow 2NO(g) + O_2(g)$$

The rate constant at 273 K is 2.3×10^{-12} L mol^{-1} s^{-1}, and the activation energy is 111 kJ/mol. How long will it take for the concentration of $NO_2(g)$ to decrease from an initial partial pressure of 2.5 atm to 1.5 atm at 500. K? Assume ideal gas behavior.

105. The activation energy for a certain uncatalyzed biochemical reaction is 50.0 kJ/mol. In the presence of a catalyst at 37°C, the rate constant for the reaction increases by a factor of 2.50×10^3 as compared with the uncatalyzed reaction. Assuming that the frequency factor *A* is the same for both the catalyzed and uncatalyzed reactions, calculate the activation energy for the catalyzed reaction.

106. For the reaction

$$2N_2O_5(g) \longrightarrow 4NO_2(g) + O_2(g)$$

the following data were collected, where

$$\text{Rate} = -\frac{d[N_2O_5]}{dt}$$

t (s)	T = 338 K $[N_2O_5]$	T = 318 K $[N_2O_5]$
0	1.00×10^{-1} *M*	1.00×10^{-1} *M*
100.	6.14×10^{-2} *M*	9.54×10^{-2} *M*
300.	2.33×10^{-2} *M*	8.63×10^{-2} *M*
600.	5.41×10^{-3} *M*	7.43×10^{-2} *M*
900.	1.26×10^{-3} *M*	6.39×10^{-2} *M*

Calculate E_a for this reaction.

107. Experiments have shown the average frequency of chirping of individual snowy tree crickets (*Oecanthus fultoni*) to be 178 min^{-1} at 25.0°C, 126 min^{-1} at 20.3°C, and 100. min^{-1} at 17.3°C.
a. What is the apparent activation energy of the reaction that controls the chirping?
b. What chirping rate would be expected at 15.0°C?
c. Compare the observed rates and your calculated rate from part b to the rule of thumb that the Fahrenheit temperature is 42 plus 0.80 times the number of chirps in 15 s.

108. Experiments during a recent summer on a number of fireflies (small beetles, *Lampyridae photinus*) showed that the average interval between flashes of individual insects was 16.3 s at 21.0°C and 13.0 s at 27.8°C.
a. What is the apparent activation energy of the reaction that controls the flashing?
b. What would be the average interval between flashes of an individual firefly at 30.0°C?
c. Compare the observed intervals and the one you calculated in part b to the rule of thumb that the Celsius temperature is 54 minus twice the interval between flashes.

109. The compound NO_2Cl is thought to decompose to NO_2 and Cl_2 by the following mechanism:

$$NO_2Cl \underset{k_{-1}}{\overset{k_1}{\rightleftharpoons}} NO_2 + Cl$$

$$NO_2Cl + Cl \xrightarrow{k_2} NO_2 + Cl_2$$

Derive the rate law for the production of Cl_2 using the steady-state approximation.

110. Many biochemical reactions are catalyzed by large protein molecules called enzymes. A typical mechanism for the conversion of a biochemical substrate (S) to product (P) catalyzed by an enzyme (E) involves the following steps:

$$E + S \underset{k_{-1}}{\overset{k_1}{\rightleftharpoons}} ES$$

$$ES \xrightarrow{k_2} P$$

The rate-determining step is the decomposition of the intermediate enzyme–substrate complex (ES) to products (P). Under these conditions, show that the overall rate of product formation is

$$\text{Rate} = \frac{d[P]}{dt} = \frac{k_1k_2[E]_T[S]}{k_{-1} + k_2 + k_1[S]}$$

where $[E]_T$ equals the total enzyme concentration:

$$[E]_T = [E] + [ES]$$

Challenge Problems

111. Consider the following reaction:

$$CH_3X + Y \longrightarrow CH_3Y + X$$

At 25°C the following two experiments were run, yielding the following data:

Experiment 1: $[Y]_0 = 3.0$ M

$[CH_3X]$	Time (h)
7.08×10^{-3} *M*	1.0
4.52×10^{-3} *M*	1.5
2.23×10^{-3} *M*	2.3
4.76×10^{-4} *M*	4.0
8.44×10^{-5} *M*	5.7
2.75×10^{-5} *M*	7.0

Experiment 2: $[Y]_0 = 4.5$ M

$[CH_3X]$	Time (h)
4.50×10^{-3} *M*	0
1.70×10^{-3} *M*	1.0
4.19×10^{-4} *M*	2.5
1.11×10^{-4} *M*	4.0
2.81×10^{-5} *M*	5.5

Experiments were also run at 85°C. The value of the rate constant at 85°C was found to be 7.88×10^8 (with the time in units of hours), where $[CH_3X]_0 = 1.0 \times 10^{-2}$ *M* and $[Y]_0 = 3.0$ *M*.

a. Determine the rate law and the value of k for this reaction at 25°C.
b. Determine the half-life at 85°C.
c. Determine E_a for the reaction.
d. Given that the C—X bond energy is known to be about 325 kJ/mol, suggest a mechanism that explains the results in parts a and c.

112. The following data were collected in two studies of the reaction

$$2A + B \longrightarrow C + D$$

where

$$\text{Rate} = \frac{-d[A]}{dt}$$

Time (s)	Experiment 1 [A] (*M*) $\times 10^{-2}$	Experiment 2 [A] (*M*) $\times 10^{-2}$
0	10.0	10.0
20.	6.67	5.00
40.	5.00	3.33
60.	4.00	2.50
80.	3.33	2.00
100.	2.86	1.67
120.	2.50	1.43

In experiment 1, $[B]_0 = 5.0$ *M*. In experiment 2, $[B]_0 = 10.0$ *M*.

a. Why is [B] much greater than [A]?
b. Give the rate law and value for k for this reaction.
c. Which of the following mechanisms could be correct for this reaction? Justify your choice.
   i. $A + B \rightleftharpoons E$ (fast equilibrium)
      $E + B \longrightarrow C + D$ (slow)
   ii. $A + B \rightleftharpoons E$ (fast equilibrium)
      $E + A \longrightarrow C + D$ (slow)
   iii. $A + A \longrightarrow E$ (slow)
      $E + B \longrightarrow C + D$ (fast)

113. Consider a reaction of the type aA → products, in which the rate law is found to be rate $= k[A]^3$ (termolecular reactions are improbable but possible). If the first half-life of the reaction is found to be 40. s, what is the time for the second half-life? *Hint:* Using your calculus knowledge, derive the integrated rate law from the differential rate law for a termolecular reaction:

$$\text{Rate} = \frac{-d[A]}{dt} = k[A]^3$$

114. For the reaction

$$2A + B \longrightarrow \text{products}$$

a friend proposes the following mechanism:

$$A + B \rightleftharpoons M$$

$$A + M \longrightarrow \text{products}$$

a. Assuming that the second step is the rate-determining step and the first step is a fast equilibrium step, determine the rate law. Represent the rate constant in terms of k_1, k_{-1}, and k_2.
b. Using the steady-state approximation, determine the rate law.
c. Under what conditions of [A] and [B] do you get the same rate law in parts a and b?

115. Consider the hypothetical reaction

$$A + B + 2C \longrightarrow 2D + 3E$$

In a study of this reaction, three experiments were run at the same temperature. The rate is defined as $-d[B]/dt$.

Experiment 1:

$[A]_0 = 2.0$ *M* $[B]_0 = 1.0 \times 10^{-3}$ *M* $[C]_0 = 1.0$ *M*

[B] (mol/L)	Time (s)
2.7×10^{-4}	1.0×10^5
1.6×10^{-4}	2.0×10^5
1.1×10^{-4}	3.0×10^5
8.5×10^{-5}	4.0×10^5
6.9×10^{-5}	5.0×10^5
5.8×10^{-5}	6.0×10^5

Experiment 2:

$[A]_0 = 1.0 \times 10^{-2}\ M \quad [B]_0 = 3.0\ M \quad [C]_0 = 1.0\ M$

[A] (mol/L)	Time (s)
8.9×10^{-3}	1.0
7.1×10^{-3}	3.0
5.5×10^{-3}	5.0
3.8×10^{-3}	8.0
2.9×10^{-3}	10.0
2.0×10^{-3}	13.0

Experiment 3:

$[A]_0 = 10.0\ M \quad [B]_0 = 5.0\ M \quad [C]_0 = 5.0 \times 10^{-1}\ M$

[C] (mol/L)	Time (s)
0.43	1.0×10^{-2}
0.36	2.0×10^{-2}
0.29	3.0×10^{-2}
0.22	4.0×10^{-2}
0.15	5.0×10^{-2}
0.08	6.0×10^{-2}

Write the rate law for this reaction, and calculate the rate constant.

116. A reaction represented by the equation

$$3O_2(g) \longrightarrow 2O_3(g)$$

was studied at a specific temperature, and the following data were collected:

Time (s)	Total Pressure (atm)
0	1.000
46.89	0.9500
98.82	0.9033
137.9	0.8733
200.0	0.8333
286.9	0.7900
337.9	0.7700
511.3	0.7233

a. Determine the rate law for this reaction.
b. Determine the value of the rate constant (including units).
c. Calculate the time it would take for the total pressure to be 0.7133 atm.

117. The gas-phase decomposition $2N_2O_5 \rightarrow 4NO_2 + O_2$ is first order but not unimolecular. A possible mechanism is

$$M + N_2O_5 \underset{k_{-1}}{\overset{k_1}{\rightleftharpoons}} NO_3 + NO_2 + M$$

$$NO_3 + NO_2 \xrightarrow{k_2} NO + O_2 + NO_2$$

$$NO_3 + NO \xrightarrow{k_3} 2NO_2$$

Apply the steady-state approximation to the concentrations of the intermediates NO_3 and NO, and derive the rate law for the decomposition of N_2O_5.

118. You are studying the kinetics of the reaction $H_2(g) + F_2(g) \rightarrow 2HF(g)$ and you wish to determine a mechanism for the reaction. You run the reaction twice by keeping one reactant at a much higher pressure than the other reactant (this lower-pressure reactant begins at 1.000 atm). Unfortunately, you neglect to record which reactant was at the higher pressure, and you forget it later. Your data for the first experiment are as follows:

Pressure of HF (atm)	Time (min)
0	0
0.300	30.0
0.600	65.8
0.900	110.4
1.200	169.1
1.500	255.9

When you run the second experiment (in which the higher-pressure reactant is run at a much higher pressure), you determine the values of the apparent rate constants to be the same. It also turns out that you find data taken from another person in the lab. This individual found that the reaction proceeds 40.0 times faster at 55°C than at 35°C. You also know, from the energy-level diagram, that there are three steps to the mechanism, and the first step has the highest activation energy. You look up the bond energies of the species involved and they are (in kJ/mol): H—H (432), F—F (154), and H—F (565).

a. Sketch an energy-level diagram (qualitative) that is consistent with the one described.
b. Develop a reasonable mechanism for the reaction. Support your answer and explain the significance of each piece of information.
c. Which reactant was limiting in the experiments?

119. Hydrogen peroxide and the iodide ion react in acidic solution as follows:

$$H_2O_2(aq) + 3I^-(aq) + 2H^+(aq) \longrightarrow I_3^-(aq) + 2H_2O(l)$$

The kinetics of this reaction were studied by following the decay of the concentration of H_2O_2 and constructing plots of $\ln[H_2O_2]$ versus time. All the plots were linear and all solutions had $[H_2O_2]_0 = 8.0 \times 10^{-4}$ mol/L. The slopes of these straight lines depended on the initial concentrations of I^- and H^+. The results follow:

$[I^-]_0$ (mol/L)	$[H^+]_0$ (mol/L)	Slope (min^{-1})
0.1000	0.0400	−0.120
0.3000	0.0400	−0.360
0.4000	0.0400	−0.480
0.0750	0.0200	−0.0760
0.0750	0.0800	−0.118
0.0750	0.1600	−0.174

The rate law for this reaction has the form

$$\text{Rate} = \frac{-d[H_2O_2]}{dt} = (k_1 + k_2[H^+])[I^-]^m[H_2O_2]^n$$

a. Specify the orders of this reaction with respect to $[H_2O_2]$ and $[I^-]$.
b. Calculate the values of the rate constants k_1 and k_2.
c. What reason could there be for the two-term dependence of the rate on $[H^+]$?

Marathon Problem

120. Consider the following hypothetical data collected in two studies of the reaction

$$2A(g) + 2B(g) \longrightarrow C(g) + 2D(g)$$

Time (s)	Experiment 1 [A] (mol/L)	Experiment 2 [A] (mol/L)
0.	1.0×10^{-2}	1.0×10^{-2}
10.	8.4×10^{-3}	5.0×10^{-3}
20.	7.1×10^{-3}	2.5×10^{-3}
30.	?	1.3×10^{-3}
40.	5.0×10^{-3}	6.3×10^{-4}

In experiment 1, [B] = 10.0 *M*. In experiment 2, [B] = 20.0 *M*.

$$\text{Rate} = \frac{-d[A]}{dt}$$

a. Use the concentration versus time data to determine the rate law for the reaction.

b. Solve for the rate constant *k* for the reaction. Include units.

c. Calculate the concentration of A in experiment 1 at $t = 30.$ s.

d. The following three mechanisms are proposed for this reaction:

i.	$2B \rightleftharpoons B_2$	(Fast equilibrium)
	$B_2 + A \longrightarrow E + D$	(Slow)
	$E + A \longrightarrow C + D$	(Fast)
ii.	$A + B \longrightarrow D + F$	(Slow)
	$F + B \longrightarrow C + G$	(Fast)
	$G + A \longrightarrow D$	(Fast)
iii.	$A + 2B \longrightarrow E + D$	(Slow)
	$E + A \longrightarrow C + D$	(Fast)

Choose the best mechanism(s). Include an explanation of which mechanism(s) you exclude and why. If you believe two (or even all three) mechanisms are equally good, explain why.

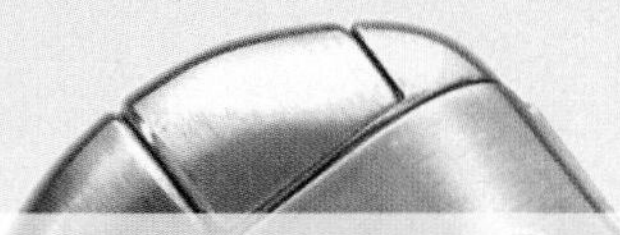

Liquids and Solids

16

chapter

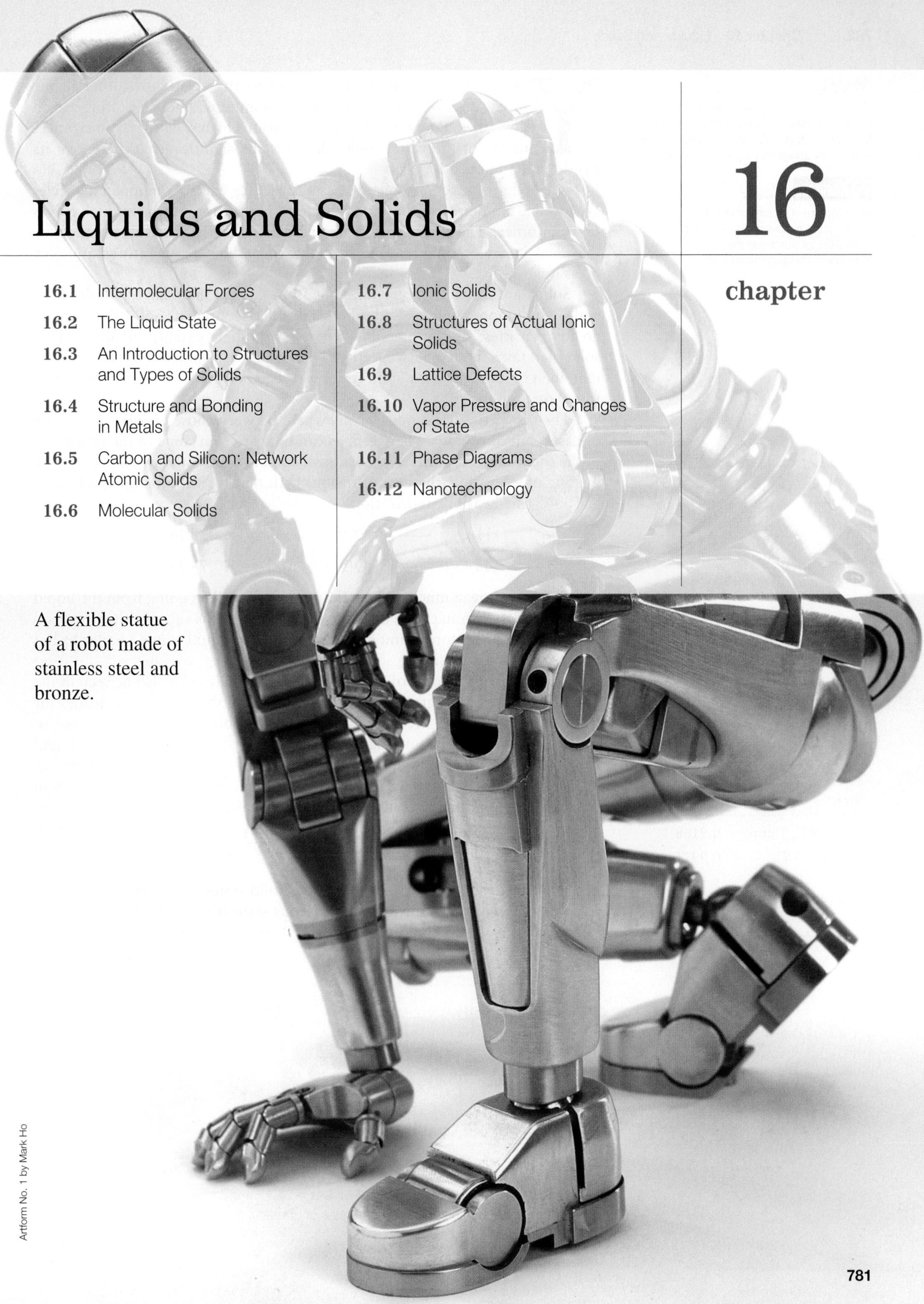

A flexible statue of a robot made of stainless steel and bronze.

Artform No. 1 by Mark Ho

OWL

Sign in to OWL at www.cengage.com/owl to view tutorials and simulations, develop problem-solving skills, and complete online homework assigned by your professor.

go Chemistry

Download mini lecture videos for key concept review and exam prep from OWL or purchase them from www.cengagebrain.com.

You have only to think about water to appreciate how different the three states of matter are. Flying, swimming, and ice skating are all done in contact with water in its various forms. Clearly, the arrangements of the water molecules must be significantly different in its gas, liquid, and solid forms.

Recall that a gas can be pictured as a substance whose component particles are far apart and are in rapid, random motion, exerting relatively small forces on each other. The kinetic molecular model was constructed to account for the ideal behavior that real gases approach at high temperatures and low pressures.

Solids are obviously very different from gases. Gases have low densities and high compressibilities and completely fill a container. Solids have much greater densities, are compressible only to a very slight extent, and are rigid—a solid maintains its shape irrespective of its container. These properties indicate that the components of a solid are close together and exert large attractive forces on each other.

The properties of liquids lie somewhere between those of solids and those of gases, but not midway between, as can be seen from some of the properties of the three states of water. For example, compare the enthalpy change for the melting of ice at 0°C (the heat of fusion) with that for vaporizing liquid water at 100°C (the heat of vaporization):

$$H_2O(s) \longrightarrow H_2O(l) \qquad \Delta H^\circ_{fus} = 6.02 \text{ kJ/mol}$$

$$H_2O(l) \longrightarrow H_2O(g) \qquad \Delta H^\circ_{vap} = 40.7 \text{ kJ/mol}$$

These values show a much greater change in structure in going from the liquid to the gas than in going from the solid to the liquid. This suggests that there are extensive attractive forces among the molecules in liquid water, similar to but not as strong as those in the solid state.

The relative similarity of the liquid and solid states can also be seen in the densities of the three states of water. As shown in Table 16.1, the densities for liquid and solid water are quite close. Compressibilities can also be used to explore the relationship among water's states. At 25°C the density of liquid water changes from 0.99707 g/cm^3 at a pressure of 1 atm to 1.046 g/cm^3 at 1065 atm. Given the large change in pressure, this is a very small variation in the density. Ice also shows little variation in density with increased pressure. On the other hand, at 400°C the density of gaseous water changes from 3.26×10^{-4} g/cm^3 at 1 atm pressure to 0.157 g/cm^3 at 242 atm—a huge variation.

The conclusion is clear. The liquid and solid states show many similarities and are strikingly different from the gaseous state (Fig. 16.1). We must bear this in mind as we develop models for the structures of solids and liquids.

Table 16.1

Densities of the Three States of Water

State	Density (g/cm^3)
Solid (0°C, 1 atm)	0.9168
Liquid (25°C, 1 atm)	0.9971
Gas (400°C, 1 atm)	3.26×10^{-4}

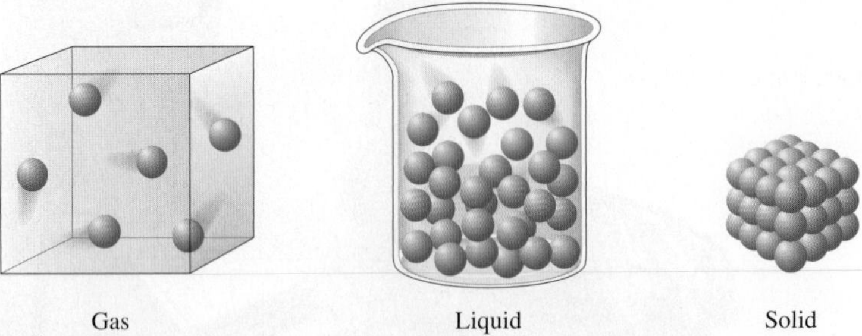

Figure 16.1
Schematic representations of the three states of matter.

16.1 Intermolecular Forces

Recall that atoms can form stable units called molecules by sharing electrons. This is called *intramolecular* (within the molecule) bonding. In this chapter we will consider the properties of the **condensed states of matter** (liquids and solids) and the forces that cause the aggregation of the components of a substance to form a liquid or a solid. These forces may involve covalent or ionic bonding, or they may involve weaker interactions usually called **intermolecular forces** (because they occur between, rather than within, molecules).

It is important to recognize that when a substance like water changes from solid to liquid to gas, *the molecules remain intact.* The changes of state are caused by changes in the forces *among* the molecules rather than those *within* the molecules. In ice, as we will see later in this chapter, the molecules are virtually locked in place, although they can vibrate about their positions. If energy is added, the motions of the molecules increase, and they eventually achieve the greater movement and disorder characteristic of liquid water; the ice has melted. As more energy is added, the gaseous state is eventually reached, where the individual molecules are far apart and thus interacting relatively little. However, the gas still consists of water molecules. It would take much more energy to overcome the covalent bonds and decompose the water molecules into their component atoms. This can be seen by comparing the energy needed to vaporize 1 mole of liquid water (40.7 kJ) with that needed to break the O—H bonds in 1 mole of water molecules (934 kJ).

Remember that temperature is a measure of the random motions of the particles in a substance.

Dipole–Dipole Forces

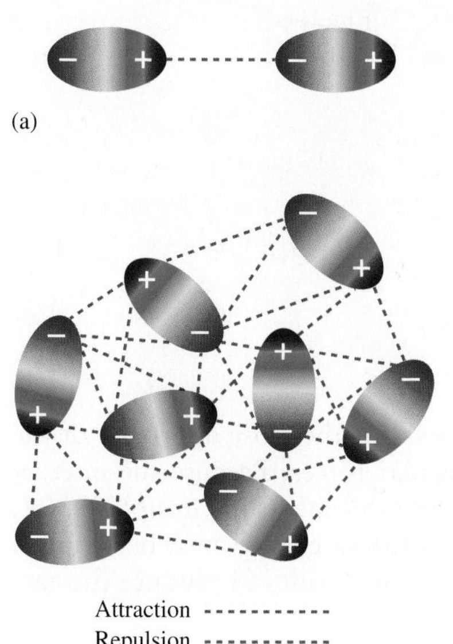

Figure 16.2
(a) The electrostatic interaction of two polar molecules. (b) The interaction of many dipoles in a condensed state.

Recall that molecules with polar bonds often behave in an electric field as if they had a center of positive charge and a center of negative charge; that is, they exhibit a dipole moment. Molecules with dipole moments can attract each other electrostatically by lining up so that the positive and negative ends are close to each other, as shown in Fig. 16.2(a). This is called a **dipole–dipole attraction.** In a condensed state such as a liquid, the dipoles find the best compromise between attraction and repulsion, as shown in Fig. 16.2(b).

Dipole–dipole forces are typically only about 1% as strong as covalent or ionic bonds, and they rapidly become weaker as the distance between the dipoles increases. At low pressures in the gas phase, where the molecules are far apart, these forces are relatively unimportant.

Particularly strong dipole–dipole forces are seen among molecules in which hydrogen is bound to a highly electronegative atom, such as nitrogen, oxygen, or fluorine. Two factors account for the strengths of these interactions: the great polarity of the bond and the close approach of the dipoles, allowed by the very small size of the hydrogen atom. Because dipole–dipole attractions of this type are so unusually strong, they are given a special name—**hydrogen bonding.** Figure 16.3 shows hydrogen bonding among water molecules.

Hydrogen bonding has a very important effect on various physical properties. For example, the boiling points for the covalent hydrides of the elements in Groups 4A, 5A, 6A, and 7A are shown in Fig. 16.4. Note that the nonpolar tetrahedral hydrides of Group 4A show a steady increase in boiling point with molar mass (that is, in going down the group), whereas for the other groups the lightest member has an unexpectedly high boiling point. Why? The answer lies in the especially large hydrogen-bonding interactions that exist among the smallest molecules with the most polar X—H bonds. These unusually strong hydrogen-bonding forces are due primarily to two factors. One factor is the relatively large electronegativity values of the lightest elements in each group, which leads to especially polar X—H bonds. The second factor is the small size

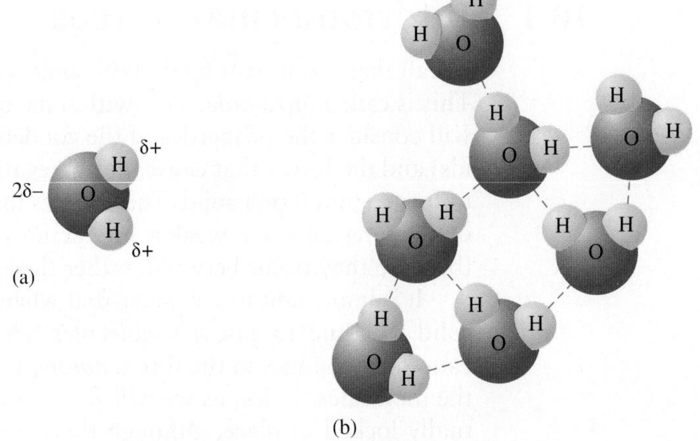

Figure 16.3
(a) The polar water molecule. (b) Hydrogen bonding among water molecules. Note that the small size of the hydrogen atom allows for close interactions.

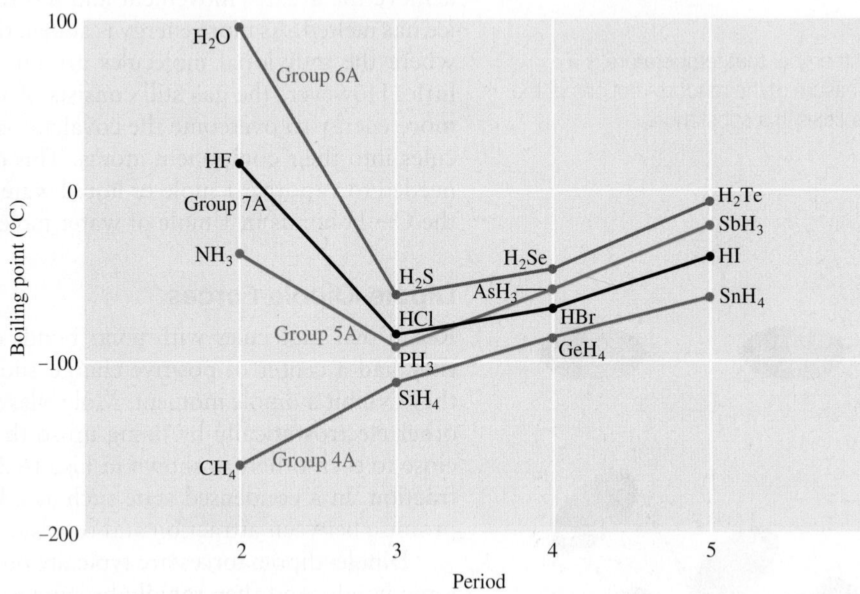

Figure 16.4
The boiling points of the covalent hydrides of elements in Groups 4A, 5A, 6A, and 7A.

of the first element of each group. This allows for the close approach of the dipoles, further strengthening the intermolecular forces. Because the interactions among the molecules containing the lightest elements in Groups 5A, 6A, and 7A are so strong, an unusually large quantity of energy must be supplied to overcome these interactions and separate the molecules to produce the gaseous state—hence the very high boiling points.

Boiling point will be defined precisely in Section 16.10.

London Dispersion Forces

The dipole–dipole and London dispersion forces are sometimes called van der Waals forces after Johannes van der Waals. See Section 5.10.

Even molecules without dipole moments must exert forces on each other. We know they do because all substances—even the noble gases—exist in the liquid and solid states under certain conditions. The relatively weak forces that exist among noble gas atoms and nonpolar molecules are called **London dispersion forces.** To understand the origin of these forces, let's consider a pair of noble gas atoms. Although we usually assume that the electrons of an atom are uniformly distributed about the nucleus, this is apparently not true at every instant. Atoms can develop a momentary nonsymmetrical electron distribution that produces a temporary dipolar arrangement of charge. This *instantaneous dipole* can then *induce* a similar dipole in a neighboring atom, as shown in Fig. 16.5(a). This phenomenon leads to an interatomic attraction that is both weak

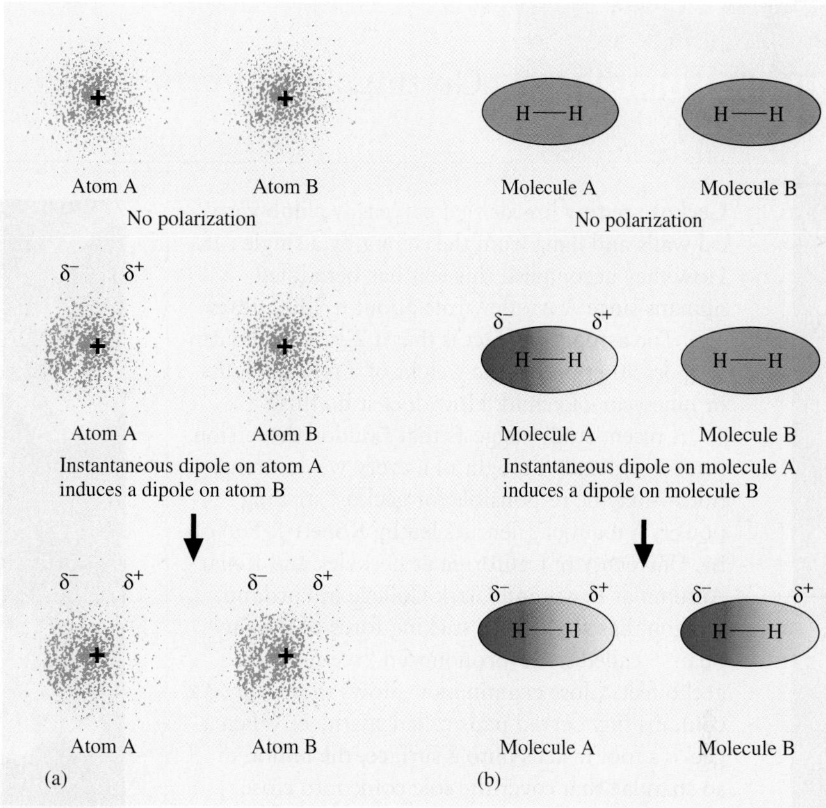

Figure 16.5
(a) An instantaneous polarization can occur on atom A, creating an instantaneous dipole. This dipole creates an induced dipole on neighboring atom B. (b) Nonpolar molecules such as H_2 can develop instantaneous and induced dipoles.

Table 16.2

The Freezing Points of the Group 8A Elements

Element	Freezing Point (°C)
Helium*	−269.7
Neon	−248.6
Argon	−189.4
Krypton	−157.3
Xenon	−111.9

*Helium is the only liquid that does not freeze when the temperature is lowered at 1 atm. It will freeze only if more pressure is applied.

and short-lived but that can be very significant for large atoms, as we will discuss later. For these interactions to become strong enough to produce a solid, the motions of the atoms must be greatly reduced. This explains, for instance, why the noble gas elements have such low freezing points (Table 16.2).

Note from Table 16.2 that the freezing point rises going down the group. The principal cause for this trend is that as the mass (and the atomic number) increases, the number of electrons increases, so there is an increased chance of the occurrence of momentary dipoles. We say that large atoms with many electrons exhibit a higher *polarizability* than small atoms. Thus the importance of London dispersion forces greatly increases as atomic size increases.

The dispersion forces in molecules with large atoms are quite significant and are often actually more important than dipole–dipole forces.

These same ideas also apply to nonpolar molecules such as H_2, CH_4, CCl_4, and CO_2 [Fig. 16.5(b)]. Since none of these molecules has a permanent dipole moment, their principal means of attraction for each other is through London dispersion forces.

16.2 The Liquid State

Liquids and liquid solutions are vital to our lives. Of course, water is the most important liquid. Besides being essential to life, it provides a medium for food preparation, for transportation, for cooling in many types of machines and industrial processes, for recreation, for cleaning, and for a myriad of other uses.

Liquids exhibit many characteristics that help us to understand their natures. We have already mentioned their low compressibility, lack of rigidity, and high density compared with gases. Many of the properties of liquids give us direct information about the forces that exist among the particles. For example, when a liquid is poured onto a solid surface, it tends to bead as drop-

Chemical Insights — Getting a Grip

Geckos are tiny lizards that can easily climb vertical walls and hang from the ceiling by a single toe. How they accomplish this feat has befuddled humans since Aristotle wrote about it 2300 years ago. The astonishing fact is that a 2-inch gecko can support 90 pounds—the weight of a typical eight- or nine-year-old child. How does it do this?

A recent study suggests that London dispersion forces—typically thought of as very weak interactions—may be responsible for geckos' sticking power. A team of scientists led by Robert J. Full of the University of California at Berkeley and Kellar Autumn at Lewis and Clark College in Portland, Oregon, has studied the sticking force of the tiny "hairs" called *setae* (pronounced "see-tee") on gecko feet. Close examination shows that each seta contains tiny curved pads called spatulae. When a gecko's foot fastens onto a surface, the billion or so spatulae that cover the sole come into close contact with the surface. The scientists' calculations show that the dispersion forces from these billions of interactions can explain the amazing gecko grip. Gecko adhesion also works in a vacuum, which eliminates the possibility that "suction" is involved in the process.

The West Coast team was tipped off to the nature of the adhesion forces by the unusual way in which a gecko walks—by setting its toes down like tongues uncurling and lifting its toes as if peeling up adhesive tape. The downward step seems to both press the setae against the surface and tug them parallel to the surface, ensuring that maximum contact is made. Tests show that the setae hold fast until the angle between them and the surface reaches about 30 degrees. To remove its toes from the surface, the gecko presumably raises the setae beyond the critical angle of 30 degrees.

Figuring out how geckos stick is not just an issue of pure scientific interest but could lead to products such as super sticky tape and robots that could climb any surface. The gecko's adhesive system never leaves a residue, never gets dirty, never wears out, and works on any surface. Many useful products can be visualized if scientists can find a way to artificially replicate gecko adhesion. It might even take all of the danger (and fun) out of rock climbing!

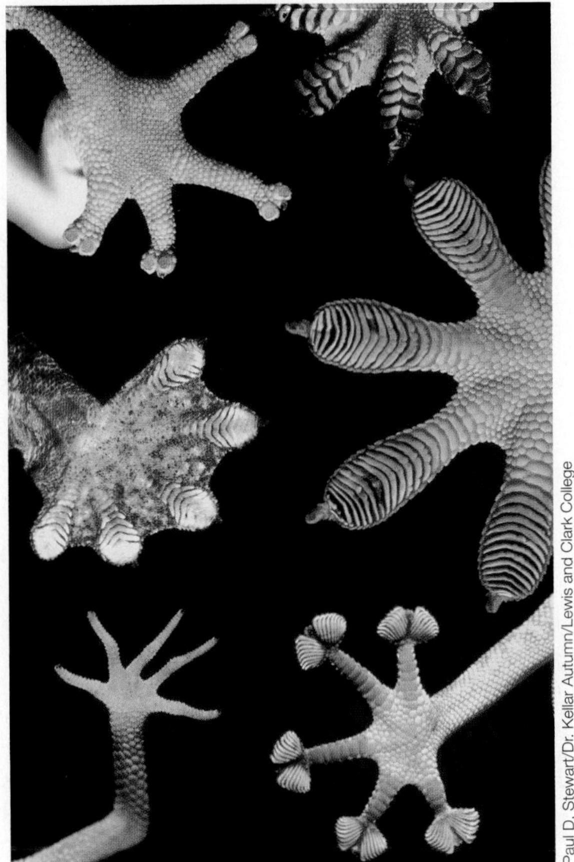

Paul D. Stewart/Dr. Kellar Autumn/Lewis and Clark College

The feet and toes of a variety of gecko lizards.

lets, a phenomenon that depends on the intermolecular forces. Although molecules in the interior of the liquid are completely surrounded by other molecules, those at the liquid's surface are subject to attractions only from the side and from below (Fig. 16.6). The effect of this uneven pull on the surface molecules tends to draw them into the body of the liquid and causes a droplet of liquid to assume the shape that has the minimum surface area—a sphere.

For a given volume, a sphere has a smaller surface area than any other shape.

To increase a liquid's surface area, molecules must move from the interior of the liquid to the surface. This requires energy, since some intermolecular

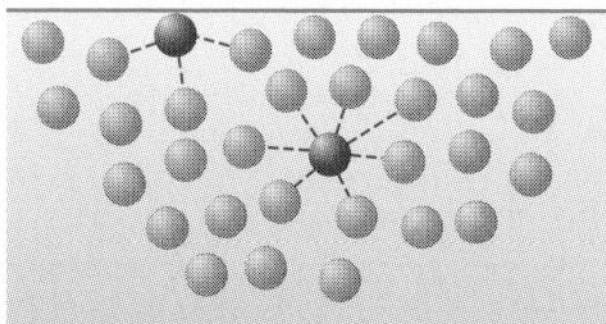

Figure 16.6
A molecule in the interior of a liquid is attracted to the molecules surrounding it, whereas a molecule at the surface of a liquid is attracted only by molecules below it and on each side of it.

The composition of glass is discussed in Section 16.5.

forces must be overcome. The resistance of a liquid to an increase in its surface area is called the **surface tension** of the liquid. As we would expect, liquids with relatively large intermolecular forces tend to have relatively high surface tensions.

Polar liquids also exhibit **capillary action,** the spontaneous rising of a liquid in a narrow tube. Two different types of forces are responsible for this property: *cohesive forces,* the intermolecular forces among the molecules of the liquid, and *adhesive forces,* the forces between the liquid molecules and their container. We have already seen how cohesive forces operate among polar molecules. Adhesive forces between a polar liquid and a given surface are strongest when the surface is made of a substance that has polar bonds. For example, glass contains many oxygen atoms with partial negative charges that are attractive to the positive end of a polar molecule such as water. This ability of water to "wet" glass makes it creep up the walls of the tube where the water surface touches the glass. This, however, tends to increase the surface area of the water, which is opposed by the cohesive forces that try to minimize the surface area. Thus, because water has both strong cohesive (intermolecular) forces and strong adhesive forces to glass, it "pulls itself" up a glass capillary tube (a tube with a small diameter) to a height where the weight of the column of water just balances the water's tendency to be attracted to the glass surface. The concave shape of the meniscus (Fig. 16.7) shows that water's adhesive forces toward the glass are stronger than its cohesive forces. A nonpolar liquid such as mercury shows a convex meniscus in a glass tube. This behavior is characteristic of a liquid in which the cohesive forces are stronger than the adhesive forces toward the glass.

Another property of liquids that is strongly dependent on intermolecular forces is **viscosity,** a measure of a liquid's resistance to flow. As might be expected, liquids with large intermolecular forces tend to be highly viscous. For example, glycerol, whose structure is

$$
\begin{array}{c}
\mathrm{H} \\
| \\
\mathrm{H{-}C{-}O{-}H} \\
| \\
\mathrm{H{-}C{-}O{-}H} \\
| \\
\mathrm{H{-}C{-}O{-}H} \\
| \\
\mathrm{H}
\end{array}
$$

has an unusually high viscosity mainly because of its high capacity to form hydrogen bonds.

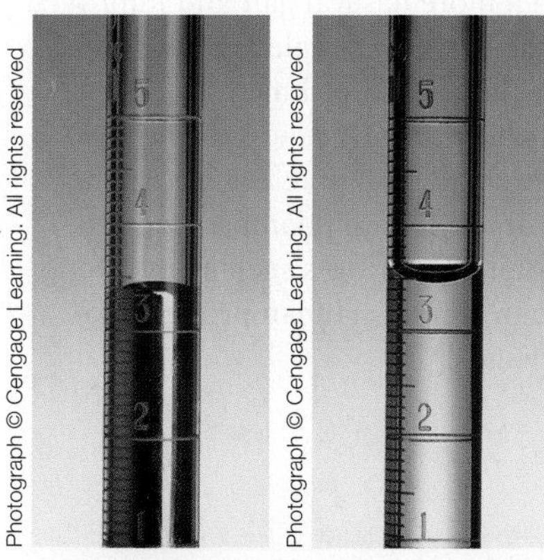

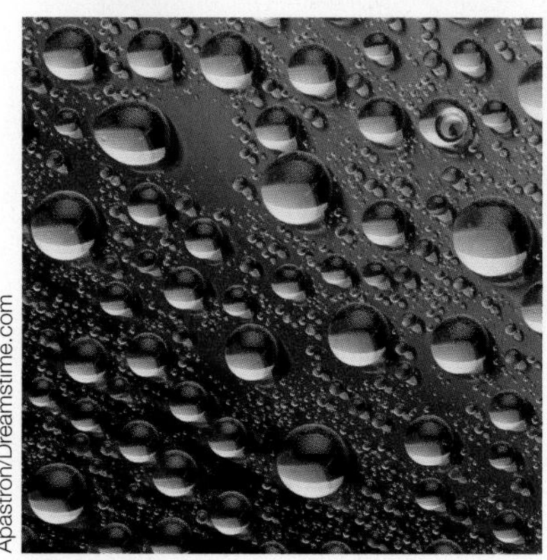

Figure 16.7
(top left) Nonpolar liquid mercury forms a convex meniscus in a glass tube. (top right) Polar water forms a concave meniscus. (bottom) Beads of water on a waxed car finish.

Molecular complexity also leads to higher viscosity because very large molecules can become entangled with each other. For example, nonviscous gasoline contains molecules of the type $CH_3—(CH_2)_n—CH_3$, where n varies from about 3 to 8. However, grease, which is very viscous, contains much larger molecules in which n varies from 20 to 25.

In many respects the development of a structural model for liquids presents greater challenges than the development of such a model for the other two states of matter. In the gaseous state, the particles are so far apart and are moving so rapidly that intermolecular forces are negligible under most circumstances. This means we can use a relatively simple model for gases. In the solid state, although the intermolecular forces are large, the molecular motions are minimal, and fairly simple models are again possible. The liquid state, however, has both strong intermolecular forces *and* significant molecular motions. Such a situation precludes the use of really simple models for liquids. Recent advances in spectroscopy—the study of the manner in which substances interact with electromagnetic radiation—make it possible to follow the very rapid

Chemical Insights Smart Fluids

Matter seems to be getting smarter these days. Increasingly, we have discovered materials that can remember their initial shape after being deformed or can sense and respond to their environment. In particular, valuable new materials have been formulated whose properties can be changed instantly by applying a magnetic or electric field.

One example of such a substance is a fluid whose flow characteristics (rheology) can be changed from free flowing to almost solid in about 0.01 second by the application of an electromagnetic field. This magnetorheological (MR) fluid was developed by Lord Corporation. Working in collaboration with Delphi Corporation, the company is applying the fluid in suspension control of General Motors automobiles such as Cadillacs and Corvettes. The so-called Magneride system has sensors that monitor the road surface and provide information about what suspension damping is needed. In response, a message is instantly sent to an electromagnetic coil in the shock absorbers, which adjusts the viscosity of the MR fluid to provide continuously variable damping. The result: an amazingly smooth ride and unerring road-holding ability.

The MR fluid is composed of a synthetic oil in which particles of an iron-containing compound are suspended. When the magnetic field is turned off,

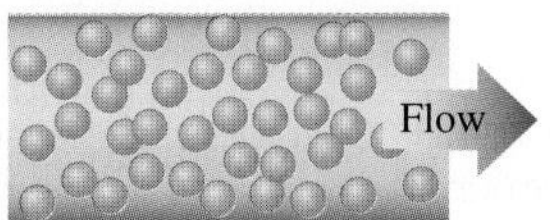

Magnetic field off
Magnetic particles flow randomly

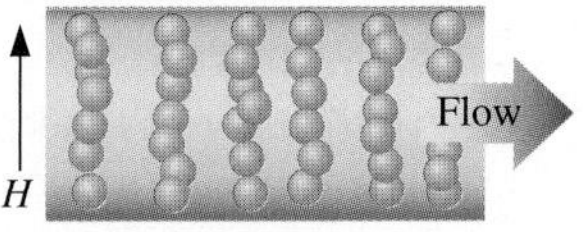

Magnetic field on
Applied field (H) creates structure that increases viscosity

these particles flow freely in all directions (see the figure above). When the field is turned on, the particles aggregate into chains that line up perpendicular to the flow of the fluid, thereby increasing its viscosity in proportion to the strength of the applied field.

Many other applications of MR fluids besides auto suspensions are under development. For example, this technology is being used in a prosthesis (see below) for above-the-knee amputees, which gives them a more natural gait and improves stair climbing. One very large-scale application is in Japan's National Museum of Emerging Science and Innovation, where an MR fluid is being used in dampers to protect the building against earthquake damage. Large MR fluid dampers are also being used for stabilizing bridges such as the Dong Ting Lake Bridge in China's Hunan province to steady it in high winds.

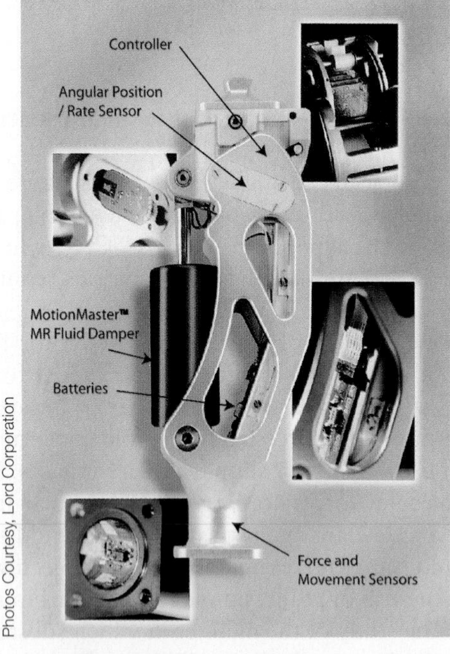

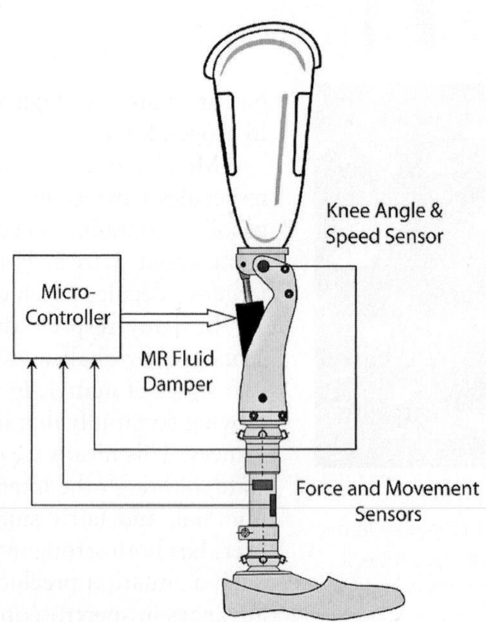

This High Intelligence Prosthesis for the knee uses an MR fluid damper to provide motion that closely duplicates the natural movement of the knee joint.

changes that occur in liquids. As a result, our models of liquids are becoming more accurate. As a starting point, a typical liquid might best be viewed as containing a large number of regions where the arrangements of the components are similar to those found in the solid, but with more disorder. In addition, a smaller number of regions exist in the liquid where gaps ("holes") occur among the molecules. The situation is highly dynamic, with rapid fluctuations occurring in both types of regions.

16.3 An Introduction to Structures and Types of Solids

There are many ways to classify solids, but the broadest categories are **crystalline solids,** those with a highly regular arrangement of their components, and **amorphous solids,** those with considerable disorder in their structures.

The regular arrangement of the components of a crystalline solid at the microscopic level produces the beautiful, characteristic shapes of crystals, such as those shown in Fig. 16.8. The positions of the components in a crystalline solid are usually represented by a **lattice,** a three-dimensional array of points designating the centers of the components (atoms, ions, or molecules) that shows the repetitious pattern of the components. The *smallest repeating unit* of

Figure 16.8
Several crystalline solids. (clockwise from upper left) Rhodochrosite, fluorite, pyrite, amethyst.

the lattice is called the **unit cell.** Thus a particular lattice can be generated by repeating (translating) the unit cell in all three dimensions to form the extended structure. Three common unit cells and their lattices are shown in Fig. 16.9.

Although we will concentrate on crystalline solids in this book, there are many important noncrystalline (amorphous) materials. An example is common glass, which is best pictured as a solution whose components are "frozen in place" before they can achieve an ordered arrangement. Although glass is a solid (it has a rigid shape), a great deal of disorder exists in its structure.

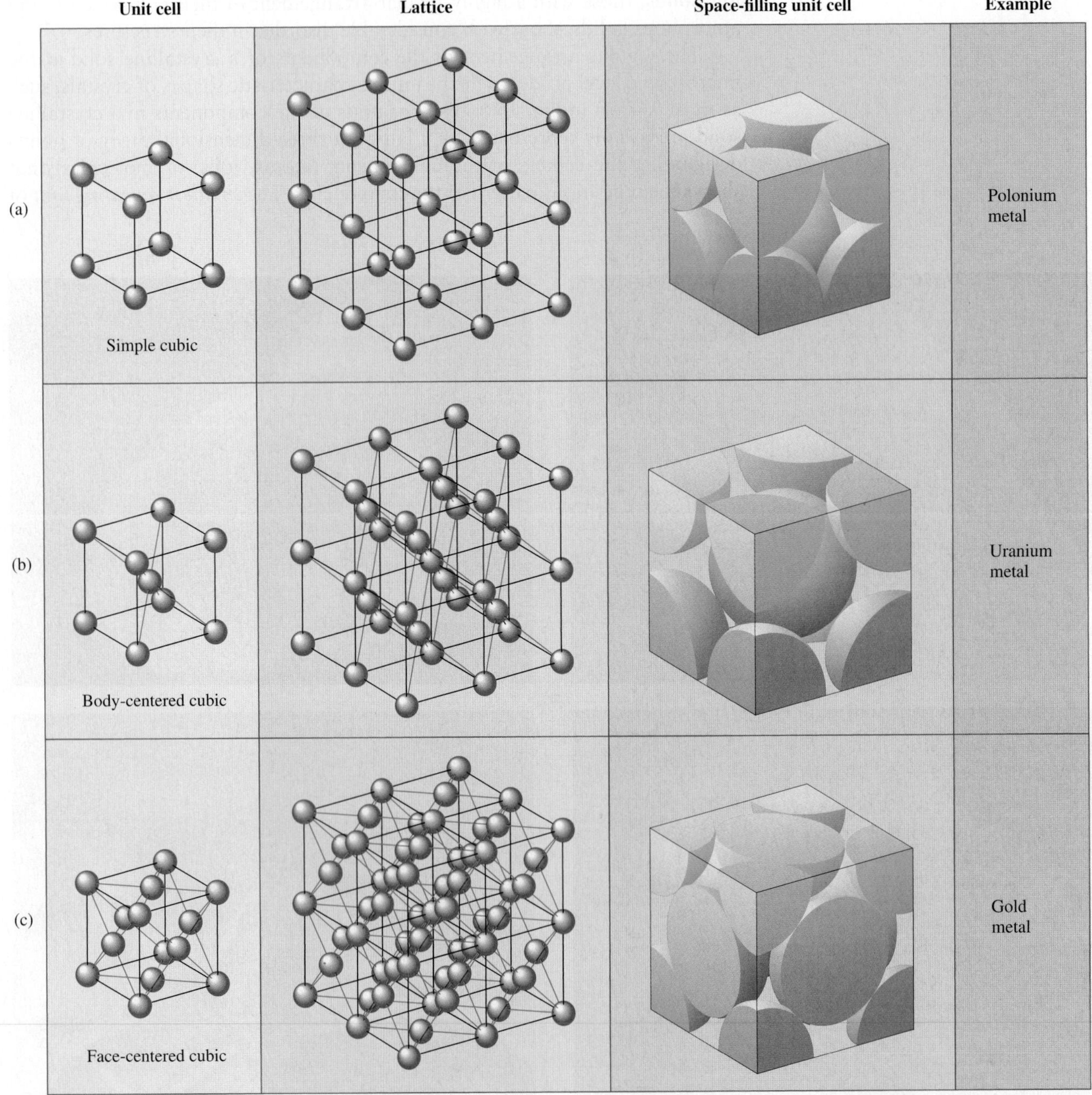

Figure 16.9
Three cubic unit cells and the corresponding lattices. Note that only parts of spheres on the corners and faces of the unit cells reside inside the unit cell, as shown by the cutoff versions.

Conch Clues

Chemical Insights

Nature is an excellent source of ideas for making materials that are useful in the human world. A recent example involves the shell of the giant pink queen conch (*Strombus gigas*). The shell of the Caribbean conch (pronounced "conk") is 100 to 1000 times stronger than the aragonite mineral that makes up 99% of the shell. Since aragonite is a form of calcium carbonate, which in its pure form breaks like chalk, why is the conch shell so tough?

Arthur Heuer and his colleagues at Case Western Reserve University in Cleveland have used microscopy to show that the secret of the conch shell's strength lies in the protein that surrounds strands of aragonite in the shell. These strands are bundled into larger strands, which are in turn packed into still larger structures. At each level of organization, adjacent structures are oriented perpendicular to one another. This arrangement allows fractures to spread without shattering the shell.

The conch shell, which is 99% calcium carbonate, is therefore really a composite material in which proteins provide the toughness that makes the shell so resistant to shattering. Although the conch shell itself is not useful (except to the conch), it illustrates how composites may offer much greater strength than the individual substances involved.

Scientists are studying many other natural materials* that are especially strong yet very light. Examples include the silica sea sponge, which has an intricate cylindrical glass skeleton approximately 20 cm long and a few centimeters in diameter. Another interesting natural structure is the 20-cm-long beak of the Toco toucan, which makes up one-third of the bird's length but accounts for only 1/20 of its mass.

Lessons learned from these natural materials may well lead to high-performance materials that may prove useful in aerospace vehicles, bridges, and construction materials.

Joanna Aizenberg

Silica sea sponge.

Kevin Schafer/Vireo

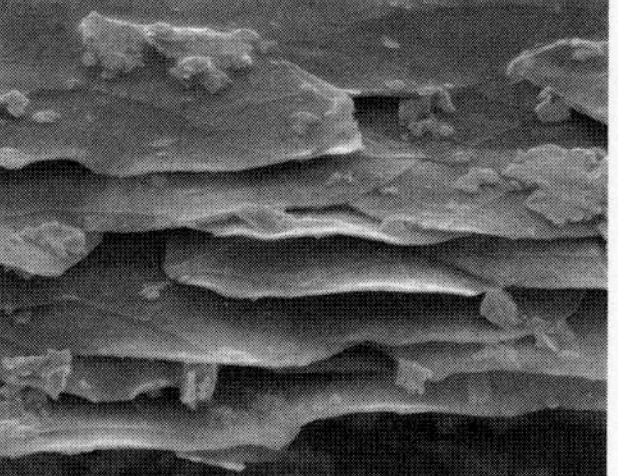

Courtesy, Dr. Marc A. Meyers/University of California, San Diego

(top) Toco toucan. (bottom) Scanning electron microscope (SEM) image of the layered and staggered keratin tiles that make up the outer shell of a Toco toucan's beak.

*For more information, see "Making the Most of It," *Sci. News* **169** (2006): 184.

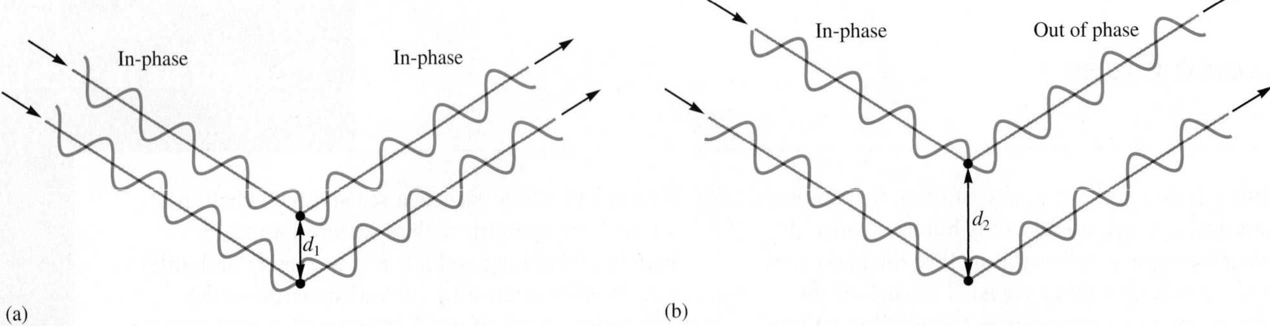

Figure 16.10
X rays scattered from two different atoms may reinforce (constructive interference) or cancel (destructive interference) one another. (a) Both the incident rays and the reflected rays are also in phase. In this case d_1 is such that the difference in the distances traveled by the two rays is a whole number of wavelengths. (b) The incident rays are in phase, but the reflected rays are exactly out of phase. In this case d_2 is such that the difference in distances traveled by the two rays is an odd number of half wavelengths.

X-Ray Analysis of Solids

The structures of crystalline solids are most commonly determined by **X-ray diffraction.** Diffraction occurs when beams of light are scattered from a regular array of points or lines where the spacings between the components are comparable to the wavelength of the light. Diffraction is caused by constructive interference when the waves of parallel beams are in phase and by destructive interference when the waves are out of phase.

When X rays of a single wavelength are directed at a crystal, a diffraction pattern is obtained, as we saw in Fig. 12.7. The light and dark areas on the photographic plate occur because the waves scattered from various atoms may reinforce or cancel each other (see Fig. 16.10). The key to whether the waves reinforce or cancel is the difference in distance traveled by the waves after they strike the atoms. The waves are in phase before they are reflected; thus, if the difference in distance traveled after reflection is an *integral number of wavelengths,* the waves will still be in phase when they meet again.

Since the distance traveled after reflection depends on the distance between the atoms, the diffraction pattern can be used to determine the interatomic spacings. The exact relationship can be formulated using the diagram in Fig. 16.11, which shows two in-phase waves being reflected by atoms in two different layers of a crystal. The extra distance traveled by the lower wave is the sum of the distances xy and yz. The waves will be in phase after reflection if

$$xy + yz = n\lambda \tag{16.1}$$

where n is an integer and λ is the wavelength of the X rays. Using trigonometry (Fig. 16.11), we can show that

$$xy + yz = 2d \sin \theta \tag{16.2}$$

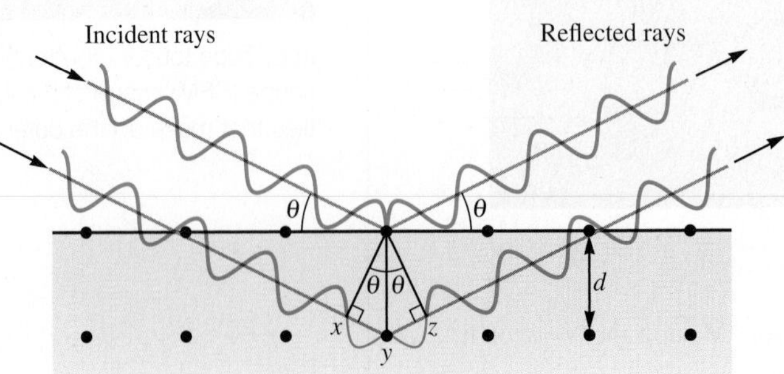

Figure 16.11
Reflection of X rays of wavelength λ from a pair of atoms in two different layers of a crystal. The lower wave travels an extra distance equal to the sum of xy and yz. If this distance is an integral number of wavelengths ($n = 1, 2, 3, \ldots$), the waves will reinforce each other when they exit the crystal.

where d is the distance between the atoms and θ is the angle of incidence and reflection. Combining Equation (16.1) and Equation (16.2) gives

$$n\lambda = 2d \sin \theta \tag{16.3}$$

Equation (16.3) is called the **Bragg equation,** after William Henry Bragg (1862–1942) and his son William Lawrence Bragg (1890–1972). They shared the Nobel Prize in physics in 1915 for their pioneering work in X-ray crystallography.

A diffractometer is a computer-controlled instrument used for carrying out the X-ray analysis of crystals. It rotates the crystal with respect to the X-ray beam and collects the data produced by the scattering of the X rays from the various planes of atoms in the crystal. The results are then analyzed by computer. The techniques for crystal structure analysis have reached a level of sophistication that allows the determination of very complex structures, such as those important in biological systems. Using X-ray diffraction, we can gather data on bond lengths and angles and, in doing so, can test the predictions of our models of molecular geometry.

Types of Crystalline Solids

There are many different types of crystalline solids. For example, although both sugar and salt dissolve readily in water, the properties of the resulting solutions are quite different. The salt solution readily conducts an electric current, but the sugar solution does not. This behavior arises from the nature of the components of these two solids. Common salt (NaCl) is an ionic solid; it contains Na^+ and Cl^- ions. When solid sodium chloride dissolves in the polar water, sodium and chloride ions are distributed throughout the resulting solution and are free to conduct an electric current. Table sugar (sucrose), on the other hand, is composed of neutral molecules that are dispersed throughout the water when the solid dissolves. Since no ions are present in the solid, the resulting solution does not conduct electricity. These examples illustrate two important types of solids: **ionic solids,** represented by sodium chloride, and **molecular solids,** represented by sucrose.

A third type of solid is illustrated by elements such as graphite, diamond, and buckminsterfullerene (all pure carbon), boron, silicon, and all metals. These substances all have atoms occupying the lattice points; we will call them **atomic solids.** Examples of the three types of solids are shown in Fig. 16.12.

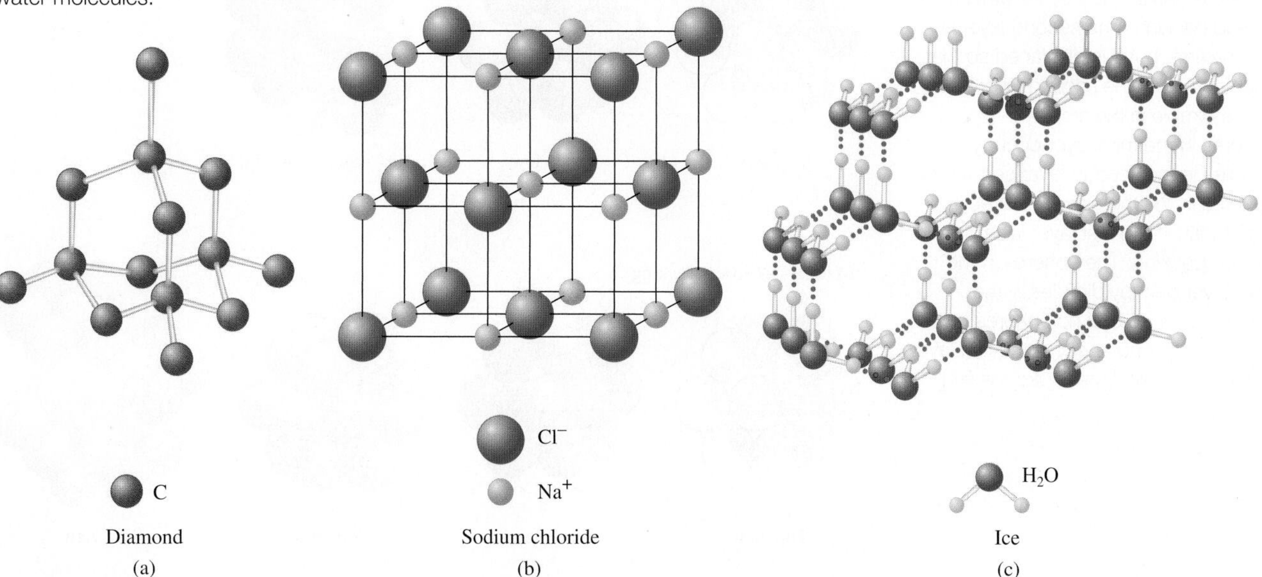

Figure 16.12 Examples of three types of crystalline solids. Only part of the structure is shown in each case. (a) An atomic solid. (b) An ionic solid. (c) A molecular solid. The dashed lines show the hydrogen bonding among the polar water molecules.

The internal forces in a solid determine the properties of the solid.

The properties of a solid are determined primarily by the nature of the forces that hold the solid together. For example, although argon, copper, and diamond all form atomic solids, they have strikingly different properties. Argon has a very low melting point (−189°C), whereas diamond and copper melt at high temperatures (about 3500°C and 1083°C, respectively). Copper is an excellent conductor of electricity, but argon and diamond are both insulators. Copper can be easily changed in shape; it is both malleable (can form thin sheets) and ductile (can be pulled into a wire). Diamond, on the other hand, is the hardest natural substance known. The marked differences in properties among these three atomic solids are caused by bonding differences. We will explore the bonding in solids in the next two sections.

16.4 Structure and Bonding in Metals

The closest packing model for metallic crystals assumes that metal atoms are uniform, hard spheres.

Metals are characterized by high thermal and electrical conductivity, malleability, and ductility. As we will see, these properties can be traced to the nondirectional covalent bonding found in metallic crystals.

A metallic crystal can be pictured as containing spherical atoms packed together and bonded to each other equally in all directions. We can model such a structure by packing uniform, hard spheres in a manner that most efficiently uses the available space. Such an arrangement is called **closest packing** (Fig. 16.13). The spheres are packed in layers in which each sphere is surrounded by six others. In the second layer, the spheres do not lie directly over those in the first layer. Instead, each one occupies an indentation (or dimple) formed by three spheres in the first layer. In the third layer, the spheres can occupy the dimples of the second layer in two possible ways. They can occupy positions so that each sphere in the third layer lies directly over a sphere in the first layer (the *aba* arrangement), or they can occupy positions so that no sphere in the third layer lies over one in the first layer (the *abc* arrangement).

The *aba* arrangement has the *hexagonal* unit cell shown in Fig. 16.14, and the resulting structure is called the **hexagonal closest packed (hcp) structure.** The *abc* arrangement has a *face-centered cubic* unit cell, as shown in Fig. 16.15,

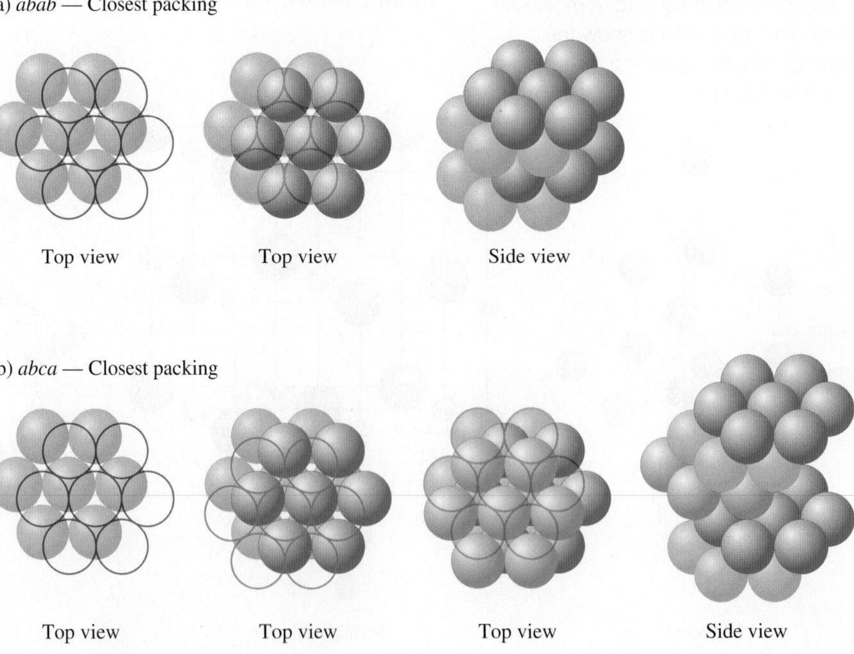

Figure 16.13
The closest packing arrangement of uniform spheres. In each layer a given sphere is surrounded by six others. (a) *aba* packing: The second layer is like the first, but it is displaced so that each sphere in the second layer occupies a dimple in the first layer. The spheres in the third layer occupy dimples in the second layer so that the spheres in the third layer lie directly over those in the first layer (*aba*). (b) *abc* packing: The spheres in the third layer occupy dimples in the second layer so that no spheres in the third layer lie above any in the first layer (*abc*). The fourth layer is like the first.

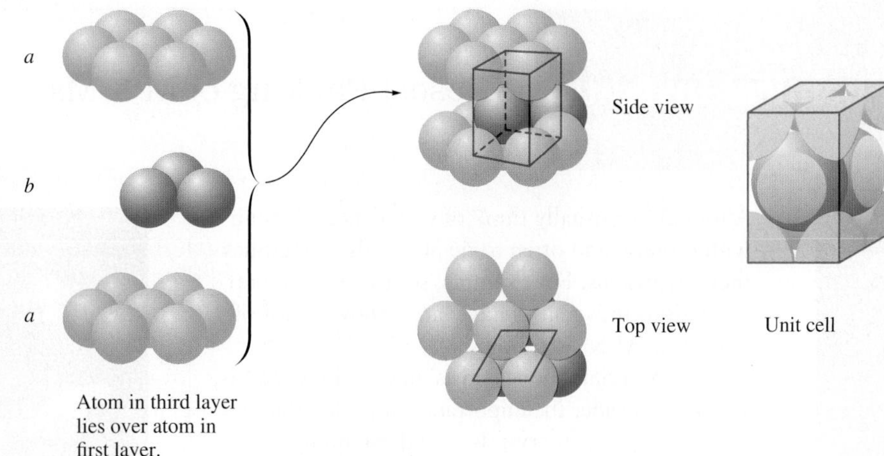

Figure 16.14
When spheres are closest packed so that the spheres in the third layer are directly over those in the first layer (*aba*), the unit cell is the hexagonal prism illustrated here in red.

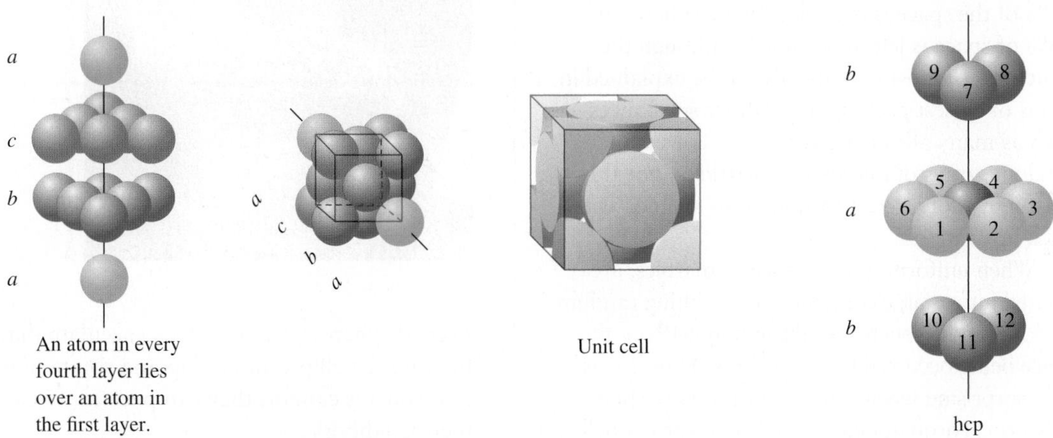

Figure 16.15
When spheres are packed in the *abc* arrangement, the unit cell is face-centered cubic. To make the cubic arrangement easier to see, the vertical axis has been tilted as shown.

Figure 16.16
The indicated sphere has 12 equivalent nearest neighbors.

and the resulting structure is called the **cubic closest packed (ccp) structure.** Note that in the hcp structure the spheres in every other layer occupy the same vertical position (*ababab* . . .), whereas in the ccp structure the spheres in every fourth layer occupy the same vertical position (*abcabca* . . .). A characteristic of both structures is that each sphere has 12 equivalent nearest neighbors: 6 in the same layer, 3 in the layer above, and 3 in the layer below (that form the dimples). This is illustrated for the hcp structure in Fig. 16.16.

Knowing the *net* number of spheres (atoms) in a particular unit cell is important for many applications involving solids. To illustrate the procedure for finding the net number of spheres in a unit cell, let's consider a face-centered cubic unit cell (Fig. 16.17). Note that this unit cell is defined by the *centers* of the spheres on the cube's corners. Thus 8 cubes share a given sphere, so $\frac{1}{8}$ of this sphere lies inside each unit cell. Since a cube has 8 corners, there are $8 \times \frac{1}{8}$ pieces, or enough to put together 1 whole sphere. The spheres at the center of each face are shared by 2 unit cells, so $\frac{1}{2}$ of each lies inside a particular unit cell. Since the cube has 6 faces, we have $6 \times \frac{1}{2}$ pieces, or enough to construct 3 whole spheres. Thus the net number of spheres in a face-centered cubic unit cell is

$$\left(8 \times \frac{1}{8}\right) + \left(6 \times \frac{1}{2}\right) = 4$$

Chemical Insights — Closest Packing of M & Ms

Although we usually think of scientists as dealing with esoteric and often toxic materials, sometimes they surprise us. For example, scientists at several prestigious universities have lately shown a lot of interest in M & M candies.

To appreciate the scientists' interest in M & Ms, we must consider the importance of packing atoms, molecules, or microcrystals in understanding the structures of solids. The most efficient use of space is the closest packing of uniform spheres, where 74% of the space is occupied by the spheres and 26% of space is left unoccupied. Although the structures of most pure metals can be explained in terms of closest packing, most other substances—such as many alloys and ceramics—consist of random arrays of microscopic particles. For this reason, it is of interest to study how such objects pack in a random way.

When uniform spheres, such as marbles, are poured into a large container, the resulting random packing of the spheres results in only 64% of the space being occupied by the spheres. Thus it was very surprising when Princeton University chemist Salvatore Torquato and his colleagues at Cornell and North Carolina Central Universities discovered that when the ellipsoidal-shaped M & Ms are poured into a large container, the candies occupy 73.5% of the available space. In other words, the randomly packed M & Ms occupy space with almost the same efficiency as closest packed spheres do.

Why do randomly packed ellipsoids occupy space so much more efficiently than randomly packed spheres? The scientists speculate that because the ellipsoids can tip and rotate in ways that spheres cannot, they can pack more closely to their neighbors.

Denise Applewhite/Princeton University

According to Torquato, these results are important because they will help us to better understand the properties of disordered materials ranging from powders to glassy solids. He also says that M & Ms make ideal test objects because they are inexpensive and uniform and "you can eat the experiment afterward."

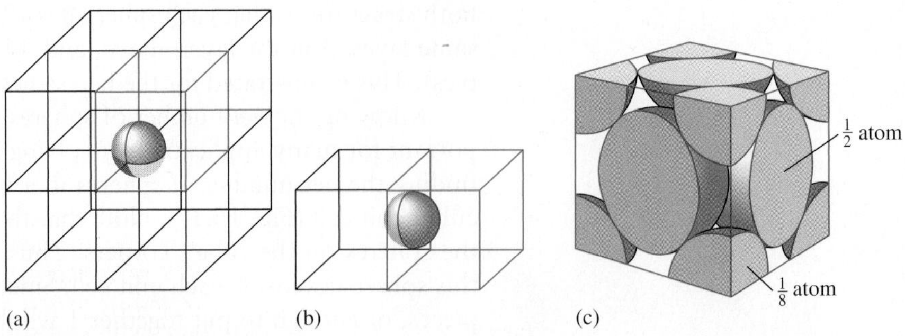

Figure 16.17
The net number of spheres in a face-centered cubic unit cell. (a) Note that the sphere on a corner of the colored cell is shared with 7 other unit cells. Thus $\frac{1}{8}$ of such a sphere lies within a given unit cell. Since there are 8 corners in a cube, there are 8 of these $\frac{1}{8}$ pieces, the equivalent of 1 net sphere. (b) The sphere on the center of each face is shared by two unit cells, and thus each unit cell has $\frac{1}{2}$ of each of these types of spheres. There are 6 of these $\frac{1}{2}$ spheres, yielding 3 net spheres. (c) Thus the face-centered cubic unit cell contains 4 net spheres.

Seething Surfaces

Chemical Insights

When we picture a solid, we think of the particles as being packed closely together with relatively little motion. The particles are thought to vibrate randomly about their positions but stay in nearly the same place. Recent research, however, indicates surface particles are a great deal more mobile than was previously thought. Independent teams of scientists from the University of Leiden in the Netherlands and Sandia National Laboratory in New Mexico have found a surprising amount of atom swapping occurring on the surface of a copper crystal.

Dutch scientist Raoul van Gastel and his colleagues used a scanning tunneling microscope (STM) to study the surface of a copper crystal containing indium atom impurities. They noted that a given patch of surface would stay the same for several scans and then, suddenly, the indium atoms would appear at different places. Surprisingly, the indium atoms seemed to make "long jumps," moving as many as five atom positions between scans. The most likely explanation for these movements is a "hole" created by a copper atom escaping the surface. This hole moves around as other atoms shift to fill it in succession (see accompanying figure). The best analogy to the movement of the hole is the toy slide puzzle with 15 numbered pieces and one missing piece in a 4 × 4 array. The object of the game is to slide a piece into the hole and then repeat the process until the numbers appear in order.

The hole on the copper surface moves very fast—up to 100 million times per second—shuffling copper atoms and allowing the indium atoms to change positions. Van Gastel believes that all of the observed motion results from just a few fast-moving holes. In fact, he suggests that just one in 6 billion copper atoms is missing at a given time, analogous to one person in the entire earth's population. Its absence causes a given atom on the surface to move every 30 or 40 seconds. Brian Swartzentruber of Sandia National Laboratories came to similar conclusions using an STM to track the movement of palladium atoms on a copper surface.

These results have important implications. For example, metal surfaces are often used to speed up particular reactions. The motions on the metal surface could significantly influence the way that reactants interact with the surface. Also, a lot of effort is now being expended to construct tiny "machines" (called nanoscale devices) by assembling individual atoms on a solid surface. These devices could be literally torn apart by excess surface motions.

A toy slide puzzle.

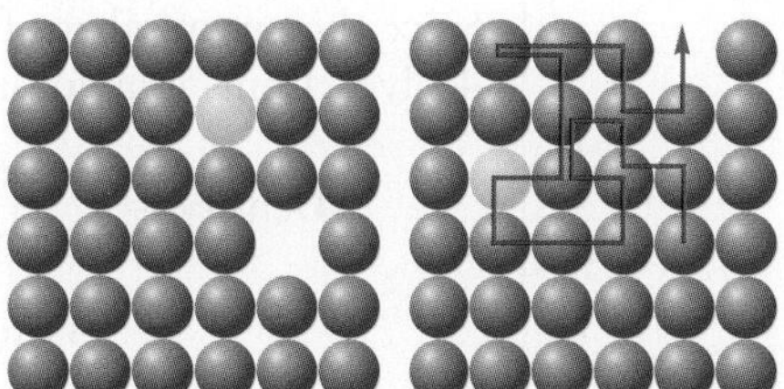

A section of a surface containing copper atoms (red) and an indium atom (yellow). A hole due to a missing copper atom is shown on the left. The blue line on the right shows the movement of this hole. As an atom moves to fill the hole, the hole moves as well. In the process, the indium atom jumps to a new position.

Illustration from R. van Gastel, et al., *Physical Review Letters,* Feb. 19, 2001, Vol. 86, Issue 5, pp. 1562–1565. Copyright 2001 by the American Physical Society.

In Example 16.1 we will see how the density of a closest packed solid can be calculated.

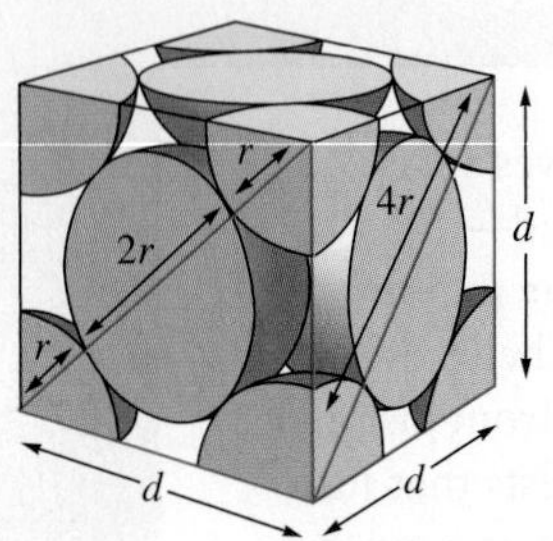

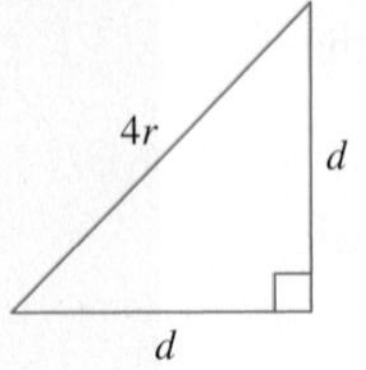

OWL INTERACTIVE EXAMPLE 16.1

Silver crystallizes in a cubic closest packed structure. The radius of a silver atom is 1.44 Å (144 pm). Calculate the density of solid silver.

Solution

■ What are we trying to solve?

In this problem we are asked to solve for the density of solid silver. Density is the mass per unit volume, so we will need to determine the mass of silver atoms that occupy a given volume in the crystal. This will require us to determine the number of silver atoms in this volume. With the number of atoms of silver and the atomic mass of silver, we can determine the mass of silver.

The structure is cubic closest packed, which means the unit cell is face-centered cubic, as shown in the figure at left.

We must find the volume of this unit cell and the net number of atoms it contains. Note that in this structure the atoms touch along the diagonals of each face and not along the edges of the cube. Thus the length of the diagonal is $r + 2r + r$, or $4r$, where r represents the radius of a silver atom. We next find the length of the edge of the cube by the Pythagorean theorem:

$$e^2 + e^2 = (4r)^2$$

$$2e^2 = 16r^2$$

$$e^2 = 8r^2$$

$$e = \sqrt{8r^2} = r\sqrt{8}$$

Since $r = 1.44$ Å for a silver atom,

$$e = (1.44\ \text{Å})(\sqrt{8}) = 4.07\ \text{Å}$$

The volume of the unit cell is e^3, which is $(4.07\ \text{Å})^3$ or $67.4\ \text{Å}^3$. We convert this to cubic centimeters as follows:

$$67.4\ \text{Å}^3 \times \left(\frac{1.00 \times 10^{-8}\ \text{cm}}{\text{Å}}\right)^3 = 6.74 \times 10^{-23}\ \text{cm}^3$$

Since the net number of atoms in a face-centered cubic unit cell is 4, there are 4 silver atoms in a volume of $6.74 \times 10^{-23}\ \text{cm}^3$. Therefore,

$$\text{Density} = \frac{\text{mass}}{\text{volume}}$$

$$= \frac{(4\ \text{atoms})(107.9\ \text{g/mol})(1\ \text{mol}/6.022 \times 10^{23}\ \text{atoms})}{6.74 \times 10^{-23}\ \text{cm}^3}$$

$$= 10.6\ \text{g/cm}^3$$

Closest packing describes the most efficient method for arranging uniform spheres. We can calculate the fraction of the space actually occupied by the spheres (f_v),

$$f_v = \frac{\text{volume occupied by spheres in the unit cell}}{\text{volume of the unit cell}}$$

by using ideas we have developed in the preceding discussion.

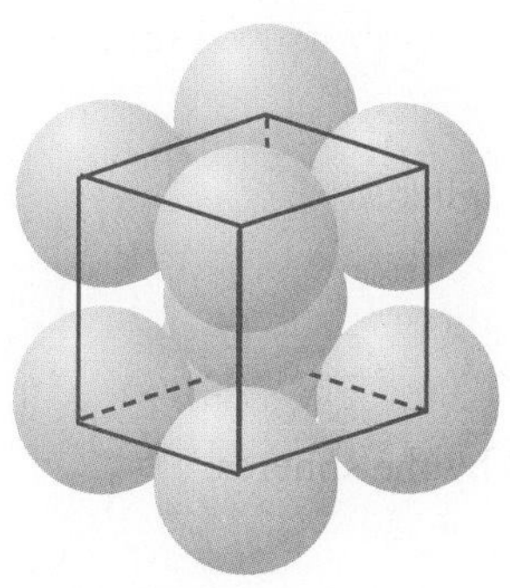
Figure 16.18
In the body-centered cubic unit cell, the spheres touch along the body diagonal.

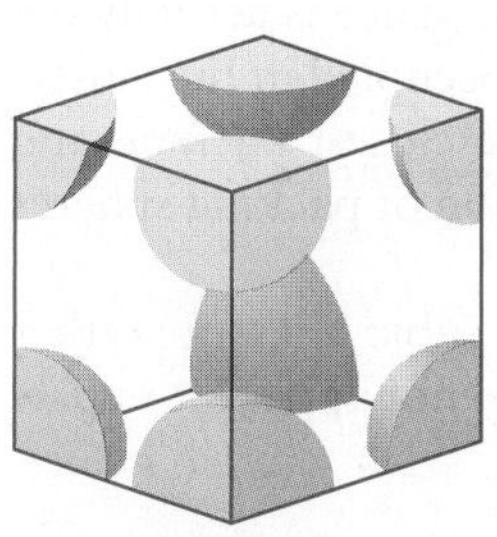
Figure 16.19
The body-centered cubic unit cell with the center sphere deleted.

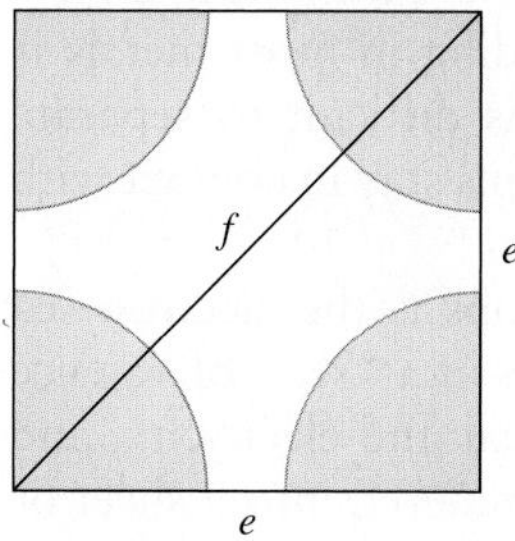

Figure 16.20
One face of the body-centered cubic unit cell. By the Pythagorean theorem, $f^2 = e^2 + e^2$.

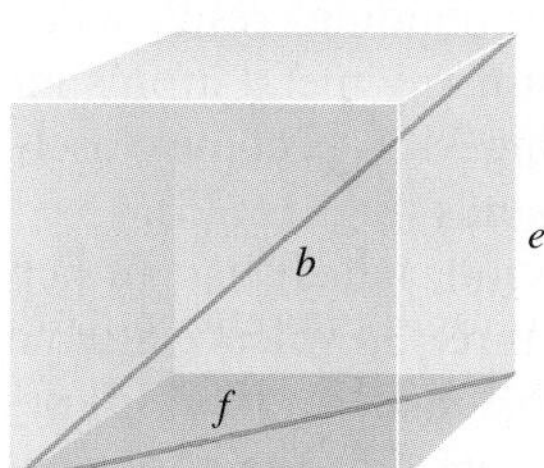

Figure 16.21
The relationship of the body diagonal (b) to the face diagonal (f) and edge (e) for the body-centered cubic unit cell.

Recall that in the cubic closest packing arrangement, the unit cell is face-centered cubic and contains four net spheres. Thus the volume occupied by spheres in the unit cell is four times the volume of each sphere:

$$4 \times \tfrac{4}{3}\pi r^3$$

In Example 16.1 we found that the edge of the unit cell (e) is related to the radius of the packed spheres (r) as follows:

$$e = r\sqrt{8}$$

Therefore, the volume of the unit cell is e^3, where

$$e^3 = (r\sqrt{8})^3$$

and the fraction of space occupied by the spheres is

$$f_v = \frac{\overbrace{4 \times \tfrac{4}{3}\pi r^3}^{\text{Volume of spheres}}}{\underbrace{(r\sqrt{8})^3}_{\text{Volume of unit cell}}} = 0.740$$

This result means that in a cubic closest packed solid 74.0% of the space is occupied by spheres (26.0% is open space). This result also applies to hexagonal closest packing, although we will not prove it here.

In contrast, for simple cubic packing (spheres stacked on top of each other in successive layers), the spheres occupy only 52.4% of the space (verify this for yourself).

Examples of metals that are cubic closest packed are aluminum, iron, copper, cobalt, and nickel. Magnesium and zinc exhibit hexagonal closest packing. Calcium and certain other metals can crystallize in either structure.

Body-Centered Cubic Packing

Although most metals assume one of the closest packed structures in the solid state, some metallic elements have structures that are not closest packed. For example, the structures of the alkali metals are characterized by a **body-centered cubic (bcc) unit cell** (Fig. 16.9). In this structure each sphere has 8 nearest neighbors (count the number of atoms around the atom at the center of the unit cell), as compared with 12 in the closest packed structures.

Note that in body-centered cubic packing the unit cell is a cube with one sphere at its center. In this structure the spheres touch along the body diagonal (Fig. 16.18). For clarity, the cube for this unit cell is shown in Fig. 16.19 with the center sphere deleted.

The body-centered arrangement of spheres is not a closest packed structure, as we can show by calculating the fraction of space occupied by the spheres. To express the volume of the cube in terms of the radius of the packed spheres, we must use the Pythagorean theorem twice. First, we express the face diagonal f in terms of the edge e (Fig. 16.20):

$$f^2 = e^2 + e^2 = 2e^2$$

Because the spheres touch along the body diagonal, the length of the body diagonal b is $4r$. We can relate the body diagonal, the face diagonal, and the cube edge as follows (Fig. 16.21):

$$b^2 = (4r)^2 = e^2 + f^2$$

and

$$(4r)^2 = e^2 + 2e^2 = 3e^2$$

Thus $$e = \frac{4r}{\sqrt{3}}$$

The body-centered cubic unit cell contains 2 net spheres:

$$\underset{\text{Corners}}{8(\tfrac{1}{8})} + \underset{\text{Center}}{1} = 2$$

Thus we can compute the fraction of space occupied by the spheres:

$$\frac{2 \times \frac{4}{3}\pi r^3}{\left(\frac{4r}{\sqrt{3}}\right)^3} = \frac{\sqrt{3}\,\pi}{8} = 0.680$$

(Numerator: Volume occupied by spheres; denominator: Cube volume)

So in the body-centered cubic arrangement, 68.0% of the space is actually occupied by spheres. This is somewhat less than the space occupied in the closest packed structures (74.0%). That is, in contrast to cubic closest packing and hexagonal closest packing, the body-centered cubic method of packing spheres does not represent a closest packed structure.

The reasons why some metals adopt a body-centered cubic structure rather than a closest packed structure are complex and are often not apparent in particular cases.

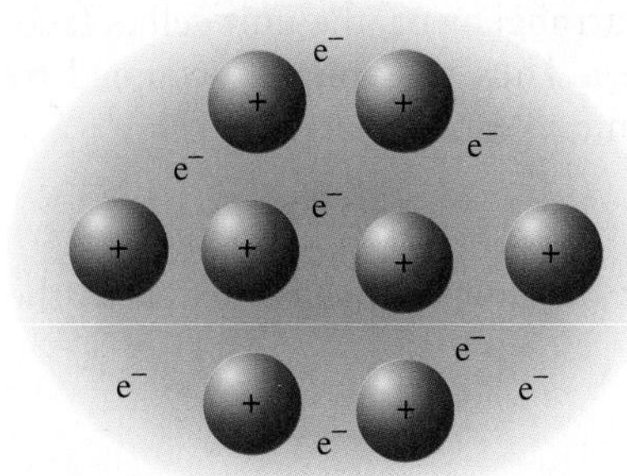

(a)

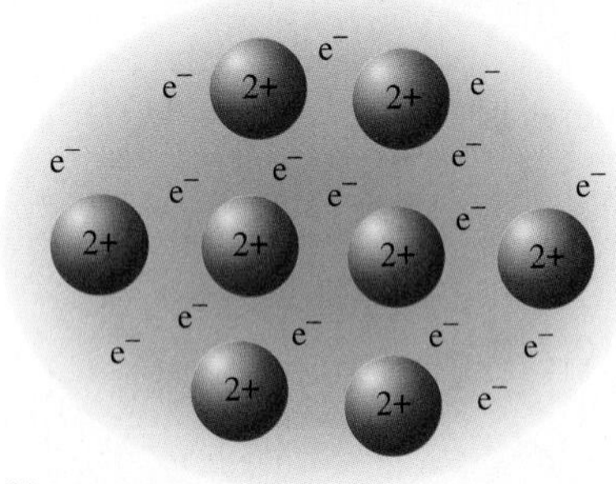

(b)

Figure 16.22
The electron sea model for metals postulates a regular array of cations in a "sea" of valence electrons. (a) Representation of an alkali metal (Group 1A) with one valence electron. (b) Representation of an alkaline earth metal (Group 2A) with two valence electrons.

Bonding in Metals

Any successful bonding model for metals must account for the typical physical properties of metals: malleability, ductility, and the efficient and uniform conduction of heat and electricity in all directions. Although the shapes of most pure metals can be changed relatively easily, most metals are durable and have high melting points. These facts indicate that the bonding in most metals is both *strong* and *nondirectional.* That is, although it is difficult to separate metal atoms, it is relatively easy to move them, if the atoms stay in contact with each other.

The simplest picture that explains these observations is the **electron sea model,** which envisions a regular array of metal cations in a "sea" of valence electrons (Fig. 16.22). The mobile electrons conduct heat and electricity, and the cations are easily moved around as the metal is hammered into a sheet or pulled into a wire.

A related model that gives a more detailed view of the electron energies and motions is the **band model,** or the molecular orbital (MO) model, for metals. In this model the electrons are assumed to travel around the metal crystal in MOs formed from the valence atomic orbitals of the metal atoms (Fig. 16.23).

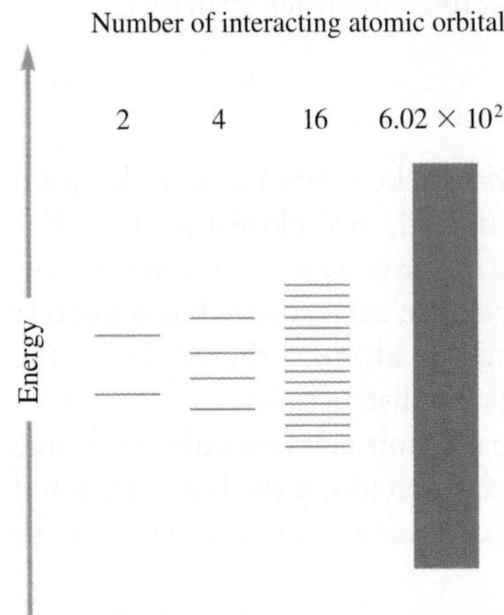

Figure 16.23
The molecular orbital energy levels produced when various numbers of atomic orbitals interact. Note that for two atomic orbitals two rather widely spaced energy levels result. (Recall the description of H_2 in Section 14.2.) As more atomic orbitals become available to form MOs, the resulting energy levels become more closely spaced, finally producing a band of very closely spaced orbitals.

Recall that in the MO model for the gaseous Li_2 molecule (Section 14.3), two widely spaced MO energy levels (bonding and antibonding) result when two identical atomic orbitals interact. However, when many metal atoms interact, as in a metal crystal, the large number of resulting MOs become closely spaced, forming a virtual continuum of levels, called **bands** (Fig. 16.23).

As an illustration, picture a magnesium metal crystal, which has an hcp structure. Since each magnesium atom has one 3*s* and three 3*p* valence atomic orbitals, a crystal with *n* magnesium atoms has available *n*(3*s*) and 3*n*(3*p*) orbitals to form the MOs, as illustrated in Fig. 16.24. Note that the core electrons are localized, as shown by their presence in the energy "well" around each magnesium atom produced by its nuclear charge. However, the valence electrons occupy closely spaced MOs, which are only partially filled.

Crystals of magnesium grown from a vapor.

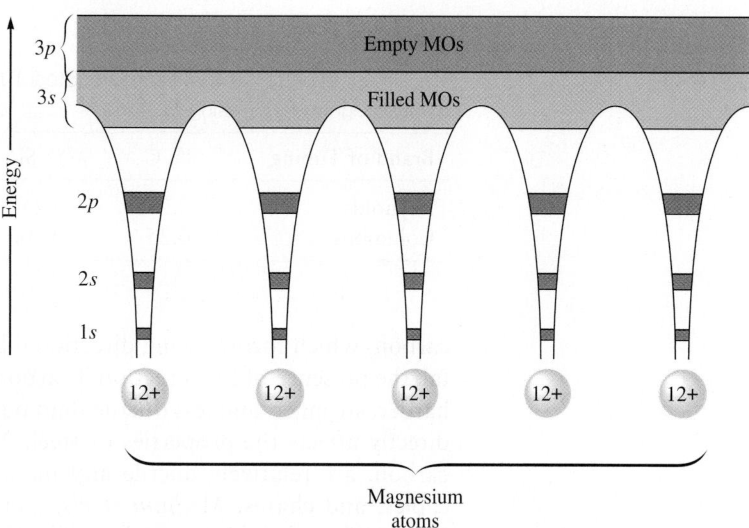

Figure 16.24
A representation of the energy levels (bands) in a magnesium crystal. The electrons in the 1*s*, 2*s*, and 2*p* orbitals are close to the nuclei and thus are localized on each magnesium atom as shown. However, the 3*s* and 3*p* valence orbitals overlap and mix to form MOs. Electrons in these energy levels can travel throughout the crystal.

The existence of empty MOs close in energy to filled MOs explains the thermal and electrical conductivity of metal crystals. Metals conduct electricity and heat very efficiently because of the availability of highly mobile electrons. For example, when an electrical potential is placed across a strip of metal, for current to flow, electrons must be free to move from the negative to the positive areas of the metal. In the band model for metals, mobile electrons are furnished when electrons in filled MOs are excited into empty ones. The conduction electrons are free to travel throughout the metal crystal as dictated by the potential imposed on the metal. The MOs occupied by these conducting electrons are called **conduction bands.** These mobile electrons also account for the efficiency of the conduction of heat through metals. When one end of a metal rod is heated, the mobile electrons can rapidly transmit the thermal energy to the other end.

Metal Alloys

Because of the nature of the structure and bonding of metals, other elements can be introduced into a metallic crystal relatively easily to produce substances called **alloys.** An alloy is best defined as *a substance that contains a mixture of elements and has metallic properties.* Alloys can be conveniently classified into two types.

In a **substitutional alloy,** some of the host metal atoms are *replaced* by other metal atoms of similar size. For example, in brass about one-third of the atoms in the host copper metal are replaced by zinc atoms, as shown in Fig. 16.25(a). Sterling silver (93% silver and 7% copper), pewter (85% tin, 7% copper, 6% bismuth, and 2% antimony), and plumber's solder (67% lead and 33% tin) are other examples of substitutional alloys.

An **interstitial alloy** is formed when some of the interstices (holes) in the closest packed metal structure are occupied by small atoms, as shown in Fig. 16.25(b). Steel, the best known interstitial alloy, contains carbon atoms in the holes of an iron crystal. The presence of the interstitial atoms changes the properties of the host metal. Pure iron is relatively soft, ductile, and malleable because of the absence of strong directional bonding. The spherical metal atoms can be rather easily moved with respect to each other. However, when

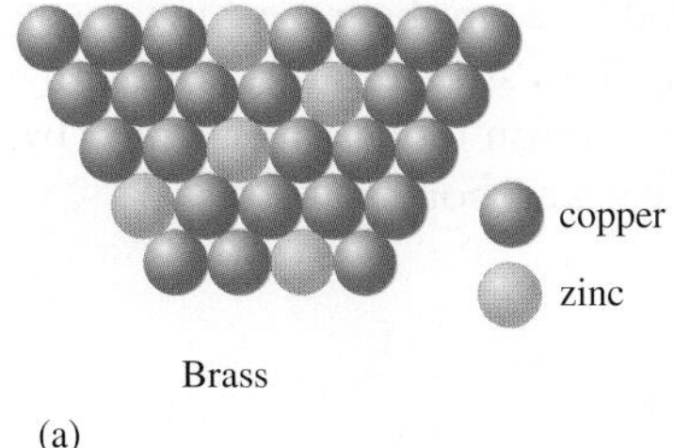

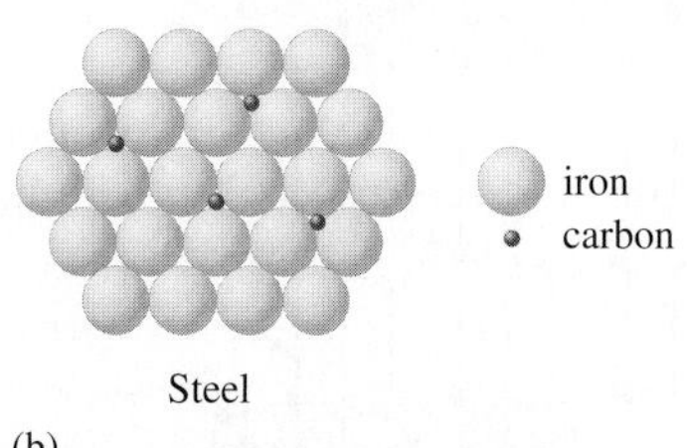

Figure 16.25
Two types of alloys.

Table 16.3

The Composition of the Two Brands of Steel Tubing Most Commonly Used to Make Lightweight Racing Bicycles

Brand of Tubing	% C	% Si	% Mn	% Mo	% Cr
Reynolds	0.25	0.25	1.3	0.20	—
Columbus	0.25	0.30	0.65	0.20	1.0

carbon, which forms strong directional bonds, is introduced into an iron crystal, the presence of the directional carbon–iron bonds makes the resulting alloy harder, stronger, and less ductile than pure iron. The amount of carbon present directly affects the properties of steel. *Mild steels,* containing less than 0.2% carbon, are relatively ductile and malleable. These steels are used for nails, cables, and chains. *Medium steels,* containing 0.2–0.6% carbon, are harder than mild steels and are used in rails and structural steel beams. *High-carbon steels,* containing 0.6–1.5% carbon, are tough and hard and are used for springs, tools, and cutlery.

Many types of steel contain elements in addition to iron and carbon. Such steels are often called *alloy steels* and can be viewed as being mixed interstitial (carbon) and substitutional (other metals) alloys. Bicycle frames, for example, are constructed from a wide variety of alloy steels. The compositions of the two brands of steel tubing used in high-quality racing bicycles are given in Table 16.3.

16.5 Carbon and Silicon: Network Atomic Solids

Many atomic solids contain strong directional covalent bonds. We will call these substances **network solids.** In contrast to metals, these materials are typically brittle and do not efficiently conduct heat or electricity. To illustrate network solids, in this section we will discuss two very important elements, carbon and silicon, and some of their compounds.

Carbon occurs in the *allotropes* (different forms) diamond, graphite, and the fullerenes. The fullerenes are molecular solids (see Section 16.6), but diamond and graphite are typically network solids. In diamond, the hardest naturally occurring substance, each carbon atom is surrounded by a tetrahedral arrangement of other carbon atoms, as shown in Fig. 16.26(a). This structure is stabilized by covalent bonds, which, in terms of the localized electron model, are formed by the overlap of sp^3 hybridized atomic orbitals on each carbon atom.

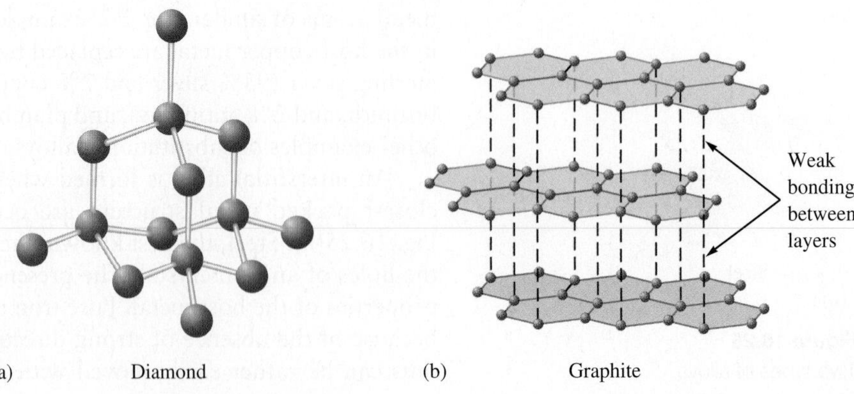

Figure 16.26
The structures of (a) diamond and (b) graphite. In each case only a small part of the entire structure is shown.

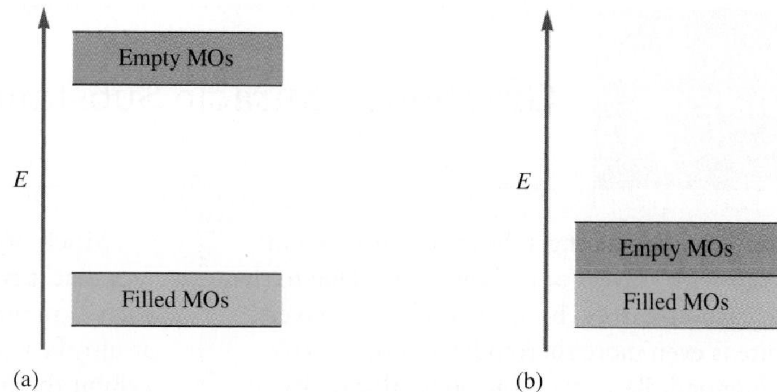

Figure 16.27
Partial representation of the MO energies in (a) diamond and (b) a typical metal.

It is also useful to consider the bonding among the carbon atoms in diamond in terms of the MO model. Energy-level diagrams for diamond and a typical metal are given in Fig. 16.27. Recall that the conductivity of metals can be explained by postulating that electrons are excited from filled levels into the very near empty levels—the conduction bands. However, note that in the energy-level diagram for diamond there is *a large gap between the filled and the empty levels.* This means that electrons cannot be easily transferred to the empty conduction bands. As a result, diamond is not expected to be a good electrical conductor. In fact, this prediction of the model agrees exactly with the observed behavior of diamond, which is known to be an electrical *insulator*—it does not conduct an electric current.

Graphite is very different from diamond. Diamond is hard, basically colorless, and an insulator; graphite is slippery, black, and a conductor. These differences, of course, arise from the differences in bonding in the two types of solids. In contrast to the tetrahedral arrangement of carbon atoms in diamond, the structure of graphite is based on layers of carbon atoms arranged in fused six-membered rings, as shown in Fig. 16.26(b). Each carbon atom in a particular layer of graphite is surrounded by three other carbon atoms in a trigonal planar arrangement with 120-degree bond angles. The localized electron model predicts sp^2 hybridization in this case. The three sp^2 orbitals on each carbon are used to form σ bonds to three other carbon atoms. One $2p$ orbital remains unhybridized on each carbon and is perpendicular to the plane of carbon atoms, as shown in Fig. 16.28. These orbitals combine to form a group of closely spaced π MOs that are important in two ways. First, they contribute significantly to the stability of the graphite layers because of the π bonding. Second,

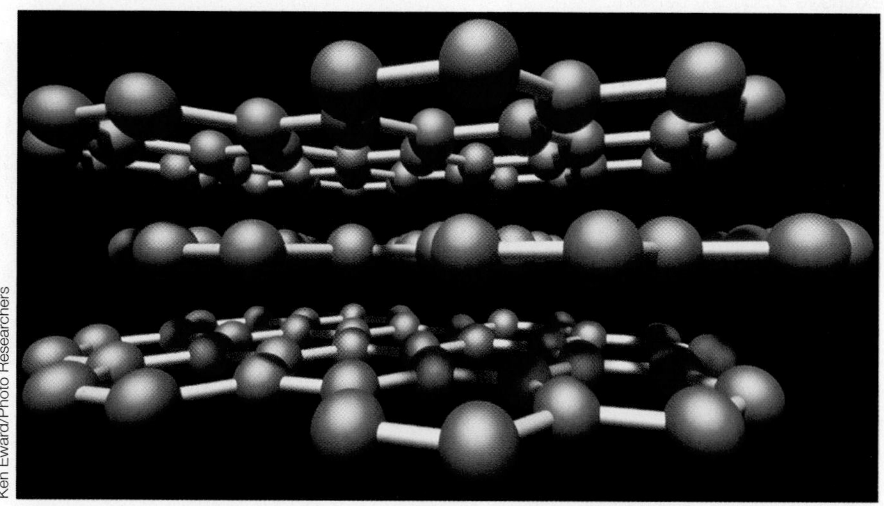

Graphite consists of layers of carbon atoms.

Chemical Insights Graphene—Miracle Substance?

Graphite is a fascinating substance consisting of "chicken-wire" layers of carbon atoms. Due to the strong carbon–carbon bonds in each of its layers, graphite is even more thermodynamically stable than diamond. Because of the unusual stability of graphite, scientists have long wondered whether the individual chicken-wire layers could exist independently. This question was answered only recently by Andre Gein and Konstantin Novoselov, two scientists working at the University of Manchester in the United Kingdom, when they were able to pull layers a single atom thick from graphite using Scotch tape. This work was deemed so important that Gein and Novoselov were awarded the Nobel Prize in physics in 2010.

Since the isolation of sheets of graphene by Gein and Novoselov, research in the properties and possible uses of this material has virtually exploded. Graphene has truly amazing properties. For example, it is believed to be the strongest material known, more than 100 times stronger than steel. Also, it is stiffer than diamond, yet it can be stretched like rubber.

Much of the interest in graphene centers on the fact that it is the best conductor of heat and electricity known. Electrons travel through graphene at ultrafast speeds. In fact, electrons in graphene exhibit the fractional quantum Hall effect: The electrons act collectively as if they are particles with only a fraction of the charge of an electron. Because of its exceptional thermal and electrical conductivities, graphene appears to be an ideal candidate for future electronic devices. Scientists at IBM have already built graphene-based transistors that can switch on and off 26 billion times per second, far faster than conventional silicon-based devices.

One problem that has hampered the development of graphene-based devices is the difficulty in producing large sheets of graphene. Byung Hee Hong and his team, from South Korea, have recently reported making rectangular sheets of graphene measuring 30 inches along the diagonal. It appears that we are very close to making commercial electronic devices based on the amazing properties of graphene.

the π MOs with their delocalized electrons account for the electrical conductivity of graphite. These closely spaced orbitals are exactly analogous to the conduction bands found in metal crystals.

Graphite is often used as a lubricant in locks (where oil is undesirable because it collects dirt). The characteristic slipperiness of graphite can be explained by noting that graphite has very strong bonding *within* the layers of

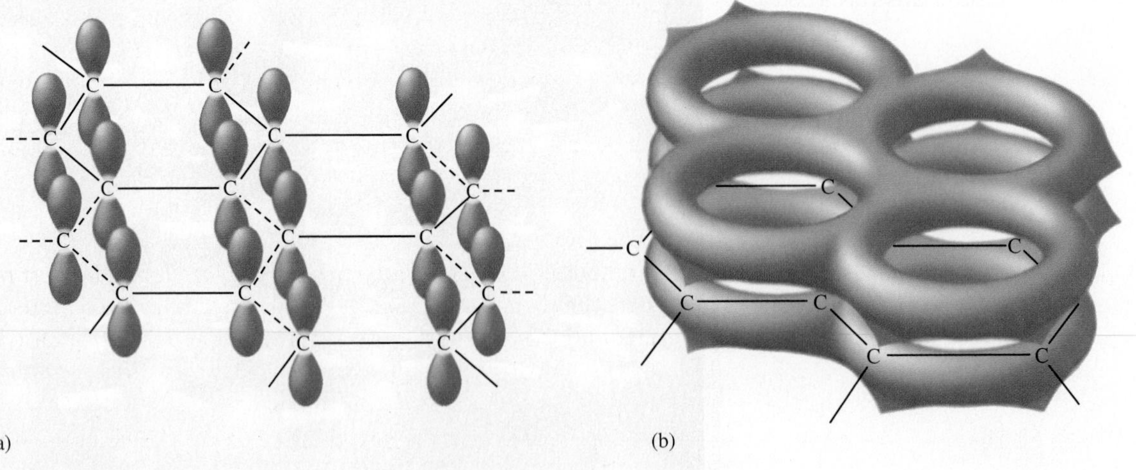

Figure 16.28
The *p* orbitals (a) perpendicular to the plane of the carbon ring system in graphite can combine to form (b) an extensive π bonding network.

carbon atoms but little bonding *between* the layers (the valence electrons are all used to form σ or π bonds among carbons within a given layer). This arrangement allows the layers to slide past one another quite readily. Graphite's layered structure is quite obvious when viewed with a high-magnification electron microscope. This structure is in contrast to that of diamond, which has uniform bonding in all directions in the crystal.

Artificial diamonds are discussed in more detail in Section 16.11.

Because of their extreme hardness, diamonds are extensively used in industrial cutting implements. Thus it is desirable to convert inexpensive graphite to diamond. As we might expect from the high density of diamond (3.5 g/cm^3) compared with that of graphite (2.2 g/cm^3), this transformation can be accomplished by applying very high pressures to graphite. The application of 150,000 atm of pressure at 2800°C converts graphite virtually completely to diamond. (The high temperature is required to break the strong bonds in graphite so that the rearrangement can occur.)

Silicon is an important constituent of the compounds that make up the earth's crust. In fact, silicon is to geology what carbon is to biology. Just as carbon compounds are the basis for most biologically significant systems, silicon compounds are fundamental to most of the rocks, sands, and soils found in the earth's crust. Although carbon and silicon are next to each other in Group 4A of the periodic table, the carbon-based compounds of biology and the silicon-based compounds of geology have markedly different structures. Carbon compounds typically contain long strings of carbon–carbon bonds, but the most stable silicon compounds involve chains with silicon–oxygen bonds. *The most important silicon compounds contain silicon and oxygen.*

The fundamental silicon–oxygen compound is **silica,** which has the empirical formula SiO_2. Knowing the properties of the apparently similar compound carbon dioxide, one might expect silica to be a gas that contains discrete SiO_2 molecules. In fact, nothing could be further from the truth—quartz and some types of sand are typical of the materials composed of silica. What accounts for this difference? The answer lies in the bonding.

The bonding in the CO_2 molecule was described in Section 14.1.

Recall that the Lewis structure for CO_2 is

$$:\ddot{O}=C=\ddot{O}:$$

and that each C=O bond is described as a combination of a σ bond, involving a carbon *sp* hybrid orbital, and a π bond, involving a carbon 2*p* orbital. On the contrary, silicon cannot use its valence 3*p* orbitals to form strong π bonds to oxygen; the larger size of the silicon atom and its orbitals results in a less effective overlap with the smaller oxygen orbitals. Therefore, instead of forming π bonds, the silicon atom satisfies the octet rule by forming single bonds to four oxygen atoms, as shown in the representation of the structure of quartz in Fig. 16.29. Note that in this structure each silicon atom is at the center of a tetrahedral arrangement of oxygen atoms, which are shared with other silicon atoms. Although the empirical formula for quartz is SiO_2, the structure is based on a *network* of SiO_4 tetrahedra with shared oxygen atoms rather than discrete SiO_2 molecules. Thus the differing abilities of carbon and silicon to form π bonds with oxygen have profound effects on the structures and properties of CO_2 and SiO_2.

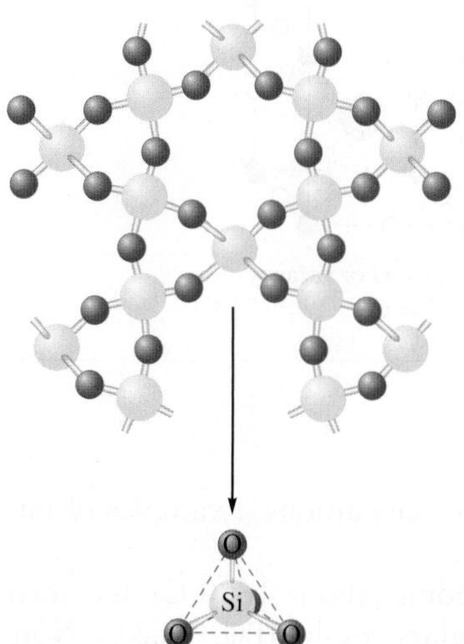

Figure 16.29
(Top) The structure of quartz (empirical formula SiO_2). Quartz contains chains of SiO_4 tetrahedra (bottom) that share oxygen atoms.

Compounds closely related to silica that are found in most rocks, soils, and clays are the **silicates.** Like silica, the silicates are based on interconnected SiO_4 tetrahedra. However, in contrast to silica, where the O:Si ratio is 2:1, silicates have O:Si ratios greater than 2:1 and contain silicon–oxygen *anions.* This means that for the formation of the neutral solid silicates, cations are needed to balance the excess negative charge. In other words, silicates are salts con-

Chemical Insights

Superconductivity

Although metals such as copper and aluminum are good conductors of electricity, up to 20% of the total energy is wasted in the transmission of a current by resistance heating of the wires. However, this loss of energy may be avoidable. Certain kinds of materials undergo a remarkable transition as they are cooled; their electrical resistance changes virtually to zero. These substances, called **superconductors,** conduct electricity with no wasted heat energy.

There is, of course, a catch. In the past superconductivity has been observed only at very low temperatures and is thus very expensive to maintain. In 1911 mercury was observed to exhibit superconductivity, but only at approximately 4 K, the boiling point of liquid helium. In the next 75 years, several niobium alloys showed superconductivity at temperatures as high as 23 K. However, in 1986 researchers realized that certain metallic oxides are superconductors at temperatures well above 23 K. The graph shows how rapidly our knowledge of superconductors is increasing.

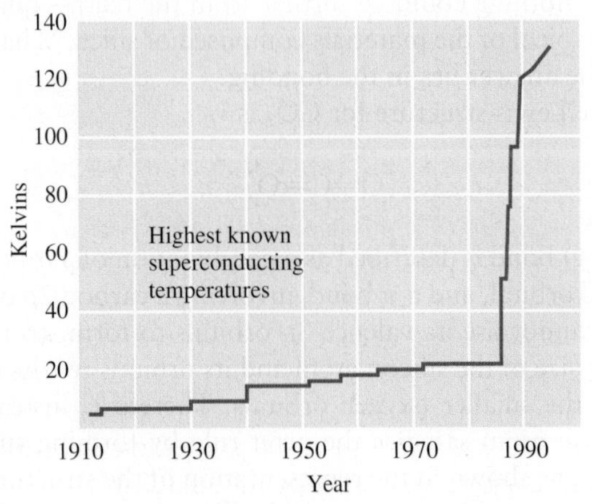

The current class of high-temperature superconductors is called *perovskites,* a well-known class of compounds that contain copper, an alkaline earth metal, a lanthanide metal, and oxygen. In the unit cell for the compounds with formula $YBa_2Cu_3O_x$, in which $x = 6.527$, the small open circles (see figure) indicate the expected oxygen positions, but some of these positions are vacant in the known materials—thus the variable oxygen content.

In addition to exhibiting superconductivity at temperatures above the boiling point of liquid nitrogen (77 K), these ceramic materials can sustain very high currents, an important characteristic necessary for any large-scale application. More recent results have shown that materials containing thallium and calcium ions in place of the lanthanide ions superconduct at 125 K. In early 1993 researchers in Zurich, Switzerland, discov-

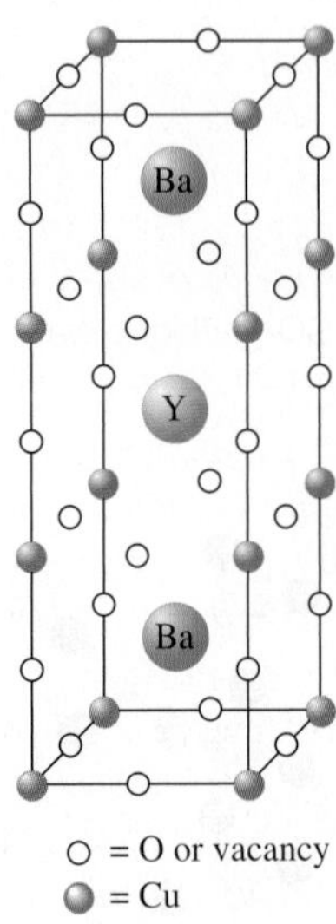

taining metal cations and polyatomic silicon–oxygen anions. Examples of important silicate anions are shown in Fig. 16.30.

When silica is heated above its melting point (about 1600°C) and then cooled rapidly, an amorphous solid called a **glass** results (Fig. 16.31). Note that a glass contains a good deal of disorder, in contrast to the crystalline nature of quartz. Glass more closely resembles a very viscous solution than a crystalline solid. Common glass results when substances such as Na_2CO_3 are added to the silica melt, which is then cooled. The properties of glass can be varied greatly by varying the additives. For example, addition of B_2O_3 produces a glass (called borosilicate glass) that expands and contracts little under

ered a mercury-containing compound that is superconducting at 133 K.

What does all this mean for the future? Superconducting materials could easily cause a revolution similar to that brought about by solid-state electronics. For example, one can envision electrical power transmitted with virtually no loss of energy. Also, new generations of superconducting electromagnets could lead to the operation of particle accelerators at unprecedented energies, the development of nuclear fusion power, the ability to produce incredibly accurate three-dimensional images of the human body for diagnosis of disease, and the development of superfast, magnetically levitated trains.

After a great surge of progress in the late 1980s, the pace slowed greatly in the early 1990s as theory tried to catch up with the results from experiments. We need to develop an understanding of why these ceramic materials are superconducting so that we can design better materials that superconduct at even higher temperatures. And there are many practical problems. For example, metal conductors can be easily bent into coils to allow construction of electromagnets, but anyone who has ever dropped a plate knows how brittle ceramic materials are. Another question concerns how current can be conducted over long distances by superconducting materials. However, there seems to be little doubt that ingenious solutions eventually will be found to the problems. The answers are already beginning to appear.

For example, American Superconductor (ASC) of Westborough, Massachusetts, in conjunction with Pirelli Cable of Milan, Italy, has produced a prototype high-temperature superconducting "wire" that exceeds the current-carrying threshold required for commercial underground power transmission cables. The 1-m-long conductor, based on a bismuth–strontium–calcium–copper oxide superconductor, carries over 2300 amps of direct current at liquid nitrogen temperatures (77 K). This is more than twice the current-carrying capability of conventional copper cables.

Yoav Levy/Phototake

A magnet is levitated over a superconducting ceramic immersed in liquid nitrogen.

ASC manufactures superconducting cables by pouring powdered reactants into a silver tube, drawing the tube into a wire, and then heating to produce the superconducting ceramic. Tens to hundreds of these fine conductors are then combined to form superconducting cable. Although ASC is now capable of producing individual superconducting wire filaments that are 1 km in length, much work remains to find a way to automate the multifilament cable manufacturing process.

large temperature changes. Thus it is useful for labware and cooking utensils. The most common brand name for this glass is Pyrex. The addition of K_2O produces an especially hard glass that can be ground to the precise shapes needed for eyeglasses and contact lenses. The compositions of several types of glass are shown in Table 16.4.

Ceramics

Ceramics are typically made from clays (which contain silicates) that are hardened by firing at high temperatures. Ceramics are a class of nonmetallic materials that are strong, brittle, and resistant to heat and attack by chemicals.

Chemical Insights Greenhouse Glass

We have emphasized in this section how the different bonding tendencies of carbon and silicon with oxygen lead to very different types of covalent oxides. Carbon and oxygen form CO_2, a gas under normal conditions, and silicon and oxygen form silica, a network solid. However, Professors Mario Santoro and Federico Gorelli of the University of Florence in Italy have discovered that when carbon dioxide is subjected to extreme temperatures and pressures, it forms a glass very similar to that formed by SiO_2. Using a diamond anvil press inside a furnace, the Italian scientists found that at 640,000 atm and 700 kelvins a solid forms that is 10 times as hard as quartz but softer than diamond. This substance, which they named "carbonia," is the hardest amorphous substance known.

Carbonia is transparent and similar in structure to amorphous silica in windows but much harder. Another nonmolecular form of carbon dioxide, called carbon dioxide 5, was made by different researchers several years ago.

Carbonia and carbon dioxide 5 are mostly of theoretical interest now because they immediately disappear when the pressure is decreased. It is very interesting that although carbon and silicon show very different bonding properties to oxygen under normal conditions, under pressure carbon becomes much more like silicon.

Like glass, ceramics are based on silicates, but with that the resemblance ends. Glass can be melted and remelted as often as desired, but once a ceramic has been hardened, it is resistant to extremely high temperatures. This behavior results from the very different structures of glasses and ceramics. A glass is a *homogeneous*, noncrystalline "frozen solution," whereas a ceramic is *heterogeneous*. A ceramic contains two phases: minute crystals of silicates that are suspended in a glassy cement.

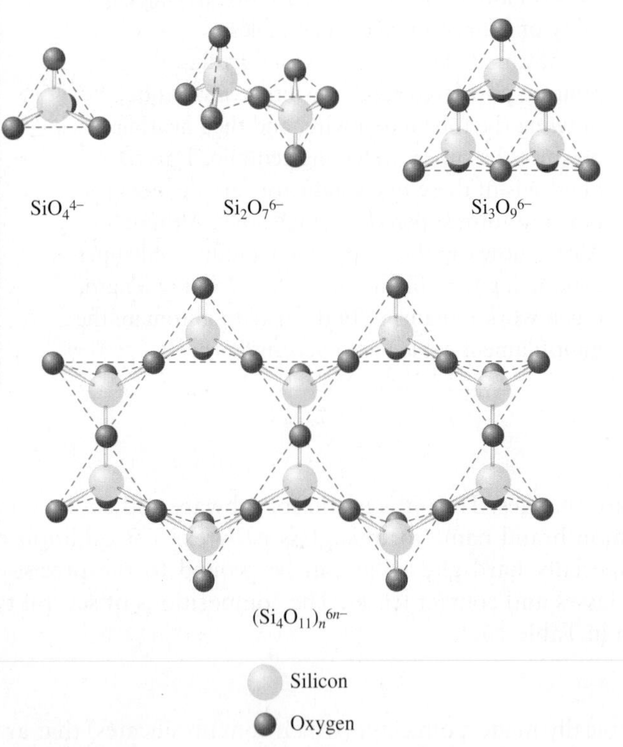

Figure 16.30
Examples of silicate anions, all of which are based on SiO_4^{4-} tetrahedra.

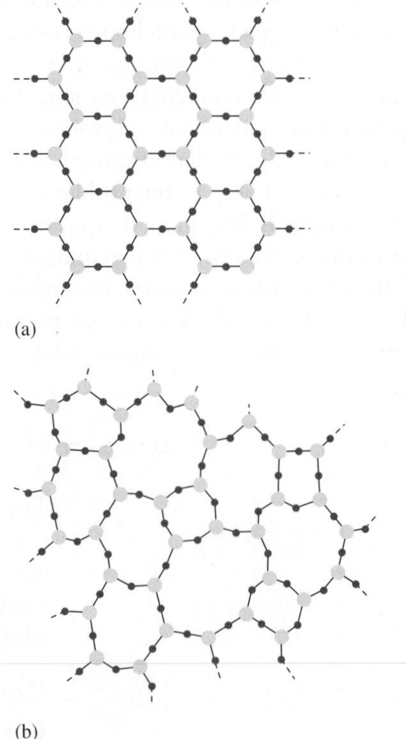

Figure 16.31
Two-dimensional representations of (a) a quartz crystal and (b) a quartz glass.

Explosive Sniffer

Chemical Insights

These days security is at the top of everyone's list of important concerns, especially for the people who are responsible for the safety of our transportation systems. In particular, airports need speedy and sensitive detectors for explosives. Plastic explosives are especially tricky to detect because they do not respond to metal detectors, and they can be shaped into innocent-looking objects to avoid X-ray detection. However, a team of scientists at Oak Ridge National Laboratory led by Thomas Thundat has just published a description of an inexpensive device that is extremely sensitive to two N-containing compounds found in plastic explosives. The key part of this detection device is a tiny (180-micrometer), V-shaped cantilever made of silicon. The cantilever is shown in the accompanying photo next to a human hair for size comparison.

The upper surface of the cantilever was first coated with a layer of gold and then a one-molecule-thick layer of an acid that binds to each of the two N-containing molecules to be detected: pentaerythritol tetranitrate (PETN) and hexahydro-1,3,5-triazine (RDX). When a stream of air containing tiny amounts of PETN or RDX passes over the cantilever, these molecules bind to the cantilever, causing it to bend "like a diving board." This bending is not due to the added mass of the attached PETN and RDX. Rather, the deformation occurs because the area of the cantilever surface where binding takes place stretches relative to the unbound areas. A laser pointed at the cantilever detects the bending motion when PETN or RDX (or both) is present. The device's sensitivity is quite remarkable: 14 parts per trillion of PETN and 30 parts per trillion of RDX.

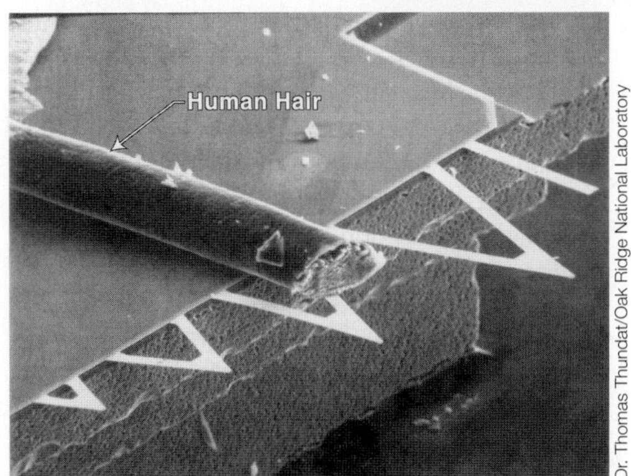

When explosive compounds bind to these V-shaped cantilevers, the microscopic structures, which are about the width of a hair, bend and produce a signal.

All in all, this device appears very promising for detecting plastic explosives in luggage. The cantilevers are inexpensive to construct (approximately $1), and the entire device is about the size of a shoe box. Also, the Oak Ridge team can fabricate thousands of cantilevers in one device. By putting different coatings on the cantilever arms, it should be possible to detect many other types of chemicals and possible biological agents.

This detector looks like a very promising addition to our arsenal of security devices.

Table 16.4

Compositions of Some Common Types of Glass

	Percentages of Various Components						
Type of Glass	SiO_2	CaO	Na_2O	B_2O_3	Al_2O_3	K_2O	MgO
Window (soda-lime glass)	72	11	13	—	0.3	3.8	—
Cookware (aluminosilicate glass)	55	15	—	—	20	—	10
Heat-resistant (borosilicate glass)	76	3	5	13	2	0.5	—
Optical	69	12	6	0.3	—	12	—

Chemical Insights Gallium Arsenide Lasers

Lasers are an essential part of modern life, providing the most efficient way to carry telephone communications by cable, the best-quality music using compact disc players, and the most accurate method for reading grocery prices. The laser (the word *lase* is an acronym for "light amplification by stimulated emission") is a device for producing an intense, coherent beam of light that results when excited electrons release photons as they change to a lower energy state. Excited electrons can return to a lower energy level by spontaneous, random emission of light. However, there is another mechanism in which the emission process is actually stimulated by a passing photon that has the same energy as the emitted photon. This is the phenomenon on which the laser is based. We will have more to say about this presently.

Gallium arsenide is a prominent member of the so-called 3–5 class of semiconductors, those containing elements from Groups 3A and 5A of the periodic table. An n-type semiconductor results when the arsenic-to-gallium ratio is greater than 1. A p-type semiconductor results when gallium is in excess. By carefully controlling the conditions under which a Ga–As semiconductor is made, one can design p–n junctions into the material. At each p–n junction, electrons from the n-type region are available to "fall into" holes in the p-type region, producing photons (see the accompanying figure). As the emitted photons travel through the solid, they *stimulate* the emission of additional photons. A mirror at the edge of the solid reflects these photons back through the substance, where they stimulate even more emission events. All these stimulated emissions produce photons with their waves in phase. Thus this process leads to a large number of coherent (in-phase) photons, which proceed out the other end of the solid where a "half-mirror" is placed. This mirror reflects some of the photons and allows the rest to leave as an intense, coherent "laser beam" of light. A current driven through the substance by an applied potential difference provides a continuous supply of electrons (and removes electrons to make new holes) so that the process occurs continuously while the power is on.

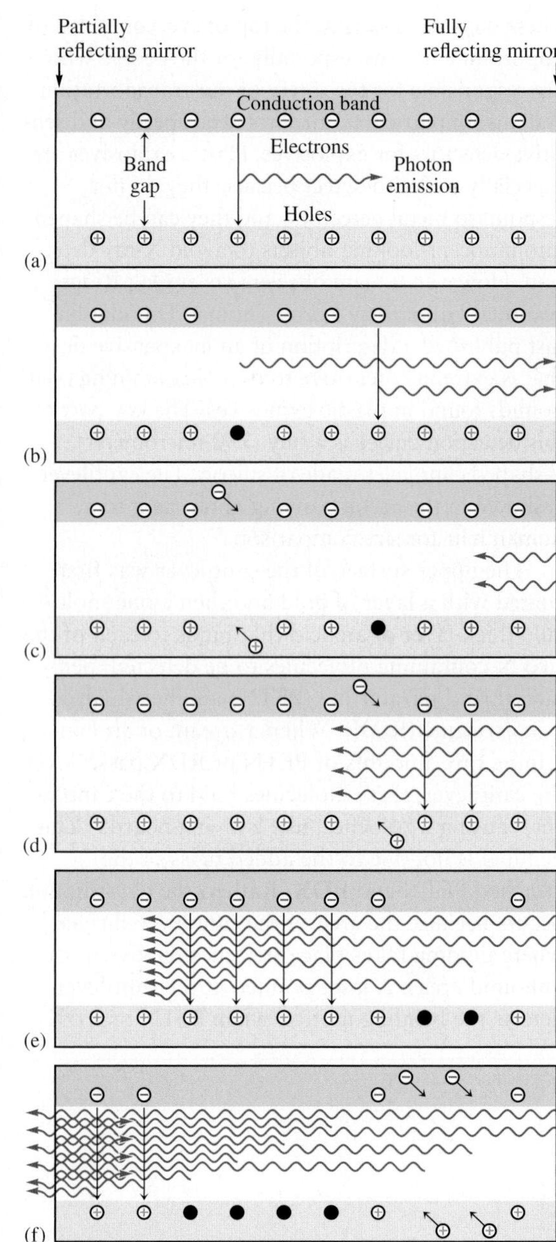

Suggested Reading

Ron Cowen, "Inventing the Light Fantastic," *Sci. News* May 8 (2010): 18.

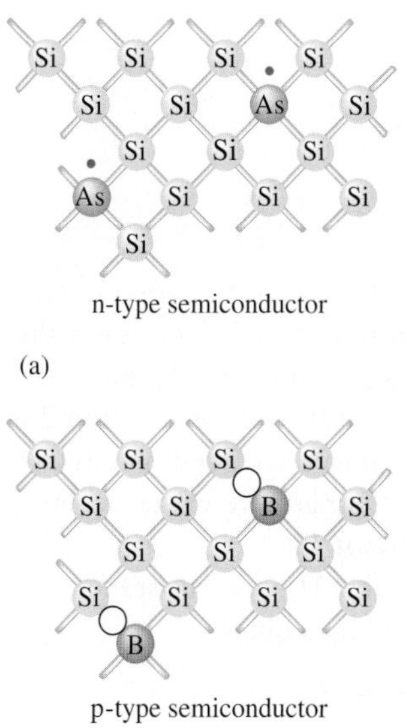

Figure 16.32
(a) A silicon crystal doped with arsenic, which has one extra valence electron. (b) A silicon crystal doped with boron, which has one less electron than silicon.

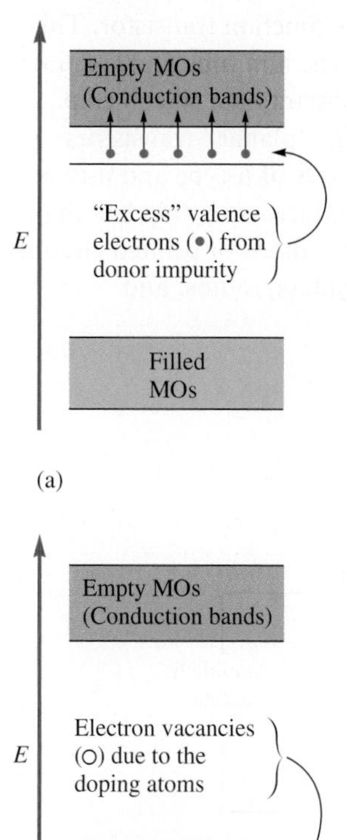

Figure 16.33
Energy-level diagrams for (a) an n-type semiconductor and (b) a p-type semiconductor.

To understand how ceramics harden, one must know something about the structure of clays. Clays are formed by the weathering action of water and carbon dioxide on the mineral feldspar, which is a mixture of silicates with empirical formulas such as $K_2O \cdot Al_2O_3 \cdot 6SiO_2$ and $Na_2O \cdot Al_2O_3 \cdot 6SiO_2$. Feldspar is really an **aluminosilicate** in which aluminum as well as silicon atoms are part of the oxygen-bridged polyanion. The weathering of feldspar produces kaolinite, consisting of tiny thin platelets with the empirical formula $Al_2Si_2O_5(OH)_4$. When dry, the platelets cling together, but when water is present, they can slide over one another, giving clay its plasticity. As clay dries, the platelets begin to interlock again. When the remaining water is driven off during firing, the silicates and cations form a "glass" that binds the tiny crystals of kaolinite.

Ceramics have a very long history. Rocks, which are natural ceramic materials, served as the earliest tools. Later, clay vessels dried in the sun or baked in fires served as containers for food and water. These early vessels were crude and quite porous. With the discovery of glazing, which probably occurred about 3000 B.C. in Egypt, pottery became more serviceable as well as more beautiful. Prized porcelain is essentially the same material as crude earthenware, but specially selected clays and glazings are used for porcelain, which is also fired at a very high temperature.

Semiconductors

Elemental silicon has the same structure as diamond, as might be expected from its position in the periodic table (in Group 4A directly under carbon). Recall that in diamond there is a large energy gap between the filled and empty MOs (Fig. 16.27). This gap prevents excitation of electrons to the empty MOs (conduction bands) and makes diamond an insulator. In silicon the situation is similar but the energy gap is smaller. A few electrons can cross the gap at 25°C, making silicon a *semiconducting element,* or **semiconductor.** In addition, at higher temperatures, where more energy is available to excite electrons into the conduction bands, the conductivity of silicon increases. This is typical behavior for a semiconducting element and is in contrast to the behavior of metals. In metals, the conductivity decreases with increasing temperature.

The small conductivity of silicon can be enhanced at normal temperatures if the silicon crystal is *doped* with certain other elements. For example, when a small fraction of silicon atoms is replaced by arsenic atoms, each having *one more* valence electron than silicon, extra electrons become available for conduction, as shown in Fig. 16.32(a). This produces an **n-type semiconductor,** a substance whose conductivity is increased by doping it with atoms having more valence electrons than the atoms in the host crystal. These extra electrons lie close in energy to the conduction bands and can easily be excited into these levels, where they can conduct an electric current [Fig. 16.33(a)].

We can also enhance the conductivity of silicon by doping the crystal with an element such as boron, which has only three valence electrons, one *less* than silicon. Because boron has one less electron than is required to form the bonds to the surrounding silicon atoms, an electron vacancy, or *hole,* is created, as shown in Fig. 16.33(b). As an electron moves to fill this hole, it leaves a new hole. As this process is repeated, the hole advances through the crystal in a direction opposite to the movement of the electrons jumping to fill the hole. Another way of thinking about this phenomenon is that in pure silicon each atom has four valence electrons, so the low-energy MOs are exactly filled. Replacing silicon atoms with boron atoms leaves vacancies in these MOs, as shown in Fig. 16.33(b). This means that there is only one electron in some of the MOs, and these unpaired electrons can function as conducting electrons.

Chemical Insights Transistors and Integrated Circuits

Transistors have had an immense impact on the technology of electronic devices for which signal amplification is needed, such as communications equipment and computers. Before the invention of the transistor at Bell Laboratories in 1947, amplification was provided exclusively by vacuum tubes, which were both bulky and unreliable. The first electronic digital computer, ENIAC, built at the University of Pennsylvania, had 19,000 vacuum tubes and consumed 150,000 watts of electricity. Because of the discovery and development of the transistor and the printed circuit, a hand-held calculator run by a small battery has the same computing power as ENIAC.

A *junction transistor* is made by joining n-type and p-type semiconductors to form an n–p–n or a p–n–p junction. The former type is shown in the diagram below. The input signal (to be amplified) occurs in circuit 1, which has a small resistance and a forward-biased n–p junction (junction 1). As the voltage of the input signal to this circuit varies, the current in the circuit varies. This means there is a change in the number of electrons crossing the n–p junction. Circuit 2 has a relatively large resistance and is under reverse bias. The key to the operation of the transistor is that current flows in circuit 2 only when electrons crossing junction 1 also cross junction 2 and travel to the positive terminal. Since the current in circuit 1 determines the number of electrons crossing junction 1, the number of electrons available to cross junction 2 is also directly proportional to the current in circuit 1. The current in circuit 2 therefore varies depending on the current in circuit 1.

The voltage (V), current (I), and resistance (R) in a circuit are related by the equation

$$V = IR$$

Since circuit 2 has a large resistance, a given current in circuit 2 produces a larger voltage than the same current in circuit 1, which has a small resistance. Thus a signal of variable voltage in circuit 1, such as might be produced by a human voice on a telephone, is reproduced in circuit 2, but with much greater voltage changes. That is, the input signal has been *amplified* by the junction transistor. This device replaces the large vacuum tube and is a tiny component of a printed circuit on a silicon chip.

Silicon chips are really "planar" transistors constructed from thin layers of n-type and p-type regions connected by conductors. A chip less than 1 cm wide can contain hundreds of printed circuits and can be used in computers, radios, and televisions.

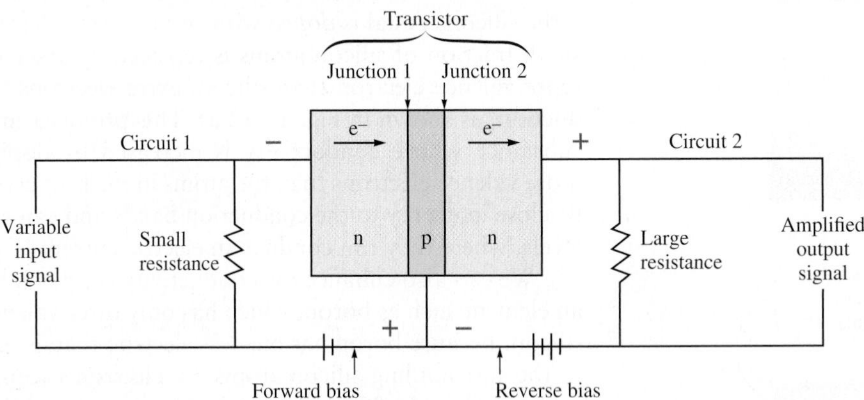

A schematic of two circuits connected by a transistor. The signal in circuit 1 is amplified in circuit 2.

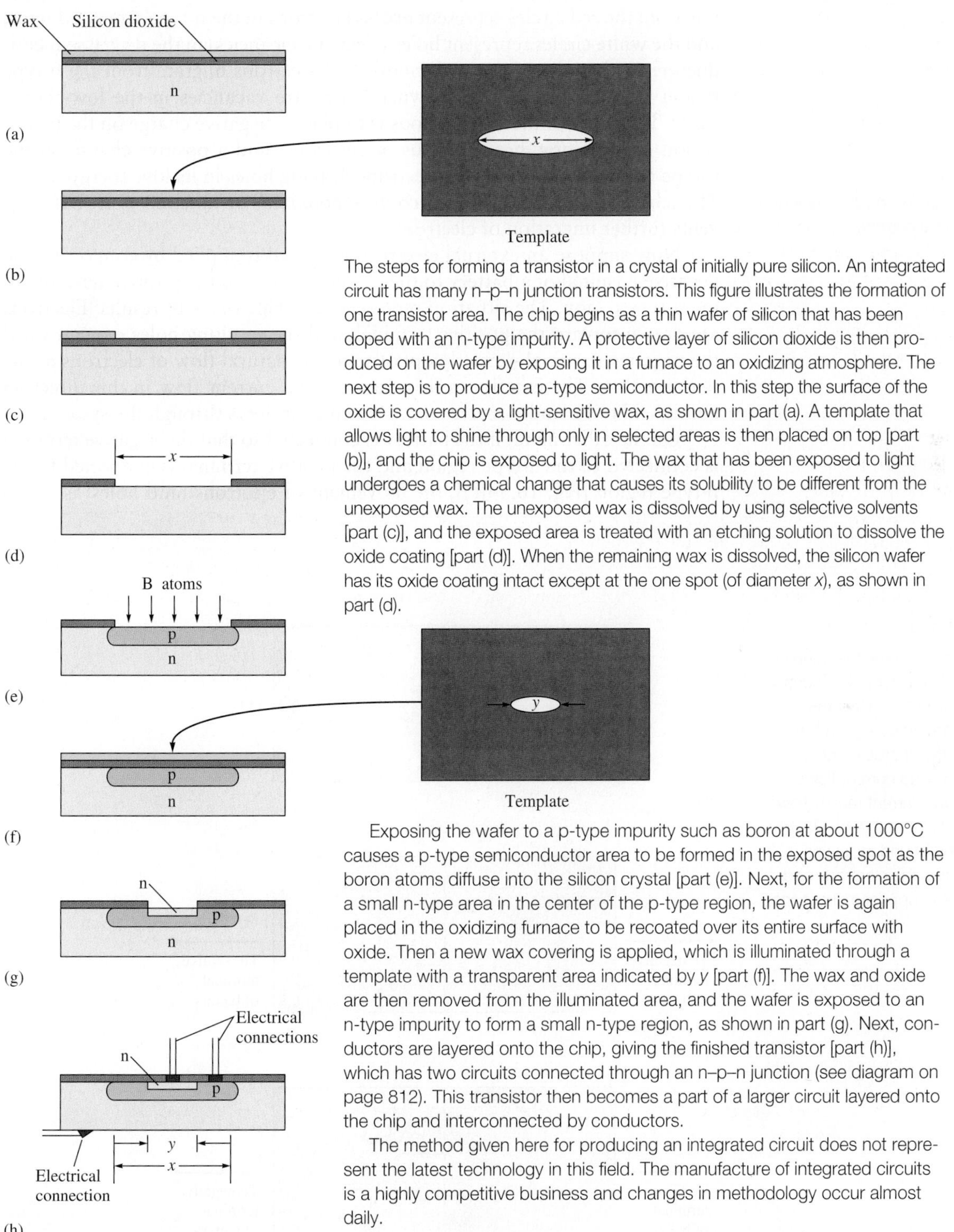

The steps for forming a transistor in a crystal of initially pure silicon. An integrated circuit has many n–p–n junction transistors. This figure illustrates the formation of one transistor area. The chip begins as a thin wafer of silicon that has been doped with an n-type impurity. A protective layer of silicon dioxide is then produced on the wafer by exposing it in a furnace to an oxidizing atmosphere. The next step is to produce a p-type semiconductor. In this step the surface of the oxide is covered by a light-sensitive wax, as shown in part (a). A template that allows light to shine through only in selected areas is then placed on top [part (b)], and the chip is exposed to light. The wax that has been exposed to light undergoes a chemical change that causes its solubility to be different from the unexposed wax. The unexposed wax is dissolved by using selective solvents [part (c)], and the exposed area is treated with an etching solution to dissolve the oxide coating [part (d)]. When the remaining wax is dissolved, the silicon wafer has its oxide coating intact except at the one spot (of diameter x), as shown in part (d).

Exposing the wafer to a p-type impurity such as boron at about 1000°C causes a p-type semiconductor area to be formed in the exposed spot as the boron atoms diffuse into the silicon crystal [part (e)]. Next, for the formation of a small n-type area in the center of the p-type region, the wafer is again placed in the oxidizing furnace to be recoated over its entire surface with oxide. Then a new wax covering is applied, which is illuminated through a template with a transparent area indicated by y [part (f)]. The wax and oxide are then removed from the illuminated area, and the wafer is exposed to an n-type impurity to form a small n-type region, as shown in part (g). Next, conductors are layered onto the chip, giving the finished transistor [part (h)], which has two circuits connected through an n–p–n junction (see diagram on page 812). This transistor then becomes a part of a larger circuit layered onto the chip and interconnected by conductors.

The method given here for producing an integrated circuit does not represent the latest technology in this field. The manufacture of integrated circuits is a highly competitive business and changes in methodology occur almost daily.

Thus the substance becomes a better conductor. When semiconductors are doped with atoms having fewer valence electrons than the atoms of the host crystal, they are called **p-type semiconductors,** so named because the positive holes can be viewed as the charge carriers.

Most important applications of semiconductors involve the connection of a p-type and an n-type to form a **p–n junction.** Figure 16.34(a) shows a typical junction; the red circles represent excess electrons in the n-type semiconductor, and the white circles represent holes (electron vacancies) in the p-type semiconductor. At the junction a small number of electrons migrate from the n-type region into the p-type region, where there are vacancies in the low-energy MOs. The effect of these migrations is to place a negative charge on the p-type region (since it now has a surplus of electrons) and a positive charge on the n-type region (since it has lost electrons, leaving holes in its low-energy MOs). This charge buildup, called the *contact potential,* or *junction potential,* prevents further migration of electrons.

Electrons must be in singly occupied MOs to conduct a current.

Now suppose an external electrical potential is applied by connecting the negative terminal of a battery to the p-type region and the positive terminal to the n-type region. The situation represented in Fig. 16.34(b) results. Electrons are drawn toward the positive terminal, and the resulting holes move toward the negative terminal—exactly opposite to the natural flow of electrons at the p–n junction. The junction resists the imposed current flow in this direction and is said to be under *reverse bias.* No current flows through the system.

On the other hand, if the battery is connected so that the negative terminal is connected to the n-type region and the positive terminal is connected to the p-type region [Fig. 16.34(c)], the movement of electrons (and holes) is in the

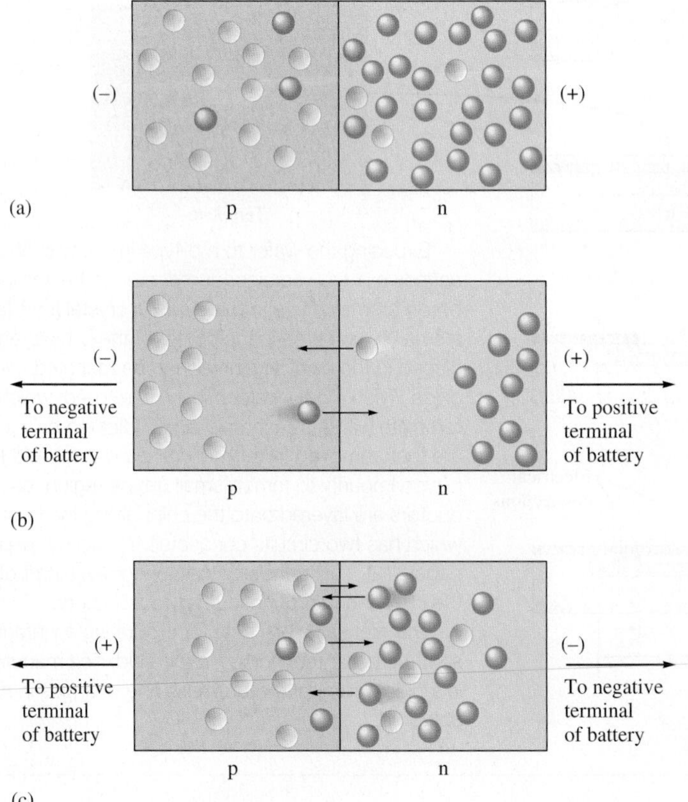

Figure 16.34
The p–n junction involves the contact of a p-type and an n-type semiconductor. (a) The charge carriers of the p-type region are holes (●). In the n-type region, the charge carriers are electrons (●). (b) No current flows (reverse bias). (c) Current readily flows (forward bias). Note that each electron that crosses the boundary leaves a hole behind. Thus the electrons and the holes move in opposite directions.

A new IBM microchip featuring silicon on a "blanket" of insulating material to protect it from temperature changes.

favored direction. The junction has low resistance, and a current flows easily. The junction is said to be under *forward bias*.

A p–n junction makes an excellent *rectifier*, a device that produces direct current (flows in one direction) from an alternating current (flows in both directions alternately). When placed in a circuit where the potential is constantly reversing, a p–n junction transmits current only under forward bias, thus converting the alternating current to a direct current. Radios, computers, and other electrical devices formerly used bulky, unreliable vacuum tubes as rectifiers. The p–n junction has revolutionized electronics; modern solid-state components contain p–n junctions in printed circuits.

16.6 Molecular Solids

So far we have considered solids in which atoms occupy the lattice positions. In most cases such crystals can be considered to consist of one giant molecule. However, there are many types of solids that contain discrete molecular units at each lattice position. A common example is ice, where the lattice positions are occupied by water molecules [see Fig. 16.12(c)]. Other examples are dry ice (solid carbon dioxide), some forms of sulfur that contain S_8 molecules [Fig. 16.35(left)], certain forms of phosphorus that contain P_4 molecules [Fig. 16.35(right)], and the fullerenes, which are all carbon-containing molecules such as C_{60}, C_{70}, and others. These substances are characterized by

Figure 16.35
(left) Sulfur crystals (yellow) contain S_8 molecules. (right) White phosphorus contains P_4 molecules. It is so reactive with the oxygen in air that it must be stored under water.

Table 16.5

Comparison of Atomic Separations Within Molecules (Covalent Bonds) and Between Molecules (Intermolecular Interactions)

Solid	Distance Between Atoms in Molecule*	Closest Distance Between Molecules in the Solid
P_4	2.20 Å	3.8 Å
S_8	2.06 Å	3.7 Å
Cl_2	1.99 Å	3.6 Å

*The shorter distances within the molecules indicate stronger bonding.

strong covalent bonding *within* the molecules but relatively weak forces *between* the molecules. For example, it takes only 6 kJ of energy to melt 1 mole of solid water (ice) because only intermolecular (H_2O—H_2O) interactions must be overcome. However, 470 kJ of energy is required to break a mole of covalent O—H bonds. The differences between the covalent bonds within the molecules and the forces between the molecules are apparent from the comparison of the interatomic and intermolecular distances in solids shown in Table 16.5.

The forces that exist among the molecules in a molecular solid depend on the nature of the molecules. Many molecules such as CO_2, I_2, P_4, and S_8 have zero dipole moments, and the intermolecular forces are London dispersion forces. Because these forces are usually small, we might expect all these substances to be gaseous at 25°C, as is the case for carbon dioxide. However, as the size of the molecules increases, the London forces become quite large, causing many of these substances to be solids at 25°C.

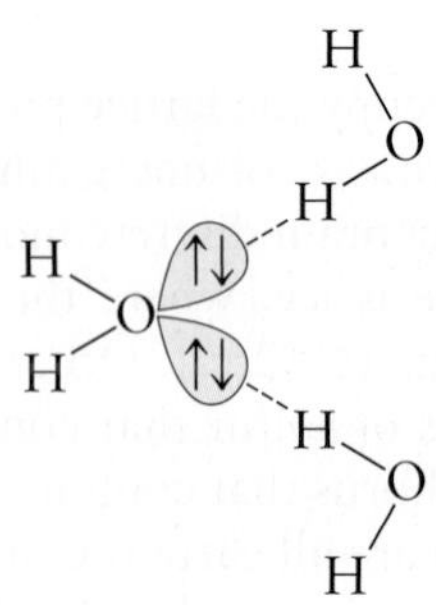

When molecules do have dipole moments, their intermolecular forces are typically greater, especially when hydrogen bonding is possible. Water molecules are particularly well suited to interact with each other because each molecule has two polar O—H bonds and two lone electron pairs on the oxygen atom. This can lead to the association of four hydrogen atoms with each oxygen: two by covalent bonds and two by dipole forces, as shown in the figure in the margin. Note the two relatively short covalent oxygen–hydrogen bonds and the two longer oxygen–hydrogen dipole interactions that can be seen in the ice structure in Fig. 16.12(c).

16.7 Ionic Solids

Ionic solids are stable, high-melting-point substances held together by the strong electrostatic forces that exist between oppositely charged ions. The principles governing the structures of ionic solids were introduced in Section 13.5. In this section we will review and extend these principles.

The structures of most binary ionic solids, such as sodium chloride, can be explained by the closest packing of spheres. Typically, the large ions, which are usually the anions, are packed in one of the closest packing arrangements (hcp or ccp), whereas the smaller cations fit into holes among the closest packed anions. The packing is done in a way that maximizes the electrostatic attractions among oppositely charged ions while minimizing the repulsions among ions with like charges.

There are three types of holes in closest packed structures:

1. Trigonal holes are formed by three spheres in the same layer [Fig. 16.36(a)].
2. Tetrahedral holes are formed when a sphere sits in the dimple of three spheres in an adjacent layer [Fig. 16.36(b)].

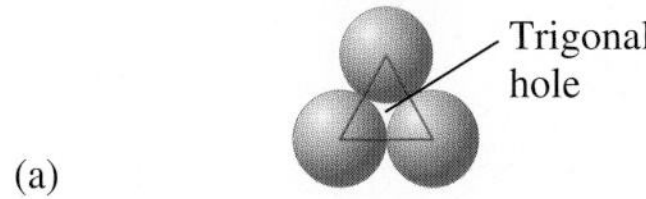

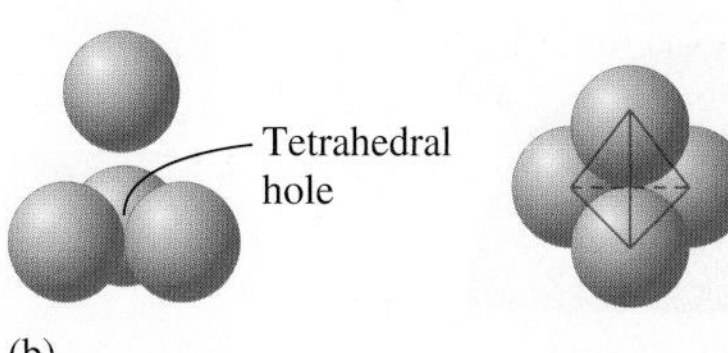

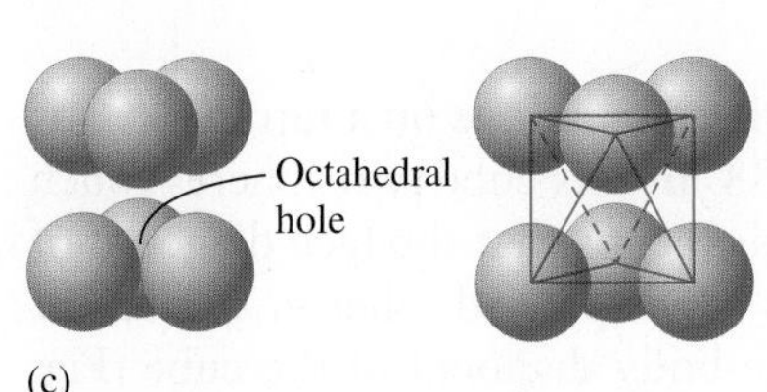

Figure 16.36
The holes that exist among closest packed uniform spheres. (a) The trigonal hole formed by three spheres in a given plane. (b) The tetrahedral hole formed when a sphere occupies a dimple in an adjacent layer. (c) The octahedral hole formed by six spheres in two adjacent layers.

3. Octahedral holes are formed between two sets of three spheres in adjoining layers of the closest packed structures [Fig. 16.36(c)].

   For spheres of a given diameter, the holes increase in size as follows:

$$\text{Trigonal} < \text{tetrahedral} < \text{octahedral}$$

In fact, trigonal holes are so small that they are never occupied in binary ionic compounds. Whether the tetrahedral or octahedral holes in a given binary ionic solid are occupied depends mainly on the *relative* sizes of the anion and cation. Next, we will determine the sizes of the octahedral and tetrahedral holes and consider guidelines for their occupation by ions.

Octahedral Holes

An octahedral hole lies at the center of six equidistant spheres whose centers define an octahedron. Three of these six spheres lie in one closest packed layer and three lie in the adjacent layer [Fig. 16.36(c)]. These six spheres can be rotated to show the octahedron more clearly (Fig. 16.37).

Since the edges of all six spheres are exactly the same distance from the center of the octahedral hole, we can calculate the radius of this hole most easily by focusing on the four spheres whose centers form a square (Fig. 16.37). Note from Fig. 16.37 that R is the radius of the packed spheres, r is the radius of the octahedral hole, and d is the length of the diagonal of the square. From the Pythagorean theorem,

$$(2R)^2 + (2R)^2 = d^2$$

where $$d = R + 2r + R = 2R + 2r = 2(R + r)$$

So we have $$8R^2 = d^2$$

and $$d = \sqrt{8}R = 2\sqrt{2}\,R = 2(R + r)$$

Solving for r yields

$$r = \sqrt{2}\,R - R = 1.414R - R = 0.414R$$

This result shows that an octahedral hole in a closest packed structure has a radius that is 0.414 times the radius of the packed spheres (ions).

Tetrahedral Holes

A tetrahedral hole lies at the center of four spheres whose centers form a tetrahedron [Fig. 16.37(b)]. Three of the spheres are in a given closest packed layer, and the fourth is in the next layer, nestling into the dimple formed by the other three.

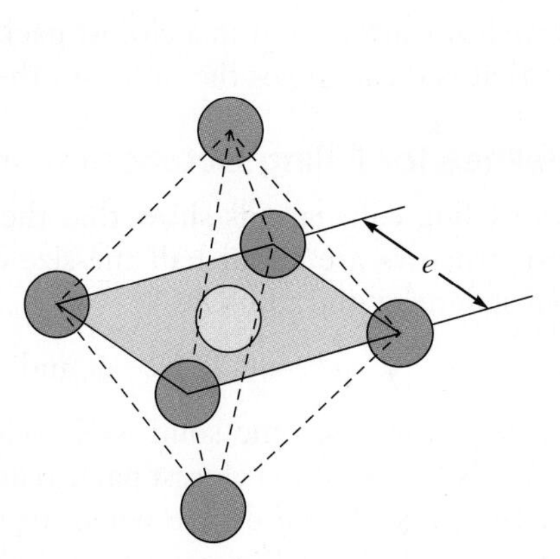

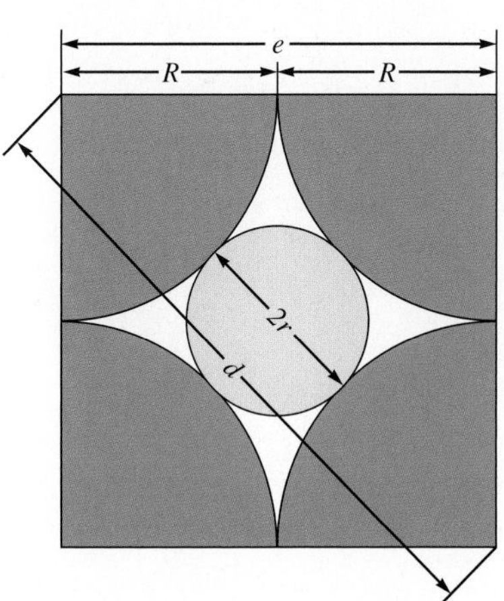

Figure 16.37
(a) The octahedral hole (shown in yellow) lies at the center of six spheres that touch along the edge (e) of the square. The packed spheres (in blue) are shown smaller than actual size for clarity. (b) The diagonal of the square (d) equals $R + 2r + R$, where r is the radius of the octahedral hole and R is the radius of the packed spheres.

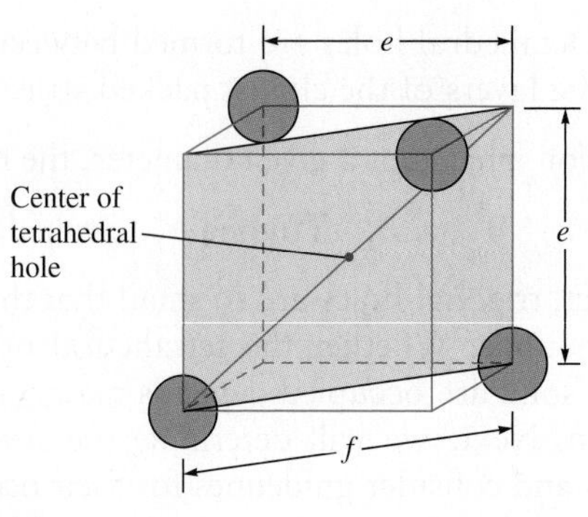

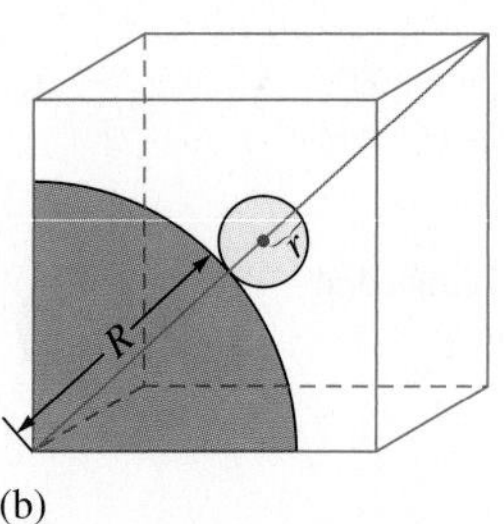

Figure 16.38
The tetrahedral hole. (a) The four spheres around a tetrahedral hole are shown inscribed in a cube. The spheres are shown much smaller than actual size. They actually touch along the face diagonal f. (b) The center of the tetrahedral hole (shown in red) is at the center of the body diagonal b (shown in purple).

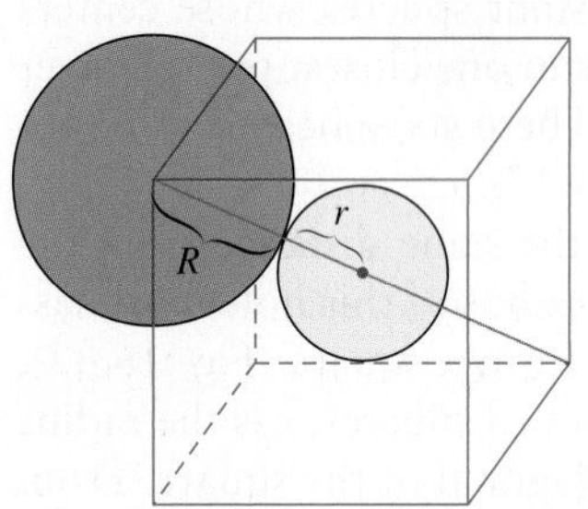

Figure 16.39
One packed sphere and its relationship to the tetrahedral hole. Note that (body diagonal)/2 $= R + r$.

The most convenient way to do geometric calculations on a tetrahedron is to inscribe it in a cube, as shown in Fig. 16.38. In this cube two spheres touch along the face diagonal of any cube face. This means that the face diagonal of the cube has length $2R$, where R is the radius of the packed spheres. The center of the tetrahedral hole is at the center of the body diagonal of the cube (Fig. 16.39). Now we will use the Pythagorean theorem twice to express the length of the body diagonal in terms of R. First, we express the face diagonal f in terms of the edge of the cube e:

$$f^2 = e^2 + e^2 = (2R)^2$$

Thus

$$e = \sqrt{2}\,R$$

Now we express the body diagonal b in terms of f and e:

$$b^2 = f^2 + e^2 = (2R)^2 + (\sqrt{2}\,R)^2$$

which leads to

$$b = \sqrt{6}\,R$$

Now the distance from the center of the body diagonal to the corner of the cube ($b/2$) is

$$\underset{\substack{\uparrow \\ \text{Radius of} \\ \text{tetrahedral} \\ \text{hole}}}{r} + \underset{\substack{\nwarrow \\ \text{Radius of} \\ \text{packed} \\ \text{spheres}}}{R} = \frac{b}{2} = \frac{\sqrt{6}\,R}{2} = \sqrt{\frac{3}{2}}\,R$$

So

$$r = \sqrt{\frac{3}{2}}\,R - R = 1.225R - R = 0.225R$$

Thus we have shown that in a closest packed structure a tetrahedral hole has a radius that is 0.225 times the radius of the packed spheres.

Guidelines for Filling Octahedral and Tetrahedral Holes

The preceding calculations show that the tetrahedral holes in a given closest packed structure are about half the size of the octahedral holes. That is, for packed spheres with radius R,

$$r^{\text{tet}} = 0.225R \qquad \text{and} \qquad r^{\text{oct}} = 0.414R$$

Now picture an ionic solid MX containing M^+ cations and X^- anions, where the X^- ions form a closest packed array. Assuming that the X^- ions have radius $R-$, how do we decide where to put the smaller cations? That is, do they occupy tetrahedral or octahedral holes? To obtain the most stable solid, we want to maximize the $M^+ \cdots X^-$ interactions and minimize the $X^- \cdots X^-$

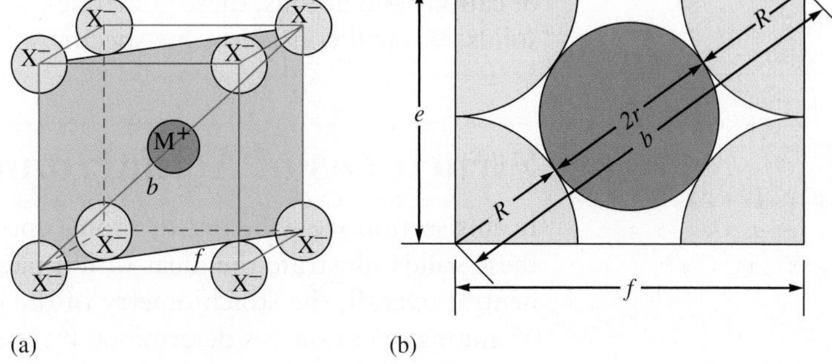

Figure 16.40
(a) A simple cubic array with X^- ions, with an M^+ ion in the center (in the cubic hole). (b) The body diagonal b equals $R + 2r + R$, since X^- and M^+ touch along this body diagonal.

In the closest packed arrangement, the X^- ions are touching. However, if M^+ ions are forced into holes in the structure that are smaller than the M^+ ions, the X^- ions are forced apart, decreasing the $X^- \cdots X^-$ repulsions. On the other hand, M^+ and X^- will be in direct contact, maximizing the $M^+ \cdots X^-$ interactions.

repulsions. We can do this by putting the M^+ ions into holes that are slightly smaller than the size of the M^+ ions. This causes the closest packed X^- ions to be pushed apart slightly. Assuming that the ions behave as hard spheres, this will result in a structure where the M^+ and X^- ions touch but where the X^- ions will no longer be touching.

With this idea in mind, we can now establish guidelines for filling the tetrahedral and octahedral holes. In terms of $r+$ (radius of M^+) and $R-$ (radius of X^-), if

$$\underset{\substack{\uparrow \\ \text{Size of} \\ \text{tetrahedral} \\ \text{holes}}}{0.225R-} < r+ < \underset{\substack{\uparrow \\ \text{Size of} \\ \text{octahedral} \\ \text{holes}}}{0.414R-}$$

then the M^+ ions are placed in the tetrahedral holes. That is, this condition ensures that the M^+ ions are larger than the tetrahedral holes in the closest packed array of X^- ions. This is what we want.

As we consider MX solids with larger and larger M^+ ions, at some point the M^+ ions become large enough to fill (and exceed the size of) the octahedral holes. Specifically, a cation with radius larger than $0.414R-$ should be placed in the octahedral holes rather than in the tetrahedral holes in the closest packed structure. As we continue to consider larger and larger M^+ ions, what happens? Is there an upper limit to the size of M^+ ions that can be forced into the octahedral holes? The answer is yes. For very large cations, the solid switches from a closest packed array of X^- ions to a simple cubic arrangement of X^- ions with an M^+ ion in the center of each cube (Fig. 16.40).

This structure can be described as M^+ occupying a *cubic hole*. Use of geometric arguments similar to those we have used for the tetrahedral and octahedral holes shows that for a cubic hole,

$$r+^{\text{cubic}} = 0.732R-$$

This means that for the solid MX, the M^+ ions occupy the octahedral holes in the range where

$$0.414R- < r+ < 0.732R-$$

For large M^+ ions ($r+ > 0.732R-$), the solid switches to the simple cubic arrangement just described.

The guidelines for filling the tetrahedral, octahedral, and cubic holes are summarized in Table 16.6.

Table 16.6

Guidelines for Filling Various Types of Holes for the Ionic Solid MX

Size of M^+	Type of Hole Filled
$0.225R- < r+ < 0.414R-$	Tetrahedral
$0.414R- < r+ < 0.732R-$	Octahedral
$0.732R- < r+$	Cubic

A few more points need to be considered. The guidelines just given assume that the ions behave as hard spheres and that only ionic forces occur. Therefore, it is not surprising that these guidelines *are not always obeyed.* They just give us a starting point for describing the structures of ionic solids. In addition, although we have considered only solids of the type MX containing a 1:1 ratio of cations and anions, these guidelines also apply to other types of binary ionic solids. We will consider some specific solids in the next section.

16.8 Structures of Actual Ionic Solids

In this section we will consider some specific binary ionic solids to show how these solids illustrate the ideas of ion packing. Because an ionic solid must be neutral overall, the stoichiometry of the compound (the ratio of the numbers of anions to cations) is determined by the ion charges. On the other hand, the structure of the compound (the placement of the ions in the solid) is determined, at least to a first approximation, by the relative sizes of the ions.

Before we consider specific compounds, we need to consider the locations and relative numbers of tetrahedral and octahedral holes in the closest packed structures. The location of the tetrahedral holes in the face-centered cubic unit cell of the ccp structure is shown in Fig. 16.41(a). Note from this figure that there are eight tetrahedral holes in the unit cell. Recall from the discussion in Section 16.4 that there are four net spheres in the face-centered cubic unit cell. Thus there are *twice as many tetrahedral holes as there are packed spheres* in the closest packed structure.

Closest packed structures contain twice as many tetrahedral holes as packed spheres. Closest packed structures contain the same number of octahedral holes as packed spheres.

The location of the octahedral holes in the face-centered cubic unit cell is shown in Fig. 16.42(a). The easiest octahedral hole to find in this structure is the one at the center of the cube. Note that this hole is surrounded by six spheres, as is required to form an octahedron. Since the remaining octahedral holes are shared with other unit cells, they are more difficult to visualize. However, it can be shown that the number of octahedral holes in the ccp structure is the *same* as the number of packed spheres.

Using these ideas, we will now consider the structures of some specific ionic solids.

The Structures of the Alkali Halides

The structure of sodium chloride, which is the prototype for most of the alkali halides, is best described as a cubic closest packed array of Cl^- ions with the Na^+ ions in all of the octahedral holes [Fig. 16.42(b)]. The relative sizes of these ions are such that $r_{Na^+} = 0.66R_{Cl^-}$, so this solid obeys the guidelines given previously. Note that the Cl^- ions are forced apart by the Na^+ ions, which are too large for the octahedral holes in the closest packed array of Cl^- ions. Since the number of octahedral holes is the same as the number of packed spheres, all the octahedral holes must be filled with Na^+ ions to achieve the required 1:1 stoichiometry. Most other alkali halides also have the sodium chloride structure. In fact, all the halides of lithium, sodium, potassium, and rubidium have this structure. Cesium fluoride has the sodium chloride struc-

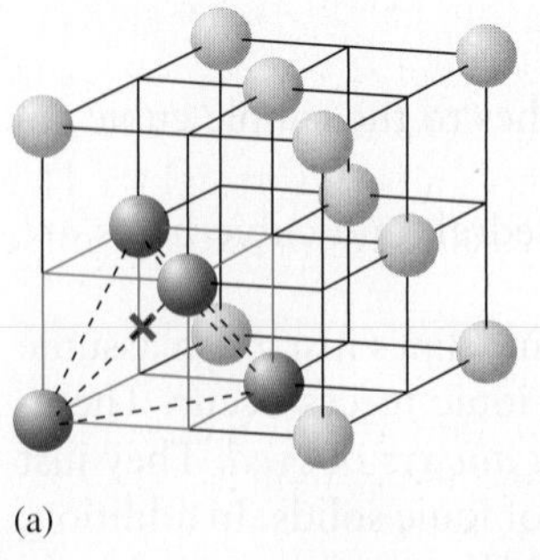
(a)

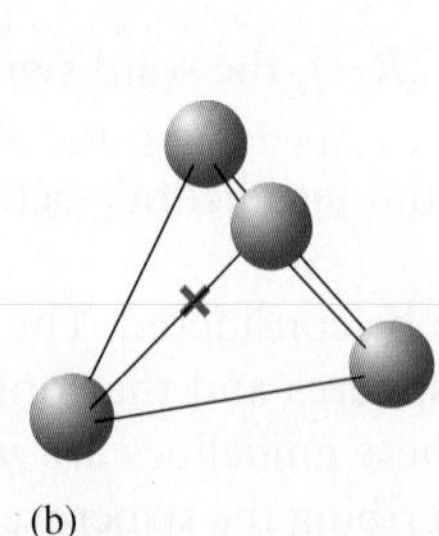
(b)

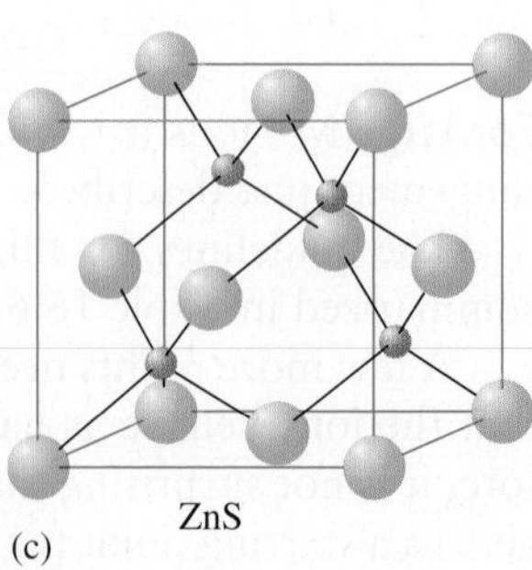

(c)

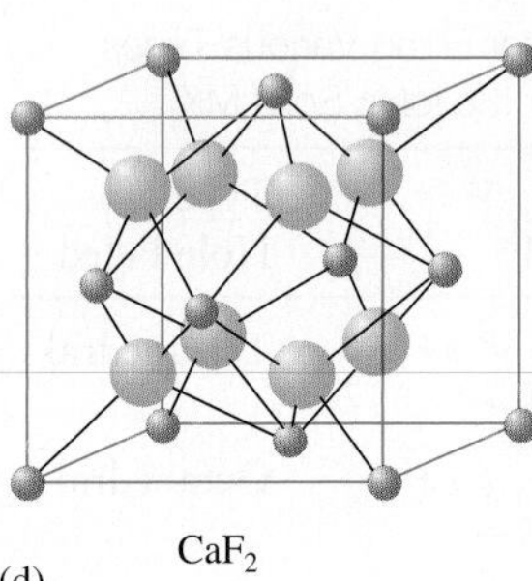

(d)

Figure 16.41
(a) The location (red ×) of a tetrahedral hole in the face-centered cubic unit cell. (b) One of the tetrahedral holes. (c) The unit cell for ZnS, where the S^{2-} ions (yellow) are closest packed, with the Zn^{2+} ions (purple) filling alternate tetrahedral holes. (d) The unit cell for CaF_2, where the Ca^{2+} ions (purple) form a face-centered cubic arrangement, with the F^- ions (yellow) in all of the tetrahedral holes.

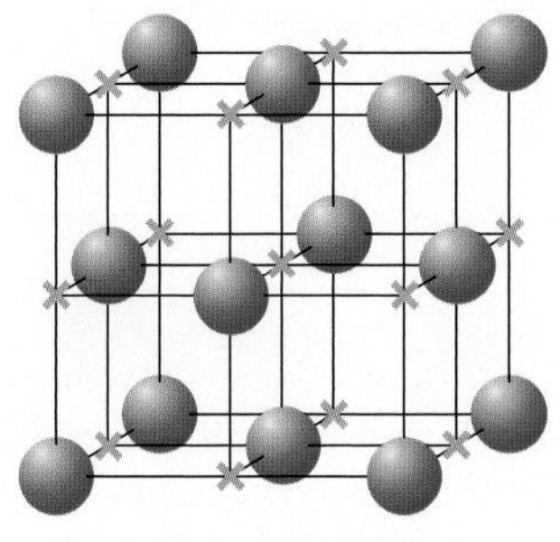

(a)

(b)

Figure 16.42
(a) The locations (gray ×) of the octahedral holes in the face-centered cubic unit cell. (b) Representation of the unit cell for solid NaCl. The Cl^- ions (green spheres) have a ccp arrangement, with Na^+ ions (gray spheres) filling all the octahedral holes. This representation shows the idealized closest packed structure of NaCl. In the actual structure, the Cl^- ions do not quite touch.

ture, but because of the large size of Cs^+ ions, in this case the Cs^+ ions form a cubic closest packed arrangement with the F^- ions in all the octahedral holes. On the other hand, cesium chloride, in which the Cs^+ and Cl^- ions are almost the same size, has a simple cubic structure of Cl^- ions, with each Cs^+ ion in the cubic hole in the center of each cube. The compounds cesium bromide and cesium iodide also have this latter structure.

The Structure of Zinc Sulfide

The compound ZnS illustrates another important general type of structure. In ZnS the ion sizes are such that $r_{Zn^{2+}} \approx 0.35R_{S^{2-}}$. According to the guidelines, this should mean that the Zn^{2+} ions occupy the tetrahedral holes among the closest packed S^{2-} ions. This is exactly what is found in ZnS. However, because there are twice as many tetrahedral holes as closest packed ions, only one-half of the tetrahedral holes are occupied in zinc sulfide, producing the required 1:1 ratio of Zn^{2+} and S^{2-} ions [Fig. 16.41(c)]. Zinc sulfide has two forms: zinc blende, where the S^{2-} ions are cubic closest packed, and wurtzite, where the S^{2-} ions are hexagonal closest packed. In both cases the Zn^{2+} ions occupy half of the tetrahedral holes. The substances ZnO and CdS also show both the zinc blende and wurtzite structures.

The Structure of Calcium Fluoride

The structure of the compound CaF_2 can be described as a face-centered cubic array of Ca^{2+} ions with the F^- ions in all the tetrahedral holes.* This gives the required 1:2 ratio of Ca^{2+} and F^- ions. This structure is called the fluorite structure [Fig. 16.41(d)] and is also observed in the compounds SrF_2, $BaCl_2$, PbF_2, and CdF_2, among others.

16.9 Lattice Defects

So far we have assumed that crystalline compounds are perfect—that is, that all the atoms, ions, or molecules are present and occupy the correct sites. Although crystalline materials are highly ordered and most of the components are where they are expected to be, all real crystals have imperfections called *lattice defects.*

Point defects refer to totally missing particles (atoms, ions, or molecules) or to cases where the particle is in a nonstandard location. A crystal with missing particles is said to have *Schottky defects* [Fig. 16.43(a)]. When ions are missing from an ionic compound, they must be missing in a way that preserves the overall electrical neutrality of the substance. For example, for every missing Ca^{2+} ion in CaF_2, there must be two missing F^- ions.

Crystals in which particles have migrated to nonstandard positions are said to exhibit *Frenkel defects* [Fig. 16.43(b)]. One group of compounds where Frenkel defects are present to the extreme is the silver halides—AgCl, AgBr, and AgI. In these compounds the anion positions are mostly those expected from closest packing ideas; however, the silver ions are distributed almost randomly in the various holes and can easily travel within the solid structure. This property is a major reason that the silver halides are so useful in photographic films.

*The structure of CaF_2 is not a closest packed structure because the Ca^{2+} ions that occupy the face-centered cubic sites are smaller than the F^- ions. An alternative description of this structure is a simple cubic array of F^- ions, with Ca^{2+} ions in half of the cubic holes.

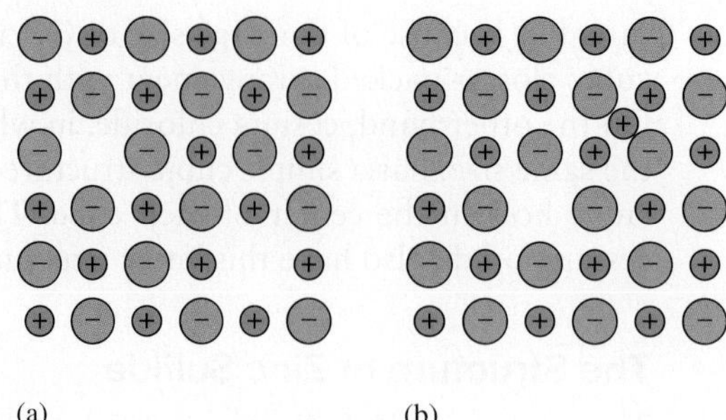

Figure 16.43
Defects in crystalline ionic solids. (a) Schottky defects, in which there are vacant sites. (b) Frenkel defects, in which an atom or an ion of either sign is present at an inappropriate site.

In some cases the crystal defects involve impurities that lead to *nonstoichiometric compounds.* For example, wüstite, which has the idealized formula FeO, actually varies from $Fe_{1.0}O_{1.0}$ to $Fe_{0.95}O_{1.0}$. These variations occur when some of the Fe^{2+} ions in the compound are replaced by Fe^{3+} ions in a way that preserves the electrical neutrality of the solid. There are many other examples of nonstoichiometric compounds. One extreme case involves TiO, which can vary from $Ti_{1.0}O_{0.7}$ to $Ti_{0.8}O_{1.0}$.

16.10 Vapor Pressure and Changes of State

Now that we have considered the general properties of the three states of matter, we can explore the processes by which matter changes state. One very familiar example of a change in state occurs when a liquid evaporates from an open container. This is clear evidence that the molecules of a liquid can escape from the liquid's surface and form a gas. Called **vaporization,** or *evaporation,* this process is endothermic because energy is required to overcome the relatively strong intermolecular forces in the liquid. The energy required to vaporize 1 mole of a liquid at a pressure of 1 atm is called the standard **heat of vaporization** or the standard **enthalpy of vaporization** and is usually symbolized as ΔH°_{vap}.

Vapor is the usual term for the gas phase of a substance that exists as a solid or liquid at 25°C and 1 atm.

The endothermic nature of vaporization has great practical significance; in fact, one of the most important roles that water plays in our world is to act as a coolant. Because of the strong hydrogen bonding among its molecules in the liquid state, water has an unusually large heat of vaporization (40.7 kJ/mol). A significant portion of the sun's energy that reaches earth is spent evaporating water from the oceans, lakes, and rivers rather than warming the earth. The vaporization of water is also crucial to the body's temperature control system through evaporation of perspiration.

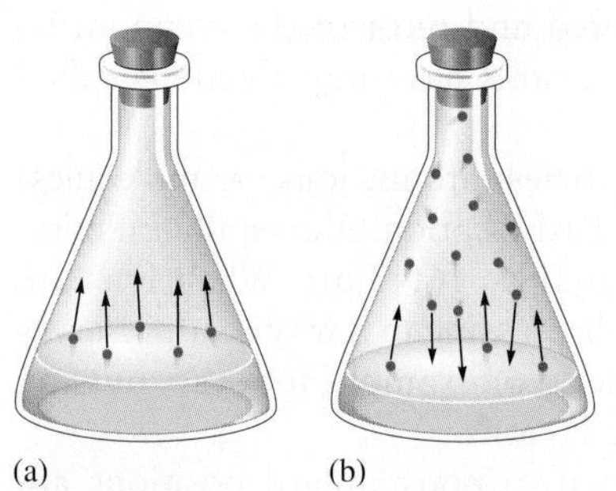

Figure 16.44
Behavior of a liquid in a closed container. (a) Initially, net evaporation occurs as molecules are transferred from the liquid to the vapor phase, so the amount of liquid decreases. (b) As the number of vapor molecules increases, the rate of return to the liquid (condensation) increases, until finally the rate of condensation equals the rate of evaporation. The system is at equilibrium, and no further changes occur in the amounts of vapor or liquid.

Vapor Pressure

When a liquid is placed in a closed container, the amount of liquid at first decreases but eventually becomes constant. The decrease occurs because there is an initial net transfer of molecules from the liquid to the vapor phase (Fig. 16.44). However, as the number of vapor molecules increases, so does the rate of return of these molecules to the liquid. The process by which vapor molecules re-form a liquid is called **condensation.** Eventually, enough vapor molecules are present above the liquid so that the rate of condensation equals the rate of evaporation (Fig. 16.45). At this point no further net change occurs in the amount of liquid or vapor because the two opposite processes exactly balance each other; the system is at equilibrium. Note that this system is highly dynamic on the molecular level—molecules are constantly escaping from and entering the liquid at a high rate. However, there is no net change because the two opposite processes just balance each other.

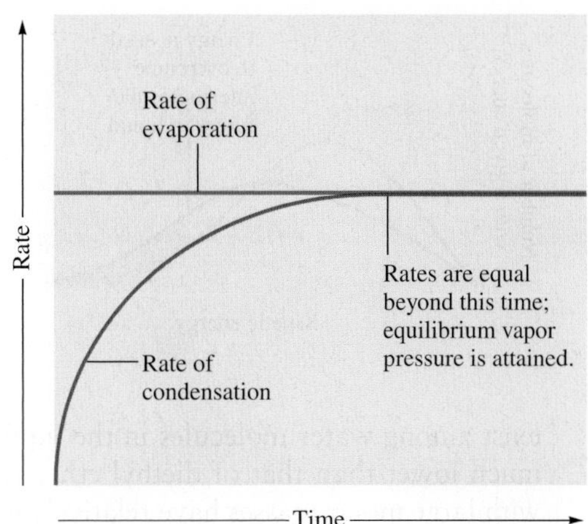

Figure 16.45
The rates of condensation and evaporation over time for a liquid sealed in a closed container. The rate of evaporation remains constant, whereas the rate of condensation increases as the number of molecules in the vapor phase increases, until the two rates become equal. At this point the equilibrium vapor pressure is attained.

The pressure of the vapor present at equilibrium is called the *equilibrium vapor pressure* or, more commonly, the **vapor pressure** of the liquid. A simple barometer can measure the vapor pressure of a liquid, as shown in Fig. 16.46. The liquid is injected at the bottom of the tube of mercury and floats to the surface because the mercury is so dense. A portion of the liquid evaporates at the top of the column, producing a vapor whose pressure pushes some mercury out of the tube. When the system reaches equilibrium, the vapor pressure can be determined from the change in the height of the mercury column, since

$$P_{atm} = P_{vapor} + P_{Hg\ column}$$

Thus

$$P_{vapor} = P_{atm} - P_{Hg\ column}$$

The vapor pressures of liquids vary widely [Fig. 16.46(b)]. Liquids with high vapor pressures are said to be *volatile*—they evaporate readily from an open dish.

The vapor pressure of a liquid is principally determined by the size of the *intermolecular forces* in the liquid. Liquids in which the intermolecular forces are large have relatively low vapor pressures because the molecules need high energies to escape to the vapor phase. For example, although water has a much lower molar mass than diethyl ether, the strong hydrogen bonding forces that

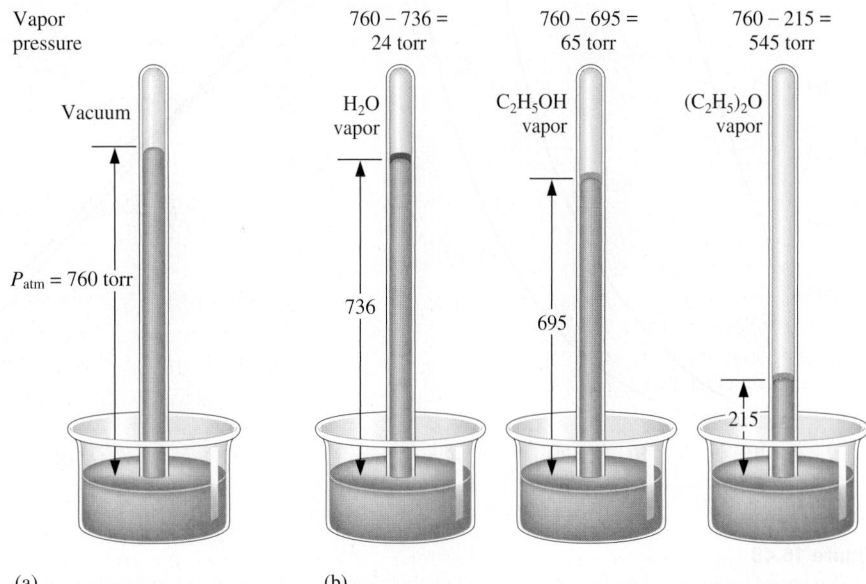

Figure 16.46
(a) The vapor pressure of a liquid can be measured easily using a simple barometer of the type shown here. (b) The three liquids, water, ethanol (C_2H_5OH), and diethyl ether [$(C_2H_5)_2O$], have quite different vapor pressures. Ether is by far the most volatile of the three. Note that in each case a little liquid remains (floating on the mercury).

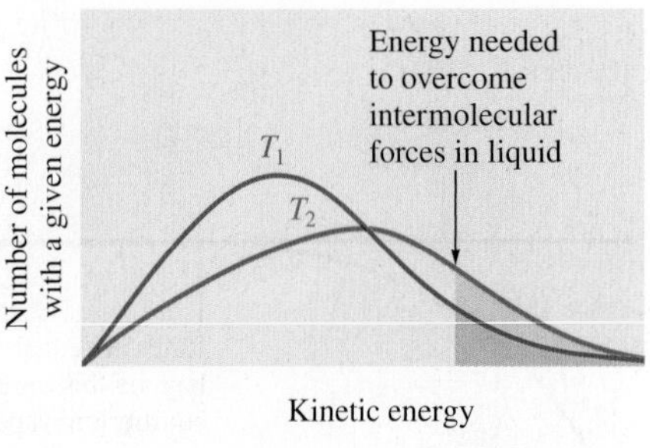

Figure 16.47
The number of molecules in a liquid with a given energy versus kinetic energy at two temperatures ($T_1 < T_2$). Note that the proportion of molecules with enough energy to escape the liquid to the vapor phase (indicated by shaded areas) increases dramatically with temperature. This causes vapor pressure to increase markedly with temperature.

exist among water molecules in the liquid cause water's vapor pressure to be much lower than that of diethyl ether [Fig. 16.46(b)]. In general, substances with large molar masses have relatively low vapor pressures mainly because of the large dispersion forces. The more electrons a substance has, the more polarizable it is, and the greater are the dispersion forces.

Measurements of the vapor pressure for a given liquid at several temperatures show that *vapor pressure increases significantly with temperature.* Figure 16.47 illustrates the distribution of molecular kinetic energies present in a liquid at two different temperatures. To overcome the intermolecular forces in a liquid, a molecule must have a minimum kinetic energy. As the temperature of the liquid is increased, the fraction of molecules having sufficient energy to overcome these forces and escape to the vapor phase increases markedly. Thus the vapor pressure of a liquid increases dramatically with temperature. The vapor pressure values for water at several temperatures are given in Table 16.7.

Table 16.7

The Vapor Pressure of Water as a Function of Temperature

T (°C)	P (torr)
0.0	4.579
10.0	9.209
20.0	17.535
25.0	23.756
30.0	31.824
40.0	55.324
60.0	149.4
70.0	233.7
90.0	525.8

The quantitative nature of the temperature dependence of vapor pressure can be illustrated graphically. Plots of vapor pressure versus temperature for water, ethanol, and diethyl ether are shown in Fig. 16.48(a). Note the nonlinear increase in vapor pressure for all the liquids as the temperature is increased.

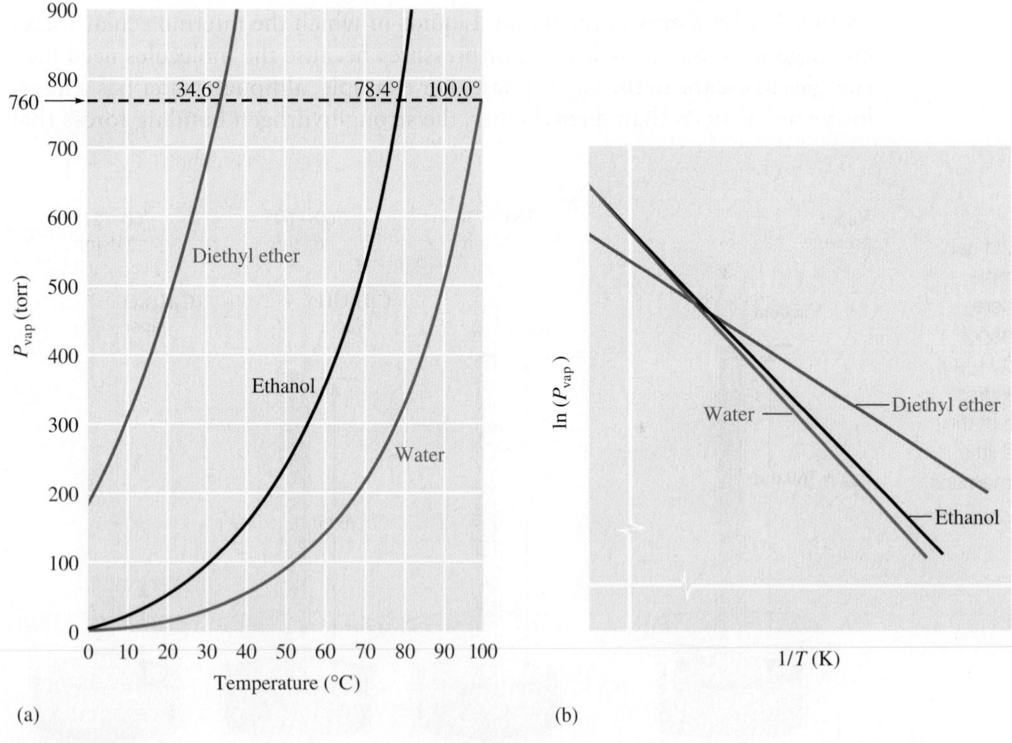

Figure 16.48
(a) The vapor pressure of water, ethanol, and diethyl ether as a function of temperature. (b) Plots of ln(P_{vap}) versus $1/T$ (Kelvin temperature) of water, ethanol, and diethyl ether.

A straight line can be obtained from these data by plotting $\ln(P_{vap})$ versus $1/T$, where T is the Kelvin temperature [Fig. 16.48(b)]. We can represent this behavior by the equation

$$\ln(P_{vap}) = -\frac{\Delta H_{vap}}{R}\left(\frac{1}{T}\right) + C \qquad (16.4)$$

where ΔH_{vap} is the enthalpy of vaporization, R is the universal gas constant, and C is a constant characteristic of a given liquid. Note that this equation is a simplified version of the equation obtained in Section 10.11 for the temperature dependence of an equilibrium constant:

$$\ln K = -\frac{\Delta H^\circ}{R}\left(\frac{1}{T}\right) + \frac{\Delta S^\circ}{R}$$

The similarity is not unexpected, since vapor pressure is an equilibrium phenomenon. Note that we can now interpret the constant C in terms of ΔS, the entropy of vaporization for a given liquid.

EXAMPLE 16.2

Using the plots in Fig. 16.48(b), determine whether water or diethyl ether has the smaller enthalpy of vaporization.

Solution When $\ln(P_{vap})$ is plotted versus $1/T$, the slope of the resulting straight line is

$$-\frac{\Delta H_{vap}}{R}$$

Note from Fig. 16.48(b) that the slopes of the lines for water and diethyl ether are both negative, as expected. Also note that the line for ether has the smaller slope. Thus ether has the smaller value of ΔH_{vap}. This makes sense because the hydrogen bonding in water causes it to have a relatively large enthalpy of vaporization.

Equation (16.4) is important for several reasons. For example, we can determine the heat of vaporization for a liquid by measuring P_{vap} at several temperatures and then evaluating the slope of a plot of $\ln(P_{vap})$ versus $1/T$. On the other hand, if we know the values of ΔH_{vap} and P_{vap} at one temperature, we can use Equation (16.4) to calculate P_{vap} at another temperature. This can be done by recognizing that the constant C does not depend on temperature. Thus, at two temperatures, T_1 and T_2, we can solve Equation (16.4) for C and then write the equality

$$\ln(P_{vap}^{T_1}) + \frac{\Delta H_{vap}}{RT_1} = C = \ln(P_{vap}^{T_2}) + \frac{\Delta H_{vap}}{RT_2}$$

Richard Megna/Fundamental Photographs

Figure 16.49
Solid iodine sublimes when heated and then recrystallizes when in contact with a cold surface.

Equation (16.5) is called the Clausius–Clapeyron equation.

which can be rearranged to

$$\ln(P_{vap}^{T_1}) - \ln(P_{vap}^{T_2}) = \frac{\Delta H_{vap}}{R}\left(\frac{1}{T_2} - \frac{1}{T_1}\right)$$

or

$$\ln\left(\frac{P_{vap}^{T_1}}{P_{vap}^{T_2}}\right) = \frac{\Delta H_{vap}}{R}\left(\frac{1}{T_2} - \frac{1}{T_1}\right) \qquad (16.5)$$

Like liquids, solids have vapor pressures. Figure 16.49 shows iodine vapor in equilibrium with solid iodine. At 25°C and 1 atm, iodine *sublimes;* that is, it goes directly from the solid to the gaseous state without passing through the liquid state. **Sublimation** also occurs with dry ice (solid carbon dioxide) under these conditions.

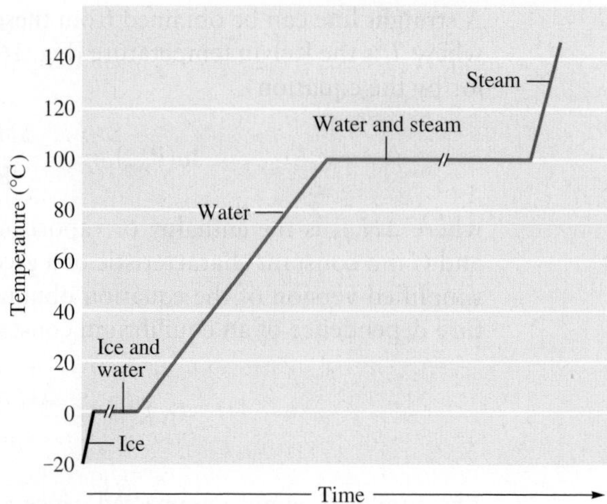

Figure 16.50
The heating curve for a given quantity of water where energy is added at a constant rate. The plateau at the boiling point is longer than the plateau at the melting point because it takes almost seven times more energy (and thus seven times the heating time) to vaporize liquid water than to melt ice. The slopes of the other lines are different because the different states of water have different molar heat capacities (the energy required to raise the temperature of 1 mole of a substance by 1°C).

Changes of State

What happens when a solid is heated? Typically, it melts to form a liquid. If the heating continues, the liquid at some point boils and forms the vapor phase. This process can be represented by a **heating curve:** a plot of temperature versus time for a process where energy is added at a constant rate.

The heating curve for water is given in Fig. 16.50. As energy flows into the ice, the random vibrations of the water molecules increase as the temperature rises. Eventually, the molecules become so energetic that they break loose from their lattice positions, and the change from solid to liquid occurs. This is indicated by a plateau at 0°C on the heating curve. At this temperature, called the **melting point,** all the added energy is used to disrupt the ice structure by breaking the hydrogen bonds, thus increasing the potential energy of the water molecules. The enthalpy change that occurs at the melting point when a solid melts is called the **heat of fusion** (or more accurately, the **enthalpy of fusion**) ΔH_{fus}. The melting points and enthalpies of fusion for several representative solids are listed in Table 16.8.

Ionic solids such as NaCl and NaF have very high melting points and very high enthalpies of fusion because of the strong ionic forces in these solids. At the other extreme is $O_2(s)$, a molecular solid containing nonpolar molecules with weak intermolecular forces.

The temperature remains constant until all the solid has changed to liquid; then it begins to increase again. At 100°C the liquid water reaches its *boiling point,* and the temperature again remains constant as the added energy is used to vaporize the liquid. When all the liquid is changed to vapor, the temperature again begins to rise. Note that changes of state are physical changes; although intermolecular forces have been overcome, no chemical bonds have been broken. If the water vapor were heated to much higher temperatures, the water mole-

Table 16.8

Melting Points and Enthalpies of Fusion for Several Representative Solids

Compound	Melting Point (°C)	Enthalpy of Fusion (kJ/mol)
O_2	−218	0.45
HCl	−114	1.99
HI	−51	2.87
CCl_4	−23	2.51
$CHCl_3$	−64	9.20
H_2O	0	6.01
NaF	992	29.3
NaCl	801	30.2

cules would break down into the individual atoms. This is a chemical change, since covalent bonds are broken. We no longer have water after it occurs.

The melting and boiling points will be defined more precisely later in this section.

The melting and boiling points for a substance are determined by the vapor pressures of the solid and liquid states. Figure 16.51 shows the vapor pressures of solid and liquid water as functions of temperature near 0°C. Note that below 0°C the vapor pressure of ice is less than the vapor pressure of liquid water. Also note that the vapor pressure of ice has a larger temperature dependence than that of the liquid. That is, the vapor pressure of ice increases more rapidly for a given rise in temperature than does the vapor pressure of water. Thus, as the temperature of the solid is increased, a point is eventually reached where the *liquid and solid have identical vapor pressures.* This is the melting point.

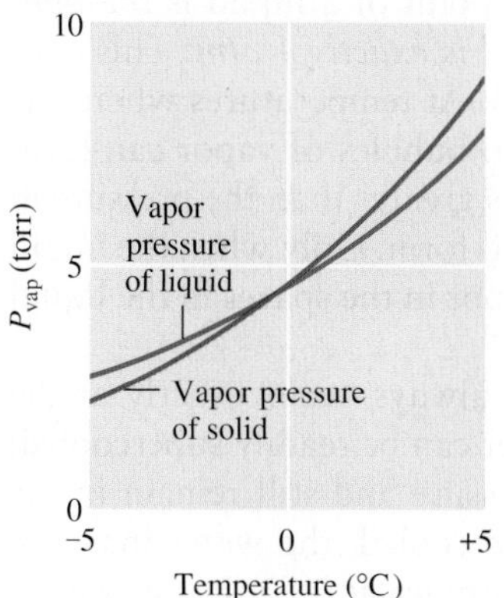

Figure 16.51
The vapor pressures of solid and liquid water as a function of temperature. The data for liquid water below 0°C are obtained from supercooled water. The data for solid water above 0°C are estimated by extrapolation of vapor pressure from below 0°C.

These concepts can be demonstrated experimentally using the apparatus illustrated in Fig. 16.52, where ice occupies one compartment and liquid water the other. Consider the following cases.

Case 1: *A temperature at which the vapor pressure of the solid is greater than that of the liquid.* At this temperature the solid requires a higher pressure of vapor than the liquid to be in equilibrium with the vapor. Thus, as vapor is released from the solid to try to achieve equilibrium, the liquid absorbs vapor in an attempt to reduce the vapor pressure to its equilibrium value. The net effect is a conversion from solid to liquid through the vapor phase. In fact, no solid can exist under these conditions. The amount of solid steadily decreases and the volume of liquid increases. Finally, there is only liquid in the right compartment, which comes to equilibrium with the water vapor, and no further changes occur in the system. The temperature for Case 1 must be *above the melting point* of ice, since only the liquid state can exist.

Case 2: *A temperature at which the vapor pressure of the solid is less than that of the liquid.* This is the opposite of the situation in Case 1. In this case the liquid requires a higher pressure of vapor than the solid to be in equilibrium with the vapor. So the liquid gradually disappears, and the amount of ice increases. Finally, only the solid remains, which achieves equilibrium with the vapor. This temperature must be *below the melting point* of ice, since only the solid state can exist.

Case 3: *A temperature at which the vapor pressures of the solid and liquid are identical.* In this case the solid and liquid states have the same vapor pressure, so they can coexist in the apparatus at equilibrium simultaneously with the vapor. This temperature represents the *melting point,* where both the solid and liquid states can exist.

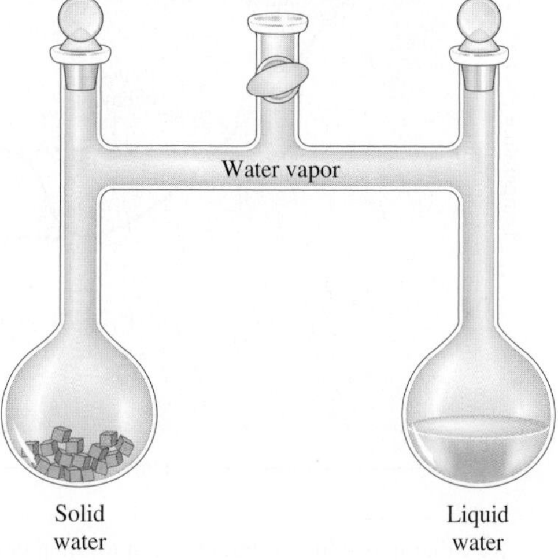

Figure 16.52
An apparatus that allows solid and liquid water to interact only through the vapor state.

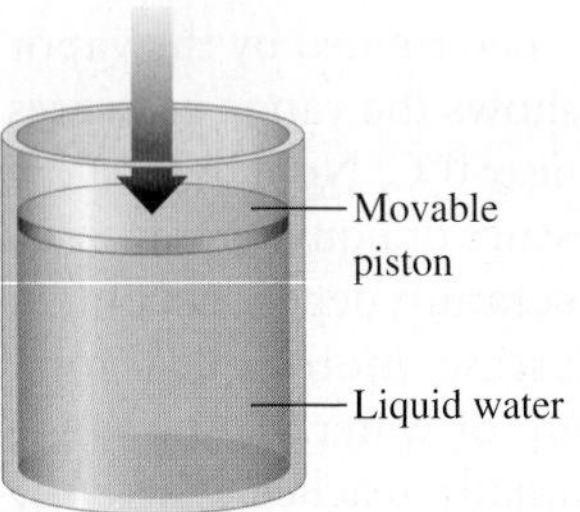

Figure 16.53
Water in a closed system with a pressure of 1 atm exerted on the piston. No bubbles can form within the liquid as long as the vapor pressure is less than 1 atm.

We can now describe the melting point of a substance more precisely. The **normal melting point** is defined as *the temperature at which the solid and liquid states have the same vapor pressure under conditions where the total pressure is 1 atm.*

Boiling occurs when the vapor pressure of a liquid becomes equal to the pressure of its environment. The **normal boiling point** of a liquid is *the temperature at which the vapor pressure of the liquid is exactly 1 atm.* This concept can be understood by considering Fig. 16.53. At temperatures where the vapor pressure of the liquid is less than 1 atm, no bubbles of vapor can form because the pressure on the surface of the liquid is greater than the pressure in any spaces in the liquid where bubbles are trying to form. Only when the liquid reaches a temperature at which the pressure of vapor in the spaces in the liquid is 1 atm can bubbles form and boiling occur.*

However, observed changes of state do not always occur exactly at the boiling point or melting point. For example, water can be readily **supercooled;** that is, it can be cooled below 0°C at 1 atm pressure and still remain in the liquid state. Supercooling occurs because, as it is cooled, the water may not achieve the degree of organization necessary to form ice at 0°C; thus it continues to exist as the liquid. At some point the correct ordering occurs and ice rapidly forms, releasing energy in the exothermic process and bringing the temperature back up to the melting point, where the remainder of the water freezes (Fig. 16.54).

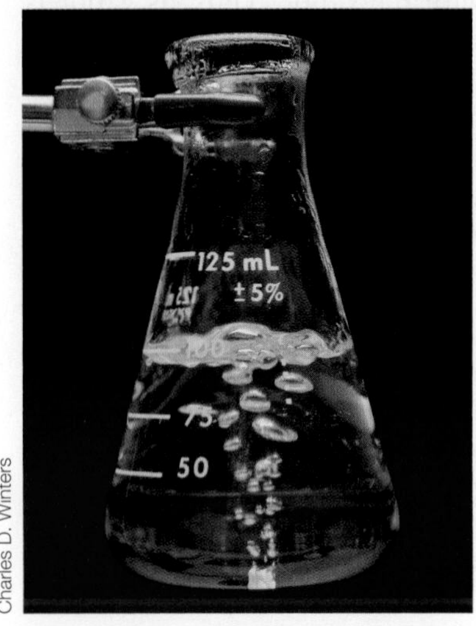

Charles D. Winters

A boiling chip releasing air bubbles acts as a nucleating agent for the large bubbles that form when water boils.

A liquid can also be **superheated,** or raised to temperatures above its boiling point, especially if it is heated rapidly. Superheating can occur because bubble formation in the interior of the liquid requires that many high-energy molecules gather in the same vicinity. This may not happen at the boiling point, especially if the liquid is heated rapidly. If the liquid becomes superheated, the vapor pressure in the liquid is greater than the atmospheric pressure. Once a bubble does form, since its internal pressure is greater than that of the atmosphere, it can burst before rising to the surface, blowing the surrounding liquid out of the container. This phenomenon is called *bumping* and has ruined many experiments. It can be avoided by adding boiling chips to the flask containing the liquid. Boiling chips are bits of porous ceramic material containing trapped air that escapes on heating, forming tiny bubbles that act as "starters" for vapor bubble formation. These starter bubbles allow a smooth onset of boiling as the boiling point is reached.

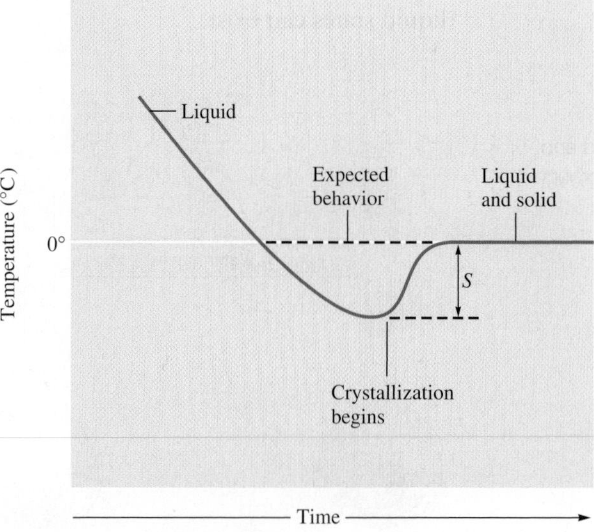

Figure 16.54
The supercooling of water. The extent of supercooling is given by *S*.

*Note that in real life the bubbles seen forming in water as it is being heated to boiling arise from the expulsion of dissolved air.

16.11 Phase Diagrams

A **phase diagram** is a convenient way of representing the phases of a substance as a function of temperature and pressure. For example, the phase diagram for water (Fig. 16.55) shows which state exists at a given temperature and pressure. It is important to recognize that a phase diagram describes conditions and events for a pure substance in a *closed* system of the type represented in Fig. 16.53, where no material can escape into the surroundings and no air is present.

To show how to interpret the phase diagram for water, we will consider heating experiments at several pressures, shown by the dashed lines in Fig. 16.56.

Experiment 1: *Pressure is 1 atm.* This experiment begins with the cylinder shown in Fig. 16.53 completely filled with ice at a temperature of −20°C and with the piston exerting a pressure of 1 atm directly on the ice (there is no air space). Since at temperatures below 0°C the vapor pressure of ice is less than 1 atm—which is the constant external pressure on the piston—no vapor is present in the cylinder. As the cylinder is heated, ice remains the only component until the temperature reaches 0°C, where the ice changes to liquid water as energy is added. This is the normal melting point of water. When all the solid has changed to liquid, the temperature again rises. At this point the cylinder contains only liquid water. *No vapor is present* because the vapor pressure of liquid water under these conditions is less than 1 atm, the constant external pressure on the piston. Heating continues until the temperature of the liquid

Below the critical temperature and pressure, the gaseous state of a substance is often referred to as a vapor.

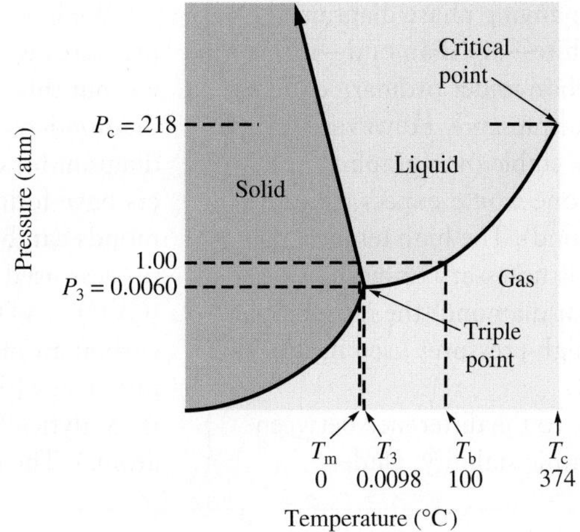

Figure 16.55
The phase diagram for water. T_m represents the normal melting point; T_3 and P_3 denote the triple point; T_b represents the normal boiling point; T_c represents the critical temperature; P_c represents the critical pressure. The negative slope of the solid/liquid line reflects the fact that the density of ice is less than that of liquid water. Note that the solid/liquid phase line continues indefinitely.

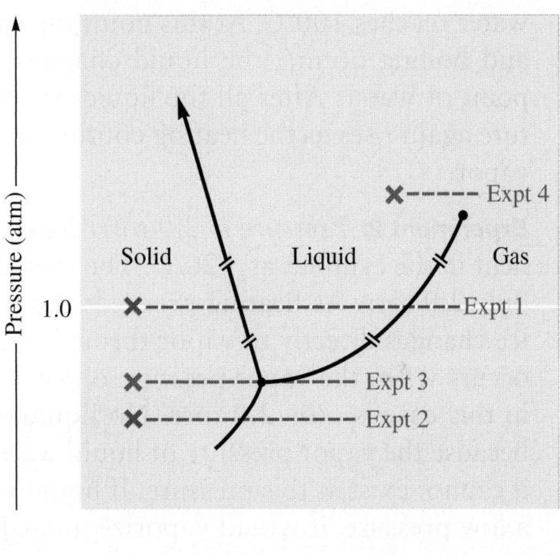

Figure 16.56
Diagrams of various heating experiments on samples of water in closed systems.

Chemical Insights

Making Diamonds at Low Pressures: Fooling Mother Nature

In 1955 Robert H. Wentorf, Jr., accomplished something that borders on alchemy—he turned peanut butter into diamonds. He and his coworkers at the General Electric Research and Development Center also changed roofing pitch, wood, coal, and many other carbon-containing materials into diamonds using a process involving temperatures of approximately 2000°C and pressures of about 100,000 atm. Although the first diamonds made by this process looked like black sand because of the impurities present, the process has now been developed to a point such that beautiful, clear, gem-quality diamonds can be produced. General Electric now has the capacity to produce 150 million carats (30,000 kg) of diamonds, virtually all of which is "diamond grit" used for industrial purposes such as abrasive coatings on cutting tools.

The high temperatures and pressures used in the GE process for making diamonds make sense if one looks at the accompanying phase diagram for carbon. Note that graphite—not diamond—is the most stable form of carbon under ordinary conditions of temperature and pressure. However, diamond becomes more stable than graphite at very high pressures (as one would expect from the greater density of diamond). The high temperature used in the GE process is necessary to disrupt the bonds in graphite so that diamond (the most stable form of carbon at the high pressures used in the process) can form.

This brings us again to the difference between kinetic and thermodynamic stability. Under normal conditions of temperature and pressure, graphite has a lower free energy than diamond (by about 2 kJ/mol). However, to get from diamond to graphite requires a large expenditure of energy. This is the energy of activation that determines the rate of a reaction at a particular temperature. As a result of this high activation energy, diamonds formed at the high pressures found deep in the earth's crust can be brought to the earth's surface by natural geological processes and continue to exist for millions of years. Although these diamonds are thermodynamically unstable (relative to graphite), they are kinetically stable. That is, the reaction to change diamond to graphite is favored by thermodynamics but is disfavored by kinetics—it is so slow that diamonds *almost* last forever.* Diamond is said to be metastable (thermodynamically unstable but kinetically stable) under normal conditions of temperature and pressure.

We have seen that diamond formed at high pressures is "trapped" in this form by slow kinetics, but this process is very expensive. Can diamond be formed at low pressures? The phase diagram for carbon says "no." However, researchers have found that under the right conditions diamonds can be "grown" at low pressures. The process used is called chemical vapor deposition (CVD). CVD uses an energy source to release carbon atoms from a compound such as methane into a steady flow of hydrogen gas (some of which is catalytically dissociated to produce hydrogen atoms). The carbon atoms then deposit as a

water reaches 100°C. At this point the vapor pressure of liquid water is 1 atm, and boiling occurs; the liquid changes to vapor. This is the normal boiling point of water. After all the liquid has been converted to steam, the temperature again rises as the heating continues. The cylinder now contains only water vapor.

Experiment 2: *Pressure is 2.0 torr.* Again we start with ice as the only component in the cylinder at −20°C. The pressure exerted by the piston in this case is only 2.0 torr. As heating proceeds the temperature rises to −10°C, where the ice changes directly to vapor through the process of sublimation. Sublimation occurs when the vapor pressure of ice is equal to the external pressure, which in this case is only 2.0 torr. No liquid water appears under these conditions because the vapor pressure of liquid water is always greater than 2.0 torr; thus it cannot exist at this pressure. If liquid water were placed in a cylinder at such a low pressure, it would vaporize immediately.

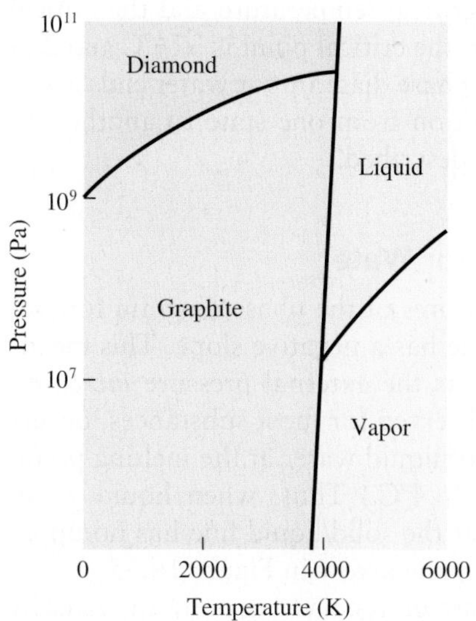

The phase diagram for carbon.

diamond film on a surface maintained at a temperature between 600°C and 900°C. Why does diamond form on this surface rather than the thermodynamically favored graphite? Nobody is sure, but it has been suggested that at these relatively high temperatures the diamond structure grows faster than the graphite structure, and so diamond is favored kinetically under these conditions. It has also been suggested that the hydrogen atoms present react much faster with graphite fragments than with diamond fragments, effectively removing any graphite from the growing film. Once it forms, of course, diamond is kinetically trapped. The major advantage of CVD is that there is no need for the extraordinarily high pressures used in the traditional process for synthesizing diamonds.

The first products with diamond films are already on the market. Audiophiles can buy tweeters that have diaphragms coated with a thin diamond film that limits sound distortion. Watches with diamond-coated crystals are planned, as are diamond-coated windows in infrared scanning devices used in analytical instruments and missile guidance systems. These applications represent only the beginning for diamond-coated products.

In addition, the CVD process is now being used to synthesize gem-quality diamonds. Companies have now been formed that can grow gem-quality diamonds as large as several carats. These synthetic diamonds are virtually indistinguishable from diamonds formed by natural processes. This has tremendous implications for the natural diamond industry, which is worth more than $60 billion per year.

*In Morocco, a 50-km-long slab called Beni Bousera contains chunks of graphite that were probably once diamonds formed in the deposit when it was buried 150 km underground. As this slab slowly rose to the surface over millions of years, the very slow reaction changing diamond to graphite had time to occur. That is, on this time scale thermodynamic control could exert itself. In the diamond-rich kimberlite deposits in South Africa, which rise to the surface much faster, kinetic control exists.

Experiment 3: *Pressure is 4.588 torr.* Again we start with ice as the only component in the cylinder at −20°C. In this case the pressure exerted on the ice by the piston is 4.588 torr. As the cylinder is heated, no new phase appears until the temperature reaches 0.0098°C. At this point, called the **triple point,** solid and liquid water have identical vapor pressures of 4.588 torr. Thus *at 0.0098°C and 4.588 torr all three states of water are present.* In fact, *only* under these conditions can all three states of water coexist.

Experiment 4: *Pressure is 225 atm.* In this experiment we start with liquid water in the cylinder at 300°C; the pressure exerted by the piston on the water is 225 atm. Liquid water can be present at this temperature because of the high external pressure. As the temperature increases, something happens that we did not observe in the first three experiments: The liquid gradually changes into a vapor but goes through an intermediate "fluid" region, which is neither true liquid nor vapor. This is quite unlike the behavior at lower temperatures and pressures, say, at 100°C and 1 atm, where the temperature remains constant while a definite phase change from liquid to vapor occurs. The unusual behavior at 300°C and 225 atm occurs because the conditions are beyond the critical point for water. The **critical temperature** can be defined as the temperature above which the vapor cannot be liquefied no matter what pressure is applied. The **critical pressure** is the pressure required to produce liquefaction *at* the critical temperature. Together, the critical temperature and the critical pressure define the **critical point.** For water, the critical point is 374°C and 218 atm. Note that the liquid/vapor line on the phase diagram for water ends at the critical point. Beyond this point the transition from one state to another involves the intermediate "fluid" region just described.

Applications of the Phase Diagram for Water

There are several additional interesting features of the phase diagram for water. Note that the solid/liquid boundary line has a negative slope. This means that the melting point of water *decreases* as the external pressure *increases.* This behavior, which is opposite to that observed for most substances, occurs because the density of ice is *less* than that of liquid water at the melting point. (The maximum density of water occurs at 4°C.) Thus, when liquid water freezes, its volume increases. Also note that the solid/liquid line has no upper limit. It extends indefinitely, as indicated by the arrow in Figure 16.55.

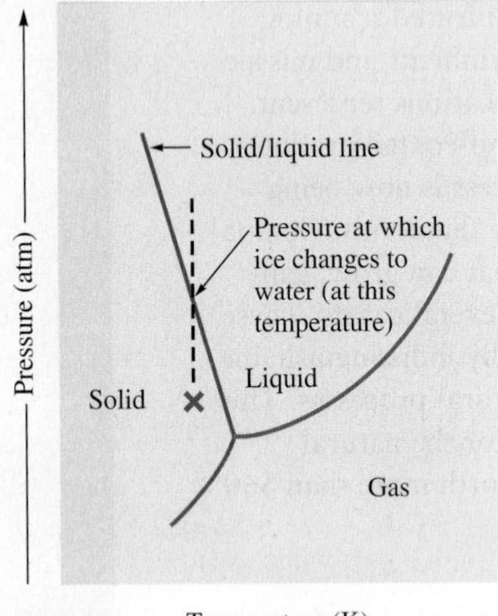

Figure 16.57
The phase diagram for water. At point × on the phase diagram, water is a solid. However, as the external pressure is increased while the temperature remains constant (indicated by the vertical dashed line), the solid/liquid line is crossed and the ice melts.

We can account for the effect of pressure on the melting point of water by using the following reasoning. At the melting point, liquid and solid water coexist—they are in dynamic equilibrium. What happens if we apply pressure to this system? When subjected to increased pressure, matter reduces its volume. This behavior is most dramatic for gases but it also occurs for condensed states. Since a given mass of ice at 0°C has a larger volume than the same mass of liquid water, the system can reduce its volume in response to the increased pressure by changing to liquid. Thus, at 0°C and an external pressure greater than 1 atm, water is liquid. In other words, the freezing point of water is less than 0°C when the pressure is greater than 1 atm.

Figure 16.57 illustrates the effect of pressure on ice. At the point × on the phase diagram, ice is subjected to increased pressure at constant temperature. Note that as the pressure is increased, the solid/liquid line is crossed, indicating that the ice melts. It has long been suggested that this is what happens in ice skating. The narrow blade of the skate exerts a large pressure, since the skater's weight is supported by the tiny area of the blade. If the pressure is high enough, this would cause the ice to melt, providing liquid water to lubricate the passing skate. However, there is now doubt that this is what happens. Recent experiments suggest that the pressure of an ice skate is not sufficient to melt the ice. In fact, new research on the structure of ice indicates that even at temperatures far below freezing the water molecules on the surface of ice vibrate faster than expected, forming a quasi-liquid with a structure intermediate between that of liquid and solid. This would make the surface of the ice slippery enough to aid in ice skating.

The motion of an ice skate blade over ice is a complex phenomenon, and other factors, such as frictional heating, may be important.

Ice's lower density has other implications. When water freezes in a pipe or an engine block, it expands and breaks the container. For this reason water pipes are insulated in cold climates and antifreeze is used in water-cooled engines. The lower density of ice also means that ice formed on rivers and lakes floats, providing a layer of insulation that helps prevent bodies of water from freezing solid in the winter. Aquatic life can therefore continue to live through periods of freezing temperatures.

A liquid boils at the temperature where the vapor pressure of the liquid equals the external pressure. Thus the boiling point of a substance, like the melt-

Table 16.9

Boiling Point of Water at Various Locations

Location	Feet Above Sea Level	P_{atm} (torr)	Boiling Point (°C)
Top of Mt. Everest, Tibet	29,028	240	70
Top of Mt. McKinley, Alaska	20,320	340	79
Top of Mt. Whitney, Calif.	14,494	430	85
Leadville, Colo.	10,150	510	89
Top of Mt. Washington, N.H.	6,293	590	93
Boulder, Colo.	5,430	610	94
Madison, Wis.	900	730	99
New York City, N.Y.	10	760	100
Death Valley, Calif.	−282	770	100.3

ing point, depends on the external pressure. This is why water boils at different temperatures at different elevations (Table 16.9), and any cooking carried out in boiling water will be affected by this variation. For example, it takes longer to hard-boil an egg in Leadville, Colorado (elevation 10,150 ft), than in San Diego (sea level), since water boils at a lower temperature, 89°C, in Leadville.

As we mentioned earlier, the phase diagram for water describes a closed system. Therefore, we must be very cautious in using the phase diagram to explain the behavior of water in a natural setting, such as on the earth's surface. For example, in dry climates (low humidity), snow and ice seem to sublime—a minimum amount of slush is produced. Wet clothes put on an outside line at temperatures below 0°C freeze and then dry while frozen. However, the phase diagram (Fig. 16.55) shows that ice should *not* be able to sublime at normal atmospheric pressures. What is happening in these cases? Ice in the natural environment is not in a closed system. The pressure is provided by the atmosphere rather than by a solid piston. This means that the vapor produced over the ice can escape from the immediate region as soon as it is formed. The vapor does not come to equilibrium with the solid, and the ice slowly disappears. Sublimation, which seems forbidden by the phase diagram, does seem to occur under these conditions, but it is not sublimation under equilibrium conditions.

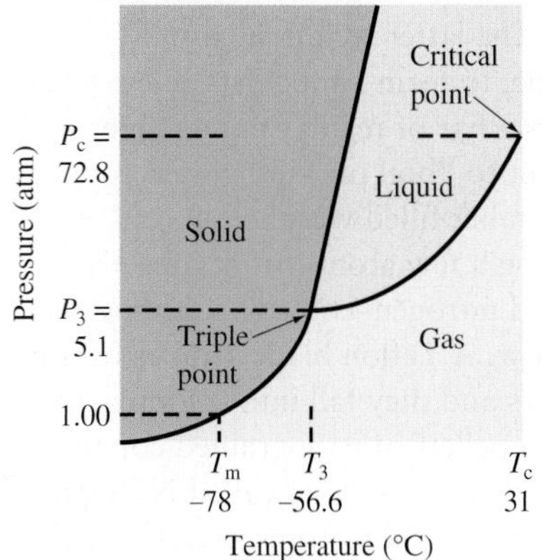

Figure 16.58
The phase diagram for carbon dioxide. The liquid state does not exist at a pressure of 1 atm. The solid/liquid line has a positive slope, since the density of solid carbon dioxide is greater than that of liquid carbon dioxide.

The Phase Diagram for Carbon Dioxide

The phase diagram for carbon dioxide (Fig. 16.58) differs significantly from that for water. The solid/liquid line has a positive slope, since solid carbon dioxide is more dense than liquid carbon dioxide. The triple point for carbon dioxide occurs at 5.1 atm and −56.6°C, and the critical point occurs at 72.8 atm and 31°C. At a pressure of 1 atm, solid carbon dioxide sublimes at −78°C, a property that leads to its common name, *dry ice*. No liquid phase occurs under normal atmospheric conditions, making dry ice a convenient coolant.

Carbon dioxide is often used in fire extinguishers, where it exists as a liquid at 25°C under high pressures. Liquid carbon dioxide released from the extinguisher into the environment at 1 atm immediately changes to a vapor. Being heavier than air, this vapor smothers the fire by keeping oxygen away from the flame. The liquid/vapor transition is highly endothermic, so cooling also results, which helps to put out the fire. The "fog" produced by a carbon dioxide extinguisher is a mixture of solid carbon dioxide and moisture frozen from the air.

Chemical Insights

Smaller Can Be Better

Because of the emphasis on crystalline materials in introductory chemistry courses, there is a tendency for students to think that most materials are single crystals—that is, that the regular array of the crystal is found without interruption throughout the entire material. But this is not true. There are some materials that are essentially single crystals—the silicon used for integrated circuits and the diamond in an engagement ring are two examples. However, most materials are polycrystalline. They contain multitudes of tiny crystalline fragments called *grains*. For example, if you examine ordinary copper metal under a 100-power microscope, you will see a patchwork quilt of grains, with no two grains the same size or shape. Each of these grains is a single crystal in which the copper atoms are arranged in a regular array. The metal is composed of grains rather than being a large single crystal because when a typical solid forms there are many nucleation sites—many single crystals start growing at about the same time. However, these microcrystals are not oriented identically in space, so when they grow large enough to contact each other, they form grain boundaries (see the figure).

Thus the solid is polycrystalline—it is composed of many grains (single crystals) of various sizes oriented in all different directions. The properties of a piece of copper, such as its hardness, color, electrical conductivity, and so on, depend on both the properties of the individual grains and the relationships and interactions among the grains. Copper is such a good electrical conductor, for example, because electrons can move easily within each grain and can jump readily from one grain to another.

Because the behavior of polycrystalline materials is so sensitive to the way the grains are packed together, the properties of these materials can be dramatically changed by altering the sizes and shapes of the grains. One person who is studying these effects is Dick Siegel of Rensselaer Polytechnic Institute in Troy, New York. Siegel has found that he can make copper that is five times harder than normal copper by lowering the grain size by a factor of 10,000—a change in size proportional to going from a boulder the size of a car to a grain of sand. Siegel has also been able to change the properties of ceramics from their characteristic brittle nature to being flexible and deformable by reducing the grain sizes. In addition, some ceramic materials that are opaque with normal grain sizes become transparent when the grain sizes are reduced.

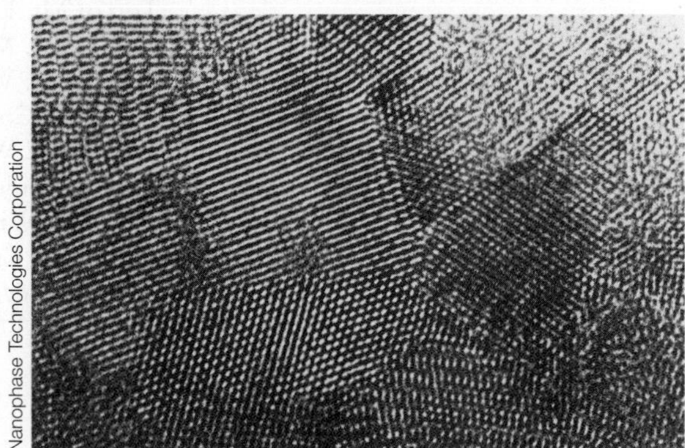

Nanophase Technologies Corporation

Grains of nanophase palladium magnified 200,000 times by an electron microscope.

Because the grain sizes in these substances are on the order of 10 to 100 nm (10^{-9} m), they are usually called "nanocrystalline" materials. The formation of nanocrystalline materials can be accomplished by mechanical grinding of the normal materials in high-speed ball mills for many hours or by controlling the growth of the grains as the solid forms. The latter technique usually works best. For example, to form nanocrystalline copper, Dick Siegel heats a bar of regular pure copper with an electric current to "boil off" individual copper atoms into a chamber filled with helium gas. As clusters containing a few atoms form, they are carried to a liquid nitrogen–cooled "cold finger" by the helium flow. A Teflon blade scrapes off the nanosized clusters and they fall into a compactor to be shaped into pellets of nanograined copper.

Siegel has formed a company called Nanophase Technologies Corporation to commercialize his work. The opportunities look bright for nanocrystalline materials. Sometimes smaller is definitely better.

16.12 Nanotechnology

A nanometer is so small that it is difficult to imagine. For example, the period at the end of this sentence has a diameter of approximately 500,000 nm. As another example, a man's beard typically grows about a nanometer in the time it takes to raise a razor to his face.

The field of nanotechnology is typically defined as dealing with particles approximately 1 to 100 nm in size. Because of the tiny size of nanoscale particles, they behave quite differently from normal-sized particles, even though the two types of particles may contain the same types of atoms. An important reason for these different behaviors is the incredible surface area present in a sample of nanosized particles. For example, about one gram of nanobeads used in a device to filter water contains an amazing 1000 square meters of surface area. This extremely high surface area of nanoparticles is very useful for catalyzing the decomposition of soil contaminants. For instance, several studies have shown that nanosized metallic particles are very efficient catalysts for the breakdown of the chlorinated solvents often found in contaminated soil.*

Another reason for the unusual properties of nanoparticles is that their tiny size allows them to penetrate into very small spaces. For example, one application of this characteristic developed at Rice University involves using gold nanoparticles ($\sim$120 nm in diameter) to penetrate the tangle of capillaries supplying blood to a cancerous tumor. The gold particles (called nanoshells) are then heated with infrared radiation, which kills the surrounding cancer cells.

The field of nanotechnology is virtually exploding today. The number of scientists involved in the field and the amount of money spent on nanoscience are increasing at an incredible rate. Although it is a multidisciplinary field, nanotechnology has its basis in chemistry, and many chemists are now working in this area.

Carbon Nanotubes

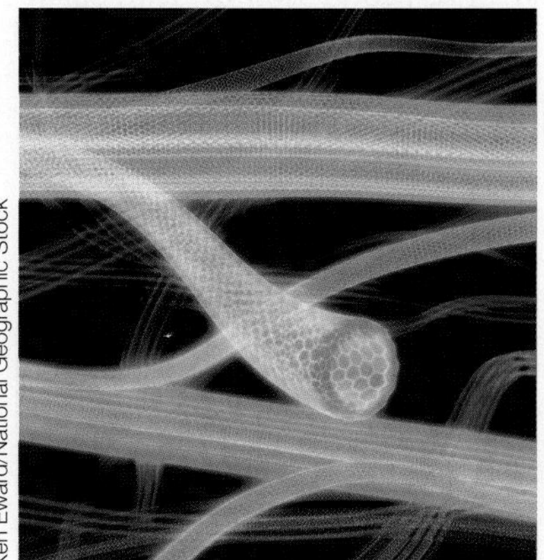

Ken Eward/National Geographic Stock

Figure 16.59
Forest of nanotubes.

The roots of the science of nanotechnology can be traced back to the discovery in 1985 of C_{60} molecules (buckminsterfullerene) by Richard Smalley and his coworkers at Rice University. Although "bucky balls" have not found many practical uses, a close relative, the carbon nanotube, has proved to have great practical potential. Carbon nanotubes contain networks of carbon atoms in interconnected six-membered rings (like "chicken wire") organized into slender tubes (Fig. 16.59). The main attraction of nanotubes is that they are over 50 times stronger than steel wire at one-sixth the weight and can carry a thousand times more electrical current than copper wire. The problem with carbon nanotubes is that when they are formed, they occur as a hodgepodge of tubes of different diameters. Sorting the tubes by size has proved to be a difficult task. However, a new technique pioneered by Michael S. Arnold and his coworkers at Northwestern University appears very promising. In this method, surfactants are added to the jumble of raw nanotubes, which makes the nanotubes water soluble. The solution of nanotubes is then added to a solution whose density increases with depth, and the mixture is centrifuged at high speeds. This treatment causes the tubes to sort themselves into bands (Fig. 16.60). So far only small amounts of sorted nanotubes have been produced, but the method appears capable of being scaled up.

Carbon nanotubes are now used in a variety of consumer products. They strengthen high-end sporting equipment such as tennis rackets, bicycle handle-

*"Special Treatment," *Sci. News* **167** (2005): 266.

Chemical Insights — Nanogenerators: Power from Motion

Many groups of researchers are working on ways to convert motions involved in everyday life into energy to power electronic devices. This new technology is based on the properties of piezoelectric nanowires and nanocrystals. Piezoelectric materials produce an electric current when they are stressed in some way, usually by bending.

One strategy being pursued is the inclusion of piezoelectric fibers into clothes to convert motions such as walking into electric currents that could power MP3 players or cell phones. One researcher working on this problem is Liwei Lin, a mechanical engineer at the University of California at Berkeley. Lin's goal is to weave long, thin fibers of piezoelectric polyvinylidene fluoride (PVDF) into clothes. Lin calculates that if a million PVDF fibers are included in a single garment, enough energy could be harvested to power an iPod or to keep a smart phone charged.

Other researchers hope to use piezoelectric nanomaterials inside the body to power sensors or even pacemakers. A group led by Zhong Lin Wang, Director of the Center for Nanostructure Characterization at the Georgia Institute of Technology, uses piezoelectric zinc oxide nanowires on a gold-coated surface, which are then attached to a platinum-coated silicon wafer. By placing these nanowires in the artery of a rat, an alternating current can be produced as the heart beats, creating a fluctuating flow of blood that moves the piezoelectric wires. Wang has also been able to boost the power of this device by connecting it to an enzyme-based "biofuel" cell that collects energy from the oxidation of glucose in the blood.

Although current power levels are much smaller than those needed to power an iPod, much work is being done to ramp up the efficiency of these nanopiezoelectric devices.

bars and frames, and golf club shafts. One of the major potential uses for nanotubes is in various electrical devices. Although carbon nanotubes can conduct prodigious amounts of current with very little production of heat, scientists have not yet figured out how to make them into practical devices. As of now, the longest electricity-conducting nanotube in existence measures only a fraction of a centimeter. Research is proceeding to find ways to make practical nanotubes.

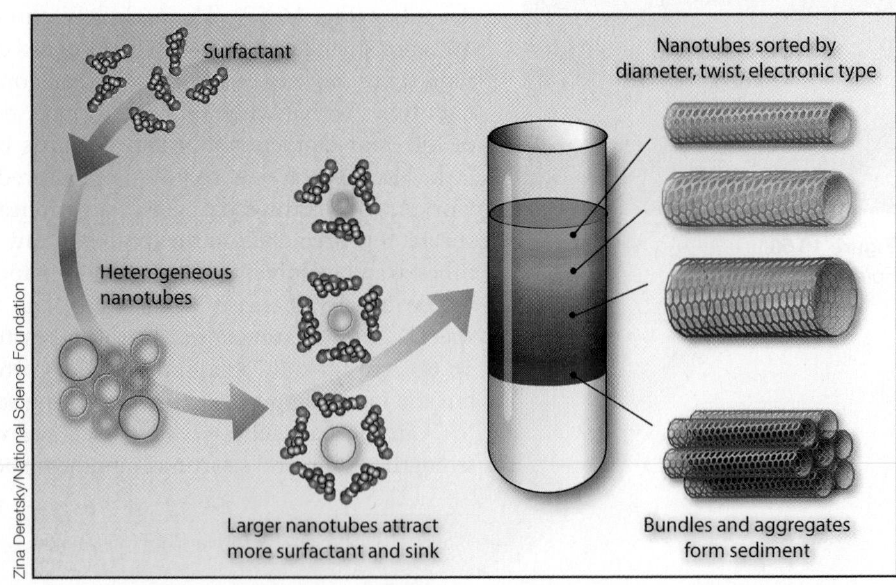

Figure 16.60
Graphic illustrating the process of sorting nanotubes.

Medical Applications of Nanotechnology

Nanoscale species are ideal for biomedical operations because they are comparable in size to many biological entities. A biomedical species smaller than about 50 nm can easily enter most cells (10,000–20,000 nm in diameter), and those smaller than 20 nm can exit blood vessels. As a result, nanoscale biomedical devices can easily interact with biomolecules inside and outside cells.

One rapidly expanding use of nanotechnology involves the transportation via nanoparticles of molecules that improve the spectroscopic detection of disease-causing cells such as cancer cells. Examples include nanosized "quantum dots" composed of cadmium selenide, zinc sulfide, and an organic polymer that can be attached to molecules. These labeled molecules then can be tracked by fluorescence techniques. Another example involves the attachment of magnetic resonance imaging (MRI) contrast agents to tiny cancer cells, making tumor cells as small as 2 mm detectable by MRI.

A further medical use of nanoscale devices is for targeting cancer cells. In this application the "nanocapsule" contains molecules lethal to tumors. The nanodevice is linked to a molecule that has a specific affinity for the receptors on the surfaces of cancer cells. Once at the site of the cancer, the nanocapsule dumps its lethal contents into the cancer cell. One method for releasing the contents of the nanocapsule is through a tiny valve developed at UCLA by Jeffrey I. Zink and his coworkers. The valves consist of dumbbell-like molecules that can be rotated by the energy from a single electron.

Nanotoxicology

The explosive growth of nanotechnology has brought with it some safety concerns about these materials.* The unique properties of nanoparticles raise the possibility of unique types of toxicity. One of the problems with studying the toxicity of nanoparticles is their almost infinite variety. For example, dozens of types of samples of carbon nanotubes are available containing different sized particles and different contaminating substances depending on how they were made. Researchers are now trying to decide how to design a systematic approach to the very complex problem of testing the possible toxicity of the great variety of nanoparticles. Some early results indicate that caution is in order as we continue to develop this technology.

*"Particular Problems," *Sci. News* **169** (2006): 280.

Key Terms

Section 16.1
condensed states of matter
intermolecular forces
dipole–dipole attraction
hydrogen bonding
London dispersion forces

Section 16.2
surface tension
capillary action
viscosity

Section 16.3
crystalline solid
amorphous solid

For Review

Sign in at **www.cengage.com/owl** to:

- View tutorials and simulations, develop problem-solving skills, and complete online homework assigned by your professor.
- Download Go Chemistry mini lecture modules for quick review and exam prep from OWL (or purchase them at **www.cengagebrain.com**)

Condensed states of matter: liquids and solids

- Held together by forces among the component molecules, atoms, or ions
- Liquids exhibit properties such as surface tension, capillary action, and viscosity that depend on the forces among the components

lattice
unit cell
X-ray diffraction
Bragg equation
ionic solid
molecular solid
atomic solid

Section 16.4

closest packing
hexagonal closest packed (hcp) structure
cubic closest packed (ccp) structure
body-centered cubic (bcc) unit cell
electron sea model
band model
bands
conduction band
alloy
substitutional alloy
interstitial alloy

Section 16.5

network solid
silica
silicate
glass
ceramic
aluminosilicate
semiconductor
n-type semiconductor
p-type semiconductor
p–n junction

Section 16.10

vaporization
heat of vaporization
enthalpy of vaporization (ΔH_{vap})
condensation
vapor pressure
sublimation
heating curve
melting point
heat of fusion
enthalpy of fusion (ΔH_{fus})
normal melting point
normal boiling point
supercooled
superheated

Section 16.11

phase diagram
triple point
critical temperature
critical pressure
critical point

Dipole–dipole forces

- Attractions among molecules with dipole moments
- Hydrogen bonding is a particularly strong form of dipole–dipole attraction
  - Occurs in molecules containing hydrogen bonded to a highly electronegative element such as nitrogen, oxygen, or fluorine
  - Produces unusually high boiling points

London dispersion forces

- Caused by instantaneous dipoles that form in atoms or nonpolar molecules

Crystalline solids

- Have a regular arrangement of components often represented as a lattice; the smallest repeating unit of the lattice is called the unit cell
- Classified by the types of components:
  - Atomic solids (atoms)
  - Ionic solids (ions)
  - Molecular solids (molecules)
- Arrangement of the components can be determined by X-ray analysis

Metals

- Structure is modeled by assuming atoms to be uniform spheres
  - Closest packing
    - Hexagonal
    - Cubic
- Metallic bonding can be described in terms of two models
  - Electron sea model: valence electrons circulate freely among the metal cations
  - Band model: electrons are assumed to occupy molecular orbitals
    - Conduction bands: closely spaced molecular orbitals with empty electron spaces
- Alloys: mixtures with metallic properties
  - Substitutional
  - Interstitial

Network solids

- Contain giant networks of atoms covalently bound together
- Examples are diamond and graphite
- Silicates are network solids containing Si—O—Si bridges that form the basis for many rocks, clays, and ceramics

Semiconductors

- Very pure silicon is "doped" with other elements
  - n-type: doping atoms typically contain five valence electrons (one more than silicon)
  - p-type: doping elements typically contain three valence electrons
- Modern electronics are based on devices with p–n junctions

Molecular solids

- Components are discrete molecules
- Intermolecular forces are typically weak, leading to relatively low boiling and melting points

Ionic solids

- Components are ions
- Interionic forces are relatively strong, leading to solids with high melting and boiling points

- Many structures consist of closest packing of the larger ions with the smaller ions in tetrahedral or octahedral holes

Phase changes

- The change from liquid to gas (vapor) is called vaporization or evaporation
- Condensation is the reverse of vaporization
- Equilibrium vapor pressure: the pressure that occurs over a liquid or solid in a closed system when the rate of evaporation equals the rate of condensation
  - Liquids whose components have high intermolecular forces have relatively low vapor pressures
  - Normal boiling point: the temperature at which the vapor pressure of a liquid equals one atmosphere
  - Normal melting point: the temperature at which a solid and its liquid have the same vapor pressure (at 1 atm external pressure)
- Phase diagram
  - Shows what state exists at a given temperature and pressure in a closed system
  - Triple point: temperature at which all three phases exist simultaneously
  - Critical point: defined by the critical temperature and pressure
    - Critical temperature: the temperature above which the vapor cannot be liquefied no matter the applied pressure
    - Critical pressure: the pressure required to produce liquefaction at the critical temperature

Discussion Questions

These questions are designed to be considered by groups of students in class. Often these questions work well for introducing a particular topic in class.

1. It is possible to balance a paper clip on the surface of water in a beaker. If you add a bit of soap to the water, however, the paper clip will sink. Explain how the paper clip can float and why it sinks when soap is added.
2. Consider a sealed container half-filled with water. Which statement best describes what occurs in the container?
   a. Water evaporates until the air is saturated with water vapor; at this point, no more water evaporates.
   b. Water evaporates until the air is overly saturated (supersaturated) with water, and most of this water recondenses; this cycle continues until a certain amount of water vapor is present, and then the cycle ceases.
   c. The water does not evaporate, because the container is sealed.
   d. Water evaporates, and then water evaporates and recondenses simultaneously and continuously.
   e. Water evaporates until it is eventually all in vapor form.

   Explain each choice, and justify the best choice. For those you did not choose, explain why they are incorrect.
3. Explain the following: You add 100 mL of water to a 500-mL round-bottom flask and heat the water until it is boiling. You remove the heat, stopper the flask, and the boiling stops. You then run cool water over the neck of the flask, and the boiling begins again. It seems as though you are boiling water by cooling it.
4. Is it possible for the dispersion forces in a particular substance to be stronger than hydrogen bonding forces in another substance? Explain your answer.
5. Does the nature of intermolecular forces change when a substance goes from a solid to a liquid or from a liquid to a gas? What causes a substance to undergo a phase change?
6. Generally, the vapor pressure of a liquid is related to (there may be more than one answer)
   a. amount of the liquid
   b. atmospheric pressure
   c. temperature
   d. intermolecular forces

   Explain.
7. What is meant by the term *vapor pressure*? Why do liquids have a vapor pressure? Do all liquids have a vapor pressure? What about solids? How does vapor pressure change with changing temperature? Why? What happens, if anything, when the vapor pressure of a liquid is equal to atmospheric pressure?

8. Consider an open beaker of water. Over time, the water evaporates. As the water is evaporating, is the vapor pressure increasing, decreasing, or staying the same? Why? What is the vapor pressure of water at 100°C? How do you know?

9. How do the following physical properties depend on the strength of intermolecular forces? Explain.
   a. surface tension
   b. viscosity
   c. melting point
   d. boiling point
   e. vapor pressure

10. Use the kinetic molecular theory to explain why a liquid in an insulated vessel gets cooler as it evaporates.

Exercises

OWL Interactive versions of these problems may be assigned in OWL.

A blue exercise number indicates that the answer to that exercise appears at the back of this book and a solution appears in the *Solutions Guide.*

Intermolecular Forces and Physical Properties

11. What are intermolecular forces? How do they differ from intramolecular forces? What are dipole–dipole forces? How do typical dipole–dipole forces differ from hydrogen-bonding interactions? In what ways are they similar? What are London dispersion forces? How do typical London dispersion forces differ from dipole–dipole forces? In what ways are they similar?

12. List the major types of intermolecular forces in order of increasing strength. Is there some overlap? That is, can the strongest London dispersion forces be greater than some dipole–dipole forces? Give examples of such instances.

13. Why is ΔH_{vap} for water much greater than ΔH_{fus}? What does this reveal concerning changes in intermolecular forces in going from solid to liquid to vapor?

14. Rationalize why chalk (calcium carbonate) has a higher melting point than motor oil (large compounds made from carbon and hydrogen), which has a higher melting point than water, which engages in relatively strong hydrogen-bonding interactions.

15. Identify the most important types of interparticle forces present in the solids of each of the following substances.
   a. Ar
   b. HCl
   c. HF
   d. $CaCl_2$
   e. CH_4
   f. CO
   g. $NaNO_3$
   h. NH_4Cl
   i. Teflon, $CF_3(CF_2CF_2)_nCF_3$
   j. polyethylene, $CH_3(CH_2CH_2)_nCH_3$
   k. $CHCl_3$
   l. NH_3
   m. NO
   n. BF_3

16. Predict which substance in each of the following pairs would have the greater intermolecular forces.
   a. CO_2 or OCS
   b. SeO_2 or SO_2
   c. $CH_3CH_2CH_2NH_2$ or $H_2NCH_2CH_2NH_2$
   d. CH_3CH_3 or H_2CO
   e. CH_3OH or H_2CO

17. In each of the following groups of substances, pick the one that has the given property. Justify your answer.
   a. highest boiling point: HBr, Kr, or Cl_2
   b. highest freezing point: H_2O, NaCl, or HF
   c. lowest vapor pressure at 25°C: Cl_2, Br_2, or I_2
   d. lowest freezing point: N_2, CO, or CO_2
   e. lowest boiling point: CH_4, CH_3CH_3, or $CH_3CH_2CH_3$
   f. highest boiling point: HF, HCl, or HBr
   g. lowest vapor pressure at 25°C: $CH_3CH_2CH_3$, $CH_3\overset{\overset{O}{\|}}{C}CH_3$, or $CH_3CH_2CH_2OH$

18. In each of the following groups of substances, pick the one that has the given property. Justify each answer.
   a. highest boiling point: CCl_4, CF_4, CBr_4
   b. lowest freezing point: LiF, F_2, HCl
   c. lowest vapor pressure at 25°C: CH_3OCH_3, CH_3CH_2OH, $CH_3CH_2CH_3$
   d. greatest viscosity: H_2S, HF, or H_2O_2
   e. greatest heat of vaporization: H_2CO, CH_3CH_3, CH_4
   f. smallest enthalpy of fusion: I_2, CsBr, or CaO

19. Rationalize the difference in boiling points for each of the following pairs of substances:
   a. *n*-pentane $CH_3CH_2CH_2CH_2CH_3$ 36.2°C
      neopentane $H_3C-\underset{\underset{CH_3}{|}}{\overset{\overset{CH_3}{|}}{C}}-CH_3$ 9.5°C
   b. HF 20°C
      HCl −85°C
   c. HCl −85°C
      LiCl 1360°C
   d. *n*-pentane $CH_3CH_2CH_2CH_2CH_3$ 36.2°C
      n-hexane $CH_3CH_2CH_2CH_2CH_2CH_3$ 69°C

20. Rationalize the following differences in physical properties in terms of intermolecular forces. Compare the first three substances with each other, compare the last three with each other, and then compare all six. Can you account for any anomalies?

Substance	bp (°C)	mp (°C)	ΔH_{vap} (kJ/mol)
Benzene, C_6H_6	80	6	33.9
Naphthalene, $C_{10}H_8$	218	80	51.5
Carbon tetrachloride	76	−23	31.8
Acetone, CH_3COCH_3	56	−95	31.8
Acetic acid, CH_3CO_2H	118	17	39.7
Benzoic acid, $C_6H_5CO_2H$	249	122	68.2

21. Consider the compounds Cl_2, HCl, F_2, NaF, and HF. Which compound has a boiling point closest to that of argon? Explain.

22. Consider the following melting point data:

Compound	NaCl	$MgCl_2$	$AlCl_3$	$SiCl_4$	PCl_3	SCl_2	Cl_2
mp (°C)	801	708	190	−70	−91	−78	−101
Compound	NaF	MgF_2	AlF_3	SiF_4	PF_5	SF_6	F_2
mp (°C)	997	1396	1040	−90	−94	−56	−220

Account for the trends in melting points for the two series of compounds in terms of interparticle forces.

23. Consider the following electrostatic potential diagrams:

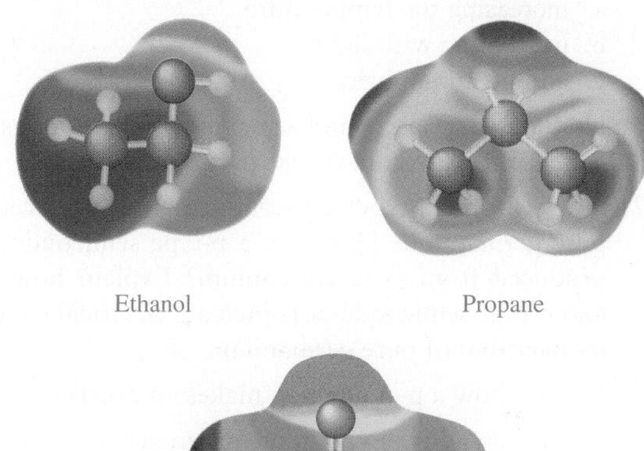
Ethanol Propane

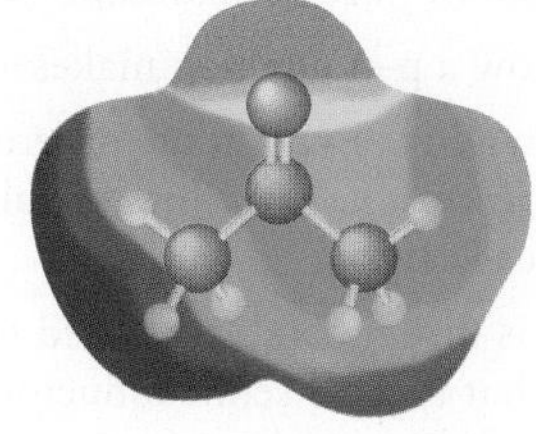
Acetone

Rank the compounds from lowest to highest boiling point, and explain your answer.

24. Consider the following enthalpy changes:

$$F^-(g) + HF(g) \longrightarrow FHF^-(g) \qquad \Delta H = -155 \text{ kJ/mol}$$

$$(CH_3)_2C{=}O(g) + HF(g) \longrightarrow (CH_3)_2C{=}O\text{---}HF(g) \qquad \Delta H = -46 \text{ kJ/mol}$$

$$H_2O(g) + HOH(g) \longrightarrow H_2O\text{---}HOH \ \text{(in ice)} \qquad \Delta H = -21 \text{ kJ/mol}$$

How do the strengths of hydrogen bonds vary with the electronegativity of the element to which hydrogen is bonded? Where in the preceding series would you expect hydrogen bonds of the following type to fall?

$$\text{—N---H—O—} \quad \text{and} \quad \text{—N---H—N}$$

25. Using the heats of fusion and vaporization for water, calculate the change in enthalpy for the sublimation of water:

$$H_2O(s) \longrightarrow H_2O(g)$$

Using the ΔH value given in Exercise 24 and the number of hydrogen bonds formed to each water molecule, estimate what portion of the intermolecular forces in ice can be accounted for by hydrogen bonding.

Properties of Liquids

26. List some physical characteristics that distinguish solids from liquids from gases.

27. Define the following terms, and describe how each depends on the strength of the intermolecular forces.
    a. surface tension
    b. viscosity
    c. melting point
    d. boiling point
    e. vapor pressure

28. The shape of the meniscus of water in a glass tube is different from that of mercury in a glass tube. Why?

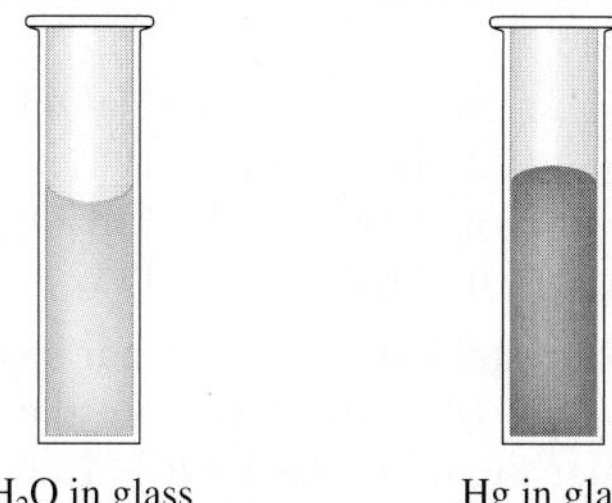
H_2O in glass Hg in glass

29. Explain why water forms into beads on a waxed car finish.

30. Some of the physical properties of H_2O and D_2O are as follows:

Property	H_2O	D_2O
Density at 20°C (g/mL)	0.997	1.108
Boiling point (°C)	100.00	101.41
Melting point (°C)	0.00	3.79
ΔH°_{vap} (kJ/mol)	40.7	41.61
ΔH°_{fus} (kJ/mol)	6.01	6.3

Account for the differences. (*Note:* D is a symbol often used for ^2H, the deuterium isotope of hydrogen.)

31. Hydrogen peroxide (H_2O_2) is a syrupy liquid with a relatively low vapor pressure and a normal boiling point of 152.2°C. Rationalize the differences between these physical properties and those of water.

32. Carbon diselenide (CSe_2) is a liquid at room temperature. The normal boiling point is 125°C, and the melting point is −45.5°C. Carbon disulfide (CS_2) is also a liquid at room temperature, with normal boiling and melting points of 46.5°C and −111.6°C, respectively. How do the strengths of the intermolecular forces vary from CO_2 to CS_2 to CSe_2? Explain your answer.

Structures and Properties of Solids

33. Distinguish between the solids in the following pairs.
    a. crystalline solid and amorphous solid
    b. ionic solid and molecular solid
    c. molecular solid and network solid
    d. metallic solid and network solid

34. What is closest packing? What is the difference between hexagonal closest packing and cubic closest packing? What is the unit cell for each closest packing arrangement?

35. Compare and contrast the structures of the following solids.
   a. diamond and graphite
   b. silica, silicates, and glass
36. Compare and contrast the structures of the following solids.
   a. $CO_2(s)$ and $H_2O(s)$
   b. $NaCl(s)$ and $CsCl(s)$
37. Will a crystalline solid or an amorphous solid give a simpler X-ray diffraction pattern? Why?
38. What type of solid will each of the following substances form?

a. CO_2	e. Ru	i. NaOH	m. GaAs
b. SiO_2	f. I_2	j. U	n. BaO
c. Si	g. KBr	k. $CaCO_3$	o. NO
d. CH_4	h. H_2O	l. PH_3	p. GeO_2

39. A topaz crystal has an interplanar spacing (d) of 1.36 Å (1 Å = 1×10^{-10} m). Calculate the wavelength of the X ray that should be used if $\theta = 15.0°$ (assume $n = 1$).
40. X rays of wavelength 2.63 Å were used to analyze a crystal. The angle of first-order diffraction ($n = 1$ in the Bragg equation) was 15.55°. What is the spacing between crystal planes, and what would be the angle for second-order diffraction ($n = 2$)?
41. X rays from a copper X-ray tube (λ = 1.54 Å) were diffracted at an angle of 14.22° by a crystal of silicon. Assuming first-order diffraction ($n = 1$ in the Bragg equation), what is the interplanar spacing in silicon?
42. In what array (simple cubic, body-centered cubic, and face-centered cubic) do atoms pack most efficiently (greatest percent of space occupied by spheres)? Support your answer mathematically.
43. Calcium has a cubic closest packed structure as a solid. Assuming that calcium has an atomic radius of 197 pm, calculate the density of solid calcium.
44. A certain form of lead has a cubic closest packed structure with an edge length of 492 pm. Calculate the value of the atomic radius and the density of lead.
45. Nickel has a face-centered cubic unit cell. The density of nickel is 6.84 g/cm^3. Calculate a value for the atomic radius of nickel.
46. You are given a small bar of an unknown metal X. You find the density of the metal to be 10.5 g/cm^3. An X-ray diffraction experiment measures the edge of the face-centered cubic unit cell as 4.09 Å (1 Å = 10^{-10} m). Identify X.
47. Titanium metal has a body-centered cubic unit cell. The density of titanium is 4.50 g/cm^3. Calculate the edge length of the unit cell and a value for the atomic radius of titanium. (*Hint:* In a body-centered arrangement of spheres, the spheres touch along the body diagonal.)
48. Barium has a body-centered cubic structure. If the atomic radius of barium is 222 pm, calculate the density of solid barium.
49. The radius of gold is 144 pm and the density is 19.32 g/cm^3. Does elemental gold have a face-centered cubic structure or a body-centered cubic structure?
50. The radius of tungsten is 137 pm and the density is 19.3 g/cm^3. Does elemental tungsten have a face-centered cubic structure or a body-centered cubic structure?
51. Use the band model to describe differences among insulators, conductors, and semiconductors. Also use the band model to explain why each of the following increases the conductivity of a semiconductor.
   a. increasing the temperature
   b. irradiating with light
   c. adding an impurity

   How do conductors and semiconductors differ as to the effect of temperature on electrical conductivity?
52. How can an n-type semiconductor be produced from pure germanium? How can a p-type semiconductor be produced from pure germanium? Explain how n-type and p-type semiconductors increase electrical conductivity over that of pure germanium.
53. Explain how a p–n junction makes an excellent rectifier.
54. What is an alloy? Explain the differences in structure between substitutional and interstitial alloys. Give an example of each type.
55. Selenium is a semiconductor used in photocopying machines. What type of semiconductor would be formed if a small amount of indium impurity were added to pure selenium?
56. The Group 3A/Group 5A semiconductors are composed of equal amounts of atoms from Group 3A and Group 5A—for example, InP and GaAs. These types of semiconductors are used in light-emitting diodes and solid-state lasers. What would you add to make a p-type semiconductor from pure GaAs? How would you dope pure GaAs to make an n-type semiconductor?
57. The band gap in aluminum phosphide (AlP) is 2.5 electron-volts (1 eV = 1.6×10^{-19} J). What wavelength of light is emitted by an AlP diode?
58. An aluminum antimonide solid-state laser emits light with a wavelength of 730. nm. Calculate the band gap in joules.
59. The structures of some common crystalline substances follow. Show that the net composition of each unit cell corresponds to the correct formula of each substance.

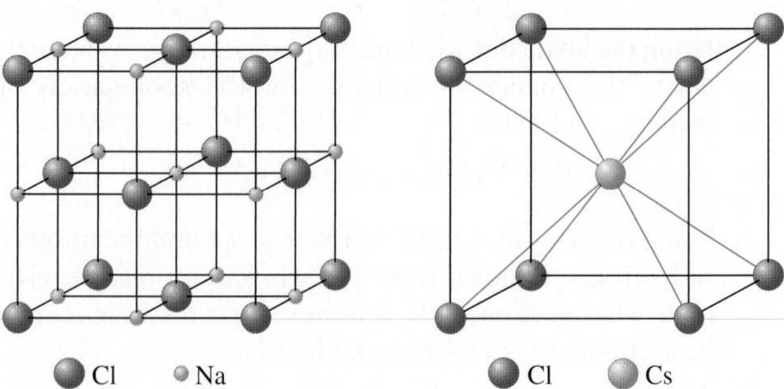

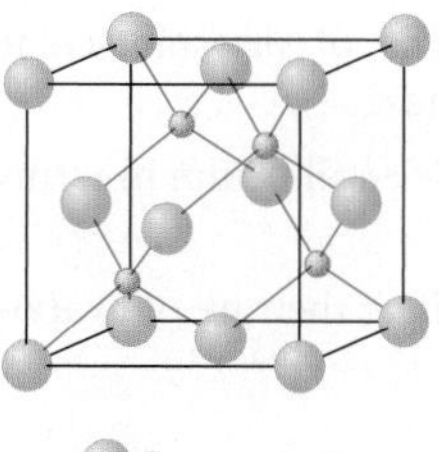

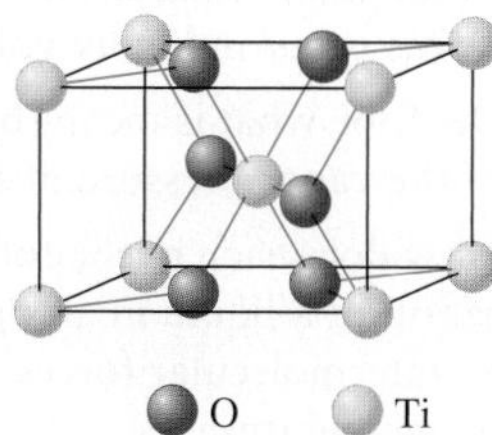

60. The unit cell for nickel arsenide is shown below. What is the formula of this compound?

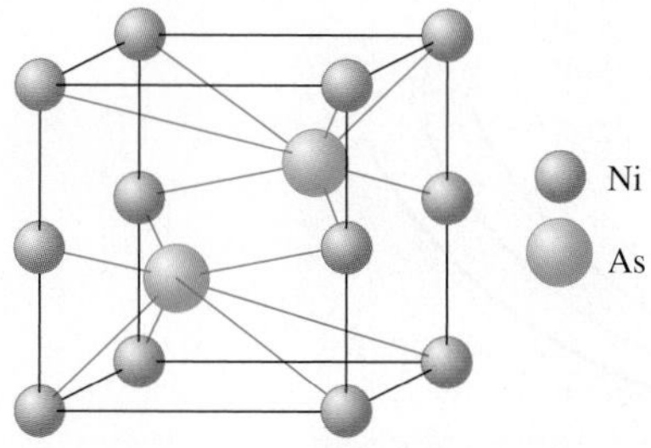

61. Describe, in general, the structures of ionic solids. Compare and contrast the structures of sodium chloride and zinc sulfide. How many tetrahedral holes and octahedral holes are there per closest packed anion? In zinc sulfide, why are only one-half of the tetrahedral holes filled with cations?

62. Assume the two-dimensional structure of an ionic compound M_xA_y is

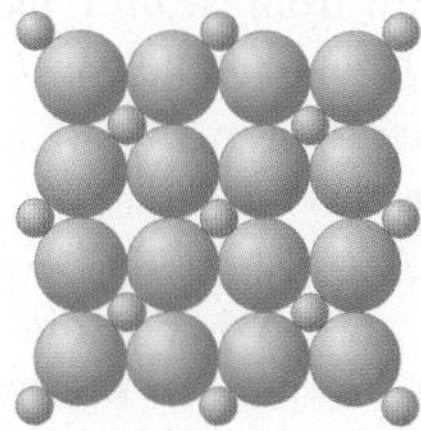

What is the empirical formula of this ionic compound?

63. Cobalt fluoride crystallizes in a closest packed array of fluoride ions, with the cobalt ions filling one-half of the octahedral holes. What is the formula of this compound?

64. The compounds Na_2O, CdS, and ZrI_4 all can be described as cubic closest packed anions with the cations in tetrahedral holes. What fraction of the tetrahedral holes is occupied for each case?

65. The structure of manganese fluoride can be described as a simple cubic array of manganese ions with fluoride ions at the center of each edge of the cubic unit cell. What is the charge of the manganese ions in this compound?

66. The unit cell of MgO is shown below.

Does MgO have a structure like that of NaCl or ZnS? If the density of MgO is 3.58 g/cm^3, estimate the radius (in centimeters) of the O^{2-} anions and the Mg^{2+} cations.

67. From the density of cesium chloride (3.97 g/cm^3), calculate the distance between the centers of adjacent Cs^+ and Cl^- ions in the solid. Compare this value with the expected distance based on the sizes of the ions (Fig. 13.8). The ionic radius of Cs^+ is 169 pm.

68. A given sample of wüstite has the formula $Fe_{0.950}O_{1.00}$. Calculate the fraction of iron ions present as Fe^{3+}. What fraction of the sites normally occupied by Fe^{2+} must be vacant in this solid?

69. Use the relative ionic radii in Fig. 13.8 to predict the structures expected for CsBr and KF. Do these predictions agree with observed structures? The ionic radius of Cs^+ is 169 pm.

70. MnO has either the NaCl type structure or the CsCl type structure. The edge length of the MnO unit cell is 4.47×10^{-8} cm, and the density of MnO is 5.28 g/cm^3.
    a. Does MnO crystallize in the NaCl type or the CsCl type structure?
    b. Assuming that the ionic radius of oxygen is 140. pm, estimate the ionic radius of manganese.
    c. Does the calculated cation-to-anion radius ratio for MnO substantiate your answer in part a? Explain.

71. Superalloys have been made of nickel and aluminum. The alloy owes its strength to the formation of an ordered phase, called the *gamma-prime phase,* in which Al atoms are at the corners of a cubic unit cell and Ni atoms are at the face centers. What is the composition (relative numbers of atoms) for this phase of the nickel–aluminum superalloy?

72. The memory metal, nitinol, is an alloy of nickel and titanium. It is called a memory metal because after being deformed, a piece of nitinol wire will return to its original shape. (See *Chem. Matters,* October 1993, pp. 4–7.) The structure of nitinol consists of a simple cubic array of Ni atoms and an inner penetrating simple cubic array of Ti atoms. In the extended lattice, a Ti atom is found at the center of a cube of Ni atoms; the reverse is also true.
    a. Describe the unit cell for nitinol.
    b. What is the empirical formula of nitinol?
    c. What are the coordination numbers (number of nearest neighbors) of Ni and Ti in nitinol?

73. A mineral crystallizes in a cubic closest packed array of oxygen ions with aluminum ions in some of the octahedral holes and magnesium ions in some of the tetrahedral holes. Deduce the formula of this mineral, and predict the fraction of octahedral holes and tetrahedral holes that are filled by the various cations.

74. Perovskite is a mineral containing calcium, titanium, and oxygen. The following diagrams represent the unit cell.

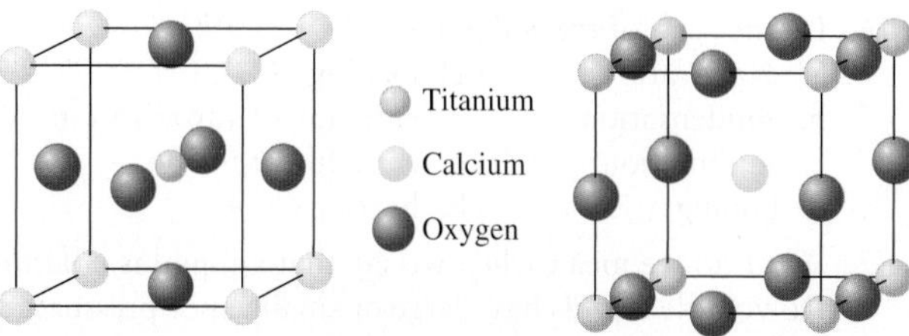

a. What is the formula of perovskite?

b. An alternative way of drawing the unit cell of perovskite has calcium at the center of each cubic unit cell. What are the positions of the titanium and oxygen atoms in this representation of the unit cell? Show that the formula for perovskite is the same for both unit cell representations.

c. How many oxygen atoms surround a given Ti atom in each representation of the unit cell?

75. Materials containing the elements Y, Ba, Cu, and O that are superconductors (electrical resistance equals zero) at temperatures above that of liquid nitrogen were recently discovered. The structures of these materials are based on the perovskite structure. Were they to have the ideal perovskite structure, the superconductors would have the structure shown in part (a) of the accompanying figure.

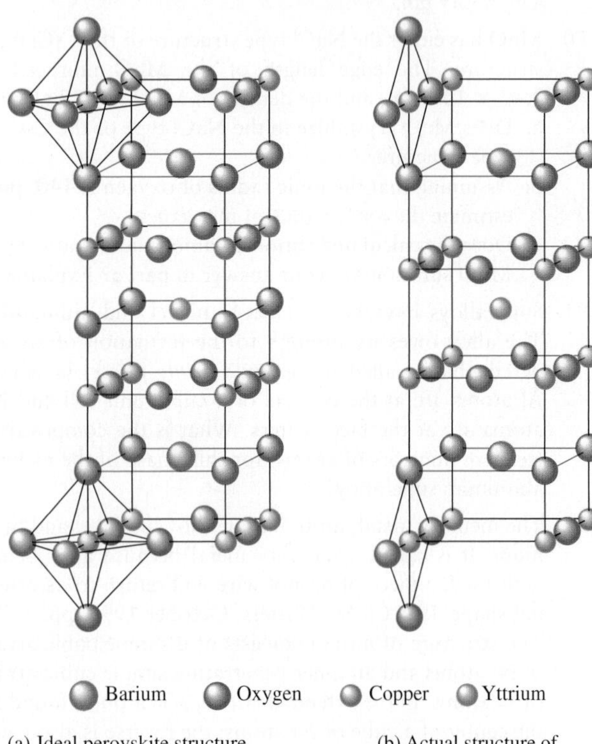

(a) Ideal perovskite structure

(b) Actual structure of superconductor

a. What is the formula of this ideal perovskite material?

b. How is this structure related to the perovskite structure discussed in Exercise 74?

These materials, however, do not act as superconductors unless they are deficient in oxygen. The structure of the actual superconducting phase appears to be that shown in part (b) of the figure.

c. What is the formula of this material?

Phase Changes and Phase Diagrams

76. Define each of the following.

a. evaporation	e. melting
b. condensation	f. enthalpy of vaporization
c. sublimation	g. enthalpy of fusion
d. boiling	h. heating curve

77. What do we mean when we say that a liquid is *volatile*? Do volatile liquids have large or small vapor pressures at room temperature? What strengths of intermolecular forces occur in highly volatile liquids?

78. Describe what is meant by dynamic equilibrium in terms of the vapor pressure of a liquid.

79. How does each of the following affect the rate of evaporation of a liquid in an open dish?

a. intermolecular forces

b. temperature

c. surface area

80. Consider the following vapor pressure versus temperature plot for three different substances A, B, and C.

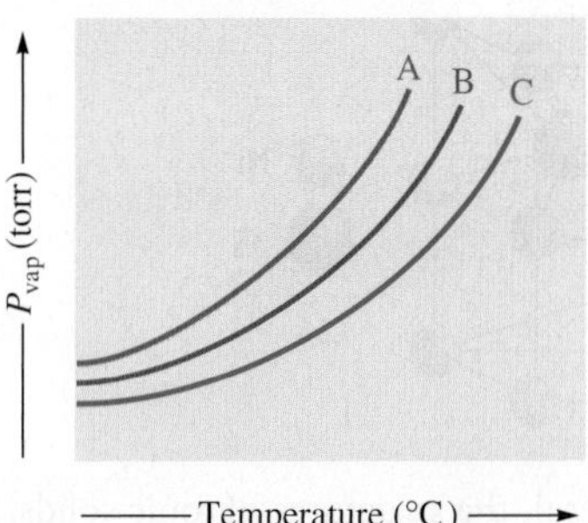

If the three substances are CH_4, SiH_4, and NH_3, match each curve to the correct substance.

81. When a person has a high fever, one therapy used to reduce the fever is an "alcohol rub." Explain how the evaporation of alcohol from a person's skin removes heat energy from the body.

82. Why is a burn from steam typically much more severe than a burn from boiling water?

83. A plot of ln (P_{vap}) versus $1/T$ (K) is linear with a negative slope. Why is this the case?

84. In Breckenridge, Colorado, the typical atmospheric pressure is 520. torr. What is the boiling point of water (ΔH_{vap} = 40.7 kJ/mol) in Breckenridge? What pressure would have to be applied to steam at 350.°C to condense the steam to liquid water?

85. Carbon tetrachloride (CCl_4) has a vapor pressure of 213 torr at 40.°C and 836 torr at 80.°C. What is the normal boiling point of CCl_4?

86. The normal boiling point for acetone is 56.5°C. At an elevation of 5300 ft, the atmospheric pressure is 630. torr. What would be the boiling point of acetone (ΔH_{vap} = 32.0 kJ/mol) at this elevation? What would be the vapor pressure of acetone at 25.0°C at this elevation?

87. Plot the following data, and from the graph determine ΔH_{vap} for magnesium and lithium. In which metal is the bonding stronger?

Vapor Pressure (mm Hg)	Temperature (°C)	
	Li	Mg
1.	750.	620.
10.	890.	740.
100.	1080.	900.
400.	1240.	1040.
760.	1310.	1110.

88. From the following data for liquid nitric acid, determine its heat of vaporization and normal boiling point.

Temperature (°C)	Vapor Pressure (mm Hg)
0	14.4
10.	26.6
20.	47.9
30.	81.3
40.	133
50.	208
80.	670.

89. A substance has the following properties:

			Specific Heat Capacities
ΔH_{vap}	20 kJ/mol	$C_{(s)}$	3.0 J g^{-1} °C^{-1}
ΔH_{fus}	5.0 kJ/mol	$C_{(l)}$	2.5 J g^{-1} °C^{-1}
bp	75°C	$C_{(g)}$	1.0 J g^{-1} °C^{-1}
mp	−15°C		

Sketch a heating curve for the substance, starting at −50°C.

90. Use the heating–cooling curve below to answer the following questions.

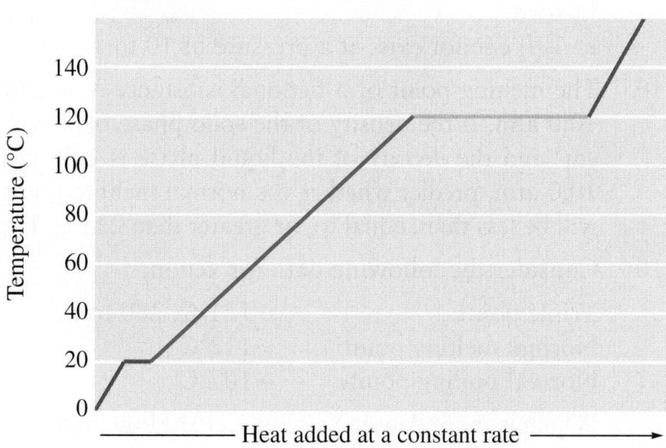

a. What is the freezing point of the liquid?
b. What is the boiling point of the liquid?
c. Which is greater, the heat of fusion or the heat of vaporization? Explain each term, and explain how the heating–cooling curve above helps you to answer the question.

91. How much energy does it take to convert 0.500 kg of ice at −20.°C to steam at 250.°C? Specific heat capacities: ice, 2.1 J g^{-1} °C^{-1}; liquid, 4.2 J g^{-1} °C^{-1}; steam, 2.0 J g^{-1} °C^{-1}; ΔH_{vap} = 40.7 kJ/mol; ΔH_{fus} = 6.01 kJ/mol.

92. Consider a 75.0-g sample of $H_2O(g)$ at 125°C. What phase or phases are present when 215 kJ of energy is removed from this sample? (See Exercise 91.)

93. A 0.250-g chunk of sodium metal is cautiously dropped into a mixture of 50.0 g of water and 50.0 g of ice, both at 0°C. The reaction is

$$2Na(s) + 2H_2O(l) \longrightarrow 2NaOH(aq) + H_2(g) \qquad \Delta H = -368 \text{ kJ}$$

Will the ice melt? Assuming the final mixture has a specific heat capacity of 4.18 J g^{-1} °C^{-1}, calculate the final temperature.

94. A 20.0-g sample of ice at −10.0°C is mixed with 100.0 g of water at 80.0°C. Calculate the final temperature of the mixture assuming no heat loss to the surroundings. The heat capacities of $H_2O(s)$ and $H_2O(l)$ are 2.08 J g^{-1} °C^{-1} and 4.18 J g^{-1} °C^{-1}, respectively, and the enthalpy of fusion for ice is 6.01 kJ/mol.

95. An ice cube tray contains enough water at 22.0°C to make 18 ice cubes, each of which has a mass of 30.0 g. The tray is placed in a freezer that uses CF_2Cl_2 as a refrigerant. The heat of vaporization of CF_2Cl_2 is 158 J/g. What mass of CF_2Cl_2 must be vaporized in the refrigeration cycle to convert all the water at 22.0°C to ice at −5.0°C? The heat capacities for $H_2O(s)$ and $H_2O(l)$ are 2.08 J g^{-1} °C^{-1} and 4.18 J g^{-1} °C^{-1}, respectively, and the enthalpy of fusion for ice is 6.01 kJ/mol.

96. In regions with dry climates, evaporative coolers are used to cool air. A typical electric air conditioner is rated at 1.00×10^4 Btu/h (1 Btu, or British thermal unit, equals the amount of energy needed to raise the temperature of 1 lb of water by 1°F). How much water must be evaporated each hour to dissipate as much heat as a typical electric air conditioner?

97. Compare and contrast the phase diagrams of water and carbon dioxide. Why doesn't CO_2 have a normal melting point and a normal boiling point, whereas water does? The slopes of the solid/liquid lines in the phase diagrams of H_2O and CO_2 are different. What do the slopes of the solid/liquid lines indicate in terms of the relative densities of the solid and liquid states for each substance? How do the melting points of H_2O and CO_2 depend on pressure? How do the boiling points of H_2O and CO_2 depend on pressure? Rationalize why the critical temperature for H_2O is greater than that for CO_2.

98. Define *critical temperature* and *critical pressure*. In terms of the kinetic molecular theory, why is it impossible for a substance to exist as a liquid above its critical temperature?

99. The critical point of NH_3 is 132°C and 111 atm, and the critical point of N_2 is −147°C and 34 atm. Which of these substances cannot be liquefied at room temperature no matter how much pressure is applied? Explain.

100. Like most substances, bromine exists in one of the three typical phases. Br_2 has a normal melting point of −7.2°C and a normal boiling point of 59°C. The triple point for Br_2 is −7.3°C and 40 torr, and the critical point is 320°C and 100 atm. Using this information, sketch a phase diagram for bromine indicating the points described above. Based on your phase diagram, order the three phases from least dense to most dense. What is the stable phase of Br_2 at room temperature and 1 atm? Under what temperature conditions can liquid bromine never exist? What phase changes occur as the temperature of a sample of bromine at 0.10 atm is increased from −50°C to 200°C?

101. Consider the following phase diagram. What phases are present at points *A* through *H*? Identify the triple point, normal boiling point, normal freezing point, and critical point. Which phase is denser, solid or liquid?

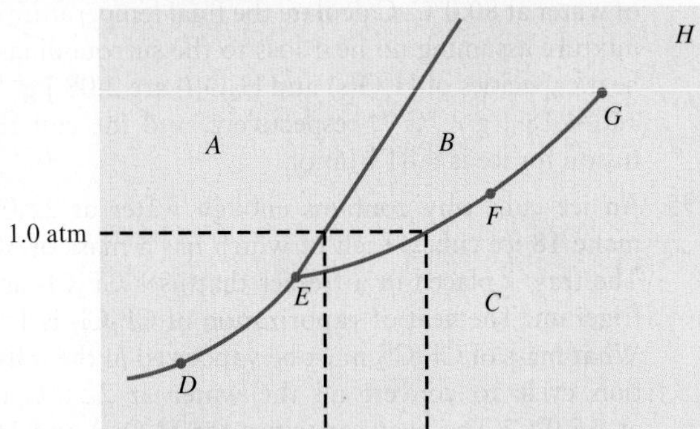

102. Use the accompanying phase diagram for sulfur to answer the following questions. (The phase diagram is not to scale.)

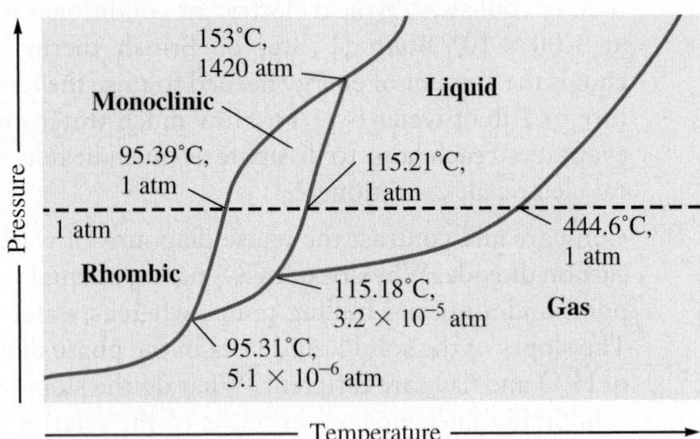

a. How many triple points are in the phase diagram?
b. What phases are in equilibrium at each of the triple points?
c. What phase is stable at room temperature and 1.0 atm pressure?
d. Can monoclinic sulfur exist in equilibrium with sulfur vapor?
e. What are the normal melting point and normal boiling point of sulfur?
f. Which is the denser solid phase, monoclinic or rhombic sulfur?

103. Use the accompanying phase diagram for carbon to answer the following questions.
a. How many triple points are in the phase diagram?
b. What phases can coexist at each triple point?
c. What happens if graphite is subjected to very high pressures at room temperature?
d. If we assume that the density increases with an increase in pressure, which is more dense, graphite or diamond?

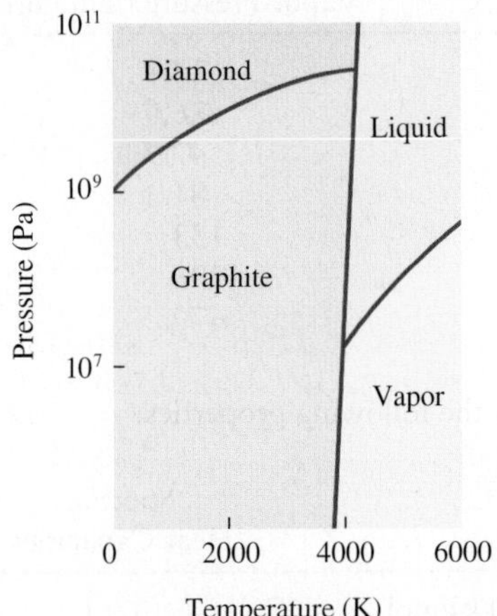

104. Iodine, like most substances, exhibits only three phases: solid, liquid, and vapor. The triple point of iodine is at 90 torr and 115°C. Which of the following statements concerning liquid I_2 must be true? Explain your answer.
a. $I_2(l)$ is more dense than $I_2(g)$.
b. $I_2(l)$ cannot exist above 115°C.
c. $I_2(l)$ cannot exist at 1 atmosphere pressure.
d. $I_2(l)$ cannot have a vapor pressure greater than 90 torr.
e. $I_2(l)$ cannot exist at a pressure of 10 torr.

105. The melting point of a fictional substance X is 225°C at 10.0 atm. If the density of the solid phase of X is 2.67 g/cm^3 and the density of the liquid phase is 2.78 g/cm^3 at 10.0 atm, predict whether the normal melting point of X will be less than, equal to, or greater than 225°C. Explain.

106. Consider the following data for xenon:

Triple point:	−121°C, 280 torr
Normal melting point:	−112°C
Normal boiling point:	−107°C

Which is more dense, Xe(*s*) or Xe(*l*)? How do the melting point and boiling point of xenon depend on pressure?

107. Some water is placed in a sealed glass container connected to a vacuum pump (a device used to pump gases from a container), and the pump is turned on. The water appears to boil and then freezes. Explain these changes by using the phase diagram for water. What would happen to the ice if the vacuum pump was left on indefinitely?

Additional Exercises

108. Oil of wintergreen, or methyl salicylate, has the following structure:

O ‖ C—OCH_3 OH

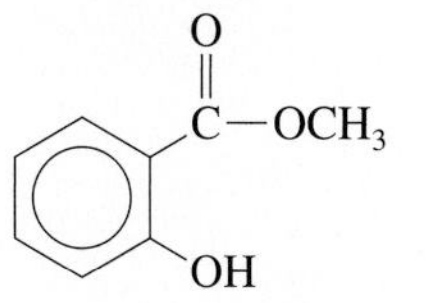

mp = −8°C

Methyl 4-hydroxybenzoate is another molecule with exactly the same molecular formula; it has the following structure:

O ‖ C—OCH_3 OH

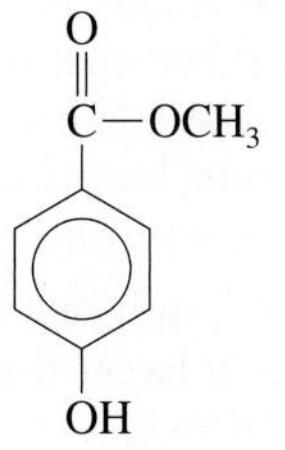

mp = 127°C

Account for the large difference in the melting points of the two substances.

109. You have three covalent compounds with three very different boiling points. All of the compounds have similar molar mass and relative shape. Explain how these three compounds could have very different boiling points.

110. Boron nitride (BN) exists in two forms. The first is a slippery solid formed from the reaction of BCl_3 with NH_3, followed by heating in an ammonia atmosphere at 750°C. Subjecting the first form of BN to a pressure of 85,000 atm at 1800°C produces a second form that is the second hardest substance known. Both forms of BN remain solids to 3000°C. Suggest structures for the two forms of BN.

111. How could you experimentally determine whether TiO_2 is an ionic solid or a network solid?

112. When wet laundry is hung on a clothesline on a cold winter day, it freezes but eventually dries. Explain.

113. Consider the following data concerning four different substances.

Compound	Conducts Electricity as a Solid	Other Properties
B_2H_6	No	Gas at 25°C
SiO_2	No	High mp
CsI	No	Aqueous solution Conducts electricity
W	Yes	High mp

Label the four substances as either ionic, network, metallic, or molecular solids.

114. A special vessel (see Fig. 16.52) contains ice and supercooled water (both at −10°C) connected by vapor space. Describe what happens to the amounts of ice and water as time passes.

115. Consider a cation in a trigonal hole. What size ion will just fit in the hole if the packed spheres have radius R?

116. Use the diagram of the unit cell for the hexagonal closest packed structure in Fig. 16.14 to determine the net number of atoms in the hcp unit cell.

117. Argon has a cubic closest packed structure as a solid. Assuming that argon has a radius of 190. pm, calculate the density of solid argon.

118. Rubidium chloride has the sodium chloride structure at normal pressures but assumes the cesium chloride structure at high pressures. What ratio of densities is expected for these two forms? Does this change in structure make sense on the basis of simple models?

119. A certain oxide of titanium is 28.31% oxygen by mass and contains a mixture of Ti^{2+} and Ti^{3+} ions. Determine the formula of the compound and the relative numbers of Ti^{2+} and Ti^{3+} ions.

120. Spinel is a mineral that contains 37.9% aluminum, 17.1% magnesium, and 45.0% oxygen, by mass, and has a density of 3.57 g/cm^3. The edge of the cubic unit cell measures 809 pm. How many of each type of atom are present in the unit cell?

121. A metallic solid with atoms in a face-centered cubic unit cell with an edge length of 392 pm has a density of 21.45 g/cm^3. Calculate the atomic mass and the atomic radius of the metal. Identify the metal.

122. The unit cell for a pure xenon fluoride compound is shown below. What is the formula of the compound?

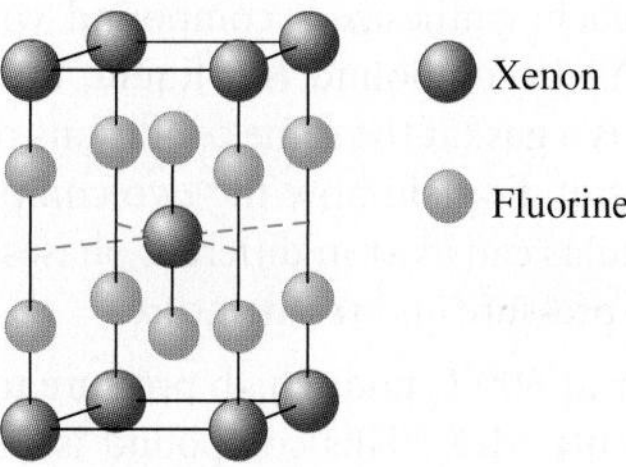

123. In solid KCl the smallest distance between the centers of a potassium ion and a chloride ion is 314 pm. Calculate the length of the edge of the unit cell and the density of KCl, assuming it has the same structure as sodium chloride.

124. The temperature inside a pressure cooker is 115°C. Calculate the vapor pressure of water inside the pressure cooker. What would be the temperature inside the pressure cooker if the vapor pressure of water was 3.50 atm?

125. Dry nitrogen gas is bubbled through liquid benzene (C_6H_6) at 20.0°C. From 100.0 L of the gaseous mixture of nitrogen and benzene, 24.7 g of benzene is condensed by passing the mixture through a trap at a temperature

where nitrogen is gaseous and the vapor pressure of benzene is negligible. What is the vapor pressure of benzene at 20.0°C?

126. A sample of dry nitrogen gas weighing 100.0 g is bubbled through liquid water at 25.0°C. The gaseous mixture of nitrogen and water vapor escapes at a total pressure of 700. torr. What mass of water has vaporized? (The vapor pressure of water at 25°C is 23.8 torr.)

127. The molar enthalpy of vaporization of water at 373 K is 41.16 kJ/mol. What fraction of this energy is used to change the internal energy of the water, and what fraction is used to do work against the atmosphere? (Assume that water vapor is an ideal gas.)

128. You are asked to help set up a historical display in the park by stacking some cannonballs next to a Revolutionary War cannon. You are told to stack them by starting with a triangle in which each side is composed of four touching cannonballs. You are to continue stacking them until you have a single ball on the top centered over the middle of the triangular base.
  a. How many cannonballs do you need?
  b. What type of closest packing arrangement is displayed by the cannonballs?
  c. The four corners of the pyramid of cannonballs form the corners of what type of regular geometric solid?

129. Consider the ionic solid A_xB_y, which has the unit cell below. The B ions are packed in a cubic arrangement, where each face has this structure:

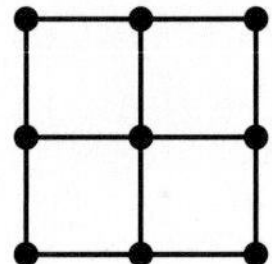

There is one B in the center of the cube. The structure can also be described in terms of three parallel planes of B's of the type shown above. The resulting structure thus contains eight intersecting cubes of B's. The A ions are found in the centers of alternate intersecting cubes (that is, four of every eight cubes have A's in the center). What is the formula of A_xB_y? In the extended structure, how many B's surround each A? What structure do the B's form?

130. Mn crystallizes in the same type of cubic unit cell as Cu. Assuming that the radius of Mn is 5.6% larger than the radius of Cu and the density of Cu is 8.96 g/cm^3, calculate the density of Mn.

Challenge Problems

131. An electron is trapped in an octahedral hole in a closest packed array of aluminum atoms (assume they behave as uniform hard spheres). In this situation the energy of the electron is quantized and the lowest-energy transition corresponds to a wavelength of 9.50 nm. Assuming that the hole can be approximated as a cube, what is the radius of a sphere that will just fit in the octahedral hole? Reference Exercise 148 in Chapter 12 for the energy equation for a particle in a cube.

132. You and a friend each synthesize a compound with the formula $XeCl_2F_2$. Your compound is a liquid, and your friend's compound is a gas (at the same conditions of temperature and pressure). Explain how the two compounds with the same formulas can exist in different phases at the same conditions of pressure and temperature.

133. A metal burns in air at 600°C under high pressure to form an oxide with formula MO_2. This compound is 23.72% oxygen by mass. The distance between the centers of touching atoms in a cubic closest packed crystal of this metal is 269.0 pm. What is this metal? What is its density?

134. Consider a perfectly insulated and sealed container. Determine the minimum volume of a container such that a gallon of water at 25°C will evaporate completely. If the container is a cube, determine the dimensions in feet. Assume the density of water is 0.998 g/cm^3.

135. The vapor pressure of water at 30.0°C is 31.824 torr; at this temperature the density of liquid water is 0.99567 g/cm^3. What is the ratio of the average distance between water molecules in the liquid and in the saturated vapor at this temperature?

136. For a simple cubic array, solve for the volume of an interior sphere (cubic hole) in terms of the radius of a sphere in the array.

137. The structures of another class of high-temperature ceramic superconductors are shown below.

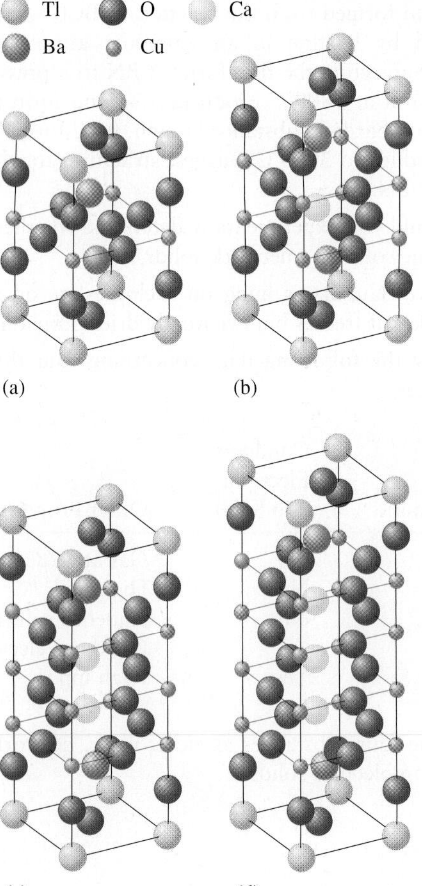

a. Determine the formula of each of these four superconductors.

b. One of the structural features that appears to be essential for high-temperature superconductivity is the presence of planar sheets of copper and oxygen atoms. As the number of sheets in each unit cell increases, the temperature for the onset of superconductivity increases. Order the four structures from the lowest to highest superconducting temperature.

c. Assign oxidation states to Cu in each structure assuming that Tl exists as Tl^{3+}. The oxidation states of Ca, Ba, and O are assumed to be +2, +2, and −2, respectively.

d. It also appears that copper must display a mixture of oxidation states for a material to exhibit superconductivity. Explain how this occurs in these materials as well as in the superconductor in Exercise 75.

138. Consider two different compounds, each with the formula C_2H_6O. One of these compounds is a liquid at room conditions and the other is a gas. Write Lewis structures consistent with this observation, and explain your answer.

139. The edge of the LiCl unit cell is 514 pm in length. Assuming that the Li^+ ions just fit in the octahedral holes of the closest packed Cl^- ions, calculate the ionic radii for the Li^+ and Cl^- ions. Compare them with the radii given in Fig. 13.8, and discuss the significance of any discrepancies.

Marathon Problem

140. The table below lists the ionic radii for the cations and anions in three different ionic compounds.

Formula	r_{cation}	r_{anion}
SnO_2	71 pm	140. pm
AlP	50. pm	212. pm
BaO	135 pm	140. pm

Each compound has either the NaCl, CsCl, or ZnS type cubic structure. Predict the type of structure formed (NaCl, CsCl, or ZnS) and the type and fraction of holes filled by the cations, and estimate the density of each compound.

Properties of Solutions

17

chapter

Water and oil are immiscible.

Charles D. Winters/Photo Researchers, Inc.

Sign in to OWL at www.cengage.com/owl to view tutorials and simulations, develop problem-solving skills, and complete online homework assigned by your professor.

Download mini lecture videos for key concept review and exam prep from OWL or purchase them from www.cengagebrain.com.

Most of the substances we encounter in daily life are mixtures: Wood, milk, gasoline, shampoo, steel, and air are all well-known examples. When the components of a mixture are uniformly intermingled—that is, when a mixture is homogeneous—it is called a *solution.* Solutions can be gases, liquids, or solids, as shown in Table 17.1. However, we will be concerned in this chapter with the properties of liquid solutions, particularly those containing water. Many essential chemical reactions occur in aqueous solutions, since water is capable of dissolving so many substances.

17.1 Solution Composition

Because a mixture, unlike a chemical compound, has a variable composition, the relative amounts of substances in a solution must be specified. The qualitative terms *dilute* (relatively little solute present) and *concentrated* (relatively large amount of solute) are often used to describe solution content. However, we need to define solution composition more precisely to perform calculations. For example, in dealing with the stoichiometry of solution reactions in Chapter 4, we found it useful to describe solution composition in terms of **molarity,** or the number of moles of solute per liter of solution.

A solute is the substance being dissolved. The solvent is the dissolving medium.

Other ways of describing solution composition are also useful. **Mass percent** (sometimes called *weight percent*) is the percent solute by mass in the solution:

When liquids are mixed, the liquid present in the largest amount is called the solvent.

$$\text{Mass percent} = \left(\frac{\text{grams of solute}}{\text{grams of solution}}\right) \times 100$$

Another way of describing solution composition is the **mole fraction** (symbolized by the Greek letter chi, χ), the ratio of the number of moles of a given component to the total number of moles of solution. For a two-component solution, where n_A and n_B represent the number of moles of the two components,

In very dilute aqueous solutions, the molality and the molarity are nearly the same.

$$\text{Mole fraction of component A} = \chi_A = \frac{n_A}{n_A + n_B}$$

Since molarity depends on the volume of the solution, it changes slightly with temperature. Molality is independent of temperature, since it depends on mass.

Still another way of describing solution composition is **molality** (symbolized by m), the number of moles of solute per *kilogram of solvent:*

$$\text{Molality} = \frac{\text{moles of solute}}{\text{kilograms of solvent}}$$

Table 17.1

Various Types of Solutions

Example	State of Solution	State of Solute	State of Solvent
Air, natural gas	Gas	Gas	Gas
Vodka in water, antifreeze	Liquid	Liquid	Liquid
Brass, steel	Solid	Solid	Solid
Carbonated water (soda)	Liquid	Gas	Liquid
Seawater, sugar solution	Liquid	Solid	Liquid
Hydrogen in platinum	Solid	Gas	Solid

17.2 The Thermodynamics of Solution Formation

Dissolving solutes in liquids is very common. We dissolve salt in the water used to cook vegetables, sugar in iced tea, stains in cleaning fluid, gaseous carbon dioxide in water to make soda water, ethanol in gasoline to make gasohol, and so on.

Solubility is important in other ways as well. For example, because the pesticide DDT is fat-soluble, it is retained and concentrated in animal tissues, where it causes detrimental effects. This is why DDT, even though it is effective for killing mosquitos, has been banned in the United States. Also, the solubilities of various vitamins are important in determining correct dosages. The insolubility of barium sulfate means it can be safely used to improve X-ray images of the gastrointestinal tract, even though Ba^{2+} ions are quite toxic.

Polar solvents dissolve polar solutes; nonpolar solvents dissolve nonpolar solutes.

What factors affect solubility? The cardinal rule of solubility is *like dissolves like.* We find that we must use a polar solvent to dissolve a polar or ionic solute and a nonpolar solvent to dissolve a nonpolar solute. Now we will attempt to see why this behavior occurs from a thermodynamic point of view. As we will see, solubility is an extraordinarily complex phenomenon, especially when water is the solvent. However, it is useful to explore some of the fundamental aspects of solubility because it has such important consequences. To simplify the discussion, we will assume that the formation of a liquid solution takes place in three distinct steps.

Step 1

Breaking up the solute into individual components (expanding the solute).

Step 2

Overcoming intermolecular forces in the solvent to make room for the solute (expanding the solvent).

Step 3

Allowing the solute and solvent to interact to form the solution.

These steps are illustrated in Fig. 17.1. Steps 1 and 2 require energy, since forces must be overcome to expand the solute and the solvent. Step 3 usually releases energy. In other words, steps 1 and 2 are endothermic, whereas step 3 is usually exothermic. The overall enthalpy change associated with the forma-

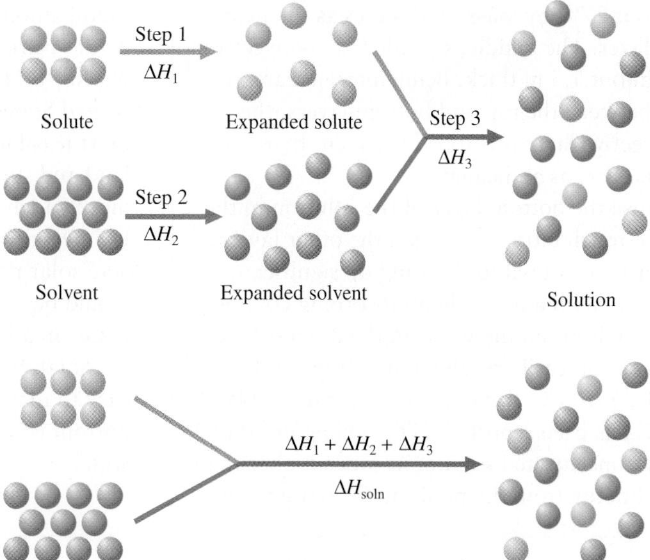

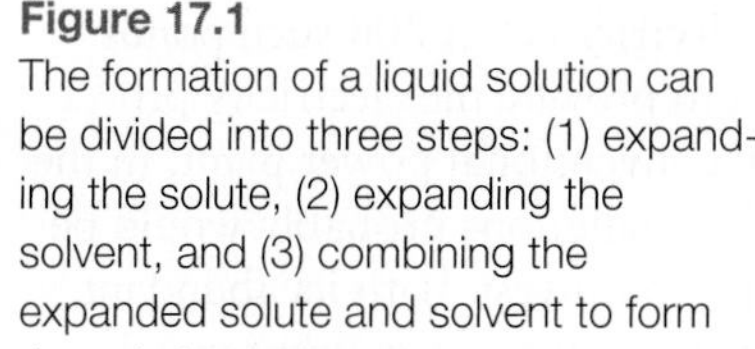

Figure 17.1
The formation of a liquid solution can be divided into three steps: (1) expanding the solute, (2) expanding the solvent, and (3) combining the expanded solute and solvent to form the solution.

Chemical Insights An Energy Solution

In today's society we face many serious problems related to energy: turmoil in the Middle East, the greenhouse effect, and polluted city air. Clearly the search for readily available, economical, and clean energy sources is of crucial importance. The solution may be in our backyards—literally.

When the sun shines on a swimming pool, the water is warmed by the solar energy. Because water is colorless (does not absorb visible light), the visible light from the sun passes through the water and strikes the bottom of the pool, where some of the light is absorbed and transformed into heat energy. This energy causes the water near the pool bottom to become warmer than the water above it. This warmer water, which has a lower density than the cooler water above it, then tends to rise toward the surface. This sets up convection currents, which circulate the warmer water to the surface, where it is cooled (mainly by evaporation). The convection currents keep the temperature of the water in the pool relatively uniform throughout.

How can this convective cooling process be stopped, thus allowing a pool of water to be an effective heat sink? The answer lies in adding common salt (NaCl). When sufficient salt is added to a pool of water 2 to 3 m deep, a salinity gradient is established. That is, the salt concentration is not uniform in the pool. In fact, three distinct layers can be identified. The top layer, called the convective layer, has a salt concentration of about 2% by mass and is only about half a meter thick. The bottom layer has a very high salt concentration (about 27% by mass) and serves as the heat storage layer. The middle, so-called nonconvective layer is about 1.5 m thick. Being intermediate in density between the top and bottom layers, the nonconvective layer is trapped between these layers and acts as an insulator.

Because the bottom layer of the solution in the pool is so much more dense than the other layers, it can get very hot without setting up significant convection. With a black, light-absorbing coating on the pool bottom, the water in the deepest layer of the pool can easily reach temperatures in the range of 90–100°C. In fact, a temperature as high as 107°C has been reported. (The boiling point of this very concentrated solution is much higher than 100°C.) Energy from the pool can be extracted if

Richard Nowitz

The solar pond in En Bokek, Israel, near the Dead Sea.

the hot brine from the storage layer is pumped through a heat exchanger to heat water or some other fluid.

A solar pond can be used to generate electricity by employing special turbines that use a working fluid other than water. Typical steam turbines cannot be used because the solar pond cannot produce temperatures high enough to generate steam efficiently. However, lower-boiling liquids can be used.

Solar ponds are now in use. A 52-acre pond at En Bokek near the Dead Sea produces 2.5 MW of electricity at peak power and could produce as much as 5 MW if more heat exchangers and turbines were installed in the powerhouse. This is a useful amount of electricity, but 200 such ponds would be required to provide the electricity produced by a typical conventional power plant. In the United States such installations probably would be most feasible in the Southwest, with its abundant land and sunshine. However, the concept may be more widely applicable. A study financed by the U.S. Department of Energy indicates that a 27,000-acre solar pond producing 600 MW of electricity would be competitive with conventional power plants in all states except Alaska.

Solar ponds are relatively inexpensive to build, easy to maintain, and nonpolluting if the pond bottom is sealed to prevent any leakage of the salt water.

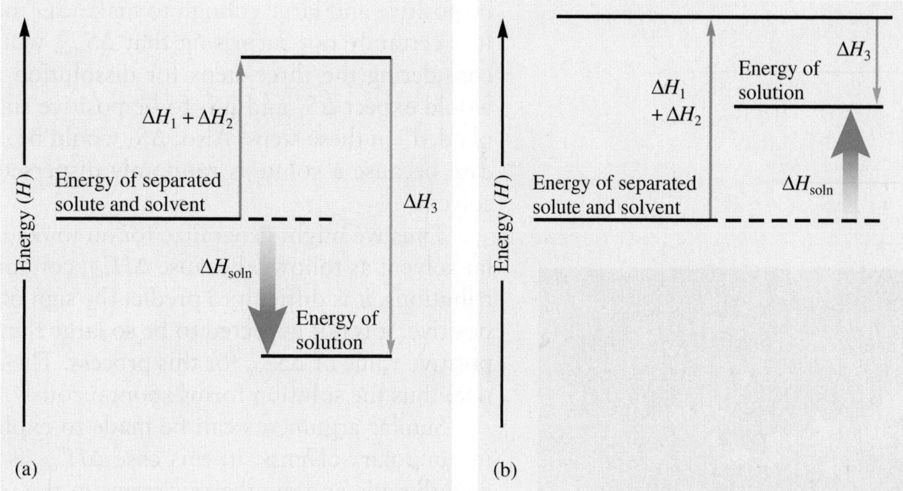

Figure 17.2
(a) The enthalpy of solution ΔH_{soln} has a negative sign (the process is exothermic) if step 3 releases more energy than is required by steps 1 and 2. (b) ΔH_{soln} has a positive sign (the process is endothermic) if steps 1 and 2 require more energy than is released in step 3. (If the energy changes for steps 1 and 2 equal that for step 3, then ΔH_{soln} is zero.)

The enthalpy of solution is the sum of energies used in expanding both the solvent and solute and the energy of solvent–solute interaction.

tion of the solution, called the **enthalpy (heat) of solution** (ΔH_{soln}), can be viewed as the sum of the ΔH values for the steps:

$$\Delta H_{soln} = \Delta H_1 + \Delta H_2 + \Delta H_3$$

where ΔH_{soln} may have a positive sign (energy absorbed) or a negative sign (energy released), as shown in Fig. 17.2.

To illustrate the importance of the various energy terms in the equation for ΔH_{soln}, we will consider a familiar case: the solubility of sodium chloride in water. It is clear that for NaCl(*s*) the term ΔH_1 is large and positive because of the strong ionic forces in the crystal that must be overcome. Also, ΔH_2 is expected to be large and positive because of the hydrogen bonds that must be broken in water. Finally, ΔH_3 is expected to be large and negative because of the strong interactions between the ions and the water molecules. In fact, the exothermic and endothermic terms essentially cancel in this case, as shown from the known values:

$$NaCl(s) \longrightarrow Na^+(g) + Cl^-(g) \qquad \Delta H_1^\circ = 786 \text{ kJ/mol}$$

$$H_2O(l) + Na^+(g) + Cl^-(g) \longrightarrow Na^+(aq) + Cl^-(aq)$$

$$\Delta H_{hyd}^\circ = \Delta H_2^\circ + \Delta H_3^\circ = -783 \text{ kJ/mol}$$

The second step shown here combines the terms ΔH_2 (for expanding the solvent) and ΔH_3 (for solvent–solute interactions) and is called the **enthalpy (heat) of hydration** (ΔH_{hyd}). This term represents the enthalpy change associated with the dispersal of a gaseous solute in water. Thus the standard enthalpy of solution for dissolving sodium chloride is the sum of ΔH_1° and ΔH_{hyd}°:

$$\Delta H_{soln}^\circ = 786 \text{ kJ/mol} - 783 \text{ kJ/mol} = 3 \text{ kJ/mol}$$

Note that ΔH_{soln}° is small but positive; the dissolving process requires a small amount of energy. Then why is NaCl so soluble in water? The answer to this question must involve the entropy change for the dissolving process. Recall from Chapter 10 that to predict whether a given process (at constant temperature and pressure) is spontaneous, we must consider the change in free energy:

$$\Delta G = \Delta H - T\Delta S$$

It is an experimental fact that NaCl(*s*) dissolves in water to form 1.0 *M* NaCl. Thus ΔG° for this process must be negative. However, the preceding calculations show that ΔH_{soln}° is positive and thus unfavorable. Therefore, ΔS_{soln}° must be positive and large enough to make $\Delta G°$ negative (through the $-T\Delta S°$ term). It is certainly not surprising that $\Delta S°_{soln}$ would be positive for this process. In considering the three steps for dissolution of a solute mentioned earlier, we would expect ΔS_1 and ΔS_2 to be positive since the solute and solvent are "expanded" in these steps. Also, ΔS_3 would be expected to be positive in a general case because a solute is randomly dispersed in the relatively large volume of solvent.

Thus we might generalize for an ionic (or polar) solute dissolving in a polar solvent as follows: Because $\Delta H°_{soln}$ contains large positive and negative contributions, it is difficult to predict the sign of $\Delta H°_{soln}$. However, even if $\Delta H°_{soln}$ is positive, it is not expected to be so large that it would overwhelm the expected positive value of $\Delta S°_{soln}$ for this process. The overall effect is to make $\Delta G°$ negative; thus the solution forms spontaneously.

Digital Globe/Getty Images

A satellite image of the 2010 oil spill in the Gulf of Mexico.

Similar arguments can be made to explain why nonpolar solutes dissolve in nonpolar solvents. In this case $\Delta H°_{soln}$ is expected to be small because the endothermic and exothermic terms in the solution process are expected to be similar in size. Thus the expected positive value of $\Delta S°_{soln}$ again would furnish the driving force for the solution process. These arguments suggest that $\Delta S°_{soln}$ provides the principal driving force for the behavior summarized by the rule "like dissolves like."

Now let's consider a process that is not spontaneous: A nonpolar solute does not dissolve in large quantities in a polar solvent. The floating oil slick that results whenever a major oil spill occurs in the ocean provides graphic evidence of this. What causes $\Delta G°$ to be positive for the dispersal of a material containing large nonpolar molecules in a polar solvent such as water? For this case $\Delta H°_{soln}$ is expected to be positive (unfavorable) because $\Delta H°_3$, the only exothermic component of $\Delta H°_{soln}$, is not expected to have a very large magnitude. On the other hand, for reasons given earlier, we expect $\Delta S°_{soln}$ to be positive (favorable). Thus we can explain the incompatibility of nonpolar and polar substances on the basis of the expected large, positive heat of solution that overwhelms the positive entropy change, thus giving a positive value for $\Delta G°$. Therefore, the solution does not form in this case.

Water as a Solvent

Because water is the most significant solvent in our world, it is especially important that we understand the solvent properties of water. As we noted in Chapter 16, water is not a typical liquid, with most of its unusual properties arising from the extensive hydrogen bonding present among the molecules. Because of water's unique nature, we must be very cautious in using simple arguments to account for the solvent properties of water.

To illustrate the unusual nature of water as a solvent, consider the values of $\Delta S°_{soln}$ listed in Table 17.2 for KCl(s), LiF(s), and CaS(s) forming aqueous solutions. Note that when KCl(s) is dissolved in water to form a 1.0 M solution, the value of $\Delta S°_{soln}$ is positive, as expected from the previous discussion. However, note that $\Delta S°_{soln}$ is *negative* for the other two salts. Why? How could the random dispersal in water of ions formerly present in a highly ordered solid produce a negative entropy change?

Table 17.2

Values of $\Delta S°_{soln}$ for Several Salts Dissolving in Water

Process	$\Delta S°$ (J K^{-1} mol^{-1})
$KCl(s) \rightarrow K^+(aq) + Cl^-(aq)$	75
$LiF(s) \rightarrow Li^+(aq) + F^-(aq)$	−36
$CaS(s) \rightarrow Ca^{2+}(aq) + S^{2-}(aq)$	−138

Obviously something must be occurring in the solution process that leads to increased order, which in some cases is large enough to dominate $\Delta S°_{soln}$. There is little doubt that this ordering effect arises from the hydration of the ions. In describing aqueous solutions containing ionic solutes in Chapter 4, we discussed the fact that the polar water molecules are attracted to the ions to form hydrated species. The assembling of a group of water molecules around the ions is an order-producing phenomenon and would be expected to make a negative contribution to $\Delta S°_{soln}$. Studies show that the more charge density an

Miracle Solvents

Chemical Insights

When a substance is heated beyond its critical temperature (T_c) and then placed under extremely high pressure, it forms a **supercritical fluid.** Although such fluids resemble liquids, they technically are not liquids. A liquid cannot exist above the critical temperature for a substance. For example, if a gaseous substance is maintained above T_c and pressure is gradually applied, there is no distinct point where the gas changes to a liquid, as would occur below T_c, but the supercritical fluid gradually forms as the pressure is increased.

In experiments at the Agriculture Department's Northern Regional Research Center in Peoria, Illinois, scientists have found that supercritical carbon dioxide behaves as a very useful nonpolar solvent for removing fat from meat. At temperatures above 31°C (T_c for CO_2) and several hundred atmospheres of pressure, the carbon dioxide fluid can dissolve virtually all the fat from samples of meat. Even more important, the fluid also will dissolve any pesticide or drug residues that may be present in the meat. When the carbon dioxide fluid is returned to normal pressures, it immediately vaporizes, and the fat, drug, and pesticide molecules come "raining" out to allow easy analysis of the types and amounts of contaminants present in the meat. Therefore, this method provides an efficient way to test meat for trace pesticide and drug residues.

Like supercritical carbon dioxide, supercritical water is a very interesting substance that has strikingly different properties from those of liquid water. For example, recent experiments have shown that supercritical (superfluid) water can behave simultaneously as both a polar and a nonpolar solvent. While the reasons for this unusual behavior remain unclear, the practical value of this behavior is very clear: It makes superfluid water a very useful reaction medium for a wide variety of substances. One extremely important application of this idea involves the environmentally sound

A scientist examines corn bran oil processed by supercritical fluid extraction.

destruction of industrial wastes. Most hazardous organic (nonpolar) substances can be dissolved in supercritical water and oxidized by dissolved O_2 in a matter of minutes. The products of these reactions are water, carbon dioxide, and possibly simple acids (which result when halogen-containing compounds are reacted). Therefore, the aqueous mixture that results from the reaction often can be disposed of with little further treatment. In contrast to the incinerators used to destroy organic waste products, a supercritical water reactor is a closed system (has no emissions).

Supercritical water has strange and wonderful properties, one of the most astonishing being that a flame can burn within the supercritical fluid during the reaction of O_2 with an organic substance. These properties promise to make supercritical water a versatile medium for chemical reactions.

ion possesses, the greater this hydration effect will be. This idea is borne out by the data in Table 17.2. For example, note that ΔS°_{soln} for KCl(s) is positive, but the value for LiF(s) is negative. This probably results from the smaller sizes (and thus larger charge densities) of Li^+ and F^- as compared with K^+ and Cl^-. The smaller ions presumably are able to bind the hydrating water molecules more firmly and thus show a more negative value for ΔS°_{soln}. The charges on the ions are also important. Note that CaS(s) exhibits a value of ΔS°_{soln} that is more negative than that for LiF(s), as might be expected for the more highly charged Ca^{2+} and S^{2-} ions.

Benzene does not dissolve in water to the extent that a 1 *M* solution is possible. The value of ΔS°_{soln} given here is extrapolated from values measured at much lower concentrations.

We have seen that ionic solutes dissolving in water can lead to negative values of ΔS°_{soln} presumably because of ion hydration effects. In certain cases the dispersal of nonpolar solute particles in water can also produce negative values of ΔS°_{soln}. For example, for benzene [a nonpolar liquid containing C_6H_6 molecules (see Section 14.5)] dissolving in water, ΔS°_{soln} has been estimated to be $-58 \text{ J K}^{-1} \text{ mol}^{-1}$. This very negative value for ΔS°_{soln} suggests that a great deal of ordering occurs when benzene is dispersed in water. This is a surprising result. What is the origin of this ordering? Clearly, the situation here is quite different from that for ionic solutes. The polar water molecules will not strongly hydrate the nonpolar benzene molecules, as they do ions.

What could cause ΔS°_{soln} to be negative for this nonpolar solute? The negative value for ΔS°_{soln} in this case seems to arise not from the hydration of the solute but from the opposite behavior. Instead of hydrating the nonpolar molecule, water seems to form a cage to isolate the nonpolar solute from the bulk water structure. This cage formation probably occurs as follows: So that a nonpolar molecule can be introduced into the water structure, a "hole" must be formed, which requires the breaking of some water–water hydrogen bonds. In response, in an effort to recover the lost hydrogen bonding, the water molecules around the edge of the hole apparently form even more hydrogen bonds than are normally found in bulk water. That is, a highly ordered cage of hydrogen-bonded water molecules forms around the nonpolar solute molecule. This ordering of the water structure makes a negative contribution to the entropy of solution and in certain cases leads to a negative value of ΔS°_{soln}. In fact, this unfavorable entropy contribution resulting from cage formation could be an important reason why nonpolar solutes are insoluble in water.

Summary

The point of this discussion has been to consider some of the thermodynamic aspects of the process that occurs when a solute is dispersed in a solvent to form a solution. Our observations tell us that "like dissolves like." However, the dissolution process is so complex that predicting whether a particular solute will dissolve in a given solvent is risky. Solubility is difficult to explain and even more difficult to predict, especially when water is the solvent. The only way to be certain about the compatibility of a given solute and solvent is to do the experiment.

17.3 Factors Affecting Solubility

Structure Effects

In the preceding section, we saw that solubility is favored if the solute and solvent have similar polarities. Since it is the molecular structure that determines polarity, there should be a definite connection between structure and solubility. Vitamins provide an excellent example of the relationship among molecular structure, polarity, and solubility.

For the last several years, there has been considerable publicity about the pros and cons of consuming large quantities of vitamins. For example, large doses of vitamin C have been advocated to combat various illnesses, including the common cold. Vitamin E has been extolled as a youth-preserving elixir and a protector against the carcinogenic (cancer-causing) effects of certain chemicals. However, there are possible detrimental effects from taking large amounts of some vitamins, depending on their solubilities.

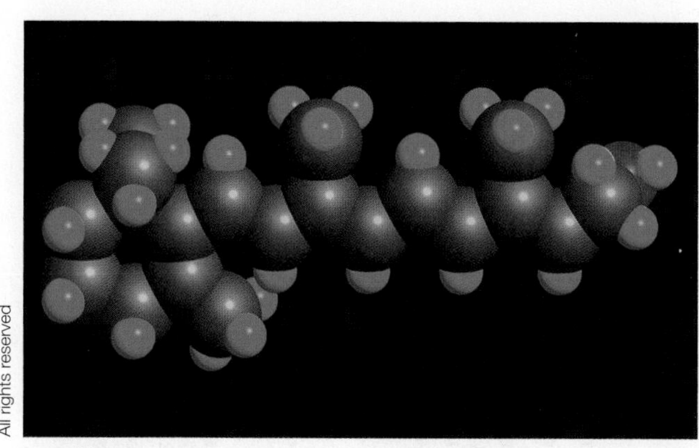

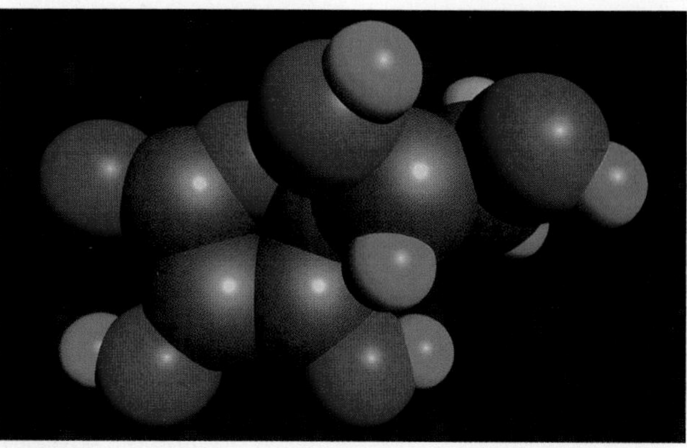

Vitamin A

Vitamin C

Figure 17.3
The molecular structures of vitamin A (nonpolar, fat-soluble) and vitamin C (polar, watersoluble). The circles in the structural formulas indicate polar bonds. Note that vitamin C contains far more polar bonds than vitamin A.

Vitamins can be divided into two classes: *fat-soluble* (vitamins A, D, E, and K) and *water-soluble* (vitamins B and C). The reason for the differing solubility characteristics can be seen by comparing the structures of vitamins A and C (Fig. 17.3). Vitamin A, composed mostly of carbon and hydrogen atoms that have similar electronegativities, is virtually nonpolar. This causes it to be soluble in nonpolar materials such as body fat, which is also largely composed of carbon and hydrogen, but not soluble in polar solvents such as water. On the other hand, vitamin C has many polar O—H and C—O bonds, making the molecule polar and thus water-soluble. We often describe nonpolar materials like vitamin A as *hydrophobic* (water-fearing) and polar substances like vitamin C as *hydrophilic* (water-loving).

Because of their solubility characteristics, the fat-soluble vitamins can build up in the fatty tissues of the body. This buildup has both positive and negative effects. Since these vitamins can be stored, the body can temporarily tolerate a diet deficient in vitamins A, D, E, and K. Conversely, if excessive amounts of these vitamins are consumed, their buildup can lead to the illness *hypervitaminosis*.

In contrast, the water-soluble vitamins are excreted by the body and therefore must be consumed regularly. This fact was first recognized when the British navy discovered that scurvy, a disease often suffered by sailors, could be prevented if the sailors regularly ate fresh limes (which are a good source of vitamin C) when aboard ship (hence the name "limey" for the British sailor).

Carbonation in a bottle of soda.

Pressure Effects

Although pressure has little effect on the solubilities of solids or liquids, it does significantly increase the solubility of a gas. Carbonated beverages, for example, are always bottled at high pressure of carbon dioxide to ensure a high concentration of carbon dioxide in the liquid. The fizzing that occurs when you open a bottle of soda results from the escape of gaseous carbon dioxide because the atmospheric pressure of CO_2 is much lower than that used in the bottling process.

Henry's law is often expressed in the form $P = kC$, where C is the concentration of the gas in mol/L. In this case k has the units L atm/mol.

The increase in gas solubility with pressure can be understood from Fig. 17.4. Figure 17.4(a) shows a gas in equilibrium with a solution; that is, the gas molecules are entering and leaving the solution at the same rate. If the pressure is suddenly increased [Fig. 17.4(b)], the number of gas molecules per unit vol-

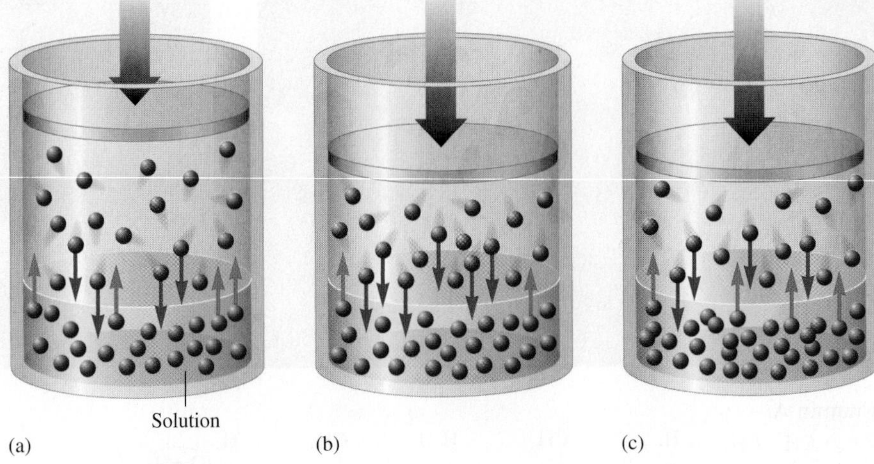

Figure 17.4
(a) A gaseous solute in equilibrium with a solution. (b) The piston is pushed in, which increases the pressure of the gas and the number of gas molecules per unit volume. This causes an increase in the rate at which the gas enters the solution, so the concentration of dissolved gas increases. (c) The greater gas concentration in the solution causes an increase in the rate of escape. A new equilibrium is reached.

ume increases; thus the gas enters the solution at a higher rate than it leaves. As the concentration of dissolved gas increases, the rate of escape of the gas also increases until a new equilibrium is reached [Fig. 17.4(c)]. At this point the solution contains more dissolved gas than before.

William Henry (1774–1836), a close friend of John Dalton, formulated his law in 1801.

The relationship between gas pressure and the concentration of a dissolved gas is given by **Henry's law:**

$$P = k_H X$$

Henry's law holds only when there is no chemical reaction between solute and solvent.

where P represents the partial pressure of the gaseous solute above the solution, X represents the mole fraction of the dissolved gas, and k_H is a constant (the Henry's law constant) characteristic of a particular solution. Henry's law states that *the amount of gas dissolved in a solution is directly proportional to the pressure of the gas above the solution.*

Henry's law is obeyed most accurately for dilute solutions of gases that do not dissociate in or react with the solvent. For example, Henry's law is obeyed by oxygen gas in water, but it does *not* correctly represent the behavior of gaseous hydrogen chloride in water because of the dissociation reaction:

$$HCl(g) \xrightarrow{H_2O} H^+(aq) + Cl^-(aq)$$

The Henry's law constants for the aqueous solutions of several gases are given in Table 17.3.

Table 17.3

The Values of Henry's Law Constants for Several Gases Dissolved in Water at 298 K

Gas	k_H (atm)
CH_4	4.13×10^2
CO_2	1.64×10^3
O_2	4.34×10^4
CO	5.71×10^4
H_2	7.03×10^4
N_2	8.57×10^4

Temperature Effects for Aqueous Solutions

Everyday experiences of dissolving substances such as sugar may lead you to think that solubility always increases with temperature. This is not the case. The dissolving of a solid occurs *more rapidly* at higher temperatures, but the amount of solid that can be dissolved may increase or decrease with increasing temperature. The effect of temperature on the solubility in water of several common solids is shown in Fig. 17.5. Note that although the solubility of most solids increases with temperature, the solubilities of some substances (such as sodium sulfate and cerium sulfate) decrease with increasing temperature.

Because ΔH°_{soln} refers to dissolving a small amount of solute in a large amount of a 1.0 *M* ideal solution, it is not necessarily relevant to the process of dissolving a solid in a saturated solution. Thus ΔH°_{soln} is of limited use in predicting the variation of solubility with temperature.

Predicting the temperature dependence of solubility is very difficult. For example, although there is some correlation between the sign of ΔH°_{soln} and the variation of solubility with temperature, important exceptions exist.* The only sure way to determine the temperature dependence of a solid's solubility is by experiment.

*For more information, see R. S. Treptow, "Le Châtelier's Principle Applied to the Temperature Dependence of Solubility," *J. Chem. Ed.* **61** (1984): 499.

Ionic Liquids?

Chemical Insights

So far in this text you have seen that ionic substances are stable solids with high melting points. For example, sodium chloride has a melting point near 800°C. One of the "hottest" areas of current chemical research is ionic liquids—substances composed of ions that are liquids at normal temperatures and pressures. This unusual behavior results from the differences in the sizes of the anions and cations in the ionic liquids. Dozens of small anions, such as BF_4^- (tetrafluoroborate) or PF_6^- (hexafluorophosphate), can be paired with thousands of large cations, such as 1-hexyl-3-methylimidazolium or 1-butyl-3-methylimidazolium (a and b, respectively, in the accompanying figure). These substances remain liquids because the bulky, asymmetrical cations do not pack together efficiently with the smaller, symmetrical anions. In contrast, in sodium chloride the ions can pack very efficiently to form a compact, orderly arrangement, leading to maximum cation–anion attractions and thus a high melting point.

The excitement being generated by these ionic liquids arises from many factors. For one thing, almost an infinite variety of ionic liquids is possible due to the large variety of bulky cations and small anions available. According to Kenneth R. Seddon, Director of QUILL (Queen's University Ionic Liquid Laboratory) in Northern Ireland, a *trillion* ionic liquids are possible. Another great advantage of these liquids is their long liquid range, typically from −100°C to 200°C.

In addition, the cations in the liquids can be designed to perform specific functions. For example, chemist James H. Davis, of the University

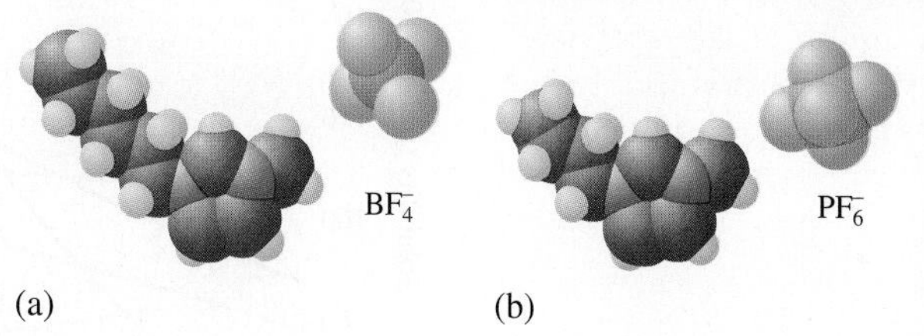

of South Alabama in Mobile, has designed various cations that will attract potentially harmful ions such as mercury, cadmium, uranium, and americium (the latter two are commonly found in nuclear waste materials) and leach them out of contaminated solutions. Davis has also developed cations that will remove H_2S (which produces SO_2 when the gas is burned) and CO_2 (which does not burn) from natural gas. Potentially, these ionic solutions might also be used to remove CO_2 from the exhaust gases of fossil-fuel–burning power plants to lessen the "greenhouse effect."

The biggest obstacle to the widespread use of ionic liquids is their cost. Normal organic solvents used in industry typically cost a few cents per liter, but ionic liquids can cost hundreds of times that amount. However, the environmentally friendly nature of ionic liquids (they produce no vapors because the ions are not volatile) and the flexibility of these substances as reaction media make them very attractive. As a consequence, efforts are under way to make their use economically feasible.

The term *ionic liquid* may have seemed like an oxymoron in the past, but these substances have a very promising future.

In contrast to the behavior of gases in water, gases typically become more soluble in most nonaqueous solvents as temperature increases.

The behavior of gases dissolving in water appears to be less complex. The solubility of a gas in water typically decreases with increasing temperature, as shown for several cases in Fig. 17.6. This temperature effect has important environmental implications because of the widespread use of water from lakes and rivers for industrial cooling. After being used as a coolant, the water is returned to its natural source at a higher-than-ambient temperature (**thermal pollution** has occurred). Because it is warmer, this water contains less than the normal concentration of oxygen and is also less dense; it tends to "float" on the colder water below, thus blocking normal oxygen absorption. This effect can be especially important in deep lakes. The warm upper layer can seriously decrease the amount of oxygen available to aquatic life in the deeper layers of the lake.

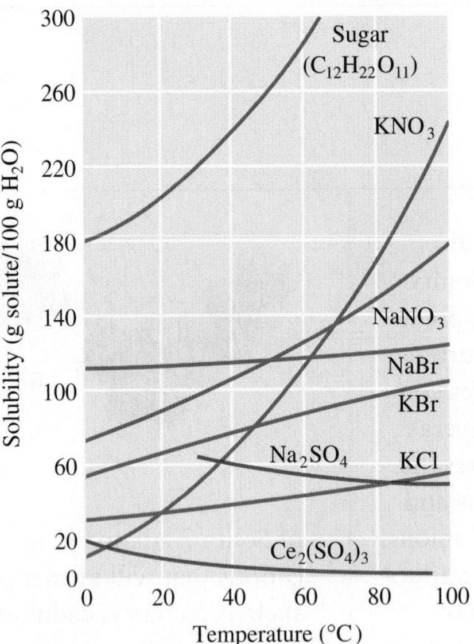

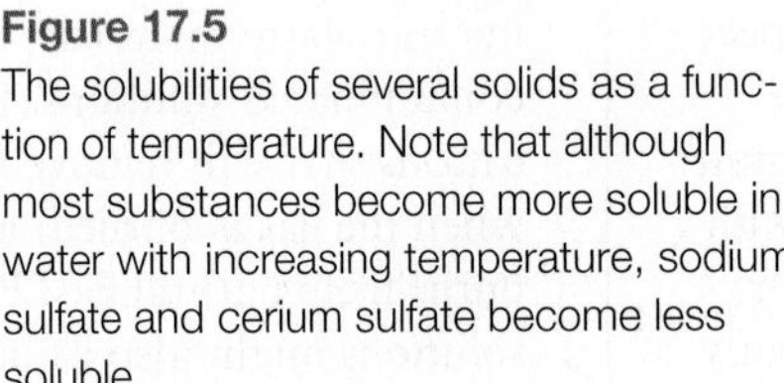

Figure 17.5
The solubilities of several solids as a function of temperature. Note that although most substances become more soluble in water with increasing temperature, sodium sulfate and cerium sulfate become less soluble.

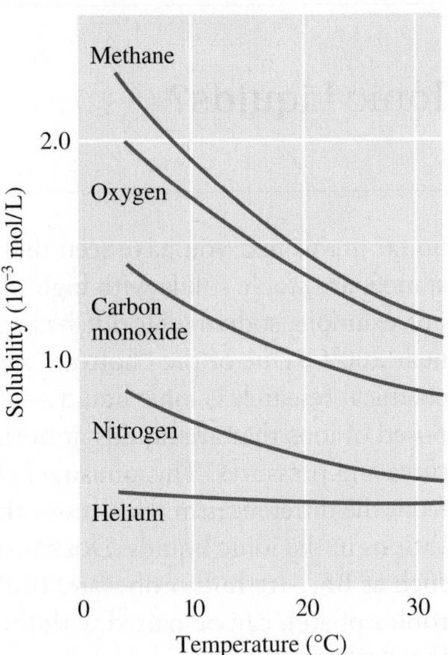

Figure 17.6
The solubilities of several gases in water as a function of temperature at a constant pressure of 1 atm of gas above the solution.

The decreasing water solubility of gases with increased temperature is also responsible for the formation of *boiler scale*. As we discussed in Chapter 7, the bicarbonate ion is formed when carbon dioxide is dissolved in water containing the carbonate ion:

$$CO_3^{2-}(aq) + CO_2(aq) + H_2O(l) \longrightarrow 2HCO_3^-(aq)$$

When the water also contains Ca^{2+} ions, this reaction is especially important—calcium bicarbonate is soluble in water, but calcium carbonate is insoluble. When the water is heated, the carbon dioxide is driven off. For the system to replace the lost carbon dioxide, the reverse reaction must occur:

$$2HCO_3^-(aq) \longrightarrow H_2O(l) + CO_2(aq) + CO_3^{2-}(aq)$$

This reaction, however, also increases the concentration of carbonate ions, causing solid calcium carbonate to form. This solid is the boiler scale that coats the walls of containers such as industrial boilers and tea kettles. Boiler scale reduces the efficiency of heat transfer and can lead to blockage of pipes (Fig. 17.7).

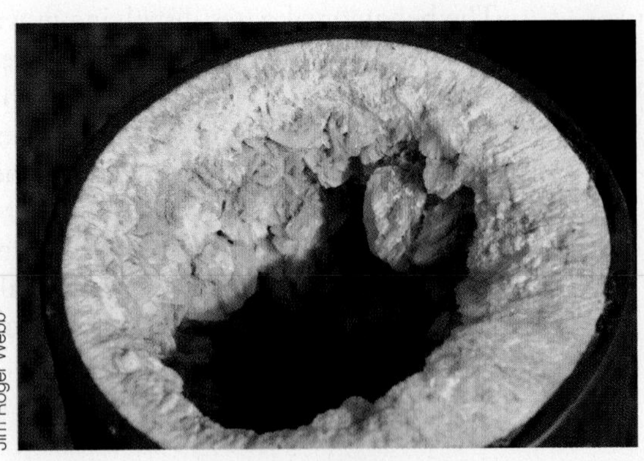

Figure 17.7
A pipe with accumulated mineral deposits.

The Lake Nyos Tragedy

Chemical Insights

On August 21, 1986, a cloud of gas suddenly boiled from Lake Nyos in Cameroon, killing nearly 2000 people. Although at first it was speculated that the gas was hydrogen sulfide, it now seems clear it was carbon dioxide. What would cause Lake Nyos to emit this huge, suffocating cloud of CO_2? Although the answer may never be known for certain, many scientists believe that the lake suddenly "turned over," bringing to the surface water that contained huge quantities of dissolved carbon dioxide. Lake Nyos is a deep lake that is thermally stratified: Layers of warm, less dense water near the surface float on the colder, denser water layers near the lake's bottom. Under normal conditions the lake stays this way; there is little mixing among the different layers. Scientists believe that over hundreds or thousands of years, carbon dioxide gas had seeped into the cold water at the lake's bottom and dissolved in great amounts because of the large pressure of CO_2 present (in accordance with Henry's law). For some reason, on August 21, 1986, the lake apparently suffered an overturn, possibly due to wind or to unusual cooling of the lake's surface by monsoon clouds. This caused water that was greatly supersaturated with CO_2 to reach the surface and release tremendous quantities of gaseous CO_2 that suffocated thousands of humans and animals before they knew what hit them—a tragic, monumental illustration of Henry's law.

T. Orban/Sygman/Corbis

Lake Nyos in Cameroon.

Since 1986, the scientists studying Lake Nyos and nearby Lake Monoun have observed a rapid recharging of the CO_2 levels in the deep waters of these lakes, causing concern that another deadly gas release could occur at any time. Apparently the only way to prevent such a disaster is to pump away the CO_2-charged deep water in the two lakes. Scientists at a conference to study this problem in 1994 recommended such a solution, but it has not yet been funded by Cameroon.

17.4 The Vapor Pressures of Solutions

Liquid solutions have physical properties significantly different from those of the pure solvent, a fact that has great practical importance. For example, we add antifreeze to the water in a car's cooling system to prevent freezing in winter and boiling in summer. We also melt ice on sidewalks and streets by spreading salt. These preventive measures work because of the solute's effect on the solvent's properties.

A nonvolatile solute has no tendency to escape from solution into the vapor phase.

To explore how a nonvolatile solute affects a solvent, we will consider the experiment represented in Fig. 17.8, in which a sealed container encloses a beaker containing an aqueous sugar solution and a beaker containing pure water. Gradually, the volume of the sugar solution increases and the volume of the pure water decreases. Why? We can explain this observation if the vapor pressure of the pure solvent is greater than that of the solution. Under these conditions, the pressure of vapor necessary to achieve equilibrium with the pure solvent is greater than that required to reach equilibrium with the solution. Thus, as the pure solvent emits vapor in an attempt to reach equilibrium, the solution absorbs vapor to try to lower the vapor pressure toward its equi-

Figure 17.8
An aqueous solution and pure water in a closed environment. (a) Initial stage. (b) After a period of time, the water is transferred to the solution.

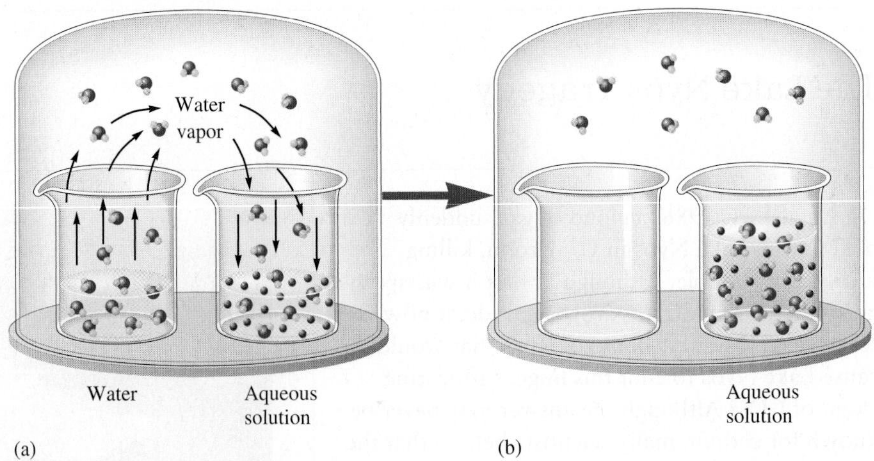

librium value. This process results in a net transfer of water from the pure liquid through the vapor phase to the solution. The system can reach an equilibrium vapor pressure only when all the water has been transferred to the solution. This experiment is just one of many observations indicating that the presence of a *nonvolatile solute lowers the vapor pressure of a solvent.*

The presence of a nonvolatile solute reduces the tendency of solvent molecules to escape.

We can account for this behavior in terms of the simple model shown in Fig. 17.9. The dissolved nonvolatile solute decreases the number of solvent molecules per unit volume. Thus it lowers the number of solvent molecules at the surface, which proportionately lowers the escaping tendency of the solvent molecules. For example, in a solution consisting of half nonvolatile solute molecules and half solvent molecules, we expect the observed vapor pressure to be half that of the pure solvent, since only half as many molecules can escape. In fact, this agrees with our observations.

Detailed studies of the vapor pressures of solutions containing nonvolatile solutes were carried out by François M. Raoult. His results are described by the equation known as **Raoult's law:**

$$P_{soln} = \chi_{solvent} P^{\circ}_{solvent}$$

where P_{soln} is the observed vapor pressure of the solution, $\chi_{solvent}$ is the mole fraction of solvent, and $P^{\circ}_{solvent}$ is the vapor pressure of the pure solvent. Note that for a solution containing half solute and half solvent molecules, $\chi_{solvent}$ is

Figure 17.9
The presence of a nonvolatile solute inhibits the escape of solvent molecules from the liquid and so lowers the vapor pressure of the solvent.

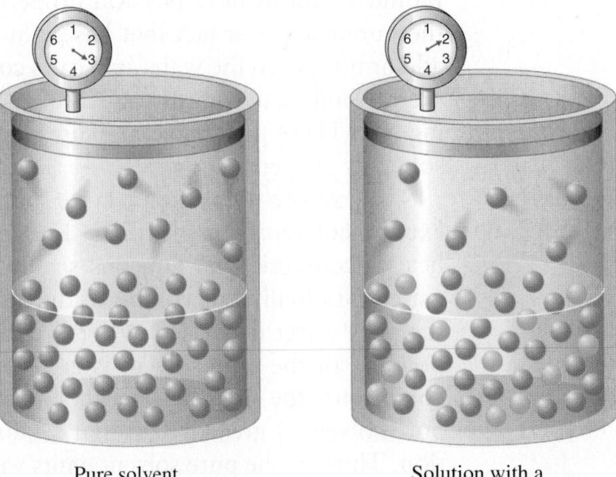

0.5, so the vapor pressure of the solution is half that of the pure solvent. On the other hand, for a solution where three-fourths of the solution molecules are solvent, $\chi_{solvent} = \frac{3}{4} = 0.75$, and $P_{soln} = 0.75P^\circ_{solvent}$. The idea is that the nonvolatile solute lowers the vapor pressure simply by diluting the solvent.

Raoult's law states that the vapor pressure of a solution is directly proportional to the mole fraction of solvent present.

Raoult's law is a linear equation of the form $y = mx + b$, where $y = P_{soln}$, $x = \chi_{solvent}$, $m = P^\circ_{solvent}$, and $b = 0$. Thus a plot of P_{soln} versus $\chi_{solvent}$ gives a straight line with a slope equal to $P^\circ_{solvent}$, as shown in Fig. 17.10.

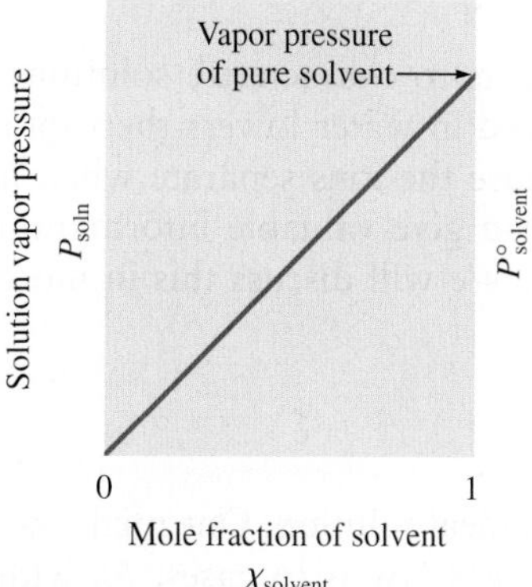

Figure 17.10
For a solution that obeys Raoult's law, a plot of P_{soln} versus $\chi_{solvent}$ yields a straight line.

The effect of the solute on the vapor pressure of a solution gives us a convenient way to "count" molecules and thus provides a means for experimentally determining molar masses. Suppose a certain mass of a compound is dissolved in a solvent, and the vapor pressure of the resulting solution is measured. Using Raoult's law, we can determine the number of moles of solute present. Since the mass of this number of moles is known, we can calculate the molar mass.

OWL INTERACTIVE EXAMPLE 17.1

A solution was prepared by adding 20.0 g of urea to 125 g of water at 25°C, a temperature at which pure water has a vapor pressure of 23.76 torr. The observed vapor pressure of the solution was found to be 22.67 torr. Calculate the molar mass of urea.

Solution

▪ What are we trying to solve?

We are trying to determine the molar mass of urea. Molar mass has the formula $\frac{\text{mass}}{\text{mole}}$, so we need to know the mass and number of moles. We have the mass as 20.0 g, so we need to determine the number of moles. We can do this by first using Raoult's law.

Raoult's law can be rearranged to give

$$\chi_{H_2O} = \frac{P_{soln}}{P^\circ_{H_2O}}$$

This form allows us to determine the mole fraction of water in the solution:

$$\chi_{H_2O} = \frac{22.67 \text{ torr}}{23.76 \text{ torr}} = 0.9541$$

However, we are interested in the urea. To find its molar mass, we must find the number of moles represented by 20.0 g. We can calculate the number of moles of urea from the definition of χ_{H_2O}:

$$\chi_{H_2O} = \frac{\text{mol } H_2O}{\text{mol } H_2O + \text{mol urea}} = \frac{n_{H_2O}}{n_{H_2O} + n_{urea}}$$

From the mass of water used to prepare the solution, we have

$$n_{H_2O} = 125 \text{ g } H_2O \times \frac{1 \text{ mol } H_2O}{18.0 \text{ g } H_2O} = 6.94 \text{ mol } H_2O$$

Since $\chi_{H_2O} = 0.9541$,

$$\chi_{H_2O} = 0.9541 = \frac{n_{H_2O}}{n_{H_2O} + n_{urea}} = \frac{6.94}{6.94 + n_{urea}}$$

or

$$0.9541(6.94 + n_{urea}) = 6.94$$

Solving for the number of moles of urea gives

$$n_{urea} = \frac{6.94 - 6.62}{0.9541} = 0.335 \text{ mol}$$

Since 20.0 g of urea was originally dissolved, 0.335 mole of urea weighs 20.0 g. Thus

$$\frac{20.0 \text{ g}}{0.335 \text{ mol}} = 59.7 \text{ g/mol}$$

The value for the molar mass of urea determined in this experiment is thus 59.7 g/mol. Urea has the formula $(NH_2)_2CO$ and a molar mass of 60.0, so the result obtained in this experiment agrees fairly well with the known value.

The lowering of vapor pressure depends on the number of solute particles present in the solution.

We can also use vapor pressure measurements to characterize solutions. For example, 1 mole of sodium chloride dissolved in water lowers the vapor pressure about twice as much as expected because the ions separate when it dissolves. Thus vapor pressure measurements can give valuable information about the nature of the solute after it dissolves. We will discuss this in more detail in Section 17.7.

Nonideal Solutions

Any solution that obeys Raoult's law is called an **ideal solution.** One might say that Raoult's law is to solutions what the ideal gas law is to gases. As with gases, ideal behavior for solutions is never perfectly achieved but is sometimes closely approached. Nearly ideal behavior is often observed when the solute–solute, solvent–solvent, and solute–solvent interactions are very similar. That is, in solutions where the solute and solvent are very much alike, the solute simply acts to dilute the solvent. However, if the solvent has a special affinity for the solute, such as if hydrogen bonding occurs, the tendency of the solvent molecules to escape will be lowered more than expected. In such cases the observed vapor pressure will be *lower* than the value predicted by Raoult's law; there is a *negative deviation from Raoult's law.*

Strong solute–solvent interaction gives a vapor pressure lower than that predicted by Raoult's law.

As mentioned, for a solution to behave ideally, the solute–solute, solvent–solvent, and solute–solvent interactions would have to be identical. This would correspond to a situation where $\Delta H_{soln} = 0$. On the other hand, when a solute and solvent release large quantities of energy in the formation of a solution—that is, when ΔH_{soln} is large and negative—we can assume that strong interactions exist between the solute and the solvent. In this case we expect a negative deviation from Raoult's law.

So far we have assumed that the solute is nonvolatile and so does not contribute to the vapor pressure over the solution. However, for liquid–liquid solutions where both components are volatile, a modified form of Raoult's law applies:

$$P_{Total} = P_A + P_B = \chi_A P_A^\circ + \chi_B P_B^\circ$$

where P_{Total} represents the total vapor pressure of a solution containing A and B, χ_A and χ_B are the mole fractions of A and B, P_A° and P_B° are the vapor pressures of pure A and pure B, and P_A and P_B are the partial pressures resulting from molecules of A and B in the vapor above the solution.

Liquid–liquid solutions obeying this form of Raoult's law are said to be *ideal.* However, as with solutions containing nonvolatile solutes, deviations from Raoult's law are often observed. These can be positive or negative, as shown in Fig. 17.11.

Again, large negative heats of solution indicate especially strong solute–solvent interactions, and such solutions are expected to show *negative* devia-

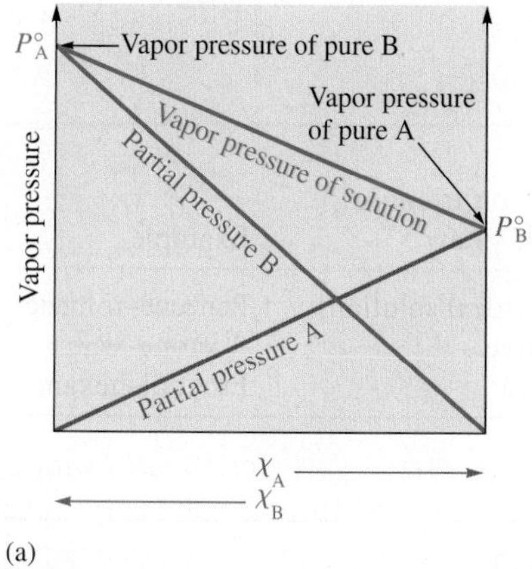

(a)

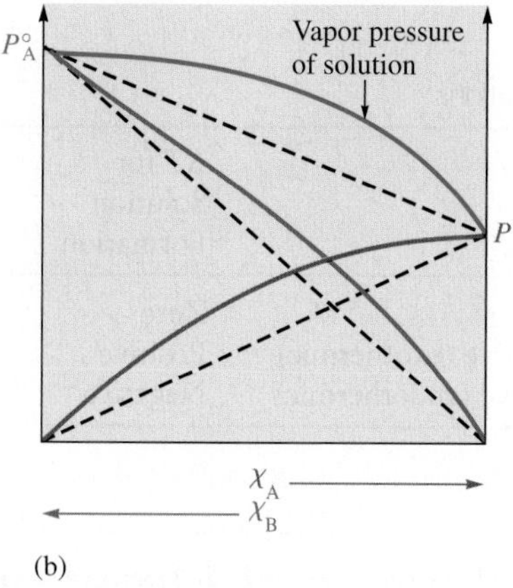

(b)

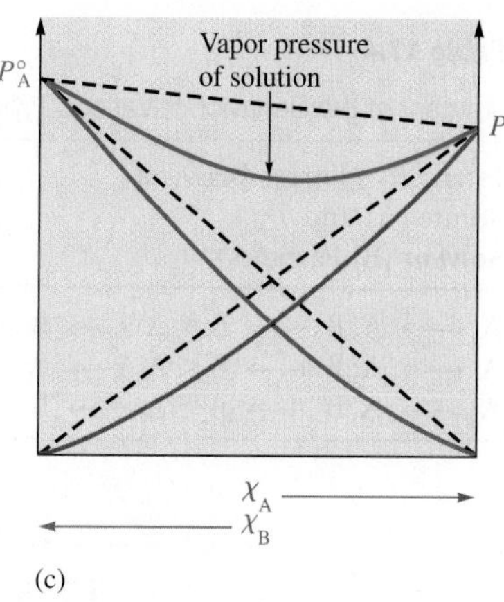

(c)

Figure 17.11
Vapor pressure for a solution of two volatile liquids. (a) The behavior predicted for an ideal liquid–liquid solution by Raoult's law. (b) A solution for which P_{Total} is larger than the value calculated from Raoult's law. This solution shows a positive deviation from Raoult's law. (c) A solution for which P_{Total} is smaller than the value calculated from Raoult's law. This solution shows a negative deviation from Raoult's law.

tions from Raoult's law. Both components have a lower escaping tendency in the solution than in the pure liquids. This behavior is illustrated by an acetone–water solution where the molecules can hydrogen-bond effectively:

$$(CH_3)_2C{=}\overset{\delta-}{O}\cdots\overset{\delta+}{H}{-}O{-}H$$

In contrast, if two liquids mix endothermically, this indicates that the solute–solvent interactions are weaker than the interactions among the molecules in the pure liquids. More energy is required to expand the liquids than is released when the liquids are mixed. In this case the molecules in the solution have a higher tendency to escape than expected, and *positive* deviations from Raoult's law are observed. An example of this case is provided by a solution of ethanol and hexane, whose Lewis structures are as follows:

$$H{-}\underset{H}{\overset{H}{C}}{-}\underset{H}{\overset{H}{C}}{-}\ddot{\underset{\cdot\cdot}{O}}{-}H \qquad H{-}\underset{H}{\overset{H}{C}}{-}\underset{H}{\overset{H}{C}}{-}\underset{H}{\overset{H}{C}}{-}\underset{H}{\overset{H}{C}}{-}\underset{H}{\overset{H}{C}}{-}\underset{H}{\overset{H}{C}}{-}H$$

Ethanol Hexane

The polar ethanol and the nonpolar hexane molecules are not able to interact effectively. Thus the enthalpy of solution is positive, as is the deviation from Raoult's law.

Finally, for a solution of very similar liquids, such as benzene and toluene,

C_6H_6 $C_6H_5CH_3$

Benzene Toluene

the enthalpy of solution is very close to zero; thus the solution closely obeys Raoult's law (ideal behavior).

A summary of the behavior of various types of solutions is given in Table 17.4.

Table 17.4

Summary of the Behavior of Various Types of Solutions

Interactive Forces Between Solute (A) and Solvent (B) Particles	ΔH_{soln}	ΔT for Solution Formation	Deviation from Raoult's Law	Example
$A \longleftrightarrow A, B \longleftrightarrow B \equiv A \longleftrightarrow B$	Zero	Zero	None (ideal solution)	Benzene–toluene
$A \longleftrightarrow A, B \longleftrightarrow B < A \longleftrightarrow B$	Negative (exothermic)	Positive	Negative	Acetone–water
$A \longleftrightarrow A, B \longleftrightarrow B > A \longleftrightarrow B$	Positive (endothermic)	Negative	Positive	Ethanol–hexane

17.5 Boiling-Point Elevation and Freezing-Point Depression

The relationships between vapor pressure and changes of state were discussed in Section 16.10.

In the preceding section, we saw how a nonvolatile solute affects the vapor pressure of a liquid solvent. Because changes of state depend on vapor pressure, the presence of a solute also affects the freezing point and boiling point of a solvent. Freezing-point depression, boiling-point elevation, and osmotic pressure (discussed in Section 17.6) are called **colligative properties.** As we will see, they are grouped together because they depend only on the number, and not on the identity, of the solute particles in an ideal solution. Because of their direct relationship to the number of solute particles, the colligative properties are very useful for characterizing the nature of a solute after it is dissolved in a solvent and for determining the molar masses of substances.

Boiling-Point Elevation

Normal boiling point was defined in Section 16.10.

The normal boiling point of a liquid is the temperature at which the vapor pressure is equal to 1 atm. We have seen that a nonvolatile solute lowers the vapor pressure of the solvent. Therefore, such a solution must be heated to a higher temperature than the boiling point of the pure solvent to reach a vapor pressure of 1 atm. This means that *a nonvolatile solute elevates the boiling point of the solvent.* Figure 17.12 shows the phase diagram for an aqueous solution containing a nonvolatile solute. Note that the liquid/vapor line for this solution is shifted to higher temperatures than that for pure water.

As you might expect, the magnitude of the boiling-point elevation depends on the concentration of the solute. The change in boiling point can be represented by the equation

$$\Delta T = K_b m_{solute}$$

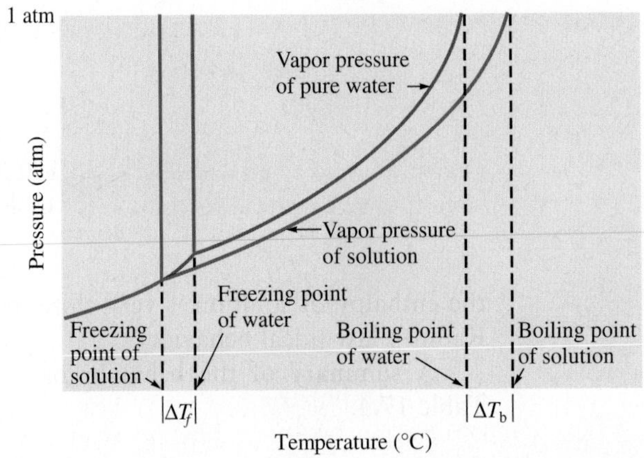

Figure 17.12 Phase diagrams for pure water (red lines) and for an aqueous solution containing a nonvolatile solute (blue lines). Note that the boiling point of the solution is higher than that of pure water. Conversely, the freezing point of the solution is lower than that of pure water. The effect of a nonvolatile solute is to extend the liquid range of a solvent.

Table 17.5

Molal Boiling-Point Elevation Constants (K_b) and Freezing-Point Depression Constants (K_f) for Several Solvents

Solvent	Boiling Point (°C)	K_b (°C kg/mol)	Freezing Point (°C)	K_f (°C kg/mol)
Water (H_2O)	100.0	0.51	0.	1.86
Carbon tetrachloride (CCl_4)	76.5	5.03	−22.99	30.
Chloroform ($CHCl_3$)	61.2	3.63	−63.5	4.70
Benzene (C_6H_6)	80.1	2.53	5.5	5.12
Carbon disulfide (CS_2)	46.2	2.34	−111.5	3.83
Ethyl ether ($C_4H_{10}O$)	34.5	2.02	−116.29	1.79
Camphor ($C_{10}H_{16}O$)	208.0	5.95	179.8	40.

where ΔT is the boiling-point elevation, or the difference between the boiling point of the solution and that of the pure solvent, K_b is a constant that is characteristic of the solvent and is called the **molal boiling-point elevation constant,** and m_{solute} is the *molality* of the solute in the solution. Values of K_b for some common solvents are given in Table 17.5.

The molar mass of a solute can be determined from the observed boiling-point elevation, as shown in Example 17.2.

OWL INTERACTIVE EXAMPLE 17.2

A solution was prepared by dissolving 18.00 g of glucose in 150.0 g of water. The resulting solution was found to have a boiling point of 100.34°C at 1 atm. Calculate the molar mass of glucose. Glucose is a molecular solid that is present as individual molecules in solution.

Solution

- What are we trying to solve?

We are trying to determine the molar mass of glucose. Molar mass has the formula $\frac{\text{mass}}{\text{mole}}$, so we need to know the mass and number of moles. We have the mass as 18.0 g, so we need to determine the number of moles. We are given the boiling point of the solution, so we make use of the equation

$$\Delta T = K_b m_{solute}$$

where $\Delta T = 100.34°\text{C} - 100.00°\text{C} = 0.34°\text{C}$

From Table 17.5, $K_b = 0.51$ for water. The molality of this solution can be calculated by rearranging the boiling-point elevation equation to give

$$m_{solute} = \frac{\Delta T}{K_b} = \frac{0.34°\text{C}}{0.51\frac{°\text{C kg}}{\text{mol}}} = 0.67 \text{ mol/kg}$$

The solution was prepared using 0.1500 kg of water. Using the definition of molality, we can find the number of moles of glucose in the solution.

$$m_{solute} = 0.67 \text{ mol/kg} = \frac{\text{mol solute}}{\text{kg solvent}} = \frac{n_{glucose}}{0.1500 \text{ kg}}$$

$$n_{glucose} = (0.67 \text{ mol/kg})(0.1500 \text{ kg}) = 0.10 \text{ mol}$$

Thus 0.10 mole of glucose weighs 18.00 g, and 1.0 mole of glucose weighs 180 g (10 × 18.00 g). The molar mass of glucose is 180 g/mol.

Freezing-Point Depression

Melting point and freezing point both refer to the temperature at which the solid and liquid coexist.

When a solute is dissolved in a solvent, the freezing point of the solution is lower than that of the pure solvent. Why? Recall that the vapor pressures of ice and liquid water are the same at 0°C. Suppose a solute is dissolved in water. The resulting solution does not freeze at 0°C because *the water in the solution has a lower vapor pressure than that of pure ice.* No ice forms under these conditions. However, the vapor pressure of ice decreases more rapidly than that of liquid water as the temperature decreases. Therefore, as the solution is cooled, the vapor pressure of the ice and that of the liquid water in the solution will eventually become equal. The temperature at which this occurs is the new freezing point of the solution and is below 0°C. The freezing point has been *depressed.*

We can account for this behavior in terms of the simple model shown in Fig. 17.13. The presence of the solute lowers the rate at which molecules in the liquid return to the solid state. Thus for an aqueous solution only the liquid state is found at 0°C. As the solution is cooled, the rate at which water molecules leave the solid ice decreases until this rate and the rate of formation of ice become equal and equilibrium is reached. This is the freezing point of the water in the solution.

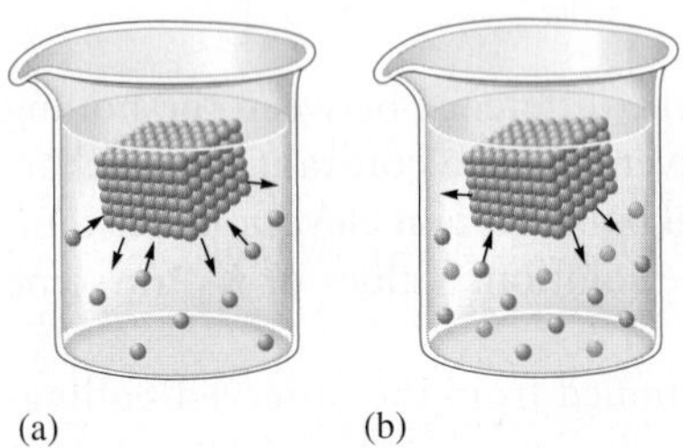

Figure 17.13
(a) Ice in equilibrium with liquid water. (b) Ice in equilibrium with liquid water containing a dissolved solute (shown in pink).

Because a solute lowers the freezing point of water, compounds such as sodium chloride and calcium chloride are often spread on streets and sidewalks to prevent ice from forming in freezing weather. Of course, if the outside temperature is lower than the freezing point of the resulting salt solution, ice forms anyway. So this procedure is not effective at extremely cold temperatures.

The solid/liquid line for an aqueous solution is shown in the phase diagram for water in Fig. 17.12. Since the presence of a solute elevates the boiling point and depresses the freezing point of the solvent, adding a solute has the effect of extending the liquid range.

The equation for freezing-point depression is analogous to that for boiling-point elevation:

$$\Delta T = K_f m_{solute}$$

where ΔT is the freezing-point depression, or the difference between the freezing point of the pure solvent and that of the solution, and K_f is a constant that is characteristic of a particular solvent and is called the **molal freezing-point depression constant.** Values of K_f for common solvents are listed in Table 17.5.

OWL INTERACTIVE EXAMPLE 17.3

What mass of ethylene glycol ($C_2H_6O_2$), the main component of antifreeze, must be added to 10.0 L of water to produce a solution for use in a car's radiator that freezes at −10.0°F (−23.3°C)? Assume that the density of water is exactly 1 g/mL.

Solution The freezing point must be lowered from 0°C to −23.3°C. To determine the molality of ethylene glycol needed to accomplish this, we can use the equation

$$\Delta T = K_f m_{solute}$$

where $\Delta T = 23.3°C$ and $K_f = 1.86$ (from Table 17.5). Solving for the molality gives

$$m_{solute} = \frac{\Delta T}{K_f} = \frac{23.3°C}{1.86 \frac{°C\ kg}{mol}} = 12.5 \text{ mol/kg}$$

This means that 12.5 moles of ethylene glycol must be added per kilogram of water. We have 10.0 L, or 10.0 kg, of water. Therefore, the total number of moles of ethylene glycol needed is

$$\frac{12.5 \text{ mol}}{\text{kg}} \times 10.0 \text{ kg} = 1.25 \times 10^2 \text{ mol}$$

The mass of ethylene glycol needed is

$$1.25 \times 10^2 \text{ mol} \times \frac{62.1 \text{ g}}{\text{mol}} = 7.76 \times 10^3 \text{ g (or 7.76 kg)}$$

Like the boiling-point elevation, the observed freezing-point depression can be used to determine molar masses and to characterize solutions.

17.6 Osmotic Pressure

Osmotic pressure, another of the colligative properties, can be understood from Fig. 17.14. A solution and pure solvent are separated by a **semipermeable membrane,** which allows *solvent but not solute* molecules to pass through. As time passes, the volume of the solution increases, whereas that of the solvent decreases. This flow of solvent into the solution through the semipermeable membrane is called **osmosis.** Eventually the liquid levels stop changing, indicating that the system has reached equilibrium. Because the liquid levels are different at this point, there is a greater hydrostatic pressure on the solution than on the pure solvent. This excess pressure is called the **osmotic pressure.** We can take another view of this phenomenon, as illustrated in Fig. 17.15. Osmosis can be prevented by applying a pressure to the solution. *The pressure that just stops the osmosis is equal to the osmotic pressure of the solution.*

A simple model to explain osmotic pressure can be constructed as shown in Fig. 17.16. The membrane allows only solvent molecules to pass through. However, the initial rates of solvent transfer to and from the solution are not the same. The solute particles interfere with the passage of solvent from the solution, so the rate of transfer is slower from the solution to the solvent than in the reverse direction. Thus there is a net transfer of solvent molecules into the solution, causing the solution volume to increase. As the solution level rises in the tube, the resulting pressure exerts an extra "push" on the solvent mole-

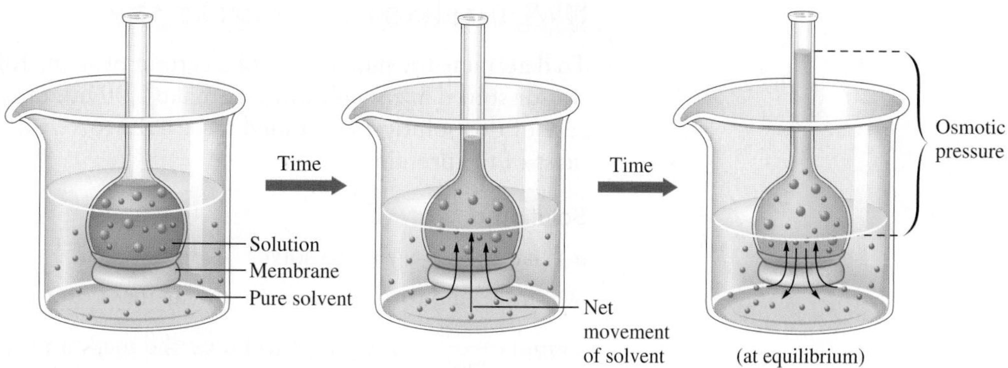

Figure 17.14
A tube with a bulb on the end is covered by a semipermeable membrane. The solution inside the tube is bathed in the pure solvent. There is a net transfer of solvent molecules into the solution until the hydrostatic pressure equalizes the solvent flow in both directions. The solvent level shows little change, because the narrow stem of the bulb has a very small volume.

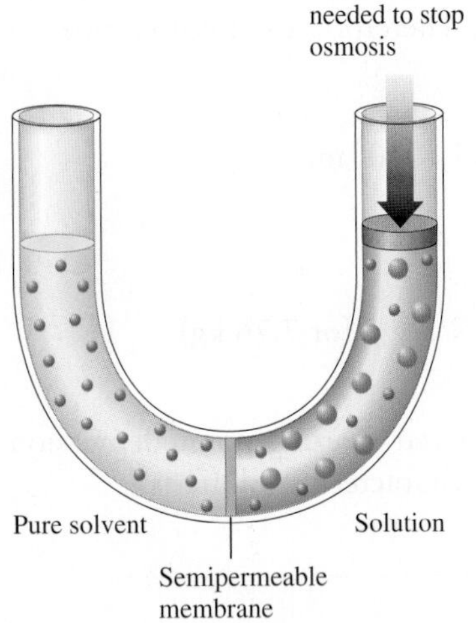

Figure 17.15
The normal flow of solvent into the solution (osmosis) can be prevented by applying an external pressure to the solution. The minimum pressure required to stop the osmosis is equal to the osmotic pressure of the solution.

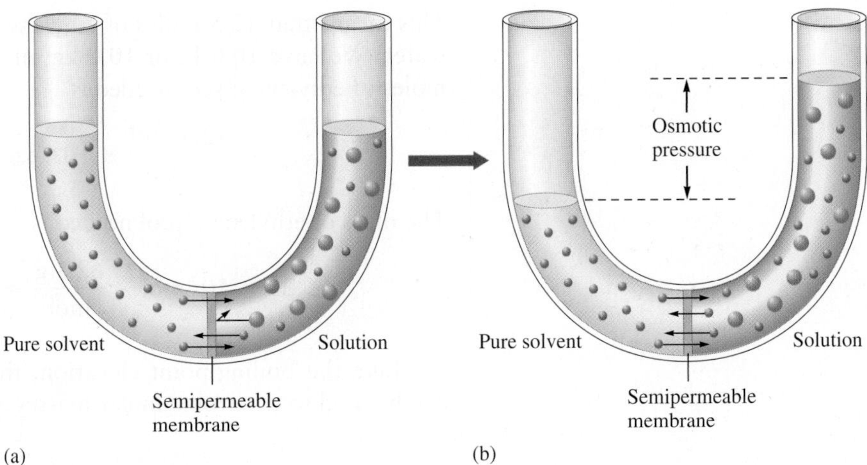

Figure 17.16
(a) A pure solvent and its solution (containing a nonvolatile solute) are separated by a semipermeable membrane through which solvent molecules (blue) can pass but solute molecules (green) cannot. The rate of solvent transfer is greater from solvent to solution than from solution to solvent. (b) The system at equilibrium, where the rate of solvent transfer is the same in both directions.

cules in the solution, forcing them back through the membrane. Eventually, enough pressure develops so that the solvent transfer becomes equal in both directions. At this point, equilibrium is achieved and the levels stop changing.

Osmotic pressure can be used to characterize solutions and to determine molar masses just as the other colligative properties can; however, osmotic pressure is particularly useful because a small concentration of solute produces a relatively large osmotic pressure.

Experiments show that the dependence of the osmotic pressure on solution concentration is represented by the equation

$$\pi = MRT$$

where π is the osmotic pressure in atmospheres, M is the molarity of the solute, R is the gas law constant, and T is the Kelvin temperature.

A molar mass determination using osmotic pressure is illustrated in Example 17.4.

OWL INTERACTIVE EXAMPLE 17.4

To determine the molar mass of a certain protein, 1.00×10^{-3} g of the protein was dissolved in enough water to make 1.00 mL of solution. The osmotic pressure of this solution was found to be 1.12 torr at 25.0°C. Calculate the molar mass of the protein.

Solution

■ What are we trying to solve?

We are trying to determine the molar mass of a protein. Molar mass has the formula $\frac{\text{mass}}{\text{mole}}$, so we need to know the mass and number of moles. We have the mass as 1.00×10^{-3} g, so we need to determine the number of moles. We are given the osmotic pressure of the solution, so we use the equation

$$\pi = MRT$$

In this case

$$\pi = 1.12 \text{ torr} \times \frac{1 \text{ atm}}{760 \text{ torr}} = 1.47 \times 10^{-3} \text{ atm}$$

$$R = 0.08206 \text{ L atm K}^{-1} \text{ mol}^{-1}$$

$$T = 25.0 + 273 = 298 \text{ K}$$

Note that the osmotic pressure must be converted to atmospheres because of the units of R.

Solving for M gives

$$M = \frac{1.47 \times 10^{-3} \text{ atm}}{(0.08206 \text{ L atm K}^{-1} \text{ mol}^{-1})(298 \text{ K})} = 6.01 \times 10^{-5} \text{ mol/L}$$

Measurements of osmotic pressure generally give much more accurate molar mass values than those from freezing-point or boiling-point changes.

Since 1.00×10^{-3} g of protein was dissolved in 1.00 mL of solution, the mass of protein per liter of solution is 1.00 g. The solution's concentration is 6.01×10^{-5} mol/L. This concentration is produced from 1.00×10^{-3} g of protein per milliliter, or 1.00 g/L. Thus 6.01×10^{-5} mole of protein has a mass of 1.00 g, and

$$\frac{1.00 \text{ g}}{6.01 \times 10^{-5} \text{ mol}} = \frac{x \text{ g}}{1.00 \text{ mol}}$$

$$x = 1.66 \times 10^{4} \text{ g/mol}$$

The molar mass of the protein is 1.66×10^4. This molar mass may seem very large, but it is relatively small for a protein.

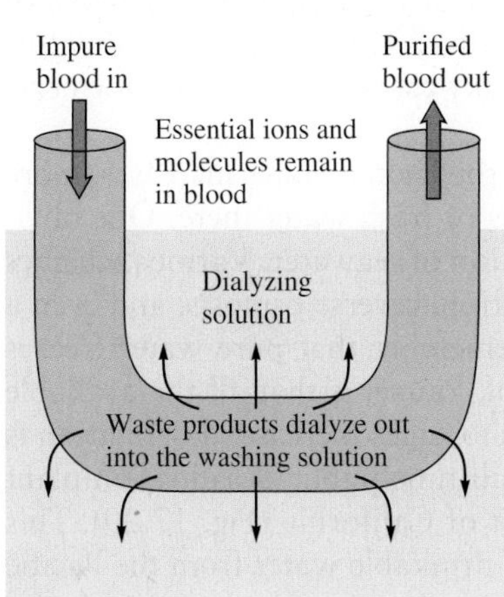

Figure 17.17
Representation of the functioning of an artificial kidney.

In osmosis a semipermeable membrane prevents transfer of *all* solute particles. A similar phenomenon called **dialysis** occurs at the walls of most plant and animal cells. However, in this case the membrane allows transfer of both solvent molecules and *small* solute molecules and ions. One of the most important applications of dialysis is the use of artificial kidney machines to purify the blood. The blood is passed through a cellophane tube, which acts as the semipermeable membrane. The tube is immersed in a dialyzing solution (Fig. 17.17). This "washing" solution contains the same concentrations of ions and small molecules as blood but has none of the waste products normally removed by the kidneys. The resulting dialysis of waste products cleanses the blood.

Solutions that have identical osmotic pressures are said to be **isotonic solutions.** Fluids administered intravenously must be isotonic with body fluids. For example, if cells are bathed in a hypertonic solution, which is a solution having an osmotic pressure higher than that of the cell fluids, the cells will shrivel because of a net transfer of water out of the cells. This phenomenon is called *crenation.* The opposite phenomenon, called *lysis,* occurs when cells are bathed in a hypotonic solution, a solution with an osmotic pressure lower than that of the cell fluids. In this case the cells rupture because of the flow of water into the cells.

We can use the phenomenon of crenation to our advantage. Food can be preserved by treating its surface with a solute that forms a solution hypertonic to bacteria cells. Bacteria on the food then tend to shrivel and die. This is why salt can be used to protect meat and sugar can be used to protect fruit.

OWL INTERACTIVE EXAMPLE 17.5

What concentration of sodium chloride in water is needed to produce an aqueous solution isotonic with blood (π = 7.70 atm at 25°C)?

Solution We can calculate the molarity of the solute from the equation

$$\pi = MRT \qquad \text{or} \qquad M = \frac{\pi}{RT}$$

$$M = \frac{7.70 \text{ atm}}{(0.08206 \text{ L atm K}^{-1} \text{ mol}^{-1})(298 \text{ K})} = 0.315 \text{ mol/L}$$

This represents the total molarity of solute particles. But NaCl gives two ions per formula unit. Therefore, the concentration of NaCl needed is 0.315/2, or 0.158 *M*. That is,

$$\underset{0.1575\,M}{\text{NaCl}} \longrightarrow \underbrace{\underset{0.1575\,M}{\text{Na}^+} + \underset{0.1575\,M}{\text{Cl}^-}}_{0.315\,M}$$

Reverse Osmosis

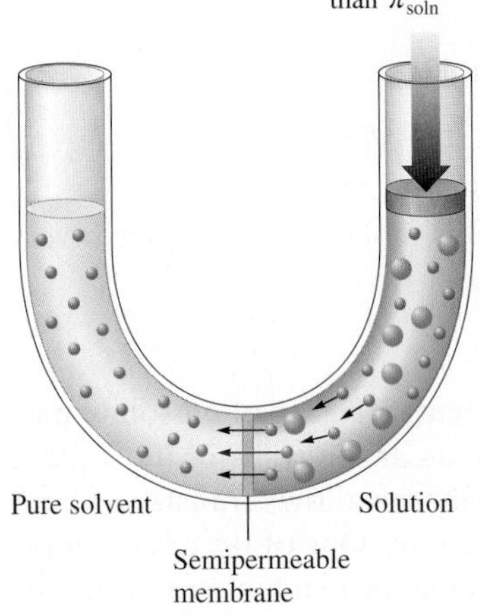

Figure 17.18
Reverse osmosis. A pressure greater than the osmotic pressure of the solution is applied, which causes a net flow of solvent molecules (blue) from the solution to the pure solvent. The solute molecules (green) remain behind.

If a solution in contact with pure solvent across a semipermeable membrane is subjected to an external pressure larger than its osmotic pressure, **reverse osmosis** occurs. The pressure will cause a net flow of solvent from the solution to the solvent, as shown in Fig. 17.18. In reverse osmosis, the semipermeable membrane acts as a "molecular filter" to remove solute particles. This fact is potentially applicable to the **desalination** (removal of dissolved salts) of seawater, which is highly hypertonic to body fluids and thus is not drinkable.

A small-scale, manually operated reverse osmosis desalinator has been developed by the U.S. Navy to provide fresh water on life rafts (Fig. 17.19). Potable water can be supplied by this desalinator at the rate of 1.25 gallons of water per hour—enough to keep 25 people alive. This compact desalinator, which weighs only 10 pounds, replaces the bulky cases of fresh water formerly stored on Navy life rafts.

As the population of the Sunbelt areas of the United States increases, more demand will be placed on the limited supplies of fresh water there. One obvious source of fresh water is from the desalination of seawater. Various schemes have been suggested, including solar evaporation, reverse osmosis, and even a plan for towing icebergs from Antarctica (remember that pure water freezes out of an aqueous solution). The problem, of course, is that all the available processes are expensive. However, as water shortages increase, desalination is becoming necessary. For example, the first full-time public desalination plant is found on Catalina Island, just off the coast of California (Fig. 17.20). This plant, which can produce 132,000 gallons of drinkable water from the Pacific

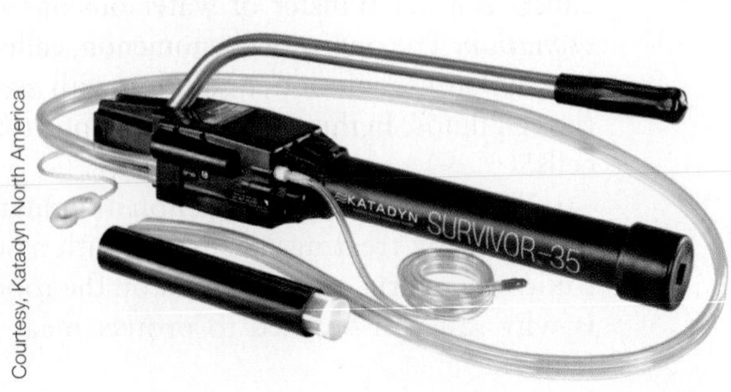

Figure 17.19
A commercially available desalinator, similar to those developed by the Navy for life rafts. The essential part of these desalinators is a cellophane-like membrane wrapped around a tube with holes in it. Seawater is forced through the tube by a hand-operated pump at pressures of about 70 atm to produce water with a salt content only 40% higher than that from a typical tap.

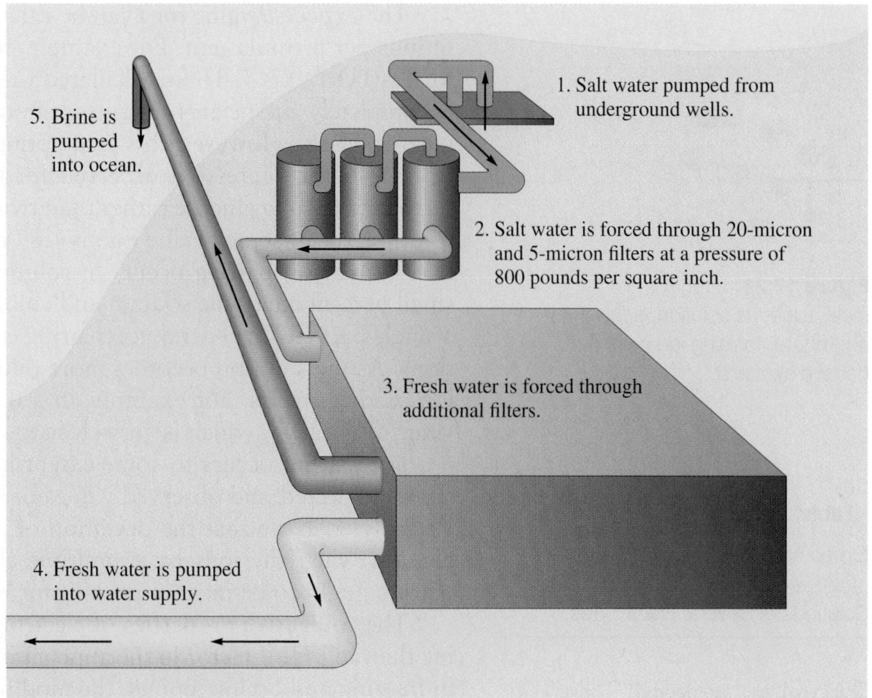

Figure 17.20
Residents of Catalina Island off the coast of southern California are benefiting from a desalination plant that can supply 132,000 gallons of drinkable water per day, one-third of the island's daily needs.

Ocean every day, operates by reverse osmosis. Powerful pumps, developing over 800 lb/in^2 of pressure, are used to force seawater through synthetic semipermeable membranes.

Catalina Island's plant may be just the beginning. The city of Santa Barbara opened a $40 million desalination plant in 1992 that can produce 8 million gallons of drinkable water per day, and the Southern California town of Carlsbad opened a plant in 2012 that can produce 50 million gallons of drinking water from seawater each day. This $320 million project is the largest desalination plant in the Western Hemisphere. Desalination plants are also in the works for Huntington Beach, California, and Camp Pendleton, a military base just north of Carlsbad.

17.7 Colligative Properties of Electrolyte Solutions

As we have seen, the colligative properties of solutions depend on the total concentration of solute particles. For example, a 0.10 *m* glucose solution shows a freezing-point depression of 0.186°C:

$$\Delta T = K_f m = (1.86°\text{C kg/mol})(0.10 \text{ mol/kg}) = 0.186°\text{C}$$

On the other hand, a 0.10 *m* sodium chloride solution should show a freezing-point depression of 0.37°C, since the solution is 0.10 *m* Na^+ ions and 0.10 *m* Cl^- ions. Therefore, because the solution is 0.20 *m* in solute particles, $\Delta T = (1.86°\text{C kg/mol})(0.20 \text{ mol/kg}) = 0.37°\text{C}$.

Dutch chemist J. H. van't Hoff (1852–1911) received the first Nobel Prize in chemistry in 1901.

The relationship between the moles of solute dissolved and the moles of particles in solution is usually expressed by the **van't Hoff factor** (*i*):

$$i = \frac{\text{moles of particles in solution}}{\text{moles of solute dissolved}}$$

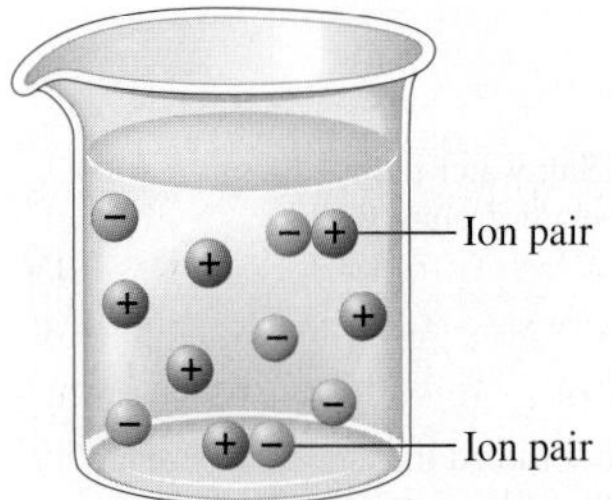

Figure 17.21
In an aqueous solution, a few ions aggregate, forming ion pairs that behave as a unit.

The *expected* value for i can be calculated for a salt by noting the number of ions per formula unit. For example, for NaCl, i is 2; for K_2SO_4, i is 3; and for $Fe_3(PO_4)_2$, i is 5. These calculated values assume that when a salt dissolves, it completely dissociates into its component ions, which then move around independently. However, this assumption is not always true. For example, the freezing-point depression observed for 0.10 m NaCl is 1.87 times that observed for 0.10 m glucose rather than twice as great. That is, for a 0.10 m NaCl solution, the observed value for i is 1.87 rather than 2. Why? The best explanation is that **ion pairing** occurs in solution (Fig. 17.21). At a given instant, a small percentage of the sodium and chloride ions are paired and thus count as a single particle. In general, ion pairing is most important in concentrated solutions. As the solution becomes more dilute, the ions are farther apart and less ion pairing occurs. For example, in a 0.0010 m NaCl solution, the observed value of i is 1.97, which is very close to the expected value.

Ion pairing occurs to some extent in all electrolyte solutions. Table 17.6 shows expected and observed values of i for a given concentration of various electrolytes. Note that the deviation of i from the expected value tends to be greatest when the ions have multiple charges. This is expected because ion pairing ought to be most important for highly charged ions.

Table 17.6

Expected and Observed Values of the van't Hoff Factor for 0.05 m Solutions of Several Electrolytes

Electrolyte	i (expected)	i (observed)
NaCl	2.0	1.9
$MgCl_2$	3.0	2.7
$MgSO_4$	2.0	1.3
$FeCl_3$	4.0	3.4
HCl	2.0	1.9
Glucose*	1.0	1.0

*A nonelectrolyte shown for comparison.

The colligative properties of electrolyte solutions are described by including the van't Hoff factor in the appropriate equation. For example, for changes in freezing and boiling points, the modified equation is

$$\Delta T = imK$$

where K represents the freezing-point depression constant or the boiling-point elevation constant for the solvent.

For the osmotic pressure of electrolyte solutions, the equation is

$$\pi = iMRT$$

OWL INTERACTIVE EXAMPLE 17.6

The observed osmotic pressure for a 0.10 M solution of $Fe(NH_4)_2(SO_4)_2$ at 25°C is 10.8 atm. Compare the expected and experimental values for i.

Solution The ionic solid $Fe(NH_4)_2(SO_4)_2$ dissociates in water to produce 5 ions:

$$Fe(NH_4)_2(SO_4)_2 \xrightarrow{H_2O} Fe^{2+} + 2NH_4^+ + 2SO_4^{2-}$$

Thus the expected value for i is 5. We can obtain the experimental value for i by using the equation for osmotic pressure:

$$\pi = iMRT \quad \text{or} \quad i = \frac{\pi}{MRT}$$

where $\pi = 10.8$ atm, $M = 0.10$ mol/L, $R = 0.08206$ L atm K^{-1} mol^{-1}, and $T = 25 + 273 = 298$ K. Substituting these values into the equation gives

$$i = \frac{\pi}{MRT} = \frac{10.8 \text{ atm}}{\left(0.10\frac{\text{mol}}{\text{L}}\right)\left(0.08206\frac{\text{L atm}}{\text{K mol}}\right)(298 \text{ K})} = 4.42$$

The experimental value for i is less than the expected value, presumably because of ion pairing.

17.8 Colloids

Mud can be suspended in water by vigorous stirring. When the stirring stops, most of the particles rapidly settle out, but even after several days some of the smallest particles remain suspended. Although undetected in normal lighting, their presence can be demonstrated by shining a beam of intense light through the suspension. The beam is visible from the side because the light is scattered by the suspended particles. In a true solution, on the other hand, the beam is invisible from the side because the individual ions and molecules dispersed in the solution are too small to scatter visible light.

The scattering of light by particles is called the **Tyndall effect** (Fig. 17.22) and is often used to distinguish between a suspension and a true solution.

A suspension of tiny particles in some medium is called a **colloidal dispersion,** or a **colloid.** The suspended particles can be single, large molecules or aggregates of molecules or ions ranging in size from 1 to 1000 nanometers. Colloids are classified according to the states of the dispersed phase and the dispersing medium. Table 17.7 summarizes various types of colloids.

What stabilizes a colloid? Why do the particles remain suspended rather than forming larger aggregates and precipitating out? The answer is compli-

Figure 17.22
The Tyndall effect. The solution on the left does not show the path of the light beam, whereas the suspension on the right clearly shows the light path.

Table 17.7
Types of Colloids

Examples	Dispersing Medium	Dispersed Substance	Colloid Type
Fog, aerosol sprays	Gas	Liquid	Aerosol
Smoke, airborne bacteria	Gas	Solid	Aerosol
Whipped cream, soap suds	Liquid	Gas	Foam
Milk, mayonnaise	Liquid	Liquid	Emulsion
Paint, clays, gelatin	Liquid	Solid	Sol
Marshmallow, polystyrene foam	Solid	Gas	Solid foam
Butter, cheese	Solid	Liquid	Solid emulsion
Ruby glass	Solid	Solid	Solid sol

Chemical Insights Organisms and Ice Formation

The ice-cold waters of the polar oceans are teeming with fish that seem immune to freezing. One might think that these fish have some kind of antifreeze in their blood. However, studies show that they are protected from freezing in a way that is very different from the way antifreeze protects our cars. As we have seen in this chapter, solutes such as sugars, salts, and ethylene glycol lower the temperature at which the solid and liquid phases of water can coexist. However, the fish could not tolerate high concentrations of solutes in their blood because of the osmotic pressure effects. Instead, they are protected by proteins in their blood. These proteins allow the water in the bloodstream to be supercooled—exist below 0°C—without forming ice. The proteins apparently coat the surface of each tiny ice crystal as soon as it begins to form, thus preventing it from growing to a size that would cause biological damage. Similar "antifreeze" proteins have been found in other organisms, including the springtail snow flea, beetles, moths, and various bacteria and plants.

Recently, teams of researchers at the University of Notre Dame and the University of Alaska–Fairbanks discovered a new type of antifreeze molecule in the Upis beetle. Instead of a protein-based antifreeze, the Upis antifreeze is a complex sugar called *xylomannan,* which appears to be equally as effective as the proteins at preventing freezing. The Upis beetle freezes at −19°F but remarkably can survive exposure to temperatures as low as about −100°F.

Although it might at first seem surprising, this research on antifreeze molecules has attracted the attention of ice cream manufacturers. Premium-quality ice cream is smooth; it does not have large ice crystals in it. The makers of ice cream would like to incorporate natural antifreeze molecules, or at least molecules that behave similarly, into ice cream to prevent growth of ice crystals during storage.

Fruit and vegetable growers have a similar interest: They also want to prevent ice formation that could damage their crops during an unusual cold wave. However, this is a very different kind of problem than keeping polar fish from freezing. Many types of fruits and vegetables are colonized by bacteria that manufacture a protein that *encourages* freezing by acting as a nucleating agent for an ice crystal. Chemists have identified the offending protein in the bacteria as well as the gene that is responsible for making it. They have learned to modify the genetic material of these bacteria in such a way that removes their ability to make the protein that encourages ice crystal formation. If testing shows that these modified bacteria have no harmful effects on the crop or the environment, the original bacteria strain will be replaced with the new form so that ice crystals will not form so readily when a cold snap occurs.

cated, but the main effect that stabilizes the colloid seems to be *electrostatic repulsion.* A colloid, like all other macroscopic substances, is electrically neutral. However, when a colloid is placed in an electric field, the dispersed particles all migrate to the same electrode; thus they must all have the same charge. How is this possible? The center of a colloidal particle (a tiny ionic crystal, a group of molecules, or a single large molecule) attracts from the medium a layer of ions, all of the same charge. This group of ions in turn attracts another layer of oppositely charged ions, as shown in Fig. 17.23. Because the colloidal particles all have an outer layer of ions with the same charge, they repel each other; thus they do not easily aggregate to form particles that are large enough to precipitate.

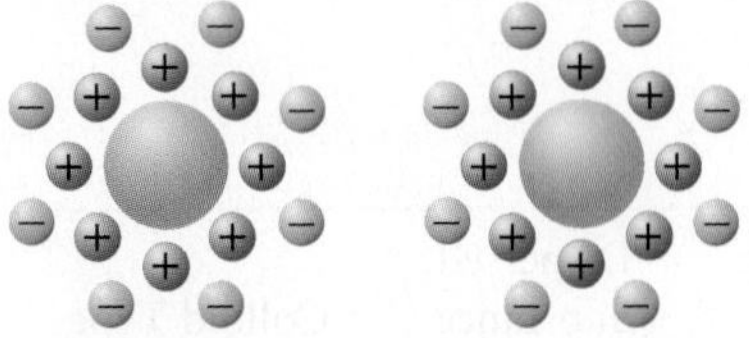

Figure 17.23
A representation of two colloidal particles. In each the center particle is surrounded by a layer of positive ions, with negative ions in the outer layer. Thus, although the particles are electrically neutral, they still repel each other because of their outer negative layer of ions.

The destruction of a colloid, called **coagulation,** can usually be accomplished either by heating or by adding an electrolyte. Heating increases the velocities of the colloidal particles, causing them to collide with enough energy so that the ion barriers are penetrated. This allows the particles to aggregate.

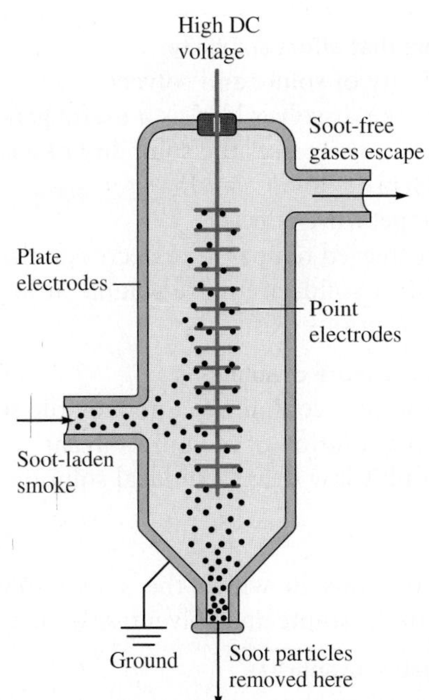

Figure 17.24
The Cottrell precipitator installed in a smokestack. The charged plates attract the colloidal particles because of their ion layers and thus remove them from the smoke.

As this is repeated many times, the particle grows to a point where it settles out. Adding an electrolyte neutralizes the adsorbed ion layers. This is why clay suspended in rivers is deposited where the river reaches the ocean, forming the deltas characteristic of large rivers like the Mississippi. The high salt content of the seawater causes the colloidal clay particles to coagulate.

The removal of soot from smoke is another example of the coagulation of a colloid. When smoke is passed through an electrostatic precipitator (Fig. 17.24), the suspended solids are removed. The use of precipitators has produced an immense improvement in the air quality of heavily industrialized cities.

Key Terms

Section 17.1
molarity
mass percent
mole fraction
molality

Section 17.2
enthalpy (heat) of solution
enthalpy (heat) of hydration
supercritical fluid

Section 17.3
Henry's law
thermal pollution

Section 17.4
Raoult's law
ideal solution

Section 17.5
colligative properties
molal boiling-point elevation constant

For Review

Sign in at **www.cengage.com/owl** to:
- View tutorials and simulations, develop problem-solving skills, and complete online homework assigned by your professor.
- Download Go Chemistry mini lecture modules for quick review and exam prep from OWL (or purchase them at **www.cengagebrain.com**)

Solution composition
- Molarity (M): moles solute per liter of solution
- Mass percent: ratio of mass of solute to mass of solution times 100%
- Mole fraction (χ): ratio of moles of a given component to total moles of all components
- Molality (m): moles solute per mass of solvent (in kg)
- Normality (N): number of equivalents per liter of solution

Enthalpy of solution (ΔH_{soln})
- The enthalpy change accompanying solution formation
- Can be partitioned into
  - The energy required to overcome the solute–solute interactions
  - The energy required to "make holes" in the solvent
  - The energy associated with solute–solvent interactions

molal freezing-point depression constant

Section 17.6
semipermeable membrane
osmosis
osmotic pressure
dialysis
isotonic solution
reverse osmosis
desalination

Section 17.7
van't Hoff factor
ion pairing

Section 17.8
Tyndall effect
colloid (colloidal dispersion)
coagulation

Factors that affect solubility

- Polarity of solute and solvent
  - "Like dissolves like" is a useful generalization
- Pressure increases the solubility of gases in a solvent
  - Henry's law: $C = kP$
- Temperature effects
  - Increased temperature decreases the solubility of a gas in water
  - Most solids are more soluble at higher temperatures but important exceptions exist

Vapor pressure of solutions

- A solution containing a nonvolatile solute has a lower vapor pressure than a solution of the pure solvent
- Raoult's law defines an ideal solution

$$P_{\text{vapor}}^{\text{soln}} = \chi_{\text{solvent}} P_{\text{vapor}}^{\text{solvent}}$$

  - Solutions in which the solute–solvent attractions differ from the solute–solute and solvent–solvent attractions violate Raoult's law

Colligative properties

- Depend on the number of solute particles present
- Boiling-point elevation: $\Delta T = K_b m_{\text{solute}}$
- Freezing-point lowering: $\Delta T = K_f m_{\text{solute}}$
- Osmotic pressure: $\Pi = MRT$
  - Osmosis occurs when a solution and pure solvent are separated by a semipermeable membrane that allows solvent molecules to pass but not solute particles
  - Reverse osmosis occurs when the applied pressure is greater than the osmotic pressure of the solution
- Because colligative properties depend on the number of particles, solutes that break into several ions when they dissolve have an effect proportional to the number of ions produced
  - The van't Hoff factor i represents the number of ions produced by each formula unit of solute

Colloids

- A suspension of tiny particles stabilized by electrostatic repulsion among the ion layers surrounding the individual particles
- Can be coagulated (destroyed) by heating or adding an electrolyte

Discussion Questions

These questions are designed to be considered by groups of students in class. Often these questions work well for introducing a particular topic in class.

1. Consider Fig. 17.8. According to the caption and picture, water is transferred from one beaker to another.
   a. Explain why this occurs.
   b. The explanation in the text uses terms such as *vapor pressure* and *equilibrium*. Explain what these terms have to do with the phenomenon. For example, what does the equilibrium involve?
   c. Does all the water end up in the second beaker?
   d. Is the water in the beaker containing the solute evaporating? If so, is the rate of evaporation increasing, decreasing, or staying constant?

   Draw pictures to illustrate your explanations.
2. Consider Fig. 17.8. Suppose that instead of having a nonvolatile solute in the solvent in one beaker, the two beakers have different volatile liquids. That is, suppose one beaker contains liquid A ($P_{\text{vap}} = 50$ torr) and the other beaker contains liquid B ($P_{\text{vap}} = 100$ torr). Explain what happens as time passes. How is this similar to the first case shown in the figure? How is it different?

3. Assume that you place a freshwater plant into a saltwater solution and examine it under a microscope. What happens to the plant cells? What if you placed a saltwater plant in pure water? Explain. Draw pictures to illustrate your explanations.
4. How does ΔH_{soln} relate to deviations from Raoult's law? Explain.
5. You have read that adding NaCl to water can both increase its boiling point and decrease its freezing point. A friend of yours explains it to you like this: "The ions prevent freezing by blocking the water molecules from joining together. The hydration of the ions also makes the water boil at a higher temperature." What do you say to your friend?
6. You drop an ice cube (made from pure water) into a saltwater solution at 0°C. Explain what happens and why.
7. Using the phase diagram for water and Raoult's law, explain why salt is spread on the roads in winter (even when the temperature is below freezing).
8. You and your friend are each drinking cola from separate 2-L bottles. Both colas are equally carbonated. You are able to drink 1 L of cola, but your friend can drink only about half a liter. You each close the bottles, and place them in the refrigerator. The next day when you each go to get the colas, whose will be more carbonated and why?
9. Is molality or molarity dependent on temperature? Explain your answer. Why is molality, and not molarity, used in the equations describing freezing-point depression and boiling-point elevation?
10. If a solution shows positive deviations from Raoult's law, would you expect it to have a higher or lower boiling point than if it were ideal? Explain.
11. Consider a beaker of salt water sitting open in a room. Over time does the vapor pressure increase, decrease, or stay the same? Explain.

Exercises

OWL Interactive versions of these problems may be assigned in OWL.

A blue exercise number indicates that the answer to that exercise appears at the back of this book and a solution appears in the *Solutions Guide*.

Solution Composition

12. The four most common ways to describe solution composition are mass percent, mole fraction, molarity, and molality. Define each of these solution composition terms. Why is molarity temperature-dependent, whereas the other three solution composition terms are temperature-independent?
13. Write equations showing the ions present after the following strong electrolytes are dissolved in water.
    a. HNO_3
    b. Na_2SO_4
    c. $Al(NO_3)_3$
    d. $SrBr_2$
    e. $KClO_4$
    f. NH_4Br
    g. NH_4NO_3
    h. $CuSO_4$
    i. NaOH
14. Calculate the sodium ion concentration when 70.0 mL of 3.0 *M* sodium carbonate is added to 30.0 mL of 1.0 *M* sodium bicarbonate.
15. An aqueous antifreeze solution is 40.0% ethylene glycol ($C_2H_6O_2$) by mass. The density of the solution is 1.05 g/cm^3. Calculate the molality, molarity, and mole fraction of the ethylene glycol.
16. Common commercial acids and bases are aqueous solutions with the following properties:

	Density (g/cm^3)	Mass Percent of Solute
Hydrochloric acid	1.19	38
Nitric acid	1.42	70.
Sulfuric acid	1.84	95
Acetic acid	1.05	99
Ammonia	0.90	28

Calculate the molarity, molality, and mole fraction of each of these reagents.

17. A solution is prepared by mixing 50.0 mL of toluene ($C_6H_5CH_3$, $d = 0.867$ g/cm^3) with 125 mL of benzene (C_6H_6, $d = 0.874$ g/cm^3). Assuming that the volumes add on mixing, calculate the mass percent, mole fraction, molality, and molarity of the toluene.
18. A bottle of wine contains 12.5% ethanol by volume. The density of ethanol (C_2H_5OH) is 0.789 g/cm^3. Calculate the concentration of ethanol in wine in terms of mass percent and molality.
19. A 1.37 *M* aqueous solution of citric acid ($H_3C_6H_5O_7$) has a density of 1.10 g/cm^3. What are the mass percent, molality, and mole fraction of the citric acid?
20. Determine the molarity and mole fraction of a 1.00 *m* solution of acetone (CH_3COCH_3) dissolved in ethanol (C_2H_5OH). (Density of acetone = 0.788 g/cm^3; density of ethanol = 0.789 g/cm^3.) Assume that the final volume equals the sum of the volumes of acetone and ethanol.
21. A solution of phosphoric acid was made by dissolving 10.0 g of H_3PO_4 in 100.00 mL of water. The resulting volume was 104 mL. Calculate the density, mole fraction, molarity, and molality of the solution. Assume water has a density of 1.00 g/cm^3.
22. In a lab you need at least 100 mL of each of the following solutions. Explain how you would proceed by using the given information.
    a. 2.0 *m* KCl in water (density of H_2O = 1.00 g/cm^3)
    b. 15% NaOH by mass in water
    c. 25% NaOH by mass in CH_3OH ($d = 0.79$ g/cm^3)
    d. 0.10 mole fraction of $C_6H_{12}O_6$ in water

Thermodynamics of Solutions and Solubility

23. What does the axiom "like dissolves like" mean? There are four types of solute–solvent combinations: polar sol-

utes in polar solvents, nonpolar solutes in polar solvents, and so on. For each type of solution, discuss the magnitude of ΔH_{soln}.

24. For each of the following solute–solvent combinations, state the sign and relative magnitudes for ΔH_1, ΔH_2, ΔH_3, and ΔH_{soln} (as defined in Fig. 17.1 of the text). Explain your answers.

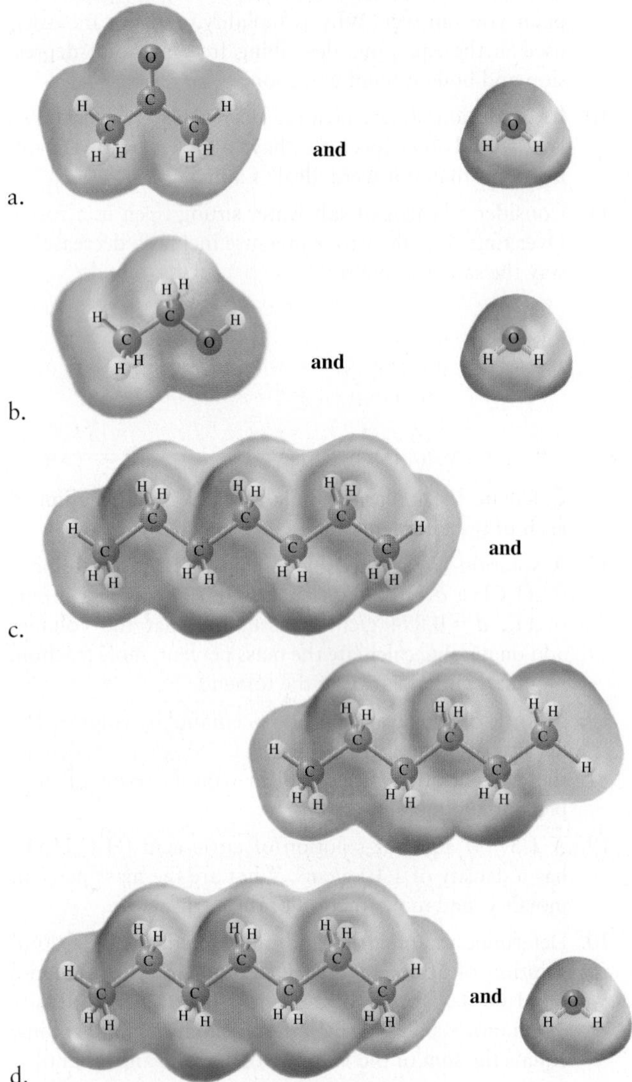

25. In order for sodium chloride to dissolve in water, a small amount of energy must be added during solution formation. This is not energetically favorable. Why is NaCl so soluble in water?

26. The high melting points of ionic solids indicate that a lot of energy must be supplied to separate the ions from one another. How is it possible that the ions can separate from one another when soluble ionic compounds are dissolved in water, often with essentially no temperature change?

27. The lattice energy* of NaI is −686 kJ/mol, and the enthalpy of hydration is −694 kJ/mol. Calculate the enthalpy of solution per mole of solid NaI. Describe the process to which this enthalpy change applies.

28. a. Use the following data to calculate the enthalpy of hydration for calcium chloride and calcium iodide.

	Lattice Energy*	ΔH_{soln}
$CaCl_2(s)$	−2247 kJ/mol	−46 kJ/mol
$CaI_2(s)$	−2059 kJ/mol	−104 kJ/mol

b. Based on your answers to part a, which ion, Cl^- or I^-, is more strongly attracted to water?

29. Although $Al(OH)_3$ is insoluble in water, NaOH is very soluble. Explain this difference in terms of lattice energies.

30. Which ion in each of the following pairs would you expect to be more strongly hydrated? Why?
   a. Na^+ or Mg^{2+}
   b. Mg^{2+} or Be^{2+}
   c. Fe^{2+} or Fe^{3+}
   d. F^- or Br^-
   e. Cl^- or ClO_4^-
   f. ClO_4^- or SO_4^{2-}

31. For each of the following pairs, predict which substance would be more soluble in water.

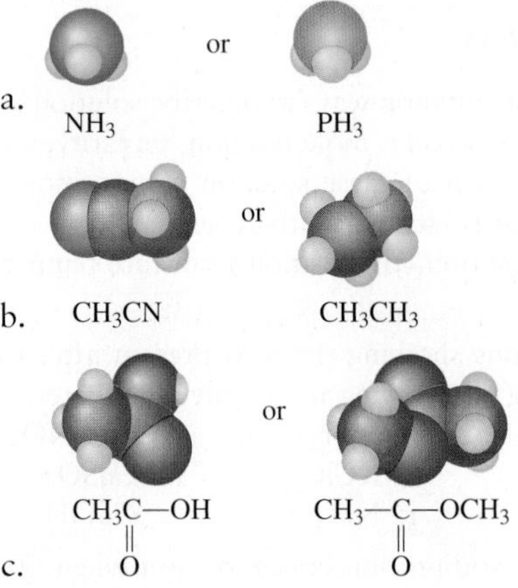

32. Which solvent, water or carbon tetrachloride, would you choose to dissolve each of the following?
   a. KrF_2
   b. SF_2
   c. SO_2
   d. CO_2
   e. MgF_2
   f. CH_2O
   g. $CH_2{=}CH_2$

*Lattice energy was defined in Chapter 13 as the energy change for the process $M^+(g) + X^-(g) \rightarrow MX(s)$.

33. Rationalize the trend in water solubility for the following simple alcohols.

Alcohol	Solubility (g/100 g H_2O at 20°C)
Methanol, CH_3OH	Soluble in all proportions
Ethanol, CH_3CH_2OH	Soluble in all proportions
Propanol, $CH_3CH_2CH_2OH$	Soluble in all proportions
Butanol, $CH_3(CH_2)_2CH_2OH$	8.14
Pentanol, $CH_3(CH_2)_3CH_2OH$	2.62
Hexanol, $CH_3(CH_2)_4CH_2OH$	0.59
Heptanol, $CH_3(CH_2)_5CH_2OH$	0.09

34. In the flushing and cleaning of columns used in liquid chromatography, a series of solvents is used. Hexane (C_6H_{14}), chloroform ($CHCl_3$), methanol (CH_3OH), and water are passed through the column in that order. Rationalize the order in terms of intermolecular forces and the mutual solubility (miscibility) of the solvents.

35. Structure, pressure, and temperature all have an effect on solubility. Discuss each of their effects. What is Henry's law? Why does Henry's law not work for $HCl(g)$? What do the terms *hydrophobic* and *hydrophilic* mean?

36. $O_2(g)$ obeys Henry's law in water but not in blood (an aqueous solution). Why?

37. The solubility of nitrogen in water is 8.21×10^{-4} mol/L at 0°C when the N_2 pressure above water is 0.790 atm. Calculate the Henry's law constant for N_2 in units of L atm/mol for Henry's law in the form $P = kC$, where C is the gas concentration in mol/L. Calculate the solubility of N_2 in water when the partial pressure of nitrogen above water is 1.10 atm at 0°C.

38. In Exercise 112 in Chapter 5, the pressure of CO_2 in a bottle of sparkling wine was calculated assuming that the CO_2 was insoluble in water. This was an incorrect assumption. Redo this problem by assuming that CO_2 obeys Henry's law. Use the data given in that problem to calculate the partial pressure of CO_2 in the gas phase and the solubility of CO_2 in the wine at 25°C. The Henry's law constant for CO_2 is 32 L atm/mol at 25°C with Henry's law in the form $P = kC$, where C is the concentration of the gas in mol/L.

39. Rationalize the temperature dependence of the solubility of a gas in terms of the kinetic molecular theory.

Vapor Pressures of Solutions

40. Glycerin ($C_3H_8O_3$) is a nonvolatile liquid. What is the vapor pressure of a solution made by adding 164 g of glycerin to 338 mL of H_2O at 39.8°C? The vapor pressure of pure water at 39.8°C is 54.74 torr and its density is 0.992 g/cm^3.

41. The vapor pressure of a solution containing 53.6 g glycerin ($C_3H_8O_3$) in 133.7 g ethanol (C_2H_5OH) is 113 torr at 40°C. Calculate the vapor pressure of pure ethanol at 40°C assuming that glycerin is a nonvolatile, nonelectrolyte solute in ethanol.

42. Which of the following will have the lowest total vapor pressure at 25°C? Which has the highest vapor pressure at 25°C? At 25°C, the vapor pressure of pure water is 23.8 torr.
    a. pure water
    b. a solution of glucose in water with $\chi_{glucose} = 0.01$
    c. a solution of sodium chloride in water with $\chi_{NaCl} = 0.01$
    d. a solution of methanol in water with $\chi_{CH_3OH} = 0.2$
    (At 25°C, the vapor pressure of pure methanol is 143 torr.)

43. The normal boiling point of methanol is 64.7°C. A solution containing a nonvolatile solute dissolved in methanol has a vapor pressure of 710.0 torr at 64.7°C. What is the mole fraction of methanol in this solution?

44. A solution is prepared by mixing 0.0300 mole of CH_2Cl_2 and 0.0500 mole of CH_2Br_2 at 25°C. Assuming the solution is ideal, calculate the composition of the vapor (in terms of mole fractions) at 25°C. At 25°C the vapor pressures of pure CH_2Cl_2 and pure CH_2Br_2 are 133 and 11.4 torr, respectively.

45. At a certain temperature, the vapor pressure of pure benzene (C_6H_6) is 0.930 atm. A solution was prepared by dissolving 10.0 g of a nondissociating, nonvolatile solute in 78.11 g of benzene at that temperature. The vapor pressure of the solution was found to be 0.900 atm. Assuming that the solution behaves ideally, determine the molar mass of the solute.

46. At 25°C the vapor in equilibrium with a solution containing carbon disulfide and acetonitrile has a total pressure of 263 torr and is 85.5 mole percent carbon disulfide. What is the mole fraction of carbon disulfide in the solution? At 25°C the vapor pressure of carbon disulfide is 375 torr. Assume that the solution and the vapor exhibit ideal behavior.

47. Pentane (C_5H_{12}) and hexane (C_6H_{14}) combine to form an ideal solution. At 25°C the vapor pressures of pentane and hexane are 511 and 150. torr, respectively. A solution is prepared by mixing 25 mL of pentane (density = 0.63 g/mL) with 45 mL of hexane (density = 0.66 g/mL).
    a. What is the vapor pressure of this solution?
    b. What is the mole fraction of pentane in the vapor that is in equilibrium with this solution?

48. Benzene and toluene form ideal solutions. Consider a solution of benzene and toluene prepared at 25°C. Assuming that the mole fractions of benzene and toluene in the vapor phase are equal, calculate the composition of the solution. At 25°C the vapor pressures of benzene and toluene are 95 and 28 torr, respectively.

49. What is the composition of a methanol (CH_3OH)–propanol ($CH_3CH_2CH_2OH$) solution that has a vapor pressure of 174 torr at 40°C? At 40°C the vapor pressures of pure methanol and pure propanol are 303 and 44.6 torr, respectively. Assume that the solution exhibits ideal behavior.

50. Match the vapor pressure diagrams with the solute–solvent combinations and explain your answers.

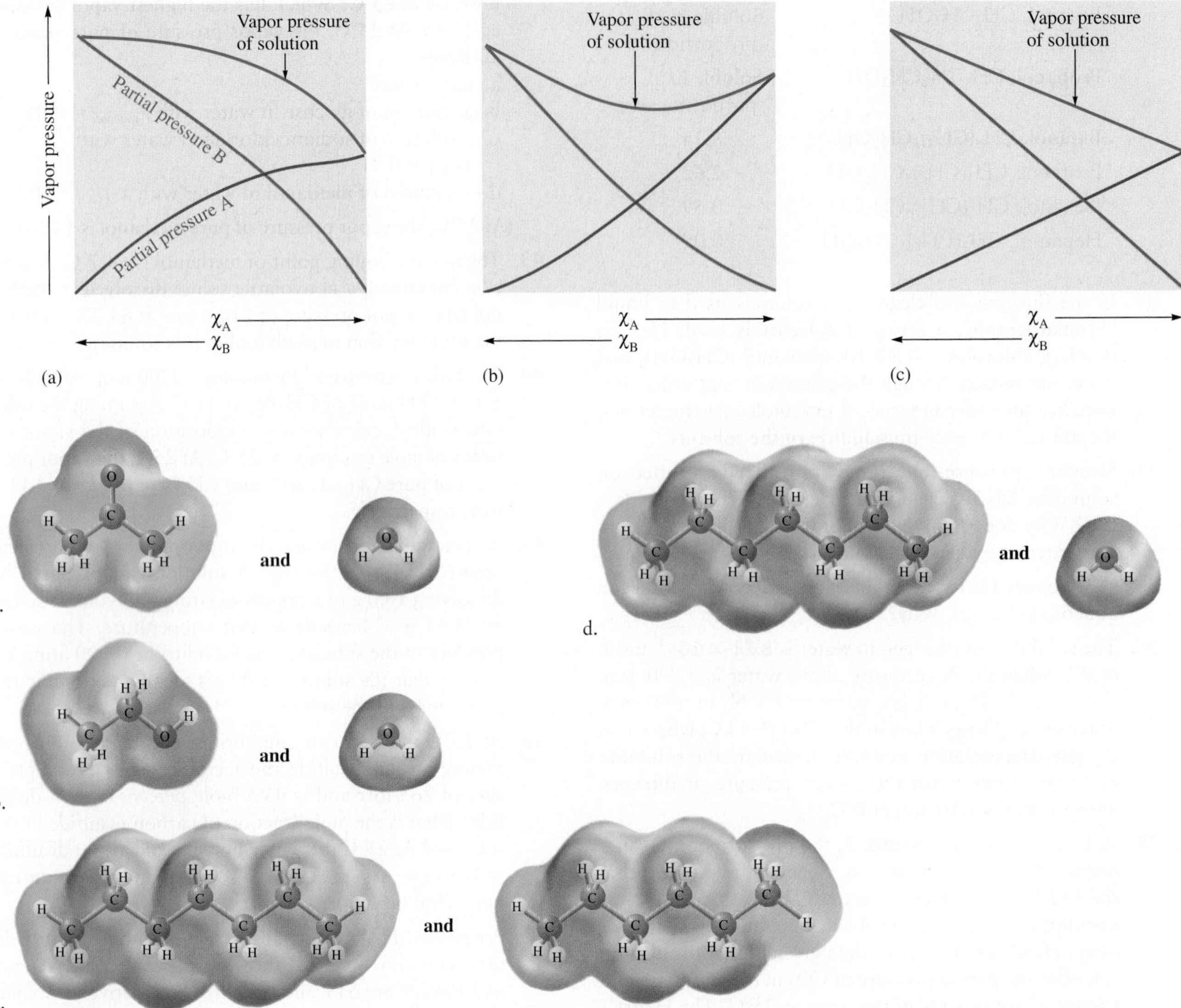

51. A solution is made by mixing 50.0 g of acetone (CH_3COCH_3) and 50.0 g of methanol (CH_3OH). What is the vapor pressure of this solution at 25°C? What is the composition of the vapor expressed as a mole fraction? Assume ideal solution and gas behavior. (At 25°C the vapor pressures of pure acetone and pure methanol are 271 and 143 torr, respectively.) The actual vapor pressure of this solution is 161 torr. Explain any discrepancies.

52. The vapor pressures of several solutions of water–propanol ($CH_3CH_2CH_2OH$) were determined at various compositions, with the following data collected at 45°C:

χ_{H_2O}	Vapor Pressure (torr)
0	74.0
0.15	77.3
0.37	80.2
0.54	81.6
0.69	80.6
0.83	78.2
1.00	71.9

a. Are solutions of water and propanol ideal? Explain.
b. Predict the sign of ΔH_{soln} for water–propanol solutions.
c. Are the interactive forces between propanol and water molecules weaker than, stronger than, or equal to the interactive forces between the pure substances? Explain.
d. Which of the solutions in the table would have the lowest normal boiling point?

53. When pure methanol is mixed with water, the solution gets warmer to the touch. Would you expect this solution to be ideal? Explain.

54. The two beakers in the sealed container illustrated below contain pure water and an aqueous solution of a volatile solute.

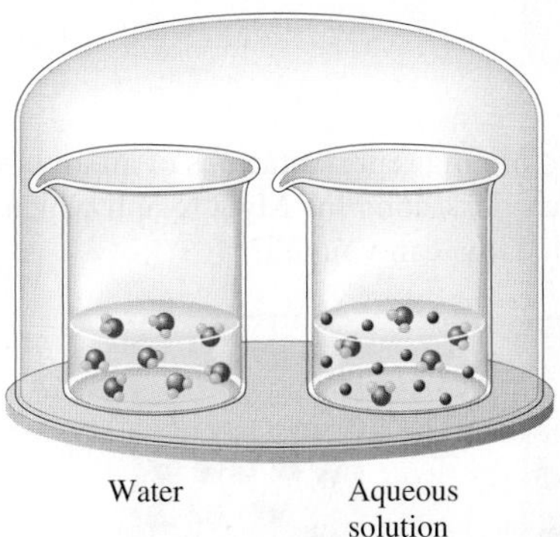

If the solute is less volatile than water, explain what will happen to the volumes in the two containers as time passes.

55. The following plot shows the vapor pressure of various solutions of components A and B at some temperature.

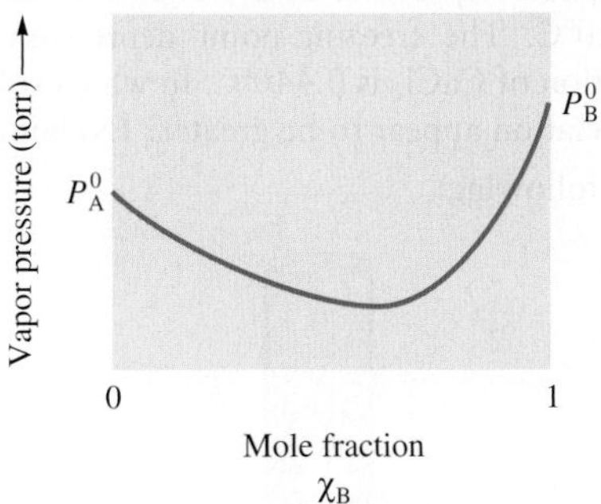

Which of the following statements is false concerning solutions of A and B?
a. The solutions exhibit negative deviations from Raoult's law.
b. ΔH_{soln} for the solutions should be exothermic.
c. The intermolecular forces are stronger in solution than in either pure A or pure B.
d. Pure liquid B is more volatile than pure liquid A.
e. The solution with $\chi_B = 0.6$ will have a lower boiling point than either pure A or pure B.

Colligative Properties

56. Vapor-pressure lowering is a colligative property, as are freezing-point depression and boiling-point elevation. What is a colligative property? Why is the freezing point depressed for a solution as compared to the pure solvent? Why is the boiling point elevated for a solution as compared to the pure solvent?

57. How would you prepare 1.0 L of an aqueous solution of sucrose ($C_{12}H_{22}O_{11}$) having an osmotic pressure of 15 atm at a temperature of 22°C? Sucrose is a nonelectrolyte.

58. Is the following statement true or false? Explain your answer. When determining the molar mass of a solute using boiling-point or freezing-point data, camphor would be the best solvent choice of all of the solvents listed in Table 17.5.

59. A solution is prepared by dissolving 27.0 g of urea $[(NH_2)_2CO]$, in 150.0 g of water. Calculate the boiling point of the solution. Urea is a nonelectrolyte.

60. What mass of glycerin ($C_3H_8O_3$), a nonelectrolyte, must be dissolved in 200.0 g of water to give a solution with a freezing point of −1.50°C?

61. Calculate the freezing point and boiling point of an antifreeze solution that is 50.0% by mass ethylene glycol ($HOCH_2CH_2OH$) in water. Ethylene glycol is a nonelectrolyte.

62. The freezing point of *t*-butanol is 25.50°C and K_f is 9.1°C kg/mol. Usually *t*-butanol absorbs water on exposure to air. If the freezing point of a 10.0-g sample of *t*-butanol is 24.59°C, how many grams of water are present in the sample?

63. The molar mass of a nonelectrolyte is 58.0 g/mol. Calculate the boiling point of a solution containing 24.0 g of this compound and 600. g of water if the barometric pressure is such that pure water boils at 99.725°C.

64. If the human eye has an osmotic pressure of 8.00 atm at 25°C, what concentration of solute particles in water will provide an isotonic eyedrop solution (a solution with equal osmotic pressure)?

65. A 2.00-g sample of a large biomolecule was dissolved in 15.0 g of carbon tetrachloride. The boiling point of this solution was determined to be 77.85°C. Calculate the molar mass of the biomolecule. For carbon tetrachloride, the boiling-point constant is 5.03°C kg/mol, and the boiling point of pure carbon tetrachloride is 76.50°C.

66. An aqueous solution of 10.00 g of catalase, an enzyme found in the liver, has a volume of 1.00 L at 27°C. The solution's osmotic pressure at 27°C is found to be 0.745 torr. Calculate the molar mass of catalase.

67. Thyroxine, an important hormone that controls the rate of metabolism in the body, can be isolated from the thyroid gland. When 0.455 g of thyroxine is dissolved in 10.0 g of benzene, the freezing point of the solution is depressed by 0.300°C. What is the molar mass of thyroxine? See Table 17.5.

68. What volume of ethylene glycol ($C_2H_6O_2$), a nonelectrolyte, must be added to 15.0 L of water to produce an antifreeze solution with a freezing point of −30.0°C? What is the boiling point of this solution? (The density of ethylene glycol is 1.11 g/cm^3, and the density of water is 1.00 g/cm^3.)

69. Calculate the freezing-point depression and osmotic pressure in torr at 25°C for an aqueous solution of 1.0 g/L of a protein (molar mass = 9.0×10^4 g/mol) if the density of the solution is 1.0 g/cm^3.

70. Considering your answer to Exercise 69, which colligative property, freezing-point depression or osmotic pressure, would be better for determining the molar masses of large molecules? Explain your answer.

71. If the fluid inside a tree is about 0.1 *M* more concentrated in solute than the groundwater that bathes the roots, how high will a column of fluid rise in the tree at 25°C? Assume that the density of the fluid is 1.0 g/cm^3. (The density of mercury is 13.6 g/cm^3.)

72. A 1.60-g sample of a mixture of naphthalene ($C_{10}H_8$) and anthracene ($C_{14}H_{10}$) is dissolved in 20.0 g of benzene (C_6H_6). The freezing point of the solution is 2.81°C. What is the composition of the sample mixture in terms of mass percent? The freezing point of benzene is 5.51°C, and K_f is 5.12°C kg/mol.

73. Before refrigeration became common, many foods were preserved by salting them heavily. Fruits were preserved by mixing them with a large amount of sugar (fruit preserves). How do salt and sugar act as preservatives?

Properties of Electrolyte Solutions

74. Calculate the freezing point and the boiling point of each of the following solutions. (Assume complete dissociation.)
    a. 5.0 g NaCl in 25 g H_2O
    b. 2.0 g $Al(NO_3)_3$ in 15 g H_2O

75. A solution of sodium chloride in water has a vapor pressure of 19.6 torr at 25°C. What is the mole fraction of NaCl in this solution? What would be the vapor pressure of this solution at 45°C? The vapor pressure of pure water is 23.8 torr at 25°C and 71.9 torr at 45°C.

76. Consider an aqueous solution containing sodium chloride that has a density of 1.01 g/mL. Assume that the solution behaves ideally. The freezing point of this solution at 1.00 atm is −1.28°C. Calculate the percent composition of this solution (by mass).

77. Consider the following solutions:

    0.010 *m* Na_3PO_4 in water
    0.020 *m* $CaBr_2$ in water
    0.020 *m* KCl in water
    0.020 *m* HF in water (HF is a weak acid.)

    a. Assuming complete dissociation of the soluble salts, which solution(s) would have the same boiling point as 0.040 *m* $C_6H_{12}O_6$ in water? ($C_6H_{12}O_6$ is a nonelectrolyte.)
    b. Which solution would have the highest vapor pressure at 28°C?
    c. Which solution would have the largest freezing-point depression?

78. From the following:

    pure water
    solution of $C_6H_{12}O_6$ ($\chi = 0.01$) in water
    solution of NaCl ($\chi = 0.01$) in water
    solution of $CaCl_2$ ($\chi = 0.01$) in water

    choose the one with the following:
    a. highest freezing point
    b. lowest freezing point
    c. highest boiling point
    d. lowest boiling point
    e. highest osmotic pressure

79. Determine the van't Hoff factor for the following ionic solute dissolved in water.

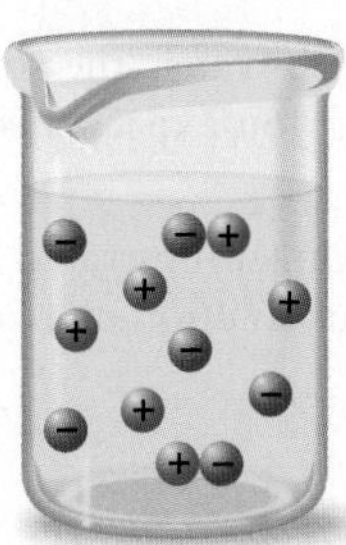

80. Consider the following representations of an ionic solute in water. Which flask contains $MgSO_4$, and which flask contains NaCl? How can you tell?

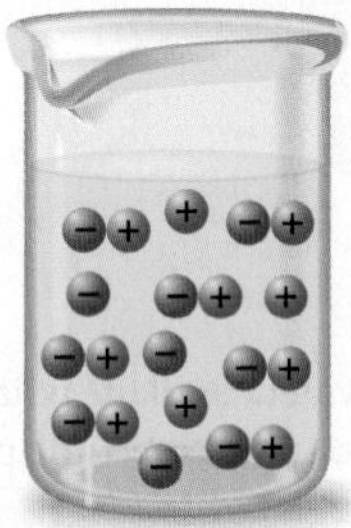

81. The freezing-point depression of a 0.091 *m* solution of CsCl is 0.320°C. The freezing-point depression of a 0.091 *m* solution of $CaCl_2$ is 0.440°C. In which solution does ion association appear to be greater? Explain.

82. Consider the following:

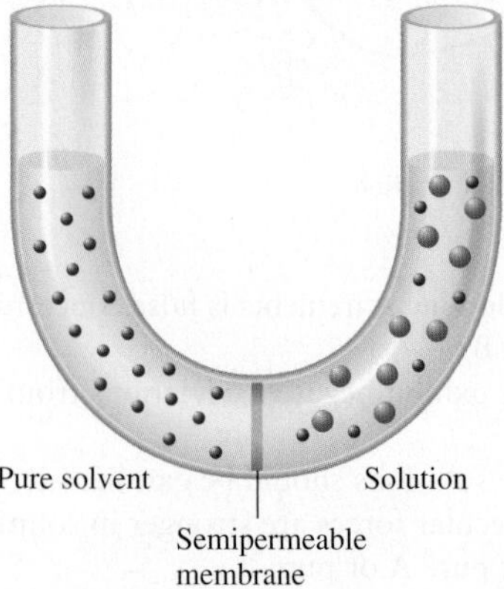

What would happen to the level of liquid in the two arms if the semipermeable membrane separating the two liquids were permeable to
a. H_2O (the solvent) only?
b. H_2O and solute?

83. A water desalination plant is set up near a salt marsh containing water that is 0.10 *M* NaCl. Calculate the

minimum pressure that must be applied at 20.°C to purify the water by reverse osmosis. Assume that NaCl is completely dissociated.

84. Calculate the freezing point and the boiling point of each of the following aqueous solutions. (Assume complete dissociation.)
   a. 0.050 *m* $MgCl_2$ b. 0.050 *m* $FeCl_3$

85. Calculate the freezing point and the boiling point of each of the following solutions using the observed van't Hoff factors in Table 17.6.
   a. 0.050 *m* $MgCl_2$ b. 0.050 *m* $FeCl_3$

86. Use the data in the following table for three different aqueous solutions of $CaCl_2$ to calculate the apparent value of the van't Hoff factor.

Molality	Freezing-Point Depression (°C)
0.0225	0.110
0.0910	0.440
0.278	1.330

87. A 0.500-g sample of a compound is dissolved in enough water to form 100.0 mL of solution. This solution has an osmotic pressure of 2.50 atm at 25°C. If each molecule of the solute dissociates into two particles (in this solvent), what is the molar mass of this solute?

88. In the winter of 1994, record low temperatures were registered throughout the United States. For example, in Champaign, Illinois, a record low of −29°F was registered. At this temperature can salting icy roads with $CaCl_2$ be effective in melting the ice?
   a. Assume that $i = 3.00$ for $CaCl_2$.
   b. Assume the average value of *i* from Exercise 86.
   (The solubility of $CaCl_2$ in cold water is 74.5 g per 100.0 g of water.)

Additional Exercises

89. The Tyndall effect is often used to distinguish between a colloidal suspension and a true solution. Explain.

90. Detergent molecules can stabilize the emulsion of oil in water as well as remove dirt from soiled clothes. A typical detergent is sodium dodecylsulfate (SDS), which has a formula of $CH_3(CH_2)_{10}CH_2SO_4{}^-Na^+$. In aqueous solution, SDS suspends oil or dirt by forming small aggregates of detergent anions called micelles. Propose a structure for micelles.

91. The destruction of a colloid is done through a process called *coagulation*. What is coagulation?

92. What stabilizes a colloidal suspension? Explain why adding heat or adding an electrolyte can cause the suspended particles to settle out.

93. The term *proof* is defined as twice the percent by volume of pure ethanol in solution. Thus a solution that is 95% (by volume) ethanol is 190 proof. What is the molarity of ethanol in a 92-proof ethanol–water solution? Assume the density of ethanol (C_2H_5OH) is 0.79 g/cm^3 and the density of water is 1.0 g/cm^3.

94. Patients undergoing an upper gastrointestinal tract laboratory test are typically given an X-ray contrast agent that aids with the radiologic imaging of the anatomy. One such contrast agent is sodium diatrizoate, a nonvolatile water-soluble compound. A 0.378-*m* solution is prepared by dissolving 38.4 g of sodium diatrizoate (NaDTZ) in 1.60×10^2 mL of water at 31.2°C (the density of water at 31.2°C is 0.995 g/mL). What is the molar mass of sodium diatrizoate? What is the vapor pressure of this solution if the vapor pressure of pure water at 31.2°C is 34.1 torr?

95. The freezing point of an aqueous solution is −2.79°C.
   a. Determine the boiling point of this solution.
   b. Determine the vapor pressure (in mm Hg) of this solution at 25°C (the vapor pressure of pure water at 25°C is 23.76 mm Hg).
   c. Explain any assumptions you make in solving parts a and b.

96. An aqueous solution containing 0.250 mole of Q, a strong electrolyte, in 5.00×10^2 g of water freezes at −2.79°C. What is the van't Hoff factor for Q? The molal freezing-point depression constant for water is 1.86°C kg/mol. What is the formula of Q if it is 38.68% chlorine by mass and there are twice as many anions as cations in one formula unit of Q?

97. The solubility of benzoic acid,

$$C_6H_5-\underset{\underset{\displaystyle O}{\|}}{C}-O-H$$

is 0.34 g/100 mL in water at 25°C and 10.0 g/100 mL in benzene (C_6H_6) at 25°C. Rationalize this solubility behavior. For a 1.0-*m* solution of benzoic acid in benzene, would the measured freezing-point depression be equal to, greater than, or less than 5.12°C? ($K_f = 5.12$°C kg/mol for benzene.)

98. Would benzoic acid be more or less soluble in a 0.1-*M* NaOH solution than it is in water?

99. In a coffee-cup calorimeter, 1.60 g of NH_4NO_3 was mixed with 75.0 g of water at an initial temperature of 25.00°C. After dissolution of the salt, the final temperature of the calorimeter contents was 23.34°C.

a. Assuming the solution has a heat capacity of 4.18 J °C^{-1} g^{-1}, and assuming no heat loss to the calorimeter, calculate the enthalpy of solution (ΔH_{soln}) for the dissolution of NH_4NO_3 in units of kJ/mol.
b. If the enthalpy of hydration for NH_4NO_3 is −630. kJ/mol, calculate the lattice energy of NH_4NO_3.

100. Specifications for lactated Ringer's solution, which is used for intravenous (IV) injections, are as follows for each 100. mL of solution:

285–315 mg Na^+
14.1–17.3 mg K^+
4.9–6.0 mg Ca^{2+}
368–408 mg Cl^-
231–261 mg lactate, $C_3H_5O_3^-$

a. Specify the amounts of NaCl, KCl, $CaCl_2 \cdot 2H_2O$, and $NaC_3H_5O_3$ needed to prepare 100. mL of lactated Ringer's solution.
b. What is the range of the osmotic pressure of the solution at 37°C, given the above specifications?

101. In the vapor over a pentane–hexane solution at 25°C, the mole fraction of pentane is equal to 0.15. What is the mole fraction of pentane in the solution? (See Exercise 47 for the vapor pressures of the pure liquids.)

102. Explain the following on the basis of the behavior of atoms and/or ions.
a. Cooking with hot water is faster in a pressure cooker than in an open pan.
b. Salt is used on icy roads.
c. Melted sea ice from the Arctic Ocean produces fresh water.
d. $CO_2(s)$ (dry ice) does not have a normal boiling point under normal atmospheric conditions, even though CO_2 is a liquid in fire extinguishers.

103. Anthraquinone contains only carbon, hydrogen, and oxygen. When 4.80 mg of anthraquinone is burned, 14.2 mg of CO_2 and 1.65 mg of H_2O are produced. The freezing point of camphor is lowered by 22.3°C when 1.32 g of anthraquinone is dissolved in 11.4 g of camphor. Determine the empirical and molecular formulas of anthraquinone.

104. An unknown compound contains only carbon, hydrogen, and oxygen. Combustion analysis of the compound gives mass percents of 31.57% C and 5.30% H. The molar mass is determined by measuring the freezing-point depression of an aqueous solution. A freezing point of −5.20°C is recorded for a solution made by dissolving 10.56 g of the compound in 25.0 g of water. Determine the empirical formula, molar mass, and molecular formula of the compound. Assume that the compound is a nonelectrolyte.

105. A forensic chemist is given a white solid that is suspected of being pure cocaine ($C_{17}H_{21}NO_4$, molar mass = 303.36 g/mol). She dissolves 1.22 ± 0.01 g of the solid in 15.60 ± 0.01 g of benzene. The freezing point is lowered by 1.32 ± 0.04°C.
a. What is the molar mass of the substance? Assuming that the percentage of uncertainty in the calculated molar mass is the same as the percentage of uncertainty in the temperature change, calculate the uncertainty in the molar mass.
b. Could the chemist unequivocally state that the substance is cocaine? For example, is the uncertainty small enough to distinguish cocaine from codeine ($C_{18}H_{21}NO_3$, molar mass = 299.37)?
c. Assuming that the absolute uncertainties in the measurements of temperature and mass remain unchanged, how could the chemist improve the precision of her results?

106. You add an excess of solid MX in 250 g of water. You measure the freezing point and find it to be −0.028°C. What is the K_{sp} of the solid?

107. A solution saturated with a salt of the type M_3X_2 has an osmotic pressure of 2.64×10^{-2} atm at 25°C. Calculate the K_{sp} value for the salt, assuming ideal behavior.

108. You take 20.0 g of a sucrose ($C_{12}H_{22}O_{11}$) and NaCl mixture and dissolve it in 1.00 L of water. The freezing point of this solution is found to be −0.426°C. Assuming ideal behavior, calculate the mass percent composition of the original mixture and the mole fraction of sucrose in the original mixture.

109. A 0.100-g sample of the weak acid HA (molar mass = 100.0 g/mol) is dissolved in 500.0 g of water. The freezing point of the resulting solution is −0.0056°C. Calculate the value of K_a for this acid. Assume molarity equals molality in this solution.

110. A solid consists of a mixture of $NaNO_3$ and $Mg(NO_3)_2$. When 6.50 g of this solid is dissolved in 50.0 g of water, the freezing point is lowered by 5.40°C. What is the composition of the solid (by mass)? Assume ideal behavior.

111. A solid mixture contains $MgCl_2$ (molar mass = 95.218 g/mol) and NaCl (molar mass = 58.443 g/mol). When 0.5000 g of this solid is dissolved in enough water to form 1.000 L of solution, the osmotic pressure at 25.0°C is observed to be 0.3950 atm. What is the mass percent of $MgCl_2$ in the solid? (Assume ideal behavior for the solution.)

112. Some nonelectrolyte solute (molar mass = 142 g/mol) was dissolved in 150. mL of a solvent (density = 0.879 g/cm^3). The elevated boiling point of the solution was 355.4 K. What mass of solute was dissolved in the solvent? For the solvent, the enthalpy of vaporization is 33.90 kJ/mol, the entropy of vaporization is 95.95 J K^{-1} mol^{-1}, and the boiling-point elevation constant is 2.5 K kg/mol.

Challenge Problems

113. Carbon tetrachloride (CCl_4) and benzene (C_6H_6) form ideal solutions. Consider an equimolar solution of CCl_4 and C_6H_6 at 25°C. The vapor above the solution is collected and condensed. Using the following data, determine the composition in mole fraction of the condensed vapor.

Substance	ΔG_f°
$C_6H_6(l)$	124.50 kJ/mol
$C_6H_6(g)$	129.66 kJ/mol
$CCl_4(l)$	−65.21 kJ/mol
$CCl_4(g)$	−60.59 kJ/mol

114. You have a solution of two volatile liquids, A and B (assume ideal behavior). Pure liquid A has a vapor pressure of 350.0 torr and pure liquid B has a vapor pressure of 100.0 torr at the temperature of the solution. The vapor at equilibrium above the solution has double the mole fraction of substance A as the solution does. What is the mole fraction of liquid A in the solution?

115. The vapor pressure of pure benzene is 750.0 torr and the vapor pressure of toluene is 300.0 torr at a certain temperature. You make a solution by pouring "some" benzene with "some" toluene. You then place this solution in a closed container and wait for the vapor to come into equilibrium with the solution. Next, you condense the vapor. You put this liquid (the condensed vapor) in a closed container and wait for the vapor to come into equilibrium with the solution. You then condense this vapor and find the mole fraction of benzene in this vapor to be 0.714. Determine the mole fraction of benzene in the original solution, assuming the solution behaves ideally.

116. Plants that thrive in salt water must have internal solutions (inside the plant cells) that are isotonic (same osmotic pressure) with the surrounding solution. A leaf of a saltwater plant is able to thrive in an aqueous salt solution (at 25°C) that has a freezing point equal to −0.621°C. You would like to use this information to calculate the osmotic pressure of the solution in the cell.
    a. In order to use the freezing-point depression to calculate osmotic pressure, what assumption must you make (in addition to ideal behavior of the solutions, which we will assume)?
    b. Under what conditions is the assumption in part a reasonable?
    c. Solve for the osmotic pressure (at 25°C) of the solution in the plant cell.
    d. The plant leaf is placed in an aqueous salt solution (at 25°C) that has a boiling point of 102.0°C. What will happen to the plant cells in the leaf?

117. In some regions of the southwest United States, the water is very hard. For example, in Las Cruces, New Mexico, the tap water contains about 560 μg of dissolved solids per milliliter. Reverse osmosis units are marketed in this area to soften water. A typical unit exerts a pressure of 8.0 atm and can produce 45 L of water per day.
    a. Assuming that all the dissolved solids are $MgCO_3$, and assuming a temperature of 27°C, what total volume of water must be processed to produce 45 L of pure water?
    b. Would the same system work for purifying seawater? (Assume that seawater is 0.60 *M* NaCl.)

118. Erythrocytes are red blood cells containing hemoglobin. In a saline solution, they shrivel when the salt concentration is high and swell when the salt concentration is low. In a 25°C aqueous solution of NaCl, whose freezing point is −0.406°C, erythrocytes neither swell nor shrink. If we want to calculate the osmotic pressure of the solution inside the erythrocytes under these conditions, what do we need to assume? Why? Estimate how good (or poor) of an assumption it is. Make this assumption and calculate the osmotic pressure of the solution inside the erythrocytes.

119. A sample containing 0.0500 mole of $Fe_2(SO_4)_3$ is dissolved in enough water to make 1.00 L of solution. This solution contains hydrated SO_4^{2-} and $Fe(H_2O)_6^{3+}$ ions. The latter behave as an acid:

$$Fe(H_2O)_6^{3+} \rightleftharpoons Fe(OH)(H_2O)_5^{2+} + H^+$$

    a. Calculate the expected osmotic pressure of this solution at 25°C if the above dissociation is negligible.
    b. The actual osmotic pressure of the solution is 6.73 atm at 25°C. Calculate K_a for the dissociation reaction of $Fe(H_2O)_6^{3+}$. (To do this calculation, you must assume that none of the ions goes through the semipermeable membrane. Actually, this is not a good assumption for the tiny H^+ ion.)

120. In a solution with carbon tetrachloride as the solvent, the compound VCl_4 undergoes dimerization:

$$2VCl_4 \rightleftharpoons V_2Cl_8$$

When 6.6834 g of VCl_4 is dissolved in 100.0 g of carbon tetrachloride, the freezing point is lowered by 5.97°C. Calculate the value of the equilibrium constant for the dimerization of VCl_4 at this temperature. (The density of the equilibrium mixture is 1.696 g/cm^3, and K_f = 29.8°C kg/mol for CCl_4.)

121. An aqueous solution is 1.00% NaCl by mass and has a density of 1.071 g/cm^3 at 25°C. The observed osmotic pressure of this solution is 7.83 atm at 25°C.
    a. What fraction of the moles of NaCl in this solution exist as ion pairs?
    b. Calculate the freezing point that would be observed for this solution.

122. Liquid A has vapor pressure x, and liquid B has vapor pressure y. What is the mole fraction of the liquid mixture if the vapor above the solution is 30.% A by moles? 50.% A? 80.% A? (Calculate in terms of x and y.) What is the mole fraction of the vapor above the solution if the liquid mixture is 30.% A by moles? 50.% A? 80.% A? (Calculate in terms of x and y.)

Marathon Problem

123.* Using the following information, identify the strong electrolyte whose general formula is

$$M_x(A)_y \cdot zH_2O$$

Ignore the effect of interionic attractions in the solution.

a. A^{n-} is a common oxyanion. When 30.0 mg of the anhydrous sodium salt containing this oxyanion (Na_nA, where $n = 1, 2,$ or 3) is reduced, 15.26 mL of 0.02313 M reducing agent is required to react completely with the Na_nA present. Assume a 1:1 mole ratio in the reaction.

b. The cation is derived from a silvery white metal that is relatively expensive. The metal itself crystallizes in a body-centered cubic unit cell and has an atomic radius of 198.4 pm. The solid, pure metal has a density of 5.243 g/cm³. The oxidation state of M in the strong electrolyte general formula is +3.

c. When 33.45 mg of the compound is present (dissolved) in 10.0 mL of aqueous solution at 25°C, the solution has an osmotic pressure of 558 torr.

*From James H. Burness, "The Use of "Marathon" Problems as Effective Vehicles for the Presentation of General Chemistry Lectures," Journal of Chemical Education, 68(11).

The Representative Elements

18

chapter

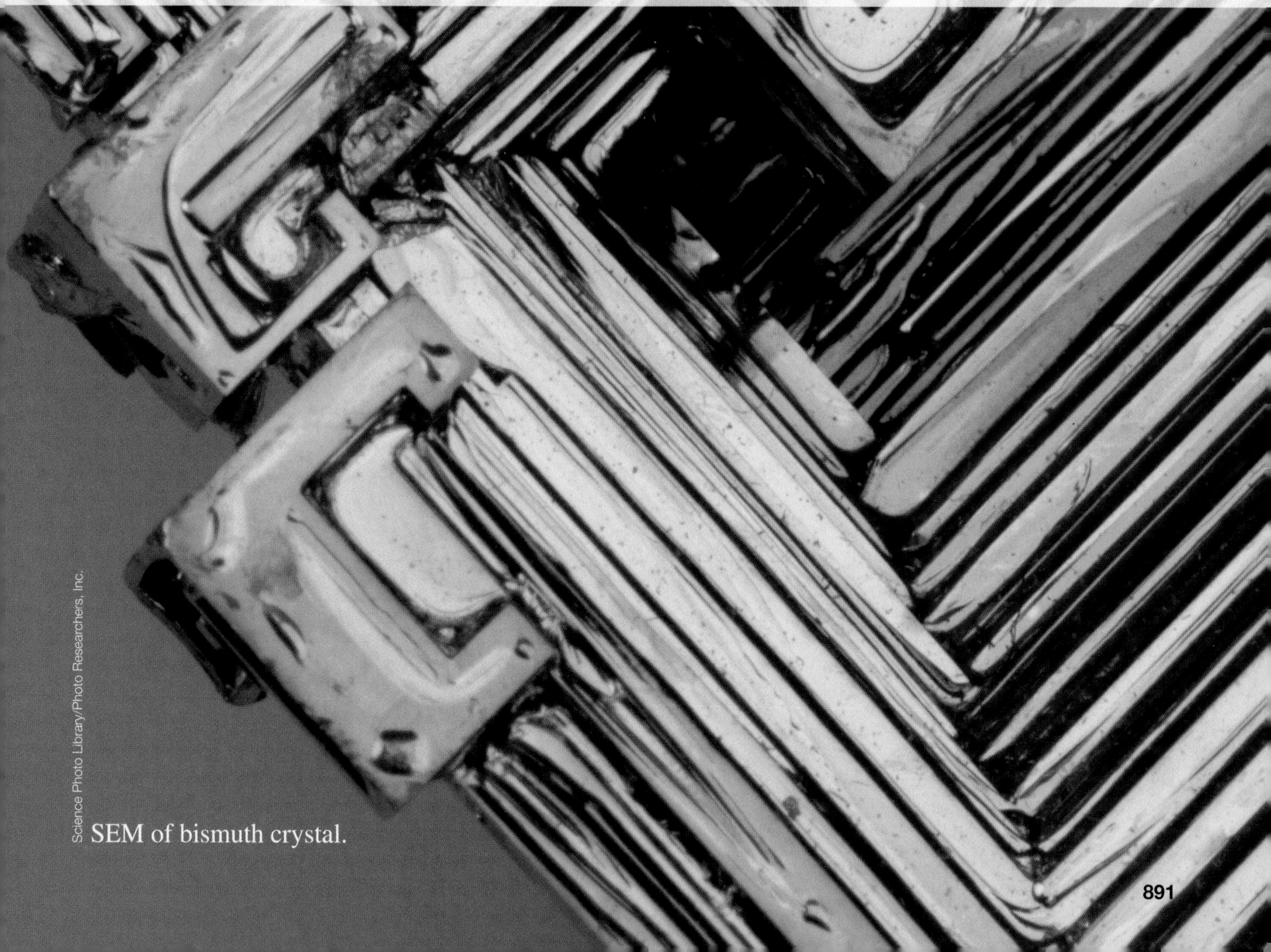

Science Photo Library/Photo Researchers, Inc.

SEM of bismuth crystal.

Sign in to OWL at www.cengage.com/owl to view tutorials and simulations, develop problem-solving skills, and complete online homework assigned by your professor.

go Chemistry

Download mini lecture videos for key concept review and exam prep from OWL or purchase them from www.cengagebrain.com.

So far in this book we have covered the major principles and explored the most important models of chemistry. In particular, we have seen that the chemical properties of the elements can be explained very successfully by the quantum mechanical model of the atom. In fact, the most convincing evidence of that model's validity is its ability to relate the observed periodic properties of the elements to the number of valence electrons in their atoms.

We have learned many properties of the elements and their compounds, but we have not discussed extensively the relationship between the chemical properties of a specific element and its position on the periodic table. In this chapter we will explore the chemical similarities and differences among the elements in the several groups of the periodic table and will try to interpret these data using the wave mechanical model of the atom. In the process we will illustrate a great variety of chemical properties and further demonstrate the practical importance of chemistry.

18.1 A Survey of the Representative Elements

The traditional form of the periodic table is shown in Fig. 18.1. Recall that the **representative elements,** whose chemical properties are determined by the valence-level *s* and *p* electrons, are designated Groups 1A through 8A. The **transition metals,** in the center of the table, result from the filling of *d* orbitals. The elements that correspond to the filling of the 4*f* and 5*f* orbitals are listed separately as the **lanthanides** and **actinides,** respectively.

The heavy black line in Fig. 18.1 separates the metals from the nonmetals, except for one case. Hydrogen, which appears on the metal side, is a nonmetal. Some elements just on either side of this line, such as silicon and germanium, exhibit both metallic and nonmetallic properties. These elements are often called **metalloids,** or **semimetals.** The fundamental chemical difference between metals and nonmetals is that metals tend to lose their valence electrons to form *cations,* which usually have the valence electron configuration of the noble gas from the preceding period. On the other hand, nonmetals tend to gain electrons to form *anions* that exhibit the electron configuration of the noble gas in the same period. Metallic character is observed to increase in going down a given group, which is consistent with the trends in ionization energy, electron affinity, and electronegativity discussed earlier (see Sections 12.15 and 13.2).

Metallic character increases going down a group in the periodic table.

1A																	8A
H	2A											3A	4A	5A	6A	7A	He
Li	Be											B	C	N	O	F	Ne
Na	Mg											Al	Si	P	S	Cl	Ar
K	Ca	Sc	Ti	V	Cr	Mn	Fe	Co	Ni	Cu	Zn	Ga	Ge	As	Se	Br	Kr
Rb	Sr	Y	Zr	Nb	Mo	Tc	Ru	Rh	Pd	Ag	Cd	In	Sn	Sb	Te	I	Xe
Cs	Ba	La	Hf	Ta	W	Re	Os	Ir	Pt	Au	Hg	Tl	Pb	Bi	Po	At	Rn
Fr	Ra	Ac	Rf	Ha	Unh	Uns	Uno	Une	Ds	Rg	Cn	Uut	Uuq	Uup		Uus	Uuo

Lanthanides	Ce	Pr	Nd	Pm	Sm	Eu	Gd	Tb	Dy	Ho	Er	Tm	Yb	Lu
Actinides	Th	Pa	U	Np	Pu	Am	Cm	Bk	Cf	Es	Fm	Md	No	Lr

Figure 18.1
The periodic table. The elements in the A groups are the representative elements. The elements shown in pink are called transition metals. The heavy black line approximately separates the nonmetals from the metals. The elements that have both metallic and nonmetallic properties (semimetals) are shaded in blue.

Atomic Size and Group Anomalies

Although the chemical properties of the members of a group have many similarities, there are also important differences. The most dramatic differences usually occur between the first and second member. For example, hydrogen in Group 1A is a nonmetal, whereas lithium is a very active metal. This extreme difference results primarily from the very large difference in the atomic radii of hydrogen and lithium, as shown in Fig. 18.2. Since the small hydrogen atom has a much greater attraction for electrons than do the larger members of Group 1A, it forms covalent bonds with nonmetals. In contrast, the other members of Group 1A lose their valence electrons to nonmetals to form 1+ cations in ionic compounds.

The effect of size is also evident in other groups. For example, the oxides of the metals in Group 2A are all quite basic except for the first member of the series; beryllium oxide (BeO) is amphoteric. The basicity of an oxide depends on its ionic character. Ionic oxides contain the O^{2-} ion, which reacts with water to form two OH^- ions. All the oxides of the Group 2A metals are highly ionic except for beryllium oxide, which has considerable covalent character. The small Be^{2+} ion can effectively polarize the electron "cloud" of the O^{2-} ion, thereby producing significant electron sharing. We see the same pattern in

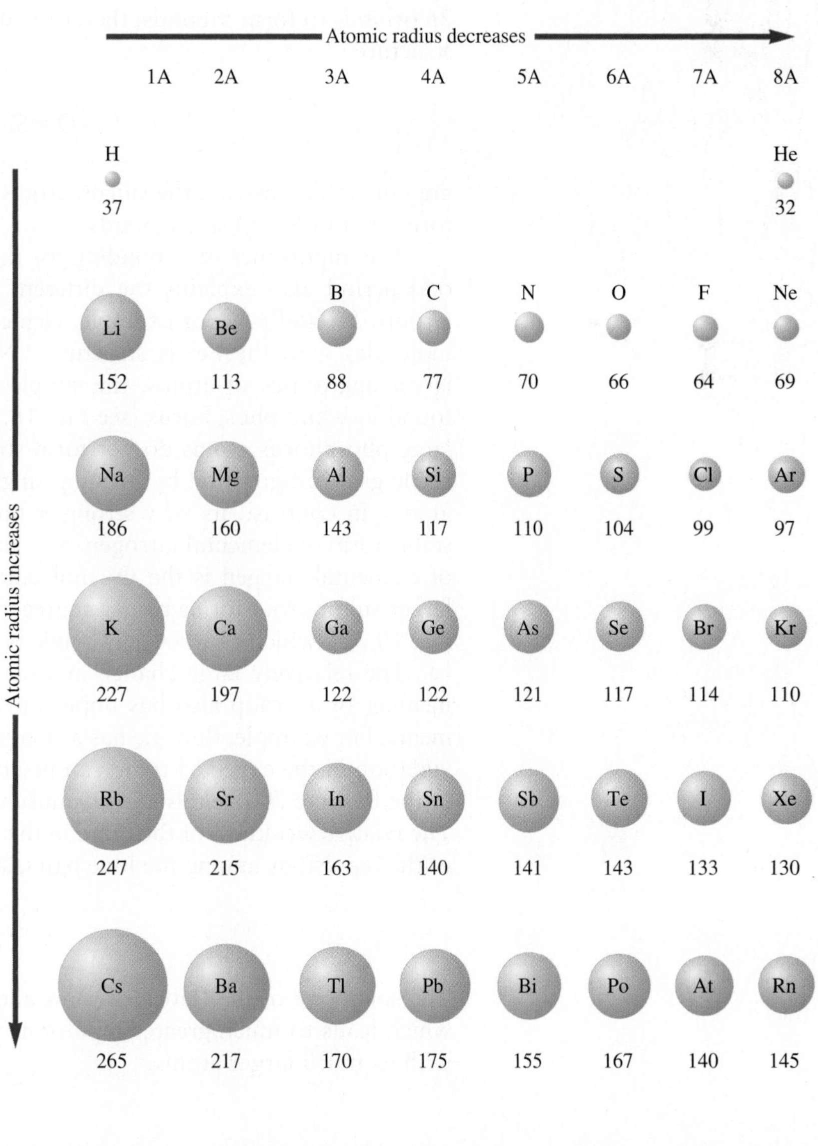

Figure 18.2
Some atomic radii (in picometers).

Atomic radius decreases →

Atomic radius increases ↓

1A	2A	3A	4A	5A	6A	7A	8A
H 37							He 32
Li 152	Be 113	B 88	C 77	N 70	O 66	F 64	Ne 69
Na 186	Mg 160	Al 143	Si 117	P 110	S 104	Cl 99	Ar 97
K 227	Ca 197	Ga 122	Ge 122	As 121	Se 117	Br 114	Kr 110
Rb 247	Sr 215	In 163	Sn 140	Sb 141	Te 143	I 133	Xe 130
Cs 265	Ba 217	Tl 170	Pb 175	Bi 155	Po 167	At 140	Rn 145

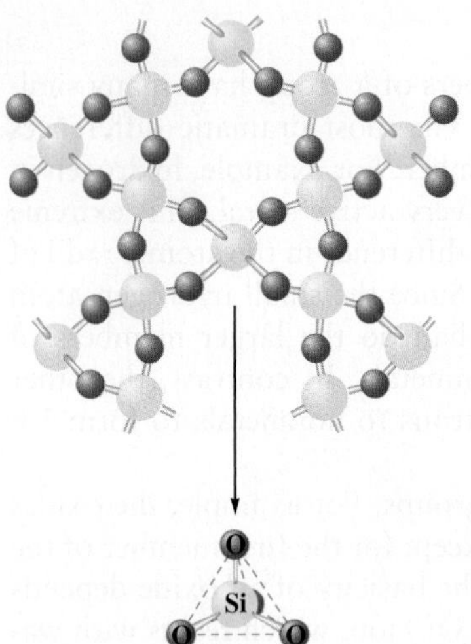

Figure 18.3
The structure of quartz, which has the empirical formula SiO_2. Note that the structure is based on interlocking SiO_4 tetrahedra, in which each oxygen atom is shared by two silicon atoms.

Group 3A, where only the small boron atom behaves as a nonmetal, or sometimes as a semimetal, whereas aluminum and the other members are active metals.

In Group 4A the effect of size is reflected in the dramatic differences between the chemical properties of carbon and silicon. The chemistry of carbon is dominated by molecules containing chains of C—C bonds, but silicon compounds mainly contain Si—O bonds rather than Si—Si bonds. Silicon does form compounds with chains of Si—Si bonds, but these compounds are much more reactive than the corresponding carbon compounds. The reasons for the difference in reactivity between the carbon and silicon compounds are quite complex but are likely related to the differences in the sizes of the carbon and silicon atoms.

Carbon and silicon also differ markedly in their abilities to form π bonds. As we discussed in Section 14.1, carbon dioxide is composed of discrete CO_2 molecules with the Lewis structure

$$\ddot{O}{=}C{=}\ddot{O}$$

where the carbon and oxygen atoms achieve the [Ne] configuration by forming π bonds. In contrast, the structure of silica (empirical formula SiO_2) is based on SiO_4 tetrahedra with Si—O—Si bridges, as shown in Fig. 18.3. The silicon $3p$ valence orbitals do not overlap very effectively with the smaller oxygen $2p$ orbitals to form π bonds; therefore, discrete SiO_2 molecules with the Lewis structure

$$\ddot{O}{=}Si{=}\ddot{O}$$

are not stable. Instead, the silicon atoms achieve a noble gas configuration by forming four Si—O single bonds.

The importance of π bonding for the relatively small elements of the second period also explains the different elemental forms of the members of Groups 5A and 6A. For example, elemental nitrogen exists as very stable N_2 molecules with the Lewis structure $:N{\equiv}N:$. Elemental phosphorus forms larger aggregates of atoms, the simplest being the tetrahedral P_4 molecules found in white phosphorus (see Fig. 18.18). Like silicon atoms, the relatively large phosphorus atoms do not form strong π bonds but prefer to achieve a noble gas configuration by forming single bonds to several other phosphorus atoms. In contrast, its very strong π bonds make the N_2 molecule the most stable form of elemental nitrogen. Similarly, in Group 6A the most stable form of elemental oxygen is the O_2 molecule with a double bond. However, the larger sulfur atom forms bigger aggregates, such as the cyclic S_8 molecule (see Fig. 18.22), which contain only single bonds.

The relatively large change in size in going from the first to the second member of a group also has important consequences for the Group 7A elements. For example, fluorine has a smaller electron affinity than chlorine. This violation of the expected trend can be attributed to the fact that the small size of the fluorine $2p$ orbitals causes unusually large electron–electron repulsions. The relative weakness of the bond in the F_2 molecule can be explained in terms of the repulsions among the lone pairs, shown in the Lewis structure:

$$:\ddot{\underset{..}{F}}{-}\ddot{\underset{..}{F}}:$$

The small size of the fluorine atoms allows close approach of the lone pairs, which leads to much greater repulsions than those found in the Cl_2 molecule with its much larger atoms.

Table 18.1

Distribution (Mass Percent) of the 18 Most Abundant Elements in the Earth's Crust, Oceans, and Atmosphere

Element	Mass Percent	Element	Mass Percent
Oxygen	49.2	Chlorine	0.19
Silicon	25.7	Phosphorus	0.11
Aluminum	7.50	Manganese	0.09
Iron	4.71	Carbon	0.08
Calcium	3.39	Sulfur	0.06
Sodium	2.63	Barium	0.04
Potassium	2.40	Nitrogen	0.03
Magnesium	1.93	Fluorine	0.03
Hydrogen	0.87	All others	0.49
Titanium	0.58		

Thus the relatively large increase in atomic radius in going from the first to the second member of a group causes the first element to exhibit properties quite different from the others.

Abundance and Preparation

Table 18.1 shows the distribution of elements in the earth's crust, oceans, and atmosphere. The major element is, of course, oxygen, which is found in the atmosphere as O_2, in the oceans as H_2O, and in the earth's crust primarily in silicate and carbonate minerals. The second most abundant element, silicon, is found throughout the earth's crust in the silica and silicate minerals that form the basis of most sand, rocks, and soil. The most abundant metals, aluminum and iron, are found in ores, in which they are combined with nonmetals, most commonly oxygen. One notable fact revealed by Table 18.1 is the small incidence of most transition metals. Since many of these relatively rare elements are assuming increasing importance in our high-technology society, it is possible that the control of transition metal ores may ultimately have more significance in world politics than will control of petroleum supplies.

The distribution of elements in living materials is very different from that found in the earth's crust. Table 18.2 shows the distribution of elements in the

Table 18.2

Abundance of Elements in the Human Body

Major Elements	Mass Percent	Trace Elements (in alphabetical order)
Oxygen	65.0	Arsenic
Carbon	18.0	Chromium
Hydrogen	10.0	Cobalt
Nitrogen	3.0	Copper
Calcium	1.4	Fluorine
Phosphorus	1.0	Iodine
Magnesium	0.50	Manganese
Potassium	0.34	Molybdenum
Sulfur	0.26	Nickel
Sodium	0.14	Selenium
Chlorine	0.14	Silicon
Iron	0.004	Vanadium
Zinc	0.003	

Frank Krahmer/Radium Images/Masterfile

Sand, such as that found in the massive sand dunes bordering the desert plain near Namib, Namibia, is composed of silicon and oxygen.

human body. Oxygen, carbon, hydrogen, and nitrogen form the basis for all biologically important molecules. The other elements, even though they are found in relatively small amounts, are often crucial for life. For example, zinc is found in over 150 different biomolecules in the human body.

Only about one-fourth of the elements occur naturally in the free state. Most are found in a combined state. The *process of obtaining a metal from its ore* is called **metallurgy.** Since the metals in ores are found in the form of cations, the chemistry of *metallurgy always involves reduction of the ions to the elemental metal (with an oxidation state of zero).* A variety of reducing agents can be used, but carbon is the usual choice because of its wide availability and relatively low cost.

Carbon is the cheapest and most readily available industrial reducing agent for metallic ions.

Electrolysis is often used to reduce the most active metals. In Chapter 11 we considered the electrolytic production of aluminum metal. The alkali metals are also produced by electrolysis, usually of their molten halide salts.

The preparation of nonmetals varies widely. Elemental nitrogen and oxygen are usually obtained from the **liquefaction** of air, which is based on the principle that a gas cools as it expands. After each expansion, part of the cooler gas is compressed, whereas the rest is used to carry away the heat of the compression. The compressed gas is then allowed to expand again. This cycle is repeated many times. Eventually, the remaining gas becomes cold enough to form the liquid state. Because liquid nitrogen and liquid oxygen have different boiling points, they can be separated by the distillation of liquid air. Both substances are important industrial chemicals, with nitrogen ranking second in terms of amount manufactured in the United States (approximately 60 billion pounds per year) and oxygen ranking third (over 40 billion pounds per year). Hydrogen can be obtained from the electrolysis of water, but more commonly it is obtained from the decomposition of the methane in natural gas. Sulfur is found underground in its elemental form and is recovered by the Frasch process (see Section 18.12). The halogens are obtained by oxidation of the anions from halide salts (see Section 18.13).

The preparation of sulfur and the halogens is discussed later in this chapter.

18.2 The Group 1A Metals

1A
H
Li
Na
K
Rb
Cs
Fr

The Group 1A elements with their ns^1 valence electron configurations are all very active metals (they lose their valence electrons very readily), except for hydrogen, which behaves as a nonmetal. We will discuss the chemistry of hydrogen in the next section. Many of the properties of the **alkali metals** have been given previously (see Section 12.16). The sources and methods of preparation of pure alkali metals are given in Table 18.3. The ionization energies, standard reduction potentials, ionic radii, and melting points for the alkali metals are listed in Table 18.4.

In Section 12.16 we saw that the alkali metals all react vigorously with water to release hydrogen gas:

$$2M(s) + 2H_2O(l) \longrightarrow 2M^+(aq) + 2OH^-(aq) + H_2(g)$$

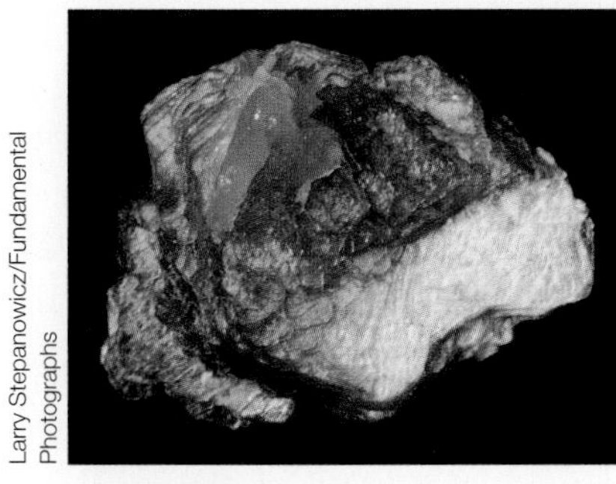

Figure 18.4
Lepidolite is mainly composed of lithium, aluminum, silicon, and oxygen, but it also contains significant amounts of rubidium and cesium.

Table 18.3

Sources and Methods of Preparation of the Pure Alkali Metals

Element	Source	Method of Preparation
Lithium	Silicate minerals such as spodumene, $LiAl(Si_2O_6)$	Electrolysis of molten LiCl
Sodium	NaCl	Electrolysis of molten NaCl
Potassium	KCl	Electrolysis of molten KCl
Rubidium	Impurity in lepidolite, $Li_2(F,OH)_2Al_2(SiO_3)_3$	Reduction of RbOH with Mg and H_2
Cesium	Pollucite ($Cs_4Al_4Si_9O_{26} \cdot H_2O$) and an impurity in lepidolite (Fig. 18.4)	Reduction of CsOH with Mg and H_2

Several properties of the alkali metals are given in Table 12.9.

We will reconsider this process briefly because it illustrates several important concepts. From the ionization energies, we might expect lithium to be the weakest of the alkali metals as a reducing agent in water. However, the standard reduction potentials indicate that it is the strongest. This reversal results mainly from the very large energy of hydration of the small Li^+ ion. Because of its relatively high charge density, the Li^+ ion very effectively attracts water molecules. A large quantity of energy is released in the process, favoring the formation of the Li^+ ion and making lithium a strong reducing agent in aqueous solution.

We also saw in Section 12.16 that lithium, although it is the strongest reducing agent, reacts more slowly with water than sodium or potassium. From the discussions in Chapters 10 and 15, we know that the *equilibrium position* for a reaction (in this case indicated by the $\mathscr{E}°$ values) is controlled by thermodynamic factors but that the *rate* of a reaction is controlled by kinetic factors. There is *no* direct connection between these factors. Lithium reacts more slowly with water than sodium or potassium because as a solid lithium has a higher melting point than either of the other elements. Since lithium does not become molten from the heat of reaction with water as sodium and potassium do, it has a smaller area of contact with the water.

Sodium reacts violently with water.

Table 18.5 summarizes some important reactions of the alkali metals.

The alkali metal ions are very important for the proper functioning of biological systems such as nerves and muscles; Na^+ and K^+ ions are present in all body cells and fluids. In human blood plasma, the concentrations are

$$[Na^+] \approx 0.15\ M \quad \text{and} \quad [K^+] \approx 0.005\ M$$

In the fluids *inside* the cells, the concentrations are reversed:

$$[Na^+] \approx 0.005\ M \quad \text{and} \quad [K^+] \approx 0.16\ M$$

Table 18.4

Selected Physical Properties of the Alkali Metals

Element	Ionization Energy (kJ/mol)	Standard Reduction Potential (V) for $M^+ + e^- \rightarrow M$	Radius of M^+ (pm)	Melting Point (°C)
Lithium	520	−3.05	60	180
Sodium	495	−2.71	95	98
Potassium	419	−2.92	133	64
Rubidium	409	−2.99	148	39
Cesium	382	−3.02	169	29

Table 18.5

Selected Reactions of the Alkali Metals

Reaction	Comment
$2M + X_2 \longrightarrow 2MX$	X_2 = any halogen molecule
$4Li + O_2 \longrightarrow 2Li_2O$	Excess oxygen
$2Na + O_2 \longrightarrow Na_2O_2$	
$M + O_2 \longrightarrow MO_2$	M = K, Rb, or Cs
$2M + S \longrightarrow M_2S$	
$6Li + N_2 \longrightarrow 2Li_3N$	Li only
$12M + P_4 \longrightarrow 4M_3P$	
$2M + H_2 \longrightarrow 2MH$	
$2M + 2H_2O \longrightarrow 2MOH + H_2$	
$2M + 2H^+ \longrightarrow 2M^+ + H_2$	Violent reaction!

Since the concentrations are so different inside and outside the cells, an elaborate mechanism involving selective ligands is needed to transport Na^+ and K^+ ions through the cell membranes.

18.3 The Chemistry of Hydrogen

Under ordinary conditions of temperature and pressure, hydrogen is a colorless, odorless gas composed of H_2 molecules. Because of its low molar mass and nonpolarity, hydrogen has a very low boiling point (−253°C) and melting point (−260°C). Hydrogen gas is highly flammable; mixtures of air containing from 18% to 60% hydrogen by volume are explosive. In a common lecture demonstration, hydrogen and oxygen gases are bubbled into soapy water. The resulting bubbles are then ignited with a candle on a long stick, producing a loud explosion.

The major industrial source of hydrogen gas is the reaction of methane with water at high temperatures (800–1000°C) and pressures (10–50 atm) in the presence of a metallic catalyst (often nickel):

$$CH_4(g) + H_2O(g) \xrightarrow[\text{Catalyst}]{\text{Heat, pressure}} CO(g) + 3H_2(g)$$

Large quantities of hydrogen are also formed as a by-product of gasoline production, when hydrocarbons with high molecular masses are broken down (or *cracked*) to produce smaller molecules more suitable for use as a motor fuel.

Very pure hydrogen can be produced by the electrolysis of water (see Section 11.7), but this method is currently not economically feasible for large-scale production because of the relatively high cost of electricity.

The major industrial use of hydrogen is in the production of ammonia by the Haber process. Large quantities of hydrogen are also used for hydrogenating unsaturated vegetable oils (those containing carbon–carbon double bonds) to produce solid shortenings that are saturated (containing carbon–carbon single bonds):

$$\sim\overset{\mathrm{H}}{\overset{|}{\mathrm{C}}}=\overset{\mathrm{H}}{\overset{|}{\mathrm{C}}}\sim \xrightarrow{H_2} \sim\underset{\mathrm{H}}{\underset{|}{\overset{\mathrm{H}}{\overset{|}{\mathrm{C}}}}}-\underset{\mathrm{H}}{\underset{|}{\overset{\mathrm{H}}{\overset{|}{\mathrm{C}}}}}\sim$$

The catalysis of this process was discussed in Section 15.9.

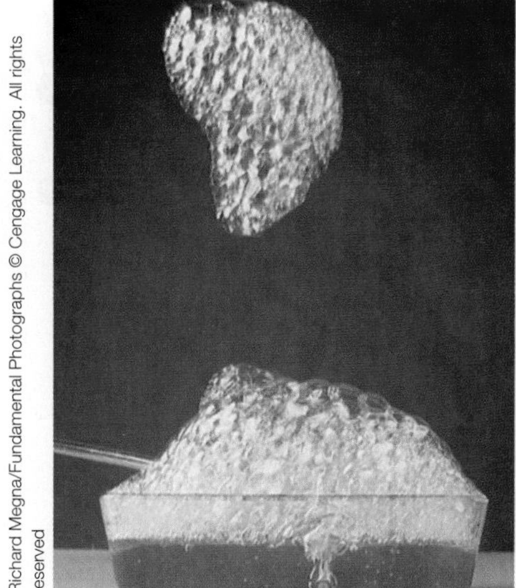

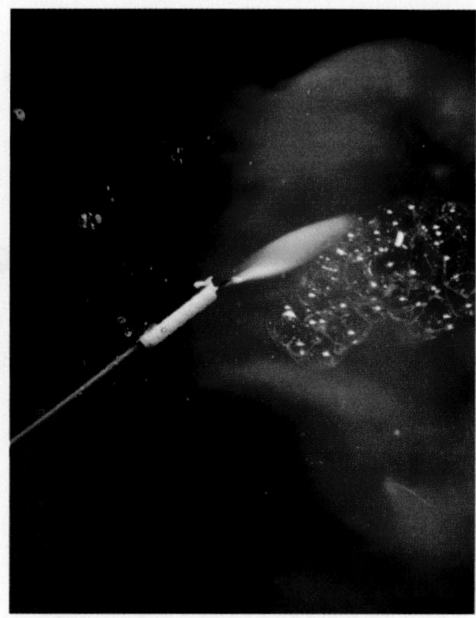

(left) Hydrogen gas being used to blow soap bubbles. (right) As the bubbles float upward, they are lighted by using a candle on a long stick. The orange flame results from the heat of the reaction between hydrogen and oxygen, which excites sodium atoms in the soap bubbles.

Chemically, hydrogen behaves as a typical nonmetal, forming covalent compounds with other nonmetals and forming salts with very active metals. Binary compounds containing hydrogen are called **hydrides,** of which there are three classes. The **ionic** (or saltlike) **hydrides** are formed when hydrogen combines with the most active metals, those from Groups 1A and 2A. Examples are LiH and CaH_2, which can best be characterized as containing hydride ions (H^-) and metal cations. Because the presence of two electrons in the small 1*s* orbital produces large electron–electron repulsions and because the nucleus has only a 1+ charge, the hydride ion is a strong reducing agent. For example, when ionic hydrides are placed in water, a violent reaction takes place. This reaction results in the formation of hydrogen gas, as seen in the equation

Boiling points of covalent hydrides were discussed in Section 16.1.

$$LiH(s) + H_2O(l) \longrightarrow H_2(g) + Li^+(aq) + OH^-(aq)$$

Covalent hydrides are formed when hydrogen combines with other nonmetals. We have encountered many of these compounds already: HCl, CH_4, NH_3, H_2O, and so on. The most important covalent hydride is water. The polarity of the H_2O molecule leads to many of water's unusual properties. Water has a much higher boiling point than is expected from its molar mass. It has a large heat of vaporization and a large heat capacity, both of which make it a very useful coolant. Water has a higher density as a liquid than as a solid because of the open structure of ice, which results from maximizing the hydrogen bonding (Fig. 18.5). Because water is an excellent solvent for ionic and polar substances, it provides an effective medium for life processes. In fact, water is one of the few covalent hydrides that is nontoxic to organisms.

The third class of hydrides is the **metallic,** or **interstitial, hydrides,** which are formed when transition metal crystals are treated with hydrogen gas. The hydrogen molecules dissociate at the metal's surface, and the small hydrogen atoms migrate into the crystal structure to occupy holes, or *interstices.* These metal–hydrogen mixtures are more like solid solutions than true compounds. Palladium can absorb about *900 times* its own volume of hydrogen gas. In fact, hydrogen can be purified by placing it under slight pressure in a vessel containing a thin wall of palladium. The hydrogen diffuses into and through the metal wall, leaving the impurities behind.

Although hydrogen can react with transition metals to form compounds such as UH_3 and FeH_6, most of the interstitial hydrides have variable compositions (often called *nonstoichiometric* compositions) with formulas such as

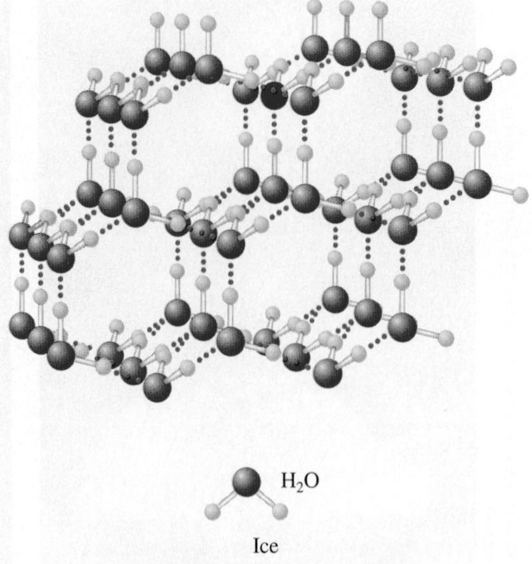

Figure 18.5
The structure of ice, showing the hydrogen bonding.

$LaH_{2.76}$ and $VH_{0.56}$. The compositions of the nonstoichiometric hydrides vary with the length of exposure of the metal to hydrogen gas.

When interstitial hydrides are heated, much of the absorbed hydrogen is lost as hydrogen gas. Because of this behavior, these materials offer possibilities for storing hydrogen for use as a portable fuel. The internal combustion engines in current automobiles can burn hydrogen gas with little modification, but storage of enough hydrogen to provide an acceptable mileage range remains a problem. One possible solution might be to use a fuel tank containing a porous solid that includes a transition metal. The hydrogen gas could be pumped into the solid to form the interstitial hydride. The hydrogen gas could then be released when the engine requires additional energy. This system is now being tested by several automobile companies.

See Section 9.8 for a discussion of the feasibility of using hydrogen gas as a fuel.

18.4 The Group 2A Elements

2A
Be
Mg
Ca
Sr
Ba
Ra

The Group 2A elements (with the valence electron configuration ns^2) are very reactive, losing their two valence electrons to form ionic compounds that contain M^{2+} cations. These elements are commonly called the **alkaline earth metals** because of the basicity of their oxides:

$$MO(s) + H_2O(l) \longrightarrow M^{2+}(aq) + 2OH^-(aq)$$

Only the amphoteric beryllium oxide (BeO) also shows some acidic properties, such as dissolving in aqueous solutions containing hydroxide ions:

$$BeO(s) + 2OH^-(aq) + H_2O(l) \longrightarrow Be(OH)_4^{2-}(aq)$$

An amphoteric oxide displays both acidic and basic properties.

The more active alkaline earth metals react with water as the alkali metals do, producing hydrogen gas:

$$M(s) + 2H_2O(l) \longrightarrow M^{2+}(aq) + 2OH^-(aq) + H_2(g)$$

Calcium, strontium, and barium react vigorously at 25°C. The less easily oxidized beryllium and magnesium show no observable reaction with water at 25°C, although magnesium reacts with boiling water. Table 18.6 summarizes various properties, sources, and methods of preparation of the alkaline earth metals.

The alkaline earth metals have great practical importance. Calcium and magnesium ions are essential for human life. Calcium is found primarily in the

Table 18.6

Selected Physical Properties, Sources, and Methods of Preparation of the Group 2A Elements

Element	Radius of M^{2+} (pm)	Ionization Energy (kJ/mol)		$\mathscr{E}°$ (V) for $M^{2+} + 2e^- \longrightarrow M$	Source	Method of Preparation
		First	Second			
Beryllium	≈300	900	1760	−1.70	Beryl ($Be_3Al_2Si_6O_{18}$)	Electrolysis of molten $BeCl_2$
Magnesium	65	735	1445	−2.37	Magnesite ($MgCO_3$), dolomite ($MgCO_3 \cdot CaCO_3$), carnallite ($MgCl_2 \cdot KCl \cdot 6H_2O$)	Electrolysis of molten $MgCl_2$
Calcium	99	590	1146	−2.76	Various minerals containing $CaCO_3$	Electrolysis of molten $CaCl_2$
Strontium	113	549	1064	−2.89	Celestite ($SrSO_4$), strontianite ($SrCO_3$)	Electrolysis of molten $SrCl_2$
Barium	135	503	965	−2.90	Baryte ($BaSO_4$), witherite ($BaCO_3$)	Electrolysis of molten $BaCl_2$
Radium	140	509	979	−2.92	Pitchblende (1 g of Ra/7 tons of ore)	Electrolysis of molten $RaCl_2$

Calcium metal reacting with water to form bubbles of hydrogen gas.

structural minerals composing bones and teeth. Magnesium (as the Mg^{2+} ion) plays a vital role in metabolism and in muscle functions. Because magnesium metal has a relatively low density and displays moderate strength, it is a useful structural material, especially if alloyed with aluminum.

Table 18.7 summarizes some important reactions involving the alkaline earth metals.

Relatively large concentrations of Ca^{2+} and Mg^{2+} ions are often found in natural water supplies. These ions in this so-called **hard water** interfere with the action of detergents and form precipitates with soap. In Section 7.6 we saw that Ca^{2+} is often removed by precipitation as $CaCO_3$ in large-scale water softening. In individual homes Ca^{2+}, Mg^{2+}, and other cations are removed by **ion exchange.** An **ion exchange resin** consists of large molecules (polymers) that have many ionic sites. A cation exchange resin is represented schemati-

Table 18.7

Selected Reactions of the Group 2A Elements

Reaction	Comment
$M + X_2 \longrightarrow MX_2$	X_2 = any halogen molecule
$2M + O_2 \longrightarrow 2MO$	Ba gives BaO_2 as well
$M + S \longrightarrow MS$	
$3M + N_2 \longrightarrow M_3N_2$	High temperatures
$6M + P_4 \longrightarrow 2M_3P_2$	High temperatures
$M + H_2 \longrightarrow MH_2$	M = Ca, Sr, or Ba; high temperatures; Mg at high pressure
$M + 2H_2O \longrightarrow M(OH)_2 + H_2$	M = Ca, Sr, or Ba
$M + 2H^+ \longrightarrow M^{2+} + H_2$	
$Be + 2OH^- + 2H_2O \longrightarrow Be(OH)_4{}^{2-} + H_2$	

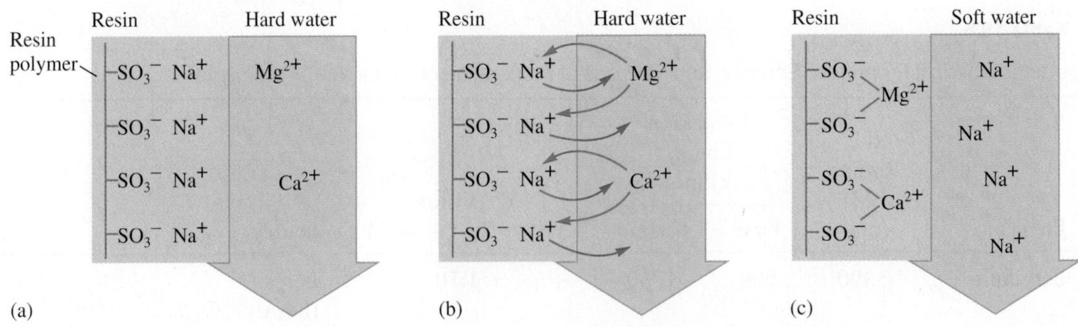

Figure 18.6
(a) A schematic representation of a typical cation exchange resin. (b) and (c) When hard water is passed over the cation exchange resin, the Ca^{2+} and Mg^{2+} bind to the resin.

cally in Fig. 18.6(a), showing Na^+ ions bound ionically to the SO_3^- groups that are covalently attached to the resin polymer. When hard water is passed over the resin, Ca^{2+} and Mg^{2+} bind to the resin in place of Na^+, which is released into the solution [Fig. 18.6(b)]. Replacing Mg^{2+} and Ca^{2+} by Na^+ [Fig. 18.6(c)] "softens" the water because the sodium salts of soap are soluble.

18.5 The Group 3A Elements

3A
B
Al
Ga
In
Tl

The Group 3A elements (valence electron configuration ns^2np^1) generally show the increase in metallic character in going down the group that is characteristic of the representative elements. Some physical properties, sources, and methods of preparation of the Group 3A elements are summarized in Table 18.8.

Boron is a typical nonmetal, and most of its compounds are covalent. The most interesting compounds of boron are the covalent hydrides called **boranes.** We might expect BH_3 to be the simplest hydride, since boron has three valence electrons to share with three hydrogen atoms. However, this compound is unstable, and the simplest known member of the series is diborane (B_2H_6), with the structure shown in Fig. 18.7(a). In this molecule the terminal B—H bonds are normal covalent bonds, each involving one electron pair. The bridging bonds are three-center bonds similar to those in solid BeH_2. Another interesting borane contains the square pyramidal B_5H_9 molecule [Fig. 18.7(b)], which

Table 18.8

Selected Physical Properties, Sources, and Methods of Preparation of the Group 3A Elements

Element	Radius of M^{3+} (pm)	Ionization Energy (kJ/mol)	$\mathscr{E}°$ (V) for $M^{3+} + 3e^- \longrightarrow M$	Source	Method of Preparation
Boron	20	798	—	Kernite, a form of borax ($Na_2B_4O_7 \cdot 4H_2O$)	Reduction by Mg or H_2
Aluminum	50	581	−1.66	Bauxite (Al_2O_3)	Electrolysis of Al_2O_3 in molten Na_3AlF_6
Gallium	62	577	−0.53	Traces in various minerals	Reduction with H_2 or electrolysis
Indium	81	556	−0.34	Traces in various minerals	Reduction with H_2 or electrolysis
Thallium	95	589	0.72	Traces in various minerals	Electrolysis

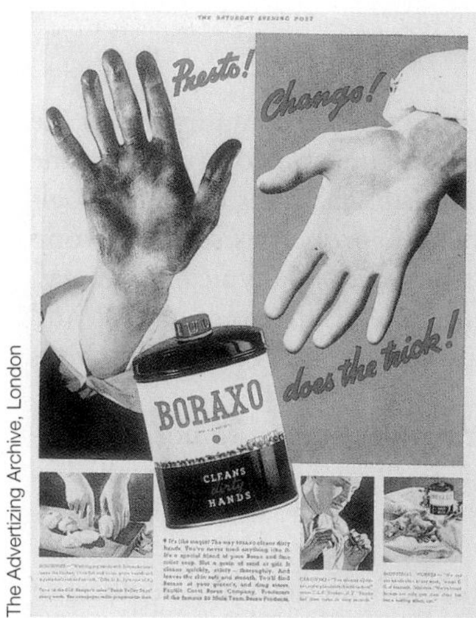

An old ad from *The Saturday Evening Post* for Boraxo, a hand-cleaning product containing sodium tetraborate (NaB_4O_7). Extensive natural deposits of borax ($Na_2B_4O_7 \cdot 10H_2O$) found in saline lakes near Death Valley, California, were hauled to a factory in wagons pulled by teams of 20 mules—hence the name 20 Mule Team Borax.

Gallium melts in the hand.

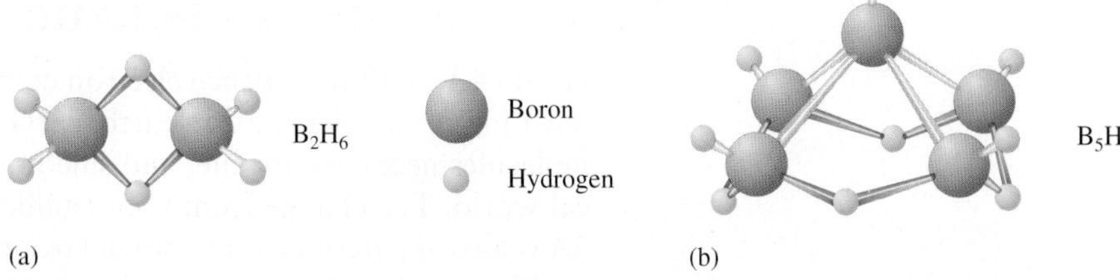

Figure 18.7
(a) The structure of B_2H_6 with its two three-center B—H—B bridging bonds and four "normal" B—H bonds. (b) The structure of B_5H_9. There are five "normal" B—H bonds to terminal hydrogens and four three-center bridging bonds around the base.

has four three-center bonds situated around the base of the pyramid. Because the boranes are extremely electron-deficient, they are highly reactive. The boranes react very exothermically with oxygen and were once evaluated as potential fuels for rockets in the U.S. space program.

Aluminum, the most abundant metal on earth, has metallic physical properties, such as high thermal and electrical conductivities and a lustrous appearance; however, its bonds to nonmetals are significantly covalent. This covalency is responsible for the amphoteric nature of Al_2O_3, which dissolves in acidic or basic solution, and for the acidity of $Al(H_2O)_6^{3+}$ (see Section 7.8):

$$Al(H_2O)_6^{3+}(aq) \rightleftharpoons Al(OH)(H_2O)_5^{2+}(aq) + H^+(aq)$$

One especially interesting property of *gallium* is its unusually low melting point of 29.8°C, which is in contrast to the 660°C melting point of aluminum. Also, since gallium's boiling point is about 2400°C, it has the largest liquid range of any metal. This makes it useful for thermometers, especially to measure high temperatures. Gallium, like water, expands when it freezes. The chemistry of gallium is quite similar to that of aluminum. For example, Ga_2O_3 is amphoteric.

The chemistry of *indium* is similar to that of aluminum and gallium except that compounds containing the 1+ ion are known, such as InCl and In_2O, in addition to those with the more common 3+ ion. The chemistry of *thallium* is completely metallic.

Table 18.9 summarizes some important reactions of the Group 3A elements.

Table 18.9

Selected Reactions of the Group 3A Elements

Reaction	Comment
$2M + 3X_2 \longrightarrow 2MX_3$	X_2 = any halogen molecule; Tl gives TlX as well, but no TlI_3
$4M + 3O_2 \longrightarrow 2M_2O_3$	High temperatures; Tl gives Tl_2O as well
$2M + 3S \longrightarrow M_2S_3$	High temperatures; Tl gives Tl_2S as well
$2M + N_2 \longrightarrow 2MN$	M = Al only
$2M + 6H^+ \longrightarrow 2M^{3+} + 3H_2$	M = Al, Ga, or In; Tl gives Tl^+
$2M + 2OH^- + 6H_2O \longrightarrow 2M(OH)_4^- + 3H_2$	M = Al or Ga

18.6 The Group 4A Elements

4A
C
Si
Ge
Sn
Pb

Group 4A (with the valence electron configuration ns^2np^2) contains two of the most important elements on earth: carbon, the fundamental constituent of the molecules necessary for life, and silicon, which forms the basis of the geological world. The change from nonmetallic to metallic properties seen in Group 3A is also apparent in going down Group 4A from carbon, a typical nonmetal, to silicon and germanium, usually considered semimetals, to the metals tin and lead. Table 18.10 summarizes some physical properties, sources, and methods of preparation of the elements in this group.

All the Group 4A elements can form four covalent bonds to nonmetals—for example, CH_4, SiF_4, $GeBr_4$, $SnCl_4$, and $PbCl_4$. In each of these tetrahedral molecules, the central atom is described as sp^3 hybridized by the localized electron model.

We have seen that carbon also differs markedly from the other members of Group 4A in its ability to form π bonds. This accounts for the completely different structures and properties of CO_2 and SiO_2. Note from Table 18.11 that C—C bonds and Si—O bonds are stronger than Si—Si bonds. This partly explains why the chemistry of carbon is dominated by C—C bonds, whereas that of silicon is dominated by Si—O bonds.

A new form of elemental carbon, the fullerenes, was discussed in Chapter 2.

Carbon occurs in the allotropic forms graphite, diamond, and fullerenes, whose structures were given in Section 16.5. The most important chemistry of carbon is organic chemistry, which is described in detail in Chapter 21.

Silicon, the second most abundant element in the earth's crust, is a semimetal found widely distributed in silica and silicates (see Section 16.5). About 85% of the earth's crust is composed of these substances. Although silicon is found in some steel and aluminum alloys, its major use is in semiconductors for electronic devices (see Chapter 16).

Germanium, a relatively rare element, is a semimetal used mainly in the manufacture of semiconductors for transistors and similar electronic devices.

Tin is a soft, silvery metal that can be rolled into thin sheets (tin foil) and has been used for centuries in various alloys such as bronze (20% Sn and 80% Cu), solder (33% Sn and 67% Pb), and pewter (85% Sn, 7% Cu, 6% Bi, and 2% Sb). Tin exists as three allotropes: *white tin,* stable at normal temperatures; *gray tin,* stable at temperatures below 13.2°C; and *brittle tin,* found at temperatures above 161°C. When tin is exposed to low temperatures, it gradually changes to powdery gray tin and crumbles away; this is known as *tin disease.*

Table 18.10

Selected Physical Properties, Sources, and Methods of Preparation of the Group 4A Elements

Element	Electronegativity	Melting Point (°C)	Boiling Point (°C)	Source	Method of Preparation
Carbon	2.6	3727 (sublimes)	—	Graphite, diamond, petroleum, coal	—
Silicon	1.9	1410	2355	Silicate minerals, silica	Reduction of K_2SiF_6 with Al, or reduction of SiO_2 with Mg
Germanium	2.0	937	2830	Germinate (mixture of copper, iron, and germanium sulfides)	Reduction of GeO_2 with H_2 or C
Tin	2.0	232	2270	Cassiterite (SnO_2)	Reduction of SnO_2 with C
Lead	2.3	327	1740	Galena (PbS)	Roasting of PbS with O_2 to form PbO_2 and then reduction with C

Beethoven: Hair Is the Story

Chemical Insights

Ludwig van Beethoven, arguably the greatest composer who ever lived, led a troubled life fraught with sickness, deafness, and personality aberrations. Now we may know the source of these difficulties: lead poisoning. Scientists have recently reached this conclusion through analysis of Beethoven's hair. When Beethoven died in 1827 at age 56, many mourners took samples of the great man's hair. In fact, it was said at the time that he was practically bald by the time he was buried. The hair that was recently analyzed consisted of 582 strands—3 to 6 inches long—bought for the Center of Beethoven Studies for $7300 in 1994 from Sotheby's auction house in London.

According to William Walsh of the Health Research Institute (HRI) in suburban Chicago, Beethoven's hair showed a lead concentration 100 times the normal levels. The scientists concluded that Beethoven's exposure to lead came as an adult, possibly from the mineral water he drank and swam in when he visited spas.

The lead poisoning may well explain Beethoven's volatile temper—the composer was subject to towering rages and sometimes had the look of a wild animal. In rare cases lead poisoning has been known to cause deafness, but the researchers remain unsure if this problem led to Beethoven's hearing loss.

According to Walsh, the scientists at HRI were originally looking for mercury, a common treatment for syphilis in the early nineteenth century, in Beethoven's hair. The absence of mercury supports the consensus of scholars that Beethoven did not have this disease. Not surprisingly, Beethoven himself wanted to know what made him so ill. In a letter to his brothers in 1802, he asked them to have doctors find the cause of his frequent abdominal pain after his death.

The Granger Collection

Portrait of Beethoven by Josef Karl Stieler.

Table 18.11

Strengths of C—C, Si—Si, and Si—O Bonds

Bond	Bond Energy (kJ/mol)
C—C	347
Si—Si	340
Si—O	368

Currently, tin is used mainly as a protective coating for steel, especially for cans used as food containers. The thin layer of tin, applied electrolytically, forms a protective oxide coating that prevents further corrosion.

Lead is easily obtained from its ore, galena (PbS). Because lead melts at such a low temperature, it may have been the first pure metal obtained from its ore. We know that lead was used as early as 3000 B.C. by the Egyptians. It was later used by the Romans to make eating utensils, glazes on pottery, and even intricate plumbing systems. The Romans also prepared a sweetener called *sapa* by boiling down grape juice in lead-lined vessels. The sweetness of this syrup was partly caused by the formation of lead(II) acetate (formerly called sugar of lead), a very sweet-tasting compound. The problem with these practices is that lead is very toxic. In fact, the Romans had so much contact with lead that it may have contributed to the demise of their civilization. Analysis of bones from that era shows significant levels of lead.

Although lead poisoning has been known since at least the second century B.C., lead continues to be a problem. For example, many children have been

Table 18.12

Selected Reactions of the Group 4A Elements

Reaction	Comment
$M + 2X_2 \longrightarrow MX_4$	X_2 = any halogen molecule; M = Ge or Sn; Pb gives PbX_2
$M + O_2 \longrightarrow MO_2$	M = Ge or Sn; high temperatures; Pb gives PbO or Pb_3O_4
$M + 2H^+ \longrightarrow M^{2+} + H_2$	M = Sn or Pb

exposed to lead by eating chips of lead-based paint. Because of this problem, lead-based paints are no longer used for children's furniture, and many states have banned lead-based paint for interior use. Lead poisoning can also occur when acidic foods and drinks leach the lead from lead-glazed pottery dishes that were improperly fired and when liquor is stored in leaded crystal decanters, producing toxic levels of lead in the drink in a relatively short time. In addition, the widespread use of tetraethyl lead [$(C_2H_5)_4Pb$] as an antiknock agent in gasoline has increased the lead levels in our environment. Concern about the effects of this lead pollution has caused the U.S. government to require the gradual replacement of the lead in gasoline with other antiknock agents. The largest commercial use of lead (over one million tons annually) is for electrodes in the lead storage batteries used in automobiles (see Section 11.5).

Table 18.12 summarizes some important reactions of the Group 4A elements.

18.7 The Group 5A Elements

5A
N
P
As
Sb
Bi

The Group 5A elements (with the valence electron configuration ns^2np^3), which are prepared as shown in Table 18.13, exhibit remarkably varied chemical properties. As usual, metallic character increases going down the group, as is apparent from the electronegativity values (Table 18.13). Nitrogen and

Table 18.13

Selected Physical Properties, Sources, and Methods of Preparation of the Group 5A Elements

Element	Electronegativity	Source	Method of Preparation
Nitrogen	3.0	Air	Liquefaction of air
Phosphorus	2.2	Phosphate rock [$Ca_3(PO_4)_2$], fluorapatite [$Ca_5(PO_4)_3F$]	$2Ca_3(PO_4)_2 + 6SiO_2 \longrightarrow 6CaSiO_3 + P_4O_{10}$ $P_4O_{10} + 10C \longrightarrow 4P + 10CO$
Arsenic	2.2	Arsenopyrite (Fe_3As_2, FeS)	Heating arsenopyrite in the absence of air
Antimony	2.1	Stibnite (Sb_2S_3)	Roasting Sb_2S_3 in air to form Sb_2O_3 and then reduction with carbon
Bismuth	2.0	Bismite (Bi_2O_3), bismuth glance (Bi_2S_3)	Roasting Bi_2S_3 in air to form Bi_2O_3 and then reduction with carbon

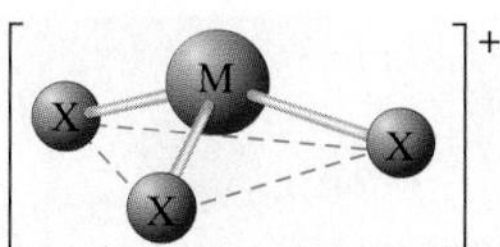

Figure 18.8
The pyramidal shape of the Group 5A MX_3 molecules.

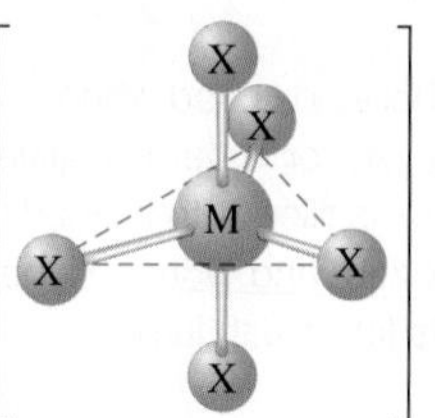

Figure 18.9
The trigonal bipyramidal shape of the MX_5 molecules.

phosphorus are nonmetals that can gain three electrons to form 3− anions in salts with active metals; examples are magnesium nitride (Mg_3N_2) and beryllium phosphide (Be_3P_2). The chemistry of these two important elements is discussed in the next two sections.

Bismuth and *antimony* tend to be metallic, readily losing electrons to form cations. Although these elements have five valence electrons, so much energy is required to remove all five that no ionic compounds containing Bi^{5+} or Sb^{5+} ions are known.

The Group 5A elements can form molecules or ions that involve three, five, or six covalent bonds to the Group 5A atom. Examples involving three single bonds are NH_3, PH_3, NF_3, and $AsCl_3$. Each of these molecules has a lone pair of electrons (and thus can behave as a Lewis base) and a pyramidal shape as predicted by the VSEPR model (Fig. 18.8).

All the Group 5A elements except nitrogen can form molecules with five covalent bonds (of general formula MX_5). Nitrogen cannot form such molecules because of its small size. The MX_5 molecules have a trigonal bipyramidal shape (Fig. 18.9) as predicted by the VSEPR model, and the central atom can be described as dsp^3 hybridized.

Although the MX_5 molecules have a trigonal bipyramidal structure in the gas phase, the solids of many of these compounds contain a 1:1 mixture of the ions MX_4^+ and MX_6^- (Fig. 18.10). The MX_4^+ cation is tetrahedral (the atom represented by M is sp^3 hybridized), and the MX_6^- anion is octahedral (the atom represented by M is d^2sp^3 hybridized). Examples are PCl_5 (which in the solid state contains PCl_4^+ and PCl_6^-) and AsF_3Cl_2 (which in the solid state contains $AsCl_4^+$ and AsF_6^-).

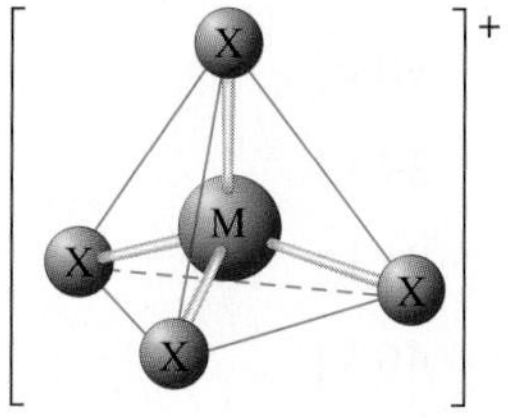

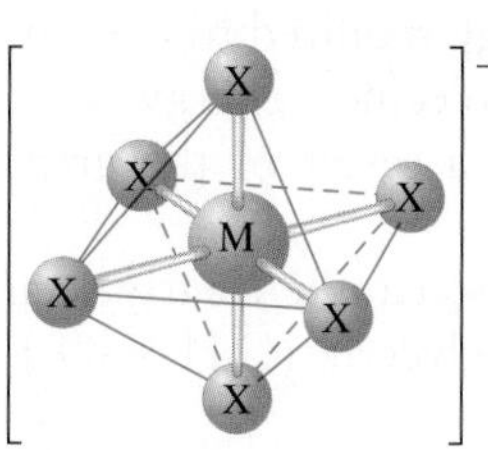

Figure 18.10
The structures of the tetrahedral MX_4^+ and the octahedral MX_6^- ions.

As discussed in Section 18.1, the ability of the Group 5A elements to form π bonds decreases dramatically after nitrogen. This explains why elemental nitrogen exists as N_2 molecules containing two π bonds, whereas the other elements in the group exist as larger aggregates containing single bonds. For example, in the gas phase the elements phosphorus, arsenic, and antimony consist of P_4, As_4, and Sb_4 molecules, respectively.

18.8 The Chemistry of Nitrogen

At the earth's surface, virtually all elemental nitrogen exists as the N_2 molecule with its very strong triple bond (941 kJ/mol). Because of this large bond strength, the N_2 molecule is so unreactive that it can coexist with most other elements under normal conditions without undergoing any appreciable reaction. This property makes nitrogen gas very useful as a medium for experiments involving substances that react with oxygen or water. Such experiments can be done using an inert atmosphere box of the type shown in Fig. 18.11.

The strength of the triple bond in the N_2 molecule is important both thermodynamically and kinetically. Thermodynamically, the great stability of the

Figure 18.11
An inert atmosphere box used when working with oxygen- or water-sensitive materials. The box is filled with an inert gas such as nitrogen, and work is done through the ports fitted with large rubber gloves.

Paul Ridgway/Lawrence Livermore National

N≡N bond means that most binary compounds containing nitrogen decompose exothermically to the elements, for example:

$$N_2O(g) \longrightarrow N_2(g) + \tfrac{1}{2}O_2(g) \qquad \Delta H° = -82 \text{ kJ}$$

$$NO(g) \longrightarrow \tfrac{1}{2}N_2(g) + \tfrac{1}{2}O_2(g) \qquad \Delta H° = -90 \text{ kJ}$$

$$NO_2(g) \longrightarrow \tfrac{1}{2}N_2(g) + O_2(g) \qquad \Delta H° = -34 \text{ kJ}$$

$$N_2H_4(g) \longrightarrow N_2(g) + 2H_2(g) \qquad \Delta H° = -95 \text{ kJ}$$

$$NH_3(g) \longrightarrow \tfrac{1}{2}N_2(g) + \tfrac{3}{2}H_2(g) \qquad \Delta H° = +46 \text{ kJ}$$

Of these compounds, only ammonia is thermodynamically more stable than its component elements. That is, only for ammonia is energy required to decompose the molecule to its elements. For the remaining molecules, energy is released when decomposition to the elements occurs, as a result of the great stability of N_2.

The importance of the thermodynamic stability of N_2 can be clearly seen in the power of nitrogen-based explosives, such as nitroglycerin ($C_3H_5N_3O_9$), which has the following structure:

```
     H          H          H
     |          |          |
  H—C—————————C—————————C—H
     |          |          |
     O          O          O
     |          |          |
     N          N          N
    / \        / \        / \
   O   O      O   O      O   O
```

When ignited or subjected to sudden impact, nitroglycerin decomposes very rapidly and exothermically:

$$4C_3H_5N_3O_9(l) \longrightarrow 6N_2(g) + 12CO_2(g) + 10H_2O(g) + O_2(g) + \text{energy}$$

An explosion occurs; that is, large volumes of gas are produced in a fast, highly exothermic reaction. Note that 4 moles of liquid nitroglycerin produce 29 (6 + 12 + 10 + 1) moles of gaseous products. This alone produces a large increase in volume. However, also note that the products, which include N_2, are very stable molecules with strong bonds. Their formation is therefore accompanied by the release of large quantities of energy as heat, which increases the gaseous volume. The hot, rapidly expanding gases produce a pressure surge and damaging shock wave.

Most high explosives are organic compounds that, like nitroglycerin, contain nitro ($—NO_2$) groups and produce nitrogen and other gases as products. Another example is *trinitrotoluene* (TNT), a solid at normal temperatures, which decomposes as follows:

$$2C_7H_5N_3O_6(s) \longrightarrow 12CO(g) + 5H_2(g) + 3N_2(g) + 2C(s) + \text{energy}$$

CH_3, NO_2, NO_2, NO_2

TNT

Note that 2 moles of solid TNT produce 20 moles of gaseous products plus energy.

The effect of bond strength on the kinetics of reactions involving the N_2 molecule is illustrated by the synthesis of ammonia from nitrogen and hydrogen, a reaction we have discussed many times before. Because a large quantity of energy is required to disrupt the $N{\equiv}N$ bond, the ammonia synthesis reaction occurs at a negligible rate at room temperature, even though the equilibrium constant is very large ($K \approx 10^8$). Of course, the most direct way to increase the rate of a reaction is to raise the temperature. However, since this reaction is very exothermic,

$$N_2(g) + 3H_2(g) \longrightarrow 2NH_3(g) \qquad \Delta H° = -92 \text{ kJ}$$

the value of K decreases significantly with a temperature increase (at 500°C, $K \approx 10^{-2}$).

Obviously, the kinetics and the thermodynamics of this reaction are in opposition. A compromise must be reached, involving high pressure to force the equilibrium to the right and high temperature to produce a reasonable rate. The **Haber process** for manufacturing ammonia represents such a compromise (Fig. 18.12). The process is carried out at a pressure of about 250 atm and a temperature of approximately 400°C. Even higher temperatures would be required if a catalyst consisting of a solid iron oxide mixed with small amounts of potassium oxide and aluminum oxide were not used to facilitate the reaction.

Impure N_2, H_2

Unwanted trace gases removed

Pure N_2, H_2

Catalytic reactors

NH_3

Cooling chamber

Liquid NH_3 (yield 20% on each cycle)

Unreacted N_2, H_2

Figure 18.12
A schematic diagram of the Haber process for the manufacture of ammonia.

Nitrogen is essential to living systems. The problem with nitrogen is not one of supply—we are surrounded by it—but rather one of changing it from the inert N_2 molecule to a form usable by plants and animals. The process of transforming N_2 to other nitrogen-containing compounds is called **nitrogen fixation.** The Haber process is one example of nitrogen fixation. The ammonia produced can be applied to the soil as a fertilizer, since plants can readily use the nitrogen in ammonia to make the nitrogen-containing biomolecules essential for their growth.

Nitrogen fixation also results from the high-temperature combustion process in automobile engines. The nitrogen in the air is drawn into the engine and reacts at a significant rate with oxygen to form nitric oxide (NO), which further reacts with oxygen from the air to form nitrogen dioxide (NO_2). This nitrogen dioxide, which contributes to photochemical smog in many urban areas (see Section 15.9), reacts with moisture in the air and eventually reaches the soil to form nitrate salts, which are plant nutrients.

Hugh Spencer/Photo Researchers

Nodules on the roots of pea plants contain nitrogen-fixing bacteria.

Nitrogen fixation also occurs naturally. For example, lightning provides the energy to disrupt N_2 and O_2 molecules in the air, producing highly reactive nitrogen and oxygen atoms. These atoms in turn attack other N_2 and O_2 molecules to form nitrogen oxides that eventually become nitrates. Although lightning has traditionally been credited with forming about 10% of the total fixed nitrogen, recent studies indicate that lightning may account for as much as half of the fixed nitrogen available on earth. Another natural nitrogen fixation process involves bacteria that reside in the root nodules of plants such as beans, peas, and alfalfa. These **nitrogen-fixing bacteria** readily allow the conversion of nitrogen to ammonia and to other nitrogen-containing compounds useful to

Chemical Insights: An Explosive Discovery

For the first time in more than 100 years, a new species containing only nitrogen atoms has been discovered. In 1999 chemists Karl O. Christe and William W. Wilson, working at Edwards Air Force Base in California, reported a compound containing the N_5^+ cation. The only other all-nitrogen species known are molecular nitrogen (N_2), isolated in 1772, and the azide ion (N_3^-), first synthesized in 1890.

The Edwards scientists synthesized the compound N_5AsF_6, which contains the N_5^+ and AsF_6^- ions, by reacting N_2FAsF_6 with hydrazoic acid (HN_3) in liquid hydrogen fluoride at $-78°C$. The resulting N_5AsF_6 is a white, powdery substance that is extremely explosive. The compound has an enthalpy of formation of $+1460$ kJ/mol, an extraordinarily large value. Because the compound is so explosive, it is treacherous to handle. Indeed, its synthesis and characterization can be carried out only using special facilities that are available at a few laboratories in the world.

The N_5^+ ion is one of the most powerful oxidizing agents ever discovered. It reacts explosively with both water and organic materials. Calculations indicate that the N_5^+ ion is V-shaped and can be represented by the Lewis structure

$$\left[:N \equiv N - \ddot{N} = N = \ddot{N}: \right]^+$$

and its accompanying resonance structures.

Christe, Wilson, and their colleagues are working to prepare other compounds containing N_5^+, including N_5SbF_6 and N_5N_3 (containing N_5^+ and N_3^-). This should be an exciting (and hazardous) search.

---

plants. The efficiency of these bacteria is intriguing: They produce ammonia at soil temperatures and 1 atm pressure, whereas the Haber process requires severe conditions of 400°C and 250 atm. For obvious reasons, researchers are studying these bacteria intensively.

When plants and animals die and decompose, the elements they consist of are returned to the environment. In the case of nitrogen, the return of the element to the atmosphere as nitrogen gas, called **denitrification,** is carried out by bacteria that change nitrates to nitrogen. The complex **nitrogen cycle** is summarized in Fig. 18.13. It has been estimated that as much as 10 million tons more nitrogen per year is currently being fixed by natural and human processes than is being returned to the atmosphere. This fixed nitrogen is accumulating in soil, lakes, rivers, and oceans, where it promotes the growth of algae and other undesirable organisms.

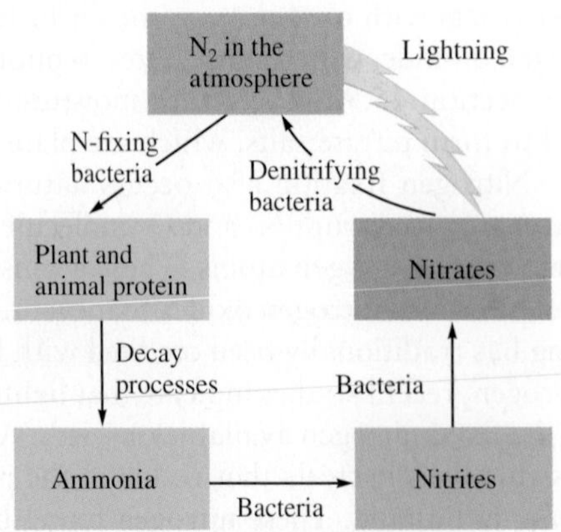

Figure 18.13
The nitrogen cycle. To be used by plants and animals, nitrogen must be converted from N_2 to nitrogen-containing compounds, such as nitrates, ammonia, and proteins. The nitrogen is returned to the atmosphere by natural decay processes.

Nitrogen Hydrides

By far the most important hydride of nitrogen is **ammonia.** A toxic, colorless gas with a pungent odor, ammonia is manufactured in huge quantities (approximately 40 billion pounds per year), mainly for use in fertilizers.

The pyramidal ammonia molecule has a lone pair of electrons on its nitrogen atom (see Fig. 18.8) and polar N—H bonds. This structure leads to a high degree of intermolecular interaction by hydrogen bonding in the liquid state, thereby producing an unusually high boiling point (−33.4°C) for a substance with such a low molar mass. Note, however, that the hydrogen bonding in liquid ammonia is clearly not as important as that in liquid water, which has about the same molar mass but a much higher boiling point. The water molecule has two polar bonds involving hydrogen and two lone pairs—the right combination for optimum hydrogen bonding—in contrast to the one lone pair and three polar bonds of the ammonia molecule.

As we saw in Chapter 7, ammonia behaves as a base, reacting with acids to produce ammonium salts. For example,

$$NH_3(g) + HCl(g) \longrightarrow NH_4Cl(s)$$

A second nitrogen hydride of major importance is **hydrazine** (N_2H_4). The Lewis structure of hydrazine

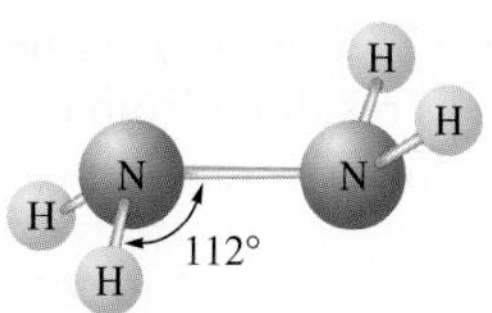

Figure 18.14
The molecular structure of hydrazine (N_2H_4). This arrangement minimizes the repulsion between the lone pairs on the nitrogen atoms by placing them on opposite sides.

```
H        H
 \.. ../
  N—N
 /    \
H      H
```

indicates that each nitrogen atom should be sp^3 hybridized with bond angles close to 109.5° (the tetrahedral angle), since the nitrogen atom is surrounded by four electron pairs. The observed structure with bond angles of 112° (Fig. 18.14) agrees reasonably well with these predictions. Hydrazine, a colorless liquid with an ammoniacal odor, freezes at 2°C and boils at 113.5°C. This boiling point is quite high for a compound with a molar mass of 32; this suggests that considerable hydrogen bonding occurs among the polar hydrazine molecules.

Hydrazine is a powerful reducing agent and has been widely used as a rocket propellant. For example, its reaction with oxygen is highly exothermic:

$$N_2H_4(l) + O_2(g) \longrightarrow N_2(g) + 2H_2O(g) \qquad \Delta H° = -622 \text{ kJ}$$

Since hydrazine also reacts vigorously with the halogens, fluorine is often used instead of oxygen as the oxidizer in rocket engines. Substituted hydrazines, where one or more of the hydrogen atoms are replaced by other groups, are also useful rocket fuels. For example, monomethylhydrazine,

```
CH3      H
   \    /
    N—N
   /    \
  H      H
```

is used with the oxidizer dinitrogen tetroxide (N_2O_4) to power the U.S. space shuttle orbiter. The reaction is

$$5N_2O_4(l) + 4N_2H_3(CH_3)(l) \longrightarrow 12H_2O(g) + 9N_2(g) + 4CO_2(g)$$

Because of the large number of gaseous molecules produced and the exothermic nature of this reaction, a very high thrust per mass of fuel is achieved. The reaction is also self-starting—it begins immediately when the fuels are mixed—which is a useful property for rocket engines that must be started and stopped frequently.

The use of hydrazine as a rocket propellant is a rather specialized application. The main industrial use of hydrazine is as a "blowing" agent in the manu-

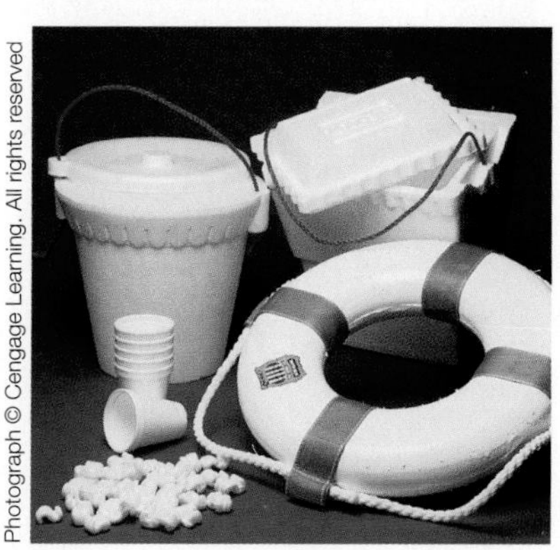

Blowing agents—such as hydrazine, which forms nitrogen gas on decomposition—are used to produce porous plastics like these polystyrene products.

facture of plastics. Hydrazine decomposes to form nitrogen gas, which causes foaming in the liquid plastic and results in a porous texture. Another major use of hydrazine is in the production of agricultural pesticides. Of the many hundreds of hydrazine derivatives (substituted hydrazines) that have been tested, 40 are used as fungicides, herbicides, insecticides, or plant growth regulators.

Nitrogen Oxides

Nitrogen forms a series of oxides in which its oxidation state ranges from +1 to +5, as shown in Table 18.14.

Dinitrogen monoxide (N_2O), more commonly called *nitrous oxide* or laughing gas, has an inebriating effect and has been used as a mild anesthetic by dentists. Because of its high solubility in fats, nitrous oxide is widely used as a propellant in aerosol cans of whipped cream. It is dissolved in the liquid inside the can at high pressure and forms bubbles that produce foaming as the liquid is released from the can. A significant amount of N_2O exists in the atmosphere, mostly produced by soil microorganisms, and its concentration appears to be gradually increasing. Because it can strongly absorb infrared radiation, nitrous oxide plays a small but probably significant role in controlling the earth's temperature in the same way that atmospheric carbon dioxide and water vapor do (see the discussion of the greenhouse effect in Section 9.7). Some scientists fear that the rapid decrease of tropical rain forests resulting from the development of countries such as Brazil will significantly affect the rate of pro-

A copper penny reacts with nitric acid to produce NO gas, which is immediately oxidized in air to give reddish brown NO_2.

Table 18.14

Some Common Nitrogen Compounds

Oxidation State of Nitrogen	Compound	Formula	Lewis Structure*
−3	Ammonia	NH_3	H—N̈—N, with H bonded below N
−2	Hydrazine	N_2H_4	H—N̈—N̈—H, with H bonded below each N
−1	Hydroxylamine	NH_2OH	H—N̈—Ö—H, with H bonded below N
0	Nitrogen	N_2	:N≡N:
+1	Dinitrogen monoxide (nitrous oxide)	N_2O	:N=N=O:
+2	Nitrogen monoxide (nitric oxide)	NO	:N=O:
+3	Dinitrogen trioxide	N_2O_3	(O)(O)N—N=O
+4	Nitrogen dioxide	NO_2	:Ö—N=O:
+5	Nitric acid	HNO_3	:Ö—N—Ö—H, with =O below N

*In some cases additional resonance structures are needed to fully describe the electron distribution.

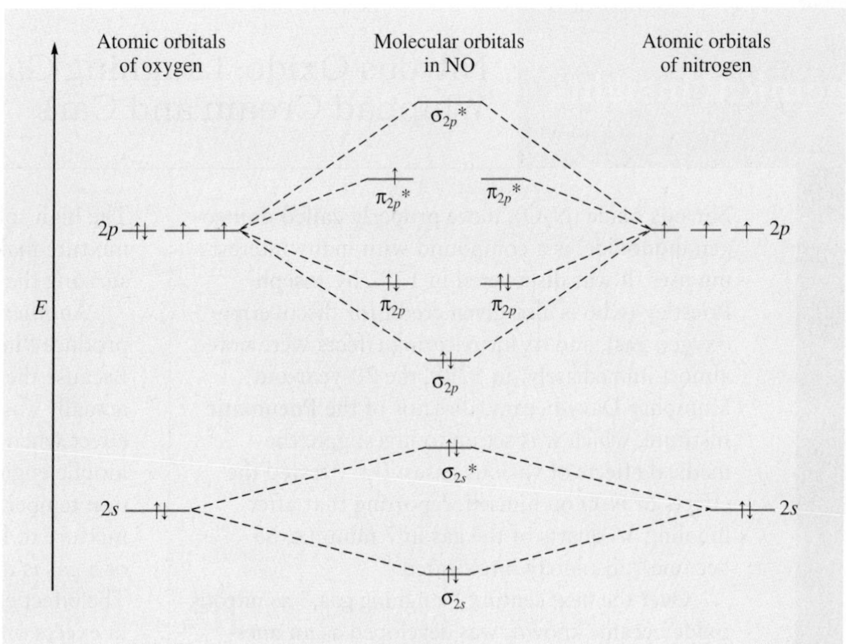

Figure 18.15
The molecular orbital energy-level diagram for nitric oxide (NO). The bond order is 2.5, or (8 − 3)/2.

duction of N_2O by soil organisms and thus will have important effects on the earth's temperature.

Nitrogen monoxide (NO), commonly called *nitric oxide*, has been found to be an important regulator in biological systems. Nitric oxide is a colorless gas under normal conditions and can be produced in the laboratory by reacting 6 *M* nitric acid with copper metal:

$$8H^+(aq) + 2NO_3^-(aq) + 3Cu(s) \longrightarrow 3Cu^{2+}(aq) + 4H_2O(l) + 2NO(g)$$

When this reaction is carried out in the air, the nitric oxide is immediately oxidized by O_2 to reddish brown nitrogen dioxide (NO_2).

Since the NO molecule has an odd number of electrons, it is most conveniently described in terms of the molecular orbital model. The molecular orbital energy-level diagram is shown in Fig. 18.15. Note that the NO molecule should be paramagnetic and have a bond order of 2.5, predictions that are supported by experimental observations. Since the NO molecule has one high-energy electron, it is not surprising that it can be rather easily oxidized to form NO^+, the *nitrosyl ion*. Because an antibonding electron is removed in going from NO to NO^+, the resulting ion should have a stronger bond (the predicted bond order is 3) than the molecule. This is borne out by experiment. The bond lengths and bond energies for nitric oxide and the nitrosyl ion are shown in Table 18.15.

Table 18.15
Comparison of the Bond Lengths and Bond Energies for Nitric Oxide and the Nitrosyl Ion

	NO	NO^+
Bond length (Å)	1.15	1.09
Bond energy (kJ/mol)	630	1020
Bond order (predicted by MO model)	2.5	3

Nitric oxide is thermodynamically unstable and decomposes to nitrous oxide and nitrogen dioxide:

$$3NO(g) \longrightarrow N_2O(g) + NO_2(g)$$

Nitrogen dioxide (NO_2), which is also an odd-electron molecule, has a V-shaped structure. The reddish brown, paramagnetic NO_2 molecule readily dimerizes to form dinitrogen tetroxide,

$$2NO_2(g) \rightleftharpoons N_2O_4(g)$$

which is diamagnetic and colorless. The value of the equilibrium constant is approximately 1 for this process at 55°C, and since the dimerization is exothermic, *K* decreases as the temperature increases.

Chemical Insights

Nitrous Oxide: Laughing Gas That Propels Whipped Cream and Cars

Nitrous oxide (N_2O), more properly called dinitrogen monoxide, is a compound with many interesting uses. It was discovered in 1772 by Joseph Priestley (who is also given credit for discovering oxygen gas), and its intoxicating effects were noted almost immediately. In 1798, the 20-year-old Humphry Davy became director of the Pneumatic Institute, which was set up to investigate the medical effects of various gases. Davy tested the effects of N_2O on himself, reporting that after inhaling 16 quarts of the gas in 7 minutes, he became "absolutely intoxicated."

Over the next century "laughing gas," as nitrous oxide became known, was developed as an anesthetic, particularly for dental procedures. Nitrous oxide is still used as an anesthetic, although it has been primarily replaced by more modern drugs.

One major use of nitrous oxide today is as the propellant in cans of "instant" whipped cream. The high solubility of N_2O in the whipped cream mixture makes it an excellent candidate for pressurizing the cans of whipping cream.

Another current use of nitrous oxide is to produce "instant horsepower" for street racers. Because the reaction of N_2O with O_2 to form NO actually absorbs heat, this reaction has a cooling effect when placed in the fuel mixture in an automobile engine. This cooling effect lowers combustion temperatures, thus allowing the fuel–air mixture to be significantly more dense (the density of a gas is inversely proportional to temperature). The effect can produce a burst of additional power in excess of 200 horsepower. Because engines are not designed to run steadily at such high power levels, the nitrous oxide is injected from a tank when extra power is desired.

The least common of the nitrogen oxides are *dinitrogen trioxide* (N_2O_3), a blue liquid that readily dissociates to gaseous nitric oxide and nitrogen dioxide, and *dinitrogen pentoxide* (N_2O_5), which under normal conditions is a solid that is best viewed as a mixture of NO_2^+ and NO_3^- ions. Although N_2O_5 molecules can exist in the gas phase, they readily dissociate to nitrogen dioxide and oxygen:

$$2N_2O_5(g) \rightleftharpoons 4NO_2(g) + O_2(g)$$

This reaction follows first-order kinetics, as discussed in Section 15.4.

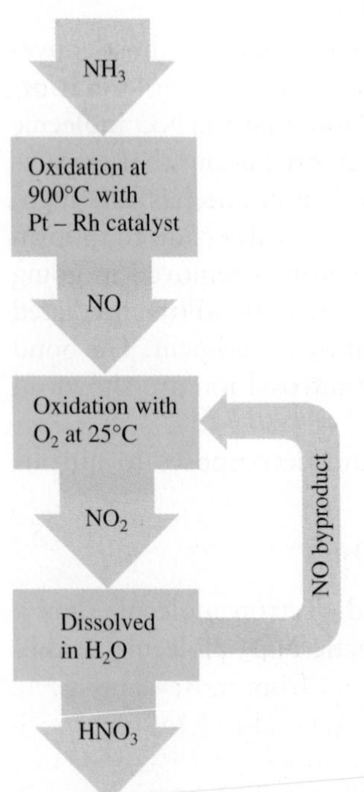

Figure 18.16
The Ostwald process.

Oxyacids of Nitrogen

Nitric acid is an important industrial chemical (approximately 8 million tons produced annually) used in the manufacture of many products, such as nitrogen-based explosives and ammonium nitrate for use as a fertilizer.

Nitric acid is produced commercially by the oxidation of ammonia in the **Ostwald process** (Fig. 18.16). In the first step of this process, ammonia is oxidized to nitric oxide:

$$4NH_3(g) + 5O_2(g) \longrightarrow 4NO(g) + 6H_2O(g) \qquad \Delta H° = -905 \text{ kJ}$$

Although this reaction is highly exothermic, it is very slow at 25°C. A side reaction occurs between nitric oxide and ammonia:

$$4NH_3(g) + 6NO(g) \longrightarrow 5N_2(g) + 6H_2O(g)$$

which is particularly undesirable because it traps the nitrogen in the very unreactive N_2 molecules. The desired reaction can be accelerated and the effects of the competing reaction can be minimized if the ammonia oxidation is carried out by using a catalyst of a platinum–rhodium alloy heated to 900°C. Under these conditions, there is a 97% conversion of the ammonia to nitric oxide.

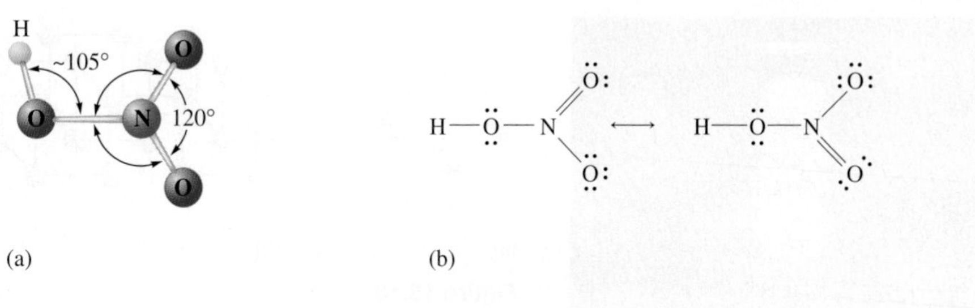

Figure 18.17
(a) The molecular structure of HNO_3.
(b) The resonance structures of HNO_3.

In the second step, nitric oxide is reacted with oxygen to produce nitrogen dioxide:

$$2NO(g) + O_2(g) \longrightarrow 2NO_2(g) \qquad \Delta H° = -113 \text{ kJ}$$

This oxidation reaction has a rate that *decreases* with increasing temperature. Because of this very unusual behavior, the reaction is carried out at approximately 25°C and is kept at this temperature by cooling with water.

The third step in the Ostwald process is the absorption of nitrogen dioxide by water:

$$3NO_2(g) + H_2O(l) \longrightarrow 2HNO_3(aq) + NO(g) \qquad \Delta H° = -139 \text{ kJ}$$

The gaseous NO produced in the reaction is recycled so that it can be oxidized to NO_2. The aqueous nitric acid from this process is about 50% HNO_3 by mass, which can be increased to 68% by distillation to remove some of the water. The maximum concentration attainable by this method is 68% because nitric acid and water form an *azeotrope* at this concentration. The solution can be further concentrated to 95% HNO_3 by treatment with concentrated sulfuric acid, which strongly absorbs water; H_2SO_4 is often used as a *dehydrating (water-removing) agent.*

An *azeotrope* is a solution that, like a pure liquid, distills at a constant temperature without a change in composition.

Nitric acid is a colorless, fuming liquid (bp = 83°C) with a pungent odor; it decomposes in sunlight by the following reaction:

$$4HNO_3(l) \xrightarrow{h\nu} 4NO_2(g) + 2H_2O(l) + O_2(g)$$

As a result, nitric acid turns yellow as it ages because of the dissolved nitrogen dioxide. The common laboratory reagent called *concentrated nitric acid* is 15.9 *M* HNO_3 (70.4% HNO_3 by mass) and is a very strong oxidizing agent. The resonance structures and molecular structure of HNO_3 are shown in Fig. 18.17. Note that the hydrogen is bound to an oxygen atom rather than to nitrogen as the formula might suggest.

Nitrous acid (HNO_2) is a weak acid,

$$HNO_2(aq) \rightleftharpoons H^+(aq) + NO_2^-(aq) \qquad K_a = 4.0 \times 10^{-4}$$

that forms pale yellow nitrite (NO_2^-) salts. In contrast to nitrates, which are often used as explosives, nitrites are quite stable even at high temperatures.

18.9 The Chemistry of Phosphorus

Although phosphorus lies directly below nitrogen in Group 5A of the periodic table, its chemical properties are significantly different from those of nitrogen. The differences arise mainly from four factors: nitrogen's ability to form much stronger π bonds, the greater electronegativity of nitrogen, the larger size of phosphorus atoms, and the potential availability of empty valence *d* orbitals on phosphorus.

White phosphorus reacts vigorously with the oxygen in air and must be stored under water. Red phosphorus is stable in air.

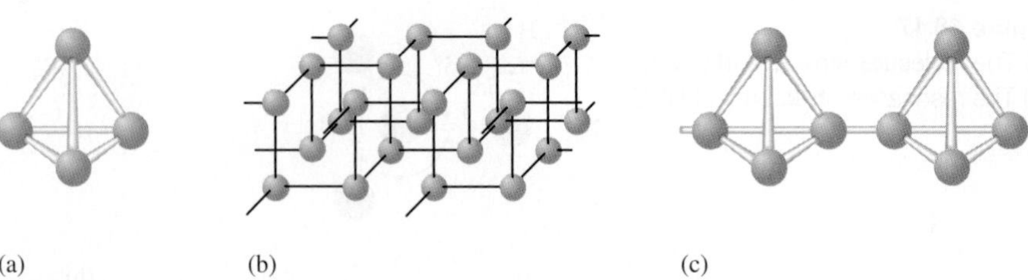

Figure 18.18
(a) The P_4 molecule found in white phosphorus. (b) The crystalline network structure of black phosphorus. (c) The chain structure of red phosphorus.

The chemical differences between nitrogen and phosphorus are apparent in their elemental forms. In contrast to the diatomic form of elemental nitrogen, which is stabilized by strong π bonds, there are several solid forms of phosphorus that all contain aggregates of atoms. *White phosphorus,* which contains discrete tetrahedral P_4 molecules [Fig. 18.18(a)], is very reactive; it bursts into flames on contact with air (it is said to be *pyrophoric*). Consequently, white phosphorus is commonly stored under water. White phosphorus is quite toxic; the P_4 molecules are very damaging to tissue, particularly the cartilage and bones of the nose and jaw. The much less reactive forms, called *black phosphorus* and *red phosphorus,* are network solids (see Section 16.5). Black phosphorus has a regular crystalline structure [Fig. 18.18(b)], but red phosphorus is amorphous and is thought to consist of chains of P_4 units [Fig. 18.18(c)]. Red phosphorus can be obtained by heating white phosphorus in the absence of air at 1 atm. Black phosphorus is obtained from either white or red phosphorus by heating at high pressures.

Even though phosphorus has a lower electronegativity than nitrogen, it will form phosphides (ionic substances containing the P^{3-} anion) such as Na_3P and Ca_3P_2. Phosphide salts react vigorously with water to produce *phosphine* (PH_3), a toxic, colorless gas:

$$2Na_3P(s) + 6H_2O(l) \longrightarrow 2PH_3(g) + 6Na^+(aq) + 6OH^-(aq)$$

Phosphine is analogous to ammonia, although it is a much weaker base ($K_b \approx 10^{-26}$) and is much less soluble in water.

Phosphine has the Lewis structure

$$\left[\begin{array}{c} \text{H}-\ddot{\text{P}}-\text{H} \\ | \\ \text{H} \end{array}\right]$$

and a pyramidal molecular structure, as we would predict from the VSEPR model. However, it has bond angles of 94° rather than 107°, as found in the ammonia molecule. The reasons for this are complex; therefore, we will simply regard phosphine as an exception to the simple version of the VSEPR model that we use.

Phosphorus Oxides and Oxyacids

Phosphorus reacts with oxygen to form oxides in which its oxidation states are +5 and +3. The oxide P_4O_6 is formed when elemental phosphorus is burned in a limited supply of oxygen, and P_4O_{10} is produced when the oxygen is in excess. Picture these oxides (shown in Fig. 18.19) as being constructed by adding oxygen atoms to the fundamental P_4 structure. The intermediate states, P_4O_7, P_4O_8, and P_4O_9, which contain one, two, and three terminal oxygen atoms, respectively, are also known.

The terminal oxygens are the nonbridging oxygen atoms.

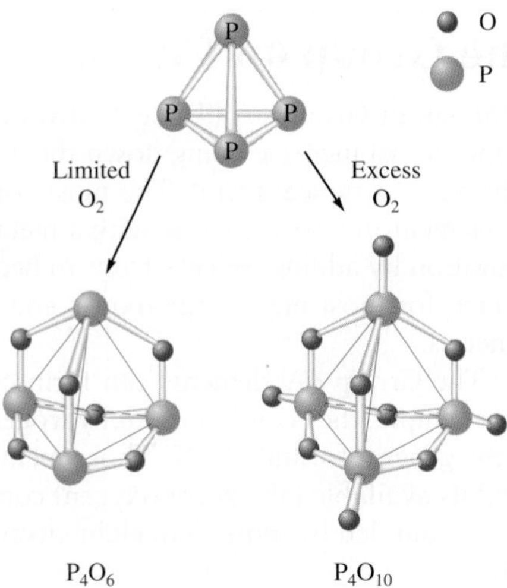

Figure 18.19
The structures of P_4O_6 and P_4O_{10}.

Tetraphosphorus decoxide (P_4O_{10}), which was formerly represented as P_2O_5 and called phosphorus pentoxide, has a great affinity for water and thus is a powerful dehydrating agent. For example, it can be used to convert HNO_3 and H_2SO_4 to their parent oxides, N_2O_5 and SO_3, respectively.

When tetraphosphorus decoxide dissolves in water, **phosphoric acid** (H_3PO_4), also called **orthophosphoric acid,** is produced:

$$P_4O_{10}(s) + 6H_2O(l) \longrightarrow 4H_3PO_4(aq)$$

Pure phosphoric acid is a white solid that melts at 42°C. Aqueous phosphoric acid is a much weaker acid ($K_{a_1} \approx 10^{-2}$) than nitric acid or sulfuric acid and is a poor oxidizing agent.

When the oxide P_4O_6 is placed in water, **phosphorous acid** (H_3PO_3) is formed [Fig. 18.20(a)]. Although the formula suggests a triprotic acid, phosphorous acid is a *diprotic* acid. The hydrogen atom bonded directly to the phosphorus atom is not acidic in aqueous solution; only those hydrogen atoms bonded to the oxygen atoms in H_3PO_3 can be released as protons.

A third oxyacid of phosphorus is *hypophosphorous acid* (H_3PO_2) [Fig. 18.20(b)], which is a monoprotic acid.

Phosphorus in Fertilizers

Phosphorus is essential for plant growth. Although most soil contains large amounts of phosphorus, it is often present in insoluble minerals, making it inaccessible to the plants. Soluble phosphate fertilizers are manufactured by treating phosphate rock with sulfuric acid to make **superphosphate of lime,** a mixture of $CaSO_4 \cdot 2H_2O$ and $Ca(H_2PO_4)_2 \cdot H_2O$. If phosphate rock is treated with phosphoric acid, $Ca(H_2PO_4)_2$, known as *triple phosphate,* is produced. The reaction of ammonia with phosphoric acid gives *ammonium dihydrogenphosphate* ($NH_4H_2PO_4$), a very efficient fertilizer that furnishes both phosphorus and nitrogen.

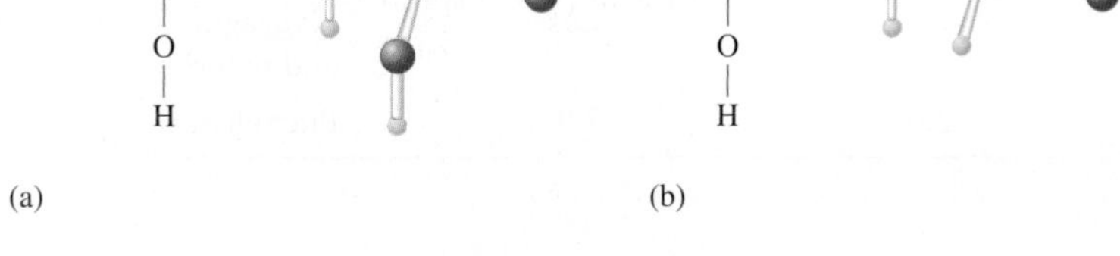

(a) H—O—P(—H)(—O)—O—H

(b) H—P(—H)(—O)—O—H

Figure 18.20
(a) The structure of phosphorous acid (H_3PO_3). (b) The structure of hypophosphorous acid (H_3PO_2).

18.10 The Group 6A Elements

6A
O
S
Se
Te
Po

Although in Group 6A (Table 18.16) there is the usual tendency for metallic properties to increase going down the group, none of the Group 6A elements behaves as a typical metal. The most common chemical behavior of a Group 6A element involves reacting with a metal to achieve a noble gas electron configuration by adding two electrons to become a 2− anion in ionic compounds. In fact, for most metals, the oxides and sulfides constitute the most common minerals.

The Group 6A elements can form covalent bonds with other nonmetals. For example, they combine with hydrogen to form a series of covalent hydrides of the general formula H_2X. Those members of the group that have valence *d* orbitals available (all except oxygen) commonly form molecules in which they are surrounded by more than eight electrons. Examples are SF_4, SF_6, TeI_4, and $SeBr_4$.

The two heaviest members of Group 6A can lose electrons to form cations. Although they do not lose all six valence electrons because of the high energies that would be required, tellurium and polonium appear to exhibit some chemistry involving their 4+ cations. However, the chemistry of these Group 6A cations is much more limited than that of the Group 5A elements bismuth and antimony.

Eyebyte/Royalty Free Image/Alamy Images

Walnuts contain trace amounts of selenium.

In recent years there has been a growing interest in the chemistry of selenium, an element found throughout the environment in trace amounts. Selenium's toxicity has long been known, but some medical studies have shown an *inverse* relationship between the incidence of cancer and the selenium levels in soil. It has been suggested that the greater dietary intake of selenium by people living in areas with relatively high selenium levels somehow furnishes protection from cancer. These studies are only preliminary, but selenium is definitely known to be physiologically important (it is involved in the activity of vitamin E and certain enzymes). Selenium (as well as tellurium) is also a semiconductor and therefore finds some application in the electronics industry.

Polonium was discovered in 1898 by Marie and Pierre Curie in their search for the sources of radioactivity in pitchblende. Polonium has 27 isotopes and is highly toxic and very radioactive. It has been suggested that the isotope ^{210}Po, a natural contaminant of tobacco and an α-particle producer (see Section 20.1), might be at least partly responsible for the incidence of cancer in smokers.

Table 18.16

Selected Physical Properties, Sources, and Methods of Preparation of the Group 6A Elements

Element	Electronegativity	Radius of X^{2-} (pm)	Source	Method of Preparation
Oxygen	3.4	140	Air	Distillation from liquid air
Sulfur	2.6	184	Sulfur deposits	Melted with hot water and pumped to the surface
Selenium	2.6	198	Impurity in sulfide ores	Reduction of H_2SeO_4 with SO_2
Tellurium	2.1	221	Nagyagite (mixed sulfide and telluride)	Reduction of ore with SO_2
Polonium	2.0	230	Pitchblende	

18.11 The Chemistry of Oxygen

US Navy photo by Photographer's Mate Airman Apprentice Adam York

A U.S. Navy test pilot in an F-14 jet using an oxygen mask.

It is hard to overstate the importance of oxygen, the most abundant element in and near the earth's crust. Oxygen is present in the atmosphere as oxygen gas and ozone; in soil and rocks in oxide, silicate, and carbonate minerals; in the oceans in water; and in our bodies in water and a myriad of other molecules. In addition, most of the energy we need to live and run our civilization comes from the exothermic reactions of oxygen with carbon-containing molecules.

The most common elemental form of oxygen (O_2) constitutes 21% of the volume of the earth's atmosphere. Since nitrogen has a lower boiling point than oxygen, nitrogen can be boiled away from liquid air, leaving oxygen and small amounts of argon, another component of air. Liquid oxygen is a pale blue liquid that freezes at $-219°C$ and boils at $-183°C$. The paramagnetism of the O_2 molecule can be demonstrated by pouring liquid oxygen between the poles of a strong magnet, where it "sticks" until it boils away (see Fig. 14.42). The paramagnetism of the O_2 molecule can be accounted for by the molecular orbital model (Fig. 14.41), which also explains its bond strength.

The other form of elemental oxygen is **ozone** (O_3), a molecule that can be represented by the resonance structures

$$O{=}O{-}O \longleftrightarrow O{-}O{=}O$$

The bond angle in the O_3 molecule is 117°, in reasonable agreement with the prediction of the VSEPR model (three effective pairs require a trigonal planar arrangement). That the bond angle is slightly less than 120° can be explained by concluding that more space is required for the lone pair than for the bonding pairs.

Ozone can be prepared by passing an electrical discharge through pure oxygen gas. The electrical energy disrupts the bonds in some O_2 molecules, thereby producing oxygen atoms, which react with other O_2 molecules to form O_3. Ozone is much less stable than oxygen at 25°C and 1 atm. For example, $K \approx 10^{-56}$ for the equilibrium

$$3O_2(g) \rightleftharpoons 2O_3(g)$$

A pale blue, highly toxic gas, ozone is a much more powerful oxidizing agent than oxygen. The strong oxidizing power of ozone makes it useful for killing bacteria in swimming pools, hot tubs, and aquariums. It is also increasingly being used in municipal water treatment and for washing produce after it comes out of the fields. One of the main advantages of using ozone for water purification is that it does not leave potentially toxic residues behind. On the other hand, chlorine, which is widely used for water purification, leaves residues of chloro compounds, such as chloroform ($CHCl_3$), which may cause cancer after long-term exposure. Although ozone effectively kills the bacteria in water, one problem with **ozonolysis** is that the water supply is not protected against recontamination, since virtually no ozone remains after the initial treatment. In contrast, for chlorination, significant residual chlorine remains after treatment.

The oxidizing ability of ozone can be detrimental, especially when it is present in the pollution from automobile exhausts (see Section 5.11).

Ozone exists naturally in the upper atmosphere of the earth. The *ozone layer* is especially important because it absorbs ultraviolet light and thus acts as a screen to prevent this radiation, which can cause skin cancer, from pene-

Scientists have become concerned that Freons and nitrogen dioxide are promoting the destruction of the ozone layer (see Section 15.9).

trating to the earth's surface. When an ozone molecule absorbs this energy, it splits into an oxygen molecule and an oxygen atom:

$$O_3 \xrightarrow{h\nu} O_2 + O$$

If the oxygen molecule and atom collide, they will not stay together as ozone unless a "third body," such as a nitrogen molecule, is present to help absorb the energy released by bond formation. The third body absorbs this energy as kinetic energy; its temperature is increased. Therefore, the energy originally absorbed as ultraviolet radiation is eventually changed to thermal energy. Thus the ozone prevents the harmful high-energy ultraviolet light from reaching the earth.

18.12 The Chemistry of Sulfur

Charles D. Winters

Cinnabar from New Almaden, California.

Sulfur is found in nature both in large deposits of the free element and in widely distributed ores, such as galena (PbS), cinnabar (HgS), pyrite (FeS_2), gypsum ($CaSO_4 \cdot 2H_2O$), epsomite ($MgSO_4 \cdot 7H_2O$), and glauberite ($Na_2SO_4 \cdot CaSO_4$).

About 60% of the sulfur produced in the United States comes from the underground deposits of elemental sulfur found in Texas and Louisiana. This sulfur is recovered by using the **Frasch process** developed by Herman Frasch in the 1890s. Superheated water is pumped into the deposit to melt the sulfur (mp = 113°C), which is then forced to the surface by air pressure (Fig. 18.21). The remaining 40% of sulfur produced in the United States either is a by-product of the purification of fossil fuels before combustion to prevent pollution or comes from the sulfur dioxide (SO_2) scrubbed from the exhaust gases when sulfur-containing fuels are burned.

In contrast to oxygen, elemental sulfur exists as S_2 molecules only in the gas phase at high temperatures. Because sulfur atoms form much stronger σ bonds than π bonds, S_2 is less stable at 25°C than larger aggregates such as S_6 and S_8 rings and S_n chains (Fig. 18.22). The most stable form of sulfur at 25°C and 1 atm is called *rhombic sulfur* [Fig. 18.23(a)], which contains stacked S_8 rings. If rhombic sulfur is melted and heated to 120°C, it forms *monoclinic sulfur* as it slowly cools [Fig. 18.23(b)]. The monoclinic form also contains S_8 rings, but the rings are stacked differently than in rhombic sulfur.

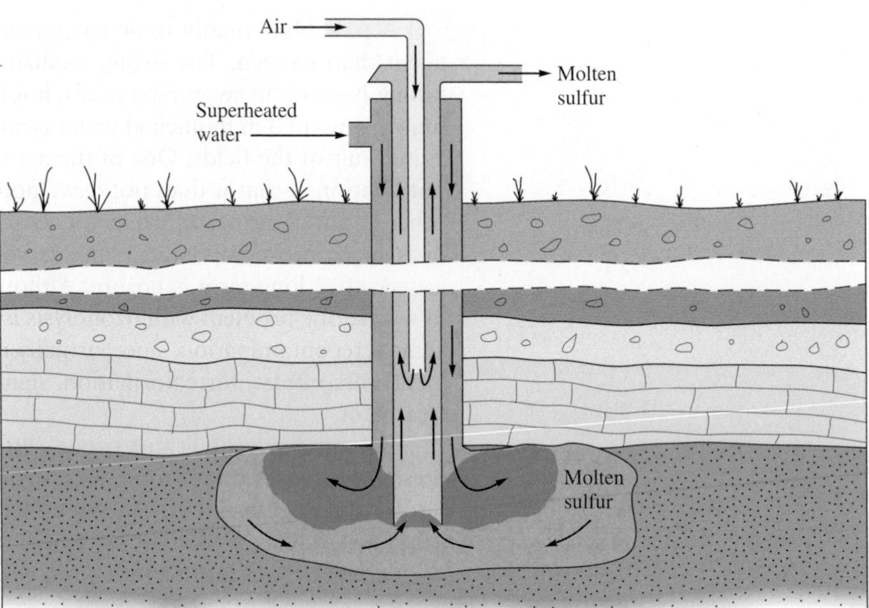

Figure 18.21
The Frasch method for recovering sulfur from underground deposits.

Sulfur piles in Vancouver, British Columbia.

Sulfur Oxides

From its position below oxygen in the periodic table, we might expect the simplest stable oxide of sulfur to have the formula SO. However, *sulfur monoxide,* which can be produced in small amounts when gaseous sulfur dioxide (SO_2) is subjected to an electrical discharge, is very unstable. The difference in the stabilities of the O_2 and SO molecules probably reflects the much stronger π bonding between oxygen atoms than between a sulfur and an oxygen atom.

The scrubbing of sulfur dioxide from exhaust gases was discussed in Section 5.11.

Sulfur burns in air with a bright blue flame to give *sulfur dioxide* (SO_2), a colorless gas with a pungent odor, which condenses to a liquid at $-10°C$ and 1 atm. Sulfur dioxide is a trigonal planar molecule, which is a very effective antibacterial agent often used to preserve stored fruit.

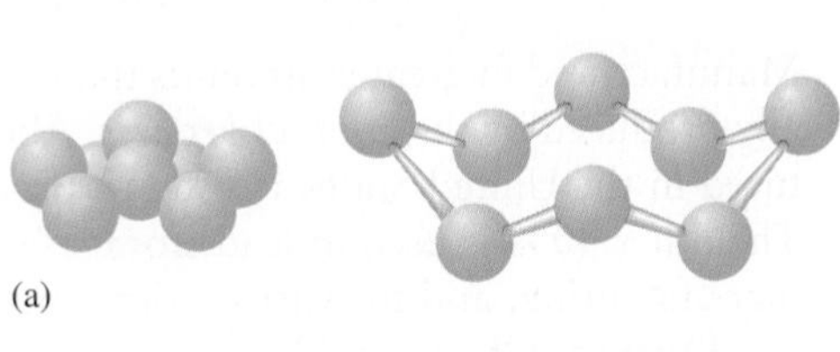
(a)

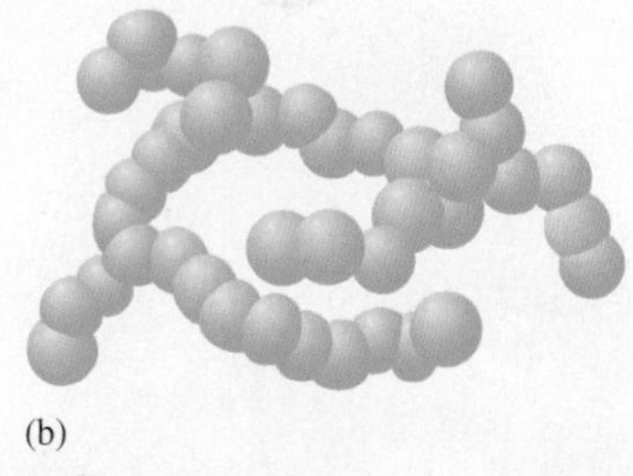
(b)

Figure 18.22
(a) The S_8 molecule. (b) Chains of sulfur atoms in viscous liquid sulfur. The chains may contain as many as 10,000 atoms.

(a)

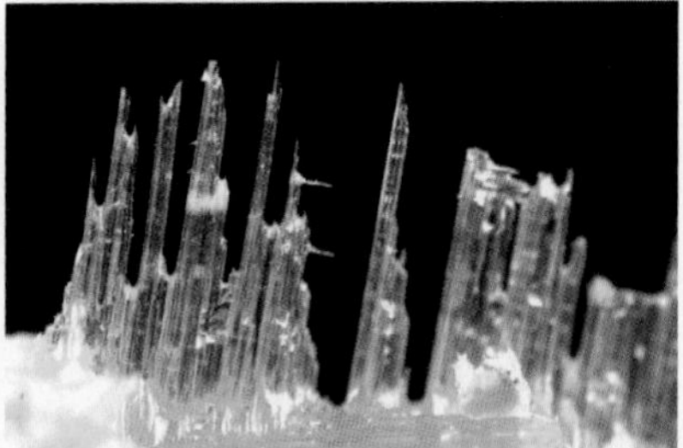

(b)

Figure 18.23
(a) Crystals of rhombic sulfur. (b) Crystals of monoclinic sulfur.

Sulfur dioxide reacts with oxygen to produce *sulfur trioxide* (SO_3):

$$2SO_2(g) + O_2(g) \longrightarrow 2SO_3(g)$$

However, this reaction is very slow in the absence of a catalyst. One of the mysteries during early research on air pollution was how the sulfur dioxide produced from the combustion of sulfur-containing fuels is so rapidly converted to sulfur trioxide in the atmosphere. It is now known that dust and other particles can act as heterogeneous catalysts for this process (see Section 15.9).

Oxyacids of Sulfur

Sulfur dioxide dissolves in water to form an acidic solution. The reaction is often represented as

$$SO_2(g) + H_2O(l) \longrightarrow H_2SO_3(aq)$$

where H_2SO_3 is called *sulfurous acid.* However, very little H_2SO_3 actually exists in the solution. The major form of sulfur dioxide in water is SO_2, and the acid dissociation equilibria are best represented as

$$SO_2(aq) + H_2O(l) \rightleftharpoons H^+(aq) + HSO_3^-(aq) \qquad K_{a_1} = 1.5 \times 10^{-2}$$

$$HSO_3^-(aq) \rightleftharpoons H^+(aq) + SO_3^{2-}(aq) \qquad K_{a_2} = 1.0 \times 10^{-7}$$

This situation is analogous to the behavior of carbon dioxide in water (see Section 7.7). Although H_2SO_3 cannot be isolated, salts of SO_3^{2-} (*sulfites*) and HSO_3^- (*hydrogen sulfites*) are well known.

Sulfur trioxide reacts violently with water to produce the diprotic acid **sulfuric acid:**

$$SO_3(g) + H_2O(l) \longrightarrow H_2SO_4(aq)$$

Figure 18.24
The reaction of H_2SO_4 with sucrose (on the left) to produce a blackened column of carbon (on the right).

Manufactured in greater amounts than any other chemical, sulfuric acid is usually produced by the *contact process.* About 60% of the sulfuric acid manufactured in the United States is used to produce fertilizers from phosphate rock. The other 40% is used in lead storage batteries, in petroleum refining, in steel manufacturing, and for various other purposes in the chemical industry.

Because sulfuric acid has a high affinity for water, it is often used as a dehydrating agent. Gases that do not react with sulfuric acid, such as oxygen, nitrogen, and carbon dioxide, are often dried by bubbling them through concentrated solutions of the acid. Sulfuric acid is such a powerful dehydrating agent that it will remove hydrogen and oxygen from a substance in a 2:1 ratio even when the substance contains no molecular water. For example, concentrated sulfuric acid reacts vigorously with common table sugar (sucrose), leaving a charred mass of carbon (Fig. 18.24):

$$\underset{\text{Sucrose}}{C_{12}H_{22}O_{11}(s)} + 11H_2SO_4(conc) \longrightarrow 12C(s) + 11H_2SO_4 \cdot H_2O(l)$$

18.13 The Group 7A Elements

7A
F
Cl
Br
I
At

In our coverage of the representative elements, we have progressed from the groups of metallic elements (Groups 1A and 2A), through groups in which the lighter members are nonmetals and the heavier members are metals (Groups 3A, 4A, and 5A), to a group containing all nonmetals (Group 6A—although some might prefer to call polonium a metal). The Group 7A elements, the **halogens** (with the valence electron configuration ns^2np^5), are all nonmetals whose properties generally vary smoothly going down the group. The only notable exceptions are the unexpectedly low value for the electron affinity of fluorine and the unexpectedly small bond energy of the F_2 molecule (see Sec-

Samples of chlorine gas, liquid bromine, and solid iodine.

Table 18.17

Trends in Selected Physical Properties of the Group 7A Elements

Element	Electronegativity	Radius of X^- (pm)	$\mathscr{E}°$ (V) for $X_2 + 2e \rightarrow 2X^-$	Bond Energy of X_2 (kJ/mol)
Fluorine	4.0	136	2.87	154
Chlorine	3.2	181	1.36	239
Bromine	3.0	195	1.09	193
Iodine	2.7	216	0.54	149
Astatine	2.2	—	—	—

tion 18.1). Table 18.17 summarizes the trends in some physical properties of the halogens.

Because of their high reactivities, the halogens are not found as free elements in nature. Instead, they are found as halide ions (X^-) in various minerals and in seawater (Table 18.18).

Although astatine is a member of Group 7A, its chemistry is of no practical importance because all its known isotopes are radioactive. The longest-lived isotope, ^{210}At, has a half-life of only 8.3 hours.

The halogens, particularly fluorine, have very high electronegativity values (Table 18.17). They tend to form polar covalent bonds with other nonmetals and ionic bonds with metals in their lower oxidation states. When a metal ion is in a higher oxidation state, such as +3 or +4, the metal–halogen bonds are polar and covalent. For example, $TiCl_4$ and $SnCl_4$ are both covalent compounds that are liquids under normal conditions.

Hydrogen Halides

The hydrogen halides can be prepared by a reaction of the elements

$$H_2(g) + X_2(g) \longrightarrow 2HX(g)$$

This reaction occurs with explosive vigor when fluorine and hydrogen are mixed. On the other hand, hydrogen and chlorine can coexist with little apparent reaction for relatively long periods in the dark. However, ultraviolet light causes an explosively fast reaction, and this is the basis of a popular lecture demonstration, the "hydrogen–chlorine cannon." Bromine and iodine also react with hydrogen, but more slowly.

Table 18.18

Some Physical Properties, Sources, and Methods of Preparation of the Group 7A Elements

Element	Color and State	Percentage of Earth's Crust	Melting Point (°C)	Boiling Point (°C)	Source	Method of Preparation
Fluorine	Pale yellow gas	0.07	−220	−188	Fluorospar (CaF_2), cryolite (Na_3AlF_6), fluorapatite [$Ca_5(PO_4)_3F$]	Electrolysis of molten KHF_2
Chlorine	Yellow-green gas	0.14	−101	−34	Rock salt (NaCl), halite (NaCl), sylvite (KCl)	Electrolysis of aqueous NaCl
Bromine	Red-brown liquid	2.5×10^{-4}	−7.3	59	Seawater, brine wells	Oxidation of Br^- by Cl_2
Iodine	Violet-black solid	3×10^{-5}	113	184	Seaweed, brine wells	Oxidation of I^- by electrolysis or MnO_2

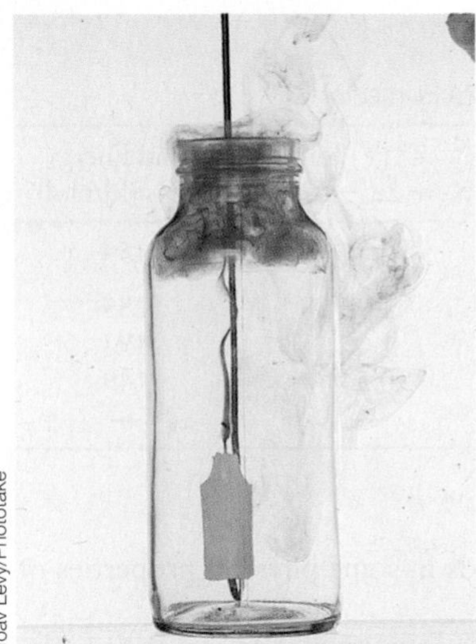

A candle burning in an atmosphere of $Cl_2(g)$. The exothermic reaction, which involves breaking C—C and C—H bonds in the wax and forming C—Cl bonds in their places, produces enough heat to make the gases in the region incandescent (a flame results).

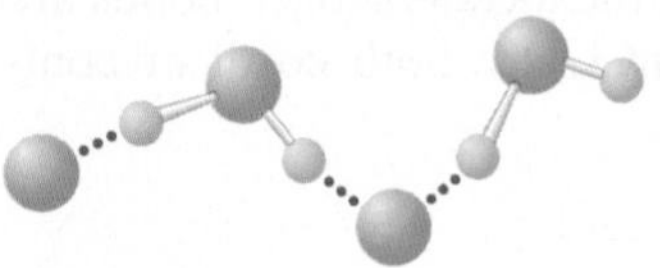

Figure 18.25
The hydrogen bonding among HF molecules in liquid hydrogen fluoride.

Table 18.19

Some Physical Properties of the Hydrogen Halides

HX	Melting Point (°C)	Boiling Point (°C)	H—X Bond Energy (kJ/mol)
HF	−83	20	565
HCl	−114	−85	427
HBr	−87	−67	363
HI	−51	−35	295

Some physical properties of the hydrogen halides are listed in Table 18.19. Note the very high boiling point for hydrogen fluoride, which results from extensive hydrogen bonding among the very polar HF molecules (Fig. 18.25). Fluoride ion has such a high affinity for protons that in concentrated aqueous solutions of hydrogen fluoride, the ion $[F{-}{-}H{-}{-}F]^-$ exists, in which an H^+ ion is centered between two F^- ions.

When dissolved in water, the hydrogen halides behave as acids, and all except hydrogen fluoride are completely dissociated. Because water is a much stronger base than Cl^-, Br^-, or I^- ion, the acid strengths of HCl, HBr, and HI cannot be differentiated in water. However, in a less basic solvent, such as glacial (pure) acetic acid, the acids show different strengths:

$$\underset{\text{Strongest acid}}{H{—}I} > H{—}Br > H{—}Cl \gg \underset{\text{Weakest acid}}{H{—}F}$$

To see why hydrogen fluoride is the only weak acid in water among the HX molecules, let's consider the dissociation equilibrium,

$$HX(aq) \rightleftharpoons H^+(aq) + X^-(aq) \qquad \text{where} \qquad K_a = \frac{[H^+][X^-]}{[HX]}$$

from a thermodynamic point of view. Recall that acid strength is reflected by the magnitude of K_a—a small K_a value means a weak acid. Also recall that the value of an equilibrium constant is related to the standard free energy change for the reaction,

$$\Delta G^\circ = -RT\ln(K)$$

As ΔG° becomes more negative, K becomes larger; a *decrease* in free energy favors a given reaction. As we saw in Chapter 10, free energy depends on enthalpy, entropy, and temperature. For a process at constant temperature,

$$\Delta G^\circ = \Delta H^\circ - T\Delta S^\circ$$

Thus, to explain the various acid strengths of the hydrogen halides, we must focus on the factors that determine ΔH° and ΔS° for the acid dissociation reaction.

What energy terms are important in determining ΔH° for the dissociation of HX in water? (Keep in mind that large, positive contributions to the value of ΔH° will tend to make ΔG° more highly positive, K_a smaller, and the acid weaker.) One important factor is certainly the H—X bond strength. Note from Table 18.19 that the H—F bond is much stronger than the other H—X bonds. This factor tends to make HF a weaker acid than the others.

Another important contribution to ΔH° is the enthalpy of hydration (see Section 17.2) of X^- (Table 18.20). As we would expect, the smallest of the halide ions, F^-, has the most negative value—its hydration is the most exother-

Table 18.20

The Enthalpies and Entropies of Hydration for the Halide Ions

$$X^-(g) \xrightarrow{H_2O} X^-(aq)$$

X^-	ΔH° (kJ/mol)	ΔS° (J/K mol)
F^-	−510	−159
Cl^-	−366	−96
Br^-	−334	−81
I^-	−291	−64

mic. This term favors the dissociation of HF into its ions more so than it does for the other HX molecules.

So far we have two conflicting factors: The large HF bond energy tends to make HF a weaker acid than the other hydrogen halides, but the enthalpy of hydration favors the dissociation of HF more than that of the others. When we compare data for HF and HCl, the difference in bond energy (138 kJ/mol) is slightly smaller than the difference in the enthalpies of hydration for the anions (144 kJ/mol). If these were the *only* important factors, HF should be a stronger acid than HCl because the large enthalpy of hydration of F^- more than compensates for the large HF bond strength.

Hydration becomes more exothermic as the charge density of an ion increases. Thus, for ions of a given charge, the smallest is most strongly hydrated.

As it turns out, the *deciding factor appears to be entropy.* Note from Table 18.20 that the entropy of hydration for F^- is much more negative than the entropy of hydration for the other halides because of the high degree of ordering that occurs as the water molecules associate with the small F^- ion. Remember that a negative change in entropy is unfavorable. Thus, although the enthalpy of hydration favors dissociation of HF, the *entropy* of hydration strongly opposes it.

When H_2O molecules cluster around an ion, an ordering effect occurs; thus ΔS°_{hyd} is negative.

When all these factors are taken into account, ΔG° for the dissociation of HF in water is positive; that is, K_a is small. In contrast, ΔG° for dissociation of the other HX molecules in water is negative (K_a is large). This example illustrates the complexity of the processes that occur in aqueous solutions and the importance of entropy effects in that medium.

Stomach acid is 0.1 *M* HCl.

In practical terms, **hydrochloric acid** is the most important of the **hydrohalic acids,** the aqueous solutions of the hydrogen halides. About 3 million tons of hydrochloric acid are produced annually for use in cleaning steel before galvanizing and in the manufacture of many other chemicals.

Hydrofluoric acid is used to etch glass by reacting with the silica in glass to form the volatile gas SiF_4:

$$SiO_2(s) + 4HF(aq) \longrightarrow SiF_4(g) + 2H_2O(l)$$

Oxyacids and Oxyanions

All the halogens except fluorine combine with various numbers of oxygen atoms to form a series of oxyacids, as shown in Table 18.21. The strengths of these acids vary in direct proportion to the number of oxygen atoms attached to the halogen, with the acid strength increasing as more oxygens are added.

The only member of the chlorine series that has been obtained in the pure state is *perchloric acid* ($HOClO_3$), a strong acid and a powerful oxidizing agent. Because perchloric acid reacts explosively with many organic materials, it must be handled with great caution. The other oxyacids of chlorine are

Table 18.21

The Known Oxyacids of the Halogens

Oxidation State of Halogen	Fluorine	Chlorine	Bromine	Iodine*	General Name of Acids	General Name of Salts
+1	HOF†	HOCl	HOBr	HOI	Hypohalous acid	Hypohalites, MOX
+3	‡	HOClO	‡	‡	Halous acid	Halites, MXO_2
+5	‡	$HOClO_2$	$HOBrO_2$	$HOIO_2$	Halic acid	Halates, MXO_3
+7	‡	$HOClO_3$	$HOBrO_3$	$HOIO_3$	Perhalic acid	Perhalates, MXO_4

*Iodine also forms $H_4I_2O_9$ (mesodiperiodic acid) and H_5IO_6 (paraperiodic acid).
†HOF oxidation state is best represented as −1.
‡Compound is unknown.

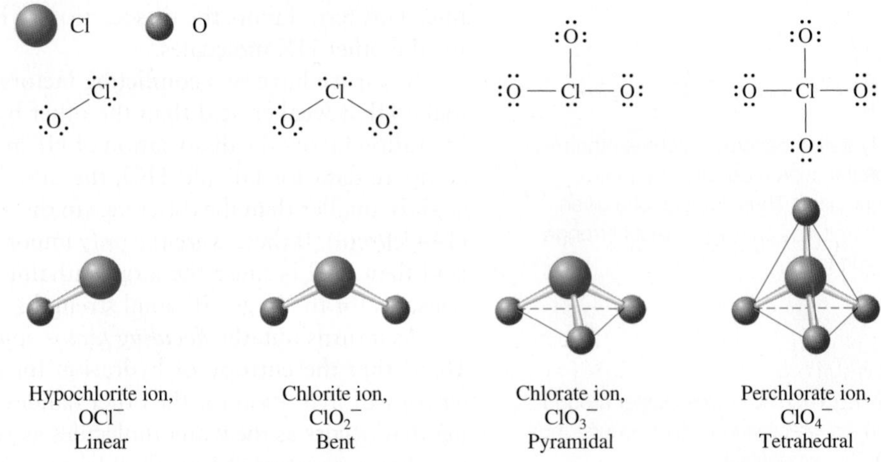

Figure 18.26
The structures of the oxychloro anions.

known only in solution, although salts containing their anions are well known (Fig. 18.26).

Hypochlorous acid (HOCl) is formed when chlorine gas is dissolved in cold water:

$$Cl_2(aq) + H_2O(l) \rightleftharpoons HOCl(aq) + H^+(aq) + Cl^-(aq)$$

Note that in this reaction chlorine is both oxidized (from 0 in Cl_2 to $+1$ in HOCl) and reduced (from 0 in Cl_2 to -1 in Cl^-). Such a reaction, *in which a given element is both oxidized and reduced,* is called a **disproportionation reaction.** Hypochlorous acid and its salts are strong oxidizing agents; their solutions are widely used as household bleaches and disinfectants.

Chlorate salts, such as $KClO_3$, are also strong oxidizing agents and are used as weed killers and as oxidizers in fireworks (see Chapter 12) and explosives.

Fluorine forms only one oxyacid, hypofluorous acid (HOF), but it forms at least two oxides. When fluorine gas is bubbled into a dilute solution of sodium hydroxide, the compound *oxygen difluoride* (OF_2) is formed:

The name for OF_2 is oxygen difluoride rather than difluorine oxide because fluorine has a higher electronegativity than oxygen and thus is named as the anion.

$$4F_2(g) + 3H_2O(l) \longrightarrow 6HF(aq) + OF_2(g) + O_2(g)$$

Oxygen difluoride is a pale yellow gas (bp $= -145°C$) that is a strong oxidizing agent. The oxide *dioxygen difluoride* (O_2F_2) is an orange solid that can be prepared by an electric discharge in an equimolar mixture of fluorine and oxygen gases:

$$F_2(g) + O_2(g) \xrightarrow[\text{discharge}]{\text{Electric}} O_2F_2(s)$$

18.14 The Group 8A Elements

8A
He
Ne
Ar
Kr
Xe
Rn

The Group 8A elements, the **noble gases,** are characterized by filled *s* and *p* valence orbitals (electron configurations of $2s^2$ for helium and ns^2np^6 for the others). Because of their completed valence shells, these elements are very unreactive. In fact, no noble gas compounds were known 50 years ago. Selected properties of the Group 8A elements are summarized in Table 18.22.

Helium was identified by its characteristic emission spectrum as a component of the sun before it was found on earth. The major sources of helium on earth are natural gas deposits, where helium was formed from the α-particle decay of radioactive elements. The α particle is a helium nucleus that can easily pick up electrons from the environment to form a helium atom. Although he-

Automatic Sunglasses

Chemical Insights

Sunglasses can be troublesome. It seems they are always getting lost or sat on. One solution to this problem for people who wear glasses is photochromic glass—glass that darkens in response to intense light. Recall that glass is a complex, noncrystalline material that is composed of polymeric silicates (see Chapter 16). Of course, glass transmits visible light—its transparency is its most useful property.

Glass can be made photochromic by adding tiny silver chloride crystals, which get trapped in the glass matrix as the glass solidifies. Silver chloride has the unusual property of darkening when struck by light—the property that makes the silver halide salts so useful for photographic films. This darkening occurs because light causes an electron transfer from Cl^- to Ag^+ in the silver chloride crystal, forming a silver atom and a chlorine atom. The silver atoms formed in this way tend to migrate to the surface of the silver chloride crystal, where they aggregate to form a tiny crystal of silver metal, which is opaque to light.

In photography the image defined by the grains of silver is fixed by chemical treatment so that it remains permanent. However, in photochromic glass this process must be reversible—the glass must become fully transparent again when the person goes back indoors. The secret to the reversibility of photochromic glass is the presence of Cu^+ ions. The added Cu^+ ions serve two important functions. First, they reduce the Cl atoms formed in the light-induced reaction. This prevents them from escaping from the crystal:

$$Ag^+ + Cl^- \xrightarrow{h\nu} Ag + Cl$$

$$Cl + Cu^+ \longrightarrow Cu^{2+} + Cl^-$$

Second, when the exposure to intense light ends (the person goes indoors), the Cu^{2+} ions migrate to the surface of the silver chloride crystal, where they accept electrons from silver atoms as the tiny crystal of silver atoms disintegrates:

$$Cu^{2+} + Ag \longrightarrow Cu^+ + Ag^+$$

The Ag^+ ions that are re-formed in this way then return to their places in the silver chloride crystal, making the glass transparent once again.

Although these glass lenses still are produced, the market for glass has decreased significantly since the development of plastic photochromic lenses in the 1990s.

The chemistry of plastic photochromic lenses is completely different from that of glass lenses. The photochromic process in plastic uses organic dyes that change structure when they absorb light. For example, the dye naphthopyran rearranges as follows when it absorbs UV light:

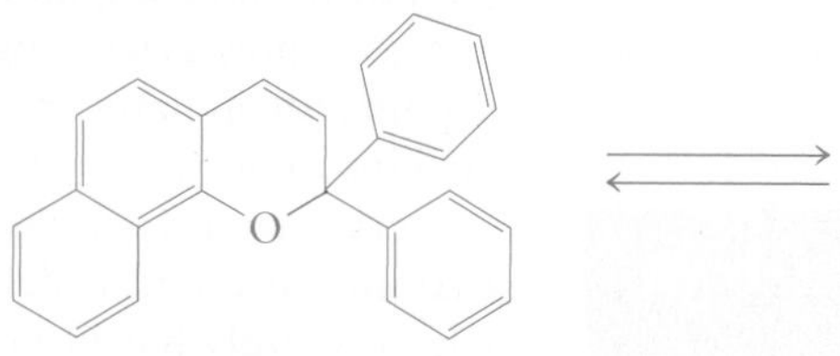

Absorbs UV light

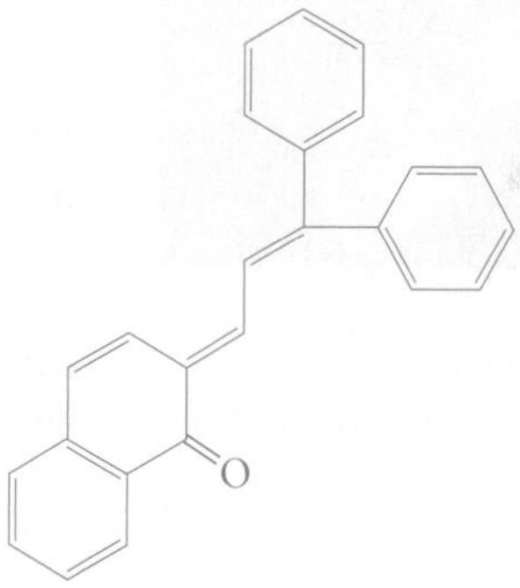

Absorbs UV and visible light

Note that the bond shown in red breaks when the molecule absorbs UV light, causing rearrangement to a molecule that absorbs visible light, which in turn causes the lens to darken. Although the dyes are central to the reversible darkening process, they do not work unless they are incorporated in special ways in the plastic lenses. For example, for lenses composed of polycarbonate, the dye is sandwiched between layers of polyurethane on the surface of the lens.

One drawback of self-darkening lenses is that they do not work well inside automobiles because automotive windshields are constructed with materials that are very effective UV light absorbers. Research is underway to find dyes that will cause darkening by absorbing more into the visible region of the spectrum to solve this problem. Until then, old-fashioned sunglasses will be required for driving.

Table 18.22

Selected Properties of Group 8A Elements

Element	Melting Point (°C)	Boiling Point (°C)	Atmospheric Abundance (% by volume)	Examples of Compounds
Helium	−270	−269	5×10^{-4}	None
Neon	−249	−246	1×10^{-3}	None
Argon	−189	−186	9×10^{-1}	HArF
Krypton	−157	−153	1×10^{-4}	KrF_2
Xenon	−112	−107	9×10^{-6}	XeF_4, XeO_3, XeF_6

lium forms no compounds, it is an important substance that is used as a coolant, as a pressurizing gas for rocket fuels, as a diluent in the gases used for deep-sea diving and spaceship atmospheres, and as the gas in lighter-than-air airships (blimps).

AP Photo/Donna McWilliam

Neon signmaker and artist Jess Baird shows off a few of the items he has made in his Weatherford, Texas, shop.

Like helium, *neon* forms no compounds, but it is a very useful element. For example, neon is widely used in luminescent lighting (neon signs). *Argon*, which recently has been shown to form chemical bonds under special circumstances, is used to provide the noncorrosive atmosphere in incandescent light bulbs, which prolongs the life of the tungsten filament.

Krypton and *xenon* have been observed to form many stable chemical compounds. The first of these was prepared in 1962 by Neil Bartlett, an English chemist who made an ionic compound that he thought had the formula $XePtF_6$. Subsequent studies indicated that the compound might be better represented as $XeFPtF_6$ and contains the XeF^+ and PtF_6^- ions.

Less than a year after Bartlett's report, a group at Argonne National Laboratory near Chicago prepared xenon tetrafluoride by reacting xenon and fluorine gases in a nickel reaction vessel at 400°C and 6 atm:

$$Xe(g) + 2F_2(g) \longrightarrow XeF_4(s)$$

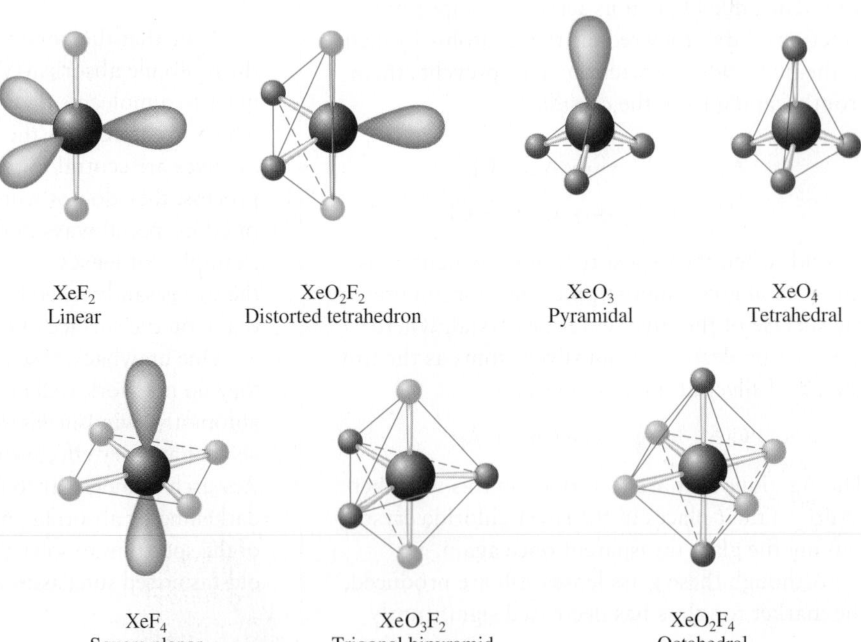

Figure 18.27
The structures of several known xenon compounds.

Xenon tetrafluoride forms stable colorless crystals. Two other xenon fluorides, XeF_2 and XeF_6, were synthesized by the group at Argonne, and a highly explosive xenon oxide (XeO_3) was also found. The xenon fluorides react with water to form hydrogen fluoride and oxycompounds. For example:

$$XeF_6(s) + 3H_2O(l) \longrightarrow XeO_3(aq) + 6HF(aq)$$

$$XeF_6(s) + H_2O(l) \longrightarrow XeOF_4(aq) + 2HF(g)$$

In the past 35 years, other xenon compounds have been prepared. Examples are XeO_4 (explosive), $XeOF_4$, $XeOF_2$, and XeO_3F_2. These compounds contain discrete molecules with covalent bonds between the xenon atom and the other atoms. A few compounds of krypton, such as KrF_2 and KrF_4, have also been observed. The structures of several known xenon compounds are shown in Fig. 18.27. Radon also has been shown to form compounds similar to those of xenon and krypton.

Key Terms

Section 18.1
representative elements
transition metals
lanthanides
actinides
metalloids (semimetals)
metallurgy
liquefaction

Section 18.2
alkali metals

Section 18.3
hydride
ionic (saltlike) hydride
covalent hydride
metallic (interstitial) hydride

Section 18.4
alkaline earth metals
hard water
ion exchange
ion exchange resin

Section 18.5
boranes

Section 18.8
Haber process
nitrogen fixation
nitrogen-fixing bacteria
denitrification
nitrogen cycle
ammonia
hydrazine
nitric acid
Ostwald process

Section 18.9
phosphoric (orthophosphoric) acid
phosphorous acid
superphosphate of lime

For Review

Representative elements

- Chemical properties are determined by their *s* and *p* valence-electron configurations
- Metallic character increases going down the group
- The properties of the first element in a group usually differ most from the properties of the other elements in the group due to a significant difference in size
  - In Group 1A, hydrogen is a nonmetal and the other members of the group are active metals
  - The first member of a group forms the strongest π bonds, causing nitrogen and oxygen to exist as N_2 and O_2 molecules

Elemental abundances on earth

- Oxygen is the most abundant element, followed by silicon
- The most abundant metals are aluminum and iron, which are found as ores

Group 1A elements (alkali metals)

- Have valence configuration ns^1
- Except for hydrogen, readily lose one electron to form M^+ ions in their compounds with nonmetals
- React vigorously with water to form M^+ and OH^- ions and hydrogen gas
- Form a series of oxides of the types M_2O (oxide), M_2O_2 (peroxide), and MO_2 (superoxide)
  - Not all metals form all types of oxide compounds
- Hydrogen forms covalent compounds with nonmetals
- With very active metals, hydrogen forms hydrides that contain the H^- ion

Group 2A (alkaline earth metals)

- Have valence configuration ns^2
- React less violently with water than alkali metals

Section 18.11
ozone
ozonolysis

Section 18.12
Frasch process
sulfuric acid

Section 18.13
halogens
hydrochloric acid
hydrohalic acids
disproportionation reaction

Section 18.14
noble gases

- The heavier alkaline earth metals form nitrides and hydrides
- Hard water contains Ca^{2+} and Mg^{2+} ions
  - Form precipitates with soap
  - Usually removed by ion-exchange resins that replace the Ca^{2+} and Mg^{2+} ions with Na^+

Group 3A

- Have valence configuration ns^2np^1
- Show increasing metallic character going down the group
- Boron is a nonmetal that forms many types of covalent compounds, including boranes, which are highly electron-deficient and thus are very reactive
- The metals aluminum, gallium, and indium show some covalent tendencies

Group 4A

- Have valence configuration ns^2np^2
- Lighter members are nonmetals; heavier members are metals
  - All group members can form covalent bonds to nonmetals
- Carbon forms a huge variety of compounds, most of which are classified as organic compounds

Group 5A

- Elements show a wide variety of chemical properties
  - Nitrogen and phosphorus are nonmetals
  - Antimony and bismuth tend to be metallic, although no ionic compounds containing Sb^{5+} and Bi^{5+} are known; the compounds containing Sb(V) and Bi(V) are molecular rather than ionic
  - All group members except N form molecules with five covalent bonds
  - The ability to form π bonds decreases dramatically after N
- Chemistry of nitrogen
  - Most nitrogen-containing compounds decompose exothermically, forming the very stable N_2 molecule, which explains the power of nitrogen-based explosives
  - The nitrogen cycle, which consists of a series of steps, shows how nitrogen is cycled in the natural environment
  - Nitrogen fixation changes the N_2 in air into compounds useful to plants
    - The Haber process is a synthetic method of nitrogen fixation
    - In the natural world, nitrogen fixation occurs through nitrogen-fixing bacteria in the root nodules of certain plants and through lightning in the atmosphere
  - Ammonia is the most important hydride of nitrogen
    - Contains pyramidal NH_3 molecules
    - Widely used as a fertilizer
  - Hydrazine (N_2H_4) is a powerful reducing agent
  - Nitrogen forms a series of oxides including N_2O, NO, NO_2, and N_2O_5
  - Nitric acid (HNO_3) is a very important strong acid manufactured by the Ostwald process
- Chemistry of phosphorus
  - Exists in three elemental forms: white (contains P_4 molecules), red, and black
  - Phosphine (PH_3) has bond angles close to 90 degrees
  - Phosphorus forms oxides including P_4O_6 and P_4O_{10} (which dissolves in water to form phosphoric acid, H_3PO_4)

Group 6A

- Metallic character increases going down the group but no element behaves as a typical metal
- The lighter members tend to gain two electrons to form X^{2-} ions in compounds with metals
- Chemistry of oxygen
  - Elemental forms are O_2 and O_3
  - Oxygen forms a wide variety of oxides
  - O_2 and especially O_3 are powerful oxidizing agents
- Chemistry of sulfur
  - The elemental forms are called rhombic and monoclinic sulfur, both of which contain S_8 molecules
  - The most important oxides are SO_2 (which forms H_2SO_3 in water) and SO_3 (which forms H_2SO_4 in water)
  - Sulfur forms a wide variety of compounds in which it shows the oxidation states +6, +4, +2, 0, and −2

Group 7A (halogens)

- All nonmetals
- Form hydrides of the type HX that behave as strong acids in water except for HF, which is a weak acid
- The oxyacids of the halogens become stronger as more oxygen atoms are present
- The interhalogens contain two or more different halogens

Group 8A (noble gases)

- All elements are monatomic gases and are generally very unreactive
- The heavier elements form compounds with electronegative elements such as fluorine and oxygen

Exercises

OWL Interactive versions of these problems may be assigned in OWL.

A blue exercise number indicates that the answer to that exercise appears at the back of this book and a solution appears in the *Solutions Guide*.

Group 1A Elements

1. Although the earth was formed from the same interstellar material as the sun, there is little hydrogen in the earth's atmosphere. How can you explain this?

2. Hydrogen is produced commercially by the reaction of methane with steam:

$$CH_4(g) + H_2O(g) \rightleftharpoons CO(g) + 3H_2(g)$$

   a. Calculate $\Delta H°$ and $\Delta S°$ for this reaction (use the data in Appendix 4).
   b. What temperatures will favor product formation assuming standard conditions and assuming that $\Delta H°$ and $\Delta S°$ do not depend on temperature?

3. The major industrial use of hydrogen is in the production of ammonia by the Haber process:

$$3H_2(g) + N_2(g) \longrightarrow 2NH_3(g)$$

   a. Using data from Appendix 4, calculate $\Delta H°$, $\Delta S°$, and $\Delta G°$ for the Haber process reaction.
   b. Is this reaction spontaneous at standard conditions?
   c. At what temperatures is the reaction spontaneous at standard conditions? Assume that $\Delta H°$ and $\Delta S°$ do not depend on temperature.

4. List two major industrial uses of hydrogen.

5. Label the following hydrides as ionic, covalent, or interstitial, and support your answer. Note: The light blue atoms are hydrogen atoms.

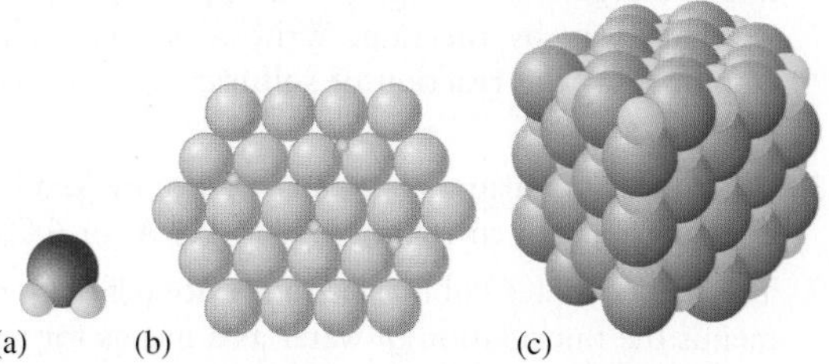

6. Many lithium salts are hygroscopic (absorb water), but the corresponding salts of the other alkali metals are not. Why are lithium salts different from the others?

7. What is the valence electron configuration for the alkali metals? List some common properties of alkali metals. How are the pure metals prepared?

8. What will be the atomic number of the next alkali metal to be discovered? How would you expect the physical properties of the next alkali metal to compare with the properties of the other alkali metals summarized in Table 18.4?

9. What evidence supports putting hydrogen in Group 1A of the periodic table? In some periodic tables hydrogen is listed separately from all of the groups. In what ways is hydrogen unlike a Group 1A element?

10. The electrolysis of aqueous sodium chloride (brine) is an important industrial process for the production of chlorine and sodium hydroxide. In fact, this process is the second largest consumer of electricity in the United States, after the production of aluminum. Write a balanced equation for the electrolysis of aqueous sodium chloride (hydrogen gas is also produced).

11. Write balanced equations describing the reaction of lithium metal with each of the following: O_2, S_8, Cl_2, P_4, H_2, H_2O, and HCl.

Group 2A Elements

12. What is the valence electron configuration for alkaline earth metals? List some common properties of alkaline earth metals. How are alkaline earth metals prepared?

13. How long will it take to produce 1.00×10^3 kg of magnesium metal by the electrolysis of molten magnesium chloride using a current of 5.00×10^4 A?

14. Suppose 10.00 g of an alkaline earth metal reacts with 10.0 L of water to produce 6.10 L of hydrogen gas at 1.00 atm and 25°C. Identify the metal, and determine the pH of the solution.

15. What ions are found in hard water? What happens when water is "softened"?

16. Write balanced equations describing the reaction of Sr with each of the following: O_2, S_8, Cl_2, P_4, H_2, H_2O, and HCl.

17. One harmful effect of acid rain is the deterioration of structures and statues made of marble or limestone, both of which are essentially calcium carbonate. The reaction of calcium carbonate with sulfuric acid yields carbon dioxide, water, and calcium sulfate. Because calcium sulfate is marginally soluble in water, part of the object is washed away by the rain. Write a balanced chemical equation for the reaction of sulfuric acid with calcium carbonate.

18. What mass of barium is produced when molten $BaCl_2$ is electrolyzed by a current of 2.50×10^5 A for 6.00 h?

19. The United States Public Health Service (USPHS) recommends the fluoridation of water as a means for preventing tooth decay. The recommended concentration is 1 mg F^-/L. The presence of calcium ions in hard water can precipitate the added fluoride. What is the maximum molarity of calcium ions in hard water if the fluoride concentration is at the USPHS recommended level? (K_{sp} for $CaF_2 = 4.0 \times 10^{-11}$.)

20. Slaked lime $[Ca(OH)_2]$ is used to soften hard water by removing calcium ions from hard water through the reaction

$$Ca(OH)_2(aq) + Ca^{2+}(aq) + 2HCO_3^-(aq) \longrightarrow 2CaCO_3(s) + 2H_2O(l)$$

Although $CaCO_3(s)$ is considered insoluble, some of it does dissolve in aqueous solutions. Calculate the molar solubility of $CaCO_3$ in water ($K_{sp} = 8.7 \times 10^{-9}$).

Group 3A Elements

21. What is the valence electron configuration for the Group 3A elements? How does metallic character change as one goes down this group? How are boron and aluminum different?

22. What are three-centered bonds?

23. Boron hydrides were once evaluated for possible use as rocket fuels. Complete and balance the following equation:

$$B_2H_6 + O_2 \longrightarrow B(OH)_3$$

24. Consider element 113. What is the expected electron configuration for element 113? What oxidation states would be exhibited by element 113 in its compounds?

25. Write equations describing the reactions of Ga with each of the following: F_2, O_2, S_8, and HCl.

26. Thallium and indium form +1 and +3 oxidation states when in compounds. Predict the formulas of the possible compounds between thallium and oxygen and between indium and chlorine. Name the compounds.

Group 4A Elements

27. What is the valence electron configuration for Group 4A elements? Group 4A contains two of the most important elements on earth. What are they, and why are they so important? How does metallic character change as one goes down Group 4A?

28. Discuss the importance of the C—C and Si—Si bond strengths and of π bonding to the properties of carbon and silicon.

29. The following illustration shows the orbitals used to form the bonds in carbon dioxide.

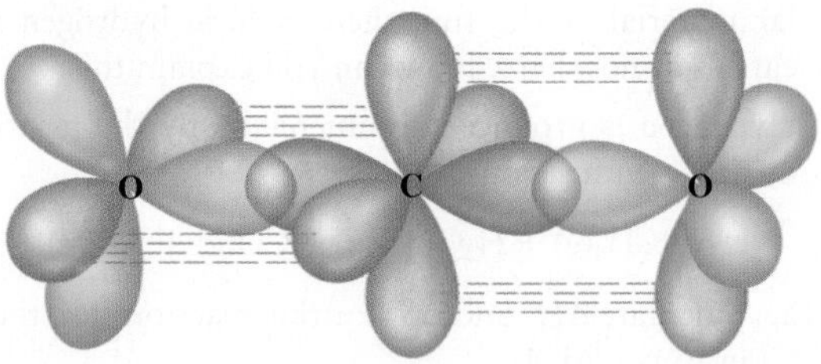

Each color represents a different orbital. Label each orbital, draw the Lewis structure for carbon dioxide, and explain how the localized electron model describes the bonding in CO_2.

30. In addition to CO_2, two additional stable oxides of carbon form. The space filling models for CO_2 and the other two stable oxides are:

What are the formulas for the two additional stable oxides of carbon? Explain the bonding in each of these forms using the localized electron model.

31. From the information on the temperature stability of white and gray tin given in this chapter, which form would you expect to have the more ordered structure (have the smaller positional probability)?

32. Silicon is produced for the chemical and electronics industries by the following reactions. Give the balanced equation for each reaction.
    a. $SiO_2(s) + C(s) \longrightarrow Si(s) + CO(g)$
    b. Silicon tetrachloride is reacted with very pure magnesium, producing silicon and magnesium chloride.
    c. $Na_2SiF_6(s) + Na(s) \longrightarrow Si(s) + NaF(s)$

33. Tin forms compounds in the +2 and +4 oxidation states. Therefore, when tin reacts with fluorine, two products are possible. Write balanced equations for the production of the two tin halide compounds and name them.

34. Write equations describing the reactions of Sn with each of the following: Cl_2, O_2, and HCl.

35. The compound Pb_3O_4 (red lead) contains a mixture of lead(II) and lead(IV) oxidation states. What is the mole ratio of lead(II) to lead(IV) in Pb_3O_4?

36. The resistivity (a measure of electrical resistance) of graphite is $(0.4 \text{ to } 5.0) \times 10^{-4}$ ohm · cm in the basal plane. (The basal plane is the plane of the six-membered rings of carbon atoms.) The resistivity is 0.2 to 1.0 ohm · cm along the axis perpendicular to the plane. The resistivity of diamond is 10^{14} to 10^{16} ohm · cm and is independent of direction. How can you account for this behavior in terms of the structures of graphite and diamond?

Group 5A Elements

37. The oxyanion of nitrogen in which it has the highest oxidation state is the nitrate ion (NO_3^-). The corresponding oxyanion of phosphorus is PO_4^{3-}. The NO_4^{3-} ion is known but is not very stable. The PO_3^- ion is not known. Account for these differences in terms of the bonding in the four anions.

38. Using data from Appendix 4, calculate $\Delta H°$, $\Delta S°$, and $\Delta G°$ for the reaction

$$N_2(g) + O_2(g) \longrightarrow 2NO(g)$$

Why does the NO formed in an automobile engine not readily decompose back to N_2 and O_2 in the atmosphere?

39. Table 18.14 lists some common nitrogen compounds having oxidation states ranging from −3 to +5. Rationalize this spread in oxidation states.

40. What is nitrogen fixation? Give some examples of nitrogen fixation.

41. Elemental nitrogen exists as N_2, whereas in the gas phase the elements phosphorus, arsenic, and antimony consist of P_4, As_4, and Sb_4 molecules, respectively. Give a possible reason for this difference between N_2 and the other Group 5A elements.

42. The compound NF_3 is quite stable, but NCl_3 is very unstable (NCl_3 was first synthesized in 1811 by P. L. Dulong, who lost three fingers and an eye studying its properties). The compounds NBr_3 and NI_3 are unknown, although the explosive $NI_3 \cdot NH_3$ is known. How do you account for the instability of these halides of nitrogen?

43. Lewis structures can be used to understand why some molecules react in certain ways. Write the Lewis structure for the reactants and products in the following reactions.
    a. Nitrogen dioxide dimerizes to produce dinitrogen tetroxide.
    b. Boron trifluoride accepts a pair of electrons from ammonia, forming BF_3NH_3.

    Give a possible explanation for why these two reactions occur.

44. Use bond energies (Table 13.6) to show that the preferred products for the decomposition of N_2O_3 are NO_2 and NO rather than O_2 and N_2O. (The N—O single-bond energy is 201 kJ/mol.) (*Hint:* Consider the reaction kinetics.)

45. Ammonia is produced by the Haber process, in which nitrogen and hydrogen are reacted directly using an iron mesh impregnated with oxides as a catalyst. For the reaction

$$N_2(g) + 3H_2(g) \rightleftharpoons 2NH_3(g)$$

equilibrium constants (K_p values) as a function of temperature are

300°C, 4.34×10^{-3}
500°C, 1.45×10^{-5}
600°C, 2.25×10^{-6}

Is the reaction exothermic or endothermic?

46. The synthesis of ammonia gas from nitrogen gas and hydrogen gas represents a classic case in which a knowledge of kinetics and equilibrium was used to make a desired chemical reaction economically feasible. Explain how each of the following conditions helps to maximize the yield of ammonia.
    a. running the reaction at an elevated temperature
    b. removing the ammonia from the reaction mixture as it forms
    c. using a catalyst
    d. running the reaction at high pressure

47. Write an equation for the reaction of hydrazine with fluorine gas to produce nitrogen gas and hydrogen fluoride gas. Estimate ΔH for this reaction, using bond energies from Table 13.6.

48. In each of the following pairs of substances, one is stable and known, whereas the other is unstable. For each pair, choose the stable substance, and explain why the other compound is unstable.
    a. NF_5 or PF_5 b. AsF_5 or AsI_5 c. NF_3 or NBr_3

49. Write balanced equations for the reactions described in Table 18.13 for the production of Bi and Sb.

50. Phosphoric acid (H_3PO_4) is a triprotic acid, phosphorous acid (H_3PO_3) is a diprotic acid, and hypophosphorous acid (H_3PO_2) is a monoprotic acid. Explain this phenomenon.

51. Trisodium phosphate (TSP) is an effective grease remover. Like many cleaners, TSP acts as a base in water. Write a balanced equation to account for this basic behavior.

52. White phosphorus is much more reactive than black or red phosphorus. Explain.

53. Arsenic reacts with oxygen to form oxides analogous to the phosphorus oxides. These arsenic oxides react with water similarly to the phosphorus oxides. Write balanced chemical equations describing the reaction of arsenic with oxygen and the reaction of the oxides with water.

54. Compare the description of the localized electron model (Lewis structure) with that of the molecular orbital model for the bonding in NO, NO^+, and NO^-. Account for any discrepancies between the two models.

55. Many oxides of nitrogen have positive values for the standard free energy of formation. Using NO as an example, explain why this is the case.

56. Draw Lewis structures for the $AsCl_4^+$ and $AsCl_6^-$ ions. What type of reaction (acid–base, oxidation–reduction, or the like) is the following?

$$2AsCl_5(g) \longrightarrow AsCl_4AsCl_6(s)$$

57. In many natural waters, nitrogen and phosphorus are the least abundant nutrients available for plant life. Some waters that become polluted from agricultural runoff or municipal sewage become infested with algae. The algae flourish, and fish life dies off as a result. Describe how these events are chemically related.

58. Nitric acid is produced commercially by the Ostwald process, represented by the following equations:

$$4NH_3(g) + 5O_2(g) \longrightarrow 4NO(g) + 6H_2O(g)$$
$$2NO(g) + O_2(g) \longrightarrow 2NO_2(g)$$
$$3NO_2(g) + H_2O(l) \longrightarrow 2HNO_3(aq) + NO(g)$$

What mass of NH_3 must be used to produce 1.0×10^6 kg HNO_3 by the Ostwald process? Assume 100% yield in each reaction, and assume that the NO produced in the third step is not recycled.

59. The space shuttle orbiter utilizes the oxidation of methylhydrazine by dinitrogen tetroxide for propulsion:

$$4N_2H_3CH_3(l) + 5N_2O_4(l) \longrightarrow 12H_2O(g) + 9N_2(g) + 4CO_2(g)$$

Calculate $\Delta H°$ for this reaction using data in Appendix 4.

60. How is phosphine's (PH_3) structure different from that of ammonia?

61. Ammonia forms hydrogen-bonding intermolecular forces resulting in an unusually high boiling point for a substance with the small size of NH_3. Can hydrazine (N_2H_4) also form hydrogen-bonding interactions?

Group 6A Elements

62. What is the valence electron configuration of Group 6A elements? What are some property differences between oxygen and polonium?

63. What are the Lewis structures for the two allotropic forms of oxygen? How can the paramagnetism of O_2 be explained using the molecular orbital model? What are the molecular structure and the bond angles in ozone?

64. Use bond energies to estimate the maximum wavelength of light that will cause the reaction

$$O_3 \xrightarrow{h\nu} O_2 + O$$

65. Ozone is desirable in the upper atmosphere but undesirable in the lower atmosphere. A dictionary states that ozone has the scent of a spring thunderstorm. How can these seemingly conflicting statements be reconciled in terms of the chemical properties of ozone?

66. Ozone is a possible replacement for chlorine in municipal water purification. Unlike chlorine, virtually no ozone remains after treatment. This has good and bad consequences. Explain.

67. Sulfur forms a wide variety of compounds in which it has +6, +4, +2, 0, and −2 oxidation states. Give examples of sulfur compounds having each of these oxidation states.

68. In large doses, selenium is toxic. However, in moderate intake, selenium is a physiologically important element. How is selenium physiologically important?

69. Write a balanced equation describing the reduction of H_2SeO_4 by SO_2 to produce selenium.

Group 7A Elements

70. What is the valence electron configuration of the halogens? Why do the boiling points and melting points of the halogens increase steadily from F_2 to I_2?

71. Draw the Lewis structure of O_2F_2. Assign oxidation states and formal charges to the atoms in O_2F_2. The compound O_2F_2 is a vigorous and potent oxidizing and fluorinating agent. Are oxidation states or formal charges more useful in accounting for these properties of O_2F_2?

72. The oxidation states of the halogens vary from −1 to +7. Identify compounds of chlorine that have −1, +1, +3, +5, and +7 oxidation states. How does the oxyacid strength of the halogens vary as the number of oxygens in the formula increases?

73. Give two reasons why F_2 is the most reactive of the halogens.

74. Explain why HF is a weak acid, whereas HCl, HBr, and HI are all strong acids.

75. Hydrazine is somewhat toxic. Use the following half-reactions to explain why household bleach (highly alkaline solutions of sodium hypochlorite) should not be

mixed with household ammonia or glass cleansers that contain ammonia.

$$ClO^- + H_2O + 2e^- \longrightarrow 2OH^- + Cl^- \quad \mathscr{E}° = 0.90\ V$$

$$N_2H_4 + 2H_2O + 2e^- \longrightarrow 2NH_3 + 2OH^- \quad \mathscr{E}° = -0.10\ V$$

76. What is a disproportionation reaction? Use the following reduction potentials,

$$ClO_3^- + 3H^+ + 2e^- \longrightarrow HClO_2 + H_2O \quad \mathscr{E}° = 1.21\ V$$

$$HClO_2 + 2H^+ + 2e^- \longrightarrow HClO + H_2O \quad \mathscr{E}° = 1.65\ V$$

to predict whether $HClO_2$ will disproportionate.

77. Photogray lenses contain small embedded crystals of solid silver chloride. Silver chloride is light-sensitive because of the reaction

$$AgCl(s) \xrightarrow{h\nu} Ag(s) + Cl(s)$$

Small particles of metallic silver cause the lenses to darken. In the lenses this process is reversible. When the light is removed, the reverse reaction occurs. However, when pure white silver chloride is exposed to sunlight, it darkens; the reverse reaction does not occur in the dark.
a. How do you explain this difference?
b. Photogray lenses do become permanently dark in time. How do you account for this?

Group 8A Elements

78. What special property of the noble gases makes them unreactive? The boiling points and melting points of the noble gases increase steadily from He to Xe. Explain.

79. Although He is the second most abundant element in the universe, it is very rare on earth. Why?

80. The noble gases were among the latest elements discovered; their existence was not predicted by Mendeleev when he published his first periodic table. Explain. In chemistry textbooks written before 1962, the noble gases were referred to as the inert gases. Why do we no longer use the term *inert gases?*

81. Using the data in Table 18.22, calculate the mass of argon at 25°C and 1.0 atm in a room 10.0 m × 10.0 m × 10.0 m. How many Ar atoms are in this room? How many Ar atoms do you inhale in one breath (approximately 2 L) of air at 25°C and 1.0 atm? Argon gas is inert, so it poses no serious health risks. However, if significant amounts of radon are inhaled into the lungs, lung cancer is a possible result. Explain the health risk differences between argon gas and radon gas.

82. For the structures of the xenon compounds in Fig. 18.27, give the bond angles exhibited and the hybridization of the central atom in each compound.

83. There is evidence that radon reacts with fluorine to form compounds similar to those formed by xenon and fluorine. Predict the formulas of these RnF_x compounds. Why is the chemistry of radon difficult to study?

Additional Exercises

84. What are the two most abundant elements by mass in the earth's crust, oceans, and atmosphere? Does this make sense? Why? What are the four most abundant elements by mass in the human body? Does this make sense? Why?

85. In most compounds the solid phase is denser than the liquid phase. Why isn't this true for water?

86. Diagonal relationships in the periodic table exist in addition to the vertical relationships. For example, Be and Al are similar in some of their properties, as are B and Si. Rationalize why these diagonal relationships hold for properties such as size, ionization energy, and electron affinity.

87. In the 1950s and 1960s, several nations conducted tests of nuclear warheads in the atmosphere. It was customary, following each test, to monitor the concentration of strontium-90 (a radioactive isotope of strontium) in milk. Why would strontium-90 tend to accumulate in milk?

88. Calculate the pH of a 0.050 M $Al(NO_3)_3$ solution. The K_a value for $Al(H_2O)_6^{3+}$ is 1.4×10^{-5}.

89. The compound with the formula TlI_3 is a black solid. Given the following standard reduction potentials,

$$Tl^{3+} + 2e^- \longrightarrow Tl^+ \quad \mathscr{E}° = +1.25\ V$$

$$I_3^- + 2e^- \longrightarrow 3I^- \quad \mathscr{E}° = +0.55\ V$$

would you formulate this compound as thallium(III) iodide or thallium(I) triiodide?

90. Atomic size seems to play an important role in explaining some of the differences between the first element in a group and the subsequent group elements. Explain.

91. The inert pair effect is sometimes used to explain the tendency of heavier members of Group 3A to exhibit +1 and +3 oxidation states. What does the inert pair effect reference? (*Hint:* Consider the valence electron configuration for Group 3A elements.)

92. How could you determine experimentally whether the compound Ga_2Cl_4 contains two gallium(II) ions or one gallium(I) ion and one gallium(III) ion? (*Hint:* Consider the electron configurations of the three possible ions.)

93. It takes 15 kWh (kilowatt-hours) of electrical energy to produce 1.0 kg of aluminum metal from aluminum oxide by the Hall–Heroult process. Compare this to the amount of energy necessary to melt 1.0 kg of aluminum metal. Why is it economically feasible to recycle aluminum cans? [The enthalpy of fusion for aluminum metal is 10.7 kJ/mol (1 W = 1 J/s).]

94. Reference Table 18.5, and give examples of the three types of alkali metal oxides that form. How do they differ?

95. Why are the tin(IV) halides more volatile than the tin(II) halides?

96. Beryllium shows some covalent characteristics in some of its compounds, unlike the other alkaline earth halides. Give a possible explanation for this phenomenon.

97. One reason suggested for the instability of long chains of silicon atoms is that the decomposition involves the transition state shown below:

$$H_3Si{-}SiH_3 \longrightarrow$$

$$\left\{ H_3Si \cdots H \cdots SiH_2 \right\} \longrightarrow SiH_4 + :SiH_2$$

The activation energy for such a process is 210 kJ/mol, which is less than either the Si—Si or the Si—H bond energy. Why would a similar mechanism not be expected to be very important in the decomposition of long carbon chains?

98. Many structures of phosphorus- and sulfur-containing compounds are drawn with some P=O and P=S bonds. These bonds are not the typical π bonds we've considered, which involve the overlap of two p orbitals. They result instead from the overlap of a d orbital on the phosphorus or sulfur atom with a p orbital on oxygen. This type of π bonding is sometimes used as an explanation for why H_3PO_3 has the first structure below rather than the second:

$$H{-}P(=O)(OH){-}OH \qquad HO{-}\ddot{P}(OH){-}OH$$

Draw a picture showing how a d orbital and a p orbital overlap to form a π bond.

99. The N_2O molecule is linear and polar.
   a. On the basis of this experimental evidence, which arrangement, NNO or NON, is correct? Explain your answer.
   b. On the basis of your answer in part a, write the Lewis structure of N_2O (including resonance forms). Give the formal charge on each atom and the hybridization of the central atom.
   c. How would the multiple bonding in

$$:N{\equiv}N{-}\ddot{\underset{\cdot\cdot}{O}}:$$

be described in terms of orbitals?

100. Bacterial digestion is an economical method of sewage treatment. The reaction

$$5CO_2(g) + 55NH_4^+(aq) + 76O_2(g) \xrightarrow{\text{Bacteria}}$$
$$\underset{\text{Bacterial tissue}}{C_5H_7O_2N(s)} + 54NO_2^-(aq) + 52H_2O(l) + 109H^+(aq)$$

is an intermediate step in the conversion of the nitrogen in organic compounds into nitrate ions. What mass of bacterial tissue is produced in a treatment plant for every 1.0×10^4 kg of wastewater containing 3.0% NH_4^+ ions by mass? Assume that 95% of the ammonium ions are consumed by the bacteria.

101. Phosphate buffers are important in regulating the pH of intracellular fluids. If the concentration ratio of $H_2PO_4^-$ to HPO_4^{2-} in a sample of intracellular fluid is 1.1:1, what is the pH of this sample of intracellular fluid?

$$H_2PO_4^-(aq) \rightleftharpoons HPO_4^{2-}(aq) + H^+(aq) \qquad K_a = 6.2 \times 10^{-8}$$

102. The unit cell for a pure xenon fluoride compound is shown below. What is the formula of the compound?

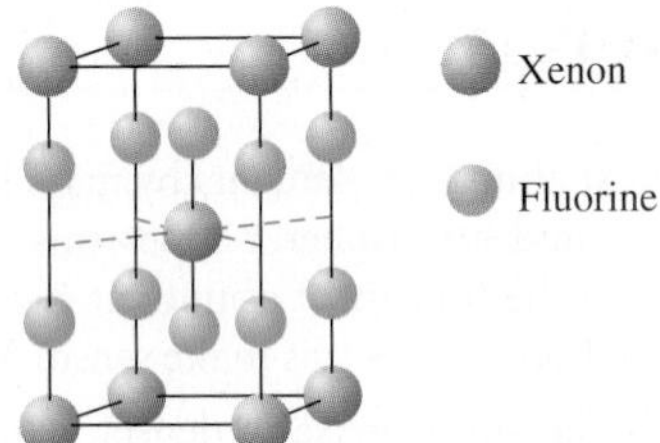

Challenge Problems

103. Indium(III) phosphide is a semiconducting material that has been frequently used in lasers, light-emitting diodes (LED), and fiberoptic devices. This material can be synthesized at 900. K according to the following reaction:

$$In(CH_3)_3(g) + PH_3(g) \longrightarrow InP(s) + 3CH_4(g)$$

   a. If 2.56 L $In(CH_3)_3$ at 2.00 atm is allowed to react with 1.38 L PH_3 at 3.00 atm, what mass of InP(s) will be produced assuming the reaction is 87% efficient?
   b. When an electric current is passed through an optoelectronic device containing InP, the light emitted has an energy of 2.03×10^{-19} J. What is the wavelength of this light, and is it visible to the human eye?
   c. The semiconducting properties of InP can be altered by doping. If a small number of phosphorus atoms are replaced by atoms with an electron configuration of $[Kr]5s^24d^{10}5p^4$, is this n-type or p-type doping?

104. While selenic acid has the formula H_2SeO_4 and thus is directly related to sulfuric acid, telluric acid is best visualized as H_6TeO_6 or $Te(OH)_6$.

a. What is the oxidation state of tellurium in $Te(OH)_6$?

b. Despite its structural differences with sulfuric and selenic acid, telluric acid is a diprotic acid with $pK_{a_1} = 7.68$ and $pK_{a_2} = 11.29$. Telluric acid can be prepared by hydrolysis of tellurium hexafluoride according to the equation

$$TeF_6(g) + 6H_2O(l) \longrightarrow Te(OH)_6(aq) + 6HF(aq)$$

Tellurium hexafluoride can be prepared by the reaction of elemental tellurium with fluorine gas:

$$Te(s) + 3F_2(g) \longrightarrow TeF_6(g)$$

If a cubic block of tellurium (density = 6.240 g/cm^3) measuring 0.545 cm on edge is allowed to react with 2.34 L fluorine gas at 1.06 atm and 25°C, what is the pH of a solution of $Te(OH)_6$ formed by dissolving the isolated $TeF_6(g)$ in 115 mL solution? Assume 100% yield in all reactions.

105. Sodium tripolyphosphate ($Na_5P_3O_{10}$) is used in many synthetic detergents. Its major effect is to soften the water by complexing Mg^{2+} and Ca^{2+} ions. It also increases the efficiency of surfactants, or wetting agents that lower a liquid's surface tension. The pK value for the formation of $MgP_3O_{10}^{3-}$ is −8.60. The reaction is $Mg^{2+} + P_3O_{10}^{5-} \rightleftharpoons MgP_3O_{10}^{3-}$. Calculate the concentration of Mg^{2+} in a solution that was originally 50. ppm Mg^{2+} (50. mg/L of solution) after 40. g of $Na_5P_3O_{10}$ is added to 1.0 L of the solution.

106. Nitrogen gas reacts with hydrogen gas to form ammonia gas (NH_3). Consider the following illustration representing the original reaction mixture (the numbers of each molecule shown are relative numbers):

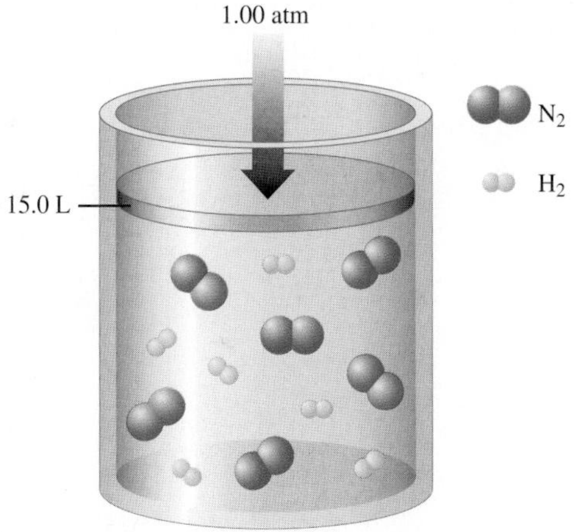

Assume this reaction mixture goes to completion. The piston apparatus allows the container volume to change in order to keep the pressure constant at 1.00 atm. Assume ideal behavior and constant temperature.

a. What is the partial pressure of ammonia in the container when the reaction is complete?

b. What is the mole fraction of ammonia in the container when the reaction is complete?

c. What is the volume of the container when the reaction is complete?

107. Using data from Appendix 4, calculate $\Delta H°$, $\Delta G°$, and K_p (at 298 K) for the production of ozone from oxygen:

$$3O_2(g) \rightleftharpoons 2O_3(g)$$

At 30 km above the surface of the earth, the temperature is about 230. K and the partial pressure of oxygen is about 1.0×10^{-3} atm. Estimate the partial pressure of ozone in equilibrium with oxygen at 30 km above the earth's surface. Is it reasonable to assume that the equilibrium between oxygen and ozone is maintained under these conditions? Explain.

108. You travel to a distant, cold planet where the ammonia flows like water. In fact, the inhabitants of this planet use ammonia (an abundant liquid on their planet) much as earthlings use water. Ammonia is also similar to water in that it is amphoteric and undergoes autoionization. The K value for the autoionization of ammonia is 1.8×10^{-12} at the standard temperature of the planet. What is the pH of ammonia at this temperature?

109. A proposed two-step mechanism for the destruction of ozone in the upper atmosphere is

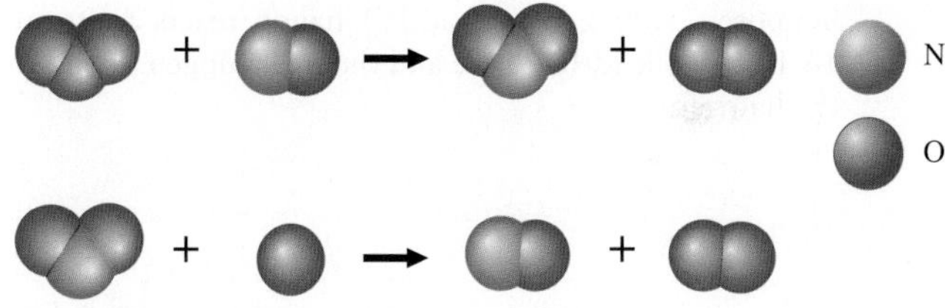

a. What is the overall balanced equation for the ozone destruction reaction?

b. Which species is a catalyst?

c. Which species is an intermediate?

d. What is the rate law derived from this mechanism if the first step in the mechanism is slow and the second step is fast?

e. One of the concerns about the use of Freons is that they will migrate to the upper atmosphere, where chlorine atoms can be generated by the reaction

$$\underset{\text{Freon-12}}{CCl_2F_2} \xrightarrow{h\nu} CF_2Cl + Cl$$

Chlorine atoms also can act as a catalyst for the destruction of ozone. The first step of a proposed mechanism for chlorine-catalyzed ozone destruction is

$$Cl(g) + O_3(g) \longrightarrow ClO(g) + O_2(g) \quad \text{Slow}$$

Assuming a two-step mechanism, propose the second step in the mechanism and give the overall balanced equation.

110. Provide a reasonable estimate for the number of atoms in a 150-lb adult human. Use the information given in Table 18.2.

111. EDTA is used as a complexing agent in chemical analysis. Solutions of EDTA, usually containing the disodium salt Na_2H_2EDTA, are also used to treat heavy metal poisoning. The equilibrium constant for the following reaction is 6.7×10^{21}:

$$Pb^{2+}(aq) + H_2EDTA^{2-}(aq) \rightleftharpoons PbEDTA^{2-}(aq) + 2H^+(aq)$$

$$EDTA^{4-} = \begin{matrix} {}^-O_2C{-}CH_2 & & & & CH_2{-}CO_2^- \\ & \diagdown & & \diagup & \\ & & N{-}CH_2{-}CH_2{-}N & & \\ & \diagup & & \diagdown & \\ {}^-O_2C{-}CH_2 & & & & CH_2{-}CO_2^- \end{matrix}$$

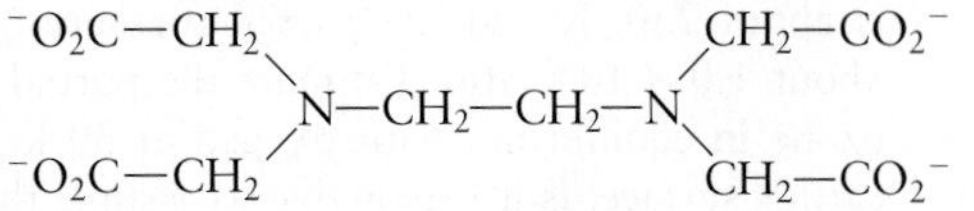

Ethylenediaminetetraacetate

Calculate $[Pb^{2+}]$ at equilibrium in a solution originally 0.0050 M in Pb^{2+}, 0.075 M in H_2EDTA^{2-}, and buffered at pH = 7.00.

112. Lead forms compounds in the +2 and +4 oxidation states. All lead(II) halides are known (and are known to be ionic). Only PbF_4 and $PbCl_4$ are known among the possible lead(IV) halides. Presumably lead (IV) oxidizes bromide and iodide ions, producing the lead(II) halide and the free halogen:

$$PbX_4 \longrightarrow PbX_2 + X_2$$

Suppose 25.00 g of a lead(IV) halide reacts to form 16.12 g of a lead(II) halide and the free halogen. Identify the halogen.

113. a. Many biochemical reactions that occur in cells require relatively high concentrations of potassium ion (K^+). The concentration of K^+ in muscle cells is about 0.15 M. The concentration of K^+ in blood plasma is about 0.0050 M. The high internal concentration in cells is maintained by pumping K^+ from the plasma. How much work must be done to transport 1.0 mole of K^+ from the blood to the inside of a muscle cell at 37°C (normal body temperature)?
b. When 1.0 mole of K^+ is transferred from blood to the cells, do any other ions have to be transported? Why or why not?
c. Cells use the hydrolysis of adenosine triphosphate, abbreviated ATP, as a source of energy. Symbolically, this reaction can be represented as

$$ATP(aq) + H_2O(l) \longrightarrow ADP(aq) + H_2PO_4^-(aq)$$

where ADP represents adenosine diphosphate. For this reaction at 37°C, $K = 1.7 \times 10^5$. How many moles of ATP must be hydrolyzed to provide the energy for the transport of 1.0 mole of K^+? Assume standard conditions for the ATP hydrolysis reaction.

Transition Metals and Coordination Chemistry

19

chapter

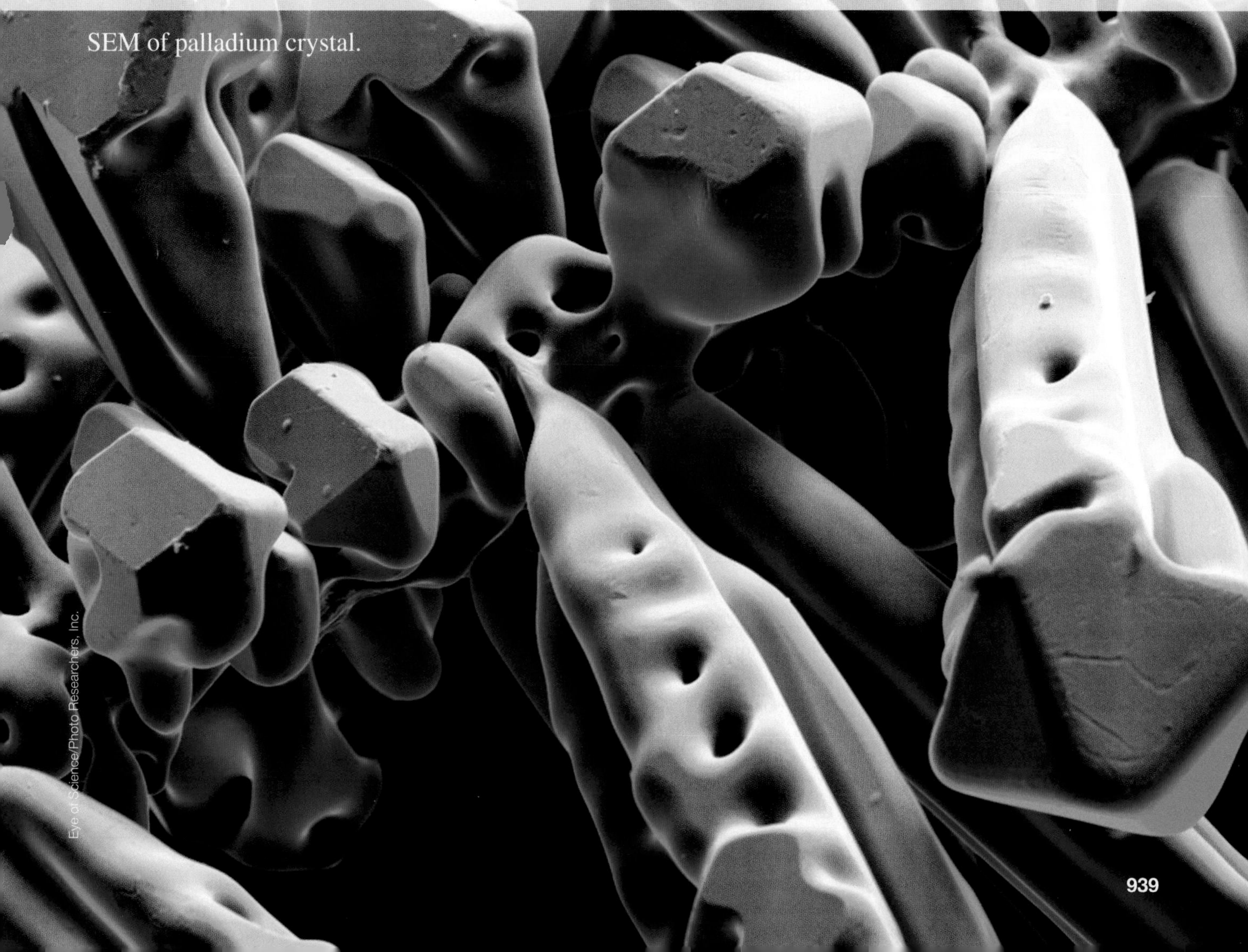

SEM of palladium crystal.

Sign in to OWL at www.cengage.com/owl to view tutorials and simulations, develop problem-solving skills, and complete online homework assigned by your professor.

Download mini lecture videos for key concept review and exam prep from OWL or purchase them from www.cengagebrain.com.

Transition metals have many uses in our society. Iron is used for steel; copper for electrical wiring and water pipes; titanium for paint; silver for photographic paper; manganese, chromium, vanadium, and cobalt as additives to steel; platinum for industrial and automotive catalysts; and so on.

One indication of the importance of transition metals is the great concern shown by the U.S. government for continuing the supply of these elements. The United States is a net importer of more than 60 "strategic and critical" minerals, including cobalt, manganese, platinum, palladium, and chromium. All these metals play a vital role in the U.S. economy and defense, but about 90% of the required amounts must be imported (Table 19.1).

In addition to being important in industry, transition metal ions play a vital role in living organisms. For example, complexes of iron provide for the transport and storage of oxygen, molybdenum and iron compounds are catalysts in nitrogen fixation, zinc is found in more than 150 biomolecules in humans, copper and iron play a crucial role in the respiratory cycle, and cobalt is found in essential biomolecules such as vitamin B_{12}.

In this chapter we explore the general properties of transition metals, paying particular attention to the bonding, structure, and properties of the complex ions of these metals.

19.1 The Transition Metals: A Survey

General Properties

One striking characteristic of the representative elements is that their chemistry changes markedly across a given period as the number of valence electrons changes. The chemical similarities occur mainly within the vertical groups. In contrast, *the transition metals show great similarities within a given period as well as within a given vertical group*. This difference occurs because the last electrons added to the transition metal elements are inner electrons: *d* electrons for the *d*-block transition metals and *f* electrons for the lanthanides and actinides. These inner *d* and *f* electrons cannot participate in bonding as readily as the valence *s* and *p* electrons can. Thus the chemistry of the transition elements is not as greatly affected by the gradual change in the number of electrons as is the chemistry of the representative elements.

Group designations are traditionally given on the periodic table for the *d*-block transition metals (Fig. 19.1). However, these designations do not relate as directly to the chemical behavior of these elements as do the designations for the representative elements (the A groups), so we will not use them.

Table 19.1

Some Transition Metals Important to the U.S. Economy and Defense

Metal	Uses	Percentage Imported
Chromium	Stainless steel (especially for parts exposed to corrosive gases and high temperatures)	~91%
Cobalt	High-temperature alloys in jet engines, magnets, catalysts, drill bits	~93%
Manganese	Steelmaking	~97%
Platinum and palladium	Catalysts	~87%

The Lanthanides: Critical Elements

Chemical Insights

The lanthanide elements are traditionally known as rare earths, but this name is misleading. Many of these elements are, in fact, quite abundant, and many of them are crucial to modern life. For example, typical hybrid cars contain about 10 kg of lanthanum in their nickel-metal hydride batteries and smaller amounts of neodymium, praseodymium, dysprosium, and terbium in their electric motors and generators. In addition, the electricity-producing windmills that seem to be springing up everywhere depend on magnets that weigh about 1 ton and contain hundreds of pounds of neodymium in the form of the neodymium-iron-boron compound $Nd_2Fe_{14}B$. These neodymium-iron-boron–based magnets are also found in today's computer disk drives.

Because these elements are so important to modern technology, they play a key role in global economics. Although China possesses only about 52% of the world's known rare-earth reserves, in 2010 nearly 100% of the world's rare-earth metals and more than 94% of rare-earth oxides came from China. China has gained this status because it produces the rare earths at a much lower cost than the rest of the world. Although the United States has rare-earth reserves estimated at 13% of the world's supply, it produces little, if any, rare earths. A plant in Mountain Pass, California, owned by Molycorp, produced ore composed of 49% cerium, 33% lanthanum, 13% praseodymium, 4% samarium, and 1% heavier rare-earth elements, but the plant was shut down in 2002 because it could not compete price-wise with China. However, China recently has begun to reduce its rare-earth exports, which has caused concern about supplies of these critical elements for the United States and other nations. As a result, Molycorp is searching for less expensive ways to produce rare earths at the Mountain Pass mine. The current plan is to reopen the Mountain Pass operation in 2012.

Although the rare-earth elements are not as economically important to the global economy as petroleum, they will become increasingly important to the global economy and world politics in the coming years.

As a class, the transition metals behave as typical metals, exhibiting metallic luster and relatively high electrical and thermal conductivities. Silver is the best conductor of heat and electrical current. However, copper is a close second, which explains copper's wide use in the electrical systems of homes and factories.

In spite of their many similarities, the properties of the transition metals do vary considerably. For example, tungsten with a melting point of 3400°C is

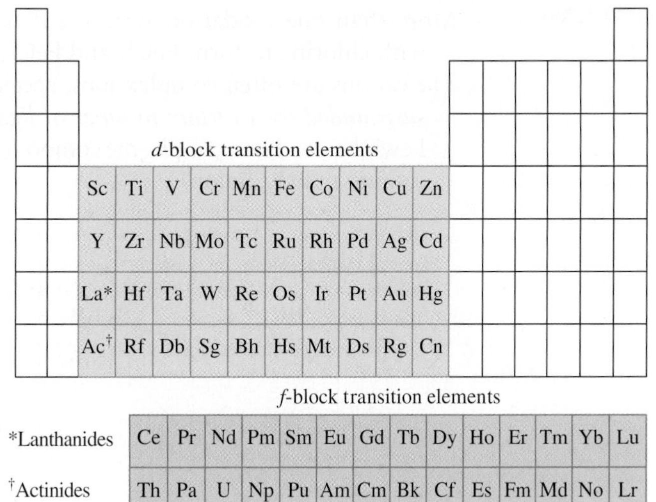

Figure 19.1
The position of the transition elements on the periodic table. The *d*-block elements correspond to filling the 3*d*, 4*d*, 5*d*, or 6*d* orbitals. The inner transition metals correspond to filling the 4*f* (lanthanides) or 5*f* (actinides) orbitals.

(clockwise from upper left) Calcite stalactites colored by traces of iron. Quartz is often colored by the presence of transition metals such as Mn, Fe, and Ni. Wulfenite contains $PbMoO_4$. Rhodochrosite is a mineral containing $MnCO_3$.

used for filaments in light bulbs; in contrast, mercury is a liquid at 25°C. Some transition metals such as iron and titanium are hard and strong and therefore make very useful structural materials; others such as copper, gold, and silver are relatively soft. The chemical reactivity of the transition metals also varies significantly. Some react readily with oxygen to form oxides. Of these metals, some, such as chromium, nickel, and cobalt, form oxides that adhere tightly to the metallic surface, thereby protecting the metal from further oxidation. Others, such as iron, form oxides that scale off, constantly exposing new metal to the corrosion process. On the other hand, the noble metals—primarily gold, silver, platinum, and palladium—do not readily form oxides.

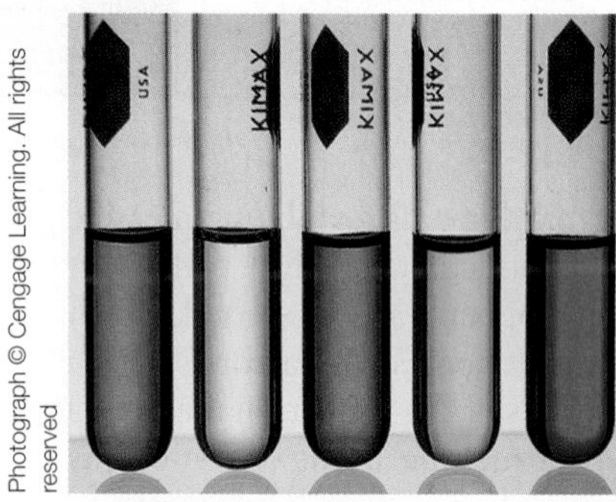

(from left to right) Aqueous solutions containing the metal ions Co^{2+}, Mn^{2+}, Cr^{3+}, Fe^{3+}, and Ni^{2+}.

In forming ionic compounds with nonmetals, the transition metals exhibit several typical characteristics:

More than one oxidation state is often found. For example, iron combines with chlorine to form $FeCl_2$ and $FeCl_3$.

The cations are often **complex ions**, species in which *the transition metal ion is surrounded by a certain number of ligands* (molecules or ions that behave as Lewis bases). For example, the compound $[Co(NH_3)_6]Cl_3$ contains $Co(NH_3)_6{}^{3+}$ cations and Cl^- anions:

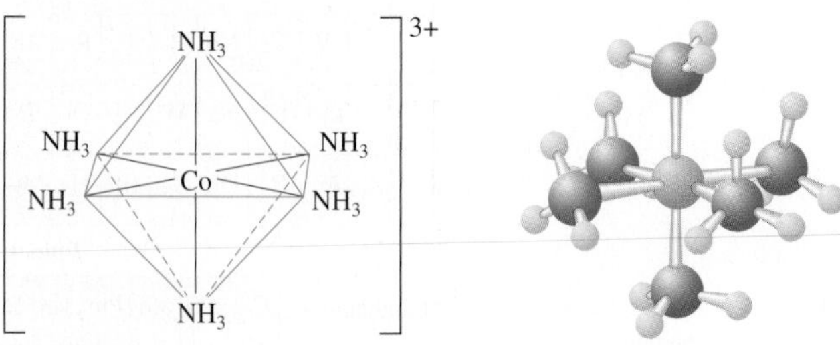

The $Co(NH_3)_6{}^{3+}$ ion

Table 19.2

Selected Properties of the First-Row Transition Metals

Property	Scandium	Titanium	Vanadium	Chromium	Manganese	Iron	Cobalt	Nickel	Copper	Zinc
Atomic number	21	22	23	24	25	26	27	28	29	30
Electron configuration*	$4s^23d^1$	$4s^23d^2$	$4s^23d^3$	$4s^13d^5$	$4s^23d^5$	$4s^23d^6$	$4s^23d^7$	$4s^23d^8$	$4s^13d^{10}$	$4s^23d^{10}$
Atomic radius† (pm)	162	147	134	130	135	126	125	124	128	138
Ionization energies (eV/atom)										
First	6.54	6.82	6.74	6.77	7.44	7.87	7.86	7.64	7.73	9.39
Second	12.80	13.58	14.65	16.50	15.64	16.18	17.06	18.17	20.29	17.96
Third	24.76	27.49	29.31	30.96	33.67	30.65	33.50	35.17	36.83	39.72
Reduction potential‡ (V)	−2.08	−1.63	−1.2	−0.91	−1.18	−0.44	−0.28	−0.23	+0.34	−0.76
Common oxidation states	+3	+2,+3, +4	+2,+3, +4,+5	+2,+3, +6	+2,+3, +4,+7	+2,+3	+2,+3	+2	+1,+2	+2
Melting point (°C)	1397	1672	1710	1900	1244	1530	1495	1455	1083	419
Density (g/cm³)	2.99	4.49	5.96	7.20	7.43	7.86	8.9	8.90	8.92	7.14
Electrical conductivity§	—	2	3	10	2	17	24	24	97	27

*Each atom has an argon inner-core configuration.
†Covalent atomic radii.
‡For the reduction process $M^{2+} + 2e^- \longrightarrow M$ (except for scandium, where the ion is Sc^{3+}).
§Compared with an arbitrarily assigned value of 100 for silver.

Most compounds are colored because the typical transition metal ion in a complex ion can absorb visible light of specific wavelengths.
Many compounds are paramagnetic (they contain unpaired electrons).

In this chapter we will concentrate on the **first-row transition metals** (scandium through zinc) because they are representative of the other transition series and because they have great practical significance. Some important properties of these elements are summarized in Table 19.2 and are discussed below.

Electron Configurations

Chromium has the electron configuration [Ar]$4s^13d^5$.

The electron configurations of the first-row transition metals were discussed in Section 12.13. The 3*d* orbitals begin to fill after the 4*s* orbital is complete—that is, after calcium ([Ar]$4s^2$). The first transition metal, *scandium,* has one electron in the 3*d* orbitals; the second, *titanium,* has two; and the third, *vanadium,* has three. We would expect *chromium,* the fourth transition metal, to have the electron configuration [Ar]$4s^23d^4$. However, the actual configuration is [Ar]$4s^13d^5$; it has a half-filled 4*s* orbital and a half-filled set of 3*d* orbitals (one electron in each of the five 3*d* orbitals). It is tempting to say that the configuration results because half-filled "shells" are especially stable. Although there are some reasons to think that this explanation might be valid, it is an oversimplification. For instance, tungsten, which is in the same vertical group as chromium, has the configuration [Xe]$6s^24f^{14}5d^4$, where half-filled *s* and *d* shells are not found. There are several similar cases that dispute the importance of half-filled shells.

The rigorous explanation of the electron configuration of chromium, which requires knowledge that is beyond the scope of an introductory course, involves the details of the electron interactions. It turns out that orbital energies are not constant for a given atom but depend on the way that the other orbitals in the atom are occupied. Thus there is no simple explanation for why chromium has the $4s^13d^5$ configuration rather than the $4s^23d^4$ configuration. Suffice it to say that for all the first-row transition metals, because the 4*s* and 3*d* orbitals have similar energies, the $4s^23d^n$ and the $4s^13d^{n+1}$ configurations have similar energies. For most elements, $4s^23d^n$ is lower in energy, but for chromium and for copper ($[Ar]4s^13d^{10}$), the $4s^13d^{n+1}$ configuration is more stable.

In contrast to the neutral transition metals, where the 3*d* and 4*s* orbitals have very similar energies, the *energy of the 3d orbitals in transition metal ions is significantly less than that of the 4s orbital.* This means that the electrons remaining after the ion is formed occupy the 3*d* orbitals, since they are lower in energy. *First-row transition metal ions do not have 4s electrons.* For example, manganese has the configuration $[Ar]4s^23d^5$, but that of Mn^{2+} is $[Ar]3d^5$. The neutral titanium atom has the configuration $[Ar]4s^23d^2$, but that of Ti^{3+} is $[Ar]3d^1$.

Oxidation States and Ionization Energies

The transition metals can form a variety of ions by losing one or more electrons. The common oxidation states of these elements are shown in Table 19.2. Note that for the first five metals the maximum possible oxidation state corresponds to the loss of all the 4*s* and 3*d* electrons. For example, the maximum oxidation state of chromium ($[Ar]4s^13d^5$) is +6. Toward the right end of the period, the maximum oxidation states are not observed; in fact, the 2+ ions are the most common. The higher oxidation states are not seen for these metals because the 3*d* orbitals become lower in energy as the nuclear charge increases, making the electrons increasingly difficult to remove. From Table 19.2 we see that ionization energy increases gradually from left to right across the period. However, the third ionization energy (corresponding to the removal of an electron from a 3*d* orbital) increases faster than the first ionization energy, clear evidence of the significant decrease in the energy of the 3*d* orbitals in going across the period (Fig. 19.2).

Standard Reduction Potentials

When a metal acts as a *reducing agent,* the half-reaction is

$$M \longrightarrow M^{n+} + ne^-$$

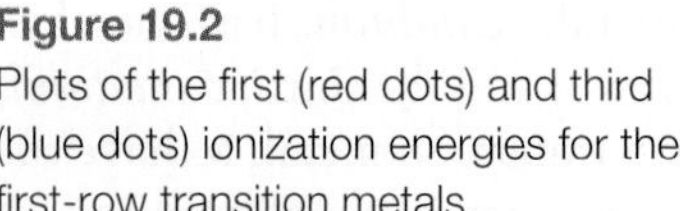
Figure 19.2
Plots of the first (red dots) and third (blue dots) ionization energies for the first-row transition metals.

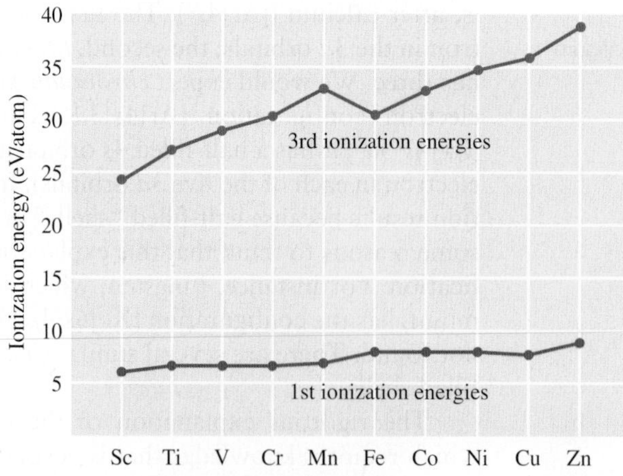

Table 19.3

Relative Reducing Abilities of the First-Row Transition Metals in Aqueous Solution

Reaction	Potential (V)
$Sc \rightarrow Sc^{3+} + 3e^-$	2.08
$Ti \rightarrow Ti^{2+} + 2e^-$	1.63
$V \rightarrow V^{2+} + 2e^-$	1.2
$Mn \rightarrow Mn^{2+} + 2e^-$	1.18
$Cr \rightarrow Cr^{2+} + 2e^-$	0.91
$Zn \rightarrow Zn^{2+} + 2e^-$	0.76
$Fe \rightarrow Fe^{2+} + 2e^-$	0.44
$Co \rightarrow Co^{2+} + 2e^-$	0.28
$Ni \rightarrow Ni^{2+} + 2e^-$	0.23
$Cu \rightarrow Cu^{2+} + 2e^-$	−0.34

(Reducing ability increases upward.)

This is the reverse of the conventional listing of half-reactions in tables. Thus, when we rank the transition metals in order of reducing ability, it is most convenient to reverse the reactions and the signs given in Table 19.2. The metal with the most positive potential is thus the best reducing agent. The transition metals are listed in order of reducing ability in Table 19.3.

Recall that $\mathscr{E}°$ is zero by definition for the process

$$2H^+ + 2e^- \longrightarrow H_2$$

Therefore, all the metals except copper can reduce H^+ ions in 1 *M* aqueous solutions of strong acid to hydrogen gas:

$$M(s) + 2H^+(aq) \longrightarrow H_2(g) + M^{2+}(aq)$$

As Table 19.3 shows, the reducing abilities of the first-row transition metals generally decrease going from left to right across the period. Only chromium and zinc do not follow this trend.

The 4*d* and 5*d* Transition Series

In comparing the 3*d*, 4*d*, and 5*d* transition series, it is instructive to consider the atomic radii of these elements (Fig. 19.3). Note that there is a general, although not regular, decrease in size going from left to right across each of the series. Also note that although there is a significant increase in radius in going from the 3*d* to the 4*d* metals, the 4*d* and 5*d* metals are remarkably similar in size. This

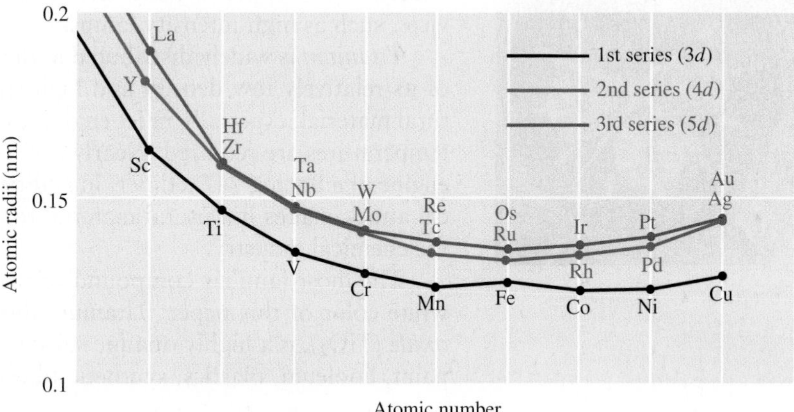

Figure 19.3
Atomic radii of the 3*d*, 4*d*, and 5*d* transition series.

latter phenomenon is the result of the **lanthanide contraction.** In the **lanthanide series,** which consists of the elements between lanthanum and hafnium (Fig. 19.1), electrons are filling the 4*f* orbitals. Since the 4*f* orbitals are buried in the interior of these atoms, the additional electrons do not add to the atomic size. In fact, the increasing nuclear charge (remember that a proton is added to the nucleus with the addition of each electron) causes the radii of the lanthanide elements to decrease significantly going from left to right. This lanthanide contraction just offsets the normal increase in size due to changing from one principal quantum level to another. Thus the 5*d* elements, instead of being significantly larger than the 4*d* elements, are almost identical to them in size. This leads to a great similarity in the chemistry of the 4*d* and 5*d* elements in a given vertical group. For example, the chemical properties of hafnium and zirconium are remarkably similar, and they always occur together in nature. Their separation, which is probably more difficult than the separation of any other pair of elements, often requires fractional distillation of their compounds.

Niobium was originally called columbium and is still occasionally referred to by that name.

In general, the differences between the 4*d* and 5*d* elements in a group increase gradually going from left to right. For example, niobium and tantalum are also quite similar, but less so than zirconium and hafnium.

Although generally less well known than the 3*d* elements, the 4*d* and 5*d* transition metals have certain very useful properties. For example, zirconium and zirconium oxide (ZrO_2) are quite resistant to high temperatures and are used, along with niobium and molybdenum alloys, for space vehicle parts that are exposed to high temperatures during reentry into the earth's atmosphere. Niobium and molybdenum are also important alloying materials for certain types of steel. Tantalum, which has a high resistance to attack by body fluids, is often used for surgical clips. The *platinum group metals*—ruthenium, osmium, rhodium, iridium, palladium, and platinum—are all quite similar and are widely used as catalysts for many types of industrial processes.

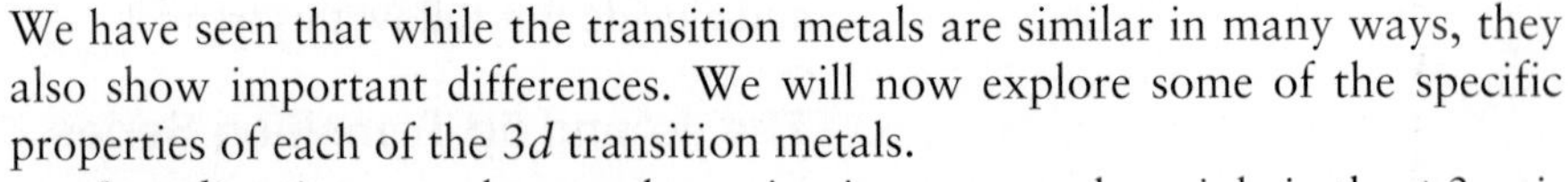

19.2 The First-Row Transition Metals

Bobby Ray Strange/Shutterstock.com

A titanium watch.

We have seen that while the transition metals are similar in many ways, they also show important differences. We will now explore some of the specific properties of each of the 3*d* transition metals.

Scandium is a rare element that exists in compounds mainly in the +3 oxidation state—for example, in $ScCl_3$, Sc_2O_3, and $Sc_2(SO_4)_3$. The chemistry of scandium strongly resembles that of the lanthanides, with most of its compounds being colorless and diamagnetic. This is not surprising, since Sc^{3+} has no *d* electrons. As we will see in Section 19.6, the color and magnetism of transition metal compounds usually arise from the *d* electrons on the metal ion. Scandium metal, which can be prepared by electrolysis of molten $ScCl_3$, is not widely used because of its rarity. However, it is found in some electronic devices, such as high-intensity lamps.

Titanium is widely distributed in the earth's crust (0.6% by mass). Because of its relatively low density and high strength, titanium is an excellent structural material, especially in jet engines, where light weight and stability at high temperatures are required. Nearly 5000 kg of titanium alloys are used in each engine of a Boeing 747 jetliner. In addition, the resistance of titanium to chemical attack makes it a useful material for pipes, pumps, and reaction vessels in the chemical industry.

The most familiar compound of titanium is no doubt responsible for the white color of this paper. Titanium dioxide, or more correctly, *titanium(IV) oxide* (TiO_2), is a highly opaque substance used as the white pigment in paper, paint, linoleum, plastics, synthetic fibers, whitewall tires, and cosmetics (sun-

Liquid titanium(IV) chloride being added to water, forming a cloud of solid titanium oxide and hydrochloric acid.

screens, for example). About 700,000 tons of TiO_2 are used annually in these and similar products. Titanium(IV) oxide is widely dispersed in nature, but the main ores are rutile (impure TiO_2) and ilmenite ($FeTiO_3$). Rutile is processed by treatment with chlorine to form volatile $TiCl_4$, which is then separated from the impurities and burned to form TiO_2:

$$TiCl_4(g) + O_2(g) \longrightarrow TiO_2(s) + 2Cl_2(g)$$

Ilmenite is treated with sulfuric acid to form a soluble sulfate,

$$FeTiO_3(s) + 2H_2SO_4(aq) \longrightarrow Fe^{2+}(aq) + TiO^{2+}(aq) + 2SO_4^{2-}(aq) + 2H_2O(l)$$

When this aqueous mixture is allowed to stand under vacuum, solid $FeSO_4 \cdot 7H_2O$ forms and is removed. The mixture remaining is then heated, and the insoluble titanium(IV) oxide hydrate ($TiO_2 \cdot H_2O$) forms. The water of hydration is driven off by heating to form pure TiO_2:

$$TiO_2 \cdot H_2O(s) \xrightarrow{\text{Heat}} TiO_2(s) + H_2O(g)$$

In its compounds titanium most often exists in the +4 oxidation state. Examples are TiO_2 and $TiCl_4$, the latter a colorless liquid (bp = 137°C) that fumes in moist air to produce TiO_2:

$$TiCl_4(l) + 2H_2O(l) \longrightarrow TiO_2(s) + 4HCl(g)$$

Titanium(III) compounds can be produced by reduction of compounds containing titanium in the +4 state. In aqueous solution Ti^{3+} exists as the purple $Ti(H_2O)_6^{3+}$ ion, which is slowly oxidized to titanium(IV) by air. Titanium(II) is not stable in aqueous solution but does exist in the solid state in compounds such as TiO and dihalides of the type TiX_2.

The manufacture of sulfuric acid was discussed in Section 3.8.

Vanadium is widely dispersed throughout the earth's crust (0.02% by mass). It is used mostly in alloys with other metals such as iron (80% of vanadium is used in steel) and titanium. Vanadium(V) oxide (V_2O_5) is used as an industrial catalyst in the production of materials such as sulfuric acid.

Pure vanadium can be obtained from the electrolytic reduction of fused salts, such as VCl_2, to produce a metal similar to titanium that is steel gray, hard, and corrosion-resistant. Often the pure element is not required for alloying. For example, *ferrovanadium,* produced by reducing a mixture of V_2O_5 and Fe_2O_3 with aluminum, is added to iron to form *vanadium steel,* a hard steel used for engine parts and axles.

The most common oxidation state for vanadium is +5.

The principal oxidation state of vanadium is +5, found in compounds such as the orange V_2O_5 (mp = 650°C) and the colorless VF_5 (mp = 19.5°C). The oxidation states ranging from +5 to +2 all exist in aqueous solution (Table 19.4). The higher oxidation states, +5 and +4, do not exist as hydrated ions of the type $V^{n+}(aq)$ because these highly charged ions cause the attached water molecules to be very acidic. The H^+ ions are lost to give the oxycations VO_2^+ and VO^{2+}. The hydrated V^{3+} and V^{2+} ions are easily oxidized and thus can function as reducing agents in aqueous solution.

Table 19.4

Oxidation States and Species for Vanadium in Aqueous Solution

Oxidation State of Vanadium	Species in Aqueous Solution
+5	VO_2^+ (yellow)
+4	VO^{2+} (blue)
+3	$V^{3+}(aq)$ (blue-green)
+2	$V^{2+}(aq)$ (violet)

Although *chromium* is relatively rare, it is a very important industrial material. The chief ore of chromium, chromite ($FeCr_2O_4$), can be reduced by carbon to give *ferrochrome,*

$$FeCr_2O_4(s) + 4C(s) \longrightarrow \underbrace{Fe(s) + 2Cr(s)}_{\text{Ferrochrome}} + 4CO(g)$$

which can be added directly to iron in the steelmaking process. Chromium metal, which is often used to plate steel, is hard and brittle. It maintains a bright surface by developing a tough, invisible oxide coating.

Chemical Insights — Titanium Makes Great Bicycles

One of the most interesting characteristics of the world of bicycling is the competition among various frame materials. Bicycle frames are now built from steel, aluminum, carbon fiber composites, and titanium, with each material having advantages and disadvantages. Steel is strong, economical, adaptable, and (unfortunately) "rustable." Aluminum is light and stiff but has relatively low fatigue limits (resistance to repeated stresses). Carbon fiber composites have amazing strength-to-mass ratios and have shock- and vibration-dampening properties superior to any metal; however, they are very expensive. Titanium has a density approximately 43% less than that of steel, a yield strength (when alloyed with metals such as aluminum and tin) that is 30% greater than that of steel, an extraordinary resistance to fatigue, and a high resistance to corrosion but is expensive and difficult to work.

Of all these materials, titanium gives the bicycle that fanatics seem to love the most (Fig. 19.4). After their first ride on a bicycle with a titanium frame, most experienced cyclists find themselves shaking their heads and searching hard for the right words to describe the experience. Typically, the word *magic* is used a great deal in the ensuing description.

Jodi Jacobson/Peter Arnold, Inc./PhotoLibrary

Figure 19.4
A bicycle with a titanium frame.

The magic of titanium results from its combination of toughness, stretchability, and resilience. A bicycle that is built stiff to resist pedaling loads usually responds by giving a harsh, uncomfortable ride. A titanium bike is very stiff against high pedaling torques, but it seems to transmit much less road shock than bikes made of competitive materials. Why titanium excels in dampening vibrations is not entirely clear. Despite titanium's significantly lower density than steel, shock waves travel more slowly in titanium than steel. Whatever the explanation for its shock-absorbing abilities, titanium provides three things that cyclists find crucial: light weight, stiffness, and a smooth ride—magic.

Titanium is quite abundant in the earth's crust, ranking ninth of all the elements and second among the transition elements. The metallurgy of titanium presents special challenges. Carbon, the reducing agent most commonly used to obtain metals from their oxide ores, cannot be used because it forms intractable interstitial carbides with titanium. These carbides are extraordinarily hard and have melting points above 3000°C. However, if chlorine gas is used in conjunction with carbon to treat the ore, volatile $TiCl_4$ is formed, which can be distilled off and then reduced with magnesium or sodium at approximately 1000°C to form a titanium "sponge." This sponge is then ground up, cleaned with aqua regia (a 1:3 mixture of concentrated HNO_3 and concentrated HCl), melted under a blanket of inert gas (to prevent reaction with oxygen), and cast into ingots. Titanium, a lustrous, silvery metal with a high melting point (1667°C), crystallizes in a hexagonal closest packed structure. Because titanium tends to become quite brittle when trace impurities such as C, N, and O are present, it must be fabricated with great care.

Titanium's unusual ability to stretch makes it hard to machine. It tends to push away even from a very sharp cutting blade, giving a rather unpredictable final dimension. Also, because titanium is embrittled by reaction with oxygen, all welding operations must be carried out under a shielding gas such as argon.

However, the bicycle that results is worth all these difficulties. One woman described a titanium bicycle as "the one God rides on Sunday."

Table 19.5

Typical Chromium Compounds

Oxidation State of Chromium	Examples of Compounds (X = halogen)
+2	CrX_2
+3	CrX_3
	Cr_2O_3 (green)
	$Cr(OH)_3$ (blue-green)
+6	$K_2Cr_2O_7$ (orange)
	Na_2CrO_4 (yellow)
	CrO_3 (red)

Chromium commonly forms compounds in which it has an oxidation state of +2, +3, or +6, as shown in Table 19.5. The Cr^{2+} (chromous) ion is a powerful reducing agent in aqueous solution. In fact, traces of O_2 in other gases can be removed from those gases by bubbling the gaseous mixture through a Cr^{2+} solution:

$$4Cr^{2+}(aq) + O_2(g) + 4H^+(aq) \longrightarrow 4Cr^{3+}(aq) + 2H_2O(l)$$

The chromium(VI) species are excellent oxidizing agents, especially in acidic solution, where chromium(VI) as the dichromate ion ($Cr_2O_7^{2-}$) is reduced to the Cr^{3+} ion:

$$Cr_2O_7^{2-}(aq) + 14H^+(aq) + 6e^- \longrightarrow 2Cr^{3+}(aq) + 7H_2O(l) \qquad \mathscr{E}° = 1.33\text{ V}$$

The oxidizing ability of the dichromate ion is strongly pH-dependent; it increases as $[H^+]$ increases, as predicted by Le Châtelier's principle. In basic solution chromium(VI) exists as the chromate ion, a much less powerful oxidizing agent:

$$CrO_4^{2-}(aq) + 4H_2O(l) + 3e^- \longrightarrow Cr(OH)_3(s) + 5OH^-(aq) \qquad \mathscr{E}° = -0.13\text{ V}$$

The structures of the $Cr_2O_7^{2-}$ and CrO_4^{2-} ions are shown in Fig. 19.5.

Red chromium(VI) oxide (CrO_3) dissolves in water to give a strongly acidic, red-orange solution:

$$2CrO_3(s) + H_2O(l) \longrightarrow 2H^+(aq) + Cr_2O_7^{2-}(aq)$$

It is possible to precipitate bright orange dichromate salts, such as $K_2Cr_2O_7$, from these solutions. When made basic, the solution turns yellow and chromate salts such as Na_2CrO_4 can be obtained. A mixture of chromium(VI) oxide and concentrated sulfuric acid, commonly called *cleaning solution,* is a powerful oxidizing medium that can remove organic materials from analytical glassware, yielding a very clean surface.

John Cunningham/Visuals Unlimited

Manganese nodules on the sea floor.

Manganese is relatively abundant (0.1% of the earth's crust), although no significant sources are found in the United States. The most common use of manganese is in the production of an especially hard steel used for rock crushers, bank vaults, and armor plate. One interesting source of manganese is *manganese nodules* found on the ocean floor. These roughly spherical "rocks" contain mixtures of manganese and iron oxides as well as smaller amounts of other metals such as cobalt, nickel, and copper. Apparently, the nodules were formed at least partly by the action of marine organisms. Because of the abundance of these nodules, there is much interest in developing economical methods for their recovery and processing.

Manganese can exist in all oxidation states from +2 to +7, although +2 and +7 are the most common. Manganese(II) forms an extensive series of salts

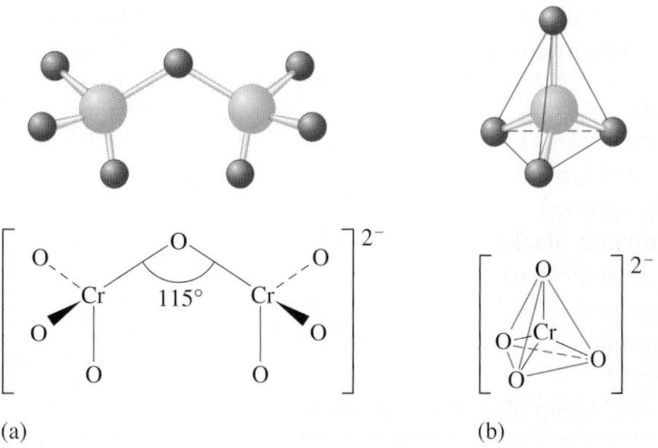

Figure 19.5

The structures of the chromium(VI) anions: (a) $Cr_2O_7^{2-}$, which exists in acidic solution; and (b) CrO_4^{2-}, which exists in basic solution.

Table 19.6

Some Compounds of Manganese in Its Most Common Oxidation States

Oxidation State of Manganese	Examples of Compounds
+2	$Mn(OH)_2$ (pink) MnS (salmon) $MnSO_4$ (reddish) $MnCl_2$ (pink)
+4	MnO_2 (dark brown)
+7	$KMnO_4$ (purple)

with all the common anions. In aqueous solution Mn^{2+} forms $Mn(H_2O)_6^{2+}$, which has a light pink color. Manganese(VII) is found in the intensely purple permanganate ion (MnO_4^-). Widely used as an analytical reagent in acidic solution, the MnO_4^- ion behaves as a strong oxidizing agent, producing Mn^{2+}:

$$MnO_4^-(aq) + 8H^+(aq) + 5e^- \longrightarrow Mn^{2+}(aq) + 4H_2O(l) \qquad \mathscr{E}° = 1.51 \text{ V}$$

Several typical compounds of manganese are listed in Table 19.6.

Iron is the most abundant heavy metal (constituting 4.7% of the earth's crust) and the most important to our civilization. It is a white, lustrous, not particularly hard metal that is very reactive toward oxidizing agents. For example, in moist air it is rapidly oxidized by oxygen to form rust, a mixture of iron oxides.

The chemistry of iron mainly involves its +2 and +3 oxidation states. Typical compounds are shown in Table 19.7. In aqueous solutions, iron(II) salts are generally light green because of the presence of $Fe(H_2O)_6^{2+}$. Although the $Fe(H_2O)_6^{3+}$ ion is colorless, aqueous solutions of iron(III) salts are usually yellow to brown in color because of the presence of $Fe(OH)(H_2O)_5^{2+}$. This latter ion results from the acidity of $Fe(H_2O)_6^{3+}$ ($K_a = 6 \times 10^{-3}$):

$$Fe(H_2O)_6^{3+}(aq) \rightleftharpoons Fe(OH)(H_2O)_5^{2+}(aq) + H^+(aq)$$

Although *cobalt* is relatively rare, it is found in ores such as smaltite ($CoAs_2$) and cobaltite ($CoAsS$) in large enough concentrations to make its production economically feasible. Cobalt is a hard, bluish white metal mainly used in alloys such as stainless steel and stellite, which is an alloy of iron, copper, and tungsten used in surgical instruments.

The chemistry of cobalt involves mainly its +2 and +3 oxidation states, although compounds containing cobalt in the 0, +1, and +4 oxidation states are known. Aqueous solutions of cobalt(II) salts contain the $Co(H_2O)_6^{2+}$ ion, which has a characteristic rose color. Cobalt forms a wide variety of coordination compounds, many of which will be discussed in later sections of this chapter. Some typical cobalt compounds are listed in Table 19.8.

Nickel ranks twenty-fourth in elemental abundance in the earth's crust. It is found in ores, where it is combined mainly with arsenic, antimony, and sulfur. Nickel metal, a silver-white substance with high electrical and thermal conductivities, is quite resistant to corrosion and is often used for plating more active metals. Nickel is also widely used in the production of alloys such as steel.

Nickel is found almost exclusively in the +2 oxidation state in its compounds. Aqueous solutions of nickel(II) salts contain the $Ni(H_2O)_6^{2+}$ ion,

Table 19.7

Typical Compounds of Iron

Oxidation State of Iron	Examples of Compounds
+2	FeO (black) FeS (brownish black) $FeSO_4 \cdot 7H_2O$ (green) $K_4Fe(CN)_6$ (yellow)
+3	$FeCl_3$ (brownish black) Fe_2O_3 (reddish brown) $K_3Fe(CN)_6$ (red) $Fe(SCN)_3$ (red)
+2, +3 (mixture)	Fe_3O_4 (black) $KFe[Fe(CN)_6]$ (deep blue, Prussian blue)

Table 19.8

Typical Compounds of Cobalt

Oxidation State of Cobalt	Examples of Compounds
+2	$CoSO_4$ (dark blue) $[Co(H_2O)_6]Cl_2$ (pink) $[Co(H_2O)_6](NO_3)_2$ (red) CoS (black) CoO (greenish brown)
+3	CoF_3 (brown) Co_2O_3 (charcoal) $K_3[Co(CN)_6]$ (yellow) $[Co(NH_3)_6]Cl_3$ (yellow)

An aqueous solution containing the Ni^{2+} ion.

Table 19.9

Typical Compounds of Nickel

Oxidation State of Nickel	Examples of Compounds
+2	$NiCl_2$ (yellow) $[Ni(H_2O)_6]Cl_2$ (green) NiO (greenish black) NiS (black) $[Ni(H_2O)_6]SO_4$ (green) $[Ni(NH_3)_6](NO_3)_2$ (blue)

which has a characteristic emerald green color. Coordination compounds of nickel(II) will be discussed later in this chapter. Some typical nickel compounds are shown in Table 19.9.

Copper, widely distributed in nature in ores containing sulfides, arsenides, chlorides, and carbonates, is valued for its high electrical conductivity and its resistance to corrosion. It is widely used for plumbing, and 50% of all copper produced annually is used for electrical applications. Copper is a major constituent in several well-known alloys (Table 19.10).

Copper roofs and bronze statues, such as the Statue of Liberty, turn green in air because $Cu_3(OH)_4SO_4$ and $Cu_4(OH)_6SO_4$ form.

Although copper is not highly reactive (it will not reduce H^+ to H_2, for example), this reddish colored metal does slowly corrode in air, producing the characteristic green *patina* consisting of basic copper sulfate,

$$3Cu(s) + 2H_2O(l) + SO_2(g) + 2O_2(g) \longrightarrow \underset{\text{Basic copper sulfate}}{Cu_3(OH)_4SO_4(s)}$$

and other similar compounds.

The chemistry of copper principally involves the +2 oxidation state, but many compounds containing copper(I) are also known. Aqueous solutions of copper(II) salts are a characteristic bright blue color because of the presence of the $Cu(H_2O)_6^{2+}$ ion. Table 19.11 lists some typical copper compounds.

Although trace amounts of copper are essential for life, this substance is quite toxic in large amounts; copper salts are used to kill bacteria, fungi, and algae. For example, paints containing copper are used on ship hulls to prevent fouling by marine organisms.

Widely dispersed in the earth's crust, *zinc* is mainly refined from sphalerite [(ZnFe)S], which often occurs with galena (PbS). Zinc is a white, lustrous, very active metal that behaves as an excellent reducing agent and that tarnishes

Table 19.10

Alloys Containing Copper

Alloy	Composition (% by mass in parentheses)
Brass	Cu (20–97), Zn (2–80), Sn (0–14), Pb (0–12), Mn (0–25)
Bronze	Cu (50–98), Sn (0–35), Zn (0–29), Pb (0–50), P (0–3)
Sterling silver	Cu (7.5), Ag (92.5)
Gold (18-karat)	Cu (5–14), Au (75), Ag (10–20)
Gold (14-karat)	Cu (12–28), Au (58), Ag (4–30)

Table 19.11

Typical Compounds of Copper

Oxidation State of Copper	Examples of Compounds
+1	Cu_2O (red) Cu_2S (black) CuCl (white)
+2	CuO (black) $CuSO_4 \cdot 5H_2O$ (blue) $CuCl_2 \cdot 2H_2O$ (green) $[Cu(H_2O)_6](NO_3)_2$ (blue)

rapidly. About 90% of the zinc produced is used for galvanizing steel. Zinc forms colorless salts containing Zn^{2+} ions.

19.3 Coordination Compounds

Keith Weller/USDA Photo

Alpine pennycress thrives on soils containing zinc and cadmium. The metals can be harvested by extracting them from the plant's shoots.

Transition metal ions characteristically form **coordination compounds,** which are usually colored and often paramagnetic. A coordination compound typically consists of a *complex ion,* a transition metal ion with its attached ligands (see Section 8.10), and **counter ions,** anions or cations needed to produce a compound with no net charge. The substance $[Co(NH_3)_5Cl]Cl_2$ is a typical coordination compound. The square brackets indicate the composition of the complex ion—in this case $Co(NH_3)_5Cl^{2+}$—and the two Cl^- counter ions are shown outside the brackets. Note that in this compound one Cl^- acts as a ligand along with the five NH_3 molecules. In the solid state, this compound consists of the large $Co(NH_3)_5Cl^{2+}$ cations with twice as many Cl^- anions, all packed together as efficiently as possible. When dissolved in water, the solid behaves like any ionic solid; the cations and anions are assumed to separate and move about independently:

$$[Co(NH_3)_5Cl]Cl_2(s) \xrightarrow{H_2O} Co(NH_3)_5Cl^{2+}(aq) + 2Cl^-(aq)$$

Coordination compounds have been known since about 1700, but their true nature was not understood until the 1890s when a young Swiss chemist named Alfred Werner proposed that transition metal ions have two types of valence (combining ability). One type of valence, which Werner called the secondary valence, refers to the ability of a metal ion to bind to Lewis bases (ligands) to form complex ions. The other type, the primary valence, refers to the ability of the metal ion to form ionic bonds with oppositely charged ions. Thus Werner explained that the compound originally written as $CoCl_3 \cdot 5NH_3$ is actually $[Co(NH_3)_5Cl]Cl_2$, in which the Co^{3+} ion has a primary valence of 3, satisfied by the three Cl^- ions, and a secondary valence of 6, satisfied by the six ligands (five NH_3 and one Cl^-). We now call the primary valence the **oxidation state** and the secondary valence the **coordination number.** The latter reflects the number of bonds formed between the metal ion and the ligands in the complex ion.

Coordination Number

The number of bonds formed by metal ions to ligands in complex ions varies from two to eight, depending on the size, charge, and electron configuration of the transition metal ion. As shown in Table 19.12, 6 is the most common co-

Table 19.12

Typical Coordination Numbers for Some Common Metal Ions

M^+	Coordination Numbers	M^{2+}	Coordination Numbers	M^{3+}	Coordination Numbers
Cu^+	2, 4	Mn^{2+}	4, 6	Sc^{3+}	6
Ag^+	2	Fe^{2+}	6	Cr^{3+}	6
Au^+	2, 4	Co^{2+}	4, 6	Co^{3+}	6
		Ni^{2+}	4, 6	Au^{3+}	4
		Cu^{2+}	4, 6		
		Zn^{2+}	4, 6		

ordination number, followed closely by 4, with a few metal ions showing a coordination number of 2. Many metal ions show more than one coordination number, and there is really no simple way to predict what the coordination number will be in a particular case. The typical geometries for the various common coordination numbers are shown in Fig. 19.6. Note that six ligands produce an octahedral arrangement around the metal ion. Four ligands can form either a tetrahedral or a square planar arrangement, and two ligands give a linear structure.

Ligands

The Lewis definition of acids and bases also encompasses the Brønsted–Lowry definition. For example,

$$H^+ + :NH_3 \longrightarrow NH_4^+$$

A **ligand** is a *neutral molecule or ion having a lone pair that can be used to form a bond to a metal ion.* In the 1920s G. N. Lewis suggested a general definition for acid–base behavior in terms of electron pairs. Lewis defined an **acid** as an **electron-pair acceptor** and a **base** as an **electron-pair donor.** Because a ligand donates an electron pair to an empty orbital on a metal ion, the formation of a metal–ligand bond can be described as the interaction between a Lewis base (the ligand) and a Lewis acid (the metal ion). This is often called a **coordinate covalent bond.**

A *ligand that can form one bond to a metal ion* is called a **monodentate ligand,** or a **unidentate ligand** (from root words meaning "one tooth"). Examples of unidentate ligands are shown in Table 19.13.

Table 19.13
Some Common Ligands

Type	Examples	
Unidentate/monodentate	H_2O NH_3 CN^- NO_2^- (nitrite)	SCN^- (thiocyanate) OH^- X^- (halides)
Bidentate	Oxalate $(-)\ddot{O}(O{=})C-C({=}O)\ddot{O}(-)$, both O bound to M	Ethylenediamine (en) H_2C-CH_2, $H_2\ddot{N}$ and $\ddot{N}H_2$ bound to M
Polydentate	Diethylenetriamine (dien) $H_2\ddot{N}-(CH_2)_2-\ddot{N}H-(CH_2)_2-\ddot{N}H_2$ Three coordinating atoms	
	Ethylenediaminetetraacetate (EDTA) $[(-):\ddot{O}-C(=O)-H_2C]_2\ddot{N}-(CH_2)_2-\ddot{N}[CH_2-C(=O)-\ddot{O}:(-)]_2$ Six coordinating atoms	

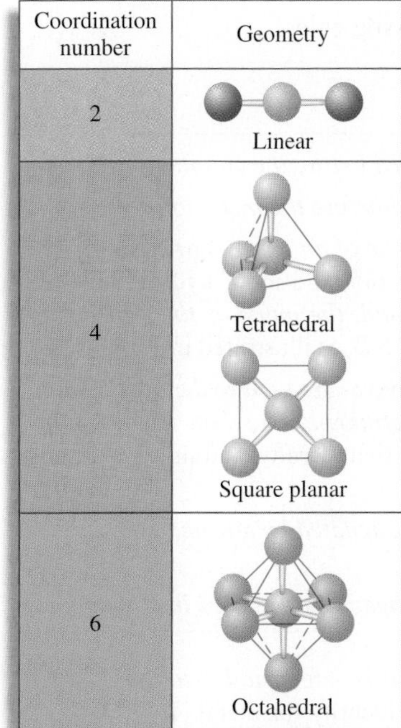

Figure 19.6
The ligand arrangements for coordination numbers 2, 4, and 6.

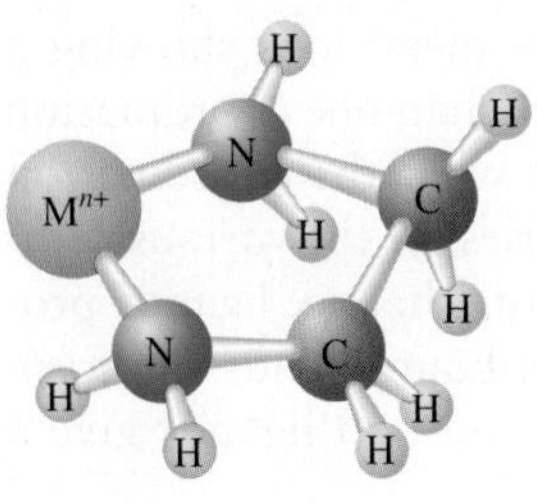

(a)

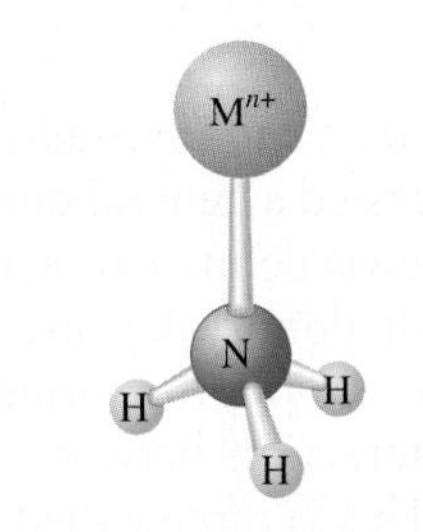

(b)

Figure 19.7
(a) The bidentate ligand ethylenediamine can bond to the metal ion through the lone pair on each nitrogen atom, thus forming two coordinate covalent bonds. (b) Ammonia is a monodentate ligand.

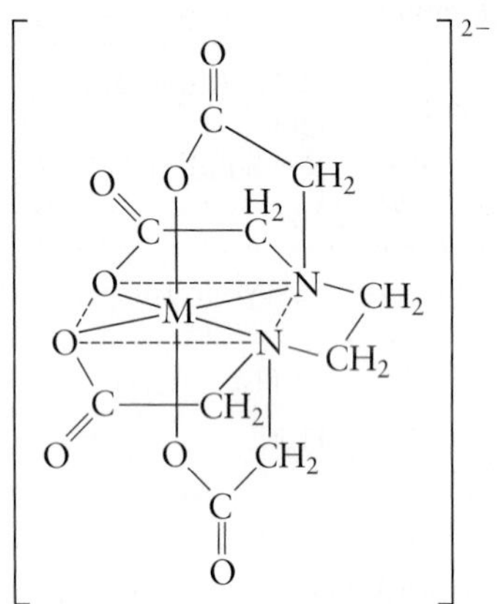

Figure 19.8
The coordination of EDTA with a 2+ metal ion.

Some ligands have more than one atom with a lone pair that can be used to bond to a metal ion. Such ligands are said to be **chelating ligands,** or **chelates** (from the Greek word *chela,* meaning "claw"). A ligand that can form two bonds to a metal ion is called a **bidentate ligand.** A very common bidentate ligand is ethylenediamine (abbreviated en), which is shown coordinating to a metal ion in Fig. 19.7(a). Note the relationship between this ligand and the unidentate ligand ammonia [Fig. 19.7(b)]. Oxalate, another typical bidentate ligand, is shown in Table 19.13.

Ligands that can form more than two bonds to a metal ion are called **polydentate ligands.** Some ligands can form as many as six bonds to a metal ion. The best-known example of such a ligand is ethylenediaminetetraacetate (abbreviated EDTA), which is shown in Table 19.13. This ligand virtually surrounds the metal ion (Fig. 19.8), coordinating through six atoms (a *hexadentate ligand*). As might be expected from the large number of coordination sites, EDTA forms very stable complex ions with most metal ions. It is therefore useful as a "scavenger" to remove toxic heavy metals such as lead from the human body. It is also used as a reagent to analyze solutions for their metal ion content. EDTA is found in countless consumer products, including soda, beer, salad dressings, bar soaps, and most cleaners. In these products EDTA ties up trace metal ions that would otherwise catalyze decomposition and produce unwanted precipitates.

Even more complicated ligands are found in biological systems, where metal ions play crucial roles in catalyzing reactions, transferring electrons, and transporting and storing oxygen. A discussion of these complex ligands will follow in Section 19.8.

Nomenclature

In Werner's lifetime no system was used to name coordination compounds. Names of the compounds were commonly based on colors and names of discoverers. As the field expanded and more coordination compounds were identified, an orderly system of nomenclature became necessary. A simplified version of this system is summarized by the following rules.

Rules for Naming Coordination Compounds

1. As with any ionic compound, *the cation is named before the anion.*
2. In naming a complex ion, *the ligands are named before the metal ion.*
3. In naming ligands, *an* o *is added to the root name of an anion.* For example, the halides as ligands are called fluoro, chloro, bromo, and iodo; hydroxide is hydroxo; and cyanide is cyano. *For a neutral ligand, the name of the molecule is used,* with the exception of H_2O, NH_3, CO, and NO, as illustrated in Table 19.14.
4. *The prefixes* mono-, di-, tri-, tetra-, penta-, *and* hexa- *are used to denote the number of simple ligands.* The prefixes *bis-, tris-, tetrakis-,* and so on are also used, especially for more complicated ligands or ones that already contain *di-, tri-,* and so on.
5. *The oxidation state of the central metal ion is designated by a Roman numeral in parentheses.*
6. *When more than one type of ligand is present, ligands are named in alphabetical order.** Prefixes do not affect the order.
7. *If the complex ion has a negative charge, the suffix* -ate *is added to the name of the metal.* Sometimes the Latin name is used to identify the metal (Table 19.15).

*In an older system, the negatively charged ligands were named first, followed by neutral ligands, with positively charged ligands named last. We will follow the newer convention in this text.

Table 19.14

Names of Some Common Unidentate Ligands

Neutral Molecules		Anions	
Aqua	H_2O	Fluoro	F^-
Ammine	NH_3	Chloro	Cl^-
Methylamine	CH_3NH_2	Bromo	Br^-
Carbonyl	CO	Iodo	I^-
Nitrosyl	NO	Hydroxo	OH^-
		Cyano	CN^-

Table 19.15

Latin Names Used for Some Metal Ions in Anionic Complex Ions

Metal	Anionic Complex Base Name
Iron	Ferrate
Copper	Cuprate
Lead	Plumbate
Silver	Argentate
Gold	Aurate
Tin	Stannate

These rules are applied in Example 19.1.

OWL INTERACTIVE EXAMPLE 19.1

Give the systematic name for each of the following coordination compounds.

a. $[Co(NH_3)_5Cl]Cl_2$ **b.** $K_3Fe(CN)_6$ **c.** $[Fe(en)_2(NO_2)_2]_2SO_4$

(top) An aqueous solution of $[Co(NH_3)_5Cl]Cl_2$.
(bottom) Solid $K_3Fe(CN)_6$.

Solution

a. To determine the oxidation state of the metal ion, we examine the charges of all ligands and counter ions. The ammonia molecules are neutral and each chloride ion has a 1− charge, so the cobalt ion must have a 3+ charge to produce a neutral compound. Since cobalt has the oxidation state +3, we use cobalt(III) in the name.

The ligands include one Cl^- ion and five NH_3 molecules. The chloride ion is designated as *chloro,* and each ammonia molecule is designated *ammine.* The prefix *penta-* indicates that there are five NH_3 ligands present. The name of the complex cation is therefore pentaamminechlorocobalt(III). Note that the ligands are named alphabetically, disregarding the prefix. Since the counter ions are chloride ions, the compound is named as a chloride salt:

pentaamminechlorocobalt(III) chloride
Cation Anion

b. First, we determine the oxidation state of the iron by considering the other charged species. The compound contains three K^+ ions and six CN^- ions. Therefore, the iron must carry a charge of 3+, giving a total of six positive charges to balance the six negative charges. Thus the complex ion present is $Fe(CN)_6^{3-}$. The cyanide ligands are each designated *cyano,* and the prefix *hexa-* indicates that six are present. Since the complex ion is an anion, we use the Latin name *ferrate.* The oxidation state is indicated by ferrate(III) in the name. The anion name is therefore hexacyanoferrate(III). The cations are K^+ ions, which are simply named potassium. Combining all this gives the name

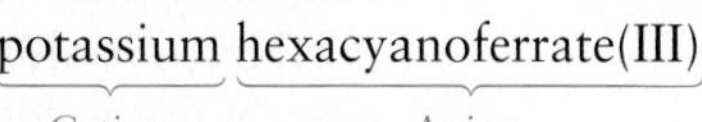

(The common name of this compound is potassium ferricyanide.)

c. We first determine the oxidation state of the iron by looking at the other charged species: four NO_2^- ions and one SO_4^{2-} ion. The ethylenediamine is neutral. Thus the two iron ions must carry a total of six positive charges to balance the six negative charges. This means that each iron has a +3 oxidation state and is designated as iron(III).

Since the name ethylenediamine already contains *di,* we use *bis-* instead of *di-* to indicate the presence of two en ligands. The name for NO_2^- as a ligand is *nitro,* and the prefix *di-* indicates the presence of two NO_2^- ligands. The anion is sulfate. Therefore, the compound's name is

$$\underbrace{\text{bis(ethylenediamine)dinitroiron(III)}}_{\text{Cation}}\ \underbrace{\text{sulfate}}_{\text{Anion}}$$

Since the complex ion is a cation in this case, the Latin name for iron is not used.

OWL INTERACTIVE EXAMPLE 19.2

Given the following systematic names, predict the formula of each coordination compound.

a. triamminebromoplatinum(II) chloride

b. potassium hexafluorocobaltate(III)

Solution

a. *Triammine* signifies three ammonia ligands, and *bromo* indicates one bromide ion as a ligand. The oxidation state of platinum is +2, as indicated by the Roman numeral II. Thus the complex ion is $[Pt(NH_3)_3Br]^+$. One chloride ion is needed to balance the 1+ charge of this cation. The formula of the compound is $[Pt(NH_3)_3Br]Cl$. Note that square brackets enclose the complex ion.

b. The complex ion contains six fluoride ligands attached to a Co^{3+} ion to give CoF_6^{3-}. Note the *-ate* ending that indicates that the complex ion is an anion. Three K^+ cations are required to balance the 3− charge on the complex ion. Thus the formula is $K_3[CoF_6]$.

19.4 Isomerism

When two or more species have the same formula but exhibit different properties, they are said to be **isomers.** Although isomers contain exactly the same types and numbers of atoms, the arrangements of their atoms differ, and this leads to different properties. We will consider two main types of isomerism: **structural isomerism,** in which the isomers contain the same atoms but one or more bonds differ, and **stereoisomerism,** in which all of the bonds in the isomers are the same but the spatial arrangements of the atoms are different. Each of these classes also has subclasses (Fig. 19.9), which we will now consider.

Structural Isomerism

The first type of structural isomerism we will consider is **coordination isomerism,** in which the composition of the complex ion varies. For example, $[Cr(NH_3)_5SO_4]Br$ and $[Cr(NH_3)_5Br]SO_4$ are coordination isomers. In the first case, SO_4^{2-} is coordinated to Cr^{3+}, whereas Br^- acts as the counter ion; in the second case, the roles of these ions are reversed.

Another example of coordination isomerism involves the $[Co(en)_3][Cr(ox)_3]$ and $[Cr(en)_3][Co(ox)_3]$ pair, where ox represents the oxalate ion, a bidentate ligand shown in Table 19.13.

In a second type of structural isomerism, **linkage isomerism,** the composition of the complex ion is the same, but the point of attachment of at least one

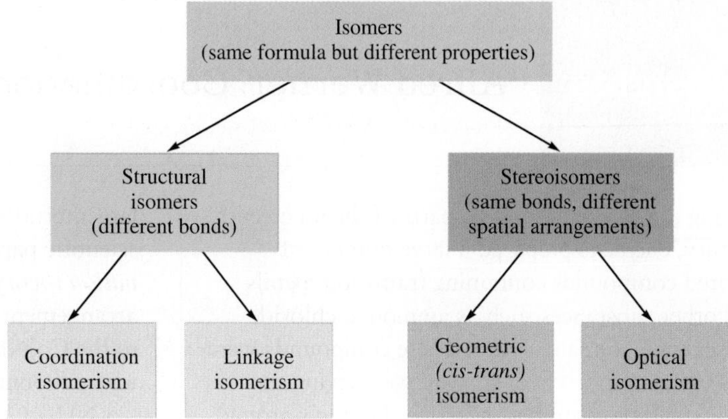

Figure 19.9
Some classes of isomers.

of the ligands differs. Two ligands that can attach to metal ions in different ways are thiocyanate, SCN^- (which can bond through lone pairs on the nitrogen or the sulfur atom), and the nitrite ion, NO_2^- (which can bond through lone pairs on the nitrogen or the oxygen atom). For example, the following two compounds are linkage isomers:

$[Co(NH_3)_4(NO_2)Cl]Cl$
Tetraamminechloronitrocobalt(III) chloride
(yellow)

$[Co(NH_3)_4(ONO)Cl]Cl$
Tetraamminechloronitritocobalt(III) chloride
(red)

In the first case, the NO_2^- ligand is called *nitro* and is attached to Co^{3+} through the nitrogen atom; in the second case, the NO_2^- ligand is called *nitrito* and is attached to Co^{3+} through an oxygen atom (Fig. 19.10).

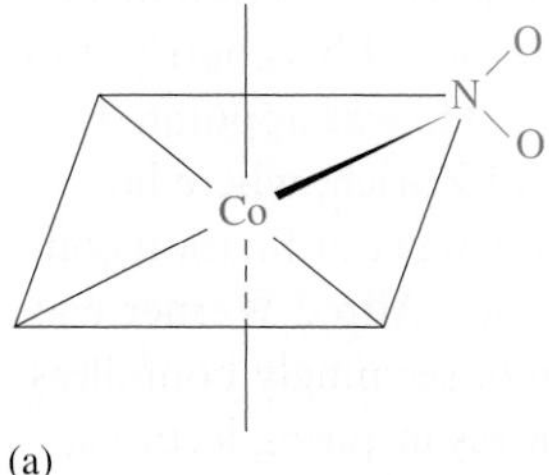

(a)

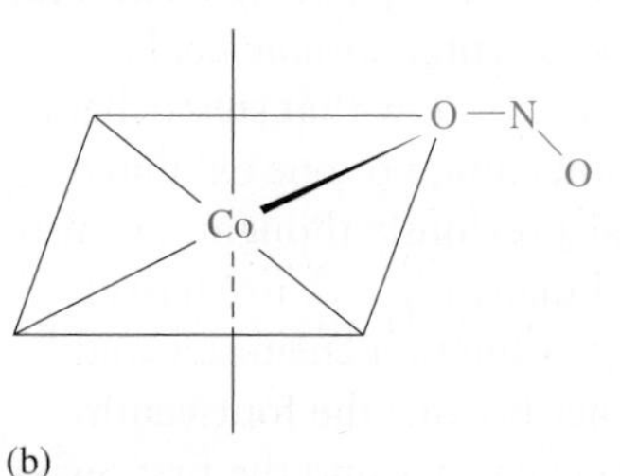

(b)

Figure 19.10
As a ligand, NO_2^- can bond to a metal ion (a) through a lone pair on the nitrogen atom or (b) through a lone pair on one of the oxygen atoms.

Stereoisomerism

Stereoisomers have the same bonds, but they exhibit different spatial arrangements of their atoms. One type, **geometrical isomerism,** or ***cis–trans*** **isomerism,** occurs when atoms or groups of atoms can assume different positions around a rigid ring or bond. An important example is the compound $Pt(NH_3)_2Cl_2$, which has a square planar structure. The two possible arrangements of the ligands are shown in Fig. 19.11. In the ***cis*** **isomer,** the ammonia molecules are next (*cis*) to each other. In the ***trans*** **isomer,** the ammonia molecules are across (*trans*) from each other.

Geometrical isomerism also occurs in octahedral complex ions. For example, the compound $[Co(NH_3)_4Cl_2]Cl$ has *cis* and *trans* isomers (Fig. 19.12).

A second type of stereoisomerism is called **optical isomerism** because the isomers have opposite effects on plane-polarized light. When light is emitted from a source such as a glowing filament, the oscillating electric fields of the

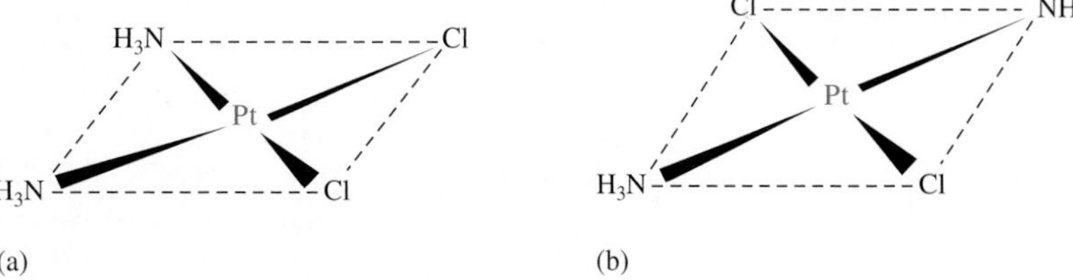

Figure 19.11
(a) The *cis* isomer of $Pt(NH_3)_2Cl_2$ (yellow).
(b) The *trans* isomer of $Pt(NH_3)_2Cl_2$ (pale yellow).

Chemical Insights — Alfred Werner: Coordination Chemist

During the early and middle parts of the nineteenth century, chemists prepared a large number of colored compounds containing transition metals and other substances such as ammonia, chloride ion, cyanide ion, and water. These compounds were very interesting to chemists who were trying to understand the nature of bonding (Dalton's atomic theory of 1808 was very new at this time), and many theories were suggested to explain these substances. The most widely accepted early theory was the *chain theory*, championed by Sophus Mads Jorgensen (1837–1914), professor of chemistry at the University of Copenhagen. The chain theory got its name from the postulate that metal–ammine* complexes contain chains of NH_3 molecules. For example, Jorgensen proposed the structure

```
   NH3—Cl
  /
Co—NH3—NH3—NH3—NH3—Cl
  \
   NH3—Cl
```

for the compound $Co(NH_3)_6Cl_3$. In the late nineteenth century, this theory was used in classrooms around the world to explain the nature of metal–ammine compounds.

However, in 1890 a young Swiss chemist named Alfred Werner, who had just obtained a Ph.D. in the field of organic chemistry, became so interested in these compounds that he apparently even dreamed about them. In the middle of one night, Werner awoke realizing that he had discovered the correct explanation for the constitution of these compounds. Writing furiously the rest of that night and into the late afternoon of the following day, he constructed a scientific paper containing his now famous *coordination theory*. This model postulates an octahedral arrangement of ligands around the Co^{3+} ion, leading to the $Co(NH_3)_6^{3+}$ complex ion with three Cl^- ions as counter ions. Thus Werner's picture of $Co(NH_3)_6Cl_3$ differed greatly from the chain theory.

In his paper on the coordination theory, Werner explained not only the metal–ammine compounds but also most of the other known transition metal compounds, and the importance of his contribution was recognized immediately. He was appointed professor at the University of Zurich, where he spent the rest of his life studying coordination compounds and refining his theory. Alfred Werner was a confident, impulsive man of seemingly boundless energy who was known for his inspiring lectures, his intolerance of incompetence (he once threw a chair at a student who performed poorly on an oral exam), and his intuitive scientific brilliance. For example, he was the first to show that stereochemistry is a general phenomenon, not one exhibited only by carbon, as was previously thought. He also recognized and named many types of isomerism.

For his work on coordination chemistry and stereochemistry, Werner became the fourteenth Nobel Prize winner in chemistry and the first Swiss chemist to be so honored. Werner's work is even more remarkable when one realizes that his ideas preceded any real understanding of the nature of covalent bonds by many years.

*Ammine is the name for NH_3 as a ligand.

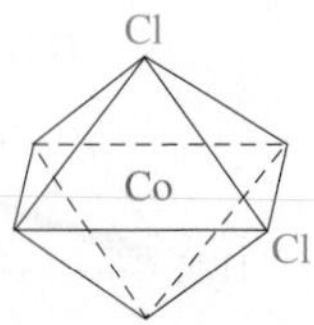

(a) (b)

Figure 19.12
(a) The *trans* isomer of $[Co(NH_3)_4Cl_2]^+$. The chloride ligands are directly across from each other. (b) The *cis* isomer of $[Co(NH_3)_4Cl_2]^+$. The chloride ligands in this case share an edge of the octahedron. Because of their different structures, the *trans* isomer of $[Co(NH_3)_4Cl_2]Cl$ is green and the *cis* isomer is violet.

photons in the beam are oriented randomly, as shown in Fig. 19.13. If this light is passed through a polarizer, only the photons with electric fields oscillating in a single plane remain, constituting *plane-polarized light*.

In 1815 French physicist Jean Biot discovered that certain crystals could rotate the plane of polarization of light. Later, scientists found that solutions of certain compounds could do the same thing (Fig. 19.14). Louis Pasteur was the first to understand this behavior. In 1848 he noted that solid sodium ammonium tartrate ($NaNH_4C_4H_4O_6$) existed as a mixture of two types of crystals, which he painstakingly separated with tweezers. Separate solutions of these two types of crystals rotated plane-polarized light in exactly opposite directions. This led to a connection between optical activity and molecular structure.

We now realize that optical activity is exhibited by molecules that have *nonsuperimposable mirror images*. Your hands are nonsuperimposable mirror images (Fig. 19.15). That is, human hands are related like an object and its

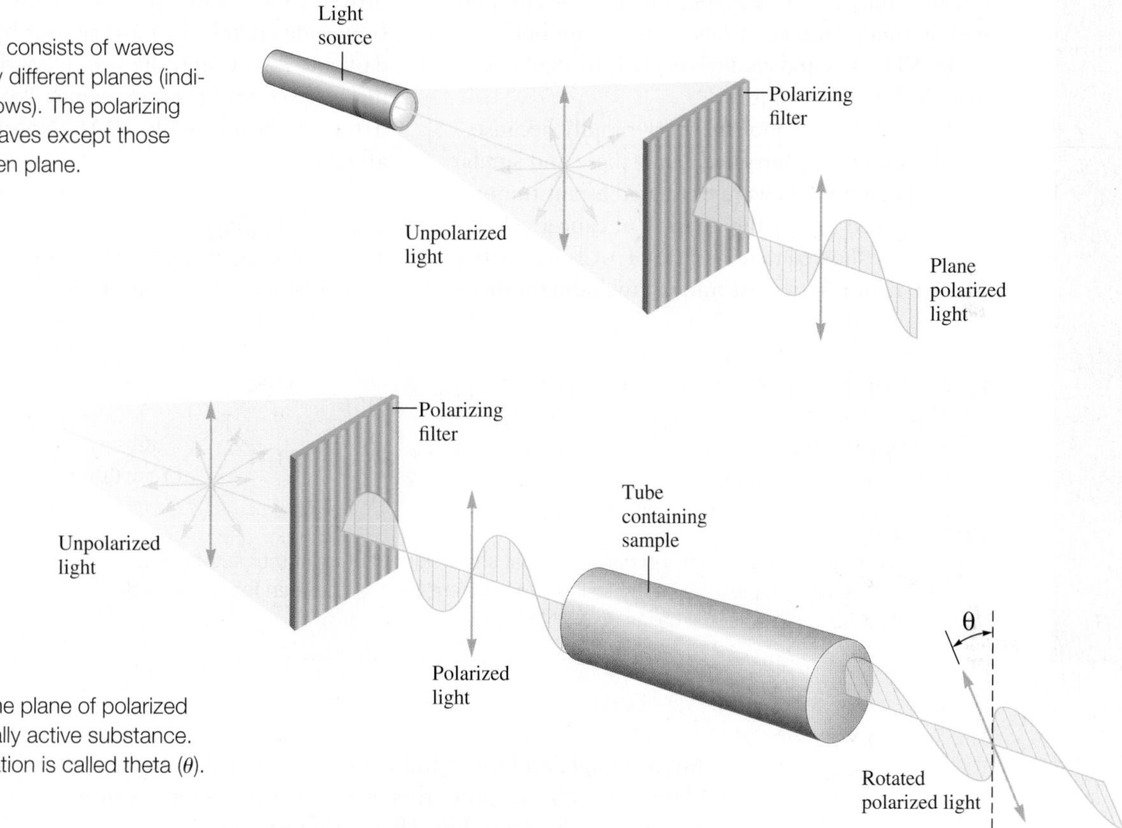

Figure 19.13
Unpolarized light consists of waves vibrating in many different planes (indicated by the arrows). The polarizing filter blocks all waves except those vibrating in a given plane.

Figure 19.14
The rotation of the plane of polarized light by an optically active substance. The angle of rotation is called theta (θ).

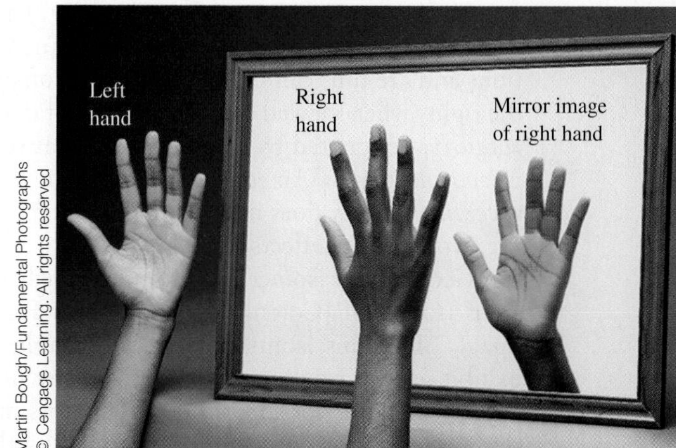

Figure 19.15
A human hand has a nonsuperimposable mirror image. Note that the mirror image of the right hand (while identical to the left hand) cannot be turned in any way to make it identical to (superimposable on) the actual right hand.

Chemical Insights: The Importance of Being *cis*

Some of the most important advancements of science are the results of accidental discoveries—for example, penicillin, Teflon, and the sugar substitutes cyclamate and aspartame. Another important chance discovery occurred in 1964, when a group of scientists using platinum electrodes to apply an electric field to a colony of *E. coli* bacteria noticed that the bacteria failed to divide but continued to grow, forming long, fibrous cells. Further study revealed that cell division was inhibited by small concentrations of the compounds *cis*-$Pt(NH_3)_2Cl_2$ and *cis*-$Pt(NH_3)_2Cl_4$ formed electrolytically in the solution.

Cancerous cells multiply very rapidly because cell division is uncontrolled. Thus these and similar platinum complexes were evaluated as antitumor agents, which inhibit the division of cancer cells. The results showed that *cis*-$Pt(NH_3)_2Cl_2$ was active against a wide variety of tumors, including testicular and ovarian tumors, which are very resistant to treatment by more traditional methods. However, although the *cis* complex showed significant antitumor activity, the corresponding *trans* complex had no effect on tumors. This illustrates the importance of isomerism in biological systems. When drugs are synthesized, great care must be taken to obtain the correct isomer.

Unfortunately, although *cis*-$Pt(NH_3)_2Cl_2$ has proved to be a valuable drug, it has some troublesome side effects, the most serious being kidney damage. As a result, the search continues for even more effective antitumor agents. Promising candidates are shown in Fig. 19.16. Note that they are all *cis* complexes.

Suggested Reading

Ann M. Thayer, "Platinum Drugs Take Their Toll," *C&EN*, June 28 (2010): 24

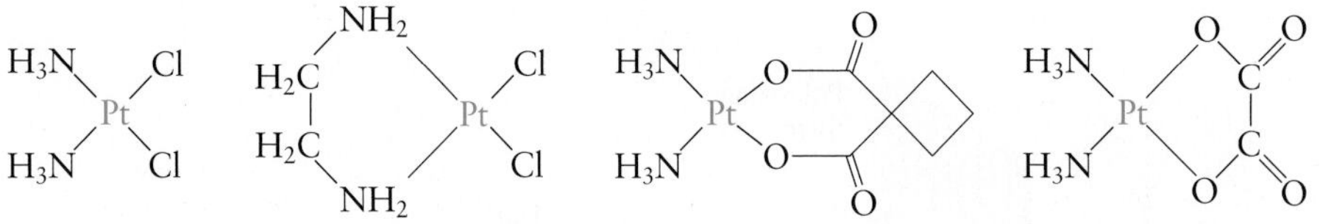

H_3N Cl Pt H_3N Cl NH_2 H_2C H_2C NH_2 Pt Cl Cl H_3N H_3N Pt O O O O H_3N H_3N Pt O O C C O O

Figure 19.16
Some *cis* complexes of platinum and palladium that show significant antitumor activity. It is thought that the *cis* complexes work by losing two adjacent ligands and then forming coordinate covalent bonds to adjacent bases on a DNA molecule.

mirror image, and one hand cannot be turned to make it identical to the other. Many molecules show this same feature—for example, the complex ion $[Co(en)_3]^{3+}$ shown in Fig. 19.17. Objects that have nonsuperimposable mirror images are said to be **chiral** (from the Greek word for hand, *cheir*).

The isomers of $[Co(en)_3]^{3+}$ (Fig. 19.17) are nonsuperimposable mirror images called **enantiomers;** they rotate plane-polarized light in opposite directions and are thus optical isomers. The isomer that rotates the plane of light to the right (when viewed down the beam of oncoming light) is said to be *dextrorotatory,* designated by *d*. The isomer that rotates the plane of light to the left is *levorotatory* (*l*). An equal mixture of the *d* and *l* isomers in solution, called a *racemic mixture,* does not rotate the plane of the polarized light at all because the two opposite effects cancel.

Geometrical isomers are not necessarily optical isomers. For instance, the *trans* isomer of $[Co(en)_2Cl_2]^+$ shown in Fig. 19.18 is identical to its mirror image. Since this isomer is superimposable on its mirror image, it does not exhibit optical isomerism and is therefore not chiral. On the other hand, *cis*-$[Co(en)_2Cl_2]^+$ is *not* superimposable on its mirror image; thus a pair of enantiomers exists for the complex ion, making the *cis* isomer chiral.

Figure 19.17
Isomers I and II of $[Co(en)_3]^{3+}$ are mirror images (the mirror image of I is identical to II) that cannot be superimposed. That is, there is no way that I can be turned in space so that it is the same as II.

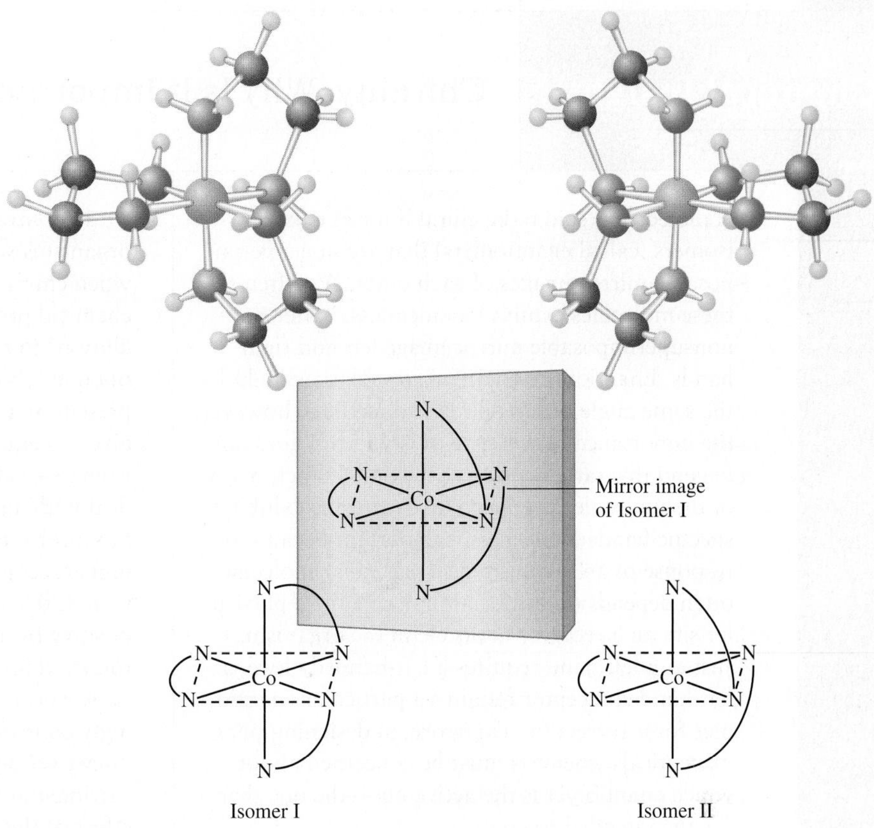

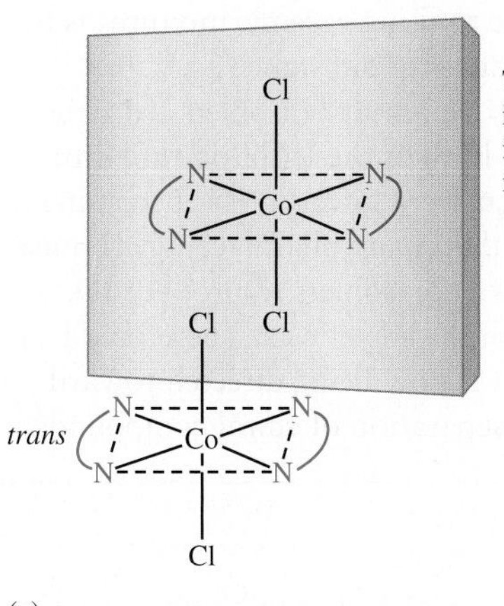

The *trans* isomer and its mirror image are identical. They are not isomers of each other.

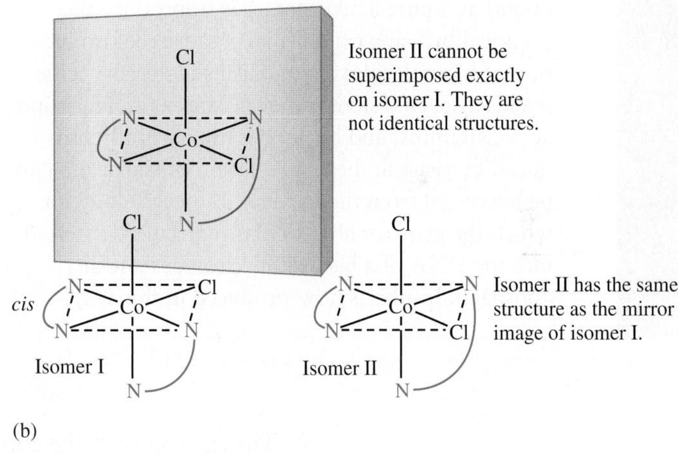

Figure 19.18
(a) The *trans* isomer of $[Co(en)_2Cl_2]^+$ and its mirror image are identical (superimposable). (b) The *cis* isomer of $[Co(en)_2Cl_2]^+$ and its mirror image are not superimposable and are thus a pair of optical isomers.

EXAMPLE 19.3

Does the complex ion $[Co(NH_3)Br(en)_2]$ exhibit geometrical isomerism? Does it exhibit optical isomerism?

Solution The complex ion exhibits geometrical isomerism, since the ethylenediamine ligands can be across from or next to each other:

Chemical Insights

Chirality: Why Is It Important?

A molecule is said to be chiral if it can exist as isomers (called enantiomers) that are nonsuperimposable mirror images of each other. We often say these molecules exhibit "handedness," after our nonsuperimposable mirror image left and right hands. Enantiomers rotate plane-polarized light by the same angle but in opposite directions; however, the importance of this type of isomerism goes far beyond this rather curious behavior. In fact, many of the molecules produced by organisms exhibit a specific handedness. This is important because the response of an organism to a particular molecule often depends on how that molecule fits a particular site on a receptor molecule in the organism. Just as a left hand requires a left-handed glove, a left-handed receptor requires a particular enantiomer for a correct fit. Therefore, in designing pharmaceuticals, chemists must be concerned about which enantiomer is the active one—the one that fits the intended receptor.

Ideally, the pharmaceutical should consist of the pure active isomer. One way to obtain the compound as a pure active isomer is to produce the chemical by using organisms, because the production of biomolecules in organisms is stereospecific (yields a specific stereoisomer). For example, amino acids, vitamins, and hormones are naturally produced by yeast in the fermentation of sugar and can be harvested from the ferment. Biotechnology, in which the gene for a particular molecule is inserted into the DNA of a bacterium, provides another approach. Insulin is now produced in this way.

In contrast to the synthesis of biomolecules by organisms where a specific isomer is produced, when chiral molecules are made by "normal" chemical procedures (reactants are mixed and allowed to react), a mixture of the enantiomers is obtained. For example, when one chiral center is present in a molecule, normal chemical synthesis gives an equal mixture of the two mirror image isomers—called a racemic mixture. How does one deal with a pharmaceutical produced as a racemic mixture? One possibility is to administer the drug in its racemic form, assuming that the inactive form (50% of the mixture) will have no effect, positive or negative. In fact, this procedure is being followed for many drugs now on the market. However, it is a procedure that is growing increasingly controversial as evidence mounts that the "inactive" form of the drug may actually produce detrimental effects often totally unrelated to the effect of the active isomer. In effect, a drug administered as a racemic mixture contains a 50% impurity, the effects of which are not well understood.

The alternative to using racemic mixtures is to find a way to produce the substance as a pure isomer or a way to separate the isomers from the racemic mixtures. Both of these options are difficult and thus expensive. However, it is becoming increasingly clear that many pharmaceuticals must be administered as pure isomers to produce the desired results with no side effects. Therefore, a great deal of effort is now being directed toward the synthesis and separation of chiral compounds.

The *cis* isomer of the complex ion also exhibits optical isomerism since its mirror images,

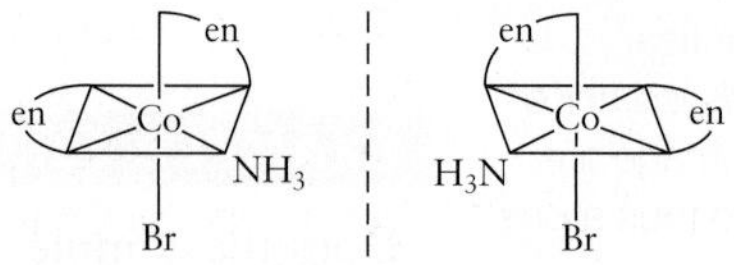

cannot be turned in any way to make them superimposable. Thus these two *cis* isomers are shown to be enantiomers that will rotate plane-polarized light in opposite directions.

Most important, since biomolecules are chiral, their reactions are highly structure-dependent. For example, a drug might have a particular effect because its molecules can bind to chiral molecules in the body. For the binding to be correct, the correct optical isomer of the drug must be administered. Just as the right hand of one person requires the right hand of another to perform a handshake, a given isomer in the body requires a specific isomer of the drug for an interaction to occur. Consequently, the syntheses of drugs, which are usually very complicated molecules, must be carried out in a way that produces the correct "handedness," a requirement that greatly adds to the difficulties in bringing these substances to market.

19.5 Bonding in Complex Ions: The Localized Electron Model

By this point in your study of chemistry, you no doubt recognize that the localized electron model, although very simple, is a very useful model for describing the bonding in molecules. Recall that a central feature of the model is the formation of hybrid atomic orbitals that are used for sharing electron pairs to form σ bonds between atoms. This same model can be used to account for the bonding in complex ions, but there are two important points to keep in mind.

1. The VSEPR model for predicting structure *does not work for complex ions*. However, we can safely assume that a complex ion with a coordination number of 6 has an octahedral arrangement of ligands and that complexes with two ligands are linear. On the other hand, complex ions with a coordination number of 4 can be either tetrahedral or square planar; there is no reliable way to predict which will occur in a particular case.
2. The interaction between a metal ion and a ligand can be viewed as a Lewis acid–base reaction, with the ligand donating a lone pair of electrons to an *empty* orbital on the metal ion to form a coordinate covalent bond:

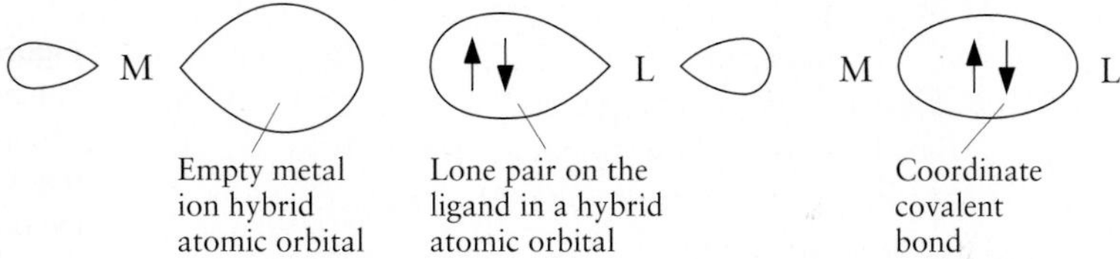

The hybrid orbitals used by the metal ion depend on the number and arrangement of the ligands. For example, accommodating the lone pair from each ammonia molecule in the octahedral $Co(NH_3)_6^{3+}$ ion requires a set of six empty hybrid atomic orbitals with an octahedral arrangement. Recall that an octahedral set of orbitals is formed by the hybridization of two *d*, one *s*, and three *p* orbitals to give six d^2sp^3 orbitals (Fig. 19.19).

The hybrid orbitals predicted for the metal ion in four-coordinate complexes depend on whether the structure is tetrahedral or square planar. For a tetrahedral arrangement of ligands, an sp^3 hybrid set is required (Fig. 19.20). For example, in the tetrahedral $CoCl_4^{2-}$ ion, the Co^{2+} is described as sp^3 hybridized. A square planar arrangement of ligands requires a dsp^2 hybrid orbital set on the metal ion (Fig. 19.20). For example, in square planar $Ni(CN)_4^{2-}$, the Ni^{2+} is described as dsp^2 hybridized.

A linear complex requires two hybrid orbitals 180° from each other. This arrangement is given by an *sp* hybrid set (Fig. 19.20). Thus in the linear $Ag(NH_3)_2^+$ ion, the Ag^+ is described as *sp* hybridized.

Although the localized electron model can account in a general way for metal–ligand bonds, it is rarely used today because it cannot predict important properties of complex ions, such as magnetism and color. Thus we will not pursue the model any further.

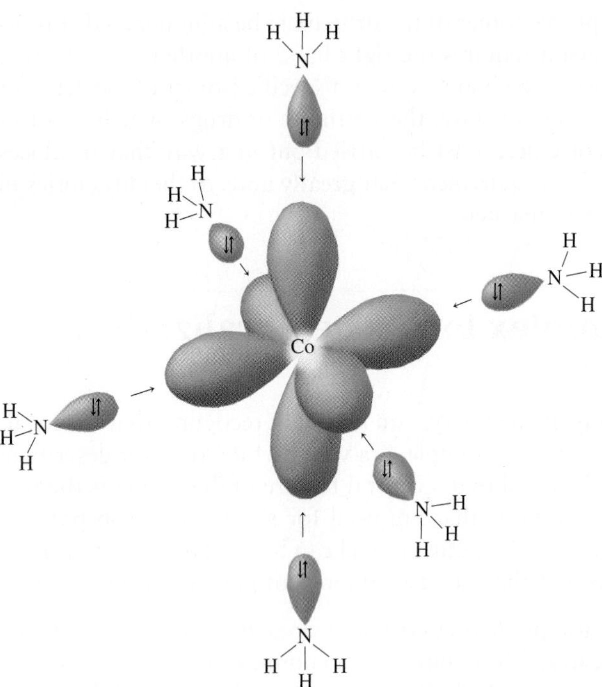

Figure 19.19
A set of six d^2sp^3 hybrid orbitals on Co^{3+} can accept an electron pair from each of six NH_3 ligands to form the $Co(NH_3)_6^{3+}$ ion.

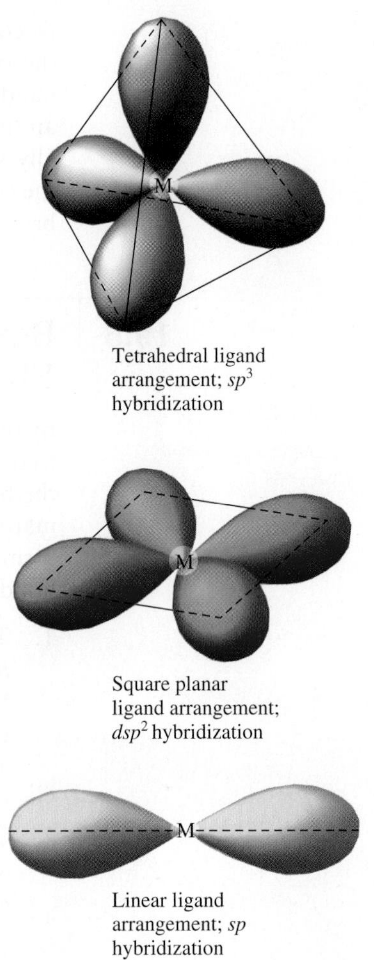

Figure 19.20
The hybrid orbitals required for tetrahedral, square planar, and linear complex ions. The metal ion hybrid orbitals are empty, so the metal ion bonds to the ligands by accepting lone pairs.

19.6 The Crystal Field Model

The main reason that the localized electron model cannot fully account for the properties of complex ions is that in its simplest form it gives no information about how the energies of the *d* orbitals are affected by complex ion formation. This is critical because, as we will see, the color and magnetism of complex ions result from changes in the energies of the metal ion *d* orbitals caused by the metal–ligand interactions.

The **crystal field model** focuses on the energies of the *d* orbitals. In fact, this model is not so much a bonding model as it is an attempt to account for the colors and magnetic properties of complex ions. In its simplest form, the crystal field model assumes that the ligands can be approximated by *negative point charges* and that metal–ligand bonding is *entirely ionic*.

Octahedral Complexes

We will illustrate the fundamental principles of the crystal field model by applying it to an octahedral complex. Figure 19.21 shows the orientation of the 3*d* orbitals relative to an octahedral arrangement of point-charge ligands. The

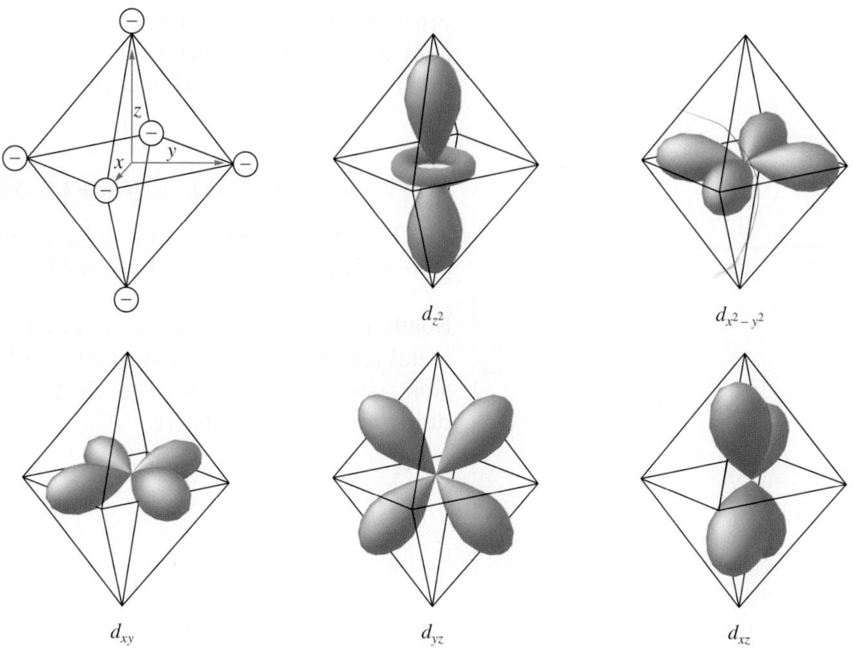

Figure 19.21
An octahedral arrangement of point-charge ligands and the orientation of the 3*d* orbitals.

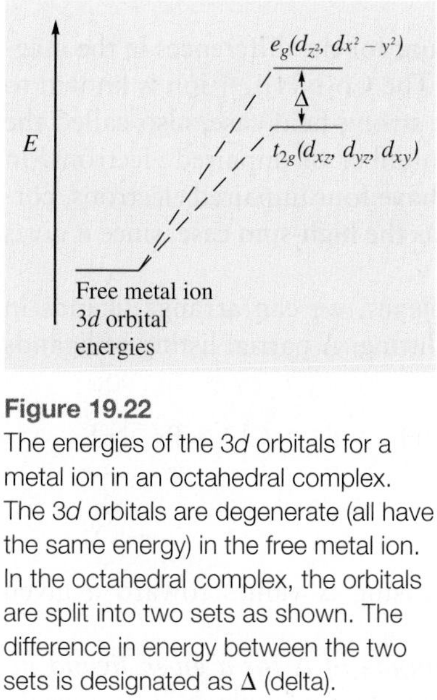

important thing to note is that two of the orbitals, d_{z^2} and $d_{x^2-y^2}$, point their lobes *directly at* the point-charge ligands, whereas three of the orbitals, d_{xz}, d_{yz}, and d_{xy}, point their lobes *between* the point charges.

To understand the effect of this difference, we need to consider which type of orbital is lower in energy. Because the negative point-charge ligands repel negatively charged electrons, the electrons will prefer the *d* orbitals farthest from the ligands. In other words, the d_{xz}, d_{yz}, and d_{xy} orbitals (called the t_{2g} set) are at a *lower energy* in the octahedral complex than are the d_{z^2} and $d_{x^2-y^2}$ orbitals (the e_g set). This is shown in Fig. 19.22. The negative point-charge ligands increase the energies of all the *d* orbitals. However, the orbitals that point at the ligands are raised in energy more than those that point between the ligands.

E
$e_g(d_{z^2}, d_{x^2-y^2})$
Δ
$t_{2g}(d_{xz}, d_{yz}, d_{xy})$
Free metal ion 3*d* orbital energies

Figure 19.22
The energies of the 3*d* orbitals for a metal ion in an octahedral complex. The 3*d* orbitals are degenerate (all have the same energy) in the free metal ion. In the octahedral complex, the orbitals are split into two sets as shown. The difference in energy between the two sets is designated as Δ (delta).

It is this **splitting of the 3*d* orbital energies** (symbolized by Δ) that explains the color and magnetism of complex ions of the first-row transition metal ions. For example, in an octahedral complex of Co^{3+} (a metal ion with six 3*d* electrons), there are two possible ways to place the electrons in the split 3*d* orbitals (Fig. 19.23). If the splitting produced by the ligands is very large, a situation called the **strong-field case,** the electrons will pair in the lower-energy t_{2g} orbitals. This gives a *diamagnetic* complex in which all electrons are paired. On the

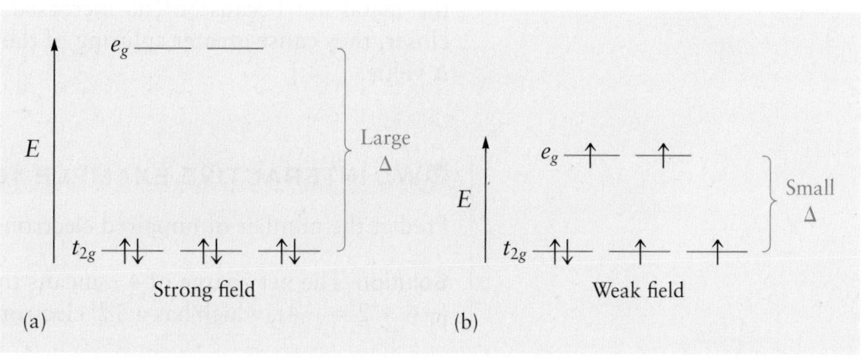

Figure 19.23
Possible electron arrangements in the split 3*d* orbitals of an octahedral complex of Co^{3+} (electron configuration $3d^6$). (a) In a strong field (large Δ value), the electrons fill the t_{2g} set first, giving a diamagnetic complex. (b) In a weak field (small Δ value), the electrons occupy all five orbitals before any pairing occurs.

other hand, if the splitting is small (the **weak-field case**), the electrons will occupy all five orbitals before pairing occurs. In this case the complex has four unpaired electrons and thus is *paramagnetic.*

OWL INTERACTIVE EXAMPLE 19.4

The $Fe(CN)_6^{3-}$ ion is known to have one unpaired electron. Does the CN^- ligand produce a strong or weak field?

Solution Since the ligand is CN^- and the overall complex ion charge is 3−, the metal ion must be Fe^{3+}, which has a $3d^5$ electron configuration. The two possible arrangements of the five electrons in the *d* orbitals split by the octahedrally arranged ligands are

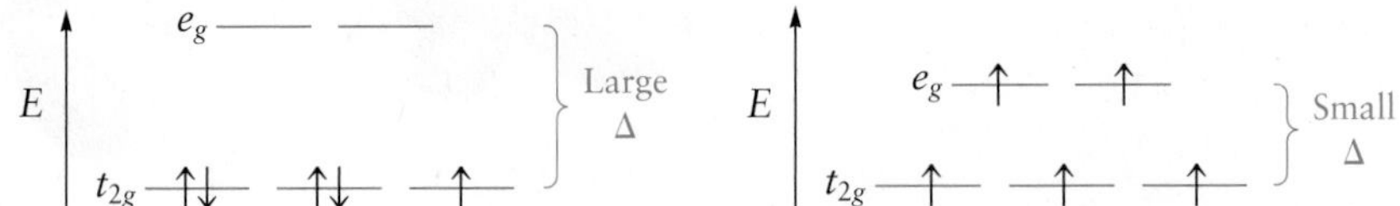

The strong-field case gives one unpaired electron, which agrees with the experimental observation. The CN^- ion is a strong-field ligand toward the Fe^{3+} ion.

The crystal field model allows us to account for the differences in the magnetic properties of $Co(NH_3)_6^{3+}$ and CoF_6^{3-}. The $Co(NH_3)_6^{3+}$ ion is known to be diamagnetic and thus corresponds to the strong-field case, also called the **low-spin case,** since it yields the *minimum* number of unpaired electrons. In contrast, the CoF_6^{3-} ion, which is known to have four unpaired electrons, corresponds to the weak-field case, also known as the **high-spin case,** since it gives the *maximum* number of unpaired electrons.

From studies of many octahedral complexes, we can arrange ligands in order of their ability to produce *d*-orbital splitting. A partial listing of ligands in this so-called **spectrochemical series** is

$$CN^- > NO_2^- > en > NH_3 > H_2O > OH^- > F^- > Cl^- > Br^- > I^-$$

Strong-field ligands (large Δ) — Weak-field ligands (small Δ)

The ligands are arranged in order of decreasing Δ values toward a given metal ion.

It has also been observed that *the magnitude of Δ for a given ligand increases as the charge on the metal ion increases.* For example, NH_3 is a weak-field ligand toward Co^{2+} but acts as a strong-field ligand toward Co^{3+}. This makes sense; as the metal ion charge increases, the ligands are drawn closer to the metal ion because of its increased charge density. As the ligands move closer, they cause greater splitting of the *d* orbitals, thereby producing a larger Δ value.

OWL INTERACTIVE EXAMPLE 19.5

Predict the number of unpaired electrons in the complex ion $[Cr(CN)_6]^{4-}$.

Solution The net charge of 4− means that the metal ion present must be Cr^{2+} $(-6 + 2 = -4)$, which has a $3d^4$ electron configuration. Since CN^- is a strong-

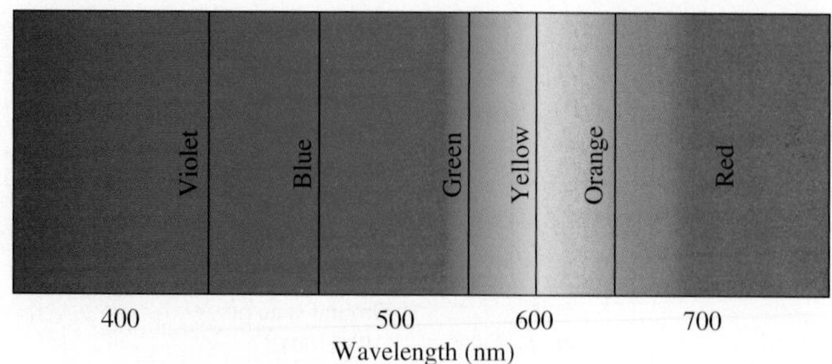

Figure 19.24
The visible spectrum.

field ligand (see the spectrochemical series), the correct crystal field diagram for $[Cr(CN)_6]^{4-}$ is

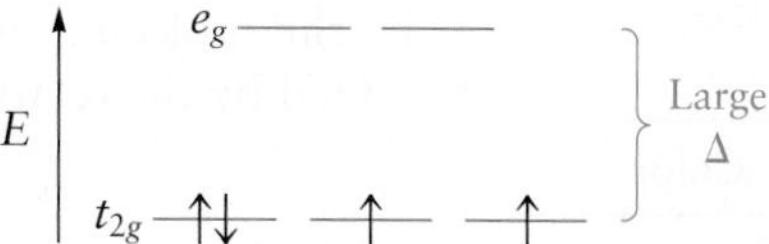

The complex ion will have two unpaired electrons. Note that the CN^- ligand produces such a large splitting that two of the electrons will be paired in the same orbital rather than forcing one electron up through the large energy gap Δ.

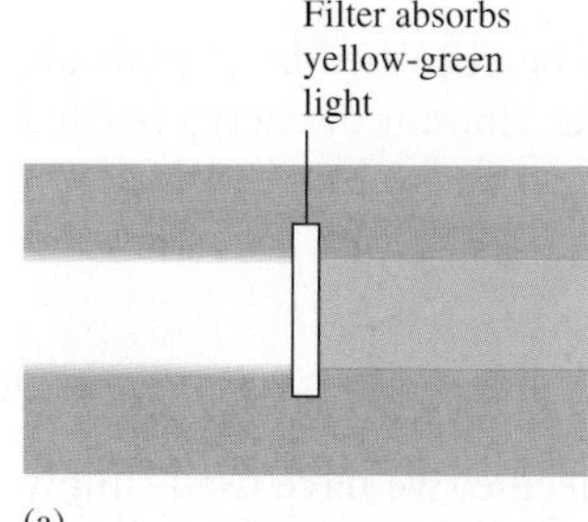

(a)

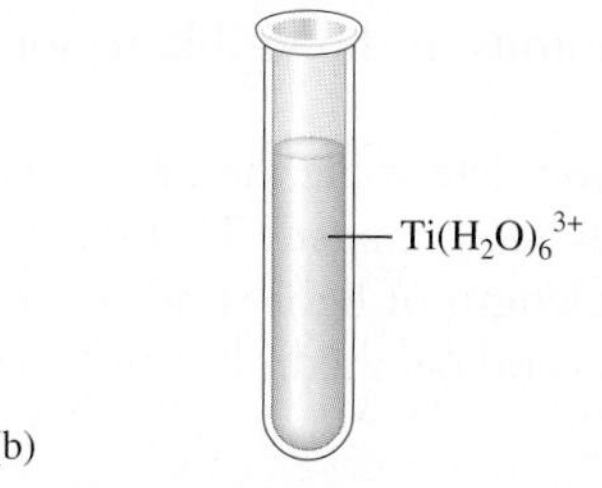

(b)

Figure 19.25
(a) When white light shines on a filter that absorbs wavelengths in the yellow-green region, the emerging light is violet. (b) Because the complex ion $Ti(H_2O)_6^{3+}$ absorbs yellow-green light, a solution of it is violet.

We have seen how the crystal field model accounts for the magnetic properties of octahedral complexes. The same model also explains the colors of these complex ions. For example, consider $Ti(H_2O)_6^{3+}$, an octahedral complex of Ti^{3+}, which has a $3d^1$ electron configuration. This complex ion is violet because it absorbs light in the middle of the visible region of the spectrum (Fig. 19.24). When a substance absorbs certain wavelengths of light in the visible region, the color of that substance is determined by the wavelengths of visible light that remain. We say that the substance exhibits the color *complementary* to those absorbed. The $Ti(H_2O)_6^{3+}$ ion is violet because it absorbs light in the yellow-green region, thus allowing red light and blue light to pass. The red and blue colors not absorbed produce the observed violet color. This effect is shown schematically in Fig. 19.25. Table 19.16 shows the general relationship between the wavelengths of visible light absorbed and the approximate color observed.

The reason that the $Ti(H_2O)_6^{3+}$ ion absorbs specific wavelengths of visible light can be traced to the transfer of the lone d electron between the split

Table 19.16

Approximate Relationship of Wavelength of Visible Light Absorbed to Color Observed

Absorbed Wavelength in nm (color)	Observed Color
400 (violet)	Greenish yellow
450 (blue)	Yellow
490 (blue-green)	Red
570 (yellow-green)	Violet
580 (yellow)	Dark blue
600 (orange)	Blue
650 (red)	Green

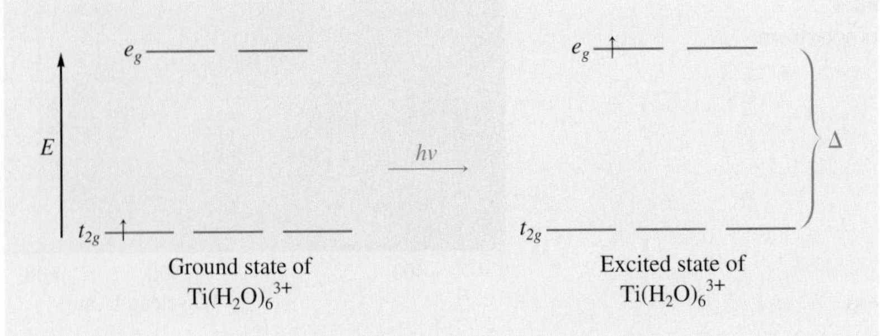

Figure 19.26
The complex ion $Ti(H_2O)_6^{3+}$ can absorb visible light in the yellow-green region to transfer the lone *d* electron from the t_{2g} to the e_g set.

d orbitals (Fig. 19.26). A given photon of light can be absorbed by a molecule only if the wavelength of the light provides the exact amount of energy needed by the molecule. In other words, whether a wavelength is absorbed is determined by the relationship

$$\Delta E = \frac{hc}{\lambda}$$

where ΔE represents the energy spacing in the molecule (we have used simply Δ in this chapter) and λ represents the wavelength of light needed to provide exactly that amount of energy. Because the *d*-orbital splitting in most octahedral complexes corresponds to the energies of photons in the visible region, octahedral complex ions are usually colored.

Since the ligands coordinated to a given metal ion determine the size of the *d*-orbital splitting, the color changes as the ligands are changed. This occurs because a change in Δ means a change in the wavelength of light is needed to transfer electrons between the t_{2g} and e_g orbitals. Several octahedral complexes of Cr^{3+} and their colors are listed in Table 19.17.

Table 19.17

Several Octahedral Complexes of Cr^{3+} and Their Colors

Isomer	Color
$[Cr(H_2O)_6]Cl_3$	Violet
$[Cr(H_2O)_5Cl]Cl_2$	Blue-green
$[Cr(H_2O)_4Cl_2]Cl$	Green
$[Cr(NH_3)_6]Cl_3$	Yellow
$[Cr(NH_3)_5Cl]Cl_2$	Purple
$[Cr(NH_3)_4Cl_2]Cl$	Violet

Other Coordination Geometries

Using the same principles developed for octahedral complexes, we will now consider complexes with other geometries. For example, Fig. 19.27 shows a tetrahedral arrangement of point charges in relation to the 3*d* orbitals of a metal ion. There are two important facts to note.

1. None of the 3*d* orbitals "points at the ligands" in the tetrahedral arrangement, as the $d_{x^2-y^2}$ and d_{z^2} orbitals do in the octahedral case. Thus the tetrahedrally arranged ligands do not differentiate the *d* orbitals as much in the tetrahedral case as in the octahedral case. That is, the difference in energy between the split *d* orbitals is significantly less in tetrahedral com-

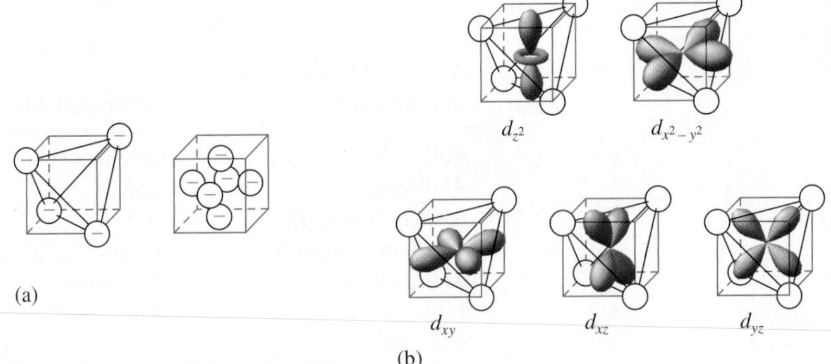

Figure 19.27
(a) Tetrahedral and octahedral arrangements of ligands shown inscribed in cubes. Note that in the two types of arrangements, the point charges occupy opposite parts of the cube: The octahedral point charges are at the centers of the cube faces, whereas the tetrahedral point charges occupy opposite corners of the cube. (b) The orientations of the 3*d* orbitals relative to the tetrahedral set of point charges.

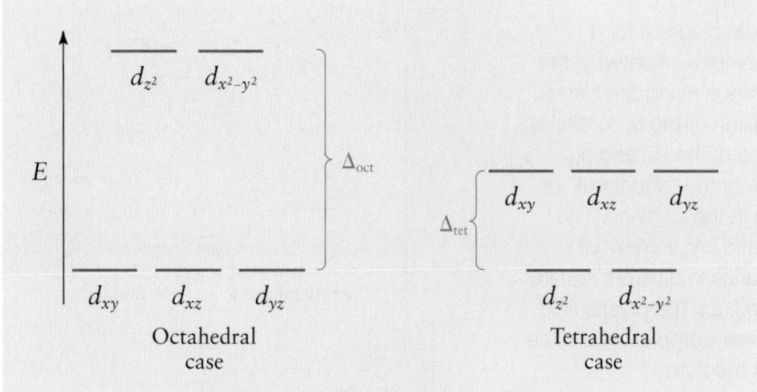

Figure 19.28
The crystal field diagrams for octahedral and tetrahedral complexes. The relative energies of the sets of *d* orbitals are reversed. For a given type of ligand, the splitting is much larger for the octahedral complex ($\Delta_{oct} > \Delta_{tet}$). This occurs because in the octahedral arrangement the d_{z^2} and $d_{x^2-y^2}$ orbitals point their lobes directly at the point charges and are thus relatively high in energy.

plexes. Although we will not derive it here, the tetrahedral splitting is $\frac{4}{9}$ that of the octahedral splitting for a given ligand and metal ion:

$$\Delta_{tet} = \tfrac{4}{9}\Delta_{oct}$$

2. Although not exactly pointing at the ligands, the d_{xy}, d_{xz}, and d_{yz} orbitals are closer to the point charges than are the d_{z^2} and $d_{x^2-y^2}$ orbitals. This means that the *d*-orbital splitting is opposite to that for the octahedral arrangement. The two arrangements are contrasted in Fig. 19.28. Because the *d*-orbital splitting is relatively small for the tetrahedral case, the weak-field case (high-spin case) *always* applies. There are no known ligands powerful enough to produce the strong-field case in a tetrahedral complex.

OWL INTERACTIVE EXAMPLE 19.6

Give the crystal field diagram for the tetrahedral complex ion $CoCl_4^{2-}$.

Solution The complex ion contains Co^{2+}, which has a $3d^7$ electron configuration. The splitting of the *d* orbitals will be small since this is a tetrahedral complex, giving the high-spin case with three unpaired electrons.

E

↑ ↑ ↑

d_{xy} d_{xz} d_{yz}

↑↓ ↑↓

d_{z^2} $d_{x^2-y^2}$

Small Δ

The crystal field model also applies to square planar and linear complexes. The crystal field diagrams for these cases are shown in Fig. 19.29. The ranking of orbitals in these diagrams can be explained by considering the relative orientations of the point charges and the orbitals. The diagram in Fig. 19.28 for the octahedral arrangement can be used to obtain these orientations. We can obtain the square planar complex by starting with an octahedral arrangement of six point charges and then removing the two point charges along the *z* axis. Removal of the two point charges on the *z* axis greatly lowers the energy of d_{z^2}. Now only the four lobes of the $d_{x^2-y^2}$ orbital point directly at the four remaining point charges. Therefore, $d_{x^2-y^2}$ is the highest-energy orbital for the square planar case. The relative energies of the remaining orbitals depend on how close their lobes are to the four point charges. The ordering of d_{z^2} and d_{xy} is not entirely obvious because d_{z^2} has a significant band of electron density centered in the *xy* plane. We will not deal with this issue here. We can obtain the linear complex from the octahedral arrangement by arbitrarily leaving the two li-

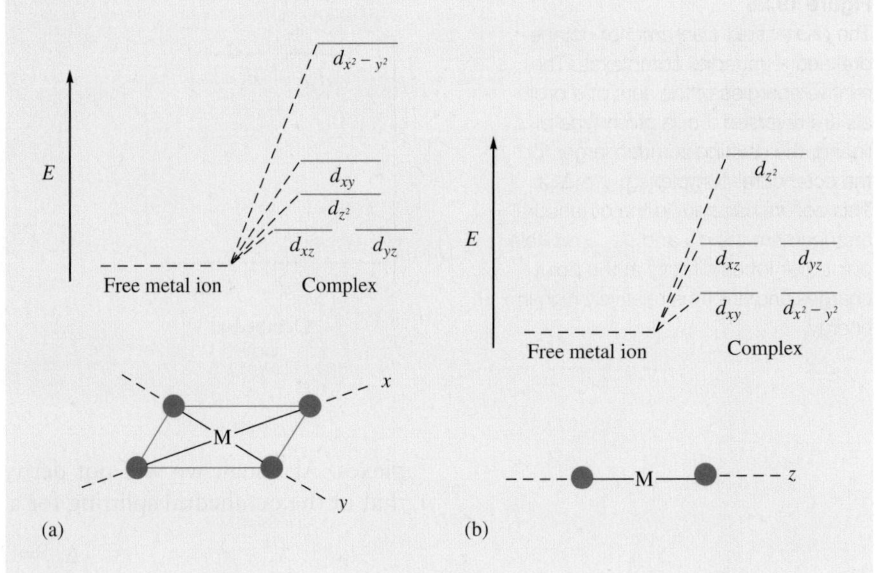

Figure 19.29
(a) The crystal field diagram for a square planar complex oriented in the *xy* plane with ligands along the *x* and *y* axes. The position of the d_{z^2} orbital is higher than those of the d_{xz} and d_{yz} orbitals because of the "doughnut" of electron density in the *xy* plane. The actual position of d_{z^2} is somewhat uncertain and varies in different square planar complexes. (b) The crystal field diagram for a linear complex where the ligands lie along the *z* axis.

gands along the *z* axis and removing the four in the *xy* plane. This means that only the d_{z^2} orbital points at the ligands and is highest in energy.

19.7 The Molecular Orbital Model

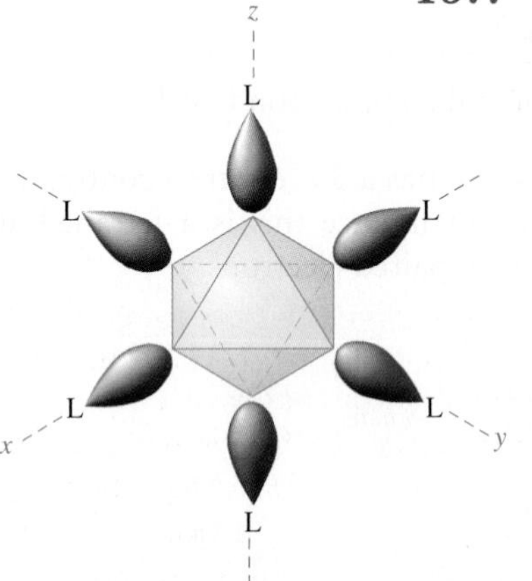

Figure 19.30
An octahedral arrangement of ligands showing their lone pair orbitals.

Although quite successful in accounting for the magnetic and spectral (light absorption) properties of complex ions, the crystal field model has limited use other than explaining those properties dependent on the *d*-orbital splitting. For example, the model gives a very crude and misleading view of the nature of the metal–ligand bonding.

Of the various models in their simplest forms, the molecular orbital (MO) model gives the most realistic view of the bonding in complex ions. Recall from our discussions in Chapter 14 that the MO model postulates that a new set of orbitals characteristic of the molecule is formed from the atomic orbitals of the component atoms. To illustrate how this model can be applied to complex ions, we will describe the MOs in an octahedral complex of general formula ML_6^{n+}. To keep things as simple as possible, we will focus only on those ligand orbitals having lone pairs that interact with the metal ion valence orbitals (3*d*, 4*s*, and 4*p*). There are two important considerations in predicting how atomic orbitals will interact to form MOs:

1. *Extent of orbital overlap.* Atomic orbitals must have a net overlap in space to form MOs. Figure 19.30 shows an octahedral arrangement of ligands with lone pair orbitals. Recall from our previous discussion that two of the metal ion's 3*d* orbitals (d_{z^2} and $d_{x^2-y^2}$) point at the ligands and thus will form MOs with the ligand lone pair orbitals. On the other hand, the d_{xy}, d_{xz}, and d_{yz} orbitals point *between* the ligands and thus will not be involved in the σ bonding with the ligands. The spherical 4*s* orbital of the metal ion overlaps with all the ligand lone pair orbitals, and each of the 4*p* orbitals overlaps with pairs of ligand lone pair orbitals on the three coordinate axes. Thus the d_{z^2}, $d_{x^2-y^2}$, 4*s*, 4_{p_x}, 4_{p_y}, and 4_{p_z} orbitals will be involved in the σ MOs in the complex ion.
2. *Relative orbital energies.* Atomic orbitals that are close in energy will interact more strongly than those widely separated in energy.

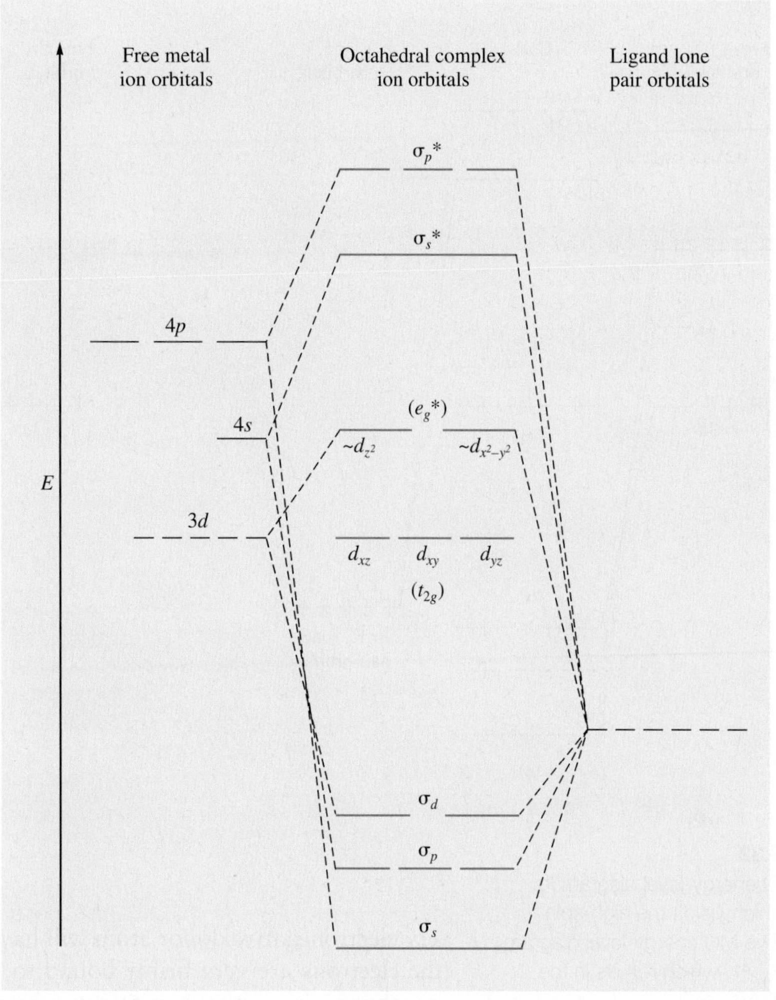

Figure 19.31
The MO energy-level diagram for an octahedral complex ion ML_6^{n+}. The e_g* MOs are essentially pure d_{z^2} and $d_{x^2-y^2}$ orbitals of the metal ion. Little mixing with the ligand orbitals occurs because of the large energy difference between the 3*d* and ligand orbitals.

The MO model was introduced in Section 14.2.

When we apply these principles to the general complex ML_6^{n+}, we obtain the energy-level diagram shown in Fig. 19.31. Note that the σ_s, σ_p, and σ_d MOs are bonding MOs; they are *lower* in energy than the ligand orbitals and metal ion orbitals that mix to form them. Electrons in these MOs have lower energies than they do in either the isolated metal ion or the ligands. These bonding electrons are mainly responsible for the stability of the complex ion.

Because the d_{xz}, d_{yz}, and d_{xy} orbitals (the t_{2g} set) of the metal ion do not overlap with the ligand orbitals, they remain unchanged. Thus they have the same energy in the complex ion as they had in the free metal ion and make no contribution to the stability of the complex. They are called *nonbonding orbitals.*

The e_g* MOs are antibonding orbitals, since they are higher in energy than the atomic orbitals that mix to form them. Since electrons in these orbitals have higher energies than they do in the free metal ion, they destabilize the complex ion relative to the separated metal ion and ligands. However, *the most important characteristic of the* e_g* *orbitals is that they are primarily composed of* d_{z^2} *and* $d_{x^2-y^2}$ *atomic orbitals,* with relatively little contribution from ligand orbitals. This lack of mixing is caused by the large energy difference between the ligand orbitals and the metal ion 3*d* orbitals. Thus we can see that the MO model predicts the same type of *d*-orbital splitting as the crystal field model, with the added advantage of giving a much more realistic picture of both the metal–ligand bonding interaction and the origin of the splitting. Because it is a more realistic physical model, we can use the MO model to explain why different ligands produce different magnitudes of splitting. In particular, a ligand with a

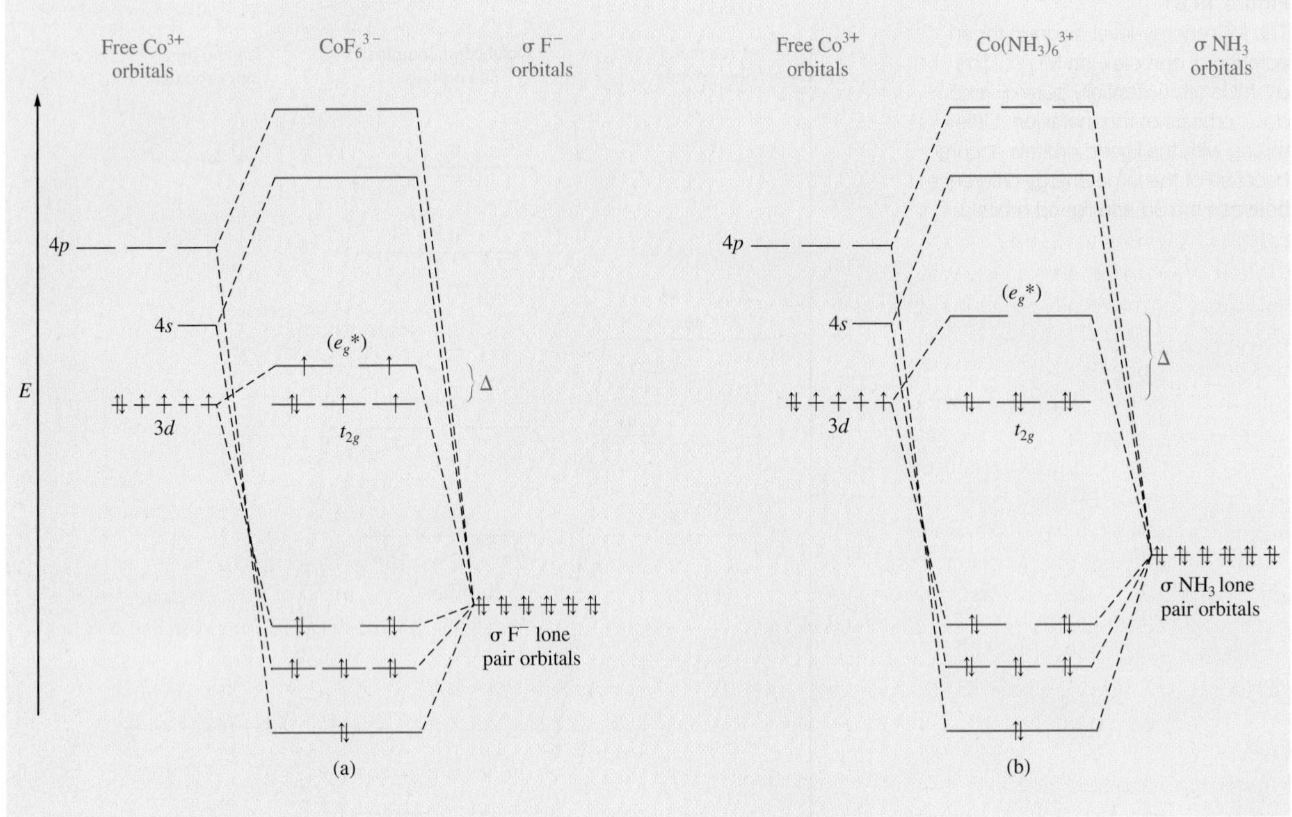

Figure 19.32
(a) The MO energy-level diagram for CoF_6^{3-}, which yields the high-spin case. (b) The MO energy-level diagram for $Co(NH_3)_6^{3+}$, which results in the low-spin case.

very electronegative donor atom will have lone pair orbitals of very low energy (the electrons are very firmly bound to the ligand); these orbitals do not mix very thoroughly with the metal ion orbitals. This will result in a small difference between the t_{2g} (nonbonding) and e_g^* (antibonding) orbitals. In other words, in this case the e_g orbitals are not much perturbed by the small amount of mixing that occurs. This means that the e_g orbital energies are not changed much by complex ion formation. For example, in CoF_6^{3-} [Fig. 19.32(a)] the low-energy orbitals of the very electronegative fluoride ion ligands mix only to a small extent with the cobalt ion orbitals. A small amount of *d*-orbital splitting results, giving the high-spin case.

On the other hand, in $Co(NH_3)_6^{3+}$ the ammonia lone pair orbitals are closer in energy to the metal ion orbitals, a situation that produces a larger degree of mixing between the two sets of orbitals. This in turn gives a relatively large amount of *d*-orbital splitting, and the low-spin case results [Fig. 19.32(b)].

We have seen that the MO model correctly predicts the *d*-orbital splitting in octahedral complexes, thereby accounting for the magnetic and spectral properties of these species. Moreover, it has a major advantage in that it accounts in a realistic way for metal–ligand bonding. However, it suffers from the disadvantage of being much more complicated to apply than the crystal field model. To take advantage of the relative strengths of the MO and crystal field models, the two have been combined to give the **ligand field model.** We will not consider the details of this model here.

The MO model can be applied to all types of complex ions, although we will not extend the model to other cases here. In addition, note that in our use of the MO model to describe octahedral complexes, we considered only σ bonding effects to keep the model as simple as possible. For a complete treatment of complex ions, π bonding interactions also would have to be considered.

Transition Metal Ions Lend Color to Gems

Chemical Insights

The beautiful pure color of gems, so valued by cultures everywhere, arises from trace transition metal ion impurities in minerals that would otherwise be colorless. For example, the stunning red of a ruby, the most valuable of all gemstones, is caused by Cr^{3+} ions, which replace about 1% of the Al^{3+} ions in the mineral corundum, which is a form of aluminum oxide (Al_2O_3) that is nearly as hard as diamond. In the corundum structure, the Cr^{3+} ions are surrounded by six oxide ions at the vertices of an octahedron. This leads to the characteristic octahedral splitting of chromium's 3*d* orbitals, such that the Cr^{3+} ions absorb strongly in the blue-violet and yellow-green regions of the visible spectrum but transmit red light to give the characteristic ruby color. (On the other hand, if some of the Al^{3+} ions in corundum are replaced by a mixture of Fe^{2+}, Fe^{3+}, and Ti^{4+} ions, the gem is a sapphire with its brilliant blue color, or if some of the Al^{3+} ions are replaced by Fe^{3+} ions, the stone is a yellow topaz.)

Emeralds are derived from the mineral beryl, which is a beryllium aluminum silicate (empirical formula $3BeO \cdot Al_2O_3 \cdot 6SiO_2$). When some of the Al^{3+} ions in beryl are replaced by Cr^{3+} ions, the characteristic green color of emerald results. In this environment the splitting of the Cr^{3+} 3*d* orbitals causes it to strongly absorb yellow and blue-violet light and to transmit green light.

A gem closely related to ruby and emerald is alexandrite, named after Alexander II of Russia. This gem is based on the mineral chrysoberyl, a beryllium aluminate with the empirical formula $BeO \cdot Al_2O_3$ in which approximately 1% of the Al^{3+} ions are replaced by Cr^{3+} ions. In the chrysoberyl environment, Cr^{3+} absorbs strongly in the yellow region of the spectrum. Alexandrite has the interesting property of changing colors depending on the light source. When the first alexandrite stone was discovered deep in a mine in the Russian Ural Mountains in 1831, it appeared to be a deep red color in the firelight of the miners' lamps. However, when the stone was brought to the surface, its color was blue. This seemingly magical color change occurs because the firelight of a miner's helmet is rich in the yellow and red wavelengths of the visible spectrum but does not contain much blue. Absorption of the yellow by the stone produces a reddish color. However, daylight has much more intensity in the blue region than firelight. Thus the extra blue in the light transmitted by the stone gives it a bluish color in daylight.

Photos Courtesy, International Colored Gemstone Association

Alexandrite is a bluish color in sunlight and changes to red in incandescent light.

Once the structure of a natural gem is known, it is usually not very difficult to make the gem artificially. For example, rubies and sapphires are made on a large scale by fusing $Al(OH)_3$ with the appropriate transition metal salts at about 1200°C to make the "doped" corundum. With these techniques gems of astonishing size can be manufactured. Rubies as large as 10 lb and sapphires up to 100 lb have been synthesized. Smaller synthetic stones produced for jewelry are virtually identical to the corresponding natural stones, and it takes great skill for a gemologist to tell the difference.

19.8 The Biological Importance of Coordination Complexes

The ability of metal ions to coordinate with and then release ligands and to easily undergo oxidation and reduction makes them ideal for use in biological systems. For example, metal ion complexes are used in humans for the transport and storage of oxygen, as electron transfer agents, as catalysts, and as drugs. Most of the first-row transition metals are essential for human health, as is summarized in Table 19.18. We will concentrate on iron's role in biological systems, since several of its coordination complexes have been studied extensively.

Iron plays a central role in almost all living cells. In mammals the principal source of energy comes from the oxidation of carbohydrates, proteins, and fats. Although oxygen is the oxidizing agent for these processes, it does not react directly with the nutrient molecules. Instead, the electrons from the breakdown of these nutrients are passed along a complex chain of molecules, called the *respiratory chain,* eventually reaching the O_2 molecule. The principal electron transfer molecules in the respiratory chain are iron-containing species called **cytochromes,** consisting of two main parts: an iron complex called **heme** and a protein. The structure of the heme complex is shown in Fig. 19.33. Note that it contains an iron ion (it can be either Fe^{2+} or Fe^{3+}) coordinated to a rather complicated planar ligand called a **porphyrin.** As a class, porphyrins all contain the same central ring structure but have different substituent groups at the edges of the rings. The various porphyrin molecules act as tetradentate ligands for many metal ions, including iron, cobalt, and magnesium. In fact, *chlorophyll,* a substance essential to the process of photosynthesis, is a magnesium–porphyrin complex of the type shown in Fig. 19.34.

A protein is a large molecule assembled from α-amino acids, which have the general structure

$$\begin{array}{c} R \\ | \\ H_2N\text{—}C\text{—}COOH \\ | \\ H \end{array}$$

where R varies.

In addition to participating in the transfer of electrons from nutrients to oxygen, iron plays a principal role in the transport and storage of oxygen in mammalian blood and tissues. Oxygen is stored using a molecule called **myoglobin,** which consists of a heme complex and a protein in a structure very

Table 19.18

The First-Row Transition Metals and Their Biological Significance

First-Row Transition Metal	Biological Function(s)
Scandium	None known
Titanium	None known
Vanadium	None known in humans
Chromium	Assists insulin in the control of blood sugar; may also be involved in the control of cholesterol
Manganese	Necessary for a number of enzymatic reactions
Iron	Component of hemoglobin and myoglobin; involved in the electron transport chain
Cobalt	Component of vitamin B_{12}, which is essential for the metabolism of carbohydrates, fats, and proteins
Nickel	Component of the enzymes urease and hydrogenase
Copper	Component of several enzymes; assists in iron storage; involved in the production of color pigments of hair, skin, and eyes
Zinc	Component of insulin and many enzymes

Figure 19.33
The heme complex, in which an Fe^{2+} ion is coordinated to four nitrogen atoms of a planar porphyrin ligand.

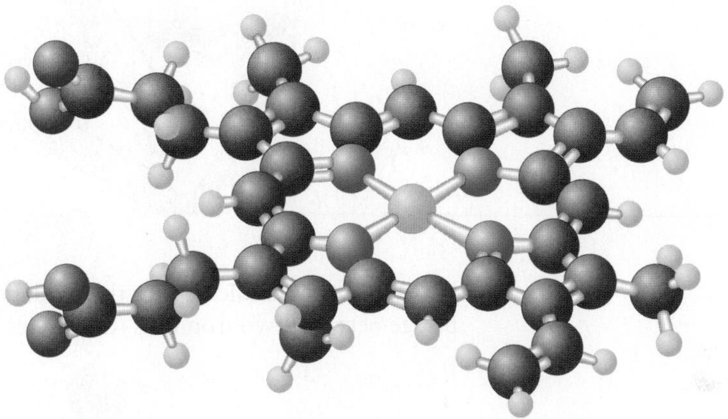

similar to that of the cytochromes. In myoglobin the Fe^{2+} ion is coordinated to four nitrogen atoms of the porphyrin ring and to one nitrogen atom of the protein chain, as shown in Fig. 19.35. Since Fe^{2+} is normally six-coordinate, this leaves one position open for attachment of an O_2 molecule.

One especially interesting feature of myoglobin is that it involves an O_2 molecule attaching directly to Fe^{2+}. However, if gaseous O_2 is bubbled into an aqueous solution containing "bare" heme (no protein attached), the Fe^{2+} is immediately oxidized to Fe^{3+}. This oxidation of the Fe^{2+} in heme does not happen in myoglobin. This fact is of crucial importance because Fe^{3+} does not form a coordinate covalent bond with O_2. Therefore, myoglobin would not function if the bound Fe^{2+} could be oxidized. Since Fe^{2+} in the "bare" heme complex can be oxidized, it must be the protein that somehow prevents the oxidation. How?

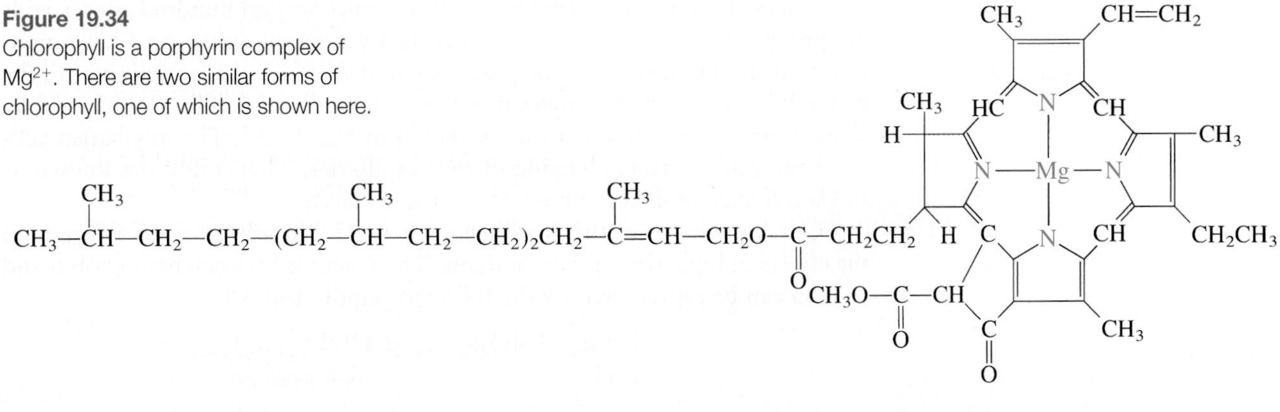

Figure 19.34
Chlorophyll is a porphyrin complex of Mg^{2+}. There are two similar forms of chlorophyll, one of which is shown here.

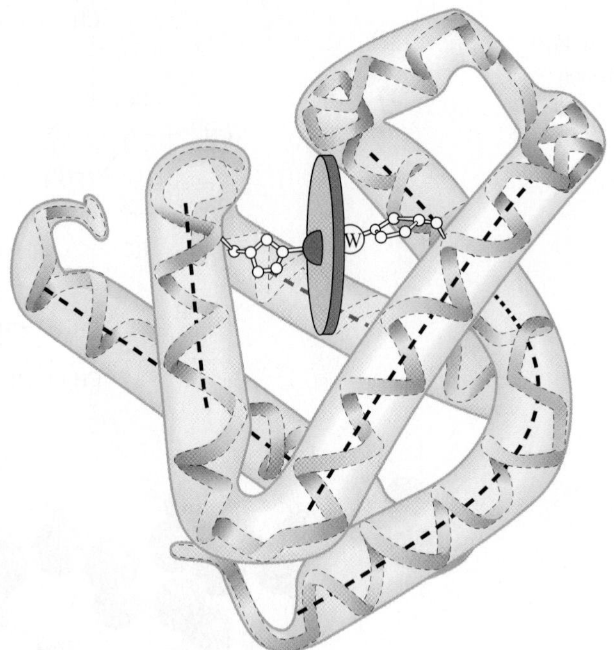

Figure 19.35
A representation of the myoglobin molecule. The Fe^{2+} ion is coordinated to four nitrogen atoms in the porphyrin of the heme (represented by the disk) and one nitrogen in the protein chain. This leaves a sixth coordination position (indicated by the W) available for an oxygen molecule.

Research results indicate that the oxidation of Fe^{2+} to Fe^{3+} involves an oxygen bridge between two iron ions (the circles indicate the ligands):

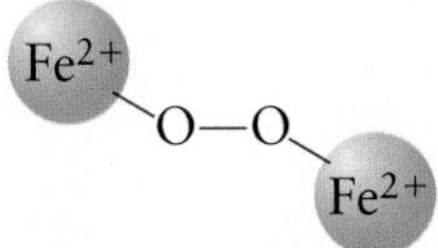

The bulky protein around the heme group in myoglobin prevents two myoglobin molecules from getting close enough to form the oxygen bridge; therefore, oxidation of the Fe^{2+} is prevented.

The transport of O_2 in the blood is carried out by **hemoglobin,** a molecule consisting of four myoglobin-like units, as shown in Fig. 19.36. Each hemoglobin can therefore bind four O_2 molecules to form a bright red diamagnetic complex. The diamagnetism indicates that oxygen is a strong-field ligand toward Fe^{2+}, which has a $3d^6$ electron configuration. When the oxygen molecule is released, a water molecule occupies the sixth coordination position around each Fe^{2+}, yielding a bluish paramagnetic complex (H_2O is a weak-field ligand toward Fe^{2+}) that gives venous blood its characteristic bluish tint.

Hemoglobin dramatically demonstrates how sensitive the function of a biomolecule is to its structure. In certain people, in the synthesis of the proteins needed for hemoglobin, an improper amino acid is inserted into the protein in two places. This may not seem very serious, since several hundred amino acids are present. However, because the incorrectly inserted amino acid has a nonpolar substituent instead of the polar one found on the proper amino acid, the hemoglobin drastically changes its shape. The red blood cells are then sickle-shaped rather than disk-shaped, as shown in Fig. 19.37. The misshapen cells can aggregate, causing clogging of tiny capillaries. This condition, known as *sickle cell anemia,* is the subject of intense research.

Our knowledge of the workings of hemoglobin allows us to understand the effects of high altitudes on humans. The reaction between hemoglobin and oxygen can be represented by the following equilibrium:

$$\underset{\text{Hemoglobin}}{Hb(aq)} + 4O_2(g) \rightleftharpoons \underset{\text{Oxyhemoglobin}}{Hb(O_2)_4(aq)}$$

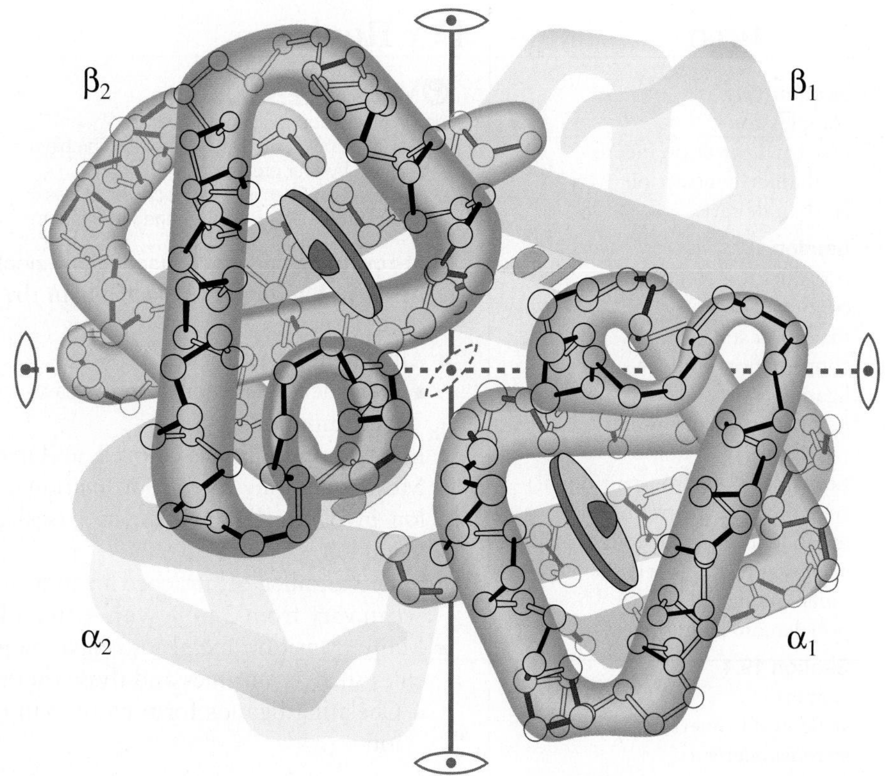

Figure 19.36
A representation of the hemoglobin structure. There are two slightly different types of protein chains (α and β). Each hemoglobin has two α chains and two β chains, each with a heme complex near the center. Thus each hemoglobin molecule can complex with four O_2 molecules.

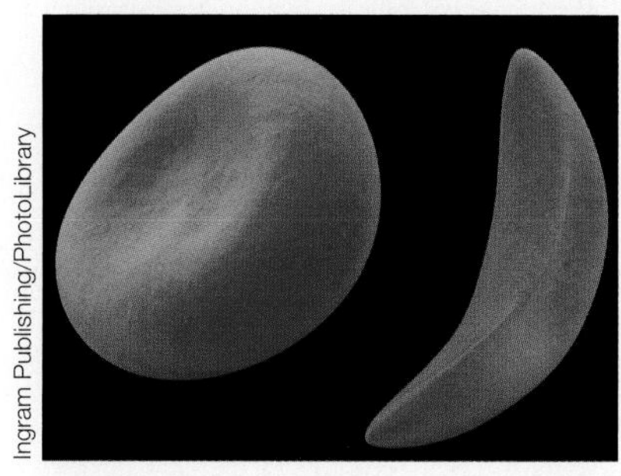

Figure 19.37
A normal red blood cell (left) and a sickle cell, both magnified 18,000 times.

At high altitudes, where the oxygen content of the air is lower than at sea level, the position of this equilibrium will shift to the left, according to Le Châtelier's principle. Because less oxyhemoglobin is formed, fatigue, dizziness, and even a serious illness called *high-altitude sickness* can result. One way to combat this problem is to use supplemental oxygen, as most high-altitude mountain climbers do. However, this is impractical for people who live at high elevations. In fact, the human body adapts to the lower oxygen concentrations by making more hemoglobin, causing the equilibrium to shift back to the right. Someone moving from Chicago to Boulder, Colorado (elevation 5300 feet), would notice the effects of the new altitude for a couple of weeks, but as the hemoglobin level increases, the effects disappear. This change is called *high-altitude acclimatization,* which explains why athletes who want to compete at high elevations should practice under such conditions for several weeks prior to the event.

Our understanding of the biological role of iron also enables us to explain the toxicities of substances such as carbon monoxide and the cyanide ion. Both CO and CN^- are very good ligands toward iron and so can interfere with the normal workings of the iron complexes in the body. For example, carbon monoxide has about 200 times the affinity for the Fe^{2+} in hemoglobin as oxygen does. The resulting stable complex, **carboxyhemoglobin,** prevents the normal uptake of O_2, thus depriving the body of needed oxygen. Asphyxiation can result if enough carbon monoxide is present in the air. The mechanism for the toxicity of the cyanide ion is somewhat different. Cyanide coordinates strongly to cytochrome oxidase, an iron-containing cytochrome enzyme that catalyzes the oxidation–reduction reactions of certain cytochromes. The coordinated cyanide thus prevents the electron transfer process, causing rapid death. Because of its behavior, cyanide is called a *respiratory inhibitor*.

Key Terms

Section 19.1
complex ion
first-row transition metals
lanthanide contraction
lanthanide series

Section 19.3
coordination compound
counter ion
oxidation state
coordination number
ligand
Lewis acid (electron-pair acceptor)
Lewis base (electron-pair donor)
coordinate covalent bond
monodentate (unidentate) ligand
chelating ligand (chelate)
bidentate ligand
polydentate ligand

Section 19.4
isomers
structural isomerism
stereoisomerism
coordination isomerism
linkage isomerism
geometrical (*cis–trans*) isomerism
cis isomer
trans isomer
optical isomerism
chiral
enantiomers

Section 19.6
crystal field model
d-orbital splitting
strong-field (low-spin) case
weak-field (high-spin) case
spectrochemical series

Section 19.7
ligand field model

Section 19.8
cytochromes
heme
porphyrin
myoglobin
hemoglobin
carboxyhemoglobin

For Review

First-row transition metals (scandium–zinc)
- All have one or more electrons in the 4*s* orbital and various numbers of 3*d* electrons
- All exhibit metallic properties
  - A particular element often shows more than one oxidation state in its compounds
- Most compounds are colored, and many are paramagnetic
- Most commonly form coordination compounds containing a complex ion involving ligands (Lewis bases) attached to a central transition metal ion
  - The number of attached ligands (called the coordination number) can vary from 2 to 8, with 4 and 6 being most common
- Many transition metal ions have major biologic importance in molecules such as enzymes and those that transport and store oxygen
  - Chelating ligands form more than one bond to the transition metal ion

Isomerism
- Isomers: two or more compounds with the same formula but different properties
  - Coordination isomerism: the composition of the coordination sphere varies
  - Linkage isomerism: the point of attachment of one or more ligands varies
  - Stereoisomerism: isomers have identical bonds but different spatial arrangements
    - Geometric isomerism: ligands assume different relative positions in the coordination sphere; examples are *cis* and *trans* isomers
    - Optical isomerism: molecules with nonsuperimposable mirror images rotate plane-polarized light in opposite directions

Spectral and magnetic properties
- Usually explained in terms of the crystal field model
- Model assumes the ligands are point charges that split the energies of the 3*d* orbitals
- Color and magnetism are explained in terms of how the 3*d* electrons occupy the split 3*d* energy levels
  - Strong-field case: relatively large orbital splitting
  - Weak-field case: relatively small orbital splitting

Metallurgy
- The processes connected with separating a metal from its ore
  - The minerals in ores are often converted to oxides (roasting) before being reduced to the metal (smelting)
- The metallurgy of iron: most common method for reduction uses a blast furnace; process involves iron ore, coke, and limestone
  - Impure product (~90% iron) is called pig iron
- Steel is manufactured by oxidizing the impurities in pig iron

Discussion Questions

These questions are designed to be considered by groups of students in class. Often these questions work well for introducing a particular topic in class.

1. You isolate a compound with the formula $PtCl_4 \cdot 2KCl$. From electrical conductance tests of an aqueous solution of the compound, you find that three ions are present, and you also notice that the addition of $AgNO_3$ does not cause a precipitate. Give the formula for this compound that shows the complex ion present. Explain your findings, and name this compound.
2. Both $Ni(NH_3)_4^{2+}$ and $Ni(CN)_4^{2-}$ have four ligands. The first is paramagnetic and the second is diamagnetic. Are the complex ions tetrahedral or square planar? Explain.
3. Which is more likely to be paramagnetic, $Fe(CN)_6^{4-}$ or $Fe(H_2O)_6^{2+}$? Explain.
4. A metal ion in a high-spin octahedral complex has two more unpaired electrons than the same ion does in a low-spin octahedral complex. Name some possible metal ions for which this arrangement would apply.
5. Discuss the advantages and disadvantages of the molecular orbital model and the crystal field model. Under which circumstances is it more useful to use the MO model? The crystal field model?

Exercises

OWL Interactive versions of these problems may be assigned in OWL.

A blue exercise number indicates that the answer to that exercise appears at the back of this book and a solution appears in the *Solutions Guide.*

Transition Metals

6. Write electron configurations for each of the following.
   a. Ti, Ti^{2+}, Ti^{4+}
   b. Re, Re^{2+}, Re^{3+}
   c. Ir, Ir^{2+}, Ir^{3+}
7. Write electron configurations for each of the following.
   a. Cr, Cr^{2+}, Cr^{3+}
   b. Cu, Cu^{+}, Cu^{2+}
   c. V, V^{2+}, V^{3+}
8. What two first-row transition metals have unexpected electron configurations? A statement in the text says that first-row transition metal ions do not have 4*s* electrons. Why not? Why do transition metal ions often have several oxidation states, whereas representative metals generally have only one?
9. What is the lanthanide contraction? How does the lanthanide contraction affect the properties of the 4*d* and 5*d* transition metals?
10. We expect the atomic radius to increase down a group in the periodic table. Can you suggest why the atomic radius of hafnium breaks this rule? (See the following data.)

Element	Atomic Radius (Å)	Element	Atomic Radius (Å)
Sc	1.57	Ti	1.477
Y	1.693	Zr	1.593
La	1.915	Hf	1.476

11. Molybdenum is obtained as a by-product of copper mining or is mined directly (primary deposits are in the Rocky Mountains in Colorado). In both cases it is obtained as MoS_2, which is then converted to MoO_3. The MoO_3 can be used directly in the production of stainless steel for high-speed tools (which accounts for about 85% of the molybdenum used). Molybdenum can be purified by dissolving MoO_3 in aqueous ammonia and crystallizing ammonium molybdate. Depending on conditions, either $(NH_4)_2Mo_2O_7$ or $(NH_4)_6Mo_7O_{24} \cdot 4H_2O$ is obtained.
    a. Give names for MoS_2 and MoO_3.
    b. What is the oxidation state of Mo in each of the compounds mentioned above?
12. Titanium dioxide, the most widely used white pigment, occurs naturally but is often colored by the presence of impurities. The chloride process is often used in purifying rutile, a mineral form of titanium dioxide.
    a. Show that the unit cell for rutile, illustrated below, conforms to the formula TiO_2. (*Hint:* Recall the discussion in Section 16.4.)

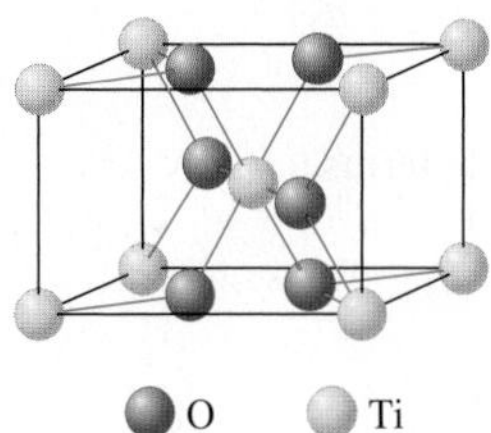

    b. The reactions for the chloride process are

$$2TiO_2(s) + 3C(s) + 4Cl_2(g) \xrightarrow{950°C} 2TiCl_4(g) + CO_2(g) + 2CO(g)$$

$$TiCl_4(g) + O_2(g) \xrightarrow{1000\text{–}1400°C} TiO_2(s) + 2Cl_2(g)$$

    Assign oxidation states to the elements in both reactions. Which elements are being reduced, and which are being oxidized? Identify the oxidizing agent and the reducing agent in each reaction.

13. The melting and boiling points of the titanium tetrahalides are given below.

	bp (°C)	mp (°C)
TiF_4	284	—
$TiCl_4$	−24	136.5
$TiBr_4$	38	233.5
TiI_4	155	377

Rationalize these data in terms of the bonding in and the intermolecular forces among these compounds.

14. Iron is present in the earth's crust in many types of minerals. The iron oxide minerals are hematite (Fe_2O_3) and magnetite (Fe_3O_4). What is the oxidation state of iron in each mineral? The iron ions in magnetite are a mixture of Fe^{2+} and Fe^{3+} ions. What is the ratio of Fe^{3+} to Fe^{2+} ions in magnetite? The formula for magnetite is usually written as $FeO \cdot Fe_2O_3$. Does this make sense? Explain.

15. Chromium(VI) forms two different oxyanions, the orange dichromate ion ($Cr_2O_7^{2-}$) and the yellow chromate ion (CrO_4^{2-}). The equilibrium reaction between the two ions is

$$Cr_2O_7^{2-}(aq) + H_2O(l) \rightleftharpoons 2CrO_4^{2-}(aq) + 2H^+(aq)$$

The following pictures show what happens when sodium hydroxide is added to a dichromate solution.

Explain what happened.

Coordination Compounds

16. Define each of the following terms.
 a. coordination compound
 b. complex ion
 c. counter ions
 d. coordination number
 e. ligand
 f. chelate
 g. bidentate

17. How would transition metal ions be classified using the Lewis definition of acids and bases? What must a ligand have to bond to a metal? What do we mean when we say that a bond is a "coordinate covalent bond"?

18. When a metal ion has a coordination number of 2, 4, or 6, what are the observed geometries and associated bond angles? For each of the following, give the correct formulas for the complex ions.
 a. linear Ag^+ complex ions having CN^- ligands
 b. tetrahedral Cu^+ complex ions having H_2O ligands
 c. tetrahedral Mn^{2+} complex ions having oxalate ligands
 d. square planar Pt^{2+} complex ions having NH_3 ligands
 e. octahedral Fe^{3+} complex ions having EDTA ligands
 f. octahedral Co^{2+} complex ions having Cl^- ligands
 g. octahedral Cr^{3+} complex ions having ethylenediamine ligands

19. The compound cisplatin, $Pt(NH_3)_2Cl_2$, has been studied extensively as an antitumor agent. The reaction for the synthesis of cisplatin is:

$$K_2PtCl_4(aq) + 2NH_3(aq) \longrightarrow Pt(NH_3)_2Cl_2(s) + 2KCl(aq)$$

Write the electron configuration for platinum ion in cisplatin. Most d^8 transition metal ions exhibit square planar geometry. With this and the name in mind, draw the structure of cisplatin.

20. What is the electron configuration for the transition metal ion(s) in each of the following compounds?
 a. $(NH_4)_2[Fe(H_2O)_2Cl_4]$
 b. $[Co(NH_3)_2(NH_2CH_2CH_2NH_2)_2]I_2$
 c. $Na_2[TaF_7]$
 d. $[Pt(NH_3)_4I_2][PtI_4]$
 Pt forms +2 and +4 oxidation states in compounds.

21. Name the following complex ions:

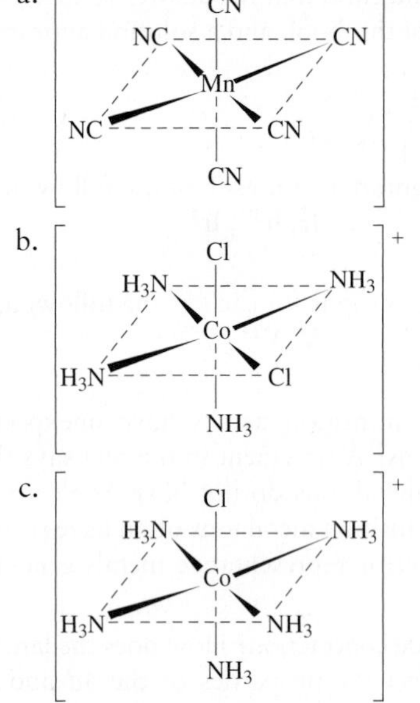

22. Name the following coordination compounds.
 a. $[Co(NH_3)_6]Cl_2$
 b. $[Co(H_2O)_6]I_3$
 c. $K_2[PtCl_4]$
 d. $K_4[PtCl_6]$
 e. $[Co(NH_3)_5Cl]Cl_2$
 f. $[Co(NH_3)_3(NO_2)_3]$

23. Name the following complex ions.
 a. $Ru(NH_3)_5Cl^{2+}$
 b. $Fe(CN)_6^{4-}$
 c. $Mn(NH_2CH_2CH_2NH_2)_3^{2+}$
 d. $Co(NH_3)_5NO_2^{2+}$

24. Name the following coordination compounds.
 a. $[Cr(H_2O)_5Br]Br_2$
 b. $Na_3[Co(CN)_6]$
 c. $[Fe(NH_2CH_2CH_2NH_2)_2(NO_2)_2]Cl$
 d. $[Pt(NH_3)_4I_2][PtI_4]$

25. Give formulas for the following.
   a. potassium tetrachlorocobaltate(II)
   b. aquatricarbonylplatinum(II) bromide
   c. sodium dicyanobis(oxalato)ferrate(III)
   d. triamminechloroethylenediaminechromium(III) iodide

26. Give formulas for the following complex ions.
   a. tetrachloroferrate(III) ion
   b. pentaammineaquaruthenium(III) ion
   c. tetracarbonyldihydroxochromium(III) ion
   d. amminetrichloroplatinate(II) ion

27. A coordination compound of cobalt(III) contains four ammonia molecules, one sulfate ion, and one chloride ion. Addition of aqueous $BaCl_2$ solution to an aqueous solution of the compound gives no precipitate. Addition of aqueous $AgNO_3$ to an aqueous solution of the compound produces a white precipitate. Propose a structure for this coordination compound.

28. When an aqueous solution of KCN is added to a solution containing Ni^{2+} ions, a precipitate forms, which redissolves upon addition of more KCN solution. No precipitate forms when H_2S is bubbled into this solution. Write reactions describing what happens in this solution. [*Hint:* CN^- is a Brønsted–Lowry base ($K_b \approx 10^{-5}$) and a Lewis base.]

29. Consider aqueous solutions of the following coordination compounds: $Co(NH_3)_6I_3$, $Pt(NH_3)_4I_4$, Na_2PtI_6, and $Cr(NH_3)_4I_3$. If aqueous $AgNO_3$ is added to separate beakers containing solutions of each coordination compound, how many moles of AgI will precipitate per mole of transition metal present? Assume that each transition metal ion forms an octahedral complex.

30. A series of chemicals was added to some $AgNO_3(aq)$. $NaCl(aq)$ was added first to the silver nitrate solution with the end result shown below in test tube 1, $NH_3(aq)$ was then added with the end result shown in test tube 2, and $HNO_3(aq)$ was added last with the end result shown in test tube 3.

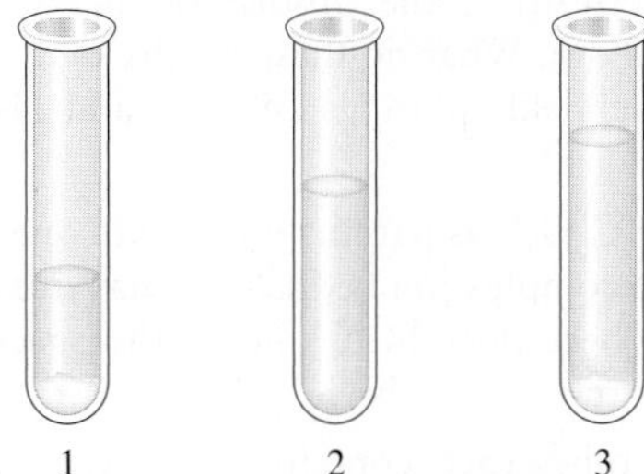

Explain the results shown in each test tube. Include a balanced equation for the reaction(s) taking place.

31. Define each of the following, and give examples.
   a. isomers
   b. structural isomers
   c. stereoisomers
   d. coordination isomers
   e. linkage isomers
   f. geometric isomers
   g. optical isomers

32. How many bonds could each of the following chelating ligands form with a metal ion?
   a. acetylacetone (acacH)

$$CH_3-\overset{\overset{\displaystyle O}{\|}}{C}-CH_2-\overset{\overset{\displaystyle O}{\|}}{C}-CH_3$$

   b. diethylenetriamine

$$NH_2-CH_2-CH_2-NH-CH_2-CH_2-NH_2$$

   c. salen

=N N=
OH HO

   d. porphine

NH N
N HN

33. Draw geometrical isomers of each of the following complex ions.
   a. $[Co(C_2O_4)_2(H_2O)_2]^-$
   b. $[Pt(NH_3)_4I_2]^{2+}$
   c. $[Ir(NH_3)_3Cl_3]$
   d. $[Cr(en)(NH_3)_2I_2]^+$

34. Draw structures for each of the following.
   a. *cis*-dichloroethylenediamineplatinum(II)
   b. *trans*-dichlorobis(ethylenediamine)cobalt(II)
   c. *cis*-tetraamminechloronitrocobalt(III) ion
   d. *trans*-tetraamminechloronitritocobalt(III) ion
   e. *trans*-diaquabis(ethylenediamine)copper(II) ion

35. Figure 19.18 shows that the *cis* isomer of $Co(en)_2Cl_2^+$ is optically active, whereas the *trans* isomer is not optically active. Is the same true for $Co(NH_3)_4Cl_2^+$? Explain.

36. For the process

$$Co(NH_3)_5Cl^{2+} + Cl^- \longrightarrow Co(NH_3)_4Cl_2^+ + NH_3$$

What would be the expected ratio of *cis* to *trans* isomers in the product?

37. Amino acids can act as ligands toward transition metal ions. The simplest amino acid is glycine ($NH_2CH_2CO_2H$). Draw a structure of the glycinate anion ($NH_2CH_2CO_2^-$), acting as a bidentate ligand. Draw the structural isomers of the square planar complex $Cu(NH_2CH_2CO_2)_2$.

38. The carbonate ion (CO_3^{2-}) can act as either a monodentate or a bidentate ligand. Draw a picture of CO_3^{2-} coordinating to a metal ion as a bidentate and as a monodentate ligand. The carbonate ion can also act as a bridge between two metal ions. Draw a picture of a CO_3^{2-} ion bridging between two metal ions.

39. Which of the following ligands are capable of linkage isomerism? Explain your answer.

$$SCN^-, N_3^-, NO_2^-, NH_2CH_2CH_2NH_2, OCN^-, I^-$$

40. Draw all geometrical and linkage isomers of $Co(NH_3)_4(NO_2)_2$.

41. Acetylacetone, abbreviated acacH, is a bidentate ligand. It loses a proton and coordinates as $acac^-$, as shown below, where M is a transition metal:

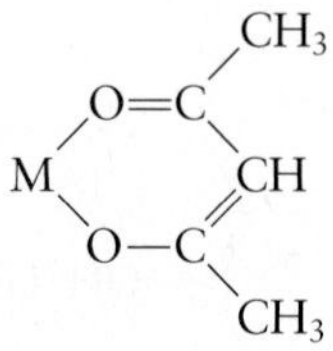

CH_3
O=C
M CH
O—C
CH_3

Which of the following complexes are optically active: *cis*-$Cr(acac)_2(H_2O)_2$, *trans*-$Cr(acac)_2(H_2O)_2$, and $Cr(acac)_3$?

42. Draw all geometrical isomers of $Pt(CN)_2Br_2(H_2O)_2$. Which of these isomers have an optical isomer? Draw the various optical isomers.

43. Draw all the geometrical isomers of $[Cr(en)(NH_3)_2BrCl]^+$. Which of these isomers also have an optical isomer? Draw the various isomers.

Bonding, Color, and Magnetism in Coordination Compounds

44. For the following crystal field diagrams, label each as low spin, high spin, or cannot tell. Explain your answers.

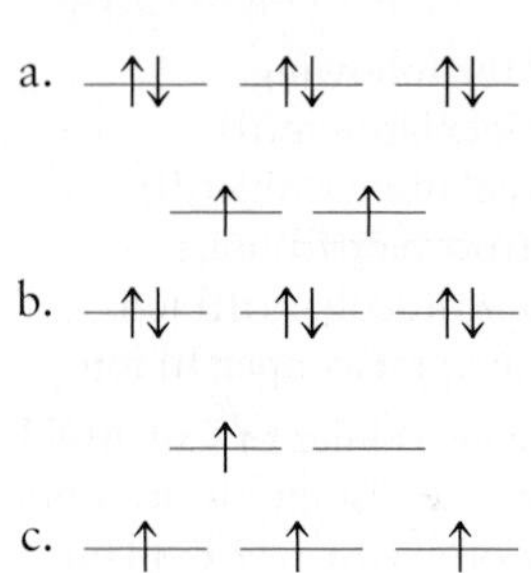

45. Define each of the following.
a. weak-field ligand
b. strong-field ligand
c. low-spin complex
d. high-spin complex

46. Compounds of copper(II) are generally colored, but compounds of copper(I) are not. Explain. Would you expect $Cd(NH_3)_4Cl_2$ to be colored? Explain.

47. Compounds of Sc^{3+} are not colored, but those of Ti^{3+} and V^{3+} are. Why?

48. Match the crystal field diagrams given below with the following complex ions.

$[Cr(NH_3)_5Cl]^{2+}$ $[Co(NH_3)_4Br_2]^+$ (assume strong field) $[Fe(H_2O)_6]^{3+}$ (assume weak field)

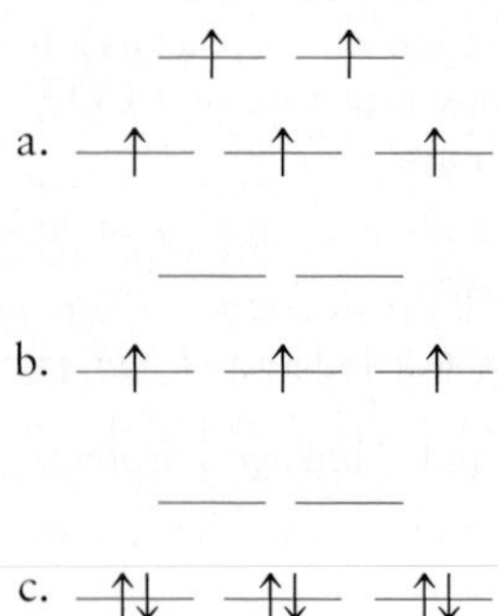

49. Match the crystal field diagrams given below with the following complex ions.

$[Fe(CN)_6]^{3-}$ $[Mn(H_2O)_6]^{2+}$

a. 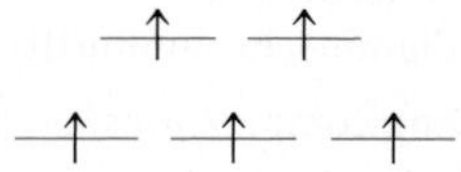 ↑ ↑ / ↑ ↑ ↑

b. 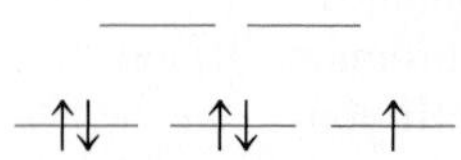 — — / ↑↓ ↑↓ ↑

50. Consider the complex ions $Co(NH_3)_6^{3+}$, $Co(CN)_6^{3-}$, and CoF_6^{3-}. The wavelengths of absorbed electromagnetic radiation for these compounds are (in no specific order) 770 nm, 440 nm, and 290 nm. Match the complex ion to the wavelength of absorbed electromagnetic radiation.

51. Draw the *d*-orbital splitting diagrams for the octahedral complex ions of each of the following.
a. Fe^{2+} (high and low spin)
b. Fe^{3+} (high spin)
c. Ni^{2+}
d. Zn^{2+}
e. Co^{2+} (high and low spin)

52. The $Co(NH_3)_6^{3+}$ ion is diamagnetic, but $Fe(H_2O)_6^{2+}$ is paramagnetic. Explain.

53. How many unpaired electrons are in the following complex ions?
a. $Ru(NH_3)_6^{2+}$ (low-spin case)
b. $Ni(H_2O)_6^{2+}$
c. $V(en)_3^{3+}$

54. Rank the following complex ions in order of increasing wavelength of light absorbed.

$Co(H_2O)_6^{3+}$, $Co(CN)_6^{3-}$, CoI_6^{3-}, $Co(en)_3^{3+}$

55. The complex ion $[Cu(H_2O)_6]^{2+}$ has an absorption maximum at around 800 nm. When four ammonias replace water, $[Cu(NH_3)_4(H_2O)_2]^{2+}$, the absorption maximum shifts to around 600 nm. What do these results signify in terms of the relative field splittings of NH_3 and H_2O? Explain.

56. The complex ion $Fe(CN)_6^{3-}$ is paramagnetic with one unpaired electron. The complex ion $Fe(SCN)_6^{3-}$ has five unpaired electrons. Where does SCN^- lie in the spectrochemical series with respect to CN^-?

57. The following test tubes each contain a different chromium complex ion.

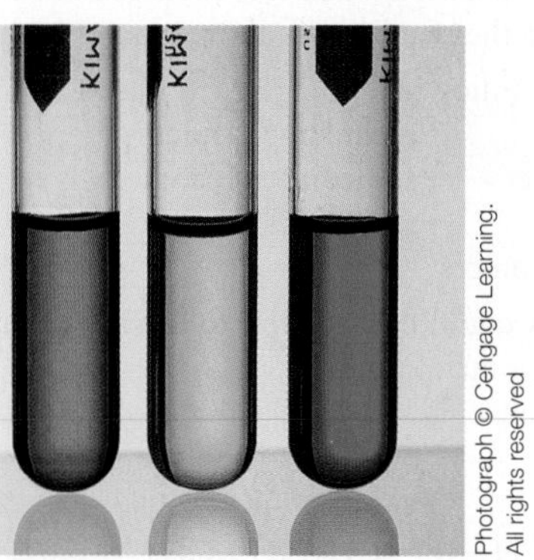

For each compound, predict the predominant color of light absorbed. If the complex ions are $Cr(NH_3)_6^{3+}$, $Cr(H_2O)_6^{3+}$, and $Cr(H_2O)_4Cl_2^{+}$, what is the identity of the complex ion in each test tube? (*Hint:* Reference the spectrochemical series.)

58. Would it be better to use octahedral Ni^{2+} complexes or octahedral Cr^{2+} complexes to determine whether a ligand is a strong-field or a weak-field ligand by determining experimentally the number of unpaired electrons? How else could the relative ligand field strengths be determined?

59. The complex ion $Ru(phen)_3^{2+}$ has been used as a probe for the structure of DNA. (Phen is a bidentate ligand.)
   a. What type of isomerism is found in $Ru(phen)_3^{2+}$?
   b. $Ru(phen)_3^{2+}$ is diamagnetic (as are all complex ions of Ru^{2+}). Draw the crystal field diagram for the *d* orbitals in this complex ion.

phen = 1,10-phenanthroline =

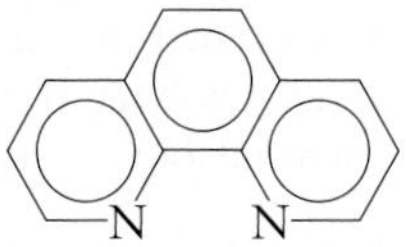

60. Tetrahedral complexes of Co^{2+} are quite common. Use a *d*-orbital splitting diagram to rationalize the stability of Co^{2+} tetrahedral complex ions.

61. Why do tetrahedral complex ions have a different crystal field diagram than octahedral complex ions? Why are virtually all tetrahedral complex ions "high spin"?

62. When concentrated hydrochloric acid is added to a red solution containing the $Co(H_2O)_6^{2+}$ complex ion, the solution turns blue as the tetrahedral $CoCl_4^{2-}$ complex ion forms. Explain this color change.

63. The wavelength of absorbed electromagnetic radiation for $CoBr_4^{2-}$ is 3.4×10^{-6} m. Will the complex ion $CoBr_6^{4-}$ absorb electromagnetic radiation having a wavelength longer or shorter than 3.4×10^{-6} m? Explain.

64. The complex ion $PdCl_4^{2-}$ is diamagnetic. Propose a structure for $PdCl_4^{2-}$.

Additional Exercises

65. When aqueous KI is added gradually to mercury(II) nitrate, an orange precipitate forms. Continued addition of KI causes the precipitate to dissolve. Write balanced equations to explain these observations. (*Hint:* Hg^{2+} reacts with I^- to form HgI_4^{2-}.) Would you expect HgI_4^{2-} to form colored solutions? Explain.

66. A certain first-row transition metal ion forms many different colored solutions. When four coordination compounds of this metal, each having the same coordination number, are dissolved in water, the colors of the solutions are red, yellow, green, and blue. Further experiments reveal that two of the complex ions are paramagnetic with four unpaired electrons and the other two are diamagnetic. What can be deduced from this information about the four coordination compounds?

67. Nickel can be purified by producing the volatile compound nickel tetracarbonyl [$Ni(CO)_4$]. Nickel is the only metal that reacts directly with CO at room temperature. What is the oxidation state of nickel in $Ni(CO)_4$?

68. Why are CN^- and CO toxic to humans?

69. What causes high-altitude sickness, and what is high-altitude acclimatization?

70. Acetylacetone (see Exercise 32, part a), abbreviated acacH, is a bidentate ligand. It loses a proton and coordinates as $acac^-$, as shown below:

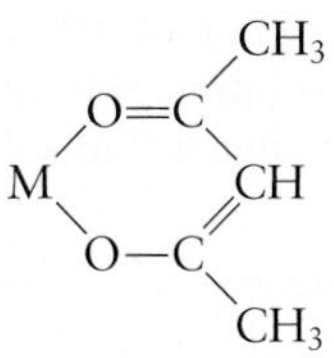

   Acetylacetone reacts with an ethanol solution containing a salt of europium to give a compound that is 40.1% C and 4.71% H by mass. Combustion of 0.286 g of the compound gives 0.112 g Eu_2O_3. Assuming the compound contains only C, H, O, and Eu, determine the formula of the compound formed from the reaction of acetylacetone and the europium salt. (Assume that the compound contains one europium ion.)

71. Ammonia and potassium iodide solutions were added to an aqueous solution of $Cr(NO_3)_3$. A solid was isolated (compound A), and the following data were collected:
   i. When 0.105 g of compound A was strongly heated in excess O_2, 0.0203 g of CrO_3 was formed.
   ii. In a second experiment, it took 32.93 mL of 0.100 *M* HCl to titrate completely the NH_3 present in 0.341 g of compound A.
   iii. Compound A was found to contain 73.53% iodine by mass.
   iv. The freezing point of water was lowered by 0.64°C when 0.601 g of compound A was dissolved in 10.00 g of H_2O (K_f = 1.86°C kg/mol).

   What is the formula of the compound? What is the structure of the complex ion present? [*Hints:* Cr^{3+} is expected to be six-coordinate with NH_3 and (possibly) I^- acting as ligands. The I^- ions will be the counter ions if needed.]

72. In the production of printed circuit boards for the electronics industry, a 0.60-mm layer of copper is laminated onto an insulating plastic board. Next, a circuit pattern made of a chemically resistant polymer is printed on the board. The unwanted copper is removed by chemical etching, and the protective polymer is finally removed by solvents. One etching reaction is

$$[Cu(NH_3)_4]Cl_2(aq) + 4NH_3(aq) + Cu(s) \longrightarrow 2[Cu(NH_3)_4]Cl(aq)$$

a. Is this reaction an oxidation–reduction process? Explain.
b. A plant needs to manufacture 10,000 printed circuit boards, each 8.0 cm × 16.0 cm in area. An average of 80.% of the copper is removed from each board (density of copper = 8.96 g/cm^3). What masses of $[Cu(NH_3)_4]Cl_2$ and NH_3 are needed to accomplish this task? Assume 100% yield.

73. BAL is a chelating agent used in treating heavy metal poisoning. It acts as a bidentate ligand. What types of linkage isomers are possible when BAL coordinates to a metal ion?

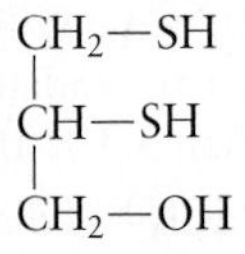

$$\begin{array}{l} CH_2{-}SH \\ | \\ CH{-}SH \\ | \\ CH_2{-}OH \end{array}$$

BAL

74. Until the discoveries of Werner, it was thought that the presence of carbon in a compound was required for it to be optically active. Werner prepared the following compound containing OH^- ions as bridging groups and then separated the optical isomers.

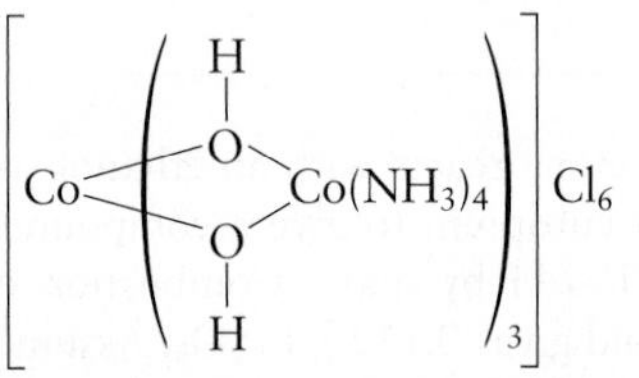

$$\left[Co \left(\begin{array}{c} H \\ | \\ O \\ \\ O \\ | \\ H \end{array} Co(NH_3)_4 \right)_3 \right] Cl_6$$

a. Draw structures of the two optically active isomers of this compound.
b. What are the oxidation states of the cobalt ions?
c. How many unpaired electrons are present if the complex is the low-spin case?

75. Chelating ligands often form more stable complex ions than the corresponding monodentate ligands form with the same donor atoms. For example,

$$Ni^{2+}(aq) + 6NH_3(aq) \rightleftharpoons Ni(NH_3)_6^{2+}(aq) \qquad K_f = 3.2 \times 10^8$$

$$Ni^{2+}(aq) + 3en(aq) \rightleftharpoons Ni(en)_3^{2+}(aq) \qquad K_f = 1.6 \times 10^{18}$$

$$Ni^{2+}(aq) + penten(aq) \rightleftharpoons Ni(penten)^{2+}(aq) \qquad K_f = 2.0 \times 10^{19}$$

where en is ethylenediamine and penten is

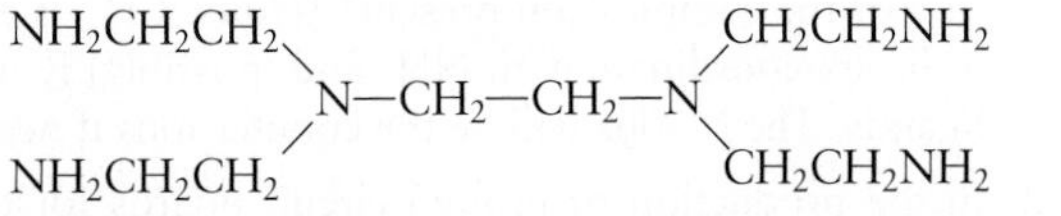

$$\begin{array}{l} NH_2CH_2CH_2 \\ \qquad\qquad N{-}CH_2{-}CH_2{-}N \\ NH_2CH_2CH_2 \end{array} \begin{array}{l} CH_2CH_2NH_2 \\ \\ CH_2CH_2NH_2 \end{array}$$

The increased stability that results is called the chelate effect. From the bond energies, would you expect the enthalpy changes for the preceding reactions to be very different? What is the order (from least favorable to most favorable) of the entropy changes for the preceding reactions? How do the values of the formation constants correlate with $\Delta S°$? How can this be used to explain the chelate effect?

76. There are three salts that contain octahedral complex ions of chromium and have the molecular formula $CrCl_3 \cdot 6H_2O$. Treating 0.27 g of the first salt with a strong dehydrating agent resulted in a mass loss of 0.036 g. Treating 270 mg of the second salt with the same dehydrating agent resulted in a mass loss of 18 mg. The third salt did not lose any mass when treated with the same dehydrating agent. Addition of excess aqueous silver nitrate to 100.0-mL portions of 0.100 *M* solutions of each salt resulted in the formation of different masses of silver chloride; one solution yielded 1430 mg AgCl; another, 2870 mg AgCl; the third, 4300 mg AgCl. Two of the salts are green and one is violet. Suggest probable structural formulas for these salts, defending your answer on the basis of the preceding observations. State which salt is most likely to be violet.

77. Henry Taube, the 1983 Nobel Prize winner in chemistry, has studied the mechanisms of oxidation–reduction reactions involving transition metal complexes. In one experiment he and his students studied the following reaction:

$$Cr(H_2O)_6^{2+}(aq) + Co(NH_3)_5Cl^{2+}(aq) \longrightarrow Cr(III)\ complexes + Co(II)\ complexes$$

Chromium(III) and cobalt(III) complexes are substitutionally inert (no exchange of ligands) under conditions of the experiment. However, chromium(II) and cobalt(II) complexes can exchange ligands very rapidly. One of the products of the reaction is $Cr(H_2O)_5Cl^{2+}$. Is this consistent with the reaction proceeding through formation of $(H_2O)_5Cr{-}Cl{-}Co(NH_3)_5$ as an intermediate? Explain your answer.

78. A compound related to acetylacetone is 1,1,1-trifluoroacetylacetone (abbreviated Htfa):

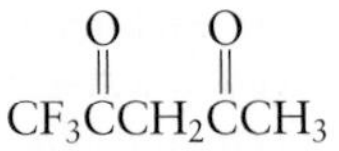

$$\begin{array}{c} \quad O \quad\; O \\ \quad \| \quad\;\; \| \\ CF_3CCH_2CCH_3 \end{array}$$

Htfa forms complexes in a manner similar to acetylacetone. (See Exercise 41.) Both Be^{2+} and Cu^{2+} form complexes with tfa^- having the formula $M(tfa)_2$. Two isomers are formed for each metal complex.

a. The Be^{2+} complexes are tetrahedral. Draw the two isomers of $Be(tfa)_2$. What type of isomerism is exhibited by $Be(tfa)_2$?
b. The Cu^{2+} complexes are square planar. Draw the two isomers of $Cu(tfa)_2$. What type of isomerism is exhibited by $Cu(tfa)_2$?

79. The equilibrium constant K_a for the reaction

$$Fe(H_2O)_6^{3+}(aq) + H_2O(l) \rightleftharpoons Fe(H_2O)_5(OH)^{2+}(aq) + H_3O^+(aq)$$

is 6.0×10^{-3}.

a. Calculate the pH of a 0.10 *M* solution of $Fe(H_2O)_6^{3+}$.
b. Will a 1.0 *M* solution of iron(II) nitrate have a higher or lower pH than a 1.0 *M* solution of iron(III) nitrate? Explain.

80. Ethylenediaminetetraacetate ($EDTA^{4-}$) is used as a complexing agent in chemical analysis with the structure shown in Fig. 19.8. Solutions of $EDTA^{4-}$ are used to treat

heavy metal poisoning by removing the heavy metal in the form of a soluble complex ion. The complex ion virtually prevents the heavy metal ions from reacting with biochemical systems. The reaction of $EDTA^{4-}$ with Pb^{2+} is

$$Pb^{2+}(aq) + EDTA^{4-}(aq) \rightleftharpoons PbEDTA^{2-}(aq) \quad K = 1.1 \times 10^{18}$$

Consider a solution with 0.010 mole of $Pb(NO_3)_2$ added to 1.0 L of an aqueous solution buffered at pH = 13.00 and containing 0.050 *M* Na_4EDTA. Does $Pb(OH)_2$ precipitate from this solution? [K_{sp} for $Pb(OH)_2 = 1.2 \times 10^{-15}$.]

81. Hemoglobin (abbreviated Hb) is a protein that is responsible for the transport of oxygen in the blood of mammals. Each hemoglobin molecule contains four iron atoms that serve as the binding sites for O_2 molecules. The oxygen binding is pH dependent. The relevant equilibrium reaction is

$$HbH_4^{4+}(aq) + 4O_2(g) \rightleftharpoons Hb(O_2)_4(aq) + 4H^+(aq)$$

Use Le Châtelier's principle to answer the following.

a. What form of hemoglobin, HbH_4^{4+} or $Hb(O_2)_4$, is favored in the lungs? What form is favored in the cells?

b. When a person hyperventilates, the concentration of CO_2 in the blood decreases. How does this affect the oxygen-binding equilibrium? How does breathing into a paper bag help to counteract this effect? (*Hint:* CO_2 reacts with water to produce carbonic acid.)

c. When a person has suffered a cardiac arrest, an injection of a sodium bicarbonate solution is given. Why is this step necessary? (*Hint:* CO_2 blood levels increase during cardiac arrest.)

82. Carbon monoxide is toxic because it binds more strongly to iron in hemoglobin (Hb) than does O_2. Consider the following reactions and approximate standard free energy changes:

$$Hb + O_2 \longrightarrow HbO_2 \quad \Delta G° = -70 \text{ kJ}$$

$$Hb + CO \longrightarrow HbCO \quad \Delta G° = -80 \text{ kJ}$$

Using these data, estimate the equilibrium constant value at 25°C for the following reaction:

$$HbO_2 + CO \rightleftharpoons HbCO + O_2$$

Challenge Problems

83. Qualitatively draw the crystal field splitting of the *d* orbitals in a trigonal planar complex ion. (Let the *z* axis be perpendicular to the plane of the complex.)

84. Consider the following complex ion, where A and B represent ligands.

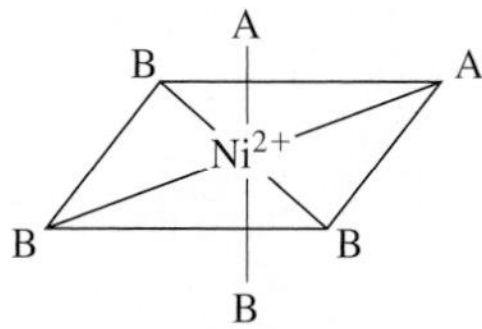

The complex is known to be diamagnetic. Do A and B produce very similar or very different crystal fields? Explain.

85. Qualitatively draw the crystal field splitting for a trigonal bipyramidal complex ion. (Let the *z* axis be perpendicular to the trigonal plane.)

86. Sketch and explain the most likely crystal field diagram for the following complex ion:

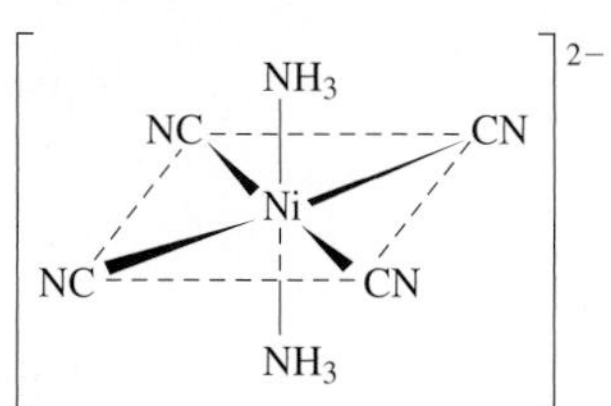

Note: The CN^- ligand produces a *much* stronger crystal field than NH_3. Assume the NH_3 ligands lie on the *z* axis.

87. Consider the following data:

$$Co^{3+} + e^- \longrightarrow Co^{2+} \quad \mathscr{E}° = 1.82 \text{ V}$$

$$Co(en)_3^{2+} \quad K_f = 1.5 \times 10^{12}$$

$$Co(en)_3^{3+} \quad K_f = 2.0 \times 10^{47}$$

where en = ethylenediamine.

a. Calculate $\mathscr{E}°$ for the half-reaction

$$Co(en)_3^{3+} + e^- \longrightarrow Co(en)_3^{2+}$$

b. Based on your answer to part a, which is the stronger oxidizing agent, Co^{3+} or $Co(en)_3^{3+}$?

c. Use the crystal field model to rationalize the result in part b.

88. Consider the following complex ion where A and B represent neutral ligands, and A produces a stronger crystal field than B.

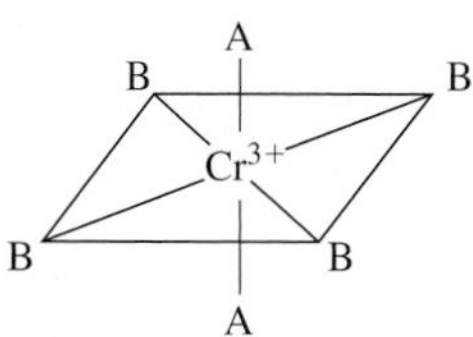

a. Draw an appropriate crystal field diagram for this complex ion (include the electrons). Assume the A ligands lie on the *z* axis.

b. Draw an approximate molecular orbital diagram for this complex ion, assuming only sigma bonding.

c. The two diagrams (parts a and b) should be slightly different with respect to *d*-orbital energies. How could the assumptions used in constructing the diagrams be modified to achieve agreement between the two?

89. a. Calculate the molar solubility of AgBr in pure water. K_{sp} for AgBr is 5.0×10^{-13}.
    b. Calculate the molar solubility of AgBr in 3.0 *M* NH_3. The overall formation constant for $Ag(NH_3)_2^+$ is 1.7×10^7.
    c. Compare the calculated solubilities from parts a and b. Explain any differences.
    d. What mass of AgBr will dissolve in 250.0 mL of 3.0 *M* NH_3?
    e. What effect does adding HNO_3 have on the solubilities calculated in parts a and b?

90. Sketch a *d*-orbital energy diagram for the following:
    a. a linear complex ion with ligands on the *x* axis
    b. a linear complex ion with ligands on the *y* axis

The Nucleus: A Chemist's View

20

chapter

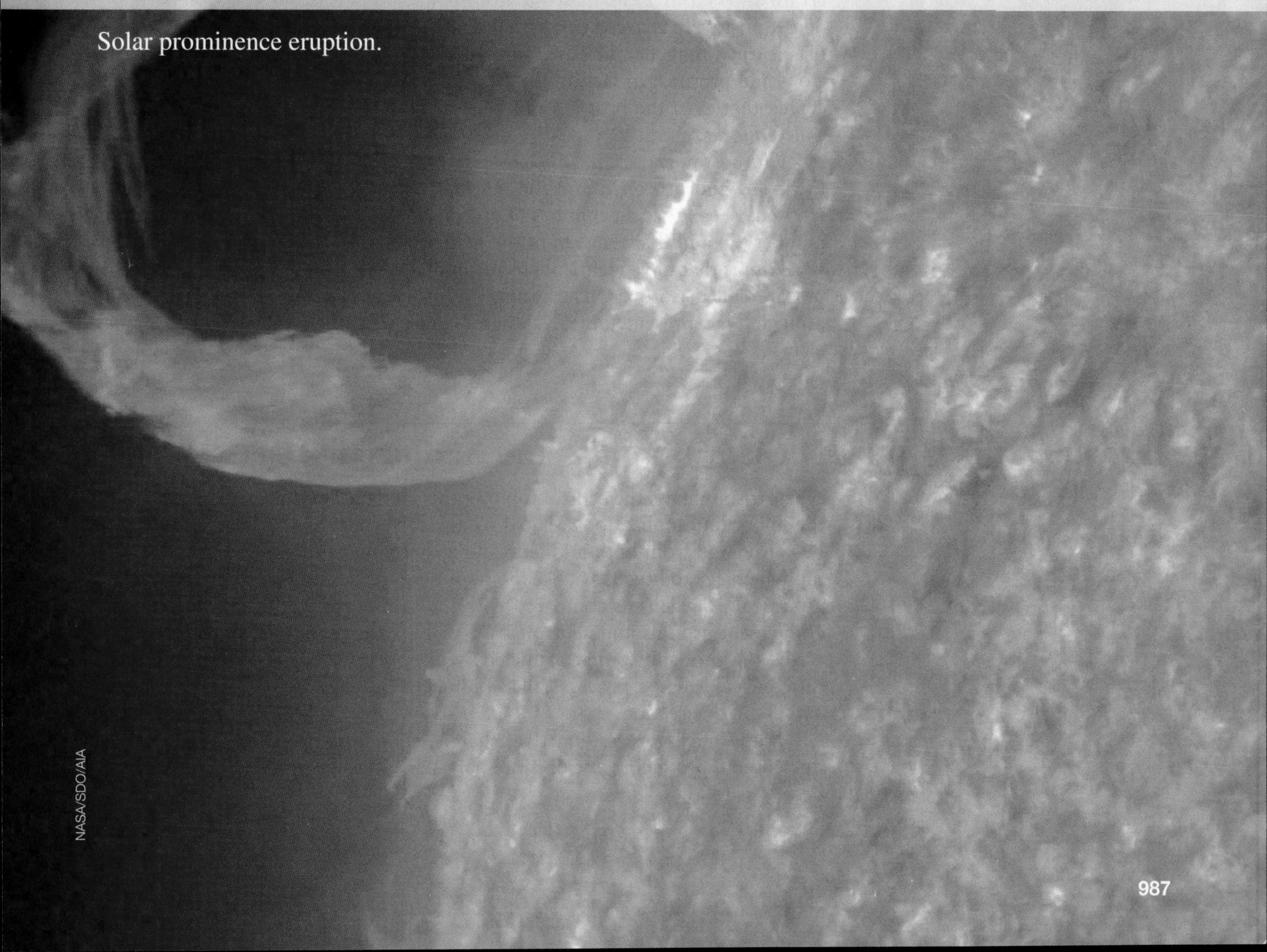

Solar prominence eruption.

NASA/SDO/AIA

OWL

Sign in to OWL at www.cengage.com/owl to view tutorials and simulations, develop problem-solving skills, and complete online homework assigned by your professor.

go Chemistry

Download mini lecture videos for key concept review and exam prep from OWL or purchase them from www.cengagebrain.com.

Since the chemistry of an atom is determined by the number and arrangement of its electrons, the properties of the nucleus are not of primary importance to chemists. In the simplest view, the nucleus provides the positive charge to bind the electrons in atoms and molecules. However, a quick reading of any daily newspaper will show you that the nucleus and its properties have an important impact on our society. This chapter considers those aspects of the nucleus about which everyone should have some knowledge.

Several aspects of the nucleus are immediately impressive: its very small size, its very large density, and the magnitude of the energy that holds it together. The radius of a typical nucleus appears to be about 10^{-13} cm. This can be compared with the radius of a typical atom, which is on the order of 10^{-8} cm. A visualization will help you appreciate the small size of the nucleus: If the nucleus of the hydrogen atom were the size of a Ping-Pong ball, the electron in the 1*s* orbital would be, on average, 0.5 kilometer (0.3 mile) away. The density of the nucleus is equally impressive—about 1.6×10^{14} g/cm^3. A sphere of nuclear material the size of a Ping-Pong ball would have a mass of *2.5 billion tons!* In addition, the energies involved in nuclear processes are typically millions of times larger than those associated with normal chemical reactions. This fact makes nuclear processes very attractive for feeding the voracious energy appetite for our civilization.

Atomos, the Greek root of the word *atom,* means "indivisible." It was originally believed that the atom was the ultimate indivisible particle of which all matter was composed. However, as we discussed in Chapter 2, Lord Rutherford showed in 1911 that the atom is not homogeneous but instead has a dense, positively charged center surrounded by electrons. Subsequently, scientists learned that the nucleus of the atom can be described as containing neutrons and protons. In fact, in the past three decades, it has become widely accepted that even the protons and neutrons are composed of smaller particles called *quarks*.

For most purposes, the nucleus can be regarded as a collection of nucleons (neutrons and protons), and the internal structures of these particles can be ignored. Recall that the number of protons in a particular nucleus is the atomic number (Z) and that the sum of the neutrons and protons is the mass number (A). Atoms that have identical atomic numbers but different mass number values are called isotopes. The general term *nuclide* is applied to each unique atom and is represented by ${}^{A}_{Z}\text{X}$, where X represents the symbol for a particular element.

20.1 Nuclear Stability and Radioactive Decay

Nuclear stability is the central topic of this chapter and forms the basis for all the important applications related to nuclear processes. Nuclear stability can be considered from both a kinetic and a thermodynamic point of view. Thermodynamic stability, as we use the term here, refers to the potential energy of a particular nucleus as compared with the sum of the potential energies of its component protons and neutrons. We will use the term *kinetic stability* to describe the probability that a nucleus will undergo decomposition to form a different nucleus—a process called radioactive decay. We will consider radioactivity in this section.

Many nuclei are radioactive; that is, they decompose, forming another nucleus and producing one or more particles. An example is carbon-14, which decays as follows:

$$ {}^{14}_{6}\text{C} \longrightarrow {}^{14}_{7}\text{N} + {}^{0}_{-1}\text{e} $$

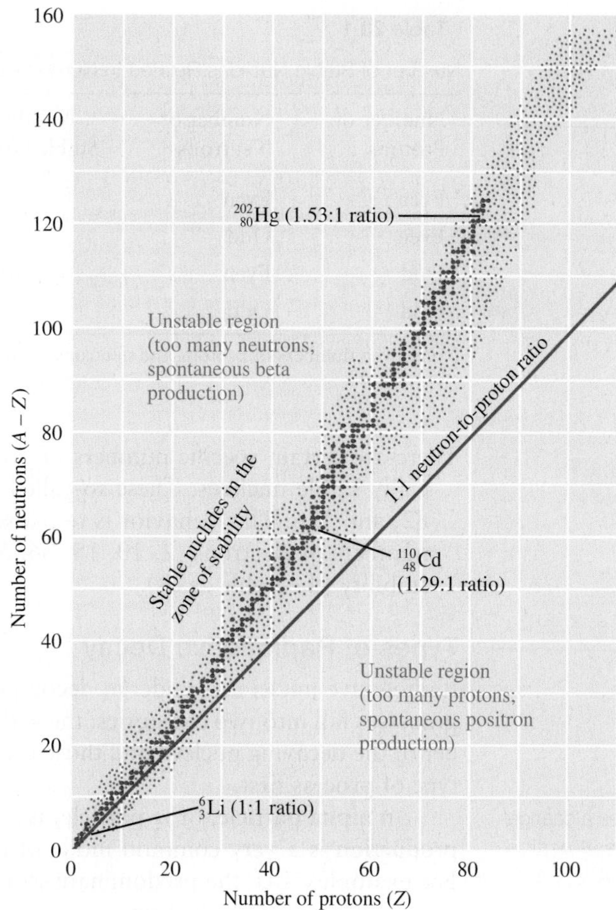

Figure 20.1
The known nuclides. The red dots indicate the nuclides that *do not* undergo radioactive decay. Note that as the number of protons (Z) in a nuclide increases, the neutron/proton ratio required for stability also increases.

where ${}_{-1}^{0}\mathrm{e}$ represents an electron, which is called a **beta particle,** or **β particle,** in nuclear terminology. This equation is typical of those representing radioactive decay in that both A and Z must be conserved. That is, the Z values must give the same sum on both sides of the equation ($6 = 7 - 1$), as must the A values ($14 = 14 + 0$).

Of the approximately 2000 known nuclides, only 279 are stable with respect to radioactive decay. Tin has the largest number of stable isotopes—10.

It is instructive to examine how the numbers of neutrons and protons in a nucleus are related to its stability with respect to radioactive decay. Figure 20.1 shows a plot of the positions of the known nuclides as a function of the number of protons (Z) and the number of neutrons ($A - Z$). The stable nuclides are said to reside in the **zone of stability.**

The following are some important observations concerning radioactive decay:

All nuclides with 84 or more protons are unstable with respect to radioactive decay.

Light nuclides are stable when Z equals $A - Z$, that is, when the neutron/proton ratio is 1. However, for heavier elements, the neutron/proton ratio required for stability is greater than 1 and increases with Z.

Certain combinations of protons and neutrons seem to confer special stability. For example, nuclides with even numbers of protons and neutrons are more often stable than those with odd numbers, as shown by the data in Table 20.1.

Table 20.1

Number of Stable Nuclides Related to Numbers of Protons and Neutrons

Number of Protons	Number of Neutrons	Number of Stable Nuclides	Examples
Even	Even	168	$^{12}_{6}C$, $^{16}_{8}O$
Even	Odd	57	$^{13}_{6}C$, $^{47}_{22}Ti$
Odd	Even	50	$^{19}_{9}F$, $^{23}_{11}Na$
Odd	Odd	4	$^{2}_{1}H$, $^{6}_{3}Li$

Note: Even numbers of protons and neutrons seem to favor stability.

There are certain specific numbers of protons or neutrons that produce especially stable nuclides. These so-called **magic numbers** are 2, 8, 20, 28, 50, 82, and 126. This behavior is reminiscent of that for atoms, where certain numbers of electrons (2, 10, 18, 36, 54, and 86) produce special chemical stability (the noble gases).

Types of Radioactive Decay

Radioactive nuclei can undergo decomposition in various ways. These decay processes fall into two categories: those that involve a change in the mass number of the decaying nucleus and those that do not. We will consider the former type of process first.

α-Particle production involves a change in A for the decaying nucleus; β-particle production has no effect on A.

An **alpha particle, or α particle,** is a helium nucleus ($^{4}_{2}He$). **Alpha-particle production** is a very common mode of decay for heavy radioactive nuclides. For example, $^{238}_{92}U$, the predominant isotope of natural uranium (99.3%), decays by α-particle production:

$$^{238}_{92}U \longrightarrow {}^{4}_{2}He + {}^{234}_{90}Th$$

Another α-particle producer is $^{230}_{90}Th$:

$$^{230}_{90}Th \longrightarrow {}^{4}_{2}He + {}^{226}_{88}Ra$$

Another decay process in which the mass number of the decaying nucleus changes is **spontaneous fission,** the splitting of a heavy nuclide into two lighter nuclides with similar mass numbers. Although this process occurs at an extremely slow rate for most nuclides, it is important in some cases. For instance, spontaneous fission is the predominant mode of decay for $^{254}_{98}Cf$.

The most common decay process in which the mass number of the decaying nucleus remains constant is **beta-particle production.** For example, the thorium-234 nuclide produces a β particle and is converted to protactinium-234:

$$^{234}_{90}Th \longrightarrow {}^{234}_{91}Pa + {}^{0}_{-1}e$$

Iodine-131 is also a β-particle producer:

$$^{131}_{53}I \longrightarrow {}^{0}_{-1}e + {}^{131}_{54}Xe$$

The β particle is assigned the mass number 0, since its mass is tiny compared with that of a proton or neutron. Because the value of Z is -1 for the β particle, the atomic number for the new nuclide is greater by 1 than for the original nuclide. Thus *the net effect of β-particle production is to change a neutron to a proton.* We therefore expect nuclides that lie above the zone of stability (those nuclides whose neutron/proton ratios are too high) to be β-particle producers.

Although the β particle is an electron, the emitting nucleus does not contain electrons. As we shall see later in this chapter, a given quantity of energy (which is best regarded as a form of matter) can become a particle (another form of matter) under certain circumstances. The unstable nuclide creates an electron as it releases energy in the decay process. The electron thus results from the decay process rather than being present before the decay occurs. Think of this process as being analogous to talking: Words are not stored inside us but are formed as we speak. Later in this chapter we will discuss in more detail this very interesting phenomenon in which matter in the form of particles and matter in the form of energy can interchange.

A **gamma ray, or γ ray,** refers to a high-energy photon. Frequently, gamma-ray production accompanies nuclear decays and particle reactions, such as in the α-particle decay of $^{238}_{92}U$:

$$^{238}_{92}U \longrightarrow {}^{4}_{2}He + {}^{234}_{90}Th + 2^{0}_{0}\gamma$$

in which two γ rays of different energies are sometimes produced in addition to the α particle. The emission of γ rays is one way a nucleus with excess energy (in an excited nuclear state) can relax to its ground state.

Positron production occurs for nuclides below the zone of stability (those nuclides whose neutron/proton ratios are too small). The positron is a particle with the same mass as the electron but opposite charge. An example of a nuclide that decays by positron production is sodium-22:

$$^{22}_{11}Na \longrightarrow {}^{0}_{1}e + {}^{22}_{10}Ne$$

Note that *the net effect of this process is to change a proton to a neutron,* causing the product nuclide to have a higher neutron/proton ratio than the original nuclide.

Besides being oppositely charged, the positron shows an even more fundamental difference from the electron: It is the **antiparticle** of the electron. When a positron collides with an electron, the particulate matter is changed to electromagnetic radiation in the form of high-energy photons:

$$^{0}_{-1}e + {}^{0}_{1}e \longrightarrow 2^{0}_{0}\gamma$$

This process, which is characteristic of matter–antimatter collisions, is called *annihilation* and is another example of the interchange of the forms of matter.

Electron capture is a process in which one of the inner-orbital electrons is captured by the nucleus, as illustrated by the process

$$^{201}_{80}Hg + \underset{\substack{\uparrow \\ \text{Inner-orbital electron}}}{^{0}_{-1}e} \longrightarrow {}^{201}_{79}Au + {}^{0}_{0}\gamma$$

This reaction would have been of great interest to the alchemists, but unfortunately, it does not occur at a rate that would make it a practical means for changing mercury to gold. The various types of radioactive decay are summarized in Table 20.2.

Table 20.2

Various Types of Radioactive Processes Showing the Changes That Take Place in the Nuclides

Process	Change in A	Change in Z	Change in Neutron/Proton Ratio	Example
β-Particle (electron) production	0	+1	Decrease	$^{227}_{89}Ac \longrightarrow {}^{227}_{90}Th + {}^{0}_{-1}e$
Positron production	0	−1	Increase	$^{13}_{7}N \longrightarrow {}^{13}_{6}C + {}^{0}_{1}e$
Electron capture	0	−1	Increase	$^{73}_{33}As + {}^{0}_{-1}e \longrightarrow {}^{73}_{32}Ge$
α-Particle production	−4	−2	Increase	$^{210}_{84}Po \longrightarrow {}^{206}_{82}Pb + {}^{4}_{2}He$
γ-Ray production	0	0	—	Excited nucleus $\longrightarrow$ ground-state nucleus + $^{0}_{0}\gamma$
Spontaneous fission	—	—	—	$^{254}_{98}Cf \longrightarrow$ lighter nuclides + neutrons

Chemical Insights — Does Antimatter Matter?

Ever since its discovery in the early twentieth century scientists have been fascinated by antimatter, the "inverse" of matter. Individual antimatter particles are well known and have been studied extensively in particle accelerators. For example, the positron, the antiparticle for the electron, has the same mass as the electron but a positive charge, and the antiproton has the same mass as the proton but a negative charge. Antiparticles are tricky to study because they undergo immediate annihilation, releasing large quantities of energy, when they meet their corresponding matter particles.

Although both the antiparticles for antihydrogen (the antiproton and the positron) can be stored for a certain time in accelerators, no one had been able to create antihydrogen until recently. In September 1995 at the European Laboratory for Particle Physics (CERN) in Geneva, a team of researchers led by Walter Oelert observed antihydrogen for the first time. Using the Low Energy Antiproton Ring (LEAR), the scientists sent a beam of antiprotons at nearly the speed of light into a stream of xenon gas. Occasionally an antiproton interacted with a xenon atom to produce a positron and an electron. In the very rare cases in which the created positron had a velocity similar to that of the antiproton beam, the positron would be captured by an antiproton to form antihydrogen.

Physicists are very interested in comparing the properties of antihydrogen with those of hydrogen, especially to determine antimatter's response to gravity. However, these antiatoms, created in flight in the accelerator, lasted for a fleeting 40-billionths of a second, too short a time for the scientists to observe their properties. Physicists are now trying to find ways to make low-energy antihydrogen so that it can be stored long enough to study it more thoroughly. They hope that these experiments will provide answers to questions such as why there is so little antimatter in the universe relative to matter—present theory indicates that equal amounts of matter and antimatter would have been created in the early universe.

CERN photo by Maximilien Brice

CERN, the world's largest particle accelerator, lies at the foot of the Jura Mountains near Geneva, Switzerland. This is a view of the CMS (compact muon solenoid).

Often a radioactive nucleus cannot reach a stable state through a single decay process. In such a case, a **decay series** occurs until a stable nuclide is formed. A well-known example is the decay series that starts with ${}^{238}_{92}\text{U}$ and ends with ${}^{206}_{82}\text{Pb}$, as shown in Fig. 20.2. Similar series exist for ${}^{235}_{92}\text{U}$:

$${}^{235}_{92}\text{U} \xrightarrow[\text{decays}]{\text{Series of}} {}^{207}_{82}\text{Pb}$$

and for ${}^{232}_{90}\text{Th}$:

$${}^{232}_{90}\text{Th} \xrightarrow[\text{decays}]{\text{Series of}} {}^{208}_{82}\text{Pb}$$

20.2 The Kinetics of Radioactive Decay

In a sample consisting of radioactive nuclides of a given type, each nuclide has a certain probability of undergoing decay. Suppose that a sample containing 1000 atoms of a certain nuclide produces 10 decay events per hour. This means

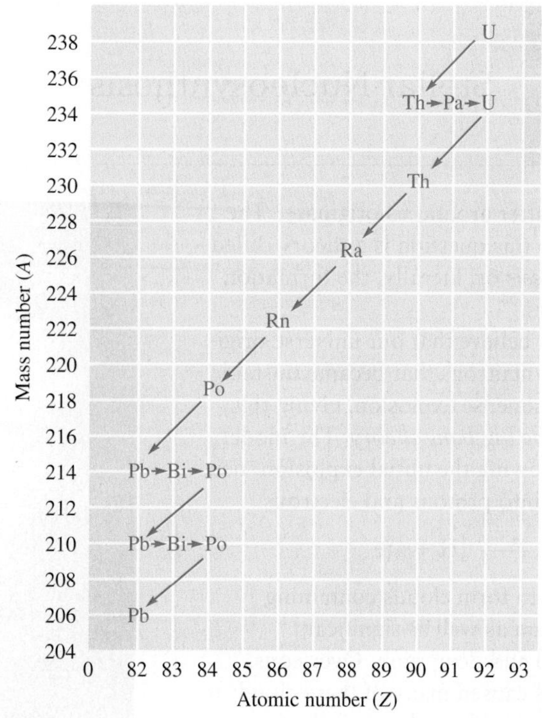

Figure 20.2
The decay series from $^{238}_{92}U$ to $^{206}_{82}Pb$. Each nuclide in the series except $^{206}_{82}Pb$ is unstable, and the successive transformations (shown by the arrows) continue until $^{206}_{82}Pb$ is finally formed. Note that horizontal arrows indicate processes where A is unchanged, whereas diagonal arrows signify that both A and Z change.

that over the span of an hour, 1 out of every 100 nuclides will decay. Assuming that this probability of decay is characteristic for this type of nuclide, we could predict that a 2000-atom sample would give 20 decay events per hour. Thus, for radioactive nuclides, the **rate of decay,** which is the negative of the change in the number of nuclides per unit time ($-\Delta N/\Delta t$), is directly proportional to the number of **nuclides** (N) in a given sample:

Rates of reaction are discussed in Chapter 15.

$$\text{Rate} = -\frac{\Delta N}{\Delta t} \propto N$$

The negative sign is included because the number of nuclides is decreasing. We now insert a proportionality constant k to give

$$\text{Rate} = -\frac{\Delta N}{\Delta t} = -\frac{dN}{dt} = kN$$

Recall from Chapter 15 that this is the rate law for a first-order process. The integrated first-order rate law is

$$\ln\left(\frac{N}{N_0}\right) = -kt$$

where N_0 represents the original number of nuclides (at $t = 0$) and N represents the number of nuclides *remaining* at time t.

Half-Life

The **half-life** ($t_{1/2}$) of a radioactive sample is defined as the time required for the number of nuclides to reach half of the original value ($N_0/2$). We can use this definition in connection with the integrated first-order rate law (see Section 15.4) to produce the following expression for $t_{1/2}$:

$$t_{1/2} = \frac{\ln(2)}{k} = \frac{0.693}{k}$$

Chemical Insights

Stellar Nucleosynthesis

How did all the matter around us originate? The scientific answer to this question is a theory called *stellar nucleosynthesis* or, literally, the formation of nuclei in the stars.

Many scientists believe that our universe originated as a cloud of neutrons that became unstable and produced an immense explosion, giving this model its name—*the big bang theory.* The model postulates that following the initial explosion, neutrons decomposed into protons and electrons,

$$^{1}_{0}\text{n} \longrightarrow {}^{1}_{1}\text{H} + {}^{\;\,0}_{-1}\text{e}$$

which recombined to form clouds containing hydrogen and helium as well as significant amounts of lithium and deuterium. Over eons, gravitational forces caused many of these clouds to contract and heat up sufficiently to reach temperatures where proton fusion was possible, releasing large quantities of energy. When the tendency to expand because of the heat from fusion and the tendency to contract because of the forces of gravity are balanced, a stable young star such as our sun can be formed.

Eventually, when the supply of hydrogen is exhausted, the core of the star again contracts with further heating until temperatures are reached where fusion of helium nuclei can occur. This leads to the formation of $^{12}_{6}\text{C}$ and $^{16}_{8}\text{O}$ nuclei. In turn, when the supply of helium nuclei is depleted, further contraction and heating occur, until the fusion of heavier nuclei takes place. This process occurs repeatedly, forming heavier and heavier nuclei until iron nuclei are formed. Because the iron nucleus is the most stable of all, energy is required to fuse iron nuclei. This endothermic fusion process cannot furnish energy to sustain the star; therefore, it cools to a small, dense *white dwarf.*

The evolution just described is characteristic of small and medium-sized stars. Much larger stars, however, become unstable during their evolution and undergo a *supernova explosion.* In the latter stages in this explosion, which is actually an extreme implosion that occurs when the fusion heat source is exhausted, some medium-mass nuclei are fused to form heavy elements. Also, some light nuclei capture neutrons. These neutron-rich nuclei then produce β particles, increasing their atomic number with each event. This eventually leads to nuclei with large atomic numbers. In fact, almost all nuclei beyond iron are thought to originate from supernova explosions. The debris of a supernova explosion thus contains a large variety of elements that are spread widely and become part of hydrogen clouds that eventually become stars, which in turn form solar systems. For example, our sun and solar system probably formed from materials that had been through two previous supernova explosions.

Although other theories for the origin of matter have been suggested, there is much evidence to support the big bang theory. It continues to be widely accepted, although the details continue to be modified.

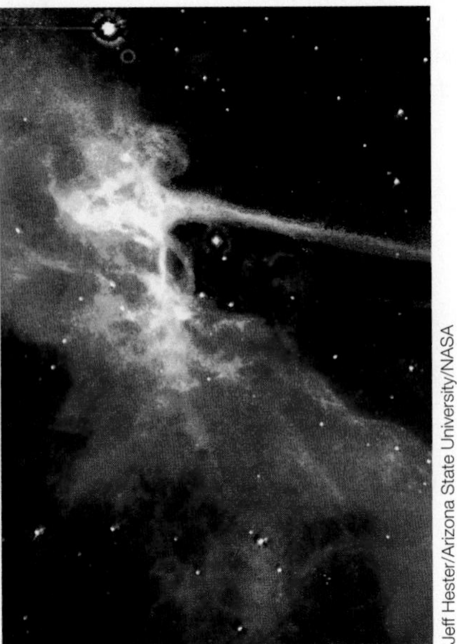

Jeff Hester/Arizona State University/NASA

Image of a portion of the Cygnus Loop supernova remnant, taken by the Hubble space telescope.

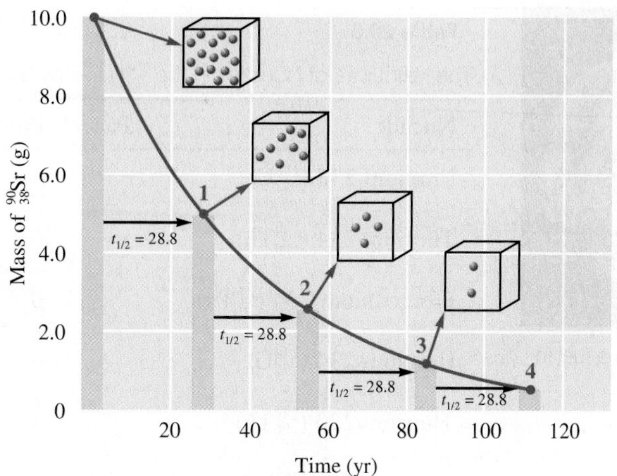

Figure 20.3
The decay of a 10.0-g sample of strontium-90 over time. Note that the half-life is a constant 28.8 years.

Thus, if the half-life of a radioactive nuclide is known, the rate constant can be easily calculated, and vice versa.

As we saw in Section 15.4, the half-life for a first-order process is constant. This is illustrated for the β-particle decay of strontium-90 in Fig. 20.3; it takes 28.8 years for each halving of the amount of $^{90}_{38}Sr$. Contamination of the environment with $^{90}_{38}Sr$ poses serious health hazards because of the similar chemistry of strontium and calcium (both are in Group 2A). Strontium-90 in grass and hay is incorporated into cow's milk along with calcium and is then passed on to humans, where it lodges in the bones. Because of its relatively long half-life, ^{90}Sr persists for years in humans, causing radiation damage that may lead to cancer.

The harmful effects of radiation will be discussed in Section 20.7.

EXAMPLE 20.1

The half-life of molybdenum-99 is 67.0 h. How much of a 1.000-mg sample of $^{99}_{42}Mo$ remains after 335 h?

Solution The easiest way to solve this problem is to recognize that 335 h represents five half-lives for $^{99}_{42}Mo$:

$$335 = 5 \times 67.0$$

We can sketch the change that occurs, as shown in Fig. 20.4. Thus, after 335 h, 0.031 mg of $^{99}_{42}Mo$ remains.

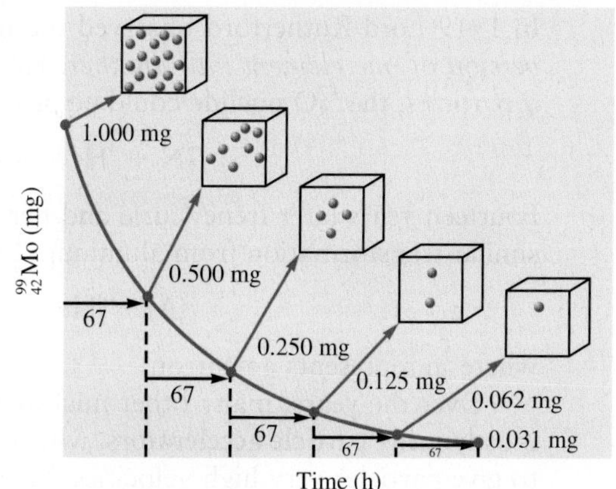

Figure 20.4
The change in the amount of $^{99}_{42}Mo$ with time ($t_{1/2} = 67$ h).

A uranium "button" for use as a fuel in a nuclear reactor.

Table 20.3

The Half-Lives of Nuclides in the $^{238}_{92}U$ Decay Series

Nuclide	Particle Produced	Half-Life
Uranium-238 ($^{238}_{92}U$) ↓	α	4.51×10^9 years
Thorium-234 ($^{234}_{90}Th$) ↓	β	24.1 days
Protactinium-234 ($^{234}_{91}Pa$) ↓	β	6.75 hours
Uranium-234 ($^{234}_{92}U$) ↓	α	2.48×10^5 years
Thorium-230 ($^{230}_{90}Th$) ↓	α	8.0×10^4 years
Radium-226 ($^{226}_{88}Ra$) ↓	α	1.62×10^3 years
Radon-222 ($^{222}_{86}Rn$) ↓	α	3.82 days
Polonium-218 ($^{218}_{84}Po$) ↓	α	3.1 minutes
Lead-214 ($^{214}_{82}Pb$) ↓	β	26.8 minutes
Bismuth-214 ($^{214}_{83}Bi$) ↓	β	19.7 minutes
Polonium-214 ($^{214}_{84}Po$) ↓	α	1.6×10^{-4} second
Lead-210 ($^{210}_{82}Pb$) ↓	β	20.4 years
Bismuth-210 ($^{210}_{83}Bi$) ↓	β	5.0 days
Polonium-210 ($^{210}_{84}Po$) ↓	α	138.4 days
Lead-206 ($^{206}_{82}Pb$)	—	Stable

The half-lives of radioactive nuclides vary over a tremendous range. For example, $^{144}_{60}Nd$ has a half-life of 5×10^{15} years, whereas $^{214}_{84}Po$ has a half-life of 2×10^{-4} second. To give perspective on this, the half-lives of the nuclides in the $^{238}_{92}U$ decay series are given in Table 20.3.

20.3 Nuclear Transformations

In 1919 Lord Rutherford observed the first **nuclear transformation,** *the conversion of one element into another.* He found that by bombarding $^{14}_{7}N$ with α particles, the $^{17}_{8}O$ nuclide could be produced:

$$^{14}_{7}N + ^{4}_{2}He \longrightarrow ^{17}_{8}O + ^{1}_{1}H$$

Fourteen years later Irene Curie and her husband Frederick Joliot observed a similar transformation from aluminum to phosphorus,

$$^{27}_{13}Al + ^{4}_{2}He \longrightarrow ^{30}_{15}P + ^{1}_{0}n$$

where $^{1}_{0}n$ represents a neutron.

Over the years, many other nuclear transformations have been achieved, mostly using **particle accelerators,** which, as the name reveals, are devices used to give particles very high velocities. Because of the electrostatic repulsion between the target nucleus and a positive ion, accelerators are needed when posi-

A cyclotron at TRIUMF, Canada's national laboratory of particle and nuclear physics.

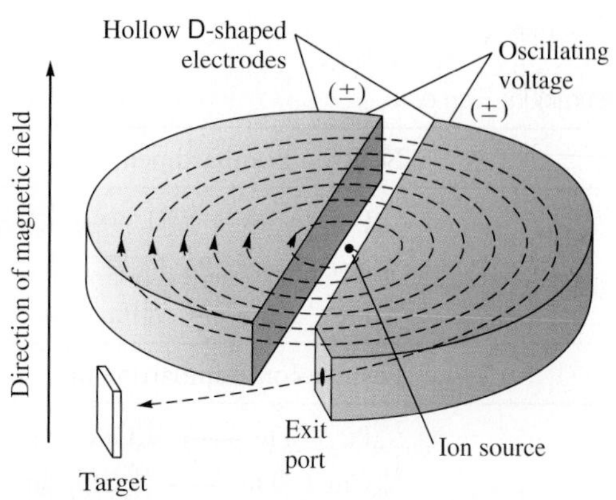

Figure 20.5
A schematic diagram of a cyclotron. The ion is introduced in the center and is then pulled back and forth between the hollow D-shaped electrodes by constant reversals of the electric field. Magnets above and below these electrodes produce a spiral path that expands as the particle velocity increases. When the particle has sufficient speed, it exits the accelerator and is directed at the target nucleus.

tive ions are used as bombarding particles. The particle, accelerated to a very high velocity, can overcome the repulsion and penetrate the target nucleus, thus effecting the transformation. A schematic diagram of one type of particle accelerator, the **cyclotron,** is shown in Fig. 20.5. The ion is introduced at the center of the cyclotron and is accelerated in an expanding spiral path by use of alternating electric fields in the presence of a magnetic field. The **linear accelerator** illustrated in Fig. 20.6 uses changing electric fields to achieve high velocities along a linear pathway.

In addition to positive ions, neutrons are often used as bombarding particles to effect nuclear transformations. Because neutrons are uncharged and thus not repelled electrostatically by a target nucleus, they are readily absorbed by many nuclei, leading to the formation of new nuclides. The most common source of neutrons for this purpose is a fission reactor (see Section 20.6).

By using neutron and positive-ion bombardment, scientists have been able to extend the periodic table. Prior to 1940, the heaviest known element was uranium ($Z = 92$). However, in 1940 neptunium ($Z = 93$) was produced by neutron bombardment of $^{238}_{92}\text{U}$. The process initially gives $^{239}_{92}\text{U}$, which then decays to $^{239}_{93}\text{Np}$ by β-particle production:

$$^{238}_{92}\text{U} + ^{1}_{0}\text{n} \longrightarrow ^{239}_{92}\text{U} \xrightarrow[t_{1/2} = 23\text{ min}]{} ^{239}_{93}\text{Np} + ^{0}_{-1}\text{e}$$

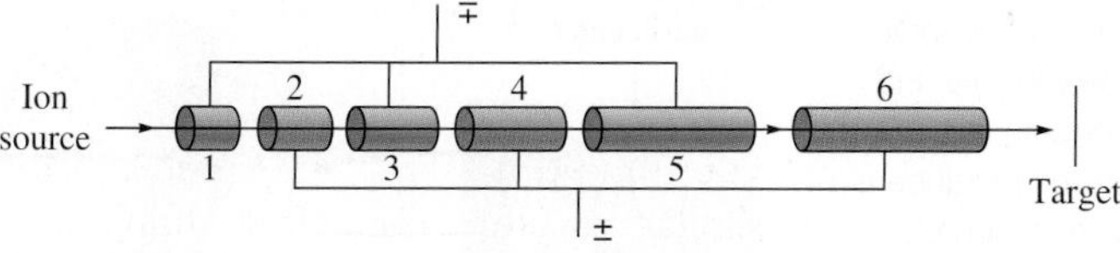

Figure 20.6
Schematic diagram of a linear accelerator, which uses a changing electric field to accelerate a positive ion along a linear path. As the ion leaves the source, the odd-numbered tubes are negatively charged, and the even-numbered tubes are positively charged. The positive ion is thus attracted into tube 1. As the ion leaves tube 1, the tube polarities are reversed. Now tube 1 is positive, repelling the positive ion, and tube 2 is negative, attracting the positive ion. This process continues, eventually producing high particle velocity.

Table 20.4

Syntheses of Some of the Transuranium Elements

Element	Neutron Bombardment	Half-Life
Neptunium (Z = 93)	$^{238}_{92}U + ^{1}_{0}n \longrightarrow ^{239}_{92}U \longrightarrow ^{239}_{93}Np + ^{0}_{-1}e$	2.35 days ($^{239}_{93}Np$)
Plutonium (Z = 94)	$^{239}_{93}Np \longrightarrow ^{239}_{94}Pu + ^{0}_{-1}e$	24,400 years ($^{239}_{94}Pu$)
Americium (Z = 95)	$^{239}_{94}Pu + 2^{1}_{0}n \longrightarrow ^{241}_{94}Pu \longrightarrow ^{241}_{95}Am + ^{0}_{-1}e$	458 years ($^{241}_{95}Am$)
Element	**Positive-Ion Bombardment**	**Half-Life**
Curium (Z = 96)	$^{239}_{94}Pu + ^{4}_{2}He \longrightarrow ^{242}_{96}Cm + ^{1}_{0}n$	163 days ($^{242}_{96}Cm$)
Californium (Z = 98)	$^{242}_{96}Cm + ^{4}_{2}He \longrightarrow ^{245}_{98}Cf + ^{1}_{0}n$ or $^{238}_{92}U + ^{12}_{6}C \longrightarrow ^{246}_{98}Cf + 4^{1}_{0}n$	44 minutes ($^{245}_{98}Cf$)
Rutherfordium (Z = 104)	$^{249}_{98}Cf + ^{12}_{6}C \longrightarrow ^{257}_{104}Rf + 4^{1}_{0}n$	
Dubnium (Z = 105)	$^{249}_{98}Cf + ^{15}_{7}N \longrightarrow ^{260}_{105}Db + 4^{1}_{0}n$	
Seaborgium (Z = 106)	$^{249}_{98}Cf + ^{18}_{8}O \longrightarrow ^{263}_{106}Sg + 4^{1}_{0}n$	

In the years since 1940, the elements with atomic numbers 93 through 112 and 114, 116, and 118, called the **transuranium elements,*** have been synthesized. Many of these elements have very short half-lives, as shown in Table 20.4. As a result, only a few atoms of some of the transuranium elements have ever been formed. This, of course, makes the chemical characterization of these elements extremely difficult.

20.4 Detection and Uses of Radioactivity

Geiger counters are often called survey meters in industry.

Various instruments measure radioactivity levels, the most familiar being the **Geiger-Müller counter,** or **Geiger counter** (Fig. 20.7). This instrument takes advantage of the fact that the high-energy particles from radioactive decay processes produce ions when they travel through matter. The probe of the Geiger counter is filled with argon gas, which can be ionized by a rapidly moving particle:

$$Ar(g) \xrightarrow[\text{particle}]{\text{High-energy}} Ar^{+}(g) + e^{-}$$

Normally, a sample of argon gas does not conduct a current when an electrical potential is applied. However, the formation of ions and electrons produced by the passage of the high-energy particle allows a momentary current to flow.

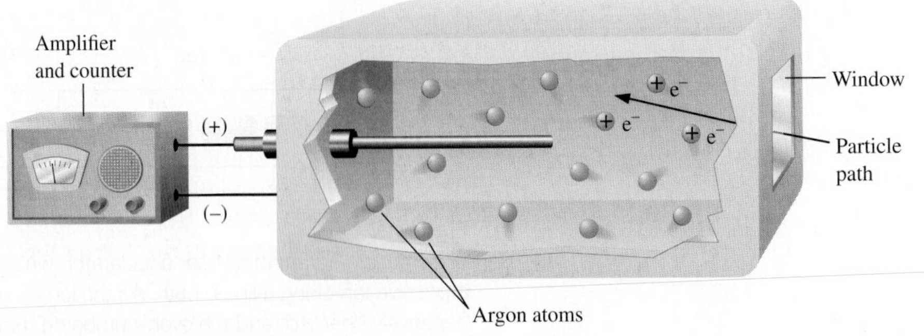

Figure 20.7
A schematic representation of a Geiger-Müller counter. The high-energy radioactive particle enters the window and ionizes argon atoms along its path. The resulting ions and electrons produce a momentary current pulse, which is amplified and counted.

*For an excellent article on the history of these elements, see G. B. Kauffman, "Beyond Uranium," *Chem. Eng. News,* November 19, 1990, p. 18.

Electronic devices detect this current flow, so the number of these events can be counted. Thus the decay rate of the radioactive sample can be determined.

Another instrument often used to detect levels of radioactivity is a **scintillation counter,** which takes advantage of the fact that certain substances, such as zinc sulfide, give off light when they are struck by high-energy radiation. A photocell senses the flashes of light that occur as the radiation strikes and thus measures the number of decay events per unit time.

Dating by Radioactivity

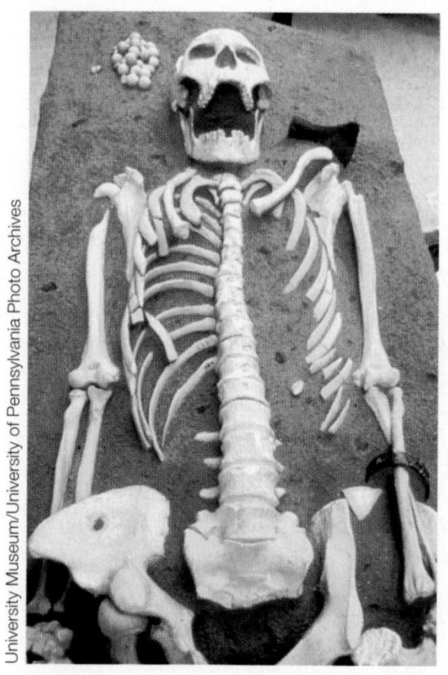
University Museum/University of Pennsylvania Photo Archives

Carbon-14 radioactivity is often used to date human skeletons found at archaeological sites.

Archaeologists, geologists, and others involved in reconstructing the ancient history of the earth rely heavily on radioactivity to provide accurate dates for artifacts and rocks. A method that has been very important for dating ancient articles made from wood or cloth is **radiocarbon dating,** or **carbon-14 dating.** This technique was originated in the 1940s by Willard Libby, an American chemist who received a Nobel Prize for his efforts in this field.

Radiocarbon dating is based on the radioactivity of the nuclide ${}^{14}_{6}C$, which decays via β-particle production:

$${}^{14}_{6}C \longrightarrow {}^{0}_{-1}e + {}^{14}_{7}N$$

Carbon-14 is produced in the atmosphere when high-energy neutrons from space collide with nitrogen-14:

$${}^{14}_{7}N + {}^{1}_{0}n \longrightarrow {}^{14}_{6}C + {}^{1}_{1}H$$

Carbon-14 is continuously produced by this process, and it continuously decomposes through β-particle production. Over the years, the rates for these two processes have become equal. Like a participant in a chemical reaction at equilibrium, the amount of ${}^{14}_{6}C$ that is present in the atmosphere remains about constant.

Carbon-14 can be used to date wood and cloth artifacts because the ${}^{14}_{6}C$, along with the other carbon isotopes in the atmosphere, reacts with oxygen to form carbon dioxide. A living plant consumes carbon dioxide in the photosynthesis process and incorporates the carbon, including ${}^{14}_{6}C$, into its molecules. As long as the plant lives, the ${}^{14}_{6}C/{}^{12}_{6}C$ ratio in its molecules remains the same as that in the atmosphere because of the continuous uptake of carbon. However, as soon as a tree is cut to make a wooden bowl or a flax plant is harvested to make linen, the ${}^{14}_{6}C/{}^{12}_{6}C$ ratio begins to decrease because of the radioactive decay of ${}^{14}_{6}C$ (the ${}^{12}_{6}C$ nuclide is stable). Since the half-life of ${}^{14}_{6}C$ is 5730 years, a wooden bowl found in an archaeological dig showing a ${}^{14}_{6}C/{}^{12}_{6}C$ ratio half that found in currently living trees must be about 5730 years old. This reasoning assumes that the current ${}^{14}_{6}C/{}^{12}_{6}C$ ratio is the same as that found in ancient times.

Dendrochronologists—scientists who date trees from annual growth rings—have used data collected from long-lived species of trees, such as bristlecone pines and sequoias, to show that the ${}^{14}_{6}C$ content of the atmosphere has changed significantly over the ages. These data have been used to derive correction factors that allow very accurate dates to be determined from the observed ${}^{14}_{6}C/{}^{12}_{6}C$ ratio in an artifact, especially for artifacts up to 10,000 years old. Recent measurements of uranium/thorium ratios in ancient coral have raised questions about the accuracy of ${}^{14}C$ dates extending back 20,000 to 30,000 years, suggesting that errors up to 3000 years may have occurred. Efforts are now being made to recalibrate the ${}^{14}C$ dates over this period.

The ${}^{14}_{6}C/{}^{12}_{6}C$ ratio is the basis for carbon-14 dating.

One drawback of radiocarbon dating is that a fairly large piece of the object (up to several grams) must be burned to form carbon dioxide, which is then analyzed for radioactivity. Another method for counting ${}^{14}_{6}C$ nuclides avoids destruction of a significant portion of a valuable artifact. This technique, requiring only about 10^{-3} gram, uses a mass spectrometer (Chapter 3), in which the carbon atoms are ionized and accelerated through a magnetic field that deflects their paths. Because of their different masses, the various ions are deflected by different amounts and can be counted separately. This allows a very accurate determination of the ${}^{14}_{6}C/{}^{12}_{6}C$ ratio in the sample.

In their attempts to establish the geological history of the earth, geologists have made extensive use of radioactivity. For example, since ${}^{238}_{92}U$ decays to the stable ${}^{206}_{82}Pb$ nuclide, the ratio of ${}^{206}_{82}Pb$ to ${}^{238}_{92}U$ in a rock can, under favorable circumstances, be used to estimate the age of the rock.

Because the half-life of ${}^{238}_{92}U$ is very long compared with those of the other members of the decay series (Table 20.3) to reach ${}^{206}_{82}Pb$, the number of nuclides in intermediate stages of decay is negligible. That is, once a ${}^{238}_{92}U$ nuclide starts to decay, it reaches ${}^{206}_{82}Pb$ relatively fast.

EXAMPLE 20.2

A rock containing ${}^{238}_{92}U$ and ${}^{206}_{82}Pb$ was examined to determine its approximate age. Analysis showed the ratio of ${}^{206}_{82}Pb$ atoms to ${}^{238}_{92}U$ atoms to be 0.115. Assuming that no lead was originally present, that all the ${}^{206}_{82}Pb$ formed over the years has remained in the rock, and that the number of nuclides in the intermediate stages of decay between ${}^{238}_{92}U$ and ${}^{206}_{82}Pb$ is negligible, calculate the age of the rock. The half-life of ${}^{238}_{92}U$ is 4.5×10^9 yr.

Solution The problem can be solved using the integrated first-order rate law:

$$\ln\left(\frac{N}{N_0}\right) = -kt = -\left(\frac{0.693}{4.5 \times 10^9 \text{ yr}}\right)t$$

where N/N_0 represents the ratio of ${}^{238}_{92}U$ atoms now found in the rock to the number present when the rock was formed. We are assuming that each ${}^{206}_{82}Pb$ nuclide present must have come from decay of a ${}^{238}_{92}U$ atom:

$${}^{238}_{92}U \longrightarrow {}^{206}_{82}Pb$$

Thus

$$\boxed{\text{Number of } {}^{238}_{92}U \text{ atoms originally present}} = \boxed{\text{number of } {}^{206}_{82}Pb \text{ atoms now present}} + \boxed{\text{number of } {}^{238}_{92}U \text{ atoms now present}}$$

$$\frac{\text{Atoms of } {}^{206}_{82}Pb \text{ now present}}{\text{Atoms of } {}^{238}_{92}U \text{ now present}} = 0.115 = \frac{0.115}{1.000} = \frac{115}{1000}$$

Think carefully about what this means. For every 1115 ${}^{238}_{92}U$ atoms originally present in the rock, 115 have been changed to ${}^{206}_{82}Pb$ and 1000 remain as ${}^{238}_{92}U$. Thus

$$\frac{N}{N_0} = \frac{\overset{\text{Now present}}{{}^{238}_{92}U}}{\underbrace{{}^{206}_{82}Pb + {}^{238}_{92}U}_{{}^{238}_{92}U \text{ originally present}}} = \frac{1000}{1115} = 0.8969$$

$$\ln\left(\frac{N}{N_0}\right) = \ln(0.8969) = -\left(\frac{0.693}{4.5 \times 10^9 \text{ yr}}\right)t$$

$$t = 7.1 \times 10^8 \text{ yr}$$

This is the approximate age of the rock. It was formed sometime during the Cambrian Period.

Medical Applications of Radioactivity

Although the rapid advances of the medical sciences in recent decades are the result of many causes, one of the most important has been the discovery and use of **radiotracers,** which are radioactive nuclides that can be introduced into

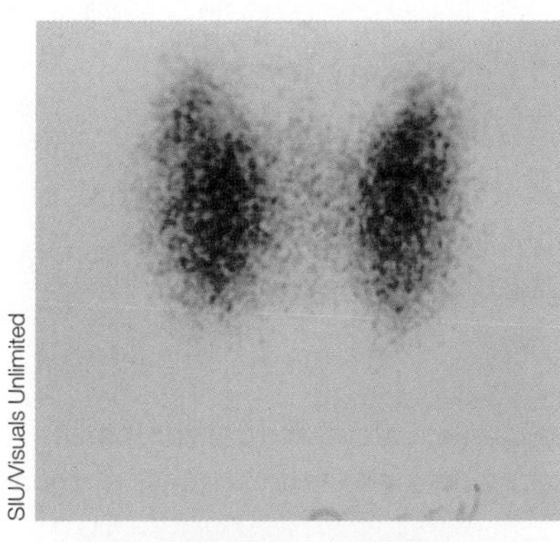
SIU/Visuals Unlimited

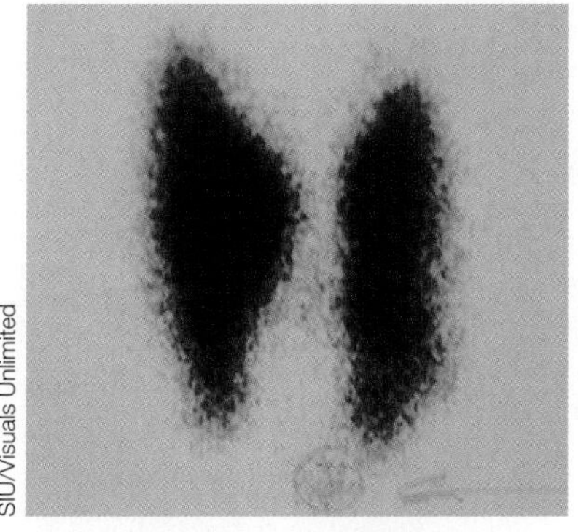
SIU/Visuals Unlimited

Figure 20.8
After consumption of $Na^{131}I$, the patient's thyroid is scanned for radioactivity levels to determine the efficiency of iodine absorption. (top) A normal thyroid. (bottom) An enlarged thyroid.

The *m* in technetium-99m designates an excited nuclear state of ^{99}Tc that decays to the ground state by γ production:

$$^{99m}Tc \longrightarrow {}^{99}Tc + \gamma$$

organisms through food or drugs and whose pathways can be *traced* by monitoring their radioactivity (Fig. 20.8). For example, the incorporation of nuclides such as $^{14}_{6}C$ and $^{32}_{15}P$ into nutrients has produced important information about metabolic pathways.

Iodine-131 has proved very useful in the diagnosis and treatment of illnesses of the thyroid gland. Patients drink a solution containing small amounts of $Na^{131}I$, and the uptake of the iodine by the thyroid gland is monitored with a scanner.

Thallium-201 can be used to assess the damage to the heart muscle in a person who has suffered a heart attack because thallium is concentrated in healthy muscle tissue. Technetium-99m is also taken up by normal heart tissue and is used for damage assessment in a similar way.

A relatively new technique that uses radioactivity to study body processes and diagnose malfunctions is commonly called *positron emission tomography* (PET). In this technique radionuclides that decay by positron emission are incorporated into compounds. For example, brain function can be studied by incorporating $^{11}_{6}C$ into glucose, which is the main source of energy for the brain. By studying how this labeled glucose is metabolized in the brain, doctors can discover abnormalities caused by diseases such as cancer, Parkinson's disease, and epilepsy.

Radiotracers provide sensitive and noninvasive methods for learning about biological systems, for detection of disease, for monitoring the action and effectiveness of drugs, and for early detection of pregnancy; their usefulness should continue to grow. Some useful radiotracers are listed in Table 20.5.

In contrast to the use of radiotracers, where the patient is exposed to only very small quantities of radiation, **radiation therapy** uses larger radiation doses to kill cancer cells. Typically the radioactive nuclides used in radiation therapy emit high-energy γ rays. The radiation can be administered external to the body or implanted in the cancerous tissue, depending on the location and nature of the cancer. For example, $^{192}_{77}Ir$ pellets (covered by a platinum shell that stops α and β particles from entering the body but is easily penetrated by γ rays) are placed directly in the tumor to be destroyed. When implants are not practical, $^{60}_{27}Co$ is often used externally to direct γ rays at the target tissue. Because virtually all the iodine that enters the body ends up in the thyroid gland, large doses of ^{131}I can be ingested to treat thyroid cancer. Because γ radiation is so highly penetrating, it is impossible to confine it to cancerous tissue in radiation therapy, and patients usually suffer nausea, hair loss, fatigue, and a weakened immune system. Despite its dangers and unpleasant side effects, however, radiation therapy is a potent tool to fight many forms of cancer.

Table 20.5

Some Radioactive Nuclides, with Half-Lives and Medical Applications as Radiotracers

Nuclide	Half-Life	Area of the Body Studied
^{131}I	8.1 days	Thyroid
^{59}Fe	45.1 days	Red blood cells
^{99}Mo	67 hours	Metabolism
^{32}P	14.3 days	Eyes, liver, tumors
^{51}Cr	27.8 days	Red blood cells
^{87}Sr	2.8 hours	Bones
^{99}Tc	6.0 hours	Heart, bones, liver, lungs
^{133}Xe	5.3 days	Lungs
^{24}Na	14.8 hours	Circulatory system

20.5 Thermodynamic Stability of the Nucleus

We can determine the thermodynamic stability of a nucleus by calculating the change in potential energy that would occur if that nucleus were formed from its constituent protons and neutrons. As an example, let's consider the hypothetical process of forming a $^{16}_{8}O$ nucleus from eight neutrons and eight protons:

$$8^1_0n + 8^1_1H \longrightarrow {}^{16}_{8}O$$

The energy change associated with this process can be calculated by comparing the sum of the masses of eight protons and eight neutrons with the mass of the oxygen nucleus:

$$\text{Mass of } (8^1_0n + 8^1_1H) = \underset{\text{Mass of } ^1_0n}{8(1.67493 \times 10^{-24}\text{ g})} + \underset{\text{Mass of } ^1_1H}{8(1.67262 \times 10^{-24}\text{ g})}$$

$$= 2.67804 \times 10^{-23}\text{ g}$$

$$\text{Mass of } ^{16}_{8}O \text{ nucleus} = 2.65535 \times 10^{-23}\text{ g}$$

The difference in mass for one nucleus is

$$\text{Mass of } ^{16}_{8}O - \text{mass of } (8^1_0n + 8^1_1H) = -2.269 \times 10^{-25}\text{ g}$$

The difference in mass for the formation of 1 mole of $^{16}_{8}O$ nuclei is therefore

$$(-2.269 \times 10^{-25}\text{ g/nucleus})(6.022 \times 10^{23}\text{ nuclei/mol}) = -0.1366\text{ g/mol}$$

Thus 0.1366 gram of mass would be lost if 1 mole of oxygen-16 were formed from the constituent protons and neutrons. Why does the mass change, and how can this information be used to calculate the energy change that accompanies this process?

The answers to these questions can be found in the work of Albert Einstein. As we discussed in Section 12.2, Einstein's theory of relativity showed that energy should be considered a form of matter. His famous equation,

Energy is a form of matter.

$$E = mc^2$$

where c is the speed of light, gives the relationship between a quantity of energy and its mass. When a system gains or loses energy, it also gains or loses a quantity of mass, given by E/c^2. Thus the mass of a nucleus is less than that of its component nucleons because the process is so exothermic.

Einstein's equation in the form

The energy changes associated with normal chemical reactions are small enough that the corresponding mass changes are not detectable.

$$\text{Energy change} = \Delta E = \Delta mc^2$$

where Δm, the change in mass (the **mass defect**), can be used to calculate ΔE for the formation of a nucleus from its component nucleons.

The thermodynamic stability of a particular nucleus is normally represented as energy released per nucleon. To illustrate how this quantity is obtained, we will continue to consider $^{16}_{8}O$. First, we calculate ΔE for 1 mole of $^{16}_{8}O$ nuclei from the equation

$$\Delta E = \Delta mc^2$$

where $c = 3.00 \times 10^8\text{ m/s}$

and $\Delta m = -0.1366\text{ g/mol} = -1.366 \times 10^{-4}\text{ kg/mol}$

as calculated above. Thus

$$\Delta E = (-1.366 \times 10^{-4}\text{ kg/mol})(3.00 \times 10^8\text{ m/s})^2 = -1.23 \times 10^{13}\text{ J/mol}$$

Next, we calculate ΔE per nucleus by dividing the molar value by Avogadro's number:

$$\Delta E \text{ per } {}^{16}_{8}\text{O nucleus} = \frac{-1.23 \times 10^{13}\text{ J/mol}}{6.022 \times 10^{23}\text{ nuclei/mol}} = -2.04 \times 10^{-11}\text{ J/nucleus}$$

In terms of a more convenient energy unit, the MeV (million electron-volts), where

$$1\text{ MeV} = 1.60 \times 10^{-13}\text{ J}$$

we have

$$\begin{aligned}\Delta E \text{ per } {}^{16}_{8}\text{O nucleus} &= \left(-2.04 \times 10^{-11}\frac{\text{J}}{\text{nucleus}}\right)\left(\frac{1\text{ MeV}}{1.60 \times 10^{-13}\text{ J}}\right)\\ &= -1.28 \times 10^{2}\text{ MeV/nucleus}\end{aligned}$$

Next, we can calculate the value of ΔE per nucleon by dividing by A, the sum of the numbers of neutrons and protons:

$$\begin{aligned}\Delta E \text{ per nucleon for } {}^{16}_{8}\text{O} &= \frac{-1.28 \times 10^{2}\text{ MeV/nucleus}}{16\text{ nucleons/nucleus}}\\ &= -8.00\text{ MeV/nucleon}\end{aligned}$$

This means that 8.00 MeV of energy per nucleon would be *released* if ${}^{16}_{8}$O were formed from free neutrons and protons. The energy required to *decompose* this nucleus into its components has the same numeric value but is positive. This is called the **binding energy** per nucleon for ${}^{16}_{8}$O.

The values of the binding energy per nucleon for the various nuclides are shown in Fig. 20.9. Note that the most stable nuclei (those requiring the largest energy per nucleon to decompose the nucleus) occur at the top of the curve. The most stable nucleus known is ${}^{56}_{26}$Fe, which has a binding energy per nucleon of 8.79 MeV.

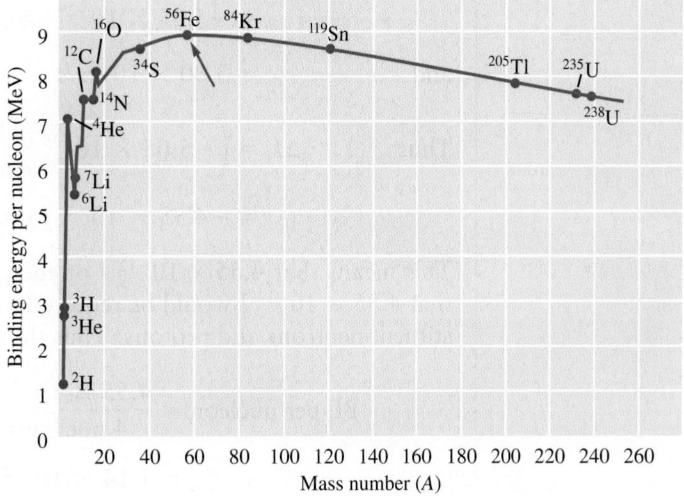

Figure 20.9
The binding energy per nucleon as a function of mass number. The most stable nuclei are at the top of the curve. The most stable nucleus is ${}^{56}_{26}$Fe.

OWL INTERACTIVE EXAMPLE 20.3

Calculate the binding energy per nucleon for the $^{4}_{2}He$ nucleus (atomic masses: $^{4}_{2}He = 4.0026$ amu, $^{1}_{1}H = 1.0078$ amu).

Solution First, we must calculate the mass defect (Δm) for $^{4}_{2}He$. Since atomic masses (which include the electrons) are given, we must decide how to account for the electron mass:

$$4.0026 = \text{mass of } ^{4}_{2}\text{He atom} = \text{mass of } ^{4}_{2}\text{He nucleus} + 2m_e$$

Electron mass

$$1.0078 = \text{mass of } ^{1}_{1}\text{H atom} = \text{mass of } ^{1}_{1}\text{H nucleus} + m_e$$

Thus, since a $^{4}_{2}He$ nucleus is "synthesized" from two protons and two neutrons, we see that

$$\Delta m = \underbrace{(4.0026 - 2m_e)}_{\text{Mass of } ^{4}_{2}\text{He nucleus}} - [2\underbrace{(1.0078 - m_e)}_{\text{Mass of } ^{1}_{1}\text{H nucleus (proton)}} + 2\underbrace{(1.0087)}_{\text{Mass of neutron}}]$$

$$= 4.0026 - 2m_e - 2(1.0078) + 2m_e - 2(1.0087)$$

$$= 4.0026 - 2(1.0078) - 2(1.0087)$$

$$= -0.0304 \text{ amu}$$

Note that in this case the electron mass cancels in taking the difference. This will always happen in this type of calculation if the atomic masses are used both for the nuclide of interest and for the $^{1}_{1}H$. Thus 0.0304 amu of mass is *lost* per $^{4}_{2}He$ nucleus formed.

The corresponding energy change can be calculated from

$$\Delta E = \Delta mc^2$$

where

$$\Delta m = -0.0304 \frac{\text{amu}}{\text{nucleus}}$$

$$= \left(-0.0304 \frac{\text{amu}}{\text{nucleus}}\right)\left(1.66 \times 10^{-27} \frac{\text{kg}}{\text{amu}}\right)$$

$$= -5.05 \times 10^{-29} \text{ kg/nucleus}$$

and

$$c = 3.00 \times 10^8 \text{ m/s}$$

Thus

$$\Delta E = \left(-5.05 \times 10^{-29} \frac{\text{kg}}{\text{nucleus}}\right)\left(3.00 \times 10^8 \frac{\text{m}}{\text{s}}\right)^2$$

$$= -4.55 \times 10^{-12} \text{ J/nucleus}$$

This means that 4.55×10^{-12} J of energy is *released* per nucleus formed and that 4.55×10^{-12} J would be required to decompose the nucleus into the constituent neutrons and protons. Thus the binding energy (BE) per nucleon is

$$\text{BE per nucleon} = \frac{4.55 \times 10^{-12} \text{ J/nucleus}}{4 \text{ nucleons/nucleus}}$$

$$= 1.14 \times 10^{-12} \text{ J/nucleon}$$

$$= \left(1.14 \times 10^{-12} \frac{\text{J}}{\text{nucleon}}\right)\left(\frac{1 \text{ MeV}}{1.60 \times 10^{-13} \text{ J}}\right)$$

$$= 7.13 \text{ MeV/nucleon}$$

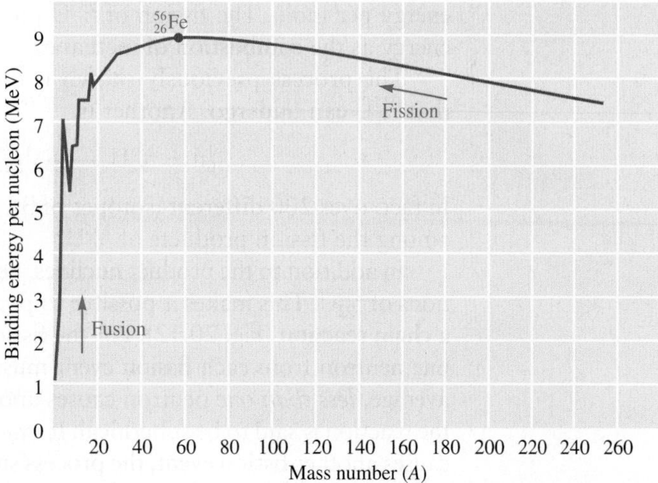

Figure 20.10
Both fission and fusion produce more stable nuclides and are thus exothermic.

20.6 Nuclear Fission and Nuclear Fusion

The graph shown in Fig. 20.9 has very important implications for the use of nuclear processes as sources of energy. Recall that energy is released—that is, ΔE is negative—when a process goes from a less stable to a more stable state. The higher a nuclide is on the curve, the more stable it is. This means that two types of nuclear processes are exothermic (Fig. 20.10):

1. Combining two light nuclei to form a heavier, more stable nucleus. This process is called **fusion.**
2. Splitting a heavy nucleus into two nuclei with smaller mass numbers. This process is called **fission.**

Because of the large binding energies involved in holding the nucleus together, both of these processes involve energy changes more than a million times larger than those associated with chemical reactions.

Nuclear Fission

Nuclear fission was discovered in the late 1930s when $^{235}_{92}$U nuclides bombarded with neutrons were observed to split into lighter elements:

$$^{1}_{0}\text{n} + {}^{235}_{92}\text{U} \longrightarrow {}^{141}_{56}\text{Ba} + {}^{92}_{36}\text{Kr} + 3\,^{1}_{0}\text{n}$$

This process shown schematically in Fig. 20.11, releases 3.5×10^{-11} J of energy per event, which translates to 2.1×10^{13} J per mole of $^{235}_{92}$U. Compare this figure with that for the combustion of methane, which releases only 8.0×10^{5} J of

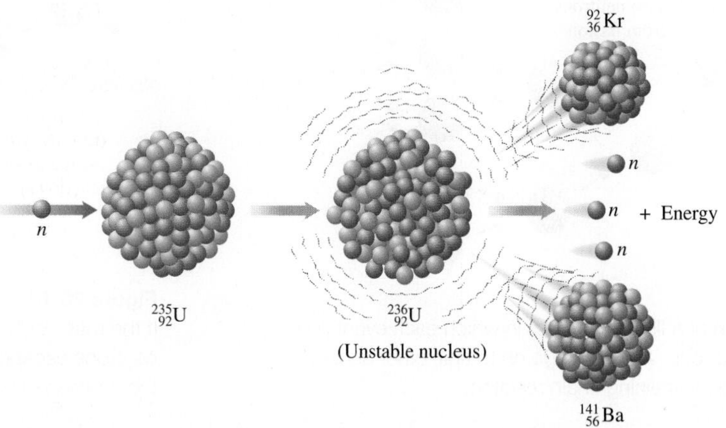

Figure 20.11
Upon capturing a neutron, the $^{235}_{92}$U nucleus undergoes fission to produce two lighter nuclides, free neutrons (typically three), and a large amount of energy.

energy per mole. The fission of $^{235}_{92}U$ produces about 26 million times as much energy as the combustion of methane.

The process previously shown is only one of the many fission processes that $^{235}_{92}U$ can undergo. Another is

$$^{1}_{0}n + ^{235}_{92}U \longrightarrow ^{137}_{52}Te + ^{97}_{40}Zn + 2^{1}_{0}n$$

In fact, over 200 different isotopes of 35 different elements have been observed among the fission products of $^{235}_{92}U$.

In addition to the product nuclides, neutrons are produced in the fission reactions of $^{235}_{92}U$. This makes it possible to produce a self-sustaining fission process—a **chain reaction** (Fig. 20.12). For the fission process to be self-sustaining, at least one neutron from each fission event must go on to split another nucleus. If, on average, *less than one* neutron causes another fission event, the process dies out; the reaction is said to be **subcritical.** If *exactly one* neutron from each fission event causes another fission event, the process sustains itself at the same level and is said to be **critical.** If *more than one* neutron from each fission event causes another fission event, the process rapidly escalates, and the heat buildup causes a violent explosion. This situation is described as **supercritical.**

The critical state requires a certain mass of fissionable material, called the **critical mass.** If the sample is too small, too many neutrons escape before they have a chance to cause a fission event; thus the process stops. This is illustrated in Fig. 20.13.

During World War II an intense research effort (the Manhattan Project) was carried out by the United States to build a bomb based on the principles of nuclear fission. This program produced the fission bombs that were used with devastating effects on the cities of Hiroshima and Nagasaki in 1945. Basically, a fission bomb operates by suddenly combining two subcritical masses of fissionable material to form a supercritical mass, thereby producing an explosion of incredible intensity.

Nuclear Reactors

Because of the tremendous energies involved, it seemed desirable after World War II to develop the fission process as an energy source for producing electricity. Therefore, reactors were designed in which controlled fission can occur.

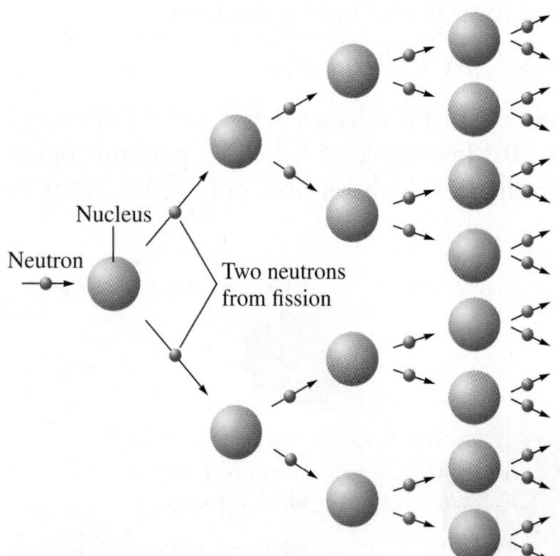

Figure 20.12
Representation of a fission process in which each event produces two neutrons, which can go on to split other nuclei, leading to a self-sustaining chain reaction.

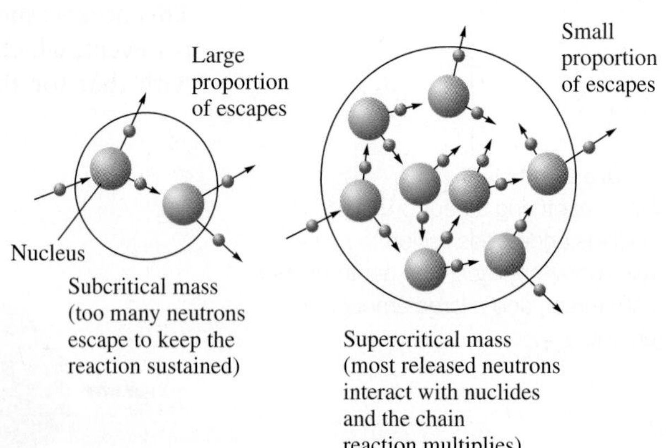

Figure 20.13
If the mass of fissionable material is too small, most of the neutrons escape before causing another fission event; thus the process dies out.

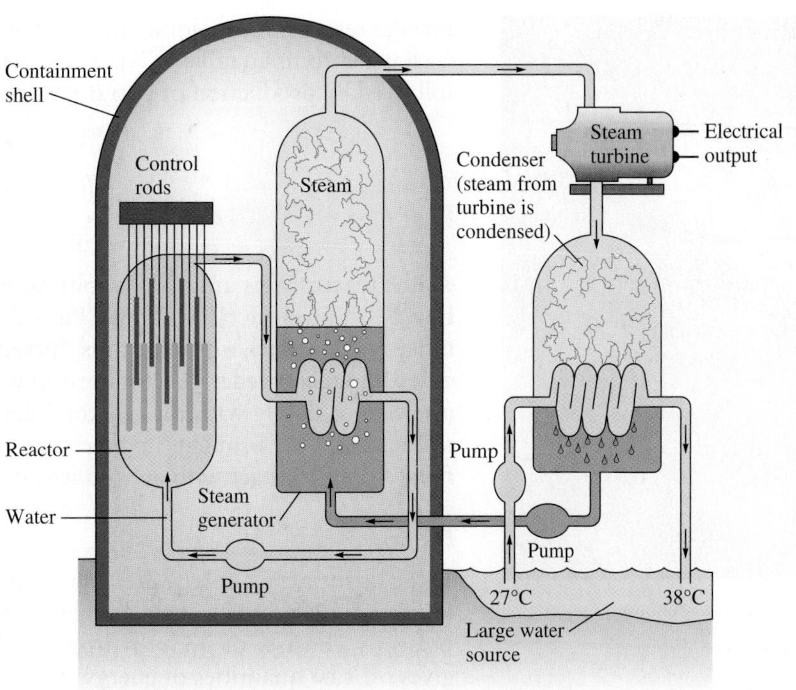

Figure 20.14
A schematic diagram of a nuclear power plant.

The resulting energy is used to heat water to produce steam to run turbine generators, in much the same way that a coal-burning power plant generates energy. A schematic diagram of a nuclear power plant is shown in Fig. 20.14.

^{235}U enrichment is described in Section 5.7.

In the **reactor core,** shown in Fig. 20.15, uranium that has been enriched to approximately 3% $^{235}_{92}$U (natural uranium contains only 0.7% $^{235}_{92}$U) is housed in cylinders. A **moderator** surrounds the cylinders to slow down the neutrons so that the uranium fuel can capture them more efficiently. **Control rods,** composed of substances that absorb neutrons, are used to regulate the power level of the reactor. The reactor is designed so that should a malfunction occur, the control rods are automatically inserted into the core to stop the reaction. A liquid (usually water) is circulated through the core to extract the heat generated by the energy of fission; the energy is then passed on via a heat exchanger to water in the turbine system.

Although the concentration of $^{235}_{92}$U in the fuel elements is not great enough to allow a supercritical mass to develop in the core, a failure of the cooling system can lead to temperatures high enough to melt the core. As a result, the building housing the core must be designed to contain the core even if a meltdown occurs. A great deal of controversy now exists about the efficiency of the safety systems in nuclear power plants. Accidents such as the one at the Three Mile Island facility in Pennsylvania in 1979 and the one in Chernobyl,* USSR, in 1986 have led many people to question the wisdom of continuing to build fission-based power plants.

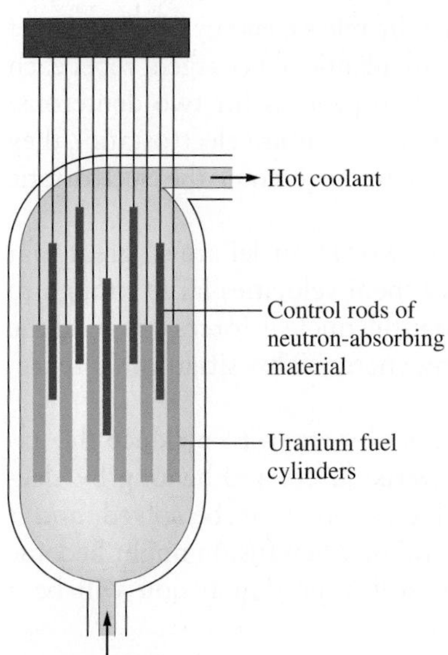

Figure 20.15
A schematic of a reactor core. The position of the control rods determines the level of energy production by regulating the amount of fission taking place.

Breeder Reactors

One potential problem facing the nuclear power industry is the supply of $^{235}_{92}$U. Some scientists have suggested that we have nearly depleted those uranium deposits rich enough in $^{235}_{92}$U to make production of fissionable fuel economically feasible. Because of this possibility, **breeder reactors** have been developed, in which fissionable fuel is actually produced while the reactor runs. In a

*For a detailed account of this incident, see C. A. Atwood, "Chernobyl—What Happened?" *J. Chem. Ed.* **65** (1988): 1037.

A core of a nuclear reactor.

breeder reactor, the major component of natural uranium, nonfissionable $^{238}_{92}U$, is changed to fissionable $^{239}_{94}Pu$. The reaction involves absorption of a neutron, followed by production of two β particles:

$$^{1}_{0}n + ^{238}_{92}U \longrightarrow ^{239}_{92}U$$

$$^{239}_{92}U \longrightarrow ^{239}_{93}Np + ^{0}_{-1}e$$

$$^{239}_{93}Np \longrightarrow ^{239}_{94}Pu + ^{0}_{-1}e$$

As the reactor runs and $^{235}_{92}U$ is split, some of the excess neutrons are absorbed by $^{238}_{92}U$ to produce $^{239}_{94}Pu$. The $^{239}_{94}Pu$ is then separated out and used to fuel another reactor. Such a reactor thus "breeds" nuclear fuel as it operates.

Although breeder reactors are now used in France, the United States is proceeding slowly with the reactors' development because of their controversial nature. One problem involves the hazards in handling plutonium, which flames upon contact with air and is very toxic.

Fusion

Large quantities of energy are also produced by the fusion of two light nuclei. In fact, stars produce their energy through nuclear fusion. Our sun, which presently consists of 73% hydrogen, 26% helium, and 1% other elements, gives off vast quantities of energy from the fusion of protons to form helium:

$$^{1}_{1}H + ^{1}_{1}H \longrightarrow ^{2}_{1}H + ^{0}_{1}e$$

$$^{1}_{1}H + ^{2}_{1}H \longrightarrow ^{3}_{2}He$$

$$^{3}_{2}He + ^{3}_{2}He \longrightarrow ^{4}_{2}He + 2^{1}_{1}H$$

$$^{3}_{2}He + ^{1}_{1}H \longrightarrow ^{4}_{2}He + ^{0}_{1}e$$

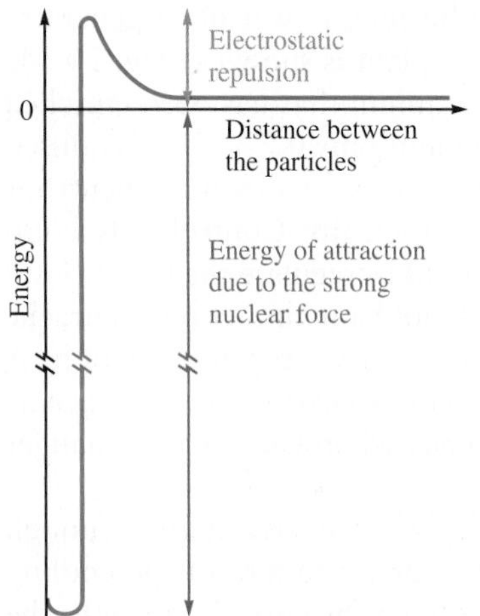

Figure 20.16
A plot of energy versus the separation distance for two $^{2}_{1}H$ nuclei. The nuclei must have sufficient velocities to get over the electrostatic repulsion "hill" and get close enough for the nuclear binding forces to become effective, thus "fusing" the particles into a new nucleus and releasing large quantities of energy. The binding force is at least 100 times the electrostatic repulsion.

Intense research is under way to develop a feasible fusion process because of the ready availability of many light nuclides (deuterium, $^{2}_{1}H$, in seawater, for example) that can serve as fuel in fusion reactors. The major stumbling block is the high temperatures required to initiate fusion. The forces that bind nucleons together to form a nucleus are effective only at *very small* distances ($\approx 10^{-13}$ cm). Thus, for two protons to bind together and thereby release energy, they must get very close together. But protons, because they are identically charged, repel each other electrostatically. This means that to get two protons (or two deuterons) close enough to bind together (the nuclear binding force is *not* electrostatic), they must be "shot" at each other at speeds high enough to overcome the electrostatic repulsion.

The electrostatic repulsion forces between two $^{2}_{1}H$ nuclei are so great that a temperature of 4×10^{7} K is required to give them velocities large enough to cause them to collide with sufficient energy that the nuclear forces can bind the particles together and thus release the binding energy. This situation is represented in Fig. 20.16.

Currently, scientists are studying two types of systems to produce the extremely high temperatures required: high-powered lasers and heating by electric currents. At present, many technical problems remain to be solved, and it is not clear which method will prove more useful or when fusion might become a practical energy source. However, there is still hope that fusion will be a major energy source in the twenty-first century.

20.7 Effects of Radiation

Everyone knows that being hit by a train is very serious. The problem is the energy transfer involved. In fact, any source of energy is potentially harmful to organisms. Energy transferred to cells can break chemical bonds and cause

Nuclear Physics: An Introduction

Chemical Insights

Nuclear physics is concerned with the fundamental nature of matter. The central focuses of this area of study are the relationship between a quantity of energy and its mass, given by $E = mc^2$, and the fact that matter can be converted from one form (energy) to another (particulate) in particle accelerators. Collisions between high-speed particles have produced a dazzling array of new particles—hundreds of them. These events can best be seen as conversions of kinetic energy into particles. For example, a collision of sufficient energy between a proton and a neutron can produce four particles: two protons, one antiproton, and a neutron:

$$^{1}_{1}\text{H} + ^{1}_{0}\text{n} \longrightarrow 2^{1}_{1}\text{H} + ^{\ 1}_{-1}\text{H} + ^{1}_{0}\text{n}$$

where $^{\ 1}_{-1}\text{H}$ is the symbol for an *antiproton,* which has the same mass as a proton but the opposite charge. This process is a little like throwing one baseball at a very high speed into another and having the collision produce four baseballs.

The results of such accelerator experiments have led scientists to postulate the existence of three types of forces important in the nucleus: the *strong force,* the *weak force,* and the *electromagnetic force.* Along with the *gravitational force,* these forces are thought to account for all types of interactions found in matter. These forces are believed to be generated by the exchange of particles between the interacting pieces of matter. For example, gravitational force is thought to be carried by particles called *gravitons.* The electromagnetic force (the classical electrostatic force between charged particles) is assumed to be exerted through the exchange of *photons.* The strong force, not charge-related and effective only at very short distances ($\approx 10^{-13}$ cm), is postulated to involve the exchange of particles called *gluons.* The weak force is 100 times weaker than the strong force and seems to be exerted by the exchange of two types of large particles, the W (has a mass 70 times the proton mass) and the Z (has a mass 90 times the proton mass).

The particles discovered have been classified into several categories. Three of the most important classes are as follows:

1. *Hadrons* are particles that respond to the strong force and have internal structure.
2. *Leptons* are particles that do not respond to the strong force and have no internal structure.
3. *Quarks* are particles with no internal structure that are thought to be the fundamental constituents of hadrons. Neutrons and protons are hadrons that are thought to be composed of three quarks each.

Courtesy, Fermilab Visual Media/Batavia, IL

The accelerator tunnel at Fermilab, a high-energy particle accelerator in Batavia, Illinois.

The world of particle physics appears mysterious and complicated. For example, particle physicists have discovered new properties of matter they call "color," "charm," and "strangeness" and have postulated conservation laws involving these properties. This area of science is extremely important because it should help us to understand the interactions of matter in a more elegant and unified way. For example, the classification of forces into four categories is probably necessary only because we do not understand the true nature of forces. All forces may be special cases of a single, all-pervading force field that governs all of nature. In fact, Einstein spent the last 30 years of his life looking for a way to unify gravitational and electromagnetic forces—without success. Physicists may now be on the verge of accomplishing what Einstein failed to do.

Although the practical aspects of much of the work in nuclear physics are not yet totally apparent, a more fundamental understanding of the way nature operates could lead to presently undreamed-of devices for energy production, communication, and so on, which could revolutionize our lives.

malfunctioning of the cell systems. This fact is behind the concern about the ozone layer in the earth's upper atmosphere, which screens out high-energy ultraviolet radiation from the sun. Radioactive elements, which are sources of high-energy particles, are also potentially hazardous, although the effects are usually quite subtle. The reason for the subtlety of radiation damage is that even though high-energy particles are involved, the quantity of energy actually deposited in tissues *per event* is quite small. However, the resulting damage is no less real, although the effects may not be apparent for years.

The α particles, β particles, and γ rays produced by radioactive nuclei have energies far greater than the bond energies and the ionization energies of biological molecules. Therefore, nuclear radiation fragments and ionizes biomolecules, generating reactive species that can cause further damage. For example, γ rays can knock an electron from a water molecule to form H_2O^+ ions, which can react with another water molecule as follows:

$$H_2O^+ + H_2O \longrightarrow H_3O^+ + OH$$

The OH fragment produced in this reaction is a *free radical,* a species with one or more unpaired electrons. Because it has an unpaired electron, the OH molecule readily attacks other molecules, forming more free radicals that in turn can react with additional molecules. Thus the formation of a single OH radical by a γ ray can lead to extensive damage in a cell.

Radiation damage to organisms can be classified as somatic or genetic damage. **Somatic damage** is damage to the organism itself, resulting in sickness or death. The effects may appear almost immediately if a massive dose of radiation is received; for smaller doses, damage may appear years later, usually in the form of cancer. **Genetic damage** is damage to the genetic machinery, which produces malfunctions in the offspring of the organism.

The biological effects of a particular source of radiation depend on several factors:

1. *The energy of the radiation.* The higher the energy content of the radiation, the more damage it can cause by formation of free radicals and ions. The official SI unit for radiation doses is the *Gray* (1 J of energy deposited per kg of tissue), although the **rad** (10^{-2} J/kg tissue) is still commonly used.
2. *The penetrating ability of the radiation.* The particles and rays produced in radioactive processes vary in their abilities to penetrate human tissue: γ rays are highly penetrating; β particles can penetrate about 1 cm; and α particles are stopped by the skin.
3. *The chemical properties of the radiation source.* When a radioactive nuclide is ingested into the body, its effectiveness in causing damage depends on its residence time. For example, ${}^{85}_{36}Kr$ and ${}^{90}_{38}Sr$ are both β-particle producers. However, since krypton is chemically inert, it passes through the body quickly and does not have much time to do damage. Strontium, being chemically similar to calcium, can collect in bones, where it may cause leukemia and bone cancer.

Because of the differences in the behavior of the particles and rays produced by radioactive decay, both the energy dose of the radiation and its effectiveness in causing biological damage must be taken into account. The **rem** (which is short for *r*oentgen *e*quivalent for *m*an) is defined as follows:

$$\text{Number of rems} = (\text{number of rads}) \times \text{RBE}$$

where RBE represents the relative effectiveness of the radiation in causing biological damage.

Table 20.6 shows the physical effects of short-term exposure to various doses of radiation, and Table 20.7 gives the sources and amounts of radiation

Table 20.6

Effects of Short-Term Exposures to Radiation

Dose (rem)	Clinical Effect
0–25	Nondetectable
25–50	Temporary decrease in white blood cell counts
100–200	Strong decrease in white blood cell counts
500	Death of half the exposed population within 30 days after exposure

Table 20.7

Typical Radiation Exposures for a Person Living in the United States (1 millirem = 10^{-3} rem)

Source	Exposure (millirems/year)
Radon	200
Cosmic radiation	27
From the earth	28
From building materials	3
In human tissues	39
Inhalation of air	5
Total from natural sources	302
X-ray diagnosis	50
Radiotherapy	10
Internal diagnosis/therapy	1
Nuclear power industry	0.2
TV tubes, industrial wastes, etc.	2
Radioactive fallout	4
Total from human activities	67
Total	369

exposure for a typical person in the United States. Note that natural sources contribute four to five times as much as human activities to the total exposure. However, although the nuclear industry contributes only a small percentage of the total exposure, the major controversy associated with nuclear power plants is the *potential* for radiation hazards. These arise mainly from two sources: accidents allowing the release of radioactive materials and improper disposal of the radioactive products in spent fuel elements. The radioactive products of the fission of $^{235}_{92}U$, although constituting only a small percentage of the total products, have half-lives of several hundred years and remain dangerous for a long time. Various schemes have been advanced for the disposal of these wastes. The one that seems to hold the most promise is the incorporation of the wastes into ceramic blocks and the burial of these blocks in geologically stable formations. At present, however, no disposal method has been accepted, and nuclear wastes continue to accumulate in temporary storage facilities.

Note from Table 20.7 that exposure to radon is by far the most significant source of natural radiation. $^{222}_{86}Rn$, a decay product of $^{238}_{92}U$ (Table 20.3), is continuously generated in the earth's crust. This gaseous radon emerges from the ground and is trapped in the basements of houses and other buildings, where it is circulated by heating and cooling systems. Because radon is an α-particle producer,

$$^{222}_{86}Rn \longrightarrow {}^{218}_{84}Po + {}^{4}_{2}He$$

with a short half-life of 3.82 days, it causes biological damage when inhaled. In addition, the $^{218}_{84}Po$ formed, a solid that can be trapped in lung tissue, is an even more potent α-particle producer ($t_{1/2}$ = 3.11 min). Because α particles have a high RBE (because of their ability to produce a very high concentration of ions in affected tissue), radon is particularly dangerous and there is great concern that this exposure can lead to a significant incidence of lung cancer.

Key Terms

Section 20.1
beta (β) particle
zone of stability
magic numbers
alpha (α) particle
alpha-particle production
spontaneous fission
beta-particle production
gamma (γ) ray
positron production
antiparticle

For Review

OWL and Go Chemistry

Sign in at **www.cengage.com/owl** to:

- View tutorials and simulations, develop problem-solving skills, and complete online homework assigned by your professor.
- Download Go Chemistry mini lecture modules for quick review and exam prep from OWL (or purchase them at **www.cengagebrain.com**)

Radioactivity

- Certain nuclei decay spontaneously into more stable nuclei
- Types of radioactive decay:
  - α-particle ($^{4}_{2}He$) production
  - β-particle ($^{0}_{-1}e$) production

electron capture decay series **Section 20.2** rate of decay nuclides half-life **Section 20.3** nuclear transformation particle accelerator cyclotron linear accelerator transuranium elements **Section 20.4** Geiger–Müller counter (Geiger counter) scintillation counter radiocarbon dating (carbon-14 dating) radiotracers radiation therapy **Section 20.5** mass defect binding energy **Section 20.6** fusion fission chain reaction subcritical reaction critical reaction supercritical reaction critical mass reactor core moderator control rods breeder reactor **Section 20.7** somatic damage genetic damage rad rem	■ Positron ($^{0}_{1}e$) production ■ γ rays are usually produced in a radioactive decay event ■ A decay series involves several radioactive decays to finally reach a stable nuclide ■ Radioactive decay follows first-order kinetics ■ Half-life of a radioactive sample: the time required for half of the nuclides to decay ■ The transuranium elements (those beyond uranium in the periodic table) can be synthesized by particle bombardment of uranium or heavier elements ■ Radiocarbon dating employs the $^{14}_{6}C/^{12}_{6}C$ ratio in an object to establish its date of origin **Thermodynamic stability of a nucleus** ■ Compares the mass of a nucleus to the sum of the masses of its component nucleons ■ When a system gains or loses energy, it also gains or loses mass as described by the relationship $E = mc^2$ ■ The difference between the sum of the masses of the component nucleons and the actual mass of a nucleus (called the mass defect) can be used to calculate the nuclear binding energy **Nuclear energy production** ■ Fusion: the process of combining two light nuclei to form a heavier, more stable nucleus ■ Fission: the process of splitting a heavy nucleus into two lighter, more stable nuclei ■ Current nuclear power reactors employ controlled fission to produce energy **Radiation damage** ■ Radiation can cause direct (somatic) damage to a living organism or genetic damage to the organism's offspring ■ The biologic effects of radiation depend on the energy, the penetrating ability, the ionizing ability of the radiation, and the chemical properties of the nuclide producing the radiation

Exercises

OWL Interactive versions of these problems may be assigned in OWL.

A blue exercise number indicates that the answer to that exercise appears at the back of this book and a solution appears in the *Solutions Guide.*

Radioactive Decay and Nuclear Transformations

1. Define or illustrate the following terms.
   a. thermodynamic stability
   b. kinetic stability
   c. radioactive decay
   d. β-particle production
   e. α-particle production
   f. positron production
   g. electron capture
   h. γ-ray emissions
2. Which type of radioactive decay has the net effect of changing a neutron into a proton? Which type of decay has the net effect of turning a proton into a neutron?
3. Supply the missing particle, and state the type of decay for each of the following nuclear processes.
   a.

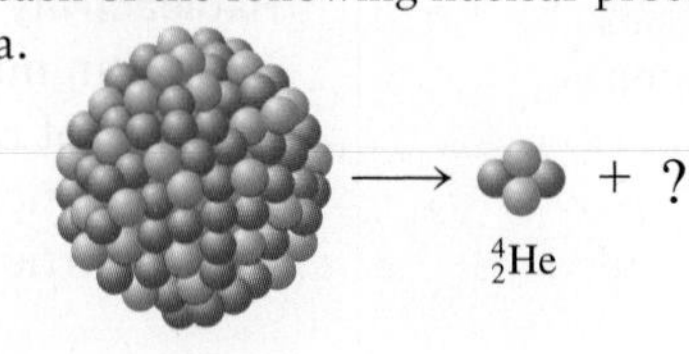

b.

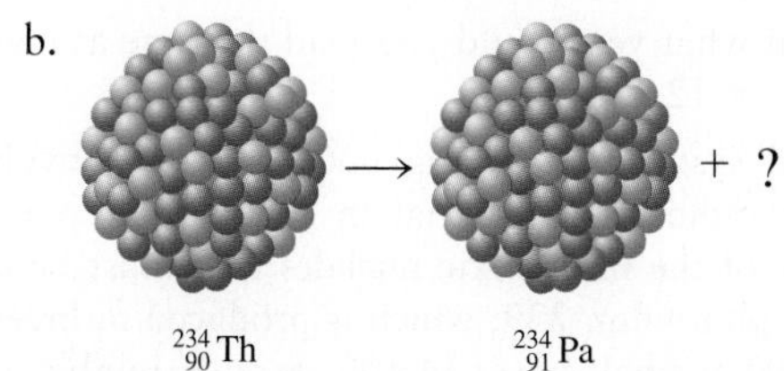

4. Write an equation describing the radioactive decay of each of the following nuclides. (The particle produced is shown in parentheses, except for electron capture, where an electron is a reactant.)
   a. $^{3}_{1}H$ (β)
   b. $^{8}_{3}Li$ (β followed by α)
   c. $^{7}_{4}Be$ (electron capture)
   d. $^{8}_{5}B$ (positron)

5. Write balanced equations for each of the following processes.
   a. Chromium-51, which targets the spleen and is used as a tracer in studies of red blood cells, decays by electron capture.
   b. Iodine-131, used to treat hyperactive thyroid glands, decays by producing a β particle.
   c. Phosphorus-32, which accumulates in the liver, decays by β-particle production.

6. In each of the following radioactive decay processes, supply the missing particle.
   a. $^{73}Ga \rightarrow {}^{73}Ge + ?$
   b. $^{192}Pt \rightarrow {}^{188}Os + ?$
   c. $^{205}Bi \rightarrow {}^{205}Pb + ?$
   d. $^{241}Cm + ? \rightarrow {}^{241}Am$
   e. $^{60}Co \rightarrow {}^{60}Ni + ?$
   f. $^{97}Tc + ? \rightarrow {}^{97}Mo$
   g. $^{99}Tc \rightarrow {}^{99}Ru + ?$
   h. $^{239}Pu \rightarrow {}^{235}U + ?$

7. Write an equation describing the radioactive decay of each of the following nuclides. (The particle produced is shown in parentheses, except for electron capture, where an electron is a reactant.)
   a. ^{68}Ga (electron capture)
   b. ^{62}Cu (positron)
   c. ^{212}Fr (α)
   d. ^{129}Sb (β)

8. One type of commercial smoke detector contains a minute amount of radioactive americium-241 (^{241}Am), which decays by α-particle production. The α particles ionize molecules in the air, allowing it to conduct an electric current. When smoke particles enter, the conductivity of the air changes, and the alarm buzzes.
   a. Write the equation for the decay of $^{241}_{95}Am$ by α-particle production.
   b. The complete decay of ^{241}Am involves successively α, α, β, α, α, β, α, α, α, β, α, and β production. What is the final stable nucleus produced in this decay series?
   c. Identify the 11 intermediate nuclides.

9. The radioactive isotope ^{247}Bk decays by a series of α-particle and β-particle productions, taking ^{247}Bk through many transformations to end up as ^{207}Pb. In the complete decay series, how many α particles and β particles are produced?

10. There are four stable isotopes of iron with mass numbers 54, 56, 57, and 58. There are also two radioactive isotopes: iron-53 and iron-59. Predict modes of decay for these two isotopes, and write a nuclear reaction for each. (See Table 20.2.)

11. The only stable isotope of fluorine is fluorine-19. Predict possible modes of decay for fluorine-21, fluorine-18, and fluorine-17.

12. Predict whether each of the following nuclides is stable or unstable (radioactive). If the nuclide is unstable, predict the type of radioactivity you would expect it to exhibit.
   a. $^{45}_{19}K$
   b. $^{56}_{26}Fe$
   c. $^{20}_{11}Na$
   d. $^{194}_{81}Tl$

13. In 1994 it was proposed (and eventually accepted) that element 106 be named seaborgium (Sg) in honor of Glenn T. Seaborg, discoverer of the transuranium elements.
   a. ^{263}Sg was produced by the bombardment of ^{249}Cf with a beam of ^{18}O nuclei. Complete and balance an equation for this reaction.
   b. ^{263}Sg decays by α-particle emission. What is the other product resulting from the α decay of ^{263}Sg?

14. Many elements have been synthesized by bombarding relatively heavy atoms with high-energy particles in particle accelerators. Complete the following nuclear reactions, which have been used to synthesize elements.
   a. $\_\_\_\_ + {}^{4}_{2}He \longrightarrow {}^{243}_{97}Bk + {}^{1}_{0}n$
   b. $^{238}_{92}U + {}^{12}_{6}C \longrightarrow \_\_\_\_ + 6{}^{1}_{0}n$
   c. $^{249}_{98}Cf + \_\_\_\_ \longrightarrow {}^{260}_{105}Db + 4{}^{1}_{0}n$
   d. $^{249}_{98}Cf + {}^{10}_{5}B \longrightarrow {}^{257}_{103}Lr + \_\_\_\_$

Kinetics of Radioactive Decay

15. Americium-241 is widely used in smoke detectors. The radiation released by this element ionizes particles that are then detected by a charged-particle collector. The half-life of ^{241}Am is 433 years, and it decays by emitting alpha particles. How many alpha particles are emitted each second by a 5.00-g sample of ^{241}Am?

16. Krypton consists of several radioactive isotopes, some of which are listed in the following table.

Isotope	Half-life
Kr-73	27 s
Kr-74	11.5 min
Kr-76	14.8 h
Kr-81	2.1×10^5 yr

Which of these isotopes is most stable, and which isotope is "hottest"? How long does it take for 87.5% of each isotope to decay?

17. Radioactive copper-64 decays with a half-life of 12.8 days.
   a. What is the value of k in s^{-1}?
   b. A sample contains 28.0 mg ^{64}Cu. How many decay events will be produced in the first second? Assume that the atomic mass of ^{64}Cu is 64.0.
   c. A chemist obtains a fresh sample of ^{64}Cu and measures its radioactivity. She then determines that to do an experiment, the radioactivity cannot fall below 25% of the initial measured value. How long does she have to perform the experiment?

18. The curie (Ci) is a commonly used unit for measuring nuclear radioactivity: 1 curie of radiation is equal to 3.7 ×

10^{10} decay events per second (the number of decay events from 1 g of radium in 1 s).

a. What mass of $Na_2{}^{38}SO_4$ has an activity of 10.0 mCi? Sulfur-38 has an atomic mass of 38.0 and a half-life of 2.87 h.

b. How long does it take for 99.99% of a sample of sulfur-38 to decay?

19. The first atomic explosion was detonated in the desert north of Alamogordo, New Mexico, on July 16, 1945. What percentage of the strontium-90 ($t_{1/2}$ = 28.9 yr) originally produced by that explosion still remains as of July 16, 2012?

20. Iodine-131 is used in the diagnosis and treatment of thyroid disease and has a half-life of 8.0 days. If a patient with thyroid disease consumes a sample of $Na^{131}I$ containing 10. μg of ^{131}I, how long will it take for the amount of ^{131}I to decrease to 1/100 of the original amount?

21. Phosphorus-32 is a commonly used radioactive nuclide in biochemical research, particularly in studies of nucleic acids. The half-life of phosphorus-32 is 14.3 days. What mass of phosphorus-32 is left from an original sample of 175 mg of $Na_3{}^{32}PO_4$ after 35.0 days? Assume that the atomic mass of ^{32}P is 32.0.

22. A chemist wishing to do an experiment requiring $^{47}Ca^{2+}$ (half-life = 4.5 days) needs 5.0 μg of the nuclide. What mass of $^{47}CaCO_3$ must be ordered if it takes 48 h for delivery from the supplier? Assume that the atomic mass of ^{47}Ca is 47.0.

23. Explain the theory behind carbon-14 dating. What assumptions must be made and what problems arise when using carbon-14 dating?

24. The decay of uranium-238 to lead-206 is also used to estimate the age of objects. Specifically, ^{206}Pb-to-^{238}U ratios allow dating of rocks. Why is the ^{238}U decay to ^{206}Pb useful for dating rocks but useless for dating objects 10,000 years old or younger? Similarly, why is carbon-14 dating useful for dating objects 10,000 years old or younger but useless for dating rocks?

25. At a flea market you've found a very interesting painting done in the style of Rembrandt's "Dark Period" (1642–1672). You suspect that you really do not have a genuine Rembrandt, so you take it to the local university for testing. Living wood shows a carbon-14 activity of 15.3 counts per minute per gram. Your painting showed a carbon-14 activity of 15.1 counts per minute per gram. Could it be a genuine Rembrandt? (For ^{14}C, $t_{1/2}$ = 5730 years.)

26. A living plant contains about the same fraction of carbon-14 as atmospheric carbon dioxide. The observed rate of decay of carbon-14 from a living plant is 15.3 counts per minute per gram of carbon. How many counts per minute per gram of carbon will be measured from a 15,000-year-old sample? Will radiocarbon dating work well for small samples of 10 mg or less? (For ^{14}C, $t_{1/2}$ = 5730 years.)

27. During World War II, tritium (^3H) was a component of fluorescent watch dials and hands. Assume you have such a watch that was made in January 1944. If 17% or more of the original tritium was needed to read the dial in dark places, until what year could you read the time at night? (For ^3H, $t_{1/2}$ = 12.3 years.)

28. A proposed system for storing nuclear wastes involves storing the radioactive material in caves or deep mine shafts. One of the most toxic nuclides that must be disposed of is plutonium-239, which is produced in breeder reactors and has a half-life of 24,100 years. A suitable storage place must be geologically stable long enough for the activity of plutonium-239 to decrease to 0.1% of its original value. How long is this period for plutonium-239?

29. A rock contains 0.688 mg of ^{206}Pb for every 1.000 mg of ^{238}U present. Assuming that no lead was originally present, that all the ^{206}Pb formed over the years has remained in the rock, and that the number of nuclides in intermediate stages of decay between ^{238}U and ^{206}Pb is negligible, calculate the age of the rock. (For ^{238}U, $t_{1/2} = 4.5 \times 10^9$ years.)

30. The mass ratios of ^{40}Ar to ^{40}K can also be used to date geological materials. Potassium-40 decays by two processes:

$$^{40}_{19}K + {}^{0}_{-1}e \longrightarrow {}^{40}_{18}Ar \ (10.7\%)$$

$$t_{1/2} = 1.27 \times 10^9 \text{ yr}$$

$$^{40}_{19}K \longrightarrow {}^{40}_{20}Ca + {}^{0}_{-1}e \ (89.3\%)$$

a. Why are $^{40}Ar/^{40}K$ ratios rather than $^{40}Ca/^{40}K$ ratios used to date materials?

b. What assumptions must be made in using this technique?

c. A sedimentary rock has a $^{40}Ar/^{40}K$ ratio of 0.95. Calculate the age of the rock.

d. How will the measured age of a rock compare with the actual age if some ^{40}Ar has escaped from the sample?

Energy Changes in Nuclear Reactions

31. The sun radiates 3.9×10^{23} J of energy into space every second. What is the rate at which mass is lost from the sun?

32. The earth receives 1.8×10^{14} kJ/s of solar energy. What mass of solar material is converted to energy over a 24-h period to provide the daily amount of solar energy to the earth? What mass of coal would have to be burned to provide the same amount of energy? Coal releases 32 kJ of energy per gram when burned.

33. The most stable nucleus in terms of binding energy per nucleon is ^{56}Fe. If the atomic mass of ^{56}Fe is 55.9349 amu, calculate the binding energy per nucleon for ^{56}Fe.

34. Calculate the binding energy per nucleon for ^2_1H and ^3_1H. The atomic masses are ^2_1H, 2.01410 amu, and ^3_1H, 3.01605 amu.

35. Calculate the binding energy in J/nucleon for carbon-12 (atomic mass 12.00000) and uranium-235 (atomic mass 235.0439). The atomic mass of ^1_1H is 1.00782 amu, and the mass of a neutron is 1.00866 amu. The most stable nucleus known is ^{56}Fe (see Exercise 33). Would the binding energy per nucleon for ^{56}Fe be larger or smaller than that for ^{12}C or ^{235}U? Explain.

36. The mass defect for a lithium-6 nucleus is −0.03434 g/mol. Calculate the atomic mass of lithium-6.

37. The binding energy per nucleon for magnesium-27 is 1.326×10^{-12} J/nucleon. Calculate the atomic mass of magnesium-27.

38. A positron and an electron annihilate each other upon colliding, thereby producing energy:

$$^{0}_{-1}e + ^{0}_{+1}e \longrightarrow 2\,^{0}_{0}\gamma$$

Assuming that both γ rays have the same energy, calculate the wavelength of the electromagnetic radiation produced.

39. The easiest fusion reaction to initiate is

$$^{2}_{1}H + ^{3}_{1}H \longrightarrow ^{4}_{2}He + ^{1}_{0}n$$

Calculate the energy released per nucleus of $^{4}_{2}He$ produced and per mole of $^{4}_{2}He$ produced. The atomic masses are as follows: $^{2}_{1}H$, 2.01410 amu; $^{3}_{1}H$, 3.01605 amu; and $^{4}_{2}He$, 4.00260 amu. The masses of the electron and neutron are 5.4858×10^{-4} amu and 1.00866 amu, respectively.

40. Calculate the amount of energy released per gram of hydrogen nuclei reacted for the following reaction. The atomic masses are $^{1}_{1}H$, 1.00782 amu, and $^{2}_{1}H$, 2.01410 amu. (*Hint:* Think carefully about how to account for the electron mass.)

$$^{1}_{1}H + ^{1}_{1}H \longrightarrow ^{2}_{1}H + ^{0}_{+1}e$$

Detection, Uses, and Health Effects of Radiation

41. The typical response of a Geiger-Müller tube is shown below. Explain the shape of this curve.

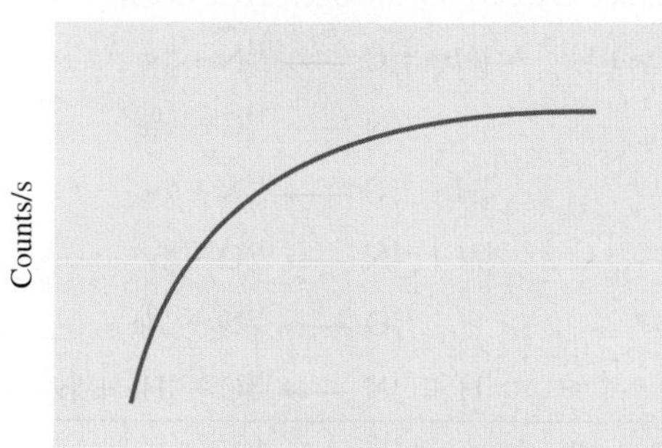

42. When using a Geiger-Müller counter to measure radioactivity, one must maintain the same geometrical orientation between the sample and the Geiger-Müller tube to compare different measurements. Why?

43. Define *fission* and *fusion*. Fusion processes are more likely to occur for lighter elements, whereas fission processes are more likely to occur for heavier elements. Explain.

44. Why are elevated temperatures necessary to initiate fusion reactions but not fission reactions?

45. What are the purposes of the moderator and control rods in a fission reactor?

46. Much of the research on controlled fusion focuses on the problem of how to contain the reacting material. Magnetic fields appear to be the most promising mode of containment. Why is containment such a problem? Why must one resort to magnetic fields for containment?

47. During the research that led to production of the two atomic bombs used against Japan in World War II, different mechanisms for obtaining a supercritical mass of fissionable material were investigated. In one type of bomb, a "gun" shot one piece of fissionable material into a cavity containing another piece of fissionable material. In the second type of bomb, the fissionable material was surrounded with a high explosive that, when detonated, compressed the fissionable material into a smaller volume. Discuss what is meant by critical mass, and explain why the ability to achieve a critical mass is essential to sustaining a nuclear reaction.

48. There is a trend in the United States toward using coal-fired power plants to generate electricity rather than building new nuclear fission power plants. Is the use of coal-fired power plants without risk? Make a list of the risks to society from the use of each type of power plant.

49. How could a radioactive nuclide be used to demonstrate that chemical equilibrium is a dynamic process?

50. Consider the following reaction to produce methyl acetate:

$$CH_3OH + CH_3\overset{\overset{O}{\|}}{C}OH \longrightarrow \underset{\text{Methyl acetate}}{CH_3\overset{\overset{O}{\|}}{C}OCH_3} + H_2O$$

When this reaction is carried out with CH_3OH containing radioactive oxygen-18, the water produced is not radioactive. Explain.

51. Photosynthesis in plants can be represented by the following overall reaction:

$$6CO_2(g) + 6H_2O(l) \xrightarrow{\text{Light}} C_6H_{12}O_6(s) + 6O_2(g)$$

Algae grown in water containing some radioactive ^{18}O (in $H_2{}^{18}O$) evolve oxygen gas with the same isotopic composition as the oxygen in the water. When algae growing in water containing only ^{16}O were furnished with carbon dioxide containing ^{18}O, no ^{18}O was found to be evolved from the oxygen gas produced. What conclusions about photosynthesis can be drawn from these experiments?

52. Radiotracers are used in the medical sciences to learn about metabolic pathways. What are radiotracers? Explain why ^{14}C and ^{32}P radioactive nuclides would be very helpful in learning about metabolic pathways.

53. The biological effects of a particular source of radiation depend on several factors. List some of these factors. Even though ^{85}Kr and ^{90}Sr are both β-particle emitters, the dangers associated with the decay of ^{90}Sr are much greater than those linked to ^{85}Kr. Why?

54. Although γ rays are far more penetrating than α particles, the latter are more likely to cause damage to an organism. Why? Which type of radiation is more effective at promoting the ionization of biomolecules?

55. Consider the following information:
    i. The layer of dead skin on our bodies is sufficient to protect us from most α-particle radiation.
    ii. Plutonium is an α-particle producer.
    iii. The chemistry of Pu^{4+} is similar to that of Fe^{3+}.
    iv. Pu oxidizes readily to Pu^{4+}.

Why is plutonium one of the most toxic substances known?

Additional Exercises

56. Consider the following graph of binding energy per nucleon as a function of mass number.

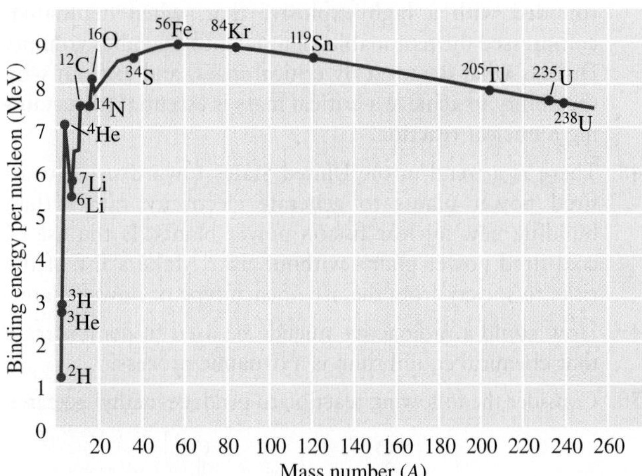

a. What does this graph tell us about the relative half-lives of the nuclides? Explain your answer.
b. Which nuclide shown is the most thermodynamically stable? Which is the least thermodynamically stable?
c. What does this graph tell us about which nuclides undergo fusion and which undergo fission to become more stable? Support your answer.

57. The mass percent of carbon in a typical human is 18%, and the mass percent of ^{14}C in natural carbon is $1.6 \times 10^{-10}\%$. Assuming a 180-lb person, how many decay events per second occur in this person due exclusively to the β-particle decay of ^{14}C? (For ^{14}C, $t_{1/2} = 5730$ years.)

58. Using the kinetic molecular theory (Section 5.6), calculate the root mean square velocity and the average kinetic energy of ^2_1H nuclei at a temperature of 4×10^7 K. (See Exercise 39 for the appropriate mass values.)

59. A small atomic bomb releases energy equivalent to the detonation of 20,000 tons of TNT; a ton of TNT releases 4×10^9 J of energy when exploded. Using 2×10^{13} J/mol as the energy released by fission of ^{235}U, about what mass of ^{235}U undergoes fission in this atomic bomb?

60. When nuclei undergo nuclear transformations, γ rays of characteristic frequencies are observed. How does this fact, along with other information in the chapter on nuclear stability, suggest that a model similar to the quantum mechanics used for atoms may apply to the nucleus?

61. A chemist studied the reaction mechanism for the reaction

$$2NO(g) + O_2(g) \longrightarrow 2NO_2(g)$$

by reacting $N^{16}O$ with $^{18}O_2$. If the reaction mechanism is

$$NO + O_2 \rightleftharpoons NO_3 \text{ (fast equilibrium)}$$

$$NO_3 + NO \longrightarrow 2NO_2 \text{ (slow)}$$

what distribution of ^{18}O would you expect in the NO_2? Assume that N is the central atom in NO_3, assume only $N^{16}O^{18}O_2$ forms, and assume stoichiometric amounts of reactants are combined.

62. Define "third-life" in a similar way to "half-life," and determine the "third-life" for a nuclide that has a half-life of 31.4 years.

63. A 0.10-cm^3 sample of a solution containing a radioactive nuclide (5.0×10^3 counts per minute per milliliter) is injected into a rat. Several minutes later, 1.0 cm^3 of blood is removed. The blood shows 48 counts of radioactivity per minute. What is the volume of blood in the rat? What assumptions must be made in performing this calculation?

64. In addition to the process described in the text, a second process called the carbon–nitrogen cycle occurs in the sun:

$$^1_1H + ^{12}_6C \longrightarrow ^{13}_7N + ^0_0\gamma$$
$$^{13}_7N \longrightarrow ^{13}_6C + ^0_{+1}e$$
$$^1_1H + ^{13}_6C \longrightarrow ^{14}_7N + ^0_0\gamma$$
$$^1_1H + ^{14}_7N \longrightarrow ^{15}_8O + ^0_0\gamma$$
$$^{15}_8O \longrightarrow ^{15}_7N + ^0_{+1}e$$
$$^1_1H + ^{15}_7N \longrightarrow ^{12}_6C + ^4_2He + ^0_0\gamma$$

$$\text{Overall reaction: } 4\,^1_1H \longrightarrow ^4_2He + 2\,^0_{+1}e$$

a. What is the catalyst in the above scheme?
b. What nucleons are intermediates?
c. How much energy is released per mole of hydrogen nuclei reacted in the overall reaction? (See Exercises 39 and 40 for the appropriate mass values.)

Challenge Problems

65. Naturally occurring uranium is composed mostly of ^{238}U and ^{235}U, with relative abundances of 99.28% and 0.72%, respectively. The half-life for ^{238}U is 4.5×10^9 years, and the half-life for ^{235}U is 7.1×10^8 years. Assuming that the earth was formed 4.5 billion years ago, calculate the relative abundances of the ^{238}U and ^{235}U isotopes when the earth was formed.

66. The curie (Ci) is a commonly used unit for measuring nuclear radioactivity: 1 curie of radiation is equal to 3.7×10^{10} decay events per second (the number of decay events from 1 g radium in 1 s). A 1.7-mL sample of water containing tritium was injected into a 150-lb person. The total activity of radiation injected was 86.5 mCi. After some time to allow the tritium activity to equally distribute throughout the body, a sample of blood plasma containing 2.0 mL water at an activity of 3.6 μCi was removed. From these data, calculate the mass percent of water in this 150-lb person.

67. Estimate the temperature needed to achieve the fusion of deuterium to make an α particle. The energy required can

be estimated from Coulomb's law [use the form $E = 9.0 \times 10^9 (Q_1Q_2/r)$, using $Q = 1.6 \times 10^{-19}$ C for a proton and $r = 2 \times 10^{-15}$ m for the helium nucleus; the unit for the proportionality constant in Coloumb's law is J · m/C^2].

68. Radioactive cobalt-60 is used to study defects in vitamin B_{12} absorption because cobalt is the metallic atom at the center of the vitamin B_{12} molecule. The nuclear synthesis of this cobalt isotope involves a three-step process. The overall reaction is iron-58 reacting with two neutrons to produce cobalt-60 along with the emission of another particle. What particle is emitted in this nuclear synthesis? What is the binding energy in J per nucleon for the cobalt-60 nucleus (atomic masses ^{60}Co = 59.9338 amu, ^1H = 1.00782 amu)? What is the de Broglie wavelength of the emitted particle if it has a velocity equal to 0.90*c*, where *c* is the speed of light?

69. To determine the K_{sp} value of Hg_2I_2, a chemist obtained a solid sample of Hg_2I_2 in which some of the iodine is present as radioactive ^{131}I. The count rate of the Hg_2I_2 sample is 5.0×10^{11} counts per minute per mole of I. An excess amount of $Hg_2I_2(s)$ is placed in some water, and the solid is allowed to come to equilibrium with its respective ions. A 150.0-mL sample of the saturated solution is withdrawn and the radioactivity measured at 33 counts per minute. From this information, calculate the K_{sp} value for Hg_2I_2.

$$Hg_2I_2(s) \rightleftharpoons Hg_2^{2+}(aq) + 2I^-(aq) \qquad K_{sp} = [Hg_2^{2+}][I^-]^2$$

70. The most significant source of natural radiation is radon-222. ^{222}Rn, a decay product of ^{238}U, is continuously generated in the earth's crust allowing gaseous Rn to seep into the basements of buildings. Because ^{222}Rn is an α-particle producer with a relatively short half-life of 3.82 days, it can cause biological damage when inhaled.
    a. How many α particles and β particles are produced when ^{238}U decays to ^{222}Rn? What nuclei is produced when ^{222}Rn decays?
    b. Radon is a noble gas so one would expect it to pass through the body quickly. Why is there a concern over inhaling ^{222}Rn?
    c. Another problem associated with ^{222}Rn is that the decay of ^{222}Rn produces a more potent α-particle producer ($t_{1/2}$ = 3.11 min) that is a solid. What is the identity of the solid? Give the balanced equation of this species decaying by α-particle production. Why is the solid a more potent α-particle producer?
    d. The U.S. Environmental Protection Agency (EPA) recommends that ^{222}Rn levels not exceed 4 pCi per liter of air (1 Ci = 1 curie = 3.7×10^{10} decay events per second; 1 pCi = 1×10^{-12} Ci). Convert 4.0 pCi per liter of air into concentration units of ^{222}Rn atoms per liter of air and mole of ^{222}Rn per liter of air.

71. For uranium to be useful as a nuclear fuel, the relative amount of ^{235}U to ^{238}U must be increased from about 0.7% ^{235}U in naturally occurring uranium to about 3%. The process of gas diffusion enrichment utilizes the velocity differences between $^{238}UF_6$ and $^{235}UF_6$ to accomplish the desired enrichment. Since the mass differences between $^{238}UF_6$ and $^{235}UF_6$ are relatively small, several steps are required in the diffusion process to enrich a natural uranium sample to the desired 3% ^{235}U. (For a complete discussion of this process, see Section 5.7.)
    a. Which molecule, $^{238}UF_6$ or $^{235}UF_6$, has the greater average velocity at a certain temperature? Explain.
    b. In theory, how many steps are required to enrich a uranium sample from 0.700% ^{235}U to 3.00% ^{235}U using the UF_6 multistage diffusion process? The molar masses of $^{235}UF_6$ and $^{238}UF_6$ are 349.03 g/mol and 352.05 g/mol, respectively.
    c. A certain sample of uranium is reacted with fluorine to form a mixture of $^{235}UF_6(g)$ and $^{238}UF_6(g)$. After 100 diffusion steps, the gas contains 1526 $^{235}UF_6$ molecules per 1.000×10^5 total number of molecules in the gas ($^{235}UF_6$ + $^{238}UF_6$). What is the ratio of ^{235}U to ^{238}U atoms in the original sample of uranium?

72. Zirconium is one of the few metals that retains its structural integrity upon exposure to radiation. The fuel rods in most nuclear reactors therefore are often made of zirconium. Answer the following questions about the redox properties of zirconium based on the half-reaction

$$ZrO_2 \cdot H_2O + H_2O + 4e^- \longrightarrow Zr + 4OH^- \qquad \mathscr{E}° = -2.36 \text{ V}$$

    a. Is zirconium metal capable of reducing water to form hydrogen gas at standard conditions?
    b. Write a balanced equation for the reduction of water by zirconium.
    c. Calculate $\mathscr{E}°$, $\Delta G°$, and *K* for the reduction of water by zirconium metal.
    d. The reduction of water by zirconium occurred during the accident at Three Mile Island in 1979. The hydrogen produced was successfully vented and no chemical explosion occurred. If 1.00×10^3 kg of Zr reacts, what mass of H_2 is produced? What volume of H_2 at 1.0 atm and 1000.°C is produced?
    e. At Chernobyl in 1986, hydrogen was produced by the reaction of superheated steam with the graphite reactor core:

$$C(s) + H_2O(g) \longrightarrow CO(g) + H_2(g)$$

    It was not possible to prevent a chemical explosion at Chernobyl. In light of this, do you think it was a correct decision to vent the hydrogen and other radioactive gases into the atmosphere at Three Mile Island? Explain.

Organic and Biochemical Molecules

21

chapter

A rainbow eucalyptus tree in Maui, Hawaii.

Ron Dahlquist/Pacific Stock/PhotoLibrary

Sign in to OWL at www.cengage.com/owl to view tutorials and simulations, develop problem-solving skills, and complete online homework assigned by your professor.

Download mini lecture videos for key concept review and exam prep from OWL or purchase them from www.cengagebrain.com.

Two Group 4A elements, carbon and silicon, form the basis of most natural substances. Silicon, with its great affinity for oxygen, forms chains and rings containing Si—O—Si bridges to produce the silica and silicates that form the basis for most rocks, sands, and soils. What silicon is to the geological world, carbon is to the biological world. Carbon has the unusual ability of bonding strongly to itself to form long chains or rings of carbon atoms. In addition, carbon forms strong bonds to other nonmetals such as hydrogen, nitrogen, oxygen, sulfur, and the halogens. Because of these bonding properties, there are a myriad of carbon compounds; several million are now known, and the number continues to grow rapidly. Among these many compounds are the **biomolecules,** those responsible for maintaining and reproducing life.

The study of carbon-containing compounds and their properties is called **organic chemistry.** Although a few compounds involving carbon, such as its oxides and carbonates, are considered to be inorganic substances, the vast majority are organic compounds that typically contain chains or rings of carbon atoms.

Originally, the distinction between inorganic and organic substances was based on whether a compound was produced by living systems. For example, until the early nineteenth century, it was believed that organic compounds had some sort of "life force" and could be synthesized only by living organisms. This view was dispelled in 1828 when German chemist Friedrich Wöhler (1800–1882) prepared urea from the inorganic salt ammonium cyanate by simple heating:

$$\underset{\text{Ammonium cyanate}}{NH_4OCN} \xrightarrow{\text{Heat}} \underset{\text{Urea}}{H_2N{-}\underset{\underset{O}{\|}}{C}{-}NH_2}$$

Urea is a component of urine, so it is clearly an organic material, yet here was clear evidence that it could be produced in the laboratory as well as by living things.

Organic chemistry plays a vital role in our quest to understand living systems. Beyond that, the synthetic fibers, plastics, artificial sweeteners, and drugs that are such an accepted part of modern life are products of industrial organic chemistry. In addition, the energy on which we rely so heavily to power our civilization is based mostly on the organic materials found in coal and petroleum.

Because organic chemistry is such a vast subject, we can provide only a brief introduction to it in this book. We will begin with the simplest class of organic compounds, the hydrocarbons, and then show how most other organic compounds can be considered to be derivatives of hydrocarbons.

21.1 Alkanes: Saturated Hydrocarbons

As the name indicates, **hydrocarbons** are compounds composed of carbon and hydrogen. Those compounds whose carbon–carbon bonds are all single bonds are said to be **saturated** because each carbon is bound to four atoms, the maximum number. Hydrocarbons containing carbon–carbon multiple bonds are described as being **unsaturated,** since the carbon atoms involved in a multiple

Chemistry in the Garden

Chemical Insights

As we learn more about the biological world, we are finding that humble plants are often master chemists. For example, certain plants in the Amazon rain forests use chemistry in a very clever way to protect themselves. When attacked by herbivores such as spider mites or caterpillars, these plants produce and release compounds into the air that attract certain mites that are natural enemies of the attacking herbivores. These "bodyguard" mites assist the plant in fending off the attackers.

Although some plants can defend themselves using chemistry to attract "bodyguards," most cannot do this. As a result, scientists are trying to modify important food plants such as tomato plants so that they will emit chemicals to attract protective mites. For example, scientists from the Netherlands and Israel have begun a joint venture to genetically engineer *Arabidopis thaliava* plants that produce mite-attracting chemicals [*Science* **309** (2005): 2070].

Another example of the cooperation of plants and insects occurs in the so-called devil's gardens, in which a single species of tree predominates (*D. hirsute*) in large areas of the Amazon rain forests. Local legend attributes this strange phenomenon to an evil forest spirit that prevents other types of trees from growing. However, scientists at the University of Colorado, Denver, have found that colonies of ants that make nests in the stems of the *D. hirsute* trees produce formic acid, which prevents growth of other plants in the area. It is clear that plants and insects can team up to make great chemistry together.

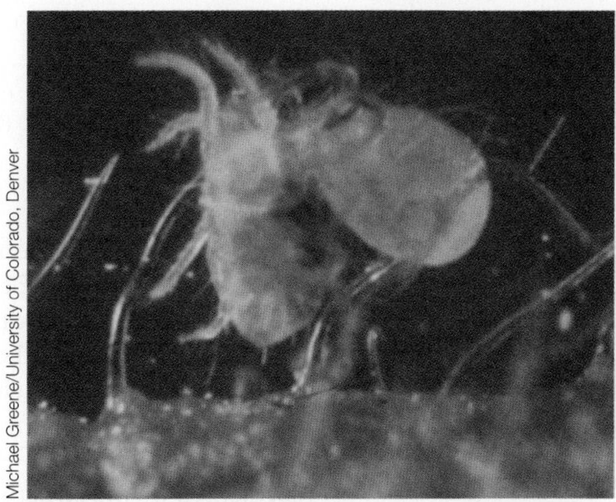

Michael Greene/University of Colorado, Denver

A "bodyguard" mite (red) attacks a spider mite.

bond can react with additional atoms, as shown by the *addition* of hydrogen to ethylene:

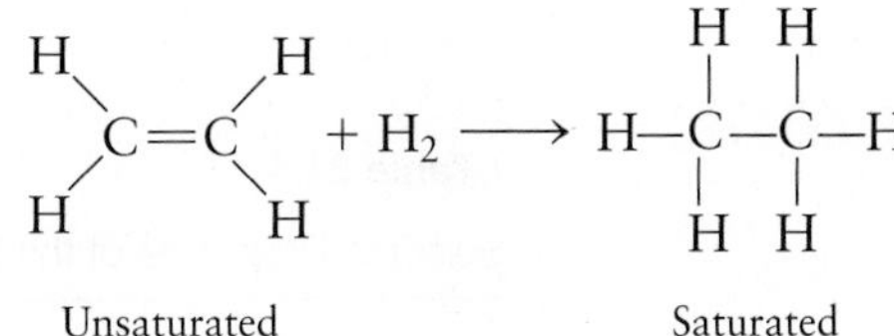

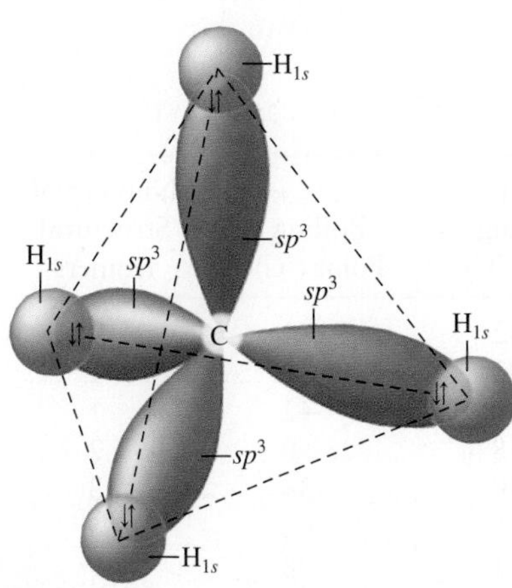

Figure 21.1
The C—H bonds in methane.

Note that each carbon in ethylene is bonded to three atoms (one carbon and two hydrogens) but that each can bond to one additional atom if one bond of the carbon–carbon double bond is broken.

The simplest member of the saturated hydrocarbons, which are also called the **alkanes,** is *methane* (CH_4). As discussed in Section 14.1, methane has a tetrahedral structure and can be described in terms of a carbon atom using an sp^3 hybrid set of orbitals to bond to the four hydrogen atoms (Fig. 21.1). The next alkane, the one containing two carbon atoms, is *ethane* (C_2H_6), as shown in Fig. 21.2. Each carbon in ethane is surrounded by four atoms and thus adopts a tetrahedral arrangement and sp^3 hybridization, as predicted by the localized electron model.

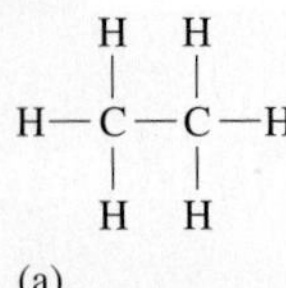

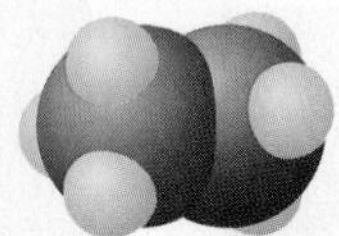

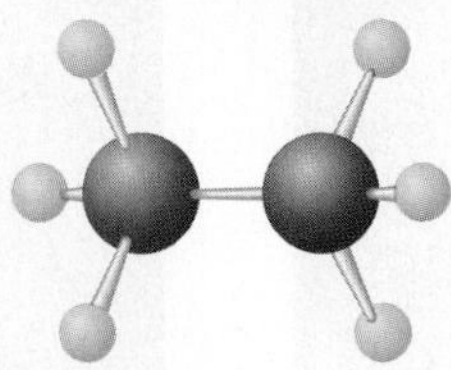

Figure 21.2
(a) The Lewis structure of ethane (C_2H_6). (b) The molecular structure of ethane represented by space-filling and ball-and-stick models.

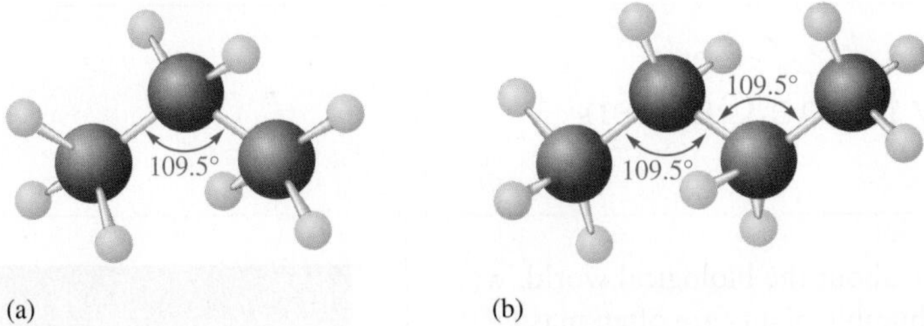

Figure 21.3
The structures of (a) propane ($CH_3CH_2CH_3$) and (b) butane ($CH_3CH_2CH_2CH_3$). Each angle shown in red is 109.5°.

The next two members of the series are *propane* (C_3H_8) and *butane* (C_4H_{10}), shown in Fig. 21.3. Again, each carbon is bonded to four atoms and is described as sp^3 hybridized.

Alkanes in which the carbon atoms form long "strings" or chains are called **normal, straight-chain,** or **unbranched hydrocarbons.** As can be seen from Fig. 21.3, the chains in normal alkanes are not really straight but zig-zag, since the tetrahedral C—C—C angle is 109.5°. The normal alkanes can be represented by the structure

$$H_3C-\left(\begin{array}{c} H \\ | \\ C \\ | \\ H \end{array}\right)_n-CH_3$$

where n is an integer. Note that each member is obtained from the previous one by inserting a *methylene* (CH_2) group. We can condense the structural formulas by omitting some of the C—H bonds. For example, the general formula for normal alkanes shown on the previous page can be condensed to

$$CH_3-(CH_2)_n-CH_3$$

The first 10 normal alkanes and some of their properties are listed in Table 21.1. Note that all alkanes can be represented by the general formula C_nH_{2n+2}. For example, nonane, which has nine carbon atoms, is represented by $C_9H_{(2\times9)+2}$, or C_9H_{20}. Also note from Table 21.1 that the melting points and boiling points increase as the molar masses increase, as we would expect.

Table 21.1

Selected Properties of the First 10 Normal Alkanes

Name	Formula	Molar Mass	Melting Point (°C)	Boiling Point (°C)	Number of Structural Isomers
Methane	CH_4	16	−182	−162	1
Ethane	C_2H_6	30	−183	−89	1
Propane	C_3H_8	44	−187	−42	1
Butane	C_4H_{10}	58	−138	0	2
Pentane	C_5H_{12}	72	−130	36	3
Hexane	C_6H_{14}	86	−95	68	5
Heptane	C_7H_{16}	100	−91	98	9
Octane	C_8H_{18}	114	−57	126	18
Nonane	C_9H_{20}	128	−54	151	35
Decane	$C_{10}H_{22}$	142	−30	174	75

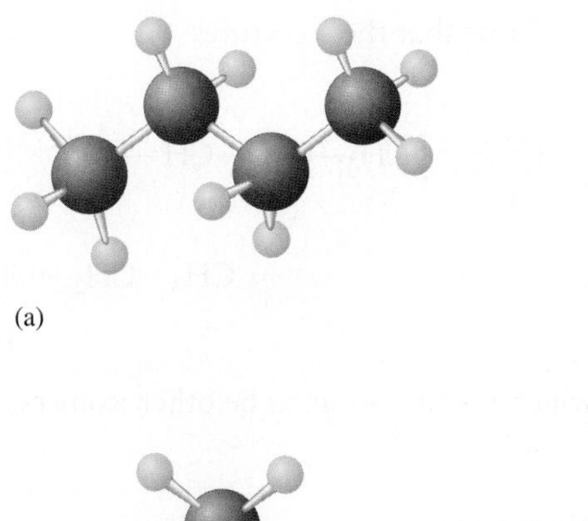
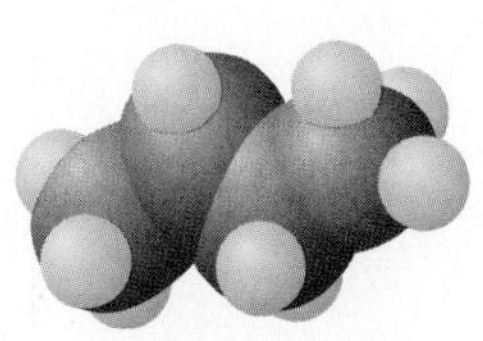

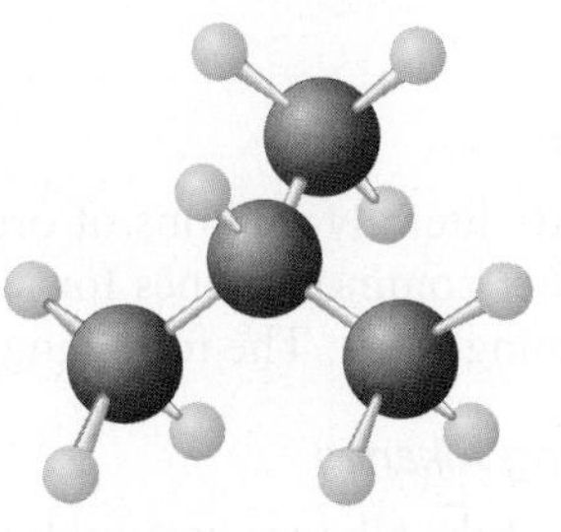
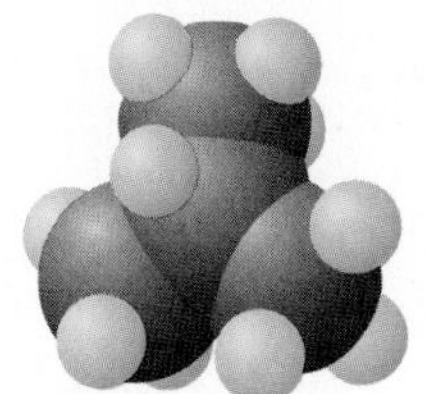

Figure 21.4
(a) Normal butane (abbreviated *n*-butane). (b) The branched isomer of butane (called isobutane).

Isomerism in Alkanes

Butane and all succeeding members of the alkanes exhibit **structural isomerism.** Recall from Section 19.4 that structural isomerism occurs when two molecules have the same atoms but different bonds. For example, butane can exist as a straight-chain molecule (normal butane, or *n*-butane) or with a branched-chain structure (called isobutane), as shown in Fig. 21.4. Because of their different structures, these molecules exhibit different properties. For example, the boiling point of *n*-butane is $-0.5°C$, whereas that of isobutane is $-12°C$.

EXAMPLE 21.1

Draw the isomers of pentane.

Solution Pentane (C_5H_{12}) has the following isomeric structures:

1.
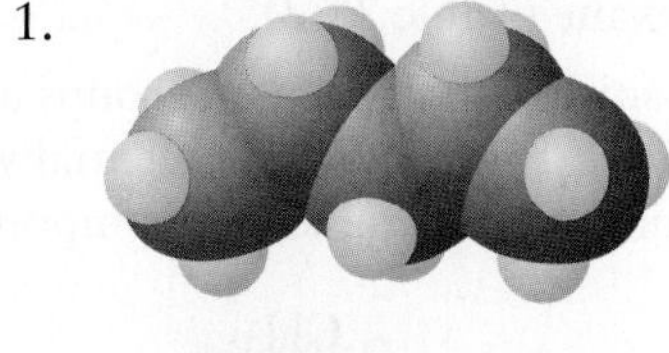

$CH_3—CH_2—CH_2—CH_2—CH_3$

n-Pentane

2.

$$\begin{array}{c} \quad CH_3 \\ \quad | \\ CH_3—CH—CH_2—CH_3 \end{array}$$

Isopentane

3.

$$\begin{array}{c} CH_3 \\ | \\ CH_3—C—CH_3 \\ | \\ CH_3 \end{array}$$

Neopentane

Note that the structures

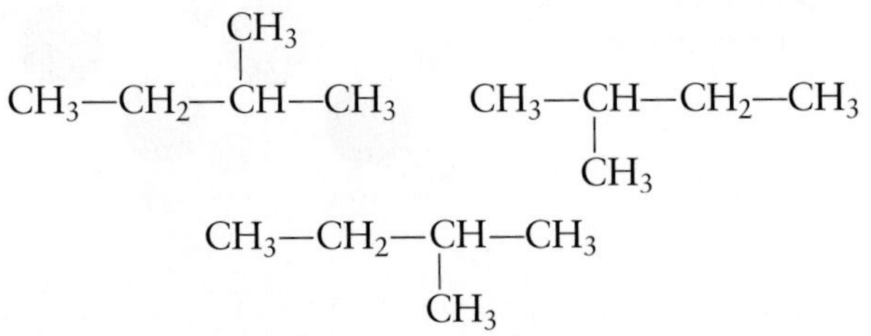

$$\begin{array}{c}CH_3\\ |\\ CH_3—CH_2—CH—CH_3\end{array}\qquad \begin{array}{c}CH_3—CH—CH_2—CH_3\\ |\\ CH_3\end{array}$$

$$\begin{array}{c}CH_3—CH_2—CH—CH_3\\ |\\ CH_3\end{array}$$

which might appear to be other isomers, are actually identical to structure 2.

Nomenclature

Because there are literally millions of organic compounds, it would be impossible to remember common names for all of them. We must have a systematic method for naming them. The following rules are used in naming alkanes.

Rules for Naming Alkanes

1. The names of the alkanes beyond butane are obtained by adding the suffix *-ane* to the Greek root for the number of carbon atoms (*pent-* for five, *hex-* for six, and so on). For a branched hydrocarbon, the longest continuous chain of carbon atoms determines the root name for the hydrocarbon. For example, in the alkane

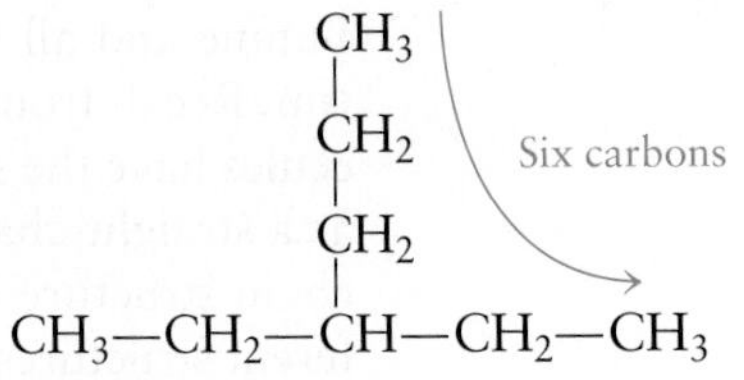

$$\begin{array}{c}CH_3\\ |\\ CH_2\\ |\\ CH_2\\ |\\ CH_3—CH_2—CH—CH_2—CH_3\end{array}$$

Six carbons

the longest chain contains six carbon atoms, and this compound is named as a hexane.

2. When alkane groups appear as substituents, they are named by dropping the *-ane* and adding *-yl*. For example, $—CH_3$ is obtained by removing a hydrogen from methane and is called *methyl*, $—C_2H_5$ is called *ethyl*, $—C_3H_7$ is called *propyl*, and so on. The compound above is therefore an ethylhexane (Table 21.2).

3. The positions of substituent groups are specified by numbering the longest chain of carbon atoms sequentially, starting at the end closest to the branching. For example, the compound

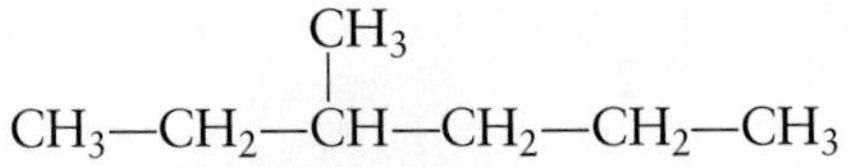

$$\begin{array}{cccccc} & & CH_3 & & & \\ & & | & & & \\ CH_3— & CH_2— & CH— & CH_2— & CH_2— & CH_3 \\ 1 & 2 & 3 & 4 & 5 & 6 \\ 6 & 5 & 4 & 3 & 2 & 1 \end{array}\quad \begin{array}{l}\\ \\ \\ \text{Correct numbering}\\ \text{Incorrect numbering}\end{array}$$

is called 3-methylhexane. Note that the top set of numbers is correct because the left end of the molecule is closest to the branching, and this gives the smallest number for the position of the substituent. Also note that a hyphen is placed between the number and the substituent name.

4. The location and name of each substituent are followed by the root alkane name. The substituents are listed in alphabetical order, and the prefixes *di-*, *tri-*, and so on are used to indicate multiple, identical substituents.

Table 21.2

The Most Common Alkyl Substituents and Their Names

Structure*	Name†
$—CH_3$	Methyl
$—CH_2CH_3$	Ethyl
$—CH_2CH_2CH_3$	Propyl
$\begin{array}{c}|\\ CH_3CHCH_3\end{array}$	Isopropyl
$—CH_2CH_2CH_2CH_3$	Butyl
$\begin{array}{l}|\\ CH_3CHCH_2CH_3\end{array}$	*sec*-Butyl
$\begin{array}{c}H\\ |\\ —CH_2—C—CH_3\\ |\\ CH_3\end{array}$	Isobutyl
$\begin{array}{c}CH_3\\ |\\ —C—CH_3\\ |\\ CH_3\end{array}$	*tert*-Butyl

*The bond with one end open shows the point of attachment of the substituent to the carbon chain.

†For the butyl groups, *sec-* indicates attachment to the chain through a secondary carbon, a carbon atom attached to *two* other carbon atoms. The designation *tert-* signifies attachment through a tertiary carbon, a carbon attached to *three* other carbon atoms.

EXAMPLE 21.2

Draw the structural isomers for the alkane C_6H_{14}, and give the systematic name for each one.

Solution We will proceed systematically, starting with the longest chain and then rearranging the carbons to form the shorter, branched chains.

1. $CH_3CH_2CH_2CH_2CH_2CH_3$ Hexane

   Note that although a structure such as

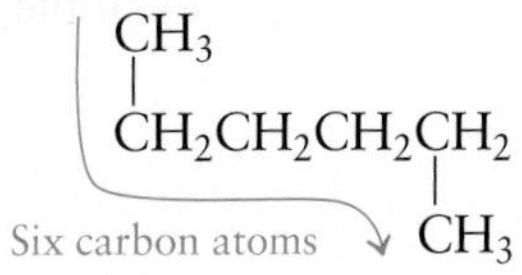

   $$\begin{array}{l} CH_3 \\ | \\ CH_2CH_2CH_2CH_2 \\ \qquad\qquad\quad | \\ \qquad\qquad\quad CH_3 \end{array}$$

   Six carbon atoms

   may look different, it is still hexane, since the longest carbon chain has six atoms.

2. We now take one carbon out of the chain and make it a methyl substituent.

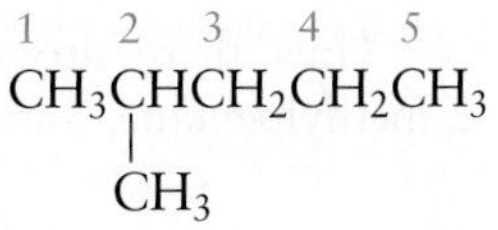

   $$\begin{array}{l} \overset{1}{CH_3}\overset{2}{C}H\overset{3}{C}H_2\overset{4}{C}H_2\overset{5}{C}H_3 \\ \quad\;\; | \\ \quad\;\; CH_3 \end{array}$$

   2-Methylpentane

   Since the longest chain consists of five carbons, this is a substituted pentane: 2-methylpentane. The 2 indicates the position of the methyl group on the chain. Note that if we numbered the chain from the right end, the methyl group would be on carbon 4. Because we want the smallest possible number, the numbering shown is correct.

3. The methyl substituent can also be on carbon 3 to give

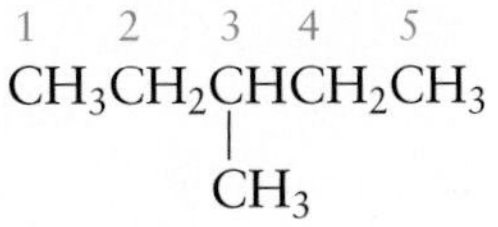

   $$\begin{array}{l} \overset{1}{CH_3}\overset{2}{C}H_2\overset{3}{C}H\overset{4}{C}H_2\overset{5}{C}H_3 \\ \qquad\quad\; | \\ \qquad\quad\; CH_3 \end{array}$$

   3-Methylpentane

   Note that we have now exhausted all possibilities for placing a single methyl group on pentane.

4. Next, we can take two carbons out of the original six-member chain:

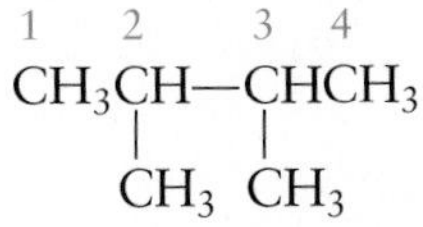

   $$\begin{array}{l} \overset{1}{CH_3}\overset{2}{C}H{-}\overset{3}{C}H\overset{4}{C}H_3 \\ \quad\;\; | \qquad | \\ \quad\;\; CH_3 \; CH_3 \end{array}$$

   2,3-Dimethylbutane

   Since the longest chain now has four carbons, the root name is butane. Since there are two methyl groups, we use the prefix *di-*. The numbers denote that the two methyl groups are positioned on the second and third carbons in the butane chain. Note that when two or more numbers are used, they are separated by a comma.

5. The two methyl groups can also be attached to the same carbon atom as shown here:

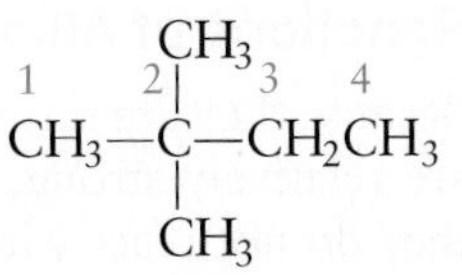

   $$\begin{array}{l} \qquad\quad CH_3 \\ \qquad\quad\; | \\ \overset{1}{CH_3}{-}\overset{2}{C}{-}\overset{3}{C}H_2\overset{4}{C}H_3 \\ \qquad\quad\; | \\ \qquad\quad CH_3 \end{array}$$

   2,2-Dimethylbutane

We might also try ethyl-substituted butanes, such as

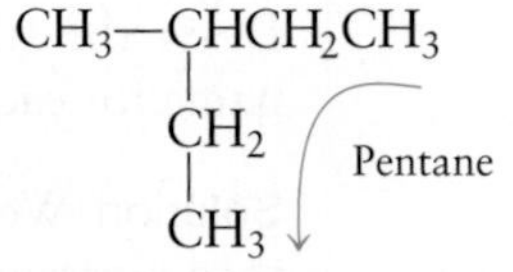

$$\begin{array}{l} CH_3\text{—}CHCH_2CH_3 \\ \qquad\quad | \\ \qquad\quad CH_2 \quad \text{Pentane} \\ \qquad\quad | \\ \qquad\quad CH_3 \end{array}$$

However, note that this is instead a pentane (3-methylpentane), since the longest chain has five carbon atoms. Thus this is not a new isomer. Trying to reduce the chain to three atoms provides no further isomers either. For example, the structure

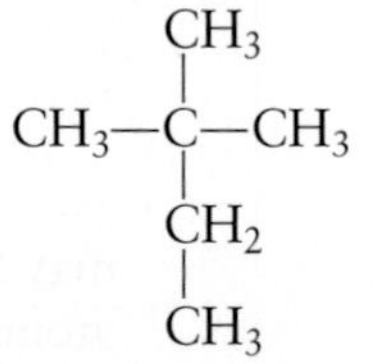

$$\begin{array}{ccc} & CH_3 & \\ & | & \\ CH_3\text{—} & C & \text{—}CH_3 \\ & | & \\ & CH_2 & \\ & | & \\ & CH_3 & \end{array}$$

is actually 2,2-dimethylbutane.

Thus there are only five distinct structural isomers of C_6H_{14}: hexane, 2-methylpentane, 3-methylpentane, 2,3-dimethylbutane, and 2,2-dimethylbutane.

EXAMPLE 21.3

Determine the structure for each of the following compounds.

a. 4-ethyl-3,5-dimethylnonane **b.** 4-*tert*-butylheptane

Solution

a. The root name *nonane* signifies a nine-carbon chain. Thus we have

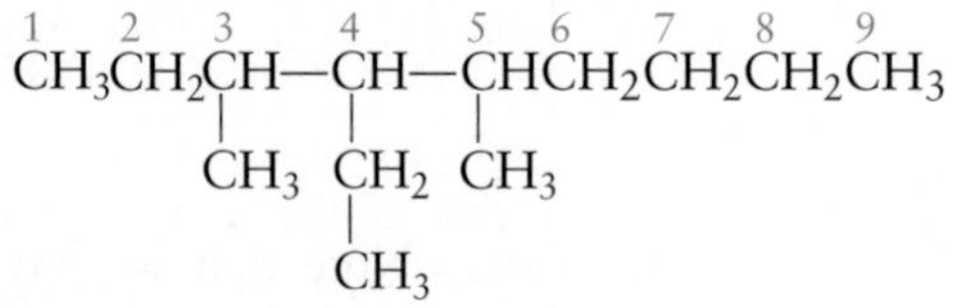

$$\begin{array}{lllll} {}^{1\ \ 2\ \ 3} & & {}^{4} & & {}^{5\ \ 6\ \ 7\ \ 8\ \ 9} \\ CH_3CH_2CH & \text{—} & CH & \text{—} & CHCH_2CH_2CH_2CH_3 \\ \qquad\quad | & & | & & | \\ \qquad\quad CH_3 & & CH_2 & & CH_3 \\ & & | & & \\ & & CH_3 & & \end{array}$$

b. Heptane signifies a seven-carbon chain, and the *tert*-butyl group is

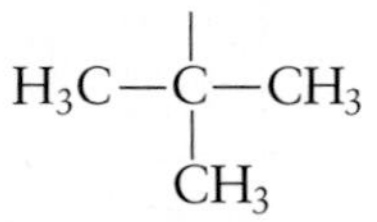

$$\begin{array}{ccc} & | & \\ H_3C\text{—} & C & \text{—}CH_3 \\ & | & \\ & CH_3 & \end{array}$$

Thus we have

$$\begin{array}{c} {}^{1\ \ \ 2\ \ \ 3\ \ \ 4\ \ \ 5\ \ \ 6\ \ \ 7} \\ CH_3CH_2CH_2CHCH_2CH_2CH_3 \\ | \\ H_3C\text{—}C\text{—}CH_3 \\ | \\ CH_3 \end{array}$$

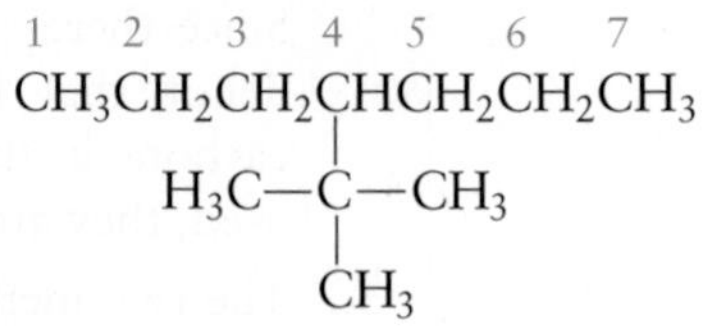

Reactions of Alkanes

Because they are saturated compounds and because the C—C and C—H bonds are relatively strong, the alkanes are fairly unreactive. For example, at 25°C they do not react with acids, bases, or strong oxidizing agents. This chemical inertness makes them valuable as lubricating materials and as the backbone for structural materials such as plastics.

At a sufficiently high temperature, alkanes do react vigorously and exothermically with oxygen, and these **combustion reactions** are the basis for their widespread use as fuels. For example, the reaction of butane with oxygen is

$$2C_4H_{10}(g) + 13O_2(g) \longrightarrow 8CO_2(g) + 10H_2O(g)$$

The alkanes can also undergo **substitution reactions,** primarily where halogen atoms replace hydrogen atoms. For example, methane can be successively chlorinated as follows:

The $h\nu$ above the arrow represents ultraviolet light.

$$CH_4 + Cl_2 \xrightarrow{h\nu} \underset{\text{Chloromethane}}{CH_3Cl} + HCl$$

$$CH_3Cl + Cl_2 \xrightarrow{h\nu} \underset{\text{Dichloromethane}}{CH_2Cl_2} + HCl$$

$$CH_2Cl_2 + Cl_2 \xrightarrow{h\nu} \underset{\substack{\text{Trichloromethane} \\ \text{(chloroform)}}}{CHCl_3} + HCl$$

$$CHCl_3 + Cl_2 \xrightarrow{h\nu} \underset{\substack{\text{Tetrachloromethane} \\ \text{(carbon tetrachloride)}}}{CCl_4} + HCl$$

Note that the products of the last two reactions have two names; the systematic name is given first, followed by the common name in parentheses. (This format will be used throughout this chapter for compounds that have common names.) Also note that ultraviolet light ($h\nu$) furnishes the energy to break the Cl—Cl bond to produce chlorine atoms:

$$Cl_2 \longrightarrow Cl\cdot + Cl\cdot$$

A chlorine atom has an unpaired electron, as indicated by the dot, which makes it very reactive and able to attack the C—H bond.

As we mentioned before, substituted methanes with the general formula CF_xCl_{4-x} containing both chlorine and fluorine as substituents are called chlorofluorocarbons (CFCs) and are also known as *Freons.* These substances are very unreactive and have been extensively used as coolant fluids in refrigerators and air conditioners. Unfortunately, their chemical inertness allows Freons to remain in the atmosphere so long that they eventually reach altitudes where they are a threat to the protective ozone layer (see Section 15.9), and the use of these compounds is being rapidly phased out.

Alkanes can also undergo **dehydrogenation reactions** in which hydrogen atoms are removed and the product is an unsaturated hydrocarbon. For example, in the presence of chromium(III) oxide at high temperatures, ethane can be dehydrogenated, yielding ethylene:

$$CH_3CH_3 \xrightarrow[500°C]{Cr_2O_3} \underset{\text{Ethylene}}{CH_2{=}CH_2} + H_2$$

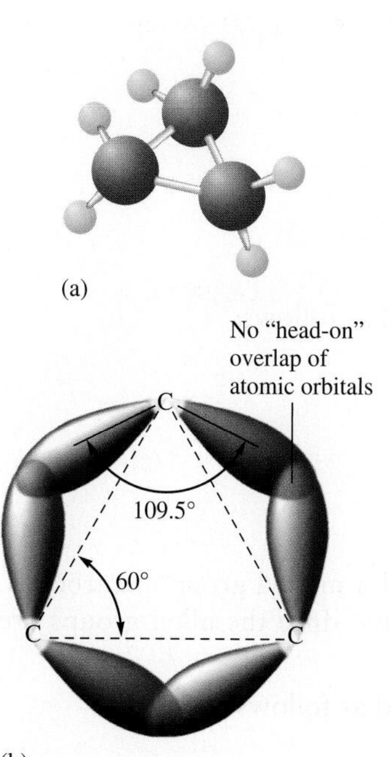

Figure 21.5
(a) The molecular structure of cyclopropane (C_3H_6). (b) The overlap of the sp^3 orbitals that form the C—C bonds in cyclopropane.

Cyclic Alkanes

Besides forming chains, carbon atoms also form rings. The simplest of the **cyclic alkanes** (general formula C_nH_{2n}) is cyclopropane (C_3H_6), shown in Fig. 21.5(a). Since the carbon atoms in cyclopropane form an equilateral triangle with 60° bond angles, their sp^3 hybrid orbitals do not overlap head-on as in normal alkanes [Fig. 21.5(b)]. This results in unusually weak, or *strained,* C—C bonds; thus the cyclopropane molecule is much more reactive than straight-chain propane. The carbon atoms in cyclobutane (C_4H_8) form a square with 90° bond angles, and cyclobutane is also quite reactive.

The next two members of the series, cyclopentane (C_5H_{10}) and cyclohexane (C_6H_{12}), are quite stable because their rings have bond angles very close to tetrahedral angles, which allows the sp^3 hybrid orbitals on adjacent carbon atoms to overlap head-on and form normal C—C bonds, which are quite strong. To attain tetrahedral angles, the cyclohexane ring must "pucker"—that is, become nonplanar. Cyclohexane can exist in two forms, the *chair* and the *boat* forms, as shown in Fig. 21.6. The two hydrogen atoms above the ring in the boat form are quite close to each other, and the resulting repulsion between these atoms causes the chair form to be preferred. At 25°C more than 99% of cyclohexane exists in the chair form.

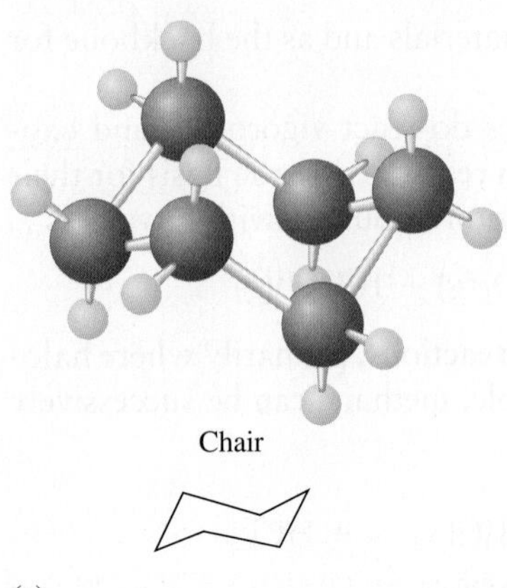

(a)

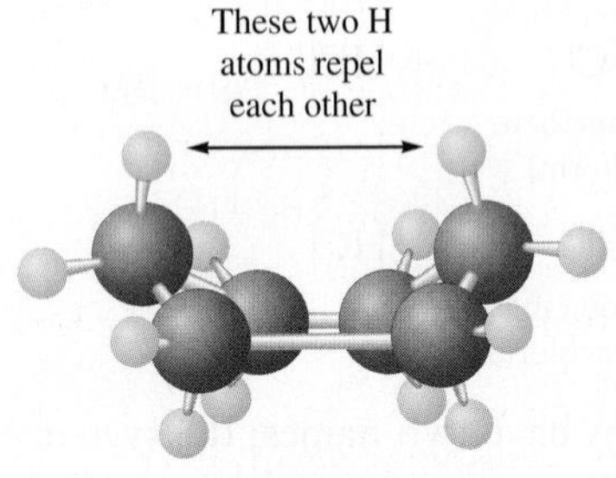

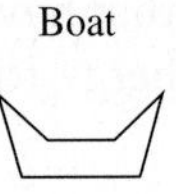

(b)

Figure 21.6
The (a) chair and (b) boat forms of cyclohexane.

For simplicity, the cyclic alkanes are often represented by the following structures:

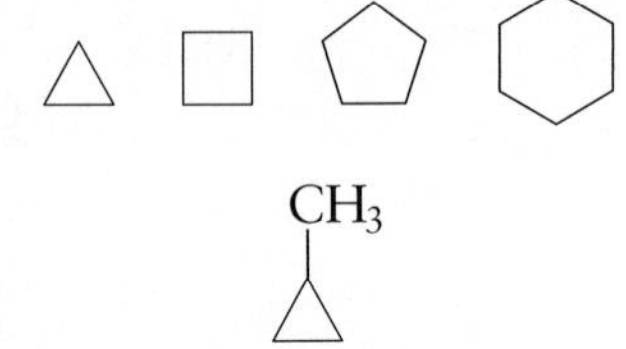

Thus the structure

CH_3 (on a triangle)

represents methylcyclopropane.

The nomenclature for cycloalkanes follows the same rules as for the other alkanes except that the root name is preceded by the prefix *cyclo-*. The ring is numbered to yield the smallest substituent numbers possible.

OWL INTERACTIVE EXAMPLE 21.4

Name each of the following cyclic alkanes.

a. CH_3—CH—CH_3 attached to a cyclohexane ring bearing a CH_3 group

b. Cyclobutane ring bearing CH_2CH_3 and $CH_2CH_2CH_3$

Solution

a. The six-carbon cyclohexane ring is numbered as follows:

CH_3—CH—CH_3 at carbon 1 (ring numbered 1–6), CH_3 at carbon 3

There is an isopropyl group at carbon 1 and a methyl group at carbon 3. The name is 1-isopropyl-3-methylcyclohexane, since the alkyl groups are named in alphabetical order.

b. This is a cyclobutane ring, which is numbered as follows:

CH_2CH_3 at carbon 1, $CH_2CH_2CH_3$ at carbon 2 (ring numbered 1–4)

The name is 1-ethyl-2-propylcyclobutane.

21.2 Alkenes and Alkynes

Multiple carbon–carbon bonds result when hydrogen atoms are removed from alkanes. Hydrocarbons that contain at least one carbon–carbon double bond are called **alkenes** and have the general formula C_nH_{2n}. The simplest alkene (C_2H_4), commonly known as *ethylene,* has the Lewis structure

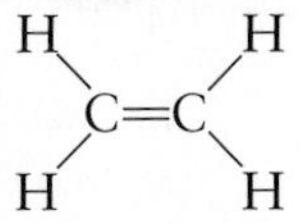

```
H       H
 \     /
  C=C
 /     \
H       H
```

As discussed in Section 14.1, each carbon in ethylene can be described as sp^2 hybridized. The C—C σ bond is formed by sharing an electron pair between sp^2 orbitals, and the π bond is formed by sharing a pair of electrons between p orbitals (Fig. 21.7).

The systematic nomenclature for alkenes is quite similar to that for alkanes.

1. The root hydrocarbon name ends in *-ene* rather than *-ane*. Thus the systematic name for C_2H_4 is *ethene* and the name for C_3H_6 is *propene.*
2. In alkenes containing more than three carbon atoms, the location of the double bond is indicated by the lowest-numbered carbon atom involved in the bond. Thus $CH_2{=}CHCH_2CH_3$ is called 1-butene, and $CH_3CH{=}CHCH_3$ is called 2-butene.

Note from Fig. 21.7 that the p orbitals on the two carbon atoms in ethylene must be lined up (parallel) to allow formation of the π bond. This prevents rotation of the two CH_2 groups relative to each other at ordinary temperatures, in contrast to alkanes, where free rotation is possible (Fig. 21.8). The restricted rotation around doubly bonded carbon atoms means that alkenes exhibit ***cis–trans* isomerism.** For example, there are two stereoisomers of 2-butene (Fig. 21.9). Identical substituents on the same side of the double bond are designated *cis,* and those on opposite sides are labeled *trans.*

Alkynes are unsaturated hydrocarbons containing at least one triple carbon–carbon bond. The simplest alkyne is C_2H_2 (commonly called *acety-*

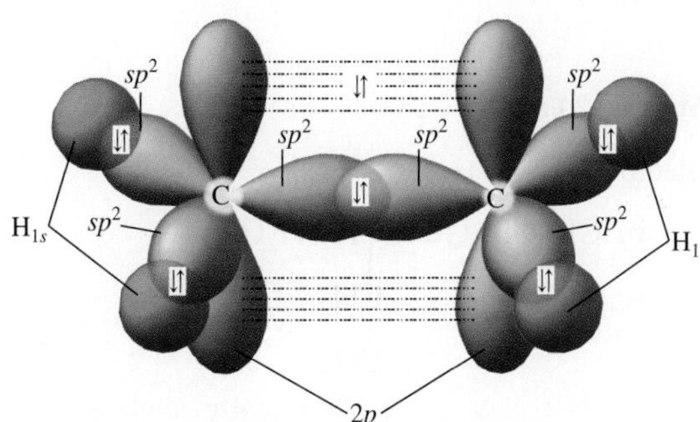

Figure 21.7
The bonding in ethylene.

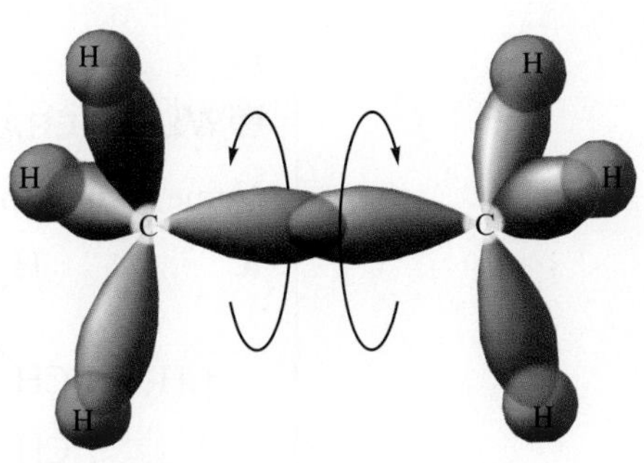

Figure 21.8
The bonding in ethane.

Figure 21.9
The two stereoisomers of 2-butene: (a) *cis*-2-butene and (b) *trans*-2-butene.

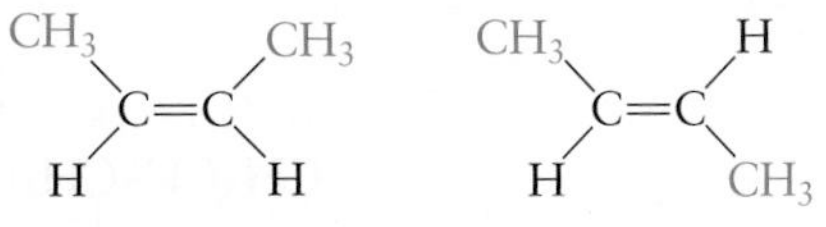

```
CH3       CH3        CH3       H
   \     /              \     /
    C=C                  C=C
   /     \              /     \
 H         H          H         CH3

(a)                  (b)
```

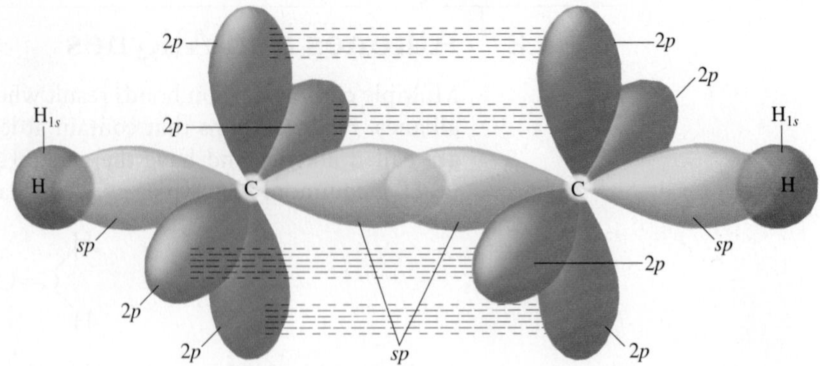

Figure 21.10
The bonding in acetylene.

lene), which has the systematic name *ethyne*. As discussed in Section 14.1, the triple bond in acetylene can be described as one σ bond between two *sp* hybrid orbitals on the two carbon atoms and two π bonds involving two 2*p* orbitals on each carbon atom (Fig. 21.10).

The nomenclature for alkynes involves the use of *-yne* as a suffix to replace the *-ane* of the parent alkane. Thus the molecule $CH_3CH_2C{\equiv}CCH_3$ has the name 2-pentyne.

Like alkanes, unsaturated hydrocarbons can exist as ringed structures, for example,

For cyclic alkenes, number through the double bond toward the substituent.

```
        CH2
   H2C       CH
    |        ||        or   [ring]
   H2C       CH
        CH2
```

Cyclohexene

```
        CH3
        |
        CH
   H2C       CH2       or   [ring with CH3]
     HC=CH
```

4-Methyl-cyclopentene

OWL INTERACTIVE EXAMPLE 21.5

Name each of the following molecules.

a.

```
            H         CH3
             \       /
              C = C
             /       \
    CH3CH2CH          H
           |
          CH3
```

b.

```
CH3CH2C≡CCHCH2CH3
           |
           CH2
           |
           CH3
```

Solution

a. The longest chain, which contains six carbon atoms, is numbered as follows:

```
                          1
             H  3     2  CH3
              \         /
  6  5  4      C  =  C
  CH3CH2CH              \
         |               H
         CH3
```

Thus the hydrocarbon is a 2-hexene. Since the hydrogen atoms are located on opposite sides of the double bond, this molecule corresponds to the *trans* isomer. The name is 4-methyl-*trans*-2-hexene.

b. The longest chain, consisting of seven carbon atoms, is numbered as shown (giving the triple bond the lowest possible number):

$$\overset{1}{C}H_3\overset{2}{C}H_2\overset{3}{C}\equiv\overset{4}{C}\underset{\displaystyle \underset{\displaystyle CH_3}{|} }{\overset{5}{C}H}\overset{6}{C}H_2\overset{7}{C}H_3$$

(with a CH_2CH_3 group attached to carbon 5: $-CH_2-CH_3$)

The hydrocarbon is a 3-heptyne. The full name is 5-ethyl-3-heptyne, where the position of the triple bond is indicated by the lower-numbered carbon atom involved in this bond.

Reactions of Alkenes and Alkynes

Because alkenes and alkynes are unsaturated, their most important reactions are **addition reactions.** In these reactions, π bonds, which are weaker than the C—C σ bonds, are broken, and new σ bonds are formed to the atoms being added. For example, **hydrogenation reactions** involve the addition of hydrogen atoms:

$$\underset{\text{1-Propene}}{CH_2{=}CHCH_3} + H_2 \xrightarrow{\text{Catalyst}} \underset{\text{Propane}}{CH_3CH_2CH_3}$$

For this reaction to proceed rapidly at normal temperatures, a catalyst of platinum, palladium, or nickel is used. The catalyst serves to help break the relatively strong H—H bond, as was discussed in Section 15.9. Hydrogenation of alkenes is an important industrial process, particularly in the manufacture of solid shortenings, where unsaturated fats (fats containing double bonds), which are generally liquid, are converted to solid saturated fats.

Halogenation of unsaturated hydrocarbons involves addition of halogen atoms. For example,

$$\underset{\text{1-Pentene}}{CH_2{=}CHCH_2CH_2CH_3} + Br_2 \longrightarrow \underset{\text{1,2-Dibromopentane}}{CH_2BrCHBrCH_2CH_2CH_3}$$

Another important reaction involving certain unsaturated hydrocarbons is **polymerization,** a process in which many small molecules are joined together to form a large molecule. Polymerization will be discussed in Section 21.5.

21.3 Aromatic Hydrocarbons

A special class of cyclic unsaturated hydrocarbons is known as the **aromatic hydrocarbons.** The simplest of these is benzene (C_6H_6), which has a planar ring structure, as shown in Fig. 21.11(a). In the localized electron model of the bonding in benzene, resonance structures of the type shown in Fig. 21.11(b) are used to account for the known equivalence of all the carbon–carbon bonds. But as we discussed in Section 14.5, the best description of the benzene molecule assumes that sp^2 hybrid orbitals on each carbon are used to form the C—C and C—H σ bonds, whereas the remaining $2p$ orbital on each carbon is used

(a)

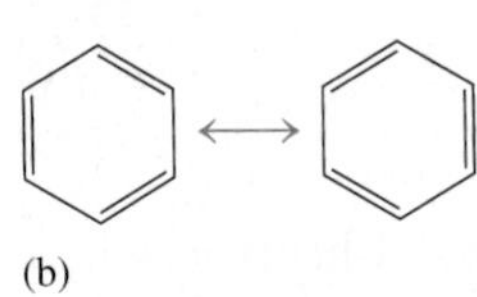

(b)

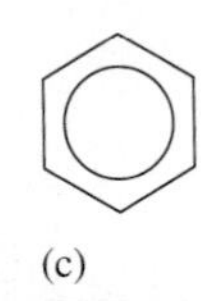

(c)

Figure 21.11
(a) The structure of benzene, a planar ring system in which all bond angles are 120°. (b) Two of the resonance structures of benzene. (c) The usual representation of benzene. The circle represents the electrons in the delocalized π system. All C—C bonds in benzene are equivalent.

to form π molecular orbitals. The delocalization of these π electrons is usually indicated by a circle inside the ring [Fig. 21.11(c)].

The delocalization of the π electrons makes the benzene ring behave quite differently from a typical unsaturated hydrocarbon. As we have seen previously, unsaturated hydrocarbons generally undergo rapid addition reactions. However, benzene does not. Instead, it undergoes substitution reactions in which *hydrogen atoms are replaced by other atoms.* For example,

$$C_6H_6 + Cl_2 \xrightarrow{FeCl_3} C_6H_5Cl + HCl$$

Chlorobenzene

$$C_6H_6 + HNO_3 \xrightarrow{H_2SO_4} C_6H_5NO_2 + H_2O$$

Nitrobenzene

$$C_6H_6 + CH_3Cl \xrightarrow{AlCl_3} C_6H_5CH_3 + HCl$$

Toluene

In each case the substance shown over the arrow is needed to catalyze these substitution reactions.

Substitution reactions are characteristic of saturated hydrocarbons, and addition reactions are characteristic of unsaturated ones. The fact that benzene reacts more like a saturated hydrocarbon indicates the great stability of the delocalized π electron system.

The nomenclature of benzene derivatives is similar to the nomenclature for saturated ring systems. If there is more than one substituent present, numbers are used to indicate substituent positions. For example, the compound

(1,2-dichlorobenzene structure, ring positions numbered 1–6, Cl at 1 and 2)

is named 1,2-dichlorobenzene. Another nomenclature system uses the prefix *ortho-* (*o*-) for two adjacent substituents, *meta-* (*m*-) for two substituents with one carbon between them, and *para-* (*p*-) for two substituents opposite each other. When benzene is used as a substituent, it is called the **phenyl group.** Examples of some aromatic compounds are shown in Fig. 21.12.

Benzene is the simplest aromatic molecule. More complex aromatic systems can be viewed as consisting of a number of "fused" benzene rings. Some examples are given in Table 21.3.

Figure 21.12
Some selected substituted benzenes and their names. Common names are given in parentheses.

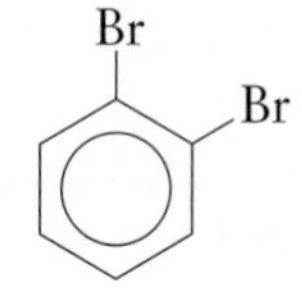

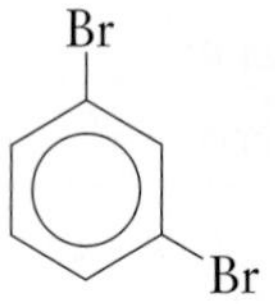

Table 21.3
More Complex Aromatic Systems

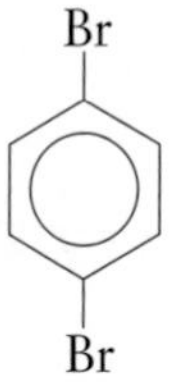

Structural Formula	Name	Use or Effect
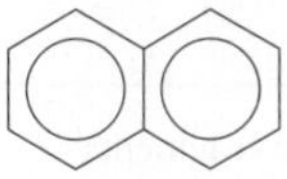	Naphthalene	Formerly used in mothballs
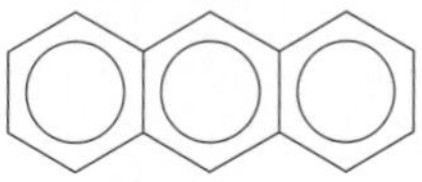	Anthracene	Dyes
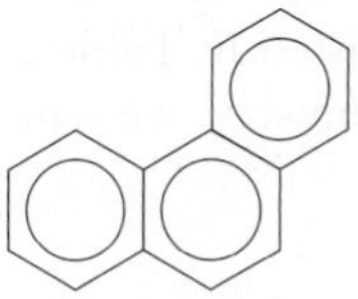	Phenanthrene	Dyes, explosives, and synthesis of drugs
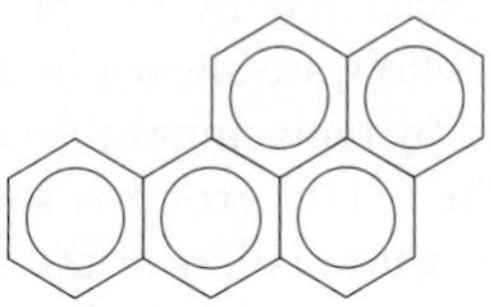	3,4-Benzpyrene	Active carcinogen found in smoke and smog

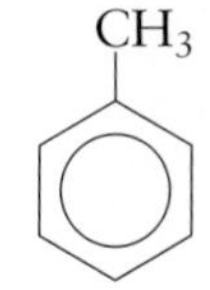

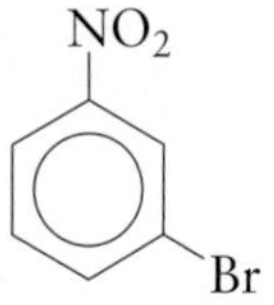

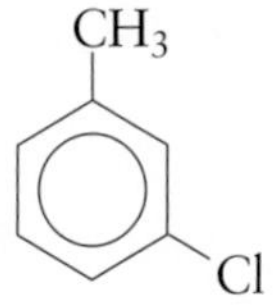

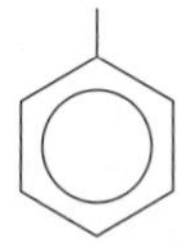

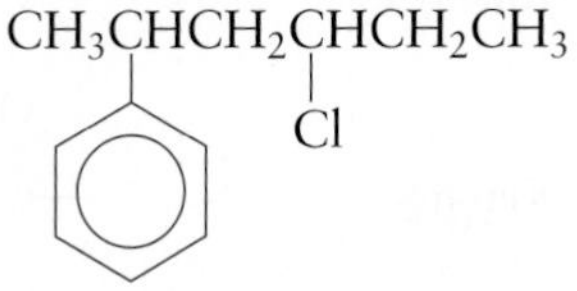

21.4 Hydrocarbon Derivatives

The vast majority of organic molecules contain elements in addition to carbon and hydrogen. However, most of these substances can be viewed as **hydrocarbon derivatives,** molecules that are fundamentally hydrocarbons but that have

Table 21.4

The Common Functional Groups

Class	Functional Group	General Formula*	Example
Halohydrocarbons	—X (F, Cl, Br, I)	R—X	CH_3I Iodomethane (methyl iodide)
Alcohols	—OH	R—OH	CH_3OH Methanol (methyl alcohol)
Ethers	—O—	R—O—R′	CH_3OCH_3 Dimethyl ether
Aldehydes	—C(=O)—H	R—C(=O)—H	CH_2O Methanal (formaldehyde)
Ketones	—C(=O)—	R—C(=O)—R′	CH_3COCH_3 Propanone (dimethyl ketone or acetone)
Carboxylic acids	—C(=O)—OH	R—C(=O)—OH	CH_3COOH Ethanoic acid (acetic acid)
Esters	—C(=O)—O—	R—C(=O)—O—R′	$CH_3COOCH_2CH_3$ Ethyl acetate
Amines	$—NH_2$	$R—NH_2$	CH_3NH_2 Aminomethane (methylamine)

*R and R′ represent hydrocarbon fragments.

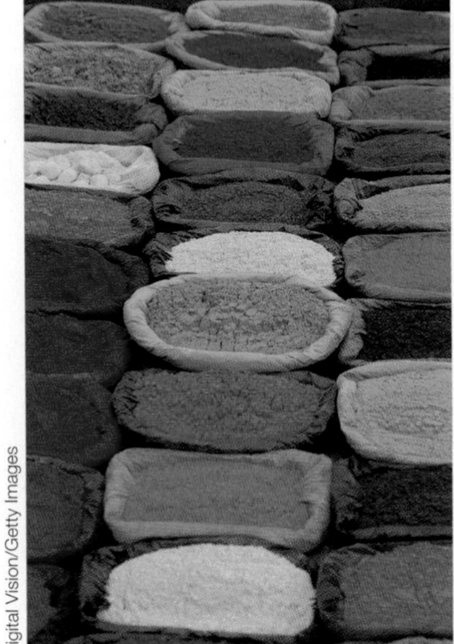

Digital Vision/Getty Images

Compounds containing aromatic rings are often used in dyes, such as these for sale in a market in Nepal.

additional atoms or groups of atoms called **functional groups.** The common functional groups are listed in Table 21.4. Because each functional group exhibits characteristic chemistry, we will consider the groups separately.

Alcohols

Alcohols are characterized by the presence of the hydroxyl group (—OH). Some common alcohols are shown in Table 21.5. The systematic name for an alcohol is obtained by replacing the final *-e* of the parent hydrocarbon with *-ol.* The position of the —OH group is specified by a number (where necessary) chosen so that it is the smallest of the substituent numbers. Alcohols are classified according to the number of hydrocarbon fragments bonded to the carbon where the —OH group is attached,

$R—CH_2OH$	$(R)(R')CHOH$	$(R)(R')(R'')COH$
Primary alcohol (one R group)	*Secondary* alcohol (two R groups)	*Tertiary* alcohol (three R groups)

Table 21.5

Some Common Alcohols

Formula	Systematic Name	Common Name
CH_3OH	Methanol	Methyl alcohol
CH_3CH_2OH	Ethanol	Ethyl alcohol
$CH_3CH_2CH_2OH$	1-Propanol	*n*-Propyl alcohol
$CH_3CH(OH)CH_3$	2-Propanol	Isopropyl alcohol

where R, R′, and R″ (which may be the same or different) represent hydrocarbon fragments.

Alcohols usually have much higher boiling points than might be expected from their molar masses. For example, both methanol and ethane have a molar mass of 30, but the boiling point for methanol is 65°C, whereas that for ethane is −89°C. This difference can be understood if we consider the types of intermolecular attractions that occur in these liquids. Ethane molecules are nonpolar and exhibit only weak London dispersion interactions. However, the polar —OH group of methanol produces extensive hydrogen bonding similar to that found in water (see Section 16.1), which results in the relatively high boiling point.

Although there are many important alcohols, the simplest ones, methanol and ethanol, have the greatest commercial value. Methanol, also known as *wood alcohol* because it was formerly obtained by heating wood in the absence of air, is prepared industrially (approximately 4 million tons annually in the United States) by the hydrogenation of carbon monoxide:

$$CO + 2H_2 \xrightarrow[ZnO/Cr_2O_3]{400^\circ C} CH_3OH$$

Methanol is used as a starting material for the synthesis of acetic acid and for many types of adhesives, fibers, and plastics. It is also used (and such use may increase) as a motor fuel. Methanol is highly toxic to humans and can lead to blindness and death if ingested.

Ethanol is the alcohol found in beverages such as beer, wine, and whiskey; it is produced by the fermentation of glucose in corn, barley, grapes, and so on:

$$\underset{\text{Glucose}}{C_6H_{12}O_6} \xrightarrow{\text{Yeast}} \underset{\text{Ethanol}}{2CH_3CH_2OH} + 2CO_2$$

The reaction is catalyzed by the enzymes found in yeast. This reaction can proceed only until the alcohol content reaches about 13% (the percentage found in most wines), at which point the yeast can no longer survive. Beverages with higher alcohol content are made by distilling the fermentation mixture.

Ethanol, like methanol, can be burned in the internal combustion engines of automobiles and is now commonly added to gasoline to form gasohol (see Section 9.8). It is also used in industry as a solvent and for the preparation of acetic acid. The commercial production of ethanol (500,000 tons per year in the United States) is carried out by reaction of water with ethylene:

$$CH_2{=}CH_2 + H_2O \xrightarrow[\text{Catalyst}]{\text{Acid}} CH_3CH_2OH$$

AP Photo/Seth Perlman

A fuel consisting of 85% ethanol and 15% gasoline is now widely available in the United States.

Many polyhydroxyl (more than one —OH group) alcohols are known, the most important being *1,2-ethanediol* (ethylene glycol),

$$\begin{array}{l} H_2C{-}OH \\ \;\;\;| \\ H_2C{-}OH \end{array}$$

a toxic substance that is the major constituent of most automobile antifreeze solutions.

The simplest aromatic alcohol is

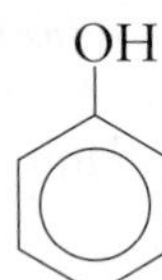

which is commonly called **phenol.** Most of the 1 million tons of phenol produced annually in the United States are used to make polymers for adhesives and plastics.

OWL INTERACTIVE EXAMPLE 21.6

For each of the following alcohols, give the systematic name, and specify whether the alcohol is primary, secondary, or tertiary.

a. $CH_3\underset{\displaystyle OH}{\underset{|}{C}}HCH_2CH_3$

b. $ClCH_2CH_2CH_2OH$

c. $CH_3\underset{\displaystyle OH}{\underset{|}{\overset{\displaystyle CH_3}{\overset{|}{C}}}}CH_2CH_2CH_2CH_2Br$

Solution

a. The chain is numbered as follows:

$$\overset{1}{CH_3}\underset{\displaystyle OH}{\underset{|}{\overset{2}{C}}}H\overset{3}{C}H_2\overset{4}{C}H_3$$

The compound is called 2-butanol, since the —OH group is located at the number 2 position of a four-carbon chain. Note that the carbon to which the —OH is attached also has $—CH_3$ and $—CH_2CH_3$ groups attached:

$$\underbrace{CH_3}_{R}-\overset{\overset{\displaystyle H}{|}}{\underset{\underset{\displaystyle OH}{|}}{C}}-\underbrace{CH_2CH_3}_{R'}$$

Therefore, this is a *secondary* alcohol.

b. The chain is numbered as follows:

$$\overset{3}{Cl-CH_2}-\overset{2}{CH_2}-\overset{1}{CH_2}-OH$$

The name is 3-chloro-1-propanol. This is a *primary* alcohol:

$$\underbrace{Cl-CH_2CH_2}-\overset{\overset{\displaystyle H}{|}}{\underset{\underset{\displaystyle H}{|}}{C}}-OH$$

One R group attached to the carbon with the —OH group

c. The chain is numbered as follows:

$$\overset{1}{CH_3}-\overset{\overset{\displaystyle CH_3}{|}}{\underset{\underset{\displaystyle OH}{|}}{\overset{2}{C}}}-\overset{3}{CH_2}-\overset{4}{CH_2}-\overset{5}{CH_2}-\overset{6}{CH_2Br}$$

The name is 6-bromo-2-methyl-2-hexanol. This is a *tertiary* alcohol since the carbon where the —OH is attached also has three other R groups attached.

Cinnamaldehyde produces the characteristic odor of cinnamon.

Aldehydes and Ketones

Aldehydes and ketones contain the **carbonyl group,**

$$\diagdown\!\!\!\!\diagup C=O$$

In **ketones** this group is bonded to two carbon atoms, as in acetone,

$$CH_3-\underset{\underset{\displaystyle O}{\|}}{C}-CH_3$$

In **aldehydes** the carbonyl group is bonded to at least one hydrogen atom, as in formaldehyde,

$$H-\underset{\underset{\displaystyle O}{\|}}{C}-H$$

or acetaldehyde,

$$CH_3-\underset{\underset{\displaystyle O}{\|}}{C}-H$$

Figure 21.13
Some common ketones and aldehydes. Note that since the aldehyde functional group always appears at the end of a carbon chain, carbon is assigned the number 1 when the compound is named.

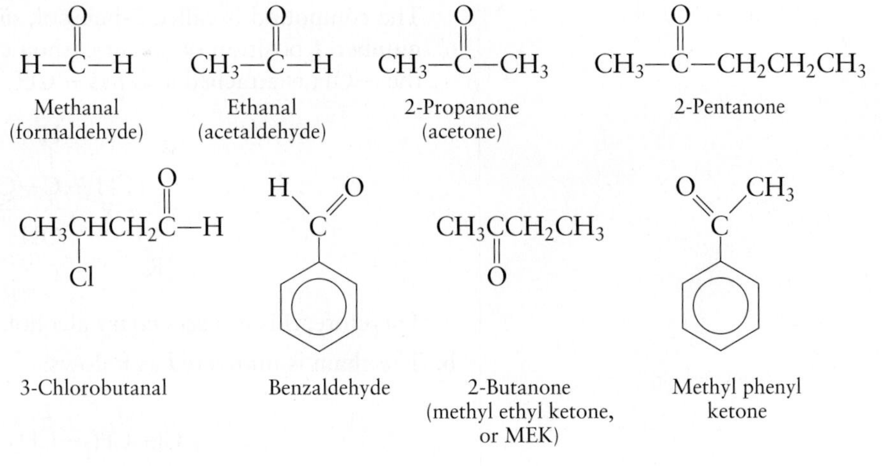

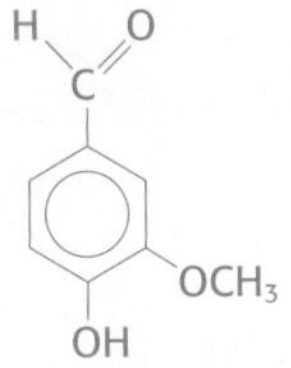

The systematic name for an aldehyde is obtained from the parent alkane by removing the final *-e* and adding *-al*. For ketones the final *-e* is replaced by *-one*, and a number indicates the position of the carbonyl group where necessary. Examples of common aldehydes and ketones are shown in Fig. 21.13. Note that since the aldehyde functional group always occurs at the end of the carbon chain, the aldehyde carbon is assigned the number 1 when substituent positions are listed in the name.

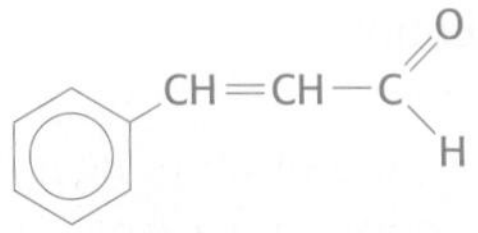

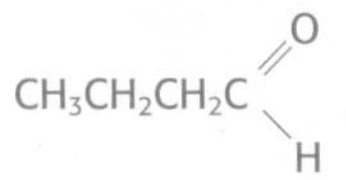

Ketones often have useful solvent properties (acetone is found in nail polish remover, for example) and are frequently used in industry for this purpose. Aldehydes typically have strong odors. Vanillin is responsible for the pleasant odor in vanilla beans; cinnamaldehyde produces the characteristic odor of cinnamon. On the other hand, the unpleasant odor in rancid butter arises from the presence of butyraldehyde.

Aldehydes and ketones are most often produced commercially by the oxidation of alcohols. For example, oxidation of a *primary* alcohol yields the corresponding aldehyde:

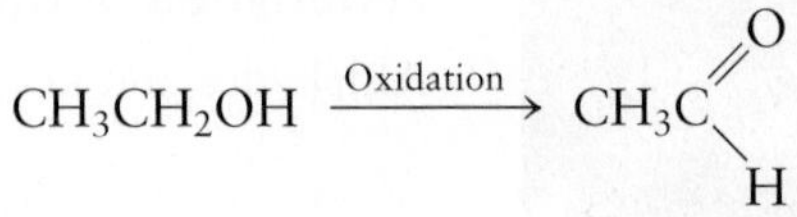

$$CH_3CH_2OH \xrightarrow{\text{Oxidation}} CH_3C(=O)H$$

Oxidation of a *secondary* alcohol results in a ketone:

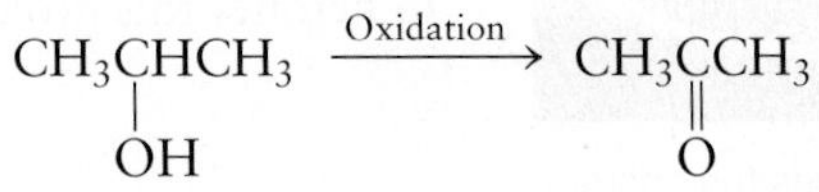

$$CH_3CH(OH)CH_3 \xrightarrow{\text{Oxidation}} CH_3C(=O)CH_3$$

Carboxylic Acids and Esters

Carboxylic acids are characterized by the presence of the **carboxyl group**

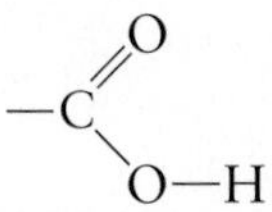

$$-C(=O)-O-H$$

that gives an acid of the general formula RCOOH. Typically, these molecules are weak acids in aqueous solution (see Section 7.5). Organic acids are named from the parent alkane by dropping the final *-e* and adding *-oic*. Thus CH_3COOH, commonly called acetic acid, has the systematic name ethanoic

$CH_3CH_2CH_2COOH$

Butanoic acid

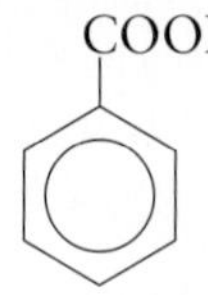

COOH (on benzene ring)

Benzoic acid

$CH_3CHCH_2CH_2COOH$ (Br on C-4)

4-Bromopentanoic acid

$Cl_3C{-}COOH$

Trichloroethanoic acid (trichloroacetic acid)

Figure 21.14
Some carboxylic acids.

acid, since the parent alkane is ethane. Other examples of carboxylic acids are shown in Fig. 21.14.

Many carboxylic acids are synthesized by oxidizing primary alcohols with a strong oxidizing agent. For example, ethanol can be oxidized to acetic acid by using potassium permanganate:

$$CH_3CH_2OH \xrightarrow{KMnO_4(aq)} CH_3COOH$$

A carboxylic acid reacts with an alcohol to form an **ester** and a water molecule. For example, the reaction of acetic acid with ethanol produces ethyl acetate and water:

$$CH_3\overset{O}{\overset{\|}{C}}{-}OH \quad H{-}OCH_2CH_3 \longrightarrow CH_3\overset{O}{\overset{\|}{C}}{-}OCH_2CH_3 + H_2O$$

React to form water

Esters often have a sweet, fruity odor that is in contrast to the often pungent odors of the parent carboxylic acids. For example, the odor of bananas is caused by *n*-amyl acetate,

$$CH_3C(=O)OCH_2CH_2CH_2CH_2CH_3$$

and that of oranges is caused by *n*-octyl acetate,

$$CH_3\underset{O}{\underset{\|}{C}}{-}OC_8H_{17}$$

The systematic name for an ester is formed by changing the *-oic* ending of the parent acid to *-oate*. The parent alcohol chain is named first with a *-yl* ending. For example, the systematic name for *n*-octyl acetate is *n*-octylethanoate (from ethanoic acid).

A very important ester is formed from the reaction of salicylic acid and acetic acid:

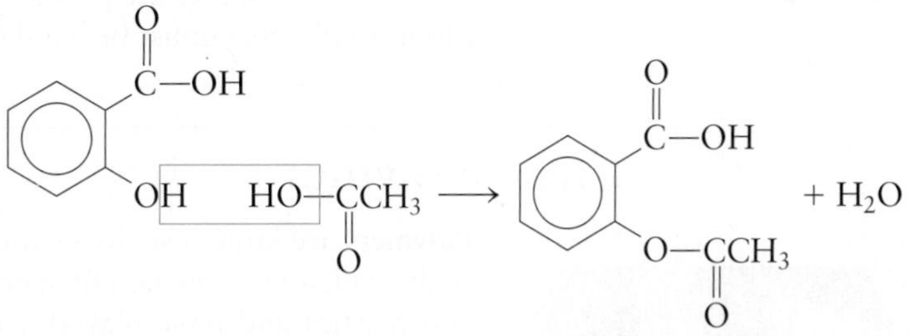

$$C_6H_4(COOH)(OH) + HO{-}\underset{O}{\underset{\|}{C}}CH_3 \longrightarrow C_6H_4(COOH)(O{-}\underset{O}{\underset{\|}{C}}CH_3) + H_2O$$

Salicylic acid — Acetic acid — Acetylsalicylic acid

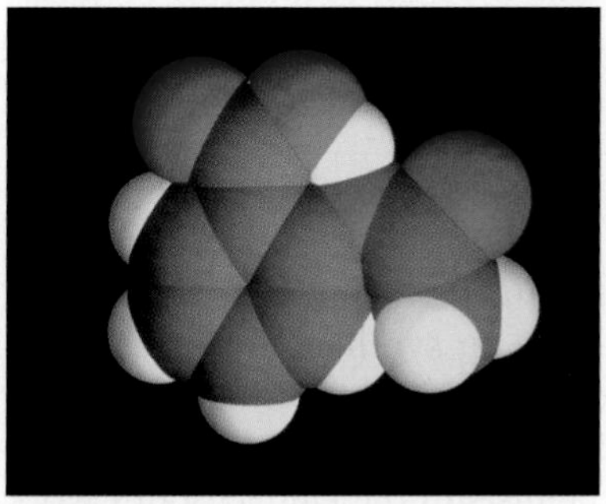

Computer-generated space-filling model of acetylsalicylic acid (aspirin).

The product is acetylsalicylic acid, commonly known as *aspirin,* which is used in huge quantities as an analgesic (painkiller).

Amines

Amines are probably best viewed as derivatives of ammonia in which one or more N—H bonds are replaced by N—C bonds. The resulting amines are classified as *primary* if one N—C bond is present, *secondary* if two N—C bonds are present, and *tertiary* if all three N—H bonds in NH_3 have been replaced by

Table 21.6

Some Common Amines

Formula	Common Name	Type
CH_3NH_2	Methylamine	Primary
$CH_3CH_2NH_2$	Ethylamine	Primary
$(CH_3)_2NH$	Dimethylamine	Secondary
$(CH_3)_3N$	Trimethylamine	Tertiary
NH_2 (on benzene ring)	Aniline	Primary
H–N (bonded to two benzene rings)	Diphenylamine	Secondary

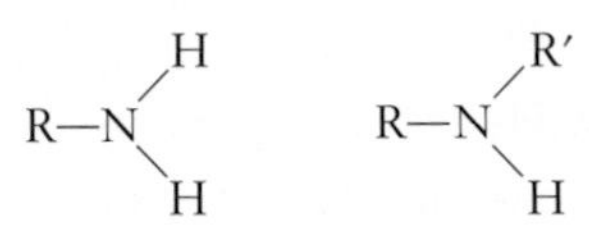

R—N R′ R″

Tertiary amine

Figure 21.15
The general formulas for primary, secondary, and tertiary amines. R, R′, and R″ represent carbon-containing substituents.

N—C bonds (Fig. 21.15). Examples of some common amines are given in Table 21.6.

Common names are often used for simple amines; the systematic nomenclature for more complex molecules uses the name *amino-* for the —NH_2 functional group. For example, the molecule

$$CH_3\underset{\displaystyle NH_2}{\underset{|}{C}}HCH_2CH_3$$

is named 2-aminobutane.

Many amines have unpleasant "fishlike" odors. For example, the odors associated with decaying animal and human tissues are caused by amines such as putrescine ($H_2NCH_2CH_2CH_2NH_2$) and cadaverine ($H_2NCH_2CH_2CH_2CH_2CH_2NH_2$).

Aromatic amines are primarily used to make dyes. Since many of them are carcinogenic, they must be handled with great care.

21.5 Polymers

Polymers are large, usually chainlike molecules that are built from small molecules called *monomers*. Polymers form the basis for synthetic fibers, rubbers, and plastics and have played a leading role in the revolution that has been brought about in daily life by chemistry during the past 50 years. It has been estimated that about 50% of the industrial chemists in the United States work in some area of polymer chemistry, a fact that illustrates just how important polymers are to our economy and standard of living.

AP Photo/Indianapolis Star/Karen Ducey

The soybeans on the left are coated with a red acrylic polymer to delay soybean emergence. This allows farmers to plant their crops more efficiently.

The Development and Properties of Polymers

The development of the polymer industry provides a striking example of the importance of serendipity in the progress of science. Many discoveries in polymer chemistry arose from accidental observations that scientists followed up.

The age of plastics might be traced to a day in 1846 when Christian Schoenbein, a chemistry professor at the University of Basel in Switzerland, spilled a flask containing nitric and sulfuric acids. In his hurry to clean up the spill, he grabbed his wife's cotton apron, which he then rinsed out and hung up in front of a hot stove to dry. Instead of drying, the apron flared and burned.

Very interested in this event, Schoenbein repeated the reaction under more controlled conditions and found that the new material, which he correctly concluded to be nitrated cellulose, had some surprising properties. As he had experienced, nitrated cellulose is extremely flammable and, under certain circumstances, highly explosive. In addition, he found that it could be molded at moderate temperatures to give objects that were, upon cooling, tough but elastic. Predictably, the explosive nature of the substance was initially of more interest than its other properties, and cellulose nitrate rapidly became the basis for smokeless gun powder. Although Schoenbein's discovery cannot be described as a truly synthetic polymer (because he simply found a way to modify the natural polymer cellulose), it formed the basis for a large number of industries that grew up to produce photographic films, artificial fibers, and molded objects of all types.

Phil Nelson

A radio from the 1930s made of Bakelite.

The first synthetic polymers were produced as by-products of various organic reactions and were regarded as unwanted contaminants. Thus the first preparations of many of the polymers now regarded as essential to our modern lifestyle were thrown away in disgust. One chemist who refused to be defeated by the "tarry" products obtained when he reacted phenol with formaldehyde was Belgian-American chemist Leo H. Baekeland (1863–1944). Baekeland's work resulted in the first completely synthetic plastic (called Bakelite), a substance that when molded to a certain shape under high pressure and temperature cannot be softened again or dissolved. Bakelite is a **thermoset polymer.** In contrast, cellulose nitrate is a **thermoplastic polymer;** that is, it can be remelted after it has been molded.

The discovery of Bakelite in 1907 spawned a large plastics industry, producing telephones, billiard balls, and insulators for electrical devices. During the early days of polymer chemistry, there was a great deal of controversy over the nature of these materials. Although the German chemist Hermann Staudinger speculated in 1920 that polymers were very large molecules held together by strong chemical bonds, most chemists of the time assumed that these materials were much like colloids, in which small molecules are aggregated into large units by forces weaker than chemical bonds.

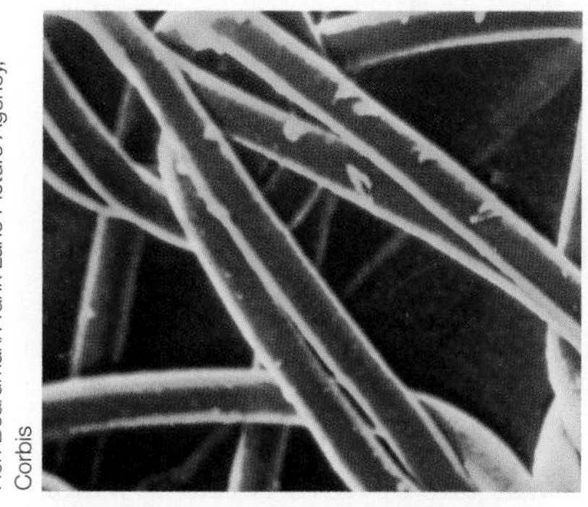

Ron Boardman/Frank Lane Picture Agency, Corbis

Nylon netting magnified 62 times.

One chemist who contributed greatly to the understanding of polymers as giant molecules was Wallace H. Carothers of the DuPont Chemical Company. Among his accomplishments was the preparation of nylon. The nylon story further illustrates the importance of serendipity in scientific research. When nylon is first prepared, the resulting product is a sticky material with little structural integrity. Because of this, it was initially put aside as having no apparently useful characteristics. However, Julian Hill, a chemist in the Carothers research group, one day put a small ball of this nylon on the end of a stirring rod and drew it away from the remaining sticky mass, forming a string. He noticed the silky appearance and strength of this thread and realized that nylon could be drawn into useful fibers.

The reason for this behavior of nylon is now understood. When nylon is first formed, the individual polymer chains are oriented randomly, like cooked spaghetti, and the substance is highly amorphous. However, when drawn out into a thread, the chains tend to line up (the nylon becomes more crystalline), which leads to increased hydrogen bonding between adjacent chains. This increase in crystallinity, along with the resulting increase in hydrogen-bonding interactions, leads to strong fibers and thus to a highly useful material. Commercially, nylon is produced by forcing the raw material through a *spinneret,* a plate containing small holes, which forces the polymer chains to line up.

Another property that adds strength to polymers is **crosslinking,** the existence of covalent bonds between adjacent chains. The structure of Bakelite is highly crosslinked, which accounts for the strength and toughness of this polymer. Another example of crosslinking occurs in the manufacture of rubber. Raw natural rubber consists of chains of the type

$$\sim\sim CH_2-CH_2-CH=\underset{\displaystyle CH_3}{\underset{|}{C}}-CH_2-CH_2-CH=\underset{\displaystyle CH_3}{\underset{|}{C}}-CH_2\sim\sim$$

Charles Goodyear tried for many years to change natural rubber into a useful product. In 1839 he accidentally dropped some rubber containing sulfur on a hot stove. Noting that the rubber did not melt as expected, Goodyear pursued this lead and developed vulcanization.

and is a soft, sticky material unsuitable for tires. However, in 1839 Charles Goodyear (1800–1860), an American chemist, accidentally found that if sulfur is added to rubber and the resulting mixture is heated (a process called **vulcanization**), the resulting rubber is still elastic (reversibly stretchable) but is much stronger. This change in character occurs because sulfur atoms become bonded between carbon atoms on different chains. These sulfur atoms form bridges between the polymer chains, thus linking the chains together.

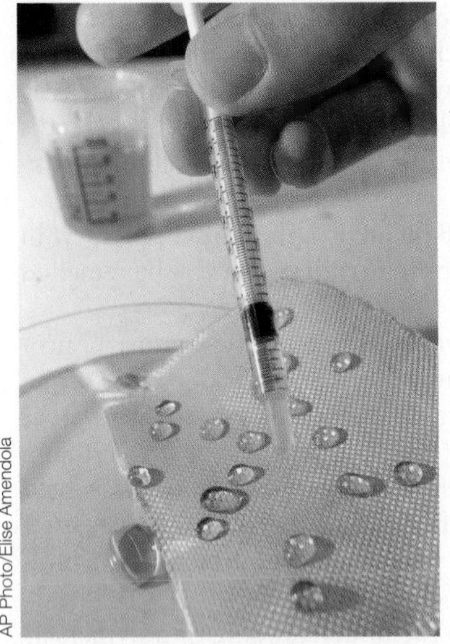
AP Photo/Elise Amendola

Colored water drops are shown beading on Kevlar fabric treated with a nanoscale water-resistant coating.

Types of Polymers

The simplest and one of the best-known polymers is *polyethylene,* which is constructed from ethylene monomers:

$$nCH_2=CH_2 \xrightarrow{\text{Catalyst}} \left(\begin{array}{c} H\ \ H \\ |\ \ \ | \\ -C-C- \\ |\ \ \ | \\ H\ \ H \end{array}\right)_n$$

where n represents a large number (usually several thousand). Polyethylene is a tough, flexible plastic used for piping, bottles, electrical insulation, packaging films, garbage bags, and many other purposes. Its properties can be varied by using substituted ethylene monomers. For example, when tetrafluoroethylene is the monomer, the polymer Teflon is obtained:

$$n\left(\begin{array}{c} F\ \ \ \ \ \ \ F \\ \diagdown\ \ \diagup \\ C=C \\ \diagup\ \ \diagdown \\ F\ \ \ \ \ \ \ F \end{array}\right) \longrightarrow \left(\begin{array}{c} F\ \ F \\ |\ \ \ | \\ -C-C- \\ |\ \ \ | \\ F\ \ F \end{array}\right)_n$$

Tetrafluoroethylene Teflon

The discovery of Teflon, a very important substituted polyethylene, is another illustration of the role of chance in chemical research. In 1938 a DuPont chemist named Roy Plunkett was studying the chemistry of gaseous tetrafluoroethylene. He synthesized about 100 pounds of the chemical and stored it in steel cylinders. When one of the cylinders failed to produce perfluoroethylene gas when the valve was opened, the cylinder was cut open to reveal a white powder. This powder turned out to be a polymer of perfluoroethylene, which was eventually developed into Teflon. Because of the resistance of the strong C—F bonds to chemical attack, Teflon is an inert, tough, and nonflammable material widely used for electrical insulation, nonstick coatings on cooking utensils, and bearings for low-temperature applications.

Other polyethylene-type polymers are made from monomers containing chloro, methyl, cyano, and phenyl substituents, as summarized in Table 21.7. In each case the double carbon–carbon bond in the substituted ethylene monomer becomes a single bond in the polymer. The different substituents lead to a wide variety of properties.

Table 21.7

Some Common Synthetic Polymers, Their Monomers, and Applications

Monomer		Polymer		
Name	**Formula**	**Name**	**Formula**	**Uses**
Ethylene	$H_2C{=}CH_2$	Polyethylene	$-\!(CH_2{-}CH_2)_n\!-$	Plastic piping, bottles, electrical insulation, toys
Propylene	$H_2C{=}C(H)(CH_3)$	Polypropylene	$-\!(CH(CH_3){-}CH_2{-}CH(CH_3){-}CH_2)_n\!-$	Film for packaging, carpets, lab wares, toys
Vinyl chloride	$H_2C{=}C(H)(Cl)$	Polyvinyl chloride (PVC)	$-\!(CH_2{-}CH(Cl))_n\!-$	Piping, siding, floor tile, clothing, toys
Acrylonitrile	$H_2C{=}C(H)(CN)$	Polyacrylonitrile (PAN)	$-\!(CH_2{-}CH(CN))_n\!-$	Carpets, fabrics
Tetrafluoro-ethylene	$F_2C{=}CF_2$	Teflon	$-\!(CF_2{-}CF_2)_n\!-$	Cooking utensils, electrical insulation, bearings
Styrene	$H_2C{=}C(H)(C_6H_5)$	Polystyrene	$-\!(CH_2CH(C_6H_5))_n\!-$	Containers, thermal insulation, toys
Butadiene	$H_2C{=}C(H){-}C(H){=}CH_2$	Polybutadiene	$-\!(CH_2CH{=}CHCH_2)_n\!-$	Tire tread, coating resin
Butadiene and styrene	(See above.)	Styrene-butadiene rubber	$-\!(CH(C_6H_5){-}CH_2{-}CH_2{-}CH{=}CH{-}CH_2)_n\!-$	Synthetic rubber

The polyethylene polymers illustrate one of the major types of polymerization reactions, called **addition polymerization,** in which the monomers simply "add together" to produce the polymer. No other products are formed. The polymerization process is initiated by a **free radical** (a species with an unpaired electron) such as the hydroxyl radical (HO ·). The free radical attacks and breaks the π bond of an ethylene molecule to form a new free radical,

$$H_2C{=}CH_2 + \cdot OH \longrightarrow H{-}C(H)(OH){-}C(H)(H)\cdot$$

which is then available to attack another ethylene molecule:

$$\mathrm{HO{-}CH_2{-}CH_2\cdot} + \mathrm{CH_2{=}CH_2} \longrightarrow \mathrm{HO{-}CH_2{-}CH_2{-}CH_2{-}CH_2\cdot}$$

Repetition of this process thousands of times creates a long-chain polymer. Termination of the growth of the chain occurs when *two radicals* react to form a bond, a process that consumes two radicals without producing any others.

Another common type of polymerization is **condensation polymerization,** in which a small molecule, such as water, is formed for each extension of the polymer chain. The most familiar polymer produced by condensation is *nylon.* Nylon is a **copolymer,** since two different types of monomers combine to form the chain; a **homopolymer** is the result of polymerizing a single type of monomer. One common form of nylon is produced when hexamethylenediamine and adipic acid react by splitting out a water molecule to form a C—N bond:

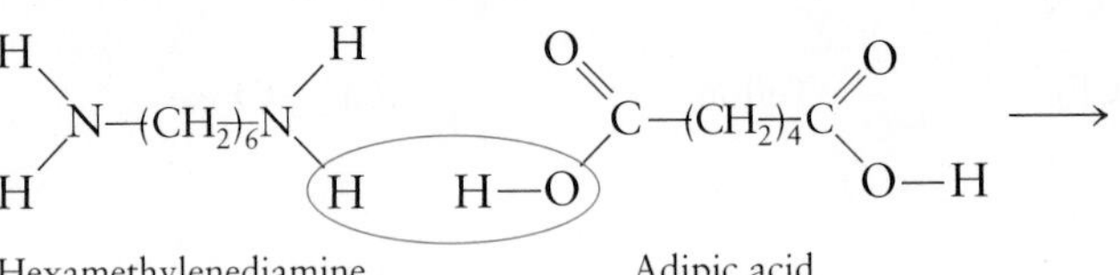

Hexamethylenediamine Adipic acid

$$\mathrm{H_2N{-}(CH_2)_6{-}N(H){-}C({=}O){-}(CH_2)_4{-}C({=}O){-}OH} + \mathrm{H_2O}$$

The molecule formed, called a **dimer** (two monomers joined), can undergo further condensation reactions since it has an amino group at one end and a carboxyl group at the other. Thus both ends are free to react with another monomer. Repetition of this process leads to a long chain of the type

$$\left(\mathrm{{-}N(H){-}(CH_2)_6{-}N(H){-}C({=}O){-}(CH_2)_4{-}C({=}O){-}}\right)_n$$

which is the basic structure of nylon. The reaction to form nylon occurs quite readily and is often used as a lecture demonstration (Fig. 21.16). The properties of nylon can be varied by changing the number of carbon atoms in the chain of the acid or amine monomer.

More than 1 million tons of nylon are produced annually in the United States for use in clothing, carpets, rope, and so on. Many other types of condensation polymers are also produced. For example, Dacron is a copolymer formed from the condensation reaction of ethylene glycol (a dialcohol) and *p*-terephthalic acid (a dicarboxylic acid):

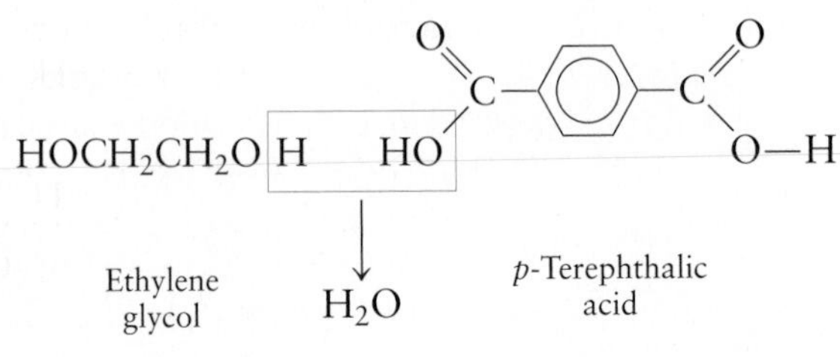

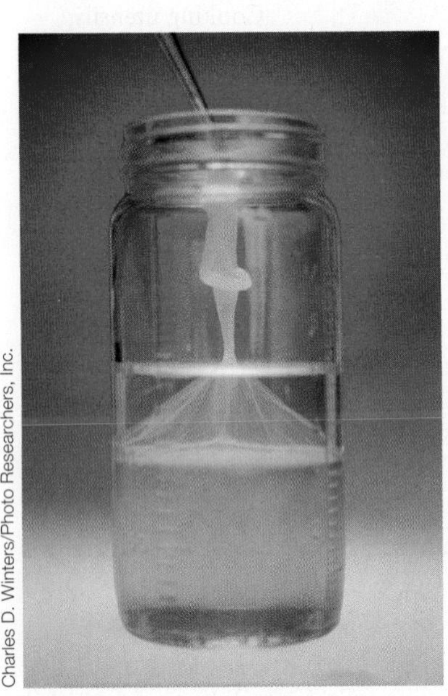

Charles D. Winters/Photo Researchers, Inc.

Figure 21.16
The reaction to form nylon can be carried out at the interface of two immiscible liquid layers in a beaker. The bottom layer contains adipoyl chloride,

$$\mathrm{Cl{-}C({=}O){-}(CH_2)_4{-}C({=}O){-}Cl}$$

dissolved in CCl_4, and the top layer contains hexamethylenediamine,

$$\mathrm{H_2N{-}(CH_2)_6{-}NH_2}$$

dissolved in water. A molecule of HCl is formed as each C—N bond forms.

The repeating unit of Dacron is

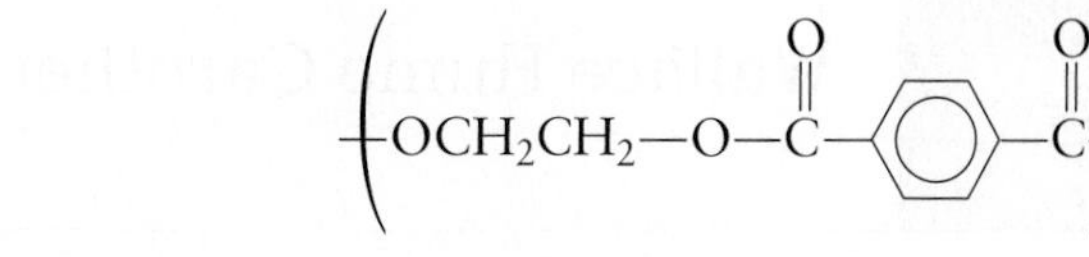

$$\left(-OCH_2CH_2-O-\overset{\overset{\displaystyle O}{\|}}{C}-C_6H_4-\overset{\overset{\displaystyle O}{\|}}{C}- \right)_n$$

Note that this polymerization involves a carboxylic acid and an alcohol forming an ester group:

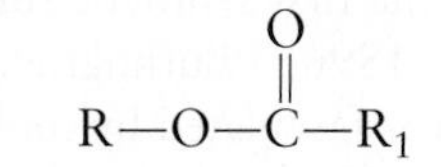

$$R-O-\overset{\overset{\displaystyle O}{\|}}{C}-R_1$$

Thus Dacron is called a **polyester.** By itself or blended with cotton, Dacron is widely used in fibers for the manufacture of clothing.

Polymers Based on Ethylene

A large section of the polymer industry involves the production of macromolecules from ethylene or substituted ethylenes. As discussed previously, ethylene molecules polymerize by addition after the double bond has been broken by some initiator:

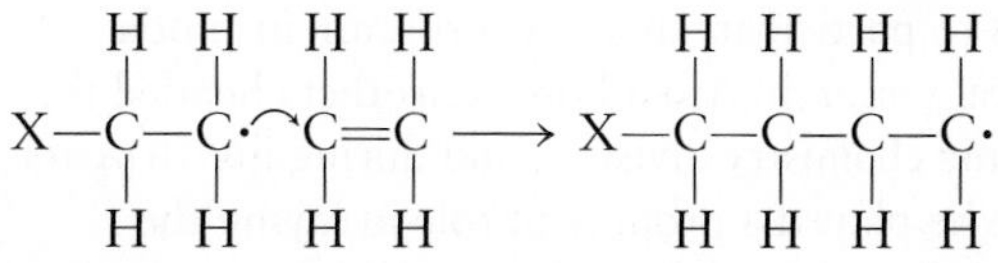

$$X-CH_2-CH_2\cdot + CH_2{=}CH_2 \longrightarrow X-CH_2-CH_2-CH_2-CH_2\cdot$$

This process continues by adding new ethylene molecules to eventually give polyethylene, a thermoplastic material.

There are two forms of polyethylene: low-density polyethylene (LDPE) and high-density polyethylene (HDPE). The chains in LDPE contain many branches and thus do not pack as tightly as those in HDPE, which consist of mostly straight-chain molecules.

psi is the abbreviation for pounds per square inch: 15 psi ≈ 1 atm.

Traditionally, LDPE has been manufactured under conditions of high pressure (≈20,000 psi) and high temperature (500°C). These severe reaction conditions require specially designed equipment, and for safety reasons, the reaction usually has been run behind a reinforced concrete barrier. More recently, lower reaction pressures and temperatures have become possible through the use of catalysts. One catalytic system using triethylaluminum $[Al(C_2H_5)_3]$ and titanium(IV) chloride was developed by Karl Ziegler in Germany and Giulio Natta in Italy. Although this catalyst is very efficient, it catches fire on contact with air and must be handled very carefully. A safer catalytic system was developed at Phillips Petroleum Company. It uses a chromium(III) oxide (Cr_2O_3) and aluminosilicate catalyst and has mainly taken over in the United States. The product of the catalyzed reaction is highly linear (unbranched) and is often called *linear low-density polyethylene.* It is very similar to HDPE.

The major use of LDPE is in the manufacture of the tough, transparent film that is used in packaging so many consumer goods. Two-thirds of the approximately 10 billion pounds of LDPE produced annually in the United States are used for this purpose. The major use of HDPE is for blow-molded products, such as bottles for consumer products (Fig. 21.17).

Molecular weight (not molar mass) is the common terminology in the polymer industry.

The useful properties of polyethylene are due primarily to its high molecular weight (molar mass). Although the strengths of the interactions between specific points on the nonpolar chains are quite small, the chains are so long that these small attractions accumulate to a very significant value so that the chains stick together very tenaciously. There is also a great deal of physical tangling of the lengthy chains. The combination of these interactions gives the

Chemical Insights — Wallace Hume Carothers

Wallace H. Carothers, a brilliant organic chemist who was principally responsible for the development of nylon and the first synthetic rubber (Neoprene), was born in 1896 in Burlington, Iowa. As a youth, Carothers was fascinated by tools and mechanical devices and spent many hours experimenting. In 1915 he entered Tarkio College in Missouri. Carothers so excelled in chemistry that even before his graduation he was made a chemistry instructor.

Carothers eventually moved to the University of Illinois at Urbana–Champaign, where he was appointed to the faculty when he completed his Ph.D. in organic chemistry in 1924. He moved to Harvard University in 1926 and then to DuPont in 1928 to participate in a new program in fundamental research. At DuPont, Carothers headed the organic chemistry division, and during his 10 years there he played a prominent role in laying the foundations of polymer chemistry.

By the age of 33, Carothers had become a world-famous chemist whose advice was sought by almost everyone working in polymers. He was the first industrial chemist to be elected to the prestigious National Academy of Sciences.

Carothers was an avid reader of poetry and a lover of classical music. Unfortunately, he also suffered from severe bouts of depression that finally led to his suicide in 1937 in a Philadelphia hotel room, where he drank a cyanide solution. He was 41 years old. Despite the brevity of his career, Carothers was truly one of the finest American chemists of all time. His great intellect, his love of chemistry, and his insistence on perfection produced his special genius.

Courtesy, Dupont

Wallace H. Carothers

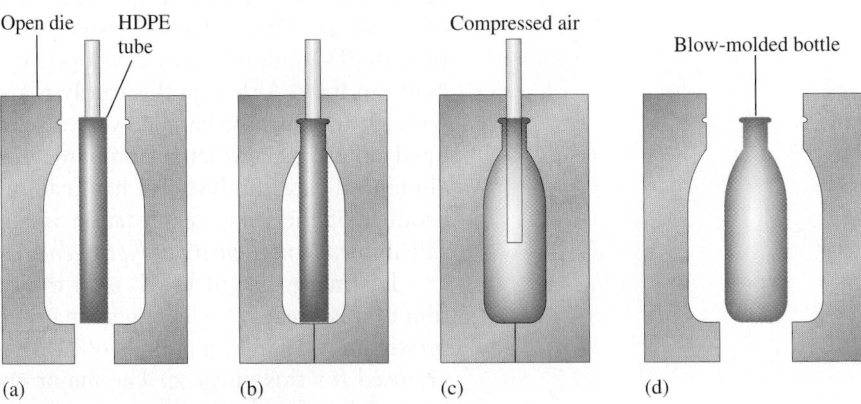

Figure 21.17 A major use of HDPE is for blow-molded objects such as bottles for soft drinks, shampoos, bleaches, and so on. (a) A tube composed of HDPE is inserted into the mold (die). (b) The die closes, sealing the bottom of the tube. (c) Compressed air is forced into the warm HDPE tube, which then expands to take the shape of the die. (d) The molded bottle is removed from the die.

polymer strength and toughness. However, a material like polyethylene can be melted and formed into a new shape (thermoplastic behavior) because in the melted state the molecules can readily flow past one another.

Since a high molecular weight gives a polymer useful properties, one might think that the goal would be to produce polymers with chains as long as possible. However, this is not the case—polymers become much more difficult to process as the molecular weights increase. Most industrial operations require that the polymer flow through pipes as it is processed. But, as the chain lengths increase, viscosity also increases. In practice, the upper limit of a polymer's molecular weight is set by the flow requirements of the manufacturing process. Thus the final product often reflects a compromise between the optimal properties for the application and those needed for ease of processing.

Although many polymer properties are greatly influenced by molecular weight, some other important properties are not. For example, chain length does not affect a polymer's resistance to chemical attack. Physical properties such as color, refractive index, hardness, density, and electrical conductivity are also not greatly influenced by molecular weight.

We have already seen that one way of altering the strength of a polymeric material is to vary the chain length. Another method for modifying polymer behavior involves varying the substituents. For example, if we use a monomer of the type

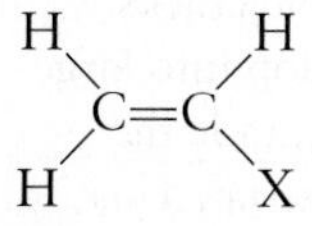

H H
C=C
H X

the properties of the resulting polymer depend on the identity of X. The simplest example is polypropylene, whose monomer is

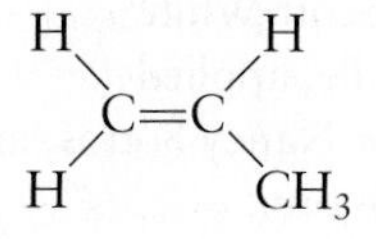

H H
C=C
H CH_3

and that has the form

H CH_3 H CH_3 H CH_3 H CH_3 H CH_3
C C C C C C C C C C
H H H H H H H H H H

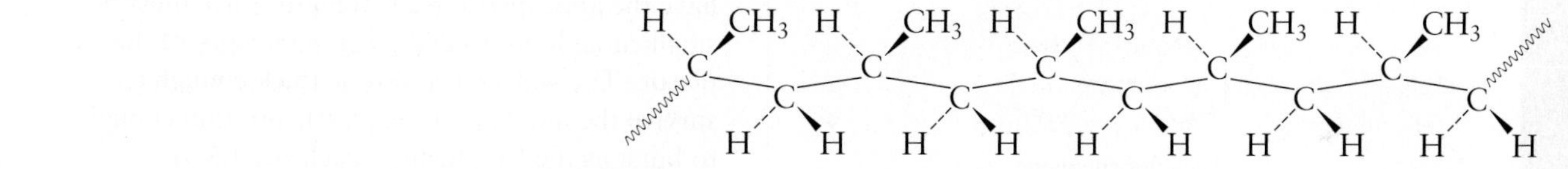

The CH_3 groups can be arranged on the same side of the chain (called an **isotactic chain**) as shown above, can alternate (called a **syndiotactic chain**) as shown below,

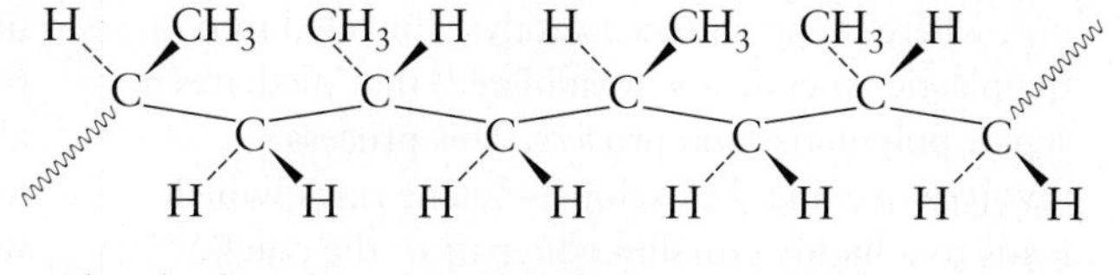

H CH_3 CH_3 H H CH_3 CH_3 H
C C C C C C C C
H H H H H H H H

or can be randomly distributed (called an **atactic chain**).

The chain arrangement has a significant effect on the polymer's properties. Most polypropylene is made using the Ziegler-Natta catalyst $[Al(C_2H_5)_3 \cdot TiCl_4]$, which produces highly isotactic chains that pack together quite closely. As a result, polypropylene is more crystalline and therefore stronger and harder than polyethylene. The major uses of polypropylene are for molded parts (40%), fibers (35%), and packaging films (10%). Polypropylene fibers are especially useful for athletic wear because they do not absorb water from perspiration, as cotton does. Rather, the moisture is drawn away from the skin to the surface of the polypropylene garment, where it can evaporate. The annual U.S. production of polypropylene is about 7 billion pounds.

Chemical Insights Heal Thyself

One major problem with structural materials is that they crack and weaken as they age. The human body has mechanisms for healing itself if the skin is cut or a bone is broken. However, inanimate materials have had no such mechanisms—until now. Scientists at the University of Illinois at Urbana–Champaign (UIUC) have invented a plastic that automatically heals microscopic cracks before they can develop into large cracks that would degrade the usefulness of the material. This accomplishment was achieved by an interdisciplinary team of scientists including aeronautical engineering professors Scott White and Philippe Geubelle, applied mechanics professor Nancy Sottos, and chemistry professor Jeffrey Moore.

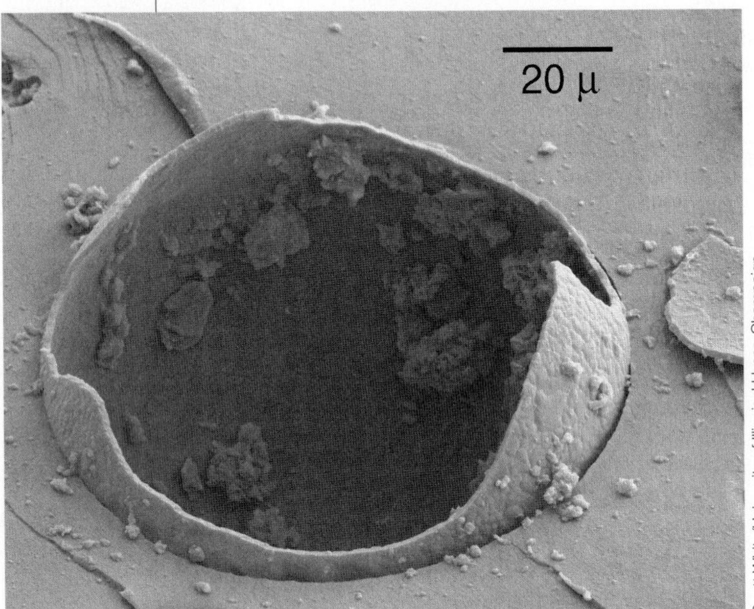

Scott White/University of Illinois, Urbana-Champaign

A scanning electron microscope image showing the fractured plane of a self-healing material with a ruptured microcapsule in a thermosetting matrix.

The self-healing system is based on microcapsules containing liquid dicyclopentadiene

Dicyclopentadiene

that are blended into the plastic. When a microscopic crack develops, it encounters and breaks a microcapsule. The dicyclopentadiene then leaks out, where it encounters a catalyst (blended into the plastic when it was formulated) that mediates a repair polymerization process. This process involves opening the cyclopentadiene rings, which leads to a highly crosslinked repair of the crack.

The trickiest part of the repair mechanism is to get the microcapsules to be the correct size and to have the appropriate wall strength. They must be small enough not to degrade the strength of the plastic. The walls must also be thick enough to survive the molding of the plastic but thin enough to burst as the lengthening crack reaches them.

Self-healing materials should have many applications. The U.S. Air Force, which partially funded the research at UIUC, is interested in using the materials in tanks that hold gases and liquids under pressure. The current materials used for these tanks are subject to microcracks that eventually grow, causing the tanks to leak. Self-healing materials would also be valuable in situations where repair is impossible or impractical, such as electronic circuit boards, components of deep space probes, and implanted medical devices.

Another related polymer, **polystyrene,** is constructed from the monomer styrene,

$$\mathrm{H_2C{=}CH{-}C_6H_5}$$

Pure polystyrene is too brittle for many uses, so most polystyrene-based polymers are actually *copolymers* of styrene and butadiene,

$$\mathrm{H_2C{=}CH{-}CH{=}CH_2}$$

thus incorporating bits of butadiene rubber into the polystyrene matrix. The resulting polymer is very tough and is often used as a substitute for wood in furniture.

Another polystyrene-based product is acrylonitrile-butadiene-styrene (ABS), a tough, hard, and chemically resistant plastic used for pipes and for items such as radio housings, telephone cases, and golf club heads, for which shock resistance is an essential property. Originally, ABS was produced by copolymerization of the three monomers:

$$\underset{\text{Acrylonitrile}}{\mathrm{H_2C{=}CH{-}CN}} + \underset{\text{Styrene}}{\mathrm{H_2C{=}CH{-}C_6H_5}} + \underset{\text{Butadiene}}{\mathrm{H_2C{=}CH{-}CH{=}CH_2}} \longrightarrow \left(\mathrm{CH_2{-}\underset{CH}{CH}{-}CH_2{-}\underset{C_6H_5}{CH}{-}CH_2{-}\underset{CH_2=CH}{CH}}\right)_n$$

It is now prepared by a special process called *grafting,* in which butadiene is polymerized first, and then the cyanide and phenyl substituents are added chemically.

Another high-volume polymer, **polyvinyl chloride (PVC),** is constructed from the monomer vinyl chloride,

$$\mathrm{H_2C{=}CHCl}$$

21.6 Natural Polymers

Proteins

We have seen that many useful synthetic materials are polymers. Thus it should not be surprising that a great many natural materials are also polymers: starch, hair, silicate chains in soil and rocks, silk and cotton fibers, and the cellulose in woody plants, to name only a few.

In this section we consider a class of natural polymers, the **proteins,** which make up about 15% of our bodies and have molecular weights (molar masses) that range from about 6000 to over 1,000,000 grams per mole. Proteins perform many functions in the human body. **Fibrous proteins** provide structural integrity and strength for many types of tissue and are the main components of muscle, hair, and cartilage. Other proteins, usually called **globular proteins** because of their roughly spherical shape, are the "worker" molecules of the body. These proteins transport and store oxygen and nutrients, act as catalysts for the thousands of reactions that make life possible, fight invasion by foreign objects, participate in the body's many regulatory systems, and transport electrons in the complex process of metabolizing nutrients.

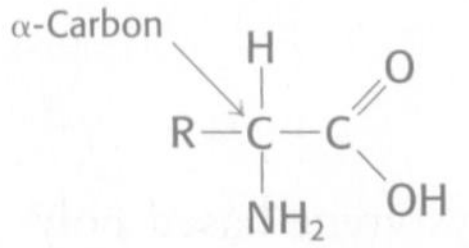

α-Carbon

$$R-\overset{H}{\underset{NH_2}{C}}-C(=O)OH$$

The building blocks of all proteins are the **α-amino acids,** where R may represent H, CH_3, or a more complex substituent. These molecules are called α-amino acids because the amino group ($-NH_2$) is always attached to the α-carbon, the one next to the carboxyl group ($-CO_2H$). The 20 amino acids most commonly found in proteins are shown in Fig. 21.18.

At the pH in biological fluids, the amino acids shown in Fig. 21.18 exist in a different form, with the proton of the —COOH group transferred to the $-NH_2$ group. For example, glycine would be in the form $H_3{}^+NCH_2COO^-$.

Note from Fig. 21.18 that the amino acids are grouped into polar and nonpolar classes, determined by the R groups, or **side chains.** Nonpolar side chains contain mostly carbon and hydrogen atoms, whereas polar side chains contain large numbers of nitrogen and oxygen atoms. This difference is important because polar side chains are *hydrophilic* (water-loving), but nonpolar side chains are *hydrophobic* (water-fearing), and this characteristic greatly affects the three-dimensional structure of the resulting protein.

The protein polymer is built by condensation reactions between amino acids. For example,

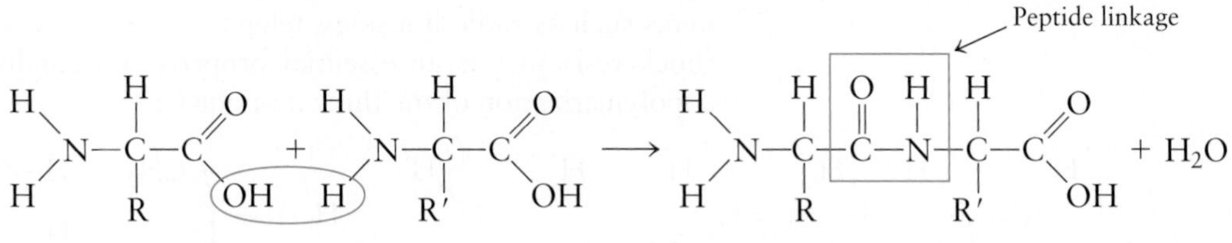

$$H_2N-\overset{H}{\underset{R}{C}}-C(=O)OH + H-\overset{H}{N}-\overset{H}{\underset{R'}{C}}-C(=O)OH \longrightarrow H_2N-\overset{H}{\underset{R}{C}}-\overset{O}{\overset{\|}{C}}-\overset{H}{N}-\overset{H}{\underset{R'}{C}}-C(=O)OH + H_2O$$

Peptide linkage

The product shown above is called a **dipeptide.** This name is used because the structure

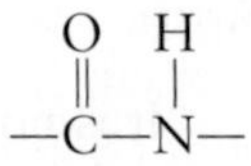

$$-\overset{O}{\overset{\|}{C}}-\overset{H}{\overset{|}{N}}-$$

is called a **peptide linkage** by biochemists. (The same grouping is called an *amide* by organic chemists.) Additional condensation reactions lengthen the chain to produce a **polypeptide,** eventually yielding a protein.

The peptide linkage is also found in nylon (see Section 21.5).

You can imagine that with 20 amino acids, which can be assembled in any order, there is essentially an infinite variety possible in the construction of proteins. This flexibility allows an organism to tailor proteins for the many types of functions that must be carried out.

The order, or sequence, of amino acids in the protein chain is called the **primary structure,** conveniently indicated by using three-letter codes for the amino acids (Fig. 21.18), where it is understood that the terminal carboxyl group is on the right and the terminal amino group is on the left. For example, one possible sequence for a tripeptide containing the amino acids lysine, alanine, and leucine is

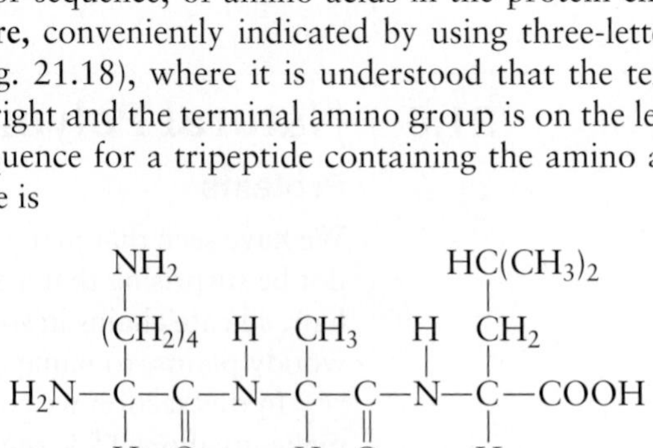

$$H_2N-\underset{H}{\overset{(CH_2)_4NH_2}{C}}-\overset{O}{\overset{\|}{C}}-\overset{H}{N}-\underset{H}{\overset{CH_3}{C}}-\overset{O}{\overset{\|}{C}}-\overset{H}{N}-\underset{H}{\overset{CH_2HC(CH_3)_2}{C}}-COOH$$

Lysine Alanine Leucine

which is represented in the shorthand notation by

lys-ala-leu

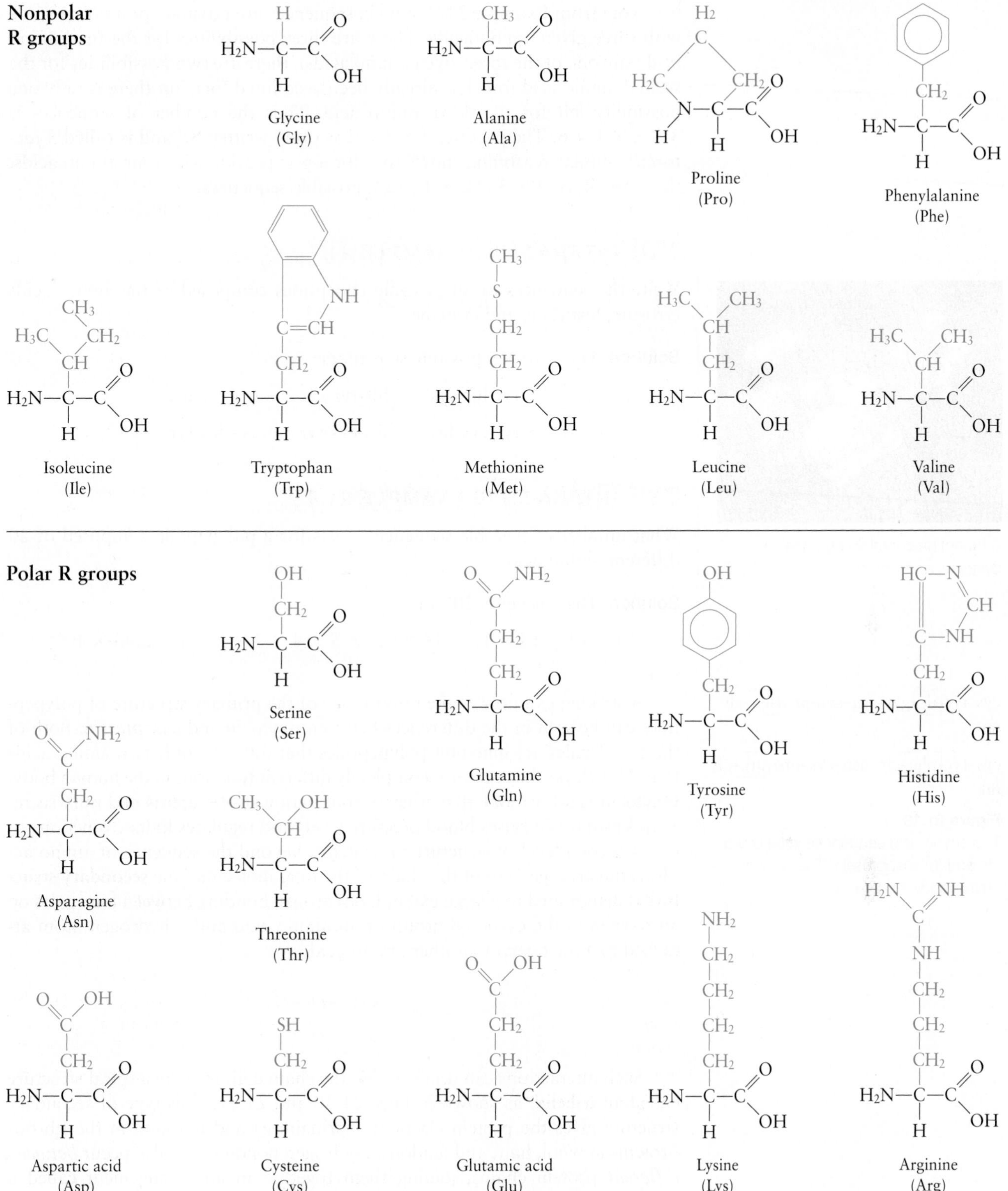

Figure 21.18
The 20 α-amino acids found in most proteins. The R group is shown in color.

Note from Example 21.7 that six sequences are possible for a polypeptide with three given amino acids. There are three possibilities for the first amino acid (any one of the three given amino acids), there are two possibilities for the second amino acid (one has already been accounted for), but there is only one possibility left for the third amino acid. Thus the number of sequences is $3 \times 2 \times 1 = 6$. The product $3 \times 2 \times 1$ is often written 3! (and is called *3 factorial*). Similar reasoning shows that for a polypeptide with four amino acids, there are 4!, or $4 \times 3 \times 2 \times 1 = 24$, possible sequences.

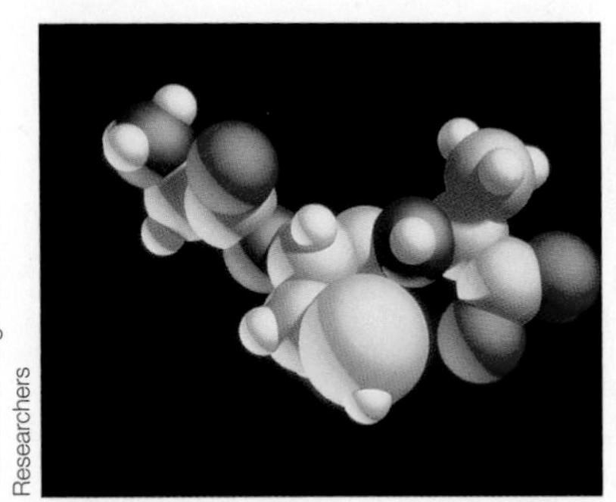

Ken Eward/Biografx/Science Source/Photo Researchers

A tripeptide containing glycine, cysteine, and alanine.

OWL INTERACTIVE EXAMPLE 21.7

Write the sequences of all possible tripeptides composed of the amino acids tyrosine, histidine, and cysteine.

Solution There are six possible sequences:

tyr-his-cys	his-tyr-cys	cys-tyr-his
tyr-cys-his	his-cys-tyr	cys-his-tyr

OWL INTERACTIVE EXAMPLE 21.8

What number of possible sequences exists for a polypeptide composed of 20 different amino acids?

Solution The answer is 20!, or

$$20 \times 19 \times 18 \times 17 \times 16 \times \ldots \times 5 \times 4 \times 3 \times 2 \times 1 = 2.43 \times 10^{18}$$

cys–tyr–ile–gln–asn–cys–pro–leu–gly
(a)

cys–tyr–phe–gln–asn–cys–pro–arg–gly
(b)

Figure 21.19
The amino acid sequences in (a) oxytocin and (b) vasopressin. The differing amino acids are boxed.

A striking example of the importance of the primary structure of polypeptides can be seen in the differences between *oxytocin* and *vasopressin*. Both of these molecules are nine-unit polypeptides that differ by only two amino acids (Fig. 21.19), yet they perform completely different functions in the human body. Oxytocin is a hormone that triggers contraction of the uterus and milk secretion. Vasopressin raises blood pressure levels and regulates kidney function.

A second level of structure in proteins, beyond the sequence of amino acids, is the arrangement of the chain of the long molecule. The **secondary structure** is determined to a large extent by hydrogen bonding between lone pairs on an oxygen in the carbonyl group of an amino acid and a hydrogen atom attached to a nitrogen of another amino acid:

$$\mathrm{C{=}\ddot{O}{:}\cdots H{-}N} \qquad \delta- \quad \delta+$$

Such interactions can occur *within* the chain coils to form a spiral structure called an **α-helix,** as shown in Figs. 21.20 and 21.21. This type of secondary structure gives the protein elasticity (springiness) and is found in the fibrous proteins in wool, hair, and tendons. Hydrogen bonding can also occur *between different* protein chains, joining them together in an arrangement called a **pleated sheet,** as shown in Fig. 21.22. Silk contains this arrangement of proteins, making its fibers flexible yet very strong and resistant to stretching. The pleated sheet is also found in muscle fibers. The hydrogen bonds in the α-helical protein are called *intrachain* (within a given protein chain), and those in the pleated sheet are said to be *interchain* (between protein chains).

As you might imagine, a molecule as large as a protein has a great deal of flexibility and can assume a variety of overall shapes. The specific shape that a protein assumes depends on its function. For long, thin structures, such as hair,

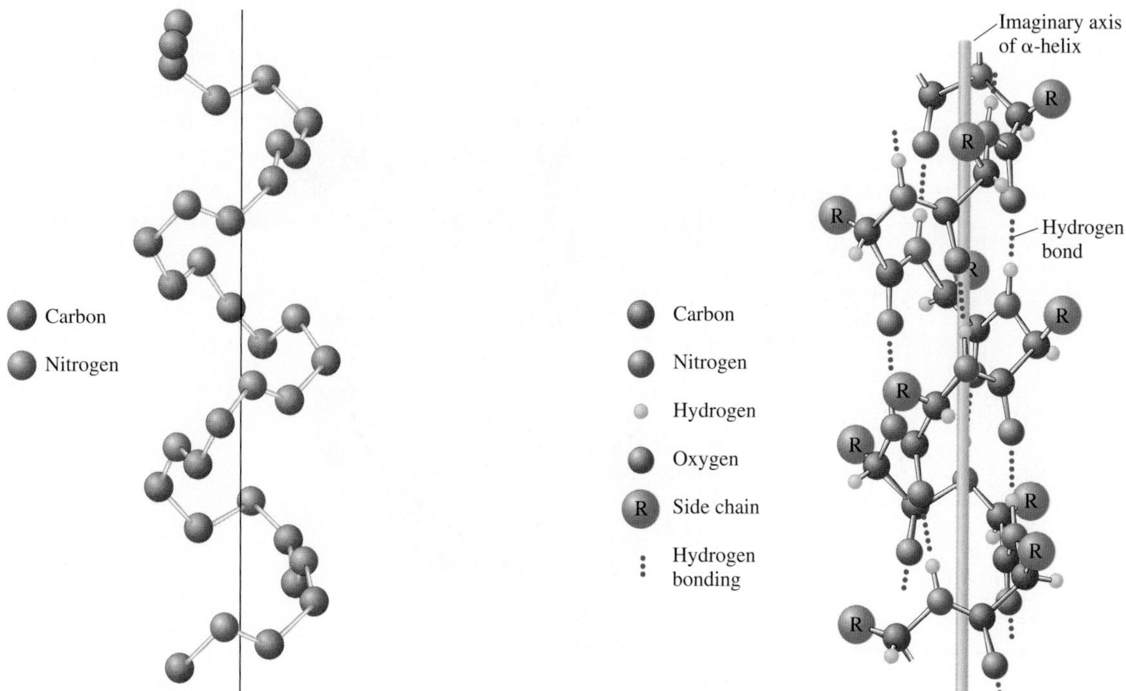

Figure 21.20
Hydrogen bonding within a protein chain causes it to form a stable helical structure called an α-helix. Only the main atoms in the helical backbone are shown here. The hydrogen bonds are not shown.

Figure 21.21
Ball-and-stick model of a portion of a protein chain in the α-helical arrangement, showing the hydrogen-bonding interactions.

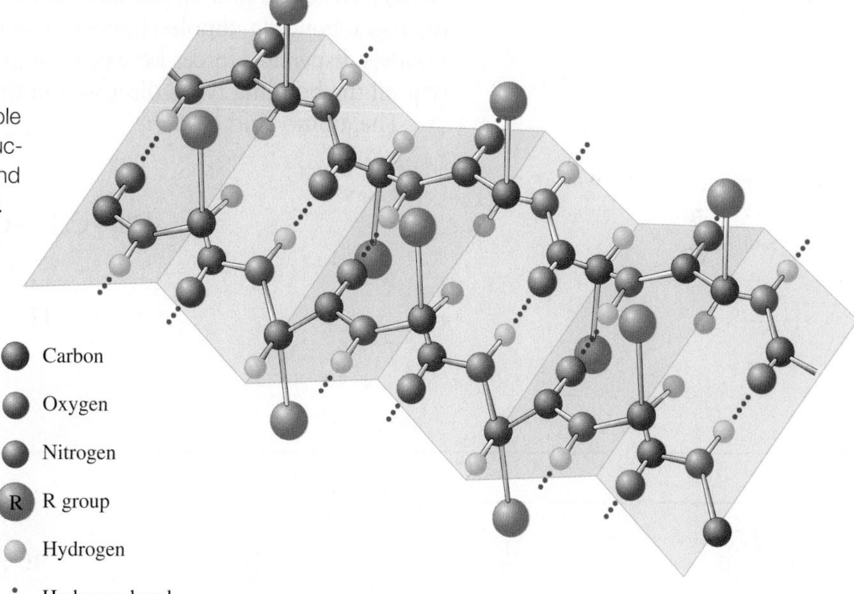

Figure 21.22
When hydrogen bonding occurs between protein chains rather than within them, a stable structure (the pleated sheet) results. This structure contains many protein chains and is found in natural fibers, such as silk, and in muscles.

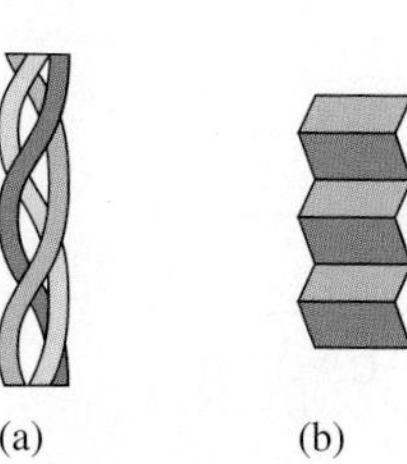

Figure 21.23
(a) Collagen, a protein found in tendons, consists of three protein chains (each with a helical structure) twisted together to form a superhelix. The result is a long, relatively narrow protein. (b) The pleated-sheet arrangement of many proteins bound together to form the elongated protein found in silk fibers.

wool and silk fibers, and tendons, an elongated shape is required. This may involve an α-helical secondary structure, as found in the protein α-keratin in hair and wool or in the collagen found in tendons [Fig. 21.23(a)], or it may involve a pleated-sheet secondary structure, as found in silk [Fig. 21.23(b)]. Many of the proteins in the body having nonstructural functions are globular, such as myoglobin (Fig. 21.24). Note that the secondary structure of myoglo-

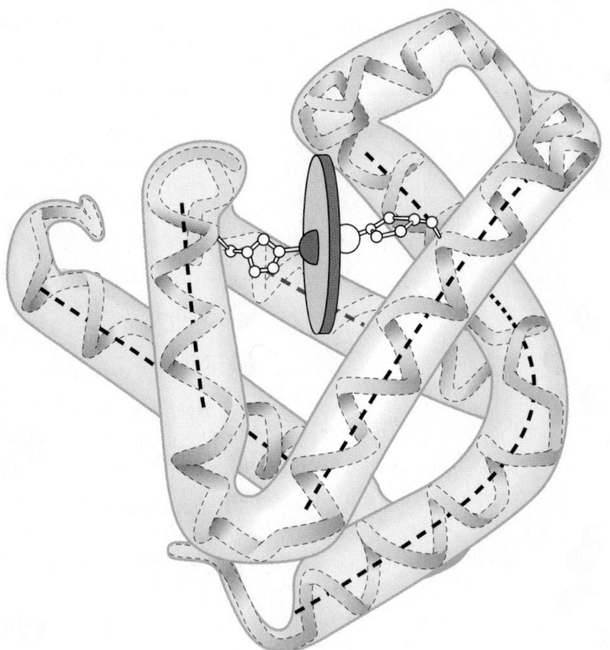

Figure 21.24
The protein myoglobin.

bin is basically α-helical. However, in the areas where the chain bends to give the protein its compact globular structure, the α-helix breaks down to give a secondary configuration known as the **random-coil arrangement.**

The overall shape of the protein, long and narrow or globular, is called its **tertiary structure** and is maintained by several different types of interactions: hydrogen bonding, dipole–dipole interactions, ionic bonds, covalent bonds, and London dispersion forces between nonpolar groups. These bonds, which represent all the bonding types discussed in this text, are summarized in Fig. 21.25.

The amino acid *cysteine*

$$\begin{array}{c} \quad\quad\quad\quad\quad\quad \text{H} \;\; \text{O} \\ \quad\quad\quad\quad\quad\quad | \;\;\; \| \\ \text{HS—CH}_2\text{—C—C—OH} \\ \quad\quad\quad\quad | \\ \quad\quad\quad\quad \text{N} \\ \quad\quad\quad\quad \text{H} \;\; \text{H} \end{array}$$

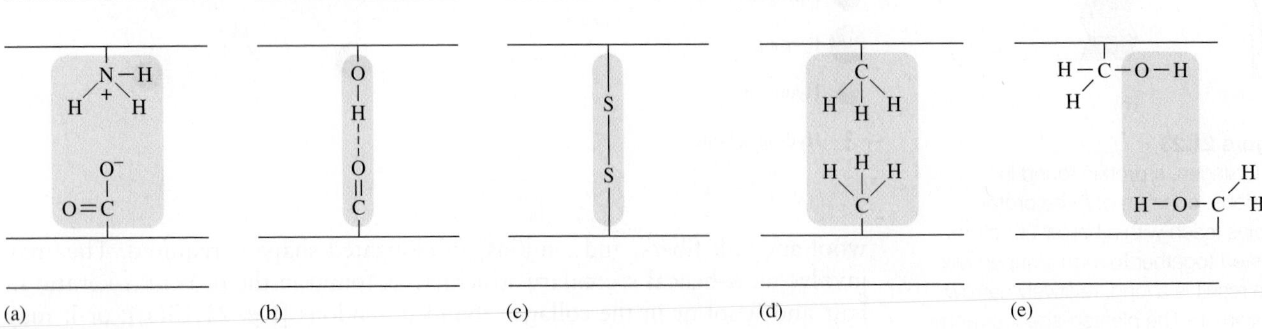

Figure 21.25
Summary of the various types of interactions that stabilize the tertiary structure of a protein: (a) ionic, (b) hydrogen bonding, (c) covalent, (d) London dispersion, and (e) dipole–dipole.

plays a special role in stabilizing the tertiary structure of many proteins because the —SH groups on two cysteines can react in the presence of an oxidizing agent to form a S—S bond called a **disulfide linkage:**

$$\text{C}-\text{CH}_2-\text{S}-\text{H} + \text{H}-\text{S}-\text{CH}_2-\text{C} \longrightarrow \text{C}-\text{CH}_2-\boxed{\text{S}-\text{S}}-\text{CH}_2-\text{C}$$

A practical application of the chemistry of disulfide bonds is permanent waving of hair, as summarized in Fig. 21.26. The S—S linkages in the protein of hair are broken by treatment with a reducing agent. The hair is then set in curlers to change the tertiary protein structure to the desired shape. Then treatment with an oxidizing agent causes new S—S bonds to form, which allow the hair protein to retain the new structure.

The three-dimensional structure of a protein is crucial to its function. The process of breaking down this structure is called **denaturation** (Fig. 21.27). For example, the denaturation of egg proteins occurs when an egg is cooked. Any source of energy can cause denaturation of proteins and is thus potentially dangerous to living organisms. For example, ultraviolet and X-ray radiation or nuclear radioactivity can disrupt protein structure, which may lead to cancer or genetic damage. Protein damage is also caused by chemicals like benzene, trichloroethane, and 1,2-dibromoethane (called EDB). The metals lead and mercury, which have a very high affinity for sulfur, cause protein denaturation by disrupting disulfide bonds between protein chains.

The tremendous flexibility in the various levels of protein structure allows the tailoring of proteins for a wide range of specific functions. Proteins are the "workhorse" molecules of living organisms.

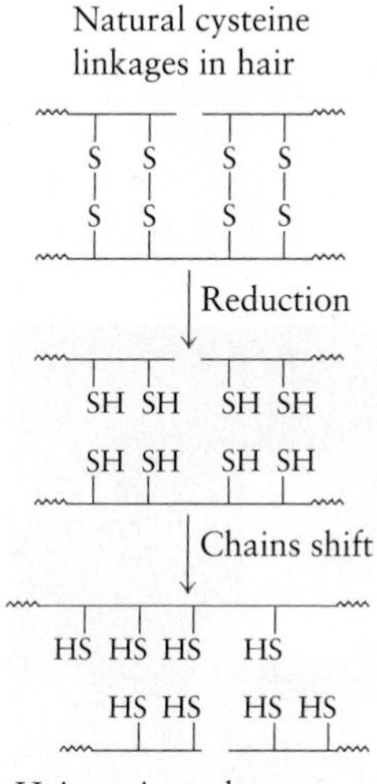

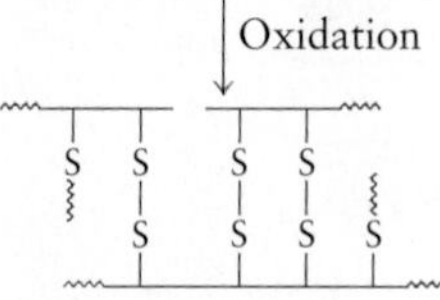

New cysteine linkages in waved hair

Figure 21.26
The permanent waving of hair.

General Name of Sugar	Number of Carbon Atoms
Triose	3
Tetrose	4
Pentose	5
Hexose	6
Heptose	7
Octose	8
Nonose	9

Carbohydrates

Carbohydrates form another class of biologically important molecules. They serve as a food source for most organisms and as a structural material for plants. Because many carbohydrates have the empirical formula CH_2O, it was originally believed that these substances were hydrates of carbon, thus accounting for the name.

Most important carbohydrates, such as starch and cellulose, are polymers composed of monomers called **monosaccharides,** or **simple sugars.** The monosaccharides are polyhydroxy ketones and aldehydes. The most important con-

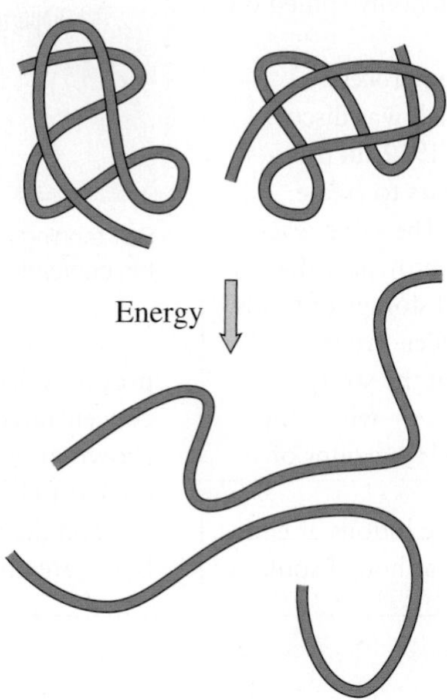

Figure 21.27
A schematic representation of the thermal denaturation of a protein.

Chemical Insights — Tanning in the Shade

Among today's best-selling cosmetics are self-tanning lotions. Many light-skinned people want to look like they have just spent a vacation in the Caribbean, but they recognize the dangers of too much sun—it causes premature aging and may lead to skin cancer. Chemistry has come to the rescue in the form of lotions that produce an authentic-looking tan. All of these lotions have the same active ingredient: dihydroxyacetone (DHA). DHA, which has the structure

$$\begin{array}{ccccccc} & & \mathrm{H} & & \mathrm{O} & & \mathrm{H} \\ & & & & \| & & \\ \mathrm{H}-\mathrm{O}- & & \mathrm{C} & - & \mathrm{C} & - & \mathrm{C}-\mathrm{OH} \\ & & \mathrm{H} & & & & \mathrm{H} \end{array}$$

is a nontoxic, simple sugar that occurs as an intermediate in carbohydrate metabolism in higher-order plants and animals. The DHA used in self-tanners is prepared by bacterial fermentation of glycerine,

$$\begin{array}{ccccccc} & & & & \mathrm{H} & & \\ & & & & | & & \\ & & & & \mathrm{O} & & \\ & & \mathrm{H} & & | & & \mathrm{H} \\ \mathrm{H}-\mathrm{O}- & & \mathrm{C} & - & \mathrm{C} & - & \mathrm{C}-\mathrm{O}-\mathrm{H} \\ & & \mathrm{H} & & | & & \mathrm{H} \\ & & & & \mathrm{H} & & \end{array}$$

The tanning effects of DHA were discovered by accident in the 1950s at Children's Hospital at the University of Cincinnati, where DHA was being used to treat children with glycogen storage disease. When the DHA was accidentally spilled on the skin, it produced brown spots.

The mechanism of the browning process involves the Maillard reaction, which was discovered by Louis-Camille Maillard in 1912. In this process amino acids react with sugars to create brown or golden-brown products. The same reaction is responsible for much of the browning that occurs during the manufacture and storage of foods. It is also the reason that beer is golden brown.

The browning of skin occurs in the stratum corneum—the outermost, dead layer—where the DHA reacts with free amino (—NH_2) groups of the proteins found there.

DHA is present in most tanning lotions at concentrations between 2% and 5%, although some products designed to give a deeper tan are more concentrated. Because the lotions themselves turn brown above pH 7, the tanning lotions are buffered at pH 5.

Thanks to these new products, tanning is now both safe and easy.

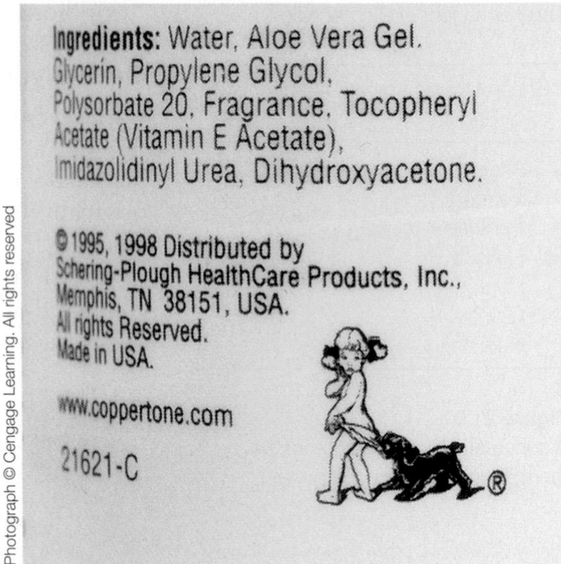

Self-tanning products and a close-up of a label showing the contents.

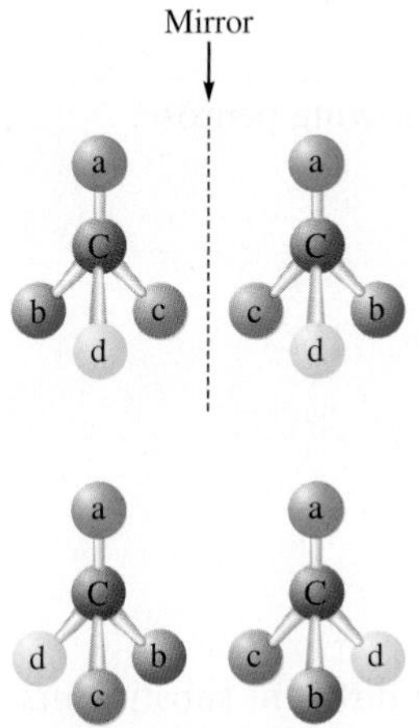

Figure 21.28
When a tetrahedral carbon atom has four different substituents, there is no way that its mirror image can be superimposed. The lower two forms show other possible orientations of the molecule. Compare these with the mirror image and note that they cannot be superimposed.

tain five carbon atoms (**pentoses**) or six carbon atoms (**hexoses**). One important hexose is *fructose*, a sugar found in honey and fruit. Its structure is

```
      CH2OH
      |
      C=O
      |
HO—*C—H
      |
 H—*C—OH
      |
 H—*C—OH
      |
      CH2OH
```

Fructose

where the asterisks indicate chiral carbon atoms. In Section 19.4 we saw that molecules with nonsuperimposable mirror images exhibit optical isomerism. A carbon atom with four *different* groups bonded to it in a tetrahedral arrangement *always* has a nonsuperimposable mirror image (Fig. 21.28), which gives rise to a pair of optical isomers. For example, the simplest sugar, glyceraldehyde,

```
  H   O
   \ //
    C
    |
H—*C—OH
    |
    CH2OH
```

which has one chiral carbon, has two optical isomers, as shown in Fig. 21.29.

In fructose each of the three chiral carbon atoms satisfies the requirement of being surrounded by four different groups. This leads to a total of 2^3, or 8, isomers that differ in their ability to rotate polarized light. The particular isomer whose structure is given above is called D-fructose. Generally, monosaccharides have one isomer that is more common in nature than the others. The most important pentoses and hexoses are shown in Table 21.8.

Figure 21.29
The mirror-image optical isomers of glyceraldehyde. Note that these mirror images cannot be superimposed.

Table 21.8
Some Important Monosaccharides

Pentoses

D-Ribose	D-Arabinose	D-Ribulose
CHO	CHO	CH_2OH
H—C—OH	HO—C—H	C=O
H—C—OH	H—C—OH	H—C—OH
H—C—OH	H—C—OH	H—C—OH
CH_2OH	CH_2OH	CH_2OH

Hexoses

D-Glucose	D-Mannose	D-Galactose	D-Fructose
CHO	CHO	CHO	CH_2OH
H—C—OH	HO—C—H	H—C—OH	C=O
HO—C—H	HO—C—H	HO—C—H	HO—C—H
H—C—OH	H—C—OH	HO—C—H	H—C—OH
H—C—OH	H—C—OH	H—C—OH	H—C—OH
CH_2OH	CH_2OH	CH_2OH	CH_2OH

OWL INTERACTIVE EXAMPLE 21.9

Determine the number of chiral carbon atoms in the following pentose:

H   O
 \\ //
  C
  |
H—C—OH
  |
H—C—OH
  |
H—C—OH
  |
  CH_2OH

Solution We must look for carbon atoms that have four different substituents. The top carbon has only three substituents and thus cannot be chiral. The three carbon atoms shown in blue each have four different groups attached to them:

H   O
 \\ //
  C
  |
H—C—OH
  |
H—C—OH
  |
H—C—OH
  |
  CH_2OH

H—C=O
  |
H—C—OH
  |
H—C—OH
  |
H—C—OH
  |
  CH_2OH

H—C=O
  |
H—C—OH
  |
H—C—OH
  |
H—C—OH
  |
  CH_2OH

Since the fifth carbon atom has only three types of substituents (it has two hydrogen atoms), it is not chiral.

Thus the three chiral carbon atoms in this pentose are those shown in blue:

H   O
 \\ //
  C
  |
H—C—OH
  |
H—C—OH
  |
H—C—OH
  |
  CH_2OH

Note that D-ribose and D-arabinose, shown in Table 21.8, are two of the eight isomers of this pentose.

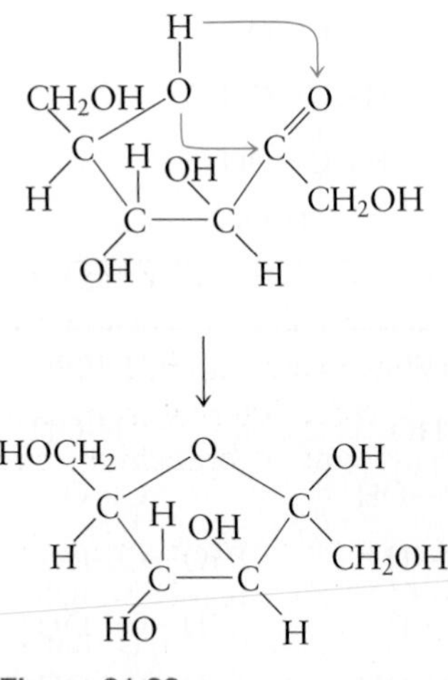

Figure 21.30
The cyclization of D-fructose.

Although we have so far represented the monosaccharides as straight-chain molecules, they usually cyclize, or form a ring structure, in aqueous solution. Figure 21.30 shows this reaction for fructose. Note that a new bond is formed between the oxygen of the terminal hydroxyl group and the carbon of the ketone group. In the cyclic form, fructose is a five-membered ring containing a C—O—C bond. The same type of reaction can occur between a hydroxyl group and an aldehyde group, as shown for D-glucose in Fig. 21.31. In this case a six-membered ring is formed.

More complex carbohydrates are formed by combining monosaccharides. For example, **sucrose,** common table sugar, is a **disaccharide** formed from glucose and fructose by elimination of water to form a C—O—C bond between the rings, which is called a **glycoside linkage** (Fig. 21.32). When sucrose is consumed in food, the preceding reaction is reversed. An enzyme in saliva catalyzes the breakdown of this disaccharide.

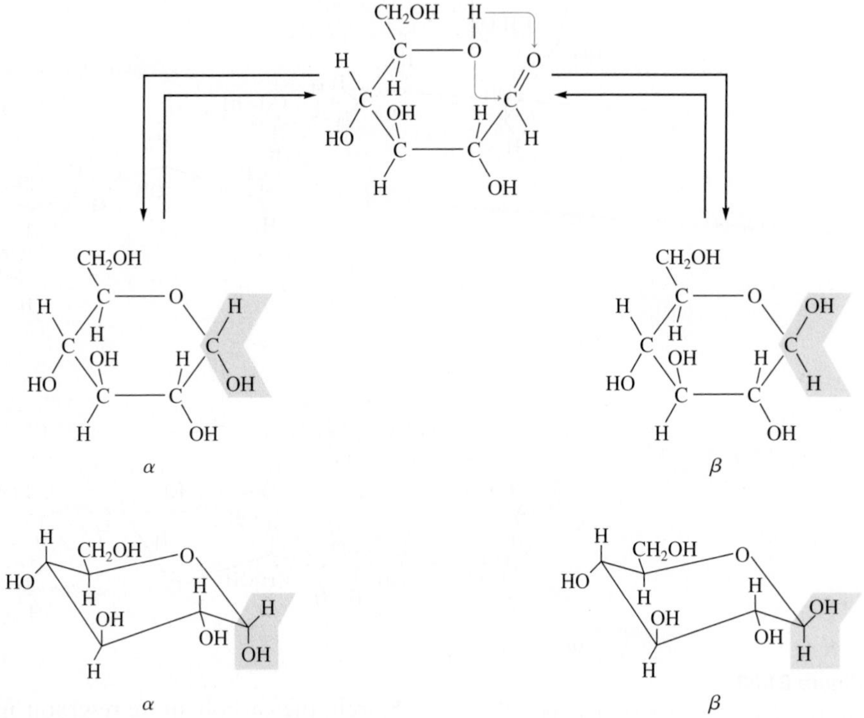

Figure 21.31
The cyclization of glucose. Two different rings are possible; they differ in the orientation of the hydroxyl group and hydrogen on one carbon, as indicated. The two forms are designated α and β and are shown here in two representations.

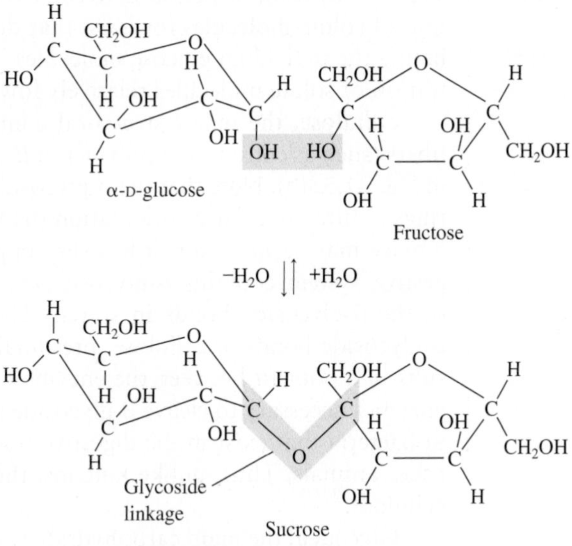

Figure 21.32
Sucrose is a disaccharide formed from α-D-glucose and fructose.

Large polymers consisting of many monosaccharide units, called polysaccharides, can form when each ring forms two glycoside linkages, as shown in Fig. 21.32. Three of the most important of these polymers are starch, cellulose, and glycogen. All these substances are polymers of glucose, differing from each other in the nature of the glycoside linkage, the amount of branching, and molecular weight (molar mass).

Starch, a polymer of α-D-glucose, consists of two parts: *amylose,* a straight-chain polymer of α-glucose [Fig. 21.33(a)], and *amylopectin,* a highly branched polymer of α-glucose with a molecular weight that is 10 to 20 times that of amylose. Branching occurs when a third glycoside linkage attaches a branch to the main polymer chain.

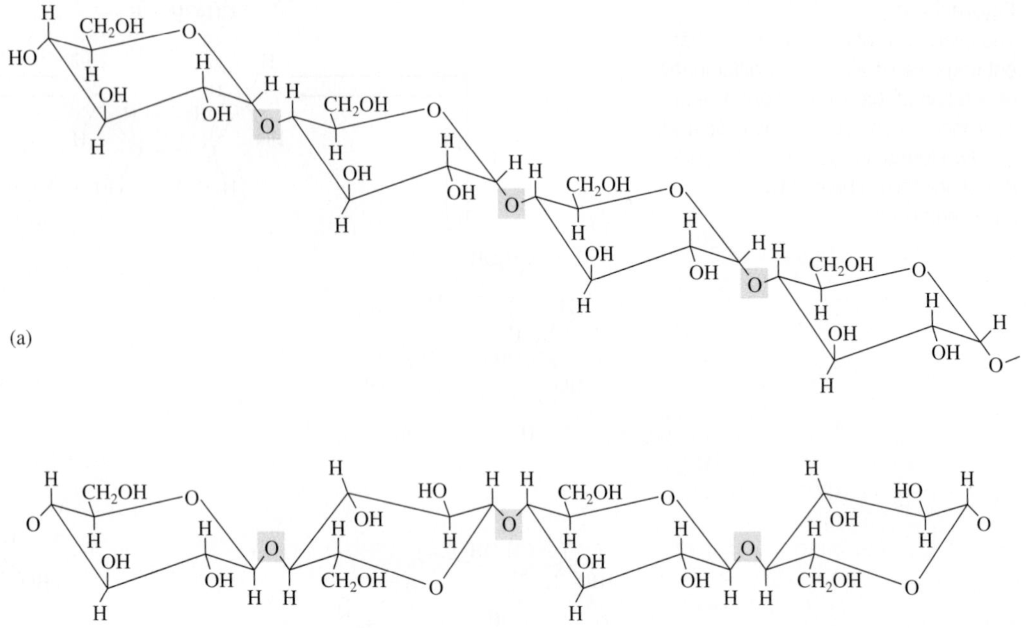

(a)

(b)

Figure 21.33
(a) The polymer amylose is a major component of starch and is made up of α-D-glucose monomers.
(b) The polymer cellulose, which consists of β-D-glucose monomers.

Starch, the carbohydrate reservoir in plants, is the form in which glucose is stored by the plant for later use as cellular fuel. Glucose is stored in this high-molecular-weight form because it results in less stress on the plant's internal structure by osmotic pressure. Recall from Section 17.6 that it is the concentration of solute molecules (or ions) that determines the osmotic pressure. Combining the individual glucose molecules into one large chain keeps the concentration of solute molecules relatively low, minimizing the osmotic pressure.

Cellulose, the major structural component of woody plants and natural fibers (such as cotton), is a polymer of β-D-glucose and has the structure shown in Fig. 21.33(b). Note that the β-glycoside linkages in cellulose give the glucose rings a different relative orientation than is found in starch. Although this difference may seem minor, it has very important consequences. The human digestive system contains α-glycosidases, enzymes that can catalyze breakage of the α-glycoside bonds in starch. These enzymes are not effective on the β-glycoside bonds of cellulose, presumably because the different structure results in a poor fit between the enzyme's active site and the carbohydrate. The enzymes necessary to cleave β-glycoside linkages, the β-glycosidases, are found in bacteria that exist in the digestive tracts of termites, cows, deer, and many other animals. Thus, unlike humans, these animals can derive nutrition from cellulose.

Glycogen, the main carbohydrate reservoir in animals, has a structure similar to that of amylopectin but with more branching. It is this branching that is thought to facilitate the rapid breakdown of glycogen into glucose when energy is required.

Nucleic Acids

Life is possible only because each cell, when it divides, can transmit the vital information about how it works to the next generation. It has been known for a long time that this process involves the chromosomes in the nucleus of the cell. Only since 1953, however, have scientists understood the molecular basis of this intriguing cellular "talent."

The substance that stores and transmits the genetic information is a polymer called **deoxyribonucleic acid (DNA),** a huge molecule with a molecular weight as high as several billion grams per mole. Together with other similar nucleic acids called the **ribonucleic acids (RNA),** DNA is also responsible for

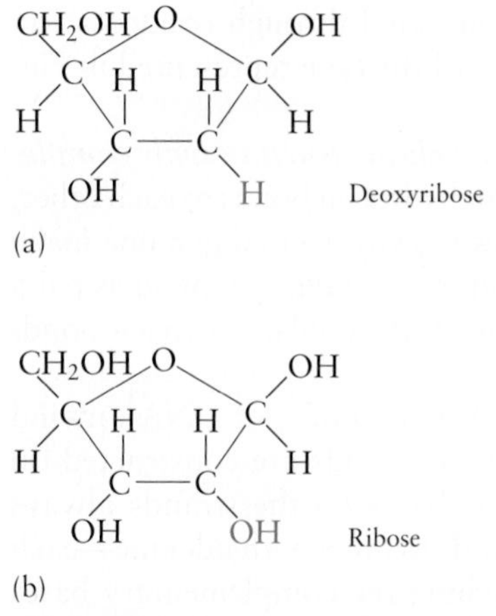

Figure 21.34
The structure of the pentoses (a) deoxyribose and (b) ribose. Deoxyribose is the sugar molecule present in DNA; ribose is found in RNA.

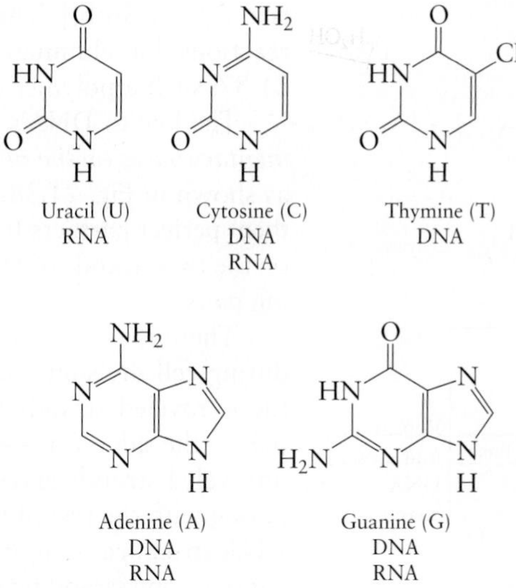

Figure 21.35
The organic bases found in DNA and RNA.

the synthesis of the various proteins needed by the cell to carry out its life functions. The RNA molecules, which are found in the cytoplasm outside the nucleus, are much smaller than DNA polymers, with molecular weights of only 20,000 to 40,000 grams per mole.

The monomers of the nucleic acids, called **nucleotides,** are composed of three distinct parts:

1. A *five-carbon sugar,* deoxyribose in DNA and ribose in RNA (Fig. 21.34)
2. A *nitrogen-containing organic base* of the type shown in Fig. 21.35
3. A *phosphoric acid molecule* (H_3PO_4)

The base and the sugar combine as shown in Fig. 21.36(a) to form a unit that in turn reacts with phosphoric acid to create the nucleotide, which is an ester

Figure 21.36
(a) Adenosine is formed by the reaction of adenine with ribose. (b) The reaction of phosphoric acid with adenosine to form the ester adenosine 5-phosphoric acid, a nucleotide. (At biological pH, the phosphoric acid would not be fully protonated as is shown here.)

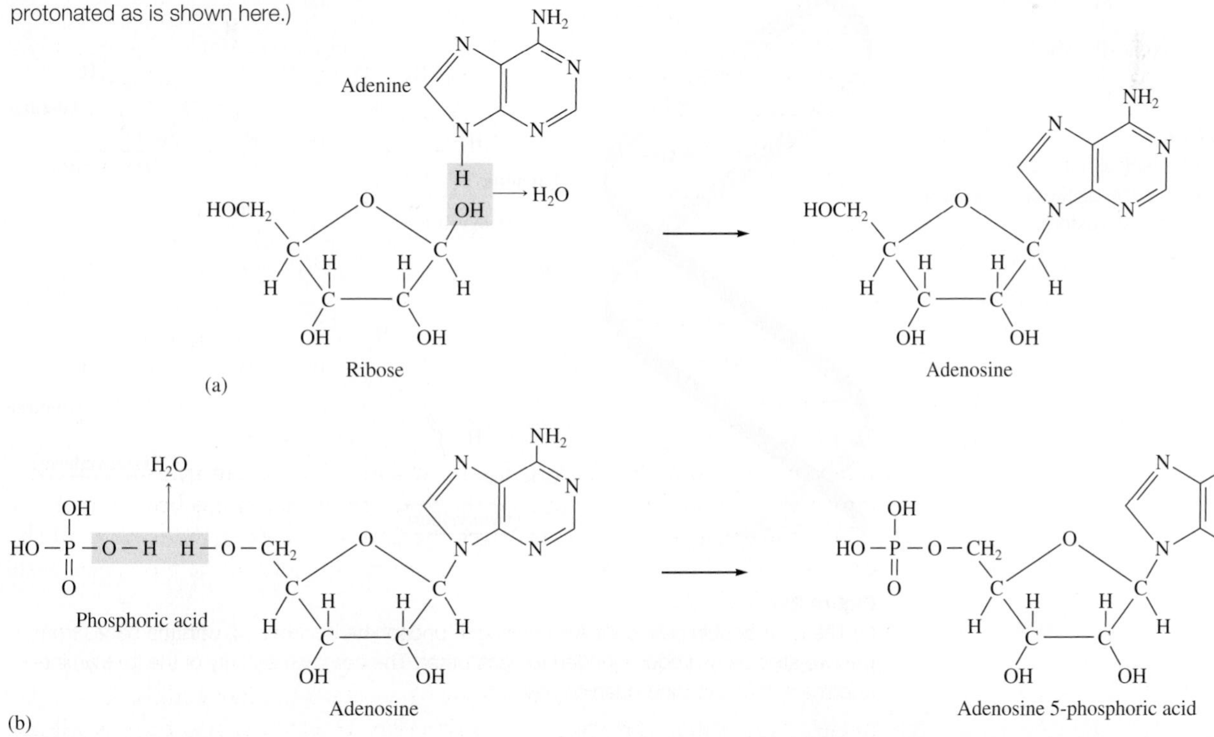

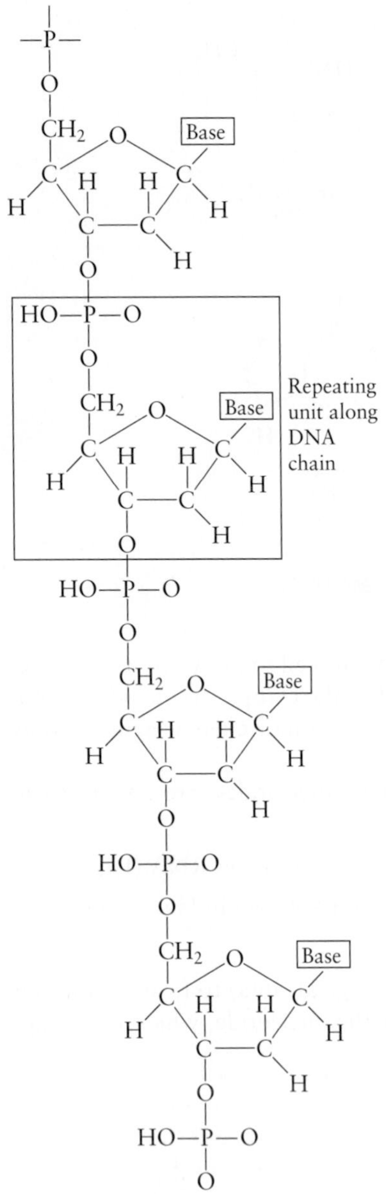

[see Fig. 21.36(b)]. The nucleotides become connected through condensation reactions that eliminate water to give a polymer of the type represented in Fig. 21.37; such a polymer can contain a *billion* units.

The key to DNA's functioning is its *double-helical structure with complementary bases on the two strands.* The bases form hydrogen bonds to each other, as shown in Fig. 21.38. Note that the structures of cytosine and guanine make them perfect partners for hydrogen bonding, and they are *always* found as pairs on the two strands of DNA. Thymine and adenine form similar hydrogen-bonding pairs.

There is much evidence to suggest that the two strands of DNA unwind during cell division and that new complementary strands are constructed on the unraveled strands (Fig. 21.39). Because the bases on the strands always pair in the same way—cytosine with guanine and thymine with adenine—each unraveled strand serves as a template for attaching the complementary bases (along with the rest of the nucleotide). This process results in two double-helix DNA structures that are identical to the original one. Each new double strand contains one strand from the original DNA double helix and one newly synthesized strand. This replication of DNA allows for the transmission of genetic information as the cells divide.

The other major function of DNA is **protein synthesis.** A given segment of the DNA, called a **gene,** contains the code for a specific protein. These codes transmit the primary structure of the protein (the sequence of amino acids) to the construction "machinery" of the cell. There is a specific code for each amino acid in the protein, which ensures that the correct amino acid will be inserted as the protein chain grows. A code consists of a set of three bases called a **codon.**

DNA stores the genetic information, whereas RNA molecules are responsible for transmitting this information to the ribosomes, where protein synthesis actually occurs. This complex process involves, first, the construction of a

Repeating unit along DNA chain

Figure 21.37
A portion of a typical nucleic acid chain. Note that the backbone consists of sugar–phosphate esters.

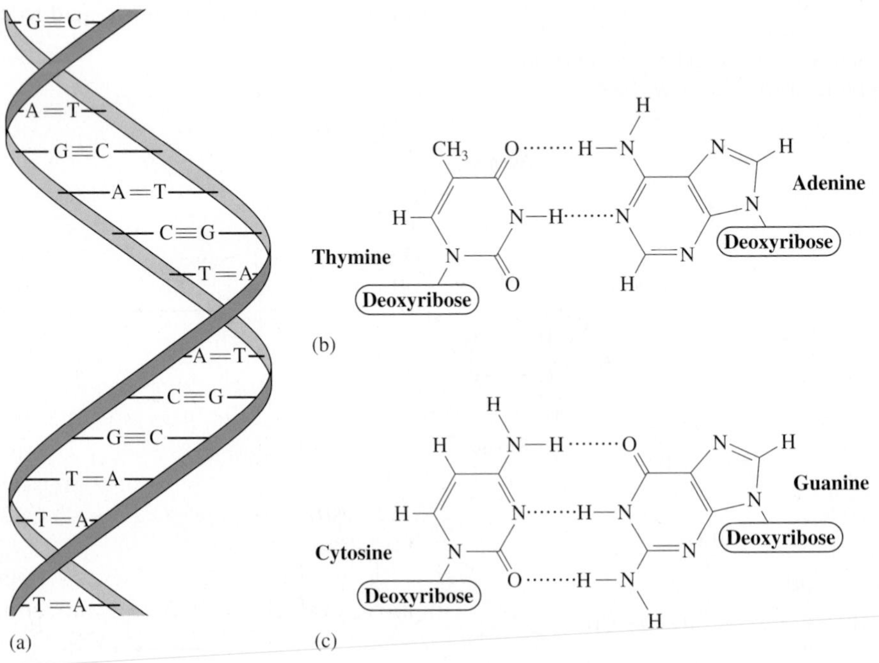

Figure 21.38
(a) The DNA double helix contains two sugar–phosphate backbones, with the bases from the two strands hydrogen-bonded to each other. The complementarity of the (b) thymine-adenine and (c) cytosine-guanine pairs.

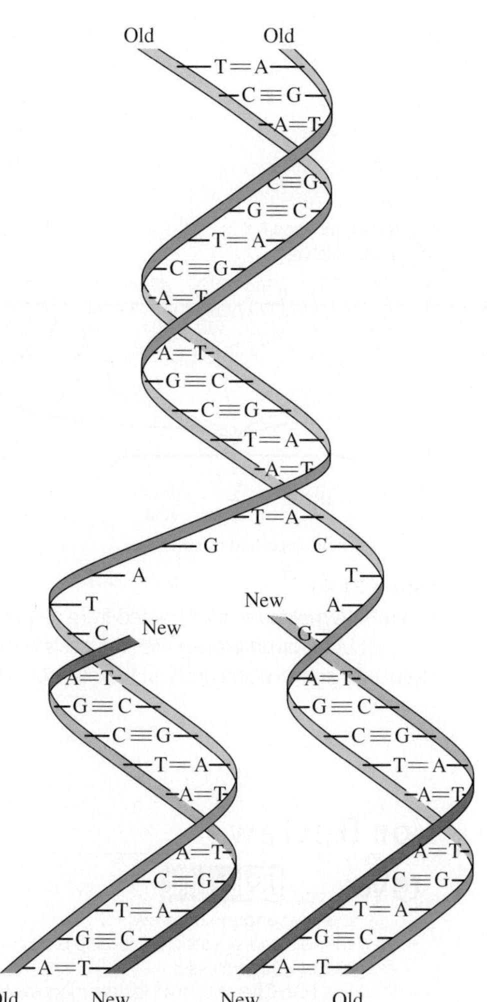

Figure 21.39
During cell division, the original DNA double helix unwinds, and new complementary strands are constructed on each original strand.

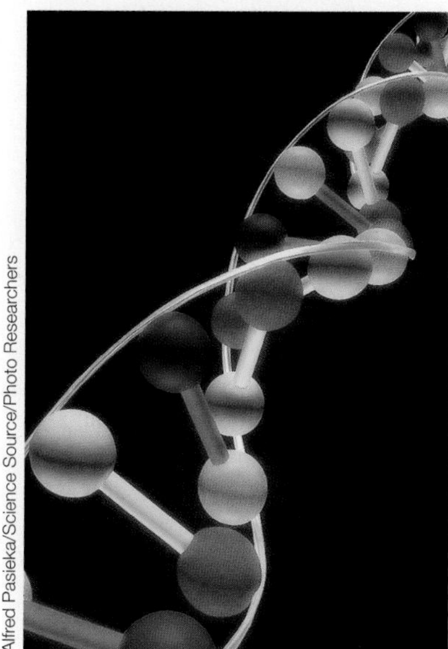
Alfred Pasieka/Science Source/Photo Researchers

A computer image of the base pairs of DNA. The blue lines represent the sugar–phosphate backbone, and the colored bars represent the hydrogen bonding between the base pairs.

special RNA molecule called **messenger RNA (mRNA).** The mRNA is built in the cell nucleus on the appropriate section of DNA (the gene); the double helix is "unzipped," and the complementarity of the bases is used in a process similar to that used in DNA replication. The mRNA then migrates into the cytoplasm of the cell, where, with the assistance of the ribosomes, the protein is synthesized.

Small RNA fragments, called **transfer RNA (tRNA),** are tailored to find specific amino acids and then to attach them to the growing protein chain as dictated by the codons in the mRNA. Transfer RNA has a lower molecular weight than messenger RNA. It consists of a chain of 75 to 80 nucleotides, including the bases adenine, cytosine, guanine, and uracil, among others. The chain folds back onto itself in various places as the complementary bases along the chain form hydrogen bonds. The tRNA decodes the genetic message from the mRNA, using a complementary triplet of bases called an **anticodon.** The nature of the anticodon governs which amino acid will be brought to the protein under construction.

The protein is built in several steps. First, a tRNA molecule brings an amino acid to the mRNA [the anticodon of the tRNA must complement the codon of the mRNA (Fig. 21.40)]. Once this amino acid is in place, another tRNA moves to the second codon site of the mRNA with its specific amino acid. The two amino acids link via a peptide bond, and the tRNA on the first codon breaks away. The process is repeated down the chain, always matching the tRNA anticodon with the mRNA codon.

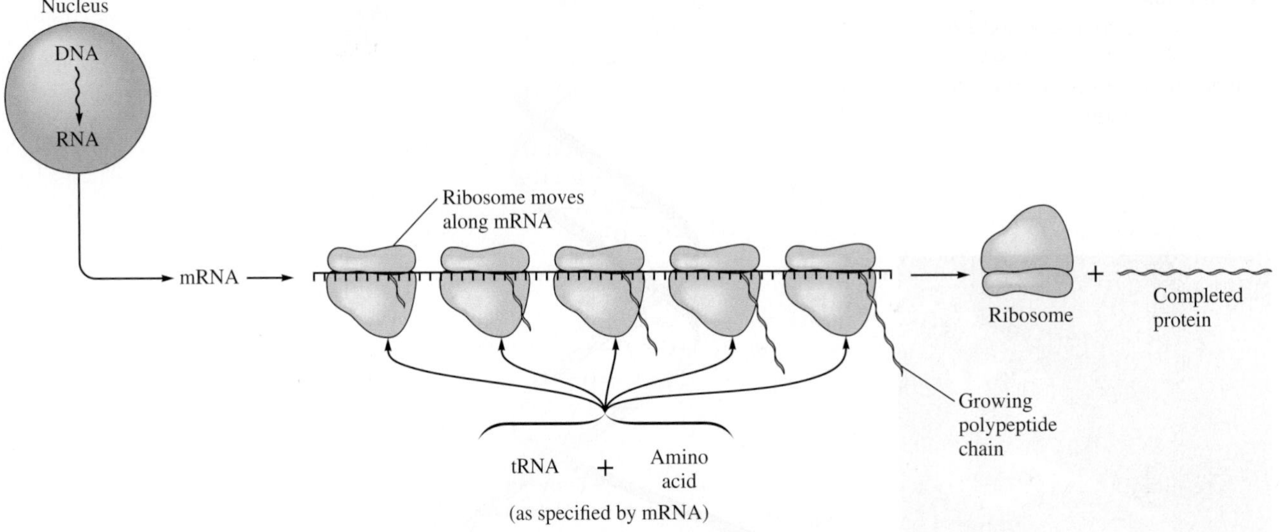

Figure 21.40
The mRNA molecule, constructed from a specific gene on the DNA, is used as the pattern to construct a given protein with the assistance of ribosomes. The tRNA molecules attach to specific amino acids and put them in place as called for by the codons on the mRNA.

Key Terms

biomolecule
organic chemistry

Section 21.1
hydrocarbons
saturated
unsaturated
alkanes
normal (straight-chain or unbranched) hydrocarbons
structural isomerism
combustion reaction
substitution reaction
dehydrogenation reaction
cyclic alkanes

Section 22.2
alkenes
cis–trans isomerism
alkynes
addition reaction
hydrogenation reaction
halogenation
polymerization

Section 22.3
aromatic hydrocarbons
phenyl group

Section 22.4
hydrocarbon derivatives
functional group
alcohols
phenol
carbonyl group

For Review

OWL and go Chemistry

Sign in at **www.cengage.com/owl** to:

- View tutorials and simulations, develop problem-solving skills, and complete online homework assigned by your professor.
- Download Go Chemistry mini lecture modules for quick review and exam prep from OWL (or purchase them at **www.cengagebrain.com**)

Hydrocarbons

- Compounds composed of mostly carbon and hydrogen atoms that typically contain chains or rings of carbon atoms
- Alkanes
  - Contain compounds with only C—C single bonds
  - Can be represented by the formula C_nH_{2n+2}
  - Are said to be saturated because each carbon present is bonded to the maximum number of atoms (4)
  - The carbon atoms are described as being sp^3 hybridized
  - Their structural isomerism involves the formation of branched chains
  - React with O_2 to form CO_2 and H_2O (called a combustion reaction)
  - Undergo substitution reactions
- Alkenes
  - Contain one or more C═C double bonds
  - Simplest alkene is C_2H_4 (ethylene), which is described as containing sp^2 hybridized carbon atoms
  - Restricted rotation about the C═C bonds in alkenes can lead to *cis–trans* isomerism
  - Undergo addition reactions
- Alkynes
  - Contain one or more C≡C triple bonds
  - Simplest example is C_2H_2 (acetylene), described as containing *sp*-hybridized carbon atoms
  - Undergo addition reactions

ketones
aldehydes
carboxylic acids
carboxyl group
ester
amine

Section 22.5
polymer
thermoset polymer
thermoplastic polymer
crosslinking
vulcanization
addition polymerization
free radical
condensation polymerization
copolymer
homopolymer
dimer
polyester
isotactic chain
syndiotactic chain
atactic chain
polystyrene
polyvinyl chloride (PVC)

Section 22.6
proteins
fibrous proteins
globular proteins
α-amino acids
side chains
dipeptide
peptide linkage
polypeptide
primary structure
secondary structure
α-helix
pleated sheet
random-coil arrangement
tertiary structure
disulfide linkage
denaturation
carbohydrates
monosaccharides (simple sugars)
pentoses
hexoses
sucrose
disaccharide
glycoside linkage
starch
cellulose
glycogen
deoxyribonucleic acid (DNA)
ribonucleic acid (RNA)
nucleotides
protein synthesis
gene
codon
messenger RNA (mRNA)
transfer RNA (tRNA)
anticodon

- Aromatic hydrocarbons
  - Contain rings of carbon atoms with delocalized π electrons
  - Undergo substitution reactions rather than addition reactions

Hydrocarbon derivatives
- Contain one or more functional groups
- Alcohols: contain the —OH group
- Aldehydes: contain a $\underset{\text{H}}{\overset{\diagdown}{\diagup}}$C=O group
- Ketones: contain the $\overset{\diagdown}{\diagup}$C=O group
- Carboxylic acids: contain the $-\text{C}\overset{\text{O}}{\underset{\text{OH}}{\lessgtr}}$ group

Polymers
- Large molecules formed from many small molecules (called monomers)
  - Addition polymerization: monomers add together by a free radical mechanism
  - Condensation polymerization: monomers connect by splitting out a small molecule, such as water

Proteins
- A class of natural polymers with molar masses ranging from 600 to 1,000,000
- Fibrous proteins form the structural basis of muscle, hair, and cartilage
- Globular proteins perform many biologic functions, including transport and storage of oxygen, catalysis of biologic reactions, and regulation of biologic systems
- Building blocks of proteins (monomers) are α-amino acids, which connect by a condensation reaction to form a peptide linkage
- Protein structure
  - Primary: the order of amino acids in the chain
  - Secondary: the arrangement of the protein chain
    - α-helix
    - pleated sheet
  - Tertiary structure: the overall shape of the protein

Carbohydrates
- Contain carbon, hydrogen, and oxygen
- Serve as food sources for most organisms
- Monosaccharides are most commonly five-carbon and six-carbon polyhydroxy ketones and aldehydes
  - Monosaccharides combine to form more complex carbohydrates, such as sucrose, starch, and cellulose

Genetic processes
- When a cell divides, the genetic information is transmitted via deoxyribonucleic acid (DNA), which has a double helical structure
  - During cell division, the double helix unravels and a new polymer forms along each strand of the original DNA
  - The genetic code is carried by organic bases that hydrogen-bond to each other in specific pairs in the interior of the DNA double helix

Exercises

OWL Interactive versions of these problems may be assigned in OWL.

A blue exercise number indicates that the answer to that exercise appears at the back of this book and a solution appears in the *Solutions Guide.*

Hydrocarbons

1. What is a hydrocarbon? What is the difference between a saturated hydrocarbon and an unsaturated hydrocarbon? Distinguish between normal and branched hydrocarbons.
2. In the shorthand notation for cyclic alkanes, the hydrogens are usually omitted. How do you determine the number of hydrogens bonded to each carbon in a ring structure?
3. Why are cyclopropane and cyclobutane so reactive?
4. What are aromatic hydrocarbons? Give a detailed description of the bonding in benzene. The π electrons in benzene are delocalized, whereas the π electrons in simple alkenes and alkynes are localized. Explain the difference.
5. Draw all the structural isomers for C_8H_{18} that have the following root name (longest carbon chain). Name the structural isomers.
   a. heptane c. pentane
   b. hexane d. butane
6. The normal (unbranched) hydrocarbons are often referred to as the straight-chain hydrocarbons. To what does this name refer? Does this mean that all carbon atoms in a straight-chain hydrocarbon actually have a linear arrangement? Explain your answer.
7. A general rule for a group of hydrocarbon isomers is that as the amount of branching increases, the boiling point decreases. Explain why this would be true.
8. Name the five structural isomers of C_6H_{14}.
9. Draw the structural formula for each of the following.
   a. 3-isobutylhexane
   b. 2,2,4-trimethylpentane, also called *isooctane*. This substance is the reference (100 level) for octane ratings.
   c. 2-*tert*-butylpentane
   d. The names given in parts a and c are incorrect. Give the correct names for these hydrocarbons.
10. Draw the structure for 4-ethyl-2,3-diisopropylpentane. This name is incorrect. Give the correct systematic name.
11. Name each of the following.

a. 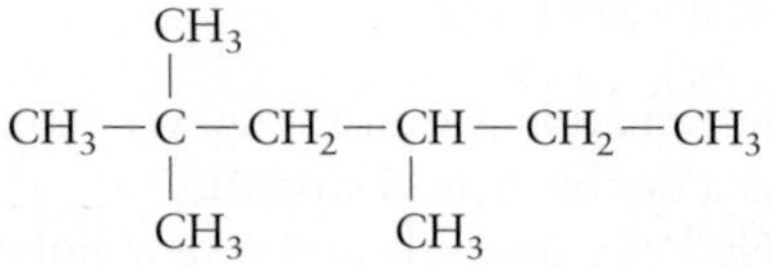
```
        CH3
        |
CH3—C—CH2—CH—CH2—CH3
        |         |
        CH3       CH3
```

b.
```
CH2—CH2—CH2—CH—CH2—CH2—CH2
|            |           |
CH3          CH3         CH3
```

c.
```
        CH3       CH3
        |         |
CH3—C—CH2—C—CH3
        |         |
        CH3       CH3
```

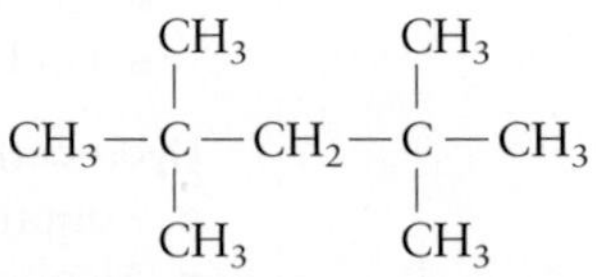

d. 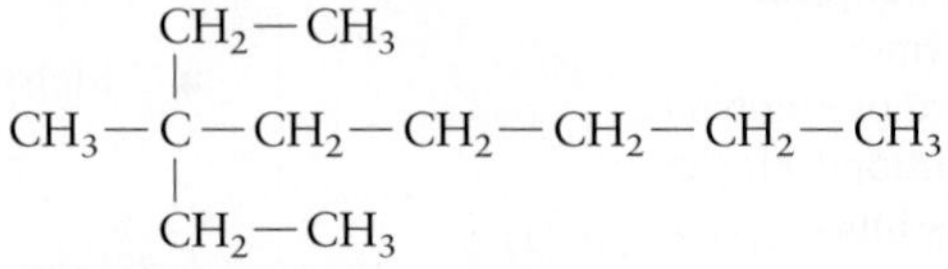
```
        CH2—CH3
        |
CH3—C—CH2—CH2—CH2—CH2—CH3
        |
        CH2—CH3
```

12. Name each of the following cyclic alkanes, and indicate the formula of the compound.

a. 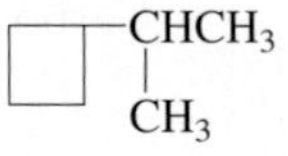 cyclobutane ring bearing $-CHCH_3$, with CH_3 below the CH

b. 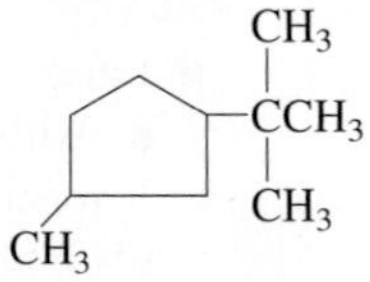 cyclopentane ring with CH_3 on one carbon and, on another carbon, $-CCH_3$ bearing CH_3 above and CH_3 below

c. 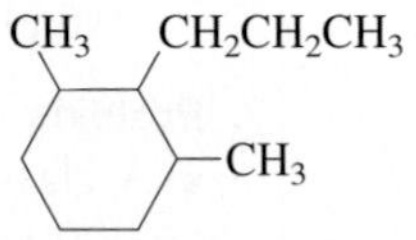 cyclohexane ring with adjacent substituents CH_3, $CH_2CH_2CH_3$, and $-CH_3$

13. Name each of the following alkenes or alkynes.

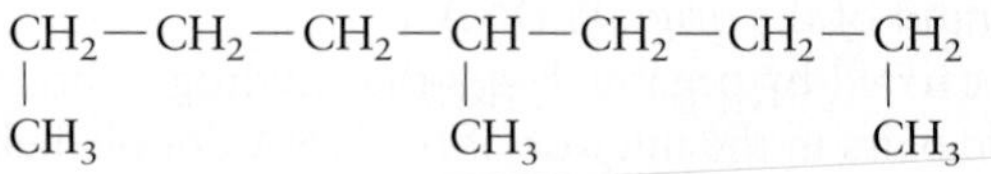

a. $CH_2{=}CH{-}CH_2{-}CH_3$

b. 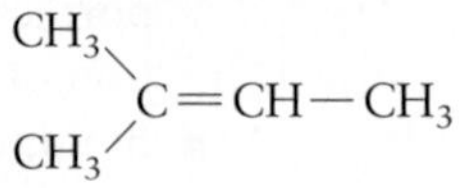
```
CH3
    \
     C=CH—CH3
    /
CH3
```

c.
```
          CH3                CH3
          |                 /
CH3CH2CH—CH=CH—CH
                            \
                             CH3
```

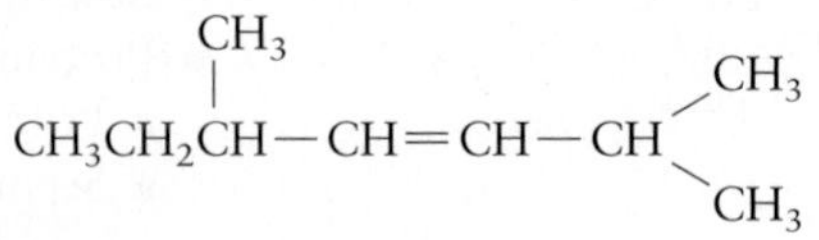

d. 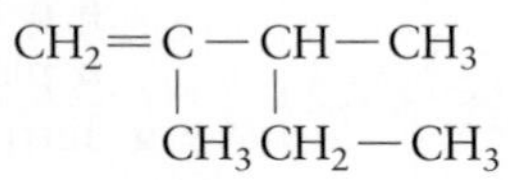
```
CH2=C—CH—CH3
       |   |
     CH3 CH2—CH3
```

e. 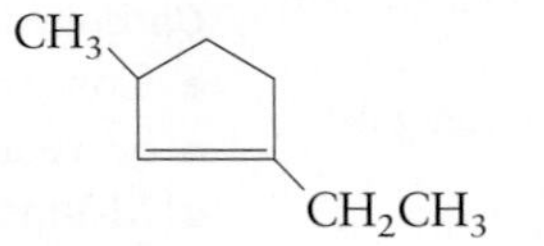 cyclopentene ring with CH_3 and CH_2CH_3 substituents

f. 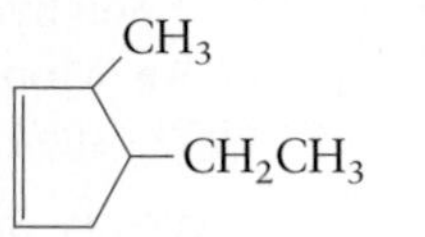 cyclopentene ring with CH_3 and $-CH_2CH_3$ substituents

g. 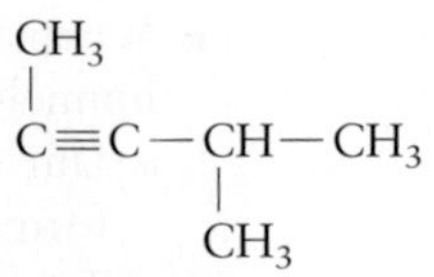
```
         CH3
         |
C≡C—CH—CH3
         |
         CH3
```

14. Give the structure for each of the following.
    a. 3-hexene
    b. 2,4-heptadiene
    c. 2-methyl-3-octene
    d. 4-methyl-1-pentyne

15. Name each of the following.

a. $Cl-CH_2-CH_2-CH(Cl)-CH_3$

b. $CH_3CH_2CH_2CCl_3$

c. $(CH_3)_2CCl-CH(Cl)-CH(CH_3)(CH_2CH_3)$

d. CH_2FCH_2F

e. $CH_3CH(I)CH=CH_2$

f. benzene ring with adjacent CH_3 and Br substituents

g. cyclohexane ring with adjacent CH_3 and Br substituents

h. cyclohexene ring with adjacent CH_3 and Br substituents

16. Give the structure for each of the following aromatic hydrocarbons.

a. *o*-ethyltoluene
b. *p*-di-*tert*-butylbenzene
c. *m*-diethylbenzene
d. 1-phenyl-2-butene

17. Cumene is the starting material for the industrial production of acetone and phenol. The structure of cumene is

$C_6H_5-CH(CH_3)-CH_3$

Give the systematic name for cumene.

Isomerism

18. Distinguish between isomerism and resonance. Distinguish between structural and geometric isomerism.

19. A confused student was doing an isomer problem and listed the following six names as different structural isomers of C_7H_{16}.

a. 1-*sec*-butylpropane
b. 4-methylhexane
c. 2-ethylpentane
d. 1-ethyl-1-methylbutane
e. 3-methylhexane
f. 4-ethylpentane

How many different structural isomers are actually present in these six names?

20. There is only one compound that is named 1,2-dichloroethane, but there are two distinct compounds that can be named 1,2-dichloroethene. Why?

21. Which of the compounds in Exercise 13 exhibit *cis–trans* isomerism?

22. Which of the compounds in Exercise 14 exhibit *cis–trans* isomerism?

23. Consider the following four structures:

(i) $H_3C(H)C=C(H)-C(H)=CH_2$

(ii) $H(H_3C)C=C(H)-C(H)=CH_2$

(iii) $H_3C(H)C=C(H)-C(H)=CH_2$

(iv) $H_3C(H)C=C(H)-C(H)=CH_2$

a. Which of these compounds have the same physical properties (melting point, boiling point, density, and so on)?
b. Which of these compounds are *trans* isomers?
c. Which of these compounds do not exhibit *cis–trans* isomerism?

24. Alkenes and cycloalkanes are structural isomers of each other. Give an example of each using C_4H_8. Another common feature of alkenes and cycloalkanes is that both have restricted rotation about one or more bonds in the compound, so both can exhibit *cis–trans* isomerism. What is required for an alkene or cycloalkane to exhibit *cis–trans* isomerism?

25. *Cis–trans* isomerism is also possible in molecules with rings. Draw the *cis* and *trans* isomers of 1,2-dimethylcyclohexane.

26. Draw the following.

a. *cis*-2-hexene
b. *trans*-2-butene
c. *cis*-2,3-dichloro-2-pentene

27. Draw all the structural and geometric (*cis–trans*) isomers of C_3H_5Cl.

28. Draw all the structural and geometric (*cis–trans*) isomers of C_4H_7F. (*Hint:* Refer to Exercise 25.)

29. Draw all the structural isomers of C_5H_{10}. Ignore any cyclic isomers. Which of the structural isomers exhibit *cis–trans* isomerism?

30. Draw all the structural and geometric (*cis–trans*) isomers of bromochloropropene.

31. Name the following compounds.

a. $(CH_3)(H)C=C(Br)(H)$

b. $(CH_3)(CH_3CH_2)C=C(CH_2CH_3)(CH_2CH_2CH_3)$

c. 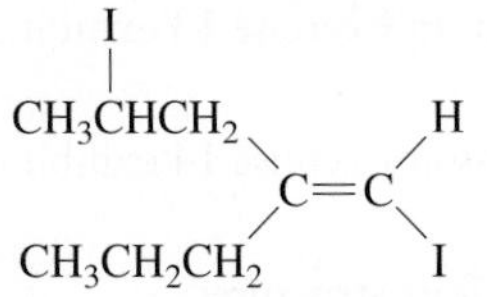 (structure: CH_3CHICH_2 and $CH_3CH_2CH_2$ on one carbon of C=C; H and I on the other)

$CH_3\underset{|}{C}HCH_2$ (I on C2) — C=C — H, I; $CH_3CH_2CH_2$

32. If one hydrogen in a hydrocarbon is replaced by a halogen atom, the number of isomers that exist for the substituted compound depends on the number of types of hydrogen in the original hydrocarbon. Thus there is only one form of chloroethane (all hydrogens in ethane are equivalent), but there are two isomers of propane that arise from the substitution of a methyl hydrogen or a methylene hydrogen. How many isomers can be obtained when one hydrogen in each of the compounds named below is replaced by a chlorine atom?
    a. *n*-pentane
    b. 2-methylbutane
    c. 2,4-dimethylpentane
    d. methylcyclobutane

33. There are three isomers of dichlorobenzene, one of which has now replaced naphthalene as the main constituent of mothballs.
    a. Identify the *ortho*, *meta*, and *para* isomers of dichlorobenzene.
    b. Predict the number of isomers for trichlorobenzene.
    c. The presence of one chlorine atom on a benzene ring will cause the next substituent to add *ortho* or *para* to the first chlorine atom on the benzene ring. What does this tell you about the synthesis of *m*-dichlorobenzene?
    d. Which of the isomers of trichlorobenzene will be the hardest to prepare?

34. Polychlorinated dibenzo-*p*-dioxins, or PCDDs, are highly toxic substances that are present in trace amounts as byproducts of some chemical manufacturing processes. They have been implicated in a number of environmental incidents—for example, the chemical contamination at Love Canal and the herbicide spraying in Vietnam. The structure of dibenzo-*p*-dioxin, along with the customary numbering convention, is

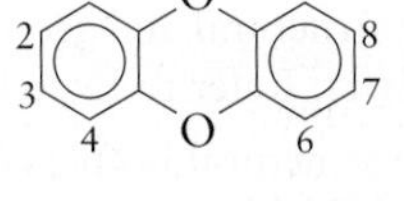

    (structure of dibenzo-*p*-dioxin with positions 1, 2, 3, 4, 6, 7, 8, 9 and two O atoms)

    The most toxic PCDD is 2,3,7,8-tetrachloro-dibenzo-*p*-dioxin. Draw the structure of this compound. Also draw the structures of two other isomers containing four chlorine atoms.

Functional Groups

35. Carboxylic acids are often written as RCOOH. What does —COOH indicate, and what does R indicate? Aldehydes are sometimes written as RCHO. What does —CHO indicate?

36. Alcohols and ethers are structural isomers of each other, as are aldehydes and ketones. Give an example of each to illustrate. Which functional group in Table 21.4 can be structural isomers of carboxylic acids?

37. Identify each of the following compounds as a carboxylic acid, ester, ketone, aldehyde, or amine.
    a. anthraquinone, an important starting material in the manufacture of dyes:

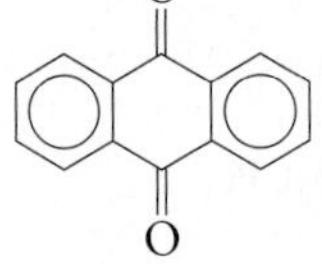

    (structure of anthraquinone)

    b. C_6H_5—CHO (benzene ring—$\overset{O}{\overset{\|}{C}}H$)

    c. HO—$\overset{O}{\overset{\|}{C}}$—$CH_2CHCH_3$ with CH_3 on CH

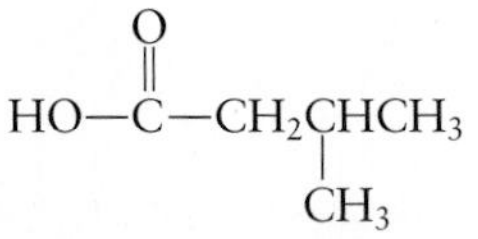

    d. 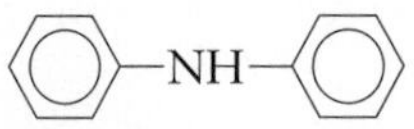 C_6H_5—NH—C_6H_5

38. Identify the functional groups present in the following compounds.
    a. 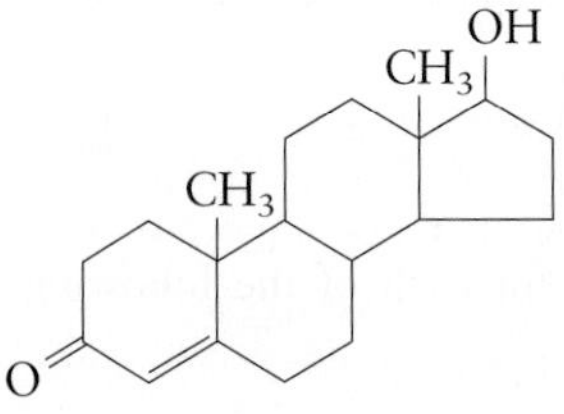 (structure of testosterone: OH, CH_3, CH_3, O)

    Testosterone

    b. 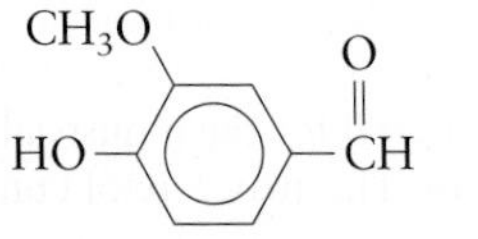 HO—(benzene ring with CH_3O)—$\overset{O}{\overset{\|}{C}}H$

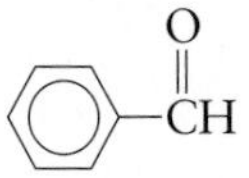

    Vanillin

    c. 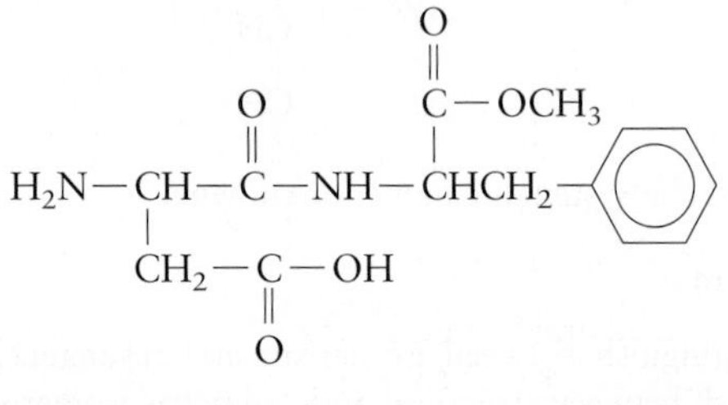 H_2N—CH—$\overset{O}{\overset{\|}{C}}$—NH—CH$CH_2$—$C_6H_5$, with CH_2—$\underset{\|}{C}$(=O)—OH on the first CH and $\overset{O}{\overset{\|}{C}}$—$OCH_3$ on the second CH

    Aspartame

39. Mimosine is a natural product found in large quantities in the seeds and foliage of some legume plants. It has been shown to cause inhibition of hair growth as well as hair loss in mice.

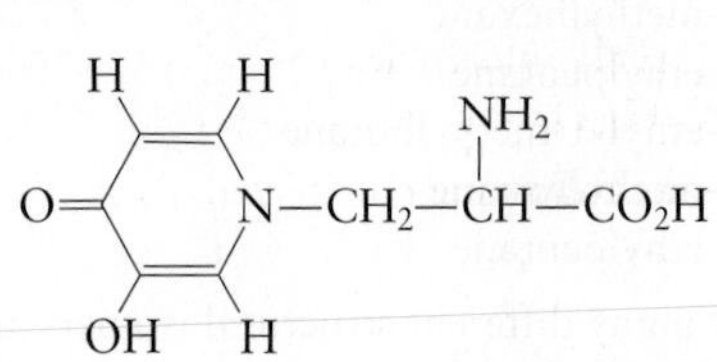

    (structure: pyridinone ring with H, H, O=, OH, H; N—CH_2—CH(NH_2)—CO_2H)

    Mimosine ($C_8H_{10}N_2O_4$)

a. What functional groups are present in mimosine?
b. Give the hybridization of the eight carbon atoms in mimosine.
c. How many σ and π bonds are found in mimosine?

40. Minoxidil ($C_9H_{15}N_5O$) is a compound produced by the Pharmacia & Upjohn Company that has been approved as a treatment for some types of male pattern baldness.

a. Would minoxidil be more soluble in acidic or basic aqueous solution? Explain.
b. Give the hybridization of the five nitrogen atoms in minoxidil.
c. Give the hybridization of each of the nine carbon atoms in minoxidil.
d. Give approximate values for the bond angles marked *a*, *b*, *c*, *d*, *e*, and *f*.
e. Including all the hydrogen atoms, how many σ bonds exist in minoxidil?
f. How many π bonds exist in minoxidil?

41. For each of the following alcohols, give the systematic name and specify whether the alcohol is primary, secondary, or tertiary.

a. $CH_3CH(Cl)CH_2CH_2OH$

b. $CH_3C(CH_2CH_2CH_3)(OH)CH_2CH_3$

c. cyclopentane ring bearing $-CH_3$ and $-OH$ on the same carbon

42. Draw structural formulas for each of the following alcohols. Indicate whether the alcohol is primary, secondary, or tertiary.
a. 1-butanol
b. 2-butanol
c. 2-methyl-1-butanol
d. 2-methyl-2-butanol

43. Name all the alcohols that have the formula $C_5H_{12}O$. How many ethers have the formula $C_5H_{12}O$?

44. Name all the aldehydes and ketones that have the formula $C_5H_{10}O$.

45. Name the following compounds.

a. $CH_3CH(Cl)CH(Cl)C(=O)CH_2CH_3$

b. $HC(=O)CH(CH_2CH_3)CH(CH_3)CH_3$

c. benzene ring with CH_3 and $-CH(=O)$ substituents (meta)

46. Name the following compounds.

a. $Cl-C_6H_4-C(=O)-OH$ (para)

b. $CH_3CH_2CH(CH_3)CH(CH_2CH_2CH_3)-C(=O)-OH$

c. $HCOOH$

47. Draw the structural formula for each of the following.
a. formaldehyde (methanal)
b. 4-heptanone
c. 3-chlorobutanal
d. 5,5-dimethyl-2-hexanone

48. Draw a structural formula for each of the following.
a. 3-methylpentanoic acid
b. ethyl methanoate
c. methyl benzoate
d. 3-chloro-2,4-dimethylhexanoic acid

49. Draw the isomer(s) specified. There may be more than one possible isomer for each part.
a. a cyclic compound that is an isomer of *trans*-2-butene
b. an ester that is an isomer of propanoic acid
c. a ketone that is an isomer of butanal
d. a secondary amine that is an isomer of butylamine
e. a tertiary amine that is an isomer of butylamine
f. an ether that is an isomer of 2-methyl-2-propanol
g. a secondary alcohol that is an isomer of 2-methyl-2-propanol

50. Which of the following statements is (are) false? Explain why the statement(s) is (are) false.
a. $CH_3CH_2CH_2C(=O)OCH_3$ is a structural isomer of pentanoic acid.
b. $HC(=O)CH_2CH_2CH(CH_3)CH_3$ is a structural isomer of 2-methyl-3-pentanone.
c. $CH_3CH_2OCH_2CH_2CH_3$ is a structural isomer of 2-pentanol.
d. $CH_2{=}CHCH(OH)CH_3$ is a structural isomer of 2-butenal.
e. Trimethylamine is a structural isomer of $CH_3CH_2CH_2NH_2$.

51. The following organic compounds cannot exist. Why?
    a. 2-chloro-2-butyne
    b. 2-methyl-2-propanone
    c. 1,1-dimethylbenzene
    d. 2-pentanal
    e. 3-hexanoic acid
    f. 5,5-dibromo-1-cyclobutanol

52. Mycomycin is a naturally occurring antibiotic produced by the fungus *Nocardia acidophilus*. The molecular formula of the substance is $C_{13}H_{10}O_2$, and its systematic name is 3,5,7,8-tridecatetraene-10,12-diynoic acid. Draw the structure of mycomycin.

Reactions of Organic Compounds

53. Distinguish between substitution and addition reactions. Give an example of each type.

54. Alkanes and aromatics are fairly stable compounds. To make them react, a special catalyst must be present. What catalyst must be present when reacting Cl_2 with an alkane or with benzene? Adding Cl_2 to an alkene or alkyne does not require a special catalyst. Why are alkenes and alkynes more reactive than alkanes and aromatic compounds?

55. The following are some other organic reactions covered in Section 21.4. Give an example to illustrate each type of reaction.
    a. Adding H_2O to an alkene (in the presence of H^+) yields an alcohol.
    b. Primary alcohols are oxidized to aldehydes, which can be further oxidized to carboxylic acids.
    c. Secondary alcohols are oxidized to ketones.
    d. Reacting an alcohol with a carboxylic acid (in the presence of H^+) produces an ester.

56. In the presence of light, chlorine can substitute for one (or more) of the hydrogens in an alkane. For the following reactions, draw the possible monochlorination products.
    a. 2,2-dimethylpropane + $Cl_2 \xrightarrow{h\nu}$
    b. 1,3-dimethylcyclobutane + $Cl_2 \xrightarrow{h\nu}$
    c. 2,3-dimethylbutane + $Cl_2 \xrightarrow{h\nu}$

57. Complete the following reactions.
    a. $CH_3CH{=}CHCH_3 + H_2 \xrightarrow{Pt}$
    b. $CH_2{=}CHCH(CH_3)CH{=}CH(CH_3) + 2Cl_2 \longrightarrow$ (CH₃ groups drawn on the third and fifth carbons)
    c. 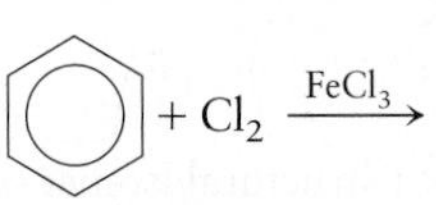 benzene + $Cl_2 \xrightarrow{FeCl_3}$
    d. $CH_3C(CH_3){=}CH_2 + O_2 \xrightarrow{Spark}$

58. Reagents such as HCl, HBr, and HOH (H_2O) can add across carbon–carbon double and triple bonds, with H forming a bond to one of the carbon atoms in the multiple bond and Cl, Br, or OH forming a bond to the other carbon atom in the multiple bond. In some cases, two products are possible. For the major organic product, the addition occurs so that the hydrogen atom in the reagent attaches to the carbon atom in the multiple bond that already has the greater number of hydrogen atoms bonded to it. With this rule in mind, draw the structure of the major product in each of the following reactions.
    a. $CH_3CH_2CH{=}CH_2 + H_2O \xrightarrow{H^+}$
    b. $CH_3CH_2CH{=}CH_2 + HBr \longrightarrow$
    c. $CH_3CH_2C{\equiv}CH + 2HBr \longrightarrow$
    d. 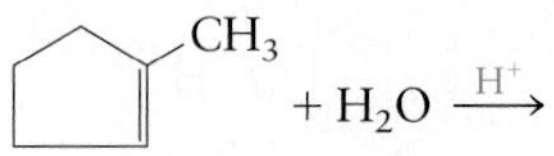 1-methylcyclopentene (ring with CH_3 on the double-bond carbon) $+ H_2O \xrightarrow{H^+}$
    e. $(CH_3CH_2)(CH_3)C{=}C(CH_3)(H) + HCl \longrightarrow$

59. Why is it preferable to produce chloroethane by the reaction of HCl(*g*) with ethene than by the reaction of $Cl_2(g)$ with ethane? (See Exercise 58.)

60. Give an example reaction that would yield the following products. Name the organic reactant and product in each reaction.
    a. alkane
    b. monohalogenated alkane
    c. dihalogenated alkane
    d. tetrahalogenated alkane
    e. monohalogenated benzene
    f. alkene

61. Using appropriate reactants, alcohols can be oxidized into aldehydes, ketones, and/or carboxylic acids. Primary alcohols can be oxidized into aldehydes, which can then be oxidized into carboxylic acids. Secondary alcohols can be oxidized into ketones. Tertiary alcohols do not undergo this type of oxidation. Give the structure of the product(s) resulting from the oxidation of each of the following alcohols.
    a. 3-methyl-1-butanol
    b. 3-methyl-2-butanol
    c. 2-methyl-2-butanol
    d. 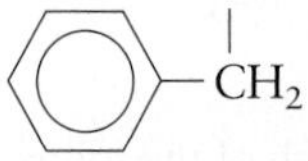 benzene ring–CH_2–OH
    e. 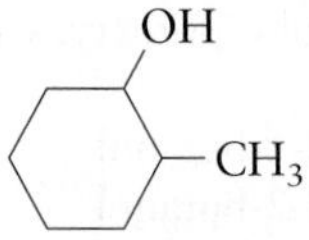 cyclohexane ring bearing OH and an adjacent CH_3
    f. 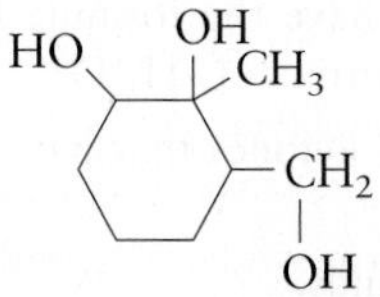 cyclohexane ring bearing HO, a carbon with OH and CH_3, and CH_2–OH

62. Oxidation of an aldehyde yields a carboxylic acid:

$$R{-}\overset{\overset{\displaystyle O}{\|}}{C}H \xrightarrow{[ox]} R{-}\overset{\overset{\displaystyle O}{\|}}{C}{-}OH$$

Draw the structures for the products of the following oxidation reactions.

a. propanal $\xrightarrow{[ox]}$

b. 2,3-dimethylpentanal $\xrightarrow{[ox]}$

c. 3-ethylbenzaldehyde $\xrightarrow{[ox]}$

63. Three different organic compounds have the formula C_3H_8O. Only two of these isomers react with $KMnO_4$ (a strong oxidizing agent). What are the names of the products when these isomers react with excess $KMnO_4$?

64. Give an example reaction that would yield the following products as major organic products. See Exercises 58 and 61 for some hints. For oxidation reactions, just write *oxidation* over the arrow and don't worry about the actual reagent.

a. primary alcohol
b. secondary alcohol
c. tertiary alcohol
d. aldehyde
e. ketone
f. carboxylic acid
g. ester

65. How would you synthesize each of the following?

a. 1,2-dibromopropane from propene
b. acetone (2-propanone) from an alcohol
c. *tert*-butyl alcohol (2-methyl-2-propanol) from an alkene (see Exercise 58)
d. propanoic acid from an alcohol

66. What tests could you perform to distinguish between the following pairs of compounds?

a. $CH_3CH_2CH_2CH_3$, $CH_2{=}CHCH_2CH_3$

b. $CH_3CH_2CH_2COOH$, $CH_3CH_2\overset{O}{\overset{\|}{C}}CH_3$

c. $CH_3CH_2CH_2OH$, $CH_3\overset{O}{\overset{\|}{C}}CH_3$

d. $CH_3CH_2NH_2$, CH_3OCH_3

67. Complete the following reactions.

a. $CH_3CO_2H + CH_3OH \rightarrow$

b. $CH_3CH_2CH_2OH + HCOOH \rightarrow$

68. How would you synthesize the following esters?

a. *n*-octylacetate

b. $CH_3CH_2CH_2CH_2CH_2CH_2O{-}\overset{O}{\overset{\|}{C}}CH_2CH_3$

Polymers

69. Define and give an example of each of the following.

a. addition polymer
b. condensation polymer
c. copolymer
d. homopolymer
e. polyester
f. polyamide

70. What is polystyrene? The following processes result in a stronger polystyrene polymer. Explain why in each case.

a. addition of catalyst to form syndiotactic polystyrene
b. addition of 1,3-butadiene and sulfur
c. producing long chains of polystyrene
d. addition of a catalyst to make linear polystyrene

71. Answer the following questions regarding the formation of polymers.

a. What structural features must be present in a monomer in order to form a homopolymer polyester?
b. What structural features must be present in the monomers in order to form a copolymer polyamide? (*Hint:* Nylon is an example of a polyamide. When the monomers link together to form nylon, an amide functional group results from each linkage.)
c. What structural features must be present in a monomer that can form both an addition polymer and a condensation polymer?

72. What monomer(s) must be used to produce the following polymers?

a. 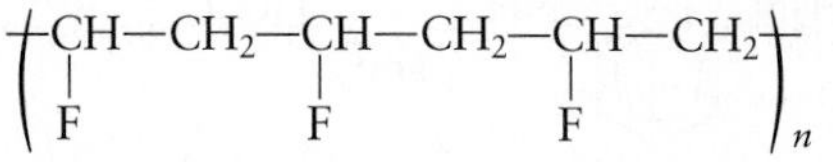 $\left(\overset{}{\underset{F}{\underset{|}{CH}}}{-}CH_2{-}\underset{F}{\underset{|}{CH}}{-}CH_2{-}\underset{F}{\underset{|}{CH}}{-}CH_2\right)_n$

b. $\left(O{-}CH_2{-}CH_2{-}\overset{O}{\overset{\|}{C}}{-}O{-}CH_2{-}CH_2{-}\overset{O}{\overset{\|}{C}}\right)_n$

c. $\left(\overset{H}{\overset{|}{N}}{-}CH_2{-}CH_2{-}\overset{H}{\overset{|}{N}}{-}\overset{O}{\overset{\|}{C}}{-}CH_2{-}CH_2{-}\overset{O}{\overset{\|}{C}}\right)_n$

d. $\left(\overset{CH_3}{\overset{|}{\underset{C_6H_5}{\underset{|}{C}}}}{-}CH_2{-}\overset{CH_3}{\overset{|}{\underset{C_6H_5}{\underset{|}{C}}}}{-}CH_2{-}\overset{CH_3}{\overset{|}{\underset{C_6H_5}{\underset{|}{C}}}}{-}CH_2\right)_n$

e. $\left(\underset{C_6H_5}{\underset{|}{CH}}{-}\underset{CH_3}{\underset{|}{CH}}{-}\underset{C_6H_5}{\underset{|}{CH}}{-}\underset{CH_3}{\underset{|}{CH}}\right)_n$

f. $\left(O\underset{H}{\underset{|}{\overset{H}{\overset{|}{C}}}}{-}C_6H_{10}{-}\underset{H}{\underset{|}{\overset{H}{\overset{|}{C}}}}O\overset{O}{\overset{\|}{C}}{-}C_6H_4{-}\overset{O}{\overset{\|}{C}}O\underset{H}{\underset{|}{\overset{H}{\overset{|}{C}}}}{-}C_6H_{10}{-}\underset{H}{\underset{|}{\overset{H}{\overset{|}{C}}}}O\overset{O}{\overset{\|}{C}}{-}C_6H_4{-}\overset{O}{\overset{\|}{C}}\right)_n$

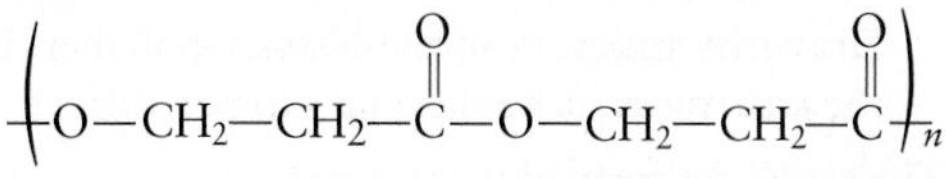

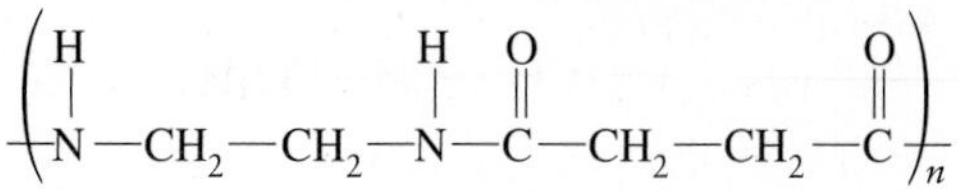

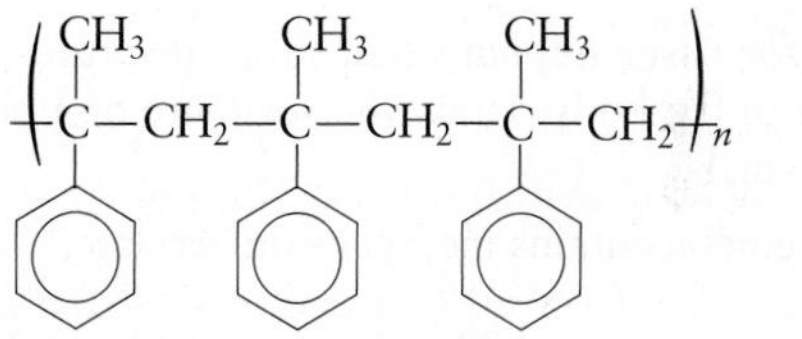

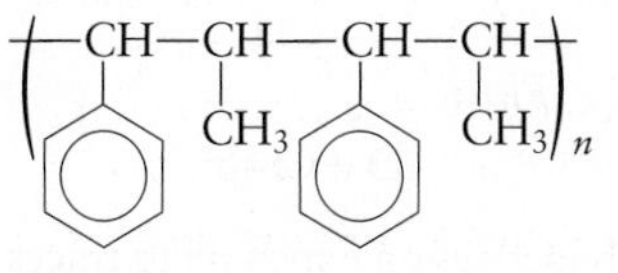

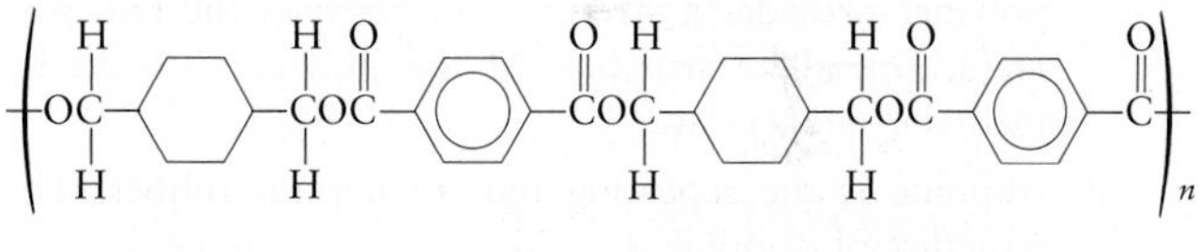

(This polymer is Kodel, used to make fibers of stain-resistant carpeting.)

Classify these polymers as condensation or addition polymers. Which are copolymers?

73. Kel-F is a polymer with the structure

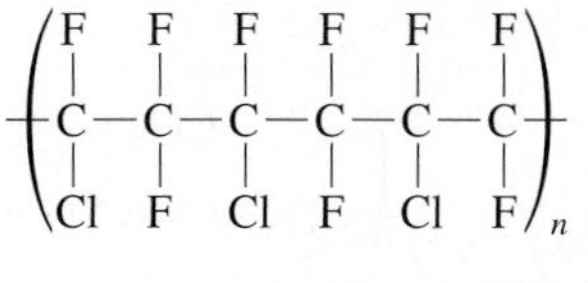

$$\left(\overset{F}{\overset{|}{\underset{Cl}{\underset{|}{C}}}}{-}\overset{F}{\overset{|}{\underset{F}{\underset{|}{C}}}}{-}\overset{F}{\overset{|}{\underset{Cl}{\underset{|}{C}}}}{-}\overset{F}{\overset{|}{\underset{F}{\underset{|}{C}}}}{-}\overset{F}{\overset{|}{\underset{Cl}{\underset{|}{C}}}}{-}\overset{F}{\overset{|}{\underset{F}{\underset{|}{C}}}}\right)_n$$

What is the monomer for Kel-F?

74. The polymer nitrile is a copolymer made from acrylonitrile and butadiene; it is used to make automotive hoses and gaskets. Draw the structure of nitrile. (*Hint:* See Table 21.7.)

75. Polyaramid is a term applied to polyamides containing aromatic groups. These polymers were originally made for use as tire cords but have since found many other uses.
    a. Kevlar is used in bulletproof vests and many high-strength composites. The structure of Kevlar is

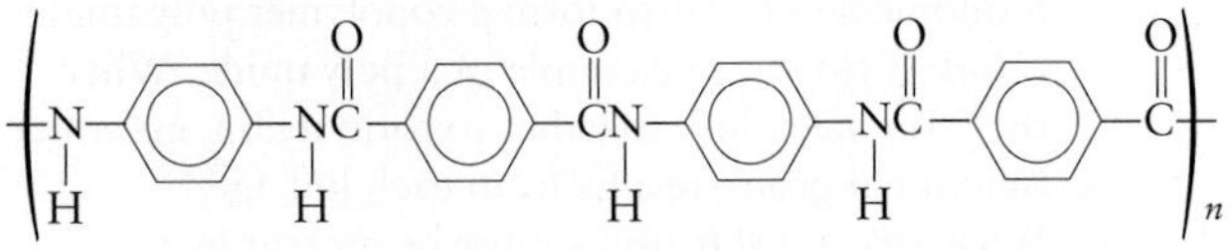

    $$\left[-\underset{\text{H}}{\text{N}}-C_6H_4-\underset{\text{H}}{\text{N}}\overset{\text{O}}{\text{C}}-C_6H_4-\overset{\text{O}}{\text{C}}\underset{\text{H}}{\text{N}}-C_6H_4-\underset{\text{H}}{\text{N}}\overset{\text{O}}{\text{C}}-C_6H_4-\overset{\text{O}}{\text{C}}- \right]_n$$

    Which monomers are used to make Kevlar?
    b. Nomex is a polyaramid used in fire-resistant clothing. It is a copolymer of

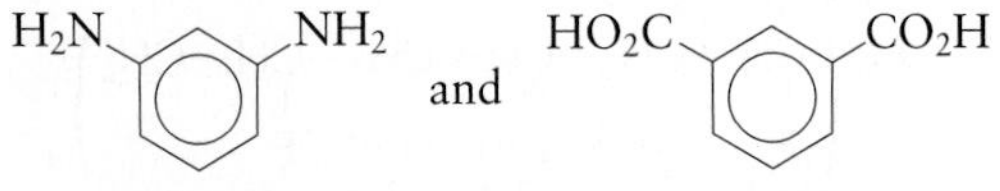

    H_2N–C_6H_4–NH_2 (meta) and HO_2C–C_6H_4–CO_2H (meta)

    Draw the structure of the Nomex polymer. How do the structures of Kevlar and Nomex differ?
76. The polyester formed from lactic acid,

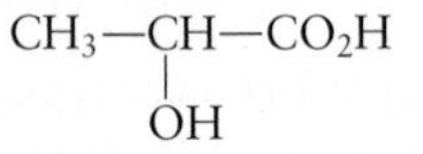

    $$CH_3-\underset{\text{OH}}{CH}-CO_2H$$

    is used for tissue implants and surgical sutures that will dissolve in the body. Draw the structure of a portion of this polymer.
77. "Super glue" contains methyl cyanoacrylate,

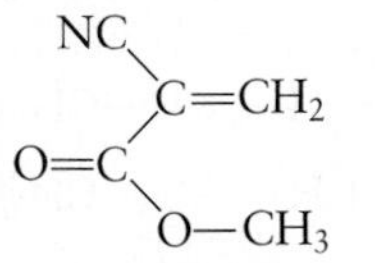

    $$\text{NC}-\underset{\text{O}=\text{C}-\text{O}-CH_3}{\text{C}}=CH_2$$

    which readily polymerizes on exposure to traces of water or alcohols on the surfaces to be bonded together. The polymer provides a strong bond between the two surfaces. Draw the structure of the polymer formed by methyl cyanoacrylate.
78. Isoprene is the repeating unit in natural rubber. The structure of isoprene is

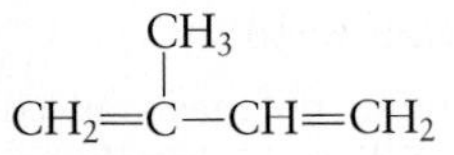

    $$CH_2=\overset{CH_3}{C}-CH=CH_2$$

    a. Give a systematic name for isoprene.
    b. When isoprene is polymerized, two polymers of the form

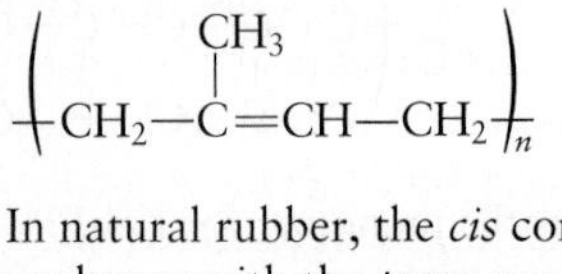

    $$\left(CH_2-\overset{CH_3}{C}=CH-CH_2 \right)_n$$

    are possible. In natural rubber, the *cis* configuration is found. The polymer with the *trans* configuration around the double bond is called gutta percha and was once used in the manufacture of golf balls. Draw the structure of natural rubber and gutta percha showing three repeating units and the configuration around the carbon–carbon double bonds.
79. Polystyrene can be made more rigid by copolymerizing styrene with divinylbenzene,

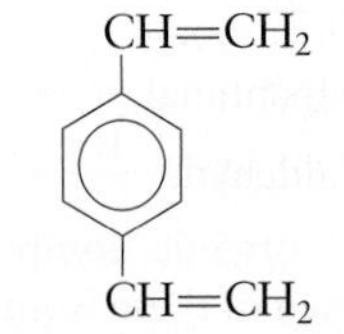

    $$CH=CH_2-C_6H_4-CH=CH_2 \text{ (para)}$$

    What purpose does the divinylbenzene serve? Why is the copolymer more rigid?
80. Polyimides are polymers that are tough and stable at temperatures up to 400°C. They are used as a protective coating on the quartz fibers in fiber optics. What monomers are used to make the following polyimide?

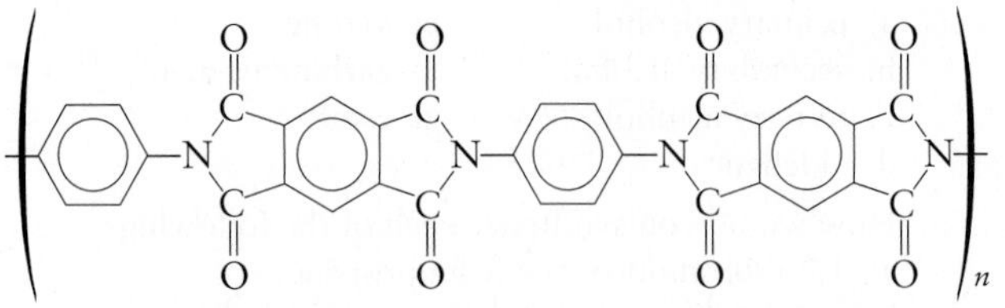

    $$\left[-C_6H_4-N\langle(\text{C}=\text{O})_2\rangle C_6H_2\langle(\text{C}=\text{O})_2\rangle N-C_6H_4-N\langle(\text{C}=\text{O})_2\rangle C_6H_2\langle(\text{C}=\text{O})_2\rangle N- \right]_n$$

81. Polycarbonates are a class of thermoplastic polymers that are used in the plastic lenses of eyeglasses and in the shells of bicycle helmets. A polycarbonate is made from the reaction of bisphenol A (BPA) with phosgene ($COCl_2$):

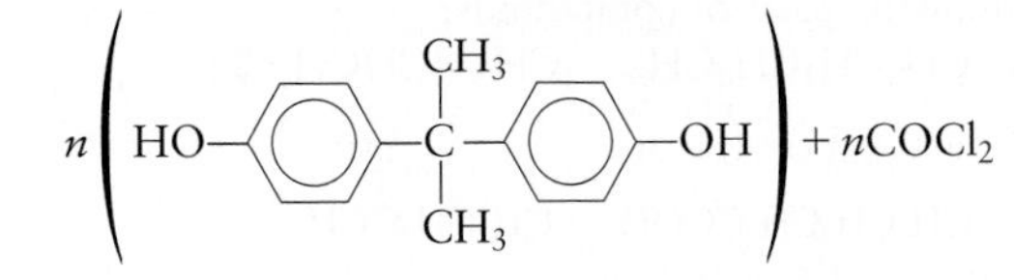

    $$n\left(HO-C_6H_4-\underset{CH_3}{\overset{CH_3}{C}}-C_6H_4-OH \right) + nCOCl_2 \xrightarrow{\text{Catalyst}} \text{polycarbonate} + 2nHCl$$

    BPA

    Phenol (C_6H_5OH) is used to terminate the polymer (stop its growth).
    a. Draw the structure of the polycarbonate chain formed in the above reaction.
    b. Is this reaction a condensation or an addition polymerization?
82. In which polymer, polyethylene or polyvinyl chloride, would you expect to find the stronger intermolecular forces, assuming that the average chain lengths are equal?
83. When acrylic polymers are burned, toxic fumes are produced. For example, in many airplane fires, more passenger deaths have been caused by breathing toxic fumes than by the fire itself. Using polyacrylonitrile as an example, what would you expect to be one of the most toxic, gaseous combustion products created in the reaction?
84. Polyesters containing double bonds are often crosslinked by reacting the polymer with styrene. This type of reaction is common in the manufacture of fiberglass.
    a. Draw the structure of the copolymer of $HO-CH_2CH_2-OH$ and $HO_2C-CH=CH-CO_2H$
    b. Draw the structure of the crosslinked polymer (after the polyester has been reacted with styrene).

85. Another way of producing highly crosslinked polyesters (see Exercise 84) is to use glycerol. Alkyd resins are a polymer of this type. The polymer forms very tough coatings when baked onto a surface and is used in paints for automobiles and large appliances. Draw the structure of the polymer formed from the condensation of

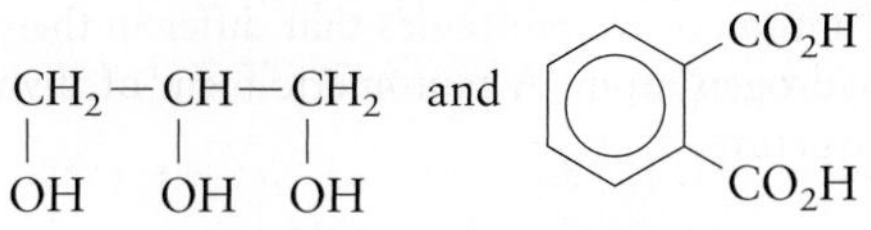

$$\underset{\text{Glycerol}}{\underset{|\quad\ \ |\quad\ \ |}{CH_2-CH-CH_2}\atop{OH\quad OH\quad OH}}\quad\text{and}\quad\underset{\text{Phthalic acid}}{C_6H_4(CO_2H)_2}$$

Explain how individual polymer chains could be crosslinked.

Natural Polymers

86. In Section 21.6 three important classes of biologically important natural polymers are discussed. What are the three classes, what are the monomers used to form the polymers, and why are they biologically important?

87. Give the general formula for an amino acid. Some amino acids are labeled hydrophilic and some are labeled hydrophobic. What do these terms refer to?

88. Distinguish between the primary, secondary, and tertiary structures of a protein. Give examples of the types of forces that maintain each type of structure.

89. Describe how denaturation affects the function of a protein.

90. Aqueous solutions of amino acids are buffered solutions. Why?

91. Which of the amino acids in Fig. 21.18 contain the following functional groups in their R group?
   a. alcohol
   b. carboxylic acid
   c. amine
   d. amide

92. When pure crystalline amino acids are heated, decomposition generally occurs before the solid melts. Account for this observation. (*Hint:* Crystalline amino acids exist as $H_3\overset{+}{N}CRHCOO^-$, called zwitterions.)

93. Aspartame, the artificial sweetener marketed under the name NutraSweet, is a methyl ester of a dipeptide. The structure of aspartame is

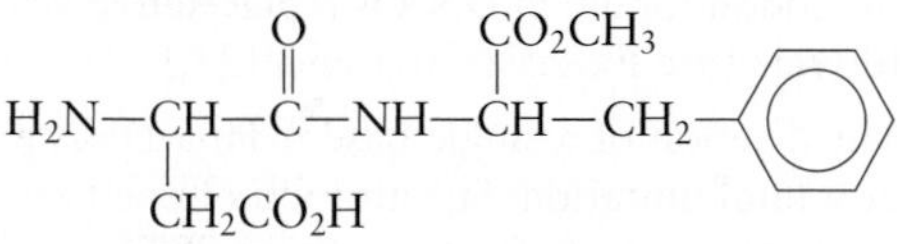

$$H_2N-\underset{\underset{CH_2CO_2H}{|}}{CH}-\overset{\overset{O}{\|}}{C}-NH-\overset{\overset{CO_2CH_3}{|}}{CH}-CH_2-C_6H_5$$

   a. What two amino acids are used to prepare aspartame?
   b. There is concern that methanol may be produced by the decomposition of aspartame. From what portion of the molecule can methanol be produced? Write an equation for this reaction.

94. Glutathione, a tripeptide found in virtually all cells, functions as a reducing agent. The structure of glutathione is

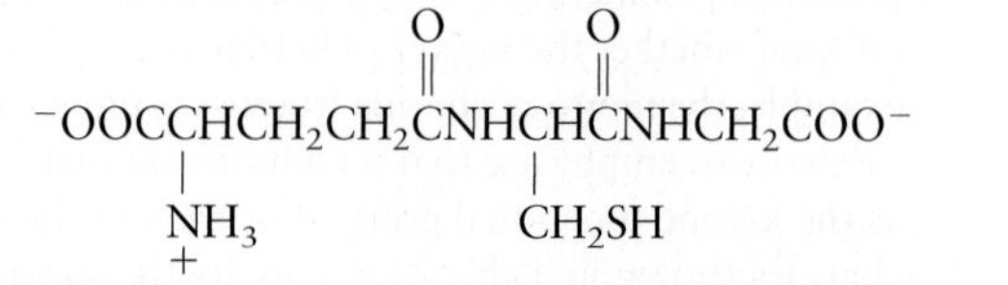

$$^-OOC\underset{\underset{\overset{+}{NH_3}}{|}}{C}HCH_2CH_2\overset{\overset{O}{\|}}{C}NH\underset{\underset{CH_2SH}{|}}{C}H\overset{\overset{O}{\|}}{C}NHCH_2COO^-$$

What amino acids make up glutathione?

95. Draw the structures of the two dipeptides that can be formed from serine and alanine.

96. Draw the structures of the tripeptides gly-ala-ser and ser-ala-gly. How many other tripeptides are possible using these three amino acids?

97. Write the sequence of all possible tetrapeptides composed of the following amino acids.
   a. two phenylalanines and two glycines
   b. two phenylalanines, glycine, and alanine

98. How many different pentapeptides can be formed using five different amino acids?

99. What types of interactions can occur between the side chains of the following amino acids that would help maintain the tertiary structure of a protein?
   a. cysteine and cysteine
   b. glutamine and serine
   c. glutamic acid and lysine
   d. proline and leucine

100. Give examples of amino acids that could give rise to the interactions pictured in Fig. 21.25 that maintain the tertiary structures of proteins.

101. Oxygen is carried from the lungs to tissues by the protein hemoglobin in red blood cells. Sickle cell anemia is a disease resulting from abnormal hemoglobin molecules in which a valine is substituted for a single glutamic acid in normal hemoglobin. How might this substitution affect the structure of hemoglobin?

102. All amino acids have at least two functional groups with acidic or basic properties. In alanine the carboxylic acid group has $K_a = 4.5 \times 10^{-3}$ and the amino group has $K_b = 7.4 \times 10^{-5}$. Three ions of alanine are possible when alanine is dissolved in water. Which of these ions would predominate in a solution with $[H^+] = 1.0\ M$? in a solution with $[OH^-] = 1.0\ M$?

103. Monosodium glutamate (MSG) is commonly used as a flavoring in foods. Draw the structure of MSG.

104. Over 100 different kinds of mutant hemoglobin molecules have been detected in humans. Unlike sickle cell anemia (see Exercise 101), not all of these mutations are as serious. In one nonlethal mutation, glutamine substitutes for a single glutamic acid in normal hemoglobin. Rationalize why this substitution is nonlethal.

105. Draw cyclic structures for D-ribose and D-mannose.

106. Indicate the chiral carbon atoms found in the monosaccharides D-ribose and D-mannose.

107. In addition to using *numerical* prefixes in the general names of sugars to indicate how many carbon atoms are present, we sometimes use the prefixes *keto-* and *aldo-* to indicate whether the sugar is a ketone or an aldehyde. For example, the monosaccharide fructose is often called a *ketohexose* to emphasize that it contains six carbons as well as the ketone functional group. For each of the monosaccharides shown in Table 21.8, classify the sugars as aldohexoses, aldopentoses, ketohexoses, or ketopentoses.

108. Glucose can occur in three forms: two cyclic forms and one open-chain structure. In aqueous solution, only a tiny fraction of the glucose is in the open-chain form. Yet tests for the presence of glucose depend on reaction with the aldehyde group, which is found only in the open-chain form. Explain why these tests work.

109. What is a *disaccharide?* What monosaccharide units make up the dissacharide sucrose? What is the bond called that forms between the monosaccharide units?

110. Cows can digest cellulose, but humans can't. Why not?

111. What are the structural differences between α- and β-glucose? Each of these two cyclic forms of glucose is the building block to form two different polymers. Explain.

112. What is optical isomerism? What do you look for to determine whether an organic compound exhibits optical isomerism? 1-bromo-1-chloroethane is optically active, whereas 1-bromo-2-chloroethane is not optically active. Explain.

113. Which of the amino acids in Fig. 21.18 contain more than one chiral carbon atom? Draw the structures of these amino acids, and indicate all chiral carbon atoms.

114. Why is glycine not optically active?

115. Which of the noncyclic isomers of bromochloropropene is (are) optically active?

116. How many chiral carbon atoms does the following structure have?

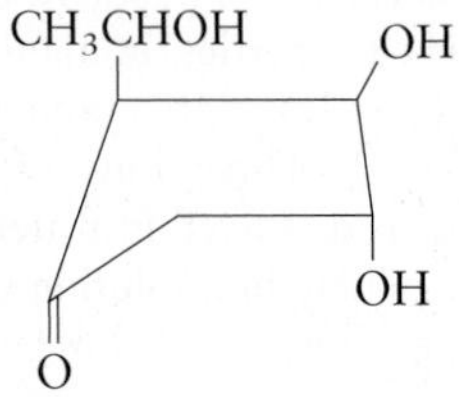

CH_3CHOH OH

OH

O

117. The compounds adenine, guanine, cytosine, and thymine are called the nucleic acid bases. What structural features in these compounds make them bases?

118. Describe the structural differences between DNA and RNA. Describe the complementary base pairing between the two individual strands of DNA that forms the overall double-helical structure. How is complementary base pairing involved in the replication of the DNA molecule during cell division?

119. Part of a certain DNA sequence is G-G-T-C-T-A-T-A-C. What is the complementary sequence?

120. The codons (words) in DNA that identify which amino acid should be in a protein are three bases long. How many such three-letter words can be made from the four bases adenine, cytosine, guanine, and thymine?

121. Which base will hydrogen-bond with uracil within an RNA molecule? Draw the structure of this base pair.

122. Tautomers are molecules that differ in the position of a hydrogen atom. A tautomeric form of thymine has the structure

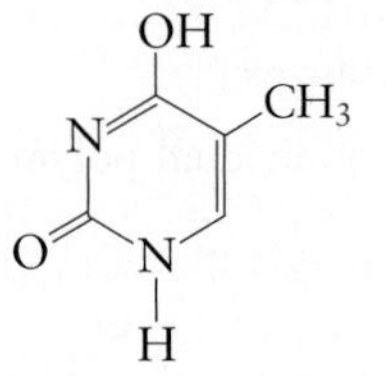

OH

N CH_3

O N

H

If this tautomer, rather than the stable form of thymine, were present in a strand of DNA during replication, what would be the result?

123. The base sequences in mRNA that code for certain amino acids are

Glu: GAA, GAG
Val: GUU, GUC, GUA, GUG
Met: AUG
Trp: UGG
Phe: UUU, UUC
Asp: GAU, GAC

These sequences are complementary to the sequences in DNA.

a. Give the corresponding sequences in DNA for the amino acids listed above.
b. Give a DNA sequence that would code for the peptide trp-glu-phe-met.
c. How many different DNA sequences can code for the peptide in part b?
d. What is the peptide that is produced from the DNA sequence T-A-C-C-T-G-A-A-G?
e. What other DNA sequences would yield the same tripeptide as in part d?

124. The change of a single base in the DNA sequence for normal hemoglobin can encode for the abnormal hemoglobin, giving rise to sickle cell anemia. Which base in the codon for glu in DNA is replaced to give the codon(s) for val? (See Exercises 101 and 123.)

125. The deletion of a single base from a DNA molecule can be a fatal mutation. Substitution of one base for another is often not as serious a mutation. Why?

126. The average molar mass of one base pair of nucleotides in DNA is approximately 600 g/mol. The spacing between successive base pairs is about 0.34 nm, and a complete turn in the helical structure of DNA occurs about every 3.4 nm. If a DNA molecule has a molar mass of 4.5×10^9 g/mol, approximately how many complete turns exist in the DNA α-helix structure?

Additional Exercises

127. Is octanoic acid more soluble in 1 *M* HCl, 1 *M* NaOH, or pure water? Explain. Drugs such as morphine ($C_{17}H_{19}O_3N$) are often treated with strong acids. The most commonly used form of morphine is morphine hydrochloride ($C_{17}H_{20}O_3NCl$). Why is morphine treated in this way? (*Hint:* Morphine is an amine.)

128. Sorbic acid is used to prevent mold and fungus growth in some food products, especially cheeses. The systematic name for sorbic acid is 2,4-hexadienoic acid. Draw structures for the four geometric isomers of sorbic acid.

129. When toluene ($C_6H_5CH_3$) reacts with chlorine gas in the presence of an iron(III) catalyst, the product is a mixture of the *ortho* and *para* isomers of $C_6H_4ClCH_3$. However, when the reaction is light-catalyzed with no Fe^{3+} catalyst present, the product is $C_6H_5CH_2Cl$. Explain.

130. Consider the reaction to produce the ester methyl acetate:

$$CH_3OH + CH_3\overset{\overset{O}{\|}}{C}OH \longrightarrow \underset{\text{Methyl acetate}}{CH_3\overset{\overset{O}{\|}}{C}OCH_3} + H_2O$$

When this reaction is carried out with CH_3OH containing radioactive oxygen-18, the water produced does not contain oxygen-18. Explain the results of this radioisotope tracer experiment.

131. A compound containing only carbon and hydrogen is 85.63% C by mass. Reaction of this compound with H_2O produces a secondary alcohol as the major product and a primary alcohol as the minor product (see Exercise 58). If the molar mass of the hydrocarbon is between 50 g/mol and 60 g/mol, name the compound.

132. Poly(lauryl methacrylate) is used as an additive in motor oils to counter the loss of viscosity at high temperature. The structure is

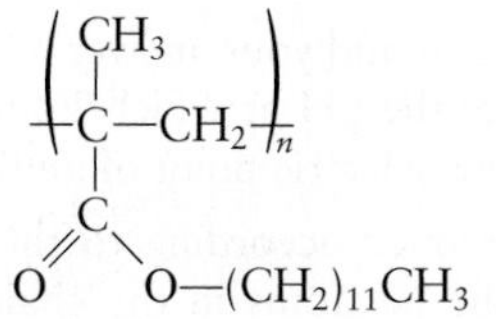

$$\left(\begin{array}{c} CH_3 \\ | \\ -C-CH_2- \\ | \\ C(=O)-O-(CH_2)_{11}CH_3 \end{array}\right)_n$$

The long hydrocarbon chain of poly(lauryl methacrylate) makes the polymer soluble in oil (a mixture of hydrocarbons with mostly 12 or more carbon atoms). At low temperatures the polymer is coiled into balls. At higher temperatures the balls uncoil and the polymer exists as long chains. Explain how this helps control the viscosity of oil.

133. Ethylene oxide,

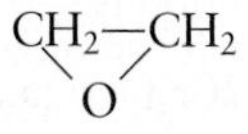

$$\begin{array}{c} CH_2-CH_2 \\ \diagdown \; \diagup \\ O \end{array}$$

is an important industrial chemical. Although most ethers are unreactive, ethylene oxide is quite reactive. It resembles C_2H_4 in its reactions in that many addition reactions occur across the C—O bond.

a. Why is ethylene oxide so reactive? (*Hint:* Consider the bond angles in ethylene oxide as compared with those predicted by the VSEPR model.)
b. Ethylene oxide undergoes addition polymerization, forming a polymer used in many applications requiring a non-ionic surfactant. Draw the structure of this polymer.

134. The Amoco Chemical Company has successfully raced a car with a plastic engine. Many of the engine parts, including piston skirts, connecting rods, and valve-train components, are made of a polymer called *Torlon:*

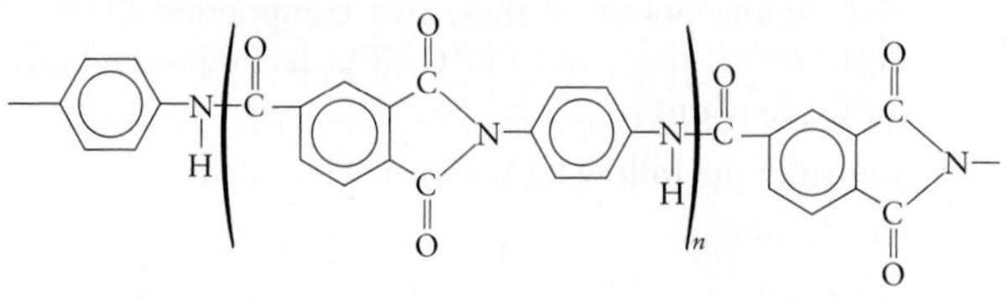

[Structure of Torlon: repeating unit of a benzene ring bonded to N(H)–C(=O)– attached to a phthalimide ring system (C=O, N, C=O), with N bonded to the next benzene ring; subscript *n*]

What monomers are used to make this polymer?

135. A urethane linkage occurs when an alcohol adds across the carbon–nitrogen double bond in an isocyanate:

$$\underset{\text{Alcohol}}{R-O-H} + \underset{\text{Isocyanate}}{O=C=N-R'} \longrightarrow \underset{\text{Urethane}}{RO-\overset{\overset{O}{\|}}{C}-\underset{\underset{H}{|}}{N}-R'}$$

Polyurethanes are formed from the copolymerization of a diol with a diisocyanate. Polyurethanes are used in foamed insulation and a variety of other construction materials. What is the structure of the polyurethane formed by the following reaction?

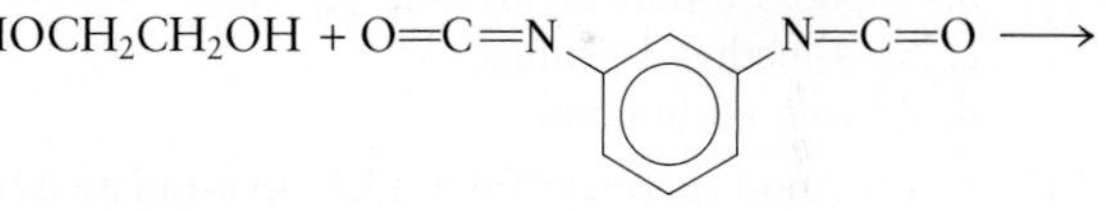

$$HOCH_2CH_2OH + O=C=N-C_6H_4-N=C=O \longrightarrow$$

(the two N=C=O groups are attached at the 1,3-positions of a benzene ring)

136. Stretch a rubber band while holding it gently to your lips. Then slowly let it relax while still in contact with your lips.
a. What happens to the temperature of the rubber band on stretching?
b. Is the stretching an exothermic or endothermic process?
c. Explain the above result in terms of intermolecular forces.
d. What is the sign of ΔS and ΔG for stretching the rubber band?
e. Give the molecular explanation for the sign of ΔS for stretching.

137. Consider the compounds butanoic acid, pentanal, *n*-hexane, and 1-pentanol. The boiling points of these compounds (in no specific order) are 69°C, 103°C, 137°C, and 164°C. Match the boiling points to the correct compound.

138. Consider the following five compounds.
   a. $CH_3CH_2CH_2CH_2CH_3$
   b.  $CH_3CH_2CH_2CH_2OH$ (OH on the terminal carbon)
   c. $CH_3CH_2CH_2CH_2CH_2CH_3$
   d.  $CH_3CH_2CH_2CH{=}O$ (aldehyde)
   e. 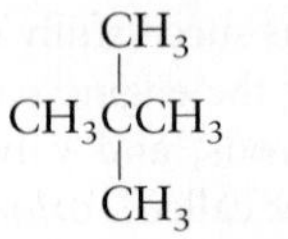 $CH_3C(CH_3)_2CH_3$

   The boiling points of these five compounds are 9.5°C, 36°C, 69°C, 76°C, and 117°C. Which compound boils at 36°C? Explain.

139. Consider the following polymer:

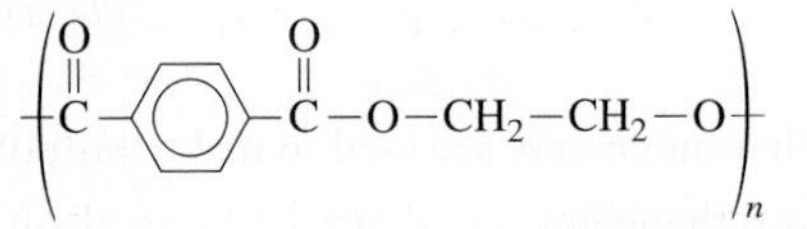

$$\left[-\overset{O}{\overset{\|}{C}}-C_6H_4-\overset{O}{\overset{\|}{C}}-O-CH_2-CH_2-O- \right]_n$$

   Is this polymer a homopolymer or a copolymer, and is it formed by addition polymerization or condensation polymerization? What is (are) the monomer(s) for this polymer?

140. For the following formulas, what types of isomerism could be exhibited? For each formula, give an example that illustrates the specific type of isomerism. The types of isomerism are structural, geometric, and optical.
   a. C_6H_{12} b. $C_5H_{12}O$ c. $C_6H_4Br_2$

141. What is wrong with the following names? Give the correct name for each compound.
   a. 2-ethylpropane
   b. 5-iodo-5,6-dimethylhexane
   c. *cis*-4-methyl-3-pentene
   d. 2-bromo-3-butanol

142. a. Use bond energies (Table 13.6) to estimate ΔH for the reaction of two molecules of glycine to form a peptide linkage.
   b. Would you predict ΔS to favor the formation of peptide linkages between two molecules of glycine?
   c. Would you predict the formation of proteins to be a spontaneous process?

143. The reaction to form a phosphate–ester linkage between two nucleotides can be approximated as follows:

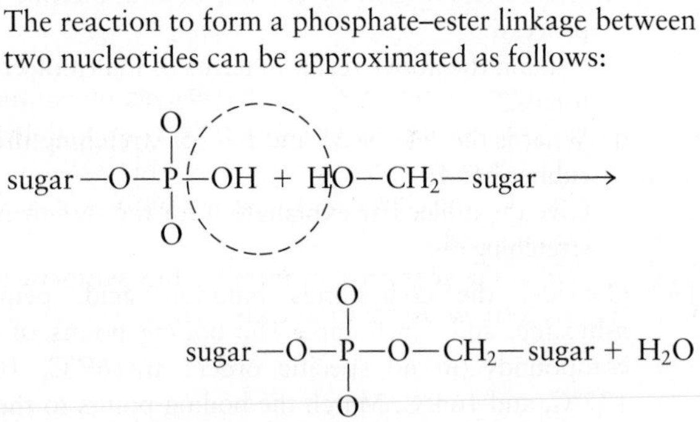

$$\text{sugar}-O-\underset{O}{\overset{O}{P}}-OH + HO-CH_2-\text{sugar} \longrightarrow$$

$$\text{sugar}-O-\underset{O}{\overset{O}{P}}-O-CH_2-\text{sugar} + H_2O$$

   Would you predict the formation of a dinucleotide from two nucleotides to be a spontaneous process?

144. Considering your answers to Exercises 142 and 143, how can you justify the existence of proteins and nucleic acids in light of the second law of thermodynamics?

145. The structure of tartaric acid is

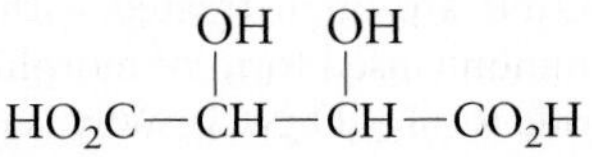

$$HO_2C-\underset{}{\overset{OH}{CH}}-\overset{OH}{CH}-CO_2H$$

   a. Is the form of tartaric acid pictured below optically active? Explain.

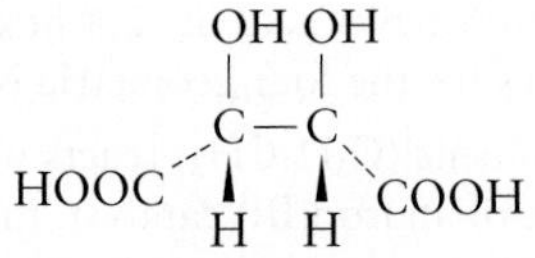

$$HOOC-\overset{OH}{\underset{H}{C}}-\overset{OH}{\underset{H}{C}}-COOH$$

   Note: The dashed lines show groups directed behind the plane of the page. The wedges show groups directed in front of the plane of the page.
   b. Draw the optically active forms of tartaric acid.

146. In glycine, the carboxylic acid group has $K_a = 4.3 \times 10^{-3}$ and the amino group has $K_b = 6.0 \times 10^{-5}$. Use these equilibrium constant values to calculate the equilibrium constants for the following.
   a. $^+H_3NCH_2CO_2^- + H_2O \rightleftharpoons H_2NCH_2CO_2^- + H_3O^+$
   b. $H_2NCH_2CO_2^- + H_2O \rightleftharpoons H_2NCH_2CO_2H + OH^-$
   c. $^+H_3NCH_2CO_2H \rightleftharpoons 2H^+ + H_2NCH_2CO_2^-$

147. The isoelectric point of an amino acid is the pH at which the molecule has no net charge. For glycine, that would be the pH at which virtually all glycine molecules are in the form $^+H_3NCH_2CO_2^-$. If we assume that the principal equilibrium is

$$2\,^+H_3NCH_2CO_2^- \rightleftharpoons H_2NCH_2CO_2^- + {}^+H_3NCH_2CO_2H \quad \text{(i)}$$

   then at equilibrium

$$[H_2NCH_2CO_2^-] = [^+H_3NCH_2CO_2H] \quad \text{(ii)}$$

   Use this result and your answer to part c of Exercise 146 to calculate the pH at which Equation (ii) is true. This will be the isoelectric point of glycine.

148. Nylon is named according to the number of C atoms between the N atoms in the chain. Nylon-46 has 4 C atoms then 6 C atoms, and this pattern repeats. Nylon-6 always has 6 atoms in a row. Speculate as to why nylon-46 is stronger than nylon-6. (*Hint:* Consider the strengths of interchain forces.)

149. A chemical "breathalyzer" test works because ethyl alcohol in the breath is oxidized by the dichromate ion (orange) to form acetic acid and chromium(III) ion (green). The balanced reaction is

$$3C_2H_5OH(aq) + 2Cr_2O_7^{2-}(aq) + 2H^+(aq) \longrightarrow 3HC_2H_3O_2(aq) + 4Cr^{3+}(aq) + 11H_2O(l)$$

   You analyze a breathalyzer test in which 4.2 mg of $K_2Cr_2O_7$ was reduced. Assuming the volume of the breath was 0.500 L at 30.°C and 750. mm Hg, what was the mole percent alcohol of the breath?

Challenge Problems

150. Alcohols are very useful starting materials for the production of many different compounds. The following conversions, starting with 1-butanol, can be carried out in two or more steps. Show the steps (reactants/catalysts) you would follow to carry out the conversions, drawing the formula for the organic product in each step. For each step, a major product must be produced.

See Exercise 58. (*Hint:* In the presence of H^+, an alcohol is converted into an alkene and water. This is the exact reverse of the reaction of adding water to an alkene to form an alcohol.)

a. 1-butanol $\longrightarrow$ butane
b. 1-butanol $\longrightarrow$ 2-butanone

151. Consider a sample of a hydrocarbon at 0.959 atm and 298 K. Upon combusting the entire sample in oxygen, you collect a mixture of gaseous carbon dioxide and water vapor at 1.51 atm and 375 K. This mixture has a density of 1.391 g/L and occupies a volume four times as large as that of the pure hydrocarbon. Determine the molecular formula of the hydrocarbon and name it.

152. Using one of the Lewis structures for benzene (C_6H_6), estimate ΔH_f° for $C_6H_6(g)$ using bond energies and given that the standard enthalpy of formation of C(*g*) is 717 kJ/mol. The experimental ΔH_f° value for $C_6H_6(g)$ is 83 kJ/mol. Explain the discrepancy between the experimental value and the calculated ΔH_f° value for $C_6H_6(g)$.

153. Consider the following reactions. For parts b–d, reference Exercise 58.

a. When C_5H_{12} is reacted with $Cl_2(g)$ in the presence of ultraviolet light, four different monochlorination products form. What is the structure of C_5H_{12} in this reaction?
b. When C_4H_8 is reacted with H_2O, a tertiary alcohol is produced as the major product. What is the structure of C_4H_8 in this reaction?
c. When C_7H_{12} is reacted with HCl, 1-chloro-1-methylcyclohexane is produced as the major product. What are the two possible structures for C_7H_{12} in this reaction?
d. When a hydrocarbon is reacted with H_2O and the major product of this reaction is then oxidized, acetone (2-propanone) is produced. What is the structure of the hydrocarbon in this reaction?
e. When $C_5H_{12}O$ is oxidized, a carboxylic acid is produced. What are the possible structures for $C_5H_{12}O$ in this reaction?

154. ABS plastic is a tough, hard plastic used in applications requiring shock resistance. (See Section 21.5 of the text.) The polymer consists of three monomer units: acrylonitrile (C_3H_3N), butadiene (C_4H_6), and styrene (C_8H_8).

a. Draw two repeating units of ABS plastic assuming the three monomer units react in a 1:1:1 mole ratio and react in the same order as the monomers previously listed.
b. A sample of ABS plastic contains 8.80% N by mass. It took 0.605 g of Br_2 to react completely with a 1.20-g sample of ABS plastic. What is the percent by mass of acrylonitrile, butadiene, and styrene in this polymer sample?
c. ABS plastic does not react in a 1:1:1 mole ratio between the three monomer units. Using the results from part b, determine the relative numbers of the monomer units in this sample of ABS plastic.

155. Consider the titration of 50.0 mL of 1.0 *M* glycine hydrochloride [$(H_3NCH_2COOH)Cl$], with 1.0 *M* NaOH. For $^+H_3NCH_2COOH$, K_a for the carboxylic acid group is 4.3×10^{-3} and K_b for the amino group is 6.0×10^{-5}.

a. Calculate the pH after 25.0 mL, 50.0 mL, and 75.0 mL of NaOH has been added. *Hint:* Refer to Section 8.7 of the text, and treat the titration of $^+H_3NCH_2COOH$ as a diprotic acid titration.
b. Sketch the titration curve and indicate the major amino acid species present after 0.0 mL, 25.0 mL, 50.0 mL, 75.0 mL, and 100.0 mL of 1.0 *M* NaOH have reacted completely. Assume the initial pH is 1.2 and the pH at the second equivalence point is 11.7.
c. At what pH do the majority of amino acid molecules have a net charge of zero? This pH is called the isoelectric point.
d. At what pH is the net charge of the major amino acid species present equal to $+\frac{1}{2}$? $-\frac{1}{2}$?

156. In 1994 chemists at Texas A&M University reported the synthesis of a non-naturally occurring amino acid (*C & E News,* April 18, 1994, pp. 26–27):

$$\begin{array}{c} H_2N \quad CH_2 \quad H \\ \diagdown \;\diagup\;\diagdown\; \diagup \\ C—C \\ \diagup \qquad\quad \diagdown \\ CO_2H \qquad CH_2SCH_3 \end{array}$$

a. To which naturally occurring amino acid is this compound most similar?
b. A tetrapeptide, phe-met-arg-phe—NH_2, is synthesized in the brains of rats addicted to morphine and heroin. (The —NH_2 indicates that the peptide ends in $-\overset{\overset{\displaystyle O}{\|}}{C}-NH_2$ instead of —CO_2H.) The TAMU scientists synthesized a similar tetrapeptide, with the synthetic amino acid above replacing one of the original amino acids. Draw a structure for the tetrapeptide containing the synthetic amino acid.
c. Indicate the chiral carbon atoms in the synthetic amino acid.
d. Draw the geometric isomers for this synthetic amino acid. (*Hint:* Refer to Exercise 25.)

Appendixes

Appendix One Mathematical Procedures

A1.1 Exponential Notation

The numbers characteristic of scientific measurements are often very large or very small; thus it is convenient to express them by using powers of 10. For example, the number 1,300,000 can be expressed as 1.3×10^6, which means multiply 1.3 by 10 six times:

$$1.3 \times 10^6 = 1.3 \times \underbrace{10 \times 10 \times 10 \times 10 \times 10 \times 10}_{10^6 = 1 \text{ million}}$$

Note that each multiplication by 10 moves the decimal point one place to the right, and the easiest way to interpret the notation 1.3×10^6 is that it means move the decimal point in 1.3 to the right six times.

In this notation the number 1985 can be expressed as 1.985×10^3. Note that the usual convention is to write the number that appears before the power of 10 as a number between 1 and 10. Some other examples are given below.

Number	Exponential Notation
5.6	5.6×10^0 or 5.6×1
39	3.9×10^1
943	9.43×10^2
1126	1.126×10^3

To represent a number smaller than 1 in exponential notation, start with a number between 1 and 10 and *divide* by the appropriate power of 10:

$$0.0034 = \frac{3.4}{10 \times 10 \times 10} = \frac{3.4}{10^3} = 3.4 \times 10^{-3}$$

Division by 10 moves the decimal point one place to the *left*. Thus the number 0.00000014 can be written as 1.4×10^{-7}.

To summarize, we can write any number in the form

$$N \times 10^{\pm n}$$

where N is between 1 and 10 and the exponent n is an integer. If the sign preceding n is positive, it means the decimal point in N should be moved n places to the right. If a negative sign precedes n, the decimal point in N should be moved n places to the left.

Multiplication and Division

When two numbers expressed in exponential notation are multiplied, the initial numbers are multiplied and the exponents of 10 are added:

$$(M \times 10^m)(N \times 10^n) = (MN) \times 10^{m+n}$$

For example,

$$(3.2 \times 10^4)(2.8 \times 10^3) = 9.0 \times 10^7$$

When the numbers are multiplied, if a result greater than 10 is obtained for the initial number, the number is adjusted to conventional notation:

$$(5.8 \times 10^2)(4.3 \times 10^8) = 24.9 \times 10^{10} = 2.49 \times 10^{11} = 2.5 \times 10^{11}$$

Division of two numbers expressed in exponential notation involves normal division of the initial numbers and *subtraction* of the exponent of the divisor from that of the dividend. For example,

$$\frac{4.8 \times 10^8}{\underbrace{2.1 \times 10^3}_{\text{Divisor}}} = \frac{4.8}{2.1} \times 10^{(8-3)} = 2.3 \times 10^5$$

Addition and Subtraction

When we add or subtract numbers expressed in exponential notation, *the exponents of the numbers must be the same*. For example, to add 1.31×10^5 and 4.2×10^4, rewrite one number so that the exponents of both are the same:

$$\begin{array}{r} 13.1 \times 10^4 \\ +\ \ 4.2 \times 10^4 \\ \hline 17.3 \times 10^4 \end{array}$$

In correct exponential notation, the result is expressed as 1.73×10^5.

Powers and Roots

When a number expressed in exponential notation is taken to some power, the initial number is taken to the appropriate power and the exponent of 10 is multiplied by that power:

$$(N \times 10^n)^m = N^m \times 10^{m \cdot n}$$

For example,*

$$(7.5 \times 10^2)^3 = 7.5^3 \times 10^{3 \cdot 2} = 422 \times 10^6 = 4.22 \times 10^8$$
$$= 4.2 \times 10^8 \text{ (rounded to 2 significant figures)}$$

When a root is taken of a number expressed in exponential notation, the root of the initial number is taken and the exponent of 10 is divided by the number representing the root:

$$\sqrt{N \times 10^n} = (N \times 10^n)^{1/2} = \sqrt{N} \times 10^{n/2}$$

For example, $(2.9 \times 10^6)^{1/2} = \sqrt{2.9} \times 10^{6/2} = 1.7 \times 10^3$

Because the exponent of the result must be an integer, we may sometimes have to change the form of the number so that the power divided by the root equals an integer; for example,

$$\sqrt{1.9 \times 10^3} = (1.9 \times 10^3)^{1/2} = (0.19 \times 10^4)^{1/2}$$
$$= \sqrt{0.19} \times 10^2 = 0.44 \times 10^2$$
$$= 4.4 \times 10^1$$

*Refer to the instruction booklet for your calculator for directions concerning how to take roots and powers of numbers.

The same procedure is followed for roots other than square roots; for example,

$$\sqrt[3]{4.6 \times 10^{10}} = (4.6 \times 10^{10})^{1/3} = (46 \times 10^{9})^{1/3}$$

$$= \sqrt[3]{46} \times 10^{3} = 3.6 \times 10^{3}$$

A1.2 Logarithms

A logarithm is an exponent. Any number N can be expressed as follows:

$$N = 10^{x}$$

For example,

$$1000 = 10^{3}$$

$$100 = 10^{2}$$

$$10 = 10^{1}$$

$$1 = 10^{0}$$

The common, or base 10, logarithm of a number is the power to which 10 must be taken to yield that number. Thus, since $1000 = 10^{3}$,

$$\log 1000 = 3$$

Similarly,

$$\log 100 = 2$$

$$\log 10 = 1$$

$$\log 1 = 0$$

For a number between 10 and 100, the required exponent of 10 will be between 1 and 2. For example, $65 = 10^{1.8129}$; that is, $\log 65 = 1.8129$. For a number between 100 and 1000, the exponent of 10 will be between 2 and 3. For example, $650 = 10^{2.8129}$ and $\log 650 = 2.8129$.

A number N greater than 0 and less than 1 can be expressed as follows:

$$N = 10^{-x} = \frac{1}{10^{x}}$$

For example,

$$0.001 = \frac{1}{1000} = \frac{1}{10^{3}} = 10^{-3}$$

$$0.01 = \frac{1}{100} = \frac{1}{10^{2}} = 10^{-2}$$

$$0.1 = \frac{1}{10} = \frac{1}{10^{1}} = 10^{-1}$$

Thus

$$\log 0.001 = -3$$

$$\log 0.01 = -2$$

$$\log 0.1 = -1$$

Although common logs are often tabulated, the most convenient method for obtaining such logs is to use a calculator.

Since logs are simply exponents, they are manipulated according to the rules for exponents. For example, if $A = 10^{x}$ and $B = 10^{y}$, then their product is

$$A \cdot B = 10^{x} \cdot 10^{y} = 10^{x+y}$$

and

$$\log AB = x + y = \log A + \log B$$

For division we have

$$\frac{A}{B} = \frac{10^{x}}{10^{y}} = 10^{x-y}$$

and $$\log \frac{A}{B} = x - y = \log A - \log B$$

For a number raised to a power, we have

$$A^n = (10^x)^n = 10^{nx}$$

and $$\log A^n = nx = n \log A$$

It follows that $$\log \frac{1}{A^n} = \log A^{-n} = -n \log A$$

or for $n = 1$, $$\log \frac{1}{A} = -\log A$$

When a common log is given, to find the number it represents, we must carry out the process of exponentiation. For example, if the log is 2.673, then $N = 10^{2.673}$. The process of exponentiation is also called taking the antilog, or the inverse logarithm, and is easily carried out by using a calculator.

A second type of logarithm, the natural logarithm, is based on the number 2.7183, which is referred to as *e*. In this case a number is represented as $N = e^x = 2.7183^x$. For example,

$$N = 7.15 = e^x$$

$$\ln 7.15 = x = 1.967$$

If a natural logarithm is given, to find the number it represents, we must carry out exponentiation to the base *e* (2.7183) by using a calculator.

A1.3 Graphing Functions

In the interpretation of the results of a scientific experiment, it is often useful to make a graph. It is usually most convenient to graph the function in a form that gives a straight line. The equation for a straight line (a *linear equation*) can be represented by the general form

$$y = mx + b$$

where y is the *dependent variable*, x is the *independent variable*, m is the *slope*, and b is the *intercept* with the y axis.

As an illustration of the characteristics of a linear equation, the function $y = 3x + 4$ is plotted in Fig. A1.1. For this equation, $m = 3$ and $b = 4$. Note that the y intercept occurs when $x = 0$. In this case the intercept is 4, as can be seen from the equation ($b = 4$).

The slope of a straight line is defined as the ratio of the rate of change in y to that in x:

$$m = \text{slope} = \frac{\Delta y}{\Delta x}$$

For the equation $y = 3x + 4$, y changes three times as fast as x (since x has a coefficient of 3). Thus the slope in this case is 3. This can be verified from the graph. For the triangle shown in Fig. A1.1:

$$\text{Slope} = \frac{\Delta y}{\Delta x} = \frac{24}{8} = 3$$

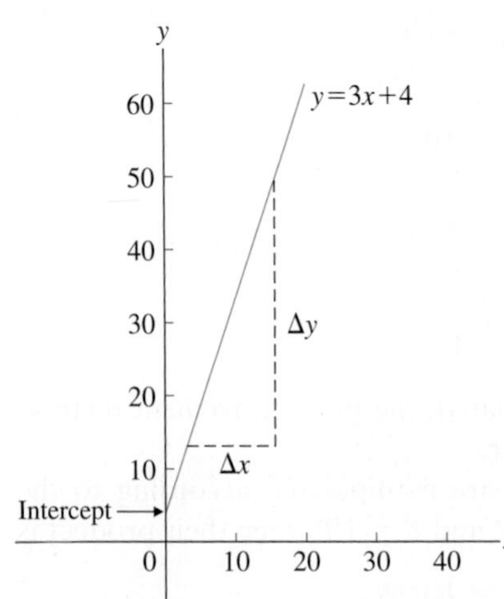

Figure A1.1
Graph of the linear equation $y = 3x + 4$.

Sometimes an equation that is not in standard form can be changed to the form $y = mx + b$ by rearrangement or mathematical manipulation. An example is the equation $k = Ae^{-E_a/RT}$, where A, E_a, and R are constants, k is the

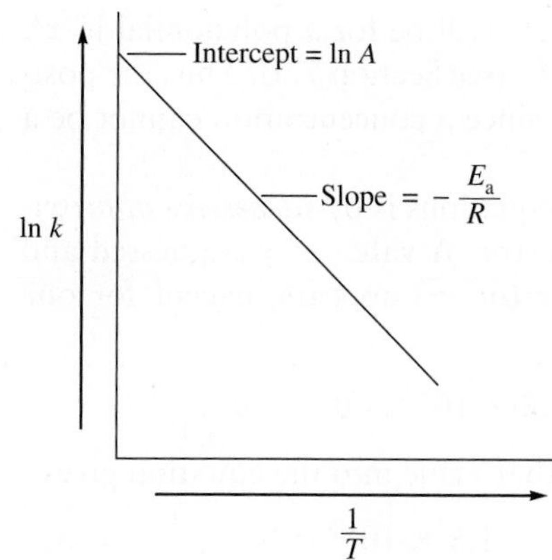

Figure A1.2
Graph of ln k *versus* $1/T$.

dependent variable, and $1/T$ is the independent variable. This equation can be changed to standard form by taking the natural logarithm of both sides,

$$\ln k = \ln Ae^{-E_a/RT} = \ln A + \ln e^{-E_a/RT} = \ln A - \frac{E_a}{RT}$$

noting that the log of a product is equal to the sum of the logs of the individual terms and that the natural log of $e^{-E_a/RT}$ is simply the exponent $-E_a/RT$. Thus in standard form the equation $k = Ae^{-E_a/RT}$ is written

$$\underbrace{\ln k}_{y} = \underbrace{-\frac{E_a}{R}}_{m}\underbrace{\left(\frac{1}{T}\right)}_{x} + \underbrace{\ln A}_{b}$$

A plot of ln k versus $1/T$ (see Fig. A1.2) gives a straight line with slope $-E_a/R$ and intercept ln A.

Of course, many relationships that arise from the description of natural systems are nonlinear, and the "slope" of a curve is continuously changing. In this case the instantaneous slope is given by the tangent to the curve at that point, which is described by a new function obtained by taking the derivative of the original function. For example, for the function in x, $f = ax^2$, the derivative (df/dx) is $2ax$. Thus the slope at each point on the curve defined by the function ax^2 is given by $2ax$.

A1.4 Solving Quadratic Equations

A *quadratic equation,* a polynomial in which the highest power of x is 2, can be written as

$$ax^2 + bx + c = 0$$

One method for finding the two values of x that satisfy a quadratic equation is to use the *quadratic formula:*

$$x = \frac{-b \pm \sqrt{b^2 - 4ac}}{2a}$$

where a, b, and c represent the coefficients of x^2 and x and the constant, respectively. For example, in the determination of $[H^+]$ in a solution of 1.0×10^{-4} M acetic acid, the following expression arises:

$$1.8 \times 10^{-5} = \frac{x^2}{1.0 \times 10^{-4} - x}$$

which yields $$x^2 + (1.8 \times 10^{-5})x - 1.8 \times 10^{-9} = 0$$

where $a = 1$, $b = 1.8 \times 10^{-5}$, and $c = -1.8 \times 10^{-9}$. Using the quadratic formula, we have

$$x = \frac{-b \pm \sqrt{b^2 - 4ac}}{2a}$$

$$= \frac{-1.8 \times 10^{-5} \pm \sqrt{3.24 \times 10^{-10} - (4)(1)(-1.8 \times 10^{-9})}}{2(1)}$$

and $$x = \frac{6.9 \times 10^{-5}}{2} = 3.5 \times 10^{-5}$$

or $$x = \frac{-10.5 \times 10^{-5}}{2} = -5.2 \times 10^{-5}$$

Note that there are two roots, as there always will be for a polynomial in x^2. In this case x represents a concentration of H^+ (see Section 7.5). Thus the positive root is the one that solves the problem, since a concentration cannot be a negative number.

A second method for solving quadratic equations is by *successive approximations,* a systematic method of trial and error. A value of x is guessed and substituted into the equation everywhere x (or x^2) appears, except for one place. For example, for the equation

$$x^2 + (1.8 \times 10^{-5})x - 1.8 \times 10^{-9} = 0$$

we might guess $x = 2 \times 10^{-5}$. Substituting that value into the equation gives

$$x^2 + (1.8 \times 10^{-5})(2 \times 10^{-5}) - 1.8 \times 10^{-9} = 0$$

or $$x^2 = 1.8 \times 10^{-9} - 3.6 \times 10^{-10} = 1.4 \times 10^{-9}$$

Thus $$x = 3.7 \times 10^{-5}$$

Note that the guessed value of x (2×10^{-5}) is not the same as the value of x that is calculated (3.7×10^{-5}) after inserting the estimated value. This means that $x = 2 \times 10^{-5}$ is not the correct solution, and we must try another guess.

We take the calculated value (3.7×10^{-5}) as our next guess:

$$x^2 + (1.8 \times 10^{-5})(3.7 \times 10^{-5}) - 1.8 \times 10^{-9} = 0$$

$$x^2 = 1.8 \times 10^{-9} - 6.7 \times 10^{-10} = 1.1 \times 10^{-9}$$

Thus $$x = 3.3 \times 10^{-5}$$

Now we compare the two values of x again:

Guessed: $x = 3.7 \times 10^{-5}$

Calculated: $x = 3.3 \times 10^{-5}$

These values are closer but still not identical.

Next, we try 3.3×10^{-5} as our guess:

$$x^2 + (1.8 \times 10^{-5})(3.3 \times 10^{-5}) - 1.8 \times 10^{-9} = 0$$

$$x^2 = 1.8 \times 10^{-9} - 5.9 \times 10^{-10} = 1.2 \times 10^{-9}$$

Thus $$x = 3.5 \times 10^{-5}$$

Compare:

Guessed: $x = 3.3 \times 10^{-5}$

Calculated: $x = 3.5 \times 10^{-5}$

Next, we guess $x = 3.5 \times 10^{-5}$, which leads to

$$x^2 + (1.8 \times 10^{-5})(3.5 \times 10^{-5}) - 1.8 \times 10^{-9} = 0$$

$$x^2 = 1.8 \times 10^{-9} - 6.3 \times 10^{-10} = 1.2 \times 10^{-9}$$

Thus $$x = 3.5 \times 10^{-5}$$

Now the guessed value and the calculated value are the same; we have found the correct solution. Note that this agrees with one of the roots found with the quadratic formula in the first method above.

To further illustrate the method of successive approximations, we will solve Example 7.9 by using this procedure. In solving for $[H^+]$ for 0.010 *M* H_2SO_4, we obtain the following expression:

$$1.2 \times 10^{-2} = \frac{x(0.010 + x)}{0.010 - x}$$

which can be rearranged to give

$$x = (1.2 \times 10^{-2})\left(\frac{0.010 - x}{0.010 + x}\right)$$

We will guess a value for x, substitute it into the right side of the equation, and then calculate a value for x. In guessing a value for x, we know it must be less than 0.010, since a larger value would make the calculated value for x negative and the guessed and calculated values will never match. We start by guessing $x = 0.005$.

The results of the successive approximations are shown in the following table:

Trial	Guessed Value for x	Calculated Value for x
1	0.0050	0.0040
2	0.0040	0.0051
3	0.00450	0.00455
4	0.00452	0.00453

Note that the first guess was close to the actual value and that there was oscillation between 0.004 and 0.005 for the guessed and calculated values. For trial 3, an average of these values was used as the guess, and this led rapidly to the correct value (0.0045 to the correct number of significant figures). Also note that it is useful to carry extra digits until the correct value is obtained, which is then rounded off to the correct number of significant figures.

The method of successive approximations is especially useful for solving polynomials containing x to a power of 3 or higher. The procedure is the same as for quadratic equations: Substitute a guessed value for x into the equation for every x term but one, and then solve for x. Continue this process until the guessed and calculated values agree.

A1.5 Uncertainties in Measurements

The number associated with a measurement is obtained by using some measuring device. For example, consider the measurement of the volume of a liquid in a buret, as shown in Fig. A1.3, where the scale is greatly magnified. The volume is about 22.15 mL. Note that the last number must be estimated by interpolating between the 0.1-mL marks. Since the last number is estimated, its value may vary depending on who makes the measurement. If five different people read the same volume, the results might be as follows:

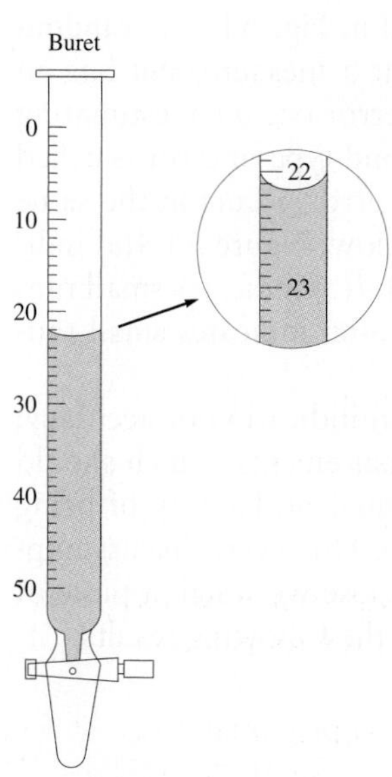

Figure A1.3
Measurement of volume using a buret. The volume is read at the bottom of the liquid curve (called the meniscus).

Person	Result of Measurement
1	22.15 mL
2	22.14 mL
3	22.16 mL
4	22.17 mL
5	22.16 mL

Note from these results that the first three numbers (22.1) remain the same regardless of who makes the measurement; these are called certain digits. However, the digit to the right of the 1 must be estimated and thus varies; it is called an uncertain digit. We customarily report a measurement by recording all the certain digits plus the *first* uncertain digit. In our example it would not make any sense to try to record the volume to thousandths of a milliliter because the value for hundredths of a milliliter must be estimated when using the buret.

It is very important to realize that a *measurement always has some degree of uncertainty*. The uncertainty of a measurement depends on the precision of the measuring device. For example, using a bathroom scale, you might estimate that the mass of a grapefruit is about 1.5 pounds. Weighing the same grapefruit on a highly precise balance might produce a result of 1.476 pounds. In the first case the uncertainty occurs in the tenths of a pound place; in the second case the uncertainty occurs in the thousandths of a pound place. Suppose we weigh two similar grapefruit on the two devices and obtain the following results:

	Bathroom Scale	Balance
Grapefruit 1	1.5 lb	1.476 lb
Grapefruit 2	1.5 lb	1.518 lb

Do the two grapefruits have the same mass? The answer depends on which set of results you consider. Thus a conclusion based on a series of measurements depends on the certainty of those measurements. For this reason, it is important to indicate the uncertainty in any measurement. This is done by always recording the certain digits and the first uncertain digit (the estimated number). These numbers are called the **significant figures** of a measurement.

The convention of significant figures automatically gives an indication of the uncertainty in a measurement. The uncertainty in the last number (the estimated number) is usually assumed to be ± 1 unless otherwise indicated. For example, the measurement 1.86 kilograms can be interpreted to mean 1.86 ± 0.01 kilograms.

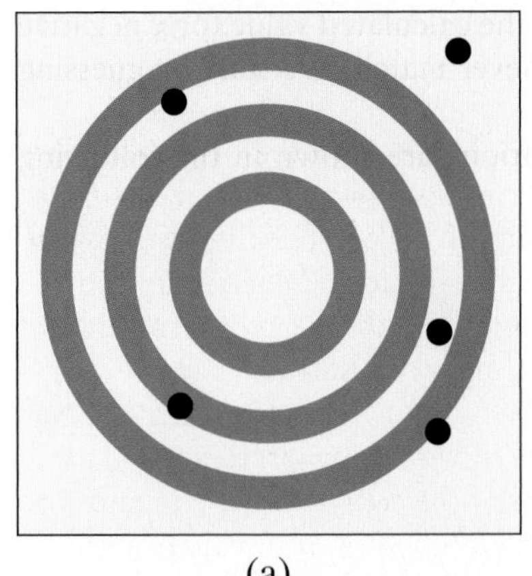

(a)

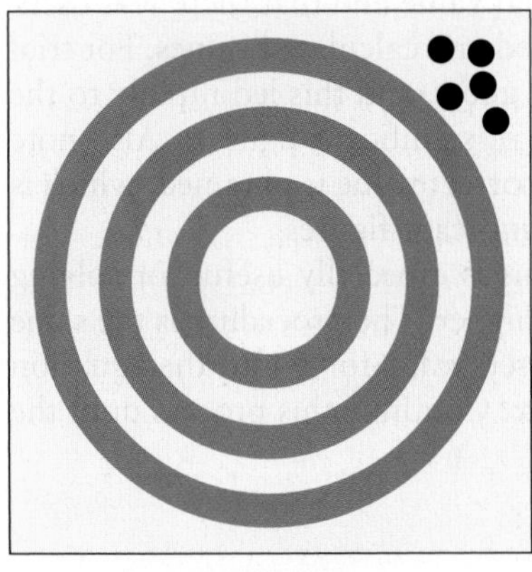

(b)

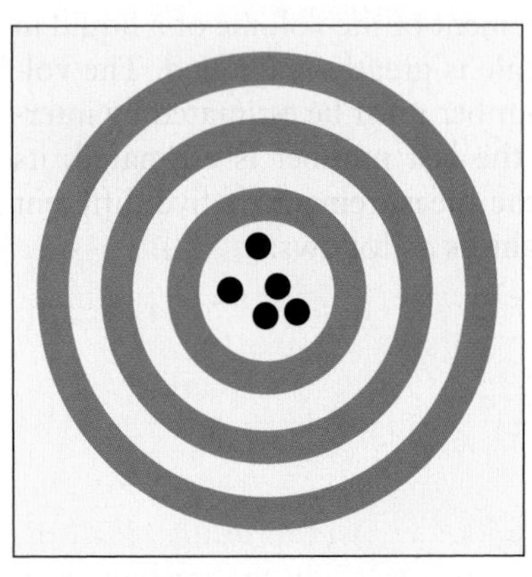

(c)

Figure A1.4
Shooting targets show the difference between *precise* and *accurate*. (a) Neither accurate nor precise (large random errors). (b) Precise but not accurate (small random errors, large systematic error). (c) Bull's-eye! Both precise and accurate (small random errors, no systematic error).

Precision and Accuracy

Two terms often used to describe uncertainty in measurements are *precision* and *accuracy*. Although these words are frequently used interchangeably in everyday life, they have different meanings in the scientific context. **Accuracy** refers to the agreement of a particular value with the true value. **Precision** refers to the degree of agreement among several measurements of the same quantity. Precision reflects the *reproducibility* of a given type of measurement. The difference between these terms is illustrated by the results of three different target practices shown in Fig. A1.4.

Two different types of errors are also introduced in Fig. A1.4. A **random error** (also called an indeterminate error) means that a measurement has an equal probability of being high or low. This type of error occurs in estimating the value of the last digit of a measurement. The second type of error is called **systematic error** (or determinate error). This type of error occurs in the same direction each time; it is either always high or always low. Figure A1.4(a) indicates large random errors (poor technique). Figure A1.4(b) indicates small random errors but a large systematic error, and Fig. A1.4(c) indicates small random errors and no systematic error.

In quantitative work precision is often used as an indication of accuracy; we assume that the *average* of a series of precise measurements (which should "average out" the random errors because of their equal probability of being high or low) is accurate, or close to the "true" value. However, this assumption is valid only if systematic errors are absent. Suppose we weigh a piece of brass five times on a very precise balance and obtain the following results:

Weighing	Result
1	2.486 g
2	2.487 g
3	2.485 g
4	2.484 g
5	2.488 g

Normally, we would assume that the true mass of the piece of brass is very close to 2.486 grams, which is the average of the five results. However, if the balance has a defect causing it to give a result that is consistently 1.000 gram too high (a systematic error of +1.000 gram), then 2.486 grams would be seriously in error. The point here is that high precision among several measurements is an indication of accuracy *only* if you can be sure that systematic errors are absent.

Expression of Experimental Results

The accuracy of a measurement refers to how close it is to the true value. An inaccurate result occurs as a result of some flaw (systematic error) in the measurement: the presence of an interfering substance, incorrect calibration of an instrument, operator error, and so on. The goal of chemical analysis is to eliminate systematic error, but random errors can only be minimized. In practice, an experiment is almost always done in order to find an unknown value (the true value is not known—someone is trying to obtain that value by doing the experiment). In this case the precision of several replicate determinations is used to assess the accuracy of the result. The results of the replicate experiments are expressed as an average (which we assume is close to the true value) with an error limit that gives some indication of how close the average value may be to the true value. The error limit represents the uncertainty of the experimental result.

To illustrate this procedure, consider a situation that might arise in the pharmaceutical industry. Assume that the specification for a commercial 500-mg acetaminophen (the active painkiller in Tylenol) tablet is that each batch of tablets must contain 450 to 550 mg of acetaminophen per tablet. Suppose that chemical analysis gave the following results for a batch of acetaminophen tablets: 428, 479, 442, and 435 mg. How can these results be used to decide whether the batch of tablets meets the specification? Although the details of how to draw such conclusions from measured data are beyond the scope of this discussion, we will consider some aspects of this process. We will focus here on the types of experimental uncertainty, the expression of experimental results, and a simplified method for estimating experimental uncertainty when several types of measurements contribute to the final result.

There are two common ways of expressing an average: the mean and the median. The mean ($\bar{x}$) is the arithmetic average of the results, or

$$\text{Mean} = \bar{x} = \sum_{i=1}^{n} \frac{x_i}{n} = \frac{x_1 + x_2 + \cdots + x_n}{n}$$

where Σ means take the sum of the values. The mean is equal to the sum of all the measurements divided by the number of measurements. For the acetaminophen results given previously, the mean is

$$\bar{x} = \frac{428 + 479 + 442 + 435}{4} = 446 \text{ mg}$$

The median is the value that lies in the middle among the results. Half of the measurements are above the median and half are below the median. For results of 465, 485, and 492 mg, the median is 485 mg. When there is an even number of results, the median is the average of the two middle results. For the acetaminophen results, the median is

$$\frac{442 + 435}{2} = 439 \text{ mg}$$

There are several advantages to using the median. If a small number of measurements is made, one value can greatly affect the mean. Consider the results for the analysis of acetaminophen: 428, 479, 442, and 435 mg. The mean is 446 mg, which is larger than three of the four results. The median is 439 mg, which lies near the three values that are relatively close to one another.

In addition to expressing an average value for a series of results, we must also express the uncertainty. This usually means expressing either the precision of the measurements or the observed range of the measurements. The range of a series of measurements is defined by the smallest value and the largest value. For the analytical results on the acetaminophen tablets, the range is from 428 to 479 mg. Using this range, we can express the results by saying that the true value lies between 428 and 479 mg. That is, we can express the amount of acetaminophen in a typical tablet as 446 ± 33 mg, where the error limit is chosen to give the observed range (approximately).

The most common way to specify precision is by the standard deviation s, which for a small number of measurements is given by the formula

$$s = \left[\frac{\sum_{i=1}^{n} (x_i - \bar{x})^2}{n - 1} \right]^{1/2}$$

where x_i is an individual result, $\bar{x}$ is the average (either mean or median), and n is the total number of measurements. For the acetaminophen example, we have

$$s = \left[\frac{(428 - 446)^2 + (479 - 446)^2 + (442 - 446)^2 + (435 - 446)^2}{4 - 1} \right]^{1/2} = 23$$

Thus we can say that the amount of acetaminophen in a typical tablet in the batch of tablets is 446 mg with a sample standard deviation of 23 mg. Statistically, this means that any additional measurement has a 68% probability (68 chances out of 100) of being between 423 mg (446 − 23) and 469 mg (446 + 23). Thus the standard deviation is a measure of the precision of a given type of measurement.

In scientific calculations it is also useful to be able to estimate the precision of a procedure that involves several measurements by combining the precisions of the individual steps. That is, we want to answer the following question: How do the uncertainties propagate when we combine the results of several different types of measurements? There are many ways to deal with the propagation of uncertainty. We will discuss one simple method below.

Worst-Case Method for Estimating Experimental Uncertainty

To illustrate this method, we will consider the determination of the density of an irregularly shaped solid. In this determination we make three measurements. First, we measure the mass of the object on a balance. Next, we must obtain the volume of the solid. The easiest method for doing this is to partially fill a graduated cylinder with a liquid and record the volume. Then we add the solid and record the volume again. The difference in the measured volumes is the volume of the solid. We can then calculate the density of the solid from the equation

$$D = \frac{M}{V_2 - V_1}$$

where M is the mass of the solid, V_1 is the initial volume of liquid in the graduated cylinder, and V_2 is the volume of liquid plus solid. Suppose we get the following results:

$$M = 23.06 \text{ g}$$
$$V_1 = 10.4 \text{ mL}$$
$$V_2 = 13.5 \text{ mL}$$

The calculated density is

$$\frac{23.06 \text{ g}}{13.5 \text{ mL} - 10.4 \text{ mL}} = 7.44 \text{ g/mL}$$

Now suppose that the precision of the balance used is ±0.02 g and that the volume measurements are precise to ±0.05 mL. How do we estimate the uncertainty of the density? We can do this by assuming a worst case. That is, we assume the largest uncertainties in all measurements, and we see what combinations of measurements will give the largest and smallest possible results (the greatest range). Since the density is the mass divided by the volume, the largest value of the density will be that obtained by using the largest possible mass and the smallest possible volume:

Largest possible mass = 23.06 + 0.02

$$D_{\text{max}} = \frac{23.08}{13.45 - 10.45} = 7.69 \text{ g/mL}$$

Smallest possible V_2 (13.45); Largest possible V_1 (10.45)

The smallest value of the density is

Smallest possible mass

$$D_{\text{min}} = \frac{23.04}{13.55 - 10.35} = 7.20 \text{ g/mL}$$

Largest possible V_2 (13.55); Smallest possible V_1 (10.35)

Thus the calculated range is from 7.20 to 7.69, and the average of these values is 7.45. The error limit is the number that gives the high and low range values when added and subtracted from the average. Therefore, we can express the density as 7.45 ± 0.25 g/mL, which is the average value plus or minus the quantity that gives the range calculated by assuming the largest uncertainties.

Analysis of the propagation of uncertainties is useful in drawing qualitative conclusions from the analysis of measurements. For example, suppose that we obtained the preceding results for the density of an unknown alloy and we want to know if it is one of the following alloys:

Alloy A:	$D = 7.58$ g/mL
Alloy B:	$D = 7.42$ g/mL
Alloy C:	$D = 8.56$ g/mL

We can safely conclude that the alloy is not C. But the values of the densities for alloys A and B are both within the inherent uncertainty of our method. To distinguish between A and B, we need to improve the precision of our determination. The obvious choice is to improve the precision of the volume measurement.

The worst-case method is useful for estimating the maximum uncertainty expected when the results of several measurements are combined to obtain a result. We assume the maximum uncertainty in each measurement and then calculate the minimum and maximum possible results. These extreme values describe the range and thus the maximum error limit associated with a particular determination.

Table A1.1

Values of t for 90% and 95% Confidence Levels

	Values of t for Confidence Intervals	
n	90%	95%
2	6.31	12.7
3	2.92	4.30
4	2.35	3.18
5	2.13	2.78
6	2.02	2.57
7	1.94	2.45
8	1.90	2.36
9	1.86	2.31
10	1.83	2.26

Confidence Limits

A more sophisticated method for estimating the uncertainty of a particular type of determination involves the use of confidence limits. A confidence limit is defined as

$$\text{Confidence limit} = \pm \frac{ts}{\sqrt{n}}$$

where t = a weighting factor based on statistical analysis

s = the standard deviation

n = the number of experiments carried out

In this context an experiment may refer to a single type of measurement (for example, weighing an object) or to a procedure that requires various types of measurements to obtain a given final result (for example, obtaining the percentage of iron in a particular sample of iron ore). Some representative values of t are listed in Table A1.1.

A 95% confidence level means that the true value (the *average* obtained if the experiment were repeated an *infinite* number of times) will lie within $\pm ts/\sqrt{n}$ of the *observed* average (obtained from n experiments) with a 95% probability (95 of 100 times). Thus the factor $\pm ts/\sqrt{n}$ represents an error limit for a given set of results from a particular type of experiment. Thus we might represent the result of n determinations as

$$\bar{x} \pm \frac{ts}{\sqrt{n}}$$

where $\bar{x}$ is the average of the results from the n experiments. This type of error limit is expected to be considerably smaller than that obtained from a worst-case analysis.

A1.6 Significant Figures

Calculating the final result for an experiment usually involves adding, subtracting, multiplying, or dividing the results of various types of measurements. Thus it is important to be able to estimate the uncertainty in the final result. In the previous section we have considered this process in some detail. A closely related matter concerns the number of digits that should be retained in the result of a given calculation. In other words, how many of the digits in the result are significant (meaningful) relative to the uncertainty expected in the result? From statistical analyses of how uncertainties accumulate when arithmetic operations are carried out, rules have been developed for determining the correct number of significant figures in a final result. First, we must consider how to count the number of significant figures (digits) represented in a particular number.

Rules for Counting Significant Figures (Digits)

1. *Nonzero integers.* Nonzero integers always count as significant figures.
2. *Zeros.* There are three classes of zeros:
   a. *Leading zeros* are zeros that *precede* all the nonzero digits. They do not count as significant figures. In the number 0.0025 the three zeros simply indicate the position of the decimal point. This number has only two significant figures.
   b. *Captive zeros* are zeros *between* nonzero digits. They always count as significant figures. The number 1.008 has four significant figures.

c. *Trailing zeros* are zeros at the *right end* of the number. They are significant only if the number contains a decimal point. The number 100 has only one significant figure, whereas the number 1.00×10^2 has three significant figures. The number one hundred written as 100. also has three significant figures.

3. *Exact numbers.* Many times calculations involve numbers that were not obtained by using measuring devices but were determined by counting: 10 experiments, 3 apples, 8 molecules. Such numbers are called *exact numbers.* They can be assumed to have an infinite number of significant figures. Other examples of exact numbers are the 2 in $2\pi r$ (the circumference of a circle) and the 4 and the 3 in $\frac{4}{3}\pi r^3$ (the volume of a sphere). Exact numbers can also arise from definitions. For example, one inch is defined as exactly 2.54 centimeters. Thus, in the statement 1 in = 2.54 cm, neither the 2.54 nor the 1 limits the number of significant figures when used in a calculation.

The following rules apply for determining the number of significant figures in the result of a calculation.

Rules for Significant Figures in Mathematical Operations*

1. For *multiplication or division* the number of significant figures in the result is the same as the number in the least precise measurement used in the calculation. For example, consider this calculation:

$$4.56 \times 1.4 = 6.38 \xrightarrow{\text{Corrected}} 6.4$$

↑ Limiting term has two significant figures — Two significant figures

The correct product has only two significant figures, since 1.4 has two significant figures.

2. *For addition or subtraction* the result has the same number of decimal places as the least precise measurement used in the calculation. For example, consider the following sum:

12.11
18.0 ← Limiting term has one decimal place
1.013
———
$31.123 \xrightarrow{\text{Corrected}} 31.1$

↑ One decimal place

The correct result is 31.1, since 18.0 has only one decimal place.

Note that for multiplication and division significant figures are counted. For addition and subtraction the decimal places are counted.

In most calculations you will need to round off numbers to obtain the correct number of significant figures. The following rules should be applied for rounding.

*Although these rules work well for most cases, they can give misleading results in certain cases. For a discussion of this, see L. M. Schwartz, "Propagation of Significant Figures," *J. Chem. Ed.* **62** (1985):693.

Rules for Rounding

1. In a series of calculations, carry the extra digits through to the final result, *then* round off.*
2. If the digit to be removed†
   a. is less than 5, the preceding digit stays the same. For example, 1.33 rounds to 1.3.
   b. is equal to or greater than 5, the preceding digit is increased by 1. For example, 1.36 rounds to 1.4.

When rounding, use only the first number to the right of the last significant figure. Do not round off sequentially. For example, the number 4.348 when rounded to two significant figures is 4.3, not 4.4.

Appendix Two Units of Measurement and Conversions Among Units

A2.1 Measurements

Making observations is fundamental to all science. A quantitative observation, or **measurement,** always consists of two parts: a *number* and a scale (a *unit*). Both parts must be present for the measurement to be meaningful.

The two most widely used systems of units are the *English system* used in the United States and the *metric system* used by most of the rest of the industrialized world. This duality obviously causes a good deal of trouble; for example, parts as simple as bolts are not interchangeable between machines built using the different systems. As a result, the United States has begun to adopt the metric system.

For many years, most scientists worldwide have used the metric system. In 1960 an international agreement established a system of units called the *International System* (*le Système International* in French), abbreviated **SI.** This system is based on the metric system and the units derived from the metric system. The fundamental SI units are listed in Table A2.1.

Because the fundamental units are not always convenient (expressing the mass of a pin in kilograms is awkward), the SI system uses prefixes to change the size of the unit. These prefixes are listed in Table A2.2.

Table A2.1

The Fundamental SI Units

Physical Quantity	Name of Unit	Abbreviation
Mass	kilogram	kg
Length	meter	m
Time	second	s
Temperature	Kelvin	K
Electric current	ampere	A
Amount of substance	mole	mol
Luminous intensity	candela	cd

*This practice will not usually be followed in the examples in this text because we want to show the correct number of significant figures in each step. However, in the answers to the end-of-chapter exercises, only the final answer is rounded.

†This procedure is consistent with the operation of calculators.

Table A2.2

The Prefixes Used in the SI System

Prefix	Symbol	Meaning	Exponential Notation*
exa	E	1,000,000,000,000,000,000	10^{18}
peta	P	1,000,000,000,000,000	10^{15}
tera	T	1,000,000,000,000	10^{12}
giga	**G**	**1,000,000,000**	$\mathbf{10^{9}}$
mega	**M**	**1,000,000**	$\mathbf{10^{6}}$
kilo	**k**	**1000**	$\mathbf{10^{3}}$
hecto	h	100	10^{2}
deka	da	10	10^{1}
—	—	1	10^{0}
deci	**d**	**0.1**	$\mathbf{10^{-1}}$
centi	**c**	**0.01**	$\mathbf{10^{-2}}$
milli	**m**	**0.001**	$\mathbf{10^{-3}}$
micro	$\boldsymbol{\mu}$	**0.000001**	$\mathbf{10^{-6}}$
nano	**n**	**0.000000001**	$\mathbf{10^{-9}}$
pico	p	0.000000000001	10^{-12}
femto	f	0.000000000000001	10^{-15}
atto	a	0.000000000000000001	10^{-18}

*The most common notations are shown in bold. See Appendix A1.1 if you need a review of exponential notation.

One physical quantity that is very important in chemistry is *volume,* which is not a fundamental SI unit; it is derived from length. A cube with dimensions of 1 m on each edge has a volume of $(1\ \text{m})^3 = 1\ \text{m}^3$. Then, recognizing that there are 10 decimeters (dm) in a meter, the volume of the cube is $(10\ \text{dm})^3 = 1000\ \text{dm}^3$. A cubic decimeter, dm^3, is commonly called a liter (L), which is a unit of volume slightly larger than a quart. Similarly, since 1 dm equals 10 centimeters (cm), the liter $(1\ \text{dm})^3$ contains $1000\ \text{cm}^3$, or 1000 milliliters (mL).

A2.2 Unit Conversions

It is often necessary to convert results from one system of units to another. The most common way of converting units is by the *unit factor method,* more commonly called **dimensional analysis.** To illustrate the use of this method, we will look at a simple unit conversion.

Consider a pin measuring 2.85 cm in length. What is its length in inches? To solve this problem, we must use the equivalence statement

$$2.54\ \text{cm} = 1\ \text{in} \qquad \text{(exactly)}$$

If we divide both sides of this equation by 2.54 cm, we get

$$\frac{2.54\ \text{cm}}{2.54\ \text{cm}} = 1 = \frac{1\ \text{in}}{2.54\ \text{cm}}$$

Note that the expression 1 in/2.54 cm equals 1. This expression is called a **unit factor.** Since 1 in and 2.54 cm are exactly equivalent, multiplying any expression by this unit factor will not change its value.

The pin has a length of 2.85 cm. Multiplying this length by the unit factor gives

$$2.85\ \cancel{\text{cm}} \times \frac{1\ \text{in}}{1.54\ \cancel{\text{cm}}} = \frac{2.85}{2.54}\ \text{in} = 1.12\ \text{in}$$

Note that the centimeter units cancel to give inches for the result. This is exactly what we wanted to accomplish. Note also that the result has three significant figures, as required by the number 2.85. Recall that the 1 and 2.54 in the conversion factor are exact numbers by definition.

STEPS

Converting from One Unit to Another

1 To convert from one unit to another, use the equivalence statement that relates the two units.

2 Derive the appropriate unit factor by noting the direction of the required change (to cancel the unwanted units).

3 Multiply the quantity to be converted by the unit factor to give the quantity with the desired units.

In dimensional analysis your verification that everything has been done correctly is that the correct units are obtained in the end. *In doing chemistry problems, you should always include the units for the quantities used.* Always check to see that the units cancel to give the correct units for the final result. This provides a very valuable check, especially for complicated problems.

Appendix Three Spectral Analysis

Although volumetric and gravimetric analyses are still commonly used, spectroscopy is the technique most often used for modern chemical analysis. *Spectroscopy* is the study of electromagnetic radiation emitted or absorbed by a given chemical species. Since the quantity of radiation absorbed or emitted can be related to the quantity of the absorbing or emitting species present, this technique can be used for quantitative analysis. There are many spectroscopic techniques, since electromagnetic radiation spans a wide range of energies to include microwaves, X rays, and ultraviolet, infrared, and visible light, to name a few of its familiar forms. However, we will consider here only one procedure, which is based on the absorption of visible light.

If a liquid is colored, it is because some component of the liquid absorbs visible light. In a solution the greater the concentration of the light-absorbing substance, the more light is absorbed, and the more intense is the color of the solution.

The quantity of light absorbed by a substance can be measured by a *spectrophotometer,* shown schematically in Fig. A3.1. This instrument consists of a source that emits all wavelengths of light in the visible region (wavelengths of $\approx$400–700 nm); a monochromator, which selects a given wavelength of light; a sample holder for the solution being measured; and a detector, which compares the intensity of incident light I_0 with the intensity of light after it has passed through the sample I. The ratio I/I_0, called the *transmittance,* is a measure of the fraction of light that passes through the sample. The amount of light absorbed is given by the *absorbance* A, where

$$A = -\log \frac{I}{I_0}$$

The absorbance can be expressed by the *Beer-Lambert law:*

$$A = \epsilon l c$$

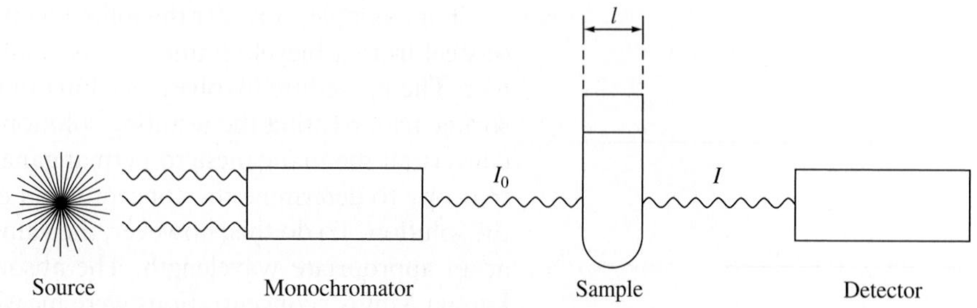

Figure A3.1
A schematic diagram of a simple spectrophotometer. The source emits all wavelengths of visible light, which are dispersed by using a prism or grating and then focused, one wavelength at a time, onto the sample. The detector compares the intensity of the incident light (I_0) with the intensity of the light after it has passed through the sample (I).

where ϵ is the molar absorptivity or the molar extinction coefficient (in L mol^{-1} cm^{-1}), l is the distance the light travels through the solution (in cm), and c is the concentration of the absorbing species (in mol/L). The Beer-Lambert law is the basis for using spectroscopy in quantitative analysis. If ϵ and l are known, determining A for a solution allows us to calculate the concentration of the absorbing species in the solution.

Suppose we have a pink solution containing an unknown concentration of $Co^{2+}(aq)$ ions. A sample of this solution is placed in a spectrophotometer, and the absorbance is measured at a wavelength where ϵ for $Co^{2+}(aq)$ is known to be 12 L mol^{-1} cm^{-1}. The absorbance A is found to be 0.60. The width of the sample tube is 1.0 cm. We want to determine the concentration of $Co^{2+}(aq)$ in the solution. This problem can be solved by a straightforward application of the Beer-Lambert law,

$$A = \epsilon l c$$

where

$$A = 0.60$$

$$\epsilon = \frac{12 \text{ L}}{\text{mol cm}}$$

$$l = \text{light path} = 1.0 \text{ cm}$$

Solving for the concentration gives

$$c = \frac{A}{\epsilon l} = \frac{0.60}{\left(12 \dfrac{\text{L}}{\text{mol cm}}\right)(1.0 \text{ cm})} = 5.0 \times 10^{-2} \text{ mol/L}$$

To obtain the unknown concentration of an absorbing species from the measured absorbance, we must know the product ϵl, since

$$c = \frac{A}{\epsilon l}$$

We can obtain the product ϵl by measuring the absorbance of a solution of *known* concentration, since

$$\epsilon l = \frac{A}{c}$$

Measured using a spectrophotometer (A)

Known from making up the solution (c)

However, a more accurate value of the product ϵl can be obtained by plotting A versus c for a series of solutions. Note that the equation $A = \epsilon l c$ gives a straight line with slope ϵl when A is plotted against c.

For example, consider the following typical spectroscopic analysis. A sample of steel from a bicycle frame is to be analyzed to determine its manganese content. The procedure involves weighing out a sample of the steel, dissolving it in strong acid, treating the resulting solution with a very strong oxidizing agent to convert all the manganese to permanganate ion (MnO_4^-), and then using spectroscopy to determine the concentration of the intensely purple MnO_4^- ions in the solution. To do this, however, the value of ϵl for MnO_4^- must be determined at an appropriate wavelength. The absorbance values for four solutions with known MnO_4^- concentrations were measured to give the following data:

Solution	Concentration of MnO_4^- (mol/L)	Absorbance
1	7.00×10^{-5}	0.175
2	1.00×10^{-4}	0.250
3	2.00×10^{-4}	0.500
4	3.50×10^{-4}	0.875

A plot of absorbance versus concentration for the solutions of known concentration is shown in Fig. A3.2. The slope of this line (change in A/change in c) is 2.48×10^3 L/mol. This quantity represents the product ϵl.

A sample of the steel weighing 0.1523 g was dissolved, and the unknown amount of manganese was converted to MnO_4^- ions. Water was then added to give a solution with a final volume of 100.0 mL. A portion of this solution was placed in a spectrophotometer, and its absorbance was found to be 0.780. We can use these data to calculate the percent manganese in the steel. The MnO_4^- ions from the manganese in the dissolved steel sample show an absorbance of 0.780. Using the Beer-Lambert law, we calculate the concentration of MnO_4^- in this solution:

$$c = \frac{A}{\epsilon l} = \frac{0.780}{2.48 \times 10^3 \text{ L/mol}} = 3.15 \times 10^{-4} \text{ mol/L}$$

However, there is a more direct way for finding c. Using a graph such as that in Fig. A3.2 (often called a Beer's law plot), we can read the concentration that corresponds to $A = 0.780$. This interpolation is shown by dashed lines on

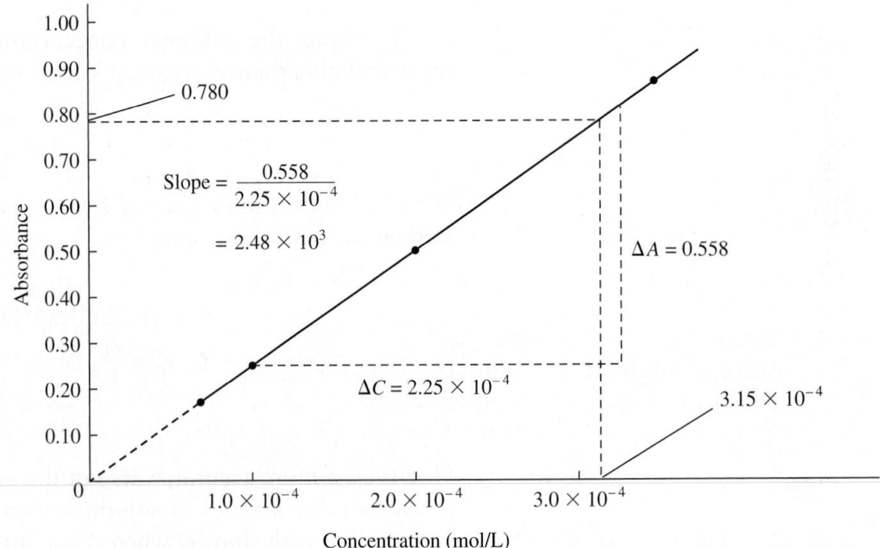

Figure A3.2
A plot of absorbance versus concentration of MnO_4^- in a series of solutions of known concentration.

the graph. By this method, $c = 3.15 \times 10^{-4}$ mol/L, which agrees with the value obtained above.

Recall that the original 0.1523-g steel sample was dissolved, the manganese was converted to permanganate, and the volume was adjusted to 100.0 mL. We now know that the $[MnO_4^-]$ in that solution is 3.15×10^{-4} *M*. Using this concentration, we can calculate the total number of moles of MnO_4^- in that solution:

$$\text{Mol of } MnO_4^- = 100.0 \text{ mL} \times \frac{1 \text{ L}}{1000 \text{ mL}} \times 3.15 \times 10^{-4} \frac{\text{mol}}{\text{L}}$$

$$= 3.15 \times 10^{-5} \text{ mol}$$

Each mole of manganese in the original steel sample yields a mole of MnO_4^-. That is,

$$1 \text{ mol of Mn} \xrightarrow{\text{Oxidation}} 1 \text{ mol of } MnO_4^-$$

so the original steel sample must have contained 3.15×10^{-5} mol of manganese. The mass of manganese present in the sample is

$$3.15 \times 10^{-5} \text{ mol of Mn} \times \frac{54.938 \text{ g of Mn}}{1 \text{ mol of Mn}} = 1.73 \times 10^{-3} \text{ g of Mn}$$

Since the steel sample weighed 0.1523 g, the percent manganese in the steel is

$$\frac{1.73 \times 10^{-3} \text{ g of Mn}}{1.523 \times 10^{-1} \text{ g of sample}} \times 100\% = 1.14\%$$

This example illustrates a typical use of spectroscopy in quantitative analysis. The steps commonly involved are as follows:

1. Preparation of a calibration plot (a Beer's law plot) from the measured absorbance values of a series of solutions with known concentrations.
2. Measurement of the absorbance of the solution of unknown concentration.
3. Use of the calibration plot to determine the unknown concentration.

Appendix Four | Selected Thermodynamic Data*

Substance and State	ΔH_f° (kJ/mol)	ΔG_f° (kJ/mol)	S° (J K^{-1} mol^{-1})	Substance and State	ΔH_f° (kJ/mol)	ΔG_f° (kJ/mol)	S° (J K^{-1} mol^{-1})
Aluminum				Bromine			
$Al(s)$	0	0	28	$Br_2(l)$	0	0	152
$Al_2O_3(s)$	−1676	−1582	51	$Br_2(g)$	31	3	245
$Al(OH)_3(s)$	−1277	—	—	$Br_2(aq)$	−3	4	130
$AlCl_3(s)$	−704	−629	111	$Br^-(aq)$	−121	−104	82
Barium				$HBr(g)$	−36	−53	199
$Ba(s)$	0	0	67	Cadmium			
$BaCO_3(s)$	−1219	−1139	112	$Cd(s)$	0	0	52
$BaO(s)$	−582	−552	70	$CdO(s)$	−258	−228	55
$Ba(OH)_2(s)$	−946	—	—	$Cd(OH)_2(s)$	−561	−474	96
$BaSO_4(s)$	−1465	−1353	132	$CdS(s)$	−162	−156	65
Beryllium				$CdSO_4(s)$	−935	−823	123
$Be(s)$	0	0	10	Calcium			
$BeO(s)$	−599	−569	14	$Ca(s)$	0	0	41
$Be(OH)_2(s)$	−904	−815	47	$CaC_2(s)$	−63	−68	70

*All values are assumed precise to at least ±1.

(continued)

Appendix Four (continued)

Substance and State	ΔH_f° (kJ/mol)	ΔG_f° (kJ/mol)	S° (J K^{-1} mol^{-1})
$CaCO_3(s)$	−1207	−1129	93
$CaO(s)$	−635	−604	40
$Ca(OH)_2(s)$	−987	−899	83
$Ca_3(PO_4)_2(s)$	−4126	−3890	241
$CaSO_4(s)$	−1433	−1320	107
$CaSiO_3(s)$	−1630	−1550	84
Carbon			
$C(s)$ (graphite)	0	0	6
$C(s)$ (diamond)	2	3	2
$CO(g)$	−110.5	−137	198
$CO_2(g)$	−393.5	−394	214
$CH_4(g)$	−75	−51	186
$CH_3OH(g)$	−201	−163	240
$CH_3OH(l)$	−239	−166	127
$H_2CO(g)$	−116	−110	219
$HCOOH(g)$	−363	−351	249
$HCN(g)$	135.1	125	202
$C_2H_2(g)$	227	209	201
$C_2H_4(g)$	52	68	219
$CH_3CHO(g)$	−166	−129	250
$C_2H_5OH(l)$	−278	−175	161
$C_2H_6(g)$	−84.7	−32.9	229.5
$C_3H_6(g)$	20.9	62.7	266.9
$C_3H_8(g)$	−104	−24	270
$C_2H_4O(g)$ (ethylene oxide)	−53	−13	242
CH_2 P $CHCN(g)$	185.0	195.4	274
$CH_3COOH(l)$	−484	−389	160
$C_6H_{12}O_6(s)$	−1275	−911	212
$CCl_4(l)$	−135	−65	216
Chlorine			
$Cl_2(g)$	0	0	223
$Cl_2(aq)$	−23	7	121
$Cl^-(aq)$	−167	−131	57
$HCl(g)$	−92	−95	187
Chromium			
$Cr(s)$	0	0	24
$Cr_2O_3(s)$	−1128	−1047	81
$CrO_3(s)$	−579	−502	72
Copper			
$Cu(s)$	0	0	33
$CuCO_3(s)$	−595	−518	88
$Cu_2O(s)$	−170	−148	93
$CuO(s)$	−156	−128	43
$Cu(OH)_2(s)$	−450	−372	108
$CuS(s)$	−49	−49	67
Fluorine			
$F_2(g)$	0	0	203
$F^-(aq)$	−333	−279	−14
$HF(g)$	−271	−273	174
Hydrogen			
$H_2(g)$	0	0	131
$H(g)$	217	203	115
$H^+(aq)$	0	0	0
$OH^-(aq)$	−230	−157	−11
$H_2O(l)$	−286	−237	70
$H_2O(g)$	−242	−229	189
Iodine			
$I_2(s)$	0	0	116
$I_2(g)$	62	19	261
$I_2(aq)$	23	16	137
$I^-(aq)$	−55	−52	106
Iron			
$Fe(s)$	0	0	27
$Fe_3C(s)$	21	15	108
$Fe_{0.95}O(s)$ (wustite)	−264	−240	59
$FeO(s)$	−272	−255	61
$Fe_3O_4(s)$ (magnetite)	−1117	−1013	146
$Fe_2O_3(s)$ (hematite)	−826	−740	90
$FeS(s)$	−95	−97	67
$FeS_2(s)$	−178	−166	53
$FeSO_4(s)$	−929	−825	121
Lead			
$Pb(s)$	0	0	65
$PbO_2(s)$	−277	−217	69
$PbS(s)$	−100	−99	91
$PbSO_4(s)$	−920	−813	149
Magnesium			
$Mg(s)$	0	0	33
$MgCO_3(s)$	−1113	−1029	66
$MgO(s)$	−602	−569	27
$Mg(OH)_2(s)$	−925	−834	64
Manganese			
$Mn(s)$	0	0	32
$MnO(s)$	−385	−363	60
$Mn_3O_4(s)$	−1387	−1280	149
$Mn_2O_3(s)$	−971	−893	110
$MnO_2(s)$	−521	−466	53
$MnO_4^-(aq)$	−543	−449	190
Mercury			
$Hg(l)$	0	0	76
$Hg_2Cl_2(s)$	−265	−211	196
$HgCl_2(s)$	−230	−184	144
$HgO(s)$	−90	−59	70
$HgS(s)$	−58	−49	78
Nickel			
$Ni(s)$	0	0	30
$NiCl_2(s)$	−316	−272	107
$NiO(s)$	−241	−213	38
$Ni(OH)_2(s)$	−538	−453	79
$NiS(s)$	−93	−90	53
Nitrogen			
$N_2(g)$	0	0	192
$NH_3(g)$	−46	−17	193
$NH_3(aq)$	−80	−27	111
$NH_4^+(aq)$	−132	−79	113
$NO(g)$	90	87	211

Appendix Four (continued)

Substance and State	ΔH_f° (kJ/mol)	ΔG_f° (kJ/mol)	S° (J K^{-1} mol^{-1})
$NO_2(g)$	34	52	240
$N_2O(g)$	82	104	220
$N_2O_4(g)$	10	98	304
$N_2O_4(l)$	−20	97	209
$N_2O_5(s)$	−42	134	178
$N_2H_4(l)$	51	149	121
$N_2H_3CH_3(l)$	54	180	166
$HNO_3(aq)$	−207	−111	146
$HNO_3(l)$	−174	−81	156
$NH_4ClO_4(s)$	−295	−89	186
$NH_4Cl(s)$	−314	−203	96
Oxygen			
$O_2(g)$	0	0	205
$O(g)$	249	232	161
$O_3(g)$	143	163	239
Phosphorus			
$P(s)$ (white)	0	0	41
$P(s)$ (red)	−18	−12	23
$P(s)$ (black)	−39	−33	23
$P_4(g)$	59	24	280
$PF_5(g)$	−1578	−1509	296
$PH_3(g)$	5	13	210
$H_3PO_4(s)$	−1279	−1119	110
$H_3PO_4(l)$	−1267	—	—
$H_3PO_4(aq)$	−1288	−1143	158
$P_4O_{10}(s)$	−2984	−2698	229
Potassium			
$K(s)$	0	0	64
$KCl(s)$	−436	−408	83
$KClO_3(s)$	−391	−290	143
$KClO_4(s)$	−433	−304	151
$K_2O(s)$	−361	−322	98
$K_2O_2(s)$	−496	−430	113
$KO_2(s)$	−283	−238	117
$KOH(s)$	−425	−379	79
$KOH(aq)$	−481	−440	9.20
Silicon			
$SiO_2(s)$ (quartz)	−911	−856	42
$SiCl_4(l)$	−687	−620	240
Silver			
$Ag(s)$	0	0	43
$Ag^+(aq)$	105	77	73
$AgBr(s)$	−100	−97	107
$AgCN(s)$	146	164	84
$AgCl(s)$	−127	−110	96
$Ag_2CrO_4(s)$	−712	−622	217
$AgI(s)$	−62	−66	115
$Ag_2O(s)$	−31	−11	122
$Ag_2S(s)$	−32	−40	146
Sodium			
$Na(s)$	0	0	51
$Na^+(aq)$	−240	−262	59
$NaBr(s)$	−360	−347	84
$Na_2CO_3(s)$	−1131	−1048	136

Substance and State	ΔH_f° (kJ/mol)	ΔG_f° (kJ/mol)	S° (J K^{-1} mol^{-1})
$NaHCO_3(s)$	−948	−852	102
$NaCl(s)$	−411	−384	72
$NaH(s)$	−56	−33	40
$NaI(s)$	−288	−282	91
$NaNO_2(s)$	−359	—	—
$NaNO_3(s)$	−467	−366	116
$Na_2O(s)$	−416	−377	73
$Na_2O_2(s)$	−515	−451	95
$NaOH(s)$	−427	−381	64
$NaOH(aq)$	−470	−419	50
Sulfur			
$S(s)$ (rhombic)	0	0	32
$S(s)$ (monoclinic)	0.3	0.1	33
$S^{2-}(aq)$	33	86	−15
$S_8(g)$	102	50	431
$SF_6(g)$	−1209	−1105	292
$H_2S(g)$	−21	−34	206
$SO_2(g)$	−297	−300	248
$SO_3(g)$	−396	−371	257
$SO_4^{2-}(aq)$	−909	−745	20
$H_2SO_4(l)$	−814	−690	157
$H_2SO_4(aq)$	−909	−745	20
Tin			
$Sn(s)$ (white)	0	0	52
$Sn(s)$ (gray)	−2	0.1	44
$SnO(s)$	−285	−257	56
$SnO_2(s)$	−581	−520	52
$Sn(OH)_2(s)$	−561	−492	155
Titanium			
$TiCl_4(g)$	−763	−727	355
$TiO_2(s)$	−945	−890	50
Uranium			
$U(s)$	0	0	50
$UF_6(s)$	−2137	−2008	228
$UF_6(g)$	−2113	−2029	380
$UO_2(s)$	−1084	−1029	78
$U_3O_8(s)$	−3575	−3393	282
$UO_3(s)$	−1230	−1150	99
Xenon			
$Xe(g)$	0	0	170
$XeF_2(g)$	−108	−48	254
$XeF_4(s)$	−251	−121	146
$XeF_6(g)$	−294	—	—
$XeO_3(s)$	402	—	—
Zinc			
$Zn(s)$	0	0	42
$ZnO(s)$	−348	−318	44
$Zn(OH)_2(s)$	−642	—	—
$ZnS(s)$ (wurtzite)	−193	—	—
$ZnS(s)$ (zinc blende)	−206	−201	58
$ZnSO_4(s)$	−983	−874	120

Appendix Five Equilibrium Constants and Reduction Potentials

Table A5.1

K_a for Some Common Monoprotic Acids

Name	Formula	Value of K_a
Hydrogen sulfate ion	HSO_4^-	1.2×10^{-2}
Chlorous acid	$HClO_2$	1.2×10^{-2}
Monochloracetic acid	$HC_2H_2ClO_2$	1.35×10^{-3}
Hydrofluoric acid	HF	7.2×10^{-4}
Nitrous acid	HNO_2	4.0×10^{-4}
Formic acid	HCO_2H	1.8×10^{-4}
Lactic acid	$HC_3H_5O_3$	1.38×10^{-4}
Benzoic acid	$HC_7H_5O_2$	6.4×10^{-5}
Acetic acid	$HC_2H_3O_2$	1.8×10^{-5}
Hydrated aluminum(III) ion	$[Al(H_2O)_6]^{3+}$	1.4×10^{-5}
Propanoic acid	$HC_3H_5O_2$	1.3×10^{-5}
Hypochlorous acid	$HOCl$	3.5×10^{-8}
Hypobromous acid	$HOBr$	2×10^{-9}
Hydrocyanic acid	HCN	6.2×10^{-10}
Boric acid	H_3BO_3	5.8×10^{-10}
Ammonium ion	NH_4^+	5.6×10^{-10}
Phenol	HOC_6H_5	1.6×10^{-10}
Hypoiodous acid	HOI	2×10^{-11}

Table A5.2

Stepwise Dissociation Constants for Several Common Polyprotic Acids

Name	Formula	K_{a_1}	K_{a_2}	K_{a_3}
Phosphoric acid	H_3PO_4	7.5×10^{-3}	6.2×10^{-8}	4.8×10^{-13}
Arsenic acid	H_3AsO_4	5×10^{-3}	8×10^{-8}	6×10^{-10}
Carbonic acid	H_2CO_3	4.3×10^{-7}	4.8×10^{-11}	
Sulfuric acid	H_2SO_4	Large	1.2×10^{-2}	
Sulfurous acid	H_2SO_3	1.5×10^{-2}	1.0×10^{-7}	
Hydrosulfuric acid	H_2S	1.0×10^{-7}	$\sim 10^{-19}$	
Oxalic acid	$H_2C_2O_4$	6.5×10^{-2}	6.1×10^{-5}	
Ascorbic acid (vitamin C)	$H_2C_6H_6O_6$	7.9×10^{-5}	1.6×10^{-12}	
Citric acid	$H_3C_6H_5O_7$	8.4×10^{-4}	1.8×10^{-5}	4.0×10^{-6}

Table A5.3

K_b for Some Common Weak Bases

Name	Formula	Conjugate Acid	K_b
Ammonia	NH_3	NH_4^+	1.8×10^{-5}
Methylamine	CH_3NH_2	$CH_3NH_3^+$	4.38×10^{-4}
Ethylamine	$C_2H_5NH_2$	$C_2H_5NH_3^+$	5.6×10^{-4}
Diethylamine	$(C_2H_5)_2NH$	$(C_2H_5)_2NH_2^+$	1.3×10^{-3}
Triethylamine	$(C_2H_5)_3N$	$(C_2H_5)_3NH^+$	4.0×10^{-4}
Hydroxylamine	$HONH_2$	$HONH_3^+$	1.1×10^{-8}
Hydrazine	H_2NNH_2	$H_2NNH_3^+$	3.0×10^{-6}
Aniline	$C_6H_5NH_2$	$C_6H_5NH_3^+$	3.8×10^{-10}
Pyridine	C_5H_5N	$C_5H_5NH^+$	1.7×10^{-9}

Table A5.4

Values of K_{sp} at 25°C for Common Ionic Solids

Ionic Solid	K_{sp} (at 25°C)	Ionic Solid	K_{sp} (at 25°C)	Ionic Solid	K_{sp} (at 25°C)
Fluorides		Chromates (*continued*)		Hydroxides (*continued*)	
BaF_2	2.4×10^{-5}	Hg_2CrO_4*	2×10^{-9}	$Co(OH)_3$	2.5×10^{-16}
MgF_2	6.4×10^{-9}	$BaCrO_4$	8.5×10^{-11}	$Ni(OH)_2$	1.6×10^{-16}
PbF_2	4×10^{-8}	Ag_2CrO_4	9.0×10^{-12}	$Zn(OH)_2$	4.5×10^{-17}
SrF_2	7.9×10^{-10}	$PbCrO_4$	2×10^{-16}	$Cu(OH)_2$	1.6×10^{-19}
CaF_2	4.0×10^{-11}			$Hg(OH)_2$	3×10^{-26}
		Carbonates		$Sn(OH)_2$	3×10^{-27}
Chlorides		$NiCO_3$	1.4×10^{-7}	$Cr(OH)_3$	6.7×10^{-31}
$PbCl_2$	1.6×10^{-5}	$CaCO_3$	8.7×10^{-9}	$Al(OH)_3$	2×10^{-32}
$AgCl$	1.6×10^{-10}	$BaCO_3$	1.6×10^{-9}	$Fe(OH)_3$	4×10^{-38}
Hg_2Cl_2*	1.1×10^{-18}	$SrCO_3$	7×10^{-10}	$Co(OH)_3$	2.5×10^{-43}
		$CuCO_3$	2.5×10^{-10}		
Bromides		$ZnCO_3$	2×10^{-10}	Sulfides	
$PbBr_2$	4.6×10^{-6}	$MnCO_3$	8.8×10^{-11}	MnS	2.3×10^{-13}
$AgBr$	5.0×10^{-13}	$FeCO_3$	2.1×10^{-11}	FeS	3.7×10^{-19}
Hg_2Br_2*	1.3×10^{-22}	Ag_2CO_3	8.1×10^{-12}	NiS	3×10^{-21}
		$CdCO_3$	5.2×10^{-12}	CoS	5×10^{-22}
Iodides		$PbCO_3$	1.5×10^{-15}	ZnS	2.5×10^{-22}
PbI_2	1.4×10^{-8}	$MgCO_3$	1×10^{-15}	SnS	1×10^{-26}
AgI	1.5×10^{-16}	Hg_2CO_3*	9.0×10^{-15}	CdS	1.0×10^{-28}
Hg_2I_2*	4.5×10^{-29}			PbS	7×10^{-29}
		Hydroxides		CuS	8.5×10^{-45}
Sulfates		$Ba(OH)_2$	5.0×10^{-3}	Ag_2S	1.6×10^{-49}
$CaSO_4$	6.1×10^{-5}	$Sr(OH)_2$	3.2×10^{-4}	HgS	1.6×10^{-54}
Ag_2SO_4	1.2×10^{-5}	$Ca(OH)_2$	1.3×10^{-6}		
$SrSO_4$	3.2×10^{-7}	$AgOH$	2.0×10^{-8}	Phosphates	
$PbSO_4$	1.3×10^{-8}	$Mg(OH)_2$	8.9×10^{-12}	Ag_3PO_4	1.8×10^{-18}
$BaSO_4$	1.5×10^{-9}	$Mn(OH)_2$	2×10^{-13}	$Sr_3(PO_4)_2$	1×10^{-31}
		$Cd(OH)_2$	5.9×10^{-15}	$Ca_3(PO_4)_2$	1.3×10^{-32}
Chromates		$Pb(OH)_2$	1.2×10^{-15}	$Ba_3(PO_4)_2$	6×10^{-39}
$SrCrO_4$	3.6×10^{-5}	$Fe(OH)_2$	1.8×10^{-15}	$Pb_3(PO_4)_2$	1×10^{-54}

*Contains Hg_2^{2+} ions. $K_{sp} = [Hg_2^{2+}][X^-]^2$ for Hg_2X_2 salts.

Table A5.5

Standard Reduction Potentials at 25°C (298 K) for Many Common Half-Reactions

Half-Reaction	$\mathscr{E}°$ (V)	Half-Reaction	$\mathscr{E}°$ (V)
$F_2 + 2e^- \longrightarrow 2F^-$	2.87	$O_2 + 2H_2O + 4e^- \longrightarrow 4OH^-$	0.40
$Ag^{2+} + e^- \longrightarrow Ag^+$	1.99	$Cu^{2+} + 2e^- \longrightarrow Cu$	0.34
$Co^{3+} + e^- \longrightarrow Co^{2+}$	1.82	$Hg_2Cl_2 + 2e^- \longrightarrow 2Hg + 2Cl^-$	0.27
$H_2O_2 + 2H^+ + 2e^- \longrightarrow 2H_2O$	1.78	$AgCl + e^- \longrightarrow Ag + Cl^-$	0.22
$Ce^{4+} + e^- \longrightarrow Ce^{3+}$	1.70	$SO_4^{2-} + 4H^+ + 2e^- \longrightarrow H_2SO_3 + H_2O$	0.20
$PbO_2 + 4H^+ + SO_4^{2-} + 2e^- \longrightarrow PbSO_4 + 2H_2O$	1.69	$Cu^{2+} + e^- \longrightarrow Cu^+$	0.16
$MnO_4^+ + 4H^+ + 3e^- \longrightarrow MnO_2 + 2H_2O$	1.68	$2H^+ + 2e^- \longrightarrow H_2$	0.00
$IO_4^- + 2H^+ + 2e^- \longrightarrow IO_3^- + H_2O$	1.60	$Fe^{3+} + 3e^- \longrightarrow Fe$	−0.036
$MnO_4^+ + 8H^+ + 5e^- \longrightarrow Mn^{2+} + 4H_2O$	1.51	$Pb^{2+} + 2e^- \longrightarrow Pb$	−0.13
$Au^{3+} + 3e^- \longrightarrow Au$	1.50	$Sn^{2+} + 2e^- \longrightarrow Sn$	−0.14
$PbO_2 + 4H^+ + 2e^- \longrightarrow Pb^{2+} + 2H_2O$	1.46	$Ni^{2+} + 2e^- \longrightarrow Ni$	−0.23
$Cl_2 + 2e^- \longrightarrow 2Cl^-$	1.36	$PbSO_4 + 2e^- \longrightarrow Pb + SO_4^{2-}$	−0.35
$Cr_2O_7^{2-} + 14H^+ + 6e^- \longrightarrow 2Cr^{3+} + 7H_2O$	1.33	$Cd^{2+} + 2e^- \longrightarrow Cd$	−0.40
$O_2 + 4H^+ + 4e^- \longrightarrow 2H_2O$	1.23	$Fe^{2+} + 2e^- \longrightarrow Fe$	−0.44
$MnO_2 + 4H^+ + 2e^- \longrightarrow Mn^{2+} + 2H_2O$	1.21	$Cr^{3+} + e^- \longrightarrow Cr^{2+}$	−0.50
$IO_3^- + 6H^+ + 5e^- \longrightarrow \frac{1}{2}I_2 + 3H_2O$	1.20	$Cr^{3+} + 3e^- \longrightarrow Cr$	−0.73
$Br_2 + 2e^- \longrightarrow 2Br^-$	1.09	$Zn^{2+} + 2e^- \longrightarrow Zn$	−0.76
$VO_2^+ + 2H^+ + e^- \longrightarrow VO^{2+} + H_2O$	1.00	$2H_2O + 2e^- \longrightarrow H_2 + 2OH^-$	−0.83
$AuCl_4^- + 3e^- \longrightarrow Au + 4Cl^-$	0.99	$Mn^{2+} + 2e^- \longrightarrow Mn$	−1.18
$NO_3^- + 4H^+ + 3e^- \longrightarrow NO + 2H_2O$	0.96	$Al^{3+} + 3e^- \longrightarrow Al$	−1.66
$ClO_2 + e^- \longrightarrow ClO_2^-$	0.954	$H_2 + 2e^- \longrightarrow 2H^-$	−2.23
$2Hg^{2+} + 2e^- \longrightarrow Hg_2^{2+}$	0.91	$Mg^{2+} + 2e^- \longrightarrow Mg$	−2.37
$Ag^+ + e^- \longrightarrow Ag$	0.80	$La^{3+} + 3e^- \longrightarrow La$	−2.37
$Hg_2^{2+} + 2e^- \longrightarrow 2Hg$	0.80	$Na^+ + e^- \longrightarrow Na$	−2.71
$Fe^{3+} + e^- \longrightarrow Fe^{2+}$	0.77	$Ca^{2+} + 2e^- \longrightarrow Ca$	−2.76
$O_2 + 2H^+ + 2e^- \longrightarrow H_2O_2$	0.68	$Ba^{2+} + 2e^- \longrightarrow Ba$	−2.90
$MnO_4^- + e^- \longrightarrow MnO_4^{2-}$	0.56	$K^+ + e^- \longrightarrow K$	−2.92
$I_2 + 2e^- \longrightarrow 2I^-$	0.54	$Li^+ + e^- \longrightarrow Li$	−3.05
$Cu^+ + e^- \longrightarrow Cu$	0.52		

Glossary

Note to the Student: The Glossary includes brief definitions of some of the fundamental terms used in chemistry. It does not include complex concepts that require detailed explanation for understanding. Please refer to the appropriate sections of the text for complete discussion of particular topics or concepts.

Accuracy: the agreement of a particular value with the true value. (A1.5)

Acid: a substance that produces hydrogen ions in solution; a proton donor. (4.2)

Acid–base indicator: a substance that marks the endpoint of an acid–base titration by changing color. (8.6)

Acid dissociation constant (K_a): the equilibrium constant for a reaction in which a proton is removed from an acid by H_2O to form the conjugate base and H_3O^+. (7.1)

Acid rain: a result of air pollution by sulfur dioxide. (5.11)

Actinide series: a group of 14 elements following actinium in the periodic table, in which the 5*f* orbitals are being filled. (12.13; 18.1)

Activated complex (transition state): the arrangement of atoms found at the top of the potential energy barrier as a reaction proceeds from reactants to products. (15.8)

Activation energy: the threshold energy that must be overcome to produce a chemical reaction. (15.8)

Addition polymerization: a type of polymerization in which the monomers simply add together to form the polymer, with no other products. (21.5)

Addition reaction: a reaction in which atoms add to a carbon–carbon multiple bond. (21.2)

Adiabatic process: a process that occurs without the transfer of energy as heat. (10.14)

Adsorption: the collection of one substance on the surface of another. (15.9)

Air pollution: contamination of the atmosphere, mainly by the gaseous products of transportation and production of electricity. (5.11)

Alcohol: an organic compound in which the hydroxyl group is a substituent on a hydrocarbon. (21.4)

Aldehyde: an organic compound containing the carbonyl group bonded to at least one hydrogen atom. (21.4)

Alkali metal: a Group 1A metal. (2.8; 18.2)

Alkaline earth metal: a Group 2A metal. (2.8; 18.4)

Alkane: a saturated hydrocarbon with the general formula C_nH_{2n+2}. (21.1)

Alkene: an unsaturated hydrocarbon containing a carbon–carbon double bond. The general formula is C_nH_{2n}. (21.2)

Alkyne: an unsaturated hydrocarbon containing a triple carbon–carbon bond. The general formula is C_nH_{2n-2}. (21.2)

Alloy: a substance that contains a mixture of elements and has metallic properties. (16.4)

Alloy steel: a form of steel containing carbon plus other metals such as chromium, cobalt, manganese, and molybdenum. (19.2)

Alpha (α) particle: a helium nucleus. (20.1)

Alpha-particle production: a common mode of decay for radioactive nuclides in which the mass number changes. (20.1)

Amine: an organic base derived from ammonia in which one or more of the hydrogen atoms are replaced by organic groups. (7.6; 21.4)

α-Amino acid: an organic acid in which an amino group and an R group are attached to the carbon atom next to the carboxyl group. (21.6)

Amorphous solid: a solid with considerable disorder in its structure. (16.3)

Ampere: the unit of electric current equal to one coulomb of charge per second. (11.7)

Amphoteric substance: a substance that can behave either as an acid or as a base. (7.2)

Angular momentum quantum number (ℓ): the quantum number relating to the shape of an atomic orbital, which can assume any integral value from 0 to $n - 1$ for each value of n. (12.9)

Anion: a negative ion. (2.7)

Anode: the electrode in a galvanic cell at which oxidation occurs. (11.1)

Antibonding molecular orbital: an orbital higher in energy than the atomic orbitals of which it is composed. (14.2)

Aqueous solution: a solution in which water is the dissolving medium or solvent. (4)

Aromatic hydrocarbon: one of a special class of cyclic unsaturated hydrocarbons, the simplest of which is benzene. (21.3)

Arrhenius concept: a concept postulating that acids produce hydrogen ions in aqueous solution, whereas bases produce hydroxide ions. (7.1)

Arrhenius equation: the equation representing the rate constant as $k = Ae^{-E_a/RT}$ where A represents the product of the collision frequency and the steric factor, and $e^{-E_a/RT}$ is the fraction of collisions with sufficient energy to produce a reaction. (15.8)

Atmosphere: the mixture of gases that surrounds the earth's surface. (5.11)

Atomic mass (average): the weighted average mass of the atoms in a naturally occurring element. (2.3)

Atomic number: the number of protons in the nucleus of an atom. (2.6)

Atomic radius: half the distance between the nuclei in a molecule consisting of identical atoms. (12.15)

Atomic solid: a solid that contains atoms at the lattice points. (16.3)

Aufbau principle: the principle stating that as protons are added one by one to the nucleus to build up the elements, electrons are similarly added to hydrogenlike orbitals. (12.13)

Autoionization: the transfer of a proton from one molecule to another of the same substance. (7.2)

Avogadro's law: equal volumes of gases at the same temperature and pressure contain the same number of particles. (5.2)

Avogadro's number: the number of atoms in exactly 12 grams of pure ^{12}C, equal to 6.022×10^{23}. (3.2)

Ball-and-stick model: a molecular model that distorts the sizes of atoms, but shows bond relationships clearly. (2.7)

Band model: a molecular model for metals in which the electrons are assumed to travel around the metal crystal in molecular orbitals formed from the valence atomic orbitals of the metal atoms. (16.4)

Barometer: a device for measuring atmospheric pressure. (5.1)

Base: a substance that produces hydroxide ions in aqueous solution, a proton acceptor. (7.2)

Base dissociation constant (K_b): the equilibrium constant for the reaction of a base with water to produce the conjugate acid and hydroxide ion. (7.6)

Basic oxide: an ionic oxide that dissolves in water to produce a basic solution. (18.4)

Battery: a group of galvanic cells connected in series. (11.5)

Beta (β) particle: an electron produced in radioactive decay. (20.1)

Beta-particle production: a decay process for radioactive nuclides in which the mass number remains constant and the atomic number changes. The net effect is to change a neutron to a proton. (20.1)

Bidentate ligand: a ligand that can form two bonds to a metal ion. (19.3)

Bimolecular step: a reaction involving the collision of two molecules. (15.6)

Binary compound: a two-element compound. (2.9)

Binding energy (nuclear): the energy required to decompose a nucleus into its component nucleons. (20.5)

Biomolecule: a molecule responsible for maintaining and/or reproducing life. (22)

Bond energy: the energy required to break a given chemical bond. (13.1)

Bond length: the distance between the nuclei of the two atoms connected by a bond; the distance where the total energy of a diatomic molecule is minimal. (13.1)

Bond order: the difference between the number of bonding electrons and the number of antibonding electrons, divided by two. It is an index of bond strength. (14.2)

Bonding molecular orbital: an orbital lower in energy than the atomic orbitals of which it is composed. (14.2)

Bonding pair: an electron pair found in the space between two atoms. (13.9)

Borane: a covalent hydride of boron. (18.5)

Boyle's law: the volume of a given sample of gas at constant temperature varies inversely with the pressure. (5.2)

Breeder reactor: a nuclear reactor in which fissionable fuel is produced while the reactor runs. (20.6)

Brønsted–Lowry definition (model): a model proposing that an acid is a proton donor, and a base is a proton acceptor. (7.1)

Buffer capacity: the ability of a buffered solution to absorb protons or hydroxide ions without a significant change in pH; determined by the magnitudes of [HA] and [A^-] in the solution. (8.4)

Buffered solution: a solution that resists a change in its pH when either hydroxide ions or protons are added. (8.2)

Calorimetry: the science of measuring heat flow. (9.4)

Capillary action: the spontaneous rising of a liquid in a narrow tube. (16.2)

Carbohydrate: a polyhydroxyl ketone or polyhydroxyl aldehyde or a polymer composed of these. (21.6)

Carboxyhemoglobin: a stable complex of hemoglobin and carbon monoxide that prevents normal oxygen uptake in the blood. (19.8)

Carboxyl group: the —COOH group in an organic acid. (7.2; 21.4)

Carboxylic acid: an organic compound containing the carboxyl group; an acid with the general formula RCOOH. (21.4)

Catalyst: a substance that speeds up a reaction without being consumed. (15.9)

Cathode: the electrode in a galvanic cell at which reduction occurs. (11.1)

Cathode rays: the "rays" emanating from the negative electrode (cathode) in a partially evacuated tube; a stream of electrons. (2.5)

Cathodic protection: a method in which an active metal, such as magnesium, is connected to steel to protect it from corrosion. (11.6)

Cation: a positive ion. (2.7)

Cell potential (electromotive force): the driving force in a galvanic cell that pulls electrons from the reducing agent in one compartment to the oxidizing agent in the other. (11.1)

Ceramic: a nonmetallic material made from clay and hardened by firing at high temperature; it contains minute silicate crystals suspended in a glassy cement. (16.5)

Chain reaction (nuclear): a self-sustaining fission process caused by the production of neutrons that proceed to split other nuclei. (20.6)

Charge balance: the positive and negative charges carried by the ions in an aqueous solution must balance. (7.9)

Charles's law: the volume of a given sample of gas at constant pressure is directly proportional to the temperature in kelvins. (5.2)

Chelating ligand (chelate): a ligand having more than one atom with a lone pair that can be used to bond to a metal ion. (19.3)

Chemical bond: the energy that holds two atoms together in a compound. (2.7)

Chemical equation: a representation of a chemical reaction showing the relative numbers of reactant and product molecules. (3.6)

Chemical equilibrium: a dynamic reaction system in which the concentrations of all reactants and products remain constant as a function of time. (6)

Chemical formula: the representation of a molecule in which the symbols for the elements are used to indicate the types of atoms present and subscripts are used to show the relative numbers of atoms. (2.7)

Chemical kinetics: the area of chemistry that concerns reaction rates. (15)

Chemical stoichiometry: the calculation of the quantities of material consumed and produced in chemical reactions. (3)

Chirality: the quality of having nonsuperimposable mirror images. (19.4)

Chlor-alkali process: the process for producing chlorine and sodium hydroxide by electrolyzing brine in a mercury cell. (11.8)

Coagulation: the destruction of a colloid by causing particles to aggregate and settle out. (17.8)

Codons: organic bases in sets of three that form the genetic code. (21.6)

Colligative properties: properties of a solution that depend on the number, and not on the identity, of the solute particles. (17.5)

Collision model: a model based on the idea that molecules must collide to react; used to account for the observed characteristics of reaction rates. (15.8)

Colloid: a suspension of particles in a dispersing medium. (17.8)

Combustion reaction: the vigorous and exothermic reaction that takes place between certain substances, particularly organic compounds, and oxygen. (21.1)

Common ion effect: the shift in an equilibrium position caused by the addition or presence of an ion involved in the equilibrium reaction. (8.1)

Complete ionic equation: an equation that shows all substances that are strong electrolytes as ions. (4.6)

Complex ion: a charged species consisting of a metal ion surrounded by ligands. (8.9; 19.1)

Compound: a substance with constant composition that can be broken down into elements by chemical processes. (2.7)

Concentration cell: a galvanic cell in which both compartments contain the same components, but at different concentrations. (11.4)

Condensation: the process by which vapor molecules reform a liquid. (16.10)

Condensation polymerization: a type of polymerization in which the formation of a small molecule, such as water, accompanies the extension of the polymer chain. (21.5)

Condensed states of matter: liquids and solids. (16.1)

Conduction bands: the molecular orbitals that can be occupied by mobile electrons, which are free to travel throughout a metal crystal to conduct electricity or heat. (16.4)

Conjugate acid: the species formed when a proton is added to a base. (7.1)

Conjugate acid–base pair: two species related to each other by the donating and accepting of a single proton. (7.1)

Conjugate base: what remains of an acid molecule after a proton is lost. (7.1)

Continuous spectrum: a spectrum that exhibits all the wavelengths of visible light. (12.3)

Control rods: rods in a nuclear reactor composed of substances that absorb neutrons. These rods regulate the power level of the reactor. (20.6)

Coordinate covalent bond: a metal–ligand bond resulting from the interaction of a Lewis base (the ligand) and a Lewis acid (the metal ion). (19.3)

Coordination compound: a compound composed of a complex ion and counter ions sufficient to give no net charge. (19.3)

Coordination isomerism: isomerism in a coordination compound in which the composition of the coordination sphere of the metal ion varies. (19.4)

Coordination number: the number of bonds formed between the metal ion and the ligands in a complex ion. (19.3)

Copolymer: a polymer formed from the polymerization of more than one type of monomer. (21.5)

Core electron: an inner electron in an atom; one not in the outermost (valence) principal quantum level. (12.13)

Corrosion: the process by which metals are oxidized in the atmosphere. (11.6)

Coulomb's law: $E = 2.31 \times 10^{-19} (Q_1Q_2/r)$, where E is the energy of interaction between a pair of ions, expressed in joules; r is the distance between the ion centers in nm; and Q_1 and Q_2 are the numerical ion charges. (13.1)

Counter ions: anions or cations that balance the charge on the complex ion in a coordination compound. (19.3)

Covalent bonding: a type of bonding in which electrons are shared by atoms. (2.7; 13.1)

Critical mass: the mass of fissionable material required to produce a self-sustaining chain reaction. (20.6)

Critical point: the point on a phase diagram at which the temperature and pressure have their critical values; the endpoint of the liquid–vapor line. (16.11)

Critical pressure: the minimum pressure required to produce liquefaction of a substance at the critical temperature. (16.11)

Critical reaction (nuclear): a reaction in which exactly one neutron from each fission event causes another fission event, thus sustaining the chain reaction. (20.6)

Critical temperature: the temperature above which vapor cannot be liquefied, no matter what pressure is applied. (16.11)

Crosslinking: the existence of bonds between adjacent chains in a polymer, thus adding strength to the material. (21.5)

Crystal field model: a model used to explain the magnetism and colors of coordination complexes through the splitting of the *d* orbital energies. (19.6)

Crystalline solid: a solid with a regular arrangement of its components. (16.3)

Cubic closest packed (ccp) structure: a solid modeled by the closest packing of spheres with an *abcabc* arrangement of layers; the unit cell is face-centered cubic. (16.4)

Cyclotron: a type of particle accelerator in which an ion introduced at the center is accelerated in an expanding spiral path by use of alternating electric fields in the presence of a magnetic field. (20.3)

Cytochromes: a series of iron-containing species composed of heme and a protein. Cytochromes are the principal electron-transfer molecules in the respiratory chain. (19.8)

Dalton's law of partial pressures: for a mixture of gases in a container, the total pressure exerted is the sum of the pressures that each gas would exert if it were alone. (5.5)

Degenerate orbitals: a group of orbitals with the same energy. (12.9)

Dehydrogenation reaction: a reaction in which two hydrogen atoms are removed from adjacent carbons of a saturated hydrocarbon, giving an unsaturated hydrocarbon. (21.1)

Delocalization: the condition where the electrons in a molecule are not localized between a pair of atoms but can move throughout the molecule. (13.9)

Denaturation: the breaking down of the three-dimensional structure of a protein resulting in the loss of its function. (21.6)

Denitrification: the return of nitrogen from decomposed matter to the atmosphere by bacteria that change nitrates to nitrogen gas. (18.8)

Deoxyribonucleic acid (DNA): a huge nucleotide polymer having a double-helical structure with complementary bases on the two strands. Its major functions are protein synthesis and the storage and transport of genetic information. (21.6)

Desalination: the removal of dissolved salts from an aqueous solution. (17.6)

Dialysis: a phenomenon in which a semipermeable membrane allows transfer of both solvent molecules and small solute molecules and ions. (17.6)

Diamagnetism: a type of magnetism, associated with paired electrons, that causes a substance to be repelled from the inducing magnetic field. (14.3)

Differential rate law: an expression that gives the rate of a reaction as a function of concentrations; often called the rate law. (15.2)

Diffraction: the scattering of light from a regular array of points or lines, producing constructive and destructive interference. (12.2)

Diffusion: the mixture of gases. (5.7)

Dilution: the process of adding solvent to lower the concentration of solute in a solution. (4.3)

Dimer: a molecule formed by the joining of two identical monomers. (21.5)

Dipole–dipole attraction: the attractive force resulting when polar molecules line up so that the positive and negative ends are close to each other. (16.1)

Dipole moment: a property of a molecule whose charge distribution can be represented by a center of positive charge and a center of negative charge. (13.3)

Disaccharide: a sugar formed from two monosaccharides joined by a glycoside linkage. (21.6)

Disproportionation reaction: a reaction in which a given element is both oxidized and reduced. (18.13)

Disulfide linkage: a S—S bond that stabilizes the tertiary structure of many proteins. (21.6)

Double bond: a bond in which two pairs of electrons are shared by two atoms. (13.8)

Downs cell: a cell used for electrolyzing molten sodium chloride. (11.8)

Dry cell battery: a common battery used in calculators, watches, radios, and portable audio players. (11.5)

Dual nature of light: the statement that light exhibits both wave and particulate properties. (12.2)

$E = mc^2$: Einstein's equation proposing that energy has mass; E is energy, m is mass, and c is the speed of light. (12.2)

Effective nuclear charge: the apparent nuclear charge exerted on a particular electron, equal to the actual nuclear charge minus the effect of electron repulsions. (12.11)

Effusion: the passage of a gas through a tiny orifice into an evacuated chamber. (5.7)

Electrical conductivity: the ability to conduct an electric current. (4.2)

Electrochemistry: the study of the interchange of chemical and electrical energy. (11)

Electrolysis: a process that involves forcing a current through a cell to cause a nonspontaneous chemical reaction to occur. (11.7)

Electrolyte: a material that dissolves in water to give a solution that conducts an electric current. (4.2)

Electrolytic cell: a cell that uses electrical energy to produce a chemical change that would otherwise not occur spontaneously. (11.7)

Electromagnetic radiation: radiant energy that exhibits wavelike behavior and travels through space at the speed of light in a vacuum. (12.1)

Electron: a negatively charged particle that moves around the nucleus of an atom. (2.5)

Electron affinity: the energy change associated with the addition of an electron to a gaseous atom. (12.15)

Electron capture: a process in which one of the inner-orbital electrons in an atom is captured by the nucleus. (20.1)

Electron sea model: a model for metals postulating a regular array of cations in a "sea" of electrons. (16.4)

Electron spin quantum number: a quantum number representing one of the two possible values for the electron spin; either $+\frac{1}{2}$ or $-\frac{1}{2}$. (12.10)

Electronegativity: the tendency of an atom in a molecule to attract shared electrons to itself. (13.2)

Element: a substance that cannot be decomposed into simpler substances by chemical or physical means. (2.1)

Elementary step: a reaction whose rate law can be written from its molecularity. (15.6)

Empirical formula: the simplest whole number ratio of atoms in a compound. (3.5)

Enantiomers: isomers that are nonsuperimposable mirror images of each other. (19.4)

Endpoint: the point in a titration at which the indicator changes color. (4.9)

Endothermic: refers to a reaction where energy (as heat) flows into the system. (9.1)

Energy: the capacity to do work or to cause heat flow. (9.1)

Enthalpy: a property of a system equal to $E + PV$, where E is the internal energy of the system, P is the pressure of the system, and V is the volume of the system. At constant pressure, where only PV work is allowed, the change in enthalpy equals the energy flow as heat. (9.2)

Enthalpy of fusion: the enthalpy change that occurs to melt a solid at its melting point. (16.10)

Entropy: a thermodynamic function that measures randomness or disorder. (10.1)

Enzyme: a large molecule, usually a protein, that catalyzes biological reactions. (15.9)

Equilibrium (thermodynamic definition): the position where the free energy of a reaction system has its lowest possible value. (10.11)

Equilibrium constant: the value obtained when equilibrium concentrations of the chemical species are substituted in the equilibrium expression. (6.2)

Equilibrium expression: the expression (from the law of mass action) obtained by multiplying the product concentrations and dividing by the multiplied reactant concentrations, with each concentration raised to a power represented by the coefficient in the balanced equation. (6.2)

Equilibrium position: a particular set of equilibrium concentrations. (6.2)

Equivalence point (stoichiometric point): the point in a titration when enough titrant has been added to react exactly with the substance in solution being titrated. (4.9; 8.4)

Exothermic: refers to a reaction where energy (as heat) flows out of the system. (9.1)

Exponential notation: expresses a number as $N \times 10^M$, a convenient method for representing a very large or very small number and for easily indicating the number of significant figures. (A1.1)

Faraday: a constant representing the charge on one mole of electrons; 96,485 coulombs. (11.3)

First law of thermodynamics: the energy of the universe is constant; same as the law of conservation of energy. (9.1)

Fission: the process of using a neutron to split a heavy nucleus into two nuclei with smaller mass numbers. (20.6)

Formal charge: the charge assigned to an atom in a molecule or polyatomic ion derived from a specific set of rules. (13.12)

Formation constant (stability constant): the equilibrium constant for each step of the formation of a complex ion by the addition of an individual ligand to a metal ion or complex ion in aqueous solution. (8.9)

Fossil fuel: coal, petroleum, or natural gas; consists of carbon-based molecules derived from decomposition of once-living organisms. (9.7)

Frasch process: the recovery of sulfur from underground deposits by melting it with hot water and forcing it to the surface by air pressure. (18.12)

Free energy: a thermodynamic function equal to the enthalpy (H) minus the product of the entropy (S) and the kelvin temperature (T); $G = H - TS$. Under certain conditions the change in free energy for a process is equal to the maximum useful work. (10.7)

Free radical: a species with an unpaired electron. (21.5)

Frequency: the number of waves (cycles) per second that pass a given point in space. (12.1)

Fuel cell: a galvanic cell for which the reactants are continuously supplied. (11.5)

Functional group: an atom or group of atoms in hydrocarbon derivatives that contains elements in addition to carbon and hydrogen. (21.4)

Fusion: the process of combining two light nuclei to form a heavier, more stable nucleus. (20.6)

Galvanic cell: a device in which chemical energy from a spontaneous redox reaction is changed to electrical energy that can be used to do work. (11.1)

Galvanizing: a process in which steel is coated with zinc to prevent corrosion. (11.6)

Gamma (γ) ray: a high-energy photon. (20.1)

Geiger-Müller counter (Geiger counter): an instrument that measures the rate of radioactive decay based on the ions and electrons produced as a radioactive particle passes through a gas-filled chamber. (20.4)

Gene: a given segment of the DNA molecule that contains the code for a specific protein. (21.6)

Geometrical (*cis-trans*) isomerism: isomerism in which atoms or groups of atoms can assume different positions around a rigid ring or bond. (19.4; 21.2)

Glass: an amorphous solid obtained when silica is mixed with other compounds, heated above its melting point, and then cooled rapidly. (16.5)

Glass electrode: an electrode for measuring pH from the potential difference that develops when it is dipped into an aqueous solution containing H^+ ions. (11.4)

Glycosidic linkage: a C—O—C bond formed between the rings of two cyclic monosaccharides by the elimination of water. (21.6)

Graham's law of effusion: the rate of effusion of a gas is inversely proportional to the square root of the mass of its particles. (5.7)

Gravimetric analysis: a method for determining the amount of a given substance in a solution by precipitation, filtration, drying, and weighing. (4.8)

Greenhouse effect: a warming effect exerted by the earth's atmosphere (particularly CO_2 and H_2O) due to thermal energy retained by absorption of infrared radiation. (9.7)

Ground state: the lowest possible energy state of an atom or molecule. (12.4)

Group (of the periodic table): a vertical column of elements having the same valence electron configuration and showing similar properties. (2.8)

Haber process: the manufacture of ammonia from nitrogen and hydrogen, carried out at high pressure and high temperature with the aid of a catalyst. (3.9; 6.1; 18.8)

Half-life (of a radioactive sample): the time required for the number of nuclides in a radioactive sample to reach half of the original value. (20.2)

Half-life (of a reaction): the time required for a reactant to reach half of its original concentration. (15.4)

Half-reactions: the two parts of an oxidation–reduction reaction, one representing oxidation, the other reduction. (4.11; 11.1)

Halogen: a Group 7A element. (2.8; 18.13)

Halogenation: the addition of halogen atoms to unsaturated hydrocarbons. (21.2)

Hard water: water from natural sources that contains relatively large concentrations of calcium and magnesium ions. (18.4)

Heat: energy transferred between two objects caused by a temperature difference between them. (9.1)

Heat capacity: the amount of energy required to raise the temperature of an object by one degree Celsius. (9.4)

Heat of fusion: the enthalpy change that occurs to melt a solid at its melting point. (16.10)

Heat of hydration: the enthalpy change associated with placing gaseous molecules or ions in water; the sum of the energy needed to expand the solvent and the energy released from the solvent–solute interactions. (17.2)

Heat of solution: the enthalpy change associated with dissolving a solute in a solvent; the sum of the energies needed to expand both solvent and solute in a solution and the energy released from the solvent–solute interactions. (17.2)

Heat of vaporization: the energy required to vaporize one mole of a liquid at a pressure of one atmosphere. (16.10)

Heating curve: a plot of temperature versus time for a substance where energy is added at a constant rate. (16.10)

Heisenberg uncertainty principle: a principle stating that there is a fundamental limitation to how precisely both the position and momentum of a particle can be known at a given time. (12.5)

Heme: an iron complex. (19.8)

Hemoglobin: a biomolecule composed of four myoglobin-like units (proteins plus heme) that can bind and transport four oxygen molecules in the blood. (19.8)

Henderson–Hasselbalch equation: an equation giving the relationship between the pH of an acid–base system and the concentrations of base and acid

$$pH = pK_a + \log\left(\frac{[\text{base}]}{[\text{acid}]}\right). \quad (8.2)$$

Henry's law: the amount of a gas dissolved in a solution is directly proportional to the pressure of the gas above the solution. (17.3)

Hess's law: in going from a particular set of reactants to a particular set of products, the enthalpy change is the same whether the reaction takes place in one step or in a series of steps; in summary, enthalpy is a state function. (9.5)

Heterogeneous equilibrium: an equilibrium involving reactants and/or products in more than one phase. (6.5)

Hexagonal closest packed (hcp) structure: a structure composed of closest packed spheres with an *ababab* arrangement of layers; the unit cell is hexagonal. (16.4)

Homogeneous equilibrium: an equilibrium system where all reactants and products are in the same phase. (6.5)

Homopolymer: a polymer formed from the polymerization of only one type of monomer. (21.5)

Hund's rule: the lowest-energy configuration for an atom is the one having the maximum number of unpaired electrons allowed by the Pauli exclusion principle in a particular set of degenerate orbitals, with all unpaired electrons having parallel spins. (12.13)

Hybrid orbitals: a set of atomic orbitals adopted by an atom in a molecule different from those of the atom in the free state. (14.1)

Hybridization: a mixing of the native orbitals on a given atom to form special atomic orbitals for bonding. (14.1)

Hydration: the interaction between solute particles and water molecules. (4.1)

Hydride: a binary compound containing hydrogen. The hydride ion, H^-, exists in ionic hydrides. The three classes of hydrides are covalent, interstitial, and ionic. (18.3)

Hydrocarbon: a compound composed of carbon and hydrogen. (23.1)

Hydrocarbon derivative: an organic molecule that contains one or more elements in addition to carbon and hydrogen. (21.4)

Hydrogen bonding: unusually strong dipole–dipole attractions that occur among molecules in which hydrogen is bonded to a highly electronegative atom. (16.1)

Hydrogenation reaction: a reaction in which hydrogen is added, with a catalyst present, to a carbon–carbon multiple bond. (21.2)

Hydrohalic acid: an aqueous solution of a hydrogen halide. (18.13)

Hydronium ion: the H_3O^+ ion; a hydrated proton. (7.1)

Hypothesis: one or more assumptions put forth to explain the observed behavior of nature. (1.3)

Ideal gas: a gas that obeys the equation, $PV = nRT$. (5.2)

Ideal gas law: an equation of state for a gas, where the state of the gas is its condition at a given time; expressed by $PV = nRT$, where P = pressure, V = volume, n = moles of the gas, R = the universal gas constant, and T = absolute temperature. This equation expresses behavior approached by real gases at high T and low P. (5.3)

Ideal solution: a solution whose vapor pressure is directly proportional to the mole fraction of solvent present. (17.4)

Indicator: a chemical that changes color and is used to mark the endpoint of a titration. (4.9; 8.5)

Inert pair effect: the tendency for the heavier Group 3A elements to exhibit the +1 as well as the expected +3 oxidation states, and Group 4A elements to exhibit the +2 as well as the +4 oxidation states. (18.5)

Integrated rate law: an expression that shows the concentration of a reactant as a function of time. (15.2)

Intermediate: a species that is neither a reactant nor a product but that is formed and consumed in the reaction sequence. (15.6)

Intermolecular forces: relatively weak interactions that occur between molecules. (16.1)

Internal energy: a property of a system that can be changed by a flow of work, heat, or both; $\Delta E = q + w$, where ΔE is the change in the internal energy of the system, q is heat, and w is work. (9.1)

Ion: an atom or a group of atoms that has a net positive or negative charge. (2.7)

Ion exchange (water softening): the process in which an ion-exchange resin removes unwanted ions (for example, Ca^{2+} and Mg^{2+}) and replaces them with Na^+ ions, which do not interfere with soap and detergent action. (18.4)

Ion pairing: a phenomenon occurring in solution when oppositely charged ions aggregate and behave as a single particle. (17.7)

Ion-product constant (K_w): the equilibrium constant for the autoionization of water; $K_w = [H^+][OH^-]$. At 25°C, K_w equals 1.0×10^{-14}. (7.2)

Ion-selective electrode: an electrode sensitive to the concentration of a particular ion in solution. (11.4)

Ionic bonding: the electrostatic attraction between oppositely charged ions. (2.7; 13.1)

Ionic compound (binary): a compound that results when a metal reacts with a nonmetal to form a cation and an anion. (13.1)

Ionic solid: a solid containing cations and anions that dissolves in water to give a solution containing the separated ions, which are mobile and thus free to conduct electric current. (16.3)

Ionization energy: the quantity of energy required to remove an electron from a gaseous atom or ion. (12.15)

Irreversible process: any real process. When a system undergoes the changes State 1 → State 2 → State 1 by any real pathway, the universe is different than before the cyclic process took place in the system. (10.2)

Isoelectronic ions: ions containing the same number of electrons. (13.4)

Isomers: species with the same formula but different properties. (19.4)

Isothermal process: a process in which the temperature remains constant. (10.2)

Isotonic solutions: solutions having identical osmotic pressures. (17.6)

Isotopes: atoms of the same element (the same number of protons) with different numbers of neutrons. They have identical atomic numbers but different mass numbers. (2.6)

Ketone: an organic compound containing the carbonyl group

$$\left(\begin{array}{c} \diagdown\, C \,\diagup \\ \| \\ O \end{array}\right)$$

bonded to two carbon atoms. (21.4)

Kinetic energy ($\frac{1}{2}\,mv^2$)**:** energy resulting from the motion of an object; dependent on the mass of the object and the square of its velocity. (9.1)

Kinetic molecular theory: a model that assumes that an ideal gas is composed of tiny particles (molecules) in constant motion. (5.6)

Lanthanide contraction: the decrease in the atomic radii of the lanthanide series elements, going from left to right in the periodic table. (19.1)

Lanthanide series: a group of 14 elements following lanthanum in the periodic table, in which the 4*f* orbitals are being filled. (12.13; 18.1; 19.1)

Lattice: a three-dimensional system of points designating the positions of the centers of the components of a solid (atoms, ions, or molecules). (16.3)

Lattice energy: the energy change occurring when separated gaseous ions are packed together to form an ionic solid. (13.5)

Law of conservation of energy: energy can be converted from one form to another but can be neither created nor destroyed. (9.1)

Law of conservation of mass: mass is neither created nor destroyed. (2.2)

Law of definite proportion: a given compound always contains exactly the same proportion of elements by mass. (2.2)

Law of mass action: a general description of the equilibrium condition; it defines the equilibrium constant expression. (6.2)

Law of multiple proportions: when two elements form a series of compounds, the ratios of the masses of the second element that combine with one gram of the first element can always be reduced to small whole numbers. (2.2)

Lead storage battery: a battery (used in cars) in which the anode is lead, the cathode is lead coated with lead dioxide, and the electrolyte is a sulfuric acid solution. (11.5)

Le Châtelier's principle: if a change is imposed on a system at equilibrium, the position of the equilibrium will shift in a direction that tends to reduce the effect of that change. (6.8)

Lewis acid: an electron-pair acceptor. (19.3)

Lewis base: an electron-pair donor. (19.3)

Lewis structure: a diagram of a molecule showing how the valence electrons are arranged among the atoms in the molecule. (13.10)

Ligand: a neutral molecule or ion having a lone pair of electrons that can be used to form a bond to a metal ion; a Lewis base. (19.3)

Lime-soda process: a water-softening method in which lime and soda ash are added to water to remove calcium and magnesium ions by precipitation. (7.6)

Limiting reactant (limiting reagent): the reactant that is completely consumed when a reaction is run to completion. (3.9)

Line spectrum: a spectrum showing only certain discrete wavelengths. (12.3)

Linear accelerator: a type of particle accelerator in which a changing electric field is used to accelerate a positive ion along a linear path. (20.3)

Linkage isomerism: isomerism involving a complex ion where the ligands are all the same but the point of attachment of at least one of the ligands differs. (19.4)

Liquefaction: the transformation of a gas into a liquid. (18.1)

Localized electron (LE) model: a model that assumes that a molecule is composed of atoms that are bound together by sharing pairs of electrons using the atomic orbitals of the bound atoms. (13.9)

London dispersion forces: the forces, existing among noble gas atoms and nonpolar molecules, that involve an accidental dipole that induces a momentary dipole in a neighbor. (16.1)

Lone pair: an electron pair that is localized on a given atom; an electron pair not involved in bonding. (13.9)

Magnetic quantum number (m_ℓ)**:** the quantum number relating to the orientation of an orbital in space relative to the other orbitals with the same ℓ quantum number. It can have integral values between ℓ and $-\ell$, including zero. (12.9)

Main-group (representative) elements: elements in the groups labeled 1A, 2A, 3A, 4A, 5A, 6A, 7A, and 8A in the periodic table. The group number gives the sum of valence *s* and *p* electrons. (12.13; 18.1)

Major species: the components present in relatively large amounts in a solution. (7.4)

Manometer: a device for measuring the pressure of a gas in a container. (5.1)

Mass defect: the change in mass occurring when a nucleus is formed from its component nucleons. (20.5)

Mass number: the total number of protons and neutrons in the atomic nucleus of an atom. (2.6)

Mass percent: the percent by mass of a component of a mixture (17.1) or of a given element in a compound. (3.4)

Mass spectrometer: an instrument used to determine the relative masses of atoms by the deflection of their ions in a magnetic field. (3.1)

Matter: the material of the universe.

Mean free path: the average distance a molecule in a given gas sample travels between collisions with other molecules. (5.6; 5.9)

Measurement: a quantitative observation. (A1.5)

Messenger RNA (mRNA): a special RNA molecule built in the cell nucleus that migrates into the cytoplasm and participates in protein synthesis. (21.6)

Metal: an element that gives up electrons relatively easily and is lustrous, malleable, and a good conductor of heat and electricity. (2.8)

Metalloids (semimetals): elements along the division line in the periodic table between metals and nonmetals. These elements exhibit both metallic and nonmetallic properties. (12.16; 18.1)

Metallurgy: the process of separating a metal from its ore and preparing it for use. (18.1)

Millimeters of mercury (mm Hg): a unit of pressure, also called a torr; 760 mm Hg = 760 torr = 101,325 Pa = 1 standard atmosphere. (5.1)

Mixture: a material of variable composition that contains two or more substances.

Model (theory): a set of assumptions put forth to explain the observed behavior of matter. The models of chemistry usually involve assumptions about the behavior of individual atoms or molecules. (1.3)

Moderator: a substance used in a nuclear reactor to slow down the neutrons. (20.6)

Molal boiling-point elevation constant: a constant characteristic of a particular solvent that gives the change in boiling point as a function of solution molality; used in molecular weight determinations. (17.5)

Molal freezing-point depression constant: a constant characteristic of a particular solvent that gives the change in freezing point as a function of the solution molality; used in molecular weight determinations. (17.5)

Molality: the number of moles of solute per kilogram of solvent in a solution. (17.1)

Molar heat capacity: the energy required to raise the temperature of one mole of a substance by one degree Celsius. (9.3; 9.4)

Molar mass: the mass in grams of one mole of molecules or formula units of a substance; also called molecular weight. (3.3)

Molar volume: the volume of one mole of an ideal gas; equal to 22.42 liters at STP. (5.4)

Molarity: moles of solute per volume of solution in liters. (4.3; 17.1)

Mole (mol): the number equal to the number of carbon atoms in exactly 12 grams of pure ^{12}C; Avogadro's number. One mole represents 6.022×10^{23} units. (3.2)

Mole fraction: the ratio of the number of moles of a given component in a mixture to the total number of moles in the mixture. (5.5; 17.1)

Mole ratio (stoichiometry): the ratio of moles of one substance to moles of another substance in a balanced chemical equation. (3.8)

Molecular equation: an equation representing a reaction in solution showing the reactants and products in undissociated form, whether they are strong or weak electrolytes. (4.6)

Molecular formula: the exact formula of a molecule, giving the types of atoms and the number of each type. (3.5)

Molecular orbital (MO) model: a model that regards a molecule as a collection of nuclei and electrons, where the electrons are assumed to occupy orbitals much as they do in atoms, but having the orbitals extend over the entire molecule. In this model the electrons are assumed to be delocalized rather than always located between a given pair of atoms. (14.2)

Molecular orientations (kinetics): orientations of molecules during collisions, some of which can lead to a reaction and some of which cannot. (15.8)

Molecular solid: a solid composed of neutral molecules at the lattice points. (16.3)

Molecular structure: the three-dimensional arrangement of atoms in a molecule. (13.13)

Molecular weight: the mass in grams of one mole of molecules or formula units of a substance; also called molar mass. (3.3)

Molecularity: the number of species that must collide to produce the reaction represented by an elementary step in a reaction mechanism. (15.6)

Molecule: a bonded collection of two or more atoms of the same or different elements. (2.7)

Monodentate (unidentate) ligand: a ligand that can form one bond to a metal ion. (19.3)

Monoprotic acid: an acid with one acidic proton. (7.2)

Monosaccharide (simple sugar): a polyhydroxy ketone or aldehyde containing from three to nine carbon atoms. (21.6)

Myoglobin: an oxygen-storing biomolecule consisting of a heme complex and a protein. (19.8)

Natural law: a statement that expresses generally observed behavior. (1.3)

Nernst equation: an equation relating the potential of an electrochemical cell to the concentrations of the cell components

$$\mathscr{E} = \mathscr{E}^\circ - \frac{0.0591}{n}\log(Q) \text{ at } 25^\circ\text{C}. \quad (11.4)$$

Net ionic equation: an equation for a reaction in solution, where strong electrolytes are written as ions, showing only those components that are directly involved in the chemical change. (4.6)

Network solid: an atomic solid containing strong directional covalent bonds. (16.5)

Neutralization reaction: an acid–base reaction. (4.9)

Neutron: a particle in the atomic nucleus with mass virtually equal to the proton's but with no charge. (2.6)

Nitrogen cycle: the conversion of N_2 to nitrogen-containing compounds, followed by the return of nitrogen gas to the atmosphere by natural decay processes. (18.8)

Nitrogen fixation: the process of transforming N_2 to nitrogen-containing compounds useful to plants. (18.8)

Nitrogen-fixing bacteria: bacteria in the root nodules of plants that can convert atmospheric nitrogen to ammonia and other nitrogen-containing compounds useful to plants. (18.8)

Noble gas: a Group 8A element. (2.8; 18.14)

Node: an area of an orbital having zero electron probability. (12.9)

Nonelectrolyte: a substance that, when dissolved in water, gives a nonconducting solution. (4.2)

Nonmetal: an element not exhibiting metallic characteristics. Chemically, a typical nonmetal accepts electrons from a metal. (2.8)

Normal boiling point: the temperature at which the vapor pressure of a liquid is exactly one atmosphere. (16.10)

Normal melting point: the temperature at which the solid and liquid states have the same vapor pressure under conditions where the total pressure on the system is one atmosphere. (16.10)

Normality: the number of equivalents of a substance dissolved in a liter of solution. (17.1)

Nuclear atom: an atom having a dense center of positive charge (the nucleus) with electrons moving around the outside. (2.5)

Nuclear transformation: the change of one element into another. (20.3)

Nucleon: a particle in an atomic nucleus, either a neutron or a proton. (2.6)

Nucleotide: a monomer of the nucleic acids composed of a five-carbon sugar, a nitrogen-containing base, and phosphoric acid. (21.6)

Nucleus: the small, dense center of positive charge in an atom. (2.5)

Nuclide: the general term applied to each unique atom; represented by ${}^{A}_{Z}X$, where X is the symbol for a particular element. (20.2)

Octet rule: the observation that atoms of nonmetals tend to form the most stable molecules when they are surrounded by eight electrons (to fill their valence orbitals). (13.10)

Optical isomerism: isomerism in which the isomers have opposite effects on plane-polarized light. (19.4)

Orbital: a specific wave function for an electron in an atom. The square of this function gives the probability distribution for the electron. (12.5)

***d*-Orbital splitting:** a splitting of the *d* orbitals of the metal ion in a complex such that the orbitals pointing at the ligands have higher energies than those pointing between the ligands. (19.6)

Order (of reactant): the positive or negative exponent, determined by experiment, of the reactant concentration in a rate law. (15.2)

Organic acid: an acid with a carbon-atom backbone; often contains the carboxyl group. (7.2)

Organic chemistry: the study of carbon-containing compounds (typically chains of carbon atoms) and their properties. (21)

Osmosis: the flow of solvent into a solution through a semipermeable membrane. (17.6)

Osmotic pressure (π): the pressure that must be applied to a solution to stop osmosis; $= MRT$. (17.6)

Ostwald process: a commercial process for producing nitric acid by the oxidation of ammonia. (18.8)

Oxidation: an increase in oxidation state (a loss of electrons). (4.10; 11.1)

Oxidation–reduction (redox) reaction: a reaction in which one or more electrons are transferred. (4.4; 4.10; 11.1)

Oxidation states: a concept that provides a way to keep track of electrons in oxidation–reduction reactions according to certain rules. (4.10)

Oxidizing agent (electron acceptor): a reactant that accepts electrons from another reactant. (4.10; 11.1)

Oxyacid: an acid in which the acidic proton is attached to an oxygen atom. (7.2)

Ozone: O_3, the form of elemental oxygen in addition to the much more common O_2. (18.11)

Paramagnetism: a type of induced magnetism, associated with unpaired electrons, that causes a substance to be attracted into the inducing magnetic field. (14.3)

Partial pressures: the independent pressures exerted by different gases in a mixture. (5.5)

Particle accelerator: a device used to accelerate nuclear particles to very high speeds. (20.3)

Pascal: the SI unit of pressure; equal to newtons per meter squared. (5.1)

Pauli exclusion principle: in a given atom no two electrons can have the same set of four quantum numbers. (12.10)

Penetration effect: the effect whereby a valence electron penetrates the core electrons, thus reducing the shielding effect and increasing the effective nuclear charge. (12.14)

Peptide linkage: the bond resulting from the condensation reaction between amino acids; represented by

$$-\overset{\overset{\displaystyle O}{\|}}{C}-\overset{\overset{\displaystyle H}{|}}{N}- \qquad (22.6)$$

Percent dissociation: the ratio of the amount of a substance that is dissociated at equilibrium to the initial concentration of the substance in a solution, multiplied by 100. (7.5)

Percent yield: the actual yield of a product as a percentage of the theoretical yield. (3.9)

Periodic table: a chart showing all the elements arranged in columns with similar chemical properties. (2.8)

pH curve (titration curve): a plot showing the pH of a solution being analyzed as a function of the amount of titrant added. (8.5)

pH scale: a log scale based on 10 and equal to $-\log[H^+]$; a convenient way to represent solution acidity. (7.3)

Phase diagram: a convenient way of representing the phases of a substance in a closed system as a function of temperature and pressure. (16.11)

Phenyl group: the benzene molecule minus one hydrogen atom. (21.3)

Photochemical smog: air pollution produced by the action of light on oxygen, nitrogen oxides, and unburned fuel from auto exhaust to form ozone and other pollutants. (5.11)

Photon: a quantum of electromagnetic radiation. (12.2)

Physical change: a change in the form of a substance, but not in its chemical composition; chemical bonds are not broken in a physical change.

Pi (π) bond: a covalent bond in which parallel *p* orbitals share an electron pair occupying the space above and below the line joining the atoms. (14.1)

Planck's constant: the constant relating the change in energy for a system to the frequency of the electromagnetic radiation absorbed or emitted; equal to 6.626×10^{-34} J s. (12.2)

Polar covalent bond: a covalent bond in which the electrons are not shared equally because one atom attracts them more strongly than the other. (13.1)

Polar molecule: a molecule that has a permanent dipole moment. (4.1)

Polyatomic ion: an ion containing a number of atoms. (2.7)

Polyelectronic atom: an atom with more than one electron. (12.11)

Polymer: a large, usually chainlike molecule built from many small molecules (monomers). (21.5)

Polymerization: a process in which many small molecules (monomers) are joined together to form a large molecule. (21.2)

Polypeptide: a polymer formed from amino acids joined together by peptide linkages. (21.6)

Polyprotic acid: an acid with more than one acidic proton. It dissociates in a stepwise manner, one proton at a time. (7.7)

Porous disk: a disk in a tube connecting two different solutions in a galvanic cell that allows ion flow without extensive mixing of the solutions. (11.1)

Porphyrin: a planar ligand with a central ring structure and various substituent groups at the edges of the ring. (19.8)

Positional probability: a type of probability that depends on the number of arrangements in space that yield a particular state. (10.1)

Positron production: a mode of nuclear decay in which a particle is formed having the same mass as an electron but opposite charge. The net effect is to change a proton to a neutron. (20.1)

Potential energy: energy resulting from position or composition. (9.1)

Precipitation reaction: a reaction in which an insoluble substance forms and separates from the solution. (4.5)

Precision: the degree of agreement among several measurements of the same quantity; the reproducibility of a measurement. (A1.5)

Primary structure (of a protein): the order (sequence) of amino acids in the protein chain. (21.6)

Principal quantum number: the quantum number relating to the size and energy of an orbital; it can have any positive integer value. (12.9)

Probability distribution: the square of the wave function indicating the probability of finding an electron at a particular point in space. (12.8)

Product: a substance resulting from a chemical reaction. It is shown to the right of the arrow in a chemical equation. (3.6)

Protein: a natural high-molecular-weight polymer formed by condensation reactions between amino acids. (21.6)

Proton: a positively charged particle in an atomic nucleus. (2.6; 20)

Qualitative analysis: the separation and identification of individual ions from a mixture. (4.7; 8.9)

Quantization: the concept that energy can occur only in discrete units called quanta. (12.2)

Rad: a unit of radiation dosage corresponding to 10^{-2} J of energy deposited per kilogram of tissue (from *r*adiation *a*bsorbed *d*ose). (20.7)

Radioactive decay (radioactivity): the spontaneous decomposition of a nucleus to form a different nucleus. (20.1)

Radiocarbon dating (carbon-14 dating): a method for dating ancient wood or cloth based on the rate of radioactive decay of the nuclide ${}^{14}_{6}C$. (20.4)

Radiotracer: a radioactive nuclide, introduced into an organism for diagnostic purposes, whose pathway can be traced by monitoring its radioactivity. (20.4)

Random error: an error that has an equal probability of being high or low. (A1.5)

Raoult's law: the vapor pressure of a solution is directly proportional to the mole fraction of solvent present. (17.4)

Rate constant: the proportionality constant in the relationship between reaction rate and reactant concentrations. (15.2)

Rate of decay: the change in the number of radioactive nuclides in a sample per unit time. (20.2)

Rate-determining step: the slowest step in a reaction mechanism, the one determining the overall rate. (15.6)

Rate law (differential rate law): an expression that shows how the rate of reaction depends on the concentration of reactants. (15.2)

Reactant: a starting substance in a chemical reaction. It appears to the left of the arrow in a chemical equation. (3.6)

Reaction mechanism: the series of elementary steps involved in a chemical reaction. (15.6)

Reaction quotient: a quotient obtained by applying the law of mass action to initial concentrations rather than to equilibrium concentrations. (6.6)

Reaction rate: the change in concentration of a reactant or product per unit time. (15.1)

Reactor core: the part of a nuclear reactor where the fission reaction takes place. (20.6)

Reducing agent (electron donor): a reactant that donates electrons to another substance to reduce the oxidation state of one of its atoms. (4.10; 11.1)

Reduction: a decrease in oxidation state (a gain of electrons). (4.10; 11.1)

Rem: a unit of radiation dosage that accounts for both the energy of the dose and its effectiveness in causing biological damage (from *r*oentgen *e*quivalent for *m*an). (20.7)

Resonance: a condition occurring when more than one valid Lewis structure can be written for a particular molecule. The actual electronic structure is not represented by any one of the Lewis structures but by the average of all of them. (13.11)

Reverse osmosis: the process occurring when the external pressure on a solution causes a net flow of solvent through a semipermeable membrane from the solution to the solvent. (17.6)

Reversible process: a cyclic process carried out by a hypothetical pathway, which leaves the universe exactly the same as it was before the process. No real process is reversible. (10.2)

Ribonucleic acid (RNA): a nucleotide polymer that transmits the genetic information stored in DNA to the ribosomes for protein synthesis. (21.6)

Root mean square velocity: the square root of the average of the squares of the individual velocities of gas particles. (5.6)

Salt: an ionic compound. (7.8)

Salt bridge: a U-tube containing an electrolyte that connects the two compartments of a galvanic cell, allowing ion flow without extensive mixing of the different solutions. (11.1)

Scientific method: the process of studying natural phenomena, involving observations, forming laws and theories, and testing of theories by experimentation. (1.3)

Scintillation counter: an instrument that measures radioactive decay by sensing the flashes of light produced in a substance by the radiation. (20.4)

Second law of thermodynamics: in any spontaneous process, there is always an increase in the entropy of the universe. (10.5)

Secondary structure (of a protein): the three-dimensional structure of the protein chain (for example, α-helix, random coil, or pleated sheet). (21.6)

Selective precipitation: a method of separating metal ions from an aqueous mixture by using a reagent whose anion forms a precipitate with only one or a few of the ions in the mixture. (4.7; 8.8)

Semiconductor: a substance conducting only a slight electric current at room temperature, but showing increased conductivity at higher temperatures. (16.5)

Semipermeable membrane: a membrane that allows solvent but not solute molecules to pass through. (17.6)

Shielding: the effect by which the other electrons screen, or shield, a given electron from some of the nuclear charge. (12.14)

SI units: International System of units based on the metric system and units derived from the metric system. (A2.1)

Side chain (of amino acid): the hydrocarbon group on an amino acid represented by H, CH_3, or a more complex substituent. (21.6)

Sigma (σ) bond: a covalent bond in which the electron pair is shared in an area centered on a line running between the atoms. (14.1)

Significant figures: the certain digits and the first uncertain digit of a measurement. (A1.5)

Silica: the fundamental silicon–oxygen compound, which has the empirical formula SiO_2, and forms the basis of quartz and certain types of sand. (16.5)

Silicates: salts that contain metal cations and polyatomic silicon–oxygen anions that are usually polymeric. (16.5)

Single bond: a bond in which one pair of electrons is shared by two atoms. (13.8)

Solubility: the amount of a substance that dissolves in a given volume of solvent at a given temperature. (4.2)

Solubility product constant: the constant for the equilibrium expression representing the dissolving of an ionic solid in water. (8.8)

Solute: a substance dissolved in a liquid to form a solution. (4.2; 17.1)

Solution: a homogeneous mixture. (17)

Solvent: the dissolving medium in a solution. (4.2)

Somatic damage: radioactive damage to an organism resulting in its sickness or death. (20.7)

Space-filling model: a model of a molecule showing the relative sizes of the atoms and their relative orientations. (2.7)

Specific heat capacity: the energy required to raise the temperature of one gram of a substance by one degree Celsius. (9.4)

Spectator ions: ions present in solution that do not participate directly in a reaction. (4.6)

Spectrochemical series: a listing of ligands in order based on their ability to produce *d*-orbital splitting. (19.6)

Spectroscopy: the study of the interaction of electromagnetic radiation within matter. (14.7)

Spontaneous fission: the spontaneous splitting of a heavy nuclide into two lighter nuclides. (20.1)

Spontaneous process: a process that occurs without outside intervention. (10.1)

Standard atmosphere: a unit of pressure equal to 760 mm Hg. (5.1)

Standard enthalpy of formation: the enthalpy change that accompanies the formation of one mole of a compound at 25°C from its elements, with all substances in their standard states at that temperature. (9.6)

Standard free energy change: the change in free energy that will occur for one unit of reaction if the reactants in their standard states are converted to products in their standard states. (10.9)

Standard free energy of formation: the change in free energy that accompanies the formation of one mole of a substance from its constituent elements with all reactants and products in their standard states. (10.9)

Standard hydrogen electrode: a platinum conductor in contact with 1 M H^+ ions and bathed by hydrogen gas at one atmosphere. (11.2)

Standard reduction potential: the potential of a half-reaction under standard state conditions, as measured against the potential of the standard hydrogen electrode. (11.2)

Standard solution: a solution whose concentration is accurately known. (4.3)

Standard state: a reference state for a specific substance defined according to a set of conventional definitions. (9.6)

Standard temperature and pressure (STP): the condition 0°C and 1 atm of pressure. (5.4)

Standing wave: a stationary wave as on a string of a musical instrument; in the wave mechanical model, the electron in the hydrogen atom is considered to be a standing wave. (12.5)

State function: a property that is independent of the pathway. (9.1)

States of matter: the three different forms in which matter can exist: solid, liquid, and gas. (5)

Stereoisomerism: isomerism in which all the bonds in the isomers are the same but the spatial arrangements of the atoms are different. (19.4)

Steric factor: the factor (always less than one) that reflects the fraction of collisions with orientations that can produce a chemical reaction. (15.8)

Stoichiometric quantities: quantities of reactants mixed in exactly the correct amounts so that all are used up at the same time. (3.9)

Strong acid: an acid that completely dissociates to produce a H^+ ion and the conjugate base. (4.2; 7.2)

Strong base: a metal hydroxide salt that completely dissociates into its ions in water. (4.2; 7.6)

Strong electrolyte: a material that, when dissolved in water, gives a solution that conducts an electric current very efficiently. (4.2)

Structural formula: the representation of a molecule in which the relative positions of the atoms are shown and the bonds are indicated by lines. (2.7)

Structural isomerism: isomerism in which the isomers contain the same atoms but one or more bonds differ. (19.4; 21.1)

Subcritical reaction (nuclear): a reaction in which less than one neutron causes another fission event and the process dies out. (20.6)

Sublimation: the process by which a substance goes directly from the solid to the gaseous state without passing through the liquid state. (16.10)

Subshell: a set of orbitals with a given angular momentum quantum number. (12.9)

Substitution reaction (hydrocarbons): a reaction in which an atom, usually a halogen, replaces a hydrogen atom in a hydrocarbon. (21.1)

Supercooling: the process of cooling a liquid below its freezing point without its changing to a solid. (16.10)

Supercritical reaction (nuclear): a reaction in which more than one neutron from each fission event causes another fission event. The process rapidly escalates to a violent explosion. (20.6)

Superheating: the process of heating a liquid above its boiling point without its boiling. (16.10)

Superoxide: a compound containing the O_2^- anion. (18.2)

Surface tension: the resistance of a liquid to an increase in its surface area. (16.2)

Surroundings: everything in the universe surrounding a thermodynamic system. (9.1)

Syngas: synthetic gas, a mixture of carbon monoxide and hydrogen, obtained by coal gasification. (9.8)

System (thermodynamic): that part of the universe on which attention is to be focused. (9.1)

Systematic error: an error that always occurs in the same direction. (A1.5)

Termolecular step: a reaction involving the simultaneous collision of three molecules. (15.6)

Tertiary structure (of a protein): the overall shape of a protein, long and narrow or globular, maintained by different types of intramolecular interactions. (21.6)

Theoretical yield: the maximum amount of a given product that can be formed when the limiting reactant is completely consumed. (3.9)

Theory: a set of assumptions put forth to explain some aspect of the observed behavior of matter. (1.3)

Thermal pollution: the oxygen-depleting effect on lakes and rivers of using water for industrial cooling and returning it to its natural source at a higher temperature. (17.3)

Thermodynamic stability (nuclear): the potential energy of a particular nucleus as compared with the sum of the potential energies of its component protons and neutrons. (20.1)

Thermodynamics: the study of energy and its interconversions. (9.1)

Third law of thermodynamics: the entropy of a perfect crystal at 0 K is zero. (10.8)

Titration: a technique in which one solution is used to analyze another. (4.9)

Torr: another name for millimeter of mercury (mm Hg). (5.1)

Transfer RNA (tRNA): a small RNA fragment that finds specific amino acids and attaches them to the protein chain as dictated by the codons in mRNA. (21.6)

Transition metals: several series of elements in which inner orbitals (*d* or *f* orbitals) are being filled. (12.13; 18.1)

Transuranium elements: the elements beyond uranium that are made artificially by particle bombardment. (20.3)

Triple bond: a bond in which three pairs of electrons are shared by two atoms. (13.8)

Triple point: the point on a phase diagram at which all three states of a substance are present. (16.11)

Tyndall effect: the scattering of light by particles in a suspension. (17.8)

Uncertainty (in measurement): the characteristics that any measurement involves estimates and cannot be exactly reproduced. (A1.5)

Unimolecular step: a reaction step involving only one molecule. (15.6)

Unit cell: the smallest repeating unit of a lattice. (16.3)

Unit factor: an equivalence statement between units used for converting from one unit to another. (A2.2)

Universal gas constant: the combined proportionality constant in the ideal gas law; 0.08206 L atm/K mol or 8.3145 J/K mol. (5.3)

Valence electrons: the electrons in the outermost principal quantum level of an atom. (12.13)

Valence shell electron-pair repulsion (VSEPR) model: a model whose main postulate is that the structure around a given atom in a molecule is determined principally by minimizing electron-pair repulsions. (13.13)

van der Waals's equation: a mathematical expression for describing the behavior of real gases. (5.10)

van't Hoff factor: the ratio of moles of particles in solution to moles of solute dissolved. (17.7)

Vapor pressure: the pressure of the vapor over a liquid at equilibrium. (16.10)

Vaporization: the change in state that occurs when a liquid evaporates to form a gas. (16.10)

Viscosity: the resistance of a liquid to flow. (16.2)

Volt: the unit of electrical potential defined as one joule of work per coulomb of charge transferred. (11.1)

Voltmeter: an instrument that measures cell potential by drawing electric current through a known resistance. (11.1)

Volumetric analysis: a process involving titration of one solution with another. (4.9)

Wave function: a function of the coordinates of an electron's position in three-dimensional space that describes the properties of the electron. (12.5)

Wave mechanical model: a model for the hydrogen atom in which the electron is assumed to behave as a standing wave. (12.7)

Wavelength: the distance between two consecutive peaks or troughs in a wave. (12.1)

Weak acid: an acid that dissociates only slightly in aqueous solution. (4.2; 7.2)

Weak base: a base that reacts with water to produce hydroxide ions to only a slight extent in aqueous solution. (4.2; 7.6)

Weak electrolyte: a material that, when dissolved in water, gives a solution that conducts only a small electric current. (4.2)

Weight: the force exerted on an object by gravity. (2.3)

Work: force acting over a distance. (9.1)

X-ray diffraction: a technique for establishing the structures of crystalline solids by directing X rays of a single wavelength at a crystal and obtaining a diffraction pattern from which interatomic spaces can be determined. (16.3)

Zone of nuclear stability: the area encompassing the stable nuclides on a plot of their positions as a function of the number of protons and the number of neutrons in the nucleus. (20.1)

Answers to Selected Exercises

The answers listed here are from the *Complete Solutions Guide,* in which rounding is carried out at each intermediate step in a calculation in order to show the correct number of significant figures for that step. Therefore, an answer given here may differ in the last digit from the result obtained by carrying extra digits throughout the entire calculation and rounding at the end (the procedure you should follow).

Chapter 2

19. ClF_5 **21.** NH_3; Avogadro's hypothesis (law) implies that volume ratios are equal to molecule ratios at constant temperature and pressure. Here, 1 volume of N_2 reacts with 3 volumes of H_2 to produce 2 volumes of the gaseous product. Or in terms of molecule ratios, 1 molecule of N_2 reacts with 3 molecules of H_2 to produce 2 molecules of a product. In order for the equation to be balanced, the product must be NH_3. **23.** All the masses of hydrogen in these three compounds can be expressed as simple whole-number ratios. The g H/g N in hydrazine, ammonia, and hydrogen azide are in the ratios 6:9:1. **25.** O, 7.94; Na, 22.8; Mg, 11.9; O and Mg are incorrect by a factor of ≈2; correct formulas are H_2O, Na_2O, and MgO. **27.** d(nucleus) = 3×10^{15} g/cm^3; d(atom) = 0.4 g/cm^3 **29.** Since all charges are whole-number multiples of 6.40×10^{-13} zirkombs, then the charge on one electron could be 6.40×10^{-13} zirkombs. However, 6.40×10^{-13} zirkombs could be the charge of two electrons (or three electrons, etc.). All one can conclude is that the charge of an electron is 6.40×10^{-13} zirkombs or an integer fraction of 6.40×10^{-13}. **31.** If the plum pudding model were correct (a diffuse positive charge with electrons scattered throughout), then α particles should have traveled through the thin foil with very minor deflections in their path. This was not the case because a few of the α particles were deflected at very large angles. Rutherford reasoned that the large deflections of these α particles could be caused only by a center of concentrated positive charge that contains most of the atom's mass (the nuclear model of the atom). **33.** The atomic number of an element is equal to the number of protons in the nucleus of an atom of that element. The mass number is the sum of the number of protons plus neutrons in the nucleus. The atomic mass is the actual mass of a particular isotope (including electrons). As we will see in Chapter 3, the average mass of an atom is taken from a measurement made on a large number of atoms. The average atomic mass value is listed in the periodic table. **35.** a. The noble gases are He, Ne, Ar, Kr, Xe, and Rn (helium, neon, argon, krypton, xenon, and radon). Radon has only radioactive isotopes. In the periodic table, the whole number enclosed in parentheses is the mass number of the longest-lived isotope of the element. b. promethium (Pm) and technetium (Tc) **37.** a. Cl; halogen; b. Be; alkaline earth metal; c. Eu; lanthanide metal; d. Hf; transition metal; e. He; noble gas; f. U; actinide metal; g. Cs; alkali metal **39.** For lighter, stable isotopes, the number of protons in the nucleus is about equal to the number of neutrons. When the number of protons and neutrons is equal to each other, the mass number (protons + neutrons) will be twice the atomic number (protons). Therefore, for lighter isotopes, the ratio of the mass number to the atomic number is close to 2. For example, consider ^{28}Si, which has 14 protons and (28 − 14 =) 14 neutrons. Here, the mass number to atomic number ratio is 28/14 = 2.0. For heavier isotopes, there are more neutrons than protons in the nucleus. Therefore, the ratio of the mass number to the atomic number increases steadily upward from 2 as the isotopes get heavier and heavier. For example, ^{238}U has 92 protons and (238 − 92 =) 146 neutrons. The ratio of the mass number to the atomic number for ^{238}U is 238/92 = 2.6. **41.** a. 12 p, 12 n, 12 e; b. 12 p, 12 n, 10 e; c. 27 p, 32 n, 25 e; d. 27 p, 32 n, 24 e; e. 27 p, 32 n, 27 e; f. 34 p, 45 n, 34 e; g. 34 p, 45 n, 36 e; h. 28 p, 35 n, 28 e; i. 28 p, 31 n, 26 e **43.** $^{151}_{63}Eu^{3+}$; $^{118}_{50}Sn^{2+}$ **45.** a. Lose 2 e$^-$ to form Ra^{2+}; b. Lose 3 e$^-$ to form In^{3+}; c. Gain 3 e$^-$ to form P^{3-}; d. Gain 2 e$^-$ to form Te^{2-}; e. Gain 1 e$^-$ to form Br^-; f. Lose 1 e$^-$ to form Rb^+ **47.** $AlCl_3$, aluminum chloride; $CrCl_3$, chromium(III) chloride; ICl_3, iodine trichloride; $AlCl_3$ and $CrCl_3$ are ionic compounds following the rules for naming ionic compounds. The major difference is that $CrCl_3$ contains a transition metal (Cr) that generally exhibits two or more stable charges when in ionic compounds. We need to indicate which charged ion we have in the compound. This is generally true whenever the metal in the ionic compound is a transition metal. ICl_3 is made from only nonmetals and is a covalent compound. Predicting formulas for covalent compounds is extremely difficult. Because of this, we need to indicate the number of each nonmetal in the binary covalent compound. The exception is when there is only one of the first species present in the formula; when this is the case, *mono-* is not used (it is assumed). **49.** a. sulfur difluoride; b. dinitrogen tetroxide; c. iodine trichloride; d. tetraphosphorus hexoxide **51.** a. copper(I) iodide; b. copper(II) iodide; c. cobalt(II) iodide; d. sodium carbonate; e. sodium hydrogen carbonate or sodium bicarbonate; f. tetrasulfur tetranitride; g. selenium tetrabromide; h. sodium hypochlorite; i. barium chromate; j. ammonium nitrate **53.** a. SO_2; b. SO_3; c. Na_2SO_3; d. $KHSO_3$; e. Li_3N; f. $Cr_2(CO_3)_3$; g. $Cr(C_2H_3O_2)_2$; h. SnF_4; i. NH_4HSO_4 (composed of NH_4^+ and HSO_4^- ions); j. $(NH_4)_2HPO_4$; k. $KClO_4$; l. NaH; m. HBrO; n. HBr **55.** a. lead(II) acetate; b. copper(II) sulfate; c. calcium oxide; d. magnesium sulfate; **e.** magnesium hydroxide; f. calcium sulfate; g. dinitrogen monoxide or nitrous oxide (common) **57.** a. nitric acid, HNO_3; b. perchloric acid, $HClO_4$; c. acetic acid, $HC_2H_3O_2$; d. sulfuric acid, H_2SO_4; e. phosphoric acid, H_3PO_4 **59.** The equation for the reaction would be $2\ Na(s) + Cl_2(g) \rightarrow 2\ NaCl(s)$. The sodium reactant exists as singular sodium atoms packed together very tightly and in a very organized fashion. This type of packing of atoms represents the solid phase. The chlorine reactant exists as Cl_2 molecules. In the picture of chlorine, there is a lot of empty space. This occurs only in the gaseous phase. When sodium and chlorine react, the ionic compound NaCl forms. NaCl exists as separate Na^+ and Cl^- ions. Because the ions are packed very closely together and are packed in a very organized fashion, NaCl is depicted in the solid phase. **61.** 299 g **63.** a. True; b. False. The isotope has 34 protons; c. False. The isotope has 45 neutrons; d. False. The identity is selenium, Se. **65.** Ra; 142 n **67.** SeO_4^{2-}: selenate; SeO_3^{2-}: selenite; TeO_4^{2-}: tellurate; TeO_3^{2-}: tellurite **69.** InO, atomic mass of In = 76.54; In_2O_3, atomic mass of In = 114.8

71. $SbCl_3$; antimony(III) chloride **73.** chlorine; 18 electrons **75.** a. The compounds have the same number and types of atoms (same formula), but the atoms in the molecules are bonded together differently. Therefore, the two compounds are different compounds with different properties. The compounds are called isomers of each other. b. When wood burns, most of the solid material in wood is converted to gases, which escape. The gases produced are most likely CO_2 and H_2O. c. The atom is not an indivisible particle, but is instead composed of other smaller particles—electrons, neutrons, and protons. d. The two hydride samples contain different isotopes of either hydrogen or lithium. Although the compounds are composed of different isotopes, their properties are similar because different isotopes of the same element have similar properties (except, of course, their mass). **77.** The ratio of the masses of R that combine with 1.00 g Q is 3:1, as expected by the law of multiple proportions. R_3Q **79.** C:H ratio = 8:18 or 4:9

Chapter 3

23. 47.88 amu; Ti **25.** 108.9 amu **27.** 185 amu **29.** There are three peaks in the mass spectrum, each two mass units apart. This is consistent with two isotopes, differing in mass by two mass units. The peak at 157.84 corresponds to a Br_2 molecule composed of two atoms of the lighter isotope. This isotope has mass equal to 157.84/2 or 78.92, which corresponds to ^{79}Br. The second isotope is ^{81}Br with mass equal to 161.84/2 = 80.92. The peaks in the mass spectrum correspond to $^{79}Br_2$, $^{79}Br^{81}Br$, and $^{81}Br_2$, in order of increasing mass. The intensities of the highest and lowest masses tell us the two isotopes are present at about equal abundance. The actual abundance is 50.68% ^{79}Br and 49.32% ^{81}Br. **31.** GaAs can be either $^{69}GaAs$ or $^{71}GaAs$. The mass spectrum for GaAs will have 2 peaks at 144 (69 + 75) and 146 (71 + 75) with intensities in the ratio of 60:40 or 3:2. Ga_2As_2 can be $^{69}Ga_2As_2$, $^{69}Ga^{71}GaAs_2$, or $^{71}Ga_2As_2$. The mass spectrum will have 3 peaks at 288, 290, and 292 with intensities in the ratio of 36:48:16 or 9:12:4. **33.** a. 180.158 g/mol; b. 2.78×10^{-3} mol; 1.67×10^{21} molecules **35.** a. 1.03×10^{-4} mol; b. 4.52×10^{-3} mol; c. 3.41×10^{-2} mol **37.** 4.0 g He < 1.0 mol F_2 < 44.0 g CO_2 < 4.0 g H_2 < 146 g SF_6 **39.** a. 165.39 g/mol; b. 3.023 mol; c. 3.3 g; d. 5.5×10^{22} atoms; e. 1.6 g; f. 1.373×10^{-19} g **41.** 71.40% C; 8.689% H; 5.648% F; 14.26% O **43.** 13.35% Y, 41.22% Ba, 28.62% Cu, and 16.81% O **45.** Only NO is 46.7% N by mass, so NO could be this species. Any other compound having NO as an empirical formula could also be the compound. **47.** 6.54×10^4 g/mol **49.** a. 40.002% C, 6.7135% H, 53.285% O; b. 40.002% C, 6.7136% H, 53.284% O; c. 40.002% C, 6.7135% H, 53.285% O; All three compounds have the same empirical formula, CH_2O, but different molecular formulas. The composition of all three in mass percent is also the same (within rounding differences). Therefore, elemental analysis will give us only the empirical formula. **51.** a. $S_4N_4H_4$; b. $N_3P_3Cl_6$; c. $Co_2C_8O_8$; d. S_4N_4 **53.** HgO and Hg_2O **55.** $C_7H_5N_3O_6$ **57.** $P_5N_5Cl_{10}$ **59.** $C_3H_4O_3$; $C_6H_8O_6$ **61.** Only one product is formed in this representation. This product has two Ys bonded to an X. The other substance present in the product mixture is just the excess of one of the reactants (Y). The best equation has smallest whole numbers. Here, answer c would be this smallest whole number equation ($X + 2Y \rightarrow XY_2$). Answers a and b have incorrect products listed, and for answer d, an equation includes only the reactants that go to produce the product; excess reactants are not shown in an equation. **63.** a. $C_6H_{12}O_6(s) + 6O_2(g) \rightarrow 6CO_2(g) + 6H_2O(g)$; b. $Fe_2S_3(s) + 6HCl(g) \rightarrow 2FeCl_3(s) + 3H_2S(g)$; c. $CS_2(l) + 2NH_3(g) \rightarrow H_2S(g) + NH_4SCN(s)$ **65.** a. $16Cr(s) + 3S_8(s) \rightarrow 8Cr_2S_3(s)$; b. $2NaHCO_3(s) \rightarrow Na_2CO_3(s) + CO_2(g) + H_2O(g)$; c. $2KClO_3(s) \rightarrow 2KCl(s) + 3O_2(g)$; d. $2Eu(s) + 6HF(g) \rightarrow 2EuF_3(s) + 3H_2(g)$; e. $2C_6H_6(l) + 15O_2(g) \rightarrow 12CO_2(g) + 6H_2O(g)$ **67.** 4355 g **69.** 21.5 g Fe_2O_3; 7.26 g Al; 13.7 g Al_2O_3 **71.** 2.8 days **73.** 150 g **75.** $2NO(g) + O_2(g) \rightarrow 2NO_2(g)$; NO is limiting. **77.** 0.301 g H_2O_2; 3.6×10^{-2} g HCl **79.** 1.20 metric tons **81.** 2.81×10^6 g HCN; 5.63×10^6 g H_2O **83.** 99.8 g F_2 **85.** 4.30×10^{-2} mol; 2.50 g **87.** M is yttrium, and X is chlorine. Yttrium(III) chloride; 1.84 g **89.** 5 **91.** Al_2Se_3 **93.** 42.8% **95.** 83.40% **97.** 86.2% **99.** 86.92 amu **101.** I. NH_3; II. N_2H_4; III. HN_3; If we set the atomic mass of H equal to 1.008, then the atomic mass of N is 14.01. **103.** 87.8 amu **105.** $C_{20}H_{30}O$ **107.** The gas mixture consists of $^{16}O^{16}O$, $^{16}O^{18}O$, and ^{40}Ar. The isotope composition is 42.82% ^{16}O, 8.6×10^{-2}% ^{18}O, and 57.094% ^{40}Ar. **109.** 207 amu; Pb **111.** Ge **113.** 184 amu **115.** 1.05 mol **117.** 10.% La^{2+}, 90.% La^{3+} **119.** 32.9% **121.** 0.48 mol

Chapter 4

11. Solution A has the most molecules per unit volume so solution A is most concentrated. This is followed by solution D, then solution C. Solution B has the fewest molecules per unit volume, so solution B is least concentrated. **13.** a. *Polarity* is a term applied to covalent compounds. Polar covalent compounds have an unequal sharing of electrons in bonds that results in unequal charge distribution in the overall molecule. Polar molecules have a partial negative end and a partial positive end. These are not full charges as in ionic compounds but are charges much smaller in magnitude. Water is a polar molecule and dissolves other polar solutes readily. The oxygen end of water (the partial negative end of the polar water molecule) aligns with the partial positive end of the polar solute, whereas the hydrogens of water (the partial positive end of the polar water molecule) align with the partial negative end of the solute. These opposite-charge attractions stabilize polar solutes in water. This process is called *hydration.* Nonpolar solutes do not have permanent partial negative and partial positive ends; nonpolar solutes are not stabilized in water and do not dissolve. b. KF is a soluble ionic compound, so it is a strong electrolyte. KF(*aq*) actually exists as separate hydrated K^+ ions and hydrated F^- ions in solution: $C_6H_{12}O_6$ is a polar covalent molecule that is a nonelectrolyte. $C_6H_{12}O_6$ is hydrated as described in part a. c. RbCl is a soluble ionic compound, so it exists as separate hydrated Rb^+ ions and hydrated Cl^- ions in solution. AgCl is an insoluble ionic compound, so the ions stay together in solution and fall to the bottom of the container as a precipitate. d. HNO_3 is a strong acid and exists as separate hydrated H^+ ions and hydrated NO_3^- ions in solution. CO is a polar covalent molecule and is hydrated as explained in part a. **15.** a. picture iv; b. picture ii; c. picture iii; d. picture i; **17.** a. Place 20.0 g NaOH in a 2-L volumetric flask; add water to dissolve the NaOH, and fill to the mark with water, mixing several times along the way. b. Add 500. mL of 1.00 *M* NaOH stock solution to a 2-L volumetric flask; fill to the mark with water, mixing several times along the way. c. Similar to the solution made in part a, instead using 38.8 g K_2CrO_4. d. Similar to the solution made in part b, instead using 114 mL of the 1.75 *M* K_2CrO_4 stock solution. **19.** 4.00 g **21.** 100.0 mL of 0.30 *M* $AlCl_3$ contains the most moles of Cl^- ions. **23.** 4.5 *M* **25.** 5.95×10^{-8} *M* **27.** a. 2.5×10^{-8} *M*; b. 8.4×10^{-9} *M*; c. 1.33×10^{-4} *M*; d. 2.8×10^{-7} *M* **29.** Bromides: NaBr, KBr, and NH_4Br (and others) would be soluble, and AgBr, $PbBr_2$, and Hg_2Br_2 would be insoluble. Sulfates: Na_2SO_4, K_2SO_4, and $(NH_4)_2SO_4$ (and others) would be soluble, and $BaSO_4$, $CaSO_4$, and $PbSO_4$ (or Hg_2SO_4) would be insoluble. Hydroxides: NaOH, KOH, $Ca(OH)_2$ (and others) would be soluble, and $Al(OH)_3$, $Fe(OH)_3$, and $Cu(OH)_2$ (and others) would be insoluble. Phosphates: Na_3PO_4, K_3PO_4, $(NH_4)_3PO_4$ (and others) would be soluble, and Ag_3PO_4, $Ca_3(PO_4)_2$, and $FePO_4$ (and others) would be insoluble. Lead: $PbCl_2$, $PbBr_2$, PbI_2, $Pb(OH)_2$, $PbSO_4$, and PbS (and others) would be insoluble. $Pb(NO_3)_2$ would be a soluble Pb^{2+} salt. **31.** a. precipitate = Hg_2SO_4; b. no precipitate forms; c. precipitate = $MgCO_3$; d. precipitate = $Al_2(CrO_4)_3$ **33.** a. $(NH_4)_2SO_4(aq) + Ba(NO_3)_2(aq) \rightarrow 2NH_4NO_3(aq) + BaSO_4(s)$; $2NH_4^+(aq) + SO_4^{2-}(aq) + Ba^{2+}(aq) + 2NO_3^-(aq) \rightarrow 2NH_4^+(aq) + 2NO_3^-(aq) + BaSO_4(s)$; $Ba^{2+}(aq) + SO_4^{2-}(aq) \rightarrow BaSO_4(s)$;

b. $Pb(NO_3)_2(aq) + 2NaCl(aq) \rightarrow PbCl_2(s) + 2NaNO_3(aq)$; $Pb^{2+}(aq) + 2NO_3^-(aq) + 2Na^+(aq) + 2Cl^-(aq) \rightarrow PbCl_2(s) + 2Na^+(aq) + 2NO_3^-(aq)$; $Pb^{2+}(aq) + 2Cl^-(aq) \rightarrow PbCl_2(s)$; c. No reaction occurs since all possible products are soluble. d. No reaction occurs since all possible products are soluble. e. $CuCl_2(aq) + 2NaOH(aq) \rightarrow Cu(OH)_2(s) + 2NaCl(aq)$; $Cu^{2+}(aq) + 2Cl^-(aq) + 2Na^+(aq) + 2OH^-(aq) \rightarrow Cu(OH)_2(s) + 2Na^+(aq) + 2Cl^-(aq)$; $Cu^{2+}(aq) + 2OH^-(aq) \rightarrow Cu(OH)_2(s)$ **35.** a. When $CuSO_4(aq)$ is added to $Na_2S(aq)$, the precipitate that forms is $CuS(s)$. Therefore, Na^+ (the gray spheres) and SO_4^{2-} (the bluish green spheres) are the spectator ions. $CuSO_4(aq) + Na_2S(aq) \rightarrow CuS(s) + Na_2SO_4(aq)$; $Cu^{2+}(aq) + S^{2-}(aq) \rightarrow CuS(s)$ b. When $CoCl_2(aq)$ is added to $NaOH(aq)$, the precipitate that forms is $Co(OH)_2(s)$. Therefore, Na^+ (the gray spheres) and Cl^- (the green spheres) are the spectator ions. $CoCl_2(aq) + 2NaOH(aq) \rightarrow Co(OH)_2(s) + 2NaCl(aq)$; $Co^{2+}(aq) + 2OH^-(aq) \rightarrow Co(OH)_2(s)$ c. When $AgNO_3(aq)$ is added to $KI(aq)$, the precipitate that forms is $AgI(s)$. Therefore, K^+ (the red spheres) and NO_3^- (the blue spheres) are the spectator ions. $AgNO_3(aq) + KI(aq) \rightarrow AgI(s) + KNO_3(aq)$; $Ag^+(aq) + I^-(aq) \rightarrow AgI(s)$ **37.** From the solubility rules in Table 4.1, the possible cations could be Ba^{2+} and Ca^{2+}. **39.** 2.9 g AgCl; 0.050 *M* Cl^-; 0.10 *M* NO_3^-; 0.075 *M* Ca^{2+} **41.** 0.607 g **43.** 16.2% **45.** 39.49 mg/tablet; 67.00% **47.** 23 amu; Na **49.** a. Perchloric acid reacted with potassium hydroxide is a possibility. $HClO_4(aq) + KOH(aq) \rightarrow H_2O(l) + KClO_4(aq)$; b. Nitric acid reacted with cesium hydroxide is a possibility. $HNO_3(aq) + CsOH(aq) \rightarrow H_2O(l) + CsNO_3(aq)$; c. Hydroiodic acid reacted with calcium hydroxide is a possibility. $2HI(aq) + Ca(OH)_2(aq) \rightarrow 2H_2O(l) + CaI_2(aq)$ **51.** a. 50.0 mL; b. 25.0 mL; c. 8.33 mL; d. 33.3 mL; e. 25.0 mL; f. 8.33 mL **53.** The acid is a diprotic acid (H_2A), meaning that it has two H^+ ions in the formula to donate to a base. The reaction is $H_2A(aq) + 2NaOH(aq) \rightarrow 2H_2O(l) + Na_2A(aq)$, where A^{2-} is what is left over from the acid formula when the two protons (H^+ ions) are reacted. For the HCl reaction, the base has the ability to accept two protons. The most common examples are $Ca(OH)_2$, $Sr(OH)_2$, and $Ba(OH)_2$. A possible reaction would be $2HCl(aq) + Ca(OH)_2(aq) \rightarrow 2H_2O(l) + CaCl_2(aq)$. **55.** 4.7×10^{-2} *M* **57.** a. 0.8393 *M*; b. 5.010% **59.** The resulting solution is not neutral. 5.9×10^{-3} *M* OH^- **61.** 2.0×10^{-2} *M* OH^- **63.** a. The species reduced is the element that gains electrons. The reducing agent causes reduction to occur by itself being oxidized. The reducing agent generally refers to the entire formula of the compound/ion that contains the element oxidized. b. The species oxidized is the element that loses electrons. The oxidizing agent causes oxidation to occur by being reduced itself. The oxidizing agent generally refers to the entire formula of the compound/ion that contains the element reduced. c. For simple binary ionic compounds, the actual charge on the ions are the oxidation states. For covalent substances, nonzero oxidation states are imaginary charges the elements would have if they were held together by ionic bonds (assuming the bond is between two different nonmetals). Nonzero oxidation states for elements in covalent compounds are not actual charges. Oxidation states for covalent compounds are a bookkeeping method to keep track of electrons in a reaction. **65.** a. K, +1; O, −2; Mn, +7; b. Ni, +4; O, −2; c. Fe, +2; d. H, +1; O, −2; N, −3; P, +5; e. P, +3; O, −2; f. O, −2; Fe, $+\frac{8}{3}$; g. O, −2; F, −1; Xe, +6; h. S, +4; F, −1; i. C, +2; O, −2; j. C, 0; H, +1; O, −2 **67.** a. Sr, +2; O, −2; Cr, +6; b. Cu, +2; Cl, −1; c. O, 0; d. H, +1; O, −1; e. Mg, +2; O, −2; C, +4; f. Ag, 0; g. Pb, +2; O, −2; S, +4; h. O, −2; Pb, +4; i. Na, +1; O, −2; C, +3; j. O, −2; C, +4; k. H, +1; N, −3; O, −2; S, +6; Ce, +4; l. O, −2; Cr, +3 **69.** a. $2Al(s) + 6HCl(aq) \rightarrow 2AlCl_3(aq) + 3H_2(g)$; H is reduced, and Al is oxidized. b. $CH_4(g) + 4S(s) \rightarrow CS_2(l) + 2H_2S(g)$; S is reduced, and C is oxidized. c. $C_3H_8(g) + 5O_2(g) \rightarrow 3CO_2(g) + 4H_2O(l)$; O is reduced, and C is oxidized. d. $Cu(s) + 2Ag^+(aq) \rightarrow 2Ag(s) + Cu^{2+}(aq)$; Ag is reduced, and Cu is oxidized. **71.** a. $3Cu(s) + 8H^+(aq) + 2NO_3^-(aq) \rightarrow 3Cu^{2+}(aq) + 2NO(g) + 4H_2O(l)$; b. $14H^+(aq) + Cr_2O_7^{2-}(aq) + 6Cl^-(aq) \rightarrow 3Cl_2(g) + 2Cr^{3+}(aq) + 7H_2O(l)$; c. $Pb(s) + 2H_2SO_4(aq) + PbO_2(s) \rightarrow 2PbSO_4(s) + 2H_2O(l)$; d. $14H^+(aq) + 2Mn^{2+}(aq) + 5NaBiO_3(s) \rightarrow 2MnO_4^-(aq) + 5Bi^{3+}(aq) + 5Na^+(aq) + 7H_2O(l)$; e. $8H^+(aq) + H_3AsO_4(aq) + 4Zn(s) \rightarrow 4Zn^{2+}(aq) + AsH_3(g) + 4H_2O(l)$; f. $7H_2O(l) + 4H^+(aq) + 3As_2O_3(s) + 4NO_3^-(aq) \rightarrow 4NO(g) + 6H_3AsO_4(aq)$; g. $16H^+(aq) + 2MnO_4^-(aq) + 10Br^-(aq) \rightarrow 5Br_2(l) + 2Mn^{2+}(aq) + 8H_2O(l)$; h. $8H^+(aq) + 3CH_3OH(aq) + Cr_2O_7^{2-}(aq) \rightarrow 2Cr^{3+}(aq) + 3CH_2O(aq) + 7H_2O(l)$ **73.** a. $8HCl(aq) + 2Fe(s) \rightarrow 2HFeCl_4(aq) + 3H_2(g)$; b. $6H^+(aq) + 8I^-(aq) + IO_3^-(aq) \rightarrow 3I_3^-(aq) + 3H_2O(l)$; c. $97Ce^{4+}(aq) + 54H_2O(l) + Cr(NCS)_6^{4-}(aq) \rightarrow 97Ce^{3+}(aq) + Cr^{3+}(aq) + 6NO_3^-(aq) + 6CO_2(g) + 6SO_4^{2-}(aq) + 108H^+(aq)$; d. $64OH^-(aq) + 2CrI_3(s) + 27Cl_2(g) \rightarrow 54Cl^-(aq) + 2CrO_4^{2-}(aq) + 6IO_4^-(aq) + 32H_2O(l)$; e. $258OH^-(aq) + Fe(CN)_6^{4-}(aq) + 61Ce^{4+}(aq) \rightarrow Fe(OH)_3(s) + 61Ce(OH)_3(s) + 6CO_3^{2-}(aq) + 6NO_3^-(aq) + 36H_2O(l)$ **75.** 1.622×10^{-2} *M* **77.** 34.6% **79.** 49.4 mL **81.** 173 mL **83.** a. 14.2%; b. 8.95 mL **85.** a. 24.8% Co, 29.7% Cl, 5.09% H, 40.4% O; b. $CoCl_2 \cdot 6H_2O$; c. $CoCl_2 \cdot 6H_2O(aq) + 2AgNO_3(aq) \rightarrow 2AgCl(s) + Co(NO_3)_2(aq) + 6H_2O(l)$, $CoCl_2 \cdot 6H_2O(aq) + 2NaOH(aq) \rightarrow Co(OH)_2(s) + 2NaCl(aq) + 6H_2O(l)$, $4Co(OH)_2(s) + O_2(g) \rightarrow 2Co_2O_3(s) + 4H_2O(l)$ **87.** 72.4% KCl; 27.6% NaCl **89.** 2.00 *M* **91.** 0.0785 ± 0.0002 *M* **93.** three acidic hydrogens **95.** 3.442% **97.** a. 31.3%; b. 6.00 *M* **99.** 14.6 g Zn, 14.4 g Ag **101.** 77.1% KCl; 22.9% KBr **103.** a. $\dfrac{\text{mass of AgCl}}{\text{mass of PCB}} = \dfrac{143.4\,n}{154.20 + 34.44\,n}$ or $\text{mass}_{AgCl}\,(154.20 + 34.44\,n) = \text{mass}_{PCB}\,(143.4\,n)$; b. 7.097 **105.** a. 5.35×10^4 L/s; b. 4.25 ppm; c. 1.69×10^6 g; d. 10.3 ppm **107.** a. $YBa_2Cu_3O_{6.5}$: +2; only Cu^{2+} present; $YBa_2Cu_3O_7$: +2.33; two Cu^{2+} and one Cu^{3+} present; $YBa_2Cu_3O_8$: +3; only Cu^{3+} present; b. $2Cu^{2+}(aq) + 5I^-(aq) \rightarrow 2CuI(s) + I_3^-(aq)$; $Cu^{3+}(aq) + 4I^-(aq) \rightarrow CuI(s) + I_3^-(aq)$; $2S_2O_3^{2-}(aq) + I_3^-(aq) \rightarrow 3I^-(aq) + S_4O_6^{2-}(aq)$; c. $YBa_2Cu_3O_{7.25}$; +2.50 **109.** 24.99% $AgNO_3$; 40.07% $CuCl_2$; 34.94% $FeCl_3$

Chapter 5

21. 47.5 torr; 6.33×10^3 Pa; 6.25×10^{-2} atm **23.** 1.01×10^5 Pa; 10.3 m **25.** a. 3.6×10^3 mm Hg; b. 3.6×10^3 torr; c. 4.9×10^5 Pa; d. 71 psi **27.** The decrease in temperature causes the balloon to contract (*V* and *T* are directly related). Because weather balloons do expand, the effect of the decrease in pressure must be dominant. **29.** $P_{H_2} = 317$ torr; $P_{N_2} = 50.7$ torr; $P_{Total} = 368$ torr **31.** 309 g **33.** 3.08 atm; $P_{CO_2} = 3.08$ atm, $P_{Total} = 4.05$ atm **35.** 4.44×10^3 g He; 2.24×10^3 g H_2 **37.** 7.0×10^2°C **39.** 12.5 mL **41.** $n_2/n_1 = 0.921$ **43.** a. The He flask has 1.5 times as many atoms of gas present as the Ne flask, so the pressure in the He flask will be 1.5 times greater (assuming a constant temperature). b. Because the flask volumes are the same, your drawing should have the various atoms equally distributed between the two flasks. So each flask should have 3 He atoms and 2 Ne atoms. c. $P_{final} = \frac{5}{6}P_{He,\,initial} = \frac{5}{4}P_{Ne,\,initial}$; d. $P_{He,\,final} = \dfrac{P_{He,\,initial}}{2}$; $P_{Ne,\,final} = \dfrac{P_{Ne,\,initial}}{2}$ **45.** 3.69 L **47.** a. $\chi_{CH_4} = 0.412$, $\chi_{O_2} = 0.588$; b. 0.161 mol; c. 1.06 g CH_4, 3.03 g O_2 **49.** $P_{N_2} = 1.1$ atm; $P_{O_2} = 0.10$ atm; $P_{NH_3} = 0.18$ atm; $P_{total} = 1.4$ atm **51.** Rigid container: pressure increases, density is constant; flexible container: pressure is constant, density decreases **53.** N_2H_4 **55.** The calculated molar masses are 209 g/mol from data set I and 202 g/mol from data set II. These values are close to the expected molar mass (207 g/mol) for the divalent metal compound, $Be(C_5H_7O_2)_2$. **57.** 1.16% **59.** 1.5×10^7 g Fe; 2.6×10^7 g 98% H_2SO_4 **61.** 46.5% **63.** $P_{N_2} = 0.74$ atm; $P_{Total} = 2.2$ atm **65.** 18.0% **67.** 13.3% **69.** 0.333 atm **71.** $2NH_3(g) \rightarrow N_2(g) + 3H_2(g)$: As reactants are converted into products, we go from 2 mol of gaseous reactants to 4 mol of gaseous products (1 mol N_2 + 3 mol H_2). Because the moles of gas double as reactants are converted into products, the volume of the gases will double (at constant P and T). Pressure is directly related to n at constant T and V. As the reaction occurs, the moles of gas will double, so the pressure will double. Because 1 mol of N_2 is produced for every 2 mol of NH_3 reacted, $P_{N_2} = \frac{1}{2}P^\circ_{NH_3}$. Due to the 3:2 mole ratio in the balanced equation, $P_{H_2} = \frac{3}{2}P^\circ_{NH_3}$. Note: $P_{Total} = P_{H_2} + P_{N_2} = \frac{3}{2}P^\circ_{NH_3} + \frac{1}{2}P^\circ_{NH_3} = 2P^\circ_{NH_3}$. As said earlier, the total pressure will double from the initial pressure of NH_3 as the reactants are completely converted into products. **73.** The number of gas particles is constant, so at constant moles of gas, either a temperature change or a pressure change results in the smaller volume. If the temperature is constant, an increase in the external pressure would cause the volume to decrease. Gases are mostly empty space, so gases are easily compressible. If the pressure is constant, a decrease in temperature would cause the volume to decrease. As the temperature is lowered, the gas particles move with a slower average velocity and don't collide with the container walls as frequently and as forcefully. As a result, the internal pressure decreases. In order to keep the pressure constant, the volume of the container must decrease in order to increase the gas particle collisions per unit area. **75.** a. Both gas samples have the same number of molecules present (n is constant); b. Since T is constant, $(KE)_{avg}$ must be the same for both gases $[(KE)_{avg} = \frac{3}{2}RT]$; c. The lighter gas A molecules will have the faster average velocity; d. The heavier gas B molecules do collide more forcefully, but gas A molecules, with the faster average velocity, collide more frequently. The end result is that P is constant between the two containers. **77.** 3.40×10^3 J/mol = 5.65×10^{-21} J/molecule (for each gas at 273 K); 6.81×10^3 J/mol = 1.13×10^{-20} J/molecule (for each gas at 546 K) **79.** No; there is a distribution of energies with only the average kinetic energy equal to $\frac{3}{2}RT$. Similarly, there is always a distribution of velocities for a gas sample at some temperature. **81.** a. All the same; b. Flask C; c. Flask A **83.** CF_2Cl_2 **85.** 63.7 g/mol **87.** a. 12.24 atm; b. 12.13 atm; c. The ideal gas law is high by 0.91%. **89.** The kinetic molecular theory assumes that gas particles do not exert forces on each other and that gas particles are volumeless. Real gas particles do exert attractive forces on each other, and real gas particles do have volumes. A gas behaves most ideally at low pressures and high temperatures. The effect of attractive forces is minimized at high temperatures since the gas particles are moving very rapidly. At low pressure, the container volume is relatively large (P and V are inversely related), so the volume of the container taken up by the gas particles is negligible. **91.** The pressure measured for real gases is too low compared to ideal gases. This is due to the attractions gas particles do have for each other; these attractions "hold" them back from hitting the container walls as forcefully. To make up for this slight decrease in pressure for real gases, a factor is added to the measured pressure. The measured volume is too large. A fraction of the space of the container volume is taken up by the volume of gas of the molecules themselves. Therefore, the actual volume available to real gas molecules is slightly less than the container volume. A term is subtracted from the container volume to correct for the volume taken up by real gas molecules. **93.** CO_2 since it has the largest a value. **95.** $u_{rms} = 667$ m/s; $u_{mp} = 545$ m/s; $u_{avg} = 615$ m/s **97.** Impact force (H_2)/impact force (He) = 0.7097 **99.** The change in momentum per impact is 2.827 times larger for O_2 molecules than for He atoms. There are 2.827 times as many impacts per second for He as compared with those for O_2. **101.** 1.0×10^9 collisions/s; 1.3×10^{-6} m **103.** a. 0.19 torr; b. 6.6×10^{21} molecules/m^3; c. 6.6×10^{15} molecules/cm^3 **105.** Benzene: 9.47×10^{-3} ppmv; 2.31×10^{11} molecules/cm^3; toluene: 1.37×10^{-2} ppmv; 3.33×10^{11} molecules/cm^3 **107.** Processes a, c, and d will all result in a doubling of the pressure. Process a has the effect of halving the volume, which would double the pressure (Boyle's law). Process c doubles the pressure because the absolute temperature is doubled (from 200. K to 400. K). Process d doubles the pressure because the moles of gas are doubled (28 g N_2 is 1 mol of N_2). Process b won't double the pressure because the absolute temperature is not doubled (303 K to 333 K). **109.** 46 mL **111.** at 25°C: 2.00 atm; at 125°C: 4.00 atm **113.** $MnCl_4$ **115.** BrF_3 **117.** 1.61×10^3 g **119.** 0.990 atm; 0.625 g Zn **121.** $P_{He} = 50.0$ torr; $P_{Ne} = 76.0$ torr; $P_{Ar} = 90.0$ torr; $P_{Total} = 216.0$ torr **123.** a. 78.0%; b. 0.907 L **125.** 1490 **127.** 60.6 kJ **129.** 7.00 mL **131.** a. $2CH_4(g) + 2NH_3(g) + 3O_2(g) \rightarrow 2HCN(g) + 6H_2O(g)$ b. 15.6 g/s **133.** 30.% **135.** 29.0% **137.** $dT = \dfrac{P(\text{molar mass})}{R}$ = constant so $d = \text{constant}\left(\dfrac{1}{T}\right)$; −272.6°C **139.** 16.03 g/mol **141.** From Figure 5.16 of the text, as temperature increases, the probability that a gas particle has the most probable velocity decreases. Since the probability of the gas particle with the most probable velocity decreased by one-half, the temperature must be higher than 300. K. The temperature is 1.20×10^3 K. **143.** 1.3 L **145.** $\chi_{CO} = 0.291$; $\chi_{CO_2} = 0.564$; $\chi_{O_2} = 0.145$ **147.** a. 8.7×10^3 L air/min; b. $\chi_{CO} = 0.0017$, $\chi_{CO_2} = 0.032$, $\chi_{O_2} = 0.13$, $\chi_{N_2} = 0.77$, $\chi_{H_2O} = 0.067$; c. $P_{CO} = 0.0017$ atm, $P_{CO_2} = 0.032$ atm, $P_{O_2} = 0.13$ atm, $P_{N_2} = 0.77$ atm, $P_{H_2O} = 0.067$ atm **149.** 2.1×10^2 stages **151.** a. A given volume of air at a given set of conditions has a larger density than helium at those conditions. We need to heat the air to greater than 25°C to lower the air density (by driving air out of the hot-air balloon) until the density is the same as that for helium (at 25°C and 1.00 atm). b. 2150 K **153.** C_3H_8 is possible.

Chapter 6

11. $2NOCl(g) \rightleftharpoons 2NO(g) + Cl_2(g)$; $K = 1.6 \times 10^{-5}$
The expression for K is the product concentrations divided by the reactant concentrations. When K has a value much less than 1, the product concentrations are relatively small and the reactant concentrations are relatively large.

$$2NO(g) \rightleftharpoons N_2(g) + O_2(g); K = 1 \times 10^{31}$$

When K has a value much greater than 1, the product concentrations are relatively large and the reactant concentrations are relatively small. In both cases, however, the rate of the forward reaction equals the rate of the reverse reaction at equilibrium (this is a definition of equilibrium). **13.** No, it doesn't matter in which direction the equilibrium position is reached. Both experiments will give the same equilibrium position since both experiments started with stoichiometric amounts of reactants or products. **15.** When equilibrium is reached, there is no net change in the amount of reactants and products present since the rates of the forward and reverse reactions are equal. The first diagram has 4 A_2B molecules, 2 A_2 molecules, and 1 B_2 molecule present. The second diagram has 2 A_2B molecules, 4 A_2 molecules, and 2 B_2 molecules. The first diagram cannot represent equilibrium because there was a net change in reactants and products. Is the second diagram the equilibrium mixture? That depends on whether there is a net change between reactants and products when going from the second diagram to the third diagram. The third diagram contains the same number and type of molecules as the second diagram, so the second diagram is the first illustration that represents equilibrium. The reaction container initially contained only A_2B. From the first diagram, 2 A_2 molecules and 1 B_2 molecule are present (along with 4 A_2B molecules). From the balanced reaction, these 2 A_2 molecules and 1 B_2 molecule were formed when 2 A_2B molecules decomposed. Therefore, the initial number of A_2B molecules present equals 4 + 2 = 6 A_2B molecules. **17.** K and K_p are equilibrium constants as determined by the law of mass action. For K, concentration units of mol/L are used, and for K_p, partial pressures in units of atm are used (generally). Q is called the reaction quotient. Q has the exact same form as K or K_p, but instead of equilibrium concentrations, initial concentrations are used to calculate the Q value. The use of Q is when it is compared to the K value. When $Q = K$ (or when $Q_p = K_p$), the reaction is at equilibrium. When $Q \neq K$, the reaction is not at equilibrium, and one can deduce the net change that must occur for the system to get to equilibrium.

19. a. $K = \frac{[H_2O]}{[NH_3]^2[CO_2]}$; $K_p = \frac{P_{H_2O}}{P_{NH_3}{}^2 \times P_{CO_2}}$; b. $K = [N_2][Br_2]^3$; $K_p = P_{N_2} \times P_{Br_2}{}^3$; c. $K = [O_2]^3$; $K_p = P_{O_2}{}^3$; d. $K = \frac{[H_2O]}{[H_2]}$; $K_p = \frac{P_{H_2O}}{P_{H_2}}$ **21.** 1.7×10^{-5} **23.** 4.6×10^3 **25.** 4.08×10^8; yes, this set of concentrations represents a system at equilibrium because the calculated value of K using these concentrations gives 4.08×10^8. **27.** 0.16 mol **29.** 4.07 **31.** 0.72; 0.017 **33.** 0.056 **35.** a. Reaction shifts left to reach equilibrium. b. At equilibrium; c. Reaction shifts left to reach equilibrium. **37.** a. decrease; b. will not change; c. will not change; d. increase **39.** a. $[HOCl] = 9.2 \times 10^{-3}\ M$, $[Cl_2O] = 1.8 \times 10^{-2}\ M$, $[H_2O] = 5.1 \times 10^{-2}\ M$; b. $[HOCl] = 0.07\ M$, $[Cl_2O] = [H_2O] = 0.22\ M$ **41.** $P_{SO_2} = 0.38$ atm, $P_{O_2} = 0.44$ atm, $P_{SO_3} = 0.12$ atm **43.** The assumption comes from the value of K being much less than 1. For these reactions, the equilibrium mixture will not have a lot of products present; mostly reactants are present at equilibrium. If we define the change that must occur in terms of x as the amount (molarity or partial pressure) of a reactant that must react to reach equilibrium, then x must be a small number because K is a very small number. We want to know the value of x in order to solve the problem, so we don't assume $x = 0$. Instead, we concentrate on the equilibrium row in the ICE table. Those reactants (or products) that have equilibrium concentrations in the form of $0.10 - x$ or $0.25 + x$ or $3.5 - 3x$, etc., is where an important assumption can be made. The assumption is that because $K << 1$, x will be small ($x << 1$) and when we add x or subtract x from some initial concentration, it will make little or no difference. That is, we assume that $0.10 - x \approx 0.10$ or $0.25 + x \approx 0.25$ or $3.5 - 3x \approx 3.5$; we assume that the initial concentration of a substance is equal to the equilibrium concentration. This assumption makes the math much easier and usually gives a value of x that is well within 5% of the true value of x (we get about the same answer with a lot less work). When the 5% rule fails, the equation must be solved exactly or by using the method of successive approximations (see Appendix A1.4). **45.** $[CO_2] = 0.39\ M$; $[CO] = 8.6 \times 10^{-3}\ M$; $[O_2] = 4.3 \times 10^{-3}\ M$ **47.** 66.0% **49.** a. 1.5×10^8; b. $P_{CO} = P_{Cl_2} = 1.8 \times 10^{-4}$ atm; $P_{COCl_2} = 5.0$ atm **51.** Only statement d is correct. Addition of a catalyst has no effect on the equilibrium position; the reaction just reaches equilibrium more quickly. Statement a is false for reactants that are either solids or liquids (adding more of these has no effect on the equilibrium). Statement b is false always. If temperature remains constant, then the value of K is constant. Statement c is false for exothermic reactions where an increase in temperature decreases the value of K. **53.** a. no effect; b. shifts left; c. shifts right **55.** $H^+ + OH^- \rightarrow H_2O$; sodium hydroxide (NaOH) will react with the H^+ on the product side of the reaction. This effectively removes H^+ from the equilibrium, which will shift the reaction to the right to produce more H^+ and $CrO_4{}^{2-}$. Since more $CrO_4{}^{2-}$ is produced, the solution turns yellow. **57.** a. right; b. right; c. no effect; d. left; e. no effect **59.** a. left; b. right; c. left; d. no effect; e. no effect; f. right **61.** An endothermic reaction, where heat is a reactant, will shift right to products with an increase in temperature. The amount of $NH_3(g)$ will increase as the reaction shifts right, so the smell of ammonia will increase. **63.** a. 2×10^3 molecules/cm³ b. There is more NO in the atmosphere than expected from the value of K. The reason is the slow rate (kinetics) of the reaction at low temperatures. Nitric oxide is produced in high-energy or high-temperature environments. In nature some NO is produced by lightning, and the primary man-made source is automobiles. The production of NO is endothermic. At high temperatures K will increase and the rates of the reaction will also increase, resulting in a higher production of NO. Once the NO gets into a more normal temperature environment, it doesn't go back to N_2 and O_2 because of the slow rate of the reaction. **65.** 2.6×10^{81} **67.** $[H_2]_0 = 11.0\ M$, $[N_2]_0 = 10.0\ M$ **69.** a. $P_{PCl_3} = P_{Cl_2} = 0.2230$ atm; $P_{PCl_5} = 0.0259$ atm; $K_p = 1.92$; b. $P_{PCl_3} = 0.0650$ atm; $P_{Cl_2} = 5.44$ atm; $P_{PCl_5} = 0.1839$ atm **71.** $P_{NO_2} = 0.704$ atm; $P_{N_2O_4} = 0.12$ atm **73.** a. $P_{CO_2} = P_{H_2O} = 0.50$ atm; b. 7.5 g $NaHCO_3$, 1.6 g Na_2CO_3; c. 3.9 L **75.** 192 g NH_4HS; 1.3 atm **77.** 6.74×10^{-6}. **79.** a. 134; b. $P_{NO} = 0.052$ atm; $P_{Br_2} = 0.18$ atm; $P_{NOBr} = 0.25$ atm **81.** $P_{P_4} = 0.73$ atm, $P_{P_2} = 0.270$ atm; 0.16 dissociated **83.** 1.5 **85.** 0.23 **87.** a. $P_{PCl_5} = 0.137$ atm; $P_{PCl_3} = P_{Cl_2} = 0.191$ atm; b. 39.4 g **89.** 71 atm **91.** 0.63 **93.** See the *Solutions Guide* for the plot. The data for the plot follow. At $P_{Total} = 1.0$ atm, $P_{NH_3} = 0.024$ atm; at $P_{Total} = 10.0$ atm, $P_{NH_3} = 1.4$ atm; at $P_{Total} = 100.$ atm, $P_{NH_3} = 32$ atm; at $P_{Total} = 1000.$ atm, $P_{NH_3} = 440$ atm; notice that as P_{Total} increases larger fractions of both N_2 and H_2 are converted to NH_3, i.e., as P_{Total} increases, (V decreases), the reaction shifts further to the right, as predicted by Le Châtelier's principle. **95.** 4.81 g/L; 5.5×10^{-3}

Chapter 7

17. a. The first equation is for the reaction of some generic acid, HA, with H_2O. HA is the proton donor (the acid), and H_2O is the proton acceptor (the base). In the reverse reaction, H_3O^+ is the proton donor (the acid) and A^- is the proton acceptor (the base). The second equation is for some generic base, B, with some generic acid, HX. Note that B has three hydrogens bonded to it. B is the proton acceptor (the base), and HX is the proton donor (the acid). When B accepts a proton, the central atom goes from having 3 bonded hydrogens to 4 bonded hydrogens. In the reverse reaction, BH^+ is the proton donor (the acid) and X^- is the proton acceptor (the base). b. Arrhenius acids produce H^+ in solution. So HA in the first equation is an Arrhenius acid. However, in the second equation, H^+ is not a product, so HX is not an Arrhenius acid. Both HA in the first equation and HX in the second equation are proton donors, so both are considered Brønsted–Lowry acids. For the bases in the two equations, H_2O and B, neither of them produce OH^- in their equations, so neither of them are Arrhenius bases. Both H_2O and B accept protons, so both are Brønsted–Lowry bases.

19.	Acid	Base	Conjugate Base of Acid	Conjugate Acid of Base
a.	H_2CO_3	H_2O	HCO_3^-	H_3O^+
b.	$C_5H_5NH^+$	H_2O	C_5H_5N	H_3O^+
c.	$C_5H_5NH^+$	HCO_3^-	C_5H_5N	H_2CO_3

21. a. $HC_2H_3O_2(aq) \rightleftharpoons H^+(aq) + C_2H_3O_2^-(aq)$; $K_a = \frac{[H^+][C_2H_3O_2^-]}{[HC_2H_3O_2]}$; b. $Co(H_2O)_6{}^{3+}(aq) \rightleftharpoons H^+(aq) + Co(H_2O)_5(OH)^{2+}(aq)$; $K_a = \frac{[H^+][Co(H_2O)_5(OH)^{2+}]}{[Co(H_2O)_6{}^{3+}]}$; c. $CH_3NH_3^+(aq) \rightleftharpoons H^+(aq) + CH_3NH_2(aq)$; $K_a = \frac{[H^+][CH_3NH_2]}{[CH_3NH_3^+]}$ **23.** The beaker on the left represents a strong acid in solution; the acid HA is 100% dissociated into the H^+ and A^- ions. The beaker on the right represents a weak acid in solution; only a little bit of the acid HB dissociates into ions, so the acid exists mostly as undissociated HB molecules in water. a. HNO_2: weak acid beaker; b. HNO_3: strong acid beaker; c. HCl: strong acid beaker; d. HF: weak acid beaker; e. $HC_2H_3O_2$: weak acid beaker **25.** $HClO_4 > HClO_2 > NH_4^+ > H_2O$ **27.** a. HCl; b. HNO_2; c. HCN **29.** a. H_2O and $CH_3CO_2^-$; b. An acid–base reaction can be thought of as a competition between two opposing bases. Since this equilibrium lies far to the left ($K_a < 1$), then $CH_3CO_2^-$ is a stronger base than H_2O. c. The acetate ion is a better base than water and produces basic solutions in water. When we put acetate ion into solution as the only major basic species, the reaction is

$$CH_3CO_2^- + H_2O \rightleftharpoons CH_3CO_2H + OH^-$$

Now the competition is between $CH_3CO_2^-$ and OH^- for the proton. Hydroxide ion is the strongest base possible in water. The above equilibrium lies far to the left, resulting in a K_b value less than 1. Those species we specifically call weak bases ($10^{-14} < K_b < 1$) lie between H_2O and OH^- in base strength. Weak bases are stronger than water but are weaker bases than OH^-. **31.** a. weak acid; b. strong acid; c. weak base; d. strong base; e. weak base; f. weak acid; g. weak acid; h. strong base; i. strong acid **33.** a. $[H^+] = [OH^-] = 1.71 \times 10^{-7}$ *M*; b. 6.767; c. 12.54 **35.** a. $[H^+] = 6.7 \times 10^{-15}$ *M*; basic; pH = 14.18; pOH = −0.18; b. $[H^+] = 2.8$ *M*; acidic; pH = −0.44; pOH = 14.44; c. $[H^+] = 1.0 \times 10^{-7}$ *M*; neutral; pH = pOH = 7.00; d. $[H^+] = 1.4 \times 10^{-11}$ *M*; basic; pH = 10.86; pOH = 3.14 **37.** a. $[H^+] = 4.0 \times 10^{-8}$ *M*; $[OH^-] = 2.5 \times 10^{-7}$ *M*; basic; b. $[H^+] = 5 \times 10^{-16}$ *M*; $[OH^-] = 20$ *M*; basic; c. $[H^+] = 10$ *M*; $[OH^-] = 1 \times 10^{-15}$ *M*; acidic; d. $[H^+] = 6.3 \times 10^{-4}$ *M*; $[OH^-] = 1.6 \times 10^{-11}$ *M*; acidic; e. $[H^+] = 1 \times 10^{-9}$ *M*; $[OH^-] = 1 \times 10^{-5}$ *M*; basic; f. $[H^+] = 4.0 \times 10^{-5}$ *M*; $[OH^-] = 2.5 \times 10^{-10}$ *M*; acidic **39.** a. 1.00; b. −0.70; c. 7.00 **41.** Use 4.2 mL of 12 *M* HCl with enough water added to make 1600 mL of solution. **43.** a. HNO_2 and H_2O; 2.00; b. CH_3CO_2H and H_2O; 2.68 **45.** $[H^+] = [F^-] = 3.5 \times 10^{-3}$ *M*; $[OH^-] = 2.9 \times 10^{-12}$ *M*; [HF] = 0.017 *M*; pH = 2.46 **47.** a. HA is a weak acid. Most of the acid is present as HA molecules; only one set of H^+ and A^- ions is present. In a strong acid, all of the acid would be dissociated into H^+ and A^- ions. b. $K_a = 2.2 \times 10^{-3}$; exactly 10% dissociated **49.** 2.02 **51.** $[C_6H_5CO_2H] = 4.1 \times 10^{-3}$ *M*; $[C_6H_5CO_2^-] = [H^+] = 5.1 \times 10^{-4}$ *M*; $[OH^-] = 1.9 \times 10^{-11}$ *M*; 3.29 **53.** 2.68 **55.** a. 1.00; b. 1.30 **57.** 0.033 **59.** 3.5×10^{-4} **61.** 6×10^{-2} *M* **63.** $NH_3 > C_5H_5N > H_2O > NO_3^-$ **65.** a. $C_6H_5NH_2$; b. $C_6H_5NH_2$; c. OH^-; d. CH_3NH_2 **67.** a. 13.00; b. 7.00; c. 14.30 **69.** 1.6×10^{-4} *M* **71.** $[OH^-] = [H_2NNH_3^+] = 2.4 \times 10^{-3}$ *M*; $[H_2NNH_2] = 2.0$ *M*; $[H^+] = 4.2 \times 10^{-12}$ *M*; pH = 11.38 **73.** Neutrally charged organic compounds containing at least one nitrogen atom generally behave as weak bases. The nitrogen atom has an unshared pair of electrons around it. This lone pair of electrons is used to form a bond to H^+. **75.** 12.00 **77.** 9.59 **79.** a. 1.3%; b. 4.2%; c. 6.4% **81.** 1.0×10^{-9} **83.** $H_3C_6H_5O_7(aq) \rightleftharpoons H_2C_6H_5O_7^-(aq) + H^+(aq)$;

$$K_{a_1} = \frac{[H_2C_6H_5O_7^-][H^+]}{[H_3C_6H_5O_7]};\ H_2C_6H_5O_7^-(aq) \rightleftharpoons$$

$$HC_6H_5O_7^{2-}(aq) + H^+(aq);\ K_{a_2} = \frac{[HC_6H_5O_7^{2-}][H^+]}{[H_2C_6H_5O_7^-]};$$

$$HC_6H_5O_7^{2-}(aq) \rightleftharpoons C_6H_5O_7^{3-}(aq) + H^+(aq);$$

$$K_{a_3} = \frac{[C_6H_5O_7^{3-}][H^+]}{[HC_6H_5O_7^{2-}]}$$

85. $[H^+] = 3 \times 10^{-2}$ *M*, $[OH^-] = 3 \times 10^{-13}$ *M*, $[H_3AsO_4] = 0.17$ *M*, $[H_2AsO_4^-] = 3 \times 10^{-2}$ *M*; $[HAsO_4^{2-}] = 8 \times 10^{-8}$ *M*, $[AsO_4^{3-}] = 2 \times 10^{-15}$ *M* **87.** 3.00 **89.** −0.30 **91.** a. These are strong acids such as HCl, HBr, HI, HNO_3, H_2SO_4, and $HClO_4$. b. These are salts of the conjugate acids of the bases in Table 7.3. These conjugate acids are all weak acids. NH_4Cl, $CH_3NH_3NO_3$, and $C_2H_5NH_3Br$ are three examples. Note that the anions used to form these salts (Cl^-, NO_3^-, and Br^-) are conjugate bases of strong acids; this is so because they have no acidic or basic properties in water (with the exception of HSO_4^-, which has weak acid properties). c. These are strong bases such as LiOH, NaOH, KOH, RbOH, CsOH, $Ca(OH)_2$, $Sr(OH)_2$, and $Ba(OH)_2$. d. These are salts of the conjugate bases of the neutrally charged weak acids in Table 7.2. The conjugate bases of weak acids are weak bases themselves. Three examples are $NaClO_2$, $KC_2H_3O_2$, and CaF_2. The cations used to form these salts are Li^+, Na^+, K^+, Rb^+, Cs^+, Ca^{2+}, Sr^{2+}, and Ba^{2+} since these cations have no acidic or basic properties in water. Notice that these are the cations of the strong bases you should memorize. e. There are two ways to make a neutral salt. The easiest way is to combine a conjugate base of a strong acid (except for HSO_4^-) with one of the cations from a strong base. These ions have no acidic/basic properties in water so salts of these ions are neutral. Three examples are NaCl, KNO_3, and SrI_2. Another type of strong electrolyte that can produce neutral solutions are salts that contain an ion with weak acid properties combined with an ion of opposite charge having weak base properties. If the K_a for the weak acid ion is equal to the K_b for the weak base ion, then the salt will produce a neutral solution. The most common example of this type of salt is ammonium acetate ($NH_4C_2H_3O_2$). For this salt, K_a for $NH_4^+ = K_b$ for $C_2H_3O_2^- = 5.6 \times 10^{-10}$. This salt at any concentration produces a neutral solution. **93.** a. HI < HF < NaI < NaF; b. HBr < NH_4Br < KBr < NH_3; c. $HNO_3 < C_6H_5NH_3NO_3 < HOC_6H_5 < NaNO_3 < C_6H_5NH_2 < KOC_6H_5 < NaOH$ **95.** a. Neutral; Sr^{2+} and NO_3^- have no acidic or basic properties; b. Basic; K_b for $CN^- > K_a$ for $C_2H_5NH_3^+$ (CN^- is a better base than $C_2H_5NH_3^+$ is as an acid); c. Acidic; K_a for $C_5H_5NH^+ > K_b$ for F^-; d. Neutral; K_a for $NH_4^+ = K_b$ for $C_2H_3O_2^-$; e. Basic; HCO_3^- is a stronger base than an acid because K_b for $HCO_3^- > K_a$ for HCO_3^-. **97.** a. $[OH^-] = [H^+] = 1.0 \times 10^{-7}$ *M*; pH = 7.00; b. $[OH^-] = 3.7 \times 10^{-6}$ *M*; $[H^+] = 2.7 \times 10^{-9}$ *M*; pH = 8.57 **99.** 3.66 **101.** a. 8.23; b. 10.56; c. 4.82 **103.** NaF; this was determined by calculating K_b for F^-. **105.** 3.00 **107.** 8.37 **109.** 7.4; when an acid is added to water, the pH cannot be basic. Must account for the autoionization of water. **111.** 6.24 **113.** 6.15 **115.** a. 1.8×10^9; b. 2.5×10^3; c. 3.1×10^{-5}; d. 1.0×10^{14}; e. 5.6×10^4; f. 4.0×10^{10} **117.** 990 mL **119.** a. In the lungs there is a lot of O_2, and the equilibrium favors $Hb(O_2)_4$. In the cells there is a lower concentration of O_2, and the equilibrium favors HbH_4^{4+}. b. CO_2 is a weak acid. Removing CO_2 essentially decreases $[H^+]$. $Hb(O_2)_4$ is then favored, and O_2 is not released by hemoglobin in the cells. Breathing into a paper bag increases CO_2 in the blood, thus increasing $[H^+]$, which shifts the equilibrium left. c. CO_2 builds up in the blood, and it becomes too acidic, which drives the equilibrium to the left. Hemoglobin can't bind O_2 as strongly in the lungs. Bicarbonate ion acts as a base in water and neutralizes the excess acidity. **121.** NH_4Cl **123.** 11.77 **125.** 3.36 **127.** a. 1.66; b. −0.78; c. Because of the lower charge, $Fe^{2+}(aq)$ will not be as strong an acid as $Fe^{3+}(aq)$. A solution of iron(II) nitrate will be less acidic (have a higher pH) than a solution with the same concentration of iron(III) nitrate. **129.** 0.022 *M* **131.** 0.022 *M* **133.** a. 2.80; b. 1.1×10^{-3} *M* **135.** 1.96 **137.** a. 1.1×10^{-4}; b. $[H_2CO_3] = [CO_3^{2-}]$;

c. $pH = \dfrac{pK_{a_1} + pK_{a_2}}{2}$ (see *Solutions Guide* for derivation);

d. 8.35 **139.** 20.0 g **141.** K_a for HX = 1.0×10^{-5} **143.** 2.492 **145.** 6.16 L **147.** 10.00 **149.** 7.20 **151.** 6.72

Chapter 8

15. A buffer solution is one that resists a change in its pH when either hydroxide ions or protons (H^+) are added. Any solution that contains a weak acid and its conjugate base or a weak base and its conjugate acid is classified as a buffer. The pH of a buffer depends on the [base]/[acid] ratio. When H^+ is added to a buffer, the weak base component of the buffer reacts with the H^+ and forms the acid component of the buffer. Even though the concentrations of the acid and base components of the buffer change some, the ratio of [base]/[acid] does not change that much. This translates into a pH that doesn't change much. When OH^- is added to a buffer, the weak acid component is converted into the base component of the buffer. Again, the ratio of [base]/[acid] does not change a lot (unless a large quantity of OH^- is added), so the pH does not change much. $H^+(aq) + CO_3^{2-}(aq) \rightarrow HCO_3^-(aq)$; $OH^-(aq) + HCO_3^-(aq) \rightarrow CO_3^{2-}(aq) + H_2O(l)$ **17.** Only the third beaker represents a buffer solution. A weak acid and its conjugate base must both be present in large quantities in order to have a buffer solution. This is only the case in the third beaker. The first beaker represents a beaker full of strong acid, which is 100% dissociated. The second beaker represents a weak acid solution. In a weak acid solution, only a small fraction of the acid is dissociated. In this representation, 1/10 of the weak acid has dissociated. The only B^- present in this beaker is from the dissociation of the weak acid. A buffer solution has B^- added from another source. **19.** When [weak acid] > [conjugate base], pH < pK_a. When [conjugate base] > [weak acid], pH > pK_a. **21.** a. 2.96; b. 8.94; c. 7.00; d. 4.89 **23.** a. 4.29; b. 12.30; c. 12.30; d. 5.07 **25.** 3.37 **27.** 3.48; 3.14 **29.** 8.18; 0.20 mol NaOH **31.** 4.37 **33.** a. 0.19; b. 0.59; c. 1.0; d. 1.9 **35.** a. 0.50 mol; b. 0.78 mol; c. 0.36 mol **37.** 15 g **39.** a. 1.2 *M*; b. 4.32 **41.** a. 1.1 $\approx$ 1; b. A best buffer has approximately equal concentrations of weak acid and conjugate base so that pH $\approx pK_a$ for a best buffer. The pK_a value for a $H_3PO_4/H_2PO_4^-$ buffer is $-\log(7.5 \times 10^{-3}) = 2.12$. A pH of 7.1 is too high for a $H_3PO_4/H_2PO_4^-$ buffer to be effective. At this high pH, there would be so little H_3PO_4 present that we could hardly consider it a buffer; this solution would not be effective in resisting pH changes, especially when a strong base is added. **43.** The $[H_2CO_3]:[HCO_3^-]$ concentration ratio must increase from 0.093 to 0.10 in order for the onset of acidosis to occur. **45.** Only mixture c results in a buffered solution. **47.** HOCl; there are many possibilities. One possibility is a solution with [HOCl] = 1.0 *M* and [NaOCl] = 0.35 *M*. **49.** 7.0×10^{-7} *M* **51.** 6.89 **53.** a. Let's call the acid HB, which is a weak acid. When HB is present in the beakers, it exists in the undissociated form, which makes it a weak acid. A strong acid would exist as separate H^+ and B^- ions. b. beaker c → beaker a → beaker e → beaker b → beaker d; c. pH = pK_a when a buffer solution is present that has equal concentrations of the weak acid and conjugate base. This is beaker e. d. The equivalence point is when just enough OH^- has been added to exactly react with all of the acid present initially. This is beaker b. e. Past the equivalence, the pH is dictated by the concentration of excess OH^- added from the strong base. We can ignore the amount of hydroxide added by the weak conjugate base that is also present. This is beaker d.

55.

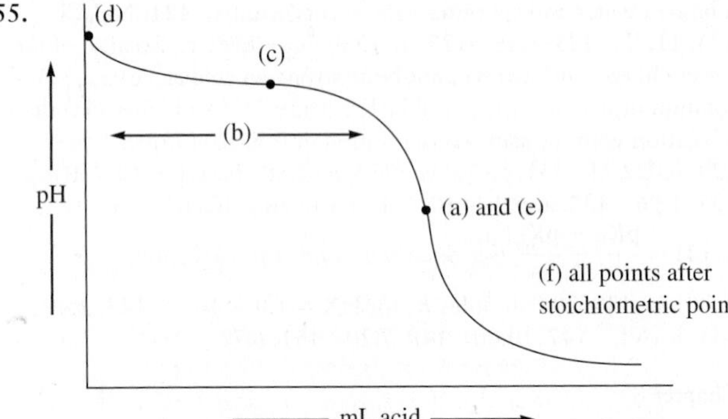

$B + H^+ \rightarrow BH^+$; added H^+ from the strong acid converts the weak base, B, into its conjugate acid, BH^+. Initially, before any H^+ is added (point d), B is the dominant species present. After H^+ is added, both B and BH^+ are present and a buffered solution results (region b). At the equivalence point (points a and e), exactly enough H^+ has been added to convert all of the weak base present initially into its conjugate acid, BH^+. Past the equivalence point (region f), excess H^+ is present. For the answer to part b, we included almost the entire buffer region. The maximum buffer region is around the halfway point to equivalence (point c) where [B] = [BH^+]. Here, pH = pK_a, which is a characteristic of a best buffer. **57.** a. all the same; b. i < iv < iii < ii; c. i < iv < iii < ii; d. iii < ii = i < iv; the only different answer would be part c. The ordering would be i < iii < iv < ii. **59.** a. f; b. a; the best point to look at to differentiate a strong acid from a weak acid titration (if initial concentrations are not known) is the equivalence point pH. If the pH = 7.00, the acid titrated is a strong acid; if the pH is greater than 7.00, the acid titrated is a weak acid. c. d **61.** a. 0.699; b. 0.854; c. 1.301; d. 7.00; e. 12.15 **63.** a. 2.72; b. 4.26; c. 4.74; d. 5.22; e. 8.79; f. 12.15

65.

Volume (mL)	pH
0.0	2.43
4.0	3.14
8.0	3.53
12.5	3.86
20.0	4.46
24.0	5.24
24.5	5.6
24.9	6.3
25.0	8.28
25.1	10.3
26.0	11.30
28.0	11.75
30.0	11.96

See *Solutions Guide* for pH plot.

67.

Volume (mL)	pH
0.0	11.11
4.0	9.97
8.0	9.58
12.5	9.25
20.0	8.65
24.0	7.87
24.5	7.6
24.9	6.9
25.0	5.28
25.1	3.7
26.0	2.71
28.0	2.24
30.0	2.04

See *Solutions Guide* for pH plot.

69. a. 4.19, 8.45; b. 10.74, 5.96; c. 0.89, 7.00 **71.** 6.6×10^{-7} **73.** a. yellow; b. 8.0; c. blue **75.** The pH is between 5 and 8. **77.** Bromthymol blue or phenol red are possible indicators for Exercise 61, and *o*-cresolphthalein or phenolphthalein are possible indicators for Exercise 63. **79.** Phenolphthalein is a possible indicator for Exercise 65, and bromcresol green is a possible indicator for Exercise 67. **81.** Methyl red changes color in a pH range of about pH = $pK_a \pm 1 = 5.3 \pm 1$. Therefore, methyl red is a useful indicator at pH values approximately between 4.3 and 6.3. In titrating a weak acid with a base, we start off with an acidic solution with pH < 4.3 so the color would change from red to reddish-orange at pH ~ 4.3. In titrating a weak base with an acid, the color change would be from yellow to yellowish-orange at pH ~ 6.3. Only a weak base–strong acid titration would have an acidic pH at the equivalence point, so only in this type of titration would the color change of methyl red indicate the approximate endpoint. **83.** a. 200.0 mL; b. i. H_2A and H_2O are the major species; ii. H_2A, HA^-, H_2O, and Na^+; iii. HA^-, H_2O, and Na^+; iv. HA^-, A^{2-}, H_2O, and Na^+; v. A^{2-}, H_2O, and Na^+; vi. OH^-, A^{2-}, H_2O, and Na^+; c. $K_{a_1} = 1 \times 10^{-4}$; $K_{a_2} = 1 \times 10^{-8}$ **85.** a. 2.18; b. 2.55; c. 3.00; d. 5.00; e. 6.40; f. 7.00; g. 9.50; h. 11.63; i. 12.00; j. 12.34 **87.** a. $K_{a_1} = 1.5 \times 10^{-4}$; $K_{a_2} = 2.8 \times 10^{-7}$; $K_{a_3} = 3.6 \times 10^{-10}$; b. The pH at the third halfway point to equivalence (60.5 mL of NaOH added) will be equal to $pK_{a_3} = 9.44$. The pH at 59.0 mL NaOH added should be slightly less than 9.44. c. 9.34 **89.** a. Na^+ is present in all solutions. A. CO_3^{2-}, H_2O; B. CO_3^{2-}, HCO_3^-, H_2O, Cl^-; C. HCO_3^-, H_2O, Cl^-; D. HCO_3^-, $CO_2(H_2CO_3)$, H_2O, Cl^-; E. $CO_2(H_2CO_3)$, H_2O, Cl^-; F. H^+ (excess), $CO_2(H_2CO_3)$, H_2O, Cl^-; b. A, 11.66; B, 10.32; C, 8.35; D, 6.37; E, 3.92 **91.** The representations all show 1:1 salts (i.e., the formula of the solid contains 1 cation for every 1 anion [either +1 and −1, or +2 and −2, or +3 and −3]). The solution with the largest number of ions (largest $[M^{n+}]$ and $[X^{n-}]$) will have the largest K_{sp} value. From the representations, the second beaker has the largest number of ions present, so this salt has the largest K_{sp} value. Conversely, the third beaker, with the fewest number of hydrated ions, will have the smallest K_{sp} value. **93.** a. 1.6×10^{-5} mol/L; 6.7×10^{-3} g/L; b. 9.3×10^{-5} mol/L; 9.3×10^{-3} g/L; c. 6.5×10^{-7} mol/L; 3.1×10^{-4} g/L **95.** a. 2.3×10^{-9}; b. 8.20×10^{-19} **97.** 5.3×10^{-12} **99.** a. CaF_2 has the smallest molar solubility since it has the smaller K_{sp} value. b. $FePO_4$ has the smallest molar solubility (must be calculated). **101.** a. 4×10^{-17} mol/L; b. 4×10^{-11} mol/L; c. 4×10^{-29} mol/L **103.** 1.5×10^{-19} g **105.** If the anion in the salt can act as a base in water, then the solubility of the salt will increase as the solution becomes more acidic. Added H^+ will react with the base, forming the conjugate acid. As the basic anion is removed, more of the salt will dissolve to replenish the basic anion. The salts with basic anions are Ag_3PO_4, $CaCO_3$, $CdCO_3$, and $Sr_3(PO_4)_2$. Hg_2Cl_2 and PbI_2 do not have any pH dependence since Cl^- and I^- are terrible bases (the conjugate bases of strong acids). $Ag_3PO_4(s) + H^+(aq) \rightarrow 3Ag^+(aq) + HPO_4^{2-}(aq) \xrightarrow{\text{Excess } H^+} 3Ag^+(aq) + H_3PO_4(aq)$; $CaCO_3(s) + H^+(aq) \xrightarrow{\text{Excess } H^+} Ca^{2+}(aq) + HCO_3^-(aq) \xrightarrow{\text{Excess } H^+} Ca^{2+}(aq) + H_2CO_3(aq)$ $[H_2O(l) + CO_2(g)]$; $CdCO_3(s) + H^+(aq) \rightarrow Cd^{2+}(aq) + HCO_3^-(aq) \rightarrow Cd^{2+}(aq) + H_2CO_3(aq)$ $[H_2O(l) + CO_2(g)]$; $Sr_3(PO_4)_2(s) + 2H^+(aq) \rightarrow 3Sr^{2+}(aq) + 2HPO_4^{2-} \xrightarrow{\text{Excess } H^+} 3Sr^{2+}(aq) + 2H_3PO_4(aq)$ **107.** a. AgF; b. $Pb(OH)_2$; c. $Sr(NO_2)_2$; d. $Ni(CN)_2$ **109.** $[Ba^{2+}] = 6.0 \times 10^{-5}$ *M*; $[Br^-] = 1.2 \times 10^{-4}$ *M*; $[K^+] = 4.8 \times 10^{-4}$ *M*; $[C_2O_4^{2-}] = 2.4 \times 10^{-4}$ *M* **111.** Precipitation of $Al(OH)_3(s)$ will begin when pH > 3.7. **113.** When $[AgNO_3]_0$ is greater than 5.6×10^{-5} *M*, then $Ag_3PO_4(s)$ will precipitate. **115.** See the *Solutions Guide* for the flowchart for each separation. A possible order of chemicals necessary to separate the ions follows. a. NaCl(*aq*) followed by $NH_3(aq)$ followed by $H_2S(aq)$; b. NaCl(*aq*) followed by $Na_2SO_4(aq)$ followed by the basic addition of $H_2S(aq)$; c. $AgNO_3(aq)$ followed by $NH_3(aq)$ followed by $Na_2S_2O_3(aq)$; d. Na_2SO_4 followed by the basic addition of $H_2S(aq)$ **117.** S^{2-} is a very basic anion and reacts significantly with H^+ to form HS^- ($S^{2-} + H^+ \rightleftharpoons HS^-$). The actual concentration of S^{2-} in solution depends on the amount of H^+ present. In basic solutions, little H^+ is present, which shifts the above reaction to the left. In basic solutions, the S^{2-} concentration is relatively high. So, in basic solutions, a wider range of sulfide salts will precipitate. However, in acidic solutions, added H^+ shifts the equilibrium to the right, which results in a lower S^{2-} concentration. In acidic solutions, only the least soluble sulfide salts will precipitate out of solution. **119.** $Hg^{2+}(aq) + 2I^-(aq) \rightarrow HgI_2(s)$, orange ppt.; $HgI_2(s) + 2I^-(aq) \rightarrow HgI_4^{2-}(aq)$, soluble complex ion **121.** 3.3×10^{-32} *M* **123.** $[F^-] = 4.0$ *M*; $[Be^{2+}] = 2.6 \times 10^{-20}$ *M*; $[BeF^+] = 8.2 \times 10^{-15}$ *M*; $[BeF_2] = 1.9 \times 10^{-10}$ *M*; $[BeF_3^-] = 4.6 \times 10^{-7}$ *M*; $[BeF_4^{2-}] = 5.0 \times 10^{-5}$ *M* **125.** a. 1.6×10^{-6}; b. 0.056 mol/L **127.** 42 g **129.** Test tube 1: added Cl^- reacts with Ag^+ to form a silver chloride precipitate. The net ionic equation is $Ag^+(aq) + Cl^-(aq) \rightarrow AgCl(s)$. Test tube 2: added NH_3 reacts with Ag^+ ions to form a soluble complex ion, $Ag(NH_3)_2^+$. As this complex ion forms, Ag^+ is removed from the solution, which causes the AgCl(*s*) to dissolve. When enough NH_3 is added, all of the silver chloride precipitate will dissolve. The equation is $AgCl(s) + 2NH_3(aq) \rightarrow Ag(NH_3)_2^+(aq) + Cl^-(aq)$. Test tube 3: added H^+ reacts with the weak base, NH_3, to form NH_4^+. As NH_3 is removed from the $Ag(NH_3)_2^+$ complex ion, Ag^+ ions are released to solution and can then react with Cl^- to re-form AgCl(*s*). The equations are $Ag(NH_3)_2^+(aq) + 2H^+(aq) \rightarrow Ag^+(aq) + 2NH_4^+(aq)$ and $Ag^+(aq) + Cl^-(aq) \rightarrow AgCl(s)$. **131.** a. 8.1; b. pH = 7.00: 0.083; pH = 9.00: 8.3; c. 8.08; 7.95 **133.** a. potassium fluoride + HCl; b. benzoic acid + NaOH; c. sodium acetate + acetic acid; d. HOCl + NaOH or ammonium chloride + sodium acetate; e. ammonium chloride + NaOH **135.** 99.5%; for a strong base–strong acid titration, the equivalence point occurs at pH = 7.0. Bromthymol blue is a good indicator choice since it changes color at pH ~ 7 (from base color to acid color). **137.** Since the equivalence point occurs at pH = 8.9, phenolphthalein would be a good indicator choice because it changes color at pH ~ 9 (from acid color to base color). **139.** Precipitation of $CaF_2(s)$ will begin to occur when just more than 9.0×10^{-6} g $Ca(NO_3)_2$ has been added. **141.** a. 6.7×10^{-6} *M*; b. 1.2×10^{-13} *M*; c. 2.3×10^{-19} *M*; no, since *Q* (= 2.3×10^{-21}) is less than the K_{sp} value. **143.** 65 mL **145.** 4.92 **147.** 180. g/mol; 3.3×10^{-4} **149.** K_{a_3} is so small (4.8×10^{-13}) that a break is not seen at the third stoichiometric point. **151.** 49 mL **153.** a. See the *Solutions Guide* for the derivation. b. ammonium formate, 6.50; ammonium acetate, 7.00; ammonium bicarbonate, 7.81; c. $NH_4^+(aq) + OH^-(aq) \rightarrow NH_3(aq) + H_2O(l)$, $C_2H_3O_2^-(aq) + H^+(aq) \rightarrow HC_2H_3O_2(aq)$ **155.** a. 0.33 mol/L; b. 0.33 *M*; c. 4.8×10^{-3} *M* **157.** a. 5.8×10^{-4} mol/L; b. Greater; since F^- is a weak base, the F^- concentration is lowered by reaction with water, which causes more $SrF_2(s)$ to dissolve. c. 3.5×10^{-3} mol/L **159.** Any pH less than 8.7 is a correct answer to this problem. **161.** 5.7×10^{-2} mol/L **163.** 3.9 L **165.** a. 2.21; b. 2.30×10^{-2} *M* **167.** 12.34 **169.** pH = 9.50: 10.0 mL; pH = 4.00: 4.55 mL

Chapter 9

15. KE = 78 J; PE = 118 J **17.** Path-dependent functions for a trip from Chicago to Denver are those quantities that depend on the route taken. One can fly directly from Chicago to Denver or one could fly from Chicago to Atlanta to Los Angeles and then to Denver. Some path-dependent quantities are miles traveled, fuel consumption of the airplane, time traveling, airplane snacks eaten, etc. State functions are path-independent; they depend only on the initial and final states. Some state functions for an airplane trip from Chicago to Denver would be longitude change, latitude change, elevation change, and overall time zone change. **19.** 70. J **21.** $q = 30.9$ kJ; $w = -12.4$ kJ; $\Delta E = 18.5$ kJ **23.** −37.56 kJ **25.** $\Delta H = \Delta E + P\Delta V$ at constant *P*; from the strict definition of enthalpy, the difference between ΔH and ΔE is the quantity $P\Delta V$. Thus, when a system at constant *P* can do pressure-volume work, then $\Delta H \neq \Delta E$. When the system cannot do *PV* work, then $\Delta H = \Delta E$ at constant pressure. An important way to differentiate ΔH from ΔE is to concentrate on *q*, the heat flow; the heat flow by a system at constant pressure equals ΔH, and the heat flow by a system at constant volume equals ΔE. **27.** Since the reaction is exothermic (heat is a product), one should provide cooling for the reaction mixture to prevent $H_2SO_4(aq)$ from boiling. **29.** a. 1650 kJ of heat released; b. 826 kJ of heat released; c. 7.39 kJ of heat released; d. 34.4 kJ of heat released **31.** When a liquid is converted into gas, there is an increase in volume. The 2.5 kJ/mol quantity is the work done by the vaporization process in pushing back the atmosphere. **33.** Constant *V*: $\Delta E = q = 74.3$ kJ, $w = 0$, $\Delta H = 88.1$ kJ; constant *P*: $\Delta H = q = 88.1$ kJ, $w = -13.8$ kJ, $\Delta E = 74.3$ kJ **35.** Pathway one: step 1: $q = 30.4$ kJ, $w = -12.2$ kJ, $\Delta E = 18.2$ kJ, $\Delta H = 30.4$ kJ; step 2: $q = -28.1$ kJ, $w = 21.3$ kJ, $\Delta E = -6.8$ kJ, $\Delta H = -11$ kJ; total: $q = 2.3$ kJ, $w = 9.1$ kJ, $\Delta E = 11.4$ kJ, $\Delta H = 19$ kJ; pathway two: step 3: $q = 6.84$ kJ, $w = 0$, $\Delta E = 6.84$ kJ, $\Delta H = 11.40$ kJ; step 4: $q = 7.6$ kJ, $w = -3.0$ kJ, $\Delta E = 4.6$ kJ, $\Delta H = 7.6$ kJ; total: $q = 14.4$ kJ, $w = -3.0$ kJ, $\Delta E = 11.4$ kJ, $\Delta H = 19.0$ kJ; state functions are independent of the particular pathway taken between two states; path functions are dependent on the particular pathway. In this problem, the overall values of ΔH and ΔE for the two pathways are the same; hence, ΔH and ΔE are state functions. The overall values of *q* and *w* for the two pathways are different; hence, *q* and *w* are path functions. **37.** In calorimetry, heat flow is determined into or out of the surroundings. Because $\Delta E_{univ} = 0$ by the first law of thermodynamics, $\Delta E_{sys} = -\Delta E_{surr}$; what happens to the surroundings is the exact opposite of what happens to the system. To determine heat flow, we need to know the heat capacity of the surroundings, the mass of the surroundings that accepts/donates the heat, and the change in temperature. If we know these quantities, q_{surr} can be calculated and then equated to q_{sys} ($-q_{surr} = q_{sys}$). For an endothermic reaction, the surroundings (the calorimeter contents) donate heat to the system. This is accompanied by a decrease in temperature of the surroundings. For an exothermic reaction, the system donates heat to the surroundings (the calorimeter) so temperature increases. $q_p = \Delta H$; $q_v = \Delta E$; a coffee cup calorimeter is at constant (atmospheric) pressure. The heat released or gained at constant pressure is ΔH. A bomb calorimeter is at constant volume. The heat released or gained at constant volume is ΔE. **39.** $H_2O(l)$, 2.30×10^3 J; $Hg(l)$, 140°C **41.** 75.0 g **43.** 28 g **45.** −66 kJ/mol **47.** 170 J/g; 20. kJ/mol **49.** 39.2°C **51.** a. 31.5 kJ/°C; b. -1.10×10^3 kJ/mol **53.** a. $C_{12}H_{22}O_{11}(s) + 12O_2(g) \rightarrow 12CO_2(g) + 11H_2O(l)$; b. −5630 kJ/mol; c. −5630 kJ/mol **55.** 1268 kJ; since the reaction is very endothermic (requires a lot of heat), high energy costs would make it an impractical way of making ammonia. **57.** 226 kJ **59.** −713 kJ **61.** −158 kJ **63.** −202.6 kJ **65.** $Na(s) + \frac{1}{2}Cl_2(g) \rightarrow NaCl(s)$; $H_2(g) + \frac{1}{2}O_2(g) \rightarrow H_2O(l)$; 6 C(graphite, s) + 6 $H_2(g)$ + 3 $O_2(g) \rightarrow C_6H_{12}O_6(s)$; $Pb(s) + S(s) + 2\,O_2(g) \rightarrow PbSO_4(s)$ **67.** a. −940. kJ; b. −265 kJ; c. −176 kJ; d. −1235 kJ; e. −320. kJ; f. −37 kJ **69.** −832 kJ; −368 kJ; −133 kJ; in both cases sodium metal reacts with the "extinguishing agent." Both reactions are exothermic and each reaction produces a flammable gas—H_2 and CO, respectively. **71.** −4594 kJ **73.** a. 632 kJ; b. Since $3C_2H_2(g)$ is higher in energy than $C_6H_6(l)$, acetylene will release more energy per gram when burned in air. **75.** −169 kJ/mol **77.** 3.97 g **79.** −129 kJ **81.** −22.7 kJ/g versus −29.67 kJ/g for ethanol. Ethanol has a higher fuel value than methanol. **83.** 25 J **85.** 282.1 kJ **87.** $C_{H_2O} = 4.18$ kJ/°C; $C_{cal} = 6.66$ kJ/°C **89.** −306 kJ/mol **91.** 4.2 kJ of heat released **93.** When $\Delta V > 0$ ($\Delta n > 0$), then $w < 0$ and the system does work on the surroundings (c and e). When $\Delta V < 0$ ($\Delta n < 0$), then $w > 0$ and the surroundings do work on the system (a and d). When $\Delta V = 0$ ($\Delta n = 0$), then $w = 0$ (b). **95.** 16.8 kJ/mol **97.** a. $2Al(s) + \frac{3}{2}O_2(g) \rightarrow Al_2O_3(s)$; b. $C_2H_5OH(l) + 3O_2(g) \rightarrow 2CO_2(g) + 3H_2O(l)$; c. $Ba(OH)_2(aq) + 2HCl(aq) \rightarrow 2H_2O(l) + BaCl_2(aq)$ d. 2C(graphite) + $\frac{3}{2}H_2(g) + \frac{1}{2}Cl_2(g) \rightarrow C_2H_3Cl(g)$; e. $C_6H_6(l) + \frac{15}{2}O_2(g) \rightarrow 6CO_2(g) + 3H_2O(l)$ (Note: ΔH_{comb} values generally assume 1 mol of compound combusted); f. $NH_4Br(s) \rightarrow NH_4^+(aq) + Br^-(aq)$ **99.** 6.02 kJ/mol **101.** 43.58 kJ/mol; from Appendix 4 data, $\Delta H° = 44$ kJ/mol. The ΔH values agree to two significant figures (as they should). **103.** 32 m^2 **105.** For an isothermal expansion of an ideal gas, $\Delta T = 0$, so $\Delta E = 0$ (and $\Delta H = 0$); therefore, $q = -w = P\Delta V$. As long as the gas expands against a nonzero external pressure, $q \neq 0$ because $w \neq 0$.

Chapter 10

13. Possible arrangements for one molecule:

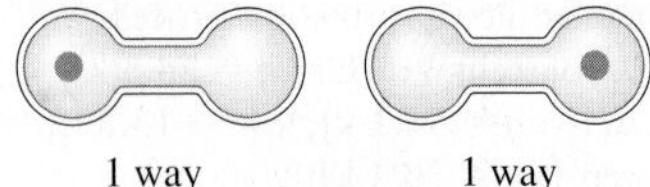

1 way 1 way

Both are equally probable.

Possible arrangements for two molecules:

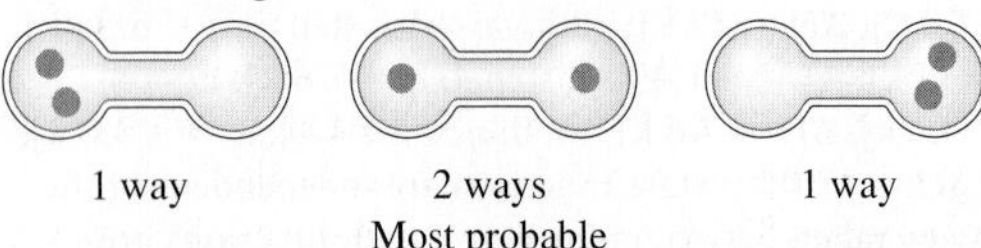

1 way 2 ways Most probable 1 way

Possible arrangement for three molecules:

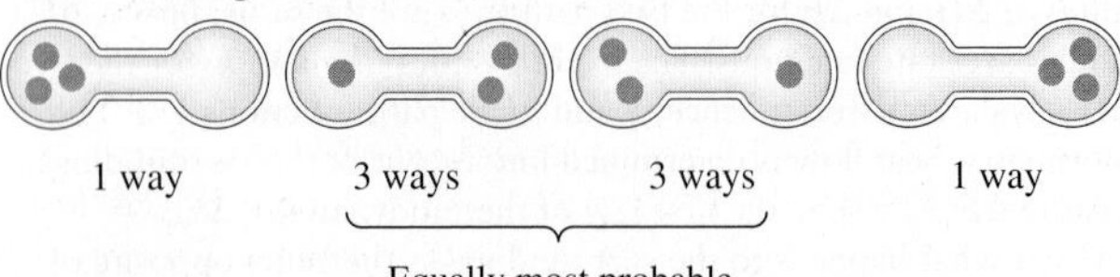

1 way 3 ways 3 ways 1 way

Equally most probable

15.

2 kJ	___	___	X	___	X	XX
1 kJ	___	X	___	XX	X	___
0 kJ	XX	X	X	___	___	___
Total E =	0 kJ	1 kJ	2 kJ	2 kJ	3 kJ	4 kJ

The most likely total energy is 2 kJ.

17. c, e, f **19.** a. Positional probability increases; there is a greater volume accessible to the randomly moving gas molecules, which increases positional probability. b. The positional probability does not change. There is no change in volume and thus no change in the numbers of positions of the molecules. c. Positional probability decreases; volume decreases. **21.** There are six ways to get a seven, more than any other number. The seven is not favored by energy; rather it is favored by probability. To change the probability we would have to expend energy (do work). **23.** $q_v = \Delta E = 71.9$ kJ; $q_p = \Delta H = 85.3$ kJ; $\Delta E = 71.9$ kJ for both constant-volume and constant-pressure processes. **25.** 77.0°C **27.** $w = q = 0$; $q_{rev} = 350$ J **29.** a. constant V, 1.51 kJ; constant P, 1.94 kJ; b. 219.63 J K^{-1} mol^{-1}; c. 218.30 J K^{-1} mol^{-1} **31.** 2.50×10^4 J **33.** 2.9 J/K **35.** Living organisms need an external source of energy to carry out these processes. Green plants use the energy from sunlight to produce glucose from carbon dioxide and water by photosynthesis. In the human body, the energy released from the metabolism of glucose helps drive the synthesis of proteins. For all processes combined, ΔS_{univ} must be greater than zero (second law). **37.** ΔS_{surr} is primarily determined by heat flow. This heat flow into or out of the surroundings comes from the heat flow out of or into the system. In an exothermic process ($\Delta H < 0$), heat flows into the surroundings from the system. The heat flow into the surroundings increases the random motions in the surroundings and increases the entropy of the surroundings ($\Delta S_{surr} > 0$). This is a favorable driving force for spontaneity. In an endothermic reaction ($\Delta H > 0$), heat is transferred from the surroundings into the system. This heat flow out of the surroundings decreases the random motions in the surroundings and decreases the entropy of the surroundings ($\Delta S_{surr} < 0$). This is unfavorable. The magnitude of ΔS_{surr} also depends on the temperature. The relationship is inverse; at low temperatures, a specific amount of heat exchange makes a larger percent change in the surroundings than the same amount of heat flow at a higher temperature. The negative sign in the $\Delta S_{surr} = -\Delta H/T$ equation is necessary to get the signs correct. For an exothermic reaction where ΔH is negative, this increases ΔS_{surr}, so the negative sign converts the negative ΔH value into a positive quantity. For an endothermic process where ΔH is positive, the sign of ΔS_{surr} is negative and the negative sign converts the positive ΔH value into a negative quantity. **39.** a. 7.45×10^3 J/K; b. −376 J/K **41.** a. negative; b. positive; c. negative; d. negative; e. negative; f. positive **43.** 262 J K^{-1} mol^{-1} **45.** 184 J K^{-1} mol^{-1} **47.** 629.7 K **49.** a. Yes, NH_3 will melt since $\Delta G < 0$. b. 196 K **51.** 43.7 K **53.** −16 kJ/mol **55.** −5.40 kJ; 328.6 K; $\Delta G°$ is negative below 328.6 K where the favorable $\Delta H°$ term dominates. **57.** a. 464 kJ; b. Since $\Delta G°$ is positive, this reaction is not spontaneous at standard conditions at 298 K. c. This reaction will be spontaneous at standard conditions ($\Delta G° < 0$) at $T > 2890$ K where the favorable entropy term will dominate. **59.** $CH_4(g) + CO_2(g) \rightarrow CH_3CO_2H(l)$, $\Delta H° = -16$ kJ, $\Delta S° = -240.$ J/K, $\Delta G° = 56$ kJ; $CH_3OH(g) + CO(g) \rightarrow CH_3CO_2H(l)$, $\Delta H° = -173$ kJ, $\Delta S° = -278$ J/K, $\Delta G° = -90.$ kJ; the second reaction is preferred at standard conditions since it will be spontaneous at high enough temperatures so that the rate of the reaction should be reasonable. It should be run at temperatures below 622 K. **61.** Enthalpy is not favorable, so ΔS must provide the driving force for the change. Thus ΔS is positive. There is an increase in positional probability, so the original enzyme has the more ordered structure. **63.** Since there are more product gas molecules than reactant gas molecules ($\Delta n > 0$), ΔS will be positive. From the signs of ΔH and ΔS, this reaction is spontaneous at all temperatures. It will cost money to heat the reaction mixture. Since there is no thermodynamic reason to do this, the purpose of the elevated temperature must be to increase the rate of the reaction, i.e., kinetic reasons. **65.** The sign of ΔG (positive or negative) tells us which reaction is spontaneous (the forward or reverse reaction). If $\Delta G < 0$, then the forward reaction is spontaneous and if $\Delta G > 0$, then the reverse reaction is spontaneous. If $\Delta G = 0$, then the reaction is at equilibrium (neither the forward reaction nor the reverse reaction is spontaneous). $\Delta G°$ gives the equilibrium position by determining K for a reaction utilizing the equation $\Delta G° = -RT \ln K$. $\Delta G°$ can be used to predict spontaneity only when all reactants and products are present at standard pressures of 1 atm and/or standard concentrations of 1 M. **67.** a. shifts right; b. no shift (at equilibrium); c. shifts left **69.** $\Delta H° = -92$ kJ, $\Delta S° = -199$ J/K, $\Delta G° = -34$ kJ, $K = 9.1 \times 10^5$; a. $\Delta G = -67$ kJ; b. $\Delta G = -68$ kJ; c. $\Delta G = -85$ kJ; d. $\Delta G = -41$ kJ **71.** $-90.$ kJ; 5.9×10^{15}; because there is a decrease in the number of moles of gaseous particles, $\Delta S°$ is negative. Because $\Delta G°$ is negative, $\Delta H°$ must be negative. The reaction will be spontaneous at low temperatures (the favorable $\Delta H°$ term dominates at low temperatures). **73.** To determine K at a temperature other than 25°C, one needs to know $\Delta G°$ at that temperature. We assume $\Delta H°$ and $\Delta S°$ are temperature-independent and use the equation $\Delta G° = \Delta H° - T\Delta S°$ to estimate $\Delta G°$ at the different temperature. For $K = 1$, we want $\Delta G° = 0$, which occurs when $\Delta H° = T\Delta S°$. Again, assume $\Delta H°$ and $\Delta S°$ are temperature-independent, and then solve for T ($= \Delta H°/\Delta S°$). At this temperature, $K = 1$ because $\Delta G° = 0$. This works only for reactions where the signs of $\Delta H°$ and $\Delta S°$ are the same (either both positive or both negative). When the signs are opposite, K will always be greater than one (when $\Delta H°$ is negative and $\Delta S°$ is positive) or K will always be less than one (when $\Delta H°$ is positive and $\Delta S°$ is negative). When the signs of $\Delta H°$ and $\Delta S°$ are opposite, K can never equal one. **75.** 725 K **77.** -71 kJ/mol **79.** -4.1 kJ/mol **81.** 140 kJ/mol **83.** 60 **85.** At 25.0°C, $K = 8.72$; at 100.0°C, $K = 0.0789$ **87.** a. $\Delta H° = 57.5$ kJ, $\Delta S° = -75.6$ J/K; b. 106.4 kJ **89.** 310 K; -310 J K^{-1} mol^{-1} **91.** 12.8 L; 3.84×10^3 J **93.** a. $q = 4.01 \times 10^3$ J; $w = -4.01 \times 10^3$ J; $\Delta E = 0$; b. $q = 1.99 \times 10^3$ J; $w = -1.99 \times 10^3$ J; $\Delta E = 0$; c. $q = 0$; $w = \Delta E = -2.6 \times 10^3$ J **95.** The light source for the first reaction is necessary for kinetic reasons. Kinetics is the study of the rate of reactions. The first reaction is just too slow to occur unless a light source is available. The kinetics of a reaction are independent of the thermodynamics of a reaction. Even though the first reaction is more favorable thermodynamically (assuming standard conditions), it is unfavorable for kinetic reasons. The second reaction has a negative $\Delta G°$ value and is a fast reaction, so the second reaction, which occurs very quickly, is favored both kinetically and thermodynamically. When considering if a reaction will occur, thermodynamics and kinetics must both be considered. **97.** All are positive. **99.** $w_{max} = \Delta G$; when ΔG is negative, the magnitude of ΔG is equal to the maximum possible useful work obtainable from the process (at constant T and P). When ΔG is positive, the magnitude of ΔG is equal to the minimum amount of work that must be expended to make the process spontaneous. Due to waste energy (heat) in any real process, the amount of useful work obtainable from a spontaneous process is always less than w_{max}, and for a nonspontaneous reaction, an amount of work greater than w_{max} must be applied to make the process spontaneous. **101.** $\Delta S_{sys} = 93.8$ J K^{-1}; $\Delta S_{surr} = -93.8$ J K^{-1}; $\Delta S_{univ} = 0$ **103.** a. $\Delta G = \Delta G° = 1.8 \times 10^4$ J/mol, shifts left; b. $\Delta G = 0$, at equilibrium; c. $\Delta G = -1.1 \times 10^4$ J/mol, shifts right; d. $\Delta G = 0$, at equilibrium; e. $\Delta G = 2 \times 10^3$ J/mol, shifts left **105.** $\Delta S°$ will be negative because 2 mol of gaseous reactants forms 1 mol of gaseous product. For $\Delta G°$ to be negative, $\Delta H°$ must be negative (exothermic). For this sign combination, K decreases as T increases. Therefore, the ratio of the partial pressure of PCl_5 to the partial pressure of PCl_3 will decrease when T is raised. **107.** Using Le Châtelier's principle: A decrease in pressure (volume increases) will favor the side with the greater number of particles. Thus 2I(g) will be favored at low pressure. Looking at ΔG: $\Delta G = \Delta G° + RT \ln(P_I^2/P_{I_2})$; $\ln(P_I^2/P_{I_2}) > 0$ for $P_I = P_{I_2} = 10$ atm and ΔG is positive (not spontaneous). But at $P_I = P_{I_2} = 0.10$ atm, the logarithm term is negative. If $|RT \ln(Q)| > \Delta G°$, then ΔG becomes negative, and the reaction is spontaneous. **109.** 6 M **111.** ΔS is more favorable for reaction two than for reaction one. In reaction one, seven particles in solution are forming one particle in solution. In reaction two, four particles are forming one particle, which results in a smaller decrease in positional probability than for reaction one and a larger equilibrium constant value for reaction two. **113.** Note that these substances are not in the solid state but are in the aqueous state; water molecules are also present. There is an apparent increase in ordering (decrease in positional probability) when these ions are placed in water as compared to the separated state. The hydrating water molecules must be in a highly ordered arrangement when surrounding these anions. **115.** 1.6×10^6 **117.** $\Delta G = 11.5$ kJ; $\Delta H = 0$; $\Delta S = -38.3$ J/K **119.** a. $\Delta E = 0$, $\Delta H = 0$, $\Delta S = 7.62$ J/K, $\Delta G = -2320$ J, $w = -1500$ J, $q = 1500$ J; b. Since ΔS_{univ} (2.7 J/K) is positive, the process is spontaneous. **121.** a. 226 K; b. 239 K **123.** a. -2.87 J/K; b. 2.87 J/K **125.** 88 J/K **127.** greater than 7.5 torr **129.** $\Delta H° = 286$ kJ, $\Delta G° = 326$ kJ, $K = 7.22 \times 10^{-58}$; $P_{O_3} = 3.3 \times 10^{-41}$ atm; this partial pressure represents one molecule of ozone per 9.5×10^{17} L of air. Equilibrium is probably not maintained under these conditions since the concentration of ozone is not large enough to maintain equilibrium. **131.** a. 0.333; b. $P_A = 1.50$ atm, $P_B = 0.50$ atm; c. $\Delta G = 2722$ J + (8.3145)(298) ln(0.50/1.50) = 2722 J − 2722 J = 0 (carrying extra significant figures)

133.

	q	w	ΔE	ΔS	ΔH	ΔG
Step 1	1480 J	−1480 J	0	7.56 J/K	0	−2250 J
Step 2	1240 J	−1240 J	0	5.77 J/K	0	−1720 J
Step 3	1250 J	−1250 J	0	5.81 J/K	0	−1730 J
Total	3970 J	−3970 J	0	19.14 J/K	0	-5.70×10^3 J

135. a. 1.24×10^{90}; b. -154 J/K **137.** $\Delta S = 29$ J K^{-1} mol^{-1} **139.** $\Delta S_{sys} = -34.3$ J/K, $\Delta S_{surr} = 33.7$ J/K

Chapter 11

15. Electrochemistry is the study of the interchange of chemical and electrical energy. A redox (oxidation–reduction) reaction is a reaction in which one or more electrons are transferred. In a galvanic cell, a spontaneous redox reaction occurs that produces an electric current. In an electrolytic cell, electricity is used to force a nonspontaneous redox reaction to occur. **17.** a. Zn is oxidized in the galvanic cell, so Zn is the reducing agent. Cu^{2+} is reduced in the galvanic cell, so Cu^{2+} is the oxidizing agent. Electrons will flow from the zinc compartment (the anode) to the copper compartment (the cathode). b. 1.10 V; c. The copper electrode will increase in mass as Cu^{2+} is reduced to Cu. The zinc electrode will decrease in mass as Zn is oxidized to Zn^{2+}. **19.** See Fig. 11.2 for a typical galvanic cell. The anode compartment contains the oxidation half-reaction compounds/ions, and the cathode compartment contains the reduction half-reaction compounds/ions. The electrons flow from the anode to the cathode. In the salt bridge, cations flow to the cathode and anions flow to the anode. For each of the following answers, all solutes are 1.0 M and all gases are at 1.0 atm. a. $7H_2O(l) + 2Cr^{3+}(aq) + 3Cl_2(g) \rightarrow Cr_2O_7^{2-}(aq) + 6Cl^-(aq) + 14H^+(aq)$; $\mathscr{E}°_{cell} = 0.03$ V; cathode: Pt electrode; Cl_2 bubbled into solution, Cl^- in solution; anode: Pt electrode; Cr^{3+}, H^+, and $Cr_2O_7^{2-}$ in solution; b. $Cu^{2+}(aq) + Mg(s) \rightarrow Cu(s) + Mg^{2+}(aq)$; $\mathscr{E}°_{cell} = 2.71$ V; cathode: Cu electrode; Cu^{2+} in solution; anode: Mg electrode; Mg^{2+} in solution; c. $12H^+(aq) + 2IO_3^-(aq) + 10Fe^{2+}(aq) \rightarrow 10Fe^{3+}(aq) + I_2(s) + 6H_2O(l)$; $\mathscr{E}°_{cell} = 0.43$ V; cathode: Pt electrode; IO_3^-, I_2, and H_2SO_4

(H^+ source) in solution; anode: Pt electrode; Fe^{2+} and Fe^{3+} in solution; d. $Zn(s) + 2Ag^+(aq) \rightarrow 2Ag(s) + Zn^{2+}(aq)$; $\mathscr{E}° = 1.56$ V; cathode: Ag electrode; Ag^+ in solution; anode: Zn electrode; Zn^{2+} in solution **21.** See Exercise 19 for a description of a galvanic cell. For each of the following answers, all solutes are 1.0 *M* and all gases are at 1.0 atm. a. $Cl_2(g) + 2Br^-(aq) \rightarrow Br_2(aq) + 2Cl^-(aq)$; $\mathscr{E}° = 0.27$ V; cathode: Pt electrode; $Cl_2(g)$ bubbled in, Cl^- in solution; anode: Pt electrode; Br_2 and Br^- in solution; b. $3H_2O(l) + 5IO_4^-(aq) + 2Mn^{2+}(aq) \rightarrow 5IO_3^-(aq) + 2MnO_4^-(aq) + 6H^+(aq)$; $\mathscr{E}° = 0.09$ V; cathode: Pt electrode; IO_4^-, IO_3^-, and H_2SO_4 (as a source of H^+) in solution; anode: Pt electrode; Mn^{2+}, MnO_4^-, and H_2SO_4 in solution; c. $2H_2O_2(aq) \rightarrow 2H_2O(l) + O_2(g)$; $\mathscr{E}°_{cell} = 1.10$ V; cathode: Pt electrode; H_2O_2 and H^+ in solution; anode: Pt electrode; $O_2(g)$ bubbled in, H_2O_2 and H^+ in solution; d. $2Fe^{3+}(aq) + 3Mn(s) \rightarrow 2Fe(s) + 3Mn^{2+}(aq)$; $\mathscr{E}°_{cell} = 1.14$ V; cathode: Fe electrode; Fe^{3+} in solution; anode: Mn electrode; Mn^{2+} in solution **23.** a. Pt | Br^- (1.0 *M*), Br_2 (1.0 *M*) ‖ Cl_2 (1.0 atm) | Cl^- (1.0 *M*) | Pt; b. Pt | Mn^{2+} (1.0 *M*), MnO_4^- (1.0 *M*), H^+ (1.0 *M*) ‖ IO_4^- (1.0 *M*), IO_3^- (1.0 *M*), H^+ (1.0 *M*) | Pt; c. Pt | H_2O_2 (1.0 *M*), H^+ (1.0 *M*) | O_2 (1.0 atm) ‖ H_2O_2 (1.0 *M*), H^+ (1.0 *M*) | Pt; d. Mn | Mn^{2+} (1.0 *M*) ‖ Fe^{3+} (1.0 *M*) | Fe **25.** a. no; b. yes; c. yes; d. no **27.** $F^- < H_2O < I_2 < Cu^+ < H^- < K$ **29.** a. $Cr_2O_7^{2-}$, O_2, MnO_2, IO_3^-; b. $PbSO_4$, Cd^{2+}, Fe^{2+}, Cr^{3+}, Zn^{2+}, H_2O **31.** Since $\mathscr{E}°_{cell}$ is positive for this reaction, at standard conditions ClO^- can spontaneously oxidize NH_3 to the somewhat toxic N_2H_4. **33.** An extensive property is one that depends directly on the amount of substance. The free energy change for a reaction depends on whether 1 mole of product is produced or 2 moles of product are produced or 1 million moles of product are produced. This is not the case for cell potentials, which do not depend on the amount of substance. The equation that relates ΔG to $\mathscr{E}$ is $\Delta G = -nF\mathscr{E}$. It is the *n* term that converts the intensive property $\mathscr{E}$ into the extensive property ΔG. *n* is the number of moles of electrons transferred in the balanced reaction that ΔG is associated with. **35.** -0.829 V; the two values agree to two significant figures (-0.83 V in Table 11.1). **37.** a. -388 kJ; b. $-270.$ kJ **39.** a. $\Delta G° = -52$ kJ, $K = 1.4 \times 10^9$; b. $\Delta G° = -90$ kJ, $K = 2 \times 10^{15}$; c. $\Delta G° = -212$ kJ, $K = 1.68 \times 10^{37}$; d. $\Delta G° = -660.$ kJ, $K = 5.45 \times 10^{115}$ **41.** a. $3Mn(s) + 8H^+(aq) + 2NO_3^-(aq) \rightarrow 2NO(g) + 4H_2O(l) + 3Mn^{2+}(aq)$; $5IO_4^-(aq) + 2Mn^{2+}(aq) + 3H_2O(l) \rightarrow 5IO_3^-(aq) + 2MnO_4^-(aq) + 6H^+(aq)$; b. reaction one: $\mathscr{E}°_{cell} = 2.14$ V, $\Delta G° = -1240$ kJ, $K \approx 10^{217}$; reaction two: $\mathscr{E}°_{cell} = 0.09$ V, $\Delta G° = -90$ kJ, $K = 2 \times 10^{15}$ **43.** $\mathscr{E}° = \dfrac{T\Delta S°}{nF} - \dfrac{\Delta H°}{nF}$; if we graph $\mathscr{E}°$ vs. *T*, we should get a straight line ($y = mx + b$). The slope of the line is equal to $\Delta S°/nF$ and the *y*-intercept is equal to $-\Delta H°/nF$. $\mathscr{E}°$ will have a small temperature dependence when $\Delta S°$ is close to zero. **45.** a. spontaneous; -35 kJ; 1.2×10^6; b. not spontaneous; c. spontaneous; -85 kJ; 7.8×10^{14} **47.** 6.5×10^{20} **49.** -0.19 V **51.** pH = 0.05 **53.** a. $Au^{3+}(aq) + 3Tl(s) \rightarrow Au(s) + 3Tl^+(aq)$; $\mathscr{E}°_{cell} = 1.84$ V; b. $\Delta G° = -533$ kJ; $K = 2.52 \times 10^{93}$; c. 2.04 V **55.** 1.50 V **57.** a. 2.12 V; b. 1.98 V; c. 2.05 V; d. Cell potential decreases with decreasing *T* (also, oil becomes more viscous). **59.** Concentration cell: a galvanic cell in which both compartments contain the same components but at different concentrations. All concentration cells have $\mathscr{E}°_{cell} = 0$ because both compartments contain the same contents. The driving force for the cell is the different ion concentrations at the anode and cathode. The cell produces a voltage as long as the ion concentrations are different. The lower ion concentration is always at the anode. The magnitude of the cell potential depends on the magnitude of the differences in ion concentrations between the anode and cathode. The larger the difference in ion concentrations, the more negative is the log *Q* term, and the more positive is the cell potential. Thus, as the difference in ion concentrations between the anode and cathode compartments increase, the cell potential increases. This can be accomplished by decreasing the ion concentration at the anode and/or by increasing the ion concentration at the cathode. **61.** Electron flow is always from the anode to the cathode. For the cells with a nonzero cell potential, we will identify the cathode, which means the other compartment is the anode. a. 0.0 V; b. 8.9×10^{-3} V; compartment with $[Ni^{2+}] = 2.0$ *M* is the cathode; c. 0.030 V; compartment with $[Ni^{2+}] = 1.0$ *M* is the cathode; d. 0.13 V; compartment with $[Ni^{2+}] = 1.0$ *M* is the cathode; e. 0.0 V **63.** a. 0.23 V; b. 0.16 V (the reverse reaction from part a occurs); c. 1.2×10^{-5} *M*; d. A graph of $\mathscr{E}$ versus $\log[Cu^{2+}]$ will yield a straight line with slope equal to 0.0296 V or 29.6 mV. **65.** a. 1.4×10^{-14} *M*; b. 2.9×10^{13} **67.** 0.99 V **69.** a. 16 g; b. 25 g; c. 71 g; d. 4.9 g **71.** 554 s **73.** a. cathode: $Ni^{2+} + 2e^- \rightarrow Ni$, anode: $2Br^- \rightarrow Br_2 + 2e^-$; b. cathode: $Al^{3+} + 3e^- \rightarrow Al$, anode: $2F^- \rightarrow F_2 + 2e^-$; c. cathode: $Mn^{2+} + 2e^- \rightarrow Mn$, anode: $2I^- \rightarrow I_2 + 2e^-$; d. cathode: $Ni^{2+} + 2e^- \rightarrow Ni$, anode: $2Br^- \rightarrow Br_2 + 2e^-$; e. cathode: $2H_2O + 2e^- \rightarrow H_2 + 2OH^-$, anode: $2H_2O \rightarrow O_2 + 4H^+ + 4e^-$; f. cathode: $2H_2O + 2e^- \rightarrow H_2 + 2OH^-$, anode: $2I^- \rightarrow I_2 + 2e^-$ **75.** Au(*s*) will plate out first since it has the most positive reduction potential, followed by Ag(*s*), which is followed by Ni(*s*), and finally Cd(*s*) will plate out last since it has the most negative reduction potential of the metals listed. **77.** Yes, since the range of potentials for plating out each metal do not overlap, we should be able to separate the three metals. The order of plating will be Ir(*s*) first, followed by Pt(*s*), and finally Pd(*s*) as the potential is gradually increased. **79.** $MgCl_2$ **81.** 9.12 L F_2 at the anode; 29.2 g K at the cathode **83.** 6.00 L **85.** a. The spoon is where Cu^{2+} is reduced to Cu, so the spoon will be the cathode. The anode will be the copper strip where Cu is oxidized to Cu^{2+}. b. cathode reaction: $Cu^{2+} + 2\,e^- \rightarrow Cu$; anode reaction: $Cu \rightarrow Cu^{2+} + 2\,e^-$ **87.** 0.98 V **89.** 1.14 V **91.** -212 kJ **93.** a. Paint: covers the metal surface so no contact occurs between the metal and air. This works only as long as the painted surface is not scratched. b. Durable oxide coatings: covers the metal surface so no contact occurs between the metal and air. c. Galvanizing: coating steel with zinc; Zn forms an effective oxide coating over steel; also, zinc is more easily oxidized than the iron in the steel. d. Sacrificial metal: a more easily oxidized metal attached to an iron surface; the more active metal is preferentially oxidized instead of iron. e. Alloying: adding chromium and nickel to steel; the added Cr and Ni form oxide coatings on the steel surface. f. Cathodic protection: a more easily oxidized metal is placed in electrical contact with the metal we are trying to protect. It is oxidized in preference to the protected metal. The protected metal becomes the cathode electrode, thus, "cathodic protection." **95.** It is easier to oxidize Zn than Fe, so the Zn will be oxidized, protecting the iron of the *Monitor*'s hull. **97.** The potential oxidizing agents are NO_3^- and H^+. Hydrogen ion cannot oxidize Pt under either condition. Nitrate cannot oxidize Pt unless there is Cl^- in the solution. Aqua regia has both Cl^- and NO_3^-. The overall reaction is $12Cl^-(aq) + 3Pt(s) + 2NO_3^-(aq) + 8H^+(aq) \rightarrow 3PtCl_4^{2-}(aq) + 2NO(g) + 4H_2O(l)$; $\mathscr{E}°_{cell} = 0.21$ V **99.** a. ± 0.02 pH units; $\pm 6 \times 10^{-6}$ *M* H^+; b. ± 0.001 V **101.** a. decrease; b. increase; c. decrease; d. decrease; e. no effect **103.** osmium(IV) nitrate; copper, Cu **105.** a. 1.14 V; b. 1.09 V; c. 78 g Zn, 319 g Cu; d. 13.4 h **107.** a. 5.77×10^{-10}; b. 1.9 *M* **109.** 2.39×10^{-7} **111.** a. 0.16 V; b. 8.6 mol **113.** 0.64 V **115.** a. yes; b. $3H_2O(l) + Zr(s) \rightarrow 2H_2(g) + ZrO_2 \cdot H_2O(s)$; c. $\mathscr{E}° = 1.53$ V; $\Delta G° = -590.$ kJ; $K \approx 10^{104}$; d. 4.42×10^4 g H_2; 2.3×10^6 L; e. Probably yes; less radioactivity overall was released by venting the H_2 than what would have been released if the H_2 had exploded inside the reactor (as happened at Chernobyl). Neither alternative is pleasant, but venting the radioactive hydrogen is the less unpleasant of the two alternatives.

Chapter 12

21. 3.0×10^{10} s^{-1}; 2.0×10^{-23} J/photon; 12 J/mol **23.** 107.1 MHz electromagnetic radiation is FM radio-waves, 2.12×10^{-10} m electromagnetic radiation is X rays, and the 3.97×10^{-19} J/photon electromagnetic radiation is visible (green) light. The order of increasing photon energy and increasing frequency is: FM radiowaves < visible (green) light < X rays. **25.** a. 5.0×10^{-6} m; b. infrared; c. 4.0×10^{-20} J/photon; 2.4×10^4 J/mol; d. less **27.** 427.7 nm **29.** 4.36×10^{-19} J **31.** The photoelectric effect refers to the phenomenon in which electrons are emitted from the surface of a metal when light strikes it. The light must have a certain minimum frequency (energy) in order to remove electrons from the surface of a metal. Light having a frequency below the minimum results in no electrons being emitted, whereas light at or higher than the minimum frequency does cause electrons to be emitted. For light having a frequency higher than the minimum frequency, the excess energy is transferred into kinetic energy for the emitted electron. Albert Einstein explained the photoelectric effect by applying quantum theory. **33.** a. 1.32×10^{-13} m; b. 5.3×10^3 m/s **35.** Ca **37.** a. For hydrogen (Z = 1), the energy levels in units of joules are given by the equation $E_n = -2.178 \times 10^{-18}(1/n^2)$. Because the differences between $1/n^2$ values for consecutive energy levels decrease as n increases, the energy levels get closer together as n increases. b. In the diagram, the red line is for the $n_i = 3$ to $n_f = 2$ transition. The calculated wavelength for this emission using the Bohr model is 656.7 nm. From Figure 12.3, λ = 656.7 nm is red light, so the diagram is correct for the red line. The green line is for the $n_i = 4$ to $n_f = 2$ transition. The calculated wavelength for this emission is 486.4 nm. From Figure 12.3, λ = 486.4 nm is green-blue light. The diagram is consistent with this line. The blue line is for the $n_i = 5$ to $n_f = 2$ transition. The calculated wavelength for this emission is 434.3 nm. From Figure 12.3, λ = 434.3 nm is blue or blue-violet light. The diagram is consistent with this line also. **39.** 10 **41.** a. False; It takes less energy to ionize an electron from $n = 3$ than from the ground state. b. True; c. False; The energy difference between $n = 3$ and $n = 2$ is smaller than the energy difference between $n = 3$ and $n = 1$. Thus the wavelength of light emitted is longer for the $n = 3$ to $n = 2$ electronic transition than for the $n = 3$ to $n = 1$ transition. d. True; e. False; $n = 2$ is the first excited state and $n = 3$ is the second excited state. **43.** $n = 7$ **45.** $n = 4$ **47.** $n = 4$ **49.** a. 5.79×10^{-4} m; b. 3.64×10^{-33} m; the diameter of an H atom is roughly 2×10^{-8} cm. The uncertainty in the position of the electron is much larger than the size of the atom, whereas the uncertainty in the position of the baseball is insignificant as compared to the size of a baseball. **51.** At $x = 0$, the value of the square of the wave function must be zero. The particle must be inside the box. For $\psi = A\cos(Lx)$, at $x = 0$, $\cos(0) = 1$ and $\psi^2 = A^2$. This violates the boundary condition. **53.** 3.50 nm **55.** As L increases, E_n will decrease and the spacing between energy levels will also decrease. **57.** Since E_n is inversely proportional to L^2, the electron in the larger box (10^{-6} m) has the lowest ground-state energy. **59.** The $2p$ orbitals differ from each other in the direction in which they point in space. The $2p$ and $3p$ orbitals differ from each other in their size, energy, and number of nodes. A nodal surface in an atomic orbital is a surface in which the probability of finding an electron is zero. **61.** $1p$, $3f$, and $2d$ are all incorrect designations. **63.** a. For $n = 3$, $\ell = 3$ is not possible. d. m_s cannot equal -1. e. ℓ cannot be a negative number. f. For $\ell = 1$, m_ℓ cannot equal 2. **65.** $1p$, 0 electrons ($\ell \neq 1$ when $n = 1$); $6d_{x^2-y^2}$, 2 electrons (specifies one atomic orbital); $4f$, 14 electrons (7 orbitals have $4f$ designation); $7p_y$, 2 electrons (specifies one atomic orbital); $2s$, 2 electrons (specifies one atomic orbital); $n = 3$, 18 electrons ($3s$, $3p$, and $3d$ orbitals are possible; there are one $3s$ orbital, three $3p$ orbitals, and five $3d$ orbitals) **67.** The diagrams of the orbitals in the text give only 90% probabilities of where the electron may reside. We can never be 100% certain of the location of the electrons due to Heisenberg's uncertainty principle. **69.** $\theta = 0 : 2.46 \times 10^{28}$; $\theta = 90° : 0$ **71.** He: $1s^2$; Ne: $1s^22s^22p^6$; Ar: $1s^22s^22p^63s^23p^6$; each peak in the diagram corresponds to a sublevel with different values of n. Corresponding sublevels are closer to the nucleus for heavier elements because of the increased nuclear charge. **73.** Valence electrons are the electrons in the outermost principal quantum level of an atom (those electrons in the highest n value orbitals). The electrons in the lower n value orbitals are all inner core or just core electrons. The key is that the outermost electrons are the valence electrons. When atoms interact with each other, it will be the outermost electrons that are involved in these interactions. In addition, how tightly the nucleus holds these outermost electrons determines atomic size, ionization energy, and other properties of atoms. Elements in the same group have similar valence electron configurations and, as a result, have similar chemical properties. **75.** a. 32; b. 8; c. 25; d. 10; e. 6; f. 0; g. 1; h. 9; i. 0; j. 2 **77.** Sc: $1s^22s^22p^63s^23p^64s^23d^1$; Fe: $1s^22s^22p^63s^23p^64s^23d^6$; P: $1s^22s^22p^63s^23p^3$; Cs: $1s^22s^22p^63s^23p^64s^23d^{10}4p^65s^24d^{10}5p^66s^1$; Eu: $1s^22s^22p^63s^23p^64s^23d^{10}4p^65s^24d^{10}5p^66s^24f^65d^1$ (actual: $[Xe]6s^24f^7$); Pt: $1s^22s^22p^63s^23p^64s^23d^{10}4p^65s^24d^{10}5p^66s^24f^{14}5d^8$ (actual: $[Xe]6s^14f^{14}5d^9$); Xe: $1s^22s^22p^63s^23p^64s^23d^{10}4p^65s^24d^{10}5p^6$; Br: $1s^22s^22p^63s^23p^64s^23d^{10}4p^5$ **79.** Cr, Cu, Nb, Mo, Tc, Ru, Rh, Pd, Ag, Pt, Au **81.** Cr: $1s^22s^22p^63s^23p^64s^13d^5$; Cr has 6 unpaired electrons.

1s	2s	2p	3s	3p
↑↓	↑↓	↑↓ ↑↓ ↑↓	↑↓	↑↓ ↑↓ ↑↓

4s		4s	3d		3d
↑	or	↓	↑ ↑ ↑ ↑ ↑	or	↓ ↓ ↓ ↓ ↓

Cu: $1s^22s^22p^63s^23p^64s^13d^{10}$; Cu has 1 unpaired electron.

1s	2s	2p	3s	3p
↑↓	↑↓	↑↓ ↑↓ ↑↓	↑↓	↑↓ ↑↓ ↑↓

4s		4s	3d
↑	or	↓	↑↓ ↑↓ ↑↓ ↑↓ ↑↓

83. a. 2 valence electrons; $4s^2$; b. 6 valence electrons; $2s^22p^4$; c. 7 valence electrons; $7s^27p^5$; d. 3 valence electrons; $5s^25p^1$; e. 8 valence electrons; $3s^23p^6$; f. 5 valence electrons; $6s^26p^3$ **85.** a. 32; b. 28; c. 23; d. 56 or 59 **87.** Li (1 unpaired electron), N (3 unpaired electrons), Ni (2 unpaired electrons), and Te (2 unpaired electrons) are paramagnetic. **89.** H, Li, Na, K, B, Al, Ga, F, Cl, Br, Sc, and Cu all have one unpaired electron in the ground state. **91.** O, 2; O^+, 3; O^-, 1; Os, 4; Zr, 2; S, 2; F, 1; Ar, 0 **93.** Ionization energy: $P(g) \rightarrow P^+(g) + e^-$; electron affinity: $P(g) + e^- \rightarrow P^-(g)$ **95.** As successive electrons are removed, the net positive charge on the resultant ion increases. This increase in positive charge binds the remaining electrons more firmly, and the ionization energy increases. The electron configuration for Si is $1s^22s^22p^63s^23p^2$. There is a large jump in ionization energy when going from the removal of valence electrons to the removal of core electrons. For silicon, this occurs when the fifth electron is removed since we go from the valence electrons in $n = 3$ to the core electrons in $n = 2$. There should be another big jump when the thirteenth electron is removed (i.e., when a $1s$ electron is removed). **97.** a. S < Se < Te; b. Br < Ni < K; c. F < Si < Ba; d. Be < Na < Rb; e. Ne < Se < Sr; f. O < P < Fe **99.** a. Ba; b. K; c. O; d. S^{2-}; e. Cs **101.** As: $[Ar]4s^23d^{10}4p^3$; Se: $[Ar]4s^23d^{10}4p^4$; The general ionization energy trend predicts that Se should have a higher ionization energy than As. Se is an exception to the general ionization energy trend. There are extra electron–electron repulsions in Se because two electrons are in the same $4p$ orbital, resulting in a lower ionization energy for Se than predicted. **103.** Size also decreases going across a period. Sc and Ti, along with Y and Zr, are adjacent elements. There are 14 elements (the lanthanides) between La and Hf, making Hf considerably smaller. **105.** a. $[Rn]7s^25f^{14}6d^{10}7p^5$; b. At; c. NaUus, $Mg(Uus)_2$, $C(Uus)_4$, $O(Uus)_2$; d. $UusO^-$, $UusO_2^-$, $UusO_3^-$, $UusO_4^-$ **107.** In each case something energetically unfavorable occurs when an electron is added. For Be, the added electron must go into a higher-energy $2p$ atomic orbital because the $2s$ orbital is full. In N, the added electron must pair up with another electron in one of the $2p$ atomic orbitals; this adds electron–electron repulsions. In Ne, the added electron must be added to a much higher-energy $3s$ atomic orbital because the $n = 2$ orbitals are full. **109.** Electron–electron repulsions are much greater in O^- than in S^- because the electron goes into a smaller $2p$ orbital versus the larger $3p$ orbital in sulfur. This results in a more favorable (more exothermic) electron affinity for sulfur. **111.** a. −1445 kJ/mol; b. −580 kJ/mol; c. 348.7 kJ/mol; d. 1255 kJ/mol; e. −1255 kJ/mol **113.** Yes; ionization energy generally decreases down a group, and atomic radius generally increases down a group. The data in Table 12.9 confirm both of these general trends. **115.** a. $6Li(s) + N_2(g) \rightarrow 2Li_3N(s)$; b. $2Rb(s) + S(s) \rightarrow Rb_2S(s)$; c. $2Cs(s) + 2H_2O(l) \rightarrow 2CsOH(aq) + H_2(g)$; d. $2Na(s) + Cl_2(g) \rightarrow 2NaCl(s)$ **117.** For 589.0 nm: $5.090 \times 10^{14}\ s^{-1}$, 3.373×10^{-19} J/photon, 203.1 kJ/mol; for 589.6 nm: $5.085 \times 10^{14}\ s^{-1}$, 3.369×10^{-19} J/photon, 202.9 kJ/mol **119.** 119: $[Rn]7s^2 5f^{14} 6d^{10} 7p^6 8s^1$ **121.** a. n; b. n and ℓ **123.** Size decreases from left to right and increases going down the periodic table. So going one element right and one element down would result in a similar size for the two elements diagonal to each other. The ionization energies will be similar for the diagonal elements since the periodic trends also oppose each other. Electron affinities are harder to predict, but atoms with similar size and ionization energy should also have similar electron affinities. **125.** 200 s (about 3 minutes) **127.** greenish yellow light **129.** When the p and d orbital functions are evaluated at various points in space, the results sometimes have positive values and sometimes have negative values. The term "phase" is often associated with the + and − signs. For example, a sine wave has alternating positive and negative phases. This is analogous to the positive and negative values (phases) in the p and d orbitals. **131.** a. 146 kJ; b. 407 kJ; c. 1117 kJ; d. 1524 kJ **133.** alkaline earth metal family **135.** a. 24; b. 6; c. 12; d. 2; e. 26; f. 24.9 g; g. $1s^2 2s^2 2p^6 3s^2 3p^6 4s^1 3d^5$ **137.** a. 4; b. four elements in the first period and 16 elements in the second period; c. 20; d. 28 **139.** S-type cone receptors detect 400–500 nm light; from Figure 12.3 in the text, this is violet to green light. M-type cone receptors detect 450–630 nm light, which is blue to orange light. L-type cone receptors detect 500–700 nm light, which is green to red light. **141.** a. Line A: 6 → 3; Line B: 5 → 3; b. 121.6 nm **143.** X = carbon; $m = 5(C^{5+})$ **145.** a. The quantum numbers are:

ground state (E_{11}) → $n_x = 1, n_y = 1$
first excited state (E_{21}) → $n_x = 2, n_y = 1$
second excited state (E_{12}) → $n_x = 1, n_y = 2$

b. 4.5×10^{-5} m **147.** a. 2.2×10^{-9}; b. 1.5×10^{-9}; c. 2.9×10^{-10}; d. 1.9×10^{-4}; e. 1×10^{-3} **149.** 2470 nm **151.** a. 43.33; b. Z_{eff} is slightly less than Z. Electrons in other orbitals can penetrate the 1s orbital. Thus a 1s electron can be slightly shielded from the nucleus, giving a Z_{eff} close to but less than Z.

Chapter 13

11. Electronegativity is the ability of an atom in a molecule to attract electrons to itself. Electronegativity is a bonding term. Electron affinity is the energy change when an electron is added to a substance. Electron affinity deals with isolated atoms in the gas phase. A covalent bond is a sharing of an electron pair in a bond between two atoms. An ionic bond is a complete transfer of electrons from one atom to another to form ions. The electrostatic attraction of the oppositely charged ions is the ionic bond. A pure covalent bond is an equal sharing of a shared electron pair in a bond. A polar covalent bond is an unequal sharing. Ionic bonds form when there is a large difference in electronegativity between the two atoms bonding together. This usually occurs when a metal with a small electronegativity is bonded to a nonmetal having a large electronegativity. A pure covalent bond forms between atoms having identical or nearly identical electronegativities. A polar covalent bond forms when there is an intermediate electronegativity difference. In general, nonmetals bond together by forming covalent bonds, either pure covalent or polar covalent. Ionic bonds form due to the strong electrostatic attraction between two oppositely charged ions. Covalent bonds form because the shared electrons in the bond are attracted to two different nuclei, unlike the isolated atoms, in which electrons are attracted to only one nuclei. The attraction to another nuclei overrides the added electron–electron repulsions.

13.

Halogen	(IE − EA)	(IE − EA)/502	EN(text)
F	2006 kJ/mol	4.0	4.0
Cl	1604 kJ/mol	3.2	3.0
Br	1463 kJ/mol	2.9	2.8
I	1302 kJ/mol	2.6	2.5

2006/502 = 4.0; The values calculated from IE and EA show the same trend as (and agree fairly closely) with the values given in the text. **15.** a. C < N < O; b. Se < S < Cl; c. Sn < Ge < Si; d. Tl < Ge < S; e. Rb < K < Na; f. Ga < B < O **17.** a. C (2.5) < N (3.0) < O (3.5), same as predicted; b. Se (2.4) < S (2.5) < Cl (3.0), same; c. Si (1.8) = Ge (1.8) = Sn (1.8), different; d. Tl (1.8) = Ge (1.8) < S (2.5), different; e. Rb (0.8) = K (0.8) < Na (0.9), different; f. Ga (1.6) < B (2.0) < O (3.5), same; Most polar bonds in Exercise 16 using actual EN values: a. Si—F and Ge—F (Ge—F predicted); b. P—Cl (same as predicted); c. S—F (same as predicted); d. Ti—Cl (same as predicted); e. Si—H and Sn—H (Sn—H predicted); f. Al—Br (Tl—Br predicted) **19.** Br—Br < N—O < C—F < Ca—O < K—F **21.** $(NH_4)_2SO_4$ and $Ca_3(PO_4)_2$ are the compounds with both ionic and covalent bonds. **23.** Anions are larger than the neutral atom, and cations are smaller than the neutral atom. For anions, the added electrons increase the electron–electron repulsions. To counteract this, the size of the electron cloud increases, placing the electrons further apart from one another. For cations, as electrons are removed, there are fewer electron–electron repulsions, and the electron cloud can be pulled closer to the nucleus. Isoelectronic: same number of electrons. Two variables, the number of protons and the number of electrons, determine the size of an ion. Keeping the number of electrons constant, we have to consider only the number of protons to predict trends in size. The ion with the most protons attracts the same number of electrons most strongly, resulting in a smaller size. **25.** a. $Cu > Cu^+ > Cu^{2+}$; b. $Pt^{2+} > Pd^{2+} > Ni^{2+}$; c. $O^{2-} > O^- > O$; d. $La^{3+} > Eu^{3+} > Gd^{3+} > Yb^{3+}$; e. $Te^{2-} > I^- > Cs^+ > Ba^{2+} > La^{3+}$ **27.** Rb^+: $[Ar]4s^2 3d^{10} 4p^6$; Ba^{2+}: $[Kr]5s^2 4d^{10} 5p^6$; Se^{2-}: $[Ar]4s^2 3d^{10} 4p^6$; I^-: $[Kr]5s^2 4d^{10} 5p^6$ **29.** a. Cs_2S is composed of Cs^+ and S^{2-}. Cs^+ has the same electron configuration as Xe, and S^{2-} has the same configuration as Ar. b. SrF_2; Sr^{2+} has the Kr electron configuration and F^- has the Ne configuration. c. Ca_3N_2; Ca^{2+} has the Ar electron configuration and N^{3-} has the Ne configuration. d. $AlBr_3$; Al^{3+} has the Ne electron configuration and Br^- has the Kr configuration. **31.** Se^{2-}, Br^-, Rb^+, Sr^{2+}, Y^{3+}, and Zr^{4+} are some ions that are isoelectronic with Kr (36 electrons). In terms of size, the ion with the most protons will hold the electrons tightest and will be the smallest. The size trend is

$$Zr^{4+} < Y^{3+} < Sr^{2+} < Rb^+ < Br^- < Se^{2-}$$

smallest → largest

33. a. Al_2S_3, aluminum sulfide; b. K_3N, potassium nitride; c. $MgCl_2$, magnesium chloride; d. CsBr, cesium bromide **35.** −411 kJ/mol **37.** 161 kJ/mol **39.** a. From the data given, less energy is required to produce $Mg^+(g) + O^-(g)$ than to produce $Mg^{2+}(g) + O^{2-}(g)$. However, the lattice energy for $Mg^{2+}O^{2-}$ will be much more exothermic than for Mg^+O^- (because of the greater charges in $Mg^{2+}O^{2-}$). The favorable lattice energy term will dominate and $Mg^{2+}O^{2-}$ forms. b. Mg^+ and O^- both have unpaired electrons. In Mg^{2+} and O^{2-}, there are no unpaired electrons. Hence, Mg^+O^- would be paramagnetic; $Mg^{2+}O^{2-}$ would be diamagnetic. Paramagnetism can be detected by measuring the mass of a sample in the presence and absence of a magnetic field. The apparent mass of a paramagnetic substance will be larger in a magnetic field because of the force between the unpaired electrons and the field. **41.** Ca^{2+} has greater charge than Na^+, and Se^{2-} is smaller than Te^{2-}. The effect of charge on the lattice energy is greater than the effect of size. We expect the trend from most exothermic to least exothermic to be

$$\underset{(-2862)}{CaSe} > \underset{(-2721)}{CaTe} > \underset{(-2130)}{Na_2Se} > \underset{(-2095\ kJ/mol)}{Na_2Te}$$

43. a. −183 kJ; b. −109 kJ; c. −158 kJ; d. −1169 kJ **45.** −42 kJ **47.** −1228 kJ **49.** −5681 kJ **51.** Since both reactions are highly exothermic, the high temperature is not needed to provide energy. It must be necessary for some other reason. This will be discussed in Chapter 15 on kinetics. **53.** a. 1549 kJ; b. 1390. kJ; c. 1312 kJ; d. 1599 kJ **55.** D_{calc} = 389 kJ/mol as compared with 391 kJ/mol in the table.

57. a. H—C≡N: b. H—P(H)—H c. :Cl—C(H)(Cl)—Cl: d. [H—N(H)(H)—H]$^+$

e. H₂C=O f. :F—Se—F: g. O=C=O h. O=O i. H—Br:

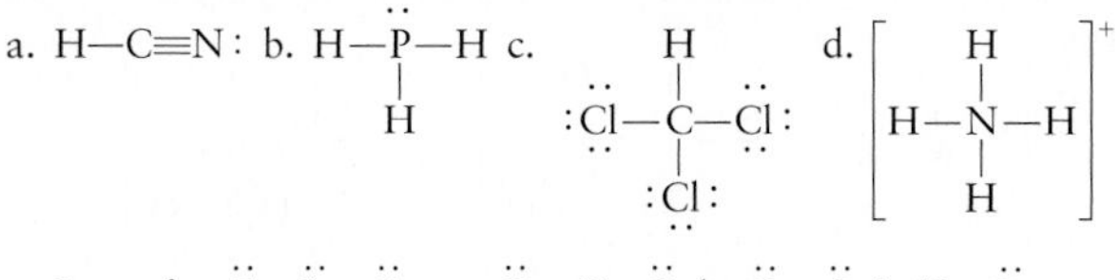

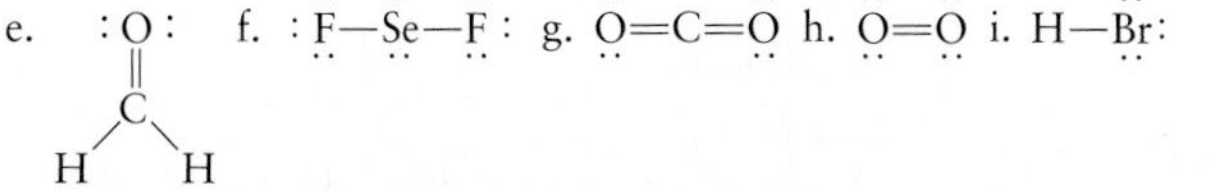

59. Molecules or ions that have the same number of valence electrons and the same number of atoms will have similar Lewis structures.

61. O=O—O: ⟷ :O—O=O

O=S—O: ⟷ :O—S=O

(O=)S(O)(O) ⟷ ⟷ (three resonance structures of SO_3)

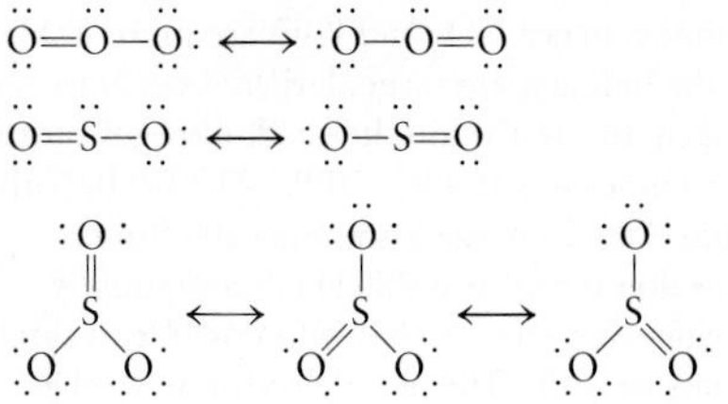

63. H—C(H)(H)—N=C=O ⟷ H—C(H)(H)—N≡C—O:

⟷ H—C(H)(H)—N—C≡O:

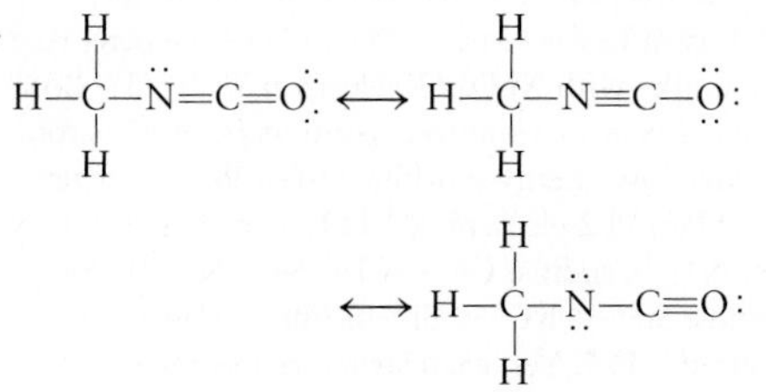

65. (two resonance structures of benzene, C_6H_6) ⟷

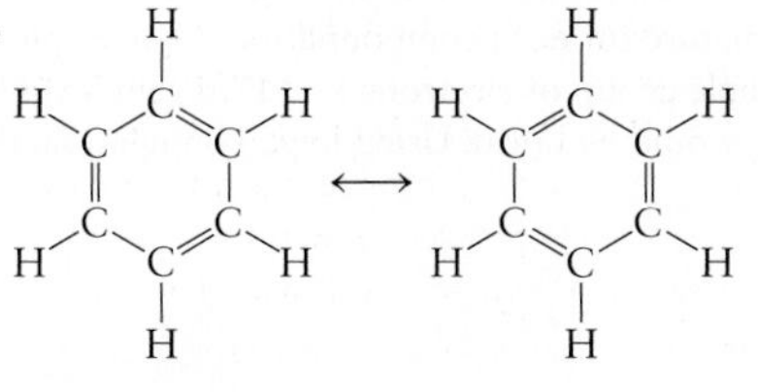

67. (two resonance structures of borazine, $B_3N_3H_6$) ⟷

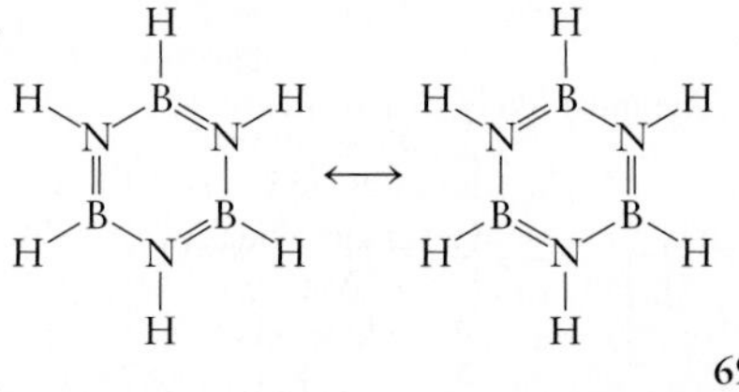

69. Statements a and c are true. For statement b, SF_4 has 5 electron pairs around the sulfur in the best Lewis structure; it is an exception to the octet rule. Because OF_4 has the same number of valence electrons as SF_4, OF_4 would also have to be an exception to the octet rule. However, Row 2 elements like O never have more than 8 electrons around them, so OF_4 does not exist. For statement d, two resonance structures can be drawn for ozone:

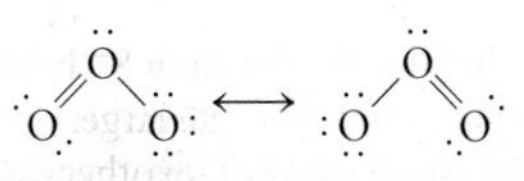

O=O—O: ⟷ :O—O=O

When resonance structures can be drawn, the actual bond lengths and strengths are all equal to each other. Even though each Lewis structure implies the two O—O bonds are different, this is not the case in real life. In real life, both of the O—O bonds are equivalent. When resonance structures can be drawn, you can think of the bonding as an average of all of the resonance structures.

71. PF_5 (P bonded to five F) ClF_3 (Cl bonded to three F)

SF_4 (S bonded to four F) Br_3^- [:Br—Br—Br:]$^-$

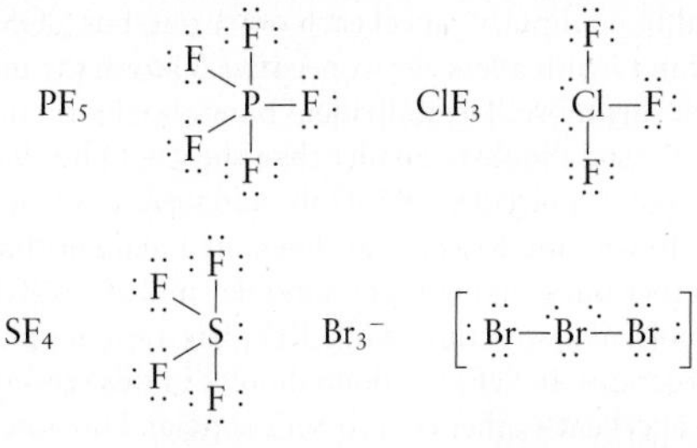

Row 3 and heavier nonmetals can have more than 8 electrons around them when they have to. Row 3 and heavier elements have empty *d* orbitals which are close in energy to the valence *s* and *p* orbitals. These empty *d* orbitals can accept extra electrons. **73.** Three resonance structures can be drawn for CO_3^{2-}. The actual structure for CO_3^{2-} is an average of these three resonance structures. That is, the three C—O bond lengths are all equivalent, with a length somewhere between a single bond and a double bond. The actual bond length of 136 pm is consistent with this resonance view of CO_3^{2-}. **75.** $NO^+ < N_2O < NO_2^- < NO_3^- < H_2NOH$

77. 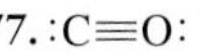 :C≡O:

In the CO Lewis structure, carbon has a formal charge of −1 and oxygen has a formal charge of +1. Electronegativity predicts the opposite polarization. The two opposing effects seem to partially cancel to give a much less polar molecule than expected. **79.** a–f and h all have similar Lewis structures:

:Y—X(Y)(Y)—Y: g. ClO_3^- [:O—Cl(O)—O:]$^-$

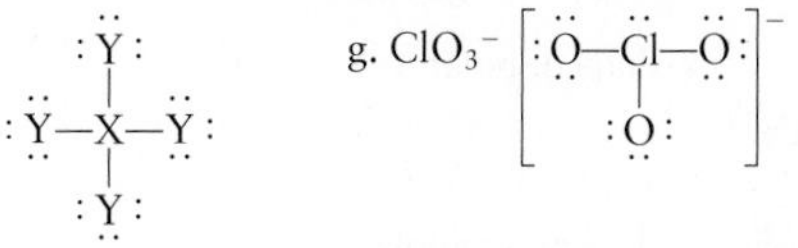

Formal charges: a. +1; b. +2; c. +3; d. +1; e. +2; f. +4; g. +2; h. +1

81. 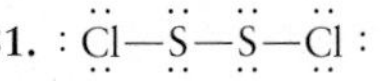 :Cl—S—S—Cl:

83. :F—O—O—F:

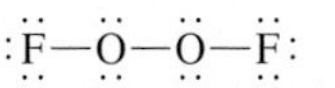

	F	O	O	F
Formal charge:	0	0	0	0
Oxidation state:	−1	+1	+1	−1

Oxidation states are more useful when accounting for the reactivity of O_2F_2. We are forced to assign +1 as the oxidation state for oxygen due to the bonding of fluorine. Oxygen is very electronegative, and +1 is not a stable oxidation state for this element. **85.** a. V-shaped or bent; b. see-saw; c. trigonal pyramid; d. trigonal bipyramid; e. tetrahedral **87.** [57] a. linear, 180°; b. trigonal pyramid, < 109.5°; c. tetrahedral, 109.5°; d. tetrahedral, 109.5°; e. trigonal planar, 120°; f. V-shaped, < 109.5°; g. linear, 180°; h. and i. linear, no bond angle in diatomic molecules; [58] a. All are tetrahedral, 109.5°; b. All are trigonal pyramid, < 109.5°; c. All are V-shaped, < 109.5°; [60] a. NO_2^-: V-shaped, ≈ 120°; NO_3^-: trigonal planar, 120°; N_2O_4: trigonal planar about both nitrogens, 120°; b. All are linear; 180° **89.** Br_3^- is linear; ClF_3 is T-shaped; SF_4 is see-saw. **91.** a. linear, 180°; b. T-shaped, ≈ 90°; c. see-saw, ≈ 120° and ≈ 90°; d. trigonal bipyramid, 120° and 90° **93.** SeO_2, PCl_3, and SCl_2 have net dipole moments (are polar). **95.** The two general requirements for a polar molecule are (1) polar bonds and (2) a structure such that the bond dipoles of the polar

bonds do not cancel. In CF_4, the fluorines are symmetrically arranged about the central carbon atom. The net result is for all of the individual C—F bond dipoles to cancel each other out, giving a nonpolar molecule. In XeF_4 the 4 Xe—F bond dipoles are also symmetrically arranged and XeF_4 is also nonpolar. The individual bond dipoles cancel out when summed together. In SF_4 we also have 4 polar bonds. But in SF_4 the bond dipoles are not symmetrically arranged and they do not cancel each other out. SF_4 is polar. It is the positioning of the lone pair that disrupts the symmetry in SF_4. CO_2 is nonpolar because the individual bond dipoles cancel each other out, but COS is polar. By replacing an O with a less electronegative S atom, the molecule is not symmetric any more. The individual bond dipoles do not cancel since the C—S bond dipole is smaller than the C—O bond dipole, resulting in a polar molecule. **97.** Only statement c is true. The bond dipoles in CF_4 and KrF_4 are arranged in a manner that they all cancel each other out, making them nonpolar molecules (CF_4 has a tetrahedral molecular structure, while KrF_4 has a square planar molecular structure). In SeF_4 the bond dipoles in this see-saw molecule do not cancel each other out, so SeF_4 is polar. For statement a, all the molecules have either a trigonal planar geometry or a trigonal bipyramidal geometry, both of which have 120-degree bond angles. However, $XeCl_2$ has three lone pairs and two bonded chlorine atoms around it. $XeCl_2$ has a linear molecular structure with a 180-degree bond angle. With three lone pairs, we no longer have a 120-degree bond angle in $XeCl_2$. For statement b, SO_2 has a V-shaped molecular structure with a bond angle of about 120 degrees. CS_2 is linear with a 180-degree bond angle, and SCl_2 is V-shaped but with an approximately 109.5-degree bond angle. The three compounds do not have the same bond angle. For statement d, central atoms adopt a geometry to minimize electron repulsions, not maximize them.
99. Element E must belong to the Group 6A elements because it has 6 valence electrons. E must also be a Period 3 or heavier element since this ion has more than 8 electrons around the central E atom (Period 2 elements never have more than 8 electrons around them). Some possible identities for E are S, Se, and Te. The ion has a T-shaped molecular structure with bond angles of $\approx 90°$.

101. a. OCl_2 — V-shaped, polar

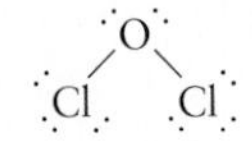

KrF_2 : F—Kr—F : — Linear, nonpolar

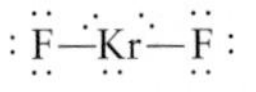

BeH_2 H—Be—H — Linear, nonpolar

SO_2 — V-shaped, polar

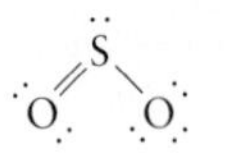

(One other resonance structure possible)

b. SO_3 — Trigonal planar, nonpolar

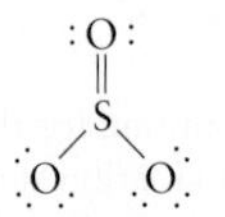

(Two other resonance structures possible)

NF_3 — Trigonal pyramid, polar

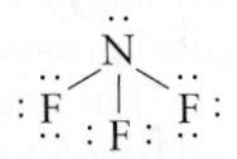

IF_3 — T-shaped, polar

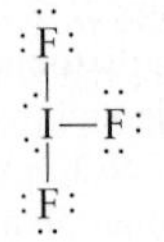

c. CF_4 — Tetrahedral, nonpolar

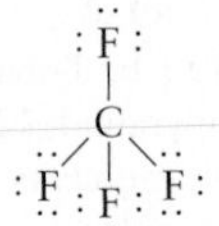

SeF_4 — See-saw, polar

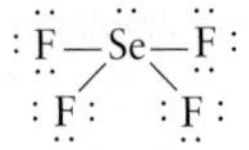

KrF_4 — Square planar, nonpolar

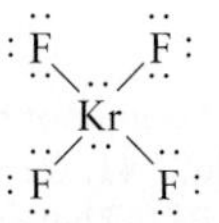

d. IF_5 — Square pyramid, polar

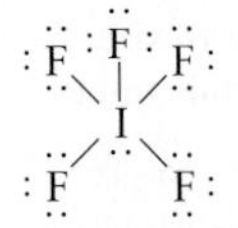

AsF_5 — Trigonal bipyramid, nonpolar

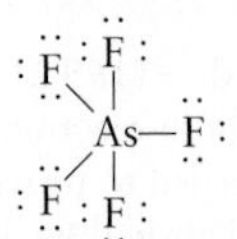

103. All these molecules have polar bonds that are symmetrically arranged around the central atom. In each molecule, the individual bond dipoles cancel to give no net overall dipole moment.

105. [Resonance structures of CO_3^{2-}, HCO_3^-, and H_2CO_3]

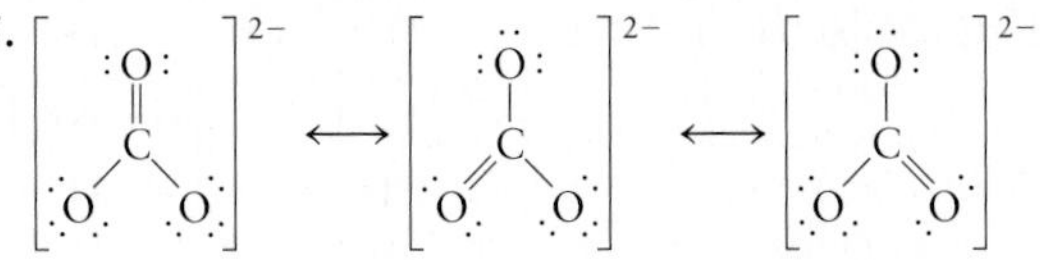

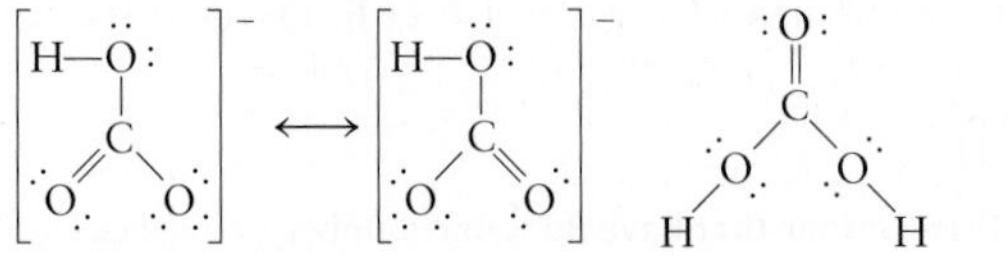

$\Delta H = -83$ kJ; the carbon–oxygen double bond is stronger than two carbon–oxygen single bonds; hence CO_2 and H_2O are more stable than H_2CO_3. **107.** As the halogen atoms get larger, it becomes more difficult to fit three halogen atoms around the small nitrogen atom, and the NX_3 molecule becomes less stable. **109.** a. NaBr: In $NaBr_2$ the sodium ion would have a +2 charge assuming each bromine has a −1 charge. Sodium doesn't form stable Na^{2+} compounds.
b. ClO_4^-: ClO_4 has 31 valence electrons so it is impossible to satisfy the octet rule for all atoms in ClO_4. The extra electron from the −1 charge in ClO_4^- allows for complete octets for all atoms.
c. XeO_4: We can't draw a Lewis structure that obeys the octet rule for SO_4 (30 electrons), unlike with XeO_4 (32 electrons). d. SeF_4: Both compounds require the central atom to expand its octet. O is too small and doesn't have low-energy *d* orbitals to expand its octet (which is true for all Period 2 elements). **111.** a. radius: $N^+ < N < N^-$; IE: $N^- < N < N^+$; b. radius: $Cl^+ < Cl < Se < Se^-$; IE: $Se^- < Se < Cl < Cl^+$; c. radius: $Sr^{2+} < Rb^+ < Br^-$; IE: $Br^- < Rb^+ < Sr^{2+}$
113. I; square pyramid **115.** Yes; each structure has the same number of effective pairs around the central atom, giving the same predicted molecular structure for each compound/ion. (A multiple bond is counted as a single group of electrons.) **117.** From VSEPR, the molecular structure would be linear. Using hyperconjugation, the resonance structures are

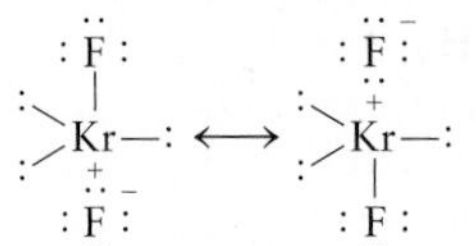

[Resonance structures of KrF_2]

119. −562 kJ **121.** a. The most likely structures are

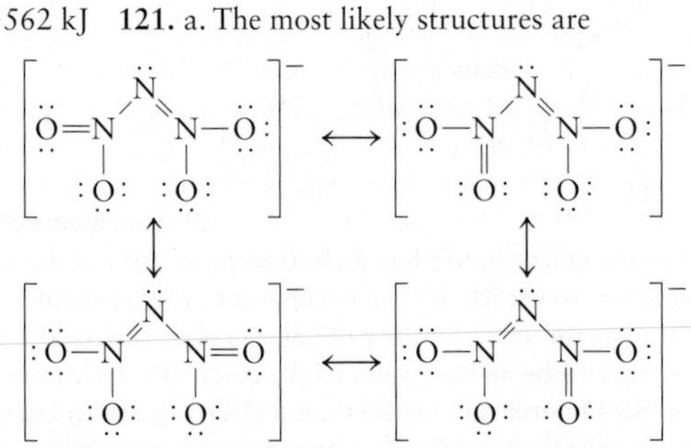

[Resonance structures]

There are other possible resonance structures, but these are most likely. b. The NNN, ONN, and ONO bond angles should be about

120°. c. $NH_4N(NO_2)_2(s) \rightarrow 2N_2(g) + 2H_2O(g) + O_2(g)$, $\Delta H = -893$ kJ; d. To estimate ΔH, we ignored the ionic interactions between NH_4^+ and $N(NO_2)_2^-$. In addition, we assumed the bond energies in Table 13.6 applied to the $N(NO_2)^-$ bonds in any one of the resonance structures above. This is a bad assumption since molecules that exhibit resonance generally have stronger overall bonds than predicted. All of these assumptions give an estimated ΔH value that is too negative. **123.** a. i. −2636 kJ; ii. −3471 kJ; iii. −3543 kJ; b. reaction iii, −8085 kJ/kg **125.** 17 kJ/mol

Chapter 14

9. The valence orbitals of the nonmetals are the *s* and *p* orbitals. The lobes of the *p* orbitals are 90 degrees and 180 degrees apart from each other. If the *p* orbitals were used to form bonds, then all bond angels should be 90 degrees or 180 degrees. This is not the case. In order to explain the observed geometry (bond angles) that molecules exhibit, we need to make up (hybridize) orbitals that point to where the bonded atoms and lone pairs are located. We know the geometry; we hybridize orbitals to explain the geometry. Sigma bonds have shared electrons in the area centered on a line joining the atoms. The orbitals that overlap to form the sigma bonds must overlap head to head or end to end. The hybrid orbitals about a central atom always are directed at the bonded atoms. Hybrid orbitals will always overlap head to head to form sigma bonds. **11.** We use *d* orbitals when we have to; that is, we use *d* orbitals when the central atom on a molecule has more than eight electrons around it. The *d* orbitals are necessary to accommodate more than eight electrons. Row 2 elements never have more than eight electrons around them so they never hybridize *d* orbitals. We rationalize this by saying there are no *d* orbitals close in energy to the valence 2*s* and 2*p* orbitals (2*d* orbitals are forbidden energy levels). However, for Row 3 and heavier elements, there are 3*d*, 4*d*, 5*d*, etc. orbitals, which will be close in energy to the valence *s* and *p* orbitals. It is Row 3 and heavier nonmetals that hybridize *d* orbitals when they have to. For sulfur, the valence electrons are in 3*s* and 3*p* orbitals. Therefore, 3*d* orbitals are closest in energy and are available for hybridization. Arsenic would hybridize 4*d* orbitals to go with the valence 4*s* and 4*p* orbitals, whereas iodine would hybridize 5*d* orbitals since the valence electrons are in $n = 5$.

13. H_2O: O bonded to H and H (Lewis structure)

H_2O has a tetrahedral arrangement of the electron pairs around the O atom that requires sp^3 hybridization. Two of the sp^3 hybrid orbitals are used to form bonds to the two hydrogen atoms, and the other two sp^3 hybrid orbitals hold the two lone pairs on oxygen. The two O—H bonds are formed from overlap of the sp^3 hybrid orbitals from oxygen with the 1*s* atomic orbitals from the hydrogen atoms. **15.** For ethane, the carbon atoms are sp^3 hybridized. The six C—H sigma bonds are formed from overlap of the sp^3 hybrid orbitals on C with the 1*s* atomic orbitals from the hydrogen atoms. The carbon–carbon sigma bond is formed from overlap of an sp^3 hybrid orbital on each C atom. For ethanol, the two C atoms and the O atom are sp^3 hybridized. All bonds are formed from overlap with these sp^3 hybrid orbitals. The C—H and O—H sigma bonds are formed from overlap of sp^3 hybrid orbitals with hydrogen 1*s* atomic orbitals. The C—C and C—O sigma bonds are formed from overlap of the sp^3 hybrid orbitals on each atom. **17.** [57] a. *sp*; b. sp^3; c. sp^3; d. sp^3; e. sp^2; f. sp^3; g. *sp*; h. each O is sp^2 hybridized; i. Br is sp^3 hybridized [58] a. All are sp^3 hybridized. b. All are sp^3 hybridized. c. All are sp^3 hybridized. [60] a. NO_2^-, sp^2; NO_3^-, sp^2; N_2O_4, sp^2 for both nitrogens; b. All are *sp* hybridized. **19.** Each of the four N—H sigma bonds are formed from overlap of an sp^3 hybrid orbital from nitrogen with a 1*s* orbital from hydrogen. Each of the two N—C sigma bonds are formed from an sp^3 hybrid orbital from N with an sp^2 hybrid orbital from carbon. The double bond between carbon and oxygen consists of one σ and one π bond. The σ bond in the double bond is formed from overlap of a carbon sp^2 hybrid orbital with an oxygen sp^2 hybrid orbital. The π bond in the double bond is formed from overlap of the unhybridized *p* atomic orbitals. Carbon and oxygen each have one unhybridized *p* atomic orbital, and they are assumed to be parallel to each other. When two parallel *p* atomic orbitals overlap side to side, a π bond results.

21. a. Tetrahedral, 109.5°, sp^3, nonpolar (Lewis structure of CF_4)

b. Trigonal pyramid, <109.5°, sp^3, polar (Lewis structure of NF_3)

c. V-shaped, <109.5°, sp^3, polar (Lewis structure of OF_2)

d. Trigonal planar, 120°, sp^2, nonpolar (Lewis structure of BF_3)

e. Linear, 180°, *sp*, nonpolar

H—Be—H

f. (TeF_4 structure) $a \approx 120°$, see-saw, dsp^3, polar; $b \approx 90°$

g. (AsF_5 structure) $a = 90°$, trigonal bipyramid, dsp^3, nonpolar; $b = 120°$

h. :F̤—Kr—F̤: Linear, 180°, dsp^3, nonpolar

i. (KrF_4 structure) Square planar, 90°, d^2sp^3, nonpolar

j. (SeF_6 structure) Octahedral, 90°, d^2sp^3, nonpolar

k. (IF_5 structure) Square pyramid, ≈90°; d^2sp^3, polar

l. (IF_3 structure) T-shaped, ≈90°, dsp^3, polar

23. For the *p* orbitals to properly line up to form the π bond, all six atoms are forced into the same plane. If the atoms are not in the same plane, then the π bond could not form since the *p* orbitals would no longer be parallel to each other.

25. Biacetyl (structure with carbonyl carbons labeled sp^2 and methyl carbons labeled sp^3)

All CCO angles are 120°. The six atoms are not forced to lie in the same plane because of free rotation around the carbon–carbon single (σ) bonds. 11 σ and 2 π bonds.

Acetoin

[structure of acetoin: labels sp^2, a, b, sp^3, sp^3]

Angle a = 120°, angle b = 109.5°, 13 σ bonds and 1 π bond **27.** To complete the Lewis structure, add lone pairs to complete octets for each atom. a. 6; b. 4; c. The center N in —N=N=N group; d. 33 σ; e. 5 π; f. 180°; g. ≈109.5°; h. sp^3
29. a.

[structures of azodicarbonamide (angles a–d) and methyl cyanoacrylate (angles e–h)]

Note: NH_2, H_2C, and CH_3 are shorthand for nitrogen or carbon atoms singly bonded to hydrogen atoms. b. In azodicarbonamide, the two carbon atoms are sp^2 hybridized, the two nitrogen atoms with hydrogens attached are sp^3 hybridized, and the other two nitrogens are sp^2 hybridized. In methyl cyanoacrylate, the CH_3 carbon is sp^3 hybridized, the carbon with the triple bond is sp hybridized, and the other three carbons are sp^2 hybridized. c. Azodicarbonamide contains three π bonds, and methyl cyanoacrylate contains four π bonds. d. a) ≈109.5°; b) 120°; c) ≈120°; d) 120°; e) 180°; f) 120°; g) ≈109.5°; h) 120°

31. CO :C≡O: CO_2 O=C=O C_3O_2 O=C=C=C=O

There is no molecular structure for the diatomic CO molecule. The carbon in CO is sp hybridized. CO_2 is a linear molecule, and the central carbon atom is sp hybridized. C_3O_2 is a linear molecule with all of the central carbon atoms exhibiting sp hybridization.

33. a. σ_p

When p orbitals are combined head-to-head and the phases are the same sign (the orbital lobes have the same sign), a sigma bonding molecular orbital is formed.

b.

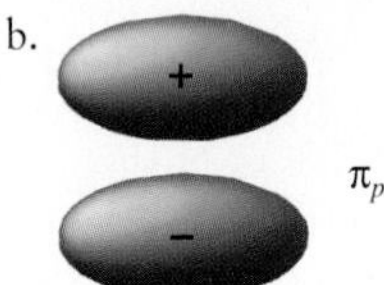

π_p

When parallel p orbitals are combined in-phase (the signs match up), a pi bonding molecular orbital is formed.

c.

σ_p^*

When p orbitals are combined head-to-head and the phases are opposite, a sigma antibonding molecular orbital is formed.

d.

π_p^*

When parallel p orbitals are combined out-of-phase (the orbital lobes have opposite signs), a pi antibonding molecular orbital is formed. **35.** From experiment, B_2 is paramagnetic. If the σ_{2p} MO is lower in energy than the two degenerate π_{2p} MOs, the electron configuration for B_2 would have all electrons paired. Experiment tells us we must have unpaired electrons. Therefore, the MO diagram is modified to have the π_{2p} orbitals lower in energy than the σ_{2p} orbitals. This gives two unpaired electrons in the electron configuration for B_2, which explains the paramagnetic properties of B_2. The model allowed for s and p orbitals to mix, which shifted the energy of the σ_{2p} orbital to above that of the π_{2p} orbitals. **37.** a. H_2 has two valence electrons to put in the MO diagram whereas He_2 has four valence electrons.

H_2: $(\sigma_{1s})^2$ Bond order = B.O. = (2 − 0)/2 = 1
He_2: $(\sigma_{1s})^2(\sigma_{1s}{}^*)^2$ B.O. = (2 − 2)/2 = 0

H_2 has a nonzero bond order, so MO theory predicts it will exist. The H_2 molecule is stable with respect to the two free H atoms. He_2 has a bond order of zero, so it should not form. The He_2 molecule is not more stable than the two free He atoms. b. See Fig. 14.41 for the MO energy-level diagrams of B_2, C_2, N_2, O_2, and F_2. B_2 and O_2 have unpaired electrons in their electron configurations, so they are predicted to be paramagnetic. C_2, N_2, and F_2 have no unpaired electrons in their MO diagrams; they are all diamagnetic. c. From the MO energy diagram in Fig. 14.41, N_2 maximizes the number of electrons in the lower-energy bonding orbitals and has no electrons in the antibonding $2p$ molecular orbitals. N_2 has the highest possible bond order of 3, so it should be a very strong (stable) bond. d. NO^+ has 5 + 6 − 1 = 10 valence electrons to place in the MO diagram and NO^- has 5 + 6 + 1 = 12 valence electrons. The MO diagram for these two ions is assumed to be the same as that used for N_2.

NO^+: $(\sigma_{2s})^2(\sigma_{2s}{}^*)^2(\pi_{2p})^4(\sigma_{2p})^2$ B.O. = (8 − 2)/2 = 3
NO^-: $(\sigma_{2s})^2(\sigma_{2s}{}^*)^2(\pi_{2p})^4(\sigma_{2p})^2(\pi_{2p}{}^*)^2$ B.O. = (8 − 4)/2 = 2

NO^+ has a larger bond order than NO^-, so NO^+ should be more stable than NO^-.
39.

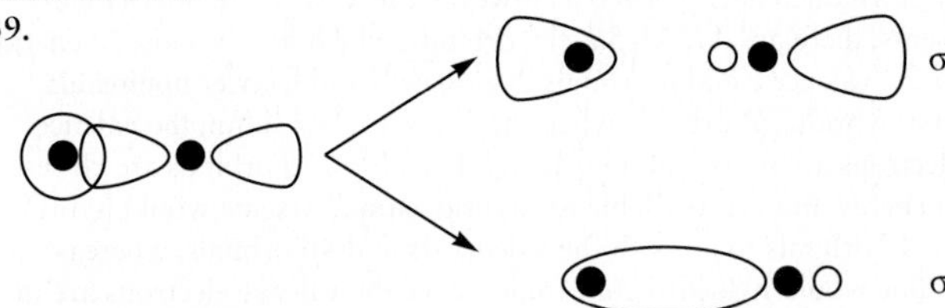

These MOs are σ MOs since the electron density is cylindrically symmetric around the internuclear axis. **41.** $N_2{}^+$ and $N_2{}^-$ both have a bond order of 2.5.
43. CN: $(\sigma_{2s})^2(\sigma_{2s}{}^*)^2(\pi_{2p})^4(\sigma_{2p})^1$;
NO: $(\sigma_{2s})^2(\sigma_{2s}{}^*)^2(\pi_{2p})^4(\sigma_{2p})^2(\pi_{2p}{}^*)^1$;
$O_2{}^{2+}$: $(\sigma_{2s})^2(\sigma_{2s}{}^*)^2(\sigma_{2p})^2(\pi_{2p})^4$;
$N_2{}^{2+}$: $(\sigma_{2s})^2(\sigma_{2s}{}^*)^2(\pi_{2p})^4$; If the added electron goes into a bonding orbital, the bond order would increase, making the species more stable and more likely to form. Between CN and NO, CN would most likely form CN^- since the bond order increases (unlike NO^-, in which the added electron goes into an antibonding orbital). Between $O_2{}^{2+}$ and $N_2{}^{2+}$, $N_2{}^+$ would most likely form since the bond order increases (unlike when $O_2{}^+$ forms). **45.** The π bonds between S atoms and between C and S atoms are not as strong. The atomic orbitals do not overlap with each other as well as the smaller atomic orbitals of C and O overlap. **47.** π molecular orbital **49.** a. The electron density would be closer to F on the average. The F atom is more electronegative than the H atom, and the $2p$ orbital of F is lower in energy than the 1s orbital of H; b. The bonding MO would have more fluorine $2p$ character because it is closer in energy to the fluorine $2p$ atomic orbital; c. The antibonding MO would place more electron density closer to H and would have a greater contribution from the higher-energy hydrogen 1s atomic orbital. **51.** Molecules that exhibit resonance have delocalized π bonding. This is a fancy way of saying that the π electrons are not permanently stationed between two specific atoms but instead can roam about over the surface of a molecule. We use the concept of delocalized π electrons to explain why molecules that exhibit resonance have equal

bonds in terms of strength. Because the π electrons can roam about over the entire surface of the molecule, the π electrons are shared by all of the atoms in the molecule, giving rise to equal bond strengths. The classic example of delocalized π electrons is benzene (C_6H_6). Figure 14.50 shows the π molecular orbital system for benzene. Each carbon in benzene is sp^2 hybridized, leaving one unhybridized p atomic orbital. All six of the carbon atoms in benzene have an unhybridized p orbital pointing above and below the planar surface of the molecule. Instead of just two unhybridized p orbitals overlapping, we say all six of the unhybridized p orbitals overlap, resulting in delocalized π electrons roaming about above and below the entire surface of the benzene molecule.

O=S—O ⟷ O—S=O

In SO_2 the central sulfur atom is sp^2 hybridized. The unhybridized p atomic orbital on the central sulfur atom will overlap with parallel p orbitals on each adjacent O atom. All three of these p orbitals overlap together, resulting in the π electrons moving about above and below the surface of the SO_2 molecule. With the delocalized π electrons, the S—O bond lengths in SO_2 are equal (and not different as each individual Lewis structure indicates). **53.** The Lewis structures for CO_3^{2-} (24 e^-) are:

$[\text{O=C(O)O}]^{2-} \longleftrightarrow [\text{O–C(O)=O}]^{2-} \longleftrightarrow [\text{O–C(=O)O}]^{2-}$

In the localized electron view, the central carbon atom is sp^2 hybridized; the sp^2 hybrid orbitals are used to form the three sigma bonds in CO_3^{2-}. The central C atom also has one unhybridized p atomic orbital that overlaps with another p atomic orbital from one of the oxygen atoms to form the π bond in each resonance structure. This localized π bond moves (resonates) from one position to another. In the molecular orbital model for CO_3^{2-}, all four atoms in CO_3^{2-} have a p atomic orbital that is perpendicular to the plane of the ion. All four of these p orbitals overlap at the same time to form a delocalized π bonding system where the π electrons can roam above and below the entire surface of the ion. The π molecular orbital system for CO_3^{2-} is analogous to that for NO_3^-, which is shown in Fig. 14.51 of the text. **55.** 411.7 N m^{-1} **57.** a. 113 pm; b. 3.45×10^{11} s^{-1} **59.** a. quartet in iv; b. singlet in i; c. triplet in iii; d. doublet in ii

61. a. XeO_3 — Trigonal pyramid; sp^3 b. XeO_4 — Tetrahedral; sp^3

c. $XeOF_4$ (or $XeOF_4$ with O in base) — Square pyramid; d^2sp^3 d. $XeOF_2$: O—Xe with two F (or F—Xe with F and O) — T-shaped; dsp^3

e. XeO_3F_2 (three arrangements) — Trigonal bipyramid; dsp^3

63. a. No, some atoms are in different places. Thus these are not resonance structures; they are different compounds. b. For the first Lewis structure, all nitrogens are sp^3 hybridized and all carbons are sp^2 hybridized. In the second Lewis structure, all nitrogens and carbons are sp^2 hybridized. c. The first structure with the carbon–oxygen double bonds is slightly more stable.

65.

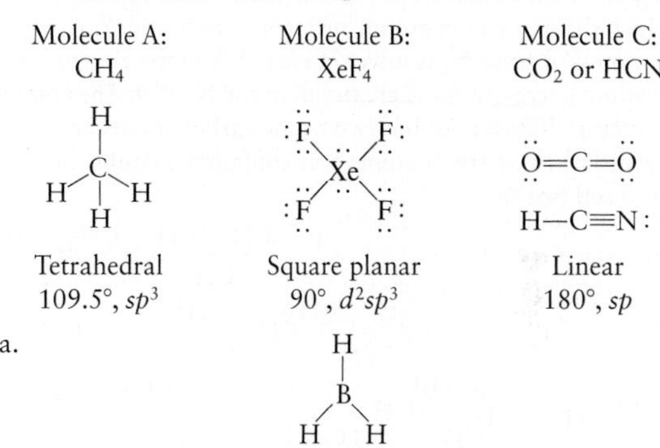

The two orbitals overlap side to side, so when the orbitals are in phase, a π bonding molecular orbital would form. **67.** Molecule A has a tetrahedral arrangement of electron pairs because it is sp^3 hybridized. Molecule B has six electron pairs about the central atom, so it is d^2sp^3 hybridized. Molecule C has two σ and two π bonds to the central atom, so it either has two double bonds to the central atom (as in CO_2) or one triple bond and one single bond (as in HCN). Molecule C is consistent with a linear arrangement of electron pairs exhibiting sp hybridization. There are many correct possibilities for each molecule; an example of each is:

Molecule A: CH_4	Molecule B: XeF_4	Molecule C: CO_2 or HCN
H–C(H)(H)–H	F₂XeF₂	O=C=O; H—C≡N:
Tetrahedral 109.5°, sp^3	Square planar 90°, d^2sp^3	Linear 180°, sp

69. a. BH_3 — Trigonal planar; nonpolar; 120°; sp^2

b. F—N=N—F (cis) Can also be: F—N=N—F (trans)

Polar Nonpolar

V-shaped about both N atoms; ≈120°; sp^2.
These are distinctly different molecules.

c. H—C(H)=C(H)—C(H)=C—H — Trigonal planar about each carbon; nonpolar; 120°; sp^2 for all carbons

d. ICl_3 — T-shaped; polar; ≈90°; dsp^3

71. (caffeine Lewis structure)

Six C and N atoms are sp^2 hybridized, six C and N atoms are sp^3 hybridized, and none of the C and N atoms are sp hybridized. Caffeine has 25 σ bonds and four π bonds. **73.** a. The NNO structure is correct. From the Lewis structures we would predict both NNO and NON to be linear, but NON would be nonpolar. NNO is polar.

b. $:\!\ddot{\text{N}}\!=\!\text{N}\!=\!\ddot{\text{O}}\!: \longleftrightarrow :\!\text{N}\!\equiv\!\text{N}\!-\!\ddot{\underset{..}{\text{O}}}\!: \longleftrightarrow :\!\ddot{\underset{..}{\text{N}}}\!-\!\text{N}\!\equiv\!\text{O}\!:$

$-1 \quad +1 \quad 0 \qquad 0 \quad +1 \quad -1 \qquad -2 \quad +1 \quad +1$ Formal charges

The central N is sp hybridized. We can probably ignore the third resonance structure on the basis of formal charge. c. sp hybrid orbitals from the center N overlap with atomic orbitals (or hybrid orbitals) from the other two atoms to form the two sigma bonds. The remaining p orbitals from the center N overlap with p orbitals from the other N to form the two π bonds. **75.** a. 116 kJ/mol;

b. F_2: $(\sigma_{2s})^2(\sigma_{2s}{}^*)^2(\sigma_{2p})^2(\pi_{2p})^4(\pi_{2p}{}^*)^4$ B.O. $= (8-6)/2 = 1$
$F_2{}^-$: $(\sigma_{2s})^2(\sigma_{2s}{}^*)^2(\sigma_{2p})^2(\pi_{2p})^4(\pi_{2p}{}^*)^4(\sigma_{2p}{}^*)^1$ B.O. $= (8-7)/2 = 0.5$

MO theory predicts that F_2 should have a stronger bond than $F_2{}^-$ because F_2 has the larger bond order. The calculated $F_2{}^-$ bond energy is indeed smaller than the F_2 bond energy (154 kJ/mol). **77.** a. N_2, $N_2{}^+$, N, and N^+ will all be present, assuming excess N_2. b. Light with wavelengths in the range of 85.33 nm $< \lambda \le$ 127 nm will produce N but no ions. c. Consider the MO energy-level diagram for N_2 (see Figs. 14.40 and 14.41). The electron removed from N_2 is in the σ_{2p} MO, which is lower in energy than the $2p$ atomic orbital from which the electron in atomic nitrogen is removed. Since the electron removed from N_2 is lower in energy than the electron in N, the ionization energy of N_2 is greater than for N. **79.** The complete Lewis structure follows. All but two of the carbon atoms are sp^3 hybridized. The two carbon atoms that contain the double bond are sp^2 hybridized (see*).

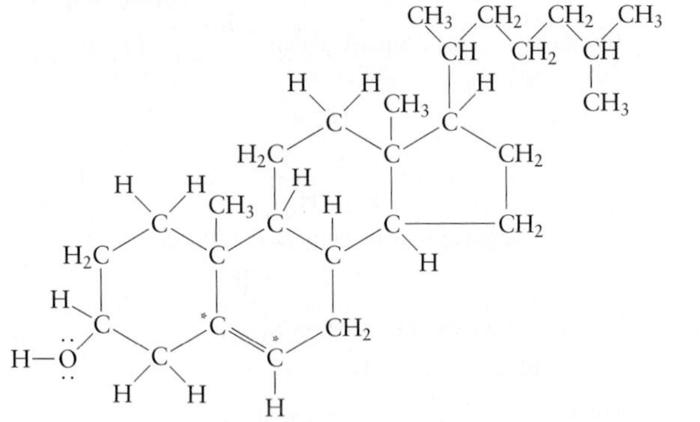

```
                                CH3  CH2   CH2  CH3
                                   \/   \ /   \/
                                   CH   CH2   CH
                      H    H        |          |
                       \  /  CH3    H         CH3
                        C     |    /
               H2C     /  \   C -- C
                  \  /      /       \
          H   H    CH3   H  C        CH2
           \ /      \ /  |  |        |
            C        C   H  C ------ CH2
       H2C /  \    /  \    /  \
          |    C           C    H
     H    |    |           |
      \  C     *C ===  *C  CH2
  H-O:   /  \  /        |
     ..  C     C        H
        /  \
       H    H
```

No; most of the carbons are not in the same plane since a majority of carbon atoms exhibit a tetrahedral structure. *Note:* CH, CH_2, H_2C, and CH_3 are shorthand for carbon atoms singly bonded to hydrogen atoms. **81.** a. The CO bond is polar with the negative end around the more electronegative oxygen atom. We would expect metal cations to be attracted to and bond to the oxygen end of CO on the basis of electronegativity. b. $:\!\text{C}\!\equiv\!\text{O}\!:$ FC (carbon) $= -1$; FC (oxygen) $= +1$. From formal charge, we would expect metal cations to bond to the carbon (with the negative formal charge). c. In MO theory, only orbitals with proper symmetry overlap to form bonding orbitals. The metals that form bonds to CO are usually transition metals, all of which have outer electrons in the d orbitals. The only MOs of CO that have proper symmetry to overlap with d orbitals are the $\pi_{2p}{}^*$ orbitals, whose shape is similar to the d orbitals (see Fig. 14.36). Since the antibonding MOs have more carbon character, one would expect the bond to form through carbon. **83.** $\Delta H_f^\circ =$ 237 kJ/mol (calculated); there is a 154 kJ discrepancy. Benzene is more stable (lower in energy) by 154 kJ than we expect from bond energies. Two equivalent Lewis structures can be drawn for benzene. The π bonding system implied by each Lewis structure consists of three localized π bonds. This, however, does not adequately explain the six equivalent C—C bonds in benzene. To explain the equivalent C—C bonds, the π electrons are said to be delocalized over the entire surface of the molecule (see Section 14.5). The large discrepancy between the $\Delta H^\circ{}_f$ values is due to the effect of a delocalized π bonding system that was not considered when calculating ΔH_f°. The extra stability associated with benzene can be called resonance stabilization. **85.** The two isomers having the formula C_2H_6O are

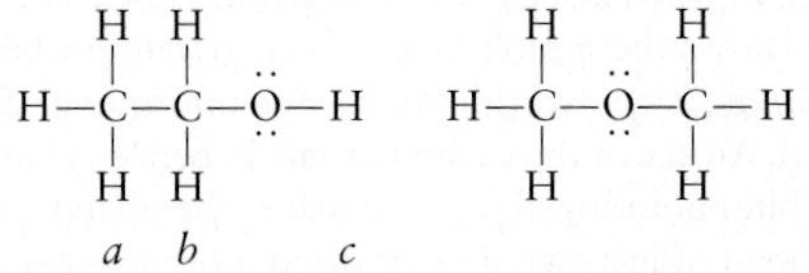

```
   H  H                H     H
   |  |   ..           |  .. |
H—C—C—O—H         H—C—O—C—H
   |  |   ..           |  .. |
   H  H                H     H
   a  b    c
```

The first structure has three types of hydrogens (a, b, and c), so the signals should be seen in three different regions of the NMR spectrum. The overall relative areas of the three signals should be in a 3:2:1 ratio due to the number of a, b, and c hydrogens. The signal for the a hydrogens will be split into a triplet signal due to the two neighboring b hydrogens. The signal for the b hydrogens should be split into a quintet signal due to the three neighboring a hydrogens plus the neighboring c hydrogen (four total protons). The c hydrogen should be split into a triplet signal due to the two neighboring b hydrogens. In practice, however, the c hydrogen bonded to the oxygen does not behave "normally." This O—H hydrogen generally behaves as if it is more than three sigma bonds apart from the b hydrogens. Therefore, the spectrum will most likely have a triplet signal for the a hydrogens, a quartet signal for the b hydrogens, and a singlet signal for the c hydrogen. In the second structure, all 6 hydrogens are equivalent, so only one signal will appear in the NMR spectrum. This signal will be a singlet signal because the hydrogens in the two —CH_3 groups are more than three sigma bonds apart from each other (no splitting occurs).

Chapter 15

11. A possible balanced equation is 2A + 3B $\rightarrow$ 4C.
13. $\dfrac{d[NH_3]}{dt} = -\dfrac{2}{3}\dfrac{d[H_2]}{dt}$ **15.** $L^{1/2}\ mol^{-1/2}\ s^{-1}$
17. a. Rate $= k[NO]^2[Cl_2]$; b. $k_{mean} = 1.8 \times 10^2\ L^2\ mol^{-2}\ min^{-1}$ **19.** a. Rate $= k[NOCl]^2$; b. $k_{mean} = 6.6 \times 10^{-29}\ cm^3\ molecules^{-1}\ s^{-1}$; c. $4.0 \times 10^{-8}\ L\ mol^{-1}\ s^{-1}$ **21.** a. Rate $= k[ClO_2]^2[OH^-]$; $k_{mean} = 2.30 \times 10^2\ L^2\ mol^{-2}\ s^{-1}$; b. 0.594 mol $L^{-1}\ s^{-1}$
23. Rate $= k[H_2SeO_3][H^+]^2[I^-]^3$; $k_{mean} = 5.2 \times 10^5\ L^5\ mol^{-5}\ s^{-1}$
25. Rate $= \dfrac{k[I^-][OCl^-]}{[OH^-]}$; $k_{mean} = 60.\ s^{-1}$ **27.** The first-order half-life is independent of concentration, the zero-order half-life is directly related to the concentration, and the second-order half-life is inversely related to concentration. For a first-order reaction, if the first half-life equals 20. s, the second half-life will also be 20. s because the half-life for a first-order reaction is concentration-independent. The second half-life for a zero-order reaction will be 1/2(20.) = 10. s. This is so because the half-life for a zero-order reaction has a direct relationship with concentration (as the concentration decreases by a factor of 2, the half-life decreases by a factor of 2). Because a second-order reaction has an inverse relationship between $t_{1/2}$ and $[A]_0$, the second half-life will be 40. s (twice the first half-life value). **29.** a. Rate $= k[A]$; $\ln[A] = -kt + \ln[A]_0$; $k = 2.97 \times 10^{-2}\ min^{-1}$; b. 23.3 min; c. 69.9 min **31.** $\ln[H_2O_2] = -kt + \ln[H_2O_2]_0$; Rate $= k[H_2O_2]$; $k = 8.3 \times 10^{-4}\ s^{-1}$; 0.037 M
33. $\dfrac{1}{[NO_2]} = kt + \dfrac{1}{[NO_2]_0}$; Rate $= k[NO_2]^2$; $k = 2.08 \times 10^{-4}$ L $mol^{-1}\ s^{-1}$; 0.131 M **35.** The rate law and integrated rate law are Rate $= k = 1.7 \times 10^{-4}$ atm/s; $P_{C_2H_5OH} = -kt + 250.\ \text{torr}\left(\dfrac{1\ \text{atm}}{760\ \text{torr}}\right)$ $= -kt + 0.329$ atm; At 900. s, $P_{C_2H_5OH} = 130$ torr **37.** a. The reaction is first order with respect to NO and first order with respect to O_3. b. Rate $= k[NO][O_3]$; c. $k' = 1.8\ s^{-1}$; $k'' = 3.6\ s^{-1}$; d. $1.8 \times 10^{-14}\ cm^3\ molecules^{-1}\ s^{-1}$ **39.** 1.6 L $mol^{-1}\ s^{-1}$ **41.** 150. s **43.** 12.5 s **45.** a. 160. s for both; b. 532 s **47.** a. 0.07 M; b. 34.7 min **49.** 4.2×10^{-5} mol $L^{-1}\ h^{-1}$ **51.** a. 1.15×10^2

$L^3\ mol^{-3}\ s^{-1}$; b. 87.0 s; c. [B] = 1.00 *M*; [A] = 1.27×10^{-5} *M* **53.** In a unimolecular reaction, a single-reactant molecule decomposes to products. In a bimolecular reaction, two molecules collide to give products. The probability of the simultaneous collision of three molecules with enough energy and proper orientation is very small, making termolecular steps very unlikely. **55.** a. Rate = $k[CH_3NC]$; b. Rate = $k[O_3][NO]$; c. Rate = $k[O_3]$; d. Rate = $k[O_3][O]$; e. Rate = $k[^{14}_{6}C]$ or rate = kN where N = the number of $^{14}_{6}C$ atoms **57.** Rate = $k[C_4H_9Br]$; $C_4H_9Br + 2H_2O \rightarrow C_4H_9OH + Br^- + H_3O^+$; the intermediates are $C_4H_9^+$ and $C_4H_9OH_2^+$ **59.** mechanism b **61.** Rate = $\frac{k_3k_2k_1}{k_{-2}k_{-1}}[Br^-][BrO_3^-][H^+]^2 =$ $k[Br^-][BrO_3^-][H^+]^2$ **63.** Rate = $\frac{k_2k_1[I^-][OCl^-]}{k_{-1}[OH^-]} = \frac{k[I^-][OCl^-]}{[OH^-]}$ **65.** a. $MoCl_5^-$; b. Rate = $\frac{d[NO_2^-]}{dt} = \frac{k_1k_2[NO_3^-][MoCl_6^{2-}]}{k_{-1}[Cl^-] + k_2[NO_3^-]}$ **67.** a. Rate = $\frac{d[E]}{dt} = \frac{k_1k_2[B]^2}{k_{-1}[B] + k_2}$; b. When $k_2 << k_{-1}[B]$, then the reaction is first order in B. c. Collisions between B molecules only transfer energy from one B to another. This occurs at a much faster rate than the decomposition of an energetic B molecule (B*). **69.** Two reasons are: a. The collision must involve enough energy to produce the reaction; that is, the collision energy must be equal to or exceed the activation energy. b. The relative orientation of the reactants when they collide must allow formation of any new bonds necessary to produce products. **71.** a. $T_2 > T_1$; as temperature increases, the distribution of collision energies shifts to the right. That is, as temperature increases, there are fewer collision energies with small energies and more collisions with large energies. b. As temperature increases, more of the collisions have the required activation energy necessary to convert reactants into products. Hence, the rate of the reaction increases with increasing temperature. **73.** $9.5 \times 10^{-5}\ L\ mol^{-1}\ s^{-1}$ **75.** 51°C **77.** a. 91.5 kJ/mol; b. $3.54 \times 10^{14}\ s^{-1}$; c. $3.24 \times 10^{-2}\ s^{-1}$ **79.** A plot of ln(k) versus $\frac{1}{T}$ gives a straight line with negative slope equal to $-E_a/R$ (see *Solutions Guide* for plot). E_a = 11.2 kJ/mol

81.

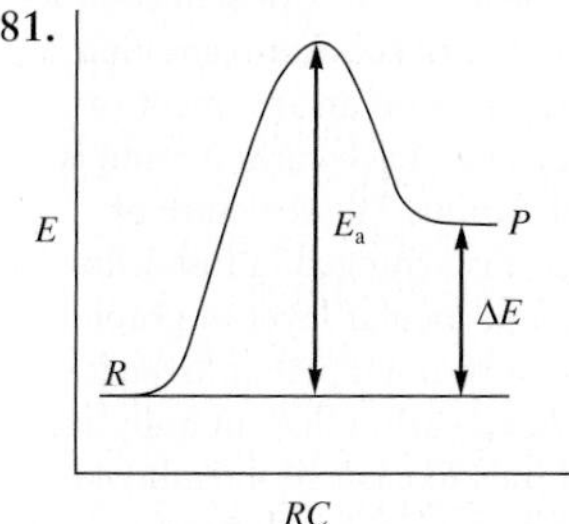

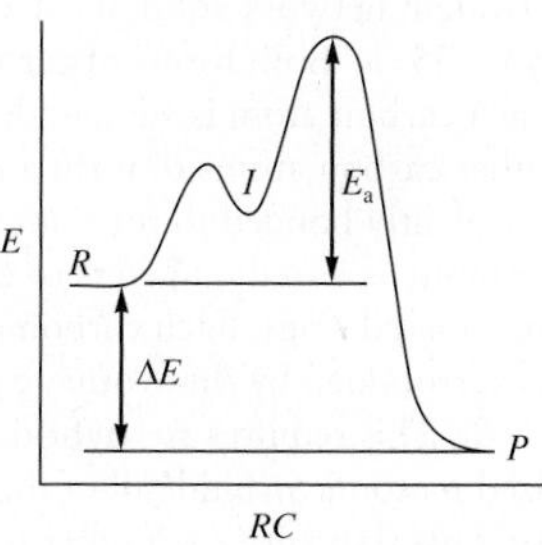

d. See plot for the intermediate plateau, which represents the energy of the intermediate. The two steps in the mechanism are $R \rightarrow I$ and $I \rightarrow P$. In a mechanism, the rate of the slowest step determines the rate of the reaction. The activation energy for the slowest step will be the largest energy barrier that the reaction must overcome. Since the second hump in the diagram is at the highest energy, then the second step ($I \rightarrow P$) has the largest activation energy and will be the rate-determining step (the slow step). **83.** 139 kJ/mol **85.** a. The blue plot is the catalyzed pathway. The catalyzed pathway has the lower activation. This is why the catalyzed pathway is faster. b. ΔE_1 represents the activation energy for the uncatalyzed pathway. c. ΔE_2 represents the energy difference between the reactants and products. Note that ΔE_2 is the same for both the catalyzed and the uncatalyzed pathways. It is the activation energy that is different for a catalyzed pathway versus an uncatalyzed pathway. d. Because the products have a higher total energy as compared to reactants, this is an endothermic reaction. **87.** a. W, because it has a lower activation energy than the Os catalyst. b. The W-catalyzed reaction is approximately 10^{30} times faster than the uncatalyzed reaction. c. Because $[H_2]$ is in the denominator of the rate law, the presence of H_2 decreases the rate of the reaction. For the decomposition to occur, NH_3 molecules must be adsorbed on the surface of the catalyst. If H_2 is also adsorbed on the catalyst surface, then there are fewer sites for NH_3 molecules to be adsorbed, and the rate decreases. **89.** Since the chlorine atom–catalyzed reaction has a lower activation energy, then the Cl-catalyzed rate is faster. Hence, Cl is a more effective catalyst. **91.** At high [S], the enzyme is completely saturated with substrate. Once the enzyme is completely saturated, the rate of decomposition of ES can no longer increase and the overall rate remains constant. **93.** ~1 second **95.** The most common method to experimentally determine the differential rate law is the method of initial rates. Once the differential rate law is determined experimentally, the integrated rate law can be derived. However, sometimes it is more convenient and more accurate to collect concentration versus time data for a reactant. When this is the case, then we do "proof" plots to determine the integrated rate law. Once the integrated rate law is determined, the differential rate law can be determined. Either experimental procedure allows determination of both the integrated and the differential rate; which rate law is determined by experiment and which is derived is usually decided by which data is easiest and most accurately collected. **97.** The experimentally determined rate law is Rate = $k[DNA][CH_3I]$. Mechanism I is possible because the derived rate law from the mechanism (Rate = $k[DNA][CH_3I]$) agrees with the experimentally determined rate law. The derived rate law for Mechanism II will equal the rate of the slowest step. This is step 1 in the mechanism, giving a derived rate law that is Rate = $k[CH_3I]$. Because this rate law does not agree with experiment, Mechanism II would not be a possible mechanism for the reaction. **99.** a. untreated: k = 0.465 day^{-1}; deacidifying agent: k = 0.659 day^{-1}; antioxidant: k = 0.779 day^{-1}; b. No, the silk degrades more rapidly with the additives; c. untreated: $t_{1/2}$ = 1.49 day; deacidifying agent: $t_{1/2}$ = 1.05 day; antioxidant: $t_{1/2}$ = 0.890 day **101.** Carbon cannot form the fifth bond necessary for the transition state because of the small atomic size of carbon and because carbon doesn't have low-energy *d* orbitals available to expand the octet. **103.** a. Second order; b. 17.1 s ≈ 20 s; c. 34.6 s ≈ 30 s; d. 847 s; e. 31 kJ/mol **105.** 29.8 kJ/mol **107.** a. 53.9 kJ/mol; b. 83 chirps per minute per insect;

c. T (°C)	T (°F)	k (min^{-1})	42 + 0.80 (k/4)
25.0	77.0	178	78°F
20.3	68.5	126	67°F
17.3	63.1	100.	62°F
15.0	59.0	83	59°F

The rule of thumb appears to be fairly accurate, about ±1°F. **109.** Rate = $\frac{d[Cl_2]}{dt} = \frac{k_1k_2[NO_2Cl]^2}{k_{-1}[NO_2] + k_2[NO_2Cl]}$ **111.** a. Rate = $k[CH_3X]$, k = 0.93 h^{-1}; b. 8.80×10^{-10} h; c. 3.0×10^2 kJ/mol; d. The activation energy is close to the C—X bond energy. A plausible mechanism is

$$CH_3X \longrightarrow CH_3 + X \quad \text{(slow)}$$
$$CH_3 + Y \longrightarrow CH_3Y \quad \text{(fast)}$$

113. 160 s **115.** Rate = $k[A][B]^2$, $k = 1.4 \times 10^{-2}\ L^2\ mol^{-2}\ s^{-1}$ **117.** Rate = $\frac{-d[N_2O_5]}{dt} = \frac{2k_1k_2[M][N_2O_5]}{k_{-1}[M] + 2k_2}$ **119.** a. Both are first order. b. k_1 = 0.82 $L\ mol^{-1}\ min^{-1}$; k_2 = 9.5 $L^2\ mol^{-2}\ min^{-1}$ c. There are two pathways, one involving H^+ with rate = $k_2[H^+][I^-][H_2O_2]$ and another pathway not involving H^+ with rate = $k_1[I^-][H_2O_2]$. The overall rate of reaction depends on which of these two pathways dominates, and this depends on the H^+ concentration.

Chapter 16

11. Intermolecular forces are the relatively weak forces between molecules that hold the molecules together in the solid and liquid phases. Intramolecular forces are the forces within a molecule. These are the covalent bonds in a molecule. Intramolecular forces (covalent bonds) are much stronger than intermolecular forces. Dipole forces are the forces that act between polar molecules. The electrostatic attraction between the partial positive end of one polar molecule and the partial negative end of another is the dipole force. Dipole forces are generally weaker than hydrogen bonding. Both of these forces are due to dipole moments in molecules. Hydrogen bonding is given a separate name from dipole forces because hydrogen bonding is a particularly strong dipole force. Any neutral molecule that has a hydrogen covalently bonded to N, O, or F exhibits the relatively strong hydrogen-bonding intermolecular forces. London dispersion forces are accidental or induced dipole forces. Like dipole forces, London dispersion forces are electrostatic in nature. Dipole forces are the electrostatic forces between molecules having a permanent dipole. London dispersion forces are the electrostatic forces between molecules having an accidental or induced dipole. All covalent molecules (polar and nonpolar) have London dispersion forces, but only polar molecules (those with permanent dipoles) exhibit dipole forces. **13.** *Fusion* refers to a solid converting to a liquid and *vaporization* refers to a liquid converting to a gas. Only a fraction of the hydrogen bonds are broken in going from the solid phase to the liquid phase. Most of the hydrogen bonds are still present in the liquid phase and must be broken during the liquid-to-gas transition. The enthalpy of vaporization is much larger than the enthalpy of fusion because more intermolecular forces are broken during the vaporization process. **15.** a. LD (London dispersion); b. dipole, LD; c. hydrogen bonding, LD; d. ionic; e. LD; f. dipole, LD; g. ionic; h. ionic; i. LD mostly; C—F bonds are polar, but polymers like Teflon are so large that LD forces are the predominant intermolecular forces. j. LD; k. dipole, LD; l. hydrogen bonding, LD; m. dipole, LD; n. LD **17.** a. HBr; HBr is polar, whereas Kr and Cl_2 are nonpolar. HBr has dipole forces unlike Kr and Cl_2. So HBr has the stronger intermolecular forces and the higher boiling point. b. NaCl; ionic forces are much stronger than the intermolecular forces for molecular substances. c. I_2; all are nonpolar, so the largest molecule (I_2) will have the strongest LD (London Dispersion) forces and the lowest vapor pressure. d. N_2; nonpolar and smallest, so it has the weakest intermolecular forces. e. CH_4; smallest, nonpolar molecule, so it has the weakest LD (London Dispersion) forces. f. HF; HF can form relatively strong H-bonding interactions unlike the others. g. $CH_3CH_2CH_2OH$; H-bonding unlike the others, so it has strongest intermolecular forces. **19.** a. Neopentane is more compact than *n*-pentane. There is less surface area contact between neopentane molecules. This leads to weaker LD (London Dispersion) forces and a lower boiling point. b. HF is capable of hydrogen bonding; HCl is not. c. LiCl is ionic and HCl is a molecular solid with only dipole forces and LD forces. Ionic forces are much stronger than the forces for molecular solids. d. *n*-Hexane is a larger molecule so it has stronger LD forces. **21.** Ar exists as individual atoms that are held together in the condensed phases by London dispersion forces. The molecule that will have a boiling point closest to Ar will be a nonpolar substance with about the same molar mass as Ar (39.95 g/mol); this same size nonpolar substance will have about the equivalent strength of London dispersion forces. Of the choices, only Cl_2 (70.90 g/mol) and F_2 (38.00 g/mol) are nonpolar. Because F_2 has a molar mass closest to that of Ar, one would expect the boiling point of F_2 to be close to that of Ar. **23.** The electrostatic potential diagrams indicate that ethanol and acetone are polar substances and that propane is a nonpolar substance. Ethanol, with the O—H covalent bond, will exhibit relatively strong hydrogen bonding intermolecular forces in addition to London dispersion forces. The polar acetone will exhibit dipole forces in addition to London dispersion forces, and the nonpolar propane will exhibit only London dispersion forces. Because all three compounds have about the same molar mass, the relative strengths of the London dispersion forces should be about the same. Therefore, ethanol (with the H-bonding capacity) should have the highest boiling point, with polar acetone having the next highest boiling point, and the nonpolar propane, with the weakest intermolecular forces, will have the lowest boiling point. **25.** 46.7 kJ/mol; 90.% **27.** a. Surface tension: the resistance of a liquid to an increase in its surface area. b. Viscosity: the resistance of a liquid to flow. c. Melting point: the temperature (at constant pressure) where a solid converts entirely to a liquid as long as heat is applied. A more detailed definition is the temperature at which the solid and liquid states have the same vapor pressure under conditions where the total pressure is constant. d. Boiling point: the temperature (at constant pressure) where a liquid converts entirely to a gas as long as heat is applied. The detailed definition is the temperature at which the vapor pressure of the liquid is exactly equal to the external pressure. e. Vapor pressure: the pressure of the vapor over a liquid at equilibrium. As the strengths of intermolecular forces increase, surface tension, viscosity, melting point, and boiling point increase, while vapor pressure decreases. **29.** Water is a polar substance and wax is a nonpolar substance. A molecule at the surface of a drop of water is subject to attractions only by molecules below it and to each side. The effect of this uneven pull on the surface molecules tends to draw them into the body of the liquid and causes the droplet to assume the shape that has the minimum surface area, a sphere. **31.** The structure of H_2O_2 is H—O—O—H, which produces greater hydrogen bonding than in water. **33.** a. crystalline solid: regular, repeating structure; amorphous solid: irregular arrangement of atoms or molecules; b. ionic solid: made up of ions held together by ionic bonding; molecular solid: Made up of discrete covalently bonded molecules held together in the solid phase by weaker forces (LD, dipole, or hydrogen bonds); c. molecular solid: discrete, individual molecules; covalent network solid: no discrete molecules; A covalent network solid is one large molecule. The interparticle forces are the covalent bonds between atoms. d. metallic solid: completely delocalized electrons, conductor of electricity (ions in a sea of electrons); covalent network solid: localized electrons; insulator or semiconductor **35.** a. Both forms of carbon are network solids. In diamond, each carbon atom is surrounded by a tetrahedral arrangement of other carbon atoms to form a huge molecule. Each carbon atom is covalently bonded to four other carbon atoms. The structure of graphite is based on layers of carbon atoms arranged in fused six-membered rings. Each carbon atom in a particular layer of graphite is surrounded by three other carbons in a trigonal planar arrangement. This requires sp^2 hybridization. Each carbon has an unhybridized p atomic orbital; all of these p orbitals in each six-membered ring overlap with each other to form a delocalized π electron system. b. Silica is a network solid having an empirical formula of SiO_2. The silicon atoms are singly bonded to four oxygens. Each silicon atom is at the center of a tetrahedral arrangement of oxygen atoms that are shared with other silicon atoms. The structure of silica is based on a network of SiO_4 tetrahedra with shared oxygen atoms rather than discrete SiO_2 molecules. Silicates closely resemble silica. The structure is based on interconnected SiO_4 tetrahedra. However, in contrast to silica, where the O/Si ratio is 2:1, silicates have O/Si ratios greater than 2:1 and contain silicon–oxygen anions. To form a neutral solid silicate, metal cations are needed to balance the charge. In other words, silicates are salts containing metal cations and polyatomic silicon–oxygen anions. When silica is heated above its melting point and cooled rapidly, an amorphous (disordered) solid called glass results. Glass more closely resembles a very viscous solution than it does a crystalline solid. To affect the properties of glass, several different additives are thrown into the mixture. Some of these additives are Na_2CO_3, B_2O_3, and K_2O, with each compound serving a specific purpose relating to the properties of the glass. **37.** A crystalline solid will have the simpler diffraction pattern because a regular, repeating arrangement is necessary to produce planes of atoms that will diffract the X rays in regular patterns. An amorphous solid does not have a regular repeating arrangement and will produce a complicated diffraction pattern. **39.** 0.704 Å **41.** 3.13 Å = 313 pm **43.** 1.54 g/cm^3 **45.** 136 pm **47.** edge length of unit cell = 328 pm; radius = 142 pm **49.** The measured density is consistent with the calculated density, assuming a face-centered cubic structure. **51.** Conductor: The energy difference between the filled and unfilled molecular orbitals is minimal. We call this energy difference the band gap. Because the band gap is minimal, electrons can easily move into the conduction bands (the unfilled molecular orbitals). Insulator: Large band gap; electrons do not move from the filled molecular orbitals to the conduction bands since the energy difference is large. Semiconductor: Small band gap; the energy difference between the filled and unfilled molecular orbitals is smaller than in insulators, so some electrons can jump into the conduction bands. The band gap, however, is not as small as with conductors, so semiconductors have intermediate conductivity. a. As the temperature is increased, more electrons in the filled molecular orbitals have sufficient kinetic energy to jump into the conduction bands (the unfilled molecular orbitals). b. A photon of light is absorbed by an electron, which then has sufficient energy to jump into the conduction bands. c. An impurity either adds electrons at an energy near that of the conduction bands (n-type) or creates holes (unfilled energy levels) at energies in the previously filled molecular orbitals (p-type). Both n-type and p-type semiconductors increase conductivity by creating an easier path for electrons to jump from filled to unfilled energy levels. In conductors, electrical conductivity is inversely proportional to temperature. Increases in temperature increase the motions of the atoms, which gives rise to increased resistance (decreased conductivity). In a semiconductor, electrical conductivity is directly proportional to temperature. An increase in temperature provides more electrons with enough kinetic energy to jump from the filled molecular orbitals to the conduction bands, increasing conductivity. **53.** A rectifier is a device that produces a current that flows in one direction from an alternating current, which flows in both directions. In a p–n junction, a p-type and an n-type semiconductor are connected. The natural flow of electrons in a p–n junction is for the excess electrons in the n-type semiconductor to move to the empty energy levels (holes) of the p-type semiconductor. Only when an external electric potential is connected so that electrons flow in this natural direction will the current flow easily (forward bias). If the external electric potential is connected in reverse of the natural flow of electrons, no current flows through the system (reverse bias). A p–n junction transmits a current only under forward bias, thus converting the alternating current to direct current. **55.** In has fewer valence electrons than Se; thus Se doped with In would be a p-type semiconductor. **57.** 5.0×10^2 nm **59.** Sodium chloride structure:

$$8 \text{ corners} \times \frac{1/8\ \text{Cl}^-}{\text{corner}} + 6 \text{ faces} \times \frac{1/2\ \text{Cl}^-}{\text{face}} = 4\ \text{Cl}^- \text{ ions}$$

$$12 \text{ edges} \times \frac{1/4\ \text{Na}^+}{\text{edge}} + 1\ \text{Na}^+ \text{ at body center} = 4\ \text{Na}^+ \text{ ions;}$$

NaCl is the formula.

Cesium chloride structure: 1 Cs^+ ion at body center;

$$8 \text{ corners} \times \frac{1/8\ \text{Cl}^-}{\text{corner}} = 1\ \text{Cl}^- \text{ ion; CsCl is the formula.}$$

Zinc sulfide structure: There are four Zn^{2+} ions inside the cube.

$$8 \text{ corners} \times \frac{1/8\ \text{S}^{2-}}{\text{corner}} + 6 \text{ faces} \times \frac{1/2\ \text{S}^{2-}}{\text{face}} = 4\ \text{S}^{2-} \text{ ions;}$$

ZnS is the formula.

Titanium oxide structure: $8 \text{ corners} \times \frac{1/8\ \text{Ti}^{4+}}{\text{corner}} + 1\ \text{Ti}^{4+}$ at body center = 2 Ti^{4+} ions

$$4 \text{ faces} \times \frac{1/2\ \text{O}^{2-}}{\text{face}} + 2\ \text{O}^{2-} \text{ inside cube} = 4\ \text{O}^{2-} \text{ ions;}$$

TiO_2 is the formula.

61. The structures of most binary ionic solids can be explained by the closest packing of spheres. Typically, the larger ions, usually the anions, are packed in one of the closest packing arrangements and the smaller cations fit into holes among the closest packed anions. There are different types of holes within the closest packed anions that are determined by the number of spheres that form them. Which of the three types of holes are filled usually depends on the relative size of the cation to the anion. Ionic solids will always try to maximize electrostatic attractions among oppositely charged ions and minimize the repulsions among ions with like charges. The structure of sodium chloride can be described in terms of a cubic closest packed array of Cl^- ions with Na^+ ions in all of the octahedral holes. An octahedral hole is formed between 6 Cl^- anions. The number of octahedral holes is the same as the number of packed ions. So, in the face-centered unit cell of sodium chloride, there are 4 net Cl^- ions and 4 net octahedral holes. Because the stoichiometry dictates a 1:1 ratio between the number of Cl^- anions and Na^+ cations, all of the octahedral holes must be filled with Na^+ ions. In zinc sulfide, the sulfide anions also occupy the lattice points of a cubic closest packing arrangement. But instead of having the cations in octahedral holes, the Zn^{2+} cations occupy tetrahedral holes. A tetrahedral hole is the empty space created when four spheres are packed together. There are twice as many tetrahedral holes as packed anions in the closest packed structure. Therefore, each face-centered unit cell of sulfide anions contains 4 net S^{2-} ions and 8 net tetrahedral holes. For the 1:1 stoichiometry to work out, only half the tetrahedral holes are filled with Zn^{2+} ions. This gives 4 S^{2-} ions and 4 Zn^{2+} ions per unit cell for an empirical formula of ZnS. **63.** CoF_2 **65.** +3 **67.** From density data, 358 pm; from ionic radii data, 350. pm. The two values differ by less than 2.5%. **69.** CsBr: $r_+/r_- = 0.867$; From the radius ratio, Cs^+ should occupy cubic holes. The structure should be the CsCl structure. The actual structure is the CsCl structure. KF: $r_+/r_- = 0.978$; Again, we would predict a structure similar to CsCl, i.e., cations in the middle of a simple cubic array of anions. The actual structure is the NaCl structure. The rule fails for KF. Exceptions are common for crystal structures. **71.** $AlNi_3$ **73.** The formula is Al_2MgO_4. Half of the octahedral holes are filled with Al^{3+} ions, and one-eighth of the tetrahedral holes are filled with Mg^{2+} ions. **75.** a. $YBa_2Cu_3O_9$; b. The structure of this superconductor material follows the alternative perovskite structure described in Exercise 74b. The $YBa_2Cu_3O_9$ structure is three of these cubic perovskite unit cells stacked on top of each other. The oxygen atoms are in the same places, Cu takes the place of Ti, two of the Ca atoms are replaced by two Ba atoms, and one Ca is replaced by Y. c. $YBa_2Cu_3O_7$ **77.** A volatile liquid is one that evaporates relatively easily. Volatile liquids have large vapor pressures because the intermolecular forces that prevent evaporation are relatively weak. **79.** a. As the strength of the intermolecular forces increases, the rate of evaporation decreases. b. As temperature increases, the rate of evaporation increases. c. As surface area increases, the rate of evaporation increases. **81.** $C_2H_5OH(l) \rightarrow C_2H_5OH(g)$ is an endothermic process. Heat is absorbed when liquid ethanol vaporizes; the internal heat from the body provides this heat, which results in the cooling of the body. **83.** If one plots $\ln(P_{vap})$ versus $1/T$ with temperature in kelvin, the slope of the straight line is $-\Delta H_{vap}/R$. Because ΔH_{vap} is always positive, the slope of the straight line will be negative. **85.** 77°C **87.** The plots of $\ln(P_{vap})$ versus $1/T$ are linear with negative slope equal to $-\Delta H_{vap}/R$ (see *Solutions Guide* for plot). For Li, ΔH_{vap} =

158 kJ/mol; for Mg, $\Delta H_{vap} = 139$ kJ/mol. The bonding is stronger in Li since ΔH_{vap} is larger for Li.

89.

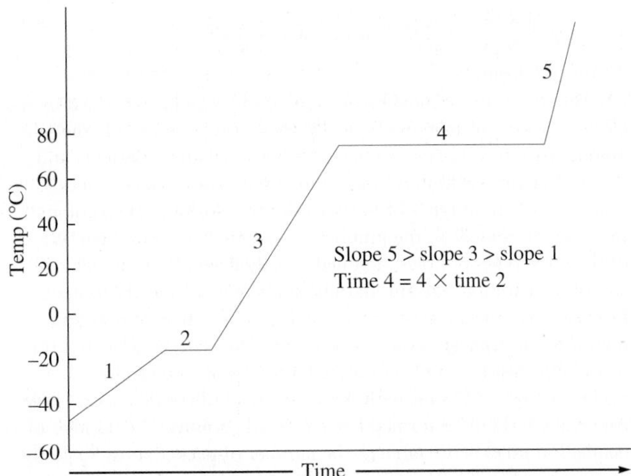

91. 1680 kJ **93.** The reaction doesn't release enough heat to melt all the ice. The temperature will remain at 0°C. **95.** 1490 g **97.** See Figs. 16.55 and 16.58 for the phase diagrams of H_2O and CO_2. Most substances exhibit only three different phases: solid, liquid, and gas. This is true for H_2O and CO_2. Also typical of phase diagrams is the positive slopes for both the liquid/gas equilibrium line and the solid/gas equilibrium line. This is also true for both H_2O and CO_2. The solid/liquid equilibrium line also generally has a positive slope. This is true for CO_2 but not for H_2O. In the H_2O phase diagram, the slope of the solid/liquid line is negative. The determining factor for the slope of the solid/liquid line is the relative densities of the solid and liquid phases. The solid phase is denser than the liquid phase in most substances; for these substances, the slope of the solid/liquid equilibrium line is positive. For water, the liquid phase is denser than the solid phase, which corresponds to a negative-sloping solid/liquid equilibrium line. Another difference between H_2O and CO_2 is the normal melting points and normal boiling points. The term "normal" just dictates a pressure of 1 atm. H_2O has a normal melting point (0°C) and a normal boiling point (100°C), but CO_2 does not. At 1 atm pressure, CO_2 only sublimes (goes from the solid phase directly to the gas phase). There are no temperatures at 1 atm for CO_2 where the solid and liquid phases are in equilibrium or where the liquid and gas phases are in equilibrium. There are other differences, but those discussed above are the major ones. The relationship between melting point and pressure is determined by the slope of the solid/liquid equilibrium line. For most substances (CO_2 included), the positive slope of the solid/liquid line shows a direct relationship between the melting point and pressure. As pressure increases, the melting point increases. Water is just the opposite since the slope of the solid/liquid line in water is negative. Here the melting point of water is inversely related to the pressure. For boiling points, the positive slope of the liquid/gas equilibrium line indicates a direct relationship between the boiling point and pressure. This direct relationship is true for all substances including H_2O and CO_2. The critical temperature for a substance is defined as the temperature above which the vapor cannot be liquefied no matter what pressure is applied. The critical temperature, like the boiling-point temperature, is directly related to the strength of the intermolecular forces. Since H_2O exhibits relatively strong hydrogen-bonding interactions and CO_2 exhibits only London dispersion forces, one would expect a higher critical temperature for H_2O than for CO_2. **99.** The critical temperature is the temperature above which the vapor cannot be liquefied no matter what pressure is applied. Since N_2 has a critical temperature below room temperature (~22°C), it cannot be liquefied at room temperature. NH_3, with a critical temperature above room temperature, can be liquefied at room temperature. **101.** A: solid; B: liquid; C: vapor; D: solid + vapor; E: solid + liquid + vapor (triple point); F: liquid + vapor; G: liquid + vapor (critical point); H: vapor; the first dashed line (at the lower temperature) is the normal melting point, and the second dashed line is the normal boiling point. The solid phase is denser because of the positive slope of the solid/liquid equilibrium line. **103.** a. two; b. Higher-pressure triple point: graphite, diamond, and liquid; lower-pressure triple point: graphite, liquid, and vapor; c. It is converted to diamond (the more dense solid form). d. Diamond is more dense, which is why graphite can be converted to diamond by applying pressure. **105.** Because the density of the liquid phase is greater than the density of the solid phase, the slope of the solid/liquid boundary line is negative (as in H_2O). With a negative slope, the melting points increase with a decrease in pressure, so the normal melting point of X should be greater than 225°C.

107.

As P is lowered, we go from a to b on the phase diagram. The water boils. The evaporation of the water is endothermic and the water is cooled ($b \rightarrow c$), forming some ice. If the pump is left on, the ice will sublime until none is left. This is the basis of freeze drying. **109.** The strength of intermolecular forces determines relative boiling points. The types of intermolecular forces for covalent compounds are London dispersion forces, dipole forces, and hydrogen bonding. Because the three compounds are assumed to have similar molar mass and shape, the strength of the London dispersion forces will be about equal among the three compounds. One of the compounds will be nonpolar, so it has only London dispersion forces. The other two compounds will be polar, so they have additional dipole forces and will boil at a higher temperature than the nonpolar compound. One of the polar compounds probably has an H covalently bonded to either N, O, or F. This gives rise to the strongest type of covalent intermolecular forces, hydrogen bonding. The compound that hydrogen bonds will have the highest boiling point, whereas the polar compound with no hydrogen bonding will boil at a temperature in the middle of the other compounds. **111.** If TiO_2 conducts electricity as a liquid, then it is an ionic solid; if not, then TiO_2 is a network solid. **113.** B_2H_6, molecular solid; SiO_2, network solid; CsI, ionic solid; W, metallic solid **115.** The cation must have a radius that is 0.155 times the radius of the spheres to just fit into the trigonal hole. **117.** 1.71 g/cm^3 **119.** The formula is $TiO_{1.182}$ or $Ti_{0.8462}O$. 63.7% of the titanium is Ti^{2+} and 36.3% is Ti^{3+}. **121.** 194 amu; 1.39×10^{-8} cm; from the periodic table, the best choice for the metal is platinum. **123.** 6.28×10^{-8} cm; 2.00 g/cm^3 **125.** 57.8 torr **127.** 92.47% of the energy goes to increase the internal energy of the water. The remainder of the energy (7.53%) goes to do work against the atmosphere. **129.** The empirical formula is AB_2. Each A atom is in a cubic hole of B atoms so 8 B atoms surround each A atom. This will also be true in the extended lattice. The structure of B atoms in the unit cell is a cubic arrangement with B atoms at every face, edge, corner, and center of the cube. **131.** 46.5 pm **133.** Rh; 12.42 g/cm^3 **135.** distance (liquid)/distance (vapor) = 0.03123 **137.** a. structure (a), $TlBa_2CuO_5$; structure (b), $TlBa_2CaCu_2O_7$; structure (c), $TlBa_2Ca_2Cu_3O_9$; structure (d), $TlBa_2Ca_3Cu_4O_{11}$; b. (a) < (b) < (c) < (d); c. structure (a), only Cu^{3+} is present; structure (b), each formula unit contains 1 Cu^{2+} ion and 1 Cu^{3+} ion; structure (c), each formula unit contains 2 Cu^{2+} ions and 1 Cu^{3+} ion; structure (d), each formula unit contains 3 Cu^{2+} ions and 1 Cu^{3+} ion; d. This superconductor material achieves variable copper oxidation states by varying the numbers of Ca, Cu, and O in each unit cell. The mixtures of copper oxidation states are discussed in part c. The superconductor material in Exercise 75 achieves variable copper oxidation states by omitting oxygen at various sites in the lattice. **139.** Calculated Li^+ radius = 75 pm; calculated Cl^- radius = 182 pm; From Fig. 13.8, the Li^+ radius is 60 pm and the Cl^- radius is 181 pm. The Li^+ ion is much smaller than calculated. This probably means that the ions are not actually in contact with each other. The octahedral holes are larger than the Li^+ ions.

Chapter 17

13. a. $HNO_3(l) \rightarrow H^+(aq) + NO_3^-(aq)$; b. $Na_2SO_4(s) \rightarrow 2Na^+(aq) + SO_4^{2-}(aq)$; c. $Al(NO_3)_3(s) \rightarrow Al^{3+}(aq) + 3NO_3^-(aq)$; d. $SrBr_2(s) \rightarrow Sr^{2+}(aq) + 2Br^-(aq)$; e. $KClO_4(s) \rightarrow K^+(aq) + ClO_4^-(aq)$; f. $NH_4Br(s) \rightarrow NH_4^+(aq) + Br^-(aq)$; g. $NH_4NO_3(s) \rightarrow NH_4^+(aq) + NO_3^-(aq)$; h. $CuSO_4(s) \rightarrow Cu^{2+}(aq) + SO_4^{2-}(aq)$; i. $NaOH(s) \rightarrow Na^+(aq) + OH^-(aq)$ **15.** 10.7 mol/kg; 6.77 mol/L; 0.162 **17.** 28.5%; 0.252; 4.32 mol/kg; 2.69 mol/L **19.** 23.9%; 1.6 mol/kg; 0.028 **21.** 1.06 g/cm^3; 0.0180; 0.981 mol/L; 1.02 mol/kg **23.** "Like dissolves like" refers to the nature of the intermolecular forces. Polar solutes and ionic solutes dissolve in polar solvents because the types of intermolecular forces present in solute and solvent are similar. When they dissolve, the strengths of the intermolecular forces in solution are about the same as in pure solute and pure solvent. The same is true for nonpolar solutes in nonpolar solvents. The strengths of the intermolecular forces (London dispersion forces) are about the same in solution as in pure solute and pure solvent. In all cases of like dissolves like, the magnitude of ΔH_{soln} is either a small positive number (endothermic) or a small negative number (exothermic). For polar solutes in nonpolar solvents and vice versa, ΔH_{soln} is a very large, unfavorable value (very endothermic). Because the energetics are so unfavorable, polar solutes do not dissolve in nonpolar solvents and vice versa. **25.** Although the enthalpy change is endothermic when NaCl dissolves in water, the magnitude of ΔH_{soln} will be a value close to zero ($\Delta H_{soln} \approx 0$). This is typical for soluble ionic compounds. So energy is not the reason why ionic solids like NaCl are so soluble in water. The answer lies in nature's tendency toward the higher probability of the mixed state; solutions form due to an increase in entopy ($\Delta S_{soln} > 0$). In the mixed state, the Na^+ and Cl^- ions have access to a larger volume and therefore have more positions available to them. When a solution forms, positional probability increases, which translates into an increase in entropy. The positive ΔS_{soln} furnishes the driving force for NaCl to dissolve. **27.** $NaI(s) \rightarrow Na^+(aq) + I^-(aq)$, $\Delta H_{soln} = -8$ kJ/mol **29.** Both $Al(OH)_3$ and NaOH are ionic compounds. Since the lattice energy is proportional to the charge of the ions, the lattice energy of aluminum hydroxide is greater than that of sodium hydroxide. The attraction of water molecules for Al^{3+} and OH^- cannot overcome the larger lattice energy and $Al(OH)_3$ is insoluble. For NaOH, the favorable hydration energy is large enough to overcome the smaller lattice energy and NaOH is soluble. **31.** a. NH_3; b. CH_3CN; c. CH_3CO_2H **33.** As the length of the hydrocarbon chain increases, the solubility decreases. The —OH end of the alcohols can hydrogen-bond with water. The hydrocarbon chain, however, is basically nonpolar and interacts poorly with water. As the hydrocarbon chain gets longer, a greater portion of the molecule cannot interact with the water molecules and the solubility decreases; i.e., the effect of the —OH group decreases as the alcohols get larger. **35.** Structure effects refer to solute and solvent having similar polarities in order for solution formation to occur. Hydrophobic solutes are mostly nonpolar substances that are "water-fearing." Hydrophilic solutes are mostly polar or ionic substances that are "water-loving." Pressure has little effect on the solubilities of solids or liquids; it does significantly affect the solubility of a gas. Henry's law states that the amount of a gas dissolved in a solution is directly proportional to the pressure of the gas above the solution ($C = kP$). The equation for Henry's law works best for dilute solutions of gases that do not dissociate in or react with the solvent. $HCl(g)$ does not follow Henry's law because it dissociates into $H^+(aq)$ and $Cl^-(aq)$ in solution (HCl is a strong acid). For O_2 and N_2, Henry's law works well since these gases do not react with the water solvent. An increase in temperature can either increase or decrease the solubility of a solid solute in water. It is true that a solute dissolves more rapidly with an increase in temperature, but the amount of solid solute that dissolves to form a saturated solution can either decrease or increase with temperature. The temperature effect is difficult to predict for solid solutes. However, the temperature effect for gas solutes is easier to predict as the solubility of a gas typically decreases with increasing temperature. **37.** 962 L atm/mol; 1.14×10^{-3} mol/L **39.** As the temperature increases, the gas molecules will have a greater average kinetic energy. A greater fraction of the gas molecules in solution will have kinetic energy greater than the attractive forces between the gas molecules and the solvent molecules. More gas molecules escape to the vapor phase and the solubility of the gas decreases. **41.** 136 torr **43.** 0.9342 **45.** 3.0×10^2 g/mol **47.** a. 290 torr; b. 0.69 **49.** $\chi_{meth} = \chi_{prop} = 0.500$ **51.** $P_{ideal} = 188.6$ torr; $\chi_{acetone} = 0.512$, $\chi_{methanol} = 0.488$; since the actual vapor pressure of the solution is smaller than the ideal vapor pressure, this solution exhibits a negative deviation from Raoult's law. This occurs when solute–solvent attractions are stronger than for the pure substances. **53.** No, the solution is not ideal. For an ideal solution, the strength of the intermolecular forces in the solution is the same as in pure solute and pure solvent. This results in $\Delta H_{soln} = 0$ for an ideal solution. ΔH_{soln} for methanol/water is not zero. Since $\Delta H_{soln} < 0$, this solution exhibits a negative deviation from Raoult's law. **55.** Solutions of A and B have vapor pressures less than ideal (see Fig. 17.11), so this plot shows negative deviations from Raoult's law. Negative deviations occur when the intermolecular forces are stronger in solution than in pure solvent and solute. This results in an exothermic enthalpy of solution. The only statement that is false is e. A substance boils when the vapor pressure equals the external pressure. Since $\chi_B = 0.6$ has a lower vapor pressure at the temperature of the plot than either pure A or pure B, one would expect this solution to require the highest temperature for the vapor pressure to reach the external pressure. Therefore, the solution with $\chi_B = 0.6$ will have a higher boiling point than either pure A or pure B. (Note that since $P_B^\circ > P_A^\circ$, then B is more volatile than A.) **57.** Dissolve 210 g of sucrose in some water and dilute to 1.0 L in a volumetric flask. To get 0.62 ±0.01 mol/L, we need 212 ±3 g sucrose. **59.** 101.5°C **61.** −29.9°C; 108.2°C **63.** 100.08°C **65.** 498 g/mol **67.** 776 g/mol **69.** 2.0×10^{-5} °C; 0.20 torr **71.** ~30 m **73.** With addition of salt or sugar, the osmotic pressure inside the fruit cells (and bacteria) is less than outside the cell. Water will leave the cells, which will dehydrate any bacteria present, causing them to die. **75.** 0.0880; 59.2 torr **77.** a. 0.010 *m* Na_3PO_4 and 0.020 *m* KCl; b. 0.020 *m* HF; c. 0.020 *m* $CaBr_2$ **79.** 1.67 **81.** In $CaCl_2$, 20.% ion association occurs, and in CsCl, 10.% ion association occurs. The ion association is greater in the $CaCl_2$ solution. **83.** A pressure greater than 4.8 atm should be applied. **85.** a. −0.25°C; 100.069°C; b. −0.32°C; 100.087°C **87.** 97.8 g/mol **89.** Both solutions and colloids have suspended particles in some medium. The major difference between the two is the size of the particles. A colloid is a suspension of relatively large particles as compared with a solution. Because of this, colloids will scatter light, whereas solutions will not. The scattering of light by a colloidal suspension is called the Tyndall effect. **91.** Coagulation is the destruction of a colloid by the aggregation of many suspended particles to form a large particle that settles out of solution. **93.** 7.9 *M* **95.** a. 100.77°C; b. 23.1 mm Hg; c. We assumed ideal behavior in solution formation and assumed $i = 1$ (no ions form). **97.** Benzoic acid is capable of hydrogen bonding, but a significant part of benzoic acid is the nonpolar benzene ring, which is composed of only carbon and hydrogen. In benzene, a hydrogen-bonded dimer forms:

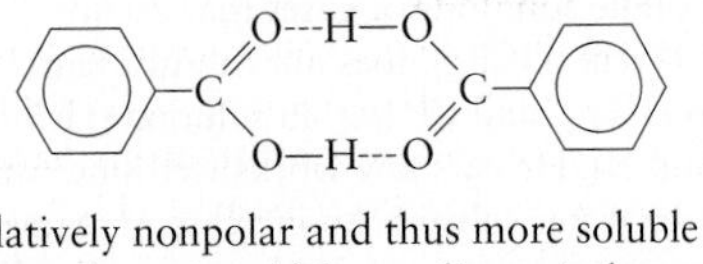

```
      O--H—O
  ⌬—C        C—⌬
      O—H--O
```

The dimer is relatively nonpolar and thus more soluble in benzene than in water. Since benzoic acid forms dimers in benzene, the effective solute particle concentration will be less than 1.0 molal. Therefore, the freezing-point depression would be less than 5.12°C ($\Delta T_f = K_f m$). **99.** a. 26.6 kJ/mol; b. −657 kJ/mol **101.** 0.050 **103.** C_7H_4O; $C_{14}H_8O_2$ **105.** a. 303 ± 9 g/mol; b. No, codeine could not be eliminated since its molar mass is in the possible range including the uncertainty; c. We would like the uncertainty to be ±1 g/mol. We need the freezing-point depression to be about 10 times what it was in this problem. Two possibilities are (1) make the solution ten times more concentrated (may be solubility problem) or (2) use a solvent with a larger K_f value, e.g., camphor. **107.** 5.08×10^{-17} **109.** 1.0×10^{-3} **111.** 72% $MgCl_2$ **113.** $\chi_{CCl_4} = 0.554$, $\chi_{C_6H_6} = 0.446$ **115.** 0.286 **117.** a. If we assume $MgCO_3$ does not dissociate, 46 L of water must be processed. If $MgCO_3$ does dissociate, then 47 L of water must be processed. b. No; A reverse osmosis system that applies 8.0 atm can only purify water with a solute concentration less than 0.32 mol/L. Salt water has a solute concentration of 2(0.60 *M*) = 1.2 mol/L ions. The solute concentration of salt water is much too high for this reverse osmosis unit to work. **119.** a. 6.11 atm; b. 8.3×10^{-3} **121.** a. 0.25 or 25%; b. −0.562°C

Chapter 18

1. The gravity of the earth is not strong enough to keep H_2 in the atmosphere. **3.** a. $\Delta H° = -92$ kJ; $\Delta S° = -199$ J/K; $\Delta G° = -33$ kJ; b. yes; c. $T < 460$ K **5.** The first illustration is an example of a covalent hydride like H_2O. Covalent hydrides are just binary covalent compounds formed between hydrogen and some other nonmetal and exist as individual molecules. The middle illustration represents interstitial (or metallic) hydrides. In interstitial hydrides, hydrogen atoms occupy the holes of a transition metal crystal. These hydrides are more like solid solutions than true compounds. The third illustration represents ionic (or salt-like) hydrides like LiH. Ionic hydrides form when hydrogen reacts with a metal from Group 1A or 2A. The metals lose electrons to form cations and the hydrogen atoms gain electrons to form the hydride anions (H^-). These are just ionic compounds formed between a metal and hydrogen. **7.** Alkali metals have an ns^1 valence shell electron configuration. Alkali metals lose this valence electron with relative ease to form M^+ cations when in ionic compounds. They all are easily oxidized. Therefore, in order to prepare the pure metals, alkali metals must be produced in the absence of materials (H_2O, O_2) that are capable of oxidizing them. The method of preparation is electrochemical processes, specifically, electrolysis of molten chloride salts and reduction of alkali salts with Mg and H_2. In all production methods, H_2O and O_2 must be absent. **9.** Hydrogen forms many compounds in which the oxidation state is +1, as do the Group 1A elements. Consider, for example, H_2SO_4 and HCl compared with Na_2SO_4 and NaCl. On the other hand, hydrogen forms diatomic H_2 molecules and is a nonmetal, while the Group 1A elements are metals. Hydrogen also forms compounds with a −1 oxidation state, which is not characteristic of Group 1A metals, e.g., NaH. **11.** $4Li(s) + O_2(g) \rightarrow 2Li_2O(s)$; $16Li(s) + S_8(s) \rightarrow 8Li_2S(s)$; $2Li(s) + Cl_2(g) \rightarrow 2LiCl(s)$; $12Li(s) + P_4(s) \rightarrow 4Li_3P(s)$; $2Li(s) + H_2(g) \rightarrow 2LiH(s)$; $2Li(s) + 2H_2O(l) \rightarrow 2LiOH(aq) + H_2(g)$; $2Li(s) + 2HCl(aq) \rightarrow 2LiCl(aq) + H_2(g)$ **13.** 44.1 hours **15.** The alkaline earth ions that give water the hard designation are Ca^{2+} and Mg^{2+}. These ions interfere with the action of detergents and form unwanted precipitates with soaps. Large-scale water softeners remove Ca^{2+} by precipitating out the calcium ions as $CaCO_3$. In homes, Ca^{2+} and Mg^{2+} (plus other cations) are removed by ion exchange. See Fig. 18.6 for a schematic of a typical cation exchange resin. **17.** $CaCO_3(s) + H_2SO_4(aq) \rightarrow CaSO_4(aq) + H_2O(l) + CO_2(g)$ **19.** $CaF_2(s)$ will precipitate when $[Ca^{2+}]_0 > 2 \times 10^{-2}$ *M*. Therefore, hard water should have a calcium ion concentration of less than 2×10^{-2} *M* to avoid precipitate formation. **21.** The valence electron configuration of Group 3A elements is ns^2np^1. The lightest Group 3A element, boron, is a nonmetal because most of its compounds are covalent. Aluminum, although commonly thought of as a metal, does have some nonmetallic properties because its bonds to other nonmetals have significant covalent character. The other Group 3A elements have typical metal characteristics; their compounds formed with nonmetals are ionic. From this discussion, metallic character increases as the Group 3A elements get larger. As mentioned previously, boron is a nonmetal in both properties and compounds formed. However, aluminum has physical properties of metals such as high thermal and electrical conductivities and a lustrous appearance. The compounds of aluminum with other nonmetals, however, do have some nonmetallic properties because the bonds have significant covalent character. **23.** $B_2H_6(g) + 3O_2(g) \rightarrow 2B(OH)_3(s)$ **25.** $2Ga(s) + 3F_2(g) \rightarrow 2GaF_3(s)$; $4Ga(s) + 3O_2(g) \rightarrow 2Ga_2O_3(s)$; $16Ga(s) + 3S_8(s) \rightarrow 8Ga_2S_3(s)$; $2Ga(s) + 6HCl(aq) \rightarrow 2GaCl_3(aq) + 3H_2(g)$ **27.** The valence electron configuration of Group 4A elements is ns^2np^2. The two most important elements on earth are both Group 4A elements. They are carbon, found in all biologically important molecules, and silicon, found in most of the compounds that make up the earth's crust. They are important because they are so prevalent in compounds necessary for life and the geologic world. As with Group 3A, Group 4A shows an increase in metallic character as the elements get heavier. Carbon is a typical nonmetal, silicon and germanium have properties of both metals and nonmetals, so they are classified as semimetals, whereas tin and lead have typical metallic characteristics. **29.** O=C=O

The darker green orbitals about carbon are *sp* hybrid orbitals. The lighter green orbitals about each oxygen are sp^2 hybrid orbitals, and the gold orbitals about all of the atoms are unhybridized *p* atomic orbitals. In each double bond in CO_2, one sigma and one π bond exists. The two carbon–oxygen sigma bonds are formed from overlap of *sp* hybrid orbitals from carbon with a sp^2 hybrid orbital from each oxygen. The two carbon–oxygen π bonds are formed from side-to-side overlap of the unhybridized *p* atomic orbitals from carbon with an unhybridized *p* atomic orbital from each oxygen. These two π bonds are oriented perpendicular to each other as illustrated in the figure. **31.** White tin is stable at normal temperatures. Gray tin is stable at temperatures below 13.2°C. Thus for the phase change Sn(gray) → Sn(white), ΔG is negative at $T > 13.2$°C and ΔG is positive at $T < 13.2$°C. This is possible only if ΔH is positive and ΔS is positive. Thus gray tin has the more ordered structure (has the smaller positional probability). **33.** $Sn(s) + 2F_2(g) \rightarrow SnF_4(s)$, tin(IV) fluoride; $Sn(s) + F_2(g) \rightarrow SnF_2(s)$, tin(II) fluoride. **35.** In each formula unit of Pb_3O_4, two atoms are Pb(II) and one atom is Pb(IV) (2:1 mole ratio). **37.** Both NO_4^{3-} and PO_4^{3-} have 32 valence electrons, so both have similar Lewis structures. From the Lewis structure for NO_4^{3-}, the central N atom has a tetrahedral arrangement of electron pairs. N is small. There is probably not enough room for all 4 oxygen atoms around N. P is larger; thus PO_4^{3-} is stable. PO_3^- and NO_3^- both have 24 valence electrons, so both have similar Lewis structures. From the Lewis structure for PO_3^-, PO_3^- has a trigonal arrangement of electron pairs around the central P atom (two single bonds and one double bond). P=O bonds are not particularly stable, whereas N=O bonds are stable. Thus NO_3^- is stable. **39.** N: $1s^22s^22p^3$; the extremes of the oxidation states for N can be rationalized by examining the electron configuration of N. Nitrogen is three electrons short of the stable Ne electron configuration of $1s^22s^22p^6$. Having an oxidation state of −3 makes sense. The +5 oxidation state corresponds to N "losing" its 5 valence electrons. In compounds with oxygen, the N—O bonds are polar covalent, with N having the partial positive end of the bond dipole. In the world of oxidation states, electrons in polar covalent bonds are assigned to the more electronegative atom; this is oxygen in N—O bonds. N can form enough bonds to oxygen to give it a +5 oxidation state. This loosely corresponds to losing all of the valence electrons. **41.** This is due to nitrogen's ability to form strong π bonds, whereas heavier Group 5A elements do not form strong π bonds. Therefore, P_2, As_2, and Sb_2 do not form since two π bonds are required to form these diatomic substances.

43. a.

```
      N                 O         O
     / \\                \\      /
    O    O                N — N
                         /      \
                        O         O
```

plus other resonance structures; plus other resonance structures;

b.

```
     :F:           N              :F:      H
      |          / | \              \     /
      B         H  H  H        :F — B — N — H
     / \                            /     \
   :F:  :F:                       :F:      H
```

In reaction a, NO_2 has an odd number of electrons so it is impossible to satisfy the octet rule. By dimerizing to form N_2O_4, the odd electrons on two NO_2 molecules can pair up, giving a species whose Lewis structure can satisfy the octet rule. In general odd-electron species are very reactive. In reaction b, BF_3 can be considered electron-deficient. Boron has only six electrons around it. By forming BF_3NH_3, the boron atom satisfies the octet rule by accepting a lone pair of electrons from NH_3 to form a fourth bond. **45.** exothermic **47.** $N_2H_4(l) + 2F_2(g) \longrightarrow 4HF(g) + N_2(g); \Delta H° = -1169$ kJ

49. $2Bi_2S_3(s) + 9O_2(g) \longrightarrow 2Bi_2O_3(s) + 6SO_2(g);$
$2Bi_2O_3(s) + 3C(s) \longrightarrow 4Bi(s) + 3CO_2(g)$

$2Sb_2S_3(s) + 9O_2(g) \longrightarrow 2Sb_2O_3(s) + 6SO_2(g);$
$2Sb_2O_3(s) + 3C(s) \longrightarrow 4Sb(s) + 3CO_2(g)$

51. TSP = Na_3PO_4; PO_4^{3-} is the conjugate base of the weak acid HPO_4^{2-} ($K_a = 4.8 \times 10^{-13}$). All conjugate bases of weak acids are effective bases ($K_b = K_w/K_a = 1.0 \times 10^{-14}/4.8 \times 10^{-13} = 2.1 \times 10^{-2}$). The weak-base reaction of PO_4^{3-} with H_2O is $PO_4^{3-}(aq) + H_2O(l) \rightleftharpoons HPO_4^{2-}(aq) + OH^-(aq)$ $K_b = 2.1 \times 10^{-2}$.

53.
$4As(s) + 3O_2(g) \longrightarrow As_4O_6(s); 4As(s) + 5O_2(g) \longrightarrow As_4O_{10}(s);$
$As_4O_6(s) + 6H_2O(l) \longrightarrow 4H_3AsO_3(aq); As_4O_{10}(s) + 6H_2O(l) \longrightarrow 4H_3AsO_4(aq)$

55. $\frac{1}{2}N_2(g) + \frac{1}{2}O_2(g) \rightarrow NO(g), \Delta G° = \Delta G°_{f,NO} = 87$ kJ/mol; By definition, $\Delta G°_f$ for a compound equals the free energy change that would accompany the formation of 1 mol of that compound from its elements in their standard states. NO (and some other oxides of nitrogen) have weaker bonds as compared to the triple bond of N_2 and the double bond of O_2. Because of this, NO (and some other oxides of nitrogen) have higher (positive) free energies of formation as compared to the relatively stable N_2 and O_2 molecules. **57.** The pollution provides sources of nitrogen and phosphorus nutrients so the algae can grow. The algae consume oxygen, which decrease the dissolved oxygen levels below that required for other aquatic life to survive, and the fish die. **59.** −4594 kJ **61.** Hydrazine also can hydrogen bond because it has covalent N—H bonds as well as a lone pair of electrons on each N. The high boiling point for hydrazine's relatively small size supports this. **63.** The two allotropic forms of oxygen are O_2 and O_3.

```
   O2                  O3
  :O=O:        O                  O
             // \       <-->     / \\
           O     O:           :O     O
```

The MO electron configuration of O_2 has two unpaired electrons in the degenerate pi antibonding ($\pi_{2p}{}^*$) orbitals. A substance with unpaired electrons is paramagnetic (see Fig. 14.41). Ozone has a V-shape molecular structure with a bond angle of 117°, slightly less than the predicted 120° trigonal planar bond angle. **65.** In the upper atmosphere, O_3 acts as a filter for UV radiation: $O_3 \xrightarrow{h\nu} O_2 + O$; O_3 is also a powerful oxidizing agent. It irritates the lungs and eyes, and at high concentration it is toxic. The smell of a "spring thunderstorm" is O_3 formed during lightning discharges. Toxic materials don't necessarily smell bad. For example, HCN smells like almonds. **67.** +6 oxidation state: SO_4^{2-}, SO_3, SF_6; +4 oxidation state: SO_3^{2-}, SO_2, SF_4; +2 oxidation state: SCl_2; 0 oxidation state: S_8 and all other elemental forms of sulfur; −2 oxidation state: H_2S, Na_2S **69.** $H_2SeO_4(aq) + 3SO_2(g) \rightarrow Se(s) + 3SO_3(g) + H_2O(l)$

71. $:\ddot{F}—\ddot{O}—\ddot{O}—\ddot{F}:$

	F	O	O	F
Formal Charge	0	0	0	0
Oxidation State	−1	+1	+1	−1

Oxidation states are more useful. We are forced to assign +1 as the oxidation state for oxygen. Oxygen is very electronegative and +1 is not a stable oxidation state for this element. **73.** Fluorine is the most reactive of the halogens because it is the most electronegative atom and the bond in the F_2 molecule is very weak. **75.** $ClO^-(aq) + 2NH_3(aq) \rightarrow Cl^-(aq) + N_2H_4(aq) + H_2O(l)$, $\mathscr{E}°_{cell} = 1.00$ V; Since $\mathscr{E}°_{cell}$ is positive for this reaction, then at standard conditions ClO^- can spontaneously oxidize NH_3 to the somewhat toxic N_2H_4. **77.** a. $AgCl(s) \xrightarrow{h\nu} Ag(s) + Cl$; the reactive chlorine atom is trapped in the crystal. When light is removed, Cl reacts with silver atoms to re-form AgCl; that is, the reverse reaction occurs. In pure AgCl, the Cl atoms escape, making the reverse reaction impossible. b. Over time, chlorine is lost, and the dark silver metal is permanent. **79.** Helium is unreactive and doesn't combine with any other elements. It is a very light gas and would easily escape the earth's gravitational pull as the planet was formed **81.** 2×10^4 g; 2×10^{26} atoms in room; 5×10^{20} atoms in one breath; Since Ar and Rn are both noble gases, both species will be relatively unreactive. However, all nuclei of Rn are radioactive, unlike most nuclei of Ar. The radioactive decay products of Rn can cause biological damage when inhaled. **83.** One would expect RnF_2 and RnF_4 to form in fashion similar to XeF_2 and XeF_4. The chemistry of radon is difficult to study because all radon isotopes are radioactive. The hazards of dealing with radioactive materials are immense. **85.** Solids have stronger intermolecular forces than liquids. In order to maximize the hydrogen bonding in the solid phase, ice is forced into an open structure. This open structure is why $H_2O(s)$ is less dense than $H_2O(l)$. **87.** Strontium and calcium are both alkaline earth metals, so both have similar chemical properties. Since milk is a good source of calcium, strontium could replace some calcium in milk without much difficulty. **89.** In solution Tl^{3+} can oxidize I^- to I_3^-. Thus we expect TlI_3 to be thallium(I) triiodide. **91.** The inert pair effect refers to the difficulty of removing the pair of *s* electrons from some of the elements in the fifth and sixth periods of the periodic table. As a result, multiple oxidation states are exhibited for the heavier elements of Groups 3A and 4A. In^+, In^{3+}, Tl^+, and Tl^{3+} oxidation states are all important to the chemistry of In and Tl. **93.** 5.4×10^4 kJ (Hall–Heroult process) versus 4.0×10^2 kJ; it is feasible to recycle Al by melting the metal because, in theory, it takes less than 1% of the energy required to produce the same amount of Al by the Hall–Heroult process. **95.** The bonds in SnX_4 compounds have a large covalent character. SnX_4 acts as discrete molecules held together by weak London dispersion forces. SnX_2 compounds are ionic and are held in the solid state by strong ionic forces. Because the intermolecular forces are weaker for SnX_4 compounds, they are more volatile (have a lower boiling point). **97.** Carbon cannot form the fifth bond necessary for the transition state since carbon doesn't have low-energy *d* orbitals available to expand the octet. **99.** a. The NNO structure is correct. From the Lewis structures we would predict both NNO and NON to be linear. However, we would predict NNO to be polar and NON to be nonpolar. Since experiments show N_2O to be polar, then NNO is the correct structure.

b. $:\ddot{N}=N=\ddot{O}: \longleftrightarrow :N\equiv N-\ddot{\underset{..}{O}}: \longleftrightarrow :\ddot{\underset{..}{N}}-N\equiv O:$

−1 +1 0   0 +1 −1   −2 +1 +1

The formal charges for the atoms in the various resonance structures appear below each atom. The central N is *sp* hybridized in all of the resonance structures. We can probably ignore the third resonance structure on the basis of the relatively large formal charges as compared to the first two resonance structures. c. The *sp* hybrid orbitals on the center N overlap with atomic orbitals (or hybrid orbitals) on the other two atoms to form the two σ bonds. The remaining two unhybridized *p* orbitals on the center N overlap with two *p* orbitals on the peripheral N to form the two π bonds. **101.** 7.17 **103.** a. 7.1 g InP; b. 979 nm; this wavelength is not visible to humans; it is in the infrared region of the electromagnetic radiation spectrum. c. n-type doping **105.** 4.8×10^{-11} *M* **107.** $\Delta H° =$ 286 kJ; $\Delta G° = 326$ kJ; $K_p = 7.22 \times 10^{-58}$; $P_{O_3} = 3.3 \times 10^{-41}$ atm; The volume occupied by one molecule of ozone at 230. K is 9.5×10^{17} L. Equilibrium is probably not maintained under these conditions. When only two ozone molecules are in a volume of 9.5×10^{17} L, the reaction is not at equilibrium. Under these conditions, $Q > K$ and the reaction shifts to the left. But with only two ozone molecules in this huge volume, it is extremely unlikely that they will collide with each other. Under these conditions, the concentration of ozone is not large enough to maintain equilibrium. **109.** a. $O_3(g) + O(g) \rightarrow 2\ O_2(g)$; b. NO; c. NO_2; d. Rate = $k[O_3][NO]$;

e.

$O_3(g) + Cl(g) \rightarrow O_2(g) + ClO(g)$	slow step 1
$ClO(g) + O(g) \rightarrow O_2(g) + Cl(g)$	fast step 2
$O_3(g) + O(g) \rightarrow 2\ O_2(g)$	overall equation

111. 1.1×10^{-37} *M* **113.** a. 8.8 kJ; b. Other ions will have to be transported to maintain electroneutrality. Either anions must be transported into the cells or cations (Na^+) in the cell must be transported to the blood. The latter is what happens: $[Na^+]$ in blood is greater than $[Na^+]$ in cells as a result of this pumping. c. 0.28 mol.

Chapter 19

7. a. Cr: $[Ar]4s^13d^5$, Cr^{2+}: $[Ar]3d^4$, Cr^{3+}: $[Ar]3d^3$; b. Cu: $[Ar]4s^13d^{10}$, Cu^+: $[Ar]3d^{10}$, Cu^{2+}: $[Ar]3d^9$; c. V: $[Ar]4s^23d^3$, V^{2+}: $[Ar]3d^3$, V^{3+}: $[Ar]3d^2$ **9.** The lanthanide elements are located just before the 5*d* transition metals. The lanthanide contraction is the steady decrease in the atomic radii of the lanthanide elements when going from left to right across the periodic table. As a result of the lanthanide contraction, the sizes of the 4*d* and 5*d* elements are very similar (see the following exercise). This leads to a greater similarity in the chemistry of the 4*d* and 5*d* elements in a given vertical group. **11.** a. molybdenum(IV) sulfide; molybdenum(VI) oxide; b. MoS_2, +4; MoO_3, +6; $(NH_4)_2Mo_2O_7$, +6; $(NH_4)_6Mo_7O_{24} \cdot 4H_2O$, +6 **13.** TiF_4: Ionic compound containing Ti^{4+} ions and F^- ions. $TiCl_4$, $TiBr_4$, and TiI_4: Covalent compounds containing discrete, tetrahedral TiX_4 molecules. As these molecules get larger, the bp and mp increase because the London dispersion forces increase. TiF_4 has the highest bp since the interparticle forces are stronger in ionic compounds as compared with those in covalent compounds. **15.** $H^+ + OH^- \rightarrow H_2O$; Sodium hydroxide (NaOH) will react with the H^+ on the product side of the reaction. This effectively removes H^+ from the equilibrium, which will shift the reaction to the right to produce more H^+ and CrO_4^{2-}. Since more CrO_4^{2-} is produced, the solution turns yellow. **17.** Because transition metals form bonds to species that donate lone pairs of electrons, transition metals are Lewis acids (electron-pair acceptors). The Lewis bases in coordination compounds are the ligands, all of which have an unshared pair of electrons to donate. The coordinate covalent bond between the ligand and the transition metal just indicates that both electrons in the bond originally came from one of the atoms in the bond. Here, the electrons in the bond come from the ligand. **19.** Pt^{2+}: $[Xe]4f^{14}5d^8$.

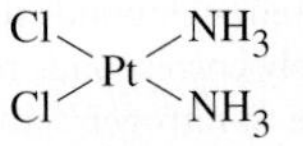

Cl, Cl, NH_3, NH_3 bonded to Pt (Cl and Cl on left; NH_3 and NH_3 on right)

21. a. hexacyanomanganate(II) ion; b. *cis*-tetraamminedichlorocobalt(III) ion; c. pentaamminechlorocobalt(II) ion **23.** a. pentaamminechlororuthenium(III) ion; b. hexacyanoferrate(II) ion; c. tris(ethylenediamine)manganese(II) ion; d. pentaamminenitrocobalt(III) ion **25.** a. $K_2[CoCl_4]$; b. $[Pt(H_2O)(CO)_3]Br_2$; c. $Na_3[Fe(CN)_2(C_2O_4)_2]$; d. $[Cr(NH_3)_3Cl(H_2NCH_2CH_2NH_2)]I_2$

27. $[Co(NH_3)_4(SO_4)]^+$ Cl^- (structure: Co with NH_3 above and below, two H_3N on left, bidentate sulfate O–S–O on right)

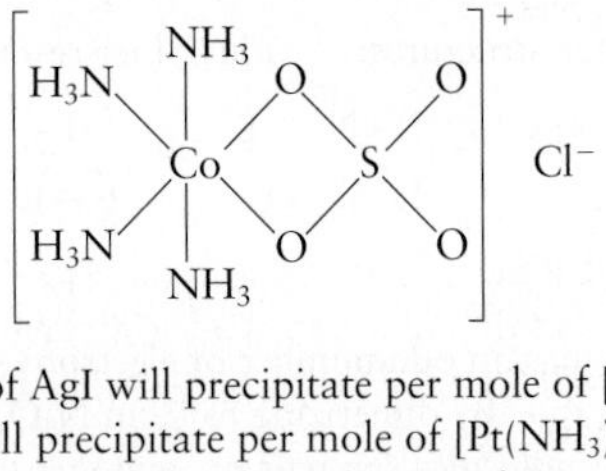

29. Three moles of AgI will precipitate per mole of $[Co(NH_3)_6]I_3$, 2 moles of AgI will precipitate per mole of $[Pt(NH_3)_4I_2]I_2$, 0 moles of AgI will precipitate per mole of $Na_2[PtI_6]$, and 1 mole of AgI will precipitate per mole of $[Cr(NH_3)_4I_2]I$. **31.** a. Isomers: Species with the same formulas but different properties. See text for examples of the following types of isomers. b. Structural isomers: Isomers that have one or more bonds that are different. c. Stereoisomers: Isomers that contain the same bonds but differ in how the atoms are arranged in space. d. Coordination isomers: Structural isomers that differ in the atoms that make up the complex ion. e. Linkage isomers: Structural isomers that differ in how one or more ligands are attached to the transition metal. f. Geometric isomers: (*cis–trans* isomerism); stereoisomers that differ in the positions of atoms with respect to a rigid ring, bond, or each other. g. Optical isomers: Stereoisomers that are nonsuperimposable mirror images of each other; that is, they are different in the same way that our left and right hands are different.

33. a. $[Co(H_2O)_2(C_2O_4)_2]^-$: *cis* (two H_2O adjacent) and *trans* (two OH_2 opposite)

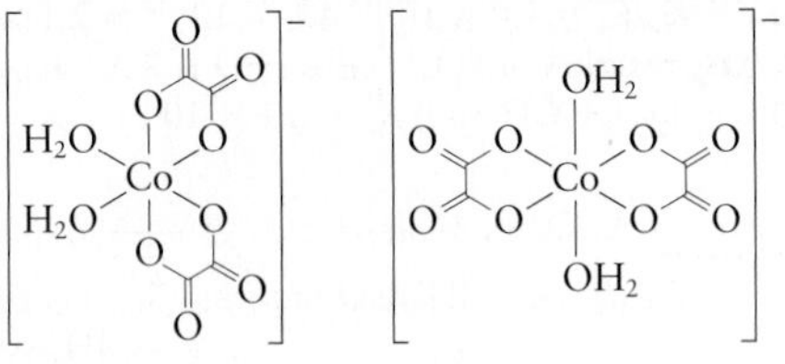

cis   *trans*

b. $[Pt(NH_3)_4I_2]^{2+}$: *cis* (ligands: I, I, H_3N, H_3N, NH_3, NH_3; two I adjacent) and *trans* (two I opposite)

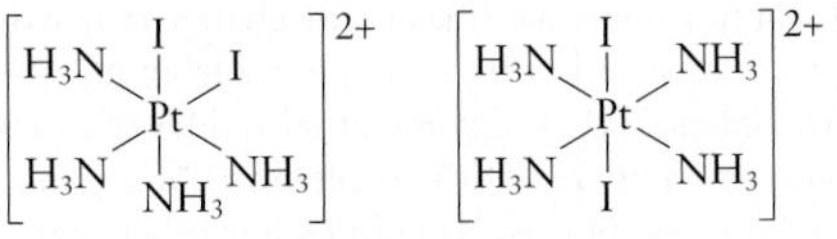

cis   *trans*

c. $Ir(NH_3)_3Cl_3$: *cis* and *trans* (ligands: Cl, Cl, Cl, H_3N, H_3N, NH_3)

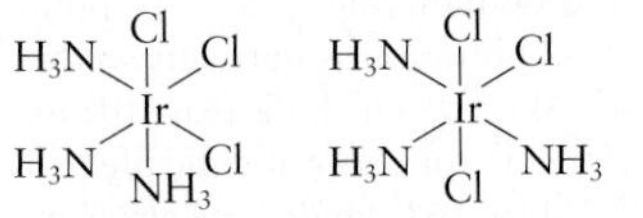

cis   *trans*

d. $[Cr(en)(NH_3)_2I_2]^+$: three isomers (ligands: N–N, H_3N, NH_3, I, I)

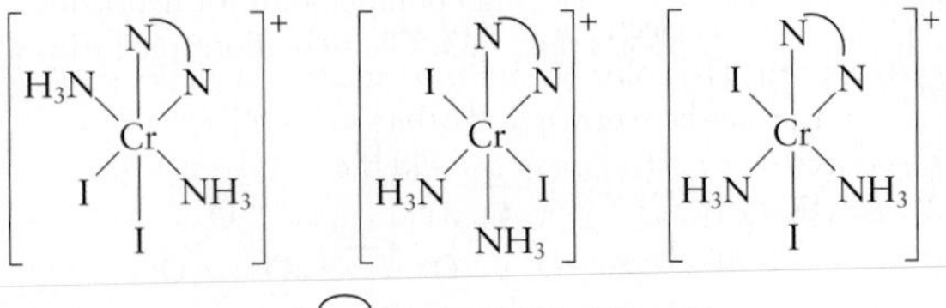

en = N⌒N = $NH_2CH_2CH_2NH_2$

35. No; both the *trans* and the *cis* forms of $Co(NH_3)_4Cl_2^+$ have mirror images that are superimposable. For the *cis* form, the mirror image needs only a 90° rotation to produce the original structure. Hence neither the *trans* nor *cis* form is optically active.

37.

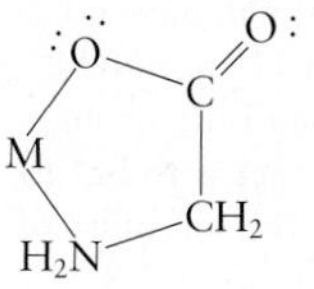

M = transition metal ion

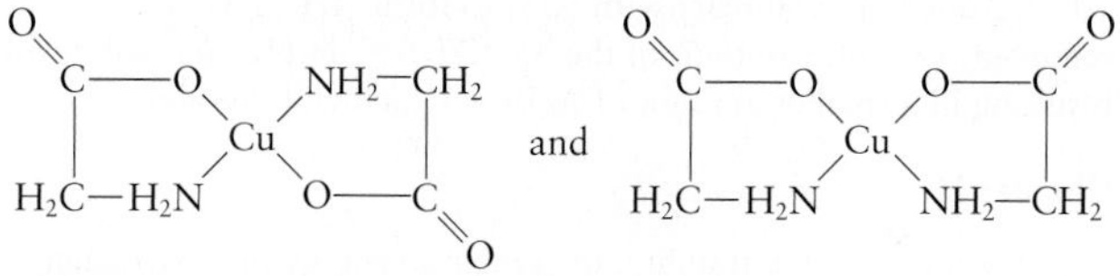

and

39. Linkage isomers differ in the way that the ligand bonds to the metal. SCN^- can bond through the sulfur or through the nitrogen atom. NO_2^- can bond through the nitrogen or through the oxygen atom. OCN^- can bond through the oxygen or through the nitrogen atom. N_3^-, $NH_2CH_2CH_2NH_2$, and I^- are not capable of linkage isomerism. **41.** $Cr(acac)_3$ and *cis*-$Cr(acac)_2(H_2O)_2$ are optically active. **43.** There are four geometrical isomers (labeled i-iv). Isomers iii and iv are optically active, and the nonsuperimposable mirror images are shown.

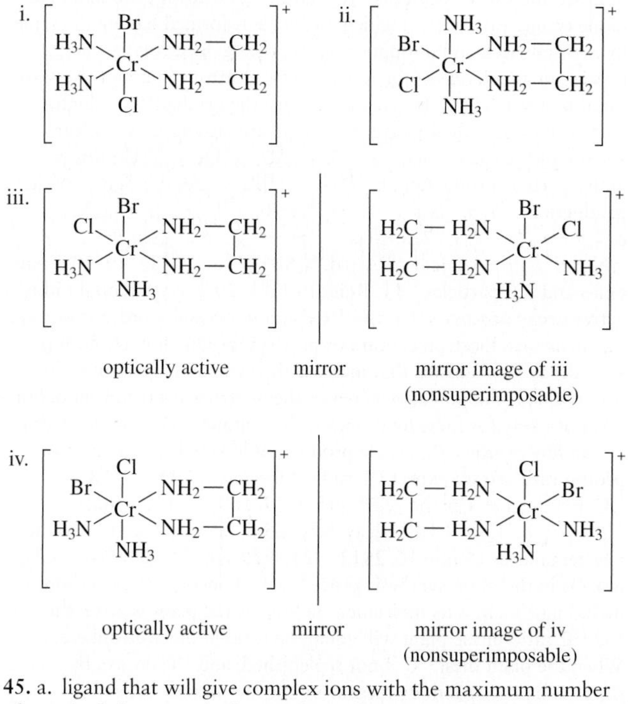

i. ii.

iii. optically active mirror mirror image of iii (nonsuperimposable)

iv. optically active mirror mirror image of iv (nonsuperimposable)

45. a. ligand that will give complex ions with the maximum number of unpaired electrons; b. ligand that will give complex ions with the minimum number of unpaired electrons; c. complex with a minimum number of unpaired electrons (low spin = strong field); d. complex with a maximum number of unpaired electrons (high spin = weak field) **47.** Sc^{3+} has no electrons in *d* orbitals. Ti^{3+} and V^{3+} have *d* electrons present. The color of transition metal complexes results from electron transfer between split *d* orbitals. If no *d* electrons are present, no electron transfer can occur and the compounds are not colored. **49.** $[Mn(H_2O)_6]^{2+}$ would have the weak-field diagram in a and $[Fe(CN)_6]^{3-}$ would have the strong-field diagram in b.

51. a. Fe^{2+} High spin Low spin

b. Fe^{3+} High spin c. Ni^{2+}

d. Zn^{2+}

e. Co^{2+} High spin Low spin

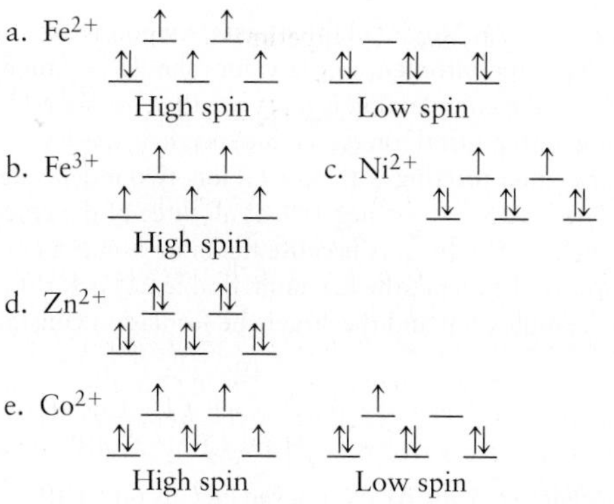

53. a. 0; b. 2; c. 2 **55.** Replacement of water ligands by ammonia ligands resulted in shorter wavelengths of light being absorbed. Energy and wavelength are inversely related, so the presence of the NH_3 ligands resulted in a larger *d*-orbital splitting (larger Δ). Therefore, NH_3 is a stronger field ligand than H_2O. **57.** The violet complex ion absorbs yellow-green light ($\lambda \approx 570$ nm) and is $Cr(H_2O)_6^{3+}$. The yellow complex ion absorbs blue light ($\lambda \approx 450$ nm) and is $Cr(NH_3)_6^{3+}$. The green complex ion absorbs red light ($\lambda \approx 650$ nm) and is $Cr(H_2O)_4Cl_2^+$.

59. a. optical isomerism; b.

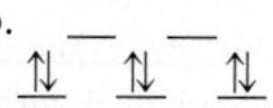

61. The crystal field diagrams are different because the geometries of where the ligands point is different. The tetrahedrally oriented ligands point differently in relationship to the *d*-orbitals than do the octahedrally oriented ligands. Plus, there are more ligands in an octahedral complex. The *d*-orbital splitting in tetrahedral complexes is less than one-half the *d*-orbital splitting in octahedral complexes. There are no known ligands powerful enough to produce the strong-field case; hence all tetrahedral complexes are weak-field or high-spin. **63.** $CoBr_6^{4-}$ has an octahedral structure and $CoBr_4^{2-}$ has a tetrahedral structure (as do most Co^{2+} complexes with four ligands). Coordination complexes absorb electromagnetic radiation (EMR) of energy equal to the energy difference between the split *d* orbitals. Since the tetrahedral *d*-orbital splitting is less than one-half of the octahedral *d*-orbital splitting, tetrahedral complexes will absorb lower-energy EMR, which corresponds to longer-wavelength EMR ($E = hc/\lambda$). Therefore, $CoBr_6^{4-}$ will absorb EMR having a wavelength shorter than 3.4×10^{-6} m. **65.** $Hg^{2+}(aq) + 2I^-(aq) \rightarrow HgI_2(s)$, orange ppt.; $HgI_2(s) + 2I^-(aq) \rightarrow HgI_4^{2-}(aq)$, soluble complex ion; Hg^{2+} is a d^{10} ion. Color is the result of electron transfer between split *d* orbitals. This cannot occur for the filled *d* orbitals in Hg^{2+}. Therefore, we would not expect Hg^{2+} complex ions to form colored solutions. **67.** zero (carbon monoxide is a neutrally charged ligand) **69.** At high altitudes, the oxygen content of air is lower, so less oxyhemoglobin is formed, which diminishes the transport of oxygen in the blood. A serious illness called high-altitude sickness can result from the decrease of O_2 in the blood. High-altitude acclimatization is the phenomenon that occurs with time in the human body in response to the lower amounts of oxyhemoglobin in the blood. This response is to produce more hemoglobin and, hence, increase the oxyhemoglobin in the blood. High-altitude acclimatization takes several weeks to take hold for people moving from lower altitudes to higher altitudes. **71.** $Cr(NH_3)_5I_3$ is the empirical formula. Cr(III) forms octahedral complexes. So compound A is made of the octahedral $[Cr(NH_3)_5I]^{2+}$ complex ion and two I^- ions; that is, $[Cr(NH_3)_5I]I_2$.

73.

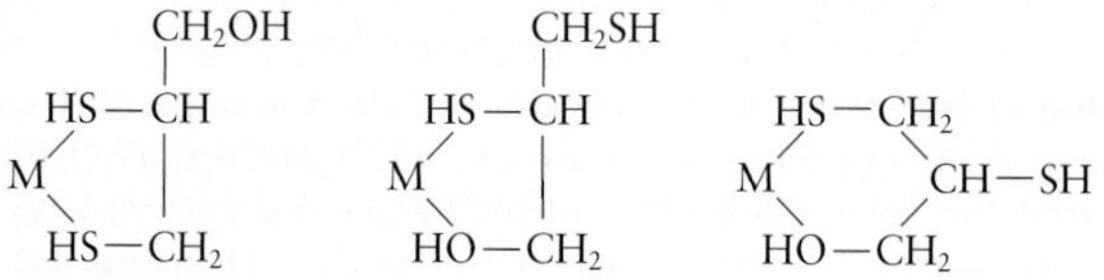

where M = metal ion **75.** No; in all three cases, six bonds are formed between Ni^{2+} and nitrogen, so ΔH values should be similar. $\Delta S°$ for formation of the complex ion is most negative for six NH_3 molecules reacting with a metal ion (seven independent species become one). For penten reacting with a metal ion, two independent species become one, so $\Delta S°$ is least negative of all three of the reactions. Thus the chelate effect occurs because the more bonds a chelating agent can form to the metal, the less unfavorable $\Delta S°$ is for the formation of the complex ion and the larger the formation constant.

$$\textbf{77.}\ (H_2O)_5\overset{II}{Cr} - Cl - \overset{III}{Co}(NH_3)_5 \rightarrow (H_2O)_5\overset{III}{Cr} - Cl - \overset{II}{Co}(NH_3)_5 \rightarrow Cr(H_2O)_5Cl^{2+} + Co(II)\ \text{complex}$$

Yes, this is consistent. After the oxidation, the ligands on Cr(III) won't exchange. Since Cl^- is in the coordination sphere, then it must have formed a bond to Cr(II) before the electron transfer occurred (as proposed through the formation of the intermediate). **79.** a. 1.66; b. Because of the lower charge, $Fe^{2+}(aq)$ will not be as strong an acid as $Fe^{3+}(aq)$. A solution of iron(II) nitrate will be less acidic (have a higher pH) than a solution with the same concentration of iron(III) nitrate. **81.** a. In the lungs, there is a lot of O_2, and the equilibrium favors $Hb(O_2)_4$. In the cells, there is a lower concentration of O_2, and the equilibrium favors HbH_4^{4+}. b. CO_2 is a weak acid in water, $CO_2 + H_2O \rightleftharpoons HCO_3^- + H^+$. Removing CO_2 essentially decreases H^+. $Hb(O_2)_4$ is then favored, and O_2 is not released by hemoglobin in the cells. Breathing into a paper bag increases $[CO_2]$ in the blood, thus increasing $[H^+]$ and shifting the reaction to the left. c. CO_2 builds up in the blood, and it becomes too acidic, driving the equilibrium to the left. Hemoglobin can't bind O_2 as strongly in the lungs. Bicarbonate ion acts as a base in water and neutralizes the excess acidity.

83. — — $d_{x^2-y^2}$, d_{xy}
— d_{z^2}
— — d_{xz}, d_{yz}

The $d_{x^2-y^2}$ and d_{xy} orbitals are in the plane of the three ligands and should be destabilized the most. The amount of destabilization should be about equal when all the possible interactions are considered. The d_{z^2} orbital has some electron density in the xy plane (the doughnut) and should be destabilized a lesser amount as compared to the $d_{x^2-y^2}$ and d_{xy} orbitals. The d_{xz} and d_{yz} orbitals have no electron density in the plane and should be lowest in energy.

85. — d_{z^2}
— — $d_{x^2-y^2}$, d_{xy}
— — d_{xz}, d_{yz}

The d_{z^2} orbital will be destabilized much more than in the trigonal planar case (see Exercise 19.83). The d_{z^2} orbital has electron density on the z-axis directed at the two axial ligands. The $d_{x^2-y^2}$ and d_{xy} orbitals are in the plane of the three trigonal planar ligands and should be destabilized a lesser amount as compared to the d_{z^2} orbital; only a portion of the electron density in the $d_{x^2-y^2}$ and d_{xy} orbitals is directed at the ligands. The d_{xz} and d_{yz} orbitals will be destabilized the least since the electron density is directed between the ligands. **87.** a. −0.26 V; b. The stronger oxidizing agent is the species more easily reduced. From the reduction potentials, Co^{3+} ($\mathscr{E}° = 1.82$ V) is a much stronger oxidizing agent than $Co(en)_3^{3+}$ ($\mathscr{E}° = -0.26$ V). c. In aqueous solution, Co^{3+} forms the hydrated transition metal complex $Co(H_2O)_6^{3+}$. In both complexes, $Co(H_2O)_6^{3+}$ and $Co(en)_3^{3+}$, cobalt exists as Co^{3+}, which has 6 d electrons. Assuming a strong-field case, the d-orbital splitting diagram for each is

— — e_g
⇅ ⇅ ⇅ t_{2g}

When each complex gains an electron, the electron enters the higher-energy e_g orbitals. Since en is a stronger field ligand than H_2O, the d-orbital splitting is larger for $Co(en)_3^{3+}$ and it takes more energy to add an electron to $Co(en)_3^{3+}$ than to $Co(H_2O)_6^{3+}$. Therefore, it is more favorable for $Co(H_2O)_6^{3+}$ to gain an electron than for $Co(en)_3^{3+}$ to gain an electron. **89.** a. 7.1×10^{-7} mol/L; b. 8.7×10^{-3} mol/L; c. The presence of NH_3 increases the solubility of AgBr. Added NH_3 removes Ag^+ from solution by forming the complex ion $Ag(NH_3)_2^+$. As Ag^+ is removed, more AgBr(s) will dissolve to replenish the Ag^+ concentration. d. 0.41g; e. Added HNO_3 will have no effect on the AgBr(s) solubility in pure water. Neither H^+ nor NO_3^- reacts with Ag^+ or Br^- ions. Br^- is the conjugate base of the strong acid HBr, so it is a terrible base. Added H^+ will not react with Br^- to any great extent. However, added HNO_3 will reduce the solubility of AgBr(s) in the ammonia solution. NH_3 is a weak base ($K_b = 1.8 \times 10^{-5}$). Added H^+ will react with NH_3 to form NH_4^+. As NH_3 is removed, a smaller amount of the $Ag(NH_3)_2^+$ complex ion will form, resulting in a smaller amount of AgBr(s) which will dissolve.

Chapter 20

1. a. Thermodynamic stability: the potential energy of a particular nucleus as compared to the sum of the potential energies of its component protons and neutrons. b. Kinetic stability: the probability that a nucleus will undergo decomposition to form a different nucleus. c. Radioactive decay: a spontaneous decomposition of a nucleus to form a different nucleus. d. β-particle production: a decay process for radioactive nuclides where an electron is produced; the mass number remains constant and the atomic number changes. e. α-particle production: a common mode of decay for heavy radioactive nuclides where a helium nucleus is produced causing the atomic number and the mass number to change. f. Positron production: a mode of nuclear decay in which a particle is formed having the same mass as an electron but opposite in charge. g. Electron capture: a process in which one of the inner-orbital electrons in an atom is captured by the nucleus. h. γ-ray emissions: the production of high-energy photons called γ rays that frequently accompany nuclear decays and particle reactions. **3.** a. ${}^{238}_{92}U \rightarrow {}^{4}_{2}He + {}^{234}_{90}Th$; this is alpha-particle production. b. ${}^{234}_{90}Th \rightarrow {}^{234}_{91}Pa + {}^{0}_{-1}e$; this is β-particle production. **5.** a. ${}^{51}_{24}Cr + {}^{0}_{-1}e \rightarrow {}^{51}_{23}V$; b. ${}^{131}_{53}I \rightarrow {}^{0}_{-1}e + {}^{131}_{54}Xe$; c. ${}^{32}_{15}P \rightarrow {}^{0}_{-1}e + {}^{32}_{16}S$ **7.** a. ${}^{68}_{31}Ga + {}^{0}_{-1}e \rightarrow {}^{68}_{30}Zn$; b. ${}^{62}_{29}Cu \rightarrow {}^{0}_{+1}e + {}^{62}_{28}Ni$; c. ${}^{212}_{87}Fr \rightarrow {}^{4}_{2}He + {}^{208}_{85}At$; d. ${}^{129}_{51}Sb \rightarrow {}^{0}_{-1}e + {}^{129}_{52}Te$ **9.** 10 α particles and 5 β particles **11.** Refer to Table 20.2 for potential radioactive decay processes. ^{17}F and ^{18}F contain too many protons or too few neutrons. Electron capture or positron production are both possible decay mechanisms that increase the neutron-to-proton ratio. α-particle production also increases the neutron-to-proton ratio, but it is not likely for these light nuclei. ^{21}F contains too many neutrons or too few protons. β-particle production lowers the neutron-to-proton ratio, so we expect ^{21}F to be a β-emitter. **13.** a. ${}^{249}_{98}Cf + {}^{18}_{8}O \rightarrow {}^{263}_{106}Sg + 4{}^{1}_{0}n$; b. ${}^{259}_{104}Rf$ **15.** 6.35×10^{11} **17.** a. $6.27 \times 10^{-7}\ s^{-1}$; b. 1.65×10^{14} decays/s; c. 25.6 days **19.** 20.0% of the ^{90}Sr remains as of July 16, 2012. **21.** 6.22 mg ^{32}P **23.** Plants take in CO_2 in the photosynthesis process, which incorporates carbon, including ^{14}C, into its molecules. As long as the plant is alive, the $^{14}C/^{12}C$ ratio in the plant will equal the ratio in the atmosphere. When the plant dies, ^{14}C is not replenished, and ^{14}C decays by β-particle production. By measuring the ^{14}C activity today in the artifact and comparing this to the assumed ^{14}C activity when the plant died to make the artifact, an age can be determined for the artifact. The assumptions are that the ^{14}C level in the atmosphere is constant or that the ^{14}C level at the time the plant died can be calculated. A constant ^{14}C level is a pure assumption, and accounting for variation is complicated. Another problem is that some of the material must be destroyed to determine the ^{14}C level. **25.** No; from ^{14}C dating, the painting was produced during the early 1900s. **27.** 1975 **29.** 3.8×10^9 yr **31.** 4.3×10^6 kg/s **33.** 1.408×10^{-12} J/nucleon **35.** ^{12}C, 1.230×10^{-12} J/nucleon; ^{235}U, 1.2154×10^{-12} J/nucleon; Since ^{26}Fe is the most stable known nucleus, the binding energy per nucleon for ^{56}Fe (1.41×10^{-12} J/nucleon) will be larger than that for ^{12}C or ^{235}U (see Fig. 20.9). **37.** 26.9830 amu

39. -2.820×10^{-12} J/nucleus; -1.698×10^{12} J/mol **41.** The Geiger-Müller tube has a certain response time. After the gas in the tube ionizes to produce a "count," some time must elapse for the gas to return to an electrically neutral state. The response of the tube levels because at high activities, radioactive particles are entering the tube faster than the tube can respond to them. **43.** Fission: Splitting of a heavy nucleus into two (or more) lighter nuclei. Fusion: Combining two light nuclei to form a heavier nucleus. The maximum binding energy per nucleon occurs at Fe. Nuclei smaller than Fe become more stable by fusing to form heavier nuclei closer in mass to Fe. Nuclei larger than Fe form more stable nuclei by splitting to form lighter nuclei closer in mass to Fe. **45.** The moderator slows the neutrons to increase the efficiency of the fission reaction. The control rods absorb neutrons to slow or halt the fission reaction. **47.** In order to sustain a nuclear chain reaction, the neutrons produced by the fission must be contained within the fissionable material so that they can go on to cause other fissions. The fissionable material must be closely packed together to ensure that neutrons are not lost to the outside. The critical mass is the mass of material in which exactly one neutron from each fission event causes another fission event so that the process sustains itself. A supercritical situation occurs when more than one neutron from each fission event causes another fission event. In this case the process rapidly escalates and the heat build-up causes a violent explosion. **49.** A nonradioactive substance can be put in equilibrium with a radioactive substance. The two materials can then be checked to see whether all the radioactivity remains in the original material or if it has been scrambled by the equilibrium. **51.** All evolved oxygen in O_2 comes from water and not from carbon dioxide. **53.** Some factors for the biological effects of radiation exposure are as follows: a. The energy of the radiation. The higher the energy, the more damage it can cause. b. The penetrating ability of radiation. The ability of specific radiation to penetrate human tissue and cause damage must be considered. c. The ionizing ability of the radiation. When biomolecules are ionized, their function is usually disturbed. d. The chemical properties of the radiation source. Specifically, the radioactive substance either is readily incorporated into the body or is inert chemically and passes through the body relatively quickly. ^{90}Sr will be incorporated into the body by replacing calcium in the bones. Once incorporated, ^{90}Sr can cause leukemia and bone cancer. Krypton is chemically inert so it will not be incorporated into the body. **55.** (i) and (ii) mean that Pu is not a significant threat outside the body. Our skin is sufficient to keep out the α particles. If Pu gets inside the body, it is easily oxidized to Pu^{4+} (iv), which is chemically similar to Fe^{3+} (iii). Thus Pu^{4+} will concentrate in tissues where Fe^{3+} is found, including the bone marrow, where red blood cells are produced. Once inside the body, α particles cause considerable damage. **57.** 3800 decays/s **59.** ~900 g ^{235}U (assuming that all ^{235}U present undergoes fission) **61.** $\frac{1}{9}$ of the NO_2 is $N^{16}O_2$, $\frac{4}{9}$ of the NO_2 is $N^{18}O_2$, and $\frac{4}{9}$ of the NO_2 is $N^{16}O^{18}O$. **63.** Assuming that (1) the radionuclide is long-lived enough that no significant decay occurs during the time of the experiment, and (2) the total activity is uniformly distributed only in the rat's blood; $V = 10.$ mL. **65.** 77% ^{238}U and 23% ^{235}U **67.** 5×10^9 K **69.** 4.3×10^{-29} **71.** a. $^{235}UF_6$; At constant temperature, average velocity is proportional to $(1/M)^{1/2}$. Therefore, the lighter the molecule, the faster the average velocity. b. In theory, 345 stages (steps) are required. c. $^{235}U/^{238}U = 1.01 \times 10^{-2}$

Chapter 21

1. A hydrocarbon is a compound composed of only carbon and hydrogen. A saturated hydrocarbon has only carbon–carbon single bonds in the molecule. An unsaturated hydrocarbon has one or more carbon–carbon multiple bonds but may also contain carbon–carbon single bonds. A normal hydrocarbon has one chain of consecutively bonded carbon atoms, with each carbon atom in the chain bonded to one or two other carbon atoms. A branched hydrocarbon has at least one carbon atom in the structure that forms bonds to three or four other carbon atoms; the structure is not one continuous chain of carbon atoms. **3.** In order to form, cyclopropane and cyclobutane are forced to form bond angles much smaller than the preferred 109.5° bond angles. Cyclopropane and cyclobutane easily react in order to obtain the preferred 109.5° bond angles.

5. a.

$$\begin{array}{l} \quad CH_3 \\ \quad | \\ CH_3CHCH_2CH_2CH_2CH_2CH_3; \end{array} \qquad \begin{array}{l} \qquad\quad CH_3 \\ \qquad\quad | \\ CH_3CH_2CHCH_2CH_2CH_2CH_3; \end{array}$$

2-Methylheptane 3-Methylheptane

$$\begin{array}{l} \qquad\qquad CH_3 \\ \qquad\qquad | \\ CH_3CH_2CH_2CHCH_2CH_2CH_3; \end{array}$$

4-Methylheptane

b.

$$\begin{array}{l} \quad CH_3 \\ \quad | \\ CH_3CCH_2CH_2CH_2CH_3; \\ \quad | \\ \quad CH_3 \end{array} \qquad \begin{array}{l} \quad CH_3 \\ \quad | \\ CH_3CHCHCH_2CH_2CH_3; \\ \qquad | \\ \qquad CH_3 \end{array}$$

2,2-Dimethylhexane 2,3-Dimethylhexane

$$\begin{array}{l} \quad CH_3 \\ \quad | \\ CH_3CHCH_2CHCH_2CH_3; \\ \qquad\quad | \\ \qquad\quad CH_3 \end{array} \qquad \begin{array}{l} \quad CH_3 \\ \quad | \\ CH_3CHCH_2CH_2CHCH_3; \\ \qquad\qquad\quad | \\ \qquad\qquad\quad CH_3 \end{array}$$

2,4-Dimethylhexane 2,5-Dimethylhexane

$$\begin{array}{l} \qquad CH_3 \\ \qquad | \\ CH_3CH_2CCH_2CH_2CH_3; \\ \qquad | \\ \qquad CH_3 \end{array} \qquad \begin{array}{l} \qquad CH_3 \\ \qquad | \\ CH_3CH_2CHCHCH_2CH_3; \\ \qquad\quad | \\ \qquad\quad CH_3 \end{array}$$

3,3-Dimethylhexane 3,4-Dimethylhexane

$$\begin{array}{l} \qquad CH_2CH_3 \\ \qquad | \\ CH_3CH_2CHCH_2CH_2CH_3; \end{array}$$

3-Ethylhexane

c.

$$\begin{array}{l} \quad\;\; H_3C\;\; CH_3 \\ \qquad\; |\qquad | \\ CH_3-C-CH-CH_2-CH_3; \\ \qquad\; | \\ \quad\;\; CH_3 \end{array} \qquad \begin{array}{l} \qquad\; CH_3 \qquad\qquad CH_3 \\ \qquad\; | \qquad\qquad\quad | \\ CH_3-C-CH_2-CH-CH_3; \\ \qquad\; | \\ \qquad\; CH_3 \end{array}$$

2,2,3-Trimethylpentane 2,2,4-Trimethylpentane

$$\begin{array}{l} \qquad CH_3\; CH_3 \\ \qquad |\qquad | \\ CH_3-CH-C-CH_2-CH_3; \\ \qquad\qquad | \\ \qquad\qquad CH_3 \end{array} \qquad \begin{array}{l} \qquad CH_3\;\; CH_3\;\; CH_3 \\ \qquad |\qquad\; |\qquad\; | \\ CH_3-CH-CH-CH-CH_3; \end{array}$$

2,3,3-Trimethylpentane 2,3,4-Trimethylpentane

$$\begin{array}{l} \qquad CH_3\;\; CH_2CH_3 \\ \qquad |\qquad | \\ CH_3-CH-CH-CH_2-CH_3; \end{array} \qquad \begin{array}{l} \qquad\qquad\qquad CH_2CH_3 \\ \qquad\qquad\qquad | \\ CH_3-CH_2-C-CH_2-CH_3; \\ \qquad\qquad\qquad | \\ \qquad\qquad\qquad CH_3 \end{array}$$

3-Ethyl-2-methylpentane 3-Ethyl-3-methylpentane

d.

$$\begin{array}{l} \qquad CH_3\; CH_3 \\ \qquad |\qquad | \\ CH_3-C-C-CH_3 \\ \qquad |\qquad | \\ \qquad CH_3\; CH_3 \end{array}$$

2,2,3,3-Tetramethylbutane

7. London dispersion (LD) forces are the primary intermolecular forces exhibited by hydrocarbons. The strength of the LD forces depends on the surface area contact among neighboring molecules.

As branching increases, there is less surface area contact among neighboring molecules, leading to weaker LD forces and lower boiling points.

9. a.

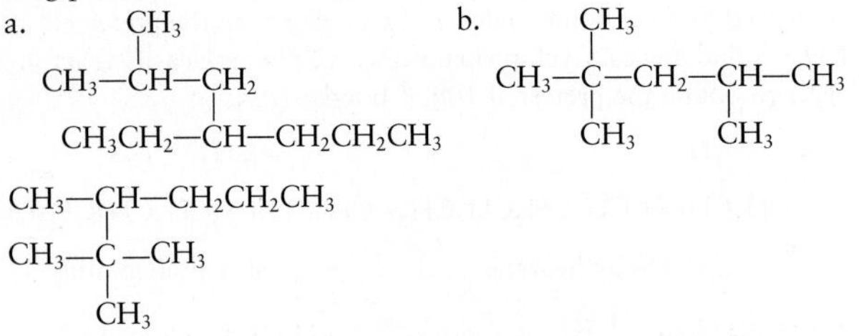

```
     CH3
     |
CH3—CH—CH2
        |
   CH3CH2—CH—CH2CH2CH3
```

b.

```
        CH3
        |
CH3—C—CH2—CH—CH3
        |        |
        CH3      CH3
```

c.

```
CH3—CH—CH2CH2CH3
     |
CH3—C—CH3
     |
     CH3
```

d. 4-ethyl-2-methylheptane, 2,2,3-trimethylhexane **11.** a. 2,2,4-trimethylhexane; b. 5-methylnonane; c. 2,2,4,4-tetramethylpentane; d. 3-ethyl-3-methyloctane **13.** a. 1-butene; b. 2-methyl-2-butene; c. 2,5-dimethyl-3-heptene; d. 2,3-dimethyl-1-pentene; e. 1-ethyl-3-methylcyclopentene; f. 4-ethyl-3-methylcyclopentene; g. 4-methyl-2-pentyne **15.** a. 1,3-dichlorobutane; b. 1,1,1-trichlorobutane; c. 2,3-dichloro-2,4-dimethylhexane; d. 1,2-difluoroethane; e. 3-iodo-1-butene; f. 2-bromotoluene (or 1-bromo-2-methylbenzene); g. 1-bromo-2-methylcyclohexane; h. 4-bromo-3-methylcyclohexene **17.** isopropylbenzene or 2-phenylpropane **19.** All six of these compounds are the same. They differ from each other only by rotations about one or more carbon–carbon single bonds. Only one isomer of C_7H_{16} is present in all of these names, 3-methylhexane. **21.** Compounds c and f exhibit *cis–trans* isomerism. See Exercise 25 for an example of *cis–trans* isomerism in ring compounds. **23.** a. Compounds ii and iii are identical compounds, so they would have the same physical properties. b. Compound i is a *trans* isomer because the bulkiest groups bonded to the carbon atoms in the $C_3{=}C_4$ double bond are as far apart as possible. c. Compound iv does not have carbon atoms in a double bond that each have two different groups attached. Compound iv does not exhibit *cis–trans* isomerism.

25.

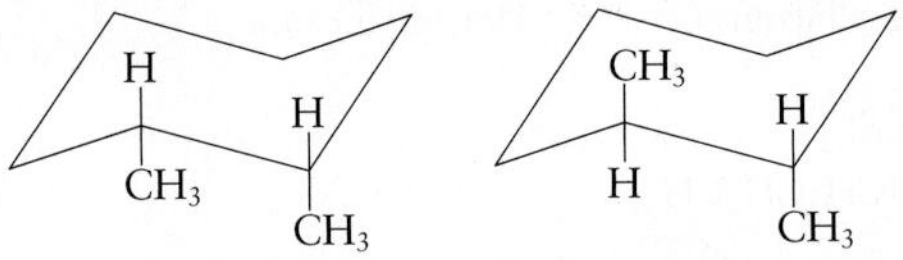

```
   H         H                 CH3        H
  (chair)                     (chair)
   CH3       CH3               H          CH3
```

cis *trans*

27.

```
Cl      CH3        H       CH3
  \    /            \     /
   C=C               C=C
  /    \            /     \
H        H         Cl       H

CH2=C—CH3      CH2=CH—CH2       Cl
    |                  |         |
    Cl                 Cl       (cyclopropane ring)
```

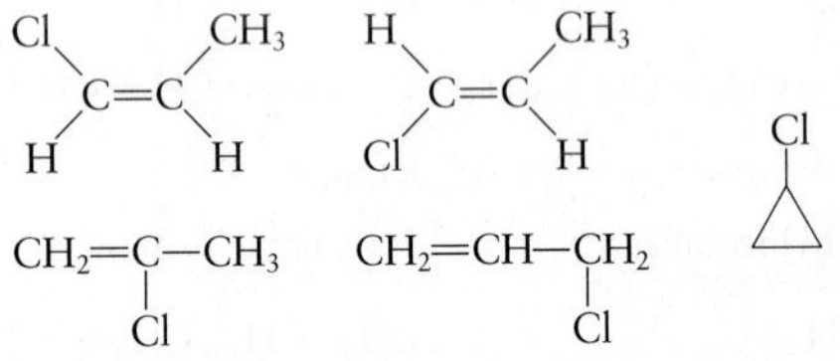

29. $CH_2{=}CHCH_2CH_2CH_3$ $CH_3CH{=}CHCH_2CH_3$

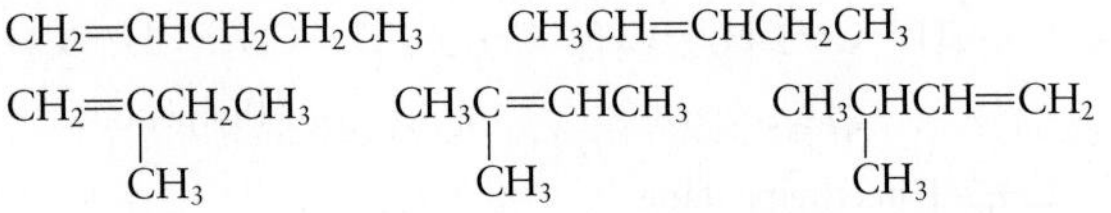

```
CH2=CCH2CH3     CH3C=CHCH3     CH3CHCH=CH2
    |               |              |
    CH3             CH3            CH3
```

Only 2-pentene exhibits *cis–trans* isomerism. The isomers are

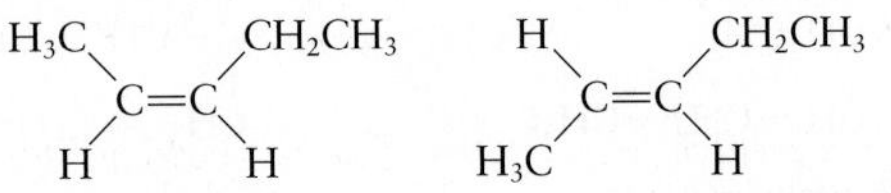

```
H3C      CH2CH3        H        CH2CH3
   \    /               \      /
    C=C                  C=C
   /    \               /      \
H         H          H3C         H
```

cis *trans*

31. a. *cis*-1-bromo-1-propene; b. *cis*-4-ethyl-3-methyl-3-heptene; c. *trans*-1,4-diiodo-2-propyl-1-pentene

33. a. (benzene rings with Cl substituents: 1,2-dichloro; 1,3-dichloro; 1,4-dichloro)

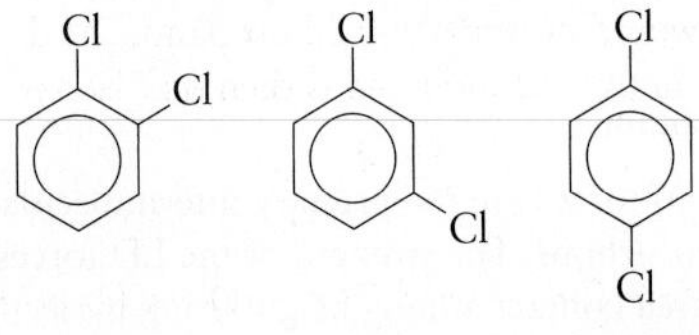

ortho *meta* *para*

b. There are three trichlorobenzenes (1,2,3-trichlorobenzene, 1,2,4-trichlorobenzene, and 1,3,5-trichlorobenzene). c. The *meta* isomer will be very difficult to synthesize. d. 1,3,5-trichlorobenzene will be the most difficult to synthesize since all Cl groups are *meta* to each other in this compound.

35.

```
Carboxylic acid        Aldehyde
     :O:                 :O:
     ||                  ||
R—C—Ö—H               R—C—H

   RCOOH                RCHO
```

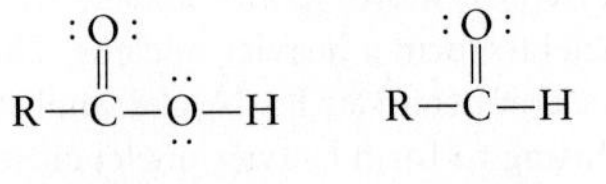

The R designation refers to the rest of the organic molecule beyond the specific functional group indicated in the formula. The R group may sometimes be a hydrogen, but is usually a hydrocarbon fragment. The major point in the R group designation is that if the R group is an organic fragment, then the first atom in the R group is a carbon atom. What the R group has after the first carbon is not important to the functional group designation. **37.** a. ketone; b. aldehyde; c. carboxylic acid; d. amine

39. a.

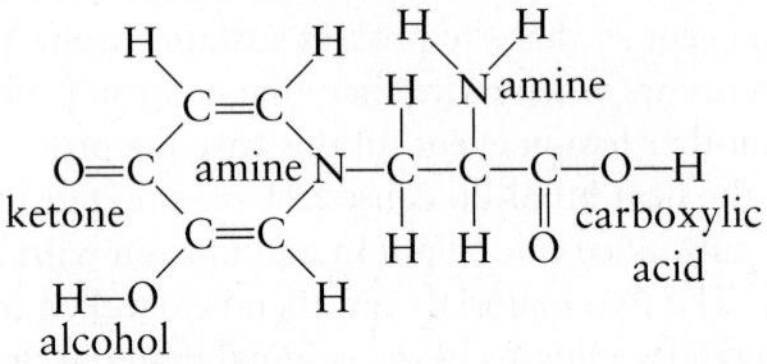

```
         H      H      H   H
          \    /        \ /
           C=C       H   N amine
          /    \     |   |
ketone O=C      amine N—C—C—C—O—H
          \    /         |  |  ||  carboxylic
           C=C           H  H  O   acid
          /    \
      H—O       H
   alcohol
```

b. 5 carbons in ring and the carbon in $—CO_2H$: sp^2, the other two carbons: sp^3; c. 24 σ bonds, 4 π bonds **41.** a. 3-chloro-1-butanol, primary alcohol; b. 3-methyl-3-hexanol, tertiary alcohol; c. 2-methylcyclopentanol, secondary alcohol

43.

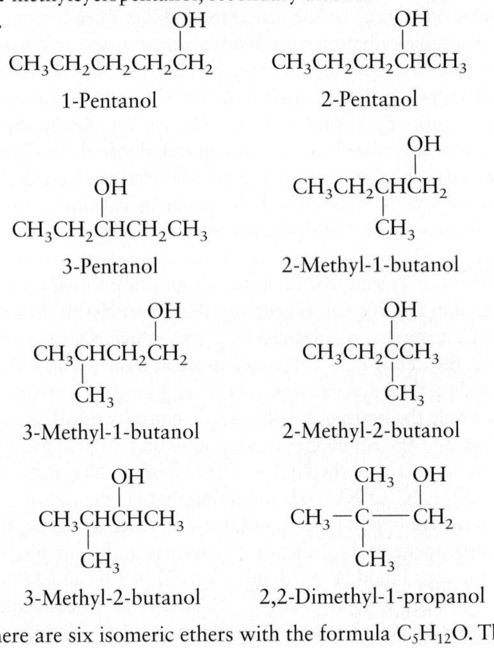

```
              OH                      OH
              |                       |
  CH3CH2CH2CH2CH2        CH3CH2CH2CHCH3
     1-Pentanol               2-Pentanol

                                       OH
         OH                            |
         |                     CH3CH2CHCH2
  CH3CH2CHCH2CH3                    |
                                   CH3
     3-Pentanol            2-Methyl-1-butanol

             OH                        OH
             |                         |
  CH3CHCH2CH2                  CH3CH2CCH3
     |                                 |
     CH3                               CH3
  3-Methyl-1-butanol       2-Methyl-2-butanol

         OH                      CH3 OH
         |                        |   |
  CH3CHCHCH3                CH3—C—CH2
     |                            |
     CH3                          CH3
  3-Methyl-2-butanol       2,2-Dimethyl-1-propanol
```

There are six isomeric ethers with the formula $C_5H_{12}O$. The structures follow.

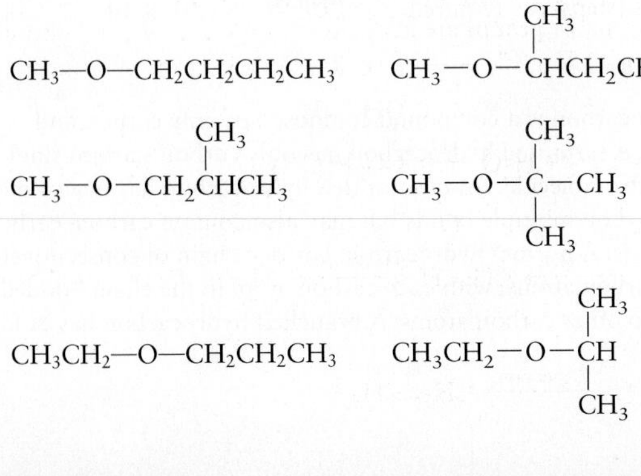

```
                                          CH3
                                          |
CH3—O—CH2CH2CH2CH3              CH3—O—CHCH2CH3

          CH3                             CH3
          |                               |
CH3—O—CH2CHCH3                  CH3—O—C—CH3
                                          |
                                          CH3

                                          CH3
                                          |
CH3CH2—O—CH2CH2CH3              CH3CH2—O—CH
                                          |
                                          CH3
```

45. a. 4,5-dichloro-3-hexanone; b. 2,3-dimethylpentanal; c. 3-methylbenzaldehyde or *m*-methylbenzaldehyde

47. a. $H{-}\overset{O}{\overset{\|}{C}}{-}H$ b. $CH_3CH_2CH_2\overset{O}{\overset{\|}{C}}CH_2CH_2CH_3$

c. $H{-}\overset{O}{\overset{\|}{C}}CH_2\underset{Cl}{\underset{|}{C}}HCH_3$ d. $CH_3\overset{O}{\overset{\|}{C}}CH_2CH_2\overset{CH_3}{\overset{|}{\underset{CH_3}{\underset{|}{C}}}}CH_3$

49. a. (cyclobutane, with eight H atoms) or (methylcyclopropane: cyclopropane ring bearing H, H / H, CH_3 / H, H)

b. $CH_3\overset{O}{\overset{\|}{C}}{-}O{-}CH_3$ or $H\overset{O}{\overset{\|}{C}}{-}O{-}CH_2CH_3$

c. $CH_3CH_2\overset{O}{\overset{\|}{C}}CH_3$

d. $CH_3{-}\underset{CH_2CH_2CH_3}{\underset{|}{N}}{-}H$ or $CH_3{-}\underset{CH_3CHCH_3}{\underset{|}{N}}{-}H$ or

$CH_3CH_2{-}\underset{CH_2CH_3}{\underset{|}{N}}{-}H$

e. $CH_3{-}\underset{CH_2CH_3}{\underset{|}{N}}{-}CH_3$

f. $CH_3{-}O{-}CH_2CH_2CH_3$ or $CH_3{-}O{-}\overset{CH_3}{\overset{|}{\underset{CH_3}{\underset{|}{C}}}}H$ or

$CH_3CH_2{-}O{-}CH_2CH_3$

g. $CH_3\overset{OH}{\overset{|}{C}}HCH_2CH_3$

51. a. 2-Chloro-2-butyne would have 5 bonds to the second carbon. Carbon never expands its octet. b. 2-Methyl-2-propanone would have 5 bonds to the second carbon. c. Carbon-1 in 1,1-dimethylbenzene would have 5 bonds. d. You cannot have an aldehyde functional group bonded to a middle carbon in a chain. Aldehyde groups can only be at the beginning and/or the end of a chain of carbon atoms. e. You cannot have a carboxylic acid group bonded to a middle carbon in a chain. Carboxylic groups must be at the beginning and/or the end of a chain of carbon atoms. f. In cyclobutanol, the 1 and 5 positions refer to the same carbon atom. 5,5-Dibromo-1-cyclobutanol would have five bonds to carbon-1. This is impossible; carbon never expands its octet. **53.** Substitution: An atom or group is replaced by another atom or group; e.g., H in benzene is replaced by Cl. $C_6H_6 + Cl_2 \xrightarrow{\text{Catalyst}} C_6H_5Cl + HCl$
Addition: Atoms or groups are added to a molecule; e.g., Cl_2 adds to ethene. $CH_2{=}CH_2 + Cl_2 \longrightarrow CH_2Cl{-}CH_2Cl$

55. a. $CH_2{=}CH_2 + H_2O \xrightarrow{H^+} \overset{OH}{\overset{|}{C}}H_2{-}\overset{H}{\overset{|}{C}}H_2$

b. $CH_3\overset{OH}{\overset{|}{C}}H_2 \xrightarrow{\text{Oxidation}} CH_3\overset{O}{\overset{\|}{C}}H \xrightarrow{\text{Oxidation}} CH_3\overset{O}{\overset{\|}{C}}{-}OH$

c. $CH_3\overset{OH}{\overset{|}{C}}HCH_3 \xrightarrow{\text{Oxidation}} CH_3\overset{O}{\overset{\|}{C}}CH_3$

d. $CH_3{-}O{-}H + HO{-}\overset{O}{\overset{\|}{C}}CH_3 \longrightarrow CH_3{-}O{-}\overset{O}{\overset{\|}{C}}CH_3 + H_2O$

57. a. $CH_3\overset{H}{\overset{|}{C}}H{-}\overset{H}{\overset{|}{C}}HCH_3$ b. $\overset{Cl}{\overset{|}{C}}H_2{-}\overset{Cl}{\overset{|}{C}}H\underset{CH_3}{\underset{|}{\overset{Cl}{\overset{|}{C}}}}H{-}\underset{CH_3}{\underset{|}{\overset{Cl}{\overset{|}{C}}}}H$

c. (benzene ring)$-Cl + HCl$

d. $C_4H_8(g) + 6O_2(g) \longrightarrow 4CO_2(g) + 4H_2O(g)$

59. When $CH_2{=}CH_2$ reacts with HCl, there is only one possible product, chloroethane. When Cl_2 is reacted with CH_3CH_3 (in the presence of light), there are six possible products because any number of the six hydrogens in ethane can be substituted for by Cl. The light-catalyzed substitution reaction is very difficult to control; hence it is not a very efficient method of producing monochlorinated alkanes.

61. a. $H{-}\overset{O}{\overset{\|}{C}}{-}CH_2\underset{CH_3}{\underset{|}{C}}HCH_3 + HO{-}\overset{O}{\overset{\|}{C}}{-}CH_2\underset{CH_3}{\underset{|}{C}}HCH_3$

b. $CH_3{-}\overset{O}{\overset{\|}{C}}{-}\underset{CH_3}{\underset{|}{C}}HCH_3$ c. no reaction

d. (benzene ring)$-\overset{O}{\overset{\|}{C}}{-}H +$ (benzene ring)$-\overset{O}{\overset{\|}{C}}{-}OH$

e. (cyclohexanone ring with C=O and adjacent $-CH_3$)

f. (cyclohexanone ring bearing OH, CH_3 and $-\underset{O}{\underset{\|}{C}}{-}H$) + (cyclohexanone ring bearing OH, CH_3 and $-\underset{O}{\underset{\|}{C}}{-}OH$)

63. The products of the reactions with excess $KMnO_4$ are 2-propanone and propanoic acid. **65.** a. $CH_3CH{=}CH_2 + Br_2 \longrightarrow CH_3CHBrCH_2Br$ (Addition reaction of Br_2 with propene)

b. $CH_3{-}\overset{OH}{\overset{|}{C}}H{-}CH_3 \xrightarrow{\text{Oxidation}} CH_3{-}\overset{O}{\overset{\|}{C}}{-}CH_3$

Oxidation of 2-propanol yields acetone (2-propanone).

c. $CH_2{=}\overset{CH_3}{\overset{|}{C}}{-}CH_3 + H_2O \xrightarrow{H^+} \underset{H}{\underset{|}{C}}H_2{-}\overset{CH_3}{\overset{|}{\underset{OH}{\underset{|}{C}}}}{-}CH_3$

Addition of H_2O to 2-methylpropene would yield *tert*-butyl alcohol (2-methyl-2-propanol) as the major product.

d. $CH_3CH_2CH_2OH \xrightarrow{KMnO_4} CH_3CH_2\overset{O}{\overset{\|}{C}}{-}OH$

Oxidation of 1-propanol would eventually yield propanoic acid. Propanal is produced first in this reaction and is then oxidized to propanoic acid.

67. a. $CH_3\overset{O}{\overset{\|}{C}}{-}OH + HO{-}CH_3 \longrightarrow CH_3\overset{O}{\overset{\|}{C}}{-}O{-}CH_3 + H_2O$

b.

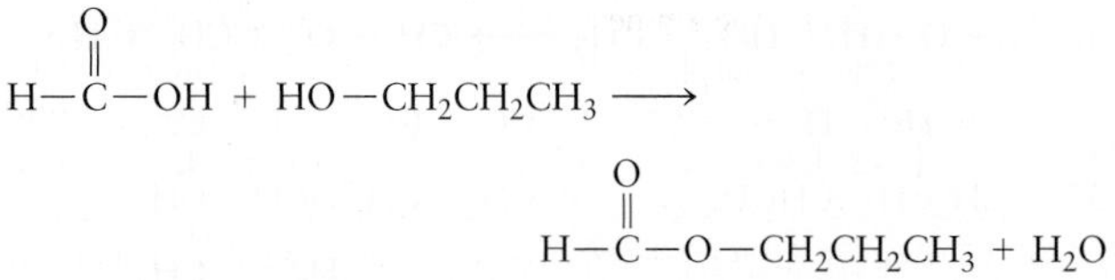

$$\text{H}-\overset{\overset{\displaystyle O}{\|}}{C}-\text{OH} + \text{HO}-\text{CH}_2\text{CH}_2\text{CH}_3 \longrightarrow \text{H}-\overset{\overset{\displaystyle O}{\|}}{C}-\text{O}-\text{CH}_2\text{CH}_2\text{CH}_3 + \text{H}_2\text{O}$$

69. a. Addition polymer: a polymer that forms by adding monomer units together (usually by reacting double bonds). Teflon, polyvinyl chloride, and polyethylene are examples of addition polymers. b. Condensation polymer: a polymer that forms when two monomers combine by eliminating a small molecule (usually H_2O or HCl). Nylon and Dacron are examples of condensation polymers. c. Copolymer: a polymer formed from more than one type of monomer. Nylon and Dacron are copolymers. d. Homopolymer: a polymer formed from the polymerization of only one type of monomer. Polyethylene, Teflon, and polystyrene are examples of homopolymers. e. Polyester: a condensation polymer whose monomers link together by formation of the ester functional group. Dacron is a polyester. f. Polyamide: a condensation polymer whose monomers link together by formation of the amide functional group. Nylon is a polyamide as are proteins in the human body. **71.** a. A polyester forms when an alcohol functional group reacts with a carboxylic acid functional group. The monomer for a homopolymer polyester must have an alcohol functional group and a carboxylic acid functional group present within the structure of the monomer. b. A polyamide forms when an amine functional group reacts with a carboxylic acid functional group. For a copolymer polyamide, one monomer would have at least two amine functional groups present, and the other monomer would have at least two carboxylic acid functional groups present. For polymerization to occur, each monomer must have two reactive functional groups present. c. To form a typical addition polymer, a carbon–carbon double bond must be present. To form a polyester, the monomer would need the alcohol and carboxylic acid functional groups present. To form a polyamide, the monomer would need the amine and carboxylic acid functional groups present. The two possibilities are for the monomer to have a carbon–carbon double bond, an alcohol functional group, and a carboxylic acid functional group present or to have a carbon–carbon double bond, an amine functional group, and a carboxylic acid functional group present.

73. $F(Cl)C{=}CF_2$ (F and Cl on one carbon; F and F on the other)

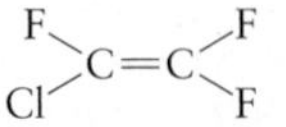

75. a. $H_2N-C_6H_4-NH_2$ (para) and $HO_2C-C_6H_4-CO_2H$ (para)

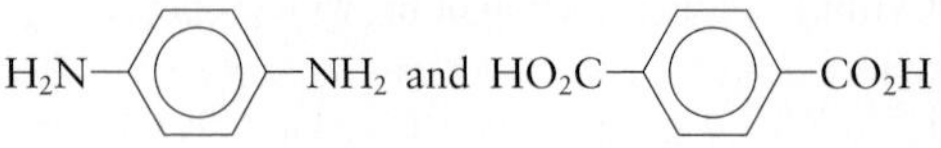

b. Repeating unit: $\left(\text{HN}-C_6H_4-\text{NHC}(=O)-C_6H_4-\text{C}(=O)\right)_n$ (meta-substituted rings)

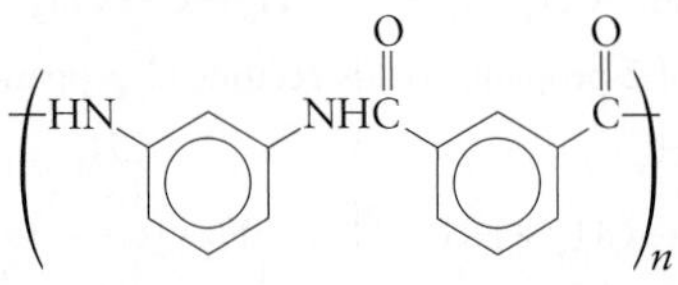

The two polymers differ in the substitution pattern on the benzene rings. The Kevlar chain is straighter, and there is more efficient hydrogen bonding between Kevlar chains than between Nomex chains.

77. $\left(\text{C}(\text{CN})(\text{C(=O)OCH}_3)-\text{CH}_2-\text{C}(\text{CN})(\text{C(=O)OCH}_3)-\text{CH}_2\right)_n$

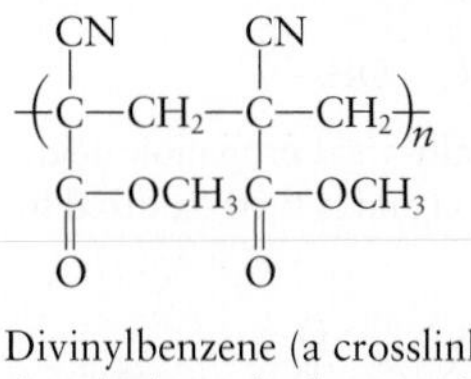

79. Divinylbenzene (a crosslinking agent) has two reactive double bonds, which are both reacted when divinylbenzene inserts itself into two adjacent polymer chains. The chains cannot move past each other because the crosslinks bond adjacent polymer chains together making the polymer more rigid.

81. a. $\left(\text{O}-C_6H_4-\text{C}(\text{CH}_3)_2-C_6H_4-\text{O}-\text{C}(=O)-\text{O}-C_6H_4-\text{C}(\text{CH}_3)_2-C_6H_4-\text{O}-\text{C}(=O)\right)_n$

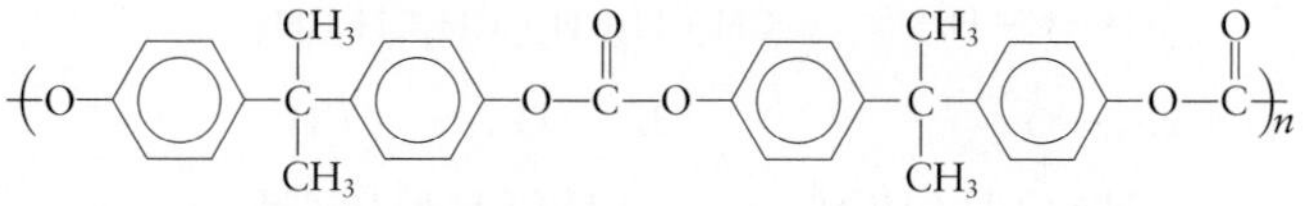

b. Condensation; HCl is eliminated when the polymer bonds form.

83. Polyacrylonitrile: $\left(\text{CH}_2-\text{CH}(\text{C}{\equiv}\text{N})\right)_n$

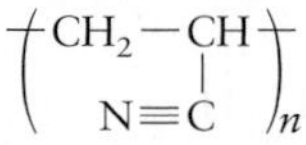

The CN triple bond is very strong and will not easily break in the combustion process. A likely combustion product is the toxic gas hydrogen cyanide [HCN(*g*)].

85. $\left(\text{OCH}_2\text{CH(OH)CH}_2\text{OC(=O)}-C_6H_4-\text{C(=O)OCH(CH}_2\text{OH)CH}_2\text{OC(=O)}-C_6H_4-\text{C(=O)}\right)_n$

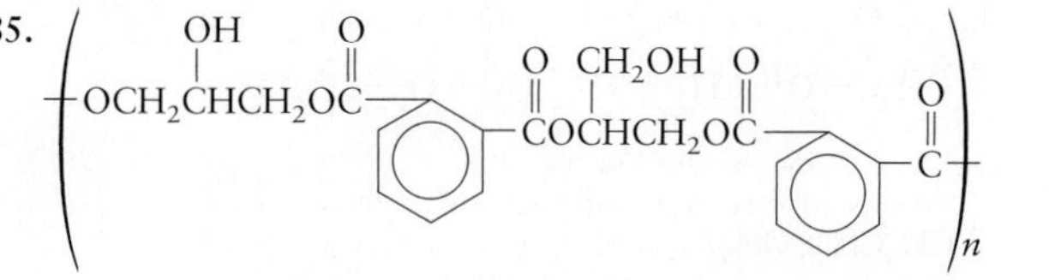

Two linkages are possible with glycerol. A possible repeating unit with both types of linkages is shown above. With either linkage, there are free OH groups on the polymer chains. These unreacted OH groups on adjacent polymer chains can react with the acid groups of phthalic acid to form crosslinks between various polymer chains.

87. $\text{R}-\text{C}(\text{H})(\text{NH}_2)-\text{C}(=O)-\text{OH}$

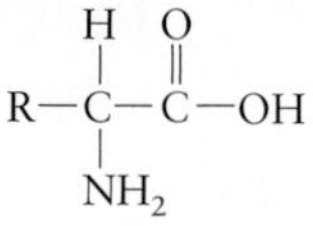

Hydrophilic (water-loving) and hydrophobic (water-fearing) refer to the polarity of the R groups. When the R group consists of a polar group, then the amino acid is hydrophilic. When the R group consists of a nonpolar group, then the amino acid is hydrophobic. **89.** Denaturation changes the three-dimensional structure of a protein. Once the structure is affected, the function of the protein will also be affected. **91.** a. serine, tyrosine, and threonine; b. aspartic acid and glutamic acid; c. histidine, lysine, arginine, and tryptophan; d. glutamine and asparagine **93.** a. aspartic acid and phenylalanine; b. Aspartame contains the methyl ester of phenylalanine. This ester can hydrolyze to form methanol, $R-CO_2CH_3 + H_2O \rightleftharpoons RCO_2H + CH_3OH$.

95. $\text{H}_2\text{NCH(CH}_2\text{OH)C(=O)}-\text{NHCH(CH}_3\text{)CO}_2\text{H}$ (ser-ala); $\text{H}_2\text{NCH(CH}_3\text{)C(=O)}-\text{NHCH(CH}_2\text{OH)CO}_2\text{H}$ (ala-ser)

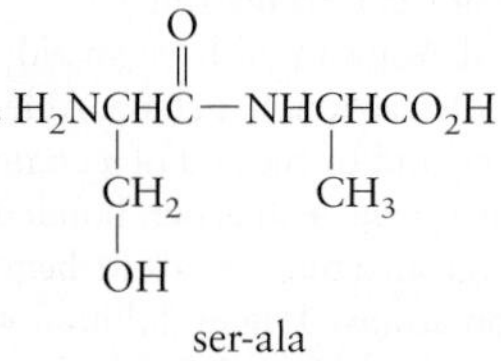

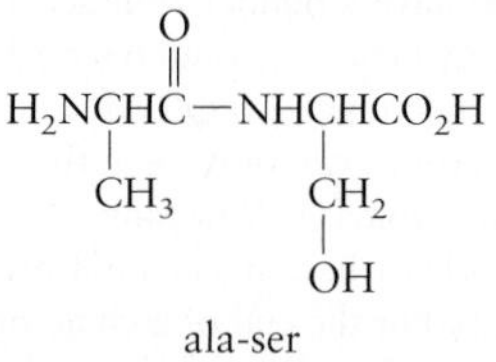

97. a. Six tetrapeptides are possible. From NH_2 to CO_2H end: phe-phe-gly-gly, gly-gly-phe-phe, gly-phe-phe-gly, phe-gly-gly-phe, phe-gly-phe-gly, gly-phe-gly-phe; b. Twelve tetrapeptides are possible. From NH_2 to CO_2H end: phe-phe-gly-ala, phe-phe-ala-gly, phe-gly-phe-ala, phe-gly-ala-phe, phe-ala-phe-gly, phe-ala-gly-phe, gly-phe-phe-ala, gly-phe-ala-phe, gly-ala-phe-phe, ala-phe-phe-gly, ala-phe-gly-phe, ala-gly-phe-phe **99.** a. covalent; b. hydrogen bonding; c. ionic; d. London dispersion **101.** Glutamic acid: R = $-CH_2CH_2CO_2H$; Valine: R = $-CH(CH_3)_2$; A polar side chain is replaced by a nonpolar side chain. This could affect the tertiary structure of hemoglobin and the ability of hemoglobin to bind oxygen.

103. $H_2N—CH(CH_2CH_2CO_2^-Na^+)—CO_2H$ or $H_2N—CH(CH_2CH_2CO_2H)—CO_2^-Na^+$

The first structure is MSG, which is impossible for you to predict.

105. [structures: D-Ribose, D-Mannose]

D-Ribose D-Mannose

107. The aldohexoses contain 6 carbons and the aldehyde functional group. Glucose, mannose, and galactose are aldohexoses. Ribose and arabinose are aldopentoses since they contain 5 carbons with the aldehyde functional group. The ketohexose (6 carbons + ketone functional group) is fructose, and the ketopentose (5 carbons + ketone functional group) is ribulose. **109.** A disaccharide is a carbohydrate formed by bonding two monosaccharides (simple sugars) together. In sucrose, the simple sugars are glucose and fructose, and the bond formed between these two monosaccharides is called a glycoside linkage. **111.** The α and β forms of glucose differ in the orientation of a hydroxyl group on one specific carbon in the cyclic forms (see Fig. 21.31). Starch is a polymer composed of only α-D-glucose, and cellulose is a polymer composed of only β-D-glucose.

113. Isoleucine: $H_3C—C^*H—CH_2CH_3$ attached to $H_2N—C^*H—CO_2H$; Threonine: $H_3C—C^*H—OH$ attached to $H_2N—C^*H—CO_2H$

Isoleucine Threonine

The chiral carbons are marked with asterisks.

115. $Br—C^*H(Cl)—CH{=}CH_2$

is optically active. The chiral carbon is marked with an asterisk. **117.** They all contain nitrogen atoms with lone pairs of electrons. **119.** C-C-A-G-A-T-A-T-G **121.** Uracil will hydrogen-bond to adenine.

[structure: Uracil—Adenine hydrogen bonding; labels: Uracil, Sugar, Adenine, Sugar]

123. a. glu: CTT, CTC; val: CAA, CAG, CAT, CAC; met: TAC; trp: ACC; phe: AAA, AAG; asp: CTA, CTG; b. ACC-CTT-AAA-TAC;
or CTC (for CTT) or AAG (for AAA)
c. Due to glu and phe, there is a possibility of four different DNA sequences; d. met-asp-phe; e. TAC-CTA-AAG; TAC-CTA-AAA; TAC-CTG-AAA **125.** A deletion may change the entire code for a protein, thus giving an entirely different sequence of amino acids. A substitution will change only one single amino acid in a protein. **127.** $CH_3CH_2CH_2CH_2CH_2CH_2CH_2COOH + OH^- \rightarrow CH_3{-}(CH_2)_6{-}COO^- + H_2O$; Octanoic acid is more soluble in 1 *M* NaOH. Added OH^- will remove the acidic proton from octanoic acid, creating a charged species. As is the case with any substance with an overall charge, solubility in water increases. When morphine is reacted with H^+, the amine group is protonated, creating a positive charge on morphine ($R_3N + H^+ \rightarrow R_3N^+H$). By treating morphine with HCl, an ionic compound results that is more soluble in water and in the bloodstream than is the neutral covalent form of morphine. **129.** To substitute for the benzene ring hydrogens, an iron(III) catalyst must be present. Without this special iron catalyst, the benzene ring hydrogens are unreactive. To substitute for an alkane hydrogen, light must be present. For toluene, the light-catalyzed reaction substitutes a chlorine for a hydrogen in the methyl group attached to the benzene ring. **131.** 1-butene **133.** a. The bond angles in the ring are about 60°. VSEPR predicts bond angles close to 109°. The bonding electrons are closer together than they prefer, resulting in strong electron–electron repulsions. Thus ethylene oxide is unstable (reactive). b. The ring opens up during polymerization; monomers link together through the formation of O—C bonds.

$$\text{—}(O{—}CH_2CH_2{—}O{—}CH_2CH_2{—}O{—}CH_2CH_2)_n\text{—}$$

135. [structure: $\left(-OCH_2CH_2OC(=O)NH{-}C_6H_4{-}NHC(=O)OCH_2CH_2OC(=O)NH{-}C_6H_4{-}NHC(=O)-\right)_n$]

137. *n*-hexane, 69°C; pentanal, 103°C; 1-pentanol, 137°C; butanoic acid, 164°C; the strength of the intermolecular forces increases when going from *n*-hexane to pentanal to 1-pentanol to butanoic acid. Hence the boiling points will increase in the same order. **139.** This is a condensation copolymer. The monomers are:

$HO—C(=O)—C_6H_4—C(=O)—OH$ $HO—CH_2—CH_2—OH$

141. a. The longest chain is 4 carbons long. The correct name is 2-methylbutane. b. The longest chain is 7 carbons long, and we would start the numbering system at the other end for lowest possible numbers. The correct name is 3-iodo-3-methylheptane. c. This compound cannot exhibit *cis–trans* isomerism because one of the double-bonded carbons has the same two groups (CH_3) attached. The numbering system should also start at the other end to give the double bond the lowest possible number. 2-Methyl-2-pentene is correct. d. The OH functional group gets the lowest number. 3-Bromo-2-butanol is correct. **143.** For the reaction, we break a P—O and O—H bond and form a P—O and O—H bond, so $\Delta H \approx 0$ based on bond dissociation energies. ΔS for this process should be negative (unfavorable) because positional probability decreases. Thus ΔG should be positive due to the unfavorable ΔS term, and the reaction is not expected to be spontaneous. **145.** a. No; the mirror image is superimposable.

b. [structures: two tartaric acid forms with OH, OH, H, CO_2H, H, CO_2H and OH, OH, HO_2C, H, H, CO_2H, separated by a mirror]

mirror

These two forms of tartaric acid are nonsuperimposable.
147. pH = 6.07 = isoelectric point **149.** 0.11% **151.** C_2H_6; ethane

153. a. $CH_3CH(CH_3)CH_2CH_3$

b. $CH_2{=}C(CH_3)CH_3$

c. [cyclohexene structure with CH_3 on a double-bonded ring carbon, remaining ring positions bearing H]; [methylenecyclohexane structure with $=CH_2$ on a ring carbon, remaining ring positions bearing H]

d. $CH_2{=}CHCH_3$

e. $CH_2(OH)CH_2CH_2CH_2CH_3$ (OH on the first carbon of $CH_2CH_2CH_2CH_2CH_3$); $CH_2CH(OH)CH_2CH_3$ bearing CH_3 on C-2 (CH_3 above, OH below the CH)

$CH_3CH(CH_3)CH_2CH_2OH$ (CH_3 above the CH, OH below the final CH_2); $CH_2(OH){-}C(CH_3)_2{-}CH_3$ (OH below the CH_2, CH_3 above and below the C)

155. a. 25.0 mL, 2.37; 50.0 mL, 6.07; 75.0 mL, 9.77

b.

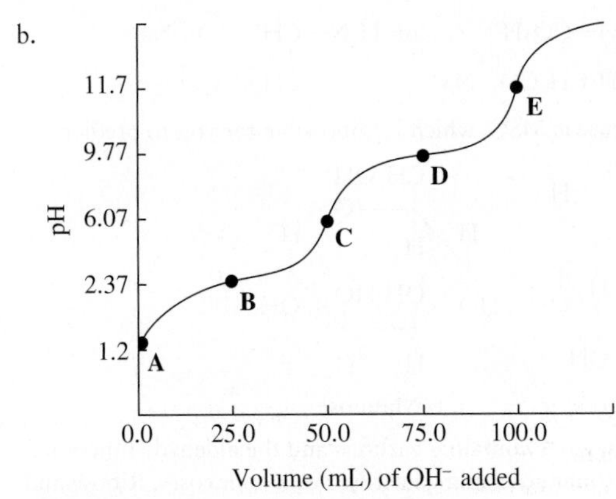

The major amino acid species present are:

point A (0.0 mL OH^-): $H_3\overset{+}{N}CH_2COOH$

point B (25.0 mL OH^-): $H_3\overset{+}{N}CH_2COOH$, $H_3\overset{+}{N}CH_2COO^-$

point C (50.0 mL OH^-): $H_3\overset{+}{N}CH_2COO^-$

point D (75.0 mL OH^-): $H_3\overset{+}{N}CH_2COO^-$, $H_2NCH_2COO^-$

point E (100.0 mL OH^-): $H_2NCH_2COO^-$

c. 6.07; d. $+\frac{1}{2}$, 2.37; $-\frac{1}{2}$, 9.77

Index

Ion(s), 31–33
  electrolysis of mixtures of, 504, 506
  in electrolytes, 97–100
  energy of interaction of, 597
  independence in solution, 96, 106
  naming of, 36–39
  oxidation states of, 126
  polyatomic, 32, 39, 614
  separating mixtures of, 112–114, 338–342, 346–347
  sizes of, 608–609
  spectator, 111–112
  See also Complex ions; Hydration

Page Numbers of Some Important Tables

Physical Constants

Constant	Symbol	Value
Atomic mass unit	amu	1.66054×10^{-27} kg
Avogadro's number	N	6.02214×10^{23} mol^{-1}
Bohr radius	a_0	5.292×10^{-11} m
Boltzmann's constant	k	1.38066×10^{-23} J K^{-1}
Charge of an electron	e	1.60218×10^{-19} C
Faraday's constant	F	96,485 C mol^{-1}
Gas constant	R	8.31451 J K^{-1} mol^{-1}
		0.08206 L atm K^{-1} mol^{-1}
Mass of an electron	m_e	9.10939×10^{-31} kg
		5.48580×10^{-4} amu
Mass of a neutron	m_n	1.67493×10^{-27} kg
		1.00866 amu
Mass of a proton	m_p	1.67262×10^{-27} kg
		1.00728 amu
Planck's constant	h	6.62608×10^{-34} J s
Speed of light	c	2.99792458×10^{8} m s^{-1}